OBSERVATION METHODS

Methods and methodology underpin all the social sciences. Published since 2001 and now exceeding 60 four-volume sets, the **SAGE Benchmarks in Social Research Methods** series has proven itself the definitive reference collection on methods available today. From ethnography to measurement, the series continues to systematically map the history of thought on the vast range of quantitative and qualitative methods in the social sciences. Edited by leaders in their fields, each set presents a careful selection of the key historical and contemporary works – classics and previously inaccessible works alike – and includes an authoritative introduction by the editor.

Barry Smart is Professor of Sociology at the University of Portsmouth and in his career has taught a wide range of sociology course units, including units on social theory, methodology, and social research, as well as contributing to courses on social research methods at both undergraduate and graduate level. Barry has extensive editorial experience and has edited and co-edited a number of academic texts and reference volumes, including *Michel Foucault: Critical Assessments I*, 3 volumes (Routledge, 1994); *Michel Foucault Critical Assessments II*, 4 volumes (Routledge, 1995); *Resisting McDonaldization* (SAGE, 1999); *Handbook of Social Theory* (edited with George Ritzer, SAGE, 2001); and *Post-Industrial Society*, 4 Volumes (SAGE, 2010). He is on the editorial advisory boards of a number of international journals including *Journal of Classical Sociology; European Journal of Social Theory; Theory, Culture and Society* and *International Journal of Japanese Sociology*.

Kay Peggs is Principal Lecturer in Sociology at the University of Portsmouth. Her teaching includes undergraduate and postgraduate courses on a range of qualitative and quantitative research methods. She is particularly interested in using complementary methods and her work *Identity and Repartnering after Separation* (with Richard Lampard, Palgrave, 2007) uses a combination of quantitative and qualitative social research techniques. Her current research centres on the complex inequalities associated with species-related oppressions. In this work she has turned to discourse analysis and papers have appeared in journals such as *Sociology, Sociological Research Online*, and *Society and Animals*. Her most recent book *Animals and Sociology* (2012) is published by Palgrave Macmillan. Her forthcoming book *Experiments, Animal Bodies and Human Values* will be published by Ashgate. She has gained substantial editorial experience through her membership of the editorial boards of journals such as *Sociology and Sociological Research Online*.

Joseph Burridge is a Senior Lecturer in Sociology at the University of Portsmouth, where he teaches research methods across the curriculum. Joseph has considerable editorial experience having co-edited a special issue of the *Social Semiotics* (Vol 18, Issue 3, 2008), which was re-published as an edited book (*Analysing Media Discourses*, Routledge, 2011). He has also organised and edited a special issue of the journal *Food and Foodways* (Vol 20, Issue 1, 2012). He is an Associate Editor of *The Poster* (published by Intellect), and has acted as a referee for the following international journals: *Sociology, British Journal of Sociology, Social Problems, Sexualities, Critical Discourse Studies; The British Journal of Social Psychology; Studies in Language and Capitalism* and the *Journal of Community and Applied Social Psychology*. He has also acted for the ESRC as a referee for large grant applications. His main research interests lie in the areas of food and culture, the sociology of culture, media representations, and discursive methods.

SAGE BENCHMARKS IN SOCIAL RESEARCH METHODS

OBSERVATION METHODS

VOLUME I

Edited by

Barry Smart, Kay Peggs and Joseph Burridge

SAGE

Los Angeles | London | New Delhi
Singapore | Washington DC

SAGE

Los Angeles | London | New Delhi
Singapore | Washington DC

SAGE Publications Ltd
1 Oliver's Yard
55 City Road
London EC1Y 1SP

SAGE Publications Inc.
2455 Teller Road
Thousand Oaks, California 91320

SAGE Publications India Pvt Ltd
B 1/I 1, Mohan Cooperative Industrial Area
Mathura Road
New Delhi 110 044

SAGE Publications Asia-Pacific Pte Ltd
3 Church Street
#10-04 Samsung Hub
Singapore 049483

First published 2013

Typeset by Arete Publishing Private Limited, Delhi

Printed on paper from sustainable resources

Printed by MPG Printgroup, UK

MIX
Paper from responsible sources
FSC
www.fsc.org
FSC® C018575

Library of Congress Control Number: 2012942348

British Library Cataloguing in Publication Data

A catalogue record for this book is available from the British Library

ISBN: 978-1-4462-0811-3 (set of four volumes)

Contents

Volume II

B. Reflections on the Practice of Observation (*Continued*)

C. Ethics, Risk and Observation

Volume III

D. Participant Observation

Volume IV

G. Observing Workplaces and Workers

H. Studying Up: Observing the Unobserved

'Serious differences among social scientists occur not between those who would observe without thinking and those who would think without observing; the differences have rather to do with what kinds of thinking, what kinds of observing, and what kinds of links, if any, there are between the two'.

Wright Mills, C. (1970 [1959]). *The Sociological Imagination*. Harmondsworth: Penguin: 42.

Appendix of Sources

All articles and chapters have been reproduced exactly as they were first published, including textual cross-references to material in the original source.

Grateful acknowledgement is made to the following sources for permission to reproduce material in this book.

1. 'The Bucket and the Searchlight: Two Theories of Knowledge', *Karl R. Popper*
   Objective Knowledge: An Evolutionary Approach (Oxford: Oxford University Press, 1979), pp. 341–361.
   Reprinted with permission of University of Klagenfurt / Karl Popper Library.

2. 'Revolutions as Changes of World View', *Thomas Kuhn*
   The Structure of Scientific Revolutions (3rd Edition) (Chicago: The University of Chicago Press, 1996), pp. 111–135.
   Published by the University of Chicago. Reprinted with permission from University of Chicago Press.

3. 'Techniques of the Observer', *Jonathan Crary*
   October, 45 (1988): 3–35.
   © 1988 by October Magazine, Ltd. and the Massachusetts Institute of Technology. Reprinted with permission from MIT Press Journals.

4. 'Interpretation: Observer Effects', *William Thompson*
   Allan Jamieson and Andre A. Moenssens (eds), *Wiley Encyclopaedia of Forensic Science* (Chichester: John Wiley & Sons Ltd, 2009), pp. 1–4.
   Reproduced with permission of John Wiley & Sons Ltd.

5. 'Seeing and Knowing', *Michel Foucault*
   A.M. Sheridan Smith (tr), *The Birth of the Clinic: An Archaeology of Medical Perception* (London: Routledge, 2003), pp. 131–151.
   Copyright © 1973 by Tavistock Publications, Ltd. Used by permission of Pantheon Books, a division of Random House, Inc. and Taylor & Francis Books UK.

6. 'Rules for the Observation of Social Facts', *Emile Durkheim*
   Steven Lukes (ed.) and W. D. Halls (tr), *The Rules of Sociological Method and Selected Texts on Sociology and Its Method* (London: Macmillan, 1982), pp. 60–84.
   Copyright © 1982 by Steven Lukes. Translation, Copyright © 1982 by Macmillan Press Ltd., UK. All rights reserved. Reprinted with the permission of Free Press, a Division of Simon & Schuster, Inc. and Palgrave Macmillan.

7. 'Weber's *Verstehen* and the History of Qualitative Research: The Missing Link', *Jennifer Platt*
   The British Journal of Sociology, XXXVI(3) (1985): 448–466.
   Reproduced with permission of Blackwell Publishing Ltd.

8. 'The Definitions of Sociology and of Social Action', *Max Weber*
   Talcott Parsons (ed.) and A.M. Henderson and Talcott Parsons (trs), *Max Weber: The Theory of Social and Economic Organization* (London: The Free Press, 1964), pp. 88–104.

9. 'Social Relationships between Contemporaries and Indirect Social Observation', *Alfred Schutz*
   George Walsh and Fredrick Lehnert (trs), *The Phenomenology of the Social World* (Evanston: Northwestern University Press, 1967), pp. 204–207.

10. 'Some Basic Problems of Interpretive Sociology', *Alfred Schutz*
    George Walsh and Fredrick Lehnert (trs), *The Phenomenology of the Social World* (Evanston: Northwestern University Press, 1967), pp. 215–250.

11. 'Unexpected Interactions: Georg Simmel and the Observation of Nature', *Matthias Gross*
    Journal of Classical Sociology, 1(3) (2001): 395–414.
    Published by SAGE Publications Ltd. Reprinted with permission.

12. 'Scopic Regimes of Modernity', *Martin Jay*
    Hal Foster (ed.), *Vision and Visuality* (Seattle: Bay Press, 1988), pp. 3–23.

13. 'Foucault's Art of Seeing', *John Rajchman*
    October, 44 (1988): 88–117.
    Published by MIT Press. Reprinted with permission from John Rajchman.

14. 'Excerpt from *The Observation of Savage Peoples*', *Joseph-Marie Baron de Gérando*
    F.C.T. Moore (tr), *The Observation of Savage Peoples* (London: Routledge and Kegan Paul, 1969), pp. 65–70.
    Reproduced by permission of Taylor & Francis Books UK.

15. 'Roles in Sociological Field Observations', *Raymond L. Gold*
Social Forces, 36(3) (1958): 217–223.
Reprinted with permission of Oxford University Press on behalf of University of North Carolina Chapel Hill Department of Sociology.

16. 'Performing Ethnography and the Ethnography of Performance', *Paul Atkinson*
British Journal of Sociology of Education, 25(1) (2004): 107–114.

17. 'Accounts, Interviews and Observations', *Robert Dingwall*
Gale Miller and Robert Dingwall (eds), *Context and Method in Qualitative Research* (London: SAGE, 1997), pp. 51–64.
Published by SAGE Publications Ltd. Reprinted with permission.

18. 'Observational Field Work', *Robert M. Emerson*
Annual Review of Sociology, 7 (1981): 351–378.

19. 'Everett C. Hughes and the Development of Fieldwork in Sociology', *Jean-Michel Chapoulie*
Journal of Contemporary Ethnography (formerly *Urban Life*), 15(3/4) (1987): 259–297.
Published by SAGE Publications, Inc. Reprinted with permission.

20. 'The Chicago School and Firsthand Data', *Jennifer Platt*
History of the Human Sciences, 7(1) (1994): 57–80.
Published by SAGE Publications Ltd. Reprinted with permission.

21. 'Mass-Observation: Social Research or Social Movement?', *Penny Summerfield*
Journal of Contemporary History, 20 (1985): 439–452.
Published by SAGE Publications Ltd. Reprinted with permission

22. 'A Problem of Sociological Praxis: The Case for Interventive Observation in Field Work', *Y. Michal Bodemann*
Theory and Society, 5(3) (1978): 387–420.

23. 'Benefits of "Observer Effects": Lessons from the Field', *Torin Monahan and Jill A. Fisher*
Qualitative Research, 10(3) (2010): 357–376.
Published by SAGE Publications Ltd. Reprinted with permission.

24. 'Can There Be a Feminist Ethnography?', *Judith Stacey*
Women's Studies International Forum, 11(1) (1988): 21–27.
© 1988 Pergamon Journals Ltd. Reprinted with permission from Elsevier via Copyright Clearance Center's RightsLink service.

25. 'On Tricky Ground: Researching the Native in the Age of Uncertainty', *Linda Tuhiwai Smith*
Norman K. Denzin and Yvonna S. Lincoln (eds), *The Landscape of Qualitative Research* (3rd Edition) (Thousand Oaks, CA: SAGE, 2008), pp. 113–143.
Published by SAGE Publications, Inc. Reprinted with permission.

26. 'Ethnographic Showcases, 1870–1930', *Raymond Corbey*
Cultural Anthropology, 8(3) (1993): 338–369.
Reproduced by permission of the American Anthropological Association. Not for sale or further reproduction.

27. 'Why Look at Animals?', *John Berger*
About Looking (NY: Pantheon Books, 1980), pp. 27–37.
About Looking (London: Bloomsbury Publishing Plc, 2009), pp. 16–28.
Copyright © 1980 by John Berger. Used by permission of Pantheon Books, a division of Random House, Inc. and Bloomsbury Publishing Plc.

28. 'Ethical Challenges in Participant Observation: A Reflection on Ethnographic Fieldwork', *Jun Li*
The Qualitative Report, 13(1) (2008): 100–115.
Published by Nova Southeastern University, School of Humanities and Social Sciences. Reprinted with permission.

29. 'The Risk of "Going Observationalist": Negotiating the Hidden Dilemmas of Being an Insider Participant Observer', *Robert V. Labaree*
Qualitative Research, 2(1) (2002): 97–122.
Published by SAGE Publications Ltd. Reprinted with permission.

30. 'Informed Consent, Anticipatory Regulation and Ethnographic Practice', *Elizabeth Murphy and Robert Dingwall*
Social Science & Medicine, 65(11) (2007): 2223–2234.
© 2007 Elsevier Ltd. All rights reserved. Reprinted with permission from Elsevier via Copyright Clearance Center's RightsLink service.

31. 'The Art and Politics of Covert Research: Doing "Situated Ethics" in the Field', *David Calvey*
Sociology, 42(5) (2008): 905–918.
Published by SAGE Publications Ltd. Reprinted with permission.

32. 'Covert Participant Observation: On Its Nature and Practice', *Richard A. Hilbert*
Journal of Contemporary Ethnography (formerly *Urban Life*), 9(1) (1980): 51–77.
Published by SAGE Publications, Inc. Reprinted with permission.

33. 'Between Overt and Covert Research: Concealment and Disclosure in an Ethnographic Study of Commercial Hospitality', *Peter Lugosi*
Qualitative Inquiry, 12(3) (2006): 541–561.
Published by SAGE Publications, Inc. Reprinted with permission.

34. 'Ethical Covert Research', *Paul Spicker*
Sociology, 45(1) (2011): 118–133.
Published by SAGE Publications Ltd. Reprinted with permission.

35. 'Lone Researchers at Sea: Gender, Risk and Responsibility', *Helen Sampson and Michelle Thomas*
Qualitative Research, 3(2) (2003): 165–189.
Published by SAGE Publications Ltd. Reprinted with permission.

36. 'When Is Disguise Justified? Alternatives to Covert Participant Observation', *Martin Bulmer*
Qualitative Sociology, 5(4) (1982): 251–264.
© 1982 Human Sciences Press. Reprinted with kind permission from Springer Science and Business Media via Copyright Clearance Center's RightsLink service.

37. 'A Comment on Disguised Observation in Sociology', *Kai T. Erikson*
Social Problems, 14(4) (1967): 366–373.
© 1967 by the Society for the Study of Social Problems. Published by the University of California Press. Reprinted with permission.

38. 'Methods: The Sociologist as Voyeur', *Laud Humphreys*
Tearoom Trade (Enlarged Edition) (New Jersey: Transaction, 1975), pp. 16–44.
Copyright © by Aldine Publishers. Reprinted by permission of AldineTransaction, a division of Transaction Publishers.

39. 'Controversies Surrounding Laud Humphreys' Tearoom Trade: An Unsettling Example of Politics and Power in Methodological Critiques', *Michael Lenza*
International Journal of Sociology and Social Policy, 24(3/4/5) (2004): 20–31.

40. 'Working in Hostile Environments', *Nigel Fielding*
Clive Seale, Giampietro Gobo, Jaber F. Gubrium and David Silverman (eds), *Qualitative Research Practice* (London: SAGE, 2004), pp. 236–248.
Published by SAGE Publications Ltd. Reprinted with permission.

41. '"Dangerous Fieldwork" Re-examined: The Question of Researcher Subject Position', *Pamela Nilan*
Qualitative Research, 2(3) (2002): 363–386.
Published by SAGE Publications Ltd. Reprinted with permission.

42. 'Doing Participant Observation in a Psychiatric Hospital – Research Ethics Resumed', *Christine Oeye, Anne Karen Bjelland and Aina Skorpen*
Social Science & Medicine, 65(11) (2007): 2296–2306.

43. 'The Researcher as Hooligan: Where "Participant" Observation Means Breaking the Law', *Geoff Pearson*
International Journal of Social Research Methodology, 12(3) (2009): 243–255.

44. 'Ethnographic Intimacy: Thinking through the Ethics of Social Research in Sex Worlds', *Maria Pérez-y-Pérez and Tony Stanley*
Sociological Research Online, 16(2) 13 (2011).
http://www.socresonline.org.uk/16/2/13.html
Published by University of Surrey. Reprinted with permission.

45. 'The Con Man as Model Organism: The Methodological Roots of Erving Goffman's Dramaturgical Self', *Michael Pettit*
History of the Human Sciences, 24(2) (2011): 138–154.
Published by SAGE Publications Ltd. Reprinted with permission.

46. 'A Note on Participant Observation', *Colin Bell*
Sociology, 3(3) (1969): 417–418.
Published by SAGE Publications Ltd. Reprinted with permission.

47. 'A Contribution to the Theory of Participant Observation', *Jiri Kolaja*
    Social Forces, 35(2) (1956): 159–163.
    Reprinted with permission of Oxford University Press on behalf of University of North Carolina Chapel Hill Department of Sociology.

48. 'Problems of Inference and Proof in Participant Observation', *Howard S. Becker*
    American Sociological Review, 23(6) (1958): 652–660.
    Republished with permission of SAGE Publications Inc. Journals; permission conveyed through Copyright Clearance Center, Inc.

49. 'Part of the Action, or "Going Native"? Learning to Cope with the "Politics of Integration"', *Duncan Fuller*
    Area, 31(3) (1999): 221–227.
    © Royal Geographical Society (with The Institute of British Geographers) 1999. Reproduced with permission of Blackwell Publishing Ltd.

50. 'The Participant Observer and "Over-Rapport"', *S.M. Miller*
    American Sociological Review, 17(1) (1952): 97–99.
    Republished with permission of SAGE Publications Inc. Journals; permission conveyed through Copyright Clearance Center, Inc.

51. 'Role Boundaries and Paying Back: "Switching Hats" in Participant Observation', *Jacqueline E. Wade*
    Anthropology & Education Quarterly, 15(3) (1984): 211–224.
    Reproduced by permission of the American Anthropological Association. Not for sale or further reproduction.

52. 'Deep Play: Notes on the Balinese Cockfight', *Clifford Geertz*
    The Interpretation of Cultures (New York: Basic Books, 1973), pp. 412–453.
    Copyright © 1973 Clifford Geertz. Reprinted by permission of Basic Books, a member of the Perseus Books Group via Copyright Clearance Center's RightsLink service.

53. 'Participant Observation as a Tool for Understanding the Field of Safety and Security', *Frédéric Diaz*
    Champ Pénal/Penal Field, II (2005).
    http://champpenal.revues.org/471
    Published by *Champ Pénal/Penal Field*. Reprinted with permission.

54. 'Participant Observation in Prison', *James B. Jacobs*
    Journal of Contemporary Ethnography (formerly *Urban Life and Culture*), 3(2) (1974): 221–240.
    Published by SAGE Publications, Inc. Reprinted with permission.

55. 'A Spy, a Shill, a Go-between, or a Sociologist: Unveiling the "Observer" in Participant Observer', *Susan B. Murray*
Qualitative Research, 3(3) (2003): 377–395.
Published by SAGE Publications Ltd. Reprinted with permission.

56. 'On Writing Fieldnotes: Collection Strategies and Background Expectancies, *Nicholas H. Wolfinger*
Qualitative Research, 2(1) (2002): 85–95.
Published by SAGE Publications Ltd. Reprinted with permission.

57. 'Thick Description: Toward an Interpretive Theory of Culture', *Clifford Geertz*
The Interpretation of Cultures (New York: Basic Books, 1973), pp. 3–30.
Copyright © 1973 Clifford Geertz. Reprinted by permission of Basic Books, a member of the Perseus Books Group via Copyright Clearance Center's RightsLink service.

58. 'Thinking through Fieldwork', *Judith Okely*
Alan Bryman and Robert G. Burgess (eds), *Analyzing Qualitative Data* (London: Routledge, 1994), pp. 18–34.
Reproduced by permission of Taylor & Francis Books UK.

59. 'On the Analysis of Observational Data: A Discussion of the Worth and Uses of Inductive Techniques and Respondent Validation', *M. Bloor*
Sociology, 12(3) (1978): 545–552.
Published by SAGE Publications Ltd. Reprinted with permission.

60. 'The Presentation of Everyday Life: Some Textual Strategies for "Adequate Ethnography"', *Kenneth Stoddart*
Journal of Contemporary Ethnography (formerly *Urban Life*), 15(1) (1986): 103–121.
Published by SAGE Publications, Inc. Reprinted with permission.

61. 'Representation, Legitimation, and Autoethnography: An Autoethnographic Writing Story', *Nicholas L. Holt*
International Journal of Qualitative Methods, 2(1) (2003): 18–28.
Published quarterly as a web-based journal by the International Institute for Qualitative Methodology at the University of Alberta. This is an open access journal.

62. 'Looking Emotionally: Photography, Racism and Intimacy in Research', *Mónica G. Moreno Figueroa*
History of the Human Sciences, 21(4) (2008): 68–85.
Published by SAGE Publications Ltd. Reprinted with permission.

63. 'Using CCTV to Study Visitors in The New Art Gallery, Walsall, UK', *Ela Beaumont*
Surveillance & Society, 3(2/3) (2005): 251–269.
Published by Surveillance Studies Network. Reprinted with permission.

64. 'Ethnographic Approaches to the Internet and Computer-Mediated Communication', *Angela Cora Garcia, Alecea I. Standlee, Jennifer Bechkoff and Yan Cui*
Journal of Contemporary Ethnography, 38(1) (2009): 52–83.
Published by SAGE Publications, Inc. Reprinted with permission.

65. 'Ethnography, the Internet, and Youth Culture: Strategies for Examining Social Resistance and "Online-Offline" Relationships', *Brian Wilson*
Canadian Journal of Education, 29(1) (2006): 307–328.
Published by Canadian Society for the Study of Education. Reprinted with permission.

66. 'Social Access in the Workplace: Are Ethnographers Gossips?', *Simon Carmel*
Work, Employment & Society, 25(3) (2011): 551–560.
Published by SAGE Publications Ltd. Reprinted with permission.

67. 'The Sweat-Shop in Summer', *Annie Marion MacLean*
American Journal of Sociology, IX(3) (1903): 289–309.
Published by the University of Chicago Press. Reprinted with permission via Copyright Clearance Center's RightsLink service.

68. 'On Doctor Watching: Fieldwork in Medical Settings', *Sandra K. Danziger*
Journal of Contemporary Ethnography (formerly *Urban Life*), 7(4) (1979): 513–532.
Published by SAGE Publications, Inc. Reprinted with permission.

69. 'Two Weeks in Department Stores', *Annie Marion MacLean*
American Journal of Sociology, IV(6) (1899): 721–741.
Published by the University of Chicago Press. Reprinted with permission via Copyright Clearance Center's RightsLink service.

70. 'An Observational Study of Shoplifting', *Abigail Buckle and David P. Farrington*
The British Journal of Criminology: An International Review of Crime and Society, 24(1) (1984): 63–73.
Reprinted by permission of Oxford University Press via Copyright Clearance Center's RightsLink service.

71. 'Glimpses at the Mind of a Waitress', *Amy E. Tanner*
American Journal of Sociology, 13(1) (1907): 48–55.
Published by the University of Chicago Press. Reprinted with permission via Copyright Clearance Center's RightsLink service.

72. 'Extracts from "Living the Kitchen Life" and "Appendix: Ethnography in the Kitchen"', *Gary Alan Fine*
Kitchens: The Culture of Restaurant Work (2nd Edition) (London: University of California Press, 2009), 'Living the Kitchen Life', pp. 17–45 and 'Appendix: Ethnography in the Kitchen', pp. 233–240.
© 2009 by the Regents of the University of California. Published by the University of California Press. Reprinted with permission.

73. 'Up the Anthropologist – Perspectives Gained from Studying Up', *Laura Nader*
Dell Hymes (ed.), *Reinventing Anthropology* (New York: Pantheon Books, 1972), pp. 284–311.
Copyright © 1972 by Random House, Inc. Used by permission of Pantheon Books, a division of Random House, Inc.

74. 'Ethnography in/of the World System: The Emergence of Multi-Sited Ethnography', *George E. Marcus*
Annual Review of Anthropology, 24 (1995): 95–117.
Copyright © 1995 by Annual Reviews Inc. All rights reserved. Republished with permission of Annual Reviews, Inc.; permission conveyed through Copyright Clearance Center, Inc.

75. 'Studying Up Revisited', *Hugh Gusterson*
PoLAR: Political and Legal Anthropology Review, 20(1) (1997): 114–119.
Reproduced by permission of the American Anthropological Association. Not for sale or further reproduction.

76. 'After Method? Ethnography in the Knowledge Economy', *David Mills and Richard Ratcliffe*
Qualitative Research, 12(2) (2012): 147–164.
Published by SAGE Publications Ltd. Reprinted with permission.

77. 'Fast Capitalism: Para-Ethnography and the Rise of the Symbolic Analyst', *Douglas R. Holmes and George E. Marcus*
Melissa S. Fisher and Greg Downey (eds), *Frontiers of Capital: Ethnographic Reflections on the New Economy* (Durham, NC: Duke University Press, 2006), pp. 33–57.
Copyright, 2006, Duke University Press. All rights reserved. Reprinted by permission of Duke University Press. www.dukeupress.edu

78. 'Introduction: Anthropology Goes to Wall Street', *Karen Ho*
Liquidated: An Ethnography of Wall Street (London: Duke University Press, 2009), pp. 1–38.

79. 'Researching Police Deviance: A Personal Encounter with the Limitations and Liabilities of Field-Work', *Maurice Punch*
The British Journal of Sociology, 40(2) (1989): 177–204.

80. 'Potential Sources of Observer Bias in Police Observational Data', *Richard Spano*
Social Science Research, 34(3) (2005): 591–617.

81. 'Observing the Observers: Researching Surveillance and Counter-Surveillance on "Skid Row"', *Thomas Kemple and Laura Huey*
Surveillance & Society, 3(2/3) (2005): 139–157.

82. 'Whistle-Blower Disclosures and Management Retaliation: The Battle to Control Information about Organization Corruption', *Joyce Rothschild and Terance D. Miethe*
Work and Occupations, 26(1) (1999): 107–128.

Editors' Introduction: Observation Methods and Social Research

Kay Peggs, Joseph Burridge and Barry Smart

Observation as a deliberate, organised, and systematic way of 'looking' or 'watching' is integral to all scientific inquiry. It is a practice that is guided by rational principles and assumptions and is motivated by an interest in obtaining data on occurrences, events, processes, reactions, forms of conduct, and relationships. Observation may take many forms and it occupies a particularly significant, if not paramount, place within the range of research methods employed within the social sciences. In line with this central importance, in this set of volumes we draw together key contributions on the observation methods deployed in social research. In so doing we give consideration to the historical development of observational methods and techniques and provide analytic reflection on the issues involved in the scientific practice of observation. The volumes also include examples of results obtained by studies now considered to be social science classics along with some less well known studies, demonstrating the rich diversity of observational approaches and associated innovations, and providing a comprehensive coverage of central themes and issues.

Observational methods have yielded some of the most interesting and exciting research in the social sciences. Early observations gave insights into factory work (Maclean, 1903) and waitress work (Tanner, 1907); more recent observations have centred on internet relationships (Wilson, 2006) and sex worlds (Pérez-y-Pérez and Stanley, 2011). We would not know as much as we do about criminal gangs, the workings of the police, the tensions among nightclub 'bouncers', or staff-patient interactions in psychiatric wards without the, at times daring, studies undertaken by observational researchers. Likewise, 'studying up' (Nader, 1972) has allowed observational researchers to find out more, but as yet not enough, about the institutions, organisations, and groups that have the power to affect our everyday lives (Williams, 2012). Of course, observational techniques are not limited to the social sciences; observation as a deliberate, organised, and systematic form of 'looking' or 'watching' is integral to all scientific inquiry (Daston and

Lunbeck, 2011). In the eighteenth and early nineteenth centuries observation and experiment were understood to work in partnership as observation was used to suggest conjectures that could then be tested and verified by experiment, giving rise to new observations for further testing (Daston and Lunbeck, 2011). However, in the 1820s philosophers of science began to oppose the use of observation because they saw it as involving the mere *description* of data rather than the required *analysis* of it. One of the major concerns was the lack of objectivity that was felt to be inherent in the approach, which rested in the view that science was theory laden, leading the observer to see only what they wanted to see (Daston and Lunbeck, 2011). But such concerns have been largely rejected and observation in some shape or form is now integral to the evidential foundations of all scientific practice (Maxwell, 2009).

While observation is a common feature of all scientific fields, as arguably is the 'observer effect', there are different understandings of what is involved in 'observation' in these diverse fields. For example, the debate in Germany in the late nineteenth century over social scientific methodology (*Methodenstreit*) led Wilhelm Windelband to contrast nomothetic approaches, which seek to emulate the logic and methodology of the physical sciences and to generate causal and law-like statements from scientific observation and research (*erklaren*), and ideographic approaches where emphasis is placed upon the historical particularity of phenomena and the importance of interpretation and understanding (*verstehen*) (Lamiell, 1998; Outhwaite, 2000). Wilhelm Dilthey drew a broadly similar distinction between the physical sciences (*Naturwissenschaften*) and the social and cultural sciences (*Geisteswissenschaften*). He argued that in the social and cultural sciences observation should be based in interpretation and understanding, which he saw as quite different from the natural scientific requirements that observe cause and effect (Outhwaite, 2000; Spiro, 1986).

Observation is not limited to research-specific scientific work; it is fundamental to our everyday lives. Because observation is ubiquitous and everyday this added to the problems that scientists saw with it as a methodological technique, as it came to be seen as too similar to lay methods to be appropriate for use in scientific research. This returns us to the aforementioned concerns of nineteenth century philosophers of science. William Herschel (1987 [1830]) remarked that observation could be done by amateurs or 'foot soldiers' (Daston and Lunbeck, 2011). However, twentieth century philosophers of science often held a different view. For example, in 1917 Russell Conwell noted that 'Millions and millions had seen the lid of a teakettle bobbing up and down over the boiling water before that Scotchman, Watt, observed it while making watches. . . Then he applied this power to the steam-engine' (1917: 66). In the 1970s Thomas Kuhn (1970) remarked that the trained observer sees differently to the layperson, though Karl Popper cautioned that observers should understand properly the assumptions underpinning their observations, including the 'horizon of expectations' that are brought to bear upon the observations that are made (1979: 346). Such ideas are encapsulated in John Berger's assertion that '[t]he way we see things is affected by what we know or what we believe' (1972: 8). Clearly seeing is influenced by a range of factors, including assumptions about social class, gender, and 'race' and a number of critical observational approaches have emerged to challenge, for example, ethnocentric and male-centred ways of proceeding (e.g. see Visweswaran, 1997). But are seeing and observing equivalent? Do researchers 'see' or 'observe'?

In the 1990s Jack Sanger argued that 'there is a world of difference between seeing and observing' (1996: 1). He explains that *seeing* and *observing* are not just different 'they are conflicted' because, as laypersons, we see without too much thought, whereas scientific observation 'is brought on by the stimulus to be necessarily aware' because 'we observe for an audience' (Sanger, 1996: 3–4). Scientific observation, conducted in accordance with the methodological requirements of each scientific discipline, effectively 'discovers the world anew' (Daston and Lunbeck, 2011: 1). But in order to achieve this, observation as a research approach 'requires training of the body and mind, material props, techniques of description and visualization, networks of communication and transmission, canons of evidence, and specialized forms of reasoning' (Daston and Lunbeck, 2011: 3). However, as Bronislaw Malinowski (1922: 6) argues, more than training is required, for the observer 'must possess real scientific aims and know the values and criteria of modern ethnography'.

In the social sciences, the observational approach has its origins in ethnographic research like that undertaken by Malinowski, which is most clearly associated with social anthropology and, from a critical perspective, the history of colonialism and the exploration (and construction) of cultural 'Others'.[1] The ethnographic method often involves direct observation of a small society or organisation (e.g. Mead, 1937) and a range of social scientists have drawn on this method to provide insights into naturally occurring social activity. It is a research technique that is especially useful when the researcher wants to study people's meanings in particular situations or in specific settings, and wants to examine the processes by which those involved construct such meanings. There are innumerable opportunities for undertaking such research. Social researchers have observed juvenile gangs, religious communities, the operation of workplaces, virtual cafes, cockfighting, interactions among members of their own families, and their own behaviour. The opportunities for observational research are as varied as the participants and situations that individuals inhabit. Although observations are often limited to small-scale social settings, this is not to say that the objective of such observations is necessarily small in scale. For example, Aviad Raz's (1999) objective in his observation of Tokyo Disneyland was to use observation as a way of viewing Japanese culture. It therefore served as what he called a 'doorway into Japan' more generally (Raz, 1999: 22–23).

Observational methods assume a number of forms. Distinctions are often drawn in terms of the researcher's relationship to the events and behaviours subjected to observation. These range from complete involvement, where the observer is a full member of the context/group that is under observation, through forms of participant observation, to completely detached observation, and finally to forms of self- or auto-observation (Adler and Adler, 1998; also see Gold, 1958) and auto-ethnography (Reed-Danahay, 1997). Distinctions also contrast forms of 'direct observation by the researcher' with 'indirect observation where the researcher collects self-observations from respondents' (Wilson, 2002). Indirect observation can be transmitted via questionnaire responses and in interview situations. Classic examples include the Mass Observation Studies that began in the UK in 1937. These studies do not usually include a 'professional' researcher who observes social phenomena in the real world and records events as they happen, rather the studies use 'lay' observers who report on daily happenings in the world around them in diaries and in questionnaires (Summerfield, 1985). In both direct and indirect observation the observational data ends up being structured, as a consequence of the categories employed in data collection and recording processes or when structure is articulated in and through the process of analysis itself.

The articles contained within these volumes mainly centre on direct observation and participant observation. This returns us to the distinctions that relate to the researcher's relationship to the observational setting. Direct observation involves a detached, watchful, and focused observer who employs a range of means by which to observe, and these may include video, one-way mirrors, and CCTV (e.g. see Corsaro, 1981). Thus the observer is not a participant in the situation or scene being studied. In contrast, the participant observer is there, present, and *participating* (either overtly or covertly) in the setting in which observation is simultaneously being conducted (e.g. Gans, 1979). But whatever approach is to be employed, observers must consider ways of gaining appropriate access to the forms of social life to be studied. In addition to addressing the critical issue of gaining access or getting into the field observers have to consider closely associated matters such as the recording of data and the writing up of results. In turn, there are a range of other related matters, notably ethical issues, as well as concerns about accuracy that have to be addressed. It is aspects such as these that are essential to the subject matter of these volumes.

Observing Others

For social researchers the elemental role of the observational fieldworker is that of studying how others view and experience the world. This often involves observers immersing themselves in worlds that are very different from their own and frequently demands that researchers learn to hear, speak, think, and act in ways very different from those with which they are familiar (Arluke and Sanders, 1996: 19). What observers seek to do is to examine people's behaviour and interactions and to explore these by considering how the subjects of their research interpret the meaning of events, the objects within them, and the situations in which they occur (Arluke and Sanders, 1996: 19). Clifford Geertz (1974: 26) calls this taking the 'native's point of view', which effectively means achieving 'anthropological knowledge of the way natives think, feel, and perceive'. Associated with this is the generation of a 'thick description', that is one that describes human behaviour in the specific cultural context in which it occurs, and which therefore makes it possible for the behaviour concerned to become intelligible and meaningful to someone who is not inside the group (Geertz, 1973). To achieve such a description, the observational researcher must become immersed in the sociocultural settings of the observed others and must strive to achieve an empathetically meaningful understanding of their behaviour (Schutz, 1972).

Observational research is generally conducted on independently occurring social events, processes, and settings, for the most part ongoing, routine, everyday happenings and associated forms of behaviour. However, covert forms of observational research have also been conducted under quasi-experimental conditions. What have come to be known as Harold Garfinkel's (1984 [1967]) 'breaching experiments' provide good examples as they were designed to observe the reactions of participants when behavioural norms were broken in naturalistic social settings. The point of the breaching experiment is to observe people's reactions to a disruption of routine, mundane, or taken-for-granted aspects of everyday life, to an experimenter (participant) who deliberately breaks a small, everyday rule or norm. In one example Garfinkel (1984 [1967]) encouraged his students to act as covert experimenters in their homes by asking for clarification from their family members when one of them might casually remark 'I am tired'. The behaviour of the students, requesting clarification of the meaning of such mundane utterances, led to anger, argument, and social sanctions being applied by family members – indicating that underlying norms and assumptions were being violated by, what was experienced by family members exposed to such questioning as, an intrusive and unnecessary request for clarification.

Experimental observation can also lead to unanticipated forms of self-observation and reflection on the part of experimenting observers. In an experiment conducted by Stanley Milgram on the New York City subway in 1972, 'experimental' observers, without providing any explanation, asked those who were seated on a crowded train to give up their seats (Luo, 2004). The principal objective was to observe passenger reactions in order to reveal the web of unwritten rules that govern behaviour on the subway. However, the experiment also produced another order of observations as the covert researchers involved in the experiment reported being troubled by undertaking this minor violation of a social norm. Such studies have led Patricia Adler and Peter Adler (1998) to extend paradigms of observational research to include dramaturgical constructions of reality, ethnomethodology, and auto-observation. Such observational techniques as these make it possible to explore how social interaction 'works', how forms of sociality are constituted, maintained, and transformed. However, insofar as subjects exposed to such forms of research are unaware they are being observed, and therefore have not had an opportunity to consent to any involvement, serious ethical questions are raised.

Accessing the Field

As with any form of social research, in practice the fieldwork begins before the researcher actually enters the field. Much of the early work will involve reviewing the literature on the substantive topic in question and the literature on the methodological approach to be employed. Personal reflection and recollection may also preface the actual fieldwork as the researcher's previous personal experiences might have influenced their chosen area of study (Burgess, 1984). For example, following arrest on a number of occasions for political activity, Miklos Harastzi (1978) got a job in the Red Star Tractor Factory and used his personal observations and reflections on his experiences as the basis of his account of working in a socialist state. Other social researchers have chosen to do research on communities in which they were previously members. For example, Simon Holdaway (1983) was a police officer before embarking on his observational study of a police force in Britain. Of course there are a number of factors that might lead a researcher to become interested in a particular field of inquiry, including intellectual background and study, personal experiences, political involvement(s), career calculations, and increasingly research funding opportunities.

There are several issues associated with gaining research access to the field, but addressing and attempting to resolve any problematic issues can be very useful when trying to understand the group(s) to be observed. Hammersley and Atkinson (1995) argue that addressing access problems can provide important insights into how groups respond to strangers and how open they are to outsiders; how aware the group is of its boundaries, and what methods are used to police them. For example, the civil servant Robin Page (1971) undertook an observational approach in his study of how people cope with living in abject poverty. His entry into homeless society in Brighton (UK) (1971) revealed the generous spirit of the homeless people he encountered, for example, one homeless man invited him to make himself at home when he arrived for his first night sleeping rough on the beach.

In contrast, Edward Hansen (1977) gained insights into the tight structure of the Catalonian village that he observed through the obstacles he had to negotiate. In order to gain access he had to accept aristocratic sponsorship and was required to work within the aristo-peasant hierarchy that operated within the village. Although there are clear differences, what these two studies show is that, in entering the setting, observational researchers often have to gain access and permission from insiders. In Hansen's case, this involved several meetings and negotiations with 'gatekeepers', the powerful figures in the group.

Some social situations and settings are more difficult to access. In some instances, access is restricted to particular categories of people and in such circumstances the help of a gatekeeper is vital. For example, to conduct a study of young men in prison Michael Little (1990) had to obtain permission from the UK Home Office 'gatekeeper'. But gatekeepers are not always easy to identify, in consequence the researcher must garner a good deal of social knowledge about the social setting or group that is the intended focus of research. For example, William Foote Whyte's (1955) access to the gang culture in the ghetto in Boston (USA) depended on him being befriended by 'Doc', a gang leader, who subsequently vouched for him. Ned Polsky (1976 [1971]) employed his billiard playing skills to establish contacts within the pool halls he wanted to study. On rare occasions individuals who would subsequently act as gatekeepers have sought out researchers. For example, James Patrick (1973) did not intend to study a juvenile gang at all. During college vacations he worked in one of Scotland's approved schools and one of the inmates, a gang leader, befriended him and asked him to join the gang so that he could observe what went on. However, in some situations it may not be possible to find a gatekeeper, or access might be denied completely. For example, as a woman, Barbara Rogers (1988) found it at times impossible to enter men's only clubs such as the freemasons, private boys' schools, and male-dominated sports clubs. In consequence, in her research she had to rely for information on male observers and members who acted as her informants.

Amongst other things what the above shows is that observational research can rarely be wholly covert. If a gatekeeper is involved, they will of course know the identity of the researcher and that research is being conducted. This can put a strain on both the researcher and the gatekeeper, both of whom must conceal what is actually occurring throughout the period of research observation. As observation usually lasts for a considerable period of time, access issues are effectively present and continually encountered throughout the duration of the research. Once initial access is secured the researcher usually finds that, if it is to be maintained, it must be continually renegotiated, in effect it constitutes an ongoing feature of the research process. For Howard Newby (1977) access to farm workers could only be secured through the agreement of the farmer and this had to be renegotiated recurrently during the whole period of observation. In contemporary ethnographic contexts, maintaining informed consent is an ethically sensitive process and constitutes an ongoing practical matter (Murray, 2003: 390).

Observational Roles

The role of the observer can take many forms. The observers can be obtrusive or unobtrusive, their presence in the research setting may be overt or covert, but these very broad differences obscure further degrees of complexity. Moreover, such distinctions do not align themselves neatly with a distinction between ethical and unethical practice(s) in the way that is sometimes assumed.

Unobtrusive research encompasses a wide range of possible 'nonreactive' methods of inquiry, methods in which the presence of the researcher is not a factor, and in and through which the data collected is not derived from information directly elicited from those who are the subjects of the research (Lee, 2000). In the context of our consideration of observation methods unobtrusive research is ultimately exemplified by forms of nonparticipant observation. In such forms of research the observer is anxious to avoid 'contaminating' in any way the research setting subject to observation. In contrast obtrusive forms of research intrude in varying degrees into the research setting and involve, in some shape or form, the participation of the researchers. The participation of researchers may be overt, in which case the subjects of research are aware that they are being observed, or covert, in which case research subjects will be unaware of their observed status (Brewer, 2003).

Covert observation is preferred by participant observers who want to minimise the effects of their research presence on the setting and on the people involved. Some researchers argue that it is often the only option if research access is to be achieved. Nigel Fielding (1981) takes this view as he argues that his observation of the National Front (an extreme right-wing political group once active in the UK) would not have been possible if his researcher role had been known (also see Billig, 1978). This created problems for Fielding (1981) as he had to set aside his hostility to the politics of the organisation and adopt the role of the potential convert. Leon Festinger, Henry Riecken, and Stanley Schachter (1956) and William Shaffir (1985) make a similar argument about the importance of a convert position in their observational research of religious groups. Following his covert observational study of organised gambling, William Chambliss (1975) concluded that other sociologists should follow his example, present themselves as legitimate members of groups and social settings to be observed and engage in covert study, as this would lead to results, insights, and understandings of a quality that could not be achieved through overt methods. In contemporary research this point of view is controversial because the emphasis increasingly is placed on the informed consent of research subjects and, in consequence, transparency and a more overt observational research approach.[2] But the requirement to obtain informed consent, and all that is associated with that, significantly limits the scope of social research. Indeed it has been argued that the restrictions placed on the use of covert research 'threaten the very nature of academic enquiry' (Spicker, 2011: 131).

Four main observational roles have been identified, and each of the roles has a particular connection with the issue of reactivity, that is with the potential of those being observed to be affected by the observational process itself (Gold, 1958; Junker, 1960). The *Complete Participant* usually uses a wholly covert approach and participates fully in the group. This role has many attractions. The researcher (e.g. Fielding, above) argues that they can often generate a complete knowledge of the group being observed and, without the use of such an approach such observational research might not take place at all. There is a potential problem of reactivity, but it is reduced to a minimum if all goes to plan, as the observed will not be aware of the presence of the researcher engaged in observing them. The *Complete Observer* is a researcher who is at the other extreme. In this role, the researcher avoids interaction with group members. For example, in his unobtrusive, nonparticipant, observational study of preschool children in the classroom, William Corsaro (1981) used a one-way mirror to observe their behaviour. The *Participant as Observer* is the researcher who is both obtrusive and overt, known to the research subjects, who in turn are fully aware that they are part of a field study. Here greater emphasis is placed on participation and interaction over observing, one key aim being to promote relationships of rapport and trust between researcher and observed subjects. In this instance potential problems of reactivity may be outweighed by the prospect of positive benefits flowing from establishment of a close rapport between researcher(s) and subject(s), allowing researchers to develop friendships with members who may then be willing to share their knowledge of the setting subject to observation (Pearsall, 1970). The fourth role, *Observer as Participant* has the reverse emphasis, with the focus being on observation at the expense of participation. In this instance it is argued that the observer aims to maintain intellectual detachment and social distance from research subjects, the intention being 'not [to] cross into the friendship domain' (Adler and Adler, 1998: 380). Here evident problems of reactivity may well be accompanied by diminished, inaccurate, or inappropriate understanding of the activities of the participants as a direct consequence of a lack of immersion in the research setting.

In practice, observational research and the roles assumed by observers are more complex than the fourfold typology outlined above allows (see Adler and Adler, 1998). For example, complete participant observers can get 'too involved', in some cases at the risk of compromising their research as they almost leave their research role behind and virtually become full members of the group. In his observational study in Bristol, Ken Pryce (1979) reported that, at times, he became so absorbed in his interactions with the group that he lost sight of the direction of his research. One way of countering the possibility of this happening might be to become a *partial participant* as Ned Polsky (1976 [1971]) did in his classic study. Polsky observed poolroom hustlers in their own 'game' setting but, in order to maintain some distance, he did not spend time with the hustlers in other social contexts. The drawback is that additional information could not be gathered from outside the poolroom. When operating in this way the researcher can only become associated with the group rather than becoming a member of it. In some cases the personal characteristics of the researcher may limit observational role possibilities. For example, Paul Willis (1977) wanted to find out why 'working class kids get working class jobs', but he clearly could not become one of the working class 'lads' in the town in England's West Midlands where he undertook his research. Instead he hung around with them and was able to observe them both in and out of school. Similarly as a 22-year-old residential community youth worker at a youth centre for teenage 'street kids' from Liverpool in the UK, Howard Parker (1974) could be no more than an outsider observer. However, he was accepted by the boys as an 'ok outsider' because he showed that he 'knew the score' about their exploits. Furthermore, even in ostensibly unobtrusive forms of observational research where the researcher assumes the role of *complete observer*, it is possible that subjects being observed may become aware of the observer's presence and in consequence, to some degree, may modify their behaviour, a manifestation of the reactivity problem already noted above.

Reactivity is a problem in most observational research and, of course, an observer cannot be present in a setting without being involved in it in some way, especially when participants are fully aware of the research focus. Because of this, participant observers often argue that it is important to be as neutral as possible so as to minimise the effects on the setting and the participants in it. The problem of reactivity is clearly in evidence in Whyte's (1955) study when Doc, the leader of the gang and Whyte's gatekeeper, revealed to Whyte that he was having an effect on Doc's behaviour. Even though Whyte had no involvement in the decisions taken by the gang leader Doc said that, since Whyte had emerged on the scene, he had begun to consider what Whyte would do in any of the circumstances involved before making a decision. Although Whyte had made a conscious decision to remain neutral, his mere presence had nevertheless been influential, had changed the situation and the behaviour of at least one of the observed subjects.

Many observers fear such 'contamination' of their research setting. However, not all researchers are so concerned, particularly if the form the 'contamination' assumes might be considered potentially progressive in social and political terms. For example, in the covert observational study Sallie Westwood (1984) conducted she took on the role of a packer in a factory and built up a good rapport with the women with whom she worked. She empathised strongly with them and became outraged by sexist comments to which they were subjected by men at social events, which she also attended, and she chose to make her feelings perfectly clear. Many of the women with whom she worked (and observed) said they were affected by her reactions and they subsequently began to question and challenge the sexist treatment to which they were exposed. Understandably Westwood did not see this unanticipated emergent quasi-participatory action research dimension to her project as a problem at all.

One way of attempting to overcome the potential problem of reactivity or contamination affecting the research field is for the observing researcher to be completely frank about their research interests and open about their involvement in the research field, the assumption being that any effect on the situation that their presence may have will diminish over time. For example, in her open observational study of girl gangs, Anne Campbell (1984) quickly recognised that she would not be able to pass off as a member or be able to maintain a covert observer role. Fortunately she found that the gang members soon got used to her being around and while she did not, and could not, become one of them she was able to generate insightful observational data. However, she did suspect that some acts, especially those of a criminal nature, were frequently hidden from her.

In addition to the range of observational roles considered so far there is an approach to observational research that effectively combines autobiography and ethnography, an approach in which researchers draw not only on methodological tools and studies within their specialist field, but also on their own personal experiences, to produce highly personalised accounts that extend understanding of a particular social setting by making aspects of it familiar to both cultural insiders and outsiders (Reed-Danahay, 1997). This approach differs from other observational roles as there is a combination of study of others and reflection on the researching observational self's interactions with others, with the focus being on the observer's subjective experience, which in turn serves to give an explicit voice to the researcher (Wall, 2006). Auto-ethnographic research is 'usually written in the first person and feature(s) dialogue, emotion, and self-consciousness as relational and institutional stories affected by history, social structure, and culture' (Holt, 2003: 3). For example, the researcher Andrew Sparkes (1996) published an auto-ethnographic account of his passage from elite athlete to a man whose participation in sport was curtailed by inflammatory back disease. Such 'inside auto-ethnography' need not focus on the self alone; such research can also explore ties that observers have with participants in the research setting.

Auto-ethnographic observational research also has its critics. Although an auto-ethnographic researcher himself, Sparkes (2000) laments that auto-ethnographic accounts can, at best, only sit at the margins of academic research because they are difficult to evaluate in terms of conventional social scientific criteria (i.e. validity, reliability, and objectivity) used to assess qualitative studies. Such criticisms generally centre on 'the problematic nature of self-as-the-only-data-source in autoethnography' (Holt, 2003: 15), but by way of contrast it has been argued that conducted appropriately auto-ethnography involves reflexivity and is not reducible to 'self-absorbed digression' (Anderson, 2006: 385).

Notwithstanding such critical concerns there is a considerable body of literature that endorses the value of a reflexive dimension in observational research. For example, Susan Krieger (1985) argues that successful observation lies in the researcher having a full sense of self, because it is only through a sense of self that an understanding of the members of social settings can be achieved. In short, '[w]e see others as we know ourselves' (Krieger, 1985: 320). In a broadly comparable manner Arnold Arluke and Clinton Sanders (1996: 25) argue that in observational research projects '[r]esearchers gain the richest data by examining their own experiences'. In such instances relatively little, if any, consideration is given to the idea that observers should try to ensure objectivity; rather they are encouraged to look inward to their feelings about the fieldwork. In writing up, observers often report on their involvement with the participants in the observational studies they have conducted (e.g. Krieger, 1985; Whyte, 1955). But observational research does not always engender positive feelings and reflections on the part of researchers. In the course of their research observers frequently have to engage in 'emotion work' (Hochschild, 1983), and the necessity for a degree of such work might give rise to discomfiting differences, if not contradictions, between what the observer believes they should be feeling and what they actually do feel. Some researchers say they 'love' the social setting they are observing (e.g. Fjellman, 1992), while others say they 'hate' it (e.g. Raz, 1999). In turn, other researchers record how they have been transformed in the course of the research process, in some instances directly, in a bodily sense, by the experience of having taken their degree of participation to an extreme (e.g. Wacquant, 2004). In some instances researchers admit to having felt troubled and upset by what they have observed, and others report witnessing sides of their 'selves' that they find distressing (e.g. Arluke and Hafferty, 1996).

Gaining Acceptance

Once access has been secured the success of observational research depends upon the quality of the relationship between the observer and the observed. In participant observation a major issue centres on how to pass as a member/participant and what social and cultural knowledge or common sense understandings are required of the researcher as participant. For instance, unless the observer is already a member of the group being observed they will be forced to confront the problem of being not *of* the world that is under investigation (Humphreys, 1975: 27) and of having to negotiate the boundary of member/stranger (Atkinson, 1990: 157). The researcher must 'fit in', which involves engagement in considerable impression management. Ascribed characteristics, such as gender (e.g. Rogers, 1988) might prohibit the granting of access to and/or acceptance within a research setting. Although granted access to a cargo vessel, Helen Sampson and Michelle Thomas (2003) refer to the way in which their research was affected by the hostility they experienced from male sailors. Observers can often go to great lengths to ensure that they fit in. For his covert study of a Glasgow gang, James Patrick (1973) dressed as a gang member and tried to immerse himself (as much as possible) in their activities. However, his nonappearance at gang fights soon meant that his 'days in the gang were numbered'. Patrick duly lost the trust of the participants in the setting. Comparable problems can arise in overt observational studies insofar as those subject to observation might have little trust in the researcher from the outset. For example, Irene Kaplan (1991) noted that the fishermen she observed initially thought she was a government inspector and were in consequence wary of her. Similarly Samson and Thomas (2003) were often snubbed by the sailors on board the ship because they were suspicious of them, thinking they were spying for the company.

The ways in which researchers collect observations and write up field notes are likely to have an effect on their acceptance by those being observed. In his study of the making of a television documentary series Philip Elliot (1972) decided that virtually everything was potentially relevant and so he traced the programme through the inception of the idea, the recruitment of the team, to the final filming in the studio. This meant he had to take copious notes, which he took openly as he did not want to miss anything. Although this might work for an overt observer, a covert participant observer needs to do things differently. In his ethnography of 'pilfering' in a bakery, Jason Ditton (1977) initially adopted a covert approach but this made note-taking highly impractical, and his regular trips to the toilet – where he made his notes – were eventually noticed and became an issue for his bakery 'colleagues'. In contrast, Whyte (1955) had a logistically much easier time of making research notes because he took on the role of a secretary in the Italian Community Club and this served to legitimate his scribbling things down during club meetings.[3]

The eventual analysis and writing up of observational data can also affect relationships established in the field. For instance, Whyte (1955) had a very good relationship with Doc the gang leader, but Doc felt betrayed when Whyte published his research because he felt let down by the ways that Whyte reported the gang's activities. Similarly, Nell Keddie (1971) was welcomed by the teachers in the school setting in which she did her research, but when she published her work she contradicted the teachers' claims that they did not stream pupils, and she was roundly condemned by the teachers.

Ethics and Risk

As with all social research, there are various ethical issues associated with observational research concerning, for example, personal privacy, anonymity, and informed consent. Furthermore, some forms of observational research may become, in some respects at least, virtually indistinguishable from, or at least may closely resemble, surveillance technologies. Covert research in particular raises a number of critical issues and is a source of considerable continuing debate. Fielding (1981) argued that the only way of gaining data on the National Front was to undertake a covert participant observation approach. The organisation was known for its racist policies and it might be argued that the questionable moral character of participants in the organisation reduced their right to be able to choose whether or not to be participants in the research. The crux of the argument is the possible value to the rest of society of such covert research. Can its value be such that it outweighs the rights of the members of the group whose consent has not been sought? Current practice is that informed consent is usually demanded by governing bodies, professional associations (such as the British Sociological Association, 2002), and, increasingly, publishers.

Even in situations where researchers do not intend to conceal their identity, gaining informed consent can be complicated. For example, it was clearly impossible for Raz (1999) to gain informed consent given the very large number of participants who took part in the tours he used as observational settings for his research in Disneyland in Tokyo. Beyond such practical difficulties, the idea of informed consent itself is potentially problematic as researchers may not be able to give sufficient information to participants to enable them to make an informed decision about their involvement. For example, although David Morgan's observations of female workers in a factory offered anonymity to the participants, the British press subsequently exposed the research site and reinterpreted his results in a way that mocked and trivialised the women's lives (Burgess 1984: 165). The problematic effects on the lives of research subjects are perhaps most obviously apparent in Stanley Milgram's (1974) controversial covert observational study of the extent to which participants were prepared to administer what they (wrongly) believed to be electric shocks to other participants. This had long-term effects on some participants as it led them to confront the ways in which their moral codes could be affected, indeed undermined, by their evident willingness to obey authority.

In online contexts researcher's virtual ethnography generates its own ethical concerns. These concerns cover a range of elements including what constitutes informed consent in a cyberspace context (Paccagnella, 1997). The ethics of virtual ethnography or *netnography* are complex, not least because it is often assumed that postings on the Internet are public knowledge and thus are freely available for research purposes. However, Robert Kozinets (2010) argues that we cannot and should not assume that online individuals and communities are happy for their postings to be used for research purposes. He cites a range of instances where online users were hostile to their postings being accessed for research purposes, including research by James Hudson and Amy Bruckman (2004) in which chat room users are shown to be very disgruntled by the presence of researchers in their chat rooms. Only four of the 766 chat room users subsequently consented to being part of the research project when they were given the opportunity to choose to be involved (Kozinets, 2010: 138).

Ethical issues associated with members have been covered a great deal in the literature on observational techniques, but what have not been considered in such depth are the risks to observers, the dangers faced by some researchers in threatening research settings (Dillman, 1977; Sluka, 1990). In the course of an observational research project conducted in Belfast in Northern Ireland during 'The Troubles', Jeffrey Sluka (1990) was involved in intense communications with guerrilla fighters. Reflecting later on the project he notes that 'it was dangerous just being with them' and concludes that measures should be taken for managing some of the dangers inherent in conducting fieldwork in violent or dangerous social contexts (Sluka, 2012: 286; see also Lee Treweek and Linkogle, 2000). It is important to recognise that sometimes the ethical demands of professionalism themselves can lead to personal losses on the part of the researcher. For example, during his observational research on ecological movements police insisted that Rik Scarce (1999) reveals details of the activists he met. He refused, citing his unwillingness to breach the American Sociological Association code of ethics on anonymity and confidentiality. As a result of that refusal, he was imprisoned for five months.

It is debatable whether contemporary ethical governance frameworks inhibit the potential for participant observation more generally. Problems in this area arise from the fact that university research ethics committees effectively require researchers to be able to specify the precise character of their results in advance of conducting the research. Given the often open-ended form of ethnographic research in particular, this makes these processes particularly tricky for ethnographers to negotiate, and may discourage researchers from proposing or pursuing the use of such methods. Moreover, although critical investigative journalism offers a more appropriate parallel, it is quasi-biomedical ethical structures that have been applied to the social sciences, eliding fundamental differences between the two forms of research and discouraging the proposal of research that does not fit with the prevalent ethical paradigm (Dingwall, 2006; Hammersley, 2009). Ethnographic research, and especially those forms proposed in the context of health-related studies, is undermined by such highly 'anticipatory' forms of regulation (see Murphy and Dingwall, 2007).

Observation, Interpretation, and Understanding

There are additional significant matters concerning the interpretation and presentation of observational results that need to be addressed. These range from description(s) of what has been observed, where processes of interpretation and associated assumptions about meanings in play in social situations are taken-for-granted, to where such meanings are recognised to be more open and themselves become subject to analytic reflection and interrogation. In the latter case the process of description of observations is recognised to be more problematic and inferential and meanings and interpretive processes are engaged with reflexively by the observer-researcher as processes of seeing/observing are analysed as indexical or meaningfully contextualised.

A major criticism of observational methods arises from scepticism about the accuracy of the results. All research methods can be quarry for scepticism over accuracy, but observational methods are amongst those that receive the most criticisms, on the grounds of impressionism, 'subjectivity', or lack of reliability, validity, and generalisability. This means that an important consideration in terms of knowledge generation is the adequate writing, accounting, or 'telling' of any instance of observation, that is, how analysis and writing up are carried out so as to produce particular effects (Stoddart, 1986: 119). Although irony has been considered a 'master trope' of ethnography, another key aim has been the production of 'authenticity', that is, finding a form that will make the observations seem vivid, valid, and 'real' (Atkinson, 1990: 158: 168; also see Adler and Adler, 1998; Dingwall, 1997).

The interpretation of observational data lies in the constructs of both observer and member. In his discussion of anthropology, Kenneth Pike (1954) defines 'emic' constructs as those that centre in member's interpretations (or at least what it feels like to be a member), whereas 'etic' constructs refer to those that are from outside the culture or setting that is being studied. Members should be able to recognise themselves in emic constructs, but would not be able to do so in etic constructs (Arluke and Hafferty, 1996). The fate of any tension between these differing sets of constructs will perhaps be an interesting feature of a research culture, which places an increasing emphasis upon direct engagement with members, and promotes the sharing of results with the groups studied as a way to facilitate 'impact'.

Concluding Remarks

Observational research is wide-ranging, creative, and challenging, and on occasion controversial. The articles included in the various sections in the four volumes reflect the variety of approaches employed in observational research and draw on the range of issues that have been and need to be addressed by researchers. Consideration is given to the origins, development, and refinement of observational methods, as well as aspects of the theoretical and philosophical understandings that underpin observation as a process. Practical matters involved in the stages of preparing for, engaging in, and analysing observations are also addressed. In addition, there is material from classic studies using observational methods, as well as critiques of observation as an approach.

With the increasing proliferation of screens and cameras, particularly following the development of information technology and digitalisation, more and more aspects of people's lives are potentially subject to being observed, in turn, people are spending an increasing amount of time observing via screens (e.g. smart phones, pads, laptops, and desktop PCs). Screens in one form or another increasingly intrude into and/or constitute many aspects of human experience and there have been developments in observational methods that utilise the possibilities afforded by 'observational screens' in creative ways. We have included contributions that address some of these matters. A penultimate section of articles focuses selectively on the ways in which observation methods have been employed in studies of workplaces and workers and the set of volumes conclude with a series of contributions that draw attention to some of the institutional problems and limits to which observational social research has been subject. These articles provide a compelling case for the beneficial insights and understandings to be gained from 'studying up', from observing those all-too-frequently invisible-to-scrutiny and account elite strata and levels of institutions and organisations where, for the most part, power is exercised, responsibility lies, and key decisions are ultimately taken.

Notes

1. See for example the philosopher Jean Marie Degérando's (1969 [1800]) early and uneven attempt to problematise the colonialist gaze, albeit while repeatedly employing the term 'savages'.
2. See for example the overt observational study of a religious group conducted by Eileen Barker (1984) following an invitation to do so and Roger Grimshaw and Tony Jefferson's (1987) nonparticipant and overt observational study of police work.
3. As Emerson, Fretz, and Shaw make clear, field notes do more than simply record observations 'they constitute a way of life through the very writing choices the ethnographer makes and the stories she tells . . . [such writing constitutes] "an interpretive process" that "creates a world"' (1995: 16).

References

Adler, P. A. and Adler, P. (1998). 'Observational Techniques', in N. Denzin and Y. S. Lincoln (Eds.) *Collecting and Interpreting Qualitative Materials*. London: SAGE: 377–391.

Arluke, A. and Hafferty, F. (1996). 'From Apprehension to Fascination with 'Dog Lab': The Use of Absolutions by Medical Students'. *Journal of Contemporary Ethnography* 25: 191–209.

Arluke, A. and Sanders, C. (1996). *Regarding Animals*. Philadelphia: Temple University Press.

Anderson, L. (2006). 'Analytic Autoethnography'. *Journal of Contemporary Ethnography* 35(4): 373–399.

Atkinson, P. (1990). *The Ethnographic Imagination*. London: Routledge.

Barker, E. (1984). *The Making of the Moonie: Brainwashing or Choice*. Oxford: Blackwell.

Berger, J. (1972). *Ways of Seeing*. Harmondsworth: Penguin.

Billig, M. G. (1978). *Fascists: A Social Psychological View of the National Front*. London: Harcourt Brace.

Brewer, J. D. (2003). 'Observation, Overt and Covert', in R. L. Miller and J. D. Brewer (Eds.) *The A-Z of Social Research*. London: SAGE.

British Sociological Association (2002). *Statement of Ethical Practice*, available online at: www.britsoc.co.uk/equality/Statement+Ethical+Practice.htm.

Burgess, R. (1984) *In the Field: An Introduction to Field Research*. London: Unwin Hyman.

Campbell, A. (1984). *The Girls in the Gang*. Oxford: Blackwell.

Chambliss, W. (1975). 'On the Paucity of Original Research on Organized Crime – a Footnote to Galliher and Cain'. *American Sociologist* 10: 36–39.

Conwell, R. H. (1917). *Observation: Every Man His Own University*. New York and London: Harper and Brothers.

Corsaro, W. A. (1981). 'Entering the Child's World: Research Strategies for Field Entry and Data Collection in a Pre-school Setting', in J. L. Green and C. Wallats (Eds.) *Ethnography and Language in Education Setting*. Norwood, NJ: Ablex.

Daston, L. and Lunbeck, A. (2011). 'Introduction: Observation Observed', in L. Daston and A. Lunbeck (Eds.) *Histories of Scientific Observation*. Chicago and London: University of Chicago Press.

Degérando, J.-M. (1969 [1800]). *The Observation of Savage People (Trans. F. G. T. Moore)*. London: Routledge and Kegan Paul.

Dillman, C. M. (1977). 'Ethical Problems in Social Science Research Peculiar to Participant Observation'. *Human Organization* 36(4): 405–407.

Dingwall, R. (1997). 'Accounts, Interviews and Observations', in G. Miller and R. Dingwall (Eds.) *Context and Method in Qualitative Research*. London: Sage: 51–65.

Dingwall, R. (2006). 'Confronting the Anti-Democrats: The Unethical Nature of Ethical Regulation in Social Science'. *Medical Sociology Online* 1: 51–58.

Ditton, J. (1977). *Part-Time Crime: An Ethnography of Fiddling and Pilferage*. London: Macmillan.

Elliot, P. (1972). *The Making of a Television Series*. London: Constable.

Emerson, R. M., Fretz, R. I. and Shaw, L. L. (1995). *Writing Ethnographic Fieldnotes* (2nd edn). Chicago: University of Chicago Press.

Festinger, L., Riecken, N. and Schachter, P. (1956). *When Prophecy Fails*. New York: Harper.

Fielding, N. (1981). *The National Front*. London: Routledge & Kegan Paul.

Fjellman, S. J. (1992). *Vinyl Leaves: Walt Disney World and America*. New York: Westview Press.

Gans, H. J. (1979). *Deciding What's News: A Study of CBS Evening News, NBC Nightly News, Newsweek, and Time*. London: Random House.

Garfinkel, H. (1984 [1967]). *Studies in Ethnomethodology*. Cambridge: Polity.

Geertz, C. (1974). 'From the Native's Point of View': On the Nature of Anthropological Understanding'. *Bulletin of the American Academy of Arts and Sciences* 28: 27–45.

Geertz, C. (1973). *Interpretation of Cultures: Selected Essays*. New York: Basic Books.

Gold, R. L. (1958). 'Roles in Sociological Field Observations'. *Social Forces*, 36(3): 217–223.

Grimshaw, R. and Jefferson, T. (1987). *Interpreting Policework: Policy and Practice in Forms of Beat Policing*. London: Allen and Unwin.

Hammersley, M. and Atkinson, P. (1995). *Ethnography: Principles in Practice*. London: Routledge.

Hammersley, M. (2009) 'Against the Ethicists: On the Evils of Ethical Regulation'. *International Journal of Social Research Methodology* 12(3): 211–225.

Hansen, E. C. (1977). *Rural Catalonia Under the Franco Regime*. Cambridge: Cambridge University Press.

Harastzi, M. (1978). *A Worker in a Worker's State*. Harmondsworth: Penguin.

Herschel, J. (1987 [1830]). *A Preliminary Discourse on the Study of Philosophy*. Chicago: University of Chicago Press.

Hochschild, A. (1983). *The Managed Heart*. Berkeley: California University Press.

Holdaway, S. (1983). *Inside the British Police Force: A Force at Work*. Oxford: Blackwell.

Holt, N. L. (2003). 'Representation, Legitimation, and Autoethnography: An Autoethnographic Writing Story'. *International Journal of Qualitative Methods* 2(1): 1–21.

Hudson, J. M. and Bruckman, A. (2004). '"Go Away". Participant Objections to Being Studied'. *The Information Society* 20(2): 127–139.

Humphreys, L. (1975). *Tea Room Trade (Enlarged Edition)*. New Jersey: Transaction.

Junker, B. H. (1960). *Field Work*. Chicago: Chicago University Press.

Kaplan, I. M. (1991). 'Gone Fishing, Be Back Later', in W. B. Shaffir and R. Stebbins (Eds.) *Experiencing Fieldwork*. London: SAGE: 232–237.

Keddie, N. (1971). 'Classroom Knowledge', in M. Young (Ed.) *Knowledge and Control*. London: Macmillan: 133–160.

Kozinets, R.V. (2010). *Nethnography: Doing Ethnographic Research Online*. London: SAGE.

Krieger, S. (1985). 'Beyond Subjectivity: The Use of the Self in Social Science'. *Qualitative Sociology* 8: 309–324.

Kuhn, T. (1970). *Structure of Scientific Revolutions*. Chicago: University of Chicago Press.

Lamiell, J. T. (1998). '"Nomothetic" and "Idiographic": Contrasting Windelband's Understanding with Contemporary Usage'. *Theory and Psychology* 8(1): 23–38.

Lee, R. M. (2000). *Unobtrusive Methods in Social Research*. Buckingham: Open University Press.

Lee-Treweek, S. and Linkogle, G. (2000). (Eds.) *Danger in the Field*. London: Routledge.

Little, M. (1990). *Young Men in Prison*. Aldershot: Dartmouth.

Luo, M. (2004). '"Excuse Me. May I Have Your Seat?" Revisiting a Social Experiment, and the Fear That Goes with It. *The New York Times*. September 14, 2004.

Maclean, A. M. (1903). 'The Sweat-Shop in Summer'. *American Journal of Sociology* 9: 289–309.

Malinowski, B. (1922). *Argonauts of the Western Pacific*. New York: Dutton & Co.

Mass Observation (1937 onwards) available at: http://www.massobs.org.uk/original_publications.htm.

Maxwell, G. (2009). 'The Ontological Status of Theoretical Entities', in T. McGrew, M. Alspector-Kelly, and F. Allhoff (Eds.) *Philosophy of Science: An Historical Anthology*. Chichester: Wiley Blackwell: 453–457.

Mead, M. (1937). *Coming of Age in Samoa: A Psychological Study of Primitive Youth for Western Civilisation*. New York: American Museum of Natural History.

Milgram, S. (1974). *Obedience to Authority*. New York: Harper and Row.

Murphy, E. and Dingwall, R. (2007). 'Informed Consent, Anticipatory Regulation and Ethnographic Practice'. *Social Science and Medicine* 65(11): 2223–2234.

Murray, S. B. (2003). 'A Spy, a Shill, a Go-Between, or a Sociologist: Unveiling the "Observer" in Participant Observation'. *Qualitative Research* 3(3): 377–395.

Nader, L. (1972). 'Up the Anthropologist – Perspectives Gained from Studying Up', in D. Hymes (Ed.) *Reinventing Anthropology*. New York: Pantheon Books: 284–311.

Newby, H. (1977). *The Deferential Worker*. London: Allen Lane.

Outhwaite, W. (2000). 'The Philosophy of Social Science', in B. Turner (Ed.) *The Blackwell Companion to Social Theory. Oxford*: Blackwell: 47–71.

Paccagnella, L. (1997). 'Getting the Seats of Your Pants Dirty: Strategies for Ethnographic Research on Virtual Communities'. *Journal of Computer-Mediated Communication* 3(1): available at: http://www.ascusc.org/jcmc/vol3/issue1/paccagnella.html.

Page, R. (1971). *Down Among the Dossers*. London: Davis Poynter.

Parker, H. (1974). *View From the Boys*. Newton Abbot: David and Charles.

Patrick, J. (1973). *A Glasgow Gang Observed*. London: Eyre Methuen.

Pearsall, M. (1970). 'Participant Observation as Rote and Method in Behavioral Research'. In W. J. Filstead (Ed.) *Qualitative Methodology: Firsthand Involvement with the Social World*. Chicago: Markham: 340–352.

Pérez-y-Pérez, M. and Stanley, T. (2011). 'Ethnographic Intimacy: Thinking Through the Ethics of Social Research in Sex Worlds'. *Sociological Research Online*, 16(2) available at: http://www.socresonline.org.uk/16/2/13.html retrieved 31/03/12.

Pike, K. (1954). *Language in Relation to a Unified Theory of Structure of Human Behaviour*. The Hague: Mouton.

Polsky, N. (1976 [1971). *Hustlers, Beats and Others*. Harmondsworth: Penguin.

Popper, K. R. (1979). *Objective Knowledge: An Evolutionary Approach* (2nd edn). Oxford: Clarendon Press.

Pryce, K. (1979). *Endless Pressure*. Penguin: Harmondsworth.

Raz, A. E. (1999). *Riding the Black Ship: Japan and Tokyo Disneyland*. Harvard: Harvard University Asia Center.

Reed-Danahay, D. (1997). *Auto/Ethnography*. New York: Berg.

Rogers, B. (1988) *Men Only*. London: Pandora.

Sampson, H. and Thomas, M. (2003). 'Lone Researchers at Sea: Gender, Risk and Responsibility'. *Qualitative Research* 3(2): 165–189.

Sanger, J. (1996). *The Compleat Observer? A Field Research Guide to Observation*. London: Falmer Press.

Scarce, R. (1999). 'Good Faith, Bad Ethics: When Scholars Go the Distance and Scholarly Associations Do Not [Commentary]'. *Law & Social Inquiry* 24(4): 977–986.

Schutz, A. (1972). *The Phenomenology of the Social World*. London: Heinemann.

Shaffir, W. B. (1985). 'Some Reflections on Approaches to Fieldwork in Hassidic Communities'. *Jewish Journal of Sociology* 27(2): 115–134.

Sluka, J. A. (1990). 'Participant Observation in Violent Social Contexts'. *Human Organization* 49(2): 114–126.

Sluka, J. A. (2012). 'Reflections on Managing Danger in Fieldwork: Dangerous Anthropology in Belfast', in A. C. G. M. Robben and J. A. Sluka (Eds.) *Ethnographic Fieldwork: An Anthropological Reader* (2nd edn). Chichester: Wiley Blackwell: 283-296.

Sparkes, A. C. (1996). 'The Fatal Flaw: A Narrative of the Fragile Body-Self'. *Qualitative Inquiry* 2(4): 463–494.

Sparkes, A. C. (2000). 'Autoethnography and Narratives of Self: Reflections on Criteria in Action'. *Sociology of Sport* 17:21–41.

Spicker, P. (2011). 'Ethical Covert Research'. *Sociology* 45(1): 118–133.

Spiro, M. E. (1986). 'Cultural Relativism and the Future of Anthropology'. *Cultural Anthropology* 1(3): 259–286.

Stoddart, K. (1986). 'The Presentation of Everyday Life: Some Textual Strategies for "Adequate Ethnograph"'. *Journal of Contemporary Ethnography* 15(1): 103–121.

Summerfield, P. (1985). 'Mass-Observation: Social Research or Social Movement?'. *Journal of Contemporary History* 20: 439–452.

Tanner, A. (1907). 'Glimpses at the Mind of a Waitress'. *American Journal of Sociology* 13: 48–55.

Thomas, N. (1991). 'Against Ethnography'. *Cultural Anthropology* 6(3): 306–322.

Visweswaran, K. (1997). 'Histories of Feminist Ethnography'. *Annual Review of Anthropology* 26: 591–621.

Wacquant, L. J. D. (2004). *Body and Soul: Notebooks of an Apprentice Boxer*. Oxford: Oxford University Press.

Wall, S. (2006). 'An Autoethnography on Learning about Autoethnography'. *International Journal of Qualitative Methods* 5(2): 1–12.

Westwood, S. (1984). *All Day Every Day: Factory and Family in the Making of Women's Lives*. London: Pluto.

Whyte, W. F. (1955). *Street Corner Society: The Social Structure of an Italian Slum*. Chicago: University of Chicago Press.

Williams, C. (2012). *Researching Power, Elites, and Leadership*. London: Sage.

Willis, P. (1977). *Learning to Labour*. Aldershot: Gower.

Wilson, T. D. (2002). ISIC4 – Fourth International Conference on Information Seeking in Context, Universidade Lusiada, Lisbon, Portugal, September 11 to 13.

Wilson, B. (2006). 'Ethnography, the Internet, Youth Culture: Strategies for Examining Social Resistance and "Online-Offline" Relationships'. *Canadian Journal of Education* 29(1): 307–328.

A. Observation: Philosophy, Science and Art

1

The Bucket and the Searchlight: Two Theories of Knowledge

Karl R. Popper

The purpose of this paper is to criticize a widely held view about the aims and methods of the natural sciences, and to put forward an alternative view.

I

I shall start with a brief exposition of the view I propose to examine, which I will call '*the bucket theory of science*' (or '*the bucket theory of the mind*'). The starting point of this theory is the persuasive doctrine that before we can know or say anything about the world, we must first have had perceptions – sense experiences. It is supposed to follow from this doctrine that our knowledge, our experience, consists either of accumulated perceptions (naïve empiricism) or else of assimilated, sorted, and classified perceptions (a view held by Bacon and, in a more radical form, by Kant).

The Greek atomists had a somewhat primitive notion of this process. They assumed that atoms break loose from the objects we perceive, and penetrate our sense organs, where they become perceptions; and out of these, in the course of time, our knowledge of the external world fits itself together [like a self-assembling jigsaw puzzle]. According to this view, then, our mind resembles a container – a kind of bucket – in which perceptions and knowledge

Source: *Objective Knowledge: An Evolutionary Approach* (Oxford: Oxford University Press, 1979), pp. 341–361.

accumulate. (Bacon speaks of perceptions as 'grapes, ripe and in season' which have to be gathered, patiently and industriously, and from which, if pressed, the pure wine of knowledge will flow.)

Strict empiricists advise us to interfere as little as possible with this process of accumulating knowledge. True knowledge is pure knowledge, uncontaminated by those prejudices which we are only too prone to add to, and mix with, our perceptions; these alone constitute experience pure and simple. The result of these additions, of our disturbing and interfering with the process of accumulating knowledge, is error. Kant opposes this theory: he denies that perceptions are ever pure, and asserts that our experience is the result of a process of assimilation and transformation – the combined product of sense perceptions and of certain ingredients added by our minds. The perceptions are the raw material, as it were, which flows from outside into the bucket, where it undergoes some (automatic) processing – something akin to digestion, or perhaps to systematic classification – in order to be turned in the end into something not so very different from Bacon's 'pure wine of experience'; let us say, perhaps, into fermented wine.

I do not think that either of these views suggests anything like an adequate picture of what I believe to be the actual process of acquiring experience, or the actual method used in research or discovery. Admittedly, Kant's view might be so interpreted that it comes much nearer to my own view than does pure empiricism. I grant, of course, that science is impossible without experience (but the notion of 'experience' has to be carefully considered). Though I grant this, I nevertheless hold that perceptions do not constitute anything like the raw material, as they do according to the 'bucket theory', out of which we construct either 'experience' or 'science'.

II

In science it is *observation* rather than perception which plays the decisive part. But observation is a process in which we play an intensely *active* part. An observation is a perception, but one which is planned and prepared. We do not 'have' an observation [as we may 'have' a sense experience] but we 'make' an observation. [A navigator even 'works' an observation.] An observation is always preceded by a particular interest, a question, or a problem – in short, by something theoretical.[1] After all, we can put every question in the form of a hypothesis or conjecture to which we add: 'Is this so? Yes or no?' Thus we can assert that every observation is preceded by a problem, a hypothesis (or whatever we may call it); at any rate by something that interests us, by something theoretical or speculative. This is why observations are always selective, and why they presuppose something like a principle of selection.

Before elaborating these points any further I shall try to introduce, as a digression, a few remarks of a biological nature. Although these are not meant to constitute a basis or even an argument for the main thesis which I intend to propose later, they may perhaps be helpful in getting over, or in circumventing, certain objections to it, and in this way facilitate later its understanding.

III

We know that all living things, even the most primitive, react to certain stimuli. These reactions are specific; that is to say, for each organism (and for each type of organism) the number of possible reactions is limited. We can say that every organism possesses a certain innate set of possible reactions, or a certain disposition to react in this or that way. This set of dispositions may change with the advancing age of the organism (partly perhaps under the influence of sense impressions or perceptions) or it may remain constant; however this may be, at any instant in the life of the organism it is, we may assume, invested with such a set of possibilities and dispositions to react, and this set constitutes what may be called its [momentary] inner state.

Upon this inner state of the organism will depend how it will react to its external environment. This is why physically identical stimuli may at different times produce different reactions, whilst physically different stimuli may result in identical reactions.[2]

Now we shall say that an organism '*learns from experience*' only if its dispositions to react change in the course of time, and if we have reason to assume that these changes do not depend merely on innate [developmental] changes in the state of the organism but also on the changing state of its external environment. (This is a necessary condition, though not a sufficient one, for saying that the organism learns from experience.) In other words, we shall regard the process by which the organism learns as a certain kind of change, or modification, in its dispositions to react, and not, as would the bucket theory, as an (ordered or classified or associated) accumulation of memory traces, left over by perceptions that are past.

These modifications in the organism's disposition to react, which go to make up the processes of learning, are closely connected with the important notion of an '*expectation*', and also with that of a '*disappointed expectation*'. We may characterize an expectation as a *disposition to react, or as a preparation for a reaction*, which is adapted to [or which anticipates] a state of the environment yet to come about. This characterization seems to be more adequate than one that describes an expectation in terms of states of consciousness; for we become conscious of many of our expectations only when they are disappointed, owing to their being unfulfilled. An example would be the encountering of an unexpected step in one's path: it is the unexpectedness of the step which may make us conscious of the fact that we expected to encounter an even surface. Such disappointments force us to *correct* our system of expectations. The process of learning consists largely in such corrections; that is, in the elimination of certain [disappointed] expectations.

IV

Let us now return to the problem of observation. An observation always presupposes the existence of some system of expectations. These expectations can be formulated in the form of queries; and the observation will be used to obtain either a confirming or a correcting answer to expectations thus formulated.

My thesis that the question, or the hypothesis, must precede the observation may at first have seemed paradoxical; but we can see now that it is not at all paradoxical to assume that expectations – that is, dispositions to react – must precede every observation and, indeed, every perception: for certain dispositions or propensities to react are innate in all organisms whereas perceptions and observations clearly are not innate. And although perceptions and, even more, observations, play an important part in the process of *modifying* our dispositions or propensities to react, some such dispositions or propensities must, of course, be present first, or they could not be modified.

These biological reflections are by no means to be understood as implying my acceptance of a behaviourist position. I do not deny that perceptions, observations, and other states of consciousness occur, but I assign to them a role very different from the one they are supposed to play according to the bucket theory. Nor are these biological reflections to be regarded as forming in any sense an assumption on which my arguments will be based. But I hope that they will help towards a better understanding of these arguments. The same may be said of the following reflections, which are closely connected with these biological ones.

At every instant of our pre-scientific or scientific development we are living in the centre of what I usually call a *'horizon of expectations'*. By this I mean the sum total of our expectations, whether these are subconscious or conscious, or perhaps even explicitly stated in some language. Animals and babies have also their various and different horizons of expectations though no doubt on a lower level of consciousness than, say, a scientist whose horizon of expectations consists to a considerable extent of linguistically formulated theories or hypotheses.

The various horizons of expectations differ, of course, not only in their being more or less conscious, but also in their content. Yet in all these cases the horizon of expectations plays the part of a frame of reference: only their setting in this frame confers meaning or significance on our experiences, actions, and observations.

Observations, more especially, have a very peculiar function within this frame. They can, under certain circumstances, destroy even the frame itself, if they clash with certain of the expectations. In such a case they can have an effect upon our horizon of expectations like a bombshell. This bombshell may force us to reconstruct, or rebuild, our whole horizon of expectations; that is to say, we may have to correct our expectations and fit them together again into something like a consistent whole. We can say that in this way our horizon of expectations is raised to and reconstructed on a higher level, and that we reach in this way a new stage in the evolution of our experience; a stage in which those expectations which have not been hit by the bomb are somehow incorporated into the horizon, while those parts of the horizon which have suffered damage are repaired and rebuilt. This has to be done in such a manner that the damaging observations are no longer felt as disruptive, but are integrated with the rest of our expectations. If we succeed in this rebuilding, then we shall have created what is usually known as an *explanation* of those observed events [which created the disruption, the problem].

As to the question of the temporal relation between observation on the one hand and the horizon of expectations or theories on the other, we may well admit that a new explanation, or a new hypothesis, is generally preceded in time by *those* observations which destroyed the previous horizon of expectations and thus were the stimulus to our attempting a new explanation. Yet this must not be understood as saying that observations generally precede expectations or hypotheses. On the contrary, each observation is preceded by expectations or hypotheses; by those expectations, more especially, which make up the horizon of expectations that lends those observations their significance; only in this way do they attain the status of real observations.

The question, 'What comes first, the hypothesis (*H*) or the observation (*O*)?' reminds one, of course, of that other famous question: 'What came first, the hen (*H*) or the egg (*O*)?' Both questions are soluble. The bucket theory asserts that [just as a primitive form of an egg (*O*), a unicellular organism, precedes the hen (*H*)] observation (*O*) always precedes every hypothesis (*H*); for the bucket theory regards the latter as arising from observations by generalization, or association, or classification. By contrast, we can now say that the hypothesis (or expectation, or theory, or whatever we may call it) precedes the observation, even though an observation that refutes a certain hypothesis may stimulate a new (and therefore a temporally later) hypothesis.

All this applies, more especially, to the formation of scientific hypotheses. For we learn only from our hypotheses what kind of observations we ought to make: whereto we ought to direct our attention; wherein to take an interest. Thus it is the hypothesis which becomes our guide, and which leads us to new observational results.

This is the view which I have called the '*searchlight theory*' (in contradistinction to the '*bucket theory*'), [According to the searchlight theory, observations are secondary to hypotheses.] Observations play, however, an important role as *tests* which a hypothesis must undergo in the course of our [critical] examination of it. If the hypothesis does not pass the examination, if it is falsified by our observations, then we have to look around for a new hypothesis. In this case the new hypothesis will come after those observations which led to the falsification or rejection of the old hypothesis. Yet what made the observations interesting and relevant and what altogether gave rise to our undertaking them in the first instance, was the earlier, the old [and now rejected] hypothesis.

In this way science appears clearly as a straightforward continuation of the pre-scientific repair work on our horizons of expectations. Science never starts from scratch; it can never be described as free from assumptions; for at every instant it presupposes a horizon of expectations – yesterday's horizon of expectations, as it were. Today's science is built upon yesterday's science [and so it is the result of yesterday's searchlight]; and yesterday's science, in turn, is based on the science of the day before. And the oldest scientific theories are built on pre-scientific myths, and these, in their turn, on still older expectations. Ontogenetically (that is, with respect to the development of the individual organism) we thus regress to the state of the expectations of a newborn child; phylogenetically (with respect to the evolution of the race, the phylum) we get to the state of expectations of unicellular organisms. (There is no danger here of a vicious infinite regress – if for no other reason than that every organism is born with *some* horizon of expectations.) There is, as it were, only one step from the amoeba to Einstein.

Now if this is the way science evolves, what can be said to be the characteristic step which marks the transition from pre-science to science?

V

The first beginnings of the evolution of something like a scientific method may be found, approximately at the turn of the sixth and fifth centuries B.C., in ancient Greece. What happened there? What is new in this evolution? How do the new ideas compare with the traditional myths, which came from the East and which, I think, provided many of the decisive suggestions for the new ideas?

Among the Babylonians and the Greeks and also among the Maoris in New Zealand – indeed, it would seem, among all peoples who invent cosmological myths – tales are told which deal with the beginning of things, and which try to understand or explain the structure of the Universe in terms of the story of its origin. These stories become traditional and are preserved in special schools. The tradition is often in the keeping of some separate or chosen class, the priests or medicine men, who guard it jealously. The stories change only little by little – mainly through inaccuracies in handing them on, through misunderstandings, and sometimes through the accretion of new myths, invented by prophets or poets.

Now what is new in Greek philosophy, what is newly added to all this, seems to me to consist not so much in the replacement of the myths by something more 'scientific', as in a *new attitude towards the myths*. That their character then begins to change seems to me to be merely a consequence of this new attitude.

The new attitude I have in mind is *the critical attitude. In the place of a dogmatic handing on of the doctrine* [in which the whole interest lies in the preservation of the authentic tradition] *we find a critical discussion of the doctrine*. Some people begin to ask questions about it; they doubt the trustworthiness of the doctrine: its *truth*.

Doubt and criticism certainly existed before this stage. What is new, however, is that doubt and criticism now become, in their turn, part of the tradition of the school. A tradition of a higher order replaces the traditional preservation of the dogma: in the place of traditional theory – in place of the myth – we find the tradition of criticizing theories (which at first themselves are hardly more than myths). It is only in the course of this critical discussion that observation is called in as a witness.

It can hardly be a mere accident that Anaximander, the disciple of Thales, developed a theory which explicitly and consciously diverges from that of his master, and that Anaximenes, the disciple of Anaximander, diverges just as consciously from his master's doctrine. The only explanation seems to be that the founder of the school himself challenged his disciples to criticize his theory, and that they turned this new critical attitude of their master's into a new tradition.

It is interesting that this happened only once, so far as I know. The earlier Pythagorean school was almost certainly a school of the old kind: its tradition does not embrace the critical attitude but is confined to the task of preserving the doctrine of the master. It was undoubtedly only the influence of the critical school of the Ionians which later loosened the rigidity of the Pythagorean school tradition and so paved the way leading to the philosophical and scientific method of criticism.

The critical attitude of ancient Greek philosophy can hardly be better exemplified than by the famous lines of Xenophanes:

> Yet if cattle or horses or lions had hands and could draw
> And could sculpture like men, then the horses would draw their gods
> Like horses; and cattle like cattle; and each would then shape
> Bodies of gods in the likeness, each kind, of its own.

This is not only a critical challenge – it is a statement made in the full consciousness and mastery of a critical methodology.

Thus it seems to me that it is the tradition of criticism which constitutes what is new in science, and what is characteristic of science. On the other hand it seems to me that the task which science sets itself [that is, the explanation of the world] and the main ideas which it uses, are taken over without any break from prescientific mythmaking.

VI

What is the task of science? With this question I have ended my preliminary examination of biological and historical trends, and I now come to the logical analysis of science itself.

The task of science is partly theoretical – *explanation* – and partly practical – *prediction and technical application*. I shall try to show that these two aims are, in a way, two different aspects of one and the same activity.

I will first examine the idea of an explanation.

One often hears it said that an explanation is the reduction of the unknown to the known; but we are rarely told how this is to be done. At any rate, this notion of explanation is not one that was ever used in the actual practice of explanation in science. If we look at the history of science in order to see what kinds of explanation were used and accepted as satisfactory at one time or another, then we find a very different notion of explanation in practical use.

I gave a short sketch of this history (I do not mean the history of the concept of explanation but the history of the practice of explanation) to the philosophical seminar this morning.[3] Unfortunately, time prevents me from dealing with this question here again at length. Yet I should mention here one general result. In the course of the historical development of science many different methods and kinds of explanation have been regarded as acceptable; but they all have one aspect in common: the various methods of explanation all consist of a *logical deduction*; a deduction whose conclusion is the *explicandum* – a statement of the thing to be explained – and whose premisses consist of the *explicans* [a statement of the explaining laws and conditions]. The main changes that have occurred in the course of the history of science consist in the silent abandonment of certain implicit demands regarding the character of the *explicans* (that it can be intuitively grasped, that it is to be self-evident, etc.); demands which turn out not to be reconcilable with certain other demands whose crucial significance becomes more and more obvious as time goes on; in particular the demand for the independent testability of the *explicans* [which forms the premisses and thus the very heart of the explanation].

Thus an explanation is always the deduction of the *explicandum* from certain premisses, to be called the *explicans*.

Here is a somewhat gruesome example, just for the purpose of illustration.[4]

A dead rat has been discovered and we wish to know what has happened to it. The *explicandum* may be stated thus: 'This rat here has died recently.' This *explicandum* is definitely known to us – the fact lies before us in stark reality. If we want to explain it, we must try out some conjectural or hypothetical explanations (as the authors of detective stories do); that is to say, explanations which introduce something *unknown*, or at any rate much less known, to us. Such a hypothesis may be, for instance, that the rat died of a large dose of rat poison. This is useful as a hypothesis in so far as, firstly, it helps us to formulate an *explicans* from which the *explicandum* can be deduced; secondly, it suggests to us a number of independent tests – tests of the *explicans* which are quite independent of whether the *explicandum* is true or not.

Now the *explicans* – which is our hypothesis – does not only consist of the sentence 'This rat has eaten some bait containing a large dose of rat poison', for from this statement alone one cannot validly deduce the *explicandum*. Rather, we shall have to use, as *explicans*, two different kinds of premisses – *universal laws* and *initial conditions*. In our case the universal law might be put like this: 'If a rat eats at least eight grains of rat poison it will die within five minutes.' The (singular) initial condition (which is a singular statement) might be: 'This rat ate at least eighteen grains of rat poison, more than five minutes ago.' From these two premisses together we may now indeed deduce that this rat recently has died [that is, our *explicandum*].

Now all this may seem somewhat obvious. But consider one of my theses – the thesis, namely, that what I have called the '*initial conditions*' [the conditions pertaining to the individual case] never suffice by themselves as an explanation, and that we always need a general law as well. Now this thesis is by no means obvious; on the contrary, its truth is often not admitted. I even suspect that most of you would be inclined to accept a remark like 'this rat has eaten rat poison' as quite sufficient to explain its death, even if no explicit statement of the universal law regarding the effects of rat poison is added. But suppose for a moment that we were living in a world in which anybody (and also any rat) who eats a lot of that chemical called 'rat poison' will feel especially well and happy for a week to come and more lively than ever before. If a universal law like this were valid, could the statement 'This rat has eaten rat poison' still be acceptable as an explanation of death? Obviously not.

Thus we have reached the important result, often overlooked, that any explanation that utilizes the singular initial conditions alone would be incomplete, and that *at least one universal law* is needed besides, even though this law is, in some cases, so well known that it is omitted as if it were redundant.

To sum up this point. We have found that an explanation is a deduction of the following kind:

U	(Universal Law)	Premisses
I	(Specific Initial Conditions)	(constituting the *Explicans*)
$\bar{E}$	[*Explicandum*)	Conclusion

VII

But are all explanations of this structure *satisfactory*? Is, for instance, our example (which explains the death of the rat by reference to rat poison) a satisfactory explanation? We do not know: the tests may show that whatever the rat may have died of, it was not rat poison.

If some friend should be sceptical of our explanation and ask, 'How do you know that this rat ate poison?', it will obviously not be sufficient to answer, 'How can you doubt it, seeing that it is dead?'. Indeed, any reason which we may state in support of any hypothesis must be other than, and independent of, the *explicandum*. If we can only adduce the *explicandum* itself as evidence, we feel that our explanation is circular, and therefore quite *unsatisfactory*. If, on the other hand, we are able to reply, 'Analyse the contents of its stomach, and you will find a lot of poison', and if this prediction (which is new – that is, not entailed by the *explicandum* alone) proves true, we shall at least consider our explanation a fairly good hypothesis.

But I have to add something. For our sceptical friend may also question the truth of the universal law. He may say, for instance, 'Granted, this rat has eaten a certain chemical; but why should it have died of it?'. Again, we must not answer: 'But don't you see it is dead? That just shows you how dangerous it is to eat this chemical.' For this again would make our explanation circular and unsatisfactory. In order to make it satisfactory we should have to submit the universal law to test cases which are independent of our *explicandum*.

With this, my analysis of the formal schema of explanation may be regarded as concluded, but I shall add some further remarks and analyses to the general schema I have outlined.

First, an observation about the ideas of cause and effect. The state of affairs described by the singular *initial conditions* can be called the '*cause*', and the one described by the *explicandum* the '*effect*'. I feel, however, that these terms, encumbered as they are with associations from their history, are better avoided. If we still want to use them, we should always remember that they acquire a meaning only relative to a theory or a universal law. It is the theory or the law which constitutes the *logical link* between cause and effect, and the statement '*A* is the cause of *B*' should be analysed thus: 'There is a theory *T* which can be, and has been, independently tested, and from which, in conjunction with an independently tested description *A*, of a specific situation, we can logically deduce a description, *B*, of another specific situation.' (That the existence of such a *logical* link between 'cause' and 'effect' is presupposed in the very use of these terms has been overlooked by many philosophers, including Hume.)[5]

VIII

The task of science is not confined to searching for purely theoretical explanations; it also has its practical sides: prediction-making as well as technical applications. Both of these can be analysed by means of the same logical schema which we introduced to analyse explanation.

(1) *The derivation of predictions*. Whereas in the search for an explanation the *explicandum* is given – or known – and a suitable *explicans* has to be found, the derivation of predictions proceeds in the opposite direction. Here the theory is given, or assumed to be known (perhaps from textbooks), and so are the specific initial conditions (they are known, or assumed to be known, by observation). What remain to be found are the logical consequences: certain logical conclusions which are not yet known to us from observation. These are the *predictions*. In this case, the prediction *P* takes the place of the *explicandum E* in our logical schema.

(2) *Technical application*. Consider the task of building a bridge which has to comply with certain practical requirements, laid down in a list of specifications. What we are given are the specifications, *S*, which describe a certain required state of affairs – the bridge to be built. [*S* are the customer's specifications, which are given prior to, and are distinct from, the architect's specifications.) We are given, further, the relevant physical theories (including certain rules of thumb). What are to be found are certain initial conditions which may be realized technically and which are of such a nature that the specifications may be deduced from them, together with the theory. So in this case, *S* takes the place of *E* in our logical schema.[6]

This makes it clear how, from a logical point of view, both the derivation of predictions and the technical application of scientific theories may be regarded as mere inversions of the basic schema of scientific explanation.

The use of our schema, however, is still not exhausted: it may also serve to analyse the *procedure of testing our explicans*. The testing procedure consists in the derivation from the explicans of a prediction, *P*, and in comparing it with an actual, observable, situation. If a prediction does not agree with the observed situation, then the *explicans* is shown to be false; it is falsified. In this case we still do not know whether it is the universal *theory* which is false, or whether the *initial conditions* describe a situation which does not correspond with the real situation – so that the initial conditions are false. [Of course, it may well be that the theory *and* the initial conditions are false.]

The falsification of the prediction shows that the *explicans* is false, yet the reverse of this does not hold: it is incorrect and grossly misleading to think that we can interpret the 'verification' of the prediction as 'verifying' the *explicans* or even a part of it. For a true prediction may easily have been validly deduced from an *explicans* that is false. It is even quite misleading to regard *every* 'verification' of a prediction as something like a practical *corroboration* of the *explicans*: it would be more correct to say that only such 'verifications' of predictions which are 'unexpected' [without the theory under examination] may be regarded as corroborations of the *explicans*, and so of the theory. This means that a prediction can be used to corroborate a theory only if its comparison with observations might be regarded as a serious attempt at testing the *explicans* – a serious attempt at refuting it. A ['risky'] prediction of this kind may be called 'relevant to a test of the theory'.[7] After all, it is fairly obvious that the passing of an examination can give an idea of the qualities of the student only if the examination which he passes is sufficiently severe, and that an examination can be designed which even the weakest student will pass easily.[8]

In addition to all this, our logical scheme permits us, finally, to analyse the difference between the tasks of a *theoretical* and of a *historical* explanation.

The theoretician is interested in finding, and testing, universal laws. In the course of testing them he uses other laws, of the most diverse kinds (many of them quite unconsciously) as well as diverse specific initial conditions.

The historian on the other hand, is interested in finding descriptions of states of affairs in certain finite, specific spatio-temporal regions – that is to say, what I have called specific initial conditions – and in testing or checking their adequacy or accuracy. In this kind of testing he uses, in addition to other specific initial conditions, universal laws of all kinds – usually rather obvious ones – which belong to his horizon of expectations, though, as a rule, he is not conscious that he uses them. In this he resembles the theoretician. [Their difference, however, is very marked: it lies in the difference between their various interests, or problems: in the difference of what each regards as problematic.]

In a logical schema [similar to our previous ones] the procedure of the theoretician may be represented in the following manner:

$$\begin{array}{cccc} U_0 & U_0 & U_0 & \dots \\ U_1 & U_2 & U_3 & \dots \\ \underline{I_1} & \underline{I_2} & \underline{I_3} & \dots \\ P_1 & P_2 & P_3 & \dots \end{array}$$

U_0 is here the universal law, the universal hypothesis, which is under examination. It is kept constant throughout the tests, and used, together with various other laws U_1, U_2, . . . and various other initial conditions I_1, I_2, . . . in order to derive various predictions P_1, P_2, . . . which may then be compared with observable actual facts.

The procedure of the historian may be represented by the following schema:

$$\begin{array}{cccc} U_1 & U_2 & U_3 & \dots \\ I_1 & I_2 & I_3 & \dots \\ \underline{I_0} & \underline{I_0} & \underline{I_0} & \dots \\ P_1 & P_2 & P_3 & \dots \end{array}$$

Here, I_0 is the historical hypothesis, the historical description, which is to be examined or tested. It is kept constant throughout the tests; and it is combined with various (mostly obvious) laws, U_1, U_2, . . . and with corresponding initial conditions, I_1, I_2, . . . for deriving various predictions, P_1, P_2, etc.

Both our schemata are, of course, highly idealized and oversimplified.

IX

Earlier I have tried to show that an explanation will be *satisfactory* only if its universal laws, its theory, can be tested independently of the *explicandum*. But this means that any satisfactory explanatory theory must always assert *more* than what was already contained in the *explicanda* which originally led us to put it forward. In other words, satisfactory theories must, as a matter of principle, transcend the empirical instances which gave rise to them; otherwise they would, as we have seen, merely lead to explanations which are circular.

Here we have a methodological principle which stands in direct contradiction to all positivistic and naïvely empiricist [or inductivist] tendencies. It is a principle which demands that we should dare to put forward bold hypotheses that open up, if possible, new domains of observations, rather than those careful generalizations from 'given' observations which have remained [ever since Bacon] the idols of all naïve empiricists.

Our view that it is the task of science to put forward explanations, or (what leads essentially to the same logical situation)[9] to create the theoretical foundations for predictions and other applications – this view has led us to the methodological demand that our theories should be testable. Yet there are *degrees of testability*. Some theories are *better* testable than others. If we strengthen our methodological demand and aim at *better and better testable* theories, then we arrive at a methodological principle – or a statement of the task of science – whose [unconscious] adoption in the past would rationally explain a great number of events in the history of science: it would explain them as steps towards carrying out the task of science. (At the same time it gives us a statement of the task of science, telling us what should in science be regarded as *progress*; for in contrast to most other human activities – art and music in particular – there really is, in science, such a thing as progress.)

An analysis and comparison of the degrees of testability of different theories shows that the testability of a theory grows with its *degree of universality* as well as with its *degree of definiteness, or precision*.

The situation is fairly simple. Along with the degree of universality of a theory goes an increase in the range of those events about which the theory can make predictions and thereby also the domain of possible falsifications. But a theory which is more easily falsified is at the same time one that is better testable.

We find a similar situation if we consider the degree of definiteness or precision. A precise statement can be more easily refuted than a vague one, and it can therefore be better tested. This consideration also allows us to explain the demand that qualitative statements should if possible be replaced by quantitative ones by our principle of increasing the degree of testability of our theories. (In this way we can also explain the part played by *measurement* in the testing of theories; it is a device which becomes increasingly important in the course of scientific progress, but which should not be used [as it often is] as a characterizing feature of science, or the formation of theories, in general. For we must not overlook the fact that measuring procedures began to be used only at a fairly late stage in the development of some of the sciences, and that they are even now not used in all of them; and we must also not overlook the fact that all measurement is dependent on theoretical assumptions.)

X

A good example from the history of science that may be used to illustrate my analysis is the transition from the theories of Kepler and Galileo to the theory of Newton.

That this transition has nothing whatever to do with induction, and that Newton's theory cannot be regarded as anything like a generalization of those two earlier theories may be seen from the undeniable [and important] fact that Newton's theory *contradicts* them. *Thus Kepler's laws cannot be deduced from Newton's* [although it has been often asserted that they can be so deduced, and even that Newton's can be deduced from Kepler's]: Kepler's laws can be obtained from Newton's only *approximately*, by making the [false] assumption that the masses of the various planets are negligible compared with the mass of the sun. Similarly, Galileo's law of free falling bodies cannot be deduced from Newton's theory: on the contrary, it contradicts it. Only by making the [false] assumption that the total length of all falls is negligible compared with the length of the radius of the earth can we obtain Galileo's law *approximately* from Newton's theory.

This shows, of course, that Newton's theory cannot be a generalization obtained by induction [or deduction] but that it is a new hypothesis which can irradiate the way to the falsification of the old theories: it can irradiate, and point the way to those domains in which, according to the new theory, the old theories fail to yield good approximations. (In Kepler's case this is the domain of the theory of perturbations, and in Galileo's case it is the theory of variable accelerations, since according to Newton gravitational accelerations vary inversely with the square of the distance.)

Had Newton's theory achieved no more than the union of Kepler's laws with Galileo's, it would have been only a *circular explanation of these laws* and therefore unsatisfactory as an explanation. Yet its power of illumination and its power of convincing people consisted just in its power to throw light on the way to independent tests, leading us to [successful] predictions which were incompatible with the two older theories. It was the way to new empirical discoveries.

Newton's theory is an example of an attempt to explain certain older theories of a lower degree of universality, which not only leads to a kind of unification of these older theories but at the same time to their falsification (and so to their correction by restricting or determining the domain within which they are, in good approximation, valid).[10] A case which occurs perhaps more often is this: an old theory is first falsified; and the new theory arises later, as an attempt to explain the partial success of the old theory as well as its failure.

XI

In connection with my analysis of the notion (or rather the practice) of explanation a further point seems significant. From Descartes [and perhaps even from Copernicus] to Maxwell, most physicists tried to explain all newly discovered relations by means of *mechanical models*; that is to say, they tried to reduce them to laws of push or pressure with which we are acquainted from handling everyday physical things – things belonging to the realm of 'physical bodies of medium size'. Descartes made this into a kind of programme for all the sciences; he even demanded that we should restrict ourselves to models that work merely by push or pressure. This programme suffered its first defeat with the success of Newton's theory; but this defeat (which was a serious affliction to Newton and his generation) was soon forgotten, and gravitational attraction was admitted into the programme on equal terms with push and pressure. Maxwell, too, first tried to develop his theory of the electromagnetic field in the form of a mechanical model of the ether; but in the end he gave up the attempt. With this, the mechanical model lost most of its significance: only the equations which originally were meant to describe the mechanical model of the ether remained. [They were interpreted as describing certain non-mechanical properties of the ether.]

With this transition from a mechanical to an *abstract theory* a stage is reached in the evolution of science at which in practice no more is demanded of explanatory theories than that they can be tested independently; we are ready to work with theories which can be intuitively represented by diagrams such as pictures [or by 'picturable' or 'visualizable' mechanical models], if they are obtainable: this yields 'concrete' theories; or else, if these are not obtainable, we are ready to work with 'abstract' mathematical theories [which, however, may be quite 'understandable' in a sense I have analysed elsewhere].[11]

Our general analysis of the notion of explanation is of course unaffected by the failures of any particular picture or model. It applies to all kinds of abstract theories in the same manner as it applies to mechanical and other models. In fact, models are, from our point of view, nothing but attempts to explain new laws in terms of old laws which have already been tested [together with assumptions about typical initial conditions, or the occurrence of a typical structure – that is to say, the model in a narrower sense]. Models often play important parts in the extension and elaboration of theories; but it is necessary to distinguish a new model in a setting of old theoretical assumptions from a new theory – that is, from a new system of theoretical assumptions.

XII

I hope that some of my formulations which at the beginning of this lecture may have seemed to you far-fetched or even paradoxical will now appear less so.

There is no road, royal or otherwise, which leads of necessity from a 'given' set of specific facts to any universal law. What we call 'laws' are hypotheses or conjectures which always form a part of some larger system of theories [in fact, of a whole horizon of expectations] and which, therefore, can never be tested in isolation. The progress of science consists in trials, in the elimination of errors, and in further trials guided by the experience acquired in the course of previous trials and errors. No particular theory may ever be regarded as absolutely certain: every theory may become problematical, no matter how well corroborated it may seem now. No scientific theory is sacrosanct or beyond criticism. This fact has often been forgotten, particularly during the last century, when we were impressed by the often repeated and truly magnificent corroborations of certain mechanical theories, which eventually came to be regarded as indubitably true. The stormy development of physics since the turn of the century has taught us better; and we have now come to see that it is the task of the scientist to subject his theory to ever new tests, and that no theory must be pronounced final. Testing proceeds by taking the theory to be tested and combining it with all possible kinds of initial conditions as well as with other theories, and then comparing the resulting predictions with reality. If this leads to disappointed expectations, to refutations, then we have to rebuild our theory.

The disappointment of some of the expectations with which we once eagerly approached reality plays a most significant part in this procedure. It may be compared with the experience of a blind man who touches, or runs into, an obstacle, and so becomes aware of its existence. *It is through the falsification of our suppositions that we actually get in touch with 'reality'.* It is the discovery and elimination of our errors which alone constitute that 'positive' experience which we gain from reality.

It is of course always possible to save a falsified theory by means of supplementary hypotheses [like those of epicycles]. But this is not the way of progress in the sciences. The proper reaction to falsification is to search for new theories which seem likely to offer us a better grasp of the facts. Science is not interested in having the last word if this means shutting off our minds from falsifying experiences, but rather in learning from our experience; that is, in learning from our mistakes.

There is a way of formulating scientific theories which points with particular clarity to the possibility of their falsification: we can formulate them in the form of prohibitions [or *negative existential statements*] such as, for example, 'There does not exist a closed physical system, such that energy changes in one part of it without compensating changes occurring in another part' (first law of thermodynamics). Or, 'There does not exist a machine which is 100 per cent efficient' (second law). It can be shown that universal statements and negative existential statements are logically equivalent. This makes it possible to formulate all universal laws in the manner indicated; that is to say, as prohibitions. However, these are prohibitions intended only for the technicians and not for the scientist. They tell the former how to proceed if he does not want to squander his energies. But to the scientist they are a challenge to test and to falsify; they stimulate him to try to discover those states of affairs whose existence they prohibit, or deny.

Thus we have reached a point from which we can see science as a magnificent adventure of the human spirit. It is the invention of ever new theories, and the indefatigable examination of their power to throw light on experience. The principles of scientific progress are very simple. They demand that we give up the ancient idea that we may attain certainty [or even a high degree of 'probability' in the sense of the probability calculus] with the propositions and theories of science (an idea which derives from the association of science with magic and of the scientist with the magician): the aim of the scientist is not to discover absolute certainty, but to discover better and better theories [or to invent more and more powerful searchlights] capable of being put to more and more severe tests [and thereby leading us to, and illuminating for us, ever new experiences]. But this means that these theories must be falsifiable: it is through their falsification that science progresses.

Notes

1. By the word 'theoretical' I do not mean here the opposite of 'practical' (since our interest might very well be a practical one); it should rather be understood in the sense of 'speculative' [as with a speculative interest in a pre-existing problem] in contrast to 'perceptive'; or 'rational' as opposed to 'sensual'.

2. Compare F. A. von Hayek, 'Scientism and the Study of Society', *Economica*, N.S. 9, 10, and 11 (1942, 1943, and 1944); [now also in his *The Counter-Revolution of Science*, 1952].
3. (Added in the translation.) Part of the fuller story will be found (though somewhat condensed and with reduced emphasis upon what has been in actual practice accepted as an explanation) in my Venice lecture: 'Philosophy and Physics: Theories of the Structure of Matter', *Atti del XII Congresso Internazionale di Filosofia*, 2, Sansoni, Florence, 1961, pp. 367–74. Other parts are to be found in the first half of my *Conjectures and Refutations*, especially chapters 6, 3, and 4. (This last chapter will be found to overlap with, and expand, some parts of the present lecture.)
4. I have made the example slightly less gruesome in the translation.
5. (Added in translation.) I made these comments on the notions 'cause' and 'effect' first in section 12 of my *Logik der Forschung* (*The Logic of Scientific Discovery*). See also my *Poverty of Historicism*, pp. 122 f.; my *Open Society and Its Enemies*, especially note 9 to chapter 25; and 'What can Logic do for Philosophy?', *Aristotelian Society, Supplementary Volume*, 22, 1948, pp. 145 ff.
6. (Added in the translation.) This analysis must not be interpreted as implying that the technologist or the engineer is concerned only with '*applying*' theories which are supplied by the pure scientist. On the contrary, the technologist and the engineer are constantly faced with *problems to be solved*. These problems are of various degrees of abstraction, but are usually, at least in part, of a theoretical character; and in trying to solve them, the technologist or engineer uses, like everybody else, the method of conjecture, or trial, and testing, or refutation, or error elimination. This is well explained on p. 43 of J. T. Davies, *The Scientific Approach*, 1965, a book in which many good applications and illustrations of the searchlight theory of science can be found.
7. A relevant prediction corresponds, in a certain sense, to an acid test, or to an '*experimentum crucis*'; for in order that a prediction P may he relevant to a test of a theory T, it must be possible to state a prediction P' which does not contradict the initial condition and the remainder of our horizon of expectations for the time being, other than T (assumptions, theories, etc.), and which, combined with the initial conditions and the remainder of the horizon of expectations, contradicts P. This is what is meant if we say that $P(= E)$ ought to be (without T) 'unexpected'.
8. Experienced examiners will feel that the word 'easily' is somewhat unrealistic. As the President of a Governmental Board of Examiners in Vienna sometimes said musingly: 'If a student, in answering the examination question "How much is 5 plus 7?" puts down "eighteen", then we give him a pass. But if he answers "green", I then sometimes think afterwards that we really ought to have ploughed him.'
9. (Added in the translation.) I have in later years (from 1950 on) made a sharper distinction between the theoretical or explanatory and the practical or 'instrumental' tasks of science, and I have stressed the logical priority of the theoretical task over the instrumental task. I have tried to stress, more especially, that predictions have not only an instrumental aspect, but also, and mainly, a theoretical one, as they play a decisive role in testing a theory (as shown earlier in the present lecture). See my *Conjectures and Refutations*, especially chapter 3.
10. (Added in the translation.) The incompatibility of Newton's theory with that of Kepler was stressed by Pierre Duhem, who wrote of Newton's '*principles of universal gravity*' that it is '*very far from being derivable by generalisation and induction from the observational laws of Kepler*' in that it '*formally contradicts these laws. If Newton's theory is correct, Kepler's laws are necessarily false.*' (The quotation is from p. 193 of P. P. Wiener's translation of Duhem's *The Aim and Structure of Physical Theory*, 1954. The term '*observational*' applied here to the '*laws of Kepler*' should be taken with a good grain of salt: Kepler's laws were wild conjectures, just as much as Newton's were:

they cannot be induced from Tycho's observations – no more than Newton's can from Kapler's laws.) Duhem's analysis is based on the fact that our solar system contains *many* heavy planets for whose mutual attraction allowance has to be made in accordance with Newton's theory of perturbation. We can, however, go beyond Duhem: even if we take Kepler's laws as holding for a set of *two-body systems*, each of them containing a central body of the mass of the sun and *one* planet (of varying mass and distance in the various different systems belonging to the set), even then Kepler's third laws fails if Newton's laws are true, as I have shown briefly in *Conjectures and Refutations*, note 28 to chapter 1 (p. 62) and in some detail in my paper 'The Aim of Science', (1957), now Chapter 5 of the present volume, also in *Theorie und Realität*, edited by Hans Albert, 1964, chapter 1, pp. 73 ff., especially pp. 82 f. In this paper I say a little more about explanations which *correct their (apparently 'known' or 'given') explicanda while explaining them approximately*. This is a view which I have developed fairly fully in my lectures since 1940 (first in a course of lectures given to the Christchurch branch of the Royal Society of New Zealand: cp. the footnote on pp. 134 f. of my *Poverty of Historicism*).

11. (Added in the translation.) A fuller analysis of 'understanding' is given in Chapter 4 of the present volume.

2

Revolutions as Changes of World View

Thomas Kuhn

Examining the record of past research from the vantage of contemporary historiography, the historian of science may be tempted to exclaim that when paradigms change, the world itself changes with them. Led by a new paradigm, scientists adopt new instruments and look in new places. Even more important, during revolutions scientists see new and different things when looking with familiar instruments in places they have looked before. It is rather as if the professional community had been suddenly transported to another planet where familiar objects are seen in a different light and are joined by unfamiliar ones as well. Of course, nothing of quite that sort does occur: there is no geographical transplantation; outside the laboratory everyday affairs usually continue as before. Nevertheless, paradigm changes do cause scientists to see the world of their research-engagement differently. In so far as their only recourse to that world is through what they see and do, we may want to say that after a revolution scientists are responding to a different world.

It is as elementary prototypes for these transformations of the scientist's world that the familiar demonstrations of a switch in visual gestalt prove so suggestive. What were ducks in the scientist's world before the revolution are rabbits afterwards. The man who first saw the exterior of the box from above later sees its interior from below. Transformations like these, though usually more gradual and almost always irreversible, are common concomitants of scientific training. Looking at a contour map, the student sees lines on paper, the cartographer a picture of a terrain. Looking at a bubble-chamber photograph, the student sees confused and broken lines, the physicist a

---

Source: *The Structure of Scientific Revolutions* (3rd Edition) (Chicago: The University of Chicago Press, 1996), pp. 111–135.

record of familiar subnuclear events. Only after a number of such transformations of vision does the student become an inhabitant of the scientist's world, seeing what the scientist sees and responding as the scientist does. The world that the student then enters is not, however, fixed once and for all by the nature of the environment, on the one hand, and of science, on the other. Rather, it is determined jointly by the environment and the particular normal-scientific tradition that the student has been trained to pursue. Therefore, at times of revolution, when the normal-scientific tradition changes, the scientist's perception of his environment must be re-educated – in some familiar situations he must learn to see a new gestalt. After he has done so the world of his research will seem, here and there, incommensurable with the one he had inhabited before. That is another reason why schools guided by different paradigms are always slightly at cross-purposes.

In their most usual form, of course, gestalt experiments illustrate only the nature of perceptual transformations. They tell us nothing about the role of paradigms or of previously assimilated experience in the process of perception. But on that point there is a rich body of psychological literature, much of it stemming from the pioneering work of the Hanover Institute. An experimental subject who puts on goggles fitted with inverting lenses initially sees the entire world upside down. At the start his perceptual apparatus functions as it had been trained to function in the absence of the goggles, and the result is extreme disorientation, an acute personal crisis. But after the subject has begun to learn to deal with his new world, his entire visual field flips over, usually after an intervening period in which vision is simply confused. Thereafter, objects are again seen as they had been before the goggles were put on. The assimilation of a previously anomalous visual field has reacted upon and changed the field itself.[1] Literally as well as metaphorically, the man accustomed to inverting lenses has undergone a revolutionary transformation of vision.

The subjects of the anomalous playing-card experiment discussed in Section VI experienced a quite similar transformation. Until taught by prolonged exposure that the universe contained anomalous cards, they saw only the types of cards for which previous experience had equipped them. Yet once experience had provided the requisite additional categories, they were able to see all anomalous cards on the first inspection long enough to permit any identification at all. Still other experiments demonstrate that the perceived size, color, and so on, of experimentally displayed objects also varies with the subject's previous training and experience.[2] Surveying the rich experimental literature from which these examples are drawn makes one suspect that something like a paradigm is prerequisite to perception itself. What a man sees depends both upon what he looks at and also upon what his previous visual-conceptual experience has taught him to see. In the absence of such training there can only be, in William James's phrase, "a bloomin' buzzin' confusion."

In recent years several of those concerned with the history of science have found the sorts of experiments described above immensely suggestive. N. R. Hanson, in particular, has used gestalt demonstrations to elaborate some of the same consequences of scientific belief that concern me here.[3] Other colleagues have repeatedly noted that history of science would make better and more coherent sense if one could suppose that scientists occasionally experienced shifts of perception like those described above. Yet, though psychological experiments are suggestive, they cannot, in the nature of the case, be more than that. They do display characteristics of perception that *could* be central to scientific development, but they do not demonstrate that the careful and controlled observation exercised by the research scientist at all partakes of those characteristics. Furthermore, the very nature of these experiments makes any direct demonstration of that point impossible. If historical example is to make these psychological experiments seem relevant, we must first notice the sorts of evidence that we may and may not expect history to provide.

The subject of a gestalt demonstration knows that his perception has shifted because he can make it shift back and forth repeatedly while he holds the same book or piece of paper in his hands. Aware that nothing in his environment has changed, he directs his attention increasingly not to the figure (duck or rabbit) but to the lines on the paper he is looking at. Ultimately he may even learn to see those lines without seeing either of the figures, and he may then say (what he could not legitimately have said earlier) that it is these lines that he really sees but that he sees them alternately *as* a duck and *as* a rabbit. By the same token, the subject of the anomalous card experiment knows (or, more accurately, can be persuaded) that his perception must have shifted because an external authority, the experimenter, assures him that regardless of what he *saw*, he was *looking at* a black five of hearts all the time. In both these cases, as in all similar psychological experiments, the effectiveness of the demonstration depends upon its being analyzable in this way. Unless there were an external standard with respect to which a switch of vision could be demonstrated, no conclusion about alternate perceptual possibilities could be drawn.

With scientific observation, however, the situation is exactly reversed. The scientist can have no recourse above or beyond what he sees with his eyes and instruments. If there were some higher authority by recourse to which his vision might be shown to have shifted, then that authority would itself become the source of his data, and the behavior of his vision would become a source of problems (as that of the experimental subject is for the psychologist). The same sorts of problems would arise if the scientist could switch back and forth like the subject of the gestalt experiments. The period during which light was "sometimes a wave and sometimes a particle" was a period of crisis – a period when something was wrong – and it ended only with the development of wave mechanics and the realization that light was a self-consistent entity different from both waves and particles. In the sciences, therefore, if perceptual switches accompany paradigm changes, we may not expect scientists to attest to these changes directly. Looking at the moon, the convert to Copernicanism does not say, "I used to see a planet, but now I see a satellite." That locution would imply a sense in which the Ptolemaic system had once been correct. Instead, a convert to the new astronomy says, "I once took the moon to be (or saw the moon as) a planet, but I was mistaken." That sort of statement does recur in the aftermath of scientific revolutions. If it ordinarily disguises a shift of scientific vision or some other mental transformation with the same effect, we may not expect direct testimony about that shift. Rather we must look for indirect and behavioral evidence that the scientist with a new paradigm sees differently from the way he had seen before.

Let us then return to the data and ask what sorts of transformations in the scientist's world the historian who believes in such changes can discover. Sir William Herschel's discovery of Uranus provides a first example and one that closely parallels the anomalous card experiment. On at least seventeen different occasions between 1690 and 1781, a number of astronomers, including several of Europe's most eminent observers, had seen a star in positions that we now suppose must have been occupied at the time by Uranus. One of the best observers in this group had actually seen the star on four successive nights in 1769 without noting the motion that could have suggested another identification. Herschel, when he first observed the same object twelve years later, did so with a much improved telescope of his own manufacture. As a result, he was able to notice an apparent disk-size that was at least unusual for stars. Something was awry, and he therefore postponed identification pending further scrutiny. That scrutiny disclosed Uranus' motion among the stars, and Herschel therefore announced that he had seen a new comet! Only several months later, after fruitless attempts to fit the observed motion to a cometary orbit, did Lexell suggest that the orbit was probably planetary.[4] When that suggestion was accepted, there were several fewer stars and one more planet in the world of the professional astronomer. A celestial body that had been observed off and on for almost a century was seen differently after 1781 because, like an anomalous playing card, it could no longer be fitted to the perceptual categories (star or comet) provided by the paradigm that had previously prevailed.

The shift of vision that enabled astronomers to see Uranus, the planet, does not, however, seem to have affected only the perception of that previously observed object. Its consequences were more far-reaching. Probably, though the evidence is equivocal, the minor paradigm change forced by Herschel helped to prepare astronomers for the rapid discovery, after 1801, of the numerous minor planets or asteroids. Because of their small size, these did not display the anomalous magnification that had alerted Herschel. Nevertheless, astronomers prepared to find additional planets were able, with standard instruments, to identify twenty of them in the first fifty years of the nineteenth century.[5] The history of astronomy provides many other examples of paradigm-induced changes in scientific perception, some of them even less equivocal. Can it conceivably be an accident, for example, that Western astronomers first saw change in the previously immutable heavens during the half-century after Copernicus' new paradigm was first proposed? The Chinese, whose cosmological beliefs did not preclude celestial change, had recorded the appearance of many new stars in the heavens at a much earlier date. Also, even without the aid of a telescope, the Chinese had systematically recorded the appearance of sunspots centuries before these were seen by Galileo and his contemporaries.[6] Nor were sunspots and a new star the only examples of celestial change to emerge in the heavens of Western astronomy immediately after Copernicus. Using traditional instruments, some as simple as a piece of thread, late sixteenth-century astronomers repeatedly discovered that comets wandered at will through the space previously reserved for the immutable planets and stars.[7] The very ease and rapidity with which astronomers saw new things when looking at old objects with old instruments may make us wish to say that, after Copernicus, astronomers lived in a different world. In any case, their research responded as though that were the case.

The preceding examples are selected from astronomy because reports of celestial observation are frequently delivered in a vocabulary consisting of relatively pure observation terms. Only in such reports can we hope to find anything like a full parallelism between the observations of scientists and those of the psychologist's experimental subjects. But we need not insist on so full a parallelism, and we have much to gain by relaxing our standard. If we can be content with the everyday use of the verb 'to see,' we may quickly recognize that we have already encountered many other examples of the shifts in scientific perception that accompany paradigm change. The extended use of 'perception' and of 'seeing' will shortly require explicit defense, but let me first illustrate its application in practice.

Look again for a moment at two of our previous examples from the history of electricity. During the seventeenth century, when their research was guided by one or another effluvium theory, electricians repeatedly saw chaff particles rebound from, or fall off, the electrified bodies that had attracted them. At least that is what seventeenth-century observers said they saw, and we have no more reason to doubt their reports of perception than our own. Placed before the same apparatus, a modern observer would see electrostatic repulsion (rather than mechanical or gravitational rebounding), but historically, with one universally ignored exception, electrostatic repulsion was not seen as such until Hauksbee's large-scale apparatus had greatly magnified its effects. Repulsion after contact electrification was, however, only one of many new repulsive effects that Hauksbee saw. Through his researches, rather as in a gestalt switch, repulsion suddenly became *the* fundamental manifestation of electrification, and it was then attraction that needed to be explained.[8] The electrical phenomena visible in the early eighteenth century were both subtler and more varied than those seen by observers in the seventeenth century. Or again, after the assimilation of Franklin's paradigm, the electrician looking at a Leyden jar saw something different from what he had seen before. The device had become a condenser, for which neither the jar shape nor glass was required. Instead, the two conducting coatings – one of which had been no part of the original device – emerged to prominence. As both written discussions and pictorial representations gradually attest, two metal plates with a non-conductor between them had become the prototype for the class.[9] Simultaneously, other inductive effects received new descriptions, and still others were noted for the first time.

Shifts of this sort are not restricted to astronomy and electricity. We have already remarked some of the similar transformations of vision that can be drawn from the history of chemistry. Lavoisier, we said, saw oxygen where Priestley had seen dephlogisticated air and where others had seen nothing at all. In learning to see oxygen, however, Lavoisier also had to change his view of many other more familiar substances. He had, for example, to see a compound are where Priestley and his contemporaries had seen an elementary earth, and there were other such changes besides. At the very least, as a result of discovering oxygen, Lavoisier saw nature differently. And in the absence of some recourse to that hypothetical fixed nature that he "saw differently," the principle of economy will urge us to say that after discovering oxygen Lavoisier worked in a different world.

I shall inquire in a moment about the possibility of avoiding this strange locution, but first we require an additional example of its use, this one deriving from one of the best known parts of the work of Galileo. Since remote antiquity most people have seen one or another heavy body swinging back and forth on a string or chain until it finally comes to rest. To the Aristotelians, who believed that a heavy body is moved by its own nature from a higher position to a state of natural rest at a lower one, the swinging body was simply falling with difficulty. Constrained by the chain, it could achieve rest at its low point only after a tortuous motion and a considerable time. Galileo, on the other hand, looking at the swinging body, saw a pendulum, a body that almost succeeded in repeating the same motion over and over again ad infinitum. And having seen that much, Galileo observed other properties of the pendulum as well and constructed many of the most significant and original parts of his new dynamics around them. From the properties of the pendulum, for example, Galileo derived his only full and sound arguments for the independence of weight and rate of fall, as well as for the relationship between vertical height and terminal velocity of motions down inclined planes.[10] All these natural phenomena he saw differently from the way they had been seen before.

Why did that shift of vision occur? Through Galileo's individual genius, of course. But note that genius does not here manifest itself in more accurate or objective observation of the swinging body. Descriptively, the Aristotelian perception is just as accurate. When Galileo reported that the pendulum's period was independent of amplitude for amplitudes as great as 90°, his view of the pendulum led him to see far more regularity than we can now discover there.[11] Rather, what seems to have been involved was the exploitation by genius of perceptual possibilities made available by a medieval paradigm shift Galileo was not raised completely as an Aristotelian. On the contrary, he was trained to analyze motions in terms of the impetus theory, a late medieval paradigm which held that the continuing motion of a heavy body is due to an internal power implanted in it by the projector that initiated its motion. Jean Buridan and Nicole Oresme, the fourteenth-century scholastics who brought the impetus theory to its most perfect formulations, are the first men known to have seen in oscillatory motions any part of what Galileo saw there. Buridan describes the motion of a vibrating string as one in which impetus is first implanted when the string is struck; the impetus is next consumed in displacing the string against the resistance of its tension; tension then carries the string back, implanting increasing impetus until the mid-point of motion is reached; after that the impetus displaces the string in the opposite direction, again against the string's tension, and so on in a symmetric process that may continue indefinitely. Later in the century Oresme sketched a similar analysis of the swinging stone in what now appears as the first discussion of a pendulum.[12] His view is clearly very close to the one with which Galileo first approached the pendulum. At least in Oresme's case, and almost certainly in Galileo's as well, it was a view made possible by the transition from the original Aristotelian to the scholastic impetus paradigm for motion. Until that scholastic paradigm was invented, there were no pendulums, but only swinging stones, for the scientist to see. Pendulums were brought into existence by something very like a paradigm-induced gestalt switch.

Do we, however, really need to describe what separates Galileo from Aristotle, or Lavoisier from Priestley, as a transformation of vision? Did these men really *see* different things when *looking at* the same sorts of objects? Is there any legitimate sense in which we can say that they pursued their research in different worlds? Those questions can no longer be postponed, for there is obviously another and far more usual way to describe all of the historical examples outlined above. Many readers will surely want to say that what changes with a paradigm is only the scientist's interpretation of observations that themselves are fixed once and for all by the nature of the environment and of the perceptual apparatus. On this view, Priestley and Lavoisier both saw oxygen, but they interpreted their observations differently; Aristotle and Galileo both saw pendulums, but they differed in their interpretations of what they both had seen.

Let me say at once that this very usual view of what occurs when scientists change their minds about fundamental matters can be neither all wrong nor a mere mistake. Rather it is an essential part of a philosophical paradigm initiated by Descartes and developed at the same time as Newtonian dynamics. That paradigm has served both science and philosophy well. Its exploitation, like that of dynamics itself, has been fruitful of a fundamental understanding that perhaps could not have been achieved in another way. But as the example of Newtonian dynamics also indicates, even the most striking past success provides no guarantee that crisis can be indefinitely postponed. Today research in parts of philosophy, psychology, linguistics, and even art history, all converge to suggest that the traditional paradigm is somehow askew. That failure to fit is also made increasingly apparent by the historical study of science to which most of our attention is necessarily directed here.

None of these crisis-promoting subjects has yet produced a viable alternate to the traditional epistemological paradigm, but they do begin to suggest what some of that paradigm's characteristics will be. I am, for example, acutely aware of the difficulties created by saying that when Aristotle and Galileo looked at swinging stones, the first saw constrained fall, the second a pendulum. The same difficulties are presented in an even more fundamental form by the opening sentences of this section: though the world does not change with a change of paradigm, the scientist afterward works in a different world. Nevertheless, I am convinced that we must learn to make sense of statements that at least resemble these. What occurs during a scientific revolution is not fully reducible to a reinterpretation of individual and stable data. In the first place, the data are not unequivocally stable. A pendulum is not a falling stone, nor is oxygen dephlogisticated air. Consequently, the data that scientists collect from these diverse objects are, as we shall shortly see, themselves different. More important, the process by which either the individual or the community makes the transition from constrained fall to the pendulum or from dephlogisticated air to oxygen is not one that resembles interpretation. How could it do so in the absence of fixed data for the scientist to interpret? Rather than being an interpreter, the scientist who embraces a new paradigm is like the man wearing inverting lenses. Confronting the same constellation of objects as before and knowing that he does so, he nevertheless finds them transformed through and through in many of their details.

None of these remarks is intended to indicate that scientists do not characteristically interpret observations and data. On the contrary, Galileo interpreted observations on the pendulum, Aristotle observations on falling stones, Musschenbroek observations on a charge-filled bottle, and Franklin observations on a condenser. But each of these interpretations presupposed a paradigm. They were parts of normal science, an enterprise that, as we have already seen, aims to refine, extend, and articulate a paradigm that is already in existence. Section III provided many examples in which interpretation played a central role. Those examples typify the overwhelming majority of research. In each of them the scientist, by virtue of an accepted paradigm, knew what a datum was, what instruments might be used to retrieve it, and what concepts were relevant to its interpretation. Given a paradigm, interpretation of data is central to the enterprise that explores it.

But that interpretive enterprise – and this was the burden of the paragraph before last – can only articulate a paradigm, not correct it. Paradigms are not corrigible by normal science at all. Instead, as we have already seen, normal science ultimately leads only to the recognition of anomalies and to crises. And these are terminated, not by deliberation and interpretation, but by a relatively sudden and unstructured event like the gesalt switch. Scientists then often speak of the "scales falling from the eyes" or of the "lightning flash" that "inundates" a previously obscure puzzle, enabling its components to be seen in a new way that for the first time permits its solution. On other occasions the relevant illumination comes in sleep.[13] No ordinary sense of the term 'interpretation' fits these flashes of intuition through which a new paradigm is born. Though such intuitions depend upon the experience, both anomalous and congruent, gained with the old paradigm, they are not logically or piecemeal linked to particular items of that experience as an interpretation would be. Instead, they gather up large portions of that experience and transform them to the rather different bundle of experience that will thereafter be linked piecemeal to the new paradigm but not to the old.

To learn more about what these differences in experience can be, return for a moment to Aristotle, Galileo, and the pendulum. What data did the interaction of their different paradigms and their common environment make accessible to each of them? Seeing constrained fall, the Aristotelian would measure (or at least discuss – the Aristotelian seldom measured) the weight of the stone, the vertical height to which it had been raised, and the time required for it to achieve rest. Together with the resistance of the medium, these were the conceptual categories deployed by Aristotelian science when dealing with a falling body.[14] Normal research guided by them could not have produced the laws that Galileo discovered. It could only – and by another route it did – lead to the series of crises from which Galileo's view of the swinging stone emerged. As a result of those crises and of other intellectual changes besides, Galileo saw the swinging stone quite differently. Archimedes' work on floating bodies made the medium non-essential; the impetus theory rendered the motion symmetrical and enduring; and Neoplatonism directed Galileo's attention to the motion's circular form.[15] He therefore measured only weight, radius, angular displacement, and time per swing, which were precisely the data that could be interpreted to yield Galileo's laws for the pendulum. In the event, interpretation proved almost unnecessary. Given Galileo's paradigms, pendulum-like regularities were very nearly accessible to inspection. How else are we to account for Galileo's discovery that the bob's period is entirely independent of amplitude, a discovery that the normal science stemming from Galileo had to eradicate and that we are quite unable to document today. Regularities that could not have existed for an Aristotelian (and that are, in fact, nowhere precisely exemplified by nature) were consequences of immediate experience for the man who saw the swinging stone as Galileo did.

Perhaps that example is too fanciful since the Aristotelians recorded no discussions of swinging stones. On their paradigm it was an extraordinarily complex phenomenon. But the Aristotelians did discuss the simpler case, stones falling without uncommon constraints, and the same differences of vision are apparent there. Contemplating a falling stone, Aristotle saw a change of state rather than a process. For him the relevant measures of a motion were therefore total distance covered and total time elapsed, parameters which yield what we should now call not speed but average speed.[16] Similarly, because the stone was impelled by its nature to reach its final resting point, Aristotle saw the relevant distance parameter at any instant during the motion as the distance *to* the final end point rather than as that *from* the origin of motion.[17] Those conceptual parameters underlie and give sense to most of his well-known "laws of motion." Partly through the impetus paradigm, however, and partly through a doctrine known as the latitude of forms, scholastic criticism changed this way of viewing motion. A stone moved by impetus gained more and more of it while receding from its starting point; distance from rather than distance to therefore became the revelant parameter. In addition, Aristotle's notion of speed was bifurcated by the scholastics into concepts that soon after Galileo became our average speed and instantaneous speed. But when seen through the paradigm of which these conceptions were a part, the falling stone, like the pendulum, exhibited its governing laws almost on inspection. Galileo was not one of the first men to suggest that stones fall with a uniformly accelerated motion.[18] Furthermore, he had developed his theorem on this subject together with many of its consequences before he experimented with an inclined plane. That theorem was another one of the network of new regularities accessible to genius in the world determined jointly by nature and by the paradigms upon which Galileo and his contemporaries had been raised. Living in that world, Galileo could still, when he chose, explain why Aristotle had seen what he did. Nevertheless, the immediate content of Galileo's experience with falling stones was not what Aristotle's had been.

It is, of course, by no means clear that we need be so concerned with "immediate experience" – that is, with the perceptual features that a paradigm so highlights that they surrender their regularities almost upon inspection. Those features must obviously change with the scientist's commitments to paradigms, but they are far from what we ordinarily have in mind when we speak of the raw data or the brute experience from which scientific research is reputed to proceed. Perhaps immediate experience should be set aside as fluid, and we should discuss instead the concrete operations and measurements that the scientist performs in his laboratory. Or perhaps the analysis should be carried further still from the immediately given. It might, for example, be conducted in terms of some neutral observation-language, perhaps one designed to conform to the retinal imprints that mediate what the scientist sees. Only in one of these ways can we hope to retrieve a realm in which experience is again stable once and for all – in which the pendulum and constrained fall are not different perceptions but rather different interpretations of the unequivocal data provided by observation of a swinging stone.

But is sensory experience fixed and neutral? Are theories simply man-made interpretations of given data? The epistemological viewpoint that has most often guided Western philosophy for three centuries dictates an immediate and unequivocal, Yes! In the absence of a developed alternative, I find it impossible to relinquish entirely that viewpoint. Yet it no longer functions effectively, and the attempts to make it do so through the introduction of a neutral language of observations now seem to me hopeless.

The operations and measurements that a scientist undertakes in the laboratory are not "the given" of experience but rather "the collected with difficulty." They are not what the scientist sees – at least not before his research is well advanced and his attention focused. Rather, they are concrete indices to the content of more elementary perceptions, and as such they are selected for the close scrutiny of normal research only because they promise opportunity for the fruitful elaboration of an accepted paradigm. Far more clearly than the immediate experience from which they in part derive, operations and measurements are paradigm-determined. Science does not deal in all possible laboratory manipulations. Instead, it selects those relevant to the juxtaposition of a paradigm with the immediate experience that that paradigm has partially determined. As a result, scientists with different paradigms engage in different concrete laboratory manipulations. The measurements to be performed on a pendulum are not the ones relevant to a case of constrained fall. Nor are the operations relevant for the elucidation of oxygen's properties uniformly the same as those required when investigating the characteristics of dephlogisticated air.

As for a pure observation-language, perhaps one will yet be devised. But three centuries after Descartes our hope for such an eventuality still depends exclusively upon a theory of perception and of the mind. And modern psychological experimentation is rapidly proliferating phenomena with which that theory can scarcely deal. The duck-rabbit shows that two men with the same retinal impressions can see different things; the inverting lenses show that two men with different retinal impressions can see the same thing. Psychology supplies a great deal of other evidence to the same effect, and the doubts that derive from it are readily reinforced by the history of attempts to exhibit an actual language of observation. No current attempt to achieve that end has yet come close to a generally applicable language of pure percepts. And those attempts that come closest share one characteristic that strongly reinforces several of this essay's main theses. From the start they presuppose a paradigm, taken either from a current scientific theory or from some fraction of everyday discourse, and they then try to eliminate from it all non-logical and non-perceptual terms. In a few realms of discourse this effort has been carried very far and with fascinating results. There can be no question that efforts of this sort are worth pursuing. But their result is a language that – like those employed in the sciences – embodies a host of expectations about nature and fails to function the moment these expectations are violated. Nelson Goodman makes exactly this point in describing the aims of his *Structure of Appearance*: "It is fortunate that nothing more [than phenomena known to exist] is in question; for the notion of 'possible' cases, of cases that do not exist but might have existed, is far from clear."[19] No language thus restricted to reporting a world fully known in advance can produce mere neutral and objective reports on "the given." Philosophical investigation has not yet provided even a hint of what a language able to do that would be like.

Under these circumstances we may at least suspect that scientists are right in principle as well as in practice when they treat oxygen and pendulums (and perhaps also atoms and electrons) as the fundamental ingredients of their immediate experience. As a result of the paradigm-embodied experience of the race, the culture, and, finally, the profession, the world of the scientist has come to be populated with planets and pendulums, condensers and compound ores, and other such bodies besides. Compared with these objects of perception, both meter stick readings and retinal imprints are elaborate constructs to which experience has direct access only when the scientist, for the special purposes of his research, arranges that one or the other should do so. This is not to suggest that pendulums, for example, are the only things a scientist could possibly see when looking at a swinging stone. (We have already noted that members of another scientific community could see constrained fall.) But it is to suggest that the scientist who looks at a swinging stone can have no experience that is in principle more elementary than seeing a pendulum. The alternative is not some hypothetical "fixed" vision, but vision through another paradigm, one which makes the swinging stone something else.

All of this may seem more reasonable if we again remember that neither scientists nor laymen learn to see the world piecemeal or item by item. Except when all the conceptual and manipulative categories are prepared in advance – e.g., for the discovery of an additional transuranic element or for catching sight of a new house – both scientists and laymen sort out whole areas together from the flux of experience. The child who transfers the word 'mama' from all humans to all females and then to his mother is not just learning what 'mama' means or who his mother is. Simultaneously he is learning some of the differences between males and females as well as something about the ways in which all but one female will behave toward him. His reactions, expectations, and beliefs – indeed, much of his perceived world – change accordingly. By the same token, the Copernicans who denied its traditional title 'planet' to the sun were not only learning what 'planet' meant or what the sun was. Instead, they were changing the meaning of 'planet' so that it could continue to make useful distinctions in a world where all celestial bodies, not just the sun, were seen differently from the way they had been seen before. The same point could be made about any of our earlier examples. To see oxygen instead of dephlogisticated air, the condenser instead of the Leyden jar, or the pendulum instead of constrained fall, was only one part of an integrated shift in the scientist's vision of a great many related chemical, electrical, or dynamical phenomena. Paradigms determine large areas of experience at the same time.

It is, however, only after experience has been thus determined that the search for an operational definition or a pure observation-language can begin. The scientist or philosopher who asks what measurements or retinal imprints make the pendulum what it is must already be able to recognize a pendulum when he sees one. If he saw constrained fall instead, his question could not even be asked. And if he saw a pendulum, but saw it in the same way he saw a tuning fork or an oscillating balance, his question could not be answered. At least it could not be answered in the same way, because it would not be the same question. Therefore, though they are always legitimate and are occasionally extraordinarily fruitful, questions about retinal imprints or about the consequences of particular laboratory manipulations presuppose a world already perceptually and conceptually subdivided in a certain way. In a sense such questions are parts of normal science, for they depend upon the existence of a paradigm and they receive different answers as a result of paradigm change.

To conclude this section, let us henceforth neglect retinal impressions and again restrict attention to the laboratory operations that provide the scientist with concrete though fragmentary indices to what he has already seen. One way in which such laboratory operations change with paradigms has already been observed repeatedly. After a scientific revolution many old measurements and manipulations become irrelevant and are replaced by others instead. One does not apply all the same tests to oxygen as to dephlogisticated air. But changes of this sort are never total. Whatever he may then sea, the scientist after a revolution is still looking at the same world. Furthermore, though he may previously have employed them differently, much of his language and most of his laboratory instruments are still the same as they were before. As a result, postrevolutionary science invariably includes many of the same manipulations, performed with the same instruments and described in the same terms, as its prerevolutionary predecessor. If these enduring manipulations have been changed at all, the change must lie either in their relation to the paradigm or in their concrete results. I now suggest, by the introduction of one last new example, that both these sorts of changes occur. Examining the work of Dalton and his contemporaries, we shall discover that one and the same operation, when it attaches to nature through a different paradigm, can become an index to a quite different aspect of nature's regularity. In addition, we shall see that occasionally the old manipulation in its new role will yield different concrete results.

Throughout much of the eighteenth century and into the nineteenth, European chemists almost universally believed that the elementary atoms of which all chemical species consisted were held together by forces of mutual affinity. Thus a lump of silver cohered because of the forces of affinity between silver corpuscles (until after Lavoisier these corpuscles were themselves thought of as compounded from still more elementary particles). On the same theory silver dissolved in acid (or salt in water) because the particles of acid attracted those of silver (or the particles of water attracted those of salt) more strongly than particles of these solutes attracted each other. Or again, copper would dissolve in the silver solution and precipitate silver, because the copper-acid affinity was greater than the affinity of acid for silver. A great many other phenomena were explained in the same way. In the eighteenth century the theory of elective affinity was an admirable chemical paradigm, widely and sometimes fruitfully deployed in the design and analysis of chemical experimentation.[20]

Affinity theory, however, drew the line separating physical mixtures from chemical compounds in a way that has become unfamiliar since the assimilation of Dalton's work. Eighteenth-century chemists did recognize two sorts of processes. When mixing produced heat, light, effervescence or something else of the sort, chemical union was seen to have taken place. If, on the other hand, the particles in the mixture could be distinguished by eye or mechanically separated, there was only physical mixture. But in the very large number of intermediate cases – salt in water, alloys, glass, oxygen in the atmosphere, and so on – these crude criteria were of little use. Guided by their paradigm, most chemists viewed this entire intermediate range as chemical, because the processes of which it consisted were all governed by forces of the same sort. Salt in water or oxygen in nitrogen was just as much an example of chemical combination as was the combination produced by oxidizing copper. The arguments for viewing solutions as compounds were very strong. Affinity theory itself was well attested. Besides, the formation of a compound accounted for a solution's observed homogeneity. If, for example, oxygen and nitrogen were only mixed and not combined in the atmosphere, then the heavier gas, oxygen, should settle to the bottom. Dalton, who took the atmosphere to be a mixture, was never satisfactorily able to explain oxygen's failure to do so. The assimilation of his atomic theory ultimately created an anomaly where there had been none before.[21]

One is tempted to say that the chemists who viewed solutions as compounds differed from their successors only over a matter of definition. In one sense that may have been the case. But that sense is not the one that makes definitions mere conventional conveniences. In the eighteenth century mixtures were not fully distinguished from compounds by operational tests, and perhaps they could not have been. Even if chemists had looked for such tests, they would have sought criteria that made the solution a compound. The mixture-compound distinction was part of their paradigm – part of the way they viewed their whole field of research – and as such it was prior to any particular laboratory test, though not to the accumulated experience of chemistry as a whole.

But while chemistry was viewed in this way, chemical phenomena exemplified laws different from those that emerged with the assimilation of Dalton's new paradigm. In particular, while solutions remained compounds, no amount of chemical experimentation could by itself have produced the law of fixed proportions. At the end of the eighteenth century it was widely known that *some* compounds ordinarily contained fixed proportions by weight of their constituents. For some categories of reactions the German chemist Richter had even noted the further regularities now embraced by the law of chemical equivalents.[22] But no chemist made use of these regularities except in recipes, and no one until almost the end of the century thought of generalizing them. Given the obvious counterinstances, like glass or like salt in water, no generalization was possible without an abandonment of affinity theory and a reconceptualization of the boundaries of the chemist's domain. That consequence became explicit at the very end of the century in a famous debate between the French chemists Proust and Berthollet. The first claimed that all chemical reactions occurred in fixed proportion, the latter that they did not. Each collected impressive experimental evidence for his view. Nevertheless, the two men necessarily talked through each other, and their debate was entirely inconclusive. Where Berthollet saw a compound that could vary in proportion, Proust saw only a physical mixture.[23] To that issue neither experiment nor a change of definitional convention could be relevant. The two men were as fundamentally at cross-purposes as Galileo and Aristotle had been.

This was the situation during the years when John Dalton undertook the investigations that led finally to his famous chemical atomic theory. But until the very last stages of those investigations, Dalton was neither a chemist nor interested in chemistry. Instead, he was a meteorologist investigating the, for him, physical problems of the absorption of gases by water and of water by the atmosphere. Partly because his training was in a different specialty and partly because of his own work in that specialty, he approached these problems with a paradigm different from that of contemporary chemists. In particular, he viewed the mixture of gases or the absorption of a gas in water as a physical process, one in which forces of affinity played no part. To him, therefore, the observed homogeneity of solutions was a problem, but one which he thought he could solve if he could determine the relative sizes and weights of the various atomic particles in his experimental mixtures. It was to determine these sizes and weights that Dalton finally turned to chemistry, supposing from the start that, in the restricted range of reactions that he took to be chemical, atoms could only combine one-to-one or in some other simple whole-number ratio.[24] That natural assumption did enable him to determine the sizes and weights of elementary particles, but it also made the law of constant proportion a tautology. For Dalton, any reaction in which the ingredients did not enter in fixed proportion was *ipso facto* not a purely chemical process. A law that experiment could not have established before Dalton's work, became, once that work was accepted, a constitutive principle that no single set of chemical measurements could have upset. As a result of what is perhaps our fullest example of a scientific revolution, the same chemical manipulations assumed a relationship to chemical generalization very different from the one they had had before.

Needless to say, Dalton's conclusions were widely attacked when first announced. Berthollet, in particular, was never convinced. Considering the nature of the issue, he need not have been. But to most chemists Dalton's new paradigm proved convincing where Proust's had not been, for it had implications far wider and more important than a new criterion for distinguishing a mixture from a compound. If, for example, atoms could combine chemically only in simple whole-number ratios, then a re-examination of existing chemical data should disclose examples of multiple as well as of fixed proportions. Chemists stopped writing that the two oxides of, say, carbon contained 56 per cent and 72 per cent of oxygen by weight; instead they wrote that one weight of carbon would combine either with 1.3 or with 2.6 weights of oxygen. When the results of old manipulations were recorded in this way, a 2:1 ratio leaped to the eye; and this occurred in the analysis of many well-known reactions and of new ones besides. In addition, Dalton's paradigm made it possible to assimilate Richter's work and to see its full generality. Also, it suggested new experiments, particularly those of Gay-Lussac on combining volumes, and these yielded still other regularities, ones that chemists had not previously dreamed of. What chemists took from Dalton was not new experimental laws but a new way of practicing chemistry (he himself called it the "new system of chemical philosophy"), and this proved so rapidly fruitful that only a few of the older chemists in France and Britain were able to resist it.[25] As a result, chemists came to live in a world where reactions behaved quite differently from the way they had before.

As all this went on, one other typical and very important change occurred. Here and there the very numerical data of chemistry began to shift. When Dalton first searched the chemical literature for data to support his physical theory, he found some records of reactions that fitted, but he can scarcely have avoided finding others that did not. Proust's own measurements on the two oxides of copper yielded, for example, an oxygen weight-ratio of 1.47:1 rather than the 2:1 demanded by the atomic theory; and Proust is just the man who might have been expected to achieve the Daltonian ratio.[26] He was, that is, a fine experimentalist, and his view of the relation between mixtures and compounds was very close to Dalton's. But it is hard to make nature fit a paradigm. That is why the puzzles of normal science are so challenging and also why measurements undertaken without a paradigm so seldom lead to any conclusions at all. Chemists could not, therefore, simply accept Dalton's theory on the evidence, for much of that was still negative. Instead, even after accepting the theory, they had still to beat nature into line, a process which, in the event, took almost another generation. When it was done, even the percentage composition of well-known compounds was different. The data themselves had changed. That is the last of the senses in which we may want to say that after a revolution scientists work in a different world.

Notes

1. The original experiments were by George M. Stratton, "Vision without Inversion of the Retinal Image," *Psychological Review*, IV (1897), 341–60, 463–81. A more up-to-date review is provided by Harvey A. Carr, *An Introduction to Space Perception* (New York, 1935), pp. 18–57.
2. For examples, see Albert H. Hastorf, "The Influence of Suggestion on the Relationship between Stimulus Size and Perceived Distance," *Journal of Psychology*, XXIX (1950), 195–217; and Jerome S. Bruner, Leo Postman, and John Rodrigues, "Expectations and the Perception of Color," *American Journal of Psychology*, LXIV (1951), 216–27.
3. N. R. Hanson, *Patterns of Discovery* (Cambridge, 1958), chap. i.
4. Peter Doig, *A Concise History of Astronomy* (London, 1950), pp. 115–16.
5. Rudolph Wolf, *Geschichte der Astronomic* (Munich, 1877), pp. 513–15, 683–93. Notice particularly how difficult Wolf's account makes it to explain these discoveries as a consequence of Bode's Law.
6. Joseph Needham, *Science and Civilization in China*, III (Cambridge, 1959), 423–29, 434–36.
7. T. S. Kuhn, *The Copernican Revolution* (Cambridge, Mass., 1957), pp. 206–9.
8. Duane Roller and Duane H. D. Roller, *The Development of the Concept of Electric Charge* (Cambridge, Mass., 1954), pp. 21–29.
9. See the discussion in Section VII and the literature to which the reference there cited in note 9 will lead.
10. Galileo Galilei, *Dialogues concerning Two New Sciences*, trans. H. Crew, and A. de Salvio (Evanston, Ill., 1946), pp. 80–81, 162–66.
11. *Ibid.*, pp. 91–94, 244.
12. M. Clagett, *The Science of Mechanics in the Middle Ages* (Madison, Wis., 1959), pp. 537–38, 570.
13. [Jacques] Hadamard, *Subconscient intuition, et logique dans la recherche scientifique (Conférence faite au Palais de la Découverte le 8 Décembre 1945* [Alençon, n.d.]), pp. 7–8. A much fuller account, though one exclusively restricted to mathematical innovations, is the same author's *The Psychology of Invention in the Mathematical Field* (Princeton, 1949).

14. T. S. Kuhn, "A Function for Thought Experiments," in *Mélanges Alexandre Koyré*, ed. R. Taton and I. B. Cohen, to be published by Hermann (Paris) in 1963.
15. A. Koyré, *Etudes Galiléennes* (Paris, 1939), I, 46–51; and "Galileo and Plato," *Journal of the History of Ideas*, IV (1943), 400–428.
16. Kuhn, "A Function for Thought Experiments," in *Mélanges Alexandre Koyré* (see n. 14 for full citation).
17. Koyré, *Etudes* . . ., II, 7–11.
18. Clagett, *op. cit.*, chaps, iv, vi, and ix.
19. N. Goodman, *The Structure of Appearance* (Cambridge, Mass., 1951), pp. 4–5. The passage is worth quoting more extensively: "If all and only those residents of Wilmington in 1947 that weigh between 175 and 180 pounds have red hair, then 'red-haired 1947 resident of Wilmington' and '1947 resident of Wilmington weighing between 175 and 180 pounds' may be joined in a constructional definition. . . . The question whether there 'might have been' someone to whom one but not the other of these predicates would apply has no bearing . . . once we have determined that there is no such person. . . . It is fortunate that nothing more is in question; for the notion of 'possible' cases, of cases that do not exist but might have existed, is far from clear."
20. H. Metzger, *Newton, Stahl, Bocrhaave et la doctrine chimique* (Paris, 1930), pp. 34–68.
21. *Ibid.*, pp. 124–29, 139–48. For Dalton, see Leonard K. Nash, *The Atomic-Molecular Theory* ("Harvard Case Histories in Experimental Science," Case 4; Cambridge, Mass., 1950), pp. 14–21.
22. J. R. Partington, *A Short History of Chemistry* (2d ed.; London, 1951), pp. 161–63.
23. A. N. Meldrum, "The Development of the Atomic Theory: (1) Berthollet's Doctrine of Variable Proportions," *Manchester Memoirs*, LIV (1910), 1–16.
24. L. K. Nash, "The Origin of Dalton's Chemical Atomic Theory," *Isis*, XLVII (1956), 101–16.
25. A. N. Meldrum, "The Development of the Atomic Theory: (6) The Reception Accorded to the Theory Advocated by Dalton," *Manchester Memoirs*, LV (1911), 1–10.
26. For Proust, see Meldrum, "Berthollet's Doctrine of Variable Proportions," *Manchester Memoirs*, LIV (1910), 8. The detailed history of the gradual changes in measurements of chemical composition and of atomic weights has yet to be written, but Partington, *op. cit.*, provides many useful leads to it.

3

Techniques of the Observer

Jonathan Crary

Near the opening of Goethe's *Color Theory* (1810) we find the following account:

> Let a room be made as dark as possible; let there be a circular opening in the window shutter about three inches in diameter, which may be closed or not at pleasure. The sun being suffered to shine through this on a white surface, let the spectator from some little distance fix his eyes on this bright circle thus admitted.[1]

Goethe, following a long established practice, has made a camera obscura the site of his optical studies. The dark room, of course, had been a crucial feature of the experiments detailed by Newton in his *Opticks* (1704), where it established categorical relations between interior and exterior, between light source, aperture, and screen, and between observer and representation. And as the work of Descartes, Leibniz, Locke, and others attests, the significance of the camera obscura went well beyond the domain of natural science. For nearly two hundred years the camera stood as a sovereign metaphor for describing the status of an observer and as a model, in both rationalist and empiricist thought, of how observation leads to truthful inferences about the world.[2]

But as Goethe continues his recitation, he abruptly and stunningly abandons the order of the camera obscura:

> *The hole being then closed* [Man schliesse darauf die Offnung], let him look towards the darkest part of the room; a circular image will now be

Source: *October*, 45 (1988): 3–35.

> seen to float before him. The middle of the circle will appear bright, colorless, or somewhat yellow, but the border will appear red.
>
> After a time this red, increasing towards the centre, covers the whole circle, and at last the bright central point. No sooner, however, is the whole circle red than the edge begins to be blue, and the blue gradually encroaches inwards on the red. When the whole is blue the edge becomes dark and colorless. The darker edge again slowly encroaches on the blue till the whole circle appears colorless.[3]

Goethe's instruction to seal the hole announces a disordering and negation of the camera obscura as both an optical system and epistemological principle. The closing off of the opening dissolves the distinction between inner and outer space on which the very functioning of the camera (as paradigm and apparatus) depended. But it is now not simply a question of an observer repositioned in a sealed interior to view its particular contents; the optical experience described here by Goethe presents a notion of vision which the classical model was incapable of encompassing.

The colored circles that seem to float, undulate, and undergo a sequence of chromatic transformations have no correlative either within or without the dark room; as Goethe explains at length, they are "physiological colors" belonging entirely to the body of the observer, and they are "the necessary conditions of vision."

> Let the observer look steadfastly on a small colored object and let it be taken away after a time while his eyes remain unmoved; the spectrum of another color will then be visible on the white plane . . . it arises from an image which now *belongs to the eye*.[4]

The corporal subjectivity of the observer, which was *a priori* excluded from the camera obscura, suddenly becomes the site on which an observer is possible. The human body, in all its contingency and specificity, generates "the spectrum of another color," and thus becomes the active producer of optical experience.

The ramifications of Goethe's color theory are manifold and have little to do with the empirical "truth" of his assertions or the "scientific" character of his experiments. Contained within the unsystematized accumulation of statements and findings is a key delineation of subjective vision, a post-Kantian notion that is both a product and constituent of modernity. What is crucial about Goethe's account of subjective vision is the inseparability of two models usually presented as distinct and irreconcilable: a physiological subject who will be described in increasing detail by the empirical sciences in the nineteenth century, and an observer posited by various "romanticisms" and early modernisms as the active, autonomous producer of his or her own content. This essay seeks to describe some of the features of this new kind of observer and to suggest how his or her formation in the nineteenth century was immanent to the elaboration of new empirical knowledge of vision and techniques of the visible.

Clearly Kant's "Copernican revolution" (*Drehung*) of the spectator, proposed in the preface to the second edition of the *Critique of Pure Reason* (1787), is a definitive sign of a new organization and positioning of the subject. For Kant, continuing the use of optical figures, it is "a change in point of view," such that "our representation of things, as they are given, does not conform to these things as they are in themselves, but that these objects as appearances, conform to our mode of representation."[5] William Blake put it more simply: "As the eye, such the object."[6] Michel Foucault makes clear that vision in the classical era was precisely the opposite of Kant's subject-centered epistemology, that it was then a form of immediate knowing, "a perceptible knowledge." For example:

> Natural history [in the eighteenth century] is nothing more than the nomination of the visible. Hence its apparent simplicity, and that air of naiveté it has from a distance, so simple does it appear and so obviously *imposed by the things themselves*.[7]

In the aftermath of Kant's work there is an irreversible clouding over of the transparency of the subject-as-observer. Vision, rather than a privileged form of knowing, becomes itself an object of knowledge, of observation. From the beginning of the nineteenth century a science of vision will tend to mean increasingly an interrogation of the makeup of the human subject, rather than of the mechanics of light and optical transmission. It is a moment when the visible escapes from the timeless incorporeal order of the camera obscura and becomes lodged in another apparatus, within the unstable physiology and temporality of the human body.

When Goethe's experiments repeatedly call for either a darkened room or, perhaps more significantly, the closed eye, he is not simply privileging an experience of being severed from contact with an external world. On the one hand he is indicating his conviction that color is always the product of an admixture of light and shadow: "Color itself is a degree of darkness; hence Kircher is perfectly right in calling it *lumen opacatum*."[8] On the other hand he is also posing conditions in which the inescapable physiological components of vision can be artificially isolated and made observable by themselves. For Goethe, and for Schopenhauer soon after, vision is always an irreducible complex of elements belonging to the observer's body and of external data. Thus the kind of separation between interior representation and exterior reality implicit in the camera obscura becomes in Goethe's work a single surface of affect on which interior and exterior have few of their former meanings and positions. Color, as the primary object of vision, is now atopic, cut from any spatial referent.

Goethe insistently cites experiences in which the subjective contents of vision are dissociated from an objective world, in which the body itself produces phenomena that have no external correlate. Notions of correspondence, of reflection on which classical optics and theories of knowledge were based, although retained elsewhere by Goethe, have lost their centrality and necessity. The priority previously accorded to Lockean primary qualities over secondary ones becomes inverted. Perhaps most important is his designation of opacity as a crucial and productive component of vision. If discourses of the visible in the seventeenth and eighteenth centuries repressed and concealed whatever threatened the transparence of an optical system, Goethe signals a reversal, and instead poses the opacity of the observer as the necessary condition for the appearance of phenomena.[9] Pure light and pure transparence are now beyond the limits of the visible.[10]

The articulation of subjective vision in the early nineteenth century is part of a shift which Foucault calls "the threshold of our modernity." When the camera obscura was the dominant model of observation it was as "a form of representation which made knowledge in general possible." But at the beginning of the nineteenth century

> the site of analysis is no longer representation but man in his finitude. . . . [It was found] that knowledge has anatomo-physiological conditions, that it is formed gradually within the structures of the body, that it may have a privileged place within it, but that its forms cannot be dissociated from its peculiar functioning; in short, that there is a nature of human knowledge that determines its forms and that at the same time can be manifest to it in its own empirical contents.[11]

Goethe's affirmation of both the subjective and the physiological in perception obviously is echoed in a wide range of well-known philosophical, scientific, and literary developments. One less familiar but significant parallel is the contemporary work of Maine de Biran. During the first decade of the century, the latter outlined a science of the *sens intime* in an attempt to understand more accurately the nature of inward experience. But in seeking to grasp the density and the immediacy of the *sens intime* Maine de Biran began to blur the identity of the very inwardness that was his original object. He employed the term *coenesthèse* to describe "one's immediate awareness of the presence of the body in perception" and "the simultaneity of a composite of impressions inhering in different parts of the organism."[12] Visual perception, for example, is inseparable from the muscular movements of the eye and the physical effort involved in focusing on an object. In a complete reversal of the classical model of the apparatus as a neutral device of pure transmission, the viewer's sensory equipment now is inextricably mixed with whatever object it beholds. Seven years before Goethe published the *Farbenlehre*, Maine de Biran discussed how perception of color was determined by the body's tendency to fatigue (by a physiological modulation over time) and that the very process of becoming tired was in fact perception.[13] For both of them, the absolute values accorded to color by Newtonian theory are displaced by an insistence on color's transient unfolding within the human subject.

Maine de Biran unravels the assumptions of Condillac and others about the composition of perception. Condillac's notion of sensation as a simple unit, a building block out of which clear perceptions were assembled, is no longer adequate to a new multilayered and temporally dispersed perception which Maine de Biran details, making impossible "a soul reduced to pure receptivity." For both Goethe and Maine de Biran, subjective observation is not the inspection of an inner space or a theater of representations. Instead, observation is increasingly exteriorized; the viewing body and its objects begin to constitute a single field on which inside and outside are confounded. Perhaps most importantly, both observer and what he sees are subject to the same modes of empirical study. For Georges Canguilhem, the reorganization of human knowledge at the beginning of the nineteenth century signals an end to the idea of a qualitatively different human order, and he cites the major discovery by Maine de Biran that since "the soul is necessarily incarnated there is no psychology without biology."[14]

The models of subjective vision outlined by Goethe and Schopenhauer (which were brought to fulfillment in the physiological optics of Helmholtz in the 1860s), also must be seen against the profound changes which took place in theories of the nature of light. The shift from emission and corpuscular theories to undulatory or wave motion explanations have a major significance for nineteenth-century culture as a whole.[15] The wave theory of light made obsolete any notion of a rectilinear propagation of light rays on which classical optics was based and, in part, the science of perspective. All the modes of representation derived from Renaissance and later models of perspective no longer had the legitimation of a science of optics. The verisimilitude associated with perspectival construction obviously persisted into the nineteenth century, but it was severed from the scientific base that had once authorized it, and it could no longer have the same meanings it had when either Aristotelian or Newtonian optics held sway. Dominant theories of vision, whether of Alberti, Kepler, or Newton (Huygens is the obvious exception), all described in their own fashion how a beam of isolated light rays traversed an optical system, with each ray taking the shortest possible route to reach its destination.[16] The camera obscura is inextricably wedded to this point-to-point epistemological setup. At the same time it must be stressed how deeply theological was the notion that light was radiant (composed of rays) and emanative.

The work of Augustin Jean Fresnel has come to stand for the paradigm shift.[17] By 1821 Fresnel had concluded that the vibrations of which light consisted were entirely *transverse*, which led him and those who followed to build mechanical models of an ether which transmitted such tranverse waves rather than longitudinal rays or waves. Fresnel's work participates in the destruction of classical mechanics and the eventual dominance of modern physics. What had been a discrete domain of optics in the seventeenth and eighteenth centuries now became merged with the study of other physical phenomena, i.e., electricity and magnetism. Above all it is a moment when light loses its ontological privilege; and during the nineteenth century the very identity of light as an independent entity became increasingly problematic. Goethe's color theory, with its proposal of a qualitative difference between light and color, had implicitly suggested such developments. More importantly here, however, as light began to be seen as an electromagnetic phenomenon, it had less and less to do with the realm of the visible and with the description of human vision. So it is at this moment in the early nineteenth century, when physical optics (the study of light and the forms of its propagation) merges with physics, that physiological optics (the study of the eye and its sensory capacities) comes to dominate the study of vision.

*

The retinal afterimage is perhaps the most important optical phenomenon discussed by Goethe in his chapter on "Physiological Colors" in his *Color Theory*. Though preceded by others in the late eighteenth century, his treatment of the topic was by far the most thorough up to that moment.[18] Subjective visual phenomena such as afterimages had been recorded since antiquity, but only as events outside the domain of optics, and they were relegated to the category of the "spectral" or mere appearance. But in the early nineteenth century, particularly with Goethe, such experiences attain the status of optical "truth." They are no longer deceptions that obscure a "true" perception; rather they begin to constitute an irreducible component of human vision. For Goethe and the physiologists who followed him there was no such thing as optical illusion: whatever the healthy corporal eye experienced was in fact optical truth.

The implications of the new "objectivity" accorded to subjective phenomena are several. First, the privileging of the afterimage allowed the thought of sensory perception cut from any necessary link with an external referent. The afterimage – the presence of sensation in the absence of a stimulus – and its subsequent modulations posed a theoretical and empirical demonstration of autonomous vision, of an optical experience that was produced by and within the subject. Secondly, and equally crucial for the rest of the nineteenth century, is the introduction of temporality as an inescapable component of observation.[19] Most of the phenomena described by Goethe in the *Color Theory* involve an unfolding over time: "the edge begins to be blue . . . the blue gradually encroaches inward . . . the image then becomes gradually fainter."[20] The virtual instantaneity of optical transmission (whether intromission or extromission) was an unquestioned foundation of classical optics and theories of perception from Aristotle to Locke. And the simultaneity of the camera obscura image with its exterior object was never questioned.[21] But as observation is increasingly tied to the body in the early nineteenth century, temporality and vision become inseparable. The shifting processes of one's own subjectivity experienced in time became synonymous with the act of seeing, dissolving the Cartesian ideal of an observer completely focused on an object.

But the problem of the afterimage and the temporality of subjective vision is lodged within larger epistemological issues in the nineteenth century. On one hand the attention given to the afterimage by Goethe and others parallels contemporary philosophical discourses (such as Maine de Biran) which describe perception and cognition as essentially temporal processes dependent upon a dynamic amalgamation of past and present. In the preface to his *Phenomenology* (1807), Hegel makes a sweeping repudiation of Lockean perception and situates perception within an unfolding that is temporal and historical. While attacking the apparent certainty of sense perception, Hegel implicitly refutes the model of the camera obscura. "It must be pointed out that truth is not like stamped coin issued ready from the mint, and so can be taken up and used."[22] Although referring to the Lockean notion of ideas "imprinting" themselves on passive minds, Hegel's remark has a precocious applicability to photography, which, like coinage, offered another mechanically and mass-produced form of "truth." Hegel's dynamic, dialectical account of perception, in which appearance negates itself to become something other, finds an echo in Goethe's discussion of afterimages:

> The eye cannot for a moment remain in a particular state determined by the object it looks upon. On the contrary, it is forced to a sort of opposition, which, in contrasting extreme with extreme, intermediate degree with intermediate degree, at the same time combines these opposite impressions, and thus ever tends to be whole, whether the impressions are successive or simultaneous and confined to one image.[23]

Goethe and Hegel, each in his own way, pose observation as the play and interaction of forces and relations, rather than as the orderly contiguity of discrete stable sensations conceived by Locke or Condillac.[24]

Other writers in the first decade also began to delineate perception as a continuous process, a flux of temporally dispersed contents. The physicist André-Marie Ampère in his epistemological writings used the term *concrétion* to describe how any perception always blends with a preceding or remembered perception. The words *mélange* and *fusion* occur frequently in his attack on classical notions of "pure" isolated sensations. Perception, as he wrote to his friend Maine de Biran, was fundamentally "*une suite de différences successives*."[25] The dynamics of the afterimage are also implied in the work of Johann Friedrich Herbart, who undertook one of the earliest attempts to quantify the movement of cognitive experience. Although his ostensible aim was to demonstrate and preserve Kant's notion of the unity of the mind, Herbart's formulation of mathematical laws governing mental experience in fact make him "a spiritual father of stimulus-response psychology."[26] If Kant gave a positive account of the mind's capacity for synthesizing and ordering experience, Herbart (Kant's successor at Königsberg) detailed how the subject wards off and prevents internal incoherence and disorganization. Consciousness, for Herbart, begins as a stream of potentially chaotic input from without. Ideas of things and events in the world were never copies of external reality but rather the outcome of an interactional process within the subject in which ideas (*Vorstellungen*) underwent operations of fusion, fading, inhibition, and blending (*Verschmelzungen*) with other previous or simultaneously occurring ideas or "presentations." The mind does not reflect truth but rather extracts it from an ongoing process involving the collision and merging of ideas.

> Let a series a, b, c, d, be given by perception; then, from the first movement of the perception and during its continuance, *a* is exposed to an arrest from other concepts already in consciousness. In the meantime, *a*, already partially sunken in consciousness, became more and more obscured when *b* came to it. This *b* at first unobscured, blended with the sinking *a*; then followed *c*, which itself unobscured, fused with *b*, which was becoming obscured. Similarly followed *d*, to become fused with *a*, *b*, and *c*, in different degrees. From this arises a law for each of these concepts. . . . It is very important to determine by calculation the degree of strength which a concept must attain in order to be able to stand beside two or more stronger ones exactly on the threshold of consciousness.[27]

All the processes of blending and opposition that Goethe described phenomenally in terms of the afterimage are for Herbart statable in differential equations and theorems. He specifically discusses color perception to describe the mental mechanisms of opposition and inhibition.[28] Once the operations of cognition become fundamentally measurable in terms of duration and intensity, it is thereby rendered both predictable and controllable. Although Herbart was philosophically opposed to empirical experimentation or any physiological research, his convoluted attempts to mathematize perception were important for the later quantitative sensory work of Johannes Müller, Gustav Fechner, Ernst Weber, and Wilhelm Wundt. He was one of the first to recognize the potential crisis of meaning and representation implied by an autonomous subjectivity, and to propose a framework for its disciplining and control. Herbart clearly was attempting a quantification of *cognition*, but he nonetheless prepared the ground for attempts to measure the magnitude of sensations, and such measurements required sensory experience that was durational. The afterimage was to become a crucial means by which observation could be quantified, by which the intensity and duration of retinal stimulation could be measured.

Also it is important to remember that Herbart's work was not simply abstract epistemological speculation, but was directly tied to his pedagogical theories, which were influential in Germany and elsewhere in Europe during the mid-nineteenth century.[29] Herbart believed that his quantification of psychological processes held the possibility for controlling and determining the sequential input of ideas into young minds, and in particular had the potential of instilling disciplinary and moral ideas. Obedience and attentiveness were central goals of Herbart's pedagogy. Just as new forms of factory production demanded more precise knowledge of a worker's attention span, so the management of the classroom, another disciplinary institution, demanded similar information.[30] In both cases the subject in question was measurable and regulated in time.

By the 1820s the quantitative study of afterimages was occurring in a wide range of scientific research throughout Europe. Working in Germany, the Czech Jan Purkinje continued Goethe's work on the persistence and modulation of afterimages: how long they lasted, what changes they went through, and under what conditions.[31] His empirical research and Herbart's mathematical methods were to come together in the next generation of psychologists and psychophysicists, when the threshold between the physiological and the mental became one of the primary objects of scientific practice. Instead of recording afterimages in terms of the lived time of the body as Goethe had generally done, Purkinje was the first to study them as part of a comprehensive quantification of the irritability of the eye.[32] He provided the first formal classification of different types of afterimages and his drawings of them are a striking indication of the paradoxical objectivity of the phenomena of subjective vision. Were we able to see the original drawings in color we would have a more vivid sense of their unprecedented overlapping of the visionary and the empirical, of "the real" and the abstract.

Although working with relatively imprecise instruments, Purkinje timed how long it takes the eye to become fatigued, how long dilation and contraction of the pupil take, and measured the strength of eye movements. For Purkinje the physical surface of the eye itself became a field of statistical information: he demarcated the retina in terms of how color changes hue depending on where it strikes the eye, describing the extent of the area of visibility, and quantified the distinction between direct and indirect vision, and also gave a highly precise account of the blind spot.[33] The discourse of dioptrics, of the transparency of mechanical optical systems in the seventeenth and eighteenth centuries, has given way to a mapping of the human eye as a productive territory with varying zones of efficiency and aptitude.

Purkinje's research, along with that of Johannes Müller and others, inaugurated the comprehensive physiology of vision in the nineteenth century. Part of this study involved establishing quantitative and statistical *norms* of the sense of sight. As Foucault and others have shown, the "scientific" assessment of "normality" in medicine, physiology, and other fields became

Jan Purkinje. Afterimages. 1823.

central to the accumulation of knowledge about individuals – whether in medicine, psychiatry, child psychology, the rationalization of labor and education – and thus essential to the exercise of power. My concern here is how the individual as observer became an object of investigation, a locus of knowledge in the first half of the 1800s, and how the nature of vision was thus modified. One feature of this period is the widespread effort by researchers from a variety of fields to establish the bounds of "normative" vision and to quantify forms of optical and other sensory response. The pervasive preoccupation with optical illusions is part of the exploration of the limits and pathology of human vision, defining ever more sharply the shape of the normal. But as Georges Canguilhem has shown, the indication of a norm is not a neutral activity; it never occurs without the specification of technical means to correct, or to produce normativeness.[34]

What is of immediate concern here is how some of the optical devices that spawned a new mass visual culture in the nineteenth century are inseparable from the new normative sciences of the observer and of the seeing body. The comprehensive articulation of subjective vision, which included the conceptual severing of visual experience from referent (in Müller's famous theory of specific nerve energies), and the quantification and study of afterimages, of persistence of vision, peripheral and binocular vision, and thresholds of attention all were directly part of the creation of a new vast domain of visual culture. On one hand there is a new abstraction and mobility of images, on the other is an inverse disciplining of the observer in terms of rigidly fixed relations to image and apparatus, particularly with the stereoscope, the kaleidoscope, the phenakistiscope, and even the diorama.

*

Beginning in the mid-1820s, the experimental study of afterimages was intertwined with the invention of a number of related optical devices and techniques. The boundary separating their use for purposes of scientific observation and as forms of popular entertainment is indistinct. Common to them all was the notion that perception was not instantaneous, and the notion of a disjunction between eye and object. Research on afterimages had suggested that some form of blending or fusion occurred when sensations were perceived in quick succession, and it was the duration involved in seeing that allowed its modification and regulation. The control of time becomes synonymous with new modalities of power.

The details and background of these devices and inventors have been well documented and discussed at length elsewhere, but almost exclusively in the service of a history of cinema.[35] Film scholars position them as the initial forms in an evolutionary technological development leading to the emergence of a single dominant form at the end of the century. Their fundamental characteristic is prematurity; they are not-yet cinema, thus nascent

Thaumatropes. c.1825.

and incomplete forms. Obviously there is a connection between cinema and these machines of the 1830s, but it is often a dialectical relation of inversion and opposition in which features of these earlier devices were negated or concealed. At the same time there is a tendency to conflate all optical devices in the nineteenth century as equally implicated in a vague collective drive to higher and higher standards of verisimilitude. Such teleological approaches most often neglect entirely how these devices were expressions of nonveridical models of perception.

One of the earliest was the Thaumatrope (literally, "wonder-turner"), first demonstrated in London by Dr. John Paris in 1825. It was a small circular disc with a drawing on either side and strings attached so that it could be twirled with a spin of the hand. The image of a bird on one side and an empty cage on the other would produce when spun the appearance of the bird in the cage. The simplicity of this "philosophical toy" made unequivocally clear the hallucinatory and fabricated nature of the image and the absolute rupture between perception and its object.

Also in 1825, Peter Mark Roget, an English mathematician and the author of the first Thesaurus, published an account of his observations of railway

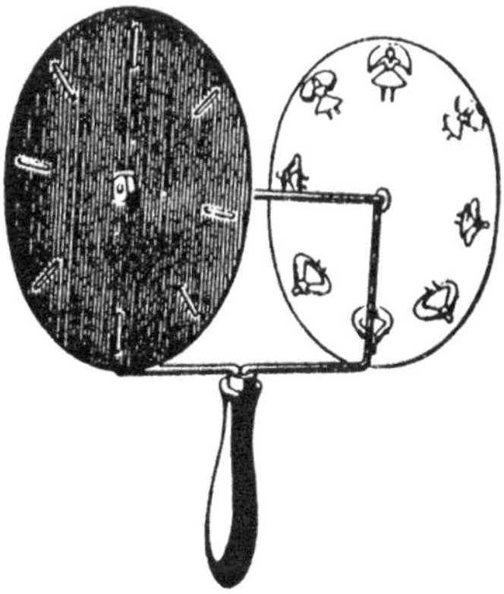

Left: Feraday wheel. 1830; Right: Phenakistiscope. 1832.

train wheels seen through the vertical bars of a fence. Roget pointed out the illusions that occurred under these circumstances in which the spokes of the wheels seemed to be either motionless or to be turning backwards. Roget's observations suggested to him how the location of an observer in relation to an intervening screen could exploit the durational properties of retinal afterimages to create various effects of motion. The physicist Michael Faraday explored similar phenomena, particularly the experience of rapidly turning wheels that *appeared* to be moving slowly. In 1831, the year of his discovery of electromagnetic induction, he produced his own device, later called the Faraday wheel, consisting of two spoked wheels mounted on the same axis. By varying the relation between the spokes of the two wheels relative to the eye of the viewer, the apparent motion of the farther wheel could be modulated. Thus the experience of temporality itself is made susceptible to a range of external technical manipulations.

During the late 1820s the Belgian scientist Joseph Plateau also conducted a wide range of experiments with afterimages, some of which cost him his eyesight due to staring directly into the sun for extended periods. By 1828 he had worked with a Newton color wheel, demonstrating that the duration and quality of retinal afterimage varied with the intensity, color, time, and direction of the stimulus. He also made a rough calculation of the average

Zellotrope. Mid-1830s.

time that such sensations lasted – about a third of a second. What is more, his research seemed to confirm the earlier speculations of Goethe and others that retinal afterimages do not simply dissipate uniformly, but go through a number of positive and negative states before vanishing. He made one of the most influential formulations of the theory of "persistence of vision."

> If several objects which differ sequentially in terms of form and position are presented one after the other to the eye in very brief intervals and sufficiently close together, the impressions they produce on the retina will blend together without confusion and one will believe that a single object is gradually changing form and position.[36]

In the early 1830s, Plateau constructed the Phenakistiscope (literally "deceptive view") which incorporated his own research and that of Roget, Faraday, and others. At its simplest it consisted of a single disc, divided into sixteen equal segments, each of which contained a small, slitted opening and a figure representing one position in a sequence of movement. The side with figures drawn on it was faced toward a mirror while the viewer stayed immobile as the disc turned. When an opening passed in front of the eye, it allowed one to see the reflected figure on the disc very briefly. The same effect occurs with each of the slits. The images then appear to be in continuous motion before the eye. By 1833, commercial models were being sold in London; by 1834 two similar devices appeared, the Stroboscope invented by the German mathematician Stampfer and the Zootrope or "wheel of life" of William G. Horner.

The empirical truth of the notion of "persistence of vision" as an explanation for the illusion of motion is irrelevant here.[37] What is important are the conditions and circumstances that allowed it to operate as an explanation and the historical subject/observer that it presupposed. The idea of persistence of vision is linked to two different sorts of studies. One is the kind of self-observation conducted first by Goethe, then by Purkinje, Plateau, Fechner, and others, in which the changing conditions of the observer's own retina was (or was then believed to be) the object of investigation. The other source was the often accidental observation of new industrial forms of movement, in particular mechanized wheels moving at high speeds. Purkinje and Roget both derived some of their ideas from noting the appearance of train wheels in motion or regularly spaced forms seen from a fast moving train. Faraday indicates that his experiments were suggested by a visit to a factory: "Being at the magnificent lead mills of Messrs. Maltby, two cog-wheels were shown me moving with such velocity that if the eye were . . . standing in such a position that one wheel appeared behind the other, there was immediately the distinct though shadowy resemblance of cogs moving slowly in one direction."[38] Like the study of afterimages, new experiences of speed and machine movement disclosed an increasing divergence between appearances and their external causes.

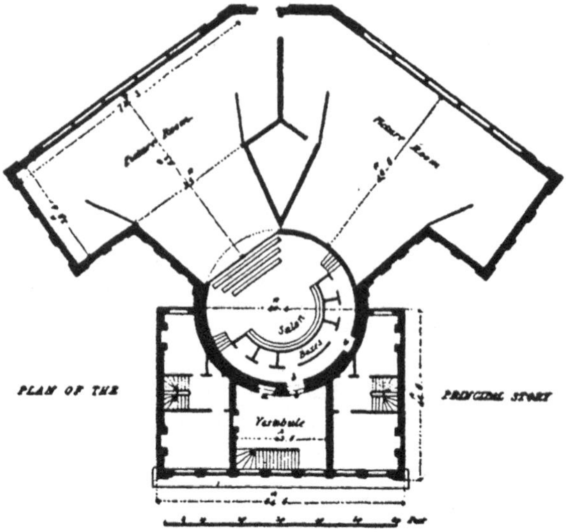

(*a*) Ground plan of the Diorama building, London, by A. Pugin and J. Morgan, 1823

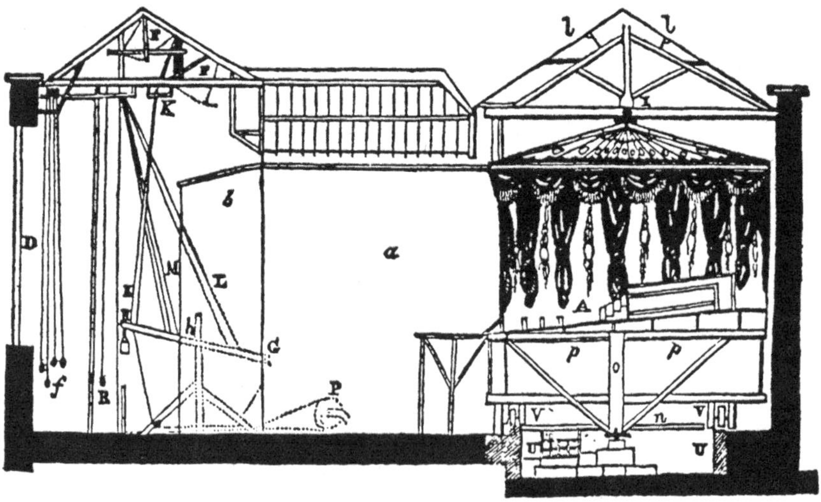

THE DIORAMA.

(*b*) Cross-section of the auditorium and picture emplacement of the Diorama, London

The Phenakistiscope substantiates Walter Benjamin's contention that in the nineteenth century "technology has subjected the human sensorium to a complex kind of training."[39] While it was of course a mode of popular entertainment, a leisure-time commodity purchasable by an expanding urban middle class, it nonetheless paralleled the format of the scientific devices used by Purkinje, Plateau, and others for the scientific study of subjective vision. That is, a form with which a new public consumed images of an illusory "reality" was isomorphic to the apparatuses used to accumulate knowledge about an observer. In fact, the very physical position required of the observer by the Phenakistiscope bespeaks a confounding of three modes: a body at once a spectator, a subject of empirical research and observation, and an element of machine production. In all three cases it is a question of a body aligned with and operating an assemblage of turning and regularly moving wheeled parts. The imperatives that generated a rational organization of time and movement in production simultaneously pervaded multiple spheres of social activity. A need for knowledge of the capacities of the eye and its regimentation pervaded many of them.

Another phenomenon that corroborates this change in the position of the observer is the Diorama, given its definitive form by Louis J. M. Daguerre in the early 1820s. Unlike the static panorama painting which first appeared in the 1790s, the Diorama is based on the incorporation of an *immobile* observer into a mechanical apparatus and a subjection to a predesigned temporal unfolding of optical experience.[40] The circular or semi-circular panorama painting clearly broke with the localized point of view of perspective painting or the camera obscura, allowing the spectator an ambulatory ubiquity. One was compelled at the least to turn one's head (and eyes) to see the entire work. The multi-media Diorama removed that autonomy from the observer, situating the audience on a circular platform that was slowly moved, permitting views of different scenes and shifting light effects. Like the Phenakistiscope or the Zootrope, the Diorama was a machine of wheels in motion and of which the observer was an inflexible component. For Marx, one of the great technical innovations of the nineteenth century was the way in which the body was made adaptable to "the few main fundamental forms of motion."[41] But if the remaking of the observer involved the adaptation of the eye to rationalized forms of movement, such a change coincided with and was possible only because of an increasing abstraction of optical experience from a stable referent. Thus a precondition of modernization was the "uprooting" of vision from the delimited and static relations of the camera obscura.

Consider also the kaleidoscope, invented in 1815 by Sir David Brewster.[42] With all the luminous possibilities suggested by Baudelaire, and later by Proust, the kaleidoscope seems radically unlike the rigid and disciplinary structure of the Phenakistiscope, with its sequential repetition of regulated representations. But, for Brewster, the justification for making the kaleidoscope was productivity and efficiency. He saw it as mechanical means for the reformation of art according to an industrial paradigm. Since symmetry was the basis of beauty in nature and visual art, he declared, the kaleidoscope was aptly suited to produce art through "the inversion and multiplication of simple forms."

kaleidoscope. Mid-19th century.

> If we reflect further on the nature of the designs thus composed, and on the methods which must be employed in their composition, the Kaleidoscope will assume the character of the highest class of machinery, which improves at the same time that it abridges the exertions of individuals. There are few machines, indeed, which rise higher above the operations of human skill. It will create in an hour, what a thousand artists could not invent in the course of a year; and while it works with such unexampled rapidity, it works also with a corresponding beauty and precision.[43]

Brewster's proposal of infinite serial production seems far removed from Baudelaire's image of the dandy as "a kaleidoscope gifted with consciousness." But the abstraction necessary for Brewster's industrial delirium is made possible by the same forces of modernization that allowed Baudelaire to use the kaleidoscope as a model for the kinetic experience of "the multiplicity of life itself and the flickering grace of all its elements."[44]

*

The most significant form of visual imagery in the nineteenth century, with the exception of photography, was the stereoscope.[45] It is easily forgotten now how pervasive was the experience of the stereoscope and how for decades it defined a principal mode of consuming photographically produced images. This is again a form whose history has thus far been confounded with another phenomenon, in this case, photography. Yet conceptually, structurally, and initially, historically it is independent of the now dominant medium. Clearly, the stereoscope utilized photographic imagery but its invention preceded photography and *in no way required* photographic procedures. Although distinct from the optical devices which represented the illusion of movement, the stereoscope is nonetheless part of the same reorganization of the observer that those devices implied.

Of concern here is the period during which the technical and theoretical principles of the stereoscope were elaborated, rather than its effects once it was distributed in a social-cultural field. Only after 1850 did the wide commercial diffusion throughout North America and Europe of the stereoscope occur.[46] The origins of the stereoscope are intertwined with research in the 1820s and 1830s on subjective vision and more generally within the field of nineteenth-century physiology already discussed. The two figures most closely associated with its invention, Charles Wheatstone and Sir David Brewster, had already written extensively on optical illusions, color theory, afterimages, and other visual phenomena. Wheatstone was in fact the translator of Purkinje's major 1823 dissertation on afterimages and subjective vision, published in English in 1830. A few years later Brewster had summarized available research on optical devices and subjective vision.

The stereoscope is also inseparable from early nineteenth-century debates about the perception of space, which were to continue unresolved indefinitely. Was space an innate form, or was it something recognized through the learning of cues after birth? The Molyneux problem had been transposed to a different century for different solutions.[47] In the eighteenth century, regardless of how the problem was ultimately answered, whether the claim was nativist or empiricist, the testimony of the senses constituted a common surface of order. The question was how the passage from one domain of sense perception to another occurred, how did the senses "reconvene," that is, come together in a body.[48] But those whose answers to Molyneux were negative – a blind man suddenly restored with sight would *not* immediately recognize the objects before him – and these included Locke, Berkeley, Diderot, Reid, and Condillac, share little with the researchers in the nineteenth century who also, with greater scientific authority, answered negatively. By insisting that knowledge, and specifically knowledge of space and depth, is built up out of an orderly accumulation and cross-referencing of perceptions on a plane independent of the viewer, eighteenth-century thought could know nothing of ideas of pure visibility which would arise in the nineteenth century. From Descartes to Berkeley to Diderot is a continued insistence that vision is grounded in the sense of touch.[49] We could not be further removed from Berkeley's theory of how distance is perceived than with the science of the stereoscope. The quintessentially nineteenth-century device, with which distance (or relief) is perceptible solely through an organization of *optical* cues, eradicates the field on which eighteenth-century knowledge arranged itself.

The question that troubled the nineteenth century had never really been a problem before. Binocular disparity, the self-evident fact that each eye sees a slightly different image, had been a familiar phenomenon since antiquity, Only in the 1830s does it become crucial for scientists to define the seeing body as essentially binocular, to quantify precisely the angular differential of the optical axis of each eye, and to specify the physiological basis for disparity. What preoccupied researchers was this: given that an observer perceives with each eye a different image, *how* are they experienced as single or unitary. Before 1800, even when the question was asked it was more as a curiosity, never a central problem. Two alternative explanations had been offered for centuries: one proposed that we never saw anything except with one eye at a time; the other was a projection theory articulated by Kepler, and proposed as late as the 1750s, which asserted that each eye projects an object to its actual location.[50]

By the late 1820s physiologists were seeking anatomical evidence in the structure of the optical chiasma, the point behind the eyes where the nerve fibers leading from the retina to the brain cross each other, carrying half of the nerves from each retina to each side of the brain.[51] But such physiological evidence was relatively inconclusive at that time. Wheatstone's conclusions in 1833 came out of the successful measurement of binocular parallax, or the degree to which the angle of the axis of each eye differed when focused on the same point. The human organism, he claimed, had the capacity under most conditions to synthesize retinal disparity into a single unitary image. While this seems obvious from our own standpoint, Wheatstone's work marked a major break from (or often disregard of) the older explanations of the binocular body.

The form of the stereoscope is linked to some of Wheatstone's initial findings: his research concerned the visual experience of objects relatively close to the eye.

> When an object is viewed at so great a distance that the optic axes of both eyes are sensibly parallel when directed towards it, the perspective projections of it, seen by each eye separately, and the appearance to the two eyes is precisely the same as when the object is seen by one eye only.[52]

Instead Wheatstone was preoccupied with objects close enough to the observer so that the optic axes had *different* angles.

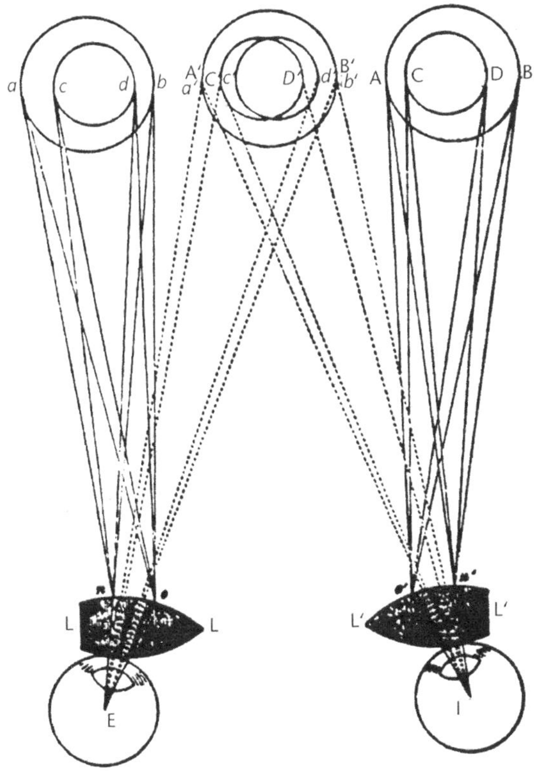

David Brewster's lenticular stereoscope. 1849.

> . . . when the object is placed so near the eyes that to view it the optic axes must converge . . . a different perspective projection of it is seen by each eye, and these perspectives are more dissimilar as the convergence of the optic axes becomes greater.[53]

Thus physical proximity brings binocular vision into play as an operation of reconciling disparity, of making two distinct views appear as one. This is what links the stereoscope with other devices in the 1830s like the Phenakistiscope. Its "realism" presupposes perceptual experience to be essentially an apprehension of differences. The relation of the observer to the object is not one of identity but an experience of disjunct or divergent images. Helmholtz's epistemology was based on such a "differential hypothesis."[54] Both Wheatstone and Brewster indicated that the fusion of pictures viewed in a stereoscope took place over time and that their convergence might not actually be secure. According to Brewster

> . . . the relief is not obtained from the mere combination or superposition of the two dissimilar pictures. The superposition is effected by turning each eye upon the object, but the relief is given by the play of the optic axes in uniting, in rapid *succession*, similar points of the two pictures. . . . Though the pictures apparently coalesce, yet the relief is given by the subsequent play of the optic axes varying themselves *successively* upon, and unifying, the similar points in each picture that correspond to different distances from the observer.[55]

Brewster then confirms there never really is a stereoscopic image, that it is a conjuration, an effect of the observer's experience of the differential between two other images.

In devising the stereoscope, Wheatstone aimed to simulate the actual presence of a physical object or scene, not to discover another way to exhibit a print or drawing. Painting had been an adequate form of representation, he asserts, but only for images of objects at a great distance. When a landscape is presented to a viewer, "if those circumstances which would disturb the illusion are excluded," we could mistake the representation for reality. He declares that up to this point in history it is impossible for an artist to give a faithful representation of any *near* solid object.

> When the painting and the object are seen with both eyes, in the case of the painting two similar objects are projected on the retina, in the case of the solid object the pictures are dissimilar; there is therefore an essential difference between the impressions on the organs of sensation in the two cases, and consequently between the perceptions formed in the mind; the painting therefore cannot be confounded with the solid object.[56]

What he seeks, then, is a complete equivalence of stereoscopic image and object. Not only will the invention of the stereoscope overcome the deficiencies of painting but also those of the Diorama, which Wheatstone singles out. The Diorama, he believed, was too bound up in the techniques of painting, which depended for their illusory effects on the depiction of distant subjects. The stereoscope, on the contrary, provided a form in which "vividness" of effect increased as the object represented appeared closer to the viewer, and the impression of three-dimensional solidity became greater as the optic axes of each eye diverged. Thus the desired effect of the stereoscope was not simply likeness, but immediate, apparent *tangibility*. But it is a tangibility that has been transformed into a purely visual experience of a kind that Diderot could never have imagined. Even as sophisticated a student of vision as Helmholtz could write, in the 1850s:

> These stereoscopic photographs are so true to nature and so lifelike in their portrayal of material things, that after viewing such a picture and recognizing in it some object like a house, for instance, we get the impression, when we actually do see the object, that we have already seen it before and are more or less familiar with it. In cases of this kind, the actual view of the thing itself does not add anything new or more accurate to the previous apperception we got from the picture, so far at least as mere form relations are concerned.[57]

No other form of representation in the nineteenth century had so conflated the real with the optical, an object with its image.

The stereoscope as a means of representation was inherently *obscene*. It shattered the *scenic* relationship between viewer and object that was intrinsic to the fundamentally theatrical setup of the camera obscura. The very functioning of the stereoscope depended on the visual priority of the objects closest to the viewer and on the absence of any mediation between eye and objects viewed.[58] It was a fulfillment of what Walter Benjamin saw as part of the visual culture of modernity: "Day by day the need becomes greater to take possession of the object – from the closest proximity – in an image and the reproduction of the image."[59] It is no coincidence that the stereoscope became increasingly synonymous with varieties of pornographic imagery in the course of the nineteenth century. The very effects of tangibility that Wheatstone had sought from the beginning were quickly turned into a mass form of ocular possession. Some have speculated that the close association of the stereoscope with pornography was in part responsible for its social demise; around the turn of the century sales of the device supposedly dwindled because of its link with 'indecent" subject matter. Although the reasons for the collapse of the stereoscope lie elsewhere, as I will suggest shortly, the simulation of tangible three-dimensionality hovers uneasily at the limits of acceptable verisimilitude.[60]

If photography preserved an ambivalent (and superficial) relation to the codes of monocular space and geometrical perspective, the relation of the stereoscope to these older forms was one of annihilation, not compromise.

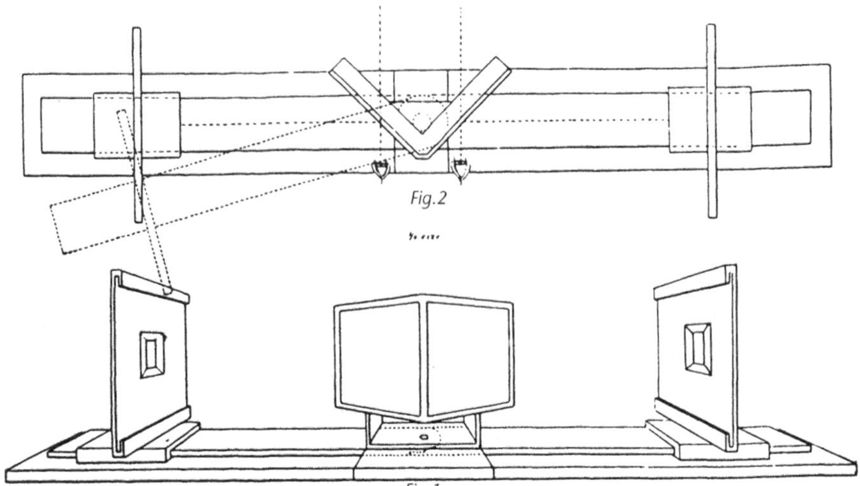

Diagram of the operation of the Wheatstone stereoscope.

Charles Wheatstone's question in 1838 was, "What would be the visual effect of simultaneously presenting to each eye, instead of the object itself, its projection on a plane surface as it appears to that eye?" The stereoscopic spectator sees neither the identity of a copy nor the coherence guaranteed by the frame of a window. Rather, what appears is the technical reconstitution of an already reproduced world fragmented into *two* nonidentical models, models that precede any experience of their subsequent perception as unified or tangible. It is a radical repositioning of the observer's relation to visual representation. The institutionalization of this de-centered observer and the stereoscope's dispersed and multiplied sign severed from a point of external reference indicate a greater break with a classical observer than that which occurs later in the century in modernist painting. The stereoscope signals an eradication of "the point of view" around which, for several centuries, meanings had been assigned reciprocally to an observer and the object of his or her vision. Perspective is not even a possibility under the terms of this technique of beholding. An observer no longer sees an image that has an intelligible or quantifiable location in space, but rather a hallucinatory composite of two dissimilar images whose positions refer to the anatomical structure of the observer's body.

To fully appreciate the rupture signified by the stereoscope, it is important to consider one of its earliest forms, the so-called Wheatstone stereoscope. In order to view images with this device, an observer placed his eyes directly in front of two plane mirrors set 90 degrees to one another. The images to be viewed were held in slots on either side of the observer, and thus were spatially completely separated from each other. Unlike the Brewster stereoscope, invented in the late 1840s, or the familiar Holmes viewer, invented in 1861, the Wheatstone model made clear the atopic nature of the perceived stereoscopic image, the disjunction between experience and its cause. The later models allowed the viewer to believe that he was looking forward *at* something "out there." But the Wheatstone model left the hallucinatory and fabricated nature of the experience undisguised. It did not support what Roland Barthes called "the referential illusion."[61] There simply was nothing "out there." The illusion of relief and of depth was thus a subjective event and the observer coupled with the apparatus was the agent of synthesis or fusion.

Like the Phenakistiscope and other nonprojective optical devices, the stereoscope also required the corporal adjacency and immobility of the observer. They are part of a nineteenth-century modulation in the relation between eye and optical apparatus. During the seventeenth and eighteenth centuries, that relationship had been essentially metaphoric: the eye and the camera obscura or the eye and the telescope or microscope were allied by a conceptual similarity, in which the authority of the eye remained unchallenged.[62] Beginning in the nineteenth century, the relation between eye and optical apparatus becomes one of metonymy: both were now contiguous instruments on the same plane of operation, with varying capabilities and features.[63] The limits and deficiencies of one will be complemented by the capacities of the other and vice versa. The optical apparatus undergoes a shift comparable to that of the tool as described by Marx: "From the moment that the tool proper is taken from man, and fitted into a mechanism, a machine takes the place of a mere implement."[64] In the older, handcraft-based work, he explained, a workman "makes use of a tool," that is, the tool had a metaphoric relation to the innate powers of the human subject.[65] In the factory, Marx contended, the machine makes use of man by subjecting him to a relation of contiguity, of part to other parts, and of exchangeability.[66] Georges Canguilheim makes a crucial distinction between eighteenth-century Utilitarianism, which derived its idea of utility from its definition of man as toolmaker, and the instrumentalism of the human sciences in the nineteenth century, which is based on "one implicit postulate: that the nature of man is to be a tool, that his vocation is to be set in his place and to be set to work."[67] Although "set to work" may sound inappropriate in a discussion of optical devices, the apparently passive observers of the stereoscope and Phenakistiscope were in fact made into producers, by virtue of specific physical capacities, of forms of verisimilitude.

A crucial feature of these optical devices of the 1830s and 1840s is the undisguised nature of their operational structure and the form of subjection they entail. Even though they provide access to "the real," they make no claim that the real is anything other than a mechanical production. The optical experiences they manufacture are clearly disjunct from the images used in the device. They refer as much to the functional interaction of body and machine as they do to external objects, no matter how "vivid" the quality of

Phantasmagoria.

the illusion. So when the Phenakistiscope and the stereoscope eventually disappeared, it was not as part of a smooth process of invention and improvement but rather because these earlier forms were no longer adequate to current needs and uses.

One reason for their obsolescence was that they were insufficiently "phantasmagoric," a word which Adorno, Benjamin, and others have used to describe forms of representation after 1850. *Phantasmagoria* was a name for a specific type of magic lantern performance in the 1790s and early 1800s, which used back projection to keep an audience unaware of the lanterns. Adorno takes the word to indicate

> the occultation of production by means of the outward appearance of the product . . . this outer appearance can lay claim to the status of being. Its perfection is at the same time the perfection of the illusion that the work of art is a reality *sui generis* that constitutes itself in the realm of the absolute without having to renounce its claim to image the world.[68]

But the effacement or mystification of a machine's operation was precisely what David Brewster hoped to overcome with his kaleidoscope and stereoscope. He optimistically saw the spread of scientific ideas in the nineteenth century undermining the possibility of phantasmagoric effects, and he presented a cursory history of civilization in terms of the development of technologies of illusion and apparition.[69] For Brewster, a Scottish Calvinist, the maintenance of barbarism, tyranny, and popery had always been founded on closely guarded knowledge of optics and acoustics, the secrets by which priestly and higher castes ruled. But his implied program, the democratization and mass dissemination of techniques of illusion, simply collapsed that older model of power onto a single human subject, transforming each observer into simultaneously the magician and the deceived.

Even in the later Holmes stereoscope the "concealment of the process of production" did not really occur.[70] Clearly the stereoscope was dependent on a physical engagement with the apparatus which became increasingly inconvenient and unacceptable, but more importantly, the abstract and synthetic nature of the stereoscopic image could never be fully effaced. An apparatus openly based on a principle of disparity, on an awkward "binocular" body, and on an illusion patently derived from the binary referent of the stereoscopic card of paired images, inevitably would give way to a form that preserved the referential illusion more fully than anything before it. Photography also defeated the stereoscope as a mode of visual consumption because it recreated and perpetuated the fiction that the "free" subject of the camera obscura was still viable. Photographs seemed to be a continuation of older "naturalistic" pictorial codes but only because their dominant conventions were restricted to a narrow range of technical possibilities (that is, shutter speeds and lens openings that rendered elapsed time invisible).[71] But photography had already abolished the inseparability of observer and camera obscura, bound together by a single point of view, and made the new camera an apparatus fundamentally independent of the spectator, yet which masqueraded as a transparent and incorporeal intermediary between observer and world. The prehistory of the spectacle *and* the "pure perception" of modernism are lodged in the newly discovered territory of a fully

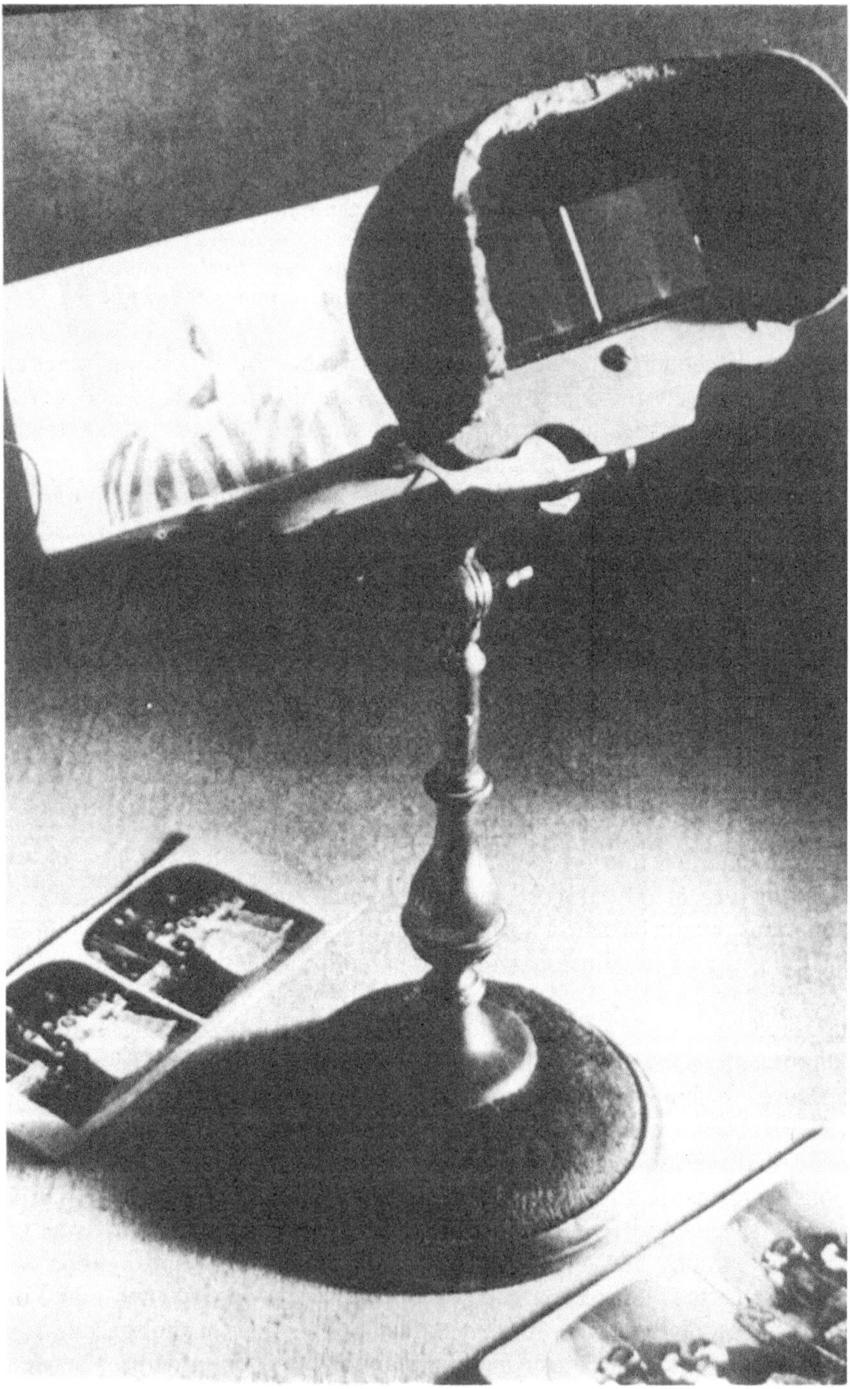

Holmes stereoscope. 1860s.

embodied viewer, but the eventual triumph of both depends on the denial of the body, its pulsings and phantasms, as the ground of vision.[72]

Notes

1. Johann Wolfgang von Goethe, *Theory of Colours*, trans. Charles Eastlake, Cambridge, Mass., MIT Press, 1970, pp. 16–17; *Gedenkausgabe der Werke, Briefe, und Gespräche*, ed. Ernst Beutler, Zurich, 1949, vol. 16, pp. 35–36.
2. The present essay is adapted from a book forthcoming from the MIT Press on the making of the observer in the nineteenth century in which I discuss the paradigm of the camera obscura at length. I argue that the camera obscura must be understood as part of a larger organization of representation, cognition, and subjectivity in the seventeenth and eighteenth centuries (common to *all* of Europe, not just the North as some have suggested) which is fundamentally discontinuous with a nineteenth-century observer. Thus I contend that the camera obscura and photography, as historical objects, are radically dissimilar. See my "Modernizing Vision," in *Vision and Visuality: Discussion in Contemporary Culture*, ed. Hal Foster, Seattle, Bay Press, 1988.
3. Goethe, *Theory of Colours*, p. 17. (Emphasis added.)
4. *Ibid.*, p. 21. (Emphasis added.) See Ernst Cassirer, *Rousseau, Kant, and Goethe*, trans. James Gutmann, Princeton, Princeton University Press, 1945, pp. 81–82: In his color theory Goethe aimed "to include nothing but the world of the eye, which contains only form and color."
5. Immanuel Kant, *Critique of Pure Reason*, trans. Norman Kemp Smith, New York, St. Martin's Press, 1965, pp. 24–25.
6. William Blake, "Annotations to Reynolds" [c. 1808], in *Complete Writings*, ed. Geoffrey Keynes, Oxford, Oxford University Press, 1972, p. 456.
7. Michel Foucault, *The Order of Things*, New York, Pantheon, 1970, p. 132. (Emphasis added.)
8. "Die Farbe selbst is ein Schattiges; deswegen Kircher volkommen recht hat, sie *Lumen opacatum* zu nennen" (Goethe, *Gedenkausgabe*, p. 45). The reference to Athanasius Kircher (1602–1680), the Jesuit turned magic-lantern showman, recalls a counter-use of classical optical systems. Kircher's career as a conjurer was based on the subversion and making opaque of the "transparency" of the camera obscura and other optical instruments.
9. The thematic of repression is central to Jean-François Lyotard's discussion of Renaissance representation in *Discours, Figure*, Paris, Klincksieck, 1978, esp. pp. 163–189.
10. This point is made in Elaine Escoubas, "L'Oeil du teinturier," *Critique*, vol. 37, no. 418 (March 1982), pp. 231–242.
11. Foucault, *The Order of Things*, p. 319.
12. Maine de Biran, *Considerations sur les principes d'une division des faits psychologiques et physiologiques*, in *Oeuvres des Maine de Biran*, vol. 13, ed. P. Tisserand, Paris, Presses Universitaires de France, 1949, p. 180. An important study is Michel Henry, *Philosophie et phénoménologie du corps: essay sur l'ontologie biranienne*, Paris, Presses Universitaires de France, 1965.
13. Maine de Biran, *Influence de l'habitude sur la faculté de penser* [1803], ed. P. Tisserand, Paris, 1953, pp. 56–60: "When the eye fixes itself on a single color, for a certain length of time, in its manner of becoming fatigued there follows a mixed form of this color and several others, and over time the original color will no longer be contained in this new mixture."

14. Georges Canguilhem, "Qu'est-ce que la psychologie," *Études d'histoire et de philosophie des sciences*, Paris, Librarie philosophique de J. Vrin, 1968, pp. 374–375.
15. See P. M. Harman, *Energy, Force, and Matter: The Conceptual Development of Nineteenth-Century Physics*, Cambridge, Cambridge University Press, 1982, pp. 19–26; and Thomas S. Kuhn, *The Structure of Scientific Revolutions*, 2nd ed., Chicago, University of Chicago Press, 1970, pp. 73–74.
16. See, for example, David C. Lindberg, *Theories of Vision from Al-Kindi to Kepler*, Chicago, University of Chicago Press, 1976; and Gerard Simon, *Le regard, l'être et l'apparence dans l'optique de l'antiquité*, Paris, Seuil, 1988.
17. See Edward Frankel, "Corpuscular Optics and the Wave Theory of Light: the Science and Politics of a Revolution in Physics," *Social Studies of Science*, vol. 6, 1976, pp. 141–184; G. N. Cantor, *Optics After Newton*, Manchester, Manchester University Press, 1983, esp. pp. 150–159; and R. H. Silliman, "Fresnel and the Emergence of Physics as a Discipline," *Historical Studies in the Physical Sciences*, vol. 4, 1974, pp. 137–162.
18. Goethe identifies some of these earlier researchers, including Robert W. Darwin (1766–1848), the father of Charles, and the French naturalist Buffon (1707–1788). See *Theory of Colours*, pp. 1–2. See also E. G. Boring, *A History of Experimental Psychology*, New York, Appleton-Century-Crofts, 1950, pp. 102–104.
19. Nineteenth-century science suggested "the idea of a reality which endures inwardly, which is duration itself" (Henri Bergson, *Creative Evolution*, trans. Arthur Mitchell, New York, Random House, 1944, p. 395).
20. Goethe, *Theory of Colours*, pp. 16–17.
21. On the instantaneity of perception see, for example, Lindberg, *Theories of Vision*, pp. 93–94.
22. G. W. F. Hegel, *The Phenomenology of Mind*, trans. J. B. Baillie, New York, Harper and Row, 1967, p. 98.
23. Goethe, *Theory of Colours*, p. 13.
24. It should be noted, however, that Hegel, in an 1807 letter to Schelling, criticized Goethe's color theory for being "restricted completely to the empirical" (*Briefe von und an Hegel*, vol. 1, ed. Karl Hegel, Leipzig, 1884, p. 94, cited in Karl Löwith, *From Hegel to Nietzsche: The Revolution in Nineteenth-Century Thought*, trans. David E. Green, New York, Holt, Rinehart and Winston, 1964, p. 13).
25. André-Marie Ampère, "Lettre à Maine de Biran" [1809], in *Philosophie des Deux Ampéres*, ed. J. Barthélemy-Saint-Hilaire, Paris, Didier, 1866, p. 236.
26. Benjamin B. Wolman, "The Historical Role of Johann Friedrich Herbart," in *Historical Roots of Contemporary Psychology*, ed. Benjamin B. Wolman, New York, Harper and Row, 1968, p. 33. See also David E. Leary, "The Historical Foundations of Herbart's Mathematization of Psychology," *Journal of the History of the Behavioral Sciences*, vol. 16 (1980), pp. 150–163.
27. Johann Friedrich Herbart, *A Textbook in Psychology: An Attempt to Found the Science of Psychology on Experience, Metaphysics, and Mathematics*, trans. Margaret Smith, New York, Appleton, 1891, pp. 21–22.
28. See Johann Friedrich Herbart, *Psychologie als Wissenschaft*, vol. 1, Königsberg, August Unzer, 1824, pp. 222–224.
29. For Herbart's theories of education, see Harold B. Dunkel, *Herbart and Herbartism: An Educational Ghost Story*, Chicago, University of Chicago Press, 1970, esp. pp. 63–96.
30. See Nikolas Rose, "The Psychological Complex: Mental Measurement and Social Administration," *Ideology and Consciousness*, no. 5 (Spring 1979), pp. 5–70; and Didier Deleule and François Guéry, *Le corps productif*, Paris, Mame, 1973, pp. 72–89.
31. Purkinje wrote in Latin, which was translated into Czech. For relevant English translations, see "Visual Phenomena," trans. H. R. John, in *History of Psychology: A Source Book in Systematic Psychology*, ed. William Sahakian, Itasca, Ill., F. E. Peacock, 1968,

pp. 101–108; and "Contributions to a Psychology of Vision," trans. Charles Wheatstone, *Journal of the Royal Institution*, vol. 1 (1830), pp. 101–117, reprinted in *Brewster and Wheatstone on Vision*, ed. Nicholas Wade, London, Academic Press, 1983, pp. 248–262.

32. Goethe provides a telling account of the subjectivity of the afterimage in which the physiology of the attentive (male heterosexual) eye and its operation are inseparable from memory and desire: "I had entered an inn towards evening, and, as a well-favored girl, with a brilliantly fair complexion, black hair, and a scarlet bodice, came into the room, I looked attentively at her as she stood before me at some distance in half shadow. As she presently afterwards turned away, I saw on the white wall, which was now before me, a black face surrounded with a bright light, while the dress of the perfectly distinct figure appeared of a beautiful sea green" (*Theory of Colours*, p. 22).
33. It should be noted that Purkinje, in 1823, formulated the first classification system for human fingerprints, another technique of producing and regulating human subjects. See Vlasilav Krutz, "Purkinje, Jan Evangelista," *Dictionary of Scientific Biography*, vol. 11, New York, 1975, pp. 213–217.
34. See Georges Canguilhem, *Le normal et le pathologique*, Paris, Presses Universitaires de France, 1966, trans, as *The Normal and the Pathological* by Carolyn Fawcett, New York, Zone Books, forthcoming.
35. See, for example, C. W. Ceram, *Archaeology of the Cinema*, New York, Harcourt, Brace and World, 1965; Michael Chanan, *The Dream that Kicks: The Prehistory and Early Years of Cinema in Britain*, London, Routledge and Kegan Paul, 1980, esp. pp. 54–65; Jean-Louis Comolli, "Technique et idéologic," *Cahiers du cinéma*, no. 229 (May–June 1971), pp. 4–21; Jean Mitry, *Histoire du cinéma*, vol. 1, Paris, Editions Universitaires, 1967, pp. 21–27; and Georges Sadoul, *Histoire générale du cinéma*, vol. 1, Paris, Denoël, 1973, pp. 15–43. See also the brief geneaology in Gilles Deleuze, *Cinema 1: The Movement-Image*, trans. Hugh Tomlinson and Barbara Habberjam, Minneapolis, University of Minnesota Press, 1986, p. 5.
36. Joseph Plateau, *Dissertation sur quelques propriétés des impressions*, thesis submitted at Liège, May 1829. Quoted in Georges Sadoul, *Histoire générate du cinéma*, p. 25.
37. Some recent studies have discussed the "myth" of persistence of vision. They tell us, not surprisingly, that current neurophysiological research shows nineteenth-century explanations of fusion or blending of images to be an inadequate explanation for the perception of illusory movement. See Joseph and Barbara Anderson, "Motion Perception in Motion Pictures," and Bill Nichols and Susan J. Lederman, "Flicker and Motion in Film," both in *The Cinematic Apparatus*, ed. Teresa de Lauretis and Stephen Heath, London, Macmillan, 1980, pp. 76–95 and 96–105.
38. Quoted in Chanan, *The Dream that Kicks*, p. 61.
39. Walter Benjamin, *Charles Baudelaire: A Lyric Poet in the Era of High Capitalism*, trans. Harry Zohn, London, NLB, 1972, p. 126.
40. An important study on the relation between the panorama and the Diorama is Eric de Kuyper and Emile Poppe, "Voir et regarder," *Communications*, no. 34 (1981), pp. 85–96. Other works include Stephan Oettermann, *Das Panorama*, Munich, Syndikat, 1980; Heinz Buddemeier, *Panorama, Diorama, Photographie: Entstehung und Wirkung neuer Medien im 19. Jahrhundert*, Munich, H. Fink, 1970; and Helmut and Alison Gernsheim, *L. J. M. Daguerre: The History of the Diorama and the Daguerreotype*, New York, Dover, 1968.
41. Karl Marx, *Capital*, vol. 1, trans. Samuel Moore and Edward Aveling, New York, International, 1967, p. 374.
42. For more on this device, see my "Notes on the Kaleidoscope and Stereoscope," *Journal*, no. 5 (Autumn 1985), pp. 38–41.

43. Sir David Brewster, *The Kaleidoscope: Its History, Theory, and Construction* [1819], rpt. London, John Murray, 1858, pp. 134–136.
44. Charles Baudelaire, "Le peintre de la vie moderne," *Oeuvres complètes*, Paris, Gallimard, 1961, p. 1161.
45. There are few serious cultural or historical studies of the stereoscope. Some helpful works are: Edward Earle, ed., *Points of View: The Stereoscope in America: A Cultural History*, Rochester, Visual Studies Workshop, 1979; A.T. Gill, "Early Stereoscopes," *The Photographic Journal*, no. 109 (1969), pp. 546–599, 606–614, 641–651; and Rosalind Krauss, "Photography's Discursive Spaces: Landscape/View," *Art Journal*, no. 42 (Winter 1982), pp. 311–319.
46. By 1856, two years after its founding, the London Stereoscopic Company alone had sold over half a million viewers. See Helmut and Alison Gernsheim, *The History of Photography*, London, Thames and Hudson, 1969, p. 191.
47. The best known formulation of the Molyneux problem is John Locke, *An Essay Concerning Human Understanding*, ed. Alexander C. Fraser, New York, Dover, vol. II, no. ix, p. 8.
48. See Ernst Cassirer, *The Philosophy of the Enlightenment*, trans. Fritz C. A. Koelln and James Petlegrove, Princeton, Princeton University Press, 1951, p. 108.
49. See Michel Serres, *Hermès ou la communication*, Paris, Minuit, 1968, pp. 124–125; and Maurice Merleau-Ponty, *The Primacy of Perception*, Evanston, Ill., Northwestern University Press, 1964, pp. 169–172.
50. See, for example, William Porterfield, *A Treatise on the Eye, the Manner and Phenomena of Vision*, Edinburgh, Hamilton and Balfour, 1759, p. 285.
51. See R. L. Gregory, *Eye and Brain: The Psychology of Seeing*, 3rd. ed., New York, McGraw-Hill, 1979, p. 45.
52. Charles Wheatstone, "Contributions to the Physiology of Vision," in *Brewster and Wheatstone on Vision*, p. 65.
53. *Ibid.*
54. Hermann von Helmholtz, "The Facts in Perception," in *Epistemological Writings*, ed. Paul Hertz and Moritz Schlick, Boston, Boston Studies in the Philosophy of Science, 1977, p. 133: ". . . our acquaintance with the visual field can be acquired by observation of the images during the movements of our eyes, provided only that there exists, between otherwise qualitatively alike retinal sensations, some or other perceptible difference corresponding to the difference between distinct places on the retina."
55. Sir David Brewster, *The Stereoscope: Its History, Theory, and Construction*, London, John Murray, 1856, p. 53. (Emphasis in original.)
56. Charles Wheatstone, "Contributions to the Physiology of Vision," p. 66.
57. Hermann von Helmholtz, *Handbook of Physiological Optics*, vol. 3, trans. George T. Ladd, New York, Dover, 1962, p. 303.
58. See Florence de Mèridieu, "De l'obscénité photographique," *Traverses*, no. 29 (October 1983), pp. 86–94.
59. Walter Benjamin, "A Short History of Photography," *Artforum*, no. 15 (February 1977), p. 49.
60. The ambivalence with which twentieth-century audiences have received 3-D movies and holography suggests the enduring problematic nature of such techniques.
61. See Roland Barthes, "L'effet de réel," *Communications*, no. 11 (1968), pp. 84–89; trans, as "The Reality Effect" by Richard Howard, in *The Rustle of Language*, New York, Hill and Wang, 1986, pp. 141–148.
62. On the telescope as metaphor in Galileo, Kepler, and others, see Timothy J. Reiss, *The Discourse of Modernism*, Ithaca, N.Y., Cornell University Press, 1980, pp. 25–29.
63. "In Metonymy, phenomena are implicitly apprehended as bearing relationships to one another in the modality of part-part relationships, on the basis of which one can

effect a *reduction* of one of the parts to the status of an aspect or function of the other" (Hayden White, *Metahistory: The Historical Imagination in Nineteenth-Century Europe*, Baltimore, Johns Hopkins University Press, 1973, p. 35).

64. Karl Marx, *Capital*, vol. 1, p. 374. In this sense, other optical instruments of the seventeenth and eighteenth centuries, like peep shows, Claude glasses, and print viewing boxes had the status of tools.
65. *Ibid.*, p. 422. J. D. Bernal has noted that the instrumental capacities of the telescope and microscope remained remarkably undeveloped during the seventeenth and eighteenth centuries. Until the nineteenth century, the microscope "remained more amusing and instructive, in the philosophical sense, than of scientific and practical value" (*Science in History, Vol. 2: The Scientific and Industrial Revolutions*, Cambridge, Mass., MIT Press, 1971, pp. 464–469).
66. Marx again indicates the new metonymic status of the human subject: "As soon as man, instead of working with an implement on the subject of his labor, becomes merely the motive power of an implement-machine, it is a mere accident that motive power takes the disguise of human muscle; and it may equally well take the form of wind, water, or steam" (*Capital*, vol. 1, p. 375).
67. Canguilhem, "Qu'est-ce que la psychologie" p. 378.
68. Theodor Adorno, *In Search of Wagner*, trans. Rodney Livingstone, London, Verso, 1981, p. 85. On Adorno, Wagner, and phantasmagoria, see Andreas Huyssen, *After the Great Divide: Modernism, Mass Culture, Postmodernism*, Bloomington, University of Indiana Press, 1986, pp. 34–42.
69. Sir David Brewster, *Letters on Natural Magic*, New York, J. J. Harper, 1832, pp. 15–21.
70. This device is described by its inventor in Oliver Wendell Holmes, "The Stereoscope and the Stereograph," *Atlantic Monthly*, vol. 3, no. 20 (June 1859), pp. 740–752.
71. For the disruptive effect of Muybridge and Marey on nineteenth-century codes of "naturalistic" representation, see Noël Burch, "Charles Baudelaire Versus Doctor Frankenstein," *Afterimage*, nos. 8–9 (Spring 1981), pp. 10–13.
72. On the problem of modernism, vision and the body, see the recent work of Rosalind Krauss, for example, "Antivision," *October*, no. 36 (Spring 1986), pp. 147–154; and "Where's Poppa," forthcoming in, *Marcel Duchamp Centennial Conference at Nova Scotia School of Art*, Thierry de Duve, ed., Cambridge, Mass., MIT Press.

4

Interpretation: Observer Effects

William Thompson

Introduction

The term *observer effect* generally describes circumstances in which the results of an observation are affected by the observer. In physics, the term has been used to describe circumstances in which the act of observation changes the phenomenon being observed, as where measurement of electrical current in a circuit changes the current flow. By contrast, in the social sciences and in forensic science, the term is used to describe circumstances in which the observer's preconceptions or motives influence conclusions drawn from data [1]. The preconceptions and motives are thought to influence the *perception and interpretation* of the evidence rather than changing the evidence itself. Observer effects are sometimes also called *examiner bias* [2, 3] although it important to note that the "bias" entailed in the phenomenon may occur without the observer intending or even being aware of it [4, 5].

Observer effects are closely related to *context effects* and the two terms are sometimes used synonymously [4]. The term *context effect* originated in psychology and has been used to describe circumstances in which the perception of a stimulus is affected by the surrounding context, as where a gray object looks lighter against a dark background than against a light background (see, e.g., http://web.mit.edu/persci/people/adelson/checkershadow_illusion.html). In forensic science, the term *context effect* has been used more broadly to describe situations in which the results of a forensic analysis

Source: Allan Jamieson and Andre A. Moenssens (eds), *Wiley Encyclopaedia of Forensic Science* (Chichester: John Wiley & Sons Ltd, 2009), pp. 1–4.

are affected by the circumstances in which it is performed, and particularly by the information available to the analyst, as when an analyst becomes more likely to identify a latent print as that of a suspect when told that another analyst has already made the identification or when told that other evidence indicates the suspect made the print. The "other evidence" might be said to provide a "context" that changes the analyst's interpretation of the data contained in the prints. Alternatively, the "other information" might be said to have changed the analyst's expectations about the data and hence to have induced an *observer effect*.

The same phenomenon is sometimes described as *confirmation bias*. In psychology and cognitive science, the term *confirmation bias* refers to a human tendency to evaluate evidence in a manner that supports or confirms one's preconceptions, such as a tendency to search out and give more weight to evidence that supports a favored hypothesis than to evidence that contradicts it [5]. In forensic science, the term *confirmation bias* has been used in a manner that is roughly interchangeable with the terms *observer effect* and *context effect* [6, 7].

Underlying Psychological Phenomenon

Underlying all of these terms is a basic phenomenon of human psychology – the tendency of observers to interpret data in a manner consistent with their expectations and desires. The existence of this phenomenon is well established – it has been called *one of the most venerable ideas of traditional epistemology* as well as *one of the better demonstrated findings of twentieth century psychology* [8]. An early discussion of the phenomenon can be found in the writings of Francis Bacon, who commented in 1620 that

> The human understanding when it has once adopted an opinion draws all things else to support and agree with it. And though there be a greater number and weight of instances to be found on the other side, yet these it either neglects and despises, or else by some distinction sets aside and rejects, in order that by this great and pernicious predetermination the authority of its former conclusions may remain inviolate [9].

Even earlier, Julius Caesar famously noted men's tendency to "believe quite readily that which they wish to be true" [10]. In the twentieth century, psychologists confirmed the existence and strength of these phenomena with innumerable experiments in which people's interpretation of a variety of stimuli, data, and forms of evidence were shown to be influenced by preconceptions and desires. The effects are greatest when the underlying data are somewhat ambiguous and when observers are influenced by strongly held expectations and motives. Extensive reviews of this literature are found in [1, 4, 5, 8, 11, 12].

The most relevant studies relate to the interpretation of scientific data. Like everyone else, scientists have a tendency to interpret data in a manner that supports theories that they favor. The tendency of academic scientists to cling to pet theories, well past the point at which such adherence could be justified by the evidence, has been widely noted [13–15]. In his iconic book *Galileo's Revenge: Junk Science in the Courtroom* [16], Peter Huber traced several false scientific theories to misinterpretation of data arising from uncontrolled observer effects. A scientist who is committed to a pet theory inevitably (and unconsciously) interprets data in a manner consistent with that theory. By Huber's account, uncontrolled observer effects are one of the hallmarks of junk science (*see* **Interpretation: Legal Perspective**).

Minimizing Observer Effects

Allowing preconceptions to influence the interpretation of data is said to be the "cardinal sin for the formal, pure scientist" [8]. Accordingly, when scientists must rely on subjective judgment to interpret the results of an experiment, they routinely take careful steps to mask or shield the person interpreting data from extraneous information that might improperly influence the interpretation. For example, scientists in most fields use "blind" or "double-blind" procedures when relying on subjective judgment to interpret data. Blind procedures are also widely used for peer-review of scientific articles, for grading of written examinations, and for other functions for which it is important to minimize observer effects.

The field of forensic science has been criticized for failing to take adequate steps to minimize observer effects [1, 4]. Through communications with police, lawyers, and other experts, forensic scientists often are exposed to information that may influence their expectations and perhaps even their hopes and desires about what a particular examination might reveal [1]. Moreover, when performing examinations forensic scientists often rely in part on subjective judgment to evaluate potentially ambiguous data [17]. Even seemingly objective procedures like DNA testing sometimes require analysts to use subjective judgment to resolve crucial ambiguities and hence could potentially be influenced by observer effects [3, 18–21]. Yet forensic scientists rarely take steps to shield themselves from extraneous information – i.e., information unnecessary for making a scientific assessment – when making comparisons or interpreting test results [1] (*see also*, **Threat Assessment: School; DNA: Degraded Samples**).

The difference between forensic science and other scientific fields may stem in part from uncertainty about what types of information are necessary and relevant to a forensic science assessment. In some forensic disciplines consideration of nonscientific evidence is not only accepted but required by professional norms. For example, fire investigators are trained to consider whether a suspect had a motive for setting a fire when deciding whether to classify a fire as arson [22] (*see also* **Fire and Explosion Investigations: Overview; Fire: Scene Investigation**). There have been reports of forensic scientists relying on nonscientific evidence in other disciplines as well, such as a DNA analyst who defended a "match" that implicated the defendant in a rape case by saying: "I know I am right. They found the victim's purse in [the defendant's] apartment" [21]. In 1997 the US Justice Department's Office of Inspector General reported that FBI explosives experts had relied on extraneous evidence when making key scientific determinations in a number of important cases [23]. In the first World Trade Center bombing in 1993, for example, FBI examiners had relied on the fact that the suspects had access to urea nitrate to reach the conclusion that urea nitrate had been used to make the bomb. The Inspector General condemned this reasoning as circular and biased. The FBI laboratory management agreed and pledged to take steps to ensure that the problem never happened again. Academic commentators have been unanimous in condemning the use of such nonscientific evidence as a basis for scientific conclusions [1, 6, 7, 17, 18]. They argue that the role of the forensic scientist is to offer conclusions derived from a scientific discipline, not to offer conclusions based in part on analysis of other evidence in the case (*see* **Fire and Explosion Investigations: Overview**).

Forensic scientists have become more aware of the importance of addressing observer effects as a result of a high-profile error by the FBI's fingerprint identification unit in the Madrid train bombing case [6, 7, 24]. Three highly trained FBI latent print examiners, and a well regarded independent examiner, all identified a latent print associated with the Madrid bombing as having come from Brandon Mayfield, a resident of Portland, Oregon. The error came to light when Spanish authorities determined the latent print actually matched an Algerian suspected terrorist. The FBI acknowledged the error and apologized to Mr Mayfield. In a subsequent analysis, the agency attributed the error in part to *confirmation bias*. After an initial erroneous identification, each subsequent examiner was fully aware that the latent print had previously been matched to Mayfield. This knowledge may have led them to place too much weight on similarities between the two prints while ignoring or discounting discrepancies.

Empirical Studies of Observer Effects in Forensic Science

Empirical studies have confirmed that observer effects can influence the interpretation of latent prints, leading to errors [25–28]. For example, in one clever study, psychologists asked five highly experienced latent print examiners to compare pairs of prints [26]. The examiners were not told that they had previously individualized these pairs (i.e., found them to match) during casework. Each examiner was instead told (incorrectly) that the prints were those that the FBI had erroneously matched in the Madrid train bombing case. Although they were instructed to "ignore all the contextual information and focus solely on the actual prints", three of the five examiners changed their previous judgment of "match" to "no match" and a fourth changed from "match" to "cannot tell". Only one of the five examiners consistently maintained that the prints were a "match". In other words, after exposure to "extraneous information" suggesting that the prints should *not* match, four of five examiners reached a different conclusion than they had reached previously when comparing the same prints.

Studies in other forensic science domains have also found evidence of observer effects on examiners' judgments [29]. Research on the precise mechanisms by which observer effects influence interpretation are just beginning to emerge [30]. Additional research in this important area is clearly needed.

Case Managers and Sequential Unmasking

There are a number of procedures for minimizing observer effects [1, 4, 31]. One method is to separate various laboratory functions and assign them to different people. A *case manager* who is fully informed of the facts of the case decides what to test and how to test it; a *laboratory analyst* who is "blind" to extraneous information analyzes and interprets the test results. To the extent possible, the analyst remains "blind" to extraneous information, such as information about the legal consequences of the judgment and non-scientific evidence in the case, when deciding whether samples "match". Once the analyst interprets and records the test results, the *case manager* (who is aware of the broader facts of the case) is then responsible for placing the test results in context and assessing the compatibility of forensic observations with various theories of what occurred.

Another approach is simply to perform the analysis and evaluation of various samples in a sequence so that analysis and interpretation of early samples cannot be influenced by knowledge of the later samples. Evidentiary samples, which are generally more difficult evaluate, are analyzed first, before the analyst knows the features of the reference samples. For example, fingerprint examiners can decide whether a latent print is interpretable and which features of the latent print constitute reliable data before knowing whether those features are consistent or inconsistent with a reference print. Similarly, DNA analysts can decide which "alleles" in an evidentiary sample profile are real and which are spurious before knowing whether those alleles match up with a suspect's. Information required to draw an ultimate conclusion is "unmasked" when needed but the analyst performs as much work as possible while "blind" to unnecessary facts [31].

References

[1] Risinger, D.M., Saks, M.J., Thompson, W.C. & Rosenthal, R. (2002). The Daubert/ Kumho implications of observer effects in forensic science: hidden problems of expectation and suggestion, *California Law Review* **90**(1), 1–56.

[2] Beckham, J.C., Annis, L.V. & Gustafson, D.J. (1989). Decision making and examiner bias in forensic expert recommendations for not guilty by reason of insanity, *Law and Human Behavior* **13**(1), 79–87.

[3] Thompson, W.C. (1998). Examiner bias in forensic RFLP analysis, *Scientific Testimony: An Online Journal*, http://www.scientific.org/case-in-point/articles/thompson/ thompson.html.

[4] Saks, M.J., Risinger, D.M., Rosenthal, R. & Thompson, W.C. (2003). Context effects in forensic science, *Science and Justice* **43**(2), 77–90.

[5] Plous, S. (1993). *The Psychology of Judgment and Decision Making*, McGraw-Hill, New York.

[6] Stacey, R.B. (2004). Report on the erroneous fingerprint individualization in the Madrid train bombing case, *Journal of Forensic Identification* **54**(6), 706–718.

[7] Office of the Inspector General U.S. Department of Justice (2006). *A Review of the FBI's Handling of the Brandon Mayfield Case*, Office of the Inspector General U.S. Department of Justice, Washington, DC, pp. 1–330.

[8] Nisbett, R. & Ross, L. (1980). *Human Inference*, Prentice Hall, Englewood Cliffs, p. 67.

[9] Bacon, F. (1620). *Novum Organum, book I*, 109, point 46, reprinted, in Hutchins, R.M. (ed) (1952) *Great Books of the Western World*, Vol. 30, Britannica Publishing, New York.

[10] Caesar, G.J. (1873). *Commentaries on the Gallic war 155 (51 BCE)* John Wiley & Sons.

[11] Schneider, D.J., Hastorf, A.H. & Ellsworth, P.C. (1979). *Person Perception*, 2nd Edition, Addison-Wesley Publishing, Reading.

[12] Gilovich, T. (1991). *How We Know What isn't So: The Fallibility of Human Reason in Everyday Life*, The Free Press, New York.

[13] Barber, B. (1952). *Science and the Social Order*, Collier, New York.

[14] Kuhn, T.S. (1962). *The Structure of Scientific Revolutions*, University of Chicago Press, Chicago.

[15] Mahoney, M.J. (1976). *Scientist as Subject: The Psychological Imperative*, Ballinger, Cambridge.

[16] Huber, P.W. (1991). *Galileo's Revenge: Junk Science in the Courtroom*, Basic Books, New York.

[17] Thompson, W.C., Cole, S.A. (2007). Psychological aspects of forensic identification evidence, in *Expert Psychological Testimony for the Courts*, M. Costanzo, D. Krauss & K. Pezdek, eds, Lawrence Erlbaum and Associates, New York.

[18] Thompson, W.C. (1995). Subjective interpretation, laboratory error and the value of forensic DNA evidence: three case studies, *Genetica* **96**, 153–168.

[19] Thompson, W.C. (1997). A sociological perspective on the science of forensic DNA testing, *UC Davis Law Review* **30**(4), 1113–1136.

[20] Thompson, W.C. (1997). Accepting lower standards: The National Research Council's second report on forensic DNA evidence, *Jurimetrics Journal* **37**(4), 405–424.

[21] Thompson, W.C., Ford, S., Doom, T., Raymer, M. & Krane, D. (2003). Evaluating forensic DNA evidence: essential elements of a competent defense review: part 1, *The Champion* **27**(3), 16–25.

[22] Lentini, J. (2006). *Scientific Protocols for Fire Investigation*, CRC Press, Boca Raton.
[23] Office of the Inspector General, U.S. Department of Justice (1997). *The FBI Laboratory: An Investigation Into Laboratory Practices and Alleged Misconduct In Explosives-Related and Other Cases*, Office of the Inspector General, U.S. Department of Justice, Washington, DC.
[24] Thompson, W.C. & Cole, S.A. (2005). Lessons from the Brandon Mayfield case, *The Champion* **29**, 32–34.
[25] Dror, I.E., Peron, A., Hind, S.L. & Charlton, D. (2005). When emotions get the better of us: the effect of contextual top-down processing on matching fingerprints, *Applied Cognitive Psychology* **19**(6), 799–809.
[26] Dror, I.E., Charlton, D. & Peron, A. (2006). Contextual information renders experts vulnerable to making erroneous identifications, *Forensic Science International* **156**, 74–78.
[27] Dror, I.E. & Charlton, D. (2006). Why experts make errors, *Journal of Forensic Identification* **56**(4), 600–616.
[28] Dror, I.E. & Rosenthal, R. (2008). Meta-analytically quantifying the reliability and biasability of forensic experts, *Journal of Forensic Sciences* **53**(4), 900–903.
[29] Miller, L.S. (1987). Procedural bias in forensic examination of hair, *Law and Human Behavior* **11**(2), 157–163.
[30] Schiffer, B. & Champod, C. (2007). The potential (negative) influence of observational biases at the analysis stage of fingermark individualization, *Forensic Science International* **167**, 116–120.
[31] Krane, D.E., Ford, S., Gilder, J.R., Inman, K., Jamieson, A., Koppl, R., Kornfield, I.L., Risinger, D.M., Rudin, N., Taylor, M.S. & Thompson, W.C. (2008). Sequential unmasking: a means of minimizing observer effects in forensic DNA interpretation (letter), *Journal of Forensic Sciences* **53**(4), 1006–1007.

5

Seeing and Knowing

Michel Foucault

'Hippocrates applied himself only to observation and despised all systems. It is only by following in his footsteps that medicine can be perfected'.[1] But the privileges that the clinic had recently recognized in observation were much more numerous than the prestige accorded it by tradition and of a quite different nature. They were at the same time the privileges of a pure gaze, prior to all intervention and faithful to the immediate, which it took up without modifying it, and those of a gaze equipped with a whole logical armature, which exorcised from the outset the naïvety of an unprepared empiricism. We must now describe the concrete exercise of such a perception.

The observing gaze refrains from intervening: it is silent and gestureless. Observation leaves things as they are; there is nothing hidden to it in what is given. The correlative of observation is never the invisible, but always the immediately visible, once one has removed the obstacles erected to reason by theories and to the senses by the imagination. In the clinician's catalogue, the purity of the gaze is bound up with a certain silence that enables him to listen. The prolix discourses of systems must be interrupted: 'All theory is always silent or vanishes at the patient's bedside';[2] and the suggestions of the imagination – which anticipate what one perceives, find illusory relations, and give voice to what is inaccessible to the senses – must also be reduced: 'How rare is the accomplished observer who knows how to await, in the silence of the imagination, in the calm of the mind, and before forming his judgement, the relation of a sense actually being exercised!'[3] The

Source: A.M. Sheridan Smith (tr), *The Birth of the Clinic: An Archaeology of Medical Perception* (London: Routledge, 2003), pp. 131–151.

gaze will be fulfilled in its own truth and will have access to the truth of things if it rests on them in silence, if everything keeps silent around what it sees. The clinical gaze has the paradoxical ability to *hear a language* as soon as it *perceives a spectacle*. In the clinic, what is manifested is originally what is spoken. The opposition between clinic and experiment overlays exactly the difference between the language we hear, and consequently recognize, and the question we pose or, rather, impose: 'The observer . . . reads nature, he who experiments questions'.[4] To this extent, observation and experiment are opposed but not mutually exclusive: it is natural that observation should lead to experiment, provided that experiment should question only in the vocabulary and within the language proposed to it by the things observed; its questions can be well founded only if they are answers to an answer itself without question, an absolute answer that implies no prior language, because, strictly speaking, it is the first word. It is this privilege of possessing an unsupersedable (*indépassable*) origin that the Double expresses in terms of causality: 'observation must not be confused with experience; the latter is the result or effect, the former the means or cause; observation leads naturally to experience'.[5] The observing gaze manifests its virtues only in a double silence: the relative silence of theories, imaginings, and whatever serves as an obstacle to the sensible immediate; and the absolute silence of all language that is anterior to that of the visible. Above the density of this double silence things seen can be heard at last, and heard solely by virtue of the fact that they are seen.

This gaze, then, which refrains from all possible intervention, and from all experimental decision, and which does not modify, shows that its reserve is bound up with the strength of its armature. To be what it must be, it is not enough for it to exercise prudence or scepticism; the immediate on which it opens states the truth only if it is at the same time its origin, that is, its starting point, its principle and law of composition; and the gaze must restore as truth what was produced in accordance with a genesis: in other words, it must reproduce in its own operations what has been given in the very movement of composition. It is precisely in this sense that it is 'analytic'. Observation is logic at the level of perceptual contents; and the art of observing seems to be

> a logic for those meanings which, more particularly, teach their operations and usages. In a word, it is the art of being in relation with relevant circumstances, of receiving impressions from objects as they are offered to us, and of deriving inductions from them that are their correct consequences. Logic is . . . the basis of the art of observing, but this art might be regarded as one of the parts of Logic whose object is more dependent on meanings.[6]

One can, therefore, as an initial approximation, define this clinical gaze as a perceptual act sustained by a logic of operations; it is analytic because it restores the genesis of composition; but it is pure of all intervention insofar as this genesis is only the syntax of the language spoken by things themselves in an original silence. The gaze of observation and the things it perceives communicate through the same Logos, which, in the latter, is a genesis of totalities and, in the former, a logic of operations. Clinical observation involves two necessarily united domains; the hospital domain and the teaching domain.

The hospital domain is that in which the pathological fact appears in its singularity as an event and in the series surrounding it. Not long ago the family still formed the natural locus in which truth resided unaltered. Now its double power of illusion has been discovered: there is a risk that disease may be masked by treatment, by a regime, by various actions tending to disturb it; and it is caught up in the singularity of physical conditions that make it incomparable with others. As soon as medical knowledge is defined in terms of frequency, one no longer needs a natural environment; what one now needs is a neutral domain, one that is homogeneous in all its parts and in which comparison is possible and open to any form of pathological event, with no principle of selection or exclusion. In such a domain everything must he possible, and possible in the same way.

> What a source of instruction is provided by two infirmaries of 100 to 150 patients each! . . . What a varied spectacle of fevers or phlegmasias, malign or benign, sometimes highly developed in strong constitutions, sometimes in a slight, almost latent, condition, together with all the forms and modifications that age, mode of life, seasons, and more or less energetic moral affections can offer![7]

The old objection that the hospital causes modifications that are both pathological disorders and disorderings of pathological forms is neither dismissed nor ignored but rigourously annulled, since the modifications in question are uniformly valid for all events; it is possible, therefore, to isolate them by analysis and to treat them separately; by setting aside modifications due to locality, season, and nature of treatment 'one can succeed in introducing into the hospital clinic and general medical practice a degree of foresight and precision'.[8] The clinic is not, therefore, that mythical landscape in which diseases appear of their own accord, completely revealed; it makes possible the integration, in experience, of the hospital modification in a constant form. What the medicine of species called *nature* is shown to be merely the discontinuity of heterogeneous and artificial conditions; the 'artificial' diseases of the hospital permit pathological events to be reduced to the homogeneous; the hospital domain is no doubt not pure transparency to truth, but the refraction that is proper to it makes possible, through its constancy, the analysis of truth.

By means of the endless play of modifications and repetitions, the hospital clinic makes possible, therefore, the setting aside of the extrinsic. But this same play makes possible the summation of the essential in knowledge: in fact, variations cancel each other out, and the effect of the repetition of constant phenomena outlines spontaneously the fundamental conjunctions. By showing itself in a repetitive form, the truth indicates the way by which it may be acquired. It offers itself to knowledge by offering itself to recognition. 'The student . . . cannot familiarize himself overmuch with the repeated sight of alterations of all kinds, whose particular practice might later show him the picture'.[9] The genesis of the manifestation of truth is also the genesis of the knowledge of truth. There is, therefore, no difference in nature between the clinic as science and the clinic as teaching. A group is thus formed consisting of the master and his pupils, in which the act of recognition and the effort to know find fulfillment in a single movement. In its structure and in its two aspects as manifestation and acquisition, medical experience now has a collective subject; it is no longer divided between those who know and those who do not; it is made up, as one entity, of those who unmask and those before whom one unmasks. The statement is the same; the disease speaks the same language to both.

The *collective* structure of medical experience, the *collective* character of the hospital field – the clinic is situated at the meeting point of the two totalities; the experience that defines it traverses the surface of their confrontation and of their reciprocal boundary. There it derives not only its inexhaustible richness but also its sufficient, enclosed form. It is the carving up of the infinite domain of events by the intersection of the gaze and mutual questions. At the Edinburgh clinic, observation consisted of four series of questions: the first concerned the patient's age, sex, temperament, and occupation; the second, his symptoms; the third, the origin and development of the disease; and the fourth, more distant causes and earlier accidents.[10] Another method – one used at Montpellier – consisted of a general examination of all the visible modifications of the organism: 'first, the alterations of the body in general; second, those in the matter excreted; third, those denoted by the exercise of the functions'.[11] Pinel levelled the same criticism at both forms of investigation: they were unlimited. To the first, he objected: 'How, in the midst of this profusion of questions . . . can one grasp the essential, specific features of the disease?' and to the second, in corresponding fashion: 'What an immense enumeration of symptoms . . . ! Will this not throw us back into a new chaos?'[12] The questions to be asked are innumerable; the things to be seen infinite. If the clinical domain is open only to the tasks of language or to the demands of the gaze, it will have no limits and, therefore, no organization. There is boundary, form, and meaning only if interrogation and examination are connected with each other, defining at the level of fundamental structures the 'meeting place' of doctor and patient. In its initial form, the clinic seeks to determine this locus by three means:

1. The Alternation of Spoken Stages and Perceived Stages in an Observation

In the schema of the ideal investigation sketched by Pinel, the general indication of the first stage is visual: one observes the present state in its manifestations. But the questionnaire already guarantees the place of language within this examination; the symptoms that first strike the senses of the observer are noted, but immediately afterwards the patient is questioned as to the pains he feels, and lastly, by observation, the state of the most important physiological functions is described. The second stage is dominated by language as well as by time, memory, developments, and successive incidents. First what, at a given moment, was perceptible must be recognized (recalling the forms of invasion, the succession of symptoms, the appearance of their present characteristics, and the medicaments already applied). Then the patient or his entourage must be questioned as to his general appearance, his occupation, his past life. The third stage of observation is again one of perception; a day-by-day account is kept of the progress of the disease under four headings: evolution of the symptoms, possible appearance of new phenomena, state of the secretions, and effect of medicaments used. The final stage is reserved to speech: the prescription of the regime during convalescence.[13] In the event of death, most clinicians – but, as we shall see, Pinel less readily than others – reserved to the gaze the final, most decisive authority, namely, the anatomy of the body. In this regular alternation of speech and gaze, the disease gradually declares its truth, a truth that it offers to the eye and ear, whose theme, although possessing only one *sense*, can be restored, in its indubitable totality, only by two *senses*: that which sees and that which listens. This is why the questionnaire without the examination and the examination without the interrogation were doomed to an endless task: it belongs to neither to fill the gaps within the province of the other.

2. The Effort to Define a Statutory Form of Correlation between the Gaze and Language

The theoretical and practical problem confronting the clinicians was to know whether it would be possible to introduce into a spatially legible and conceptually coherent representation that element in the disease that belongs to a visible symptomatology and that which belongs to a verbal analysis. This problem was revealed in a technical difficulty that was very revealing of the demands of clinical thinking: the *picture*. Is it possible to integrate into a picture, that is, into a structure that is at the same time visible and legible, spatial and verbal, that which is perceived on the surface of the body by the clinician's eye, and that which is heard by that same clinician in the essential language of the disease? Perhaps the most naïve attempt was made by Fordyce: he included in the x axis all the notations concerning the climate, the seasons, prevalent diseases, the patient's temperament, idiosyncrasy, appearance, age, and previous accidents; and he classified in the y axis the symptoms according to the organ or function in which they were manifested (pulse, skin, temperature, muscles, eyes, tongue, mouth, breathing, stomach, intestines, urine).[14] It is clear that this functional distinction between visible and expressible (*énonçable*), and their correlation in the myth of an analytic geometry, could be of no use in the work of clinical thought; such an effort is significant only of the data of the problem and of the terms to be correlated. The pictures drawn up by Pinel seem simpler: their conceptual structure is in fact more subtle. As in Fordyce, the y axis Includes the symptomatic elements that the disease offers to perception; but in the x axis, he indicates the significant values that these symptoms may assume. In an acute fever, a painful sensitivity in the pit of the stomach, a headache, and a violent thirst are to be included in a gastric symptomatology; on the other hand, prostration and abdominal tension have an adynamic meaning; lastly, pain in the limbs, a dry tongue, rapid breathing, a paroxysm, especially one occurring in the evening, are signs of both gastricity and adynamism.[15] Thus each visible segment assumes a significant value, and the picture certainly serves an analytical function in clinical knowledge. But it is obvious that the analytical structure is neither produced nor revealed by the picture itself; the analytical structure preceded the picture, and the correlation between each symptom and its symptomatological value was fixed once and for all in an essential a priori; beneath its apparently analytical function, the picture's only role is to divide up the visible within an already given conceptual configuration. The task is not, therefore, one of correlation, but, purely and simply, of redistribution of what was given by a perceptible extent in a conceptual space defined in advance. It makes nothing known; at most, it makes possible recognition.

3. The Ideal of an Exhaustive Description

The arbitrary or tautological appearance of these pictures led clinical thought towards another form of correlation between the visible and the expressible, namely, the continuous correlation of an entirely – that is, doubly – faithful description; in relation to its object it must be, in effect, without any gap; and in language describing the object it must allow no deviation. Descriptive *rigour* will be the result of *precision* in the statement and of *regularity* in the designation: which, according to Pinel, is 'the method now followed in all other parts of natural history'.[16] Thus language is charged with a dual function: by its value as precision, it establishes a correlation between each sector of the visible and an expressible element that corresponds to it as accurately as possible; but this expressible element operates, within its role as description, a denominating function which, by its articulation upon a constant, fixed vocabulary, authorizes comparison, generalization, and establishment within a totality. By virtue of this dual function, the work of description ensures 'a prudent reserve in rising to general views without lending reality to abstract terms' and 'a simple, regular distribution, invariably based on the relations of structure or the organic functions of the parts'.[17]

It is in this exhaustive and complete passage from the *totality of the visible* to the *over-all structure of the expressible* (*structure d'ensemble de l'énonçable*) that is fulfilled at last that significative analysis of the perceived that the naïvely geometric architecture of the picture failed to provide. It is description, or, rather, the implicit labour of language in description, that authorizes the transformation of symptom into sign and the passage from patient to disease and from the individual to the conceptual. And it is there that is forged, by the spontaneous virtues of description, the link between the random field of pathological events and the pedagogical domain in which they formulate the order of their truth. To describe is to follow the ordering of the manifestations, but it is also to follow the intelligible sequence of their genesis; it is to see and to know at the same time, because by saying what one sees, one integrates it spontaneously into knowledge; it is also to learn to see, because it means giving the key of a language that masters the visible. The well-made language, which Condillac and his successors saw as the ideal of scientific knowledge, must not therefore be sought, as do certain over-hasty doctors,[18] on the side of a language of calculation, but on the side of that *measured language* that has the measure of both the things that it describes and the language in which it describes them. For the dream of an arithmetical structure of medical language must be substituted, therefore, the search for a certain internal measurement consisting of fidelity and fixity, of primary and absolute openness to things and rigour in the considered use of semantic values. 'The art of describing facts is the supreme art in medicine: everything pales before it'.[19]

Over all these endeavours on the part of clinical thought to define its methods and scientific norms hovers the great myth of a pure Gaze that would be pure Language: a speaking eye. It would scan the entire hospital field, taking in and gathering together each of the singular events that occurred within it; and as it saw, as it saw ever more and more clearly, it would be turned into speech that states and teaches; the truth, which events, in their repetitions and convergence, would outline under its gaze, would, by this same gaze and in the same order, be reserved, in the form of teaching, to those who do not know and have not yet seen. This speaking eye would be the servant of things and the master of truth.

It is understandable that, after the revolutionary dream of an absolutely open science and practice, a certain medical esotericism could be revived around these themes: one now sees the visible only because one knows the language; things are offered to him who has penetrated the closed world of words; and if these words communicate with things, it is because they obey a rule that is intrinsic to their grammar. This new esotericism is different in structure, meaning, and use from that which made Molière's doctors speak in Latin: then it was simply a matter of not being understood and of preserving at the level of linguistic formulation the corporate privileges of a profession; now operational mastery over things is sought by accurate syntactic usage and a difficult semantic familiarity with language. Description, in clinical medicine, does not mean placing the hidden or the invisible within reach of those who have no direct access to them; what it means is to give speech to that which everyone sees without seeing – a speech that can be understood only by those initiated into true speech. 'Whatever precepts are given about so delicate a matter, it will always remain beyond the reach of the multitude'.[20] Here, at the level of theoretical structures, we encounter once again the theme of initiation, the outline of which is already to be found in the institutional forms of the same period:[21] we are at the heart of the clinical experience – a form of the *manifestation* of things in their truth, a form of *initiation* into the truth of things. It was this that Bouillaud was to declare as a self-evident banality some forty years later: 'The medical clinic may be regarded either as a science or as a way of teaching medicine'.[22]

A hearing gaze and a speaking gaze: clinical experience represents a moment of balance between speech and spectacle. A precarious balance, for it rests on a formidable postulate: that all that is *visible* is *expressible*, and that it is *wholly visible* because it is *wholly expressible*. A postulate of such scope could permit a coherent science only if it was developed in a logic that was its rigorous outcome. But the reversibility, without residue, of the visible in the expressible remained in the clinic a requirement and a limit rather than an original principle. Total *description* is a present and ever-withdrawing horizon; it is much more the dream of a thought than a basic conceptual structure.

There is a simple historical reason for this: Condillac's logic did not allow a science in which the visible and the describable were caught up in a total adequation. Condillac's philosophy gradually shifted from an analysis of the original impression to an operational logic of signs, then from this logic to the constitution of a knowledge that would be both language and calculation: used at these three levels, and each time with different meanings, the notion of *element* sustained throughout this reflexion an ambiguous continuity, but one without a defined, coherent logical structure. Condillac never derived a universal logic from the element – whether this element was perceptual, linguistic, or calculable; he never ceased to hesitate between two logics of operations: of genesis and of calculation. Hence the dual definition of analysis: reduce complex ideas 'to the simple ideas of which they are made up and follow the progress of their generation';[23] and seek the truth 'by a kind of calculation, that is, by composing and decomposing notions and comparing them in the most favourable way with the discoveries that one has in view'.[24]

This ambiguity had its effect on clinical method, but this method followed a conceptual 'slope' that was the exact opposite of Condillac's development: the term by term reversal of the point of origin and the point of culmination. It redescended from the exigency of calculation to the primacy of genesis; after seeking to define the postulate of equation of the visible with the expressible by a *universal*, rigorous calculability, it gave that postulate the meaning of total, exhaustive *description*. The essential operation was no longer combinative but a matter of syntactic transcription. Nothing is more typical of this movement – which takes up again, in the opposite direction, Condillac's whole approach – than Cabanis's thought, and this is particularly apparent if we compare it with Brulley's analysis. Brulley wished 'to regard certainty as a whole divisible into as many probabilities as one may wish'. 'A probability is therefore a degree, a part of certainty from which it differs as the part differs from the whole';[25] medical certainty must thus be obtained by a combination of probabilities; after laying down the rules of such a combination Brulley declares that he will go no further, that he must leave to a more celebrated doctor the task of elucidating this subject – a task that he would have great difficulty in carrying out.[26] In all probability, it was Cabanis to whom he referred. For in *Les Révolutions de la médecine* the certain form of science is not defined by a type of calculation but by an organization whose values are essentially expressive; it is not a question of drawing up a calculation to proceed from the probable to the certain, but of determining a syntax in order to proceed from the element of the perceived to the coherence of discourse: 'the theoretical part of a science must, therefore, be the simple statement of the sequence of classification and of the relationship of all the facts which make up this science; it must, so to speak, be its summary expression'.[27] And if Cabanis finds room for the calculation of probabilities in the construction of medicine, it is only as one element among others in the total construction of scientific discourse. Brulley tried to place himself at the level of *La Langue des calculs*; although Cabanis cited this text, his thought is structurally on a footing with the *Essai sur l'origine des connaissances*.

It might be thought – and all the clinicians of that generation thought so – that things would rest there and that an unproblematic equilibrium was possible at that level between the composition of the visible and the syntactic rules of the expressible. But this was to be no more than a brief period of euphoria, a golden age with no future, in which seeing, saying, and learning to see by saying what one saw communicated in an immediate transparence: experience was rightfully science; and 'knowing' was in step with 'learning'. The gaze saw sovereignty in a world of language whose clear speech it gathered up effortlessly in order to restore it in a secondary, identical speech: given by the visible, this speech, without changing anything, made it possible to see. In its sovereign exercise, the gaze took up once again the structures of visibility that it had itself deposited in its field of perception.

But this generalized form of transparence leaves opaque the status of the language that must be its foundation, its justification, and its delicate instrument.

Such a deficiency, which also occurs in Condillac's logic, opens up the field to a number of epistemological myths that are destined to mask it. But these myths are already engaging the clinic in new spatial figures, in which visibility thickens and becomes cloudy, in which the gaze is confronted by obscure masses, by impenetrable shapes, by the black stone of the body.

1. The First of These Epistemological Myths Concerns the Alphabetical Structure of Disease

At the end of the eighteenth century, the alphabet appeared to grammarians to be the ideal schema of analysis and the ultimate form of the decomposition of a language; by that very fact it constituted the way in which that language was learnt. This alphabetical image was transposed essentially unaltered into the definition of the clinical gaze. The smallest possible observable segment, that from which one must set out and beyond which one cannot go back, is the singular impression one receives of a patient, or, rather, of a symptom of that patient; it signifies nothing in itself, but assumes meaning and value and begins to speak if it blends with other elements:

> Particular, isolated observations are to science what letters and words are to discourse; discourse is founded only on the concourse and coming together of letters and words whose mechanism and value must have been studied and reflected upon before correct and practical use was made of them; the same may be said of observations.[28]

This alphabetical structure of disease ensures not only that one can always return to the 'unsupersedable' (*indépassable*) element; it also ensures that the number of these elements will be finite and even small. It is not first impressions that are diverse and apparently infinite, but their combination within a single disease: just as the small number of 'modifications designated by the grammarians under the name of consonants' is enough to give 'to the expression of feeling the precision of thought', so, for pathological phenomena, 'with each new case, one might think that one is presented with new facts, whereas they are merely new combinations of facts. In the pathological state, there is never more than a small number of principal phenomena. . . . The order in which they appear, their importance, and their various relations are enough to give birth to every variety of disease'.[29]

2. The Clinical Gaze Effects a Nominalist Reduction On the Essence of the Disease

Composed as they are of letters, diseases have no other reality than the order of their composition. In the final analysis, their varieties refer to those few simple individuals, and whatever may be built up with them and above them is merely Name. And name in a double sense: in the sense in which the Nominalists use it when they criticize the substantial reality of abstract, general beings; and in another sense, one closer to a philosophy of language, since the form of composition of the being of the disease is of a linguistic type. In relation to the individual, concrete being, disease is merely a name; in relation to the isolated elements of which it is made up, it has all the rigorous architecture of a verbal designation. To ask what is the essence of a disease is like 'asking what is the nature of the essence of a word'.[30] A man coughs; he spits blood; he has difficulty in breathing; his pulse is rapid and hard; his temperature is rising; these are all so many immediate impressions, so many letters, as it were. Together, they form a disease, pleurisy: 'But what, then, is pleurisy?. . . It is the concourse of the accidents that constitute it. The word pleurisy merely retraces them in a more abbreviated manner.' 'Pleurisy' has no more being than the word itself; it 'expresses an abstraction of the mind'; but, like the word, it is a well-defined structure, a multiple figure 'in which all or almost all the accidents are combined. If one or more are lacking, it is no longer pleurisy, or at least not real pleurisy'.[31] Disease, like the word, is deprived of being, but, like the word, it is endowed with a configuration. The nominalist reduction of existence frees a constant truth. That is why:

3. The Clinical Gaze Operates on Pathological Phenomena a Reduction of a Chemical Type

Until the end of the eighteenth century the gaze of the nosographers was a gardener's gaze; one had to recognize the specific essence in the variety of appearances. At the beginning of the nineteenth century another model emerged: that of the chemical operation, which, by isolating the component elements, made it possible to define the composition, to establish common points, resemblances, and differences with other totalities, and thus to found a classification that was no longer based on specific types but on forms of relations: 'Instead of following the example of the botanists, should not the nosologists have, rather, taken as their model the systems of the chemist-mineralogists, that is, be content to classify the elements of diseases and their more frequent combinations?'[32] The notion of analysis in which, applied to the clinic, we have already recognized a quasi-linguistic sense and a quasi-mathematical sense[33] will now move towards a chemical signification: it will have as its horizon the isolation of pure bodies and the depiction of their combinations. One has passed from the theme of the combinative to that of syntax and finally to that of combination.

And, by reciprocity, the clinician's gaze becomes the functional equivalent of fire in chemical combustion; it is through it that the essential purity of phenomena can emerge: it is the separating agent of truths. And just as combustions reveal their secret only in the very vividness of fire, and it would be useless to ask, once the flame was extinguished, what can remain in the inert powders, the *caput mortuum*, so it is in the act of voice and the brightness that it sheds over phenomena that truth is revealed: 'It is not the remains of the morbid combustion that the doctor should know, but the species of the combustion'.[34] The clinical gaze is a gaze that burns things to their furthest truth. The attention with which it observes and the movement by which it states are in the last resort taken up again in this paradoxical act of consuming. The reality, whose language it spontaneously reads in order to restore it as it is, is not as adequate to itself as might be supposed: its truth is given in a decomposition that is much more than a reading since it involves the freeing of an implicit structure. One can now see that the clinic no longer has simply to read the visible; it has to discover its secrets.

4. The Clinical Experience Is Identified with a Fine Sensibility

The clinical gaze is not that of an intellectual eye that is able to perceive the unalterable purity of essences beneath phenomena. It is a gaze of the concrete sensibility, a gaze that travels from body to body, and whose trajectory is situated in the space of sensible manifestation. For the clinic, all truth is sensible truth; 'theory falls silent or almost always vanishes at the patient's bedside to be replaced by observation and experience; for on what are observation and experience based if not on the relation of our senses? And where would they be without these faithful guides?'[35] And if this knowledge, at the level of the immediate use of the senses, is not attained at the outset, if it can acquire depth and mastery, it is not a shift in level that enables it to accede to something other than itself, it is a sovereignty that is entirely internal to its own domain; it only acquires depth at its own level, which is that of pure sensory perception; for sense can only spring from sense. What, then, is

> the doctor's glance, which so often involves such vast erudition and such solid instruction, if not the result of the frequent, methodical, and accurate exercise of the senses, from which derive that facility of application, that alertness to relations, that confidence of judgement that is sometimes so rapid that all these acts seem to occur simultaneously, and are comprised together under the name of 'touch'?[36]

Thus this sensory knowledge – which nevertheless implies the conjunction of a hospital domain and a pedagogic domain, the definition of a field of probability and a linguistic structure of the real – is reduced to praise of the immediate sensibility.

The whole dimension of analysis is deployed only at the level of an aesthetic. But this aesthetic not only defines the original form of all truth, it also prescribes rules of exercise, and it becomes, at a secondary level, aesthetic in that it prescribes the norms of an art. The sensible *truth* is now open, not so much to the senses themselves, as to a *fine* sensibility. The whole complex structure of the clinic is summarized and fulfilled in the prestigious rapidity of an art: 'Since everything, or nearly everything, in medicine is dependent on a glance or a happy instinct, certainties are to be found in the sensations of the artist himself rather than in the principles of the art'.[37] The technical armature of the medical gaze is transformed into advice about prudence, taste, skill: what is required is 'great sagacity', 'great attention', 'great precision', 'great skill', 'great patience'.[38]

At this level, all structures are dissolved, or, rather, those that constituted the essence of the clinical *gaze* are gradually, and in apparent disorder, replaced by those that are to constitute the *glance*. And they are very different. In fact, the gaze implies an open field, and its essential activity is of the successive order of reading; it records and totalizes; it gradually reconstitutes immanent organizations; it spreads out over a world that is already the world of language, and that is why it is spontaneously related to hearing and speech; it forms, as it were, the privileged articulation of two fundamental aspects of *saying* (what is said and what one says). The glance, on the other hand, does not scan a field: it strikes at one point, which is central or decisive; the gaze is endlessly modulated, the glance goes straight to its object. The glance chooses a line that instantly distinguishes the essential; it therefore goes beyond what it sees; it is not misled by the immediate forms of the sensible, for it knows how to traverse them; it is essentially demystifying. If it strikes in its violent rectitude, it is in order to shatter, to lift, to release appearance. It is not burdened with all the abuses of language. The glance is silent, like a finger pointing, denouncing. There is no statement in this denunciation. The glance is of the non-verbal order of *contact*, a purely ideal contact perhaps, but in fact a more *striking* contact, since it traverses more easily, and goes further beneath things. The clinical eye discovers a kinship with a new sense that prescribes its norm and epistemological structure; this is no longer the ear straining to catch a language, but the index finger palpating the depths. Hence that metaphor of 'touch' (*le tact*) by which doctors will ceaselessly define their glance.[39]

And by that very fact, clinical experience sees a new space opening up before it: the tangible space of the body, which at the same time is that opaque mass in which secrets, invisible lesions, and the very mystery of origins lie hidden. The medicine of symptoms will gradually recede, until it finally disappears before the medicine of organs, sites, causes, before a clinic wholly ordered in accordance with pathological anatomy. The age of Bichat has arrived.

Notes

1. Clifton, *État de la médecine ancienne et moderne* (Paris, 1742, preface by the translator, unpaged).

2. Corvisart, Preface to the French translation of Auenbrugger, *Nouvelle méthode pour reconnaître les maladies internes de la poitrine* (Paris, 1808, p. vii).
3. *Ibid.*, p. viii.
4. Roucher-Deratte, *Leçons sur l'art d'observer* (Paris, 1807, p. 14).
5. Double, *Séméiologie générale*, vol. I, p. 80.
6. Sénebier, *Essai sur l'art d'observer et de faire des expériences* (2nd edn., Paris, 1802, Vol. I, p. 6).
7. Ph. Pinel, *Médecine clinique* (Paris, 1815, introduction, p. ii).
8. *Ibid.*, p. i.
9. Maygrier, *Guide de l'étudiant en médecine* (Paris, 1818, pp. 94–5).
10. Pinel, *op. cit.*, p. 4.
11. *Ibid.*, p. 3.
12. *Ibid.*, pp. 5 and 3.
13. *Ibid.*, p. 57.
14. Fordyce, *Essai d'un nouveau plan d'observations médicales* (Fr. trans., Paris, 1811).
15. Pinel, *op. cit.*, p. 78.
16. Pinel, *Nosographie philosophique*, introduction, p. iii.
17. *Ibid.*, pp. iii–iv.
18. Cf. above, Chapter 6.
19. Amard, *Association intellectuelle* (Paris, 1821, vol. I, p. 64).
20. *Ibid.*, p. 65.
21. Cf. above, Chapter 5.
22. Bouillaud, *Philosophie médicale* (Paris, 1831, p. 244).
23. Condillac, *Origine des connaissances humaines*, p. 162.
24. *Ibid.*, p. 110.
25. C.-A. Brulley, *Essai sur l'art de conjecturer en médecine*, pp. 26–7.
26. *Ibid.*
27. Cabanis, *Coup d'oeil sur les Révolutions et la Réforme de la médecine* (Paris, 1804, p. 271).
28. Double, *op. cit.*, p. 79.
29. Cabanis, *Du degré de certitude* (3rd edn., Paris, 1819, p. 86).
30. *Ibid.*, p. 66.
31. *Ibid.*, p. 66.
32. Demorcy-Delettre, *Essai sur l'analyse appliquée au perfectionnement de la médecine* (Paris, 1818, p. 135).
33. Cf. above, Chapter 6.
34. Amard, *op. cit.*, vol. II, p. 389.
35. Corvisart, *op. cit.*, p. vii.
36. *Ibid.*, p. x.
37. Cabanis, *op. cit.*, p. 126.
38. Roucher-Deratte, *op. cit.*, pp. 87–99.
39. Corvisart, *op. cit.*

Bibliography

I. Nosology

Alibert, J.-L., *Nosologie naturelle* (Paris, 1817).
Boissier de Sauvages, Fr., *Nosologie méthodique* (Fr. trans., Lyons, 1772, 10 vols.).
Capuron, J., *Nova medicinae elementa* (Paris, 1804).
Ch . . . , J.-J., *Nosographiae compendium* (Paris, 1816).

Chaussier, Fr. *Table générale des méthodes nosologiques* (Paris, n.d.).
Cullen, W., *Apparatus ad nosologiam methodicam* (Amsterdam, 1775).
———, *Institutions de médecine pratique* (Fr. trans., Paris, 1785, 2 vols.).
Dupont, J.-Ch., *Y a-t-il de la différence dans les systèmes de classification dont on se sert avec avantage dans l'étude de l'histoire naturelle et ceux qui peuvent être profitable à la connaissance des maladies?* (Bordeaux, 1803).
Duret, F.-J.-J., *Tableau d'une classification générale des maladies* (Paris, 1813).
Fercoq, G.A., *Synonymie ou concordance de la nomenclature de la Nosographie philosophique du P' Pinel avec les anciennes nosologies* (Paris, 1812).
Frank, J. P., *Synopsis nosologiae methodicae* (Ticini, 1790).
Latour, F.-D., *Nosographie synoptique* (Paris, 1810, 1st vol. only).
Linnaeus, C., *Genera morborum* (Fr. trans., Sauvages, cf. above).
Pinel, Ph., *Nosographie philosophique* (Paris, year VI).
Sagar, J. B. M., *Systema morborum systematicum* (Vienna, 1771).
Sydenham, Th., *Médecine pratique* (Fr. trans., Paris, 1784).
Voulonne, *Déterminer les maladies dans lesquelles la médecine agissante est préférable à l'expectante* (Avignon, 1776).

II. Medical Administration and Geography

Audin-Rouviére, J.-M., *Essai sur la topographie physique et médicale de Paris* (Paris, Year II).
Bacher, A., *De la médecine considérée politiquement* (Paris, Year IX).
Banau and Turben, *Mémoires sur les épidémies du Languedoc* (Paris, 1766).
Barberet, D., *Mémoire sur les maladies épidémiques des bestiaux* (Paris, 1766).
Bienville, J.-D.-T., *Traité des erreurs populaires sur la médecine* (The Hague, 1775).
Cattet, J.-J. and Gardet, J.-B., *Essai sur la contagion* (Paris, Year II).
Cerveau, M., *Dissertation sur la médecine des casernes* (Paris, 1803).
Clerc, *De la contagion* (St. Petersburg, 1771).
Colombier, J., *Préceptes sur la santé des gens de guerre* (Paris, 1775).
———, *Code de médecine militaire* (5 vols., Paris, 1772).
Daignan, G., *Ordre du service des hôpitaux militaires* (Paris, 1785).
———, *Tableau des variétés de la vie humaine* (2 vols., Paris, 1786).
———, *Centuries médicales du XIX*[e] siècle (Paris, 1807–1808).
———, *Conservatoire de Santé* (Paris, 1802).
Desgenettes, R.-N., *Histoire médicale de l'armée d'Orient* (Paris, 1802).
———, *Opuscules* (Cairo, n.d.).
Fouquet, H., *Observations sur la constitution des six premiers mois de l'an V à Montpellier* (Montpellier, Year VI).
Frank, J.-P., *System einer vollständigen medizinischen Polizei* (4 vols., Mannheim, 1779–1790).
Frier, F., *Guide pour la conservation de l'homme* (Grenoble, 1789).
Gachet, L.-E., *Problème médico-politique pour ou contre les arcanes* (Paris, 1791).
Gachet, M., *Tableau historique des événements présents relatif à leur influence sur la santé* (Paris, 1790).
Ganne, A., *L'homme physique et moral* (Strasbourg, 1791).
Guindant, T., *La nature opprimée par la médecine moderne* (Paris, 1768).
Guyton-Morveau, L.-B., *Traité des moyens de désinfecter l'air* (Paris, 1801).
Hautesierck, F.-M., *Recueil d'observations de médecine des hôpitaux militaires* (2 vols., Paris, 1766–1772).
Hildenbrand, J.-V., *Du typhus contagieux* (Fr. trans., Paris, 1811).
Home, D. R. de, *Mémoire sur quelques objets qui intéressent plus particulièrement la salubrité de la ville de Paris* (Paris, 1788).

Instruction sur les moyens d'entretenir la salubrité et de purifier l'air des salles dans le hôpitaux militaires (Paris, Year II).
Jacquin, A.-P., *De la Santé* (Paris, 1762).
Lafon, J.-B., *Philosophie médicale* (Paris, 1796).
Lanthenas, F., *De l'influence de la liberté sur la santé, la morale et le bonheur* (Paris, 1798).
Laugier, E.-M., *L'art de faire cesser la peste* (Paris, 1784).
Lebègue de Presle, *Le conservateur de Santé* (Paris, 1772).
Lebrun, *Traité théorique sur les maladies épidémiques* (Paris, 1776).
Lepecq de La Cloture, L., *Collection d'observations sur les maladies et constitutions épidémiques* (2 vols., Rouen, 1778).
Lioult, P.-J., *Les charlatans dévoilés* (Paris, Year VIII).
Mackenzie, J., *Histoire de la santé et de l'art de la conserver* (The Hague, 1759).
Maret, M., *Quelle influence les mœurs des Français ont sur leur santé* (Amiens, 1772).
Médecine militaire ou Traité des maladies tant internes qu'externes auxquelles les militaires sont exposés pendant la paix ou la guerre (6 vols., Paris, 1778).
Menuret, J.-J., *Essai sur l'action de l'air dans les maladies contagieuses* (Paris, 1781).
———, *Essai sur l'histoire médico-topographique de Paris* (Paris, 1786).
Murat, J.-A., *Topographie médicale de la ville de Montpellier* (Montpellier, 1810).
Nicolas, P.-F., *Mémoires sur les maladies épidémiques qui ont régné dans la province de Dauphiné* (Grenoble, 1786).
Petit, M.-A., *Sur l'influence de la Révolution sur la santé publique* (1796).
———, in *Essai sur la médecine du cœur* (Lyon, 1806).
Pichler, J.-F.-C., *Mémoire sur les maladies contagieuses* (Strasbourg, 1786).
Préceptes de santé ou Introduction au Dictionnaire de Santé (Paris, 1772).
Quatroux, Fr., *Traité de la peste* (Paris, 1771).
Razoux, J., *Tables nosologiques et météorologiques dressées à l'Hôtel-Dieu de Nîmes* (Basel, 1767).
Réflexions sur le traitement et la nature des épidémies lues à la Société royale de Médecine le 27 mai 1785 (Paris, 1785).
Roy-Desjoncades, A., *Les lois de la nature applicables aux lois physiques de la médecinty* (2 vols., Paris, 1788).
Rochard, C.-C.-T., *Programme de cours sur les maladies épidémiques* (Strasbourg, Year XIII).
Ruette, F., *Observations cliniques sur une maladie épidémique* (Paris, n.d.).
Salverte, E., *Des rapports de la médecine avec la politique* (Paris, 1806).
Souquet, *Essai sur l'histoire topographique médico-physique du district de Boulogne* (Boulogne, Year II).
Tallavignes, J.-A., *Dissertation sur la médecine où l'on prouve que l'homme civilisé est plus sujet aux maladies graves* (Carcassonne, 1821).
Thiery, F., *Vœux d'un patriots sur la médecine en France* (Paris, 1789).

III. Reform of Practice and Teaching

Baraillon, J.-F., *Rapport sur la partie de police qui tient à la médecine, 8 germ. an VI* (Paris, Year VI).
———, *Opinion sur le projet de la commission d'Instruction publique relatif aux Ecoles de Médecine, 7 germ, an VI* (Paris, Year VI).
Baumes, J.-B.-J., *Discours sur la nécessité des sciences dans une nation libre* (Montpellier, Year III).
Cabanis, P.-J.-C., *Œuvres* (2 vols., Paris, 1956).
Calès, J.-M., *Projet sur les Ecoles de santé, 12 prairial an V* (Paris, Year V).
———, *Opinion sur les Ecoles de Médecine, 17 germinal an VI* (Paris, Year VI).
Cantin, D.-M.-J., *Projet de réforme adressé à l'Assemblée Nationale* (Paris, 1790).

Caron, J.-F.-C., *Réflexions sur l'exercice de la médecine* (Paris, 1804).
________, *Projet de règlement sur l'art de guérir* (Paris, 1801).
Chambon de Montaux, *Moyens de rendre les hôpitaux utiles et de perfectionner la médecine* (Paris, 1787).
Colon de Divol, *Réclamations des malades de Bicêtre* (Paris, 1790).
Coqueau, C.-P., *Essai sur l'établissement des hôpitaux dans les grandes villes* (Paris, 1787).
Daunou, P.-C., *Rapports sur les Ecoles spéciales* (Paris, Year V).
Demangeon, J.-B., *Tableau d'un triple établissement réuni en un seul hospice à Copenhague* (Paris, Year VII).
________, *Des moyens de perfectionner la médecine* (Paris, 1804).
Desmonceaux, A., *De la bienfaisance nationals* (Paris, 1787).
Duchanoy, *Projet d'organisation médicale* (n.p., n.d.).
Du Laurens, J., *Moyens de rendre les hôpitaux utiles et de perfectionner les médecins* (Paris, 1787).
Dupont de Nemours, P., *Idées sur les secours à donner aux pauvres malades dans une grande ville* (Paris, 1786).
Ehrmann, J.-F., *Opinion sur le projet de Vitet, 14 germinal an VI* (Paris, Year VI).
Essai sur la réformation de la société dite de médecine (Paris, Year VI).
Etat actuel de l'Ecole de Santé (Paris, Year VI).
Fourcroy, A. F., *Discours sur le projet de loi relatif à l'exercice de la médecine, 19 ventôse an XI* (Paris, Year XI).
________, *Exposé des motifs du projet de loi relatif à l'exercice de la médecine* (Paris, n.d.).
________, *Rapport sur les Ecoles de Médecine, frimaire an III* (Paris, Year III).
________, *Rapport sur l'enseignement libre des sciences et des arts* (Paris, Year II).
Fourot, *Essai sur les concours en médecine* (Paris, 1786).
Callot, J.-G., *Vues générales sur la restauration de l'art de guérir* (Paris, 1790).
Géraud, M., *Projet de décret à rendre sur l'organisation civile des médecins* (Paris, 1791).
Guillaume, J., *Procès-verbaux du Comité d'Instruction publique* (Paris, 1899).
Guillemardet, F.-P., *Opinion sur les Ecoles spéciales de Santé, 14 germinal an VI* (Paris, Year VI).
Imbert, J., *Le droit hospitalier de la Révolution et de l'Empire* (Paris, 1954).
Instituta facultatis medicae Vidobonensis, curante Ant. Storck (Vienna, 1775).
Jadelot, N., *Adresse à Nos Seigneurs de l'Assemblée Nationale sur la nécessité et les moyens de perfectionner l'enseignement de la médecine* (Nancy, 1790).
Lefévre, J., *Opinion sur le projet de Vitet, 16 germinal an VI* (Paris, Year VI).
Lespagnol, N.-L., *Projet d'établir trois médecins par district pour le soulagement des gens de la campagne* (Charleville, 1790).
Marquais, J.-Th., *Rapport au Roi sur l'état actuel de la médecine en France* (Paris, 1814).
Menuret, J.-J., *Essai sur les moyens de former de bons médecins* (Paris, 1791).
Motif de la réclamation de la Faculté de Médecine de Paris contre l'établissement de la Société royale de Médecine (n.p., n.d.; the author is Vacher de La Feutrie).
Observations sur les moyens de perfectionner l'enseignement de la médecine en France (Montpellier, Year V).
Pastoret, C.-E., *Rapport sur un mode provisoire d'examen pour les officiers de Santé (19 thermidor an V)* (Paris, Year V).
Petit, M. A., *Projet de réforme sur l'exercise de la médecine en France* (Paris, 1791).
________, *Sur la meilleure manière de construire un hôpital* (Paris, 1774).
Plan de travail présenté à la Société de Médecine de Paris (Paris, Year V).
Plan général d'enseignement dans l'Ecole de Santé de Paris (Paris, Year III).
Porcher, G.-C., *Opinion sur la résolution du 19 fructidor an V, 16 vendémiaire an VI* (Paris, Year VI).
Précis historique de l'établissement de la Société royale de Médecine (n.p., n.d.).

Prieur de La Côte-d'Or, C.-A., *Motion relative aux Ecoles de Santé* (Paris, Year VI).
Programme de la Société royale de Médecine surles cliniques (Paris, 1792).
Programme des cours d'enseignement dans l'Ecole de Santé de Montpellier (Paris, Year III).
Prunelle, Cl.-V., *Des Ecoles de Médecine, de leurs connexions et de leur methodologie* (Paris, 1816).
Recueil de discours prononcés à la Faculté de Montpellier (Montpellier, 1820).
Régnault, J.-B., *Considérations sur l'état de la médecine en France depuis la Révolution jusqu'à nos jours* (Paris, 1819).
Retz, N., *Exposé succinct à l'Assemblée Nationale sur les Facultés et Sociétés de Médecine* (Paris, 1790).
Royer, P.-F., *Bienfaisance médicale et projet financier* (Provins, Year IX).
________, *Bienfaisance médicale rurale* (Troyes, 1814).
Sabarot de L Avernière, *Vue de législation médicale adressée aux Etats généraux* (n.p., 1789).
Tissot, S.-A.-D., *Essai sur les moyens de pefectionner les études de médecine* (Lausanne, 1785).
Vicq d'Azyr, F., *Œuvres* (6 vols., Paris, 1805).
Vitet, L., *Rapport sur les Ecoles de Santé, 17 ventôse an VI* (Paris, Year VI).
Würtz, *Mémoire sur l'établissement des Ecoles de Médecine pratique* (Paris, 1784).

IV. Methods

Amard, L.-V.-F., *Association intellectuelle* (2 vols., Paris, 1821).
Amoreux, P.-J., *Essai sur la médecine des Arabes* (Montpellier, 1805).
Audibert-Caille, J.-M., *Mémoire sur l'utilité de l'analogie en médecine* (Montpellier, 1814).
Auenbrugger, *Nouvelle méthode pour reconnoître les maladies internes* (Fr. trans, in Rozière de La Chassaigne, *Manuel des pulmoniques*, Paris, 1763).
Beullac, J.-P., *Nouveau guide de l'étudiant en médecine* (Paris, 1824).
Bordeu, Th., *Recherches sur le pouls* (4 vols., Paris, 1779–1786).
Bouillaud, J., *Dissertation sur les généralités de la clinique* (Paris, 1831).
Broussonnet, J.-L.-V., *Tableau élémentaire de séméiotique* (Montpellier, Year VI).
Brulley, C.-A., *Essai sur l'art de conjecturer en médecine* (Paris, Year X).
Bruté, S.-G.-G., *Essai sur l'histoire et les avantages des institutions cliniques* (Paris, 1803).
Chomel, J.-B.-L., *Essai historique sur la médecine en France* (Paris, 1762).
Clos de Sorèze, J.-A., *De l'analyse en médecine* (Montpellier, Year V).
Corvisart, J.-N., *Essai sur les maladies et lésions du cœur et des gros vaisseaux* (Paris, 1806).
Dardonville, H., *Réflexions pratiques sur les dangers des systèmes en médecine* (Paris, 1818).
Demorcy-Delettre, J.-B.-E., *Essai sur l'analyse appliquée au perfectionnement de la médecine* (Paris, 1818).
Double, F.-J., *Séméiologie générate ou Traité des signes et de leur valeur dans les maladies* (3 vols., Paris, 1811–1822).
Duvivier, P.-H., *De la médecine considérée comme science et comme art* (Paris, 1826).
Essyg, *Traité du diagnostic médical* (Fr. trans., Paris, Year XII).
Fabre, *Recherche des vrais principes de l'art de guérir* (Paris, 1790).
Fordyce, G., *Essai d'un nouveau plan d'observations médicales* (Fr. trans., Paris, 1811).
Fouquet, H., *Discours sur la clinique* (Montpellier, Year XI).
Frank, J.-P., *Ratio instituti clinici Vicinensis* (Vienna, 1797).
Gilbert, N.-P., *Les théories médicales modernes comparées entre elles* (Paris, Year VII).
Girbal, A., *Essai sur l'esprit de la clinique médicale de Montpellier* (Montpellier, 1857).
Goulin, J., *Mémoires sur l'histoire de la médecine* (Paris, 1779).
Hélian, M., *Dictionnaire de diagnostic ou l'art de connaître les maladies* (Paris, 1771).
Hildenbrand, J., *Médecine pratique* (Fr. trans., Paris, 1824, 2 vols.).

Landré-Beauvais, A.-J., *Séméiotique ou traité des signes des maladies* (Paris, 1810).
Leroux, J.-J., *Cours sur les généralités de la médecine* (Paris, 1818).
________, *Ecole de Médecine. Clinique interne* (Paris, 1809).
Lordat, J., *Conseils sur la manière d'étudier la physiologie de l'homme* (Montpellier, 1813).
________, *Perpétuité de la médecine* (Montpellier, 1837).
Mahon, P.-A.-O., *Histoire de la médecine clinique* (Paris, Year XII).
Martinet, L., *Manuel de clinique* (Paris, 1825).
Maygrier, J.-P., *Guide de l'étudiant en médecine* (Paris, 1807).
Menuret, J.-J., *Traité du pouls* (Paris, 1798).
Moscati, P., *De l'emploi des systèmes dans la médecine pratique* (Strasbourg, Year III).
Petit, M.-A., *Collection d'observations cliniques* (Lyons, 1815).
Pinel, Ph., *Médecine clinique* (Paris, 1802).
Piorry, P. A., *Tableau indiquant la manière d'examiner et d'interroger le malade* (Paris, 1832).
Rostan, L., *Traité élémentaire de diagnostic, de pronostic, d'indications thérapeutiques* (6 vols., Paris, 1826).
Roucher-Deratte, Cl., *Leçons sur l'art d'observer* (Paris, 1807).
Selle, Ch.-G., *Médecine clinique* (Fr. trans., Montpellier, 1787).
________, *Introduction à l'étude de la nature et de la médecine* (trad., Montpellier, Year III).
Sénebier, J., *Essai sur l'art d'observer et de fairs des expériences* (3 vols., 1802).
Thiery, F., *La médecine expérimental* (Paris, 1755).
Vaidy, J.-V.-F., *Plan d'études médicates à l'usage des aspirants* (Paris, 1816).
Zimmermann, G., *Traité de l'experience en médecine* (Fr. trans., Paris, 1774, 3 vols.).

V. Morbid Anatomy

Baillie, M., *Anatomie pathologique des organes les plus importants du corps humain* (Fr. trans., Paris, 1815).
Bayle, G.-L., *Recherches sur la phtisie pulmonaire* (Paris, 1810).
Bichat, X., *Anatomie générate apphquée à la physiologie et à médecine* (3 vols., Paris, 1801).
________, *Anatomie pathologique* (Paris, 1825).
________, *Recherches physiologiques sur la vie et la mort* (Paris, Year VIII).
________, *Traité des membranes* (Paris, 1807).
Bonet, Th., *Sepulchretum* (3 vols., Lyons, 1700).
Breschet, G., *Répertoire général d'anatomie et de physiologie pathologiques* (6 vols., Paris, 1826–1828).
Cailliot, L., *Eléments de pathologie et de physiologie pathologique* (2 vols., Paris, 1819).
Chomel, A.-F., *Eléments de pathologie générate* (Paris, 1817).
Cruveilhier, J., *Essai sur l'anatomie pathologique en général* (2 vols., Paris, 1816).
Dezeimeris, J.-E., *Aperçu rapide des découvertes en anatomie pathologique* (Paris, 1830).
Guillaume, A., *De l'influence de l'anatomie pathologique sur les progrès de la médecine* (Dôle, 1834).
Laënnec, R., *Traité de l'auscultation médiate* (2 vols., Paris, 1819).
________, *Traité inédit de l'anatomie pathologique* (Paris, 1884).
Lallemand, F., *Recherches anatomo-pathologiques sur l'encéphale et ses dépendances* (2 vols., Paris, 1820).
Morgagni, J.-B., *De sedibus et causis morborum* (Venice, 1761).
Portal, A., *Cours d'anatomie médicale* (5 vols., Paris, Year XII).
Prost, P.-A., *La médecine éclairée par l'observation et l'ouverture des corps* (2 vols., Paris, Year XII).
Rayer, P., *Sommaire d'une histoire abrégée de l'anatomie pathologique* (Paris, 1818).

Ribes, Fr., *De l'anatomie pathologique considérée dans ses vrais rapports avec la science des maladies* (2 vols., Paris, 1828–1834).
Richerand, B.-A., *Histoire des progrès récents de la chirurgie* (Paris, 1825).
Saucerotte, C., *De l'influence de l'anatomie pathologique sur les progrès de la médecine* (Paris, 1834).
Tacheron, C.-F., *Recherches anatomo-pathologiques sur la médecine pratique* (3 vols., Paris, 1823).

VI. Fevers

Barbier, J.-B.-G., *Réflexions sur les fièvres* (Paris, 1822).
Boisseau, F.-G., *Pyrétologie physiologique* (Paris, 1823).
Bompart, A., *Description de la fièvre adynamique* (Paris, 1815).
Bouillaud, J., *Traité clinique ou expérimental des fièvres dites essentielles* (Paris, 1830).
Broussais, F.-J.-V., *Catéchisme de médecine physiologique* (Paris, 1824).
———, *Examen des doctrines médicates* (Paris, 1821).
———, *Histoire des phlegmasies ou inflammations chroniques* (Paris, 1808, 2 vols.).
———, *Leçons sur la phlegmasie gastrique* (Paris, 1819).
———, *Mémoire sur l'influence que les travaux des médecins physiologistes ont exercée sur l'état de la médecine* (Paris, 1832).
———, *Traité de physiologie appliquée à la pathologie* (2 vols., 1822–1823).
Caffin, J.-F., *Quelques mots de réponse à un ouvrage de M. Broussais* (Paris, 1818).
Castel, L., *Réfutation de la nouvelle doctrine médicale de M. le Dr. Broussais* (Paris, 1824).
Chambon de Montaux, *Traité de la fièvre maligne simple et des fièvres compliquées de malignité* (4 vols., Paris, 1787).
Chauffard, H., *Traité sur les fièvres prétendues essentielles* (Paris, 1825).
Chomel, A. F., *De l'existence des fièvres* (Paris, 1820).
———, *Des fièvres et des maladies pestilentielles* (Paris, 1821).
Collineau, J.-C., *Peut-on mettre en doute l'existence des fièvres essentielles* (Paris, 1823).
Dagoumer, Th., *Précis historique de la fièvre* (Paris, 1831).
Dardonville, H., *Mémoire sur les fièvres* (Paris, 1821).
Ducamp, Th., *Réfexions critiques sur les écrits de M. Chomel* (Paris, 1821).
Fodéra, M., *Histoire de quelques doctrines médicales comparées à celles de M. Broussais* (Paris, 1818).
Fournier, M., *Observations sur les fièvres putrides et malignes* (Dijon, 1775).
Gérard, M., *Peut-on mettre en doute l'existence des fièvres essentielles?* (Paris, 1823).
Giannini, *De la nature des fièvres* (Fr. trans., Paris, 1808).
Giraudy, Ch., *De la fièvre* (Paris, 1821).
Grimaud, M. de, *Cours complet ou Traité des fièvres* (3 vols., Montpellier, 1791).
Hernandez, J.-F., *Essai sur le typhus* (Paris, 1816).
Hoffmann, F., *Traité des fièvres* (Fr. trans., Paris, 1746).
Hufeland, C.-W., *Observations sur les fièvres nerveuses* (Fr. trans., Berlin, 1807).
Huxham, J., *Essai sur les différentes espèces de fièvres* (Fr. trans., Paris, 1746).
Larroque, J.-B. de, *Observations cliniques opposées à l'examen de la nouvelle doctrine* (Paris, 1818).
Leroux, F.-M., *Opposition aux erreurs de la science médicale* (Paris, 1817).
Lesage, L.-A., *Danger et absurdité de la doctrine physiologique* (Paris, 1823).
Monfalcon, J.-B., *Essai pour servir à l'histoire des fièvres adynamiques* (Lyons, 1823).
Mongellaz, P.-J., *Essai sur les irritations intermittentes* (2 vols., Paris, 1821).
Pascal, Ph., *Tableau synoptique du diagnostic des fièvres essentielles* (Paris, 1818).
Petit, M.-A., *Traité de la fièvre entéro-mésentérique* (Paris, 1813).
Petit-Radel, Ph., *Pyrétologie médicale* (Paris, 1812).

Quitard-Piorry, H.-H., *Traité sur la non-existence des fièvres essentieiles* (Paris, 1830).
Roche, L.-Ch., *Réfutation des objections faites à la nouvelle doctrine des fièvres* (Paris, 1821).
Roederer and Wagler, *Tractatus de morbo mucoso* (Göttingen, 1783).
Roux, G., *Traité des fièvres adynamiques* (Paris, 1812).
Selle, Ch.-G., *Eléments de pyrétologie méthodique* (Fr. trans., Lyons, Year IX).
Stoll, M., *Aphorismes sur la connaissance et la curation des fièvres* (Fr. trans., Paris, Year V).
Tissot, S.-A.-D., *Dissertation sur les fièvres bilieuses* (Fr. trans., Paris, Year VIII).

6

Rules for the Observation of Social Facts

Emile Durkheim

The first and most basic rule is *to consider social facts as things.*

I

At the moment when a new order of phenomena becomes the object of a science they are already represented in the mind, not only through sense perceptions, but also by some kind of crudely formed concepts. Before the first rudiments of physics and chemistry were known, men already possessed notions about physical and chemical phenomena which went beyond pure perception alone. Such, for example, are those to be found intermingled with all religions. This is because reflective thought precedes science, which merely employs it more methodically. Man cannot live among things without forming ideas about them according to which he regulates his behaviour. But, because these notions are closer to us and more within our mental grasp than the realities to which they correspond, we naturally tend to substitute them for the realities, concentrating our speculations upon them. Instead of observing, describing and comparing things, we are content to reflect upon our ideas, analysing and combining them. Instead of a science which deals with realities, we carry out no more than an ideological analysis. Certainly this analysis does not rule out all observation. We can appeal to the facts to corroborate these notions or the conclusions drawn from them. But then the facts intervene only secondarily, as examples

Source: Steven Lukes (ed.) and W.D. Halls (tr), *The Rules of Sociological Method and Selected Texts on Sociology and Its Method* (London: Macmillan, 1982), pp. 60–84.

or confirmatory proof. Thus they are not the subject matter of the science, which therefore proceeds from ideas to things, and not from things to ideas.

It is clear that this method cannot yield objective results. These notions or concepts – however they are designated – are of course not legitimate surrogates for things. The products of common experience, their main purpose is to attune our actions to the surrounding world; they are formed by and for experience. Now a representation can effectively perform this function even if it is theoretically false. Several centuries ago Copernicus dispelled the illusions our senses experienced concerning the movements of the heavenly bodies, and yet it is still according to these illusions that we commonly regulate the distribution of our time. For an idea to stimulate the reaction that the nature of a thing demands, it need not faithfully express that nature. It is sufficient for it to make us perceive what is useful or disadvantageous about the thing, and in what ways it can render us service or disservice. But notions formed in this way can only present a roughly appropriate practicality, and then only in the general run of cases. How often are they both dangerous and inadequate! It is therefore not by elaborating upon them, however one treats them, that we will ever succeed in discovering the laws of reality. On the contrary, they are as a veil interposed between the things and ourselves, concealing them from us even more effectively because we believe to be more transparent.

Such a science can only be a stunted one, for it lacks the subject matter on which to feed. It has hardly come into existence, one might say, before it vanishes, transmuted into an art. Allegedly its notions contain all that is essential to reality, but this is because they are confused with the reality itself. From then onwards they appear to contain all that is needful for us not only to understand what is, but also to prescribe what should be done and the means of implementation, for what is good is in conformity with the nature of things. What goes against nature is bad, and the means of attaining the good and eluding the bad both derive from that same nature. Thus if we have already comprehended the reality from the first, to study it has no longer any practical interest. Since it is this interest which is the reason for our study, there is henceforth no purpose to it. Our reflective thought is thus induced to turn away from what is the true subject matter of the science, namely the present and the past, and in one fell swoop to proceed to the future. Instead of seeking to understand the facts already discovered and acquired, it immediately undertakes to reveal new ones, more in accord with the ends that men pursue. If men think they know what is the essence of matter, they immediately embark on the quest for the philosopher's stone. This encroachment of art upon science, which hinders the latter's development, is made easy also by the very circumstances which determine the awakening of scientific reflection. For, since this reflection comes into being only to satisfy vital needs, it is quite naturally directed towards practical matters. The needs which it is called upon to assuage are always pressing ones, and consequently urge it to arrive at conclusions. Remedies, not explanations, are required.

This procedure is so much in accordance with the natural inclination of our mind that it is even to be found in the beginnings of the physical sciences. It is what characterises alchemy as distinct from chemistry, and astrology from astronomy. It is how Bacon characterises the method followed by the scholars of his day – one which he fought against. Indeed the notions just discussed are those *notiones vulgares*, or *praenotiones*,[1] which he points out as being at the basis of all the sciences,[2] in which they take the place of facts.[3] It is these *idola* which, resembling ghost-like creatures, distort the true appearance of things, but which we nevertheless mistake for the things themselves. It is because this imagined world offers no resistance that the mind, feeling completely unchecked, gives rein to limitless ambitions, believing it possible to construct – or rather reconstruct – the world through its own power and according to its wishes.

If this has been true for the natural sciences, how much more had it to be true for sociology. Men did not wait on the coming of social science to have ideas about law, morality, the family, the state or society itself, for such ideas were indispensable to their lives. It is above all in sociology that these preconceptions, to employ again Bacon's expression, are capable of holding sway over the mind, substituting themselves for things. Indeed, social things are only realised by men: they are the product of human activity. Thus they appear to be nothing save the operationalising of ideas, which may or may not be innate but which we carry within us, and their application to the various circumstances surrounding men's relationships with one another. The organisation of the family, of contracts, or repression, of the state and of society seems therefore to be a simple development of the ideas we have about society, the state, justice, etc. Consequently these and similar facts seem to lack any reality save in and through the ideas which engender them and which, from then on, become the subject matter proper of sociology.

The apparent justification for this view derives from the fact that since the details of social life swamp the consciousness from all sides, it has not a sufficiently strong perception of the details to feel the reality behind them. Lacking ties that are firm enough or close enough to us, this all produces the impression upon us that it is clinging to nothing and floating in a vacuum, consisting of matter half unreal and infinitely malleable. This is why so many thinkers have seen in the social organisation mere combinations which are artificial and to some degree arbitrary. But if the details and the special concrete forms elude us, at least we represent to ourselves in a rough, approximate way the most general aspects of collective existence. It is precisely these schematic, summary representations which constitute the prenotions that we employ in our normal way of life. Thus we cannot visualise their existence being called into question, since we see it at the same time as we see our own. Not only are they within us, but since they are the product of repeated experiences, they are invested with a kind of ascendancy and authority, by dint of repetition and the habit which results from it. We feel their resistance when we seek to free ourselves from them, and we cannot fail to regard as real something which pits itself against us. Thus everything conspires to make us see in them the true social reality.

And indeed up to now sociology has dealt more or less exclusively not with things, but with concepts. It is true that Comte proclaimed that social phenomena are natural facts, subject to natural laws. In so doing he implicitly recognised their character as things, for in nature there are only things. Yet when, leaving behind these general philosophical statements, he tries to apply his principle and deduce from it the science it contained, it is ideas which he too takes as the object of his study. Indeed, what constitutes the principal subject matter of his sociology is the progress over time of humanity. His starting point is the idea that the continuous evolution of the human species consists of an ever-growing perfection of human nature. The problem with which he deals is how to discover the sequence of this evolution. Yet, even supposing this evolution exists, its reality can only be established when the science has been worked out. Thus the evolution cannot be made the subject of research unless it is postulated as a conception of the mind, and not a thing. In fact, so much is this a wholly subjective idea, this progress of humanity does not exist. What do exist, and what alone are presented to us for observation, are particular societies which are born, develop and die independently of one another. If indeed the most recent societies were a continuation of those which had preceded them, each superior type might be considered merely as the repetition of the type at the level immediately below it, with some addition. They could all then be placed end-on, so to speak, assimilating together all those at the same stage of development; the series thus formed might be considered representative of humanity. But the facts do not present themselves with such extreme simplicity. A people which takes the place of another is not merely a prolongation of the latter with some new features added. It is different, gaining some extra properties, but having lost others. It constitutes a new individuality, and all such distinct individualities, being heterogeneous, cannot be absorbed into the same continuous series, and above all not into one single series. The succession of societies cannot be represented by a geometrical line; on the contrary, it resembles a tree whose branches grow in divergent directions. Briefly, in his consideration of historical development, Comte has taken his own notion of it, which is one that does not differ greatly from that commonly held. It is true that, viewed from a distance, history does take on somewhat neatly this simple aspect of a series. One perceives only a succession of individuals all moving in the same direction, because they have the same human nature. Moreover, since it is inconceivable that social evolution can be anything other than the development of some human idea, it appears entirely natural to define it by the conception that men have of it. But if one proceeds down this path one not only remains in the realm of ideology, but assigns to sociology as its object a concept which has nothing peculiarly sociological about it.

Spencer discards this concept, but replaces it with another which is none the less formed in the same way. He makes societies, and not humanity, the object of his study, but immediately gives to societies a definition which causes the thing of which he speaks to disappear and puts in its place the preconception he has of them. In fact he states as a self-evident proposition that 'a society is formed only when, besides juxtaposition, there is co-operation'; it is solely in this way that the union of individuals becomes a society proper.[4] Then, starting from this principle, that co-operation is the essence of social life, he divides societies into two classes according to the nature of the predominant mode of co-operation. 'There is', he states, 'a spontaneous co-operation which grows up without thought during the pursuit of private ends; and there is a co-operation which, consciously devised, implies distinct recognition of public ends',[5] The first category he dubs industrial societies, the latter military societies. One may say of this distinction that it is the seminal idea for his sociology.

But this initial definition enunciates as a thing what is only a mental viewpoint. It is presented as the expression of a fact that is immediately apparent, one sufficiently ascertained by observation, since it is formulated from the very beginning of the science as an axiom. Yet from mere inspection it is impossible to know whether co-operation really is the mainspring of social life. Such an assertion is only scientifically justified if at first all the manifestations of collective life have been reviewed and it has been demonstrated that they are all various forms of co-operation. Thus once again a certain conception of social reality is substituted for that reality.[6] What is defined in this way is not society but Spencer's idea of it. If he feels no scruples in proceeding in this fashion it is because for him also society is only, and can be only, the realisation of an idea, namely that very idea of co-operation by which he defines society.[7] It would be easy to show, in each of the particular problems that he tackles, that his method remains the same. Also, although he has an air of proceeding empirically, because the facts accumulated in his sociology are used to illustrate analyses of notions rather than to describe and explain things, they seem indeed to be there to serve as arguments. All that is really essential in his doctrine can be directly deduced from his definition of society and the different forms of co-operation. For if we have only the choice between co-operation tyranically imposed and one that is free and spontaneous, it is plainly the latter which is the ideal towards which humanity does and ought to strive.

These common notions are not to be encountered only at the basis of the sciences, but are also to be found constantly as the arguments unravel. In our present state of knowledge we do not know exactly what the state is, nor sovereignty, political freedom, democracy, socialism, communism, etc.

Thus our method should make us forswear any use of these concepts so long as they have not been scientifically worked out. Yet the words that express them recur continually in the discussions of sociologists. They are commonly used with assurance, as if they corresponded to things well known and well defined, while in fact they evoke in us only confused notions, an amalgam of vague impressions, prejudices and passions. Today we mock at the strange ratiocinations that the doctors of the Middle Ages constructed from their notions of heat and cold, humidity and dryness, etc. Yet we do not perceive that we continue to apply the selfsame method to an order of phenomena which is even less appropriate for it than any other, on account of its extreme complexity.

In the specialised branches of sociology this ideological character is even more marked.

It is particularly so in the case of ethics. It may in fact be asserted that there is not a single system which does not represent it as the simple development of an initial idea which enshrines it potentially in its entirety. Some believe that men possess this idea complete at birth; on the other hand, others believe that it has grown up at a varying rate in the course of history. But for both empiricists and rationalists this is all that is truly real about morality. As for detailed legal and moral rules, these would have, in a manner of speaking, no existence *per se*, being merely applications of the basic notion to the particular circumstances of living, and varying according to different cases. Hence the subject matter of morality cannot be this unreal system of precepts, but the idea from which the precepts derive and which is interpreted differently according to cases. Thus all the questions that ethics normally raises relate not to things but to ideas. We must know what constitutes the ideas of law and morality and not what is the nature of morality and law considered in their own right. Moralists have not yet even grasped the simple truth that, just as our representations of things perceived by the senses spring from those things themselves and express them more or less accurately, our representation of morality springs from observing the rules that function before our very eyes and perceives them systematically. Consequently it is these rules and not the cursory view we have of them which constitute the subject matter of science, just as the subject matter of physics consists of actual physical bodies and not the idea that ordinary people have of it. The outcome is that the basis of morality is taken to be what is only its superstructure, namely, the way in which it extends itself to the individual consciousness and makes its impact upon it. Nor is it only for the more general problems of science that this method is followed; it is not modified even for more specialised questions. From the essential ideas that he studies at the outset the moralist passes on to the examination of second-order ideas, such as family, country, responsibility, charity and justice – but it is always to ideas that his thinking is applied.

The same applies to political economy. John Stuart Mill states that its subject matter is the social facts which arise principally or exclusively with a view to the acquisition of wealth.[8] But, in order for the facts defined in this way to be submitted to the scrutiny of the scientist as things, at the very least it should be possible to indicate the means whereby those which satisfy this condition can be recognised. With a new science one is no position to affirm that the facts exist, and even less to know what they are. In any kind of investigation it is only when the explanation of the facts is fairly well advanced that it is possible to establish that they have a goal and what that goal is. There is no problem more complex or less likely to be resolved at the very beginning. We therefore lack any prior assurance that a sphere of social activity exists where the desire for wealth really plays this predominant role. Consequently the subject matter of economics so conceived is made up not of realities which may be precisely pointed to, but merely of possible ones, pure conceptions of the mind. They are facts which the economist *conceives* of as relating to the purpose under consideration, and facts as he conceives them. If, for example, he embarks on a study of what he terms production, he believes it possible immediately to spell out and review the principal agencies which assist it. This means therefore that he has not ascertained their existence by studying on what conditions depends the thing that he is studying. If he had, he would have begun by setting out the operations from which he drew that conclusion. If, in summary terms, at the beginning of his researches he proceeds to make such a classification, it is because he has arrived at it by mere logical analysis. He starts from the idea of production and as he dissects it he finds that it logically entails ideas of natural forces, of work, of tools or capital and he then goes on to treat in the same way these ideas which he has derived.[9]

The most basic economic theory of all, that of value, has clearly been built up according to the same method. If value were studied as a fact having reality should be, the economist would show how the thing so designated could be identified; he would then classify its various kinds, testing by methodical inductions how these vary according to different causes, and finally comparing the various results in order to arrive at a general formulation. A theory could therefore only emerge when the science was fairly well advanced. Instead it is met with at the very beginning. To do this the economist contents himself with his own reflective thinking, evoking his idea of value, namely that of an object capable of being exchanged. He finds that this implies the ideas of utility and scarcity, etc., and it is from these fruits of his analysis that he constructs his definition. He doubtless backs it up with a few examples. But, reflecting on the countless facts which such a theory must explain, how can one concede the slightest validity of proof to the necessarily very few facts which are cited at random as they suggest themselves to him?

Thus in political economy, as in ethics, the role of scientific investigation is extremely limited, and that of art is preponderant. The theoretical part of ethics is reduced to a few discussions on the ideas of duty, goodness and right. But such abstract speculations do not strictly speaking constitute a science, since their purpose is not to determine what is, in fact, supreme moral law, but what ought to be. Likewise, what economists dwell on most in their researches is the problem of knowing, for example, whether society *should be* organised on individualistic or socialist lines; whether *it is better* for the state to intervene in industrial and commercial relations or abandon them entirely to private initiative; whether the monetary system *should be* based on monometallism or bimetallism, etc. Laws properly so called are very few; even those which by custom we call laws do not generally merit the term, but are merely maxims for action, or in reality practical precepts. For example, the celebrated law of supply and demand has never been established inductively as an expression of economic reality. Never has any experiment or methodical comparison been instituted to establish whether, *in fact*, it is according to this law that economic relations are regulated. All that could be done, and has been done, has been to demonstrate by dialectical argument that individuals should act in this way if they perceive what is in their best interest; any other course of action would be harmful to them, and if they followed it would indeed constitute an error of logic. It is logical that the most productive industries should be the most prized, and that those who hold goods most in demand and most scarce should sell them at the highest price. But this entirely logical necessity in no way resembles the one that the true laws of nature reveal. These express the relationships whereby facts are linked together in reality, and not the way in which it would be good for them to be linked.

What we state about this law can be repeated for all those that the orthodox school of economists term 'natural' and which, moreover, are scarcely more than special cases of this first law. They may be said to be natural in the sense that they enunciate the means which are, or may appear to be, natural to employ in order to reach some assumed goal. But they should not be termed so if by a natural law is understood any inductively verified mode of existence of nature. All in all, they are mere counsels of practical wisdom. If it has been possible to present them to a more or less plausible extent as a clear expression of reality, it is because, rightly or wrongly, the assumption has been that these counsels were effectively those followed by most men and in the majority of cases.

Yet social phenomena are things and should be treated as such. To demonstrate this proposition one does not need to philosophise about their nature or to discuss the analogies they present with phenomena of a lower order of existence. Suffice to say that they are the sole *datum* afforded the sociologist. A thing is in effect all that is given, all that is offered, or rather forces itself upon our observation. To treat phenomena as things is to treat them as *data*, and this constitutes the starting point for science. Social phenomena unquestionably display this characteristic. What is given is not the idea that men conceive of value, because that is unattainable; rather is it the values actually exchanged in economic transactions. It is also not some conception or other of the moral ideal; it is the sum total of rules that in effect determine behaviour. It is not the idea of utility or wealth; it is all the details of economic organisation. Social life may possibly be merely the development of certain notions, but even if this is assumed to be the case, these notions are not revealed to us immediately. They cannot therefore be attained directly, but only through the real phenomena that express them. We do not know a *priori* what ideas give rise to the various currents into which social life divides, nor whether they exist. It is only after we have traced the currents back to their source that we will know from where they spring.

Social phenomena must therefore be considered in themselves, detached from the conscious beings who form their own mental representations of them. They must be studied from the outside, as external things, because it is in this guise that they present themselves to us. If this quality of externality proves to be only apparent, the illusion will be dissipated as the science progresses and we will see, so to speak, the external merge with the internal. But the outcome cannot be anticipated, and even if in the end social phenomena may not have all the features intrinsic to things, they must at first be dealt with as if they had. This rule is therefore applicable to the whole of social reality and there is no reason for any exceptions to be made. Even those phenomena which give the greatest appearance of being artificial in their arrangement should be considered from this viewpoint. *The conventional character of a practice or an institution should never be assumed in advance*. If, moreover, we are allowed to invoke personal experience, we believe we can state with confidence that by following this procedure one will often have the satisfaction of seeing the apparently most arbitrary facts, after more attentive observation, display features of constancy and regularity symptomatic of their objectivity.

In general, moreover, what has been previously stated about the distinctive features of the social fact gives us sufficient reassurance about the nature of this objectivity to demonstrate that it is not illusory. A thing is principally recognisable by virtue of not being capable of modification through a mere act of the will. This is not because it is intractable to all modification. But to effect change the will is not sufficient; it needs a degree of arduous effort because of the strength of the resistance it offers, which even then cannot always be overcome. We have seen that social facts possess this property of resistance. Far from their being a product of our will, they determine it from without. They are like moulds into which we are forced to cast our actions. The necessity is often ineluctable. But even when we succeed in triumphing, the opposition we have encountered suffices to alert us that we are faced with something independent of ourselves. Thus in considering facts as things we shall be merely conforming to their nature.

In the end, the reform that must be introduced into sociology is identical in every respect to that which has transformed psychology over the last thirty years. Just as Comte and Spencer declare that social facts are facts of nature, but nevertheless refuse to treat them as things, the different empirical schools had long recognised the natural character of psychological phenomena, while continuing to apply to them a purely ideological method. Indeed the empiricists, no less than their opponents, proceeded exclusively by introspection. But the facts observable in ourselves are too few, too fleeting and malleable, to be able to impose themselves upon the corresponding notions that habit has rooted in us and to prevail over them. Thus when these notions are not subject to some other control, no countervailing force exists; consequently they take the place of facts and constitute the subject matter of the science. Thus neither Locke nor Condillac considered physical phenomena objectively. It is not sensation they study, but a certain idea of it. This is why, although in certain respects they were its forerunners, scientific psychology arose only much later. It arose after it had been finally established that states of consciousness can and must be studied externally and not from the perspective of the individual consciousness which experiences them. This is the great revolution that has been accomplished in this field of study. All the special procedures and new methods which have enriched this science are only various expedients for realising more fully this basic idea. Such an advance remains to be accomplished in sociology, which must pass from the subjective stage, beyond which it has hardly progressed, to the objective stage.

This transition, moreover, is less difficult to accomplish in sociology than in psychology. Psychical facts naturally appertain to states of the individual, from whom they do not even appear to be separable. Internal by definition, such states cannot seemingly be treated as external save by doing violence to their nature. Not only is an effort of abstraction necessary, but a whole gamut of procedures and artifices as well, for them to be considered successfully from the external viewpoint. Social facts, on the other hand, display much more naturally and immediately all the characteristics of a thing. Law is enshrined in legal codes, the events of daily life are registered in statistical figures and historical monuments, fashions are preserved in dress, taste in works of art. By their very nature social facts tend to form outside the consciousnesses of individuals, since they dominate them. To perceive them in their capacity as things it is therefore not necessary to engage in an ingenious distortion. From this viewpoint sociology has significant advantages over psychology which have hitherto not been perceived, and this should accelerate its development. Its facts are perhaps more difficult to interpret because they are more complex, but they are more readily accessible. Psychology, on the other hand, has not only difficulty in specifying its facts, but also in comprehending them. Thus one may legitimately believe that as soon as this principle of sociological method has been universally acknowledged and is put into practice, sociology will be seen to progress at a speed that its present slow rate of development would scarcely allow one to suppose, even making up the lead of psychology, which it owes solely to its prior historical place[10].

II

But our predecessors' experience has shown us that, in order to realise in practice the truth just established, it is not enough to demonstrate it theoretically or even to absorb it oneself. The mind has such a natural disposition to fail to recognise it that inevitably we will relapse into past errors unless we submit ourselves to a rigorous discipline. We shall formulate the principal rules for this discipline, all of which are corollaries of the previous rule.

(1) The first of these corollaries is: *One must systematically discard all preconceptions*. Special proof of this rule is unnecessary: it follows from all that we have stated above. Moreover, it is the basis of all scientific method. Descartes' method of doubt is in essence only an application of it. If at the very moment of the foundation of science Descartes prescribed a rule for himself to question all the ideas he had previously accepted, it is because he wished to use only concepts which had been scientifically worked out, that is, constructed according to the method that he devised. All those of another origin had therefore to be rejected, at least for the time being. We have seen that Bacon's theory of the idols has the same significance. The two great doctrines, so often placed in contradiction to each other, agree on this essential point. Thus the sociologist, either when he decides upon the object of his research or in the course of his investigations, must resolutely deny himself the use of those concepts formed outside science and for needs entirely unscientific. He must free himself from those fallacious notions which hold sway over the mind of the ordinary person, shaking off, once and for all, the yoke of those empirical categories that long habit often makes tyrannical. If necessity sometimes forces him to resort to them, let him at least do so in full cognisance of the little value they possess, so as not to assign to them in the investigation a role which they are unfit to play.

What makes emancipation from such notions peculiarly difficult in sociology is that sentiment so often intervenes. We enthuse over our political and religious beliefs and moral practices very differently from the way we do over the objects of the physical world. Consequently this emotional quality is transmitted to the way in which we conceive and explain our beliefs. The ideas that we form about them are deeply felt, just as are their purposes, thereby taking on such authority that they brook no contradiction. Any opinion which is embarrassing is treated as hostile. For example, a proposition may not accord with our view of patriotism or personal dignity. It is therefore denied, whatever may be the proofs advanced. We cannot allow it to be true. It is rejected, and our strong emotions, seeking a justification for so doing, have no difficulty in suggesting reasons which we find readily conclusive. These notions may even be so prestigious that they will not tolerate scientific examination. The mere fact of subjecting them, as well as the phenomena they express, to cold, dry analysis is repugnant to certain minds. The sociologist who undertakes to study morality objectively as an external reality seems to such sensitive souls bereft of moral sense, just as the vivisectionist seems to the ordinary person devoid of normal feelings. Far from admitting that these sentiments are subject to science, it is believed that it is to them one should address onself in order to construct the science of things to which they relate. 'Woe', writes an eloquent historian of religions, 'Woe to the scientist who approaches the things of God without having in the depths of his consciousness, in the innermost indestructible parts of his being, in which sleep the souls of his ancestors, an unknown sanctuary from which at times there arises the fragrance of incense, a verse of a psalm, a cry of sorrow or triumph that as a child, following his brothers' example, he raised to heaven, and which suddenly joins him once again in communion with the prophets of yore!'[11]

One cannot protest too strongly against this mystical doctrine which – like all mysticism, moreover – is in essence only a disguised empiricism, the negation of all science. Feelings relating to social things enjoy no pride of place over other sentiments, for they have no different origin. They too have been shaped through history. They are a product of human experience, albeit one confused and unorganised. They are not due to some transcendental precognition of reality, but are the result of all kinds of disordered impressions and emotions accumulated through chance circumstance, lacking systematic interpretation. Far from bringing enlightenment of a higher order than the rational, they are composed exclusively of states of mind which, it is true, are strong but also confused. To grant them such a predominant role is to ascribe to the lower faculties of the intelligence supremacy over superior ones and to condemn oneself more or less to a rhetorical logomachy. A science constituted in this way can only satisfy those minds who prefer to think with their sensibility rather than their understanding, who prefer the immediate and confused syntheses of sensation to the patient, illuminating analyses of the reason. Feeling is an object for scientific study, not the criterion of scientific truth. But there is no science which at its beginnings has not encountered similar resistances. There was a time when those feelings relating to the things of the physical world, since they also possessed a religious or moral character, opposed no less violently the establishment of the physical sciences. Thus one can believe that, rooted out from one science after another, this prejudice will finally disappear from sociology as well, its last refuge, and leave the field clear for the scientist.

(2) But the above rule is entirely negative. It teaches the sociologist to escape from the dominance of commonly held notions and to direct his attention to the facts, but does not state how he is to grasp the facts in order to study them objectively.

Every scientific investigation concerns a specific group of phenomena which are subsumed under the same definition. The sociologist's first step must therefore be to define the things he treats, so that we may know – he as well – exactly what his subject matter is. This is the prime and absolutely indispensable condition of any proof or verification. A theory can only be checked if we know how to recognise the facts for which it must account. Moreover, since this initial definition determines the subject matter itself of the science, that subject matter will either consist of a thing or not, according to how this definition is formulated.

To be objective the definition clearly must express the phenomena as a function, not of an idea of the mind, but of their inherent properties. It must characterise them according to some integrating element in their nature and not according to whether they conform to some more or less ideal notion. When research is only just beginning and the facts have not yet been submitted to any analysis, their sole ascertainable characteristics are those sufficiently external to be immediately apparent. Those less apparent are doubtless more essential. Their explanatory value is greater, but they remain unknown at this stage of scientific knowledge and cannot be visualised save by substituting for reality some conception of the mind. Thus it is among the first group of visible characteristics that must be sought the elements for this basic definition. Yet it is clear that the definition will have to include, without exception or distinction, all the phenomena which equally manifest these same characteristics, for we have no reason nor the means to discriminate between them. These properties, then, are all that we know of reality. Consequently they must determine absolutely how the facts should be classified. We possess no other criterion which can even partially invalidate the effect of this rule. Hence the following rule: *The subject matter of research must only include a group of phenomena defined beforehand by certain common external characteristics and all phenomena which correspond to this definition must be so included.* For example, we observe that certain actions exist which all possess the one external characteristic that, once they have taken place, they provoke on the part of society that special reaction known as punishment. We constitute them as a group *sui generis* and classify them under a single heading: any action that is punished is termed a crime and we make crime, so defined, the subject matter of a special science of criminology. Likewise we observe within all known societies the existence of a smaller society outwardly recognisable because it is formed for the most part of individuals linked by a blood relationship and joined to each other by legal ties. From the relevant facts we constitute a special group to which we assign a distinctive name: phenomena of domestic life. We term every aggregate of this kind a family and make the family, so defined, the subject matter of a specific investigation which has not yet received a special designation in sociological terminology. When we later pass on from the family in general to the different types of family, the same rule should be applied. For example, embarking upon a study of the clan, or the maternal or patriarchal family, we should begin by defining them according to the same method. The subject matter of each topic, whether general or specialised, should be constituted according to the same principle.

By proceeding in this way from the outset the sociologist is immediately grounded firmly in reality. Indeed, how the facts are classified does not depend on him, or on his own particular cast of mind, but on the nature of things. The criterion which determines whether they are to be grouped in a particular category can be demonstrated and generally accepted by everybody, and the observer's statements can be verified by others. It is true that a notion built up in this way does not always chime – or does not generally even chime at all – with the notion commonly held. For example, it is evident that acts relating to freedom of thought or lapses in etiquette which are so regularly and severely punished in many societies, from the viewpoint of common sense are not regarded as crimes when people consider those societies. In the same way a clan is not a family in the usual sense of the word. But this is of no consequence, for it is not simply a question of how we can discover with a fair degree of accuracy the facts to which the words of common parlance refer and the ideas that they convey. What has to be done is to form fresh concepts *de novo*, ones appropriate to the needs of science and expressed by the use of a special terminology. It is certainly not true that the commonly held concept is useless to the scientist. It serves as a benchmark, indicating to him that somewhere there exists a cluster of phenomena bearing the same name and which consequently are likely to possess common characteristics. Moreover, since the common concept is never without some relationship to the phenomena, it occasionally points to the approximate direction in which they are to be discovered. But as the concept is only crudely formulated, it is quite natural for it not to coincide exactly with the scientific concept which it has been instrumental in instituting.[12]

However obvious and important this rule is, it is scarcely observed at present in sociology. Precisely because sociology deals with things which are constantly on our lips, such as the family, property, crime, etc., very often it appears useless to the sociologist initially to ascribe a rigorous definition to them. We are so accustomed to using these words, which recur constantly in the course of conversation, that it seems futile to delimit the meaning being given to them. We simply refer to the common notion of them, but this is very often ambiguous. This ambiguity causes us to classify under the same heading and with the same explanation things which are in reality very different. From this there arises endless confusion. Thus, for example, there are two kinds of monogamous unions: the ones that exist in fact, and those that exist legally. In the first kind the husband has only one wife, although legally he may have several; in the second kind polygamy is legally prohibited. Monogamy is met with *de facto* in several animal species and certain societies at a lower stage of development, not sporadically, but indeed with the same degree of generality as if it had been imposed by law. When a tribe is scattered over a wide area the social bond is very loose and consequently individuals live isolated from each other. Hence every man naturally seeks a female mate, but only one, because in his isolated state it is difficult for him to secure several. Compulsory monogamy, on the other hand, is only observed in societies at the highest stage of development. These two kinds of conjugal union have therefore very different significance, and yet the same word serves to described them both. We commonly say that certain animals are monogamous, although in their case there is nothing remotely resembling a legal tie. Spencer, embarking on his study of marriage, uses the term monogamy, without defining it, in its usual and equivocal sense. Consequently for him the development of marriage appears to present an incomprehensible anomaly, since he thinks he can observe the higher form of sexual union from the very earliest stages of historical development, while it apparently tends to disappear in the intermediate period, only to reappear again later. He concludes from this that there is no consistent relationship between social progress in general and the progressive advance towards a perfect type of family life. A definition at the appropriate time would have obviated this error.[13]

In other cases great care is taken to define the subject matter of the research but instead of including in the definition and grouping under the same heading all phenomena possessing the same external properties, a selection is made. Certain phenomena, a kind of elite, are chosen as those considered to have the sole right to possess these characteristics. The others are held to have usurped these distinctive features and are disregarded. It is easy to envisage that, using this procedure, only a subjective and partial notion can be obtained. Such a process of elimination can in fact only be made according to a preconceived idea, since at the beginnings of a science no research would have been able to establish whether such a usurpation was real, even assuming it to be possible. The phenomena selected can only have been chosen because, more than the others, they conformed to the ideal conception that had already been formed of that kind of reality. For example, Garofalo, at the beginning of his *Criminologie*, demonstrates extremely well that the point of departure for that science should be 'the sociological notion of crime'.[14] Yet, in order to build up this notion, he does not compare indiscriminately all the actions which in different types of society have been repressed by regular punishment, but only certain of them, namely those which offend the normal and unchangeable elements in the moral sense. As for those moral sentiments which have disappeared as a result of evolution, for him they were apparently not grounded in the nature of things for the simple reason that they did not succeed in surviving.

Consequently the acts which have been deemed criminal because they violated those sentiments seemed to him to have merited this label only through chance circumstances of a more or less pathological kind. But he proceeds to make this elimination by virtue of a very personal conception of morality. He starts from the idea that moral evolution, considered at the source or its close proximity, carries along with it all sorts of deposits and impurities which it then progressively eliminates; only today has it succeeded in ridding itself of all the extraneous elements which at the beginning troubled its course. But this principle is neither a self-evident axiom nor a demonstrated truth: it is only a hypothesis, which indeed nothing justifies. The variable elements of the moral sense are no less founded in the nature of things than those that are immutable; the variations through which the former elements have passed evidence the fact that the things themselves have varied. In zoology those forms peculiar to the lower species are not considered any less natural than those which recur at all levels on the scale of animal development. Similarly, those actions condemned as crimes by primitive societies, but which have since lost that label, are really criminal in relation to those societies just as much as those we continue to repress today. The former crimes correspond to the changing conditions of social life, the latter to unchanging conditions, but the first are no more artificial than the rest.

More can be added to this: even if these acts had wrongly assumed a criminal character, they neverthless should not be drastically separated from the others. The pathological forms of a phenomenon are no different in nature from the normal ones, and consequently it is necessary to observe both kinds in order to determine what that nature is. Sickness is not opposed to health; they are two varieties of the same species and each throws light on the other. This is a rule long recognised and practised both in biology and psychology, and one which the sociologist is no less under an obligation to respect. Unless one allows that the same phenomenon can be due first to one cause and then to another – which is to deny the principle of causality – the causes which imprint upon an action, albeit abnormally, the distinctive mark of a crime, cannot differ in kind from those which normally produce the same effect. They are distinguishable only in degree, or because they are not operating in the same set of circumstances. The abnormal crime therefore continues to be a crime and must consequently enter into the definition of crime. But what happens? Thus Garofalo takes for the *genus* what is only the species or merely a simple variation. The facts to which his formulation of criminality are applicable represent only a tiny minority among those which should be included. His formulation does not fit religious crimes, or crimes against etiquette, ceremonial or tradition, etc., which, although they have disappeared from our modern legal codes, on the contrary almost entirely fill the penal law of past societies.

The same error of method causes certain observers to deny to savages any kind of morality.[15] They start from the idea that our morality is *the* morality. But it is either clearly unknown among primitive peoples or exists only in a rudimentary state, so that this definition is an arbitrary one. If we apply our rule all is changed. To decide whether a precept is a moral one or not we must investigate whether it presents the external mark of morality. This mark consists of a widespread, repressive sanction, that is to say a condemnation by public opinion which consists of avenging any violation of the precept. Whenever we are confronted with a fact that presents this characteristic we have no right to deny its moral character, for this is proof that it is of the same nature as other moral facts. Not only are rules of this kind encountered in more primitive forms of society, but in them they are more numerous than among civilised peoples. A large number of acts which today are left to the discretion of individuals were then imposed compulsorily. We perceive into what errors we may fall if we omit to define, or define incorrectly.

But, it will be claimed, to define phenomena by their visible characteristics, is this not to attribute to superficial properties a kind of preponderance over more fundamental qualities? Is this not to turn the logical order upside down, to ground things upon their apex and not their base? Thus when crime is defined by punishment almost inevitably one runs the risk of being accused of wanting to derive crime from punishment, or, to cite a well known quotation, to see the source of shame in the scaffold rather than in the crime to be expiated. But the reproach is based upon a confusion. Since the definition, the rule for which we have just enunciated, is made at the beginnings of the science its purpose could not be to express the essence of reality; rather is it intended to equip us in order to arrive at this essence later. Its sole function is to establish the contact with things, and since these cannot be reached by the mind save from the outside, it is by externalities that it expresses them. But it does not thereby explain them; it supplies only an initial framework necessary for our explanations. It is not of course punishment that causes crime, but it is through punishment that crime, in its external aspects, is revealed to us. And it is therefore punishment that must be our starting point if we wish to understand crime.

The objection referred to above would be well founded only if these external characteristics were at the same time merely accidental, that is, if they were not linked to the basic properties of things. In these conditions science, after having pointed out the characteristics, would indeed lack the means of proceeding further. It could not penetrate deeper into reality, since there would be no connection between the surface and the depths. But, unless the principle of causality is only empty words, when clearly determined characteristics are to be found identically and without exception in all phenomena of a certain order, it is assuredly because they are closely linked to the nature of these phenomena and are joined indissolubly to them. If any given set of actions similarly presents the peculiarity of having a penal sanction attached to it, it is because there exists a close link between the punishment and the attributes constituting those actions. Consequently, however superficial these properties may be, provided they have been methodically observed, they show clearly to the scientist the path that he must follow in order to penetrate more deeply into the things under consideration. They are the prime, indispensable link in the sequence later to be unfolded by science in the course of its explanations.

Since it is through the senses that the external nature of things is revealed to us, we may therefore sum up as follows: in order to be objective science must start from sense-perceptions and not from concepts that have been formed independently from it. It is from observable data that it should derive directly the elements for its initial definition. Moreover, it is enough to call to mind what the task of scientific work is to understand that science cannot proceed otherwise. It needs concepts which express things adequately, as they are, and not as it is useful in practical living to conceive them. Concepts formed outside the sphere of science do not meet this criterion. It must therefore create new concepts and to do so must lay aside common notions and the words used to express them, returning to observations, the essential basic material for all concepts. It is from sense experience that all general ideas arise, whether they be true or false, scientific or unscientific. The starting point for science or speculative knowledge cannot therefore be different from that for common or practical knowledge. It is only beyond this point, in the way in which this common subject matter is further elaborated, that divergences will begin to appear.

(3) But sense experience can easily be subjective. Thus it is a rule in the natural sciences to discard observable data which may be too personal to the observer, retaining exclusively those data which present a sufficient degree of objectivity. Thus the physicist substitutes for the vague impressions produced by temperature or electricity the visual representation afforded by the rise and fall of the thermometer or the voltmeter. The sociologist must needs observe the same precautions. The external characteristics whereby he defines the object of his research must be as objective as possible.

In principle it may be postulated that social facts are more liable to be objectively represented the more completely they are detached from the individual facts by which they are manifested.

An observation is more objective the more stable the object is to which it relates. This is because the condition for any objectivity is the existence of a constant, fixed vantage point to which the representation may be related and which allows all that is variable, hence subjective, to be eliminated. If the sole reference points given are themselves variable, continually fluctuating in relationship to one another, no common measure at all exists and we have no way of distinguishing between the part of those impressions which depends on what is external and that part which is coloured by us. So long as social life has not succeeded in isolating itself from the particular events which embody it, in order that it may constitute itself a separate entity, it is precisely this difficulty which remains. As these events do not take on the same appearance each time nor from one moment to another and as social life is inseparable from them, they communicate to it their own fluctuating character. Thus social life consists of free-ranging forces which are in a constant process of change and which the observer's scrutinising gaze does not succeed in fixing mentally. The consequence is that this approach is not open to the scientist embarking upon a study of social reality. Yet we do know that social reality possesses the property of crystallising without changing its nature. Apart from the individual acts to which they give rise, collective habits are expressed in definite forms such as legal or moral rules, popular sayings, or facts of social structure, etc. As these forms exist permanently and do not change with the various applications which are made of them, they constitute a fixed object, a constant standard which is always to hand for the observer, and which leaves no room for subjective impressions or personal observations. A legal rule is what it is and there are no two ways of perceiving it. Since, from another angle, these practices are no more than social life consolidated, it is legitimate, failing indications to the contrary,[16] to study that life through these practices.

Thus when the sociologist undertakes to investigate any order of social facts he must strive to consider them from a viewpoint where they present themselves in isolation from their individual manifestations. It is by virtue of this principle that we have studied elsewhere social solidarity, its various forms and their evolution, through the system of legal rules whereby they are expressed.[17] In the same way, if an attempt is made to distinguish and classify the different types of family according to the literary descriptions imparted by travellers and sometimes by historians, we run the risk of confusing the widely differing species and of linking types extremely dissimilar. If, on the other hand, we take as the basis of classification the legal constitution of the family, and more especially the right of succession, we have an objective criterion which, although not infallible, will nevertheless prevent many errors.[18] If we aim at a classification of different kinds of crime, the attempt must be made to reconstitute the various modes of living and the 'professional' customs in vogue in the different worlds of crime. As many criminological types will be identified as there are organisational forms. To penetrate the customs and popular beliefs we will turn to the proverbs and sayings which express them. Doubtless by such a procedure we leave outside science for the time being the concrete data of collective life. Yet, however changeable that life may be, we have no right to postulate *a priori* its incomprehensibility. But in order to proceed methodically we must establish the prime bases of the science on a solid foundation, and not on shifting sand. We must approach the social domain from those positions where the foothold for scientific investigation is the greatest possible. Only later will it be feasible to carry our research further and by progressive approaches gradually capture that fleeting reality which the human mind will perhaps never grasp completely.

Notes

1. Bacon, *Novum Organum*, I, p.26.
2. Ibid., I, p.17.
3. Ibid., I, p.36.
4. H. Spencer, *The Principles of Sociology*, II, p.244 (London, Williams & Norgate, 1882).
5. Ibid., II, p.245.
6. This is moreover a conception which is controversial (cf. *Division du travail social*, II, 2, ss. 4).
7. Spencer, op. cit., II, p.244: 'Cooperation, then, is at once that which cannot exist without a society, and that for which a society exists.'
8. J.S. Mill, *A System of Logic*, vol. 2, book VI. ch. IX, p.496 (London, Longmans, Green Reader & Dyer, 1872): 'Political economy shows mankind as occupied solely in acquiring and consuming wealth.'
9. This trait emerges from the very expressions used by economists. They continually talk of ideas, of the ideas of utility, savings, investment and cost. (Cf. C. Gide, *Principes de l'économie politique* book III, ch. 1, ss. 1; ch. 2, ss. 1; ch. 3, ss. 1 [First edition, Paris, 1884].)
10. It it true that the greater complexity of social facts renders the science that relates to them more difficult. But, as compensation, precisely because sociology is the latest arrival on the scene, it is in a position to benefit from the progress realised by the lesser sciences and to learn from them. This use of previous experience cannot fail to hasten its development.
11. J. Darmsteter, *Les Prophètes d'Israël* (Paris, 1892) p.9.
12. It is in practice always the common concept and the common term which are the point of departure. Among the things that in a confused fashion this term denotes, we seek to discover whether any exist which present common external characteristics. If there are any, and if the concept formed by grouping the facts brought together in this way coincides, if not entirely (which is rare) but at least for the most part, with the common concept, it will be possible to continue to designate the former by the same common term, retaining in the science the expression used in everyday parlance. But if the difference is too considerable, if the common notion mixes up a number of different notions, the creation of new and special terms becomes a necessity.
13. It is the same absence of definition which has sometimes caused it to be stated that democracy occurred both at the beginning and the end of history. The truth is that primitive and present-day democracy are very different from each other.
14. R. Garofalo, *Criminologie* (Paris, 1888) p.2 (trans. by the author from the Italian).
15. J. Lubbock, *Origins of Civilization*, ch. VIII. More generally still, it is stated, no less inaccurately, that ancient religions are amoral or immoral. The truth is that they have their own morality. [Durkheim may have read Sir John Lubbock's work in translation. It was published in French translation by E. Barbier in 1873. Two further editions in French followed in 1877 and 1881.]
16. For example, one should have grounds to believe that, at a given moment, law no longer expressed the real state of social relationships for this substitution to be invalid.
17. Cf. *Division du travail social*, I, 1.
18. Cf. Durkheim, 'Introduction à la sociologie de la famile', *Annales de la Faculté des Lettres de Bordeaux*, 1889.

7

Weber's *Verstehen* and the History of Qualitative Research: The Missing Link

Jennifer Platt

The name of Weber is commonly invoked in connection with qualitative research methods emphasizing understanding, of which participant observation tends to be seen as the archetype.[1] This paper argues that this is a historically misleading association, and that the importance now attached to Weber's work has blotted out both the much lesser importance attached to it at the relevant times and places, and the availability of functionally alternative theorists whose work actually was influential. Many of the points made will not be novel to Weber specialists; but it may, none the less, be useful to bring them together in relation to this issue.

Detailed documentation of the extent to which the association with Weber has been made will not be attempted; it is a theme which is sufficiently 'in the air' to make this unnecessary. A few sources will, however, be drawn on to show some of the ways in which the connection is made. Bruyn in his important book on participant observation invokes Weber as a key representative of the tradition on which he focuses.[2] McCall and Simmons, in the introduction to the first section of their useful reader on participant observation,[3] see controversy over the value of *verstehen* as important in the decade 1920–30; Filstead in the introduction to his similar reader[4] equates the tradition of *verstehen* with qualitative methodology and cites Weber as his first reference on it. Easthope,[5] and Ackroyd and Hughes,[6] in their more strictly historical works, both convey the impression that Weber was relied on at a later period to a much greater extent than he really was through a

Source: *The British Journal of Sociology*, XXXVI(3) (1985): 448–466.

simple historical error. Each correctly quotes an article by Howard P. Becker[7] as strongly emphasizing the importance of Weberian method, but wrongly imputes it to the well-known advocate and theorist of participant observation Howard S. Becker (no relation to the senior Becker).[8] John Hughes in his methods textbook[9] invokes Weber as articulating the rationale for participant observation. Plummer, in his book on the life history,[10] describes the tradition which emphasizes the necessity of grasping subjects' own perspectives as extending 'all the way at least from Weber's notion of *verstehen*. . .', and then immediately mentions as a central figure in clarifying this position Znaniecki in *The Polish Peasant*.[11] In such sources, the line between retrospectively providing a rationale and historically having given the inspiration is easily blurred, and even when they make the distinction clearly one has the impression that it is the collocation rather than its precise context which sticks in the notes and minds of most students.

Crucial steps in the emergence of qualitative research methods are generally agreed to have taken place at the University of Chicago in the 1920s and 1930s, and then more specifically in the elaboration of participant observation as a distinct method in the decade following World War II, and classic exemplars were produced there; the argument will, therefore, focus mainly on that body of work, and on the extent to which Weber's '*verstehende soziologie*' was known to those producing it, drawn on by them in their thinking about it, and was necessary to produce the research outcomes.

Knowledge of Weber

For Weber to have been influential, he must first have been known. For anyone who did not read German, his major relevant works were only directly accessible after World War II. The dates of the English translations are: *The Protestant Ethic and the Spirit of Capitalism*, 1930;[12] *From Max Weber: Essays in Sociology*, 1946;[13] *The Theory of Social and Economic Organization*, 1947;[14] *The Methodology of the Social Sciences*, 1949.[15] (Indirect access through secondary sources is, of course, also possible; this is discussed below.) Some American sociologists could read German, but certainly not all of them, and not all of those would have been able to grapple with Weber's notoriously difficult language. No systematic data are available on this point, but some suggestive statements by knowledgeable people may be cited:

(i) Edward Shils says that in the 1920s some of the older generation had studied in Germany, but the younger had not – and later on American academics could commonly read no language but their own.[16]
(ii) Robert Angell, a graduate student at Michigan in the early 1920s, reports that, although French and German were required, what sociologists writing in those languages had to say was generally learned from

secondary sources; he himself went to Heidelberg in 1938 to read Weber in the original, feeling that he could only grapple with the difficulties in such a conducive environment.[17,18]

(iii) Parsons and Barber, writing in 1948, say: 'Max Weber's work, which is more recent in its influence in this country, has been handicapped by the fact that hitherto only fragments have been available in English translation, and few American scholars are competent to read his difficult German'.[19]

It would no doubt also have helped Weber's influence within sociology, without being a necessary condition for it, if he had been generally regarded as a sociologist – rather than as an economic historian or an economist – which, in the earlier period, he was not. The very earliest translation into English was that by Frank H. Knight, an institutional economist, of the *General Economic History* in 1927. Of the items in the Gerth bibliography (see note 49) not by authors named below which appeared before 1930 the great majority are in non-sociological journals or, if books, by non-sociologists. (Apart from those which are obituaries or reviews of recent German work, the majority relate specifically to *The Protestant Ethic and the Spirit of Capitalism*.)

But, despite this, some sociologists in America did read Weber and did publicize his views; who were they, and what did they do? One of the earliest interpreters was Theodore Abel. He came across Weber quite accidentally: a fellow-student when he was a graduate student at Columbia was, for some reason, given the works of Weber, could not read German himself, and so passed them on to Abel who, as a Pole of German origins who came to America only as a graduate student, spoke it fluently. Abel was deeply impressed by what he read and incorporated a section on Weber into his thesis, published in 1929 as *Systematic Sociology in Germany*, as well as writing later articles on Weber.[20] Alexander von Schelting, the author of a study in German of Weber's thought very warmly reviewed by Parsons,[21] came to Columbia in the late 1920s[22] and presumably taught about Weber, though he does not appear to have written much on him in English. Sorokin, a Russian refugee and then at Minnesota, gave an extended and respectful discussion of Weber's ideas in his *Contemporary Sociological Theories*[23] in 1928. Howard P. Becker was of German origin, although born in America, and studied towards his Ph.D. at Cologne with Paul Honigsheim in the later 1920s; he taught for much of his career at Wisconsin, and published a lot on Weber (and other German sociologists, especially von Wiese).[24] Honigsheim was German, and took his Ph.D. at Heidelberg, where he was a member of Weber's inner circle; he reached the USA, at Michigan State University, in 1938, and continued to write and teach about Weber.[25] Talcott Parsons went to Heidelberg in 1925 as a graduate student, and discovered Weber's work there;[26] on his return he both translated and wrote about Weber's work. Albert Salomon was another German who took his Ph.D. at Heidelberg, in

1921; he came to America in the New School for Social Research in the mid-1930s publishing some important articles on Weber in its journal *Social Research* in 1934–5.[27] Carl Mayer, another German refugee and member of the New School, did not publish much but is described in a memorial issue of *Social Research* as 'the lifelong bearer of Weber's sociology to successive generations of students at the New School for Social Research.'[28] Hans Speier, who wrote a number of articles on Weber, took his Ph.D. at Heidelberg in 1928 and was by 1933 at the New School. Hans Gerth, who translated and wrote about Weber, was born in Germany, studied at Heidelberg in 1927–9, reached the USA and a teaching post at Illinois in 1939–40 and then went to Wisconsin. Reinhard Bendix left Germany in 1938; a friend gave him Weber's *Economy and Society* as a going-away present, and in consequence he discovered Weber after reaching America.[29] The discovery led him to become an important Weber scholar.[30]

The general pattern is clear: the proponents of Weber spoke German well, had usually done graduate work in Germany (especially at Weber's own university, Heidelberg – see Parsons[31] on the 'Weber circle' in Heidelberg even after his death), and were most often of German origin.[32] They wrote about Weber in English from the later 1920s onwards, but it was not until the late 1930s that more than a handful of them were active in the USA. They taught at Harvard, Columbia, Wisconsin and the New School for Social Research, all well-regarded graduate schools but not known for their qualitative empirical research. The centre of work on Weber was in New York, at Columbia – the traditional centre of quantitative work, as contrasted with Chicago's qualitative style – and the New School, heavily theoretical in its concerns. A majority of them were theorists rather than researchers who would themselves produce exemplary research work that might have served as models. Honigsheim[33] and Shils[34] are clearly right when they suggest that before the war it was primarily those who had some special interest in German sociology who gave close attention to Weber. But their works were published and they taught students, so they must have diffused knowledge of him more widely.

Weber's Influence

How well known and influential did Weber's work become as a result of the advocacy and exposition of these men? Three kinds of evidence can be brought to bear on the question: statements by participants or informed commentators, archival data, and references in published work. First, some statements by participants in qualitative work. In interviews with a number of those who were in the Chicago department or involved with qualitative research in the 1920s and 1930s, or involved in the later efflorescence of participant observation, I raised the question of Weber's significance to them, and received the following answers. Ruth Shonle Cavan (Ph.D. Chicago

1926) said she was probably not aware of Weber as a student, and did not think the idea of *verstehen* had any influence in relation to the case-study method.[35] L. S. Cottrell (Ph.D. Chicago 1933) says that Weber was mentioned as one among other theorists, but without special emphasis or attention to his methodology.[36] Herbert Blumer (Ph.D. Chicago 1928) confirms this impression, saying that although people in the department knew the work they did not really understand the ideas, and were not methodologically sophisticated.[37] Faris, in his participant's book on the Chicago department, does not mention Weber, although Durkheim is referred to.[38] Philip Hauser (Ph.D. Chicago 1938) gives a somewhat different impression, but there is reason to believe that this testimony blurs periods whose differences are important for this paper.[39] Hauser, too, was primarily a demographer, as contrasted with the others working in more qualitative styles, and to the extent that we are concerned with actual rather than potential influences their perceptions are more relevant for our purposes. (Two graduates of Minnesota, a methodologically self-conscious but very quantitatively oriented department, give a yet stronger impression of the lack of interest in Weber there. C. A. Anderson (Ph.D. 1932) saw all systematic rationale for qualitative method as the postwar product of the influx of German refugees;[40] William Sewell (Ph.D. 1939) says he did not read a word of Weber as a graduate student.[41] Anderson 1984; Sewell 1984.)

Coming to the later period, William F. Whyte said that he did not read any Weber until he had finished the fieldwork on *Streetcorner Society*,[42] and ideas on *verstehen* were not salient to him in relation to participant observation.[43] Anselm Strauss (Ph.D. Chicago 1946), Howard Becker (Ph.D. Chicago 1951) and Robert Habenstein (Ph.D. Chicago 1954), all committed to qualitative work and participant observation, agree that when they were graduate students they knew of Weber's translated work and the word *verstehen* but it did not make any particular impact. *The Structure of Social Action* on Weber was used as a cribbing device to pass theory exams,[44] and recent Weber translations were used more as ammunition when required in argument than as a significant influence.[45] Herbert Gans (M.A. Chicago 1950) said Weber was taught in theory, and thought of as concerned with capitalism, class and power and value-free sociology, but he has never read him on *verstehen* and does not think he ever heard of that then.[46] Murray Wax (Ph.D. Chicago 1959) does not recall significant reference to verstehen in the classes of Blumer or Hughes in 1951.[47] Raymond Gold (Ph.D. Chicago 1954), another committed qualitative researcher, on the other hand, says that he was taught about Weber on fieldwork courses.[48] The Chicago department was large and varied enough for it to be possible for people passing through it at the same time to have divergent experiences, but this difference seems more likely to be a personal one. It seems safe to conclude that, although by the postwar period Weberian ideas were much more current than before, *verstehen* was by no means an important part of the inspiration of participant observation and fieldwork.

Among those with more theoretical interests, Weber was not necessarily better known in the earlier period. Parsons, when he went to Heidelberg as a graduate student in 1925, had never heard of Weber.[49] Abel's book on Weber and three other German sociologists says that 'the works of the writers discussed, excepting Simmel's, are comparatively unknown in this country. . .'[50] Re-analysis of data collected by L. L. Bernard in 1927 on who American sociologists claimed to have been influenced by showed none mentioning Weber.[51] Well-informed commentators agree that it was only later that he became generally influential even among those with theoretical interests: Bendix and Roth say that Weber's reception did not become significant until the 1940s;[52] Roscoe and Gisela Hinkle refer to Weber as arousing interest in the late 1930s;[53] Bierstedt says that Weber was not a great name in the USA until Parsons gave him fame and that 'the history of American sociological theory up to Parsons is as independent of Durkheim and Weber as it is of Marx. To most of our Americans the holy trinity is neither holy nor a trinity';[54] Shils also argues that Weber had been largely unknown to indigenous American sociologists until Parsons took him up.[55]

If we turn now to references in published work, it is obviously impossible to undertake a complete review, but some strategic observations may be made. First, no reference to Weber has been found in the classic Chicago qualitative empirical studies.[56,57,58] Park and Burgess' famous *Introduction to the Science of Sociology* (1924)[59] has only three references to Weber, none of them involving discussion in the text (Gumplowicz and Ratzenhofer are mentioned more often); Bernard's (1934) compendium reviewing the field[60] has references to Weber only in the chapter by H. P. Becker, despite the obvious opportunities provided by other chapters on 'The field and problems of the sociology of religion', 'The sources and methods of the sociology of religion' and 'The sources and methods of historical sociology'.

Weber appears in Appendix E of Rice's 1931 compendium *Methods in Social Science*[61] where books and articles of importance in sociological method and standpoint are listed, but it is his *Gesammelte Aufsätze zur Religionssoziologie* which is picked out, under the subheading 'History of Social Institutions'. A series of important textbooks on research methods, chosen to include the major ones of the period paying special attention to qualitative methods,[62,63,64,65] have no index references to Weber – even though Odum and Jocher have specific chapters on economic and historical approaches, and Bogardus refers to Weber elsewhere as a social thinker.[66]

A sequence of publications listing the most widely cited authors in general textbooks, and sometimes also other sources, has been examined, with the rollowing results:

(i) Hart, investigating who were regarded as the most important social thinkers, did not find Weber mentioned enough to be even included on his list.[67]

(ii) Odum does not have Weber on his list of authors mentioned 20 or more times in textbooks published from 1883 to 1951.[68]
(iii) Levine *et al.* find no reference to Weber at all in 5 selected sociology texts published over the period 1924–33.[69]
(iv) In three consecutive postwar studies of the authors mentioned in 10 recent leading introductory textbooks Weber appears in these positions: mentioned 50 times in four texts, with those most mentioned scoring 108[70]; mentioned 115 times in nine texts, ranking second only to Merton[71]; mentioned 98 times in seven texts, after Parsons, Merton and Lipset.[72]
(v) In a study of citations in the *American Sociological Review* and *American Journal of Sociology* in 1955 and 1970 Weber does not appear among the 26 most cited authors in 1955, while he is fifth on the list with 17 citations in 1970.[73] An analysis of the extremely thorough Weber bibliography produced by Gerth and Gerth in 1949 shows only 17 works in English before the 1930s, as compared with 183 in German.[74]

Thus these sources too confirm the impression that Weber's influence was not significantly diffused before the 1940s, which is at least consistent with the suggestion made by several writers that Parsons, and particularly his *The Structure of Social Action* (1937), played a crucial role in making Weber's ideas salient to the general sociological public in America.[75,76,77,78]

Pure ignorance of Weber's work no doubt made a significant contribution to this neglect, but it does not alone provide a sufficient explanation. Some central figures who were *not* ignorant of his work, but were important in inspiring qualitative research, did not make any use of his ideas in that connection. Robert Park, the central figure guiding the early Chicago research, worked at Harvard in 1898 when Munsterberg, a friend and former colleague of Weber's at Freiburg who was responsible for inviting him to the U.S.A. in 1904,[79] was there in the same department, and took his Ph.D. at Heidelberg in 1904 when Weber was there. He thus had some of the personal contacts, as well as the linguistic skills, which might have led him to take an active interest in Weber's work – but clearly they did not do so to more than a very limited extent. Robert Angell, who was Cooley's nephew as well as graduate student, reports that although Cooley could read German well, and was responsible for ordering *The Protestant Ethic and the Spirit of Capitalism* for the University of Michigan library, he did not suggest to his students that they study Weber: 'I really believe that he did not see them [the statement also covers Durkheim and Pareto] as central figures in the development of sociology'.[80]

One might speculate that, for writers to become considered really important, it is among the necessary conditions that their work be relatively easily available and that a few people in the field actively expound and publicize it, but that to complete the set of conditions jointly sufficient to produce the end result something else, which might weakly be labelled '*Zeitgeist*' or 'readiness', is required. What that might consist of in this case, which was present by the 1940s but not earlier, is a fascinating question, but beyond the scope of this paper. (However, an alternative explanation of the observed outcome could be simply that it takes time for exposition and publicity to have its effect, and that it was only by the later 1940s that the efforts of Weber's early supporters had had the time.)

Influence of Alternative Theorists

Why, when it might seem that there was such a good fit between Weber's methodological ideas and the intellectual interests and needs of qualitative researchers, was there so little interest in them among this group? I suggest that one prime reason is the existence of alternative and more familiar theorists who satisfied those intellectual needs. (This implies that to explain the neglect of Weber in terms of the *content* of his relevant ideas is unsatisfactory to the extent that they overlapped, or were felt to overlap, with those of these other theorists.) There were home-grown theorists, well-known to all American sociologists before the war and particularly to those at Chicago, who provided whatever theoretical inspiration and justification was needed for qualitative research: Dewey, Mead, Thomas, and perhaps especially Cooley. Dewey, Mead and Thomas were all members of the Chicago faculty, and Cooley was at nearby Michigan (where Dewey and Mead were for a short time his colleagues[81]) and in close contact with Chicago.[82]

If we return to the sources of evidence used earlier, we may trace the greater significance of Cooley's influence in the earlier period. Faris refers frequently to him, along with Dewey, Mead and Thomas.[83] Park and Burgess have 26 index references to Cooley, of which 5 are to passages directly quoted.[84] Odum found Cooley mentioned in more textbooks than any other author, and second only to Park and Burgess in the total number of references.[85]

Levine *et al.*[86] report that Cooley was the author most often mentioned as influencing them by Bernard's sample, with 56 mentions from 258 sociologists, and that he was referred to from 7 to 49 times in the textbooks studied. Hart did not find Cooley in a leading position, but he scored 37 per cent of the maximum possible on the index used.[87] (Top scorers were Spencer, Comte, Ward, Ross, Giddings.) Bain found Cooley mentioned 68 times in 8 different texts.[88] Only in the studies of work published from 1963 onwards[89,90,92] does Weber take the lead – a very marked one – over Cooley. (Thomas figures in none of these lists.)

The Bernard *Fields and Methods* volume cites Cooley 13 times in five different chapters;[92] among the methods texts, Cooley is referred to by all but one: by Odum and Jocher, nine times;[93] by Young, 13 times;[94] and, most significantly, by Palmer, the one arising directly from the Chicago work, only

Table I: Mentions of weber and cooley in various sources

Sources and date(s) referred to	*Weber*	*Cooley*
Key Chicago textbook:		
Park & Burgess 1924	3	26
Studies of textbooks:		
Hart c. 1927	below minima for inclusion on lists	37% of maximum
Odum 1883–1951		434 in 37 texts
Levine *et al.*, 1924–33	0	114
Bain 1958–1962	50	68
Oromaner 1963–1967	115	46
Swatos & Swatos 1968–1972	98	41
Influences mentioned by sociologists:		
Levine *et al.*, 1927	0	56
Compendium reviewing field:		
Bernard 1934	5 (all in one chapter)	13 (in 5 chapters)
Important methods textbooks:		
Palmer 1928	0	3
Odum & Jocher 1929	0	9
Bogardus 1936	0	0
Young 1939	0	13
Study of articles in *ASR* and *AJS*:		
Oromaner: 1955	0	0
1970	17	0

three times, but always very precisely in the context of the case study.[95] (Table I summarizes these data for Weber and Cooley.) The earliest source found which explicitly cites Weber on *verstehen* in relation to a discussion of concrete methods of qualitative empirical research is House in 1936, and this is only in a footnote.[96]

In the interviews from the prewar Chicago department, Ruth Cavan mentions *The Polish Peasant* and Cooley as methodologically important,[97] Cottrell mentions Thomas and Mead,[98] Hauser cites Cooley and Mead as among those authors one was expected to read[99] and Herbert Blumer sees Thomas and Park as influential on method, with the general ethos being one of 'getting out and getting close to what one wished to study' rather than of theorizing about refinements of method.[100] Perhaps surprisingly, the interviews from the postwar department still give the same impression: Howard Becker and Robert Habenstein agreed that Cooley and other writers in that tradition were more important than Weber,[101] and Anselm Strauss said

> *Verstehen* was a word we knew, and there were some articles on it, but I think that stuff didn't make any particular impact on it, except that it was like Cooley's sympathetic introspection, Park, Mead and the whole business – it didn't mean anything special. We did read Dilthey.[102]

Gans and Gold give more empirically oriented lists, with an emphasis on community studies.[103,104] William Whyte mentioned only anthropological writers as influential on his fieldwork, along with the 'interactionist' (n.b. *not* 'symbolic interactionist') ideas then being developed at Harvard.[105]

It must be concluded that the native American tradition was of much greater practical importance than Weberian ideas in the emergence of qualitative empirical research.[106] Several writers have pointed out that Cooley had independently reached a methodological position similar to Weber's on *verstehen* by the early 1900s.[107,108,109] His works expounding that position appeared from 1918–1928, and his ideas were of course also transmitted by informal networks and exemplified in his substantive publications.[110,111,112,113,114]

Their general gist was to advocate the importance of sympathetic insight or understanding, and of a holistic and qualitative case-study approach involving close observation and the use of life histories. His introduction of the concept of 'primary group', and emphasis on its importance, was obviously useful and congenial to empirical research. W. I. Thomas, although his only major work on method was the famous 'Methodological Note' written with Znaniecki in *The Polish Peasant*[115] provided several practical exemplars of empirical work, all of which were influential. There were also, of course, other American writers in the same tradition, both theoretical and empirical. If we look at the actual methods used, they are the 'case study', the life-history and other 'personal documents', and direct observation, as advocated and/or practiced by Cooley and Thomas, rather than the historical work in which Weber exemplified his own principles and for which he is and was best known.[116] Thus even though it was in principle possible for the work of Weber to have been drawn on for inspiration, it was in practice unnecessary in the sense that adequate functional alternatives nearer to home were readily to hand.[117] Strauss, quoted above, is not alone in having seen this aspect of Weber's though primarily in terms of its similarity to Cooley. Read Bain (a theorist, and well known for the breadth of his reading) was reading Weber in German in 1930, and wrote in a letter to a friend 'As nearly as I can get the drift of the first 10 pages or so, his verstehende soziologie is something like Cooley's sympathetic insight. . .'.[118] Anyone already familiar with Cooley's work was likely to detect the similarity, to see the ideas as not altogether novel, and possibly for that reason not to give them great attention.

Discussion

Whether it is in some sense intellectually essential for appropriate theoretical/methodological positions to have been elaborated to provide a foundation for empirical work is not obvious. It may help, but abstract methodological programmes do not invariably bear fruit, and there is some tendency for theoretical/methodological writing to be done by different people from empirical work and for each group to direct its main attention inwards rather than outwards. (I have been struck, in interviewing a number of members of the early Chicago department, by the extent to which this group, commonly seen by others as representing a methodological school, do not themselves treat method as a very salient aspect of what was going on; the main focus of interest was more often on the substantive problem.) One could construct a plausible argument which saw the methods of early Chicago qualitative research as inspired largely by Park's experience as a newspaperman, combined with the influence of existing social-work practices and the large body of data they made available – and later empirical works as inspired by the earlier ones. This point would hold whether or not theorists are actually invoked in the public work; that can be obeisance to academic convention, or striving for retrospective legitimation, as much as an indication of real intellectual foundations.

The neglect of Cooley in Britain, and the enormous attention paid to Weber, are part of the mysterious processes by which a discipline selects (and reselects) who to regard as its founding fathers, as well as a simple matter of parochialism. This selection is a matter of normative choice, not just of historical description, and as such tends to interfere with accurate historical description.[119] There is always the danger of projecting current preoccupations backwards, looking for what one knows is 'important', and taking intellectual compatibility of standpoint as *prima facie* evidence of influence. (It seems curious that no one working in this style has yet explored the thesis that Cooley was a crucial influence on Weber.)[120,121] This goes together with a tendency to use key names as labels for any position that can be identified as having some resemblance to theirs – as when, for instance, anyone writing on stratification who takes status into account is referred to as Weberian. Such ascriptions provide convenient classificatory devices and mnemonics for teaching, but do not contribute to our understanding of the social processes of intellectual history.

It seems possible that another reason for the British neglect of an important American tradition in this connection is the fact that there is no British equivalent to the American sociological social psychology, which was exceptionally strong at Chicago but by no means confined to that department; over the years, members of the American Sociological Association when invited to specify their main interests have regularly put 'social psychology' in the lead.[122,123] All the influential figures referred to above are commonly described in American sources as 'social psychologists', although their formal affiliations were with departments of sociology (Cooley, Thomas), philosophy (Mead, Dewey) or education (Dewey). Against that background, the idea of understanding motives and subjective meanings is automatically associated with 'social psychology.' (It is hard to imagine a British thesis, like that of Bensman, entitled 'Max Weber as a social psychologist: a critical study.'[124]) British sociology, when theoretically inclined, has had a more structural and macroscopic orientation which makes Weber seem a much more natural reference point.

The material reviewed in this paper might suggest a more general point about the ways in which intellectual influences are transmitted. It has been shown that almost all of the active early proponents of Weber in America had some kind of social connection, direct or indirect, with Weber himself and/or with Germany. It has also been shown that there were social connections, such as membership of the same university, among the researchers influenced and those who influenced them. It seems likely that such social contacts and connections, whatever the precise mechanisms of their operation, certainly facilitate influence, though they do not ensure it. At any given period there are a large number of potential influences, present or past, many of which may overlap in the possibilities they suggest, but not all of which are equally accessible or likely to become actual influences in a particular case. Several different patterns of influence could be causally sufficient to have produced the observed outcome. To understand influence processes we must look not only at intellectual content, but also at social relations.[125] To look only at similarities in the content of ideas always has weaknesses as a method of study of influence, but these are particularly striking when there are alternative sources available for similar intellectual content.

Conclusion

In sum, it is concluded that there is no evidence that Weber's ideas on method had significant influence on those involved at formative stages in the development of qualitative research methods in the USA. His work was not widely known until the 1940s; before then, those who took an active interest in it at all did so as part of a special interest in Germany, and most of them were primarily known as general theorists. Others who knew his work, especially those relatively close to empirical research, did not treat it as especially important or as standing out from the work of many others now hardly mentioned. At the same time there were other writers on methodology and exemplars of research, of a kind adequate and sufficient to inspire and legitimate qualitative research, who were much more widely known, cited and defined as important and relevant. The link with Weber is missing.

Notes

1. This perception is anachronistic for the earlier period, when 'participant observation' in the modern sense had not emerged as a distinct entity, the observation done by participants did not lay special stress on empathy or understanding, and the concepts used for the qualitative methods then current were 'case study' and 'personal documents'. For detailed discussion of this, see Jennifer Platt, 'The development of the "participant observation" method in sociology: origin myth and history', *Journal of the History of the Behavioural Sciences*, vol. 19, no. 4, October 1983, pp. 379–93.
2. Severyn T. Bruyn, *The Human Perspective in Sociology*, Englewood Cliffs, N.J., Prentice-Hall, 1966, pp. 2–9.

3. George J. McCall and J. L. Simmons, *Issues in Participant Observation*, Reading, Mass., Addison-Wesley, 1969, p. 2.
4. William J. Filstead, *Qualitative Methodology*, Chicago, Rand McNally, 1970, p. 4.
5. Gary Easthope, *History of Social Research Methods*, London, Longman, 1974, p. 99.
6. S. Ackroyd and J. A. Hughes, *Data Collection in Context*, London, Longman, 1981, p. 115.
7. Howard P. Becker, 'Interpretive sociology and constructive typology' in G. Gurvitch and W. E. Moore, *Twentieth Century Sociology*, New York, Philosophical Library, 1945, pp. 70–95.
8. The younger Becker had in 1945 not yet got his BA, and his style is very dissimilar from that of the older Becker, so the error is in some ways surprising. It is interesting how such a confusion can lead to over-interpretation of a text. Gary Easthope (op. cit.) imputes to the older Becker a specific point about the superiority of participant observation, to which I am unable to detect any reference. The passage which he probably has in mind (H. P. Becker, op. cit. note 8, pp. 72–4) appears to refer, in so far as it is clear on the matter, to external observation and the exercise of empathy ('the interpreter puts himself in the place of the actor as best he can. . .').
9. John A. Hughes, *Sociological Analysis: Methods of Discovery*, London, Nelson, 1976, p. 116.
10. K. Plummer, *Documents of Life*, London, Allen & Unwin, 1983, p. 66.
11. W. I. Thomas and F. Znaniecki, *The Polish Peasant in Europe and America*, Chicago, University of Chicago Press (vols. I, II), Boston, Badger, (vols III–V), 1918–20.
12. M. Weber, *The Protestant Ethic and the Spirit of Capitalism*, tr. Parsons, London, Allen & Unwin, 1930.
13. H. H. Gerth and C. W. Mills, *From Max Weber: Essays in Sociology*, New York, Oxford University Press, 1946.
14. M. Weber, *The Theory of Social and Economic Organisation*, tr. Henderson & Parsons, Glencoe, Ill., Free Press and The Falcon's Wing Press, 1947.
15. M. Weber, *The Methodology of the Social Sciences*, trs. Shils & Finch, Glencoe, Ill., Free Press, 1949.
16. E. Shils, *The Calling of Sociology*, Chicago, University of Chicago Press, 1980, p. 102.
17. R. C. Angell, Autobiography, 'Science and values in a sociologist's career', box 3, Robert Angell papers, Michigan Historical Collections, Bentley Historical Library, University of Michigan, 1970, p. 9.
18. R. C. Angell, 'The joys of modest success', *Society*, vol. 18, 1980, pp. 72–81.
19. T. Parsons and B. Barber, 'Sociology 1941–6', *American Journal of Sociology*, vol. 53, no. 4, January 1948, pp. 235–57; p. 254.
20. T. Abel, interview, September 1983.
21. T. Parsons, review of A. von Schelting, *Max Webers Wissenschaftslehre, American Sociological Review*, vol. 1, no. 4, August 1936, pp. 675–81.
22. See note 21.
23. P. Sorokin, *Contemporary Sociological Theories*, New York, Harper, 1928, pp. 673–84.
24. H. H. Gerth, obituary of H. P. Becker, *American Sociological Review*, vol. 25, no. 5, October 1960, pp. 743–5.
25. J. A. Beegle and R. H. Schulze, obituary of P. Honigsheim, *American Sociological Review*, vol. 28, no. 4, August 1963, pp. 629–30.
26. T. Parsons, 'The circumstances of my encounter with Max Weber', in R. K. Merton and M. W. Riley, *Sociological Traditions from Generation to Generation*, Norwood, N.J., Ablex, 1980, pp. 37–43.
27. A. J. Salomon, 3 articles on the thought of Max Weber, *Social Research*, vol. 1, no. 2, May 1934, pp. 147–68; vol. 2, no. 1, February 1935, pp. 60–73; vol. 2, no. 3, August 1935, pp. 368–84.

28. A. J. Vidich, 'Editor's introduction', *Social Research*, vol. 42, Winter 1975, pp. 567–9.
29. R. Bendix, 'Western Europe as the object and source of social science research', *Social Research*, vol. 46, Winter 1979, pp. 787–804, p. 787.
30. These names were chosen because they were among the translators of Weber, and/or are referred to by others as significant in propagating knowledge of his works. Edward Shils should also be included on the list, but is excluded by the absence of any information about his career in the usual works of reference. Louis Wirth, whose name some might expect to appear on it, has been excluded on the ground of testimony by a colleague from personal acquaintance in the 1930s, and of the lack of reference to Weber in his teaching materials at appropriate points. E.g. his question for an exam in History of Sociology set in 1928: 'State the chief contributions to sociology of 6 of the following: Wundt, Ratzel, Ratzenhofer, Gumplowicz, Simmel, Von Wiese, Tönnies, Durkheim, Tarde, Le Bon, De Grey, Wallas, Hobhouse, Ward, Cooley, Ross, Giddings, Thomas.'! (Park papers, Special Collections Department, Joseph Regenstein Library, University of Chicago: Addenda, Box 1, folder 4). (cf. also note 40.) Sources of biographical information used include, in addition to those cited in the text, *American Men of Science: The Social and Behavioral Sciences*, The Jacques Cattell Press, R. R. Bowker, New York, vols. 10 (1962), and 11 (1968); David L. Sills (ed.), *International Encyclopaedia of the Social Sciences* and *Biographical Supplement*, vol. 18, Macmillan and Free Press, New York, 1968 and 1979; *Who's Who in America 1980–81*; and the membership lists of the American Sociological Association.
31. Op. cit., note 27, p. 40.
32. Cf. Levine *et al.* 'Simmel's influence on American sociology', *American Journal of Sociology*, vol. 81, no. 4, January, 1976, pp. 813–44, p. 820) on the American sociologists who promoted Simmel.
33. P. Honigsheim, *On Max Weber*, New York, Free Press, 1968, p. 145.
34. Shils, op. cit., note 17, p. 154.
35. R. S. Cavan, interview, September 1982.
36. L. S. Cottrell, interview, September 1982.
37. H. Blumer, interview, September 1982.
38. Robert E. L. Faris, *Chicago Sociology 1920–32*, Chicago, University of Chicago Press, 1967.
39. Philip M. Hauser, interview, September 1982. Hauser's own papers are deposited in the University of Chicago's archives (Hauser Papers, Special Collections Department, Joseph Regenstein Library) and these include many from his time as a student. The negative evidence there includes: (i) Notes on the history of sociology, with no mention of Weber (Box 21, folder 10); (ii) Notes on the history of sociology, in which Weber is described as 'primarily economist', and the total of three-quarters of a page devoted to him contains no reference to method (box 22, folder 1); (iii) in a set of sheets each listing some key names under a heading, the following: 'Methodology // Savigny, Maine, Niebuhr, Ranke, Menger, Comte, Spencer, Simmel, Durkheim, Dilthey, Windelband, Rickert, Tönnies, Pareto; 'Verein für Sozialpolitik // Scholler [sic], Wagner, Knies.' The dates in the bibliography suggest that this course was taught c.1932 (Box 22, folder 1); (iv) an apparent term paper, on the question 'What seem to you to be the important European ancestors of American sociology?', which refers to other German sociologists but not Weber (Box 22, folder 2); long and detailed typed notes on Wirth's 1932 course 'History of Sociology in the US', referring to the general importance of European influences but not mentioning Weber (Box 22, folder 2). Weber on *verstehen* appears on the reading list for Park's 1932 course 'Methods of Social Research' – though low down – but not anywhere that I could detect in the notes from it (Box 23, folder 7). However, in response to a draft copy of this paper Professor Hauser has sent me a copy of some notes on Weber (not

deposited in the archives), which he made in 1932–5 when working towards his Ph.D. exam. These do mention *verstehen*, and stress the emphasis on the understanding of meaning as one of the defining characteristics of Weber's position. (At two places his view is compared with Cooley's.) This document, unlike the others, suggests that Weber was taken seriously as a leading theorist, though it throws no light on whether his views were treated as relevant to current research. In all relevant reading lists etc. from *after* the war Weber appears and is treated as important.

This detail is laboured, both because of the importance of taking evidence counter to one's thesis seriously, and because it illustrates well a point relevant to that thesis. Those of us with shorter professional careers may invent a plausible history which looks right from our present standpoint; those who experienced the history personally may still rework it unconsciously to fit later intellectual development, perhaps especially on points not central to our own concerns.

40. C. A. Anderson, interview, February 1984.
41. W. Sewell, oral presentation at Midwest Sociological Society meeting, April 1984, on graduate school experience in the 1930s.
42. W. F. Whyte, *Streetcorner Society*, Chicago, University of Chicago Press, 1943.
43. W. F. Whyte, interview, September 1982.
44. A. Strauss, interview, September 1982.
45. H. S. Becker and R. Habenstein, joint interview, September 1982.
46. H. Gans, interview, May 1984.
47. M. Wax, personal letter, May 1984.
48. R. L. Gold, interview, February 1984.
49. Parsons, op. cit., note 27, p. 39.
50. Theodore Abel, *Systematic Sociology in Germany*, 1929, reprinted New York, Octagon Books, 1965.
51. D. E. Sutherland, 'Who now reads European sociology? Reflections on the relationship between European and American sociology', *Journal of the History of Sociology*, vol. 2, no. 1, 1978, pp. 35–66.
52. R. Bendix and G. Roth, *Scholarship and Partisanship*, Berkeley, University of California Press, 1971, p. 48.
53. R. and G. Hinkle, *The Development of Modern Sociology*, New York, Random House, 1954, p. 53.
54. Robert Bierstedt, *American Sociological Theory: a Critical History*, New York, Academic Press, pp. 91, 499.
55. Shils, op. cit., note 17, pp. 122–3.
56. Nels Anderson, *The Hobo*, Chicago, University of Chicago Press, 1923.
57. Paul G. Cressey, *The Taxi Dance Hall*, Chicago, University of Chicago Press, 1932.
58. Frederic M. Thrasher, *The Gang*, Chicago, University of Chicago Press, 1927.
59. Robert E. Park and Ernest W. Burgess, *Introduction to the Science of Sociology*, Chicago, University of Chicago Press, 1924.
60. Luther L. Bernard, *The Fields and Methods of Sociology*, New York, Farrar & Rinehart, 1934.
61. Stuart A. Rice, *Methods in Social Science*, Chicago, University of Chicago Press, 1931.
62. Emory S. Bogardus, *Introduction to Social Research*, Los Angeles, Suttonhouse, 1936.
63. Howard W. Odum and K. Jocher, *Introduction to Social Research*, New York, Henry Holt, 1929.
64. Vivien Palmer, *Field Studies in Sociology*, Chicago, University of Chicago Press, 1928.
65. Pauline V. Young, *Scientific Social Surveys and Research*, Englewood Cliffs, N.J., Prentice-Hall, 1939.
66. E. S. Bogardus, *The Development of Social Thought*, New York, Longmans, Green, 1940.

67. Hornell Hart, 'A history of social thought: a consensus of American opinion', *Social Forces*, vol. 6, December 1927.
68. Howard W. Odum, *American Sociology*, New York, Greenwood, 1969 (reprinted from original 1951 edition).
69. Levine *et al.*, op. cit., note 33, pp. 840, 842.
70. Read Bain, 'The most important sociologists?', *American Sociological Review*, vol. 27, no. 5, October 1962, pp. 746–8.
71. Mark J. Oromaner, 'The most cited sociologists: an analysis of introductory text citations', *The American Sociologist*, vol. 3, no. 2, May, 1968, pp. 124–6.
72. William H. and Priscilla L. Swatos, 'Name citations in introductory sociology texts: a note on further research', *The American Sociologist*, vol. 9, no. 4, November 1974, pp. 225–8.
73. Mark J. Oromaner, 'Influentials' in sociological textbooks and journals, 1955 and 1970', *The American Sociologist*, vol. 15, no. 3, pp. 169–74.
74. H. H. Gerth and H. I. Gerth, 'Bibliography on Max Weber', *Social Research*, vol. 16, o. 1, March 1949, pp. 70–89. One article in English, and four in German, were excluded from the count by the absence of a date in the reference given.
    The complete figures are:

	up to 1920s	1930s	1940s
English	17	43	54
German	183	88	34

    The pattern of these figures obviously confirms the interpretations made in the text. It should be noted that in a number of cases there are articles by the same people in German and in English, usually the former first followed by the latter after emigration. Only 2 of the 20 authors of items before 1930 who are not named in the text appear not to be of German origin. Were it not for Hitler, the pattern would certainly be different.
75. Odum op. cit., note 69, passim.
76. Honigsheim, op. cit., note 34, p. 145.
77. Harry E. Barnes, 'Historical sociology', in J. S. Roucek, *Contemporary Sociology*, London, Peter Owen, 1959.
78. Shils, op. cit., note 17, pp. 118–123.
79. Gerth and Mills, op. cit., note 14, pp. 14, 21.
80. Angell, op. cit., note 18, pp. 9–10. Perhaps this is less surprising than it might intitially seem in the light of Schroeter's carefully-argued claim that Weber did not have any significant influence even on German sociology in the period 1920–33, because his definitions and ideas were not congenial to the intellectual climate of the time. (G. Schroeter, 'Max Weber as outsider: his normal influence on German sociology in the twenties', *Journal of the History of the Behavioral Sciences*, vol. 16, no. 4, October 1980, pp. 317–32.) Shils goes so far as to suggest that his ideas eventually became influential in Germany 'in consequence of his elaboration and deciphering in America'. (Op. cit., note 17, p. 187), and Vidich supports this (op. cit., note 29.)
81. Lewis Coser, *Masters of Sociological Thought*, New York, Harcourt Brace Jovanovich, 1971, pp. 326–7.
82. I refrain deliberately from using the term 'symbolic interactionism' here where it might be expected because, though it serves useful purposes, it was only invented after the relevant period. It is interesting that Mead, like Weber, appears to have been the subject of retrospective overestimation of his influence at an earlier period as postwar interest in his work increased (Coser, op. cit., note 82, p. 345; J. D. Lewis & R. L. Smith, *American Sociology and Pragmatism*, Chicago, University of Chicago Press, 1980.)

83. Faris, op. cit., note 39.
84. Park and Burgess, op. cit., note 60.
85. Odum, op. cit., note 69.
86. Levine *et al.*, op. cit., note 33, pp. 840, 842.
87. Hart, op. cit., note 68.
88. Bain, op. cit., note 71.
89. Oromaner, op. cit., note 72.
90. Swatos and Swatos, op. cit., note 73.
91. Oromaner, op. cit., note 74.
92. Bernard, op. cit., note 61.
93. Odum and Jocher, op. cit., note 64.
94. Young, op. cit., note 66.
95. Palmer, op. cit., note 65.
96. Floyd N. House, *The Development of Sociology*, New York, McGraw-Hill, 1936.
97. Cavan, op. cit., note 36.
98. Cottrell, op. cit., note 37.
99. Hauser, interview, September 1982.
100. Blumer, op. cit., note 38.
101. Becker and Habenstein, op. cit., note 46.
102. Strauss, op. cit., note 45. This last comment draws attention to the fact that Weber is not the only possible source of ideas on *verstehen*; it is likely that Dilthey and other German authors have also been relatively underestimated as a result of the growing emphasis on Weber.
103. Gans, op. cit., note 47.
104. Gold, op. cit., note 49.
105. Whyte, op. cit., note 44. On the intellectual background to this, see Platt, op. cit., note 2, p. 385.
106. Cf. Roscoe Hinkle, 'Antecedents of the action orientation in American sociology before 1935', *American Sociological Review*, vol. 28, no. 5, October 1963, pp. 705–15.
107. Coser, op. cit., note 82, p. 310.
108. Hinkle, op. cit., note 107, p. 709.
109. Angell, op. cit., note 18, p. 10a.
110. Charles H. Cooley, 'Social Science', in Cooley, *Social Process*, Carbondale, Ill., Southern Illinois Press, 1918, reprinted 1966.
111. Charles H. Cooley, 'Case study of small institutions as a method of research', reprinted in Cooley, *Sociological Theory and Social Research*, New York, Kelley, 1927.
112. Charles H. Cooley, 'The roots of social knowledge', *American Journal of Sociology*, vol. 32, no. 1, pp. 59–79.
113. Charles H. Cooley, 'The life study method as applied to rural social research', *Publications of the American Sociological Society*, vol. 23, 1928, pp. 248–54.
114. Charles H. Cooley, 'Sumner and Methodology', *Sociology of Social Research*, vol. 12, no. 4, 1928, pp. 303–6.
115. Thomas and Znaniecki, op. cit., note 12.
116. Weber did actually have some involvement with work using questionnaires, and may even have done something approaching participant observation (Rosalie H. Wax, *Doing Fieldwork: Warnings and Advice*, Chicago, University of Chicago Press, 1971, p. 27) but he is not generally known for this. See R. C. Angell, 'Cooley's heritage to social research', *Social Forces*, vol. 8, March 1930, pp. 340–7, for a summary of Cooley's methodological precepts as they were seen at the time.
117. It may also be noted that, insofar as German theorists were of practical importance in this context, Simmel is a more promising candidate. Park was actively interested

in him, and did much to diffuse his work (Levine *et al.*, op. cit., note 33.) At least one of the classic qualitative researchers drew inspiration for participant observation from his ideas on the stranger (Martin Bulmer, 'The methodology of *The Taxi-Dance Hall, Urban Life*, vol. 12, no. 1, April 1983, pp. 95–120). His micro-sociological orientation is of more obvious direct bearing on the issues of face-to-face fieldwork, and it could be argued that 'the power to reproduce in one's own imagination . . . the thoughts and motives of another' (Park on *verstehen*, from an unpublished work on method, ch. 1, p. 12; Park papers (on which see note 19) Addenda, Box 5, folder 4) is redundant when one has direct access to observational data or, as in many of the early Chicago qualitative studies, is drawing on personal experience.

118. Read Bain papers, Michigan Historical Collections, Bentley Historical Library, University of Michigan: Correspondence, letter to 'Josiphilus', 11 February 1930.
119. Cf. Augustine Brannigan, *The Social Basis of Scientific Discoveries*, Cambridge, Cambridge University Press, 1981, ch. 6, on the retrospective creation of Mendel as the founder of genetics.
120. Cf. Robert K. Merton, 'On the history and systematics of sociological theory', in Merton, *On Theoretical Sociology*, New York, Free Press, 1967, pp. 1–37.
121. Quentin Skinner, 'Meaning and understanding in the history of ideas', *History and Theory*, vol. 8, no. 1, 1969, pp. 3–53.
122. G. A. Lundberg, 'The interests of members of the American Sociological Society, 1930', *American Journal of Sociology*, vol. 37, no. 3, November 1931, pp. 458–60.
123. Nico N. Stehr and L. E. Larson, 'The rise and decline of areas of specialization', *The American Sociologist*, vol. 7, no. 7, August 1972, pp. 5–6.
124. Joseph Bensman, 'Max Weber as a social psychologist: a critical study', M.A. thesis, University of Wisconsin, 1947. (Listed in Gerth and Gerth, op. cit., note 75.)
125. For further discussion of related issues in other contexts, see: Jennifer Platt, 'Whatever happened to the case study?', University of Sussex, xeroxed, 1981; Platt, 'Stouffer & Lazarsfeld: patterns of influence', in ed. H. Kuklick, *Knowledge and Society*, forthcoming 1985; Platt, op. cit., note 2.

8

The Definitions of Sociology and of Social Action

Max Weber

1. Sociology (in the sense in which this highly ambiguous word is used here) is a science which attempts the interpretive understanding of social action in order thereby to arrive at a causal explanation of its course and effects. In 'action' is included all human behaviour when and in so far as the acting individual attaches a subjective meaning to it. Action in this sense may be either overt or purely inward or subjective; it may consist of positive intervention in a situation, or of deliberately refraining from such intervention or passively acquiescing in the situation. Action is social in so far as, by virtue of the subjective meaning attached to it by the acting individual (or individuals), it takes account of the behaviour of others and is thereby oriented in its course.[1]

(a) The Methodological Foundations of Sociology[2]

1. 'Meaning' may be of two kinds. The term may refer first to the actual existing meaning in the given concrete case of a particular actor, or to the average or approximate meaning attributable to a given plurality of actors; or secondly to the theoretically conceived *pure type*[3] of subjective meaning attributed to the hypothetical actor or actors in a given type of action. In no case does it refer to an objectively 'correct' meaning or one which is 'true' in

---

Source: Talcott Parsons (ed.) and A.M. Henderson and Talcott Parsons (trs), *Max Weber: The Theory of Social and Economic Organization* (London: The Free Press, 1964), pp. 88–104.

some metaphysical sense. It is this which distinguishes the empirical sciences of action, such as sociology and history, from the dogmatic disciplines in that area, such as jurisprudence, logic, ethics, and esthetics, which seek to ascertain the 'true' and 'valid' meanings associated with the objects of their investigation.

2. The line between meaningful action and merely reactive behaviour to which no subjective meaning is attached, cannot be sharply drawn empirically. A very considerable part of all sociologically relevant behaviour, especially purely traditional behaviour, is marginal between the two. In the case of many psychophysical processes, meaningful, i.e. subjectively understandable, action is not to be found at all; in others it is discernible only by the expert psychologist. Many mystical experiences which cannot be adequately communicated in words are, for a person who is not susceptible to such experiences, not fully understandable. At the same time the ability to imagine one's self performing a similar action is not a necessary prerequisite to understanding; 'one need not have been Caesar in order to understand Caesar.' For the verifiable accuracy[4] of interpretation of the meaning of a phenomenon, it is a great help to be able to put one's self imaginatively in the place of the actor and thus sympathetically to participate in his experiences, but this is not an essential condition of meaningful interpretation. Understandable and non-understandable components of a process are often intermingled and bound up together.

3. All interpretation of meaning, like all scientific observation, strives for clarity and verifiable accuracy of insight and comprehension (*Evidenz*). The basis for certainty in understanding can be either rational, which can be further subdivided into logical and mathematical, or it can be of an emotionally empathic or artistically appreciative quality. In the sphere of action things are rationally evident chiefly when we attain a completely clear intellectual grasp of the action-elements in their intended context of meaning. Empathic or appreciative accuracy is attained when, through sympathetic participation, we can adequately grasp the emotional context in which the action took place. The highest degree of rational understanding is attained in cases involving the meanings of logically or mathematically related propositions; their meaning may be immediately and unambiguously intelligible. We have a perfectly clear understanding of what it means when somebody employs the proposition $2 \times 2 = 4$ or the Pythagorean theorem in reasoning or argument, or when someone correctly carries out a logical train of reasoning according to our accepted modes of thinking. In the same way we also understand what a person is doing when he tries to achieve certain ends by choosing appropriate means on the basis of the facts of the situation as experience has accustomed us to interpret them. Such an interpretation of this type of rationally purposeful action possesses, for the understanding of the choice of means, the highest degree of verifiable certainty. With a lower degree of certainty, which is, however, adequate for most purposes of explanation,

we are able to understand errors, including confusion of problems of the sort that we ourselves are liable to, or the origin of which we can detect by sympathetic self-analysis.

On the other hand, many ultimate ends or values toward which experience shows that human action may be oriented, often cannot be understood completely, though sometimes we are able to grasp them intellectually. The more radically they differ from our own ultimate values, however, the more difficult it is for us to make them understandable by imaginatively participating in them. Depending upon the circumstances of the particular case we must be content either with a purely intellectual understanding of such values or when even that fails, sometimes we must simply accept them as given data. Then we can try to understand the action motivated by them on the basis of whatever opportunities for approximate emotional and intellectual interpretation seem to be available at different points in its course. These difficulties apply, for instance, for people not susceptible to the relevant values, to many unusual acts of religious and charitable zeal; also certain kinds of extreme rationalistic fanaticism of the type involved in some forms of the ideology of the 'rights of man' are in a similar position for people who radically repudiate such points of view.

The more we ourselves are susceptible to them the more readily can we imaginatively participate in such emotional reactions as anxiety, anger, ambition, envy, jealousy, love, enthusiasm, pride, vengefulness, loyalty, devotion, and appetites of all sorts, and thereby understand the irrational conduct which grows out of them. Such conduct is 'irrational,' that is, from the point of view of the rational pursuit of a given end. Even when such emotions are found in a degree of intensity of which the observer himself is completely incapable, he can still have a significant degree of emotional understanding of their meaning and can interpret intellectually their influence on the course of action and the selection of means.

For the purposes of a typological scientific analysis it is convenient to treat all irrational, affectually determined elements of behaviour as factors of deviation from a conceptually pure type of rational action. For example a panic on the stock exchange can be most conveniently analysed by attempting to determine first what the course of action would have been if it had not been influenced by irrational affects; it is then possible to introduce the irrational components as accounting for the observed deviations from this hypothetical course. Similarly, in analysing a political or military campaign it is convenient to determine in the first place what would have been a rational course, given the ends of the participants and adequate knowledge of all the circumstances. Only in this way is it possible to assess the causal significance of irrational factors as accounting for the deviations from this type. The construction of a purely rational course of action in such cases serves the sociologist as a type ('ideal type') which has the merit of clear understandability and lack of ambiguity. By comparison with this it is possible to understand the ways in which actual action is influenced by irrational factors of all sorts, such as affects[5] and errors, in that they account for the deviation from the line of conduct which would be expected on the hypothesis that the action were purely rational.

Only in this respect and for these reasons of methodological convenience, is the method of sociology 'rationalistic.' It is naturally not legitimate to interpret this procedure as involving a 'rationalistic bias' of sociology, but only as a methodological device. It certainly does not involve a belief in the actual predominance of rational elements in human life, for on the question of how far this predominance does or does not exist, nothing whatever has been said. That there is, however, a danger of rationalistic interpretations where they are out of place naturally cannot be denied. All experience unfortunately confirms the existence of this danger.

4. In all the sciences of human action, account must be taken of processes and phenomena which are devoid of subjective meaning,[6] in the role of stimuli, results, favouring or hindering circumstances. To be devoid of meaning is not identical with being lifeless or non-human; every artifact, such as for example a machine, can be understood only in terms of the meaning which its production and use have had or will have for human action; a meaning which may derive from a relation to exceedingly various purposes. Without reference to this meaning such an object remains wholly unintelligible.[7] That which is intelligible or understandable about it is thus its relation to human action in the role either of means or of end; a relation of which the actor or actors can be said to have been aware and to which their action has been oriented. Only in terms of such categories is it possible to 'understand' objects of this kind. On the other hand processes or conditions, whether they are animate or inanimate, human or non-human, are in the present sense devoid of meaning in so far as they cannot be related to an intended purpose. That is to say they are devoid of meaning if they cannot be related to action in the role of means or ends but constitute only the stimulus, the favouring or hindering circumstances.[8] It may be that the incursion of the Dollart at the beginning of the twelfth century[9] had historical significance as a stimulus to the beginning of certain migrations of considerable importance. Human mortality, indeed the organic life cycle generally from the helplessness of infancy to that of old age, is naturally of the very greatest sociological importance through the various ways in which human action has been oriented to these facts. To still another category of facts devoid of meaning belong certain psychic or psychophysical phenomena such as fatigue, habituation, memory, etc.; also certain typical states of euphoria under some conditions of ascetic mortification; finally, typical variations in the reactions of individuals according to reaction-time, precision, and other modes. But in the last analysis the same principle applies to these as to other phenomena which are devoid of meaning. Both the actor and the sociologist must accept them as data to be taken into account.

It is altogether possible that future research may be able to discover non-understandable uniformities underlying what has appeared to be specifically meaningful action, though little has been accomplished in this direction thus far. Thus, for example, differences in hereditary biological constitution, as of 'races,' would have to be treated by sociology as given data in the same way as the physiological facts of the need of nutrition or the effect of senescence on action. This would be the case if, and in so far as, we had statistically conclusive proof of their influence on sociologically relevant behaviour. The recognition of the causal significance of such factors would naturally not in the least alter the specific task of sociological analysis or of that of the other sciences of action, which is the interpretation of action in terms of its subjective meaning. The effect would be only to introduce certain non-understandable data of the same order as others which, it has been noted above, are already present, into the complex of subjectively understandable motivation at certain points. Thus it may come to be known that there are typical relations between the frequency of certain types of teleological orientation of action or of the degree of certain kinds of rationality and the cephalic index or skin colour or any other biologically inherited characteristic.

5. Understanding may be of two kinds: the first is the direct observational understanding[10] of the subjective meaning of a given act as such, including verbal utterances. We thus understand by direct observation, in this sense, the meaning of the proposition $2 \times 2 = 4$ when we hear or read it. This is a case of the direct rational understanding of ideas. We also understand an outbreak of anger as manifested by facial expression, exclamations or irrational movements. This is direct observational understanding of irrational emotional reactions. We can understand in a similar observational way the action of a woodcutter or of somebody who reaches for the knob to shut a door or who aims a gun at an animal. This is rational observational understanding of actions.

Understanding may, however, be of another sort, namely explanatory understanding. Thus we understand in terms of *motive* the meaning an actor attaches to the proposition twice two equals four, when he states it or writes it down, in that we understand what makes him do this at precisely this moment and in these circumstances. Understanding in this sense is attained if we know that he is engaged in balancing a ledger or in making a scientific demonstration, or is engaged in some other task of which this particular act would be an appropriate part. This is rational understanding of motivation, which consists in placing the act in an intelligible and more inclusive context of meaning.[11] Thus we understand the chopping of wood or aiming of a gun in terms of motive in addition to direct observation if we know that the woodchopper is working for a wage or is chopping a supply of firewood for his own use or possibly is doing it for recreation. But he might also be 'working off' a fit of rage, an irrational case. Similarly we understand the motive of a person aiming a gun if we know that he has been commanded to shoot as a member of a firing squad, that he is fighting against an enemy, or that he is doing it for revenge. The last is affectually determined and thus in a certain sense irrational. Finally we have a motivational understanding of the outburst of anger if we know that it has been provoked by jealousy, injured pride, or an insult. The last examples are all affectually determined and hence derived from irrational motives. In all the above cases the particular act has been placed in an understandable sequence of motivation, the understanding of which can be treated as an explanation of the actual course of behaviour. Thus for a science which is concerned with the subjective meaning of action, explanation requires a grasp of the complex of meaning in which an actual course of understandable action thus interpreted belongs.[12] In all such cases, even where the processes are largely affectual, the subjective meaning of the action, including that also of the relevant meaning complexes, will be called the 'intended' meaning.[13] This involves a departure from ordinary usage, which speaks of intention in this sense only in the case of rationally purposive action.

6. In all these cases understanding involves the interpretive grasp of the meaning present in one of the following contexts: (a) as in the historical approach, the actually intended meaning for concrete individual action; or (b) as in cases of sociological mass phenomena the average of, or an approximation to, the actually intended meaning; or (c) the meaning appropriate to a scientifically formulated pure type (an ideal type) of a common phenomenon. The concepts and 'laws' of pure economic theory are examples of this kind of ideal type. They state what course a given type of human action would take if it were strictly rational, unaffected by errors or emotional factors and if, furthermore, it were completely and unequivocally directed to a single end, the maximization of economic advantage. In reality, action takes exactly this course only in unusual cases, as sometimes on the stock exchange; and even then there is usually only an approximation to the ideal type.[14]

Every interpretation attempts to attain clarity and certainty, but no matter how clear an interpretation as such appears to be from the point of view of meaning, it cannot on this account alone claim to be the causally valid interpretation. On this level it must remain only a peculiarly plausible hypothesis. In the first place the 'conscious motives' may well, even to the actor himself, conceal the various 'motives' and 'repressions' which constitute the real driving force of his action. Thus in such cases even subjectively honest self-analysis has only a relative value. Then it is the task of the sociologist to be aware of this motivational situation and to describe and analyse it, even though it has not actually been concretely part of the conscious 'intention' of the actor; possibly not at all, at least not fully. This is a borderline case of the interpretation of meaning. Secondly, processes of action which seem to an observer to be the same or similar may fit into exceedingly various complexes of motive in the case of the actual actor. Then even though the situations appear superficially to be very similar we must actually understand them or interpret them as very different, perhaps, in terms of meaning, directly opposed.[15] Third, the actors in any given situation are often subject to opposing and conflicting impulses, all of which we are able to understand. In a large number of cases we know from experience it is not possible to arrive at even an approximate estimate of the relative strength of conflicting motives and very often we cannot be certain of our interpretation. Only the actual outcome of the conflict gives a solid basis of judgment.

More generally, verification of subjective interpretation by comparison with the concrete course of events is, as in the case of all hypotheses, indispensable. Unfortunately this type of verification is feasible with relative accuracy only in the few very special cases susceptible of psychological experimentation. The approach to a satisfactory degree of accuracy is exceedingly various, even in the limited number of cases of mass phenomena which can be statistically described and unambiguously interpreted. For the rest there remains only the possibility of comparing the largest possible number of historical or contemporary processes which, while otherwise similar, differ in the one decisive point of their relation to the particular motive or factor the role of which is being investigated. This is a fundamental task of comparative sociology. Often, unfortunately, there is available only the dangerous and uncertain procedure of the 'imaginary experiment' which consists in thinking away certain elements of a chain of motivation and working out the course of action which would then probably ensue, thus arriving at a causal judgment.[16]

For example, the generalization called Gresham's Law is a rationally clear interpretation of human action under certain conditions and under the assumption that it will follow a purely rational course. How far any actual course of action corresponds to this can be verified only by the available statistical evidence for the actual disappearance of under-valued monetary units from circulation. In this case our information serves to demonstrate a high degree of accuracy. The facts of experience were known before the generalization, which was formulated afterwards; but without this successful interpretation our need for causal understanding would evidently be left unsatisfied. On the other hand, without the demonstration that what can here be assumed to be a theoretically adequate interpretation also is in some degree relevant to an actual course of action, a 'law,' no matter how fully demonstrated theoretically, would be worthless for the understanding of action in the real world. In this case the correspondence between the theoretical interpretation of motivation and its empirical verification is entirely satisfactory and the cases are numerous enough so that verification can be considered established. But to take another example, Eduard Meyer has advanced an ingenious theory of the causal significance of the battles of Marathon, Salamis, and Platea for the development of the cultural peculiarities of Greek, and hence, more generally, Western, civilization.[17] This is derived from a meaningful interpretation of certain symptomatic facts having to do with the attitudes of the Greek oracles and prophets towards the Persians. It can only be directly verified by reference to the examples of the conduct of the Persians in cases where they were victorious, as in Jerusalem, Egypt, and Asia Minor, and even this verification must necessarily remain unsatisfactory in certain respects. The striking rational plausibility of the hypothesis must here necessarily be relied on as a support. In very many cases of historical interpretation which seem highly plausible, however, there is not even a possibility of the order of verification which was feasible in this case. Where this is true the interpretation must necessarily remain a hypothesis.

7. A motive is a complex of subjective meaning which seems to the actor himself or to the observer an adequate ground for the conduct in question. We apply the term 'adequacy on the level of meaning'[18] to the subjective interpretation of a coherent course of conduct when and in so far as, according to our habitual modes of thought and feeling, its component parts taken in their mutual relation are recognized to constitute a 'typical' complex of meaning. It is more common to say 'correct.' The interpretation of a sequence of events will on the other hand be called *causally* adequate in so far as, according to established generalizations from experience, there is a probability that it will always actually occur in the same way. An example of adequacy on the level of meaning in this sense is what is, according to our current norms of calculation or thinking, the correct solution of an arithmetical problem. On the other hand, a causally adequate interpretation of the same phenomenon would concern the statistical probability that, according to verified generalizations from experience, there would be a correct or an erroneous solution of the same problem. This also refers to currently accepted norms but includes taking account of typical errors or of typical confusions. Thus causal explanation depends on being able to determine that there is a probability, which in the rare ideal case can be numerically stated, but is always in some sense calculable, that a given observable event (overt or subjective) will be followed or accompanied by another event.

A correct causal interpretation of a concrete course of action is arrived at when the overt action and the motives have both been correctly apprehended and at the same time their relation has become meaningfully comprehensible. A correct causal interpretation of typical action means that the process which is claimed to be typical is shown to be both adequately grasped on the level of meaning and at the same time the interpretation is to some degree causally adequate. If adequacy in respect to meaning is lacking, then no matter how high the degree of uniformity and how precisely its probability can be numerically determined, it is still an incomprehensible statistical probability, whether dealing with overt or subjective processes. On the other hand, even the most perfect adequacy on the level of meaning has causal significance from a sociological point of view only in so far as there is some kind of proof for the existence of a probability[19] that action in fact normally takes the course which has been held to be meaningful. For this there must be some degree of determinable frequency of approximation to an average or a pure type.

Statistical uniformities constitute understandable types of action in the sense of this discussion, and thus constitute 'sociological generalizations,' only when they can be regarded as manifestations of the understandable subjective meaning of a course of social action. Conversely, formulations of a rational course of subjectively understandable action constitute sociological types of empirical process only when they can be empirically observed with a significant degree of approximation. It is unfortunately by no means the case that the actual likelihood of the occurrence of a given course of overt action is always directly proportional to the clarity of subjective interpretation. There are statistics of processes devoid of meaning such as death rates, phenomena of fatigue, the production rate of machines, the amount of rainfall, in exactly the same sense as there are statistics of meaningful phenomena. But only when the phenomena are meaningful is it convenient to speak of sociological statistics. Examples are such cases as crime rates, occupational distributions, price statistics, and statistics of crop acreage. Naturally there are many cases where both components are involved, as in crop statistics.

8. Processes and uniformities which it has here seemed convenient not to designate as (in the present case) sociological phenomena or uniformities because they are not 'understandable,' are naturally not on that account any the less important. This is true even for sociology in the present sense which restricts it to subjectively understandable phenomena – a usage which there is no intention of attempting to impose on anyone else. Such phenomena, however important, are simply treated by a different method from the others; they become conditions, stimuli, furthering or hindering circumstances of action.

9. Action in the sense of a subjectively understandable orientation of behaviour exists only as the behaviour of one or more *individual* human beings. For other cognitive purposes it may be convenient or necessary to consider the individual, for instance, as a collection of cells, as a complex of bio-chemical reactions, or to conceive his 'psychic' life as made up of a variety of different elements, however these may be defined. Undoubtedly such procedures yield valuable knowledge of causal relationships. But the behaviour of these elements, as expressed in such uniformities, is not subjectively understandable. This is true even of psychic elements because the more precisely they are formulated from a point of view of natural science, the less they are accessible to subjective understanding. This is never the road to interpretation in terms of subjective meaning. On the contrary, both for sociology in the present sense, and for history, the object of cognition is the subjective meaning-complex of action. The behaviour of physiological entities such as cells, or of any sort of psychic elements may at least in principle be observed and an attempt made to derive uniformities from such observations.

It is further possible to attempt, with their help, to obtain a causal explanation of individual phenomena, that is, to subsume them under uniformities. But the subjective understanding of action takes the same account of this type of fact and uniformity as of any others not capable of subjective interpretation. This is true, for example, of physical, astronomical, geological, meteorological, geographical, botanical, zoological, and anatomical facts and of such facts as those aspects of psycho-pathology which are devoid of subjective meaning or the facts of the natural conditions of technological processes.

For still other cognitive purposes as, for instance, juristic, or for practical ends, it may on the other hand be convenient or even indispensable to treat social collectivities, such as states, associations, business corporations, foundations, as if they were individual persons. Thus they may be treated as the subjects of rights and duties or as the performers of legally significant actions. But for the subjective interpretation of action in sociological work these collectivities must be treated as *solely* the resultants and modes of organization of the particular acts of individual persons, since these alone can be treated as agents in a course of subjectively understandable action. Nevertheless, the sociologist cannot for his purposes afford to ignore these collective concepts derived from other disciplines. For the subjective interpretation of action has at least two important relations to these concepts. In the first place it is often necessary to employ very similar collective concepts, indeed often using the same terms, in order to obtain an understandable terminology. Thus both in legal terminology and in everyday speech the term 'state' is used both for the legal concept of the state and for the phenomena of social action to which its legal rules are relevant. For sociological purposes, however, the phenomenon 'the state' does not consist necessarily or even primarily of the elements which are relevant to legal analysis; and for sociological purposes there is no such thing as a collective personality which 'acts.' When reference is made in a sociological context to a 'state,' a 'nation,' a 'corporation,' a 'family,' or an 'army corps,' or to similar collectivities, what is meant is, on the contrary, *only* a certain kind of development of actual or possible social actions of individual persons. Both because of its precision and because it is established in general usage the juristic concept is taken over, but is used in an entirely different meaning.

Secondly, the subjective interpretation of action must take account of a fundamentally important fact. These concepts of collective entities which are found both in common sense and in juristic and other technical forms of thought, have a meaning in the minds of individual persons, partly as of something actually existing, partly as something with normative authority. This is true not only of judges and officials, but of ordinary private individuals as well. Actors thus in part orient their action to them, and in this role such ideas have a powerful, often a decisive, causal influence on the course of action of real individuals. This is above all true where the ideas concern a recognized positive or negative normative pattern.[20] Thus, for instance, one of the important aspects of the 'existence' of a modern state, precisely as a complex of social interaction of individual persons, consists in the fact that the action of various individuals is oriented to the belief that it exists or should exist, thus that its acts and laws are valid in the legal sense. This will be further discussed below. Though extremely pedantic and cumbersome it would be possible, if purposes of sociological terminology alone were involved, to eliminate such terms entirely, and substitute newly-coined words. This would be possible even though the word 'state' is used ordinarily not only to designate the legal concept but also the real process of action. But in the above important connexion, at least, this would naturally be impossible.

Thirdly, it is the method of the so-called 'organic' school of sociology[21] to attempt to understand social interaction by using as a point of departure the 'whole' within which the individual acts. His action and behaviour are then interpreted somewhat in the way that a physiologist would treat the role of an organ of the body in the 'economy' of the organism, that is from the point of view of the survival of the latter.[22] How far in other disciplines this type of functional analysis of the relation of 'parts' to a 'whole' can be regarded as definitive, cannot be discussed here; but it is well known that the bio-chemical and bio-physical modes of analysis of the organism are on principle opposed to stopping there. For purposes of sociological analysis two things can be said. First this functional frame of reference is convenient for purposes of practical illustration and for provisional orientation. In these respects it is not only useful but indispensable. But at the same time if its cognitive value is overestimated and its concepts illegitimately 'reified,'[23] it can be highly dangerous. Secondly, in certain circumstances this is the only available way of determining just what processes of social action it is important to understand in order to explain a given phenomenon.[24] But this is only the beginning of sociological analysis as here understood. In the case of social collectivities, precisely as distinguished from organisms, we are in a position to go beyond merely demonstrating functional relationships and uniformities. We can accomplish something which is never attainable in the natural sciences, namely the subjective understanding of the action of the component individuals. The natural sciences on the other hand cannot do this, being limited to the formulation of causal uniformities in objects and events and the explanation of individual facts by applying them. We do not 'understand' the behaviour of cells, but can only observe the relevant functional relationships and generalize on the basis of these observations. This additional achievement of explanation by interpretive understanding, as distinguished from external observation, is of course attained only at a price – the more hypothetical and fragmentary character of its results. Nevertheless, subjective understanding is the specific characteristic of sociological knowledge.

Notes

1. In this series of definitions Weber employs several important terms which need discussion. In addition to *Verstehen*, which has already been commented upon, there are four important ones: *Deuten, Sinn, Handeln*, and *Verhalten*. *Deuten* has generally been translated as 'interpret.' As used by Weber in this context it refers to the interpretation of subjective states of mind and the meanings which can be imputed as intended by an actor. Any other meaning of the word 'interpretation' is irrelevant to Weber's discussion. The term *Sinn* has generally been translated as 'meaning'; and its variations, particularly the corresponding adjectives, *sinnhaft, sinnvoll, sinnfremd*, have been dealt with by appropriately modifying the term meaning. The reference here again is always to features of the content of subjective states of mind or of symbolic systems which are ultimately referable to such states of mind.

   The terms *Handeln* and *Verhalten* are directly related. *Verhalten* is the broader term referring to any mode of behaviour of human individuals, regardless of the frame of reference in terms of which it is analysed. 'Behaviour' has seemed to be the most appropriate English equivalent. *Handeln*, on the other hand, refers to the concrete phenomenon of human behaviour only in so far as it is capable of 'understanding,' in Weber's technical sense, in terms of subjective categories. The most appropriate English equivalent has seemed to be 'action.' This corresponds to the editor's usage in *The Structure of Social Action* and would seem to be fairly well established. 'Conduct' is also closely similar and has sometimes been used. *Deuten, Verstehen*, and *Sinn* are thus applicable to human behaviour only in so far as it constitutes action or conduct in this specific sense. – Ed.
2. Weber's text is organized in a somewhat unusual manner. He lays down certain fundamental definitions and then proceeds to comment upon them. The definitions themselves are in the original printed in large type, the subsidiary comments in smaller type. For the purposes of this translation it has not seemed best to make a distinction in type form, but the reader should be aware that the numbered paragraphs which follow a definition or group of them are in the nature of comments, rather than the continuous development of a general line of argument. This fact accounts for what is sometimes a relatively fragmentary character of the development and for the abrupt transition from one subject to another. Weber apparently did not intend this material to be 'read' in the ordinary sense, but rather to serve as a reference work for the clarification and systematization of theoretical concepts and their implications. While the comments under most of the definitions are relatively brief, under the definitions of Sociology and of Social Aciion, Weber wrote what is essentially a methodological essay. This makes sec. 1 out of proportion to the other sections of this and the following chapters. It has, however, seemed best to retain Weber's own plan for the subdivision of the material. – Ed.
3. Weber means by 'pure type' what he himself generally called and what has come to be known in the literature about his methodology as the 'ideal type.' The reader may be referred for general orientation to Weber's own Essay (to which he himself refers below), *Die Objektivität sozialwissenschaftlicher Erkenntnis*; to two works of Dr. Alexander von Schelting, 'Die logische Theorie der historischen Kulturwissenschaften von Max Weber' (*Archiv fuer Sozialwissenschajt*, vol. xlix), and *Max Webers Wissenschaftslehre*; and to the editor's *Structure of Social Action*, chap. xvi. A somewhat different interpretation is given in Theodore Abel, *Systematic Sociology in Germany*, chap. iv. – Ed.
4. This is an imperfect rendering of the German term *Evidenz*, for which, unfortunately, there is no good English equivalent. It has hence been rendered in a number of different ways, varying with the particular context in which it occurs. The primary meaning refers to the basis on which a scientist or thinker becomes satisfied of the

certainty or acceptability of a proposition. As Weber himself points out, there are two primary aspects of this. On the one hand a conclusion can be 'seen' to follow from given premises by virtue of logical, mathematical, or possibly other modes of meaningful relation. In this sense one 'sees' the solution of an arithmetical problem or the correctness of the proof of a geometrical theorem. The other aspect is concerned with empirical observation. If an act of observation is competently performed, in a similar sense one 'sees' the truth of the relevant descriptive proposition. The term *Evidenz* does not refer to the process of observing, but to the quality of its result, by virtue of which the observer feels justified in affirming a given statement. Hence 'certainty' has seemed a suitable translation in some contexts, 'clarity' in others, 'accuracy' in still others. The term 'intuition' is not usable because it refers to the process rather than to the result – Ed.

5. A term now much used in psychological literature, especially that of Psychoanalysis. It is roughly equivalent to 'emotion' but more precise. – Ed.
6. The German term is *sinnfremd*. This should not be translated by 'meaningless,' but interpreted in the technical context of Weber's use of *Verstehen* and *Sinndeutung*. The essential criterion is the impossibility of placing the object in question in a complex of relations on the meaningful level. – Ed.
7. *Unverstehbar*.
8. Surely this passage states too narrow a conception of the scope of meaningful interpretation. It is certainly not *only* in terms such as those of the rational means-end schema, that it is possible to make action understandable in terms of subjective categories. This probably can actually be called a source of rationalistic bias in Weber's work. In practice he does not adhere at all rigorously to this methodological position. For certain possibilities in this broader field, see the editor's *Structure of Social Action*, chaps, vi and xi. – Ed.
9. A gulf of the North Sea which broke through the Netherlands coast, flooding an area. – Ed.
10. Weber here uses the term *aktuelles Verstehen*, which he contrasts with *erklärendes Verstehen*. The latter he also refers to as *motivationsmaessig*. 'Aktuell' in this context has been translated as 'observational.' It is clear from Weber's discussion that the primary criterion is the possibility of deriving the meaning of an act or symbolic expression from immediate observation without reference to any broader context. In *erklärendes Verstehen*, on the other hand, the particular act must be placed in a broader context of meaning involving facts which cannot be derived from immediate observation of a particular act or expression. – Ed.
11. The German term is *Sinnzusammenhang*. It refers to a plurality of elements which form a coherent whole on the level of meaning. There are several possible modes of meaningful relation between such elements, such as logical consistency, the esthetic harmony of a style, or the appropriateness of means to an end. In any case, however, a *Sinnzusammenhang* must be distinguished from a system of elements which are causally interdependent. There seems to be no single English term or phrase which is always adequate. According to variations in the context, 'context of meaning,' 'complex of meaning,' and sometimes 'meaningful system' have been employed. – Ed.
12. On the significance of this type of explanation for causal relationship. See para. 6, pp. 96 ff. below in the present section.
13. The German is *gemeinter Sinn*. Weber departs from ordinary usage not only in broadening the meaning of this conception. As he states at the end of the present methodological discussion, he does not restrict the use of this concept to cases where a clear self-conscious awareness of such meaning can be reasonably attributed to every individual actor. Essentially, what Weber is doing is to formulate an operational concept. The question, is not whether in a sense obvious to the ordinary person such an

intended meaning 'really exists,' but whether the concept is capable of providing a logical framework within which scientifically important observations can be made. The test of validity of the observations is not whether their object is immediately clear to common sense, but whether the results, of these technical observations can be satisfactorily organized and related to those of others in a systematic body of knowledge. – ED.

14. The scientific functions of such construction have been discussed in the author's article in the *Archiv für Sozialwissenschaft*, vol. xix, pp. 64 ff.
15. Simmel, in his *Probleme der Geschichtsphilosophie*, gives a number of examples.
16. The above passage is an exceedingly compact statement of Weber's theory of the logical conditions of proof of causal relationship. He developed this most fully in his essay *Die Objektivität sozialwissenschaftlicher Erkenntnis*, op. cit. It is also discussed in certain of the other essays which have been collected in the volume, *Gesammelte Aufsätze zur Wissenschaftslehre*. The best and fullest secondary discussion is to be found in Von Schelting's book, *Max Webers Wissenschaftslehre*. There is a briefer discussion in chap, xvi of the editor's *Structure of Social Action*. – ED.
17. See Edvard Meyer, *Geschichte des Altertums*, Stuttgart, 1901, vol. iii, pp. 420, 444 ff.
18. The expression, *sinnhafte Adäquanz* is one of the most difficult of Weber's technical terms to translate. In most places the cumbrous phrase 'adequacy on the level of meaning' has had to be employed. It should be clear from the progress of the discussion that what Weber refers to is a satisfying level of knowledge for the particular purposes of the subjective state of mind of the actor or actors. He is, however, careful to point out that *causal* adequacy involves in addition to this a satisfactory correspondence between the results of observations from the subjective point of view and from the objective; that is, observations of the overt course of action which can be described without reference to the state of mind of the actor. For a discussion of the methodological problem involved here, see *Structure of Social Action*, chaps, ii and v. – ED.
19. This is the first occurrence in Weber's text of the term *Chance* which he uses very frequently. It is here translated by 'probability,' because he uses it as interchangeable with *Wahrscheinlichkeit*. As the term 'probability' is used in a technical mathematical and statistical sense, however, it implies the possibility of numerical statement. In most of the cases where Weber uses *Chance* this is out of the question. It is, however, possible to speak in terms of higher and lower degrees of probability. To avoid confusion with the technical mathematical concept, the term 'likelihood' will often be used in the translation. It is by means of this concept that Weber, in a highly ingenious way, has bridged the gap between the interpretation of meaning and the inevitably more complex facts of overt action. – ED.
20. By a negative normative pattern, Weber means one which prohibits certain possible modes of action. – ED.
21. A classical example is Schäffle's brilliant work, *Bau und Leben des sozialen Körpers*.
22. One of the most illuminating treatments of physiological problems from such a functional point of view, which is readily understandable to the layman, is W. B. Cannon: *The Wisdom of the Body*, second edition, 1938. The point of reference on this physiological level is not primarily survival value to the species in the sense of the Darwinian theory of evolution, but rather the maintenance of the individual organism as a 'going concern' in carrying through its typical life cycle. What is the life cycle, is to the physiologist essentially a matter of empirical observation. – ED.
23. The term 'reification' as used by Professor Morris Cohen in his book, *Reason and Nature*, seems to fit Weber's meaning exactly. A concept or system of concepts, which critical analysis can show to be abstract, is 'reified' when it is used naively as though it provided an adequate total description of the concrete phenomenon in question.

The fallacy of 'reification' is virtually another name for what Professor Whitehead has called 'the fallacy of misplaced concreteness.' See his *Science and the Modern World*. – Ed.

24. Compare the famous dictum of a well-known physiologist: 'sec. 10. The spleen. Of the spleen, gentlemen, we know nothing. So much for the spleen.' Actually, of course, he 'knew' a good deal about the spleen – its position, size, shape, etc.; but he could say nothing about its function, and it was his inability to do this that he called 'ignorance.'

9

Social Relationships between Contemporaries and Indirect Social Observation

Alfred Schutz

In the world of direct social experience there is a radical difference between *participation* and *observation*. This difference disappears when we get into the world of contemporaries. The reason is that in the latter we never encounter real living people at all. In that world, whether we are participants or observers, we are dealing only with ideal types. Our whole experience is in the mode of the "They." Nevertheless, the ideal type of an observer in the world of contemporaries necessarily differs from the ideal type of a participant in that same world. For, as we have noted, the ideal type varies with the interests of the person who constructs it. The latter's aim is always to visualize a certain objective meaning-context, which he already grasps, as someone else's subjective context of meaning. Now, the total context of experience with which the observer approaches the other person differs from that of the participant. Likewise his interests are radically different. His ideal type can be more or less detailed, more concrete or more formalized, of a greater or lesser degree of anonymity. Whatever the case, it will always be different.

Now, it may be that what is above all interesting to the observer of a social relationship among contemporaries is the conscious experiences of the two participants. Or it may be the course of the relationship. If the former is the case, the observer will either construct or draw from his past experience an

Source: George Walsh and Fredrick Lehnert (trs), *The Phenomenology of the Social World* (Evanston: Northwestern University Press, 1967), pp. 204–207.

ideal type equipped with those conscious experiences which anyone in such a relationship would necessarily observe in himself. The observer then "identifies" himself with this ideal type; he lives it out, imagining himself involved in just this situation. He can then imagine himself having all those experiences which are by definition proper to the ideal type in question. He can also make definite statements about the nature of the relationship he is observing and about the interrelations between the corresponding ideal types that are involved. He can do this quite easily because, as a human being, he is more than just an observer since he himself has in the past been involved in innumerable social relationships, direct and indirect. He may indeed have had such relationships with the very persons he is now observing. Indeed, he may even now be involved in a direct Thou-orientation with one of these persons. Such cases as the last are especially frequent.

Observation of the social behavior of another involves the very real danger that the observer will naïvely substitute his own ideal types for those in the minds of his subject. The danger becomes acute when the observer, instead of being directly attentive to the person observed, thinks of the latter as a "case history" of such and such an abstractly defined type of conduct. Here not only may the observer be using the wrong ideal type to understand his subject's behavior, but he may never discover his error because he never confronts his subject as a real person. Social observation thus tends to develop into second-order ideal-typical construction: the observed actor is himself an ideal type of the first order, and the presumed ideal type in terms of which the actor understands his partner is an ideal type of the second order. Both of these are logical constructions of the observer and are determined by his point of view.

This situation is very significant from the standpoint of every empirical social science involving indirect observation. Its ideal-typical concept formation underlies the principles of meaning-adequacy and causal adequacy which we have yet to discuss. Interpretive sociology, however, must go beyond this. It must construct personal ideal types for social actors that are compatible with those constructed by the latter's partners. This aim may be regarded as a postulate for interpretive sociology. Upon closer scrutiny, it reduces to a more basic principle – the postulate of meaning-adequacy. This postulate states that, given a social relationship between contemporaries, the personal ideal types of the partners and their typical conscious experiences must be congruent with one another and compatible with the ideal-typical relationship itself.

A good example of the type of clarification that is required lies in the field of legal sociology. This discipline encounters great difficulties when it seeks to formulate descriptions of legal relationships between various partners, e.g., legislator and interpreter of the law, executor and subject of the law. Legal sociology seeks to interpret these relationships in terms of the subjective meanings of the persons in question. But, in doing this, it confuses the ideal types in terms of which each of the persons imagines his real partner with the sociologist's own ideal types of the partner. There are only two possible ways to remedy this situation and make possible a genuine descriptive concept of the kind desired by legal sociologists. The first would be to fix from the beginning the standpoint from which the type is to be constructed. This would mean that the legal sociologist would identify himself with one of the actors, postulating as invariant not only the latter's acts but also his interpretive schemes of his partners. The sociologist would then have to regard the ideal-typical concepts so constructed as binding upon himself. If this were the procedure adopted, the kind of sociological concept used would be directly derived from the field of law itself: legislator, judge, lawyer, partner, verdict, execution, etc. The alternative would be to come up with a principle according to which these more general ideal types can be transformed into the individual ideal types which the partners have of each other in concrete situations.

In Chapter 5 we shall deal with the special systematic problems which arise when indirect social observation develops into social *science* as such.

10

Some Basic Problems of Interpretive Sociology

Alfred Schutz

Summary of Our Conclusions to This Point

The results so far achieved are sufficient to allow us to state precisely and in conclusive form our theory of the understanding of meaning. We began by demonstrating the lack of clarity inherent in Max Weber's concept of intended meaning. We saw that, so long as action itself remains undefined, one cannot speak intelligibly of the intended meaning "which the actor attaches to his action." In order to reach a satisfactory definition of action, we found it necessary to make a detailed and exhaustive analysis of its constituting processes. We came, finally, to the conclusion that action is (1) a lived experience that is (2) guided by a plan or project arising from the subject's spontaneous activity and (3) distinguished from all other lived experiences by a peculiar Act of attention. We then saw that, on the basis of this definition, the formula "the actor attaches a meaning to his action" must be interpreted metaphorically. For the meaning is merely the special way in which the subject attends to his lived experience; it is this which elevates the experience into an action. It is incorrect, then, to regard meaning as if it were some kind of predicate which could be "attached" to an action. We further distinguished between the action (*actio; Handeln*) as an experience in process and the completed act (*actum; Handlung*), and we described the peculiar mode of constitution of the projected act, according to which it is anticipated in its own project in the future perfect tense.

Source: George Walsh and Fredrick Lehnert (trs), *The Phenomenology of the Social World* (Evanston: Northwestern University Press, 1967), pp. 215–250.

Our next step was to formulate a preliminary definition of meaning applicable to every kind of lived experience. We said that the "meaning" of a lived experience can be reduced to a turning of the attention to an already elapsed experience, in the course of which the latter is lifted out of the stream of consciousness and identified as an experience constituted in such and such a way and in no other. Meaning in this initial sense is prepredicative and pertains to prephenomenal experience. We found it necessary to enlarge upon and enrich the concept in order to make it coincide with the object of our investigation, namely, the *specific meaning* which the actor "attaches" to his experience when he acts. It is this which is meant by "intended meaning." In order to analyze this concept of meaning, we examined the series of polythetically constructed Acts, which, according to a fundamental principle of phenomenology, can be taken in by a single glance of attention. We saw that every such series stands in a context of meaning, and we analyzed the constitution of the world of experience (*Erfahrungswelt*) as a total structure made up of different arrangements of such meaning-contexts. We then explained (1) the concept of schemes of experience (*Schemata der Erfahrung*), (2) the concept of the lower strata (*Unterstufen*) that are constituted as taken for granted, and (3) the concept of the "stock of knowledge at hand" (*Erfahrungsvorrat*). By studying the theory of attentional modifications we learned that it is *the interests of the subject* and his particular vantage point which defines the borderline between that which he takes for granted and that which is problematic for him. We thus gave the pragmatic element in thinking its due. As for the concept of action itself, we established (1) that a course of action is a polythetically constructed series of Acts (*Akte*) upon which, after their completion, one can direct one's attention in one "single-rayed" or concentrated shaft of attention within which they are seen as a deed or act (*Handlung*); and (2) that therefore action is itself a complex of meaning or meaning-context. At the same time we recognized that the specific meaning-context of an action is dependent upon the scope of the project which constitutes it as a single action. Therefore, if one is in earnest about seeking the subjective meaning of an action, one will find it in that which is the action's own principle of unity. The latter is always determined subjectively and only subjectively. It is methodologically inadmissible to interpret a given series of acts objectively as a unified sequence without any reference to a project and then ascribe to them a subjective meaning. We saw that Weber failed to distinguish the projected act from the completed act, leading him to confuse the meaning of an action with its motives. We, on the other hand, concluded that a series of complicated meaning-structures is already pregiven to the motive. We saw that the motive is really a context of meaning connecting that which motivates with that which is motivated. We drew the important distinction between the in-order-to motive and the genuine because-motive. Then, within the in-order-to motives of an action we pointed out various strata, and we showed that the in-order-to motive of an action is nothing more nor less than the act itself projected in the future perfect tense. We saw that it is for the sake of this act that the action is carried out step by step. We established all these points while our study was as yet limited to the stream of consciousness of the solitary Ego. We brought that study to a conclusion by introducing the concept of the "self-understanding" or self-interpretation of one's own act and of one's own action, which we saw to be an Act of synthetic recognition, the identification and ordering under mutually consistent schemes of experience of that which has already been grasped in the previous Act of attention.

We then turned to an analysis of the social world. Here the ego, as we saw, lights upon the alter ego, a being which, like itself, has consciousness and duration and which, also like itself, interprets its own lived experiences. But the fact that my partner *is* another self, the fact that he performs such and such acts whose external manifestations I can see – this is something I recognize solely by ordering and classifying my own perceptions of him within the total context of my knowledge. *But this is not yet knowledge of the other person as such*. All I am doing so far is ordering and classifying my own experiences of the social world just as if they were experiences of the natural world. But I can at any moment abandon this whole approach and adopt a new one. I can turn my attention away from the objective meaning-context into which I have ordered *my* experiences of the other person's experiences. For the perceived course of the other person's act which I perceive stands for him also within a context of meaning. This is because he sees in one glance of attention the polythetic phases which have gone to make up the whole act. It is only when I begin to grasp the other person's point of view as such, or, in our terminology, only when I make the leap from the objective to the subjective context of meaning, that I am entitled to say that I understand him.

We can attend to the subjective meaning-context of all sorts of human products and cultural objects, which can always be interpreted as evidence for what went on in the minds of their creators. Now, we have already seen that all knowledge of the subjective experiences of others must be obtained signitively. Among all the different kinds of products and indications we have considered, we have singled out for special attention signs, which stand in a context of meaning that is on the one hand an expressive scheme for the sign-user and on the other an interpretive scheme for the sign-interpreter. Both of these schemes can be interpreted as objective contexts of meaning if they are first abstracted from the living Acts and actions in which these signs were used and if the interpretation is confined to the signs themselves. However, we can start out from the external sign itself and, regarding it as a product, trace it back to the original actions and subjective experiences of its inventor or user. This is how, within the world of signs, the transition is made from the objective to the subjective context of meaning. The word *Verstehen* is generally used for the interpretation of both the subjective and objective meaning-contexts of products. This situation conceals the essential problem of knowledge of the social world. Only when the equivocation is identified does the problem come to light: that the meaning of one's own experiences is radically different from the meaning of the experiences of someone else and that consequently it is one thing to interpret one's own experience and quite another to interpret the experiences of someone else. It then becomes clear that the meaning attributed to a product, in contrast to the meaning attributed to a natural object, amounts precisely to this: that the product not only stands in a meaning-context for me – for the interpreter – but that it is also evidence for the further meaning-context in which it stands in the mind of you – its creator. It should here be emphasized that I, the interpreter, do not interpret alone and that your product as a thing in the world belongs not only to my private world but to the one intersubjective world common to us all. In this sense, the term "meaningful world" (*die sinnhafte Welt*), in contrast to the term "the natural world," carries within it an implicit reference to "the Other" who originated this thing that is meaningful. For, being-a-natural object and being-a-meaningful object are both conceptualized to the same degree in objective meaning-contexts by us. the interpreters, since we classify all our experiences in schemes of knowledge.

Our study of the social relationship and of observation in the social world has shown that we can have insight into the inner life of the other person by looking at it as subjective context of meaning. We found that all understanding of the Other is based on Acts of self-explication, that the objective meaning (*Sinn*) of a sign contains within it both actual and occasional meanings (*Bedeutungen*). We saw that we could distinguish the meaning-function (*Bedeutungsfunktion*) from the expressive function (*Ausdrucksfunktion*) of signs and that we could describe – even though only in outline – the special method by means of which the Other's interpretive schemes are comprehended. Our study of the context of motive in the social sphere showed us that all establishment of meaning was for the sake of interpretation and that all interpretation led back to the Act of meaning-establishment. At this point we finally gained access to the fields of "social action" and "social relationship."

An analysis of Weber's concept of social action revealed the nature of Other-orientation and of affecting-the-Other. This led in turn to the problems of orientation-relationship and of social interaction. We examined the general formal structure of these, for both participants and observers. In the process it became clear to us that the concepts of social action and social relationship undergo many modifications, depending on whether the object of the Other-orientation is an alter ego of the world of directly experienced social reality, the world of mere contemporaries, the world of predecessors, or the world of successors. We then turned to an analysis of these regions of the social world. Here we found that only in the direct social relationship as such can we have immediate awareness of the Thou's stream of subjective experiences in its living and present actuality. In contrast, we found that our own present subjective experiences at the moment of self-interpretation are in fact inaccessible to us and that the only experiences of our own that are open to self-interpretation are past ones. We analyzed the genuine We-relationship (*Wirbeziehung*), a relationship within which you and I can grasp each other's living stream of consciousness simultaneously and in one undivided glance. This We-relationship, too, is subject to manifold shadings: it occurs in different levels of actualization and concretization and can include lived experiences which lie close to, or further away from, the intimately grasped Thou of the face-to-face situation – in other words, lived experiences of greater or lesser proximity. On the other hand, we saw that in the world of mere contemporaries the other person is not given to me directly and bodily, but only indirectly. To a certain extent the Other has now become anonymous; we may even say that he has been replaced by an ideal type that has been constructed out of previously given experiences of certain courses of action. This ideal type, again, can be more or less removed from a real Thou, more or less concrete and full of content. We saw that the ego is oriented to the alter ego of the world of mere contemporaries in a special way: we called this the They-orientation (*Ihreinstellung*) because its object is not the thusness (*Sosein*) – or immediately apprehended qualities – of another person but rather his whatness (*Wie-sein*) – his being of such and such a general type (*Gleichsam-sein*). We have also analyzed the stratification of the They-relationships and have shown that they form a continuous series of ever increasing anonymization, beginning with the ideal type "my friend N" and culminating in the most general ideal type "one" or "someone" (*Man*), the originator of artifacts and objective sign systems. Hand in hand with the increasing anonymization of the previously given other person, there occurs for me a greater and greater self-distancing from his living personality. The more anonymous my partner, the less direct and personal the relationship and the more conceptualized must my dealings with him be. And the more I conceptualize my partner, the less can I regard him as a free agent. When I am face to face with someone, I immediately grasp him as a spontaneous and freely acting being: His future action is as yet open and undecided, and I can only hazard a guess as to what he is going to do. The ideal type, on the other hand, is, when rightly conceived, without any freedom; he cannot transcend his type without ceasing to be a mere contemporary and becoming a consociate of mine in direct experience. As for the world of predecessors, it is completely lacking in freedom. The world of successors, on the other hand, is free. We saw in section 11 that the problem of freedom, when rightly understood, is a *time problem*. Following our argument out to its conclusion, we now see that meaning in the social world is itself conditioned by time, a point that we had already proved with respect to the individual consciousness (see Chapter 2).

All the points made above we applied not only to the participant within the social relationship but to the observer as well. In the latter's case we found that the same basic distinctions can be drawn, namely, between observation of one's fellow men in direct experience, observation of those who are merely one's contemporaries, and observation of one's predecessors.

Indirect Social Observation and the Problem of Knowledge in the Social Sciences

We shall now attempt to draw further conclusions about the modifications which Other-orientation and social relationship undergo in the four regions of the social world. Up to this point we have been concerned chiefly with the problem of how the man in the natural attitude, the man actually living in the social world, comprehends this world and interprets it. Only now and then, in connection with special points, did we make reference to the unique problem of social scientific knowledge. What is this problem? It consists in the fact that, although the social sciences start out from, and take for granted, the same social world in which we live from day to day, yet their methods of gathering knowledge are quite different from those of everyday life. For the social scientist organizes and classifies his data into quite different contexts of meaning and works them up in quite different ways.

In our introduction to Chapter 4, we referred to the relation between the knowledge we gain in everyday life and the knowledge we gain in the social sciences, and we showed how difficult it is to draw a sharp boundary between them. When in everyday life I think conceptually about my fellow man, I am actually taking up toward him the attitude of a social scientist. On the other hand, when I am engaging in social research, I am still a human being among human beings; in fact, it pertains to the very nature of science that it is not science for me only, but for everyone. Science always presupposes the experiences (*Erfahrungen*) of a whole scientific community, the experiences of others who, like me, with me, and for me, are carrying on scientific work.[1] And so the problem of the social sciences is already present in the prescientific sphere, and social science itself is only possible and conceivable within the general sphere of life in the social world. This is by no means to say that the social scientist may characterize as scientific that knowledge which he picks up in everyday life and in his ordinary associations. Our point has been merely to indicate the region where we must first look if we are to engage in a critique of the methodology of the social sciences.

What, then, is the specific attitude of social science to its object, the social world? Fundamentally, it is the same as the attitude of the indirect social observer toward his contemporaries. It is different, however, in one respect: no directly experienced social reality is pregiven to social science as such. The world of social science is simply not identical with the world of the social scientist, who is also a man living in the social world. But the world of predecessors is indeed pregiven to social science, and only this is pregiven to history. The whole context of knowledge of social science is therefore necessarily different from that of the indirect observer in everyday life.

With his usual acuity, Max Weber has also seen this problem. In his controversy with Munsterberg he speaks about the basic difference between scientific psychology and the psychology of the *Menschenkenner* or intuitive observer of others. In so doing, he sets himself in opposition to Münsterberg's assertion that the *Menschenkenner* either knows the whole man or does not know him at all. He replies: all he knows about him is what is *relevant for his immediate purposes* and nothing more.

> What is significant in a human being from one particular and limited point of view cannot for *logical* reasons serve as the basis of a pure psychological theory which seeks to express itself in the form of general laws. Actually, however, intuitive understanding takes into consideration the endless variety of human feeling and behavior – something which no theory can absorb into its "presuppositions."[2]

One must not allow oneself to be confused here by Weber's terminology. The distinction he draws applies, not only to intuitive versus scientific psychology, but more generally and fundamentally to everyday knowledge as opposed to scientific knowledge.

It is to Husserl that one must turn in order to find a definitive statement of the distinction in question. As he showed in his *Formal and Transcendental Logic*, all scientific judgment has as its goal knowledge of the world with a maximum of explicit clarity and distinctness. In scientific judgment no presupposition nor any pregiven element can be accepted as simply "at hand" without need of any further explanation. On the contrary, when I act as a scientist, I subject to a detailed step-by-step analysis everything taken from the world of everyday life: my own judgments, the judgments of others which I have previously accepted without criticism, indeed everything that I have previously taken as a matter of belief or have even thought in a confused fashion. Every social science, including interpretive sociology, therefore, sets as its primary goal the greatest possible clarification of what is thought about the social world by those living in it. Weber undertook the task of analyzing the processes of meaning-establishment as they occur in the social world – occur, of course, in a manner still lacking complete clarity. In so doing, he took as his basic theme the "intended meaning of human action," presupposing, at the same time, that the implicit meanings of everyday judgments in the social world can be rendered explicit by scientific means.

We shall be dealing shortly with the difficult question of how many social sciences there are and how their areas are to be marked off from one another. First, let us, with the aid of Weber's sociology, deal with the consequences resulting from the attitude of every social science to its object, in other words, from the attitude of whoever observes the world of mere contemporaries or the world of predecessors.

We have seen that the world of mere contemporaries is not given to the observer in any immediate and direct fashion and that the ego comprehends the contemporary alter ego only as an ideal type. When such ideal types are being constructed, the selection of their fixed and essential elements depends on the point of view of the observer at the moment of interpretation. It depends on his stock of knowledge at hand and upon the modifications of his attention to his knowledge of the world in general and of the social world in particular. Even the construction of scientific ideal types depends on the total context of scientific knowledge or, what is the same thing, on the total context of clear and distinct judgments about the world. All these judgments, however, insofar as they are scientific, must be ordered into those highest contexts of meaning which, to employ an image of Husserl's,[3] comprehend in one expression all the axioms, fundamental principles, theorems, and deductions of a science. The interpretive schemes used by the observer in the social world to understand those who are his mere contemporaries are necessarily different from those of the social scientist, however. The indirect observer's knowledge is determined by his own direct experience regardless of whether what he knows is comprehended in Acts of judgment that are positional or neutralizing, explicit or vague, or are merely in the prepredicative "having" stage of the social world. This is true because of the living intentionality of the Acts (*Akte*) *in* which this individual lives. On the other hand, the complex of knowledge of the social sciences is based exclusively on explicit positional Acts of judgment, on constituted ideal objectifications, that is to say, on conclusions of thought, and never on prepredicative Acts of laying hold on (*in Selbsthabe erlebte Erfassungen*) the other person himself. Social science is through and through an explicit knowledge of either mere contemporaries or predecessors; it nowhere refers back to the face-to-face experience. Moreover, it must be recognized that scientific experience (*Erfahrung*) embraces the conclusions of all the sciences of the world and that the interpretive schemes of the social sciences must be compatible not merely with experience of the social world but with scientific experience as a whole. The original and fundamental scheme of science, the expressive scheme of its propositions, and the interpretive scheme of its explanations is, therefore, essentially that of *formal logic*. Accordingly, science is always an objective context of meaning, and the theme of all sciences of the social world is *to constitute an objective meaning-context either out of subjective meaning-contexts generally or out of some particular subjective meaning-contexts*. The problem of every social science can, therefore, be summarized in the question: *How are sciences of subjective meaning-context possible?*[4]

Our analysis of the social world of contemporaries has already partly answered this question. The fact that subjective meaning-contexts can be comprehended in objectivating and anonymizing constructions can be shown and described with the aid of the personal ideal types of the worlds of contemporaries and predecessors that are built up in the naïve natural point of view of everyday life. Since every social science starts out by taking for granted a social world which it sees as either a world of mere contemporaries or a world of predecessors, it can comprehend this world only by the method of ideal types, whether course-of-action types or personal types. In other words, the social world is pregiven to each social science only indirectly and never with the immediacy of living intentionality. Now, since it is typifying experience, social science is an objective meaning-context, whose object, however, is subjective meaning-contexts (to be precise, the typical subjective processes of personal ideal types).

We shall now describe the modifications which the laws of type-formation take on in the social sciences (as opposed to everyday life) due to the absence of direct social experience and the presence of the over-all scientific picture of the world.

We have seen that the observer of the world of contemporaries may, in order to understand the other person, appropriately construct only such ideal types as are in agreement with his past experience. The position of the social scientist is a parallel one. His ideal types must not only be compatible with the established conclusions of all the sciences but must explain in terms of motivations the very subjective experiences which they cover. To put the point in Weber's terminology, the ideal types constructed by social science, and, above all, by interpretive sociology, must possess at the same time both causal adequacy and meaning-adequacy. We shall presently examine the role played by these two concepts in interpretive sociology.

The Function of the Ideal Type in Weber's Sociology

In analyzing a few of the categories of Weber's sociology, our source will be his great and, regrettably, unfinished work, *Wirtschaft und Gesellschaft*. Weber's views on methodology changed in important respects in the course of the years, as indeed one would expect in the case of a man of such outstanding intellectual integrity. We shall forego tracing these changes, first of all in order not to burden ourselves too heavily, and second because there are already a number of excellent works on this subject.[5]

Let us begin with a few quotations from Weber's main work:

> Sociology is a science which attempts the interpretive understanding of social action.[6]
>
> Sociology seeks to formulate type concepts and generalized uniformities of empirical processes. This distinguishes it from history, which is oriented to causal analysis and explanation of individual actions, structures and personalities possessing cultural significance.[7]
>
> Among the various bases on which its concepts are formulated and its generalizations worked out, is an attempt to justify its important claim to be able to make a contribution to the causal explanation of some historically and culturally important phenomenon. As in the case of every generalizing science, the abstract character of the concepts of sociology is responsible for the fact that, compared with actual historical reality, they are relatively lacking in fullness of concrete content. To compensate

> for this disadvantage, sociological analysis can offer a greater precision of concepts. This precision is obtained by striving for the highest possible degree of adequacy on the level of meaning. . . . This can be realized in the case of concepts and generalizations which formulate rational processes. But sociological investigation attempts to include in its scope various irrational phenomena, as well as prophetic, mystic and affectual modes of action, formulated in terms of theoretical concepts which are adequate on the level of *meaning*. In *all* cases, rational or irrational, sociological analysis both abstracts from reality and at the same time helps us to understand it, in that it shows with what degree of *approximation* a concrete historical phenomenon can be subsumed under one or more of these concepts. . . . In order to give *precise* meaning to these terms it is necessary for the sociologist to formulate pure ideal types of the corresponding forms of action which in each case involve the highest possible degree of logical integration by virtue of their complete adequacy on the level of *meaning*. But precisely because this is true, it is probably seldom if ever that a real phenomenon can be found that corresponds exactly to one of these ideally constructed pure types. The case is similar to a physical reaction which has been calculated on the assumption of an absolute vacuum.[8]
>
> Meaning (as the term is used in interpretive sociology) may be of two kinds. The term may refer first to the actually existing meaning in the given concrete case of a particular actor, or to the average or approximate meaning attributed to a given plurality of actors; or, secondly, to a theoretically conceived *pure type* of subjectively intended meaning attributed to the hypothetical actor or actors in a given type of action.[9]
>
> In all these cases, understanding involves the interpretive grasp of the meaning present in one of the following contexts: (a) as in the historical approach, the actually intended meaning for concrete individual action; or (b) as in cases of sociological mass phenomena, the average of, or an approximation to, the actually intended meaning; or (c) the meaning appropriate to a scientifically formulated pure type (an ideal type) of a common phenomenon. The concepts and "laws" of pure economic theory are examples of this ideal type. They state what course a given type of human action would take if it were strictly rational, unaffected by errors or emotional factors, and if, furthermore, it were completely and unequivocally directed to a single end, the maximization of economic advantage. In reality, action takes exactly this course only in unusual cases, as sometimes on the stock exchange; and even then there is usually only an approximation to the ideal type.[10]

The foregoing quotations are sufficient to give an adequate idea of the function within interpretive sociology of the ideal types as their creator conceived them. Since we have up to this point so often gone beyond the concepts of Max Weber and often found cause to disagree with them, the tremendous significance for all the social sciences of Weber's achievement cannot be sufficiently stressed. Again and again Weber refers to the problem of the ideal type as the central problem of all the social sciences. Our studies have shown how well founded this conception is. For the world of contemporaries and the world of predecessors can only be comprehended in an ideal-typical way. The individual episodes and events in this world are already abstracted from the concrete other person encountered in the face-to-face relationship. They are more or less anonymous and belong to typical courses of consciousness, which are to be found in all degrees of concreteness and richness of content, running from the type of an individual to the type of "someone."

Weber believes that he has taken care of all these manifold variations by dividing intended meanings into three classes: (a) the meaning intended by an individual actor in a historically given case, (b) the meaning intended on the average by a given group of several actors, and (c) the meaning of an ideal-typical actor. Weber's motives are clear; he is distinguishing the method of history and the method of statistics, on the one hand, from the method of interpretive sociology, on the other. The distinction is abundantly justified as long as the sciences in question are understood as sciences dealing with objective meaning-context and therefore with the external course of the act quite apart from any concern with the conscious experiences of the actor. This premise being granted, history (at least in Weber's conception of it) does have to do with the particular actions of individuals, statistics with the average actions of masses, and sociology with the action of a pure ideal type.[11] However, Weber's distinction turns out to be without foundation to the extent that these sciences are interpreted in terms of intended meaning – to the extent, that is, to which attention is turned away from the external course of action to the subjective context of meaning. For, since what is thematically pregiven to sociology and every other social science is the social reality which is indirectly experienced (never immediate social reality) – a social reality which can only be comprehended in the They-relationship and therefore typically – it follows that *even when social science is dealing with the action of a single individual, it must do so in terms of types*. Weber saw that clearly enough, for he expressly allows that all three methods of understanding meaning (*Sinn-Verstehens*) are valid for interpretive sociology. However, if we look more deeply into these distinctions, we will find that to each of the three modes of understanding there corresponds a different degree of verifiability of the external behavior. The boundaries, of course, are fluid. However, it can be said that the ideal type of an individual's behavior, say that of the behavior of a friend of mine, is derived from a very intimate acquaintance with his personal characteristics, that acquaintance which I enjoy in the We-relationship. This acquaintance is much greater than is the case with a personal ideal type constructed to fit a given piece of behavior. We have already explained this phenomenon in the preceding chapter. He who lives in the social world is a free being: his acts proceed from spontaneous activity. Once the action has transpired, once it is over and done with, it has become an act and is no longer free but closed and determinate in character. Nevertheless, it was free at the time the action took place; and if the question concerning the intended meaning refers, as it does in Max Weber's case, to the point in time before the completion of the act, then the answer must be that the actor always acts freely, and this is true even though I am able to know him only indirectly and in ideal-typical fashion. On the other hand, the personal *ideal type* that is correctly constructed, that is, one which is nontype-transcendent, is essentially unfree. This is true whether his action is regarded as now occurring or whether (after its occurrence) it is interpreted as an already constituted ideal objectivity.

Let us remember the important distinction between the construction of the ideal type and the application of this type as an interpretive scheme to real concrete actions. Let us take a case of interpreting a *future* action by means of an ideal type. Our ideal type will be defined as having definite and invariant motives, and from these motives we will be able to deduce invariant acts and sequences of acts. Suppose that our ideal type is that of a bureaucrat. Applying the type to a concrete person, I can say, "*N* is a typical bureaucrat; therefore we may expect him to be visiting our office regularly." Or else, "*N* has just performed action *a; a* corresponds to ideal type *A; a′* is also characteristic of *A*; we may, therefore, expect *N* also to perform action *a′*." Now, how reliable are such judgments? Since action *a′* is still in the future and therefore free, I cannot be certain that *N* will perform it. The application of a personal ideal type to a future action of another person is something that can only be done with the assumption that it is *probably* correct. If the person does not act as predicted, we must assume that we have applied the wrong ideal type to the person in question. We will therefore look around for *another* personal ideal type which *will* make his action comprehensible. This principle will hold regardless of whether *N* is immediately experienced or is himself known only as a type. Now, the more freedom *N* has, the less anonymous he is, the closer to the We-relationship he stands, the less likelihood will there be that he will behave "according to ideal type." But if *N* himself is no more than an ideal type, if his actions are controlled by his observer, then the ideal type must *always* receive positive verification, must always "come out right" insofar as it was constructed according to a correct methodology, that is to say, in a manner that is both adequate on the level of meaning and causally adequate.

Now, what does it mean to construct an ideal type according to both these criteria? In Weber's thought this does not mean the applicability of the ideal type to future actions. Rather it means the selection of certain acts of one or more persons as typically relevant. An act is defined as "typically relevant" if it originates from motives that can be established as constant and invariant in the actor in question. But this means merely that the act is repeatable, in other words, that the ideal type derived from it has the ideality of the "and so forth," of the "again and again." Therefore, the concept of adequacy on the level of meaning and causal adequacy applies to the correct choice of motive and only implicitly to the acts that are postulated as following from these motives. And, in fact, it is above all the in-order-to motive which is thus postulated as constant. For the search for the genuine because-motive takes place, so to speak, in the pluperfect mode, on the basis of in-order-to motives posited as already given.

But how can we know the motives of another person? According to Weber, a motive is a context of meaning which appears either to the actor or to the observer as the meaningful ground (*als sinnhafter Grund*) of behavior. Now, we have already shown that this definition fails to distinguish between two quite different situations. In direct social observation, the observer assumes that the meaningful ground of the action was the project which was fulfilled by the already completed act. In this case the observer starts out by tacitly assuming that the action was really projected or planned. But he can, any time he wishes, simply question the actor and possibly discover that the latter intended to do something quite different. In other words, the actor can tell the observer just what the "span of his project" was. It is precisely this span which the observer cannot determine *merely* by observing. But in indirect social observation the situation is quite different. Here there is no distinction between the meaning-context of the observer and that of the actor. The reason is simple: if there is a real person corresponding to the observer's postulated ideal type, then he will by definition intend what the observer has in mind. However – and this is the basic postulate of social *science* – the motives ascribed to the ideal type must be both causally adequate and adequate on the level of meaning.

Before we proceed further, let us try to obviate a confusion that may arise from our own terminology. When sociology undertakes to interpret a concrete action, it has the act already given to it as a datum. From the act it tries to draw inferences about the motives that would be typical of a person acting in this way. In the process, recourse is had to a personal ideal type. For purposes of abbreviation, we shall not in the following paragraphs refer to the personal type as such, but merely to the typical motive. However, it should be quite clearly understood that by "typical motive" we mean the motive of an individual person who is comprehended via the ideal-typical method.

Let us now find out exactly what Weber means by the two terms "causal adequacy" and "adequacy on the level of meaning."

Causal Adequacy

Weber makes the distinction between the two concepts quite clear at an early point in his *Wirtschaft und Gesellschaft*:

> We apply the term *adequacy on the level of meaning* to the subjective interpretation of a coherent course of conduct, when, and insofar as

> according to our habitual modes of thought and feeling, its component parts, taken in their mutual relation, are recognized to constitute a typical complex of meaning. It is more common to say "correct." The interpretation of a sequence of events will, on the other hand, be *causally adequate* insofar as, according to established generalizations from *experience*, there is a probability that it will always actually occur in the same way. An example of *adequacy on the level of meaning* in this sense is what is, according to our current *norms* of calculation or thinking, the correct solution to an arithmetical problem. On the other hand, a causally adequate interpretation of the same phenomenon would concern the statistical probability that, according to verified generalizations from experience, there would be a "correct" or "erroneous" solution of the same problem. This also refers to currently accepted norms, but includes taking account of typical errors or of typical confusions. Thus causal explanation depends on being able to determine that there is a probability, which, in the rare ideal case, can be numerically stated, but is always in some sense calculable, that a given observable event (overt or subjective) will be followed or accompanied by another event.
>
> A *correct* causal interpretation of a concrete course of action is arrived at when the overt action and the motives have both been correctly apprehended, and at the same time their relation has become *meaningfully* comprehensible. A correct causal interpretation of typical action means that the process which is claimed to be typical is shown to be both adequately grasped on the level of meaning and at the same time the interpretation is to some degree causally adequate. If adequacy in respect to meaning is lacking, then no matter how high the degree of uniformity and how precisely its probability can be numerically determined, it is still an *incomprehensible statistical* probability, whether dealing with overt or subjective processes. On the other hand, even the most perfect adequacy on the level of meaning has causal significance from the sociological point of view only insofar as there is some kind of proof for the existence of a *probability* that the action in *fact* normally *takes* the course which has been held to be meaning-adequate. For this there must be some degree of determinable frequency of approximation to an average or a pure type.
>
> Statistical uniformities constitute understandable types of action in the sense of this discussion, and thus constitute "sociological generalizations," only when they can be regarded as manifestations of the understandable subjective meaning of a course of social action. Conversely, formulations of a rational course of subjectively understandable action constitute sociological types of empirical processes only when they can be empirically observed with a significant degree of approximation. It is unfortunately by no means the case that the actual likelihood of the occurrence of a given course of overt action is *always* directly proportional to the clarity of subjective interpretation.[12]

We shall now seek to bring these remarks of Weber into accord with the requirements of our own theory. Let us begin with the concept of causal adequacy. A sequence of events is causally adequate to the degree that experience teaches us it will probably happen again. The concept of causal adequacy relates, therefore, to that objective context of meaning which is social science itself. That certain acts are followed by certain other acts is a generalization founded (1) in everyday life in my interpretation of my own experiences and (2) in social science, in a scientific complex of knowledge. In both cases the generalization is achieved through a synthesis of recognition. But this should in nowise be identified with knowledge of the conscious experience of the other person or with knowledge of the "intended meaning" of his action. A sequence of events is, therefore, causally adequate if it is in accord with past experience. Here it is immaterial whether the events in question add up to a human action or whether they are nothing more than a series of happenings in the world of nature. As a matter of fact, the concept of causal adequacy was first advanced by the physiologist Johannes von Kries[13] in connection with certain problems involved in the calculating of probabilities. His aim was to make a contribution to the theory of legal accountability in criminal law, but he introduced the idea as a general concept independent of any specific application. There are weighty objections against the use of the word "causal" in sociological discourse. For when we formulate judgments of causal adequacy in the social sciences, what we are really talking about is not causal necessity in the strict sense but the so-called "causality of freedom," which pertains to the end-means relation. Therefore, one cannot really speak of a causal relation in the general sense postulated by Kries[14] so long as one confines oneself to the external event, the objective context of meaning, and so forth. However, if one interprets the concept in Weber's sense, then the postulate of causal adequacy is identical with what we have previously called "the postulate of the coherence of experience." A type construct is causally adequate, then, if it is probable that, according to the rules of experience, an act will be performed (it does not matter by whom or in what context of meaning) in a manner corresponding to the construct.

But this formulation is still lacking in precision. If I start out from a real action as my datum, then every ideal-typical construct that I base on it will already be in itself causally adequate. This is because the objective meaning-context of the act with which I start itself discloses the typical subjective meaning-context which corresponds to the act or, more strictly speaking, *can* correspond to it. Therefore, if I am going to construct a personal ideal type in a scientifically correct manner, it is not enough that the action in question probably take place. Rather, what is required in addition to this is that the action be *repeatable* and that the postulate of its repeatability not be inconsistent with the whole body of our scientific knowledge. This is a good time to repeat our previous observation that Weber starts out with an external action and seeks to connect with it an intended meaning without accounting for the fact that even the concept of the unity of the action presupposes a subjective foundation once we ask what the intended meaning is. However, this error turns out to be harmless if we follow his further train of thought. Causal adequacy is for him above all a category of the *social sciences*; hence, only sociological and historical understanding is bound by it. However, such understanding takes place via the construction of personal ideal types derived from a course of external behavior that has been isolated arbitrarily by the social scientist. If we set it down as a requirement that such constructs be derived only from acts occurring with a certain known frequency, then what we really have here is a heuristic principle based on the economy of thought. It means simply that a construct is appropriate and to he recommended only if it derives from acts that are not isolated but have a certain probability of repetition or frequency. If the postulate of causal adequacy is conceived in this way, then it is by no means a principle essential to all the social sciences. It would be binding upon sociology only and not upon history, since it derives from the basic approach of the sociologist toward his problems. But this would then leave everyone at liberty to decide whether he wanted to carry on the scientific study of the social world qua sociologist or qua historian.

But Weber's postulate of causal adequacy means something more than this. For reasons we have yet to discuss, the sociologist prefers the interpretive scheme of the *rational* action (specifically, either that of the action oriented to an ordinary purpose, or the action oriented to an absolute value[15]) to all other interpretive schemes. Every ordinary purposive action takes place within the means-end relationship. Establishing the pattern of such an action simply means seeking out how typical ends and typical means are related. In other words, the actor's choice of goals, his in-order-to projects, is determined via ideal-typical construction. Once this is done – that is, once the actor's goal is defined – it is only a matter of selecting those means for him that experience has shown to be appropriate. We can now interpret Weber's postulate of causal adequacy in the following way: In a type construct of ordinary purposive action, the means must be, in the light of our past experience, appropriate to the goal. Later, when we discuss rational action and rational method, we shall explain in detail what we mean by this second concept of causal adequacy.

An ideal-typical construct is said to be causally adequate when it turns out to predict what actually happens, in accord with all the rules of frequency. But this does not mean that what it predicts must always happen. Weber himself gives as an example the probability of a typical error in calculation. Let us suppose that we wish to multiply a given number by a two-digit number. Then, instead of placing the second partial product one point to the left of the first, we place it one point to the right. It would be causally adequate to conclude that we are going to come out with the wrong answer. But this conclusion would not be correct for all cases in which the above procedure was employed; for instance, if the two digits of the multiplier are the same, it does not matter whether the second partial product is moved to the left or to the right of the first. Here, as a matter of fact, we have Weber's ideal case of numerically assignable probability, for, out of ten such operations, nine will be incorrect and one correct. However, if we look more closely, we shall see that causal adequacy, or agreement with past experience, is based on typically comprehended meaning-adequate relations, in this case the laws of arithmetic and number theory as applied to the operation of multiplication. We can even go further and make the general statement that all causal adequacy which pertains to human action is based on principles of meaning-adequacy of some kind or other. For such causal adequacy means the consistency of the type construct of a human action with the total context of our past experience. Furthermore, we can come to know a human action only by ordering it within a meaning-context, whether objective or subjective. Causal adequacy, then, insofar as it is a concept applying to human behavior, is only a special case of meaning-adequacy.[16]

Our position on this point will immediately become more intelligible as we proceed to our analysis of the nature of meaning-adequacy.

Meaning-Adequacy

According to Weber, a continuous course of behavior is meaning-adequate, or adequate on the level of meaning, to the degree that the relation of its constituent parts is affirmed by us as a typical meaning-context in accordance with average habits of thought and feeling. Here again we encounter the paradox that dominates Weber's whole philosophy of social science. He postulates as the task of social science the discovery of intended meaning – indeed, the intended meaning of the actor. But this "intended meaning" turns out to be a meaning which is given to the observer and not to the actor. In our terminology, Weber is saying that an action is meaning-adequate when it can be ordered under an objective context of meaning. We have already shown that such objective interpretation is quite a different thing from discovering what the actor himself has in mind. Our next question must therefore be whether meaning-adequacy is attained through objective interpretation or whether we have to go further and show without contradiction how the actor could himself have subjectively intended a certain meaning. We shall have to decide in favor of the second alternative, as we shall see.[17]

This distinction is by no means irrelevant for Weber's theory of meaning-adequacy. For him, behavior is meaning-adequate if it is in accord with "*average* habits of thought and feeling." What he means by this afterthought is not at all clear. For average habits of thought and feeling are a matter of causally adequate, not meaning-adequate, interpretation. It seems contradictory to set up the sociologist as judge of what is meaning-adequate, unless we mean by "knowledge of average habits of thought and feeling" the knowledge the social sciences have of all conceivable subjective experiences whatsoever. It is enough for the meaningful interpretation of another's behavior that I assume that my ideal construct stands in a context of meaning *for him*. This suffices even if such a meaning-context clashes with my own knowledge. For instance, I can regard the totemistic interpretation of the behavior of a primitive tribe as meaning-adequate even though the whole totemistic way of thinking is foreign to the "average habits of thought and feeling" of our culture, or at least of the sociologists of our culture. But that is not at all what Max Weber means. For he is very conscious of the fact that these "average habits of thought and feeling" refer back to given personal ideal types. He knows also that it is a matter of our experience, that is, of the experience of the social sciences, whether certain meaning-contexts can be ordered under a definite personal ideal type in a way that is typically adequate rather than type-transcendent. And so our attempt to discover a criterion for what is meaning-adequate has come down to this: we are back once again at the subjective meaning-context and the personal ideal type, which in turn have to be constructed in terms of the postulate of causal adequacy.

On the other hand, we can regard an ideal-typical construct as adequate for a given action if the corresponding subjective meaning-context can really be ascribed to the actor in question without contradicting what else we know about him. Of course this person whose subjective experiences we are interpreting may appear to us as more or less determinate depending on how well we know him. Thus understood, the problem of meaning-adequacy pertains only to the interpretation of a concrete action via already constituted ideal types. On the other hand, the sociologist would have a completely free hand in the construction of a personal ideal type, because he so equips the latter's ideal consciousness that it is quite capable of having the subjective experiences appropriate to the typical behavior in question.

Our analysis has thus shown that, so far as Max Weber is concerned, the two concepts of causal adequacy and meaning-adequacy are convertible. Any interpretation which is meaning-adequate must also be causally adequate, and vice versa. The two postulates really require that there be no contradiction to previous experience. As soon as one assumes that there is a definite stock of such experience at hand – in other words, as soon as only one person is making the interpretation, and from only one point of view – then either both of the postulates will be fulfilled or neither of them will. If it appears otherwise, that is only because a number of interpreters are introduced or because a number of temporal vantage points are assumed, in which, for instance, one interpretation already meaning-adequate in itself conflicts causally with another, later one.

For even where a given instance of behavior seems incomprehensible to the observer, for instance, behavior which is on the one hand causally adequate but on the other seems lacking in meaning-adequacy, meaning-adequacy may well exist, even in such cases, from the point of view of the actor himself.

Suppose, for instance, that an observer who is quite ignorant of the use of linguistic statistics in historical research comes on a man counting the frequency of certain words in the works of Plato. In terms of "average habits of thought and feeling" he will simply not know what to make of such behavior. He will begin to make sense out of the man's actions only when it is explained to him that in different periods of his life a person shows a preference for certain words and that, therefore, by studying the frequency of given words in his writings one will have made a start toward establishing a chronology for them. What was merely causally adequate then becomes meaning-adequate as well and therefore fully intelligible. We shall presently see how Weber's concept of meaning-adequacy really derives from the in-order-to motive of rational action and how his concept of intelligibility (*Verstehbarkeit*) is closely bound up with his notion of an action oriented to an ordinary purpose.

At this point we must add a remark on the situation underlying the distinction between causal adequacy and meaning-adequacy. The postulate that an ideal-typical construct must be both causally adequate and adequate on the level of meaning implies that it must be formulated as a *pure* construct without any admixture of type-transcending behavior.[18] Furthermore, it must be compatible with our experience of the world in general and therefore with our experience of other people in general and of the particular person in general whose acts we are seeking to understand by means of the construct. Another demand of the postulate is that the construct be based only on repeatable behavior. So much for the demands of the postulate of adequacy insofar as it deals with the *formation* of ideal-typical constructs. What are its requirements so far as the application of these types to concrete acts is concerned? Here the postulate of adequacy states that the type must *be sufficient to explain the action without contradicting previous experience*. But an action is sufficiently explained via an ideal type only when its motives are understood as typical ones; the explanation must, therefore, be meaning-adequate. To say that the motives must be causally adequate means merely that the motives could have brought about this action and, more strictly, that they probably did so. We must now examine the concept of probability.

Objective and Subjective Probability

Weber distinguishes between two kinds of probability, subjective and objective. Objective probability consists in the fact that certain behavior can be conceived with both causal adequacy and adequacy on the level of meaning without regard to the subjective experiences of the actor. *Objective* probability is, therefore, a category of *interpretation*. Subjective probability, on the other hand, is predicated only of the subjective meaning-context, in other words, of the "intended meaning." Subjective probability looks forward into the future from the vantage point of the actor. It pictures something in that future as already over and done with, or, as we have expressed it, in the future perfect tense. Subjective probability is synonymous with expectation in the broadest sense; above all, therefore, it is predicated of the project and of the protentions directed toward the latter's goal. Accordingly, for any actor, only a subjective probability can be attributed to each project as he plans the action to carry it out. Furthermore, all in-order-to motives have subjective probability, a fact which is already implied when we say that all actions are "oriented."

On the other hand, a genuine because-motive can have only objective probability. That is just another way of saying that the because-motive can be regarded as operative only if it has been so constructed in terms of meaning-adequacy and causal adequacy that it could have been operative. Here the position of the external observer and the position of the actor are in principle the same. The actor can discover his genuine because-motives only through a process of self-observation. Either his completed act or his in-order-to motive is then taken as datum, and he tries to picture in the pluperfect tense just what lived experience still further back in the past could have led him to plan such and such or do such and such. In search for such an experience, he uses, of course, the criterion of adequacy, which involves an objective meaning-context. It is, then, a question of objective probability whether the real because-motive of the act has been identified.

As for subjective probability, when we say that it is predicable of the in-order-to motive, we mean that every projected act calls for fulfillment through real action and that the actor counts on such fulfillment. But this assumption that the act is going to be carried through is based on the knowledge available to the actor at the time he formulates his project. This knowledge is largely a matter of his past experience as to whether things like this "can be done." He thus brings his planned action into a context of meaning, namely, certain previous projects of a similar nature that are now imagined in the pluperfect tense, judged as to whether they were successfully carried through and thus made the standard of "adequacy."

We must center our attention on the relation between the concept of probability and those ideal types that are constructed scientifically, that is, according to the postulates of adequacy.

As far as the construction of the ideal type is concerned, the postulate of adequacy required that it be probable that a real person would behave in the manner specified by the type. Objective probability and adequacy with respect to type are, therefore, correlative as far as behavior that has already transpired is concerned. But if the type construct is to be applied to future action, then the criterion of meaning-adequacy is different. Here the observer must postulate an action of such a nature that the *actor* would *think* its performance probable. In short, what is here called for is subjective probability. Projects have a positive degree of subjective probability if those who formulate them expect to be able to carry them out.

Probability, whether objective or subjective, involves potentiality. Now, we have known ever since Husserl's *Ideas*[19] that potentiality can originate from two different sources. First of all, it can derive from positionality, in other words, be the result of thetic positing acts. Second, it can result from the transformation into potentiality of neutralized contents of consciousness.[20] The concept of objective and subjective probability, when applied here, embraces both these categories. The difference is this: in the case of thetic potentiality, the judgments are made with relative explicitness and in clarity, whereas, in the case of neutralized contents of consciousness transformed into potentiality, the probability of their proving true remains from the subjective standpoint quite undecided or else is taken for granted. But all this depends on the actor's original mode of attention, which is antecedent to all contexts of meaning. If, however, his mode of attention is properly established as a typical and invariant way in which the Ego regards its own lived experiences, the subjective probability originating from the neutralizing experiences of consciousness can then be left out of account, and the probability originating from thetic positionality can be brought to light. It can then be assumed that the actor makes a series of positive judgments (*thetische Setzungsakte*) about his goal, its possibility of attainment, and the means available to him. Since these judgments are explicit and clear, he is said to be acting *rationally*.

The clarification of this notion of rational action will be the final step in our study of Weber's basic concepts.

Interpretive Sociology's Preference for Rational Action Types

Let us recall once again our definition of action. Action is behavior based on an antecedent project. Since every project has an "in-order-to" or "for-the-sake-of-which" structure, it follows that every action is rational. Without such a project, one does not "act"; one merely "behaves" or "has experiences." Every action can, in its turn, be placed in a higher context of meaning, within which it is merely a means to a further end. Now, this end or higher goal may be clearly pictured, while the action leading up to it is carried out in a confused and uncertain manner. Or conversely, the goal may be vaguely conceived, while the action leading to it is well thought out. An example of the first situation would be a direction like "The post office is that way," in contrast to "Take your first right, then after two blocks a left." An example of the second situation would be a chemist carrying out careful experiments on a newly discovered substance whose nature is as yet unknown. Both of these situations are alien to the kind of ideal type that is constructed in the social sciences and, as a matter of fact, in all indirect knowledge of social reality. The ideal type proper to such indirect social experience is one in which both ends and means are clearly conceived. For, since within these types the in-order-to motive is fixed and invariant, the corresponding ends and means must be assumed to have a maximum of meaning-adequacy and the action itself a maximum chance of being carried out. An action type of this kind is, according to Weber, a *rational* action.[21] It does not matter whether the rational action is oriented to an ordinary purpose or to an absolute value. This latter distinction really pertains to the genuine because motive which can be coordinated to the typical in-order-to motive. Whether an act is oriented to an ordinary purpose or to an absolute value depends upon the interest of the actor; the same may be said for the problems he sets up for himself and the experiences he selects as relevant to their solution.[22]

This means-end relation can be thought of in an objective context of meaning and its objective probability estimated. With a suitable choice of type, the objective meaning-context of the means-end relation can be treated as a subjective meaning-context and the objective probability as a subjective probability. This will be true the more universal are the problem situations which are the genuine because-motives corresponding to the typical in-order-to motives in question. For this reason interpretive sociology – but in this it is by no means alone – prefers rational action types. Irrational action (namely, action whose ends or means are confused or uncertain) is interpreted as a variant function of rational action. This is done by postulating a rational action type and then making certain changes in its in-order-to motives; the result is a deviant type. We must keep in mind the fact that sociology is concerned primarily with social interactions and that the latter involve reciprocal orientations in which the calculation of means and ends plays a large role. It is precisely because of the centrality of this calculation that rational action is such an important concept for interpretive sociology. But this does not by any means imply that interpretive sociology neglects irrational action. Weber has again and again stressed that the latter is part of the subject matter of sociology. His works on the sociology of religion, for instance, make exemplary use of categories of irrational, emotional, and traditional action.

This preference for rational action types we must very sharply distinguish from the so-called "rational method" of interpretive sociology. Sociology can claim no monopoly on rational method. The methodologies of all true sciences are rational, involving, as they do, the use of formal logic and interpretive schemes. All true sciences demand the maximum of clarity and distinctness for all their propositions. There is no such thing as an irrational science. We must never cease reiterating that the method of Weber's sociology is a rational one and that the position of interpretive sociology should in no way be confused with that of Dilthey, who opposes to rational science another, so-called "interpretive" science based on metaphysical presuppositions and incorrigible "intuition."

It is true that the postulate of such an interpretive science arose historically from the necessity of breaking through the barriers that were erected between the rational special sciences and the understanding of living human experience. But it was forgotten by those proposing this new approach that life and thought are two different things and that science remains a matter of thought even when its subject matter is life. It cannot, therefore, base itself on some vague and confused empathy or on value presuppositions or on descriptions lacking in intellectual rigor. It was this point and nothing else that lay at the center of Weber's insistence on the objectivity of the knowledge attained in the social sciences. And it was Weber who first raised interpretive sociology to the rank of a science.

Objective and Subjective Meaning in the Social Sciences

Having completed our analysis of the most important basic concepts of interpretive sociology, we must now try to answer the questions we formulated in section 43 concerning the relationship between the meaning-endowing acts of everyday life and their interpretation by the social sciences. Our answer is this: *All social sciences are objective meaning-contexts of subjective meaning-contexts*. We shall now try to clarify what we mean by this statement.

All scientific knowledge of the social world is indirect. It is knowledge of the world of contemporaries and the world of predecessors, never of the world of immediate social reality. Accordingly, the social sciences can understand man in his everyday social life not as a living individual person with a unique consciousness, but only as a personal ideal type without duration or spontaneity. They can understand him only as existing within an impersonal and anonymous objective time which no one ever has, or ever can, experience. To this ideal type are assigned only such conscious experiences as are required to accompany motives already formally postulated. We have already outlined the methodology involved in this postulation. We have seen that it must take place in a manner that is both meaning-adequate and causally adequate. This means that there must be constant recourse to pregiven knowledge of the social world and of the world in general. It means that the motives postulated must not be incompatible with those of the observer's previously constructed ideal types.

Since the social sciences qua social sciences never actually encounter real people but deal only in personal ideal types, it can hardly be their function to understand the subjective meaning of human action in the sense that one person understands another's meaning when he is directly interacting with him. However, we saw that the nature of subjective meaning itself changes with the transition from direct to indirect social experience. In the process of ideal-typical construction, subjective meaning-contexts that can be directly experienced are successively replaced by a series of objective meaning-contexts.

These are constructed gradually, each one upon its predecessor, and they interpenetrate one another in Chinese-box fashion, so that it is difficult to say where one leaves off and the other begins. However, it is precisely this process of construction which makes it possible for the social scientist, or indeed for any observer, to understand what the actor means; for it is this process alone which gives a dimension of objectivity to his meaning. Of course, this process of constitution can only be disclosed to the interpreter by means of his own typifying method. What he will thus come to know is only a conceptual model, not a real person.

We have already seen that there can be personal ideal types of all degrees of anonymity or concreteness. By studying a given cultural product we can gain some insight into what its creator had in mind, regardless of the anonymity of the ideal type we are employing. Accordingly, the different social sciences deal with subject matter of very different degrees of anonymity and concreteness. This should be obvious enough when we consider that the social sciences include, according to our own concept, such widely separated disciplines as individual biography, jurisprudence, and pure economics. And here we should add that not all the social sciences have as their goal the interpretation of the subjective meaning of products by means of personal ideal types. Some of them are concerned with what we have called course-of-action types. Examples of such social sciences are the history of law, the history of art, and political science. This latter group of disciplines simply takes for granted the lower stages of meaning-establishment and pays no attention to them. Their scientific goal is not to study the process of meaning-establishment but rather the cultural products which are the result of that meaning-establishment. These products are then regarded as meaningful in themselves (*als sinnhafte Erzeugnisse*) and are classified into course-of-action types.

At this point an obvious objection will be raised. It will be pointed out that the existence of the so-called law-constructing (or nomothetic) social sciences contradicts our assertion that all social sciences are type-constructing in nature. These law-constructing social sciences, it will be said, are able to provide us with universally valid knowledge prior to all experience. Let us look closely at these sciences and their attitude toward the subjective and objective meaning of the social world, using as our example pure economics.

The Austrian marginal utility school, the Anglo-American scholars working along similar lines, and the mathematical economists as well all claim to have an exact theoretical science, the principles of which are universally valid for all situations in which economic activity occurs. Among the more recent writers of this orientation, Mises can be regarded as the most significant advocate of the pure a priori character of economics. In his treatise "Soziologie und Geschichte," which we have already quoted repeatedly, he takes up a position opposed to that of Weber on the problem of the contrast between theoretical and historical social science. For Mises economics is only a part of sociology, though, to be sure, the most highly developed part. In his polemic against Weber, Mises asks "whether the concepts of economics actually have the logical character of ideal types." His conclusion is:

> This question must be answered quite flatly in the negative. In fact, our theoretical concepts "can be empirically discovered nowhere in reality in their pure conceptual form." Concepts can never be encountered in reality; they belong not to the realm of reality, but to that of thought. They are the intellectual means with which we seek to grasp reality on the level of thought. But one cannot say of these economic concepts that they are formed "by the one-sided accentuation of one or more points of view and by the synthesis of a great many diffuse, discrete, more or less present and occasionally absent *concrete individual* phenomena, which are arranged according to those one-sidedly emphasized viewpoints into a unified *analytical* construct (*Gedankenbild*)."[23] Rather, they are acquired by means of abstraction, which aims at selecting for conceptualization certain aspects of each of the individual phenomena under consideration.[21]
>
> Max Weber's basic error lies in his misunderstanding of what is meant by saying that the sociological principle is universally valid. The economic principle, the fundamental laws of the formation of rates of exchange, the law of profit, the law of population and all other such propositions are valid always and everywhere when the conditions presupposed by them are present.[25]

No doubt Mises' criticism is valid against Weber's earliest formulations of the concept of ideal type, and it is these to which Mises is here referring. According to this earliest view of Weber's, the ideal types would in principle be applicable only to historical data. They would stand in contrast to the concepts of theoretical sociology derived by abstraction from aspects of *each* of the individual phenomena under consideration. However, the theory of ideal types which I have set forth in the present work – a method which is, in my opinion, already foreshadowed in Weber's later works[26] – is an entirely different one, so far as its deduction is concerned. According to our view, ideal types are constructed by postulating certain motives as fixed and invariant within the range of variation of the actual self-interpretation in which the Ego interprets its own action as it acts. To be sure, this postulation of certain motives as invariant does refer back to previous "experience" (*Erfahrung*). But this is not the "experience" of shallow empiricism. It is rather the immediate prepredicative encounter which we have with any direct object of intuition.[27] The ideal type may, therefore, be derived from many kinds of "experiences" and by means of more than one kind of constituting process. Both "empirical" and eidetic ideal types may be constructed. By empirical we mean "derived from the senses," and by eidetic we mean "'derived from essential insight."[28] The manner of construction may be abstraction, generalization, or formalization, the principle of meaning-adequacy always, of course, being observed. Our own theory of ideal types, therefore, covers the concepts and propositions of the theoretical social sciences, including those of pure economics. For even the examples cited by Mises – the economic principle, the basic laws of price formation, and so forth – are in our sense ideal types. Of course these principles must be based upon a thoroughgoing formalization and generalization of material that has already been postulated as fixed and invariant. It is this formalization and generalization which give the ideal types universal validity.[29] Such ideal types do not refer to any individual or spatio-temporal collection of individuals. They are statements about anyone's action, about action or behavior considered as occurring in complete anonymity and without any specification of time or place. They are precisely for that reason lacking in concreteness.[30] Mises[31] is right when he criticizes Weber for interpreting the marginal utility theory in too narrow a fashion, so that it appears to describe an economy run entirely according to the calculations of entrepreneurs. He justly remarks that Weber is here confusing the marginal utility model with that of classical political economy. The latter, he points out, has in mind a more concrete and less anonymous concept of "economic man." Modern theoretical economics,[32] on the other hand, starts not from the behavior of the businessman but from that of the consumer, in other words, from the behavior of anyone and everyone. Such behavior, of course, can serve as the basis of an ideal type of a higher degree of anonymity. It is because of this, in turn, that the principles of catallactics possess a higher degree of generality. Here, as Mises repeatedly emphasizes, is to be found the basis of the objectivism and objectivity of the propositions of catallactics.[33] But this "objectivity" of Mises is, therefore, the same as the concept of objectivity we ourselves put forward in our discussion of the objective and subjective contexts of meaning. The law of marginal utility, then, turns out to be a stipulation that merely marks out the fixed boundaries of the only area within which economic acts can by definition take place.[34]

In our view, pure economics is a perfect example of an objective meaning-complex about subjective meaning-complexes, in other words, of an objective meaning-configuration stipulating the typical and invariant subjective experiences of anyone who acts within an economic framework. Of course the word "typical" takes on a special meaning here, as Mises comes to admit when he emphasizes that an action running contrary to the "principle of marginal utility" (and therefore in our sense "atypical") is inconceivable. But that holds true only so long as one conceives the principle of marginal utility is a definition of the purely formal action as such. Excluded from such a scheme would have to be any consideration of the uses to which the "goods" are to be put after they are acquired.[35] But once we do turn our attention to the subjective meaning of a real individual person, leaving the anonymous "anyone" behind, then of course it makes sense to speak of behavior that is atypical – atypical in relation to standardized economic goals. To be sure, such behavior is irrelevant from the point of view of economics, and it is in this sense that economic principles are, in Mises' words, "not a statement of what usually happens, but of what necessarily must happen."[36]

Mises' criticism therefore does not rule out the applicability of ideal types as such to economic activity. For how could ideal types be excluded from this area, since all scientific knowledge is essentially ideal-typical in character? On the contrary, Mises' argument really turns out to be a defense against the intrusion of ideal types of too great concreteness and too little anonymity into economics. And with this we must agree. At the same time, we must state that the very objectivity of economic knowledge consists in the ordering of subjective meaning-contexts (such as subjective valuations) into the objective meaning-context of scientific knowledge.

Let us now see how the contrast between objective and subjective meaning exhibits itself in a science that is methodologically of a quite different character, namely, the "pure jurisprudence" of Hans Kelsen. Here we find our problem cropping up in the following way:

> Is a constitution republican, for instance, merely because it announces itself as such? Is a state federal merely because its constitution calls it such? Since legal acts usually have a verbal form, they can say something about their own meaning. This fact alone betrays an important difference between the subject matter of jurisprudence, indeed of the *social sciences* as such, and the subject matter of the natural sciences. We need not fear, for instance, that a stone will ever announce itself to be an animal. On the other hand, one cannot take the declared legal meaning of certain human acts at their face value; to do so is simply to beg the question of whether such declared meaning is really the objective legal meaning. For whether these acts are really legal acts at all, if they are, what their place is in the legal system, what significance they have for other legal acts – all these considerations will depend on the *basic norm* by means of which the scheme that interprets them is produced.[37]
>
> Jurisprudence must pronounce that certain acts standing at the outer boundary of the legal system are, contrary to their own claim, *invalid acts*. The root of the problem is that the human acts which are the subject matter of jurisprudence have their own immanent subjective meaning which may or may not coincide with the objective meaning that accrues to them in the legal system to which they belong, *and by the basic norm postulated by the theory governing the system*.[38]

It would be hard to find a more penetrating formulation of the true relation of the social sciences to their subject matter, which we have defined as the ordering of subjective meaning-contexts within an objective meaning-context. According to Kelsen, the subjective meaning which the individual legal acts have for those enacting or performing them must be ordered within an objective meaning-context by means of what *we* should call ideal-typical constructions on the part of the interpreting science of jurisprudence. The ideal-typical construction that we find in jurisprudence is carried out through formalization and generalization, just as in pure economics. In pure economics the principle of marginal utility is the defining principle of the whole field and presents a highest interpretive scheme which alone makes possible the scientific systematization of the subjective meaning-contexts of individual economic acts. Correspondingly, in the realm of pure jurisprudence, as Kelsen himself clearly recognizes, application of a presupposed basic norm determines the area of invariance for all those subjective meaning-contexts of legal acts which are relevant for jurisprudence or which, to use technical terminology, bear the mark of positivity.[39] In another work Kelsen formulates this thought in the following way:

> While positivism means that only that is law which has been created by constitutional procedure, it does not mean that everything which has been thus created is acceptable as law, or that it is acceptable as law in the sense which it attributes to itself. The assumption of a basic norm which establishes a supreme authority for the purpose of law-making is the ultimate presupposition which enables us to consider as "law" only those materials which have been fashioned by a certain method. The above described interpretation of legal material has actually long been in use by legal science. If it is correct, and if this imputation of an objective meaning is possible (without which there can be no legal science), then it must be the basic norm itself which gives the significance of law to material produced by a certain procedure. It must, moreover, be possible to ascertain from this basic norm which part of the material is valid "law," and also the objective meaning of the legal material, which actually may conflict with its own subjective meaning. The hypothesis of the basic norm simply expresses the assumptions necessary for legal cognition.[40]

There is nothing to add to these ideas from the standpoint of the theory being advocated here. Kelsen quite clearly indicates that his basic norm is the principle by which are constructed those ideal-typical schemes which alone make it possible to interpret subjective meaning-contexts as objective meaning-contexts of law.

In these two examples we have shown how the two most advanced "theoretical" social sciences – pure economics and jurisprudence – make use of ideal-typical constructs (in our sense) in order to delimit their subject areas and establish an objective context of meaning. What is true for the "theoretical" social sciences is generally true for all the social sciences.[41] Subjective meaning-contexts are comprehended by means of a process in which that which is scientifically relevant in them is separated from that which is irrelevant. This process is made possible by an antecedently given highest interpretive scheme which defines once and for all the nature of the constructs which may be used.

It would require a treatise in itself to define the specific problems of each social science – especially the historical disciplines – and the methods peculiar to each of them and then, on the basis of these determinations, to attempt a classification of the sciences in question. As the principle of classification we should, first of all, put forward the degree of anonymity of the ideal constructs used in each social science, in other words, the fundamental attitude of each science to the subjective meaning-context with which it deals. Furthermore, the social sciences fall into two classes. First, they can be *pure theories of the form* of the social world, which deal with the constitution of social relationships and social patterns, the act-objectivities and artifacts in the conscious processes of individuals who live in the social world, meanwhile comprehending all these things by a purely descriptive method. However, the social sciences can also take as their subject matter the *real-ontological content* of the social world as already constituted and study the relationships and patterns in themselves – the already given historical or social acts and the artifacts as objects independent of the subjective experiences in which they were constituted.

There is still a word to be said about the field and method of interpretive sociology. The primary task of this science is to describe the processes of meaning-establishment and meaning-interpretation as these are carried out by individuals living in the social world. This description can be empirical or eidetic; it can take as its subject matter the individual or the typical; it can be performed in concrete situations of everyday life or with a high degree of generality. But, over and above this, interpretive sociology approaches such cultural objects and seeks to understand their meaning by applying to them the interpretive schemes thus obtained.

Conclusion: A Glance at Further Problems

We have now arrived at the end of our study, which, of course, could deal with only one aspect of the complex problem of the interpretation of meaning in the social world. There are many other tasks lying ahead for a sociology which is based on phenomenological principles and which is willing to take as its starting point our analysis of duration and of the latter's connection with meaning. The group of problems which held the center of our attention in this work was that which was concerned with the *sociological person*. We have by no means clarified to a sufficient degree the questions which we discussed under the heading of the Thou-orientation and the They-orientation, the We-relationship and the They-relationship, of the physically present alter ego and the personal ideal type. A crucial question which we never faced up to was that of interpretive sociology's right to make valid assertions about the form of social relationships, regardless of whether the entities involved in such relationships were one or more individuals, a personal ideal type, or a social collective. If we recall what we said about the relation of the individual to the ideal type, we shall find that even here the boundaries are fluid throughout, that the individual in his physical reality can be just as easily comprehended in terms of the content of all the possible ideal types of him as, on the other hand, each ideal type can be comprehended as a statement about an individual considered under an anonymous function. Every statement about the action of a personal ideal type removes the latter to some extent from the They-relationship and places it within a We-relationship. Every statement about an individual removes him from the direct We-relationship out of his concrete thusness into a like-thatness, into a typical relationship in the world of contemporaries.

A second group of problems reaches far beyond the boundaries of the subject matter of the social sciences. It is the whole *problem of relevance*, which has kept cropping up again and again in the present study. The definitive clarification of this problem will be possible only through an over-all phenomenological analysis, which nevertheless can be begun within the field of the social sciences. Whether we take our departure from the ideal type, from the existence of in-order-to and because-motives, from the "projected" character of the act, from the possibility of reproduction, even from the mere distinguishability of our lived experiences, repeatedly we come up against the same problem. This is the question of why these facts and precisely these are selected by thought from the totality of lived experience and regarded as relevant. The settling of this question is of crucial importance for all categories of social science which are based on the tacit assumption that the observer's interest situation and the statement of the problem which is determined by it were already satisfactorily deduced through the clarification of the problem of relevance.

A third group of problems would include the *constitution of the Thou* as such, the illumination of the intersubjective structure of all thinking, and the constitution of the transcendental alter ego from the transcendental ego. Hand in hand with the solution of this problem will come the solution of the intersubjective validity of our experience of the world in general. In his *Formal and Transcendental Logic*, Husserl has already laid the foundation for the solution of this problem. He has announced a future work which will make the whole question the center of attention, a question whose conclusion solution will probably give us for the first time an ontology of the human being on a phenomenological basis.[42]

The other two basic problems, namely, the problem of the sociological person and the problem of relevance in the social world, can be taken on by interpretive sociology working closely along the lines originally set down by Max Weber.

Notes

1. Cf. Husserl's *Formale und transzendentale Logik*, pp. 29 f. and 206.
2. Weber, "Roscher und Knies und die logischen Probleme der historischen Nationalökonomie," *Gesammelte Aufsätze zur Wissenschaftslehre* (1904), p. 81, obs. 3.

3. *Logik*, p. 13.
4. Cf. below, sec. 49, pp. 275 ff. ["Wie sind Wissenschaften vom Subjektiven Sinnzusammenhang überhaupt möglich?" Cf. Kant, *Critique of Pure Reason*, B 20: "How is pure science of nature possible?" ("Wie ist reine Naturwissenschaft möglich?")]
5. Walther, "Max Weber als Soziologe," *Jahrbuch für Soziologie*, II, 1–65; Schelting, "Die logische Theorie der historischen Kulturwissenschaft von Max Weber und im besonderen sein Begriff des Idealtypus," *Archiv für Sozialwissenschaften und Sozialpolitik*, XLIX (1922), pp. 623–752; Hans Oppenheimer, "Die Logik der sozialwissenschaftlichen Begriffsbildung mit besonderer Berücksichtigung von Max Weber," *Heidelberger Abhandlungen zur Philosophie*, V (1925); Freyer, *Soziologie als Wirklichkeitswissenschaft*, pp. 145 ff., 175 ff., etc. For Weber's personal development see Voegelin, "Über Max Weber," *Deutsche Vierteljahrsschrift für Literaturwissenschaft und Geisteswissenschaft*, III, 177 ff., and, by the same author, "Gedenkrede auf Max Weber," *Kölner Vierteljahrshefte für Soziologie*, IX, 1 ff.; and, finally, the extensive and very important work of Marianne Weber, *Max Weber, ein Lebensbild* (Tübingen, 1926). [The reader is also referred to the above-named Alexander von Schelting's *Max Webers Wissenschaftslehre* (Tübingen: J. C. B. Mohr [P. Siebeck], 1934).]
6. *Wirtschaft und Gesellschaft*, p. 1 [E.T., p. 88].
7. *Ibid.*, p. 9 [E.T., p. 109].
8. *Ibid.*, pp. 9, 10 [E.T., pp. 109–110].
9. *Ibid.*, p. 1 [E.T., p. 89; "subjectively intended meaning" has been substituted for "subjective meaning" in the Henderson-Parsons translation].
10. *Ibid.*, p. 4 [E.T., p. 96].
11. With respect to Weber's concept of history and statistics, see Mises, "Soziologie und Geschichte," *Archiv für Sozialwissenschaften und Sozialpolitik*, LXI, 465–512.
12. Weber, *Wirtschaft und Gesellschaft*, pp. 5–6 [E.T., pp. 99–100].
13. "Über den Begriff der objektiven Möglichkeit und einige Anwendungen desselben," *Vierteljahrsschrift für wissenschaftliche Philosophie* (1888), pp. 180 ff.; on the concept of causal adequacy, see esp. pp. 201 f. With respect to Max Weber's concept, cf. the essay devoted to this theme in *Gesammelte Aufsätze zur Wissenschaftslehre*, pp. 78 ff.
14. A critique of this concept, for which we do not have the space here, would show that its universal validity is quite doubtful. Cf., with respect to its usefulness in criminal law, Felix Kaufmann, *Die philosophischen Grundprobleme der Lehre von der Strafrechtsschuld* (Leipzig and Vienna, 1929), pp. 78 ff.
15. [See *Wirtschaft und Gesellschaft*, p. 12; E.T., p. 115.]
16. But not, of course, in the case of the natural sciences. The phenomena of nature are in principle beyond interpretive understanding and have no "meaning," since they fall outside man's consciousness and belong to an objective spatiotemporal order. This is not the place to investigate more deeply the distinction between the natural and the cultural sciences.
17. Since only a conscious experience can be meaningful (*sinnhaft*), we need not, in speaking of meaning-adequacy, distinguish between its application to cultural and its application to natural objects, as we did in the case of causal adequacy.
18. With respect to this concept, see above, p. 191.
19. Pp. 228 ff. [E.T., pp. 313 ff.]
20. With respect to this concept cf. sec. 11, p. 67.
21. For an analysis of the concept of rational action see Hermann J. Grab's valuable monograph *Der Begriff des Rationalen in der Soziologie Max Webers* (Karlsruhe, 1927). Needless to say, my agreement with Grab can only be partial, since he presupposes Scheler's concept of objective values.

22. For the derivation of the two types of action in question see Mises, "Soziologie und Geschichte," p. 479.
23. Quoted from Weber's "Die Objektivität sozialwissenschaftlicher und sozial-politischer Erkenntnis," *Gesammelte Aufsätze zur Wissenschaftslehre* (1904), p. 191. [Cf. *Max Weber on the Methodology of the Social Sciences*, trans, and ed. by Edward A. Shils and Henry A. Finch (Gleneoe, Ill., 1949), p. 90, from which our translation of the above quotation is taken.]
24. Mises, "Soziologie und Geschichte," p. 474.
25. *Ibid.*, p. 480.
26. Max Weber's well-known formulation of the concept of ideal type, made in 1904, which he himself calls "sketchy and therefore perhaps partially incorrect" is indeed fragmentary because it has in mind chiefly the ideal type of his theory of history. It must be strongly emphasized that once Weber's thought makes the transition to sociology, the conception of the ideal type itself undergoes a thorough change. Unfortunately, this fact is only hinted at in a few statements in *Wirtschaft und Gesellschaft*, e.g., on page 10 [E.T., p. 110], Cf. Walther, "Max Weber als Soziologe," *Jahrbuch für Soziologie*, II, 1–65.
27. [See the first chapter of Husserl's *Ideas*.]
28. [See *ibid.*, esp. § 3 (E.T., p. 54).]
29. These two points are merely expressed in a slightly different way by Mises when he says that the theoretical propositions are universally valid *under the stipulated conditions*.
30. In the sense of our discussion in sec. 39; see above, p. 195.
31. *Op. cit.*, p. 486.
32. [The reference here is to the marginal utility school emanating from Jevons, Menger, and Böhm-Bawerk.]
33. *Op. cit.*, pp. 482, 486.
34. Cf. Felix Kaufmann, "Logik und Wirtschaftswissenschaft," *Archiv für Sozialwissenschaften*, LIV, 614–56, esp. 650.
35. We need not here pursue the problem of the reduction of the concept "economic good" to less anonymous and more concrete psychological concepts. Cf. Mises, *op. cit.*, p. 476; also Kaufmann, "Logik und Wirtschaftswissenschaft," p. 628.
36. Mises, *op. cit.*, p. 484.
37. Kelsen, *Allgemeine Staatslehre* (Berlin, 1925), p. 129; italics ours.
38. *Ibid.*, p. 278.
39. [Cf. Kelsen, *General Theory of Law and State* (Cambridge, Mass., 1945), pp. 114 ff.: "Law is always positive law and its positivity lies in the fact that it is created and annulled by acts of human beings, thus being independent of morality and similar norm systems" (E.T., Anders Wedberg).]

    For a discussion of the concept of "basic norm" see Felix Kaufmann, "Juristischer und soziologischer Rechtsbegriff," in the anniversary volume for Hans Kelsen, *Gesellschaft, Staat und Recht: Untersuchungen zur reinen Rechtslehre* (Vienna, 1931), pp. 14–41, esp. pp. 19 ff. and 30 f.
40. Kelsen, "Die philosophischen Grundlagen der Naturrechtslehre und des Rechtspositivismus," *Philosophische Vorträge der Kantgesellschaft* (Charlottenburg, 1928), pp. 24 f. [E.T., "Natural Law Doctrine and Legal Positivism," by Wolfgang Kraus in Kelsen, *General Theory of Law and the State*.]
41. Cf. the discussion in sec. 28.
42. Husserl's *Cartesian Meditations* have already in part fulfilled his promise.

11

Unexpected Interactions: Georg Simmel and the Observation of Nature

Matthias Gross

Introduction

The material environment remains at best on the fringe of sociology, sometimes acknowledged, but rarely theorized. However, in recent years the physical and symbolic properties of society have been debated among an increasing number of sociologists, especially in the field of environmental sociology and in some areas of science studies.[1] Current debates on the ontological status of the physical world of human society and the sociological approaches to it can be roughly divided into two familiar camps: the 'social constructivist' and the 'realist'. Those who have chosen a social constructivist route place people and their interpretation of the natural world at the center of their research. In the other camp are those who claim that the reality of the material environment needs to be taken into sociological theorizing as direct information, but theoretically the latter stream has not gone beyond this mere claim.[2] Of course, one will find few realists without an interest in cultural constructions and few constructivists who deny the significance of natural powers for social processes. However one puts it though, even if a constructivist approach does not explicitly say that nature does not have a voice, it is nevertheless unclear to what extent cultural patterns do determine the view of nature and the extent to which nature has a say. Thus, the theoretical debate between both camps is stuck in an either/or impasse. In recent years a third stream has emerged, which may be seen as

Source: *Journal of Classical Sociology*, 1(3) (2001): 395–414.

a way of circumventing the shortcomings of the other two streams – the so-called symmetric anthropology of the actor-network theory (ANT) of authors like Bruno Latour (1993). This is a metaphorical midway position that can be used to look at both sides simultaneously via a declared a priori symmetry between humans and their material environment.

One of the few things that is true for most proponents of all three streams is their conviction that their findings are new, a claim they uphold by positing that their approaches are superior to the classical tradition of sociological scholarship, which is said to have excluded a sociological concept of the natural world; by contrast, their respective approaches are tackling recent social developments, including environmental problems, the green movement, ecological awareness, and the declaration of a generally changing relationship between society and the natural world.

By way of opposing this view, I examine some of the reflections of classical sociologist Georg Simmel (1858–1918).[3] I aim to show that Simmel outlined a way of moving back and forth between the natural and the cultural, neither reducing society to nature, nor nature to society, but instead exploring their interactions via a switch between two levels of observation. This, I argue, is not only a move beyond worn-out debates on 'nature as physical' versus 'nature as symbolic', but is also a promising starting point for sociological observation, avoiding some of the pitfalls of the streams already mentioned. That is to say, I use Simmel's reflections on the human place in nature as a guiding line for sociological observation of humans' interaction with the non-social.[4]

In the following section, I briefly discuss this background to Georg Simmel's reflections on 'the nature of society'. Then I introduce Simmel's reflections on society and nature in order to subsequently discuss his conception of social development in its natural environment and its affinities with 21st-century ecological practice. Finally the theoretical potential of Simmel's take on the nature–society interaction – the switch of observation between the social and the natural environments – is illustrated. To show how enlightening Simmel's theoretical ideas can be for the sociological description of contemporary areas, I have chosen examples that stem from today's ecological field experiments. In this final section of the article, I present what I believe Simmel would have done, had he been a fieldworker in fields of interest to today's environmental sociology.

Reciprocal Effects: Simmel's Core Concept of Sociology

Simmel rejected organicist, or natural determinist ideas, as well as the culturalist or idealist ideas of what we would call today the realist and constructivist sociological approaches. He did not see society as an entity or an organism, nor merely as a convenient label for something without a real existence (Simmel, 1992: 13 ff.). Simmel's solution was to picture society as a web of reciprocal interactions (*Wechselwirkungen*) between people and people as well as between people and the material world. Simmel's core notion of *Wechselwirkung* can be translated as both 'reciprocal interaction' and, with its reference to causality, as 'reciprocal effect' and also 'reciprocal causation'. This is demonstrated by his statement on the study of *Wechselwirkungen*:

> Because all social life is reciprocal interaction (*Wechselwirkung*), it is a unity; because what else can unity mean, if not that everything (*das Viele*) is connected with everything else and the destiny of every element does not leave another element untouched. (1912: 53)[5]

Furthermore, in his approach, Simmel also included the idea that a *material object* is also 'in *Wechselwirkungen* with all that floods or persists (*flutet oder beharrt*) around it' (1986: 111). His concept of *Wechselwirkung*, then, can be understood as a breakdown of the linear dichotomy between nature and society. In fact, this is a definition that is so obvious, if not a truism, in terms of an everyday understanding and, at the same time, so broad, theoretically and methodologically, that not much is gained. As a first step to clarification, one could use the following statement:

> The important point here is to find out what constellation is scientifically functional, where the interaction (*Wechselwirkung*) between beings is robust enough for their isolated observation of interactions with other beings to make an optimal explanation. The main thing in this respect is whether the combination under consideration is a frequent one or not, so that the understanding of it may be typical. (Simmel, 1989: 130)[6]

Simmel subsequently defined the 'matter' of sociology not as the study of society, community, or even of groups, but the study of the *modes* or *forms* of 'sociation' or more literally of 'societalization' (*Vergesellschaftung*). This, he believed, ought to be the focus of sociology. Simmel explicitly included in his notion of *Wechselwirkung* the 'reception of action' (1992: 18) and effect (*Wirkung*). This perspective at least includes the opportunity of going beyond a fixed idea of what belongs to society and what does not, or of deciding what the original causes of the social are. Thus, the term *Wechselwirkung* in the sense understood by Simmel strictly speaking cannot be translated into the English term *interaction* without sacrificing a very important aspect of it.

Throughout his work, Simmel considered social actions not in themselves but in relation to particular structures and processes. For him sociology is the science that studies the processes that lead to 'sociation'. This idea of sociology was a response to the critiques from the historical sciences as well as from psychology that a science of society would never become established, because it would fail to have a field of its own. Thus the sociologist's proper object of study should not be an entity called society but the intricate connections that give rise to society. Society, for Simmel, consists of nothing more than reciprocal effects and interactions (*Wechselwirkungen*), almost like a network of the most varied relations between persons as well as between humans and the material world. This society is characterized by urbanization and individualization as well as modern techniques and technologies.

Natural Interactions

With the versatile notion of *Wechselwirkung* at his disposal, Simmel proceeded to reflect on the human place in nature and outlined a sociological perspective that can be used to observe natural and social changes that is virtually forgotten today.[7] Of further significance are his reflections and theoretical fragments on the dynamics that nature can *develop*, with and without human interference. Simmel's approach was to think of a sociological observation where the natural world – that is, the wind, storms, heat, the natural growth of plants and animals – engages with human society. Simmel often speaks of the 'surprise', 'amazement', 'admiration', or even the 'fanatical fascination' felt about a certain landscape (1993b), about a tree that has been grafted (1986: 197 ff.), or about a building that is about to turn into a ruin and hence is being taken over by nature again, thus representing a natural–social entity (p. 118 ff.). In Simmel's words:

> Nature around us excites us sometimes to aesthetic enjoyment, sometimes to fright and horror, and gives us a feeling for the sublimity of her force (*Übergewalt*) . . . sometimes its harsh reality scares us into analyzing our basic feelings, something that I would only be able to describe as shock (*Erschütterung*). (1912: 19)

Furthermore:

> That the violation (*Vergewaltigung*) of a work of human will by natural powers can have an aesthetic effect at all is founded on the precondition that nature has never completely relinquished her rightful claim to this work, however much it may be formed by the human spirit (*Geist*). In its material, its given condition (*Gegebenheit*), it has always remained nature, and if now nature becomes once more completely mistress over it, she is merely carrying out a right which has remained silent until now but which, in a way, she has never renounced. (1986: 121)

In Simmel's view, nature has not necessarily given up its rights – as he called them – even if it has been 'invented' by human hand. For Simmel, it is even possible that a product made by human hands 'becomes more beautiful by chemical and physical means' (1986: 121). The fascination lies in the fact that after a certain amount of time 'a work made by humans appears entirely as a product of nature' (p. 120), because

> . . . as soon as the human-made work is completed (*dasteht*), it not only has an objective being and an individual existence (*Eigenleben*) independent of humans, but it also holds in its being (*Selbstsein*) . . . strengths and weaknesses, components and significances (*Bedeutsamkeiten*), that we are completely innocent of and which often take us by surprise. (p. 213)

In Simmel's metaphorical language: 'Humans can make themselves accomplices (*Mitschuldige*) of nature and of her inherent tendencies (*Wirkungsrichtung*), which may be opposed to their own interests' (1986: 119). He even speaks up for nature, tries to take the perspective of nature: 'A new form arises that is, from the standpoint of nature, entirely meaningful, comprehensible (*begreiflich*), and differentiated' (p. 120). But although a natural process can start off on the surface of human-made nature and can completely take it over against humans' own interests, it follows that what is made by human skill can also become 'unintentionally and unenforceably something obviously new, that is often even more beautiful and more consistent than what has been intended' (p. 121). In his concept of *Wechselwirkung*, Simmel allows natural powers to move freely. Unlike his contemporary Durkheim (1982), Simmel did not try to incorporate nature into society, but allowed both sides to have a voice of their own. Simmel did not perceive the material environment of society as merely something that is distinguishable by the degree of societal consolidation, as Durkheim proposed in his *Rules of Sociological Method* (1982).[8] Departing from the Aristotelean example of the duality between nature and technology, Simmel instead shifted the notion of nature towards the possibility of human-made nature. In contrast to Aristotle, Simmel clarifies the degrees and suggests a distinction between *culturalized* and *cultivated* nature.

> A garden fruit that has been raised (*gezogen*) by the work of a gardener from a 'woody' and inedible fruit, we will call cultivated. Alternatively, we might say that the wild tree has been cultivated into a garden fruit-tree. But if a ship's mast were produced from the same tree, with even the same purposive work expended upon it, we would not regard the tree as being cultivated into a mast. This linguistic nuance makes explicit that, although the fruit could not have grown without human effort, the tree developed from its indigenous powers and only fulfilled the potentialities which were already sketched in its own natural tendencies. (Simmel, 1986: 197)

Simmel distinguishes human-made things, such as *buildings* or a mast, from human-made *nature*, such as a cultivated tree. Human-made nature is still natural when it is able to fulfill potentials which do not work against its natural tendencies. Although in this example nature has given way to culture, the final product is completely natural. Speaking from a contemporary perspective, in this conception the natural ecological dynamics are initiated by humans, but thereafter nature, so to speak, designs itself. This 'self-design' can generate numerous potential outcomes owing to the complexity and unpredictability of ecological dynamics.

In a way, the implementation of cultivation overcomes the subject–object distinction between society and nature. Cultivation resembles the nexus between the members of a human society who have their own ideas about, and constructions of, the rest of nature that surrounds humans. Cultivation is thus the term that best grasps the interface of the reciprocal interaction between humans and the natural world. With the notion of 'cultivation', Simmel's versatile concept of *Wechselwirkung* finds a more succinct definition. The crucial point here seems to be that a human-made thing can develop 'naturally', that is with an independent existence, extricating and dissociating itself from human planning. The cultural performances of human morphing of nature then do become an independent natural power, which leads to a natural–cultural interaction process.

Furthermore, in Simmel's reflections on nature–society interactions, aesthetics is clearly of vital importance. Turning to the example of a ruin, Simmel points out that, although a ruin was originally a work of architecture (1986: 118 ff.), both forms of material environment – a *cultivated* tree as well as a *cultural* building – can develop a 'natural' life of their own and hence can turn into something that transmutes into a product of nature, indistinguishable for humans from 'untouched' nature. Art and nature then merge into 'a new whole' (p. 118). Here *cultivation* for Simmel means the invention of a 'new nature' that involved the development of a latent natural power as well as culture for its expression. In his definition, cultivation does not mean that it is subject to manipulation in all stages of its development. Cultivation in Simmel's usage does not mean the design of an intentionally produced or engineered ecosystem. It is in fact precisely because it is unplanned and independent of human involvement that the achievement is so amazing and delightful. In this regard, nature is also perceived as a meaningful actor: 'Nature has transformed the work of art into material for her own expression (*Formung*), as she had previously served as material for art' (p. 120). The idea of *Wechselwirkung* between the human and the natural also becomes the focus for the development of an aesthetic landscape.

The cultural achievements of human societies can, so to speak, turn into something natural with their own – so far, latent – natural powers. 'Cultivated' nature for Simmel is still something natural, although it has been invented by humans. If humans leave nature enough space to unfold according to its own rules – as with the grafted apple tree – Simmel still regards it as natural. For Simmel 'the work of the gardener really only develops potentialities inherent in a tree, but unawakened in its inner core' (1993a: 85). The gardener *completes* nature and brings it 'to a fulfillment determined by the essential and fundamental tendencies' (p. 85). In other words, the gardener also 'awakes' nature. This completion emerges through the 'interaction of nature with new teleological interventions which follow the inherent proclivities (*Anlageeinrichtungen*) that could be called *its* culture' (p. 85). Concurrently with the gardener's intervention, nature also invents its own natural culture. When nature is culturalized, as in his example of the mast made out of the same apple tree, Simmel does not immediately speak of the rights of nature. The form of a mast made out of a tree 'is given to nature purely from the outside, under the impact of a purposive system that is foreign to its own fundamental tendencies' (p. 85). Here Simmel implicitly differentiates between other forms of landscaping, modeling, farming and manipulation where a predetermined endpoint is the desired stage.

The Development of Society in Nature

Although Simmel's outlook for the future of modern society is sometimes quite a pessimistic one, his general take on the environment is less negative. He simply shows that humans subjectively change their environments, but these environments become objectified and thus change social life. Concurrently Simmel also describes the development of large cities quite naturalistically. In his essay 'The Term and the Tragedy of the Culture' (first published in 1911) he claimed the city develops 'after no fixed plan, but after the coincidental needs and inclinations of particular individuals' (1986: 211). However, Simmel nevertheless describes the city as 'a whole that is meaningful, and expressively closed in itself, like an organic being (*Gebilde*)'. Furthermore:

> Just as a human being does not end with the limits of his/her physical body or the personal belongings with which he or she is immediately confronted, but rather, with the totality of meaningful effects (*Wirkungen*) which emanate from him/her temporally and spatially, in that way a city exists only in the totality of its effects (*Wirkungen*), which transcend their immediate sphere (*Unmittelbarkeit*). This is the real context in which its being (*Sein*) is expressed. (1903: 200)

Departing from his general concept of *Wechselwirkung*, Simmel becomes aware that the sociological observation of the totality of the effects of the human – nature interaction is faced with a time-honored problem. In his chapter dealing with the question 'How is society possible?', Simmel reflected on the tension between nature and society and the problem of human freedom, which – as discussed later in this article – can be interpreted as a solution towards the switch between the two spheres of society and the natural world. As Simmel wrote in his *Soziologie*:

> On the one hand we know ourselves as part (*eingegliedert*) of nature, as one of her products, which stands alongside every other as an equal

> among equals, as a point which nature's substance and energies reach and leave, just as they circle through flowing water and blossoming plants. (1992: 54)

This points to one side of the observation, to humans as a part of the natural order that directly influences their social world. However, Simmel continues:

> And yet the mind (*Seele*) has the feeling of something with an independent existence, something to which we assign the logically unsound designation 'freedom', offering opposition and resistance to all that energy, an element of which we still retain, which leads to the radicalism which culminates (*aufgipfelt*) in the formula: nature is only a figment of the human imagination. (p. 54)

This appears as the other side of the sociological observation, the side that is abbreviated to the term 'social constructivism' or the 'typical' or 'pure' sociological perspective today. Yet, one also has to pay attention to the fact that the human ego 'with all its freedom and "being itself" (*Fürsichsein*), with its juxtaposition to "mere nature", is still a member or limb of nature; this is precisely the all-encompassing connection of nature' (p. 54). Simmel summarized the tension of the idea of *Wechselwirkung* and highlighted 'that the individual desires to be whole (*ein Ganzes*), but the all-embracing whole (*das übergreifende Ganze*) allows him or her only to be in the position of a chain' (1912: 68).

Thus the individual is given the opportunity to be a whole and at the same time a piece of the whole, to build an order beyond individuality with full individual freedom' (p. 68). The unity of nature and society is revealed by Simmel only from a bird's eye view. Viewed from this perspective, Simmel understands nature as meaning 'the endless connection of the things, the continuous bearing and destroying of forms, the flooding unit of what happens (*des Geschehens*) that expresses itself in the continuity of the temporal and spatial existence' (1993b: 130). But also with Simmel there is a hierarchy: 'In the layering of nature and spirit nevertheless, following their cosmic order, nature presents itself as if it were the substructure, the material or the half product (*Halbprodukt*), the spirit presents itself as the definitive form, the figuration and crowning' (1986: 120). From a quasi-evolutionary perspective, Simmel understood social development as coming from nature. The development of nature simply leads to the point where nature is replaced by cultural development (1993a: 84). Culture is therefore an advancement of natural history: 'The point at which this separation of the forces of development takes place marks the boundary of the natural state against the cultural status' (p. 84). This boundary, as we have seen, can be grasped in terms of what Simmel called cultivation: that is, the interaction between the natural and the social. The development of culture is thus understood as evolutionary, whereby the evolutionary process originates from the subjective ideas of individuals. Simmel stresses that culture is to be derived 'causally from its naturally developing conditions' (p. 84).

With the example of the ruin, Simmel makes clear how 'the equation between nature and spirit, which represented the building, shifts in favor of nature' (1993a: 84). A ruin for Simmel is the meeting point or the interface of human action and the action of nature. He called it 'the special attraction (*Reiz*) of the ruin that an object invented by human skill is finally felt like a natural product' (p. 120). Simmel's points of observation thus shift and switch, move back and forth between the observation of individual subjective human goals, like the building of a house or the grafting of a tree, and the objective – non-foreseeable and unexpected – natural action that he describes in terms of his understanding of nature, that is, his sensing of nature via his experiences stemming from his life world, like the description of a house turning into a ruin or a grafted apple tree that nevertheless follows its own inherent and fundamental tendencies.

To await the 'natural' reaction in the process of reciprocal interaction, the sociological observer in Simmel's descriptions moves to the 'objective' side of the natural world, that is, the 'independent existence' that is 'autonomous from humans', but is 'unintentionally and unenforceably something obviously new' (1986: 121). Here Simmel moves the sociological eye (or alternatively, the tongue, ear or nose) from the ideas and understanding of humans to the actions of nature until it is 'felt like a natural product' (p. 120). When humans point to the enigma of nature that they are completely unconnected with and which takes them by surprise (p. 213), Simmel sees this as the time to switch to the 'objective' natural environment, like the deteriorating building or the grafted apple tree.

Simmel's Two-Level Observation: Experiences from Ecological Practice

Before proceeding any further, we need to examine some issues relating to recent developments in some streams of ecology, since they relate to Simmel's reflections. In ecology, dating at least from the 1970s, the ideal of a balance of nature or an equilibrium has been superseded by a paradigm of nature that could be called non-linear, chaotic, or unstable. If these ecological findings were to be taken seriously, there would be no reliable state that sociologists could treat as a platform for social action. Practitioners in areas of ecological fieldwork like ecological restoration (e.g. Jordan et al., 1987; Packard and Mutel, 1997) or adaptive management (e.g. Gunderson et al., 1995; Holling, 1998) have long accepted the fact of an ever-changing nature that interacts – often violently – with human action. For example, these ecologists are designing and restoring landscapes that seem to have a mind of their own and do not follow the suggestions made by humans; natural action simply differs from human plans. Thus there can be no master plan of action, since ecologists today know that the view of a balance of nature is wrong at local, regional and global levels. All the field practitioners in ecological management – ecological restoration or ecological landscaping, for instance – do is 'adaptive management'. In adaptive management, nature is seen as a constantly moving target. Practitioners go out and try something and then observe the results, modifying what they do next on the basis of the natural action that ensues. Compared to the ideal type notion of a balance of nature, the practitioners' knowledge of nature in adaptive management is admittedly minuscule, since there is no single ideal type.[9] In the words of ecologist Daniel Botkin: 'Chance arises from truly inherent randomness in nature' (1990: 122). The assumption that human activities are to a certain degree caused by these natural factors, which also influence societal development, is challenging. Thus, sociology also has to move beyond the conception of nature as static, seeing it instead as a dynamic engagement with natural processes. The natural in sociology should be treated as something that is always changing. In a Simmelian manner, even Carolyn Merchant makes it unmistakably clear that human inventions – here, buildings etc. – are also subject to natural changes:

> In the mechanistic framework of classical physics, nature was rendered passive and inert, subject to prediction and control through linear differential equations. Within that framework, suspension bridges, tunnels, skyscrapers are engineering triumphs because mechanical systems are considered to be closed, spatially defined, and subject to the classical law of statics and equilibrium dynamics. Yet even these stable structures can be vulnerable to chaotic forces created by unusual weather patterns or geological events generated from outside the system. (1998: 70)

After this short focus on contemporary ecological practice and its understanding of nature, let us now consider an example from experiments in field ecology in order to illustrate the fruitfulness of Simmel's approach. When humans invent and restore a prairie, agency is being attributed to the behavior of nature (i.e. to certain plants, animals, thunder, or a prairie fire) as a result and expression of uncertainty and unpredictability. In other words, it is a different way of saying that natural agency is the observed capacity of a part of nature, as Simmel (1986: 213 ff.) has put it, to surprise its human observers. In doing so, the observing sociologist tries to develop *a feeling for nature* as ecological practitioners do in order to understand the ecosystem from the inside (Helford, 1999). Thus, this approach means following the human actors describing nature in order to know as a sociologist when to switch to the level of observing natural action. The 'action' of nature I have in mind is an acting of the natural world that is to some degree *independent of the planning* of the ecological practitioners. This presupposes the sociological investment in 'listening', that is, in *allowing nature an independent*

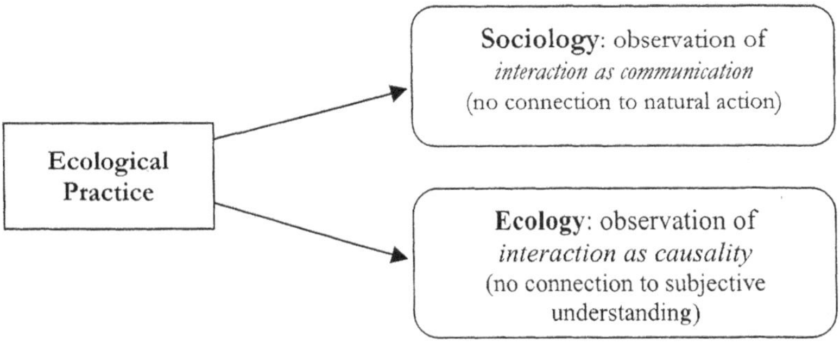

Figure 1: Current division of the social and the ecological: Sociological observation with no connection to natural action

voice, although the starting point of observation focuses on the social. This brings us closer to the possibility of grasping the natural world (a piece of land, an urban park, or a restored ecosystem) and its inherited powers from a traditional sociological perspective. Although authors in the post World War II tradition of sociology have mostly excluded the material world of social interaction, Simmel did not. One reason for this is that he did not ask *why* nature is doing certain things, but *how* nature does things in response to people's action and then how (and why) people again do things differently, supposedly because of the way nature has done things before.

Viewed that way, a codified approach that only talks about interpretations, symbolic meanings, or social constructions, would be very one-sided and incapable of providing a sociological explanation and understanding, since social constructivism – implicitly – presupposes that the natural world is arbitrarily moldable (see Figure 1).

In the following section, I outline the methodological guidelines and theoretical conclusions one can obtain by reading Simmel's ontological reflections against the background of recent trends in ecological field practice. Simmel teaches us that the focus on intentional interpretations should be the starting point of observation. This approach involves an understanding or a feeling from the point of view of the human actors themselves. If left merely at that, though, the limitation here would be that one cannot observe causal connections between humans and material occurrences, although there seems to be nothing to prevent an observer from determining behavioral events and treating them in this fashion. However, to the extent that sociological observers adopt this strategy, they sacrifice the ability to penetrate people's interpretation of their interaction with nature. Thus, we need to determine when the switch from interpretation to natural occurrences can be made (relatively) safely. I suggest that the self-descriptions of the human participants in an ecological field experiment are taken seriously so that nature as an actor can be introduced into the equation *later*; in order to move back and forth between the two sets of observation and to deliver a description of patterns and typologies in the (inter-)action of both sides – patterns of interacting, of trying out and learning in the experimental performance of humans *in* and *with* nature.

Simmel's interpretations of the objective character of a human-invented nature show that, as soon as it has been constructed, it 'has an objective being and an individual existence independent of humans' (Simmel, 1986: 213). Humans are only initially involved in this existence. Hence the analysis of humans acting in and with nature needs to be undertaken via a *permanent switching* between the interpretations and utterances of humans and the features that can be 'factually' observed, e.g. growth and movement, the heat and light of a prairie fire, or the blossoming of certain plants in reaction to people's interpretations. It is the surprise, puzzlement, disorientation, or – at its most extreme – the helplessness of human action in nature with regard to natural action, that will be observed in order to switch to the origin of the surprise and amazement.[10] The observing sociologist has to get close to humans while interacting with the natural world to acquire an understanding or a similar feeling for humans *in* the landscape.

What are the theoretical implications of this and how is the proposed switch from here to the landscape, that is, from 'ideas to nature', possible? In the case of ecological field experiments, this *does not mean giving the authority of observation to natural scientists* (Collins and Yearley, 1992).[11] Why is this so? Given the fact that ecological practitioners (scientific or not) at certain points do not know, according to their self-descriptions, what is happening or how nature is going to react to the societal suggestion – what Simmel has called amazement and surprise and which in today's ecology of chaos and non-linear systems can simply be called puzzlement, or disorientation – the position of the observing sociologist is on the same level as that of the other human actors in the field.[12] The sociological descriptions might be called 'pre-scientific' but so are those coming from the 'scientists', if we take as our point of departure the idea of naturally changing environments, or of a new ecology for the 21st century (Botkin, 1990), which suggests that changes are natural and lack of 'scientific' control is natural too. To speak of 'naturally changing environments' includes an understanding of nature where it eludes common and expert knowledge and therefore is experienced by people (scientific ecologists, hands-on practitioners *and* observing sociologists) as surprise, puzzlement, or uncertainty.[13] Admitting that both natural–scientific and lay knowledge are needed in a successful practice of ecological management, and that at certain times there is no clear difference between the two spheres, is a point that occasionally places the observing sociologist in a similar position to the practicing ecologists and their interactions with the natural world. Phrased differently, this Simmelian approach to the natural environment can be regarded a starting point which acknowledges the practitioners' self-descriptions. It can subsequently be completed with a switch to a *nature-does-act* perspective.

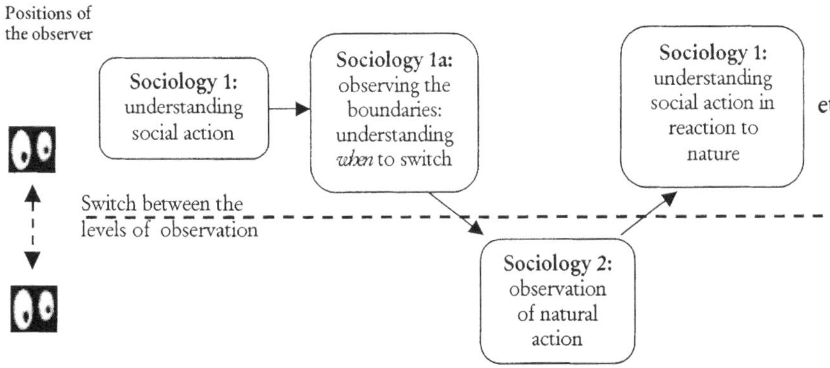

Figure 2: 'The simmelian switch': Situative coordination between the levels of observation

According to Simmel, the observer has to start out by relying on the descriptions and reports of the practitioners and will only turn to his or her own sense of nature (the nature-*does*-act approach) when the self-descriptions of the observed people indicate puzzlement and uncertainty; that is, when the social order in terms of its (scientific) consensus regarding the natural environment has ceased to function, the natural or ecological order comes into the picture and directly determines social processes. It therefore follows that the ecological element not only changes its importance for the practitioners, but will also become accessible for sociological observation. That means the *life-world* and the *everyday nature 'sensings' of the sociological observer* will be allowed to move into the picture of observation.[14] This is borrowed from Simmel's view of the observed wildness and unpredictability of nature, where nature can develop an independent objective that humans are unconnected with. As soon as this stage of observation is arrived at, nature can be described in what Simmel understood as 'objective' terms. The social scientist then does not a priori pretend to find nature sitting or moving objectively and interacting with the social, but has 'direct' access to nature by following the self-descriptions of people who lead or point the observer to the possibility of facing and understanding the non-understanding 'objectively' (Sociology 1a in Figure 2). This includes an *observation of the boundaries* of the self-descriptions of ecological practitioners. Thus, for the observing sociologist, nature becomes an independent actor as part of the human community (Sociology 2 in Figure 2). The switch back to the level of departure of observation, the subjective meanings of restoration practitioners (Sociology 1 in Figure 2) will depend on the explanation of these practitioners of the non-explanation of non-linear natural action. Is there an explanation? Or will it be taken as it comes? How will it be understood?

In the methodological approach of switching between the two levels of observation, in a way between the two environmental sociologies referred to in the introduction to this article, none of the metaphorical syncretism prevalent in many of the more recent streams that claim to acknowledge nature – that is, among others, the talk of cyborgs (Haraway, 1991), or of hybrids and quasi-objects (Latour, 1993) – need to be used either.[15]

In addition, especially noteworthy is the fact that in the previously mentioned streams of ecology, the invention of plans and goals for the morphing of nature includes the plan that these goals will never be accomplished. The conversion into something completely different is sometimes even desired because, to use Simmel's metaphoric language once again, what is made by human skill can also become 'unintentionally and unenforceably something obviously new, that is often even more beautiful and more consistent than what was intended' (Simmel, 1986: 121). *This* natural action – the everyday nature 'sensings' of the sociological observer – is the focus of attention in the practice (Sociology 2 in Figure 2) of ecological work. At this point Simmel thus proposes a *direct*, but *not* a *naive* realism. The observer studies *the natural powers that can be exercised by ecological practitioners and sociologists alike*. Yet, in order to do this, the sociological observer in Sociology 2 needs to go along with the human actors. To switch back to Sociology 1, the *impact* of the observed natural actions – the self descriptions – on society needs to be focused on.

Outlook

In order to understand the practice of ecological field practice from a sociological point of view, the observing sociologist (Sociology 1) has to switch his or her focus to the self-acknowledged reason for the human practitioner's puzzlement: nature or parts of nature (a fire, certain plants or animals, or a human-made ecosystem) as independent members of society. This is the point of the switch of the levels of observation from Sociology 1 to Sociology 2, when the life-world of the sociological observer is *forced* to move into the picture, since no new information on nature can be gained from the social actors, but only from natural action itself: for instance, from the heat of a fire out of control (that is, beyond the ecologists' understanding), from falling trees, or from creeping sand-dunes. However, to fully understand the impact of the observed natural action on society, the observer needs to switch back to Sociology 1: the starting point of the observation of the *turn-taking process* between society and nature.

In conclusion, it needs to be said – although Simmel's theoretical reflections are quite different from the ideas of actor–network theory (ANT) – that a perspective that includes natural action does not leave the terrain of classical sociology (as purists in the field might hastily want to add, if they feared their discipline was being questioned). Read this way, the debate between actor–network theoreticians and 'traditional' humanistic sociologists who argue (1) that actor–network theory is non-sociological (as claimed by traditionalists like Collins and Yearley, 1992; Bloor, 1999; or Breslau, 2000); and (2) that actor–network–theory is something new that pushes sociology to new horizons (as claimed by proponents of ANT like Latour, 1993, 1999; or Pickering, 1995) has been proven premature. Put another way, the Simmelian conception of the material environment of society as dynamic, and hence often beyond the control of humans, outlines a context of society or of the performance of 'societalization' that goes beyond a cultural determinist as well as a static perspective of nature. Almost 100 years ago, Simmel had already rejected misleading assumptions about the natural world predominant in most of contemporary sociology, including so-called environmental sociology.

A Simmelian perspective means that the observing sociologist has to employ the analysis of social processes in interdependence with the processes of the external world. Nothing more, but certainly nothing less. Hence, contrary to acclaimed midway positions, natural action will be involved in what is said about it and what is observable at that moment (examined in the way introduced in this article). It is part of Simmel's theory of the endless process of interaction between humans and humans, and humans and the natural world that leads to what he has termed societalization (*Vergesellschaftung*). Simmel *switches* his perspective of observation from the subjective plane, the intentions, feelings, attitudes and behavior of humans, to the objective action of nature that often differs from human intentions. He also indicates the moment when the switch should be performed: when humans are taken by surprise and are puzzled by natural action. Again, humans change their plans depending on nature's actions. In Simmel, one can find an emphasis on the dynamism of nature by which new properties, materials and forms of societalization are created. His theoretical reflections make it possible to emphasize how human activities modify nature and how nature's *responses* provide cause for subsequent individual action and social organization, without losing the 'social'.

Notes

1. To name but a few: Bell (1998), Buttel (1996), Dunlap and Catton (1979), Goldman and Schurman (2000), Irwin (2001), Knorr Cetina (1997), Latour (1993), Pickering (1995) and the contributions in Redclift and Woodgate (1997) as well as Spaargaren et al. (2000).
2. There is, of course, much more to 'social constructivism' and 'realism' than I am able to indicate in this brief sketch, but its crude outline should help to throw into relief its main contours. A fuller discussion is provided in Gross (2001b). See also Hacking (1999).
3. For a discussion of some other classics of sociology and their perspectives on the material environment of human society, see, for instance, Foster (1999), Gross (2000, 2001a, 2002), Mitchell (2001) and West (1994).

4. I am aware that it is difficult to fully understand Simmel's take on non-human nature without considering his seemingly archaic notion of 'Leben' that was influenced by – among others – Friedrich Nietzsche. Thus I merely use Simmel's writings on nature as an impulse and stimulus for further theory development. For a discussion on Simmel's *Lebensphilosophie* and its importance for the understanding of his ideas on culture and landscapes see the excellent treatise by Solies (1998: esp. chap. 5).
5. Due to the difficulty of Simmel's language, as well as the dearth or imprecision of available translations, this and the following quotes from Simmel are my own translations from the German.
6. Simmel's most detailed discussions of *Wechselwirkung* can be found in Simmel (1992: 13–41) and especially in Simmel (1989: 115–38). For a general introduction to the etymology of the German term *Wechselwirkung* from the 18th century to the time of Simmel, see Christian (1978: 110–33).
7. It needs to be noted that most interpretations of Simmel up to now have picked out what he wrote about everyday social life or about the philosophy of history and not about his reflections on nature. Even the enlightening work of Gerald Young (e.g. Young, 1996), who tried to use the idea of connective processes that link all objects of ecological interest as the foundational concept for human ecology in order to connect ecological and sociological ideas, never refers to Simmel.
8. For a focused discussion and comparison of Durkheim's and Simmel's reflections on nature see Gross (2000). On Durkeim, see also Järvikoski (1996).
9. A detailed discussion of the challenges of new streams in ecology for sociology can be found in Gross (2001b).
10. This can be seen as a more consistent and systematic realization of Durkheim's plea not only to 'proceed from ideas to things', but also from 'things to ideas' (1982: 60).
11. In Collins and Yearley's (1992) critique of 'nonhumans as actors' approaches (most notably of Bruno Latour's *symmetric anthropology*, e.g. 1993), where the material world and nature are given the a priori status of *actants*, the determination of that action is given back to scientists and technologists, who are supposedly detached from the rest of society and do deliver objective knowledge on the natural world.
12. So far, this is similar to Pickering's (1995) take on the observation of the practice of science where he stresses the importance of the temporal emergence of practice. The example of Simmel's work nevertheless shows that the analysis of what Pickering calls a real-time perspective has always been – though forgotten in the meantime – a part of classical sociology.
13. In fields like ecological restoration, the traditional differentiation between scientists and lay actors seems to be moving one step forward and – at least currently – coalescing into a new whole. Understood in this way – and only in this way – a sociological observation in the particular field of ecological restoration is not fundamentally different from the observation undertaken by natural scientists. Taking this as a point of departure, sociological studies of ecological field practice are easier to undertake as regards the problem of scientific authority (see Gross, 2001b).
14. The concept of the life-world (*Lebenswelt*) has been introduced into sociology via the work of Alfred Schütz. For Schütz the chief characteristic of the life-world is that it is taken for granted. The routine character of the life-world, in Schütz's understanding, is to be contrasted with, for example, the world of dreams, of science, or of religion (see Grathoff, 1989). The observation switch I introduce here, thus, in a sense switches between two worlds: the world of the science of society (sociologist) and the world of everyday life and commonsense knowledge about the social and natural world: that is, the life-world of the observing sociologist.
15. I hasten to add that there is no denying the fruitfulness of actor-network theory (ANT) in a lot of other areas. It has in many respects stimulated my own thinking and

research in environmental sociology and ecological field experiments. However, in terms of the a priori decision of a symmetry between the natural and the social, Simmel's reflections serve as a better theoretical starting point for framing ecological field practice. One could also say that Simmel also believed in networks, but has a different take on observing the interactions of the entities in the network than ANT. Furthermore, Simmel's reflections – his focus on aesthetics as well as his thoughts on an ever-changing natural landscape – also include important aspects not dealt with in ANT.

References

Bell, Michael M. (1998) *An Invitation to Environmental Sociology*. Thousand Oaks, CA: Pine Forge Press.

Bloor, David (1999) 'Anti-Latour', *Studies in the History and Philosophy of Science* 30(1): 81–112.

Botkin, Daniel B. (1990) *Discordant Harmonies: A New Ecology for the Twenty-First Century*. New York: Oxford University Press.

Breslau, Daniel (2000) 'Sociology after Humanism: A Lesson from Contemporary Science Studies', *Sociological Theory* 18(2): 289–307.

Buttel, Frederick H. (1996) 'Environmental and Resource Sociology: Theoretical Issues and Opportunities for Synthesis', *Rural Sociology* 61(1): 56–67.

Christian, Petra (1978) *Einheit und Zwiespalt: Zum hegelianischen Denken in der Philosophie und Soziologie Georg Simmels*. Berlin: Duncker & Humblot.

Collins, Harry M. and Steven Yearley (1992) 'Epistemological Chicken', in Andrew Pickering (ed.) *Science as Practice and Culture*, pp. 302–26. Chicago: University of Chicago Press.

Dunlap, Riley E. and William R. Catton Jr (1979) 'Environmental Sociology: A Framework for Analysis', in Timothy O'Riordan and Ralph C. D'Arge (eds) *Progress in Resource Management and Environmental Planning*, Vol. 1, pp. 57–85. Chichester: Wiley.

Durkheim, Émile (1982[1895]) *The Rules of Sociological Method and Selected Texts on Sociology and its Method*, trans. W.D. Halls. New York: Free Press.

Foster, John B. (1999) 'Marx's Theory of Metabolic Rift: Classical Foundations for Environmental Sociology', *American Journal of Sociology* 105(2): 366–405.

Goldman, Michael and Rachel A. Schurman (2000) 'Closing the "Great Divide": New Social Theory on Society and Nature', *Annual Review of Sociology* 26: 563–84.

Grathoff, Richard (1989) *Milieu und Lebenswelt*. Frankfurt: Suhrkamp.

Gross, Matthias (2000) 'Classical Sociology and the Restoration of Nature: The Relevance of Émile Durkheim and Georg Simmel', *Organization & Environment* 13(3): 277–91.

Gross, Matthias (2001a) *Die Natur der Gesellschaft: Eine Geschichte der Umweltsoziologie*. Weinheim: Juventa Verlag.

Gross, Matthias (2001b) 'Inventing Nature: Sociological Investigations into Ecological Restoration', PhD thesis, Department of Sociology, Bielefeld University.

Gross, Matthias (2002) 'When Ecology and Sociology Meet: The Contributions of Edward A. Ross', *Journal of the History of the Behavioral Sciences* 38(1): 1–16.

Gunderson, Lance, H., Crawford S. Holling and Stephen S. Light (eds) (1995) *Barriers and Bridges to the Renewal of Ecosystems and Institutions*. New York: Columbia University Press.

Hacking, Ian (1999) *The Social Construction of What*? Cambridge, MA: Harvard University Press.

Haraway, Donna (1991) *Simians, Cyborgs, and Women: The Reinvention of Nature*. London: Free Association Books.

Helford, Reid M. (1999) 'Rediscovering the Presettlement Landscape: Making the Oak Savanna Ecosystem "Real"', *Science, Technology, & Human Values* 24(1): 55–79.

Holling, Crawford S. (1998) 'Surprise for Science, Resilience for Ecosystems, and Incentives for People', *Perspectives on Ecosystem Management* 6(3): 733–5.

Irwin, Alan (2001) *Sociology and the Environment: A Critical Introduction to Society, Nature, and Knowledge*. Oxford: Polity Press.

Järvikoski, Timo (1996) 'The Relation of Nature and Society in Marx and Durkheim', *Acta Sociologica* 39(1): 73–86.

Jordan, William R. III, Michael E. Gilpin and John D. Aber (eds) (1987) *Restoration Ecology: A Synthetic Approach to Ecological Research*. Cambridge: Cambridge University Press.

Knorr Cetina, Karin (1997) 'Sociality with Objects: Social Relations in Postsocial Knowledge Societies', *Theory, Culture & Society* 14(4): 1–30.

Latour, Bruno (1993[1991]) *We Have Never Been Modern*, trans. C. Porter. Cambridge, MA: Harvard University Press.

Latour, Bruno (1999) 'For David Bloor . . . and Beyond: A Reply to David Bloor's "Anti-Latour"', *Studies in the History and Philosophy of Science* 30(1): 113–29.

Merchant, Carolyn (1998) 'Partnership with Nature', *Landscape Journal* 17(2): 69–71.

Mitchell, Ross E. (2001) 'Thorstein Veblen: Pioneer in Environmental Sociology', *Organization & Environment* 14(4): 389–408.

Packard, Stephen and Cornelia F. Mutel (eds) (1997) *The Tallgrass Restoration Handbook: For Prairies, Savannas, and Woodlands*. Washington, DC: Island Press.

Pickering, Andrew (1995) *The Mangle of Practice: Time, Agency, and Science*. Chicago: University of Chicago Press.

Redclift, Michael and Graham Woodgate (eds) (1997) *The International Handbook of Environmental Sociology*. Cheltenham: Elgar.

Simmel, Georg (1903) 'Die Großstädte und das Geistesleben', *Jahrbuch der Gehestiftung* 9: 187–206.

Simmel, Georg (1912[1906]) *Die Religion*. Frankfurt: Rütten & Loening.

Simmel, Georg (1986[1923]) *Philosophische Kultur: Gesammelte Essais*. Berlin: Wagenbach.

Simmel, Georg (1989[1890]) *Über Sociale Differenzierung* (Gesamtausgabe 2). Frankfurt: Suhrkamp.

Simmel, Georg (1992[1908]) *Soziologie: Untersuchungen über die Formen der Vergesellschaftung*. Frankfurt: Suhrkamp.

Simmel, Georg (1993a[1908]) 'Vom Wesen der Kultur', in *Das Individuum und die Freiheit*, pp. 84–91. Frankfurt: Fischer.

Simmel, Georg (1993b[1913]) 'Philosophie der Landschaft', in *Das Individuum und die Freiheit*, pp. 130–9. Frankfurt: Fischer.

Solies, Dirk (1998) *Natur in der Distanz: Zur Bedeutung von Georg Simmel's Kulturphilosophie für die Landschaftsästhetik*. St Augustin: Gardez Verlag.

Spaargaren, Gert, Arthur P.J. Mol and Frederick H. Buttel (2000) *Environment and Global Modernity*. London: Sage.

West, Patrick C. (1994) 'Natural Resources and the Persistance of Rural Poverty in America: A Weberian Perspective on the Role of Power, Domination, and Natural Resource Bureaucracy', *Society and Natural Resources* 7(4): 415–27.

Young, Gerald L. (1996) 'Interaction as Concept Basic to Human Ecology: An Exploration and Synthesis', in Lee Freese (ed.) *Advances in Human Ecology*, Vol. 5, pp. 157–211. Greenwich, CT: JAI Press.

12

Scopic Regimes of Modernity

Martin Jay

The modern era, it is often alleged,[1] has been dominated by the sense of sight in a way that set it apart from its premodern predecessors and possibly its postmodern successor. Beginning with the Renaissance and the scientific revolution, modernity has been normally considered resolutely ocularcentric. The invention of printing, according to the familiar argument of McLuhan and Ong,[2] reinforced the privileging of the visual abetted by such inventions as the telescope and the microscope. "The perceptual field thus constituted," concludes a typical account, "was fundamentally nonreflexive, visual and quantitative."[3]

Although the implied characterization of different eras in this generalization as more favorably inclined to other senses should not be taken at face value,[4] it is difficult to deny that the visual has been dominant in modern Western culture in a wide variety of ways. Whether we focus on "the mirror of nature" metaphor in philosophy with Richard Rorty or emphasize the prevalence of surveillance with Michel Foucault or bemoan the society of the spectacle with Guy Debord,[5] we confront again and again the ubiquity of vision as the master sense of the modern era.

But what precisely constitutes the visual culture of this era is not so readily apparent. Indeed, we might well ask, borrowing Christian Metz's term, is there one unified "scopic regime"[6] of the modern or are there several, perhaps competing ones? For, as Jacqueline Rose has recently reminded us, "our previous history is not the petrified block of a single visual space since, looked at obliquely, it can always be seen to contain its moment of unease."[7]

Source: Hal Foster (ed.), *Vision and Visuality* (Seattle: Bay Press, 1988), pp. 3–23.

In fact, may there possibly be several such moments, which can be discerned, if often in repressed form, in the modern era? If so, the scopic regime of modernity may best be understood as a contested terrain, rather than a harmoniously integrated complex of visual theories and practices. It may, in fact, be characterized by a differentiation of visual subcultures, whose separation has allowed us to understand the multiple implications of sight in ways that are now only beginning to be appreciated. That new understanding, I want to suggest, may well be the product of a radical reversal in the hierarchy of visual subcultures in the modern scopic regime.

Before spelling out the competing ocular fields in the modern era as I understand them, I want to make clear that I am presenting only very crude ideal typical characterizations, which can easily be faulted for their obvious distance from the complex realities they seek to approximate. I am also not suggesting that the three main visual subcultures I single out for special attention exhaust all those that might be discerned in the lengthy and loosely defined epoch we call modernity. But, as will soon become apparent, it will be challenging enough to try to do justice in the limited space I have to those I do want to highlight as most significant.

Let me begin by turning to what is normally claimed to be the dominant, even totally hegemonic, visual model of the modern era, that which we can identify with Renaissance notions of perspective in the visual arts and Cartesian ideas of subjective rationality in philosophy. For convenience, it can be called Cartesian perspectivalism. That it is often assumed to be equivalent to the modern scopic regime per se is illustrated by two remarks from prominent commentators. The first is the claim made by the art historian William Ivins, Jr., in his *Art and Geometry* of 1946 that "the history of art during the five hundred years that have elapsed since Alberti wrote has been little more than the story of the slow diffusion of his ideas through the artists and peoples of Europe."[8] The second is from Richard Rorty's widely discussed *Philosophy and the Mirror of Nature*, published in 1979; "in the Cartesian model the intellect *inspects* entities modeled on retinal images. . . . In Descartes' conception – the one that became the basis for 'modern' epistemology – it is *representations* which are in the 'mind.'"[9] The assumption expressed in these citations that Cartesian perspectivalism is *the* reigning visual model of modernity is often tied to the further contention that it succeeded in becoming so because it best expressed the "natural" experience of sight valorized bv the scientific world view. When the assumed equivalence between scientific observation and the natural world was disputed, so too was the domination of this visual subculture, a salient instance being Erwin Panofsky's celebrated critique of perspective as merely a conventional symbolic form.[10]

But for a very long time Cartesian perspectivalism was identified with the modern scopic regime *tout court*. With full awareness of the Schematic nature of what follows, let me try to establish its most important characteristics.

There is, of course, an immense literature on the discovery, rediscovery, or invention of perspective – all three terms are used depending on the writer's interpretation of ancient visual knowledge – in the Italian Quattrocento. Brunelleschi is traditionally accorded the honor of being its practical inventor or discoverer, while Alberti is almost universally acknowledged as its first theoretical interpreter. From Ivins, Panofsky, and Krautheimer to Edgerton, White, and Kubovy,[11] scholars have investigated virtually every aspect of the perspectivalist revolution, technical, aesthetic, psychological, religious, even economic and political.

Despite many still disputed issues, a rough consensus seems to have emerged around the following points. Growing out of the late medieval fascination with the metaphysical implications of light – light as divine *lux* rather than perceived *lumen* – linear perspective came to symbolize a harmony between the mathematical regularities in optics and God's will. Even after the religious underpinnings of this equation were eroded, the favorable connotations surrounding the allegedly objective optical order remained powerfully in place. These positive associations had been displaced from the objects, often religious in content, depicted in earlier painting to the spatial relations of the perspectival canvas themselves. This new concept of space was geometrically isotropic, rectilinear, abstract, and uniform. The *velo* or veil of threads Alberti used to depict it conventionalized that space in a way that anticipated the grids so characteristic of twentieth-century art, although, as Rosalind Krauss has reminded us, Alberti's veil was assumed to correspond to external reality in a way that its modernist successor did not.[12]

The three-dimensional, rationalized space of perspectival vision could be rendered on a two-dimensional surface by following all of the transformational rules spelled out in Alberti's *De Pittura* and later treatises by Viator, Dürer, and others. The basic device was the idea of symmetrical visual pyramids or cones with one of their apexes the receding vanishing or centric point in the painting, the other the eye of the painter or the beholder. The transparent window that was the canvas, in Alberti's famous metaphor, could also be understood as a flat mirror reflecting the geometricalized space of the scene depicted back onto the no less geometricalized space radiating out from the viewing eye.

Significantly, that eye was singular, rather than the two eyes of normal binocular vision. It was conceived in the manner of a lone eye looking through a peephole at the scene in front of it. Such an eye was, moreover, understood to be static, unblinking, and fixated, rather than dynamic, moving with what later scientists would call "saccadic" jumps from one focal point to another. In Norman Bryson's terms, it followed the logic of the Gaze rather than the Glance, thus producing a visual take that was eternalized, reduced to one "point of view," and disembodied. In what Bryson calls the "Founding Perception" of the Cartesian perspectivalist tradition,

> *the gaze of the painter arrests the flux of phenomena, contemplates the visual field from a vantage-point outside the mobility of duration, in an eternal moment of disclosed presence; while in the moment of viewing, the viewing subject unites his gaze with the Founding Perception, in a moment of perfect recreation of that first epiphany.*[13]

A number of implications followed from the adoption of this visual order. The abstract coldness of the perspectival gaze meant the withdrawal of the painter's emotional entanglement with the objects depicted in geometricalized space. The participatory involvement of more absorptive visual modes was diminished, if not entirely suppressed, as the gap between spectator and spectacle widened. The moment of erotic projection in vision – what St. Augustine had anxiously condemned as "ocular desire"[14] – was lost as the bodies of the painter and viewer were forgotten in the name of an allegedly disincarnated, absolute eye. Although such a gaze could, of course, still fall on objects of desire – think, for example, of the female nude in Dürer's famous print of a draftsman drawing her through a screen of perspectival threads[15] – it did so largely in the service of a reifying male look that turned its targets into stone. The marmoreal nude drained of its capacity to arouse desire was at least tendentially the outcome of this development. Despite important exceptions, such as Caravaggio's seductive boys or Titian's *Venus of Urbino*, the nudes themselves fail to look out at the viewer, radiating no erotic energy in the other direction. Only much later in the history of Western art, with the brazenly shocking nudes in Manet's *Déjeuner sur l'herbe* and *Olympia*, did the crossing of the viewer's gaze with that of the subject finally occur. By then the rationalized visual order of Cartesian perspectivalism was already coming under attack in other ways as well.

In addition to its de-croticizing of the visual order, it had also fostered what might be called de-narrativization or de-textualization. That is, as abstract, quantitatively conceptualized space became more interesting to the artist than the qualitatively differentiated subjects painted within it, the rendering of the scene became an end in itself. Alberti, to be sure, had emphasized the use of perspective to depict *istoria*, ennobling stories, but in time they seemed less important than the visual skill shown in depicting them. Thus the abstraction of artistic form from any substantive content, which is part of the clichéd history of twentieth-century modernism, was already prepared by the perspectival revolution five centuries earlier. What Bryson in his book *Word and Image* calls the diminution of the discursive function of painting, its telling a story to the unlettered masses, in favor of its figural function,[16] meant the increasing autonomy of the image from any extrinsic purpose, religious or otherwise. The effect of realism was consequently enhanced as canvases were filled with more and more information that seemed unrelated to any narrative or textual function. Cartesian perspectivalism was thus in league with a scientific world view that no longer hermeneutically read the world as a divine text, but rather saw it as situated in a mathematically regular spatio-temporal order filled with natural objects that could only be observed from without by the dispassionate eye of the neutral researcher.

It was also complicitous, so many commentators have claimed, with the fundamentally bourgeois ethic of the modern world. According to Edgerton, Florentine businessmen with their newly invented technique of double-entry bookkeeping may have been "more and more disposed to a visual order that would accord with the tidy principles of mathematical order that they applied to their bank ledgers."[17] John Berger goes so far as to claim that more appropriate than the Albertian metaphor of the window on the world is that of "a safe let into a wall, a safe in which the visible has been deposited."[18] It was, he contends, no accident that the invention (or rediscovery) of perspective virtually coincided with the emergence of the oil painting detached from its context and available for buying and selling. Separate from the painter and the viewer, the visual field depicted on the other side of the canvas could become a portable commodity able to enter the circulation of capitalist exchange. At the same time, if philosophers like Martin Heidegger are correct, the natural world was transformed through the technological world view into a "standing reserve" for the surveillance and manipulation of a dominating subject.[19]

Cartesian perspectivalism has, in fact, been the target of a widespread philosophical critique, which has denounced its privileging of an ahistorical, disinterested, disembodied subject entirely outside of the world it claims to know only from afar. The questionable assumption of a transcendental subjectivity characteristic of universalist humanism, which ignores our embeddedness in what Maurice Merleau-Ponty liked to call the flesh of the world, is thus tied to the "high altitude" thinking characteristic of this scopic regime. In many accounts, this entire tradition has thus been subjected to wholesale condemnation as both false and pernicious.

Looked at more closely, however, it is possible to discern internal tensions in Cartesian perspectivalism itself that suggest it was not quite as uniformly coercive as is sometimes assumed. Thus, for example, John White distinguishes between what he terms "artificial perspective," in which the mirror held up to nature is flat, and "synthetic perspective," in which that mirror is presumed to be concave, thus producing a curved rather than planar space on the canvas. Here, according to White, Paolo Uccello and Leonardo da Vinci were the major innovators, offering a "spherical space which is homogeneous, but by no means simple, and which possesses some of the qualities of Einstein's finite infinity."[20] Although artificial perspective was the dominant model, its competitor was never entirely forgotten.

Michael Kubovy has recently added the observation that what he calls the "robustness of perspective"[21] meant that Renaissance canvases could be successfully viewed from more than the imagined apex of the beholder's visual pyramid. He criticizes those who naively identify the rules of perspective established by its theoretical champions with the actual practice of the artists themselves. Rather than a procrustean bed, they were practically subordinated to the exigencies of perception, which means that denunciations of their failings are often directed at a straw man (or at least his straw eye).

Equally problematic is the subject position in the Cartesian perspectivalist epistemology. For the monocular eye at the apex of beholder's pyramid, could be construed as transcendental and universal – that is, exactly the same for any human viewer occupying the same point in time and space – or contingent – solely dependent on the particular, individual vision of distinct beholders, with their own concrete relations to the scene in front of them. When the former was explicitly transformed into the latter, the relativistic implications of perspectivalism could be easily drawn. Even in the nineteenth century, this potential was apparent to thinkers like Leibniz, although he generally sought to escape its more troubling implications. These were not explicitly stressed and than praised until the late nineteenth century by such thinkers as Nietzsche. If everyone had his or her own camera obscura with a distinctly different peephole, he gleefully concluded, then no transcendental world view was possible.[22]

Finally, the Cartesian perspectivalist tradition contained a potential for internal contestation in the possible uncoupling of the painter's view of the scene from that of the presumed beholder. Interestingly, Bryson identifies this development with Vermeer, who represents for him a second state of perspectivalism even more disincarnated than that of Alberti. "The bond with the viewer's physique is broken and the viewing subject," he writes, "is now proposed and assumed, as a notional point, a non-empirical Gaze."[23]

What makes this last observation so suggestive is the opening it provides for a consideration of an alternative scopic regime that may be understood as more than a subvariant of Cartesian perspectivalism. Although I cannot pretend to be a serious student of Vermeer able to quarrel with Bryson's interpretation of his work, it might be useful to situate the painter in a different context from the one we have been discussing. That is, we might include him and the Dutch seventeenth-century art of which he was so great an exemplar in a visual culture very different from that we associate with Renaissance perspective, one which Svetlana Alpers has recently called *The Art of Describing*.[24]

According to Alpers, the hegemonic role of Italian painting in art history has occluded an appreciation of a second tradition, which flourished in the seventeenth-century Low Countries. Borrowing Georg Lukács's distinction between narration and description, which he used to contrast realist and naturalist fiction, she argues that Italian Renaissance art, for all its fascination with the techniques of perspective, still held fast to the storytelling function for which they were used. In the Renaissance, the world on the other side of Alberti's window, she writes, "was a stage in which human figures performed significant actions based on the texts of the poets. It is a narrative art."[25] Northern art, in contrast, suppresses narrative and textual reference in favor of description and visual surface. Rejecting the privileged, constitutive role of the monocular subject, it emphasizes instead the prior existence of a world of objects depicted on the flat canvas, a world indifferent to the beholder's position in front of it. This world, moreover, is not contained entirely within the frame of the Albertian window, but seems instead to extend beyond it. Frames do exist around Dutch pictures, but they are arbitrary and without the totalizing function they serve in Southern art. If there is a model for Dutch art, it is the map with its unapologetically flat surface and its willingness to include words as well as objects in its visual space. Summarizing the difference between the art of describing and Cartesian perspectivalism, Alpers posits the following oppositions:

> *attention to many small things versus a few large ones; light reflected off objects versus objects modeled by light and shadow; the surface of objects, their colors and textures, dealt with rather than their placement in a legible space; an unframed image versus one that is clearly framed; one with no clearly situated viewer compared to one with such a viewer. The distinction follows a hierarchical model of distinguishing between phenomena commonly referred to as primary and secondary: objects and space versus the surfaces, forms versus the textures of the world.*[26]

If there is a philosophical correlate to Northern art, it is not Cartesianism with its faith in a geometricalized, rationalized, essentially intellectual concept of space but rather the more empirical visual experience of observationally oriented Baconian empiricism. In the Dutch context Alpers identifies it with Constants Huygens. The nonmathematical impulse of this tradition accords well with the indifference to hierarchy, proportion, and analogical resemblances characteristic of Cartesian perspectivalism. Instead, it casts its attentive eye on the fragmentary, detailed, and richly articulated surface of a world it is content to describe rather than explain. Like the microscopist of the seventeenth century – Leeuwenhoeck is her prime example – Dutch art savors the discrete particularity of visual experience and resists the temptation to allegorize or typologize what it sees, a temptation to which she claims Southern art readily succumbs.

In two significant ways, the art of describing can be said to have anticipated later visual models, however much it was subordinated to its Cartesian perspectivalist rival. As we have already noted, a direct filiation between Alberti's *velo* and the grids of modernist art is problematic because, as Rosalind Krauss has argued, the former assumed a three-dimensional world out there in nature, whereas the latter did not. A more likely predecessor can thus be located in the Dutch art based on the mapping impulse. As Alpers notes,

> *Although the grid that Ptolemy proposed, and those that Mercator later imposed, share the mathematical uniformity of the Renaissance perspective grid, they do not share the positioned viewer, the frame, and the definition of the picture as a window through which an external viewer looks. On these accounts the Ptolemaic grid, indeed cartographic grids in general, must be distinguished from, not confused with, the perspectival grid. The projection is, one might say, viewed from nowhere. Nor is it to be looked through. It assumes a flat working surface.*[27]

Secondly, the art of describing also anticipates the visual experience produced by the nineteenth-century invention of photography. Both share a number of salient features: "fragmentariness, arbitrary frames, the immediacy that the first practitioners expressed by claiming that the photograph gave Nature the power to reproduce herself directly unaided by man."[28] The parallel frequently drawn between photography and the anti-perspectivalism of impressionist art, made for example by Aaron Scharf in his discussion of Degas,[29] should thus be extended to include the Dutch art of the seventeenth century. And if Peter Galassi is correct in *Before Photography*, there was also a tradition of topographical painting – landscape sketches of a fragment of reality – that resisted Cartesian perspectivalism and thus prepared the way both for photography and the impressionist return to two-dimensional canvases.[30] How widespread or self-consciously oppositional such a tradition was I will leave to experts in art history to decide. What is important for our purposes is simply to register the existence of an alternative scopic regime even during the heyday of the dominant tradition.

Alpers's attempt to characterize it is, of course, open to possible criticisms. The strong opposition between narration and description she posits may seem less firm it we recall the de-narrativizing impulse in perspectival art itself mentioned above. And if we can detect a certain fit between the exchange principle of capitalism and the abstract relational space of perspective, we might also discern a complementary fit between the valorization of material surfaces in Dutch art and the fetishism of commodities no less characteristic of a market economy. In this sense, both scopic regimes can be said to reveal different aspects of a complex but unified phenomenon, just as Cartesian and Baconian philosophies can be said to be consonant, if in different ways, with the scientific world view.

If, however, we turn to a third model of vision, or what can be called the second moment of unease in the dominant model, the possibilities for an even more radical alternative can be discerned. This third model is perhaps best identified with the baroque. At least as eatly as 1888 and Heinrich Wöfflin's epochal study, *Renaissance and Baroque*, art historians have been tempted to postulate a perennial oscillation between two styles in both painting and architecture.[31] In opposition to the lucid, linear, solid, fixed, planimetric, closed form of the Renaissance, or as Wölfflin later called it, the classical style, the baroque was painterly, recessional, soft-focused, multiple, and open. Derived, at least according to one standard etymology, from the Portuguese word for an irregular, oddly shaped pearl, the baroque connoted the bizarre and peculiar, traits which were normally disdained by champions of clarity and transparency of form.

Although it may be prudent to confine the baroque largely to the seventeenth century and link it with the Catholic Counter Reformation or the manipulation of popular culture by the newly ascendant absolutist state – as has, for example, the Spanish historian José Antonio Maravall[32] – it may also be possible to see it as a permanent, if often repressed, visual possibility throughout the entire modern era. In the recent work of the French philosopher Christine Buci-Glucksmann, *La raison baroque* of 1984 and *La folie du voir* of 1986,[33] it is precisely the explosive power of baroque vision that is seen as the most significant alternative to the hegemonic visual style we have called Cartesian perspectivalism. Celebrating the dazzling, disorienting, ecstatic surplus of images in baroque visual experience, she emphasizes its rejection of the monocular geometricalization of the Cartesian tradition, with its illusion of homogeneous three-dimensional space seen with a God's-eye-view from afar. She also tacitly contrasts the Dutch art of describing, with its belief in legible surfaces and faith in the material solidity of the world its paintings map, with the baroque fascination for opacity, unreadability, and the indecipherability of the reality it depicts.

For Buci-Glucksmann, the baroque self-consciously revels in the contradictions between surface and depth, disparaging as a result any attempt to reduce the multiplicity of visual spaces into any one coherent essence. Significantly, the mirror that it holds up to nature is not the flat reflecting glass that commentators like Edgerton and White see as vital in the development of rationalized or "analytic" perspective, but rather the anamorphosistic mirror, either concave or convex, that distorts the visual image – or, more precisely, reveals the conventional rather than natural quality of "normal" specularity by showing its dependence on the materiality of the medium of reflection. In fact, because of its greater awareness of that materiality – what a recent commentator, Rodolphe Gasché, has drawn attention to as the "tain of the mirror"[34] – baroque visual experience has a strongly tactile or haptic quality, which prevents it from turning into the absolute ocularcentrism of its Cartesian perspectivalist rival.

In philosophical terms, although no one system can be seen as its correlate, Leibniz's pluralism of monadic viewpoints,[35] Pascal's meditations on paradox, and the Counter Reformation mystics' submission to vertiginous experiences of rapture might all be seen as related to baroque vision. Moreover, the philosophy it favored self-consciously eschewed the model of intellectual clarity expressed in a literal language purified of ambiguity. Instead, it recognized the inextricability of rhetoric and vision, which meant that images were signs and that concepts always contained an irreducibly imagistic component.

Baroque vision, Buci-Glucksmann also suggests, sought to represent the unrepresentable and, necessarily failing, produced the melancholy that Walter Benjamin in particular saw as characteristic of the baroque sensibility. As such, it was closer to what a long tradition of aesthetics called the sublime, in contrast to the beautiful, because of its yearning for a presence that can never be fulfilled. Indeed, desire, in its erotic as well as metaphysical forms, courses through the baroque scopic regime. The body returns to dethrone the disinterested gaze of the disincarnated Cartesian spectator. But unlike the return of the body celebrated in such twentieth-century philosophies of vision as Merleau-Ponty's, with its dream of meaning-laden imbrication of the viewer and the viewed in the flesh of the world, here it generates only allegories of obscurity and opacity. Thus it truly produces one of those "moments of unease" which Jacqueline Rose sees challenging the petrification of the dominant visual order (the art of describing seeming in fact far more at case in the world).

A great deal more might be said about these three ideal typical visual cultures, but let me conclude by offering a few speculations, if I can use so visual a term, on their current status. First, it seems undeniable that we have witnessed in the twentieth century a remarkable challenge to the hierarchical order of the three regimes. Although it would be foolish to claim that Cartesian perspectivalism has been driven from the field, the extent to which it has been denaturalized and vigorously contested, in philosophy as well as in the visual arts, is truly remarkable. The rise of hermeneutics, the return of pragmatism, the profusion of linguistically oriented structuralist and poststructuralist modes of thought have all put the epistemological tradition derived largely from Descartes very much on the defensive. And, of course, the alternative of Baconian observation, which periodically resurfaces in variants of positivistic thought, has been no less vulnerable to attack, although one might argue that the visual practice with which it had an elective affinity has shown remarkable resilience with the growing status of photography as a nonperspectival art form (or, if you prefer, counter-art form). There are as well contemporary artists like the German Jewish, now Israeli painter Joshua Neustein, whose fascination with the flat materiality of maps has recently earned a comparison with Alpers's seventeenth-century Dutchmen.[36]

Still, if one had to single out the scopic regime that has finally come into its own in our time, it would be the "madness of vision" Buci-Glucksmann identifies with the baroque. Even photography, if Rosalind Krauss's recent work on the Surrealists is any indication,[37] can lend itself to purposes more in line with this visual impulse than the art of mere describing. In the postmodern discourse that elevates the sublime to a position of superiority over the beautiful, it is surely the "palimpsests of the unseeable,"[38] as Buci-Glucksmann calls baroque vision, that seem most compelling. And if we add the current imperative to restore rhetoric to its rightful place and accept the irreducible linguistic moment in vision and the equally insistent visual moment in language, the timeliness of the baroque alternative once again seems obvious.

In fact, if I may conclude on a somewhat perverse note, the radical dethroning of Cartesian perspectivalism may have gone a bit too far. In our haste to denaturalize it and debunk its claims to represent vision per se, we may be tempted to forget that the other scopic regimes I have quickly sketched are themselves no more natural or closer to a "true" vision. Glancing is not somehow innately superior to gazing; vision hostage to desire is not necessarily always better than casting a cold eye; a sight from the situated context of a body in the world may not always see things that are visible to a "high-altitude" or "God's-eye-view." However we may regret the excesses of scientism, the Western scientific tradition may have only been made possible by Cartesian perspectivalism or its complement, the Baconian art of describing. There may well have been some link between the absence of such scopic regimes in Eastern cultures, especially the former, and their general lack of indigenous scientific revolutions. In our scramble to scrap the rationalization of sight as a pernicious reification of visual fluidity, we need to ask what the costs of too uncritical an embrace of its alternatives may be. In the case of the art of describing, we might see another reification at work, that which makes a fetish of the material surface instead of the three-dimensional depths. Lukács's critique of naturalist description in literature, unmentioned by Alpers, might be applied to painting as well. In the case of baroque vision, we might wonder about the celebration of ocular madness, which may produce ecstasy in some, but bewilderment and confusion in others. As historians like Maravall have darkly warned, the phantasmagoria of baroque spectacle was easily used to manipulate those who were subjected to it. The current vision of "the culture industry," to use the term Maravall borrows from Horkheimer and Adorno in his account of the seventeenth century, does not seem very threatened by postmodernist visual experiments in "la folie du voir." In fact, the opposite may well be the case.

Rather than erect another hierarchy, it may therefore be more useful to acknowledge the plurality of scopic regimes now available to us. Rather than demonize one or another, it may be less dangerous to explore the implications, both positive and negative, of each. In so doing, we won't lose entirely the sense of unease that has so long haunted the visual culture of the West, but we may learn to see the virtues of differentiated ocular experiences. We may learn to wean ourselves from the fiction of a "true" vision and revel instead in the possibilities opened up by the scopic regimes we have already invented and the ones, now so hard to envision, that are doubtless to come.

Notes

1. See, for example, Lucien Febvre, *The Problem of Unbelief in the Sixteenth Century: The Religion of Robelais*, trans. Beatrice Gottlieb (Cambridge, Mass.: Harvard University Press, 1982) and Robert Mandrou, *Introduction to Modern France, 1500–1640: An*

Essay in Historical Psychology, trans. R. E. Hallmark (New York: Holmes & Meier, 1975).

2. Marshall McLuhan, *Understanding Media: The Extensions of Man* (New York: McGraw-Hill, 1964); Walter J. Ong, *The Presence of the Word: Some Prolegomena for Cultural and Religious History* (New Haven: Yale University Press, 1967); see also Elizabeth I. Eisenstein, *The Printing Press as an Agent of Change: Communications and Cultural Transformations in Early Modern Europe*, 2 vols. (Cambridge: Cambridge University Press, 1979).
3. Donald M. Lowe, *History of Bourgeois Perception* (Chicago: University of Chicago Press, 1982), p. 26.
4. For an account of the positive attitude towards vision in the medieval church, see Margaret R. Miles, *Image as Insight: Visual Understanding in Western Christianity and Secular Culture* (Boston: Beacon Press, 1985). Contrary to the argument of Febvre and Mandrou, which has been very influential, she shows the extent to which sight was by no means widely demeaned in the Middle Ages.
5. Richard Rorty, *Philosophy and the Mirror of Nature* (Princeton: Princeton University, Press, 1979); Michel Foucault, *Discipline and Punish: The Birth of the Prison*, trans. Alan Sheridan (New York: Vintage Books, 1979); Guy Debord, *Society of the Spectacle*, rev. ed. (Detroit: Black and Red, 1977).
6. Christian Metz, *The Imaginary Signifier: Psychoanalysis and the Cinema*, trans. Celia Britton et al. (Bloomington: Indiana University Press, 1982), p. 61.
7. Jacqueline Rose, *Sexuality in the Field of Vision* (London: Verso, 1986), pp. 232–23.
8. William M. Ivins, Jr., *Art and Geometry: A Study in Space Intuitions* (Cambridge, Mass.: Harvard University Press, 1946), p. 81.
9. Rorty, p. 45.
10. Erwin Panofsky, "Die Perspektive als 'symolischen Form,'" *Vorträge der Bibliothek Warburg* 4 (1924–1925): 258–331.
11. William M. Ivins, Jr., *On the Rotionalization of Sight* (New York: Metropolitan Museum of Art, 1938); Panofsky, "Die Perspektive als 'symbolischen Form'"; Richard Krautheimer, "Brunelleschi and Linear Perspective," in *Brunelleschi in Perspective*, comp. Isabelle Hyman (Englewood Cliffs, N.J.: Prentice-Hall, 1974); Samuel Y. Edgerton, Jr., *The Renaissance Discovery of Linear Perspective* (New York: Basic Books, 1975); John White, *The Birth and Rebirth of Pictorial Space*, 3rd ed. (Cambridge, Mass.: Belknap Press, 1987); Michael Kubovy, *The Psychology of Perspective and Renaissance Art* (Cambridge: Cambridge University Press, 1986).
12. Rosalind F. Krauss, *The Originality of the Avant-Garde and Other Modernist Myths* (Cambridge, Mass.: The MIT Press, 1985), p. 10.
13. Norman Bryson, *Vision and Painting: The Logic of the Gaze* (New Haven: Yale University Press, 1983), p. 94.
14. Augustine discusses ocular desire in Chapter 35 of the *Confessions*.
15. For a discussion of the gender implications of this work, see Svetlana Alpers, "Art History and its Exclusions," in *Feminism and Art History: Questioning the Litany*. ed. Norma Broude and Mary D. Garrard (New York: Harper and Row, 1982), p. 187.
16. Norman Bryson, *Word and Image: French Painting of the Ancient Régime* (Cambridge: Cambridge University Press, 1981), chapter 1.
17. Edgerton, p. 39.
18. John Berger, *Ways of Seeing* (London: BBC, 1972), p. 109.
19. Martin Heidegger, "The Question Concerning Technology," in *The Question Concerning Technology, and Other Essays*, trans. William Lovitt (New York: Harper and Row 1977). Heidegger's most extensive critique of Cartesian perspectivalism can be found in his essay "The Age of the World Picture," in the same volume.
20. White, p. 208.

21. Kubovy, chapter IV.
22. Sarah Kofman, *Camera Obscura, de l'idéologie* (Paris: Éditions Galilée, 1973), treats this theme in Nietzsche.
23. Bryson, *Vision and Painting*, p. 112.
24. Svetlana Alpers, *The Art of Describing: Dutch Art in the Seventeenth Century* (Chicago: University of Chicago Press, 1983).
25. Ibid., p. XIX.
26. Ibid., p. 44.
27. Ibid., p. 138.
28. Ibid., p. 43.
29. Aaron Scharf, *Art and Photography* (London: Allen Lane, 1968; reprint, New York: Penguin Books, 1986), chapter VIII.
30. Peter Galassi, *Before Photography: Painting and the Invention of Photography* (New York: Museum of Modern Art, 1981).
31. Heinrich Wölfflin, *Renaissance and Baroque*, trans. Kathrin Simon (Ithaca, N.Y.: Cornell University Press, 1966). See also the systematic development of the contrast in *Principles of Art History: The Problem of the Development of Style in Later Art*, trans. M. D. Hottinger (London: G. Bell & Sons. Ltd., 1932).
32. José Antonio Maravall, *Culture of the Baroque: Analysis of a Historical Structure*. trans. Terry Cochran (Minneapolis: University of Minnesota Press, 1986).
33. Christine Buci-Glucksmann, *la raison haroque: de Baudelaire à Benjamin* (Paris: Éditions Galilée, 1984) and *La folic du voir: de l'esthétique baroque* (Paris Éditions Galilée, 1986).
34. Rodolphe Gasché, *The Tain of the Mirror: Derrida and the Philosophy of Reflection* (Cambridge, Mass.: Harvard University Press, 1986).
35. As Buci-Glucksman recognizes, Leibnizian pluralism retains a faith in the harmonizing of perspectives that is absent from the more radically Nietzschean impulse in the baroque. See *La folie du voir*, p. 80, where she identifies that impulse with Gracián and Pascal.
36. See Irit Rogoff, "Mapping Out Strategies of Dislocation," in the catalogue for Neustein's October 24–November 26, 1987 show at the Exit Art gallery in New York.
37. Krauss, "The Photographic Conditions of Surrealism," in *The Originality of the Avant-Garde*. See also her work with Jane Livingston, *l'amour fou: Photography and Surrealism* (New York: Abbeville Press, 1985).
38. Buci-Glucksmann, *La folie du voir*, chapter VI.

13

Foucault's Art of Seeing

John Rajchman

In his book called *Foucault*, Gilles Deleuze says of Michel Foucault that he was a great seer, a *voyant*. He declares that Foucault's seeing, and his discussion of seeing, are a constant and central part not only of his histories but also of his thought. He says that those who fail to take this part of his thought into account "mutilate" it to the point where it becomes comparable to analytic philosophy, something "with which it does not have much in common."[1]

Deleuze attributes many things to the visual part of Foucault's thought. The territory of the visual spans knowledge, art, ethics, and politics, and so it illustrates why Foucault had no difficulties in dealing with "the relations of science and literature, or the imaginary and the scientific, or the known and the lived."[2] The visual is also central to the way Foucault's thought would develop. It is the other component, along with "discourse," of what Deleuze sees as Foucault's "neo-Kantianism," and so it is linked to the theme of the "transcendental imagination" in Kant, and to the attempts on the part of Merleau-Ponty and Heidegger to go beyond intentionality to the "opening" of Being. But Deleuze also applies to Foucault the categories of the Danish semiologist Louis Hjelmslev that he had found useful in his study of film. He says that Foucault was a great "audiovisual" thinker, who was "singularly close to contemporary film."[3]

I think Deleuze is the first to "see" this side of Foucault's thought and to demonstrate its general importance in his work.[4] I will not follow all the intricate paths Deleuze gathers together in his analysis. I will try to present what I think Deleuze had in mind in somewhat different terms. I start with Foucault's art of historical depiction.

Source: *October*, 44 (1988): 88–117.

Wood engraving of a ship of fools from a 1498 Latin translation of Sebastian Brant's work.

Historical Pictures

Foucault was an exceedingly *visual* historian. His histories are filled with vivid pictures that stick in the mind. Visualizing events or historical depiction is, of course, an art which itself has a history. Events have not always been visualized in the same way or under the same description. Michelet might be one example. So would a whole aspect of the "new history" with which Foucault associates his work in the Introduction to the *Archeology*, where an attempt would be made to "turn documents into monuments" – the preoccupation of the new historians with the "spaces" in which people lived, and the reconstruction of *tableaux de moeurs* – the sort of thing useful in making "period films."[5] But Foucault's pictures are more than such *tableaux*. They are puzzles that call for analysis. They form part of a philosophical exercise in which seeing has a part.

A frequent device in Foucault's writing is before-and-after pictures. One is shown a picture from one period and then one from another. Thus the question of how one passed from one system of thought to another is visualized. The device occurs throughout, but is particularly prominent in the two "birth" books, the birth of the prison and the birth of the clinic.

In *Discipline and Punish*, one is shown the picture of the excruciating execution of Damiens, regicide, and then a timetable of observed activities. In the *Birth of the Clinic* one is shown Pomme's bathing cure of a hysteric in which the "heat" of her nervous system is "dried out." And then one is shown Bayle's careful examination of the lesions in the brain, that "dingy-looking pulp."

In both cases we have pictures not simply of what things looked like, but how things were *made* visible, how things were *given* to be seen, how things were "*shown*" to knowledge or to power – two ways in which things became *seeable*. In the case of the prison, it is a question of two ways crime was made visible in the body, through "spectacle" or through "surveillance." In the case of the clinic, it is a question of two ways of organizing "the space in which bodies and eyes meet." With Bayle, the eye acquires depth, and the body, volume; in examining the brain he is looking into the depths of the individual body where disease is located. Pomme was still looking for that general "portrait" of the disease which allows for the classification of fevers.

In both instances Foucault links the two techniques of making things visible to a larger conception of seeing in the period. This is one theme in Deleuze – what he calls *visibilités*. There is a history not simply of what was seen, but of what could be seen, of what was *seeable*, or visible. A "visualization," a scheme through which things are given to be seen, belongs to the "positivity" of knowledge and power of a time and place.

But there is a second feature of Foucault's before-and-after pictures: the one which proposes a philosophical exercise in seeing. For, at the end of the analysis of the passage from before to after, one is led to "see" the depicted events in a new light, or in a different way – in the light of their underlying, unseen concepts. Thus, after reading *Discipline and Punish*, it is hard not to "see" annular prison construction in a new light, hard not to be surprised that "prisons resemble factories, schools, barracks, hospitals, which all resemble prisons."[6] This is the aspect of Foucault's historical depiction Deleuze calls *évidences*. Foucault would be a "seer" because of the way "visibility" and "evidence" are linked in his history, and in his thought.

Visibility

Foucault finds in what he calls "the classical age" a whole range of ways of seeing, and of letting things be seen, which would have been unthinkable in the preceding period in which the eye was linked to the ear in the deciphering of "resemblances": in the classificatory tables of its forms of knowledge; in the primacy it accords to perceptual evidence; in its conception of madness as "dazzled reason"; in its conception of painting; in its utopian literature of the transparent society; in its natural histories as well as in its way of "displaying" mad people, so different from the "ship of fools." Foucault tried to determine the deep conceptual organization which gathered these seeings together into a form of "visibility" different from others.

Foucault's hypothesis was that there exists a sort of "positive unconscious" of vision which determines not what is seen, but what *can* be seen. His idea is that not all ways of visualizing or rendering visible are possible at once. A period only lets some things be seen and not others. It "illuminates" some things and so casts others in the shade. There is much more regularity, much more *constraint*, in what we can see than we suppose. To see is always to think, since what is seeable is part of what "structures thought in advance." And conversely to think is always to see.

What makes the visual intelligible is itself unseen. It is an anonymous body of practice spread out in different places. As Deleuze puts it, "*visibilités* are neither the acts of a seeing subject nor the givens of a visual sense."[7] In the *Archeology of Knowledge* Foucault discusses "enunciative modalities" as properties of discourse. But in his histories he also discusses "modalities of seeing" as properties of visual intelligence: who sees what or whom and where are integral features of the visual thinking of a period and not an independent fact about its contexts. And this visual thought is rooted in a specific sort of "material existence" – the spaces in which it is exercised (such as hospital, prison, museum or home), and the techniques through which its images are reproduced and circulated (such as printing, markets, and so forth).

In one sense, it is "the subject" which is given in the forms of "visibility." Foucault finds that the same organization a period assigns to inner or psychological processes recurs in external "public" ones such as making maps or illustrating scholarly works. Thus the scheme of Renaissance resemblances is placed, by the classical period, within the imagination as a source of error to be cleared up by proper observation. And, with the birth of the clinic, the "visionary space" in which disease had been discussed is put into the head of the patient. Visualization belongs to the great internalizing or psychologizing practices Foucault associates with modernity. Thus, Freud's idea of dreaming as a way of showing to oneself one's innermost desires, contrasts with Artemidorus's scheme where dreams are ways of making visible one's fortune in a hierarchical society.[8]

The visual thinking of a period would thus have a positive organization. But that organization is not rooted in keeping something concealed. As Foucault came to realize, the "classical" way of making madness visible was not based on the repression or concealment of the true way of seeing it. The conceptual scheme that determines what can be seen is, in the phrase of the *Archeology*, "invisible but not hidden." The visibility of a period may be invisible to it, but not as something hidden or kept from sight. What is invisible is just the light which illuminates things or makes them visible.

In short, visibility is a matter of a positive, material, anonymous body of practice. Its existence shows that we are much less free in what we see than we think, for we do not see the constraints of thought in what we can see. But it also shows that we are much *more* free than we think, since the element of visibility is also something that opens seeing to historical change or transformation.[9] That is the problem of *évidence*.

Evidence

Evidence, in both English and French, comes from *videre*, to see. In the course of its history, the word acquires the senses of proof, testimony, and clarity or indubitability to the mind. Ian Hacking has studied one change in the concept of evidence in his history of the "emergence of probability."[10] It is the one that made Hume possible.

There is one sense of the French word *évidence* which is particularly marked in Foucault's visual idiom – the one English translates as "self-evidence": what is taken for granted or accepted without question. Foucault introduces this concept of evidence into his historical depiction in a new way.

In a discussion with historians, Foucault explains that his way of seeing the birth of the prison was an attempt to see "events" behind self-evident entities and continuities, and so to "event-alize" history. One starts with a *rupture d'évidence*, a break with self-evidence, with "those *évidences* on which our knowledges, our agreements, our practices, rest."[11] And then one asks how such "*évidences*" arose and took form. Where there is self-evidence, one tries to uncover the singular formation of an event unseen.

Evidence is used in this sense in both birth books. Thus, in the *Birth of the Clinic*, Foucault says that "the exact superposition of the 'body' of disease and the body of the sick man . . . is *self-evident* only for us."[12] And, in *Discipline and Punish* he refers to the "self-evident character (*le caractère d'évidence*) that the prison form soon assumed."[13] There is both a legal and an economic self-evidence to the prison form; together they explain why, despite the fact that the prison was not doing what it was meant to do, "one could not 'see' how to replace it."[14]

"Seeing" in this sense is part of doing. We cannot see what to do because we are "prisoners" of the self-evidence of one *way* of seeing what to do. We participate, we do our bit, in the practices which make that way of seeing self-evident to us – a participation or acceptance we can refuse. Thus in Foucault's idiom, *évidence* is related to the *acceptability* of a practice. It is part of what makes a "strategy of power" *tolerable*, despite its difficulties. Thus, to see the events through which things become self-evident is to be able to see in what ways they may be *intolerable* or *unacceptable*. It is to try to see how we might *act* on what cannot yet be seen in what we do. It is, in short, a "critical" art, and it is in exercising it that Foucault would be, in Deleuze's term, a seer or *voyant*.

For Deleuze, a seer is not basically, nor in the first instance, someone who can depict future events. Nor is he necessarily the sort of "visionary" or "utopian" who looks forward to the place where everything that ought to be is finally made transparent to all.[15] "A seer," says Deleuze, "is someone who sees something not seen."[16] Foucault's art of seeing is an art of exposing the unseen *évidences* that make the things we in fact do acceptable or tolerable to us.

Deleuze finds this sort of seeing in the work Foucault did for the Group for Information about Prisons, of which Deleuze was a member. A kind of "public space" of discussion with prisoners was opened up. It was in this space that Foucault "saw something which at the time no one else saw." This act of seeing required a *rupture d'évidence*: the gap between the self-evidence of the economical and legal conception of the prison, and what was actually going on. In this gap one could start to see something intolerable in those practices, which opened up a question for historical analysis: an analysis that would initiate new ways of seeing and thinking not simply about French penal institutions, but also about the strategic organization of power in modern societies, its relations to forms of knowledge and modes of living. Foucault's seeing would lie in this critical opening in thinking. Says Deleuze:

> He saw things, and like all people who know how to see something and see it deeply, he found what he saw to be intolerable. For him to think meant to react to the intolerable, to something intolerable that he had seen. And the intolerable was never the visible, it was something more. . . .[17]

One sense of "evidence" in the study of history is the sense of the "eye-witness" to actual events as distinct from the "eye" that reads forthcoming events. Foucault's idea of the events of evidence has to do with the eye of *historians*. In *fictional* depiction, Foucault found a similar aim of making visible the unseen spaces of seeing. In Maurice Blanchot's depiction of the "space" in which encounters transpire and words are exchanged, he saw, in fiction, an attempt

> not to show (*faire voir*) the invisible, but to show the extent to which the invisibility of the visible is invisible. Hence [fiction] bears a profound kinship with space. . . .[18]

In several interviews, Foucault also describes his *own* histories as fiction. It is not that these histories lack the validity that would distinguish them from fiction. It is rather that they share an *aim* with fiction: the aim not of explanation, or of showing how our ways of seeing and doing are historically necessitated, but, on the contrary, of showing how things might be otherwise, beyond our self-evidences. That is why the history of the "evidences" of the way things are seen includes the "evidences" in the thinking of historians. To "see" is to open history to new domains and new questions, "to do this history of the 'objectification' of those events historians take as objectively given."[19] When Foucault says that he writes works of history that are more than works of a historian,[20] it is in part because of this other aim of seeing, which the philosopher would share with the writer.

Seeing is important in Foucault's work as philosopher and historian in this sense as an art of trying to see what is unthought in our seeing, and to open as yet unseen ways of seeing. A peculiar idiom of space and sight unfolds in Foucault's writing that moves in many directions. I would now like to bring various facets of this idiom into focus, to look at seeing in knowing, seeing in doing, seeing in thinking, and in living.

Seeing in Knowing

Foucault did not see knowledge as simply built up from ordinary perceptual evidence through a logic of inference, inductive or deductive. He was concerned with the ways seeing in knowledge has been itself conceptually constructed. In his idiom a *savoir* requires, and sets up, a way of *spatializing* itself, a sort of "technology of the visible." Foucault wanted to get away from what might be called our modern philosophical obsession with what we call "observation" in knowledge – a piece of philosophical "self-evidence" he found in different forms in both phenomenology and positivism. Our philosophical conception of observation is a recent one, and it prevents us from seeing how knowledge is *in fact* "spatialized" or "visualized." In science, seeing is more than meets the eye.

"Ocular metaphors," it has often been observed, occupy a central place in our vocabulary of knowing: truth is something we say we see. But these metaphors have not always worked in the same way. Foucault thought the changes were in part due to the *actual* ways people invented to "spatialize" their knowledge, the *actual* role of seeing in knowledge. Richard Rorty reviews the analytic literature which shows how the Cartesian idea of perception differs from the "hylomorphic" seeing in Scholastic thought; nature is mirrored in a new way.[21] But in Foucault's archeology of the visual, the emergence of the Cartesian privilege of perception, with its idea of evidence as transparency to the mind, is a rather more complicated one.

In the "Discourse on Language," there is talk of a general change in seeing that would arise in Britain of the sixteenth and seventeenth centuries, captured in the precept "look rather than read, verify rather than comment." It involved a whole "scheme of possible, observable, measurable, classifiable objects,"[22] a scheme which preceded the actual collection of the "contents" of knowledge. This change matches with what, in the *Order of Things*, Foucault calls a "reorganization of culture in which we are still caught," where the eye no longer deciphers the "prose of the world," and where, therefore, "the eye was . . . destined to see and only see, and the ear to hear and only hear."[23] There thus arises the doctrine of separation of the senses central to the emergence of the new discipline of "aesthetics" in Lessing and Diderot.[24]

In knowledge, one place we see the change is in the "spatialization" of the natural histories of the classical period. In the classifications of Linnaeus, plants were studied, without microscopes, in terms of their visual "character," the colorless, odorless, "primary qualities" of a planar space. The principle of classification of elements and their arrangement in this space was based on a sort of "optics" of plant morphology, one which could be shown through the illustrations the new printing techniques made possible, and one which figured in the account of the "reproduction" of the plants themselves. It is this conceptual reorganization or "spatialization" which made natural history "nothing more than the nomination of the visible."[25] The "natural histories of the classical period," Foucault argues, "did not become possible when men looked harder and more closely,"[26] but when what they saw was organized in this new way.

The *Birth of the Clinic* records another change in seeing that occurs in French medicine at the end of the eighteenth century. Foucault is again at pains to dispute the view that "men looked harder and more closely." He disputes the account of the change in medicine in which the eye would move from fantastic imaginings to the careful observations of things. This is part of his general quarrel with the dichotomy between the imaginary and the epistemic, or the ideological and the scientific, in the history of knowledge.[27]

In fact, he argues, the "visionary space" in which doctors, physiologists, and patients discussed disease was itself a quite regular and coherent form of "spatialization," one based on recognizing the "portrait" of the disease in the body. And, what was involved in the "birth of the clinic" was a change in the whole idea of what it is to be "seen" by a doctor – where, with what instruments, and under what concepts.

Foucault argues that the change cannot be explained by "thematic contents," or "logical modalities," or the use of quantitative methods alone. It is the "space" of disease itself which changes, the place in the individual body where it is located, and the institutional "space" in which such localization occurs. Moreover, there was nothing inevitable about the change; one had to wait decades to get cures. The explanation for the change turns on institutional factors which emerge through the new programs of the French Revolution.

In this way, Foucault tries to show that it was the complex "spatialization" of disease which accounted for the role of observation in the new medicine, and not the primacy of observation which accounted for its new conception of illness. Such processes of "spatialization," however, are not the same thing as "theory-dependence." Foucault is not saying that medicine started to use a new theoretical vocabulary with which its "observation-reports" were "laden." It is rather a matter of the construction of a "space" in which not just observation, but also theory, becomes possible.

Foucault might be said in this regard to extend the distinction Georges Canguilhem had developed in his study of the reflex between "the history of theories" and the "history of concepts."[28] There is a history of the *concepts* through which things were given to be seen, which is separate from the history of *theories* about them. In particular, Foucault was interested in the history of how the concepts of visualization came to be embedded in institutional practices, or what he calls "tertiary spatialization." Thus, in *Discipline and Punish*, he goes on to explain that

> One of the essential conditions for the epistemological "thaw" of medicine at the end of the 18th century was the organization of the hospital as an "examining apparatus."[29]

For through this "apparatus" was "established over individuals the visibility by which they were differentiated and judged."[30]

The spatial "scheme" of a form of knowledge is not only distinct from the theories which occur within it; it often precedes and makes them possible. Thus the singular manner in which the general hospital gave mad people to be seen precedes the elaboration of the classical theory of madness, and the architectural reorganization of prisons precedes the new theory of crime. The relation between theory and visualization in knowledge is not fixed or given as in the Kantian idea of a "schematism" locked in the recesses of the human soul. It is rather a matter of contingent historical configuration.

In asking how such entities as "madness," "illness," or "crime" were made visible in the knowledge of different periods, Foucault thus focused on practices of "spatialization" that were more complex, and more deeply embedded in external processes, than the mere exercise of the naked eye aided with a theoretical vocabulary. In his archeology of seeing in knowing, what he "excavated" was how the eye of observation was oriented in such practices, in a manner that does not simply derive from theory.

He thus revived an old philosophical debate about seeing and *reality*. From the fact that disease is "spatialized," does it follow that it is not "real"? Does it follow that when a physician sees a patient he is not seeing something real, but only the phantom of the discourse of his time? One source of such questions is the old idea that the real is what is observable.

Ian Hacking takes on just this idea in his discussion of the question, in the philosophy of the natural sciences, as to whether or not abstract or theoretical entities are real.[31] Hacking argues that it is the obsession with "observation" in the philosophy of the natural sciences that has obscured the recognition of the role of complex experimental apparatuses in what one might call the "visualization" of theoretical entities in physics or genetics. Natural science also has its "modes of spatialization"; there is a whole natural "technology of the visual" as there is a human one: observatories, microscopes, cyclotrons. And experimentation is central to them.

For Hacking, "observation" is a misleading inheritance of logical positivism. In apparent allusion to Foucault, he says that phenomenology and positivism both descend from a "change in seeing" that occurs around 1800. Then would be forged the link between what is observable and what is real. Thus, in Hacking's example, both phenomenologists and positivists can agree that while meatballs are real, mesons are not.

In particular Hacking says that the positivist preoccupation with observation has led to two philosophical themes which have conjointly obscured the role of experimentation in natural science: Willard Quine's idea of semantic assent, and Norman Hanson's idea of theory-laden observation. Together they have led to the obtuse idea that, in physical science, seeing is saying. In fact, verbal "observations reports" that test theories are quite rare in physics. What matters is rather that engineering that evokes or constructs entities in highly artificial conditions. The relation between experimental engineering and theory is a complex and variable one. It is a matter of history; it is not given in a "logic" of inferring theory from perception. To understand experiment is to understand the question of what makes modern science modern; for the link between theory and experiment is part of a larger history of the link between technology and science, which helped to determine the very sense we give to "technology."

Hacking writes in praise of Francis Bacon, philosopher of experiment. Bacon's idea not of observation, but of "prerogative instances" gives a better picture than Carnap's of the way theoretical entities are made visible. Thus Hacking proposes to substitute an experimental for an observational realism. As far as he is concerned, if you can spray electrons, then they are real – just as real as meatballs. He thinks that it is the Hanson-Quine idea that seeing is saying that has induced philosophical doubts as to whether or not there exist real entities outside the verbalizations of science. Experimentation is a corrective; it offers a better way of understanding the sense in which theoretical entities can be said to be real.

Foucault's archeology of the "spatializations" of madness or illness, while it disputes simple observational realism, nevertheless does not lead, in a parallel way, to the sort of "pragmatic realism" which says: if you can cure a patient, then the illness you have seen in him is real. On the contrary, it leads in the opposite direction of a sort of nominalism; and in the *Archeology of Knowledge*, for example, Foucault talks of "de-presentifying" the very things of which he writes the "archeology."[32]

Our knowledge is, of course, such that we can "do" things to illness or madness just as we can to electrons or genes. But seeing and doing are not related in the same way. In the case of illness or madness, the construction of seeing, and the way it fits in institutions and comes to be related to other fields, never loses contact with the way certain "real" social problems are seen.

There is one sense in which Foucault's question about how we "see" psychotics is a different *sort* of question from that of what we should do about atoms or genes – even if our best *theory* about psychosis should turn out to be a genetic one. For he is not asking what to do with the psychoses our knowledge lets us see, but whether we can or want to refuse the "evidences" of the way they are given to be seen in a whole range of practices, and invent other ways of seeing/dealing with them. It is this sort of interconnection between seeing, doing, and practical self-evidence to which he turns in *Discipline and Punish*. In that book he draws a distinction between disciplinary and Baconian ways of seeing.

A central topic is "normalization." "Normality" as a fundamental category of our behavior, and even of our identity, becomes "visible" through an expanding network of practices in the nineteenth century. One of the basic things our knowledge makes visible to us is *abnormalities* both of persons and of societies.

One source is precisely medicine, the change in what it means to be "seen" by a doctor. There had been no independent way of identifying illness except as the deprivation of the state of health in the whole body. Then physiological anatomy introduced criteria for being a diseased organ independent of the health of the whole person. "Normality" could be defined as the absence of pathological symptoms in the organs. Abnormality started to be related to degeneracy. All this was part of the change in the medical gaze: when a doctor "saw" a patient he began to ask not "what is the matter with you?" but "where does it hurt?"

But this new rationality of the normal came to be applied in other places – for example, in Durkheim's attempt to distinguish "normal" from "pathological" states of societies, or to specify the "degenerate" portion of a population. The art of seeing "abnormality" fit within a network of practice. And it is the organization of that network that was rather different from the one that allows us to spray electrons.

In *Discipline and Punish*, the term "technology" enters Foucault's methodological lexicon. Discipline involves a new "technology of the visual." There is a comparison with experimental devices, and with Bacon, philosopher of experiment. As telescopes, microscopes, and prisms helped transform not simply what physics could see, but the place of "seeing" in it, so the techniques of surveillance and examination (that "microscope of behavior") not simply made such things as the "abnormal" or "criminal" personality visible; they also helped change the place of the visual in knowledge and power. These "observatories of human multiplicities," writes Foucault, introduced "an obscure art of light and the visible" which was "secretly preparing a whole new knowledge of man."[33]

And yet there is a basic difference between the two types of visual technology, or of the place of the visual "technique" in knowledge and power. "Another knowledge, another power," says Foucault. Then he refers to Bacon:

> On the threshold of the classical age, Bacon, lawyer and statesman, tried to develop a methodology for the empirical sciences. What Great Observer will produce the methodology of examination in the human sciences?[34]

But he quickly adds:

> Unless, of course, such a thing is not possible. For, although it is true that, in becoming a technique for the empirical sciences, the investigation has detached itself from the inquisitorial procedure, in which it was historically rooted, the examination has remained close to the disciplinary power that shaped it. It has always been and still is an intrinsic element of the disciplines.[35]

Bacon's conception of experiment may have roots in Inquisitorial procedure – putting Nature to the rack to extract her secrets – but the technique has long been cut loose from the sorts of problems the Inquisition was designed to deal with. In the case of discipline and its technology of seeing, its "art of light and the visible," by contrast, we see a process through which it multiplied and complexified its links to the problems it was designed to deal with; it spread out in a range of institutions where it retains the rationality of a "technical matrix."

In the difficulties that arose from the implementation of "hard" technologies such as the steam engine, the electric plant, or the television, the focus of the "problematization" was not on the "evidence" of seeing steam, electricity, or electrons. By contrast just what Foucault found important in anti-psychiatric protest or prison revolts, was the way it questioned the very "technical matrix" of the disciplines which made the madman or the criminal visible; the way it exposed the very "evidence" through which those practices are accepted.

Thus the "philosophical" problem of "seeing electrons" is not at issue in the great questions of what to *do* with them, make war or energy. But the problem of "seeing psychoses" *is* involved in the questioning of what we should do about *them*. Seeing and doing are related in different ways. Another knowledge, another power. That is why the philosophical attitude towards one can be realistic, and towards the other, nominalistic. The realist of experiments and the nominalist of disciplines can agree that seeing in knowledge is a more complex matter than inferring from perception. For their differences lie in the way knowledge comes to be "visualized" or "spatialized."

Spaces of Constructed Visibility

"Space" is a constant topic in Foucault's histories and in his thinking. As already indicated, it plays a prominent role in his study of medicine, and then it is taken up in a different way and generalized in his study of penal practices. The "spaces" we call the "territories" of states also become central in his study of the "police science" which helped introduce a new administrative sort of rationality, and a "geo-political" orientation and organization of war and diplomacy.

In the historical study of space, Foucault was impressed by the work of such social historians as Bloch, Braudel, and Ariès. He thought their work might serve as a corrective to one tendency in the philosophy of time in Bergson, Heidegger, and Sartre – the tendency of putting "space" on the side of the "pratico-inert," while reserving for time the great questions of project and history.[36]

A significant portion of Foucault's discussion of "space" is devoted to the problem of *visibility* – how spaces were designed to make things seeable, and seeable in a specific way. In his histories of the visual unthought, the construction of space plays a key role.

During the course of an interview devoted to space, Foucault declares:

> I think it is somewhat arbitrary to try to dissociate the effective practice of freedom by people, the practice of social relations, and the spatial distributions in which they find themselves. If they are separated, they become impossible to understand. Each can only be understood through the other.[37]

"The spatial distributions in which (people) find themselves" – that names a recurrent topos in Foucault's work: hospitals, poorhouses, museums, public baths, schools, homes, asylums are all spaces in which one can reconstitute the rationality of an elaborate construction of what can be seen. They are spaces of constructed visibility.

We are surrounded by spaces which help form the evidences of the ways we see ourselves and one another. Where we "dwell," how we are housed, helps in this way to determine who and what we think we are – and so they involve our freedom. We are beings who are "spatialized" in various ways; there is a historical spatialization of ourselves as subjects.

Foucault's analysis of "spaces of constructed visibility" brings out how they serve to "constitute the subject," the way they serve to construct the spatialization of the subject or his "being in space." "The art of light and the visible," which such spaces are designed to deploy, is one which makes certain *kinds* of properties of ourselves stand out as self-evident.

Foucault suggests it is just this link between visibility and constructed space which would make a "technological" history of the art of architecture possible. For the art of building is, among other things, an art of rendering visible, and so discovers one of its central interconnections with power. Architecture helps "visualize" power in other ways than simply manifesting it. It is not simply a matter of what a building shows "symbolically" or "semiotically," but also of what it makes visible about us and within us. Châteaux and churches may do this through the way they manifest divinity, sovereignty, and might. Before the museum, they may supply, as Malraux stressed, the central "imaginary" spaces which secure the categories through which art was given to be seen. But in Ledoux's salt mine Foucault finds another relation between power, visibility, and constructed space, one connected to new problems of poverty and work, one to which Bentham would give the name "panopticism." The construction of buildings is involved in the new "art of light and the visible," which doesn't look up to the glory of those who possess or embody power, but looks down to the marginalized anonymous mass that escapes it.

The art or technique of the visual in "panoptic" architecture is not exhausted by the wondrous contrivances that make the constant surveillance of the inmates of its enclosed space invisible to them. Panoptic establishments also inscribe in cellular stone the new sorts of classifications designed to deal with wayward populations. They are constructed to facilitate the introduction of the "examination" procedures that rank and judge people according to their "visible" characteristics. This spatialization makes the new classifications, unlike contemporary botanical ones, "disciplinary" – what makes a person classifiable submits him to an "individualizing" control. Thus Foucault says, where natural taxonomy links category and character, disciplinary tactics link the singular and the multiple; they give attention to each and every member of a multiplicity individually. They make the categories into which the "character" of people are slotted "visible" in them; they create in people an "individuality" that is endowed with certain "essential" or evident sorts of properties. Thus it is not simply that the "eye of power" looks down rather than up. What it sees is no longer heroic acts, but dysfunctional personalities. It focuses light not on illegal acts, but on behavioral deficiencies. It "spatializes" this new thing which is the "personality" of an individual.

Bentham's panoptic scheme is related to the "self-evidence" of his great moral principle of the rational calculability of the good of individuals. For one's good to be tabulated one must be "seen" in a certain way, or under certain categories, just the sort of categories architecture would serve to make visible. An art of spatializing human multiplicities would then be central to the formulation of utilitarian ethics.

Ludwig Wittgenstein tried to show that "looking within" or "introspection" was nothing more than a rule-governed art of language. Foucault's analysis of the rules that govern the art of space shows that when we look within we often see not so much our Cartesian minds as the worrisome sources of our behavioral deficiencies, but that in this, we are no less participating in a practice which makes the sort of thing we see seem self-evident to us.

Seeing in Power

"Spatialization" is thus one technique in the exercise of power. That is why it cannot be separated from the "effective practice" of our freedom, or our relations with one another. There is a political history of the visual unthought: a history of the way forms of power "visualize" themselves. A principle of this history is that "visibility" is one of the great "self-evidences" of the workings of power. Power becomes acceptable or tolerable through its spatialization or the way it is given to be seen.

Thus, in the *History of Sexuality*, Foucault says, "Power is *tolerable* only on the condition that it mask a substantial part of itself . . . would it be *accepted* if it were entirely cynical?"[38] One way it masks itself is to get people to "see" it in a certain way. Power conceals itself by visualizing itself. Its workings become acceptable because one sees of it only what it lets one see, only what it makes visible.

We are fascinated by the pomp, the "ostentatious signs" of power. That contributes to the "self-evidence" of our idea that it is owned or possessed, where in fact it is being anonymously exercised. One reason we don't see discipline as the form of power that it is that we don't see how it makes us visible. The inspector in the tower doesn't possess or embody, doesn't *see*, the power he implements. The sort of "visibility" the disciplines introduce conceals not simply how they work, but what they are. We don't "see" discipline as power, because we don't "see" power as strategy.

In particular the techniques for visualizing power as sovereignty, nobility, and law have prevented us from seeing it as anonymous technique. In the *History of Sexuality*, Foucault suggests how this proposition might be applied to the analysis of fascism. Something as the neo-Classical facades of panoptic prison architecture concealed the strategic construction of visibility within, so the great fascist "premodern" rituals of the ostentation of sovereignty, law, and blood concealed the very "modern" way its power in fact was working, and so helped to make it tolerable.

Seeing through Desire

In the *History of Sexuality*, Foucault introduces another theme: the history of the sort of *pleasure* we take in what we see. He came to think that it is because it belongs to what we conceive of as our "sexuality" that we are *fascinated* by it, that we want to *expose* (or *exhibit*) it, a fascination and an exposition which are linked to *knowing* it, or to the sort of truth it might tell us about ourselves: "We have invented" declared Foucault,

> a peculiar pleasure in knowing that truth, in discovering it and exposing (or exhibiting) it, the fascination of seeing it and telling it.[39]

It has not always been this way. What has been taken to be most glorious or problematic about our sexual experience has not always been this thing that fascinates us, and which we must expose; the "spaces" and "techniques" which give it to be seen in this way have not always been with us. We have not always been fascinated by our sexual *desire*, this dangerous thing Freud would be the first to have the courage to look at straight on.

Foucault wanted to determine how the "self-evidence" of this sexuality we must show and see arose in our knowledge, our agreements, and our practices. One place he looked was in what might be called the "voyeurism" of the nineteenth-century medicine of sexual deviancy: its special curiosity, its prurient preoccupation with its object – a "fascination" structurally not so unlike the "pornographic" search of one's "secret life." Foucault takes this interest as a historical property of a medical discourse and practice, not as a quirk of the doctors taken individually. But it was also a property of the "spaces" they worked in, a fact about a medical art of making sexuality visible, an incitement to see and show its dangerous truth.

One example is Jean-Marie Charcot's La Salpêtrière, where Freud made his "discovery" of the unconscious. That space was not simply a space of constructed visibility, but also a space of the *incitement* to see. "It was an enormous apparatus for observation, with its examinations, interrogations and experiments." But it was also:

> a machinery for incitement, with its public presentations, its theatre of ritual crises, carefully staged with the help of amil nitrate, its interplay of dialogues, palpitations, laying on of hands, postures which doctors elicited or obliterated with a gesture or a word, its hierarchy of personnel who kept watch, organized, provoked, monitored, and reported, and who accumulated an immense pyramid of observations and dossiers.[40]

And yet it refused to *name* what it thus incited to see: sexuality.

Charcot was said to be, and called himself, *un grand visuel*. And in his obituary, Freud says of Charcot that he was an artistically gifted seer, that the chaos of symptoms was set in order by the eye of his spirit, that he talked incessantly about the merits and difficulties of seeing in the ward of the sick, in which he said he found his greatest *satisfaction*. But, adds Freud, while he was a seer, a *visuel*, he was not a thinker or a "brooder." He could not *name* theoretically what he saw. In this obituary we see the beginnings of what, for Freud, it would mean to "observe" the processes of unconscious desire.

Charcot invented a differential diagnostic scheme for the various "types" of *la grande hystérie*, as they were exhibited in the scenographic tableaux he staged. It was this typology which allowed him to connect hysteria to witchcraft, as it was depicted in paintings. But the explanatory part of his theory connected the bodily "poses" of the hysterics only to mysterious "organic lesions" of the cortex; sexuality was not a causal factor.

In this manner, he overcame the confusion of symptoms which had made hysteria emblematic of those mental disorders whose symptoms were a form of dissimilation or lying. But the staging of hysterical poses opened the possibility of a sort of counter-movement on the part of the hysterics: the possibility of disrupting the clear space of visibility, by introducing a sexual body. In this way, sexuality got into the picture. "I believe," Foucault said at the Collège de France in 1974, "that there was a *battle* of hysteria. . . . Hysteria was the set of phenomena of *struggle* that unfolded around this new medical machinery which was the neurological clinic."[41]

In the face of this sexual body, Charcot was obliged to turn away his "admirable gaze." Babinski gave up the theatricalization of the sickness and invented as an explanation "pithiatism," or the aptitude to let oneself be theatricalized. But Freud resolved to say, and to say theoretically, just what this space gave to be seen. The origin of hysteria, he said, lay in a woman's relation to her *sexuality*. And, around the theme of sexuality, Freud began to devise ways to connect what the hysteric *showed* with what she *said* in a new sort of space: the space of psychoanalysis.

With this sort of analysis, Foucault hoped to account for one feature that would distinguish the fascination and exhibition of sexuality in our society; one he sees Freud as having basically extended: its *medicalization*. What it is that we still want to see and to show about our sexual experience is its abnormality, its perversion, its sickness. And that fascination is an integral part of the pleasure we take in *knowing* about it.

One technology employed in Charcot's demonstrations was the new art of photography, and the photographs of the hysterical postures were published by the surrealists, who, as is well known, took a particular interest in the relation of seeing to mental disorder.

The Eye of Thought

What is it then to see the events in what is unthought in our thought? One visual image Foucault offers is that of surrounding the event with a sort of "polyhedron of intelligibility," the sides of which would extend indefinitely in many directions.[42] It is to multiply the things associated with its "intelligibility," and the ways they are associated with it. The greater and more specific the internal conceptual analysis of the event, the greater the external processes with which its "invisible" or "evident" intelligibility is linked. Thus in Foucault's analysis of the event of the prison-form, he finds an intelligibility that connects it to pedagogical practices, professional armies, British empiricism, the new division of labor, and the invention of gunpowder, through the transference of technical models to other domains, new applications of theories, or new strategies to deal with local problems.

Foucault thus starts with the idea that there may be no such thing, no such "essence" as *the* visual, something that might be described by a "phenomenology of perception" or a "theory of the gaze," something which, following Martin Jay, Foucault would be against. Rather, history presents us with many different singular sorts of visual intelligibility, ways of seeing and making seeable, the unity of which is not to be found in the nature of the eye, empirical or transcendental, or in "the imaginary order." Where and when "the visual" acquires the status of essence, universality, and necessity, the philosopher's eye must look for the singular and contingent processes that make it self-evident, and so acceptable.

Something as in the philosophical task captured in Wittgenstein's famous precept "don't think, look!", the philosopher's eye, the eye of thought, is neither contemplative nor introspective. It does not look up to see the forms it has forgotten, or look within to see the point from which action should proceed, or the self-certainty from which knowledge should be derived.[43] It looks *out* to those events in thought through which things are given to be seen. It looks out in order to *change* its way of seeing. Thus in 1981, Foucault declared:

> Each time I have attempted to do theoretical work it has been on the basis of elements from my experience – always in relations I *saw* taking place around me. It is in fact because I thought I recognized something cracked, dully jarring or dysfunctioning in the things I *saw*, in the institutions with which I dealt, in my relations with others, that I undertook a particular piece of work, several fragments of an autobiography.[44]

When a philosopher "sees" something problematic or dysfunctioning around him, he doesn't turn his eye around to the ideals in whose light the problems appear as imperfections or counter-instances; he doesn't turn it within to see the true or authentic self in whose light the problems figure as distortions or mystifications. His seeing starts a form of theoretical work which, in analyzing how the problems arose and were conceived, transforms his way of seeing them – and so his way of *living*.

Ethos, Beauty, Danger

"Fragments of an autobiography" – Foucault's art of seeing is also a philosophical art of living. His theoretical work is "autobiographical" not because it is a way of transcribing his experiences, but because it forms an exercise in which to periodically question the given conceptions of his experience, and so look for new ones. Autobiography in this sense is not the attempt to provide an image or picture of who one was or how one should be seen, but a form of work to change oneself by changing one's way of seeing.

Foucault's writings do not unfold as a single doctrine or theory of things. At particular moments he changed his mind as to his aims, objects, and methods. As in the case of Nietzsche, Wittgenstein, or Heidegger, his thought is punctuated by transformations in the way he conceived of his own philosophical task. In the Introduction to the volumes published shortly before his death, he pictures this process as an exercise or *ascesis* of disengaging himself from himself in his work (*se déprendre de soi-même*), through "essays" that try to alter his way of seeing things. But he adds, there is an irony in this process. It is precisely such efforts to free oneself from oneself that makes one's work one's own; one finds who one has been by always getting away from oneself.

> Such is the irony of the efforts one makes to change one's way of seeing (*façon de voir*). . . . Have they in fact led to thinking in another way (*penser autrement*)? Maybe at best they have allowed one to think in another way what one already thought, and to see what one has done from a different vantage point and in a clearer light. . . .[45]

And the new light under which Foucault saw what he had been doing in his previous work was the light of "problematizations":

> I now feel I better perceive how . . . I was proceeding in this enterprise of a history of truth: to analyze . . . the *problematizations* through which being gives itself as what can be, and what must be, thought, and the *practices* through which these *problematizations* are formed.[46]

There may be no final "enlightened" solution to our relations with pain, illness, crime, madness, and death. And yet it is a historical fact that there arose various forms of intelligibility or rationality in the way people actually came to see such things, and the way they consequently erected around themselves forms of knowledge and action, and modes of living. Those experiences have not always been seen as raising the same sorts of problems. To analyze their history is to see the specific sorts of danger or problem which led to the evidence of a particular way of conceiving and dealing with them. Thus, according to his new way of seeing, what he would have been studying, in his previous work, is how people *saw* what is dangerous in being mad, ill, or criminal, how they envisaged those dangers and made them "visible" or "spatialized" them in knowledge and action.

Foucault proposed to view his histories of madness, of illness, or of crime as histories of just what it was in the experience of madness, illness, or crime which was so problematic as to become something which both could be, and *had* to be thought. And he began to associate such problematizations of experience with particular modes of living or being, modes of being a person of a particular sort. Thus he would have been asking what are the problematizations and consequent practices "by which man proposes to think his own nature when he perceives himself to be mad; when he considers himself to be ill; when he conceives of himself as a living, speaking, laboring being; when he judges himself as a criminal?"[47]

In the *Birth of the Clinic*, Foucault had looked for something more basic than the "mindless phenomenologies" of the "encounter" between doctor and patient, and the "so-called 'liberal'" notion of a contract or pact between two individuals. He had tried to see a "major event in the relations of man to himself and to the language of things."[48] The new "spatialization" of illness in the pathologies of the individual organs would have fundamentally changed the relations people had to themselves and to one another in "being ill." It would introduce a whole new sort of "ethical" intelligibility of the problems of doctor, patient, and pathologist.

Similarly, Foucault's history of madness may be read as an examination of how "being mad" was seen as a source of danger in society and in the individual. Thus there arose a new way of seeing the problem of madness. The class of inmates in the General Hospital may seem heteroclite to us. It answered to a perfectly intelligible way of seeing a problem, a "sensibility" for which the central danger to society and self had become *idleness*. This way of seeing derived in part from a theological promotion of idleness over vanity as the cardinal sin. It also arose from a new conception of labor and poverty that was to become the target of a new administrative sort of rationality. And it was central in the thinking through which the hysteria of idle women could enter medical discourse.

But as the techniques of moral assignation in the "enlightened" asylums of Tuke and Pinel show, there arose a question not simply of dangers from without, or in social relations, but also from within, or in one's relation to oneself, and so to others. Guilt, shame, irresponsibility, weakness, or the will as forms of relations to one's self arise within different spaces and under different conceptions; they are conceived in relation to particular dangers or problems.[49]

The new light of problematization thus brought into focus the *ethical* concerns of his previous histories. It offered Foucault a way of thinking about the origins and the changes in the very conception of ethics: its basic questions, what it supposed to be true about us, the sorts of relations it was thought to have with knowledge, law, or politics.

To be sure, Foucault had not been worried about the justification of the principles of ethical action; rather, he had been studying how ethical thought and practice had seen and responded to certain sorts of problems or dangers: the ways *it* had conceived of the obstacles one must overcome to be good or do right, the ways *it* had rationalized a way of dealing with what it saw as wrong, sinful, or evil.

The possibility of a history of ethics arose that would study the specific sorts of dangers or problems it was designed to overcome. But, in studying this history, we should not look "for the solution of a current problem in the solution another problem raised at another moment by other people."[50] The history of the problematizations of "ways of being" in ethics is not a nostalgic one. "History," Foucault said, "protects us from historicism – from the historicism that calls on the past to respond to questions of the present."[51]

Rather, Foucault says, the analysis of problematizations is itself "dangerous." In the *Order of Things*, Foucault had already said that when, in modern thinking, the "cosmological" form of moral thought is no longer possible, it is thought itself which becomes dangerous – a "perilous act."[52] And, speaking of his "pessimistic activism" in a late interview, Foucault declared:

> I would like to do the genealogy of problems, of *problématiques*. My point is not that everything is bad, but that everything is dangerous, which is not the same thing as bad. If everything is dangerous, then we always have something to do.[53]

In his last work, and in his last conception of his work, Foucault connected his art of seeing to the "ethico-political choice" one makes of "determining which is the real danger." The choice of trying to see just what it is that we have to struggle against in order to free ourselves (and free ourselves *from* ourselves). And this freedom is dangerous, since we can never have in advance a determinant or complete picture of it.

For, as a thinker and seer, Foucault was concerned with a situation, prior to the possibility of deductive normative reasoning, where one sees something must be done without yet knowing what. A space not of deduction but of questioning and analysis is thus opened up between the choice one makes and what one does, in which one tries to conceive what the danger is which one does not yet fully see, but in relation to which one must take action. It is one's responsibility to this thing that troubles, but which one can't yet describe or name, that requires one to work to change oneself. One's work is the attempt to change one's way of seeing and living in relation to those specific dangers one does not yet know what to do about.

Foucault thought that this kind of relation between seeing, living, and action might be conceived as an "aesthetics of existence" which contrasted with the one he tried to reconstruct in the ancient ethical schools of *savoir-vivre*.

In Foucault's reconstruction of ancient ethical thought, what in sex was seen as dangerous enough to become the object of a whole practice of self-transformation was not yet the sins of the flesh or the odd desires that fascinate us and that lurk in the pathogenic recesses of our heads. It was the excessive activity that threatened a loss of self-mastery in those meant to rule – a danger to the *ethos*, the fitting mode of being, of the free adult male. But the *ethos* of the "civic man" in his health, home, and courtship, around which the ancient ethical practices turned, was rather different from the ethos of the Christian "inward man," and the Augustinian problem that all sexual desire is tinged by the Fall, and is the model of sin in general, or the "democracy of shame" that linked the Christian community to the problem of poverty. Foucault sought to analyze these great changes in the problems of "being sexual" in terms of the ethical practices in which one is asked to transform oneself under a particular conception of one's experience, aims, and obligations.

In the analysis of the ancient arts of being virtuous, seeing and being seen played a significant role. As has often been observed, the ancient conception of "beauty" – beauty in the ethical sense – was a particularly *visual* one. And yet such activities of "presentation of self" were "problematized" and conceived in a rather different manner from the questions of identity, authenticity and engagement Sartre associated with "the gaze." If the proud virile body of the master was a "lived body," it was lived in a different way than the one Sartre had envisaged in his account of our "concrete relations" with ourselves and one another. Seeing oneself and one's mode of living had a different sort of intelligibility in this ethics.

Self-mastery was like the visualized evidence of one's fitness to rule. One had to be able to show, and to show *truly*, one's state of temperate control of soul and body as something noble and beautiful to be glorified for posterity. The aim or telos of the self-forming practices was thus a beautiful accord or harmony between who one is and what one does and says. Beauty was an ethical category; seeing and showing oneself were thus part of an etho-poetics, an aesthetics of existence. And the link between visible beauty and mode of being took place and was formed in certain "spaces," the *oikos*, and the *agora*. The *oikos* was a "space of constructed visibility." Discussion of marital duties was a discussion of the distribution of roles and natures in the "space" of the *oikos*, whose "roof" divided what was inside from outside. Thus there was the problem of makeup and paint in a debate about pleasures and truth: makeup can have no place in the beautiful way the mistress of the house must "carry herself," in which "standing and walking will give her body that certain demeanor, that carriage (*allure*) which, in the eyes of the Greeks, characterized the figure (*la plastique*) of the free individual."[54] In this idea of beauty, *mimesis* was a category at once of one's relation to oneself, or one's "figure," and of the relation of the "figure" of a work to what it makes appear.

If Hegel saw in this Greek experience of beauty a first "moment" of that to which spirit returns in the work of history which passes through art and religion to the state, for Foucault, the Greek experience is rather a "lost evidence," a solution to a danger no longer ours. The art of making oneself in the image of an active freedom no longer has the same central and self-evident place in our ethical thought. We have lost the "ethopoetics" that made existence the object of an aesthetics. We see other dangers, and deal with them in other ways. Our disciplined "docile and useful" bodies are not the "self-possessed" bodies of the ancient master. The Benthamite spaces which helped give the calculability of what is good for us its central ethical importance are rather different from the spaces in which the ancient question of the wisdom of the good life was raised.

For there to be a modern "aesthetics of existence," the very concept of beauty in living must change. This is the change Foucault associated with Baudelaire, and with the modern principle that "the subject is not given." For us the danger is not that we might fail to become what we are meant to be, but that we might only be what we can see ourselves to be. In ancient thought, freedom was something beautiful to be made visible in soul and body; and the loss of freedom could thus be seen as something ugly. But when "the subject is not given," our freedom ceases to have an image. It is that which we can never yet see. Its beauty lies in this danger. For us, beauty no longer resides in the perfection of a living harmony between ourselves and the "spaces" in which we can become what it is natural, given, and possible for us to be, but in what cannot yet be seen or named about us in the spaces we inhabit.

Beauty in living would lie in a dissonance or disharmony between one's given nature and one's possibilities of existence, between one's identity and what one can see in oneself and in the processes going on around one. The danger of beauty thus gives rise to an *oeuvre de soi* in which to see what one should do is to change one's mode of living, and in which changing one's mode of living involves changing one's way of seeing, one's *façon de voir*. One changes oneself as one comes to see what is dangerous in one's existence, and comes to see what is dangerous by changing oneself.

Foucault thus came to think of the art of seeing he had been practicing as an "aesthetics of existence," an art of living. In this art, the events one sees going on around one interrupt one's sense of one's self and cause one to think and to rethink. Foucault's exercise in seeing and thinking would then be the exercise for a modern *ethos* of freedom – the freedom for which we never yet have the image and the ethos in which beauty lies in coming to see the real dangers we must face.

Seeing Outside

This attempt to move out from oneself in the "essays" one writes is not so unlike the "attitude to modernity" Foucault finds in Baudelaire. In the course of a late discussion of Kant's paper on enlightenment, he says that Baudelaire's "exercise on himself" as a writer would transpire in "another, different place," outside the realm of society and politics – the place Baudelaire called "art."[55] How, then, is Foucault's "art of seeing" related to the way he saw this "art"? And, in particular, what does this "art of seeing" have to do with the conception of "modern" art and literature, which he himself advanced in a series of essays in the '60s, but from which he later sought to distance himself? It is important for Deleuze's reading to establish a continuity throughout Foucault's work based in his early conception of work or oeuvre.

In the last pages of *Madness and Civilization*, one can already find the germs of this question in the discussion of *absence d'oeuvre*, a post-Romantic idea with roots in Sade and Hölderlin, in which one's relation to madness would be linked to one's oeuvre not as its expressive content, but as the unsayable or unspeakable source from which the work would emerge, and into which it would again disappear.[56] "*Absence d'oeuvre*" doesn't mean there is no work or that one is out of work, as the English translations oddly suggest.[57] Rather through one's work one tries to say something as yet unsayable, or to see something as yet invisible, and so one opens out a space of a sort of rhythmic "disappearance" of oneself in and through one's work. The "being of language" (as distinct from its regulated use in "discourse") would offer the occasion and the condition of this modern "attitude" to oneself in one's work. In the '60s Foucault argued that this was a conception basic to the work of such writers as Klossowski, Bataille, and Blanchot.

English-language discussion has mostly ignored this theme in Foucault's own oeuvre. And yet it is the one Derrida admired in his critical review of Foucault. It is also the one which Deleuze invokes in another way in his presentation of Foucault as seer. In effect, he sees Foucault as bringing the exercise of writing as *désoeuvrement* into the fields of history, politics, and epistemology. Deleuze says the "being of language," which would be the condition and occasion of literature is also the condition and occasion of Foucault's archeology of discourse. And he says the opening of visibility, or "the being of light," which would be the chance of the visual arts, is also the condition of Foucault's art of seeing.[58]

That is one reason Deleuze can say that Foucault is "singularly close to film." For the central concept in Deleuze's own analysis of the role of thought in film is the concept of *désoeuvrement*. In response to Godard's prognostication of an end to film theory, he says "the concepts of film are not given in film."[59] There is a sort of "filmic unthought" from which film tries constantly to free itself and so open itself to other ways of thinking and showing. Thought in film is a *tout ouvert*, a conceptual ensemble open to transformation. As such it can be analyzed as a great art of conceiving of light, movement, time, and space, of conceptualizing the visuality of the "spaces" through which we are given to be seen. Deleuze contests the attempt of Christian Metz to make narrative the central question around which filmic thought must turn, and the resulting choice one would have to make between the good abstract or theoretical film, and the bad commercial, ideological, narrative one. The visuality films make intelligible is not that of the physical medium theory must purify of all *récit*; it is a tacit conceptual organization which links film to the way space, subjectivity, and time are philosophically understood.

And yet, after '68, or with his "political" turn, one hears a good deal less about the "being of language" (or of light) in Foucault's work, and the figures of Nietzsche, Hölderlin, and Bataille cease to haunt the margins of his institutional histories. As I say in my book, Foucault changes his mind about Sade and Bataille. And in a posthumously published interview from 1975, Foucault himself says as much.[60] He says that in his essays on such figures as Blanchot and Bataille in the '60s, it was not the idea of literature itself which was important to him. The reference to such figures in his histories, he says, was a matter of simple *constat*, something he would have noted in passing as though on a walk. What was important to him in Blanchot was the attempt to get out of a certain "Hegelianism" that assigned literature a privileged expressive role in history, and to ask instead about the singular place a society would accord to the writing it calls literary. Blanchot was one way out of a certain style of philosophy; it offered the possibility of another way of thinking in which one tries constantly to see outside the bounds of seeing and think outside the bounds of thinking. But at the same time, Foucault says he came to adopt a "negative attitude" towards the "sacralization" of this new conception of literature which took hold in the university, a sacralization, which, paradoxically, this new literature had first intended to resist. He says there arose an "ultra-rationalist" and "ultra-lyrical" idea of a literature radical in referring only to itself, in which writing would acquire inviolable rights to "subversion," and in which the more involuted one could make one's writing, the more "revolutionary" one would become. Foucault saw this idea as a form of "political blockage," and presents his own books on Roussel and Rivière as his way out of the new academic sanctification of literature.

Nevertheless, one can argue that his early ideas about the oeuvres of modernity did not altogether disappear from his thinking. If *absence d'oeuvre* ceases to be an object of his histories, it comes to supply something of the *ethos* of his work as historian. Rather than being the obscure hero of his histories, Nietzsche would be someone he would put to a new *use*; and he would continue Baudelaire's "attitude to modernity" in his own work in another way. It is thus that Foucault would extend the place Baudelaire called "art" to a particular ethic of thinking, seeing, and living.

In his 1966 essay on Blanchot, Foucault says that "absence" is not inside a work but *outside* it; modernist literature is not the literature that turns in on itself, but the one that opens itself outside itself; it does not make beautiful forms appear within, but it takes those forms outside of what they have been. Perhaps this sort of "absence" became part of Foucault's view of his own oeuvre: what we cannot yet see in the forms of knowledge, action, and experience through which we are given to ourselves. Foucault's philosophy was about possibility: what we *can* think and what we *can* change in what we think. He wanted to do a history not of what is true or false, but of what can be; not of what to do, but of what can be done; not of how to live but of the possibilities of living. He thus discovered a kind of "impossibility" that was not logical but historical: not the impossibility of a round square or a non-existent god, but of what is no longer or not yet possible to think; not what is meaningless, but what is not yet or no longer meaningful. Such "absence" was a matter of a historical constraint on thought, which it would be the work of philosophy to bring to light.

In this sense Foucault's art of seeing might be said to be the art of seeing outside ourselves, or seeing the "absence" in our work. Not to look within to a true or authentic self; not to master one's time by holding it in one's thought; not to find a place for oneself within society or state, but to look out from oneself, to open one's time to what has not yet been seen, to transform or displace one's instituted, assigned identity at a time and place. In this sense Foucault's art of seeing is an art of looking out, which would "give new impetus, as far and wide as possible, to the undefined work of freedom."[61]

Notes

1. Gilles Deleuze, *Foucault*, Paris, Editions Minuit, 1986, p. 57; English translation forthcoming from the University of Minnesota Press. One might trace an original view of "vision" in Deleuze's readings of philosophers. Thus, for example, in the '60s, he also presents Spinoza, philosopher and lens-polisher, as a *vivant-voyant*. Spinoza said that the geometrical demonstrations of his *Ethics* were as "the eyes of the soul"; Deleuze sees a vital optical method of rectifying those sad passions that ruin life, a way of polishing the glass for an inspired free vision. Deleuze's latest book is about Leibniz and the Baroque.
2. *Ibid.*, p. 59.
3. *Ibid.*, p. 72.
4. In an essay titled "In the Empire of the Gaze: Foucault and the Denigration of Vision in Twentieth-Century French Thought" (in *Foucault: A Critical Reader*, ed. David Cousins Hoy, London, Basil Blackwell, 1986), Martin Jay provides a useful inventory of some of the places where Foucault discusses matters visual. But I am not convinced that he succeeds in understanding what is involved in that discussion. His basic idea is that Foucault, like many of his compatriots, was "against vision." Yet it is not clear what he could mean by this. To say that Panoptic surveillance is a "diagram" of a form of power, or that it contributes to making this form of power self-evident and so acceptable, is not to be against vision, or even to make vision central ("oculo-centrism"). Jay seems to start with the hypothesis that a host of diverse French thinkers were united in a sort of conspiracy to "denigrate" the visual, and that, across the Rhine, in German sociology, more "optimistic" views are to be found. If one replaces "the visual" with "the rational" in this formulation, one finds a familiar pattern of disqualification of contemporary French thought, expounded in a more shrill manner by Apel than by Habermas. For Jay really to join this polemic, he would have to show that the French

thinkers in question identified the visual with the rational, or were opposed to the one because opposed to the other. I think this would considerably compound the difficulties or incoherences in the original charge of irrationalism. Failing this, Jay owes some account of what he means by "the visual," and what it would be to "denigrate" or be against it. I discuss Habermas's views in a forthcoming essay for *New German Critique*.

5. It would be interesting to study in which ways *Annales* historians have come in fact to contribute to period films. In "Anti-Rétro" (*Cahiers du Cinéma* [July 1974]), Foucault discusses such films as *Lacombe Lucien* and *The Night Porter* in relation to the "rétro-style" in clothes and home decoration. His analysis of the return to previous styles is neither that of "simulation" or empty recycling, nor that of anamnesis.
6. Michel Foucault, *Discipline and Punish*, trans. Alan Sheridan, New York, Pantheon, 1977, p. 228.
7. Deleuze, *Foucault*, p. 65.
8. I contrast Freud's view of dreaming with that of Artemidorus, as discussed by Foucault in *The Care of the Self* (John Rajchman "Ethics after Foucault," *Socialtext* [Winter 1985]). In his early introduction to his translation of Binswanger's *Dream and Existence (Review of Existential Psychology and Psychiatry*, vol. XIX, no. 1 [1984–5]), Foucault objects to Freud that he reduces the dream to the dream-report. But in his mature work the visuality of dreaming is understood in historical rather than existential terms.
9. See the remarks in *Technologies of the Self: A Seminar with Michel Foucault* (Amherst, University of Massachusetts Press, 1988, p. 14): "What I react against is the fact that there is a breach between social history and the history of ideas. Social historians are supposed to describe how people act without thinking, and historians of ideas are supposed to describe how people think without acting. Everybody both acts and thinks. The way people act or react is linked to a way of thinking, and, of course, thinking is related to tradition."

   The idea of *visibilités* is that how people act and react when they see something is made possible by a particular way of thinking related to tradition. One might understand in this light the project Foucault announced in the last pages of the *Archeology of Knowledge* to study painting as a "discursive practice" rather than seeing it as "pure vision that must be inscribed into the materiality of space," or as "naked gesture," or as "always a way of saying" (Michel Foucault, *The Archeology of Knowledge*, trans. A. M. Sheridan Smith, New York, Harper & Row, 1976, pp. 193–194). Painting would be "shot through with positivity"; the self-evident character of its "visuality" would derive from a materially rooted way of thinking. Thought would make one particular kind of visuality seem natural or essential to painting. To study such "positivity" would consist in asking how the concepts were brought together under which paintings could be seen, where, by what means, by whom, and so forth. Through what "system of thought" were the "objects" of painting, the class of things that could be painted, specified? And how was this delimitation connected to the spaces in which it could be seen (church or château, gallery or museum) and those in which it came to be made (studio, academy, etc.) and so to the legal and economic rules that governed its ownership and its circulation? In what way was the mode of "being a painter" conceived? How did technological innovations become part of the rationality or intelligibility of the "techniques" open to painting? How did the "materiality" of painting become more than the "context" in which it was seen or made, part of the way it was conceived (as when Foucault says that Manet was the first museum painter)? And in which ways was this construction of the conceptual space of painting linked to other or associated fields in the thinking of the age? To study thus the "events of the visual" in the history of painting would suppose that people are much less free to paint than they think, that there is much more conceptual "regularity" in the practice of painting than they imagine; but since that regularity is also what opens painting to change and transformation, people are much *more* free to paint than they imagine.

10. Ian Hacking, *The Emergence of Probability*, Cambridge, England, Cambridge University Press, 1975.
11. *L'impossible prison*, ed. Michelle Perrot, Paris, Seuil, 1980, p. 44.
12. Michel Foucault, *The Birth of the Clinic*, trans. A. M. Sheridan Smith, New York, Vintage Books, 1975, p. 3.
13. Foucault, *Discipline and Punish*, p. 232.
14. *Ibid.* In *The Use of Pleasure*, Foucault also says, "I wanted first to dwell on that quite recent and banal notion of 'sexuality': to stand detached from it, *to get around its familiar self-evidence (contourner son évidence familière)* . . . it was a matter of *seeing* how an 'experience' came to be constituted in modern Western societies . . ." (*The Use of Pleasure*, trans. Robert Hurley, New York, Random House, 1985, pp. 3–4 [translation revised and emphasis added]).
15. In the "Eye of Power," referring to the work of Jean Starobinski, Foucault briefly alludes to the Rousseauistic dream of a transparent society (in *Power/Knowledge*, New York, Pantheon, 1980, pp. 152–153).
16. "An Interview with Gilles Deleuze," *History of the Present* (Spring 1986), p. 1.
17. *Ibid.*
18. *Foucault/Blanchot*, New York, Zone Books, 1987, p. 24.
19. Foucault, *The Use of Pleasure*, p. 9.
20. Richard Rorty, *Philosophy and the Mirror of Nature*, Princeton, Princeton University Press, 1979.
21. Foucault, *The Archeology of Knowledge*, p. 218.
22. Michel Foucault, *The Order of Things*, New York, Pantheon, 1970, p. 43.
23. It would be the same "separation of the eye" that characterizes the classical primacy of observation in knowledge, which would characterize the classical primacy of perception in painting. The tradition of *ut pictura poesis* would be challenged by saying that painting employs a different sort of sign, or has a different form, from poetry; it would be given to the eye alone, not to the ear. There would emerge an examination of this form through which painting presents itself to the perceiving eye, which would permit the "criticism" of paintings to be distinguished from their "commentary." This classical distinction between criticism and commentary, or between form and content, would then initiate a long debate Foucault here sees Mallarmé as changing (Foucault, *The Order of Things*, pp. 78–87).
24. Clement Greenberg's "formalism" is "classical" in this sense. He even cites Lessing in his attempt to find the principle of the turn to abstraction in modern painting in a process by which each of the "classical" arts would turn to the specific problems of its "medium." Such would be the secret of the isolation of "form" central to the avant-garde's attempt to preserve the "value" of art in an age of kitsch and socialist realism. In thus making the physical medium the privileged object of visual intelligence, Greenberg thought he had discovered the *essence* of the visuality or opticality of the visual arts. Historically, this essence was rooted in the classical self-evidence of painting as a perceived object; and there is a self-declared *positivism* in Greenberg's account of abstraction in modern art. What now seems invisible in Greenberg's conception of the essence of the visual is precisely the famous "eye" of the formalist critics that has learned to see only forms, and the way that eye was transferred to painter or sculptor as an obligation to "purify" his visual intelligence by seeing his object in "purely" formal terms. What that eye could not "see" was the other *conception* of the visual found in Duchamp, dadaism and surrealism. On this point, see Rosalind Krauss

on the optical unconscious ("In the Blink of an Eye," forthcoming). Thierry de Duve contrasts the essentiality of Greenberg's conception of the avant-garde with the "pictorial nominalism" of the Duchampian avant-garde, where the question of how the visual is itself to be named or conceived becomes a central artistic problem (in *Le nominalism pictural*, Paris, Editions Minuit, 1984).

25. Foucault, *The Order of Things*, p. 43.
26. *Ibid.*
27. See the discussion of science and ideology in *The Archeology of Knowledge*, pp. 184–186.
28. In his *Michel Foucault and the History of Reason* (forthcoming), Gary Gutting presents a clear and detailed account of Canguilhem's distinction between the history of concepts and the history of theories. His book is a good corrective to the view that Foucault was against objectivity or rationality.
29. Foucault, *Discipline and Punish*, p. 185.
30. *Ibid.*, p. 184.
31. Ian Hacking, *Representing and Intervening*, Cambridge, England, Cambridge University Press, 1983.
32. Foucault, *Archeology of Knowledge*, p. 47.
33. Foucault, *Discipline and Punish*, p. 171.
34. *Ibid.*, p. 226.
35. *Ibid.*
36. Foucault, "The Eye of Power," *Knowledge/Power*, pp. 149–150.
37. Hoy, ed., *Foucault Reader*, p. 86.
38. Michel Foucault, *The History of Sexuality*, trans. Robert Hurley, New York, Pantheon, 1978, p. 86.
39. *Ibid.*, p. 71.
40. *Ibid.*, pp. 55–57.
41. Quoted in Jacques Lagrange, "Versions de la psychanalyse dans le texte de Foucault," in *Psychanalyse à l'université* (April 1987), p. 263. In "The Ethics of Care for the Self as a Fractice of Freedom" (*Philosophy and Social Criticism*, vol. XII [Summer 1987], p. 122), Foucault says that hysteria strikes him as the "very illustration" of a struggle of being constituted as mad: "It is not altogether a coincidence that the important phenomena of hysteria have been studied exactly where there was a maximum of coercion to compel individuals to consider themselves mad."
42. Deleuze says Foucault's "conception of the visible seems pictorial, close to Delaunay, for whom light . . . created its own forms and its own movements. Delaunay said: Cézanne broke the fruit-dish, and we should not glue it together again, as the cubists do." (Cf. also *Image-Temps*, where Deleuze has similar things to say about the non-expressionist use of light in the films of Rivette [*L'Image-Temps*, Paris, Editions Minuit, 1985, pp. 20ff.].) This is of course not the only way of connecting Foucault's art of seeing to modern pictorial practices.
43. Foucault briefly discusses Plato's "famous metaphor of the eye," in Hoy, ed., *Foucault Reader*, pp. 367–368.
44. "Est-il done important de penser?" *Libération*, May 30–31, 1981.
45. Foucault, *The Use of Pleasure*, p. 11 (translation revised).
46. *Ibid.*, (translation revised).
47. *Ibid.*, p. 7.
48. Foucault, *The Birth of the Clinic*, p. xiv.
49. In his *L'Etat Providence* (Paris, Grasset, 1986), François Ewald argues that accidents constituted a new sort of danger or problem, which, for example, does not figure in the sort of catalogue of the evils of the eighteenth century one finds in Voltaire's *Candide*. With the insurance strategy used to administer the risks of accidents emerged a new sort of "juridical experience" and a new class of rights.

50. Hoy, ed., *Foucault Reader*, p. 343.
51. *Ibid.*, p. 250.
52. Foucault, *The Order of Things*, p. 328.
53. Hoy, ed., *Foucault Reader*, p. 343.
54. Foucault, *The Use of Pleasure*, p. 162. In "The Ethics of Care for the Self as a Practice of Freedom," Foucault says, "The Greeks, in fact, considered . . . the freedom of the individual as an ethical problem. But ethical in the sense that the Greeks could understand. *Ethos* was the deportment and the way to behave. It was the subject's mode of being and a certain manner of acting *visible to others*. One's *ethos was seen* by his dress, by his bearing, by his gait, by the poise with which he reacts to events, etc. For them that is the concrete expression of liberty. That is the way they 'problematized' their freedom" (p. 117 [emphasis added]).
55. Hoy, ed., *Foucault Reader*, p. 42.
56. Michel Foucault, *Madness and Civilization*, trans. Richard Howard, New York, Vintage Books, 1973, pp. 229ff.
57. Richard Howard's "the absence of the work of art" would translate *l'absence de l'oeuvre* not *l'absence d'oeuvre*. Alan Sheridan thinks the term means "an unproductive idleness, outside human achievement" (in Alan Sheridan, *Foucault: The Will to Truth*, London, Tavistock, 1980, p. 15).
58. Deleuze tends to pair "discursivity" with words and concepts and "visibility" with things and intuitions or sensibilities. But, according to another reading of Foucault's idiom, one can "say" things with images and space just as one can "show" things with words or sentences. Language is one way of "spatializing" or "visualizing," and there are "evidences" of discourses just as there are of "sensibilities." Conversely spaces or images can make statements. Thus, for example, just as much as the letters on the typewriter keyboard, it might count as an *énoncé* that, during the colonial period, maps were made with Britain at the center. Perhaps the relation between ways of seeing and ways of saying in Foucault lies in the histories of the tacit thinking that underlies them, and does not easily match with the traditional distinctions between concept and intuition, or word and thing.
59. Deleuze, *L'Image-Temps*, p. 365.
60. "Foucault, passe-frontières de la philosophic," *Le Monde*, September 6, 1986.
61. Hoy, ed., *Foucault Reader*, p. 46.

B. Reflections on the Practice of Observation

14

Excerpt from *The Observation of Savage Peoples*

Joseph-Marie Baron de Gérando

Let us review the main faults of the observations on savage man made by these explorers, and the gaps that they have left in their accounts. When we realize what they have not done, we shall see better what remains to be done.

The first fault that we notice in the observations of explorers on savages is their incompleteness; it was only to be expected, given the shortness of their stay, the division of their attention, and the absence of any regular tabulation of their findings. Sometimes, confining themselves to the study of some isolated individuals, they have given us no information on their social condition, and have thus deprived us of the means of estimating the influence which these social relations might have on individual faculties. Sometimes, pausing on the smallest details of the physical life of the savages, they have given us scarcely any details of their moral customs. Sometimes, describing the customs of grown men, they have failed to find out about the kind of education received in childhood and youth: and above all, preoccupied almost entirely with the external and overt characteristics of a people, of its ceremonies and of its dress, they have generally taken too little care to be initiated in the far more important circumstances of its theoretical life, of its needs, its ideas, its passions, its knowledge, its laws. They have described forms rather than given instructive reports; they have marked certain effects, and explained scarcely any causes.

Source: F.C.T Moore (tr), *The Observation of Savage Peoples* (London: Routledge and Kegan Paul, 1969), pp. 65–70.

Further, such insufficient observations have not always been very certain or authentic, whether because they have sometimes been too particular, and explorers have wished to judge a society by a few of its members, a character by a few actions, or because they have sometimes confined themselves to hearsay, to the stories of the Savages whom they met, and who perhaps were not properly understood, perhaps not well-versed in what was asked, and perhaps had no interest in telling the truth, or at least in making it known in its entirety.

We should add that these observations have been badly ordered, and even in many cases quite without order. The explorers had not enough understood that there is a natural connection between the various facts that one gathers about the condition and character of societies, that this order is necessary to the precision of the individual facts, and that often some of them should serve as preparation for the others. We should study effects before trying to go back to first principles; observe individuals before trying to judge the society; become acquainted with domestic relations inside families before examining the political relations of society; and above all we should aim at full mutual understanding when we speak to men before basing certain conclusions on the accounts that we claim to have received.

Often explorers have based the accounts that they bring us on incorrect or at least on dubious hypotheses. For example, they habitually judge the customs of Savages by analogies drawn from our own customs, when in fact they are so little related to each other. Thus, given certain actions, they suppose certain opinions or needs because among us such actions ordinarily result from these needs or opinions. They make the Savage reason as we do, when the Savage does not himself explain to them his reasoning. So they often pronounce excessively severe judgments on a society accused of cruelty, theft, licentiousness, and atheism. It were wiser to gather a large number of facts, before trying to explain them, and to allow hypothesis only after exhausting the light of experience.

In the case of the accounts of explorers there is another cause of uncertainty, a fault of language rather than of imperfect observation, namely that the terms used to pass on to us the results of their observations are often in our own language of vague and ill-determined meaning. Consequently, we are in danger of taking their accounts in a way which they did not intend. This happens particularly when they try to record the religious, moral, and political beliefs of a people. It happens too when instead of giving a detailed and circumstantial account of what they have actually seen, they limit themselves to summary descriptions of the impressions which they received, and of the general judgments which they inferred on the character of peoples. Yet this drawback could easily have been avoided by making it a policy either to describe things without judging them, or to choose expressions whose sense is more agreed, or to give a precise stipulation of the sense in which one intends their use.

This is not the place to enumerate the inaccuracies springing from a lack of impartiality in explorers, from prejudices imposed by their particular opinions, from the interests of vanity or the impulse of resentment. The character of the worthy men today devoting themselves to this noble undertaking is a sufficient guarantee that such a stamp will never shape their accounts. But explorers with the purest and most honest intentions have often been led into error about the character of peoples by the behaviour they meet with. They have inferred too lightly from the circumstances of their reception, conclusions about the absolute and ordinary character of the men among whom they have penetrated. They have failed to consider sufficiently that their presence was bound to be a natural source of fear, defiance, and reserve; that reasons of policy might exaggerate this unusual circumspection; that the memory of former attacks might have left dark prejudices in the mind of such peoples; that a community might be gentle and sociable, and yet believe itself in a state of natural war with strangers whose intentions are unknown; and finally that for a just estimate of the character of a tribe, one should first leave time for the reactions of astonishment, terror, and anxiety bound to arise in the beginning to be dispelled, and secondly one should be able to be initiated into the ordinary relations which the members of the community have with each other.

But of all the regrets left by the accounts of the explorers who went before you, the strongest is their failure to tell us of the language of the peoples visited. In the first place, the scanty information which they do give lacks precision and exactness, whether because they fail to record how they went about questioning the Savages, or because they themselves have often taken little care to pose the questions properly. The demonstrative and natural gestures which they have used to ask the Savages the names of objects were often themselves liable to considerable uncertainty; one cannot know if those who were questioned understood the gestures in the same way as the explorers who were using them, and so whether they were giving proper replies to their questions. Further, to provide us with some useful and positive ideas of the idioms of savage peoples, it was wrong for explorers to limit themselves as they did to taking at random names of various objects with scarcely any relation between them; at least a family of analogous ideas should have been followed up, when it was impossible to make a record of the whole language, so that some judgment could be made on the generation of terms, and on the relations between them; it was not enough to be content with some detached words; but it would have been sensible to record whole sentences to give some idea of the construction of discourse. Further, one should have discovered whether these words were simple, or composite, as their length would often lead us to suppose; whether they were qualified by any articles or particles; and finally whether they were inflected or remained in the absolute, and whether they were liable to any kind of grammatical laws.

Failing to acquaint themselves thoroughly with the idiom of savage peoples, explorers have been powerless to draw on perhaps the most interesting ideas that could have been available. They have been unable to pass on the traditions that such peoples may preserve of their origin, of the changes that they have undergone, and of the various details of their history; traditions which perhaps would have thrown great light on the important question of how the world was peopled, and on the various causes of the present state of these societies. They have been unable to explain the significance of a mass of ceremonies and customs which are probably no more than allegorical; they have given us bizarre descriptions which tickle the idle curiosity of the many, but which offer no useful instruction to the philosophically minded. Lacking the means to carry on connected conversation with such peoples, they have been able to form only very hazardous and vague ideas of their opinions and notions; finally, they have been unable to provide us with these data, as revealing as they are abundant, that the language of a society presents on its way of seeing and feeling, and on the most intimate and essential features of its character.

15

Roles in Sociological Field Observations

Raymond L. Gold

Buford Junker has suggested four theoretically possible roles for sociologists conducting field work.[1] These range from the complete participant at one extreme to the complete observer at the other. Between these, but nearer the former, is the participant-as-observer; nearer the latter is the observer-as-participant. As a member of Junker's research team, I shared in the thinking which led to conceptualization of these research roles. After the work of the team was completed, I continued the search for insight regarding processes of interaction learning in field observation in a special study of my own.[2] A considerable portion of this study was devoted to exploration of the dimensions of Junker's role-conceptions and their controlling effects on the product of field study.

My aim in this paper is to present extensions of Junker's thinking growing out of systematic interviews with field workers whose experience had been cast in one or more of these patterns of researcher-subject relationship. All of these field workers had gathered data in natural or nonexperimental settings. I would like in this paper to analyze generic characteristics of Junker's four field observer roles and to call attention to the demands each one places on an observer, as a person and as a sociologist plying his trade.

Every field work role is at once a social interaction device for securing information for scientific purposes and a set of behaviors in which an observer's self is involved.[3] While playing a field work role and attempting to take the role of an informant, the field observer often attempts to master hitherto strange or only generally understood universes of discourse relating to many attitudes and behaviors. He continually introspects, raising endless questions

Source: *Social Forces*, 36(3) (1958): 217–223.

about the informant and the developing field relationship, with a view to playing the field work role as successfully as possible. A sociological assumption here is that the more successful the field worker is in playing his role, the more successful he must be in taking the informant's role. Success in both role-taking and role-playing requires success in blending the demands of self-expression and self-integrity with the demands of the role.

It is axiomatic that a person who finds a role natural and congenial, and who acts convincingly in it, has in fact found how to balance role-demands with those of self. If need be he can subordinate self-demands in the interest of the role and role-demands in the interest of self whenever he perceives that either self or role is in any way threatened. If, while playing the role, someone with whom he is interacting attacks anything in which he has self-involvement, he can point out to himself that the best way to protect self at the moment is to subordinate (or defer) self-expression to allow successful performance in the role. In other words, he uses role to protect self. Also, when he perceives that he is performing inadequately in the role he can indicate to himself that he can do better by changing tactics. Here he uses self as a source of new behaviors to protect role. The case of using role to protect self from perceived threat is one of acute self-consciousness, a matter of diminishing over-sensitivity to self-demands by introspectively noting corresponding demands of role. The case of using self to protect role from perceived threat is one of acute role-consciousness, a matter of diminishing over-sensitivity to role-demands by introspectively indicating that they are disproportionately larger than those of self. Both cases represent situations in which role-demands and self-demands are out of balance with each other as a result of perceived threat, and are then restored to balance by appropriate introspection.

Yet, no matter how congenial the two sets of demands seem to be, a person who plays a role in greatly varied situations (and this is especially true of a sociologist field observer) sometimes experiences threats which markedly impair his effectiveness as an interactor in the situation. When attempting to assess informational products of field work, it is instructive to examine the field worker's role-taking and role-playing in situations of perceived, but unresolved, threat. Because he defines success in the role partly in terms of doing everything he can to remain in even threatening situations to secure desired information, he may find that persevering is sometimes more heroic than fruitful.

The situation may be one in which he finds the informant an almost intolerable bigot. The field worker decides to stick it out by attempting to subordinate self-demands to those of role. He succeeds to the extent of refraining from "telling off" the informant, but fails in that he is too self-conscious to play his role effectively. He may think of countless things he would like to say and do to the informant, all of which are dysfunctional to role-demands, since his role requires taking the role of the other as an informant, not as a bigot. At the extreme of nearly overwhelming self-consciousness, the field worker may still protect his role by getting out of the situation while the getting is good. Once out and in the company of understanding colleagues, he will finally be able to achieve self-expression (i.e., finally air his views of the informant) without damaging the field role.[4]

Should the situation be such that the field worker finds the informant practically inscrutable (i.e., a "bad" informant), he may decide to persevere despite inability to meet role-taking and role-playing demands. In this situation he becomes acutely role-conscious, since he is hypersensitive to role-demands, hyposensitive to self. This partial breakdown of his self-process thwarts his drawing on past experiences and current observations to raise meaningful questions and perceive meaningful answers. At the extreme, a role-conscious field worker may play his role so mechanically and unconvincingly that the informant, too, develops role-and-self problems.

The following discussion utilizes these conceptions of role and self to aid in analyzing field work roles as "master roles" for developing lesser role-relationships with informants.[5] While a field worker cannot be all things to all men, he routinely tries to fit himself into as many roles as he can, so long as playing them helps him to develop relationships with informants in his master role (i.e., participant-as-observer, etc.).

Complete Participant

The true identity and purpose of the complete participant in field research are not known to those whom he observes. He interacts with them as naturally as possible in whatever areas of their living interest him and are accessible to him as situations in which he can play, or learn to play, requisite day-to-day roles successfully. He may, for example, work in a factory to learn about inner-workings of informal groups. After gaining acceptance at least as a novice, he may be permitted to share not only in work activities and attitudes but also in the intimate life of the workers outside the factory.

Role-pretense is a basic theme in these activities. It matters little whether the complete participant in a factory situation has an upper-lower class background and perhaps some factory experience, or whether he has an upper-middle class background quite divorced from factory work and the norms of such workers. What really matters is that he knows that he is pretending to be a colleague. I mean to suggest by this that the crucial value as far as research yield is concerned lies more in the self-orientation of the complete participant than in his surface role-behaviors as he initiates his study. The complete participant realizes that he, and he alone, knows that he is in reality other than the person he pretends to be. He must pretend that his real self is represented by the role, or roles, he plays in and out of the factory situation in relationships with people who, to him, are but informants, and this implies an interactive construction that has deep ramifications. He must bind the mask of pretense to himself or stand the risk of exposure and research failure.

In effect, the complete participant operates continually under an additional set of situational demands. Situational role-and-self demands ordinarily tend to correspond closely. For this reason, even when a person is in the act of learning to play a role, he is likely to believe that pretending to have achieved this correspondence (i.e., fourflushing) will be unnecessary when he can actually "be himself" in the role. But the complete observer simply cannot "be himself"; to do so would almost invariably preclude successful pretense. At the very least, attempting to "be himself" – that is, to achieve self-realization in pretended roles – would arouse suspicion of the kind that would lead others to remain aloof in interacting with him. He must be sensitive to demands of self, of the observer role, and of the momentarily pretended role. Being sensitive to the set of demands accompanying role-pretense is a matter of being sensitive to a large variety of overt and covert mannerisms and other social cues representing the observer's pretended self. Instead of being himself in the pretended role, all he can be is a "not self," in the sense of perceiving that his actions are meaningful in a contrived role.

The following illustration of the pretense of a complete participant comes from an interview with a field worker who drove a cab for many months to study big-city cab drivers. Here a field worker reveals how a pretended role fosters a heightened sense of self-awareness, an introspective attitude, because of the sheer necessity of indicating continually to himself that certain experiences are merely part of playing a pretended role. These indications serve as self-assurance that customers are not really treating *him* as they seem to do, since he is actually someone else, namely, a field worker.

Well, I've noticed that the cab driver who *is* a cab driver acts differently than the part-time cab drivers, who don't think of themselves as real cab drivers. When somebody throws a slam at men who drive only part of the year, such as, "Well, you're just a goddamn cab driver!," they do one of two things. They may make it known to the guy that they are not a cab driver; they are something else. But as a rule, that doesn't work out, because the customer comes back with, "Well, if you're not a cab driver what the hell are you driving this cab for?" So, as a rule, they mostly just rationalize it to themselves by thinking, "Well, this is not my role or the real me. He just doesn't understand. Just consider the source and drop it." But a cab driver who *is* a cab driver, if you make a crack at him, such as, "You're just a goddamn cab driver!" he's going to take you out of the back seat and whip you.

Other complete participant roles may pose more or less of a challenge to the field worker than those mentioned above. Playing the role of potential convert to study a religious sect almost inevitably leads the field worker to feel not only that he has "taken" the people who belong to the sect, but that he has done it in ways which are difficult to justify. In short, he may suffer severe qualms about his mandate to get information in a role where he pretends to be a colleague in moral, as well as in other social, respects.

All complete participant roles have in common two potential problems; continuation in a pretended role ultimately leads the observer to reckon with one or the other. One, he may become so self-conscious about revealing his true self that he is handicapped when attempting to perform convincingly in the pretended role. Or two, he may "go native," incorporate the role into his self-conceptions and achieve self-expression in the role, but find he has so violated his observer role that it is almost impossible to report his findings. Consequently, the field worker needs cooling-off periods during and after complete participation, at which times he can "be himself" and look back on his field behavior dispassionately and sociologically.

While the complete participant role offers possibilities of learning about aspects of behavior that might otherwise escape a field observer, it places him in pretended roles which call for delicate balances between demands of role and self. A complete participant must continually remind himself that, above all, he is there as an observer: this is his primary role. If he succumbs to demands of the pretended role (or roles), or to demands of self-expression and self-integrity, he can no longer function as an observer. When he can defer self-expression no longer, he steps out of the pretended role to find opportunities for congenial interaction with those who are, in fact, colleagues.

Participant-as-Observer

Although basically similar to the complete observer role, the participant-as-observer role differs significantly in that both field worker and informant are aware that theirs is a field relationship. This mutual awareness tends to minimize problems of role-pretending; yet, the role carries with it numerous opportunities for compartmentalizing mistakes and dilemmas which typically bedevil the complete participant.

Probably the most frequent use of this role is in community studies, where an observer develops relationships with informants through time, and where he is apt to spend more time and energy participating than observing. At times he observes formally, as in scheduled interview situations; and at other times he observes informally – when attending parties, for example. During early stages of his stay in the community, informants may be somewhat uneasy about him in both formal and informal situations, but their uneasiness is likely to disappear when they learn to trust him and he them.

But just when the research atmosphere seems ripe for gathering information, problems of role and self are apt to arise. Should field worker and informant begin to interact in much the same way as ordinary friends, they tend to jeopardize their field roles in at least two important ways. First, the informant may become too identified with the field worker to continue functioning as merely an informant. In this event the informant becomes too much of an observer. Second, the field worker may over-identify with the informant and start to lose his research perspective by "going native." Should this occur the field worker may still continue going through the motions of observing, but he is only pretending.

Although the field worker in the participant-as-observer role strives to bring his relationship with the informant to the point of friendship, to the point of intimate form, it behooves him to retain sufficient elements of "the stranger" to avoid actually reaching intimate form. Simmel's distinction between intimate content and intimate form contains an implicit warning that the latter is inimical to field observation.[6] When content of interaction is intimate, secrets may be shared without either of the interactors feeling compelled to maintain the relationship for more than a short time. This is the interaction of sociological strangers. On the other hand, when form of interaction is intimate, continuation of the relationship (which is no longer merely a field relationship) may become more important to one or both of the interactors than continuation of the roles through which they initiated the relationship.

In general, the demands of pretense in this role, as in that of the complete participant, are continuing and great; for here the field worker is often defined by informants as more of a colleague than he feels capable of being. He tries to pretend that he is as much of a colleague as they seem to think he is, while searching to discover how to make the pretense appear natural and convincing. Whenever pretense becomes too challenging, the participant-as-observer leaves the field to re-clarify his self-conceptions and his role-relationships.

Observer-as-Participant

The observer-as-participant role is used in studies involving one-visit interviews. It calls for relatively more formal observation than either informal observation or participation of any kind. It also entails less risk of "going native" than either the complete participant role or the participant-as-observer role. However, because the observer-as-participant's contact with an informant is so brief, and perhaps superficial, he is more likely than the other two to misunderstand the informant, and to be misunderstood by him.

These misunderstandings contribute to a problem of self-expression that is almost unique to this role. To a field worker (as to other human beings), self-expression becomes a problem at any time he perceives he is threatened. Since he meets more varieties of people for shorter periods of time than either the complete participant or the participant-as-observer, the observer-as-participant inclines more to feel threatened. Brief relationships with numerous informants expose an observer-as-participant to many inadequately understood universes of discourse that he cannot take time to master. These frustratingly brief encounters with informants also contribute to mistaken perceptions which set up communication barriers the field worker may not even be aware of until too late. Continuing relationships with apparently threatening informants offer an opportunity to redefine them as more congenial partners in interaction, but such is not the fortune of a field worker in this role. Consequently, using his prerogative to break off relationships with threatening informants, an observer-as-participant, more easily than the other two, can leave the field almost at will to regain the kind of role-and-self balance that he, being who he is, must regain.

Complete Observer

The complete observer role entirely removes a field worker from social interaction with informants. Here a field worker attempts to observe people in ways which make it unnecessary for them to take him into account, for they do not know he is observing them or that, in some sense, they are serving as his informants. Of the four field work roles, this alone is almost never the dominant one. It is sometimes used as one of the subordinate roles employed to implement the dominant ones.

It is generally true that with increasingly more observation than participation, the chances of "going native" become smaller, although the possibility of ethnocentrism becomes greater. With respect to achieving rapport in a field relationship, ethnocentrism may be considered a logical opposite of "going native." Ethnocentrism occurs whenever a field worker cannot or will not interact meaningfully with an informant. He then seemingly or actually rejects the informant's views without ever getting to the point of understanding them. At the other extreme, a field worker who "goes native" passes the point of field rapport by literally accepting his informant's views as his own. Both are cases of pretending to be an observer, but for obviously opposite reasons. Because a complete observer remains entirely outside the observed interaction, he faces the greatest danger of misunderstanding the observed. For the same reason, his role carries the least chance of "going native."

The complete observer role is illustrated by systematic eavesdropping, or by reconnaissance of any kind of social setting as preparation for more intensive study in another field role. While watching the rest of the world roll by, a complete observer may feel comfortably detached, for he takes no self-risks, participates not one whit. Yet, there are many times when he wishes he could ask representatives of the observed world to qualify what they have said, or to answer other questions his observations of them have brought to mind. For some purposes, however, these very questions are important starting points for subsequent observations and interactions in appropriate roles. It is not surprising that reconnaissance is almost always a prelude to using the participant-as-observer role in community study. The field worker, feeling comfortably detached, can first "case" the town before committing himself to casing *by* the town.

Conclusions

Those of us who teach field work courses or supervise graduate students and others doing field observations have long been concerned with the kinds of interactional problems and processes discussed above. We find such common "mistakes" as that of the beginner who over-identifies with an informant simply because the person treats him compassionately after others have refused to grant him an interview. This limited, although very real, case of "going native" becomes much more understandable to the beginner when we have analyzed it for him sociologically. When he can begin utilizing theory of role and self to reflect on his own assets and shortcomings in the field, he will be well on the way to dealing meaningfully with problems of controlling *his* interactions with informants.

Beyond this level of control, sophistication in field observation requires manipulating informants to help them play their role effectively. Once a field worker learns that a field relationship in process of being structured creates role-and-self problems for informants that are remarkably similar to those he has experienced, he is in a position to offer informants whatever kinds of "reassurances" they need to fit into their role. Certainly a field worker has mastered his role only to the extent that he can help informants to master theirs. Learning this fact (and doing something about it!) will eliminate nearly all excuses about "bad" or "inept" informants, since, willy-nilly, an informant is likely to play his role only as fruitfully or as fruitlessly as a field worker plays his.[7]

Experienced field workers recognize limitations in their ability to develop relationships in various roles and situations. They have also discovered that they can maximize their take of information by selecting a field role which permits them to adjust their own role-repertories to research objectives. Objectively, a selected role is simply an expedient device for securing a given level of information. For instance, a complete participant obviously develops relationships and frames of reference which yield a somewhat different perspective of the subject matter than that which any of the other field work roles would yield. These subjective and objective factors come together in the fact that degree of success in securing the level of information which a field role makes available to a field worker is largely a matter of his skill in playing and taking roles.

Each of the four field work roles has been shown to offer advantages and disadvantages with respect to both demands of role and self and level of information. No attempt has been made in this report to show how a sociological conception of field work roles can do more than provide lines of thought and action for dealing with problems and processes of field interaction. Obviously, however, a theory of role and self growing out of study of field interaction is in no sense limited to that area of human activity. Learning to take and play roles, although dramatized in the field, is essentially the same kind of social learning people engage in throughout life.

In any case, the foregoing discussion has suggested that a field worker selects and plays a role so that he, being who he is, can best study those aspects of society in which he is interested.

Notes

1. Buford Junker, "Some Suggestions for the Design of Field Work Learning Experiences," in Everett C. Hughes, et al, Cases on Field Work (hectographed by The University of Chicago, 1952), Part III-A.
2. Raymond L. Gold, Toward a Social Interaction Methodology for Sociological Field Observation, unpublished Ph. D. dissertation, University of Chicago, 1954.
3. To simplify this presentation, I am assuming that the field worker is an experienced observer who has incorporated the role into his self-conceptions. Through this incorporation, he is self-involved in the role and feels that self is at stake in it. However, being experienced in the role, he can balance role-demands and self-demands in virtually all field situations, that is, all except those to be discussed shortly.
4. An inexperienced field worker might "explode" on the spot, feeling that role and self are not congenial in this *or any other* situation. But an experienced field worker would leave such a situation as gracefully as possible to protect the role, feeling that role and self are not congenial in *this* situation only.
5. Lesser role-relationships include all achieved and ascribed roles which the field worker plays in the act of developing a field relationship with an informant. For example, he may become the "nice man that old ladies can't resist" as part of his over-all role-reportoire in a community study. Whether he deliberately sets out to achieve such relationships with old ladies or discovers that old ladies ascribe him "irresistible" characteristics, he is still a participant-as-observer who interacts with local old ladies as a "nice man." Were he not there to study the community, he might choose *not* to engage in this role-relationship, especially if being irresistible to old ladies is not helpful in whatever master role(s) brought him to town. (Cf. any experienced community researcher.)
6. In other words, intimacy is not based on the *content* of the relationship. . . . Inversely, certain external situations or moods may move us to make very personal statements and confessions, usually reserved for our closest friends only, to relatively strange people. But in such cases we nevertheless feel that this 'intimate' *content* does not yet make the relation an intimate one. For in its basic significance, the whole relation to these people is based only on its general, unindividual ingredients. That 'intimate' content, although we have perhaps never revealed it before and thus limit it entirely to this particular relationship, does nevertheless not become the basis of its form, and thus leaves it outside the sphere of intimacy." K. H. Wolff (ed.), *The Sociology of Georg Simmel* (Glencoe, Illinois: The Free Press, 1950), p. 127.

7. In a recent article on interviewing, Theodore Caplow also recognizes the key role played by the field worker in structuring the field relationship. He concludes, "The quality and quantity of the information secured probably depend far more upon the competence of the interviewer than upon the respondent." "The Dynamics of Information Interviewing," *American Journal of Sociology*, LXII (September 1956), 169. Cf. also the studies by Junker and Gold, *op. cit.*

16

Performing Ethnography and the Ethnography of Performance

Paul Atkinson

Norman Denzin's (2003) new book is part of a personal programme of publications for a revised interactionist sociology, and for a cycle of revisionist accounts of research methods (Denzin, 2001a, b, c, 2002a, b). This latest chapter in his evolving intellectual career is devoted to the overall theme of performance and performance ethnography. It is, in turn, a contribution to the performative turn in the social and cultural disciplines.

Denzin's account relates to the work of a relatively small but highly creative and productive group of American sociologists. They are affiliated through the Society for the Study of Symbolic Interaction, although they do not confine themselves to either classical or second Chicago-school symbolic interactionism. They promote innovative approaches to interactionist sociology, and novel approaches to the work of ethnographic research. In particular, they are engaged in experimental writing exercises. In addition to the publication of ethnographic texts, this group are also committed to the spoken presentation of their work – which takes the form of autobiographical narratives and poems. Sites of performance include special sessions in conferences such as the annual Couch-Stone symposium and the Society for the Study of Symbolic Interaction annual meetings. There is, therefore, the potential for a reflexive relationship between the subject-matter of sociological scrutiny – everyday performance – and the mode of its representation – performance ethnography. Such performative innovations are, moreover,

Source: *British Journal of Sociology of Education*, 25(1) (2004): 107–114.

claimed to have a political as well as a methodological significance. Indeed, in this current formulation of ethnographic and other qualitative research, the personal, the political and the performance are inseparable aspects of the research act. The core members of this set include Laurel Richardson and Patricia Clough (see Clough, 1998, 2000; Richardson, 2000, 2002a). The forthcoming third edition of the *Handbook of qualitative research*, edited by Denzin and Yvonna Lincoln, will reflect this new approach. The group have strong links with the network centred on Carolyn Ellis and Arthur Bochner (for example, Ellis & Bochner, 1996). The Sage journal edited by Denzin and Lincoln, *Qualitative Inquiry*, contains many papers in that vein. The *New Ethnography* series edited by Ellis and Bochner, published by AltaMira Press, also reflects similar tendencies. These new personal, textual and performance approaches to ethnographic and interpretative research exert a disproportionately large influence on the literature of qualitative methodology.

For various reasons, the potential interest for sociologists of education in these developments, and in other approaches to the performative aspects of social action, are considerable. In the course of this essay I shall present a brief overview of Denzin's own contribution, comment on some of his collaborators, make reference to other accounts of performance with which they do not explicitly engage and suggest some possible areas of significance for readers of this journal.

Denzin's approach to performance ethnography derives from several sources. First, there is the classic pragmatist interactionism of George Herbert Mead. This tradition stresses the active work of exploration and engagement in any form of social or natural inquiry, and Mead's version of the Act is an essentially performative one. Second, there is a tradition of American race theory and writing that stresses performance: from the revolutionary work of Du Bois, in his call for a black theatre that captured black life and culture, to contemporary work, like that of bell hooks, that address a black performance aesthetic. Third, it reflects the calls of Henry Giroux and others for performative acts that recuperate possible meanings for people's lives in the face of cynicism and impotence. Fourth, it also draws on the tradition of testimonio and memoria, first-person accounts with political implications, that provide witness to social injustice and exclusion; they include enactments of chicano/chicana experience, and are themselves drawn on by critical race theorists.

These contemporary aspects of cultural and critical scholarship are all concerned with the enactment of cultural experience. In the hands of Denzin and his network, they are aligned with the autoethnographic methodological turn. Autoethnography has a variety of related connotations. They all imply a radical weakening or obliteration of the distinction between the observer and the observed, the researcher and the topic of inquiry. The ethnographer herself/himself becomes as much the topic of reflection as the research resource. The autoethnographic turn therefore implies a thoroughgoing reflexivity, obliterating any distinction between the personal and the public, between research and experience. It thus brings into conjunction reflexive ethnography and the contemporary interest in auto/biographical research strategies.

Denzin is by no means alone in giving prominence to the performative aspects of the everyday. A number of strands in cultural studies, sociology and anthropology have treated performance as a guiding principle. John Tulloch (1999) synthesizes research on a variety of topics into a conceptual framework that engages with performative analysis. His own analysis ranges over 'high' and 'low' culture, 'expert' and 'lay' cultural productions. He draws on the literatures of anthropology and theatre studies in stressing the local and situated performance of mundane and expert understandings. The anthropological preoccupation with performativity is brought together in the collection of essays edited by Felicia Hughes-Freeland (1998), with analyses of ritual, dance, spectacles and everyday social encounters. Schieffelin's essay in that edited collection provides a useful overview. He draws on Goffman and the symbolic interactionists to suggest that performance is inherent in human conduct and in reality construction. The only limitation with such accounts is that they treat performance as so pervasive in cultural affairs that there remains little to say specifically about the performing arts or the extra-ordinary performances of art, music, opera and so on. The specificities of performance and its social organization are in danger of being lost if 'performance' is treated as a fundamental and generic process of social life. The notion has become prominent among social anthropologists, for instance. The current emphasis on performance among anthropologists represents in part a corrective to the analysis of rituals and other collective observances in terms of their systems of signification. Structuralist and semiotic approaches to ritual had little or nothing to say about rites in action, and had equally little to say about the practical engagement of social actors (Schieffelin, 1998). In addressing the actions of performers themselves, the current anthropological emphasis is also a corrective to analyses of culture that pay exclusive attention to the audiences and the consumption of cultural artifacts.

The performative move that Denzin and his collaborators make is the creation of performance works and texts. In other words, the ethnographic text is the basis for a performance in its own right. This reflects the textual experimentations that Richardson, Clough and others have promoted among American sociologists. There have been several approaches to the development of unconventional and experimental texts for the reportage of interpretative ethnography in recent years. The characteristic that they share is the use of autobiographical 'literary' modes – first-person pieces of autobiography, or poems of personal reflection and experience. The second half of Denzin's book therefore consists of 'performance' pieces that are intended to be read aloud – although they can obviously be read privately just as well. These are autobiographical fragments. They are occasioned by a series of biographical events: the death of Denzin's parents, the death of Carl Couch (a close friend and intellectual sparring partner in the interactionist movement), the purchase of a cabin in rural Montana. The latter brings Denzin in touch with a literature of the American West, and a personal response to the landscape. In Denzin's own words (p. 133) these pieces are 'part memoir, part essay, part autoethnography'. This is an attempt to embed himself in memories of the mundane, in order to find what is 'sacred' within it.

So Denzin writes and performs fragmentary texts of memory. In a short piece he intertwines a childhood admiration for Mickey Mantle, a baseball hero, and an episode that reveals the violence of his father, who drank heavily. The recollection is framed by the recent death of Mantle as well as his parents. The baseball star had a troubled life too. It is all remembered in Montana. 'Performing Montana' is a mélange of reflections on celebratory events: the Red Lodge Festival, Finn Day (at one time a substantial proportion of the State's population were of Finnish origin) and Montana Day (when the people of Red Lodge have the chance to dress up as cowboys and Indians), and the Red Lodge Mountain Man Rendezvous, an annual event that re-creates Montana of the 1840s 'authentically'. Denzin also reflects on trout fishing in Montana, an activity that connects him simultaneously with the memory of Carl Couch, the ecstatic moment of catching a trout, and yet another way of performing Montana. (A way that expensively kitted-out holiday-makers and weekenders enact their own sense of the 'authentic' outdoors.) These and other pieces in the book are, therefore, shot through with personal reminiscence, reflections on the cultural codes of performed authenticity, spectacles of heritage, and a strong sense of the local.

It is difficult to quarrel with these texts, or with texts like them. Whether read off the printed page, or heard as performance pieces (I have heard Denzin read one of the texts) they are highly engaging. They combine autobiographical reflections with more sociologically imbued understandings of what they describe. Denzin also has a strong sense of various literary traditions, especially those of Western literature, that help to shape his own sensibilities. His pieces have, therefore, an intertextual quality. In addition to literary sources, he also refers to and quotes the work of fellow social scientists such as Laurel Richardson.

In general, however, I find some problems with this genre of (auto) ethnographic interpretative writing. I want to suggest that in the pursuit of a performative ethnography, Denzin, Richardson, Clough, Ellis, Bochner and others are reflecting far too narrow a sense of performance. The scope of their texts seems unnecessarily restricted. They insist that the personal is political, but then remain in the realm of the personal and relegate the political to an economy of personal experience and emotional response. I also want to suggest that, in the great scheme of things, their shared sense of the performative is too narrowly ethnocentric.

I am not alone in thinking that the 'autoethnographic' mode carries with it dangers as well as intriguing possibilities. I do not think that modesty is a great virtue, and that false modesty is definitely a vice. But as a social scientist, I am convinced that I am not the most interesting or important topic for inquiry. Indeed, I am perfectly convinced that taking seriously and finding interest in my fellow men and women is the main point of being a social scientist. It is certainly what inspires the spirit of ethnographic research. There is a terrible danger, it seems to me, of collapsing the social world into one's own lifeworld. There is, therefore a terrible risk of trivializing complex and serious social issues. To take but one example that Denzin himself cites, consider the performance text by Laurel Richardson (2002b). It is a personal reflection on the World Trade Centre attack. It begins:

> When I hear of the airplanes and the towers, my first thoughts are – the children . . . What will the children be told.
>
> And then I see that the children are being told, as the adults are through television cameras and media voices. The children are seeing the airplane and the second tower, and the airplane/tower, airplane/tower over and over until it's All Fall Down. And All Fall Down again and again.

It continues in this vein. It seems to be about Richardson's family and her immediate response to the televised events. As a contribution to sociology it seems to me to have no value whatsoever. What to tell the children is not the most important point to make now or then. Richardson is good at seeing things through children's eyes, and elsewhere writes rather well in the person of herself as a child, but the infantilism of this particular piece is revealing. This is, incidentally, just one example of an entire genre of post-9/11 reflections and autoethnographic pieces by American sociologists. A recent issue of Qualitative Inquiry was devoted to the topic. While we may all want to agree that in many ways the 'personal is political', that does not mean that the political is reducible to the personal. The social and the political in this context risk being transmuted into matters of experience. Indeed, a great deal of current ethnographic and interpretative sociology in the United States is being justified in terms of capturing and evoking people's experience. Now one cannot deny the social and cultural significance of memory, the emotions, sentimental work, autobiography and biography. Indeed, in many ways these have been at the forefront of research agendas since the emergence of strategies like the life-history. But 'experience' has to be recognized as a cultural artifact, organized in terms of socially shared cultural categories.

Furthermore, the forms of these autoethnographic and performance pieces are reflections of a rather restricted genre. I am repeatedly struck, when reading across the growing corpus of such work, that there is an odd uniformity in the writing styles. There is remarkably little textual experimentation here. While it may still be unconventional for academic work to be written in autobiographical and literary styles, a number of different styles of textual presentation have been essayed – and they go beyond the autobiographical. Intellectually more radical were the experiments by various sociologists of science and technology – largely inspired by Michael Mulkay – in which one-act plays, dialogues, encyclopaedia entries and other forms were used. These were deployed in order to develop specific epistemological and methodological points (see Mulkay, 1985; Ashmore *et al.*, 1989). They were not indulged in just in order to create personal reflections. Their version of reflexivity was a more theoretically sophisticated than the experiential texts of autoethnography that have become widespread, and that Denzin's work contributes to.

There are, after all, many forms of performance that can be explored ethnographically. The ethnography of performance is potentially more rich and varied – and of potentially greater analytic significance – than performance texts derived from ethnography. The creation of performance pieces leaves unexplored the nature of performance itself. It would, I think, deepen and widen Denzin's understanding of performance were he and his colleagues to take a broader perspective on performance itself. There is a growing literature on performance and the spectacular, from studies of spectacular displays of science and culture (Karp & Levine, 1991; Macdonald, 2002) or the collective memorialization of tragedy (Tota, 2003), to the anthropology of the bull-fight (Marvin, 1988; Pink, 1998), to praise-singers in West Africa (Ebron, 2002), song and national identity in Tanzania (Askew, 2002), the enactment of Welsh identity (Trosset, 1993; Davies, 1998), the performance of capoeira in Brazil (Lewis, 1992), shadow plays in Java (Keeler, 1987), Andalusian-carnival song (Mintz, 1997), deep flamenco song (Washabaugh, 1996), Tango (Savigliano, 1994), blues (Oliver, 1997) and jazz (Kahn, 2000). There is, moreover, an anthropology of such embodied performing arts as ballet and other dance forms (Buckland, 1999; Wulff, 2000; Dyck & Archetti, 2003) or the body-performative texts of manners and gender (Foster, 2000). It is undeniable that contemporary emphases on the consumption of culture leaves its production and performance relatively neglected. There is, nevertheless, a corpus of material available. Moreover, recent anthropological interest, building on the work of Victor Turner (1987) means that there is a growing volume of ethnographic material and conceptual reflection (Schieffelin, 1976; Napier, 1992; Stokes, 1994; Hughes-Freeland, 1998; Washabaugh, 1998). While there is clearly an important difference between the ethnography of performance and the performance ethnography that Denzin is exemplifying and advocating here, there are also significant links to be drawn. At the very least, we need a broader sense of the range of cultural performances and their analytic treatment.

There are several senses in which readers of this journal should interest themselves in this work, and reflect on it. Pedagogy is performative. This is not just a reflection of the dramaturgical skills of the teacher in the classroom

– although this has received an oddly muted treatment on the part of most sociologists and anthropologists, perhaps because of an unconscious fear that it may render pedagogical display inauthentic. The performance of culture, through spoken, theatrical, musical or spectacular actions, and display has pedagogic functions entirely beyond the classroom – embodying collective identities of history and heritage, nationhood and ethnicity, place and region, personhood and gender. The ethnography of performance, and performance ethnography, should therefore be fundamental to a sociological understanding of cultural and social reproduction. Furthermore, the work of the dramaturg, the film or theatrical director, the operatic producer, the choregos and the orchestral conductor is pedagogic. The ethnography of such direction should be thoroughly illuminating for an understanding of pedagogy in its broadest sense. Goffman (1959, 1974) remains a guide to pedagogy as does Stanislavski's (1967) work on theatre.

Some sociologists and anthropologists of education have become increasingly preoccupied with performances of identity and resistance. The genre of testimonio has been widely documented in the intellectual space defined by critical race theory and critical pedagogy. The journal *Qualitative Studies in Education* has carried many exemplars and discussions. It is, however, analytically valuable to expand the ethnographic horizons of ethnographic inquiry. Pedagogy and performance are to be found in a diverse array of cultural contexts and are expressed through many genres and media. Performance media extend well beyond the autobiographical accounts that Denzin here produces and celebrates. They also extend beyond the genres of American culture that predominate. Denzin's previous work on cinematic culture is remarkably valuable. What is needed is a sense of performance that expands the performative and cultural horizons.

References

Ashmore, M., Mulkay, M. & Pinch, T. (1989) *Health and efficiency: a sociology of health economics* (Buckingham, Open University Press).

Askew, K. M. (2002) *Performing the nation: Swahili music and cultural politics in Tanzania* (Chicago, IL, University of Chicago Press).

Buckland, T. J. (Ed.) (1999) *Dance in the field: theory, methods and issues in dance ethnography* (London, Macmillan).

Clough, P. T. (1998) *The end(s) of ethnography: from realism to social criticism* (2nd edn) (New York, Peter Lang).

Clough, P. T. (2000) Comments on setting criteria for experimental writing, *Qualitative Inquiry*, 6, 278–291.

Davies, C. A. (1998) A oes Heddwch? Contesting meanings and identities in the Welsh National Eisteddfod, in: F. Hughes-Freeland (Ed.) *Ritual, performance, media* (London, Routledge).

Denzin, N. K. (2001a) *Interpretive interactionism* (2nd edn) (Thousand Oaks, CA, Sage).

Denzin, N. K. (2001b) The reflexive interview and a performative social science, *Qualitative Research*, 1, 23–46.

Denzin, N. K. (2001c) Confronting ethnography's crisis of representation, *Journal of Contemporary Ethnography*, 31, 482–507.
Denzin, N. K. (2002a) The call to performance, *Symbolic Interaction*, 25, 327–346.
Denzin, N. K. (2002b) The cinematic society and the reflexive interview, in: J. F. Gubrium & J. A. Holstein (Eds) *The handbook of interview research: context and method* (Thousand Oaks, CA, Sage).
Denzin, N. K. (2003) Performance ethnography: critical pedagogy and the politics of culture (Thousand Oaks, CA, Sage).
Dyck, N. & Archetti, E. P. (Eds) (2003) *Sport, dance and embodied identities* (Oxford, Berg).
Ebron, P. A. (2002) *Performing Africa* (Princeton, NJ, Princeton University Press).
Ellis, C. & Bochner, A. P. (Eds) (1996) *Composing ethnography: alternative forms of qualitative writing* (Walnut Creek, CA, AltaMira).
Foster, G. A. (2000) *Troping the body: gender, etiquette, and performance* (Carbondale, IL, Southern Illinois University Press).
Goffman, E. (1959) *The presentation of self in everyday life* (Garden City, NY, Doubleday).
Goffman, E. (1974) *Frame analysis: an essay on the organization of experience* (Boston, MA, Northeastern University Press).
Hughes-Freeland, F. (Ed.) (1998) *Ritual, performance, media* (London, Routledge).
Kahn, A. (2000) Kind of blue: the making of the Miles Davis masterpiece (London, Granta Books).
Karp, I. & Levine, S. D. (Eds) (1991) *Exhibiting cultures: the poetics and politics of museum display* (Washington, DC, Smithsonian Institution).
Lewis, J. L. (1992) *Ring of liberation: deceptive discourse in Brazilian capoeira* (Chicago, IL, University of Chicago Press).
Marvin, G. (1988) *Bullfight* (Oxford, Blackwell).
Mintz, J. R. (1997) *Carnival song and society: gossip, sexuality and creativity in Andalusia* (Oxford, Berg).
Keeler, W. (1987) *Javanese shadow plays, Javanese selves* (Princeton, NJ, Princeton University Press).
Macdonald, S. (2002) *Behind the scenes at the science museum* (Oxford, Berg).
Mulkay, M. (1985) *The word and the world* (London, George Allen and Unwin).
Napier, A. D. (1992) *Foreign bodies: performance, art, and symbolic anthropology* (Berkeley, CA, University of California Press).
Oliver, P. (1997) *Conversation with the blues* (2nd edn) (Cambridge, Cambridge University Press).
Pink, S. (1998) From ritual sacrifice to media commodity: anthropological and media constructions of the Spanish bullfight and the rise of women performers, in: F. Hughes-Freeland (Ed.) *Ritual, performance, media* (London, Routledge).
Richardson, L. (2000) Writing: a method of inquiry, in: N. K. Denzin & Y. S. Lincoln (Eds) *Handbook of qualitative research* (2nd edn) (Thousand Oaks, CA, Sage).
Richardson, L. (2002a) Poetic representation of interviews, in: J. F. Gubrium & J. A. Holstein (Eds) *Handbook of interview research: context and method* (Thousand Oaks, CA, Sage).
Richardson, L. (2002b) Small world, *Cultural Studies ↔ Critical Methodologies*, 2, 23–25.
Savigliano, M. E. (1994) *Tango and the political economy of passion* (Boulder, CO, Westview Press).
Schieffelin, E. L. (1976) *The sorrow of the lonely and the burning of the dancers* (New York, St. Martin's Press).
Schieffelin, E. L. (1998) Problematizing performance, in: F. Hughes-Freeland (Ed.) *Ritual, performance, media* (London, Routledge).
Stanislavski, C. (1967) *On the art of the stage* (London, Faber and Faber).

Stokes, M. (Ed.) (1994) *Ethnicity, identity and music: the musical construction of place* (Oxford, Berg).
Tota, A. L. (2003) *La città ferita* (Bologna, Il Mulino).
Trosset, C. (1993) *Welshness performed: Welsh concepts of person and society* (Tucson, AZ, University of Arizona Press).
Tulloch, J. (1999) *Performing culture: stories of expertise and the everyday* (London, Sage).
Turner, V. (1987) *The anthropology of performance* (New York, PAJ Publications).
Washabaugh, W. (1996) *Flamenco: passion, politics and popular culture* (Oxford, Berg).
Washabaugh, W. (Ed.) (1998) *The passion of music and dance: body, gender and sexuality* (Oxford, Berg).
Wulff, H. (2000) Access to a closed world: methods for a multilocale study on ballet as a career, in: V. Amit (Ed.) *Constructing the field: ethnographic fieldwork in the contemporary world* (London, Routledge).

17

Accounts, Interviews and Observations

Robert Dingwall

When I began my career, as a graduate student in 1971, the UK had relatively little tradition of formal training in research methods, apart from fairly elementary statistics. (I suspect that I may be a member of the last generation to use log tables and slide rules!) I am not even particularly sure that I consciously intended to be a qualitative researcher: I simply happened to be offered a scholarship by a research group whose work inclined that way.

To the extent that I received any training, it was through being encouraged to read what my mentors considered to be the great books of the 1960s, many of which happened to have used participant observation. My own PhD was an occupational socialization study in medical sociology (Dingwall 1977). The models were defined for me as *Boys in White* (Becker et al. 1961), *Asylums* (Goffman 1961a) and *The Silent Dialogue* (Olesen and Whittaker 1968). The people around me had various other empirical interests. I see from the thesis bibliography that I also read *The Social Organization of Juvenile Justice* (Cicourel 1968; my own title was a deliberate echo of Cicourel's), *The Racing Game* (Scott 1968) and *Identity and Community in the Gay World* (Warren 1974).

As one of my graduate students reminded me later, I wrote an (unpublished) section of my thesis about the PhD itself as an experience in occupational socialization that could be examined in the same way as my health visitors. Occupational socialization is the organizational production of competence, not the internalization of some fixed set of knowledge and values.

Source: Gale Miller and Robert Dingwall (eds), *Context and Method in Qualitative Research* (London: SAGE, 1997), pp. 51–64.

Although I only fully realized it later, the focus of the study should have been the staff teaching the health visitor training programme and their methods for recognizing and certifying competence in their students (Dingwall 1986). In the same way, the qualitative researchers who taught me affirmed the priority of observation in a way that made it the heart of the studies that their students undertook. To be recognized by them as a competent qualitative sociologist it was essential to be a competent participant observer.

I have done rather less of this kind of work in recent years. I had attributed this to advancing age and to the contingencies of academic careers. When you have to sit on committees, teach classes, supervise PhD students and worry about trying to take some share of domestic and child-care responsibilities, it is genuinely more difficult to organize your life in ways that can accommodate the unpredictable schedules of the field. Researchers cannot make the field fit their lives, but it is not easy to make your life fit the field. My most recent studies have focused on professionals involved in personal injury litigation and on divorce mediators. The first had the attractions of doing interviews by appointment, wearing a suit and sitting in plush law firm or insurance offices drinking tea from porcelain cups. The second was even better: the mediators did not like the idea of being directly observed but were very happy to tape their work and mail it in. This seemed a brilliant scam: lots of data without having to set foot outside the office! Conversation analysis has a great deal to recommend it to the middle-aged. However, having finally secured a post with tenure in 1990 and begun to train my own PhD students, I was surprised to find one of them comment to me that actually there did not seem to be much participant observation going on anywhere any more.

While I think that this is a slight exaggeration, it does seem to be true that the balance has tipped. The dominant kind of qualitative study appears to be one in which the investigator carries out a bunch of semi-structured interviews which are then taped and transcribed. The results are thrown into a qualitative data management package and a few themes dragged out in ways that seem rather like what we used to call 'data dredging'. This involved loading the results of a survey into the mainframe, correlating everything with everything else and seeing what came out. The problem, of course, is that the laws of probability require that a certain number of correlations will be significant purely as a matter of chance: having found them, we still have to determine whether the result has anything more than a statistical significance. This can be done only in the light of a well-thought out theoretical question and in the context of other investigators' findings. Alternatively, the interview study seems to have become an excuse for introspection, an occasion on which to contemplate the sociological navel, and to find it pierced in contemporary fashion. In effect, the researcher is simply looking for some good quotes to illustrate a previously determined position on some personal or political issue.

Both of these tendencies seem to me regrettable and this chapter is directed to challenging them. It seeks, first, to re-establish the primacy of observation and, secondly, to assert that it can still be used to say something interesting about the world rather than about the observer.

Two (or Possibly Three) Methods of Social Research

I once heard a distinguished anthropologist say something that I have shamelessly plagiarized ever since. 'There are', he declared, 'only two basic methods of social research. One is called "asking questions" and the other is called "hanging out".' Since anthropologists traditionally study non-literate societies, sociologists might want to add a third method: 'reading the papers'. All human beings have used pretty much the same kinds of methods to find out about the social world around them since our species came down from the trees.

'Hanging out' has experienced less systematic refinement than the others. Although it was recognized as a fairly obvious way to investigate certain categories of people, especially low-life, who might not answer questions truthfully, and tribal peoples, who might not understand the modern game of interviewing and take it seriously, it has remained a rather haphazard business. A number of early social investigators used observation as a means of studying the poor. Walter Wyckoff (1971), for example, who was then an assistant professor in political economy at Princeton, published an account in 1901 of a walking tour of the Midwest in 1891, during which he posed as an itinerant labourer, which was partly designed to investigate the reasons for imperfection in the labour market: why did people prefer to remain unemployed in Chicago rather than move to Iowa where the farmers were desperate for labour? However, his description takes the form of a diary or a traveller's tale, although he was clearly quite systematic in observing the behaviour of other itinerants and the local communities. Beatrice Webb worked as a seamstress as part of a study of sweated labour in the 1880s and wrote up her experiences as an influential piece of social journalism, which led to some of her early public appearances in front of Parliamentary Committees (Webb 1946: 312–345). Although she and her husband, Sidney, used observation a great deal in their studies of local government organization, they produced only a formal account of this method in 1932, when they devoted a chapter of their textbook *Methods of Social Study* to 'Watching the Institution at Work'. They asserted: 'An indispensable part of the study of any social institution, wherever this can be obtained, is deliberate and sustained personal observation of its actual operation' (Webb and Webb 1975: 158).

Participant observation became established as the dominant method for anthropological study of tribal peoples during the 1920s and 1930s as students read Bronislaw Malinowski's accounts of his researches in New Guinea during the First World War and tried to copy what they thought he had done. Actually, Malinowski's own diary shows that observation provided a context for the statistical data that he collected and for the structured questioning of informants, both of which were the methods of his predecessors (Kuper 1973).

> In working out the rules and regularities of native custom, and in obtaining a precise formula for them from the collection of data and native statements, we find that this very precision is foreign to real life, which never adheres rigidly to any rules. It must be *supplemented* by the observation of the manner in which a given custom is carried out, of the behaviour of the natives in obeying the rules so exactly formulated by the ethnographer, of the very exceptions which in sociological phenomena almost always occur. (Malinowksi 1922; my italics)

Pat Barker's reworking of W.H.R. Rivers's notes from his 1911 fieldwork in Melanesia for her (1995) novel, *The Ghost Road*, suggests that, in fact, the movement from questioning to observing was already under way before Malinowski's time. In Rivers's own words:

> A typical piece of intensive work is one in which the worker lives for a year or more among a community of perhaps four or five hundred people and studies every detail of their life and culture; in which he comes to know every member of the community personally; in which he is not content with generalized information, but studies every feature of life and custom in concrete detail and by means of the vernacular language . . . It is only by such work that it is possible to discover the incomplete and even misleading character of much of the vast mass of survey work which forms the existing material of anthropology. (Rivers 1913: 7)

Traditionally, sociologists tell the story of participant observation from the conjunction of Robert Park and W.I. Thomas at the University of Chicago around the end of the First World War. However, modern scholarship has reminded us that the work that they inspired was quite diverse and far from being uniquely observational. Platt (1994, 1996) argues that the dominant Chicago method of the 1920s and 1930s may be better described as one of case studies, with a strong emphasis on the kind of life histories and personal documents collected by case-workers in the course of social work with the marginal groups of the metropolis. Although some observation went on, she suggests that there was a relative indifference to specific questions of method; Park simply encouraged students to go out and find the story;

> one thing more is needful: first-hand observation. Go and sit in the lounges of the luxury hotels and on the doorsteps of the flop-houses, sit on the Gold Coast settees and on the slum shakedowns; sit in the Orchestra Hall and in the Star and Garter Burlesk. In short, gentlemen,

> go get the seat of your pants dirty in real research. (Bulmer 1984: 97; cf. Carey 1975: 155; Platt 1996: 52)

It must be said that Platt tends to reinforce the stock picture of Park as a journalist who happened to wander into sociology, a picture rather at odds with the range and subtlety of his theoretical writing. Remarks like this seem to have been a Chicago style: Atkinson (1977: 32) recounts the story of Howard Becker, on a visit to the University of Manchester, being asked by a graduate student how to choose a paradigm and being told to stop worrying, 'get in there and see what's going on.' As Platt herself notes, though, Becker was one of the major contributors to the articulation of participant observation as a self-conscious method within sociology during the late 1940s and to systematic published exposition during the 1950s and 1960s. Rather than being created in opposition to positivism and the survey, she suggests, observation was another manifestation of a widely shared attempt to create a more scientific and professional way to do social research. Paul Lazarsfeld, the high priest of the survey, and Everett Hughes, the guru of the post-war Chicago School of occupational ethnographers, shared a vision of a methodologically self-conscious discipline, even if they tried to operationalize this in competing ways.

This puts a rather different slant on my generation's occasional bafflement about observational methods. Platt notes that the first textbook treatment of participant observation was Junker (1960), based on Everett Hughes's graduate methods course at Chicago. Before this, anybody doing an observational study would have had either to be trained at a centre where such work was going on or to have invented an approach for themselves. While there was a flurry of relevant texts in the US over the next decade, the lag in transmission meant that my cohort in the UK were being trained by supervisors who were themselves pretty vague about what participant observation research in sociology might look like. The main exceptions were those supervised by anthropologists who had already inspired some work on British field sites through the community studies tradition (see Frankenberg 1966). However, these were not necessarily relevant models for the kind of organizational studies that our American exemplars had carried out, although Chicago ethnography had closer ties with anthropology than many of the later sociological accounts acknowledge: both Malinowski and A.R. Radcliffe-Brown were regular visitors in the 1930s and the intellectual influence of Robert Redfield, Park's son-in-law, and Lloyd Warner on developments after the Second World War was arguably as great as that of Everett Hughes.

In recent years, observation has suffered from the absence of powerful external sponsors. The pressures on the resources available to graduate students and rising expectations about their training requirements have encouraged the – cheaper and quicker – use of interview methods, which have themselves found an external constituency. The informal interview has become a routine tool of commercial research as well as of social science. Interviews, like surveys, seem to represent transferable skills, with an obvious market value. While the world of commerce can worry about its own standards, however, the world of social science professes a concern for the integrity of its conclusions which sits uncomfortably with the neglect of observational research. Other qualitative methods generate problems of validity and reliability which are so fundamental that the neglect of observation, and its proxies in direct audio- and video-recording, fatally undermines many of the conclusions that are alleged to have been drawn.

Accounting and Social Life

Scientific revolutions are preceded by the accumulation of small anomalies that ultimately pull down the paradigm that has framed them. Given this, it may be difficult to single out the snowflake whose movement ultimately caused the avalanche. The basic anomaly was present from the earliest formulations of interactionist sociology and the notion that humans were fundamentally reflexive beings. During the 1920s and 1930s, G.H. Mead and Alfred Schutz, in slightly different ways, reflecting their different intellectual contexts, restated two propositions that had first been advanced by the Stoics two thousand years before: that our understanding of our actions was derived from others' responses and that our action-projects were designed in the light of the expected responses of those others. These theorists had a limited influence until after the Second World War but their impact was eventually revolutionary. It was felt first in the doing of sociology, in the post-war Chicago studies of work and organizations, with their abandonment of the formal analysis of scientific management for more fluid images of negotiation, uncertainty and conflict and in Erving Goffman's development of strategic models of human action.

The line of argument that eventually emerged in the 1960s at the intersection between symbolic interactionism and ethnomethodology went something like this. Social order is constituted through interaction by a dance of expectations. I produce my actions in the expectation that you will understand them in a particular way. Your understanding reflects your expectations of what would be a proper action for me in these particular circumstances which, in turn, becomes the basis of your response which, itself, reflects your expectations of how I will respond. And so on. At any point, there may be a disjuncture between actions, responses and expectations which requires that the parties engage in some sort of repair work. The whole process is underpinned by what Goffman (1983) came to call 'Felicity's Condition', that we act in such a way as not to disconfirm the assumption of our sanity by those around us. In effect, we are obliged to participate in the everydayness of everyday life or be regarded as incompetent, deranged, disordered and generally unfit to be around right-thinking people.

Once interaction came to be analysed as a matter of impression management in support of personal goals and structured by the expectations of others, however, it could only be a matter of time before the analysis was applied to the tools of sociology. The key text was Aaron Cicourel's *Method and Measurement in Sociology*, published in 1964. Cicourel's book is usually cited for its attack on quantitative social science. It is often forgotten, though, that Cicourel also discussed the practice of participant observation (or field research as he tends to call it) and of interviewing. The chapter on interviewing is remarkable for the clarity with which it lays out the problem. If the interview is a social encounter, then, logically, it must be analysed in the same way as any other social encounter. The products of an interview are the outcome of a socially situated activity where the responses are passed through the role-playing and impression management of both the interviewer and the respondent. This is not the technical problem that survey researchers assume, which might be resolved by better interviewer training, more elaborate scripts or whatever. The interview is an artefact, a joint accomplishment of interviewer and respondent. As such, its relationship to any 'real' experience is not merely unknown but in some sense unknowable.

The implications for the research interview are quite dramatic. The research interview is, above all, an occasion for the elicitation of accounts. The term 'account' needs some explanation because it has been employed in' two slightly different but related ways. It was first used in an ethnomethodological context by Harold Garfinkel (1967) as he laid out his manifesto in *Studies in Ethnomethodology*. While Cicourel had been clearer about his criticisms of conventional sociological methods than about his proposed alternative, Garfinkel formulated an alternative programme. In true Kuhnian style, Garfinkel asserted that what he thought should be done was incommensurable with the paradigm of traditional sociology. Its object of study would be 'the activities whereby members produce and manage settings or organized everyday affairs [which] are identical with members' procedures for making those settings "accountable"' (Garfinkel 1967: 1). What the world is is the way we call it into existence through talk. But this is not just *any* talk. It is talk that shapes a world that others will recognize and for which they will hold us responsible. 'Accounting' is how we build a stable order in social encounters and in society.

However, the idea gained a wider currency in the more restricted sense given to it by Scott and Lyman (1968). They focused on the way people responded to questions about the normality or the rationality of their behaviour. How did we deal with challenges which imply that our behaviour has not been in accord with the expectations of the observer? How do we *account* for ourselves? Scott and Lyman argued that 'accounts' fell into two categories which they called *justifications* and *excuses*. This work has had a wide impact on the sociology of deviance (see Scott and Lyman 1970). John Eekelaar, Topsy Murray and I used it, for example, to examine how parents accused of maltreating their children tried to explain their actions in court hearings to consider the removal of their children or some other compulsory measure by child protection services (Dingwall et al. 1983). Parents tried to justify their actions by denying that they exceeded the boundaries of reasonable behaviour, by questioning the right of child protection workers to judge them, especially if the worker were not a parent, by telling 'sad tales' about their own upbringing or by appealing to higher virtues, as in the case of religious zealots who claimed a biblical justification for physical punishment. They tried to excuse their behaviour by pointing to some impairment of their capacity (by drugs, alcohol or mental disorder), to some biological drive (particularly in the case of sexual abusers), or by pointing to characteristics of the victim (especially if the child were handicapped in some way).

Scott and Lyman (1968: 61) acknowledged that the production of accounts was a socially constrained process:

> We want to know how the actors take bits and pieces of words and appearances and put them together to produce a perceivedly normal (or abnormal) state of affairs. This kind of inquiry crucially involves a study of background expectations. On the basis of such investigations, the analyst should be able to provide a set of instructions on 'how to give an account' that would be taken by other actors as 'normal'. These instructions would specify how different categories of statuses affect the honoring of an account and what categories of statuses can use what kinds of accounts.

In our study of child protection, we looked at the differential response of the courts to the various categories of justification or excuse. Basically, it was only the tellers of 'sad tales' or those with some principled libertarian objection to the agencies' judgements on their standards of child care who had any prospect of persuading the court to rule in their favour and then only if they could demonstrate what Emerson has called their 'absolute rectitude' (Emerson 1969: 167).

This restriction on the sense of accounting has troubled a number of more ethnomethodologically inclined writers. Atkinson and Drew (1979: 138–141), for instance, show that accounts may be offered without specific charges being made: they use the example of the rejection of invitations where this, dispreferred, action seems to be routinely accompanied by some explanation without that explanation actually being asked for. They note also that people do not necessarily give reasons in such contexts but rather describe situations in such a way as to imply reasons. In analysing the response of a police witness to the Scarman Inquiry into civil disorder in Northern Ireland in 1969, for instance, they note that the witness's inaction is provided for as much through a description of the scene – the petrol bombing of a newsagent's shop where the witness notes that the window was already broken and the premises alight – as through the stated reason that is provided subsequently, 'we were under gunfire at the time'. The witness expects that the tribunal will appreciate that there is no point in him taking the risk of entering a building which is already damaged and burning and from which anyone injured is already likely to have left or to have been evacuated. A possible sequence of questions has been projected and answered. They also note that the difference between justification and excuse is abstract rather than practical. Witnesses use both in connection with the same events and weave around to find whichever is most persuasive in the local context. Similar points have been made by Buttney (1987: 1), who notes the way in which human action 'is all too susceptible to error, unfulfilled goals and unintended consequences'. Accounts are the way in which the fabric of order is restored at the points where it comes under stress.

Interviews as Accounts

An interview is a point at which order is deliberately put under stress. It is a situation in which respondents are required to demonstrate their competence in the role in which the interview casts them. At one level, this is the role of respondent: hence some of the difficulties that Isobel Bowler describes in Chapter 5 in dealing with people who do not have a concept of the interview. Her respondents are not members of what Silverman (1997) has called 'the Interview Society', where the self is established as an object of narration. Her Pakistani women do not share her concept of the propriety of talking about themselves and their personal experiences of maternity care. Adequate performance of the role of respondent rests on the essential feature of interviews that people are put on notice to talk about *something*. A key feature of the framing of interviews is that the interviewer defines what the parties are going to talk about and what will count as relevant. The interview is a turn-taking system that requires that the interviewer proposes topics and that the respondent seeks to produce locally acceptable answers. This is true even of so-called unstructured interviews. The sequence may be flexible; the question wording may be flexible; it may be dressed up like a conversation between friends. But an interview is not a conversation. It is a deliberately created opportunity to talk about something that the interviewer is interested in and that may or may not be of interest to the respondent. If the interviewer refuses to propose topics, the respondent is obliged to guess what might be relevant until the interviewer gives some indication that he or she is happy with the line being taken.

An interview is never an occasion for a respondent to tell whatever story comes into, his or her head. Try it sometime. Sit someone down in front of a tape recorder and ask him/her to talk about his/her life. You will almost invariably find that the response is, 'What shall I talk about?' The initial talk is punctuated by uncertainty, checks on relevance, requests for confirmation of the direction and so on. What you have done is to carry out a breaching experiment. As Garfinkel (1967) showed, such fundamental challenges to the interaction order can be quite devastating in their impact, although they also provoke quite remarkable displays of ingenuity by those involved as they try to make sense of what is happening. The classic example of this process is McHugh's (1968) discussion of an experiment where people were told that they would receive counselling from a counsellor hidden behind a screen who would only answer 'Yes' or 'No' to their questions. The experimenter's confederate selected the answers at random but the recordings of the recipients show that they treated even radically contradictory responses as indicators of some stable underlying order. The random answers were treated as answers-to-their-questions.

Whether of interest or not, the respondent is still concerned to bring the occasion off in a way that demonstrates his or her competence as a member of whatever community is invoked by the interview topic. This is an inescapable constraint on face-to-face interaction. The consequence is that the data produced by interviews are social constructs, created by the self-presentation of the respondent and whatever interactional cues have been given off by the interviewer about the acceptability or otherwise of the accounts being presented. Once again, it must be emphasized that this is a critique of *all* interviews. It subsumes the standard critiques of the forced-choice survey interview for boxing up experience in ways that do violence to the respondent's meanings: indeed, there may be some reason to suppose that the more rigorous training of interviewers on such studies to minimize their reactions to any response may make the results somewhat less artefactual than in informal interviews where the interviewer takes an active role in dialogue with the respondent. At least with a formal and heavily structured interview, we have a good idea what the interviewer effects *are*! It also cuts across the so-called public and private distinctions: all experience is expressed through the social medium of language and there is no prior medium. Public accounts and private accounts satisfy the requirements of competence on different occasions. The private account is not the product of a more intimate understanding between interviewer and respondent. Private experience can be expressed only through a social medium in a social setting where the expression will immediately be evaluated for its fit to the range of versions of the self that are sanctioned in that environment. Thus feminist researchers elicit feminist accounts; Black researchers elicit Black accounts; and sociological researchers elicit. . . what?

Interview data cannot offer us literal descriptions of the respondents' reality. This is not to say that no use can be made of them, although a detailed consideration of this lies outside the scope of the present chapter.

However, work like that of Baruch (1981), Moore (1974) and Voysey (1975) illustrates how interviews can be analysed for what they can say about the kind of accounts that are treated as legitimate in a particular setting. As Baruch (1981) and Voysey (1975) note, what parents with a handicapped child say in an interview does not tell us what it is like to live in such a family. Ultimately, that experience is unknowable. What they do tell us quite a lot about is the work of doing being a normal family. The everyday lives of these families are natural breaching experiments which reveal the order that would otherwise be taken for granted. The ways in which they seek to restore that order tell us both about its moral force and about the kinds of accounts that are honoured in our society. Moore's (1974) progressive clergymen have a line on the marginalization of the church that nevertheless allows them to demonstrate that they are reasonable men doing a sensible job rather than, say, naïve victims of a delusion who are propping up a bankrupt and irrelevant institution. It may also be that some aspects of the respondent's reality can be glimpsed through the accounts: the selection of details, the choice of 'facts' in the narrative, perhaps. In real life we recognize that the accounts we receive every day contain some mix of the real and the representation and there seems no good reason why the accounts we receive as sociologists should be essentially different. My point is merely that interview data are fraught with problems because of the activity of the interviewer in producing them. At the same time, data are never *merely* accounts or versions, such that any reading is as good as any other. I shall return to this point.

Observation and Its Virtues

While observation does not entirely avoid some of these difficulties, it does potentially strike out interviewer variation as a first-order problem. Where interviewers *construct* data, observers *find* it. The case must not be overstated: clearly observers also select from the universe of sensations to which they are exposed in any given setting and may exercise some impact on that setting. Nevertheless, the fundamental virtue of observation, whether direct or via the proxies of audio- or video-recording, is that it enables us to document members accounting to each other in natural settings. It is the difference between the experiment on the laboratory animal and the animal in the wild. This gives a quite different purchase on the production of social order. No longer is it a matter of members trying to make themselves appear rational to us: now it is a question of how they appear rational to each other. How do they make the world a reasonably ordered, stable and reliable experience? Alternatively, in some settings, how do they deliberately set out to destabilize it, whether by means of dope or postmodernism? The observer is a part of the audience but no longer the most important part. He or she will be here today and gone tomorrow, while the other actors are potentially far more consequential. In recording their dealings with each other, we come closer to understanding the production of everyday life in a much wider range of environments. Interviews tell us about the construction of mundane reality in the interview. They are documents of the researcher–researched relationship. Observation is a document of the transactions between members themselves.

If our objective is to understand the foundations of social order, the constitution of society, the organization of settings or any of the classic questions of sociology, observation must be the method of preference. If our intention is to change the world, the same is true. As Harvey Sacks frequently observed, there can be many great events in the world that have remarkably little impact on the order of everyday life (Sacks 1992: II, 215–221). Conversely, the obduracy of everyday life can be a massive obstacle to desired change. Consider, for example, the struggle between agricultural and industrial time in the transition to capitalism (Thompson 1967) or the catastrophic failures of technical systems that are ill matched to the working practices of human operators. Chernobyl, perhaps, where the operators sought to liven up a dull shift by experimenting with the safety devices? Or, at a more mundane level, the hitches in the IBM systems used for the Atlanta Olympics. A company spokesman was reported as admitting: 'Probably the biggest lesson we learned was that technology in itself is not the solution – in this case a lot of the work has to be done by human beings' (*The Guardian Online*, 8 August 1996).

Observation shows us everyday life being brought into being. It does not, of course, show us what is real. It does not tell us what is going on inside the heads of the people who are making the world real for each other. But, since we are not a telepathic species, this is not just our problem. Members have the same difficulties as sociologists in making sense of what is going on, deciding what is real and what is fantasy, deciding what is true and what is false, deciding what is sincerity and what is deceit. But they have solved these problems, for all practical purposes, partly by creating a shared set of interactional practices, knowledge, presuppositions or whatever, and partly by creating a set of social institutions that legitimate closure. Professions and their associated workplaces can, for example, be seen as devices for managing uncertainty: medicine orders the variety of human anatomy and physiology; engineering orders the elemental forces of nature; law produces determinacy in the face of ambiguity and dispute. These are not real controls but they allow others to get on with their lives with a working assumption of boundedness and stability. Observation tells us about the set of solutions that have been produced, how people solidify and stabilize their social environment and how, on occasion, they play with it and test it.

Of course, some settings are harder to penetrate than others. Interviews may be a way to glimpse something of what goes on within these boundaries.

However, the analysis of interview data must always be reconcilable with what is known about everyday life from observation. We may not be able to observe Cabinet discussions of national security issues: when we interview ministers about them, we must keep in mind what we know of group processes and committee work. A Cabinet meeting is one of a class of occasions, some of which are very accessible, rather than something *sui generis*. If we hear something inconsistent with this, then it is a prompt to ask why our informant is putting a particular spin on to his or her story. There are, of course, some public settings that adopt the interview format: the journalist's interview or the chat show, for example. Sometimes we may have a deliberate reason for eliciting an on-the-record account. In the child protection study, we interviewed agency directors particularly in order to get an 'official version' of the agency. How did the organization represent itself to the public? This gave us some important clues to the tensions within an interagency system, where agencies were caught between their separate charters and missions (see Chapter 10) and the unavoidability of cooperation and interdependence. How could members both assert their differences and build trust in each other? The point is that we knew we were creating a particular kind of account, which is well rehearsed by agency directors who frequently have to explain their actions to journalists, politicians, lawyers and the like, but we also understood the context in which it had to be read. It was not a process of triangulation but a recognition that these were different pieces of the jigsaw, that members might have the problem of reconciling what had to be said at one time and place with what was said or done at another. This is not the classic observers irony: 'Look these people say one thing and do something different.' It is an acknowledgement that different settings and interactions exact different self-presentations but that, from time to time, these differences cannot be kept apart and must be managed.

What Is It All Worth?

In a previous paper (Dingwall 1992), I discussed some of the problems of validity in qualitative research and proposed three tests that a reader might apply: can she distinguish clearly between data and analysis? Can she see how the study has looked for contradictory or negative evidence and set out to test statements proposed on theoretical grounds or reported from previous studies? Can she see how it reflects the interactive character of social life and deals even-handedly with the people being studied? These tests are designed to screen out those writings which simply explore the emotional or intuitive response of the investigator, which fail to contribute to a cumulative science of society or which are exposé journalism masquerading as scholarship. I want to return to these issues and conclude the present chapter by restating the case against certain readings of the social constructionist argument.

Dingwall (1992) formed part of an emerging critique of the Romantic movement in ethnography (see, for example, Hammersley 1992a,b; Silverman 1989; Strong and Dingwall 1989). That movement rests on a set of assumptions about the nature of the sociological project that are fundamentally untenable. Three of them are important here. The first may be called the assumption of *authenticity*, the idea that it is possible to locate a real self. This is the mistake that qualitative interview studies frequently make: the idea that greater informality gets the interviewer nearer some ultimate truth about the respondent which is obscured by the pre-coding and structuring of more formal methods. As we have seen, the fact that language is a social medium and the interview is a social situation means that the self presented to the interviewer is an artefact of the encounter. The same, of course, is true of observation of members interacting with each other: their self-presentations, their accounts of the self, are also artefacts.

To the extent that this has been accepted by some Romantics, however, they have drifted towards the second assumption, namely that of *plasticity*. If there is no 'real self', then perhaps there is no 'real world': we can make ourselves and our worlds any way we choose. This is sometimes identified with the 'postmodern turn' in ethnography. It takes much of its inspiration from Michel Foucault's later work, where he came to see us as prisoners in Max Weber's iron cage, where the only chance of freedom and authenticity came through play, irony and transformation. It finds a metaphor in the argot of the Internet, where the speed and impersonality of communication allows us to be anybody we choose in any part of cyberspace that we happen to occupy. I can join a usegroup as a 17-year-old Mormon virgin and be that person for as long as I can sustain the role. Undoubtedly, the Internet offers many interesting issues for social science, although I am reminded of Sacks's discussion of the resilience of interactional forms with the development of the telephone. The technology creates new possibilities but these tend to be framed through existing social forms.

> There's a funny kind of thing, in which each new object becomes the occasion for seeing again what we can see anywhere; seeing people's nastinesses or goodnesses and all the rest, when they do this initially technical job of talking over the phone. This technical apparatus is, then, being made at home with the rest of our world. And that's a thing that's routinely being done, and it's the source for the failures of technocratic dreams that if we only introduced some fantastic new communication machine the world will be transformed. Where what happens is that the object is made at home in the world that has whatever organization it already has. (Sacks 1992: II, 548–549)

In Schutz's terminology, the Internet may offer a new province of meaning, alongside dreams, fantasy, science or whatever. However, it is not the primary reality. We can order food on the Internet; we need physically to consume it. We may play at sex on the Internet; reproduction requires materiality, whether in a physical act or in a laboratory. When our Internet connection goes down because our hard disks have crashed, we need a human being with a screwdriver to fix it. The fundamental conditions of our everyday existence, which are, in turn, the prerequisites and the motives for social organization, can be satisfied only in a material world. Cyberspace may be a new kind of time-out: it does not displace everyday life.

Within the everyday life-world, the possibilities are not indefinite. Our talk is organized to produce order and determinacy. That is what language is about – sharing, coordinating, restricting private meaning in the accomplishment of sociality. We expect and are expected to produce the simulacrum of a 'real self'. Indeed, to deny this may be to disconfirm 'Felicity's Condition'. Inconsistency is an accountable matter. And the plausibility of the account is the audience's decision not the speaker's. The sum of this argument is that we cannot make the world any way we choose. We are always constrained by the material conditions of our biological existence and by the responses of those around us.

Which brings us to the third assumption, which I have called *integrity*. Romantics do not need to worry whose side they are on – they know they are right. Since there are no constraints on the self, on the construction of the world or whatever, then we may characterize it in any way we please. Silverman (1997) has noted his aversion to the results in terms of second-rate poetry, drama and home videos and called for a return to the spirit of Wittgenstein's plea for clarity and rigour of expression. It is the difference between trying to convey a feeling, for which the creative arts may well be an appropriate vehicle, and a fact, for which the formality of science is not mere rhetoric but a well-honed tool. The intersubjective construction of reality through accounting is a social fact. It is there; it is observable; it is reportable. There is nothing occult about the processes involved in its analysis. It has the relentless logic of forensic science. But only the corrupt scientist becomes more concerned with proving guilt than with the truth of his or her analysis. The result of that corruption is the sort of miscarriage of justice that our Romantic would unhesitatingly condemn at the same time as reproducing it in his or her own work. The only protection in the end is the discipline of science, of a concern for evidence, for standards of procedure, for objectivity in the face of interests. The real world of scientific practice may fall short; this does not negate the importance of the ideals as standards of judgement. Unfortunately, this sometimes means that we do not produce politically acceptable conclusions. But what is the value of a scholarly enterprise that is more concerned with being 'right on' than with being right?

Observation is the most fundamental discipline for the sociologist. In an interview study, we can pick and choose the messages that we hear and that we elicit. In observation, we have no choice but to listen to what the world is telling us.

References

Atkinson, J.M. (1977) 'Coroners and the categorization of deaths as suicides: changes in perspective as features of the research process', in C. Bell and H. Newby (eds), *Doing Sociological Research*. London: Allen and Unwin. pp. 31–46.

Atkinson, J.M. and Drew, P. (1979) *Order in Court*. London and Atlantic Highlands, NY: Macmillan and Humanities Press.

Baruch, G. (1981) 'Moral tales: parents' stories of encounters with the health professions', *Sociology of Health and Illness,* 3: 275–296.

Becker, H.S., Geer, B., Hughes, E.C. and Strauss, A.L. (1961) *Boys in White*. Chicago: University of Chicago Press.

Bulmer, M. (1984) *The Chicago School of Sociology: Institutionalization, Diversity and the Rise of Socoiological Research*. Chicago: University of Chicago Press.

Buttney, R. (1987) 'Sequence and practical reasoning in accounts episodes', *Communication Quarterly,* 35: 67–83.

Carey, J.T. (1975) *Sociology and Public Affairs: The Chicago School*. Beverly Hills, CA: Sage.

Cicourel, A.V. (1968) *The Social Organization of Juvenile Justice*. New York: John Wiley.

Dingwall, R. (1977) *The Social Organisation of Health Visitor Training*. London: Croom Helm.

Dingwall, R. (1986) 'The certification of competence: assessment in occupational socialization', *Urban Life,* 15: 367–393.

Dingwall, R. (1992) '"Don't mind him – he's from Barcelona"; qualitative methods in health studies', in J. Daly, I. McDonald and E. Wallis (eds), *Researching Health Care: Designs, Dilemmas, Disciplines*. London: Routledge. pp. 161–175.

Dingwall, R., Eekelaar, J.M. and Murray, T. (1983) *The Protection of Children: State Intervention and Family Life*. Oxford: Basil Blackwell.

Emerson, R.M. (1969) *Judging Delinquents*. Chicago: Aldine.

Frankenberg, R. (1966) *Communities in Britain*. Harmondsworth: Penguin.

Garfinkel H. (1967) *Studies in Ethnomethodology*. Englewood Cliffs, NJ: Prentice Hall.

Goffman, E. (1961a) *Asylums*. New York: Anchor.

Goffman, E. (1983) 'Felicity's condition', *American Journal of Sociology,* 83: 1–53.

Hammersley, M. (1992a) *What's Wrong with Ethnography: Methodological Explorations*. London: Routledge.

Hammersley, M. (1992b) 'On feminist methodology', *Sociology,* 26: 187–206.

Junker, B. (1960) *Field Work*. Chicago: University of Chicago Press.

Kuper, A. (1973) *Anthropologists and Anthropology: the British School 1922–1972*. Harmondsworth: Penguin.

McHugh, P. (1968) *Defining the Situation: the Organization of Meaning in Social Interaction*. Indianapolis: Bobbs-Merrill.

Malinowski, B. (1922) *Argonauts of the Western Pacific*. London: Routledge and Kegan Paul.

Moore, M.W.B. (1974) 'Demonstrating the rationality of an occupation'. *Sociology,* 8: 111–124.

Olesen, V.L. and Whittaker, E.W. (1968) *The Silent Dialogue*. San Francisco: Jossey-Bass.

Platt, J. (1994) 'The Chicago School and firsthand data'. *History of the Human Sciences,* 7: 57–80.

Platt, J. (1996) *The History of Sociological Research Methods in America 1920–1960*. Cambridge: Cambridge University Press.

Rivers, W.H.R. (1913) 'Report on anthropological research outside America', in W.H.R. Rivers, A.E. Jenks and S.G. Morley (eds), *The Present Condition and Future Needs of the Science of Anthropology*. Washington DC: Gibson Brothers.

Sacks, H. (1992) *Lectures on Conversation*. Oxford: Basil Blackwell.

Scott, M.B. (1968) *The Racing Game*. Chicago: Aldine.
Scott, M.B. and Lyman, S.M. (1968) 'Accounts', *American Sociological Review*, 33: 46–62.
Scott, M.B. and Lyman, S.M. (1970) 'Accounts, deviance and social order', in J. D. Douglas (ed.), *Deviance and Respectability*. New York: Basic Books, pp. 89–119.
Silverman, D. (1989) 'The impossible dreams of reformism and romanticism', in J. F. Gubrium and D. Silverman (eds), *The Politics of Field Research: Sociology beyond Enlightenment*. London: Sage. pp. 30–48.
Silverman, D. (1997) 'Towards an aesthetics of research', in D. Silverman (ed.), *Qualitative Research: Issues of Theory and Method*. London: Sage.
Strong, P.M. and Dingwall, R. (1989) 'Romantics and Stoics', in J.F. Gubrium and D. Silverman (eds), *The Politics of Field Research: Sociology beyond Enlightenment*. London: Sage. pp. 49–69.
Thompson, E.P. (1967) 'Time, work-discipline and industrial capitalism', *Past and Present*, 38: 56–97.
Voysey, M. (1975) *A Constant Burden: the Reconstitution of Family Life*. London: Routledge and Kegan Paul.
Warren, C.A.B. (1974) *Identity and Community in the Gay World*. New York: John Wiley.
Webb, B. (1946) *My Apprenticeship*, 2nd edn. London: Longman (1st edn, 1926).
Webb, S. and Webb, B. (1975) *Methods of Social Study*. Cambridge: Cambridge University Press (first published 1932).
Wyckoff, W.A. (1971) *A Day with a Tramp and Other Days*. New York, Benjamin Blom (first published 1901).

18

Observational Field Work

Robert M. Emerson

Introduction

Over the past decade the theoretical and methodological literature on field methods has increased enormously in volume and sophistication. In this literature old problems recur: generating theory, validity and reliability, the ethical dilemmas inevitably confronting the field worker. And new problems have surfaced: the implications of naturalism and methodological self-consciousness, field work as a social and personal as well as a research enterprise.

Here I consider the recent developments in field research carried out in natural social settings through direct, continuing interaction with people in those settings. Observation through regular *participation* in the naturally occurring activities of the social groupings being studied provides the key characteristic of such research (M. Wax 1980): Such participation both generates the richness of observation and produces the distinctive methodological and theoretical problems of field research.

In this review I do not treat research that employs structured or unstructured interviewing as the primary method of data collection. Such research is not observational in any strong sense. Furthermore, while there are close parallels between some social processes in interviews and those in field studies (Cicourel 1964), participation in the former is limited to several encounters at most and hence involves qualitatively different kinds of research relations.

Source: *Annual Review of Sociology*, 7 (1981): 351–378.

Nor do I deal here in any detail with field work using preestablished schedules of observational categories. [Delamont & Hamilton (1976) and McCall (1978) review recent developments using such field methods in studying classroom education and criminal justice interactions, respectively.] This exclusion reflects basic differences between prestructured observation and more open field methods: (*a*) Rather than developing relevant categories and analytic distinctions over the course of the study, these approaches use preestablished observational schedules in order to collect data. (*b*) The data collected are conceptualized in terms of "the frequency distribution of behavior events" that are assumed to have objective meaning that can be grasped by the outside observer (Schwartz & Merten 1971:279). (*c*) Prestructured observation studies employ what is essentially a survey design, with prespecified procedures for when and what to observe or sample. (*d*) The use of these sorts of procedures narrows and restricts the observer's participation in the setting. In prestructured observations, field interactions are seen as irrelevant (or even a hazard) to the methodology and findings. Participation is unnecessary because there is no concern with trying to enter the worlds of meaning of those studied. Except for its practical impossibility in most natural settings, concealed observation would be the recommended procedure in prestructured observation settings.

I do not review here analyses concerned primarily with discrete segments of audio- or video-taped interactions collected in natural settings (e.g. Cicourel et al 1974; Bellman & Jules-Rosette 1977), or with observational studies conducted through one-way mirrors; again, in such studies issues of participation are distinctively different from those arising in more classic field work. Finally, only occasional reference will be made to the often parallel issues arising in anthropological field work, although at many places there is sufficient overlap between sociological and anthropological concerns to require some treatment of the latter.

In sum, in this essay I review developments in sociological field work during the past ten years where the researcher directly observes and interacts with those in some natural setting over a relatively long period.

Trends in Field Research

Despite the inevitable simplifications entailed, it is useful to identify significant trends in the general development of observational field work over the past decade.

First, field work has become a more clearly recognized and self-identified methodology within sociology. One indication of this consolidation as a specialty is the publication of a number of monographs and readers devoted to observational field methods, including: McCall & Simmons (1969), Filstead (1970), Habenstein (1970), Lofland (1971, 1976), Bogdan (1972), Douglas

(1972, 1976), Glazer (1972), Schatzman & Strauss (1973), Bogdan & Taylor (1975), Weppner (1977), Fiedler (1978), McCall (1978), Schwartz & Jacobs (1979), and Shaffir et al (1980). A second indicator is the appearance of several journals explicitly devoted to field research, notably *Urban Life and Culture* (first published in 1972 and renamed *Urban Life* in 1975), and *Qualitative Sociology*.

Second, field-work proponents have sought in more concerted and less defensive ways than in the past to develop distinctively qualitative methodological principles for conducting field research. Field researchers have become less apologetic about their methods and put less emphasis on correcting or systematizing field methods to bring them in line with quantitative ideals (see Douglas 1976; Lofland 1976). Increasingly, field-work practitioners have challenged notions that their studies are merely intuitive, necessarily exploratory and unsystematic, insightful but incapable of offering disciplined theorizing or solid evidence. Glaser & Strauss' explorations of field methods as a procedure for generating theory (1967) and Becker's (1970) and Katz' (1982) attempts to identify the distinctive rigor of field studies exemplify such efforts to highlight and sharpen the distinctive qualitative features of field work compared with other methods.

But even more critical have been efforts to elaborate a distinctive epistemological basis for field research, usually by challenging notions that social actions have "objective" meanings. Instead, field workers have begun to emphasize what has been termed an interpretative paradigm (Overington 1979), which understands "facts" and "data" not as objective entities, but rather in terms of the social meanings attributed by social actors (including the field researcher) in interaction with others. This tendency in field work leads to a conceptualization of "subjective," "situated," or "constructed" meanings as the object of the research enterprise. Such an intent is incompatible with the assumption that events are objectively meaningful, an assumption that underlies many quantitative approaches and associated standards of validity and reliability.

Finally, over the past decade field research has shown a tendency for greater "self-consciousness," to use the term Nash & Weintrob (1972) employed to describe a similar movement in anthropological field work. While unevenly implemented and perhaps even declining since the peaking of the counter-culture in the late 1960s and early 1970s, this trend is reflected in such developments as the publication of a number of monograph-length, personally reflective accounts of field-work studies (R. Wax 1971; Johnson 1975), greater concern with the personal dimensions and consequences of field work (Douglas 1976; Clarke 1975; Georges & Jones 1980), and increasing awareness of the strengths and limitations of particular field work roles. These developments and their implications are considered more closely in the following pages.

In the next sections of this review I address a number of focal concerns apparent in contemporary field work, including ethnography and the study of actors' perspectives, ethnomethodological concerns in field work, field work and theory, and approaches to validity and reliability in field studies. In the concluding sections I discuss the movement toward self-consciousness of the social processes involved in carrying out field studies, including the nature of field relations and roles, appreciation of the personal dimensions of field research, and the political and ethical implications of field work.

Issues and Concerns in Recent Field Work

Contemporary field work is marked by widely varying emphases on just what concerns are primary for field studies, and what are the appropriate procedures for pursuing them. Some field researchers see description of the subjectively meaningful worlds of others as the primary contribution of field studies, while others see such description as only preliminary to the development of sociological theory. Some field workers hold closely to the standard notions of validity and reliability honored in the profession at large, while others seek to assess validity and reliability while taking into account the distinctive qualities of field studies. Finally, the field-work tradition has been enriched by the critical insights of ethnomethodology, insights giving new accent to some traditional field-work concerns and taking others in totally new directions.

Field Work, Description, and Actor's Perspectives

Much field work takes as its primary task the discovery and description of the distinctive "subjective meanings" or "perspectives" of particular groups of social actors. Such research traces its origins to the *verstehen* tradition in European sociology and has strong roots in the American school of symbolic interactionism.[1] Blumer argues that "in order to treat and analyze social action one has to observe the process by which it is constructed" (1969:56). Similar statements abound in the field-work literature. Lofland (1971:7), for example, argues: "In order to capture the participants 'in their own terms' one must learn their analytic ordering of the world, their categories for rendering explicable and coherent the flux of raw reality. That, indeed, is the first principle of qualitative analysis." Schatzman & Strauss locate a similar concern at the center of their discussion of naturalism in field work (1973:14): ". . . the naturalist as researcher will not, in advance, presuppose more than the barest rudiments of social order and social value. What he will do is to maximize the possibilities of discovering these as they are developed by people within the situation."

To the extent that the initial (and sometimes the primary, or sole) task of field work is seen as discovering the actors' perspectives or subjective meanings, field research takes on a strongly *descriptive* emphasis. This emphasis has been most sharply articulated in the advocacy of *ethnographic* field work. Spradley, for example, defines ethnography as "the work of describing a culture" (1979:3), and locates its distinctive task in identifying "the different ways of making sense out of life" employed by any of the multiplicity of cultures and subcultures found in contemporary society (1977:14). Cavan (1972:16) argues that such ethnographic research requires description of both "the routine social practices members collectively engage in" and the social knowledge and beliefs that make these practices meaningful and coherent. She then focuses her description of the hippie community in San Francisco on what members of this community "do on a day-to-day basis in order to organize the diverse courses of action that are salient facts of the life of the community into a coherent whole" (1972:24). [For additional illustrations and treatments of ethnographic field work, see Rubinstein (1973), Weppner (1977), and Akins & Beschner (1980).]

Wiseman's work (1970, 1974) provides a specific example of this descriptive approach. Wiseman's concern is to "apprehend," "reconstruct," or "decipher" "the actor's perspective" for two distinct social groups found in skid row locales: skid row problem drinkers and agents of social control (1970:269ff). These perspectives are to be inferred from close, continuing talk with and comprehensive observations of the activities of these two groups, especially in interaction with each other. As Wiseman summarizes the method, the field worker "observes everyday action and attempts to reconstruct the definition of the situation on which 'natives' appear to base their activities," and "asks questions of 'the natives' to find out 'what is going on' and what sort of action is expected to ensue as a result" (270). Echoing Becker's concern with socially based "hierarchies of credibility" (1967), the processes by which each group maintains its distinctive world view in dealing with the other comprise the heart of the study.

Gould et al (1974) further elaborate this methodological focus on the actors' perspectives. These field workers seek to describe the worlds of street drug use – "to construct an account of 'what's going on out there'" (1974:xvii). Furthermore, "at the descriptive level, the only reality we need concern ourselves with is the reality that is experienced by the research subjects" (1974:xix), not the "objective truth about events." As an example, they suggest that while pharmacologists generally view methadone and heroin as essentially similar and contend that street addicts cannot tell the difference between them under double-blind conditions, addicts contend that there are fundamental differences between the two drugs. Methadone for addicts is a "fall-back" drug, used to prevent sickness or cut down on a habit, while heroin is used to get high. Because the addicts' beliefs about the drugs, rather than existing scientific knowledge of their effects, are critical, the pharmacological reality is irrelevant to understanding the reality of street drug worlds.

The research approach that seeks to describe events as they are understood or experienced by social actors is not without its problems. First, a number of field workers note that social actors have knowledge and experience that they cannot articulate (Douglas 1976). Under these conditions description from a member's point of view courts distortion and oversimplification.

Further problems arise around the process of getting actors to identify what is meaningful to them. The actors' perspectives cannot be determined simply by asking them to explain what they are doing. To do so produces what Gould et al term "perspectives *of* action" (1974:xxiv), that is, accounts intended "to make the system meaningful to an outsider." Such a perspective differs from a "perspective *in* action," which consists of explanations or accounts arising in naturally occurring interactions "which were intended to make the action meaningful to the actors themselves or to other actors in the same situation" (1974:xxv). Perspective in this latter sense is thus "the background information employed by a participant in rendering situations meaningful enough to act intentionally within the parameters of that situation" (1974:xxvii). Yet this approach raises several questions. While it admittedly creates a new situation, does asking actors about what is meaningful to them necessarily preclude learning something about how they see and understand the initial situation? Second, how specifically are perspectives to be inferred from the observation of naturally occurring actions? Such inferences presuppose some working "theory of the actor" (Cicourel, unpublished) with which to assess what is meant by, and meaningful to, the actors involved.

Finally, and perhaps most telling, some have asked whether it is possible in principle to describe events from the actor's perspective. Bittner (1973) has most forcefully developed this argument, suggesting that field workers who seek to grasp the actor's perspective generate an irremediable contradiction: They treat as a "perspective" what actors on most occasions view as the way the world is. The field worker, then, does not produce a description from the actor's point of view, but a description of the actor's point of view from the point of view of a sociological observer. This is true even if the observer seeks to empathize closely with actors' concerns and meanings. As a consequence, field-work descriptions tend to depict social life as *perceived* events and meanings, ignoring or distorting the lived reality of actors' worlds, worlds characterized by "traits of depth, stability and necessity that people recognize as actually inherent in the circumstances of their existence" (Bittner 1973:123). Douglas (1976), however, suggests that even while the observer's understanding of the actor's perspective differs from that actor's perspective, this understanding is useful, and even critical, for most sociological purposes.

Ethnomethodological Concerns in Field Work

Field researchers concerned with the actor's perspective and subjective meanings usually trace their theoretical roots back to the Chicago school and symbolic interactionism. During the 1970s, however, the concerns of ethnomethodology have influenced some Chicago-school field workers (Cavan 1972; Gould et al 1974) and have even produced some collaboration between these two approaches (e.g. Schwartz & Jacobs 1979). While there are significant conceptual and methodological differences between neo-Chicago-school field work and ethnomethodology, there are also important continuities.

1. Ethnomethodological approaches attribute greater significance to the field worker's *methods* than is characteristic of neo-Chicago-school field work, arguing that the "findings" of field research are products of the specific methods employed and cannot be understood independently of these methods (Garfinkel 1967). Methodological and substantive issues are thus inseparable: The "problems" of the field-work role are in fact salient social features of the setting under study (Cicourel 1964) and must be described and analyzed in detail as a central rather than an optional part of any field study. Wieder's (1974) field work on the "convict code" in a halfway house for ex-cons provides a revealing example: That Wieder was unable to form close, "trusting" relations with the residents of the house highlighted provisions of the code that prohibited confiding in those associated with staff.

2. Methods in most neo-Chicago-school field work refer primarily to the course and character of field relations (conventionally, "field-work role"). Such discussions of field methods lay out the processes of establishing and maintaining field relations, as these affect the kinds of access and data obtained. Ethnomethodologists have addressed several additional dimensions of method. First, ethnomethodologists call attention to the more minute, ongoing processes whereby the field worker comes to interpret the actions of others, requiring the field worker to explicate the commonsense knowledge (Leiter 1980:86–91) and interpretive procedures (Cicourel 1968; 1974a:30–41) he/she used to understand others' behaviors. For the ethnomethodologists, these interpretive practices imply a "model of the actor" (Cicourel 1974a:27) that informs and determines the attribution of meanings to specific incidents. In a similar vein, Bittner criticizes field work centered on describing the actor's perspective for its failure to make the field worker's own point of view "either implicitly obvious or explicitly explained" (1973:118).

Second, Cicourel (unpublished paper) in particular has begun to examine the complex informational processing whereby researchers compile their major data base of written field notes. Inevitably field workers comprehend, interpret, and code in memory ongoing social activities in ways that depend upon their presuppositions, general cultural knowledge, prior experiences in and particular knowledge of the setting. Written field notes involve subsequent "summarizations" of recalled exchanges and hence reflect the structure of memory processes. Cicourel has examined these processes by comparing the transcripts of medical interviews with physicians' written summaries (1974b). More precise field data can be collected, he maintains, through "the explicit study of interpretational and summarization procedures, as they occur during interaction, interviews, and taking notes or elaborating on them."

3. Ethnomethodologically oriented field work is more concerned with actors' practices and practical concerns (Zimmerman 1970) than with their perspectives and cognitive categories. Actors attend to the often mundane "tasks at hand," attempting to realize immediate "practical purposes" that tend to be logically contradictory when viewed as parts of a single rational plan. Moreover, the field worker is seen as confronting this same essential need to get on with the practical tasks of making his way into and through the setting (Hilbert 1980). "Perspective" is both too cognitive and too subjective in its connotations: It leads the analyst to slip "in and out of points of view," treating "meanings of objects as more or less freely conjured" (Bittner 1973:121–22).

4. Ethnomethodological studies insist on the reflexive character of all descriptions of social order. As Schwartz & Jacobs (1979:51) summarize this principle, "descriptions about some aspects of the social world are simultaneously within (part of) the very world that they describe." Thus, actors' expressions to field workers are not factual data or objective reports. Rather, such talk provides "accounts" (Garfinkel 1967) that simultaneously (*a*) instruct the researcher in how to appreciate the situationally specific meaning and order recognized in the setting and (*b*) create this very meaning and order. Such accounts are radically situational in character, unavoidably tied to the background knowledge and "normal form typifications" that prevail in particular settings (Cicourel 1974a:53–4). Wieder's (1974) analysis of the social uses of the convict code in a halfway house exemplifies this approach and merits deeper consideration than it has received from most field researchers. Wieder first noted the tendency of both sociologists and halfway house personnel to explain resident behavior as caused by the code. He then offers an ironic reinterpretation, emphasizing that the code was used by participants – residents, staff, and researcher alike – to find order in, and make sense of, a variety of otherwise disparate acts and events.

Field Work and Sociological Theory

A number of field researchers emphasize the relevance of field studies for sociological theories and direct attention to the processes of theorizing from field data. The *grounded theory* approach of Glaser & Strauss (1967) and

Schatzman & Strauss (1973:108–27) remains the most prominent such effort. These works examine the processes of generating theory that is appropriate to (grounded in) activities observed in particular settings. They insist that the formulation and elaboration of theory form an enterprise distinct from the testing or validation of theory. Formulation has a logic and a method fundamentally different from those appropriate to verification. Rather than establishing the relation between a few key variables, theory comprises a rich set of analytic propositions that identifies and relates many diverse variables. The generation of such innovative theory is more likely if the field researcher minimizes commitment to preconceived theory, treating preconceptions only as "sensitizing concepts" (Schwartz & Jacobs 1979:26–28) to guide initial lines of inquiry. In the collection, coding, and analysis of field data, initial presuppositions are constantly reevaluated until all the researcher's "presumed classes are displaced by those based upon observations" (Schatzman & Strauss 1973:113). Analysis begins with the identification of conceptual classes and their properties. Propositions are then developed by exploring the variety of linkages among these classes and their properties. With "an ever-increasing density of linkages," and with the focus provided by a "key linkage" (Schatzman & Strauss 1973:111), these propositions can be organized into an integrated theory.

At the core of these analytic procedures lies the method of "constant comparison," in which the researcher incorporates already observed differences and variations into the evolving theory. In addition, grounded theorists urge that future observations maximize opportunities for comparison. Hence they recommend the procedure of "theoretical sampling," in which new observations are selected explicitly to pursue analytically relevant distinctions rather than to establish the frequency or distribution of phenomena.

Over the past decade, a number of studies have cited grounded theory as a broad justification for field research methods, but few substantive studies applying the approach have appeared (but see Cavan 1974) beyond those by the concept's originators (recent studies include Glaser 1976; Fagerhaugh & Strauss 1977; Strauss 1978). Furthermore, as Katz (1982) points out, grounded-theory arguments depend on a sharp dichotomy between discovery and verification that is difficult to sustain. In addition, this rigid divorce between discovery and verification lends support to the critique of field work as insightful but unrigorous.

Two further considerations of theorizing in field work merit review. First, Lofland (1971) has emphasized the desirability of raising the levels of abstraction and generalization in field work analyses while continuing to base such studies on close observation of particular groups and settings. His recent work (1976) addresses the task of conceptually integrating and synthesizing empirically rich case studies. Lofland contends that such cumulation and consolidation proceeds by elaboration of some specific conceptual framework. His work attempts to integrate and consolidate a variety of field studies explicitly or implicitly organized around Goffmanesque concepts of strategy and situation. Despite close affinity with grounded theory, Lofland is insistent on accumulating a rich and varied body of specific field studies before attempting theoretical consolidation, warning of the dangers of "short-circuiting" (1976:93–94) by generalizing from insufficient data. He is also critical of short-term, superficial excursions into the field, and stresses comparison of a number of intensively conducted field studies rather than the grounded-theory procedure of ongoing theoretical sampling.

Second, Katz (1982) has sought to revitalize and extend the procedure of analytic induction as a distinctively qualitative yet rigorous way to analyze field data. Criticizing grounded theory's constant comparison procedure as vague, Katz considers the search for negative cases and the modification of analytic propositions to incorporate them to be the essential processes in theorizing with field data. Katz contends that such procedures are implicit in the recurrent "pursuit of data that are related but qualitatively different" in field research, and can be made more systematic if employed more self-consciously and without claims of "universal explanation."

Approaches to Validity and Reliability

During the past decade proponents of field work have developed a number of approaches to validity and reliability different from those routinely employed in survey and experimental research.

Most researchers argue that analyses of field data are closely tied to empirical observation and hence have high internal validity (Denzin 1970: 201–5). Becker (1970) has identified several features of field methods that promote this result. First, field data tend to be less artifactual than most other sorts of research data: "[T]he people the field worker observes are ordinarily constrained to act as they would have in his absence, by the very social constraints whose effects interest him; he therefore has little chance, compared to practitioners of other methods, to influence what they do, for more potent forces are operating" (1970:43). Second, since large numbers of observations are collected over an extended period of time through multiple and varied procedures, findings are recurrently revised and tested (1970:51–62). Similar assurances are provided by making diverse comparisons and by continually searching for negative cases (Katz 1982).

Greater problems are acknowledged in validly generalizing field findings to settings and groups other than those studied, although Katz (1982) emphasizes that at least some such assurances are built into most field research: "By finding negative cases and then qualifying the analysis, theory is made generalizable to more types of situations." Denzin (1970:200) notes the importance of demonstrating that the cases studied are "representative of the class of units to which generalizations are made."

Standard measures to ensure reliability (fixed designs for collecting and coding data, retests and replications) have few direct analogs in field research. Becker (1970:44–51) proposes two considerations for assessing the reliability of field data: first, that the observed do not consider the observer significant relative to the factors that naturally constrain their actions; second, that data are obtained by observing interactions between group members rather than from researcher-member contacts. Team field work and multiple observers are also felt to enhance the reliability of field data (Douglas 1976). Replication is possible only in a very loose way, through comparison of different studies of similar groups. Becker (1970:40–2) argues, for example, that two field workers studying the same setting but with "different theories and different questions in mind" inevitably report different findings. Such findings may still be *compatible* if "the conclusions of one study do not implicitly or explicitly contradict those of the other" (1970:41–42).

In order to assess the adequacy of their findings, field researchers increasingly seek procedures that do not rely on the standard notions of reliability and validity derived from quantitative models of research. One such approach recommends identifying explicitly the procedures, analytic assumptions, and interpretive devices used to collect, make sense of, and communicate field reports. While Becker (1970) recommends a "natural history" of the field project along these lines (see also Glaser & Strauss 1967:230–31), Cicourel (1964, unpublished paper) has most explicitly advanced this proposal. As an alternative to replication, he urges detailing the procedures and working theories used to assemble and interpret findings.

A second recommendation, particularly advocated by field researchers concerned with the subjectively meaningful worlds of actors ("members," "hosts"), calls for member validation (Gould et al 1974), "member tests of validity" (Douglas 1976), or "host verification" (Schatzman & Strauss 1973). Since findings are intended to capture the subjective meanings of members, those findings can be taken back to members whose responses can then be used to determine the adequacy of the study. Thus, the field researcher seeks to determine "if the members recognize, understand and accept one's descriptions of the setting" (Douglas 1976:131; see also Wiseman 1970:280).

A variety of difficulties confront such proposals. First, in seeking "member validation," field workers obtain additional equivocal evidence; additional situationally influenced statements must then be interpreted, just as the earlier statements were, in order to establish a version of the member's perspective. Under these conditions, taking findings back to the field is not a test but an opportunity for reflexive elaboration, paralleling the logic behind what Cicourel (1974:124–27) has termed "indefinite triangulation." These procedures seek to elicit as many different members' accounts of a particular event (including further accounts of first accounts) as possible, not in order to determine "what really happened" but to reveal the diversity and depth of members' understandings of this event. Similarly, taking findings back to the field may deepen and extend understanding and analysis by providing the field researcher with new, revealing situations for interpretation.

A final procedure for evaluating the adequacy of field studies involves tests of "interaction effectiveness" (Douglas 1976:123), in which the ability to "pass as a member" documents the field worker's presumptive grasp of appropriate behavior in the setting. In a similar vein, a number of field workers have proposed that descriptions of actors' perspectives should serve as adequate *instructions* to outsiders on how to act as a competent member of that social world (Frake 1964, 1975). Wiseman (1970:280–81) reports, for example, that she "tested" her findings on skid row life by having several novices read her descriptions and then "successfully" pass as skid row drunks for short periods. Yet it may be possible to pass as a member without actually performing as a competent insider (owing to the politeness or face-saving graces of members, for example). Furthermore, while it may be useful to communicate descriptions in an instructional format (Schwartz & Jacobs 1979:256–9), such efforts encounter the difficulties inherent in any effort to obey rules: The procedures for applying rules to specific situations can never be completely specified by the rules themselves. Competent performance may thus result more from the novice's reliance on generally available commonsense knowledge than from his or her use of the instructions specifically produced by the field study.

Social Processes in Field Work

As field work findings come to be seen as products of the methods employed, the question of just what these methods entail – concretely and interactionally – has received greater attention. Research published during the 1970s has increasingly and self-consciously examined the social processes involved in field work itself, including: the actual character and course of ongoing field relations as part of the natural history of the research; the personal dimensions and consequences of field work; and the political and ethical implications and issues of field work. In many ways, this self-consciousness represents a change of style rather than of practice. Intensive participation in the daily lives of those studied has always generated recurring personal, moral, and methodological decisions and dilemmas. What has changed is the willingness to address these issues and processes publically, explicitly, and more profoundly.

Field Work Relations

While the humanistic tradition of field work has produced a number of first-person accounts of the actual process of field study [see, for example, Vidich, Bensman & Stein (1964)], the last decade has seen several extensive treatments of the nature and course of field relations. Most notable among these are R. Wax's (1971) detailed analyses of three extended field studies, Glazer's (1972) examination of the wider problems and contexts of field work, and Johnson's (1975) reflective account of his methods and personal experiences in studying social casework in two public welfare offices. These and other studies [e.g. those collected by Bell & Newby (1977)] have raised and extended analysis of a number of major issues in field-work relations, including the nature and variety of field-work roles, the development of trust and rapport, the implications of reactivity, and processes of involvement and detachment.

While Gold's (1958) observation-participation continuum continues to enjoy currency in the analysis of field-work roles (e.g. Denzin 1970; Roy 1970), a number of researchers have contended that ongoing field relations show greater variation and flexibility than these categories imply. R. Wax (1971:52) holds that the traditional role imagery of field work is overly rigid and narrow, as "many of the researcher's most useful roles are spontaneously invented and developed by the combined efforts of his respondents and himself." Field relations vary from situation to situation, person to person, and over time (Johnson 1975), variations not fully captured by the participant-observer distinction. Schatzman & Strauss (1973) talk not about mutually exclusive field-work roles but distinct modes of participation in ongoing situations (1973:58–63), each mode providing a tactical choice available for different times and occasions. Finally, standard accounts of field relations have been criticized for their rationalistic imagery, depicting field work as a narrow process of applying impersonal techniques to define researcher role and purpose and neglecting its personal and experiential dimensions (Johnson 1975; Clarke 1975).

Variations in field-work roles and in the character and course of field relations have also received greater attention. Consideration of the relevance of such ascribed attributes as gender, age, and race on field relations has been undertaken by Daniels (1967), Liebow (1967), Golde (1970), Johnson (1975), Easterday et al (1977), Warren & Rasmussen (1977), Yancey & Rainwater (1970), R. Wax (1979), Moore et al (1978), and Fine & Glassner (1979). These accounts typically ask how such attributes promote or inhibit access to certain sorts of social relations and hence to certain kinds of data (i.e. what the relevant group knows or does). Somewhat less attention has been devoted to the more subtle effects and consequences of these attributes on the process of content of field interactions, although Liebow's reflections on the persistent relevance of race in his study of black streetcorner men (1967:248–51) provides a worthy model.

One response to exclusion as an outsider is to use or train members of settings under study as field observers. Walker & Lidz (1977) examined the use of "indigenous observers" to gain access to the worlds of black street addicts otherwise closed to them, and similar procedures have been employed in other settings by Douglas (1976:212–15), Moore et al (1978), and Heilman (1976, 1980). While Walker & Lidz's observers contributed as informants (although relying too heavily on personal knowledge and contacts) and as evaluators of in-process analyses, they failed to become member ethnographers of the black drug scene. In contrast, Moore reports that her field work on gangs and drugs with Chicano *pintos* (ex-convicts) emerged as a more collaborative enterprise with at least partial if stressful resolution of the deep strains between academic and *pinto* researchers.

Underlying and indeed propelling many of these considerations of field-work relations is a deepening concern with assessing the field worker's effects on the particular setting studied. This concern derives in large part from a reconceptualization of the problem of *reactivity*. Classical formulations of field-work roles advocated passive, nonobtrusive presence and modes of participation in order to avoid changing or even creating the phenomena under study. Many field workers now reject this approach, seeing the solution more in the direction of becoming aware of one's effects on the field rather than of trying to avoid or minimize such effects.[2] Cassell & Wax (1980:261) have outlined the conceptual basis of this emerging approach: "In another view, the investigator is conceptualized as part of the reality being studied. Here, it is taken for granted that the observer alters that which is observed; but these alterations are the subject of study . . . in the first [view], the alterations, resulting from the effect of the observer upon that which is observed, are interference; in the second, they are data." Thus, field work is unavoidably interactional in character, and the field worker is necessarily consequential for those studied. Solutions to reactivity thus lie not in restricting, cutting off, or regularizing field interactions, but in trying to become sensitive to and perceptive of how one is perceived and treated by others. The researcher, in Clarke's phrase, is viewed as "a source for the result, not as a contaminant of it" (1975:99).

These issues have important implications for the recurrent concern with field-work trust and rapport. Trust is often touted as a solution to the problems of reactivity: Once fully trusted, it is assumed, the field worker has reliable access to that hidden, valid information ordinarily available to insiders only. Johnson has challenged the notion that a "field researcher establishes trust with the members of a setting *as such*" emphasizing that even those individuals most liked and most trusting shield or hide various sorts of information (1975:121, 138–39). Trust is not a mechanism allowing the field worker to transcend the limitations of specific role arrangements and their related patterns for conveying and withholding information. Even extreme "intimacy" provides only situated access, as one field worker found on assuming the role of "boyfriend" in studying masseuses (Warren & Rasmussen 1977). This very intimacy ("trust"), while providing some sorts of information, inevitably cut him off from others, for masseuses were evasive or reluctant to talk to boyfriends about their sexual activities in the parlors (354). (See also Rasmussen & Kuhn 1976.) The implication, Johnson argues, is that methodological reflection becomes particularly critical *after* rapport has been established and some secrets have been revealed, because reactive effects persist.

Reconceptualization of the nature of reactivity also has important implications for evaluating covert as opposed to overt field work. Traditionally, covert roles have been assumed to minimize reactive effects, although constraining the researcher to act within the limits of the role assumed (Lofland 1971:93–101). But recognition that the reactions to the known observer are not different in kind from "natural" reactions to others in the setting has blunted this apparent advantage (Bodemann 1978:393). [Wieder's (1974) study is exemplary, showing how halfway house residents treated the field worker in just the ways they treated the house staff.] Moreover, elaborating Erikson's (1967) earlier argument, Cassell (1980:80–81) contends that covert field work inhibits the researcher's sensitivity to reactive effects: "The investigator who presents a false self is affecting both the ability to judge what is going on and the course of interaction by the need for constant monitoring of one's 'act' or 'front'." These concerns also produce an exaggeratedly manipulative view of those studied. Hilbert (1980) offers a more positive evaluation, suggesting that where the goal of field work is to discover what it takes to act as a competent member of the setting under study, the covert role offers special insights: In learning to pass as a member, secret field workers not only "must make use of their knowledge of 'what typical members do' in order to protect the secret" (55), but are also specifically instructed by members in how to do this.

A further significant change in field methods is increasing advocacy of more highly involved, fully participatory roles in carrying out field research. Criticizing the ideal of the passive, noninvolved and nonobtrusive observer honored in most earlier field work, a number of field workers advocate action as a means for understanding critical social processes in field situations (Bodemann 1978, 1979; Whyte 1979; Schensul 1980). Whyte, invoking the tradition in action anthropology (Tax 1975; Jacobs 1974) that the field worker can and indeed should aid those she or he is studying, argues that intervention in stable settings will reveal "the forces that are keeping the social system in place," which would otherwise remain invisible (1979:63). Bodemann seeks to highlight the relevance of the Marxian concept of *praxis* for participant observation, urging the field worker (*a*) to participate "fully, freely and self-critically in the setting," and, on the basis of theoretically grounded understandings generated in this way, (*b*) to intervene actively by presenting options or by returning descriptive and theoretical findings to those studied (1978:410–11).

Another strand in the advocacy of more active field-work roles emphasizes not intervention so much as *participation* in the worlds of those studied in order to attain understanding of those worlds. Douglas (1976), for example, views "immersion in the natural situation" as the proper goal of field research, because only such immersion provides the field worker with the rich "direct experience" of everyday life that is the basis for reliable data. Douglas allows for such immersion through participation as a member, but, aside from the gains in access, seems more comfortable with the distance and detachment of a clearcut researcher role. Similarly, Henslin (1972:40) discusses the distinctive insights he gained by actually working as a cabbie, but suggests that they were equally available through standard observation and interviewing techniques. Others insist that it is crucial to go beyond this sort of empathetic yet ultimately detached relationship to some fuller form of committed membership. Bodemann (1978) provides a strong programmatic statement of this position, Jules-Rosette (1975) a substantive illustration. Having become a member in good faith of the African sect she was studying, Jules-Rosette emphasizes the subtle insights and understandings of their ritual available only through this experience. (See also Thorne 1979.) Such procedures need not be covert (both Thorne and Jules-Rosette openly pursued their research concerns as well as their natural activities): Emphasis is on participating as a member participates, experiencing events in "good faith" as members do, not on whether members identify the field worker as authentic.

Highly active field-work roles, whether interventionist or experiential in emphasis, assume high degrees of collaboration, cooperation, and sympathy with those studied. Yet field relations have been proposed and conducted on very different premises. Davis (1973) suggests that taking the stance and attitude of the convert in field work is only one option; many have adopted instead the detached attitude that takes nothing for granted. Invoking the imagery of investigative reporting, Douglas (1976) argues that field-work methods must be adapted to conditions of conflict as well as cooperation in social life. Toward this end he outlines the strategies and tactics of an "investigative field research" in which the field worker's aim is to penetrate the various fronts, lies, evasions and self-deceptions groups and individuals employ to cloak the truth. Proposals for a "conflict methodology" (Lehmann & Young 1974; Lundman & McFarlane 1976; Christie 1976) rest on similar adversarial premises, highlighting the inherent conflict between dominant institutions' control of information and researchers' needs to obtain data. Hessler et al (1980), however, suggest that the sharp dichotomy between conflict and consensual methods is simplistic when applied to the reality of field work in community settings, and M. Wax (1980) has pointed out several dilemmas confronting such "antagonistic fieldwork" arising from the need to empathize with those studied.

Those field workers embracing reactivity and advocating fuller participation confront a number of unresolved problems. First, even if participation and hence reactive effects are inevitable, different degrees and kinds of participation may still have different implications and effects. In practice, most field workers tend to act in nonobtrusive ways, if only out of personal preference or a commitment to being instructed by members. Controlled, minimal participation may still be preferable to full, total participation – especially, for example, if restrained participation maximizes the likelihood of identifying particular observer effects.

Second, how does the involved and committed field worker handle the need for detachment from the events and people studied? Douglas questions his own earlier dichotomy between empathy (feeling or seeing things from the standpoint or perspective of those studied) and sympathy ("identify[ing] with or act[ing] from this standpoint") as overly simple (1976:106). But this concern does point up the fact that some sort of detachment is necessary for field work; this is harder to achieve the more involved is the field worker. The best recommendation to date is withdrawal from the field and then detached analysis after the period of high involvement (R. Wax 1971; Glazer 1972).

Third, how can self-consciousness be promoted and maximized? One standard procedure is to note direct and indirect statements and actions that suggest one's perception by, and effects on, those studied. Johnson (1975:136) suggests such materials can often be augmented by directly inviting those observed to reflect and comment on the impact of the field worker's presence in specific situations. Sanders (1980:168) urges further that field notes regularly include "feelings of discomfort, threat, and self-consciousness" so that the field worker can monitor and later review his or her subjective experiences, possible biases, and perception by others in the field. Douglas (1976:189–225) has proposed that team field work, using multiple observers in the same setting, provides a method for getting a variety of readings on field-worker effects.

Personal Dimensions of Field Work

In most published accounts the personal dimensions of field research have been regarded as irrelevant; they have been either omitted altogether or relegated to brief mention in introductions or appendixes. Reflecting "a formal hiatus between the researcher and his facts" (Clarke 1975:118), descriptions of how methodological procedures produce data typically excluded consideration of the field worker's involvement as a person. Within the past decade this practice has changed significantly, as personal aspects and dimensions of field work have received increasing acknowledgment. Field researchers have explored the dynamics and consequences of personal involvement and identification with those studied (Wolff 1964, 1976; R. Wax 1971; Jules-Rosette 1975; Thorne 1979), the relation of personal and subjective motives and feelings to research efforts (Seeley 1964; Johnson 1975; Douglas 1976; Georges & Jones 1980), and the central relevance of personal characteristics of the field worker to those studied (Gans 1968; Johnson 1975; Warren & Rasmussen 1977). Such studies insist that field work is not solely or even primarily a research enterprise. Rather, field work prototypically involves a total human and personal experience, of which research is but one part. Field work is a process of "relearning and resocialization" (R. Wax 1971:20) that engenders deep feelings of insecurity, anxiety, loneliness, frustration, and confusion (Gans 1968; R. Wax 1971; Johnson 1975; Sanders 1980). Personal values and beliefs tend to be challenged and questioned (Clarke 1975) and attitudes fundamentally changed as the result of the contact (Johnson 1975; Jules-Rosette 1975).

A number of these accounts address the tendency for personal involvement to produce bias or distortion in observation and interpretation. R. Wax examines the changing course of her strong identification with several ideological factions in relocation camps (1971:118–62), yet suggests that continual exposure to different groups in the field gradually corrected initial biases (141). Douglas urges field workers to avoid studying issues "too close to their hearts" or in which they have intense and unresolved emotional conflicts (1976:93–103). Gans (1968), extending Schwartz & Schwartz's earlier analysis (1955) of the distortive effects of intense anxiety on participant observation, identifies a number of subjective processes specific to field relations. Such processes, including the experiences of irremediable marginality and of guilt about the subtly deceptive stance assumed toward those studied, lead to over-identification with one's subjects.

Another trend insists that although subjective experience is a source of bias, it is also a source of insight and understanding, both of those studied and of oneself. Sanders (1980:168) observes that field notes on personal feelings "provide a foundation for understanding the discomfort normally experienced by other regular actors in the setting." Douglas (1976) insists that introspection provides a critical method for grasping the meanings and worlds of others. Wolff invokes the concept of surrender – comprised of "total involvement, suspension of received notions, pertinence of everything, identification, and risk of being hurt" (1964:237) – to analyze his deepening understanding of the New Mexico community he studied. Hughes (1971:571) sounds a similar note in pointing to the close link between field workers' research concerns and a personal or "sympathetic emancipation" from some significant aspect of their pasts.

In an excellent review, Clarke (1975) points out that such analyses draw attention to processes of conversion or resocialization in field work, processes that fundamentally change the field worker through "the impact of a new way of thought and life upon their own" (1975:119). These processes and experiences are critical for understanding field work findings, since "the data only sees the light of day after it has gone through the self of the investigator" (1975:104–5). As Clarke succinctly and forcefully summarizes this argument (1975:118):

> We must accept that social scientific research involves the researcher relating to those he investigates, and that the result is the outcome of their relationship, a relationship which, like all relationships, will change both parties. The knowledge thus gained from these relationships, not only changes the knower, it becomes part of the knower. We must look then to the knower as much as to his field if we would understand what he is saying, and recognise that he has a valid part to play as a person, not just as manipulator of techniques, in the acquisition of knowledge.

Clark suggests two remedies (1975:119–20): "to make what counts as 'research,' 'data,' and so on, more flexible, and/or to develop the scrutiny of the experiences and insights in question to such a point as to be able to incorporate them into the public and verifiable research process."

A number of unresolved issues remain. First, to what extent is full disclosure of the personal experiences of field work a realistic possibility? Lofland (1971:132), while highly sympathetic to personal dimensions of field work, suggests that field workers, particularly the young and professionally unestablished, will inevitably shy away from full disclosure. At best, methodological accounts describe "the second worst things that happened." Johnson (1975:167), despite his own revelations of personal reactions and feelings, similarly suggests that one should expect only "limited hangout" and not full disclosure.

Second, to what extent is full disclosure necessary? Clarke warns against the dangers of "vulgarization and distortion" posed by highly personal accounts, and advises against including "every passing emotion" in field reports (1975:120). One minimal criterion might restrict inclusion to those personal and emotional experiences that significantly affect the researcher's understanding of the setting in general or of critical activities within it. In this regard, Thorne (1979) provides a useful model for linking intense personal experience with wider sociological understanding. Reflecting on the recurrent strains between simultaneously participating in and observing the draft resistance movement, she highlights a personal "moment of truth" when publically called upon to join a group providing sanctuary to two draft resisters and hence risk arrest (1979:80). She felt tremendous pressure to join, but stayed seated, in great inner turmoil. While her reluctance to join accentuated her sense of participating as a researcher and not a full member, her very retreat from this form of political action made her aware of the distinctive commitments and epistemological outlook of Resistance activists, including a "demand for complete involvement based on a totalizing world view; a sense of crisis and apocalypse; and emphasis on collectivity and control by the group" (1979:82).[3]

Political and Ethical Issues

The political directions and implications of field-work studies have increasingly been brought into question. Noting the relatively abundant field work literature on "the poor, the ethnic groups, the disadvantaged," and the comparative neglect of the upper and even middle classes, Nader (1969:289) called for a reorientation of field work toward "studying up." Yet during the past decade the marginal and the subordinate still comprise the subjects of the bulk of field studies, although with a growing sense of unease. Many field workers question the political implications of revealing the worlds and perspectives of the deviant and the powerless, especially where such worlds are protected by secrecy (Liazos 1972; Warwick 1973; Warren 1980). Yet the call to study up has produced uneven results at best. Several studies of federal agencies have appeared (Weaver 1977; Kagan 1978), and there is a growing body of field research on the mass media (Altheide 1976; Tuchman 1978; Fishman 1980) and corporate executives (Kantor 1977). With few exceptions, however, this redirection remains subject to Gouldner's (1968) critique that the activities of lower-level officials rather than of top policy makers receive disproportionate attention. Problems of access and cooperation frequently arise when such field research is attempted (Daniels 1974; Hoffman 1980), leading some to suggest that direct participant observation may not be feasible with elites (Nader 1969; Galliher 1980; but see Lofland 1976:18–24). Indeed, many such studies of the powerful (e.g. Domhoff 1975) necessarily rely on interviews with key informants and documentary evidence rather than long-term observation and participation.

The last decade has also seen increasing consideration of various ethical issues and implications of field work. Although the extension of federal regulations seeking to protect human subjects in scientific research (DHEW 1974, 1978) has occasioned and focused much of the literature on these issues, field workers' concerns here both predate and extend beyond the federal requirements.

For field research, the main import of DHEW regulations derives from the requirement that "subjects at risk" ("risk" broadly defined as "the possibility of physical, psychological or social injury") give "legally effective informed consent," ordinarily by means of signed forms, to participation in research. These regulations are administered by granting agencies and university review boards as a condition for funding or sponsoring research. Largely informal reports (Cassell 1978) indicate that the actual application of these provisions to field research varies widely. Some institutional review boards exempt field research entirely from review, presumably on the grounds that no risks are involved, while others routinely waive written informed consent requirements under the provision that "the research present no more than minimal risk of harm to subjects." Still other IRBs strictly enforce written consent provisions on all field research projects. The response of field workers also varies, though this process has not been systematically studied. There are some reported instances of outright fabrication of written consent forms (Cassell 1978). Evaluations of the actual processes and consequences of obtaining written informed consent in field research are rare (but see the case discussed by Thorne 1980:289).

The response of field workers to these regulations has been sharply critical (M. Wax 1977, 1980; Cassell 1978, 1980; Olesen 1979; Thorne 1980). Two main components of the regulations have received particular criticism:

THE RISK-BENEFIT CALCULUS The federal regulations provide that the risks to subjects be weighed against the benefits that derive from the research, thus requiring some utilitarian calculation of risks and harms in relation to benefits. Critics have pointed out that the harms and benefits of field work not only are "less immediate, measurable, and serious" than those associated with other forms of research (Cassell 1980), but also are difficult to predict in advance of the actual study or even of publication. As a result, cost-benefit calculation tends to become a ritualistic "exercise in creativity" and provides no substantive criteria to guide the recurrent ethical decisions field workers encounter (Klockars 1977). May (1980) links these problems to the underlying limitations of utilitarian ethics – its individualistic assumptions, disregard of issues of distributive justice, and insensitivity to collective and communal values – when applied to the conduct of field work.

INFORMED CONSENT PROCEDURES Critics note that federal informed consent requirements, designed for biomedical and experimental research generally, embody assumptions that are incompatible with standard field methods (Duster et al 1979; Cassell 1980; M. Wax 1977, 1980). Field work relations are emergent, based on reciprocity and relatively equal power, continuous, and usually long-term. Such features contrast with the short-term, essentially contractual encounter presupposed by informed consent procedures, an encounter between strangers possessing grossly disparate amounts of power and knowledge. Furthermore, in most field situations, the researcher neither possesses great power as perceived by those studied, nor exercises unilateral control over the setting, conduct, or interactional contours of the research encounter (Cassell 1980). Informed consent then becomes not only formalistic, but also patronizing (M. Wax 1977; Klockars 1977), assuming "ignorant, powerless, innocent subjects who become the victims of the sly maneuvers of crafty professors" (M. Wax 1977:324). The informed consent model assumes also that the researcher confronts a series of individual subjects, ignoring the way field work proceeds within the "contours of the natural groups and settings" that offer few "formal choice points" for participation in the study (Thorne 1980:293). Finally, the informed consent model assumes that the researcher relates to those studied within narrow, researcher-subject role relations, ignoring the multiple roles in field work and their complex and often conflicting responsibilities, obligations, and expectations (Klockars 1977:218).

Field workers have begun to explore and propose alternative ethical principles for conducting field work. Cassell (1980:35) suggests abandoning the utilitarian in favor of a Kantian ethic, in which "persons [will] be treated at all times as ends in themselves, never merely as means." Cassell holds that this standard would legitimate everyday evasiveness and deceit in field work (e.g. not regularly reminding those studied that they are being observed), but would not condone covert research, which violates the autonomy of those studied or treats them as means rather than ends. Klockars (1977) seems to propose a standard close to this one in emphasizing that "respect" characterize the relation between field worker and those studied, although he tends to see this respect as something arising in normal social relations and hence not particularly needing statement as a principle. Thorne (1980:287) suggests that field workers can and should share research purposes and procedures – including "research goals, frameworks, methods, patterns of sponsorship, and expected reporting" – as completely and fully as possible with those studied. Despite the expected practical impediments to implementation, such procedures would be more effective than federal informed consent requirements in placing those studied in a position to assess the gains and risks for themselves and to decide whether they want to participate. Thorne also suggests that such consent should be renewed as the conditions and goals of field work change.

Other field workers, however, seem less dubious about covert field work. Some challenge the claim that any significant harm or risk has been done to those studied by disguised methods (Humphreys 1975; Douglas 1976:xiv–xv). Others, including most conflict methodologists, argue that informed consent requirements ignore the system of stratification and the actual differences in power in American society; according to this argument, elite groups neither need nor warrant these or other protections against covert research (Christie 1976; Galliher 1980).

Ironically, federal regulations do not provide one of the most significant protections that field workers could offer those studied – protection against the state's power of subpoena (Nejelski & Lerman 1971; Freidson 1976; Bond 1978). Often the major risks posed by field work derive from breaches of confidentiality and privacy (Warren 1980). While a number of ingenious dodges have been proposed, and while many field workers claim that they would burn notes or go to jail rather than have field notes made public, without a legal right to testimonial privilege such proposals and assurances may well be empty (Humphreys 1975; Trend 1980). Freidson recommends "the routine destruction of identifiers the minute they are no longer necessary for planned research" (1978:159), but recognizes even such a procedure is a stop-gap measure.

One of the deeper implications of these issues is the increasing role of the state and professional groups in regulating and reviewing field-work studies. Douglas (1978) decries these regulations in principle as serious governmental intrusion into the free conduct of research. Critics also warn of the dangers of bureaucratic review and accreditation of research, anticipating nearly as much threat from colleagues as from official agencies (Lofland 1976:21–22).

Conclusion

Over the past decade a major part of the field-work tradition has tied itself to the emerging interpretive paradigm and actively sought to separate and distinguish itself from the dominant quantitative models of research in sociology. Sharp criticism both of the claimed rigor of quantitative procedures (Cicourel 1964) and of its deterministic assumptions (Blumer 1960) and "passivist" implications (Lofland 1976) accompanied this separation, which produced growing appreciation of the absolute strengths of field methods: Flexible rather than prespecified design produced not faulty but contextually grounded findings; first-hand, "intimate familiarity" with the empirical world might purchase more than impressionistic evidence and might help to generate theory. Qualitative field work has established its distinctive grounding, but considerable effort still lies ahead in specifically working through the diverse methodological and theoretical implications of an interpretive paradigm.

Notes

1. A number of contemporary field researchers similarly oriented toward actors' subjective meanings also identify their work with the tradition of phenomenology. Warren (1974:10), for example, characterizes her field work findings as phenomenological in that they present "the meanings of the gay world as the members experienced it." See also Douglas & Johnson (1977).
2. Gussow (1964:231), acknowledging his debt to Anselm Strauss, provided one of the first and clearest statements of this position: "Ordinarily, in good fieldwork, researchers are not greatly concerned about whether they have disturbed the natural field or not, provided that they can analyze how they affected it structurally. Indeed, by affecting it, they often get to know better what it is they are studying."
3. In contrast, Johnson (1975) describes moments of intense personal anguish but often fails to link them to his deepening understanding and analysis of the setting (see especially pp. 157–60).

Literature Cited

Akins, C., Beschner, G., eds. 1980. *Ethnography: A Research Tool for Policymakers in the Drug and Alcohol Fields*. US Dept. Health Human Serv. Publ. No. (ADM) 80–946

Altheide, D. L. 1976. *Creating Reality: How TV News Distorts Events*. Beverly Hills: Sage. 221 pp.

Becker, H. S. 1967. Whose side are we on? *Soc. Probl.* 14:239–48

Becker, H. S. 1970. *Sociological Work: Method and Substance.* Chicago: Aldine. 358 pp.

Bell, C., Newby, H., eds. 1977. *Doing Sociological Research.* NY: Free Press. 186 pp.

Bellman, B. L., Jules-Rosette, B. 1977. *A Paradigm for Looking: Cross-Cultural Research with Visual Media.* Norwood, NJ: Ablex. 211 pp.

Bittner, E. 1973. Objectivity and realism in sociology. In *Phenomenological Sociology: Issues and Applications,* ed. G. Psathas, pp. 109–25. NY: Wiley

Blumer, H. 1969. *Symbolic Interactionism: Perspective and Method.* Englewood Cliffs, NJ: Prentice-Hall. 208 pp.

Bodemann, Y. M. 1978. A problem of sociological praxis: The case for interventive observation in field work. *Theor. Soc.* 5:387–420

Bodemann, Y. M. 1979. The fulfillment of fieldwork in Marxist praxis. *Dialect. Anthropol.* 4:155–61

Bogdan, R. 1972. *Participant Observation in Organizational Settings.* Syracuse, NY: Syracuse Univ. Div. Spec. Educ. Rehab. 106 pp.

Bogdan, R., Taylor, S. J. 1975. *Introduction to Qualitative Research Methods: A Phenomenological Approach to the Social Sciences.* NY: Wiley. 266 pp.

Bond, K. 1978. Confidentiality and the protection of human subjects in social science research: A report on recent developments. *Am. Sociol.* 13:144–52

Cassell, J. 1978. Risk and benefit to subjects of fieldwork. *Am. Sociol.* 13:134–43

Cassell, J. 1980. Ethical principles for conducting fieldwork. *Am. Anthropol.* 82:28–41

Cassell, J., Wax, M. L. 1980. Editorial introduction: toward a moral science of human beings. *Soc. Probl.* 27:259–64

Cavan, S. 1972. *Hippies of the Haight.* St. Louis: New Critics Press

Cavan, S. 1974. Seeing social structure in a rural setting. *Urb. Life & Cult.* 3:329–46

Christie, R. M. 1976. Comment on conflict methodology: a protagonist position. *Sociol. Q.* 17:513–19

Cicourel, A. V. 1964. *Method and Measurement in Sociology.* NY: Free Press

Cicourel, A. V. 1968. *The Social Organization of Juvenile Justice.* NY: Wiley. 345 pp.

Cicourel, A. V. 1974a. *Cognitive Sociology: Language and Meaning in Social Interaction.* NY: Free Press. 189 pp.

Cicourel, A. V. 1974b. Interviewing and memory. In *Pragmatic Aspects of Human Communications,* ed. C. Cherry, pp. 51–82. Dordrecht, Holland: D. Reidel

Cicourel, A. V., Jennings, K. H., Jennings, S. H. M., Leiter, K. C. W., MacKay, R., Mehan, H., Roth, D. R. 1974. *Language Use and School Performance.* NY: Academic. 368 pp.

Clarke, M. 1975. Survival in the field: implications of personal experience in field work. *Theor. Soc.* 2:95–123

Daniels, A. K. 1967. The low-caste stranger in social research. In *Ethics, Politics, and Social Research,* ed. G. Sjoberg, pp. 267–96. Cambridge, Mass: Schenkman

Daniels, A. K. 1974. Getting in and getting on: the sociology of infiltration and ingratiation. Presented at the Pac. Sociol. Assoc. Meet., San Jose, Calif.

Davis, F. 1973. The Martian and the convert: ontological polarities in social research. *Urb. Life & Cult.* 2:333–43

Delamont, S., Hamilton, D. 1976. Classroom research: a critique and a new approach. In *Explorations in Classroom Observation,* ed. M. Stubbs, S. Delamont, pp. 3–20. NY: Wiley

Denzin, N. K. 1970. Participant observation: varieties and strategies of the field method. In *The Research Act: A Theoretical Introduction to Sociological Methods,* pp. 185–218. Chicago: Aldine

Department of Health, Education and Welfare. 1974. Protection of human subjects. *Fed. Reg.* (39 FR 18914)

Department of Health, Education and Welfare. 1978. Protection of human subjects: institutional review boards. *Fed. Reg.* (43 FR 56174)
Domhoff, G. W. 1975. *The Bohemian Grove and Other Retreats*. NY: Harper and Row. 250 pp.
Douglas, J. D., ed. 1972. *Research on Deviance*. NY: Random House. 270 pp.
Douglas, J. D. 1976. *Investigative Social Research: Individual and Team Field Research*. Beverly Hills: Sage. 229 pp.
Douglas, J. D., Johnson, J. M. 1977. *Existential Sociology*. Cambridge: Cambridge Univ. Press. 327 pp.
Douglas, J. D. 1978. Letter to the *Am. Sociol.* 13:157–58
Duster, T., Matza, D., Wellman, D. 1979. Field work and the protection of human subjects. *Am. Sociol.* 14:136–42
Easterday, L., Papdemas, D., Schorr, L., Valentine, C. 1977. The making of a female researcher: role problems in field work. *Urb. Life* 6:333–48
Erikson, K. T. 1967. A comment on disguised observation in sociology. *Soc. Probl.* 14:366–73
Fagerhaugh, S., Strauss, A. 1977. *Politics of Pain Management: Staff-Patient Interaction*. Reading, Mass: Addison-Wesley. 323 pp.
Fiedler, J. 1978. *Field Research: A Manual for Logistics and Management of Scientific Studies in Natural Settings*. San Francisco: Jossey-Bass. 188 pp.
Filstead, W. J., ed. 1970. *Qualitative Methodology: Firsthand Involvement with the Social World*. Chicago: Markham. 352 pp.
Fine, G. A., Glassner, B. 1979. Participant observation with children: promise and prospects. *Urb. Life* 8:153–74
Fishman, M. 1980. *Manufacturing the News*. Austin: Univ. Austin Press. 180 pp.
Frake, C. O. 1964. How to ask for a drink in Subanun. *Am. Anthropol.* 66:127–32
Frake, C. O. 1975. How to enter a Yakan house. In *Socioculture Dimensions of Language Use*, ed. M. Sanchez, B. Blount, pp. 25–40. NY: Academic
Freidson, E. 1976. The legal protection of social research: criteria for definition. In *Social Research in Conflict with Law and Ethics*, ed. P. Nejelski, pp. 123–37. Cambridge, Mass: Ballinger
Freidson, E. 1978. Letter. *Am. Sociol.* 13:159–60
Galliher, J. F. 1980. Social scientists' ethical responsibilities to superordinates: looking upward meekly. *Soc. Probl.* 27:298–308
Gans, H. J. 1968. The participant-observer as a human being: Observations on the personal aspects of field work. In *Institutions and the Person: Papers Presented to Everett C. Hughes*, ed. H. S. Becker et al. Chicago: Aldine
Garfinkel, H. 1967. *Studies in Ethnomethodology*. Englewood Cliffs, NJ: Prentice-Hall. 288 pp.
Georges, R. A., Jones, M. O. 1980. *People Studying People: The Human Element in Fieldwork*. Berkeley: Univ. Calif. Press. 178 pp.
Glaser, B. G. 1976. *The Patsy and the Sub-contractor*. New Brunswick, NJ: Transaction Books
Glaser, B. G., Strauss, A. 1967. *The Discovery of Grounded Theory: Strategies for Qualitative Research*. Chicago: Aldine. 271 pp.
Glazer, M. 1972. *The Research Adventure: Promise and Problems of Field Work*. NY: Random House. 203 pp.
Gold, R. L. 1958. Roles in sociological field observations. *Soc. Forces* 36:217–23
Golde, P., ed. 1970. *Women in the Field: Anthropological Experiences*. Chicago: Aldine. 343 pp.
Gould, L., Walker, A. L., Crane, L. E., Lidz, C. W. 1974. *Connections: Notes from the Heroin World*. New Haven: Yale Univ. Press. 236 pp.

Gouldner, A. 1968. The sociologist as partisan: sociology and the welfare state. *Am. Sociol.* 3:103–16
Gussow, Z. 1964. The observer-observed relationship as information about structure in small-group research: a comparative study of urban elementary school classrooms. *Psychiatry* 27:230–47
Habenstein, R. W., ed. 1970. *Pathways to Data: Field Methods for Studying Ongoing Social Organizations*. Chicago: Aldine. 276 pp.
Heilman, S. C. 1976. *Synagogue Life: A Study in Symbolic Interaction*. Chicago: Univ. Chicago Press. 306 pp.
Heilman, S. C. 1980. Jewish sociologist: native-as-stranger. *Am. Sociol.* 15:100–8
Henslin, J. M. 1972. Studying deviance in four settings: research experiences with cabbies, suicides, drug users, and abortionees. In *Research on Deviance*, ed. J. D. Douglas, pp. 35–79. NY: Random House
Hilbert, R. A. 1980. Covert participant observation: on its nature and practice. *Urb. Life* 9:51–78
Hoffman, J. E. 1980. Problems of access in the study of social elites and boards of directors. In *Fieldwork Experience: Qualitative Approaches to Social Research*, ed. W. B. Shaffir, R. A. Stebbins, A. Turowetz, pp. 45–56. NY: St. Martin's Press
Hughes, E. C. 1971. *The Sociological Eye: Selected Papers*. Chicago: Aldine. 584 pp.
Humphreys, L. 1975. *Tearoom Trade: Impersonal Sex in Public Places*. Chicago: Aldine. Enlarged Edition. 238 pp.
Jacobs, S. 1974. Action and advocacy anthropology. *Hum. Org.* 33:209–15
Johnson, J. M. 1975. *Doing Field Research*. NY: Free Press. 225 pp.
Jules-Rosette, B. 1975. *Vision and Realities: Aspects of Ritual and Conversion in an African Church*. Ithaca: Cornell Univ. Press. 302 pp.
Kagan, R. 1978. *Regulatory Justice: Implementing a Wage-Price Freeze*. NY: Russell Sage. 200 pp.
Kantor, R. M. 1977. *Men and Women of the Corporation*. NY: Basic Books. 348 pp.
Katz, J. 1982. A theory of qualitative methodology: the social system of analytic fieldwork. In *Poor People's Lawyers in Transition*. New Brunswick, NJ: Rutgers Univ. Press. In press
Klockars, C. B. 1977. Field ethics for the life history. In *Street Ethnography*, ed. R. S. Weppner, pp. 201–26. Beverly Hills: Sage
Lehmann, T., Young, T. R. 1974. From conflict theory to conflict methodology: an emerging paradigm for sociology. *Soc. Inquiry* 44:15–28
Leiter, K. 1980. *A Primer on Ethnomethodology*. NY: Oxford Univ. Press. 252 pp.
Liazos, A. 1972. The poverty of the sociology of deviance: nuts, sluts, and preverts. *Soc. Probl.* 20:103–20
Liebow, E. 1967. *Tally's Corner: A Study of Negro Streetcorner Men*. Boston: Little, Brown & Co. 260 pp.
Lofland, J. 1971. *Analyzing Social Settings: A Guide to Qualitative Observation and Analysis*. Belmont, Calif: Wadsworth. 136 pp.
Lofland, J. 1976. *Doing Social Life: The Qualitative Study of Human Interaction in Natural Settings*. NY: Wiley. 328 pp.
Lundman, R. J., McFarlane, P. T. 1976. Conflict methodology: an introduction and preliminary assessment. *Sociol. Q.* 17:503–12
May, W. 1980. Doing ethics: the bearing of ethical theories on fieldwork. *Soc. Probl.* 27:358–70
McCall, G. J. 1978. *Observing the Law: Field Methods in the Study of Crime and the Criminal Justice System*. NY: Free Press. 259 pp.
McCall, G. J., Simmons, J. L. 1969. *Issues in Participant Observation: A Text and Reader*. Reading, Mass: Addison-Wesley. 359 pp.

Moore, J. W., Garcia, R., Garcia, C., Cerda, L., Valencia, F. 1978. *Homeboys: Gangs, Drugs, and Prison in the Barrios of Los Angeles*. Philadelphia: Temple Univ. Press. 239 pp.
Nader, L. 1969. Up the anthropologist – perspectives gained from studying up. In *Reinventing Anthropology*, ed. D. Hymes, pp. 284–311. NY: Vintage
Nash, D., Weintrob, R. 1972. The emergence of self-consciousness in ethnography. *Curr. Anthropol.* 13:527–42
Nejelski, P., Lerman, L. M. 1971. A researcher-subject testimonial privilege: what to do before the subpoena arrives. *Wis. Law Rev.* 1971, pp. 1085–148
Olesen, V. 1979. Federal regulations, institutional review boards and qualitative social science research: comments on a problematic era. In *Federal Regulations: Ethical Issues and Social Research*, ed. M. L. Wax, J. Cassells, pp. 103–18. Boulder, Colo: Westview
Overington, M. A. 1979. Doing what comes rationally: some developments in meta-theory. *Am. Sociol.* 14:2–12
Rasmussen, P. K., Kuhn, L. L. 1976. The new masseuse: play for pay. *Urb. Life* 5:271–92
Roy, D. 1970. The study of Southern labor union organizing campaigns. In *Pathways to Data: Field Methods for Studying Ongoing Social Organizations*, ed. R. W. Habenstein, pp. 216–44. Chicago: Aldine
Rubinstein, J. 1973. *City Police*. NY: Ballantine. 462 pp.
Sanders, C. R. 1980. Rope burns: impediments to the achievement of basic comfort early in the field research experience. In *Fieldwork Experience: Qualitative Approaches to Social Research*, ed. W. B. Shaffir, R. A. Stebbins, A. Turowetz, pp. 158–71. NY: St. Martin's Press
Schatzman, L., Strauss, A. L. 1973. *Field Research: Strategies for a Natural Sociology*. Englewood Cliffs, NJ: Prentice-Hall. 149 pp.
Schensul, S. L. 1980. Anthropological fieldwork and sociopolitical change. *Soc. Probl.* 27:309–19
Schwartz, G., Merten, D. 1971. Participant observation and the discovery of meaning. *Philos. Soc. Sci.* 1:279–98
Schwartz, H., Jacobs, J. 1979. *Qualitative Sociology: A Method to the Madness*. NY: Free Press. 458 pp.
Schwartz, M. S., Schwartz, C. G. 1955. Problems in participant observation. *Am. J. Sociol.* 60:343–54
Seeley, J. R. 1964. Crestwood Heights: intellectual and libidinal dimensions of research. In *Reflections on Community Studies*, ed. A. J. Vidich, J. Bensman, M. R. Stein, pp. 157–206. NY: Wiley
Shaffir, W. B., Stebbins, R. A., Turowetz, A. 1980. *Fieldwork Experience: Qualitative Approaches to Social Research*. NY: St. Martin's Press. 329 pp.
Spradley, J. P. 1977. Foreward. In *Street Ethnography: Selected Studies of Crime and Drug Use in Natural Settings*, ed. R. S. Weppner, pp. 13–16. Beverly Hills: Sage
Spradley, J. P. 1979. *The Ethnographic Interview*. NY: Holt, Rinehart & Winston. 247 pp.
Strauss, A. 1978. *Negotiations: Varieties, Contexts, Processes, and Social Order*. San Francisco: Jossey-Bass. 275 pp.
Tax, S. 1975. Action anthropology. *Curr. Anthropol.* 16:514–17
Thorne, B. 1979. Political activist as participant observer: conflicts of commitment in a study of the draft resistance movement of the 1960's. *Symbol. Interact.* 2:73–88
Thorne, B. 1980. "You still takin' notes?" Fieldwork and problems of informed consent. *Soc. Probl.* 27:284–97
Trend, M. G. 1980. Applied social research and the government: notes on the limits of confidentiality. *Soc. Probl.* 27:342–49
Tuchman, G. 1978. *Making News: A Study in the Construction of Reality*. NY: Free Press. 244 pp.

Vidich, A. J., Bensman, J., Stein, M. R., eds. 1964. *Reflections on Community Studies*. NY: Wiley. 359 pp.
Walker, A. L., Lidz, C. W. 1977. Methodological notes on the employment of indigenous observers. In *Street Ethnography*, ed. R. S. Weppner, pp. 103–23. Beverly Hills: Sage
Warren, C. A. B. 1974. *Identity and Community in the Gay World*. NY: Wiley. 191 pp.
Warren, C. A. B. 1977. Fieldwork in the gay world: issues in phenomenological research. *J. Soc. Iss*. 33(4):93–107
Warren, C. A. B. 1980. Data presentation and the audience: Responses, ethics, and effects. *Urb. Life* 9:282–308
Warren, C. A. B., Rasmussen, P. K. 1977. Sex and gender in field research. *Urb. Life* 6:349–70
Warwick, D. P. 1973. Tearoom trade: means and ends in social research. *Hastings Cent. Stud*. 1:27–38
Wax, M. L. 1977. On fieldworkers and those exposed to fieldwork: federal regulations and moral issues. *Hum. Org*. 36:321–28
Wax, M. L. 1980. Paradoxes of "consent" to the practice of fieldwork. *Soc. Probl*. 27:272–83
Wax, R. H. 1971. *Doing Fieldwork: Warnings and Advice*. Chicago: Univ. Chicago Press. 395 pp.
Wax, R. H. 1979. Gender and age in fieldwork and fieldwork education: no good thing is done by any man alone. *Soc. Probl*. 26:509–22
Weaver, S. 1977. *Decision to Prosecute: Organization and Policy in the Antitrust Division*. Cambridge, Mass: MIT Press. 196 pp.
Weppner, R. A., ed. 1977. *Street Ethnography: Selected Studies of Crime and Drug Use in Natural Settings*. Beverly Hills: Sage. 288 pp.
Whyte, W. F. 1979. On making the most of participant observation. *Am. Sociol*. 14:56–66
Wieder, D. L. 1974. *Language and Social Reality: The Case of Telling the Convict Code*. The Hague: Mouton. 236 pp.
Wiseman, J. P. 1970. *Stations of the Lost: The Treatment of Skid Row Alcoholics*. Englewood Cliffs, NJ: Prentice-Hall. 346 pp.
Wiseman, J. P. 1974. The research web. *Urb. Life & Cult*. 3:317–28
Wolff, K. H. 1964. Surrender and community study: the study of Loma. In *Reflections of Community Studies*, ed. A. J. Vidich, J. Bensman, M. R. Stein, pp. 233–63. NY: Wiley
Wolff, K. H. 1976. *Surrender and Catch: Experience and Inquiry Today*. Dordrecht, Holland/Boston: Reidel. 410 pp.
Yancey, W. L., Rainwater, L. 1970. Problems in the ethnography of the urban underclass. In *Pathways to Data: Field Methods for Studying Ongoing Social Organizations*, ed. R. W. Habenstein, pp. 245–69. Chicago: Aldine
Zimmerman, D. H. 1970. The practicalities of rule use. In *Understanding Everyday Life*, ed. J. D. Douglas, pp. 221–38. Chicago: Aldine

19

Everett C. Hughes and the Development of Fieldwork in Sociology

Jean-Michel Chapoulie

> To make progress in our job, we need to give full and comparative attention to the not-yets, the didn't-quite-make-its, the not quite respectable, the unremarked, and the openly "anti" goings-on in our society [E. C. Hughes 1971: 53].

> The method of observation has a sure procedure: it gathers facts to compare them, and compares them to know them better. The natural sciences are in a way no more than a series of comparisons. As each particular phenomenon is ordinarily the result of the combined action of several causes, it would be only a mystery for us if we considered it on its own; but if it is compared with analogous phenomena, they throw light on each other. The particular action of each cause we see as distinct and independent, and general laws are the result. Good observation requires analysis; now, one carries out analysis in philosophy by comparisons, as in chemistry by the play of chemical affinities (J.-M. De Gerando, 1800, 1969).

The history of the social sciences is not just one of "theories" or successive "paradigms"; it is also a history in which new areas of empirical research are opened up and new methods of data gathering are worked out simultaneously. In general, the first attempts at empirical research in social science dealt with subjects that were distant from the researcher in time or in social space: either with "primitive" populations or with the working class and peasants in countries undergoing industrialization. Empirical

Source: *Journal of Contemporary Ethnography* (formerly *Urban Life*), 15(3/4) (1987): 259–297.

researchers did not discover the middle class and its institutions for many years, until after 1940 in the United States. It took another twenty years, and Goffman's work, before they included everyday behavior among the subjects of research (before that, it was a domain reserved for etiquette books), and ten years more for them to include the work activities of researchers in the natural and social sciences.

Social scientists' prolonged avoidance of subjects closely related to them shows how difficult they found it to be objective about the social reality around them; we see the same phenomenon in their choice of methods, since they had similar problems integrating the simplest method of all – observation *in situ* by the researcher – into the panoply of ordinary social science methods. It took ethnographers, for example, more than a century to put into practice the recommendations formulated by De Gerando at the beginning of the nineteenth century, substituting direct and prolonged observation for the collection of uncertain testimony in unknown conditions.[1] Even Malinowski (who was both the initiator and propagandist of this method) did not actually use it, although he described it in the *Argonauts*.[2]

In studies of their own societies, the first researchers to use empirical methods favored the use of official statistics or, sometimes, legal texts, even if those trained as physicians or naturalists – such as Villermé, Parent-Duchâtelet, or LePlay in France – did not look down on the collection of evidence by direct observation.[3] Those who used these methods worked outside the academic framework, while the first university sociologists either failed in their attempts at empirical research (Weber, for example)[4] or, like Durkheim, developed epistemological arguments that dissuaded people from collecting data by direct observation. Such data, suspected of depending on the subjectivity of researchers were not considered useful for scientific purposes. As a result, they were left to people who used them routinely, like journalists and administrators on inspection visits, or to the moving spirits of various social movements, like LePlay's disciples, who became observers of the working class.[5]

The Department of Sociology at the University of Chicago was the first academic group to develop an empirical research enterprise making much use of observation *in situ*, along with other sources of data. At first, researchers – most of them from rural middle-class Protestant families in the Midwest – used this method primarily to study populations socially very distant from them. They used Simmel's description of "the stranger" as an analogy when they tried to explain the role of the observer. (See, for instance, Paul G. Cressey, 1983.)

Beginning in the 1940s in the United States, studies of the middle class (or at least, of subjects involving it) showed how inadequate this analogy was, when sociologists found themselves in the role of "native" rather than "stranger." The first research on subjects connected to the middle class was done, in midcareer, by sociologists born around the turn of the century: Parsons, Lazarsfeld, Merton, and Hughes. Among them, Everett C. Hughes and, in another way, Herbert Blumer, became the principal proponents of observation *in situ* as a method of data gathering, through both their publications and their teaching at the University of Chicago, widely considered "influential," although it is not always clear just how that influence was exercised.[6] In this article I will analyze, from a historical point of view, the context in which the use of observation *in situ* and, more generally, fieldwork, developed in sociology. I take as my central point of reference the research style Hughes represented and as my principal example of that style in action the research in the sociology of work he sponsored at the University of Chicago.[7]

> The vocabulary used to designate research based, at least in part, on observation is confusing – like the English language itself – because the terms are used in so many different ways and, even more, because there are so many different perspectives in social science on the use of observation as a method of data gathering. I have used a terminology derived from Hughes's own; he used "observation *in situ*" or "direct observation" to designate the activities of researchers who observe personally and for an extended time some situation or behavior that interests them and who are not therefore reduced to understanding these situations or behaviors only through the categories used by participants. I have used the term "fieldwork" to designate the research design that includes observation *in situ*, plus collecting interviews and the use of archives. I have, on the other hand, avoided the term "participant observation" because its meaning is ambiguous: sometimes it simply means fieldwork (see, for example, McCall and Simmons, 1969), and sometimes *only* cases where researchers become members, in fact or in name only, of the social groups they study (see, for example, Wax, 1968).

Hughes and the "Chicago Tradition"

Although Hughes has been called a typical representative of the "Chicago School of sociology," that claim is as ambiguous as the idea of "schools" in social science generally.[8] The concept of a Chicago School of sociology is among the most ambiguous, and to link Hughes with the Chicago School, we must stipulate precisely what we mean by that expression, which has been used in so many different ways. Although he was Park's student and the editor of his works, Hughes was not really connected with the theory of urban development we associate with Park, Burgess, and MacKenzie. Nor was he connected with the theory of social disorganization actively supported by Park, Burgess, and Wirth.[9] Likewise, Hughes's references to the familiar ideas of George Herbert Mead are rare and cursory, and one of his principal empirical studies is based on the analysis of biographies à la Thomas and Znaniecki. Nevertheless, Hughes's affiliation with the Chicago School is clear if we use the expression to refer to the multiform enterprise of empirical research that developed in the Department of Sociology at the University of Chicago under the impetus of Thomas and later Park, and produced a series of monographs, including some of the "minor classics" in the discipline.[10]

Sociologists who are the more or less distant heirs of the Chicago tradition often use the term "Chicago tradition" or "Chicago style," instead of Chicago School, which appeared only after 1940. (See, for example, Fisher and Strauss, 1978, or the remarks by Hughes and Blumer in Lofland, 1980.) Since the early 1960s and the publication of Rose's (1962) edited volume, the term "interactionism" (or "symbolic interactionism"), coined by Blumer before 1940, has frequently been used to identify research belonging to that tradition: the work of sociologists known as students or associates of Hughes and Blumer (although the same terms are applied to the more "positivist" followers of the social psychologist Manfred Kuhn). More recently, since the founding of the Society for the Study of Symbolic Interaction, these same labels have been applied to a much wider range of research, including work by sociologists who adopt a "phenomenological" point of view (usually that of Alfred Schutz).

I have used the terms "Chicago tradition" or "interactionism" to refer to what is common in the approaches to social reality of several successive generations of researchers. We can refer, for examples, to the work of Park and Ellsworth Faris in the first generation, Hughes and Blumer in the second generation, and Strauss, Becker, Freidson, and, for his early work, Goffman in the third generation. Of course, we cannot expect the body of research defined by these terms to have the same homogeneity and coherence as each one has, taken in isolation.

Like Park, whose most direct follower he seems to be in this regard, Hughes was primarily interested in the study of contemporary social reality and eclectic in his choice of research methods. His monograph on the effects of industrialization on a French Canadian town (Hughes, 1943), for example, relied on the same kinds of evidence as the monographs of Wirth (1928) or Thrasher (1927): a combination of official statistics, archives and journals, interviews he gathered himself, and, finally, on more or less systematic observations of the ordinary life of the community.

Nevertheless, like Park – or perhaps this is better seen as an echo of the pragmatism Ellsworth Faris introduced into the Chicago department – Hughes always preferred to observe the phenomena he studied *in situ*. This technique has an affinity with a sociology that puts at the center of its program of study, not "facts" constituted in the manner of Durkheim, but collective actions and social processes that can be understood, in part, through direct interaction, and whose meaning for those involved is neither given nor negligible.

Here, I must insist, against the most frequent interpretation, that the term "interaction" was used neither by Park, nor later in the Chicago tradition, to mean only face-to-face interactions among individuals. According to

> the terminology in the Park and Burgess textbook, it first referred to relations of competition among groups that did not imply actual contact. (On the other hand, Park and Burgess used "conflict" to mean antagonisms manifested in direct contacts.) The emphasis on direct interaction in many interactionist studies results primarily from an almost exclusive use of observation *in situ* and interviews as data, and not from an a priori refusal to analyze what other traditions call "objective facts." For instance, in critical remarks about studies of industrial sociology during the 1940s, Blumer insisted on the importance of chronic conflict between management and trade unions and on the nature of these two types of organization; he observed that local "relations between management and workers tend to lose their autonomous character" and to be determined by the "policies . . . projects, and strategies of central organizations," that is, national unions and industrial federations (Blumer, 1947: 273).

Hughes also retained, from the Chicago tradition of the 1930s, an attachment to the unity of the social sciences – or, one might say, an indifference to academic boundaries – that was relatively rare in that period, when each discipline was concerned with establishing itself and defending its specialty. His frequent references to anthropological research served as antidotes to the ethnocentrism of American sociology and as a source of examples furnishing points of comparison for the construction of analytic categories. Furthermore, Hughes insisted over and over (for example, apropos the division of labor) that empirical research must use a long temporal perspective, like the one historians develop during their training (Hughes, 1971: 292–294).

Finally, even more than Park, Hughes's familiarity with German social science from around the turn of the century – with the work of Weber but even more than that of Simmel – convinced him that the analysis of social reality should be based on the construction of abstract and general categories.

If the continuity between Hughes's and Park's conceptions of social science is always evident,[11] notable differences nevertheless exist in the research styles each helped to promote in the different sociopolitical, university, and intellectual contexts in which they operated.

With Hughes, sociology first diverged from social work, which, as is well known, was very much present when the Department of Sociology at Chicago was first organized (Diner, 1975; Carey, 1975). In Park's time, the boundary between these two activities had already developed, but Burgess remained in contact with charitable institutions, and numerous studies between 1920 and 1940 reflected the maintenance of privileged ties between sociology and social work: for example, those of Shaw and McKay, Thrasher, and Wirth, who had been a part-time social worker.[12] At the opposite extreme, Hughes's use of the comparative method to construct his analytic categories reveals a religious, moral, and social relativism far removed from the reformism (and the ethnocentrism) latent in most of the research done between 1930 and 1950. This relativism, as we shall see, led him to apply analytic categories that had proved pertinent in studies of professionally marginal groups of lesser status to studies of the medical profession.

The growing distance between sociology and social work coincided with a broadening of research topics in sociology, itself correlated with the diversification of topics for which one could obtain financing for research. Hughes thus participated in a study of the socialization of medical doctors[13] at about the same time as that of Merton at Columbia,[14] and in another study of the profession of nursing (Hughes et al., 1958), two types of projects without equivalence in the 1930s.[15]

Finally, Hughes's insistence on observation *in situ* was not derived, as Park's was, from journalistic experience (or a knowledge of journalists' working methods), but from a great familiarity with the studies, the problems, and the methods of anthropology in the post-Malinowski era. Except that his field experience was relatively closer to home, with respect to his relation to the population studied, his study of a French Canadian town resembles studies done at the same time by anthropologists like Redfield.

As Jennifer Platt remarked in her study of the history of participant observation, sociologists of the 1930s interpreted their research methods in terms of an opposition between case studies and statistical studies.[16] "Case studies" included the collection of biographical accounts produced at the request of the researcher (see, for example, Shaw, 1930); descriptions of a milieu or an activity, collected from people involved in them (for example, *The Professional Thief* by Sutherland and Conwell, 1937); the collection of talk in a situation; and observations carried out by the researcher himself or herself.[17] The distinctions were not always clearly recognized. It was not until the end of the 1930s that the outlines of fieldwork, with its insistence on direct observation, were a little more defined. After 25 years that design was already familiar to Anglo-Saxon ethnographers, of course, but they had hardly developed any analysis of it – nor, it seems, any specific instruction in it – and, in any case, its transfer to the study of researchers' own societies encountered other obstacles. The University of Chicago provided an intellectual and institutional setting favorable to the rapprochement of anthropology and sociology: the two disciplines were only separated administratively after 1928, and even then they remained interdependent as far as teaching was concerned. But above all there were anthropologists close to sociology at the University of Chicago – like Park's son-in-law, Robert Redfield; Conrad Arensberg, who was then working on his study of a rural community in Ireland; or Lloyd Warner, who was then studying in Newburyport (Yankee City), as well as sociologists close to anthropology, like Hughes,[18] and those who developed a perspective in affinity with the anthropologists' perspective, like Blumer. If the latter did not himself do fieldwork, he nevertheless made a large contribution to this type of research: His essays and his teaching furnished a framework and epistemological justifications for "naturalistic" studies where the subjects of research are studied in their natural environment. Blumer's contribution was fundamental all through the period in which fieldwork research had not yet gained an assured scientific status.

Collaboration between sociologists and anthropologists was primarily carried on by interdisciplinary research committees. The Committee on Human Relations in Industry, for example, led by Warner and Hughes among others, provided an institutional framework for the occupational research that Hughes and his associates carried on.

The recognition of fieldwork as a legitimate research design in sociology did not depend only on the example of ethnography. It was also a product of the Chicago School's reaction to research based on the collection and statistical analysis of questionnaires, developed after 1940 by Stouffer, Lazarsfeld, and Merton.

In comparison with the questionnaire method, the Chicago research of the 1930s – a combination of not very systematic direct observations, analyses of official statistics, and collections of biographical documents – seemed to belong to the past. In 1948 Edward Shils – a former student of Park's, and Wirth's research assistant – published an influential book on the state of sociology. Laid out like a balance sheet, it crystallized negative judgments of the research conducted at Chicago during Park's time. Shils stressed the absence of systematic (that is, statistical) verification of explicitly formulated propositions in the Chicago monographs of the 1930s, implicitly taking as his model of the scientific method the one sociologists then attributed to the physical sciences, which questionnaire research seemed to resemble most closely. This very general critique was accompanied by a relative devaluation of direct observation as a method of data gathering: It was judged less rigorous because it seemed to depend on the individual characteristics of researchers and because it was harder to conduct according to the principles of systematic sampling. Observational data, considered suggestive at best, were relegated to the initial stages of research, when one prepares to gather the kind of data susceptible to statistical treatment. This use of observation was stressed in the well-known article by Barton and Lazarsfeld (1955), for example.

> Throughout the 1950s, sociologists in the Chicago tradition had very little public debate with the detractors of observation *in situ* as a primary method of data gathering. Their most influential spokesman, Blumer, spent less time defending fieldwork than pointing out the weaknesses in research based on questionnaires and standardized interviews. Behind Blumer's criticisms – which rested on the illusory character of standardization, on the arbitrariness of the categories used as variables, and on the uncertain relationship between behavior in real situations and responses gathered in interview situations – it is easy to glimpse a defense of fieldwork. But he did not provide an argument justifying fieldwork until much later (in the introduction to his 1969 volume). As Richard LaPiere's retrospective testimony suggests, arguments by the detractors of questionnaire

> research carried less weight in the 1940s and 1950s than arguments by the adversaries of observation.[19] Looking back, it is possible to see that the controversy in fact concerned the criteria according to which one measures the rigor of a method, criteria that determine the weight accorded to this or that argument.[20]

If one relies on Chicago monographs of the 1930s and 1940s, some of the criticisms formulated against data gathering by observation are, based on the evidence, partially founded. At first, the conditions of data gathering and especially the dealings researchers had with the people they studied, were shadowy, confirming the impression that they did not try to exercise systematic control over the "facts" they reported and that interpretations of those facts were therefore dependent, insofar as it is possible to estimate, on the subjectivity of the researchers and on the conditions in which they worked.[21] Besides, even if some "model" monographs (Thrasher's on adolescent gangs and especially Cressey's on dance halls) reproduced numerous extracts from observation notes, these consisted of imprecise descriptions (especially Thrasher's) interspersed with miscellaneous value judgments. Furthermore, these extracts were used exactly as extracts from interviews and documents were used: for the information they explicitly contained and not for the elaboration of analytic categories.[22]

On these different points, the studies based on fieldwork after 1940 are much less criticizable, probably due to the indirect influence of questionnaire research.[23] Beginning early in the 1950s, fieldwork researchers published detailed accounts of the conditions in which they carried out their research and of the practical problems they encountered in getting access to their subjects and gathering their data. These provided their readers with some rudimentary estimates of the validity of their analyses: the appendix W. F. Whyte added to the 1955 edition of *Street Corner Society* is one of the best and most often cited. Starting with the 1960s, almost every monograph based on fieldwork contains such an account.[24] At the same time, various articles examined three groups of problems that confront field-workers: first, the appropriate mode of relations with the population studied, at the beginning, during, or at the end of research; second, the problems posed by the construction of analytic categories; and finally, problems of the validity of the data gathered.[25]

Accounts of research based on fieldwork also became more precise. Instead of briefly summarizing fieldnotes and remarks gathered by interviewing, or even simply referring to these data, many of these accounts cited them word for word, which forced researchers to construct finer categories of analysis and to explain their interpretations of remarks and behavior in more detail. For the most part, the occupational studies carried on under Hughes's patronage are examples.[26] This change in the direction of greater rigor and precision, in the norms of composition and in the collection of observations, seems to have been encouraged by the previously mentioned critiques by partisans of questionnaire research. One can measure the extent of the change by comparing Hughes's study of a city in Quebec with the one he did 20 years later on the socialization of medical students.

Although one finds very few published debates among sociological fieldworkers, except the one concerning the moral problems posed by clandestine observation, the mass of studies based on field research between 1945 and the middle 1960s does not constitute a unified whole with regard to conceptions of the fieldwork design. Certainly, all the studies list, among their objectives, understanding the meaning people give to their activities. In this, they are conspicuously different from all those forms of research that consider such meanings to be obvious, quasi-obvious, or, most frequently, negligible and superficial aspects of social reality.[27] However, that is the end of fieldworkers' agreement about objectives. One can compare at least two contrasting conceptions of fieldwork, clearly recognizable in programmatic essays, although accounts of actual research are ranged along a continuum between them.[28]

The first conception of fieldwork attributes to it an essentially descriptive purpose. Advocates of this conception expect researchers to describe the "culture" of the population studied, from the point of view of those who have experienced it. Among them there is often an emphatic insistence on "participation" by the researcher as the unavoidable means of gathering data.

This insistence is principally found in the work of researchers who belong to the tradition of cultural anthropology – for example, Spradley (1980) or Wax (1968) – and also in certain essays by Blumer, in which he argues against positivist research in social science. In an article first published in 1962 and often cited, Blumer asserts, for example:

> To catch the process [of interpretation], the student must take the role of the acting unit whose behavior he is studying. Since the interpretation is being made by the acting unit in terms of the objects designated and appraised, meanings acquired, and decisions made, the process has to be seen from the standpoint of the acting unit. . . . To try to catch the interpretive process by remaining aloof as a so-called "objective" observer and refusing to take the role of the acting unit is to risk the worst kind of subjectivism – the objective observer is likely to fill in the process of interpretation with his own surmises in place of catching the process as it occurs in the experience of the acting unit which uses it (Blumer, 1969: 86).

In fact, outside of programmatic essays, the term "participation" often seems to mean that the researcher has to undergo a kind of socialization in the milieu studied and not that the researcher must have done the activities he or she gives an account of, which would often be absurd anyway, given the subjects studied.

The accounts that illustrate this first conception of fieldwork always include, in a mixture of social science language and the language of everyday life, descriptions of the group studied (its locale, the circumstances its members act in, and so on), and explanations of members' "definition of their situation," which constitute a kind of translation of these accounts into the experience of supposed readers, the educated public. These accounts do not use elaborate analytic ideas and are limited in general to making behavior appear coherent and relating the categories explicitly used by actors and the tacit categories underlying their behavior.[29]

A second group of researchers consider participant observation research to be a means of gathering evidence for the objectification of the activities and lived experiences of some actors. For these researchers, description of the actors' subjectivity ceases to be an end in itself, whenever it does not appear as a contradiction in terms,[30] and knowledge of the categories of language and action used by the "subjects" studied is merely the basic material from which sociologists construct analytic categories. It is not considered a necessity or even an advantage in every case to participate as closely as possible in the life of the people studied; in principle, the degree of participation is determined by the characteristics of the subject being studied and of the researcher (his or her individual characteristics, mode of access into the milieu, and so on).

The best, and most convincing, illustration of this second conception of fieldwork is made up of the occupational studies carried out by Hughes and some of his students and associates.[31] Hughes's methodological essays, published toward the end of his career, and the ideas he developed throughout his publications, provide an overview of that research enterprise, an enterprise that combined empirical work and the construction of a system of analytic categories not reducible to the categories used by the actors themselves.

According to his own testimony (see, for example, his interview with Weiss, 1981), Hughes's thoughts on fieldwork were based in large part on his teaching at the University of Chicago. The belated publication of his methodological essays partially explains why recent analyses (see Douglas, 1976) contend that Hughes used fieldwork in a nonreflexive manner, just as the authors of Chicago monographs in the 1930s did. In his 1969 book on interactionism, which in fact focuses on studies of the 1960s, Rock attributes to Hughes and to all sociologists of his time an analogous conception of fieldwork, although he regards as a merit what Douglas considers a failing. In my judgment, Rock exaggerates the importance of direct knowledge and has not examined the fieldwork practiced by Hughes and his associates closely (or critically) enough. On the contrary, Hughes's originality and, for example, Anselm Strauss's in the second generation, lie in the care they took to explain the different stages of fieldwork and to maintain a reflexive attitude with regard to these stages, instead of praising the observer's direct subjective experience as do many field researchers, particularly ethnographers.

Studies in the Sociology of Work from 1940–1960 and the Maturing of Fieldwork Research

After 1945, studies in the sociology of work were an appreciable part of fieldwork research at the University of Chicago. These studies are distinguishable from the research done there twenty years earlier, not only, as we have seen, because the data gathering is more systematic and the research accounts more explicit, but also because their focus is more restricted. The published articles generally center on a theme that is narrow enough to allow for examination of a circumscribed social situation: for example, the way workers control their work situations, the institution of norms of production, or the types of relations between a category of workers and those who use their services.[32] In addition, these articles use almost the same set of ideas as Hughes presented and discussed in several articles he published in the same period of time.[33] But the relationship between these empirical studies and Hughes's more general essays is not simply that of a theory to its "applications" or "illustrations"; it is rather a process of elaboration in which empirical studies and more abstract analyses are conducted abreast.[34]

Some of these empirical studies – maybe the majority, if one believes certain suggestions by Hughes – were done by researchers already familiar, by virtue of their own biographies and even before beginning their research, with the occupations or institutions they studied. For example, Roy worked on an assembly line before undertaking his study of restriction of production, Becker was a jazz pianist before he did the study republished in *Outsiders*, Dalton was employed as a manager in a business concern, Davis drove a taxi to pay for his education, and so on.[35]

Their familiarity with the milieu or institution studied, acquired by acting in another role than that of "disinterested" observer, put these researchers in a radically different position vis-à-vis their subject than ethnographers or, with very rare exceptions, the Chicago sociologists who published monographs during the 1930s.[36] For those earlier researchers, access to the subject of study was often a critical problem and, having resolved it, they had to find a way of behaving that did not block the collection of data. In the 1930s, for example, when sociologists from the rural middle classes studied the behavior of members of the urban working classes, they had to bracket the moral judgments these behaviors aroused in them and also participate according to established norms in the domain of relations between middle and working classes. In contrast, researchers who study populations or situations with which they are familiar have to substitute, for the attitude of immediate participation they had before they started doing research, the point of view of the "disinterested" observer, unconstrained by the norms and judgments that accompany the position they occupied.[37] This connection with the subject studied, "a subtle equilibrium between detachment and participation" (Hughes, 1971: 420), cannot be established by a simple decision in principle, but proceeds through a prolonged process that Hughes called, in various descriptions he gave of it, "emancipation."[38] This process varies with the sociohistorical context and with the social, cultural, and religious characteristics of the researcher. Therefore, according to Hughes, it was in part due to certain peculiarities in their social trajectories that some of the young sociologists participating in research on the sociology of work at the University of Chicago after 1945 could adopt an attitude of objectivity vis-à-vis their subjects.[39]

Within the limits of their own sociohistorical context, field researchers must depend upon two kinds of resources to establish adequate rapport with the subjects they study: first, knowledge of other researchers' experiences and a reflexive attitude vis-à-vis their own research acts; and, second, the use of a comparative method in a much larger framework than the one ordinarily used by sociologists.

Hughes's methodological essays furnish examples of a reflexive attitude toward research activities without equivalence at that time. He applied to observation and interviewing the analytic schemes unearthed in studies of other types of interaction[40] or, to take another point of view, analytic categories borrowed from the sociology of work[41] (since, after all, the work of sociologists is only one kind of work among many). Hughes also questioned the relations among the three languages sociologists must use: that of the people whom they study, that of their colleagues, and that of the public that reads their published accounts (Hughes, 1962).

These thoughts on sociology as an activity obviously did not emerge from a body of rules to be followed in all circumstances when gathering data by observation or interviewing, like the rules one finds in methods texts. Unlike those who put the accent on researchers' participation in the social life they study, Hughes insisted that the roles from which researchers can observe a social activity can only be controlled within narrow limits. We learn from the experience of past research what the basic elements are from which one can evaluate what this or that position of observation implies. Hence, Hughes made it his project, which Junker's (1960) work realized in part, to gather and compare various accounts of research experiences, particularly accounts by students at the University of Chicago.

But the comparative method not only permits field researchers to take an objective point of view toward their own activities and thus exercise a certain control over them, it also allows them to avoid established representations of the subjects they study, especially those associated with their own familiar, everyday points of view. Thus, the originality of Hughes's analysis of the division of labor is derived from his unconventional and irreverent (in the context of sociology and American society at that time) use of the comparative method.

Since the middle of the 1930s, social science studies of middle-class occupations in the United States had inevitably encountered the ideas of "profession" and "professionalism" as elaborated by members of occupations ambitious to reach the status and privileges of medicine, the model of a profession in the Anglo-Saxon meaning of that term.[42] After 1940 Hughes was convinced of the sterility of the approach – used by his colleagues and, earlier, by Parsons, Merton, and their students – which took its problematic from the profession being studied and depended upon comparisons between concrete examples of occupations and an "ideal type" abstracted from the supposed characteristics of medicine (Hughes, 1951b; 1970b; 1971: 339–340). Almost alone at that time, he was interested in the concrete processes through which the division of labor is established and in the appropriation by some workers of the tasks of other categories of workers. Thus Hughes broke with the idea, unanimously but tacitly accepted, that professional groups are categories naturally defined by the technical division of labor.[43] Furthermore, starting with the commonplace idea "that all sorts of occupations belong in the same series, regardless of their place in hierarchies of prestige or of moral value" (Hughes, 1951a; 1971: 316), he concluded that one could expect to find many characteristics of professional work in "humble" occupations: among plumbers and prostitutes as well as doctors and psychiatrists.

From the first studies of marginal professions and lower-status occupations, Hughes concluded that these constitute a privileged field of observation, because their members are less capable than high-status professions of maintaining a facade and opposing their own value symbols ("professional ethics" or "scientific knowledge") to those that researchers by virtue of their own social position, are inclined to adhere (Hughes, 1951b; 1971: 342). Therefore, Hughes and his associates sought, through research that was the inverse of that done by students of Parsons and Merton, to study professions of high standing with reference to conclusions they unearthed in studying occupations of lower standing. Then, by enlarging the range of professions and work situations they compared, they progressively elaborated a framework of questions that proved fruitful for studying the processes by which the division of labor is established and transformed, the way groups of workers try to exercise control over their own activities, and the unfolding of professional careers. Hughes thus used the comparative method, not for demonstrating proofs, but to construct analytic categories free from value judgments and from the categories used by workers themselves (or by their employers) for practical purposes. This is why the comparisons Hughes made were rarely of a global character and, as a general rule, only dealt with particular aspects of situations and work behaviors – those related to his favorite notions of *dirty work, routine and emergency, restriction of production* or *career line*.

Although the studies that used these ideas in the 1940s and 1950s focused on relations among workers or between workers and others with whom they interacted directly, Hughes did not overlook the importance of relationships between occupations and the surrounding society: several of his essays elaborate analytic categories (license, mandate, and so forth) capable of accounting for complex transactions between professional groups, which always develop their own conceptions of their activities, and groups that, in title or otherwise, use the products of professionals' work.[44] In this sense, one finds in Hughes's work the point of departure for Freidson's later analyses of physicians' control of their medical practices and the relations between the culture of patients and the culture of practitioners (Freidson, 1971).

The occupational studies of Hughes and his associates did not emerge from a definite theory (in the sense of a body of propositions claiming to establish and account for a series of "facts"), but from a framework of investigation capable of guiding empirical work without reducing it to the simple verification of previously formulated hypotheses. Moreover, these studies are consistent with the interactionist conception of sociology: that the final product of research is a set of ideas – what Blumer called sensitizing concepts – and a list of questions, not a catalogue of definitive assertions about the relations between this or that aspect of social reality. According to their perspective, these relations are affected by social actors' uninterrupted interpretations. Thus, as Daniels noted (1972), the remark Hughes made about the "classic" authors of sociology applies perfectly to his own essays: "What you get from . . . [Durkheim, Park, Weber, and Simmel] . . . is a statement of general problems, a set of concepts and sometimes some fruitful ideas and hypotheses with which to enrich your own thinking and to suggest methods for solving the problems with which you are concerned" (Hughes, 1961; 1971: 565).

Conclusion: Fieldwork Studies in the United States after 1960

The continuing process of explanation, and therefore control, of fieldwork research was not completed by the beginning of the 1960s, although Hughes's career at the University of Chicago ended and so did the collective research enterprise in the sociology of work, which he had begun.[45] Thereafter, the interactionist tradition was enriched by numerous studies of either deviance or organizations (hospitals, for example), and by Glaser and Strauss's (1967) attempt to define a way of using the comparative method to construct analytic categories, a stage of research that had seemed, in Hughes's formulation, to depend on the individual imagination or the individual virtuosity of researchers.

During the 1960s there was a renewed interest in fieldwork, as shown by the publication of textbooks and the creation of doctoral programs. By the end of that decade, however, sociologists had created a new division of work, which led to the establishment of two separate domains, each with publications specific to it: "quantitative studies" (that is, studies using statistical instruments) and "qualitative studies" (those that rely on field research, autobiographies, and so on). In this new context, research in the interactionist tradition was labeled "qualitative" research, along with two other kinds of field research. One is made up of monographs written by anthropologists who were converted, by necessity, to the study of their own society; most often these studies are direct intellectual heirs of the cultural anthropology of the 1950s, and do not show any noteworthy innovation in the collection or interpretation of data.[46] The second is the product of the conjunction of ideas borrowed from Schutz, Garfinkel, and phenomenology more generally, with the point of view of fieldwork in the interactionist style.

> In the matter of fieldwork research, the relationship between the interactionist tradition and phenomenology seems complex, and cannot be understood completely without explicitly describing the network of relationships among sociologists in the 1960s and 1970s.[47] But, without anticipating the results of this analysis, one may suggest that some of the young sociologists in the 1960s who were most interested in phenomenology were also closely connected with sociologists in the interactionist tradition; and notice that the differences in research practice were certainly far less radical than the differences in vocabulary. On the other hand, no clues of a direct contact with phenomenology can be found in the published work of interactionists before 1965. One never finds references to philosophical works in Hughes's essays and Blumer's are almost always presented as refinements in the philosophy of G. H. Mead, the only author Blumer cites. Nevertheless, several articles by Schutz were published in the *American Journal of Sociology* during the time Blumer was its editor, and there is also evidence (see Bruyn, 1967) of similarities between the methods of field research as practiced by interactionists and the method of phenomenology. However, it was not until the middle of the 1960s – and principally through the intermediary of Garfinkel – that the thematic of phenomenology found a rather large audience among sociologists who did empirical research. Several remarks suggest that, at least in the beginning, phenomenology chiefly furnished a language that allowed some young sociologists (for example, Manning, 1973, and Douglas, 1976) to express their misgivings about the structural-functionalist perspective and research based on the use of questionnaires, and some sociologists of earlier generations (for example, Deutscher, 1973, and Davis, 1973) to reformulate their familiar themes.

One contribution the researchers who called themselves phenomenologists made to the explanation of fieldwork lies in the attention they paid to the social construction, in concrete interactions, of what sociologists ordinarily treat as data: that is, categorizations of behavior and persons in social life, the statistics that follow from these categorizations, and even interviews about them, produced at the initiative of a researcher (Cicourel, 1968, 1974; Sudnow, 1967). Many fieldwork studies influenced by phenomenology also analyzed the effective use of formal rules in organizations (see Bittner, 1967; Cicourel, 1968; Emerson, 1969), whereas this subject of investigation was only peripheral in interactionist studies. Methodologically, some of these studies focused on the relation between the observer and his or her subject, abandoning the "absolutist" conception of description – that is, the idea that descriptions are independent of the researcher's point of view (Douglas, 1976; Manning, 1973). From there, they went on to analyze the implicit models around which researchers organize their accounts and researchers' methods of composition.

* * *

I will not develop this examination of changes in the conception of fieldwork in American sociology after 1960 any further here. These brief remarks are only intended to show the originality of Hughes's work – and the originality of the Chicago tradition – in the context of the whole, and his contribution to the rational use of data produced by field research.

By drawing attention to details of the circumstances in which observation *in situ* was first used systematically as a source of data for sociological analyses, I intended to accomplish two other more immediate objectives. First, I have tried to show that one can find, in essays by the first generation of academic sociologists to conduct empirical research, most of the themes around which later field researchers' thoughts on their activities were arranged, so that the originality of most recent publications on the subject consists almost entirely of formulations the first generation proposed. Thus, knowledge of the history of sociology gives the discipline, on one hand, access to a certain cumulative knowledge (or at least, knowledge about) and, on the other hand, discourages the periodic treatment of reports of recurrent research problems as if they were original discoveries. Also, when referring intellectual works to their conditions of production, historical studies of sociology favored a more detached appraisal of research enterprises, refraining from interpreting these works as if they were holy texts free of worldly entrenchment (exactly the misfortune of the founding fathers of sociology like Marx, Durkheim, and the rest).

More implicitly, I have also tried to show the mutual relationships among three aspects of the development of social science, usually considered separately: the elaboration and transformation of methods of gathering and using data; the conditions of researchers' work and their position in the social structure; and the intellectual influences that are always blurred, only partially and with difficulty perceptible, when one does not limit oneself to the examination of systems of general ideas in which researchers formulate their programs of study a priori or a posteriori and in which they tend always to accentuate their affiliations with some and their differences with others.

Appendix: The Chicago Tradition, Everett Hughes, and French Sociology

Analyzing the way an intellectual tradition is introduced or not introduced in this or that foreign country, the interpretations it is given, and the influence it exerts makes us aware of some important characteristics of that tradition that are too obvious to be noticed by those who belong to it, and reveals certain properties of the same discipline in the recipient country. I want to consider here, briefly, the case of the Chicago tradition and of Hughes in relation to French sociology. Perhaps this will help the English reader understand the point of view I have adopted in this study.

Although Hughes had numerous personal contacts with French sociologists after 1945 and even though *French Canada in Transition* was translated into French as early as 1945 (for a Canadian publisher), although he had a continuing interest in research by French sociologists, anthropologists, historians, and geographers, and even though, up to the end of his life he published numerous reviews of works about France, French sociologists otherwise ignored, or at least did not understand, Hughes's publications. This lack of understanding went beyond Hughes's case and in fact encompassed all of the Chicago tradition, particularly the conception of social science research characteristic of that tradition.

An article published in 1932 by Maurice Halbwachs seems to have been, for a long time, the principal medium through which French sociologists knew the Chicago School. That article focused on the development of the city of Chicago compared to that of Paris, which Halbwachs had studied twenty years before. Only a few works by Park, Burgess, and their students were mentioned and they were slanted to the article's focus. What interested Halbwachs and later French sociologists after 1945 – Chombart de Lauwe (1952) and Castells (1972) – were general schemes, frequently called "theories," for analyzing urban development. By contrast, the multiform enterprise of research at Chicago, Park's emphasis on intimate knowledge of the populations studied, and the studies of contacts between races and cultures and of delinquency were almost completely ignored. Certainly, French researchers knew only about the most general texts, like *The City*, the volume Park and Burgess edited: in 1980, not one of the great Parisian libraries held an even moderately complete collection of the classic monographs in the series Park and Burgess directed to the University of Chicago Press.

Unknown or not understood in France, Park's research enterprise could not serve as an introduction to Hughes's. On the only occasion when Hughes addressed an audience of French researchers, he was introduced by the historian Lucien Febvre as a typical representative of American sociology, as if it were a homogeneous entity (Gurvitch, 1949: xiii). And the discussions at that symposium revealed the lack of fit between French researchers' concerns and Hughes's. Nevertheless, Hughes maintained regular relations with the French sociologist Georges Friedmann, himself a sociologist of work. Some of the students and researchers in Friedmann's circle visited the University of Chicago during the 1950s but – with the exception of Jean-Rene Treanton, who wrote an article on "the concept of career" for the first issue of the *Revue Française de Sociologie* in 1960 and a review of *Men and Their Work* for the first issue of *Sociologie du Travail* in 1959 – none of Friedmann's followers seem to have been convinced of the interest of Hughes's research enterprise. (Friedmann himself did not cite Hughes in the general article that opened the first issue of *Sociologie du Travail*). Information I have gathered informally since the French version of this article was published suggests that French sociologists who met Hughes considered him an "important" sociologist because of his official position (chairman of the sociology department at Chicago, editor of the *American Journal of Sociology*) or a "brilliant" intellectual because of his conversation; they did not, it seems, realize Hughes was the leader of a living research enterprise.

One can find various reasons for the difficulties in the diffusion of the Chicago tradition in France. In a general way, the disrepute of pragmatism among French philosophers – most notably James's work and Dewey's – no doubt influenced sociologists who remembered the philosophical background of the Chicago tradition. This certainly did not help them understand Park's and Hughes's point of view on research in the social sciences, with their insistence on direct knowledge in the manner of ethnographers and on the idea that one of the principal products of research should be the elaboration of analytic categories for use in later investigations. Because they have been influenced by Durkheimism or by this or that interpretation of Marxism, French sociologists, to the contrary, profess a pronounced distrust of the experience of those whom they study and cling to the idea of a radical boundary between "scientific" knowledge and the "practical" knowledge of social actors. Statistics and the mysterious manipulation of statistics seem (from the point of view of their literary backgrounds) to be the instrument par excellence for the "rupture" between these two forms of knowledge, whereas ethnographic methods can only seem suspect to them. And it seems to them that the objective of research ought to be establishing a "theory," which is to say, establishing general relations among classes of phenomena in the manner of Marx, of Durkheim, or even occasionally of Weber.

Even if there had not been institutional reasons, the research style represented by Columbia University sociologists, with its dual insistence on "theories of the middle range" and the statistical analysis of questionnaires, would have satisfied the diffuse expectations of French sociologists after 1945 much better than the research style Hughes represented in the same period. Although some research on French society before the middle of the 1960s used ethnographic methods, this research developed totally independent of research in the Chicago tradition. Therefore, French fieldworkers did not benefit from the Chicago sociologists' accumulated experience in collecting data; nor did they benefit from reflections on this type of research, of which certain essays by Hughes furnish some of the better examples. For lack of a group engaged in common projects (like the research on work and, later, the research on delinquency in the United States) the cumulative process that is needed for research reports to become more explicit and more analytical has not emerged in France.

Notes

1. The story of De Gerando's report seems typical. It was written for the *Société des Observateurs de l'Homme*, whose members were preparing an expedition to Australia with, among other things, scientific aims. De Gerando's report did not have any influence on the expedition, for evident reasons. Though republished in 1883, the report remained almost unknown among French anthropologists and sociologists. It was rediscovered by English historians of anthropology in the 1960s, and in 1969, translated and republished in English by F.C.T. Moore, with a preface by E. E. Evans-Pritchard. The report was later published in French in Jean Copans et al. (1978). A young philosopher when he wrote the report, De Gerando became a high official in the empire and, for the rest of his life, was involved in various philanthropic enterprises, most notably in the diffusion of elementary education. He died in 1842.
2. The recent publication of Malinowski's *Journal* made the slight differences between his method and the account he gave of it in the Introduction to the *Argonauts* (1922) more apparent. In particular, he relied more on informants than on his own observation, without doubt because of the social position he occupied in the village he studied: on this point, see M. Wax (1972). One will find in concise form some elements of a history of observation in the social sciences in R. Wax (1971).
3. Despite his arguments in favor of observation, in his first book of family monographs (1855), LePlay himself relied principally on interviews with interested parties as a means of displaying the budgets that he attributed to families; indeed, he seems to have occasionally taken some pains to distinguish between what he could verify himself and the "facts" reported by interviewees. Among other early users of observation are the researchers associated with Charles Booth's investigations in a working class area of London and, so it would seem, with other investigations of the working class being done at the same time in England and the United States. In Germany, one of the first and foremost users of observation was Paul Gohre, a student of theology who published in 1891 a study of the workers in an industrial enterprise (see Obershall, 1965).
4. Obershall's study (1965) contains a precise description of these unsuccessful attempts, particularly the research on agricultural workers that Weber attempted in association with Gohre.
5. The distrust that the first university specialists of social science maintained to the end of the nineteenth century with regard to any method based on a researcher's own collection of data was in part a result of the situation they found themselves in. Because their place in the university was poorly defined – somewhere between the classic humanities and the physical and natural sciences – they deemed it necessary (judging from the rhetoric they used) to justify their activities as conforming to what they considered the "scientific model" or, at least, to insist on the rigorous, critical examination of documents as was done by the French historians Langlois and Seignobos in their book on the historical critique (1897); in fact, this book was used as a methods textbook by certain departments of sociology in American universities at the beginning of the twentieth century.

6. Everett C. Hughes (1897–1983), born in Ohio, was the son of a minister (which those familiar with his essays could not overlook), and was a student at the University of Chicago in the same doctoral class as Louis Wirth, Robert Redfield, and Herbert Blumer. He taught at McGill University in Montreal (1927–1938), then at the University of Chicago (1938–1961), afterwards ending his career at Brandeis, followed by Boston College. He was editor of the *American Journal of Sociology* from 1952 to 1960.

   Herbert Blumer (born in 1901) remained at the University of Chicago as a professor after obtaining his doctorate in 1928. He was the representative of George Herbert Mead's teaching after Mead's death. He was editor of the *American Journal of Sociology* from 1941 to 1952, when he left for a position at Berkeley. Blumer's principal essays on methodological problems were republished in his 1969 collection.

   One can find a first appraisal of their influence in two collections of papers: one in honor of Hughes (Becker, Geer, Riesman and Weiss, 1968) and one in honor of Blumer (Shibutani, 1970).
7. Hughes's principal articles on the sociology of work have been republished in his 1958 and 1971 collections of articles. The second collection also contains his principal essays on methodological problems, with the exception of one article published in 1974, and essays on institutions and race relations.

   Recall, too, that Hughes's *oeuvre* also includes numerous publications on cultural contacts, the central theme of his doubtless most finished work (Hughes and Hughes, 1952), and a monograph on a little town in French Canada.
8. These ambiguities result, in part, from the fact that the history of sociology is done mainly by sociologists with a normative point of view and an interest in the current state of the discipline. They search for intellectual unity where there are only enterprises in universities and research institutions; they use equivocal folk concepts without any critical analysis (see, for example, the frequent use of the concept "student" to describe relations between sociologists of two generations). I will develop these points in a paper to be published about the recent history of the Chicago School of sociology.
9. On Hughes's distance from the theory of social disorganization, see the testimony of Whyte (1981/1943: 354–358). Wirth seems to have been a more constant defender of these theories, which one finds in certain of Park's essays, than Hughes and as a result he is frequently seen as the more direct follower of Park.
10. Since the book by Ellsworth Faris's son, Robert Faris, was published (1962), there have been many analyses of different aspects of the Chicago School of sociology: Carey (1975); Bulmer (1984); Raushenbush (1979), a biography of Park largely inspired by Hughes; and, in my opinion the best, Matthews (1977). See also the recent special issue of *Urban Life* (January 1982) and many testimonies: Anderson (1961, 1982, 1983), Cavan (1983), Hughes (1981), Hughes (1980–1981), Lofland (1980), and Shils (1981).

    The most continuously cited monographs are probably Anderson (1923) on hoboes, Thrasher (1927) on juvenile gangs, Wirth (1928) on the Jewish ghetto in Chicago, Zorbaugh (1929) on "natural areas" of Chicago, and Cressey (1932) on taxi dance halls; to this list one should also add life histories like the one Shaw (1930) gathered from a young delinquent. In fact, the enterprise was much broader, with other monographs published later – for example, Hayner (1936), and various research reports, such as *The Negro in Chicago* (Chicago Commission on Race Relations, 1922) and Landesco (1929). Hughes himself wrote a monograph on real estate agents in Chicago, but it was not typical of the Chicago School monographs of the 1930s and was not published until 1979.

11. Faught (1980) presents evidence of the similarities between Park's approach and Hughes's, drawn from their most general and programmatic texts. Such an exercise is patently misleading because it rests on statements of principle and not on finished products, that is, published accounts of research. One doubts that the examination of these accounts would lead to the same conclusions.
12. About Shaw and McKay, see Snodgrass (1976). On the role of Burgess as intermediary between sociology and social work, see Hughes's testimony in Raushenbusch (1979). The closeness of Wirth's point of view to that of social work, always evident to outside observers, has been noted by Hughes (1981).
13. See Becker et al. (1961) for the final published account of the first part of that research, begun in 1956; see Becker, Geer, and Hughes (1968) for the second part. (In both cases, the principal authors of the empirical work and of the published accounts of it were Becker and Geer.)
14. See Merton et al. (1957) for a first report of this research, undertaken in 1952, which has until now only been published in scattered reports.
15. Recall, however, Parson's attempt to study medicine at the end of the 1930s, using observation and interviews, which never produced a final published report. (On this point, see the testimony of the author in Parsons, 1970).
16. See Platt (1983). The two foremost examples of use of the expression "participant observation" with the meaning common at the present time are a rather obscure article by Lohman (1937), a student of Park's who taught at the University of Chicago after 1939, and a still often-cited article by anthropologist Kluckhohn (1940), based on her research experience in a Hispanic village in New Mexico.
17. In fact, this distinction between testimony and direct observation has been acquired only with difficulty (if at all) by the rest of the social sciences, as is suggested by the examples given from LePlay and Malinowski. To neglect it obviously leads to ignoring the problem of the construction of categories by the researcher or those he or she interviews.
18. At the level of teaching, Hughes began (when he arrived in Chicago in 1938) a beginning course in fieldwork (see Hughes, 1960, 1981), which according to many people played a large part in orienting many members of the next generation to the use of this method: see Becker and Debro (1970), Polsky (1969), and Becker, Geer, Riesman, and Weiss (1968). One can get a precise idea of the influence on Hughes of the ethnographic tradition (or, perhaps better, of how he relied equally and without discriminating on both sociological and anthropological researches) by comparing the book by Junker (1960) to the similar manual used by Park's students (Palmer, 1928) in which references to anthropology are absent.
19. The first of these is located in Deutscher (1966); the second is located in LaPiere (1969).
20. On the other hand, this intellectual disdain for ethnographic methods did not prevent several sociologists trained in other methods (for example, at Harvard or Columbia) from using it: see, for example, Whyte (a little before 1940) and Gouldner in the 1950s.
21. Several authors of classic Chicago School monographs have since written analyses of their methods of research: see, for example, Anderson (1961, 1982, 1983) or the recently discovered article by Cressey (1983). Anthropologists began publishing concrete descriptions of their field research earlier, maybe because they had to insist on the differences that separated their analyses from those of other travelers – missionaries, colonial administrators, and even traders – with whom they had ambivalent relationships. The foremost example is the Introduction to *Argonauts of the Western Pacific*, published in 1922.

22. Indeed, Thrasher's writing is much more typical of Chicago monographs of the 1930s than Cressey's who, with his numerous and doubtless literal citations from interviews and his effort to construct analytic schemes, anticipated the style of reporting research that was later adopted. However, this originality seems not to have been perceived at the time.
23. My interpretation on this point is a rejoinder to one by Wax (1971: 40).
24. One of the best examples in sociology of a detailed description of the conditions of data gathering in a fieldwork study is, to my knowledge, the one Stein wrote for Gouldner's book (1954). The methodological appendix by Dalton (1959), a more developed version of which was published in 1964, helped to establish the convention of publishing an account of the research process. After 1970, paradoxically, several descriptions of research projects were published without analyses of their results, interest in the research method having replaced interest in substantive results.
25. Among the best examples of this kind of article are Miller (1952), Vidich and Bensman (1954), Vidich (1955), Gusfield (1955), Schwartz and Schwartz (1955), Becker and Geer (1957), Becker (1958), and Gold (1958). To these one may add several contributions and research notes published in *Human Organization* and, in some cases, reprinted in Adams and Preiss (1960).
26. For examples, see the articles on jazz musicians published in Becker (1963). Of course, one also finds research accounts, written by sociologists in the Chicago tradition, that do not give explicit documentation (for example, Goffman's).
27. It follows from this interest in actors' lived meanings that fieldwork studies cannot be described as verifications of precise propositions formulated in advance; instead, their themes are almost always established in the course of collecting data. See, for example, Schatzman and Strauss (1973: 7) and Becker (1958: 653). Thus, the accusation of "empiricism," often made against these studies, is a complete misunderstanding of fieldwork research.
28. In making this distinction, analogous to the one Emerson (1981) applies to some recent studies, I only intend to make clear where Hughes's research style fits in the realm of fieldwork studies. Other equally pertinent distinctions exist and cannot be reduced to this one.
29. For a good example of these monographs, see Liebow (1967).
30. See, for example, Bittner's incisive remarks (1973).
31. Another illustration of these ideas about fieldwork, inspired more by phenomenology than by interactionism, is an article by Manning (1979).
32. Some of these articles originated as doctoral dissertations or even master's theses. Among the best known are several that have become the kind of classic republished in numerous collections: Becker's on jazz musicians (reprinted in Becker, 1963), Roy's on assemblyline work (1952, 1954), Dalton's on managers (reprinted in his 1959 volume), and Gold's on janitors (1952). To these one can add Davis's on taxicab drivers (1959) and Westley's on the police (1953), although they are not so closely related to Hughes's analyses; studies that rely less on observation than on a series of interviews like Hall's on physicians (1949) and Becker's on school teachers (reprinted in Becker, 1970); and various studies that deal more with the functioning of an institution than with the problems of a category of workers, like that of Roth (1963).
33. Hughes's 1958 volume was based on these articles, although it lacks his suggestive analysis of the division of labor based on studies of the occupation of nursing. All the principal themes in Hughes's essays on the sociology of work are presented in an excellent article by Daniels (1972).
34. Explicit references to these concrete studies by his students are rather infrequent in Hughes's essays, which abound instead with allusions to published articles. As a remark of Platt's suggested to me, some of the studies Hughes frequently referred to

belong to the body of industrial research sponsored by Warner. Treating the studies Hughes frequently cited as a relatively coherent whole must not, therefore, be interpreted to mean that Hughes influenced all of them, exclusively or preponderantly. Indeed, it is clearly impossible to isolate the influences of Hughes, Blumer, Warner, or Wirth on each generation of sociologists. Or to say it better, the idea of influence (like that of "student") is profoundly ambiguous since it is merely a convention researchers use to recognize or deny this or that form of originality in their own work and the work of others. This article is centered on Hughes because his essays contain the most inclusive and the most lucid discussions of fieldwork research in the Chicago tradition.

35. See Becker and Debro (1970), Dalton (1964), and the suggestive description later written by Davis (1974). Examination of a much larger range of monographs based on fieldwork suggests that their previous familiarity with their research settings was not unusual: see, for example, Roth (1963 and 1974b), Liebow (1967), Polsky (1969), and Humphreys (1970).
36. The only one of these well-known monographs of the 1930s done by a researcher who was not a "stranger" seems to be Anderson's, and even he spent a period of time tracking down the existence of hoboes before undertaking the principal study. Wirth seems to have been less familiar with his subject even though he came from a little town in Rhenamie and had lived in rural areas of the United States from 1911 until his arrival in Chicago. In some cases the authors of monographs written at Chicago in the 1930s discovered their research subjects through their social work activities, but this was not an advantage because of the difficulty they had in abandoning that point of view.
37. Dubin's article (1983), which examines the reports of research on dance halls carried out in 1920 by four students (among them, Saul Alinsky and Paul G. Cressey), illustrates the variety of problems fieldworkers encountered, depending upon their relationships to the subjects they studied. The poorest accounts are those whose authors locate themselves at the two extremes of intense participation and moral condemnation of dance hall activities.
38. Hughes developed the theme of emancipation – as opposed to alienation – in discussions of the sociology of work (1940–1960), the sociology of religion (1955), the several generations of students whom he taught (1970a), and even his own biography (1981).
39. Among them were members of several minority ethnic groups and former GIs returning to their studies on scholarships, who were emancipated from their ethnic, working-class, or provincial origins.
40. See, for example, Hughes (1952), where one finds certain of the themes later developed by Goffman, who studied with Hughes (among others) at Chicago.
41. This point of view has since been developed by Roth (1956, 1966), another "student" of Hughes's.
42. On this point, often discussed in studies of the professions after 1970, see Roth (1974a) or Chapoulie (1973).
43. See Hughes (1959 and 1971: 292–294). Several years later, some sociological studies of deviance – those done by Becker, by Goffman, and even, in one case, by Cicourel – adopted a perspective like the one Hughes adopted vis-à-vis the division of labor.
44. See, for example, chapter 6 of *Men and Their Work* (Hughes, 1958). This chapter is the only previously unpublished one in that work.
45. The beginning of the 1960s was without doubt the period when American sociologists were least interested in fieldwork and interactionism was most discredited, but it was also the period when the field research for several studies that became famous at the end of the 1960s was carried out (studies by Becker, Goffman, Strauss, and others).

During the same period one also discerns a relative lack of interest in the sociology of work and a fortiori in the essays of Hughes. These found a much larger audience at the end of the 1960s, as suggested by numerous replications and references to them in research reports.

46. Although ethnographers have carried out studies in their own societies almost continuously since 1945 – primarily studies of communities – only a few of them (to whom I have not referred here) have shown any familiarity with interactionist works. Since the beginning of the 1960s these studies have been more numerous. In the same period ethnographers have studied more limited segments of social life – hospitals and schools in particular; they have also written essays on the methods discussed by Hughes and others in the interactionist tradition, without referring to the research experience accumulated in that tradition. See, as examples, Jones (1970), Ablon (1977), and Cassel (1977).
47. I use here, among other testimonies, the still unpublished transcription of a conversation (in Berkeley, probably in 1981 or 1982) between Troy Duster, Howard Becker, Aaron Cicourel, John Kitsuse, Edwin Lemert, and David Matza.

References

Ablon, J. (1977) "Field methods in working with middle class Americans: new issues of values, personality, and reciprocity." Human Organization 36: 69–72.

Adams, R. N. and J. J. Preiss (1960) Human Organization Research. Homewood, IL: Dorsey Press.

Anderson, N. (1923) The Hobo: The Sociology of the Homeless Man. Chicago: Univ. of Chicago Press.

Anderson, N. (1961) "Introduction," in The Hobo: The Sociology of the Homeless Man. Chicago: Univ. of Chicago Press.

Anderson, N. (1982) "Sociology has many faces: part II." J. of the History of Sociology 3: 1–19.

Anderson, N. (1983) "A stranger at the gate: reflections on the Chicago school of sociology." Urban Life 11: 396–406.

Barton, A. H. and P. F. Lazarsfeld (1955) "Some functions of qualitative analysis in social research." Frankfurter Bieträge zur Soziologie 1: 321–361.

Becker, H. S. (1958) "Problems of inference and proof in participant observation." Amer. Soc. Rev. 23: 652–660.

Becker, H. S. (1963) Outsiders: Studies in the Sociology of Deviance. New York: Free Press.

Becker, H. S. (1970) Sociological Work: Method and Substance. Chicago: Aldine.

Becker, H. S. and J. Debro (1970) "Dialogue." Issues in Criminology 5: 159–179.

Becker, H. S. and B. Geer (1957) "Participant observation and interviewing: a comparison." Human Organization 16: 28–32.

Becker, H. S., B. Geer, and E. C. Hughes (1968) Making the Grade: The Academic Side of College Life. New York: John Wiley.

Becker, H. S., B. Geer, E. C. Hughes, and A. L. Strauss (1961) Boys in White: Student Culture in Medical School. Chicago: Univ. of Chicago Press.

Becker, H. S., B. Geer, D. Riesman, and R. S. Weiss (1968) Institutions and the Person: Essays Presented to Everett C. Hughes. Chicago: Aldine.

Bittner, E. (1967) "The police on skid-row: a study of peace-keeping." Amer. Soc. Rev. 32: 699–715.

Bittner, E. (1973) "Objectivity and realism in sociology," pp. 109–125 in G. Psathas (ed.) Phenomenological Sociology: Issues and Applications. New York: John Wiley.

Blumer, H. (1947) "Sociological theory in industrial relations." Amer. Soc. Rev. 12: 271–278.

Blumer, H. (1962) "Society as symbolic interaction," pp. 179–192 in A. M. Rose (ed.) Human Behavior and Social Processes: An Interactionist Approach. London: Routledge & Kegan Paul.

Blumer, H. (1969) Symbolic Interactionism: Perspective and Method. Englewood Cliffs, NJ: Prentice-Hall.

Bruyn, S. T. (1967) "The new empiricists: the participant observer and phenomenologist." Sociology and Social Research 51: 317–322.

Bulmer, M. (1984) The Chicago School of Sociology. Chicago: Univ. of Chicago Press.

Carey, J. T. (1975) Sociology and Public Affairs: The Chicago School. Beverly Hills, CA: Sage.

Cassel, J. (1977) "The relationship of the observer to the observed in peer group research." Human Organization 35: 412–416.

Castells, M. (1972) La question urbaine. Paris: François Maspero.

Cavan, R. (1983) "The Chicago school of sociology." Urban Life 11: 407–420.

Chapoulie, J.-M. (1973) "Sur l'analyse sociologique des groupes professionnels." Revue Française de Sociologie 16: 86–114.

Chicago Commission on Race Relations (1922) The Negro in Chicago: A Study of Race Relations and a Race Riot. Chicago: Univ. of Chicago Press.

Chombart de Lauwe, P.-H. (1952) Paris et l'agglomération parisienne. Paris: Presses Universitaires de France.

Cicourel, A. V. (1968) The Social Organization of Juvenile Justice. New York: John Wiley.

Cicourel, A. V. (1974) Theory and Method in a Study of Argentine Fertility. New York: John Wiley.

Cressey, P. G. (1932) The Taxi-Dance Hall. Chicago: University of Chicago Press.

Cressey, P. G. (1983) "A comparison of the roles of the 'sociological stranger' and the 'anonymous stranger' in field research." Urban Life 12: 102–120.

Dalton, M. (1959) Men Who Manage. New York: John Wiley.

Dalton, M. (1964) "Preconceptions and methods in *Men Who Manage*," pp. 50–95 in P. Hammond [ed.] Sociologists at Work. New York: Basic Books.

Daniels, A. K. (1972) "The irreverent eye." Contemporary Sociology 1: 402–409.

Davis, F. (1959) "The cabdriver and his fare: facets of a fleeting relationship." Amer. J. of Sociology 65: 158–165.

Davis, F. (1973) "The martian and the convert: ontological polarities in social research." Urban Life and Culture 2: 333–343.

Davis, F. (1974) "Stories and sociology." Urban Life and Culture 3: 310–316.

de Gerando, J.-M. (1800) Considérations sur les diverses méthodes à suivre dans l'observations des peuples sauvages. Paris: Société des Observateurs de l'homme. Republished in J. Copans and J. Jamin [eds.] (1978) Aux origines de l'anthropologie française. Paris: Le Sycomore.

de Gerando, J.-M. (1969) The Observation of Savage Peoples (F.C.T. Moore, ed. and trans.). Berkeley: Univ. of California Press.

Deutscher, I. (1966) "Words and deeds: social science and social policy." Social Problems 13: 235–254.

Deutscher, I. (1973) What We Say/What We Do: Sentiments and Acts. Glenview, IL: Scott, Foresman.

Diner, S. J. (1975) "Department and discipline: the department of sociology at the University of Chicago (1892–1920)." Minerva 13: 514–553.

Douglas, J. D. (1976) Investigative Social Research: Individual and Team Field Research. Beverly Hills, CA: Sage.

Dubin, S. C. (1983) "The moral continuum of deviance research: Chicago sociologists and the dance hall." Urban Life 12: 75–94.

Emerson, R. M. (1969) Judging Delinquents: Context and Process in Juvenile Court. Chicago: Aldine.

Emerson, R. M. (1981) "Observational field research". Ann. Rev. of Sociology 7: 351–378.

Faris, R.E.L. (1962) Chicago Sociology: 1920–1932. Chicago: University of Chicago.

Faught, J. (1980) "Presuppositions of the Chicago school in the work of Everett C. Hughes." Amer. Sociologist 15: 72–82.

Fisher, B. M. and A. L. Strauss (1978) "Interactionism," pp. 457–498 in T. Bottomore and R. N. Nisbet [eds.] A History of Sociological Analysis. New York: Basic Books.

Freidson, E. (1971) Profession of Medicine: A Study in the Sociology of Applied Knowledge. New York: Dodd, Mead.

Glaser, B. G. and A. L. Strauss (1967) The Discovery of Grounded Theory: Strategies for Qualitative Research. Chicago: Aldine.

Gohre, P. (1891) Drei Monate Fabrikarbeiter und Hardwerksbursche. Leipzig: Grünow.

Gold, R. L. (1952) "Janitor versus tenant: a status-income dilemma." Amer. J. of Sociology 57: 486–493.

Gold, R. L. (1958) "Roles in sociological field observations." Social Forces 36: 217–233.

Gouldner, A. W. (1954) Patterns of Industrial Bureaucracy. Glencoe, IL: Free Press.

Gurvitch, G. [ed.] (1949) Industrialisation et technocratie. Paris: Armand Colin.

Gusfield, J. R. (1955) "Fieldwork reciprocities in studying a social movement." Human Organization 14: 29–34.

Halbwachs, M. (1932) "Chicago, expérience ethnique." Annales d'histoire économique et sociale 4: 11–49.

Hall, O. (1948) "The stages of a medical career." Amer. J. of Sociology 53: 327–336.

Hall, O. (1949) "Types of medical careers." Amer. J. of Sociology 55: 243–253.

Hayner, N. (1936) Hotel Life. Chapel Hill: University of Carolina Press.

Hughes, E. C. (1931/1979) The Growth of an Institution: The Chicago Real Estate Board. Chicago: Arno Press.

Hughes, E. C. (1943) French Canada in Transition. Chicago: University of Chicago Press. [French trans. J. C. Falardeau, "Rencontre de deux mondes: la crise d'industrialisation du Canada Français." Montréal: Les Editions du Boreal Express.]

Hughes, E. C. (1951a) "Mistakes at work." Canadian J. of Economics and Political Sci. 17: 320–327.

Hughes, E. C. (1951b) "Work and self," pp. 313–323 in J. H. Rohrer and M. Sherif (eds.) Social Psychology at the Crossroads. New York: Harper & Row.

Hughes, E. C. (1952) "The sociological study of work: an editorial foreword." Amer. J. of Sociology 57: 423–426.

Hughes, E. C. (1955) "The early and contemporary study of religion." Amer. J. of Sociology 60: i–iv.

Hughes, E. C. (1956) "Of sociology and the interview: editorial preface." Amer. J. of Sociology 62: 137–142.

Hughes, E. C. (1958) Men and Their Work. Glencoe, IL: Free Press.

Hughes, E. C. (1959) "The study of occupations," pp. 442–458 in R. K. Merton et al. (eds.) Sociology Today. New York: Basic Books.

Hughes, E. C. (1960) "The place of field work in social science," pp. iii–xiii in B. H. Junker (ed.) Field Work: An Introduction to the Social Sciences. Chicago: University of Chicago Press.

Hughes, E. C. (1961) "Tarde's Psychologie Economique: an unknown classic by a forgotten sociologist." Amer. J. of Sociology 66: 553–559.

Hughes, E. C. (1962) "Sociologists and the public." Transactions of the Fifth Congress of Sociology 1.Washington, DC: International Sociological Association.

Hughes, E. C. (1970a) "Teaching as fieldwork." Amer. Sociologist 5: 13–18.

Hughes, E. C. (1970b) "The humble and the proud: comparative study of occupations." Sociological Q. 11: 147–156.

Hughes, E. C. (1971) The Sociological Eye: Selected Papers. Chicago: Aldine.
Hughes, E. C. (1974) "Who studies whom?" Human Organization 33: 327–334.
Hughes, E. C. (1981) Interview by R. S. Weiss. (Unpublished manuscript)
Hughes, E. C. and H. M. Hughes (1952) Where Peoples Meet: Racial and Ethnic Frontiers. Glencoe, IL: Free Press.
Hughes, E. C., H. M. Hughes, and I. Deutscher (1958) Twenty Thousand Nurses Tell Their Story. Philadelphia: Lippincott.
Hughes, H. M. (1980–1981) "On becoming a sociologist." J. of the History of Sociology 3: 27–39.
Humphreys, L. (1970) Tearoom Trade. Chicago: Aldine.
Jones, D. J. (1970) "Towards a native anthropology." Human Organization 29: 251–259.
Junker, B. H. (1960) Field Work: An Introduction to the Social Sciences. Chicago: Univ. of Chicago Press.
Kluckhohn, F. R. (1940) "The participant observer technique in small communities." Amer. J. of Sociology 46: 331–343.
Landesco, J. (1929) Organized Crime in Chicago: Part III, Illinois Crime Survey. Chicago: Univ. of Chicago Press.
Langlois, C. and C. Seignobos (1897) Introduction aux études historiques. Paris: Hachette.
LaPiere, R. T. (1969) "Comment on Irwin Deutscher's Looking Backward." Amer. Sociologist 4: 41–42.
LePlay, F. (1855) Les Ouvriers Européens. Paris: Imprimerie Impériale.
Liebow, E. (1967) Tally's Corner. Boston: Little, Brown.
Lofland, L. [ed.] (1980) "Reminiscences of classic Chicago: the Blumer-Hughes talk." Urban Life 9: 251–281.
Lohman, J. D. (1937) "The participant observer in community studies." Amer. Soc. Rev. 2: 890–898.
Malinowski, B. (1922) Argonauts of the Western Pacific. Prospect Heights, IL: Waveland Press.
McCall, G. J. and J. L. Simmons (1969) Issues in Participant Observation: A Text and Reader. Reading, MA: Addison-Wesley.
Manning, P. K. (1973) "Existential sociology." Sociological Q. 14: 200–225.
Manning, P. K. (1979) "Metaphors of the field: varieties of organizational discourse." Administrative Sci. Q. 24: 660–671.
Matthews, F. H. (1977) Quest for an American Sociology: Robert E. Park and the Chicago School. Montreal: McGill-Queen's Univ. Press.
Merton, R. K., G. Reader, and P. L. Kendall (1957) The Student-Physician: Introductory Studies in the Sociology of Medical Education. Cambridge, MA: Harvard Univ. Press.
Miller, S. M. (1952) "The participant observer and over-rapport." Amer. Soc. Rev. 17: 97–99.
Obershall, A. R. (1965) Empirical Social Research in Germany (1848–1914). Amsterdam: Mouton.
Palmer, V. (1928) Field Studies in Sociology: A Student's Manual. Chicago: Univ. of Chicago Press.
Park, R. E. and E. W. Burgess (1921) Introduction to the Science of Sociology. Chicago: Univ. of Chicago Press.
Park, R. E. and E. W. Burgess (1925) The City. Chicago: Univ. of Chicago Press.
Parsons, T. (1970) "On building social system theory: a personal history." Daedalus 99: 826–881.
Platt, J. (1983) "The development of the 'participant observation' method in sociology: origin myth and history." J. of the History of the Behavioral Sciences 19: 379–393.
Polsky, N. (1969) Hustlers, Beats and Others. Garden City, NY: Doubleday.

Raushenbush, W. (1979) Robert E. Park: Biography of a Sociologist. Durham, NC: Duke Univ. Press.
Rock, P. (1979) The Making of Symbolic Interactionism. London: Macmillan.
Rose, A. M. [ed.] (1962) Human Behavior and Social Processes: An Interactionist Approach. London: Routledge & Kegan Paul.
Roth, J. A. (1956) "The status of interviewing." Midwest Sociologist 19: 8–11.
Roth, J. A. (1963) Timetables: Structuring the Passage of Time in Hospital Treatment and Other Careers. Indianapolis, IN: Bobbs-Merrill.
Roth, J. A. (1966) "Hired hand research." Amer. Sociologist 1: 190–196.
Roth, J. A. (1974a) "Professionalism: the sociologist's decoy." Sociology of Work and Occupations 1: 6–23.
Roth, J. A. (1974b) "Turning adversity to account." Urban Life and Culture 3: 347–361.
Roy D. (1952) "Quota restriction and goldbricking in a machine shop." Amer. J. of Sociology 57: 427–442.
Roy, D. (1954) "Efficiency and 'the fix': informal intergroup relations in a piecework machine shop." Amer. J. of Sociology 60: 255–266.
Schatzman, L. and A. L. Strauss (1973) Field Research: Strategies for a Natural Sociology. Englewood Cliffs, NJ: Prentice-Hall.
Schwartz, M. S. and C. G. Schwartz (1955) "Problems in participant observation." Amer. J. of Sociology 60: 343–354.
Shaw, C. A. (1930) The Jackroller: A Delinquent Boy's Own Story. Chicago: Univ. of Chicago Press.
Shibutani, T. [ed.] (1970) Human Nature and Collective Behavior: Papers in Honor of Herbert Blumer. Englewood Cliffs, NJ: Prentice-Hall.
Shils, E. (1948) The Present State of American Sociology. Glencoe, IL: Free Press.
Shils, E. (1981) "Some academics, mainly in Chicago." Amer. Scholar 50: 179–196.
Snodgrass, J. (1976) "Clifford R. Shaw and Henry D. McKay: Chicago criminologists." British J. of Criminology 16: 1–19.
Spradley, J. P. (1980) Participant Observation. New York: Holt, Rinehart & Winston.
Stein, M. R. (1954) "Field work procedures: the social organization of a student research team," pp. 247–269 in A. W. Gouldner, (ed.) Patterns of Industrial Bureaucracy. Glencoe, IL: Free Press.
Sudnow, D. (1967) Passing On: The Social Organization of Dying. Englewood Cliffs, NJ: Prentice-Hall.
Sutherland, E. H. and C. Conwell (1937) The Professional Thief. Chicago: Univ. of Chicago Press.
Thrasher, F. M. (1927) The Gang: A Study of 1313 Gangs in Chicago. Chicago: Univ. of Chicago Press.
Vidich, A. J. (1955) "Participant observation and the collection and interpretation of data." Amer. J. of Sociology 55: 354–360.
Vidich, A. J. and J. Bensman (1954) "The validity of field data." Human Organization 13: 20–27.
Wax, M. L. (1972) "Tenting with Malinowski." Amer. Sociological Rev. 37: 1–13.
Wax, R. H. (1968) "Participant observation." Intl. Encyclopedia of the Social Sciences 11: 238–241.
Wax, R. H. (1971) Doing Fieldwork: Warnings and Advice. Chicago: Univ. of Chicago Press.
Westley, W. A. (1953) "Violence and the police." Amer. J. of Sociology 59: 34–41.
Whyte, W. F. (1981) Street Corner Society: The Social Structure of an Italian Slum, (originally published in 1943) Chicago: Univ. of Chicago Press.
Wirth, L. (1928) The Ghetto. Chicago: Univ. of Chicago Press.
Zorbaugh, H. H. (1929) The Gold Coast and the Slum. Chicago: Univ. of Chicago Press.

20

The Chicago School and Firsthand Data

Jennifer Platt

It is generally recognized that the 'Chicago school' of sociology was of crucial historical importance in the 1920s and 1930s.[1] Its importance has encouraged an element of myth-making in the treatment of its contribution, and also a secondary industry of debunking the myths (see, for example, Bulmer, 1984; Deegan, 1988; Harvey, 1987). This article is part of that industry in that it questions common interpretations, but not part of it in the sense that it argues that different accounts may be appropriate to their own terms of reference (even if such accounts are often inappropriately read without consideration of what those terms of reference were). The main focus is on the research methods used, and especially on the extent to which the data deployed can be regarded as 'firsthand'. Many of the Chicago data were collected by other people, whose contribution has been played down. Were 'firsthand' data as important to the Chicago sociologists as later writers have suggested?[2] Indeed, were the Chicago sociologists as innovative in other respects as is commonly assumed? The answers given to these questions lead to consideration of the social and intellectual processes which have led to anachronistic accounts becoming dominant both in the generally accepted history of sociology and in most of the more specialized revisionist versions.

The standard version of history places the University of Chicago at the centre of things. Why was it so important? Because it pioneered in the collection of systematic empirical data, with a special emphasis on firsthand acquaintance with social life, inventing the new methods that were needed;

Source: *History of the Human Sciences*, 7(1) (1994): 57–80.

because it adopted a scientific rather than a merely reformist approach, and initiated novel topics of research. In consequence, it produced research of a new and superior kind. In this article the validity of all these claims is questioned, although the emphasis is on issues to do with research methods, and other aspects are dealt with primarily to give fuller documentation of processes relevant to methods. Each is considered in turn below, although material mentioned under one head is often also relevant under another.

Pioneering Systematic Empirical Data?

Did the University of Chicago Department of Sociology pioneer in the collection of systematic empirical data? It certainly engaged in large-scale data collection, and maintained its collection of material on the city as a central resource; many of the students also gathered data from a variety of sources on a smaller scale for their dissertations. But how far was this innovative? Several authors have pointed out that the women associated with Hull-House collected systematic data about Chicago before the department had produced any work of that kind (Deegan, 1988:33–69; Fish, 1985; Sklar, 1985) and, although Hull-House residents played a leading role in the welfare and reform activities of the time, they were by no means the only people outside the university or outside sociology to do so. The 'social survey movement' at an earlier period had led to the widespread collection of factual information about local communities (Converse, 1987:11–53, Young, 1939:18–44); this is sufficiently well known not to require documentation here. *Hull-House Maps and Papers* (1895) was in this tradition. The great movement of action and feeling among the educated middle class represented by the Progressive Party led to a widespread concern with social reform which was often associated with the collection of data (Diner, 1980). Jane Addams of Hull-House remarked of the Progressive Party that 'One subcommittee would seem to be like a session of the National Conference of Charities and Corrections; another like a session of the American Sociological Society' (Addams, 1930:28–9). The idea of 'social science' included both these strands (see, for example, Leash, 1981:316–22). Social workers also frequently carried out research, which was seen as an important part of their training; the 'field work' which later came to mean social work action had a strong research orientation, and its sociological character was encouraged by the conception at the time of 'social work' as something more like social engineering than individual case work (cf. Turner, 1992:12). In Chicago alone, the Chicago Commons settlement house undertook an investigation of a different aspect of its neighbourhood each year (Taylor, 1936:216); the Chicago School of Civics and Philanthropy which grew out of it had a research department which regularly undertook investigation into topics such as housing and truancy, teaching students research methods and then sending them out under the direction of the staff (Wade, 1964:172); the University of Chicago settlement run by Mary McDowell undertook 'surveys' every year (Wilson, 1928:66–8). The Juvenile Protective Association (JPA) was also responsible for the collection of large amounts of data, employing investigators for the purpose. The JPA was part of the nexus of institutions which led to the setting-up in Chicago of what has been claimed as the first juvenile court, and the first child guidance clinic under Dr William Healy; his many published case studies of children, such as *The Individual Delinquent* (Healy, 1915), were drawn on by innumerable other writers, including many sociologists. Miriam van Waters, a pioneering California-based social worker with delinquent girls who was in close touch with Chicago reformers, presented many case studies and an interpretive framework in her *Youth in Conflict* (van Waters, 1926). This list could easily be extended.

Inventing New Methods?

Did the university invent new methods? Many of the classic Chicago monographs give little or no account of their methods, but a certain amount of what they did can be inferred. Nowadays they are particularly noted for their use of participant observation and intensive case studies (Platt, 1991b) as well as the collection and mapping of quantitative ecological data.[3] (How far they did use direct observation is to some extent questioned below.) The reader is invited to guess which of the following extracts come from the work of university sociologists:

> In order that the investigation might not be limited to those who had failed in the profession and had been obliged to fall back on charity, another investigator (a man) was employed to visit . . . restaurants and music halls, and by treating the women to drinks to get them to tell their own stories . . . investigators spent some time watching prostitutes, either at their professional haunts or at common lodging houses and other places where they talked among themselves when off duty.
>
> Investigators . . . mingled with the men and girls, sat in the saloons, danced in the halls, and talked with manager, employees and patrons. Their observations were carefully noted on cards prepared for the purpose and filed daily.
>
> he devoted the first 6 weeks of his time to personal observance of saloon life. . . . He extended this personal inquiry to many other districts and the saloons of all types and characters. Then by extensive correspondence, personal interviews and experience in various ways he ascertained what was being attempted by fraternal organisations, trade unions, culture clubs, public schools and educational institutions.

It may perhaps have been guessed that none of these extracts is from the work of sociologists. The first comes from Royden (1916:13–14), the second from Bowen (1917:3) and the third from a description by Graham Taylor, of the Chicago Commons settlement, of work done there by R.L. Melendy around 1900 (Taylor, 1936:174). Maude Royden was a British feminist and preacher; she reports the work of several contributors to an inquiry originally begun 'for social workers specially interested in discovering how far economic causes contribute to the making of prostitutes' (Royden, 1916:12). (British and American workers in these fields were closely in touch at this period [Platt, 1991a].) Louise de Koven Bowen was a Chicago society lady, the daughter of a banker, who in the 1890s started to be involved in many of the activities of Hull-House and the JPA, and was also active on many civic issues (Bowen, 1926; Sicherman and Green, 1980). In the quotation she is reporting work done by the JPA. Both Royden's and Bowen's inquiries also collected quantitative data on a large scale, in addition to the systematic participant observation reported above. Rodgers (1978:74) cites a number of cases where middle-class social investigators participated in working life to find out more about it. Jane Addams' book, on prostitution, also drawing on material provided by the JPA, included, in addition to a number of special investigations, the personal histories of '200 department store girls, of 200 factory girls, of 200 immigrant girls, of 200 office girls, and of girls employed in over one hundred hotels and restaurants' (Addams, 1912:x). Mrs Rippin, chief probation officer in Philadelphia, may not have published her work but a summary of 1,000 cases is mentioned in correspondence between her and Mrs Ethel Sturges Dummer (Dummer papers 24:379, 22 February 1919). Mrs Dummer was another of the extraordinarily active society ladies of Chicago, and played an important, though insufficiently acknowledged, role in the sociology of the 1920s (Platt, 1992).[4]

Emily Greene Balch, a social scientist who did not primarily identify herself as a sociologist although much of her work could be described as sociological, taught at Wellesley for a number of years from 1898, and her pre-Chicago school methods sound very familiar:

> A great deal of attention seems to have been given to field work in the early days of the department. We hear of Miss Balch's students in 1901–2 mapping the North End of Boston. . . . Each student handed in a final paper describing the section from personal observation, together with a 'social map' showing centres of social significance – schools, hospitals, saloons, churches and settlement houses. (Randall, 1964:104, 110)

She also conducted a study of Slavic immigrants for which

> . . . I spent the greater part of the year 1905 in Austria-Hungary, studying emigration on the spot, and over a year visiting Slavic colonies in the United States. . . . One autumn was spent as a boarder in the family of a Bohemian workingman in New York City. (Balch, 1910)

The plotting of quantitative data on maps had ample further precedent, most obviously in the work of Booth (1892–7), which was well known in Chicago, and in *Hull-House Maps and Papers*.

It is not clear that the university added anything to the repertoire of methods of data collection and presentation already in use elsewhere, although they made a useful intellectual contribution to rationalizing and naming the methods in use and so cultivating a valuable self-consciousness about them. (See, for example, Palmer [1928], a departmental methods text.) One of the areas in which the department played an active role in methodological discussion was 'case-study method', seen at the time as the key qualitative approach; it subsumed 'life history and personal documents as types of data. The cases used were, however, very frequently those first collected by social workers for their own purposes, and it seems likely that the conception of the 'case' was strongly affected by this origin. (In this field, too, there were published social work precedents. Miner [1916:xi] describes how many of her stories of prostitutes are given in their own words, and in her first chapter presents a series of cases of different types: 'an old offender', 'a parlour house girl', and so on.) This leads us to the next issue to be raised: to what extent did the work of the department rest on a direct firsthand acquaintance with social life?

Using Firsthand Data?

It has become traditional to quote Park in this connection:

> . . . one thing more is needful: first-hand observation. Go and sit in the lounges of the luxury hotels and on the doorsteps of the flop-houses, sit on the Gold Coast settees and on the slum shakedowns; sit in the Orchestra Hall and in the Star and Garter Burlesk. In short, gentlemen, go get the seat of your pants dirty in real research. (Bulmer, 1984:97)

This approach is connected, by Bulmer and others, with Park's background as a journalist, and Park himself drew an analogy between his role as a supervisor and his previous experience as a city editor directing reporters (Bulmer, 1984:91).[5] This is a stirring manifesto and a romantic analogy; every sociologist can probably vaguely recall Chicago work which they seem to fit well – but memory is often deceptively selective.

A close reading of the classic monographs suggests that the extent to which they rely on firsthand data collected by their authors has been exaggerated. Most of them contain large quantities of secondary data, and many of the data which look firsthand turn out to be so at one remove – collected firsthand by others. Its frequently vivid and naive style, the apparently direct recording of speech, and the exposure of exotic and disreputable modes of life, give a misleading sense of directness. For secondary data, particularly prominent sources are newspaper stories and court or coroners records, and census materials are often used as background. The 'firsthand' data provided by others most often come from the records of social work agencies; the same names appear again and again in the prefaces and footnote acknowledgements. It is also quite common for unpublished data collected by other authors, sometimes but not always fellow members of the department, to be used. More ambiguous is the status of the frequent written accounts solicited for the purpose of the research; sometimes these are accounts of their own lives by the subjects of the research, but there are also accounts of the subjects written by third parties. The only unequivocally firsthand material is the interviews or observations done by the author or his collaborators.

These general assertions will be supported by a detailed examination of four of the more famous monographs to show the types of data used. These examples cannot be treated as quantitatively representative of all the empirical work of the school but, since they are drawn from those which probably seem most clearly firsthand, should provide a sufficient basis for an *a-fortiori* argument. We start with Cressey's *The Taxi-Dance Hall* (1932), the study which comes nearest to relying on planned participant observation carried out by the author.

The Taxi-Dance Hall is somewhat more explicit than many of the others about its methods, and what it says can fortunately be supplemented from an archival source discussing methods, now published (Bulmer, 1983), although this deals more with general principles than with the practices followed. In the study observers, of whom the author was one and others were fellow graduate students, took part in the life of the dance halls and those involved with them, and conducted informal interviews with those they met there. Cressey sometimes identified himself as a researcher; he recruited informants to write life histories for him, visited families to collect data on the family adjustment of taxi dancers, or overtly interviewed such people as the proprietors of halls. Many quotations from numbered 'cases' appear in the text, and these are usually first-person accounts by patrons or dancers from the halls. At least one of them is referred to in the text as a 'document', which may imply that it was a written life history. Others, however, were written up by whomever collected the data. Philip Hauser has described how he took notes at the earliest opportunity on what girls had told him, and reports that many of the cases used in the book were his (Hauser, 1982). It is not clear how these are distinguished from some other first-person quotations which are labelled 'record of an investigator', but it is likely that the distinction is between a relatively extensive 'case study' of an individual and single observations. The term 'investigator' has some ambiguity, because it was also used to describe the role played by people employed by the JPA, who both collected data for its purposes and acted as social workers and/or moral reformers. In his preface, Cressey (1932:xvii) refers to reports from 'observers and investigators', which implies a distinction. Cressey was in fact employed by the JPA as a case worker and investigator, and it was the JPA which initially asked him to look at taxi-dance halls (Cressey, 1932: ix, foreword by Jessie F. Binford of the JPA). He drew on the case records of the JPA (and other social agencies) in the book, and a fair number of the book's data are footnoted as from JPA records; some of them may have been collected by himself in his JPA role, but it seems evident that not all of them were. It sounds as though the common 'case study' practice of supplementing material collected directly from the individuals studied with information from the records of social agencies was followed, though the wording used (Cressey, 1932:xviii) does not make it absolutely clear that this was what was done with the agency material. Not all the data collected are directly cited in the book, so it cannot be assumed that the proportions there from different sources precisely represent the sources used in reaching the conclusions.

We may conclude that an unknown, though quite large, proportion of the total data used in the book was indeed the product of firsthand observation by the author or the colleagues he recruited for the purpose. A fair proportion, however, was secondhand, most of it taken from agency records compiled for other purposes, and a small part from various other documentary sources such as newspapers. Although the modern reader is most likely to identify the method as 'participant observation' or 'ethnography', it is evident that to Cressey the relevant methodological category was 'case study'; for this data could be collected by many means (Platt, 1983). Two interesting features may be noted. First, the difficulty of identifying just what was done suggests that this was not regarded as important. Second, the complete data set was a product of cooperation both within the department and between a member of the department and outside agencies. Both these features recur, and their significance is discussed below.

The second book to be considered is Thrasher's *The Gang* (1927). In this he identified 1313 'gangs' in Chicago, and reached general conclusions about gangs. It is clear that he was actively involved in the collection of observational data, though the process is not described in the book; some hints are given in a later article (Thrasher, 1928). The fullest available published account, however, is given in the summary of his thesis published in Burgess and Bogue (1964:658–9):

> For statistical purposes, a census was taken of the gangs in Chicago . . . . A spot map was constructed to show the distribution of the gangs. . . .
>
> A study of the typical gang areas revealed in the census was undertaken. This tended to develop along both descriptive and historical lines.
>
> The collection of case studies of gangs and gang boys was a very important part of the investigation. The linkages and ramifications of the gang with the other problems were studied largely by means of the case method. . . . Extensive interviewing tended to supplement the lack of printed sources. Interviews were had with most of the workers in both

> private and public agencies which deal with boys. These agencies also served as important sources of data by having their workers fill out questionnaires. Their printed bulletins and reports and their unpublished records were invaluable . . .
>
> The most valuable material for the study was obtained from boys who were or had been members of gangs. Many boys were interviewed on the streets and in the neighborhood institutions. More than one hundred boys were interviewed in the Chicago Cook County School for Boys and twenty-eight in the Chicago Parental School. The technique followed in getting the boy's own story may be described as one of sympathetic approach. Every effort was made . . . to penetrate the boy's own social world, and to speak in terms of his own universe of discourse. . .
>
> Direct observation of gang behavior was undertaken by the investigator at settlements and similar institutions where neighborhood gangs were accustomed to meet for club activities. The gang was studied in its native habitat of streets and alleys, vacant lots. . . . On occasion, the investigator would fall in with a gang on the street. At times he would penetrate the mysteries of the hangout. . . . The physical and social environments of the gangs were also studied first-hand. . . .
>
> The census of the gangs was made by gathering data on as many different gangs as possible from every available source. Case studies were built up by combining the results of the boys' own stories with personal observations, interviews with local residents and officials, documents and records wherever these were available. In so far as possible, no source of information about gangs in the city was neglected.

Thus a wide range of types of data was available to be drawn upon, several from material directly collected by the author for the research. But how were the data used? In the text of the book, almost all the specific data are presented as 'documents', numbered in the order of their appearance. Only 9 of the 268 documents report Thrasher's own observations, 15 give extracts from interviews with gang members or others reporting on their own behaviour, and 67 give accounts – presumably solicited – by gang boys, some of which sound as if they were given in writing rather than orally. There are, however, 61 reports of interviews with informants such as railway police, talking about gangs rather than about themselves (and thus in that sense secondhand); there are also 68 items from published studies or newspaper stories, 24 uses of unpublished material provided by others, 8 extracts from social work agency records, and 5 passages from manuscripts by other non-professionals (for the remaining items no source is given).[6] Of these latter items, some refer to particular individuals or incidents while others make generalizations. A majority of the 'documents', therefore, are not ones providing firsthand data collected by the author. He does not seem interested in this potential distinction. Each document appears to be treated as a statement of fact, and no attention is given to the possible differences in the status of a generalization and an account of a single episode. Very few of the generalizations made are supported by general evidence; the function of the evidence overtly provided seems to be to act as illustrations or examples for assertions which are grounded on wider, but unspecified, data or experience.

The third book to be considered is *The Unadjusted Girl* (1923), by W. I. Thomas, a leader in the older generation of the department (though by 1923 he had left it after a scandal) and seen as an important pioneer in empirical work. All the book's data are presented in the form of 93 numbered 'documents', some very brief and others several pages long. The content of these 'documents' ranges from generalizations from academic books, through letters sent to newspapers, to case histories from social work records. It is striking, however, to what extent the data are inherently secondhand or, while primary, were not collected by the author. A maximum of 5 of the documents appear to report firsthand data collected by Thomas – and one of these is from an autobiographical account which had already appeared in *The Polish Peasant in Europe and America* (Thomas and Znaniecki, 1918–19) and was collected by his collaborator Znaniecki. The largest categories of source are published books or articles, letters written to newspapers, and court or social work records. Although the book itself gives little information about Thomas' methods, we may learn more from his correspondence with Mrs Ethel Sturges Dummer. (Mrs Dummer played a very important role in relation to this book in particular [Platt, 1992]. That is mentioned further below, but here we need only note that she was very active in supplying him with cases to use for the book, most of them collected from the social workers with whom she had extensive contacts.) Some quotations are suggestive:

> At any time you can get any cases, Mrs Dummer, or the loan of them, don't omit it . . . it is necessary to have a lot of them. What I shall need more than the court cases is life histories, showing what processes lead up to a given condition. I also want cases of reform. (W. I. Thomas to Mrs Dummer, 4 March 1920)
>
> I have not begun to call for materials on our study systematically. I am picking them up, getting material from the library, visiting courts, and formulating the problem. (W. I. Thomas to Mrs Dummer, 30 March 1920)
>
> [I am] going over about 500 cases which I have already in manuscript – transcripts from records of juvenile courts etc. . . . I have planned to get the life histories of several mature and stable women. (W. I. Thomas to Mrs Dummer, 12 April 1920)
>
> I am now getting some data from 3 young women who were attendants in the Pennsylvania home for delinquent girls. (W. I. Thomas to Mrs Dummer, 11 December 1920)
>
> [All the above quotations are from the Dummer papers, box 36, folder 785.]

There are also several references in the correspondence to 'cases' being exchanged with other people, usually research-minded social workers. It is clear that Thomas was happy to get 'cases' in any way he could, except that this did not normally include interviewing or other means by which he might have personally collected data on the subjects of the research by any form of observation. There is, however, occasional brief anecdotal material from his own observation, as when he refers to one case he saw in a juvenile court (Thomas, 1923:108), but that might have been merely the accidental by-product of 'case' collection. No statistics of his own are provided, although those of others are quoted, but there is a passing reference in support of one point to having examined about 3000 cases (Thomas, 1923:109). It is clear that this general pattern of data was a deliberate methodological choice – although the potential alternatives implicitly rejected may not be those we would think of today – as is shown by the discussion in the final chapter.

> A social institution can be understood . . . only if we do not limit ourselves to the study of its formal organisation but analyse the way in which it appears in the personal experience of various members of the group and follow the influence which it has had on their lives. And an individual can be understood only if we . . . determine the means by which a certain life-organisation is *developed* . . . as we accumulate records of personal evolution, with indications of the means by which the wishes seek expression and of the conditions of their normal satisfaction, we shall be in a better position to measure the influence of particular institutions. . .
>
> The 'human document', prepared by the subject . . . is one means of measuring social influence. It is capable of presenting life as a connected whole . . . by the multiplication and analysis of life records we may expect to determine typical lines of the genesis of character as related to types of influence. . . . Even fictitious representations are significant when viewed as showing the tendency at a given moment to idealise certain schematizations of life. The autobiography has a more positive value for the student . . . one of the largest and most important bodies of spontaneous material for the study of the personality and the wishes passes through the mails . . .
>
> Another type of behavior record is now being prepared by those social workers, psychologists and psychiatrists who have to handle the problems of maladjustment. . . . Under the pressure of practical needs they . . . are studying the evolution of personal life-organisation and making the record as complete as possible. (Thomas, 1923:244–51)

It is interesting, too, that Thomas in writing about *Old World Traits Transplanted* (Park and Miller, 1921) (of which he was the author, although his name did not appear on it), which was put together in a similar way, says to Mrs Dummer: 'Probably the documentation will seem strange to you . . . but I believe it is possible to bring out points in that way better than by description' (W. I. Thomas to Mrs Dummer, 4 March 1920, as above). The placement of the 'documents' in *The Unadjusted Girl*'s argument suggests that they are being used as illustrations of points on which the conclusions may have been reached using other materials, rather than as data. To the extent that this is so, the text of the book does not directly represent the research process. No doubt the same could be said of many books. In the end we have to treat the published work as a definitive product, even if there is more to the whole story.

The final book we look at is Harvey W. Zorbaugh, *The Gold Coast and the Slum* (1929); Albert Hunter (1983:461) describes this as having been 'the model or exemplar of participant observer field research which characterised the Chicago School of urban community studies'. Once again, however, the range of its types of data is striking. It contains, like most of the other Chicago monographs, a variety of maps, which show a creative use of documentary sources, plotting the distribution of the addresses of the United Charities' contributors and relief cases, or of persons buried from the morgue when their bodies were not identified or claimed, or of those listed in the *Social Register*. One chapter, on the 'Gold Coast', is based on documents written by its residents for the research; many of these give general descriptions of its social circles and their way of life. Bulmer (1984:100) reports that the records of the university's Local Research Committee show that

> Persons interviewed included old residents, influential people in the area, representatives of civic and social organisations, and people with useful specialised information. In addition, informal contacts were made with newspaper men, the night court, the nurses of the free ward of the hospital, and others familiar with the life of the area.

We may note that this list suggests respondents suitable to be treated as informants about the lives of others. Bulmer (1984:103) also reproduces formal schedules used to collect information about housing conditions and demographic and financial information about families; some data from them appear in Zorbaugh's text. Once again most of the specific data cited are in numbered 'documents' which include both interviews with named individuals and quotations from published reports, as well as a few instances which sound like field notes but are not identified as such. Extensive use is made of social workers' case histories, and Document 48 is 'a summary of family schedules on file at the Lower North Community Council'. More than one diary is quoted. Document 51 is several hundred papers written by children about their communities, and Document 60 is 'five hundred themes written by eighth-grade children in the schools of the Near North Side'; we are not told how these originated. Much use is also made of data or written accounts provided by other members of the department; 14 of them are thanked in the Acknowledgements for placing their research results at his disposal. In addition to all these identified data sources, however, there are many passages for which no source is given but which might well come from field observation. Some of these are physical descriptions of the area, and those which are more social in content are often put in generalized form; for example:

> On a warm summer evening one may find a crowd of anarchists, socialists, proletarians. . . (90)
>
> As one watches these types merge and mingle . . . one is struck by the fact that in Towertown nearly everyone plays a role, wears a masque. (96)
>
> The police and the social agency alike meet with little cooperation or response on the part of the population. Those who see an opportunity for profit give a nominal or calculating cooperation. Others are indifferent, suspicious, resentful or in open opposition. (197)

Whatever direct observation Zorbaugh has done, it is scarcely if at all explicitly acknowledged. The issue is also somewhat confused by his frequently highly literary and popular style, rich with metaphors, picturesque adjectives and moral judgements; there is little sense of any felt responsibility to distinguish data from interpretation.

These analyses show that Bulmer (1984) was certainly right to emphasize the diversity of the sources of data used, which is truly impressive. However, the extent to which it can be described as firsthand is less impressive. Indeed, if a journalist is the model it often seems more like one who turns first to the cuttings file rather than to fearless firsthand investigation. (No special merit is here attached to being 'firsthand'; documentary sources, reports by informants or data already collected by others may often have been the best or the only source of the information wanted.)

We have in effect explored the meanings which can be given to the idea of 'firsthand' data. There are three central themes: were the data actually collected by the researcher, did the researcher have direct contact with the subjects of the research, and were the data collected directly observed instances of the behaviour studied, or only reports on it? It has been shown that, although the proportions varied in different books, the data used were often not originally collected by that researcher, or even by a sociological colleague. It follows that for many of the items used the author did not have direct contact with the subjects; this applies both where someone else had direct contact and may, for instance, have elicited a life history, and in the many cases of use of material such as census data or membership lists where 'direct contact' was hardly relevant or appropriate. Where there was some sort of direct contact, it was quite often used to get reports on third parties from informants; the police are, for instance, interviewed about gangs rather than about police behaviour towards gangs. The idea that direct observation of behaviour by the researcher has special intrinsic merit was probably only developed in the polemic for participant observation, as opposed to the survey, in the 1950s (cf. Platt, 1983 and forthcoming). A key rationale for the use of life histories was that they provided access to personal meanings and showed the viewpoint of the subject (Bulmer, 1984:55). However, their users often in practice seem not to have resisted the temptation to use them simply as a factual account of the course of events.

The frequent absence of any clear indication of the sources of the data, and of whether or not they are firsthand, can also be taken to have a significant substantive meaning. It suggests that the distinction was not one of much interest to the authors. 'Documents' used in similar ways in the argument can contain anything from a quotation from an academic source to an extract from participant observation field notes (cf. Platt, 1983:382), data collected by others both for research and for non-research purposes are freely used, and 'firsthand' data collected in various ways often use the respondent as an informant at secondhand on others whose lives are the object of interest. It seems to be regarded as relatively unimportant who obtained the material, whether it was originally written or oral, and whether it reported specific incidents or generalizations. It also seems to have been felt to be important that data, of whatever origin, should have objective existence in writing independent of the publication in which they were used (cf. Bennett, 1981:140), and so even observational material appears as a 'document'. The descriptions provided of the data thus almost systematically blur the distinction between first- and secondhand. This suggests either that the emphasis on the firsthand was not a very salient methodological theme in the writers' thinking, or that practice diverged from some of their rhetoric – or both.

Descriptions apart, the question of who actually did what in the compilation of the data also arises. Chicago University authors draw freely on data collected by others; some of the others were also members of the Department of Sociology, but many were not. Those who were not were most frequently members of the network of social workers and civic reformers described above. In the case of *The Unadjusted Girl*, the great bulk of the data appear to have come from such sources. As Deegan (1988:309–42) has pointed out, some of the most important women involved were originally regarded as sociologists at the time, but became redefined and marginalized in fields such as social work, home economics, or domestic science; subsequent histories have then neglected their contributions because they were 'not sociology'. (Similar preconceptions have also led to the tendency to treat the reformist involvements of the university sociologists as outside their professional lives, or at most instrumental to professional sociological goals, because they were not activities of the kind that nowadays we call sociology. Why should we not assume that these were an integral part of a total life pattern of social concern, expressed in what at the time were often not regarded as separate spheres?) Cressey worked for the JPA before he chose his thesis topic (see above), and Nels Anderson was employed by them as an investigator after his MA thesis which became *The Hobo* (Anderson, 1923). Juvenile court records, or JPA associated data, were used in the monographs of Cressey (1932), Frazier (1932), Park and Miller (1921), Reckless (1933), Shaw et al. (1929), Thomas (1923), Thomas and

Znaniecki (1918–19), Thrasher (1927) and Zorbaugh (1929) (cf. Deegan and Burger, 1981). A similar and related network of connections could be traced for research and social work concern with prostitution, with many data and much employment stemming from the prominent citizens' Chicago Vice Commission and Committee of 15 (cf. Diner, 1975). Clifford Shaw was a part-time parole officer and then a probation officer for the juvenile court while a graduate student, as well as initially living on a settlement (Snodgrass, 1976:4–5). It was commonplace for graduate students to be employed in such activities, sometimes simply to earn money but sometimes also getting their dissertations from them. Thus in many ways the boundaries between purely 'sociological' and other data collection were blurred.

In addition, the choice of the city as the focus of the department's research meant that there was an active programme of general data accumulation independent of particular projects; quantitative and qualitative data about the city as a whole were collected, often from social agencies, and mapped, and were then available as a central resource (Burgess and Bogue, 1964:6). Many graduate students involved in this. This centralized activity must have encouraged the high sense of departmental community, one manifestation of which was both the sharing of data already gathered and the preparedness to do field work for each other. The same items several times appear in more than one book by different authors. For instance, the diary of Miriam Donaven bulks large in both Mowrer (1927) and Cavan (1928). Thus both the internal and the external relations of the department seem to have promoted a sense of data as community rather than individual property (cf. Carey, 1975:158).

The blurring of the boundaries between 'sociological' and other research and data leads on to our final questions: did the university sociologists address novel topics, and was their approach to them distinguished by being more scientific and less reformist?

Initiating Novel Topics?

Much of the material already presented shows that university researchers were working on some of the same topics as the reformers and the social workers, even in conjunction with them. Burgess (Burgess and Bogue, 1964:4) pointed out that a number of studies of slums, immigrants and other aspects of urban life in Chicago had been carried out at Hull-House and by the women associated with it before the well-known university work. Juvenile delinquency had also already been studied by Breckinridge and Abbott (1912). A report on the dance halls of the city had already been published by Bowen (1917) of the JPA long before Cressey started work on the subject. Prostitution and promiscuity was a hackneyed topic by the time Thomas came to it; Addams (1912) may be taken as the book-length example most closely associated with Chicago. I have not found any other work on hobos done in Chicago before that of Anderson (1923), although the JPA had a pamphlet on 'The Juvenile and the Tramp' (1922), but Parker's *The Casual Laborer and Other Essays* (1920:78–9, 121) describes some of the same social patterns and gives the same picturesque slang terms associated with them. (Rodgers [1978:285] cites several participative and journalistic studies of tramps from the early 1900s.) Finally, there is even a near-predecessor to *The Polish Peasant* (1919) and *Old World Traits Transplanted* in Balch's *Our Slavic Fellow Citizens* (1910), as Deegan (1983:104) has pointed out.

This lack of novelty is not at all surprising when the research is set in its social context. The daily facts of Chicago life could hardly fail to draw attention to some of the issues and, more importantly, some of the work was initiated outside the university. The most striking instance is *The Unadjusted Girl*, which was not just funded by Mrs Dummer but commissioned by her to carry forward her existing interests (Platt, 1992). This was done against the background of an extensive reformist concern with 'vice' and 'social hygiene' (cf. Connelly, 1980), reflected in the formation of Chicago's Vice Commission and similar movements in other cities as well as in a growing feminist concern about practices which penalized the women involved but not the men. *Old World Traits Transplanted* (Park and Miller, 1921) and The *Immigrant Press and its Control* (Park, 1922) were both part of the Carnegie Corporation 'Americanization' programme (Bulmer, 1984:246–7). *The Gold Coast and the Slum* appears to originate at least in part from the concerns of the Lower North Community Council. Much of the work on delinquency was directly supported by local social agencies and, like *The Taxi-Dance Hall*, arose from their concerns. However, a concern with policy issues is prominent even when the work was not initiated outside sociology; as many authors have pointed out, most of the topics chosen are in one sense or another 'social problems'.[7] Final chapters with recommendations are common, and value-judgements are unselfconsciously explicit and taken for granted in both (cf. Dubin, 1983). In this respect the work of the sociologists is hardly distinguished from that of the social workers. Nor is it clearly separated by its theoretical orientation or its scientific approach.[8] General theorizing, like that of Thomas on 'the four wishes', does indeed appear, but the model is commonly that described by Matthews (1977:118):

> The books published under his [Park's] direction are often club sandwiches, with slices of theoretical suggestions surrounding thick slabs of direct quotation from newspapers, informants and public documents.

No later book can compete with *The Polish Peasant* for the sheer bulk of its unanalysed data, but characteristically very extensive data are only loosely organized under headings. If the practice was to formulate guiding hypotheses before collecting data, we are not told so; where so many data were secondhand, it might not have provided much practical guidance if this had been done. The process by which the data led to the conclusions is not normally described, so that it is often unclear whether the materials presented in the text are meant merely as illustrations, or as the evidence to ground the conclusions. On the other hand, work done by non-sociologists often contained extensive empirical data with reasonable claims to be at least as systematic as that of the sociologists, and with as much or more quantitative sophistication (Deegan, 1988:45–7).

Discussion

This article does not argue that the work of the non-sociologists was better, more scientific, or more truly sociological than that of the sociologists, merely that on key points there was considerable overlap between the two – and this was not usually because the sociologists got there first and took the lead. A riposte to my claims for the social workers and reformers could well be made by arguing that the sociologists were also involved in many of their civic activities. That is true, although it does not follow that the sociologists necessarily led or dominated them, as some sociological writers appear to assume. It is simply not at all easy to distinguish between the contributions made by the two groups, and indeed to treat them as two separate groups could be seen as misleading. However, the history of sociology has generally been written as the history of 'sociology', and this has distorted the actual course of events.

It is possible to imagine a reasonably plausible account which reverses the picture. In this alternative history, the work of the sociologists would appear minor and dependent: carrying on established traditions of concern, funded by external agencies with their own agendas, often done at the behest of and carried out under the supervision of non-sociologists, drawing many of its data from what was already available, and feeding back into practical action a modest stream of useful results to justify its existence.[9] Some of those on the other side did indeed sometimes see the matter rather like that. Jane Addams says that

> *The Polish Peasant in Europe and America, The Race Relations Study of the Pacific Coast,* the investigations made by the Immigrants' Protective League of Chicago in connection with the School of Social Service Administration at the University of Chicago, are all examples of an attempt to keep the social worker informed. (Addams, 1930:303)

Mrs Dummer's contract with W. I. Thomas specified that the work was to be done

> . . . primarily to gather and interpret such data as shall lead to less unjust and futile treatment than is at present accorded to so-called delinquent women. (Dummer papers, Box 36, Folder 785, 27 January 1920)

Karpf (1926:219–20), a social worker, describes the Americanization studies of the Carnegie Corporation – which included the books mentioned above by Park and Thomas – as undertaken to help social workers and others involved with practical problems. I do not argue here that this alternative picture is the right one, but it does need to be taken into account as part of the whole. Its potential is suggestive about the nature of the accounts more generally current among sociologists. Why have these been so partial?

One reason is a problem generally recognized in writing the history of ideas: to define the relevant past as those parts of it which can most directly be related to current concerns is often to leave out much that was significant at the time (cf. Kragh, 1987). However, this line of thought does not take us very far, because so much of what is omitted is intellectually extremely similar to what is included, and so equally related to current concerns. Deegan (1988, and see also 1986) offers an explanation in terms of sexism. The great majority of those whose work is now forgotten were women, and she shows how they became redefined as social workers even when, like Jane Addams and others, they had been initially recognized as sociologists; within universities, the departments to which they were relegated became the distinct and inferior sphere for women outside sociology. This is a persuasive line of argument, but to account for our problem it needs something more: the history of sociology needs to have been written to exclude any reference to what is not now defined as sociology. This is indeed what has happened, in a process of retrospective professionalization. An anachronistic picture has thus been drawn. One consequence of this is to throw into spurious relief the contributions of the Chicago department; against the broader background of their time, what they accomplished in empirical research was excellent, but not so strikingly novel.

Different versions of history have emerged for different purposes. The official volumes about social science research at the university (Smith and White, 1929; Wirth, 1940) give due credit to city predecessors, but define their work as an earlier stage on which more systematic and scientific university work has built; current connections with civic reformers and social agencies are treated as a source of funding, sometimes for work done as public service rather than academic priority. (Park's chapter [Smith and White, 1929:13–14] which sketches some of the relevant history, however, refers to the work of the Juvenile Court, the Institute for Juvenile Research, Healy and Adler without mentioning at whose initiative they took place, thus implicitly coopting them for the university in a way which does little justice to their history.) Within the university, however, the volumes take a strongly interdisciplinary perspective which does not isolate the special contribution of sociology but shows how it fitted into a larger enterprise. Accounts written from within sociology as history of sociology generally treat both other disciplines and groups outside the academy as parts of the background, instrumental to the main ends or introducing distortions into the natural or appropriate course of pure sociological development. Accounts which focus specifically on the 'Chicago school' have, as Harvey (1987:11–13) points out, often selected particular parts of the whole to create a picture which legitimates current activities claiming to be associated with it.[10] In relation to research methods, the accident of more recent developments has meant that this has led to an exaggeration of the emphasis on 'qualitative' work, and a writing back into the past of categories and meanings (especially those associated with participant observation) which were not salient at the time (Platt, forthcoming). Thus a stress laid in the 1920s on 'firsthand' observation, which distinguished the empirically oriented Chicago department from the others whose work was done in the armchair and library, has been misleadingly taken to have the meanings recent writers need to distinguish their preferred styles of work from the survey. This tendency has been encouraged by the availability of some attractive programmatic statements, like the one by Park quoted above, which have often been rather naively treated as literal descriptions of practice. It is commonplace for manifestos and statements of methodological principle to have a rather poor correspondence to practice, but the convenience of a capsule formulation can tempt the historian to rely on them rather than to face the more laborious task of analysing the details of practice.

Conclusion

It has been shown that the Chicago department has been treated by historians of sociology as pioneering many aspects of research which were actually initiated outside sociology, or outside what is now regarded as sociology.[11] Because the focus of interest has been on sociology as we now define it, developments outside this have not been noticed, or have been treated as unimportant. Professional sociology is probably insulated from the outside world to varying degrees at different times and places; at Chicago in the 1920s the boundary was highly permeable, with ideas, practices and people (or at least men) moving freely across it. The production of data about the city and its problems could be seen as a communal enterprise, in which the department did not always play the leading role. But if this detracts from the reputation of the department for 'firsthand' data, that need not be seen as inconsistent with its ethos at the time; its textual practices suggest that whether or not data were 'firsthand' in the sense implied by modern writers was not then seen as important. The accounts of the department given by its members have, like all other accounts, been influenced by the uses for which they were intended and the audiences to which they were addressed. Some which were manifestos or declarations of intent have been treated by later readers more as if they were direct descriptions of current practice. For an audience of sociologists, it is priority within sociology that counts, and so other claims to priority were neglected. The accounts of the department given by outsiders often serve non-historical purposes; they redefine the aspects of history which interest their authors in terms of the categories appropriate to their current concerns, rather than to the shared meanings of the time and place described. It is likely that similar comments could also be made about other versions of disciplinary history.

Notes

1. Support for work on which this article draws came from the Radcliffe Research Support Program; this support is acknowledged with gratitude.
2. 'Firsthand' is placed in quotation marks both because its meanings are critically explored below, and because the idea has been used by many other writers. Some, but not all, have actually used the word to express the idea. For instance, Bulmer (1984:5) contrasts Chicago sociologists with earlier writers who did not see 'first- or even secondhand acquaintance with the contemporary world . . . as a necessary requirement for fruitful sociological generalisation', and describes the Chicago group as part of a movement to give greater significance to 'firsthand empirical research'. Matza (1969:37) refers to 'the Chicago school's stress on direct observation and field work'. Other writers, more concerned with the Chicago group as ancestors for their own practices, have raised similar themes in the name of 'ethnography' or participant observation, as Hammersley and Atkinson (1983) do. Phrases such as their 'the ethnographer participates . . .' (1983:2) take personal involvement for granted, even though they later (1983:126–9) concede a role to documents, and recognize that the Chicago researchers whom they repeatedly invoke used them heavily. (Interestingly, they redefine some documents as participant observation by the writer, thus attempting to retain the aura of the firsthand while referring to what in our terms is secondhand.) The group associated with *Urban Life* (Lofland, 1980) have associated themselves with similar meanings.

   For the purposes of this article the issue has been couched in terms of 'firsthand' data' rather than 'observation', for reasons which should become clear later on.
3. The extent to which the department's quantitative work was innovative is not discussed, because the focus is on images of the department's work and this aspect has not been much emphasized. However, as Deegan (1988:46–7) has pointed out, the departmental tradition of mapping was only building on earlier work.
4. The participant observation done by Cressey and his colleagues for *The Taxi-Dance Hall* (Cressey, 1932) may not appear quite so novel and intrepid when we read in one of Mrs Dummer's letters, written when she was 55, that

   > At a committee meeting last evening I met Miss Campbell, going with her later to a closed dance hall recently established here for the Filipino boys in the navy. (Dummer, 25:403, 30 May 1921)

   Mrs Dummer was married to a prominent Chicago banker, and moved in elite social circles. At that time and place it was not unusual for members of their class to be heavily involved in social welfare and reform activities, as Mrs Dummer was, often in active cooperation with the network of women and men for whom it was their profession. From around the turn of the century she became extraordinarily active on a wide range of issues; she initiated discussions and activities among academics and practitioners, funded research and herself collected data and diffused results. For fuller details, see Platt (1992).

The Dummer papers are held at the Schlesinger Library, Radcliffe College; I am most grateful to the Schlesinger Library for the use of its excellent facilities. All references to the papers give first the box and then the folder number.

5. A somewhat different perspective on Robert Park's involvement with the city is suggested by an anecdote in the Dummer papers. A social worker describes meeting Park at a dinner party, and remarks:

> These men at the University are so far away really from the things that are happening it was interesting Saturday to see how very much interested he was in each little human story that was told. Each story seemed to be such a revelation to him. (Jessie Binford to Mrs Dummer, 5 April 1922, Dummer papers, 25:404)

This perhaps reflects the differing spheres of interest of men and women (representing sociology and social work), although it might also indicate that insofar as Park was himself directly involved in observation it was of a kind which gave access only to some sorts of material.

6. These figures suggest a slightly false precision, although the general picture they give is not misleading. Some apparently multiple sources have been classified by what seemed predominant, and ambiguous cases (such as 'interview' when the text does not make clear who it was with) have been allocated alternately between categories (cf. Madge, 1963:116 for another attempt at quantifying these sources of data).
7. When Mrs Dummer wrote to Zorbaugh complaining of the lack of counterbalance in his book to the 'vice and vapidity' he showed among her neighbours, he wrote back:

> I am frank to confess it portrays but one side of the life of the Near North Side. . . I did not attempt to portray the life of that fascinating area in its entirety. I attempted merely to show the types of social problems which arise. . . . *The Gold Coast and the Slum* is my last piece of muck-raking. (Dummer papers, 39:873, letters of 10 June 1929 and 1 July 1929)

8. One social worker saw social work practice as in some ways *more* scientific than that of sociology, though presumably he was not thinking specifically of Chicago sociology:

> How much of sociology can be said to have been subjected to the tests to which any hypothesis should be subjected before it can be said to be a law? In this respect the social worker's method . . . is superior to the sociologist's because. . . the methods which are found not to work are gradually discarded. . . . True that comparatively few generalisations have thus far been made in social work, but the material is there ready to be classified for the purpose of deducing such laws and principles as are indicated. This is the inductive process which is very largely the method of science. With sociology, the process is reversed. First we have some keen observer or analyst who observes and analyses human behavior as he sees it or reads about it. Next he formulates principles which he thinks govern such behavior, and he writes a book which he may call the 'Principles of Human Behavior'. . . . Some other sociologist . . . if he likes the promulgator of the principles and if the principles themselves happen to agree with his own pet notions, he promptly accepts them. . . . But in most instances the case method is not used in either the formulation, confirmation or rejection of the principles. (Karpf, 1925:422)

9. Diner (1980:141–4), writing about the period 1892–1919, rather cynically suggests that the work of sociologists was used only when it confirmed, or seemed likely to confirm, plans already made on other grounds by the interested parties. If that were so at the later period it would confirm the general line of argument here, though requiring some modification of the wording of the alternative account suggested.
10. Bulmer's important book (1984) is an exception to this, and is valuable for its emphases both on the cross-disciplinary context of the Department of Sociology within the University of Chicago and on the department's methodological eclecticism. However, it still gives an account which distinguishes sharply between sociological research and the work of the social activists, and treats the latter as an earlier stage in the development of the academic work on which the book focuses.
11. It is interesting to note how some chronologically prior work, whether inside or outside formal sociology, becomes defined as early or primitive, while other work is defined as that of forefathers; it is not always obvious that historically the latter has been a more important ancestor of lasting or recent developments.

Bibliography

Addams, Jane (1912) *A New Conscience and an Ancient Evil*. New York: Macmillan.

Addams, Jane (1930) *The Second Twenty Years at Hull-House*. New York: Macmillan.

Anderson, Nels (1923) *The Hobo*. Chicago: University of Chicago Press. Reprinted 1961.

Balch, Emily Greene (1910) *Our Slavic Fellow Citizens*. New York: Charities Publication Committee.

Bennett, James (1981) *Oral History and Delinquency*. Chicago: University of Chicago Press.

Booth, Charles (1892–7) *Life and Labour of the People in London*. London: Macmillan.

Bowen, Louise de Koven (1917) *The Public Dance Halls of Chicago*. Chicago: Juvenile Protective Association.

Bowen, Louise de Koven (1926) *Growing Up with a City*. New York: Macmillan.

Breckinridge, Sophonisba and Abbott, Edith (1912) *The Delinquent Child and the Home*. New York: Russell Sage Foundation.

Bulmer, Martin (1983) 'The Methodology of *The Taxi-Dance Hall*', *Urban Life* 12:95–120.

Bulmer, Martin (1984) *The Chicago School of Sociology*. Chicago: University of Chicago Press.

Burgess, Ernest W. and Bogue, Donald J. (1964) *Contributions to Urban Sociology*. Chicago: University of Chicago Press.

Carey, James T. (1975) *Sociology and Public Affairs*. Beverly Hills, CA: Sage.

Cavan, Ruth S. (1928) *Suicide*. Chicago: University of Chicago Press. Reprinted (1965) New York: Russell & Russell.

Connelly, Mark T. (1980) *The Response to Prostitution in the Progressive Era*. Chapel Hill: University of North Carolina Press.

Converse, Jean M. (1987) *Survey Research in the US: Roots and Emergence 1890–1960*. Berkeley: University of California Press.

Cressey, Paul G. (1932) *The Taxi-Dance Hall*. Chicago: University of Chicago Press.

Deegan, Mary Jo (1983) 'Sociology at Wellesley College: 1900–1919', *Journal of the History of Sociology* 5:91–115.

Deegan, Mary Jo (1986) 'The Clinical Sociology of Jessie Taft', *Clinical Sociology Review* 4:30–45.

Deegan, Mary Jo (1988) *Jane Addams and the Men of the Chicago School*. New Brunswick, NJ: Transaction.

Deegan, Mary Jo and Burger, Jon S. (1981) 'W. I. Thomas and Social Reform: his Work and Writings', *Journal of the History of the Behavioral Sciences* 17:114–25.
Diner, Steven J. (1975) 'Department and Discipline: The Department of Sociology at the University of Chicago, 1892–1920', *Minerva* 13:514–53.
Diner, Steven J. (1980) *A City and its Universities: Public Policy in Chicago 1892–1919*. Chapel Hill: University of North Carolina Press.
Dubin, Steven C. (1983) 'The Moral Continuum of Deviancy Research: Chicago Sociologists and the Dance Hall', *Urban Life* 12: 75–94.
Fish, Virginia K. (1985) 'Hull House: Pioneer in Urban Research during its Creative Years', *History of Sociology* 6:33–54.
Frazier, E. Franklin (1932) *The Negro Family in Chicago*. Chicago: University of Chicago Press.
Hammersley, Martyn and Atkinson, Paul (1983) *Ethnography: Principles in Practice*. London: Routledge.
Harvey, Lee (1987) *Myths of the Chicago School of Sociology*. Aldershot, Hants: Avebury, Gower.
Hauser, Philip M. (1982) personal interview.
Healy, William (1915) *The Individual Delinquent*. Boston, MA: Little, Brown.
Hull-House, residents of (1895) *Hull-House Maps and Papers*. New York: Crowell.
Hunter, Albert (1983) 'The Gold Coast and the Slum Revisited', *Urban Life* 11:461–76.
Karpf, Maurice J. (1925) 'The Relation between Sociology and Social Work', *Social Forces* 3:419–27.
Karpf, Maurice J. (1926) 'The Development of the Relationship between Sociology and Social Work', *Publications of the American Sociological Society* 21:213–22.
Kragh, H. (1987) *An Introduction to the Historiography of Science*. Cambridge: Cambridge University Press.
Leach, William (1981) *True Love and Perfect Union*. London: Routledge & Kegan Paul.
Lofland, Lyn H. (1980) 'Reminiscences of Classic Chicago', *Urban Life* 9:251–81.
Madge, John (1963) *The Origins of Scientific Sociology*. London: Tavistock.
Matthews, Fred H. (1977) *Quest for an American Sociology: Robert E. Park and the Chicago School*. Montreal: McGill-Queen's University Press.
Matza, David (1969) *Becoming Deviant*. Englewood Cliffs, NJ: Prentice-Hall.
Miner, Maude E. (1916) *Slavery of Prostitution: a plea for emancipation*. New York: Macmillan.
Mowrer, Ernest (1927) *Family Disorganisation*. Chicago: University of Chicago Press.
Palmer, Vivien (1928) *Field Studies in Sociology*. Chicago: University of Chicago Press.
Park, Robert E. (1922) *The Immigrant Press and its Control*. New York: Harper.
Park, Robert E. and Miller, H. A. (1921) *Old World Traits Transplanted*. New York: Harper. Reprinted (1969) New York: Arno Press.
Parker, Carleton H. (1920) *The Casual Laborer and Other Essays*. New York: Harcourt Brace & Howe.
Platt, Jennifer (1983) 'The Development of the "Participant Observation" Method in Sociology: Origin Myth and History', *Journal of the History of the Behavioral Sciences* 19:379–93.
Platt, Jennifer (1991a) 'Anglo-American Contacts in the Development of Research Methods before 1945', in Martin Bulmer, Kevin Bales and Kathryn Kish Sklar (eds) *The Social Survey in Historical Perspective*. Cambridge: Cambridge University Press, 340–58.
Platt, Jennifer (1992) 'Acting as a Switchboard: Mrs Ethel Sturges Dummer's Role in Sociology', *The American Sociologist* 23:23–36.
Platt, Jennifer (forthcoming) 'Research Methods and the Second Chicago School', in Gary A. Fine (ed.) *A Second Chicago School*? Publication pending.

Randall, Mercedes M. (1964) *Improper Bostonian: Emily Greene Balch*. New York: Twayne.

Reckless, Walter C. (1933) *Vice in Chicago*. Chicago: University of Chicago Press.

Rodgers, Daniel T. (1978) *The Work Ethic in Industrial America 1850–1920*. Chicago: University of Chicago Press.

Royden, A. Maude (1916) *Downward Paths*. London: Bell.

Shaw, Clifford R. et al. (1929) *Delinquency Areas*. Chicago: University of Chicago Press.

Sicherman, Barbara and Green, Carol H., eds (1980) *Notable American Women*. Cambridge, MA: Harvard University Press.

Sklar, Kathryn Kish (1985) 'Hull House in the 1890s: A Community of Women Reformers', *Signs* 10:658–77.

Smith, T. V. and White, Leonard D., eds (1929) *Chicago: An Experiment in Social Science Research*. Chicago: University of Chicago Press.

Snodgrass, Jon (1976) 'Clifford R. Shaw and Henry D. McKay: Chicago Criminologists', *British Journal of Criminology* 16:1–19.

Taylor, Graham (1936) *Chicago Commons Through Forty Years*. Chicago: University of Chicago Press.

Thomas, William I. (1923) *The Unadjusted Girl*. Boston, MA: Little, Brown. Reprinted (1969) Montclair, NJ: Patterson Smith.

Thomas, William I. and Znaniecki, Florian (1918–19) *The Polish Peasant in Europe & America*. Boston, MA: Badger. Reprinted (1958) New York: Dover.

Thrasher, Frederic M. (1927) *The Gang*. Chicago: University of Chicago Press. Abridged edn 1963.

Thrasher, Frederic M. (1928) 'How to Study the Boys' Gang in the Open', *Journal of Educational Sociology* 1:244–54.

Turner, Stephen (1992) 'The Pittsburgh Survey and its Intellectual Context', paper presented at American Sociological Association meetings.

van Waters, Miriam (1926) *Youth in Conflict*. London: Methuen.

Wade, Louise C. (1964) *Graham Taylor: Pioneer for Social Justice*. Chicago: University of Chicago Press.

Wilson, Howard E. (1928) *Mary McDowell Neighbor*. Chicago: University of Chicago Press.

Wirth, Louis, ed. (1940) *Eleven Twenty-Six*. Chicago: University of Chicago Press.

Young, Pauline M. (1939) *Scientific Social Surveys and Research*. New York: Prentice-Hall.

Zorbaugh, Harvey W. (1929) *The Gold Coast and the Slum*. Chicago: University of Chicago Press.

21

Mass-Observation: Social Research or Social Movement?[1]

Penny Summerfield

Historians of the recent past are turning in growing numbers to Mass-Observation as a source of information about social life between 1937 and 1948.[2] The volume, depth and diversity of this vast collection of material, deposited at Sussex University in 1970, provokes something like culture shock in researchers familiar with sources less rich in intimate details about people's lives, thoughts and feelings, but more tidily sorted and thoroughly catalogued, such as government social surveys, opinion polls and newspapers. Using the Mass-Observation archive presents its own set of problems,[3] but at least some of them cannot be resolved without an understanding of Mass-Observation itself. This paper is an attempt to explore possible interpretations of Mass-Observation as an organization in a specific historical and sociological context.

Mass-Observation could be understood, as most contemporary accounts of the development of sociology saw it, as an organization pioneering a particular type of social research which some, e.g. Bronislaw Malinowski, saw as a vital new departure in scientific research, and others, e.g. Mark Abrams, wrote off as misguided.[4] But it might be more appropriate to regard it as recent historians, like Tom Jeffery, have tended to see it, as a social movement with quasi-political objectives and an active and diverse following. Further to this, Mass-Observation as originally founded, came to an end in 1949. Its demise could be interpreted as the result of the research methods adopted, or of the stresses common to social movements, or as the peculiar

Source: *Journal of Contemporary History*, 20 (1985): 439–452.

product of trying to combine two different objectives: academically respectable research and the creation of social change.

The idea of the social research originated with two men, Tom Harrisson and Charles Madge. Tom Harrisson was a somewhat larger-than-life ornithologist-cum-anthropologist, who left his public school to go on an Arctic expedition and left Cambridge, before gaining a degree, to go on a tropical expedition. He spent the years 1931–36 in Central Borneo and the New Hebrides islands of the Western Pacific, in particular Malekula, and published his anthropological findings on the cannibals of the New Hebrides in a book called *Savage Civilisation* in 1937. On returning to Britain he was struck by the thought that the things he had been doing in the Pacific had not been done within these shores, and he moved to Bolton to study the Lancastrians in the same way he had studied the Malekulans. Simultaneously, Charles Madge, at the time a *Daily Mirror* reporter, published a letter in the *New Statesman* announcing the formation of a group of poets, painters and film-makers in Blackheath committed to social documentation. Harrisson and Madge joined forces in a project to develop what they called a 'science of ourselves' through the organization they named Mass-Observation.

Of course, social research was not entirely new. Before the first world war Booth, Rowntree and Bowley had undertaken research into conditions in various towns, and these surveys were repeated along with others in the context of the economic depression of the 1920s and 1930s.[5] But the objectives and methods of Mass-Observation were entirely different. Whereas the poverty surveys concentrated on carefully constructed statistical sampling of indices of social conditions such as housing, health, employment, and income, and paid relatively little attention to what people thought and felt, Mass-Observation's mission was to liberate 'facts' about what people did and said in order to 'add to the social consciousness of the time'. A major part of Madge and Harrisson's own motivation was that they thought that this social consciousness was being stifled or distorted by those with power in the 1930s, (particularly political, commercial, and media power), and that its release would lead to change.

A brief look at Mass-Observation's very first enterprise illustrates these themes and shows how they dictated the research methods which remained central to the organization. Mass-Observation began with a study of a single national event, the coronation of George VI on 12 May 1937, in the hope of discovering what (if anything) such an event meant in the lives of 'ordinary people', an objective springing from the belief of both Madge and Harrisson that the official interpretation of such occasions, reported in the media, was at odds with what people really thought and felt. The research methods used were to document the day from subjective accounts on the one hand, and observations on the other. The first came from a group of volunteers who were prepared to write down everything they did on 12 May, a method referred to as the 'day survey'. And the observations were recorded by twelve observers who moved around on the day, noting down what they saw and heard.[6]

These two methods became the twin pillars of Mass-Observation's research methodology. By 1940 the organization had a team of full-time observers working in Bolton to watch people, to record overheard comments, to engage in conversations which were recorded as indirect interviews and more rarely to interview directly. In all this, observers were not supposed to involve their own reactions. Mass-Observation also had a panel of volunteers, to whom it sent monthly 'directives' consisting of a list of questions to which subjective answers were expected. What was wanted from them were detailed, candid, personal reactions: as Harrisson put it, 'material which could not be obtained by interviewing or by any contact made between the investigator and strangers'. They were regarded as informants, in the anthropological sense.[7] The number of voluntary panellists has been exaggerated, not least by the organization itself, which repeatedly claimed to have a panel of 2,000. Nick Stanley, who has subjected the figures to rigorous analysis, throws some light on the claim. During the period 1937 to 1945, a total of 2,847 individuals replied to at least one directive, but less than half this number, 1,095, replied to more than one.[8] In addition, 200 of these volunteers kept diaries, originally just on the twelfth of the month, following the coronation initiative, but during the war diarists were asked to send in full monthly diaries. As in the case of replies to the directives, commitment varied and few diarists submitted instalments every month, but during the six years of war, Mass-Observation had a total of 500 diarists on its books.[9]

Who was prepared to undertake this sort of self-observation on a purely voluntary basis and why? Samuel Hynes in his book *The Auden Generation* describes them as young members of the provincial lower-middle-class, 'the lonely bored livers of unexciting lives'. Closer analysis by Tom Jeffery confirms that a sizeable minority was indeed lower-middle-class but Jeffery argues that the idea that their lives were empty is a misrepresentation.[10] Many had won scholarships from elementary to secondary school, but few had gone to university. Typical occupations of men and women were clerk and schoolteacher. Many male panellists in 1939 were shopkeepers, journalists, scientists or students, though a rising number joined the forces during the war. Throughout, a substantial minority of men, though not women, were manual workers in manufacturing industry. The vast majority of women classified themselves as housewives, though of course they may have been in paid employment prior to marriage.[11] Many experienced unemployment between the wars. Most read widely and were keen to go beyond their limited schooling.

This educative urge was an important part of their motivation in joining Mass-Observation. Few belonged to any political party, yet many were 'left of centre'. The desire for self-education was coupled to a desire for a better understanding of current events. This is reflected in the fact that many Mass-Observation members were also members of the Left Book Club. This was a publishing enterprise started by Gollancz in 1936 as an outlet for broadly left-wing publications which most booksellers were at the time refusing to take. The co-membership of the Left Book Club and Mass-Observation is indicative of the socio-political motivation of many of Mass-Observation's participants. The Left Book Club's objective was to inform people of the rise of fascism and its horrific consequences, to introduce them to the socialist experiments in Russia and China, and to draw attention to the social conditions at home, notably mass unemployment and poverty, which the government, currently appeasing the dictators, was apparently content to ignore.[12] Jeffery writes that the majority of Mass-Observation's panellists 'volunteered to work for Mass-Observation because they wanted to be of some use in the fight against fascism and against official neglect of ordinary people'.[13] They could clearly identify with the key ideas of Mass-Observation's introductory pamphlet of February 1937, which stressed the need to investigate popular fears of the coming war, and the way that the press manipulated ordinary people. The invitation to send in their most intimate thoughts and feelings and what they knew of those of others around them must have given them a sense of being listened to which was itself a counter to their powerlessness. This may have been particularly important for women, who in spite of having won the vote, were still largely deprived of a 'voice' in traditional party politics and trade union organization between the wars. In addition, gossip as well as news, home life as well as work, sex as well as politics, were considered important, an acknowledgement that 'the personal is political'. As many women as men kept Mass-Observation diaries, and possibly rather more women kept full diaries on a sustained basis.

This is not to say that *all* Mass-Observation's members were lower-middle-class autodidacts. There were middle-class members, particularly among the full-time observers, some of whom were university friends or acquaintances of Madge and Harrisson. As Jeffery says, members of this group, who included Humphrey Jennings, Humphrey Spender, John Sommerfield, Celia Fremlin and Woodrow Wyatt, stand out by their wider-ranging experience, in terms of friendship networks, university education, foreign travel, etc. In addition there were working-class members, mostly from the artisan stratum, for whom the key difference from the lower-middle-class was that their hours of work were longer, so they had less time to devote to Mass-Observation. Whichever group one looked at, however, the numbers of women with small children were very small, presumably because the time constraints of child care are even greater than those of manual work. Women Mass-Observers tended to be either single or older married women with grown-up children.

Mass-Observation's leaders referred to their recruits as 'men and women of goodwill', by which was meant both goodwill towards Mass-Observation, which they were supplying with data free of charge, and goodwill towards society in general – for surely a science of ourselves must benefit us all?

The way in which Mass-Observation defined its social research objectives determined the methods it adopted, and these involved the participation of a loose organization of an unlimited number of members whose voluntary commitment depended upon a sense of shared objectives. Even though Mass-Observation was definitely never a mass movement, it certainly bears resemblance to a social movement, for all that it was small. To put it in perspective, the most generous aggregation can credit Mass-Observation with no more than 2,847 individuals on its panel of volunteers, compared with a high point of 57,000 Left Book Club members.[14] (Many would query whether the Left Book Club itself should be seen as a 'movement'.) However, Mass-Observation's influence extended beyond its panellists, since an essential concomitant of collecting information was its dissemination for the purpose of augmenting 'the social consciousness of the time'.

The organization produced over twenty books between 1937 and 1950, most of which were written rapidly, with the overt intention of returning information to the 'masses'. Their style was not dissimilar from that of documentary film: they tended to present a series of 'clips', each consisting of a word picture, some faithfully recorded dialogue, and a short commentary written in a direct, journalistic style on various aspects of the topic in hand, be it Bolton pubs, the phoney war, wartime production, post-war prospects, or religious belief.[15] Further research would be necessary to discover the circulation, let alone the impact, of these books. However, the fact that Mass-Observation's main publishers were Gollancz (of Left Book Club fame), Penguin Special (trying to do something similar to the Left Book Club) and John Murray (another 'progressive' publisher) indicates that circulation was likely to be within the same social milieu as that from which Mass-Observation's active participants were drawn.

In addition to publishing books, Harrisson and other full-timers, notably H. D. Willcock, produced numerous newspaper and journal articles, and made frequent broadcasts. It should be noted that their intention was to communicate the findings of Mass-Observation to a much wider audience than an academic one. Mass-Observation's relationship with academia was rather like that with the press. It had friends in both camps (e.g. Bronislaw Malinowski, Professor of Anthropology at the University of London, and Tom Driberg of the *Daily Express*). But in general it saw both the press and academia as rivals, and as detractors of Mass-Observation's projects. This led to long-term mutual antagonism, for all that Tom Harrisson was eventually appointed to a personal chair at Sussex University on the occasion of depositing the Mass-Observation archive there.

We shall now consider three problems encountered by Mass-Observation between 1937 and 1948: how to continue its existence in wartime, its relationship with government, and its relationship with business, and shall argue that because of the particularly close integration between the social research and the movement upon which it depended, these problems eventually destroyed the organization as originally conceived.

The war presented Mass-Observation with problems of attrition, as its members were gradually called up, and their freedom to observe and report was hampered by censorship. The war put an end to Harrisson's Bolton project (though some of the data collected was published as *The Pub and the People* in 1943), and Harrisson himself was called up in 1942. Within two years he was airlifted into Borneo, to remain in the Far East for twenty-five years, apart from brief home visits. But in some ways Mass-Observation did not suffer as much as other allied groupings, like the Left Book Club, from the advent of war. The Left Book Club's membership fell steadily from 1939, whereas record numbers wrote replies to Mass-Observation directives in 1939 and after a fall from this peak in 1940, numbers climbed again. 1942 and 1943 were Mass-Observation's best years in terms of the frequency of responses from its panellists.[16] Mass-Observation may have benefited from its concentration on Britain, which freed it from association with a commitment to Russia as a model of anti-fascism, an association which caused deep rifts in many left-wing organizations, including the Left Book Club, on the occasion of the Nazi-Soviet Anti-Aggression Pact of August 1939, which looked rather like socialist appeasement of fascism and sowed seeds of doubt and disillusion. The war had been anticipated by Mass-Observation and it brought a new challenge: it was now more urgent than ever to defeat fascism, and Mass-Observation directed its energies in wartime towards producing efficiency in government and industry. Mass-Observation's interpretation of 'efficiency' was securing the maximum effort from the people through willing co-operation. Of course, leaders in government and industry would, by its analysis, fail to do so if they ignored what people really thought, felt and wanted. It maintained a critique of these leaders for living at a distance from those they were supposed to lead, and redoubled its energies to collect information which would connect them with social reality.

Many of Mass-Observation's members moved on account of the war into new contexts, such as the forces, war industry and voluntary organizations like the Home Guard and the Women's Voluntary Service, which brought them into contact with new social groups, especially of working-class people. Even though censorship as well as the rigours of service life affected what service-men and women could send in to Mass-Observation, some continued to correspond with the organization, which was able to amass collection which affords considerable insights into wartime forces' life. Equally, the direction of people, especially women, into types of industrial work that were new to them, gave fresh scope for participant observation. Mass-Observers were quick to pick up the kinds of matters which would interest their organization. For example, a woman welding trainer sent in all the letters her trainees had written to her during 1942, and one of the diarists gave long verbatim reports of the proceedings of a Women's Parliament held in January 1942.[17] Mass-Observation published its wartime findings rapidly. For example, *War Begins at Home*, a study of such things as government propaganda, class divisions within evacuation and fears of air attack, came out early in 1940, and *People in Production*, researched between October 1941 and March 1942, was written and published before the end of that year. It examined aspects of production such as divisions in the work-place, poor organization of processes and people, and neglect of the interrelation of home and work, especially as far as women were concerned. Mass-Observation's message, expressed in such publications, was that it was no good relying on national unity and patriotic resolution: these things could only be achieved by genuinely bringing the objectives of the war and the methods of waging it into line with the aspirations and preferences of 'the masses'. In this Mass-Observation was not alone, though the milieu of which it was a part was rather different from the pre-war one. A. C. H. Smith in *Paper Voices* has depicted a wartime movement embracing George Orwell, J. B. Priestley, the Commonwealth Party and the *Daily Mirror*, to which could be added the Army Bureau of Current Affairs and army education more generally, as waging a 'war to win the war'.[18] Mass-Observation belonged in this context.

However, the very success with which Mass-Observation brought information about ordinary people to the attention of the nation's leaders during the war, caused tensions within Mass-Observation's leadership and changed its relationship with its followers.

The occasion for an irredeemable split between the two founding fathers, Madge and Harrisson, was the employment of Mass-Observation by the Ministry of Information in the spring of 1940. The 'Home Intelligence' section of the MOI was run by an old friend of Harrisson's, Mary Adams, and she commissioned Mass-Observation to send in regular 'morale reports'. With Arnold Plant she also set up an official investigative body, the Wartime Social Survey, which eventually became a rival to Mass-Observation. Madge did not approve of Mass-Observation's link with officialdom, and after six months of quarrelling left the organization. The split is sometimes seen in terms of personality. Jeffery writes of Tom Harrisson: 'You either got on with him or you gave up. Madge gave up.'[19] Under J. M. Keynes' patronage, Madge did a socio-economic survey of wartime patterns of saving and spending in Bolton, and later became Professor of Sociology at Birmingham University. But Madge's objections were important. Mass-Observation's intention had been to stand as an independent critic of government, rather than as an organ of propaganda in itself. How could it work against manipulation of the masses if it was co-opted by the Ministry officially engaged in manipulation?

Harrisson clung to the idea that by refusing to sign the Official Secrets Act he maintained the independence of Mass-Observation. But it is clear that he increasingly saw his work for the state as a means to the end of changing the government's orientation – one of the objectives, after all, of Mass-Observation as a social movement.

The impact of the Morale Reports which recorded reactions to the news, is itself indicative of the unwillingness of many members of government to accept evidence of what people were thinking when it came from sources other than those they were used to, i.e. the press, the police, and social worthies like JPs, clergymen and doctors. The Labour members of the wartime coalition, e.g. Bevin and Morrison, were among the worst in this respect, and a parliamentary row blew up in early summer 1940, in which Mass-Observation personnel were slated as 'Cooper's Snoopers', after the Minister of Information Duff Cooper. There was further tension over Mass-Observation's reports on the effects of the blitz of autumn 1940–41. Mass-Observation started by producing reports for the Admiralty on morale in bombed ports, but was soon covering all the major blitzed towns. In November 1940, it criticized inadequate government emergency provisions for the distressed inhabitants of Coventry, and misleading press statements to the effect that Coventry people were crying out for merciless reprisals on German cities. In contrast, it suggested, Coventry people were in a state of shock; they lacked both welfare and information and further bombing would have produced panic. These reports, hotly contested as 'unreliable' in government, did lead to better provisions for Coventry, in a sense vindicating Harrisson. But they also had more sinister consequences at the hands of the Admiralty and Service chiefs, who concluded from them that the saturation bombing of German civilian populations would be the shortest way to victory. In a sense this vindicates Madge's doubts.[20]

The wartime relationship with government was seen by Harrisson as consistent with Mass-Observation's objectives. The same was true of Mass-Observation's relationship with business concerns at this time. Mass-Observation undertook at least one work-place study on behalf of its management. The works manager at one of the factories of an industrial enterprise, Ekco Ltd., which was producing radar equipment, was concerned about what he saw as slackness among his workers, principally women. Mass-Observation sent a full-time observer incognito to discover, through participant observation, the causes of the problem. This study, written up as *War Factory* and published in 1943, remains a superb picture of the responses of a group of women, mainly unaccustomed to industrial work, to the processes and work-place relationships of an engineering factory, and the prejudices with which they were confronted by men and management. With the usual combination of autocracy and a genuine democratic urge for all to learn from such findings, Harrisson got Gollancz to publish the book and disseminated it amongst the Ekco work-force, rather to the surprise of the management.[21] But in spite of Tom Harrisson's *post hoc* openness, the spirit behind the research was manipulative. The company was the paymaster, the research into its work-force was conducted in secrecy on its behalf, and the conclusions were intended to suggest how it could extract higher productivity from its women workers.

The question of how the 'science of ourselves' was to be used, depended to a large extent on who was paying for it. The Advertising Service Guild, an association of eight advertising agents, funded an increasing number of Mass-Observation studies during and after the war. It should not be imagined that this meant that Mass-Observation immediately became narrowly focused on market research, since the Advertising Service Guild's wartime aim was to 'assist the war effort' by promoting 'independent surveys into fundamental problems of war activity and organization'. Under this heading came Mass-Observation's clothes rationing, propaganda, savings and homes post-war plans and a large-scale investigation into war production generally.[22] After the war, however, Mass-Observation did increasingly conduct narrowly focused surveys for commercial companies with products to sell, such as its surveys on washing habits, on the domestic use of paint and on the public taste for cosmetics, custard powder, baked beans and frozen fish.[23]

In these projects Mass-Observation used the sources described earlier – on the one hand, the observations of full-time, largely incognito observers, and on the other hand, the replies to its 'directives' and the diaries of its panel of volunteers. Many of the projects, even *People in Production*, which was almost evangelical in its advocacy of 'human factor management' as the solution to wartime inefficiency, could be seen as consistent with the original aim of Mass-Observation, which was to bring to the attention of national leaders the social reality experienced by ordinary people, for the purpose of changing the balance of power within society. Yet it is also possible to see in this material a redefinition of this objective, along the lines that Mass-Observation's information would simply offer guidance on how to wield power more effectively, by telling leaders how to lead, managers how to manage and salesmen how to sell.

Possibly the original objective of the movement was so vague that it was inevitable that its leaders would eventually produce some sort of sharper working definition. After all, the original objective amounted to no more than the notion that 'information is power'. There was no analysis of power relations – between men and women, managers and workers, press and readers, government and people. In fact Tom Harrisson, a committed empiricist, shied away from theory which could have led to such analysis.[24] But since he was determined to use Mass-Observation's information to produce change, he ended up devoted to changing the style with which power was wielded, rather than the *structure* of relationships which maintained that power.

It could be argued that the distinction we are making is too fine, and that too much is being read into the changes which occurred in the course of Mass-Observation's history. The organization inevitably had to attract funds, and the objectives of the social research and the movement on which it depended necessarily had to evolve. After all, Mass-Observation's information did lead to some improvements in the way people were both governed and managed during the war, and marketing in itself is not immoral! But we shall conclude by arguing that these changes were of fundamental importance to Mass-Observation, both as social research, and as a social movement.

After the war, the social research methods undertaken by Mass-Observation were pushed increasingly in the direction of statistical sampling. This was popular with market research because it rendered simplistic results rapidly. It was also favoured in the academic community, by economists, political scientists and especially sociologists, because it was seen as more scientific than Mass-Observation's anthropological technique of using informants. There was what amounted to an onslaught against Mass-Observation's methods after the publication of *People in Production* in 1942. Before the war, rude but memorable comments had been made such as: 'Facts simply multiply like maggots in a cheese and leave no shape' and references to Mass-Observers as 'busybodies of the left . . . scientifically they are about as valuable as the chimpanzees' tea party at the zoo'. During and after the war numerous academic criticisms appeared in serious academic journals, culminating in Mark Abrams' hostile critique in *Social Surveys and Social Action*, published in 1951.[25] The attack concentrated both on the absence of a clear framework within which data was collected, and the use of indirect interviews and informants. The structured questionnaire and the representative statistical sample were the favoured methods.

Under pressure from these criticisms, and in order to retain credibility with clients, Mass-Observation introduced more statistics into its publications, even though its published statistics were in many cases useless, because the size and composition of the respondent group from which percentages were calculated were not given. Of course this led to further academic attacks. H. D. Willcock eventually gave up and went to the Government Social Survey, originally founded as the Wartime Social Survey by Harrisson's friend Mary Adams, to do statistical surveys. However recent research, including the author's and that of Di Parkin, has found a strong consistency between Mass-Observation's findings and those of sample-based questionnaire research, such as that done by the Wartime Social Survey. The main divergence is that Mass-Observation's data was so much deeper that it was able to give a fuller picture. For example, Mass-Observation discovered not just that a sizeable minority of women were unwilling to undertake war work, a fact which the Wartime Social Survey reported too, but also that under certain conditions women were prepared to enter war work and that these conditions were, specifically, collective provisions for domestic work, reduction of hours, offers of better pay and more interesting industrial work. Similarly, Mass-Observation's findings on attitudes to Churchill supported those of The British Institute of Public Opinion to the effect that he was popular, but enabled Mass-Observation to add that in spite of his personal popularity Labour would win the next General Election.[26] However, Mass-Observation did not have the courage of its convictions after the war, and endeavoured, damagingly, to appear respectably quantitative.

As far as Mass-Observation as a social movement is concerned, more research needs to be done on the fortunes of the organization after 1945. As far as we know, there is not yet a statistical analysis of recruitment to and composition of the panel in the years 1945–49, of the calibre of Stanley's work on the earlier period, which indicates a decline in the numbers of panellists replying to directives in 1944 and 1945. A superficial perusal of directive replies and diary instalments suggests a continuing shrinkage thereafter. As to causes, one possible explanation is that the decline was the result of the betrayal of trust in selling the intimate confessions of voluntary informants to cosmetics manufacturers. A further hypothesis is that there may have been a feeling among those running the organization themselves as well as outside, that the old objective had been realized by the Labour victory in the General Election of 1945. Now the aspirations of ordinary people would be fulfilled, so no further effort was necessary. Obviously a more thoroughgoing analysis of power would have pointed towards the need for continued effort to impress upon the Labour government what 'people of good will' were keen to see achieved through post-war reconstruction. As it was, Tate and Lyle, the British Medical Association, and other particular interest groups had the field to themselves after the war, claiming, for example, that no one wanted nationalization, or a wholly public as opposed to a partly private health service.[27]

Gradually, during the years after the war, the organization moved further towards market research and away from a 'science of ourselves' aimed at the creation of social change. In 1949 it was superceded by Mass-Observation Ltd., run by two ex-observers, Leonard England and Mollie Tarrant. This was a market research organization, using principally the 'representative sample' and the questionnaire, rather than the subjective report of the voluntary informant. Tom Harrisson exchanged his rights in the organization for control of all the material collected by Mass-Observation prior to 1949. He commented on this material that over 90 per cent of it was never used in the compilation of reports and publications. It now sits in boxes in the Sussex University Library, an amazingly rich source whose depths historians are barely beginning to plumb.

Notes

1. This paper is intended as a discussion piece, and does not pretend to be based on in-depth research on Mass-Observation, for which the reader should see Tom Jeffery, 'Mass-Observation – A Short History' (Centre for Contemporary Cultural Studies Occasional Paper, Birmingham 1978); N. S. Stanley, '"The Extra Dimension": a study and assessment of the methods employed by Mass-Observation in its first period 1937–40', (CNAA PhD 1981); Angus Calder, *The Mass-Observers* (Cape, forthcoming).

2. As a member of this group, I have used the Mass-Observation Archive as a source of information on education in the Armed Forces and, more extensively, the wartime employment of women; see P. Summerfield, 'Education and Politics in the British Armed Forces in the Second World War', *International Review of Social History*, XXVI (1981), Part 2; idem, 'Women Workers in the Second World War' (University of Sussex DPhil 1982); idem, 'Women, Work and Welfare: a study of child care and shopping in Britain in the Second World War', *Journal of Social History*, 17, 2 (Winter 1983); idem, *Women Workers in the Second World War, Production and Patriarchy in Conflict* (London 1984).
3. Some of these were discussed by myself and others at a session entitled 'Using the Mass-Observation Archive' at the British Association for the Advancement of Science Conference held at Sussex University, August 1983.
4. Essay by Bronislaw Malinowski in Mass-Observation, *First Year's Work 1937–38* (London 1938); Mark Abrams, *Social Surveys and Social Action* (London 1951).
5. See *inter alia* Charles Booth, *Life and Labour of the People of London* (London 1892–1900); B. S. Rowntree, *Poverty: A study of town life* (London 1901); A. L. Bowley, *Has Poverty Diminished?* (London 1905); Sir H. L. Smith, *The New Survey of London Life and Labour* (London 1930–35); B. S. Rowntree, *Poverty and Progress* (London 1941); John Hilton, *Rich Man, Poor Man* (London 1944); Pilgrim Trust, *Men without Work* (Cambridge 1938).
6. Charles Madge and Humphrey Jennings (eds), *May 12th* (London 1937).
7. Mass-Observation, *War Begins at Home* (London 1940), 16–22.
8. Stanley, op. cit., 155–156.
9. D. Sheridan, *The Mass-Observation Records 1937–1949, A Guide for Researchers* (Mass-Observation Archive, University of Sussex 1982), 15–16; idem, 'Mass Observing the British', *History Today*, 34 (July 1984), 45.
10. S. Hynes, *The Auden Generation* (London 1976), 282; Jeffery (1978), 29.
11. Stanley, op. cit., Table 7 opposite p. 167.
12. Stuart Samuels, 'The Left Book Club', *Journal of Contemporary History*, 1, 2 (April 1966); John Lewis, *The Left Book Club: An Historical Record* (London 1970).
13. Jeffery, op. cit., 29.
14. Samuels, op. cit., 71.
15. These topics were dealt with in the following Mass-Observation publications: *The Pub and the People* (London 1943); *War Begins at Home* (London 1940); *People in Production* (London 1942); *The Journey Home* (London 1944); *Puzzled People* (London 1947).
16. Stanley, op. cit., 155 and Table 2, p. 158.
17. Mass-Observation Archive, TC32, 'Women in Wartime', Box 3, File F; Diary No. 5390.
18. A. C. H. Smith with E. Immirzi and T. Blackwell, *Paper Voices, The Popular Press and Social Change 1935–1965* (London 1975).
19. Jeffery, op. cit., 40.
20. Tom Harrisson, *Living through the Blitz* (London 1976), chapter x.
21. M. Lipman, *The War, a Memoir* (unpublished ms. 1975). Lipman was works manager at Ekco, Malmesbury, and describes how the War Factory project was set up and carried out. Mass-Observation, *War Factory* (London 1943).
22. Mass-Observation, *People in Production* (London 1942), iii, foreword by Advertising Service Guild. See also The Advertising Service Guild's Bulletins, entitled 'Change', in which Mass-Observation reported on the other topics mentioned here.
23. Mass-Observation Archive, File Report 2315, Survey of Laundry Usage, January 1946; File Report 2546, The Application of Face Cream, 1947; File Report 2547, Custard Powder, 1947; File Report 3143, Washing Habits, 1949; Mass-Observation, *People and Paint* (ICI Publications, 1949); Mass-Observation Bulletin No. 43, 'Shopping

at Department Stores', Jan./Feb. 1952; File Report 3646, Baked Beans, 1957; File Report 3652, Frozen Fish, 1957; File Report 3888, Instant Coffee, 1959.

24. See, for example, Mass-Observation, *War Begins at Home* (London 1940), 2–3, and Tom Harrisson, *Britain Revisited* (London 1961), 19.
25. Harrisson quotes examples in *Britain Revisited*, 16–17. See also Margaret Bunn, 'Mass-Observation: A comment on People in Production', *Manchester School of Economic and Social Studies*, vol. xiii, 1944; Abrams, op. cit.
26. P. Summerfield, *Women Workers in the Second World War*, (London 1984), 40–41; Di Parkin, 'Using Mass-Observation', paper presented to BAAS conference, August 1983.
27. P. Addison, *The Road to 1945, British Politics and the Second World War* (London 1975), 273; R. Miliband, *Parliamentary Socialism* (London 1972), 301–302 especially footnote 3, p. 301, on Tate and Lyle's anti-nationalization propaganda.

OBSERVATION METHODS

SAGE BENCHMARKS IN SOCIAL RESEARCH METHODS

OBSERVATION METHODS

VOLUME IV

Edited by

Barry Smart, Kay Peggs and Joseph Burridge

SAGE

Los Angeles | London | New Delhi
Singapore | Washington DC

SAGE

Los Angeles | London | New Delhi
Singapore | Washington DC

SAGE Publications Ltd
1 Oliver's Yard
55 City Road
London EC1Y 1SP

SAGE Publications Inc.
2455 Teller Road
Thousand Oaks, California 91320

SAGE Publications India Pvt Ltd
B 1/I 1, Mohan Cooperative Industrial Area
Mathura Road
New Delhi 110 044

SAGE Publications Asia-Pacific Pte Ltd
3 Church Street
#10-04 Samsung Hub
Singapore 049483

Typeset by Arete Publishing Private Limited, Delhi

Printed on paper from sustainable resources

Printed by MPG Printgroup, UK

MIX
Paper from responsible sources
FSC® C018575

First published 2013

Library of Congress Control Number: 2012942348

British Library Cataloguing in Publication Data

A catalogue record for this book is available from the British Library

ISBN: 978-1-4462-0811-3 (set of four volumes)

Contents

Volume IV

'What if . . . [we] were to study the colonizers rather than the colonized, the culture of power rather than the powerless, the culture of affluence rather than the culture of poverty?

Studying "up" as well as "down" would lead us to ask many "common sense" questions in reverse. Instead of asking why some people are poor, we would ask why other people are so affluent? How on earth would a social scientist explain the hoarding patterns of the American rich and middle class? How can we explain the fantastic resistance to change among those whose options appear to be so many?'

Nader, L. (1972). 'Up the Anthropologist – Perspectives Gained from Studying Up', in D. Hymes (Ed.) *Reinventing Anthropology*. New York: Pantheon Books: 289.

G. Observing Workplaces and Workers

66

Social Access in the Workplace: Are Ethnographers Gossips?

Simon Carmel

Introduction

This article investigates what it means to gain and retain 'social access' in workplace ethnography, in particular examining the role of gossip. Several difficulties encountered in attempting to gain such access within a health care setting are analysed. The analysis reveals the importance of participation in workplace gossip and the length of time spent in the field as facilitators of social access. The research on which this article is based involved periods of non-participant observation in intensive care units, studying these well-defined organizational units as places of work. Three intensive care units were studied, each for a period of between three weeks and four months. The data presented in this article, relating to social access, were collected during periods of observation and include conversational interviews in the field. Ethnographic notes were generally recorded openly, contemporaneously and at a low level of inference. These notes were reviewed daily and later transcribed for data analysis.

'Access' in Ethnography

The term 'access' is to be preferred over 'entry' or 'rapport', since access 'focuses attention on the social scientific goal of ethnography: access to information' (Harrington, 2003: 599). While 'gaining access' is widely acknowledged

Source: *Work, Employment & Society*, 25(3) (2011): 551–560.

as vital to ethnography, depending on the school of ethnography espoused, analytical separation of 'access' from other ethnographic data could be seen as contrary to the spirit of the enterprise. For example, in much anthropology the primary aim is cultural interpretation (Fetterman, 1989: 28). In this case the very processes of gaining access are likely to demonstrate aspects of the culture being studied, and analysis of access processes is rightly included as part of the analysis of the culture itself. Indeed, where organizational research is avowedly anthropological (i.e. researching 'organisational culture'), the access process is regarded as a potentially 'rich source of data' (Schwartzman, 1993: 48).

On the other hand, in organizational ethnography more generally, gaining access tends to be portrayed as a particular *phase* of a research project which is assumed to be historically prior to – and analytically distinct from – the phase of collecting data for the substantive matter under investigation. Burgess (1984: 45), for example, refers to gaining access as 'an essential phase . . . a prerequisite; a precondition'. This phase has been given a fair degree of attention in organizational research – in part because it has proved so difficult for many organizational researchers to actually *get in* to organizations (Bryman, 1989: 161; Smith, 2001).

In contrast to the phase of 'getting in' to an organization, which we may term physical access, social access ('getting along') is recognized as an *on-going process*:

> The problem of obtaining access to the data . . . is often at its most acute in initial negotiations to enter a setting . . . the problem persists, to one degree or another, throughout the data collection process. (Hammersley and Atkinson, 1995: 54)

To gain and retain social access entails the active creation and maintenance of personal relationships. There is a need, therefore, to recognize what Coffey (1999: 57) calls the 'interpersonal and emotional craft work' inherent in ethnography. This emotional labour can be intense, since for ethnographers friendly relationships in the field are necessary for our work; we may well not have chosen the particular relationships in the field apart from the work that we do. It is, to be blunt, in the researcher's interest to develop and maintain friendly relations, even where the researcher may not actually *like* field-work participants. Although we are 'interested in people' and their social relations, the interest we have is as social scientists: fundamentally, we are interested in people only for the data they can provide.

Such problematic characteristics of inquiry have been identified in feminist critiques of social research, especially in the conduct of interviews (Oakley, 1981). Duncombe and Jessop (2002), for example, describe their ambivalence about establishing 'rapport' in qualitative interviews – that this seems to be manipulative and a false form of friendship. However, it is worth noting that the relevant issues, for example, asymmetry in power relationships between researcher and participants, are slightly different in ethnographic inquiry to those in qualitative interviews. Most notably, the length of time researchers spend to get to know participants is generally much longer in ethnographic research, and this has practical, ethical and political implications. For example, we can note that while the reasons ethnographers form a relationship can be construed as cynical, we can only speculate on participants' reasons (Coffey, 1999: 42). In fact, a participant's excessive interest in one's research can even be restrictive to ethnographic inquiry – an example of which is analysed below.

The issue of social access also brings out the importance of flexibility in ethnography:

> Gaining access is a political rather than a normative problem, so that 'playing it by ear' is an inevitable component of the negotiations, with the researcher frequently having to make instant decisions and think on his or her feet. (Hornsby-Smith, 1993: 54–5)

The term 'political' here, in contradistinction to 'normative', emphasizes that there are few *general* rules, norms, standards or models to follow. The flexibility of ethnography as a research method applies not only to discovering and developing themes of analytical interest, but also applies to dealing with the inevitable political issues and the improvisational skills consequently required (Humphreys et al., 2003).

A general comment to be made about organizational ethnography is that we are researching fractured fields. That is, where the site of inquiry is a workplace, there are likely to be particularly pressing points of conflict arising out of power relations and occupational competition; the fractured character of the field is a more salient point in ethnography of work than in other ethnographic inquiries. For while any ethnographic inquiry will be able to identify subcultures and even insurgent groups, in a workplace there is, at least notionally, an idea that all workers should be contributing towards super-ordinate organizational goals. That said, perhaps this notion is more pertinent in health care organizations. For the intensive care units where I conducted my research, it was clear that in general all workers wanted to contribute to the overall aim of treatment, care and appropriate discharge of severely ill patients, despite apparently divergent occupational perspectives (Carmel, 2006a, 2006b). Nevertheless, there is a lack of voluntarism in most workplaces and this might be considered an important factor in precipitating interpersonal tensions – an example of which is gossip.

Gossip

Early social scientific accounts of gossip identified gossip's social functions. For example, Gluckman (1963) identified gossip as a hallmark of group membership. Paine (1967) and Handelman (1973) saw gossip rather more individualistically as information management, particularly related to self-interest. More recent social scientific inquiry has undoubtedly been useful in developing properly empirical accounts (Tebbutt and Marchington, 1997) but sometimes questions the traditional negative moral connotations by, for example, emphasizing the 'important personal and social functions' (Goodman, 1994: 1) that gossip fulfils. However, since gossip can be defined as the 'process of passing value-laden information' (Noon and Delbridge, 1993), the moral connotations of gossip cannot be ignored. In principle gossip might be benevolent, but as Taylor (1994: 39) avers, 'as a matter of fact, instances of gossip are more likely to be malicious than benevolent'. Gossip can be seen as an exercise in power, since it often aims to denigrate (Collins, 1994: 108), for example as a process of 'othering' (Davies, 2003). It can also be seen as 'idle talk', in that it requires 'no abstract thought or the working out of political, ideological or moral views' (Taylor, 1994: 34). Overall despite some articulation of a contrary view, 'gossip' is regarded in the literature as morally problematic, even if we allow that on some occasions it may be positive in intent.

Two aspects of gossip, as noted by Taylor (1994), are particularly pertinent to the topic of this research note. Most relevant for ethnographic inquiry is that 'those who gossip, critically or not, are curious about and take an interest in the activities and relationships of other human beings' (Taylor, 1994: 42). Thus gossip shares with ethnography at least one positive attribute (unless curiosity itself is deemed morally problematic, which would write off almost all academic inquiry). In addition, Taylor comments that it is the attitude and level of detachment of the gossiper which is most important in determining the morality of a particular instance of gossip. Such an observation might have the potential to alleviate any moral qualms a gossiping ethnographer might face.

While ethnographers have no claim to be especially honest or candid (Fine, 1993), we do need to acknowledge the moral, ethical and political issues we face (Punch, 1986). To this end, four situations related to gaining social access are now presented. In the first example below, social access seemed straightforward; it was only on reflection that I realized that this 'successful' social access had actually only been with one particular group in the workplace, to the possible exclusion of others.

Straightforward Access?

In this case formal 'gatekeeper' approval and physical access to the organization had been granted by a senior manager who was known to me and had indicated his willingness to participate in the study. When I turned up on the first day I was introduced by the senior manager to his staff. Later, I had plenty of opportunity to assure his staff that I was not undertaking a formal evaluation or assessment.

Early on in the observational period, I joined in with a junior doctor undertaking work at very unsocial hours. I 'shadowed' David,[1] who appeared to be on friendly terms with many members of staff. I suspect that by this I gained credibility – or at least, signalled that I was interested in the whole work of the organization rather than just senior staff or just work during office hours. I felt happy when I arrived at the workplace on my following visit:

> I arrived at 8.45am. At the entrance I announced myself as 'Simon Carmel, here to do a research study.' The reply was 'Come on in.' This was Ellie – she later told me that she recognised me on the monitor. When I went through the door I said 'Hello' straight away to those I could see waiting; I took off my coat and said hello to the others who were there . . . I had a feeling of being welcome – smiles, eye contact etc. I recognised most of the faces from earlier visits.

That I *was* welcome, and not just *felt* welcome, is supported by the fact that later that morning I was invited to join two junior doctors on their coffee break. I was sure that David would have told these two that I had accompanied him throughout his unsocial working hours. This then is a positive aspect of gossip – discussion about me, behind my back, but which placed me in a favourable light.

There was, though, a deeper methodological lesson to be drawn. In reviewing the theme of 'social access' I noticed that there was more data on my acceptance by certain groups of worker (e.g. junior doctors) than others (e.g. nurses). This was a function, at least in part, of David's role as a key member of that particular group. In more self-critical mode, I realized that I found this group easy to get along with. The problem is that while social access was facilitated within this group (both by gossip and, presumably, aspects of a shared social identity), there remains the problem that they were not the only group in the workplace. Social access to one group is no guarantee of successful social access to another.

Dangers of Over-interest

Formal access to another intensive care unit had been greatly facilitated by a senior, influential manager, Peter. However, I found that once this had occurred, this individual's involvement in the research had to be managed very carefully – basically, he was *too* interested in my research. It is worth noting here that Peter had some knowledge of social research methodology and told me that what I was doing reminded him of a research methods course. On some occasions he would pointedly flag up what I should be observing, using a phrase such as 'This is an interesting vignette for you', although he did later apologize ('Sorry, I shouldn't be trying to put these thoughts into your head'). To begin with, I was prepared to accept this attempted influence as a corollary of Peter's interest, and a price worth paying for the speedy physical access he had negotiated on my behalf, but soon I felt I had to interpose some distance. Two incidents confirmed this realization for me.

The first was after I had witnessed a very heated discussion in a meeting and Peter suggested that I should not discuss this argument further (nor the incident which had precipitated it) with anyone but him. By this point in my research though I was on friendly terms with other individuals, such that they were talking to me about this matter of their own volition. The second incident was when he suggested that I should record that the unit is 'quieter than normal' because of a particularly serious medical situation. As I had not actually observed any difference in the ambience or noise levels of the unit I became worried that my powers of observation were inadequate. Later that day I had the opportunity to ask some of his staff about this 'serious medical situation'. They indicated to me that as far as they were concerned the day had been normal, so at least I was reassured that my ethnographic skills were not totally amiss.

What is important for the purposes of this research note is that the interaction with further members of staff had been independent of my 'key sponsor' himself and, indeed, both sets of interactions had been behind his back. In order to gain the confidence of various groups of workers, and to gain access to fuller information about the workplace, I had engaged in gossip which had been critical of Peter. At the risk of being disloyal (Peter had, after all, been instrumental in expediting my physical access to the organization), I resolved to spend more time with other workers; individuals I hoped would be sufficiently interested in my research to talk to me about *their* work but not so interested that they wanted to influence mine. To extend Lofland and Lofland's (1995) typology of ethnographic access, gaining access had become not only a question of 'getting in' and 'getting along' but also 'getting *around*'.

A Manager Worried about Other Agendas

The importance of 'getting around' (i.e. in terms of increasing the number of social contacts) can be extended to the notion of *being* around – that is in terms of the time an ethnographer spends in the field. This can be seen in my third example. At this workplace, the initial process of gaining access had been quite formal and proceeded in a top-down fashion (Burgess, 1984: 196). I met the manager, Liz, together with her boss. I promised them I would anonymize the data and consult with them about how I would describe the unit in any report I might produce that would be seen by senior organizational managers. Liz explained to me that recently she had been managing another department, which had had 'big problems'; it had been under close observation by a group reporting to senior management. She was concerned that senior managers had 'misread certain lines in the report', which had placed her in a difficult position. For this reason she wanted to make sure that any report I wrote should be filtered through her and her immediate boss (rather than going to senior managers directly). I was happy to comply with this request.

Over two months later, when the research study was well established at this site, one of her staff told me that Liz had originally been very worried about me being around. She had told her staff to 'be careful what you say' when I was around. This was corroborated by some written evidence, where Liz had written about me on the date of my first visit:

> Simon Carmel will be starting soon to look at 'organisational and management' issues, 'so be professional and be wise'.

Although she had initially been worried about my presence, some time later, Liz seemed to have become reassured about what I was doing:

> Liz says to me, 'Are you still writing notes? I keep telling people – no lewd comments [when Simon's around]'. SC: 'Does it work?' Liz: 'No'.

So although Liz had originally been worried about my presence, and especially what I might report back to senior managers, after I had been around for several months she was able to joke about her earlier concerns. Here, apparently, was a manager who had originally attempted to prevent too much gossip. I had been unaware of how concerned Liz had been about my research, until much later in the research study. It is difficult to identify any particular factors related to Liz's change of perception, other than the length of time I had spent with her team such that my presence became an accepted feature of the working environment. The importance of being around, in particular the time spent with people, was an important way to alleviate managerial concerns.

An Unpopular Manager

The importance of managerial influence is also seen in my final example, where I began to suspect that one manager, John, although trying to be helpful, unintentionally impeded my social access. For example, he introduced some staff members to me as potential 'informants', who then subsequently claimed not to have time to speak to me and even seemed hostile. There are many possible causes of ethnographers' paranoia (Hammersley and Atkinson, 1995: 79), but I developed a strong suspicion that people were avoiding me. It transpired that there were two parallel issues. One was the way that John had described the research (which had connotations of a managerial audit). Another was that John, unfortunately, had less than ideal working relationships with his staff.

Much later in the research, as I got to know other staff and chatted to them, it was confirmed to me that I had, indeed, inherited some animosity. In a series of conversations, participants moaned to me about John's behaviour. One of these conversations is what led directly to the topic of this research note. As John walked into the room Jane, one of his staff, had been complaining about him to her colleague and to me. Taken by surprise, I was flustered and embarrassed at almost being 'caught in the act' and left the room. Later that day I reflected that

> I need to develop a suitable cover for 'gossip', and the confidence to use it – isn't 'gossip' what this research method thrives on?

I am sure that many of us have been in the situation where the target of gossip comes within earshot. In fact, apart from the obvious feeling of embarrassment, I felt bad partly because I had been inveigled into this discussion by a research participant, merely undertaking (so I would like to think) the normal social interactive work of an ethnographer. My tactic for exploring workplace relationships up to this point had been a 'softly-softly' approach, aiming to be sensitive and discreet, yet enabling others to express their points of view. However, Jane had, in effect, thwarted any attempt at discretion as I had become implicated in workplace gossip.

Following this incident, I realized that in order to gain social access to non-managerial workers, I would need to be more direct and allow other people to be as straightforward in their complaints as Jane had been. This was necessary to get beyond social niceties of surface interactions, perhaps becoming a catalyst of gossip – and a gossiper myself.

Discussion

Problematic instances of social access were resolved through the use of workplace gossip with various members of staff. Even in the apparently successful instances of social access the positive presence of gossip can be inferred, since it appeared that the participants were talking about the researcher. Gossipy conversations were an important resource in understanding the workplace, yet at the same time they demonstrated a potentially seedy side to ethnography. Other ethnographers, like myself, might remain uneasy about the extent to which gossip appears to be necessary to successful ethnographic inquiry. The advantages, even necessity, of participating in covert discussions and workplace gossip are apparent: these conversations were an important resource in enabling me to become more accepted as a member of particular social groups, as well as gaining deeper understanding of the character of some workplace relationships (for example, that some individuals did not get on especially well with others). All this, it seems to me, is vital to the ethnographic endeavour.

Nevertheless, it remains the case that in this research I participated in talk behind people's backs – a morally problematic characteristic of gossip (Taylor, 1994). Yet Taylor's analysis also provides a way through this issue for those who, like me, feel morally compromised by gossip. She highlights both that gossip is related to curiosity about and interest in human beings (also a characteristic of ethnography), and also that it is the attitude of the gossiper that is key in determining moral culpability. While I am not fully persuaded that this is sufficient for asserting the moral worth of ethnographic gossip, the problem is lessened somewhat in recognizing both the importance of attitude and the link between gossip and curiosity. Furthermore, we may ask how could one deal morally and ethically with malevolent gossip? The ideal of being an honest and candid ethnographer (Fine, 1993) is noble, but even taking gossip back to the subject is likely to violate ethical principles such as non-maleficence and the avoidance of 'uncalled for self-knowledge of participants' (British Sociological Association, 2004 [2002]: Section 27). All this, then, reinforces the tenet that as a method where the researcher is the 'instrument' and the 'objects' of research are human beings – as in ethnographic inquiry – ethical and political issues are inextricably linked (Murphy and Dingwall, 2001; Punch, 1986).

There is a point of practical relevance to ethnographic research on work and organizations. While problems of access can reveal much about the social structure of a social group (and, by extension, a workplace), the social structure of a group may not necessarily be the main topic of inquiry. The data *about* access, interesting as they are, may only tangentially impact on the primary research question. That said, a generality about social access in the workplace, at least for the hierarchical organizations represented in this article, is the potentially profound influence of managerial staff. Managers can (perhaps contrary to their intention) hinder social access; they can also intend to influence the direction of research and they can, through their own worries and concerns, attempt to influence the behaviour of their staff.

Therefore we need to manage the research participants' level of interest in our research. While some interest is undoubtedly beneficial, there can be occasions where participants are overly interested in the ethnographer's research and may be too concerned that the ethnographer notes certain events or interprets them in a particular way. There can be a danger of over-reliance on a few individuals in attempting to gain social access, partly because other social relations in the field may be at risk. An important principle in ethnographic research, therefore, is not only to 'get in' and 'get along' but also to 'get around'.

Finally, this article has identified the importance of ethnographers spending time and 'being around' research participants. On this note, it is appropriate to return to the feminist critique of social research and the power relations between researcher and researched. It is interesting to note that the critique of 'rapport' has been most forcefully expressed in terms of qualitative interviews (Duncombe and Jessop, 2002) rather than observational or ethnographic research. The 'false friendship' inherent in much qualitative research, while not absent from ethnographic inquiry, is a bigger issue in standalone interviews than in longer term participant-observation. One dimension of the ethical and political issues in research is that of the length of time the researcher has in getting to know his or her participants. This is a dimension that, it seems to me, has been lacking in ethical discussion of qualitative methodology. The time spent in getting to know participants *as people* is much greater in ethnographic research than in research that comprises qualitative interviews only. This time is generally beneficial in safeguarding against overly manipulative relations between researcher and research participant.

Note

1. All names of participants are pseudonyms.

References

British Sociological Association (2004 [2002]) *Statement of Ethical Practice for the British Sociological Association – March 2002*. [Appendix updated 2004] Available (consulted 29August 2006) at: http://www.britsoc.co.uk/user_doc/Statement%20of%20 Ethical%20Practice.pdf

Bryman A (1989) *Research Methods and Organization Studies*. London: Routledge.

Burgess RG (1984) *In the Field: An Introduction to Field Research*. London: Routledge.

Carmel S (2006a) Boundaries obscured and boundaries reinforced: incorporation as a strategy of occupational enhancement for intensive care. *Sociology of Health and Illness* 28(2): 154–77.

Carmel S (2006b) Health care practices, professions and perspectives: a case study in intensive care. *Social Science and Medicine* 62(8): 2079–90.

Coffey A (1999) *The Ethnographic Self: Fieldwork and the Representation of Identity*. London: Sage.

Collins L (1994) Gossip: a feminist defence. In: Goodman RF and Ben-Ze-ev A (eds) *Good Gossip*. Lawrence, KS: University Press of Kansas, 106–14.

Davies C (2003) Workers, professions and identity. In: Henderson J and Atkinson D (eds) *Managing Care in Context*. London: Routledge, 189–210.

Duncombe J and Jessop J (2002) 'Doing rapport' and the ethics of 'faking friendship'. In: Mauthner M, Birch M, Jessop J and Miller T (eds) *Ethics in Qualitative Research*. London: Sage, 107–21.

Fetterman DM (1989) *Ethnography: Step by Step*. London: Sage.

Fine GA (1993) Ten lies of ethnography. *Journal of Contemporary Ethnography* 22(3): 267–94.

Gluckman M (1963) Gossip and scandal. *Current Anthropology* 4(3): 307–16.

Goodman RF (1994) Introduction. In: Goodman RF and Ben-Ze-ev A (eds) *Good Gossip*. Lawrence, KS: University Press of Kansas, 1–8.

Hammersley M and Atkinson P (1995) *Ethnography: Principles in Practice,* 2nd Edition. London: Routledge.

Handelman D (1973) Gossip in encounters: the transmission of information in a bounded social setting. *Man* 8(2): 210–27.

Harrington B (2003) The social psychology of access in ethnographic research. *Journal of Contemporary Ethnography*, 32(5): 592–625.

Hornsby-Smith M (1993) Gaining access. In: Gilbert GN (ed.) *Researching Social Life*. London: Sage, 52–67.

Humphreys M, Brown AD and Hatch MJ (2003) Is ethnography jazz? *Organization* 10(1): 5–31.

Lofland J and Lofland LH (1995) *Analyzing Social Settings: A Guide to Qualitative Observation and Analysis*. London: Wadsworth.

Murphy E and Dingwall R (2001) The ethics of ethnography. In: Atkinson P, Coffey A, Delamont S, Lofland J and Lofland LH (eds) *Handbook of Ethnography*. London: Sage, 339–51.

Noon M and Delbridge R (1993) News from behind my hand: gossip in organizations. *Organization Studies* 14(1): 23–36.

Oakley A (1981) Interviewing women: a contradiction in terms. In: Roberts H (ed.) *Doing Feminist Research*. London: Routledge, 30–61.

Paine R (1967) What is gossip about? An alternative hypothesis. *Man* 2(2): 278–85.

Punch M (1986) *The Politics and Ethics of Fieldwork*. London: Sage.

Schwartzman H (1993) *Ethnography in Organizations*. London: Sage.

Smith V (2001) Ethnographies of work and the work of ethnographers. In: Atkinson P, Coffey A, Delamont S, Lofland J and Lofland LH (eds) *Handbook of Ethnography*. London: Sage, 220–33.

Taylor G (1994) Gossip as moral talk. In: Goodman RF and Ben-Ze-ev A (eds) *Good Gossip*. Lawrence, KS: University Press of Kansas, 34–46.

Tebbutt M and Marchington M (1997) 'Look before you speak': gossip and the insecure workplace. *Work, Employment and Society* 11(4): 713–35.

67

The Sweat-Shop in Summer[1]

Annie Marion MacLean

The exploration of fields of industrial labor by those who from circumstances belong outside of the manual laboring class has become so common that one feels constrained to apologize to the long-suffering public for thrusting upon it still another "experience." The only excuse for this, as for many another, is the hope that it may rouse to thought and action some people who have heretofore been apathetic or listless, and this time in regard to a vigorous relic of an earlier industrial system – the sweat-shop.

From the time of Paul Göhre[2] and Frau Dr. Minna Wettstein-Adelt[3] to Walter Wycoff,[4] many educated people have been actively interested in the hardships of some phases of industry, and from time to time have thrown light upon actual conditions by experiencing the toil. Others again, in a dilettante fashion, have gone down and mixed with the so-called dregs of society long enough to focus a figurative kodak on little groups of workers here and there, for the purpose of ekeing out "copy." To the serious student of society such work is valueless, to the general reader it may have a morbid sort of interest, and to the toiler it is an insult; whereas an honest portrayal of the conditions under which the people work may be of inestimable value to all. To the student, the philanthropist, and the legislator it suggests a rational ground for constructive action; to the general public it may serve to awaken a sense of personal responsibility; while to the toiler himself it may mean hope in the future.

It is inevitable that the mighty changes constantly taking place in modern industry should push to oppression some who can but poorly stand the strain. The responsibility is not theirs; circumstances and the consumer are

Source: *American Journal of Sociology*, IX(3) (1903): 289–309.

to blame. Then let us, as consumers, accept the dictum of fate and conduct ourselves accordingly. Duty may lead me to endure the hardships of the worker in the interests of amelioration; she just as surely leads others to assist in lifting the burden when once it has been pointed out.

The field of investigation included in my present task is difficult of delimitation, owing to the widely differing conceptions in men's minds; but in this study I use the term "sweat-shop" as synonymous with "tenement-house workshop," the same in which it is used by factory inspectors in their reports. The term itself appeared in England during the troublous times of 1847–48, when the working people were in the direst straits and commenced taking work home for a mere pittance rather than sit quietly awaiting starvation. "In England and America alike the sweater is simply a sub-contractor who, at home or in small workshops, undertakes to do work which he in turn sublets to other contractors, or has done under his own eyes."[5] An inquiry into the sweating system conducted by a committee of the House of Lords in 1888–90 defined sweating as "no particular method of remuneration, no particular form of industrial organization, but certain conditions of employment, viz., unusually low rates of wages, excessive hours of labor, and unsanitary work-places."

The work of which I am to speak was undertaken in all seriousness with the hope that it might throw some light upon the evolution of the ready-made clothing industry, and thus incidentally aid the Consumers' League in its crusade against sweated garments, and in this way to awaken in the minds of buyers an appreciation of the danger lurking near them, when they unthinkingly purchase goods made under unsanitary, not to say demoralizing, conditions. The average person, it must be admitted, knows little or nothing of the conditions and processes of garment-making. He knows equally little of the manufacture of the butter and the production of the milk he buys; but his ignorance does not save him from inoculation with fever germs, neither does it excuse him when he communicates disease to others by lack of precaution. City authorities hold him responsible in the observance of quarantine regulations; and so in this case ignorance of the conditions should be no excuse.

It may appear to some that the summer is not an auspicious season for sweat-shop study. It is true that it is a slack time; yet work is being done, and under more trying conditions, in many respects, than in the cold-weather period. First of all, the slackness of the work renders life more precarious, and thus the sweater is tempted to use the services of all members of his family. Children are home from school, and they must justify their existence some way; or, as one man said to me: "What are our children for, if not to help support us?" I must confess that this was to me a rather new view of family obligation. The duty, I should think, is on the side of the parents, who alone are responsible for the existence of their offspring. One feels this keenly when he sees slum streets swarming with small bits of humanity, festering in the summer sultriness. It is all very well to talk of man's duty to the state in reproducing himself. The chief trouble with the poor in the great cities is that they reproduce themselves too many times. Seven or eight little children in a home where there is only enough bread for two is a monstrous outrage against the innocent victims, and it should be condemned by enlightened sentiment everywhere.

But it is not my intention to offer a panacea for such a lamentable social condition, but only to call attention to an existing horror which fosters the survival of an industrial anachronism – the tenement-house workshop. To the ignorant, harassed parent sewing at home the temptation to utilize childish activity is great. I have no quarrel with those who decry the modern tendency to "race-suicide;" they are sincere, and their protest, no doubt, is necessary in some quarters; but to one who has spent more or less time during the summer season in the slums of London, New York, and Chicago the crying need of "race-limitation" must be apparent. The high rate of infant mortality saves the slums. This opinion may savor of brutishness, but it is born of common-sense. Once I saw a little baby die in a small room in a rear tenement, where the mother, her sister, and two older children were finishing Ascot ties at two and a half cents a dozen, and the mercury stood at ninety-six degrees. The mother had the baby on her lap, and she was working half an hour before its little life ebbed out. She screamed with grief for a few minutes, and then she said: "Thank God it's gone! I couldn't take care of it." In a few months there was another, and the struggle still goes on. All sweaters, however, are not in such dire straits as these, but they are working under the worst possible conditions.

My interest in this particular phase of industry was first actively aroused during the summer of 1900, when I visited some fifty or sixty sweat-shops in New York for the Tenement House Commission.[6] My work was insignificant compared with that of others, but full of significance to me in that it revealed a hitherto undreamed of condition of toil for thousands of unskilled workers. My official visits were to the necktie workers entirely, and each morning before starting out I carefully scanned the board of health reports so that I might avoid the neighborhood of contagious diseases; but, in spite of my vigilance, on more than one occasion I found measles and scarlet-fever patients in or very near the domestic workshop; yet the sewing went on as usual. Now, someone bought those neckties bearing the germs of disease, and so an innocent victim of the greed for gain may have paid the penalty.

But all of this is not pleasant to think upon. Factory inspectors' reports have told the tale of the spread of loathsome diseases by means of garments made in the home workshop, and notable among these is the special report on smallpox by Mrs. Florence Kelley, of Illinois, in 1894. But a very small proportion of the people who buy ready-made clothing read state reports; and so they go blindly on, encouraging the perpetuation of a system wretched in the extreme – a system which, in spite of the law, utilizes the service of children and invalids.

My insight into the neckwear industry led me to explore unofficially another phase of sweating, known as the "knee pants," or, more correctly, the garment-workers' trade. One who lives for a time on the lower East Side of New York is bound to be impressed by the continual morning and evening procession of people bearing on their heads and shoulders great bundles of these unfinished garments. I used to wonder if the making of knee "pants" was the staple industry of the country. I suppose the small boy's ability to transform whole cloth into rags in the twinkling of an eye is at the root of this great industry, with its concomitant hardships; but as, in the old tales, the blacksmith's horse went unshod, so the children of the people who wear their lives out making "pants" rarely had more than an apology for such a garment to cover their nakedness.

At last I determined to see for myself just the conditions under which these garments were made, and so learn by experience how lucrative it proved. I went some blocks away from my abiding place to a region where I was not known, to seek work as a sweater. Many weary blocks I walked in the scorching sun, and many weary stairs I climbed in what proved to be a fruitless effort to find employment. No one wanted me, because the season was dull and there was not enough work to warrant the hiring of a "green" hand. In piece-work the race certainly is to the swift and skilled. Thus I was obliged to retire from the field to swell the ranks of the unemployed. But my plan to work as a sweater was not abandoned, only postponed.

My next attempt to get employment was in Chicago, several years later, where my efforts were crowned with success, and this in the hard heat of the city summer.

One day I went down to the region where many sweat-shops flourish. The street I selected first was near the river – narrow, dirty, ill-smelling, with treacherous board walks at the sides, and walled in by rickety houses, reeking with stale smells; and through that street I walked seeking employment. I may mention incidentally that I covered my more than usually shabby clothes with a long coat, which I deposited in a pawn-shop before seeking work. Sweat-shops were everywhere. I groped my way to the third or fourth floor of many a house, asking if they needed a "hand," and as many times I was turned away, and mostly with a savage look or word. One man said, with brutal frankness, and in broken English, that I was holding my head too high, and that he did not want my kind about. I retreated, communed with myself a little, and gave myself a few lessons in humility of spirit and practiced a hang-dog position of the head. Another was skeptical because my fingers were not needle-pricked, while still another objected to me on the ground that I was an American and likely to be lazy! That seemed highly amusing. Is the native-born worker really discriminated against in his own

Scene in sweat-shop

country? I found that these foreign sweaters – and there are more than three thousand in Chicago – favored foreign employees, and I was unable to determine whether the reason was due to jealousy or fear.

At last I found work in a rear tenement on the second floor, where eight men and two women were engaged in making knee pants. A third woman who worked there was ill that day, and the owner told me I might take her place, my wages to depend on the amount of work I did. I eagerly took up my task, which was sewing the pockets for boys' pants. It was machine-work entirely. I was told to take a seat at a big foot-power machine and go to work on a stack of pockets a yard high. After a few brief instructions I was able to run the machine; but I fear I was rather slow, for I was barely able to keep a man who sat at a machine next to me, putting on the facings, employed; and he had an enormous pile to start with. Fortunately, he had some basting to do, and that helped me a little. After an hour or two I became more of an expert, and turned off a constant stream of pockets. And just here I wish to enter a protest against putting pockets in the trousers of small boys. It is not necessary, and they are so hard to make. If anyone imagines it is an easy thing to work a sewing machine all day, let him try it for awhile. Abounding health and strength kept me from being prostrated at the end of my first day; a weaker woman unused to toil could not have endured the strain. It is said that strong men who have worked foot-power machines from youth are worn out at thirty-five, oftentimes wholly or partially paralyzed from hips down. It is a gloomy outlook, indeed, for the worker, who realizes that he will be a disabled old man while still young in years. In the modern factory foot-power has virtually passed away and electricity has been substituted. There one may stitch all day, using the foot only when it is necessary to push a button to start or stop the machine.

I cannot now describe the utter weariness that possessed me when my first day's work was over. I sewed from nine in the morning to six in the evening, with a brief halt for lunch; the others were at work much earlier. I know, of course, that the newness of the task made it doubly hard for me. My companions were not so tired as I.

But let us pass over the physical feeling and look at my surroundings. As was stated before, we were in a rear tenement, which was reached through an alley paved with tin cans and broken bottles, and the outside stairway leading to the house was rickety beyond belief. Each step groaned and trembled as I crept carefully up to the door, where I had to stoop a little to enter a small room lighted by one window. In this room were six people and five machines, the sixth man being a presser. Opening off the first room was a second, smaller and darker, where the head of the establishment and two others worked. The whole place was filthy, and was rendered even more unpleasant by unsavory odors from the living-rooms adjoining, where the enterprising wife boarded the workers. The presser was an unsightly object, clad in one brace and a pair of trousers, and he literally boiled over his work. The perspiration dropped continually from his short brown beard to the steaming clothes he was ironing. The irons were heated on an ill-smelling oil stove, and all of this made the room horribly unpleasant. But these things made no difference to the men and women, whose whole beings were absorbed in the various processes of making knee pants, for each was paid according to the number of pieces he turned off.[7]

The sweaters are not unionized, and so there is no minimum wage. The state factory law theoretically regulates the number of hours in a working day. I say theoretically, because in actual practice the sweater and his family work as many hours as they see fit. I found in New York women, and even children under the legal age, working till nearly midnight in the busy season. Very young children can be utilized advantageously, pulling out basting threads. Their small fingers do the work well enough, and much time is saved to the adult members of the family. This is appalling to those of us whose childhood's memories carry only long days of delightful play out of doors with a twilight-hour bedtime. And this is what belongs to every child in the state; and cursed, I say, be the industrial system that defrauds the children of this God-given right!

One of the saddest sights I ever saw was a group of three little girls under ten years of age sitting on a tenement-room floor pulling bastings out of neckties the mother and aunt were finishing. The little things were piling up the threads and pretending they were balls which they would play with

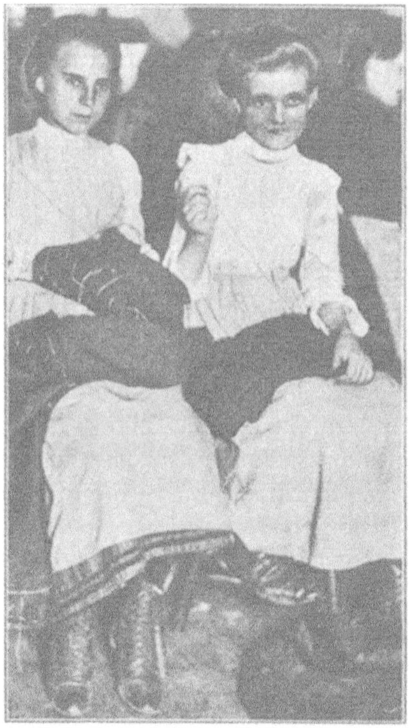

Girls of fourteen and fifteen years

when the night's toil was over. But the little tired eyes grew sleepy long before they were free to play as they had planned.

My employer had two children who worked in the shop during the daytime, and I have every reason to believe that they performed some services after we had gone home.[8]

The second morning I was at my place at seven o'clock, a worn and weary creature, to face the hardships of the day, but glad to be allowed to return. I was told that my work was not of a high order, but as I was swifter than many beginners I was regarded as hopeful. Two days' work making pockets quite reconciled me to belong to the pocketless sex.

The two women who were my companions were stolid German Jews with expressionless faces. They were both married and had several children at home. During the lunch hour of the second day I asked them if they were very tired, and they both agreed that they would rather die than go on much longer at that work; their backs ached incessantly from constant running of the machines. I asked why they did not try factory life. One said that the near-by factories could not take non-union people, and she could not join a union for two reasons: first, her husband would not allow her, because he had been nearly killed in a strike riot once when he was a "scab" worker; and, second, she did not know the trade well enough. The other woman said that one had no independence in a factory, which, being interpreted, really meant that she did not care to be a very little cog in a very big wheel. She thought she was the wheel in Herr G.'s establishment; and I learned later that her thoughts were quite correct, for the rest were afraid of her, and when she told the presser to take his unsightly form into the next room and out of the presence of the "new lady," he sheepishly slunk away. This was at lunch time, when we were trying to eat the scraps the manager's wife had brought us. I could not bring myself to the point of eating. My honest conviction is that I would have elected starvation first, but fortunately I was not called upon to make the choice. My appetite was gone. The food may have been wholesome enough, but, to put it mildly, it was not attractive.

At the close of my second day I was told that the woman for whom I was substituting would return on the morrow, and so my services would not be required. This news came as a relief, for my whole soul hated the big machine I had to work all day. I felt as if I were in bondage and shuddered at the thought of the treadmill existence.

Three youthful sweaters

I learned but little from the workers. The men were non-committal, even suspicious of the stranger. The women were weary and unresponsive, and we were together only two days. But I saw much. I saw a group of human beings working under conditions not fit for human beings; I saw boys' pants made in filth too odious to describe; and I saw the mere pittance handed to me for two days' toil. I learned that those garments when finished were sent to dealers all over the state, and were sold so cheap that even country mothers bought them for their children. Good people are amazed that clothes can be sold at such low prices, and well they may wonder. They should stand aghast at the conditions that made them cheap. Knee pants are finished for from fifty cents to a dollar and a quarter per dozen pairs, or an average of six or seven cents a pair.

No reasonable person condemns the legitimate cheapening of clothing, but it is the demand for cheapness at all hazards that should be subject to condemnation; and we must not lose sight of the fact that the cheapening of ready-made garments has brought relief to many an overburdened mother whose days are all too short to give time for the family sewing.

My experience in the shop of Herr G. forced upon me the conclusion that no money would induce me to buy knee pants without finding out first of all, if possible, where they were made. It is the duty of the consumer to acquaint himself with the conditions of production. And we who buy need not soothe our consciences with the belief that we are helpless in the matter, while the people who give out the work are the only sinners. This may be comfortable, but it is criminal. The contractor gives us what we are willing to take. Sufficient protest from us will induce him to change his method, if it be objectionable.

My second employer was a frail-looking woman with a Polish name. She and her daughter were working on women's cotton wrappers. When I appeared asking for work, they were skeptical about needing help, but finally told me I could stay, and if I proved useful they would give me fifty cents a day. I evidently lived up to their expectation, for they kept me four days, when they dismissed me. We commenced to work about eight in the morning, and worked steadily until about half-past six, with a few minutes for lunch. I carried a lunch with me and ate it alone in the bedroom where we sewed; the others ate in the little kitchen, the only remaining room in the tenement flat. I found that the daughter slept in the kitchen, while the father and mother occupied the bedroom. The father was an unskilled laborer and worked at odd jobs. Thus the living- and sleeping- and workrooms were one – a distinct violation of the state factory and workshops act.

This condition exists in places too numerous to mention, and the fact that it does exist casts no odium on the factory inspectors, who are conscientious and efficient, but in numbers wholly inadequate to cope with the situation. There are but nineteen for Illinois, while New York has fifty, Massachusetts thirty, and Pennsylvania twenty-six. It takes eight deputies three months to inspect the sweat-shops in Chicago once. It is clear that an annual inspection is insufficient. The inspectors visit the tenement-house workshop in the autumn – the busy season, which lasts from September to December. During the remainder of the year the sweaters do as they please. When one remembers that nineteen people inspect upward of twenty thousand establishments, employing about half a million people during the year, the herculean task of the inspectors becomes evident. There are in Chicago alone nearly six thousand[9] garment-makers' shops, employing twenty-five thousand women and two thousand or more girls under sixteen years of age, besides twenty-one thousand men and five hundred little boys.

For graphic statement the following tabulation is adopted as a recapitulation of the foregoing facts:

Factory inspectors in Illinois	19
Annual inspections	17,219
Total employees	484,172
GARMENT WORKERS (CHICAGO).	
Establishments	5,313
Total employees	50,417
Women	25,573
Girls under sixteen years of age	2,417
Men	21,759
Boys under sixteen years of age	569

And this is the great army whose life and conditions of work interested me, and should interest everyone who buys clothing. One cannot quiet his conscience by announcing that he goes to high-priced clothiers for his garments, and so cannot come in contact with sweated goods; hence he is free from responsibility in the matter. Such soothing syrup may prove fatal in the end. In the first place, the mere fact of buying expensive clothing does not exempt one from the danger of tenement-house goods. The tailor who charges fancy prices is quite liable to let his work out by contract, and the original contractor, though not a sweater himself, may sublet the work to one who is; and so one's hundred-dollar coat may repose on the bed of a scarlet-fever patient before it is delivered ready for use. Costliness alone is no guarantee that a garment is made under decent conditions.

On the other hand, even though we are certain that our clothes are made in a wholesome place by workers who receive fair wages, our responsibility to society does not end. The fact that any garments at all are manufactured and sold with the germs of disease and the life-blood of the workers upon them should be sufficient incentive to all people to demand that the horror cease.

"What can we do?" some may ask. The answer is clear: Insist upon a guarantee on every article of clothing you buy. Dealers are only too eager to

Sweat-shop coat carriers

satisfy their patrons. If we accept without question what they offer, why should they change? Their only desire is to suit their customers – the consumers.

The tenement-house workshop should be exterminated, and there is but one sure way of driving it out of existence; that is, by united action on the part of those who buy clothes. There is an organization whose primary aim is the rousing to action of lethargic persons who disregard social and moral responsibility, and in addition to this the Consumers' League aims to endorse only clothes made under wholesome conditions. The Consumers' League label has become quite a potent factor in the mercantile world, and it bids fair to become even more powerful as the organization enlists more and more of the intelligent sympathy of the community. It may be noted here that the Women's Union Label League stands for the same principle, but its indorsement is a trades-union label. This is creating a public conscience in a class but little affected by Consumers' League logic, and so is doing an excellent work for society.

Another employer of mine was a red-bearded Jew who made shirt waists. He had three women and one young boy working for him, and agreed to take me at sixty cents a day. I went to work immediately finishing cuffs. The waists I worked on came from an ultra-fashionable shirtmaker – a woman – who charges eight dollars for the mere making of a garment. I learned her identity from the boy who carried the parcels back and forth. A few days later I called on that woman and inquired her prices. She quoted them as I have stated. I then asked her if the waists were made in her own shop. With a bewitching little smile she answered: "Oh yes, we make everything right in the back rooms. That is why I charge good prices and take only a limited amount of work from select people!" I told her I thought I would not leave an order then, and passed out, not caring to belong to the "select." She added the insult of a lie to the injury of sweated waists. Her waists were carefully cut and fitted, but were just as poorly made as much cheaper ready-to-wear ones.

A sojourn among the garment workers certainly reveals some strange facts about the making of fashionable clothing. Now I never see certain more or less fashionable establishments without thinking at the same time of certain tenement attics bound to them by an invisible tie. My lady who scatters smiles through slumdom little dreams that the stylish clothes she wears may have been partly made beyond the ill-smelling alley she will not enter. It may be her first visit there but her clothes should feel at home!

Work in the shop of the red-bearded Jew was hard for me, because the days were insufferably warm, and the light was poor, as the workroom was on a court. I got along rather slowly, and I had a feeling that possibly I was not worth more than sixty cents a day. But my pride suffered a frightful shock when my employer told me that I was not worth anything, and he did not propose to pay me at all. This decision was precipitated upon me at the end of the second day, when I asked for my wages, as I did not intend to return.

A young toiler

I learned but little from my companions there. The women were up in years and uncommunicative. One of them had an ugly sore on her hand which she tried to keep covered. I would not "finish" for her, and so incurred her dislike. I said nothing, but always took work from someone else in preference.

Another place, where I worked but half a day, was more interesting to me because a number of young girls were there and they kept the place lively all the time. I asked one of them if she supposed we could get work in a factory, and she said: "Law, yes, if yer wants to join the union!" As for herself, she claimed that anything was better than "livin' by rule." The others declared that she was not skilled enough to belong to a union, and they grew quite merry in the dispute, so I could not determine just what her real objection to the factory was. These were the first happy sweaters I found. The six girls chewed gum, and laughed while they sewed. They ran the machines in a care-free way that almost made my hair stand on end. I got fifteen cents for my half-day's work. I was not needed longer, so I had to seek other scenes of labor.

I did not want long engagements in the sweat-shops. They were not necessary for my purpose. I simply wished to verify my knowledge regarding actual conditions of work. I had visited over a hundred of these places, and I already knew the aspect they present to the observer; but a few minutes' visit can never teach one the hardships of the workers. We may gasp when we are told of women who toil twelve or fourteen hours for a mere pittance, but, after all, it is without meaning until one has experienced the weary eyes and dizzy head and aching back caused by a long day's sewing in a badly ventilated and poorly lighted room. My poor cramped shoulders made me understand the feelings of the woman who sang the "Song of the Shirt;" and only too many of our present-day toilers know what it means to "stitch, stitch, stitch, in poverty, hunger, and dirt." And who makes them do it? We do. Our mad craze for cheapness has cheapened life.

But the tenement-house workshop must go; and the Christianized consumer must make it go.

Still another employer was a woman engaged in making corset covers. I toiled part of a day and received thirty-five cents for my time. It was handwork entirely – the kind that appears on expensive *lingerie*. The sewing itself was rather pleasant, as the materials and lace were soft and fine; but before the day was over my inexperienced fingers were pricked past recognition. The others, accustomed to needlework, did themselves no such bodily violence.

The workroom was clean and bright and cheery, and looked out on the street from the fifth floor of a rather poor tenement. Our employer was a sharp-featured American spinster, whose contact with the industrial world had not softened her as she advanced in years. She appeared to have two missions on earth – one, the making of underwear; the other, inveighing against matrimony; and she threw herself with equal zeal into each. I believe she made a good living from her sewing. She took work from a fashionable trousseau-maker and employed as many helpers as she needed from time to time. Her method was to engage girls for a week and then discharge them at night if she wished. Her reason, of course, for agreeing to give a week's work was that few would consider a proposition for one day only. I did, it is true, but a self-supporting sewing-woman would hesitate to stop for such an engagement. Our employer told me that she usually paid fifty cents a day to a smart seamstress, but, as I stated before, she gave me only thirty-five cents, claiming that I did not work a full day. I actually sewed nine and one-half hours, while the five regulars sewed twelve hours. The granite spinster believed in grinding the soul of her employees. Her philosophy of life was that women are too idle and too prone to sentiment. The first evil trait she seeks to eradicate from all who become a part of her industrial machine; while she harangues on the second whenever her mouth is free from pins and buttons. "Beware of men!" is the slogan that makes the needles glitter through the muslin and lace. She even gives free lectures after hours on this important subject, and yet she is unwilling to pay her women wages that would render them financially independent of men. I was not with her long enough to learn her justification of such inconsistency.

Hasty marriage is usually the working-girl's last protest against a wage-earning system that pushes her to the wall. It is not a hope of bettering her condition so much as a desire to escape immediate wretchedness that leads her to plunge into what often proves the infernal fire of matrimony. Thus the supposedly ideal state frequently turns into purgatory, from which the divorce court or death offers the only means of escape. Some such thoughts as these were evidently surging through the mind of my employer, who mingled muslin with matrimonial misery in her lamentations and warnings. She was a unique type of sweater.

Such garments as we made there sell at five dollars apiece, and my inexperienced hands finished one that day. I could buy the materials at retail for seventy-five cents in any good store, so there was a profit of nearly four dollars on one after paying for my work, and this on the false assumption that the goods were bought at retail. The fashionable modiste took orders from her stylish customers for hand-made underwear "almost at cost," for the sake of accommodating them, and then sent the work out to a sweater who lived in an unhealthful quarter of the city. It may be as well that we are not omniscient! Still there was not the serious question of filth in that establishment that frequently faces one in a tenement-house workshop. But the equally serious one of long hours with insufficient wages was there demanding an answer from the conscientious buyer. And this, after all, is the vital issue.

Unwholesome physical conditions may cause loss of life, but the other stunts the mental, moral, and spiritual growth of all those engaged in the fight for bread. When prosperity is flaunting itself before the people's gaze, it seems incongruous that thousands upon thousands of citizens, nominally free, are in the thraldom of slavery more harassing than that banished forever by the Civil War. The black slaves were sure of food and shelter, while the white industrial slaves lead a most precarious existence. It is not pleasant to live with famine staring one in the face, and, in spite of the ingenious schemes for living on five cents a day originated by people with satisfied appetites and warm rooms, it is a serious matter for a woman alone in the city to try to live on two or three dollars a week. Such unfortunates who earn this pittance in the sweat-shop usually give up the struggle to maintain a decent appearance, and go about with the slouching gait and dull-eyed discontent that indicate a hopeless heart. The joy of service is unknown to them, and one cannot wonder. It is a simple thing to grow enthusiastic over the zealous delight that should be in the soul of the one who works with his hands, but the actual emotion is a dread travesty on the idealist's dream. Joyful thoughts are a mockery when one sews all day long in a dismal, unsanitary room and is only half fed. The sweater is fathoms below the factory operator in all that goes to make life more than mere physical existence.

I observed much while I toiled among the sweaters. It was no light task for one unaccustomed to machine- and hand-sewing to sit for hours and stitch, stitch, stitch, with every nerve tense in the effort to accomplish as much as possible; it was no childish prank to endure the aching bones and dizzy head and weary eyes that are a part of such work. But I have purposely ignored much of this for two reasons. In the first place, I claim no heroism in enduring hardships, because they were self-imposed and for a purpose; and, in the second place, it would be necessary to charge much of my discomfort to inexperience. Any unusual physical strain brings with it undue weariness. So, while there is no significance attaching to my fatigue, it should be of keenest moment that there are thousands of women victims of a system vocal with the sobbing of children and the groans of weary women, the adequacy of whose compensation bears no fitting relation to the labor contributed to society. Such a condition is fatal to progress in its widest sense.

This may apply with equal truth to other branches of industry, but that does not weaken its potency as an argument for change in this particular occupation. Sweated goods are a constant menace to society, both in the injury done to the workers owing to the bad conditions under which they work, and in the liability from exposure to contagious diseases; and popular opinion should support this verdict. Theoretically this is indorsed by all intelligent people, but an indorsement to be of any avail must be capable of practical demonstration. In this case it must mean an utter rejection of garments thus made, and this implies insistence upon a label assuring the buyer that the article in question either was or was not made in a sweat-shop. In either case he could make his own choice.

As each buyer cannot become a volunteer inspector, the most rational method of attaining his end is to unite his protest to an organized one and

Girl fifteen years of age who runs a machine all day

thus increase its potency. The Consumers' League[10] is a rational and effective crusade in favor of goods made under wholesome conditions, and its guarantee carries weight. It seems therefore a simple thing for the purchaser of ready-made clothing to aid in the effort to crystallize public opinion into definite action along the lines suggested by this organization; and those who are not informed would do well to acquaint themselves with its principles. It stands for fair treatment of employees in mercantile establishments; it stands for sanitary workshops; and it stands for an attempt to place responsibility where it belongs. It indorses those merchants whose standards are high.

I believe the sweat-shop evil could be eradicated if thinking people would lend their aid. The trades-organizations are working to this end; and when all these forces become sufficiently emphatic in their denunciation of this deplorable system, it will vanish away, as negro slavery vanished from the civilized world.

Notes

1. Photographs of Chicago sweaters furnished by courtesy of the Chicago *American*.
2. *Three Months in a Workshop* (1895).
3. *3½ Monate Fabrik-Arbeiterin* (1897).
4. *The Workers* (1898).
5. Helen Campbell, *Prisoners of Poverty Abroad*, p. 34.
6. Acting under appointment of Governor Roosevelt.

7. An exception was extra help hired and paid by the day.
8. Illinois child-labor law in force after July 1, 1903: "*Section 1. Child under fourteen years.* – Be it enacted by the people of the state of Illinois, represented in the general assembly: That no child under the age of fourteen years shall be employed, permitted or suffered to work at any gainful occupation in any theater, concert hall, or place of amusement where intoxicating liquors are sold, or in any mercantile institution, store, office, hotel, laundry, manufacturing establishment, bowling alley, passen-gen or freight elevator, factory or workshop, or as messenger or driver therefor, within this state. That no child under fourteen years of age shall be employed at any work performed for wages or other compensation, to whomsoever payable, during any portion of any month when the public schools of the town, township, or village or city in which he or she resides are in session, nor be employed at any work before the hour of seven o'clock in the morning or after the hour of six o'clock in the evening. *Provided*, that no child shall be allowed to work more than eight hours in any one day."
9. Round numbers only are used here.
10. National headquarters in New York city.

68

On Doctor Watching: Fieldwork in Medical Settings

Sandra K. Danziger

The conduct of qualitative inquiry is traditionally subject to descriptive rather than analytic treatment. In research articles and monographs, authors frequently report their methods of collecting field data, their style of carrying out participant observation, and their attempts to overcome the biases inherent in settings or in pressures from participants. Methods textbooks assert that these problems exist in every research situation and are more or less amenable to resolution by the individual investigator. Thus, future researchers can learn from strategies devised by their predecessors.

Problems of field strategy and how they are resolved, however, are shaped to some extent by the specific context of the field situation. An analysis of the researcher's attempts to overcome these difficulties can then demonstrate key sociological features of the setting. In this paper I provide a personal account of doing fieldwork in medical settings to document certain organizational features of medical care. I shall argue that the problem of fieldwork practice are of theoretical significance and deserve a more central place in the reporting of research findings.

Background of the Research Project

In a study of doctor-patient interaction during pregnancy and childbirth (Danziger, 1978), I analyzed some of the ways in which professionals exercise social control over their patients and documented some adverse effects

Source: *Journal of Contemporary Ethnography* (formerly *Urban Life*), 7(4) (1979): 513–532.

of these practices. Through qualitative observational data, I examined the sequence of medical encounters over the course of childbearing as a career and focused specifically on how legitimation of the expert's power occurs interactionally. The data suggest that one of the goals of everyday routine medical work – the maximizing of the professional's authoritative control – leads to defining the welfare of patients in terms of the welfare of staff members, physicians, and nurses. This results from interaction between parties of unequal status; asymmetry is a key feature of the situation. My focus on interactional effects of such status discrepancies is shifted in this paper to how members of medical settings respond to a fieldworker's presence. Emphasizing the impact of this status differentiation, I describe the methods used to obtain legitimacy in clinics and hospitals and to carve a niche for myself as a conditional participant-member. It is important that we understand how such contextual constraints impinge on fieldworker role relations in order to compare findings obtained in dissimilar ways or from various vantage points. For sociology to operate as a cumulative discipline, such contrasts of the researcher's perspective and contextual base are essential.

Three classic problems in participant observation strategy require special accommodation due to the predominance of elites in the setting: (a) obtaining access to medical settings, (b) establishing rapport with highly stratified classes of participants, and (c) avoiding "going-going-gone native" (Freilich, 1970: 530). My effort in each area is analyzed with respect to the contextual backdrop of preexisting role relationships. Following this, I draw a parallel between the expectant father's role during hospital labor and delivery and that of the participant observer, so as to substantiate the case made for situational shaping of fieldwork choices.

Obtaining Access

> Heroic literature is replete with tales in which the hero must pass a series of tests and questions before he is permitted to gain the prize. A similar experience, though less ritualized and unaccompanied by operatic music, awaits the sociologist in the hospital [Mauksch, 1970: 188].

Nine months elapsed between the time I began to look for a research site and my first days in the field. A good part of the troubles and delays evolved from the firm resistance of doctors to an outside researcher. Medical settings are organized to protect the physicians' autonomy and to prevent their accountability (Freidson, 1970, 1972, 1975). The care physicians take to circumscribe scrutiny by others is usually defended on grounds of ethical concerns for the confidentiality of patients. Indeed, most institutions now have a vast set of bureaucratic guidelines for protecting the rights of patients from researchers. Before one even reaches the point of going through the bureaucracy's review channels, however, the opposition of the profession to exposure of its activities is encountered.

The success of doctors' resistance – their self-protection in the name of patient protection – is demonstrated in the pessimism that other health care researchers expressed about the feasibility of my project. My early talks with professors in sociology, hospital administration, nursing, psychology, and social psychiatry were dismal at best. They were all convinced that even if my research ideas were good, the physicians would never allow it. The intimidation produced by medicine's reputation for opposition succeeded in dissuading me from directly approaching any doctors themselves for some time.

One very special credential, however, served to minimize my threatening outsider status with respect to the medical profession. I had a personal connection in that my father is a practicing obstetrician gynecologist. Not only did I have my first opportunity to observe on labor and delivery units where he practiced or where he was a friend of the chief of staff, but I also had a letter of introduction from him, which proved to be a door opener. It meant that I was officially a member of the fraternal order of colleagues. This was an enormous advantage in overcoming the classic double-bind of (1) not getting in because you lack prior experience and (2) never getting the first break of entering in order to obtain experience.

My new-found legitimacy allowed me to be a novice spectator for a short period of time and to learn my way around the highly mysterious occupational world of baby-delivering. This glimpse into the network of role relations and coordination of activities greatly enhanced my later attempts at integration of the observer role. In addition, I used my letter of reference when I encountered institutional resistance in other settings. The letter put my research proposal on the physicians' desks. One other maneuver insured their reading it. I consulted who had completed a study involving observation on hospital maternity wards, and she gave me some important advice that linked me into the occupational network. She felt that I needed feedback from the head nurses at the maternity units. If they were interested, she said, they would help push it and would tell me how to go about receiving permission.

This proved to be a remarkably wise strategy. The head nurses were extremely receptive to my ideas and were flattered that I was seeking their counsel before that of the doctors. It was at this point that I began to build some quite contradictory alliances. I claimed membership in the fraternal order of the elite and concurrently sought the help of the second order of authority to win over the first; similar problems of conflicting allegiances plagued me later in the field. In order to demonstrate their own position of power, the nurses would have to be successful in helping me win over the doctors. They were being counted on to get this project under way; as a result, I obtained an insider advocate for my research. Each of these women gave me hints as to how to "sell" the proposal to "their" group of physicians.

This dual strategy started the bureaucracy's wheels in motion in two instances and opened the door without a further hitch in a third. In one of the former, I again met the resistance to exposure of the elite, exemplified in a fiery phone call to me at 8:00 A.M. one weekend morning. The doctor was obviously quite busy; that he could take an odd moment to bother himself with "one of these people" was something for which I could never, I suppose, express enough gratitude. His concern was that I might not be objective, that my motives were somehow irregular. He said that he had had experiences with patients' rights advocates in the past that had resulted in a lot of "unnecessary" trouble.

When I could get a word in edgewise, I argued that my purpose was to understand the perspective of the staff as well as the patients, that I was interested in what problems doctors and nurses saw from their involvement on a routine day-to-day basis. I said that the problems of health care delivery systems cited in the literature were probably irrelevant to the concerns of the practicing clinician and that understanding this view was the object of my study. He replied, "Well, now that's another story and it's a damned worthwhile thing to document." I had argued the merits of the case study method, and, after what must have been three-fourths of an hour, I had won him over. I knew that access was no longer a question of legitimacy, but now only a matter of wading through the bureaucratic quagmire of committees and paperwork. I had learned to flush out the opposition. I had also gained invaluable knowledge about the structure of medical care settings. Bogdan and Taylor (1975: 32) take note of the importance of such information.

> The getting-in stage is also a time in which the researcher will gain insight into how other clients and other outsiders are processed by the organization. *The best way to learn about the structure and hierarchy of an organization is to be handed around through it.*

Establishing Rapport

Having gained access to the field, I was now faced with the difficulties of "learning the ropes," of establishing myself as legitimate and acceptable to others *on the scene*. Learning the ropes is defined by Geer et al. (1968: 209) as:

> The process of initial learning by the newcomer in any social situation – his attempts to master where things and people are, the niceties of rank and privilege, who expects him to do what, at what time, for how long; what the rules are – which ones can or must be broken, which followed to the letter.

I soon learned that patients generally defer to staff members and that if the nurse and/or doctor treats an observer as appropriate and welcome, the patient is likely to follow suit. Patients at clinics and in labor and delivery units act like guests or company in someone else's home, while the settings are like homes to the host workers, doctors, and nurses. As an observer I fell somewhere in between and in the presence of patients was usually categorized like a worker/host. When I was alone with staff members I tended to ingratiate myself to the doctors and nurses. What is a more befitting demeanor in this hierarchical system with strict rules of cross-occupational conduct? While settings vary in the flexibility with which doctors and nurses behave toward each other, the former usually make rules which the latter carry out. A norm basic to medical settings is that one set of participants performs the other's dirty work, i.e., nurses, "bow and scrub"[1] for doctors. Thus, I chose to conduct myself in a serviceable manner, to be not just untroublesome but even a little useful, handy for the staff to have around.

Millman (1975) notes the difficulties of maintaining the favor of the doctors. She describes the observer's position in sexual metamorphical terms as that of having to penetrate the intimate borders of their world. She found that one has to cajole them and humor them and yet also remain aloof and sociologically or analytically distant (Millman, 1975: 619). Physicians may not be the only group who zealously guard their own autonomy to the extent of excluding other groups from achieving independence, but they certainly excel in this craft. They constantly impose their standards and values on others and glorify their own position.

Several doctors displayed a "defending-the-reigning-title" type of attitude. One doctor, when told that I was a student of medical sociology, quipped, "Oh, what are you trying to socialize?" Another ousted me from observing a cesarian section delivery on the grounds that he preferred no one be in the room "who wasn't contributing," despite the fact that a medical student was permitted to observe. Because this feeling that the outside observer is an intruder is not uncommon, I often stooped to tying a doctor's sterile gown or running to find a nurse or paging someone or doing whatever trivial task was requested. Sometimes I would do anything to keep from being attacked or confronted.

Occasionally, I was tested by a staff member to see how "on the ball" I was. Although I took offense at their thinking that one's technical medical knowledge is the only measure of his or her worth, I never minded claiming ignorance and encouraged them to lecture about this or that aspect of medicine whenever they so desired. If I asked something about the patient's progress or about a procedure, I usually got a didactic explanation, similar to what would be given any medical student. Although technical expertise was not my primary interest, I often asked medical questions out of a sense of politeness.

Nurses were occasionally intrigued by what I was writing "in my little notebook." Some commented amongst themselves that I made them nervous. I was enormously sensitive to the precariousness of my being in the covetous territory of the medical exam room. Because my aim was to view all aspects of staff-patient interaction, however, I broke unstated rules: I did not leave when husbands were asked to do so; nor did I always wait until the nurses went into the patient's room to begin observations of doctor-patient interaction. Sometimes I recorded conversations of staff at the nurses' station or at "report" during a shift change.

One major drawback of a setting where the potential for crisis and the "on call" attitude pervade the environment is that, as "mere spectator," I felt compelled to avoid discussion with staff members and patients that would focus attention on the research. I was hypersensitive to the fact that my security in the setting depended on minimizing the obtrusiveness of my role. In order to prevent anyone from reproaching me with "For heaven's sake, can't you see I'm busy now?", I tried to act either innocuous, invisible, or else supportive of participants at all times. Adopting this role limitation worked in that my conduct rarely offended anyone, although I was perhaps excessively shy and may have passed up opportunities to obtain information by engaging in informal conversation.

I went to great lengths to obtain legitimacy in the form of staff confidence in my role. I wanted to avoid hostilities toward researchers "who come and go whenever they please." To this end, I began fieldwork by observing on full work shifts at one hospital and full days of seeing patients at the doctor's offices. I suffered the whole onslaught of the work routine so as to sensitize myself to what it was like for them. And suffer I did on those long and dreary 11-to-7 night shifts! At the expense of writing up good fieldnotes, I immersed myself in their world of work, so as to be in a more comfortable, familiar status when I would later shorten my observation time and focus my attention on specific aspects of the work setting. This soothed my own conscience because I was bending over backward, catering on their behalf to the structure of the work place. Hopefully, I can proceed with the self-imposed discipline of fieldwork more rapidly in the future, having prior experience with what Hughes (1971) calls the marginal status of the participant-observer role. In other words, one's legitimacy as a member of the setting is always more or less conditional, i.e., tenuous; the researcher may remain somewhat on the outside throughout fieldwork. If experience teaches, one's skills in adjusting to marginality and in acquiring a niche in the setting should improve.

Rapport with the patients was never as much of a problem for me as was acceptance by staff members. The method by which I obtained access to the settings virtually guaranteed the patients' acceptance. One woman did refuse me permission by claiming that she preferred not to be a "guinea pig," and several asked me not to appear at a particular session after I had observed their previous visits. There were also many who were largely apathetic to my presence. Perhaps if some of these women had been more assertive or less intimidated by the institutional setting, they may have denied my access. But largely they never questioned what appeared to be staff policy. I asked eighteen women for permission to follow their pregnancy through labor and delivery, and only five – all of whom were having their first baby – turned down the request. One claimed her husband refused; the other four women gave the same explanation – that they did not want another person watching them deliver. They expected birth to be an intimate, private experience between the parents, the doctor, and the baby, and they chose to have no one else present. Little did they know about the hustle and bustle of the institutional setting and its lack of regard for personal privacy. I was tempted to warn them, but decided against suggesting that their refusal was based on a rose-colored expectation.

I developed quite positive, warm relationships with the women I accompanied on prenatal visits and during their births. They were generally excited and happy about the coming event and enjoyed my interest in their experiences. In my role as observer, I found myself performing many functions for these women. I offered support and encouragement, especially during labor when one's physical and emotional endurance is often tested beyond imagined limits. I sat with them and helped pass the time while we waited to see doctors or waited out the next labor contraction. I called to see how they were holding out when they were overdue or when a problem arose. I relieved their husbands during long hours of labor and coached their Lamaze breathing techniques, massaged their backs, cushioned them in a pushing position, held the bedpan, held the emesis pan, fed them ice chips, wiped their foreheads, retrieved nurses for them.

Although I could act neither as a medical agent nor as an involved, personal advocate on behalf of each patient, I found that the many little services I provided helped cushion the experience of pregnancy and birth. Just the presence of a familiar, friendly face proved to be a source of comfort to many of the women and their spouses. One commented to me that her prenatal sessions were more informative, that the doctor seemed to spend more time with her, whenever I was along. After delivery, several husbands took my picture along with baby's and the nurse's and thanked me profusely for my "help." I received several gifts of gratitude, which embarrassed me but also indicated that my presence was perceived by the patient as beneficial.

By working to maintain good fieldworker rapport, I became a person outside of the family who offered individualized, undivided attention to these people. Staff members only provided such interest and support when they had one single case at hand (a situation that occurred in only two hospital instances and at the home births). Where doctors and nurses have competing priorities and routines to manage, the patient is processed accordingly; she is "another one of those cases" to which staff are obliged to apply the normal standards of routine treatment. Although staff members differ vastly in style and personality – i.e., some seem more concerned and interested than others – they all possess normal routines for managing the more typical or common aspects of their cases. In contrast, the attention I gave these women was special by virtue of its marginal, extrasystemic, nonroutine nature.

A final note on my acceptance in these settings relates to the style of data collection I used. I wrote up most of my data after I left the field, relying on my memory and eye to fill in the context of skimpy notes. I found that when I became too absorbed in my own data-recording work during a course of observation, my availability for helping others suffered. My rapport in the field was based on being viewed as somewhat *useful* to all three types of participants – doctors, nurses, and parents/patients. I have since had the opportunity to collect more systematic, codified observations in labor and delivery, and I find that my rapport with both staff and patients during that period suffered and became more tenuous, less congenial. The flexibility of field-note-taking allowed me to be more serviceable and responsive to participants on the scene. Ironically, "the "scientific" appeal of precoded, quantifiable, hypothesis-testing research guarantees that it will be approved more easily by research review boards, so that access to clinical settings is less often prohibited. However, participants in those settings are likely to be more hostile to observers whose prestructured approach prevents them from reducing their obtrusiveness. Thus, increasing the observer's legitimacy as a scientist can influence his/her role to the detriment of the research. It was important for the substantive nature of my interests that I prevent potential hostility and maintain a low-status, aide-like position vis-à-vis others in the field setting.

Avoiding Going Native

The attempt to integrate oneself in the field is not without negative consequences. The problem here has to do with over-empathizing with one group's perspective on the setting. This can result in the observer's uncritical acceptance of or identification with that viewpoint to the exclusion of others. To prevent this bias, one must remain delicately balanced between several interest groups and positions. One must also carefully separate what one does to maintain legitimacy in the sense of the personal favor of members of the setting from the way one writes about it in the constructed fieldnotes.

These medical settings engendered a feeling of lack of trust in myself as a fieldworker. I was worried that bias might result from behaving in a deferential manner. Was I carving too low a niche for myself in the hierarchy or was this my only safe course of action? It seemed too easy to say or do something that would challenge the rigid order of things and therefore endanger my position. My hypersensitivity paid off, however, in that I struck an alliance with members of every group – doctors, nurses, patients.

This meant that in the long run I developed a much deeper understanding of the different perspectives and roles that comprise the situation. However, at first I was so immersed in my own favor-seeking that I was blinded to the actual dynamics of interactional control. Indeed, my initial field behavior is a good indicator of the extent to which a medical setting's operations are closed off to observability.

Other aspects of my role relationships in the field bear on the types of information that were and were not accessible to me. In particular, I am certain that my being female offered certain advantages and liabilities. I could "pass," in a formal sense, as a student nurse or nurse's aide, a figure of respectable but *low* status. Such a person threatens no one and is, in fact, of potential *use* to everyone. I doubt that a man could have followed female patients through the sequence of medical encounters and could have expressed the level of personal interest in them that I did without arousing great suspicion. Perhaps a man would have been able to forge a stronger alliance with the physicians in these settings, but his rapport with the nurses and patients might have suffered. The preexisting pressures to adopt the medical perspective and to go native certainly would increase in such a situation.[2] In general, I overcame the pressures to go native by capitalizing on and maintaining cognizance of the role expectations that all others in the setting had for me. This enabled me to distance myself from all participant perspectives and not overidentify with any one group of participants.[3]

The Case of the Expectant Father's Role

Apart from how one resolves the specific problems of obtaining access, achieving rapport, and avoiding going native, the choices of behavior styles in which a person can conduct herself or himself are somewhat limited by the context of a field setting. One does not assume a specified fieldwork demeanor in a multiplicity of situations; one adapts his or her manners, bearing, presence, and so on to the particular site. In hospitals one behaves differently than in bars, according to cultural proscriptions. In addition to the fact that observational research demands a heightened sensitivity to the situational accommodations that people make,[4] hospitals and clinics themselves elicit a guarded style of comportment.

Thus, my own preoccupation with "fitting in" and presenting myself as "appropriate" may be compared with the ways expectant fathers act in obstetrical encounters. They were overwhelmingly occupied with minimizing their intrusion on the scene, which is itself organized to *not* placate their discomfort. A person absorbed in not making a nuisance of himself or herself is too busy to see how well this behavior serves to maintain or reproduce the hierarchical role relationships. Questioning the behavior of important figures in the setting is difficult if one believes one's own legitimacy to be on the line. Thus, such pre-occupations are compatible with the interests of the staff in pursuing their work concerns. They are not solely derived from the motivations of the observer or the fathers, but are instead socially constructed or situationally produced.

Two instances from the data illustrate such timidity on the part of parents. The fathers in the cases of patients Ursula and Susan were both very quiet throughout labor, having remained undecided about whether to stay with their wives during the delivery. Each tried hard not to bother anyone, i.e., they moved quickly out of the way when the nurse wanted to check the woman's blood pressure and when the doctor entered the room. Neither of them initiated conversation with the staff. I asked one how he was doing at one point in the evening after we had spent most of the day there with his wife. "Oh, you just don't think about yourself when you see something like this," he responded quickly – which was to say, "Never mind about me." He attempted to be completely subservient to the situation; he did not try to achieve a sense of control or work out his own perspective on the situation. He acted as if he could only be troublesome to others and he could not have displayed more appreciation or gratitude to the staff for letting him be present with his wife throughout the course of events.

The other husband sat quietly throughout most of active labor, while the nurse stayed with them and at one particular point proceeded to give his wife an injection. He never raised any questions to the nurse, but instead waited until she left the room to direct his desire for information to me. As observer, I clearly had nothing to do with implementing any decisions, nor had I any authority in the situation. "What's this shot all about?" was how he phrased his puzzlement. I answered him the way I had heard many nurses explain the use of analgesic drugs. Then he asked me another question often directed to staff, "How much longer is it going to be?" I gave him the classic answer I had heard many times, "I don't think they can really tell; no one knows for sure; it depends on so much." His questions differed little from those of others, but he rarely voiced them, and even then he waited to direct them to one whose status was marginal, who presumably had minimal functions in the setting.

In both these cases the fathers did not participate in childbirth education classes, and their wives displayed great difficulty in tolerating hard labor. Several other factors might account for their relative timidity. First, having not received adequate preparation for the situation, they may have felt they did not understand enough even to pose questions to the staff members. Second, they may have been quite afraid for their wives who were experiencing great pain. Their timidity, even if a product of such fear of the unknown, however, is received by staff members as largely something to be ignored or avoided. It is not problematic for staff because the husbands made themselves untroublesome. Staff members encourage their deferential behavior; they do not object to or deal with passive shyness insofar as it makes their routine easier to accomplish.

The parallels between this behavior of husbands and that of the fieldworker are clear. My own deferential manner made my life on the wards far more pleasant as I coped with the ever-present fear of disrupting a scene, incurring the wrath of a staff person, and possibly being booted out of the hospital. This behavior, then, reflects the complicity and perhaps encouragement of the staff members as much as it does the stance taken by the fathers or observers themselves. It is a reaction to what seems appropriate, which is a moral order jointly produced by members of the setting. Those who are regular, full-fledged members do a greater share of this constructing than those who pass through as marginal members or occasional guests, i.e., as strangers. Birthing women and the fathers (like observers) primarily adapt to the implicit and explicit rules and regulations of staff members who go about their work.

However, the differences between the father's role as involved participant and that of the outside observer are also noteworthy. Because the role of supportive family member/labor coach is becoming increasingly institutionalized in many hospitals, each particular father faces adjusting himself to a structured position in the setting, rather than having to carve out a totally new niche for himself. Given his position in the preestablished network of participant roles, he is far less likely to perceive and question this social order than is the observer who attempts to make sense of and ameliorate the contradictory demands placed on her or the confusion generated by her presence. Thus, although each takes on a passive behavioral stance in deference to staff authority, the observer and expectant father yield to different pressures. Each is threatened with loss of being in the staff member's favor, but this loss has important distinguishable consequences. The observer risks losing the delicate balance of information sources. The father/husband risks incurring the wrath of the persons in whom the health of his wife and child are temporarily entrusted. The *moral* pressures on the parent to acquiesce are quite strong. For the observer, then, such pressure becomes a crucial analytical feature of the contextual structure of the setting.

A good bit of confirmatory evidence of the weight of this moral pressure is revealed in the experience of parents who do not acquiesce. The patients and expectant fathers who were not deferential and who expressed dissident views are treated as if they are violating taken-for-granted norms of behavior. Husbands who express reservations about procedural actions on the part of staff are met with hostility. Patients who make requests for special nonroutine variations on procedure are barely acknowledged, given some vague assurances, then disregarded. The usual form of hostility doled out in reply to suspicious reactions to medical plans is one in which the doctor invokes his superior ability to judge based on his years of experience in obstetrics, as in the following example.

In the case of patient Deanna, the father presented his objection to the procedure of pitocin stimulation of labor to the nurse, who replied, "Well, she's putzying around." She further explained that when a patient has ruptured membranes, "we like to get the baby out within 24 hours." The doctor later answered the father by referring to his ten years of being in the field of obstetrics. He used his tenure to give greater weight to his judgment that the risks of complications from waiting and not implementing the induction were greater than those posed by the procedure itself. He equated the father's objection with neglect of the patient, a shirking of medical responsibility, by portraying the possible negative consequences of not following through with his order.

This is a classic strategy in which the professional invalidates the lay person's perspective by claiming privileged access to information. Thus, the patient has no option but to place his or her complete trust in the autonomy of the staff. In fact, encouraging people to trust the staff blindly is considered to promote a therapeutic milieu. As with the bus driver who asks that "you leave the driving to us" and the insurance agent who claims "you're in good hands," the prescription offered in labor is, "let us do the worrying; the less anxiety you have, the better off you'll be; let us handle everything."

This suggestion is the implicit normative rule provided in the patterns of restricted communication between staff and patients in these settings. It allows staff members to proceed in unhampered fashion, with "cooperation." The way in which they define and structure ongoing events and conduct their work, then, is dependent on their ability to provide this selective feedback containing behavioral prescriptions. For the observer, this is all grist for the mill, data to be collected and analyzed. His/her interests by definition conflict with those of the staff member, because s/he must view the situation from beyond the practitioner's perspective. The interests of the parent, on the other hand, at various points converge with the staff member's. They typically conform to suggested behavior patterns without further question.

Summary and Conclusions

In my case, legitimacy in the role I had developed necessitated that I, too, conform in a way that minimized the conflict that I might generate across participant and occupational groups. I did what many researchers in similar situations have done – I adopted one approach while in the field and another while recording observations away from the setting. Thus, although fathers may have adapted in ways that were nondisruptive of staff work routines, I had to master behaving in deferential ways while conducting critical and therefore potentially disruptive investigations. I carefully censored every question that occurred to me in the course of observation and played off the political alliances I had forged, so as to "fit in" with most of the participants, to the fullest extent possible.

This variation in field strategy is not an uncommon one, nor is it one that has been neglected in the methods literature. That one's strategy reflects the *context* of the research and reveals structural features of the setting, however, has received scant attention. It is as if these choices are made on strictly theoretical or methodological grounds, or that one's adoption of a field strategy is arbitrary. It is precisely because of the tenuously taken-for-granted nature of social reality that we, as observers, have such difficulty negotiating through the everyday world. The process of finding a place for oneself from which to study a setting reveals, then, a great deal about the contextual features of interaction.

For example, in terms of role strategy, Oleson and Whittaker (1967) describe changes that occurred over the course of their research project. They found that the amount of interpersonal intimacy they displayed, as defined by the number and variety of roles in which they interacted with subjects, expanded. Scott (1963) discusses the relative status of the observer/researcher vis-à-vis other participants in a work setting. He does not address the status attainment process that the researcher undergoes, nor does he view this process as having analytic significance, as providing data on the setting's background organizational features.

Cavan (1974), in describing how she studied the Hippies of the Redwood Forest, reports her daily rounds of observation and refers to how her own activities meshed with those of the rural Hippie life. In this sense she, unlike the above others, does exemplify the relationship between observer strategy and situational factors, thus revealing much about the lifestyle of her subjects. She does not in this essay analyze her relationship with various classes of Hippies, with male versus female members of the community, for example, which would have enriched her treatment of the research context.

In contrast to these authors, my aim has been to demonstrate the theoretical insight gained through empirical negotiation of fieldworker role stategy. What we learn about the social world as a result of inserting ourselves as strangers into it should be brought back to the conceptualization stage of sociological research. The moral order of status hierarchy in medical settings has been amply documented in other works with perhaps greater passion (see Duff and Hollingshead, 1968; Shaw, 1974; and Millman, 1976), but deserves further consideration on various planes in our discipline.

Taken as a problem for fieldwork practice, how contextual features are perceived and resolved by different researchers has significance for the utility of ethnographic studies. By sharing the insights we obtain *in situ*, we may compare our *perceptions* of social worlds, rather than simply comparing the worlds we study and assuming the validity or accuracy of our sociological perceptions. When we contrast two ethnographies, we tend to that discrepant findings refer to differences in the situations or settings of study. Yet the variation may be related to different approaches and field maneuvers that open and close alternative information channels. By presenting the whole spectrum of strategic options that s/he perceived in the setting and analyzing his/her choices, the researcher offers more data from which to evaluate the work. We learn *how* observations are generated, not just *what* was observed.

Notes

1. A clinic nurse used the phrase "bow and scrub" to describe the work on a particular labor and delivery ward. She viewed this ward negatively, as a place where nurses were treated as handmaidens for carrying out doctors' orders rather than as possessors of intrinsically valuable, separate skills.
2. The importance of the sexual status and gender of the participant observer as it affects field strategy choices and as in part determined by contextual features of settings is addressed in a recent article by Easterday et al. (1977).
3. A classic article by Miller (1952) discusses the implications of overrapport for the type of data that one may collect.
4. The future of a research project often hangs on the fieldworker's maintenance of legitimacy in the setting, no matter what the status level of his/her role. Whatever accommodations seem necessary to allow the person to continue collecting data must be well understood and strictly followed.

References

Barton, A. H. and P. F. Lazarsfeld (1961) "Some functions of qualitative analysis in social research," pp. 95–122 in S. Lipset and N. Smelser (eds.) Sociology: The Progress of a Decade. Englewood Cliffs, NJ: Prentice-Hall.

Bogdan, R. and S. J. Taylor (1975) Introduction to Qualitative Research Methods. New York: John Wiley.

Cavan, S. (1974) "Seeing social structure in a rural setting." Urban Life and Culture 3: 329–346.

Danziger, S. K. (1978) "The medical context of childbearing: a study of social control in doctor-patient interactions." Boston: Boston University Department of Sociology, (unpublished)

Duff, R. S. and A. B. Hollingshead (1968) Sickness and Society. New York: Harper & Row.

Easterday, L., D. Papademas, L. Schorr, and C. Valentine (1977) "The making of a female researcher: role problems in field work." Urban Life 6: 333–348.

Freidson, E. (1975) Doctoring Together: A Study of Professional Social Control. New York: Elsevier.

________ (1972) Profession of Medicine: A Study of the Sociology of Applied Knowledge. New York: Dodd, Mead.

________ (1970) Professional Dominance: The Social Structure of Medical Care. New York: Atherton.

Freilich, M. (1970) Marginal Natives: Anthropologists At Work. New York: Harper & Row.

Geer, B., J. Haas, C. Vivona, S. Miller, C. Woods, and H. S. Becker (1968) "Learning the ropes: situational learning in four occupational training programs," in I. Deutscher and E. Thompson (eds.) Among the People: Encounters with the Poor. New York: Basic Books.

Glazer, M. (1972) The Research Adventure: Promise and Problems of Field Work. New York: Random House.
Gold, R. L. (1958) "Roles in sociological field observations." Social Forces 36: 217–223.
Hughes, E. C. (1971) The Sociological Eye: Selected Papers on Work, Self, and the Study of Society. Chicago: Aldine.
Johnson, J. M. (1975) Doing Field Research. New York: Free Press.
Lofland, J. (1971) Analyzing Social Settings. Belmont, CA: Wadsworth.
Mauksch, H. O. (1970) "Studying the hospital," pp. 185–203 in R. W. Habenstein (ed.) Pathways To Data. Chicago: Aldine.
Miller, S. M. (1952) "The participant observer and "over-rapport". Amer. Soc. Rev. 17: 97–99.
Millman, M. (1976) The Unkindest Cut: Life in the Backrooms of Medicine. New York: William Morrow.
——— (1975) "Feature review essay of The Courage To Fail." Contemporary Sociology 4: 617–619.
Oleson, V. and E. Whittaker (1967) "Role-making in participant observation: processes in the researcher-actor relationship." Human Organization 26: 273–281.
Roth, J. (1974) "Turning adversity to account." Urban Life and Culture 3: 347–361.
Scott, R. W. (1963) "Fieldwork in a formal organization: some dilemmas in the role of observer." Human Organization 22: 162–168.
Shaw, N. S. (1974) Forced Labor: Maternity Care in the United States. New York: Pergamon.

69

Two Weeks in Department Stores[1]

Annie Marion MacLean

It is so common for those who purchase goods to think nothing at all about the clerk in attendance, or the conditions under which the goods were produced, that it seems timely just now, when the Consumers' League[2] has started upon a crusade of educating the public, to give a true picture of some conditions existing in Chicago.

The necessity for a thorough investigation of the work of women and children in the large department stores in the city was apparent, and the difficulties manifold. With a view to ascertaining some things which could be learned only from the inside, the investigation which is to form the subject-matter of this paper was undertaken. It seemed evident that valuable information could be obtained if someone were willing to endure the hardships of the saleswoman's life, and from personal experience be able to pass judgment upon observed conditions. The urgency of the need, coupled with an enthusiastic interest in the work for which the Consumers' League stands, led me to join the ranks of the retail clerks for two weeks during the rush of the holiday trade. It may be urged that just judgments could not be formed at a time when conditions must be abnormal. It is true that conditions were abnormal, but the importance of knowing to what extent cannot be overestimated. The consumer should know how far his Christmas shopping works hardship for the clerks. Moreover, he should concern himself with the question as to whether the abnormal conditions he has helped to create are in part mitigated by adequate payment for the work exacted. The law in Illinois[3] prohibits the employment of children under fourteen years, and

Source: *American Journal of Sociology*, IV(6) (1899): 721–741.

limits the working day of those between the ages of fourteen and sixteen to ten hours in manufacturing and mercantile establishments, and it should be a matter of concern to the purchaser if his persistence in late shopping leads the merchant to break, or at least evade, the law. It is admittedly a menace to the social weal to have children and young girls working late at night, and thus exposed to the dangers of city streets at a time when physical and moral safety demand that they be at home. One of the objects of this investigation was to find the amount of overtime exacted, and the compensation, if any, that was given. Employers are always ready to tell the best conditions that exist; it remains to others to find the worst. And the Consumers' League utterly refuses to indorse stores that do not live up to its standard all the time.

And yet some will argue that any effort in behalf of the employés in the great stores is unnecessary. Many objections were urged against factory legislation in the early days of that reform. The champions of the movement in England met with strenuous opposition, but finally their frightful revelations of actual conditions overcame their opponents, and a wave of enthusiastic reform set in. The history in this country is similar. From 1830 to 1874 agitation for the protection of women and children in the factories was kept up, till finally, at the latter date, the Massachusetts Act became a reality. Then other states followed the example set, until, at the present time, almost all the states having large manufacturing interests have very good factory laws. Illinois is a notable exception.[4] Such, in a word, has been the history of the factory laws. We are just on the eve of an agitation for the amelioration of the conditions under which a vast army of saleswomen and cash children work. Thoughtful people all over the country have already recognized the necessity for this; but the whole body of the people must be awakened. And to help, in a small way, the educative movement here my labor was undertaken.

The difficulty of finding employment was not so great as might be supposed. Owing to the holiday rush, and the consequent need of large reinforcements to the original help, the employers were not insistent on experience as a requisite for the successful applicant. However, it was not until several visits had been made that I was promised a position at three dollars a week. Work was to begin the following Monday, which would give me just two weeks of the Christmas trade. Employment being promised, it seemed desirous to engage board in some home for working women; for the environment which such a place would provide gave promise of the best results. I was fortunate in finding a most satisfactory place not far from the heart of the city, and there I went as a working-woman. This home is deserving of more than passing mention. It provides board and lodging, together with the use of pleasant parlors and library, to working-women under thirty years of age for two dollars and a half a week, if they are content to occupy a single bed in a dormitory. These dormitories are thoughtfully planned, and accommodate from ten to fifteen each. A large proportion of the sixty-five residents were saleswomen, and they, in the course of conversation, gave me much useful information. All classes of girls were there, and most of them received very low wages. A few entries in the house register are here inserted to show the nature of the records kept, and the way in which the girls fill in the columns.

Name	*Age*	*Nationality*	*Occupation*	*Wages per week*
________	18	American	Saleslady	$4.00
________	27	Virginian	Stenographer	6.00
________	24	American	Clerk	4.50
________	23	American	Clerk	3.00
________	29	German	Cashier	6.50
________	23	Irish	Saleswoman	6.00
________	28	American	Fur worker	5.00
________	20	American	Saleslady	3.00

This, then, was the place from which I started out to work on the appointed Monday morning. The hurried breakfast, the rush out into the street thronged with a lunch-carrying humanity hastening to the down-town district, and the cars packed with pale-faced, sleepy-eyed men and women, made the working world seem very real. Hurrying workers filled the heart of the city; no one else was astir. I reached my destination promptly at eight, the time of opening. Then I had to stand in line at the manager's office awaiting my more definite appointment, which was received in due time. But the manager had changed his mind about wages, and said he would give me two dollars a week plus 5 per cent. commission on sales, instead of the regular salary he had mentioned in our former interview. I was then given a number, and by "424" I was known during my stay there. I was sent to the toy department, where I found sixty-seven others who were to be my companions in toil. The place was a dazzling array of all kinds of toys, from a monkey beating a drum to a doll that said "mamma," and a horse whose motor force was to be a small boy. Our business was first to dust and condense[5] the stock, and then to stand ready for customers. We all served in the double capacity of floorwalkers and clerks, and our business was to see that no one escaped without making a purchase. The confusion can be readly imagined. As soon as the elevators emptied themselves on the floor, there was one mad rush of clerks with a quickly spoken, "What would you like, madam?" or, "Something in toys, sir?" And the responses to these questions were indicative of the characters of the people making them. The majority were rude, some amused, and a few alarmed at the urgency of the clerks. One young boy, on being assailed by half a dozen at once, threw up his hands in horror, and said: "For God's sake, let me get out of here!" and fled down the stairs, not even waiting for the elevator. The cause of such watchful activity on the part of so many employés was the 5 per cent. commission which was to eke out the two or three dollars a week salary. Those who were experienced received the latter sum. And the extra nickels earned meant so much to many of them. Most of the girls in that department lived at home or with relatives, but in many cases the necessity for money was most urgent.

One of the difficult things at first was keeping track of the prices, for they were frequently changed during the day, and the penalty for selling under price was immediate discharge, while selling above price met with no disapproval.

Every morning there were special sales. Sometimes articles that had sold for one dollar would be reduced to ninety-eight cents, with much blowing of trumpets, while, again, twenty-five cent articles would be offered at a bargain for *forty cents* "today only." But we soon learned what things were to be "leaders" from day to day, and the manager's brief instructions each morning were sufficient to keep us posted on the bargains. The charms of the bargain counter vanish when one has been behind the scenes and learned something of its history. The humor of it seemed to impress the clerks, for often knowing winks would be exchanged when some unwary customer was being victimized.

Oh, the weariness of that first morning! The hours seemed days. "Can I possibly stand up all day?" was the thought uppermost in my mind, for I soon learned from my companions that abusive language was the share of the one who was found sitting down. Later in the week I found this to be true. One of the girls who was well-nigh exhausted sat a moment on a little table that was for sale – there was not a seat of any kind in the room, and the only way one could get a moment's rest was to sit on the children's furniture that was for sale on one part of the floor. The manager came along and found the poor girl resting. The only sympathy he manifested was to call out in rough tones: "Get up out of that, you lazy huzzy, I don't pay you to sit around all day!" Under such circumstances it is small wonder that the stolen rests were few. By night the men as well as the women were limping wearily across the floor, and many sales were made under positive physical agony.

How well I remember my first service there! The people were slow in coming that morning; in fact, they were every morning. We scarcely ever had any business worth mentioning till eleven o'clock, and the greatest rush came about six. From half-past twelve to two was a busy time also. People seemed determined to shop when we ought to be getting our meals. My first two customers were of a type that abounds. First an angular woman with a business-like expression came to me and in peremptory tones demanded that I show her building blocks. They were dutifully shown, but proved unsatisfactory. Then dolls' buggies, boys' sleds, laundry sets, and skates were examined in slow succession, and I was catechized in a thoroughly pedagogical manner regarding the prices and merits of the same. When the last skate had been critically examined, she fixed a patronizing gaze upon me and said: "I do not intend to buy today; I merely wished to examine your goods." "Was she a revenue officer?" was the first thought that came to my mind. Oh, no! in the language of the shop, she was only a "rubber-neck." I afterward estimated the distance walked with her, and found it to be about one-twelfth of a mile, and still I had not a sale on my book. She took half an hour of my time.

The next customer who fell to my lot was a man of vinegary mien who wanted a boy's sled at a cost of one dollar and a half. Now, we had none at that particular price, but we had them at one dollar and thirty-five, and one dollar and sixty-five cents, either of which I thought would suit him. But I was mistaken, for he turned upon me a look of utter scorn, and then proceeded to denounce me for advertising things we did not have in stock. I meekly suggested that I was not responsible for the advertisements which appeared in the morning papers, but he was not at all mollified, and left in high dudgeon. I felt rather blue, but the comforting voice of a little cash girl said: "Don't yer mind him, he's only a cheap skate." Thus reassured I started out on another venture. This time it was a small boy who wanted to buy, and the bright-faced little fellow did me good. He had eighty cents, he said, and he wanted presents for the baby, and Tom, and Freda, and cousin Jack, and several others. I suggested one thing after another, till finally he had spent his money; so I made out my first check and looked at it with pride. It read thus:

	X. Y. Z. HOUSE.		
	Sold by		*Am't rec'd*
	424		*.80*
1	"Dewey" bank		05
2	Sets dishes	15	30
1	Laundry set		15
1	Mother Goose ladder		12
1	Rubber ball		10
2	Bb'ls clothespins	04	08
			80
Cash no.			*Amount*
127			.80
			.80

The boy was happy, and so was I. I looked admiringly at the eighty cents set down on my index sheet. It meant that I had earned four cents. After that the sales came frequently. They were all small, of course, and amounted to only $14.98 for the day. But this was more than I sold any succeeding day. It has often been noticed that new clerks do better at first than they do later.

With me, freshness and interest in the novelty helped to banish weariness and invite sales.

My first day ended at half-past six. Through some oversight, a supper ticket was not given to me, and so I was allowed to go home. I went wearily to the cloak-room and more wearily to my boarding place. When I arrived there, I could only throw myself upon my little white cot in the dormitory and wildly wonder if it would be all right for a working girl to cry. Presently I was dreaming that blows from an iron mallet were falling fast upon me; and in a little while it was morning, and another day was begun. Hundreds of clerks in the city were starting out for work just as weary as I, but with them there was not the knowledge that labor could be ended at will.

It must be understood that "our house" was open every evening till about ten o'clock, and the only compensation given for the extra work was a supper, the market value of which was about fifteen cents. That, like the lunch, had to be eaten in great haste. The maximum time allowed, in either case, was thirty minutes, but our instructions were to "hurry back." That half an hour was wholly inadequate one can readily imagine. It sometimes took ten or fifteen minutes to get a simple order filled in the crowded restaurants near by. The lunch outside meant from ten to fifteen cents a day out of our small earnings, but the breath of even the smoky outdoor air was worth that to us. The air inside was always foul, and the continual noise was fairly maddening. We were obliged to eat our supper in the store, where it was provided. The second day I partook of what the management magnanimously called the "free supper." We were fed in droves and hurried away before the last mouthful was swallowed. The menu consisted of a meat dinner and an oyster stew, the latter of which I always elected with the lingering hope that it had not been made of scraps left from the regular café dinner earlier in the day. The said stew consisted of a bowl of hot milk, in the bottom of which lurked *three* oysters, except on that memorable day when I found *four*.

The days in the store were much the same, with their endless fatigue. At times the rush would be great; then again we would have nothing to do but stand around and talk. Thus we became surprisingly well acquainted in a short time. We talked about our wages and compared index sheets on every possible occasion. Some sold very little and at the end of the week had no more than three dollars.[6] The mental anguish of some of the girls when they saw at night how small their sales had been is impossible to describe. One may elect to become a worker, and endure the hardships of the toil, and live the life of the laborer, and receive the same starvation wages, but he can never experience the abject wretchedness of not knowing where to turn when the last dollar is gone. Three dollars a week to a girl alone in the city means starvation or shame.

The fourth day of the week was one I remember well. There had been special sales the day before, and everyone was more tired than usual;

consequently those in charge were more than usually harsh and discourteous. One girl was ill. She should not have left home, but she feared losing her place if she remained away. She found after an hour or two that she could not work, so she asked permission to go home. The answer given was that she need not return if she left then. The floorwalker, who had a spark of humanity in his breast, told her that she could go to the toilet-room to lie down, if she would come out to her place once in a while to show that she was there. That poor girl spent the day on the rough, dirty floor, with a cash girl's apron for a pillow. At intervals she dragged herself out to her place in the department, only to crawl back more wretched than before. We wondered sometimes why there was no large chair or couch provided for an emergency case of that kind. There were comforts in the customers' waiting-rooms, but discharge was the fate of the employe who dared go in there.

A shop girl might die on the bare, hard floor, while easy chairs and couches in another room were unoccupied. Surely it would not be unreasonable to require that suitable rest-rooms be provided for the employés. Undue advantage could not well be taken of such a thing, for we could not leave the floor without asking the floorwalker – a man – for a pass, and his injunction always was, "Don't stay long." The unpleasantness of asking for a pass was sometimes overcome by girls slipping away in the crowd without permission. We thought some woman might be commissioned to grant such requests. We had to endure so many unnecessary hardships.

The cloak-, toilet-, and lunch-rooms were the gloomiest and filthiest it was ever my misfortune to enter. The cobwebs and dirt-besmeared floors looked "spooky" under the flickering glare of insufficient gaslight. The only ventilation came through a foul basement, and there the little girl attendants stayed all day and late into the night. And that was where the girls who brought lunches had to eat them. A few rough board tables and chairs in a more or less advanced state of ruin were provided, and scores of hungry girls sat around and ate lunches from newspaper parcels and drank coffee from tin cans.[7] It was not a healthful atmosphere, either physically or morally, and yet it was typical of the poorer class of stores. The slang of the streets, interspersed with oaths, formed the staple medium of communication. A young and innocent newcomer could not fail to feel shocked at what she heard. But the surroundings were not conducive of elevated thoughts. Refinement of thought and speech would soon disappear in such an environment. I never saw a clean towel in the toilet-room. Several hundred pairs of hands were wiped on the coarse, filthy piece of crash each day, and there was no woman in attendance to see that things were kept in a sanitary condition. Two little girls were in the cloak-room, but they had nothing to do with the adjoining places. The rooms were merely narrow hallways. The wretchedness of all these appointments was forced upon me the day my fellow-worker was so ill. It was so hard to get our wraps at night, for then all the employés were there pushing their way to the front. One night a young girl in the line was rather restless, and one of the store officials charged her with crowding, and jerked her out of line so that she struck against a counter on the right. He then shoved her back with such force that she fell against another on the left. She was badly hurt, and the uproar which followed was mob-like in its intensity. The boys were going to shoot the offender, they said, but he only smiled, secure in the justness of his attack. The case was afterward reported to the managers, but no reparation was ever made. The girl was unable to work the next day on account of the soreness of her back. In addition to the physical discomfort she had to endure, she lost a day's wages. From that warlike atmosphere we went forth into the night, and many of us had to go alone. That night I felt timid; so I asked if anyone was going my way. A little cash girl of only thirteen years spoke up and said : "I'll go wid yez." She had eight blocks to walk after she left me. The only mitigating circumstance was her total lack of fear. She was used to sights and sounds to which I was a stranger. There were always men on the street corners ready to speak to a girl alone, and one hesitating step meant danger. Almost every morning the girls had some story to tell of encounters with men of that class; and that they were not exaggerating was proved satisfactorily to me by an experience of my own. I stepped from the car one night after midnight, and soon found that I was being followed. The chase continued for two blocks, when I staggered breathless into my doorway, with my pursuer not five feet away. My terror had given me power to outrun him.[8]

I always pitied the cash children. Many of them were too young to be working, but the sin was at their parents' door. They placed on file the required affidavits,[9] and the employer asked no questions. One little girl confessed to me that she was *not quite twelve years old*, but she told me not to tell anyone, because her mother told her to say she was fourteen. This burst of childish confidence came when I was pitying her because she had the toothache. The poor little things always had the toothache. There seems to be something about enforced work that brings on that malady in a child! But their trouble was probably more real than imagined. They often carried some cheap candy in one apron pocket and a little vial of toothache drops in the other, but they thought of no relationship existing between the two. The little girls frequently responded to the cry of, "Cash! Here cash! "with tears streaming down their faces; and the cause was always the one just mentioned, or sore feet. They got tired, of course. They were only children, and the instinct for play was strong. They would kiss the dolls and trundle the carts they were taking to the wrapping-room. A change of any kind was hailed with delight, as for instance the "running" of a C. O. D. check, which occasioned a trip to another floor. The added labor entailed was not considered. There were a great many extra ones employed at that time, and there was a continual war between the regulars and extras. The latter were engaged every morning at thirty-three cents a day, while the former were paid two dollars a week. The "enunciator," or one who had charge of the others, received two dollars and a half a week. In that particular store the little girls looked down on the boys, of whom there were comparatively few. The latter were supposed to clear away rubbish from under the counters, and on one occasion, when no boy was in sight, the floorwalker told a girl to carry away some waste papers, and she replied with a toss of her head: "You bloke you, I ain't no cash boy!" Childlike, they had their favorites among the clerks, and the fortunes of those they watched with much interest. One day the manager of the store appeared on our floor, and in ringing tones called out "424!" As I was starting to answer the summons, my young friend threw her arms around me and said : "Don't you mind Tom Jones,[10] he can't hurt you. Tell him you're a new girl, if he scolds you; and if he's ugly, tell him to go to h–." I did not do any of those things, and I got away unhurt. He had no grievance against me, but he had such a rude way of addressing the clerks that they were all afraid of him.

We had our troubles with the manager and other officials, but they were not all. Some of the customers were so hard to please and so uncivil; and they made us feel like criminals because of our inability to do what apparently could not be done. Then there was the well-meaning buyer who persisted in asking us how much wages we got. Just why saleswomen should be subjected to such rudeness by seemingly intelligent people is difficult to see. One rather independent girl, on being asked this question for the fifth time one day, replied that she got ten dollars a week, and added: "How much do you get?" The questioner was a gentleman of clerical appearance, and he replied: "My dear young woman, I am afraid your surroundings are corrupting your good manners!" Then he passed on, doubtless feeling very righteous over his reproval.

On the whole, the week there passed quickly, and on Saturday night I decided to leave and try my fortune elsewhere. I thought that one week each in two stores would be better than the whole time spent in one. I told the manager that I wanted my pay because I was going to leave. He was rather abusive and said: "What do you want to leave for? You are making good money; you girls want the earth." I left that night with my two dollars in my pocket; my commission could not be obtained till the following week. Wages are always paid weekly there.

I was "out of a job," and trusted to luck to find another.

Sunday in the home was a quiet day. Everybody was tired and discouraged. There had been extra work, but no extra pay, and there were so many Christmas things to be bought. Sunday had to be the general mending day, and that day many were making little gifts for the friends at home. Most of the girls were sensible about dress, and they guarded their small earnings carefully. I guided my expenditure by theirs and kept an accurate account of my expenses for the week. The items are here presented:

Board for one week	$2.50
Car fare, 6 days,[11] @ 10c.	.60
Lunch, 5 days, @15c.	.75
" 1 day, @10c.	.10
For charity dinner[12]	.13
Paper, 3 nights	.06
Postal cards	.05
Candy[13]	.10
Stamps	.10
Oranges[13]	.09
Present for table girl	.05
" " matron	.10
Laundry	.16
Total expense	$4.79

What I earned for the week was as follows:

Wages	$2.00
Commission	3.25
	$5.25
Less fines[14]	30
Total earnings	$4.95

Thus I had a balance of sixteen cents after my bills were paid, and that was as much as many had. At that rate it would take a long time to earn enough to buy a pair of boots.

The next week I started out again to look for a place, and I found one where I most wished to work. When I first sought employment I was an unskilled laborer, but the next time I was an experienced saleswoman, and as such I was engaged at a salary of four dollars a week plus 1 per cent. commission on sales. This time my work was selling dolls, and there were four of us at the one counter. I realized at once that this was a much better place than the first one. The managers and floorwalkers were gentlemanly and kind, and the work was carried on in a thoroughly business-like way. I breathed freely when I found that no one would swear at me. There it was no crime to sit down, and behind each counter could be found one or two little boxes which the girls used for seats. They were awkward things, though, and very much in the way when we were moving around, waiting on customers. A hinged seat that could be swung under the counter would be such a boon.

The hours were very long. We worked from eight in the morning till eleven at night, with the exception of Christmas eve, when we worked until twelve. Half an hour was the time allowed for each meal. The only extra pay given was thirty cents each night for supper money. There was a very good cheap restaurant in the store, and there we bought our suppers for from twenty to thirty cents. Many of the clerks ate two cold lunches a day in order to save the money, while others were quite reckless and bought what they considered dainties. One day a girl who had a very bad headache went to lunch with me, and this is what she ordered:

Plum pudding with wine sauce	$0.10
Swiss cheese sandwiches	.05
Chocolate ice cream	.05
Strong coffee	.05

My astonishment was too great for words.

The work in this store was in many ways not so difficult as in the first. Our work was confined to one counter, and then we could sit down for a moment once in a while; but the customers were just as hard to suit and equally regardless of our feelings. And how long the days were! It seemed to me that my thoughts were always centered on my feet! Our arms got tired, too; we had to reach a good deal for stock. A man made me open and take the dolls from nineteen boxes to see if I could not find him one with black eyes and yellow hair. I told him they were all gone, at the price he desired, but he wanted me to verify my statement. As if it would matter to his two-year-old baby whether the doll had black or green eyes! He was evidently buying one for his own delectation. That is only one instance of the many exacting customers we met.

There the sanitary conditions were good, lunch- and cloakroom accommodation ample, and the treatment kind and courteous; but the wages were woefully insufficient. From four to five dollars a week was the average. The commission given was only temporary, and designed to give an extra impetus to the sale of the holiday goods. One girl who had worked there for seven years told me that she had never received more than five dollars a week; and she had to keep up a respectable appearance. It was an openly acknowledged fact among the girls there that the paths of dishonor were traversed to supplement their small incomes. Some of them did not hesitate to advise newcomers of this lucrative employment. They viewed the matter solely from a commercial standpoint, and justified their conduct by the urgency of the need. The girls themselves said that more than a third of them were leading lives of shame. It was common to hear such expressions as this uttered in agonized seriousness: "If I don't get more wages I'll have to go bad. But I'd hate to disgrace my family." Lecherous men were always around ready to offer aid. They came, professedly, to buy, but it was not the wares of the store they wanted. The young and pretty girls yielded most easily. They would weep, sometimes, and say : "Good people look down on us. But they don't know – they don't know. *We have to earn our living.*"

Surely any effort which is being made to bring the saleswoman's wages up to a point where she can live without the wages of sin is worthy of the most respectful consideration. Whatever is done in this direction is manifestly a social good. And, moreover, the best interests of society demand that thinking people should consider this matter seriously. All the hardships of the shop girl's life fade into insignificance before this grave danger she has to face. Adequate support is the first necessity. Improved sanitary conditions and opportunity for rest may well take a second place. They can be secured by legislation; the other must come from united action on the part of the buyers, and the organization of the saleswomen themselves. The trades-union spirit should be fostered, and the working-women taught the power of united effort.

Many merchants in this city do give living wages, but there are others who do not. I know from actual experience, and I know from reliable testimony.

My earnings for the first week have already been presented, and those of the second are here given:

Salary	$4.00
Commission on sales	1.53
Supper money	1.80
Total	$7.33
Less fines[15]	40
Week's wages	$6.93

My expenses for the week were as given below:

Board	$2.50
Car fare, 6 days, @ 10c.	60
Lunch, 4 days, @ 15c.	60
" 2 days, @ 10c.	20
Supper, 6 days, @ 25c.	1.50
Paper, 3 days, @ 2c.	06
Stamps	04
Toy dog for cook's baby	11
Bananas[16]	10
Witch hazel	10
Chewing gum[17]	06
Laundry	18
	$6.05

Thus my balance was eighty-eight cents. One girl at the home had only two cents left when her bills were paid. And she it was who said in answer to someone's expressed wish one night that "tomorrow would be Sunday": "I wish there wouldn't be any tomorrow." So many times they were overcome by the utter hopelessness of the future. They have to grapple with the wages problem in a most practical way.

It is true that the present rate of women's wages has been brought about by forces over which the public has or can have but little control; yet it is equally true that a conscientious investigation of the whole subject in this city could not be valueless. While it is impossible to improve matters at once, or perhaps ever, it is yet certainly worthy of an attempt. That women in other vocations may be in dire straits does not preclude the legitimacy of centering public efforts on one occupation.

In the two weeks I was employed I worked one hundred and seventy-five hours and received eleven dollars and eighty-eight cents, or a little more than six cents an hour. Under normal conditions the hours would be about one hundred and twenty for the same length of time. This, of course, would be exclusive of Sunday work, which is required all the year, at least in some stores. A certain number of clerks are needed for two or more hours during the day. In my first place no remuneration of any kind was given for this; in the second, car fare was always given, and lunch if the duties did not end before noon. The cash children in those two stores earned three and one-third cents an hour. When kept overtime, as they were in the first store in which I worked, they earned not quite two and one-third cents an hour. It must be said, however, that the managers did not insist upon children who worked at night being there promptly at eight in the morning. Sometimes they did not go until nine.

I am going to present just here in tabulated form some information I gathered from my fellow-workers regarding themselves. The statements are made for the month of December, and refer only to employés in the big down-town stores. In some cases I have had a dozen or more names followed by identical information, so I have inserted in the table but one to serve as a type. As a rule the working-women object to telling the wages they receive, particularly if they are low. They like to give the impression that they receive much more than they do, and this, I suppose, is not characteristic of any one particular class in society. Any figures based on a canvass made from the outside are almost sure to be misleading. The cases here cited have been carefully examined, and from intimate acquaintance with the individuals I believe them to be correct:

The organization that is attempting to mitigate the evils connected with life in mercantile establishments has most laudable aims and methods. The ameliorative movement on the part of consumers is a rational one. It is representative of the most enlightened forces in society, and rests on a sound

Name	*Employment*	*Hours*	*Overtime*	*Weekly wage*	*For extra work*	*Cost of living per week*	*Conjugal condition*	*Health*	*Remarks*
		A.M. P.M.							
A	saleswoman	8:00–6:00	none	$ 6.00		$2.50	single	fair	
B	"	8:00–6:30	evenings till 10	3.00		2.50	"	"	
C	"	8:00–6:00	none	5.00		3.00	"	"	same place 3 years
D	inspector	8:00–6:30	till 10 or 11 P.M.	3.50		lived home	"	bad	
E	wrapper	8:00–6:00	till 7 sometimes	4.00		" "	"	fair	
F	buyer	8:00–6:00	none	10.00		$4.75	separated	bad	
G	enunciator	8:00–6:30	till 10 or 11	2.50	supper	lived home	single	good	
H	saleswoman	8:00–6:30	" "	3.50	"	$2.50	married	fair	
I	cashier	8:00–6:30	" "	6.50	50c per week	3.00	single	"	
J	saleswoman	8:00–6:30	till 11	3.00		2.50	married	"	husband and 4 children
K	"	8:00–6:00	till 10	5.00	35c for supper	2.50	widow	good	one child
L	cash girl	9:30–4:30	none	3.25		lived home	single	"	
M	saleswoman	8:00–6:30	evenings	5% commission	supper	" "	"	fair	
N	cash girl	8:00–6:30	"	$2.00	"	" "	"	"	
O	saleswoman	8:00–6:30	"	$2.00+5% com.	"	$2.50	"	good	
P	"	8:00–6:30	"	$3.00	cup of coffee	2.50	"		
Q	sewer	8:30–5:30	none	6.50		2.50	widow	"	one child
R	saleswoman	8:00–6:00	evenings	5.00	30c for supper	2.50	single	fair	same place 7 years
S	cash girl	8:00–6:00	none	2.00		lived home	"	"	
T	inspector	8:00–6:00	evenings	2.50	30c for supper	" "	"	"	
U	saleswoman	8:00–6:00	"	$4.50+1% com.	" "	$1.50 at home	"	bad	mother to help
V	"	8:00–6:00	"	4.00+1% com.	" "	$2.50	"	fair	
W	"	8:00–6:30	"	2.00+5% com.	supper	$2.00 at home	"	"	
X	shirt-maker	8:00–6:00	none	$40\frac{3}{4}$c per doz.		lived home	married	bad	husband had no work
Y	wrapper-maker	8:00–6:00	"	$12\frac{1}{2}$c each		" "	"	fair	" ill; 2 children
Z	saleswoman	8:30–5:30	"	$7.00		$3.75	single	good	

basis. So long as the consumer will patronize bad stores, so long will they exist; so long as people will buy clothing produced under inhuman conditions, so long will they continue to be produced under just those conditions. Has the public no duty in the matter ? Women and children are in the industrial world, and it is useless to wrangle over the expediency of their filling the places they do. They are there, and as the weaker members of society they need protection. Inhuman and demoralizing conditions must be removed. Some of the evils here could be speedily remedied by legislation and faithful inspection. Those who have not already considered the matter would do well to peruse carefully the Consumers' standard of a fair house, and ask themselves whether or not they can do something to lessen the hardships of the salespeoples' lives.

CONSUMERS' LEAGUE OF ILLINOIS.

CONSUMERS' STANDARD.

Children. – A standard house is one in which no child is allowed to work after six o'clock in the evening, and the requirements of the child-labor law are all complied with.

Wages. – A standard house is one in which equal pay is given for work of equal value, irrespective of sex. In the departments where women only are employed the minimum wages are $6 per week for adult workers of six months' experience, and fall in few instances below $8.

In which wages are paid weekly or fortnightly.

In which fines, if imposed, are paid into a fund for the benefit of the employés.

In which the minimum wages of cash girls and boys are $2.25 per week, with the same conditions regarding weekly payments and fines.

Hours. – A standard house is one in which the hours from 8 A. M. to 6 P. M. (with not less than three quarters of an hour for lunch) constitute the working day, and a general half holiday is given on one day of each week during the summer months.

In which a vacation of not less than one week is given, with pay, during the summer season to employés of six months' standing.

In which all overtime is compensated for.

Physical conditions. – A standard house is one in which work-, lunch-, and retiring rooms are apart from each other and are in good sanitary condition.

In which seats are provided for saleswomen and the use of seats permitted.

Other conditions. – A standard house is one in which humane and considerate behavior toward employés is the rule.

In which fidelity and length of service meet with the consideration which is their due.

It is a comparatively easy matter to enlist the sympathy of intelligent and educated people, and through them reform must be brought about. The great body of buyers who regularly patronize the cheap stores will take no interest in the matter. Some may feel that they have done their duty when they cease buying at stores where evils exist; but that is a dwarfed conception of social obligation. We should not rest until the bad stores improve or go out of business.

Notes

1. It should be distinctly stated that the two department stores in which the material for this paper was collected are not the establishments which have the best reputation of their class in Chicago.
2. The Consumers' League of Illinois was organized by the collegiate alumnæ of this city in February, 1897, when a standard was adopted and a provisional constitution drawn up. A permanent organization, with Mrs. Charles Henrotin as president, was effected at a meeting held in Hull-House, November 30, 1898. The league at the present time has about eight hundred members.
3. Child-labor law of Illinois, February, 1897.
4. The supreme court declared the law of 1893 unconstitutional.
5. This meant to pile like things together in as small space as possible.
6. On Saturday night all those whose sales averaged less than five dollars a day were discharged.
7. Coffee was supplied to employés at the rate of two cups for five cents.
8. This happened during my second week.
9. The child-labor law of this state requires all children under sixteen to file affidavits sworn to before a notary public.
10. This was the way the manager was spoken of by everybody. The name here is fictitious.
11. Many of the girls walked as far as two miles to save car fare.
12. The matron asked for contributions from two cents up. Every girl in the home responded.
13. These articles were for a "treat."
14. A fine of ten cents was imposed for each tardiness, unless over half an hour; then twenty-five cents was charged.
15. These were unavoidable owing to the crowds of employés using the elevators, and the shortness of the time allowed for meals.
16. For a "treat".
17. This investment was to enable me to respond affirmatively to the oft-repeated question: "Got any chewin's on you?"

70

An Observational Study of Shoplifting

Abigail Buckle and David P. Farrington

Our knowledge about the nature and incidence of offending and about factors influencing it is severely limited by our methods of measurement. For many years, the official criminal statistics and records were the major sources of information about offending. However, these have many well-known deficiencies (see, *e.g.* Farrington, 1979b). For example, official statistics reflect the behaviour of official agencies as well as that of offenders, and these two aspects are difficult to disentangle.

The widespread dissatisfaction with official statistics has led to the increasing use of interview or self-completion methods of measuring offending, notably self-report and victim surveys. However, these are also rather biased and indirect measures of offending. Unfortunately, they often have to be validated against official records as an external criterion (see, *e.g.* Farrington, 1973; Sparks, Genn and Dodd, 1977).

More valid information about offending could be obtained if more direct methods of measurement were used. In particular, our knowledge about the nature and incidence of offending would be increased greatly if more research projects were carried out in which offences were observed as they occurred. Direct, systematic observation of offending is not easy to arrange. As McCall (1975) pointed out, offences occur with low predictability and low probability, observers may have reactive effects in deterring potential offenders, and observers may be in physical danger. Furthermore, offenders often try to commit offences in such a way that they are not observed.

Source: *British Journal of Criminology*, 24(1) (1984): 63–73.

While direct, systematic observation of offending is not easy, the argument here is that more efforts should be made by criminologists to use this method. There are a number of unsystematic participant observation studies in the literature (*e.g.* Parker, 1974; Gill, 1977), but systematic non-participant observation of offending is rare. It is likely to be most feasible with offences which are relatively frequent and which are committed in public. As a recent example, Graham (1981) studied institutional vandalism by counting broken windows twice a day. However, he did not observe these offences actually being committed.

One possible solution to the problem of observing offending is for researchers to provide systematic, controlled opportunities for offending to members of the public. For example, Farrington and Kidd (1977) carried out research in which members of the public were given opportunities to claim dishonestly coins which had apparently been dropped on the pavement. Farrington and Knight (1979, 1980) gave people opportunities to steal money found in apparently lost letters, and Farrington, Knapp, Erickson and Knight (1980) gave them opportunities to steal money during a coin-sorting task. However, these studies raise ethical issues (*e.g.* of encouraging offending) and can be carried out only with rather minor acts of dishonesty (for a fuller discussion, see Farrington, 1979a).

Observing Shoplifting

It would clearly be better to use systematic observation with more naturally occurring offending. One type which lends itself to this approach is shoplifting. Three well-known observational studies have been carried out (Astor, 1971; Group 4, 1972; Marks, 1975) in which random samples of shoppers were followed through stores by security officers and store detectives. The results of these studies are shown in Table 1, and they have been reviewed by Mayhew (1977) and referred to in standard textbooks (*e.g.* Feldman, 1977, p.7).

The most striking result is the much higher shoplifting rate in the Irish (5.5 per cent.) and American (6.6 per cent. overall) research than in the

Table 1: Observational studies of shoplifting

		% Shoplifting (N)		
Study	*Description and location of shops*	*All*	*Men*	*Women*
Astor (1971)	New York department store 1	8.4 (500)	6.4 (156)	9.2 (344)
	New York department store 2	5.2 (361)	5.7 (135)	5.3 (226)
	Boston department store	4.4 (404)	2.6 (149)	5.4 (255)
	Philadelphia department store	7.8 (382)	6.0 (132)	8.8 (250)
Group 4 (1972)	U.K. department stores	0.8 (524)	1.9 (158)	0.3 (366)
	U.K. supermarkets	2.0 (494)	2.3 (131)	1.9 (363)
Marks (1975)	5 Dublin department and convenience stores	5.5 (567)	4.4 (180)	5.9 (387)

English study (1.4 per cent. overall). Mayhew (1977, p.560) suggested that this might be because, in the English study, "the store detectives who followed shoppers were instructed to apprehend any shoplifter they saw. Thus, the requirement of proof of shoplifting may have been at a much higher level than it was in the other two exercises. In the American and Irish studies, the followers were separate from the detectives employed by the stores involved, and virtually none of the shoplifters was apprehended (1 out of 109 in the American studies, and 0 out of 31 in Dublin).

Another result to which Mayhew drew attention was the similarity between the shoplifting rates of men and women. She suggested that, if opportunity was held constant, men and women might be equally likely to commit offences. It was interesting that, in the American shoplifting studies, women were more likely to steal than men, and this difference was almost statistically significant overall (7.4 per cent. of 1,075 as opposed to 5.1 per cent. of 572; $\chi^2 = 3.03$, 1 d.f., $p < 0.10$).

Unfortunately, none of the observational studies of shoplifting was carried out by social scientists or published in a scholarly journal. It is impossible to be sure in all cases that the shoppers followed were selected at random from all those entering a store. Furthermore, the reports provide little information about such things as how people shoplifted, amounts stolen or bought, and the average time spent in stores by shoplifters and other shoppers. A time measure is useful in trying to derive an index of opportunity.

In all the observational studies, it appears that only one observer followed each shopper from the moment of entering to the moment of leaving a store. However, only a small amount of pilot work convinced us that it is extremely difficult, in most shops, for one observer to keep a shopper under constant, unobtrusive surveillance. In order to be sure that someone has taken an item, it is vital that the person's hand and arm movements should be visible at all times. This requires at least two observers watching from two different directions.

Because only one observer was used in the existing studies, it may be that they missed a number of shoplifting incidents, and hence provide under-estimates of the true incidence. Alternatively (at least in the American and Irish studies), it may be that they tended to count a number of doubtful cases where the field of vision was obscured, and hence over-estimated shoplifting. We thought that the best way of minimising the observational problems noted here was to have two observers, one man and one woman, working together to observe each subject.

The Present Study

The main aim of the present research was to develop an observational methodology suitable for investigating the nature and incidence of shoplifting. The research was essentially exploratory, and was severely limited by the small amount of funding available for it.

Shoplifting was operationally defined as taking an item from a store without authority and without paying for it. In legal terms, this establishes the *actus reus* of theft but not the *mens rea*, or intention to steal. Only the courts can adjudicate on legal questions of intent and hence decide whether any given act was technically a criminal offence. The definition of shoplifting used here was behavioural rather than legal (see also Farrington, 1983).

The store used in the present research was a small department store, part of a national chain. It was similar to many other department stores in having a wide range of goods on display which were attractive to people of different sexes and ages. It was also similar in being partly self-service and partly counter-service. An important reason for selecting the store (as with much criminological research) was that the controlling company and staff working in it were willing to co-operate with the researchers. Another factor was that, unlike larger department stores, customers did not spend much time in it. In addition, it was not unusually difficult or easy to observe customers in this store. In order to get a reasonable estimate of the incidence of shoplifting, it was thought essential to observe at least 500 people. Within the constraints of the present research, this would not have been possible if each person spent an average of 45 minutes in the store (as was the case in a large West End department store which was initially studied).

The store was described by senior security personnel of the company as being typical of those in the chain. The inventory loss, or "stock shrinkage" figure, was five per cent, for the year 1980, the year before the research was carried out. This figure was obtained independently from two sources within the company and was said to be about average for stores in the chain. The store claimed that it did not operate any regular exclusion practices (*e.g.* of vagrants, groups of unaccompanied children or known shoplifters), and the researchers did not observe any person being refused admission.

The store was located in one of the main shopping streets of a city in the South East of England. The city was not a tourist centre, and its unemployment rate was near the national average. The store's customers were drawn mainly from the city and the surrounding rural areas, and the majority were white, British and local in origin. On the basis of their dress, accents, manners and general behaviour, most could be crudely designated as belonging to the lower-middle or working classes.

The study was carried out during three consecutive weeks in August 1981. This month might have been an unusual one, in that the schools were on holiday. The sample consisted of 503 shoppers, randomly chosen from those entering the store. Before starting the research, a list of random numbers was compiled by tossing a dice, and this list determined which shopper was chosen as the subject. The observers took up their starting positions, signalled to each other that they were ready, and then followed the Nth person (N between 1 and 6) through the store. The starting positions were systematically varied, from door to door. The research was carried out on all days of the week and on all hours of the day during which the store was open.

The store had two entrances, a wide front entrance which opened to the main shopping street, and a relatively infrequently used rear entrance. A count taken between 10 a.m. and 11 a.m. on a Tuesday morning in August 1981 showed that 574 people entered the store by the front entrance and 29 by the back entrance. Tuesday was thought to be an averagely busy day, and 10–11 a.m. an averagely busy hour. Since the number of people who entered by the front entrance was about 20 times as great as those who came in at the back, it was decided to select 25 subjects from those coming in at the rear entrance. The remaining 478 who were observed entered at the front.

As indicated above, it was thought desirable to have two observers in this study, one male and one female. They were both English, white and in their mid-thirties. They dressed conventionally, and wore dull rather than bright-coloured clothing, so that their dress would blend in with the surroundings. The observers communicated with each other using a system of hand signals. One always kept ahead of the subject and the other always behind.

Both observers were trained as psychologists and skilled and experienced in observing children and adults in classroom situations. One had accompanied store detectives around large department stores for five months, and had been trained by them in methods of unobtrusively observing customers. In addition, both researchers had completed 27 hours of practice observation and data collection in this and other stores before beginning this study.

There was one store detective permanently allocated to the store. She performed other tasks in addition to detective work (*e.g.* collecting till rolls), and was known to the staff and regular customers. The store's policy was to prosecute all apprehended shoplifters, and notices were displayed stating that "thieves will be prosecuted." The store detective (like the store manager and security officer) knew that the aim of the research was to observe shoplifting. She initially followed the researchers at a distance, but soon gave this up and did not hinder the observations in any way. The store assistants were told that the researchers were studying shopping behaviour.

For each person followed, the observers recorded the date and day, the time of entry into and exit from the store, the total cost of all purchases made, and the sex, race and estimated age of the person and of any companions. In addition, of course, they made a detailed record of the behaviour of anyone who shoplifted.

Results

The major results of this research are summarised in Table 2. Nine of the 503 customers followed (1.8 per cent.) took at least one item without paying for it. None of the shoplifters was apprehended by the store detective,

Table 2: Summary of results

	All	*Males*	*Females*	*Aged over 55*	*Aged 55 or less*
Number followed	503	142	361	102	401
Percentage shoplifting	1.8	2.8	1.4	4.9	1.0
Average amount bought	£1.82	£2.38	£1.60	£1.30	£1.95
Value of goods stolen as percentage of value of goods taken out of store	0.9	1.2	0.6	2.3	0.6
Average time in store (min.)	6.9	5.7	7.3	7.3	6.7
Number of items stolen (per 10 customer-hours)	2.1	5.2	1.1	4.0	1.6
Value of items stolen (per 10 customer-hours)	£1.37	£3.18	£0.81	£2.46	£1.06

possibly because she was keeping out of the way of the researchers. Four of the 142 males shoplifted (2.8 per cent.) in comparison with five of the 361 females (1.4 per cent.). These figures are not too dissimilar from those obtained in the Group 4 survey (combining department stores and supermarkets: 1.4 per cent. overall, 2.1 per cent. of males, 1.1 per cent. of females).

Shoplifting was most frequent among those estimated to be aged over 55 (4.9 per cent. took items, in comparison with 1.0 per cent. of the remainder; $p < 0.04$, two-tailed, on a Fisher exact probability test). None of the 24 thought to be juveniles took anything. As far as the researchers could tell, none of the shoppers noticed that they were being observed. The observations of eight customers (not included in the above 503) had to be abandoned, because both researchers lost sight of the subject's hands or failed to observe the total amount charged for items purchased. There is no reason to suppose that any of these subjects shoplifted.

Three of the nine shoplifters (all men) stole two items each, while the other six stole one item each, making a total of 12 items stolen. The total value of these 12 items was £7.86, or £0.66 each on average. The purchases of these 503 shippers totalled £915.31, or £1.82 per customer on average. The value of the items shoplifted was only 0.9 per cent. of the total value of the items taken out of the store by these customers. Table 2 shows that shoplifting accounted for 1.2 per cent. (by value) of goods taken out of the store by males, and for 2.3 per cent. by value of goods taken out of the store by those over 55.

Assuming that the stock shrinkage figure is five per cent., the total purchases (£915.31) represent 95 per cent, of the stock, and the corresponding value of stock lost is therefore £48.17. The value of items shoplifted (£7.86) therefore represents about 16 per cent. of stock loss. This estimate is of the same order as that reported by Gibbens (1981), based on data from a security firm. He stated that 25 per cent. of stock loss was attributable to shoplifting, 35 per cent. to short deliveries, 25 per cent. to theft by shop assistants, and 15 per cent. to shopsoiled goods.

Only 59 of the 503 shoppers (11.7 per cent. did not purchase anything (9.2 per cent. of the males and 12.7 per cent. of the females). Similarly, only one of the nine shoplifters did not purchase anything. For the remainder, the value of the items purchased was greater than that of the items stolen in seven out of eight cases. The average amount spent in the store by the nine shoplifters was much greater than the average value of the items which they stole (£3.00 as opposed to £0.87).

Table 2 shows that the average time spent in the store was 6.9 minutes, with females spending a little longer than males (7.3 minutes as opposed to 5.7). The shoplifters spent an average of 11.0 minutes in the store. The average rate of theft was 2.1 items (average value £1.37) per 10 customer-hours. Males had a considerably higher shoplifting rate (5.2 items per 10 customer-hours), as did those aged over 55 (4.0).

There are problems in generalising from a sample to a population when a relatively infrequent event is being studied. Assuming random sampling, the 95 per cent. confidence limits for a shoplifting rate of 1.8 per cent. in a sample of 503 are ± 1.2 per cent. In other words, we can be 95 per cent. certain that the shoplifting rate in the population is between 0.6 and 3.0 per cent., but this seems rather a wide range. Assuming a sample shoplifting rate of about two per cent., it would require a sample size of about 3,000 to reduce the confidence limits to ± 0.5 per cent.

Another problem in estimating the shoplifting rate is to establish how many customers are at risk. Shoplifting acts can be committed by two people jointly, as when one person takes an item from a shelf and places it in a bag held open by another. It would be reasonable to count this as one act committed by two people. Subjects who were accompanied by companions were three times as likely to be involved in shoplifting as those who were alone (3.1 per cent. of 195 accompanied subjects as opposed to 1.0 per cent. of 308 alone). The subjects were accompanied by a total of 222 companions, so it might perhaps be more realistic to relate the 12 items stolen to 725 people at risk rather than 503. The major problem with this is that the 222 companions were not continuously under surveillance, so shoplifting acts which they committed and which did not involve the subjects might have escaped the attention of the researchers. Therefore, the rate of 12 items stolen by 725 people is probably too low, just as the rate of 12 items stolen by 503 people is probably too high.

This discussion is important in regard to estimating the weekly rate of shoplifting from the store. Taking the previous figure of about 600 people entering per hour, and the lower estimate of 12 items stolen by 725 people, leads to an estimate of about 10 items per hour stolen by customers from the store. Since the store was open for 52 hours per week, it can be estimated that over 500 items were being stolen per week. This is, of course, a rough estimate, but it does give some idea of the order of magnitude of shoplifting in this store.

It is instructive to compare this estimate with the number of shoplifting crimes recorded by the police. For the whole police force area containing this city, just over 7,000 crimes of shoplifting were recorded for the year 1981, or about 135 per week. The police division including this city accounted for about one-third of all recorded crimes in the area, or for an estimated 45 shopliftings per week. It can be seen that the estimated number of over 500 shopliftings in one store per week is more than 10 times the estimated number recorded by the police in the whole of the city and its surrounding rural area. Assuming that the total number of shopliftings actually committed in this police division is between 10 and 100 times the number in this particular store, it follows that the police are recording between 1 in 100 and 1 in 1,000 shoplifting incidents.

Of the 7,000 crimes of shoplifting recorded by the police in 1981, nearly 97 per cent, were classified as detected, no doubt because a shoplifting was only recorded when a suspect was apprehended. Therefore, the likelihood of a shoplifter being apprehended and reported to the police for any given shoplifting incident is similar to the likelihood of a shoplifting being recorded. The likelihood of a shoplifter being caught by a store detective is almost certainly higher than this.

Results obtained in the survey by West and Farrington (1977) also suggest that the probability of being apprehended by the police for shoplifting is less than one per cent. The 389 youths who were interviewed at age 18–19 reported that they had committed a total of 1,214 shopliftings in the previous three years. According to police records, these youths had been convicted of only five offences of shoplifting committed during this period. Therefore, this survey suggests that the probability of being convicted for any given shoplifting offence is 0.4 per cent.

Discussion

The results obtained in this research agree with those reported by Group 4 (1972) showing that between one and two per cent. of customers entering a store shoplift and that men are proportionally twice as likely to shoplift as women. In addition, in agreement with results obtained by West and Farrington (1977), this research suggests that the probability of any given shoplifting leading to apprehension and official police action is less than one per cent. This study also suggests that the majority of stock shrinkage is not attributable to shoplifting, and that customers aged over 55 are especially likely to shoplift.

This research also draws attention to a neglected dimension in criminology – time. The idea that offending is a dichotomous variable, and that a minority are offenders to be compared with the majority of law-abiding people, was abandoned years ago. It has been replaced with the idea that people vary in their frequency and seriousness of offending. This study suggests that offending *per unit time* might be a feasible dependent variable. The time measure might be regarded as an index of opportunity. Furthermore, it might be possible to draw conclusions about causes and effects, or to derive equations specifying the effects of independent variables on dependent variables, by studying changes in the rate of offending over time. The concepts of acceleration or deceleration of offending might come into use in criminology, and in turn equations specifying cause and effect might include time variables.

The present research has many limitations, of course. The limited funds meant that the number of shoplifters who could be observed was very small. As already explained, the margin of error in generalising the shoplifting rate from the sample to the population is relatively high. In particular, how far the male-female difference in shoplifting would hold up in a larger sample is unknown. Again, how far similar results would be obtained in other stores, other areas or at other times of the year are essentially empirical questions to which we do not yet know the answers. It would be desirable to study shoplifting in a random sample of shops, but this might be difficult to arrange in practice, because of the problem of securing co-operation. It would also be desirable to follow shoppers through all the shops in which they went, but this would also be difficult to achieve in practice.

This research involved non-participant observation, but not necessarily non-reactive observation. We believe that no shopper was aware of being observed, but we cannot be certain of this. The fact that each customer spent a very short time in the store helped the observation to remain unobtrusive. American research on the reporting of staged shoplifting incidents suggests that shoppers are extremely unobservant. For example, Steffensmeier and Terry (1973) reported that the shoplifting had to be "blatant and aggressive" before anyone noticed it. However, the mere presence of the observers may have deterred some people from shoplifting. It might be thought that shoplifting could be observed in a completely non-reactive way using close-circuit television cameras, but a large number of cameras would be needed to identify all shoplifting incidents with certainty. The use of television cameras would be more feasible in studying shoplifting in one particular area of a store.

Some kinds of theft did not occur in this research, notably price-tag switching, using goods on display (*e.g.* using cosmetics or eating sweets) and collusion between shoppers and store employees such as till operators. The first and second of these, at least, may be rare. The third may have been inhibited in this study, since the till operators knew that the amounts which they rang up were being observed and recorded.

The shoplifters in this research seemed very concerned to minimise their likelihood of being detected, by (a) looking round carefully to check that nobody was watching them, (b) stealing small, low-cost items, and (c) buying goods as well as stealing. It is reasonable to argue that shoplifting could be reduced if the subjective probability of detection could be increased. The effect of such devices as mirrors, television cameras or prominent warning signs on shoplifting is essentially unknown, because of the difficulty of measuring shoplifting.

In order to test any hypothesis about shoplifting, it is desirable to carry out a randomised experiment with observed shoplifting as the dependent variable. This raises a number of problems. Perhaps the major one is to secure co-operation from a store or a chain of stores, since shops may be understandably worried about frightening away customers or occupying staff time. Another problem is that the rate of shoplifting seems so low that it would require very large samples to detect a reduction in it. This difficulty might be overcome either by restricting the observation to "high-risk" shoppers (*e.g.* persons estimated to be over 55) or by discovering shops with relatively high shoplifting rates. An alternative method of studying shoplifting would be to take an inventory of specified items every day (see, *e.g.* McNees, Egli, Marshall, Schnelle and Risley, 1976; McNees, Kennon, Schnelle, Kirchner and Thomas, 1980). However, direct observation would be preferable, since it would be difficult to determine from inventory checks whether items were taken by customers or employees.

The observational method could be used with other kinds of crimes which do not require permission (because they occur on the street) and which are likely to occur relatively frequently. An obvious example is exceeding the speed limit. This could be observed and recorded using equipment already in use by the police. If filming cars without their owners' or drivers' consent is considered ethical, it would also be possible to link up subjects observed at one time with those observed at another (by means of number plates). It is surprising that more studies involving the observation of motoring offences have not been carried out (see, *e.g.* Konecni, Ebbesen and Konecni, 1976; Jonah, Dawson and Smith, 1982).

This research is an example of the study of offending using direct, systematic observation. The increased use of this method can lead to a great increase in criminological knowledge. It is not argued that other methods should not be used. On the contrary, there is considerable merit in studying a phenomenon using a variety of methods, and it would be useful to study shoplifting using both interviews and observation, for example. The argument here is that direct, systematic observation has rarely been used, and should be used more. The challenge to criminologists is to overcome the difficulties and develop it as a method of studying offending.

References

Astor, S. D. (1971). "Shoplifting survey." *Security World*, **8** (3), 34–35.

Farrington, D. P. (1973). "Self-reports of deviant behaviour: predictive and stable?" *Journal of Criminal Law and Criminology*, **64**, 99–110.

Farrington, D. P. (1979a). "Experiments on deviance with special reference to dishonesty." In L. Berkowitz (ed.), *Advances in Experimental Social Psychology*, Vol. 12. New York: Academic Press.

Farrington, D. P. (1979b). "Longitudinal research on crime and delinquency." In N. Morris and M. Tonry (eds.), *Crime and Justice: An Annual Review of Research*, Vol. 1. Chicago: University of Chicago Press.

Farrington, D. P. (1983). "Measuring the natural history of delinquency and crime." In R. A. Glow (ed.), *Advances in the Behavioural Measurement of Children*, Vol. 1. Greenwich, Connecticut: JAI Press, in press.

Farrington, D. P. and Kidd, R. F. (1977). "Is financial dishonesty a rational decision?" *British Journal of Social and Clinical Psychology*, **16**, 139–146.

Farrington, D. P., Knapp, W. S., Erickson, B. E. and Knight, B. J. (1980). "Words and deeds in the study of stealing." *Journal of Adolescence*, **3**, 35–49.

Farrington, D. P. and Knight, B. J. (1979). "Two non-reactive field experiments on stealing from a 'lost' letter." *British Journal of Social and Clinical Psychology*, **18**, 277–284.

Farrington, D. P. and Knight, B. J. (1980). "Stealing from a 'lost' letter: effects of victim characteristics." *Criminal Justice and Behaviour*, **7**, 423–436.

Feldman, M. P. (1977). *Criminal Behaviour: A Psychological Analysis*. London: Wiley.

Gibbens, T. C. N. (1981). "Shoplifting." *British Journal of Psychiatry*, **138**, 346–347.

Gill, O. (1977). *Luke Street*. London: Macmillan.

Graham, F. (1981). "Probability of detection and institutional vandalism." *Brit. J.Criminol.*, **21**, 361–365.

Group 4 (1972). "Are Britons four times more honest than Yankees?" Paper submitted to the Home Office Working Party on Internal Shop Security.

Jonah, B. A., Dawson, N. E. and Smith, G. A. (1982). "Effects of a selective traffic enforcement programme on seat belt usage." *Journal of Applied Psychology*, **67**, 89–96.

Konecni, V. J., Ebbesen, E. B. and Konecni, D. K. (1976). "Decision processes and risk taking in traffic: driver response to the onset of yellow light." *Journal of Applied Psychology*, **61**, 359–367.

Marks, D. A. (1975). "Retail store security in Ireland." *Top Security* (September), 204–206.

Mayhew, P. (1977). "Crime in a man's world." *New Society*, **40**, 560 (June 16).

McCall, G. J. (1975). *Observing the Law*. Rockville, Maryland: National Institute of Mental Health.

McNees, M. P., Egli, D. S., Marshall, R. S., Schnelle, J. F. and Risley, T. R. (1976), "Shoplifting prevention: providing information through signs." *Journal of Applied Behaviour Analysis*, **9**, 399–405.

McNees, M. P., Kennon, M., Schnelle, J. F., Kirchner, R. E. and Thomas, M. M. (1980). "An experimental analysis of a programme to reduce retail theft." *American Journal of Community Psychology*, **8**, 379–385.

Parker, H. J. (1974). *View from the Boys*. Newton Abbot: David and Charles.

Sparks, R. F., Genn, H. G. and Dodd, D. J. (1977). *Surveying Victims*. Chichester: Wiley.

Steffensmeier, D. J. and Terry, R. M. (1977). "Deviance and respectability: an observational study of reactions to shoplifting." *Social Forces*, **51**, 417–426.

West, D. J. and Farrington, D. P. (1977). *The Delinquent Way of Life*. London: Heinemann.

71

Glimpses at the Mind of a Waitress

Amy E. Tanner

Last June a friend and I, two teachers with a commingled yearning for adventure and philanthropy, resolved to submerge ourselves still further into the masses than we are by nature. For some time we vibrated between work in a laundry, a saloon, and a restaurant, but at last we took positions as waitresses. We secured them in a fashionable apartment house much frequented by army and navy officers, serving in the café connected with the house. Here we earned fifteen dollars a month and board and room. We had to do our laundry ourselves or pay for it, and we were required to wear white shirtwaist and aprons and to have them always clean. If we were not spotless our mistress, whom we will call Mrs. Jones for convenience sake, had no scruples about telling us so. Considering that she neither did the laundry nor paid the bills, this was but natural.

Our duties consisted in setting the tables, waiting on them, clearing them, and washing the silver and glass. After we had finished this work, we were given extras, such as polishing silver. Usually there were also trays to be prepared, taken to rooms, and later brought back and cleared. This fell to the waitress at whose table the person sat, and she was expected to prepare anything not on the bill of fare, such as toast or eggs.

We were expected to be in the dining-room about 7:30 in the morning, and at night, working as rapidly as possible, we would be through about nine o'clock. Every other afternoon two of us were relieved about three o'clock, and came up again about a quarter to six, while the others stayed to prepare the salad for dinner, answer the bell, and do anything that turned

Source: *American Journal of Sociology*, 13(1) (1907): 48–55.

up. Sometimes we had an hour to ourselves in the morning, but more often not. That is, we had as a rule a thirteen-hour day, with a break of three hours in it every other day. We had no afternoon off and we worked seven days in the week. This is literal. No time ought to be taken out for meals, for we had no table to sit down at, and ate on the run, so to speak. We filled a plate with what we could find, and perched on a stool, and gulped the food down amid the sights and smells of dishwashing, with dirty dishes all about us, and the pile of scraps growing bigger every minute as the dish-washer pursued his merry way. Under such conditions, fifteen minutes is a liberal estimate for the time spent in a meal.

For food we had what was left – which was not much. As soon as the last guest was served, cake, fruit, and luxuries that would keep, were locked away. I should not object to this if we had had plenty of other things, but Mrs. Jones never counted in her nine servants when she bought, and so as a rule some or all of us had meager pickings. For breakfast we could have an egg if we chose, and usually some bacon was left, so that we did not suffer then. At lunch, however, we had literally nothing but bread and butter most of the time. For dinner, there were usually some kind of meat, potato, and vegetables, all cold because the cook would not take the trouble to keep them hot for us. Sometimes we had dessert, sometimes not, according to its character.

As a matter of fact, we were not very hungry, for we were not allowed to eat until we were through serving, and by that time the sight and smell of so much food had destroyed our appetites. We did not really get enough to eat for the work we were doing.

The work itself soon made us lame and bruised from head to foot. The bruises came as the direct result of carrying trays about five hours a day, pushing forcibly through swinging doors, and knocking ourselves against table corners and other pugnacious articles of furniture. All of us could display choice collections of black and blue spots, especially on the right side, since we turned to the right to push through the doors. Our arms ached from finger-tips to shoulders, and our backs and necks were lame from the strain of lifting the trays. Our feet were sore, swollen, and in some cases blistered, from being on them so many hours a day. These things were true of all of us, not simply of the two teachers.

What was the effect of this bodily condition upon our mind? Our first waking consciousness was an ache. By degrees this localized itself, and we cautiously tried raising our arms, closing our hands, and so on, to see just how much hurt we were. Then it would occur to us that we ought to rise, and we would crawl out resignedly or profanely, according to our temperament. If sufficiently desperate, we went to sleep again, creating confusion in the dining-room as others tried to do our work.

All day long the back-ground of living was an ache. This showed itself in an indisposition to do anything more than was absolutely necessary, to sit down at every opportunity, to stand laxly in the dining-room, instead of being brisk and alert – in short to act like the typical shiftless servant. We soon were that. We were not on the lookout for work. On the contrary, we came to have a vague feeling of resentment against our mistress, and enjoyed taking advantage of her. The one who did the chamber work filched black- and white-headed pins, hairpins, and other small articles, and kept us extravagantly supplied with towels. We ordered desserts for guests who were not there, and ate them ourselves. We supplied ourselves with plenty of ice-water for our rooms. We took every opportunity of proving to our own satisfaction that though our mistress could work us for thirteen hours a day, we could even the score in the end. Still, this puts the matter too definitely, for only the teachers so framed the case to themselves, and that humorously. Rather, the truth was that with this constant ache in the background, our minds became dulled, the inhibitions which usually prevent such belittling tricks were removed, and the ethical tone was lowered.

In just the same way all of us were careless about personal cleanliness. Mrs. Jones kept us to a certain standard of appearances, but we were too tired and indifferent to do more. Even we teachers were, after a few days, sore put to it to force ourselves to a bath at night, though we had a nicely appointed bathroom. During all the time we were there, no one used the tub but ourselves. I can understand, too, why the others did not. They were too tired, and they did not realize that the bath would rest them. Then too, they had no kimonos to wear and could not go through the extra mental and physical effort involved in partially dressing after undressing. When one's senses are somewhat dulled, it does not take long to become accustomed to the results of infrequent bathing.

The same general effect was noticeable in other ways. Bear in mind that for thirteen hours a day all our living was concerned with the paraphernalia of eating. We saw, heard, and touched nothing else. When our hour of leisure came, we had no time to go down town and were too tired to walk, so that usually we lay on our beds and dozed instead of getting new sights and sounds. The effect upon me, and I think it was much the same with the rest, was that my mind became more and more engrossed in the present. In dozing off to sleep, I saw endless processions of knives and forks and smelt the reek of hot dishwater. In the daytime, my thoughts of outside interests, my friends, my books, even my family, all such thoughts became far away and uninteresting. They lost their tang. I was too confused in mind and too dulled to care whether my friends objected to my neglect, and I was far too tired to anticipate any pleasure from seeing or writing to them. I became a creature ruled chiefly by sensations.

The wider bearing of this is easily seen. Such a girl is not one who can improve her condition. All of us talked of getting better places. One of us, who had been a stenographer, actually rented a typewriter and began to acquire speed. But the average girl came, stayed until she was desperate, and departed without knowing what she would do next. Some impulse of freedom would seize her one day, and before night she would be gone, conscious only that she must go.

How, indeed, can a girl in such a place be expected to retain anything of mental freshness or moral poise? What incentive has she to use her mind on her work? We were worked by the day. As soon as we finished one thing, we were given another, and a time never came when Mrs. Jones could not find more for us to do. At first my friend and I did about twice as much work as the others, for we were quick and willing. The only result was the satisfaction to our mistress. We had more and more work, with no increase of wages or shortening of hours. Tired as we were, with no chance of promotion, we would have been fools if we had not become malingerers as soon as possible. What was it to us if the silver had a slight extra polish? An additional ache in our arms! So much for the virtue of conscientiousness in work!

But how many of the sterner virtues ought we to expect from a girl who looks forward to such work all her life? Fortunately, most of them do not. There is the ubiquitous "gentleman friend" the only topic of conversation outside of the dining-room interests. Naturally, for he is the golden bow of promise. He is their only avenue of escape. It is small wonder that they take the first one who comes, and are quite satisfied if they only have a chance to "lie around." No wonder, too, that many of them prefer cheap boarding-houses to housekeeping, after years of such work as ours. The young man is the ideal, the hope of better things, to such girls, and to leave him out is an impossibility.

What will happen, then, if a girl has no place in which to see him, as was the case with us? We had not even a kitchen, for ours was in use until nine o'clock or later, and no girl will receive company under a fire of stares from her companions. Nor will she give up seeing him, so that either she must take him to her own room, or go out with him to walk or to some place of amusement, as is usually the case. Either alternative is bad. The two cannot get acquainted under simple, open conditions, and if marriage does follow, they may find each other's real character far different from the one displayed in the unusual circumstances of their meeting. Still more, if you imagine the wild craving for excitement that sometimes surges up in a girl bound down to excessive and monotonous work, you will understand how easily she might yield to any form of temptation. It is surprising that so few girls of this class use liquor.

If the girl has no "fellow," her condition is even worse, unless she has some other strong interest. Genuine religious belief saves her, since she looks for a reward in the life to come. Some one dependent on her saves her. But if she chances to be alone, if she has only drudgery to expect, it would seem nearly inevitable that she should feel so damned already that nothing else could be worse. Such girls become reckless, if they have any fire of youth. They are the ones who fall by the way.

How far is the mistress blameworthy? Mrs. Jones was not a harsh or cruel woman. She was soft-spoken and in many respects fair-minded. At times she found fault, but upon legitimate occasions and in a proper way, while at other times she showed patience and forbearance under considerable provocation. Once in a while she "nagged" somewhat, but only when she was unusually worried, and I think none of us laid it up against her. We are all human. Her fault did not lie in any of these things, but in a total failure to put herself in our place. She honestly did not know that we worked thirteen hours a day until I told her. She was openly incredulous of it until I counted up the hours. I have no doubt that she would have denied it with indignation if anyone unable to prove it had made the assertion, and quite probably she has forgotten it long since, she was so unwilling to accept it.

I think she never tried to imagine the state of feeling which work for so many hours brings about, nor could she have done so if she had tried. That background of ache and lassitude so changes the values of life that I believe something of a gulf will always exist between the woman who has and the woman who has not possessed it. But Mrs. Jones did not even try to imagine it. She came among us physically fresh, alert, brisk, bright-eyed, and could not understand how we could be bleary, slouchy, and slow. After our day of work, she would ask one of us to go to her room to massage her, and would keep her there rubbing her for an hour, because she was "so completely exhausted." She was, for she had much mental worry – but in what condition did she suppose we were?

If she had stopped to think of it, probably she would have said, "Yes, the hours are long, but the work is easy." I grant that it is unskilled, that it requires comparatively little mental ability; but let us see about the easiness. For about five hours, we had to load and carry trays weighing about twelve pounds. The rest of the time, our hands and bodies were exercised in all the movements of washing and wiping dishes, putting them away, and so on. The occupation is about paralleled by light gymnastics with half-pound dumb bells, with occasional walking. How would a person feel after eight hours of such gymnastics with five of heavier work coming at intervals?

Further, as our mistress lacked ability to imagine our bodily feelings, she lacked it in understanding our great need for variety, for something besides dishes and food. Here again her attitude was most naïve. When we gave notice that we were going, she asked under what sort of people we would work, with utter simplicity, "for you know, you will not often have a home of such culture and refinement as this." She must have thought that sweetness and light somehow oozed through the walls of her private apartment into the kitchen. I cannot account for her idea on any other basis, for we never went into any rooms except the kitchen and dining-room, and she never conversed with us on any subjects save those connected with eating.

This mistress, then, sinned chiefly in inability to imagine; and it is the sin that I have found in most of the mistresses with whom I have talked before and since this experience. She had herself done housework and therefore felt that she could criticize freely, but either she had never done as much as she expected of us, or had forgotten how it felt. She needed to live in the kitchen again.

Some will ask, "Why did not the girls complain of the work?" I wonder too, for she was not entirely unamenable to reason. But I think that they only showed a trait common to most people. How many teachers will bring a complaint to the school board or even to the principal? How many employees will carry a grievance to their employer? All of them will grumble to their fellows, but a combination of motives holds them back from complaint to the employer, a fear of losing their position, dislike to being thought a grumbler, and most of all, probably, the inertia coming from the overwork itself, which holds the generality of mankind in the "station to which it has pleased God to call them."

This same inertia makes it wellnigh impossible for such girls to organize themselves so as to command an eight-hour day and better wages. Any woman who should undertake this would need leisure, or a tremendous amount of vitality in order to do it after her long working-day. In either case, she would need strong personal qualities, to draw these tired girls out to anything so stupid as a meeting. As a class, they seem unable to better themselves and the greatest consolation that I find in looking at them is that for most of them the young man is lurking in the background.

But I would not forget that there is another side to the story. We were irresponsible and neglectful of our mistress' interests. It was not unusual for a waitress who had been engaged not to come at all, while many who came left without telling Mrs. Jones that they were going. In the course of a day they would turn up for the money owing them, and always received it. The average servant, as we saw her, has little feeling that anything is due from her to her mistress. I had none myself. Is it not possible that many of the faults of which mistresses complain, aside from ignorance, rise from the half unconscious feeling in the servant's mind that, after all, the mistress has taken all that she can grab, and that she can easily go without the rest? Is not this feeling justified to some extent by their indefinite hours of work?

What is the solution? It would be premature for me to attempt an answer on the basis of a single experience. I have been giving only a snap shot at reality. But I would quote Jane Addams. She believes that the most essential thing is stated hours for work, and my past aches and pains urge me to add that the hours stated should be eight in number. What reasons can be given for making the average servant's day from one-third to one-half longer than anyone's else, especially when she works seven days in the week? An eight-hour day and six days in the week! What will become of dinner? I don't know – nor would I care if I were the waitress or cook!

72

Extracts from 'Living the Kitchen Life' and 'Appendix: Ethnography in the Kitchen'

Gary Alan Fine

Heaven sends us good meat, but the devil sends cooks.

– *David Garrick*

The day begins slowly. Entering an empty, clean kitchen on a cool summer morning, one has little sense of the blistering tornado of action to come. That the room has no air-conditioning or windows hardly matters when the door to the dining room and the backdoor are left open. Slowly workers arrive to prepare for lunch. Mel, the day cook, enters at about 9:00. The maître d' slightly after. Some busboys arrive early to prepare the dining room. Later a pantry worker, another cook, a potman, half a dozen servers, and a bartender show up. Phil, the owner, and Paul, the head chef, appear shortly before lunch.

Mel begins by checking that the restaurant has sufficient ingredients for lunch. He and Paul have already determined what specials will be offered. Since the special is ivory salmon with a beurre blanc sauce, he checks the fish for freshness. He tastes the beef stock that has been slowly simmering for two days and casually tosses in some vegetable scraps. Denise, the pantry worker, is asked to clean the newly arrived asparagus, peel potatoes and carrots, and boil some eggs. If they fall on the floor, no matter, they will be boiled. Mel and Denise prepare anything that once completed can keep.

Source: *Kitchens: The Culture of Restaurant Work* (2nd Edition) (London: University of California Press, 2009), 'Living the Kitchen Life', pp. 17–45 and 'Appendix: Ethnography in the Kitchen', pp. 233–240.

The goal is to be prepared by 11:30 for the first orders. At 11:10, a supplier brings in tomorrow's walleyed pike, and Paul, dressed casually in chinos and a checkered work shirt, examines the fish and signs for them. He has had problems with this company, which is in conflict with the local Teamsters union, and which had recently delivered tenderloin instead of rib-eye steaks. But today the pike is fresh and good. Later that afternoon the fish will be filleted for dinner. Slightly before noon, Jon, a second cook, arrives; he has been told to be in later than usual because the restaurant hopes to save on labor costs and does not expect a large number of customers for lunch.

Geri, a veteran server, hands Mel the first ticket at about 11:45. Normally the orders are to be ready in about twenty minutes, but because there is no competition for his attention, Mel begins work quickly, and the order is ready shortly before noon. Paul samples the buerre blanc sauce with his finger and approves. It sits on the counter for a few minutes before Geri's customer is ready for it. Little by little the tempo (and temperature) heats up, and Mel and Jon soon find themselves snowed under – perhaps there is a convention in town, perhaps everyone wants to eat out, but whatever the reason, the kitchen is swamped with orders: a real lunch rush. Some twenty orders are waiting at any given moment. One steak falls on the stove and is wiped off and placed back on the plate. The situation is so desperate that Paul pitches in even though he was planning to work on the books. There is much banging of pans and anger when a server takes the wrong order, and the cooks have to scramble to prepare another. Jon prepares the vegetables, and Mel, the fish. The dish is ready, but not before the server has been abused for her incompetence. The kitchen is sweltering, smoky, and greasy from the large number of salmon and London broil served that day. Paradise has become hell: a communal one. Finally at 1:10 the orders let up, and by 1:30 there are only three orders left to prepare. The cooks survived lunch. The owner strolls in to congratulate and tease Paul. Together the three cooks have served over 120 diners in about ninety minutes. The servers made good tips, which will not be reported to the IRS.

Now the dynamics change. By 2:00, Mel has left to play a round of golf. Two evening cooks, Bruce and Larry, arrive by 4:00. Paul goes to his books and later slices the pike that arrived that morning, and Jon begins to prepare for dinner, readying vegetables, reducing the stock to a beef glaze, and checking the storeroom. Life is easy as Eddie, the bartender, smokes an illegal cigarette in the hall near the cooler; Larry jokes about spiking the drink of Ray, the mildly retarded potman. Those few servers who remain congregate near the door of the kitchen, joking with the kitchen staff about their romantic lives and teasing them about the tips they received.

Not until 5:30 do workers begin thinking about the dinner to come. Today is Friday, and the restaurant has reservations for several large parties.

Paul asks Jon to stay late to prepare for a crowd of nearly one hundred customers. No one expects much business at six o'clock, but at seven the restaurant is still nearly empty, and the cooks stand around chatting. Suddenly Roy, the maître d', enters the kitchen cursing, a party of eight has suddenly canceled, and a party of ten is fifteen minutes late. Kitchen life is pathetic.

At 8:00, things are so slow that Jon is sent home. By nine only thirty customers have arrived, far fewer than expected. Cooks and servers stand outside the backdoor of the restaurant making sarcastic comments about the customers and their own idiocy in becoming involved in such madness. Everyone is frustrated and bored. By ten, the nightly kitchen cleanup is nearly finished. Only one further affront awaits. Minutes before the 11:00 P.M. closing time, a regular arrives and waits to be served. Larry remains, cooking, banging pots, and grumbling about the inconsiderateness of diners. A day that began with hope ends with frustration. Emotions and sauces have been spilled. Friendships have grown and been rended. The community survived. (Abstracted from field notes, Owl's Nest)

* * *

Consider the life of the cook, who faces enormous challenges, toiling in an environment less pastoral than infernal. Cooks must ready the kitchen several hours before customers arrive, not knowing precisely how many to expect. Preparation must permit flexibility, depending on the walk-in trade and last-minute reservations. They must then be ready to cook numerous dishes, simultaneously and without warning, with sufficient speed that those with whom they must deal – servers and ultimately diners – do not become frustrated. Cooks have several masters. Restaurants are both service and production units, and, so, cooks work simultaneously for customers and management (Gross 1958). While also part of the burgeoning population of service workers (Fuchs 1968), cooks remain part of the diminishing manufacturing segment of the American workforce.

Customers can legitimately return food to the kitchen for additional work and have, as their agents, servers whose economic interests (though not necessarily their social loyalties) are with the customers.[1] Servers do not have authority over cooks, but they can and do make demands to which cooks must respond. Cooks and servers experience different pressures. Cooks hope to have the authority to prepare dishes in an unhurried manner, whereas servers need to maximize the satisfaction of the customers, who are their immediate source of income. Since servers do not share tips, cooks have little invested – in the short term – in insuring that customers are optimally satisfied (Paules 1991, p. 108).

The second source of control is management. Management wants to limit costs while maximizing customer satisfaction. As a result, management hopes to employ as few cooks as possible, demanding efficient performance, and to use inexpensive ingredients and limit waste. In practice this means that backstage workers will be the cheapest labor available; the restaurant industry is known for hiring undocumented aliens and the mentally impaired.

As a consequence, cooks operate under heavy constraints and feel a lack of autonomy, leading to occupational dissatisfaction. This lack is compounded by the hierarchy within the kitchen. Most restaurants employ a chef, or head cook, one of whose responsibilities is to manage the other cooks. The distinction between cook and chef is real, and may provoke friction. Beyond this occupational division, cooks (or chefs) have different responsibilities and degrees of power and autonomy.

This chapter explores how, despite these forms of social control, cooks make their lives tolerable, and how they define the satisfactions and dissatisfactions of their work lives. Specifically I examine the routine grounds of cooking and how personal organization, shortcuts, tricks of the trade, approximations, dirty work, and a negotiated division of labor among cooks affects the production of food. I further describe those elements that cooks see as characterizing their occupational status, both positively and negatively.

The Routine Grounds of Restaurant Cooking

Like all workers, cooks attempt to "get by." They do not demand paradise but strive for a passably smooth routine. Yet, routine has its dangers. Cooking can be both difficult and boring (Molstad 1986). Cooks wish to transform a potentially oppressive environment into a regime in which they can live, and from which they gain a measure of satisfaction. Formal rules and demands are secondary to the practical doing of food preparation. The classic account of this problem is that of George Orwell in his memorable, disturbing *Down and Out in Paris and London* (1933, pp. 80–81):

> It is not a figure of speech, it is a mere statement of fact to say that a French cook will spit in the soup – that is, if he is not going to drink it himself. He is an artist, but his art is not cleanliness. To a certain extent he is even dirty because he is an artist, for food, to look smart, needs dirty treatment. When a steak, for instance, is brought up for the head cook's inspection, he does not handle it with a fork. He picks it up with his fingers and slaps it down, runs his thumb round the dish and licks it to taste the gravy, runs it round and licks again, then steps back and contemplates the piece of meat like an artist judging a picture, then presses it lovingly into place with his fat, pink fingers, every one of which he has licked a hundred times that morning. When he is satisfied, he takes a cloth and wipes his fingerprints from the dish, and hands it to the waiter. . . . Whenever one pays more than, say, ten francs for a dish of meat in Paris, one may be certain that it has been fingered in this manner. . . . A customer orders, for example, a piece of toast. Somebody, pressed with work in a cellar deep underground, has to prepare it. How can he stop and say to himself, "This toast is to be eaten – I must make it eatable"? All he

> knows is that it must look right and must be ready in three minutes. Some large drops of sweat fall from his forehead onto the toast. Why should he worry? Presently the toast falls among the filthy sawdust on the floor. Why trouble to make a new piece? It is much quicker to wipe the sawdust off. On the way upstairs the toast falls again, butter side down. Another wipe is all it needs. And so with everything.

While American restaurants – at least those I observed – are not blessed by the same standards of sanitary "care," Orwell is correct in attributing to cooks the desire to have the food look and taste right without excess concern about the *process* by which it becomes right. Workers do what they must within the reality of the structure of the restaurant.

Personal Organization as Coping

When one asks cooks what is essential to help them get through the day, they frequently point to personal organization – organizing those projects that comprise the arc of work (Strauss 1991, p. 72). Workers with numerous unpredictably arrayed tasks find that it is not the work but the *preparation* for that work that is critical. I asked one cook at the Owl's Nest what he considered the most demanding part of his job: "The job is as easy as you make it. If you get the stuff lined up, it's easy. There's nothing hard once you have a system. You know what you're going to do and when" (Field notes, Owl's Nest). Crucial to culinary success is to segment projects and to know their proper order. Without this ordering, what is doable becomes disastrous. The challenge of cooking (and much work) is less what is done than the *relationship* among acts: "Things seem to fall together really easy for me. . . . When I have twenty-five different things that I have to prep up, I can usually. I know how to organize things" (Personal interview, Owl's Nest). After my first day in a restaurant kitchen I wrote: "Each action in the kitchen requires but a few seconds. It is almost as though the cooks are working on twenty assembly lines simultaneously – each requires a different action. It also requires remarkable coordination among cooks" (Field notes, Owl's Nest). The skill is to order multiple tasks under intense pressure – even if they are unable to specify the rules for what is to be done when. Each task is relatively unproblematic if provided sufficient time, but the sum is nearly impossible for the inexperienced. The nearly impossible is routine because cooks are experienced enough to adjust their speed and sequencing to meet demands of the arc of work – the totality of tasks. Perhaps the greatest challenge for cooks is when they fall behind or lose track of their tasks. The arc of work assumes detailed behavioral monitoring. The finely tuned system can fall apart, to which anyone who has had their focused concentration disturbed can attest. Cooking under pressure demands attention to an internal agenda. When I asked a cook at the steakhouse about his greatest frustration, he shared concrete instances that confirm the salience of concentration: "Falling behind on your backup supplies like your sour cream and your tartar sauce. Just not having the time or the manpower to recuperate" (Field notes, Stan's). The desire to keep pace means that cooks attempt, whenever possible, to "get ahead," incorporating slack time into the process. Particularly when dealing with cold food (e.g., salads, sandwich fixings, or desserts) that does not spoil, cooks may prepare more than actually ordered (e.g., Whyte 1948, p. 3) – they have the luxury to overproduce for later use.

One means of facilitating this organization of work is to limit the options available to customers and, hence, the degree of organization needed by workers. This is output control of kind, not quantity. To control the work pace, restaurants may provide limited menus or incorporate the same elements in a large selection of dishes (the latter practice is common in Asian establishments). Restaurants with extensive menus have either simple preparations or a large staff. Repeatedly preparing the same items is easier to organize than offering a wide range of choices: flexibility can go too far in an industrial workplace. As a result, large parties are given restricted menus to ease the chores for the cooks: "A party of seventy-five will arrive for dinner at 7:30 P.M. They are given a choice of two items. Charles, the manager, tells me: 'We'll sheet pan the steaks.[2] We'll seer the steak, then bake it. We must be restrictive with them. That's how all restaurants which serve parties do it. . . . It comes out nice.' Charles admits that he can taste the difference" (Field notes, Stan's). A limited range of selections effectively controls the enormity of the task. This limitation, however, may provoke dissatisfaction among clients, who if they do not find choices to their liking may patronize other establishments.

Easing the Way

Every occupation has informal, sub-rosa procedures that make work tolerable: techniques labeled "the underside of work." Despite the "official" practices that workers are expected to follow, the practical accomplishment of the job encourages other techniques that lighten the burden of work.

I classify these sub-rosa techniques into three classes: (1) approximations, (2) shortcuts, and (3) tricks of the trade. Approximations are techniques that deny the primacy of formal rules, suggesting that workers have the autonomy to make choices around a zone of acceptable practice. Every cook has the option to make decisions, and, in fact, measuring and timing devices are never so precise that approximation is absent. Professional cooks take these approximations as necessary and natural, whereas some home cooks (and novice professionals) attempt to avoid it, unsure of the effects of their choices. Shortcuts are techniques accessible to all those who know the task: options of which every cook – professional or amateur – is potentially aware. These involve making "improper" choices that bend or break the rules of production, but that save time and effort. Tricks of the trade are primarily known within the occupation, whether in an individual establishment or in the industry as a whole, and are contained within the boundaries of the occupation as subcultural knowledge. Unlike shortcuts, these need not be formally improper but are easier techniques of reaching a desired end.

These techniques differ in the degree to which they do violence to the final product – whether they affect the quality of the finished dishes. Tricks of the trade are generally less noticeable in the final outcome – thus, we label them "tricks" – than shortcuts. Approximations, depending on how approximate they are, may have little or great effect. These terms are used in professional restaurant kitchens with similar meanings although without distinguishing between tricks as knowledge held within the boundary of the occupation and shortcuts as accessible knowledge.

Approximations

Some occupations demand precision. Yet, all produce "slop" with which workers can mess. Few occupations require the microscopic precision of draftsmen or machine-tool operators, but even for these workers there are micromillimeters of choice. To permit approximation is to provide autonomy. Entering through the portals of a commercial kitchen, a home cook may notice a lack of precise measurement. The head chef at the Blakemore emphasized that he stresses conceptual, practical working knowledge:

> The basic recipes are in the book. As a matter of fact, when we were in school we didn't figure this out 'til [final] quarter. In first quarter everyone expected that this is how we make whatever it was. Salisbury steak. This is how we make it. It got to second quarter, they said no, this is how you make it, and they gave you a new recipe. Third quarter they said this is the recipe we're going to use. You thought to yourself, "What kind of education are we getting if they can't even decide how it works?" By fourth quarter we finally figured out that what they were telling us is that if you took all three recipes, the base is going to be there. Hamburgers and vegetables are your base. Whether you put in oregano, basil, tomato juice or tomato paste, that's your option.
>
> (Personal interview, Blakemore Hotel)

Although cooks have recipes, they ignore them, interpret them, and move beyond them to creative autonomy. Recipes are suggestions, not orders, although many home cooks follow them. Restaurant cooks have a different perspective:

BARBARA: I think there's a lot of common sense in knowing how to interpret a recipe.
GAF: What do you mean "interpret a recipe"?
BARBARA: It just seems to me that when I read a recipe and I see certain instructions how to fold something in, but common sense tells me that I have to do that very gently in this specific recipe because of the ingredients that are in it.
GAF: Beyond the recipe?
BARBARA: Yeah. And I've seen people make all the same recipe at the same time, and you have five different results, anywhere from disaster to marvelous. It all was dependent on whether that person was paying attention to what they were doing and whether they were concentrating or whether they had common sense to realize how that should look at a certain stage.

(Personal interview, La Pomme de Terre)

Dishes can be prepared in many ways; the skill is to decide which preparation should be used so the food is consistent and fine. Even here the knowledge is from memory and experience. When Jon asks Mel whether he has a recipe for making crepes, Mel replies: "I just mix things together." No more specific instructions are given, and Jon's crepes turn out well.

Because it is difficult to recall specifically how dishes taste, cooks work "by the seat of their pants." Much cooking involves adding approximate amounts of ingredients. While this might surprise those who imagine the cook must follow a recipe precisely to have the food meet an ideal standard, it reflects practical cooking. Adding ingredients in this way not only saves time but also allows the cook more autonomy.

Perhaps the best example of the use of approximation is the production of stock, the basis of sauces and soups. The stock has been described as the key to haute cuisine. The great chef Escoffier believed: "Stock is everything in cooking. At least in good and well-flavored cooking. Without it, nothing can be done. If one's stock is of good flavor, what remains of the work is easy; if on the other hand, flavor is lacking or merely mediocre, it is quite hopeless to expect anything approaching a satisfactory result" (Crocker 1945, p. 109). Cooks at the Owl's Nest and La Pomme de Terre are proud that they prepare their own stock, avoiding canned broth or powdered stock, but, despite its crucial quality, they do not follow a recipe. What is added to the stock-pot is a matter of convenience, rather than planning:

> Paul, the head chef, prepares beef stock for brown gravy. He tells me that he usually lets the stock cook for forty-eight hours, but Mel needs the vat tomorrow, so it will only cook for a day. . . . Later Bruce dumps egg whites and shells into the vat. Larry comments: "Sometimes it's just the garbage can." The stock vat is located right under the water faucet, so excess water falls in the vat.
>
> (Field notes, Owl's Nest)

GAF: What things in the kitchen do you think a layperson would be most shocked by?

DIANE: The stocks, how gross it looks to look in a stockpot. The first time I ever saw them I thought it was just repulsive. I think . . . for a layperson they think, "Oh, my God, don't you just throw that away."

(Personal interview, La Pomme de Terre)

Stocks and soups represent instances in which workers' choices may seem arbitrary if, indeed, they are conscious decisions. We may have confidence that we know our work, and that everything will be fine, and it usually is. The fact that the stock cooked on Monday differs from the stock on Thursday doesn't affect the evaluation of the meal – it tastes close enough for unknowing, mortal tongues, just as cars, surgical operations, and cowboy boots can pass muster despite their microdifferences. As the ingredients are approximations, so is timing (as discussed in chapter 2). When the Owl's Nest prepares gravlax (smoked salmon), they marinate the fish for anywhere between twenty-four and forty-eight hours, depending when it is needed.

Approximations are so integral in the work environment that cooks josh about the significant margin of error in their work: "Mel pours a dash of vinegar into the salad dressing and jokes to me: 'It comes out perfect every time'" (Field notes, Owl's Nest). The nice thing about many foodstuffs is that no matter what one does to them they taste "the same" to most customers. They are "forgiving." This doesn't mean, of course, that they taste identical, but our memories of flavors are not so precise as to distinguish between tastes not dramatically distinct. Although we may be more sensitive to textures, cooks can get away with a world of imprecision that would not be possible if their customers were able to engage in comparative tasting. To be sure, customers make judgments between good and bad dishes (and dishes that are better and worse), but most consumers accept the expertise of the cook and do not have sophisticated or educated palates. The evanescent character of cooking, distinguishing it from most other arts that are either material or can be captured in a written, auditory, or visual record, allows for imprecision that is not possible elsewhere. Memory is a capricious judge.

To the degree that workers use forgiving materials, they have flexibility and opportunity for error denied to others. This, for instance, gives psychiatrists an edge over anesthesiologists in malpractice suits although, as the latter practitioners are aware, bodies can stand a range of gases. One illusion that "professionals" – or those who claim the label – demand for themselves is that they work with unforgiving materials while they hide their secrets. It is trying for workers to perform before a knowing and critical audience, but even here the knowing audience may be unaware of the script, and some errors can be rescripted into the drama (Goffman 1974).

Shortcuts

While preparing meals, home cooks make many decisions outside the rules of the recipes they follow (Tomlinson 1986): do you fry bacon for a crumbled topping or just add Bac-O-Bits; should your whipped cream come from a mixer or a can? Similar culinary trade-offs characterize professional cooks. Any competent food preparer would be aware of these techniques but might not select them because of their effects on the outcome. Some shortcuts have noticeable consequences as when instant whipped cream is used instead of cream whipped by hand; others have minor effects as when a food is defrosted in a microwave, rather than at room temperature. Of course, what constitutes a significant change in sensory quality is a matter of personal judgment and collective construction, rather than objective fact. The representation of a dish is engraved in a customer's mind, but the means by which a presented dish is judged in light of this representation is complex, affected by cost, the reputation of the house, sophistication of one's palate, and the spirits consumed. The contextual understanding of objects is critical to their evaluation (Dickie 1974).

Audience awareness and demands determines what constitutes an acceptable shortcut. Each occupation has its own audience, but all are evaluated by someone. The question in each work sphere is not whether to limit quality, but how to do so. If the client will not notice the difference, does a difference exist? A difference exists in that the cook knows that he could do "better," and this affects his occupational self-esteem; yet, other pressures may make this trade-off necessary or desirable. Like all service workers, cooks have at least three audiences for their products: (1) themselves and their peers, who strive for high subcultural standards as long as they can be reasonably met with appropriate effort; (2) management, which demands profits by keeping labor, material, and fixed overhead costs low, and by having customers return to the establishment; and (3) customers, who insist on what they define as high quality, but who are possibly unaware of what quality consists of, and who also demand "good value" (low profit) in the given market niche.

The culinary challenge is to balance these demands. These are not personal standards but demands built into the structure of the setting and the expertise of those evaluating. How dishes are prepared, while grounded in interaction, is also constrained by a set of external and communal standards.

In order to ease their burdens, workers often cook a large quantity of a food at one time and then reheat the food as needed:

> The mushroom mousse is cooked halfway through. As Diane tells me: "It takes a long time to set. So we do it this way to save time." They also do not cook beef Wellington to order but reheat slices when needed.
>
> (Field notes, La Pomme de Terre)

> The cooks prepare prime rib by putting it in the hot au jus sauce to heat up, after having previously cooked it to certain degrees of doneness. Al comments: "It's not really the best way to do it, but here [he shrugs] it's what we got to do."
>
> (Field notes, Stan's)

Cooks also reuse pans to prepare most dishes, only briefly wiping it out to remove some of the previous flavor. Doors of walk-in coolers remain ajar because they are too much trouble to open dozens of times a day. Likewise, all food is cooked at the same temperature. Kitchens do not have enough stoves to vary temperatures. All food that needs to be floured is dipped in the same flour – whether shrimp, scallops, or onion rings. There is not enough staff or energy for cooks to do differently. These techniques are practiced in home kitchens and do not presuppose extensive knowledge. Anyone is capable of choosing these techniques, even though many customers have idealistic views of backstage life in a kitchen.

Convenience foods. The most compelling balancing of values and outcomes can be seen in the decision to serve convenience foods. In theory, all who work in, or are served by, a kitchen object to convenience foods. Customers desire food made from scratch, or what is the point of dining out? Likewise, managers do not want the public to know they serve convenience foods, scarring their reputation. Cooks dislike convenience foods, which diminish their role in the kitchen, transforming them from skilled craftsmen to manual laborers – culinary de-skilling. The chef at the Owl's Nest commented that his goal was for the restaurant to be a good "scratch house," preparing food from "scratch." Yet, each restaurant served some convenience foods although "more" convenience foods were served at Stan's and the Blakemore Hotel than at the Owl's Nest and La Pomme de Terre, and their staffs seemed less defensive. With the exception of La Pomme de Terre, each restaurant served instant mashed potatoes, and La Pomme de Terre used canned tomato products, which could have been made from scratch if the chef desired. The issue is not *whether* to use convenience foods but *when*. Every occupation engages in shortcuts; the question is what kind, and how will their use be justified rhetorically.

Instant mashed potatoes demonstrate the value of convenience foods. Few customers consider "real" mashed potatoes glamorous, and, thus, these spuds have little economic value. Yet preparing them is labor intensive. As a result, cooks admit the utility of instant potatoes: "Some chefs will say, 'Well, only fresh vegetables, we'll stick to that,' and that's good. And others will . . . use boxed vegetables or canned vegetables. There's a quality difference. . . . Stuff like mashed potatoes, it's unrealistic to cook off eight thousand tons of potatoes and then mash them. Instant is so much easier" (Personal interview, Blakemore Hotel). Yet, there are limits to what is legitimate. Cooks who rely too heavily on convenience foods are scorned by others. They have chosen to be de-skilled:

> One cook complains about the current chef: "They shouldn't have any canned, I'm not real big on frozen vegetables. . . . They had a chef [previously] . . . he didn't even like dry garlic powder. He was picky. Denver likes everything in cans. . . . They use too much fake stuff, too much canned stuff. They ought to use real stuff."
>
> (Personal interview, Blakemore Hotel)

> The head chef describes the adjustments that he instituted when he took the job: "Probably the hardest part of the whole thing was retraining, to reestablishing things and getting back, doing things the basic way, cutting out the shortcuts, incorporating as much as possible fresh items as I can. Getting back to a good basic cooking."
>
> (Personal interview, Owl's Nest)

Cooks resent those who use too much convenience food, but they recognize that they themselves indulge. The decision about when and where to use convenience foods is not personal but organizational, with policy set by the head chef or the manager, who responds to imagined customer demands. Rather than operating under rules shared by the industry as a whole, each restaurant has its own cooking traditions, in which the proper use of convenience items has been negotiated and then established.

Shortcuts are inevitable but are troubling reminders of ideal standards and the distance between reality and these standards. They measure what could be achieved given ideal conditions. The time-space limits of the kitchen direct the kinds of dishes that come forth.

Tricks of the Trade

Like all workers, cooks rely on techniques that make their occupational lives easier but are not widely known to the public. These tricks of the trade are subcultural in character. One cook asserted: "It takes a degree of skill to be a cook, and it takes a greater degree of skill to be a good cook. If a new man were asked to make something . . . he wouldn't even know how to cut. He would use a layman's method to cut something, not a chef's method. Also he wouldn't have the knowledge of the materials – the meat, produce, staples, and other things" (Schroedl 1972, p. 184). The novice cook must be socialized to acquire the "operational knowledge base" of the work (Bishop 1979). Just as some tasks are imagined to be easy and are not – preparing mashed potatoes or consommé – other tasks seem difficult but are easy. Preparing a tomato peel in the shape of a rose ("a tomato rose") looks complicated but with practice can be done in seconds, even by a clumsy sociologist. Omelettes have a frightening reputation but are easy to prepare:

"Denny, the day cook, prepares a mushroom omelette by cooking one side, adding the mushrooms on top, and broiling it for thirty seconds, then folding it. He pokes with his fingers to prod it into an 'ideal' omelette shape" (Field notes, La Pomme de Terre). A key skill is knowing those techniques that transform a difficult, time-consuming task into one that is easier, without a loss of quality: "Ron is preparing a dish that requires chopped orange peel. Denver, the head chef, explains that he should grate the oranges rather than chopping the peel. Ron immediately recognizes that this is more efficient. Denver replies: That's why I get the extra nickel'" (Field notes, Blakemore Hotel). To melt sticks of butter, cooks casually toss the wrapped stick in the pan. When the butter melts, they remove the paper (Field notes, Blakemore Hotel). At one restaurant, cooks wash parsley in Dreft, a dishwashing liquid. A cook explains that soapy water "perks up" the parsley (Field notes, Stan's). As a potential customer, I was shocked at this cleansing ritual; the cooks were amused at my reaction. The techniques are simple enough, but the boundary of knowledge is real. All occupations have tricks that are unknown to outsiders, and which collectively constitute socialization.

Some tricks of the trade involve misleading customers. A thorny problem is preparing meat to a requested degree of doneness. For buffets, cooks employ impression management skills to encourage the cooperation of customers: "Denver tells Ron that one technique to satisfy buffet customers is to illuminate the toast beef with red light, making the beef appear rarer than it is. If a customer doesn't want rare meat, the chef holds the sliced meat away from the light" (Field notes, Blakemore Hotel). These techniques ease the life of the cook, without, in theory, affecting the taste of the food. Of course, the customer may not receive what he or she expects, but that is the lot of the client in a mass-service organization, particularly when the client has a loose tie to the organization and none to the worker.

Restaurants sometimes sear steaks on the grill to add the distinctive grill marks and then bake them in a conventional oven. From the lack of complaints and the routine use of the technique, it seems that most customers cannot determine that their meat hasn't been grilled.

Tricks of the trade are not only used to make hard things easy but also to correct seemingly uncorrectable errors. To work is to err. Whether a doctor who misses stitches in surgery, a scholar who makes an erroneous citation, or a carpenter who places a screw poorly, every worker requires slack and the means to cope with that slack (Hughes 1971; Bosk 1979). Cooks acquire techniques for coping with inevitable mistakes. It is the ability to deal with errors, not the ability to avoid them, that characterizes the skilled worker. The following incidents are typical:

> Paul is frying eggs for hash, and an egg yolk breaks and begins to run. Paul quickly picks up the pan, holding the point of the broken egg over the flame, sealing the egg. That side of the egg will be served face down.
> (Field notes, Owl's Nest)

> Howie tells Barbara, the pastry chef, that he mashed one of her cakes when he pushed a dish into the refrigerator. Barbara isn't upset, saying "That's all right, I can cut it down to a smaller portion."
>
> (Field notes, La Pomme de Terre)

The cracks that appear when cakes and tarts are baked are hidden by covering those areas with topping or whipped cream, with customers blissfully unaware (e.g., McPhee 1979, p. 94) – a technique known to barbers and realtors. Cooks serve the more appealing side of a piece of meat or fish face up – giving them two chances on every dish, just as a photographer has two profiles from which to select. The competent cook can manage the inevitable problems by advantageously using subcultural knowledge of cooking science and customer psychology. To be "professional" is to transform a disaster into a culinary triumph.[3] As Orwell recognized, the final product is judged, not the backstage process that produced it.

Doing Dirt

When backstages become front stages, workers face a challenge of cleanliness. Production leaves little time for amenities. Kitchens, like many production lines, are dirty. We recognize this from our own kitchens, but in such settings it is personal, known dirt, under our control. In restaurant kitchens dirt is anonymous; diners wish to believe that the backstage of restaurants is as spotless as the front stage. Alas, the real kitchens of restaurants are not like the "display" kitchens that some restaurants use to entertain their customers.[4]

Coping with filth is a classic instance of what Everett Hughes (1971, p. 343) speaks of as "dirty work": "Dirty work of some kind is found in all occupations. It is hard to imagine an occupation in which one does not appear, in certain repeated contingencies, to be practically compelled to play a role of which he thinks he ought to be a little ashamed morally." To prepare food in a dirty environment is potentially identity smudging.

I asked all the cooks what in their kitchens would most upset the public. A strong plurality cited the mess and dirt:

> There's times if you don't know the business and you don't have to do [things], you don't know what it's like to get hit. I guess I'd be upset if I walked back into the kitchen, and there was meat sitting on the board and a fish over in the sink, and the cooler door was open, and there was a couple of buckets sitting on the floor. That would upset me.
>
> (Personal interview, Owl's Nest)

> DENVER: Fifty percent of the American public, if they saw what goes on inside of the kitchen, they would never eat out again.

GAF: Give me some examples of that.

DENVER: Chicken laying out for a couple of hours while you're panning it up [for a banquet]. When it sits on the table and you're in the middle of doing something else when it came out, and so it's sitting there for a while. . . . As a matter of fact, my brother-in-law was in here the other night and was absolutely appalled that someone was cutting chicken on the cutting board and didn't sanitize the cutting board again, which the health department really would get you for.

(Personal interview, Blakemore Hotel)

These seemingly candid comments echo Hughes's insight that while this dirt is "structurally" necessary, it is undesirable and seen as embarrassing by the workers. In their values workers are not so different from their customers, except they eventually must take dirt for granted. As one cook stated explicitly: "You want it to look nice, but, you know, it's so busy that you can't possibly clean it" (Personal interview, Blakemore Hotel).[5] The demands of the front stage limit sanitation: "[I]f those serving find it difficult to provide quick service and maintain standards of hygiene, it is poor hygiene which can be readily concealed. Many examples arise; for instance, reusing unwashed dishes, using spittle to clean cutlery, wiping china and cutlery with a serving cloth that is dirty through over-use, handling food to test how hot it is, and so on" (Mars and Nicod, 1984, p. 42).[6] George Orwell's observation cited above does not reflect how food is treated in these restaurants. His horrifying "traditions" have been largely erased as governmental control over health and concerns about germs have increased; still, the challenge of cooking efficiently and pleasantly while maintaining standards of hygiene is a trade-off, even if it is not always explicitly recognized.

Observing kitchens, I became inured to sanitation "problems" – from not refrigerating sauces for hours – letting bacteria grow – to using filthy towels to wipe pans to touching food with sweaty hands. Perhaps the most salient problem is what to do when a piece of food gets "dirty." People are fumblers, and food often falls from plates and pans. Food costs money and takes time and energy to prepare. While cooks do not want to waste, they prefer not to serve what they would hesitate to eat.

Among the criteria used by culinary workers in their decision to dispose of "dirty" food is whether it is prepared or "raw" (untransformed). The latter is less problematic – it is believed that heat cures all ills, particularly as the customer will never discover the mishap: "I get Bruce a dish of escargots from the freezer. One of the snails falls on the floor, and I ask Bruce: 'Can we use that one?' Bruce assures me: 'Sure. They won't know'" (Field notes, Owl's Nest). Even when prepared food lands on the floor, the cook must not be overly fastidious – wiping or reheating will solve any problem:

DIANE: Once when I was doing Sunday brunch [at another restaurant], and this was during the French toast day, [another cook] dropped a piece of French toast on the floor, and he picked it up and wiped it off and put it back on the plate. He didn't have time to do another piece.

GAF: Does that ever happen here?

DIANE: If it does, we usually wipe it off. A lot of people have dropped a piece of meat on the floor, and you pick it up and wash it off with hot water from the sink and throw it back in the pan, and it's fine. I really don't think that anything creepy-crawly got into it in that amount of time.

(Personal interview, La Pomme de Terre)

The fact that the customer never learns justifies the worker's doing what is easiest. Workers have too much work to do, and customers can rarely trace a flaw (or illness) to a hidden event. Doctors, for instance, like cooks, know that the iatrogenic illness that they cause cannot be traced.

The emotional tension of accepting sanitation standards below one's professed values is implicit in joking, grounded in a need for role distance, that takes place when food does fall on the floor and others notice:

> Al drops a steak on the floor and serves it after quickly putting it back on the grill, warming both sides. He jokes to one of his fellow cooks: "What's that saying, Gene. You'll serve it off the floor, you won't eat it off the floor. That's how we professionals do it. I'd like to meet a germ that could live off that floor."
>
> (Field notes, Stan's)

> Bruce drops an order of spaghetti on the floor and mutters: "Son of a bitch." Roy, the maître d', jokes: "Just pick it up and wash it off. Who'll know the difference?" They don't, but the remark suggests staff solidarity.
>
> (Field notes, Owl's Nest)

Dropping food on the floor is a mistake, but one that, on account of work pressures, can hardly be avoided. Cooks must make the best of what they have despite shared values with customers. Customers are partly responsible for being served dirty food because of their desire for reasonably priced food, rapidly prepared. As I describe in chapter 6 when considering the aesthetic structure of food, temporal and economic constraints affect what is served.

Workers are frustrated in responding to those who do not know the "practical accomplishment" of the job – or who pretend not to know – professional outsiders such as journalists or government regulators. In the kitchen this is evident in the attitudes of cooks toward health inspectors, who are a source of annoyance and not taken seriously:

> I ask Jon about the report of the health inspector who had been there a few days before. Jon says that he hadn't read the report, but "he always finds something to write. There's something sitting on the floor, the ceiling's dirty. They'll always find something." Inspectors regularly complain about Mel's ashtray. He smokes in the back alcove of the kitchen, even though regulations require a break room.
>
> (Field notes, Owl's Nest)

> DENVER: [The health inspectors] and I have never seen eye to eye. If you did everything their way, you wouldn't be able to run your operation at all. I went to two of their seminars, and I was really appalled by their ideas because they just aren't working. You're taking advice in the kitchen from people who have never been in the kitchen other than to scream at you for doing something wrong.
>
> GAF: For example?
>
> DENVER: Like spaghetti sauce. They want you to put it into a two-inch-deep pan to cool it. If you're making fifty gallons of spaghetti sauce, do you know how many two-inch-deep containers you would have? A lot of them, and you don't have room to put those things.
>
> (Personal interview, Blakemore Hotel)

Fortunately for cooks, but perhaps not for patrons, local governments do not enforce their rules effectively and do not constrain kitchen activity much. A smart restaurant can agree to change and then return to cooking as its culinary staff wishes. The loose structure of government oversight, in which visits are infrequent and often inconsequential, permits the cooks more leeway than would be possible with a government that took its assignment more conscientiously – and funded more inspector positions. Semiannual inspections, with options for corrections and appeals, permit kitchen workers latitude to cook as they wish. Thus, although government inspection could be a major concern, directing behavior, in reality it has little effect. The structure of government oversight permits a range of activity that might not otherwise be tolerated.

For governments as well as cooks, sanitation is a trade-off. In principle, everyone believes that kitchens should be clean, but keeping them clean may cost more than the cleanliness is worth, particularly in the absence of an immediate health threat. Epidemics of food-borne illnesses – such as hepatitis or salmonella – are infrequent. Responses to such threats occur only after rare, major, publicized food-poisoning scares. Routine poisonings, however often they may occur, are ignored by cooks, inspectors, and journalists. They are part of doing business and dining out, and rarely can be traced. Closing down an independent small business is not something that a government that embraces capitalism wishes to do. This oversight is similarly light in hospitals, nuclear plants, chemical refineries, poultry plants, and high schools, suggesting the limits to the intrusions of ideals enforced on dirty work practice by external agencies. Organizations depend on the trust of regulators and clients. This trust is typically well placed; but even when it is not, it is difficult to monitor without increased commitment.

Dividing Labor in the Kitchen

All occupational work is grounded in collective action and a division of labor (Becker 1974; Strauss 1991). Cooks in large kitchens are no exception, even when their work appears chaotic to the untrained eye. The haute cuisine French restaurant in this journalistic account differs only in degree from the restaurants I observed:

> The pressures mount to a peak. The orders are like a barrage of machine-gun fire. One has the vague feeling of a crew of white coated seamen trying to keep their ship afloat in a hurricane. The blare of noise, the figures rushing hither and thither, the irresistible chaos of enticing smells, the heat and spitting of the frying, the clang of pots, the bloomp-bloomp of chopping knives all beat down with enveloping force until one feels dizzy. Yet, in reality, everything is proceeding normally, everyone is efficiently absorbed. A boy is quickly shelling a bowl of beautiful, pink crayfish. Michel is adding a shower of bright-green sorrel to a brilliantly yellow sauce. Andre is making patterns with peach halves on a tart shell. Pierre [the head chef] watches everything and misses nothing. He could take over any job, from anyone, at any moment, and do it better. Everyone knows this and the effect is both disciplinary and exhilarating.
> (De Groot, 1972, p. 246)

The same system operates in modified form at the Owl's Nest as the head chef explains: "Everyone knows what everyone is doing. Because you work together, you begin to think alike. Three people become one person. The only time I go around the kitchen is if there's a problem or somebody calls for help. Otherwise, I work at the kitchen" (Field notes, Owl's Nest). This chaos consists of "the fitting together of lines of action" (Blumer 1969), particularly when cooks collaborate on the same plate (e.g., one preparing the meat, a second, the accompaniments, and a third, the garnish).

An ideal-typical instance of the division of labor is the preparation for a banquet, which because of the number of customers to be served simultaneously requires fine-tuned organization, overseen by authority. The head chef at the Blakemore Hotel described the banquet setup at a large hotel at which he had once worked:

> I was in the banquet kitchen. It was a fantastic setup. . . . The people who worked in the evening would come in, and they would set up the entire breakfast and the entire lunch for the next day. And put their meals out for that evening. You came in the morning, your eggs were cracked, your

> bacon was set out, everything was ready to slide in the oven, to put on the plates. And then you would set up the dinner for them in the evening. So when they came in, their vegetables were ready to be warmed up, they may even be half-cooked. It was just a matter of putting them in the steamer for six minutes and get[ting] them back to temperature and out they'd go. The steaks would all be scored off and on a tray.[7] Another example of banquet cooking: When you cook up a banquet with steaks, you go to your broiler which you have for à la carte, but instead of cooking it, you just score it, put the lines on it. You put them back on a sheet pan and put them back in the refrigerator. Then fifteen minutes before they're to be served, you slide them in the oven to finish cooking to get them back to the temperature.
>
> (Personal interview, Blakemore Hotel)

Even though some might see cooking as a solitary activity only mediated by the food itself, professional cooking, like many occupations embedded within organizations, demands teamwork and coordination, particularly in restaurants that attempt complex presentations. The work team is as much the unit of analysis as is the individual worker.

The division of labor is not a given but must be negotiated with more or less strain. Flexibility is, of course, desirable, but when interests diverge or when communication is ineffective, tension results as workers have different ideas of what is expected of them and their colleagues.

Flexibility in a Community of Interest

One means by which a division of labor becomes flexible is through explicit or implicit expectations of direct and reciprocal cooperation. Even though a division of labor exists in midsize restaurant kitchens, this is negotiable in practice. As noted in numerous descriptions of the informal organization of occupations, workers perform each other's jobs and cover for each other. They do so willingly because they assume that later this cooperation will be reciprocated. The articulated structure of the kitchen need not be repeatedly negotiated, because of an unstated assumption that others will be available for future aid. Flexibility is built into institutional relations of co-workers. Cooperation is required, and when it is not easily given, there is surprise and tension. A lack of cooperation demands an account. For an occupation to operate efficiently, a community of interest is assumed, which makes patterns of aid flow through a network without a specific debt and obligation incurred.

In smoothly functioning organizations, workers are socialized to believe that asking for help is both expected and desired. In kitchens this is explicit in occupational rhetoric, but elsewhere mutual aid may be more sub rosa. As the head chef at the Owl's Nest comments to his workers, "Don't be afraid to ask for help." A fellow cook chimes in, "It's always easier to ask for help before you get in the shits" (Field notes, Owl's Nest). When describing the kitchen as a work community, I note that this instrumental cooperation is tethered to expressive friendships and perquisites – cooks getting drinks from the bar, dishwashers being served steaks, and servers eating fancy desserts. A "favor bank" operates in most occupational worlds.

Perhaps cooperation in the kitchen is most dramatically evident in the surprising reality that cooks regularly work unpaid overtime to help peers. Day cooks often choose to finish tasks that they have not completed during their paid hours to prevent inconvenience to the evening staff. Evening workers routinely remain until everyone has finished cleaning up, even though only one of them – or none – is paid for that time. A breakfast cook at the Blakemore Hotel regularly arrived an hour early to complete his assigned work. A norm of community lightens the establishment's labor costs, but this norm can disintegrate if workers believe that management is consciously manipulating their fellowship for profit.

The Tension of Division

Although cooperation is far more frequent than the lack of it (Gross 1958, p. 387), anyone who has worked in kitchens can attest that they are not settings of eternal harmony. Yet, in my observation, emotional displays are rare, not the rule. As a result, whereas emotional outbursts in kitchens are notable when they occur, they typify the scene for outsiders.

In one tense restaurant a kitchen staff meeting diffused much interpersonal annoyance through negotiation and clarification:

> The structural positions of the cooks were clarified by the chef in response to the complaint of Larry, a cook, who claimed that "we used to know what was expected. . . . I got upset. I walked off the line. I was upset about what was happening. . . . People don't know what they should do." The problem was a function of role ambiguity: the role of the "middleman" (aka the "slouch cook," or "swing cook") – a "backup" for the other cooks. After Paul, the head chef, detailed how the middleman should collaborate with the broiler cook and the stove cook, the, tension lessened. That evening every cook made a special effort to demonstrate publicly that [he or she] could cooperate. Each man pointedly commented to me privately on how effective the meeting was. As Larry explained, by the end of that evening: "I felt a lot more relaxed and a lot more free. Jon [the cook with whom he had been annoyed] understood. . . . It's funny that's all it takes, fifteen minutes of talking and everything's different. It's the way it should be every night."
>
> (Field notes, Owl's Nest)

The point is not that cooks always work harmoniously, but that – in American kitchens (and much of American culture [Stearns 1987]) – an ideology of harmony prevails.[8] Americans believe that cooks should be able to get along with each other, and that if anger is evident and full cooperation absent, the organization is "dysfunctional" and needs help. A therapeutic model applies to organizations as well as persons. Cooperation is a central ideological tenet of the lived experience of work.

Being a Cook

To understand the experienced reality of cooking as a practical activity, we need to address how cooks and chefs see their work, how they perceive public attitudes, and how they were recruited to kitchen work. Occupational identity is tied to the pleasures and pain of work, and the imagined responses of the "other," the consuming public.

Cooking is demanding work; it is experienced as hard labor. Like athletes, cooks must "play" in pain; like a policeman, a cook only rarely has the luxury to call in sick. Those cooks employed by small organizations find their presence is required daily. One cook described his severe back pains, necessitating physical therapy, but continued cooking (Field notes. Owl's Nest). I was often told that cooks must work no matter what:

> I ask Mel if Paul will be in today. He seems surprised by the question and tells me that as far as he knows Paul will be here, adding "In this business, you don't get sick. You're either drunk or shacked up. You gotta drink or shack, so you better decide which. When I was young and working at the Lexington, I called in sick one day, and the next day the manager called me in and said that to me. Cooks don't often get sick."
>
> (Field notes, Owl's Nest)

> Paul, Jon, and Mel joke about one of Jon's absences. Mel is cracking eggs, and Paul tells him; "You should let Jon do it. He was the best breakfast cook we had [at the hotel at which they both worked] – when he would come in." This is a reference to one occasion on which Jon didn't show up; Paul jokingly accused him of shacking up with someone. The cooks then talk about the excuses that cooks use, including one who said that he was late because his mother tripped over the cord of his alarm clock. This joking colloquy has a strong element of social control.
>
> (Field notes, Owl's Nest)

This reality would surely be disconcerting to customers, who might be horrified to realize that all too often the cooks are sniffling, sneezing, exhausted, hungover, distracted, or bleeding.

The Downside

In addition to being required to be "iron men," other structural drawbacks mark cooks. They face challenges of time, pressure, working conditions, and a lack of personal satisfaction. Food preparation, although currently a trendy job for children of upper-middle-class baby boomers, will never have wide appeal.

Hours. Those in some occupations labor while their clients play – restaurant workers are among them. As the head chef at the Owl's Nest notes sarcastically, "What an exciting way to spend a Saturday evening!" (Field notes, Owl's Nest). Some cooks like least "the long hours, weekends, holidays. . . . Everybody else is out having fun, and you have to work" (Personal interview, Owl's Nest). Others most dislike "having to be here when you'd like a little time off to do some of your own things. Take time to be with your family. Things you should be doing, but you can't be. Being involved more with community things, home things, PTA meetings, kids' baseball games, and stuff like that. You have to forfeit a lot" (Personal interview, Owl's Nest).

Pressure. Even at its best, cooking is not known for its calm placidity. It can be a draining, pressured occupation – low paid, poorly regarded, and hard. (see chapter 2). As one cook explained: "It wears you. Try to cook the way I [do] now, and I'd be dead by the time I was forty" (Personal interview, La Pomme de Terre). Another emphasized he wouldn't be a cook for the rest of his life because "I don't want to be forty years old and grouchy" (Personal interview, Owl's Nest). Cooking is a young man's game.

Working Conditions. A kitchen is a hot, dirty, close place – no expansive office with flowers and big picture windows. Over time this reality affects cooks. For some the prime frustration is the ill-fitting uniforms or hair nets; for others, the odors. One told me that cooking "gets into your pores. When I go home, my kids can smell me. I'm told by a lot of people that 'you smell like vegetable soup'" (Field notes, Blakemore Hotel). Leaving the steak house, I was perfumed by cooking oil. Other cooks mentioned the stifling heat from standing over stoves and burners, and the pervasive dirt and grease. Although restaurant work is cleaner than some outdoor blue-collar jobs, it is far from the white-collar life that some desire.

Personal Dissatisfaction. Cooks feel unappreciated, which translates into a general sense of despair. One cook at the Blakemore reported a motto on a button that she found symbolically relevant to her situation: "The Torture Never Stops." She joked that being fired might be the best thing that could happen to her. Another cook commented that "my job is worthless. There's so many incompetent people there. It's like a big joke" (Personal interview, Blakemore Hotel). Although her view is not universal, it is a feeling many cooks have experienced.

Public Suspicion. However cooks may judge their own work, they must cope with a widely shared belief that the public does not respect them. They are, of course, not alone in this concern. Even such a high-paid professional as a lawyer, or a credentialed one as a doctor, must cope with what may seem a tide of public scorn. Most, if not all, occupations are challenged by outsiders. Every occupation develops strategies to cope with public attitudes. If one asks cooks, one will hear that the public, often ambivalent, does not give them the respect that they desire. The images of the drunken, ignorant chef and the artistic chef may be superficially contradictory, but they can coexist. Genius and deviance are, despite their distinct images, compatible.[9]

The public frequently sees restaurant kitchens as brutal places. One cook felt this lack of respect especially deeply:

AL: I think people should know how cooks feel. They're human beings and some people treat them like they're robots, and they have to do this and that. They should have more respect than what we get.

GAF: Do you think that if you had more respect, you might continue to be a cook?

AL: Probably. I don't know. . . . Working at a place like this, a lot of people say you do a good job, but you just don't get the respect that you want.

(Personal interview, Stan's)

Many cooks felt that their contacts just didn't see their career as suitable for someone who could get a "professional job," or who could be a success:

GAF: What do you like least about being a cook?

LARRY: Right now the feeling of not being respected as much as being a doctor. Right away when you hear someone say, "Well, he's a doctor," you think, "Well, he's got money; he's got a nice house, and a nice car, and a really nice family, and an airplane and a cabin." I'm a chef. "I bet he's got a dumpy apartment, and he's not married. His plants die on him. He's a loser. He's a high school graduate. He's just a dummy." . . . "Well, what do you do?" "I'm a cook." "What are you going to do when you get out of school." "I'm going to keep doing it." "Really? Is that all you're going to do. You're not going to go [to college]?" There are times when I feel that when people ask me what I do, I'm not going to tell them. I just say, "Well, I don't know. I'm just doing it for now, so I can get through school." I think it's a respectable job, and I'll stand up to anybody and say that I enjoy what I do.

(Personal interview, Owl's Nest)

This cook reveals the ambivalence within the occupation – the beginnings of an embarrassment bordering on self-loathing, revealing pride mixed with defensiveness. Cooks are unsure of how they appear in others' eyes – the stigmatized others are too polite to insult, like African Americans in a society that does not tolerate public racism. Cooks wonder about the thoughts beneath the veneer of toleration.

The Bright Side

Balancing the problems, satisfactions are an integral part of kitchen work. While one's reaction to work is a function of individual needs and what we label "personality," several components of cooking are frequently mentioned as benefits, including employment options, self-satisfaction, and the potential for pleasing others.

Employment Options. Throughout the 1980s, the restaurant and hospitality industry expanded rapidly. Americans increasingly ate outside the home, particularly as the upper-middle-class had more disposable income and more women were in the paid workforce. Since entry-level positions didn't require extensive training and positions were opening rapidly, opportunities to work where they wished arose for cooks. One cook explained that "you can always change [jobs]. It's so easy to find another job if you're good; if you have the skills" (Field notes, Owl's Nest). Several mentioned that kitchen work satisfied their desire to travel. They could relocate and search for a comparable position with the confidence that one would be readily available. One cook remarked that job security was no problem, even though restaurants frequently closed: "In my job hunting I wasn't that worried about finding one. I knew I would eventually, and now that I'm here in Minneapolis, and I do have a job, I think that even if La Pomme closed, I would be absorbed into another place real fast. Once you get attached to an area, established, I think it's pretty secure. One of the few jobs left that there's a need for, you always have to eat" (Personal interview, La Pomme de Terre). Job mobility permitted cooks to decide where they wanted to be. By changing restaurants, they could climb the industry status ladder.

Self-Satisfaction. Cooks are producers. They create products that can be beautiful and appealing to the senses. Anyone who can produce such things has the "right to feel proud" – to recognize his or her accomplishments. Skill is associated with an occupational identity (Grzyb 1990, p. 176). Cooks gain a sense of identity from their work, and from this they learn to identify with their occupation (Hughes 1971). These workers produce within an organization, and these organizations attempt to generate assumptions about the *proper* identity of workers – why they should feel satisfied despite limitations on autonomy, wages, and benefits (Leidner 1993). Many cooks commented that their prime satisfaction derived from pleasing customers. In the words of one cook: "That gives me a real feeling of satisfaction that I know that I've pleased someone" (Personal interview, Blakemore Hotel). For others, it is the ability to cook up to one's "internal standard" (Personal interview, La Pomme de Terre). For still others, it is the ability to know what one can do with food, and that one can control a situation that would be impossible for those outside the occupation: "It's interesting that guys come home, and there's nothing to eat in the house, and I come home, and I look around and throw all this stuff together, and I can make a really nice dinner. They don't even realize that it's possible to do that. . . . It's really an accomplishment thing. You feel like you've accomplished something when you're a cook. Like working at the [display kitchen]. Ron, when he has 160 people in there, he might be really tired out, but he can handle it. He was in control of the situation, and he could feel the accomplishment" (Personal interview, Blakemore Hotel). Through these skills and their public display, cooks persuade themselves that they matter in an institutional order that sometimes disregards them; they are worthy of self-respect and honor, achieving things of which others only dream.

Public Acknowledgement. Although cooks typically do not have *direct* contact with customers, they do on occasion; often these relations are mediated by servers who routinely inform cooks of a significant compliment. Many cooks are young men without much training or education, and it is understandable that they marvel that "it's amazing people are eating what you cook. It's really self-satisfying, but it's also amazing that people will pay eighty, ninety dollars a meal" (Field notes, Owl's Nest). When this is combined with stroking – public recognition – one's satisfaction is complete. Seeing the smile is important, but having the smile verbalized can be equally significant: "The reason that chefs don't make good food and beverage directors is that being a technician [i.e., a chef], you need stroking. You need a pat on the back. You need somebody to say, 'Hey, this is a really good meat loaf.' You need that stroking. . . . You make something, and 'That was a wonderful table,' somebody would say. 'You really outdid yourself. That was a wonderful meal. That was great.' That's what you're here for. You're not here for the money. You need the money, you want the money, but there's more to it" (Personal interview, Blakemore Hotel). Workers judge their satisfaction both internally and externally, and they need both internal and external positive feedback to be satisfied. When workers gain this satisfaction, they feel that they are making a difference, and that they are competent. This feeling increases the likelihood that they will remain in the kitchen.

Living the Kitchen Life

1. In this book I use the word *servers* to indicate those who might otherwise be termed *waiters* or *waitresses*. In most restaurants one gender or the other predominates in the service staff. I find the gender-neutral terms *waitron* or *waitperson* gangly and discordant.
2. To sheet pan a steak is to bake it on a sheet pan.
3. Craig Claiborne comments about his friend and collaborator, Pierre Franey: "I have known and interviewed countless chefs over the years

but I have never known any chef with such an extraordinary capacity to improvise and rectify when working in the kitchen as Pierre Franey. He is a veritable Merlin when it comes to changing failed sauces into triumphs, in knowing precisely how to make a culinary catastrophe into a thing of genius. It may be as seemingly simple as turning a curdled hollandaise into a masterpiece of silken homogeneity or whisking in a little cold water to revitalize and reconstitute a mayonnaise" (Claiborne 1982, p. 215).

4. Cooks on display have the challenging Goffmaniacal task of looking like cooks while cooking. As a result, they are far less efficient than their colleagues who cook behind closed doors, and often they are helped by those behind the doors.
5. One cook, who was the head chef at a health food restaurant, made the opposite case: "I have a good stomach. . . . In fact I take a definite perverse pleasure when I run into an irate customer who found a hair in the food. I tend to be, you know, diplomatic about it, but inwardly I'm laughing that he's offended by a hair" (Personal interview, Minneapolis).
6. Americans seem more fastidious about bodily fluids than are their European counterparts and do not spit on cutlery or reuse unwashed dishes; yet, in American restaurants dirty serving cloths and touching food are common.
7. "Scored off" refers to the technique whereby steaks are placed on the grill in order to put grill marks on them. After that, the steaks are baked, which is easier for cooks.
8. There is some anecdotal evidence that this same emphasis on cooperative teamwork is not found in the great European kitchens. One German chef is quoted about his work in America: "I learned a great deal in America, especially how to treat people. I learned respect for those who worked with me. In Europe there is little comprehension of what Americans call teamwork" (Wechsberg 1980, p. 36).
9. Cooking, like other occupations, is internally divided, with segments of the industry valued differently (Bucher and Stelling 1977). One goal of socializers is to direct the "right" students to the "right" section of the industry while providing some indication of the forthcoming public responses (Manning and Hearn 1969). This connects to theories of "social reproduction," as mediated by an internal occupational hierarchy. The culinary elite has always had high status, and this exalted views of chefs may be spreading among the population, particularly with the development of city centers by those with cultural capital, such as yuppies and gentrifiers (Zukin 1990, p. 1). Not all cooks agree that attitudes have changed. For some of these working-class males in the Twin Cities, for instance, cooking is viewed by friends and relatives as an occupation "for girls." In most working- and middle-class homes wives cooked, and husbands ate.

Appendix: Ethnography in the Kitchen

Issues and Cases

> Africa assailed my senses. I smelled and tasted ethnographic things and was both repelled by and attracted to a new spectrum of odors, flavors, sights, and sounds.
>
> – *Paul Stoller*, The Taste of Ethnographic Things

Every setting has quirks, tricks, surprises. That is the challenge – and the fun – of participant observation. Veteran anthropologists who have studied several non-Western cultures know this well. What one learned about "getting on" in one site scarcely serves as a model for what to do in the next. Margaret Mead's Samoans provided little guidance for later examining the Manus of Papua New Guinea. For those mining American middle-class sites, less intrepid spies, like myself are at least comforted that the shared culture serves as a guide.

In studying kitchens two themes are salient: the research takes place in an organization, and it deals with the preparation of food. The former has challenged others before me, but the latter, relatively unstudied, presents novel issues.

Watching Work

Observers of a work organization must recognize that workers, however fond they may be of what they do, are not present in an entirely voluntary capacity. Employers expect workers at certain times and on a regular schedule. Further, workers belong to a hierarchical organization that can decide, under some circumstances, to terminate the association.

Because these spaces are not public arenas, access is provided through management; as a consequence, researchers will have, even in the most optimal circumstances, a burden of trust to overcome (Burawoy 1979). Whose side are we on (Becker 1967)? As a result, who I *really* represented was an issue, although one that became muted when it grew clear that I was not reporting to management.

The most direct expression of this sentiment occurred in the kitchen at the Blakemore, which as part of a hotel chain operated under direct, external corporate control. One day early in the research one of the kitchen workers whom I had not previously met, seeing me constantly jotting in my notepad, inquired if I was conducting a time study.

More often my role emerged in the context of joking and teasing, but always with underlying concern:

> Davis, a server, jokes with me: "Who do you really work for? What hotel chain do you really work for?" Diane, one of the cooks, claims that I am a reporter for the *National Enquirer*.
>
> (Field notes, La Pomme de Terre)

> Paul, the head chef, jokes about me being a spy, adding that "he's watching to make sure we work."
>
> (Field notes, Owl's Nest)

This became salient when I observed minor deviance, especially among low-status workers, who feared my power over their careers: "Felicia, a pantry worker, eats a piece of the roast beef that had been trimmed off and giggles when she sees me watching. She says to me laughing, but nervous: 'Are you going to put this in your book?' [Felicia, I have.] Later Lee, a dishwasher, takes some of the beef and jokes to me: 'Which part will we steal today.' I am expected to legitimate their deviance, or, better, to participate in it [I do]." (Field notes, Blakemore Hotel). These workers trust that I will place myself on their side as a true, if limited, member of their group, embracing its underside. This was evident at Stan's, where cooks received beers from the bar. Once a waitress informed a cook that I noted this. He responded, "He's on our side. He's exposing the scandals of cooking" (Field notes, Stan's). As I gave them leeway, they returned the favor.

Only twice was I asked not to observe, both times during the early days of my observation. Stan's was featured on a local television show, and viewers were told that they would receive a special price on steak that weekend. The manager thought that the restaurant would be too crowded to observe. Later I was present on equally busy nights. At the Owl's Nest the head chef asked that I skip one day because a new server was being trained by observing the kitchen. However, later in the research, when another server was being trained, I was allowed to observe. I was accepted.

Although I was careful not to provide information to chefs and managers about particular workers, or even about how to improve the kitchen, these individuals had my ear when they wished and would query my general reactions. Workers might well worry about what I knew and what I might say. While I am confident that my comments did not have direct effects on individual workers or result in immediate changes in the kitchen structure, I may have oriented the concerns of managers or chefs by my presence or indirect comments.

Self-Reflection and Deprecation: Fine's Law of Shared Madness

Informants have images of themselves and their work. These self-images involve doubt, which becomes magnified when observed by an outsider. To defend themselves against the potentially critical eye of the researcher, informants typically display "role-distance" (Goffman 1961b), permitting them, in theory at least, to escape from some of the potentially negative implications of their activity. They need not fully justify their esoteric culture.

As I found among Little League baseball coaches, fantasy gamers, debate teachers, and mushroom collectors, informants suggested that they must be "crazy" to engage in their activity. This role-distance seems to be a fundamental regularity. "Fine's Law of Shared Madness" playfully proposes that in any adult research setting, informants explain that their activity is "irrational." Staff in each kitchen made similar comments:

> Howie says to me: "Everyone knows you've got to be real mental to be a cook."
>
> (Field notes, La Pomme de Terre)

> Mel, the day cook, jokes: "Did you learn that you didn't want to work in the kitchen?" A few weeks later Dan, the owner's son and the restaurant manager, jokes with me: "What have you learned? That you never want to enter the food business?"
>
> (Field notes, Owl's Nest)

> Candy, one of the waitresses, jokes about my observations on one very busy night: "Checking abnormal behavior? You'll definitely find it here. You're standing in the right place."
>
> (Field notes, Stan's)

> Irene, a former cook, asks Keith, a current cook, about me, joking: "You're giving him your memoirs?" Keith responds: "I'm the only guy who'll talk to him. The book will be *Portrait of a Cook*. I'll have my name up in lights. The movie will be filmed in an insane asylum."
>
> (Field notes, Blakemore Hotel)

This same theme is evident among informants who saw me as a "psychologist" or "psychoanalyst," judging "are [they] crazy or sane?" (Field notes, Blakemore Hotel). Throughout ethnography, particularly in middle-class settings, informants desire to know "how good are we?" "how are we doing?" and "will we pass?" (Field notes, Owl's Nest).

Some jokingly attempted to mimic the ethnographic enterprise, sometimes discussing writing books:[1] "Gordon says to Cal and Dana: 'One day I'm gonna write a book, and then you guys are just going to be starring in it' " (Field notes, Blakemore Hotel). Or they would observe me: "Mel and Dan kid me: 'Now we're gonna ask you questions. What is the most interesting thing about your job?' They joke that they will do a study of researchers" (Field notes, Owl's Nest). This partially derives from my research style: I am overt about my presence, publicly taking notes, believing that the novelty of my role will wear off quickly as I am around every day. Daily research breeds familiarity, and openness is both more ethical and leads to more precise data, since memory tricks are minimized. Yet, this strategy leads to a justifiable desire to turn the tables.

The Plimpton Effect

One way in which my ambiguous presence was handled was through a continuing stream of joking, transforming me from researcher to worker, from observer to full participant (Adler and Adler 1987): the George Plimpton effect, after the journalist-essayist who briefly participated as a member of professional sports teams. On a few occasions I performed such undemanding manual labor as peeling potatoes, de-stringing beans, stirring soup, or retrieving food from shelves or coolers.

> Denny, a cook at La Pomme de Terre, seriously suggests that "on your last day here we could let you cook some. It would be like George Plimpton."
> (Field Notes, La Pomme de Terre)

> Toward the end of my research Paul, the head chef, asks me if I would like to work the window tonight, saying "You've seen enough." He has seen the film *Paper Lion* and wants to put me into the George Plimpton role. I am tempted to agree, but it is not clear whether he is joking, and we don't pursue it.
> (Field notes, Owl's Nest)

Most often the attempts to involve me were jocular, indicating both my closeness and my distance. These remarks occurred in each restaurant:

> The owner jokes to me as he is leaving for the night: "We'll make you mop next week. Everyone gets a job if they stay around long enough."
> (Field notes, La Pomme de Terre)

> Evan jokes to me: "Why don't you take over. You've been trained in."
> (Field notes, Stan's)

> Willy, the stockroom manager, jokes to Sanjay, the food and beverage manager: "[If] he stays around long enough, we're gonna put him to work."
> (Field notes, Blakemore)

> Paul jokes to me as he is cutting meat: "You've watched me long enough, now you've got to try it."
> (Field notes, Owl's Nest)

Like the desire to involve the researcher in occupational deviance, a strong drive exists, helpful for research, to allow – or demand – the observer to participate in all aspects of the scene: in part from a destre to help make the experience transparent and in part to cope with the frustration of seeing the researcher standing around, watching, "doing nothing."

Sharing Food

One way that workers attempt to gain the allegiance of a researcher is to provide organizational perquisites: pleasant little bribes. This is an extension of the "gift" of information that informants give. To accept a gift is to remain in the giver's thrall until one reciprocates. Cooks do this with goods that technically are not theirs to give but belong to their bosses: food. Food was continually pressed on me, and, perhaps to my discredit, I rarely refused. During each month that I observed in the kitchens, I gained approximately ten pounds, which I diligently attempted to lose in the two months between observations. Even had I wished to refuse the offers, my friends were expecting me to taste their wares, and this occasionally was a learning experience. Attempts to ply me with food were most notable at La Pomme de Terre and the Owl's Nest, the better restaurants. Perhaps because steak is expensive and because the preparation was routine, I was not given as much at Stan's, although french fries and onion rings were offered, and when a steak was returned, it was shared. At the Blakemore food was more carefully monitored, and the better dishes were prepared in the "display kitchen." However, even there I was offered coffee, drinks, and an occasional "nibble."

The staff at La Pomme de Terre was particularly generous. During my first evening I was offered snapper Provençale, spicy sausage, lemon balm sorbet, elderberry sorbet, and chocolate cake. A few days later Davis, a room captain, gave me a large slice of cheesecake, an offer I felt rather embarrassed about. He started by asking me if I have a family. When I told him I do, he insisted I take it because they will throw it out otherwise. While they were being generous, the underlying assumption is that I would only refuse the food if I didn't like it *and them*.

The situation was more complex at the Owl's Nest, where food was supposed to be accounted for, and where staff members were expected to pay for their meals. While I was given food, it was often done with a wink and teasing remarks that I would be charged:

> Paul offers me salmon for dinner, adding "You must get hungry standing around." He adds: "Between you and me and the fence post," meaning that I should not mention his largesse.
>
> (Field notes, Owl's Nest)

> Bruce gives me a plate of salmon *sorbise*, and Paul jokes: "That will be ten dollars." Later he adds: "Don't steal the hash browns again."
>
> (Field notes, Owl's Nest)

These comments reflect a tension in our relationship, common in ethnography, about the researcher using and being used by informants. These cooks liked me, and I became, in some measure, a member of the group; yet, simultaneously I was and would remain an outsider, potentially affecting their lives and reputations. By bribing me and telling me about their lives, they hoped to gain my support. Because they work in scenes that they do not entirely control, and because I have access to those who control them, they were at my mercy. I, in turn, was at their mercy to the degree that I was allowed to operate, briefly and marginally, as one of them.

Ethnographic Sense

As the anthropologist Paul Stoller (1989) cogently notes, contemporary Western thought has given primacy to the visual mode, particularly in academic writing (but see Corbin 1986). For those exploring distant cultures the experience of the other senses is central: "Naked children defecating into the ditches which carried the city's sewage; clouds of aromatic smoke rising from grills on which butchers roasted mouth-watering slices of mutton; dirt roads rendered impassible by rat-infested hills of rotting garbage; gentle winds carrying the pungent smell of freshly pounded ginger; skeletal lepers thrusting their stump-hands in people's faces" (Stoller, 1989, p. 3). One not only reads social structure but also hears, touches, smells, and tastes it. Yet, because of our Western training, we find that we easily misplace these skills of understanding. We are taught what to see and, implicitly, taught what we should not sense. Stoller (1989, p. 4) reflects:

> My sensual openness, however, was shortlived. I quickly lost touch with those scenes of abject deprivation which blended into those of insensitive consumption. I soon lost scent of the nose-crinkling stench of the open sewer that gave way to the aromatic aromas of roasting meat. My ears soon deafened to the moans of a sick child that were overwhelmed by the happy laughter of a healthy one. I had become an experience-hardened Africa hand. My immersion in Niger, in Africa, had been, in short, distanciated, intellectualized – taken out of the realm of sensual sentiment. The world of ethnographic things had lost its tastes.

We become so taken with the reality of the world that we have *learned to experience* that we forget to notice the world that we *are experiencing*. For this research, my attempt to remain sensitive to the challenges of senses was critical. If not as dramatic as the sewers of Niger, the kitchens of Minnesota had their share of distinct smells, touches, sounds, and tastes. The challenge was learning from these senses: to let, in the words of one cook, "the kitchen talk to you." As I describe in chapter 2, when I discuss the timing of food preparation, these senses are critical to the competent doing of cuisine. Done food has a different smell from burned or underdone food, and these smells are clues for the ethnographer.

A second element of the ethnography of senses is to avoid imposing one's own taste on one's informants. The rejection of "proper" food can be a means of differentiating insiders from outsiders. Craig Claiborne's mother ran a rooming house in Indianola, Mississippi, where the social psychologist John Dollard happened to stay while researching his admirable *Caste and Class in a Southern Town* (1949). As Claiborne (1982, pp. 32–33) recalls Dollard's tastes, "In the beginning [Dr. Dollard] criticized the cooking of the greens, complaining that there was not a vitamin left in the lot. And as a result of his well-intentioned explanations and at the base encouragement of the other boarders, my mother willingly committed one of the most wicked acts of her life. Dr. Dollard was placed at a bridge table, covered, of course, with linen and set with sterling, and he was served a mess of raw greens that he ate with considerable and admirable composure and lack of resentment." His northern taste almost prevented him from understanding the society that he had chosen to examine. As a promiscuous eater, I did not have Dollard's problem, but occasionally I would be given something to taste, because a cook wanted to know if I judged it as he did. Workers within the kitchen would no doubt gauge its quality with certainty, but I often had to guess whether the food would be judged a success or a failure. I recall particularly a highly and oddly spiced sausage served to me at La Pomme de Terre, which could be fairly described as "distinctive," a basil-cantaloupe sorbet that sticks with me still, and a beet fettucini with herbed tomato sauce that was dubbed an abject failure. An observer should not like everything but must develop tastes similar to those whom he or she is studying.

References

Adler, Patricia and Peter Adler. 1987. *Membership Roles in Field Research*. Newbury Park, Calif.: Sage.

Becker, Howard S. 1967. "Whose Side Are We On?" *Social Problems* 14: 239–47.

Becker, Howard S. 1974. "Art as Collective Action." *American Sociological Review* 39: 767–76.

Bishop, James M. 1979. "Institutional and Operational Knowledge in Work: A Sensitizing Framework." *Sociology of Work and Occupations* 6: 328–52.

Blumer, Herbert. 1969. *Symbolic Interactionism*. Englewood Cliffs, N.J.: Prentice-Hall.

Bosk, Charles. 1979. *Forgive and Remember*. Chicago: University of Chicago Press.

Burawoy, Michael. 1979. *Manufacturing Consent*. Chicago: University of Chicago Press.

Charpentier, Henri and Boyden Sparkes. 1934. *Life à la Henri*. New York: Simon and Schuster.

Claiborne, Craig. 1982. *A Feast Made for Laughter: A Memoir with Recipes*. New York: Holt, Rinehart and Winston.

Corbin, Alain. 1986. *The Foul and the Fragrant*. Cambridge: Harvard University Press.

Crocker, E.C. 1945. *Flavor*. New York: McGraw-Hill.

De Groot, Roy. 1972 "Have I Found the Greatest Restaurant in the World?" *Playboy* (April): 107, 116, 244–49.

Dickie, George. 1974. *Art and the Aesthetic: An Institutional Analysis*. Ithaca: Cornell University Press.

Dollard, John. 1949. *Caste and Class in a Southern Town*. New York: Harper.

Fuchs, Victor R. 1968. *The Service Economy*. New York: National Bureau of Economic Research.

Goffman, Erving, 1961b. *Encounters*. Indianapolis: Bobbs-Merrill.

——— 1974. *Frame Analysis*. Cambridge: Harvard University Press.

Gross, Edward. 1958. *Work and Society*. New York: Crowell.

Grzyb, Gerard J. 1990. "Deskilling, Decollectivization, and Diesels." *Journal of Contemporary Ethnography* 19: 163–87.

Hughes, Everett C. 1971. *The Sociological Eye*. Boston: Little, Brown.

Leidner, Robin. 1993. *Fast Food, Fast Talk: Service Work and the Routinization of Everyday Life*. Berkeley: University of California Press.

Manning, Peter K. and H.L. Hearn. 1969. "Student Actresses and Their Artistry: Vicissitudes of Learning a Creative Trade." *Social Forces* 48: 202–13.

Mars, Gerald and Michael Nicod. 1984. *The World of Waiters*. London: George Allen and Unwin.

McPhee, John. 1979. "Profiles: Brigade de Cuisine." *The New Yorker* (February 19): 49–99.

Molstad, Clark. 1986. "Choosing and Coping with Boring Work." *Urban Life* 15: 215–36.

Orwell, George. 1933. *Down and Out in Paris and London*. New York: Harcourt Brace.

Paules, Greta Foff. 1991. *Dishing It Out: Power and Resistance among Waitresses in a New Jersey Restaurant*. Philadelphia: Temple University Press.

Stoller, Paul. 1989. *The Taste of Ethnographic Things*. Philadelphia: University of Pennsylvania Press.

Strauss, Anselm L. 1991. *Creating Sociological Awareness: Collective Images and Symbolic Representations*. New Brunswick, N.J.: Transaction.
Tomlinson, Graham. 1986. "Thought for Food: A Study of Written Instructions." *Symbolic Interaction* 9: 201–16.
Wechsberg, Joseph. 1980. "Three Chefs in Germany." *Gourmet* (January): 24–37.
Whyte, William Foote. 1948. *Human Relations in the Restaurant Industry*. New York: McGraw-Hill.
Wygan, Camilla. 1981. "Thanks to the Culinary Institute . . ." *Taste* 10: 20–32.
Zukin, Sharon. 1990. "New Cusine as a Landscape of Cultural Consumption." Unpublished Manuscript.

H. Studying Up: Observing the Unobserved

73

Up the Anthropologist – Perspectives Gained from Studying Up

Laura Nader

In this essay I shall describe some opportunities that anthropologists have for "studying up" in their own society, hoping to generate further discussion of why we study what we do (Nader, 1964). Anthropologists have a great deal to contribute to our understanding of the processes whereby power and responsibility are exercised in the United States. Moreover, there is a certain urgency to the kind of anthropology that is concerned with power (cf. Wolf, 1969), for the quality of life and our lives themselves may depend upon the extent to which citizens understand those who shape attitudes and actually control institutional structures. The study of man is confronted with an unprecedented situation: never before have so few, by their actions and inactions, had the power of life and death over so many members of the species. I shall present three reasons for "studying up": its energizing and integrating effect for many students; scientific adequacy; and democratic relevance of scientific work. Finally, I shall consider some frequent obstacles and objections, and try to answer them.

Indignation as Motive

Many of our brighter students look at the anthropology journals of recent times and conclude that anthropology appears to be phasing out, content to make a living for the most part by rediscovering what has been discovered

Source: Dell Hymes (ed.), *Reinventing Anthropology* (New York: Pantheon Books, 1972), pp. 284–311.

or by selling our wares to other disciplines and professions. The audience is too narrow, the nitpicking too precious. Making a living by selling one's wares is not an inappropriate way to subsist; it is however, in this case, symptomatic that a talent, the perspective of a *Mirror for Man*, is being underused.

Today we have anthropology students who are indignant about many problems affecting the future of *Homo sapiens*, but they are studying problems about which they have no "feelings." Some think this is the only appropriate stance for a science. Yet the things that students are energetic about they do not study. I think we are losing something here. The normative impulse often leads one to ask important questions about a phenomenon that would not be asked otherwise, or to define a problem in a new context. A rapid growth in civil rights studies is directly attributable to activities whereby the victims of the system made their victimization visible. By a process of contagion, this visibility spread moral indignation into the law schools and the legal profession, which in turn led to research into civil rights questions and the related area of poverty law. In anthropology we have the example of Ruth Benedict's *The Chrysanthemum and the Sword* (1946), an effort to understand opponents in war. The normative impulse here, generated by patriotism and loyalty, considered appropriate in World War II, was responsible for an insightful book and the development of new techniques for studying culture at a distance.

Looking back to an early founder of American anthropology, L. H. Morgan (the first anthropologist to become president of the AAAS), we discover that he broke new ground in science as a result of having been interested in a social problem (Resek, 1960). Throughout his career, Morgan was indignant at how American Indians were being treated, at how they were being pushed off the land. It was his initial indignation which led him to study American Indians, and his indignation preceded his curiosity about kinship systems and social structure. In Morgan's case, indignation had an energizing effect.

As Jules Henry put it:

> To think deeply in our culture is to grow angry and to anger others; and if you cannot tolerate this anger, you are wasting the time you spend thinking deeply. One of the rewards of deep thought is the hot glow of anger at discovering a wrong, but if anger is taboo, thought will starve to death. (Henry, 1963, p. 146)

I see among young students at Berkeley an energizing phenomenon in studying major institutions and organizations that affect everyday lives, such as the California Insurance Commission, the Better Business Bureau, air pollution agencies, and the like. The following extended excerpts illustrate something about what motivated these students to study what they did:

> I chose to study the insurance industry primarily because it is one of those "things" (there is no term) which is made of vast networks of people

> who have effects on many aspects of the lives of all people in California. Most are affected in a direct way, by owning an insurance policy. All are affected in other ways, such as by the vast political influence of the "insurance industry" and its tremendous economic influence over our personal lives.
>
> For example, one drives to the market in one's car, which is itself insured. The market one arrives at is insured for loss, theft, damages and liability. The food was delivered by trucks which were insured for the cargo they carry, which is also protected against various problems. The price of these coverages also affects the price of your food. Incidentally, the factories where the truck and your auto were manufactured are insured with various policies (as are all the employees). One of these, termed "pollution insurance," protects the company for liabilities incurred if they are sued for pollution damages. The price of this affects other prices, as mentioned, but the ownership of the insurance permits the factory or the network of people who control the factory to pollute the air without economic risk to themselves. To extend this further, the auto company is probably owned in large part by insurance companies. To get an idea of the tremendous wealth of the insurance industry, consider that California owners of insurance policies paid over $6 billion in insurance premiums in 1970, which is over $300 per person in the State of California. This is greater than the per capita income of most of the world's population. As a matter of fact, the annual amount of premiums taken in the U.S. by the insurance industry is greater than the gross national product of all but five nations in the world. (Serber, 1971, pp. 2–3)

Another student had the following to say about her study of the Oakland Better Business Bureau:

> In our complex society, we obtain many goods and services in a prepackaged state. Like the proverbial city child who grows up believing that milk grows in paper cartons, most consumers know little about what their purchases are made of, how they work, how to evaluate their potential before buying them, and how to repair them if they break down. This ignorance is not limited to goods but extends to services, investments, charities, to say nothing of the legal and medical professions. We rely on [Pacific Gas and Electricity] to install gas equipment, to check it for safety, even to relight the pilot if we cannot locate it when it is accidentally extinguished. We take our special garments to a dry cleaner. When the transmission on an automobile doesn't work, the car must be towed to a transmission specialist. Goods are ordered by telephone or through the mail. Appointments at a photographer's studio, cosmetics, magazine subscriptions, and investments may be sold and contributions to charities collected by door-to-door solicitors. When a carpet is desired, the consumer depends on a salesman to explain the qualities of the constituent fibers, to calculate the number of yards needed to cover a given area, and to make sure that incidentals like matting, tacks, and labor are included in the quoted installation cost. We depend upon specialists to provide services and often even to give us the criteria by

> which we are to judge their work. Many of our transactions take place infrequently, which means that the consumer may be totally inexperienced in evaluating what he pays for when he buys a large appliance, an insurance policy, or a vacation trip.
>
> Likewise he may find himself incapable of obtaining redress of his grievances when he thinks he has been misled or cheated. Most contacts with businesses are limited to the disembodied voice of the switchboard operator, to the secretary or public relations representative in the front office, or to the salesman who happens to be on the floor when the customer walks into the store. The consumer phones the company to see what it will do for him, or he writes to a newspaper complaint column or a broadcasting station's "hotline" program. The services of these expediters are also "packaged": the complainant sends in his story and waits for the machinery to grind out an answer. This situation extends even to the law enforcement and consumer aid organizations to which the irate customer may eventually refer his problem. . . .
>
> I began this project on the BBB in total ignorance of what it is, what it does and does not do, and why. Like the typical citizen, I began with the simple knowledge that there is a business-supported organization called the Better Business Bureau and that it is customarily contacted by telephone when a person has a question about the dependability (or existence!) of a firm or has a complaint against a business that has failed to give satisfaction. Few people go beyond these facts to ask who the voice on the other end of the line is, where she gets her information, or what actually happens to the complaint form which arrives, is returned, and whose results are relayed back to the consumer by mail. And yet thousands of people use the BBB every year. (Eaton, 1971, pp. 2–3)

Maybe these are attempts to get behind the facelessness of a bureaucratic society, to get at the mechanisms whereby faraway corporations and large-scale industries are directing the everyday aspects of our lives. Whatever the motivation, the studies raise important questions as to responsibility, accountability, self-regulation, or on another level, questions relating to social structure, network analysis, library research, and participant observation.

Scientific Adequacy

If we look at the literature based on field work in the United States, we find a relatively abundant literature on the poor, the ethnic groups, the disadvantaged; there is comparatively little field research on the middle class and very little firsthand work on the upper classes. Anthropologists might indeed ask themselves whether the entirety of field work does not depend upon a certain power relationship in favor of the anthropologist, and whether indeed such dominant-subordinate relationships may not be affecting the kinds of theories we are weaving. What if, in reinventing anthropology, anthropologists were to study the colonizers rather than the colonized, the culture of power rather than the culture of the powerless, the culture of affluence rather than the culture of poverty?

Studying "up" as well as "down" would lead us to ask many "common sense" questions in reverse. Instead of asking why some people are poor, we would ask why other people are so affluent? How on earth would a social scientist explain the hoarding patterns of the American rich and middle class? How can we explain the fantastic resistance to change among those whose options "appear to be many"? How has it come to be, we might ask, that anthropologists are more interested in why peasants don't change than why the auto industry doesn't innovate, or why the Pentagon or universities cannot be more organizationally creative? The conservatism of such major institutions and bureaucratic organizations probably has wider implications for the species and for theories of change than does the conservatism of peasantry.

If, in reinventing anthropology, we were principally studying the most powerful strata of urban society, our view of the ghetto might be largely in terms of those relationships larger than the ghetto. We would study the banks and the insurance industry that mark out areas of the city to which they will not sell insurance or extend credit. We would study the landlord class that "pays off" or "influences" enforcement or municipal officials so that building codes are not enforced. Slums are technically illegal; if building codes and other municipal laws were enforced, our slums would not be slums (if enforcement were successful), or they might be called by another name which would indicate that they were results of white-collar crime. One might say that if business crime is successful, it will produce street crime. With this perspective on white-collar crime, our analysis of gang delinquency might be correspondingly affected, and in developing theories of slum-gang behavior we might ask: Is it sufficient to understand gangs as products of the value systems of that subculture alone? We might study the marketing systems or the transportation system which, as in Watts, makes virtual islands of some ghetto areas. We might study the degree to which legal practices, or the kind of legal services, mold the perceptions of law that are present in the ghettos.

The consequences of not studying up as well as down are serious in terms of developing adequate theory and description. If one's pivot point is around those who have responsibility by virtue of being delegated power, then the questions change. From such a perspective, one notices different facets of culture – the ghetto may be viewed as being without law, lawless. The courts are not geared to the complaints of the poor (which would fall in the $20 to $80 range); furthermore, they are not geared for cheap and quick resolution of conflict – crucial features for the poor. From this perspective, ghetto communities may be said to be shut out of the legal system except as defendants, and indeed they are often shut off from other municipal services ranging from garbage-collecting to police protection. From this orientation, then, the question may be raised again: In our studies of delinquency, is it sufficient to understand gangs as products of the value systems of that subculture alone?

Let's ask another question: What have been the consequences of social science research on crime? By virtue of our concentration on lower-class crimes, we have aided in the public definition of the "law and order problem" in terms of lower-class or street crimes. Let's assume that the taxpaying public in a democracy, after listening to a presidential speech calling for more tax money for enforcement and protection from street crimes, decides to see for itself. No matter what library they went to, the most they could get is some information on crimes committed by the lower class. They would have no way of evaluating, given present descriptive materials, whether, in a situation of limited money, they would do better to put their money on street crime or on white-collar crime, both of which, after all, imperil the lives of all taxpayers every day in many ways.

As Clyde Mitchell has noted, it was with such problems in mind that anthropologists first introduced the concept of "social field."

> The classical anthropological study takes a unit – a "tribe" or "society" or "community" – and presents the behavior of its members in terms of a series of interlocking institutions, structures, norms, and values. It is not only anthropologists working in urban areas who have found this sort of assumption difficult to maintain, but also those who have been conducting "tribal" studies in modern Africa (and presumably also elsewhere). They have found that the effect of groups and institutions not physically present in the tribal area influences the behavior of people in it. The unit of interacting relationships, in other words, is larger than the tribe. (Mitchell, 1966, p. 56)

Lowie may have studied the Crow, Llewellyn and Hoebel the Cheyenne, as if they were "islands" unrelated to the wider society and even unrelated to the policies and actions of the Bureau of Indian Affairs, but there has raged a whole literature since the fifties challenging the limited ethnographic community view of the world, and a recognition of methodological need has been, as Mitchell noted, what has perhaps stimulated the development of network theory and the development of nation state studies (Adams, 1970).

If anthropology were reinvented to study up, we would sooner or later need to study down as well. We are not dealing with an either/or proposition; we need simply to realize when it is useful or crucial in terms of the problem to extend the domain of study up, down, or sideways. If we become interested in the determinants of family patterns (rather than the poor or the rich as such), then studying this problem across class, or at least on a vertical slice, would be a way to test hypotheses regarding whether certain aspects of lower-class or upper-class plight are somehow due to a particular kind of family pattern (serially monogamous, matrifocal, father-absent), whether poverty, for example, is generated by certain types of employment patterns or external factors. At least posing the problems in a comparative frame would help improve our chances for understanding the forces that generate excessive poverty or affluence and the origins of those forces, whether intrusive from the larger society or "determined by cultural transmission within the group." Depending on one's view of the processes that generate behavior, one would seek solutions to social problems either by a policy directed to reforming the society as a whole or by one directed to modifying the behavior of the subculture, or both (Valentine, 1969; Gladwin, 1969).

On the basis of such work in our own society,[1] we could rewrite the books on American Society, whose indexes make no mention of the advertising, insurance, banking, realty, or automobile industries, which most people on the street know have played a major role in forming modern American society. Ethnographic reports would describe the communications industries, the agencies which regulate them, the institutions that undergird the industrial sector, such as the legislative bodies, the universities and professional organizations, and such descriptions would be from the point of view of the *users* as well as the managers. It is appropriate that a reinvented anthropology should study powerful institutions and bureaucratic organizations in the United States, for such institutions and their network systems affect our lives and also affect the lives of people that anthropologists have traditionally studied all around the world.

It is particularly appropriate that anthropologists should lead the way in this work by virtue of a number of characteristics of our discipline. The study of man has had to be eclectic in its methods, broad in its vision of what it takes to understand man – his past, his present, his culture, his biology. We have specialized in understanding *whole* cultures in a cross-cultural context. We should, for example, be at home in studying the law firm as a secret society, in finding and analyzing the networks of power – which on paper may not be there – in describing those unwritten customary behaviors that are completely indispensable for understanding, for example, what makes Congress tick. The anthropologist should, above all, by virtue of his understanding of the principle of reciprocity, be able to analyze why it is that decisions of Federal Communications Commissioners may not be "rational," or the cultural dimensions involved in the failure of national programs ostensibly geared to reintegrate society. It is the anthropologist who, by virtue of his populist values, may be able to define the role of citizen-scholar – a science of man for man.

Democratic Relevance

"Studying up" seems to be one track for integrating paramount social concerns with the goals and aims of the science of man. The service function we have performed in the past could be amplified to include another service, social as well as scientific, that is, writing ethnographies for the "natives." A monograph that should be taken into account by managers for the benefit of people concerned is Colson's recent (1971) book on *The Impact of the Kariba Resettlement upon the Gwembe Tonga.*

> Massive technological development hurts. This is a fact largely ignored by economic planners, technicians, and political leaders. In planning drastic alterations in environment that uproot populations or make old adjustments impossible, they count the engineering costs but not the social costs. After all, they do not think of themselves as paying the latter. . . . This book is a study in the impact of forced change upon some of its victims.

Another example is Spradley's *You Owe Yourself a Drunk*, an ethnographic description of the interaction and the consequences of the interaction that drunks have with the legal and enforcement systems. This monograph is already serving to educate managers of that system about the consequences of specific legal decisions and procedures. This is not a novel role for social scientists to play, and unfortunately our findings have often served to help manipulate rather than aid those we study. Another role, however, is related to the concept of citizenship in a country that is to be run on a democratic framework and the control that citizens must have to harness managerial manipulation. We cannot, as responsible scientists, educate "managers" without at the same time educating those "being managed." A democratic framework implies that citizens should have access to decision-makers, institutions of government, and so on. This implies that citizens need to know something about the major institutions, government or otherwise, that affect their lives. Most members of complex societies and certainly most Americans do not know enough about, nor do they know how to cope with, the people, institutions, and organizations which most affect their lives. I believe that anthropologists would be surprisingly good at applying their descriptive and analytical tools to a major problem: How can a citizenry function in a democracy when that citizenry is woefully ignorant of how the society works and doesn't work, of how a citizen can "plug in" as a citizen, of what would happen should citizens begin to exercise rights other than voting as a way to make the "system" work for them? But first, as we know, we have to describe the bureaucracy and its culture.

Love and Eaton (1970) began their study of the Bay Area Air Pollution Control Agency with questions about the functions of the agency: How does the agency perceive itself? Who uses it? How do the users perceive the agency? Public access was a key question.

> Our approach was, at first, guarded due to our doubts. We pretended innocence, and in fact found out that we really were innocent. We then began to realize that we were "outsiders." We were the public who did not understand the professional language being spoken. The avenues we

> approached were those the public generally approached. Gradually, a picture of the agency and its position in the legal system emerged.
>
> Its structure, the personalities of the decision-makers, the limitations reality places on any ideal system, and finally the kinds of uses made of it became clearer. . . . (pp. 2–3)
>
> When the citizen goes to the agency, he is translated into statistical data which separate him from the actual procedure or use of the agency. . . . Assumed in this is the notion that since the agency is supposed to protect the public interest, the public will seek access to it. The reality of the situation is very different. The agency acts as autonomously as possible to combat air pollution and in so doing, comes into close contact with the industry officials who speak the same technical and legal language. It is industry which has the greatest access to the agency, especially at the legislative level. It is industry which makes the greatest *use* of the agency to protect *its* interests. (pp. 32–33)

This same study notes that, in the legal division of the agency, violation notices are treated like parking tickets – after so many are collected the violator is prosecuted. But what does prosecution consist of, given the intimate patterns of social interaction described above? These were not ordinary criminals.

> In the legal division, the agency lawyer emphasized that the principles of criminal law were not a solution, hence the civil fines. The type of "crime" committed does not merit the "responsible" official being put in jail with "prostitutes and muggers." . . . An interesting footnote to this procedure is that Regulation 1, which does *not* apply to most large industries but to private citizens and land developers, is treated as a misdemeanor where the violator can be put in jail with the "prostitutes and muggers." (p. 33)

Apart from being a useful report on bureaucratic culture, this thirty-seven-page report is the kind of ethnographic information that citizens need prior to an attempt to gain *access* to, or attempt to *use*, a public agency. Such reports would introduce them to the structure and culture of the subgroup in such a way as to allow them to gauge whether the cards are stacked and in what direction they are stacked in terms of real access to, and use of, a public agency.

The study of the California Department of Insurance, and in particular the processing of complaints by the Policy Services Bureau of that same agency, is another attempt to describe the workings of an organization whose acts of omission or commission affect the lives of many (Serber, 1971). Serber concludes (p. 62) that the Department of Complaints does not meet the needs of the people of the State of California because "the vast majority of the people are not aware of its existence." He adds, by means of a quote, a further insight which has been noted by other student studies of governmental agencies (and it is not much different for private agencies that purport to serve the public) and which suggests that such public institutions are not structured for public access:

> It is much worse to deal with someone in the public because you know that you are very limited in what kinds of answers you can give them and the results you can get for them. They expect more and often get impatient. With the industry, it's different: they are usually friendly and polite, at least to our faces; we always know where we stand and how far we can go. It's less stressful because I feel less responsible for the outcome of the conversations. (Insurance Officer III).

The report goes on to note that "there is a qualitative difference in the nature of the interaction between the complainants and the Insurance Officer and the representative of the industry and the Insurance Officer."

> It's not very pleasant to arrive here at a quarter of eight in the morning after battling to cross the Bay Bridge for forty-five minutes and before I can finish a cup of coffee some *hysterical* fat bitch who can hardly talk, she's so stupid and excited, will come in, and they will call me. When I catch sight of her my stomach tightens and my mouth gets dry; sometimes the burning in my pipe starts before I can even get up to the desk, and I'll have to take a sip of water. (Insurance Officer IV)

This same report makes a set of predictions as to what might happen to this Department of Complaints were access by the public easily available; the structure and function of the department would move more closely in line with a major goal of the Department of Insurance "to enforce insurance laws so as to achieve the highest possible degree of protection for the public in general and all policy holders and beneficiaries in particular" (Serber, 1971, p. 64).

The above-stated goal raises a more general question: Who is to decide what is good for the public? Eaton's paper on the Better Business Bureau of Oakland illustrates the dilemma of a value conflict.

> A major limitation in the value of the BBB to the consumer lies in the very fact that it *is* an organization designed to further the interests of legitimate business. The movement assumes that what is good for business is good for the economy and for the consumer. It assumes that the power of conscience and the power of public opinion will triumph over the unbridled profit motive, that an informed public will be able to mold the responsive market to its own desires. These assumptions may be true on some levels, but the picture is not that simple. There are areas of the society in which power is concentrated and areas where it is dispersed or absent altogether. The consumer's complaint has more weight with the locally competitive retailer than with the faraway corporation which made the product that the retailer sells. The Bureau regulates retail advertising, but the consumer is also exposed to national advertising,

> especially on television. The retailer is not a free agent: he is limited by the distributor and the supplier. . . .
>
> The Bureau is concerned with truth in advertising, but is it equally concerned about relevance in advertising? There are many things which can be said about a product which are true, but which have very little to do with its significant attributes: durability, safety, efficiency. . . . The consumer is told on the one hand that it is his responsibility to be informed and to exercise his power of choice to bring the market into line with his needs and desires. . . . On the other hand, he is assaulted by national advertising, which stresses the nonpractical attributes of products, and he is confronted with a range of products from different manufacturers which have essentially no differences between them in areas which the consumer may think are significant. He is told to understand the warranty that comes with his new car, but he is not told what he can do if he does not like its terms and finds that all warranties from all companies are just the same. As an individual, he is essentially powerless to bargain in the greater market system that characterizes the modern complex society. (Eaton, 1971, pp. 68–70)

Understanding the differences in the perceptions of producers on the one hand and consumers on the other allows a citizen to evaluate for himself any government statement about the need for government not to encroach on the self-regulatory organizations such as the BBB that are set up by business groups. Eaton's study of the BBB touches but one tiny part of the self-regulatory attempts of business. Since 1966, there has evolved a whole series of "complaint handling mechanisms," from "hot lines" to corporate ombudsmen. A comparative study of such mechanisms would be a much-needed contribution to the literature on the nature of extralegal attempts at voicing complaints and obtaining redress. It would be fascinating to know to what degree "informal law" is dominated by public relations and Madison Avenue techniques in conflict management. The use of advertising in grievance resolution may be related to an upper-class perception of upper-class law and order which says, "Cool it rather than resolve it."

Some years ago, the criminologist Edwin H. Sutherland wrote a book entitled *White-Collar Crime*. A landmark finding documented in that work was the simple fact that white-collar personnel commit crimes, a fact which should have helped combat the belief, at least among social scientists, that the poor had a monopoly on crime. If, as scientists, we are interested in understanding the determinants of crime, then the "discovery" that the rich as well as the poor commit "crimes" (something that is well known to the average citizen and most certainly known by the poor) is very important. The fact that crimes are differentially stigmatized and prosecuted according to class should lead us to disregard oversimplistic theories explaining criminal behavior. Very few sociological works of this type followed Sutherland's study, and indeed there was a long dry period between the muckraking of the turn of the century and Sutherland.

Instead, sociologists such as Lewis Coser (1968) prefer to tell us why some poor do and why some poor do not commit crimes in terms of the theory of related criminal behavior. If we look at the question in relation to a vertical slice, it is a well-known fact that the criminal law has been oriented toward individual crimes, preferably street crime, and collective criminal behavior by an agency or corporation is often dealt with in administrative agencies or in ways which leave very little stigma on those involved (Pound, 1906; Sutherland, 1949). Yet our analyzed data base is slim. Henry Ruth, Director of the National Institute of Law Enforcement and Criminal Justice, noted as late as 1970 that

> the National Institute of Law Enforcement and Criminal Justice has developed an intensive concern that so-called "white-collar crime" receives scant attention from the law enforcement and research communities. . . . The entire field of white-collar crime represents a national priority for action and research – to define the problem, to examine its many faces, to measure its impact, to look for ways in which its victims can be helped, and to determine how such crime can be prevented, deterred, and effectively prosecuted. (Edelhertz, 1970, p. iii)

With regard to benefit to citizens, it is astounding that in as legalistic a country as the United States, nowhere in the educational system does one get a *working* knowledge of the law as part of a general education. In fact, after years of studying the Zapotec legal system of Oaxaca, Mexico, I would conclude that the single most important difference between the Zapotec legal system of southern Mexico and the American legal system (from the point of view of a middle-class consumer) is that Zapotecs have access to, know how to use access to, the legal system. In the United States, most citizens do not have access to the legal system, either because they are ignorant of the workings of the system or because they cannot afford the professional (lawyer) who would have adequate knowledge of the workings of the system. In California, for example – and I imagine this is much more widespread – there are few books for citizens describing the legal system, what it is, and how it works.

This situation is representative of the larger problem of citizen education. Most of what we learn about the law we absorb vicariously from TV westerns and Perry Mason – style shows. Ethnographic works on the subject of law would be filling a scientific and descriptive need, as well as informing the native about a system which at times heavily weights the direction his life takes. For example, one student began a study of the Immigration and Naturalization Service in an effort to find out how immigration and the INS have molded and influenced the communities of third-world immigrants. The basic hypothesis was that the INS is the historical product of negative American attitudes toward non-Northern European immigrants – fear of foreigners, dislike of strange cultures, isolationism, and the like; that while there has been a major liberalization of the laws, the administration of immigration, and indeed recruitment to the agency, continue to be affected by these historical attitudes. Other anthropological studies might involve the use of personal documents – the memoirs of judges, lawyers, and corporate executives are more noticeable for their absence from the bookshelves. The Washington law firms whose lobbying functions have earned them the label of fourth branch of government would be a fascinating place to test some of Elizabeth Bott's hypotheses about networks. What shapes and functions do the networks of such firms have in an organization where, at mid-career, the majority of firm members fan out into positions about Washington, yet still maintain relations with the law firm even after they are no longer on the payroll? What kind of reciprocity is involved here?

Obstacles and Objections

But there are those who would not want to entertain any such reorientation of anthropology, and it is important to appreciate the reasons why present-day anthropologists would say "impossible," "improbable" "irrelevant," "off the mark," even "impertinent." The obstacles that are posed are many, but for our purposes here they may be discussed in terms of *access, attitudes, ethics*, and *methodology*.

Departments of anthropology have generally believed that students should do their dissertation field work in a non-Western culture. At some points in time that was a useful policy to implement, if in training anthropologists one valued the importance of culture shock and the detachment which accompanies it. For many students today, the experience of working in a Washington law firm, in a company town, or in an international industrial complex would be more bizarre than anything a student anthropologist could find in a Mexican village, or in New Guinea for that matter. We anthropologists have studied the cultures of the world only to find in the end that ours is one of the most bizarre of all cultures and one, by virtue of its world influence for "bad" or "good," in urgent need of study.

The most usual obstacle is phrased in terms of access. The powerful are out of reach on a number of different planes: they don't want to be studied; it is dangerous to study the powerful; they are busy people; they are not all in one place, and so on. As some of our students found out in their studies of corporate use of the courts:

> The belief that corporations work secretly and surreptitiously in their own interests has been somewhat verified. Their desire for secrecy, their paranoid fear of all but self-fashioned publicity, their refusal to discuss questions on their operation, and the overconscious regard of their lawyers for the confidential nature of the lawyer-client relationship

(even when the public's interests are at stake), all serve to eliminate any free flow of information which should be available to the public forum, and are reminiscent of secret societies. The stealth of the corporation is epitomized in those wily chess masters they employ to handle their cases, the corporate lawyers. (Zelf and Bush, 1970)

These difficulties are true of the people that anthropologists have studied in many different places. That problems of access are any different, or at least any more problematic, in studying up in the United States is a proposition which has not been adequately tested. Anthropologists have had problems of access everywhere they have gone; solving such problems of access is part of what constitutes "making rapport." In view of our successes among peoples of the wrorld who have been incredibly hostile, it is rather surprising that anthropologists could be so timid at home (see Riesman, 1954, pp. 440–66). Furthermore, it could be argued that access to bureaucratic organizations (such as governmental agencies) frequented by the wealthy and powerful should be open to social scientists by virtue of laws which protect public access to information affecting the public interest. In addition, there are wealthy anthropologists who would presumably have access "up." Cleveland Amory (1947) and E. Digby Baltzell (1964) have made substantial contributions to understanding the power status of the upper class, although neither one is an anthropologist. No, there must be more plausible reasons why the less powerful are more attractive for study in the United States.

It has been said that anthropologists value studying what they like and liking what they study and, in general, we prefer the underdog. Braroe and Hicks (1967), discussing the mystique of anthropology, make reference again to the traditional alienation from their own culture that characterizes anthropologists, and they explore how such alienation relates to their lack of intense commitment to social reform. This could be phrased more positively: Anthropologists have favored studying non-Western cultures as a way of fulfilling their mission to study the diverse ways of mankind; they have not had an intense commitment to social reform because of their relativistic stance and a belief that such a stance was necessary to a truly "objective, detached, scientific perspective," or because they thought that others, such as sociologists, were involved in social reform. While scientific findings may be ideally viewed as "value-free," certainly the choice of subject for scientific inquiry is most certainly not. Anthropologists of the future will have a greater responsibility for what they choose to study as well as how they study.

The ethical problems that are raised in studying up almost always appear to be confused, particularly in discussing ethics of working in one's own society. One student made the following comment:

To say that kula-ring participants don't perform in practice what they say they do has very different consequences from saying that a government

> agency is not living up to its standards. This isn't to say that the government agency shouldn't be studied, or that the fact it isn't living up to its standards shouldn't be pointed out. The question is: Can the anthropologist do a structural study and then in his role as citizen point out that the agency is screwing the American public?

The same student asked:

> How can we gain access to the same kinds of information as when we "study down" without being dishonest (i.e., a fake secretary or other role)? If we did get information without letting informants know we were social scientists, how could we publish it? It seems that the only "open" way of doing a study would end up being fairly superficial – questionnaires and formal interviews as versus what we learn by participant observation.

The problems raised by this student are ethical problems anthropologists have had to face no matter what culture they are studying. In discussing such ethical questions involved in studying up in our own society, I have the impression that confusion results depending on whether or not one recognizes the implicit double standard – is there one ethic for studying up and another for studying down? Or is it, as this student suggests, that the consequences of describing what may be systemic inadequacies may be greater for government agencies than peasant economic systems or for conflict resolution (or just plain conflict) in a small fishing village, and that therefore our subjects of study should be treated accordingly?

There is an important distinction to be recognized as to "public" and "private," even though informant anonymity may be important to both sectors. For the most part, anthropologists working in the United States can be said to have worked in the "private" sphere: we study families, small groups, those aspects of communities which are more private than public. We should not necessarily apply the same ethics developed for studying the private, and even ethics developed for studying in foreign cultures (where we are guests), to the study of institutions, organizations, bureaucracies that have a broad public impact. In reinventing anthropology, any discussion of ethics should consider the public-private dimensions as well as the home-abroad component. Furthermore, in the present anthropology, work that is considered in the objective social science mode, when carried out abroad might well be dubbed "journalistic" by the subjects. Telling it like it is may be perceived as muckraking by the subjects of study (Oscar Lewis' work on Mexico was so viewed), *or* by fellow professionals who feel more comfortable if data is presented in social science jargon which would protect the work from common consumption.

The concept of participant observation plays a determining role in *what* anthropologists choose to study. The power of participant observation as such was only discovered in the twentieth century. Malinowski and Radcliffe-Brown, among the first to do field work by the techniques of participant observation, set a new standard for ethnographic descriptions. When an anthropologist goes to study the culture of a people, he lives with them; the resultant description is rich in contextual information and is the result of the many points of view that one is opened to by virtue of "living with the natives." Hortense Powdermaker has described the components of participant observation as follows:

> The conditions for successful mutual communication include 1) physical proximity of the field worker to the people he studies, 2) knowledge of their language, and 3) psychological involvement. (1966, p. 287)

She goes on to say:

> The ability to be psychologically mobile is important in hierarchical situations where it is necessary to move easily between different levels in the power structure. Some field workers identify so completely with the underdog that they are unable to make effective contacts with those on the top level of the social (or political) hierarchy. (p. 291)

At the same time that Hortense Powdermaker has described the value of participant observations, she has also alluded to the limitations of such complete acceptance of participant observation as a distinctive feature of all social anthropological field work. When the anthropologist participant observes, he or she resides and generally partakes with the "natives." Such a method has weighed heavily in the decisions as to where anthropologists study: we prefer residential situations, whether the residence is in a primitive village or a modern hospital.

The degree to which our field choices might be determined by whether or not we can observe as participant was made clear to me when two of my students went to Washington to study a law firm that did not want to be studied (even though individual members were willing to cooperate in a limited way). How could they participant-observe if the firm wouldn't let them in the door, and *if they couldn't participant-observe, how could they do anthropology*? These questions have, of course, been raised before in anthropology, and when anthropologists thought it important enough they surmounted the problems raised. Witness the culture-at-a-distance studies that cropped up during World War II, or witness the work of Elizabeth Bott (1957) in her network study of kinship in London, which was based principally on face-to-face interviewing.

The point here is that there is a mystique about participant observation that carries points with it, yet it remains that the anthropologist's image of himself is shattered (Fischer, 1969) if he cannot participant-observe, and for the most part our students are not generally trained in the kinds of techniques that they would need to work on problems in nonresidential settings such as banks, insurance companies, government agencies, electronics industries, and the like. How many anthropologists know how to find out who owns a city? If Sol Tax is right in pointing out that anthropologists are not working on the most relevant problems of the world today – such as population, pollution, and war – because they cannot participant-observe such problems in a community, then, in reinventing anthropology, we might have to shuffle around the value placed on participant observation that leads us to forget that there are other methods (see Gussow and Tracy, 1971) more useful for some of the problems and situations we might like to investigate. The use of personal documents, memoirs, may substitute for anthropological participation in some areas of culture that take long years of participation to really understand. One student makes the following comments about field methodology:

> The principal research method of the anthropologist, participant observation, is, needless to say, not wholly applicable when one is studying a government agency or elite institution and its interaction with various people. A particular situation can he dealt with, but characteristically the data gleaned would be through observation rather than participation. We can define participation in two ways. One definition would maintain that to say the researcher is a participant means he is able to interact as a native in the situation studied and is therefore able to use himself as an informant. The other definition considers the participant in a status achieved by an outsider, since he is treated as an insider. Ideally, the more intimate the acceptance, the less the participant/observer will influence the situation he is observing and the closer he will be to the status of participant. Considering these two definitions, the term "participant/observer" could not be applicable to the types of situations that the ethnographer would want to study in large-scale institutions, unless he was actually to become a member of the group he is studying. The term "participant/observer" would not truly apply to these researcher's techniques, even in the situations where I was attempting to fulfill the ideal. In studying one's own society, especially if it is complex and highly specialized and heterogeneous, the question is to determine the levels of actual participation and the level barred from participation.
>
> If the anthropologist is going to make a contribution to the understanding of the institutions which in a complex way affect the lives of many people, he must take a methodologically eclectic approach. . . . (Serber, 1971, pp. 5–6)

Interviews of various sorts (formal/informal, face-to-face/telephone) were used by my students. Documents were used (see the *NACLA Guide*, 1970) – public relations documents for understanding the preferred self-image of the organization, internal documents on the structure and statistics of work planned and accomplished by the organization, all useful in discovering trends and what is thought of as problematic by the actors. Also important is what Marian Eaton has labeled "self-analysis" – an awareness on the part of the student of how he as a social scientist is perceived, run around, encultured, and described in the veiled and not-so-veiled encounters with informants and the members of organizations and the like whose job it is to deal with outsiders. We may have to give higher priority to traditional anthropological values such as using our knowledge of others as a mirror for ourselves and allowing questions to lead us to methodology (rather than vice versa).

We may have to reorder our conception of urgent anthropology. Surely it should be the needs of mankind for the study of man that lead the way.

Note

1. See M. N. Srinivas' book *Social Change in Modern India*, Chapter 5, "Some Thoughts on the Study of One's Own Society," for a discussion of the problems involved in such an endeavor.

References

Adams, Richard Newbold. 1970. *Crucifixion by Power*. Austin: University of Texas Press.

Amory, Cleveland. 1947. *The Proper Bostonians*. New York: E. P. Dutton & Co.

Baltzell, E. Digby. 1964. *The Protestant Establishment: Aristocracy and Caste in America*. New York: Vintage Books.

Benedict, Ruth. 1946. *The Chrysanthemum and the Sword*. Boston: Houghton Mifflin Co.

Bott, Elizabeth. 1957. *Family and Social Network: Roles, Norms and External Relationships in Ordinary Urban Families*. London: Tavistock Publications.

Braroe, Niels Winther, and George L. Hicks. 1967. "Observations on the Mystique of Anthropology." *Sociological Quarterly* 7, No. 2: 173–86.

Colson, Elizabeth. 1971. *The Impact of the Kariba Resettlement upon the Gwembe Tonga*. Manchester: University of Manchester Press.

Coser, Lewis A. 1968. "Violence and the Social Structure." In *Violence in the Streets*, ed. Shalom Endleman. Chicago: Quadrangle Books. Pp. 71–84.

Eaton, Marian. 1971. "An Ethnography of BBB Oakland: One Consumer's View." Unpublished undergraduate thesis, Department of Anthropology, University of California, Berkeley.

Edelhertz, Herbert. 1970. *The Nature, Impact, and Prosecution of White-Collar Crime*. Washington, D.C.: Government Printing Office.

Fellmeth, Robert, ed. 1971. *Power and Land in California, Preliminary Draft*. Center for the Study of Responsive Law, Washington, D.C.

Fischer, Ann. 1969. "The Personality and Subculture of Anthropologists and Their Study of U.S. Negroes." In *Concepts and Assumptions in Contemporary Anthropology*, ed. Stephen A. Tyler. Proceedings of the Southern Anthropological Society No. 3. Athens: University of Georgia Press. Pp. 12–17.

Gellhorn, Walter. 1966. *When Americans Complain*. Cambridge, Mass.: Harvard University Press.

Gladwin, Thomas. 1969. Review of *Culture and Poverty: Critique and Counter-Proposals*, by Charles Valentine. *Current Anthropology*, 10, Nos. 2–3: 185.

Gussow, Zachary, and George S. Tracy. 1971. "The Use of Archival Materials in the Analysis and Interpretation of Field Data: A Case Study in the Institutionalization of the Myth of Leprosy as 'Leper.'" *American Anthropologist* 73, No. 3: 695–709.

Hannerz, Ulf. 1969. *Soulside: Inquiries Into Ghetto Culture and Community*. New York: Columbia University Press.

Henry, Jules. 1963. *Culture Against Man*. New York: Random House.

Kluckhohn, Clyde, 1960. *Mirror for Man*. Greenwich, Conn.: Fawcett Publications.

Leacock, Eleanor Burke, ed. 1971. *The Culture of Poverty: A Critique*. New York: Simon & Schuster.

Love, Norma, and Marian Eaton. 1970. "The Bay Area Air Pollution Control District: An Anthropological Perspective." Unpublished undergraduate paper, Department of Anthropology, University of California, Berkeley.

Malinowski, Bronislaw. 1922. *Argonauts of the Western Pacific*. London: Routledge & Kegan Paul.

Mitchell, J. Clyde. 1966. "Theoretical Orientations in African Urban Studies." In *The Social Anthropology of Complex Societies*, ed. Michael Banton. Association of Social Anthropologists Monograph No. 4. London: Tavistock Publications. Pp. 37–68.

Nader, Laura. 1964. "Perspectives Gained from Field Work." In *Horizons of Anthropology*, ed. Sol Tax. Chicago: Aldine Press. Pp. 148–59.

North American Congress on Latin America. 1970. *NACLA Research Methodology Guide*. New York.

Pound, Roscoe. 1906. "The Causes of Popular Dissatisfaction with the Administration of Justice." *Reports of the American Bar Association* 29, Part 1: 395–417.

Powdermaker, Hortense. 1966. *Stranger and Friend: the Way of an Anthropologist*. New York: W. W. Norton & Co.

Resek, Carl. 1960. *Lewis Henry Morgan, American Scholar*. Chicago: University of Chicago Press.

Riesman, David. 1954. *Individualism Reconsidered and Other Essays*. Glencoe, Ill.: Free Press.

Serber, David. 1971. "A Discussion of the Policy Services Bureau of the California Department of Insurance, with Specific Reference to the Interaction with the Insured Public and the Insurance Industry." Unpublished undergraduate thesis, Department of Anthropology, University of California, Berkeley.

Spradley, J. P. 1970. *You Owe Yourself a Drunk: An Ethnography of Urban Nomads*. Boston: Little, Brown & Co.

Srinivas, M. N. 1966. *Social Change in Modern India*. Berkeley: University of California Press.

Sutherland, Edwin H. 1949. *White-Collar Crime*. New York: Dryden Press.

Valentine, Charles. 1969. Book review of his *Culture and Poverty: Critique and Counter-Proposals. Current Anthropology* 10, Nos. 2–3: 181–200.

Weakland, J. H. 1960. "'The Double Bind' Hypothesis of Schizophrenia and Three-Party Interaction." In *The Etiology of Schizophrenia*, ed. Don D. Jackson. New York: Basic Books.

Wolf, E. 1969. "American Anthropologists and American Society." In *Concepts and Assumptions in Contemporary Anthropology*, ed. Stephen A. Tyler. Proceedings of the Southern Anthropological Society No. 3. Athens: University of Georgia Press. Pp. 3–11.

Weff, David, and Peggy Bush. 1970. "Corporate Use of the Courts." Unpublished undergraduate paper, Department of Anthropology, University of California, Berkeley.

74

Ethnography in/of the World System: The Emergence of Multi-Sited Ethnography

George E. Marcus

Introduction

In the mid-1980s, I specified two modes in which ethnographic research was embedding itself within the context of an historic and contemporary world system of capitalist political economy (56, 57). The most common mode preserves the intensively-focused-upon single site of ethnographic observation and participation while developing by other means and methods the world system context. Examples of these other methods include working in archives and adapting the work of macrotheorists and other kinds of scholars as a mode of contextualizing portraiture in terms of which the predicaments of local subjects are described and analyzed. In this mode, a vital literature continues to appear on the historic (colonial) and contemporary incorporation of peoples as working classes or on the apparent reduction of local cultures by the macro-processes associated with capitalist political economy in the many forms it has taken (e.g. 10, 11, 31, 70, 74, 100). Such ethnography has produced refined examinations of resistance and accommodation – a concern with the dynamics of encapsulation, focused on the relationships, language, and objects of encounter and response from the perspectives of local and cosmopolitan groups and persons who, although in different relative power positions, experience a process of being mutually

Source: *Annual Review of Anthropology*, 24 (1995): 95–117.

displaced from what has counted as culture for each of them. This mode has shown that the heart of contemporary ethnographic analysis is not in the reclamation of some previous cultural state or its subtle preservation despite changes, but rather in the new cultural forms to which changes in colonial subaltern situations have given rise.

The other, much less common mode of ethnographic research self-consciously embedded in a world system, now often associated with the wave of intellectual capital labeled postmodern, moves out from the single sites and local situations of conventional ethnographic research designs to examine the circulation of cultural meanings, objects, and identities in diffuse time-space. This mode defines for itself an object of study that cannot be accounted for ethnographically by remaining focused on a single site of intensive investigation. It develops instead a strategy or design of research that acknowledges macrotheoretical concepts and narratives of the world system but does not rely on them for the contextual architecture framing a set of subjects. This mobile ethnography takes unexpected trajectories in tracing a cultural formation across and within multiple sites of activity that destabilize the distinction, for example, between lifeworld and system (49), by which much ethnography has been conceived. Just as this mode investigates and ethnographically constructs the lifeworlds of variously situated subjects, it also ethnographically constructs aspects of the system itself through the associations and connections it suggests among sites.

This second, still emergent mode of ethnography, upon which I focus in this review, may begin in the world system, but because of the way it evolves its object of study, this mode comes circumstantially to be of the world system as well. In particular, I focus on the various mapping strategies evident in this mode of ethnography and on the challenges that it poses for the assumptions and expectations embedded in the ethnographic method itself. Of course, the intellectual capital of so-called postmodernism has provided ideas and concepts for the emergence of multi-sited ethnography, but more importantly it arises in response to empirical changes in the world and therefore to transformed locations of cultural production (see especially 47). Empirically following the thread of cultural process itself impels the move toward multi-sited ethnography.

Research in anthropology that has embedded ethnographic subjects of study within contexts of a world system, historical political economies of colonialism, market regimes, state formation, and nation-building has developed explicitly within genres of Marxist anthropology (e.g. 16), anthropology and political economy (e.g. 79), and anthropology and history (e.g. 11, 79). Although some contemporary exemplars of multi-sited ethnography have developed within these traditional genres, many of the most striking examples have emerged in arenas of work that have not been identified with these typically world system-based contexts. These studies arise instead from anthropology's participation in a number of interdisciplinary (in fact, ideologically antidisciplinary) arenas that have evolved since the 1980s, such as media studies, feminist studies, science and technology studies, various strands of cultural studies, and the theory, culture, and society group (see 23, 50). Precisely because such interdisciplinary arenas do not share a clearly bounded object of study, distinct disciplinary perspectives that participate in them tend to be challenged. For ethnography this means that the world system is not the theoretically constituted holistic frame that gives context to the contemporary study of peoples or local subjects closely observed by ethnographers, but it becomes, in a piecemeal way, integral to and embedded in discontinuous, multi-sited objects of study. Cultural logics so much sought after in anthropology are always multiply produced, and any ethnographic account of these logics finds that they are at least partly constituted within sites of the so-called system (i.e. modern interlocking institutions of media, markets, states, industries, universities – the worlds of elites, experts, and middle classes). Strategies of quite literally following connections, associations, and putative relationships are thus at the very heart of designing multi-sited ethnographic research.

Shifts in macro-perspectives of the world system since the 1970s have accommodated well the trends of ethnography described here. Wallerstein's world system initiative (97) revived historically embedded social science generally. It provided a grand systemic narrative of world history that invited itself to be filled in and debated through the production of regional and micro-geographic social histories and ethnographies. In 1982, Wolf (101) provided an articulation of the specifically anthropological version of the grand world system narrative, which preserved, albeit on a comparative scale, the model of the ethnographic research project as single-site probing of local situations and peoples.

Successor views of the world system in the 1980s were pushed both by new sets of intellectual influences that operated against working within the frame of closed, though dynamic, systems narratives of macro-social processes and by the reflective awareness throughout the academy of massive changes afoot in the post–World War II international regimes of political economy. For those across disciplines interested in placing their specific projects of research in the unfolding of new arrangements for which past historical narratives were not fully adequate, a firm sense of a world system framework was replaced by various accounts of dissolution and fragmentation, as well as new processes – captured in concepts like post-Fordism (48), time-space compression (48), flexible specialization (48), the end of organized capitalism (51), and most recently, globalization (23, 44, 84) and transnationalism (39a,b) – none of which could be fully understood in terms of earlier macro-models of the capitalist world system. Even from within the heart of neoclassical economics, there are eloquent (and not altogether pessimistic) statements about the contemporary predicament of the loss of a firm systemic grasp of contemporary political economy. For example, as Robert Solow, the MIT Nobel laureate said in 1991, "There is not some glorious theoretical synthesis of capitalism that you can write down in a book and follow. You have to grope your way" (86a). What does such groping mean for the ethnographer?

For ethnographers interested in contemporary local changes in culture and society, single-sited research can no longer be easily located in a world system perspective. This perspective has become fragmented, indeed, "local" at its very core. With the collapse, then, of the easy distinction between system and lifeworld (49) as the mode for situating and designing ethnographic research on the contemporary world, the only alternatives have been to use various successor works of scholarship on global changes in political economy as the framing for single-site studies that are fully defined and contextualized in terms of those mostly nonethnographic works, or to pursue the more open-ended and speculative course of constructing subjects by simultaneously constructing the discontinuous contexts in which they act and are acted upon. The distinction between lifeworlds of subjects and the system does not hold, and the point of ethnography within the purview of its always local, close-up perspective is to discover new paths of connection and association by which traditional ethnographic concerns with agency, symbols, and everyday practices can continue to be expressed on a differently configured spatial canvas (see 56).

At stake here are conventional views and commitments to ethnographic method, which in recent times have not been discussed very much in methodological terms. Rather, novelty in method has been embedded in a discourse of reflexive self-presentation in contemporary ethnography in which the emphasis is on ethics, commitment, and activism. The pure, scaffold-like methodological implication of the way that multi-sited ethnography is devised in more committed language might seem to be mechanical and smack of older forms of positivism and of the disengaged positioning characteristic of value-free social science. The selection of space and sites of investigation emerge inseparably from the highly politicized way that the problem of investigation and then writing is cognized. Still, for conventional ethnography as it has been practiced in anthropology, the most interesting issues concerning emergent multi-site studies are most clearly understood in methodological terms (see especially 54a), so I have adopted such a methodological focus in this review. In the final section, however, I consider the reflexive activist persona through which this mode of ethnographic research actually articulates and designs methodological questions and research designs.

Methodological Anxieties

Among anthropologists, the move toward multi-sited ethnography might give rise to three sets of methodological anxieties: a concern about testing the limits of ethnography, a concern about attenuating the power of fieldwork, and a concern about the loss of the subaltern.

Testing the Limits of Ethnography

Ethnography is predicated upon attention to the everyday, an intimate knowledge of face-to-face communities and groups. The idea that ethnography might expand from its committed localism to represent a system much better apprehended by abstract models and aggregate statistics seems antithetical to its very nature and thus beyond its limits. Although multi-sited ethnography is an exercise in mapping terrain, its goal is not holistic representation, an ethnographic portrayal of the world system as a totality. Rather, it claims that any ethnography of a cultural formation in the world system is also an ethnography of the system, and therefore cannot be understood only in terms of the conventional single-site mise-en-scene of ethnographic research, assuming indeed it is the cultural formation, produced in several different locales, rather than the conditions of a particular set of subjects that is the object of study. For ethnography, then, there is no global in the local-global contrast now so frequently evoked. The global is an emergent dimension of arguing about the connection among sites in a multi-sited ethnography. Thus, the multi-sited ethnography is content to stipulate some sort of total world system as long as the terms of any particular macro-construct of that system are not allowed to stand for the context of ethnographic work that becomes opportunistically constituted by the path or trajectory it takes in its design of sites.

Attenuating the Power of Fieldwork

The issue then arises of whether multi-sited ethnography is possible without attenuating the kinds of knowledges and competencies that are expected from fieldwork. In other words, is multi-site fieldwork practical? One response is that the field broadly conceived and encompassed in the fieldwork experience of most standard ethnographic projects indeed already crosses many potentially related sites of work, but as research evolves, principles of selection operate to bound the effective field in line with long-standing disciplinary perceptions about what the object of study should be. Thus, fieldwork as traditionally perceived and practiced is already itself potentially multi-sited.

Furthermore, standard cultural history (e.g. 7, 38a) is very much multi-sited, but unlike in anthropology, this feature of research is unproblematic. This undoubtedly has something to so with the fragmentary, reconstructive nature of historical method, in which the composition and probing of the relationships of dispersed materials are basic. It is perhaps anthropologists' appreciation of the difficulty of doing intensive ethnography at any site and the satisfaction that comes from such work in the past when it is done well that would give them pause when the ethnographer becomes mobile and still claims to have done good fieldwork.

Indeed, something of the mystique and reality of conventional fieldwork is lost in the move toward multi-sited ethnography. But not all sites are treated by a uniform set of fieldwork practices of the same intensity. Multi-sited ethnographies inevitably are the product of knowledge bases of varying intensities and qualities. To do ethnographic research, for example, on the social grounds that produce a particular discourse of policy requires different practices and opportunities than does fieldwork among the situated communities such policy affects (see especially 22). To bring these sites into the same frame of study and to posit their relationships on the basis of first-hand ethnographic research in both is the important contribution of this kind of ethnography, regardless of the variability of the quality and accessibility of that research at different sites.

Many factors thus control for the quality of fieldwork in multi-sited research. The point is that in such research a certain valorized conception of fieldwork and what it offers wherever it is conducted threatens to be qualified, displaced, or decentered in the conduct of multi-sited ethnography. Still, what is not lost but remains essential to multi-sited research is the function of translation from one cultural idiom or language to another. This function is enhanced since it is no longer practiced in the primary, dualistic "them-us" frame of conventional ethnography but requires considerably more nuancing and shading as the practice of translation connects the several sites that the research explores along unexpected and even dissonant fractures of social location. Indeed, the persuasiveness of the broader field that any such ethnography maps and constructs is in its capacity to make connections through translations and tracings among distinctive discourses from site to site.

In this enhanced challenge of translation, literal language learning remains as important as it has been in preparing for traditional fieldwork. Just as "knowing the language" guarantees the integrity of traditional fieldwork and gives the bounded field – e.g. a people, an ethnic group, a community – its most important coherence as a culture, this skill is as important in multi-sited fieldwork and with even more exactitude. It is perhaps no accident that exemplars thus far of multi-sited fieldwork have been developed in monolingual (largely Anglo-American) contexts in which fine-grained knowledge of the language is unproblematic for native English speakers. Yet, if such ethnography is to flourish in arenas that anthropology has defined as emblematic interests, it will soon have to become as multilingual as it is multi-sited. In this sense, it conforms to (and often exceeds) the most exacting and substantive demands of traditional fieldwork.

The Loss of the Subaltern

It is not just any situated subjects that ethnography concerned with the world system focuses upon, but in this context, it habitually focuses upon subaltern subjects, those positioned by systemic domination (ultimately traceable to capitalist and colonialist political economy in its variety of forms). Although multi-sited ethnography may not necessarily forsake the perspective of the subaltern, it is bound to shift the focus of attention to other domains of cultural production and ultimately to challenge this frequently privileged positioning of ethnographic perspective. In the frame of science studies, Haraway is eloquent on this point: "A commitment to mobile positioning and to passionate detachment is dependent on the impossibility of innocent 'identity' politics and epistemologies as strategies for seeing from the standpoints of the subjugated in order to see well. One cannot 'be' either a cell or molecule – or a woman, colonized person, labourer, and so on – if one intends to see and see from these positions critically . . ." (46:192).

In yielding the ethnographic centering on the subaltern point of view, one is also decentering the resistance and accommodation framework that has organized a considerable body of valuable research (see 82) for the sake of a reconfigured space of multiple sites of cultural production in which questions of resistance, although not forgotten, are often subordinated to different sorts of questions about the shape of systemic processes themselves and complicities with these processes among variously positioned subjects.

So, it is a mistake to understand multi-sited ethnography, as it sometimes has been, as merely adding perspectives peripherally to the usual subaltern focus – e.g. adding perspectives on elites and institutions, or studying "up" (68) for mere completeness. Rather, this kind of ethnography maps a new object of study in which previous situating narratives like that of resistance and accommodation become qualified by expanding what is ethnographically "in the picture" of research both as it evolves in the field and as it is eventually written up.

Nor is multi-sited ethnography merely a different kind of controlled comparison, long a part of anthropological practice, as it has also sometimes been understood, although it does represent a revival of comparative study in anthropology. Conventional controlled comparison in anthropology is indeed multi-sited, but it operates on a linear spatial plane, whether the context is a region, a broader culture area, or the world system (see e.g. 31, 85); comparisons are generated for homogeneously conceived conceptual units (e.g. peoples, communities, locales), and such comparisons usually are developed from distinctly bounded periods or separate projects of fieldwork.

In projects of multi-sited ethnographic research, de facto comparative dimensions develop instead as a function of the fractured, discontinuous plane of movement and discovery among sites as one maps an object of study and needs to posit logics of relationship, translation, and association among these sites. Thus, in multi-sited ethnography, comparison emerges from putting questions to an emergent object of study whose contours, sites, and relationships are not known beforehand, but are themselves a contribution of making an account that has different, complexly connected real-world sites of investigation. The object of study is ultimately mobile and multiply situated, so any ethnography of such an object will have a comparative dimension that is integral to it, in the form of juxtapositions of phenomena that conventionally have appeared to be (or conceptually have been kept) "worlds apart." Comparison reenters the very act of ethnographic specification by a research design of juxtapositions in which the global is collapsed into and made an integral part of parallel, related local situations rather than something monolithic or external to them. This move toward comparison embedded in the multi-sited ethnography stimulates accounts of cultures composed in a landscape for which there is as yet no developed theoretical conception or descriptive model.

Interdisciplinary Arenas and New Objects of Study

There are several inspirations for multi-sited ethnography within the high theoretical capital associated with postmodernism: One might think, for example, of Foucault's power/knowledge and heterotopia (18), Deleuze & Guattari's rhizome (13), Derrida's dissemination (15), and Lyotard's juxtaposition by "blocking together" (78). These concepts anticipate many of the contemporary social and cultural conditions with which ethnographers and other scholars are trying to come to terms in shaping their objects of study in the absence of reliable holistic models of macroprocess for contextualizing referents of research, such as "the world system," "capitalism," "the state," "the nation," etc. However, such high theoretical capital usually is not the most proximate source for the terms by which multi-sited ethnographic research is thought through and conceived. Instead, multi-sited ethnography is intellectually constructed in terms of the specific constructions and discourses appearing within a number of highly self-conscious interdisciplinary arenas that use the diverse high theoretical capital that inspires postmodernism to reconfigure the conditions for the study of contemporary cultures and societies. This section briefly samples three such milieus in which objects of study have been evoked appropriate for composite, multimethod, mobile works of scholarship, including specifically multi-sited ethnography.

Unfortunately, there are many more concepts and visions for doing multisited ethnography than there are achieved exemplars (see next section). There is no doubt, however, that within the various interdisciplinary arenas, the following concepts for reconfigured objects of study come not from detached theoretical exercises, but from vital and active research efforts in progress, the forms of whose written and published results are yet to be established fully.

Media studies has been one important arena in which multi-sited ethnographic research has emerged. Distinct genres of research have appeared on production (especially in television and film industries), on the one hand, and on the reception of such productions, on the other. These two functions have been encompassed and related to each other within the frame of individual projects of research, thus making even more complex the trajectory of modes of ethnographic research that had already tended to be multi-sited in their construction of objects of study (77).

In anthropology, there has been a shift from older interests in ethnographic film toward a more encompassing terrain for the study of indigenous media [Ginsburg's writings have been key in this shift (35–37)]. This change has been stimulated by ethnographic study and participation in contemporary indigenous peoples' movements within and across nation-states. The control of means of mass communication and the activist role of indigenous peoples as media producers in these movements have reconfigured the space in which the ethnography of many of anthropology's traditional subjects can effectively be done; they also have made this space inherently multi-sited (see e.g. 96). The above-noted merging of production and reception sites in media studies has reinforced this tendency in the design of ethnographic research on specifically indigenous media.

The social and cultural study of science and technology (see Franklin, this volume) is another major arena in which genres of multi-sited ethnographic research have established their importance. Theorists such as Latour (52, 53) and Haraway (45, 46) have been crucially important in pushing the ethnographic dimensions of this field beyond pioneering lab studies to more complex (and multi-sited) social and cultural time-spaces. Haraway's cyborg (45) has been an especially influential construct in stimulating field researchers to think unconventionally about the juxtaposed sites that constitute their objects of study (17).

In anthropological work within the field of cultural studies of science and technology, the tendency toward multi-sited research is most prevalent in the following topical areas: the study of issues concerning reproduction and reproductive technologies (originating in an important domain of feminist research in medical anthropology) (38); epidemiological studies in medical anthropology (4a); studies of new modes of electronic communication such as the Internet (see e.g. 19, 61); and studies concerned with environmentalism and toxic disasters (e.g. 54, 88, 102). Another area is the study of the emergence of biotechnology and "big" science projects like the human genome project [of particular interest here is Rabinow's (76) work on the discovery and commodification of the polymerase chain reaction, especially related to the multi-sited style of his earlier work on French modernity (75)]. The title of a recent survey of biotechnology, *Gene Dreams, Wall Street, Academia, and the Rise of Biotechnology* (93), captures the methodological tendency toward multi-sited objects of study.

Amid the diffuse inspirations and influences of the broad interdisciplinary arena of cultural studies in the United States, the collection edited by

Grossberg et al (41) surveys the possibilities and limits of this remarkable remaking of the early and equally diffuse discussions of postmodernism during the 1970s and 1980s. Within this diffuse area of cultural studies, the Public Culture project deserves special mention because it addresses the long-standing concerns of anthropology and area studies. It was originated by Arjun Appadurai and Carol Breckenridge and developed through the independent Center for Transnational Cultural Studies in Chicago (8), with the journal *Public Culture* as its major publication. This project has constituted a major point of intersection for many diverse strands of cultural studies, broadly conceived around issues of the rethinking of ideas of culture (especially questions of trans- and cross-cultural production) in the face of contemporary world system changes. Appadurai's widely read paper (4) on the global cultural economy has provided a complex multi-sited vision for research in this transnational domain that defies older practices of "locating" culture(s) in place(s).

Theoretically rethinking concepts of space and place in ethnographic research (43, 47), for which the work of cultural geographers and sociologists (30, 86) has been a reinforcing inspiration, has stimulated the opening of established genres of anthropological research to multi-sited constructions of ethnographic research designs. For example, migration studies have become part of a much richer body of work on mobile and contingently settled populations, across borders, in exile, and in diasporas (e.g. 9, 33, 71). This work, concerned theoretically with the construction of identities in global-local frames, merges with the methods and spaces constructed by media studies (e.g. 1, 69).

Development studies are similarly being reconceived. Important critiques by Ferguson (25) and Escobar (20) of older development agencies and paradigms have been followed by a much more diverse sense of the field in which any study of development must now be evolved. For example, Escobar's study of a region in Colombia (21) draws the intersections among social movements, older development approaches, and the powerful global environmentalist doctrine of biodiversity. Again, redrawing the boundaries of topics of study here inevitably causes overlap with the terrains being established by other interdisciplinary arenas such as media studies and science and technology studies. But the most interesting and specific manifestations of these reconfigurations of perspective in overlapping interdisciplinary arenas are in the modes of constructing multi-sited spaces of investigation within individual projects of research, to which we now turn.

Modes of Construction

Powerful conceptual visions of multi-sited spaces for ethnographic research that have been especially influential in anthropology, such as Haraway's construct of the cyborg (45) and Appadurai's idea of the global cultural economy with its variety of "scapes" (4), do not also function as guides for designing the research that would exemplify and fulfill such visions. This requires a more literal discussion of methodological issues, such as how to construct the multi-sited space through which the ethnographer traverses.

Such explicitly methodological discussions are rare. An interesting exception is Strathern's (89) highly theoretical discussion of rethinking problems of relationality and connectivity in light of influential ideas within science and technology studies deriving from chaos theory (39) as well as from Haraway's notion of the cyborg. Despite the abstract character of Strathern's work, she remains close to issues of how ethnographic research is to be designed.

Multi-sited research is designed around chains, paths, threads, conjunctions, or juxtapositions of locations in which the ethnographer establishes some form of literal, physical presence, with an explicit, posited logic of association or connection among sites that in fact defines the argument of the ethnography. Indeed, such multi-sited ethnography is a revival of a sophisticated practice of constructivism, one of the most interesting and fertile practices of representation and investigation by the Russian avant-garde of momentous social change just before and after their revolution. Constructivists viewed the artist as an engineer whose task was to construct useful objects, much like a factory worker, while actively participating in the building of a new society. Film-making, especially the work of Vertov (e.g. "The Man with the Movie Camera"), was one of the most creative and de facto ethnographic media through which constructivism (72) was produced. From a methodological perspective, Vertov's work is an excellent inspiration for multi-sited ethnography.

Multi-sited ethnographies define their objects of study through several different modes or techniques. These techniques might be understood as practices of construction through (preplanned or opportunistic) movement and of tracing within different settings of a complex cultural phenomenon given an initial, baseline conceptual identity that turns out to be contingent and malleable as one traces it.

Follow the People

This technique is perhaps the most obvious and conventional mode of materializing a multi-sited ethnography. Malinowski's *Argonauts of the Western Pacific* is the archetypal account (55). The exchange or circulation of objects or the extension in space of particular cultural complexes such as ritual cycles and pilgrimages may be rationales for such ethnography, but the procedure is to follow and stay with the movements of a particular group of initial subjects. Migration studies are perhaps the most common contemporary research genre of this basic mode of multi-sited ethnography. Within this genre, a recent paper by Rouse (80) [but see also the statement by Gupta & Ferguson (42) as well as their edited collection (43)] is notable and often cited for moving migration studies (e.g. 40) into the terrain of diaspora studies, which has arisen as one of the key genres of cultural studies. Rouse follows his Mexican subjects across borders and sites in the conventional mode of migration studies, but in the spirit of contemporary, self-consciously multi-sited ethnography, he materializes a new object of study, a sense of a diasporic world independent of the mere movement of subjects from one place to another.

Willis's study (99), and Foley's study (29) of a school in Texas, inspired by the former, is a foreshortened version of "following the people" in that their strategic significance as single-site research with multiple sites evoked is their "off-stage" knowledge, so to speak, of what happens to their subjects in the other sites. The sense of "system" in their work arises from the connection between ethnographic portraits of their subjects and the posited relationship of these portraits to the fates of these same subjects in other locations.

Follow the Thing

This mode of constructing the multi-sited space of research involves tracing the circulation through different contexts of a manifestly material object of study (at least as initially conceived), such as commodities, gifts, money, works of art, and intellectual property. This is perhaps the most common approach to the ethnographic study of processes in the capitalist world system. Indeed, this technique is at the heart of Wallerstein's method for fine-grained study of process in the world system (97:4):

> The concept of commodity chain is central to our understanding of the processes of the capitalist world-economy. . . . Take any consumable product, say clothing. It is manufactured. The manufacturing process minimally involves material inputs, machinery, and labor. Material inputs are either manufactured or produced in some way. Machinery is manufactured. And labor must be recruited either locally or by immigration, and must be fed. . . . We may continue to trace each "box" further back in terms of its material inputs, machinery, land, labor. The totality constitutes a commodity chain.

Wallerstein's commodity chain is hardly laid out with a specifically ethnographic sensibility, but it is clearly a blueprint appropriate for multi-sited research.

In anthropology, Mintz's culture history of sugar (66) is an exemplar of the "follow the thing" technique, but also within a conventional political economy framework that depends on a master historical narrative of the workings of colonialism and capitalism. However, the most important and influential statement of this technique for multi-sited research on the circulation of things is Appadurai's introduction to his collection, *The Social Life of Things* (3, see also 12). In tracing the shifting status of things as commodities, gifts, and resources in their circulations through different contexts, Appadurai presumes very little about the governance of a controlling narrative of macroprocess in capitalist political economy but allows the sense of system to emerge ethnographically and speculatively by following paths of circulation. Although there are no ethnographies in the genre traditionally associated with studies of contemporary capitalist political economy that literally take a thing-oriented approach, an impressive ethnographic literature on consumption and commodities has appeared, which if not multi-sited in actual research design, is produced in the speculative, open-ended spirit of tracing things in and through contexts (see especially 65, 98).

The most explicit experimentation with multi-sited research using this technique seems to have emerged in studies of contemporary worlds of art and aesthetics (see especially 63). Notable examples include Myers's study (67) of the circulation of Pintupi acrylic paintings in Western art worlds, Savigliano's study of Tango (81a), Steiner's study (87) of the transit of African curios into Western art markets, along with Taylor & Barbash's film (92) based on Steiner's study, Silverman's study of taste in Reagan's America (83) across three intensively explored sites, and Feld's mapping (24) of "world music" and "world beat."

Finally, among some of the most influential, self-consciously multi-sited work in the arena of science and technology studies, the "follow the thing" mode of constructing the space of investigation has been prominent. Latour's work (52, 53) exemplifies this mode, albeit less so than does Haraway's, which has as much a metaphorical as a material sense of the things she traces. Latour's study (53) of the triumph of Pasteur's biology in France provocatively places, with a claim of equivalence, microbes, machines, and humans in various locations on the same plane or map of investigation.

Follow the Metaphor

When the thing traced is within the realm of discourse and modes of thought, then the circulation of signs, symbols, and metaphors guides the design of ethnography. This mode involves trying to trace the social correlates and groundings of associations that are most clearly alive in language use and print or visual media. Haraway's influential studies work primarily through this mode of constructing the object of study. In anthropology, the most fully achieved multi-sited ethnography in this mode (and in a sense, the most fully achieved and rationalized multi-site ethnography, whatever mode of construction, thus far) is Martin's *Flexible Bodies: Tracking Immunity in American Culture From the Days of Polio to the Age of AIDS* (64). Her initial interest is in ways of thinking about the body's immune system at various locations in American society – in the mass media, "on the street," in the treatment of AIDS, among alternative practitioners, and among scientists. She is interested in the variety of distinct discourses and registers concerning the immune system and in the ethnographic characteristics of their social locations. She uses a variety of methods and modes of participation for each location she probes – some in more depth than others.

Martin notes a pivotal point in her research: "One of the clearest moments of 'implosion' in my fieldwork, when elements from different research contexts seemed to collapse into one another with great force, occurred in a graduate course I was taking in immunology . . ." (64:91). With an ear for metaphor, Martin makes the association between the trope of flexibility so prominent in scientific conceptions of the immune system and the regime of flexible specialization so salient in late twentieth-century capitalism. She is then led to a fascinating exploration of complexity theory in which the trope of flexibility seems to be most systematically thought out, to theories and practices of corporate management, and to new ideologies of work and how they are inculcated in training programs in which she participates. Her provocative argument about an emergent form of post-Darwinist subjectivity in the United States depends for its persuasiveness on the multi-sited ethnographic space she has tracked by working through discovered metaphorical associations. This mode of constructing multi-sited research is thus especially potent for suturing locations of cultural production that had not been obviously connected and, consequently, for creating empirically argued new envisionings of social landscapes.

Follow the Plot, Story, or Allegory

There are stories or narratives told in the frame of single-site fieldwork that might themselves serve as an heuristic for the fieldworker constructing multi-sited ethnographic research. This has been a routine technique in the disciplinary history of Levi-Straussian myth analysis within so-called traditional societies. In the framework of modernity, the character of the stories that people tell as myth in their everyday situations is not as important to fieldworkers tracking processes and associations within the world system as is their own situated sense of social landscapes. Reading for the plot and then testing this against the reality of ethnographic investigation that constructs its sites according to a compelling narrative is an interesting, virtually untried mode of constructing multi-sited research. However, Brooks's reading for the plot (6) in classic Freudian case studies as a way of developing innovative re-envisionings of relationships in Victorian society is suggestive of the way that plots in ethnographically found stories and narratives might be used to diversify the space of an object of study in fieldwork (58).

Perhaps the one genre of work where this technique is now being used is the renewed interest among anthropologists and others in social memory. Boyarin's recent collection (5) on the remapping of memory concerns social struggles over alternative visions about the definition of collective reality. Processes of remembering and forgetting produce precisely those kinds of narratives, plots, and allegories that threaten to reconfigure in often disturbing ways versions (myths, in fact) that serve state and institutional orders. In this way, such narratives and plots are a rich source of connections, associations, and suggested relationships for shaping multi-sited objects of research.

Follow the Life or Biography

The life history, a particularly favored form of ethnographic data in recent years, is a special case of following the plot. How to produce and develop life histories as ethnography has been the subject of much reflection, but the use of biographical narrative as a means of designing multi-sited research rarely has been considered. Fischer has produced one of the few discussions (26) of the use of life history in this way, and his work with Abedi (28) is a partial implementation of a strategy of developing more systematic analysis, generalized from the story of a particular individual's life [see also his recent work (27) on scientists' autobiographies as documents that suggest more general ways to materialize rich and diverse cultural formations within the history and practices of various sciences].

Life histories reveal juxtapositions of social contexts through a succession of narrated individual experiences that may be obscured in the structural study of processes as such. They are potential guides to the delineation of ethnographic spaces within systems shaped by categorical distinctions that may make these spaces otherwise invisible. These spaces are not necessarily subaltern spaces (although they may be most clearly revealed in subaltern life histories), but they are shaped by unexpected or novel associations among sites and social contexts suggested by life history accounts.

Follow the Conflict

Finally, following the parties to conflicts defines another mode for generating a multi-sited terrain in ethnographic research. In small-scale societies, this has been an established technique ("the extended case method") in the anthropology of law. In the more complex public spheres of contemporary societies, this technique is a much more central, organizing principle for multi-sited ethnography. Beyond the context of the anthropology of law, most notable contested issues in contemporary society involve simultaneously spheres of everyday life, legal institutions, and mass media. Ethnographic study of these issues thus requires multi-sited construction, perhaps more obviously than do any of the other above modes. The collections edited by Sarat & Kearns (81) and the Amherst Seminar on Law and Society (2) are excellent samplings of work that is inherently multi-sited. Ginsburg's study (34) of the abortion controversy in a small community and Gaines's study (32) of conflict over the legal status of cultural productions as copyrighted exemplify how law- and media-focused topics of ethnographic research ramify quickly into multi-sited terrains of investigation.

The Strategically Situated (Single-Site) Ethnography

As with Paul Willis's now classic study (99) of English working-class boys at school, some ethnography may not move around literally but may nonetheless embed itself in a multi-sited context. This is different than assuming or constructing a world system context.

The sense of the system beyond the particular site of research remains contingent and not assumed. Indeed, what goes on within a particular locale in which research is conducted is often calibrated with its implication for what goes on in another related locale, or other locales, even though the other locales may not be within the frame of the research design or resulting ethnography (e.g. in Willis's work the particular kind of interest that he develops in the boys at school, on which he focuses solely, is guided by his knowledge of what happens to them on the factory floor).

This strategically situated ethnography might be thought of as a foreshortened multi-sited project and should be distinguished from the single-site ethnography that examines its local subjects' articulations primarily as subalterns to a dominating capitalist or colonial system. The strategically situated ethnography attempts to understand something broadly about the system in ethnographic terms as much as it does its local subjects: It is only local circumstantially, thus situating itself in a context or field quite differently than does other single-site ethnography.

The consideration of this foreshortened version of the multi-sited project gives us the opportunity to ask what sorts of local knowledges are distinctively probed within the sites of any multi-sited ethnography. If not the resistance and accommodation frame alone for studying subjects' articulation to larger systems, then what? The key question is perhaps: What among locally probed subjects is iconic with or parallel to the identifiably similar or same phenomenon within the idioms and terms of another related or "worlds apart" site? Answering this question involves the work of comparative translation and tracing among sites, which I suggested were basic to the methodology of multi-sited ethnography. Within a single site, the crucial issue concerns the detectable system-awareness in the everyday consciousness and actions of subjects' lives. This is not an abstract theoretical awareness such as a social scientist might seek, but a sensed, partially articulated awareness of specific other sites and agents to which particular subjects have (not always tangible) relationships. In Willis's study, it is how much the boys manifest in their talk a "knowingness" about the very specific system and set of relations in which they are caught as labor. In the more fractured, discontinuous sites of Martin's study (64), it is lay notions "on the street," so to speak, of the body's immune system, compared to notions of the immune system in the lab, compared to ideas of flexibility in the corporate boardroom. In my study of the dynastic rich (62), it is how the abstract management of wealth elsewhere subtly enters the daily lives of prominent families. In the vision of the novelist DeLillo (14), it is getting at the "white noise" in any setting that makes the ethnographic probing of either multi-sited or strategically situated research distinctive.

In iconically identifying a cultural phenomenon in one site that is reproduced elsewhere, a number of conceptual discussions are guides to how to see or ethnographically probe a "sensibility" for the system among situated subjects. Taussig's essays (90) under the governing notion of the "nervous system" are suggestive here, as is his ethnographically embedded investigation (91) of Benjamin's "mimetic faculty." Pietz's (73) discussion of Marx's notion of fetishism in the theory of capitalism makes this important concept usable as another way of thinking about the system-sensitive dimensions of the everyday articulated thoughts and actions of ethnographic subjects. Studies of the phenomenology of the ethnographically situated awareness among subjects of doubled or multiply constructed selfhood in contexts of new forms of electronic communication (95) and the inheritance of great wealth (62) provide clues to the ethnographic registering of a multi-sited sensibility within any particular site. Tsing's recent ethnography (94) might also be understood as a bold attempt to establish a new way of seeing the broader registers of rich materials arising from fieldwork in an out-of-the-way place. Finally, a primary goal of the *Late Editions* series of annuals (59–61) is to expose, under different themes, the varieties of *fin-de-siècle* consciousness and sensibilities embedded in different sites as they are articulated by interlocutors in experiments with the interview or conversation format, employed by anthropologists and other scholars who return to sites of previous work.

The most important form of local knowledge in which the multi-sited ethnographer is interested is that which parallels the ethnographer's own interest – in mapping itself. Sorting out the relationships of the local to the global is a salient and pervasive form of local knowledge that remains to be recognized and discovered in the embedded idioms and discourses of any contemporary site that can be defined by its relationship to the world system. In this cognitive and intellectual identification between the investigator and variously situated subjects in the emergent field of multi-sited research, reflexivity is most powerfully defined as a dimension of method, serving to displace or recontextualize the sort of literal methodological discussion that I have provided above. Haraway's discussion of positioning (46) is perhaps the most eloquent statement of the reflexive context and significance of multi-sited research. In contemporary multi-sited research projects moving between public and private spheres of activity, from official to subaltern contexts, the ethnographer is bound to encounter discourses that overlap with his or her own. In any contemporary field of work, there are always others within who know (or want to know) what the ethnographer knows, albeit from a different subject position, or who want to know what the ethnographer wants to know. Such ambivalent identifications, or perceived identifications, immediately locate the ethnographer within the terrain being mapped and reconfigure any kind of methodological discussion that presumes a perspective from above or "nowhere."

In practice, multi-sited fieldwork is thus always conducted with a keen awareness of being within the landscape, and as the landscape changes across sites, the identity of the ethnographer requires renegotiation. Only in the writing of ethnography, as an effect of a particular mode of publication itself, is the privilege and authority of the anthropologist unambiguously reassumed, even when the publication gives an account of the changing identities of the fieldworker in the multi-sited field.

The virtue of Haraway's discussion of positioning is that it argues persuasively for the objectivity (rather than the often presumed subjectivism) that arises from such a scrupulous, methodological practice of reflexivity. However, the qualification or effacement of the traditional privileged self-identification as ethnographer that seems inevitable in multi-sited research in favor of a constantly mobile, recalibrating practice of positioning in terms of the ethnographer's shifting affinities for, affiliations with, as well as alienations from, those with whom he or she interacts at different sites constitutes a distinctly different sense of "doing research."

Ethnographer as Circumstantial Activist

It is appropriate in conclusion to come full circle and to place the literal methodological concerns developed in this review in terms of a particular ethos of self-perception commonly evidenced in multi-sited research out of the just-mentioned experience of positioning. The conventional "how-to" methodological questions of social science seem to be thoroughly embedded in or merged with the political-ethical discourse of self-identification developed by the ethnographer in multi-sited research. The movement among sites (and levels of society) lends a character of activism to such an investigation. This is not (necessarily) the traditional self-defined activist role claimed by the left-liberal scholar for his or her work. That is, it is not the activism claimed in relation to affiliation with a particular social movement outside academia or the domain of research, nor is it the academic claim to an imagined vanguard role for a particular style of writing or scholarship with reference to a posited ongoing politics in a society or culture at a specific historic moment. Rather, it is activism quite specific and circumstantial to the conditions of doing multi-sited research itself. It is a playing out in practice of the feminist slogan of the political as personal, but in this case it is the political as synonymous with the professional persona and, within the latter, what used to be discussed in a clinical way as the methodological.

In conducting multi-sited research, one finds oneself with all sorts of cross-cutting and contradictory personal commitments. These conflicts are resolved, perhaps ambivalently, not by refuge in being a detached anthropological scholar, but in being a sort of ethnographer-activist, renegotiating identities in different sites as one learns more about a slice of the world system. For example, in Martin's *Flexible Bodies* (64), she is an AIDS volunteer at one site, a medical student at another, and a corporate trainee at a third. Politically committed though she is at the start of her research, ethnographer though she is throughout it, the identity or persona that gives a certain unity to her movement through such disjointed space is the circumstantial activism involved in working in such a variety of sites, where the politics and ethics of working in any one reflects on work in the others.

In certain sites, one seems to be working with, and in others one seems to be working against, changing sets of subjects. This condition of shifting personal positions in relation to one's subjects and other active discourses in a field that overlap with one's own generates a definite sense of doing more than just ethnography, and it is this quality that provides a sense of being an activist for and against positioning in even the most self-perceived apolitical fieldworker.

Finally, the circumstantial commitments that arise in the mobility of multi-sited fieldwork provide a kind of psychological substitute for the reassuring sense of "being there," of participant observation in traditional single-site fieldwork. One often affiliates with literal activists in the space of multi-sited research, and given anthropology's past preference for focusing on subaltern or marginal subjects, such activists are often surrogates for one's "people" of traditional research. The emerging and circumstantial sense of activism that develops among ethnographers in a multi-sited space and their close personal affiliations with cultural producers (e.g. artists, filmmakers, organizers), who themselves move across various sites of activity, thus preserve for ethnographers engaged in multi-sited research an essential link with the traditional practice of participant observation, single-site ethnography in the peripatetic, translative mapping of brave new worlds.

Literature Cited

1. Abu-Lughod L, ed. 1993. Screening politics in a world of nations. *Public Cult.* 11: 465–606 (Special Segment)
2. Amherst Seminar. 1988. Law and ideology. *Law Soc. Rev.* 22(4) (Special Issue)
3. Appadurai A, ed. 1986. *The Social Life of Things: Commodities in Cultural Perspective.* New York: Cambridge University Press
4. Appadurai A. 1990. Disjuncture and difference in the global cultural economy. *Public Cult.* 2:1–24

4a. Balshem M. 1993. *Cancer in the Community: Class and Medical Authority.* Washington, DC: Smithsonian Inst. Press

5. Boyarin J. 1994. Space, time, and the politics of memory. In *Remapping Memory: The Politics of TimeSpace*, ed. J Boyarin, p. 1–24. Minneapolis: Univ. Minn. Press
6. Brooks P. 1984. *Reading for the Plot: Design and Intention in Narrative.* New York: Knopf
7. Campbell C. 1987. *The Romantic Ethic and the Spirit of Modern Consumerism.* Oxford: Blackwell
8. Chicago Cultural Studies Group. 1992. Critical multiculturalism. *Crit. Inq.* 18(3): 530–55
9. Clifford J. 1994. Diasporas. See Ref. 47, pp. 302–38
10. Comaroff J, Comaroff JL. 1991. *Of Revelation and Revolution: Christianity, Colonialism, and Consciousness in South Africa.* Chicago: Univ. Chicago Press
11. Comaroff J, Comaroff JL. 1992. *Ethnography and the Historical Imagination.* Boulder, CO: Westview
12. Coombe RJ. 1995. The cultural life of things: globalization and anthropological approaches to commodification. *Am. J. Int. Law Polit.* 10(1): In press
13. Deleuze G, Guattari F. 1988. *A Thousand Plateaus: Capitalism and Schizophrenia.* London: Athlone
14. DeLillo D. 1984. *White Noise.* New York: Penguin
15. Derrida J. 1981. [1972]. *Dissemination.* Chicago: Univ. Chicago Press
16. Donham DL. 1990. *History, Power, Ideology: Central Issues in Marxism and History.* New York: Cambridge Univ. Press
17. Downey GL, Dumit J, Traweek S. 1995. *Cyborgs and Citadels: Anthropological Interventions in Emerging Sciences and Technologies.* Santa Fe, NM: School Am. Res. Press. In press
18. Dreyfus HL, Rabinow P. 1983. *Michel Foucault: Beyond Structuralism and Hermeneutics.* Chicago: Univ. Chicago Press
19. Escobar A. 1993. Welcome to Cyberia: notes on the anthropology of cyberculture. *Curr. Anthropol.* 35:211–31
20. Escobar A. 1994. *Encountering Development: The Making and Unmaking of the Third World.* Princeton, NJ: Princeton Univ. Press
21. Escobar A. 1994. *Cultural politics and biological diversity: state, capital, and social movements in the Pacific coast of Columbia.* Presented at Guggenheim Found. Conf. "Dissent and Direct Action in the Late Twentieth Century," Otavalo, Ecuador, June 15–19
22. Farmer P. 1992. *Aids and Accusation: Haiti and the Geography of Blame.* Berkeley: Univ. Calif. Press
23. Featherstone M, ed. 1990. *Global Culture, Nationalism, Globalism, and Modernity.* London: Sage
24. Feld S. 1994. From schizophonia to schismogenesis: on the discourses and commodification practices of "world music" and "world beat." In *Music Grooves*, by C Keil, S Feld, pp. 257–89. Chicago: Univ. Chicago Press

25. Ferguson J. 1990. *The Anti-Politics Machine: "Development", Depolitization, and Bureaucratic Power in Lesotho*. New York: Cambridge Univ. Press
26. Fischer MJ. 1991. The uses of life histories. *Anthropol. Hum. Q*. 16(1):24–27
27. Fischer MJ. 1995. (Eye) (I) ing the sciences and their signifiers (language, tropes, auto-biographers): InterViewing for a cultural studies of science & technology. See Ref. 60. In press
28. Fischer MJ, Abedi M. 1990. *Debating Muslims: Cultural Dialogues in Postmodernity and Tradition*. Madison: Univ. Wisc. Press
29. Foley DE. 1990. *Learning Capitalist Culture: Deep in the Heart of Tejas*. Philadelphia: Univ. Penn. Press
30. Friedland R, Boden D. 1994. *NowHere: Space, Time, and Modernity*. Berkeley: Univ. Calif. Press
31. Friedman J. 1994. *Cultural Identity and Global Process*. London: Sage
32. Gaines J. 1991. *Contested Culture: The Image, the Voice, and the Law*. Chapel Hill: Univ. NC Press
33. Gilroy P. 1993. *The Black Atlantic: Modernity and Double Consciousness*. Cambridge, MA: Harvard Univ. Press
34. Ginsburg F. 1989. *Contested Lives: The Abortion Debate in an American Community*. Berkeley: Univ. Calif. Press
35. Ginsburg F. 1993. Aboriginal media and the Australian imaginary. *Public Cult*. 5: 557–78
36. Ginsburg F. 1994. Embedded aesthetics: creating a discursive space for indigenous media. *Cult. Anthropol*. 9(3):365–82
37. Ginsburg F. 1996. Mediating culture: indigenous media, ethnographic film, and the production of identity. In *Fields of Vision*, ed. L Deveraux, R Hillman. Berkeley: Univ. Calif. Press. In press
38. Ginsburg F, Rapp R, eds. 1996. *Conceiving the New World Order: The Global Stratification of Reproduction*. Berkeley: Univ. Calif. Press. In press

38a. Ginzburg C. 1993. Microhistory: two or three things that I know about it. *Crit. Inq*. 20(1):10–35

39. Gleick J. 1987. *Chaos: Making a New Science*. New York: Penguin

39a. Glick Schiller N, Basch L, Blanc Szanton C. 1992. *The Transnationalization of Migration: Perspectives on Ethnicity and Race*. New York: Gordon & Breach

39b. Glick Schiller N, Fouron N. 1990. "Everywhere we go we are in danger": Ti Manno and the emergence of a Haitian transnational identity. *Am. Ethnol*. 17(2):329–47

40. Grasmuck S, Pessar P. 1991. *Between Two Islands: Dominican International Migration*. Berkeley: Univ. Calif. Press
41. Grossberg L, Nelson C, Treichler P. 1992. *Cultural Studies*. New York: Routledge
42. Gupta A, Ferguson J. 1992. Beyond 'culture': space, identity, and the politics of difference. *Cult. Anthropol*. 7:6–23
43. Gupta A, Ferguson J, eds. 1992. Space, Identity, and the Politics of Difference. Cult. Anthropol. 7(1) (Theme issue)
44. Hannerz U. 1992. *Cultural Complexity: Studies in the Social Organization of Meaning*. New York: Columbia Univ. Press
45. Haraway D. 1991. A cyborg manifesto: science, technology, and socialist-feminism in the late twentieth century. In *Simians, Cyborgs, and Women: The Reinvention of Nature*, pp. 149–82. New York: Routledge
46. Haraway D. 1991. Situated knowledges: The science question in feminism and the privilege of partial perspective. In *Simians, Cyborgs, and Women: The Reinvention of Nature*, pp. 183–202. New York: Routledge
47. Harding S, Myers F, eds. 1994. *Further Inflections: Toward Ethnographies of the Future. Cult. Anthropol*. 9(3) (Special Issue)

48. Harvey D. 1989. *The Condition of Post-modernity: An Inquiry into the Origins of Cultural Change*. Oxford: Blackwell
49. Holub RC. 1991. *Jurgen Habermas: Critic in the Public Sphere*. New York: Routledge
50. Lash S, Friedman J, eds. 1992. *Modernity and Identity*. Oxford: Blackwell
51. Lash S, Urry J. 1987. *The End of Organized Capitalism*. Madison: Univ. Wisc. Press
52. Latour B. 1987. *Science In Action*. Cambridge, MA: Harvard Univ. Press
53. Latour B. 1988. *The Pasteurization of France*. Cambridge, MA: Harvard Univ. Press
54. Laughlin K. 1995. Rehabilitating science, imagining Bhopal. See Ref. 60. In press
54a. Lindenbaum S, Lock M, eds. 1993. *Knowledge, Power and Practice*. Berkeley: Univ. Calif. Press
55. Malinowski B. 1922. *Argonauts of the Western Pacific*. New York: Dutton
56. Marcus GE. 1986. Contemporary problems of ethnography in the modern world system. In *Writing Culture: The Poetics and Politics of Ethnography*, ed. J Clifford, G Marcus, pp. 165–93. Berkeley: Univ. Calif. Press
57. Marcus GE. 1989. Imagining the whole: ethnography's contemporary efforts to situate itself. *Crit. Anthropol*. 9:7–30
58. Marcus GE. 1992. The finding and fashioning of cultural criticism in ethnography. In *Dialectical Anthropology: Essays in Honor of Stanley Diamond*, ed. CW Gailey, S Gregory, pp. 77–101. Gainseville: Fla. State Univ. Press
59. Marcus GE, ed. 1993. *Perilous States: Conversations on Culture, Politics and Nation*. Late Eds. 1: *Cultural Studies for the End of the Century*. Chicago: Univ. Chicago Press
60. Marcus GE, ed. 1995. *Techno-Scientific Imaginaries*. Late Eds. 2: *Cultural Studies for the End of the Century*. Chicago: Univ. Chicago Press
61. Marcus GE, ed. 1996. *Connected: Engagements With Media at the Century's End*. Late Eds. *3: Cultural Studies for the End of the Century*. Chicago: Univ. Chicago Press. In press
62. Marcus GE with Hall PD. 1992. *Lives in Trust: The Fortunes of Dynastic Families in Late Twentieth Century America*. Boulder, CO: Westview
63. Marcus GE, Myers F, eds. 1996. *The Traffic in Art and Culture: New Approaches to a Critical Anthropology of Art*. Berkeley: Univ. Calif. Press
64. Martin E. 1994. *Flexible Bodies: Tracing Immunity in American Culture From the Days of Polio to the Age of Aids*. Boston: Beacon
65. Miller D. 1994. *Modernity: An Ethnographic Approach*. Oxford: Berg
66. Mintz S. 1985. *Sweetness and Power: The Place of Sugar in Modern History*. New York: Viking
67. Myers F. 1992. Representing culture: The production of discourse(s) for Aboriginal acrylic paintings. In *Rereading Cultural Anthropology*, ed. GE Marcus, pp. 319–55. Durham, NC: Duke Univ. Press
68. Nader L. 1969. Up the anthropologist–perspectives gained from studying up. In *Reinventing Anthropology*, ed. D Hymes, pp. 284–311. New York: Pantheon
69. Naficy H. 1993. *The Making of Exile Cultures: Iranian Television in Los Angeles*. Minneapolis: Univ. Minn. Press
70. Ong A. 1987. *Spirit of Resistance and Capitalist Discipline: Factory Women in Malaysia*. Albany: State Univ. NY Press
71. Ong A. 1993. On the edge of empires: flexible citizenship among Chinese in diaspora. *Positions* 1:745–78
72. Petric V. 1987. *Constructivism-in-Film: "The Man With the Movie Camera," a Cinematic Analysis*. New York: Cambridge Univ. Press
73. Pietz W. 1993. Fetishism and materialism: The limits of theory in Marx. In *Fetishism as Cultural Discourse*, ed. E Apter, W Pietz, pp. 119–51. Ithaca, NY: Cornell Univ. Press
74. Pred A, Watts M. 1992. *Reworking Modernity: Capitalism and Symbolic Discontent*. New Brunswick, NJ: Rutgers Univ. Press

75. Rabinow P. 1989. *French Modern: Norms and Forms of Social Environment*. Cambridge, MA: MIT Press
76. Rabinow P. 1995. *The polymerase chain reaction*. Unpublished manuscript
77. Radway J. 1988. Reception study: ethnography and the problems of dispersed audiences and nomadic subjects. *Cult. Stud.* 2(3):359–76
78. Readings B. 1991. *Introducing Lyotard: Art and Politics*. New York: Routledge
79. Roseberry W. 1989. *Anthropologies and Histories: Essays in Culture, History, and Political Economy*. New Brunswick, NJ: Rutgers Univ. Press
80. Rouse R. 1991. Mexican migration and the social space of postmodernity. *Diaspora* 1: 8–23
81. Sarat A, Kearns TR, eds. 1993. *Law in Everyday Life*. Ann Arbor: Univ. Mich. Press

81a. Savigliano ME. 1995. *Tango and the Political Economy of Passion*. Boulder, CO: Westview

82. Scott JC. 1985. *Weapons of the Weak: Everyday Forms of Peasant Resistance*. New Haven, CT: Yale Univ. Press
83. Silverman D. 1986. *Selling Culture: Bloomingdale's, Diana Vreeland, and the New Aristocracy of Taste in Reagan's America*. New York: Pantheon
84. Sklair L. 1991. *The Sociology of the Global System*. Baltimore, MD: Johns Hopkins Univ. Press
85. Smith C. 1976. *Regional Analysis*, Vols. 1–2. New York: Academic
86. Soja EW. 1989. *Postmodern Geographies: The Reassertion of Space in Critical Social Theory*. London: Verso

86a. Solow R. 1991. *New York Times*, Sept. 29, Sect. 4, p. 1

87. Steiner CB. 1994. *African Art in Transit*. New York: Cambridge Univ. Press
88. Stewart K. 1995. Bitter faiths. See Ref. 60. In press
89. Strathern M. 1991. *Partial Connections*. Savage, MD: Rowman & Littlefield
90. Taussig M. 1990. *The Nervous System*. New York: Routledge
91. Taussig M. 1992. *Mimesis and Alterity*. New York: Routledge.
92. Taylor L, Barbash I. 1993. *In and Out of Africa*. Berkeley: Univ. Calif. Extens. Cent. Media Indep. Learn. (Video)
93. Teitelman R. 1989. *Gene Dreams, Wall Street, Academia and the Rise of Biotechnology*. New York: Basic Books
94. Tsing A. 1993. *In the Realm of the Diamond Queen: Marginality in an Out-of-the-Way Place*. Princeton, NJ: Princeton Univ. Press
95. Turkle S. 1984. *The Second Self: Computers and the Human Spirit*. London: Grenada
96. Turner T. 1991. Representing, resisting, re-thinking: historical transformations of Kayapó culture and anthropological consciousness. In *Colonial Situations: Essays on the Contextualization of Ethnographic Knowledge*, ed. GW Stocking Jr, pp. 285–313. Madison: Univ. Wisc. Press
97. Wallerstein I. 1991. *Report on an Intellectual Project: The Fernand Braudel Center, 1976–1991*. Binghamton, NY: Fernand Braudel Cent.
98. Weiner AW, Schneider J, eds. 1989. *Cloth and Human Experience*. Washington, DC: Smithsonian Inst. Press
99. Willis P. 1981. *Learning to Labour: How Working Class Kids Get Working Class Jobs*. New York: Columbia Univ. Press
100. Wilmsen EN. 1989. *Land Filled With Flies: A Political Economy of the Kalahari*. Chicago: Univ. Chicago Press
101. Wolf E. 1982. *Europe and the People Without History*. Berkeley: Univ. Calif. Press
102. Zonabend F. 1993. *The Nuclear Peninsula*. New York: Cambridge Univ. Press

75

Studying Up Revisited

Hugh Gusterson

Twenty-five years ago, in her article "Up the Anthropologist," Laura Nader appealed for a critical repatriated anthropology that would, in studying the cultures of the powerful as well as the powerless, throw new light on processes of domination in American society and help revitalize American democracy.[1] Ten years ago, in their book *Anthropology as Cultural Critique*, George Marcus and Michael Fischer renewed the call, this time with a postmodern more than a social democratic inflection. In 1995 – in a world marked by a wrenching intensification of capitalist accumulation and inequality, by the globalization of industrial and bureaucratic elites, by the enduring strength of the national security state, and by the growing power of new technoscientific elites in the electronics and biotechnology industries – these appeals for a critical repatriated anthropology are as relevant as ever and, in important respects, remain substantially unrealized. And now anthropology – revitalized by the return of Marxism, the eruption of feminism, and the infusion of Foucault's theories of power, to name just three developments – has new theoretical tools to apply to studying up. In this article I want to assess the progress we have made since the late 1960's in developing a critical repatriated anthropology, pointing out lacunae as well as advances, then discuss the methodological and writing problems inherent in studying up. The article draws in part on my own experience writing an ethnography of the Lawrence Livermore National Laboratory – the nuclear weapons laboratory in California where such weapons as the MX and the neutron bomb were designed. I see the Livermore Laboratory as emblematic

Source: *PoLAR: Political and Legal Anthropology Review*, 20(1) (1997): 114–119.

of exactly the kinds of sites of power that more anthropologists should be studying.

In her original article Nader pointed out that, although there were good reasons to study up, there were also significant obstacles in the way. These included the prevailing attitude of disdain towards those who did fieldwork at home as well as problems of access. Since the 1960's the prevailing attitudes within the profession have shifted to a large degree so that it is now increasingly permissible to do even one's first fieldwork at home. As Marcus and Fischer observed ten years ago, the reasons for this are practical as well as intellectual:

> There is less funding for social science research, especially for ethnography abroad, the practical applications of which are not apparent. Host societies, protective of their nationalisms, have complicated the acquisition of research permits. And there is indeed a growing awareness in anthropology that the functions of ethnography at home are as compelling and legitimate as they have been abroad (Marcus and Fisher 1986:113).

Still, if one looks at the kinds of repatriated ethnographies that have been written, in many cases anthropology's traditional taste for the marginal and exotic has not so much been transgressed as imported and transposed upon American society, leaving us with more studies of scientologists and crack dealers than of federal bureaucrats and corporate executives. The best-known examples of repatriated anthropology involve studying across or studying down more than studying up. Thus we have, for example, the work on reproductive and bio-politics by Robbie Davis-Floyd (1992), Faye Ginsburg (1989), Ginsburg and Rapp (1995), Emily Martin (1987, 1994), and Paul Rabinow (1996); the work on American religious culture by Susan Harding (1987) and Harriet Whitehead (1987); new studies of the poor by Philippe Bourgois (1995), Elliot Liebow (1993) and Jay Macleod (1987), and work on identity politics by Micaela di Leonardo (1984), Kath Weston (1991) and others. With the exception of George Marcus' new book *Lives In Trust*, we have very few ethnographic descriptions of how America looks to the rich. Indeed there seems to be a kind of glass ceiling for ethnographers, with the consequence that the ethnographic literature on the U.S. leaves the upper reaches of the social system largely in shade. As Renato Rosaldo (1989:201–2) has observed, "full citizenship and cultural visibility appear to be inversely related." The cultural invisibility of the rich and powerful is as much a part of their privilege as their wealth and power, and a democratic anthropology should be working to reverse this invisibility.

Laura Nader originally called for ethnographic studies of banks, realtors, law firms, insurance companies, manufacturing corporations, the communications industry, and government regulatory agencies – institutions at the heart of capitalist processes of production and stratification. There have been few takers. I would add that anthropologists, paying little attention to the cold war as a global process or to the national security state as a social institution, have ignored military elites even more than capitalist elites, despite militarism's evident importance as both a mode of social organization and a site of potent ideological production in America.

Part of the reason we have not had more studies of these elites – and this brings us to the second obstacle Laura Nader named – concerns the problems of access and methodology inherent in studying elites. As the business professor Robert Thomas writes in his article "Interviewing Important People in Big Companies":

> Large corporations . . . have gates, guards, and security devices . . . Even welcome visitors encounter inner lines of defense: public relations departments, "official spokespeople," and whole levels of management trained in how to represent the company to the outside world . . . In one case it took me nearly two years of phone calls, screening meetings with executive assistants and networking to interview two executives in a major manufacturing company (Thomas 1995:5–6).

We might call this the Roger and Me syndrome in ethnography. In my own case I had to find a way to study the relationships between antinuclear groups dispersed across several hundred square miles and a nuclear weapons laboratory shielded by armed guards and barbed wire fences. I was not allowed to enter, let alone do participant-observation, inside most parts of the laboratory.

How does an anthropologist study such institutions as weapons laboratories and corporations? In most cases participant-observation will be highly problematic, if not impossible, though some have succeeded in undertaking it. For example, I have one student, Pat Bentley, who is currently doing participant observation inside a software corporation. She is able to do this because she first took the precaution of becoming a corporate vice president of the company – a strategy beyond the reach of most of us. The sociologist Jennifer Pierce (1995, n.d.) was able to do participant observation inside a major law firm by getting a job there as a paralegal, though her decision to do her ethnographic study covertly is, from an ethical point of view, highly problematic to say the least.

These examples notwithstanding, participant observation is a research technique that does not travel well up the social structure. Participant observation was designed to facilitate the understanding of small, face-to-face societies, such as the Trobriand Islanders, where a stranger could easily be absorbed into the flow of daily life and no one was likely to tell the anthropologist that he or she was on private property and should leave. This technique may not be readily portable to elite contexts in the U.S. where ethnographic access is by permission of people with careers at stake, where loitering strangers with notebooks are rarely welcome, and where potential informants are too busy to chat. Although, as Laura Nader (1974: 306) herself worried, "the anthropologist's image of himself is shattered if he cannot participant-observe," it may be that anthropologists who want to study up will have to abandon, or at least subordinate, the research technique that has defined anthropology as a discipline and served as our own parochial rite of passage into maturity since Malinowski. If so, we will have to rely more on our canonical texts for a sense of shared identity as anthropologists and more on other research techniques for our field information.

In a forthcoming article, Mary Des Chene says that, although "the model of a sojourn of not less than nine months, and preferably more than a year, in a geographically defined fieldsite remains standard disciplinary practice, there is a growing dissonance between research projects and research sites and methods . . . [and] it may again be time [for anthropology] to become unmoored from the field." Alluding to George Marcus' notion of "multi-site ethnography," to the Comaroffs' interest in the relation of global systems to historical movements, and to anthropologists' increasing engagement in archival research, Des Chene argues that we need to "search for ways to study 'global systems and epochal movements' in a way that is not rooted in the quotidian of particular locales, yet does not surrender all sense of local variability and human agency in the manner of world systems theory." And indeed, as they search for ways to write the global and the local into one another, many anthropologists are increasingly de-emphasizing participant observation and are developing research strategies that blur the disciplinary boundaries between anthropology, sociology, political science, cultural studies, and even journalism.

Thus I would suggest that ethnographers de-emphasize participant observation in favor of what I call "polymorphous engagement." Polymorphous engagement means interacting with informants across a number of dispersed sites, not just in local communities, and sometimes in virtual form; and it means collecting data eclectically from a disparate array of sources in many different ways. Polymorphous engagement preserves the pragmatic amateurism that has characterized anthropological research, but displaces it away from a fetishistic obsession with participant observation. In my own case participant observation within the laboratory was impossible (except for meals at the cafeteria), though my fieldwork did involve a component of participant observation as I socialized informally with scientists from the laboratory in local churches, social clubs, bars, hiking groups and so on. However, polymorphous engagement also involved an eclectic mix of other research techniques: formal interviews of the kind often done by journalists and political scientists; extensive reading of newspapers and official documents, and careful attention to popular culture, for example. In my new research I am tracing the evolving fortunes not just of one nuclear weapons laboratory, but of the entire American nuclear weapons community, a community that is dispersed between California, New Mexico, and Washington DC.

I am also looking, more intermittently, at the Russian counterpart community of weapons scientists. Now I find myself doing more fieldwork by telephone and by email. As the communities we study disperse and link up across the globe, as all sorts of elites now are, and as we try to match the emergence of global cultures with a global anthropology, we must develop new research strategies. If virtual space increasingly becomes a real space of social interaction then we will need virtual anthropologists to follow our subjects there.

I want to conclude by saying a word about the difficulties of *writing* critical repatriated anthropology, which pertain to the fact that our subjects are powerful and literate, and they read what we say. Critiques of anthropological writing in the 1980s emphasized the importance of liberating our subjects from the objectifying practices of anthropologists whose ethnographic authority within the text was enmeshed with privileges over their subjects in the world outside the text. Jennifer Pierce (1995), arguing against James Clifford and in favor of what she calls "outlaw ethnography," makes the case that, in writing about those who are more powerful than us, anthropologists should draw on conventional kinds of ethnographic authority to even the balance of power. In my own case, my first attempts to write about weapons scientists were very much in this style as, writing about the laboratory as a ritual secret society and nuclear tests as rituals, I tried to draw on old conventions of anthropological discourse to make strange the world of the weapons scientists and to increase my own ethnographic authority over it. This was in keeping with the style of the early ethnographies of science, such as Sharon Traweek's *Beamtimes and Lifetimes* and Latour and Woolgar's *Laboratory Life*, two books that deploy against American physicists and biologists roughly the rhetoric that Evans-Pritchard used on the Nuer.

I began to modify this strategy at about the time that I was invited by a group of weapons scientists to present my research inside the Livemore Laboratory itself. I arrived to find that one of the Laboratory's leading weapons scientists had come to my talk wearing nothing but a loincloth and carrying a cane to which he had nailed an animal skull. He shook this at me and grunted whenever my presentation displeased him – which seemed to be quite often. As he so aptly dramatized for me, the objectifying, exoticizing language of anthropology is as objectionable at home as abroad, and one is less likely to get away with it. Over time, as the weapons scientists read drafts of my work and responded to it, we learned to enter into a dialogue around it. I have tried to build spaces into my work where my subjects speak back and they, in turn, have invited me into some of their community spaces. For example, I've spoken inside the laboratory, written columns for the local newspaper, and even given three sermons at a local church. We often disagree but, as we converse in public, the public learns more about the workings of their formerly hidden world.

In the past anthropologists wrote about people who lived far away for audiences that had to take our interpretations largely at face value. The anthropologist who studies up in the United States will have colleagues who may already have strong opinions about his or her subjects, who may even have met them, and subjects who will read and argue with what is written about them. In this situation we need to rethink both fieldwork and writing, as many already have, exploring hybrid research and writing strategies that blur the boundaries between anthropology and other disciplines. Such strategies of course play havoc with ethnographic authority, and therefore threaten to undermine the critical force of studying up itself. However, they also offer the chance to incite new conversations about power in the U.S., and this is exactly the democratizing project to which Laura Nader called us 25 years ago.

Note

1. This article is adapted from a paper given on the panel "Studying Up Revisited: Papers in Honor of Laura Nader" at the American Anthropology Association Meetings, Washington DC, November 18, 1995.

References Cited

Bourgois, Philippe 1995 *In Search of Respect: Crack in El Barrio*. New York: Cambridge University Press.

Davis-Floyd, Robbie 1992 *Birth as an American Rite of Passage*. Berkeley: University of California Press.

Des Chene, Mary n.d. Locating the Past. In James Ferguson and Akhil (eds.) *Anthropology and the* Field: Boundaries, Areas and Grounds in the *Constitution of a Discipline*. University of California Press.

di Leonardo, Micaela 1984 *The Varieties of Ethnic Experience*: Kinship, Class, and Gender Among California *Italian-Americans*. Ithaca: Cornell University Press.

Ginsburg, Faye 1989 *Contested Lives: The Abortion Debate in an American Community*. Berkeley: University of California Press.

Ginsburg, Faye and Rayna Rapp 1995 *Conceiving the New World Order: The Global Politics of Reproduction*. Berkeley: University of California Press.

Harding, Susan 1987 Convicted by the Holy Spirit: The Rhetoric of Fundamental Baptist Conversion. *American Ethnologist*. 14(1): 167–181.

Latour, Bruno and Steve Woolgar 1979 *Laboratory Life*. London: Sage.

Liebow, Elliot 1993 *Tell Them Who I am: The Lives of Homeless Women*. New York: Penguin Books.

Macleod, Jay 1987 *Ain't No Makin' It: Leveled Aspirations in a Low-Income Neighborhood*. Boulder: Westview Press.

Marcus, George 1995 *Lives In Trust*. Boulder: Westview Press.

Marcus, George and Michael Fischer 1986 *Anthropology as Cultural Critique*. Chicago: University of Chicago Press.

Martin, Emily 1987 *The Woman in the Body*. Boston: Beacon Press.
________ 1994 *Flexible Bodies*. Boston: Beacon Press.
Nader, Laura 1974(1969) Up the Anthropologist – Perspectives Gained From Studying Up. In Dell Hymes (ed.) *Reinventing Anthropology*, New York: Vintage Books, pp. 284–311.
Pierce, Jennifer 1995 Reflections on Fieldwork in a Complex Organization: Lawyers, Ethnographic Authority, and Lethal Weapons In Rosanna Herz and Jonathan Imber (eds.) *Studying Elites Using Qualitative Methods*. London: Sage Books.
n.d. *Gender Trials: Emotional Lives in Contemporary Law Firms*. Berkeley: University of California Press.
Rabinow, Paul 1996 *Making PCR*. Cambridge: MIT Press.
Rosaldo, Renato 1989 *Culture and Truth*. Boston: Beacon Press.
Thomas, Robert J. 1995 Interviewing Important People in Big Companies, in Rosanna Herz and Jonathan Imber (eds.) *Studying Elites Using Qualitative Methods*. London: Sage Books.
Traweek, Sharon 1988 *Beamtimes and Lifetimes*. Cambridge: Harvard University Press.
Weston, Kath 1991 *Families We Choose*. New York: Columbia University Press.
Whitehead, Harriet 1987 *Renunciation and Reformulation: A Study of Conversion in an American Sect*. Ithaca: Cornell University Press.

76

After Method? Ethnography in the Knowledge Economy

David Mills and Richard Ratcliffe

Introduction

The setting: a packed Oxford seminar-room. The debate: the future of ethnographic research in a knowledge economy. There are two protagonists: one is a senior professor of anthropology; the other a professor of education who has used ethnographic methods throughout her career to address educational problems. The first professes disinterest in defining ethnography, seeing it as a licence to explore the curious, the messy, and the unexpected. Her view is that ethnography is best understood as a methodological and existential sensibility to be cultivated. She argues for seeing ethnographic fieldwork as a process of discovery, a sensibility developed through one's dislocation by encounters alien and the unexpected. Her opponent insists on the importance of an explicit and rigorous set of methods and approaches for ethnography, and defends a robust, disciplined empiricism.

It quickly becomes clear that there are rather different understandings of ethnography within the room – two fields divided in their understanding of a common method. Whilst this is perhaps no surprise – methodological and epistemological allegiances have long been a proudly worn badge of academic affiliation – we were surprised at the extent of their disagreement. Did their methodological disagreements stem from different disciplinary traditions of training and socialization? Or were they also the product of the

Source: *Qualitative Research*, 12(2) (2012): 147–164.

contradictory expectations and possibilities placed upon ethnography in a knowledge economy?

In this article we explore both of these explanations, and ask how the measurement of the monetary value of social research may be changing ethnographic practice. With an increasing interest in anthropological research techniques within and beyond universities (e.g. Neyland, 2008; Westbrook, 2008), ethnographic rhetorics and imageries are being appropriated across a range of fields and contexts. We see a connection between the new fashion for ethnography beyond the academy and the growing market for qualitative research. We begin with quotes from our interviews with anthropologists about these changes, linking this to the way that within the broader economy, the value of ethnographic knowledge is now being calibrated in monetary terms. We explore how this development affects the status of fieldwork in anthropology and the influence of ethnographic methods and imaginaries beyond the discipline.

A core aspect of our argument is that social research practices have become increasingly intertwined with what Thrift calls the 'cultural circuits' of capitalism (2005). For Thrift, the critiques and models developed by business theorists (whether in the media, business schools or management consultancies) have a symbiotic relationship with academic research: their analytical practices are built from and redefine academic values of rigour and craftsmanship. We suggest that this history is key to the success of a 'knowledge economy', but also key to academic responses to it. The revaluing of social research in commodity terms is made possible by academics' beliefs in a higher purpose to their work, echoing the role of the producer's moral economy in the development of commodity production and transformation of social relations during the industrial revolution (Thompson, 1970; Willis, 1977). Whilst caught up in these shifts, disciplinary debates over method also offer a way of understanding and responding to such pressures.

Anthropologists and the Knowledge Economy

> Sadly, having the skill to produce really nice ethnography is different from being employable. Even within anthropology departments, contemporary anthropology isn't based on the old LSE anthropology. Very few departments do that sort of thing. That way of looking at things is no longer epistemologically possible, because the only way you could do with those kinds of writings was to pretend that the outside didn't impinge, and we can't pretend that any more. (NH, Senior Lecturer in Anthropology)

Conducting biographical interviews with anthropologists trained in a variety of UK departments, conversation increasingly turned to the implicit pressures they felt under to be productive researchers.[1] We have written elsewhere about the gendered and embodied aspects of these expectations (Mills and Berg, 2010). Here we want to focus instead on our interviewees views of changing anthropological research practices. The quote with which we start this section is from a senior researcher, wistful about the passing of 'really nice ethnography' inculcated during her training at the London School of Economics (LSE).

These pressures led others to talk about no longer being real anthropologists, especially if they were working as contract researchers or in multidisciplinary research teams. There was repeated attention drawn to deadlines and a lack of time to complete long-term ethnographic research. Whilst there has been much work on the intensification of the work pressures facing today's academics (e.g. Acker and Armenti, 2004; Menzies and Newson, 2007; Nelson and Watt, 2004; Walker, 2009), little of it has focused on the consequences for disciplinary methods and research practices.

Our respondents' views on their anthropological identity partly depended on their working environments, and whether they felt they had the time to conduct what they called 'proper' anthropological research. When asked whether she would call herself an anthropologist, BC, working in development studies, said 'No, I'm not sure if I want to call myself an anthropologist – it might be counter-productive. I might use something like 'a freelance development consultant'. Despite being a privately enthusiastic about her anthropological identity she suggested that others in development studies 'feel that anthropology is very narrow and not very practical – especially when development projects are time-bound'.

Yet those working outside universities sometimes use their anthropology as a passport, whilst admitting that they were not able to do what they called 'real ethnography'. KM, working in applied social research, acknowledged that 'sometimes I drop in that I'm an anthropologist. If it's like a cross-cultural thing . . . I'll use it when it suits me and a lot of the time, I won't'. Asked whether the 'skills required of applied anthropologists differ from those required of applied social researchers more broadly', she made her views very clear:

> there isn't such a thing as applied anthropologists. I think applied anthropologists are just doing social research. Well, how many applied anthropologists do you know who go off and do ethnography? I don't think they do . . . there isn't the money to do the in-depth ethnography we're all trained to do.

This repeated attention to time has a number of sources, not least the UK research and funding council expectations for research productivity. Our interest is less in time constraints per se than in the consequences of measuring academic efficiency and research utility. We suggest that this is moving research design away from exploratory and long-term fieldwork towards more tightly defined research frameworks. It leads to a preference for predictability in outcome, and away from a vision of ethnography as discovery. It led us to begin to think about the ways that, in and beyond the UK, anthropological understandings and ethnographic practices were being redefined by the expectations and values of the 'knowledge economy', a term presaged in Bell's (1973) *The Coming of Post-Industrial Society* and later popularized by business gurus and policy think-tanks (e.g. Drucker, 1993; OECD, 1996).

Valuing Research in the Knowledge Economy

> Public investment in research makes certain that the UK is able to compete in the global economy. RCUK (Research Councils UK) ensures that the UK is a world leader in a global market. RCUK confirms that our research excellence and relevance today, positions the nation strongly to seize tomorrow's opportunities. (RCUK, 2010)

An association between knowledge production and larger economic forces is nothing new. Universities have long shaped, and been shaped by, the national and transnational polities and economies in which they are located. The economic role of science, particularly in the development of industrial production, was increasingly explicit from the 19th century onwards. The two-way intellectual thoroughfare linking pure and applied research has sustained and regenerated many an academic discipline. If anything, the traffic of ideas, artefacts and value across the academic/non-academic divide is intensifying. Challenged to train knowledge workers by governments, universities are being redefined by this intense circulation of concepts and commodities. It is almost 50 years since Kerr's prescient aside that 'the university and segments of industry are becoming more alike' (1963: 90). But this is not simply to argue for the loss of university ideals under the pressures of 'academic capitalism' (Slaughter and Rhoades, 2004). As Delanty notes, even if the university is 'no longer the privileged site of particular kinds of knowledge', 'it can become a key institution in a society that is coming to depend more and more on knowledge' (2001: 7). Delanty talks not of the knowledge economy but of the 'knowledge society'. He refuses to see the university reduced to a 'passive actor drawn helplessly into the market', instead arguing that it as a 'transformer of such value systems' (Delanty, 2001: 151). Whilst we admire his vision, and accept that the term 'knowledge economy' is over-used and under-defined, we nonetheless wish to explore how the university is both transforming and being transformed through these new connections.

In *The Uses of the University*, Kerr suggested that 'as the university becomes tied into the world of work, the professor . . . takes on the characteristics of an entrepreneur' (1963: 90). Amongst others, Olssen and Peters

(2005) have convincingly shown how governments now see universities as key drivers of an entrepreneurial knowledge economy. Core to this reconfiguration have been closer links between industry and academia, whether defined in terms of a 'triple helix' (Etzkowitz and Leydesdorff, 1997) or as 'mode 2' science (Gibbons et al., 1994; Nowotny et al., 2001, 2003), with the economic contributions of the universities now measurable by governments through proxies such as 'impact', 'employability', 'engagement' and 'knowledge transfer'. Such inventive ways of measuring bring their own risks, and Barry has pointed to the 'fragility of metrological regimes' (Barry 2002). Metrics have profound and unanticipated consequences, such as the institutional 'isomorphism' (Di Maggio and Powell, 1983) that accompanies global university league tables (Espeland and Sauder, 2007; Hazelkorn, 2011), or strategically instrumentalist responses to research evaluation exercises and student experience surveys. Tuchman (2009) wittily describes how a slew of new administrator-convened one-credit courses were offered to first-year students by her university to improve its position within *US News and World Report* league tables. The new politics of measurement is simultaneously also linked to a profound, and intensely resented, reordering of academic governance (Evans, 2004; Shore and Wright, 1999).

Whilst a number of critics of what has been called 'audit culture' (Morley, 2003; Power, 1997; Strathern, 2000) have sought to describe the effects of neoliberal educational regulation on educational subjectivities, fewer have focused on its impact on research practice. In particular, we are interested in the consequences of valuing research knowledge in monetary terms. Analytical work on the 'cultural circuits of capitalism' (Thrift, 2005) shows how this conversion is made possible by the networks linking producers with the consumers. It becomes increasingly difficult to distinguish academic knowledge practices from those of business theorists – the two are increasingly interdependent (Olds and Thrift, 2005). The argument is that within a cultural economy value is created through the circulation of images and virtual commodities of various kinds. For this article, the crucial point is that knowledge has a changed role when it can be ascribed a monetary value (Simmel, 1978). This is not just to say that knowledge is being turned into commodity, a point first made by Lyotard (1984), but that the value of ethnographic and social research can increasingly be quantified in economic terms.

What are the implications of valuing research in this way? Universities can no longer resort to an ideology of detachment; their justification now has to be in terms of their economic contributions. They are expected to *add value*. If social research practices can be both measured and given a monetary value, their material contribution to this economy can be ascertained. In the UK, as elsewhere, social researchers are increasingly expected to pursue research that 'delivers impact' (ESRC, 2009) in a public policy context, though precise definitions of impact are proving elusive.

This shift in academic values would seem to be one-way. The business studies theory of the 'value chain' (Porter, 1985) becomes increasingly relevant. In this model, value is produced in a linear way – a chain of inputs that can be neatly linked to a set of defined and measurable outputs and outcomes. Within business, the concept of the value chain is now seen as common-sense, along with the expectation that it be constantly refined and improved to remove inefficiencies and inconsistencies. The aspiration is for all the different parts to become aligned and work together in the production of value. Its application to the university is perhaps less appealing. By its logic, researchers can be monitored in terms of their contribution to departmental and universities research agendas. Individual academic idiosyncrasies and research interests become a potential source of inefficiency. So too do unpredictable research methods, such as anthropological approaches to ethnography.

The value chain is all about productivity. The impetus is to increase the rate of what the bureaucrats now calls 'research outputs', and where possible to create commercial value from publicly funded research. Along the way, universities are adopting a range of other business practices. University managers are closely concerned with their institutional research rankings and with positioning themselves within a global hierarchy. As in business, there is a new belief that organizational reform drives change and development. The consequence is perpetual reform aimed at gaining a competitive advantage (Pedler et al., 1997; Thrift, 2005).

What do these shifts in value, discourse and practice mean for what we do as researchers?

One positive consequence of this expectation to continually communicate knowledge and confront other ways of thinking is enhanced communication between academic disciplines. There is a greater cross-fertilization of ideas and approaches. But we also see a number of less welcome developments.

First, the focus on the efficient exchange of ideas for value (including monetary value) begins to rework academic values. The ethos of disinterested impartiality competes with new models of academic entrepreneurialism (Marginson and Considine, 2000), participation and engagement. The push for efficiency within the circuit potentially squeezes the unpredictable, the tangential and the creative. This, we suggest, can lead to methodological instrumentalism.

Second, it leads to a loss of academic authority and community. Academia has a diminished epistemological status as gatekeeper, since the new focus on knowledge transfer and knowledge exchange marks a shifted space of knowledge's reception and validation. Osborne argues that academics and other knowledge workers are increasingly acting like 'mediators', 'bringing ideas quickly and decisively into public focus, brokering their ideas in the context of different spheres of influence' (2004: 435). Here Osborne is building on Bauman's influential distinction between knowledge 'legislators' and 'interpreters' (Bauman, 1989). Increasingly the public and the market are seen as the best arbiters of useful knowledge, via new structures that offer proxies of quality and status, such as citation indexes and university rankings. One might argue that this leads to a crisis of academic community and audience. Academic work is less engaged with intellectually, even as it circulates more widely. There are growing numbers of conferences, journals and sub-fields, but the audience for each is shrinking.

Finally, these new circuits discourage a sustained investment of thought or time by academics. They favour intellectual dexterity, quick responses to emergent funding opportunities and intellectual multi-tasking. The premium placed on planned and programmed deliverables is reworking older scholastic horizons. Academic time is increasingly defined by Gantt charts (named after Henry Gantt) rather than long-term speculative research. It is a productivity focused on predictability.

Whilst we are not the first to describe these trends, critical research has tended to overlook academic involvement in the evolution of new knowledge circuits. The knowledge economy works because it is built on established academic values and knowledge practices, even whilst it reshapes them. The new technologies for valuing research co-opt academic values of peer-review, evidence and research. As Strathern notes, 'auditors are not aliens; they are a version of ourselves' (1997: 319). Not only is 'audit culture' built on an ethos of impartiality and professionalism, the very field of accounting can be traced back to nineteenth century university examinations (Hoskin, 1993). This can make critical distance difficult. The professional principle of peer review has made many academics largely complicit with the increasingly commercial approach of scholarly journals, and in the UK allowed the reification of research within successive research assessment exercises.

Methodological Debates in the Disciplines

How are disciplinary traditions responding to these demands for impact? The picture is very mixed. Protecting the relative autonomy of an intellectual field is difficult cultural work. It depends on having institutional defences (being based in an elite university is useful), strong personal networks (preferably within the state elite), dedicated scholarly work, and the financial and academic capital to make the temptations of commissioned policy research and consultancy less compelling. Not every field has such resources, which is why these trends are uneven in their scope and impact. Given the 80-year history of social anthropology within British universities, it is a good discipline through which to explore these changes. In this section we explore how anthropologists have interpreted and practised ethnographic fieldwork over the last thirty years, and the emerging relationship between these relatively autonomous disciplinary dynamics and the broader research economy. We then discuss the growing popularity of ethnographic methods beyond the academy.

Different germinations of ethnography have led to rather different national and disciplinary methodological blooms. Even within the UK, anthropologists, sociologists and others have developed different understandings of the ethnographic method. Much of this efflorescence can only be loosely linked to the changing nature of research funding. Within the UK, Malinowski's evangelical vision for social anthropology at the LSE in the 1930s was underpinned by the central role he accorded to fieldwork, and the importance of the individual ethnographic experience. Skilled at cultivating his patrons, he was less explicit about how his methods led to his functional theories of human sociality, leading critics to accuse him of 'pseudo-science', 'dubious methods' and '*synthetic a posteriori*' theorizing (Grimshaw and Hart, 1993: 20). Nonetheless, his ability at creating a minor personality cult ensured a loyal following for his methodological manifesto. His insistence on extended fieldwork and on seeking to capture the 'native's point of view' was championed by his students and followers and gradually spread to Oxbridge. After the war Max Gluckman wrote his new department into this history, arguing that Malinowski's vision was the precursor to the Manchester school's commitment to the extended case-study (Mills, 2008). These early manifestos continue to shape the British field today.

As a result, an extended period of fieldwork has continued to be an essential disciplinary rite of passage, and a shibboleth of anthropological rigour amongst many UK anthropologists (Delamont et al., 2000). The emphasis has been on the quality of the social 'immersion' made possible through linguistic competency and the social relationships that develop over time. Sometimes time itself becomes a measure of quality, with the minutiae of method left largely unexamined. Extended fieldwork remains a source of cultural capital within the discipline and a vital empirical resource for further development and exploration over one's anthropological career.

During the past 30 years, the changing has led some US anthropologists to rethink this consensus. For an influential group of anthropologists and cultural critics, initially linked to Rice University, fieldwork was not just an empirical source for the inductive generation of theory, but also a methodological trope that deserved closer investigation. Works such as Geertz's *Works and Lives* (1989), and *Writing Culture* by Clifford and Marcus (1986) highlighted anthropologists' involvement (and even complicity) in co-constructing the world about which they wrote, and have been a conduit for a range of other theoretical debates into the discipline. For some critics, this literary turn has led to anthropology becoming a form of cultural criticism. Postmodern and postcolonial theoretical perspectives have led some to question a romantic neo-Malinowskian attachment to the insights gained from in-depth fieldwork in one isolated fieldsite. Challenging the 'spatial metaphysics' of the British colonial tradition and the accompanying fetishization of the 'local', anthropological critics such as Gupta and Ferguson (1997) sought to decentre and rethink the field, suggesting that a single-site ethnography may not capture the multiple and multi-stranded networks of contemporary social life. This challenge to the orthodoxy was also driven by an attention to the politics of the research relationship and the role of collaboration. A recent distillation was entitled *Fieldwork is Not What it Used to Be* (Faubion and Marcus, 2009).

What has been at stake in these debates over fieldwork? Its protagonists have called not for the abandonment of fieldwork, but rather for its reconstruction, as 'one element in a multi-stranded methodology', a methodological shift from 'bounded fields' to 'shifting locations' (Gupta and Ferguson, 1997: 37). This shift to what Gupta and Ferguson call 'location-work' echoes debates within feminism around the importance of 'situated knowledges' (Haraway, 1991). George Marcus has consistently been an influential anthropological advocate for new models of 'multi-sited' and collaborative ethnographic practices. Back in 1995, he labelled this trend 'mobile ethnography', the shift from 'a conventional single-site location . . . to multiple sites of observation and participation', so allowing the researcher to follow the 'circulation of cultural meanings, objects, and identities in diffuse time-space' (1995: 96). If the artefacts and meanings one seeks to understand don't stand still, why should the researcher? The methodological implication – 'multi-sited ethnography' – neatly echoed the broader emphasis within social theory (particularly influenced by thinkers such as Deleuze, Callon and Latour) on understanding and following the networks and connections that shape the social. Previous notions of the 'field' have been unpicked as it has become theory-rich, multi-sited, rather than an objective place. Ethnography increasingly began to be understood as something inter-subjective, co-constructed and mediated.

There were also more prosaic factors at work in these debates. Along with a recognition that people's social worlds extend far beyond their 'home' communities, concerns over research funding and the ethics of access, have encouraged (and sometimes forced) US anthropologists to conduct more ethnographic research 'at home' and to rethink the nature of the 'field'.

While this influential re-imagination of the field as a set of interconnected empirical and conceptual spaces has gradually diffused across the North American discipline, any erosion in the status of fieldwork continues to be contested by many in British social anthropology. The result has been a series of debates around practical questions of fieldwork length. Many British anthropologists now bemoan (in conversation, if not always in print) the misappropriation and short-cutting of what they see as their method by others, scorning high-speed 'drive-by' ethnographies and ethnographically-informed studies carried out within predefined research schedules. In a riposte to an article by Marcus (2007a) entitled 'How short can fieldwork be?' Okely defended the classic fieldwork method (Okely, 2007). Length becomes a measure of the intensity of the research relationship, even if it is potentially possible to spend three years in a field-site without developing meaningful insights. Ethnographies should be judged on their quality, not on the length of fieldwork. Indeed, Geertz calculates that Evans-Pritchard spent only four months doing the fieldwork with the Nuer that led to his classic monograph (Geertz, 1989). The very existence of the debate acknowledges the iconic status of fieldwork as shorthand for anthropological rigour and awareness, and the disciplinary capital accrued as a result of one's fieldwork experience.

The debate also reflects an unease that the space for exploratory empirical enquiry is being reduced. Anthropologists have a peculiarly intense degree of emotional and epistemological investment in their ethnographic embodiment. Whilst Ingold has provocatively argued that anthropology and ethnography are 'endeavours of quite different kinds' (2009: 69) most within the anthropological tradition would insist that the two are closely linked. For them ethnography is not simply a research tool, but rather an commitment to an open-ended, iterative, non-prescriptive vision for social research, where the researcher is encouraged to acknowledge the complexity and unpredictability of the research encounter. The power of the method comes out of the theoretical and historical traditions of anthropological practice.

Ethnographers more influenced by sociological traditions have been less influenced by these debates, or by the debates around the call for more 'public sociology' (Burawoy, 2005). Instead they have continued to articulate a systematic and coherent account of ethnographic principles and practices (e.g. Hammersley and Atkinson, 2007 [1983]), downplaying its potential messiness. Delamont and Atkinson (1980) recount this history of sociological ethnography in convincing detail. They later suggest the 'high degree of explicit reflection and sophistication concerning methodology' amongst this community has meant that 'the most successful textbooks on ethnographic fieldwork published in Britain are by authors whose empirical research includes educational settings' (Atkinson et al., 1988: 235). This 'sophistication' and attention to reflexivity reflects its applied character, and the need to communicate to audiences in Education.

This methodological explicitness reflects the different institutional and epistemological politics faced by sociologically-oriented ethnographers in the UK. After all, many anthropologists also engage closely with theoretical debates, but few write methods textbooks. Tasked with defending the validity of ethnographic approaches, particularly in more positivist Education departments, sociologists have repeatedly clarified the nature of ethnography's empirical depth and rigour, developing a methodological robustness. Eschewing the anthropologists' fashion-conscious attention to the latest theoretical ideas or explanations, and the more politicized vision of ethnography offered by Smith in her work on institutional ethnography (2006), their focus has instead been on calling theory to account, and defending inductive principles.

Reviving the Popperian project, this work foregrounds close-grained active empirical research, with theoretical ideas refined and tested in relation to findings that are checked and cross checked, rather than absorbed in some mystified non-strategic manner. Hammersley has been a key advocate of this project, and his many textbooks on ethnography underscore the importance of striving for objectivity and rigour: 'I argue that researchers should be committed primarily to the pursuit of knowledge, and should be as neutral as they can towards other values and interests in their work, in an attempt to maximize the chances of producing sound knowledge of the social world' (Hammersley, 2000: 12).

Appropriating Ethnography: Method beyond Discipline

British anthropology has a long tradition of promising utility to colonial officials and others, while constructing itself as a pure scholarly enterprise. While mid-century anthropologists questioned the credibility of applied anthropology (Mills, 2006), anthropologists in the UK, Australia and elsewhere are now confident advocates of the discipline's wider relevance (Gordon, 1998; Pink, 2006).

In the US, Applied Anthropology has long been a vibrant disciplinary subfield, with its own journal (*Human Organisation*) promoting the quality and legitimacy of applied work. Inspired by the example offered by Margaret Mead, and receptive to influences from cultural psychology, scholars in the Anthropology of Education have a similar tradition of evaluative and applied research (Eddy, 1985). An early recipient of federal grants to research educational acculturation amongst America's minority communities, the sub-field has continued to produce a range of empirical single-sited ethnographies based within schools and colleges, sometimes making contributions to broader theoretical debates.

However, in the last two decades these anthropological engagements with public policy have been overshadowed by ethnography's application and deployment in the commercial sector. In the UK and USA, there is a growing interest in corporate ethnography, and hence career opportunities for anthropologists in the business and development sector (Cefkin, 2009). A range of market research companies, consultancies and design studios employ ethnographers. Whilst this methodological migration is an indicator of the method's adaptability, in economic terms it reflects the way in which ethnography adds value to a value chain. Ethnography becomes a marketable commodity, which provides it with new opportunities and challenges. In what follows, we explore this trend through the views of more than a dozen applied anthropologists who had trained in British departments but who were now working in the development, business and health sectors.

In our interview and focus groups, we noticed that some of those who now work outside academic anthropology nevertheless espouse a strong sense of disciplinary identity: an identity which purportedly offers a unique perspective on intellectual problem-solving, a certain oblique way of seeing situations and asking questions. Yet at the same time many respondents had difficulty in actually spelling out the specific research and organizational skills they had acquired along with this sense of disciplinary belonging, as if their identity was one based on implicit socialization. Others we spoke to had given their disciplinary past very little thought, sometimes admitting that we were the first people to ask them about anthropology in some years.

Amongst this first group, we identified a spectrum of opinions about anthropology and its utility. At one extreme are those who continue to proselytize for anthropology, and are almost evangelical about the contribution anthropological methods can make. One respondent, ST, working in financial services, admitted that he often pointed to, and made use of, his doctoral background – 'I leverage it every time I apply for a new job, and in the way I do my work . . . I frankly leverage it more than I have to'. He felt that his anthropological training had acted as a useful preparation for the qualitative approaches that he used in his work, even if he felt he would now do anthropology far better, as it is 'hard for anthropology lecturers to train students in getting information out of people'.

His reflections on information gathering are reflected in the literature. Writing about the banking industry, Holmes and Marcus (2006) focus on what Reich (1992) called the new 'symbolic analysts' leading international financial organizations, and their adoption of a form of anecdotal reasoning that seeks to 'represent the social nature of the contemporary' in a way that approximates to ethnographic labour. As a result they suggest that 'the key cultural practices of expert subjects converge with our own analytical endeavours' (Holmes and Marcus, 2006: 41). Echoing Osborne's discussion of academics as mediators, this blending of research and business practices poses academic ethnographers with a number of conundrums. It underscores the end of a positivist and distanced research paradigm, blurs the 'us'/'them' distinction, and makes collaborative and participatory knowledge creation more than just a normative vision. Whilst this poses challenges for academic practice, it also opens up opportunities for what Holmes and Marcus call 'paraethnography'. Marcus has independently been an influential advocate for collaborations between ethnographers and their research subjects. Attentive to contemporary shifts in academic practice, to the realpolitik of the research economy, and to the ethical challenges that anthropology has struggled with, he argues for work that 'blurs the boundaries between the professional community of observers and those observed'. Instead, he has a vision of 'substantively collaborative projects', which he argues, somewhat romantically, was a 'humanistic ideal of anthropology all along' (Marcus, 2006: 116).

In our research we collected particularly rich material from a small number of anthropologists who have been successfully marketing their anthropological skills – especially the ethnographic method – in more commercial settings. A number were involved in market research or for independent research institutes, often in areas like IT and communications technology. In this case, it was often a rather caricatured ethnographic method that they saw as their distinctive contribution, their 'added value'. Amongst the anthropological 'missionaries', TS stands out. After setting up his own anthropological consultancy, which highlighted its ethnographic approach to market research he subsequently joined a major IT hardware provider to lead its research team. An enthusiast for an anthropological take on the world, TS suggested that, within his current company, 'anthropologists are one of the key tools we use to ensure we connect our technologies with human needs'.

One can easily see the appeal of ethnography in this commercial and public policy context. Amongst policy researchers, the method is seen as offering access to difficult-to-reach communities, accessing day-to-day life, voicing marginal or silenced perspectives, ideal for unpacking failure and operational problems. Ethnography is seen to reach the parts of social life that other methods cannot reach. It provides access to 'local' perspectives for marketers, for government, and even the military, as illustrated by the controversy surrounding anthropologists employed in Human Terrain Teams (Gonzalez, 2007).

In commercial worlds, ethnography adds value. Its success is tied into the growth of cultural explanations of how to make organizations successful, the associated rise of behavioural economics and a shift to what has been termed the 'cultural economy' (Du Gay and Pryke, 2002). The fact that many of those with anthropological training are now in senior positions within these organizations makes Marcus' account even more convincing. For example, the FT journalist Gillian Tett makes the argument that her anthropological training enabled her to foresee the impending economic crash in 2008 (Tett, 2009).

Not everyone we spoke to felt able to appropriate ethnography in this way. DI, running a major policy and advocacy NGO saw anthropology as having been 'very influential in my choice of careers', and in helping her 'understand how others could live, and a broad acceptance of the possibilities of living in different ways'. However she was highly critical of her doctoral training, and now found herself working with economists and medical scientists, and having to reconcile 'two different approaches'. She also admitted that 'I call myself a lapsed anthropologist, even when talking to other anthropologists', as she was no longer able to do the 'pure ethnography', the 'long-term research process . . . the nuances of getting to grips with a particular culture'.

Others are much more agnostic. TJ, an ex-PhD now working in the World Bank, admitted that 'there are many anthropologists in the bank, but I don't mix much with other social anthropologists – I don't feel myself as an anthropologist – I don't use the jargon . . . the infrastructural people see it as bullshit' He went on to say that 'they are sometimes too anthropological – most anthropologists are working on purely sociological issues – they don't need to translate social sciences or understand non-social issues. They never try and understand road-transport economics'.

The last comment reveals the challenge of collaboration. Ethnographers are often brought into multidisciplinary teams, often through the brokerage of anthropologists and sociologists themselves. Indeed sometimes collaboration can seem an end in itself. Within disciplines urged to promote and measure the social and economic impact of their research, collaboration becomes a new target and aspiration. Yet collaboration is never straightforward. There is an extensive literature on the complexities of scholarly collaboration when working across different analytical and disciplinary frameworks (e.g. Barry et al., 2008; Strathern, 2004). In particular, collaborations pose new ethical and professional challenges for the representation and use of ethnographic knowledge and the answers will depend on the ways that particular academic fields understand, create and engage their users.

Moreover, collaboration requires a measure of like-mindedness. As Marcus observes, 'ethnographic tales of the field' would only make sense among mixed audiences of fellow anthropologists and our epistemic partners in the same room, so to speak . . . different kinds of stories are required' (2006: 106). This requires anthropologists to adopt a different narrative style, one that can no longer take a disciplinary speech community for granted. To be useful one has to be accessible, to represent anthropology to others, and to be able to explain oneself. Again, this makes new and rather different demands on disciplinary expertise.

It produces, we suggest, a trend towards short-term studies, whose findings can be easily used by those – be they bankers, investigative journalists, financial analysts or management consultants – whose work benefits from an ethnographic sensibility. These techniques often make use of what one might call quasi-ethnographic skills, approaches that label themselves as ethnographic but that are not easily recognizable by either anthropology or sociology. Given their scope, these studies are accused by some of appropriating ethnography (e.g. Moore, 2002).

These new uses of ethnography place it in a subtly different position. Collaboration, accessibility and the democratization of knowledge practices all sound admirable virtues, but they pose other problems for ethnographic notions of rigour and sensibility. Ethnographic mediators have to be plausible and to demonstrate their research competence. They offer telling anecdotes, a bit of theory and make claim to a method. Yet the contextually rich understandings that come from an epistemological grounding are simply not possible in a drive-by approach to ethnography or in a predefined collaborative research project.

In this new world, the demand is for clarity, and for production in a form that is both simple to digest for its users, and of a content that is relevant for them to do so. There is less interest in complexity and nuance outside the terms of enquiry set by the collaboration. The ethnographer's role in collaboration can be thought-provoking, but also has to be made explicit and defined. The requirement is to mediate, to be explicit, to add value, to identify key action points for change, and to do so as efficiently as possible.

We have suggested that the market creates a new sphere for ethnographic practice. Yet, beyond the university ethnography risks being less an empirical method than a popular discourse. The danger is that it gets reduced to a marketing strategy that appeals to the vanities of dominant cultural ideologies, such as an anthropology for 'watching the English' (Fox, 2005), as a form of reality TV (Caplan, 2005), or even enlisted to advise armies or support military occupations (Gusterson, 2007). For one trenchant critic, 'the gap between critical intellectuals and simple salesmanship seems only to shrink' (Frank, 2002: 52). Ethnography risks reemerging not just as tourism, but as a kind of voyeurism, aimed at entertainment and profit. This situates it in very different kinds of meanings from those traditionally associated with ethnographic enquiry.

Herein lies the final message of our article: the vital importance of sustaining disciplinary moral economies. Academic communities and individual scholars respond in different ways to the pressures of the knowledge economy. Peoples' experience varies: concerns over disciplinary rigour at the centre can coexist with an interdisciplinary ethos of 'anything goes' at the borders. Debates over method and methodologies are of less significance to some than others. Sociologists have developed an inclusive approach to professional belonging, hosting major conferences, encouraging students to participate in disciplinary associations, and allowing a plethora of sub-fields and international networks to develop (Platt, 2003). In Britain, by contrast, anthropology has remained relatively small, maintaining a more exclusive identity and collegiality through shared theoretical conversations and intellectual legacies (Mills, 2008). But in the new knowledge economy, both inclusive and exclusive approaches to disciplinary reproduction face the same challenges: maintaining their academic capital, justifying a disciplinary logic, and competing for students, funds and impact. For all these reasons, methodological communities become increasingly important.

The keen temper of methodological battles and boundaries are illustrative of these moral communities. They matter because they offer a way in which disciplines can respond to concerns about the broader reshaping of social research. Yet maintaining methodological traditions is only half the battle. To return to the vignette with which we began, we see an intriguing parallel between those committed to an ethnographic research practice that seeks to promote rigour and minimize bias (Hammersley, 2000) and those, particularly within anthropology, who advocate for an ethnographic imagination that sees the unexpected and unpredicted as key moments of insight. Whilst seemingly in opposition, both are strong visions, both transcend the technicalities of method, and both visions emerge from robust and supportive intellectual communities that view ethnography as a mode of empirical social research. It is these 'communities of critics' (Strathern, 2006) that face threat. It is these methodological communities that merit sustaining.

Conclusion: Where Next for Ethnography?

What future awaits ethnography in the knowledge economy? Do heated debates over methodological principles – such as the status of fieldwork – matter outside the disciplines? As Ingold (2009) reminds us, anthropology is not reducible to ethnography, and there are 'pitfalls involved in identifying such research activity solely or primarily in terms of the method' (Delamont and Atkinson, 1980: 139). Nonetheless, methodological disputes, such as the tensions surrounding the 'field', are valuable prisms through which to understand the larger social forces shaping academic work.

The knowledge economy offers new opportunities, as well as eroding existing practices. As we have shown, ethnography has shown itself to be resilient and to have a lasting appeal. Its migration into other fields and multidisciplinary areas has given anthropologists and others the opportunity to popularize ethnographic approaches. But this comes with the risks of making too much – and too little – of the method. If ethnography is rethought as a 'way of seeing' (Berger, 2008) that anyone can invoke, it becomes a populist discourse rather than a method. On the other hand, reducing ethnography to a set of instrumental tools to be deployed by market researchers loses much of what characterizes the ethnographic imagination. We would argue that disciplinary pedagogies in the social sciences serve to challenge both extremes. They nurture a 'culture of craftsmanship' (Marcus, 2009), a disciplinary 'imagination' (Mills, 1959) and a methodological 'habitus' (Bourdieu, 1977). The challenge is to defend and rethink these communities and methodological cultures in new times.

Note

1. Around 40 interviews were conducted with a cross-section of anthropologists trained in a range of UK anthropology departments as part of an ESRC-funded study of 700 students who completed their doctorates in UK anthropology departments between the early 1990s and 2004 (Spencer et al., 2005). They were chosen to broadly represent a range of different academic constituencies (e.g. those in tenured posts, on fixed term contracts, in anthropology departments, etc.). Twenty of the 40 were still working in academic roles, the others employed outside universities.

References

Acker S and Armenti C (2004) Sleepless in academia. *Gender and Education* 16(1): 3–22.

Amin A and Thrift N (eds) (2003) *The Blackwell Cultural Economy Reader*. Oxford: Blackwell.

Atkinson P, Delamont S and Hammersley M (1988) Qualitative research traditions: a British response to Jacob. *Review of Educational Research* 58(2): 231–250.

Barry A (2002) The anti-political economy. *Economy and Society* 31(2): 268–284.

Barry A, Born G and Weszkalnys G (2008) Logics of interdisciplinarity. *Economy and Society* 37(1): 20–49.

Bauman Z (1989) *Legislators and Interpreters: On Modernity, Postmodernity and Intellectuals*. Cambridge: Polity.

Bell D (1973) *The Coming of Post-Industrial Society: A Venture in Social Forecasting*. New York: Basic Books.

Berger J (2008) *Ways of Seeing*. London: Penguin.

Bourdieu P (1977) *Outline of a Theory of Practice*. Cambridge: Cambridge University Press.

Burawoy M (2005) For public sociology. *American Sociological Review* 70(1): 4–28.

Caplan P (2005) In search of the exotic: a discussion of the BBC2 series Tribe. *Anthropology Today* 21(2): 3–7.

Cefkin M (ed.) (2009) *Ethnography and the Corporate Encounter: Reflections on Research In and Of Corporations*. Oxford: Berghahn.

Clifford J and Marcus G (eds) (1986) *Writing Culture: The Poetics and Politics of Ethnography*. Berkeley, CA: University of California Press.

Delamont S and Atkinson P (1980) The two traditions in educational ethnography: sociology and anthropology compared. *British Journal of Sociology of Education* 1(2): 139–152.

Delamont S, Atkinson P and Parry O (2000) *The Doctoral Experience: Success and Failure in Graduate School*. London: Falmer.

Delanty G (2001) *Challenging Knowledge: The University in the Knowledge Society*. Buckingham: SRHE/Open University Press.

Di Maggio P and Powell WW (1983) The iron cage revisited: institutional isomorphism and collective rationality in organizational fields. *American Sociological Review* 48: 147–160.

Drucker PF (1993) *Post-Capitalist Society*. Oxford: Butterworth-Heinemann.

Du Gay P and Pryke M (2002) *Cultural Economy: Cultural Analysis and Commercial Life*. London: Sage.

Eddy E (1985) Theory, research, and application in educational anthropology. *Anthropology & Education Quarterly* 16(2): 92–105.

Espeland WN and Sauder M (2007) Rankings and reactivity: how public measures recreate social worlds. *American Journal of Sociology* 113(1): 1–40.

ESRC (2009) ESRC strategic plan 2009–2014: delivering impact through social science. Swindon: ESRC.

Etzkowitz H and Leydesdorff L (1997) *Universities in the Global Economy: A Triple Helix of University Industry Government Relations*. London: Cassell.

Evans M (2004) *Killing Thinking: The Death of the Universities*. London: Continuum.

Faubion JD and Marcus GE (2009) *Fieldwork is Not What it Used to Be: Learning Anthropology's Method in a Time of Transition*. Ithaca, NY: Cornell University Press.

Fox K (2005) *Watching the English*. London: Penguin.

Frank T (2002) *New Consensus for Old: Cultural Studies from Left to Right*. Cambridge: Prickly Pear Press.

Geertz C (1989) *Works and Lives: The Anthropologist as Author*. Palo Alto, CA: Stanford University Press.

Gibbons M, Limoges C, Nowotny H, Schwartzman S, Scott P and Trow M (1994) *The New Production of Knowledge: The Dynamics of Science and Research in Contemporary Societies*. London: Sage.

Gonzalez R (2007) Towards mercenary anthropology? The new US army counterinsurgency manual FM 3–24 and the military-anthropology complex. *Anthropology Today* 23(3): 14–19.

Gordon J (1998) The resurgence of applied anthropology in a post-exotic world: an Australian perspective. *Human Organisation* 57(2): 127–133.

Grimshaw A and Hart K (1993) *Anthropology and the Crisis of the Intellectuals*. Cambridge: Prickly Pear Press.

Gupta A and Ferguson J (1997) *Anthropological Locations: Boundaries and Grounds of a Field Science*. Berkeley, CA: University of California Press.

Gusterson H (2007) Anthropology and militarism. *Annual Review of Anthropology* 36: 155–175.

Hammersley M (ed.) (2000) *Taking Sides in Social Research' Essays on Partisanship and Bias*. London: Routledge.

Hammersley M and Atkinson P (2007 [1983]) *Ethnography: Principles in Practice* (3rd Edition). London: Routledge.

Haraway D (1991) *Simians, Cyborgs, and Women: The Reinvention of Nature*. London: Free Association Books.

Hazelkorn E (2011) *Rankings and the Reshaping Higher Education: The Battle for World-Class Excellence*. Basingstoke: Palgrave Macmillan.

Holmes DR and Marcus GE (2006) Fast capitalism: para-ethnography and the rise of the symbolic analyst. In: Fisher M and Downey G (eds) *Frontiers of Capital: Ethnographic Reflections on the New Economy*. Durham: Duke University Press.

Hoskin K (1993) Education and the genesis of disciplinarity: the unexpected reversal. In: Messer-Davidow E, Shumway D and Sylvan D (eds) *Knowledges: Historical and Critical Studies in Disciplinarity*. Charlottesville, VA: University of Virginia Press.

Ingold T (2009) Anthropology is not ethnography: Radcliffe-Brown lecture in social anthropology. *Proceedings of the British Academy* 154: 69–92.

Kerr C (1963) *The Uses of the University*. Cambridge, MA: Harvard University Press.

Lyotard J-F (1984) *The Postmodern Condition: A Report on Knowledge*. Manchester: Manchester University Press.

Marcus G (1995) Ethnography in/of the world system: the emergence of multi-sited ethnography. *Annual Review of Anthropology* 24: 95–117.

Marcus G (2006) Where have all the tales of fieldwork gone? *Ethnos* 71(1): 113–122.

Marcus G (2007a) How short can fieldwork be? *Social Anthropology* 15(3): 353–357.

Marcus G (2009) Notes towards an ethnographic memoir of supervising graduate research through Anthropology's decades of transformation. In: Faubion J (ed.) *Fieldwork is Not What it Used to Be: Learning Anthropology's Method in a Time of Transition*. Cornell, NY: Cornell University Press.

Marginson S and Considine M (2000) *The Enterprise University: Power, Governance and Reinvention in Australia*. Cambridge: Cambridge University Press.

Menzies H and Newson J (2007) No time to think: academics' life in the globally wired university. *Time and Society* 15(1): 83–98.

Mills CW (1959) *The Sociological Imagination*. New York: Oxford University Press.

Mills D (2006) Dinner at Claridges? Anthropology and the captains of industry 1947–55. In: Pink S (ed.) *Applications of Anthropology: Professional Anthropology in the Twenty-First Century*. Oxford: Berghahn.

Mills D (2008) *Difficult Folk: A Political History of Social Anthropology*. Oxford: Berghahn.

Mills D and Berg M (2010) Gender, disembodiment and vocation: exploring the unmentionables of British academic life. *Critique of Anthropology* 30(4): 331–353.
Moore D (2002) Ethnography and the Australian drug field: emaciation, appropriation and multi-disciplinary myopia. *International Journal of Drug Policy* 13(4): 271–284.
Morley L (2003) *Quality and Power in Higher Education*. Buckingham: SRHE and Open University Press.
Nelson C and Watt S (2004) *Office Hours: Activism and Change in the Academy*. New York: Routledge.
Neyland D (2008) *Organisational Ethnography*. London: Sage.
Nowotny H, Scott P and Gibbons M (2001) *Re-thinking Science: Knowledge and the Public in an Age of Uncertainty*. Cambridge: Polity.
Nowotny H, Scott P and Gibbons M (2003) Mode 2 revisited: the new production of knowledge. *Minerva* 41(3): 179–194.
OECD (1996) *The Knowledge-based Economy*. Paris: The Organisation.
Okely J (2007) Response to George Marcus 'How short can fieldwork be? *Social Anthropology* 15(3): 357–361.
Olds K and Thrift N (2005) Cultures on the brink: reengineering the soul of capitalism – on a global scale. In: Ong A and Collier S (eds) *Global Assemblages: Technology, Politics, and Ethics as Anthropological Problems*. Oxford: Blackwell.
Olssen M and Peters M (2005) Neoliberalism, higher education and the knowledge economy: from the free market to knowledge capitalism. *Journal of Education Policy* 20(3): 313–345.
Osborne T (2004) On mediators: intellectuals and the ideas trade in the knowledge society. *Economy and Society* 33(4): 430–447.
Pedler M, Burgoyne J and Boydell T (1997) *The Learning Company*. London: McGraw Hill.
Pink S (ed.) (2006) *Applications of Anthropology: Professional Anthropology in the Twenty-First Century*. Oxford: Berghahn.
Platt J (2003) *The British Sociological Association: A Sociological History*. Durham: sociologypress.
Porter M (1985) *Competitive Advantage*. New York: Free Press.
Power M (1997) *The Audit Society: Rituals of Verification*. Oxford: Oxford University Press.
RCUK (2010) Strategy and Delivery Plan 2010–2011. Unpublished paper, Research Councils UK, Swindon.
Reich R (1992) *The Work of Nations: Preparing Ourselves for 21st Century Capitalism*. New York: Alfred A Knopf.
Shore C and Wright S (1999) Audit culture and anthropology: neoliberalism in British higher education. *Journal of the Royal Anthropological Institute* 5: 557–575.
Simmel G (1978) *The Philosophy of Money*. London: Routledge & Kegan Paul.
Slaughter S and Rhoades G (2004) *Academic Capitalism and the New Economy: Markets, State and Higher Education*. Baltimore, MD: Johns Hopkins University Press.
Smith DE (2006) *Institutional Ethnography as Practice*. Lanham, MD: Rowman & Littlefield.
Spencer J, Mills D, et al. (2005) *Career Paths and Training Needs of Social Anthropology Research Students: Final Report to the ESRC*. Edinburgh: University of Edinburgh, 12.
Strathern M (1997) 'Improving ratings': audit in the British University system. *European Review* 5(3): 305–321.
Strathern M (ed.) (2000) *Audit Cultures: Anthropological Studies in Accountability, Ethics and the Academy*. London: Routledge.
Strathern M (2004) *Commons and Borderlands: Working Papers on Interdisciplinarity, Accountability and the Flow of Knowledge*. Wantage: Sean Kingston Publishing.
Strathern M (2006) A community of critics? Thoughts on new knowledge. *The Journal of the Royal Anthropological Institute* 12(1): 191–210.

Tett G (2009) *Fool's Gold: How Unrestrained Greed Corrupted a Dream, Shattered Global Markets and Unleashed a Catastrophe*. London: Little, Brown.
Thompson EP (1970) *The Making of the English Working Class*. Harmondsworth: Penguin.
Thrift N (2005) Assembling the 'global schoolhouse' in Pacific Asia: the case of Singapore. In: Daniels PW, Ho KC, Hutton TA (eds) *Service Industries and Asia-Pacific Cities. New Development Trajectories*. London: Routledge, 199–215.
Thrift N (2005) *Knowing Capitalism*. London: Sage.
Tuchman G (2009) *Wannabe U: Inside the Corporate University*. Chicago: University of Chicago Press.
Walker J (2009) Time as the fourth dimension in the globalization of higher education. *The Journal of Higher Education* 80(5): 483–509.
Westbrook D (2008) *Navigators of the Contemporary: Why Ethnography Matters*. Chicago: University of Chicago Press.
Willis P (1977) *Learning to Labour: Why Working Class Kids Get Working Class Jobs*. Farnborough: Saxon House.

77

Fast Capitalism: Para-Ethnography and the Rise of the Symbolic Analyst

Douglas R. Holmes and George E. Marcus

> The social theorist's oscillation between the market as the artifact of knowledge and as evidence of its limits finds powerful expression in the events now shaping the financial world. In the aftermath of the Asian Financial Crisis and the consciousness of so-called "systemic risk" amplified, the failure of the economic models that were the source of central bankers' expertise to offer useful predictions or fine-tuned policy solutions to recent market problems, and current efforts by global banks to do away with the regulation and management of the market by state actors altogether . . . the central bankers at the Bank of Japan . . . were both amazed at what they imagined as their own creation – a global market of seemingly infinite scale and complexity – and fearful of their own powerlessness vis-à-vis their creation. In particular, these central bankers' sense of powerlessness took the form of a heightened awareness of the contradictions in their role as creators and guardians of the global market.
>
> – Riles, "Real-Time: Governing the Market after the Failure of Knowledge"

Annelise Riles's acute insight locates anxieties infiltrating the work of central bankers, but we believe these fears also imbue the outlook of actors across numerous domains of expertise. These figures share the apprehension that their distinctive analytical endeavors have been compromised, that in some deep way their knowledge is in doubt. As Riles notes,

Source: Melissa S. Fisher and Greg Downey (eds), *Frontiers of Capital: Ethnographic Reflections on the New Economy* (Durham, NC: Duke University Press, 2006), pp. 33–57.

central bankers are in a particularly important position in this regard insofar as their anxiety, their sense of powerlessness, is the direct outcome of their own analytical enterprise: the institutional shaping and management of global financial markets. This predicament, faced paradigmatically by central bankers, is the starting point for what we have termed "para-ethnography," the intellectual response to a retreat of those forms of technical knowledge – whether in arts, science, business, finance, politics, government, or law – that have been underwritten and contextualized historically by the nation-state. In the shadow of this failure of technical knowledge is an alternative form of knowledge creation that has integrity in its own right, integrity with which anthropologists have a keen familiarity.

We begin with a discussion of a series of decisions made by the Federal Open Market Committee (FOMC), the key policymaking body of the U.S. Federal Reserve System, and how these decisions reveal distinctive intellectual practices that, we believe, have unusual methodological significance. We then turn to a historical account, informed by the work of David Westbrook, that describes the creation of a global *polity* – the "City of Gold," as he terms it – integrated constitutionally by supranational markets. In this brief assessment, we look at the historical circumstance that reciprocally gave birth to this global polity and brought to an end the era of the nation-state as the preeminent institutional context defining our lives together. We do this in terms of two key figures, John Maynard Keynes and Jean Monnet, who engineered this historical transformation through the Bretton Woods and European institutions, respectively, and whose analytical endeavors help us understand the role of technocratic knowledge in shaping a radical politics of our time: a politics that is not simply congruent with or reducible to the precepts of neoliberalism. This discussion also helps us assess the emergence of "symbolic analysts," a class of technocrats and political figures who, though appointed or elected in nation-states, designed supranational institutional realms in which they became key actors (Reich, 1992).

Fast capitalism thus designates the circumstances under which knowledge is created and effaced as the communicative space of the nation-state is eclipsed and our subjects and we, too, must think and act within a communicative space mediated increasingly by supranational markets. In the broadest terms, we are interested in what kind of intellectual practices are plausible, what kind of methodological innovations are necessary under the sway of contemporary political economy. For us, a reflexive ethnography is key for addressing these challenges and for investigating global assemblages constituting themselves in our midst (Agger, 1989; S. Collier and Ong, 2005; Holmes, 2000; Holmes and Marcus, 2005b; Maurer, 2005; Rabinow, 2003; Westbrook, 2003).[1]

Between Thick and Thin

We recapitulate in this text our discussions in which we sought to work out for the first time our conceptualization of the para-ethnographic. This initial exercise has since become the basis of a far broader reappraisal of the character of anthropological ethnography in which we are currently engaged that encompasses what we have termed a "refunctioning" of ethnography (Holmes and Marcus, 2005a, 2005b). Our discussion turns on the postulation of the para-ethnographic as an orienting object of study and a key component in the construction of multisited research designs. Such a notion derives from the long-established anthropological interest in "the native point of view," with its representation being, in generic terms, the goal of ethnographic description. Of course, there have been many means and ways to this goal in the history of anthropology. In the framework of a specifically critical anthropology that is interested in the histories and processes by which distinct peoples and populations have contended with the encroachments of world historical systems, particularly Western ones, in the forms of colonialism, capitalism, and now globalization, this representation of native points of view has been as much about the resistance, accommodation, and resilience of such peoples in the face of these historic encroachments. This kind of study has developed recently along with the strong, critically reflexive examination by anthropologists of their own discipline. In this intersection is the conception that the traditional subjects of study have developed something like an ethnography of both their own predicaments and those who have encroached on them, and that their knowledge practices in this regard are in some sense parallel to the anthropologist's and deserving of more consideration than mere representation in the archive of the world's peoples that anthropologists have created. It is an idea that has been expressed from time to time but has not been fully examined, especially for its implications with regard to the forms, norms, and methods of ethnography (Marcus, 1995, 1998a, 1999a, 1999b, 1999c, 2002a, 2002b).

As we have argued, our postulation of a para-ethnographic dimension to the worlds of meaning constructed by our subjects of study is perhaps a radical investment in this not yet fully examined idea that circulates currently in the production of ethnography. We do so, however, within a fully conceived ethnography of modernity where anthropologists have become accustomed to defining objects of study that are no longer on the border between the encroaching West and the transforming traditional, but are fully located within the shared, but differently situated and located, predicaments of contemporary life. In such a frame, the affinity between the effort to define the contemporary in composing ethnography and the same effort locatable in embedded discourses and practices among subjects is not at all a radical idea, but a necessary one (Boyer, 2003; Maurer, 2002a, 2002b; Miyazaki, 2003; Miyazaki and Riles, 2005; Riles, 2004b).

Thus, the illustrative account of decision making of the FOMC that follows provides the raw material, so to speak, out of which our conception of ethnography constitutes itself by demonstrating how embedded para-ethnographic practice is found in the scene of fieldwork, logically giving rise to the multi-sited terrain and problematic of contemporary fieldwork.

The impetus for the analysis that follows grew out of a rather focused discussion on the nature of ethnographic data, in particular, the role of "thin" data in shaping ethnographic analyses in what we term cultures of expertise, professional settings in which some form of research function is integral. We were preoccupied with how highly refined, technical documents – thin data – can serve as the material, so to speak, out of which we can create contexts and preconditions for "thick" description (Holmes and Marcus, 2005a; Marcus, 1999b). Coincidently, during this discussion, Holmes was involved in the preliminary research for his current project on central banking, which entailed reviewing enormous volumes of this kind of austere technocratic data, scrutinizing precisely those artifacts that Bruno Latour (1988, 54; quoted in Riles, forthcoming) describes as "the most despised of all ethnographic subjects." In the midst of this survey he encountered a dissonant term, literally a single word that altered our discussion fundamentally. This term was "anecdotal," as in a short, informal description or narrative account. This dissonant usage recurred in transcripts when senior officials of the U.S. Central Bank invoked the notion of "anecdotal data" during crucial deliberations that became the basis of a dramatic change in monetary policy. As we pursued the significance of this usage, we found that it not only provided a means to move between thick and thin description, but, more significant, it marked a breach in technocratic knowledge that opened the way for realigning the relationship between ethnography and political economy.[2]

The Fed

On January 3, 2001, the Federal Reserve Board instigated a surprise reduction in the discount rate of a full half percent (50 basis points, from 6.5 to 6 percent). The announcement came between the Federal Reserve Board's regular Open Market Committee meetings, when these decisions are normally taken, suggesting a marked anxiety about the state of the economy at the end of 2000. Subsequent data, as we now know, confirmed these apprehensions.

The FOMC meets formally eight times a year in Washington, D.C., with telephone consultations and other meetings when needed. The Committee is composed of seven members of the Board of Governors of the Federal Reserve System and five of the twelve presidents of the Federal Reserve district banks. The president of the New York Reserve Bank is a permanent member, and the other reserve bank presidents serve one-year terms on a rotating basis. All twelve presidents of the district banks participate in the FOMC discussions, but only the five presidents who are members of the Committee participate in policy decisions. The Committee is charged under law "to oversee open market operations, the principal tool of national monetary policy" (Board of Governors of the Federal Reserve System, 1994, 12–13). This legal authority over the management of monetary policy, exercised primarily through the setting of interest rates, makes the Committee perhaps the most powerful institution governing the global economy (see Blinder, 1998, 2004; Meyer, 2004). Hence, the social behavior of its members is of unusual significance, and the events of 2000–2001 are particularly intriguing.

The year 2000 had opened strongly, but by late spring problems began to surface at the heart of the New Economy, the technology sector, along with broader problems gripping manufacturing in general. A savage, though widely predicted, correction in equity prices of Internet-related companies unfolded through midyear, and these problems leached into the pricing of chipmakers and related technology companies. The consensus opinion held by the governors and presidents of the district banks of the Federal Reserve during most of 2000 was that the economy was still strong; weakness in the technology sector, outside of Internet-based companies, was understood to involve a short-term inventory correction. By the last few days of the year, the perception had shifted markedly, suggesting that a far broader and more precipitous downturn was looming, with a growing risk of recession. This change in perception impelled the rate cut in early January 2001.

So something happened during the fifteen-day period from the December 19, 2000, regularly scheduled FOMC meeting to the unscheduled telephone conference meeting on January 3 that altered significantly the judgment of the members of the FOMC about the state of the U.S. economy (http://www.federalreserve.gov/fomc/minutes/20001219.htm#jan3minutes). What transpired during this brief period intrigued us because it reveals decisively the operation of the para-ethnographic in a culture of expertise committed to a technocratic ethos predicated on the management of vast amounts of quantitative data. The imperative here for deploying the para-ethnographic is the acute need to interpret the "contemporary" in a meaningful fashion to inform decision making.

In the expert culture of the Federal Reserve System, the structure of the contemporary has a twofold manifestation that its members and staff must act within and on. On the one hand, it takes form as a skein of information – typically, highly refined quantitative data – that is constantly being generated and analyzed. On the other, it has a keen discursive character whereby information is endowed with social perspective and meaning. It is in this latter, rather oblique form of knowledge practice, typically glossed as anecdotal, that the para-ethnographic takes form, and it is under these conditions that key cultural practices of expert subjects converge with our own analytical endeavors as ethnographers. In other words, our subjects are doing something that approximates ethnographic labor to represent the *social* nature of the contemporary, within which they and we pursue our expert practices. The figures we have been looking at summon these narratives to conceptualize the contemporary financial world for a particular purpose: to anticipate the future, to capture shifting configurations of expectations and sentiment. As Westbrook (2003, 145) eloquently puts it, making "the future a task that must be confronted now, and on the terms determined by capital," has a deeply unsettling effect.

> In helping us to secure ourselves from, or at least ameliorate, the dangers that lurk in the middle distance, finance robs us of the ability to enjoy the present. The expectation of the near future, the time of finance, is built up and yet obscures the reality of the present. As a result, we experience the present as quite literally surreal, as something somehow built on top of reality. . . . Surrealism portrays us situated among conventions that we understand to have no foundations, to be arbitrary and quite possibly impossible, and that yet stand because no other basis for tomorrow is historically or otherwise imaginable.

Thus, we have argued that the para-ethnographic is not merely an aspect of expert practice, but is inlaid in the architecture of a future-oriented contemporary; it is the means by which the contemporary is reproduced socially (Holmes and Marcus, 2005b).

Less than Perfect Knowledge

The Fed is disarmingly frank to the point of deadpan about the imperfection of its own knowledge and its power to manage the economy:

> In practice, monetary policymakers do not have up-to-the-minute, reliable information about the state of the economy and prices. Information is limited because of lags in the publication of data and because of later revisions in data. Also, policymakers have a less-than-perfect understanding of how the economy works, including knowledge of when and to what extent policy actions will affect aggregate demand. (Board of Governors of the Federal Reserve System 1994, 24)

The irony of this declaration is that the Fed's own technocratic agenda has served to accelerate the pacing of what we are calling here fast capitalism, rendering analysis of economic performance increasingly refractory. Indeed, it is precisely this conundrum that makes the U.S. Central Bank such an intriguing object of study (Miyazaki and Riles, 2005).

Alan Greenspan, former chairman of the Federal Reserve Board, whose role in creating the New Economy has been celebrated widely, illustrates some of the underlying rationale for para-ethnography. Greenspan's technical innovation centered on his conviction that the U.S. economy could grow sustainably at a higher rate, roughly in excess of 2.5 percent per year, than previously thought possible without unleashing the perils of inflation. His insight that inflationary constraints on the economy can be circumvented by growth in productivity – impelled by technological dynamism – has distinguished his management of the U.S. Central Bank. What Greenspan embraced is referred to as "the economy's productivity-enhanced potential" as underwritten by new technologies. Indeed, this is the theoretical position that has been at stake in the unfolding drama that continues to reverberate through the U.S. economy and is echoed in the meetings of the FOMC.

If we return to the January 3 event, the reportage attributed the decision almost exclusively to the personal acumen of Chairman Greenspan, in particular his "intuition," his "instinct," his "feel" about what was "really" happening in the U.S. economy. What this meant was that the enormous range of quantitative data tracking the performance of the economy available to the Fed chairman was inconclusive: it suggested that some kind of change was clearly taking place within the economy, but its significance was ambiguous, hence the need for an alternative means of judgment that draws on experience. In other words, to enter analytically "the contemporary" the chairman had to move from science to art; crucially, he needed an "ethnographic" narrative to think and act in something that approximates "real time" (see Riles, this volume). Para-ethnography, thus, first surfaces here under the ambiguous guise of "intuition," as essentially a structure of feeling. There is, however, far more intriguing evidence of this kind of knowledge practice in the transcripts of the FOMC meetings.[3]

The importance of para-ethnography was, in fact, alluded to at critical points in the discussions of the FOMC meeting of December 19. In this case, the para-ethnography appears not as "a structure of feeling" but as "anecdotal evidence." At those points in the FOMC minutes when serious concern is registered about the state of the economy, "anecdotal reports" rather than quantitative data are invoked. Here, then, the pretext for a dramatic change of Central Bank policy is established using something other than the formidable authority of the massive quantitative measures available to the FOMC (http://www.federalreserve.gov/fomc/minutes/20001219.htm#jan3minutes). Indeed, it is through these anecdotal reports, which are reported publicly only in summary form in the so-called Beige Book, that the space of para-ethnography is marked (Holmes and Marcus, 2005a).

Anecdotes

What are these "anecdotal reports"? Rather than informal observations and casual asides, as the term "anecdotal" might suggest, these reports constitute a sophisticated means of tracking and interpreting the economy and endowing it with social context and meaning. Each member on the FOMC, as well as the senior officers of the Federal Reserve's twelve district banks, cultivates highly developed social networks with human interlocutors who oversee daily transactions in strategic spheres of the economy. More formally, each of the twelve Reserve Banks has its own nine-member board of directors appointed for three-year terms who are specifically charged with gleaning precisely these kinds of finely tuned anecdotal accounts of very detailed facets of the U.S. economy and society.

Indeed, the directors are classified in terms of the character of their situated knowledge. Three directors of the nine-member district banks, designated Class A directors, represent the stockholding member banks in the region and are typically senior bank executives. The three Class B (appointed by the district bank) and three Class C (appointed by the Board of Governors of the Federal Reserve System) directors are chosen as representatives of the public, and they are *not* permitted to be officers, directors, or employees of any bank; rather, they are chosen "with due, but not exclusive, consideration of the interests of agriculture, commerce, industry, services, labor and consumers" (http://www.federalreserve.gov/generalinfo/listdirectors/about.htm). Apart from formal oversight responsibilities managing the operation of the twelve district banks, these figures serve as strategically positioned interlocutors who report on a regular basis on the conditions of key sectors and often geographically localized aspects of the economy.

The range of informants is impressive. For example, in District n-Dallas (the district that covers Texas, northern Louisiana, and northern New Mexico), directors serving in 2004 included among the B and C Classes the president of the Texas Iron Workers Union, the CEO of an architectural firm in Austin, the owner of a cattle and land company in Van Horn, a real estate developer in Houston, the founder of the USA/Mexico Alliance in El Paso, and the provost of New Mexico State University. In District 9-Minneapolis (the district covering Minnesota, Montana, northern Michigan, North Dakota, and South Dakota), the Class A directors are bank executives of a small farmers' bank in Victor, Montana, and two commercial banks in Fairfax, Minnesota, and Sioux Falls, South Dakota. The B and C directors included the executive administrator of the Twin City Pipe Trades Service Association in Saint Paul; the former director of the Museum of the Rockies in Boseman, Montana; and the CEO of a health care business in Minneapolis. The deliberations of each of these twelve boards, whether they are composed of a professor at the Harvard Medical School, the CEO of Citigroup, or the director of the Arizona Department of Housing, not only draw on their technical proficiency, but evoke the deep social nature of their diverse expertise and their interrelated predicaments.

These interlocutors are informants – bankers, manufacturers, retailers, educators, union leaders, representatives of governmental and nongovernmental agencies – who transact loans, book orders (and cancellations), track store sales, hire and terminate employees, solicit charitable contributions, and follow the price of wheat and pork bellies minute to minute, hour to hour. They operate in a continuously unfolding present, providing the closest approximation to a contemporaneous engagement with the economy. These men and women, typically senior managers, not only have access to an extraordinary range of quantitative data, but they are constantly in conversations with clients, customers, and colleagues: auto executives walk showroom floors talking to potential car purchasers; bank executives converse with prospective borrowers about the state of their businesses and their outlooks for the future; manufacturers discuss with their customers their future needs in order to plan capital expenditures; union leaders appraise the labor market as they negotiate contracts. By gleaning knowledge from these interlocutors, the Fed gains access to those profound and elusive forces guiding the economy: expectations and sentiments as they take form across an intricate, geographically diverse analytical tableau.

It is not merely that these reports provide a means to overcome the inevitable lag attendant with quantitative data; rather, it is their inherently social character that provides these anecdotal reports with an agile purchase on the contemporary. And these reports would have little force if if were not for the fact that these actors speak from an intimate sense of situated business practices and predicaments. These anecdotes are not just a different kind of supplementary data; they have cogency in their own right in that they represent "a native point of view" mediated through an implicit ethnographic imaginary. These intricate exchanges that report on the economy in something that approximates real time constitute fully formed para-ethnographies extending over a vast multisited scene (Holmes and Marcus, 2005b; Riles, 2005).

What makes these reports persuasive is the experience of these interlocutors, their judgment, their feel. Above all, what these interlocutors struggle with is the creation of a semiotics of contemporary capitalism intimately linked to the integrity of their own expert practices (and perhaps even to their own unfolding existential struggles). The accounts of these situated actors are powerful as anecdotal interventions because they subtly compensate for, or even shore up, cracks in a technocratic ethos (exposed as "less than perfect knowledge") whereby the statistical mode of analysis that sustains the authority of institutions like the Fed is compromised by the speed and the unpredictability of supranational markets.

"After the Fact"

One of the most basic reports gauging the state of the economy is the quarterly gross domestic product (GDP) statistics. The rule of thumb is that two consecutive quarters of declining GDP numbers constitutes the onset of a recession.[4] This means that the events of January 3 would gain full coherence only sometime after July 2001, at which point it would be possible to determine definitively whether or not a recession had in fact commenced. Of course, the bind here for the FOMC is that by July, effective intervention to alter any decline would in all likelihood be too late to avoid significant damage to the economy.

The contemporary thus emerges as the lacuna, as the lag during which events are fundamentally uncertain, when information has yet to confer a conventionalized empirical account of the phenomenon. This is a highly dynamic contemporary on which new information continually encroaches. With each passing week as new data are released, increasingly secure inferences about whether or not a recession has materialized may be drawn. Yet, as the data on the economy are given an increasingly precise longitudinal representation, certain forms of analysis and action are foreclosed (Miyazaki, 2003).

The minutes of the January 30–31, 2001, FOMC meeting are more measured than those of December 19 and January 3, even though the committee approved another major discount rate reduction of 50 basis points. The only reference to anecdotal reports comes when the Committee refers back to their action of January 3. The discussion opened with the tabling of a report on the consequence of the precipitous decline in the U.S. debt (which was subsequently radically reversed) on the Fed's ability and latitude for developing policy. This was followed by an overall summary of the state of the U.S. economy. Then, toward the end of the January 30–31 meeting, there is the following, rather offhand statement:

> Several members observed that the evolving nature of the domestic economy, including the ongoing improvements in inventory management and the increase in managerial flexibility to alter the level and mix of capital equipment, associated in part with the greater availability of information, appeared to have fostered relatively prompt adjustments by businesses to changing economic conditions. As a consequence monetary policy reactions to shifts in economic trends needed in this view to be undertaken more aggressively and completed sooner than in the past. (http://www.federalreserve.gov/fmoc/minutes/200110131.htm)

This aside was in all likelihood intended as an oblique justification or explanation of the action taken at the January 3 meeting. Yet there is also a theoretically significant assertion conveyed in this laconic prose. The pacing and the behavior of the economic system, which the Fed and members of the FOMC have had a major role in creating, means that their own expert practices, based on the management of enormous amounts of quantitative data, have been undercut. The world they must regulate is far more opaque and unpredictable by virtue of the new accelerated pacing of economic transformation. These experts can no longer always wait for the data to come in to calibrate their action. They can no longer take action "after the fact." They participate directly in the continuous generation of a contemporary, constructed out of information over which they have increasingly tenuous analytical purchase. Under these circumstances, which demand observation and action in real time, ethnography provides the modality of access for the expert and the anthropologist alike, access predicated on socially mediated observation.

By late spring of 2001, a formidable retrospective critique began to take hold regarding critical assumptions about the nature of the New Economy and the Fed's husbandry of it. This critique also marked an important stage by which the complex nature of the contemporary was reconciled with a theoretical narrative. Again, the fundamental issue was the role of technologically induced increases in productivity in transforming the overall economy as well as analytical techniques that we use to understand it. In the evangelical words of John T. Chambers, the chief executive of Cisco Systems, articulated just before the downturn, the technologically driven productivity gains allowed companies to "reduce costs, generate revenue in new ways, empower employees and citizens and provide the agility needed for the Internet economy's rapid pace" (quoted in Norris, 2001). Faith in the acceleration of productivity growth, as suggested earlier, had become, at least in part, the basis for the Fed's policy decisions and expectations about the economy's growth potential. Questions about the nature and the scale of productivity gains thus have far-reaching implications. There are three basic questions: Did government statistics exaggerate productivity gains? Were the gains in productivity narrowly concentrated in high-tech companies? How did faith (and/or hype) in the New Economy itself influence investment and consumption? By the late summer of 2001, these questions were given new force by the release of revised economic statistics. Productivity figures were recalculated for the period 1996 to 2000, revealing gains of 2.5 percent rather than 2.8 percent that had been previously reported (*Economist*, August 13, 2001).

Clearly, members of the Fed did act, to a greater or lesser degree, as if the world had entered a new era of economic possibility and that faith did influence policy decisions. And if it can be shown that the Fed was wrong, if it had in fact exaggerated the scale and the breadth of the productivity surge during the 1990s and its impact on the economy, then the Fed itself was in part responsible for underwriting a speculative bubble. If credit supplied to the market had not sustained balanced programs of investment in an ever more productive high-tech economy, but rather speculative excesses, then the Fed had pursued policy that was ultimately destructive to the economy. This is one of a number of stories that marks the beginning of the process by which the events of 1999–2001 were given clarity via a retrospective narrative.

But what is lost analytically through this process? What interests us is not whether this critique of the New Economy is right or wrong but how it forecloses certain kinds of analysis. This is a classic bind; the narratives that are employed to conventionalize the events are drawing on the categories, measures, perspectives, and language that are still more or less predicated on the political economy of the nation-state. By endowing events with this kind of narrative framing, we believe, the contemporary is effaced and access to inchoate processes foreclosed. By invoking these retrospective narratives, we inevitably draw on theoretical frameworks that are allied with an eclipsed political economy. The authority of the ethnographic present is typically overridden via an analytical treatment and representational techniques that tether the action of subjects to convention, to tradition, and to the past, obscuring emergent practices and processes. What we are arguing here is that by interleaving our ethnography and the para-ethnography of our subjects, we can create a means to circumvent this bind, to engage a semiotics of society interpretable from experience.

Uncertainty continued to hang over the FOMC meeting of June 26–27, 2001, as the prospect of economic troubles loomed:

> In the Committee's discussion of current and prospective economic developments, members noted that by some measures the overall economic activity remained at a reasonably high level. However, recent data indicated that growth of spending and output was quite sluggish and below the pace many members had anticipated at the time of the previous meeting. . . . Continuing softness in the expansion of economic activity was mirrored in anecdotal reports of business conditions in much of the nation. Typical regional reports referred to slowing increases in economic activity from an already reduced pace to the persistence of sluggish business activity and generally downbeat business sentiment. Manufacturing continued to display particular weakness. (http://www.federalreserve.gov/fmoc/minutes/20010627.htm)

The danger, too, was becoming clearer, though shrouded in the flat prose of the Committee's minutes. There is also a marked sense that conditions had deteriorated since the FOMC meeting just a month earlier.

> It was noted, however, that the unique characteristics of the current cyclical experience, including the heavy concentration of weakness in business expenditures and manufacturing output, increased the uncertainty that surrounded any forecast. Most of the members believed that the risks to the expansion, notably for the nearer term, remained to the downside of the current forecasts. Potential sources of shortfalls included the effect of possible further increases in unemployment or consumer and business confidence; the risks of disappointing business earning that could damp investment and, through lower equity prices, consumption; and the growing indications of weakness in foreign economies could

> limit demand for exports. (http://www.federalreserve.gov/fomc/minutes/20010627.htm)

A plausible scenario for steep economic decline was taking shape, given the data available to the Committee. What was becoming clearer was how a series of more or less discrete problems in the economy could converge to create an escalating deterioration that spread across the entire system. The likelihood that an accelerating weakness would materialize was still perceived as modest, but the threat was becoming conceivable, if not precisely measurable. There is also an added anxiety that this kind of alarming convergence was possible globally, that similar kinds of weaknesses were surfacing beyond the FOMC's statutory control in foreign economies.

The notion of "foreign" economies is, of course, analytically disingenuous. The members of the FOMC know well how the growing integration of national economies has made their jurisdiction and influence fully global. Their anxieties were, no doubt, focused on something like a repeat of the Asian financial crisis of 1997, in which a confluence of supranational market forces amplified "systemic risk" in a manner that could challenge not just the analytical authority of policymakers, but their power and their ability to intervene. These fears of some ill-defined and unexpected crisis expressed in the FOMC in the early summer were, as we know too well, fully defined on the morning of September 11.

The material world, even in Manhattan, was for the most part intact on the afternoon of September 11, but the structures of sentiment and expectation were profoundly transformed. Obscure global struggles became intimate psychological realities. The personnel of the New York District branch of the Federal Reserve, located two blocks from the Twin Towers, found themselves literally in the midst of the disaster. The research staff had just completed its summary on the state of the regional economy on September 10 intended for the regularly scheduled edition of the Beige Book. For the next two weeks, the New York District offices were closed; within hours of the disaster an expanded research staff began working from their homes, calling their networks of interlocutors to assess the impact of the horrific day on the economy. At this moment, when virtually all quantitative data were rendered irrelevant, para-ethnography became the means for gaining analytical purchase on the radically changed conditions of the world. An internal document, compiled using essentially the same method as the Beige Book, though with this larger staff, was generated within a few days and served as a starting point for assessing the disaster and thus served as the briefing document for the meeting of the FOMC on September 17 (Holmes and Marcus, 2005a).

On September 17, the FOMC responded by lowering its target for the federal funds rate by 50 basis points, and then again on October 2 the Committee lowered the target another 50 basis points to 2.5 percent while adding unusually large volumes of liquidity to the financial markets. This ninth cut in interest rates in 2001 reduced the discount rate to its lowest level in almost forty years; two more interest cuts ensued in 2002, reducing the discount rate to an emergency level of 1 percent, where it remained until June 2004. A recession was at hand. The anxieties that the Fed sought to negotiate revolved around how to restore confidence, if not optimism, to American consumers so that by virtue of their proclivity to spend they might redeem the global economy. And, as we now know, that is precisely what they did: liquidity provided by the Fed fueled heroic purchases of ever more extravagant homes, automobiles, and the like.

Constitution of Markets

We turn now to an illustrative scenario in the form of a historical outline, of the origins of the contemporary globalization in which new, critical forms of knowledge production were spawned. We do this from the standpoint of key actors – symbolic analysts – who created this system and endowed it with an abiding moral urgency and a future-oriented practice. Their decisive technical innovation of establishing supranational markets as *political* instruments is the foundational achievement defining our era (Westbrook, 2003). Among other things, this discussion also helps us understand why central banks play such a significant role in defining and regulating our lives together (Blinder, 2004).

John Maynard Keynes, working with Harry Dexter White, and Jean Monnet, working with Robert Schuman, while seeking to resolve the most fundamental questions of nation-states – war, trade, and debt – arrived at solutions that yielded two radically subversive projects with a global reach: Bretton Woods and European institutions.

Both Monnet and Keynes experienced World War II from within the bureaucracies and ministries that managed, supplied, and financed the war efforts. They were aware of one another's work; Keynes is reputed to have credited Monnet's skill in orchestrating the Allies' (supranational) logistical efforts with shortening the duration of the war by a year. Both figures understood the inner operations of the *Leviathan*; they knew how to manage the *Wealth of Nations*. Both were in fundamental ways economic nationalists and liberal internationalists, and, to say the least, they both had a critical understanding of the workings of financial markets. What they sought, however, was a deployment of markets to achieve a profound historical transformation. Through the European and Bretton Woods institutions, they sought to create an integrated world economy that, to paraphrase the now famous words of the Schuman Declaration establishing the European project, would make war between nation-states not merely unthinkable, but materially impossible. In other words, they used supranational markets as a *political* tool to eviscerate the nation-state of its economic sovereignty and hence what they understood to be its most rapacious proclivities, as personified by Adolf Hitler (Westbrook, 2003). In its place, they created a politics of a distinctive sort, a politics that for very intriguing reasons never gained expression as a wide-ranging political rhetoric; rather, it operated as a deeply subversive technocratic discourse that transformed the world in the second half of the twentieth century. Their veiled efforts were aimed at achieving, in Keynes's words, "a new order," "a realizable utopia" (Lipgens, 1985; Monnet, 1978; Skidelsky, 2000, 208).

The birth of the new order arose from a technical problem, one that was negotiated at an unlikely time and in an unlikely context. In 1941, Keynes faced a problem posed by a troublesome provision of the Lend-Lease Agreement with the United States. The famous agreement is known for allowing the Roosevelt administration to provide aid to the British in the form of aging and surplus warships against the backdrop of strong American isolationism prior to Pearl Harbor. Inserted in the agreement was a provision, Article VII, that the British found profoundly irksome, if not humiliating, and that bound the parties to an agenda that would define the future status of international trade in the postwar world, a world in which the British would be debtors and the Americans creditors. This obscure provision of a diplomatic treaty amounted to a fait accompli in which the American side betrayed the obvious: that they had the financial might to determine the rules governing the postwar economic order. But it went further, demanding implicitly an end to the economic relations that sustained the twentieth-century remains of the British Empire.

Article VII became part of wide-ranging debates within the British Treasury and the Bank of England already under way in 1941 on planning for the postwar British economy and, more broadly, the structure of international trade. Keynes played a pivotal role in these discussions, particularly in the debate on the relative merits for the British of planned trade, based on the illiberal Schachtian system of bilateral managed trade relations developed by Hitler's banker and economics minister Hjalmar Schacht and Walther Funk, who succeeded Schacht as economics minister, and the laissez-faire approach of "non-discrimination in trade" or "Hullism" espoused by the American secretary of state Cordell Hull (Skidelsky 2000, 179–232). The former arrangement, at least in theory, would have perpetuated the "imperial preferences" of the Commonwealth; the latter would have decisively ended the political economy of empire.

Over a long weekend in September 1941, Keynes drafted two papers as rejoinders to Article VII, "Post War Currency Policy" and "Proposal for International Currency Union" that Robert Skidelsky (2000, 208), Keynes's biographer, described as "the most important he ever wrote in terms of their direct influence on events." In many respects, these were incomplete, ambiguous, and rather messy documents, yet together they constituted a major "fragment of a Grand Design" that was to be fully realized in the Bretton Woods agreements in 1946 (ibid.). At the center of this technical design were the mechanisms of an International Central Bank (ICB) that would, as Keynes fatefully put it, "[make] a beginning at the construction of the future government of the world" (quoted in Skidelsky, 2000, 223). "The challenge with which Article VII faced Keynes was to devise a plan to lock the USA into a system which would maintain balance of payments equilibrium between all countries without trade discrimination but also without forcing deflation, unemployment or debt-bondage on the deficit countries. He met it with the experience of the inter-war years and his economic theories, which were largely commentaries on experience" (182).

The negotiations over this plan first within the British government and then crucially in response to the plan drafted by Harry Dexter White, a New Deal – appointed assistant to Treasury Secretary Henry Morgentheau, yielded the now famous tripartite Bretton Woods institutional arrangement for global financial integration. What is most significant for our purposes is how these complex negotiations defined the space of technocratic intervention predicated on monetary issues. Skidelsky (2000, 181) comments on the prosaic conditions that made possible this kind of technocratic intervention that circumvented conventional political scrutiny:

> It might seem odd that planning for economic disarmament started on the monetary rather than on the trade side, and made faster progress. After all, Article VII was about trade, not money. But it made sense logically and politically. World trade had broken down in the 1930s because the gold standard had disintegrated. So the way to start unblocking the channels of trade was to unblock the flow of money by which traded goods were paid for. . . . It also seemed a lot easier for Britain and America to reach agreement about money. Money was too esoteric and boring to engage interests, passions, and prejudices. So technicians had a chance to make much more progress on money before their plans attracted political attention.

At stake in these negotiations were issues of highest national concern to the two parties, inspiring potentially divisive political confrontation.

> It must not be supposed that the technicians, as they were somewhat contemptuously called, operated in a historical or political vacuum. They brought to their work different understandings of what had gone wrong with the old gold standard; and different national experiences and interests. Above all the British and American plans came from two countries which expected to be very differently situated in the post-war world. The British expected to emerge from the war stripped of trade and assets and burdened with debts; the Americans in the opposite position. Keynes's therefore was a debtor's, White's a creditors, perspective. The British wanted a scheme which would enable them to borrow without strings;

> the Americans one which would lend with strings. This difference was reinforced by another. The British plan came from a banking tradition, the American one from a legal tradition. The British wanted a scheme based on prudential maxims; the Americans one based on legal rule. (182)

Yet the translation of these highly fraught political issues into technical terms opened the way to, though did not guarantee, Keynes's "realizable utopia." What this translation entailed was a conceptualization of the supranational as a sphere of theory, analysis, and intervention.

> The technical debates were not merely a cover for national self-interest. Keynes and White were both interested in theory, and in finding technical solutions to technical problems. Economics often enables political conflicts which seem intractable if carried on in the language of politics to be bypassed by assuming (or the cynic would say pretending) that they can be reduced to the technical, and imposing a language of debate consistent with that assumption. As the technicians from both sides got to know each other better, a genuine camaraderie developed *contra mundis* – against the politicians, bankers, vested interests on both sides waiting in the wings to sabotage both the Keynes and White Plans. (Skidelsky, 2000, 182)

This conceptualization of a supranational institutional framework also defined the role of a new class of actors, symbolic analysts, whose interests were no longer fully aligned with or reducible to those vested interests that defined the nation-state. These actors were no longer constrained by its theories, methods, and histories; indeed, by virtue of their institutional projects, they sought to render the intellectual apparatus underwriting the nation-state increasingly irrelevant.

Within a decade (by May 9, 1950), Jean Monnet and Robert Schuman engineered in a remarkably similar fashion the founding of the European institutions. The European project was orchestrated procedurally through diplomatic treaties that established complex technocratic agendas that created a supranational space in which a new class of functionaries could operate. An explicit technocratic ethos and practice was embraced by the founders of the European Union that entailed a comprehensive rejection of the nation-state as an instrument for managing human affairs and the development of an expert method – an administrative science of the supranational – that would impel the political integration of Europe.

Jean Monnet made a decisive discovery about the communicative possibilities (and limitations) constituted by supranationalism, a discovery that he relentlessly exploited and built into the fabric of the European project. Commenting on the journalists covering the announcement of the Schuman Declaration, Monnet (1978, 304) notes their bafflement: "They were still uncertain about the significance of the proposal, whose technical aspects at first sight masked its political meaning." It is precisely this bafflement that allowed one of the most important political projects of the twentieth century to develop in plain sight with little, if any, serious public scrutiny. In other words, the inscrutability of the European project and, even more so, the Bretton Woods project was not merely the outcome of public inattention, but a consequence of how these projects displaced fundamental historical realities that endowed events with coherence.

Inspired by Conrad Adenauer, first postwar chancellor of the Federal Republic of Germany, drafted by Robert Schuman, French prime minister, and Jean Monnet with a small group of aides, the founding document of the European Union explicitly aimed to end the possibility of war in Europe via comprehensive and wide-ranging market integration, starting with the production of coal and steel. Monnet was struck at the press conference announcing the Schuman plan that the journalists failed to grasp its significance; the elite French intellectual community lacked a framework and language to understand what the document represented. Monnet expresses in his memoir initial disappoint with this tepid response to this remarkable declaration, but then, one guesses, he realizes what a remarkable asset this public indifference could be: he could build Europe in plain sight without serious public scrutiny. He was free to pursue this staggering project because the journalists, the intellectuals, the political elite of France and Europe would be unperturbed; they operated within intellectual purviews and critical political agendas that were seamlessly tied to histories and theories from which his supranational ambitions were largely inscrutable. Like Keynes, he had discovered the tightly defined communicative character of supranationalism: its technical language and its goal of market integration. Monnet, one of the shrewdest minds of the twentieth century, recognized almost instantaneously that with the Schuman Declaration a vast political space was created that was virtually invisible and inaccessible from the standpoint of conventional political ideology and practice.

An explicit ethos and practice, designed by Monnet, was embraced by the founders of the European Union and used to fill this technocratic space of Europe. At its philosophical core were a comprehensive rejection of the nation-state as an instrument for managing human affairs and the development of an expert method – an administrative science of the supranational – that would impel the political integration of Europe. The founders of the EU were emphatic that their goal was to escape the blighted history of the European nation-state and to create a new grammar and semiotics of society (Haas, 1964, 1968; Herzfeld, 1993, 1997; Lipgens, 1985).

Thus, the dynamism of this new regime incites change that unfolds largely outside the conceptual ambit and political control of actors within the nation-state, change that transpires largely unobserved (and unmeasurable) from the perspectives of conventionally situated politicians, journalists, academics, and intellectuals. Ulrich Beck, Anthony Giddens, and Scott Lash

(1994) have characterized the resulting condition as "reflexive modernization" by which, as the director of research at the Atlanta District branch of the Fed averred, "systems, instruments, and markets are evolving faster than the political entities can bring their various rules and regulations into harmony" (quoted in Riles, this volume). This world of fast capitalism that presents itself as structurally inscrutable is, as we have argued here, nonetheless susceptible to meaningful and to manifold scrutiny through the creative assimilation of socially mediated observation, what we have termed para-ethnography.

One additional analytical move enhances this possibility: a move by which the fundamental nature of contemporary capitalism, as shaped by Bretton Woods and European institutional transformations, is restated in explicitly political terms. David Westbrook's (2003) remarkable text, *The City of Gold*, performs this critical intellectual labor.[5]

Westbrook has persuasively argued that a particular appropriation of the market mechanism as a constitutional instrument has yielded a new type of *polity*, what he terms the City of Gold. By shifting our theoretical preoccupations about capital markets, from conceptualizing them in relationship to those commercial transactions organizing production, distribution, and consumption to conceptualizing markets as *constitutional* devices, he provides analytical purchase on the forces giving social form and cultural content to the contemporary world.[6] He distills the constitutional logic governing the City's operation and establishes criteria for a reflexive politics of the contemporary by which we can appraise, as he puts it, the truth, beauty, justice, and tyranny of our time:

> To understand the way we now live rests therefore, on a restatement of politics as it appears in the context of supranational capital, legitimated through our faith in the institutions of money and property, as opposed to the modern nation state, legitimized through the familiar mechanisms of the liberal republic. . . . The communicative space formed by financial markets is the object of political thought in our time, as the nation state was for most political thought during the time we still regard as modern, the pope, king, or emperor were the protagonists of the medieval political imagination, and the *polis* was the context of classical Greek thought. This book uses urban imagery and self-consciously revives the old Marxian personification, "capitalism" in order to solidify the understanding of financial markets as the grammar through which much contemporary political thought should take place. (2003, 12–13)

The predicament we face thus can be stated as a problem of meaning, and specifically, the shifting conditions of its social mediation.

> The market's grammar, the dialectic between property and money, does not express many things important to being human. Capitalism is therefore radically impoverished as a system of politics. Insofar as we long for

> community, we necessarily experience life in capitalism as a sort of exile. . . . The construction of markets – the creation and alienation of property rights – involves the destruction of meaning, and in longing for that meaning, we complain not only about the market before us, but about the arrangement of social affairs through markets per se. (164)

Like our subjects, we, too, are legatees of a theoretical apparatus that conventionalizes data by drawing on the categories, measures, perspectives, and language that are still aligned fundamentally with those key concepts that animated nineteenth-century political economy and social theory. We are arguing here for a method that we can creatively pursue with our subjects, which circumvents this bind to encompass a semiotics of society interpretable from experience. We have sketched out in this essay a very preliminary rendering of this method and its historical relationship to and position within the communicative space of supranationalism. The great potential of this intellectual practice is that it can position the ethnographer within a fraught contemporary, a position from which he or she can delineate and address the questions of our time.

In our version, among other possible versions of this formulation of the object of ethnography, what gives rise to the para-ethnographic are the predicaments of a world of fast capitalism that engenders a constant task to produce the contemporary and in so doing to rethink, reobserve, and represent the categories of society in order to reproduce one's own identity and familiar contexts of agency and meaning. Although this may not apply to everyone everywhere today, it certainly applies, we argue, to the production of the discourses of politics and political interest in their official and displaced forms, and certainly to many other milieus and institutional arenas that anthropologists are increasingly interested in studying these days. The para-ethnographic is that side of diverse discourses and practices that represents the social ground for specific purposes and goals and done thickly or thinly with considerable consequences for events and action. This is an ethnographic process of knowledge creation, what anthropologists do, albeit in a more explicit and committed way. The question of what anthropologists will make of this affinity, how they will shape from it a tradition of research beyond the thick description of situated native points of view, remains in development.

Notes

1. One of our students, setting out for her own research on venture capitalists in Korea, presented an interesting gloss on this alternative project of contemporary ethnography in her comments on a candidate for an anthropology faculty position, defined as a specialist in the interdisciplinary field of science and technology studies:

> In some way, he is already recognizably someone like us, existing as he does between two scholarly worlds that produce their own separate sets of expert discourses. I imagine our department to be pioneers of that kind – to re-imagine and put into practice the kind of ethnography that deals with the production of contemporary knowledge. So the emphasis is on the leaving behind of life-worlds not simply by saying that we are all "connected" somehow, but to actually look at how other knowledge makers make the common "facts" around and on which we build the routines of our understanding and shape the structures of feeling and arcs of expectations.

See Marcus (1995, 1998a, 1999a, 1999b, 1999c, 2002a, 2002b) on these manifold predicaments of contemporary ethnography.

2. This is a propitious moment to work in the anthropology of finance, central banking, and, more generally, contemporary configurations of capitalism given the vibrant scholarship that is currently defining the field. Two superb papers by Mitchel Abolafia (2004, forthcoming) demonstrate the possibilities of reconstructing the context and the processes of decision making at the Fed. We also have superb insider accounts by former senior officials of the Fed that we can draw on that have a strong para-ethnographic dimension (Blinder, 1998; Meyer, 2004). We have wonderful ethnographic work by Annelise Riles on the Bank of Japan (2004b, this volume) that not only provides acute comparative analysis but also serves as rich and creative examples of ethnographic experimentation in this formidable research setting. In addition, there are distinguished ethnographic treatments of traders and the financial markets in which they operate by Hiro Miyazaki (2003) and Caitlin Zaloom (2005, this volume) and outstanding examples, notably by Bill Maurer (1995, 1999, 2002a, 2002b), of how to address the theoretical issues at stake within distinctive configurations of global finance. Michel Callon and Koray Caliskan (2005) have compiled a comprehensive summary of the literature in economic sociology, with particular emphasis on the study of markets, as well as a provocative argument for reconstituting economic anthropology.
3. The crucial work that both legitimates intuition as a method of knowing and connects it to an essentially ethnographic, or cultural historian's, sensibility is Carlo Ginzburg's 1980 essay, "Morelli, Freud and Sherlock Holmes: Clues and Scientific Method."
4. Officially, a private organization, Business Cycle Dating Committee of the National Bureau of Economic Research, is responsible for declaring the onset and duration of a recession. See Abolafia (forthcoming).
5. We have struggled to do justice to Westbrook's magnificent text, to simply gloss what it is about, and we have repeatedly failed. Broadly, the text is about political economy and insofar as it works out the economic logic and legal conventions defining contemporary political economy, it is absolutely essential, but it is much more. Perhaps broader still, it is about the at times excruciating and at times exhilarating possibilities of intellectual life in our time. What follows is just one example of his technique of analysis and of argumentation. In this case, it is concerned with the character of social criticism available to us in the City of Gold and the difficult task of creating an "authentic," as he would put it, basis of political and moral inquiry in which we are all implicated.

> And here we come to a weakness that seems entailed in the doing of social criticism, and that therefore infects the criticism of monetization, and so this apology. Social critics depict ruin in part because they are social critics. As already suggested, social criticism attacks developments that are seen to be contrary to life as we have come to understand it, not so much our actual life

as our current ideals. Social criticism tells stories, always already begun, which if continued will destroy the meaning we have constructed. A coming must also be a going – to say that a situation has come to mean something is also to say that the same situation no longer means what it once did . . . Insofar as we are fond of the way of life in which we grew up – and because it made us who we are, we have to have a certain respect for it, whatever might be said against it from the outside – then we will tend to view such change as a bad thing, and we may even speak of dehumanization. Social criticism thus tends to rest, if only implicitly, on a sense of dislocation, a sense of losing a pattern of meaning, along with the past in which such meaning was formed. Social criticism is always, at bottom, a charge of impiety, betrayal, leaving home.

Money (like technologies and wars) dehumanizes because it changes patterns, destroys meanings that human life has built up so far. Capitalism, in its relentless commitment to the future, remakes home, for better and worse. And yet there are reasons that capital is attractive, that people choose to treat themselves and others as slaves. Any successful impiety has its attractions, which ideology presumes to state in general terms. Money provides, if in the abstract, the realization of our will, which most of us conceive to be happiness, and what could be more tempting? The dramatic expression of destructive conflicts between goods, such as old meanings and new temptations, is tragedy.

It is also important to remember, however, that humans are resilient. People make meaning all the time. Culture reconstitutes itself. While people leave and die, they also find one another and babies are born. As an intellectual matter, however, it is much harder to say much about things which have not yet come together, about meanings as yet unconceived. So the narrative structure of social criticism – the effort to understand this moment inevitably in terms of a past that we suspect we are losing (indeed, that is a condition of our mortality) – fosters the sense that things are falling apart, and that one ought to be alienated. Social criticism tends to spring from a partial vision, a formal, analytical, and ultimately tragic perspective, as opposed to an intimate, synthetic, and comic perspective. Knowing when one or the other perspective is appropriate would be wisdom. (2003, 285–286)

6. This constitutional process that defines the City can be observed in the technical negotiations over China's membership in the WTO or the recently completed negotiations by which Cyprus, the Czech Republic, Estonia, Hungary, Latvia, Lithuania, Malta, Poland, Slovakia, and Slovenia each gained membership in the EU. Through intricate diplomatic treaties, these countries became integrated into the global system of trade under the jurisdiction of supranational markets that increasingly superseded the regulative role and thereby the economic sovereignty of the nation-state.

78

Introduction: Anthropology Goes to Wall Street

Karen Ho

I first became interested in studying Wall Street on 21 September 1995.[1] AT&T had just announced that it would split into three different companies, engendering one of the largest dismantlings of a corporation in U.S. history: 77,800 managers received "buy-out offers" and 48,500 workers were downsized. Living in New Jersey at the time, I was dismayed by the extent of downsizing-induced worker trauma and even more troubled to hear that, on the first day of the announcement, AT&T stock leaped 6.125 points to 63.75, or 10.6 percent of its total value, "growing" another $9.7 billion. But what shocked me the most, upon further investigation, was that the stock prices of Wall Street investment banks also rose. What was the connection?

This seemingly counterintuitive relationship between AT&T's massive downsizing and its soaring stock price was not an isolated case. According to the *New York Times* on the day of the AT&T restructuring, during this period of increased merger and reorganization activity and in anticipation of future telecommunications industry restructurings, the stocks of the Wall Street investment banks that initiated, organized, and gave advice on these deals also rose. "Brokerage stocks were one of the session's stronger groups, with analysts citing their belief that volume growth is sustainable and merger activity will continue" (Sloane 1995). The stocks of Morgan Stanley, Merrill Lynch, and Lehman Brothers all rallied with the assumption that if AT&T, a

Source: *Liquidated: An Ethnography of Wall Street* (London: Duke University Press, 2009), pp. 1–38.

bellwether for the telecommunications industry, restructured and downsized, then other companies would follow that "economic fashion," thereby bringing in more business for Wall Street investment banks (Klein 2000, 199).[2]

For the past three decades it is precisely these kinds of inversions that have dominantly characterized the corporate landscape and the relationships among layoffs, corporate profits, and stock prices. In this period, which includes what has been proclaimed as the greatest economic boom in U.S. history (early 1990s–2000), the economy experienced *not only* record corporate profits and the longest rising stock market ever, *but also* record downsizings (O'Sullivan 2000). Research reports by Challenger, Gray, and Christmas, a Chicago outplacement firm, found that in 1994, 516,000 workers were downsized "when American corporations recorded their best profits in years; for 1995 it was 440,000 when profits were even better; and for 1996 and 1997 the totals were 447,000 and 434,000, respectively, when profits were better still" (O'Boyle 1999, 219). While the U.S. stock market, as measured by the Dow Jones Industrial Average, boomed from just above 4,000 points in February 1995 to over 7,000 in February 1997, then to 11,000 in 1999, job insecurity also spiked as corporations, on average, downsized over 3 million people per year (Oldham 1999; New York Times 1996). To give another example of this new cultural code of conducting business, in 1995, Mobil Corporation announced unprecedented earnings of $626 million for the first quarter, a reversal from a $145 million loss a year earlier, then a week later announced plans to eliminate 4,700 jobs. Wall Street analysts, reacting "enthusiastically to the news," praised Mobil's aggressiveness: they were pleasantly "surprised" when layoffs were not only higher than expected but also included refining and marketing personnel in the United States, who were paid more. Wall Street institutional investors demonstrated their confidence in Mobil by bidding up shares to a fifty-two-week high (Ritter 1995; Fiorini 1995).

What was so arresting about Wall Street's approach to corporate downsizing was its celebratory tone, its rejoicing in the very fact of corporate restructuring.[3] Throughout the mid-1990s, countless financial news articles demonstrated what seemed to be a new "structure of feeling." To continue with the case study of AT&T, a few months after it announced that it would fundamentally restructure and divide itself into three different companies, a move Wall Street analysts generally applauded, AT&T announced in January 1996 that it planned to eliminate forty thousand jobs over the next four years. According to the *Wall Street Journal*,

> The magnitude of the cuts stunned even some veteran AT&T-watchers. It broadly signaled that, for many major U.S. corporations, the eagerness and urgency for wholesale restructuring continues unabated. For AT&T, in particular, it also underscored that – even after trimming some 85,000 people in the decade since the breakup of the old Bell System empire – AT&T still employs far too many workers. . . . "This is a big number – a

> very, very big number," said Blake Bath, an analyst at Sanford Bernstein & Co. "It's a lot bigger than Wall Street had been anticipating." Wall Street responded well, sending AT&T shares up $2,625 yesterday to close at $67,375. (Keller 1996)

In fact, Wall Street was so excited about the magnitude of these layoffs that Salomon Brothers' infamous superstar telecommunications research analyst Jack B. Grubman thought it necessary to temper investor enthusiasm, cautioning that "investors shouldn't expect a huge jump in earnings from the cost cuts, as AT&T reinvests much of the savings to accelerate forays into wireless and local service." He did state, however, that "it's a good aggressive move, but the earnings impact going forward will be much, much less" (Keller 1996). In other words, AT&T, in the near future, would need to find even more ways of boosting its share price. Then, in March 1996, the company retracted its initial claim of forty thousand jobs cut, announcing it planned "only" eighteen thousand layoffs. An article in *USA Today* noted that "observers say AT&T deliberately inflated its initial layoff estimates to *impress* Wall Street, which sees job cuts as increasing profit. AT&T's stock price jumped almost 6% in the two days following the January announcement" (D. Lynch 1996, my emphasis).

While the desire for profit accumulation is certainly not new, what is clearly unique in the recent history of capitalism in the United States is the complete divorce of what is perceived as the best interests of the corporation from the interests of most employees.[4] Only twenty-five years ago, the public corporation in the United States was mainly viewed as a stable social institution involved in the steady provision of goods and services, responsible for negotiating multiple constituencies from employees to shareholders, and judged according to a longer-term time frame that went beyond Wall Street's short-term financial expectations to unlock immediate investment income (O'Sullivan 2000).[5] Today, in contrast, the primary mission of corporations is understood to be the increase of their stock prices for the benefit of their "true owners," the shareholders (that is, to create shareholder value). Employees, located *outside* the corporation's central purpose, are readily liquidated in the pursuit of stock price appreciation. Whereas, under the assumptions of post-Second World War welfare capitalism, workers struggled for and (sometimes) received their (unfair) share of corporate earnings, today even this traditional capitalist hierarchy has been largely eliminated such that employees often no longer benefit at all (or even suffer) when the corporation makes a profit. It is this new logic which I encountered on Wall Street that I investigate throughout this ethnography.[6]

In light of a celebratory Wall Street, what does the ostensible dominance of finance capital, which I would argue has centrally characterized our times, mean concretely? How might an in-depth ethnographic analysis of Wall Street investment banks, and the processes leading to their ultimate demise, provide a key to understanding the sea change occurring in corporate America? How do investment bankers actively *make markets* – that is, produce the dominant sensibilities of the stock market and Wall Street financial norms through their daily cultural practices?[7] How do Wall Street investment bankers negotiate the relationship between massive downsizing, shareholder value, and the production of market crisis, which leads not only to the overhaul of mainstream business values but ultimately to the liquidation (and reinvention) of Wall Street itself? Given my assumption that the measure of corporate health has something to do with employment – an understanding bolstered by my academic training and political beliefs and supported, I thought, by mainstream American culture – how could it be that a time of record corporate profits and soaring stock prices could also be an era of record downsizings and rampant job insecurity? More broadly, how have the severe social dislocations social scientists have usually attributed to global capitalism at large – the dismantling of corporate and governmental safety nets; the wave of corporate downsizings, mergers, and restructurings; the changing nature of what it means to be a successful worker; the growing concentration of wealth at the top; the social violence of financial booms and busts – been actualized? These questions propelled me to conduct fieldwork on Wall Street to investigate what role the stock market and investment banks played in these radical socioeconomic shifts.

Wall Street Habitus: The Cultural Production of Liquidation

My central purpose in this ethnography is to analyze both Wall Street's role in the reshaping of corporate America and its corresponding effects on market formations, *and* how Wall Street helped to instantiate these changes.[8] By Wall Street, I mean the concentration of financial institutions and actor-networks[9] (investment banks, pension and mutual funds, stock exchanges, hedge funds and private equity firms) that embody a particular financial ethos and set of practices, and act as primary spokes-people for the globalization of U.S. capitalism.[10] I examine in particular the relationship between the values and actions of investment banks, the corresponding restructuring of U.S. corporations, and the construction of markets, specifically financial market booms and busts. I ask how exactly Wall Street investment bankers and banks, at the level of the everyday, helped to culturally produce a financially dominant, though highly unstable, capitalism: what kinds of experiences and ideologies shaped investment banker actions, how they were empowered to make these shifts, and how these changes were enacted and understood to be righteous. As such, this book will contribute to contemporary anthropological understandings of the globalization of U.S. capitalism through an "on the ground" ethnographic approach that counters the widespread conception of capitalist globalization as an abstract metanarrative and homogenizing force too unwieldy for ethnographic translation.

Multiple key functions and institutions constitute Wall Street and the financial markets in the United States. Aside from investment banks there are asset management companies (hedge funds, pension and mutual funds, and private equity firms), and the securities exchanges themselves. Financial firms contain departments that deal with trading and sales, corporate finance, mergers and acquisitions (M&A), research, and investment management, each with slightly divergent goals, methods, and perspectives. From this broad landscape I chose as my primary field sites the central, iconic institutions of Wall Street, the major investment banks such as Morgan Stanley and Merrill Lynch. Within the banks I focused on those functions commonly considered investment banking proper – corporate finance and M&A – because they directly demonstrate the interconnections between financial and productive markets, between financial and corporate institutions. Through their middlemen roles as financial advisors to major U.S. corporations as well as expert evaluators of and spokespeople for the stock and bond markets, investment bankers work to transfer and exchange wealth from corporations to large shareholders (and their financial advisors), hold corporations accountable for behavior and values that generate short-term shareholder value, and generate debt and securities capital to fund these practices. It is in the activities of these corporate-finance-related departments that we most clearly see the imbrication of the productive economy, investment banks, and the financial markets. The work of Wall Street does not consist only of trading and exchange in the global financial markets; it is also linked to corporate restructurings and attendant shifts in corporate values.

It is important to understand both the connections and distinctions between Wall Street investment banks and corporate America. Although investment banks are corporations in an organizational sense with specific corporate cultures, they also possess a supplementary role as the voice of the financial markets, and claim to speak for millions of shareholders as well. They thus occupy a unique social position: investment banks represent both "the market" and the corporate entities that are subject to the market. Locating the supposedly abstract market in sites with particular institutional cultures localizes the market, demonstrates its embodiment, and shows how it is infused with the organizational strategies of investment banks.[11]

Despite the recent, much-proclaimed "death of Wall Street," it is important to understand that the financial culture that I generally denote by the term "Wall Street investment banks" does not always map neatly onto particular institutions. I use the designation to broadly signify an ethos and set of practices that continue to be deeply embedded in the intricate network of institutions, investments, and people we call high finance. Moreover, I argue that the supposed end of investment banking does not signal the permanent vanquishing of Wall Street ideologies or practices. Rather, instability and crisis fundamentally characterize this particular culture of liquidity, and signal not the decline but the influence of Wall Street values and practices. During the latest period of finance capital dominance of American business, Wall Street – the names, physical locations, and institutional identities and structures of almost all Wall Street investment banking firms – has continually transformed itself through mergers, acquisitions, bankruptcies, and failures.

This observation is methodologically and theoretically crucial to the argument of this book: it demonstrates that Wall Street investment banks have incubated, promulgated, and themselves undergone the very radical shifts they imposed and recommended for corporate America. The very particular cultural system that Wall Street has constructed and nurtured – one that promotes the volatile combination of unplanned risk-taking with the search for record profits, constant identification with the financial markets and short-term stock prices, and continual corporate downsizing – has not only been imposed on corporate America but also fundamentally characterizes and affects Wall Street itself. In fact, investment banks' self-effects are perhaps even more pronounced than their restructuring of other corporations as they consider themselves to be the incarnations of the market.

Wall Street's vanishing acts are indicative of its corporate culture; a glance at its recent history is telling. Even during the time of my research from 1996 to 1999, concurrent with the longest rising bull market ever, the institutional identities and locations of many of my central field sites radically shifted. For example, in 1997, Bankers Trust (BT) bought boutique investment bank Alex Brown and was in turn acquired by Deutsche Bank in 1998; BT was then merged out of existence. In 1993 Travelers Group, an insurance and financial conglomerate, bought Smith Barney, a brokerage and mid-size investment banking firm, then in 1997 purchased major investment bank Salomon Brothers to form Salomon Smith Barney; in 1998, Travelers merged into Citicorp to create the behemoth Citigroup, a bank holding company. In 2000, J. P. Morgan and Chase Manhattan Bank merged to create J. P. Morgan Chase. With the formation of Citigroup and J. P. Morgan Chase, major investment bank Salomon Brothers effectively disappeared, and J. P. Morgan, already a "hybrid" investment and commercial bank,[12] solidified its status as a bank holding company – which is also what Goldman Sachs and Morgan Stanley changed their status to in September 2008. Not surprisingly, the mergers and acquisitions (M&A) mania, the bull market, and the dot-com bubble of the late 1990s led, not only to the millennial crash and record downsizings, but also to the corresponding restructuring of a majority of the Wall Street investment banks I researched (see Map 1 and 2).

It is important to mention that while I argue that these dismantlings germinate out of the specific corporate culture of investment banking, massive governmental deregulation and the shareholder value revolution helped to catalyze this cultural system. The late 1990s Gramm-Leach-Bliley Act ensured the rollback of most of the stipulations of the Glass-Steagall Act, which was passed during the Great Depression to prevent deposit-taking

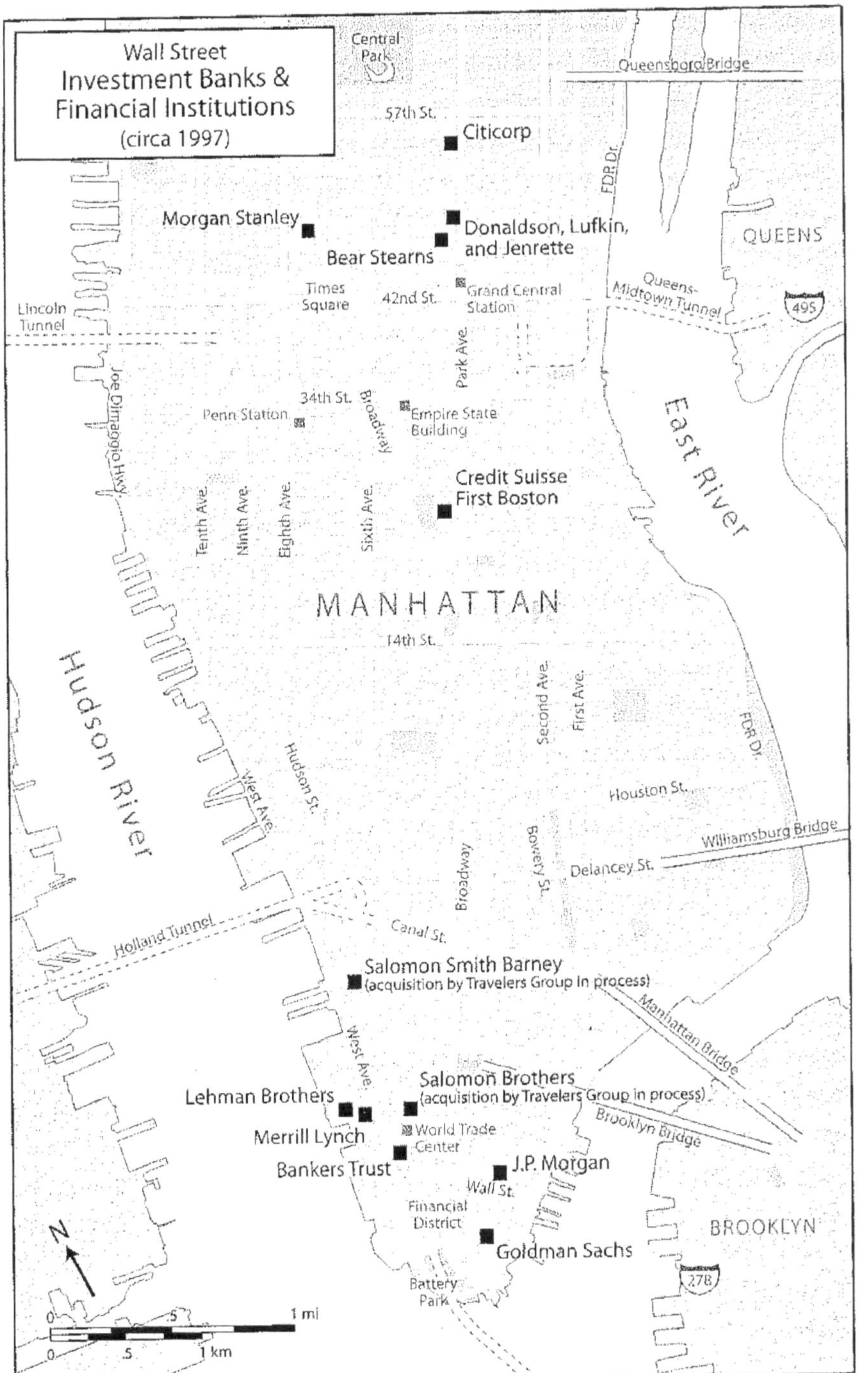

Map 1: The author's central field sites at the beginning of fieldwork. *Map by University of Minnesota Cartography Laboratory.*

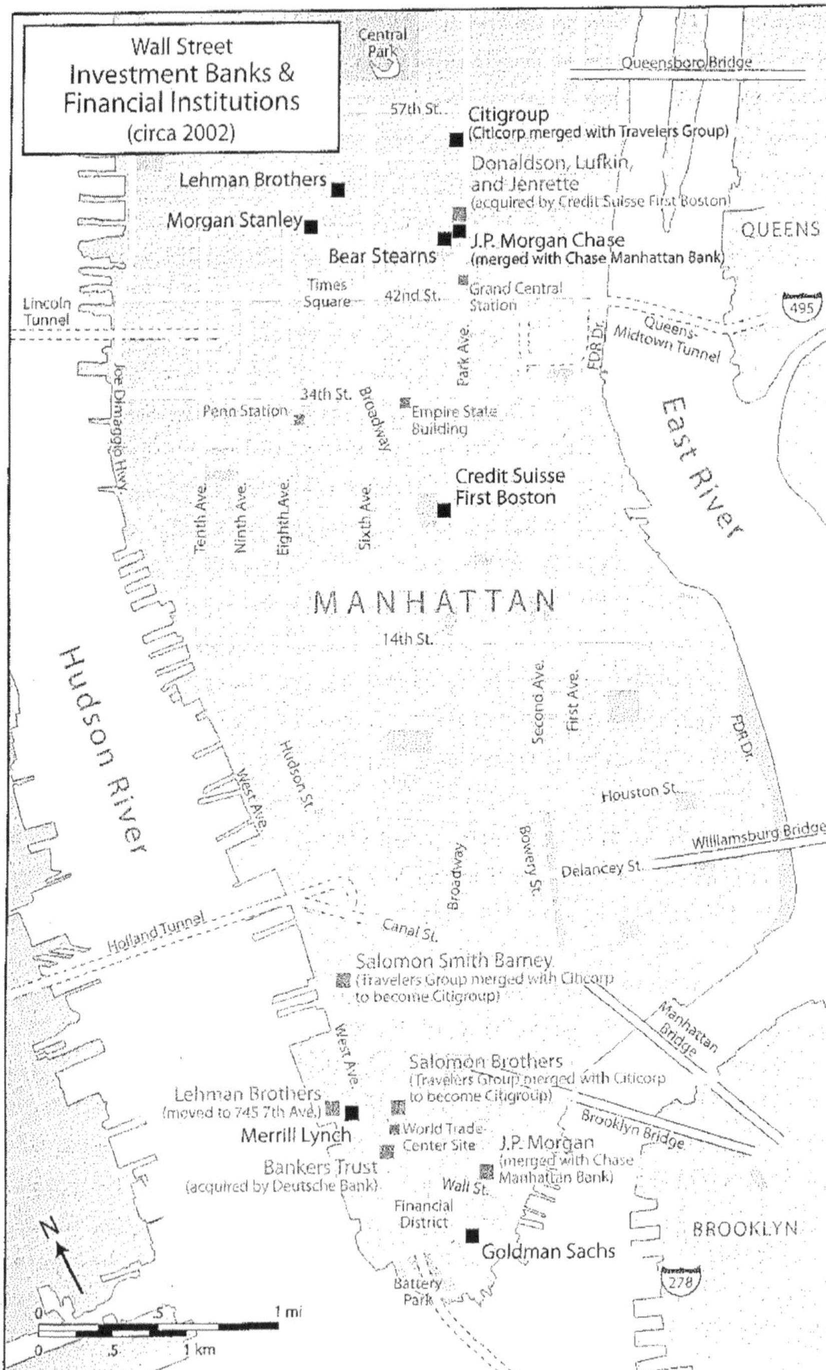

Map 2: After five years. *Map by University of Minnesota Cartography Laboratory.*

commercial banks from speculating in the financial markets. It had sought specifically to segregate investment banking, which engages in the "risky" capital markets, from the businesses of everyday banking and insurance, which are supposed to safeguard savings and provide loans. However, given the rise in stature, power, and profitability of Wall Street investment banks since the 1980s, retail and commercial banks – the titans of "low" finance – were eager to acquire investment banking capabilities and presence. Moreover, many investment banks were themselves pressured by the M&A boom to become gigantic, "one-stop shopping" institutions for corporate clients, which was deemed to increase shareholder value, not to mention the fact that senior executives received astronomical financial incentives for advising and participating in as many deals and transactions as possible. In this climate of regulation *for* the latest trends of the financial markets, investment banks participated in their own dismantling – that is, they relinquished a "purist" notion of investment banking – and financial institutions were allowed to "have it all," both commercial and investment banking ambitions.[13]

Not ironically, then, the very influence and success of Wall Street – its ability to globally market and proselytize both its products and its ethos – has generated not only record profits but also volatility, crisis, and a continual existence on the brink of annihilation (for itself as well as corporate America). For example, just as the M&A and internet bubble was the nineties analogue to the millennial subprime boom and bust, the leveraged buyout movement of the "ga-ga" eighties led, not only to the insider trading scandals, junk bond crisis, and stock market crash of 1987, but also to the bankruptcy of one of the largest investment banks of the time, Drexel Burnham Lambert, and the eventual liquidation of the investment bank Kidder Peabody. Over the past three decades, then, given the continued resurgence of Wall Street practices, I would argue that Wall Street's self-annihilation, even cannibalization, has not so much led to the disappearance of its particular cultural practices or power over American business as it has to constant financial crisis and ever widening socioeconomic inequality. Although the enormous scale and scope of what is currently being dubbed the worst financial crisis since the Great Depression – the federal bailout of Wall Street brought on by investment banks' engendering of the subprime debacle, which has in turn caused a global credit panic – is certainly extraordinary and has instigated a groundswell of reform and regulation, it remains to be seen whether Wall Street's particular investment banking ethos will disappear or will resurge in new and varied institutional forms, as it has regularly done since the 1980s. As Andy Serwer, managing editor of *Fortune* magazine, has observed: "The party is over on Wall Street – until it comes back again. . . . I've been around long enough to see that we have these cycles. These guys get their cigars and champagne. They have a great time. The whole thing blows up. But then they re-emerge years later. This one is a really, really bad one. But I don't think Wall Street is dead" (Boudreau, Fitzpatrick, and Zamost 2008).

The ostensible demise of my field site – Wall Street investment banking – in 2008, then, is not so much a radical turn of events as the fairly predictable outcome of a peculiar corporate culture. A central tenet of this ethnography is that the everyday practices and ideologies of investment bankers (which have solidified and gained currency since Wall Street achieved dominance over corporate America and global influence) continually set the stage for Wall Street's possible liquidation – and reconstitution. Yet this insight is rendered invisible precisely because Wall Street investment bankers as well as academic and popular analysts of finance often resort to an abstraction they call "the market" to explain these crises. Junk bonds, merger crazes, internet bubbles, highly leveraged housing meltdowns, and subprime debacles are mistaken for, and understood to be the organic results of, market cycles (what goes up *must* come down) with a dash of greed and hubris as human nature thrown in. As such, the construction of booms and busts are simply conflated with "the market" and are not understood as arising from the particular workplace models, corporate culture, and organizational values of Wall Street financial institutions (investment banks in particular) or the specific and personal experiences of those who work for them.

In this book, I make the case that accessing the central ideas and practices of investment banking culture allows us to unpack the very process of market making and will give us insight into the cultural workings of the so-called market. To do so, I take inspiration from Pierre Bourdieu's notions of "disposition" and "habitus," where "disposition" refers to a "way of being," "inclination," and "predisposition," often of the body, which collectively constitute the habitus, "a system of dispositions," which in turn organizes action, "produces practices," and constructs social structures and worlds (Bourdieu 1990, 73–87, 214). Specifically, I examine the structure and formation of investment bankers' habitus – how they have developed an investment banking ethos and set of experiences that frame and empower them to impose regimes of restructuring and deal making onto corporate America and, ultimately, help to engender financial market crisis. I demonstrate how, for example, the personal biographies of investment bankers play into, and converge with, job status and workplace experiences to shape a "commonsense" understanding of the righteousness of Wall Street analyses and recommendations. Recruited from elite universities and represented as "the smartest," investment bankers enter into a Wall Street workplace of rampant insecurity, intense hard work, and exorbitant "pay for performance" compensation. Forged in these experiences is a particular investment banker habitus which allows them to embrace an organizational model of "employee liquidity" and to recommend these experiences for all workers. Wall Street work environments, it turns out, are notorious for downsizing privileged investment bankers, even during bull markets, multiple times a year. To answer why constant corporate liquidation, including the downsizing of employees, has been celebrated and justified on Wall Street, it is necessary to understand how the people heralding downsizing themselves experience it.

And yet – Wall Street investment bankers understand the necessity of constantly performing in notoriously insecure work environments not as a liability, but as a challenge. Bankers, recruited as they are only from the Ivy League and a few comparable schools like MIT and Stanford, are trained to view themselves as "the best and the brightest," for whom intense deal-making through insecurity becomes a sign of their "smartness" and superiority as well as a way to cope with an anxious environment. Empowered by cultural capital, extensive elite networks, and an organizational structure of exorbitant compensation premised on numbers of transactions, investment bankers often successfully weather and negotiate (and create) crises until the next resurgence. They understand their lack of employment security as testing and developing their "mettle."[14] In this context of privilege and insecurity, investment bankers, on a practical level, are incentivized and learn to relentlessly push more deals (usually short-term transactions intended to boost stock prices) onto corporate America. By thus pressuring corporations, bankers transfer their own models of employee liquidity onto corporate America and set the stage for market crisis. Buttressed by the shareholder revolution, which I chronicle, Wall Street is empowered to shape and discipline corporate America, and such a relationship allows Wall Street to impose its own organizational practices – the very particular industry culture of banking – onto corporations at large. Instead of recognizing constant deal-making and rampant employee liquidity as their own local culture, my Wall Street informants conflated their organizational practices with their cultural roles as interpreters of the market and saw themselves, not as describing their own work and life circumstances, but as explaining "natural" market cycles and economic laws.

Bourdieu's models are prescient: these everyday practices of anxious deal-making and performance of smartness and market immediacy serve as the link between habitus and field, between the cultural structures and habits, anchored in individual's and group's bodies and minds, and the larger "fields" of social relations in the world. I thus approach the daily practices and corporate cultural values of investment bankers in the workplace as the site which links the cultural frame, dispositions, and habitus of investment bankers with broader U.S. corporate restructuring *and* the construction of financial market booms and busts (Bourdieu 1990). The very organization of this book – starting with biographies, delving into powerful ideological and institutional transformations, and ending with crises – mirrors this goal of demonstrating empirically how Wall Street's subjectivities, its specific practices, constraints, and institutional culture, dynamically intersect to constitute powerful systemic effects on U.S. corporations and financial markets, and beyond. It seeks to unpack the cultural labors necessary to construct a social and historical phenomenon – the dominance of Wall Street models and actors of finance capital.

It is not my intention, however, to assert that the Wall Street financial community, a heterogeneous site itself, single-handedly caused these massive corporate transformations, as multiple actors and institutions beyond investment banks, securities exchanges, and investment funds were critically involved, among them: governmental policies and federal "re-regulating" of corporations, a neoliberal resurgence in departments of economics and business schools, various crises in the rate of corporate profits and the choices made by corporate managers, the invention and popularity of new financial instruments from junk bonds to mutual funds, and the compromising of labor movements punctuated by racialized and gendered inequalities. It is therefore beyond the scope of this project to document all the myriad and complex conditions and practices that have enabled financial market values and actors to consolidate their influence over corporate America and have helped to solidify shareholder value as a dominant measure of business success.

Wall Street Institutional Culture: Access, Initiation, and Method

My particular strategy in "studying up," to break through the barriers of security and public relations, was based on institutional kinship.[15] To enable this research, I leveraged my socioeconomic background and connections with elite universities – the only sites from which Wall Street investment banks recruit and hire. I was first introduced to investment banking as a career option while an undergraduate at Stanford University, although admittedly Wall Street was not the destination of most of my network of friends (graduate school, nonprofits, and, of course, Silicon Valley were bigger draws). (When I began fieldwork, I also relied on Stanford alumni who worked on Wall Street as contacts and potential informants.) I then became a graduate student at Princeton University, a recruiting hotbed for investment banks. There, I had the opportunity not only to make contacts with alumni but also to participate in the job recruitment process itself. My path to Wall Street was made possible by the institutional, elite "familial" connections between particular universities and Wall Street investment banks, where alumni from prestigious universities have an inside-track into Wall Street. It is thus not far-fetched to argue that elite kinship creates a bridge or network to access finance capital. As Sylvia Yanagisako points out, framing kinship and family as dichotomous with, or external to, the very processes of capitalist formation ignores the centrality of the connections and sentiments of kinship that make capitalist production possible (Yanagisako 2002).

My particular strategy for accessing the elite lifeworlds of investment banking relied on research methods that blurred the boundaries of anthropology and sociology. I got a job at a Wall Street investment bank, a strategy with sociological precedent, while still following anthropological ethical norms, as I did not conduct ethnographic research covertly. In 1996, I participated in the spring career services recruitment process at Princeton. It was during the grueling interview process that I first recognized that anthropological discourses about globalization were not only appreciated by, but also similar to, those of Wall Street investment bankers. That June I was hired by Bankers Trust New York Corporation (also known as BT),[16] a "hybrid" investment and commercial bank, as an "internal management consultant" analyst, part of a group that acted as an "agent and advisor of change" for the different businesses within the bank.[17] Upon receiving the job offer, I told my boss (and when I began work, my colleagues) that I was taking a leave of absence from graduate studies to work at BT for two reasons. First, I was genuinely interested in learning more about the world of finance and acquiring "real world" work experience. Second, in the future I wanted to return to graduate school to study Wall Street culture. At BT, I was an employee first, friend second, and a fieldworker third; thus, while I learned much from my time there, my fieldwork mainly took place after I left my job. I did not secretly conduct fieldwork in my workplace; instead, in order to record the initial strangeness and surprises of the ethnographic encounter, the awakening into Wall Street life, I kept a journal of personal reflections and experiences, taking care not to describe in detail the thoughts and actions of my coworkers and friends who – although they knew of my research interests – were "on the job." As such, the experiences that I relate from this period are based on my observations and journal writing and not on any information that was considered private or proprietary.

As an "internal management consultant" at BT, I rotated through the many businesses of the bank to advise on a number of projects on strategy, workflow efficiency, operations, and change management. Although I did not work directly in the investment banking businesses (such as the corporate finance department), as a financial-services consultant within an investment bank, I was trained and immersed in the perspectives and mores of Wall Street financial practices. My role as a consultant allowed me to access and make contacts in multiple parts of the bank, as well as with young investment bankers throughout Wall Street. By participating intensely in conferences and financial networking events, as well as the after-work social lives of my coworkers, I also extended my contacts across multiple institutions. My goal for the year was to imbibe the general, taken-for-granted language and landscape of Wall Street, usually gathered during intense participant observation, as well as to establish a web of informants for "actual" fieldwork when I quit my job.

Living in a two-bedroom apartment share in the Cobble Hill/Carroll Gardens neighborhood of Brooklyn, New York, where I paid lower rent ($425 a month) than at Princeton, gave me the emotional sustenance to both work at BT and conduct this fieldwork. Although today my friends tell me that my old neighborhood is "the place to be" among young professionals,

Wall Streeters included (with one-bedroom apartment rents now pushing $2,000 a month), in the late 1990s it was precisely the fact that almost no one from Wall Street lived in my overwhelmingly Latino, white ethnic, and Arab American neighborhood that enabled me to put on a suit for three years and take the F train into a highly sanitized world of shareholder value proponents spouting the bull market hubris of Wall Street. I could come home to the smells of cubano sandwiches and pizza so juicy it had to be folded in half. Walking home around 8:30 p.m., I enjoyed making small talk with the porch and sidewalk crowd, who wondered why I always came home so late, looking quite forlorn, although my hours were early by Wall Street standards.

Toward the end of my time at BT, I was hitting my stress limit at work. Under constant pressure to vie for more and more projects in order to demonstrate my smartness, my capacity for hard work, and my ambition for deal-making responsibility, I was expected to demonstrate the desire for more money as evidence that I had properly imbibed these new sensibilities. As a graduate student used to living on $1000 per month, whose main purpose – seriously – was ethnographic knowledge and connection, I was reluctantly pushed into an alternate conception of space and time. Then the urgency (not to mention desperation) motivating everyone else hit me.

In January 1997, after a mere six months of employment in the Management Consulting Group (MCG) at BT, I was canned. In the midst of prefield-work, preparing myself to undertake research on investment banks' role in the downsizing of corporate America, I abruptly found myself downsized, along with many of my potential informants. The rationale given by BT for eliminating MCG was that we were a fixed expense that detracted from shareholder value. Management consultants internal to BT were on the permanent payroll, giving the bank a steady source of oversight and advice on business strategy and streamlining work processes. But key senior managers had decided that our work could be contracted out to external consultants. Our salaries were an unnecessary drag on shareholder value.[18] Two years later, BT itself was bought by the German monolith Deutsche Bank and was itself, as the lingo goes, "absorbed."

The moment appeared rife with ethnographic significance: financiers, the instigators of mass corporate restructurings throughout the United States, were downsizing themselves, BT's loss was ironically the anthropologist's gain. These seemingly mundane experiences of downsizing and job insecurity, everyday occurrences at investment banks, might yield crucial insights into the contemporary moment of financial crises and globalization. I learned that bankers' institutional culture and their own personal experiences of downsizing could reveal how the goals and practices of Wall Street not only reshaped global capitalism but also reverberated within. Investment bankers were subjected to, and suffered from, the same concepts and practices they imposed on others.

Two months after the announcement of my own downsizing, a vice president in my group, Lacey Meadows, called me into her office. She was leading "The Account Services Project" and wanted to inform me that she had chosen me to be her analyst on the project, her "right hand." "Do you want to know why I chose you?" she offered. When I shrugged, feeling slightly apprehensive about what was to come, she answered, "Because you are so nice; you're an anthropologist, and you can understand and empathize with how other people are feeling. We need someone they can talk to, and you'll be able to gain their trust." Needless to say, by the time she uttered "empathize," I was already dreading my soon-to-be role as potential collaborator, or put bluntly, her fellow axe man. I had just been called upon to streamline a "back office" department at BT that played a support role for the investment management "front office," processing trades, managing ledgers, and fielding customer accounts. Given the hierarchical structure of investment banks, front-office workers, such as investment bankers, traders, and investment managers who take credit for all profits and deals done, depend on the back office for daily support, all the while looking for ways to restructure these "cost centers." I knew immediately that I would resist; for me, this was the freedom of being a future fieldworker.

Specifically, I was asked to conduct a Taylorist time-motion study of their workday, actually charting and measuring the kinds of tasks and the time needed for completion to judge how many workers were necessary. I firmly intended not to recommend any downsizing. While interviewing the workers, who were overwhelmingly people of color and white women, I occasionally tried to assure them that my findings would bolster the necessity of their labor and protect their jobs. (Of course, I also recognized that given the role of investment bankers and consultants as agents of restructuring, this could have been viewed as trickery, as attempts to "soften them up") When talking to my team members (both of them higher up in the Wall Street hierarchy – a vice president and an associate with an MBA), I carefully asserted evidence that cast doubt on the very agenda of this project: why aren't these workers needed, who would do the job if they are "restructured," and does BT actually have a plan in place to make these changes? My fellow consultants, not to mention all the investment bankers whose reactions I solicited, were unclear as to why I defended the workers in Account Services. To say that Wall Street had little respect for back-office workers is an understatement. Although they were not openly disparaged, they were casually dubbed career nine-to-fivers; their work ethic was questioned, as was their smartness, drive, and innovation. Were they really "adding value," defined as directly boosting revenues or stock prices? The associate on the team wondered why we were wasting so much time there; in addition to being low prestige, the Account Services project was certainly not representative of the sort of financial deal-making he had hoped to be involved in.

At the same time, I was at a loss to explain why Wall Streeters, especially those in my group who were just downsized, were not more sympathetic to the concerns of these about-to-be-canned workers. It was as if these privileged bankers and consultants hadn't really lost their jobs, as if restructuring did not apply to them in the same way, as if the anxiety of constant insecurity did not cause them to doubt their financial prowess or desirability to employers. Exploring this quandary would take center stage during fieldwork. I would come to understand that, not only were investment bankers'-experiences of downsizing qualitatively different than that of most workers, but also that Wall Street investment banks' institutional culture helped to produce a model of workplace relations that was crucial to analyzing and understanding finance capital's approach to both American employment writ large *and* the making of markets. Given that recent work in social studies of finance has demonstrated that economists and financial models and theories, not only describe and analyze financial markets, but also perform and produce them (Callon 1998; MacKenzie 2006), it thus seemed a distinct possibility that studying the personal crises of investment bankers could provide a unique insight into the production of corporate restructuring and financial crises. Far from being only the perpetrators, investment bankers have themselves been subjected to the revolving-door model of employment that they recommend for other workers, although their particular mix of privilege, pedigree, compensation, and networking affords them a very different "lesson learned" from the experiences of downsizing and insecurity. By investigating Wall Street's own culture of employment, I examine how and why elite financial actors have over the past few decades radically altered the nature of American corporations, ultimately helping to shape a world of socioeconomic inequality, insecurity, and crisis.

As for my role in the Account Services project, after many sleepless nights and acne breakouts worse than a teenager's, I created a spreadsheet demonstrating that although communications channels and repetition could be reworked to flow better and some discrete tasks could be combined, there continued to be as many labor hours needed to accomplish necessary work as there were people. In other words, I concluded that the amount of labor equaled the number of people employed. Upon reviewing my spreadsheet, my vice president inquired, "How come everything fits together so neatly! Are you up to something?" I replied, "BT doesn't have its own house in order; we just got downsized, so I think it would create more inefficiency to restructure Account Services than any real long-term savings." In the end, Account Services was not downsized. I suspect it had less to do with me – though our group offered BT no justification for such action – than with Wall Street investment banks' notorious lack of planning and follow-through even in their search for short-term gains.

A year after my downsizing,[19] I undertook seventeen months of field-work, from February 1998 to June 1999, among differently positioned investment bankers working at most major Wall Street financial institutions. I drew on university alumni connections and a web of contacts that I had made during my prefieldwork job, which allowed me access to a number of investment banks as well as other fieldwork sites, from bars to outplacement agencies, conferences to panel discussions. I conducted participant observation and over one hundred interviews. Had I relied solely on my alumni contacts to create a network of informants without getting a job, my toolkit would have been mainly limited to interviews. Had I relied primarily on my job and contacts at BT, I would have learned the language and mores of finance, but my ethnography would have been contained within the walls of one investment bank and could not have addressed Wall Street as a broad occupational community.[20] Since the goal of my research was to analyze key financial agents making markets and their effects on socioeconomic inequality, I designed a methodology that combined immersion with movement, broad enough to access Wall Street worldviews and practices, yet particular enough to understand how such norms were constituted on a daily basis within particular institutions. Given that the fieldwork process constitutes and constrains one's ethnographic findings (methodology is theoretical after all), in this field space, I was able to focus my study not only on "the interior lives of experts as an elite as such, but rather to understand their frame . . . a project of tracking the global, being engaged with its dynamics from their orienting point of view" (Holmes and Marcus 2005, 248).

Partly planned and partly fortuitous, my access to Wall Street lifeworlds resulted from exploring and combining multiple sites and techniques of fieldwork, similar to what Hugh Gusterson has called "polymorphous engagement." He de-emphasizes participant observation as an often impossible method in studying up. Instead, he writes, the ethnography of the powerful needs to consist of "interacting with informants across a number of dispersed sites, not just in local communities, and sometimes in virtual form; and it means collecting data eclectically from a disparate array of sources in many different ways [such as] . . . formal interviews . . . extensive reading of newspapers and official documents . . . careful attention to popular culture," as well as informal social events outside of the actual corporate office or laboratory (Gusterson 1997, 116). This is not to say that immersion is no longer an indispensable anthropological staple in a varied toolkit; rather, such a methodology has been constituted via particular ways of imaging culture and the proper anthropological subject. The very notion of "pitching tent" at the Rockefellers' yard, in the lobby of J. P. Morgan, or on the floor of the New York Stock Exchange is not only implausible but also might be limiting and ill-suited to a study of "the power elite."[21]

Although I was able to incorporate substantial participant observation during fieldwork, the majority occurred during "prefieldwork" since I did not (nor was it my goal to) obtain investment banks' official permission to "hang out" within their workplaces. Such a strategy would have led me directly to their public relations office. During "actual" fieldwork, I relied mainly on interviews, some "shadowing," and attendance at industry conferences, panel discussions, formal networking events, and informal social events. The circumstances, connections, and affiliations that allowed (as well as circumscribed) my access to potential informants are instructive to lay out. First, because of my job at BT and subsequent experience of downsizing, I had little trouble creating a sizeable network of informants, especially through the process of direct referral. Employing this method through Wall Street insiders might normally have led me to a rather homogenous, mainly white male crowd. But, as it happened, I was a member of the internal management consulting group (MCG), one of the most diverse departments in the front and middle offices of BT. The head of MCG was an African American managing director who made a concerted effort to build a group with half women and one-third people of color. My coworkers' referrals often reflected this diversity. For example, through our head, I met Thomas Douglass, an African American managing director at BT who in turn introduced me to Corey Fisher, an African American managing director at Vanguard Investments, and Roy Allen, a white managing director at Fidelity Investments. Julie Cooper was a former white associate in my internal management consulting group, yet when we were all downsized, she secured a job in the high-yield investment banking division and introduced me to her entire team, which included John Carlton, a white managing director, Christine Chang, an Asian American vice president, and Chris Logan, a white analyst. Luckily, these connections often snowballed.

I also relied on alumni and friendship networks. While at Stanford, my social network did not include many business types, but I was actively involved in Ethnic Studies organizing as well as various Asian American student organizations. I also lived in both the Asian American and African American issues "theme" dorms. As such, I had over the years formed a handful of fairly close relationships (on my own and through friends) with a few professional Asian American and African American informants such as Joseph Tsai, associate at Donaldson, Lufkin and Jenrette (DLJ), Malinda Fan, senior vice president at Lehman Brothers, Joannie Trinh, an associate at Morgan Stanley, Raina Bennett, an analyst at Lehman Brothers, and Jason Kedd, an associate at DLJ. Many of them referred me to their bosses (most of whom were white), as I was eager to hear from more senior bankers, as well as their friends (many of whom were quite diverse). Unfortunately, at Princeton, despite the fact that almost 40 percent of the undergraduates go to work on Wall Street, as a graduate student I hardly knew any of them. Furthermore, since Princeton does not have a business school, I was unable to network via the MBA program, and since most of my Stanford acquaintances had not yet finished business school in the late 1990s, I could not tap into a potentially rich network of elite MBA graduates – although, for example, Jason Kedd did introduce me to two of his Harvard Business School classmates who worked in investment banking. There were of course acquaintances of mine from both Stanford and Princeton who knew me as a student activist and feminist studies/anthropology major and were suspicious of my research: one had established a vast network of Wall Street investment bankers, but proved reluctant to help.

My attempts at "cold-call" socializing with investment bankers while living in New York City were hit and miss. In some cases, I found some affinity with fellow Asian American women and men, especially those who were struggling with issues of what it meant to be an Asian American professional (expectations of upward mobility, relative class privilege, racial discrimination and stereotyping at work, bicultural identity formations), and they often agreed to be interviewed. In other cases, I left large events or conferences with rich, informal anecdotes gained from chatting, yet no one had been willing or able to sit down for further conversation. In one unexpected encounter that happened while volunteering for an economic justice organization in New York City, I met a white feminist minister whose partner, Jacob Carnoy, used to work on Wall Street. It turned out that he had graduated from Princeton in the 1980s and offered to introduce me to "fellow Princeton grads" such as two senior investment bankers at Goldman Sachs who, according to him, were "complaining that they only made 20 million that year." Unfortunately, I was only able to converse with them via e-mail. Finally, through my participation in SEO (Sponsors for Educational Opportunity, which focused on supporting minorities from elite campuses to enter investment banking and management consulting) conferences, I became close friends with a small network of young African American investment bankers. Through these various interconnections and chance encounters, I was able to assemble a diverse financial crowd at multiple levels of the Wall Street hierarchy. Around 40 percent of my informants were people of color, with a slim majority of them men.

As I have footnoted, all of my informants have been given pseudonyms, and although I struggled with whether or not to disguise the financial institutions themselves, I ultimately decided not to. First, this project focuses on Wall Street investment banking culture broadly conceived; as such, I am interested in general ethnographic data, not information on specific banks that might be considered proprietary. Furthermore, precisely because I seek to confront and unpack powerful globalizing institutions that also claim to speak for the markets and corporate America, it makes sense to name these institutions and call attention to their pronouncements, strategies, and influence, as I do with the speeches and announcements by Wall Street CEOs and senior management made during major events and conferences covered by the press. It is also instructive to note that providing pseudonyms for Wall Street financial institutions is practically a futile exercise given the prevalent cultural norms of the financial market where corporate names statuses, and identities constantly shift over time. For example, in 1997, Dean

Witter Discover, a retail brokerage, merged with prestigious investment bank Morgan Stanley to form Morgan Stanley Dean Witter Discover; yet, by 2001, to reclaim prestige and name recognition, the firm, which had already dropped the name "Discover," renamed itself Morgan Stanley.

Shareholder Value, Decentering Privileged Models and Histories, and the Politics of Ethnographic Representation

When I rode the subway to the field,[22] determined not to allow "the study of the stock market" to be "left to economists" (Hertz 1998, 16), I was bombarded by images and representations of the United States as a nation of stockholders and investors, and of the stock market as a populist site of economic empowerment for all Americans. President Bill Clinton gave numerous speeches about the "New Economy," locating our unprecedented national prosperity on the shoulders of the longest bull market in history and on the fact that more Americans owned stocks than in any other time (albeit largely through pension and retirement funds). The White House's figures for stock market ownership hovered around 150 million Americans, and the rising stock market was seen as a primary indicator of the improved economic lives of most Americans. A cover of *Fortune* magazine in 1999 proclaimed America "a Trader Nation": "At work, at home, all day, all night: Everybody wants a piece of the stock market." Inside, the article claims that "there's a revolution under way, and it's changing the way we invest and work and live. Our money is no longer with some broker or fund manager. Our money is with ourselves" (Serwer 1999, 116). We see an illustration of four fists, three clutching computer mice, tearing down a sign that reads, "Wall St." Such representations of "revolutions" were rampant in the late 1990s (Serwer 1999, 118).

In Wall Street's new rhetoric of market populism via shareholder value, "each mass-market success by a bank or brokerage [was declared] a victory for a democratic 'revolution'" (Frank 2000, 125). To take just one example, Thomas Frank describes how E*Trade appropriated the language and imagery of the civil rights and feminist movements. "In its 1999 annual report, entitled 'From One Revolution to the Next', E*Trade used photos of black passengers sitting in the back of a bus," with a caption reading "They Said Equality Was Only For Some of Us" to signal their role in the destruction of exclusionist and elitist Wall Street and the mass triumph of the individual investor (Frank 2000, 91). Wall Street thus allied itself with the "common people," constructing a pro-Wall Street populism and incorporating the disenfranchised into a cool new image opposed to its older, stodgier self.

In parallel, I also encountered Jesse Jackson's "Wall Street Project," a division of his Rainbow/PUSH coalition based in New York City, articulating a strategy of incorporating African-Americans and other excluded groups into the stock market. This project, founded in 1997, began as a coalition challenging Wall Street on three fronts: diversity and representation in corporate America and on Wall Street, stock market democratization, and access to capital for "inner cities" and Appalachia.[23] In his speech at the Wall Street Project Conference held at the World Trade Center in 1999, Jackson posed the question, "Why are African Americans continuing to 'invest' in the bear lotto when they need to be included as participants in the bull market? Why are our youths buying hundred dollar Nike shoes instead of Nike stock?"[24]

Perhaps not so ironically, it was precisely at the moment when Jackson advocated the incorporation of marginalized communities into the so-called shareholder value revolution that my Wall Street informants began to suspect the impending burst of the bubble. Many subscribed to the old Wall Street adage: When cab drivers start asking for investment advice and stock picks, its time to get out of the market. As Wall Streeters understand it, by the time stock market knowledge seeps to the masses, the bull market has turned into a bubble economy. This assumption only makes sense, of course, if success in the stock market depends on a delicate balance of insider knowledge, market hype, and timing.[25] Wall Street, then, views the democratization of stock market participation as a bell-wether of oversubscription and as a signal for insiders to sell, meaning "latecomers" to the market tend to bear the brunt of crashes.

Despite Wall Street's and corporate America's proclamations about putting shareholders' interests above all other constituencies of the corporation, the very practices that constitute the shareholder value repertoire do not necessarily enrich owners of corporate stock or empower shareholders to make corporate management decisions. On 25 February 2002, chronicling the continued dot-com stock market bust, *Business Week* ran a cover story on "The Betrayed Investor," which documented how the "true believers" in the stock market – the new investor class of the 1990s, comprised of "predominantly middle-class, suburban baby boomers" – had lost "$5 trillion, or 30% of their stock wealth since the spring of 2000" and were now beginning to doubt that the stock market "treats average investors fairly" (Vickers and McNamee 2002, 105).[26] The quintessential case is Enron, where shareholder value was proclaimed by Enron senior executives, Wall Street investment banks, and accounting firms such as Arthur Anderson as the central goal, yet their actions mainly benefited themselves. To keep the stock price "artificially" high, top management, who were paid via stock options (and sold their shares before the crash using inside information), worked with investment banks, who received millions in advisory and transaction fees, to find and invent new financial structures and "hypothetical transactions" to both project windfall profits and keep the massive debts off the balance sheet. When Enron imploded, not only did employees lose their jobs and savings, but the investing "public," that is, the shareholders, lost an estimated $200 billion (McClean and Elkind 2003).

If the shareholder value revolution was not sustainably enriching the average investor, who may also have been facing unemployment as his or her 401(k) appreciated a few hundred dollars (if that), then was Wall Street not increasing the stock price of corporations – their stated central mission? When I worked at BT, I certainly heard of corporations drastically cutting costs, whereupon their quarterly earnings and stock prices immediately jumped, but research and development suffered, productivity gains were negligible, and shareholder value over a longer time horizon did not increase and even declined. But it was during fieldwork that the full contradictions of shareholder value hit home.

To return to my earlier discussion of AT&T: the little-known backstory to the massive breakup of AT&T in 1995 was the disastrous acquisition of National Cash Register (NCR). This prehistory is crucial because four years prior in 1991, AT&T, also under the advisement of Morgan Stanley (a connection buried after the failure of this deal), acquired NCR in a hostile takeover for $7.4 billion (Zuckerman 1995). This aggressive act not only generated massive downsizings and insecurity in NCR's hometown of Dayton, Ohio (once a stable, thriving "company town"), but was also an utter disaster for AT&T, which "lost a half-billion dollars in the first nine months of 1995 alone" (Rimer 1996). In a desperate move to push up its stock price, CEO Robert Allen decided to spin off the ill-conceived purchase of NCR in the "trivestiture" of 1995. Yet, by 1997, less than two years after this "bold" breakup, AT&T's stock was performing dismally, trading at $33.625 a share. Allen stepped down in the face of persistent questions about the profitability of long-distance service in the new economy, which the restructurings in 1991 and 1995 did not address but exacerbated. Allen, however, continued to defend "the breakup on the grounds that AT&T unlocked almost $40 billion in shareholder value," insisting that "the whole idea was to avoid [its] destruction" (Landler 1997).

Morgan Stanley's advice, intended to bolster shareholder value, actually damaged AT&T's stock price in the long run, despite the fact that this deal making helped to generate an explosion of wealth for shareholders primed to cash out during the short-term price spikes. It is important to remember that investment bankers always receive high compensation for the deal *no matter the result*. The higher the risk, the bigger the deal, the more radical the change, the more money Wall Street makes, even though a merger or restructuring of a large company is precisely the kind of transaction that leads to a deterioration of long-term shareholder value. Many mergers have failed to deliver the expected profits due to unforeseen difficulties in integrating the companies.[27]

Despite the missteps that AT&T had taken under the advice of Morgan Stanley, in 1998, Credit Suisse First Boston (CSFB) and Goldman Sachs advised new AT&T CEO Michael Armstrong in the acquisition of TCI *and* the subsequent purchase of MediaOne, both among the largest cable television and modem companies in the United States. These deals were both named "Merger of the Year" by Wall Street trade magazine *Investment Dealer's Digest* (for 1998 and 1999, respectively). Although AT&T's acquisition spree in the late 1990s is generally blamed as much on the trigger-happy, "acquisitive" Armstrong as on his Wall Street advisors, it is crucial to note that the investment banks made over $100 million on these mergers, and that by 2000, AT&T announced that it would once again split into four separate companies, thereby breaking up and undoing the previous two years (Stokes 2000; Waters 1999). Scott Cleland, "tele-communications analyst at . . . an independent research firm," called the deals "a multibillion-dollar oops" (Hiltzik 2001). Armstrong spent over $112 billion in acquisitions and expansions, yet by 2002 had losses of $60 billion to show for it. AT&T's stock was at $13.51 in 2002, "well below their 1999 high of $49.77," and by April 2004, the once-storied AT&T was actually delisted from the Dow Jones Industrial Average (S. Lynch 2004).

A central question for this book, then, becomes: why do Wall Street investment banks continually fail to achieve their raison d'être? How can investment bankers do what they do and engage in seemingly irrational practices, such as proclaiming shareholder value while engaging in actions that not only undermine it but produce corporate and financial market crisis? How does shareholder value maintain cultural legitimacy despite the inconsistencies and failures of its champions?

Addressing such questions is complicated by dilemmas in ethnographic representation and methodology encountered when studying the powerful. For example, part of the crisis of representation and the turn toward greater anthropological reflexivity was the realization that anthropologists were unable to see "the objects of social analysis" (usually the marginalized) as "also analyzing subjects"; furthermore, ethnographers did not sufficiently locate or recognize their gaze, influence, or dominant authorship (Rosaldo 1989, 207). These historical mistakes, however, only make sense with a "typical" power configuration in mind, such as that of the Western white upper-middle-class male anthropologist studying the less-powerful "native" in the context of the colonial and postcolonial encounter. Underlying this politics of representation is the notion of the invisible expert and the marked subject. To rectify these mistakes, many anthropologists, in their attempts to "give voice" to their subjects, "run the risk of merely shifting the burden of representation from the anthropologist to the subject," making the representations of subjects the "privileged 'reality' " (Yanagisako 2002, 48). Given the historical imbalance between the native's voice and that of the ethnographer, this strategy is understandable. However, when one is studying the powerful, this approach is no longer appropriate, as it can overprivilege the informant.

My informants' representational power, influential models, historical interpretations, and prolific theorizing on capitalism, markets, and globalization rendered their voices loud and clear, even in relation to the anthropologist studying them. The task is *not* simply to "recognize" Wall Streeters as subjects acting in the world, but rather to situate and critique Wall Street worldviews in relation to other cultural models, histories, and voices. For instance, throughout much of my fieldwork, I was myself entranced by Wall Street's dominant model of shareholder value. Delving into the intricacies of bankers' proclamations and explanations of shareholder value as objects of importance in and of themselves, I did not always locate these discourses within the corporate culture through which they speak. Eventually I came to realize that Wall Street investment banks do not so much enact an ideal or model of shareholder value as perform a particular version of it, mediated through their corporate culture and experiences as investment bankers – one that in many cases actually undermines their very proclamations of shareholder value.

Similarly, I found that part of what imbues the presentist shareholder value ideal with such explanatory power is its rootedness in dominant historical narratives that legitimize Wall Street's identity as guardian of shareholder value and empower its role and practice in shaping corporate America.[28] The use of shareholder value is part and parcel of a broader project of laying claim to a restorative narrative of entitlement and succession, through which Wall Street investment bankers have been able to define their professed beneficial social contributions to our economy. Wall Street cultural legitimacy and shareholder value are naturalized through "origin myths," particular interpretations of neoclassical economic thought, and investment banking histories of shareholder rights.

Specifically, my informants viewed themselves as gatherers and purveyors of the capital that forms the foundations and enables the growth and expansion of our largest corporations and public and private works. Not surprisingly, much of the academic literature on Wall Street, in disciplines ranging from business administration to law, replicate and perpetuate investment bankers' assumptions about themselves and the history of their profession. The presumption is that investment banks financed the very creation of the U.S. corporate system and have throughout history been the primary suppliers of fresh capital to maintain and expand it. So ingrained are these notions that even the project of delineating "what Wall Street actually does" proved a much more intricate endeavor than I had realized precisely because most of the literature (popular, "native," and academic) about Wall Street is framed by this dominant, taken-for-granted narrative of its roles and responsibilities, one which relies on a particular interpretation of American corporate and financial history. Michael Jensen's idealized portrait of investment bankers in *The Financiers* (1976) is indicative of mainstream culture's approach to Wall Street as both highly secretive and esoteric as well as crucially important for the functioning of our capitalist system:

> They are the elite of Wall Street. Their offices are furnished with expensive antiques and original works of art. They dress in conservatively cut $500 suits, and are as quick to place a telephone call to Rome or Zurich or Frankfurt as most Americans are to call their next-door neighbor. . . . They engineer multi-million dollar transactions and, although they render middleman services only, enough-money remains in their hands to make them the richest wage earners in the world. They are the investment bankers of Wall Street; the men who raise billions in cash for America's giant corporations. . . . Unknown to the general public, as dissimilar to the neighborhood savings banker as Harold Geneen of ITT is to the owner of a neighborhood delicatessen, the investment banker is an invisible man. Most Americans have only the vaguest notion of what he does. . . . His art is arcane. But just as the rainmaker promised to draw from the sky the drops that nourished the farmers' crops, so these latter-day rainmakers draw from the people and institutions around them the dollars that are needed to build the nation's factories. . . . What is investment banking? Simply the art of raising money for a client. (Jensen 1976, 1–2)

One of the key ways in which Wall Street investment bankers control their present and future representation as a fountain of capital that "build[s] the nation's factories" is to strengthen their hold on the past. Such a storied role allows them to spin the pursuit of seemingly selfish goals as, in the end, a force for social good, arguing that the incentive of personal gain leads to an efficient economy, greater innovation, and better jobs.

To address the inconsistencies, failures, and ramifications of Wall Street investment banks' approach to shareholder value while paying attention to its explanatory power necessitates a multiprong approach that decenters the models, histories, and discourses of Wall Street. First, to interrogate the gaps in their dominant models of shareholder value, it is crucial to build on research in social studies of finance on the mundane cultural details and failures of finance, as well as on finance's productive effects and the (often unpredictable) relationship between financial models and their instantiation (J. Guyer 2004; Maurer 2002; Miyazaki 2003, 2006; Miyazaki and Riles 2005; MacKenzie 2006). As Jane Guyer (2004, 4) points out, "people's exchanges" are not only constructed out of "essential or archetypical" models or transactions, but also through "popular conventions" and "market experience," thus the importance of examining the interface between model and effect. As Hirokazu Miyazaki and Annelise Riles (2005, 320) observe, given the "series of high-profile financial failures" and continued examples of financial market collapse, for "market participants," "the failure of economic knowledge to predict, plan, and regulate the market seems self-evident." The question for my purposes, however, is *how and why* my informants continually construct failure, not only against their own stated values of shareholder value, but also in such a way that promotes and validates their own job insecurity.

Drawing from the scholarship on models and performativity in social studies of finance also allows for the productive analysis of both the powerful influence *and* the failures of shareholder value and other dominant models and ideals on Wall Street. Given that these theories and models are "our culture's most authoritative forms of knowledge of financial markets," it is crucial to interrogate *how* they perform and what they actually do in the world (MacKenzie 2006, 275). Donald MacKenzie and Bill Maurer point out that financial models, in my case the ideal of shareholder value, both shape the construction of markets *and* have their limitations. "Keeping performativity in mind reminds us also to ask: if the model is adopted and used widely, what will its effects be?" Further, are these models "accurate" and "what sort of a world do we want to see performed" (MacKenzie 2006, 275; Maurer 2006)? Given the limits of the shareholder value model and the gap between the ideal and the effects, my approach to shareholder value inquires what other financial cultural values and norms, beyond the "most authoritative" and dominant, shape Wall Street investment banks' restructuring of corporate America and financial markets.

Second, to directly investigate why and how these particular representations of Wall Street hold such explanatory and naturalizing power, I approach Wall Street historiographies of shareholder entitlement and Wall Street's conception of itself as fundraiser to the world as origin myths, indicative of a particular worldview and socioeconomic interest rather than objective statements of fact. Articulating the relationship between myth and social practice, Bronislaw Malinowski (1948, 96) writes that "an intimate connection exists between the word, the mythos, the sacred tales of a tribe, on the one hand, and their ritual acts, their moral deeds, their social organization, and even their practical activities, on the other." I document how Wall Street, through strategic alignments with long-standing neoclassical desires entrenched in American cultural norms, evokes nostalgia to construct a "restoration" narrative central to its "rightful" succession.

In a similar vein, to particularize Wall Street voices and global claims and decenter their realities, I analyze investment bankers' narratives of shareholder value alongside the narratives of those in the business community who have historically struggled with Wall Street's finance-centric approach to corporations.[29] Delineating the differences and contestations *between* capitalists serves as a foil to better situate and historicize investment bankers' universalizing and taken-for-granted assumptions. By examining the contestations between competing capitalist worldviews and practices of the postwar era, I write against assumptions of a singular, static, totalizing capitalist worldview, promulgated by homogenous capitalists. In her ethnography of Italian capitalist family firms, Yanagisako points out that in studies of the powerful who have considerable access to interpretive control over the terrain of meaning and self-representation, it is pivotal to go beyond their "official" versions in order to "supplement, challenge, and interpret" their dominant and authoritative discourses with other understandings.[30] At issue in the task of "describing" Wall Street is why particular narratives and social groups have come to be dominant and what is at stake in their varying visions of the world.

Finally, the politics of ethnographically representing and interpreting the powerful goes beyond decentering their models and histories; how their voices are methodologically accessed also shapes ethnographic findings. In a "studying up" context, for example, interviews are oftentimes the most accessible form of evidence: in my case, most of my written ethnographic data is in the form of interviews drawn from prearranged meetings with investment bankers, supplemented by field notes based on participant observations and my own experiences working on Wall Street. As such, much of my recorded fieldwork focused on the "talk" of investment bankers, which I both embrace and problematize. On the one hand, Michel de Certeau (1984, 77) writes, "If the art of speaking is itself an art of operating and an art of thinking, practice and theory can be present in it." He further explains that "stories" serve to "authorize" and "found" a set of social practices, to delimit and carve out "a theater of actions" (Certeau 1984, 123). Similarly, Teresa Caldeira (2000, 19–20, 78) demonstrates that the "talk of crime," through daily "repetition of histories" and stereotyping to "establish order in a universe that seems to have lost coherence," resignifies "segregation and exclusion as central logics of urban space and movement," reestablishes "a static picture of the world" that is "expressed in very simplistic terms, relying on the creation of clear-cut oppositional categories," and shapes "the scenario for social interactions." I argue similarly that Wall Street's narratives of shareholder value resignify the business landscape, creating an approach to corporate America that not only promotes socioeconomic inequality but also precludes a more democratic approach to corporate governance. Banker talk of shareholder value simplifies corporate history, limits others who may have claims on corporate profits, and forecloses a range of more equitable corporate practices. Just as the "talk of crime" Caldeira encountered was "not meant to describe the world accurately but to organize and classify it symbolically," bankers' shareholder value discourses do not reflect the complex histories of the struggles for corporate resources, but rather reorganize corporate history and values such that certain interests hold a monopoly on corporate decision-making and profits. Through circulation and repetition, these stories "delegitimate" the corporation as a social institution and "legitimate" the corporation as a private investment vehicle for the few (Caldeira 2000, 38).

On the other hand, the focus on privileged talk can be methodologically limiting, as it was often easy to detach what my informants said from what they did; such was the nature of interviewing powerful informants without a corresponding referent and constant access to participant observation. Focusing on the discourses, logics, justifications, and talk of shareholder value without a corresponding analysis of how investment bankers actually go through with it – the practices through which a particular version of shareholder value comes to have power – is not only ethnographically flat but also assumes the self-actualization of a self-serving model and discourse of the world. Recognizing that cultural effects and production are never as neat as their powerful, legitimating logics, I worked to include and juxtapose multiple observations, field notes, situations, and experiences – which I had access to due to the particular combination of my network, Wall Street job, and fieldwork – into the same analytic frame. I thus realized that embracing the interfaces, gaps, and frictions[31] constitutive of "worldviews-in-action" was crucial in examining the cultural productions and influence of Wall Street. Though admittedly intriguing and rewarding, having my hard-to-access informants "tell all" was only one piece of the puzzle, not the analytic goal. Given often unchallenged Wall Street discourses of shareholder value and market truths, the strength of what anthropology can contribute to unpacking Wall Street lies in its ability to demonstrate ethnographically how Wall Street's institutional practices produce experiences that then make discourses come alive, more than the other way around.

Countering Abstraction, Constructing Financial Markets, and Particularizing Global Capitalism

What would it mean for anthropology to resist top-down categories, distinctions, and definitions of what "markets," "capitalism," and "globalization" are or should be? What if anthropologists approach the market as a set of daily, embodied practices and models? Given the resurgence of neoliberal discourses and practices in the mid- to late 1990s, where "the market" (often referring to financial markets) occupies in mainstream, financial, and social-scientific understandings an increasingly dominant, globalizing, and normative space, this book examines the building blocks of these macro structures. Because Wall Street investment bankers are highly visible in terms of their own self-representations and claims to truth and authority, yet culturally invisible in terms of their everyday practices and assumptions, by directly accessing key agents of change on Wall Street, a site widely deemed the epitome of global capitalist markets, I attempt to localize the very actors and institutions with a world-making influence on the global economy, and thus on the livelihoods of many.

The push to explain the massive socioeconomic inequalities generated by new configurations of financial markets in this neoliberal moment has invigorated in anthropology an approach to understanding finance that could hinder the cultural analysis of influential financial actors and reproduce power-laden assumptions about the "winners" and "losers" in the global economy. By taking as central the assumption that finance capital is abstract and abstracting – that is, separated and decontextualized from concrete lived realities, in turn shaping and corroding social relations in mystifying ways – we run the risk of allowing elite players in the global economy even more space to define and decipher our socioeconomic lives (see Gregory 1998; Tsing 2000a). Of course, it is crucial to remember that anthropologists have long been at the forefront of culturalizing the economy, in particular critiquing neoclassical economic theories as narrow ideological models divorced from and unable to represent the on-the-ground complexities of economic reality, especially in "non-Western" societies (Dalton 1961; Dilley 1992; Gudeman 1986; Sahlins 1972). This legacy stems in large part from the work of Karl Polanyi, who challenged the "'economistic fallacy' of liberal economic thinking, in which market relations . . . come to be viewed as universal models of human conduct," and argued that economic practices are embedded in social networks, relations, and institutions (Slater and Tonkiss 2001, 94). The multiple dichotomies, however, upon which Polanyi and many social scientists after him rested their analyses continue to have reverberating effects. Whereas nonmarket, premodern economies are assumed to be embedded in social relations, markets in modern industrial society are frequently imagined as operating according to formal and abstract economistic models. In this formulation, binaries of concrete/abstract, embedded/disembedded, and culture/economy are implicit and perpetuated.

Recent anthropological and sociological works have challenged many of these taken-for-granted dichotomies, demonstrating that economic practices take place in complex webs of social relations, which change in form and degree over time. Just as "nonmarket" gift exchanges are characterized by a high degree of formal calculation, "market economies are more fully embedded in social networks than Polanyi's strict separation allows" (Slater and Tonkiss 2001, 101).[32] The "actual practice[s] of economies" defy top-down notions of market: "high finance is largely concerned with personalities, private perks and little interest groups, prestige, imagination, almost anything but what might be called a market" (D. Miller 2002, 224, 228). Also, anthropological investigations of "the free market" have deconstructed the concept of the market in Western culture, demonstrating how market ideology in the West is intimately tied to British and American notions of individualism, property, and neoclassical economics (Carrier 1997).

Despite these important contributions, given the rise of neoliberalism and its resurgent representations of abstracting and globalizing markets, many scholars, especially when referring to Western money and finance, revert back to our legacy of binary assumptions where the financial dominance of investment banks are often attributed to abstract, all-powerful global markets. In moments when finance and the stock market are the ruling paradigms of capitalist practice, many academic critics of market fundamentalism inadvertently take as foundational the notion that the economy has become "disembedded" from society, that financial market logics – as utopian ideals – are being used to abstractly shape social relations, leading to social violence and inequality on a global scale (Arrighi 1996; LiPuma and Lee 2004; McMichael 1998). Neoliberal actors and institutions are restructuring social worlds according to virtual economic models which privilege elite institutions and transnational corporations (Carrier and Miller 1998).[33] Taken to the extreme, these narratives of money and finance predict a homogenized and reductionist "global" world where the complexities of "local" social relations are narrowed to, and judged against, an abstract and singular bottom line, where the world is remade in the image of financial logics. I describe these academic critiques of neoliberalism as "neoliberal exceptionalism," where the confrontation with conditions of socioeconomic inequality encourages scholars to privilege distant logics over particularity and grounded cultural analysis, which (ironically) overempowers neoliberalism.

So, what is at stake in these evolutionary narratives of increasing abstraction? Overarching scripts of universalizing financial logics and capitalist globalization not only obscure the heterogeneous particularities of Wall Street practices and effects and prevent the interrogation of Wall Street investment banks' hegemonic claims, but also ironically parallel the marketing schemes and hyped representations of Wall Street capitalist promoters.[34] Recent innovative research in social studies of finance, by taking as their subjects (and objects) of study key sites in the construction and globalization of financial markets, have been central in addressing these dominant rubrics by making powerful actors in finance culturally knowable and embedded in novel forms of sociality.[35] Bill Maurer (2006, 15, 19) rightly points out that the notion of money as abstract and deracinating is a dominant "Western folk theory" and in analyzing markets, money, and finance, "anthropology . . . too often repeats the same story of the 'great transformation' from socially embedded to disembedded to abstracted economic forms." Demonstrating the naturalization of this logic in the narrative of money, Maurer writes: "The story of money is repeatedly told in venues scholarly and popular as an evolutionary tale of greater and greater distance from actual things, of greater dematerialization, in a linear trajectory from barter to metal coin, to paper backed by metal, to paper declared valuable by fiat, and finally, perhaps to complex financial entities like derivatives, with future, not anterior, backing" (Maurer 2005, 100). Similarly, Hirokazu Miyazaki and Annelise Riles (2005, 321) have observed that the scholarly preoccupation with "the mystique of finance" has prevented giving "ethnographic attention to the mundane quality of the mundane." Given that finance and money may be anthropology's "new exotic," demonstrating the quotidian particularities and insufficiencies of finance becomes all the more crucial (Maurer 2006, 18).

Put simply, allowing finance to be simply abstract lets it off the hook. I make the case that massive corporate restructurings are not caused so much by abstract financial models as by the local, cultural habitus of investment bankers, the mission-driven narratives of shareholder value, and the institutional culture of Wall Street. While I am sympathetic to explanations of corporate downsizing and rampant job insecurity as the intensification of abstraction, my intervention is to demonstrate that what seems like abstraction can actually be culturally decoded. Critics resorting to abstraction are well reminded not only of our own legacy (of binaries) but also that of neoclassical economics, where there is the "attachment of great value to detachment; in its passion for dispassionate analysis" (Nelson 1998, 78). At the level of ethnographic interpretation, it is crucial to recognize that a finance capital-led version of capitalism, which privileges downsizing, stock price, and market crisis, is perhaps not so much about disembedding as it is about power relations and unequal clashes of differently valued social domains with diverging visions of the world. But I first want to recognize that abstraction is a powerful explanatory tool in light of pressing social problems.

There seems to be little question that particular economic measures from prices to interest rates to the Dow Jones industrial averages shape our lives, that powerful knowledge producers from financial economists to corporate executives to Wall Street bankers have over the past two decades used and relied even more heavily on these indicators to make top-down decisions about jobs and policies. Tackling these profound changes, anthropologist James Carrier explains that abstract neoclassical economics, armed with greater institutional power, is engaged in "the conscious attempt to make the real world conform to the virtual image." It is precisely this "move to greater abstraction and virtualism" in economic thought that is creating a prescriptive model for reality, a "virtual reality" that is reductive, dislocating, and divorced from responsible and engaged social relationships (Carrier 1998, 2, 5, 8; Carrier and Miller 1998). Specifically, given that we now live in a business environment where corporate decisions are based less on strategic knowledge produced within the organization on the ground, but rather are dictated by financial measures, stock prices, and the expectations of Wall Street investment banks, does not abstraction seem like the appropriate diagnosis?[36] As Carrier (1998, 4) claims, "It is not too far-fetched to see the disaggregation of firms and the increasing use of outsourcing and temporary workers as a kind of disembedding, for economic activities that had occurred within the structure of the firm and the durable employment and institutional relations contained within it, move outside and are acquired through relatively more impersonal and transient market relationships." These examples do seem to demonstrate that "external" forces, "disembedded" from local and organizational contexts, with allegiances only to abstract financial markers like stock price and profit accumulation, are not only loosening social ties but also generating conditions of supreme socioeconomic inequality by obliterating any concern for the daily lives and dilemmas of everyday people.

Part of my argument here is that the construction – the feeling – of abstraction is absolutely about power and hierarchy in that powerful changes stemming from very different values, priorities, and interests "feel like" the triumph of a penetrating, alien abstraction from the point of view of the marginalized. For example, many downsized workers have argued that the stock price is just one value among many, and that other values such as a stable job and the American Dream are more important. From the stance of the disenfranchised, financial parameters like the stock price are understood as *overtaking* other values and affecting differently positioned people unequally. The unequal conflict between the priorities and agendas of the powerful versus the powerless, not to mention the dismissal, may in turn be experienced as being "turned into a dollar sign from above," yet such a phenomenon is perhaps better explained as the social effect of concrete manifestations of power relations, not abstraction.

Similarly, when one directly interrogates the powerful, what gets dubbed as abstract is about very particular values, interests, and origin myths. From my natives' point of view, their use of stock prices as the primary measure of evaluating corporations is not about heralding abstraction, but about reclaiming the "rightful" capitalist unity between ownership (of stock) and control over corporations that had been sundered during the heyday of managerial, "welfare" capitalism, which in turn fosters the values of responsibility, efficiency, and individual proprietorship. Fixation on the stock price is a mission-driven cause, not a cold abstraction, that counters and overcomes the wrongful allocation of capital. Representing a host of values, shareholder value allows bankers to translate their particular values into a number, which acts as its own explanatory force. It is a discursive strategy used by powerful financial institutions to articulate their vision of the world and fight for their elite interests by utilizing a shareholder value worldview to impose short-term, financial market-based decision-making on corporations. Simultaneously, shareholder value, Wall Street investment bankers' central worldview, is often not enacted or actualized. While I would argue that downsizing workers based on stock prices is not about liquidating local values for the sake of abstract numbers, I would also make the case that Wall Street does not downsize based only on stock prices: other models and cultural and institutional norms are at play.

Given that on Wall Street, financial models are not fully actualized in practice, the interstitial space between "virtual" models and its "real" effects is a crucial analytical site. The heated academic debate between Michel Callon and Daniel Miller is instructive here. Callon has suggested that the models and theories of financial economics are not virtual, as Miller claims, but "real" in the sense that economic practices actually perform and enact "homo economicus" on a daily basis. Economic practice socially constructs the kinds of conditions and frames which allow economic thought to be actualized. In this vein, Callon (1998) refuses the dichotomies of abstraction/real. I would argue, however, that both Callon and Miller take as their defining assumption either the distance or lack thereof between economics and markets, between theory and practice. Whereas interpretations along the lines of Callon and Miller might see the current dominance of shareholder value as evidence *either* that homo economicus now exists, as financial economics has converged with market practices, *or* that market mechanisms have abstracted social relations according to virtual, utopian capitalist fantasies, I demonstrate that economic ideals are neither wholly performed and instantiated into reality *nor* virtual substitutions for "real life complexity." Analyzing what Wall Street investment bankers actually enact necessitates attention to the interface between virtual and real, model and effect, as well as the existence of other key cultural contexts that would influence this interaction.

When Wall Street's actions in the world are framed and translated as "people got downsized because of stock prices" or shrugged off because "markets crash," investment banking practices are represented and interpreted as abstract, cordoning "the market" off from social decisions. Socioeconomic dislocations are less cases of abstraction than instances of power being experienced and enacted. When conflicts between unequal values and interests are interpreted mainly in terms of abstraction, which in turn is refracted back as a core characteristic of finance, such assumptions further obfuscate the task of grounding Wall Street actors.

This discussion of abstraction and power is incomplete without an analogy to the power of whiteness as a racial construct. Richard Dyer (1997, 38–39) has written that one of the central markers of white identity is "the attainment of a position of disinterest – abstraction, distance, separation, objectivity." Yet, while privileging the notion of universality, abstraction, and invisibility, whiteness paradoxically also claims individuality, a particular display of spirit and character, as well as a sort of privileged "race." Dyer (1997, 39) argues that the representational power of whiteness – its flexibility, productivity, covetedness, and exclusivity – stems from its ability "to be everything and nothing, literally overwhelmingly present and yet apparently absent." Not surprisingly, part of the discursive power of the financial market is precisely its representation as abstract, its seeking to be everywhere and claiming to be nowhere *coupled with* its particular mission and claims to freedom, democracy, property, and prosperity. Market power thrives on this representational flexibility: though its beneficiaries can emerge as an interest group when necessary, usually this kind of power is cloaked in its abstraction and universality. It is instructive to compare, then, Dyer's claim that the power of whiteness can be challenged by "locating and embodying it in a particular experience of being white" (Dyer 1997, 4). In this ethnography, I intervene against the flexible and productive power of markets by rendering it concrete and by demonstrating that investment banking decisions and the very experiences of the investment bankers themselves are thoroughly informed by cultural values and the social relations of race, gender, and class. I hope to portray a Wall Street shot through with embodiment, color, and particularity.

The dual work of studying ethnographically the globalization of capital markets and the values and strategies of financial actors is to both recognize their power and demonstrate their locality and instability, even their fragility.[37] On the one hand, I emphasize the location of Wall Street practices in the United States and how specific investment banking subjectivities and actions work to produce American hegemony through their brand of finance. As such, Wall Street finance is distinct from global finance, and the interests of the United States, though powerful and globalizing, should not be conflated with those of the world. On the other hand, investment bankers and banks located in the United States often do *claim* the entire globe: they effect rippling, global changes and self-represent themselves as the ultimate global actors promoting global capitalism. As Douglas Holmes and George Marcus (2005, 237) write, the "contemporary system of technocratic expertise . . . conceives and produces the idea of the global as daily practice. Central banks operate not merely under the sway of fast-capitalism; they have played a direct role in creating and mediating it."

Of course, although these actors do strongly shape social relations, part of the project of localizing Wall Street involves emphasizing the hype, miscalculations, and discrepancies of global strategies and practices, especially given the central role that boastings of global reach and power play in the self-representations of investment bankers.[38] We can acknowledge the global spread and influence of Wall Street's investment banking practices and worldviews without assuming that the actors' plans are flawlessly conceived and executed. As Michael Fischer points out, the global is "polycentric": however homogenizing, it is multiply situated, dynamic, local, and productive of odd ethnographic juxtapositions (Fischer 2003). Experts and cultures of expertise can make blunders, spinmeisters can fall prey to spin, and masters of the universe can be undermined by hubris. My approach to Wall Street investment bankers writes against this tendency to attribute infallibility and necessity to their cultural practices by investigating the fragilities of market practices and the inconsistencies of expert opinions, especially instances when encompassing global logics of shareholder value break down. It is precisely this global confidence in their capabilities that allows them the freedom to act unimpeded.

Notes

1. Over the past thirty years, what constitutes Wall Street – the number and kinds of institutions, and the amount of investment capital under management – has grown exponentially. From day trading to hedge funds (now numbering in the thousands),

Wall Street has expanded so much in both scale and scope that its values and practices have become increasingly diffuse. (Wall Street, of course, is also an actual street, a metaphor for capitalism itself, a "jungle" culture of greed; the list goes on.) My research, however, focused mainly on investment banks and on the worldviews and practices of investment bankers. Claiming to be "the oil that greases the wheels of capitalism," these bankers serve as a bridge, as financial culture brokers between the securities markets, corporations, investors, and the mass media. Because they have helped to fundamentally transform corporate America, studying their daily discourses and practices is essential to understanding the complexities of a finance-capital-oriented vision of the world.

2. From the late 1990s to the present, the telecommunications and media industries, like most industries, have engaged in a merger frenzy, downsizing workers while creating communications monopolies. For example, in 1994, Via-com acquired Blockbuster Video and Paramount Pictures and in 1998, CBS; in 1998, AT&T acquired Tele-Communications Inc. and in 1999, MediaOne Group; in 2000, AOL acquired Time Warner, which had acquired Turner Broadcasting; in 1996, WorldCom acquired MCI; in 2006 BellSouth merged with AT&T; and so on.
3. Two other examples of Wall Street's celebration of downsizing are instructive. When Sara Lee Corporation announced a complete restructuring of its business that would outsource most of its manufacturing capabilities, closing down and spinning off factories and product lines, Wall Street greeted the news with jubilation: "Sara Lee Corp. disclosed a fundamental reshaping of the consumer-products company that would move it away from making the brand-name goods it sells. . . . While the after-tax charge is much more than the $1.01 billion Sara Lee earned in the latest fiscal year . . . it seemed that on Wall Street nobody didn't like Sara Lee's idea. In New York Stock Exchange trading yesterday, Sara Lee jumped $6, or 14%, to close at a 52-week high of $48.5625" (J. Miller 1997). When Unilever, a European, multinational conglomerate, demonstrated, via downsizings, that it was willing to "emulate" the most radical restructurings much more characteristic of American companies, Wall Street "cheered." Its announcement that it would "embark on a severe program of corporate cutbacks, eliminating 25,000 jobs over five years, closing 100 factories and possibly jettisoning up to 75 percent of its products . . . surprising in its sharpness and scope, reflected a growing trend by European companies to emulate the cutbacks made by big American companies. . . . Today's announcement cheered investors" (Cowell 2000).
4. One strategy of Wall Street bankers, financial analysts, and corporate executives to divert attention from these heightened contradictions was to tout low unemployment statistics for the economy as a whole, with little regard for the fact that the majority of new jobs created were temporary, part-time, underpaid, and without benefits. Many scholars have critiqued the justification of the high employment rate, demonstrating that this new era of work is actually a "postemployment" era, where stable organizations and the notion of a long-term "career" are discarded in favor of rampant insecurity, greater wage inequality, and continuous work and retraining (Reich 2000; Sennett 1998; V. Smith 2002). That Wal-Mart Corporation and the temporary agency Manpower Temporary Services, Inc. are the largest employers of workers in the United States is indicative of the kinds and conditions of "jobs" being created today.
5. It is important to realize that most of the stable jobs in corporate America were available mainly to white, middle-class Americans. While in the past twenty years, white working and middle-class men have experienced a serious decline in earnings and job security, these experiences of downward mobility are even more pronounced among people of color and women. Unfortunately, there is the misplaced tendency to interpret the struggle for equal opportunity by previously marginalized groups as the

reason for white middle-class jab losses. As David Wellman points out, because white middle-class men have historically experienced their own monopoly on jobs and rights as "fair" and "equal," it comes as no surprise that moving toward "actual" equal opportunity is experienced by the relatively privileged as decline. What gets rendered invisible is a critical interrogation of racism and sexism and, of course, capitalism (Wellman 1997).

6. This ethnography is not so much a totalizing indictment of shareholder value as it is an examination and critique of those who speak in its name. As I will demonstrate, what defines and constitutes shareholder value is historically contingent and dependent on the larger institutional and cultural contexts in which it is embedded. While shareholder value in the past twenty-five years has come to symbolize corporate practices that boost short-term stock prices, this has not always been the case. I will also argue that many of the practices that Wall Street investment bankers construct in the name af shareholder value often result in its decline.
7. I borrow this term "making markets" from Mitchel Abolafia's pioneering ethnographic study of how traders in the stock, bond, and futures markets are embedded in socially constructed institutions (Abolafia 2001).
8. I use the term "corporate America" to reference large public corporations headquartered and operating in the United States, which may include those corporations that are also transnational in their production and delivery of services. (Many of these corporations are incorporated or chartered in the state of Delaware, though not physically located there, because of favorable corporate tax and disclosure laws.) In particular, I use "corporate America" to convey the normative business values and practices that came to characterize the modern corporation as it was codified and constructed, in conversation with American culture at large, throughout the twentieth century. Corporate America, interestingly enough, ran increasingly afoul of both U.S. corporate law and Wall Street ideology. How Wall Street in the past thirty years worked to restructure and align corporate America to its values and practices is the subject of this ethnography. For most of the twentieth century, public corporations were dominantly defined in corporate law as wholly "owned" by and responsible to their shareholders (who can be located globally, though many corporate shares are owned by institutional investors based in the United States). But in their daily practices, most workers and managers governed and enacted the corporation as a longtern social institution responsible to multiple stakeholders beyond the shareholder. In this vein, my usage of "corporate America" both refers to public corporations in the United States and emphasizes a particular imagining of those corporations. That corporate America is a contingent term demonstrates that the nature of the corporation – its very definition, meaning, and purpose – is contextual and contested. I would argue that Wall Street has transformed corporate America into a very different kind of institution, one that privileges its financial and legal form.
9. I use the concept "actor-network," derived from the work of Bruno Latour and Michel Callon, as a way to conceive and concretize the space of financial markets as constructed out of a multiplicity of human and nonhuman actors such as investment bankers and traders, financial products and technologies, which together form the complex interactions and "networks" of the market.
10. I use "U.S. capitalism" to demonstrate the cultural particularity and contingency of all economic practices and values. Given that capitalism in the U.S. is routinely conflated with "Western capitalism" or global capitalism, understood as universal, rational, and natural, and represented as the pinnacle of capitalist development, my point is that capitalist actions are always already constructed through complex, contested, and locally specific cultural ethos, motivations, and practices that change over time. All too often, it is "non-Western" capitalisms that are characterized by cultural habits

and defined as culturally different from Western capitalism, which remains normative and acultural. In this ethnography, by examining how Wall Street investment banks came to power in the late twentieth century and reshaped the culture of corporate America, I showcase capitalism-in-the-making in the United States and in so doing write against a static and acultural approach to both finance and capitalism. This is not to say that this financially dominant capitalism does not have far-reaching global influence, but rather that it grows out of a particular industry and set of institutional worldviews, subjectivities, and practices, and spreads under certain conditions and at specific moments. I contend that it is precisely this capitalism's location at a distinctly American center of financial power – Wall Street investment banking – that enables its global recognition, spread, and dominance.

11. At the same time, it is important not to homogenize or totalistically equate market making with investment banks. Markets are not reducible to their dominazit voices; they are more multiple, unwieldy, and unpredictable than that. Similarly, the daily workings and values of particular investment banks are not always commensurate with dominant Wall Street "market demands."
12. Investment banks advise major corporations on financial deals and transactions; they do not "do retail," that is, service the mass, consumer market composed of individuals. Their interaction with individuals via their private wealth management departments are limited to "high-net-worth individuals," that is, people with $2 to 5 million in assets. Commercial banks, on the other hand, are the banks with which most Americans are familiar. They have branches, take deposits, maintain accounts, and lend to both individuals and corporations. This separation of commercial and retail banking from the securities industry, of course, occurred with the spate of regulations enacted during the Great Depression when Wall Street was widely discredited and reviled, and commercial banks rebuilt their business only through legal separation and social distancing from financial markets. For most of the mid-twentieth century, commercial banks were considered more reputable institutions than investment banks, yet, since the 1980s, as Wall Street financial values rose to dominance, commercial banks have not had the same cultural cachet as investment banks, which are understood to be elite, prestigious institutions and expert spokespeople for the financial markets. Commercial bankers also receive much lower compensation. Yet, because of Reagan-, Bush-, and Clinton-era deregulations, many commercial banks, through expansions, mergers, and acquisitions, became "hybrid" commercial/investment banks like Bankers Trust, Citigroup, Deutsche Bank, and J. P. Morgan Chase, which are allowed to mix the best of both worlds: their role and expertise in capital markets with access to retail markets.
13. Certainly, concrete legal and policy changes favorable to Wall Street investment banks and the financial markets enabled the dominance of finance capital in the American economy. While mainstream academic and popular discourse describe this restructuring of rules as "deregulation," it is important to point out that in addition to the loosening and removal of controls (the lack of governmental involvement that deregulation connotes), creating "free markets" also necessitated the active re-writing of policy and the hands-on participation of government. As Karl Polanyi long pointed out, free markets were assiduously planned, not organic expressions of natural laws (Polanyi 1944).
14. During the past few decades investment bankers, especially those who worked in the most storied investment banks, were usually quickly rehired at another Wall Street firm, as these institutions all practiced revolving-door employment with constant demand and downsizings. Even if a particular downturn in the financial sector was prolonged, Wall Street dominance and cachet were influential enough such that commercial banks and most corporations relished the opportunity to hire "hot-shot"

investment bankers who ostensibly brought the know-how of high finance to the corridors of corporate America. In the current (2009) economic meltdown, Wall Street investment bankers, perhaps for the first time in thirty years, have not quickly found new jobs with comparable compensation, though many are able to depend on significant bonuses and savings to cushion their transitions. Many of my former informants have received job offers but turned them down because of the sharp decline in pay – a sign that the era of the "big payout" bonus might be over. Given the massive restructuring of the industry and the discrediting (to some extent) of Wall Street practices and incentives, it is unclear if this recession will usher in a new status quo, whether Wall Street will recuperate its standing in a few years, and how investment bankers will weather whatever shifts may occur.

15. Twenty-five years after anthropologist Laura Nader (1972, 289) first issued her call for anthropologists to "study up," to analyze "the colonizers rather than the colonized, the culture of power rather than the culture of the powerless, the culture of affluence rather than the culture of poverty," anthropologist Hugh Gusterson (1997, 114) writes that "these appeals for a critical repatriated anthropology . . . remain substantially unrealized." It is crucial to note, however, that for the past few decades, the anthropology of science and technology has been a long series of encounters of "studying up," beginning with Sharon Traweek's entry into a high energy physics laboratory (Traweek 1996). Eliding this history ignores the recent, important breakthroughs of anthropological inventiveness and dooms anthropology to continually proclaim its inability to study the powerful. As Michael Fischer importantly points out, "anthropologists need legitimation based on our competence, not just always proclaiming 'first contact'" (personal communication).

    In part because of lack of access, the "Roger and Me syndrome in ethnography," where the anthropologist is unable to get through the front lines of Fortress, Inc., "there have been few takers" of Nader's challenge to conduct ethnography in "institutions at the heart of capitalist processes of production and stratification" (Gusterson 1997, 115). At the same time, anthropologists are pioneering new strategies to garner access such as developing the requisite set of experiences, connections, and techniques, which allow them to enlist highly placed and privileged individuals as coinvestigators rather than merely informants. See, for example, Holmes and Marcus 2006.

16. Although all of my informants have been given pseudonyms, the investment banks in which they worked have not. Bankers Trust, for example, is not a pseudonym; it is the name of the bank where I worked. I changed the name of bankers' Wall Street affiliations only when their confidentiality could be compromised. It is important to remember that most of my informants are no longer working at the same banks and that many banks have merged or changed their name since the late 1990s.

17. Business divisions such as the Private Bank, Risk Management, and Investment Management hired us to do a variety of projects for them. We analyzed how much market share a group within the bank had in relation to its competitors, documented how the group could implement better "policies and procedures," and streamlined a group's "workflow" so that it could "run a tighter ship." Projects usually lasted a couple of months.

18. Shareholder value is understood by financial economists, Wall Street investment banks, and members of the investment community as a set of ideas and actions whose primary purpose is to increase the stock price of corporations to generate wealth for shareholders.

19. Although our downsizing was announced in January 1997, I did not leave BT until June 1997 because of an unprecedented deal for transition time our boss struck with the firm's most senior management. I spent the rest of the year writing grants, defending my dissertation prospectus, and garnering human subjects approval.

20. Ironically, precisely because my entire group was downsized, my coworkers dispersed throughout major investment banks, fortuitously solidifying my access to Wall Street more broadly.
21. The canonical fieldwork methodology of single-sited immersion in everyday life might not be the most generative of ethnographic data in an investigation of global financial institutions, nor should it necessarily be the primary tool of fieldwork, given rethinkings in the past two decades of what constitutes "the field," attention to the deterritorialization of culture and place, and increasing realization that cultural formations are characterized as much by change, disjuncture, and movement as by stability or persistence (Gupta and Ferguson 1992; Martin 1997).
22. I borrow this phrase from Joanne Passaro's article "'You Can't Take the Subway to the Field!': 'Village' Epistemologies in the Global Village" (Passaro 1997).
23. Jackson's mission for the Wall Street Project was to partner with large Wall Street investment banks and Fortune 500 companies to provide "access to capital" to "underserved markets, underutilized talent, and untapped capital." Jackson has long maintained that access to capital is the fourth stage of the civil rights movement after the abolition of slavery, the end of legal segregation, and the right to vote (Jackson Sr., Jackson Jr., and Gotschall 1999). Not surprisingly, of the three strategies that Jackson articulated, Wall Street was most interested in the surface multiculturalism of improving diversity.
24. During this speech, what was also striking was his observation of the stark contrast between Wall Street investment banks' approaches to "emerging markets" in Asia and Eastern Europe in relation to Wall Street's nonexistent relationship with Harlem. He questioned, for example, why Wall Street firms and institutional investors lent (short-term) billions to the "Asian Tigers" but could not even support a small business a few blocks away in upper Manhattan.
25. While a rising stock market necessitates exuberant marketing to generate buyers, it is also undermined by this exuberance as the companies the stocks represent often do not "grow" as quickly as the stock price.
26. Although most shareholders should certainly be highly critical and suspicious of Wall Street's and corporate America's breach of "trust" and rhetoric, the very notion of a betrayed investor is itself based on the problematic logic of shareholder value in the first place. The cause of the beleaguered investor, though certainly a compelling one given the extreme losses of the middle classes, often serves to obscure cuts in social programs and other effects of financial recession on the (nonshareholding) poor and detracts from a deeper interrogation of the ideology of shareholder value.
27. For example, Morgan Stanley's merger with Dean Witter, Discover in 1997 did not capitalize on the purported synergies of cross-selling Wall Street and Main Street. Daimler-Benz AG's $40 billion purchase of Chrysler Corporation in 1998 has been widely critiqued for "poor integration" and lack of cultural fit. AT&T's acquisitions of Tele-Communications and MediaOne Group in 1999 were considered not only overpriced but also a strategic waste of money given that it decided to divest its cable acquisitions to Comcast a few years later in 2002 (Lipin and Deogun 2000). In 2004, AT&T was delisted from the Dow Jones Industrial Average, due in part to its colossal mismanagement of dozens of M&A deals throughout the past fifteen years (S. Lynch 2004).
28. I found that these shareholder value origin stories are by and large shared and retold among all the investment bankers that I met and interviewed. These myths are reinforced by, intersect with, and are a part of larger narratives passed down in a variety of sites of economic knowledge production: business schools (where most investment bankers are trained), undergraduate economics classes, financial news, and mainstream media.

29. Whereas Wall Street worldviews are grounded in the practices of the stock market and the values and principles of finance, and largely intersect with neoclassical economic assumptions of ownership and property, managerial worldviews, especially those codified during the post–Second World War era, were grounded in the bureaucratic practices of the corporation as a long-term social institution and largely intersected with what might be termed a Keynesian, paternalistic, or "social entity" conception of corporate governance.
30. Yanagisako argues that "in ethnographic studies, where the subjects are endowed with sufficient financial and cultural resources to make themselves well heard, the challenge is less how to give them voice as it is to let others speak as well." Specifically, she situated the "official histories" of Italian capitalist firms as told by founder-capitalists "within the broader context of the stories that are told by other people" so that "critical methods for interpreting individuals' accounts of their lives and histories" are "part and parcel of cultural anthropology's reflexive critique of ethnographic representations" (Yanagisako 2002, 48). She analyzed founder narratives alongside those of other family members as well as archival accounts of the firm from state records.
31. I borrow this term from Anna Tsing's *Friction* (2005), an ethnography that provides an incisive anthropological model for approaching the encounters and interactions between multiple locals and universals.
32. See, for example, Carrier 1995, Granovetter 1985, J. Guyer 1995, Gratiovetter and Swedberg 2001, D. Miller 1998, and Zelizer 1997.
33. Such a premise builds on a long history of Marxist thinking about the fetishization of money and capital and its effects on social relations, and a rich anthropological and sociological literature about the disembedding (and the concurrent reembedding) of market relations from social relations (Polanyi 1944, 1957).
34. The feminist geographers J. K. Gibson-Graham warned of this potential conflation by pointing out that leftist critics of capitalism have discursively helped to create a "capitalist system" so hegemonic that it is not only omnipresent, but also the only possible economic form (Gibson-Graham 1996).
35. See Abolafia 2001, Buenza and Stark 2005, Goede 2005, Hertz 1998, Knorr Cetina and Preda 2005, MacKenzie 2001, Maurer 2005, Miyazaki 2003, and Zaloom 2006.
36. There are myriad examples of workers being downsized when corporations do not meet Wall Street quarterly stock price expectations, or small loans becoming increasingly impossible as local banks with community ties disappear and money-lending practices are dictated from national or international headquarters (Leyshon and Thrift 1997).
37. See, for example, Tsing 2000a.
38. My informants are often dubbed "masters of the universe," following Tom Wolfe's description of the Wall Street bond trader Sherman McCoy in the bestselling *Bonfire of the Vanities*, by scholars and themselves alike. To compile descriptions from Robert Reich, Rosabeth Moss Kanter, and Saskia Sassen, Wall Streeters are globe-trotting, "world class" knowledge workers; they are the investor elite, the "new monied class"; they are "the labor aristocracy of 'symbolic analysts' who spiral through the weblike structure of the elevated corporation, through workplaces connected across national boundaries, removed from local contexts"; they are the "sky workers . . . detached from those stranded on the ground, production workers and the growing sector of in-person services" (Burawoy et al. 2000, 3). In these representations, Wall Street investment bankers are hyperagents, able to marshal the resources of global institutions and the "transnational capitalist class" to recreate the world in their image and according to their values.

References

Abolafia, Mitchel. 2001. *Making Markets: Opportunism and Restraint on Wall Street*. Cambridge, Mass.: Harvard University Press.

Arrighi, Giovanni. 1996. *The Long Twentieth Century: Money, Power, and the Origins of Our Times*. London: Verso.

Boudreau, Abbie, David Fitzpatrick, and Scott Zamost. 2008. "Wall Street: Fall of the Fat Cats." *CNN.com*, 17 October, www.cnn.com (accessed 17 October 2008).

Bourdieu, Pierre. 1990. *Outline of a Theory of Practice*. Cambridge: Cambridge University Press.

Buenza, Daniel, and David Stark. 2005. "How to Recognize Opportunities: Heterarchical Search in a Trading Room." *The Sociology of Financial Markets*, edited by K. Knorr Cetina and A. Preda, 2–20. Oxford: Oxford University Press.

Burawoy, Michael, Joseph A. Blum, Sheba George, Zsuzsa Gille, Teresa Gowan, Lynne Haney, Maren Klawiter, Steven H. Lopez, Seán O. Riain, and Millie Thayer. 2000. *Global Ethnography: Forces, Connections, and Imaginations in a Postmodern World*. Berkeley: University of California Press.

Caldeira, Teresa. 2000. *City of Walls: Crime, Segregation, and Citizenship in Sao Paulo*. Berkeley: University of California Press.

Callon, Michel. 1998. "Introduction: The Embeddedness of Economic Markets in Economics." *The Laws of the Markets*, edited by M. Callon, 1–68. Oxford: Blackwell.

Carrier, James G. 1995. *Gifts and Commodities: Exchange and Western Capitalism since 1700*. London: Routledge.

———, ed. 1997. *Meanings of the Market: The Free Market in Western Culture*. Oxford: Berg.

———. 1998. "Introduction." *Virtualism: A New Political Economy*, edited by James G. Carrier and Daniel Miller, 1–24. Oxford: Berg.

Carrier, James G., and Daniel Miller, eds. 1998. *Virtualism: A New Political Economy*. Oxford: Berg.

Certeau, Michel de. 1984. *The Practice of Everyday Life*. Berkeley: University of California Press.

Dalton, George. 1961. "Economic Theory and Primitive Society." *American Anthropologist* 63, no. 1, 1–25.

Dilley, Roy. 1992. "Contesting Markets: A General Introduction to Market Ideology, Imagery, and Discourse." *Contesting Markets: Analyses of Ideology, Discourse, and Practice*, edited by R. Dilley, 1–36. Edinburgh: Edinburgh Press.

Dyer, Richard. 1997. *White*. London: Routledge.

Fiorini, Phillip. 1995. "Mobil Restructuring Pumps Shares." *USA Today*, 2 May, 3B.

Frank, Thomas. 2000. *One Market under God: Extreme Capitalism, Market Populism, and the End of Economic Democracy*. New York: Doubleday.

Gibson-Graham, J. K. 1996. *The End of Capitalism (as We Knew It): A Feminist Critique of Political Economy*. Cambridge, Mass.: Blackwell.

Goede, Marieke de. 2005. *Virtue, Fortune, and Faith: A Genealogy of Finance*. Borderlines 24. Minneapolis: University of Minnesota Press.

Granovetter, Mark. 1985. "Economic Action and Social Structures: The Problem of Embeddedness." *American Journal of Sociology* 91, no. 3, 481–510.

Granovetter, Mark, and Richard Swedberg, eds. 2001. *The Sociology of Economic Life*. Boulder, Colo.: Westview Press.

Gregory, Steven. 1998. "Globalization and the 'Place' of Politics in Contemporary Theory: A Commentary." *City and Society* 10, no. 1, 47–64.

Gudeman, Stephen. 1986. *Economics as Culture: Models and Metaphors of Livelihood*. Boston: Routledge and Kegan Paul.

Gupta, Akhil, and James Ferguson. 1992. "Beyond 'Culture': Space, Identity, and the Politics of Difference." *Cultural Anthropology* 7, no. 1, 6–23.
Gusterson, Hugh. 1997. "Studying Up Revisited." *POLAR: Political and Legal Anthropology Review* 20, no. 1, 114–19.
Guyer, Jane, ed. 1995. *Money Matters: Instability, Values, and Social Payments in the Modern History of West African Communities*. Portsmouth: Heinemann.
———. 2004. *Marginal Gains: Monetary Transactions in Atlantic Africa*. Chicago: University of Chicago Press.
Hertz, Ellen. 1998. *The Trading Crowd: An Ethnography of the Shanghai Stock Market*. Cambridge: Cambridge University Press.
Hiltzik, Michael. 2001. "Giant Cable Deal; AT&T Returning to Basics." *Los Angeles Times*, 21 December, C1.
Holmes, Douglas R., and George E. Marcus. 2005. "Cultures of Expertise and the Management of Globalization: Toward the Re-Functioning of Ethnography." *Global Assemblages: Technology, Politics, and Ethics as Anthropological Problems*, edited by A. Ong and S. J. Collier, 235–52. Maiden, Mass.: Blackwell.
Jackson, Rev. Jesse L., Sr., Jesse L. Jackson Jr., and Mary Gotschall. 1999. *It's about the Money! The Fourth Movement of the Freedom Symphony: How to Build Wealth, Get Access to Capital, and Achieve Your Financial Dreams*. New York: Random House.
Jensen, Michael. 1976. *The Financiers: The World of the Great Wall Street Investment Banking Houses*. New York: Weybright and Talley.
Keller, John. 1996. "AT&T Will Eliminate 40,000 Jobs and Take a Charge of $4 Billion." *Wall Street Journal*, 3 January, A3, A6.
Klein, Naomi. 2000. *No Logo*. New York: Picador.
Knorr Cetina, Karin, and Alex Preda, eds. 2005. *The Sociology of Financial Markets*. Oxford: Oxford University Press.
Landler, Mark. 1997. "Investing It: A Plodding Ma Bell and Her Precocious Child." *New York Times*, 13 April, F1, F7.
Leyshon, Andrew, and Nigel Thrift. 1997. *Money/Space: Geographies of Monetary Transformation*. London: Routledge.
Lipin, Steven, and Nikhil Deogun. 2000. "Deals & Deal Makers: Big Mergers of '90s Prove Disappointing to Shareholders: Poor Market Reaction Does Little to Slow Pace of Acquisitions." *Wall Street Journal*, 30 October, C1, C21.
LiPuma, Edward, and Benjamin Lee. 2004. *Financial Derivatives and the Globalization of Risk*. Durham, N.C.: Duke University Press.
Lynch, David. 1996. "AT&T Downsizes Estimate of Layoffs by 40%." *USA Today*, 15 March, B1.
Lynch, Stephen. 2004. "Index Face-Lift; Verizon In, AT&T Out as Dow Shuffles Lineup." *New York Post*, 2 April, 35.
———. 2006. *An Engine, Not a Camera: How Financial Models Shape Markets*. Cambridge, Mass.: MIT Press.
Malinowski, Bronislaw. 1948. *Magic, Science, and Religion and Other Essays*. Prospect Heights: Waveland Press.
———. 1997. "Anthropology and the Cultural Study of Science: From Citadels to String Figures." *Anthropological Locations: Boundaries and Grounds of a Field Science*, edited by A. Gupta and James Ferguson, 131–46. Berkeley: University of California Press.
———. 2002. "Repressed Futures: Financial Derivatives' Theological Unconscious." *Economy and Society* 31, no. 1, 15–36.
———. 2005. *Mutual Life, Limited: Islamic Banking, Alternative Currencies, Lateral Reason*. Princeton, N.J.: Princeton University Press.
———. 2006. "The Anthropology of Money." *Annual Review of Anthropology* 35, 15–36.

McLean, Bethany, and Peter Elkind. 2003. *The Smartest Guys in the Room: The Amazing Rise and Scandalous Fall of Enron*. New York: Portfolio.
McMichael, Philip. 1998. "Development and Structural Adjustment." *Virtualism: A New Political Economy*, edited by James G. Carrier and Daniel Miller, 95–116. Oxford: Berg.
Miller, Daniel. 1998. *A Theory of Shopping*. Ithaca, N.Y.: Cornell University Press.
———. 2002. "Turning Callon the Right Way Up." *Economy and Society* 31, no. 2, 218–33.
Miller, James. 1997. "Sara Lee to Retreat from Manufacturing." *Wall Street Journal*, 16 September, A3, A10.
Miyazaki, Hirokazu. 2003. "The Temporalities of the Market." *American Anthropologist* 105, no. 2, 255–65.
———. 2006. "Economy of Dreams: Hope in Global Capitalism and Its Critique." *Cultural Anthropology* 21, no. 2, 147–72.
Miyazaki, Hirokazu, and Annelise Riles. 2005. "Failure as Endpoint." *Global Assemblages: Technology, Politics, and Ethics as Anthropological Problems*, edited by A. Ong and S. J. Collier, 320–31. Oxford: Blackwell.
Nader, Laura. 1972. "Up the Anthropologist–Perspectives Gained from Studying Up." *Reinventing Anthropology*, edited by D. Hymes, 284–311. New York: Pantheon Books.
Nelson, Julie A. 1998. "Abstraction, Reality, and the Gender of 'Economic Man'." *Virtualism: A New Political Economy*, edited by James. G. Carrier and Daniel Miller, 75–94. Oxford: Berg.
New York Times. 1996. *The Downsizing of America*. New York: Times Books.
O'Boyle, Thomas F. 1999. *At Any Cost: Jack Welch, General Electric, and the Pursuit of Profit*. New York: Vintage.
Oldham, Jennifer. 1999. "A History of the Dow." *Los Angeles Times*, 30 March 1999, C6.
O'Sullivan, Mary. 2000. *Contests for Corporate Control: Corporate Governance and Economic Performance in the United States and Germany*. Oxford: Oxford University Press.
Passaro, Joanne. 1997. "'You Can't Take the Subway to the Field!': 'Village' Epistemologies in the Global Village." *Anthropological Locations: Boundaries and Grounds of a Field Science*, edited by A. Gupta and J. Ferguson, 147–62. Berkeley: University of California Press.
Polanyi, Karl. 1944. *The Great Transformation: the Political and Economic Origins of Our Time*. Boston: Beacon Press.
———. 1957. "The Economy as Instituted Process." *Trade and Market in the Early Empires*, edited by K. Polanyi, C. M. Arensberg, and H. W. Pearson, 243–69. Glencoe, Ill.: Free Press.
Reich, Robert. 2000. *The Future of Success: Working and Living in the New Economy*. New York: Vintage Books.
Rimer, Sara. 1996. "A Hometown Feels Less Like Home." *New York Times*, 6 March, A1, A16–17.
Ritter, Scott. 1995. "Mobil to Reduce Its Work Force by About 9.2%." *Wall Street Journal*, 2 May, A4.
Rosaldo, Renato. 1989. *Culture and Truth: The Remaking of Social Analysis*. Boston: Beacon Press.
Sahlins, Marshall. 1972. *Stone Age Economics*. New York: Aldine.
Sennett, Richard. 1998. *The Corrosion of Character: The Personal Consequences of Work in the New Capitalism*. New York: W. W. Norton.
Serwer, Andy. 1999. "A Nation of Traders." *Fortune*, 11 October, 116–20.
Slater, Don, and Fran Tonkiss. 2001. *Market Society*. Oxford: Polity Press.

Sloane, Leonard. 1995. "AT&T Dominates as Dow Climbs 25.65." *New York Times*, 21 September, D12.
Smith, Vicki. 2002. *Crossing the Great Divide: Worker Risk and Opportunity in the New Economy*. Ithaca, N.Y.: Cornell University Press.
Stokes, Jeanie. 2000. "Shareholders Dump Stock." *Rocky Mountain News* (Denver, Colo.), 26 October, 7B.
Traweek, Sharon. 1992. *Beamtimes and Lifetimes: The World of High Energy Physicists*. Cambridge, Mass.: Harvard University Press.
Tsing, Anna. 2000a. "The Global Situation." *Cultural Anthropology* 15, no. 3, 327–60.
———. 2005. *Friction: An Ethnography of Global Connection*. Princeton, N.J.: Princeton University Press.
Vickers, Marcia, and Mike McNamee. 2002. "The Betrayed Investor." *Business Week*, 25 February, 104–15.
Waters, Richard. 1999. "Hungry for More: Man in the News Michael Armstrong." *Financial Times*, 24 April, 13.
Wellman, David. 1997. "Minstrel Shows, Affirmative Action Talk, and Angry White Men: Marking Racial Otherness in the 1990s." *Displacing Whiteness: Essays in Social and Cultural Criticism*, edited by R. Frankenburg, 311–31. Durham, N.C.: Duke University Press.
———. 2002. *Producing Culture and Capital: Family Firms in Italy*. Princeton, N.J.: Princeton University Press.
Zaloom, Caitlin. 2006. *Out of the Pits: Traders and Technology from Chicago to London*. Chicago: University of Chicago Press.
Zelizer, Viviana. 1997. *The Social Meaning of Money*. Princeton, N.J.: Princeton University Press.
Zuckerman, Laurence. 1995. "The Second Breakup of AT&T: The Computer Business; The Costly Marriage to NCR Becomes a Vision That Failed." *New York Times*, 21 September, D9.

79

Researching Police Deviance: A Personal Encounter with the Limitations and Liabilities of Field-Work

Maurice Punch

(I) Introduction

Accounts of field-work occasionally reveal the stress, the deep personal involvement, the role-conflicts, and the time-consuming nature of observational studies for the researcher (e.g. Whyte, Wax, Hofland, Johnson[1]). Clarke[2] has written of the anguish of certain anthropologists, causing some even to abandon the field setting; while Van Maanen[3] has contrasted the rich oral tradition among academics, centred on what ostensibly 'really' happened in the field, with the self-censored material in publications on their research experience. That experience may appear incoherent, blurred, or stressful at the time and researchers may have considerable difficulty portraying the story behind the story of their field-work (perhaps because it is embarrassing or else it is considered unsuitable for a 'serious' publication). Reading Van Maanen's insightful work stimulated me to take out an article I had written several years ago, but which I had been unable to publish fully at the time, and to rework it in an attempt to convey in depth both the constraints encountered in the field and my reflections on the research experience long after departure. My initial account was somewhat gloomy and reeked of weary disillusionment. Now I can take a more

Source: *The British Journal of Sociology*, 40(2) (1989): 177–204.

dispassionate look at the research I conducted with the Amsterdam Police in the period 1974–1980. My purpose in this paper is to focus on three elements of that fieldwork. First, I wish to convey the strain I experienced in a demanding field setting; second, I want to communicate my acute awareness of the limitations of research in penetrating sensitive areas of institutional life; and, third, I hope to illuminate some of the ethical dilemmas encountered in the field (which one may only fully be conscious of in retrospect). And perhaps I should add that, however rough the passage, the experienced field-worker has almost a moral obligation not to discourage others but rather to incite them to get out into the field. For there is something irreducible in the mundane dilemmas of even the most unproblematic field situations that forces academics to reflect on the meaning, and boundaries, of conducting social science.

My contention is not only that observation can be a tough training in social interaction but also that it ineluctably exposes the researcher to a number of troublesome interactional, if not ethical, dilemmas. My argument is that these dilemmas arise spontaneously in 'straightforward' field research. But, crucially, they emerge in a particularly acute form in those settings where participants wish to hide their deviance. This is of vital importance for those who claim that academics should deliberately investigate, and expose, powerful institutions in our society.[4] For tackling those institutions can bring us to appreciate both the limitations of field-work as a method but also the inability to push our investigations as far as we would wish. This paper discusses precisely those personal and ethical dilemmas in research and, in my case, I am referring especially to the police organization. According to Skolnick,[5] the police organization is the most secluded part of the criminal justice system. The researcher's task then becomes how to circumvent the minefield of defences that protect the concealed reality of police work. Prolonged participant observation seems to me to be the most appropriate method for breaking through the culture and for cracking the code of policing.[6] This is particularly the case because the police organization is itself accused of routinely concealing its own deviance.[7] Research on the police documents abuses and unearths built-in malpractices[8] while official enquiries and journalism have unearthed corruption, prejudice, violence, denial of rights, and manipulation of evidence.[9]

I take this to imply that, first, observation can aid us in penetrating those areas that other sources of information alert us to; and, second, that those areas are more likely to be reached by observational techniques than by more formal styles of research methodology. It is in the nature of much deviant behaviour that it is concealed,[10] and this is particularly the case with police deviance. Ironically however, researching that deviance may present the field-worker with a number of methodological and ethical dilemmas about getting at 'the dirt' which begin to reflect precisely some of the dubious practices of the policeman's trade.[11] These dilemmas will be examined here in terms of a study of police corruption that I carried out in Amsterdam in the late seventies.[12]

(II) Background to the Research and Access

Elsewhere I have argued that *infiltration* constitutes the key technique of participant observation.[13] Although the word infiltration has a negative tone related to espionage and deception, I use it consciously to emphasize that entry and departure, confidence and trust, and attachment and desertion in the field may sponsor social and moral dilemmas that spell out a virtually *continuous* process of negotiation of the research role. The weakness of the concept is that it highlights getting in. Getting out may be equally, if not more, problematic. For in a sense you never fully abandon the field; you carry it with you as a seminal life-experience that you continue to mull over and to rework.

In my case the first steps to entering the largely concealed world of the police were taken in Britain where, although I was unable to conduct official field-work, I built up a close relationship with a number of officers who studied at the university where I taught. Later I gained informal entry to a number of police establishments through contacts built up at conferences and courses. The standard literature on the police, novels, press accounts, and personal socializing with policemen gradually helped to peel away the layers of myth and imagery that insulate the hidden reality of policing from public scrutiny. That reality was often profane, bawdy, and irreverent while deviousness was a central element in the occupational culture. One police informant revealed the ambivalence of this world when he told me, 'one thing you have to understand is that when you join the police you have to learn to break the law'.[14]

The primary insight that I gained then was that there exists a wide disparity between the public presentation of police work – as sober, legal, competent, professional and even 'sacred'[15] – and the backstage reality. Out of sight, there is another world of largely instrumental concerns, of simply getting through the day, of manipulation, violence, incompetence, humour and tomfoolery, and also of informal norms, rewards, and sanctions. My initial views on the police, then, were permeated with perceptions of in-built deviant practices which were also strongly conveyed in the standard American works of the time (Westley, Skolnick, Reiss, Wilson[16]) with their emphasis on brutality, corruption and systematic deviance in urban police work. In Britain, however, access to the police world via the unavoidable central vetting of the Home Office in London proved difficult for me. Then, in the early seventies, I began to focus increasingly on policing in The Netherlands (where

I have lived since 1975) and gained access to the Amsterdam City Police where I was able to conduct three projects between 1974–1980.[17]

(a) Project 1; 1974–1976. My first piece of research was an observational study of patrol work in the inner-city district of Amsterdam.[18] Most of my time (amounting to perhaps six months full-time participation spread over two years) was spent with patrolmen and I was struck at first by the *absence* of serious deviance. In fact it was going on around me, almost literally under my nose, but I did not see it (and, perhaps unconsciously, may not have wanted to see it).

(b) Project 2; 1977–1978. In the summer of 1976 I was extricating myself from the field when a corruption scandal involving policemen in the research station began to surface.[19] My reaction was to ignore it on the grounds that I simply could not afford the time to get involved in a new study. By 1977 the scandal was front-page news and, prompted by a number of academic colleagues, I decided to see if I could get back into the department to look at police deviance.[20] Permission was granted and the research was based primarily on interviewing policemen and on gathering documentary evidence.

(c) Project 3; 1979–1980. As I was teaching in a business school I decided to study senior police officers in terms of 'managing' a police district and returned to part-time observation in the original research station. This project brought me into closer touch with the world of senior officers, and also of detectives, and proved particularly revealing because I could witness the reverberations of the scandal which continued deeply to affect relationships within the force. I began to perceive more clearly that deviance is endemic to police work and found myself in effect still carrying out Project 2 under the guise of Project 3.

Gaining entry to the Amsterdam Police was quite easy and all three projects were readily accepted and supported by the Chief Constable.[21] Although some academics have encountered problems with officialdom in attempts to research the police,[22] and although Van Maanen states that 'antipathy and distrust of the academic researcher are endemic to most police departments',[23] I found nothing but open doors in gaining formal access in Amsterdam. This continues to be true for other researchers not only in Amsterdam but elsewhere in The Netherlands which must have one of the most open criminal justice systems in the world. The smoothness of my return to a department undergoing an unprecedented criminal investigation into corruption was related to the trust I had built up earlier, the presence of high-ranking sponsors for me within the department, and the fact that I was an individual researcher with no need to go through any gatekeeping institution for funds. The Chief Constable agreed to my research on corruption (Project 2) in May 1977 and promised me access to senior officers, to policemen suspected of corruption, and also to the corruption case dossier. Having said all this, it did prove more difficult to gain 'secondary' access.[24]

Initially, however, things went well and I began to arrange interviews. Being granted a second bite at the cherry proved fascinating. People seemed to be far more candid about the practical dilemmas in policing and about making mistakes than they had been during my first project while the word corruption was no longer taboo. I interviewed senior officers at Headquarters and in the Warmoesstraat (the name of the research station in the city-centre), plain-clothes men, detectives, investigating officers from Internal Affairs and the State Detectives, journalists and representatives of the legal authorities. There was no clear-cut end to the corruption scandal which continued to unravel and to enter new phases.[25] Perhaps pushing my luck I approached the Chief with my third proposal, to study senior 'management' in the research station, and observations began in late 1978. I entered this third phase with the second project uncompleted as the repercussions of the corruption cases were still echoing throughout the department, while an internal disciplinary investigation was dragging on interminably. To a certain extent, then, the two projects had an element of overlap which none of us had foreseen.

It was at this stage that I began to encounter resistance and deflection. During the first two projects I was not conscious of doors being closed, and even now no-one actually refused to be interviewed, but I was excluded from certain conversations between senior officers in the Warmoesstraat, discovered coolness outside of the Warmoesstraat, was explicitly not welcome at secret meetings of a clique of officers engaged in lobbying within and without the organization for a more stringent investigation, was asked not to interview the police suspects, was refused the case dossier which had been promised, encountered friction in the field situation, was not permitted to attend a week-long 'retreat' of personnel from the station (in relation to a planned change programme[26]), and was barred from sensitive confrontations between officers and detectives in the Warmoesstraat. In retrospect, this appears only natural whereas, at the time, I felt rather disappointed and let down. The resistance was partly due to the fact that in my third project I was getting much closer to powerful interests at the top of the organization and was discovering that my acceptance outside of the Warmoesstraat was tenuous; and partly because I was caught in the continuing backlash of the corruption affair which aggravated inter-rank relationships at the very time when I was observing both senior officers and detectives. To a certain extent access was continually problematic, particularly in the third project, and to some parts of the organization I was a stranger and was treated accordingly. Furthermore, I was gaining different sorts of data at different stages and, although the third project was not ostensibly about corruption, it did open my eyes to the scandal as part of a much wider struggle related to resilience to change and to the investigation within the organization.[27]

(III) Evidence

All information is managed to a greater or lesser extent and all observations are subject to interpretation and reinterpretation. The data for my study, then, have varying levels of validity and the observations, interviews, documents, and journalistic accounts on which I rely are all biassed in certain ways. This would be true of any study but it is particularly so in one concerned with deviant behaviour which, by definition, is sensitive and subject to concealment. In this section I wish to deal with the nature of the evidence collected and the extent to which respondents were open and truthful in their accounts.

In the first project, for instance, I had a close relationship with one group of patrolmen and a couple of them functioned as 'informants' whom I thought were reasonably open with me. Generally, my impressions, in contrast to what I had anticipated given my previous reading and the negative reputation of the Warmoesstraat, were of reasonably well-behaved policemen who were not noticeably violent, corrupt or crude.[28] Indeed, the first person who seriously broached the subject of corruption to me was a chief-inspector, who, in the summer of 1976, told me that he had been confronted with some practices among his plain-clothes men that amounted to corruption.[29] This was the first time that I heard the word mentioned seriously in the station.

Clearly, my informants had been less than honest and knew a great deal more than they were prepared to tell me. This ignorance of on-going deviance on my part probably had two causes. First, some of the out-of-work deviance, particularly that related to drinking, was often not visible to me because as a non-drinker I rarely took part and this was particularly true when I moved house away from Amsterdam.[30] Second, some of the work-related deviance was of such a highly incriminating sort that it would be dangerous for them to discuss it with *anyone*. And, yet in a way they did want to tell me. Policemen in general delight in talking and gossiping and even enjoy revealing glimpses of their underculture to outsiders. A number of researchers have remarked on the surprising openness of policemen when discussing their work and even in not disguising their deviance in front of observers.[31] My informants became particularly frank after I had completed the first project and after one of them had moved away from the department. Two men told me of dubious practices including fabricating statements, forging signatures on crime reports, and of one case where they 'were one thousand per cent certain' that a suspect was a drug dealer so they replaced his fake heroin with real heroin in order to get a conviction.

Also one of them explained to me later that it was impossible to trust anyone in that milieu, *including* your partner, and that you had to learn to keep your mouth shut, 'otherwise you are digging your own grave'. He then recounted an incident where he had searched a flat for firearms with a colleague, 'Tom'. It appears that Tom may have removed a revolver from the flat without reporting it

> When I heard that story later I felt really pissed off. This has now lead me to the practice of never letting a colleague out of my sight when we're engaged in a search. Pulling a stroke like that can land you in sticky situations. Which boss or judge is going to believe your story if it comes out later? 'You were with your partner weren't you?'

This revealing comment shows not only the *absence* of trust among policemen but also their vulnerability to control when engaged in risky escapades. Some of the patrolmen were clearly locked into intricate and intimate relationships of collusion and cooperation which would be almost impossible for an outsider to prize open.

Furthermore, some of the men justified drinking in the inner-city, which was explicitly against instructions, as a means of gathering information from underworld characters and some of them took off on risky escapades as a result of these 'tips'. On some occasions I accompanied them on these unofficial jaunts but was not always fully aware of what was at stake. Perhaps to a degree I was fortunate in *not* having made the breakthrough to the deviant practices going on in the area during the first project. This may sound a curious thing for a sociologist to say – who should be dutifully sniffing around for the first whiff of a free-meal or a handout – but, in retrospect, it kept *me* out of trouble. For example, one of the men, who was nicknamed 'Crazy Horse' because of his impulsive behaviour, went out on patrol with a new policeman. Late at night the old hand fired his pistol at a car with some youths in it which had driven straight at them when they had attempted to stop it. Not a soul was around and 'Crazy Horse', wishing to cover up his rather hasty decision, turned to his new colleague and said, 'You did not see anything, and you did not hear anything, o.k.?' They did not report the incident although all use of firearms must be documented for later scrutiny at Headquarters. Unfortunately for the two men the incident was reported later when the driver of the car was arrested and complained bitterly about being shot at. As a result, both men were identified and were carpeted before the station chief.

Now the whole point is that if I had been present then I just know that *I* too would have kept quiet and this might have damaged my credibility with senior officers when the incident was exposed. What would I have done if I had found out that women suspects were being sexually harassed in the station or that some men were on the take (as was the case); turn them in and destroy the research? What if I had gone out to a club and been invited to indulge my wildest erotic fantasies with a voluptuous lady paid for by a Chinese gambler (which was also happening)? Would I have shrugged my shoulders philosophically, thought that social science demands the occasional sacrifice for the sake of data, and plunged between the sheets while ethnographically absorbing as much of the situation as possible?

Knowledge of, and involvement in, deviant practices within an organization can be dangerous for the field-worker in terms both of sanctions from senior members if caught and of the ethical dilemma as to whether or not one should expose the practices at the cost of terminating the field-work.

In the two later projects no-one really emerged as a classical informant although one or two senior officers were particularly helpful and a certain marginal figure within the organization approached me at a late stage with a good deal of juicy inside information. I never had the close relationship with people that I enjoyed in the first project and yet people still seemed reasonably, and in some cases remarkably, frank. This was probably because the 'shit had hit the fan', people had taken sides, and I was seen as a chronicler who could represent their particular point of view. However, it proved impossible for some time to get in touch with the policemen under suspicion of corruption. The Chief Constable had suddenly requested that I postpone interviews with the suspects until after their trial. This may have been a seriously missed opportunity in that the suspects were suspended, were sitting at home biting their nails, were deeply frustrated by their treatment, were uncertain of the outcome of their cases, and might have been willing to talk. When I did get to talk to them almost two years had elapsed, their cases had been to court, they had told their stories to journalists, and had doubtless developed a pat version that they had been mulling over for two years and which they could trot out on request. As it was, I only managed to talk to four of the eight men who had been arrested and, significantly, all four were from the Warmoesstraat. My acceptability was plainly much more limited at Headquarters. In addition, the Chief Constable also went back on his promise to provide me with the case dossier which he said was now the responsibility of the Justice Department and, therefore, outside of his authority to give me. Fortunately, however, I managed to get my hands on a copy of the 300 page dossier, containing transcripts of interrogations of witnesses and suspects up to the trial in May 1978, with the help of a journalist to whom it had been leaked. Simply possessing these documents was illegal.[32]

To a large extent I never actually witnessed the phenomenon I was studying – 'corruption' – so that observation (and certainly *participant* observation) is something of a misnomer in relation to an activity that is basically concealed from view. This means that my major data sources are interviews and documents which are biassed in the sense that the interviews are forms of special pleading and the documents represent official versions of reality. During the second project it felt as if I was getting through to levels of information about the department which are not normally accorded to outsiders while I was under the impression that I was concerned with a number of cases in the *past* but that those cases revealed considerable current defects within the department.[33]

During the third phase of my involvement with the Amsterdam Police, however, I became increasingly aware that the continuous and unforeseen internal rumblings and external criticism surrounding the corruption issue had begun to overshadow the original intention of studying senior officers as 'managers'.[34] During this stage, moreover, material was gathered incidentally, and partly covertly, for the corruption research under the mantle of the third project. This opportunity came from the continued contact with senior officers who were closely involved in the affair and who occasionally said things pertaining to the persistence of the investigations, from interviewing the suspects, from obtaining the corruption dossier, from keeping in touch with a number of informants, and from field-work with detectives that largely by chance led to a glimpse of yet more widespread and persistent corruption than I had anticipated.

In a sense, I had moved into a new level of information which proved uncomfortable to handle because it identified highly-placed individuals who were difficult to disguise, the information itself was either given in confidence or might reveal my source, and because I had moved into a circuit of gossip and rumour which by its very nature was impossible to verify. Often policemen would intimate cryptically that they had a lot more to tell and I got rather tired of this and began to think that they really did not know anymore. Much of my data, then, was based on hearsay, gossip, and suspicions. Like a nervous girl at the doctor, enquiring about a 'friend' in trouble, respondents always pointed to someone else who had committed the act. None of the suspects ever levelled with me and simply said, 'O.K. Maurice, this is the way we went about it'. And, of course, some people were simply keeping quiet, having doubtless very sound reasons for saying as little as possible.

All the evidence presented in my work, then, needs to be treated with more than the usual dose of caution. The quality of press reporting was patchy and tended to follow events rather than take an 'investigative' strategy. The case dossier was incomplete and contains only what witnesses and suspects were prepared to reveal and what the investigators considered worth writing up. My acceptance within the organization was varied and I received different sorts of information from various parts of the organization. Indeed, most of the incidents referred to in the investigation occurred *before* my arrival, so that by 1976, when the first internal police investigation commenced, some people were already running for cover.

(IV) Dilemmas of the Research Role

Initially I approached field-work with a naive and almost insatiable enthusiasm. Toward the end, however, I increasingly experienced frustration, fatigue, despondency, and ethical qualms and these, together with other considerations, made me leave the field prematurely in the middle of my third project. These feelings were related to practical considerations (such as domestic constraints, professional commitments, pressure to move off the police area from within my institution) but my emotional reactions were also connected with certain features of the research role and the field-work situation.

(a) Factionalism and role-playing. There existed a different sort of relationship between me and the senior officers than with patrolmen and detectives.[35] The latter were amicable and jocular whereas the former were often busy in sedentary pursuits (reading, writing, phoning) that could not easily be shared, had less time or inclination for 'idle' gossip, and never invited me to take part in out-of-work socializing. In addition senior officers were less anonymous, were more easily identifiable, and had other career prospects to consider than the more anonymous members of the lower ranks and this made them more cautious in what they were prepared to reveal. Another aspect was that I was moving between officers and men, between 'upstairs' and 'downstairs', and I now had a highly visible role flitting between groups where I had to be carefeul not to say too much to either side. Once I was talking to a detective when another one called out 'Watch out, he'll tell it all to the bosses', and the first one replied, 'That's why I don't tell him everything'. It was said as a joke but the detectives were aware that I was closely involved with the 'bosses'. Clearly, field-work becomes more difficult when the researcher is dealing with more than one group within the same organization and particularly when there exists a high measure of factionalism.[36]

(b) Research fatigue. Another feature of my predicament was a certain weariness with the research role. Over a period of six years I carried out three projects in one and the same police station. Obviously as a researcher you are keeping up a front continually and, in the end, you do get tired of it, laughing and smiling and trotting out your potted biography, and never letting your guard drop ('the field-worker has to get used to being treated as a sort of fool and being laughed at and ridiculed').[37] The tiredness even became mutual to a certain extent as the station itself was experiencing 'research-fatigue' in that it had been exposed to negative publicity for some seven or eight years. Because of its notoriety it had become a favourite spot for journalists, television crews, foreign academics passing through (and even members of a Royal Commission from Britain), students doing a thesis, trainees from the Justice Department, and so on. One of the senior officers said to me,

> You know you should get away from the Warmoesstraat. It doesn't matter what you say about us because somehow it all turns out negatively. The reputation of this station is contaminated and journalists are only interested in material that confirms that reputation. If you come along and say something positive about us then they attack you in order to get

> at us. But it's still *our* reputation that gets blackened. And it keeps our name in the news. What we need is peace and quiet in order to be left alone to get on with our job. I really think it's better for us but also for you, and I don't mean this negatively – I'm just being realistic – and for your research to break with the Warmoesstraat. Go to another station in Amsterdam or go and look at a station in a completely different city.

(c) Disillusionment. An essential element of field-work is *sharing* the experience of a group, either as a full or partial member, with which the researcher to a greater or lesser extent identifies. Over-identification may involve 'going native' but practically all field-work implies that the researcher not only identifies, but actually *likes*, the subjects of his study. I found that it became difficult to believe that I was genuinely sharing the experiences of policemen, while I began to suffer a measure of disenchantment with the policeman's world. This was perhaps because you tend to be carried along in your first major project by boundless curiosity and naive enthusiasm, by the novelty of the experience, and by your involvement with, and attachment to, the group. Later, however, there was in my case no clear focus for identification while there were no heroes anymore as the seedy side of organizational life was exposed.

Other features that played a role in my disillusionment were a feeling that I had overstayed my welcome; a growing lack of stimulation in the field as I ceased to learn new things and as incidents became predictable. (I had been around longer than some of the policemen and when they start learning from *you* then maybe it is time to retire gracefully). And also there emerged a growing consciousness of the logistical and practical limitations of the research role. There was so much going on at so many different levels that I, as a part-time researcher, just could not cope with it. Now that my focus had broadened to include the whole organization, not to mention external environmental influences, what I really needed was four or five colleagues to cover the senior officers at Headquarters, the specialised detective-squads, the Mayor and the political environment, the Justice Department, and the underworld of the inner-city. In a sense I was conducting two projects at once and there was so much data available that I simply could not handle it all. Furthermore, field-work is meant to enable one to develop confidential relationships with key informants in the organization but 'there is a limit to the sheer number of people to whom he (the researcher) can relate on a basis of personal confidence'.[38]

Yet another element in this process of disenchantment came with increasing insight into the intractability of organizational mechanisms which trapped people in their roles and which were powerful enough to undermine and deflect the 'good guys' wishing to clean out the stables. Compared to the real battles for survival going on around me my research seemed almost futile and even self-indulgent. I was surrounded by rampant disarray and rancorous dissatisfaction and all I could say was that they were faced with difficult dilemmas and that my book would be out in a couple of years time! It was plain too that the corruption affair had created victims by a selective process that left some of the more culpable ones untouched. I found it extremely difficult to hide my feelings at questionable developments. For instance two internal commissions were set up to discuss the issue of corruption within the department and one was chaired by an officer whose passivity had probably helped to sponsor corruption while the other was chaired by an officer who was said to have had his finger, if not his entire fist, in the deviant pie.

In a way I became somewhat disenchanted with certain members of the Amsterdam Police. Like the detective, I became suspicious of everyone and felt that no-one was to be trusted. By this time, I had interviewed several of the suspects and had seen the effects that two years of personal turmoil – suspension, arrest, trial, disciplinary investigation, public exposure, etc. – had had on them and their families. Furthermore, their stories were diametrically opposed to the stories of other respondents including some highly-placed officers. It was clear that some people were lying to me, or were spreading misinformation, and I began to have this 'Rashomon' feeling of hearing widely varying versions from different people. I began to resent the resistance I was encountering and wearied of respondents who failed to turn up for successive appointments and who promised hidden documents which never materialized. Perhaps the researcher was being conned?[39]

A key issue here is not so much that it is difficult to recapture that initial excitement in the field but, that far more importantly, you become increasingly self-conscious of the social processes involved in gathering data. In 1975 I could convince myself that I was at one with a patrolman whereas in 1979 I could see through this pretence because I was a professor whereas my former buddy was still a patrolman. It seemed fraudulent to pretend otherwise than that we lived in two totally different social worlds. Now I was increasingly aware that relationships were transistory and that my purpose was not to build up friendships but to get data. I was going around being nice to everyone and smiling disarmingly at them and then later I would take down everything they said or did or that I had overheard. Maybe they understood this better than I and did not care particularly (and I was feeling guilty for nothing) but I could not help feeling increasingly uncomfortable, if not manipulative, in the research-role. If a latent aim of field-work is to create trust in the researcher then what was the aim of that trust? And did not the relationship involve a double betrayal: first by them of me but then by me of them?

In short, I felt that in field-work the subjects are conning you until you can gain their trust and then, once you have their confidence, *you begin conning them.*[40] In other words, I could not escape the realization that deceit and dissemblance were part of the research role and I did not feel ethically comfortable with that insight. Lies, deceit, concealment, and bending the truth are mentioned in many reports of field-work.[41] Indeed, Berreman states that 'participant observation, as a form of social interaction, always involves impression management. Therefore, as a research technique, it inevitably entails some secrecy and some dissimulation'.[42] At the time I found this all genuinely distressing and confusing. Evans Pritchard remarked that 'an anthropologist has failed unless, when he says goodbye to the natives, there is on both sides the sorrow of parting'.[43] In the end I decided to fade away discreetly and walked out of the station in early 1980. I have never been back.

(d) Breakthrough and dead-end. If there was one particular incident which finally confirmed my intention to depart then it was, paradoxically, on the night that the veil was lifted to reveal yet a deeper layer of deviance. Detectives in the district station, who normally only handle fairly mundane cases (compared to the specialised units at Headquarters), had become involved in a long drawn-out, complex series of cases (which I will call the 'big case') involving several professional criminals who had carried out burglaries where the victims were held hostage at gun-point. The detectives working on the big case were deeply engaged in interrogations, identity parades, searches, and arrests but there were no strong leads as suspects were plainly scared of talking in case they were labelled 'stool-pidgeons' and met with reprisals from the underworld.

Then one evening a young woman walked in with clean clothes for one of the prisoners concerned in the case. It turned out that she was the daughter of a senior police officer, was living with the suspect, and had a child by him. In addition, the suspect's alibi for a particular evening was that he was at the senior policeman's house with the daughter. Now clearly this was a highly delicate issue but one of the policemen, X, wanted to treat the woman as a suspect. He questioned her and at one stage used a phrase that intimated that she was 'whoring with trash'. The daughter of the senior officer was so incensed at her treatment that she complained to her father. He allegedly phoned X and gave him a dressing down hinting that he still had several years in the department and that X would have to reckon with him. The detectives believed that the senior officer would get his own back on X. Subsequently, X, who could have expected an attractive move to a specialised detective unit as the next step in his career was put back into uniform (which is in most policemen's eyes a form of demotion). This also meant that he came directly under the authority of the senior officer whose daughter he had insulted. The detectives were convinced that X had been 'screwed' and that the daughter had been avenged. The significance of this was that I could glimpse the informal system at work in terms of retaliatory sanctions and the manipulation of a policeman engaged in a criminal case involving the relative of a senior officer. But all my information on this matter came from gossip which was sometimes impossible to verify and too personal to be used without identifying both the people concerned and my sources.

By chance I was also present one evening when the detectives involved in the case came back from interrogating a key suspect. They seemed almost physically disgusted and, rather emotionally, they began to express considerable sympathy for the suspect.

> I tell you poor old Jansen (the suspect) has been screwed right and proper. If you listen to him then you open up a right old cesspit and it really turns your stomach. If we told you everything then your hair would stand on end. He was conned completely by detectives in 'Molendam' (a nearby town) and also by the Justice Department. He gave interviews to detectives in confidence but they were taping him secretly in a drawer and when he arrives at the Commissioner of Enquiry,[44] the tape is pulled out and played back to him in evidence. He swears straight up that he didn't shoot van Luijk (a policeman wounded during a traffic control at night in Amsterdam) but that they are determined to pin the blame for it on him. I'm telling you there's an underworld out there with all sorts of connections, and even with gun factories where you can order a pistol to your own requirements, that we know almost nothing about and where we have almost no control. And senior officers and Public Prosecutors and politicians all have their connections – it's incredible. You know the shooting in District East, when van Luijk was wounded, well the night before a detective from Headquarters was paying a friendly visit to Jansen (who was on the run from prison and thought to be armed)! We're telling you that you can't trust anyone, but especially at Headquarters. The Warmoesstraat is clean but other departments are not to be trusted and that includes Public Prosecutors as well.

Now this conversation took place in the detective room and the detectives concerned seemed out of their depth and were somewhat distraught. They were on to something but it was getting too big for them but also for me. A sergeant advised them to call in X and Y (senior officers) from their homes, put their cards on the table, and take the matter to one of the more trustworthy Public Prosecutors. X and Y arrived and went straight into a conclave with the detectives from which I was excluded. From the detectives' room the rest of us could look across to another part of the building and into X's room where the meeting was taking place. The remaining detectives were now also rather distraught and concerned. One of them ventilated their unease

> It's just like three years ago all over again (he is referring to the first corruption investigation of 1976 – MP). Then you would watch X and Y pacing around that room and wonder what was going to come next. Now they are going to use those detectives to get at the people they couldn't get three years ago. But those detectives will suffer because they'll become contaminated and they will suffer as a consequence. This district has already been through it once and now once again we'll be contaminated because Cees and Bert (two of the detectives on the big case) have

> started to push a case against powerful people. We don't want to spend the rest of our lives typing out statements for a stolen bike or a mugging because later when we apply for transfer to another department they'll say, 'Wait a minute, you're from the Warmoesstraat, the guys who screw their mates, no thanks, try somewhere else'. Believe you me, it'll be the ordinary detective who'll suffer here because you can't take on Headquarters and hope to get away with it. There are senior officers involved and we're convinced that things are just not right with some of them, that they are bent, but you can't expect to take them on and get away with it. They are just too powerful.

This incident hinted that key members of the organization could impose informal sanctions, particularly against the rank-and-file, which could be cloaked as rational decision-making in terms of suitability for promotion or assignments but which might be utilized against those merely *associated* with the exposure of deviance.

I was on the verge of a breakthrough to an area that promised to expose a network of connections in Amsterdam between policemen and criminals that perhaps also involved senior officers and even Public Prosecutors. And yet I decided to leave. Why did I get out at this stage?

There were a number of pressures outside of my research but the basic ones were disillusion with the research role and growing scepticism about getting to the mucky area exposed by this big case. I was not Woodward and Bernstein with unlimited time to spend on a case. I was, in effect, a part-time researcher and I baulked at the time and energy that would be required to complete the picture. But basically I felt that I could not get further within the police organization without betraying my purpose to expose the new vein of corruption and that too many powerful interests were at stake to allow me to do this. It was tantalizing to be part of the gossip circuit but more concrete evidence was needed if I was to piece together a much deeper investigation. But I just could not see how I would ever get to the level of information needed. It was clearly in no-one's interest to give it to me.

Perhaps we have to recognize that, as we move higher up the hierarchy of an organization and as we begin to encounter powerful entrenched groups and individuals who are identifiable, jealous of their reputations, prepared to fight for their survival, and powerful enough to deflect attention, that we may be attempting to research areas of institutional life which are to all intents and purposes *unresearchable*. Ironically and frustratingly for me, at any rate, the 'break-through' proved to be a dead-end.

(V) Ethical Considerations

Generally discussions of ethics in research centre on informed consent, deception, privacy, identity, confidentiality, and spoiling the field for further research.[45] Applying B.S.A. or A.S.A. Codes of Ethics in my field-work seemed ludicrously inappropriate as there was no way I could have employed them without effectively destroying my research. All three projects were formally permitted by the Amsterdam Police and were overt but in the third project I concealed my motives and interests and therefore engaged in a measure of deception. Disguising the name of the town ('Windmill West'?) or of the station scarcely seemed feasible as it is instantly recognizable to anyone in The Netherlands. This does mean, however, that when I talk of the Chief of the Warmoesstraat then that can only mean *one* man. When you move into research with the top of the organization, the people are far more easily identifiable than when researching the relatively anonymous lower ranks. At the same time it is true that many people mentioned in my book were public figures in the sense that their names, and their photos, were kept in the news over a period of years and that much of what I have to say about them was covered extensively in the press. Although I have used various techniques to disguise the identities of respondents (I employ pseudonyms for all the Dutch policemen) I can imagine that these will be fairly transparent for people inside the organization. But I see no-way around this without making my account so anodyne as to be almost worthless.

As much as possible I utilized my data in such a way that I hoped it could not be used to bring harm to anyone although there is no way that I can guarantee this completely. Most of the incidents are related in a style that precludes revealing material in a manner that might lead to a disciplinary or criminal enquiry. In the end I did begin to feel uneasy about the many confidences I was getting or was overhearing. Some information came in the form of inside gossip which was potentially highly damaging to individuals yet by its nature was impossible to verify. Finally I am not aware that I 'fouled the nest' to any great extent, although I gather that my publications are scrutinized closely at Headquarters for their degree of favourableness towards the department. Certainly there has been no sign of it being more difficult for researchers to gain access in Amsterdam (if anything the reverse is true). In retrospect, my research raised a number of ethical issues with regard to the conduct of field-work and, while I engaged in a measure of deception and misrepresentation, I did endeavour to follow convention as closely as possible in terms of avoiding identification and harm.

Clearly, academic researchers can end up in the same moral predicament as policemen and may even employ the same imagery of muddy boots and grubby hands ('in getting at the dirt one may get dirty oneself', according to Marx).[46] When I withdrew from the field I did so with a disappointed feeling that I had failed to investigate corruption satisfactorily. I had begun with the cosy cohesive world of the patrol group but ended up studying the predicament of a large, incoherent organization caught in the intense glare of publicity and ineptly endeavouring to set its house in order. In addition I was accorded a view of the nasty side of the organization with feuds, victims, taking revenge, pulling strokes, 'screwing' opponents, broken promises, and broken careers. I believe that part of my problem was that I commenced with what might be called a 'supportive' approach.[47] But I became frustrated at my inability to change the research paradigm in mid-stream to an 'investigative' or 'conflict methodology' approach (personal communication, M. Clarke). In the last resort I shrank from getting my hands too dirty.

(VI) The Sting in the Tail: Libel

Having withdrawn painfully from the field I set about exorcising my research through writing. In 1983 I approached Tavistock in London about publishing a book and in 1984 I signed a contract with them. In the summer of that year I handed over the final typescript and looked forward eagerly to publication. On signing the contract I was amused by the clauses mentioning rights related to Book Clubs, serials, strip cartoons and television but rather glossed over the 'Author's Warranty'. That clause contained the assurance that 'the work contains nothing obscene, indecent, objectionable or libellous' and that the author was responsible for any legal costs incurred by the publishers. Libel had not occurred to me because the corruption affair had been extensively reported in the press and also because libel cases are rare in The Netherlands where damages are low in such cases and where parties usually prefer to reach agreement out of court. I had not reckoned with the intimidating laws of libel in Britain.[48]

The first intimation of difficulty came in a letter from Tavistock in the autumn of 1984. The manuscript had been read in the publisher's legal department. The opinion was that it was libellous and therefore it was being passed on to a legal firm for a specialised appraisal. This was an extremely busy period for me and the editor was unable to contact me personally. She left an urgent message cryptically requesting my attendance at a crucial meeting with a lawyer in London. I was petrified.

From the offices of Tavistock in central London we walked to the Inns of Court and to a cramped and busy Dickensian office in Dr Johnson's Buildings. The editor, myself, a diminutive solicitor (who mutely and impassively earned his fee merely for being present), and a secretary crowded into a tiny, overheated room where a young, bouncy barrister opened the proceedings by cross-examining me on the meaning of symbolic interactionism. Having clearly failed to satisfy him on that score he then turned to my manuscript, and in an advice that was later written up in eighteen pages, alerted us to the danger of libel. In essence, he explained that many passages were potentially defamatory. He considered it unlikely that anyone would sue, mentioned that any case against me would be difficult to prove, and estimated that damages would be minimal. But costs might be high. Around £20,000 (currently about $12,000) was mentioned at which I nearly passed out.

The barrister maintained that it is defamatory to suggest that someone is corrupt, connived at corruption, or hindered an investigation into corruption. A reference could be actionable if any individual was identifiable. This was true even of references to policemen in published secondary sources such as newspapers and magazines. I had not appreciated this and assumed that it was safe to quote from published sources. A section in the book on London mentioned several real names of senior detectives who had been convicted of corruption in well-publicised trials (which had been documented in seemingly reputable sources such as 'The Sunday Times', 'The Observer', and 'The Economist'). It was necessary to check with the Registrar of Criminal Appeals at the Royal Courts of Justice that conviction had not been quashed on appeal. It was also necessary to contact other publishers in order to enquire if any writs were outstanding in respect of allegations made against policemen in London following publication of their material.

Apart from those steps there were two other measures imposed on me. First, I had to build up a dossier of my secondary sources on the Amsterdam scandal. If anyone then approached the publisher he could be confronted with the public documents on the affair as a warning that similar accusations had already been aired widely in the Dutch media. Second, a whole series of deletions and amendments had to be made in the text where individuals were identified or might be identified. At times it seemed that any negative reference, however innocuous, to a living person could be construed as libellous in British law. Eventually I provided the publisher with a number of assurances, compiled the dossier, and made a number of deletions. The publisher was only then prepared to give the green-light for the book to go into print.

This episode scared the daylights out of me. I was clearly ignorant of the law. I had not anticipated the extent of potentially libellous material in the book and was unaware of the crippling costs that legal action might impose on me. The cautionary message is that research on a sensitive area of a public bureaucracy may easily land the researcher in the courts. This is particularly the case where the institution and/or individuals are readily identifiable. In short, assurances on confidentiality and identifiability given to individuals in the field are not only required ethically but can also be essential legally. In England the libel laws are taken seriously, damages can be high, costs are potentially prohibitive for the author and publishers are intimidated by the mere threat of legal action. And policemen are especially likely to be aware of their rights to seek redress through the courts given their legal training. Let the unsuspecting field-worker beware; read the small print.

(VII) Conclusion

In this conclusion I wish to return to the three main issues I raised earlier and which were inspired by Van Maanen's contention that it proved extremely difficult for him to tell the field-work story as it 'really' was. There are several accounts that do reveal stress, conflict, physical and mental effort, drudgery and discomfort, frustration, disorientation and even danger in field-work.[49] For, in essence, the field-worker has to learn how to inveigle his or her way into the life of a group, build up contacts with key actors, and continually retain an emotional balance in order not to spoil acceptance and also to keep on collecting research material. Wax states that the most fundamental technique of all is 'alleviating suspicion'; Gans argues that the researcher must be 'dishonest to get honest data'; and Ditton claims that participant observation is inevitably unethical 'by virtue of being interactionally deceitful'.[50] Yet I suspect that field-workers have only the haziest conception of the ethical, and legal, dimensions of observation and merely wish to immerse themselves in the field and frequently do so without a great deal of forethought and preparation.[51] This is despite a considerable literature indicating that in the field almost inevitably a measure of mutual deception and impression management develops that borders on the edge of deception and untruth but that is doubtless interactionally unavoidable and even morally tolerable. To a greater or lesser extent people are evasive and may lie[52] and the researcher may feel that some degree of duplicity is required to penetrate fronts and to conceal his real purpose. In this paper I have revealed that I became acutely conscious of some of these dilemmas during a field-study and that eventually I found it impossible to continue with the research in question on police deviance.

It is interesting, then, and somewhat ironic, that some sociologists find themselves employing the techniques normally associated with morally polluted professions, such as policing and spying, and enjoy some of the moral ambivalence surrounding those occupations. The ironies and ambivalence are magnified when sociologists study deviants and run the danger of what Klockars calls getting 'dirty hands'; they may end up getting dirty themselves.[53] In entering polluted areas, the 'back' region where 'trade secrets' may be exposed,[54] the researcher faces moral dilemmas related both to involvement in the practices and to exposing the wrongdoing observed. This may cause a measure of discomfort if the academic reflects seriously on the moral and political dilemmas of field-work. For the observational method is suffused with irreducible ambivalence that may cause such emotional and intellectual confusion in the researcher that he or she abandons the field, refuses to write up the data, or becomes 'converted' and ceases to investigate altogether.[55] The moral imperative dictating that the researcher guarantees authenticity in accounts may, in practice, founder on the realization that the reality of the field evades description, the research process may seem fraudulent if not predatory, and the attempt to experience membership at first-hand may become almost self-defeating. Paradoxically, the sensitive field-worker may come to feel that the research experience is untranslatable and that even reflecting on it somehow betrays the intimacy and irreducibility of direct involvement.[56]

This raises an issue of over-sensitivity on the part of the researcher and even of prior matching of the personal style of research to the norms and culture of the setting. By that I mean that perhaps the demanding and volatile Amsterdam situation required a tougher and more cynical investigator than I could be in order not to be distracted and diverted by the personal and moral flak that the setting kept firing off and to which I perhaps over-reacted at the time. I can only confess that I was unable to alter my personality, and my 'supportive' research paradigm, in mid-research. As a consequence I may have dubbed the setting as virtually unresearchable as a rationalization for my own sense of defeat.

Reaching this painful insight has lead me to the position that it is incumbent on mature field-workers to reflect on the dynamics of the micro-politics and interactional ethics of research in order to shape their selective 'war-stories' into a body of knowledge available for analysis and instruction.[57] The moral career of the project should be an essential part of the methodological analysis of the field-work experience.[58] But this is a lofty ambition whereas in practice we are frequently left with the feeling that professional, and publishing, constraints mean that authors could tell us far more about the concealed micro-politics of research.[59] Frustrating as it may be we simply have to face up to the fact that institutionally it proves difficult for social scientists to be totally open about their research experience.[60]

Here, in somewhat self-conscious 'confessional' style (as opposed to the 'didactic deadpan' of many methodology accounts),[61] I have endeavoured to reflect on those limitations of field-work with which I was rudely confronted in attempting to study police corruption. From that research experience with the Amsterdam Police I evolved a number of views concerning observational studies. Initially I felt that most people carry out only one major field-study and then at a specific period in their career when abundant time is available and when domestic and professional commitments are low. Furthermore, once one becomes fully aware of the social processes involved in researching people then it becomes more difficult to go through with the almost inevitable deceit and dissimulation. Building up a relationship of simulated friendship with people can lead one to feel dishonest and the observer can become afflicted with doubts about deceit and fraudulence. This may be particularly true of the 'lone wolf'[62] and I consider team-research to be healthier (while I would argue that all field-workers, even experienced ones, should have constant 'debriefing' sessions with a close colleague as a release mechanism and as a technique for getting the field into focus).

The whole question of trust needs to be explicated too in order to make clear the extent to which observation is largely a *means* to establishing confidence so that one can make the breakthrough to interviews which tell it 'like it is'.[63] How much is observation actually used as a basis of *data* in observational studies because, as Holdaway has pointed out for the police and Dingwall *et al.* for ethnographies in general, many writers are lamentably obscure on this.[64] The dilemma of gaining trust is that one may be told much in confidence,[65] that cannot be used because it cannot be verified, can damage individual's careers, and can reveal one's sources.[66]

Observation often seems best-suited to diffuse projects of low visibility with lower-level participants in an organization. For, the higher you go up to hierarchy, the more likely it is that restrictions will be placed on the researcher. Deviance and sensitive issues in an organization can rarely be studied frontally and have to be approached obliquely while it may require pure good luck to make the breakthrough to the 'dark' regions.

Despite my own set-backs and frustrations in Amsterdam (with mixed and confused feelings of loyalty, abandonment, bad faith and betrayal) I remain a fervent supporter of observational studies. But then with the proviso that researchers appreciate that field-work is *demanding* and that it requires sound reflection on the ethical dimensions involved. Field-work may not be easy because the object of research may only emerge as the research develops[67] while at some time the necessity will eventually arise to finish and to draw a line under further commitment. Rather like corruption investigations, field-work never really ends and you can go on indefinitely peeling away successive layers of institutional reality (until exhaustion sets in or you are kicked out).

However, it is probably wise not to stay too long.[68] Experienced researchers speak of qualities such as stamina, determination, curiosity, vitality, and doggedness (and even of having to be young and in good physical condition, according to Wax[69]) but I cannot avoid the conviction that most successful field-studies exude two basic qualities – naive enthusiasm ('naive' because then you are still learning) plus more than a measure of good luck. Field-work also implies a form of personal commitment and involvement. This can make departure a hard process and perhaps my earlier gloom and pessimism was simply part of the field-worker's withdrawal symptoms. (This was coupled to a prolonged 'mid-career crisis' as I found myself locked into seemingly inescapable family commitments and turbulent and time-consuming administrative burdens at work.)

For on reflection I can now perceive two positive elements that may help encourage others to return to the field. First, the experience gained is part of the maturation process of the field-worker and the insights gained through substantial exposure should make it easier to continue in qualitative research. Perhaps prolonged immersion is something of a fetish retained from the anthropological tradition and the experienced researcher can later use time far more efficiently and chart his way through the social process more effectively. And, second, I have become increasingly conscious that advancing age and higher status are positively functional in gaining access to areas denied younger academics and this is particularly evident, for instance, in the world of management which I now encounter as a teacher in a business school (and this doubtless holds true of certain other professions, social groups, and institutions). Perhaps too there are periods where sabbaticals and/or an altered domestic situation can allow one a renewed opportunity to return to the field where prior experience and social maturity prove valuable in combating the strains and limitations of the research process. A whole new set of areas to be researched emerge if we think in terms of the advantages of promoting field-work for the over-forties.

In my case I researched an institution that is anathema to some of my colleagues but which I readily accepted. However, over a period of six years, my attitude to the police went through a cycle of ignorance, curiosity, admiration, disillusion and scepticism.[70] Initially I wrote empathetically about them but became uncomfortable when I came on negative material which could be damaging to them and could also be misused by others. Earlier I had seen my uniformed group in almost heroic terms whereas now I perceived policing as having the sad and sordid elements of a morality play ('the ethnographic hero is a fixer . . . winning daily victories over the hassles of life').[71] But then I too began to sense that my role was becoming increasingly less laudable and more manipulative while I occasionally felt in danger of becoming yet another victim of a bitter and protracted institutional struggle. The difficulty of conveying this accurately to those who have no first-hand knowledge of participant involvement in qualitative research so enraged me that I wanted to be able to physically force all academics into extended field-work. I was grimly prepared to frog-march reluctant colleagues out of their offices and to abandon them on sleazy street corners in slums, before the intimidating gates of mental hospitals, and within the walls of chilly morgues. My conviction was that even the most routine immersion in the field is potentially valuable in the social-moral development of the sociologist. This preoccupation with the salutary effects of field-work emerged from my own insights that participant observation can be demanding and stressful; that it can have limitations in terms of exploring the sensitive areas of organizational deviance; and that it can confront one with ethical dilemmas that may seem situationally almost insoluble.

The research in Amsterdam brought me face-to-face with my own personal limitations in trying to crack through the tough nut of police deviance.[72] As one reviewer of my book observed, the norms of the academic profession demanded that I press on fearlessly whereas, in practice, I followed the 'operational code' of Sociology which 'requires us to avoid being found, floating face-down, in the canals of Amsterdam'.[73] At the time getting out seemed the most expedient and even the most honest thing to do.

Notes

1. W. F. Whyte, *Street Corner Society*, Chicago, University of Chicago Press, 1955 (2nd edn). R. H. Wax, *Doing Fieldwork*, Chicago, University of Chicago Press, 1971. J. F.

Lofland, *Analyzing Social Settings*, New York, Wadsworth, 1971. J. M. Johnson, *Doing Fieldwork Research*, New York, Free Press, 1975.

2. M. Clarke, 'Survival in the Field: Implications of Personal Experience in Field-Work', *Theory and Society*, vol. 2, no. 1, pp. 95–123.
3. J. Yan Maanen, *Tales of the Field*, Chicago, University of Chicago Press, 1988.
4. Douglas represents strongly this 'conflict methodology' style of investigative research. J. D. Douglas (ed.), *Investigative Social Research*, Beverly Hills, Calif, Sage, 1976. For a critical discussion of it see M. Punch, *The Politics and Ethics of Fieldwork*, Beverly Hills, Calif, Sage, 1986.
5. J. H. Skolnick, *Justice Without Trial*, New York, Wiley, 1975 (2nd edn).
6. M. R. Chatterton, 'From Participant to Observer', *Sociologische Gids*, vol. 25, no. 6, pp. 502–16.
7. L. W. Sherman (ed.), *Police Corruption*, New York, Anchor, 1974.
8. W. A. Westley, *Violence and the Police*, Cambridge, Mass., MIT Press, 1970.
9. J. Rubinstein, *City Police*, New York, Ballantine, 1973.
10. D. Downes and P. Rock, *Understanding Deviance*, Oxford, Oxford University Press, 1982.
11. Norris' research with the police in Britain revealed some highly dubious and illegal practices and he felt himself in 'a quagmire of ethical considerations' related to informed consent, invasion of privacy, trust and deceit. C. Norris, 'Fieldwork with the Police: Personal Dilemmas and Professional Consequences', paper presented at American Society of Criminology, Atlanta, 1986.
12. M. Punch, *Conduct Unbecoming: The Social Construction of Police Deviance and Control*, London, Tavistock, 1985.
13. Punch, *op. cit.*, 1986, p. 11.
14. Van Maanen, *op. cit.*, 1988, p. 89, described himself as a 'walking violation' when on patrol with the police because of the illegal weaponry he was encouraged to carry by patrolmen.
15. P. K. Manning, *Police Work*, Cambridge, Mass., MIT Press, 1977.
16. Westley, *op. cit.* Skolnick, *op. cit.* A. J. Reiss, *The Police and the Public*, New Haven, Yale University Press, 1971. J. Q. Wilson, *Varieties of Police Behaviour*, Cambridge, Mass., Harvard University Press, 1968.
17. My research base was in the 'Warmoesstraat' station which is one of the six district stations in Amsterdam (which is a force of c. 3000 officers, policing a population of c. 700,000 inhabitants).
18. M. Punch, *Policing the Inner City*, London, Macmillan, 1979.
19. M. Punch, 'Developing Scandal', *Urban Life*, vol. 11, no. 2, pp. 201–30.
20. Amsterdam experienced in the early seventies a growth of drug-related crime connected to the import of heroin by Chinese 'Triads' (Bresler). Then between 1976–1980 the Amsterdam City Police Department became embroiled in a long drawn-out scandal related to accusations of entrapment, planting evidence, stealing from suspects, abusing female suspects, possessing illegal firearms, and transporting drugs. Previously The Netherlands had been remarked on for the *absence* of corruption and scandal of this magnitude was virtually unprecented. It revolved around the Chinese run drug trade in the inner-city and was a constant source of media attention. The specialized detective units were housed at head-quarters. The Chief Constable was answerable to the Mayor and to the Public Prosecutor. When faced with a possible criminal investigation involving his personnel the Chief must inform the Mayor who then asks the Public Prosecutor to mount an investigation. The Prosecutor can call on centrally-appointed State Detectives attached to the Amsterdam Court for assistance in conducting the enquiry. F. Bresler, *De Chinese Mafia*, Amsterdam/Brussel, Elsevier, 1981.

21. M. Punch, 'Backstage: Observing Police Work in Amsterdam', *Urban Life*, vol. 7, no. 3, pp. 309–35. The 'research bargain' (Becker) was very diffuse and while I hinted that there might be some form of feedback, of utility in coping with the issue of corruption, I made it plain that my interview material was confidential and also that I intended to publish my material in some form. No formal restrictions were placed on me by the department at all and I was not asked to sign a contract. H. S. Becker, *Sociological Work*, London, Allen Lane, 1970.
22. R. Reiner, 'Assisting with Enquiries: Problems of Research on the Police'. Paper presented to British Sociological Association Conference, Warwick University, 1979. R. V. Ericson, *Making Crime: A Study of Detective Work*, Toronto and Vancouver, Butterworth, 1981.
23. J. Van Maanen, 'On Watching the Watchers', in P. K. Manning and J. Van Maanen (eds) *Policing: A View from the Street*, Santa Monica, Calif, Goodyear Publishing Company, 1978, p. 63. Van Maanen was denied access to fourteen police departments. Also Wax, *op. cit.*, 1971, p. 47, remarks that if the researcher 'tries to obtain membership in elite groups (such as physicians, big business executives, Orthodox Jews), he will soon "observe" that most doors are firmly closed in his face'.
24. *Ibid.*, p. 61.
25. Punch, *op. cit.*, 1982.
26. M. Punch, 'Organizational Change and Police Deviance', paper delivered at American Sociological Association Conference, Toronto, 1981.
27. *Ibid.*
28. From the early seventies the Warmoesstraat had been bathed in negative publicity related particularly to violence and racial discrimination, Punch, *op. cit.*, 1979. Its officers frequently attracted complaints for abusive or aggressive behaviour.
29. This chief inspector tried to reorganize his plain-clothes squad and stumbled on deviant practices. His determination to push the matter lead to a snowball effect; plainclothes men pointed at detectives, detectives asked that attention be paid to specialised units at headquarters, and some suspects began to implicate senior officers.
30. I have never acquired a taste for alcohol. This 'handicap' (which would seem to debar me from any serious research with the police – certainly in Britain where among detectives the liquid lunch is virtually mandatory) does mean that I cannot actually join in drinking sessions but, on the other hand, I console myself with the thought that, by remaining sober when I do go along, I can continue to collect data and was even in demand as a driver for drunken policemen. I've always wondered what it must be like trying to do field-work when you are drunk particularly when you try to reconstruct events and conversations from the night before (contributions please to a special issue of 'Contemporary Ethnography').
31. Reiner, *op. cit.*, 1979, p. 12. C. P. Florez and G. L. Kelling, 'Issues in the Use of Observers in Large Scale Program Evaluation: the Hired Hand and the Lone Wolf', Unpublished paper, Kennedy School of Government, Harvard University, 1979. Reiss, *op. cit.*, 1971.
32. Eight men were arrested (four men from the research station and four from specialised units at headquarters). Six of them came to court and five were fined. They all faced a disciplinary enquiry as well as a criminal investigation. Eventually they either retired on health grounds or were taken back after their long suspension in non-detective functions. No senior officers were ever prosecuted. Technically it is illegal to possess the corruption dossier and to quote from it. But one of the suspects received a copy from his lawyer and sold it to the press. The media were not penalised for reproducing extensive extracts and my legal advice was that no-one would bother prosecuting an academic using it for research purposes providing individual identities were concealed. Rules of 'sub judice' are not enforced in The Netherlands.

33. M. Punch, *Management and Control of Organizations*, Leiden/Antwerpen, Stenfert Kroese, 1981.
34. M. Punch, 'Officers and Men: Occupational Culture, Inter-Rank Antagonism, and the Investigation of Corruption', in M. Punch (ed.), *Control in the Police Organization*, Cambridge, Mass., MIT Press, 1983, pp. 227–50.
35. M. Cain, *Society and the Policeman's Role*, London, Routledge and Kegan Paul, 1973. Ostensibly I was studying 'management' by senior officers of a police district but, as their work was confined to office-hours, I used the opportunity to spend evenings and weekends with the detectives.
36. Those features which I employed consciously in studying senior officers, such as my new status as professor at a 'respectable' business school and wearing formal clothes, distanced me from my old comrades from the uniformed branch whom I would pass amicably, but uncomfortably, on the stairs or in the canteen.
37. Wax, *op. cit.*, p. 370.
38. L. Klein, *A Social Scientist in Industry*, London, Gower, 1976.
39. Ericson, *op. cit.*, 1981, p. 232.
40. N. Fielding, 'Observational Research on the National Front' in M. Bulmer (ed.), *Social Research Ethics*, London, Macmillan, 1982, p. 370.
41. Florez and Kelling, *op. cit.*, p. 13. Van Maanen, *op. cit.*, 1978, p. 328. Wax, *op. cit.*, p. 103.
42. G. D. Berreman, *Behind Many Masks: Ethnography and Impression Management in a Himalayan Village*, Ithaca, N.Y., Society for Applied Anthropology, Monograph nr. 4, 1964, p. 18.
43. E. E. Evans-Pritchard, 'Some Reminiscences and Reflections on Field-work', *Journal of the Anthropological Society of Oxford*, no. 4, pp. 1–12.
44. A judge who does not sit on a case but who can conduct further investigations of a case in close cooperation with the Public Prosecutor.
45. C. B. Klockars and F. W. O'Connor (eds), *Deviance and Decency: The Ethics of Research with Human Subjects*, Beverly Hills, Calif., Sage, 1979. M. Bulmer (ed.), *Social Research Ethics*, London, Macmillan, 1982.
46. G. Marx, 'Notes on the Discovery, Collection and Assessment of Hidden and Dirty Data', Unpublished paper delivered at Society for the Study of Social Problems Annual Meeting, New York, 1980. Fielding, *op. cit.*, p. 96.
47. 'We fall into deep sympathy with the people we are studying', Becker, *op. cit.*, p. 124.
48. F. R. Wallis, *The Road to Total Freedom*, London, Heinemann, 1976.
49. R. G. Burgess (ed.), *Field Research: a Sourcebook and Field Manual*, London, Allen & Unwin, 1982. G. McCall and J. L. Simmons (eds), *Issues in Participant Observation*, Reading, Mass., Addison Wesley, 1969.
50. Respectively, Wax, *op. cit.*, p. 79. H. J. Gans, *The Levittowners*, London, Allen Lane, 1976, p. 46. J. Ditton, *Part-time Crime*, London, Macmillan, 1977, p. 10.
51. Anecdotes recount that Hughes encouraged students at Chicago to get out of the nest quickly and 'fly on their own' (Gans, *op. cit.*, p. 310) without the benefit of much instruction; also that Becker was almost appalled when a British student asked him what paradigm he should use in the field and Becker advised him to stop worrying and just 'to get in there and see what's going on'; M. Atkinson, 'Coroners and the Categorisation of Deaths as Suicides: Changes in Perspective as Features of the Research Process' in C. Bell and H. Newby (eds), *Doing Sociological Research*, London, Allen & Unwin, 1977, p. 32. Wax, (*op. cit.*, p. 61) reinforces this idea of diving in without elaborate preparation beforehand when stating 'like many field-workers I went into the field ill prepared, technically and intellectually'.
52. J. Van Maanen, *Qualitative Methodology*, Santa Monica, Calif., Goodyear Publishing Company, 1979, p. 539.

53. C. B. Klockars, 'Dirty Hands and Deviant Subjects' in C. B. Klockars and F. W. O'Connor (eds), *op. cit.* Marx, *op. cit.*
54. E. Goffman, *The Presentation of Self in Everyday Life*, Harmondsworth, Penguin, 1959.
55. Clarke, *op. cit.*
56. *Ibid.*
57. Van Maanen, *op. cit.*, 1988.
58. P. K. Manning and L. J. Redlinger, 'The Political Economy of Fieldwork Ethics', in C. B. Klockars and F. W. O'Connor (eds), *op. cit.* P. Hammond (ed.), *Sociologists at Work*, New York, Basic Books, 1964.
59. In my manuscript for Tavistock I tried to raise some of the points covered in this paper on the assumption that this account was essential to understanding the research. Before the question of libel arose, and caution was called for, I received comments from the reader for the publisher who wrote unfavourably of the methods chapter as 'anecdotal, over-subjective, too flippant, almost storytelling'. I had hoped that the chapter would be central to the book, and would be placed as an early chapter before the main text, but, following the libel scare, it was banished, bowdlerised and condensed to just a few pages, to a forlorn appendix.
60. C. Bell and H. Newby (eds), *op. cit.*
61. J. W. N. Watkins, 'Confession is Good for Ideas', *The Listener*, vol. 69, 18 April, pp. 667–8.
62. Florez and Kelling, *op. cit.*
63. Cain, *op. cit.*, p. 19.
64. S. Holdaway, *The Occupational Culture of Urban Policing: An Ethnographic Study*, Unpublished Ph.D., University of Sheffield, 1980. R. Dingwall, C. Payne and I. Payne, 'The Development of Ethnography in Britain', Mimeo, Oxford, Centre for Socio-Legal Studies, 1980.
65. One can perhaps be told *too* much, c.f. Marx, *op. cit.*, p. 13.
66. This was exactly my position on the 'big case'. For I had a tenuous connection with the detectives anyway and could not associate myself closely with the four detectives working on the 'big case' because it would have aroused suspicions. In addition, I myself was virtually in the position of a detective with a key informant in that any information I used could easily be traced back to that source. This was complicated by the fact that two of the detectives were suddenly attached to headquarters in order to follow-up on the case. Now they had a different boss and new loyalties and, who knows, they may have been moved to headquarters – which they themselves had condemned as untrustworthy – in order to get more control over them. I tried to keep in touch with them but discovered that they would talk far less readily now that they were part of a specialised unit at headquarters. Furthermore at one stage I was approached by a marginal figure in the organization who, out of a mixture of distrust of senior officers and sympathy with the plight of the suspects, had become involved in the affair and who had a great deal of inside knowledge that was of value to me. He almost implored my help in a bid to tackle the problems surrounding deviance in the department. But I neatly, if guiltily, side-stepped the issue in order to avoid an involvement which would have forced me to show my colours.
67. J. A. Roth, 'Comments on "Secret" Observation', *Social Problems*, vol. 9, no. 3, pp. 283–4.
68. Wax, *op. cit.*, p. 45.
69. *Ibid.*, p. 93.
70. Van Maanen, *op. cit.*, 1978, p. 334.
71. Dingwall *et al.*, *op. cit.*, p. 16.
72. Wallis muses as to whether or not his field days are over now that he has been scarred by various legal skirmishes and by a series of hassles (and highly embarrassing

harassments) related to his research on religious sects. In Amsterdam some police officers had their phones secretly tapped by suspects and also had received threats of violence to their families. Some witnesses were beaten up. Around that time a senior officer on another case had a brick thrown through his window at home. I became worried both about being called as a prosecution or defence witness, in which I might be asked to reveal the confidences of my interviews, and also about being visited by some of the suspects to frighten me off divulging anything to senior officers or officials of the Justice Department. In fact, my suburban tranquillity was never disturbed and publication sparked off an almost disconcerting silence. An English language book on a Dutch police scandal brought no response from the Dutch police, the Dutch media, or for that matter, from Dutch social science. I almost felt like throwing in the sponge and was acutely conscious of a lack of recognition of what I had been through and felt frustrated at an inability to fully convey the world I had experienced which was one of faces, smells, situations, and emotions which were for me indelible, colourful, and three-dimensional. 'Is it all worth it?' was my maudlin preoccupation of the time and I mused about writing a novel as the only way to portray effectively in print the richness and complexity of the stories behind the stories. See also, R. Wallis, 'My Secret Life: Dilemmas of Integrity in the Conduct of Field Research', unpublished paper, Queen's University, Belfast, 1986.

73. D. Smith, Review of M. Punch (1985), 'Conduct Unbecoming', *Sociology*, vol. 20, no. 3, pp. 486–7.

80

Potential Sources of Observer Bias in Police Observational Data

Richard Spano

1. Introduction

Since the discovery of police discretion, researchers have recognized the potential for bias, inequities, and prejudice in the application of the law by the police (Walker, 1992). Police researchers have documented the limitations of organizational controls on police behavior "on the street" as well as the factors that influence the behavior of beat officers (Reuss-Ianni, 1993). There is also a growing appreciation for the role of field research and observational data in the study of discretionary decision-making given the limitations of other sources of information on police behavior. Official data in the form of police records are collected on a very limited number of police behaviors that are typically related to law enforcement (e.g., the decision to arrest, use of force, traffic enforcement, etc.). More importantly, police records include a limited number of possible explanatory variables that are usually legal and/or demographic in nature (e.g., seriousness of the offense, offender's age, sex, and race), but little if any information about the officer (e.g., work orientation), the nature of his/her interaction with the suspect (e.g., suspect demeanor, the presence of a weapon), or the more general social context of the encounter (e.g., number of bystanders).

In contrast, police observational data have strengthened our understanding of the determinants of police behavior by examining explanatory variables

Source: *Social Science Research,* 34(3) (2005): 591–617.

that encompass multiple units of analysis (e.g., officer and organizational characteristics) (Mastrofski and Parks, 1990). For example, police researchers have observed officer behavior during encounters with citizens and found that police behavior is strongly influenced by the structural characteristics of the situation or "situational factors" such as, the nature of the problem, the characteristics and behavior of involved citizens, and the location of the encounter (Riksheim and Chermak, 1993).

In short, simply "being there" gives the researcher access to events, aspects of police behavior, and situational dynamics not recorded in police reports and arrest records (Emerson, 2002a, p. 2; Ferrell, 1998, p. 27). However, some researchers categorize field research as a "marginal methodology" that produces data of questionable quality (Denzin and Lincoln, 2000; Goodwin and Horowitz, 2002). The most serious challenge centers on the concept of observer bias. Since the observer is part of the context of observed behavior, observers can potentially "bias" or contaminate observational data and undermine its reliability and validity (Tedlock, 1991). Although aspects of the immediate context of encounters with citizens (i.e., situational factors) are the strongest predictors of police behavior, there has been little research on how the *observer* acts as a key part of the immediate context of police behavior and can influence police behavior.

Over three decades ago, Reiss (1968, p. 358) remarked that ". . . there are almost no systematic data with respect to observer effects in observation studies in natural settings." His comment still applies today since few criminal justice researchers have examined the methodological issues related to observational data beyond anecdotal and fragmented accounts of problems encountered while in the field (e.g., Hunt, 1989; Manning, 1976; Reiss, 1971; Skolnick, 1966; Van Maanen, 1988). Surprisingly, there have been very few in-depth analyses of the impact of observer bias on police observational data.[1] The purpose of this study is to identify potential sources of observational bias and to conduct a limited test of its impact on police observational data. First, anecdotal accounts of problems encountered in a variety of field settings are synthesized and categorized into four potential sources of observer bias.[2] One of the strengths of the ethnographic field research literature is the diversity of settings because it suggests that observer bias is a problem that transcends the narrow research interests of a typical field researcher.[3] Next, I will describe descriptive and quantitative police observational data from the Project on Policing Neighborhoods (or POPN) that will be utilized to examine the impact of observer bias in this large-scale observational study of police. Specific research questions will be developed utilizing concepts derived from the extant literature to operationalize three out of four types of observer bias (reactivity, culture shock, and burnout) and assess its impact on POPN observational data. The final section of this paper will describe more general issues not addressed in the analysis of

POPN observational data and how a semi-structured field diary can be used to increase our understanding of observer bias in future research.

2. Four Types of Observer Bias

2.1. Reactivity

One common criticism of observational data is that research subjects may react to the presence of an observer by changing his/her behavior. Two aspects of police culture suggest that police observational data are susceptible to reactivity. First, police officers feel socially isolated from a hostile public (Skolnick, 1966) and an organization that attempts to control their behavior through punitive accountability mechanisms (Brown, 1988). Second, police are secretive about their behavior (Crank, 1998). Since the police must sometimes make quick decisions with incomplete information in unpredictable situations, mistakes are inevitable. "Street cop culture" attempts to shelter its activities from scrutiny since both the public and management cannot understand the need for "shortcuts" or the prevalence of mistakes due to the nature of the work (Reuss-Ianni, 1993). Loyalty to the immediate workgroup transcends concerns about adherence to official police policies as officers attempt to protect co-workers from disciplinary action through the development of a code of secrecy. Even new recruits must display their loyalty to their colleagues (e.g., backing up an officer) before becoming accepted and reaping the benefits of mutual protection from their immediate workgroup (Van Maanen, 1974).

In short, police are not used to having their decisions scrutinized by supervisors, let alone an observer who is not only a stranger, but also an outsider. Hunt (1984) suggested that the establishment of trust or rapport is necessary to get access to "the hidden dimensions of the subjects' world" (p. 283). For example, Skolnick (1966) posited that a police researcher who only rode with police officers for a day or two would get a superficial "whitewash tour." The implication is that police officers will give socially acceptable answers when questioned and shelter observers from the more brutal aspects of police work during the early stages of fieldwork. Since police culture also suggests that police are preoccupied with danger and violence, officers may initially be reluctant to expose observers to dangerous aspects of police work because they fear that observers may: (1) criticize their decision-making; or (2) be unable to provide physical protection from disrespectful or aggressive suspects.

Given these concerns, criminal justice scholars have recognized the potential for reactivity in observational data. However, anecdotal evidence from a variety of research settings suggests that research subjects gradually become accustomed to the presence of observers and begin to act naturally and spontaneously (Emerson, 2002b; Leo, 2002; Skolnick, 1966; Westley, 1970; see Spano, 2002, 2003 for a review). Van Maanen (1988) argued that his participation in the exciting and mundane aspects of police work, his visibility as an observer, and simply his presence eroded any reactive effects on the behavior of Union City police officers. Similarly, Martin (1980, p. 225) argued that time in the field was necessary to reduce reactive effects.

2.2. Culture Shock

Immersion into an alien environment can be an intense and unsettling experience (Emerson, 2002c). Culture shock refers to feelings of disorientation and anxiety that observers feel during the early stages of field research (see DeWalt and DeWalt, 2002 for a review). Unobtrusive working relationships with research subjects must be developed quickly even though the field researcher is bombarded with an overwhelming amount of information when first entering the field (Hammersley and Atkinson, 1995). Field researchers have remarked that they feel like helpless children since they often lose their bearings and cannot effectively communicate because of the use of technical jargon by research subjects (see Hunt, 1989 for a review). As a result, researchers feel anxious about being rejected by research subjects and incapable of collecting any observational data, let alone carrying fieldwork through to completion. For example, Wax (1971) experienced understandable suspicion and hostility while studying Japanese-Americans in a relocation center after the Second World War. She felt like a failure since she obtained very little data after the first month of fieldwork. She also reacted to intense feelings of inadequacy and anxiety by crying, writing letters to friends and relatives, or overeating so much that she gained 30 pounds in three months.

In short, fieldworkers may initially feel depressed, inhibited about conducting certain activities, and paranoid about being cheated, plotted against, or laughed at by research subjects. Howell (1990) examined the extent of mental distress of anthropologists and found that 14% of the sample experienced depressive episodes and 16% reported instances of anxiety. Culture shock implies that observational data collected in the early stages of field research are less accurate than data collected at later stages (Béteille and Madan, 1975). Field researchers cannot simply discard this information because it contains unique and irreplaceable information about reactivity. As a result, observational studies always contain an unknown amount of data contaminated by culture shock.

2.3. Going Native

Gans (1982) argued that observers become sympathetic toward individuals under study for a number of reasons. One reason is that observers and research subjects work together in close quarters over the course of fieldwork. A natural consequence of the observer's task is a more sympathetic outlook toward individuals under observation. A second reason is that researchers feel a pull toward increased involvement while collecting observational data (see Adler and Adler, 1987; for an overview). Pollner and Emerson (2002) argued that research subjects could use observers as an assistant and a resource while conducting their work. This level of involvement can range from innocuous requests for observers to act as a "go-fer" to participation in more critical activities. For example, while studying drug dealers, Adler (1985) loaned drug dealers money, watched their children for extended periods of time, testified in court on their behalf and allowed them to conduct drug deals in Adler's home. While observing police officers, Norris (1993) not only made tea for the shift, but also helped to chase and arrest suspects and administered first aid following suicide attempts. Since he did not look like a police officer, Skolnick (1966) aided police by walking into a bar to find a dangerous armed robber. Skolnick also helped some officers get past a lookout by driving a disguised truck up to a building.

A more general concern is that observing research subjects in their natural settings has the potential to resocialize the researcher. This process of resocialization has been described as "overapport" and conceptualized as the problem of "going native," since observers lose sight of their role as a researcher and, in extreme cases, begin to think of themselves as members of the group under study. Observers may become involved as participants and systematically influence the phenomenon under study (Manning, 1976; Pollner and Emerson, 2002). Reiss (1968, pp. 365–366) described one instance of observer socialization that occurred during an observational study of police. In a precinct located in a high crime area populated predominantly by African-Americans, a common practice in the stationhouse was to walk down the row of cells and flush the toilets when an inmate asked for water. One evening an inmate asked for water and an observer imitated the officers by walking down the cellblock and flushing the toilets. On his last shift, one officer asked the observer what he learned over the summer. The observer shocked a fellow observer by replying "I learned to hate niggers." In extreme cases, observers could transfer their loyalty to research subjects, completely abandon data collection, and refuse to return from the field. Anthropologists who have gone native have stopped publishing material completely, rejected academia, and decided to live with indigenous people who were the topic of inquiry (Gronewold, 1972).

The values of the fieldworkers are potential manifestations of going native since the researcher's attitudes can change over the course of fieldwork as a result of socialization. Reiss (1968) found a pattern of adjustment in observers over the course of fieldwork: all of the observers became more pro-police. McCall (1978) speculated that observers who participate in police work may be more susceptible to the biasing effect of going native. Participant observation research implies a certain amount of participation since observers attempt to understand the motives and behavior of research subjects in their natural surroundings. Concerns about going native are related to broader concerns about the need to strike a critical balance between participation and observation that is implicit in all field research projects (Harrington, 2002; Liebling, 2001). The blurring of the line between observer as participant and observer as researcher is one manifestation of going native. Observers could also ignore or distort undesirable or unethical behavior, leading to collection of invalid data (Gans, 1982) since "our assumptions define and limit what we see . . . even if this involves distortion or omission" (Johnson, 1953, p. 79).

2.4. Burnout

Burnout refers to inaccurate documentation of observational data at later stages of fieldwork due to the mentally and physically demanding nature of data collection (Fine, 1993; Lofland and Lofland, 1995). Punch (1986) compares fieldwork with the strain of constantly being on stage. The mentally draining aspects of fieldwork deal with the strain of "playing the fool," being polite at all costs, and maintaining focus and fighting off lapses in concentration during fieldwork. The collection of observational data also requires a high level of physical stamina since collecting observational data is a labor intensive process. To convert detailed field notes into usable data, researchers may have to spend several hours reworking field notes for every hour spent collecting them (Sanjek, 1990). Hammersley and Atkinson (1995) warned that field researchers should set aside plenty of time to write up field notes and then at least double it since it takes longer than expected.

Since field notes are 'summarizations' of social context, social dynamics, and specific recalled exchanges, they reflect the structure of memory processes (Emerson, 1981). Fatigue or exhaustion could undermine the accuracy of an observer's memory, reducing the accuracy of observational data over time (Georges and Jones, 1980). Observers could also be motivated to cut corners if the observer was too tired to complete the work. Intense observation schedules and high levels of workload per observation session could increase worker fatigue and exacerbate the physically and mentally taxing nature of field research. The end result: observational data collected at later stages of field work could be less accurate and less comprehensive than data collected at earlier stages.

3. Data

This study utilizes POPN observational data of patrol officers collected in Indianapolis and St. Petersburg during the summers of 1996 and 1997, respectively. Shifts were selected for observation based on time of day and day of the week, but busier days (Thursday through Saturday) were over-sampled to maximize the number of police–citizen encounters. Observers rode with patrol officers in 12 of Indianapolis' 50 patrol beats and 12 of St. Petersburg's 48 patrol beats. Beats in both sites were selected to capture variation in service conditions within each jurisdiction. In addition, an attempt was made to match service conditions for beats selected from St. Petersburg and Indianapolis (Mastrofski et al., 1998). Thirty-seven observers accompanied patrol officers during 729 shifts and documented a total of 2488 encounters with suspects. Observers accompanied patrol officers over the course of a shift and were trained to unobtrusively take brief field notes on their behavior. Observers later prepared a detailed descriptive narrative account of events[4] and coded items associated with observed events on computers located at the research office. The most detailed information focused on police encounters with suspects. Before fieldwork and immediately following fieldwork, observers also completed a survey that contained their attitudes toward police, career aspirations, and demographic information. This study also draws on casual conversations with observers during data collection. Data collection protocols dictated that observer rode with officers for a shift one day and spent the next day inputting quantitative data on police behavior and writing up descriptive accounts of events observed in a computer laboratory. Informal conversations with observers provided additional anecdotal evidence of observer bias that were used to substantiate the measures of observer bias in this study.

3.1. POPN Descriptive Data on Reactivity

POPN had two types of descriptive data on reactivity prospectively built into its research design. One source of descriptive data was an assessment of ride level reactivity. After each shift, observers recorded descriptive data that gave an overall assessment of how officers reacted to the presence of the observer. Observers were instructed to watch and listen to patrol officers over the course of a shift. Observers watched for changes in an officer's attitude toward the observer (e.g., facial expressions, comments to other patrol officers, expressions of rage or acceptance, etc.). More importantly, observers tried to identify changes in officer behavior and, if possible, the motivations behind those changes. For example, the patrol officer may change his or her behavior by saying that he or she wants to show you a drug dealer. However, the motivations behind doing so are numerous: to avoid looking lazy, to look good in the eyes of the observer, or for safety.

The second source of descriptive data is accounts of activities and encounters with reactivity. For each event over the course of a shift,[5] observers were instructed to describe in the narrative what the patrol officer did or what the patrol officer said to someone else that would lead a reasonable person to presume that the behavioral change was due to the presence of the observer. Thus, while measures of ride-level reactivity are more general descriptions of an observer's assessment of reactivity over the course of a ride, activities and encounters with reactivity are more specific examples of *events within a ride* where observers felt their presence affected the behavior of the patrol officer under observation. The end result of my review of POPN descriptive data on reactivity was the creation of a coding scheme that primarily focused on identifying important aspects of interpersonal dynamics of reactivity (i.e., social cues from patrol officers that were potential manifestations of reactivity) as well as patrol officer behavior susceptible to reactivity (Spano, 2003). This coding scheme was used to provide more direct evidence based on interpersonal exchanges between officers and observers as a justification for the quantitative measures of reactivity and outcome measures utilized in this study. For example, in the descriptive data on reactivity, observers documented that police were reacting to their presence by altering their arrest and use of force behavior. As a result, these two types of officer behavior were included as outcome variables.

3.2. POPN Quantitative Data on Reactivity

POPN data collection also included a variety of coded, quantitative data on citizen and patrol officer behavior. Coded data on activities and encounters included items where observers specified if they felt the police changed their behavior because of the observer's presence. Observers coded whether they helped the police perform any tasks (e.g., gave advice or performed some physical aspect of police work). Observers also coded whether or not the citizen changed his or her behavior as a result of the presence of the observer.

3.3. Systematic Social Observation versus Small-scale Field Research Projects

Observational data are typically collected as part of small-scale participant observation or observational field research project. Typically, a single field researcher conducts interviews to compliment direct observation of research subjects. The observer is the primary data collection instrument and faces challenges that are common to all observational studies whether large or small: the researcher, a stranger, must document and describe the behavior of individuals under study while disrupting the normal flow of events as little as possible. The researcher also becomes part of the context of observation since s/he both changes and is changed by this process.

Although these small-scale studies provide an important starting point for identifying potential sources of observer bias, they have several important limitations. First, most field researchers give retrospective accounts of problems encountered in the field and observer bias. Since memory decays over time, these accounts could be distorted or exaggerated to an unknown degree (Reiss, 1968). Second, and more importantly, these retrospective accounts give only anecdotal accounts of observer bias so the exact prevalence of observer bias in observational data is unknown. Third, the lone researcher must often make a choice between direct observation and relying on second-hand accounts (Geertz, 1967). Given the limited amount of time that researchers have to collect data, the labor intensive nature of direct observation, and the breadth of the information included in published accounts, it is not surprising that at one time most anthropologists favored interviewing key informants over direct observation (Hammersley and Atkinson, 1983).

POPN observational data fall squarely in the tradition of systematic social observation (SSO) of police that draws from a tradition of anthropological field research of direct observation "in the field" (see Mastrofski et al., 1998). This form of field research (a.k.a. ethnographic field research) has been utilized in some classic studies of policing in the 1960s and 1970s (e.g., Skolnick, 1966). In SSO of police, POPN observers were instructed to accompany officers wherever they go unless the officers believe that safety is an issue. Much like ethnographers, observers make brief field notes on a small notepad during the shift to help them to reconstruct what they observe. Following each observation session, POPN observers wrote detailed narrative descriptions of the events that occurred during the shift. This descriptive data derived from direct observation is very similar to the observational data collected in small-scale field research projects. The more systematic and structured part of SSO by POPN observers included the collection of coded quantitative data on police behavior utilizing a survey instrument that was used to prompt the observer's memory and standardize the recording of the quantitative data (Mastrofski et al., 1998).

3.4. The Reliability and Validity of POPN Observational Data

As noted above, researchers have recognized the limitations of police observational data. One limitation of all observational data is that it is based on observer reports. Given that all measures are based on observer accounts, to an unknown degree, each observer could have been more or less attentive and/or contentious about recognizing and then recording descriptive data and coded data on police behavior or reactivity. The potential for observer error is implicit in all observational data whether large or small-scale and this study is no exception (Reiss, 1968; Spano, 2002). However, there are three different sources of error or bias in POPN data compared to smaller scale projects. First, the large number of POPN observers in the research team may partially mitigate the problem of going native since observers can act as a source of social support for one another. In contrast, small-scale studies are often conducted by a single researcher who has no source of social support to reinforce and remind them of their role as a researcher. Second, the additional step of entering the quantitative data on police behavior makes burnout more likely since data collection is more labor intensive. Finally, there is a related concern about potential coding errors in the quantitative data that is not applicable to smaller scale projects.

As a result, the issue of the reliability and validity of police observational data has been addressed in a number of studies. In one study, observed officers ($N = 20$) were given an anonymous survey post-observation where they were asked to assess the impact of the observer's presence on police behavior (Mastrofski and Parks, 1990). Observers also compiled descriptive data after each observation session where they documented whether their presence affected police behavior. Consistent with the observer's accounts, frequency distributions of reactivity related items across five different types of officer behavior derived from officer survey showed that about one-fourth of the officers were more proactive due to the presence of the observer. Mastrofski and Parks recognized that it is open to question whether officers are good judges of the impact of reactivity on their behavior. However, anecdotal evidence from observer's accounts provided additional support for their central finding.

Mastrofski et al. (2002) noted several reliability checks that were prospectively built into the POPN research design. Before fieldwork, observers rode along with patrol officers while completing a semester long course to become familiar with POPN instruments and data collection protocols. In addition, both before and during fieldwork, the reliability of the observers' coding was cross-checked by coding videotaped police–citizen encounters. Finally, during fieldwork, the coded observational data was checked by coding supervisors and a computerized quality control program that assessed the internal logic of the observers' coding. Observers immediately corrected any inconsistencies in the coding or unclear events.

In terms of observer bias, most studies have focused on the impact of reactivity. Mastrofski et al. (2000) noted that observers documented that there was evidence of reactivity during only one-half of 1% of encounters with the public. In a more recent study, Mastrofski et al.'s (2002) post hoc analyses revealed no statistical evidence of reactivity based on observer race, sex, and observer's attitudes about the police. In contrast, a more recent multivariate analysis by Spano (2003) showed that officers who expressed concerns about safety were less likely to arrest suspects when accompanied by female versus male observers.

4. Rationale and Description of the Analysis

The substantive thrust of the meager literature on observer bias is difficult to determine since the results are fragmented and anecdotal in nature. More importantly, the analyses from this study utilize (secondary) data originally collected for other purposes. The few studies done to date have focused on reactivity since these measures were prospectively built into POPN's research design. Although not ideal, POPN data allow for the measurement of *three out of the four types of observer bias* and therefore provide an opportunity to conduct a limited test of the impact of observer bias in this large-scale observational study of police. However, since this is a post hoc assessment, this study could best be described as a heuristic. On a more positive note, the large number of POPN observers and the large number of shifts (N = 729) and encounters with suspects (N = 2488) provide a much larger pool of examples of observer effects compared to anecdotal and often retrospective accounts by a single field researcher. In addition, the structure of POPN data collection allows for the analyses of items from observer surveys, qualitative coding based on descriptive data on reactivity, and nontraffic encounters with suspects within the larger quantitative data set from a variety of different "vantage points" or units of analysis (e.g., shift level measures of rapport [N = 729], analyses of the decision to arrest and use force that focus on nontraffic encounters with suspects [N = 2488], assessing shifts in observer attitudes toward police derived from observer surveys [N = 32]). More importantly, it is unclear whether observer bias is a systematic form of bias within the data or a more benign and randomly occurring phenomenon. Since the impact of observer bias has not been systematically examined in the literature, the presence of a systematic form of bias has important implications for the developing body of research that is based on observational data of police arrest and use of force behavior.

Two criteria are utilized to assess the impact of observer bias on police observational data. First, this analysis will make inferences about the presence of observer bias by interpreting variation between measures of observer bias and outcome measures derived from POPN observational data. For example, if there is a statistically significant relationship between measures of observer bias and arrest and use of force behavior, then observer bias is presumed to be present in the data. The second criterion is that the direction of the relationship between measures of observer bias and arrest and use of force behavior should also be consistent with hypothesized effects specified as research questions for selected types of observer bias specified below.

4.1. Research Questions[6]

4.1.1. Reactivity

This study addresses three research questions related to reactivity. First, does rapport between patrol officers and observers improve over the course of a shift? On a related note, the next two other research questions address the impact of rapport and acclimation between officers and observers on police behavior. A second research question examines the impact of negative rapport on police arrest and use of force behavior. In other words, is a lack of trust (or poor rapport) between officers and observers having an impact on arrest and/or use of force behavior? The third research question addresses the impact of acclimation between officers and observers on arrest and use of force behavior. The extant research suggests that officers are suspicious of observers and most likely to react to their presence during early stages of fieldwork. One way to provide partial substantiation of this claim is to see if officers behaving differently when these suspicions are the highest (i.e., the first time an officer rides with an observer).

4.1.2. Going Native

This study will examine two research questions related to going native. First, are observers becoming socialized by officers and developing more positive attitudes toward the police over the course of fieldwork? Second, do these shifts in observer attitudes toward police (ATP) predict whether observers help police officers and/or participate in police work?

4.1.3. Burnout

The analyses in this paper will also focus on one research question related to burnout: since the literature suggests that the effects of burnout are cumulative, are observers less likely to document most labor intensive aspects of police observational data during the later stages of fieldwork?

5. Results

5.1. Reactivity

Table 1 examines the impact of reactivity as rapport, acclimation, and behavioral change over time. At the end of each shift ($N = 729$), observers assessed the level of rapport between patrol officers and observers at the beginning and end of the shift. Panel A in Table 1 shows the patrol officer's attitude toward the observer for the first and last half hour of the shift.[7] A comparison of these two frequency distributions shows that patrol officers' attitudes toward observers tended to become less negative and more positive over the course of the shift ($p < .001$). For example, officers were nearly 6.5 times more likely to have either very negative or negative attitudes toward the observer for the first half hour of the shift compared to attitudes by the last half hour of the shift (11.6% versus 1.8%, respectively). Similarly, 63.9% of officer had either very positive or positive attitudes toward having

Table 1: Assessing the impact of observer bias on police observational data: Reactivity as rapport, acclimation, and behavioral change over time

	First 1/2 hour of shift (%)	*Last 1/2 hour of shift (%)*
(A) Rapport over time – patrol officer's attitude toward observer: frequency distribution		
Very negative	3.6**	.4
Negative	8.0	1.4
Neutral	24.6	15.0
Positive	44.4	55.4
Very positive	19.5	27.8
N (number of rides or shifts observed)	729	729
	Arrest[a]	Use of force[b]
(B) The impact of rapport and acclimation on arrest and use of force behavior: cross-tabs		
Bad rapport(both first and last 1/2 h of shift)		
Very negative or negative	.20 (51)	.00* (59)
Other	.21 (2241)	.08 (2429)
Number of times an officer rides with the same observer		
First time	.22 (1809)	.08* (1941)
Two or more times	.19 (483)	.05 (547)
N (number of nontraffic encounters with suspects)	2292	2488

[a] In column one in Panel B, arrest is measured as a dichotomous variable. Citizens were considered "under arrest" if he or she was taken into custody after being charged with a criminal offense. In addition, the citizen needed to be in custody at the end of the encounter (see Spano, 2003 for a more detailed discussion).

[b] In column two in Panel B, use of force is measured as a dichotomous variable where the officer either: (a) restrained suspects using nonpain compliance holds; or (b) restrained citizens using pain compliance holds or physically struck the suspect with either their fists, feet, or some type of weapon (e.g., baton, mace, taser, or drew and fired their firearm). Instances where officers handcuffed suspects were not included in this measure of use of force to more clearly differentiate more nonroutine instances of use of force against suspects from the more routine occurrence of handcuffing which is often the only form of "force" used against suspects (see Spano, 2003 for a more detailed discussion).

$^*p < .05$; $^{**}p < .001$ (Two-tailed significance level indicated by a χ^2 test of independence).

an observer at the beginning of the shift, but 83.2% of officers had very positive or positive attitudes toward the observer by the end of the shift.

Looking more closely at rapport serves two purposes. First, observers in Spano's (2002) analysis of descriptive data on reactivity illustrated techniques used to diffuse the anger, frustration, and hostility of very negative patrol officers at the beginning of the shift. At times, observers could "break the ice" and even make officers with very negative attitudes toward observers drastically change their mood. One observer started a conversation with an officer that completely changed his attitude. The officer admitted later in the ride that he had asked the lieutenant to reassign the observer, but he would not do it. He had planned to keep quiet and not talk for the entire shift, but the observer made him laugh and he started having fun. Another observer felt unwelcome until he asked the officer a serious question: should I become a police officer? The patrol officer then became very friendly and eagerly talked about his vast policing experiences. When an officer was angry because he did not get advance notice about having a rider, the observer said, "I realize you weren't prepared to have me along. I'm a little surprised myself." This comment defused the officer's anger and set him at ease. He started talking and, by mid-shift, was speaking freely.

A second purpose of considering rapport is to target the 1.8% of observed shifts where officers had negative or very negative attitudes toward observers during the entire shift for more in-depth analysis of reactive effects. If the establishment of rapport is critical to minimize the effects of reactivity, then observational sessions in which observers failed to diffuse overt hostility from research subjects should be prime candidates for in-depth analysis for reactive effects. Thus, Panel B in Table 1 addresses a second research question: is bad rapport affecting police arrest or use of force behavior? Panel B examines this issue from the vantage point of nontraffic encounters with suspects. Column one in Panel B shows that bad rapport is not significantly related to arrest behavior. In contrast, bad rapport is significantly related to use of force behavior in column 2 in Panel B ($p < .05$). The literature suggests that police officers will attempt to shelter observers from more "brutal" aspects of police work since police are suspicious of outsiders (Skolnick, 1966) and the findings for use of force behavior support this contention. Officers with bad rapport with observers never used force against suspects, but other officers used force against suspects during 8% of nontraffic related encounters with suspects.

Panel B in Table 1 also examines the impact of acclimation effects on police arrest and use of force behavior. One implicit assumption in the field research literature is that the more times an officer rides with an observer, the more likely it is that the level of trust will grow and suspicion will dissipate. Acclimation effects are measured by examining the number of times an officer rides with the same observer (see Panel B in Table 1). Column one in Panel B shows that the difference in arrest behavior is not statistically significantly different when comparing the first time an officer rides with an observer versus other rides. However, there is a statistically significant difference ($p < .05$) when examining use of force behavior: officers are more likely to use force against suspects the first time an observer rides with them versus other times (see column two in Panel B).[8]

5.2. Going Native

In Table 2, the impact of going native in POPN observational data is assessed by examining: (1) shifts in observer's attitudes toward police (ATP) over the course of fieldwork and (2) the probability that an observer will participate during fieldwork by helping with police work. Panel A in Table 2 addresses the first issue by examining observer surveys ($N = 32$) that were completed before the start of fieldwork (T1 – see column one) and immediately following fieldwork (T2 – see column two). Observer's ATP became more positive

Table 2: Assessing the impact of observer bias on police observational data: Going native as a shift in observer's attitudes toward police and observer's level of participation in police work (observer effects)

	Before fieldwork (T1)	*After fieldwork (T2)*
(A) Observer's attitudes toward police (ATP): mean comparison		
Do police discriminate toward certain groups? (coded 0,1 where 1 = never or rarely)	.13	.28
Would you approve if an officer struck a citizen who was attacking the officer with his/her fists? (coded 0,1 where 1 = yes)	.84	.94
Would you approve if an officer struck a citizen attempting to escape custody? (coded 0,1 where 1 = yes)	.59**,[a]	.72
Would you approve if an officer struck a citizen who said vulgar or obscene things to the officer? (coded 0,1 where 1 = yes)	.00	.00
N (number of observers who completed both T1 and T2 observer surveys)	32	32

	Observer effects	
	Observer helps officer[c]	*Observer helps with police work*[d]
(B) The relationship between change in ATP[b] and observer effects: odds ratios from logistic regression		
Observers with more positive ATP (T2 > T1)	.50*,[e]	.54
Observers with no change in ATP (T2 = T1)	–[f]	–
Observers with more negative ATP (T2 < T1)	1.25	.69
Initial ATP (before fieldwork)	2.07**	2.46***
Model χ^2	19.3***	17.0***
N (number of shifts where observers completed both T1 and T2 observer surveys)	598	598

[a] Two-tailed significance level indicated by a χ^2 test of independence.
[b] Measures of change in ATP in Panel B were derived from an additive scale of the first three observer survey questions from T1 and T2 in Panel A. This more general measure of ATP at T1 and T2 ranged from zero to three. To determine if the observer's ATP became more positive (T2 > T1), stayed the same (T2 = T1), or more negative (T2 < T1), scores on this additive scale were compared as described above.
[c] Dichotomous coding derived from descriptive data on reactivity (1 = rides where the observer helps patrol officer) (see Spano, 2003 for a more detailed description).
[d] Dichotomous coding derived from descriptive data on reactivity (1 = rides where the observer helps patrol officer with police work) (see Spano, 2003 for a more detailed description).
[e] Level of statistical significance is derived from odds ratios from multivariate logistic regression.
[f] Observers with no change in ATP is the comparison group.
*$p < .1$.
**$p < .05$.
***$p < .01$.

between T1 and T2 for the first three out of four attitudinal measures, but this change is only statistically significant at the .05 level for one out of the three measures (see Panel A). For the fourth measure, there was no change in observer ATP.

Panel B in Table 2 addresses the second issue by investigating whether observers with more positive ATP are more likely to "participate" during fieldwork by helping police officers. An additive scale was developed using the observer's first three attitudinal measures from Panel A from T1 and T2. Observer's ATP were categorized as more positive if this additive scale of ATP at T2 was greater than ATP at T1. If there was no change in ATP, then T2 and T1 ATP were equal. Finally, observer's ATP were categorized as more negative if T2 was less than T1 ATP.

Two measures of observer participation (or observer effects) were derived from a dichotomous coding scheme derived from descriptive and quantitative data on reactivity (see Spano, 2002, 2003 for a more detailed description). Column one in Panel B contains rides where observers helped police officers in any capacity. This measure includes more mundane and innocent forms of participation (e.g., the observer tells the officer that s/he left his/her lights on or offers advice when asked about whether to pull over a vehicle) to instances when observers helped officers with police work (e.g., holding handcuffs or helping to fill out paperwork, arrest and/or detain suspects) (see column one in Panel B) (Spano, 2002). Column two in Panel B is a subset of cases from the qualitative coding in Column one where observers helped the officer with police work.

Panel B examines shifts where observers completed both T1 and T2 observer surveys ($N = 598$). Observers with more positive ATP (T2 > T1), no change in ATP (T2 = T1), and more negative ATP (T2 < T1) were regressed on the two dichotomous measures of observer effects discussed above using multivariate logistic regression after controlling for the observer's initial ATP before fieldwork. Observers with more positive ATP are about 50% less likely to help police officers (see column one in Panel B). Although this relationship is marginally significant ($p < .1$), the direction of the effect is not in the expected direction. Observers with more positive ATP are not more likely to help police officers or participate in police work, which would be consistent with going native. In fact, this finding suggests exactly the opposite.

5.3. Burnout

Table 3 examines whether time in the field is related to more labor intensive aspects of POPN data collection. The structure of POPN quantitative data gives two different perspectives to examine the effects of burnout on police observational data: the number of days of fieldwork (see Panel A) and the sequence of rides from first to last (see Panel B). Both measures were dichotomized to compare the effects of observer workload for: (1) last three weeks of fieldwork versus the rest of fieldwork (see Panel A); and (2) the last 20% of shifts or rides observed versus the rest of fieldwork (see Panel B).[9]

Table 3 focuses on two possible ways that observers could "cut corners" by excluding the detailed documentation of suspect characteristics and situational dynamics that occur during encounters with suspects.[10] Table 3

Table 3: Assessing the impact of observer bias on police observational data: Burnout as time in the field and labor intensive aspects of POPN data collection

	Labor intensive aspects of POPN data collection	
Time in the field	*Number of encounters with suspects*	*Number of suspects per encounter*
(A) Number of days of fieldwork (12 weeks total)		
Last three weeks of fieldwork	2.72 (125)	3.73 (125)
Rest of fieldwork	2.94 (496)	4.07 (496)
(B) Sequence of rides from first to last		
Last 20% of rides or shifts observed	2.90 (109)	4.17* (109)
Rest of fieldwork	2.90 (512)	3.97 (512)
N (number of shifts with encounters with suspects)	621	621

*$p < .05$ (Two-tailed significance level indicated by a χ^2 test of independence.)

examines the effects of burnout from the vantage point of the number of encounters with suspects per shift ($N = 621$). Since there are two ways that observers could exclude this information, two different measures are used.[11] Observers could exclude these events by documenting fewer encounters with suspects during later stages of fieldwork (see column one). Another way observers could exclude these events is by reducing the number of suspects per encounter (see column two). The evidence presented in Table 3 does not provide evidence that suggests that burnout is present in labor intensive aspects of POPN data collection. The relationship between number of days of fieldwork and both number of encounters with suspects (see column one in Panel A) and number of suspects per encounter (see column two in Panel A) are nonsignificant. In addition, the relationship between the sequence of rides from first to last is nonsignificant for number of encounters with suspects (see column one in Panel B), but is significant at the .05 level for number of suspects per encounter (see column two in Panel B). However, the direction of the effect is not consistent with burnout: observers document more suspects per encounter during latter stages of fieldwork compared to earlier stages of fieldwork (see column two in Panel B).

6. Beyond POPN Observational Data: More General Issues Related to Observer Bias

6.1. The Relationship between Observer Bias and Time in the Field

The four types of observer bias represent four potential threats to the validity of observational data. Field researchers often use the term "time in the field" to refer to the time period when observational data are collected.

Thus, the four types of observer bias can also be categorized based on whether they undermine the validity of observational data collected during early stages of fieldwork (reactivity and culture shock) or later stages of fieldwork (going native and burnout). The timing of the different types of observer bias over the course of fieldwork is significant for two reasons. First, it makes the determination of the face validity of observational data much more complicated. One possibility is that the effects of the four types of observer bias may cancel one another out. Thus, examining one type of observer bias without looking at all four would give an incomplete assessment about the validity of observational data. Second, the effects of each type of observer bias may be gradual over the course of fieldwork. So, for example, reactivity may not just "stop" drastically after early stages of fieldwork, but may gradually dissipate over the course of fieldwork (Spano, 2003). Thus, although the literature suggests that these four types of observer bias occur at specific points in time, the simple partitioning of fieldwork into early and late may be an oversimplification. It is also possible that there could be a significant amount of overlap between the four types of observer bias.

6.2. Observers as Data Collection Instruments (or Observers as a Potential Source of Error in Data Collection)

Going native, culture shock, and burnout also illustrate the potential for human error during field research. One criticism of observational data centers on potential problems with using observers as data collection instruments. Going native is concerned with inaccurate documentation or omission of observational data due to resocialization and/or changes in the observer's belief system. On a similar note, implicit concerns related to both culture shock and burnout is the potential for observer error due to disorientation and fatigue. Observational studies that are primarily exploratory may exacerbate the problem of culture shock since observers will feel both disoriented and overwhelmed by potentially important observational data, especially at the early stages of field work. On a similar note, field research takes a physical, emotional, and psychological toll on the observer over time. Even if observer burnout does not affect the accuracy of "core" items under observation, burnout may undermine the exploratory nature of data collection at later stages of fieldwork.

6.3. Is Rapport a Problem (Going Native) or a Solution (Reactivity)?

Another contradiction in the methodological field research literature relates to rapport. The development of rapport is essential to minimize reactivity to get research subjects to feel comfortable and trust observers, and to act naturally while under observation. Reactivity suggests that, for example, too little rapport between police officers and observers can cause police officers to engage in atypical behavior and thus undermine the validity of observational data collected during the early stages of fieldwork. In contrast, going native suggests the opposite: too much rapport between observers and research subjects will affect the accuracy of data collected at later stages of fieldwork. Going native could be rationalized as a way of mitigating reactivity by establishing a high level of rapport with those being observed. However, Miller (1952) observed that friendship is based on an all-accepting outlook that can limit the depth of an observer's investigation into and criticism of the behavior of research subjects (see also Van Maanen, 1983a,b). Thus, going native could limit the scope, the depth, and the accuracy of observational data.

6.4. Incorporating Observer Bias into the Research Design: The Role of a Semi-structured Field Diary

There is a movement within ethnographic field research to become more self-conscious about the effects of field research on both the observer and the observed (Coffey, 1999). The publication of monograph-length field diaries and chapters in edited books by established researchers that specify problems encountered while in the field are one outgrowth of this movement (Emerson, 2002a). To improve the quality of police observational data, I propose making field diaries a standard part of data collection. The content of the field diaries would document problems encountered while in the field more systematically and more accurately than the most anecdotal and retrospective accounts in the literature. Given the state of our knowledge about observer bias, the proposed field diary should be partially open-ended and unstructured.

Table 4 specifies themes from the field research literature to operationalize observer bias. This table represents an attempt to systematically document and measure observer bias (see column one) and its impact on police observational data (see column two). More general concepts that should be included as a starting point for open-ended data collection in the field diary are denoted by "dashes" while more specific POPN measures used in this study or additional measures that could be incorporated into future research are denoted by "stars."

As noted above, POPN data collection included a semi-structured field diary that focused on reactivity. Observers were given a general set of instructions and then documented descriptive data on reactivity after each shift (ride level reactivity). This prospective measure of reactivity provided key insights into specific social cues that officers gave to observers (e.g., concerns about safety, self-consciousness, and explosive outbursts of anger)

Table 4: Operationalizing observer bias using a semi-structured field diary: Key concepts, POPN measures, and additional measures

Potential sources of observer bias	*Outcome measures*
Reactivity	
– Time in the field * POPN measure: early vs. later stages of fieldwork	– Police behavior susceptible to reactivity * POPN measures: (1) the decision to arrest; (2) use of force
– Rapport between officer and observer * POPN measure: observer's assessment of the officer's rapport with observer. * Additional measures: (1) observer's attitude toward officer; (2) officer assessment of rapport	
Culture shock	
– Time in the field * POPN measure: early vs. later stages of fieldwork – Observer assessment of feelings of disorientation, inadequacy, and paranoia – Observer assessment of physical and psychological symptoms of stress from the organizational behavior literature	– Observer error * POPN measures: none * Additional measures: (1) comparing police report data or calls for service data to observational data to assess the accuracy of the observer's geo-codes of the location of encounters with suspects; (2) observer assessment of the accuracy of the observational data; and (3) the observer assessment of his/her knowledge of technical jargon or slang used by research subjects.
Going native	
– Time in the field * POPN measure: later vs. early stages of fieldwork – Level of identification with police * POPN measures: 5 measures of observer's ATP (see Panel A in Table 2)	– Level of participation during fieldwork * POPN measure: observer effects – are observers helping officers do police work?
Burnout	
– Time in the field * POPN measure: later vs. early stages of fieldwork – Observer assessment of level of fatigue – Standardized and open-ended measures of job burnout from the organizational behavior literature	– Labor intensive aspects of data collection * POPN measures: (1) number of encounters with suspects; (2) number of suspects per encounter

that are potential manifestations of reactivity that have been documented for the first time (Spano, 2002, 2003). However, the analyses conducted in this paper did not have prospective measures of observer bias for the other three types of observer bias. As a result, no analyses were conducted for the impact of culture shock since no measures of feelings of the observer's feelings of inadequacy, disorientation, or paranoia were included in POPN. By developing a more extensive set of prospective measures of observer bias, future researchers could conduct more definitive analyses of its effects on police observational data. After each shift, observers could use the more general themes included in Table 4 denoted by "dashes" to direct the attention of observers to manifestations of observer bias documented in the field research literature. One promising area of future research is the application of insights and measurement models from the organizational behavior literature on job burnout (Maslach et al., 2001). This body of work views burnout as a response to chronic emotional and interpersonal stressors in the workplace that are exacerbated by job demands (e.g., too much work in too little time). More importantly, these studies have been used to develop interventions to alleviate burnout in the workplace that may be applicable to observers in future field research projects.

Since these entries would be completed after each observational session, the field diary would link field diary entries, observational data, and time in the field to determine: (1) *when* type of observer bias occurs over the course of fieldwork (early versus later stages of fieldwork); and (2) *how often* each type of observer bias occurs. The ultimate goal is to determine the prevalence of observer bias within each observational study, the strength of any observer effects, and how much of a threat observer bias poses to the face validity of observational data. By comparing the content of field diaries for different observational studies, field researchers can develop a better understanding of the face validity of observational data collected for different types of research subjects.

6.5. Limitations of the Semi-structured Field Diary

One important limitation of the semi-structured field diary is that observers would need to evaluate their own work by identifying and documenting different forms of observer bias. Observers may have a built-in incentive for not reporting the incidence of observer bias since these types of admissions could compromise the value of their work. In addition, three types of observer bias focus on potential forms of observer error (culture shock, going native, and burnout). Observers would have the most difficulty accurately documenting whether or not they "go native" since they may not be aware of: (1) the development of a more sympathetic outlook toward research subjects over the course of fieldwork; (2) drastic changes in their belief system which develop over time; (3) a tendency unquestioningly to accept statements from research subjects without probing for deeper meaning or inconsistencies; and (4) a tendency to become more involved as participants.

Although there are several reasons to doubt the value of utilizing a field diary in assessing the impact of observer bias in field research, there is also compelling evidence that suggests that it would be a valuable tool. First, as noted (and cited) above, experienced field researchers have been either formally or informally using field diaries to document problems encountered while in the field for many decades. Since this paper categorized these accounts as four types of observer bias, it suggests that observers are capable of identifying and documenting these sources of bias.

Second, evidence from a large-scale observational study of police suggests that observers are willing to document reactivity in spite of the criticisms noted above. More specifically, in Spano's (2002) analysis of descriptive data on reactivity, observers identified changes in officer behavior (e.g., the patrol officer may say that he or she wants to show you a drug dealer) and, if possible, the motivations behind those changes (i.e., to avoid looking lazy, to prevent the observer from becoming too bored, to look good in the eyes of the observer, or to demonstrate certain types of police work) in spite of having no incentive to do so (Spano, 2002).

Third, the field diary could be supplemented by other sources of information to increase our understanding of observer bias in observational data. In other words, the inherent weaknesses of the field diary can be mitigated by utilizing a multi-method approach. For example, descriptive data on reactivity collected within the field diary could be compared to an anonymous post-observation survey of research subjects to cross-check observer assessments of reactivity (Mastrofski and Parks, 1990). Observers could be surveyed several times over the course of fieldwork to check for radical shifts in attitudes toward research subjects. Mastrofski and Parks (1990) conducted weekly group debriefings and conversations with observers to discuss and share common problems in conducting observation of the police. Thus, the periodic interviewing and debriefing of *observers* over the course of field research could act as an important early warning system for going native.

7. Summary and Conclusion

Observers are a potential source of bias for two reasons. First, observers are part of the context of observed behavior. As a result, the research subject could react to the presence of the observer and behave atypically (reactivity). Second, observers are also data collection instruments. Thus, the quality of observational data could be undermined if observers: (1) get confused and make mistakes (culture shock); (2) get tired, cut corners, and/or make mistakes (burnout); or (3) become resocialized, contaminate observational data by participating in police work, and/or document inaccurate observational data (going native). However, few studies have focused on the impact of observer bias in observational data due to a due to a lack of data (Reiss, 1968). POPN descriptive and quantitative observational data provided a unique opportunity to measure and assess the impact of three out of four types of observer bias in a large-scale observational study of police (reactivity, going native, and burnout).[12] The logic of the analysis was to make inferences about the presence of observer bias by interpreting variation between measures of observer bias and outcome measures derived from POPN observational data based on two criterion. First, the relationship between, for example, measures of observer bias and arrest and use of force behavior, needed to be statistically significant. Second, the direction of the relationship between measures of observer bias and the outcome measures also needed to be consistent with hypothesized effects specified as research questions for reactivity, going native, and burnout.

The analyses of POPN descriptive and quantitative data focused on three types of observer bias (reactivity, going native, and burnout) and revealed three key findings. The first key finding is support for anecdotal evidence that rapport between officers and observer improve over time. A second key finding is that there is evidence of reactivity for use of force, but not arrest behavior. However, there is an added level of complexity since reactive effects are not uni-directional. Officers are less likely to use force against suspects when there is bad rapport for the entire shift, but are more likely to use force against suspects the first time an observer rides with an officer. This pattern of effects is not uni-directional so it is possible for reactive effects to cancel each other out. More importantly, these findings does not provide unqualified support for the notion that police are suspicious of outsiders and will react to their presence by sheltering observers from brutal aspects of police work. One possible explanation is that reactivity may be an officer or observer specific phenomenon rather than a universal effect (Spano, 2003).

A third key finding is the very limited support for observer bias in the form of going native and burnout. On the surface, observer's ATP became more positive over the course of fieldwork in terms of *absolute change*, but the change in ATP was statistically significant for only one of four measures. In addition, findings from multivariate logistic regression were marginally significant ($p < .1$) and showed that observers whose ATP became more positive over the course of fieldwork were less likely to "participate" in the form of helping officers for one out of two measures of observer effects. In contrast, the literature and logic of going native would predict that these observers would be the most likely to participate in police work.

Analyses that examined burnout assessed whether observers were "cutting corners" by being less likely to document more labor intensive aspects of POPN data collection during the latter stages of fieldwork. One out of four possible comparisons was significant at the .05 level, but the direction of the effect was inconsistent with the outcome predicted by burnout: observers documented more suspects per encounter during latter stages of fieldwork when research questions related to burnout predicted that they would document fewer. One possible explanation for this counterintuitive finding is that observer skill could improve over time and result in the more efficient and more detailed documentation of these types of events.[13]

Overall, the findings suggest that researchers should be especially conscious of how reactivity can affect police use of force behavior. However, this pattern of findings could be partially due to the fact that measures of reactivity were prospectively built into POPN's research design. Additional prospective data collected with the semi-structured field diary will provide more insight into the impact of culture shock, going native, and burnout on police observational data.

7.1. Limitations of Using POPN Observational Data to Examine Observer Bias

Although this study represents the first attempt to describe, operationalize, and analyze the impact of observer bias on police observational data, several limitations should be noted. First, there is no way to compare patrol officer behavior *without* an observer to patrol officer behavior *with* an observer. This criticism is less relevant given more general concerns about acclimation over the course of fieldwork with reactivity, the potential for resocialization over the course of fieldwork with going native, and a reduction in productivity during latter stages of fieldwork with burnout. Nevertheless, there is no baseline of officer behavior to use as a point of comparison to estimate the effect of an observer's presence on officer behavior for a more definitive examination of reactivity. Instead, this analysis is a more limited test that examined variation in patrol officer behavior during fieldwork. On a similar note, the rationale for the analysis was to interpret variation in POPN observational data that is attributable to the measures of observer bias and different aspects of police behavior (arrest and use of force) and observer behavior (e.g., observer participation and observer documentation of encounters with suspects). In other words, inferences about the validity of police observational data were made based on, for example, the statistical significance of measures of acclimation on arrest and use of force behavior.

Given these concerns, future research should include additional control variables to rule out alternative explanations for patrol officer arrest and use of force behavior. To further explore the robustness of these findings, multivariate analyses should be conducted to determine if observer bias is a significant predictor of, for example, arrest and use of force behavior net of a variety of control variables (e.g., officer work orientation, suspect characteristics, and situational factors). If the significant effect is suppressed in the "full" multivariate model, then it is presumed that observer bias has not substantively affected the quality of police observational data.

Since observer bias is inferred based on the basis of statistical significance and direction of effects in this study, a semi-structured field diary can also increase our understanding of observer bias as a process. Liebling (2001, p. 475) describes this process more generally as filling "the gap between correlation and explanation." By linking field diary entries, observational data, and time in the field, researchers can increase our understanding of the effects of observer bias in police observational data. By including all four measures in the field diary, researchers can get a much more complete picture about manifestations of observer bias in police observational data. In addition, researchers can examine observer bias by linking measures of observer bias to specific types of police behavior to determine if observer bias is a random and intermittent effect or a more systematic (and harmful) form of bias. More importantly, field diary entries could be used to demystify the process of field research since participant observation is more than simply immersing oneself in data and 'having insights' (Becker, 1958, p. 660). Thus, the field diary could also be used as a resource to improve training as well as to ensure the quality of observational data.

Notes

1. For some notable exceptions of this trend and an overview of the criminal justice research to date, see Mastrofski and Parks (1990) and Spano (2002, 2003).
2. Since field research is conducted in a number of other areas of study, such as psychology (Kidd, 2002; Rogers, 2000), organizational sociology (Morrill and Fine, 1997), and education (Metz, 2000), the significance of the findings from this study are not necessarily unique and only applicable for police field research.
3. A much more complex issue is the ending point of an inquiry into accuracy and bias in observational data (see Bloor, 2002 or Merrick, 1999 for an overview). Field research methodology has evolved over the past 30 years from a largely oral tradition (Denzin and Lincoln, 2000; Harrington, 2002; Shaffir, 1999) where observers are depicted as unobtrusive flies on the wall (Goodwin and Horowitz, 2002) to more sophisticated attempts to explicitly assess both the interpretive aspects of fieldwork as well as the impact of the observer on the observed through reflexive accounts (see Denzin and Lincoln, 2000; Emerson, 2002a; Shaffir, 1999 for an overview). While recognizing the difficulties associated with determining "bias" and "accuracy" of observational data (Hammersley, 1992, 1998; Lincoln and Guba, 2000), an *implicit objective of this paper* is to "be aware of who we are, understand our biases as much as we can, and understand and interpret our interactions with the people we study. Once we have done that, we can strive to determine whether there are regularities in human behavior" (DeWalt and DeWalt, 2002, p. 31; see also Merrick, 1999).
4. These narrative accounts were written in a story-like fashion. Observers were instructed to pretend they were writing a description of a scene for a play or movie. If a director was trying to recreate what happened, s/he would need to know who did what and in what order events occurred. The description should be detailed enough to allow the "director" to pick a "cast" of characters that accurately reflects those involved in the encounter and give them instructions to "act" appropriately.
5. See Mastrofski et al. (1998) or Spano (2002) for a more in-depth discussion of activities and encounters.
6. POPN observational data did not contain measures to address the impact of culture shock in the data. As a result, no research questions or analyses are included in this section.
7. Observers were instructed to use the following criteria to categorize an officer's attitudes toward the observer. An officer's attitude was categorized as "very negative" if the officer complained to a supervisor, peers, or the observer about having an observer

assigned or if the officer tried to get out of having an observer assigned to him/her. The officer's attitude was categorized as "negative" if the officer was reluctant to engage in conversation, ignored the observer's questions, is terse in answering questions, or attempted to exclude the observer from observing the officer's behavior. An officer's attitude was categorized as "neutral" if the officer was neither positive nor negative about being assigned an observer. The officer's attitude was categorized as "positive" if the officer was eager to engage in conversation and attentive to the observer's questions. The officer's attitude was categorized as "very positive" if the officer spoke positively to supervisors, peers, or the observer about having the observer along, made special efforts to explain things to the observer, and made special effort to ensure that the observer had access to all events.

8. Additional multivariate logistic regression analyses were run to examine the "robustness" of the finding in Panel B of Table 1 after controlling for demeanor of suspect, seriousness of the offense, number of bystanders, and number of officers as control variables (results available upon request). The nonsignificant findings for (a) bad rapport and arrest; and (b) number of times an officer rides with the same observer and arrest were replicated. The significant finding for number of times an officer rides with the same observer and use of force was also replicated in these multivariate analyses: officers were more than 100% more likely to use force against suspects if it was the first time an observe rode with an officer. However, a zero cell count in the 2×2 cross-tabulation of bad rapport and use of force prevented me from using the measure of bad rapport reported in Table 1 (bad rapport at *both the beginning and end of the shift*). As a result, I recoded this measure of bad rapport and used a more liberal measure of bad rapport where the observer reported bad rapport at *either the beginning or the end of the shift* (Menard, 1995). In contrast to the results reported in Table 1, the regression coefficient for the more liberal coding of bad rapport was not significantly related to use of force behavior. (I would like to thank the anonymous reviewer for making this suggestion.)
9. While these two dichotomized measures of time in the field are correlated ($\phi = .727$, $p < .001$), each provides a slightly different perspective on the effects of time in the field on burnout in POPN data.
10. For each citizen involved in an encounter with the police, observers wrote up detailed narrative accounts were written in a story-like fashion (see Footnote 3). Special emphasis was placed on including detailed descriptions of encounter level dynamics and temporal ordering of both police and citizen behavior to increase our understanding of factors that escalate or deescalate the likelihood of arrest and use of force behavior. In addition, observers entered up to 140 coded, quantitative items for each citizen that participated in the encounter. Based on my experience as an observer for this project, the combination of these two factors made encounters with suspects the most labor intensive aspects of POPN data collection.
11. One implicit assumption of the analyses in Table 3 is that encounters with suspects and the number of suspects per encounter are randomly distributed over the course of fieldwork.
12. On a more general note, the relevance of these findings is not limited to the field of criminology and criminal justice. Ethnographers, researchers who conduct face-to-face interviews, and child development researchers who "unobtrusively" study the interactions between parents and children (to name a few) all collect and utilize observational data to further our scientific understanding of various real-world phenomenon.
13. I would like to thank the anonymous reviewer for suggesting this alternate interpretation.

References

Adler, P.A., 1985. Wheeling and Dealing. Columbia University Press, New York.

Adler, P.A., Adler, P., 1987. Membership Roles in Field Research. Sage Publications, Newbury Park, CA.

Becker, H., 1958. Problems of inference and proof in participant observation. American Sociological Review 23/6, 652–660.

Béteille, A., Madan, T.N., 1975. Introduction. In: Béteille, A., Madan, T.N. (Eds.), Encounter and Experience: Personal Accounts of Fieldwork. University of Hawaii Press, Honolulu, Hawaii, pp. 1–12.

Bloor, M., 2002. Techniques of validation in qualitative research: a critical commentary. In: Emerson, R.M. (Ed.), Contemporary Field Research: Perspectives and Formulations, second ed. Waveland Press, Prospect Heights, IL, pp. 383–398.

Brown, M.K., 1988. Working the Street: Police Discretion and the Dilemmas of Reform. Russell Sage Foundation, New York.

Coffey, A., 1999. The Ethnographic Self: Fieldwork and the Representation of Identity. Sage Publications, Thousand Oaks, CA.

Crank, J., 1998. Understanding Police Culture. Anderson, Cincinnati, OH.

Denzin, N.K., Lincoln, Y.S., 2000. The discipline and practice of qualitative research. In: Denzin, N.K., Lincoln, Y.S. (Eds.), Handbook of Qualitative Research, second ed. Sage Publications, Thousand Oaks, CA, pp. 1–28.

DeWalt, K.M., DeWalt, B.R., 2002. Participant Observation: A Guide for Fieldworkers. AltaMira Press, New York.

Emerson, R.M., 1981. Observational field work. Annual Review of Sociology 7, 351–378.

Emerson, R.M., 2002a. Introduction: the development of ethnographic field research. In: Emerson, R.M. (Ed.), Contemporary Field Research: Perspectives and Formulations, second ed. Waveland Press, Prospect Heights, IL, pp. 1–26.

Emerson, R.M., 2002b. Producing ethnographies: theory, evidence, and representation. In: Emerson, R.M. (Ed.), Contemporary Field Research: Perspectives and Formulations, second ed. Waveland Press, Prospect Heights, IL, pp. 281–316.

Emerson, R.M., 2002c. Fieldwork Practice: issues in participant observation. In: Emerson, R.M. (Ed.), Contemporary Field Research: Perspectives and Formulations, second ed. Waveland Press, Prospect Heights, IL, pp. 113–151.

Ferrell, J., 1998. Criminologial verstehen: inside the immediacy of crime. In: Ferrell, J., Hamm, M.S. (Eds.), Ethnography at the Edge: Crime, Deviance, and Field Research. Northeastern University Press, Boston, pp. 20–42.

Fine, G.A., 1993. Ten lies of ethnography: moral dilemmas of field research. Journal of Contemporary Ethnography 22/3, 267–294.

Gans, H., 1982. The participant observer as a human being: observations on the personal aspects of fieldwork. In: Burgess, R. (Ed.), Field Research: A Sourcebook and Field Manual. George Allen and Unwin, London, pp. 53–61.

Geertz, C., 1967. Under the mosquito net. New York Review of Books 9/4, 12–13.

Georges, R., Jones, M., 1980. People Studying People: The Human Element in Fieldwork. University of California Press, Berkeley.

Goodwin, J., Horowitz, R., 2002. Introduction: the methodological strengths and dilemmas of qualitative sociology. Qualitative Sociology 25 (1), 33–47.

Gronewold, S., 1972. Did Frank Hamilton Cushing go native?. In: Kimball S., Watson, J. (Eds.), Crossing Cultural Boundaries: The Anthropological Experience. Chandler Publishing, London, pp. 33–50.

Hammersley, M., 1992. What's Wrong with Ethnography: Methodological Explorations. Routledge, New York.

Hammersley, M., 1998. Ethnographic Research: A Critical Guide, second ed. Longman, New York.

Hammersley, M., Atkinson, P., 1983. Ethnography: Principles in Practice. Tavistock Publications, New York.

Hammersley, M., Atkinson, P., 1995. Ethnography: Principles in Practice, second ed. Routledge, London.

Harrington, B., 2002. Obtrusiveness as strategy in ethnographic research. Qualitative Sociology 25 (1), 49–61.

Howell, N., 1990. Surviving Fieldwork: A Report on the Advisory Panel on Health and Safety in Fieldwork. American Anthropological Association, Washington, DC.

Hunt, J., 1984. The development of rapport through the negotiation of gender in field work among police. Human Organization 43 (4), 283–296.

Hunt, J., 1989. Psychoanalytic Aspects of Fieldwork. Sage Publications, Beverly Hills, CA.

Johnson, M.L., 1953. Seeing's Believing. New Biology 15, 60–80.

Kidd, S.A., 2002. The Role of qualitative research in psychology journals. Psychological Methods 7 (1), 126–138.

Leo, R.A., 2002. Trials and tribulations: courts, ethnography, and the need for an evidentiary privilege for academic researchers. In: Emerson, R.M. (Ed.), Contemporary Field Research: Perspectives and Formulations, second ed. Waveland Press, Prospect Heights, IL, pp. 260–279.

Liebling, A., 2001. Whose side are we on? Theory, practice, and allegiances in prison research. British Journal of Criminology 41 (3), 472–484.

Lincoln, Y.S., Guba, E.G., 2000. Paradigmatic controversies, contradictions, and emerging confluences. In: Denzin, N.K., Lincoln, Y.S. (Eds.), Handbook of Qualitative Research, second ed. Sage Publications, Thousand Oaks, CA, pp. 163–188.

Lofland, J., Lofland, L.H., 1995. Analyzing Social Settings: A Guide to Qualitative Observation and Analysis I, third ed. Wadsworth, Belmont, CA.

Manning, P., 1976. The researcher: an alien in the police world. In: Niederhoffer, A., Blumberg, A. (Eds.), The Ambivalent Force: Perspectives on Police, second ed. Dryden Press, Hinsdale, IL, pp. 103–120.

Martin, S., 1980. Breaking and Entering: Policewomen on Patrol. University of California Press, Berkeley.

Maslach, C., Schaufeli, W.B., Leiter, M.P., 2001. Job burnout. Annual Review of Psychology 52, 397–422.

Mastrofski, S.D., Parks, R.B., 1990. Improving observational studies of police. Criminology 28 (3), 475–496.

Mastrofski, S.D., Parks, R.B., Reiss, A.J., Worden, R.E., DeJong, C., Snipes, J.B., Terrill, W., 1998. Systematic observation of public police: applying field research methods to policy issues. National Institute of Justice, Washington, DC, NCJ 172859.

Mastrofski, S.D., Snipes, J.B., Parks, R.B., Maxwell, C.D., 2000. The helping hand of the law: police control of citizens on request. Criminology 38, 307–342.

Mastrofski, S.D., Reisig, M.D., McCluskey, J.D., 2002. Police disrespect toward the public: an encounter-based analysis. Criminology 40, 519–551.

McCall, G., 1978. Observing the Law: Field Methods in the Study of Crime and the Criminal Justice System. The Free Press, New York.

Menard, S., 1995. Applied Logistic Regression Analysis. Sage Publications, Thousand Oaks, CA.

Merrick, E., 1999. An Exploration of quality in qualitative research: are reliability and "validity" relevant? In: Kopola, M., Suzuki, L.A. (Eds.), Using Qualitative Methods in Psychology. Sage Publications, Thousand Oaks, CA, pp. 25–36.

Metz, M.H., 2000. Sociology and qualitative methodologies in educational research. Harvard Educational Review 70 (1), 60–74.

Miller, S.M., 1952. The participant observer and over-rapport. American Sociological Review 17 (1), 97–99.
Morrill, C., Fine, G.A., 1997. Ethnographic Contributions to Organizational Sociology. Sociological Methods and Research 25 (4), 424–451.
Norris, C., 1993. Some ethical considerations on fieldwork with police. In: Hobbs, D., May, T. (Eds.), Interpreting the Field: Accounts of Ethnography. Oxford University Press, New York, pp. 123–143.
Pollner, M., Emerson, R.M., 2002. Constructing participant/observation relations. In: Emerson, R.M. (Ed.), Contemporary Field Research: Perspectives and Formulations, second ed. Waveland Press, Prospect Heights, IL, pp. 239–259.
Punch, M., 1986. The Politics and Ethics of Fieldwork. Sage Publications, Beverly Hills, CA.
Reiss, A.J., 1968. Stuff and nonsense about social surveys and observation. In: Becker, H. (Ed.), Institutions and the Person. Aldine, Chicago, pp. 351–367.
Reiss, A.J., 1971. Systematic observation of natural social phenomenon. Sociological Methodology 3, 3–33.
Reuss-Ianni, E., 1993. Two cultures of policing: street cops and mnagement cops. Transaction, New Brunswick (USA).
Riksheim, E.C., Chermak, S.M., 1993. Causes of police behavior revisited. Journal of Criminal Justice 21 (4), 353–382.
Rogers, A.G., 2000. When methods matter: qualitative research issues in psychology. Harvard Educational Review 70 (1), 75–85.
Sanjek, R., 1990. The secret life of fieldnotes. In: Sanjek, R. (Ed.), Fieldnotes: The Makings of Anthropology. Cornell University Press, Ithaca, NY, pp. 187–272.
Shaffir, W., 1999. Doing ethnography: reflections on finding your way. Journal of Contemporary Ethnography 28 (6), 676–686.
Skolnick, J., 1966. Justice Without Trial. Wiley, New York.
Spano, R., 2002. Potential sources of observer bias in observational studies of police. PhD thesis, State University of New York at Albany.
Spano, R., 2003. Concerns about safety, observer sex, and the decision to arrest: evidence of reactivity in a large scale observational study of police. Criminology 41, 909–932.
Tedlock, B., 1991. From participant observation to the observation of participation: the emergence of narrative in ethnography. Journal of Anthropological Research 47 (1), 69–74.
Van Maanen, J., 1974. Working the street: a developmental view of police behavior. In: Jacob, H. (Ed.), The Potential for Reform of Criminal Justice. Sage Publications, Beverly Hills, CA, pp. 83–130.
Van Maanen, J., 1983a. The fact of fiction in organizational ethnography. In: Van Maanen, J. (Ed.), Qualitative Methodology. Sage Publications, Beverly Hills, CA, pp. 37–55.
Van Maanen, J., 1983b. On the ethics of fieldwork. In: Smith, R. (Ed.), An Introduction to Social Research: A Handbook of Social Science, vol. 1. Ballinger, Cambridge, MA, pp. 227–251.
Van Maanen, J., 1988. Tales from the Field. University of Chicago Press, Chicago.
Walker, S., 1992. The origins of the contemporary criminal justice paradigm: The American Bar Foundation Survey, 1953–1969. Justice Quarterly 9 (1), 47–76.
Wax, R., 1971. Doing Fieldwork: Warnings and Advice. University of Chicago Press, Chicago.
Westley, W., 1970. Violence and the Police: A Sociological Study of Law, Custom, and Morality. MIT Press, Cambridge, MA.

81

Observing the Observers: Researching Surveillance and Counter-Surveillance on 'Skid Row'

Thomas Kemple and Laura Huey

Quis custodiet ipsos custodes?
(Who shall watch the watchers? – Juvenal)

Introduction

What has historically been referred to as 'skid row' is both a highly contested physical space and an intensely conflicted moral order. Within these urban zones, drug dealers fight turf wars, prostitutes vie for corners, while 'respectable' families and local businesses launch community watches and neighborhood reclamation projects directed at removing problematic 'others.' At the same time, public police and private security agents attempt various measures to 'take back' the streets from targeted residents and community groups who actively resist such efforts. The term 'skid row' has itself long been at the center of conflicts over the naming of this space and the people who live there. These battles are fought between neighborhood organizations, municipal authorities and a variety of other stakeholders (Huey and Kemple, 2005; Hasson and Ley, 1994: 189–201). Such spatial, symbolic and political struggles render this area an important site in which to study the use of low-level surveillance and counter-surveillance

Source: *Surveillance & Society*, 3(2/3) (2005): 139–157.

measures: to the extent that disparate groups compete to stamp their social, political and/or economic agendas on the physical location, they often use surveillance and counter-surveillance techniques as components of their arsenals of domination and resistance. At the same time, the volatility of these dynamics presents the urban field researcher with a number of methodological and ethical challenges.

This essay discusses an important consideration arising from the surveillance researcher's function as an analyst of relations of power and knowledge within the contested physical, cultural and moral space of 'skid row.' In particular it concentrates on how as a consequence of conducting systematic disciplined observation of subjects in the field in order to uncover the surveillance and counter-surveillance strategies used by others, the *researcher herself* may become a subject of surveillance. Further, the intensity and proximity of the researcher's involvement in this field of observation may also lead her to be identified as an *agent* of surveillance, and thus to become a potential *target* of counter-surveillance. The methodological, theoretical and ethical significance of this phenomenon for the research process, understood as a form of social interaction and struggle in its own right, has important implications for the conduct of surveillance studies in general, and of urban field research in particular.

This paper is structure to follow two main axes of critical analysis and self-reflection. First, after offering a brief introduction to the setting and subjects of the two empirical studies from which we draw our observations, we develop a reflexive discussion of the methods employed in this research in terms of how sociological observation may be implicated in the surveillance strategy of so-called 'broken windows' policing. Second, we situate recent developments in the policing of 'skid row' by illustrating the problem of how the researcher may herself be positioned as the subject of surveillance or viewed as an agent of counter-surveillance. In each instance we offer a critical analysis of first-hand accounts from the field of how the sociological observer both studies and is immersed in techniques of surveillance and counter-surveillance. A concluding section further problematizes the methodological-epistemological and ethical-political implications of this phenomenon for field study and considers the extent to which the issues we have raised are unique to skid row research or whether they might be generalizable to other aspects of surveillance studies.

Sociological Observation and 'Broken Windows' Surveillance

> Conduct is continually monitored and reshaped by logics immanent within all networks of practice. Surveillance is 'designed in' to the flows of everyday existence. The calculated modulation of conduct according to principles of optimization of benign impulses and minimization of malign impulses is dispersed across the time and space of ordinary life (Rose, 1999: 234).

Among the several shades of meaning of the verb 'to observe' listed in *Webster's*, perhaps the two most common usages are "to conform one's action or practice" (i.e. to laws, customs, ceremonies, or rules) and "to see or sense, esp. through directed careful analytic attention" (i.e. to noted facts, evidence, knowable realities). It seems then that observation, whether casual and ordinary or disciplined and professional, is inescapably ingrained in relations of *knowledge* (especially when acquired through visual perception) and *power* (particularly involving compliance with established conventions of authority). This essay draws from two larger ethnographic studies of public and private policing in so-called 'skid row' districts that exhibit each of these dimensions of observational practice.[1] What we call 'the private security study,' was conducted in 2000 and examined the use of 'broken windows-style' policing by private security agents to regulate the presence of suspicious or disorderly individuals in a gentrified area of Vancouver's historic 'skid row' district commonly referred to as Gastown (Huey, Ericson and Haggerty, 2005). The second project, conducted in 2003 and referred to as 'the public policing study,' explored the politics of public policing in skid row districts in the San Francisco Tenderloin district, Vancouver's Downtown Eastside and Edinburgh's Cowgate and Grassmarket neighborhood (Huey, 2005). While the primary focus of both studies was on the policing of skid row districts, a principal research goal in each was more generally to examine the extent to which routine surveillance of target populations was being undertaken by public and/or private policing agencies within these communities. Although for the purposes of illustration in the present essay we draw mostly from the relevant geographical, cultural and political particularities of the Vancouver site, it is important to note that the other two sites exhibit similar characteristics with respect to the composition of the population and the use of surveillance and counter-surveillance as a means to control territory.

As surveillance is a broad concept used to describe a wide range of phenomena – from the use of infrared technology in military helicopters to the ways in which individuals routinely spy on their neighbors – we must also clarify our use of this term. Following Gary Marx (2003: 370), we define surveillance as those technologies of visibility and techniques of exposure that "seek to eliminate privacy in order to determine normative compliance or to influence the individual" for particular aims or interests. Marx (2004: 276) further provides a useful contrast between 'traditional surveillance,' defined as "close observation, especially of a suspected person,: and its new forms, which he characterizes as "scrutiny through the use of technical means to extract or create personal or group data, whether from individuals or contexts." We are reporting on studies which examined the use of relatively low-tech surveillance on skid row by police, private security, and community watch groups, on the one hand, and on the other hand, surveillance used by traditional targets of normative surveillance, including buyers and sellers in the drug, prostitution and/or stolen goods markets, and other individuals similarly defined as 'deviant others.' Here our concern is less with the extension or intensification of sophisticated surveillance techniques as such (for example, in the interests of business or government) than with the asymmetrical, anonymous, and non-reciprocal surveillance of vulnerable or suspicious social groups in public spaces. What these practices share is the use of surveillance techniques and strategies for: 1) classifying and interpellating identities within a moralized physical space, for example to distinguish 'good' residents (and/or others who use these spaces) and zones from those designated as 'bad,' and 2) fostering compliance with a particular set of normative values in that space. Rather than follow the lead of other researchers who have focused on these contested fields of surveillance and counter-surveillance themselves (Marx, 2003; Lyon, 2002; Staples, 2000), our primary aim is to examine the extent to which the research enterprise itself – through direct observation, note-taking, photography, and the subsequent accumulation and analysis of visual and recorded data – is inextricably embedded within similar and overlapping relations of power and knowledge.

A total of 161 individuals identified as stakeholders were interviewed for both studies. They were selected because they were identified as different forms of stakeholders with an interest in these spaces. This included 'street level' users of these spaces (including 46 area residents, 14 business owners and 8 tourists), front-line professionals (including 46 police officers and private security agents and 36 members of community groups), and civic officials (8 individuals). At the risk of oversimplification, we characterize these various stakeholders in terms of their location within a broader set of practices that have come to be referred to as the 'broken windows' strategy of maintaining law and order. 'Broken windows' policing was initially promoted over twenty years ago on the assumption that minor infractions, or more precisely, the visible signifiers of decay and neglect such as litter, graffiti, and broken windows tend to produce criminogenic effects including theft, prostitution, and drug-trafficking. When combined, these behaviors increase and legitimate the fears of 'respectable' citizens (Wilson and Kelling, 1982). The view that minor incivilities lead to major crimes is itself based on the assumption that visual environmental cues mark a territory as 'undefended' and therefore lacking in social controls. This approach has since been expanded from an informal policing policy to a more *generalized civic strategy of surveillance and counter-surveillance* (Johnston and Shearing, 2003: 102; Huey, Ericson and Haggerty, 2005). Public police, private security officers, and civic officials have increasingly favored the 'recruitment' of area residents, business owners, and community groups to help produce the outward appearance of order and to regulate images of disorder on skid row (Ellickson, 1996; Kelling and Coles, 1996). Beyond the institutionally sanctioned efforts of public police and private security officers, various community members have been solicited to participate in a wide range of practices of impression-management and social control. These can entail acts such as removing delinquents from public places, informing on criminals, or simply

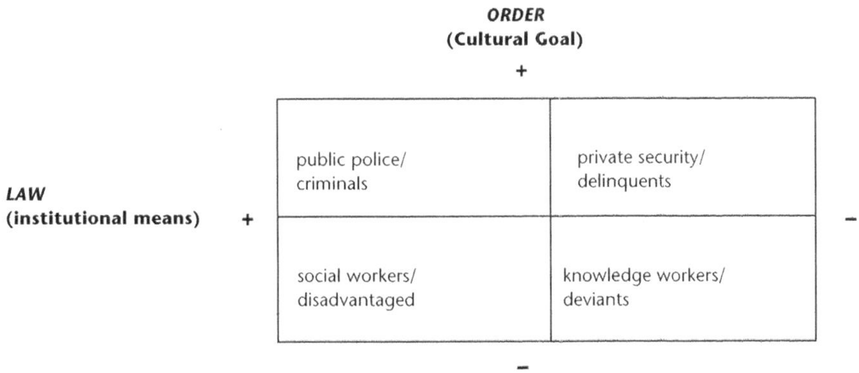

\+ = relative access to or observance of goals and/or means

– = relative lack of access to or observance of goals and/or means

Figure 1: 'Broken windows' as a strategy of (counter-)surveillance on 'skid row'

assisting the disadvantaged and keeping a watch on 'deviants' in the area. 'Broken windows' then is not just a reactive zero-tolerance policing strategy that makes use of *coercive force*; it may also (or alternatively) involve pre-emptive ceremonial displays of *symbolic power*.

Figure 1 displays these modes of adaptation and resistance imaginatively in the form of a 'window' onto the contested social field of 'skid row surveillance.' Here 'order' refers to the culturally prescribed goal of this strategy as it is normally defined in terms of middle class standards of consumption and socially dominant values of decorum, civility and respectability. 'Law' is understood as the institutionally legitimized means for achieving this goal as it is formally defined by legislative statutes that are enforced by public authorities, or as it is informally upheld by various private and other public agents who may or may not be authorized to operate in an official capacity. When arranged to suggest the coordinates of a grid (the heuristic device suggested by Figure 1), 'law' and 'order' represent a virtual frame for how the social space of 'skid row' is occupied by differently located social actors operating across a range of conflicting or complementary relationships.[2]

In documenting the activities of mobile police patrols, neighbourhood watch programs and business improvement associations, other commentators have examined the 'broken windows' ideology specifically as a tactic of zero-tolerance law enforcement, as a cornerstone of so-called 'community policing,' or as an aspect of urban gentrification (Johnston and Shearing, 2003: 102; Huey, Ericson and Haggerty, 2005). By contrast, our objective here is to highlight the broader pattern by which *observers themselves come under the gaze of observation* within the contested space of interactions between institutional and individual agents on 'skid row.'

To be implemented effectively, 'broken windows' does not necessarily have to be formally sanctioned by legal statutes or officially recognized by the police.

This method of maintaining order may also be informally enforced by security guards or merely socially approved by various social and knowledge workers, the latter term interpreted broadly to cover a range of public and private professionals (sometimes including academic researchers, a point we return to in the following section) concerned with collecting and sharing information on individuals and groups for various purposes. At the same time, understood as a multifaceted strategy of surveillance and counter-surveillance, 'broken windows' may also be honoured or observed – if only in the breach – by numerous individuals who may not acknowledge or have direct access to the dominant codes of law and order, including those individuals who are its primary targets. For example, in the interests of maintaining order (at the 'upper half' of the diagram), the police are officially charged with keeping a close watch on suspected criminals, who attempt in turn to slip under their gaze, while private security officers stand guard against the covert and unruly actions of delinquents, who likewise try to elude them. In each instance, the identities of the various actors are mutually defined or undermined by their conflictual relationships and asymmetrical interactions with one another. By contrast, at the 'lower half' of this social space, other stakeholders may have more pressing concerns than upholding law and order, such as alleviating deprivations, reducing harm, or managing dependencies. For example, social workers are typically charged with looking out for the interests of disadvantaged (who are then expected to look after themselves), while knowledge workers of various kinds are expected to gather information on deviants (which may then be put to use in rehabilitating or normalizing them), regardless of whether these individuals are engaging in criminal or disorderly conduct. These designations then should not be understood as psychological states or character roles but rather as *patterns of interaction* and as locations within a social field. In particular, they refer to relations of exposure and (in)visibility, and of domination and (in)subordination, in which individuals may be engaged simultaneously or in succession at various times and depending on the circumstances. Thus, each of the stakeholders in the space depicted here can assume a fluid range of perspectives (or 'windows') on the field. Consequently, figure 1 does not represent fixed points on a map but offers a compass for navigating through the moving equilibrium of interests and shifting practices of surveillance and counter-surveillance on 'skid row' (cf. Marx, 2004).

When the researcher enters this highly charged field of observation, she is necessarily and inextricably implicated in this grid of relationships. Fieldwork and observational research for the two ethnographic studies under consideration consisted primarily of touring the site (alone or with police or community outreach workers) or taking up a stationary position in public space to document physical and social landmarks and the ordinary lives of inhabitants and their interactions. In most instances the researcher simply sat or stood while quietly observing the people and the surroundings, sometimes taking notes or the occasional photograph. For the policing study if someone asked about her note-taking she explained the nature the research being conducted in general terms. On a few occasions, typically involving the observation of police activities, if the meaning of what she was witnessing was unclear, she would approach an officer, briefly describe her research project and ask for clarification. For the private security study she also undertook covert observation of security agents on patrol in public spaces. Thus, alongside the usual innocuous kinds of 'vernacular watching' that typically characterize urban life (often accompanied by the boredom and distraction experienced while waiting for 'something to happen') the disciplined observation of the researcher routinely entails modes of apparent voyeurism and 'suspicious looking' that often resemble the practices of surveillants and others engaged in the visual documentation of social space. For this reason, the researcher cannot avoid being positioned as a 'visual subject' in the field, that is, as an agent of sight (regardless of her physical ability to see) whose identity is constituted in large part by categories of (reciprocal) visuality, (mutual) observation, and/or (counter)-surveillance (cf. Mirzoeff, 2002).

Many of these issues are graphically displayed in Illustration 1, an otherwise unremarkable photograph taken in the course of the private security study and which depicts a security guard questioning a local resident. Considered both in the context of the present paper and in its original setting, the photograph is as interesting for what it conceals as for what it

Illustration 1: Security guard questioning a resident in Gastown, Vancouver (Photo: L. Huey)

reveals. In contrast to the activist or inflammatory intentions of photojournalism, the ethical conventions of social science research require researchers to alter the image to conceal the identities of the two individuals depicted. At the same time, the perspective assumed by the researcher/photographer also hides the technical and human conditions of the image's production, including the presence of the camera and of the photographer herself in this scene. Despite being 'off-stage' and 'out-of-frame,' these features of the image are indicative of *the visual organization of the observational relationship* with its alternation between overt activity and action incognito. Acknowledging these aspects of visual documentation also helps us to expose the paradox of photography in attempting to capture social reality by framing and immobilizing a moment – perceived as 'now/here' – within a succession of events (Becker, 1995). Thus, the particular physical and cultural boundaries of the area are 'screened' out of this image – such as the broader context of the 'broken windows' style of order-maintenance that had recently been intensified in this highly touristed and increasingly gentrified part of Vancouver's Downtown Eastside – and so too is the sequence of encounters of which this is only a frozen moment (a point we return to in the next section).

Covert observation as a research strategy requires subjects either to be unaware of the fact that they are being observed, or to be unaware of the true purpose of that observation (Bowling, 1997). In the private security study, guards were covertly observed from either fixed vantage points (sidewalk benches, café seats, bus stops, etc.) or were discretely followed while on patrol. As guard patrols took place in public spaces in a tourist district, the researcher frequently dressed in shorts and carrying a camera as a way to blend in with the crowds of vacationing tourists (a deception that was not always successful, as we discuss below). In light of the ethical questions and methodological problems raised by several critics of the use of covert observation in the field (Lauder, 2003; Elms, 1994; Bulmer, 1982; Cassell, 1982), it is reasonable to ask why discrete surveillance of security guards on patrol is worth conducting as part of the fieldwork exercise.

One answer to this question is that in contested urban spaces, especially those characterized by high levels of crime, poverty, and insecurity and which are subject to unusually high levels of social control and regulation, it is not only addicts, prostitutes, pimps, drug dealers and 'deviant others' who may wish to keep certain activities hidden from public scrutiny. For example, the private security study was initiated at the request of a local community group who wished to examine the use of security agents as a mechanism for pushing the area's poor out of the neighborhood. Community residents advised that overt conspicuous forms of surveillance were being employed by guards as a means of harassing panhandlers and other disadvantaged residents (including homeless, unemployed, physically disabled and mentally ill people), and thus to induce them to abandon their use of public space. To the extent that these activities infringe upon the rights of some of the most vulnerable or oppressed urban inhabitants, the use of covert observation can be justified as necessary to expose activities which harm or oppress disadvantaged groups (Bok, 1978; Warwick, 1973). Such scrutiny reduces the likelihood that those being observed would alter their activities and gets around the tendency for people to deny such behavior in interviews (Turnock and Gibson, 2001). Thus, in instances where dense networks of control and multiple techniques of monitoring envelop virtually anyone who enters the field of observation, covert research 'may be the only way to develop accurate descriptions that are not affected by deliberate distortions, biased recollections, or outright denial' (Reynolds, 1982: 185). It may also help develop new perspectives on repressive relations and power structures.

A further consideration with respect to the use of covert methods concerns the claim that the inability to secure informed consent when using this strategy results in an invasion of subjects' privacy, thereby intensifying the anxiety and suspicion that citizens feel in a world that is perceived as increasingly subjected to information gathering and surveillance practices (Bulmer, 1980; 1982). All observational research for both studies occurred in public space, where the activities of those observed were potentially observable by anyone in this environment. At different times this might include several or several hundred individuals. As Homan (1980) suggests, individual and group expectations of privacy are considerably lessened when they operate within public forums, and thus where observers themselves may be the objects of observation. In the following section, we turn our attention to how this situation also poses significant methodological (as well as ethical) problems for the researcher when she is positioned as a target of surveillance or identified as an agent of (counter-)surveillance. In the concluding section we return to the question of whether these concerns point to what may be a more general structural dimension of social life beyond the contingent aspects of sociological research. If, as Niklas Luhmann (2002: 87) remarks, the inherent paradox of observation is that it 'has to operate unobserved in order to be able to cut up the world,' then the observation of observation is indispensable in understanding or rethinking such divisions.

The Sociologist as Subject and Agent of (Counter-)Surveillance

> Knowing that targets of surveillance may respond in kind also can be a factor limiting or inhibiting the initial use of surveillance . . . Certainly there is greater equality in access to and use of surveillance technologies today than in much of recorded history. However, we are certainly far from equivalence here. The kind of technologies that are developed, apart from who has them, is also much affected by inequality of resources
> (Marx, 2003: 384).

In the preceding discussion, we noted some of the ways in which field observations, can involve fairly routine activities of watching and asking, as well as the more systematic labour of tracking, recording, exposing, and documenting. These may require the use of techniques of concealment, disguise and anonymity not unlike those employed by other forms of surveillance. Of course, these techniques are normally part of an assemblage of analytical and observational methods which entail varying degrees of distance and contact, intimacy and objectivity, and may include surveys, interviews and psychological tests with their attendant audio and visual recording devices. Gary Marx's comment above regarding the increasing democratization of surveillance techniques (or at least their broader dissemination) points not only to the persistence of their unequal distribution and intensification but also to the potential for their neutralization or even subversion. The employment of counter-surveillance, as practiced in the communities we discuss here as well as in other social locations, thus can represent attempts to expose power asymmetries through the tactical use or disruption of surveillance techniques (Monahan, forthcoming). Rather than focus on the full array of these tactics of resistance and subversion, here we wish to examine the implications for field research when the observer is not only positioned as an active (and anonymous) surveillant, but also identified as an object of suspicion and targeted as a possible agent of (counter-)surveillance.

The reality of fieldwork in contested spaces is that on particular occasions an apparently disengaged observer frequently becomes part of the research setting through her immersion in the dynamics of exposure and concealment that characterize the field, and may thereby become an *interactant* with those who occupy these spaces. Her embeddedness is revealed as a consequence of the fact that all strangers in such sites are frequently subjected to varying levels of surveillance aimed at determining: 1) their identity, and thus 2) their potential threat to the observer and/or to others. In order to trace the effects of these interactions on the meaning and outcome of the research process, in what follows we examine three illustrations of a researcher's interactions in the field. In view of the preceding discussion it is perhaps significant that each of these encounters happen to break off through the medium of a window: at a café, through a camera lens, and on a bus.

For skid row denizens and other residents of the street, strangers can represent a possible source of danger to their businesses and to the conduct of other everyday activities. This potential threat is compounded in those instances when unknown individuals are spotted engaging in what appears to be more than casual watching. Typically police and social workers have sought to keep watch over skid row population in attempts to normalize 'deviant others' through coercive social work or spotting (potential) criminals for punishment by the justice system. Today, however, most residents are also aware that fellow neighbors, local workers, and business owners are being recruited into becoming willing surveillants and police informers, that is, into contributing to 'the broken windows' model as 'responsibilized' citizens (O'Malley, 1992; Garland, 1996). Careful scrutiny and identification of strangers by street residents themselves thus serves as a means of self-protection from the prying eyes of state authorities and the private institutions through which these authorities increasingly work.

An incident from the field – which might otherwise be regarded as too bizarre to be relevant to the analytical aims of a research project – serves to illustrate the embeddedness of the surveillance researcher within these contexts. During the spring of 2003 the Vancouver Police Department implemented a policing program, the City-wide Enforcement Teams (CET), which resulted in a public crackdown of the open-air drug market in its skid row district (the Downtown Eastside, or DTES). Throughout the summer of 2003, CET teams created an expanded and very visible police presence throughout the DTES: walking in pairs, on horseback, on motorcycles, bicycles, and in patrol cars, police used overt surveillance to break up pockets of the local drug market. An examination of CET both as an actual surveillance strategy and as a symbolic display of panoptic power formed a part of the larger public policing study. Over the course of several afternoons, fieldwork consisted of discrete but mostly overt observation of police observing (and being observed by) area residents. The following excerpts from field notes were taken while on break at a nearby coffee shop:

> Writing up field notes in a coffee shop on Hastings [Street]. Female addict walks in and approaches. It is clear that she is 'dual-diagnosis' [addicted and mentally ill]. The conversation is strange. She has seen my notes and wants to know what I'm doing. I explain that I'm doing research on how the police treat people in her community. She becomes suspicious and angrier. She demands money. She tells me that she thinks I'm a 'narc,' and is going to tell her friends about me. She asks for money again, and when I say 'no,' she calls me a 'narc' in a loud voice (so that others in the restaurant can hear) and storms out the door. To find her friends, she says. She stares at me in the restaurant window for a minute, and then walks away. This is a serious threat. I quickly leave before she comes back
>
> (field notes, August 15, 2003).

Although the peculiar features of this incident should not be underplayed, in view of the fairly common spectacle of uniformed patrol squads in this area (as depicted in Illustration 1 for instance) and sometimes plainclothes officers taking notes about suspicious people, the woman in the coffee shop had good reason to identify the researcher as a surveillant. While the technologies employed by the researcher were fairly unsophisticated (her gaze and pen and paper) the range of meanings behind their potential use was unmistakable to any observer in this highly charged though otherwise ordinary space. Despite repeated attempts by the research to explain the nature of her project, the woman persisted in misidentifying the role of the researcher: that is, as a police informant rather than as a sociological observer. At the same time, however, the researcher was able to instantly arrive at and sustain a view of the woman (whether correctly or not) as 'a female addict' and as 'dual diagnosis,' and thus to characterize her staring through the restaurant window as a potentially 'serious threat.' Once the identities of each participant in the encounter are assigned in this way, the subsequent interaction and its outcome (the woman yelling, the researcher escaping) seem to be virtually assured. If we return to the surveillance schema represented in Figure 1 (particularly the lower right quadrant), we can see how both the identity and the security of observer and observed may be challenged or established on the basis of their (presumed or actual) mutual knowledge and ignorance of their respective social location within the space.

The sociological observer in the field ostensibly seeks to acquire knowledge of social processes. Her goal is not to affirm or support the cultural goal (order) of the setting under examination or to affirm or support the institutional means of meeting this goal (law). Rather it is to sustain *an interest in disinterestedness*, that is, to engage as much in participant observation as in *participant objectification*. Her task is to expose the ways in which human subjects are transformed into objects of power and knowledge, including how this transformation takes place through the process of research itself (Bourdieu, 2004). Whatever her moral sympathies or political commitments, the researcher does not need to have access to or endorse the goals and means (the subcultural values or official institutions) that define the field of study. In employing surveillance techniques as a strategy for gathering information for research purposes, she may nevertheless be seen to operate as a 'deviant' counter-surveillant in a relation of opposition to the established codes of law and order. In part, this perception arises from the circumstance that, as a 'knowledge worker,' her primary location in the field is not as an informant working on behalf of authorities charged with enforcing law and maintaining order in this setting. Rather, as an academic researcher (in fact, one of many who have turned their gaze on 'skid row' since the early days of urban sociology) she is also *an actor within the university field*, particularly insofar as her notes and analyses (for example) are employed for the accumulation of educational capital or in gaining symbolic recognition through publication.

The potential misidentification of the researcher as a formal agent of surveillance – as a 'narc' – encompasses a range of potentially serious methodological as well as ethical implications for the conduct and outcome of research on skid row. As in other communities, police on 'the skids' function on behalf of the goal of order. Those seen as being 'on the other side,' as one addict interviewed referred to her 'deviant' status (or in other words, as occupying the 'lower end' of the slash in each of the quadrants of Figure 1), are the usual targets of police surveillance, often leading to the loss of their livelihood, to their eventual arrest, or even to 'starlighting' (the practice of dumping a suspected criminal or disorderly person in another part of the city, or sometimes outside city limits). As a consequence, individuals identified as 'narcs' may find themselves subjected to 'preventative' or retributive treatment from other individuals in this community (Rosenfeld, Jacobs and Wright, 2003). In the course of conducting fieldwork in San Francisco's Tenderloin district, for example, an undercover police officer working a low-level drug buying operation was stabbed by a drug dealer who had identified him as a police officer. As such incidents were not uncommon in the areas where interviews and field observations are being conducted (within two or three blocks, in the San Francisco case), the researcher had to take seriously the types of accusatory threats like those posed by the woman in the coffee shop. Nevertheless, insofar as knowledge-gathering activities in 'skid row' neighborhoods (including those contributing to university based projects) are frequently viewed with confusion and distrust, it is unlikely that even a correct perception of the researcher's intentions (in this instance, as suggested perhaps in the woman's initial demand for money) would be enough to erase all suspicions.

In light of the possibility that the researcher may be mistakenly identified as an informer by skid row residents, it is somewhat ironic that, under the banner of 'broken windows,' she may also find herself treated as a potential threat to middle class community order. As noted above, 'broken windows' is deployed not just as a model for training police and security guards to scrutinize spaces for criminogenic cues, but also as a framework for maintaining the appearance of civic order as it is normatively defined by values and meanings found predominately in white middle-class neighborhoods (Huey, Ericson and Haggerty, 2005; Harcourt, 2001). In the process, a variety of civic agents are 'responsibilized' to adopt and promote this narrowly defined goal in informal ways (such as casual or conspicuous observation of one's neighbors) and through formal channels (for example, association and support for institutionalized surveillance programs, including policing and private security systems). Thus, residents seen as 'disorderly' or as threats to newly imposed community norms are likely to find themselves not only under the gaze of security guards and police officers, but also under the watchful eyes of neighbors, area workers and local businesses.

During fieldwork for the private security study, evidence of how surveillance practices can become an aspect of 'broken windows'-inspired vigilance was acquired firsthand when the researcher found herself targeted as a potential threat to community security. In this second example, she was conducting routine fieldwork by visually documenting aspects of the built environment with a camera, as described in the following field notes:

> Have located the Community Police Office. It is closed. I attempt to take a picture of it, but have difficulty due to the glare of the glass. As I

maneuver around trying to take a picture, I observe a man watching me with interest. I give up on the photo and move to the corner of Abbott and Water streets. The man follows me and continues to observe me. I stand there for a couple of minutes to see what he will do. He stands there and watches me. I decide to turn west and move on. While watching a guided tour group, I notice the same man is walking towards me from the west side of the street (he must have taken a short cut). He sees me and stops. I make deliberate eye contact. I continue walking, he continues following me. I dart into a store and look at items for a moment, check the store window and see that he is looking in. I exit and continue walking. As he is still following me, I decide to walk to an area where I can sit down and observe the length of Water Street. I sit down and take out my camera. He stops twenty or so feet away from me, directly in my line of sight. I line up a camera shot and as he attempts to move out of camera range I take his picture [Figure 2]. He then moves away . . . suspect that he is security personnel . . . the technique he used is similar to that of the security guards: overtly follow a target, but don't make contact (intimidation)

(field notes of June 17, 2000).

Illustration 2: Mystery surveillant, Gastown (photo: L. Huey)

The suspicion that this individual had received security training was supported in the course of a subsequent interview with a police officer who promotes 'broken windows' policing through his training of community patrols designed to protect area businesses. In the officer's view it was likely this individual was a local merchant seeking to prevent a potential burglary. This speculation was subsequently confirmed when the researcher recognized the mystery surveillant behind the counter of a local shop. As the officer explains,

> One of the guys that I supervised as a civilian in the Crime Watch, he helped those guys out and he created the Gastown patrol. I know that the training that he got, he did share with [area merchants]. And a lot of it was training in lower level observation. So I can see the guy circling the block. He's probably thinking about how clever he is (laughs).

This encounter allows us to appreciate how actions that otherwise appear innocuous (or ridiculous) may be viewed through the lens of 'the broken windows theory' as (potential) threats to the dominant order. The merchant in this example sees himself as working on behalf of community security. Moreover, he is using what he believes to be socially sanctioned means in support of this goal – after all, his training has come courtesy of the local police department (through the auspices of the local community watch group) which clearly has a vested interest in increasing the number of 'eyes and ears' on the street (Harcourt, 2001). That the target of his observation happened to be what the police would view as a 'legitimate' space user (in this case, a law abiding citizen engaged in authorized academic research) rather than the illegitimate target that the man suspected her to be, serves as a source of amusement for the officer (an advocate of this policing system) rather than as a cautionary tale illustrating the perils of unrestrained surveillance. At the same time, the incident inadvertently illustrates how the academic researcher herself, especially when equipped with a camera and defying intimidation, may be suspiciously viewed as a potentially subversive agent of *counter*-surveillance.

A third example of the treatment of the skid row researcher within this context further illustrates the possible implications of her identification as a *counter-surveillant*, in this case with respect to those functioning as formal agents of surveillance. As noted above, during the private security study, the surveillance activities of security guards were routinely documented in notes and occasionally in photographs. While taking the photograph discussed in the previous section depicting a security guard questioning a local resident (see Figure 1), fellow officers were alerted to the presence of the suspicious activities of the photographer/researcher who was accidentally caught in the act of observing and recording the officer's observations. The following excerpt is from field notes taken immediately after the researcher was spotted by the guard:

> I see that the guard is there. He is talking on a walky-talky. He seems to be observing me, but is not approaching . . . I have definitely been spotted. I decide to cross the street and I see the male guard is now with the female guard. I cross the street to observe them. They are standing almost directly across the street from me and are both directly watching me and taking notes. I feel that they are trying to intimidate me. I continue to watch and take notes. A bus comes and I head towards it. The female guard leaves. The male guard stays on spot, taking notes. I board the bus. He watches me through the bus window. I wave to him. No response.
>
> (field notes of May 30, 2000).

While studying the use of overt surveillance techniques by guards directed at local residents, the researcher became identified by the guards as a counter-surveillant, and therefore as an object of concern. At one level the identification was correct: in this situation, through the use of what were (up to this point) covert observation techniques, the researcher was evidently operating as a counter-surveillant. However, as in the preceding examples, her purpose was misrecognized as a potential threat to order or to their ability to carry out their mandate to uphold a commercially defined normative order in this space. As with the two preceding examples, the researcher's activities were misunderstood as a result of the suspicions that attach to the use of observation techniques by differently situated actors within this contested space. The broken windows model of upholding community order, with its emphasis on proliferating the conspicuous surveillance of activities within an area, exacerbates the underlying tensions within this community through the requirement that all activities and individuals that do not conform to dominant normative expectations be treated as objects of suspicion. Ironically, the researcher might also have appeared to each party as fulfilling a function that either would view as legitimate and potentially useful to their interests, for example, as an informant with evidence of criminal activity or an activist documenting police abuse. In any case, even when an academic observer is removed from these conflicts she may still be caught in the intersection of gazes, and thus embedded in the knowledge base and power-network of surveillance and counter-surveillance, that constitutes this space.

Conclusion: Sociological Reflexivity and the Politics of Research on 'Skid Row'

> [In control societies], individuals become '*dividuals*,' and masses become samples, data, markets, or '*banks*' . . . Disciplinary man produced energy in discrete amounts, while control man undulates, moving among a continuous range of different orbits
>
> (Deleuze 1995: 180).

Our discussion of the serious game of hide and seek that is sometimes played between a variety of stakeholders on 'skid row' needs to be placed within the context of both the broader *theoretical questions* that are raised by sociological observation, as well as some of the specific *ethical issues* posed by the unique features of this space. In one of his last comments on the 'reflexive problem' in sociology, Pierre Bourdieu (2004: 86, 89) provides a useful introduction to these questions with regard to what he refers to as the constitution of a sociological 'vision of reality' and the composition of a scientific 'point of view':

> To bring to light what is 'the hidden' par excellence, what escapes the gaze of science because it is hidden in the very gaze of the scientist, the transcendental unconscious, one has to historicize the subject of historicization, to objectivate the subject of the objectivation . . . To be able to apply to their own practice the objectivation techniques that they apply to the other sciences, sociologists have to convert reflexivity into a disposition constitutive of their scientific habitus, a *reflexivity reflex*.

Bourdieu's suggestion that the action and location of the sociological enterprise can be illuminated by the historicizing and objectivizing methods of science entails acknowledging the research situation itself as a *mode of social interaction* in its own right. That is, the sociological perspective, what he refers to as the 'principles of vision and division' of its 'scientific habitus,' consists of a distinctive disposition that is cultivated and constrained within overlapping fields of social interaction, competence, and power relations. As Luhmann (2002) might point out, the examples we have cited above suggest that the irreducible blind spot of sociology may well be the systematic organization of its own observational gaze, or at least the visual organization of the social and cultural fields it both inhabits and examines. For this reason, rather than attempting to reconcile the polarized attitudes of objectivism versus subjectivism, or even of passionate engagement versus cool detachment, our aim here has been to exercise this 'reflexivity reflex' by exposing how these tensions are often bound up with powerful patterns of revelation and concealment that are sometimes beyond our ability to fully account for or control.

As the above illustrations suggest, the limits of scientific objectification are also the limits of disciplined observation, at least as long as the methodological procedures such observation entails are carried out according to a sociological project within contemporary settings that are already thoroughly surveyed and 'sociologised.' Not only are social groups often made 'knowable' under the gaze of social science through the application of discrete classification systems, but such information can also be accumulated, checked, stored, retrieved and compared for use in assessing the risks and opportunities these groups may pose (Lyon, 2002: 3). Institutional and informal techniques of sociological observation and everyday surveillance increasingly overlap with and reinforce one another when human subjects are documented, interviewed, photographed, and profiled, often with the aim of controlling or intervening in the daily conduct of their affairs in a 'neutral' way or 'at a distance.' H. T. Wilson (2004: 296) has commented on this feature of social science research as constitutive of the institutions that support it:

The social, behavioural, and administrative/managerial sciences have become part of the *institutional structure* of the advanced societies . . . Nowadays these disciplines serve as a model for social practice rather than simply a way of studying the nature of commonsense activities that take place independently of their interests and values. In effect, they have become a veritable force of production in their own right, all the more successful in this endeavor where the fact of their influence has been muted by internalized acceptance of their commitment to observe, record, even intervene *sine ira et studio*.

The social sciences are increasingly in a position to offer legitimacy and rationality to procedures of gathering data, examining evidence, reporting information, and formulating classifications that are integral to the routine operation of *non*-academic institutions, both public and private. Such methods may be mobilized to contribute to governmental objectives of coercion and normalization, for example, or they may be redeployed collaboratively in an effort to account for and transform conventional ways of accumulating social, educational and professional capital (Wilson, 2004: 313–327; Culhane, 2003/04). In any case, neither scenario can avoid the fact that the research enterprise itself is implicated in relations of mutual observation or even in techniques of (counter-)surveillance that have ethical and not simply theoretical implications.

The insuperable challenge for researchers immersed in a precariously configured setting like 'skid row' is that many of the people and places in question are themselves subject to imperatives of mobility and transience that are enforced in large part by the police and other agents of order: "The police are primarily concerned with making manifest what is there, or rather, what is not there: 'Move along, there's nothing to see'" (Rancière, 1998: 242). At the risk of being positioned as an agent of social control or targeted as a subject of surveillance, the researcher assumes a position within this circulating flux of observers observing observers being observed. There is a sense in which modern social science and modern surveillance were each from their earliest days a means of 'keeping tabs on the mobile,' and yet today both are components in a much larger assemblage and part of the flow (Lyon, 2002: 3; Haggerty and Ericson, 2000). Among the ironies of the supposed 'democratization of surveillance' is that many of the most marginalized and mobile subjects of observation are already hypervisible and hypervigilent by virtue of their lack of access to private and protected places, and yet at the same time they remain largely unseen and unaccounted for, forgotten and unheard from (Archand, 1979). To understand the conditions that unequally affect the most vulnerable, exposed and invisible urban dwellers, and to reconfigure what there is to see and do in the spaces we share, researchers can move beyond simply observing, enumerating and documenting only by acknowledging the dialectic of visibility and invisibility which envelops the lives of each and everyone of us.

Notes

1. As the present essay is about performing research in a particular type of contested territory, our use of the morally freighted and stigmatized term 'skid row' is a deliberate choice. In using it we draw explicit attention to the historic and contemporary association of particular geographical and social locations with predominant moral attitudes that lead to the social neglect and/or harsh treatment of citizens. Also, since fieldwork for both studies was conducted by Laura Huey, in what follows we use the feminine pronoun when referring to 'the researcher' in the field, even when this term is used generically.
2. This schema is loosely inspired by Merton's (1957) classic typology of modes of individual adaptation, which he illustrates with reference to the broader American value-system that places a high premium on the goal of success through wealth. Here we narrow his frame of reference to focus specifically on the social structural strains and anomic cultural patterns on 'skid row,' which for the most part Merton characterizes in terms of the pattern of 'retreatism.' For a fuller discussion of Merton's general theory, including his displacement of the adaptation pattern of 'rebellion' (as well as social transformation and revolution) to a 'residual category,' (see Kemple, 2004: 7–10).

References

Archand, P. (1979) *Vagrancy, Alcoholism, and Social Control*. London: Macmillan.

Becker, H. S. (1995) Visual Sociology, Documentary Photography and Photojournalism: It's Almost All a Matter of Context. *Visual Sociology* 10: 5–14.

Bok, S. (1978) *Lying: Moral Choice in Public and Private Life*. New York: Pantheon.

Bourdieu, P. (2004) Why the Social Sciences Must Take Themselves as their Object. In *Science of Science and Reflexivity*. Chicago: University of Chicago Press, 85–114.

Bowling, A. (1997) *Research Methods in Health*. Buckingham, U.K.: Open University Press.

Bulmer, M. (1980) Comment on 'The Ethics of Covert Methods.' *The British Journal of Sociology* 31 (1): 59–65.

Bulmer, M. (1982) The Merits and Demerits of Covert Participant Observation." In *Social Research Ethics: An Examination of the Merits of Covert Participant Observation*, edited by M. Bulmer. New York: Holmes & Meier, 217–251.

Cassell, J. (1982) "Harm, Benefits, Wrongs, and Rights in Fieldwork. In *The Ethics of Social Research: Fieldwork, Regulation, and Publication*, edited by J. Sieber. New York: Springer-Verlag, 7–31.

Culhane, D. (2003/2004) Documenting Time and Restricted Space: University and Community Women in Downtown Eastside Vancouver. *BC Studies* 140: 91–106.

Deleuze, G. (1995) Postscript on Control Societies. In *Negotiations 1972–1990*, M. Joughin (trans.). New York: Columbia University Press, 177–182.

Ellickson, R. (1996) Controlling Chronic Misconduct in City Spaces: Of Panhandlers, Skid Rows, and Public-Space Zoning. *The Yale Law Journal* 105 (5): 1165–1248.
Elms, A. (1994) Keeping Deception Honest: Justifying Conditions for Social Scientific Research Stratagems. In *Ethical Issues in Scientific Research: An Anthology*, edited by E. Erwin, S. Gendin, and L. Kleiman. New York: Garland, 121–140.
Garland, D. (1996) The Limits of the Sovereign State: Strategies of Crime Control in Contemporary Society. *The British Journal of Criminology* 26: 445–467.
Haggerty, K.D. and R.V. Ericson (2000) The Surveillant Assemblage. *British Journal of Sociology* 51 (4): 605–622.
Harcourt, B. (2001) *Illusion of Order: The False Promise of Broken Windows Policing*. Cambridge, MA.: Harvard University Press.
Hasson, S. and D. Ley (1994) The Downtown Eastside: 'One Hundred Years of Struggle.' In *Neighborhood Organizations and the Welfare State*. Toronto: University of Toronto Press, 172–204.
Homan, R. (1980) The Ethics of Covert Methods. *The British Journal of Sociology* 31(1): 46–59.
Huey, L. (2005) *Negotiating Demands: The Politics of Skid Row Policing in Vancouver, San Francisco and Edinburgh* (unpublished manuscript).
Huey, L. and T.M. Kemple (2005) 'Let the Streets Take Care of Themselves': Making Historical, Sociological and Common Sense of 'Skid Row' (unpublished manuscript).
Huey, L., R. Ericson and K. Haggerty (2005) Policing Fantasy City. In *Re-imagining Policing in Canada*, edited by D. Cooley. Toronto: University of Toronto Press, 140–208.
Johnston, L. and C. Shearing (2003) *Governing Security: Explorations in Policing and Justice*. New York: Routledge.
Kelling, G. and C. Coles (1996) *Fixing Broken Windows: Restoring Order and Reducing Crime in Our Communities*. New York: Martin Kessler Books.
Kemple, T. M. (2004) Editor's Note to Part I: The Limits of 'Rationality'. In *The Vocation of Reason: Studies in Critical Theory and Social Science in the Age of Max Weber*, edited by T.M. Kemple. Leiden, The Netherlands: Brill, 3–14.
Lauder, M. (2003) Covert Participant Observation of a Deviant Community: Justifying the Use of Deception. *Journal of Contemporary Religion* 18 (2): 185–196.
Luhmann, N. (2002) Paradox of Observing Systems. In *Theories of Distinction: Redescribing the Descriptions of Modernity*, edited by William Rasch, translated by J. O'Neil. Stanford, CA: Stanford University Press, 79–93.
Lyon, D. (2002) Surveillance Studies: Understanding Visibility, Mobility, and the Phenetic Fix. *Surveillance & Society* 1 (1): 1–7. http://www.surveillance-and-society.org/articles1/editorial.pdf
Marx, G. T. (2003) A Tack in the Shoe: Neutralizing and Resisting the New Surveillance. *Journal of Social Issues* 59 (2): 369–390.
Marx, G. T. (2004) Surveillance and Society, in *Encyclopedia of Social Theory, 2nd ed.*, edited by George Ritzer. Thousand Oaks, CA: Sage, 275–280.
Merton, R.K. (1949) Social Structure and Anomie. In *Social Theory and Social Structure*. Glencoe, IL: The Free Press, 131–160.
Mirzoeff, N. (2002) The Subject of Visual Culture. In *The Visual Culture Reader*, edited by N. Mirzoeff. London: Routledge, 3–21.
Monahan, T. (forthcoming) Counter-Surveillance as Political Intervention? *Social Semiotics*.
O'Malley, P. (1992) Risk, Power and Crime Prevention. *Economy and Society* 21 (3): 252–275.
Rancière, J. (1998) Dix thèses sur la politique. In *Aux bords dupolitique*. Paris: Gallimard, 223–254.

Reynolds, P. (1982) Moral Judgments: Strategies for Analysis with Application to Covert Participant Observation, in *Social Research Ethics: An Examination of the Merits of Covert Participant Observation*, edited by. M. Buhner. New York: Holmes & Meier, 185–213.

Rose, N. (1999) *Powers of Freedom: Reframing Political Thought*. Cambridge, UK: Cambridge University Press.

Rosenfeld, R., B. Jacobs and R. Wright (2003) Snitching and the Code of the Street. *British Journal of Criminology* 43 (2): 291–309.

Staples, W. (2000) *Everyday Surveillance: Vigilance and Visibility in Postmodern Life*. Lanham, Md.: Rowman & Littlefield.

Turnock, C. and V. Gibson (2001) Validity in Action Research: A Discussion On Theoretical and Practical Issues Encountered whilst using Observation to Collect Data. *Journal of Advanced Nursing* 36 (4): 401–411.

Warwick, D. (1973) Survey Research and Participant Observation: A Benefit-Cost Analysis. In *Comparative Research Methods*, edited by D. Warwick and S. Osherson. Englewood Cliffs: Prentice-Hall, 189–203.

Wilson, H. T. (2004) Essential Process of Modernity: A Critical Analysis of Social Science Research Practices and an Alternative. In *The Vocation of Reason: Studies in Critical Theory and Social Science in the Age of Max Weber*, edited by T. M. Kemple. Leiden, The Netherlands: Brill, 295–329.

Wilson, J. and G. Kelling (1982) Broken Windows: The Police and Neighborhood Safety. *Atlantic Monthly* (March): 29–38.

82

Whistle-Blower Disclosures and Management Retaliation: The Battle to Control Information about Organization Corruption

Joyce Rothschild and Terance D. Miethe

Ever since Karl Marx's original analysis of the relationship between an economy and a society and how that relationship is defined by class struggle, social scientists from many disciplines have attempted to illuminate the class struggles that shape the contemporary workplace. Histories of work in the 20th century have described how a never-finer division of labor gave rise to a deskilling of the labor force. This finer division served to accumulate profit more than it did industrial efficiency (Braverman, 1974). Although contemporary scholars and researchers have recognized the workplace as a site of contention – often there is intense contestation between capital and labor – their focus tends to emphasize organized, collective expressions of this struggle, such as union-led labor resistance (Buroway, 1979; Cornfield, 1989; Edwards, 1979; Kimeldorf, 1988). However, recently scholars have begun to recognize less formal, more individual forms of worker resistance (Brower & Abolafia, 1995; Hodson, 1995; Jermier, 1988).

At the same time, commentators on our society describe the shift taking place from an economy based on manufacturing to an economy based on information processing. Most citizens of our society are aware of some of the dimensions of this shift and its implications (e.g., the downsizing of certain

Source: *Work and Occupations*, 26(1) (1999): 107–128.

occupational categories and the burgeoning of others). It is recognized by many that for this shift in the nature of our economy to be successful, educational and skill levels of the work force have to rise.

It is our contention that the information-intensive and information-driven economy – given the higher education and skill levels that it requires – has a significant consequence that has yet to be noticed by the experts or the public: That unanticipated outcome is the increasingly widespread use of "whistle-blowing." We argue that this type of disclosure represents a new and fast-growing form of worker resistance, albeit individual, that can be enormously effective at inhibiting the accumulation of profit by corporations. For example, when leaked information by a whistle-blower ended profit being made off of young people in the U.S. tobacco industry. Similar to this, in the public sector leaked information by a whistle-blower revealed the routine dynamics at work at National Aeronautics Space Administration (NASA) that underlaid the Challenger disaster. Although the business press provides frequent case examples of the power of new ideas and information in the process of production, and in the production of profits, what has been largely neglected by social scientists is workers' use of previously secret information to resist and oppose managerial directives.

Although we are not the only social scientists to write about whistle-blowers from a sympathetic perspective (see Glazer & Glazer, 1989), nor the only ones to see in their exposés of bureaucratic corruption the potential for workers to oppose bureaucratic or managerial authority (Weinstein, 1979; Perrucci, Anderson, Schendel, & Tractman, 1989), we present comprehensive, nationwide empirical data in this article that will support this perspective.

Sample Description

Various approaches to the study of whistle-blowers have been taken. Previous researchers have identified a small number of high-profile whistle-blowers from media sources and performed case studies (see Glazer and Glazer 1989). Others (see Jos, Tompkins, & Hays, 1989) have derived samples of whistle-blowers from support agencies (e.g., Government Accountability Project), and still others have drawn a sample from a single occupation (Miceli & Near, 1990). Unfortunately, however, these strategies are limited because they tend to yield atypical samples (e.g., high-profile cases) and are unable to compare different types of whistle-blowers (e.g., internal versus external whistle-blowers) with other types of employees (e.g., non-observers and silent observers).

For our purposes, we needed a unique data set. We needed to survey whistle-blowers from all walks of life and from every region of the country to yield the validity we sought. We needed to interview them in considerable depth so that we would be able to draw richly grained conclusions about the meaning of this phenomenon. Our choice of selecting multiple samples from a diverse range of employees across a variety of work settings was based on the nature of our research questions and the practical and economic constraints on drawing a large random sample of all U.S. workers.

Survey data were collected and analyzed from several sources. First, we targeted and surveyed employees in specific professions or organizations. This included workers in a petro-chemical plant (n = 66), employees in a high-security defense related manufacturing company (n = 148), employees at a large bookstore (n = 27), personnel in an administrative unit within a university (n = 18), employees in nonprofit organizations in two states (n = 108), and nurses in the state of Nevada (n = 116). Second, we conducted intensive telephone interviews with whistle-blowers and silent observers from every kind of workplace across the country (n = 292). Our complete sample involves 761 individuals in one of four groups: (a) nonobservers of misconduct (n = 149), (b) observers of misconduct who remained silent to officials within and outside their organization (n = 218 silent observers), (c) individuals who reported observed misconduct to officials within their employing organization (n = 147 internal whistle-blowers), and (d) individuals who reported misconduct to authorities outside their organization (n = 247 external whistle-blowers). Third, we reanalyzed data from the U.S. Merit System Protection Board (1993) on more than 13,000 federal employees.

Because whistle-blowing is a relatively rare event, it would be prohibitively expensive to gather enough cases to study it using a random sample of the full U.S. labor force. Thus, we constructed a conglomerated sample from a diverse range of employees and organizations, as described above. A complete description of the sample and our answers to the five major research questions that organize this study follow.

Parade Magazine Sample

The primary data source for our interviewed sample of whistle-blowers involved persons who asked for additional information about an article that appeared in *Parade Magazine* in August of 1991. The one-page article was about whistle-blowing and Integrity International – a support group for whistle-blowers. The founder of that support group provided us access to almost 1,000 persons who contacted him about their whistle-blowing experiences. The fact that this magazine is distributed every week in Sunday newspapers across the country made it extremely valuable for gaining a crosssection of whistle-blowers in various work settings. We can think of no other way to efficiently draw a national sample of self-defined whistle-blowers.

Of the phone numbers or postal addresses provided, we discovered that about one third were either outdated or disconnected. For 300 of the potential whistle-blowers who provided a postal address, we sent a self-addressed mail survey for reporting their experiences and attitudes. Of these 300 mail surveys, 83 were completed and returned. An additional 210 individuals were interviewed by telephone. The average telephone interview lasted about 90 minutes (1.5 hours).

Our in-depth telephone interviews and surveys led us to several conclusions about the appropriate methods for studying whistle-blowers. First, although many whistle-blowers had initial reservations about talking with us, this hesitancy was overcome once they realized that we were not company spies. By the end of the interviews, most expressed gratitude for this opportunity to speak openly and at length about their experience. Second, we discovered that well-designed and anonymous mail surveys could be used to study whistle-blowers. Compared to telephone interviews, mail surveys are more cost efficient and also provide anonymity. Third, we planned to study silent observers by getting the names of other employees from the whistle-blowers themselves. However, we found that whistle-blowers were generally unwilling to provide the names of others and, when they did, these other persons were typically their friends. As a result, we conducted industry-based surveys to compile a sample of silent observers.

Industry-based Surveys

For purposes of studying the correlates of whistle-blowing and identifying demographic, attitudinal, and organizational differences between whistle-blowers and nonwhistle-blowers, it is important to examine employees in similar work environments. Accordingly, we designed a mail survey that was administered to employees in several work settings.

Surveys were given to several specific groups of employees. First, 400 self-administered surveys and return envelopes were provided at the entrance and exits of a high-security manufacturing firm. A box specifying "Employee survey: Please take one" was placed at these locations. Usable surveys were returned to us directly from 148 of these employees. Second, using a similar research protocol, 66 of a possible 200 surveys were completed and returned from employees at a large, petro-chemical plant. Third, two thirds of the staff personnel (18 of 30 people) in an administrative unit within a major university completed and returned the survey instrument. Fourth, over two thirds of the workers at a large bookstore (27 of 40 people) completed the same survey. Fifth, a total of 600 mail surveys were distributed to a sample of registered nurses in Nevada and 116 usable surveys were returned.

Both the nurse sample and our interviews with whistle-blowers yielded some employees in non-profit organizations; however, additional procedures were needed to achieve a comprehensive sample of nonprofit employees. In particular, nonprofit agencies in North Carolina and in Nevada were contacted and asked to participate in our study. Participating organizations included the following: The United Way, American Cancer Society, YMCA, United Cerebral Palsy, Boys and Girls Club, Community Counseling Services, Big Brother/Big Sister, Family Religious Services, and Catholic Community Services. Across these organizations, 280 surveys were distributed. A total of 108 surveys were returned.

Compared to general labor force statistics in the U.S. (see Statistical Abstract of the United States, 1994), our industry-based sample tends to underestimate the prevalence of male workers, nonsupervisors, and young employees. There are several possible reasons why nonresponse rates might affect our yielded sample. Specifically, we expect survey respondents across all samples to be individuals with either greater personal interest or more direct experience with whistle-blowing than nonrespondents. Under these conditions, the sampling of more concerned employees would inflate overall estimates of the prevalence of observing misconduct and whistle-blowing. However, this potential sampling bias did not have any impact on our ability to accurately characterize whistle-blowers nor to compare them to other types of employees. We used a national telephone survey conducted for another project (see part 3 below) to provide us with additional data that is better able to yield accurate estimates of national rates of observing misconduct and whistle-blowing.

Extent of Whistle-Blowing

For purposes of estimating the true extent of whistle-blowing in the United States, the best sampling procedure involves drawing a random sample of employees across the country. There is no substitute for this approach for population estimations, but it is very expensive selecting such a sample (given the number of contacts necessary to find employed adults who have observed misconduct and reported it). Fortunately, we were able to accomplish the goal of estimating the national rate of whistle-blowing by "piggybacking" several questions on a national telephone survey of 1004 adults' experiences with the legal system (for details of the sample see Miethe, 1995). Through the assistance of the Center of Survey Research at Virginia Tech, we also incorporated several questions about whistle-blowing in a statewide survey of 326 adult employees in Virginia.

The specific questions about whistle-blowing in the national survey asked whether the respondent had ever observed illegal or unethical activity at work, and, if so, whether they reported it. Our national survey responses revealed that 37% of the employed adults had observed some type of misconduct at work and that 62% of those reported it. The vast majority of whistle-blowers (84%) reported the misconduct to someone within the company rather than to external authorities. By comparison, the statewide survey in Virginia indicated considerably lower rates of observing misconduct (17%) and higher levels of remaining silent (52% were silent observers) and external whistle-blowing (25% of whistle-blowers reported misconduct to persons outside the company). When combined with the results from other large-scale surveys of federal and private employees (for review see Miethe & Rothschild, 1994), our best overall estimate is that about one third of U.S. employees have observed conduct they consider to be unethical or illegal in their workplace. Over one half of these observers remain silent, whereas less than one fourth of whistle-blowers are external informants.

These findings on the extent of whistle-blowing have significant implications. First, the fact that a sizeable minority of workers observe organizational misconduct suggests that workers themselves have enormous potential to detect and control waste, fraud, and crime in the American workplace. Given the low public visibility and high technical complexity of much illegal activity in the workplace, the successful monitoring and prevention of organizational abuse rests in large part on the willingness of employees to become whistle-blowers. Second, two groups of employees (i.e., silent observers and external whistle-blowers) are especially important to study so we can gather data regarding both the resistance to and/or the inhibition of organizational misconduct. Specifically, systematic comparisons of external whistle-blowers with other types of employees enable us to identify the particular demographic, attitudinal, and organizational factors that enhance or inhibit the full disclosure of misconduct. Comparisons between silent observers and whistle-blowers provide us with a means of identifying the major individual and organizational impediments to whistle-blowing. The relatively low rates of external whistle-blowing and the high proportion of silent observers makes it all the more important to study and compare the experiences of these people.

Individual and Organizational Correlates of Whistle-Blowing

Previous studies have searched for a whistle-blower profile (Miceli & Near, 1990) that could lead managers to avoid hiring persons prone to whistle-blowing. We, too, looked for correlates that might mark those prone to blowing the whistle. Table 1 summarizes some of the quantitative results of our analyses.

One of our most important and striking findings is that almost no sociodemographic characteristics can distinguish the whistle-blower from the silent observer. Put another way, we find that there is no empirical basis for asserting that any demographic profile of the "typical" whistle-blower exists. Past research that has made claims for a demographic profile failed to include a sufficient number of whistle-blowers and silent observers to

Table 1: Comparison of average ratings

	Nonobservers	*Silent observers*	*Internal whistle-blowers*	*External whistle-blowers*	*Private organizations*	*Public organizations*	*Nonprofit organizations*
Demographic							
Female (%)	51	55	65	45	44	54	73
Less than 40 years old (%)	39	47	38	21	39	20	54
Married (%)	70	70	65	66	74	63	57
Education[a] ($\bar{x}$)	4.46	4.58	4.64	4.70	4.46	4.79	4.76
Religious[b] ($\bar{x}$)	2.17	2.13	2.14	2.16	2.19	2.09	2.15
Income[c] ($\bar{x}$)	3.39	3.50	3.48	3.06	3.48	3.17	3.22
Years employed at company ($\bar{x}$)	6.78	6.25	5.74	4.49	6.44	5.12	4.26
Supervisor position (%)	50	45	53	47	45	50	58
Promotions (0 to 3+) ($\bar{x}$)	1.49	1.37	1.37	1.37	1.49	1.38	1.10
General attitudes and/or values							
Decisions should be made on case-by-case basis (%)	73	77	73	63	77	63	72
I'm a person of worth (%)	99	97	99	98	98	98	99
What happens to me due to luck (%)	10	5	11	11	9	12	4
People should sacrifice for greater good (%)	83	77	84	80	79	78	86
People should prevent harm to others if they can (%)	98	99	96	97	99	96	97
Organizational characteristics							
Code of ethics for type of work (%)	85	82	84	71	81	77	84
Does employer have strong moral values (%)	83	62	57	29	60	31	82
Democratic decision making (%)	60	35	38	13	39	15	53
Personal opportunity to participate in decisions (%)	71	49	49	22	52	24	67
Type of misconduct observed							
Fraud (%)	0	14	22	43	15	33	16
Health and/or safety violations (%)	0	21	34	30	21	27	15
Poor production and/or service (%)	0	11	20	34	16	26	9
Total number of types of misconduct (%)	0	1.13	2.11	2.56	1.24	2.13	1.30
Sample size	149	218	147	247	385	253	150

a. 1 = some high school, to 6 = graduate school.
b. 1 = not religious, to 3 = very religious.
c. 1 ≤ \$1,000, to 6 ≥ 100,000.

make valid comparisons (Miceli & Near, 1990). The current sample covers a large number of employees from a wide-range of occupations and organizations and, therefore, is better suited for addressing possible differences.

When comparisons are made between nonobservers, silent observers, internal whistle-blowers, and external whistle-blowers (see Table 1), several findings emerge. First, there are no significant gender differences between whistle-blowers and nonwhistle-blowers. Internal whistle-blowers were far more likely to be women than men (65% vs. 35%), but an opposite trend was found for external whistle-blowers (45% vs. 55%). Second, external whistle-blowers are somewhat older (i.e., a higher proportion are older than 40), earn less income, and have shorter employment histories with the company than other employees. However, and most importantly, there are no substantial differences across the four groups in terms of their marital status, educational attainment, religiosity, number of promotions, and whether they occupied a supervisory position within the organization. Our supplemental analysis of data on 13,000 federal employees in the merit board survey also reveals few demographic differences between whistle-blowers and silent observers. The overall pattern of these results supports the conclusion that individual sociodemographic attributes do not explain differences in individuals' exposure to occupational deviance very well, nor do they explain their response to this deviance.

Attitude and Value Differences

There has been much speculation and little empirical data on whether a whistle-blowing personality exists. Previous studies using small samples, case studies, and anecdotal reports suggest that whistle-blowers are strong-willed, opinionated individuals with universalistic values and high levels of self-esteem and self-efficacy (Jos et al., 1989).

As shown in Table 1, there are some differences across groups of employees in their general attitudes and values, but these differences are not significant enough to qualify as support for a presumed whistle-blowing personality. The largest difference across groups involves "whether decisions should be made according to general rules or on a case-by-case basis" (cf. text of survey). Whistle-blowers held slightly more universalistic values than nonwhistle-blowers. This was most apparent when external reporters are compared to silent observers. Regardless of their group membership, nearly all employees had high self-esteem (i.e., agreed with the statement "I am a person of worth") and high social responsibility (i.e., "people should prevent harm to others"). There are also no major differences across employees in their perceptions of self-efficacy (i.e., "what happens to me is a matter of luck") and beliefs about altruism (i.e., "people should make sacrifices for the greater good").

Organizational Characteristics

Whistle-blowing occurs within an organizational context, and we believe that any adequate theory of its prevalence and correlates must examine this context.

Concerning organizational structure, we hypothesized that more participatory work environments should have lower levels of organizational misconduct and lower levels of external whistle-blowing. We expected these patterns because such organizations tend to have more group cohesion and agreement on goals, more cultivated norms of trust and openness in working relationships, and more personal investment in organizational success. Two patterns of results in Table 1 clearly support this hypothesis about whistle-blowing and its organizational structure. First, internal whistle-blowers are about three times more likely than external whistle-blowers to view their company as democratic in terms of its decision making. Almost two thirds (60%) of the nonobservers of misconduct thought their company was democratic compared to only 13% among external whistle-blowers. Second, external whistle-blowers were far less likely than other employees to believe that they had enough personal opportunity to participate in company decisions. Thus, more democratic and less bureaucratic organizations seem more likely to have the open channels of communication that are necessary to deal with misconduct informally.

The moral and ethical climate of the organization and the individual's specific type of job are also important correlates of whistle-blowing. For example, about 83% of the nonobservers thought their employer had strong moral values compared to only 29% of the external whistle-blowers. Although codes of ethics tended to characterize the work conducted by most employees in our sample, external whistle-blowers were less likely than other groups to report that their work activities were regulated by professional ethics.

In terms of whistle-blowing experiences, rates of external disclosure were highest among employees of public agencies and/or institutions (involving 70% of these employees), followed by workers in nonprofit organizations (18%), and an even smaller proportion of employees in private businesses (13%). Separate analysis of the nurse sample also revealed the highest level of external whistle-blowing among those working in public agencies (14% of these employees), but there were no major differences in the likelihood of external whistle-blowing for nurses in private businesses versus nonprofit organizations. The much higher rate of whistle-blowing (especially external reporting) in public institutions persists even after controlling for differences between public, private, and nonprofit organizations on other characteristics (e.g., level of occupational deviance in these organizations and perceived ethical standards of company).

To understand this finding, we looked for other important patterns. For instance, we found that nonprofit workers are far more likely than other types of workers to think their employer has strong moral values, is democratic in their decision-making processes, and provides opportunities for all employees to participate in decision making. Nonprofit employees also indicate greater feelings of self-efficacy as indicated by a lower percentage of their workers agreeing with the statement that "what happens to me is due to chance."

Overall, these findings indicate that the opportunity for whistle-blowing varies dramatically across occupational groupings and organizational structures. We attribute the higher level of whistle-blowing in public institutions to the following factors: (a) public institutions tend to be larger, more bureaucratic, and perceived as less responsive to workers' comments and suggestions; (b) private employees have less legal protection than public employees from "at will firing" and, therefore, are more prone to remain silent about observed misconduct; and (c) federal employees and employees of many state agencies are aware of special hotlines that have been set up to encourage their reporting of waste, fraud, or abuse. Employees in nonprofit organizations hold the most positive views of management; thus, they may be more likely than private employees to report infractions internally to protect their employer from future possible disgrace while still staying within the confines of their organization.

Although our investigation of the correlates of whistle-blowing has focused on organizational determinants, theories of deviance may also help to explain demographic, attitudinal, and organizational differences in whistle-blowing experiences. For example, a rational choice perspective would explain the relatively low rate of external whistle-blowing in terms of its high risks and limited benefits to the whistle-blower. In contrast, differential association theory would argue that the norms of loyalty held by management in an organizational climate that is generally antagonistic toward exposing misconduct would increase the likelihood of silence or, at most, it would increase the incidence of internal whistle-blowing while reducing the likelihood of external whistle-blowing. As we have noted elsewhere (Miethe & Rothschild, 1994), several other theories of deviance are also directly relevant to understanding the extent and distribution of whistle-blowing.

Motivations for Whistle-Blowing

Mass media presentations of whistle-blowers frequently speculate about their underlying motivations. Many present whistle-blowers as selfless and altruistic; they allegedly undertake this task only at extraordinary personal cost. Others consider whistle-blowers egoistic and spiteful, motivated solely by greed and self-interest. Our detailed interviews with whistle-blowers from a variety of work settings suggest that neither of these images adequately captures a whistle-blower's set of motivations.

To better understand why employees report wrongful or illegal acts at work, in our telephone interviews with whistle-blowers ($N = 210$) we asked, "What was the main thing that compelled or persuaded you to report this violation?" We studied and codified their sometimes long and qualitative answers to this question. We learned that at times their motivation stemmed from their job requiring them to report their observation (for example, social workers are required in many states to report abuse of patients; engineers are paid to find and report defects in the integrity of projects; accountants are supposed to report financial fraud or embezzlement). In addition, whistle-blowers in our study often cited professional norms and values as setting the stage for their disclosure. Many said that their own personal values, grounded in their religious or humane traditions, moved them. Of the sample, 79% of our interview respondents spoke of the personally held values that drove them to act. Whether these values stemmed from their job, their profession, their religion, their community, their family, or some combination of these, it was their belief that the reported behavior was wrong, harmful, or illegal. Thus, based on that belief they were moved to report. In another 11% of these cases, their motivations would have to be classified as chiefly defensive, in the sense that they feared that if they did not report and the malfeasance was eventually discovered, that it would be blamed on them. In only 2% (4 cases) did the whistle-blowers receive a promotion or raise as a result of their report. Finally, in a scant 3% of our interview sample (7 cases) the whistle-blowers said that resentment of management fueled their report.

Many individuals in our sample came to their whistle-blowing almost by accident. Typically, they witnessed something at work that troubled them. If their boss appeared to be the perpetrator of the problem, they took it to their boss's boss believing that if upper management knew what was going on, they would surely step in to correct the situation. They expected that higher officials would be grateful for the information. Only when senior officials showed that they were inert or complicit in the wrongdoing did the employee consider going to authorities outside the organization. In almost no case did the individual accurately anticipate the retaliation and severe personal consequences that would follow their report. In this sense, it would be more accurate to depict the prototypical whistle-blower as organizationally naïve; in other words, they truly believe that the organization wants its practices to be in line with its mission, rather than particularly altruistic or vengeful. As we will show, the ordeal that follows their disclosure brings an end to their illusions.

Overall, it became increasingly clear through the interviews that whistle-blowers are in the ultimate no-win situation. By speaking out they face the wrath of managers for being a squealer, but inaction makes them potentially culpable and/or easy scapegoats when and if the misconduct is discovered. In this sense, many whistle-blowers can be characterized as reluctant dissenters, moved neither by altruistic nor selfish concerns, but rather by a tide of events over which they feel they have little control.

Further, one of the most important findings from our study involves the crucial importance of the organization's response to the employee's initial concerns. It is management's response that shapes the potential whistle-blower's subsequent actions. Specifically, our interviews revealed a common pattern in which management's efforts to discredit or retaliate against the claimant become the major catalyst for the political transformation of the concerned employee into a "persistent resister" (see also Glazer & Glazer, 1989). In fact, it appears that most whistle-blowers go public with their information only after their companies begin to cover up the wrongdoing and intensify their level of retaliation against the whistle-blower. It is by virtue of the organization's retaliatory response that whistle-blowers become convinced of the moral correctness of their actions and strengthened in their conviction to fight – both to exonerate themselves and to correct the organization's wrong.

Consequences of Whistle-Blowing for the Individual and the Organization

By all indications, whistle-blowing has enormous consequences for both the individual and the organization. From the point of view of management, internal whistle-blowing may be applauded when it identifies *bad apples* within the company and allows for their removal before they cost the organization financial loss, scathing publicity, and/or litigation. On the other hand, external whistle-blowing is often condemned by management because the exposure of wrongdoing often brings adverse publicity. Also, it can bring financial ruin if the wrongdoing is far-reaching and if the company is liable for the harm involved. In contrast, whistle-blowers – even where their disclosures are found to be true and of great benefit to the employer – still face significant risk of various types of organizational retaliation, such as ostracism by coworkers, long-term economic harm, and psychological injury.

Based on data from our nationwide interviews of whistle-blowers from all occupational categories, organizational retaliation against whistle-blowers (both internal and external reporters) is severe and common. About two thirds of the whistle-blowers in our study experienced each of the following forms of retaliation from their employer or coworkers as a result of their disclosures:

- lost their job or were forced to retire (69%).
- received negative job performance evaluations (64%).
- had work more closely monitored by supervisors (68%).
- were criticized or avoided by coworkers (69%).
- were blacklisted from getting another job in their field (64%).

Retaliation against external reporters was even more common than retaliation against internal whistle-blowers. For each of the items above, the rate of retaliation was generally 10 to 15% points higher among external whistle-blowers.

As a consequence of their whistle-blowing experience, the majority of the whistle-blowers in our sample suffered intensely. The most common fallout from their whistle-blowing involved: (a) severe depression or anxiety (84%), (b) feelings of isolation or powerlessness (84%), (c) distrust of others (78%), (d) declining physical health (69%), (e) severe financial decline (66%), and (f) problems with family relations (53%). Repeatedly, respondents mentioned that whistle-blowing undermined their trust in others. These numbers are extremely high; however, the numbers cannot convey the emotional distress that we heard from so many of our respondents.

When the forms of organizational retaliation and the personal declines in the whistle-blowers' lives were considered together, it became clear to us that for many of these individuals the act of whistle-blowing had become what sociologists refer to as a "master status." One's master status is the critical bedrock of one's personal identity; it is how we first label ourselves and are recognized by others. For most of our whistle-blowers, the experience of whistle-blowing and its aftermath have been so traumatic that their "master status" is now defined by their act of whistle-blowing. Their new identity – one based on the act of whistle-blowing – defines and engulfs nearly everything in their lives. Other experiences become selectively interpreted to be consistent with the master status, and the whistle-blowing experience is repeatedly discussed. Embattled and embittered, the typical external whistle-blower becomes viewed by management and coworkers as a loose cannon and a social outcast, thereby reinforcing the salience of the whistle-blowing ordeal in the formation of the whistle-blower's personal identity and future life chances. Given the personal and financial devastation that we heard from so many whistle-blowers, it may surprise one to learn that 90% of the whistle-blowers in our sample said that they would still report misconduct if they had a chance to do things all over again. Through the suffering they have endured – but never anticipated before their disclosure – they have come to see themselves as exceedingly moral. They have distanced themselves from what they now see as the corruption of their former employer, and many begin to see themselves as possessing extraordinary integrity that they now bring to their endeavors.

Can the Individual Be Insulated from Management Retaliation?

Some of our most recent research involves examining which factors contribute most to the likelihood and severity of management retaliation and which factors, if any, may help to insulate the individual whistle-blower from reprisals. For this project, we have united our in-depth qualitative interviews with secondary analysis of survey data on federal employees conducted by the U.S. Merit Systems Protection Board (1993).

Given the ample discussions in the sociological literature that finds race, gender, age, and class are related to almost every form of discriminatory behavior, we wanted to test how these variables related to the severity and extent of retaliation. Many hypotheses seemed plausible. For example, African Americans and women might be more prone to reprisals because these groups are less likely to hold positions of power within contemporary U.S. work organizations. Similar to this, managers, high-level supervisors, and veteran employees might be expected to be insulated from retaliation for whistle-blowing because either they have the power to avoid negative reactions or they have cultivated strong interpersonal relationships with upper management that might protect them from reprisals. In contrast, the same type of high-level employees might actually be more susceptible to organizational retaliation because their reportage could be seen as a more serious violation of company loyalty norms. Also, their information and criticism often may be seen as more costly to the organization.

The most comprehensive data for examining this insulation hypothesis involves the 1992 survey of federal employees conducted by the U.S. Merit Systems Protection Board (MSPB, 1993). Of the over 13,000 federal workers who responded to this survey, slightly over 1,100 of them were either internal or external whistle-blowers. About one fourth of these whistle-blowers had personally experienced some type of reprisal – or threat of reprisal – by management for having reported misconduct. Factors associated with the likelihood of retaliation in this sample are summarized in Table 2.

Analysis of the MSPB survey data indicates that the likelihood of retaliation by management does not necessarily vary in the expected ways. As shown in Table 2, retaliation was less likely for supervisors than for nonsupervisors, but there were no major differences in the likelihood of reprisals based on the workers' gender, age, educational attainment, or years of employment in their current job. However, African American workers were nearly twice as likely as White employees to suffer organizational retaliation for whistle-blowing. Retaliation by management was also significantly more common for external whistle-blowers and when the misconduct was systemic (i.e., frequently occurring within the organization). Whistle-blowing on organizational misconduct that was both severe and frequent elicited the highest risk of retaliation in comparison to less severe or less frequent types of wrongdoing. In sum, regardless of the employee's gender, age or education, whistle-blowers who report on systemic organizational corruption or major activity (i.e., misconduct involving over $100,000 in losses and that frequently occurs) are most likely to experience organizational reprisals.

Table 2: Differences in the likelihood of organizational retaliation for whistle-blowing related to personal characteristics of the whistle-blower and situational factors ($N = 1,102$).

Characteristic	*Percentage retaliated against*
Personal characteristics	
Gender	
Male	24
Female	26
Age	
Less than 40 years old	25
40 years old or older	25
Race	
African American	40**
White	22
Education level	
Less than college graduate	22*
College graduate	26
Work position	
Nonsupervisor	27**
Supervisor	19
Length of employment at current job	
Less than 4 years	24
4 years or longer	25
Vested employee (supervisor and > 4 years at work)	
No	26*
Yes	20
Situational Factors	
Type of whistle-blowing	
Internal reporting	22**
External reporting	30
Dollar value loss from the observed misconduct	
More than $100,000	27
Less than $100,000; can't place $ value	24
Frequency of occurrence of this misconduct	
Once or rarely	11
Occasionally	19
Frequently	32
Major activity (frequent and > $100,000)	
No	24**
Yes	32

*$p < .10$. **$p < .05$.

Discussion and Conclusions

Several major findings emerge from our data that suggest that whistle-blowing can and should be understood as a new form of worker resistance; that is, as a form of worker assertion and dissent in the unending battle between labor and management to control the workplace. We believe it is a method that is especially well suited to an information-based economy that depends more than ever on proprietary information and organizational secrets for profit accumulation.

First, we saw that in 79% of our cases whistle-blowers were moved to report and dissent only after repeatedly witnessing some organizational behavior that, from the vantage point of the ethics or values that imbue their profession, their religion, their community or their family, they see as criminal or wrong. They feel they are doing the "right" thing by bringing information forward that they believe will help to correct a situation that has gone awry. In almost no case did they anticipate the retaliation that was to follow. It is only as management begins to "circle the wagons" to discredit and fire the whistle-blower that the whistle-blower turns to defense. In an arduous process of vindication, they find themselves in countless organizational skirmishes and, often, in the courtroom. Thus, it is the management-induced ordeal of reprisals and character assassination that is the catalyst for whistle-blower activism. Even the whistle-blowers who told us that they were apolitical before this event felt that being the recipient of this reign of reprisals forever changed their consciousness. They now saw this management (and often managements everywhere) as the point men for organizational cover-ups and corruption in the interest of profits. Whistle-blowers in the United States and Australia have recently formed social movement organizations that reflect their new found insights and activism. Although we cannot say from our study the effect these whistle-blower ordeals have on the many bystanders (i.e., on other employees who are not part of the whistle-blower event but who are in the workplace and observing the organizational fall-out), we speculate that there is also the potential to alter subsequent labor-management struggles. This group needs to be studied also.

Second, we have found that when the organization is seen as open – as permissive of employee voice in the decision-making process – and as essentially ethical, employees with critical observations to make will keep them internal. Employees go to external authorities with their claims only once they come to believe that internal channels are closed to them, that the organization is not moral, and that senior management is inert or complicit in the wrongdoing. This evidence tells us that employees who go to external aids (to the media, to prosecutors, etc.) with their evidence, are using their whistle-blowing as a weapon – perhaps their weapon of last resort – to make management stop engaging in whatever misconduct they see and to return to the legitimate purpose of the organization. The casual observer may not see this as worker resistance because the exposure and dissent is often individual and it is not aimed at raising the wages paid to the employees. However, the chief issue to the whistle-blower is almost always one of dignity or pride in one's work. This, our whistle-blower respondents told us, is the chief reason they want the organization to cease and desist their misconduct; they want the organization to return to what the employee sees as the reason they joined the organization in the first place: its legitimate purpose.

Third, we found in our analysis that retaliation against whistle-blowers is most likely and most severe when the observed wrongdoing is most systemic and most central to the operation of the agency (i.e., misconduct that involves losses over $100,000 and that occurs on a frequent basis generally evokes the greatest managerial retaliation). In other words, organizations are more likely to self-correct (and not to retaliate against the whistle-blower) when the misconduct that the whistle-blower has disclosed involves an isolated bad apple. However, when the misconduct in question is systemic – part of the regular way that this organization does business – management denial and retaliation are quick and, as we have found, virtually certain.

Fourth, almost no sociodemographic characteristics distinguished the whistle-blower from the silent observer. One of the qualitative themes that emerges from our in-depth telephone interviews is that our culture is in conflict as to how it views whistle-blowers. Some see them as traitorous violators of organizational loyalty norms; others see them as heroic defenders of values considered to be more important than company loyalty (e.g., the public health, truth-in-advertising, environmental respect). From the deviance point of view, this means that whistle-blowing can be viewed as either conformity or as deviance, which may help us to understand the reasons for so few sociodemographic predictors.

Lastly, and one of our most important findings, neither gender, age, race, educational attainment, nor years in the job can save you from retaliation. Indeed, even people in supervisory positions and in their current position for over four years garner only a modest amount of insulation, and this is only true if they report on less severe, nonsystemic abuses. The bottom line is that the larger and more systemic the abuse that is being exposed, the more intense will be the managerial effort to discredit and to punish the whistle-blower. Our findings in this regard squarely support a class analysis of the workplace because they show that the organization reserves its most explicit discrimination and punishment for those who block the profit accumulation process by exposing the practices that undergird this process.

As we have shown, most of the whistle-blowers in our study suffered enormously as a result of their disclosures and, as Clegg (1994) has pointed out, it is too easy for management to outflank the lone-employee whistle-blower. However, we would be remiss if we did not notice that this is not always the case. Indeed, a sizeable minority of the whistle-blowers that we interviewed – 27% of the internal whistle-blowers and 44% of the external-whistle-blowers – thought that their organization had changed its practices as a result of their disclosure.

Recently, the Government Accountability Project (GAP), a non-profit group intended to track and support legitimate whistle-blower claims, has published a how-to book intended to help would be whistle-blowers around the country make more effective disclosures of waste and abuse without having to suffer (GAP, 1997). In addition, whistle-blower hotlines have been enacted for many federal and state agencies. The U.S. Department of Defense states that their fraud hotline has saved taxpayers over $150 million.

Public confidence in business and public institutions is declining as the public becomes increasingly aware that organizations sometimes make defective products, falsify claims, pollute the environment, subject their workers to hazardous conditions, and/or abuse clients. With the recent revelations of wrongdoing at the national office of the United Way, even the nonprofit sector has aroused public suspicion that an organization's practices are not always in keeping with its purposes.

Yet, because of low public visibility, technical complexity, and explicit cover-ups of much misconduct that goes on in organizations, we would have little chance of learning about organizational wrongdoings were it not for conscientious employees who are in the best position to observe the wrongdoings firsthand. In fact, often the only people who are in a position to detect when the organization's practices have gone awry (e.g., when it produces a defective product or does something that might endanger or defraud the public) are the employees themselves. Without their aid in bringing to light abuses of lawor public trust, we would have little hope of learning or controlling what goes on inside most work organizations. Despite the public need for information of this kind and despite a number of federal and state laws that have been enacted to protect whistle-blowers, our data on reprisals suggest strongly that the laws as written are insufficient.

Our own numbers, presented earlier in this article, indicate that the extent of whistle-blowing is burgeoning. If our estimate is correct that approximately one third of U.S. employees observe workplace practices that trouble them and that they consider to be unethical or illegal, then whistle-blowing would seem to have an almost limitless future in an information-based economy. For this reason, we find it a surprising oversight that even so comprehensive a review of the history of contention between workers and management as that by Tilly & Tilly (1998) fails to notice whistle-blowing as a new weapon in the arsenal available to workers. Similar to this, Coleman's (1998) thorough review of white-collar crime fails to mention whistle-blowing as a way to identify and expose such crimes. As the burden of our study and the experience of the whistle-blowers amply show, whistle-blowing by employees who are in position to observe misconduct may be the most immediate and effective way of exposing and controlling organizational crime. In addition, the experience of the whistle-blowers points the way to an often overlooked weapon that workers can use – observation and hard evidence of wrongdoing by management – to bring more leverage to their position and more pride to their work.

References

Braverman, H. (1974). *Labor and monopoly capital*. New York: Monthly Review Press.

Brower, R., & Abolafia, M. (1995). The structural embeddledress of resistance among public managers. *Group and Organization Management, 20*, 149–166.

Buroway, M. (1979). *Manufacturing consent*. Chicago: University of Chicago Press.

Clegg, S. (1994). Power relations and the constitution of the resistant subject. In J. Jermier, D. Knights, W. Nord (Eds.), *Resistance and power in organizations* (pp. 274–325). London: Routledge.

Coleman, J. W. (1998). *The criminal elite: Understanding white-collar crime*. New York: St. Martin.

Cornfield, D. (1989). *Becoming a mighty voice: Conflict in change in the United Furniture Workers*. New York: Russell Sage.

Edwards, R. (1979). *Contested terrain*. New York: Basic Books.

Glazer, M., & Glazer, P. (1989). *The whistleblowers: Exposing corruption in government and industry*. New York: Basic Books.

Graham, J. W. (1986). Principled organizational dissent: A theoretical essay. In L. L. Cummings & B. M. Staff (Eds.), *Research in organizational behavior* (Vol. 8, pp. 1–52). Greenwich, CT: JAI.

Hodson, R. (1995). Worker resistance: An underdeveloped concept in the sociology of work. *Economic and Industrial Democracy, 16*, 79–110.

Jermier, J. (1988). Sabotage atwork. In N. DiTamaso (Ed.), *Research in the sociology of organizations* (Vol. 6, pp. 101–135). Greenwich, CT: JAI.

Jos, P., Tompkins, M. E., & Hays, S. W. (1989, November/December). In praise of difficult people: A portrait of the committed whistleblower. *Public Administration Review*, pp. 552–561.

Kimeldorf, H. (1988). *Reds or rackets? The making of radical and conservative unions on the waterfront*. Berkeley: University of California Press.

Miceli, M. P., & Near, J. P. (1990, March). *Is there a whistleblowing personality? Personal variables that may be associated with whistleblowing*. Working Paper Series 90–19). Columbus: Ohio State University, College of Business.

Miceli, M. P., & Near, J. P. (1992). *Blowing the whistle*. New York: Lexington Books.

Miethe, T. D. (1995). Predicting future litigiousness. *Justice Quarterly, 14*(3), 407–428.

Miethe, T. D., & Rothschild, J. (1994). Whistleblowing and the control of organizational misconduct. *Sociological Inquiry, 64*(3), 322–347.

Perrucci, R. M., Anderson, R. M., Schendel, D. E., & Tractman, L. E. (1989). Whistle-blowing: Professionals' resistance to organizational authority. *Social Problems, 28*, 149–164.

Statistical Abstract of the United States 1994 (114th ed.). (1994). Washington, DC: U.S. Bureau of the Census.

Tilly, C., & Tilly, C. (1998). *Work under capitalism*. Boulder, CO: Westview.

U.S. Merit Systems Protection Board. (1993). *Whistleblowing in the federal government: An update*. Report to the President and Congress by the U.S. Merit Systems Protection Board. Washington, DC: Government Printing Office

Weinstein, D. (1979). *Bureaucratic Opposition*. Elmsford, NY: Pergamon.

OBSERVATION METHODS

SAGE BENCHMARKS IN SOCIAL RESEARCH METHODS

OBSERVATION METHODS

VOLUME III

Edited by

Barry Smart, Kay Peggs and Joseph Burridge

SAGE

Los Angeles | London | New Delhi
Singapore | Washington DC

SAGE

Los Angeles | London | New Delhi
Singapore | Washington DC

SAGE Publications Ltd
1 Oliver's Yard
55 City Road
London EC1Y 1SP

SAGE Publications Inc.
2455 Teller Road
Thousand Oaks, California 91320

SAGE Publications India Pvt Ltd
B 1/I 1, Mohan Cooperative Industrial Area
Mathura Road
New Delhi 110 044

SAGE Publications Asia-Pacific Pte Ltd
3 Church Street
#10-04 Samsung Hub
Singapore 049483

Typeset by Arete Publishing Private Limited, Delhi

Printed on paper from sustainable resources

Printed by MPG Printgroup, UK

MIX
Paper from responsible sources
FSC® C018575
FSC www.fsc.org

First published 2013

Library of Congress Control Number: 2012942348

British Library Cataloguing in Publication Data

A catalogue record for this book is available from the British Library

ISBN: 978-1-4462-0811-3 (set of four volumes)

Contents

Volume III

D. Participant Observation

E. Interpretation and Presentation of Observational Data

F. Observational Screens: Photography, CCTV and Internet

'As soon as we observe the social world, we introduce in our perception of it a bias due to the fact that, to study it, to describe it, to talk about it, we must retire from it more or less completely. This *theoreticist* or *intellectualist bias* consists in forgetting to inscribe into the theory we build of the social world the fact that it is the product of a theoretical gaze, a "contemplative eye". A genuinely reflexive sociology must constantly guard itself against this epistemocentrism, or this "ethnocentrism of the scientist", which consists in ignoring everything that the analyst injects into his perception of the object by virtue of the fact that he is placed outside of the object, that he observes it from afar and from above'.

Bourdieu, P. (1992). in P. Bourdieu and L. J. D. Wacquant (Eds.) *An Invitation to Reflexive Sociology*. London: The University of Chicago Press: 69–70.

D. Participant Observation

45

The Con Man as Model Organism: The Methodological Roots of Erving Goffman's Dramaturgical Self

Michael Pettit

In a series of books and essays published in the 1950s, the sociologist Erving Goffman (1922–82) famously championed a dramaturgical perspective on the self. He argued against the common-sense distinction between the sincerity of everyday behavior and the contrived performances of actors. He portrayed humans as 'impression managers', constantly staging performances to meet and exceed the expectations of various 'audiences'. Individuals, often operating as part of a team, prepared for their roles in the 'back region' to perform a kind of play in front of the persons they encountered in their daily routines. The theatrical play provided an apt model since 'ordinary social intercourse itself is put together as a scene is put together, by the exchange of dramatically inflated actions, counteractions, and terminating replies' (Goffman, 1959: 78). His dramaturgical theory built upon his 'realization that while the performance offered by impostors or liars is quite flagrantly false and differs in this respect from ordinary performances, both are similar in the care their performers must exert in order to maintain the impression that is fostered'. Since both genuine and dishonest performances require the use of dramatic effects to render them convincing, Goffman claimed that 'we can profitably study performances that are quite false in order to learn about ones that are quite honest' (ibid.: 73). Philip Manning (1989, 1992) has argued that in between the book's 1956 and 1959 editions,

Source: *History of the Human Sciences*, 24(2) (2011): 138–154.

Goffman made changes which moved the text's center of gravity away from a view of the self as a hidden manipulator to focus instead on rituals of trust and tact. Left unaddressed in this account is the origin of this theory and why Goffman is often seen as championing a deceitful self. How did the con man come to serve as a model for human nature?

Despite the fact that Goffman remains one of the most highly cited figures in the social sciences, he has attracted little sustained interest among historians of the human sciences. Insofar as he rarely spoke about his methods, disliked being interviewed and left no personal papers, he does not make for an easy subject of historical analysis.[1] Another reason for this neglect is a preoccupation among historians of the postwar social sciences with the Cold War as the over-arching explanatory framework (Heims, 1993; Herman, 1995; Edwards, 1996; Latham, 2000; O'Connor, 2000; Solovey, 2001; Gilman, 2003; Rohde, 2009). Goffman's career does not fit neatly in a narrative organized around the growth of social scientific knowledge in service of the American state, especially its military ambitions on the world stage. Recently, historian Joel Isaac has suggested the need to attend to how the practices of cold-war era social scientific research fit within the ways of knowing developed over the longer histories of these disciplines and deployed by other sciences (Isaac, 2007).

This article moves away from a view casually mentioned in the sociology literature with some frequency that Goffman is a thinker too idiosyncratic to withstand sustained analysis. Although he preferred the artfully crafted essay over the exact analysis of data (Lofland, 1980), Goffman shared with the other postwar social scientists an interest in forms of communication within 'small groups'. This commitment was not only methodological, but ontological: these scientists conceived of society in terms of an assortment of small collectives rather than individualized atoms or an undifferentiated mass (Pooley, 2006; Cohen-Cole, 2009). Moreover, his focus on communication in 'relatively closed systems' (Goffman, 1959: 232) demonstrated common assumptions with the cybernetics movement (Edwards, 1996).[2]

My approach is to historicize Goffman's early writings from the 1950s in terms of the investigative practices he deployed. He was a social scientist who privileged ethnography in the field over the laboratory experiment, the survey questionnaire, or the mental test as a means of generating knowledge about human nature. As historian Robert Kohler has argued, laboratories and field sites are more than just geographic places. They 'are cultural domains first and foremost, where different languages, customs, materials and moral economies, and ways of life prevail' (Kohler, 2002: 5). Moreover, rather than trying to formulate universal laws Goffman focused on the models provided by richly detailed concrete examples, a style of reasoning fitting within the tradition of natural historical description.

Taking such an approach offers a different picture of Goffman's historical place than that advanced by social theorists. Alvin Gouldner argued Goffman's thinking belonged to a specific moment in the history of capitalist development, the emergence of a post-industrial service economy. Gouldner contended that the dramaturgical theory reflected the bourgeois ideology of the postwar corporate middle-manager and represented 'the transition from an older economy centered on production to a new one centered on mass marketing and promotion, including the marketing of the self' (Gouldner, 1970: 381). In contrast, Gonos (1980) argued that Goffman articulated the viewpoint of the alienated 'lumpen-bourgeoisie' or small-time business owner pushed aside by a nationally integrated economy. More recently, commentators have moved away from identifying him with a particular class, seeing him instead as advancing a theory of the 'postmodern self' (Tseëlon, 1992; Manning, 2000).

In analysing the relationship between theory and practice in Goffman's dramaturgical writings, what the cognitive psychologist Gerd Gigerenzer (1991) calls a 'tools-to-theories heuristic' provides a helpful framework.[3] Gigerenzer contends that 'scientific tools (both methods and instruments) suggest new theoretical metaphors and theoretical concepts once they are entrenched in scientific practice'. Familiarity with these tools within a disciplinary community also facilitates the reception of the theoretical metaphors they inspire (1991: 256). Reconfiguring certain distinctions central to the logical positivists' understanding of scientific progress, he contends that tools used in the context of justification loop back to provide resources in the context of discovery. This is not a deliberate, intentional process. Rather familiarity with the practical aspects of certain techniques helps make a particular understanding of human nature more thinkable. Gigerenzer focuses on tools for data analysis used in cognitive science such as inferential statistics, but his approach has broader implications. Left unexplored is the possibility that tools used for the collection of data also provide metaphors and concepts available to the scientist.

The connection between two sources for Goffman's theory of the self seems particularly worthy of consideration: his own use of participant-observation as a method and his repeated evocation of the confidence man. Goffman's 1950s essays are not narratives of his experiences of the field (Fine and Martin, 1990). Instead, I argue that the methodology of participant-observation became an unacknowledged source in Goffman's modeling of human nature. The performative nature of participant-observation resonated with his interest in the figure of the confidence man. He never undertook any field studies of these professional criminals, but he repeatedly returned to their example as a model for behavior. Goffman's depiction of the con man owed much to his own experiences as a participant-observer. Both were modern professionals who succeeded in life by cultivating their own insincerity to allay the worries of a potentially incredulous audience. The model of the participant-observer is particularly interesting because, unlike his criminal counterpart, this form of insincerity was in part virtuous.

Goffman's Natural History

In 1969, Everett C. Hughes, one of Goffman's primary mentors at the University of Chicago, made a telling analogy about his student's methodology. Hughes suggested that 'Goffman and [Konrad] Lorenz work in the same way; they observe with aggressive intensity the behavior of their subjects in situations, or occasions' (Hughes, 1969: 426).[4] At first, the comparison may strike the reader as odd as the two were polar opposites at the level of theory. The Austrian naturalist emphasized the primacy of instincts in the constitution of behavior while Goffman stressed the malleability of the self through a multitude of personae, tailored to meet local circumstances. Lorenz dealt with the deep time of evolution where Goffman concerned himself with the dynamics of the immediate situation.

Goffman had first suggested the affinities between his approach and that of the European ethologists. He claimed that the inspiration for the analysis of 'place' in *Asylums* was not the theatrical stage, but the work of Heini Hediger and Lorenz on animal territoriality (Goffman, 1961: 227). Goffman cited a paper by the Swiss-Canadian psychiatrist Henri Ellenberger which turned to ethology to interpret the patient's experience of the mental hospitals. Just as captive animals came to behave in a stereotyped fashion due to the conditions of captivity, Ellenberger suggested that some of the patient's actions might reflect the conditions of confinement more than the underlying disease (Ellenberger, 1960). The zoo offered a powerful metaphor for questioning the asylum's status as a curative space. Goffman concurred with Ellenberger that each institution constituted a 'closed community' that elicited stereotyped behaviors from its occupants.[5] In the preface to *Relations in Public* (1971), he expanded upon his appreciation of ethology. Because of its focus on social groups, ethology turned the naturalist into a student of 'face-to-face interaction'. He admired the ethologists for developing 'a field discipline that leads them to study animal conduct in very close detail and with a measure of control on preconceptions'. He also warned that the ethologist's tendency 'to apply a Darwinian frame' led to 'some very unsophisticated statements' when applied to human behavior (ibid.: xvii). In his masterful exegesis of Goffman's *oeuvre*, Burns (1992: 27) concludes that ethology contributed little to the sociologist's theoretical formulations.

Hughes never suggested that Goffman shared a theory of human nature with Lorenz, but rather he argued that they *worked* in the same fashion. The similarity resided at the level of scientific practice. They held in common a style of observation rather than any specific theory. Lorenz rejected the control of the laboratory as a scientific ideal. In its place, he stressed the importance of observing over long stretches of time the naturally occurring behaviors of various organisms. He transformed his estate in Altenberg into a scientific farm, where he raised and bred geese and jackdaws to study their developmental patterns from birth (Burkhardt, 2005). What differentiated

Goffman from the ethologist was his rejection of the ideal of a pure nature. Lorenz's science was predicated on the conviction that one can observe authentic behavior only in a relatively uncontaminated pastoral setting while Goffman reveled in the fleeting artifice and appearances characteristic of the modern. The anthropologist Lloyd Fallers recognized this way of knowing and praised Goffman for it. Where most sociologists seemed content to 'draw their data from experimental small groups' Goffman was 'a fieldworker in the anthropological tradition, an observer (whether through his own eyes or those of others) of man in his native habitat' (Fallers, 1962: 191).

Goffman had not begun his career as a human ethologist. When he commenced his graduate studies at the University of Chicago, he first worked as an employee at Social Research Inc. Although the agency was not as expansive as Paul Lazarsfeld's Bureau for Applied Social Research (Barton, 1979), the social anthropologist W. Lloyd Warner created it in 1946 to conduct market research for paying clients (Platt, 1996: 158). Under its auspices, Warner undertook a project sponsored by the Columbia Broadcasting System (CBS) to ascertain the reaction of housewives to the radio serial, *Big Sisters*. Of particular interest was connection between the audience member's socioeconomic position and her personality traits. To accomplish this task, Warner recruited 60 wives of junior executives, skilled and white-collar workers. To provide a measure of the individual housewife's personality, he turned to a leading projective technique, the Thematic Apperception Test (TAT) (Warner and Henry, 1948). Projective tests like the Rorschach and TAT materialized a specific understanding of the self (Galison, 2004). Using a fixed and tightly controlled set of evocative images whose meaning was by design ambiguous, the scientist asked subjects to craft narratives based upon what they perceived. Advocates of these tests largely adhered to the psychoanalytic conviction that such talk would uncover unconscious tendencies and personal secrets that the interviewee would not intentionally reveal.

Under Warner's supervision, Goffman set out to expand the sample to include housewives recruited from the Hyde Park neighborhood surrounding the university. As his research progressed, he grew increasingly dissatisfied with the approach taken in the CBS study. He became more interested in the embodied performances and rituals through which the women responded to and challenged the psychological test (Goffman, 1949). He argued that the transforming of the TAT from an aid to clinical judgment to a standardized tool for measurement in social research meant that its utility had been lost. He wrote: 'Bracketed statements by the interviewer concerning the pauses and reactions of the subject frequently lead to fruitful, clinical insights; in most instances, however, such statements are insufficiently precise and specific for purposes of standardization and quantification.'[6] In place of the standardized test, he came to practise a social psychology based on field observations, transforming his own camouflaged person into an instrument of observation.[7]

The Participant-Observer as Model

Gigerenzer argues that tools become heuristics for scientific theories not when they are first introduced as innovations, but once their use becomes unremarkable and diffuse within a disciplinary community. Fieldwork as a participant-observer had achieved this status among sociologists at the University of Chicago by the time of Goffman's arrival in 1948. More importantly, his dramaturgical theory closely mirrored how he and others described participant-observation as a scientific activity.

As a methodology, participant-observation began in the mid-1880s when a wide array of reform-minded journalists began practising deceptive 'class passing' as a means of investigating social phenomena supposedly out of sight of respectable society. They feigned madness to gain access to asylums, 'tramped with tramps' and waited on tables. They aimed for a total and unnoticed immersion within a given community for an extended period of time. Researchers used both observations of their surroundings and their own embodied experiences as equivalent kinds of evidence. Participant-observation was pioneered by journalists like Nellie Bly and Josiah Flynt, but academics, like the political economist Walter Wyckoff and the psychologist Amy Tanner, also adopted it (Pittenger, 1997; Higbie, 1997; Lutes, 2002; Pittenger, 2003; Pettit, 2008).

Under the guidance of Robert E. Park, the Chicago School of Sociology likewise derived its scientific authority from 'the evidence of experience' (Scott, 1991). Park and his students turned the city of Chicago into a privileged 'truth-spot' to generate models about urbanization, poverty and criminality as social processes. In the 1920s and 1930s, Chicago served as both their universalized laboratory for understanding modernization and their field-site rife with local concerns (Gieryn, 2006). Research consisted of informal interviews, the collection of official documents, and first-hand observation. When training students, Park drew upon the skills he developed as a newspaper reporter between 1891 and 1898. In part, he conceived of the sociologist as a 'superreporter' committed to conveying the long-term stories beneath the surface of the daily headlines (Bulmer, 1984: 90). The celebrated monograph series Park oversaw trafficked in the lives of those formerly deemed marginal: prostitutes, tramps, delinquents and criminal gangs. Like the naturalistic proletarian and hard-boiled novels of the era, the resulting sociological narratives depicted the activity of such persons as both the *locus* for the breakdown of social order and a defining feature of the modern, urban experience (Cappetti, 1993; Salerno, 2007).

Platt (1983) has effectively debunked the myth that participant-observation was the primary investigative practice deployed in the postwar Chicago department, but her research also demonstrates its wide acceptance as a method there. Nels Anderson's ethnography of the hobo brought participant-observation as a technique within the purview of the department's program

(Anderson, 1923). While they adopted an eclectic array of tools, the department's postwar leaders still advocated a 'you've got to touch it' empiricism, which privileged the evidence of first-hand experience. Platt also found that Chicago graduates of Goffman's generation cited the compulsory course on fieldwork as the formative rite in their education (Platt, 1995). In addition, Goffman's two main mentors, Hughes and Warner, encouraged both a methodological and an institutional rapprochement between sociology and anthropology (Abbott, 1999: 62–77).

In contrast to this continued emphasis on situated experience, Talcott Parsons organized the practice of social inquiry very differently at Harvard's Department of Social Relations. There research centered on the formulation of a highly abstract general theory of action (Isaac, 2010). The cultural anthropologist David M. Schneider made this contrast central to his positive evaluation of his friend's promotion to associate professor in 1958. He explained to the tenure committee that Goffman 'is not, like Talcott Parsons, concerned with the systematic development of a logically coherent theoretical scheme of ample proportions which may (or may not) at some later date (usually unspecified) be run against a few facts for goodness of fit and perhaps even tested'. Instead, 'Goffman's initial query is always a theoretical one, and once the question is asked, it is asked in terms of concrete observational material and organized with reference to them'.[8] In another context, Hughes claimed that 'the essence of his method is close observation and the reporting of behavior in "small" but dramatic and fateful situations, which he then extrapolates to larger systems of things in a sophisticated but not quantitative way' (Hughes, 1969: 425). The hallmark of Goffman's ethnography was his skill at generating theory out of his descriptions of minute social encounters recorded as an embedded observer.

Goffman famously resisted publishing an account of his methodology for conducting fieldwork, but hints are found in his doctoral dissertation on a formerly isolated farming community on the Shetland Islands north of mainland Scotland. His aim was the observation of the everyday rituals of communication within this community that had recently opened itself to a small tourist trade. Living among the islanders for a year and half, he struggled 'to play an unexceptional and acceptable role in community life. My real aim was to be an observant participant, rather than a participating observer' (Goffman, 1953: 2). He wanted to be present enough to observe behavior, but he also sought to absent himself from the actual communicative process. His goal was 'to observe people off their guard' (ibid.: 5), but this required a certain level of invisibility. Because of his desire to pass unnoticed and not interfere with the 'nature' of social intercourse, he felt that he had to eschew the standardizing technologies of the anthropologist.[9] He conceded that 'mechanical devices such as tape recorders and motion-picture cameras, or rigid techniques such as time-sampling, would have provided a desirable check on these recording biases', but he warned his readers that 'these corrective devices, however, were not practical for social, economic, and technical reasons' (ibid.: 4). To be a successful ethnographer, he had first and foremost to camouflage his intentions and manage people's impressions of this outsider so as not to arouse suspicion. At a time when Robert K. Merton labored to 'focus' and structure the scientific interview (Lemov, 2010), Goffman moved in the opposite direction. Rather than seeking greater control over the situation, he hoped to ensure accuracy by managing the presentation of his own presence.

In making such claims, Goffman suggested that observing humans accurately necessitated sacrificing one's sincerity. As a social scientific practice, Goffman's brand of ethnography involved a specific configuration of what the philosopher Bernard Williams calls 'the virtues of truth'. According to Williams, the cultivation of truthfulness requires a commitment to two distinct virtues, accuracy and sincerity. Adherence to these virtues mitigates the distorting influence of the will towards self-deception and deceit of others. Accuracy is a disposition towards acquiring correct beliefs about the world by struggling against the pleasing assurances of prior convictions. A commitment to sincerity guarantees that one is as transparent and forthcoming as possible when representing one's knowledge to others (Williams, 2002). Williams suggests that accuracy is the virtue that pertains to the acquisition of knowledge while sincerity governs its transmission. The example of ethnography demonstrates that this is not always the case. Measuring and recording devices that would have granted Goffman's observations greater exactitude risked making his objects of study act differently. He worried that revealing his true identity would render certain communicative situations out of reach. Too much transparency compromised the collection of data while insincerity granted him the access necessary to make accurate observations of fleeting face-to-face interactions. Insincerity was virtuous behavior because it literally got the investigator closer to the truth. When it came to the site of inquiry, he advocated a version of objectivity best described as accuracy in the absence of sincerity.

Goffman celebrated the vantage point of an insincere performer rather than the perspective of a distant and impartial recording device. In 1960, he wrote to Hughes, describing his experiences studying 'the service relations and the frayed edges of American civilization' in Las Vegas. He insisted on the necessity of observing behavior as a card dealer at the casino. He contrasted his own physical position with that of the management which took the form of 'roof mirrors behind which an invisible man watches the employees'.[10] Although the technology was readily available, Goffman rejected this disembodied, aerial-eye view in favor of the truthfulness acquired disguised on the floor. His best-known investigation of this type consisted of the ethnography he conducted on the social life of institutionalized psychiatric patients at St Elizabeth's Hospital in Washington, DC. Starting in 1954, he spent a year on the ward floor, posing 'in the role of an assistant to the athletic director' while only 'the top hospital management knew what my aims were' (Goffman, 1961: ix). He always circumscribed his position in these situations, remaining more of an observer than a fully immersed participant. He consistently chose a role adjacent to the primary action, whether that of the athletics director rather than the patient or the card dealer rather than the gambler.

He claimed that professional subterfuge served as a means of identifying members of his cohort at Chicago. He concluded his 1960 letter to Hughes on a wry note, promising to 'save my card, collect one from Howie [Becker] and members of all the other locals, and we'll present them to you on a birthday: for by their union cards you shall know the participant observers'.[11] Such prolonged work in the field semi-disguised as a member of a trade or profession became a means of establishing one's authority and credentials. Becker famously used his piano-playing skills to investigate 'the natural history of an individual's use of marihuana' in Chicago's jazz clubs (Becker, 1953: 236).

A number of Goffman's contemporaries offered ethnographies of the asylum from the viewpoint of an attendant. The year 1948 witnessed a new wave of asylum exposés including Albert Deutsch's photo-essay *The Shame of the States* alongside the novel and film versions of Mary Jane Ward's *The Snake Pit*, a classic in the confinement narrative genre. Harold Orlansky, an anthropology doctoral candidate at Yale, published a scathing account in Dwight Macdonald's magazine, *Politics*. Based upon the observations he recorded during a 14-month stint as an attendant, Orlansky concluded that 'the American asylum manifests, in embryo, some of the same social mechanisms which in Germany matured into death camps' (Orlansky, 1948: 162). A number of Chicago sociologists took up Orlansky's suggestion that the asylum attendant, endowed with arbitrary power over others, constituted a potentially dangerous occupation worthy of concern. Harold Taxel served as an orderly for 9 months while 'no one on the staff had knowledge of the research being conducted' (Taxel, 1953: 19). Robert H. Willoughby argued that 'participant-observation was thought to be the best tool for putting one's self into the same frame of reference as the attendant'. He prided himself on conducting 8 months of observation 'without the assistance of and to a great extent without the knowledge of the administration' (Willoughby, 1953: 15–16). In the early 1950s, such insincere performances were standard practice at Chicago.

More controversially, this cultivated insincerity shaped Goffman's expository style. His style of observation did fit within the long tradition of natural history, but his work did not read like traditional natural histories. Where Orlansky's article was an earnest denunciation of the dehumanizing tendencies built into the social organization of the mental hospital, Goffman never offered an outright condemnation of the treatment of patients. Instead, he carefully described the behaviors he witnessed on the ward floor, but altered their meaning by ironically framing them with the categories derived from the sociology of occupations rather than the psychopathology of mental illness

(Becker, 2003). In person and on the page, he presented himself through a distant voice saturated by sarcasm, satire and irony (Fine and Martin, 1990). Playful trickery had long formed an important part of his self-presentation. His sister, the actress Frances Bay, recalled that he was a 'real prankster as a kid' (Posner, 2008). Goffman, along with Becker, became known for their preferences for 'the offbeat to the familiar, the vivid enthnographic detail to the dull taxonomy, the sensuously expressive to dry analysis, naturalistic observation to formal questionnaires, the standpoint of the hip outsider to the square insider' (Gouldner, 1962: 208). This stance resonated with a style of mid-century dispassionate emotional restraint, what historian Peter Stearns calls 'American Cool' (Stearns, 1994). Cultivated insincerity allowed Goffman to advance biting criticisms while appearing detached from and disinterested in the matter at hand. Irony served as his surrogate for objectivity. This calculated cynicism about modern organizational structures made him a hero among many intellectuals on the left, but, as Gouldner suggested (1970), his was also a style championed in the hallways of 1960s Madison Avenue advertising agencies (Frank, 1997).

Goffman's innovation then was neither the problems he studied nor his investigative practice, but his advocacy of a model of personhood that closely resembled the well-established methodology of participant-observation. Historian Joel Isaac (2009) has documented how the practice of theory became central to the identity of the social scientist in the 1950s. His account focuses on a set of highly influential 'organization men' such as Parsons and Herbert Simon. Coming out of the University of Chicago, which had long insisted that the researcher develop an intimate knowledge of the *noir* aspects of the city (Salerno, 2007), Goffman developed a different model of the social theorist. His was an embedded and attentive yet coolly detached observer skilled at playing roles, the sociologist as urban hipster. This model of the researcher informed his satirical expository style and account of the self. It also informed his notorious 'hypermasculine' trickster persona centered on his own composure, poise and emotional control (Scheff, 2006: 13). His career reveals a 'tools-to-theory' heuristic as the methodology he deployed – manipulating others to pass unnoticed in the farming village, asylum ward, or casino – which became a model for the individual's experience of putting on various masks to fit the occasion. The similarities between this professional behavior and the activities of the confidence man may have suggested the latter as a model for human nature.

The Ethnographer and the Con Man

The normalization of insincerity formed a major and rather unique component of Goffman's account of the self. The clearest expression of this viewpoint is also the earliest, 'On Cooling the Mark out' (1952). In this essay,

Goffman made the play metaphor central to his analysis, but it did not refer to the theater. Instead, he discussed the carefully choreographed performances and management of people's identities that go into a confidence game. Gleefully adopting the criminal's slang as his own, he noted how 'the operation of any particular racket, taken through the full cycle of its steps or phases, is sometimes called a play'. Moreover, confidence games are 'practiced on private persons by talented actors who methodically and regularly build up informal social relationships just for the purpose of abusing them' (Goffman, 1952: 451).

Confidence games provided Goffman with the material to think through the conflict between one's self-image and perception by others. The successful defrauding of the mark disconfirmed his self-understanding as a savvy commercial actor. 'A mark's participation in a play, and his investment in it, clearly commit him in his own eyes to the proposition that he is a smart man' (ibid.: 1952: 452). In order to avoid legal reprimand, con men had developed techniques to mitigate the frustration and anger that derive from this disappointment. This process of 'cooling-out' became a model for all situations where individuals were confronted with disappointing outcomes that contradicted their established self-images. Through social interactions, the victim is successfully cast in a new role to perform without insult to the actor. Cooling-out is the management of disappointment when confronted with the contradictions of self-deception, when a person recognizes that his or her most cherished identity was simply a temporary role. The very lack of integration among the various social roles a person performs was the source of salvation since she or he could find refuge in other roles. The individual's capacity to be cooled out and reassigned an identity signaled to the young sociologist the malleability of human nature. Writing in the journal *Psychiatry*, Goffman contended that his theory explained the psychotherapeutic encounter. Rather than revealing hidden depths or bringing forth a person's authentic inner self, the psychotherapist operated as 'the society's cooler' by assigning a new, safer role to disappointed and frustrated clients (ibid.: 462).

Goffman never conducted fieldwork among confidence men, but rather drew heavily on the linguist David W. Maurer's ethnographic studies of the argots of the professional criminal. Maurer also saw in the figure of the criminal swindler a model for thinking about social relations in general. He insisted that 'their methods differ more in degree than in kind from those employed by more legitimate forms of business' (Maurer, 1940: 16). Trained as a scholar of Elizabethan drama at Ohio State University in the early years of the Great Depression, Maurer began his forays into the vernacular with a study of the jargon exclusive to the men working in the North Atlantic fishing trade (Maurer, 1930). This examination of 'schoonerisms' set the pattern for his subsequent career. He would record through correspondence, stenography and new audio technologies the idiosyncratic speech of various occupational subcultures.

Starting with the cant deployed among members of traveling carnivals (Maurer, 1931a), he became particularly intrigued by the language and culture of crime. Central to his linguistic studies was the belief that each criminal specialty had developed its own argot. This was because the various types of criminals constituted professions in a sociological sense. Gamblers, forgers and confidence men were not isolated rogues, but modern entrepreneurs engaged in highly organized industries with rules, apprenticeships, alliances and, most importantly, an exclusive, technical vocabulary. In academic journals, Maurer produced encyclopedic lexicons of the various trades of the underworld. He insisted that to comprehend these speech-systems as linguistic phenomena required an understanding of the behavioral patterns performed within the subcultural classes. The psycho-sociological ramifications of acquiring an argot can most readily be seen in Maurer's interpretation of prostitution, a sector of the crime world he claimed lacked its own distinct language. Maurer argued that argots did not develop out of an urge to deceptively camouflage the criminal activity from the law. Rather they emerged out of the sense of camaraderie that characterized the homosocial community of an occupation. A member acquired fluency by achieving a certain status within a professional network, an economic position never granted the prostitute, who always remained subservient to the pimp. Moreover, unlike other professional criminals who 'depend heavily on a shrewd analytical and psychological approach', Maurer concluded that prostitution did not require such mental skills (Maurer, 1939: 549). He defined argot-formation as a masculine trait that required intelligence and creativity, qualities he found wanting among the prostitutes he interviewed.

Despite his claims about abstaining from sensationalism, glamour and excitement, Maurer's project was rooted in his conviction that the argots he collected constituted elements of a dying culture. In an early study of the cant cultivated in the social space of the carnival, he stressed how criminal argots had their origin in the rough world of horse-and-wagon travel. Things had changed and 'commercialization – so some of the old-timers feel – has emasculated modern showmanship, and along with it modern circus and carnival slang. However that may be, there yet remains plenty of glamour and of grief in the road-show business and its jargon still retains a redeeming modicum of romance' (Maurer, 1931b: 327–8). Argot was a remaining auditory vestige of a more rugged, masculine era that needed to be preserved in the industrial age.

What Maurer, trained in the analysis of literature, brought to the study of swindlers was an explication of the dramaturgical structure of the confidence game. 'Big time confidence games are in reality only carefully rehearsed plays in which every member of the cast *except the mark* knows his part perfectly.' With this metaphor in place, Maurer began assigning roles:

> The insideman is the star of the cast; while the minor participants are competent actors and can learn their lines perfectly, they must look to

> the insideman for their cues; he must be not only a fine actor, but a playwright extempore as well. And he must be able to retain the confidence of an intelligent man even after that man has been swindled at his hands. (Maurer, 1940: 108)

This description of social intercourse as an inherently deceptive dramatic performance provided a rich resource for Goffman. Since the coining of the term during the 1849 trial of William Thompson (Bergman, 1969), writers had depicted the confidence man as a threat to a middle-class culture organized around the ideal of sincerity (Halttunen, 1982; Lindberg, 1982). Goffman challenged this tradition of defining the American middle class against the figure of the con man. The confidence game provided him with a social situation where the artifice of behavior and the reality of its consequences were not dichotomous. During the big con, one had to perform a deceptive role to achieve real success. It does not seem unlikely that Goffman's own professional insincerity as a participant-observer made this re-evaluation of the confidence man's import more thinkable.

Conclusion

The swindler and the participant-observer occupy an ambiguous place in *The Presentation of Self in Everyday Life* as stage-acting becomes the governing metaphor. In terms of the methods of communication, Goffman made no distinction between the operation of the con man and the everyday performers. Both 'must employ elaborate and meticulous personal fronts and often engineer meticulous social settings' in order to secure trust and reliability among the audience (Goffman, 1959: 218–19). Despite these affinities, he argued that the con man and the participant observer were not ordinary actors. These individuals occupied 'discrepant roles', such as the shill and the informer, which enabled them to move between the regions ascribed to the performer and the audience. Capable of mimicking the actions of both teams, these specialists could accrue inside information about each side and threaten the stability of the communicative system with this privileged knowledge (ibid.: 144–8). No longer the primary metaphor for human behavior, they remained central to his account. Embedded within the communication system, they were also one of the few classes of persons capable of understanding its true functioning from within its operation. The confidence man, the actor and the ordinary person all perform the life-episodes, but only the former two have the capacity to articulate the process. Indeed, social science for Goffman consisted precisely in just this articulated knowledge.

If 'the play' originally referred to the confidence man's carefully choreographed swindle, when did Goffman acquire the more literal theatrical metaphor? Following his biographer's insightful reconstruction of the various locations Goffman inhabited (Winkin, 1999), Ian Hacking suggests that

Goffman's wartime work classifying films depicting scenes of daily life for the National Film Board of Canada may have inspired such a perspective (Hacking, 2004: 289–90). In a discussion of the irrelevancy of the common-sense distinction between sincere and contrived performances added to the 1959 edition, Goffman noted 'recent uses of the "psychodrama" as a therapeutic technique' (1959: 78). During the 1930s, Jacob Levy Moreno introduced the theory of psychodrama to American psychiatrists through his private Beacon Hill Sanitarium. With the assistance of other participants, patients were expected to confront and overcome their inner conflicts by acting out a series of roles relevant to their daily lives in front of an audience and under the supervision of a therapist/director. In 1939, St Elizabeth's became the first state hospital to begin experimenting with Moreno's psychodrama technique. Two years later, Frances Herriott, a protégée of Moreno with a background in musical theater, oversaw the construction of a stage in the hospital's Hitchcock Hall to set the scene properly for therapeutic relief. She documented that Red Cross volunteers, nurses and attendants participated in her psychodrama class (Herriott, 1940; Herriott and Hagan, 1941; Herriott, 1945).[12] There is no evidence that Goffman had direct contact with the psychodramatic workers during his year at St Elizabeth's, but their shared institutional niche offers a possible explanation for his shift in metaphors during this period.

Submerged within the earliest version of Goffman's dramaturgical theory are two closely related models of the self: the participant-observer and the confidence man. His language of actors and plays originally did not refer to the theater, but to the activities of the commercial swindler. He followed Maurer in noting how with the increasingly organized nature of crime, swindlers were not all that different from the denizens of respectable society. Before advancing his dramaturgical theory, he had pursued a number of studies as an insincere participant-observer. The model of the ethnographer at the University of Chicago closely resembled Goffman's description of the con man: each was a cool, detached observer who navigates the social by practising small interpersonal hustles. The participant-observer's accuracy hinged upon success at emulating the con man's art of self-presentation. These were the professionals that Goffman most identified with the cool management of people's impressions and emotions. His adoption of the con man as model was a choice, but one which his experiences as a participant-observer facilitated. He reversed the moral valence surrounding such confidence games, whether scientific or commercial. He argued that all social interaction consisted of the constant transmission and reception of insincere signals.

Notes

1. His only explicit statements about the nature of fieldwork were published posthumously and derived from a series of spoken remarks that he requested not to be recorded (Goffman, 1989). The absence of personal or professional papers has been

partially remedied by Dmitri Shalin who edits the online Erving Goffman Archive, a site that contains a wonderful collection of interviews with Goffman's family, colleagues and students, at: http://www.unlv.edu/centers/cdclv/ega/

2. During the 1960s and 1970s, Goffman moved away from his dramaturgical theory, embracing game theory in its place. Despite this new theory's more robust mathematical foundation, it too trafficked in the idiom of confidence men with its gambling metaphors and an emphasis on the human as a strategic manipulator of others.
3. I describe this process as 'a tools-to-theory heuristic' rather than reflexivity because Goffman does not appear to *deliberately* model the self upon his scientific practices. He did not exhibit the degree of reflexivity about ethnography famously presented in Rabinow (1977).
4. On Goffman's relationship with Hughes, see Jaworski (2000).
5. This fascination was not an idiosyncratic particularity, but rather reflected a broad romance with ethology in the American behavioral sciences of the postwar era (Vicedo, 2009).
6. Erving Goffman, 'Filed Report for Soc. 301 A, 1948', Ernest W. Burgess Papers, box 131, folder 8, University of Chicago Library.
7. Goffman would incorporate unpublished findings from Social Research's market research into his account of people's 'secret consumption', which deviated from prescribed standards of taste (1959: 50–1).
8. David M. Schneider to Reinhard Bendix, 6 November 1958, David M. Schneider Papers, box 14, folder 6, University of Chicago Library.
9. Pinch (2010) discusses Goffman's attention to the materiality of the setting in his analysis of the social interaction.
10. Goffman to Hughes, 13 December 1960, Everett C. Hughes Papers, University of Chicago Library.
11. ibid.
12. Buchanan (1982) notes that psychodrama had been practised at St Elizabeth's Hospital continuously since 1939.

Bibliography

Abbott, A. (1999) *Department & Discipline: Chicago Sociology at One Hundred*. Chicago, IL: University of Chicago Press.

Anderson, N. (1923) *The Hobo: The Sociology of the Homeless Man*. Chicago, IL: University of Chicago Press.

Barton, A. H. (1979) 'Paul Lazarsfeld and Applied Social Research: Invention of the University Applied Social Research Institute', *Social Science History* 3: 4–44.

Becker, H. (1953) 'Becoming a Marihuana User', *American Journal of Sociology* 59: 235–42.

Becker, H. S. (2003) 'The Politics of Presentation: Goffman and Total Institutions', *Symbolic Interaction* 26: 659–69.

Bergman, J. D. (1969) 'The Original Confidence Man', *American Quarterly* 21: 560–77.

Buchanan, D. R. (1982) 'Psychodrama: a Humanistic Approach to Psychiatry Treatment for the Elderly', *Hospital & Community Psychiatry* 33: 220–23.

Bulmer, M. (1984) *The Chicago School of Sociology: Institutionalization, Diversity, and the Rise of Sociological Research*. Chicago, IL: University of Chicago Press.

Burkhardt, R. (2005) *Patterns of Behavior: Konrad Lorenz, Niko Tinbergen, and the Founding of Ethology*. Chicago, IL: University of Chicago Press.

Burns, T. (1992) *Erving Goffman*. London: Routledge.

Cappetti, C. (1993) *Writing Chicago: Modernism, Ethnography, and the Novel*. New York: Columbia University Press.
Cohen-Cole, J. (2009) 'The Creative American: Cold War Salons, Social Science, and the Cure for Modern Society', *Isis* 100: 219–62.
Edwards, P. (1996) *Closed World: Computers and the Politics of Discourse in Cold War America*. Cambridge, MA: MIT Press.
Ellenberger, H. F. (1960) 'Zoological Garden and Mental Hospital', *Canadian Psychiatric Association Journal* 5 (1960): 136–49.
Fallers, L. (1962) Review of 'The Presentation of Self in Everyday Life', *American Anthropologist* 64: 190–1.
Fine, G. A. and Martin, D. D. (1990) 'A Partisan View: Sarcasm, Satire, and Irony as Voices in Erving Goffman's Asylums', *Journal of Contemporary Ethnography* 19: 89–115.
Frank, T. (1997) *The Conquest of Cool: Business Culture, Counterculture, and the Rise of Hip Consumerism*. Chicago, IL: University of Chicago Press.
Galison, P. (2004) 'Image of Self', in L. Daston (ed.) *Things that Talk: Object Lessons from Art and Science*. New York: Zone Books, pp. 257–94.
Gieryn, T. F. (2006) 'City as Truth-Spot: Laboratories and Field-Sites in Urban Studies', *Social Studies of Science* 36: 5–38.
Gigerenzer, G. (1991) 'From Tools to Theories: a Heuristic of Discovery in Cognitive Psychology', *Psychology Review* 98: 254–67.
Gilman, N. (2003) *Mandarins of the Future: Modernization Theory in Cold War America*. Baltimore, MD: Johns Hopkins University Press.
Goffman, E. M. (1949) 'Some Characteristics of Response to Depicted Experience', SM thesis, University of Chicago.
Goffman, E. (1952) 'On Cooling the Mark out: Some Aspects of Adaptation to Failure', *Psychiatry* 15: 451–63.
Goffman, E. (1953) 'Communication Conduct in an Island Community', PhD dissertation, University of Chicago.
Goffman, E. (1959) *The Presentation of Self in Everyday Life*. Harmondsworth, Mx: Penguin.
Goffman, E. (1961) *Asylums: Essays on the Social Situation of Mental Patients and Other Inmates*. New York: Anchor Books.
Goffman, E. (1971) *Relations in Public*. New York: Doubleday.
Goffman, E. (1989) 'On Fieldwork', *Journal of Contemporary Ethnography* 18: 123–32.
Gonos, G. (1980) 'The Class Position of Goffman's Sociology: Social Origins of an American Structuralism', in J. Ditton (ed.) *The View from Goffman*. London: Macmillan, pp. 134–69.
Gouldner, A. W. (1962) 'Anti-Minotaur: The Myth of a Value-Free Sociology', *Social Problems* 9: 199–213.
Gouldner, A. W. (1970) *The Coming Crisis of Western Sociology*. New York: Basic.
Hacking, I. (2004) 'Between Michel Foucault and Erving Goffman: Between Discourse in the Abstract and Face-to-Face Interaction', *Economy and Society* 33: 277–302.
Halttunen, K. (1982) *Confidence Men and the Painted Women: A Study of Middle-Class Culture in America, 1830–1870*. New Haven, CT: Yale University Press.
Heims, S. J. (1993) *Constructing a Social Science for Postwar America: The Cybernetics Group, 1946–1953*. Cambridge, MA: MIT Press.
Herman, E. (1995) *The Romance of American Psychology: Political Culture in the Age of Experts*. Berkeley: University of California Press.
Herriott, F. (1940) 'Diagnostic Examination of Mental Patients on the Psychodramatic Stage', *Sociometry* 3: 383–98.
Herriott, F. (1945) 'Some Uses of Psychodrama at St. Elizabeths Hospital', *Sociometry* 8: 54–7.

Herriott, F. and Hagan, M. (1941) 'The Theatre for Psychodrama at St. Elizabeths Hospital', *Sociometry* 4: 168–76.

Higbie, T. (1997) 'Crossing Class Boundaries: Tramp Ethnographers and Narratives of Class in Progressive Era America', *Social Science History* 21: 559–92.

Hughes, E. C. (1969) Review of 'Interactive Ritual', *American Journal of Sociology* 75: 425–6.

Isaac, J. (2007) 'The Human Sciences in Cold War America', *The Historical Journal* 50: 725–46.

Isaac, J. (2009) 'Tangled Loops: Theory, History, and the Human Sciences in Modern America', *Modern Intellectual History* 6: 397–424.

Isaac, J. (2010) 'Theorist at Work: Talcott Parsons and the Carnegie Project on Theory, 1949–1951', *Journal of the History of Ideas* 71: 287–311.

Jaworski, G. D. (2000) 'Erving Goffman: The Reluctant Apprentice', *Symbolic Interaction* 23: 299–308.

Kohler, R. E. (2002) *Landscapes and Labscapes: Exploring the Lab – Field Border in Biology*. Chicago, IL: University of Chicago Press.

Latham, M. (2000) *Modernization as Ideology: American Social Science and 'Nation Building' in the Kennedy Era*. Chapel Hill: University of North Carolina Press.

Lemov, R. (2010) '"Hypothetical Machines": The Science Fiction Dreams of Cold War Social Science', *Isis* 101: 401–11.

Lindberg, G. (1982) *The Confidence Man in American Literature*. New York: Oxford University Press.

Lofland, J. (1980) 'Early Goffman: Style, Structure, Substance, Soul', in J. Ditton (ed.) *The View from Goffman*. London: Macmillan, pp. 24–51.

Lutes, J. M. (2002) 'Into the Madhouse with Nellie Bly: Girl Stunt Reporting in Late Nineteenth-Century America', *American Quarterly* 54: 217–53.

Manning, P. (1989) 'Goffman's Revisions', *Philosophy of the Social Sciences* 19: 341–3.

Manning, P. (1992) *Erving Goffman and Modern Sociology*. Stanford, CA: Stanford University Press.

Manning, P. (2000) 'Credibility, Agency, and the Interaction Order', *Symbolic Interaction* 23: 283–97.

Maurer, D. W. (1930) 'Schoonerisms: Some Speech-Peculiarities of the North-Atlantic Fishermen', *American Speech* 5: 387–95.

Maurer, D. W. (1931a) 'The Argot of the Underworld', *American Speech* 7: 99–118.

Maurer, D. W. (1931b) 'Carnival Cant: a Glossary of Circus and Carnival Slang', *American Speech* 6: 327–37.

Maurer, D. W. (1939) 'Prostitutes and Criminal Argots', *American Journal of Sociology* 44: 546–50.

Maurer, D. W. (1940) *The Big Con*. New York: BobbsMerrill.

O'Connor, A. (2000) *Poverty Knowledge: Social Science, Social Policy, and the Poor in Twentieth-Century U.S. History*. Princeton, NJ: Princeton University Press.

Orlansky, H. (1948) 'An American Death Camp', *Politics* 5: 162–7.

Pettit, M. (2008) 'The New Woman as "Tied-up Dog": Amy E. Tanner's Situated Knowledges', *History of Psychology* 11: 145–63.

Pinch, T. (2010) 'The Invisible Technologies of Goffman's Sociology: From the Merry-Go-Round to the Internet', *Technology and Culture* 51: 409–24.

Pittenger, M. (1997) 'A World of Difference: Constructing the "Underclass" in Progressive America', *American Quarterly* 49: 26–65.

Pittenger, M. (2003) 'What's on the Worker's Mind: Class Passing and the Study of the Industrial Workplace in the 1920s', *Journal of the History of the Behavioral Sciences* 39: 143–61.

Platt, J. (1983) 'The Development of "Participant Observation" Method in Sociology: Origin Myth and History', *Journal of the History of the Behavioral Sciences* 19: 379–93.

Platt, J. (1995) 'Research Methods and the second Chicago School', in G. A. Fine (ed.) *A Second Chicago School?: The Development of a Postwar American Sociology*. Chicago, IL: University of Chicago Press, pp. 82–107.

Platt, J. (1996) *A History of Sociological Research Methods in America 1920–1960*. New York: Cambridge University Press.

Pooley, J. (2006) 'Fifteen Pages that shook the Field: *Personal Influence*, Edward Shils and the Remembered History of Mass Communication Research', *The Annals of the American Academy of Political and Social Sciences* 608: 130–56.

Posner, M. (2008) 'Seinfeld's Marble Rye Lad honoured', *Globe and Mail* (6 September): R4.

Rabinow, P. (1977) *Reflections on Fieldwork in Morocco*. Berkeley: University of California Press.

Rohde, J. (2009) 'Gray Matters: Social Scientists, Military Patronage, and Democracy in the Cold War', *Journal of American History* 96: 99–121.

Salerno R. A. (2007) *Sociology Noir: Studies at the University of Chicago in Loneliness, Marginality, and Deviance, 1915–1935*. Jefferson, NC: McFarland.

Scheff, T. J. (2006) *Goffman Unbound!: A New Paradigm for Social Science*. Boulder, CO: Paradigm.

Scott, J. W. (1991) 'The Evidence of Experience', *Critical Inquiry* 17: 773–97.

Solovey, M. (2001) 'Project Camelot and the 1960s Epistemological Revolution: Rethinking the Politics-Patronage-Social Science Nexus', *Social Studies of Science* 31: 171–206.

Stearns, P. (1994) *American Cool: Constructing a Twentieth-Century Emotional Style*. New York: New York University Press.

Taxel, H. S. (1953) 'Authority Structure in a Mental Hospital Ward', AM thesis, University of Chicago.

Tseëlon, E. (1992) 'Is the Presented Self Sincere? Goffman, Impression Management and the Postmodern Self', *Theory, Culture & Society* 9: 115–28.

Vicedo, M. (2009) 'The Father of Ethology and the Foster Mother of Ducks: Konrad Lorenz as Expert on Motherhood', *Isis* 100: 263–91.

Warner, W. L. and Henry, W. E. (1948) 'The Radio Day Time Serial: a Symbolic Analysis', *Genetic Psychology Monographs* 37: 3–72.

Williams, B. (2002) *Truth and Truthfulness: An Essay in Genealogy*. Princeton, NJ: Princeton University Press.

Willoughby, R. H. (1953) 'The Attendant in the State Mental Hospital', SM thesis, University of Chicago.

Winkin, Y. (1999) 'Erving Goffman: What is Life? The Uneasy Making of Intellectual Biography', in G. Smith (ed.) *Goffman and Elements in Social Interaction*. New York: Routledge, pp. 19–41.

46

A Note on Participant Observation

Colin Bell

As sociologists most of us have some idea of what is involved in participant observation as a method of data collection. It is, however, possible to formalize this to make clearer the roles that are available to the research worker. This will provide a framework for some of the procedural and ethical decisions in participant observation.

Firstly, social systems, institutions, organizations and groups can be characterized by their degree of 'openness', by which is meant ease of access for the research worker. Communities in this country are more open than universities, universities than firms, firms than prisons and so on. 'Openness' is a relative concept but may for these purposes be dichotomized as 'open' and 'closed'. Secondly the research worker can make the choice as to whether he will work openly and overtly or secretly and covertly. This again is not in reality a true dichotomy but will vary between social situations. Combining these two dichotomies, however, gives a useful classification of styles and types of participant observation:

		SYSTEM	
		OPEN	CLOSED
ROLE	OVERT	1	2
	COVERT	3	4

The three stages that participant observation passes through, of entry, of maintaining a position and of exit, vary with the four styles. Overt entry

Source: *Sociology*, 3(3) (1969): 417–418.

into an open system involves publicity, into a closed system sponsorship; covert entry into an open system involves stealth, into a closed system guile. Community studies usually operate in boxes 1 and 3. Gans (1962 and 1967) whilst working overtly found he also played covert roles, interestingly referred to as 'performance breaks'. Industrial sociology usually operates in box 2 with the attendant problems of topdog identification after sponsorship. Dalton (1959) for this reason has argued for covert work in this field and his study falls into box 4, along with Festinger *et. al*'s (1956) study of a group that predicted the end of the world. Participant observation obliges the research worker to adopt and be adopted in, a position in some existing structure of social relations. (The technique of course is not a single activity, it varies from total participation with no observation, to observation with no participation.) In overt participant observation 'position maintenance' may involve the socialization of the members of the system that is being studied into understanding the role of 'sociological research worker', this may only apply to topdogs in box 2. In covert participant observation, the role of 'sociological research worker' must be hidden in order to maintain the position the research worker has achieved. The general statement made by Vidich (1954) that 'the social position of the observer determines what he is likely to see' applies in all four types of participant observation.

Leaving the field also varies, especially when social relations adopted for instrumental reasons by the research worker are not always understood as such by those that have been observed. Publication will surprise and shock if noticed by the subjects in boxes 3 and 4, less so in 2, but approval of topdog sponsors may be necessary. If the open field worker in an open system has acted correctly publication should be expected by his subjects, keeping confidences and confidentiality of course.

Two final problems: data collected by participant observation are often only used to illustrate not prove a hypothesis. 'For example is no proof'. We are all familiar with Merton's (1957) strictures on *ex post facto* hypothesis testing. And secondly, data may be subject to retrospective selection, if not falsification to fit the final analysis. A solution to both of these problems is to write what might be called the 'natural history' of the research: stating hypothesis in advance and what will be accepted as proof, and documenting the field workers' research positions which are important data in their own right, giving details of entry to and exit from the system studied. This vital activity, although increasingly common in the United States (Hammond 1964, Vidich, Bensman and Stein 1964 and Riesman 1958) does not seem to have begun in this country, but it is necessary if participant observation is to advance along with other techniques of data collection in sociology.

Bibliography

Dalton, Melville. 1959. *Men Who Manage*. New York: Wiley.

Festinger, Leon, Riecken, Henry W., Schachter, Stanley. 1956. *When Prophecy Fails*. Minneapolis: University of Minnesota.

Gans, Herbert J. 1962. *The Urban Villagers: Group and Class in the Life of Italian-Americans*. New York: The Free Press.

Gans, H. J. 1967. *The Levittowners*. London, Allen Lane.

Hammond, Philip E. (ed.). 1964. *Sociologists at Work*. New York: Basic Books.

Merton, Robert K. 1957. *Social Theory and Social Structure*, New York: The Free Press.

Reisman, David, 1958 'Field Report' pp. 266–370 in Paul F. Lazarsfeld and Wagner Thielans Jr. *The Academic Mind*. New York: Free Press.

Vidich, A. J. 1954. Participant Observation and the Collection and Interpretation of Data. *Amer. J. Sociol* 60. pp. 354–360.

Vidich, A. J., Bensman, J. and Stein, M. R., 1964. *Reflections on Community Studies*. New York: Wiley.

47

A Contribution to the Theory of Participant Observation

Jiri Kolaja

One of the major problems in the epistemological aspect of the method of participant observation is the question as to what degree a researcher should identify himself with his new role of observer based upon his participation in the studied group. Whyte, considering this problem, concludes that ". . . there is no one standard role for the observer of social behavior. Valuable data have been collected through the full range from totally non-participant observation to the full participation of a regular member. . . . Midway between the two extremes we find the observer who is recognized by the group members as a research man, but one who at the same time participates freely in their activities. . . ."[1] According to Goldenweiser, a researcher should attempt full participation, while Radin maintains that this is impossible.[2] Paul reports on Frank Cushing who identified himself to such a degree with the natives that he refused to publish the results of his observation.[3] Florence Kluckhohn reports that after a while she went to dances not for observation but for her own recreation.[4]

In this article we would like to attack the problem of identification or non-identification of the observer from the epistemological standpoint. By definition, the identification brings about a performance on the part of the researcher. If the observed person is afflicted with pain, under the conditions of identification the observer should go more or less through the same processes as the object. The nature of the knowledge is here different from the knowledge acquired only in terms of a concept of pain. The epistemological problem

Source: *Social Forces*, 35(2) (1956): 159–163.

is which of these two kinds of knowledge is more adequate. Should we study human behavior by performing the same or similar behavior, or is it satisfactory to grasp the relatedness between particular phases of behavior, rather to conceive than perceive the unity or similarities between the phases?

The above question is in fact a traditional problem as to the merits of knowledge by acquaintance or knowledge by description, starting with Plato and finding its echo in sociological disagreements about the "understanding" method and the "operational positivistic" method. A pupil of Dilthey, Weber,[5] Kluckhohn, Sorokin, Blumer, etc., would be inclined to subscribe to the idea that performance and frequencies of the performance are not as important as the conception-understanding of the pattern within which the phases of performance are carried out. However, as this pattern is often more a creation of the observer than a phenomenal configuration, a pupil of Solvay, Lundberg, Lazarsfeld, and Zipf, just to name a few, would maintain that a really positive knowledge is a frequency-like recording. Theoretically, the most reliable observation, from the positivistic standpoint, would be a record that would include as many phases as possible of the observed process, seen from as many different viewpoints as possible. The observed process would be almost restored in its original nature with the exception of being transferred to a symbolic level.

Thus, the problem can be posed as an alternative between the process and the concept of the process.[6] To translate this into more operational terms, let us compare participant observation with interviewing. In both instances the researcher, himself, is acting. However, while participant observation makes it possible for the observer to attend the process directly, the interviewer is informed about the event only indirectly, symbolically; he gets only a concept about the process.

Furthermore, while the observer in the participation situation is exposed to more or less the same stimulus as the other member or members of the group, both the interviewer and the interviewee serve each other as stimuli. The participation situation is, in most cases, a sort of parallel relation between members of the group, interspersed with dialectical relations when persons interact. The interview is only a dialectical relation.[7] It is worthwhile to quote Theodor Litt in this connection. Litt maintained that the dialectical relationship between I-You is not genuinely social, because it is by necessity perceived subjectively. A genuine social relation occurs with at least three persons. The first person can observe the interaction between two others from "outside" without bringing himself into the game.[8]

For the purpose of clarification, let us arrange the above differences between participant observation and interview within a fourfold diagram. (See figure 1.)

It should be noted that the interview is classified in the figure on the conceptual level only. This is done for theoretical purposes. Later we shall show that the interview is also a performance under specific conditions. The same

	Interview	Participant Observation
conceptual-symbolic	I-You	I-You You-he
performance		You-the object I-the object (I-You) (You-he)

Figure 1

holds true for symbolic interaction of participant observation. Here it should be pointed out that the purely conceptual-symbolic interaction is only that which contains extra-situational references. As soon as the verbal symbols are related to objects in the situation, they become a part of the performance. Usually they tend to lose their symbolic function, becoming more gesture-signals than symbols.[9] It is proper to refer the reader to Malinowski's notion of speech as an integrated part of the action in a specific situation.[10] In our figure this situation is expressed by the parenthesis.

It is advantageous to introduce briefly the George Herbert Mead and Max Scheler theories pertaining to the relationship between two persons. Mead's cognitive theory of the self is explicitly a dialectical theory. The "other" serves as a springboard for the knowledge of "Me."[11] Scheler, whose phenomenological analysis has not been surpassed in subtlety, is interested in the emotional identification.[12] He differentiates between dialectical, parallel, and unified relations. The latter would occur when both persons lose perception of their individualities, and totally identify themselves.[13]

Neither Mead nor Scheler, however, was in fact concerned with the epistemological process of getting to know the self of the other person. Their assumption was that the manifested behavior corresponds to the self of the other person. Scheler maintained that expression is the first thing we, in fact, understand. This understanding is a sort of an "a priori," being given to us intuitively.[14] Likewise, Mead proposed that we "carve out" the social world prior to the physical world.[15] Scheler expressed it in other words, saying that originally everything was an expression.[16]

It is evident that the interviewer or participant observer cannot take for granted the equivalence of the role-expression and the person's self. Although a society could not exist without a normative correspondence between both, in individual persons we often find a conscious or unconscious disparity between the role-appearance and the self.[17]

Returning to our comparison of the interview and participant observation, let us classify both methods within a fourfold diagram whose columns

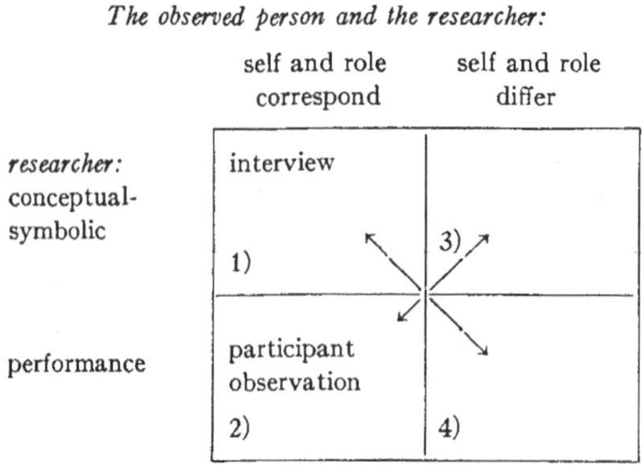

Figure 2

would show the two possible assumptions concerning the correspondence of the role and the self. (See figure 2.)

In Figure 2 we start with the assumption that the role equals the self in both interview and participant observation. Now, suppose that interaction in participant observation evokes a certain suspicion in the researcher that persons behave insincerely; or the researcher might not understand well motives and meanings shared by the other person. Then he moves from cell No. 2 into cell No. 3. He stops performing and starts thinking it over. The spontaneity of interaction, actions and gestures, has been broken. Interestingly, there is a certain reciprocity of perspectives[18] of the researcher and of the observed person. Once the researcher discovers or guesses that the other person's role differs from the other's self, he immediately loses the full identification of his self with his role. This happens to every person in the course of social interaction under similar conditions. The lover who suddenly discovers unfaithfulness of the partner, experiences a sort of shocking change from his previous spontaneous identification toward the split between his self and his role. Mead's concept of the role of the other fits here; if I assume that the other person has differentiated between his role and his self (and this realization is experienced through my imaginative taking of the role of the other), I suddenly perceive myself split between my former, or still automatically continued role, and my disappointed and doubting self.

Let us consider the interview situation. The picture is somewhat different this time. The researcher-interviewer seeks to reduce his "own performance" as much as possible. The personality of the researcher should be semantically neutralized in order that the interviewee would be exposed to symbols only. The interviewee is supposed to react to symbolically circumscribed events only. However, if he doesn't, and if there is a high degree of inhibition or evasiveness on the part of the interviewee, the interviewer starts performing himself. He seeks to establish rapport, to warm up his "subject." He moves from cell No. 1 to cell No. 4 (figure 2). While in cell No. 1 the researcher is supposed to behave as naturally and inconspicuously as possible and the respondent to relate his verbal behavior with his self, in cell No. 4 there is again a discrepancy between role and self. Here the interviewer purposively acts out his role. Thus, curiously, in both cases, i.e., in participant observation as well as in interviewing, the researcher moves across the table diagonally if he meets a lack of relatedness between the self and the role.

Theoretically, we are now in position to give an answer to the question posed at the beginning of the analysis. Is performance or concept of performance a more adequate method of gaining knowledge about human behavior?

In a perfect communication in which all phenomena are adequately translated into symbols or in which symbols refer but to "themselves" (e.g., in aesthetic products), the researcher would prefer to "understand" rather than "perform" the symbols. Performance absorbs time that could be allocated to the cognitive function. For example, the performance of writing a mathematical formula on the blackboard is irrelevant for its understanding.

However, if the symbols refer to phenomena inadequately or not exhaustively, the researcher will attempt to gain knowledge about the phenomena through his own performance-experience. For example, some emotional processes are sometimes inadequately symbolized because of the lack of terms. Theoretically, an identification-performance of the researcher himself would increase the validity of his observation. He would be better prepared to translate the experience into more adequate terms, or eventually to create some new words if necessary. It should be noted that the performance increases validity, but hardly reliability. Only symbols allow for attainment of the superpersonal level on which objectivity can be realized.[19]

Apparently, whether validity or reliability should be emphasized more depends upon the research goal – upon the strategy of the research as well as upon the nature of the studied social phenomena. Though human interaction is primarily a symbolic process, the non-symbolic elements are in certain situations of utmost importance. For example, in my research on laundry workers, the job, its physical nature and exhaustiveness, gave me an explanation as to why there were differences in symbolic interaction among the workers.[20] Those who performed relatively heavier tasks used their rest period more for physical than mental recreation. Theoretically, one could propose that performance of a job is itself irrelevant for the study of symbolic interaction if: 1) members of the group do not ascribe any special meaning to it, for example in terms of prestige; 2) its nature does not physically interfere with symbolic processes or with the physical resources of the person.

Obviously, the latter cannot be further defined. It is the result of contingencies and the research goal held by the researcher.

Summarizing: In this article we have attempted to develop two analytical categories in terms of performance and concept of performance. Both categories were cross-tabulated with the categories of correspondence and non-correspondence between the self and the role. Applying this fourfold table theoretically to the dynamic processes of participant observation and interviewing, we have found that the original correspondence between the self and the role, i.e., the performance, is interrupted under certain conditions. Then the researcher switches from the performance to the concept of performance. In the interview the correspondence between the self and the role exists first on the symbolic level and changes to the performance level under similar conditions as in participant observation.

Finally, it is of interest to realize that our explanation of the change from performance to the concept of performance is in fact a pragmatistic explanation of the emergence of thought. The switch from the conceptual level to the performance level in the interview, on the other hand, is parallel to the operationalistic tendency to translate concepts into performance of measurement. Also, theoretically, all epistemological possibilities are exhausted within our model: within the relationship of the subject-object, one can either emphasize the subject as the idealists do, or the object as the empiricists do. In our case this corresponds either to the concept-understanding approach or the process-measurement approach.

Notes

1. William Foote Whyte, *Modern Methods in Social Research*, Office of Naval Research, p. 14.
2. Benjamin D. Paul, "Interview Technique and Field Relationships," in A. L. Kroeber (ed.), *Anthropology Today* (Chicago: University of Chicago Press, 1953), p. 438.
3. *Ibid.*, p. 441.
4. Florence R. Kluckhohn, "The Participant Observer Technique in Small Communities," *The American Journal of Sociology*, 46 (November 1940), p. 342.
5. "One need not to be Caesar in order to understand Caesar." Quoted by Robert K. Merton, *Social Theory and Social Structure* (Glencoe, Ill.: The Free Press, 1951), p. 12.
6. One could, of course, also raise the question whether a concept is a process like the psychological problem as to whether perception is a behavior. For our purpose, we shall, however, not enter upon such an analysis.
7. To my knowledge, the concept of parallel and dialectical social relations was first developed by Theodor Litt, *Individuum und Gemeinschaft* (Berlin: B. G. Teubner, 1924), p. 30
8. Theodor Litt, *op. cit.*, p. 237. The analysis of the power structure in the three-member group should also be referred to. See Theodore M. Mills, "Power Relations in Three-Person Groups" in D. Cartwright and A. Zander, *Group Dynamics* (Evanston, Ill.: Row, Peterson, and Co., 1953), pp. 428–12.
9. For a definition of symbols, signs, and signals see Charles Morris, *Signs, Language and Behavior* (New York: Prentice-Hall, 1946), p. 354 ff.
10. Bronislaw Malinowski, *Magic, Science and Religion and Other Essays* (Glencoe, Ill.: The Free Press, 1948), p. 243 ff.
11. George H. Mead, *Mind, Self and Society* (Chicago: The University of Chicago Press, 1934).

12. Gordon W. Allport, "The Historical Background of Modern Social Psychology," *Handbook of Social Psychology*, G. Lindzey (ed.) (Cambridge, Mass.: Addison-Wesley, 1954), I, 21.
13. Max Scheler, *Wesen und Formen der Sympathie* (Bonn: Fr. Cohen, 1923). Scheler isolates eight categories of emotional identification according to a) whether we perceive the difference between the intensity and nature of emotions in both the observer and the other person, and b) whether the stimulus of emotions is perceived by the subject as being related to him directly, parallel to the other person, or through the other person in a dialectical way. The terms dialectical, parallel, and unified, which we use are not Scheler's terms, but our own more comprehensive categories of Scheler's eight original categories.
14. *Ibid.*, p. 288.
15. George H. Mead, *op cit.*, p. xix.
16. Max Scheler, *op. cit.*, p. 275.
17. Incidentally, this assumption concerning correspondence of the role and the self has been indirectly criticized on the ground that neither consciousness nor motality are imported from outside. See: W. H. Dismonde, Self-Actualization: Loving and Strategic, Ph.D. dissertation, Columbia University 1951, p. 32. See also Paul E. Pfuetze, *The Social Self*, (New York: Bookman Associates, 1954) p. 252 ff.
18. It is worthwhile to remember that this aspect of social relations has been stressed by many sociologists, and in addition to G. H. Mead, we would like to list G. Gurvitch, T. Parsons, et al.
19. ". . . any model procedure must somehow compromise between the requirements of reliability and validity." H. H. Hyman et al., *Interviewing in Social Research* (Chicago: University of Chicago Press, 1954), p. 27.
20. See my paper, *Ecological, Societal and Cultural Orders of a Group of Laundry Workers*, read at the annual meeting of the American Sociological Society, 1955.

48

Problems of Inference and Proof in Participant Observation

Howard S. Becker

The participant observer gathers data by participating in the daily life of the group or organization he studies.[1] He watches the people he is studying to see what situations they ordinarily meet and how they behave in them. He enters into conversation with some or all of the participants in these situations and discovers their interpretations of the events he has observed.

Let me describe, as one specific instance of observational technique, what my colleagues and I have done in studying a medical school. We went to lectures with students taking their first two years of basic science and frequented the laboratories in which they spend most of their time, watching them and engaging in casual conversation as they dissected cadavers or examined pathology specimens. We followed these students to their fraternity houses and sat around while they discussed their school experiences. We accompanied students in the clinical years on rounds with attending physicians, watched them examine patients on the wards and in the clinics, sat in on discussion groups and oral exams. We ate with the students and took night call with them. We pursued internes and residents through their crowded schedules of teaching and medical work. We stayed with one small group of students on each service for periods ranging from a week to two months, spending many full days with them. The observational situations allowed time for conversation and we took advantage of this to interview students about things that had happened and were about to happen, and about their own backgrounds and aspirations.

Source: *American Sociological Review*, 23(6) (1958): 652–660.

Sociologists usually use this method when they are especially interested in understanding a particular organization or substantive problem rather than demonstrating relations between abstractly defined variables. They attempt to make their research theoretically meaningful, but they assume that they do not know enough about the organization *a priori* to identify relevant problems and hypotheses and that they must discover these in the course of the research. Though participant observation can be used to test *a priori* hypotheses, and therefore need not be as unstructured as the example I have given above, this is typically not the case. My discussion refers to the kind of participant observation study which seeks to discover hypotheses as well as to test them.

Observational research produces an immense amount of detailed description; our files contain approximately five thousand single-spaced pages of such material. Faced with such a quantity of "rich" but varied data, the researcher faces the problem of how to analyze it systematically and then to present his conclusions so as to convince other scientists of their validity. Participant observation (indeed, qualitative analysis generally) has not done well with this problem, and the full weight of evidence for conclusions and the processes by which they were reached are usually not presented, so that the reader finds it difficult to make his own assessment of them and must rely on his faith in the researcher.

In what follows I try to pull out and describe *the basic analytic operations carried on in participant observation*, for three reasons: to make these operations clear to those unfamiliar with the method; by attempting a more explicit and systematic description, to aid those working with the method in organizing their own research; and, most importantly, in order to propose some changes in analytic procedures and particularly in reporting results which will make the processes by which conclusions are reached and substantiated more accessible to the reader.

The first thing we note about participant observation research is that analysis is carried on *sequentially*,[2] important parts of the analysis being made while the researcher is still gathering his data. This has two obvious consequences: further data gathering takes its direction from provisional analyses; and the amount and kind of provisional analysis carried on is limited by the exigencies of the field work situation, so that final comprehensive analyses may not be possible until the field work is completed.

We can distinguish three distinct stages of analysis conducted in the field itself, and a fourth stage, carried on after completion of the field work. These stages are differentiated, first, by their logical sequence: each succeeding stage depends on some analysis in the preceding stage. They are further differentiated by the fact that different kinds of conclusions are arrived at in each stage and that these conclusions are put to different uses in the continuing research. Finally, they are differentiated by the different criteria that are used to assess evidence and to reach conclusions in each stage. The three stages of field analysis are: the selection and definition of problems, concepts, and indices; the check on the frequency and distribution of phenomena; and the incorporation of individual findings into a model of the organization under study.[3] The fourth stage of final analysis involves problems of presentation of evidence and proof.

Selection and Definition of Problems, Concepts, and Indices

In this stage, the observer looks for problems and concepts that give promise of yielding the greatest understanding of the organization he is studying, and for items which may serve as useful indicators of facts which are harder to observe. The typical conclusion that his data yield is the simple one that a given phenomenon exists, that a certain event occurred once, or that two phenomena were observed to be related in one instance; the conclusion says nothing about the frequency or distribution of the observed phenomenon.

By placing such an observation in the context of a sociological theory, the observer selects concepts and defines problems for further investigation. He constructs a theoretical model to account for that one case, intending to refine it in the light of subsequent findings. For instance, he might find the following: "Medical student X referred to one of his patients as a 'crock' today."[4] He may then connect this finding with a sociological theory suggesting that occupants of one social category in an institution classify members of other categories by criteria derived from the kinds of problems these other persons raise in the relationship. This combination of observed fact and theory directs him to look for the problems in student-patient interaction indicated by the term "crock." By discovering specifically what students have in mind in using the term, through questioning and continued observation, he may develop specific hypotheses about the nature of these interactional problems.

Conclusions about a single event also lead the observer to decide on specific items which might be used as indicators[5] of less easily observed phenomena. Noting that in at least one instance a given item is closely related to something less easily observable, the researcher discovers possible shortcuts easily enabling him to observe abstractly defined variables. For example, he may decide to investigate the hypothesis that medical freshmen feel they have more work to do than can possibly be managed in the time allowed them. One student, in discussing this problem, says he faces so much work that, in contrast to his undergraduate days, he is forced to study many hours over the weekend and finds that even this is insufficient. The observer decides, on the basis of this one instance, that he may be able to use complaints about weekend work as an indicator of student perspectives on the amount of work they have to do. The selection of indicators for more abstract variables occurs in two ways: the observer may become aware of some very specific phenomenon first and later see that it may be used as an indicator of some larger class of phenomena; or he may have the larger problem in mind and search for specific indicators to use in studying it.

Whether he is defining problems or selecting concepts and indicators, the researcher at this stage is using his data only to speculate about possibilities. Further operations at later stages may force him to discard most of the provisional hypotheses. Nevertheless, problems of evidence arise even at this point, for the researcher must assess the individual items on which his speculations are based in order not to waste time tracking down false leads. We shall eventually need a systematic statement of canons to be applied to individual items of evidence. Lacking such a statement, let us consider some commonly used tests. (The observer typically applies these tests as seems reasonable to him during this and the succeeding stage in the field. In the final stage, they are used more systematically in an overall assessment of the total evidence for a given conclusion.)

The Credibility of Informants. Many items of evidence consist of statements by members of the group under study about some event which has occurred or is in process. Thus, medical students make statements about faculty behavior which form part of the basis for conclusions about faculty-student relations. These cannot be taken at face value; nor can they be dismissed as valueless. In the first place, the observer can use the statement as evidence *about the event*, if he takes care to evaluate it by the criteria an historian uses in examining a personal document.[6] Does the informant have reason to lie or conceal some of what he sees as the truth? Does vanity or expediency lead him to mis-state his own role in an event or his attitude toward it? Did he actually have an opportunity to witness the occurrence he describes or is hearsay the source of his knowledge? Do his feelings about the issues or persons under discussion lead him to alter his story in some way?

Secondly, even when a statement examined in this way proves to be seriously defective as an accurate report of an event, it may still provide useful evidence for a different kind of conclusion. Accepting the sociological proposition that an individual's statements and descriptions of events are made from a perspective which is a function of his position in the group, the observer can interpret such statements and descriptions as indications of the individual's perspective on the point involved.

Volunteered or Directed Statements. Many items of evidence consist of informants' remarks to the observer about themselves or others or about something which has happened to them; these statements range from those which are a part of the running casual conversation of the group to those arising in a long intimate tete-a-tete between observer and informant. The researcher assesses the evidential value of such statements quite differently, depending on whether they have been made independently of the observer (volunteered) or have been directed by a question from the observer. A freshman medical student might remark to the observer or to another student that he has more material to study than he has time to master; or the observer might ask, "Do you think you are being given more work than you can handle?", and receive an affirmative answer.

This raises an important question: to what degree is the informant's statement the same one he might give, either spontaneously or in answer to a question, in the absence of the observer? The volunteered statement seems likely to reflect the observer's preoccupations and possible biases less than one which is made in response to some action of the observer, for the observer's very question may direct the informant into giving an answer which might never occur to him otherwise. Thus, in the example above, we are more sure that the students are concerned about the amount of work given them when they mention this of their own accord than we are when the idea may have been stimulated by the observer asking the question.

The Observer-Informant-Group Equation. Let us take two extremes to set the problem. A person may say or do something when alone with the observer or when other members of the group are also present. The evidential value of an observation of this behavior depends on the observer's judgment as to whether the behavior is equally likely to occur in both situations. On the one hand, an informant may say and do things when alone with the observer that accurately reflect his perspective but which would be inhibited by the presence of the group. On the other hand, the presence of others may call forth behavior which reveals more accurately the person's perspective but would not be enacted in the presence of the observer alone. Thus, students in their clinical years may express deeply "idealistic" sentiments about medicine when alone with the observer, but behave and talk in a very "cynical" way when surrounded by fellow students. An alternative to judging one or the other of these situations as more reliable is to view each datum as valuable in itself, but with respect to different conclusions. In the example above, we might conclude that students have "idealistic" sentiments but that group norms may not sanction their expression.[7]

In assessing the value of items of evidence, we must also take into account the observer's role in the group. For the way the subjects of his study define that role affects what they will tell him or let him see. If the observer carries on his research incognito, participating as a full-fledged member of the group, he will be privy to knowledge that would normally be shared by such a member and might be hidden from an outsider. He could properly interpret his own experience as that of a hypothetical "typical" group member. On the other hand, if he is known to be a researcher, he must learn how group members define him and in particular whether or not they believe that certain kinds of information and events should be kept hidden from him. He can interpret evidence more accurately when the answers to these questions are known.

Checking the Frequency and Distribution of Phenomena

The observer, possessing many provisional problems, concepts, and indicators, now wishes to know which of these are worth pursuing as major foci of his study. He does this, in part, by discovering if the events that prompted their development are typical and widespread, and by seeing how these events are distributed among categories of people and organizational sub-units. He reaches conclusions that are essentially quantitative, using them to describe the organization he is studying.

Participant observations have occasionally been gathered in standardized form capable of being transformed into legitimate statistical data.[8] But the exigencies of the field usually prevent the collection of data in such a form as to meet the assumptions of statistical tests, so that the observer deals in what have been called "quasi-statistics."[9] His conclusions, while implicitly numerical, do not require precise quantification. For instance, he may conclude that members of freshmen medical fraternities typically sit together during lectures while other students sit in less stable smaller groupings. His observations may indicate such a wide disparity between the two groups in this respect that the inference is warranted without a standardized counting operation. Occasionally, the field situation may permit him to make similar observations or ask similar questions of many people, systematically searching for quasi-statistical support for a conclusion about frequency or distribution.

In assessing the evidence for such a conclusion the observer takes a cue from his statistical colleagues. Instead of arguing that a conclusion is either totally true or false, he decides, if possible, how *likely* it is that his conclusion about the frequency or distribution of some phenomenon is an accurate quasi-statistic, just as the statistician decides, on the basis of the varying values of a correlation coefficient or a significance figure, that his conclusion is more or less likely to be accurate. The kind of evidence may vary considerably and the degree of the observer's confidence in the conclusion will vary accordingly. In arriving at this assessment, he makes use of some of the criteria described above, as well as those adopted from quantitative techniques.

Suppose, for example, that the observer concludes that medical students share the perspective that their school should provide them with the clinical experience and the practice in techniques necessary for a general practitioner. His confidence in the conclusion would vary according to the nature of the evidence, which might take any of the following forms: (1) *Every* member of the group said, *in response to a direct question*, that this was the way he looked at the matter. (2) *Every* member of the group *volunteered* to an observer that this was how he viewed the matter. (3) *Some given proportion* of the group's members either *answered* a direct question or *volunteered* the information that he shared this perspective, but none of the others was asked or volunteered information on the subject. (4) Every member of the group was asked or volunteered information, but *some given proportion said* they viewed the matter from the differing perspective of a prospective specialist. (5) No one was asked questions or volunteered information on the subject, but *all members were observed to engage in behavior* or to make other statements from which the analyst *inferred* that the general practitioner perspective was being used by them as a basic, though unstated, premise. For example, all students might have been observed to complain that the University Hospital received too many cases of rare diseases that general practitioners rarely see. (6) *Some given proportion* of the group *was observed* using the general practitioner perspective as a basic premise in their activities, but *the rest of the group* was not observed engaging in such activities. (7) *Some proportion* of the group *was observed* engaged in activities implying the general practitioner perspective while *the remainder* of the group was observed engaged in activities implying the perspective of the prospective specialist.

The researcher also takes account of the possibility that his observations may give him evidence of different kinds on the point under consideration. Just as he is more convinced if he has many items of evidence than if he has a few, so he is more convinced of a conclusion's validity if he has *many kinds* of evidence.[10] For instance, he may be especially persuaded that a particular norm exists and affects group behavior if the norm is not only described by group members but also if he observes events in which the norm can be "seen" to operate – if, for example, students tell him that they are thinking of becoming general practitioners and he also observes their complaints about the lack of cases of common diseases in University Hospital.

The conclusiveness which comes from the convergence of several kinds of evidence reflects the fact that separate varieties of evidence can be reconceptualized as deductions from a basic proposition which have now been verified in the field. In the above case, the observer might have deduced the desire to have experience with cases like those the general practitioner treats from the desire to practice that style of medicine. Even though the deduction is made after the fact, confirmation of it buttresses the argument that the general practitioner perspective is a group norm.

It should be remembered that these operations, when carried out in the field, may be so interrupted because of imperatives of the field situation that they are not carried on as systematically as they might be. Where this is the case, the overall assessment can be postponed until the final stage of post-field work analysis.

Construction of Social System Models

The final stage of analysis in the field consists of incorporating individual findings into a generalized model of the social system or organization under study or some part of that organization.[11] The concept of social system is a basic intellectual tool of modern sociology. The kind of participant observation discussed here is related directly to this concept, explaining particular social facts by explicit reference to their involvement in a complex of interconnected variables that the observer constructs as a theoretical model of the organization. In this final stage, the observer designs a descriptive model which best explains the data he has assembled.

The typical conclusion of this stage of the research is a statement about a set of complicated interrelations among many variables. Although some progress is being made in formalizing this operation through use of factor analysis and the relational analysis of survey data,[12] observers usually view currently available statistical techniques as inadequate to express their conceptions and find it necessary to use words. The most common kinds of conclusions at this level include:

(1) Complex statements of the necessary and sufficient conditions for the existence of some phenomenon. The observer may conclude, for example, that medical students develop consensus about limiting the amount of work they will do because (a) they are faced with a large amount of work, (b) they engage in activities which create communication channels between all members of the class, and (c) they face immediate dangers in the form of examinations set by the faculty.

(2) Statements that some phenomenon is an "important" or "basic" element in the organization. Such conclusions, when elaborated, usually point to the fact that this phenomenon exercises a persistent and continuing influence on diverse events. The observer might conclude that the ambition to become a general practitioner is "important" in the medical school under study, meaning that many particular judgments and choices are made by students in terms of this ambition and many features of the school's organization are arranged to take account of it.

(3) Statements identifying a situation as an instance of some process or phenomenon described more abstractly in sociological theory. Theories posit relations between many abstractly defined phenomena, and conclusions of this kind imply that relationships posited in generalized form hold in this particular instance. The observer, for example, may state that a cultural norm of the medical students is to express a desire to become a general practitioner; in so doing, he in effect asserts that the sociological theory about the functions of norms and the processes by which they are maintained which he holds to be true in general is true in this case.

In reaching such types of conclusions, the observer characteristically begins by constructing models of parts of the organization as he comes in contact with them, discovers concepts and problems, and the frequency and distribution of the phenomena these call to his attention. After constructing a model specifying the relationships among various elements of this part of the organization, the observer seeks greater accuracy by successively refining the model to take account of evidence which does not fit his previous formulation;[13] by searching for negative cases (items of evidence which run counter to the relationships hypothesized in the model) which might force such revision; and by searching intensively for the interconnections *in vivo* of the various elements he has conceptualized from his data. While a provisional model may be shown to be defective by a negative instance which crops up unexpectedly in the course of the field work, the observer may infer what kinds of evidence would be likely to support or to refute his model and may make an intensive search for such evidence.[14]

After the observer has accumulated several partial-models of this kind, he seeks connections between them and thus begins to construct an overall model of the entire organization. An example from our study shows how this operation is carried on during the period of field work. (The reader will note, in this example, how use is made of findings typical of earlier stages of analysis.)

When we first heard medical students apply the term "crock" to patients we made an effort to learn precisely what they meant by it. We found, through interviewing students about cases both they and the observer had seen, that the term referred in a derogatory way to patients with many subjective symptoms but no discernible physical pathology. Subsequent observations indicated that this usage was a regular feature of student behavior and thus that we should attempt to incorporate this fact into our model of student-patient behavior. The derogatory character of the term suggested in particular that we investigate the reasons students disliked these patients. We found that this dislike was related to what we discovered to be the students' perspective on medical school: the view that they were in school to get experience in recognizing and treating those common diseases most likely to be encountered in general practice. "Crocks," presumably having no disease, could furnish no such experience. We were thus led to specify connections between the student-patient relationship and the student's view of the purpose of his professional education. Questions concerning the genesis of this perspective led to discoveries about the organization of the student body and communication among students, phenomena which we had been assigning to another part-model. Since "crocks" were also disliked because they gave the student no opportunity to assume medical responsibility, we were able to connect this aspect of the student-patient relationship with still another tentative model of the value system and hierarchical organization of the school, in which medical responsibility plays an important role.

Again, it should be noted that analysis of this kind is carried on in the field as time permits. Since the construction of a model is the analytic operation most closely related to the observer's techniques and interests he usually spends a great deal of time thinking about these problems. But he is usually unable to be as systematic as he would like until he reaches the final stage of analysis.

Final Analysis and the Presentation of Results

The final systematic analysis, carried on after the field work is completed, consists of rechecking and rebuilding models as carefully and with as many safeguards as the data will allow. For instance, in checking the accuracy of statements about the frequency and distribution of events, the researcher can index and arrange his material so that every item of information is accessible and taken account of in assessing the accuracy of any given conclusion. He can profit from the observation of Lazarsfeld and Barton that the "analysis of 'quasi-statistical data' can probably be made more systematic than it has been in the past, if the logical structure of quantitative research at least is kept in mind to give general warnings and directions to the qualitative observer."[15]

An additional criterion for the assessment of this kind of evidence is the state of the observer's conceptualization of the problem at the time the item of evidence was gathered. The observer may have his problem well worked out and be actively looking for evidence to test an hypothesis, or he may not be as yet aware of the problem. The evidential value of items in his field notes will vary accordingly, the basis of consideration being the likelihood of discovering negative cases of the proposition he eventually uses the material to establish. The best evidence may be that gathered in the most unthinking fashion, when the observer has simply recorded the item although it has no place in the system of concepts and hypotheses he is working with at the time, for there might be less bias produced by the wish to substantiate or repudiate a particular idea. On the other hand, a well-formulated hypothesis makes possible a deliberate search for negative cases, particularly when other knowledge suggests likely areas in which to look for such evidence. This kind of search requires advanced conceptualization of the problem, and evidence gathered in this way might carry greater weight for certain kinds of conclusions. Both procedures are relevant at different stages of the research.

In the post field work stage of analysis, the observer carries on the model building operation more systematically. He considers the character of his conclusions and decides on the kind of evidence that might cause their rejection, deriving further tests by deducing logical consequences and ascertaining whether or not the data support the deductions. He considers reasonable alternative hypotheses and whether or not the evidence refutes them.[16] Finally, he completes the job of establishing interconnections between partial models so as to achieve an overall synthesis incorporating all conclusions.

After completing the analysis, the observer faces the knotty problem of how to present his conclusions and the evidence for them. Readers of qualitative research reports commonly and justifiably complain that they are told little or nothing about the evidence for conclusions or the operations by which the evidence has been assessed. A more adequate presentation of the data, of the research operations, and of the researcher's inferences may help to meet this problem.

But qualitative data and analytic procedures, in contrast to quantitative ones, are difficult to present adequately. Statistical data can be summarized in tables, and descriptive measures of various kinds and the methods by which they are handled can often be accurately reported in the space required to print a formula. This is so in part because the methods have been systematized so that they can be referred to in this shorthand fashion and in part because the data have been collected for a fixed, usually small, number of categories – the presentation of data need be nothing more than a report of the number of cases to be found in each category.

The data of participant observation do not lend themselves to such ready summary. They frequently consist of many different kinds of observations which cannot be simply categorized and counted without losing some of their value as evidence – for, as we have seen, many points need to be taken into account in putting each datum to use. Yet it is clearly out of the question to publish all the evidence. Nor is it any solution, as Kluckhohn has suggested for the similar problem of presenting life history materials,[17] to publish a short version and to make available the entire set of materials on microfilm or in some other inexpensive way; this ignores the problem of how to present *proof*.

In working over the material on the medical school study a possible solution to this problem, with which we are experimenting, is a description of the natural history of our conclusions, presenting the evidence as it came to the attention of the observer during the successive stages of his conceptualization of the problem. The term "natural history" implies not the presentation of every datum, but only the characteristic forms data took at each stage of the research. This involves description of the form that data took and any significant exceptions, taking account of the canons discussed above, in presenting the various statements of findings and the inferences and conclusions drawn from them. In this way, evidence is assessed as the substantive analysis is presented. The reader would be able, if this method were used, to follow the details of the analysis and to see how and on what basis any conclusion was reached. This would give the reader, as do present modes of statistical presentation, opportunity to make his own judgment as to the adequacy of the proof and the degree of confidence to be assigned the conclusion.

Conclusion

I have tried to describe the analytic field work characteristic of participant observation, first, in order to bring out the fact that the technique consists of something more than merely immersing oneself in data and "having insights". The discussion may also serve to stimulate those who work with this and similar techniques to attempt greater formalization and systematization of the various operations they use, in order that qualitative research may become more a "scientific" and less an "artistic" kind of endeavor. Finally, I have proposed that new modes of reporting results be introduced, so that the reader is given greater access to the data and procedures on which conclusions are based.

Notes

1. There is little agreement on the specific referent of the term *participant observation*. See Raymond L. Gold, "Roles in Sociological Field Observations," *Social Forces*, 36 (March, 1958), pp. 217–223, for a useful classification of the various procedures that go by this name. Our own research, from which we have drawn our illustrations, falls under Gold's type, "participant-as-observer." The basic methods discussed here, however, would appear to be similar in other kinds of field situations.
2. In this respect, the analytic methods I discuss bear a family resemblance to the technique of *analytic induction*. Cf. Alfred Lindesmith, *Opiate Addiction* (Bloomington: Principia Press, 1947), especially pp. 5–20, and the subsequent literature cited in Ralph H. Turner, "The Quest for Universal in Sociological Research," *American Sociological Review*, 18 (December, 1953), pp. 604–611.
3. My discussion of these stages is abstract and simplified and does not attempt to deal with practical and technical problems of participant observation study. The reader should keep in mind that in practice the research will involve all these operations simultaneously with reference to different particular problems.
4. The examples of which our hypothetical observer makes use are drawn from our own current work with medical students.
5. The problem of indicators is discussed by Paul F. Lazarsfeld and Allen Barton, "Qualitative Measurement in the Social Sciences: Classification, Typologies, and Indices," in Daniel Lerner and Harold D. Lasswell, editors, *The Policy Sciences: Recent Developments in Scope and Method*, Stanford: Stanford University Press, 1951, pp. 155–192; "Some Functions of Qualitative Analysis in Sociological Research," *Sociologica*, 1 (1955), pp. 324–361 (this important paper parallels the present discussion in many places); and Patricia L. Kendall and Paul F. Lazarsfeld, "Problems of Survey Analysis," in R. K. Merton and P. F. Lazarsfeld, editors, *Continuities in Social Research*, Glencoe: Free Press, 1950, pp. 183–186.
6. Cf. Louis Gottschalk, Clyde Kluckhohn, and Robert Angell, *The Use of Personal Documents in History, Anthropology, and Sociology*, New York: Social Science Research Council, 1945, pp. 15–27, 38–47.
7. See further, Howard S. Becker, "Interviewing Medical Students," *American Journal of Sociology*, 62 (September, 1956), pp. 199–201.
8. See Peter M. Blau, "Co-operation and Competition in a Bureaucracy," *American Journal of Sociology*, 59 (May, 1954), pp. 530–535.
9. See the discussion of quasi-statistics in Lazarsfeld and Barton, "Some Functions of Qualitative Analysis . . .," *op. cit.*, pp. 346–348.
10. See Alvin W. Gouldner, *Patterns of Industrial Bureaucracy*, Glencoe, Ill.: Free Press, 1954, pp. 247–269.
11. The relation between theories based on the concept of social system and participant observation was pointed out to me by Alvin W. Gouldner. See his "Some Observations on Systematic Theory, 1945–55," in Hans L. Zetterberg, editor, *Sociology in the United States of America*, Paris: UNESCO, 1956, pp. 34–42; and "Theoretical Requirements

of the Applied Social Sciences," *American Sociological Review,* 22 (February, 1957), pp. 92–102.

12. See Alvin W. Gouldner, "Cosmopolitans and Locals: Toward an Analysis of Latent Social Roles," *Administrative Science Quarterly,* 2 (December, 1957), pp. 281–306, and 3 (March, 1958), pp. 444–480; and James Coleman, "Relational Analysis: The Study of Social Structure with Survey Methods," mimeographed.
13. Note again the resemblance to analytic induction.
14. See Alfred Lindesmith's discussion of this principle in "Comment on W. S. Robinson's 'The Logical Structure of Analytic Induction,'" *American Sociological Review,* 17 (August, 1952), pp. 492–493.
15. "Some Functions of Qualitative Analysis . . .," *op. cit.,* p. 348.
16. One method of doing this, particularly adapted to testing discrete hypotheses about change in individuals or small social units (though not in principle limited to this application), is "The Technique of Discerning," described by Mirra Komarovsky in Paul F. Lazarsfeld and Morris Rosenberg, editors, *The Language of Social Research,* Glencoe, Ill.: Free Press, 1955, pp. 449–457. See also the careful discussion of alternative hypotheses and the use of deduced consequences as further proof in Lindesmith, *Opiate Addiction, passim.*
17. Gottschalk, Kluckhohn, and Angell, *op. cit.,* pp. 150–156.

49

Part of the Action, or 'Going Native'? Learning to Cope with the 'Politics of Integration'

Duncan Fuller

Introduction

Within a discipline that has displayed sporadic interest in matters of relevance, social responsibility and academic activism (see Johnston 1997), limited attempts have been made to address the politics of integration within a 'researched community' – issues and worries that arise (Blomley 1994) when the researcher (sometimes unexpectedly, sometimes through choice) becomes actively involved within the community or group that (s)he was supposed to be 'studying'. When such involvement occurs, anxieties ensue. There is perhaps a tacit notion that to be 'committed', to be inside the group and work with it, results in the wholesale adoption of an uncritical, unquestioning position of approval in relation to that group and its actions; thus the standard of the research becomes questioned, its validity threatened (Stanley and Wise 1993). The inclusion within the research of the 'researcher as person' is interpreted as an apparent inability to distance him/herself from the events in which (s)he is participating, ultimately undermining the authority of the voice of the 'researcher as academic'.

In simple terms, such issues have at their core the process commonly referred to as 'going native'. This is indicative of the development of a sense of over-rapport between the researcher and those under study (Armstrong

Source: *Area*, 31(3) (1999): 221–227.

1993, 18). Hobbs (1988), Armstrong (1993) and May (1993) have all recounted how they were saved from reaching such a point by 'going academic'; but whilst this might be seen to be preferable to their potential new careers (as petty criminal, football hooligan and probation officer respectively), this paper argues for greater awareness of another route through such troubles. In particular, I hope to illustrate how mixing and manipulating the researcher's various identities, and the transparent and overt recognition and awareness of multiple positionality (as person, as academic, as activist) can instead benefit geographical ethnographic research (and researchers) in three main ways.

First, I want to suggest that the process of achieving such an aim, of being comfortable within the integrated position, and of being assured that the integrity of the research and the social group in question is not being compromised, is a journey laden with difficulties of both a personal and professional nature. This paper seeks to elucidate this journeying, as I believe transparency of thought and reflection will improve the design, implementation and documentation of ethnographic research within geography. I consider the political, emotional and ethical issues that have arisen through my involvement as public relations officer inside the credit union development movement in Kingston upon Hull (the 'researched community'), and the process of learning to cope with the constant repositioning of identity, reassessment of motives and multiplicity of roles necessitated by such involvement (the politics of integration). Secondly, and as a consequence of this reflexive practice, I argue that a further layer of professional accountability is added to the research project, as the researcher interacts with his/her positioning and role(s). Thirdly, I note that this critical engagement may also allow academics to play a greater role in effecting social change – a role that has perhaps been limited by the degree to which such allegiances are seen within wider academic circles to be methodologically and contextually acceptable.

Positionality, Situatedness and Relevance

During the course of my research, I became increasingly aware that 'the personal is the political' (Stanley and Wise 1993). Routledge (1996) has noted how political awareness can come through personal involvement, and that strategies of involvement and integration enable the deconstruction of the all-too-apparent barriers between the academy and 'the lives of the people it professes to represent' (Katz 1994, 73). In such a manner, academics can strive to ensure that 'scholarly work interprets and effects social change' (Katz 1994, 73).

Clearly, concern for the role, situatedness and positionality of the researcher is hardly new (see Johnston 1997, Chapter 9). The interface between geographical research and social action was explored to some extent within the radical geographies of the early 1970s with the 'revolution of relevance' (Dickinson and Clarke 1972) and the 'revolution of social responsibility' (Prince 1971). Smith (1976, 84) noted at the time how,

> [t]he real world requires involvement in social change, for we are among the 'actors' ourselves. As part of the problem, we must participate in the solution.

In many ways, however, participation (specifically that related to notions of action and activism) appeared restrained, particularly in terms of direct contact and cooperation with those deemed to be under study.

Bunge (1971; 1973a; 1973b) represents one exception to this general rule, as a member of the academy who did take the step 'onto the streets', and in so doing helped identify and embody many of the issues that can arise when such a move is made.[1] In developing a geobiography of his home area in Detroit, Bunge arguably achieved a,

> redefinition of the research problematic and intellectual commitment of the researcher away from a smug campus career, to one incorporating a dedicated community perspective which pivots around what Howe (1954), in another context, called a 'spirit of iconoclasm'. (Merrifield 1995, 57)

This redefinition was, however, an uncomfortable process. Bunge's 'critical positioning' (Merrifield 1995, 52) manifested itself through an awareness that his 'life had been spent buried in books' (Bunge 1979, 170, cited in Merrifield 1995, 53). As Merrifield highlights, the interaction between Bunge and the Detroit locals, and his integration with them, became defined in terms of survival, or 'the fragile thread binding logic, ethics and politics' (1995, 54). Through his awareness of his positionality in relation to the community, he questioned the ability of the researcher to empathize and situate him/herself within an impoverished community. This was encapsulated in his acceptance of the 'big important gaps' that would exist as a result of his inability to situate himself outside his past. In particular, the route through this survival would involve emotional difficulty, an honest political and intellectual commitment to the expedition, and dogged determination (Merrifield 1995); part of the learning was coping.

This desire to effect social change through our privileged positions as academics is revealed more recently by the works of Katz (1992), Blomley (1994), Routledge (1996) and Kitchin (1999). In calling for a greater degree of engagement between those within the academy and those on the 'outside', these works are underlain by the belief that academic disengagement is tantamount to complicity in oppression:

> To ignore our implications in the strategies of domination that inhere in all modes of representation is not only irresponsible, but ultimately disables

> political engagement with the structures of dominance and power. (Katz 1992, 496)

In order for the academic actively to counter oppression, (s)he must occupy a space in which the situatedness of our knowledges and positionalities is constantly renegotiated and critically engaged with. This space necessarily involves the removal of artificial boundaries between researcher, activist, teacher and person, and proposes instead movement between these various identities in order to facilitate engagement between and within them (Routledge 1996). This critical engagement involves a continual questioning of the researcher's social location (in terms of class, gender, ethnicity and so on), in addition to the physical location of their research, their disciplinary location and their political position and personality (Robinson 1994). Use of this 'space of betweeness' (Katz 1994) allows the framing, manipulation and intermixing of questions and work of substantive, theoretical and practical significance:

> By operating within these multiple contexts all the time, we may begin to learn not to displace or separate so as to see and speak, but to see, be seen, speak, listen and be heard in the multiply determined fields that we are everywhere, always in. In this way we can build a politics of engagement and simultaneously practice committed scholarship. (Katz 1994, 72)

As Katz notes, however, at the heart of these intentions lies a strange paradox. In our daily lives, we are constantly repositioning and renegotiating our identities and personalities in line with different situations, different spaces and different people, and we seem to do so relatively unproblematically; we are different people in different circumstances, we have different identities or roles in different spaces or places. When confronted with the seemingly straightforward task of moving between academic and activist identities or activities, however, a range of concerns seems to come to the fore. For instance, Blomley (1994) has noted that the relationship between activism and the academy continues to promote feelings of anxiety. In considering the 'unexplored bifurcation' in his political-academic life, Blomley talks of being 'increasingly embroiled in political activism outside the academy' (1994, 383). He has struggled with the linkages between the academic world and community activism, and argues that consideration of such issues 'takes us to the heart of many knotty and unsettling questions that also relate to academic inquiry and critical theory' (1994, 383). It can, however, be suggested that such feelings of anxiety are not really surprising, in that (to turn Stanley and Wise's 1993 statement around) 'the political' is unavoidably 'the personal'. Political actions, and the subsequent movement between academics' various identities, are informed by, and reflective of, a range of personal and emotional issues and consequences. These issues and the need to interact with them should be viewed as inseparable from wider methodological considerations.

However, as Katz also suggests, the ability to achieve such movement forms part of a learning process for the researcher, a task encapsulated within the following quote from Routledge (1996, 407):

> This involves what Probyn (1992) refers to as placing the experiences of our selves, and of our differences, to work where our voices may be able to reach, touch, change others, to the centrality of theorizing through one's feeling in and through another. It involves creating heterogeneous affinities with resisting others that are not merely confined to conferences, reading groups, and specialty groups, but which articulate alternative geographies (both personal and collective) in landscapes of our own choosing.

The constant renegotiation of positionality required by such engagement is not an easy experience to come to terms with personally or professionally, in either practical or theoretical terms. This is particularly the case when the researcher finds him/herself thrust into a participative or integrative position that had not been foreseen, intended or planned at the outset of the work; when this happens, the resultant effects (at least in my own experience) include the development of an almost persistent questioning of oneself. Theorizing this experience has not been met by sudden revelation (Doel 1998); instead, it is a painful, abject journey, most notably (and simply) through the continuous possibility that one may have been affecting/effecting one's own research agenda. However, whilst claims to objectivity and detachment are perhaps the easiest routes out of this conundrum ('going academic'?), an alternative route is for this journeying to be made more transparent – to be highlighted as an integral (or unavoidable) part of the research process. Such thoughts must pervade the research as it is undertaken, and continue to hover (Kristeva 1982) long after any withdrawal has been made. This is part of the learning process – as much reflection as it is action.

Put simply, the researcher needs to learn how to move between his/her various identities, and to be aware (to be able to read) the effects of this movement on his/her research as a whole. As a consequence, there is a clear need for such skills and their acquisition/development to be highlighted as an integral part of the learning and documentation of ethnographic method. From the supposition that the experiential nature of ethnographic work is part of its inherent strength as a research method, such anxieties can be understood (at least in part) as a product of the lack of inclusion of the researcher as person within such work, and a lack of awareness of the necessary learning required in order to come to terms with this inclusion. Such a process can be placed in the context of coming to terms with, or being more aware at the outset of, the implications of these roles within our more general identities as academics. There is a need for the researcher to consider his/her place within the research process, not least because inclusion in this process is both undeniable and unavoidable. Yes, there is a danger of over-stressing the importance of the researcher as person within our work. However, I would argue that such a danger is slight, compared with the false and misleading presentation of the researcher (and research itself) as inert, detached and neutral.

Consequently, the remainder of this paper briefly documents my learning to cope with the interaction and repositioning of my academic and activist identities. It seeks to develop the notion of a learning experience within the need to maintain a critical, multi-positioned (and repositioned) identity.

Becoming Part of the Action

My early interactions with the credit union community in Hull largely involved the textbook-style ethnography of the detached observer: sitting and observing, making no comment. However, as Stanley and Wise have argued, 'reasonable people behave in ordinary and everyday ways – unless they are odd or peculiar in some way' (1993, 156), and looking at me looking at them, I certainly felt out of place. Part of feeling 'odd' was that I could walk away at the end of any meeting or gathering and return to my own social and economic normality. However, the implications of my academic identity would also surface in time. Despite these obvious differences in experience, my relations with the credit union development community began to evolve over a period of months. Relationships that had commenced on a professional basis became increasingly personal. As I had got to know the people involved, I became increasingly interested in the development itself, and began to form an affinity with its aims and operations.

As a consequence, in 1996, I was formally invited to stand for a place on Hull Credit Union Forum's Committee, an umbrella organization that brings together existing and developing credit unions in the city in order to share issues of common concern and develop resources, training materials, publicity and other shared activities. At the time, I did not perceive that acceptance of this position would present any major problems for my work; indeed, I thought it would probably improve my understanding of events and views in Hull, and enable me to be closer to the action. However, after the main positions on the Committee had been elected, discussion began over the proposed new role of public relations officer, to handle the publicity aspect of credit union issues within the city. I admit I suggested the role, as it seemed a logical idea to have someone specifically charged with raising public awareness, and, after everyone had agreed that such a post would be beneficial, I was nominated and seconded to carry out the duties. This was also a task I happily accepted. My reaction was that I was now in a position to repay the community, not in terms of any lofty academic intervention, but through more relevant attributes such as writing and presentation skills and the use of links I had developed with a variety of other influential actors within the city. I could also act as a facilitator for linking wider practical and theoretical considerations from national and international development with local strategies. As such, I was pleased that I would have the opportunity to be useful to them – but certainly not in any heroic sense.

However, as time progressed, my position within the community was becoming more aligned to activism than it was to observing that activism. The 'political' had developed through the 'personal', requiring an increasing degree of reflection on the way my own thoughts and actions were incorporated within the operations of these groups. This was a complication that had not been expected or considered at the beginning of my research, and certainly had not been planned for at its outset. The growing anxiety surrounding my role(s) was epitomized in January 1997, when it was announced that the entirety of Hull City Council's Leisure Services department, including the city's two part-time credit union development workers, had been given 90 days notice in order to achieve the huge savings the council were aiming for before 1 April. One part of me was absolutely shocked, not to mention extremely annoyed that the council were apparently willing (as I perceived it) to sacrifice the interests of the credit union development community within the city (Fuller 1997). However, on hearing the news, I could not help another part of me thinking of the great 'copy' such an announcement would make – this was exactly the kind of material that would benefit my research story. These thoughts soon passed when I realized what the situation actually meant, not only in terms of its possible effects on the credit union movement in Hull, but also in terms of the effect on all the workers involved, and I was quickly ashamed of my initial thoughts.

Learning to Cope with the Politics of Integration

In terms of the methodological issues involved, I was becoming increasingly aware of the conflict between the detached observer and the unavoidable inclusion of the researcher. My initial attempts at understanding these feelings were encapsulated by the analogy of wearing different hats in different circumstances. However, as the situation became more complicated, with more of my time being spent working with and for the groups, the analogy shifted to being two 'heads', one the researcher's, the other the 'normal me', a problem vividly described by Stanley and Wise (1993, 156):

> [H]ow [can we] tell 'when we are experiencing things as a researcher' and 'when we are experiencing them as a person'[?] We are encouraged to believe that there is a difference between these two states of being – that we do different things, conduct ourselves differently, in each of them. If we fail to recognize our research as suitably 'objective', 'scholarly', 'non-directive', then we may fail to recognize when the research has 'begun'. Frequently research students doing ethnographic work report that none

> of the expected events and stages that they have read about have occurred to them while many that are taboo have (Georges and Jones 1980). 'Rapport' does not occur, 'overinvolvement' does, 'detachment' is lost sight of. And after this comes the problem of coping with yourself as a 'failed researcher' – usually at the point when your research has to be written up and presented in such a way that your credibility is maintained.

Over time, however, and as my experience with the situation grew, the 'hat and head' problem seemed to ease, as both 'heads' came together. I was learning how to position myself so as to be both committed and critical. At meetings, I became aware that I was not treated differently, as perhaps I had been before – I was no longer an outside observer, but was inside making a contribution. Views, opinions and strategies expressed at meetings and gatherings would be represented sympathetically but critically. Where I did not agree with what other members were saying, or directions they were taking, I voiced such concerns, whereas before I would perhaps have criticized in the safety of my notebook. I was also becoming aware that the other members at such meetings were undergoing a learning experience in terms of their (inter)relations with me, as my label changed from 'the researcher/academic' to 'fellow member'. The growing awareness of this enabled me to place my anxieties concerning positionality in some form of context: for me, anxieties arose through the relationship between me, the group and my research; for them, anxieties were cast in terms of the impact on their lives of my being there, and the positive or negative effects that my work could potentially represent. Indeed, initially, the impacts were largely perceived as negative. However, over time, and as my commitment to credit union development increased (essentially, as I proved myself), they became aware of the positive effects that my position and research could potentially bring.

This was exemplified through my interaction with one study group in particular. After my election as public relations officer, I found it particularly difficult (especially during the period when I was uncomfortable in my new role) to develop strong links and relationships with one of the city's study groups. This was mainly due to their belief that the forum failed to serve any beneficial role for study groups prior to their registration, and, as I had now taken on the forum label (rather than simply being an outside researcher), I was perceived as representing and endorsing the views and beliefs of the forum. This was a specific problem of access that I had not previously encountered.

However, as I became more comfortable with my position(s), and learnt to negotiate between my various roles, I undertook an examination of what this group felt the forum lacked in terms of facilitating study-group development. Through increased contact over a period of weeks, I attempted to emphasize my researcher-identity over my forum role, restoring their vision of me as the external researcher, probing their likes and dislikes and enabling them to articulate concerns that perhaps would not have been shared with another forum member. With this information, I would then gradually rearticulate their views through my forum role, allowing them to understand (and learn) how to use my differing identities to voice their concerns. Ultimately, these discussions fuelled a reassessment of the activities of the forum, which has acted as a catalyst for the evolution of a city-wide approach to credit union development within Hull (Fuller 1998).

Conclusions

From my initial, pre-thesis perception of research as 'smooth . . . neat, tidy and unproblematic' (Punch 1986, 13–14), I have gradually moved towards Fraser's assertion that critical social theory (and critical geography) should frame:

> its research program and its conceptual framework with an eye to the aims and activities of those oppositional movements with which it has a partisan, though not uncritical identification. (1989, 113, cited in Routledge 1996)

Indeed, Routledge goes further, arguing for engagement and collaboration, 'a situation of critical thought as action-oriented and engaged with the claims, goals, and actions of social movements' (1996, 406), and I would agree with such sentiments. Where feasible, geographers should attempt to make an active/direct difference, or at least not consciously to avoid being in such a position. Clearly there are implications for the relationship between researchers and those social movements or groups with which they have no apparent sympathy, or for researchers working in areas of the subject where the scope for increased synthesis of academic and activist perspectives appears more limited. However, it is better to engage with these implications than to use them uncritically in order to justify political neutrality and inactivity.

Such engagement, and the development of a committed, collaborative approach to political issues (in its broadest sense) ought to represent few problems, in that such aims are common features of many forms of normal social interaction (Katz 1994). However, during the course of this paper, I have suggested that becoming 'part of the action' within a researched community, specifically those associated with forms of social action, is instead fraught with difficulties. Such difficulties are related on a general level to the way in which such an integrated position – the combination of professional and personal identities – is arguably viewed in a negative manner within academic circles. This is indicated by attempts to steer researcher involvement away from the dreaded consequences of 'going native': the apparent loss of validity, integrity, criticality, necessary distance, formality and, ultimately, reputation. In this context, the integrated position is one of anxiety, in which the researcher is faced with the realization that, by being incorporated into the research community that (s)he was supposed to be studying, (s)he has effectively failed – as noted by Stanley and Wise (1993).

Faced with such a situation, however, I have argued that 'going academic' represents only one alternative. There is a space in which constant reassessment, renegotiation and repositioning of a researcher's various identities allows the development of a collaborative position from which 'the construction of flexible, practical relations of solidarity (Pfeil 1994, 225) [can be] constructed through various forms of dialogue and struggle' (Routledge 1996, 414) between those in the academy and those actively involved in effecting social change. However, the achievement of such an aim itself is fraught with difficulties of both a professional and personal nature, with anxieties developing as the personal and professional spheres of the researcher/activist are combined, intertwined, manipulated and recast. Specifically, it represents a learning process of coming to terms with such multiple positionality and repositioning, which is comparable to the learning of so many other integral facets associated with the effective use of ethnographic or participant observation approaches to research.

As such, my paper has sought to make awareness of such issues more transparent, through my own experiences working inside the credit union development movement in Hull, and the documentation of my own learning process. In so doing, it may also provide a route for academics to play a greater role in striving actively to effect social change, within a discipline that, as Blomley (1994) has argued, appears to be complicit in the paradox of educating students in the simultaneous existence of the political and the theoretical, when our actions (or lack of them) as academics often seem to indicate otherwise. Surely this is an ideal that is worth striving, learning and putting ourselves on the line for?

Note

1. Whilst the context here is situated in terms of the methodological insights that can be derived from his work, Bunge's relations with the academy itself can also be viewed as indicative (to some degree at least) of the perils of merging academic and activist roles.

References

Armstrong G (1993) 'Like that Desmond Morris?' in Hobbs R And May T (eds) *Interpreting the field: accounts of ethnography* (Oxford University Press, Oxford), 3–43

Blomley N (1994) 'Activism and the academy' *Environment and Planning D: Society and Space* 12, 383–5

Bunge W (1971) *Fitzgerald: geography of a revolution* (Schenkman, Cambridge, MA)

——— (1973a) 'Ethics and logic in geography' in Chorley R J (ed) *Directions in geography* (Methuen, London), 317–31

——— (1973b) 'The geography of human survival' *Annals of the Association of American Geographers* 63, 275–95

——— (1979) 'Perspective on theoretical geography' *Annals of the Association of American Geographers* 69, 169–74

Dickinson J P and Clarke C G (1972) 'Relevance and the newest geography' *Area* 3, 25–7

Doel M (1998) 'Sudden theory' Paper presented at the CGF-SCGRG (RGS-IBG) meeting on 'New critical geographies – theory into practice, practice into theory', Coventry University, 16 September (copy available from CGF website at www.mailbase.ac.uk/lists/crit-geog-forum/files/marcus.htm)

Fraser N (1989) *Unruly practices* (University of Minnesota Press, Minneapolis)

Fuller D (1997) 'Credit union development in Kingston upon Hull' *Regional Review* 7(1), 5–6

——— (1998) 'Credit union development: financial inclusion and exclusion' *Geoforum* 29(2), 145–58

Georges R and Jones M (1980) *People studying people* (University of California Press, Los Angeles)

Hobbs R (1988) *Doing the business* (Oxford University Press, Oxford)

Howe I (1954) 'The age of conformity' *Partisan Review* 21, 7–33

Johnston R J (1997) *Geography and geographers: Anglo-American human geography since 1945* (Arnold, London)

Katz C (1992) 'All the world is staged: intellectuals and the projects of ethnography' *Environment and Planning D: Society and Space* 10, 495–510

——— (1994) 'Playing the field: questions of fieldwork in geography' *Professional Geographer* 46(1), 73–80

Kitchin R (1999) 'Morals and ethics in geographical studies of disability' in Proctor J and Smith D (eds) *Geography and ethics: a journey through a moral terrain* (Routledge, London), 223–36

Kristeva J (1982) *Powers of horror* (Columbia University Press, New York)

May T (1993) 'Feelings matter: inverting the hidden equation' in Hobbs R and May T (eds) *Interpreting the field: accounts of ethnography* (Oxford University Press, Oxford), 69–98

Merrifield A (1995) 'Situated knowledge through exploration: reflections on Bunge's geographical expeditions' *Antipode* 27(1), 49–70

Pfeil F (1994) 'No basta teorizar: in-difference to solidarity in contemporary fiction, theory, and practice' in Grewal I and Kaplan C (eds) *Scattered hegemonies* (University of Minnesota Press, Minneapolis), 197–230

Prince H C (1971) 'America! America? Views on a melting pot 1: questions of social relevance' *Area* 3, 150–53

Probyn E (1992) 'Technologizing the self: a future anterior for cultural studies' in Grossberg L, Nelson C and Treichler P A (eds) *Cultural studies* (Routledge, New York), 501–11

Punch M (1986) *The politics and ethics of fieldwork* (Sage, Thousand Oaks, CA)

Robinson J (1994) 'White women researching/representing others: from anti-apartheid to postcolonialism' in Blunt A and Rose G (eds) *Writing, women and space* (Guilford Press, New York), 197–226

Routledge P (1996) 'The third space as critical engagement' *Antipode* 28(4), 399–419

Smith D (1976) 'To what extent is the geographer's world the real world – invited comment' *Area* 8(2), 83–4

Stanley L and Wise S (1993) *Breaking out again: feminist ontology and epistemology* (Routledge, London)

50

The Participant Observer and "Over-Rapport"

S.M. Miller

In making a field study involving a participant observer relationship, the writer made mistakes which could have been avoided if earlier researchers had pointed out these possible pitfalls.[1] The error in question is that of "over-rapport." This neologism has been adopted because no existing word expresses the idea that the researcher may be so closely related to the observed that his investigations are impeded. Studies of the participant observer method concentrate on such problems as how to gain entry and achieve rapport. The usual difficulties pointed to are those of insufficient rapport or under-rapport, for in such cases it is difficult to continue the study. But is it not possible to gain too much rapport?

In my study of local union leadership, I grew very close to the leaders. I was accepted by them, even liked by them, despite my academic background. Many personal things were told to me in a friend-to-friend relationship; undoubtedly I gained information because of this relationship which would not have been available to me in any other way.

On the other hand, once I had developed a close relationship to the union leaders I was committed to continuing it, and some penetrating lines of inquiry had to be dropped. They had given me very significant and delicate information about the internal operation of the local; to question closely their basic attitudes would open up severe conflict areas. To continue close rapport and to pursue avenues of investigation which appeared antagonistic to the union leaders was impossible. To shift to a lower level of rapport would

Source: *American Sociological Review*, 17(1) (1952): 97–99.

be difficult because such a change would induce considerable distance and distrust. It would reveal that the attitude of the participant observer to the leaders was not the same as the leaders' feelings of friendship for the observer. They accepted the observer as an individual, a friend, not as one playing a delimited social role. Friendship connotes an all-accepting attitude; to probe beneath the surface of long-believed values would break the friend-to-friend relationship. It may also be that development of a friend-to-friend relationship between the leaders and the participant observer was a means used by the former to limit the observer's investigations and criticisms. In a sense, the observer may be co-opted into friendship, a situation which may have prevailed in some studies of management-worker interaction.

Over-rapport had a second limiting effect of greater subtlety. We have been told of situations in which rapport with leaders may mean lack of rapport with rank and file individuals.[2] This situation does not merely mean that rank and file members may be diffident in articulating their grievances to "administration men," which is how the observer, who is friendly to the leaders, may appear to them. The neglected element is what happens to the observer: he first hears about things from the leaders with whom he has rapport; he develops their "set" toward problems; when he talks to rank and filers he readily accepts those of their statements which conform to articulated leadership attitudes, even when these statements are not deeply meaningful to the rank and file members. In short, the observer has become so attuned to the sentiments of the leaders that he is ill-attuned to the less clearly articulated feelings of the rank and file.

The researcher should not become a mere machine, but insituations involving overt and covert controversy, he should be wary of identifying himself symbolically and emotionally with a particular group.[3]

In other words, the participant observer relationship requires rapport combined with objectivity.[4] The achievement of such a relationship is difficult and complex. The researcher has to gauge how much rapport is necessary to get the cooperation required to continue the study.[5] In most situations, it should not be too difficult to develop this basic level of rapport. The second problem of preventing the rapport from growing to such an extent that it hinders the study is more difficult. When rapport does move beyond what is necessary for the study is difficult to decide, for rapport is more than a technique of acceptance. It involves a sensitive understanding of individuals so that one is able to make insightful analyses of behavior. To protect himself from developing impeding over-rapport, the researcher should ask himself: At what point does closeness to the subjects limit the research role? He should try to make clear that he is interested in a number of people in the particular situation, and that his research activities are his prime reasons for being present. In some cases, he must resolutely decide to prevent relationships from becoming more personal than is desirable for the development of insight and the maintenance of rapport. For the participant observer, the problem is not merely that of developing rapport; the question rather is what kind and quality of rapport are desirable?

Notes

1. The field study referred to in this note is my doctoral dissertation, *Union Structure and Industrial Relations: A Case Study of a Local Labor Union*, Princeton University, 1951 (microfilm).
2. Marvin K. Opler, review of Dorothy S. Thomas and Richard S. Nishimoto, *The Spoilage, in American Anthropologist*, No. 50 (April–June, 1948), pp. 307–10.
3. The amount of covert controversy is frequently underestimated by investigators, because they tend to accept the leading group's analysis of the situation. Industrial sociology studies, in particular, should avoid this error of perception and feeling.
4. This statement of the problem bears many similarities to that of "over-identification" of social workers with clients.
5. Researchers probably overestimate rather than underestimate the amount necessary for the continuation of the study. The desire to get along extremely well with his observed subjects may be in part a function of the insecurity of the researcher, particularly in relation to non-academic people. The anxiety of the researcher in his work is similar to the counter-transference of the therapist in a psychotherapeutic situation. This latter problem, interestingly enough, is becoming increasingly discussed in the psychoanalytical literature.

51

Role Boundaries and Paying Back: 'Switching Hats' in Participant Observation

Jacqueline E. Wade

Researcher role in participant observation has received considerable attention in the literature about field study methods. Most articles give useful counsel to researchers about establishing and maintaining good field relationships by detailing recurrent problems encountered by the observer, analyzing the sources of these problems, and underscoring mistakes to be avoided (see, for example, McCall and Simmons 1969). Related literature emphasizes researcher role and field relationships (during the initiation stage of the study as well as throughout the life of it) as primary determinants for what the observer will be able to learn (McCall and Simmons 1969; Riecken 1956; Schatzman and Strauss 1973; Schwartz and Schwartz 1955). Accordingly, reference works often highlight involvements between the observer and the observed that describe researcher-role dilemmas, issues of role choices, and the dynamics of field relationships. The reader becomes aware that these are paramount concerns that affect the observer's ability to create and sustain the stance of the objective scientist producing a study that is naturalistic, emergent, nonpreordinate, contextual, and holistic (Harvey 1982).

However, researcher roles and field relationships generally are described from the perspective of the "typical" observer – the *one-time* participant who initiates, proceeds with, completes the study, and then leaves the site.

Source: *Anthropology & Education Quarterly*, 15(3) (1984): 211–224.

Methodological guidance for researcher role-taking in studies conducted where the observer is employed and has a distinct, official role is rarely found. These dual roles create specialized tactical and structural features (as well as some role definition dilemmas) that have important implications for the field research process. In addition to developing and utilizing skills associated with entry, access, data collection practices, and human relationship strategies, the observer is compelled to devise appropriate means for "switching hats" in conjunction with the dual functioning in the field.

This article focuses on my experiences of switching hats in field relationships. It is based on examples from my field study of black undergraduate students at the University of Pennsylvania, "Penn," where I am employed as a student affairs administrator and am one of the few black faculty.

The purpose of my study was to describe the academic and sociocultural experiences of black students in the university. Since the mid-1960s, the university has launched a series of programs, services, and activities designed to respond to the specific needs and interests of its increasing population of black students. However, these students consistently have expressed dissatisfaction with their college experience at Penn and have persisted in pressing their grievances and demanding improvement.

University policymakers, planners, and other officials knew little about the nature of black students' problems from the students' point of view, nor did they understand these students' interactional processes within the Penn environment as they pursued their educational and social development goals. This research specifically examined these matters. Through rich description, the study sought to assist university policymakers in determining new policies, programs, and other alternatives needed to improve the quality of life for Penn's black students.

In light of the study's purpose, my role as a university official, with responsibilities related to institutional policies and practices that affected the lives of black students, meant that my role as a researcher became complex. This article describes my experience of balancing the two roles by illustrating instances of role switching from researcher to university official in response to the social developmental or academic interests and needs of some of the research subjects. Most prominent among the roles adopted were *counselor, informant and collaborator*, and *teacher*. This theme is amplified with examples from the literature of subject-initiated encounters that added unexpected dimensions to field relationships in the course of a field study.

Subject Use of the Field Relationship: Some Typical Accounts

Participant observers use several strategies for establishing field relationships and pursuing data collection by announcing sponsorship of the research, by assuming a particular role, and by selecting initial subjects. Often guided by the literature, observer field tactics usually are designed to determine precisely the kinds of data to be collected and how human relations will be managed to ensure a scientific study.

Yet several documents chronicling field study approaches reveal that subjects frequently use interactions with an observer for reasons not expressed in the purposes of the study. On their own initiative, subjects will shape and use observers to serve their own interests and needs and will adjust the research relationship to those ends (Gold 1958). From the subjects' point of view, the observer may become someone far different from the formally established role of researcher. Observers may find themselves responding to unanticipated demands and interests of their subjects that create a myriad of reactions and involvements. While such instances may foster the retrieval of substantive data unenvisioned by the observer in the intitial stages of a study, they heighten the reciprocal quality of the researcher-subject relationship.

The field relationship between observer and observed – a strong social tie – carries a profusion of investigative arrangements likely to affect both parties to some degree for the remainder of their lives. Even as observers employ all of the carefully reasoned strategies of their formally defined researcher role, their subjects will expect a reward to be forthcoming, which, in some way, makes their roles as subjects fulfilling. Observers cannot expect to be only on the receiving end of the participant observer process. They must consider straightforwardly what kinds of rewards they can offer their subjects – rewards that ordinarily will not prejudice or jeopardize the observer's overall relations with subjects, but rewards that can be viewed by all parties concerned as fairly distributed and of sufficient value to compensate them in some measurable sense for their time and cooperation (Cannell and Axelrod 1956; Gusfield 1955; Riecken 1956; Wax and Shapiro 1952). Yet, subjects' quests for extended involvement with the observer beyond the scope of the research effort may be, in fact, a quest for role reciprocity as *they* define and need it.

Harvey (1982), in her field study of an innovative public school program in Trinidad, elaborated on many perceptions of her research stance other than those of her official character as program evaluator. Because of additional perceptions of her role, Harvey found herself embarked on unforseen relationships with her subjects. For example, she bridged communication gaps with subjects who previously had no formal mechanism for sharing professional knowledge and expertise; she articulated problems and suggested solutions to subjects who expressed a desire for ways to improve professional performance in the setting; and she was a therapeutic listener to the stresses and frustrations suffered by subjects who needed "talking-it-through" experiences with an objective nonthreatening other.

Harvey's appropriate responses to her subjects and her extended presence in their lives both established a unique form of role reciprocation and facilitated the acquisition of data relevant to the study.

Olesen and Whittaker's study of student nurses discussed the expansion of their researcher roles as the "ultimate reality" presented by their subjects (1968). They became multifaceted personalities in the lives of their subjects, acting as friends, advisors, social catalysts, and "omniscient" individuals. These interactions fostered unexpected reciprocal connections with the subjects' world and spurred complementary commitments to the basic intentions of the research.

Anderson's study (1978) of low-income, poverty-class black men who frequented a bar and liquor store on the south side of Chicago revealed that his ongoing relationship with his subjects spawned role arrangements apart from his formal stance as an outsider learning about their social order. Anderson's acquiescence to the role dimensions of his researcher presence in the setting helped him attain massive amounts of data and promoted a deep sense of mutual obligation throughout the study.

Switching Hats in the Home Setting

Vidich asserts that the "perspectives and perceptions of social reality are shaped by the social positions and interests of both the observed and the observer as they live through a passing present" (1955:360). Accordingly, interactions between observer and subjects always are infused with some measure of shared responsibility for each other's needs that surface in a variety of ways. This aspect of field study method is crucial, particularly when the researcher is studying a social issue in a setting where she or he wears another hat. In such situations the researcher must create techniqes for managing this dual role. Researchers who elect to conduct a study in their home field must respond to the competing pressures and unanticipated demands arising from this relatively unusual position in a manner that best preserves the scientific and professional integrity of both role stances.

As a black researcher studying the world of black undergraduate students at the University of Pennsylvania, I soon discovered that balancing my dual status as faculty member and administrator and as researcher required considered tactics. Initially, I came to the fieldwork with some apprehension about my role duality. I was known to a few of my subjects because of my administrative and teaching responsibilities, and to others by reputation because of my student advocacy role. I clearly regarded my administrative status as having both positive and negative implications for the study's development and progress. On the one hand, I took advantage of my official relationship with some black students to inaugurate black student participation in the study and to gain entry into their world. The trust and acceptance of black student leaders was used to acquire general black student sanctioning of the study and enhanced my ability to communicate broadly the study's purpose.

On the other hand, I was concerned that my status as a member of the university administration could cause some black students to mistrust the intent of the study. I worried that the subjects would be overly cautious about being exploited as a source of data for the development of social science theories of no particular benefit to them. Moreover, there was a brief but intense period when several black students (especially student leaders) were besieged with death threats, overt racial slurs, and other forms of harassment by white student assailants. Additionally, a residence hall housing a large number of black students was the focal point of bomb scares. The building often was defaced by missiles of garbage and trash thrown from windows in an adjacent dormitory and a nearby fraternity house (both populated predominantly by white students), and was spray-painted with racial slurs.

These incidents precipitated high visibility for me as an administrator who had some official tasks associated with the investigation of these occurrences and caused me to reduce the time I planned to use in the research. The technique of switching hats from researcher to university official was used during this period in order to effect the appropriate performance of both roles. When I was in the dormitories (especially the black residence hall) and other centers of black student life, I openly and consistently had to state whether I was there as a researcher or as university administrator.

On some occasions when I entered the dormitory on an official matter, I was approached by distraught students requesting that I interview them so the story of their lives "in this hell-hole" could be told. While I sometimes was tempted to take this opportunity to record an emotionally charged account of the life situation of black students at Penn, I resisted and, maintaining my administrative role, suggested that such students meet with me in my office to discuss their fears and feelings of indignation.

Other times, when I joined a black student gathering at the student union or sat at their customary tables in a dining hall to collect data for the study, I encountered hushed discussions or guarded statements and superficial comments. I relied heavily on the trust and confidence of student supporters to help me navigate this kind of reception. On several occasions, I solicited their help to communicate that whatever I heard or saw as a researcher or as an official during this period (and at any other time, for that matter) would be handled discreetly and fairly. A student once objected to my presence at an informal discussion about a racial confrontation between a white male student and a black female student. Some of the men were making statements about their desire to accost the white student. As I was about to accede to the objection and take my leave, one of my student sponsors stated:

> Aw man, cool out. Jacqui's O.K. She hates this stuff as much as we do. Anyway, you just talkin', you ain't go' do nothing.

This was, and remains, my favorite among the comments sanctioning my participant observation in black student domains. Fortunately, I did not witness anything requiring an official intervention or report. If I had, I am confident that I would have watched the situation closely and offered what I considered more appropriate and socially acceptable alternatives (in terms of Penn's code of social conduct). That hypothetical occasion probably would have been the one time when my dual roles were almost indistinguishable.

From the first stages of this study through its completion, I was, at times, held to a series of characterizations by some of my subjects that signified my need to switch hats in the research context. The following illustrations portray some of the distinct properties of these role shifts and convey guidelines derived from them.

The Observer as Counselor

There were times, primarily during taped interviews with students, when some of the subjects' reflections and explanations about their lives at Penn highlighted a specific problem with which they were struggling. With each student I had a verbal agreement that I was then in her or his company only for certain matters and issues relevant to my study. Yet the nature of my questions and my pursuit of an indepth knowledge of the subject's background, aspirations, relationships, and other aspects of her or his sojourn at Penn sometimes engendered the respective roles of *counselor* and *counseled*. In these cases, I found myself responding instinctively to the student's need to be heard in a manner well beyond the research purpose. It became apparent that the student had seized an opportunity to try to work through some trouble that had surfaced by my probing, ethnographic interview. At such points, I was not in an interview relationship with the student as a subject of research; rather, I was someone whom the student chose to regard as a person who could help with a problem. Accordingly, I switched hats and reacted at some level of personal distress as I would have had the problem been brought to my attention in my office.

The following excerpt from an interview with a female junior exemplifies the quality of the role switch to counselor:

> Interviewer: It sounds like you've had a pretty interesting three years here at Penn.
>
> Student: Yes, it has been interesting but now I don't know how "pretty."
>
> Interviewer: Oh? Is there more?
>
> Student [in a low tearful tone of voice]: All of my life I've thought of myself as a person who judged people as individuals, and not by the color of their skin, and I've always thought that other people judged me that way, too.

> Interviewer: I hear something more there. What's the matter?
>
> Student: Well, lately with all that mess at DuBois [the black residence hall] and everything, I find myself really hating white people.
>
> Interviewer: Can you describe what you're feeling?
>
> Student: I don't know, it's really hard to put into words.
>
> Interviewer: Try? Take your time, I can stay longer.

The student began to pour out feelings of confusion, self-conflict, and fear associated with her generalized anger toward whites related to her knowledge of campus racial harassment. Although not personally the target of harassment, she revealed that she had been "struggling for weeks" with a sense of identification with the plight of harassed students. It was evident that my request for details about this student's ways of interacting in the Penn setting stimulated her need to engage in a counseling relationship to deal specifically with her new-found negative feelings. She agreed to subsequent meetings with me in my office, and was able to use my help to acquire a more appropriate campus resource for extensive counseling related to other issues important to her self-image and personal relationships.

Other shifts from researcher to counselor were prompted by a clear recognition that some revelations about the lives of my subjects were extremely delicate. Problems such as sexual preference conflicts, differences with parents about future life plans, feelings of depression, and other emotional conflicts emerged during research interviews. While some facets of these issues were important to the research, in these cases of role shifts the students capitalized on my research attention to receive immediate help with their own concerns. I was compelled to forego my interests in our interactions and to relate as they demanded.

In one case, the student abandoned the stated purpose of the interview soon after my arrival in his dorm room:

> I've been looking forward to your meeting with me all day. I have a lot to talk to you about. I really enjoyed talking with you in the Commons [a dining hall] the other night, and I know you can help me.

Seen by this student as a good adult listener with whom he could share and ventilate a pressing problem, I responded according to this projection of my observer presence in his life. Significantly, my response to this student's need for some form of social development counseling was a catalyst for his later reception of my initial research interest in interviewing him.

These illustrations show that workable field relationships sometimes call upon the researcher to use skills and knowledge not necessarily related to the research. This became a tenet underlying my dual role management. I had to be alert for indications that students might be receptive to me because of their own immediate life pressures and internal dilemmas. I yielded to my perception that their need sometimes was greater than my need for an organized exchange. I saw no reason to evade or ignore these unexpected encounters and willingly relaxed my initial fieldwork schedule to provide counseling assistance.

Observers who find themselves in this situation should feel free to become a mature, helpful, significant other in this fairly brief moment of human need. Indeed, observers can stand ready to give help in countering disturbances evident in their subjects' lives, especially if these troubles are stirred by the content of the ethnographic interview. However, neutral and unbiased reactions to whatever the subject divulges are still optimal aids for eliminating elements of observer distortions. Observers must feel their way through these events, guided by common sense and professional decorum.

My role management in this regard was hedged by awareness that my entry into some of my subjects' lives would engender independent impressions and ideas about my relationships to them and that they might attempt to exploit me to serve their own interests. I was comfortable in risking my objectivity and trusted that my switching hats was appropriate.

The Observer as Informant and Collaborator

The fact that my subjects were aware of my access to information at decision-making levels sometimes cast me in the role of *informant*. Throughout the study, I was conscious that the realities of my primary life on campus occasionally entered my research life in the eyes of my subjects, and I was drawn into providing them with knowledge about their immediate sociopolitical concerns.

An instance occurred in one of the dormitory lounges as I observed a small group of students discussing a campus speech by a prominent black national leader. I was pressed to give my "expert" data: current statistics on black faculty members, tenured and untenured, and their department locations; the ratio of black undergraduates to white undergraduates; and other facts about the black presence at Penn. Although I resisted becoming the informant in this case – by suggesting other officials and documents to be consulted (even this response was informing) – the students insisted on my direct participation, and I found myself liberally providing relevant data.

On other occasions, I was positioned as a *collaborator*. At a meeting of the Black Student League (the major black student activity group), a dispute arose over student apathy about the future direction of their organization. Aware that I held certain attitudes about this problem as part of my official duties, I tried to be unobtrusive, sitting at the back of the room behind a robust male student. My presence, however, was felt by the students; a few made eye contact with me or mentioned my name as "someone who knows."

As the discussion proceeded I was urged to give an opinion. I tried to resist official comment: "C'mon y'all, I'm just here to listen"; "Gimme a break, do you think I work twenty-four hours a day?" Still, I was lured into making a statement on the history of the organization, its problem of student participation, and pertinent information about comparable organizations in other institutions. This information helped validate for the students that the problem was generic, and seemed to lift the "pall of projected pessimism" (my characterization of the aura of this meeting in my notes) about the future of the organization. The meeting continued with my involvement in the deliberations and specific planning to elicit greater participation from black students. At subsequent meetings of the group, I noted that membership attendance increased and the tone of the discussions about the priority work of the organization and other related business seemed more focused on the external conditions that had an impact on their functioning rather than on unproductive intergroup conflict.

My role switch from observer to that of administrator providing information to black student groups was not a projection of a need on my part to exhibit that I was a superior, better informed character in the social world of black students. I had gone to great lengths to be on the alert for signs of biases or preconceptions I might have brought to the field so that they would not take precedence over my performance as an impartial researcher.

I understood that while I was watching and learning about my subjects' world from their point of view, they quite properly were watching and learning about me as researcher, black university administrator, and human being. It was imperative that I "be cool" (as some black students probably would term the notion of scientific detachment) and allow the subjects to lead me through Penn from *their* vantage points.

I am reminded of the directives in the literature that caution the observer to contain any expert knowledge on the subject's problems lest the observer's motives become suspect (Scott 1963; Goffman 1959). This stance was crucial to my image in the field where I had to present myself as a detached investigator *and* as the university administrator committed to the general well-being of black students.

Observers (especially in active and affective participation in a setting) can relax any stricture to be the "totally unaligned" researcher. The nature of participant observation calls on the investigator to react to subjects emotionally and behaviorally. To present oneself as an unalterably "neutral" character in the course of the subjects' life events courts an impression that the observer is gullible, amateurish, inane, or uncommitted (or some combination of these) and, thus, unworthy of subjects' attention and time. Furthermore, such a stance could convey to subjects that the observer has, in truth, a negative regard for their inner workings, thereby potentially causing inimical involvements in future areas of field relationships.

The informant/collaborator venture might be misconstrued as an abandonment of the objective stance that mobilizes participant observation. Such arguments are familiar in the historical development of this research approach (Paul 1953; Radin 1933; Vidich 1955). It seems apparent, however, that the participant observer who carries dual and equally important responsibilities in the study site should acknowledge the interdependence of roles and selectively use them on behalf of the subjects. In so doing, an observer can draw out other cogent data that advance her or his comprehension of the subjects' social circumstances.

Observer as Teacher

Closely related to my subjects' perceptions of my role as informant/collaborator were their notions about my educational expertise on the subject of institutional racism in America. Many black students were aware that I taught academic courses about racism and conducted workshops engaging both students and staff in learning about the historical and current existence of this social problem. Since my chief employment in the Office of Student Life required the development and implementation of a series of "racism awareness" programs for students, I had a certain visibility as the person who officially acted against the manifestations of racism on the campus.

This aspect of my role sometimes generated controversy centering on the assumptions contained in my discourses on American racism. Most black students knew of this controversy. Some came to my defense in a variety of ways: submitting letters to the editor of the campus paper, engaging in informal yet heated debates with white friends and associates, and discussing the issue among themselves. I made no public comment about the disagreements; however, I soon became aware that such arguments had an impact on some of my subjects and considerably affected the manner in which they responded to my requests for research time.

There were times during my participant observation when I was asked to explain a concept related to racial problems in America; to give an expert opinion on the merits of books, articles, or print media accounts on the issue; or to lecture on the problem as it pertained to blacks in general or to blacks at Penn. The following is excerpted from a taped interview with a student leader who was conducting a small survey for her sociology class on whether black women considered their race or their sex to be the greatest inhibitor in their lives:

> Student: This survey I'm doing is really blowing my mind. Let me show you some of the things people are telling me. Here's the questionnaire I sent around.
>
> Interviewer: Wow! I'm really impressed with this work. Looks like you're reaching for all of the important things. Have you begun to draw some inferences about the problem from what people are telling you thus far?

> Student: Yes I have. I know that most of the women think that racism is the biggest problem, but I'm not sure what that means yet because, also, a lot of women are downing [criticizing] the black man and talk about the sexism that these men have.
>
> Interviewer: Well, it's a complex situation. There's a good book I've just read. You can use some of its concepts to get at the complexity of this.

The student continued the discussion by giving an account of a workshop I had conducted, and, revealing gaps in her understanding of the concepts, asked me to elaborate. My tapes of this interview record my giving this student a minilecture on the nature of racism, suggestions for related readings, and some pointers on her research method. As we talked further, a new twist to my teacher role came about:

> Student: I know for myself, living here as the only black RA [Resident Advisor] in this building, the things that happen to me that I can say have to do with my race or my sex, it's the racism that comes out first. [Student gives many examples of this.] I imagine that you get a lot of that too.
>
> Interviewer: Yes, I know exactly what you mean. Throughout my life I've always felt that when I walk into a room where there are mostly white people, I'm first seen as a black, then as a woman.

I proceeded to describe my "racism versus sexism" experiences as well as the dilemmas I sometimes have felt in trying to determine which of the two, in certain situations, seemed more prevalent. This was a significant growth-producing exercise for me, especially when I realized that the student was taking notes as I spoke. We later joked with each other about our individual research agendas during this interview.

Early in this interview it was apparent that the student's major reason for meeting with me was to elicit my help as well as to have me participate in her own project. I did not view the student's motive in this situation as dishonest. I recognized it as one of her ways of "getting something for something."

During this research, I saw my standing as an academician unite my observer role with my subjects' educational interests. Clearly, some of the subjects brought their own meaning to our relationship as it impacted on their need for expanded intellectual enlightenment. My submission to this role shift was grounded in my recognition that subjects often need to engage in deliberate interactions with the observer in order to receive definitive information to increase their wisdom about their own cultural circumstances.

The role of participating scientist is not only an investigative one that may provide knowledge to the scientific community. It is a diversified role that can service the investigated's interests in educative and evaluative information. My switch to the teacher role, thereby giving scholastic advantage to some of my subjects, is indicative of this aspect of participant observation.

Observers rightfully should concern themselves with problems associated with objectivity, reliability, and validity in the course of their fieldwork. However, observers should not be over-concerned with these matters, lest they lose a natural responsiveness to their subjects' own initiated engagement with them. As an observer studying the nature of people and things in a setting in which I was also a member, I found it imperative to hold a view of myself as a thoughtful academic. This stance spurred my data gathering.

Conclusion

This discussion has focused on the technique of switching roles while conducting research in a setting where the researcher also has official duties. It featured the diversified nature of the participant observation process inherent in this type of researcher role management, and the appropriateness of observer responses to subject demands for role shifts. Utilizing some of the standard propositions regarding the multiple dimensions of participant observation, the discussion has described my approach to "switching hats."

The interactions that necessitated my role shift from researcher to university official highlight the uniqueness of research reciprocity in this dual-role capacity. Typically, the relationship between the observer and subject is weighted by some sense of mutual reward, both direct and indirect. This bond requires the observer to "pay dues" (i.e., reciprocate) for the intrusion into the subjects' life in the setting. To pay dues in the manner of particularized responses to the affective, cognitive, or social developmental needs of the subject is not an unreasonable expectation, given the observer's demands for information, time, and opinions or whatever forms the basic features of the study. This expectation seems especially appropriate if the researcher holds professional and academic credentials useful to the subjects.

My experience of role switching in response to black students' needs for my skills as counselor, informant/collaborator, and teacher, encompassed in my faculty-administrative functioning on the campus, was also my way of paying dues as a researcher. And why shouldn't I? After all, I was the "bladerunner," the procurer of racial justice on the campus," the "sister for the cause!"[1]

These phrases reveal the intrinsic perceptions of many of my subjects about my "real place" in the Penn environment. My presence as a black official in the institution's instructional and administrative structure implied something about my usefulness to the needs and interests of black students at Penn; and "Payback-is-a-dog" (to use the vernacular of some of my subjects). Indeed, I was compelled to switch hats: to allow ventilation of some of these students' personal difficulties in a manner leading to resolution; to help allay their interpersonal anxieties; and to be a reliable resource who could contribute to the educational and social growth of my subjects.

Finally, I must reiterate that, as a black researcher and a black official in the university, I was both within and outside my subjects' cultural milieu. It often was apparent that these role positions were viewed by a number of my subjects as inseparable. Although I was able to switch hats when necessary, neither of these role presentations were of a "colorless" neutrality. My sense of racial identification with the specialized needs and interests of my subjects was ever present in my fieldwork. The empathic quality of these relationships was the bridge that facilitated the management of my dual functioning in this field study and generated substantial data about the intriguing world of Penn's black students.

Note

1. Actually these were characterizations of me pronounced by one rather exhuberant black student leader as he introduced me to a group of his fraternity brothers visiting the campus – characterizations, I might add, that I felt were rather excessive. I said as much while shaking hands with these visitors. Besides, I know "BS" when I hear it!

References Cited

Anderson, Elijah 1978 A Place on the Corner. University of Chicago Press.

Cannell, Charles F., and Morris Axelrod 1956 The Respondent Reports on the Interview. American Journal of Sociology 62:177–181.

Goffman, Erving 1959 The Presentation of Self in Everyday Life. Garden City, NY: Doubleday Anchor Books.

Gold, Raymond L. 1958 Role in Sociological Field Observations. Social Forces 36:217–223.

Gusfield, Joseph G. 1955 Field Work Reciprocities in Studying a Social Movement. Human Organization 14(3):29–34.

Harvey, Claudia 1982 Roles and Relationships in the Ethnology of the Implementation of an Innovation. Unpublished paper presented at the third annual Ethnography in Education Research Forum, Graduate School of Education, University of Pennsylvania.

McCall, George J., and J.L. Simmons 1969 Issues in Participation Observation: A Text and Reader. Menlo Park, CA: Addison-Wesley.

Olesen, Virginia L., and Elvi Whittaker 1968 The Silent Dialogue: A Study in the Social Psychology of Professional Socialization. San Francisco, CA: Jossey-Bass.

Paul, Benjamin 1953 Interview Techniques and Field Relationships. *In* Anthropology Today. A.L. Kroeber, ed. Pp. 430–451. University of Chicago Press.

Radin, Paul 1933 The Method and Theory of Ethnology. New York: McGraw-Hill.

Riecken, Henry W. 1956 The Unidentified Interview. American Journal of Sociology 62:210–212.

Schatzman, Leonard, and Anselm Strauss 1973 Field Research. Englewood Cliffs, NJ: Prentice-Hall.

Schwartz, Morris S., and Charlotte G. Schwartz 1955 Problems in Participant Observation. American Journal of Sociology 60:343–354.

Scott, W. Richard 1963 Field Work in a Formal Organization: Some Dilemmas in the Role of the Observer. Human Organization 22:162–168.
Vidich, Arthur J. 1955 Participant Observation and the Collection and Interpretation of Data. American Journal of Sociology 60:354–360.
Wax, Murray, and Leopold J. Shapiro 1952 Repeated Interviewing. American Journal of Sociology 62:215–217.

52

Deep Play: Notes on the Balinese Cockfight

Clifford Geertz

The Raid

Early in April of 1958, my wife and I arrived, malarial and diffident, in a Balinese village we intended, as anthropologists, to study. A small place, about five hundred people, and relatively remote, it was its own world. We were intruders, professional ones, and the villagers dealt with us as Balinese seem always to deal with people not part of their life who yet press themselves upon them: as though we were not there. For them, and to a degree for ourselves, we were nonpersons, specters, invisible men.

We moved into an extended family compound (that had been arranged before through the provincial government) belonging to one of the four major factions in village life. But except for our landlord and the village chief, whose cousin and brother-in-law he was, everyone ignored us in a way only a Balinese can do. As we wandered around, uncertain, wistful, eager to please, people seemed to look right through us with a gaze focused several yards behind us on some more actual stone or tree. Almost nobody greeted us; but nobody scowled or said anything unpleasant to us either, which would have been almost as satisfactory. If we ventured to approach someone (something one is powerfully inhibited from doing in such an atmosphere), he moved, negligently but definitely, away. If, seated or leaning against a wall, we had him trapped, he said nothing at all, or mumbled what for the Balinese is the ultimate nonword – "yes." The indifference, of course,

Source: *The Interpretation of Cultures* (New York: Basic Books, 1973), pp. 412–453.

was studied; the villagers were watching every move we made, and they had an enormous amount of quite accurate information about who we were and what we were going to be doing. But they acted as if we simply did not exist, which, in fact, as this behavior was designed to inform us, we did not, or anyway not yet.

This is, as I say, general in Bali. Everywhere else I have been in Indonesia, and more latterly in Morocco, when I have gone into a new village, people have poured out from all sides to take a very close look at me, and, often an all-too-probing feel as well. In Balinese villages, at least those away from the tourist circuit, nothing happens at all. People go on pounding, chatting, making offerings, staring into space, carrying baskets about while one drifts around feeling vaguely disembodied. And the same thing is true on the individual level. When you first meet a Balinese, he seems virtually not to relate to you at all; he is, in the term Gregory Bateson and Margaret Mead made famous, "away."[1] Then – in a day, a week, a month (with some people the magic moment never comes) – he decides, for reasons I have never quite been able to fathom, that you *are* real, and then he becomes a warm, gay, sensitive, sympathetic, though, being Balinese, always precisely controlled, person. You have crossed, somehow, some moral or metaphysical shadow line. Though you are not exactly taken as a Balinese (one has to be born to that), you are at least regarded as a human being rather than a cloud or a gust of wind. The whole complexion of your relationship dramatically changes to, in the majority of cases, a gentle, almost affectionate one – a low-keyed, rather playful, rather mannered, rather bemused geniality.

My wife and I were still very much in the gust-of-wind stage, a most frustrating, and even, as you soon begin to doubt whether you are really real after all, unnerving one, when, ten days or so after our arrival, a large cockfight was held in the public square to raise money for a new school.

Now, a few special occasions aside, cockfights are illegal in Bali under the Republic (as, for not altogether unrelated reasons, they were under the Dutch), largely as a result of the pretensions to puritanism radical nationalism tends to bring with it. The elite, which is not itself so very puritan, worries about the poor, ignorant peasant gambling all his money away, about what foreigners will think, about the waste of time better devoted to building up the country. It sees cockfighting as "primitive," "backward," "unprogressive," and generally unbecoming an ambitious nation. And, as with those other embarrassments – opium smoking, begging, or uncovered breasts – it seeks, rather unsystematically, to put a stop to it.

Of course, like drinking during Prohibition or, today, smoking marihuana, cockfights, being a part of "The Balinese Way of Life," nonetheless go on happening, and with extraordinary frequency. And, as with Prohibition or marihuana, from time to time the police (who, in 1958 at least, were almost all not Balinese but Javanese) feel called upon to make a raid, confiscate the cocks and spurs, fine a few people, and even now and then expose some of them in the tropical sun for a day as object lessons which never, somehow, get learned, even though occasionally, quite occasionally, the object dies.

As a result, the fights are usually held in a secluded corner of a village in semisecrecy, a fact which tends to slow the action a little – not very much, but the Balinese do not care to have it slowed at all. In this case, however, perhaps because they were raising money for a school that the government was unable to give them, perhaps because raids had been few recently, perhaps, as I gathered from subsequent discussion, there was a notion that the necessary bribes had been paid, they thought they could take a chance on the central square and draw a larger and more enthusiastic crowd without attracting the attention of the law.

They were wrong. In the midst of the third match, with hundreds of people, including, still transparent, myself and my wife, fused into a single body around the ring, a superorganism in the literal sense, a truck full of policemen armed with machine guns roared up. Amid great screeching cries of "pulisi! pulisi!" from the crowd, the policemen jumped out, and, springing into the center of the ring, began to swing their guns around like gangsters in a motion picture, though not going so far as actually to fire them. The superorganism came instantly apart as its components scattered in all directions. People raced down the road, disappeared headfirst over walls, scrambled under platforms, folded themselves behind wicker screens, scuttled up coconut trees. Cocks armed with steel spurs sharp enough to cut off a finger or run a hole through a foot were running wildly around. Everything was dust and panic.

On the established anthropological principle, "When in Rome," my wife and I decided, only slightly less instantaneously than everyone else, that the thing to do was run too. We ran down the main village street, northward, away from where we were living, for we were on that side of the ring. About halfway down another fugitive ducked suddenly into a compound – his own, it turned out – and we, seeing nothing ahead of us but rice fields, open country, and a very high volcano, followed him. As the three of us came tumbling into the courtyard, his wife, who had apparently been through this sort of thing before, whipped out a table, a tablecloth, three chairs, and three cups of tea, and we all, without any explicit communication whatsoever, sat down, commenced to sip tea, and sought to compose ourselves.

A few moments later, one of the policemen marched importantly into the yard, looking for the village chief. (The chief had not only been at the fight, he had arranged it. When the truck drove up he ran to the river, stripped off his sarong, and plunged in so he could say, when at length they found him sitting there pouring water over his head, that he had been away bathing when the whole affair had occurred and was ignorant of it. They did not believe him and fined him three hundred rupiah, which the village raised collectively.) Seeing me and my wife, "White Men," there in the yard, the policeman performed a classic double take. When he found his voice again he asked, approximately, what in the devil did we think we were doing there. Our host of five minutes leaped instantly to our defense, producing an impassioned description of who and what we were, so detailed and so accurate that it was my turn, having barely communicated with a living human being save my landlord and the village chief for more than a week, to be astonished. We had a perfect right to be there, he said, looking the Javanese upstart in the eye. We were American professors; the government had cleared us; we were there to study culture; we were going to write a book to tell Americans about Bali. And we had all been there drinking tea and talking about cultural matters all afternoon and did not know anything about any cockfight. Moreover, we had not seen the village chief all day; he must have gone to town. The policeman retreated in rather total disarray. And, after a decent interval, bewildered but relieved to have survived and stayed out of jail, so did we.

The next morning the village was a completely different world for us. Not only were we no longer invisible, we were suddenly the center of all attention, the object of a great outpouring of warmth, interest, and most especially, amusement. Everyone in the village knew we had fled like everyone else. They asked us about it again and again (I must have told the story, small detail by small detail, fifty times by the end of the day), gently, affectionately, but quite insistently teasing us: "Why didn't you just stand there and tell the police who you were?" "Why didn't you just say you were only watching and not betting?" "Were you really afraid of those little guns?" As always, kinesthetically minded and, even when fleeing for their lives (or, as happened eight years later, surrendering them), the world's most poised people, they gleefully mimicked, also over and over again, our graceless style of running and what they claimed were our panic-stricken facial expressions. But above all, everyone was extremely pleased and even more surprised that we had not simply "pulled out our papers" (they knew about those too) and asserted our Distinguished Visitor status, but had instead demonstrated our solidarity with what were now our covillagers. (What we had actually demonstrated was our cowardice, but there is fellowship in that too.) Even the Brahmana priest, an old, grave, halfway-to-heaven type who because of its associations with the underworld would never be involved, even distantly, in a cockfight, and was difficult to approach even to other Balinese, had us called into his courtyard to ask us about what had happened, chuckling happily at the sheer extraordinariness of it all.

In Bali, to be teased is to be accepted. It was the turning point so far as our relationship to the community was concerned, and we were quite literally "in." The whole village opened up to us, probably more than it ever would have otherwise (I might actually never have gotten to that priest, and our accidental host became one of my best informants), and certainly very much faster. Getting caught, or almost caught, in a vice raid is perhaps not a very generalizable recipe for achieving that mysterious necessity of anthropological field work, rapport, but for me it worked very well. It led to a sudden and unusually complete acceptance into a society extremely difficult for outsiders to penetrate. It gave me the kind of immediate, inside-view grasp of an aspect of "peasant mentality" that anthropologists not fortunate enough to flee headlong with their subjects from armed authorities normally do not get. And, perhaps most important of all, for the other things might have come in other ways, it put me very quickly on to a combination emotional explosion, status war, and philosophical drama of central significance to the society whose inner nature I desired to understand. By the time I left I had spent about as much time looking into cockfights as into witchcraft, irrigation, caste, or marriage.

Of Cocks and Men

Bali, mainly because it is Bali, is a well-studied place. Its mythology, art, ritual, social organization, patterns of child rearing, forms of law, even styles of trance, have all been microscopically examined for traces of that elusive substance Jane Belo called "The Balinese Temper."[2] But, aside from a few passing remarks, the cockfight has barely been noticed, although as a popular obsession of consuming power it is at least as important a revelation of what being a Balinese "is really like" as these more celebrated phenomena.[3] As much of America surfaces in a ball park, on a golf links, at a race track, or around a poker table, much of Bali surfaces in a cock ring. For it is only apparently cocks that are fighting there. Actually, it is men.

To anyone who has been in Bali any length of time, the deep psychological identification of Balinese men with their cocks is unmistakable. The double entendre here is deliberate. It works in exactly the same way in Balinese as it does in English, even to producing the same tired jokes, strained puns, and uninventive obscenities. Bateson and Mead have even suggested that, in line with the Balinese conception of the body as a set of separately animated parts, cocks are viewed as detachable, self-operating penises, ambulant genitals with a life of their own.[4] And while I do not have the kind of unconscious material either to confirm or disconfirm this intriguing notion, the fact that they are masculine symbols par excellence is about as indubitable, and to the Balinese about as evident, as the fact that water runs downhill.

The language of everyday moralism is shot through, on the male side of it, with roosterish imagery. *Sabung*, the word for cock (and one which appears in inscriptions as early as A.D. 922), is used metaphorically to mean "hero," "warrior," "champion," "man of parts," "political candidate," "bachelor," "dandy," "lady-killer," or "tough guy." A pompous man whose behavior presumes above his station is compared to a tailless cock who struts about as though he had a large, spectacular one. A desperate man who makes a last, irrational effort to extricate himself from an impossible situation is likened to a dying cock who makes one final lunge at his tormentor to drag him along to a common destruction. A stingy man, who promises much, gives little, and begrudges that, is compared to a cock which, held by the tail, leaps at another without in fact engaging him. A marriageable young man still shy with the opposite sex or someone in a new job anxious to make a good impression is called "a fighting cock caged for the first time."[5] Court trials, wars, political contests, inheritance disputes, and street arguments are all compared to cockfights.[6] Even the very island itself is perceived from its shape as a small, proud cock, poised, neck extended, back taut, tail raised, in eternal challenge to large, feckless, shapeless Java.[7]

But the intimacy of men with their cocks is more than metaphorical, Balinese men, or anyway a large majority of Balinese men, spend an enormous amount of time with their favorites, grooming them, feeding them, discussing them, trying them out against one another, or just gazing at them with a mixture of rapt admiration and dreamy self-absorption. Whenever you see a group of Balinese men squatting idly in the council shed or along the road in their hips down, shoulders forward, knees up fashion, half or more of them will have a rooster in his hands, holding it between his thighs, bouncing it gently up and down to strengthen its legs, ruffling its feathers with abstract sensuality, pushing it out against a neighbor's rooster to rouse its spirit, withdrawing it toward his loins to calm it again. Now and then, to get a feel for another bird, a man will fiddle this way with someone else's cock for a while, but usually by moving around to squat in place behind it, rather than just having it passed across to him as though it were merely an animal.

In the houseyard, the high-walled enclosures where the people live, fighting cocks are kept in wicker cages, moved frequently about so as to maintain the optimum balance of sun and shade. They are fed a special diet, which varies somewhat according to individual theories but which is mostly maize, sifted for impurities with far more care than it is when mere humans are going to eat it, and offered to the animal kernel by kernel. Red pepper is stuffed down their beaks and up their anuses to give them spirit. They are bathed in the same ceremonial preparation of tepid water, medicinal herbs, flowers, and onions in which infants are bathed, and for a prize cock just about as often. Their combs are cropped, their plumage dressed, their spurs trimmed, and their legs massaged, and they are inspected for flaws with the squinted concentration of a diamond merchant. A man who has a passion for cocks, an enthusiast in the literal sense of the term, can spend most of his life with them, and even those, the overwhelming majority, whose passion though intense has not entirely run away with them, can and do spend what seems not only to an outsider, but also to themselves, an inordinate amount of time with them. "I am cock crazy," my landlord, a quite ordinary *afficionado* by Balinese standards, used to moan as he went to move another cage, give another bath, or conduct another feeding. "We're all cock crazy."

The madness has some less visible dimensions, however, because although it is true that cocks are symbolic expressions or magnifications of their owner's self, the narcissistic male ego writ out in Aesopian terms, they are also expressions – and rather more immediate ones – of what the Balinese regard as the direct inversion, aesthetically, morally, and metaphysically, of human status: animality.

The Balinese revulsion against any behavior regarded as animal-like can hardly be overstressed. Babies are not allowed to crawl for that reason. Incest, though hardly approved, is a much less horrifying crime than bestiality. (The appropriate punishment for the second is death by drowning, for the first being forced to live like an animal.)[8] Most demons are represented – in sculpture, dance, ritual, myth – in some real or fantastic animal form. The main puberty rite consists in filing the child's teeth so they will not look like animal fangs. Not only defecation but eating is regarded as a disgusting, almost obscene activity, to be conducted hurriedly and privately, because of its association with animality. Even falling down or any form of clumsiness is considered to be bad for these reasons. Aside from cocks and a few domestic animals – oxen, ducks – of no emotional significance, the Balinese are aversive to animals and treat their large number of dogs not merely callously but with a phobic cruelty. In identifying with his cock, the Balinese man is identifying not just with his ideal self, or even his penis, but also, and at the same time, with what he most fears, hates, and ambivalence being what it is, is fascinated by – "The Powers of Darkness."

The connection of cocks and cockfighting with such Powers, with the animalistic demons that threaten constantly to invade the small, cleared-off space in which the Balinese have so carefully built their jives and devour its inhabitants, is quite explicit. A cockfight, any cockfight, is in the first instance a blood sacrifice offered, with the appropriate chants and oblations, to the demons in order to pacify their ravenous, cannibal hunger. No temple festival should be conducted until one is made. (If it is omitted, someone will inevitably fall into a trance and command with the voice of an angered spirit that the oversight be immediately corrected.) Collective responses to natural evils – illness, crop failure, volcanic eruptions – almost always involve them. And that famous holiday in Bali, "The Day of Silence" (*Njepi*), when everyone sits silent and immobile all day long in order to avoid contact with a sudden influx of demons chased momentarily out of hell, is preceded the previous day by large-scale cockfights (in this case legal) in almost every village on the island.

In the cockfight, man and beast, good and evil, ego and id, the creative power of aroused masculinity and the destructive power of loosened animality fuse in a bloody drama of hatred, cruelty, violence, and death. It is little wonder that when, as is the invariable rule, the owner of the winning cock takes the carcass of the loser – often torn limb from limb by its enraged owner – home to eat, he does so with a mixture of social embarrassment, moral satisfaction, aesthetic disgust, and cannibal joy. Or that a man who has lost an important fight is sometimes driven to wreck his family shrines and curse the gods, an act of metaphysical (and social) suicide. Or that in seeking earthly analogues for heaven and hell the Balinese compare the former to the mood of a man whose cock has just won, the latter to that of a man whose cock has just lost.

The Fight

Cockfights (*tetadjen; sabungan*) are held in a ring about fifty feet square. Usually they begin toward late afternoon and run three or four hours until sunset. About nine or ten separate matches (*sehet*) comprise a program. Each match is precisely like the others in general pattern: there is no main match, no connection between individual matches, no variation in their format, and each is arranged on a completely ad hoc basis. After a fight has ended and the emotional debris is cleaned away – the bets have been paid, the curses cursed, the carcasses possessed – seven, eight, perhaps even a dozen men slip negligently into the ring with a cock and seek to find there a logical opponent for it. This process, which rarely takes less than ten minutes, and often a good deal longer, is conducted in a very subdued, oblique, even dissembling manner. Those not immediately involved give it at best but disguised, sidelong attention; those who, embarrassedly, are, attempt to pretend somehow that the whole thing is not really happening.

A match made, the other hopefuls retire with the same deliberate indifference, and the selected cocks have their spurs (*tadji*) affixed – razor-sharp, pointed steel swords, four or five inches long. This is a delicate job which only a small proportion of men, a half-dozen or so in most villages, know how to do properly. The man who attaches the spurs also provides them, and if the rooster he assists wins, its owner awards him the spur-leg of the victim. The spurs are affixed by winding a long length of string around the foot of the spur and the leg of the cock. For reasons I shall come to presently, it is done somewhat differently from case to case, and is an obsessively deliberate affair. The lore about spurs is extensive – they are sharpened only at eclipses and the dark of the moon, should be kept out of the sight of women, and so forth. And they are handled, both in use and out, with the same curious combination of fussiness and sensuality the Balinese direct toward ritual objects generally.

The spurs affixed, the two cocks are placed by their handlers (who may or may not be their owners) facing one another in the center of the ring.[9] A coconut pierced with a small hole is placed in a pail of water, in which it takes about twenty-one seconds to sink, a period known as a *tjeng* and marked at beginning and end by the beating of a slit gong. During these twenty-one seconds the handlers (*pengangkeb*) are not permitted to touch their roosters. If, as sometimes happens, the animals have not fought during this time, they are picked up, fluffed, pulled, prodded, and otherwise insulted, and put back in the center of the ring and the process begins again. Sometimes they refuse to fight at all, or one keeps running away, in which case they are imprisoned together under a wicker cage, which usually gets them engaged.

Most of the time, in any case, the cocks fly almost immediately at one another in a wing-beating, head-thrusting, leg-kicking explosion of animal fury so pure, so absolute, and in its own way so beautiful, as to be almost abstract, a Platonic concept of hate. Within moments one or the other drives home a solid blow with his spur. The handler whose cock has delivered the blow immediately picks it up so that it will not get a return blow, for if he does not the match is likely to end in a mutually mortal tie as the two birds wildly hack each other to pieces. This is particularly true if, as often happens, the spur sticks in its victim's body, for then the aggressor is at the mercy of his wounded foe.

With the birds again in the hands of their handlers, the coconut is now sunk three times after which the cock which has landed the blow must be set down to show that he is firm, a fact he demonstrates by wandering idly around the ring for a coconut sink. The coconut is then sunk twice more and the fight must recommence.

During this interval, slightly over two minutes, the handler of the wounded cock has been working frantically over it, like a trainer patching a mauled boxer between rounds, to get it in shape for a last, desperate try for victory. He blows in its mouth, putting the whole chicken head in his own mouth and sucking and blowing, fluffs it, stuffs its wounds with various sorts of medicines, and generally tries anything he can think of to arouse the last ounce of spirit which may be hidden somewhere within it. By the time he is forced to put it back down he is usually drenched in chicken blood, but, as in prize fighting, a good handler is worth his weight in gold. Some of them can virtually make the dead walk, at least long enough for the second and final round.

In the climactic battle (if there is one; sometimes the wounded cock simply expires in the handler's hands or immediately as it is placed down again), the cock who landed the first blow usually proceeds to finish off his weakened opponent. But this is far from an inevitable outcome, for if a cock can walk, he can fight, and if he can fight, he can kill, and what counts is which cock expires first. If the wounded one can get a stab in and stagger on until the other drops, he is the official winner, even if he himself topples over an instant later.

Surrounding all this melodrama – which the crowd packed tight around the ring follows in near silence, moving their bodies in kinesthetic sympathy with the movement of the animals, cheering their champions on with wordless hand motions, shiftings of the shoulders, turnings of the head, falling back en masse as the cock with the murderous spurs careens toward one side of the ring (it is said that spectators sometimes lose eyes and fingers from being too attentive), surging forward again as they glance off toward another – is a vast body of extraordinarily elaborate and precisely detailed rules.

These rules, together with the developed lore of cocks and cockfighting which accompanies them, are written down in palm-leaf manuscripts (*lontar; rontal*) passed on from generation to generation as part of the general legal and cultural tradition of the villages. At a fight, the umpire (*saja komong; djuru kembar*) – the man who manages the coconut – is in charge of their application and his authority is absolute. I have never seen an umpire's judgment questioned on any subject, even by the more despondent losers, nor have I ever heard, even in private, a charge of unfairness directed against one, or, for that matter, complaints, about umpires in general. Only exceptionally well trusted, solid, and, given the complexity of the code, knowledgeable citizens perform this job, and in fact men will bring their cocks only to fights presided over by such men. It is also the umpire to whom accusations of cheating, which, though rare in the extreme, occasionally arise, are referred; and it is he who in the not infrequent cases where the cocks expire virtually together decides which (if either, for, though the Balinese do not care for such an outcome, there can be ties) went first. Likened to a judge, a king, a priest, and a policeman, he is all of these, and under his assured direction the animal passion of the fight proceeds within the civic certainty of the law. In the dozens of cockfights I saw in Bali, I never once saw an altercation about rules. Indeed, I never saw an open altercation, other than those between cocks, at all.

This crosswise doubleness of an event which, taken as a fact of nature, is rage untrammeled and, taken as a fact of culture, is form perfected, defines the cockfight as a sociological entity. A cockfight is what, searching for a name for something not vertebrate enough to be called a group and not structureless enough to be called a crowd, Erving Goffman has called a "focused gathering" – a set of persons engrossed in a common How of activity and relating to one another in terms of that flow.[10] Such gatherings meet and disperse; the participants in them fluctuate; the activity that focuses them is discrete – a particulate process that reoccurs rather than a continuous one that endures. They take their form from the situation that evokes them, the floor on which they are placed, as Goffman puts it; but it is a form, and an articulate one, nonetheless. For the situation, the floor is itself created, in jury deliberations, surgical operations, block meetings, sit-ins, cockfights, by the cultural preoccupations – here, as we shall see, the celebration of status rivalry – which not only specify the focus but, assembling actors and arranging scenery, bring it actually into being.

In classical times (that is to say, prior to the Dutch invasion of 1908), when there were no bureaucrats around to improve popular morality, the staging of a cockfight was an explicitly societal matter. Bringing a cock to an important fight was, for an adult male, a compulsory duty of citizenship; taxation of fights, which were usually held on market day, was a major source of public revenue; patronage of the art was a stated responsibility of princes; and the cock ring, or *wantilan*, stood in the center of the village near those other monuments of Balinese civility – the council house, the origin temple, the marketplace, the signal tower, and the banyan tree. Today, a few special occasions aside, the newer rectitude makes so open a statement of the connection between the excitements of collective life and those of blood sport impossible, but, less directly expressed, the connection itself remains intimate and intact. To expose it, however, it is necessary to turn to the aspect of cockfighting around which all the others pivot, and through which they exercise their force, an aspect I have thus far studiously ignored. I mean, of course, the gambling.

Odds and Even Money

The Balinese never do anything in a simple way that they can contrive to do in a complicated one, and to this generalization cockfight wagering is no exception.

In the first place, there are two sorts of bets, or *toh*.[11] There is the single axial bet in the center between the principals (*toh ketengah*), and there is the cloud of peripheral ones around the ring between members of the audience (*toh kesasi*). The first is typically large; the second typically small. The first is collective, involving coalitions of bettors clustering around the owner; the second is individual, man to man. The first is a matter of deliberate, very quiet, almost furtive arrangement by the coalition members and the umpire huddled like conspirators in the center of the ring; the second is a matter of impulsive shouting, public offers, and public acceptances by the excited throng around its edges. And most curiously, and as we shall see most revealingly, *where the first is always, without exception, even money, the second, equally without exception, is never such*. What is a fair coin in the center is a biased one on the side.

The center bet is the official one, hedged in again with a webwork of rules, and is made between the two cock owners, with the umpire as overseer and public witness.[12] This bet, which, as I say, is always relatively and sometimes very large, is never raised simply by the owner in whose name it is made, but by him together with four or five, sometimes seven or eight, allies – kin, village mates, neighbors, close friends. He may, if he is not especially well-to-do, not even be the major contributor; though, if only to show that he is not involved in any chicanery, he must be a significant one.

Of the fifty-seven matches for which I have exact and reliable data on the center bet, the range is from fifteen ringgits to five hundred, with a mean at eighty-five and with the distribution being rather noticeably trimodal: small fights (15 ringgits either side of 35) accounting for about 45 percent of the total number; medium ones (20 ringgits either side of 70) for about 25 percent; and large (75 ringgits either side of 175) for about 20 percent, with a few very small and very large ones out at the extremes. In a society where the normal daily wage of a manual laborer – a brickmaker, an ordinary farmworker, a market porter – was about three ringgits a day, and considering the fact that fights were held on the average about every two-and-a-half days in the immediate area I studied, this is clearly serious gambling, even if the bets are pooled rather than individual efforts.

The side bets are, however, something else altogether. Rather than the solemn, legalistic pactmaking of the center, wagering takes place rather in the fashion in which the stock exchange used to work when it was out on the curb. There is a fixed and known odds paradigm which runs in a continuous series from ten-to-nine at the short end to two-to-one at the long: 10–9, 9–8, 8–7, 7–6, 6–5, 5–4, 4–3, 3–2, 2–1. The man who wishes to back the *underdog cock* (leaving aside how favorites, *kebut*, and underdogs, *ngai*, are established for the moment) shouts the short-side number indicating the odds he wants *to be given*. That is, if he shouts *gasal*, "five," he wants the underdog at five-to-four (or, for him, four-to-five); if he shouts "four," he wants it at four-to-three (again, he putting up the "three"); if "nine," at nine-to-eight, and so on. A man backing the favorite, and thus considering giving odds if he can get them short enough, indicates the fact by crying out the color-type of that cock – "brown," "speckled," or whatever.[13]

As odds-takers (backers of the underdog) and odds-givers (backers of the favorite) sweep the crowd with their shouts, they begin to focus in on one another as potential betting pairs, often from far across the ring. The taker tries to shout the giver into longer odds, the giver to shout the taker into shorter ones.[14] The taker, who is the wooer in this situation, will signal how large a bet he wishes to make at the odds he is shouting by holding a number of fingers up in front of his face and vigorously waving them. If the giver, the wooed, replies in kind, the bet is made; if he does not, they unlock gazes and the search goes on.

The side betting, which takes place after the center bet has been made and its size announced, consists then in a rising crescendo of shouts as backers of the underdog offer their propositions to anyone who will accept them, while those who are backing the favorite but do not like the price being offered, shout equally frenetically the color of the cock to show they too are desperate to bet but want shorter odds.

Almost always odds-calling, which tends to be very consensual in that at any one time almost all callers are calling the same thing, starts off toward the long end of the range – five-to-four or four-to-three – and then moves, also consensually, toward the short end with greater or lesser speed and to a greater or lesser degree. Men crying "five" and finding themselves answered only with cries of "brown" start crying "six," either drawing the other callers fairly quickly with them or retiring from the scene as their too-generous offers are snapped up. If the change is made and partners are still scarce, the procedure is repeated in a move to "seven," and so on, only rarely, and in the very largest fights, reaching the ultimate "nine" or "ten" levels. Occasionally, if the cocks are clearly mismatched, there may be no upward movement at all, or even a movement down the scale to four-to-three, three-to-two, very, very rarely two-to-one, a shift which is accompanied by a declining number of bets as a shift upward is accompanied by an increasing number. But the general pattern is for the betting to move a shorter or longer distance up the scale toward the, for sidebets, nonexistent pole of even money, with the overwhelming majority of bets falling in the four-to-three to eight-to-seven range.[15]

As the moment for the release of the cocks by the handlers approaches, the screaming, at least in a match where the center bet is large, reaches almost frenzied proportions as the remaining unfulfilled bettors try desperately to find a last-minute partner at a price they can live with. (Where the center bet is small, the opposite tends to occur: betting dies off, trailing into silence, as odds lengthen and people lose interest.) In a large-bet, well-made match – the kind of match the Balinese regard as "real cockfighting" – the mob scene quality, the sense that sheer chaos is about to break loose, with all those waving, shouting, pushing, clambering men is quite strong, an effect which is only heightened by the intense stillness that falls with instant suddenness, rather as if someone had turned off the current, when the slit gong sounds, the cocks are put down, and the battle begins.

When it ends, anywhere from fifteen seconds to five minutes later, *all bets are immediately paid*. There are absolutely no IOUs, at least to a betting opponent. One may, of course, borrow from a friend before offering or accepting a wager, but to offer or accept it you must have the money already in hand and, if you lose, you must pay it on the spot, before the next match begins. This is an iron rule, and as I have never heard of a disputed umpire's decision (though doubtless there must sometimes be some), I have also never heard of a welshed bet, perhaps because in a worked-up cockfight crowd the consequences might be, as they are reported to be sometimes for cheaters, drastic and immediate.

It is, in any case, this formal asymmetry between balanced center bets and unbalanced side ones that poses the critical analytical problem for a theory which sees cockfight wagering as the link connecting the fight to the wider world of Balinese culture. It also suggests the way to go about solving it and demonstrating the link.

The first point that needs to be made in this connection is that the higher the center bet, the more likely the match will in actual fact be an even one. Simple considerations of rationality suggest that. If you are betting fifteen ringgits on a cock, you might be willing to go along with even money even if you feel your animal somewhat the less promising. But if you are betting five hundred you are very, very likely to be loathe to do so. Thus, in large-bet fights, which of course involve the better animals, tremendous care is taken to see that the cocks are about as evenly matched as to size, general condition, pugnacity, and so on as is humanly possible. The different ways of adjusting the spurs of the animals are often employed to secure this. If one cock seems stronger, an agreement will be made to position his spur at a slightly less advantageous angle – a kind of handicapping, at which spur affixers are, so it is said, extremely skilled. More care will be taken, too, to employ skillful handlers and to match them exactly as to abilities.

In short, in a large-bet fight the pressure to make the match a genuinely fifty-fifty proposition is enormous, and is consciously felt as such. For medium fights the pressure is somewhat less, and for small ones less yet, though there is always an effort to make things at least approximately equal, for even at fifteen ringgits (five days' work) no one wants to make an even money bet in a clearly unfavorable situation. And, again, what statistics I have tend to bear this out. In my fifty-seven matches, the favorite won thirty-three times overall, the underdog twenty-four, a 1.4 : 1 ratio. But if one splits the figures at sixty ringgits center bets, the ratios turn out to be 1.1 : 1 (twelve favorites, eleven underdogs) for those above this line, and 1.6 : 1 (twenty-one and thirteen) for those below it. Or, if you take the extremes, for very large fights, those with center bets over a hundred ringgits the ratio is 1 : 1 (seven and seven); for very small fights, those under forty ringgits, it is 1.9 : 1 (nineteen and ten).[16]

Now, from this proposition – that the higher the center bet the more exactly a fifty-fifty proposition the cockfight is – two things more or less immediately follow: (1) the higher the center bet is, the greater the pull on the side betting toward the short-odds end of the wagering spectrum, and vice versa; (2) the higher the center bet is, the greater the volume of side betting, and vice versa.

The logic is similar in both cases. The closer the fight is in fact to even money, the less attractive the long end of the odds will appear and, therefore, the shorter it must be if there are to be takers. That this is the case is apparent from mere inspection, from the Balinese's own analysis of the matter, and from what more systematic observations I was able to collect. Given the difficulty of making precise and complete recordings of side betting, this argument is hard to cast in numerical form, but in all my cases the odds-giver, odds-taker consensual point, a quite pronounced mini-max saddle where the bulk (at a guess, two-thirds to three-quarters in most cases) of the bets are actually made, was three or four points further along the scale toward the shorter end for the large-center-bet fights than for the small ones, with medium ones generally in between. In detail, the fit is not, of course, exact, but the general pattern is quite consistent: the power of the center bet to pull the side bets toward its own even-money pattern is directly proportional to its size, because its size is directly proportional to the degree to which the cocks are in fact evenly matched. As for the volume question, total wagering is greater in large-center-bet fights because such fights are considered more "interesting," not only in the sense that they are less predictable, but, more crucially, that more is at stake in them – in terms of money, in terms of the quality of the cocks, and consequently, as we shall see, in terms of social prestige.[17]

The paradox of fair coin in the middle, biased coin on the outside is thus a merely apparent one. The two betting systems, though formally incongruent, are not really contradictory to one another, but are part of a single larger system in which the center bet is, so to speak, the "center of gravity," drawing, the larger it is the more so, the outside bets toward the short-odds end of the scale. The center bet thus "makes the game," or perhaps better, defines it, signals what, following a notion of Jeremy Bentham's, I am going to call its "depth."

The Balinese attempt to create an interesting, if you will, "deep," match by making the center bet as large as possible so that the cocks matched will be as equal and as fine as possible, and the outcome, thus, as unpredictable as possible. They do not always succeed. Nearly half the matches are relatively trivial, relatively uninteresting – in my borrowed terminology, "shallow" – affairs. But that fact no more argues against my interpretation than the fact that most painters, poets, and playwrights are mediocre argues against the view that artistic effort is directed toward profundity and, with a certain frequency, approximates it. The image of artistic technique is indeed exact: the center bet is a means, a device, for creating "interesting," "deep" matches, *not* the reason, or at least not the main reason, *why* they are interesting, the source of their fascination, the substance of their depth. The question of why such matches are interesting – indeed, for the Balinese, exquisitely absorbing – takes us out of the realm of formal concerns into more broadly sociological and social-psychological ones, and to a less purely economic idea of what "depth" in gaming amounts to.[18]

Playing with Fire

Bentham's concept of "deep play" is found in his *The Theory of Legislation*.[19] By it he means play in which the stakes are so high that it is, from his utilitarian standpoint, irrational for men to engage in it at all. If a man whose fortune is a thousand pounds (or ringgits) wages five hundred of it on an even bet, the marginal utility of the pound he stands to win is clearly less than the marginal disutility of the one he stands to lose. In genuine deep play, this is the case for both parties. They are both in over their heads. Having come together in search of pleasure they have entered into a relationship which will bring the participants, considered collectively, net pain rather than net pleasure. Bentham's conclusion was, therefore, that deep play was immoral from first principles and, a typical step for him, should be prevented legally.

But more interesting than the ethical problem, at least for our concerns here, is that despite the logical force of Bentham's analysis men do engage in such play, both passionately and often, and even in the face of law's revenge. For Bentham and those who think as he does (nowadays mainly lawyers, economists, and a few psychiatrists), the explanation is, as I have said, that such men are irrational – addicts, fetishists, children, fools, savages, who need only to be protected against themselves. But for the Balinese, though naturally they do not formulate it in so many words, the explanation lies in the fact that in such play, money is less a measure of utility, had or expected, than it is a symbol of moral import, perceived or imposed.

It is, in fact, in shallow games, ones in which smaller amounts of money are involved, that increments and decrements of cash are more nearly synonyms for utility and disutility, in the ordinary, unexpanded sense – for pleasure and pain, happiness and unhappiness. In deep ones, where the amounts of money are great, much more is at stake than material gain: namely, esteem, honor, dignity, respect – in a word, though in Bali a profoundly freighted word, status.[20] It is at stake symbolically, for (a few cases of ruined addict gamblers aside) no one's status is actually altered by the outcome of a cockfight; it is only, and that momentarily, affirmed or insulted. But for the Balinese, for whom nothing is more pleasurable than an affront obliquely delivered or more painful than one obliquely received – particularly when mutual acquaintances, undeceived by surfaces, are watching – such appraisive drama is deep indeed.

This, I must stress immediately, is *not* to say that the money does not matter, or that the Balinese is no more concerned about losing five hundred ringgits than fifteen. Such a conclusion would be absurd. It is because money *does*, in this hardly unmaterialistic society, matter and matter very much that the more of it one risks, the more of a lot of other things, such as one's pride, one's poise, one's dispassion, one's masculinity, one also risks, again only momentarily but again very publicly as well. In deep cockfights an owner and his collaborators, and, as we shall see, to a lesser but still quite real extent also their backers on the outside, put their money where their status is.

It is in large part *because* the marginal disutility of loss is so great at the higher levels of betting that to engage in such betting is to lay one's public self, allusively and metaphorically, through the medium of one's cock, on the line. And though to a Benthamite this might seem merely to increase the irrationality of the enterprise that much further, to the Balinese what it mainly increases is the meaningfulness of it all. And as (to follow Weber rather than Bentham) the imposition of meaning on life is the major end and primary condition of human existence, that access of significance more than compensates for the economic costs involved.[21] Actually, given the even-money quality of the larger matches, important changes in material fortune among those who regularly participate in them seem virtually non-existent, because matters more or less even out over the long run. It is, actually, in the smaller, shallow fights, where one finds the handful of more pure, addict-type gamblers involved – those who *are* in it mainly for the money – that "real" changes in social position, largely downward, are affected. Men of this sort, plungers, are highly dispraised by "true cockfighters" as fools who do not understand what the sport is all about, vulgarians who simply miss the point of it all. They are, these addicts, regarded as fair game for the genuine enthusiasts, those who do understand, to take a little money away from – something that is easy enough to do by luring them, through the force of their greed, into irrational bets on mismatched cocks. Most of them do indeed manage to ruin themselves in a remarkably short time, but there always seems to be one or two of them around, pawning their land and selling their clothes in order to bet, at any particular time.[22]

This graduated correlation of "status gambling" with deeper rights and, inversely, "money gambling" with shallower ones is in fact quite general. Bettors themselves form a sociomoral hierarchy in these terms. As noted earlier, at most cockfights there are, around the very edges of the cockfight area, a large number of mindless, sheer-chance-type gambling games (roulette, dice throw, coin-spin, pea-under-the-shell) operated by concessionaires. Only women, children, adolescents, and various other sorts of people who do not (or not yet) fight cocks – the extremely poor, the socially despised, the personally idiosyncratic – play at these games, at, of course, penny ante levels. Cockfighting men would be ashamed to go anywhere near them. Slightly above these people in standing are those who though they do not themselves fight cocks, bet on the smaller matches around the edges. Next, there are those who fight cocks in small, or occasionally medium matches, but have not the status to join in the large ones, though they may bet from time to time on the side in those. And finally, there are those, the really substantial members of the community, the solid citizenry around whom local life revolves, who fight in the larger fights and bet on them around the side. The focusing element in these focused gatherings, these men generally dominate and define the sport as they dominate and define the society. When a Balinese male talks, in that almost venerative way, about "the true cockfighter," the *bebatoh* ("bettor") or *djuru kurung* ("cage keeper"), it is this sort of person, not those who bring the mentality of the pea-and-shell game into the quite different, inappropriate context of the cockfight, the driven gambler (*potét*, a word which has the secondary meaning of thief or reprobate), and the wistful hanger-on, that they mean. For such a man, what is really going on in a match is something, rather closer to an *affaire d'honneur* (though, with the Balinese talent for practical fantasy, the blood that is spilled is only figuratively human) than to the stupid, mechanical crank of a slot machine.

What makes Balinese cockfighting deep is thus not money in itself, but what, the more of it that is involved the more so, money causes to happen: the migration of the Balinese status hierarchy into the body of the cockfight. Psychologically an Aesopian representation of the ideal/demonic, rather narcissistic, male self, sociologically it is an equally Aesopian representation of the complex fields of tension set up by the controlled, muted, ceremonial, but for all that deeply felt, interaction of those selves in the context of everyday life. The cocks may be surrogates for their owners' personalities, animal mirrors of psychic form, but the cockfight is – or more exactly, deliberately is made to be – a simulation of the social matrix, the involved system of cross-cutting, overlapping, highly corporate groups – villages, kingroups, irrigation societies, temple congregations, "castes" – in which its devotees live.[23] And as prestige, the necessity to affirm it, defend it, celebrate it, justify it, and just plain bask in it (but not, given the strongly ascriptive character of Balinese stratification, to seek it), is perhaps the central driving force in the society, so also – ambulant penises, blood sacrifices, and monetary exchanges aside – is it of the cockfight. This apparent amusement and seeming sport is, to take another phrase from Erving Goffman, "a status bloodbath."[24]

The easiest way to make this clear, and at least to some degree to demonstrate it, is to invoke the village whose cockfighting activities I observed the closest – the one in which the raid occurred and from which my statistical data are taken.

Like all Balinese villages, this one – Tihingan, in the Klungkung region of southeast Bali – is intricately organized, a labyrinth of alliances and oppositions. But, unlike many, two sorts of corporate groups, which are also status groups, particularly stand out, and we may concentrate on them, in a part-for-whole way, without undue distortion.

First, the village is dominated by four large, patrilineal, partly endogamous descent groups which are constantly vying with one another and form the major factions in the village. Sometimes they group two and two, or rather the two larger ones versus the two smaller ones plus all the unaffiliated people; sometimes they operate independently. There are also subfactions within them, subfactions within the subfactions, and so on to rather fine levels of distinction. And second, there is the village itself, almost entirely endogamous, which is opposed to all the other villages round about in its cockfight circuit (which, as explained, is the market region), but which also forms alliances with certain of these neighbors against certain others in various supravillage political and social contexts. The exact situation is thus, as everywhere in Bali, quite distinctive; but the general pattern of a tiered hierarchy of status rivalries between highly corporate but various based groupings (and, thus, between the members of them) is entirely general.

Consider, then, as support of the general thesis that the cockfight, and especially the deep cockfight, is fundamentally a dramatization of status concerns, the following facts, which to avoid extended ethnographic description

I shall simply pronounce to be facts – though the concrete evidence, examples, statements, and numbers that could be brought to bear in support of them, is both extensive and unmistakable:

1. A man virtually never bets against a cock owned by a member of his own kingroup. Usually he will feel obliged to bet for it, the more so the closer the kin tie and the deeper the fight. If he is certain in his mind that it will not win, he may just not bet at all, particularly if it is only a second cousin's bird or if the fight is a shallow one. But as a rule he will feel he must support it and, in deep games, nearly always does. Thus the great majority of the people calling "five" or "speckled" so demonstratively are expressing their allegiance to their kinsman, not their evaluation of his bird, their understanding of probability theory, or even their hopes of unearned income.
2. This principle is extended logically. If your kingroup is not involved you will support an allied kingroup against an unallied one in the same way, and so on through the very involved networks of alliances which, as I say, make up this, as any other, Balinese village.
3. So, too, for the village as a whole. If an outsider cock is fighting any cock from your village, you will tend to support the local one. If, what is a rarer circumstance but occurs every now and then, a cock from outside your cockfight circuit is fighting one inside it, you will also tend to support the "home bird."
4. Cocks which come from any distance are almost always favorites, for the theory is the man would not have dared to bring it if it was not a good cock, the more so the further he has come. His followers are, of course, obliged to support him, and when the more grand-scale legal cockfights are held (on holidays, and so on) the people of the village take what they regard to be the best cocks in the village, regardless of ownership, and go off to support them, although they will almost certainly have to give odds on them and to make large bets to show that they are not a cheapskate village. Actually, such "away games," though infrequent, tend to mend the ruptures between village members that the constantly occurring "home games," where village factions are opposed, rather than united, exacerbate.
5. Almost all matches are sociologically relevant. You seldom get two outsider cocks fighting, or two cocks with no particular group backing, or with group backing which is mutually unrelated in any clear way. When you do get them, the game is very shallow, betting very slow, and the whole thing very dull, with no one save the immediate principals and an addict gambler or two at all interested.
6. By the same token, you rarely get two cocks from the same group, even more rarely from the same subfaction, and virtually never from the same sub-subfaction (which would be in most cases one extended family)

fighting. Similarly, in outside village fights two members of the village will rarely fight against one another, even though, as bitter rivals, they would do so with enthusiasm on their home grounds.

7. On the individual level, people involved in an institutionalized hostility relationship, called *puik*, in which they do not speak or otherwise have anything to do with each other (the causes of this formal breaking of relations are many: wife-capture, inheritance arguments, political differences) will bet very heavily, sometimes almost maniacally, against one another in what is a frank and direct attack on the very masculinity, the ultimate ground of his status, of the opponent.
8. The center bet coalition is, in all but the shallowest games, *always* made up by structural allies – no "outside money" is involved. What is "outside" depends upon the context, of course, but given it, no outside money is mixed in with the main bet; if the principals cannot raise it, it is not made. The center bet, again especially in deeper games, is thus the most direct and open expression of social opposition, which is one of the reasons why both it and matchmaking are surrounded by such an air of unease, furtiveness, embarrassment, and so on.
9. The rule about borrowing money – that you may borrow *for* a bet but not *in* one – stems (and the Balinese are quite conscious of this) from similar considerations: you are never at the *economic* mercy of your enemy that way. Gambling debts, which can get quite large on a rather short-term basis, are always to friends, never to enemies, structurally speaking.
10. When two cocks are structurally irrelevant or neutral so far as *you* are concerned (though, as mentioned, they almost never are to each other) you do not even ask a relative or a friend whom he is betting on, because if you know how he is betting and he knows you know, and you go the other way, it will lead to strain. This rule is explicit and rigid; fairly elaborate, even rather artificial precautions are taken to avoid breaking it. At the very least you must pretend not to notice what he is doing, and he what you are doing.
11. There is a special word for betting against the grain, which is also the word for "pardon me" (*mpura*). It is considered a bad thing to do, though if the center bet is small it is sometimes all right as long as you do not do it too often. But the larger the bet and the more frequently you do it, the more the "pardon me" tack will lead to social disruption.
12. In fact, the institutionalized hostility relation, *puik*, is often formally initiated (though its causes always lie elsewhere) by such a "pardon me" bet in a deep fight, putting the symbolic fat in the fire. Similarly, the end of such a relationship and resumption of normal social intercourse is often signalized (but, again, not actually brought about) by one or the other of the enemies supporting the other's bird.

13. In sticky, cross-loyalty situations, of which in this extraordinarily complex social system there are of course many, where a man is caught between two more or less equally balanced loyalties, he tends to wander off for a cup of coffee or something to avoid having to bet, a form of behavior reminiscent of that of American voters in similar situation.[25]
14. The people involved in the center bet are, especially in deep fights, virtually always leading members of their group – kinship, village, or whatever. Further, those who bet on the side (including these people) are, as I have already remarked, the more established members of the village – the solid citizens. Cockfighting is for those who are involved in the everyday politics of prestige as well, not for youth, women, subordinates, and so forth.
15. So far as money is concerned, the explicitly expressed attitude toward it is that it is a secondary matter. It is not, as I have said, of no importance; Balinese are no happier to lose several weeks' income than anyone else. But they mainly look on the monetary aspects of the cockfight as self-balancing, a matter of just moving money around, circulating it among a fairly well-defined group of serious cockfighters. The really important wins and losses are seen mostly in other terms, and the general attitude toward wagering is not any hope of cleaning up, of making a killing (addict gamblers again excepted), but that of the horse-player's prayer: "Oh, God, please let me break even." In prestige terms, however, you do not want to break even, but, in a momentary, punctuate sort of way, win utterly. The talk (which goes on all the time) is about fights against such-and-such a cock of So-and-So which your cock demolished, not on how much you won, a fact people, even for large bets, rarely remember for any length of time, though they will remember the day they did in Pan Loh's finest cock for years.
16. You must bet on cocks of your own group aside from mere loyalty considerations, for if you do not people generally will say, "What! Is he too proud for the likes of us? Does he have to go to Java or Den Pasar [the capital town] to bet, he is such an important man?" Thus, there is a general pressure to bet not only to show that you are important locally, but that you are not so important that you look down on everyone else as unfit even to be rivals. Similarly, home team people must bet against outside cocks or the outsiders will accuse them – a serious charge – of just collecting entry fees and not really being interested in cockfighting, as well as again being arrogant and insulting.
17. Finally, the Balinese peasants themselves are quite aware of all this and can and, at least to an ethnographer, do state most of it in approximately the same terms as I have. Fighting cocks, almost every Balinese I have ever discussed the subject with has said, is like playing with fire only not getting burned. You activate village and kingroup rivalries and hostilities, but in "play" form, coming dangerously and entrancingly close to

> the expression of open and direct interpersonal and intergroup aggression (something which, again, almost never happens in the normal course of ordinary life), but not quite, because, after all, it is "only a cockfight."

More observations of this sort could be advanced, but perhaps the general point is, if not made, at least well-delineated, and the whole argument thus far can be usefully summarized in a formal paradigm:

The more a match is . . .

1. Between near status equals (and/or personal enemies)
2. Between high status individuals

The deeper the match.

The deeper the match . . .

1. The closer the identification of cock and man (or, more properly, the deeper the match the more the man will advance his best, most closely-identified-with cock).
2. The finer the cocks involved and the more exactly they will be matched.
3. The greater the emotion that will be involved and the more the general absorption in the match.
4. The higher the individual bets center and outside, the shorter the outside bet odds will tend to be, and the more betting there will be overall.
5. The less an "economic" and the more a "status" view of gaming will be involved, and the "solider" the citizens who will be gaming.[26]

Inverse arguments hold for the shallower the fight, culminating, in a reversed-signs sense, in the coin-spinning and dice-throwing amusements. For deep fights there are no absolute upper limits, though there are of course practical ones, and there are a great many legendlike tales of great Duel-in-the-Sun combats between lords and princes in classical times (for cockfighting has always been as much an elite concern as a popular one), far deeper than anything anyone, even aristocrats, could produce today anywhere in Bali.

Indeed, one of the great culture heroes of Bali is a prince, called after his passion for the sport, "The Cockfighter," who happened to be away at a very deep cockfight with a neighboring prince when the whole of his family – father, brothers, wives, sisters – were assassinated by commoner usurpers. Thus spared, he returned to dispatch the upstart, regain the throne, reconstitute the Balinese high tradition, and build its most powerful, glorious, and prosperous state. Along with everything else that the Balinese see in fighting cocks – themselves, their social order, abstract hatred, masculinity, demonic power – they also see the archetype of status virtue, the arrogant, resolute, honor-mad player with real fire, the ksatria prince.[27]

Feathers, Blood, Crowds, and Money

"Poetry makes nothing happen," Auden says in his elegy of Yeats, "it survives in the valley of its saying . . . a way of happening, a mouth." The cockfight too, in this colloquial sense, makes nothing happen. Men go on allegorically humiliating one another and being allegorically humiliated by one another, day after day, glorying quietly in the experience if they have triumphed, crushed only slightly more openly by it if they have not. *But no one's status really changes*. You cannot ascend the status ladder by winning cockfights; you cannot, as an individual, really ascend it at all. Nor can you descend it that way.[28] All you can do is enjoy and savor, or suffer and withstand, the concocted sensation of drastic and momentary movement along an aesthetic semblance of that ladder, a kind of behind-the-mirror status jump which has the look of mobility without its actuality.

Like any art form – for that, finally, is what we are dealing with – the cockfight renders ordinary, everyday experience comprehensible by presenting it in terms of acts and objects which have had their practical consequences removed and been reduced (or, if you prefer, raised) to the level of sheer appearances, where their meaning can be more powerfully articulated and more exactly perceived. The cockfight is "really real" only to the cocks – it does not kill anyone, castrate anyone, reduce anyone to animal status, alter the hierarchical relations among people, or refashion the hierarchy; it does not even redistribute income in any significant way. What it does is what, for other peoples with other temperaments and other conventions, *Lear* and *Crime and Punishment* do; it catches up these themes – death, masculinity, rage, pride, loss, beneficence, chance – and, ordering them into an encompassing structure, presents them in such a way as to throw into relief a particular view of their essential nature. It puts a construction on them, makes them, to those historically positioned to appreciate the construction, meaningful – visible, tangible, graspable – "real," in an ideational sense. An image, fiction, a model, a metaphor, the cockfight is a means of expression; its function is neither to assuage social passions nor to heighten them (though, in its playing-with-fire way it does a bit of both), but, in a medium of feathers, blood, crowds, and money, to display them.

The question of how it is that we perceive qualities in things – paintings, books, melodies, plays – that we do not feel we can assert literally to be there has come, in recent years, into the very center of aesthetic theory.[29] Neither the sentiments of the artist, which remain his, nor those of the audience, which remain theirs, can account for the agitation of one painting or the serenity of another. We attribute grandeur, wit, despair, exuberance to strings of sounds; lightness, energy, violence, fluidity to blocks of stone. Novels are said to have strength, buildings eloquence, plays momentum, ballets repose. In this realm of eccentric predicates, to say thai the cockfight, in its perfected cases at least, is "disquietful" does not seem at all unnatural, merely, as I have just denied it practical consequence, somewhat puzzling.

The disquietfulness arises, "somehow," out of a conjunction of three attributes of the fight: its immediate dramatic shape; its metaphoric content; and its social context. A cultural figure against a social ground, the fight is at once a convulsive surge of animal hatred, a mock war of symbolical selves, and a formal simulation of status tensions, and its aesthetic power derives from its capacity to force together these diverse realities. The reason it is disquietful is not that it has material effects (it has some, but they are minor); the reason that it is disquietful is that, joining pride to selfhood, selfhood to cocks, and cocks to destruction, it brings to imaginative realization a dimension of Balinese experience normally well-obscured from view. The transfer of a sense of gravity into what is in itself a rather blank and unvarious spectacle, a commotion of beating wings and throbbing legs, is effected by interpreting it as expressive of something unsettling in the way its authors and audience live, or, even more ominously, what they are.

As a dramatic shape, the fight displays a characteristic that does not seem so remarkable until one realizes that it docs not have to be there: a radically atomistical structure.[30] Each match is a world unto itself, a particulate burst of form. There is the matchmaking, there is the betting, there is the fight, there is the result – utter triumph and utter defeat – and there is the hurried, embarrassed passing of money. The loser is not consoled. People drift away from him, look around him, leave him to assimilate his momentary descent into nonbeing, reset his face, and return, scarless and intact, to the fray. Nor are winners congratulated, or events rehashed; once a match is ended the crowd's attention turns totally to the next, with no looking back. A shadow of the experience no doubt remains with the principals, perhaps even with some of the witnesses of a deep fight, as it remains with us when we leave the theater after seeing a powerful play well-performed; but it quite soon fades to become at most a schematic memory – a diffuse glow or an abstract shudder – and usually not even that. Any expressive form lives only in its own present – the one it itself creates. But, here, that present is severed into a string of flashes, some more bright than others, but all of them disconnected, aesthetic quanta. Whatever the cockfight says, it says in spurts.

But, as I have argued lengthily elsewhere, the Balinese live in spurts.[31] Their life, as they arrange it and perceive it, is less a flow, a directional movement out of the past, through the present, toward the future than an on-off pulsation of meaning and vacuity, an arhythmic alternation of short periods when "something" (that is, something significant) is happening, and equally short ones where "nothing" (that is, nothing much) is – between what they themselves call "full" and "empty" times, or, in another idiom, "junctures" and "holes." In focusing activity down to a burning-glass dot, the cockfight is merely being Balinese in the same way in which everything from the monadic encounters of everyday life, through the clanging pointillism of *gamelan* music, to the visiting-day-of-the-gods temple celebrations are. It is not an imitation of the punctuateness of Balinese social life, nor a depiction of it, nor even an expression of it; it is an example of it, carefully prepared.[32]

If one dimension of the cockfight's structure, its lack of temporal directionality, makes it seem a typical segment of the general social life, however, the other, its flat-out, head-to-head (or spur-to-spur) aggressiveness, makes it seem a contradiction, a reversal, even a subversion of it. In the normal course of things, the Balinese are shy to the point of obsessiveness of open conflict. Oblique, cautious, subdued, controlled, masters of indirection and dissimulation – what they call *alus*, "polished," "smooth" – they rarely face what they can turn away from, rarely resist what they can evade. But here they portray themselves as wild and murderous, with manic explosions of instinctual cruelty. A powerful rendering of life as the Balinese most deeply do not want it (to adapt a phrase Frye has used of Gloucester's blinding) is set in the context of a sample of it as they do in fact have it.[33] And, because the context suggests that the rendering, if less than a straightforward description, is nonetheless more than an idle fancy; it is here that the disquietfulness – the disquietfulness of the *fight*, not (or, anyway, not necessarily) its patrons, who seem in fact rather thoroughly to enjoy it – emerges. The slaughter in the cock ring is not a depiction of how things literally are among men, but, what is almost worse, of how, from a particular angle, they imaginatively are.[34]

The angle, of course, is stratificatory. What, as we have already seen, the cockfight talks most forcibly about is status relationships, and what it says about them is that they are matters of life and death. That prestige is a profoundly serious business is apparent everywhere one looks in Bali – in the village, the family, the economy, the state. A peculiar fusion of Polynesian title ranks and Hindu castes, the hierarchy of pride is the moral backbone of the society. But only in the cockfight are the sentiments upon which that hierarchy rests revealed in their natural colors. Enveloped elsewhere in a haze of etiquette, a thick cloud of euphemism and ceremony, gesture and allusion, they are here expressed in only the thinnest disguise of an animal mask, a mask which in fact demonstrates them far more effectively than it conceals them. Jealousy is as much a part of Bali as poise, envy as grace, brutality as charm; but without the cockfight the Balinese would have a much less certain understanding of them, which is, presumably, why they value it so highly.

Any expressive form works (when it works) by disarranging semantic contexts in such a way that properties conventionally ascribed to certain things are unconventionally ascribed to others, which are then seen actually to possess them. To call the wind a cripple, as Stevens does, to fix tone and manipulate timbre, as Schoenberg does, or, closer to our case, to picture an art critic as a dissolute bear, as Hogarth does, is to cross conceptual wires; the established conjunctions between objects and their qualities are altered, and phenomena – fall weather, melodic shape, or cultural journalism – are clothed in signifiers which normally point to other referents.[35] Similarly, to connect – and connect, and connect – the collision of roosters with the divisiveness of status is to invite a transfer of perceptions from the former to the latter, a transfer which is at once a description and a judgment. (Logically, the transfer could, of course, as well go the other way; but, like most of the rest of us, the Balinese are a great deal more interested in understanding men than they are in understanding cocks.)

What sets the cockfight apart from the ordinary course of life, lifts it from the realm of everyday practical affairs, and surrounds it with an aura of enlarged importance is not, as functionalist sociology would have it, that it reinforces status discriminations (such reinforcement is hardly necessary in a society where every act proclaims them), but that it provides a metasocial commentary upon the whole matter of assorting human beings into fixed hierarchical ranks and then organizing the major part of collective existence around that assortment. Its function, if you want to call it that, is interpretive: it is a Balinese reading of Balinese experience, a story they tell themselves about themselves.

Saying Something of Something

To put the matter this way is to engage in a bit of metaphorical refocusing of one's own, for it shifts the analysis of cultural forms from an endeavor in general parallel to dissecting an organism, diagnosing a symptom, deciphering a code, or ordering a system – the dominant analogies in contemporary anthropology – to one in general parallel with penetrating a literary text. If one takes the cockfight, or any other collectively sustained symbolic structure, as a means of "saying something of something" (to invoke a famous Aristotelian tag), then one is faced with a problem not in social mechanics but social semantics.[36] For the anthropologist, whose concern is with formulating sociological principles, not with promoting or appreciating cockfights, the question is, what does one learn about such principles from examining culture as an assemblage of texts?

Such an extension of the notion of a text beyond written material, and even beyond verbal, is, though metaphorical, not, of course, all that novel. The *interpretatio naturae* tradition of the middle ages, which, culminating in Spinoza, attempted to read nature as Scripture, the Nietszchean effort to treat value systems as glosses on the will to power (or the Marxian one to treat them as glosses on property relations), and the Freudian replacement of the enigmatic text of the manifest dream with the plain one of the latent, all offer precedents, if not equally recommendable ones.[37] But the idea remains theoretically undeveloped; and the more profound corollary, so far as anthropology is concerned, that cultural forms can be treated as texts, as imaginative works built out of social materials, has yet to be systematically exploited.[38]

In the case at hand, to treat the cockfight as a text is to bring out a feature of it (in my opinion, the central feature of it) that treating it as a rite or a pastime, the two most obvious alternatives, would tend to obscure: its use of emotion for cognitive ends. What the cockfight says it says in a vocabulary of sentiment – the thrill of risk, the despair of loss, the pleasure of triumph. Yet what it says is not merely that risk is exciting, loss depressing, or triumph gratifying, banal tautologies of affect, but that it is of these emotions, thus exampled, that society is built and individuals are put together. Attending cockfights and participating in them is, for the Balinese, a kind of sentimental education. What he learns there is what his culture's ethos and his private sensibility (or, anyway, certain aspects of them) look like when spelled out externally in a collective text; that the two are near enough alike to be articulated in the symbolics of a single such text; and – the disquieting part – that the text in which this revelation is accomplished consists of a chicken hacking another mindlessly to bits.

Every people, the proverb has it, loves its own form of violence. The cockfight is the Balinese reflection on theirs: on its look, its uses, its force, its fascination. Drawing on almost every level of Balinese experience, it brings together themes – animal savagery, male narcissism, opponent gambling, status rivalry, mass excitement, blood sacrifice – whose main connection is their involvement with rage and the fear of rage, and, binding them into a set of rules which at once contains them and allows them play, builds a symbolic structure in which, over and over again, the reality of their inner affiliation can be intelligibly felt. If, to quote Northrop Frye again, we go to see *Macbeth* to learn what a man feels like after he has gained a kingdom and lost his soul, Balinese go to cockfights to find out what a man, usually composed, aloof, almost obsessively self-absorbed, a kind of moral autocosm, feels like when, attacked, tormented, challenged, insulted, and driven in result to the extremes of fury, he has totally triumphed or been brought totally low. The whole passage, as it takes us back to Aristotle (though to the *Poetics* rather than the *Hermeneutics*) is worth quotation:

> But the poet [as opposed to the historian], Aristotle says, never makes any real statements at all, certainly no particular or specific ones. The poet's job is not to tell you what happened, but what happens: not what did take place, but the kind of thing that always does take place. He gives you the typical, recurring, or what Aristotle calls universal event. You wouldn't go to *Macbeth* to learn about the history of Scotland – you go to it to learn what a man feels like after he's gained a kingdom and lost his soul. When you meet such a character as Micawber in Dickens, you don't feel that there must have been a man Dickens knew who was exactly like this: you feel that there's a bit of Micawber in almost everybody you know, including yourself. Our impressions of human life are picked up one by one, and remain for most of us loose and disorganized. But we constantly find things in literature that suddenly coordinate and bring

> into focus a great many such impressions, and this is part of what Aristotle means by the typical or universal human event.[39]

It is this kind of bringing of assorted experiences of everyday life to focus that the cockfight, set aside from that life as "only a game" and reconnected to it as "more than a game," accomplishes, and so creates what, better than typical or universal, could be called a paradigmatic human event – that is, one that tells us less what happens than the kind of thing that would happen if, as is not the case, life were art and could be as freely shaped by styles of feeling as *Macbeth* and *David Copper-field* are.

Enacted and re-enacted, so far without end, the cockfight enables the Balinese, as, read and reread, *Macbeth* enables us to see a dimension of his own subjectivity. As he watches fight after fight, with the active watching of an owner and a bettor (for cockfighting has no more interest as a pure spectator sport than does croquet or dog racing), he grows familiar with it and what it has to say to him, much as the attentive listener to string quartets or the absorbed viewer of still life grows slowly more familiar with them in a way which opens his subjectivity to himself.[40]

Yet, because – in another of those paradoxes, along with painted feelings and unconsequenced acts, which haunt aesthetics – that subjectivity does not properly exist until it is thus organized, art forms generate and regenerate the very subjectivity they pretend only to display. Quartets, still lifes, and cockfights are not merely reflections of a pre-existing sensibility analogically represented; they are positive agents in the creation and maintenance of such a sensibility. If we see ourselves as a pack of Micawbers, it is from reading too much Dickens (if we see ourselves as unillusioned realists, it is from reading too little); and similarly for Balinese, cocks, and cockfights. It is in such a way, coloring experience with the light they cast it in, rather than through whatever material effects they may have, that the arts play their role, as arts, in social life.[41]

In the cockfight, then, the Balinese forms and discovers his temperament and his society's temper at the same time. Or, more exactly, he forms and discovers a particular facet of them. Not only are there a great many other cultural texts providing commentaries on status hierarchy and self-regard in Bali, but there are a great many other critical sectors of Balinese life besides the stratificatory and the agonistic that receive such commentary. The ceremony consecrating a Brahmana priest, a matter of breath control, postural immobility, and vacant concentration upon the depths of being, displays a radically different, but to the Balinese equally real, property of social hierarchy – its reach toward the numinous transcendent. Set not in the matrix of the kinetic emotionality of animals, but in that of the static passionlessness of divine mentality, it expresses tranquillity not disquiet. The mass festivals at the village temples, which mobilize the whole local population in elaborate hostings of visiting gods – songs, dances, compliments, gifts – assert the spiritual unity of village mates against their status inequality and project a mood of amity and trust.[42] The cockfight is not the master key to Balinese life, any more than bullfighting is to Spanish. What it says about that life is not unqualified nor even unchallenged by what other equally eloquent cultural statements say about it. But there is nothing more surprising in this than in the fact that Racine and Molière were contemporaries, or that the same people who arrange chrysanthemums cast swords.[43]

The culture of a people is an ensemble of texts, themselves ensembles, which the anthropologist strains to read over the shoulders of those to whom they properly belong. There are enormous difficulties in such an enterprise, methodological pitfalls to make a Freudian quake, and some moral perplexities as well. Nor is it the only way that symbolic forms can be sociologically handled. Functionalism lives, and so does psychologism. But to regard such forms as "saying something of something," and saying it to somebody, is at least to open up the possibility of an analysis which attends to their substance rather than to reductive formulas professing to account for them.

As in more familiar exercises in close reading, one can start anywhere in a culture's repertoire of forms and end up anywhere else. One can stay, as I have here, within a single, more or less bounded form, and circle steadily within it. One can move between forms in search of broader unities or informing contrasts. One can even compare forms from different cultures to define their character in reciprocal relief. But whatever the level at which one operates, and however intricately, the guiding principle is the same: societies, like lives, contain their own interpretations. One has only to learn how to gain access to them.

Notes

1. G. Bateson and M. Mead, *Balinese Character: A Photographic Analysis* (New York, 1942), p. 68.
2. J. Belo, "The Balinese Temper," in *Traditional Balinese Culture*, ed. J. Belo (New York, 1970) (originally published in 1935), pp. 85–110.
3. The best discussion of cockfighting is again Bateson and Mead's *Balinese Character*, pp. 24–25, 140; but it, too, is general and abbreviated.
4. Ibid., pp. 25–26. The cockfight is unusual within Balinese culture in being a single-sex public activity from which the other sex is totally and expressly excluded. Sexual differentiation is culturally extremely played down in Bali and most activities, formal and informal, involve the participation of men and women on equal ground, commonly as linked couples. From religion, to politics, to economics, to kinship, to dress, Bali is a rather "unisex" society, a fact both its customs and its symbolism clearly express. Even in contexts where women do not in fact play much of a role – music, painting, certain agricultural activities – their absence, which is only relative in any case, is more a mere matter of fact than socially enforced. To this general pattern, the cockfight, entirely of, by, and for men (women – at least *Balinese* women – do not even watch), is the most striking exception.

5. C. Hooykaas, *The Lay of the Java Prana* (London, 1958), p. 39. The lay has a stanza (no. 17) with the reluctant bridgegroom use. Jaya Prana, the subject of a Balinese Uriah myth, responds to the lord who has offered him the loveliest of six hundred servant girls: "Godly King, my Lord and Master/I beg you, give me leave to go/such things are not yet in my mind;/like a fighting cock encaged/indeed I am on my mettle/I am alone/as yet the flame has not been fanned."
6. For these, see V. E. Korn, *Het Adatrecht van Bali*, 2d ed. (The Hague, 1932), index under *toh*.
7. There is indeed a legend to the effect that the separation of Java and Bah is due to the action of a powerful Javanese religious figure who wished to protect himself against a Balinese culture hero (the ancestor of two Ksatria castes) who was a passionate cockfighting gambler. See C. Hooykaas, *Agama Tirtha* (Amsterdam, 1964), p. 184.
8. An incestuous couple is forced to wear pig yokes over their necks and crawl to a pig trough and eat with their mouths there. On this, see J. Belo, "Customs Pertaining to Twins in Bali," in *Traditional Balinese Culture*, ed. J. Belo, p. 49; on the abhorrence of animality generally, Bateson and Mead, *Balinese Character*, p. 22.
9. Except for unimportant, small-bet fights (on the question of fight "importance," see below) spur affixing is usually done by someone other than the owner. Whether the owner handles his own cock or not more or less depends on how skilled he is at it, a consideration whose importance is again relative to the importance of the fight. When spur affixers and cock handlers are someone other than the owner, they are almost always a quite close relative – a brother or cousin – or a very intimate friend of his. They are thus almost extensions of his personality, as the fact that all three will refer to the cock as "mine," say "I" fought So-and-So, and so on, demonstrates. Also, owner-handler-affixer triads tend to be fairly fixed, though individuals may participate in several and often exchange roles within a given one.
10. E. Goffman, *Encounters: Two Studies in the Sociology of Interaction* (Indianapolis, 1961), pp. 9–10.
11. This word, which literally means an indelible stain or mark, as in a birthmark or a vein in a stone, is used as well for a deposit in a court case, for a pawn, for security offered in a loan, for a stand-in for someone else in a legal or ceremonial context, for an earnest advanced in a business deal, for a sign placed in a field to indicate its ownership is in dispute, and for the status of an unfaithful wife from whose lover her husband must gain satisfaction or surrender her to him. See Korn, *Het Adatrecht van Bali*; Th. Pigeand, *Javaans-Nederlands Hand-woordenboek* (Groningen, 1938); H. H. Juynboll, *Oudjavaansche-Nederlandsche Woordenlijst* (Leiden, 1923).
12. The center bet must be advanced in cash by both parties prior to the actual fight. The umpire holds the stakes until the decision is rendered and then awards them to the winner, avoiding, among other things, the intense embarrassment both winner and loser would feel if the latter had to pay off personally following his defeat. About 10 percent of the winner's receipts are subtracted for the umpire's share and that of the fight sponsors.
13. Actually, the typing of cocks, which is extremely elaborate (I have collected more than twenty classes, certainly not a complete list), is not based on color alone, but on a series of independent, interacting, dimensions, which include – besides color – size, bone thickness, plumage, and temperament. (But *not* pedigree. The Balinese do not breed cocks to any significant extent, nor, so far as I have been able to discover, have they ever done so. The *asil*, or jungle cock, which is the basic fighting strain everywhere the sport is found, is native to southern Asia, and one can buy a good example in the chicken section of almost any Balinese market for anywhere from four or five ringgits up to fifty or more.) The color element is merely the one normally used

as the type name, except when the two cocks of different types – as on principle they must be – have the same color, in which case a secondary indication from one of the other dimensions ("large speckled" v. "small speckled," etc.) is added. The types are coordinated with various cosmological ideas which help shape the making of matches, so that, for example, you fight a small, headstrong, speckled brown-on-white cock with flat-lying feathers and thin legs from the east side of the ring on a certain day of the complex Balinese calendar, and a large, cautious, all-black cock with tufted feathers and stubby legs from the north side on another day, and so on. All this is again recorded in palm-leaf manuscripts and endlessly discussed by the Balinese (who do not all have identical systems), and a full-scale componential-cum-symbolic analysis of cock classifications would be extremely valuable both as an adjunct to the description of the cockfight and in itself. But my data on the subject, though extensive and varied, do not seem to be complete and systematic enough to attempt such an analysis here. For Balinese cosmological ideas more generally see Belo, ed., *Traditional Balinese Culture*, and J. L. Swellengrebel, ed., *Bali: Studies in Life, Thought, and Ritual* (The Hague, 1960).

14. For purposes of ethnographic completeness, it should be noted that it is possible for the man backing the favorite – the odds-giver – to make a bet in which he wins if his cock wins or there is a tie, a slight shortening of the odds (I do not have enough cases to be exact, but ties seem to occur about once every fifteen or twenty matches). He indicates his wish to do this by shouting *sapih* ("tie") rather than the cock-type, but such bets are in fact infrequent.
15. The precise dynamics of the movement of the betting is one of the most intriguing, most complicated, and, given the hectic conditions under which it occurs, most difficult to study, aspects of the fight. Motion picture recording pins multiple observers would probably be necessary to deal with it effectively. Even impressionistically – the only approach open to a lone ethnographer caught in the middle of all this – it is clear that certain men lead both in determining the favorite (that is, making the opening cock-type calls which always initiate the process) and in directing the movement of the odds, these "opinion leaders" being the more accomplished cockfighters-cum-solid-citizens to be discussed below. If these men begin to change their calls, others follow; if they begin to make bets, so do others and – though there are always a large number of frustrated bettors crying for shorter or longer odds to the end – the movement more or less ceases. But a detailed understanding of the whole process awaits what, alas, it is not very likely ever to get: a decision theorist armed with precise observations of individual behavior.
16. Assuming only binomial variability, the departure from a fifty-fifty expectation in the sixty-ringgits-and-below case is 1.38 standard deviations, or (in a one direction test) an eight in one hundred possibility by chance alone; for the below-forty-ringgits case it is 1.65 standard deviations, or about five in one hundred. The fact that these departures though real are not extreme merely indicates, again, that even in the smaller fights the tendency to match cocks at least reasonably evenly persists. It is a matter of relative relaxation of the pressures toward equalization, not their elimination. The tendency for high-bet contests to be coin-flip propositions is, of course, even more striking, and suggests the Balinese know quite well what they are about.
17. The reduction in wagering in smaller fights (which, of course, feeds on itself; one of the reasons people find small fights uninteresting is that there is less wagering in them, and contrariwise for large ones) takes place in three mutually reinforcing ways. First, there is a simple withdrawal of interest as people wander off to have a cup of coffee or chat with a friend. Second, the Balinese do not mathematically reduce odds, but bet directly in terms of stated odds as such. Thus, for a nine-to-eight bet, one man wagers nine ringgits, the other eight; for five-to-four, one wagers five,

the other four. For any given currency unit, like the ringgit, therefore, 6.3 times as much money is involved in a ten-to-nine bet as in a two-to-one bet, for example, and, as noted, in small fights betting settles toward the longer end. Finally, the bets which are made tend to be one- rather than two-, three-, or in some of the very largest fights, four- or five-finger ones. (The fingers indicate the *multiples* of the stated bet odds at issue, not absolute figures. Two fingers in a six-to-five situation means a man wants to wager ten ringgits on the underdog against twelve, three in an eight-to-seven situation, twenty-one against twenty-four, and so on.)

18. Besides wagering there are other economic aspects of the cockfight, especially its very close connection with the local market system which, though secondary both to its motivation and to its function, are not without importance. Cockfights are open events to which anyone who wishes may come, sometimes from quite distant areas, but well over 90 percent, probably over 95, are very local affairs, and the locality concerned is defined not by the village, nor even by the administrative district, but by the rural market system. Bali has a three-day market week with the familiar "solar-system"-type rotation. Though the markets themselves have never been very highly developed, small morning affairs in a village square, it is the microregion such rotation rather generally marks out – ten or twenty square miles, seven or eight neighboring villages (which in contemporary Bali is usually going to mean anywhere from five to ten or eleven thousand people) from which the core of any cockfight audience, indeed virtually all of it, will come. Most of the fights are in fact organized and sponsored by small combines of petty rural merchants under the general premise, very strongly held by them and indeed by all Balinese, that cockfights are good for trade because "they get money out of the house, they make it circulate." Stalls selling various sorts of things as well as assorted sheer-chance gambling games (see below) are set up around the edge of the area so that this even takes on the quality of a small fair. This connection of cockfighting with markets and market sellers is very old, as, among other things, their conjunction in inscriptions [R. Goris, *Prasasti Bali*, 2 vols. (Bandung, 1954)] indicates. Trade, has followed the cock for centuries in rural Bali, and the sport has been one of the main agencies of the island's monetization.
19. The phrase is found in the Hildreth translation, International Library of Psychology (1931), note to p. 106; see L. L. Fuller, *The Morality of Law* (New Haven, 1964), p. 6 ff.
20. Of course, even in Bentham, utility is not normally confined as a concept to monetary losses and gains, and my argument here might be more carefully put in terms of a denial that for the Balinese, as for any people, utility (pleasure, happiness . . .) is merely identifiable with wealth. But such terminological problems are in any case secondary to the essential point: the cockfight is not roulette.
21. M. Weber, *The Sociology of Religion* (Boston, 1963). There is nothing specifically Balinese, of course, about deepening significance with money, as Whyte's description of corner boys in a working-class district of Boston demonstrates: "Gambling plays an important role in the lives of Cornerville people. Whatever game the corner boys play, they nearly always bet on the outcome. When there is nothing at stake, the game is not considered a real contest. This does not mean that the financial element is all-important. I have frequently heard men say that the honor of winning was much more important than the money at stake. The corner boys consider playing for money the real test of skill and, unless a man performs well when money is at stake, he is not considered a good competitor." W. F. Whyte, *Street Corner Society*, 2d ed. (Chicago, 1955), p. 140.
22. The extremes to which this madness is conceived on occasion to go – and the fact that it is considered madness – is demonstrated by the Balinese folk tale *I Tuhung Kuning*. A gambler becomes so deranged by his passion that, leaving on a trip, he orders his pregnant wife to take care of the prospective newborn if it is a boy but to feed it as

meat to his fighting cocks if it is a girl. The mother gives birth to a girl, but rather than giving the child to the cocks she gives them a large rat and conceals the girl with her own mother. When the husband returns, the cocks, crowing a jingle, inform him of the deception and, furious, he sets out to kill the child. A goddess descends from heaven and takes the girl up to the skies with her. The cocks die from the food given them, the owner's sanity is restored, the goddess brings the girl back to the father, who reunites him with his wife. The story is given as "Geel komkommertje" in J. Hooykaas-van Leeuwen Boomkamp, *Sprookjes en Verhalen van Bali* (The Hague, 1956), pp. 19–25.

23. For a fuller description of Balinese rural social structure, see C. Geertz, "Form and Variation in Balinese Village Structure." *American Anthropologist* 61 (1959); pp. 94–108; "Tihingan, A Balinese Village." in R. M. Koentjaraningrat, *Villages in Indonesia* (Ithaca, 1967), pp. 210–243; and, though it is a bit off the norm as Balinese villages go, V. E. Korn, *De Dorpsrepubliek tnganan Pagringsingan* (Santpoort, Netherlands, 1933).
24. Goffman, *Encounters*, p. 78.
25. B. R. Berelson, P. F. Lazersfeld, and W. N. McPhee, *Voting: A Study of Opinion Formation in a Presidential Campaign* (Chicago, 1954).
26. As this is a formal paradigm, it is intended to display the logical, not the causal, structure of cockfighting. Just which of these considerations leads to which, in what order, and by what mechanisms, is another matter – one I have attempted to shed some light on in the general discussion.
27. In another of Hooykaas-van Leeuwen Boomkamp's folk tales ("De Gast," *Sprookjes en Verhalen van Bali*, pp. 172–180), a low caste *Sudra*, a generous, pious, and carefree man who is also an accomplished cockfighter, loses, despite his accomplishment, fight after fight until he is not only out of money but down to his last cock. He does not despair, however – "I bet," he says, "upon the Unseen World."

His wife, a good and hard-working woman, knowing how much he enjoys cockfighting, gives him her last "rainy day" money to go and bet. But, filled with misgivings due to his run of ill luck, he leaves his own cock at home and bets merely on the side. He soon loses all but a coin or two and repairs to a food stand for a snack, where he meets a decrepit, odorous, and generally unappetizing old beggar leaning on a staff. The old man asks for food, and the hero spends his last coins to buy him some. The old man then asks to pass the night with the hero, which the hero gladly invites him to do. As there is no food in the house, however, the hero tells his wife to kill the last cock for dinner. When the old man discovers this fact, he tells the hero he has three cocks in his own mountain hut and says the hero may have one of them for fighting. He also asks for the hero's son to accompany him as a servant, and, after the son agrees, this is done.

The old man turns out to be Siva and, thus, to live in a great palace in the sky, though the hero does not know this. In time, the hero decides to visit his son and collect the promised cock. Lifted up into Siva's presence, he is given the choice of three cocks. The first crows: "I have beaten fifteen opponents." The second crows, "I have beaten twenty-five opponents." The third crows, "I have beaten the king." "That one, the third, is my choice," says the hero, and returns with it to earth.

When he arrives at the cockfight, he is asked for an entry fee and replies, "I have no money; I will pay after my cock has won." As he is known never to win, he is let in because the king, who is there fighting, dislikes him and hopes to enslave him when he loses and cannot pay off. In order to insure that this happens, the king matches his finest cock against the hero's. When the cocks are placed down, the hero's flees, and the crowd, led by the arrogant king, hoots in laughter. The hero's cock then flies at the king himself, killing him with a spur stab in the throat. The hero flees. His house is encircled by the king's men. The cock changes into a Garuda, the great mythic bird of Indic legend, and carries the hero and his wife to safety in the heavens.

When the people see this, they make the hero king and his wife queen and they return as such to earth. Later their son, released by Siva, also returns and the hero-king announces his intention to enter a hermitage. ("I will fight no more cockfights. I have bet on the Unseen and won.") He enters the hermitage and his son becomes king.

28. Addict gamblers are really less declassed (for their status is, as everyone else's, inherited? than merely impoverished and personally disgraced. The most prominent addict gambler in my cockfight circuit was actually a very high caste *satria* who sold off most of his considerable lands to support his habit. Though everyone privately regarded him as a fool and worse (some, more charitable, regarded him as sick), he was publicly treated with the elaborate deference and politeness due his rank. On the independence of personal reputation and public status in Bali, see above, Chapter 14.
29. For four, somewhat variant, treatments, see S. Langer, *Feeling and Form* (New York, 1953); R. Wollheim, *Art and Its Objects* (New York, 1968); N. Goodman, *Languages of Art* (Indianapolis, 1968); M. Merleau-Ponty, "The Eye and the Mind," in his *The Primacy of Perception* (Evanston, Ill., 1964), pp. 159–190.
30. British cockfights (the sport was banned there in 1840) indeed seem to have lacked it, and to have generated, therefore, a quite different family of shapes. Most British fights were "mains," in which a preagreed number of cocks were aligned into two teams and fought serially. Score was kept and wagering took place both on the individual matches and on the main as a whole. There were also "battle Royales," both in England and on the Continent, in which a large number of cocks were let loose at once with the one left standing at the end the victor. And in Wales, the so-called Welsh main followed an elimination pattern, along the lines of a present-day tennis tournament, winners proceeding to the next round. As a genre, the cock fight has perhaps less compositional flexibility than, say, Latin comedy, but it is not entirely without any. On cockfighting more generally, see A. Ruport, *The Art of Cockfighting* (New York, 1949); G. R. Scott, *History of Cockfighting* (London, 1957); and L. Fitz-Barnard, Fighting Sports (London, 1921).
31. Above, pp. 391–398.
32. For the necessity of distinguishing among "description," "representation," "exemplification." and "expression" (and the irrelevance of "imitation" to all of them) as modes of symbolic reference, see Goodman, *Languages of Art,* pp. 61–110, 45–91, 225–241.
33. N. Frye, *The Educated Imagination* (Bloomington, Ind., 1964), p. 99.
34. There are two other Balinese values and disvalues which, connected with punctuate temporality on the one hand and unbridled aggressiveness on the other, reinforce the sense that the cockfight is at once continuous with ordinary social life and a direct negation of it: what the Balinese call *ramé*, and what they call *paling. Ramé* means crowded, noisy, and active, and is a highly sought-after social state: crowded markets, mass festivals, busy streets are all *ramé*, as, of course, is, in the extreme, a cockfight. *Ramé* is what happens in the "full" times (its opposite, *sepi*, "quiet," is what happens in the "empty" ones). *Paling* is social vertigo, the dizzy, disoriented, lost, turned-around feeling one gets when one's place in the coordinates of social space is not clear, and it is a tremendously disfavored, immensely anxiety-producing state. Balinese regard the exact maintenance of spatial orientation ("not to know where north is" is to be crazy), balance, decorum, status relationships, and so forth, as fundamental to ordered life (*krama*) and *paling,* the sort of whirling confusion of position the scrambling cocks exemplify as its profoundest enemy and contradiction. On

ramé, see Bateson and Mead, *Balinese Character*, pp. 3, 64; on *paling*, ibid., p. 11, and Belo, ed., *Traditional Balinese Culture*, p. 90 ff.

35. The Stevens reference is to his "The Motive for Metaphor" ("You like it under the trees in autumn,/Because everything is half dead./The wind moves like a cripple among the leaves/And repeats words without meaning") [Copyright 1947 by Wallace Stevens, reprinted from *The Collected Poems of Wallace Stevens* by permission of Alfred A. Knopf, Inc., and Faber and Faber Ltd.]; the Schoenberg reference is to the third of his *Five Orchestral Pieces* (Opus 16), and is borrowed from H. H. Drager, "The Concept of 'Tonal Body,'" in *Reflections on Art*, ed. S. Langer (New York, 1961), p. 174. On Hogarth, and on this whole problem – there called "multiple matrix matching" – see E. H. Gombrich, "The Use of Art for the Study of Symbols," in *Psychology and the Visual Arts*, ed. J. Hogg (Baltimore, 1969), pp. 149–170. The more usual term for this sort of semantic alchemy is "metaphorical transfer," and good technical discussions of it can be found in 'M. Black, *Models and Metaphors* (Ithaca, N.Y., 1962), p. 25 ff; Goodman, *Language as Art*, p. 44 ff; and W. Percy, "Metaphor as Mistake," *Sewanee Review* 66 (1958): 78–99.
36. The tag is from the second book of the *Organon, On Interpretation*. For a discussion of it, and for the whole argument for freeing "the notion of text . . . from the notion of scripture or writing" and constructing, thus, a general hermeneutics, see P. Ricoeur, *Freud and Philosophy* (New Haven, 1970), p. 20 ff.
37. Ibid.
38. Lévi-Strauss' "structuralism" might seem an exception. But it is only an apparent one, for, rather than taking myths, totem rites, marriage rules, or whatever as texts to interpret, Lévi-Strauss takes them as ciphers to solve, which is very much not the same thing. He does not seek to understand symbolic forms in terms of how they function in concrete situations to organize perceptions (meanings, emotions, concepts, attitudes); he seeks to understand them entirely in terms of their internal structure, *independent de tout sujet, de tout objet, et de toute contexte*. See above, Chapter 13.
39. Frye, *The Educated Imagination*, pp. 63–64.
40. The use of the, to Europeans, "natural" visual idiom for perception – "see," "watches," and so forth – is more than usually misleading here, for the fact that, as mentioned earlier, Balinese follow the progress of the fight as much (perhaps, as fighting cocks are actually rather hard to see except as blurs of motion, more) with their bodies as with their eyes, moving their limbs, heads, and trunks in gestural mimicry of the cocks' maneuvers, means that much of the individual's experience of the fight is kinesthetic rather than visual. If ever there was an example of Kenneth Burke's definition of a symbolic act as "the dancing of an attitude" [*The Philosophy of Literary Form*, rev. ed. (New York, 1957), p. 9] the cockfight is it. On the enormous role of kinesthetic perception in Balinese life. Bateson and Mead, *Balinese Character*, pp. 84–88; on the active nature of aesthetic perception in general, Goodman. *Language of Art*, pp. 241–244.
41. All this coupling of the occidental great with the oriental lowly will doubtless disturb certain sorts of aestheticians as the earlier efforts of anthropologists to speak of Christianity and totemism in the same breath disturbed certain sorts of theologians. But as ontological questions are (or should be) bracketed in the sociology of religion, judgmental ones are (or should be) bracketed in the sociology of art. In any case, the attempt to deprovincialize the concept of art is but part of the general anthropological conspiracy to deprovincialize all important social concepts – marriage, religion, law, rationality – and though this is a threat to aesthetic theories which regard certain works of art as beyond the reach of sociological analysis, it is no threat to the conviction, for which Robert Graves claims to have been reprimanded at his Cambridge tripos, that some poems are better than others.

42. For the consecration ceremony, see V. E. Korn, "The Consecration of the Priest," in Swellengrebel, ed., *Bali: Studies*, pp. 131–154; for (somewhat exaggerated) village communion, R. Goris, "The Religious Character of the Balinese Village," ibid., pp. 79–100.
43. That what the cockfight has to say about Bali is not altogether without perception and the disquiet it expresses about the general pattern of Balinese life is not wholly without reason is attested by the fact that in two weeks of December 1965, during the upheavals following the unsuccessful coup in Djakarta, between forty and eighty thousand Balinese (in a population of about two million) were killed, largely by one another – the worst outburst in the country. [J. Hughes, *Indonesian Upheaval* (New York, 1967), pp. 173–183. Hughes' figures are, of course, rather casual estimates, but they are not the most extreme.] This is not to say, of course, that the killings were caused by the cockfight, could have been predicted on the basis of it, or were some sort of enlarged version of it with real people in the place of the cocks – all of which is nonsense. It is merely to say that if one looks at Bali not just through the medium of its dances, its shadow-plays, its sculpture, and its girls, but – as the Balinese themselves do – also through the medium of its cockfight, the fact that the massacre occurred seems, if no less appalling, less like a contradiction to the laws of nature. As more than one real Gloucester has discovered, sometimes people actually get life precisely as they most deeply do not want it.

53

Participant Observation as a Tool for Understanding the Field of Safety and Security

Frédéric Diaz

"Much of the literature on controversial police topics breaks down into two categories: uncritical work by well-informed insiders, and critical work by uninformed outsiders" Gary Marx (1988)

It is because practices that persisted in being incomprehensible and "unfathomable begin to become meaningful, through patient observation, like the pieces of a puzzle that hinge together little by little, that the field worker feels rewarded for his obstinacy".[1] Although observation is mentioned in the manuals and methodology courses, it remains a minor *genre* in the range of tools available to sociologists for their studies. "Constraints of time, size and object, plus the need to quantify, verify and establish proof convincingly in the eyes of the lay person, can only lead them to prefer other research techniques."[2]

However, observation is still a tool, a method for the comprehension of the mechanisms of social interaction and of life in society.

The present article is the fruit of an encounter between sociology and ethnology, and is based on two theories of social reaction: interactionist and ethnomethodological perspectives. The former, so as to plead for going back to field work and taking the standpoint of actors rather than of institutions; the latter, so as to focus on the practices implemented by those actors.[3] "The

Source: *Champ Pénal/Penal Field,* II (2005).
http://champpenal.revues.org/471

question is (then), how actors produce their worlds, what rules generate those worlds and govern the actors' judgment."[4]

Observation of phenomena and of actors is inherent in any scientific approach. It "has asserted itself, then, as the primordial condition for the construction of knowledge in the social sciences (as well as in the pure sciences), by establishing a direct relation, but also a distance, between subject and object."[5] "While sociologists necessarily work on objectivation so as transform their empirical objects into sociological objects, the social actor too performs a similar labor, so as to interpret the surrounding world and thus, accomplish his or her actions."[6] This would mean that there are as many social realities as there are actors. Sociologists may then define their motivation as the description, accounting for and understanding of the collective dimension of human life and social relations.

The postulate that this writer would like to defend is as follows: participant observation is the most appropriate method for attempting to understand the complexity of a field[7] such as safety/security. The concept of safety/ security (henceforth designated as safety) is then understood through the demarcations set up by the actors when organizing sports and cultural events. This concept of safety will be analyzed, with attention to the practices and techniques for ensuring safety, the roles of the various private actors (private security companies, organizers of shows) and public actors (the city, police, fire department, medical emergency services) and their interactions.

In this article I will first explain the motivations behind participant observation. The low visibility of the field of safety gradually led me to go from an interview and observation scheme (in which there was little to observe if I stayed in an outsider position) to a participant observation scheme.

However, although the latter scheme definitely facilitated access to data, it nonetheless brought into the forefront the idea that there was a risk of only seeing one viewpoint (that of the organizer) and therefore a need, at some point, to put the object at a distance so as to view it as an observer rather than as an actor.

This methodological scheme is similar to the one used by some sociologists from the Chicago school, with the limits, controversies and debates raised by this approach.

The point here is to emphasize the specific contribution of the approach used, which resides in the variety of functions endorsed, which enabled us not only to define the safety problems as viewed by the actors, and the practical responses, but also, and above all, to relativize each experience.

The narrative form will be employed here to describe a research project that began in 1998,[8] and has continued over the last year at the *Centre International de Criminologie Comparée* of the University of Montreal, in what may be described as a progression from learning to a degree of "mastery" of work *in situ*. The idea is to give a detailed account of the question underlined in a collective work by Poupart, Deslauriers, Groulx, Laperrière, Mayer and Pires (1997, 212–213), which is to say, the conditions under which research is done, as well as the practical problems encountered, focusing on three key points: i) my position and role as observer, my relation to the particular fieldwork context; ii) the criteria governing the validity and accuracy of the study, ranging from the choice of the sites observed and of the target group sampled to the theoretical production and the generalization of the findings; iii) ethical questions pertaining to the relations between professional people working in a particular sphere and scientific research. Through an articulation between extant writings and the narrative of my seven-year experience in fieldwork, I will attempt to determine, firstly, the motivations and limits of the use of participant observation in a research framework. The method corresponds to the specificity of the object. Secondly, I will then explain the phases and tools involved in this original method. The method then becomes the object of the study.

A Method for an Object – On the Choice of Participant Observation: Justification, Limits and Definition

Sociologists have two qualitative methodological tools at their disposal: observation and interviews. In this debate, we will concentrate more specifically on one of the two: observation, and on one of the forms it may take in participant observation: that is, direct observation, or observation *in situ*, in the situation.[9]

But what, specifically, are the different forms taken by observation, and what is the content of what we call "participant" observation?

There is a debate within the academic community, including in sociology and anthropology, on the forms to be taken by observation. In the case of "simple" observation, the person remains outside the group, to describe it. In the case of participant observation there is an attempt to enter the group, to be a part of it so as to study it more closely.

But my choice was not a haphazard one, it was the result of a conclusion drawn after my first year of research, a year before I began my doctoral thesis. It was also based on some seminal writings whose most controversial points will be mentioned before I go on to assess what I thought I was – or was not – doing, in a reflexive attitude, at a time when I did not know what I would or would not be able to achieve.

Justification of the Shift from "Simple" Observation, External to the Object (Completed by Interviews) to Participant Observation, and Its Limits

A First Year of Research, Which Enabled Me to Define Five Features of the Field of Safety/Security

My first studies involved fieldwork in unknown territories. My only tool was my doubts, my questions and my observations. When the observer[10] first arrives in the fieldwork situation he is unfamiliar with his surroundings, and must therefore adopt a methodology which enables him to account for the complexity of the particular field.

- A complex field: many actors, many spaces

In January 1998, when I began my study,[11] I had only a vague idea of what safety represented for the organization of sports and cultural events. My point of departure was the French act on orienting and planning safety (LOPS) of January 21, 1995, which established legal liability for private organizers and distributed responsibilities between private and public actors depending on the spaces involved in the event. On the basis of that act and those consequences, the responsibilities of the different actors were structured according to the private and public spaces involved in the event. My motivation, then, was to see and analyze how that act applied, practically, in real situations.

The group targeted by my research (PhD in political sociology and comparative politics, University of Paris X-Nanterre) included twenty individuals. I had established an organization chart of the main actors intervening in the field of safety. The idea then was to try to contact them and meet them. Over an eight-month period I succeeded in meeting the main officials, both public and private, including various specialists in safety matters. I had prepared a grid for my interviews, with key questions along four lines: personal history and responsibilities, active and passive safety arrangements, determination of risks, interactions between actors. I was trying to understand the paths followed by these individuals, what had brought them to occupy their present position, and their role during the event. In addition, I tried to have as accurate as possible a vision of the arrangements made, as they were defined by the actors, on the different aspects of safety strictly speaking (and in particular, reception of the public and control of drug dealing), but also on the assistance end (especially medical aid, fire fighting and crisis management). In addition, I looked at the inside of private space as well as the outside (stadium patrols, fencing, surveillance cameras, the command post, police forces and so on). Next, I tried to understand what the main risks were, from the viewpoint of the different actors, during the competition, and what they thought about them. Last, I was intent on bringing out any interactions between the various actors involved in safety.

> "It is through a more individualized, deeper relation with the person, thanks to the interview, that the sociologist attempts to immerse himself in the sense given by actors to their practices. The challenge then consists of going beyond the mere individual pronouncements so as to bring out the more fundamental aspects contained in some particularly enlightening excerpts, which make it possible to understand the phenomenon or the social group studied.[12]

For me, interviews were the preferred tool for obtaining precise information. The idea was to identify resource people and to cross the information obtained from them with that of other actors in the field. I felt it was essential to obtain as exhaustive as possible a vision of all of the types of actors involved, through interviews.

- the discrepancy between a necessarily simplified discourse (the interview) and the reality in the field.

But these interviews made me sensitive to their limits, mostly affected by the naivety that separates the student from the professional. Those limits were also due to the simplified discourse that was served to me, owing to the time allotted to the interview, in particular.

I also became aware of the disparity between what I was told by the actors (but also their behavior, positions, attitudes, action and decisions and so on) and their practices. These actors tended to speak very positively about their arrangements and methods, and how they functioned. Rarely did they mention any problems encountered.

- *Low visibility of the field of safety: demanded by the actors to deflect attention* (although, paradoxically, there is a preventive dimension in making some arrangements visible so as to be threatening)

The limits also resided in the interviewee's interrogations about the researcher's motivations. The actor interviewed is always wondering, legitimately, how his discourse will be interpreted and analyzed, and to what end.

Officials make a strong claim to the right to protect themselves from public attention, whereas, paradoxically, the fact of publicizing the arrangements (in the mass media in particular) definitely has a preventive dimension: "exhibit yourself so as not to have to take action; conceal yourself to take action".

- No access to arrangements as long as there is no climate of trust between the actors and the student or researcher

The student, like the researcher, must therefore prove himself trustworthy, so to speak. It is the time variable, mostly through writings, meetings and experience that will make his research legitimate, and thus establish the basis for a climate of trust. As the study advances, the researcher gains experience and knowledge, thus becoming less and less alien to the milieu he is studying. His web of knowledge widens, and his questioning becomes less "naive", more relevant, and by the same token more legitimate.

- A peculiarity of this object: the difficulty in perceiving efficiency if everything goes well

Lastly, the specificity of "deviance and social control" as a field resides in the difficulty, for an outsider, in realizing what problems are encountered, particularly when the object involved involves marginal deviancy. Specialists who have paid special attention to the safety problems raised by crowds have shown that contrary to popular opinion, they usually involve very little serious violence; there is generally no major incident (Diaz, 2003a, 2003b; Fillieule, Jobard, 1998).

Seminal Texts (Particularly by the Chicago School) Criticizing or Legitimating the Choice of Participant Observation

Two main forms of criticism may be noted, based on a twofold limit: seeing only one point of view and therefore having a sort of "prejudiced view", a lack of objectivity; and the necessity of setting the object at a distance so as to see it through the eyes of an observer rather than as an actor.

Durkheim (1895) states that descriptions by actors are too vague and ambiguous to be used by scholars for scientific purposes.

According to Bourdieu (1978, 68), "the sociologist only has some chances of succeeding in the work of objectivation if, he himself an observer observed, he subjects everything to objectivation, not only everything he is, his own social conditions of production and therefore the 'limits of his mind', but also his own work of objectivation, the hidden interests invested therein and the benefits they promise."

The second limit evidenced by Bourdieu (1978, 67) is that "there is no denying the practical contradiction: we all know how difficult it is to be both caught up in interplay and an observer of it". When you play you can't do everything at once, you can't play and take the time to see what you are doing.

Conversely, several authors (and especially Chapoulie, 1985, 1996, 2001) have used the sociological tradition of the Chicago school to show the invaluable contribution of observation in sociology. This begins with Thrasher (1927) and Whyte (1943), and later participant observation with Anderson (1923) and differently, Shaw (1930, 1931), or again the interactionists with their theory of labeling and the disciples of Hughes (1996), but also Mead (1934), Blumer (1962, 1969) and later Becker (1963, 2002). Lastly, we may add the ethnomethodologists following in the footsteps of Garfinkel (1967), who look at the ways in which actors construct their reality in the course of their everyday activities.

Sociologists should cease to be strangers observing strangers to report to other strangers (Coulon, 1992). They should, henceforth, report on observations made, not as complete strangers, but as members of an in-group to some extent (Hughes, 1996).

> "To catch the process, the student must take the role of the acting unit whose behavior he is studying. Since the interpretation is being made by the acting unit in terms of objects designated and appraised, meanings acquired, and decisions made, the process has to be seen from the standpoint of the acting unit. . . To try to catch the interpretative process by remaining aloof as a so-called "objective" observer and refusing to take the role of the acting unit is to risk the worst kind of subjectivism – the objective observer is likely to fill in the process of interpretation with his own surmises in place of catching the process as it occurs in the experience of the acting unit which uses it."[13]
>
> "One would have to take the role of the actor and see his world from his standpoint. This methodological approach stands in contrast to the so-called objective approach so dominant today, namely, that of viewing the actor and his action from the perspective of an outside detached observer (. . .). The actor acts toward his world on the basis of how he sees it and not on the basis of how that world appears to the outside observer."[14]
>
> "Finding them where there are, staying with them in some role which, while acceptable to them, will allow both intimate observation of certain parts of their behavior, and reporting it in ways useful to social science but not harmful to those observed."[15]
>
> "For the sociologist was now reporting upon observations made, not in the role of the stranger, but as a full member of the little world he reported, on. He observed as a member of an in-group but, in the act of objectifying and reporting his experience, became of necessity a sort of outsider."[16]

Here we are clearly at the crux of fundamental issues as to the "role of the research worker and of his subjectivity in the production of knowledge".[17]

> In the positivist perspective, knowledge of reality necessarily involved the elimination of any possible sources of bias, including those connected with the researcher's own status and ways of thinking, in other words, if researchers are to arrive at truly objective knowledge of reality, they must be neutral and make sure that factors such as their cultural or social identity, for example, will not in any way interfere with the process of gaining knowledge. . . . In more recent writings, there is an attempt to rehabilitate the place of the researcher as social actor in the research process. The researcher's subjectivity is viewed not only as inseparable from the process of construction and production of data, but also as a resource in its own right in the knowledge-gaining process, provided, however, that researchers remain critical of their own approach and constantly exhibit a reflective attitude toward its underlying postulates.[18]

Choosing to Do "Active", "Immerged" Fieldwork, Multiplying Fieldwork Situations and Varying Activities with Different Hierarchical Positions, Alternating between Phases of Observation, Thinking and Action

On the strength of these references, I therefore chose integration in the organization of these sports and cultural events. I would study these events on two levels: everyday life and the long term. To keep pitfalls to a minimum, I had to gain acceptance by the people I met, and create regular, in-depth communication with the different actors in the field. In this case, that meant achieving direct contact with the fieldwork situation and the actors therein. The safety mechanisms, operational techniques, inter-group and interpersonal relations are so complex that one must take part in them to gain any understanding. I therefore felt it necessary to be a part of the very organization of the event. Understanding of social complexity requires direct contact with the actors and the objective study of their practices.

According to the terminology used by Patricia and Peter Adler (1987), three broad categories may be distinguished: i) a "peripheral" role: in contact but without participating; ii) an "active" role: with participation and responsibilities; iii) an "immerged" role, as natural member with the same feelings and the same goals as actors in the field. I set myself somewhere between the "active" and the "immerged" roles. "Active", since from 2000 on I began to work within groups as a salaried worker with a varying load of responsibilities and colleague-to-colleague relations; "immerged", since working on special events is more than a job, it incites people to surpass themselves, to believe in some ideas, for a common goal – that is, the show must go on.

Moreover, to reduce "prejudice", lack of objectivity and the observer's distance, I felt it relevant to attempt to be integrated in a variety of positions for each of the events studied. This change of activity also helped me to understand the viewpoints of the different actors in safety (and especially, the differences between public institutions and private organizers). It gave me some distance with respect to my own participation, by assimilating the motivations of other actors. Last, by multiplying fieldwork contexts, and moreover, for short periods (a feature specific to the special event scene), I was able to systematically extract material that refined my analysis and enabled me to very rapidly resume my distance from the object studied.

> "Other ways of solving the dialectic include being a part-time participant and part-time reporter, privately participant and publicly reporter, or publicly participant and secretly reporter."[19]

This method of observation *in situ*[20] has the advantage of reducing the social distance that may separate the student (or the researcher) from professionals in the field studied.

The difficulty, then, was to put these remarks into practice, and to conceive the conditions of achievement of participant observation.

A Method as Object: Conditions and Achievement of Participant Observation

It may seem problematic to separate the object and issues of research from the method, and certainly all the more so when the method is taken as the object of the research. The point of this second part is not to look for the theoretical questions motivating this approach and the perspectives that method may contribute, in terms of results.[21] The analysis of a field such as safety is viewed as extremely difficult (Bayley, 1994). The researcher must therefore find the means enabling him to approach the complexity of the field.

My methodology is entirely within the scope of the qualitative socio-anthropology approach.[22] This led me to choose the methods of participant observation and interviews, and later comparison, for my approach. From the outset of my thesis, the comparative dimension became essential. "It makes it possible to draw up taxonomies and sequences, then to establish correlations." (Bouvier, 2000, 56)

> During my second year of doctoral research I had to gradually change my strategy, then, as I used participant observation. This was made possible by access to fieldwork situations, and was based on three criteria: "observer's experience, awareness, and personality".[23]

"Researchers can only have access to those signifying social productions of actors, which are private phenomena, by participating, equally, as actors, in the world they propose to study . . . . Symbolic interactionism holds that it is the actors' conception of the social world which is the main object of sociological research, in the last analysis."[24]

Social significations must be viewed as "creations that are formed in and through the defining activities of people as they interact".[25] This implies that the observer who proposes to understand and analyze these significations must adopt a methodological posture which makes this analysis possible.

Private Security Jobs Are Readily Accessible, Enabling Observation to be Increasingly More Participant in the Object Studied

My research took place between 2000 and 2004, as an increasingly participant investigation. This is one of the peculiarities of that investigation. Following the phase of socializing with the milieu studied, I decided that I myself had to do some of the work I intended to study and on which I would report. This was made possible by the fact that the various private security jobs, and more generally, the positions connected with the organization of safety/security for sports and cultural events, are easily accessible.[26]

The Three Criteria Defined by Schwartz and Schwartz (1955): The Observer's Personality, Experience and Awareness

- The observer's personality and training: socialization with the milieu studied

As Wax (1971), quoted by Peretz, pointed out, all observers are not interchangeable, and the characteristics of each individual partially determine the milieus to which he has access.[27] The first step in conducting an observation depends on the worker's relations with the milieu to be observed, namely: whether or not he effectively has relations with it; the worker's own personal characteristics (sex, age, ethnic group and social class). Although, as I have emphasized, I knew very little about the field of social control and deviance, on the other hand I was very well acquainted with the sports and cultural (especially the musical) milieu, both as actor[28] and consumer.[29]

In addition, by doing fieldwork repeatedly in the same places year after year,[30] I was able to gain training in some areas[31] and to be employed in several positions. This is what makes for the peculiarity of my qualitative methodological approach: it evolved toward a participant method as my research progressed. This was not so from the outset, as was the case for "Roy (1952, 1954) [who] worked on the production line before doing his study on braking, Becker (1963) [who] was a jazz pianist before undertaking the study to be republished in Outsiders, Dalton (1959) [who] was a business executive, Davis (1959) [who] worked as a cab driver to pay for his education",[32] and more recently, Esterle-Hedibel (1995), who took a job as street educator, or Bossis (2003), as court clerk. Rather, in my case, year after year, event after event, I started with a simple observation post in jobs requiring very little skills (security guard), then went from low-skill positions to responsible positions (manager, safety coordinator).[33]

- The experience itself: criteria of validity based on changes in activity, a multiplicity of fieldwork situations, distancing between active research and theoretical work, everyday contact and the long term

I therefore felt it relevant to attempt to take on different positions for each of the events studied. Over and beyond the few experiences described here in order to give the reader a better idea of the various functions I occupied, the idea is to understand my trajectory and its evolution, the role of chance and of deliberate strategies in the choice of the positions held, as well as of the constraints imposed by these positions.

I had a deliberate strategy in the choice of fieldwork situations. It was based on four main criteria: the manifestation is susceptible of attracting a great many people; the event entails considerable stakes (for the mass media, economic, political, in sports or culture); the places and shows are viewed as references in terms of the handling of safety; the sports and cultural calendar and my own availability between 1998 and 2004. Next, I had a deliberate strategy involving the decision to hold positions touching on the problems of safety and first aid. But there is an element of chance, tied to the context of the events and the needs of the organizers. Chance certainly was a factor at the beginning of the project, in the first jobs I got, and it gradually became less important as my personality and experience and the motivations of the research project[34] gave me more freedom to choose positions (although there is no denying that these very features simultaneously induced constraints). I was concerned with being exhaustive in the functions I occupied. There was the concern with occupying those positions most apt to reveal the stakes connected with passive safety (the structures, architecture, grandstands and so on) for one event (the Bourges *Printemps* in 2000 and 2001), those connected with active safety (the human apparatus, the private security professions in different capacities and at different hierarchical levels and so on.) for another event (*Eurockéennes* in Belfort, 1999, 2000, 2001 and 2002) as well as the stakes involved in interactions between private and public actors and negotiations between those actors (Aurillac 2000 and 2001; *Vieilles Charrues* in Carhaix, 2001 and 2002). It that sense, I tried to have my safety-linked positions with private enterprises be exhaustive.

Manager of Safety Commissions, Then Site Manager at the Printemps in Bourges

I had observed the Bourges *Printemps* in 1999 as a spectator, and was well aware that it was difficult for me to have an objective vision of what transpired there, in terms of safety. I could record a number of facts, such as the names of the enterprises working there, their position at a given point in time, their attitude toward the public, etc. I could also observe the places where the police forces were active, their various interventions and so forth. But on the other hand I was unable to get an overall picture of all of the arrangements, find out the exact number of people working on the location, the problems encountered throughout the event and the interactions between all of the institutions. For that reason, when I met the technical manager at the end of the festival, I offered to work for him at the following year's festival, on the basis of my two previous years of research.

At the time, the festival was having problems with the administration over the safety commissions. Since the incidents at the Furiani stadium in Corsica on June 5, 1992, those safety commissions had been made compulsory. A commission, composed of representatives of the prefect, the mayor, the prevention and first aid agency and the *direction départementale de l'équipement* (the county-level agency of the national road and facilities directorate) inspected the festival every day throughout its duration and pointed out a number of safety-related problems. Their inspection covered the respect of regulations having to do with fire prevention, the risk of panic and accessibility for the public, especially for people with low mobility. Moreover, the commission took up much of the technical manager's time, since he had to follow it around all afternoon so as to consult with it before making the changes needed to have the place meet the safety requirements. The technical manager therefore asked me to be the liaison manager with the safety commissions, and to do everything I could to meet their demands. The position was exactly what I needed, since it enabled me to take part in the organization of the festival by working at a real job and moreover, one that was centrally located in the interactions between administrative agencies and a private organization. It gave me access to all of the information I needed, and enabled me to attend all of the preparatory and post-festival evaluation meetings. I found myself where I had hoped to be, at the heart of the event, in a role where I had to be familiar with all of the structures and all the people in charge of each location. My function was to accompany the safety commission to every location within the festival, to act as intermediary between the commission and the site manager and to be the representative of the festival, making sure that the modifications needed in order to respect the legal obligations were actually made. I held that position for two consecutive years, in 2000 and 2001.

For the 2002 festival, the *Maison de la Culture* (cultural center) of Bourges asked me to manage its two theaters (seating 1,000 and 500), including reception of the public and the ticket office, reception halls, the cafeteria, the local *France Inter* radio station for live broadcasting and so on. This role of site manager meant I was in charge of both the building in which I worked and the staff I headed. In all, there were no less than 40 people running the places under my supervision, including maintenance workers, private security guards, people monitoring access and tickets, etc. This enabled me to realize the role played by a manager in securing the safety and reception of the public and the entertainers, ensuring maintenance of the facilities, monitoring the working hours of employees, handling the circulation of people and the various problems encountered with the public. I was no longer in the strict safety framework and could look into other matters which may affect safety in one way or another. For instance, I had to oversee the different jobs done by the employees, along with their role, make sure they were properly clothed, were friendly and courteous toward the public, make sure the theaters were well kept and maintain surveillance through a system of searches and checkrooms for some sorts of objects.

Last, I met often with the main actors involved in the safety of the festival outside of that event itself, and especially with the head of the private security agency, the person in charge of first aid and prevention, the sub-prefect and the main private secretary of the prefect. These talks helped me to refine my research project and to get to the heart of the subject, making a distinction between my work during the festival and the research questions tied to my thesis, with the advantage of the experience and problems encountered during the festival.

Private Security Guard and Assistant Safety Manager for the Belfort Eurockéennes Festival

- a first interview to clear the ground

Again proceeding along the same methodological lines, I asked to meet the technical manager of the festival four months before it began. The technical manager of a festival coordinates all the on-site logistics. It is he who organizes and handles every aspect connected with the management of the equipment, the mounting of the structures and management of the personnel. In particular, he is in charge of all safety/security questions concerning both the facilities and the safety personnel. But as I had discovered previously, the limits of a "simple" interview, even one lasting three hours, were soon evident, and I therefore requested that I be allowed to analyze the safety arrangements during the festival.

- Meeting with an official in charge of the safety/security of a festival: his function and missions

I was met by the safety manager two days before the beginning of the festival. I later learned that our meeting was supposed to last only one hour, after which he could, if he chose, leave me to my own sad fate in the midst of the spectators for the rest of the festival. But things went differently, and I was lucky enough to be able to follow his every move, 20 hours a day. In exchange I served as his chauffeur, secretary and aide-mémoire, and we developed a genuine working relationship.

In five days he taught me how a safety manager handled the safety of a festival, what arrangements he made, how he handled conflicts, how he developed relations with the administration. He showed me the importance of preliminary work, with risk evaluation, participation in preparatory meetings with the public safety services, decisions as to which private security agencies would be hired and what would be demanded of them. He taught me his work as coordinator, managing 120 workers from a security guard company, spread over the campgrounds, the entrance points and throughout the site, as well as 50 first aid agents; also, his role as interface with the administration and how he coped with points of tension. Last, he allowed me to attend the various meetings held during the festival and the assessment meetings based on feedback afterwards. This gave me a comprehensive view of the safety problems encountered during a festival, from the inmost position, since he supervised everything.

- My first job as security guard: guarding a space

The following year I entered his security company as a full-fledged security guard, while playing the role of assistant after my 12 hours a day as a guard. The first thing I was asked to do was to spend three days in the woods. I saw no more than twenty people a day, mostly people in charge, and none of the spectators. I learned nothing about the function of a security guard aside, definitely, from his humility, but it is true that by accepting that role I certainly won the confidence of the safety manager.

The second year I was able to get a better full-time perspective on the mission of a security guard, since I worked at different duties in the VIP area. After my hours as security guard, I returned to my job as assistant, with interventions throughout the premises. As time went by the manager completed my training and handed more work, and more important work, over to me. That ranged from the distribution of objects (fire extinguishers) to making sure there were no "leaks" in the separation between the area open to the public and the first aid area, delivery of messages to security guards, etc. The following year I retained the same type of duty at one entrance point between the area open to the public and the VIP space while continuing to work as assistant outside of the times when the site was open to the public. My position as security guard gave me an idea of how the public regards a simple security guard on duty. In particular, I realized how difficult it is to stay calm and receptive when faced with a very diversified public.

- A different duty in 2002: the flying squad

Along with the set positions guarded on a site, there is, simultaneously, a "flying squad" function. As its name indicates, this squad moves around the area. It is composed of a group of guards (5 in 2001, 3 in 2002) who do whatever is needed and cover all sorts of missions. For example, they may be called in if some spectators try to go on stage, for a fight between two people, to reinforce coverage of an area and so forth. Their main mission is to limit drug dealing (mostly cannabis and ecstasy) by taking in any dealers found on the premises. It was in 1999 that the administration first discovered the existence of a great many dealings (selling and consumption) on the site.

The following year, in accordance with a recommendation from the prefecture, the festival safety manager suggested the creation of a team of guards in charge of taking in dealers caught in the act and putting them off premises and into the hands of the *gendarmerie* which would record the facts. This experience placed me at the center of the partnership between the police forces and the private organizer, then. I was able to analyze the roles of each actor, the modes of operation set up and the results of those procedures.

Technical Counselor for the City of Aurillac

The events described above had been observed from the viewpoint of the private organizer. It was important for me to do the same for the viewpoint of another major actor: the city. The involvement of the latter in the management of this festival was particularly strong since most of the events took place in the city center, on specially outfitted squares or sites such as school courtyards, gymnasiums, public gardens or anywhere on the streets.

The secretary-general was in charge of organizing those elements of the event for which the city was responsible. He was therefore the first person I met, for an overview of the various safety arrangements and the interactions between the various actors involved in safety. He was also my main contact throughout the festival. He suggested that I assist him during the festival, and that I serve as counselor for safety matters. I was therefore able to determine the missions of a town hall in the organization of safety, first aid and prevention, over the entire event. Those missions were of three sorts: the circulation of vehicles and pedestrians, and parking of vehicles, by the municipal police force; street-cleaning by the city cleaning department; prevention and control of drug dealing, and above all, the reception of what is known as "drifters".

Coordinator of Safety and First Aid for the Vieilles Charrues Festival in Carhaix

- A first year for defining needs and appropriate responses

I had decided to keep this festival for the end. It was the last event I covered before writing my thesis, but above all, it was the event for which I held the most responsible position and also, for which I did ongoing preparatory work throughout the year between the 2001 and the 2002 editions.

For years, up to the mid 90s, the *Eurockéennes* festival in Belfort had been the largest in terms of numbers of spectators, and then, in a matter of years, the *Vieilles Charrues* festival became the largest in France. It seemed relevant for me to participate in it, then, in order to complete my field work.

Two months before the festival I met the production manager and described my four years of research, my participation in various festivals and sports events, and my desire to be involved in that event. The person in charge of outdoor safety had recently left, and I was named to replace him, since the position was vacant.

Actually, the job was not very well defined, and my position varied with the concrete needs and the objectives I set for myself. My four years of researching had led me to identify the overall situation, how to behave with different people, how to set rules for making premises safe, the kinds of responses needed with respect to first aid and safety. I also had to define my work in connection with the existing personnel and their missions. It took me two weeks to find my niche and to show the need to name a safety coordinator.

- Safety coordinator

As coordinator I was in charge of everything having to do with safety, first aid and prevention. In addition, I had to draw up the dossier on safety and handle the interaction between the private organizer of the festival and the administration, which mostly meant the national *gendarmerie* and the county-level firefighting and first aid departments. My work therefore began early, in October 2001, with preparatory assessment meetings with the main officials from the private organization, with emphasis on all of the technical aspects. I worked on every theme; safety, parking grounds, campgrounds, logistics and technical services. Once they had all been assessed, the preparatory work for the 2002 edition could begin. The general manager also asked me to write up the safety dossier, which is the main document describing the event, given to the administration a month before it begins. The constitution of this report led me to think about all of the problems tied to the safety of an event, and enabled me to enter into close relations with some representatives of the administration, for whom I was the main contact, along with the general manager. The report was divided into twelve points: description of the festival, the entertainment program, practical information, overall organization with the various contacts for the administration, definition of the boundaries of the premises and their environment, the chain of command for the organization of safety (missions of safety coordinator, the safety manager and assistant manager, the safety headquarters, the *gendarmerie*), the chain of command for the organization of first aid, first aid resources on the premises, means of communication, private security agencies, the regulations and various plans. I also took part in selecting the private companies. There had been problems at the entrances in 2001, leading us to change the company that worked there. I recommended the company that worked at the *Printemps* in Bourges, which I thought had been very professional. Similarly, we used the same group of guards as at the Belfort *Eurockéennes* to try to limit the presence of drug dealers on the premises. The procedure used and the coordination with the administration were the same.

Through these different experiences, we see the importance of changes of position and changes of viewpoints, so as to obtain information as close as possible to the complexity of the field studied. I was not "familiar"[35] with these situations and was able to take some distance at the end of each of the events studied, none of which lasted very long. Another advantage was that I studied these events from hierarchically different positions. Some jobs, such as security guard, were close to the grass roots level, while others were more distant, such as site manager or technical counselor, or again, involved great responsibility such as security coordinator at the *Vieilles Charrues*. These positions correspond to different points of view. But one might object that active observation is caught in the trap of the possibility of making decisions (actually there is very little freedom in this respect, given the constraints connected with the duties and the number of actors involved). In this sense, we may speak of a position in which one has increasingly less freedom as one rises in the hierarchy, given the confidentiality of the information received, in particular. Now the fact of holding a position carrying weighty responsibilities does not prevent one from reflecting on that position once the event is over. Still and all, as a decision-maker and participant in negotiations, I lose one form of neutrality. I am bound by the demands of my employers. So, in order to partially avoid that trap, I resolved that in the future I would act more as a counselor and leave it to the decision-makers to act. This was the position I adopted, in particular, at the International Jazz Festival and the *Francofolies* in Montreal in 2004. In counseling one may set forth all of the possibilities for evaluating and implementing arrangements, and especially one may include the viewpoints of all participants in the negotiation. In doing so one does not take sides with any particular actor.

- Awareness: the difficulty of preserving the goal of scientific neutrality while meeting the demands of professional activity

Like any research on individuals and social groups, on actors and their practices, my work raised ethical and deontological issues. By taking interest in actors and in both private and public security institutions, and by becoming a full-fledged actor in the field of security toward the end of my research, I had to make some "compromises".

My subject, security, involved a field in which information had to remain confidential, and could not be released. It was my duty, then, to preserve secrecy. Things were clear, however, with my contacts. The articles I wrote were automatically sent to them before they were submitted for publication. The aim of my research was quite clear and well-defined, even if there was an employer/employee relationship from the second year on. I was there for the purposes of my doctoral research only, and was viewed in that perspective.

The point at which the ethical dimension was most touchy was the one raised by Tremblay (1985, 31): the fact of not taking sides. As I worked on risk management, my work produced changes in the relations between various actors, and especially between private and public actors (Sabourin, Grenier, 1993). I took part in setting up arrangements and strategies for action which had repercussions in the field.

Last, the fact of having set up an intervention squad to take in dealers on the premises raises an ethical problem, in which the participant observer has to make a clear separation between his position as researcher and that of the worker paid to find a solution to the problems encountered by private and public safety actors.

The Observation Itself: Inseparable Phases of Interaction with the Surroundings, Interviews, Observation, Note-Writing and Diversified Reading

If we apply the distinction operated by Peretz (2004, 71),[36] I did open observation of formal organizational groups. I will attempt to provide a synthetic, systematic narration, as complete as possible, of the various phases of my observation. This, as Peretz points out, seems to be lacking in most research.

- Previous relations with the milieu to be observed

I have now acquired a degree of expertise – although I dislike the term – in a field in which we see the various members of the organization function by trial and error, on the basis of their own experience. But people are rather poorly informed of what is done elsewhere, essentially owing to lack of time. Because I had worked in different places, the actors found that having the viewpoint of someone outside their organization was interesting for evaluating their work and introducing improvements.

- Entering the milieu

This made it easier for me to enter the milieu and to negotiate with officials. There is a sort of give-and-take: access to field work, to individuals and to information in exchange for advice. This supposes, then, that one be aware of which side one is on: with the private or public actors, for example, in the case of the organization of these shows.

But in contrast to my experience in France for instance, where I was a salaried worker, in Montreal I was completely independent, free to move around, meet people and make my own schedule, among other things.

Being a researcher also provides legitimacy, of a very different sort than that of a security counseling company, which may have interests (financial, for instance) in giving particular advice.

- Settling down in the organization and developing relationships: observing, writing (taking notes, recording and charting observations), analyzing and presenting findings (differences between academic and professional expectations)

First, there is a role defined for the observer by those observed. Then the field work itself begins. The idea is to try to understand the space, the actors circulating within the space, and the organization, roughly at first, then increasingly precisely. There is then a constant to-and-fro movement between observation, more or less formal interviews and the analysis of documents. The fact of having institutional legitimacy makes it possible to take notes and make recordings, and facilitates first contacts, although there may be some resentment against an outsider whose "hidden" motivations, or what they suppose are "hidden", are not necessarily known to them. Everyone must be placed on equal footing, and people's confidence must be won so as to learn more. There is no secret, it takes time and personal investment to ease relationships.

The fact of having worked for several years and in several contexts also teaches you what to look at and where. The interpretative framework I developed in the course of the thesis helped me to collect information faster. To complete your approach, you add a log book, a diary or a tape recorder.

Next comes the time to present your findings, with different objectives depending on whether you are addressing an academic milieu or a professional one, with the limit that you must respect confidentiality to avoid being excluded from the field. This is an extremely delicate time, in which the legitimacy of the research project is at stake, and which will – or will not – lay the bases for a more in-depth, long-term study, which is one of the foundations of participant observation.

This article may be reproached for not going into the analysis of ethnographic data. What does this method contribute in terms of findings? This was not my intention here, and readers are referred to other published (or forthcoming) articles in specialized journals.[37] The methodology is inseparable from the object and from the attendant research issues. The point of this research project was to advance knowledge on the management of safety and practices therein during gatherings involving large crowds. The objective was to uncover the negotiated distribution of responsibilities between the different actors (private and public) involved in ensuring the safety of these events, and to discover the technical and human arrangements set up to prevent accidents, offending and violence. The idea was to list the elements of knowledge and know-how; to determine those preventive measures that are effective and those that are not. "Above all, ethnography is an attempt at understanding, by bringing the distant closer, by making what is foreign familiar . . . By definition, the ethnographer is someone who is not content with overviews, who does not confine himself to the categories already in existence for the description of the social world (statistical categories, dominant or standardized categories of thinking). He is skeptical by principle with respect to "generalizing" analyses and pre-established ways of carving up the social world. The ethnographer claims the right to have a priori doubts about the ready-made explanations of the social order. He is always intent on going to take a closer look at social reality . . . It is this curiosity which [leads him] to push investigation farther, to observe in detail, to proceed by close-ups or zooms, when other sociologists systematically take a view from farther above and farther away."[38] For the theme of the negotiation of responsibilities and the roles of institutions and actors in the field of safety, I felt it was necessary to adopt a method that yielded an understanding of the stakes of that negotiation, as well as of the interplay of powers and interests, by immersing myself in the everyday life of those organizations, mostly through preparatory meetings.

The second limit was that in some sense my research project had led me to a degree of involvement with actors on the safety scene such that I was obliged to make decisions, using the means given to me, and as a rule on the short term, so as to meet the expectations of the organization. It was outside of the scheduled events that my research work became fully meaningful in that it obliged me to "be constantly alert to new ideas and to questions requiring verification"[39] and to work on the long term. My position and personal involvement led me to modify some elements in the management of the event itself. That was particularly true of the last two festivals on which I worked: the *Printemps* in Bourges and *Vieilles Charrues* in Carhaix. For the first time I was managing a location, with all of its dimensions (the shows, reception of the public, communication, the staff, etc.), and I was therefore able to integrate the safety dimension with the other dimensions, and to make it a factor in its own right. We had to take people into the place, offer them shared emotions during a show and, a few hours later, put them safely back on the street by which they had arrived. For four years I had had the word safety on my mind, through the elements furnished by actors in safety, and in the space of ten days I myself had to take a position in which to structure a location and make sure it would be safe. The main difficulty resided in the need to integrate theoretical issues such as the balance between freedom and safety in an operational apparatus handling real risks: the need to integrate the issue of new governance of safety with the reception of the public and prevention of potential crises; protection of the public and the negotiation of roles and responsibilities between the various actors.

> It is possible for researchers, sociologists involved in sociological action, to take on these functions. They should not stay in their ivory tower, but

should change their relations with practitioners so as to bring out responses to societal problems.

Lastly, no more than any other research method, participant observation cannot make any claim to exclusivity. Other approaches must be used. I, for instance, realize the limits of using the qualitative approach only. I see it necessary to use tools such as analysis of networks, mapping and the constitution of databases on the problems encountered, among others, in an attempt to cover the complexity of a field such as safety, and ultimately, in particular, to evaluate the arrangements implemented.

Notes

1. Coenen-Huther, 1995, 6.
2. Ibid, 6.
3. Ethnomethodology (Garfinkel, 1967) views institutions "as the outcome of an ongoing process of construction and adjustment between the various actors and logics present" (Faget, 2002, 98). "This explains why ethnomethodology, methodologically speaking, requires that sociologists be direct witnesses to the phenomena they wish to observe and analyze, whenever possible. This implies, to various extens, a position of participant observer." (Coulon, 1999, 202)
4. K. M. Van Meter, 1992, 648.
5. Jaccoud, Mayer, 1997, 211.
6. K. M. Van Meter, 1992, 648.
7. The notion of field is intended here as defined by Bourdieu (1989, 14): "the structure of a field is the state of the power relations between agents or institutions engaged in the struggle, or if we prefer, of the specific distribution of capital which, accumulated during previous struggles, orients subsequent strategies." The field of safety/security alone would then involve several forms of struggles and by the same token of alliances, such as, in particular, the struggles (and alliances) over the definitions that actors may have of the concept of safety or of risk, the struggles (and alliances) for power between the private and the public field, or again, within each field, such as within the police, for instance.
8. Diaz, 2003a. This thesis was based on sixteen different fieldwork experiences: the World Soccer Cup (1998), the Paris-Saint Germain soccer season (1999), the World Rugby Cup (1999), the European Basketball Championship (1999), concerts at the Paris Bercy *Palais Omnisports* (1999), concerts at Paris' Zenith hall (1999), the *Printemps* festival in Bourges (1999, 2000, 2001, 2002), the *Francofolies* festival (1999), the *Eurockéennes* festival in Belfort (1999, 2000, 2001, 2002), the *Astropolis* festival in Concarneau (2000), the international street theater festival in Aurillac (2000, 2001), the World Championship of road cycling (2000), the World handball championship (2001), the Roland Garros tennis tournament (2001), the *Vieilles Charrues* (Old Ploughs) festival in Carhaix (2001, 2002) and the Cannes festival (2002). For an overview, see Diaz, 2003b.
9. The expression "personal observation" was originally used by Webb (Webb, Webb, 1932, 50), and it is in the late 30s that the expression "participant observation" seems to have appeared, in its present acceptation (Platt, 1983). Hughes is the inventor of the expression "observation *in situ*" or "direct observation" to designate the study of collective action or social processes through direct interaction (Chapoulie, 1984), quoted by Jaccoud, Mayer (1997, 212).
10. Be he sociologtist, ethnologist or anthropologist.

11. An overview of this first year of research may be found in Diaz, 1998; 2001.
12. Mendras, Oberti, 2000, 133.
13. Blumer, 1962, 188.
14. Blumer, 1966, 542.
15. Hughes, 1996, 276.
16. Hughes, 1996, 276.
17. Poupart, Lalonde, Jaccoud, 1997, 330.
18. Ibid, 331.
19. Hughes, 1996, 275.
20. "It must be said that the limit between observation *in situ*, the ethnographic approach and fieldwork research is not always very clear-cut. Furthermore, some authors tend increasingly to drop the expression 'observation *in situ*', and to prefer 'fieldwork research' or 'ethnographic approach'. These expressions, it is contended, better reflect the reality of research practices which make as much use of observation as of interviews and documents, in the last analysis," Poupart, Lalonde and Jaccoud, 1997, 85–86.
21. For further developments, the reader is referred to Diaz, 2001, 2003b, 2005 and 2006 (forthcoming).
22. Bouvier, 2000.
23. Schwartz, Schwartz, 1955, 343.
24. Coulon, 1992, 14.
25. Blumer, 1969, 5.
26. This would have been much more difficult to achieve had I been studying an occupation such as doctor, lawyer, police commissioner or other.
27. Peretz, 2004, 52.
28. Rock singer for three years, with several concerts in halls accommodating 50 to 500 spectators; soccer player for 15 years, at the *département* level.
29. I have been passionately interested in all sorts of sports since my earliest childhood, an attentive reader of the sports press, and a fervent spectator of events in stadiums and entertainment halls. By the time I began work on my thesis I had attended close to 50 indoor shows and a number of festivals.
30. The *Printemps de Bourges* and the *Eurockéennes* in Belfort were studied in 1999, 2000, 2001 and 2002, the street theater in Aurillac in 2000 and 2001 and the *Vieilles Charrues* festival in Carhaix in 2001 and 2002.
31. For the technical aspects, at the *Printemps de Bourges*, as manager of the safety commissions, then site manager, which required an ERP2 level (that is, the diploma required for people in charge of places open to the public). Or again, my experience in the martial arts, where my degree (*diplôme d'Etat*) as sports educator gave me more confidence when working at the Belfort *Eurockéennes*.
32. Chapoulie, 1984, 597.
33. The analysis would be incomplete if it did not mention that one of the consequences (not one of the causes) was that this work enabled me, partially, to finance my academic work.
34. On this point, the methodology cannot be separated from the object and the academic questions tied to it.
35. "Researchers studying groups or situations that are familiar to them have to replace their prior attitude of immediate participation by that of the "disinterested" observer unfettered by the judgments and norms attached to the position they occupy . . . According to Hughes, the ability to adopt an objectivating attitude toward the subjects they study is partly due to some peculiarities of their social trajectories." Chapoulie, 1984, 597–598.

36. Peretz (2004, 53–78) distinguishes between clandestine observation in a closed setting, open observation within an informal group and open observation within a formal, institutionalized group.
37. Diaz, 2001, 2003b, 2005 and 2006 (forthcoming).
38. Beaud, Weber, 1998, 9–11.
39. Jaccoud, Mayer, 1997, 243.

Bibliographie

Adler P. A., Adler P., 1987, Membership roles in field research, qualitative research methods, vol.6, Newbury Park, Sage.

Anderson N., 1923, Le Hobo, sociologie du sans-abri, traduction par Annie Brigant, Paris, Nathan, Coll. Essais et Recherches, 1993.

Bayley D., 1994, Police for the future, New York, Oxford University Press.

Beaud S., Weber F., 1998, Guide de l'enquête de terrain, Paris, La Découverte.

Becker H., 2002, Les ficelles du métier – Comment conduire sa recherche en sciences sociales, Guides Repères, Paris, La Découverte.

Becker H., 1963, Outsiders, New York, Free Press, tr. Fr. Outsiders, Etudes de sociologie de la déviance, préface de Chapoulie J. M., Paris, A. M. Métailié, 1985.

Blumer H., 1969, Symbolic interactionism : perspective and method, Englewood Cliffs, N.J., Prentice Hall.

Blumer H., 1966, Sociological implications of the thought of George Herbert Mead, American Journal of Sociology, 71, 5, 535–544.

Blumer H., 1962, Society as symbolic interaction, in Rose A. M., Human behavior and social processes. An interactionist approach, London, Routledge and Kegan Paul, 179–192.

Bossis R., 2003, La question de la professionnalisation des greffiers, Thèse pour le doctorat de sociologie préparée sous la direction de Robert Ph., Université de Versailles Saint-Quentin en Yvelines.

Bourdieu P., 1978, Sur l'objectivation participante, Réponse à quelques objections, Actes de la recherche en sciences sociales, 23, 67–69.

Bouvier P., 2000, La socio-anthropologie, Paris, Armand Colin/HER, Coll. Sociologie.

Chapoulie J. M., 2001, La tradition sociologique de Chicago 1892–1961, Paris, éditions du Seuil.

Chapoulie J. M., 1996, Hughes, Everett Cherrington. Le regard sociologique: essais choisis, textes rassemblés et présentér par Chapoulie J.-M., Paris, Editions de l'Ecote des Hautes Etudes en Sciences Sociales.

Chapoulie J. M., 1985, « Préface », in Becker H. S., Outsiders. Etudes de sociologie de la déviance, Paris, Métailié, 9–22.

Chapoulie J. M., 1984, Everett C. Hughes et le développement du travail sur le train, Revue française de sociologie, 25, 4, 582–608.

Coenen-Huther J., 1995, Observation participante et théorie sociologique, Paris, Editions l'Harmattan, Coll. « Logiques sociales ».

Coulon A., 1999, Ethnométhodologie, in Akoun A., Ansart P., Dictionnaire de sociologie, Tours, Le Robert/Seuil, 201–202.

Coulon A., 1992, L'école de Chicago, Paris, PUF, Coll. « Que sais-je ? », n°2639.

Dalton M., 1959, Men who manage, New York, Wiley.

Davis F., 1959, The cabdriver and his fare: facets of a fleeting relationship, American journal of sociology, 65, 2, 158–165.

Diaz F., 2006, La négociation au centre des questions de sécurité des manifestations festives: logiques, stratégies et processus de partages entre acteurs privés et publics, (in collaboration with Catherine Rossi), Négotiations, (forthcoming).
Diaz F., 2005, Gouvernance de la sécurité et gestion des risques: mise en sécurité de manifestations sportives et culturelles, généralisation à tout type d'espace, in Les violences en milieu scolaire et éducatif – Connaître, prévenir, intervenir (dir. B. Gaillard), Rennes, Presses Universitaires de Rennes, Didact Education, 507–516.
Diaz F., 2003a, Coproduction de la sécurité: un nouveau mode de contrôle et de régulation sociale – La gestion des risques dans les espaces privés lors de manifestations sportives et culturelles, Thesis for a doctoral degree in sociology. Supervised by Philippe Robert, Université de Versailles Saint-Quentin-en-Yvelines, France.
Diaz F., 2003b, Coproduction de la sécurité: une nouvelle forme de l'interventionnisme étatique pour une meilleure sécurité du public? (Le cas des grands rassemblemeuts de populations en France), Déviance et Société, 27, 4, 429–458.
Diaz F., 2001, La sécurité des grands rassemblements sportifs et culturels: une gestion particulière des risques, Paris, IHESI, Etudes et Recherches.
Diaz F., 1998, Sécurité privée et sécurité publique, une nouvelle forme de contrôle social: logique et gestion des risques à l'intérieur et à l'extérieur du stade de France pendant la Coupe du monde de football 1998, Université Paris X-Nanterre, mémoire de DEA de politique comparée et de sociologie politique.
Durkheim E., 1895, Les règies de la méthode sociologique (1988), Manchecourt, Flammarion.
Esterle Hédibel M., 1995, Le rite et le risque: la culture du risque dans les bandes de jeunes de milieu populaire à travers la conduite routière, thèse de doctorat en anthropologie sociale et sociologie comparée, Université Paris V René Descartes, UFR de Sciences sociales, Sorbonne – Sciences Humaines.
Faget J., 2002, Sociologie de la délinquance et de la justice pénale, Ramonville Saint-Agne, Editions Erès.
Filleule O., Jobard F., 1998, The maintenance of order in France. Towards a model of protest policing, in Della Porta D., Reiter H., (dir.), The policing of mass demonstrations in Contemporary Democraties, University of Minnesota Press.
Garfinkel H., 1967, Studies in ethnomethodology, Englewood cliffs (N. J.), Prentice-Hall.
Hughes E. C., 1996, Le regard sociologique: essais choisis, textes rassemblés et présentérs par Chapoulie J.-M., Paris, Edition de l'Ecole des hautes études en sciences sociales.
Hughes E. C., 1960, Introduction: the place of field work in social science, in Junker B. H., Field work: an introduction to social sciences, Chicago, University of Chicago Press.
Jaccoud M., Mayer R., 1997, L'observation en situation et la recherche qualitative, in Poupart J., Groulx L-H., Deslauriers J-P., Laperriere A., Mayer R., Pires A., La recherche qualitative; enjeux épistémologiques et méthodologiques, Boucherville, Gaëtan Morin Editeur, 211–249.
Junker B. H., Field work: an introduction to social sciences, Chicago, University of Chicago Press, 1960.
Marx G. T., 1988, Undercover Police Surveillance in America, Berkeley, University of California Press.
Mead G. H., 1934, L'esprit, le soi et la société (1965), Paris, PUF.
Mendras H., Oberti M., 2000, Le sociologue et son terrain, trente recherches exemplaires, Paris, Armand Colin.
Peretz H., 2004, Les méthodes en sociologie. L'observation, Paris, La Découverte, coll. Repères, n°234.
Platt J., 1983, The development of the « participant observation » method in sociology: origin, myth and history, Journal of the history of the behavioral sciences, 19, 4, 379–383.

Poupart J., Deslauriers J-P., Groulx L-H., Laperriere A., Mayer R., Pires A., 1997, La recherche qualitative, Enjeux épistémologiques et méthodologiques, Boucherville, Canada, Gaëtan Morin éditeur.
Poupart J., Lalonde M., Jaccoud M., (avec la collaboration de Beliveau D. et Bourdages A.), 1997, De l'Ecole de Chicago au postmodernisme: trois quarts de siècle de travaux sur la méthodologie qualitative. Bibliographie alphabétique et thématique, Cap-Rouge, Canada, Les Presses Inter Universitaires, Casablanca, Maroc, Les Editions 2 Continents.
Roy D., 1954, Efficiency and 'the fix': informal intergroup relations in a piecework machine shop, American journal of sociology, 60, 3, 255–266.
Roy D., 1952, Quota restriction and goldbricking in a machine shop, American journal of sociology, 57, 5, 425–442.
Sabourin C., Grenier R., 1993, L'éthique de la recherche menée auprès des personnes assistées sociales, Revue de l'association pour la recherche qualitative, 9, 86–101.
Schwartz M. S., Schwartz C. G., 1955, Problems in participant observation, American journal of sociology, LX, 4, 343–353.
Shaw C. R., 1931, The natural history of a delinquent career, New York, Greenwood Press Publishers (2nd éd. 1968).
Shaw C. R., 1930, The Jack-Roller: a delinquent boy's own story, Chicago, University of Chicago Press (2nd éd. 1966, avec une introduction de Becker H.)
Thrasher F. M., 1927, The Gang (1963), Chicago, Phénix.
Tremblay M.A., 1985, Une première saison en Acadie ou comment je suis devenu anthropologue sans le savoir, dans Genest S. (dir.), La passion de l'échange: terrains d'anthropologues du Québec, Chicoutimi, Gaëtan Morin Editeur, 15–52.
Van Meter K. M., 1992, La sociologie, Paris, Larousse.
Wax R. H., 1971, Doing field work: warnings and advice, The University of Chicago Press, Chicago.
Webb S., Webb B., 1932, Methods of social study, Londres, Longmans, Green.
Whyte W. F., 1943, Street Corner Society: the social structure of an Italian slum.

54

Participant Observation in Prison

James B. Jacobs

Much of what we know about life within prison and other total institutions can be attributed to the reports of social scientists who have functioned as participant observers in order to gather data through intensive interaction with staff and inmates over extended periods of time. Some have entered the field as full time staff members,[1] others as involuntarily committed inmates[2] and a few as known observers without formal status in either the formal or informal organization.[3]

Among the known observers, Gresham Sykes (1966) has presented the most thorough account of life behind the walls of a maximum security prison for men. Sykes pays scant attention, however, to methodology or to the participant observer role. The brief methodological appendix in *Society of Captives* merely admonishes future students to avoid becoming identified with staff or being "corrupted" by the inmate organization. We are not told how Sykes gained entrance to the prison, how he developed informants, how he carved out a viable research role, or how he was able to gain inmate cooperation. The only full methodological report is that supplied by Rose Giallombardo (1966b) who spent a year of participant observation at Alderson, a federal maximum security penitentiary for women. Giallombardo emphasizes her conscious efforts to define for herself and others an acceptable research role. She carefully describes her entrance into the prison, her strategy for meeting and obtaining interviews with inmates and her efforts to maintain cordial but arm's length relations with staff. While Giallombardo's account provides much insight on the researcher's role playing within a

Source: *Journal of Contemporary Ethnography* (formerly *Urban Life and Culture*), 3(2) (1974): 221–240.

women's prison, it is not sufficiently attentive to the fragility and tenuousness of the researcher's position to be relevant to the extreme conditions of a men's maximum security mega prison.

The prison participant observer enters a highly unstable social setting abounding with rumor, suspicion, factionalism, and open conflict. While no participant observer can merely impose his role on those with whom he is interacting, this inability is dramatically apparent to the prison researcher who must daily negotiate the legitimacy, content and boundaries of his role with a society which is hostile to his presence.

For the field worker what is at stake in this "how" of face to face interaction is the very fate of the research. His behavior is without clear meaning to his informants until he settles on an acceptable role. The way in which his informants come to define his role as well as their own (and the way he comes to define his own role as well as theirs) will determine with whom he comes in contact, the kinds of inquiries that are permissible, the meanings assigned to the inquiries, the kinds of information communicated by informants, the meanings imputed to those communications by the researcher – ultimately, the validity and success of the project.

This paper suggests that carving out a workable research role inside prison is at once crucial and highly problematic. Qualitative prison research should be regarded as the product of an intense, dynamic, often unstable, and potentially explosive interaction between the researcher and those being observed.

Entering the Field

Through the influence of Professor Norval Morris and under the sponsorship of a Ford Foundation grant, during the summer of 1972 I carried out a participant observation study of social organization at Stateville Penitentiary, Illinois' largest and most notorious maximum security institution (Jacobs, 1974). Located in Joliet, Illinois, about thirty miles southwest of Chicago, Stateville housed approximately two thousand inmates at the time of the study. For four months, I spent five days a week and several weekends at Stateville. Reasons of prison security and personal apprehensiveness prevented observation during the night shift although that might have been useful. During the days I had complete freedom to walk about the prison unescorted and to conduct interviews and observations in the prisoners' cells, in the work areas, in solitary, and in the prison yard.

Unlike Giallombardo, whose entrance to Alderson Prison was preceded by written announcements distributed to the inmate population, I entered Stateville without a coherent definition of my role having been offered to staff or inmates. To those who inquired, I explained that I was a joint law and sociology student interested in writing a book about "the way life is" in the prison. During the first several days I was given no choice but to accept a comprehensive tour of the entire institution during which time I was thoroughly acquainted with what might be called "the institutional line." Mindful of the Sykes and Giallombardo warnings, I was anxious to keep staff contacts (especially with guards) to a minimum. However, it was necessary to maintain cordial relations with staff and administration in the face of considerable skepticism about the "advisability" of this research.

The extent of factionalism and conflict within the prison could hardly be exaggerated. The most prominent division, of course, was that between staff and inmates. The caste structure of the total institution has been extensively examined by Goffman (1961) and others. We need not belabor it here. Scarcely less significant than the caste system were the impregnable lines separating the races. At the time of the study, seventy percent of Stateville's inmates were Black, twenty percent were Caucasian and the remainder Latino. Almost the entire staff was Caucasian. Therefore, to a large extent, the caste system was reinforced by the racial divisions.

While the inmates were divided along racial lines, there also existed substantial intra-racial factionalism. About fifty percent of the Black inmates aligned themselves with three large and mutually antagonistic Chicago street gangs (Black P. Stone Nation, Vicelords, Disciples), which had carried their organizational structures and activities from Chicago's streets to Illinois' prisons. In addition, membership swelled within the prison as a result of competition for the allegiance of unaffiliated young Blacks. While the Latino population was relatively small, it was divided between Puerto Ricans and Mexican Americans as well as between members of a Chicago street gang, the Latin Kings, and unaffiliated Latinos. Among the Caucasian inmates, no formal organizational structures existed but there were divisions into various cliques. These cliques could be characterized either as protective or non-protective. Members of protective cliques were secure because of fighting ability, gangland ties, legal skills or alliance with staff members. Whites within non-protective cliques were highly vulnerable to physcial assault, rape, extortion and general harassment. The prison staff was not immune to conflict and factionalism either. Sharp hostility existed between administrators and upper echelon security officers, front line guards and treatment personnel.

Everyone within the prison has a reason for being there. People do not just wander in and out. Consequently, from the start, inmates as well as many staff members were reluctant to accept my explanation for being allowed entrance and free access to Stateville. To my knowledge I was the first individual ever to have been granted this "privilege." Rumors that I was an agent for the F.B.I. or the Chicago Police Department's Gang Intelligence Unit or an investigator for the Governor's staff recurred throughout the research and had to be dealt with daily. As Giallombardo notes, for the researcher who chooses to study prisons, it is a formidable problem to "prove" that one is who one claims to be. The first task was to dispel the notion that I was an imposter.

With respect to this problem, the importance of my association with Professor Norval Morris cannot be exaggerated. While touring a prison workshop on the second day in the field, a middle age Black clerk introduced himself to me as a member of the inmate council and probed for biographical information beyond my cursory explanation. I told B.D.,[4] who later became a key informant, that for several years I had worked closely with Professor Norval Morris at the University of Chicago and had embarked upon this research at his urging. Much to my surprise, B.D. was quite familiar with Professor Morris' "progressive" positions on prison reform. The previous spring, Professor Morris had addressed the high school graduation at Stateville on "the future of corrections" and had received a standing ovation. The connection with Professor Morris thus located my group affiliations with a well known university and suggested that I had access to "important" individuals in the Department of Corrections and in the prison reform movement. At one point in the research, Professor Morris accompanied me to Stateville and met the inmate leadership informally. This immeasurably enhanced my credibility, but even this vouchsafe of my identity did not dissolve all suspicion. After all, it was not inconceivable that both Professor Morris and myself were allied with law enforcement in some insidious venture.

Establishing Working Relationships

The problem of establishing the validity of my claim to an independent and legitimate research identity was never solved. Even at the conclusion of the research, there were inmates who shunned extensive conversation. Informants with whom I shared a friendly rapport frequently refused to answer questions which they felt too sensitive, incriminating, or secretive. I told inmates that they did not "have to" answer every question and that I appreciated their frankness in stating when they did not wish to pursue a particular subject. This in itself provided valuable data by sensitizing me to the limits of my acceptance in the setting and by illuminating the kinds of information only revealed to the most trusted insiders. Often an inmate would, several weeks later, feel more confident about my loyalty and would re-open the previously closed topic. And when it became known that I kept information confidential, inmates gradually began to speak more freely in my presence and permitted me to witness various illicit activities. One informant took me on a tour of the underground stills at Stateville. From time to time I was offered hootch (alcohol) or a joint (a Marijuana cigarette). These were politely declined.

Even after my presence was accepted at the prison and my trustworthiness reasonably well established, the success of the field work was far from assured. It was most disturbing, although understandable, to find that inmates were not interested in contributing to research or in helping a university student further his academic career. From their perspective, my presence was tolerable only if I could be helpful to them.

B.D. told me that I might be "useful" to the inmate leadership and that the inmate council executive committee would like an opportunity to counteract the "poison" given me by the administration and to "run down" the situation from their viewpoint. At the meeting held in a cramped office above the chapel without any staff member present, I was introduced to leaders from the Stones, Disciples, Vicelords and Latin Kings as well as to several articulate "off brands" (independents). For the next several hours I was sharply drilled with respect to my loyalties, contacts, and propensity to act on behalf of inmates. Throughout this "interview" I attempted to outline my research objectives and to emphasize my concern with inmate problems. I explained that I would not write an exposé or go to the press but that I would faithfully report what I saw to Professor Morris, Chairman of the Governor's Advisory Commission on the Treatment of Adult Offenders, and through him, to the Director of Corrections and other senior administrators. I said it might be possible sometime in the future to arrange a meeting between the council and these individuals.

The inmate leaders at once defined my role as prison-reform advocate. They called to my attention innumerable prison conditions that demanded immediate rectification and beseeched me to take action. I expressed concern with their situation and promised informally to investigate abuses. Throughout the research I faced the constant dilemma of appearing morally committed to the inmate "cause" while pursuing interests unrelated to reform but vital to the research.

In order to demonstrate my moral commitment to the inmate "cause" and in order to provide some incentive for highly suspicious inmates to sustain an interactive relationship it was necessary to provide many favors during my months in the field. While this kind of behavior deviates from the ideal participant observer position of neutrality and ignores the warnings of Giallombardo and Sykes concerning role corruption, reciprocities may be necessary if the terms of the interaction are not to be considered violated. With respect to this problem in field work, Lofland observes:

> At the practical level of maintaining the role and one's acceptability to the participants, it is probably necessary for one to perform services in the setting. In terms of "exchange of services" – the pure observer role involves a highly imbalanced relation to the participants. They let him watch but he does nothing for them in return. More immediate reciprocities are necessary. Indeed, in a wide range of emergent circumstances, it will seem peculiar if the observer does not volunteer his help [Lofland, 1971: 98].

Maximum security inmates require and demand a great deal of help. Consequently, I offered legal advice, contacted families and lawyers, made suggestions for strategies in dealing with administration, served as a conduit for messages, and provided mobility from one work assignment to another.

When one of the Stones' leader's daughter was critically ill, I intervened with the warden in an unsuccessful attempt to arrange a furlough. More significantly, I brought two University of Chicago professors to the prison to speak with the inmates about their grievances and took the case against dissolving the inmate council to senior administrators in the Department of Corrections. On another occasion I argued strenuously with the Assistant Director of Corrections on the prison yard concerning the inappropriateness of cancelling the popular inmate art show.

The difficulty implicit in such demonstrations of solidarity is in setting limits. Inmates urged me to carry letters out of the prison, to smuggle them joints, and to leak information to newspapers. With respect to the newspapers, I explained that the short-run shock value of an expose (if it were even printed) would not compensate for the destructive precedent against future outside observers being allowed behind the walls that would surely result from my expulsion from the setting. Despite these arguments, requests for favors persisted and the tight rope act remained precarious through the last day in the field.

At the first meeting with the inmate council executive committee, only slight indication was given of gang affiliations. Individuals were introduced by name as elected representatives of the inmate body. Within days, however, staff members approached me to confide that the inmate council was dominated by gang members and warned me against "falling in with this crowd." The inmate leaders themselves deflected questions concerning gangs, preferring to interact with me in their roles as inmates rather than as gang members. Where this definition of the situation prevailed, conversation was limited to issues of inmate concern. The meeting was only the first of continuous interactions with the gang leaders in which I attempted to cast myself in the role of researcher and to cast the leaders in the role of informants, while they attempted to cast me in the role of advocate and ally and themselves in the role of aggrieved partisans. As a consequence, compromise had continually to be hammered out and renegotiated. At times of crises (for example when the inmate council was abolished), researching had to take a back seat to advocacy; on calmer days, there was more opportunity and energy for lengthy hair-down conversation about prison life.

While the inmate leaders' definition of my role as reform-advocate and their own as inmate-spokesmen prevailed at all "meetings," "team performance" was not maintained in private conversation with each leader. During such occasions, partisan interests and antagonisms among the gangs were aired. I learned, for example, that the Disciples felt betrayed by the Stones' alleged refusal to discipline their own members according to treaty provisions.

In addition, skilled informants were able to provide far more information privately when less sophisticated inmates were not present to misinterpret what was being said. C.N., clerk for the Assistant Warden and articulate spokesman for the Latin Kings, quickly understood that my research posed no threat either to the Kings or to the other organizations. He met me at the gate each morning and personally undertook to orient me to Stateville. Over many weeks, he gradually revealed to me the complex power relationships and social structures of the inmate organization. While G.B., the younger and far less sophisticated leader of the Disciples, initially refused to enter into a relationship based upon his role as a gang leader, he eventually became an outstanding informant, providing deep insight into the organization and activities of Disciples and his own leadership position. What accounted for his acceptance of an informant role was his growing understanding of this new role set, increased confidence in my integrity, and especially, my expanding knowledge of inmate society. In the early weeks, off brand inmates had been extremely helpful in describing the extent and type of power the gangs held within the prison. Gang leaders soon realized that I knew far more about the underlife at Stateville than any staff member. Once I was credited with being an insider, the amount of information to which I was exposed increased exponentially.

Initially I preferred to keep my identity vague, presenting myself as a student of the prison, concerned with matters of prison reform and "telling it like it is" (cf. Vidich, 1969). In order to present a student identity, I carried a University of Chicago clip board, wore a beard, and dressed considerably more casually than members of the staff. At lunch I almost always sat with the clinical staff, the most tolerated members of the administration.

When it came to defining more narrowly the contours of the student role, I vacillated between the identities of lawyer and sociologist. Within the prison, the identities of lawyer and sociologist have different meanings for actors and different implications for interaction. There were two thousand inmates at Stateville at the time of the study and at least as many legal problems. Furthermore, as a law student I was accorded special prestige. Inmates accurately perceived the administration's fear that lawyers and judges would further intrude upon prison practices and procedures. As a law student I was for the inmates an individual of potential power not easily dismissed by the authorities. Thus, despite the danger of becoming full-time legal adviser to two thousand inmates, I was reluctant to reject my law student identity because of its high exchange value and because it placed me in a position to receive information on a great diversity of issues.

The identity of sociologist was more ambiguous for inmates. Most confusing, within the prison "sociologist" is a clinical position, whose incumbents are responsible for writing up diagnostic classifications and parole evaluations and toward which, inmates often feel great resentment. While hostility on the part of inmates toward prison sociologists is considerably less than that directed toward guards, the former are still defined as control agents to be duped and exploited. When asserting claim to sociologist status, it was necessary for me explicitly to disassociate myself from prison sociologists.

Excessive role clarity would have been extremely restrictive. As a law student it seemed appropriate to inmates that I be interested in disciplinary hearings and administrative discretion but not in the involvement of gangs in the distribution of contraband. As a sociologist, all inquiries were permissible but all suspect to the extent that this identity was seen to be integrated within the prison's formal organization. Neither identity was alone entirely satisfactory, but by manipulating the two it was possible to structure a research role of sufficient breadth to make legitimate my concern with a wide variety of prison matters.

Neutrality

Prison is a pervasive institution. Actors are deeply committed to their prison roles. Where individuals define their roles as crucial to their identities, it may be seen as morally impermissible for a new-comer to remain neutral and aloof. Had I assumed a "balanced" attitude toward matters of inmate concern, I would certainly have been shut off from any meaningful interaction with inmates. In this setting, to be neutral is to be allied with the enemy.

On the other hand, with respect to competing factions among inmates, the best strategy was strict neutrality, while emphasizing concern over matters pertinent to all inmates. This convenient stance was not always successful in the face of strong pressures to commit one's self.

I felt, for example, that the problem young White inmates were having with respect to sexual pressure and "rip offs" (rapes) was quite important in adequately assessing what appeared to be an increasingly anarchic situation brought about by the influx of street gangs into the prison. Despite careful explanation of my purposes within the prison, White inmates imputed to me a sentiment of racial solidarity with Whites and a motive to work on *their* behalf to improve *their* position. When they concluded that I was unwilling to take decisive action and that I would continue to maintain my relations with Black gang members, our rapport deteriorated rapidly, culminating with the following written note:

> Mr. Jacobs: Hey, you Super Liberal Piece of Shit. So far this week two white guys have been jumped on and beaten by groups of Afro-Americans.** The next time you go over to "B" house lock-up to let some jackoff ripoff artist know who told on him for raping some white kid so he can get even later on why don't you tell him about yourself too. Also instead of doing your bull shit research from an armchair, why didn't you come in as an inmate so you could find out what it's all about, you phoney cock-sucker.
>
> **Sub human nigger scum

What had been a shaky working agreement between this faction and myself had utterly dissolved. Now I was redefined as a "nigger lover" and racial traitor. No longer was there enough common agreement on social identities and social tasks to support more than the most superficial communication. Perhaps this demonstrates that neutrality itself is a role enactment subject to interpretation?

Because of the frequent tension between the Disciples and the Stones, the best strategy was to cultivate closest ties with the Latin Kings and the Vicelords. From informants within these groups, I received a valuable perspective on the nature of gang organization and activity within the prison without being pulled into a possible conflict situation. It was not long, however, before a Black inmate informed me that rumors were circulating throughout the prison that I was actually Puerto Rican – that although I tried to conceal it, I was propelled into close contact with my Latin people! Once again, imputed identification (with the Latino faction) was damaging for my "neutral" role, since the interests of the Spanish and Black gangs were not wholly coincident. It is no easy task to maintain the confidence of all partisan groups. If research is to be carried out for a long period of time in such a conflict-ridden society, it may eventually be necessary to focus one's primary attention on one faction.

Even with respect to a single faction, it is important to honor status hierarchies and local arrangements lest the researcher unwittingly get pulled into an intra-organizational dispute. Failure to observe this principle ultimately led to the loss of meaningful communication with the Black P. Stone Nation.

A cooperative clinical counselor introduced me to an inmate gang member he thought would be especially insightful. I explained to L.B. that I was interested in understanding what life was like in the prison and asked him several questions about how the presence of the gangs had "changed things" since prisons were described in decades past. L.B. told me that he was a lieutenant in the Stones and vaguely described some of the obligations which this position entailed while in the prison. Not long after this conversation I was "summoned" to the cell of one of the Stones' top leaders. There I was warned that I was to stop asking questions about the Stones' organization of anyone other than B.J. himself. B.J. reminded me of another serious breach of protocol that had occurred a week before. As I recalled that incident, three women had approached me in the prison parking lot one rainy afternoon after the day shift had been relieved. They explained that their car had broken down and asked for a ride to Chicago. During the ride I told them about my research and they volunteered that they were related to various Black P. Stone Nation members. I asked further questions about prison and the history of the Stones until they reached their destination. In the prison the next day, several Stone leaders interrogated me concerning that conversation. At the time I felt that my explanations had satisfied them.

Apparently it had not satisfied everyone, because B.J.'s insistence that I had delved too deeply into the activities of the Stones led him, on behalf of the Stones, to break off all but superficial relations with me during the remainder of the field work.

Problems of Proof and Inference

Participant observers are often concerned with the objective accuracy of informants' statements of fact and interpretation. Some, of course, have abandoned the search for objective reality in favor of "getting into the head" of the actors in order to discover the subjective reality of various positions in the social setting (Becker, 1958). The prison researcher faces this epistemological problem and a more practical one as well. In this highly balkanized and conflict-ridden society, the known observer must be especially sensitive to being "duped." This term is reserved for those instances when informants consciously falsify or distort information in order to stimulate some impression or action.

The researcher needs to be sensitive to those issues with respect to which he is likely to be duped. Once they are identified, he can then proceed to investigate the interests perceived to be at stake. By recognizing the social organizational positions of various informants, it may be possible to check and recheck information so that ideology and interpretation can be analytically distinguished.

During the summer of 1970, a melee of unclear dimensions was touched off at Stateville's ball diamond. In the course of the melee, several guards were seriously injured and the entire prison was consequently placed on "lockup" (inmates confined to their cells) for four months. In order to understand the impact of the gangs on prison social organization, I felt it crucial to reconstruct that event. Yet I was unable to do so even after several months of interviewing.

The staff adhered to the point of view that the fight between two inmates which touched off the conflict had been "staged" by the gangs in order to draw line officers into a conflict. The gang leadership denied this, claiming that the participants in the fight were unaffiliated with gangs and that the administration exploited an insignificant incident in order to lock up the entire prison and transfer gang leaders to a new super segregation unit (S.P.U.). Independent inmates were somewhat divided in their account of the incident. Some adhered to one or the other of the above interpretations while others volunteered alternate versions. After checking and counterchecking stories for weeks, I was unable to satisfy myself either as to objective facts (e.g., were the two individuals members of gangs?) or as to whether informants actually believed the interpretations they were offering.

The existence of the gangs within Illinois prisons was a sensitive issue for everyone. Staff members were committed to an anti-gang ideology and attributed all disorders at Stateville to their presence. Whoever I was and whatever use would be made of my research, the staff was interested in seeing the gang leaders defined as troublemakers and the administrative decision to segregate the leadership at S.P.U. interpreted as necessary and legitimate.

Gang leaders, of course, wanted to present their ideology and activities in the light they perceived would be most favorable to the researcher. Thus, in discussing their grievances, they constantly stressed their commitment to social and political action programs on the streets and to progressive prison reform behind the walls. When questioned closely about the racial assaults, shakedowns, and illicit activities which most Whites attributed to them, they explained that the gang leadership opposed all these activities, that accounts of them were highly exaggerated and that they were powerless to police their soldiers when they could offer them nothing in the way of administrative concessions for adherence to institutional rules.

Several of the thoughtful "old cons" at Stateville were among my most cooperative informants. They readily grasped the purpose of the research and helped introduce me to veteran inmates who could provide perspective on the changes in prison life over the past several years. Even the old cons and independents, however, were not without vested interests in the setting. The old con power structure, dominated by clerks, runners, and knowledgeable old timers, had been destroyed as the gangs solidified power. For the first time old cons no longer enjoyed a secure existence at Stateville. Not wishing to antagonize the powerful gang hierarchy and risk physical harm, these informants spoke cautiously about the gangs, although often there was implicit in their conversation disapproval of "gangbangers" and their methods.

Inmates, who often have had long experience in duping representatives of the establishment, may provide information other than that which they believe to be true even where vested interests are not perceived as threatened. Inmates may simply exaggerate in order to emphasize their own power and importance or for lesser reasons of self-aggrandizement. This was true of elite Whites who denigrated the importance of other factions. For example, informants from this group stressed their "connections" inside and outside the prison, boasting that they, as opposed to any gang leader, could obtain steak and a bottle of scotch any time they desired.

Furthermore, inmates may fabricate information in order merely to sustain an interaction that they find intrinsically satisfying. Giallombardo reports that on many occasions she had to put her research aside in order to provide a sounding board and emotional outlet for an inmate who had had a "bad day." This was no less true at Stateville. Many inmates found conversing with me in private offices or around the prison yard a pleasant break from the boredom and stress of prison life. In the researcher they found a polite and sympathetic listener. One "wise" inmate suggested to me that various individuals sought out my company only in order to spend a few moments with a young male.

The known observer should never be insensitive to the impact which he is having on the setting. Individuals may believe that there is something to gain by allying with him. For example, being seen locked in serious conversation with the researcher may serve to validate and reinforce the leadership status at the expense of potential and actual rivals. Few days passed at Stateville without at least one inmate flagging me down on the yard or interrupting an ongoing conversation in order to ask when I was coming to interview him.

Ethical Problems

The deprivations and pains of imprisonment have been documented over the years. The gravity of the prison setting was dramatically impressed upon me at the initial meeting with the executive committee of the inmate council. The field notes from that meeting state:

> Since Professor Morris' speech last spring things have been continuously deteriorating. Everything will be coming to a head soon because ________ is coming back to Stateville which means a return to the old order. This individual thinks of inmates as scum and is an admirer of the repressive system of Warden Joe Ragan. For weeks the guards have been telling inmates that the inmate council will be destroyed, that it is their turn to get even. Weapons are being stockpiled up front and the inmate leadership fears that an incident will be provoked or staged in order to carry out an Attica-like genocide. Over and over again the inmate leaders urged me to take them seriously and not to consider their fears paranoid. They are convinced that a riot is imminent and complain that they can no longer control the rank and file. All this was summed up by one gang leader who said, "The whole place could blow any day."

To be sure, prison is neither a pleasant place to live or in which to carry out field work. It is depressing to enter the din of the cell, to observe men shouting to you while gripping the bars of their steel cages. Despite one's sincere concern for those inmates one knows personally as well as for inmates in general, it is scarcely possible to avoid being hounded by the feeling that one does not care enough. The daily return home to Chicago was a constant reminder that prison for the researcher is merely a field setting, a place to stop on the road to other goals.

According to Lofland (1971: 97), one of the gravest problems in maintaining the observer role through the field work situation is the experience of marginality often felt by the observer. Although he is so close to the participants that he can empathize with their joys, pains and boredoms, he is never really one of them. Very often during the field experience it was pointed out to me that when the day ended I was going home while the deprivations attending incarceration persist inexorably. Often inmates asked why I didn't become an inmate for a week or two in order to experience the totality of their world. During such questioning it seemed to me that my self conception was severely threatened; the marginality of one's position in the field often initiates introspection into the ethics of that position. Perhaps as Vidich has said:

> In avoiding commitments to political issues, [the participant-observer] plays the role of political eunuch. He is socially marginal to the extent that he measures his society as a non-involved outsider and avoids committing his loyalties and allegiances to segments of it. This is not hypocrisy but rather, as Howe has noted of Stendhal, "it is living a ruse." Being both a participant and an observer is "the strategy of 'Deceiving the society to study it and wooing the society to live in it'" [Vidich, 1969: 81].

The known observer can never really belong. Perhaps this is best discussed in terms of role strain. When, because of faulty role enactment, moral qualms, or marginality, the gap between self and role widens, one's role playing becomes more self-conscious and the performance consequently more problematic. The individual increasingly feels more frustrated and is more likely to withdraw from the scene to repatch the damaged role. This is a tactic I found necessary throughout the research.

As one wanders about the prison pursuing research hypotheses and preparing field notes, one cannot help but wonder if this is not the part of the voyeur. Consistently deflecting comments from complaints and accounts of suffering in order to steer conversation toward tentative hypotheses, the question arises as to whether this type of research should be done at all.

My best answer is that such research must be continued. Not only is the prison an important setting in which to study individual and group behavior, but prisons too long have been isolated from the scrutiny of neutral observers. Even where the researcher's avowed purpose is not prison reform, it is important that prisons be open to social scientists who can bear witness to what occurs within. Observing the prison disciplinary committee or pursuing administrative channels to check out inmate grievances, I found the staff, conscious of my status as an attorney, anxious to follow written procedure and due process. The presence of outside observers may have the desirable effect of encouraging all actors within the setting to adhere to the rules.

It might well be argued that all field researchers have some moral debt to the individuals who have made their research possible. These obligations can perhaps be fulfilled through individual reciprocities (e.g., parole recommendations, help in finding employment) or through more diffuse contributions. Giallombardo (1966b) reports that she told inmates at Alderson that though her research could not be expected to help them personally it was aimed at helping "women in trouble everywhere."

This may be the consequence if those of us who have had the extraordinary opportunity to look at life behind the walls have the courage and desire to make known our feelings as well as our more technical research. Academics to date have played a not unimportant role in the prison reform movement and that role should become more significant in the future.

Even when we are certain that such research should be continued we cannot be insensitive to the impact which the publication of the research may have for the participants in the field. Illumination of the gang situation in Illinois prisons could, of course, result in severe repercussions for the gangs and their leadership. This is unlikely, however, since prison administrators are already aware of the gangs with whom they are in daily confrontation. The initials used in this report to refer to leaders are fictitious, and the substantive information on gang activities, no more than what an alert administrator would in any case be likely to know.

The presence of the gangs at Stateville cannot and should not be ignored. It is crucial to understand the society of captives in order to establish priorities, allocate resources and develop meaningful reforms. Perhaps this research fosters opportunity for the development of a more rational and sensitive prison administration?

In their manifesto for sociologists, Etzkowitz and Schaflender (1968) call upon field workers to get involved; to help, to work for the people working for us. They claim that the scholarship which emanates from such participation is superior to that of more passive research. In prison research, active involvement in the inmates' struggle to achieve humane penal conditions is an important and necessary contribution. In addition, it may be the only way possible for the research-observer to overcome his own sense of role strain, marginality, and anxiety.

Notes

1. Cf. D. Clemmer (1958); E. Goffman (1961).
2. Cf. B. Bettelheim (1958); J. Galtung (1961).
3. Cf. G. Sykes (1966); Morris and Pentonville (1963).
4. Initials referring to inmates are fictitious.

References

Becker, H. S. (1958) "Problems of inference and proof in participant observation." Amer. Soc. Rev. 23: 652–660.

Bettelheim, B. (1958) "Individual and mass behavior in extreme situations," pp. 300–310 in E. Maccoby et al. (eds.) Readings in Social Psychology. New York: Holt.

Clemmer, D. (1958) The Prison Community. New York: Holt.
Etzkowitz, H. and G. Schaflander (1968) "A manifesto for sociologists: institution formation – a new sociology." Social Problems 15 (Spring): 379–408.
Galtung, J. (1961) "Prison: the organization of dilemma," pp. 120–143 in D. R. Cressey (ed.) The Prison. New York: Holt, Rinehart & Winston.
Giallombardo, R. (1966a) Society of Women: A Study of a Women's Prison. New York: John Wiley.
——— (1966b) "Interviewing in the prison community." J. of Criminal Law, Criminology and Police Sci. 57: 318–324.
Goffman, E. (1961) Asylums. Garden City, N.Y.: Doubleday.
Jacobs, J. (1974) "Street gangs behind bars." Social Problems 21: 395–409.
Lofland, J. (1971) Analyzing Social Settings. Belmont, Calif.: Wadsworth.
Morris, T. and P. Pentonville (1963) A Sociological Study of an English Prison. London: Routledge & Kegan Paul.
Sykes, G. (1966) Society of Captives. New York: Random House.
Vidich, A. (1969) "Participant observation and the collection and interpretation of data,' pp. 78–86 in G. McCall and J. Simmons (eds.) Issues in Participant Observation. Reading, Mass.: Addison-Wesley.

55

A Spy, a Shill, a Go-between[1], or a Sociologist: Unveiling the 'Observer' in Participant Observer

Susan B. Murray

Introduction

Back stage excerpt from my field notes 1993:

> Phone call from Char [union organizer]. I started to take notes on this phone call but I stopped half way through. Something about our relationship crossing the line into friendship makes me uncomfortable taking notes. I also felt a twinge of this last night while talking to Sarah [another union organizer] but dismissed it and took notes anyway. It's not that I have a problem with keeping a record of what we talked about, it's the physical act of taking notes while she is talking to me – and her not knowing I'm taking notes and me not saying.

Front stage excerpt from my teaching notes 2001:

> In my office at the university talking with a graduate student about her participant-observation field study:
>
> **Graduate Student**: I feel a little strange taking notes on these people. I mean they're my friends and I believe in what they are doing.
>
> **Me**: Well you can be a part of their movement and still take notes.

Source: *Qualitative Research*, 3(3) (2003): 377–395.

Graduate Student: Sometimes I just get so caught up I forget to take notes. Or, we are out all night doing an action and I don't have the time.

Me: If you want to finish this (field study/thesis) you have to be a sociologist first. The only way to get through is to really see yourself as a sociologist – more than anything else.

What once resided solely in my field notes – my personal ethical dilemmas and debates – I am now required to reproduce as public knowledge (Edwards and Ribbens, 1998). As a thesis advisor to several graduate students conducting field research, I find myself in the difficult position of having to reveal the kinds of decisions one must make as a field researcher. Having completed two participant-observation field studies as a graduate student, I have pages and pages of field notes documenting my ethical and methodological journey alongside my observations and interpretations from the field. And while I feel personally comfortable negotiating the ethical quandaries of my own work in the field, it is quite another story to articulate those quandaries to a new generation of field researchers. To teach what I know about doing fieldwork means allowing a portion of my audience to see behind the scenes – offering them, in essence, a back stage pass to the everyday world of field research.

In my work and in this analysis, I use Goffman's (1959) dramaturgical metaphor to try and make sense out of the experience of doing fieldwork.[2] I find his analysis of 'regions' and 'region behaviour' especially instructive in sorting through the fragmented identities endemic to the work of participant observation. According to Goffman, 'a region may be defined as any place that is bounded to some degree by barriers to perception' (1959: 106). Regions, in turn, contain both 'front' and 'back' stages – the former where the performance is given, and the latter where it is routinely contradicted. These constantly shifting regions are about place and identity. Some regions are defined and delimited by physical space. Other front and back regions arise in response to who is present or not present. And finally, within the confines of an individual performance, front and back regions develop and dissolve in response to shifting identities.

Applying this framework to the experience of fieldwork, I conceptualize the acts and actions I undertake as a 'participant' as a front stage performance, and my analysis of this performance [enacted in my field notes] as a back stage performance. It is my contention that much of the process of becoming a sociologist[3] – moving from a primary to a secondary identification with this identity – occurs within these shifting spaces in the field. Up to this point my back stage performance has had no audience – no one has been granted access to my field notes.[4]

My status as 'professor' now compels me to grant partial access to the back stage arena that characterizes 'doing fieldwork'. My status as qualitative researcher necessitates that access be undertaken systematically. The article that follows is, therefore, based upon an inductive examination of my 'methodological' and 'personal' field notes. My reasons for pursuing this analysis are both practical and, at this stage in my career, morally necessary. Practically speaking, I need better answers for my students. As a qualitative researcher I know that good reflexive practice is endemic to the fieldwork process – data collection, analysis, and writing cannot proceed without reflexivity. In order to understand the structures that are generated through face-to-face interaction the researcher must become a part of those interactions. At the same time, she must be continually cognizant of the larger social structure of which that interaction is only a part. The methodological decisions that I must make in the field are made in reference to the larger structural (and political) context that frames my research. When I began teaching it came to me that, though I understand this connection, many of my students do not. The problem is that they grasp the methodological skills and techniques much more quickly than the theoretical framework that necessitates those skills and techniques. This analysis, in part, arises from my need to bridge this gap and offer my students clearer guidelines for their acts and actions in the field.

Morally, I need to feel certain that my performances in the field are guided by both professional and personal ethical standards. Two main questions guide this examination: (1) How did I negotiate my multiple identities in the field given the complex web of social relationships that characterized my field sites, and (2) What am I really, 'a spy, a shill, a go-between' or a sociologist?

After a brief discussion of both theory and method, my analysis addresses the two questions above. The negotiation process is unveiled through a close examination of several key 'situations' from my field experiences. The situations are laid out as acts in a play. Each 'scene' both builds upon the last, and moves deeper into the methodological plot. This analytic process parallels my own emergent understanding of my self as a field researcher. The question concerning my status as 'a spy, a shill, a go-between' or a sociologist is really a question about the ethics of doing fieldwork. These terms, taken from Goffman's (1959) analysis of *The Presentation of Self in Everyday Life*, characterize the issues of loyalty, keeping secrets, confidentiality, and informed consent that confront the field researcher during her daily rounds.

The Theory Behind the Method

In his chapter on 'The Methodological Position of Symbolic Interactionism', Blumer (1969:1–60) instructs the sociologist to enter the everyday empirical world that is the focus of their study. In this essay Blumer attempts to disrupt the 'nonsensical' and 'profoundly misunder[stood]' position of symbolic interaction and naturalistic inquiry within the discipline of sociology (p. 49). He is responding to the, then current, positioning of established scientific protocol and direct examination of the empirical world as opposites. Blumer decries the position of the 'detached outside observer', and instead asserts '. . . if the scholar wishes to understand the actions of people it is necessary for him to see their objects as they see them' (p. 51). From a symbolic interactionist perspective, the social world is composed of individuals continuously creating and recreating meaning through interaction. Social structure, social organizations, and indeed social stratification, are all structures-in-process upheld through interaction. To 'discover' how social life is possible within any given structure, organization, or system of stratification the sociologist must become a part of that complex web of social interactions. Through systematic observation and careful analytic reflection, the sociologist – as both objective social scientist and subjective participant – is able to re-present the social world under study.

Adding a feminist methodological voice to Blumer's convictions means paying even closer attention to the 'subjective' dimension of this methodological equation. Initial calls for a feminist revolution in sociology framed issues of subjectivity in reference to the race, class, gender, and power dynamics of both the researcher and those being researched.[5] In their demand for a deeper exploration of the knower/known relationship, Stacey and Thorne (1985) called specifically for a feminist epistemology that attends equally to race, class, sexuality, and gender (p. 311). Paying close (and systematic) attention to the researcher's own social location is, in other words, an essential component of the ethnographic process. In addition, other feminist social scientists have cautioned researchers to attend to problems of trust (Reinharz, 1992), personal involvement (Brettell, 1997; Letherby, 2000; Oakley, 1981; Stacey, 1996), reciprocity (Adler and Adler, 1991), reflexivity (Hurtado, 1996; Letherby, 2000), and the dangers of emotional involvement in the field (Kleinman, 1991; Lee-Treweek, 2000). Like Stacey (1996: 90), I took up my fieldwork armed both with these 'feminist principals' and the ethnographic methodologies of Blumer (1969), Humphreys (1970), Liebow (1967), and others. And therein lies the ethical challenge.

As a social scientist my definition of various field situations may stand in opposition to my definition as subjective participant in those situations. In the opening quote from my field notes, for example, I am – at the same time – a union organizer, a field researcher, and a friend. From the vantage point of these identities, the possible lines of action available to me may head off into two (or more) different directions. A phone call from another union organizer initiates a 'situation' that requires a line of action to be taken. As a field researcher, conducting participant-observation research on childcare union organizers, I pick up a pen and start taking field notes on the phone conversation. As a friend to the person on the phone, I hesitate in my note taking. Ultimately I stop taking notes on this particular call and later I try to explain this decision in my personal notes: 'Something about our relationship crossing the line into friendship makes me uncomfortable taking notes.'

My decision to not take field notes in *this* situation privileged my identity as 'friend' over that of sociologist. In the phone call the night before, I made the opposite choice. The question is, is one choice more ethical than the other?

These situations – these moments of ethical debate and decision-making – are typically hidden from view. They are part of the interpretive process that shapes my self as sociologist. As the 'game' (Mead, 1934: 152–157) commences and the internal dialogue begins, I take into account each of my identities and those of any other persons present. In addition, my graduate-level field methods class, the American Sociological Association code of ethics, my close reading of Babbie (1992), Blumer (1969), Bordo (1994), Denzin (1978), DeVault (1999), Emerson (1983), Edwards and Ribbens (1998), Oakley (1981), Schatzman and Strauss (1973), and Stacey (1997) all coalesce into a 'generalized other' that asserts itself in these moments of interaction. The attitude of my 'generalized other', which Mead (1934) defines as 'the attitude of the whole community' (p. 154), forms the basis for most of my actions.

All along I assume this generalized other is a feminist other, a moral other. I feel comfortable making decisions, choosing lines of action, and relentlessly pursuing meaning as it is revealed in the empirical worlds I study. And then, in the process of turning this private knowing into public knowledge, I am confronted with how this must look to the outsider – to the novice. If my students do not yet have a clear understanding of the theory (and politics) behind the method, then how are they making (ethical) sense of the methodological techniques I am teaching them? When I advise my students to be – first and foremost – sociologists in all their interactions in the field, what am I really asking them to be? What characterizes the occupational realm that I am inviting and encouraging them to join? And, am I following my own advice? The goal of this essay is to make visible – to unveil – this back stage region of my profession.

The Method Inspired by the Theory

Data for this article were collected as part of two field research projects (Murray, 1988, 1995). In the first project I spent two years as a crisis line volunteer and relief staff at a battered women's shelter. In the second, I worked as an administrative assistant, cook, janitor, and union organizer at several childcare centres. In my research, conducted primarily in Northern California between 1985 and 1995, I used a combination of participant-observation field methods, in depth interviews, and survey methods.

As a participant observer I kept systematic and detailed field notes. Following the advice of Schatzman and Strauss (1973), I divided my note taking into four categories: observational notes, methodological notes, theoretical notes and personal notes. After a day or a night in the field I would return to my computer and spend almost as much time writing field notes. My writing would switch back and forth between the four categories as my observations in the field reminded me of additional field sites I agreed to visit, of something Goffman wrote about, or I simply became aware that I needed some sleep. In the end, the observational and theoretical notes grew into dissertation chapters and journal articles (public knowledge), and the methodological and personal notes remained as private documentation of my struggles as both participant and observer. They remained, in other words, in the back stage arena of my work life – until now.

Act One: Battered Women's Shelter Intake

> **The Setting**: Three of us from the shelter drive to a local motel to do an intake on a woman who has called on the crisis line. The two other women with me are long time shelter workers (it is 5.30 pm – shift change). Normally we do not go to where a woman is directly staying. We meet her in a public place to avoid endangering ourselves. But this woman is afraid she is under surveillance and does not have a car.
>
> **Front Stage**: We knock on the motel door and a woman's voice asks us through the door who we are, then lets us in. Maria (staff) and Katie (the woman) sit on the bed. Pamela (the other worker) and I sit in chairs on either side of the bed. Katie asks us our names. We tell her and Maria explains, 'we all work in the shelter.' Maria then asks Katie to 'tell us about your situation.' Katie says that she is on SSI (Supplemental Security Income) and had been staying in transitional housing and that one of the counsellors in transitional housing had raped her. She had called the police and filed charges. Her social worker had set her up in the motel. Today – before calling the crisis line – she had received a life-threatening note from the man who had raped her. During the telling of this story both the other counsellors are nodding at Katie. At one point Maria puts her hand on Katie's arm. After about 20 minutes Maria interrupts Katie and says she has to go. The other counsellor Pamela says 'I'll walk you to your car.' I follow them out. Pamela says to Katie, 'I will be right back.' (Field notes 11/18/86)

Implicit to every interaction are front and back stage regions. There are several front stage performances being enacted during this 25 minute interaction. Maria, Pamela, and I are presented as 'all working in the shelter.' No status is given. We are presented as workers in seemingly interchangeable roles. The back stage to this presentation of status consistency is that Maria is a shelter manager, Pamela an intern, and I am a volunteer. We occupy very different positions of status and power within the organization, but present ourselves as status equals. In our performances we also exhibit a consistent message of sympathy (constantly nodding, emitting sympathetic 'ummhm's', Maria's hand on Katie's arm), and belief. 'Belief' is performed through the intentional avoidance of certain behaviours: no furrowed brows, raised eyebrows, no questions asked with an inflection of skepticism, no probing inconsistencies. From an organizational perspective we are 'peer counsellors' operating under the feminist and political assumption that women tell the truth about abuse. We serve women by believing them. Other social institutions have failed women by not believing them.

> **Back Stage**: We walked down the length of the motel porch. Pamela asks Maria what she thinks about Katie's story. Maria says 'we definitely should let her into the shelter.' Pamela agrees nodding and then continues, 'there is something troubling about her story.' Maria, 'I agree, but we need to let her in. At least for tonight.' They both look at me. I nod in agreement . . .
>
> Ten minutes later: After hugging Katie goodbye and telling her we would pick her up at 7.00 pm, Pamela and I are back in the car on our way back to the shelter: I ask Pamela what she finds 'troubling' about Katie's story. She says, 'well, she says she is on SSI and women on SSI sometimes have more problems than we can handle.' 'What do you mean?' I press. 'It's not that I don't believe her, it's just that there are certain inconsistencies in her story. It is difficult to sort out stories like hers.' (Field notes 11/18/86)

The feminist stance of unwavering loyalty and belief is confronted with the exigencies of running an emergency shelter for battered women. In this situation, the team performance of the shelter workers is immediately – though carefully – contradicted upon arrival back stage. What is '*troubling*' about Katie's 'story' is not the violence she described, but instead the veracity of her statements. On the porch and in the car – away from the performance in the motel room – the shelter workers indicate some disbelief in Katie's story. While the front stage performance is not entirely contradicted, it is also not entirely over. In the car Pamela's performance continues with me (an agency volunteer) as the audience instead of the client. Though she never says that Katie is lying ('It's not that I don't believe her . . .'), she does draw upon symbols that suggest fabrication or deceit. The use of 'story', the reference to 'inconsistencies', and the necessity of 'sorting out' her words all indicate an underlying current of disbelief.

This telling shift from front stage to back region becomes one of those defining moments for me – one of those interactional planes where I find I must choose. I am a feminist. After coming of age during 'the second wave' of feminism and after 30 hours of domestic violence training I 'buy' the peer counsellor approach. I am a feminist sociologist acutely aware of the privileged interpretive power that I possess (Acker, et al., 1983), I am an agency volunteer who is fully cognizant of the fact that I may, someday, be left 'in charge' at the shelter. There are some women who – though survivors of domestic abuse – have personal needs that are beyond the scope of shelter services. I am a symbolic interactionist. I see the contradiction (the hypocrisy) engendered by this clash of perspectives. I see the emotional danger confronting me (Lee-Treweek, 2000). My response is to slide further into my sociological self. I don't have to live in the contradiction. My job (my moral escape hatch) is to observe interaction, to write about these moments of contradiction, and try to make sociological sense of this slice of social life.

My job, in other words, is to uncover the (power-laden) structures that compel the various social actors in this interactional situation to pursue specific lines of action. Though we are 'peer' counsellors we are not peers in class, status, or situation. Though we are feminists, we are not equal in our access to privileges, resources, and power. We (the shelter workers) are women with homes and jobs, Katie is homeless and reliant on government benefits. And while the shelter began as a feminist collective, it has since been reorganized into a hierarchal agency run by a Board of Directors, an executive director, and the government agencies that fund it. Each of these strands of power form the larger context, within which the daily decisions about 'who is let into the shelter' are made. I justify my escape from the face-to-face dilemma (of deciding who gets to be safe) by viewing the dilemma as part of the larger social and political inequities that my work is attempting to address.

I am not, however, the only one making decisions about my performance. While I may choose among the various identities I occupy, others may also choose for me. Audience and performers are, of necessity, a matter of perspective. And, much of what occurs interactionally is far more complicated than any one actor's definition might take into account. What is front stage to one team of performers may also be front stage to their audience who is, in turn, giving her own performance.

> **Even Further Back Stage**: At the Shelter 7.30 pm: Pamela arrives at the shelter with Katie. As she heads to the office at the back of the shelter Pamela asks me to show Katie to her room. Katie follows me down the hall and into a small bedroom with two beds. I indicate the bed near the door for her to put her stuff on. I ask her how she is doing. Before answering me she says, 'you won't tell anyone this, will you?' and gets up and closes the door. I say 'no', and she proceeds to tell me a slightly different version of what happened to her starting with the fact that she is a lesbian and was raped by her 'lover' and not a counsellor. (Field notes 11/18/86)

Through the telling of *this* truth, Katie is trying to pull me on to her team. By revealing to me that she has, indeed, given a phoney performance she is attempting to remove me from the audience. From her perspective, the audience is both the presumably heterosexual and feminist shelter staff and the shelter organization itself. Her 'performance' was designed to meet the expectations of this audience. 'In every social establishment, there are official expectations as to what the participant owes the establishment' (Goffman, 1961: 304). In a battered women's shelter – governed by a feminist analysis of violence against women as rooted in patriarchy – heterosexual violence is the officially expected explanation for entry. From a feminist perspective male violence against women is understandable, is indeed expected; women's violence against other women is not. In telling a heterosexual story, Katie was giving the performance that she surmised would gain her entry to shelter – an assumption that proved correct. This is not to say that she wouldn't have been let in had she outed herself during the intake; she probably would have. However, the subsequent 'lines of action' chosen by the various participants might have been more difficult to 'fit' into existing feminist scripts.[6]

In this case, the back and front stage boundaries are delineated both by identity and by place. Katie assumed I was a lesbian. This assumption, in all likelihood, was based around my physical presentation at the time: short spiky haircut, jeans, boots, and black 'silence = death' tee shirt with the sleeves cut off. The 'back stage region' in this case was demarcated by an assumption of congruent sexualities and a closed door. As lesbians in a predominantly heterosexual world Katie and I (presumably) shared a world view different from other non-lesbian interactants. By outing herself to me Katie was calling upon that shared understanding. The creation of this type of identity-based back stage region is a common experience whenever gay and lesbian folks find themselves in 'situations' devoid of heterosexual people. When no straight people are present, gay and lesbian people routinely contradict front stage performances that assume heterosexuality. These 'contradictions' are really about connection; they are interactional routines designed to establish a lesbian and gay normative world view.

Katie's offer of a 'back stage' pass to her performance intensified my dilemma. After describing the above situation with Katie in my 'observational notes', I made the following entry into my 'personal notes':

> **PN**: I just didn't know what to do at this point. Really I felt pulled in several directions all at once with my various identities competing for my attention. As a lesbian my response to Katie was one of sympathy. I felt a desire to protect one of 'my own'. As a shelter advocate 'in training' I didn't feel ready to take this upon myself, and I felt a certain loyalty and responsibility to the other workers. Finally, as a sociologist I was thrilled to be taken into confidence by one of the site participants. I am poised to record the 'underlife' of this organization. (Field notes 11/18/86)

The social and personal identities that surface in this fieldwork entry provide a rich illustration of the moral dimensions of the ever-shifting front and back regions. To tell or not to tell forms the core of this dilemma. My 'participant' role as a shelter worker obliged me to 'tell' the other shelter workers the 'truth' about Katie's story. Part of this was about loyalty to that role, and part was about safety. Because Katie's batterer was a woman, the possibility existed that she [the batterer] could also gain access to the shelter and the other workers needed to know this. My front stage identity as a shelter worker obligated me to tell. My back stage identity as a sociologist compelled me to keep my mouth shut. My interpretation of these events was circumscribed by my theoretical position as a symbolic interactionist and my methodological stance as an 'observer'. To meaningfully re-construct and represent the social world that is – in part – the 'under life' of the battered women's shelter, I needed to gain the confidence of the women who stayed there. Ethically and morally these discrepant roles [participant and observer] put me into a rather untenable position. I was – in Goffman's (1959) characterization – on the brink of becoming a 'double-shill or go-between'[7]:

> The go-between learns the secrets of each side and gives each side the true impression that he will keep its secrets; but he tends to give each side the false impression that he is more loyal to it than to the other (p. 149).

Keeping secrets and impression management are necessary tools for the field researcher. Gaining the confidence of all participants in each research site is always the goal. Sometimes, however, the various participants in each setting are at odds with one another. As a sociologist my job is not to judge which side is 'correct' or 'morally wanting'; instead, my job is to understand, contextualize, and re-construct the definitions that each side holds to be true. But when competing definitions have potential to cause harm, and I am the only participant privy to both world views, I am forced to make some choices. Guided by the code of ethics for my discipline (and good 'common sense'), I weighed the import of confidentiality against 'do no harm' and quickly surmised that the latter concern was paramount. In this case I was able to convince Katie to tell the truth about her situation to the shelter staff (partly by assuring her that I would stand by her through the whole process) and was saved from having to choose sides. I was able, in other words, to maintain my rapport with staff *and* gain entry to the everyday world of shelter residents. I was able to continue doing fieldwork.

In retrospect, Goffman's depiction of the 'go-between' does resonate with the experience of doing fieldwork.[8] In all of my interactions at the shelter I fostered a sense of confidentiality with whomever I was speaking. Like all the other shelter advocates, it was my job to give a 'true impression' that I could keep secrets. But, unlike all the other shelter advocates, my ultimate loyalties were not with the shelter, with the shelter residents, or even with the battered women's movement. Though I chose violence against women as 'my cause', and I chose this field site as my feminist trench, my final goal was 'to become' a sociologist. And while giving the not-quite-true impression of being loyal to all 'sides' of this arena, my ultimate loyalty lay with my chosen profession. It is this choice that provides the key to my moral absolution. I wanted to become a sociologist and one of the requirements 'of becoming' is to conduct original research. Like any job, the job of field researcher involves impression management. In this way, it was possible to conceptualize these ongoing identity dilemmas as simply 'doing my job'.

In my current analysis of these field experiences it occurred to me that my behaviour was, perhaps, even more in line with Goffman's characterization of a 'spy'. 'The individual who all along has meant to inform on the team and originally joins only for this purpose, is sometimes called a spy.' (Goffman, 1959: 145)

Act Two: Informing on the Team

My work with the battered women's shelter ended in 1987 and my field notes from those months of participation were transformed into my Masters thesis. Though my intention was not 'to inform on the team', in the end my analysis did just that. As is the case with inductive research, the researcher never knows what she will find until she begins collecting and analysing data. My thesis, which was eventually published as a journal article (Murray, 1988), focused on the contradictions between feminist ideology and not-so-feminist practice. As a crisis line volunteer and shelter relief worker I underwent a fairly rigorous (and feminist) domestic violence training. As a part-time staff member I attended staff meetings, fund-raisers, and workshops. My analysis of these experiences led to the following findings:

> The shelter has two main goals, the first is to provide immediate and safe shelter for battered women who are in crisis situations. . . . The second and less tangible goal of the shelter is 'empowerment' of shelter residents. This goal, consistent with the feminist analysis of violence against women, is seen by staff as a need of shelter residents: the women must recognize their place within the cycle of violence and simultaneously recognize their own power to end this cycle if they so choose. (Murray, 1988: 78–9)

As a field researcher committed to 'discovering' the underlife of this social organization, I found competing goals among the shelter residents:

> In sharp contrast, shelter residents themselves say they have needs of a more practical sort: 'I need a place to live and a job.' If these needs are not met and a battered woman has no money, family, or friends to turn to, then returning to the abusive situation seems to be the only viable alternative. Since most of the women who come into the shelter are in desperate financial circumstances, the priority of these needs become apparent. This poses a conflict between the ideological goals of the shelter and the reality of the resident's needs. (Murray, 1988: 79)

In the end – after detailing the basic social processes maintaining these (and other) contradictions – I conclude, 'Shelters, while necessary and worthwhile, are a response and not a solution to the problem of violence against women in our society' (Murray, 1988: 92). Though my research was not 'news' to the women who work in shelters, it did provide an open window into a feminist space that was then available for use by both feminists and non-feminists alike. To the extent my participation on the shelter team began and ended with my field research project, it is possible to conceptualize the goal of my membership as informing on the team.[9]

In the shelter I gained permission for conducting research by meeting with the director of the larger organization that housed the shelter. In my next field project, childcare, I resolved to gain entry by asking permission of 'the team'. Though I started at the top – applying for an administrative position at a childcare centre and interviewing with the director – once offered the job I made my acceptance contingent on receiving permission from the staff to conduct research on their profession. I attended a staff meeting and explained to the teachers who were present that I was applying for the administrative assistant job, in part, so I could conduct research for my PhD dissertation on childcare workers. I explained that I would eventually be asking each of them to consent to be interviewed, and that I would be making daily observations of life at the centre. I even supplied them with copies of the journal article that my thesis had become so they could get a sense for the kind of research and analysis I had conducted in the past. I explained in full that my ultimate intention was to write a book about the working lives of childcare workers. I explained, in other words, that my intention was to 'inform on the team'. They agreed to participate, they were excited about my research, and their enthusiasm (momentarily) absolved me of any lingering doubts about my ethical status.

Act Three: The Tiny Tot Toddler Center

The Setting: In January 1990 I began working and conducting research at the Tiny Tot Toddler Centre. The centre, serving one to two-and-a-half year olds, was located in a small coastal city in Northern California. My administrative assistant position at the Tiny Tot Toddler Centre fit well with my research agenda. The job put me in an ideal position (literally and figuratively) to observe workers. My desk at this centre was located in the office along with those of the two centre directors. This office also served as the break room for the teaching staff. On their breaks, many of the teachers would come to the office and talk about their experiences 'on the floor'. The office served, in other words, as a 'back stage region' for teachers. In the office, teachers were not constrained by the presence of the children. They spoke freely about the children, the children's parents, and other co-workers who were not present. It was a situation ripe for observation.

When I was sitting at my desk in the role of administrative assistant, I was privy to a constant flow of insider information. The teachers talked about the children and parents. The directors talked about the teachers. The directors would even talk about one another when one or the other of them was out of the room. Much of this information was confidential. A good portion of the talk was the back stage bantering one would expect to hear from workers who spend most of their time on the front stage of the classroom. I did not remind workers that I was doing research each time they came into the office. I did not take notes on their conversations in front of them. I did, however, record many of these conversations in my field notes. As is evident from the following excerpt from my field notes, my status as a researcher did not inhibit those around me.

> **The Stage(s)**: Friday afternoon: I am in the office getting book orders together. [The program director – PD] is at her desk working on her computer. Mary [head teacher in training] rushes into the office. She has something in her hand. She thrusts her hand under the PD's face and says (loudly) 'Oh my god, is this lice?'. I turn to look just as the PD pulls her head back and grabs on to Mary's hand to push it away from her face. The PD says, 'Calm down and let me see. And keep your voice down.' Mary continues: 'This is so disgusting. I'm not surprised that they [the child's parents] were reported to CPS.' The PD looks at me and opens her mouth to say something when Christina [an assistant teacher] opens the office door and walks in saying 'Is it lice? She has them all over her head. How gross.' The PD says [a bit sharply] 'I can't tell in this light' and walks out the office door into the yard [presumably to see in the sunlight]. Christina looks at Mary and says 'What is her trip?'. Mary replies, 'I don't know, she seems to really want to protect this family or something.' Christina says, 'It just seems so unprofessional. They are terrible parents.' The PD walks back into the office. Mary and Christina stop talking. The PD says 'It looks like lice. I'll call her parents. You [Mary] take [the child] out in the back yard and keep her away from the other kids.' Mary and Christina leave the office. The PD sits at her desk. She sighs and says 'those two' and looks over at me. 'I can only imagine what they said while I was out of the room.' I say, 'probably what you would expect them to say.' She says: 'They are so unprofessional.' I just kind of nod at her. (Field notes 5/6/90)

This scene – one of many I experienced during my tenure as an administrative assistant – really captures the kind of impression management endemic to centre-based childcare giving. The scene begins when Mary rushes into the office with a note of hysteria in her voice '*Oh my god, is this lice*?'. Presumably, she did not respond to her initial discovery with this same emotion. Professional child caregivers are trained to respond to any kind of medical situation with calm reassurance. Mary's appearance back stage in the office enabled her to drop her professional 'front' and express her true emotional state. The program director, however, is still on front stage and she responds to Mary with the expected calm (though not quite as reassuring as one would be to a child). My presence does not register on the scene until Mary opens another back stage door: '*This is so disgusting. I'm not surprised that they [the child's parents] were reported to CPS.*' By virtue of my presence in the room Mary has crossed the line of confidentiality – though she does not recognize her own violation. Then when Christina enters the stage and the program director momentarily exits, the two teachers form a quick alliance. This alliance is established through a direct challenge to their boss's professional credibility: *Christina looks at Mary and says, 'What is her trip?'. Mary replies, 'I don't know, she seems to really want to protect this family or something.' Christina says, 'It just seems so unprofessional. They are terrible parents.*' When the program director returns, the momentary coup dissolves as the teachers are sent back to work. I am then pulled back into the scene in order to re-establish the professional authority of the program director – to repair any damage done in her absence: *She sighs and says 'those two' and looks over at me. 'I can only imagine what they said while I was out of the room.' I say,' probably what you would expect them to say.' She says: 'They are so unprofessional.' I just kind of nod at her.*

Being able to capture this type of interplay is necessary for the study of human group life:

> Human groups are seen as consisting of human beings who are engaging in action. The actions consist of the multitudinous activities that the individuals perform in their lives as they encounter one another and as they deal with the succession of situations confronting them. . . . The import of this simple and essentially redundant characterization is that fundamentally human groups or society *exists in action* and must be seen in terms of action. This picture of human society as action must be the starting point (and the point of return) for any scheme that purports to treat and analyse human society empirically (Blumer, 1969: 6).

Capturing this type of interplay as a morally neutral and seemingly unobtrusive participant observer exacerbates one of the major dilemmas of the field researcher: the issue of informed consent. Both the teachers and the directors at the Tiny Tot Toddler Centre knew I was doing research on childcare workers. Before hiring me, the teachers and directors consented to being a part of my research. I told them that my research involved observation and interviews. I told them I would be taking field notes on my experiences and observations. I did not, however, tell them when exactly I would be observing or what 'taking field notes' meant, and nobody asked me about this. Thus I did not *completely* inform those I studied about all my research interests (e.g. Thorne, 1980).

Childcare workers are used to being observed. Their jobs involve routine and daily observation by parents, student teachers, co-workers, and administrators. By agreeing to be 'observed' did they really understand that I would be describing verbatim what I saw and what I heard? Did they understand that my observations of them extended beyond the classroom, beyond their interactions with the children in their care? Did they know that when they called me at home I often made notes of our conversations *while we were talking*? Did Mary and Christina and the program director realize that their interactions in the office would someday end up in a journal article?

On the one hand my response to these questions would be 'no'. Though I clearly explained my research intentions to all of the participants in each research site, I cannot know for certain how they interpreted my explanations. I do know that some of the workers forgot that I was doing research because they would act surprised when my status as a graduate student would come up in conversation. I never tried to hide or disguise my status as a researcher but, as I have illustrated, it was not always my master status. I talked openly about my ongoing analysis. I conducted in-depth interviews with workers at my research sites the entire time I was doing fieldwork. But none of the people whom I studied had previously participated in ethnographic research and none had been trained as sociologists.

The issue of informed consent posed an ongoing ethical dilemma that served as the final catalyst in my transformation from a primary to a secondary identification with my self as sociologist.[10] In order to continue 'doing sociology' within acceptable ethical boundaries, I scoured the work of other ethnographers (feminist and non-feminist) for practical advice on doing fieldwork. I followed their suggestions, I began to internalize their normative codes, and I continued with my fieldwork. I kept my field notes confidential, and I used fictitious names in all my observations and interviews (cf. Liebow, 1967). I made up a list of aliases for each of the Tiny Tot Toddler Centre employees and kept this list secure and physically separate from my field notes (cf. Schatzman and Strauss, 1973). When using excerpts from my field notes in my writing I changed names and other personal identifying characteristics (except gender) (cf. Liebow, 1993). When certain relationships in the field crossed over an indefinable line into friendship, I stopped taking notes on our personal conversations. It felt, at some point, dishonest to take notes on people I felt emotionally close to. If our conversations centred on a subject related to my research concerns, I talked directly with my friends about my 'observations' of their comments and we discussed them openly. In some ways, these friends became my informants in the field. With them, I more readily shared my observations, explored tentative hypotheses, and verified my interpretations of events at the centre (cf. Hendry, 1992; Oakley, 1981). Through these (and other) techniques I worked to meet my obligation to adhere to the practiced ethics of informed consent. In so doing – in meeting this and other ethical concerns and dilemmas – I was able ultimately to complete my transition and secure my identity as a sociologist.

The Final Curtain Call: The Sociologist Takes a Bow

Throughout this analysis I have employed Goffman's (1959) framework for understanding the organization of social life within the confines of a concrete social establishment. Ostensibly the establishments under examination have been – in order of appearance – a battered woman's shelter circa 1985–1987, and a childcare centre circa the 1990s. Each 'act' of this article documents the generic social processes that comprise social interactions that sustain each of these establishments. In this analysis I have relied on Goffman's (1959) metaphor of front and back stage regions to illustrate current applications of impression management techniques.

In reality, the *capricious* social establishment under examination is the everyday working world of the participant observer. Again the foci of analysis are the generic social processes that comprise social interaction. Similarly, Goffman's (1959) metaphor of front and back stage regions provides the field researcher with a map for understanding the field experience. The field researcher, like all other participants in the setting, engages in social interaction by 'fitting together lines of action'. Just what action is taken in any given situation depends upon the definition the researcher holds to be relevant to that particular situation. Definitions held, and therefore actions taken, shift and change in response to the participants who are present and the various statuses and identities they invoke. These 'shifts and changes' have been herein identified as transformations from front to back stage regions.

In his depiction of these transformations, Goffman also provides the field researcher with an ethical escape hatch from the quandaries of doing fieldwork. At the end of his treatise, Goffman reveals to the reader that, in fact, all the world is not a stage. As he raises the theatrical curtain from his analysis he reveals his true focus on '. . . the structure of social encounters – the structure of those entities in social life that come into being whenever persons enter one another's immediate physical presence' (1959: 254). The field researcher, like all other *persons* who engage social encounters, is not engaged in deliberate deception, misrepresentation, or obfuscation of her intent; she is simply engaged in the maintenance of a particular definition 'in the face of a multitude of potential disruptions' (1959: 254).

The key, then, becomes my definition of what it means to be a sociologist. As a sociologist my job is to study social inequality and power. Broadly speaking, my intellectual goal is to identify structures of inequality, and my political goal is to dismantle them. My chosen methodology means that I study inequality at the level of face-to-face interaction. In so doing – by virtue of my own various social locations – I become a part of the very inequities I am researching. At times I may even end up perpetuating power inequities at the interactional level in order to continue my analysis at the structural level. In maintaining my particular definition it seems as though I have answered my own question: What am I really? I am a spy, a shill, a go-between, and (consequently) a sociologist.

Notes

1. These terms are from Goffman (1959) *The Presentation of Self in Everyday Life*. New York: Doubleday, pp. 145–51.
2. My use of Goffman's framework is not out of some undying loyalty to my sociological roots, or a dismissal of the last 50 years of sociology. Goffman's work – especially when applied to the profession of sociology – seems to parallel (and perhaps foreshadow) postmodernist discussions of epistemology and partiality (cf. Bordo, 1994; Clifford and Marcus, 1986; DeVault, 1999; Okely and Callaway, 1992; Stacey, 1996).
3. For those of us whose primary method is participant observation field research.
4. The exception to this is, of course, my observational notes. These notes – which include direct descriptions of the field, the interactants, the physical setting of various situations, and the conversations therein – have been extensively excerpted in various publications (cf. Murray, 1988, 1998, 2001).
5. For a more complete understanding of the feminist revolution in sociology see Baca Zinn, 1979; Cannon et al., 1988; Collins, 1986; Smith, 1979; Stacey and Thorne, 1985.
6. There is historical precedence for this fear on Katie's part: in the preface to the book, *Naming the Violence: Speaking Out About Lesbian Battering* (1986), the editor Lobel describes the first ever national gathering to raise awareness on domestic violence: In January 1979, the US Commission on Civil Rights held a national hearing on woman abuse in Washington DC, and out of that meeting the National Coalition on Domestic violence was formed. During the meetings 40 lesbians met unofficially in an empty meeting room to discuss homophobia in the battered women's movement and lesbian battering. Of the meeting Lobel writes: The revelations about lesbian abuse by and of our partners were shocking and frightening. Although we could not reach consensus about the causes or dimensions of lesbian violence, we quickly agreed that this discussion could not be taken to the non-lesbian women in the movement. We felt it would destroy our credibility and that it would make us even more vulnerable to homophobic attacks by our sisters and those people in the dominant culture that wanted to discredit the efforts of the battered women's movement. So we pulled a tight lid down on the subject and didn't discuss lesbian battering even among ourselves until the NCADV conference two years later. (p. 12)
7. From the perspective of either side – shelter client or staff – I was, in all probability, seen as a 'shill'. 'A shill is someone who acts as though he were an ordinary member of the audience but is in fact in league with the performers'(Goffman, 1959: 146).
8. Though, of course, in Goffman's (1959) analysis he disparages the role of the 'go-between' as an individual while, at the same time, lauding it in reference to teamwork (p. 149). In the field, however, one of my teams – the professional sociological team – is invisible.
9. In retrospect, I think this is the moment where I began to see my sociological self as 'partially feminist' (Stacey, 1996), and partially not. While I was not comfortable 'informing on the team', with my class-based analysis of resident interests versus staff interests, I was also not willing to stand down from my analysis. I agonized over my choice (to publish). The question that occurs to me now is whether this really was a feminist agony, or was it something else? As DeVault (1999) so insightfully comments:

   > Concerns about the ethics of representation, and attempts to equalize interpretive authority, are central to feminist methodological innovations. However, I wonder if they also reflect an unwitting collusion with ideological constructions of 'woman' as especially moral or caring, or perhaps, a learned discomfort with authority that many women feel. (p. 189)

10. For an important corollary to this process, see Lemert's (1967) discussion of the shift from primary to secondary deviance.

References

Acker, Joan, Barry, Kate and Esseveld, Joke (1983) 'Objectivity and Truth: Problems in Doing Feminist Research', *Women's Studies International Forum* 6(4): 423–35.

Adler, Patricia A. and Adler, Peter (1991) 'Stability and Flexibility: Maintaining Relationships Within Organized and Unorganized Groups', in W.B. Shaffir and R.A. Stebbins (eds) *Experiencing Fieldwork: An Inside View of Qualitative Research*, pp. 173–83. London: Sage Publications.

Babbie, Earl (1992) *The Practice of Social Research*. Belmont, CA: Wadsworth, Inc.

Baca Zinn, Maxine (1979) 'Field Research in Minority Communities: Ethical, Methodological, and Political Observations by an Insider', *Social Problems* 27(2): 209–19.

Blumer, Herbert (1969) *Symbolic Interactionism: Perspective and Method*. Berkeley, CA: University of California Press.

Bordo, Susan (1994) 'Feminism, Postmodernism, and Gender Skepticism', in A.C. Herrmann and A.J. Stewart (eds) *Theorizing Feminism: Parallel Trends in the Humanities and Social Sciences*, pp. 458–81. Boulder, CO: Westview Press.

Brettell, Caroline B. (1997) 'Blurred Genres and Blended Voices: Life History, Biography, Autobiography, and the Auto/Ethnography of Women's Lives', in D.E. Reed-Danahay (ed.) *Auto/Ethnography: Rewriting the Self and the Social*, pp. 223–46. Oxford: Berg.

Cannon, Lynn Weber, Higginbotham, Elizabeth and Leung, Marianne L.A. (1988) 'Race and Class Bias in Qualitative Research on Women', *Gender & Society* 2(4): 449–62.

Clifford, James and Marcus, George (eds) (1986) *Writing Culture: The Poetics and Politics of Ethnography*. Berkeley, CA: University of California Press.

Collins, Patricia Hill (1986) 'Learning From the Outsider Within: The Sociological Significance of Black Feminist Thought', *Social Problems* 33(6): S14–S32.

Denzin, Norman K. (1978) *Sociological Methods: A Sourcebook*. New York: McGraw-Hill Book Company.

DeVault, Marjorie L. (1999) *Liberating Method: Feminism and Social Research*. Philadelphia, PA: Temple University Press.

Edwards, Rosalind and Ribbens, Jane (1998) 'Living on the Edges: Public Knowledge, Private Lives, Personal Experience', in J. Ribbens and R. Edwards (eds) *Feminist Dilemmas in Qualitative Research*, pp. 1–23. London: Sage Publications.

Emerson, Robert M. (1983) *Contemporary Field Research: A Collection of Readings*. Boston, MA: Little, Brown and Company.

Goffman, Erving (1959) *The Presentation of Self in Everyday Life*. Garden City, New York: Doubleday Anchor Books.

Goffman, Erving (1961) *Asylums: Essays on the Social Situation of Mental Patients and Other Inmates*. Garden City, New York: Anchor Books.

Hendry, Joy (1992) 'The Paradox of Friendship in the Field: Analysis of a Long-Term Anglo-Japanese Relationship', in J. Okely and H. Callaway (eds) *Anthropology and Autobiography*, pp. 163–74. London: Routledge Press.

Humphreys, Laud (1970) *Tearoom Trade: Impersonal Sex in Public Places*. Chicago, IL: Aldine.

Hurtado, Aida (1996) *The Color of Privilege: Three Blasphemies on Race and Feminism*. Ann Arbor, MI: University of Michigan Press.

Kleinman, Sherryl (1991) 'Field-workers' Feelings: What We Feel, Who We Are, How We Analyze', in W.B. Shaffir and R.A. Stebbins (eds) *Experiencing Fieldwork: An Inside View of Qualitative Research*, pp. 184–95. London: Sage Publications.

Lee-Treweek Geraldine (2000) 'The Insight of Emotional Danger: Research Experiences in a Home For Older People', in G. Lee-Treweek and S. Linkogle (eds) *Danger in the Field: Risk and Ethics in Social Research*, pp. 114–31. London: Routledge.

Lemert, Edwin M. (1967) *Human Deviance, Social Problems and Social Control.* Englewood Cliffs, NJ: Prentice-Hall.
Letherby, Gayle (2000) 'Dangerous Liaisons: Auto/biography in Research and Research Writing', in G. Lee-Treweek and S. Linkogle (eds) *Danger in the Field: Risk and Ethics in Social Research*, pp. 91–113. London: Routledge.
Liebow, Eliot (1967) *Tally's Corner: A Study of Negro Streetcorner Men*. Boston, MA: Little, Brown, and Company.
Liebow, Eliot (1993) *Tell Them Who I Am: The Lives of Homeless Women*. New York: The Free Press.
Lobel, Kerry (1986) 'Introduction', in K. Lobel (ed.) *Naming the Violence: Speaking Out About Lesbian Battering*, pp. 1–7. Seattle, WA: The Seal Press.
Mead, George Herbert (1934) *Mind, Self & Society*. Chicago, IL: The University of Chicago Press.
Murray, Susan B. (1988) 'The Unhappy Marriage of Theory and Practice: An Analysis of a Battered Women's Shelter', *National Women's Studies Association Journal* 1(1): 75–92.
Murray, Susan B. (1995) Child Care Work: The Lived Experience. PhD. dissertation. University of California, Santa Cruz.
Murray, Susan B. (1998) 'Child Care Work: Intimacy in the Shadows of Family-Life', *Qualitative Sociology* 21(2): 149–68.
Murray, Susan B. (2001) '"When a Scratch Becomes a Scary Story" The Social Construction of Micro Panics in Center-Based Child Care.' *The Sociological Review* 49(4): 512–29.
Oakley, Ann (1981) 'Interviewing Women: A Contradiction in Terms', in H. Roberts (ed.) *Doing Feminist Research*, pp. 30–61. London: Routledge & Kegan Paul.
Okely, Judith and Callaway, Helen (1992) *Anthropology and Autobiography*. London: Routledge Press.
Reinharz, Shulamit (1992) *Feminist Methods in Social Research*. Oxford: Oxford University Press.
Schatzman, Leonard and Strauss, Anselm L. (1973) *Field Research: Strategies for a Natural Sociology*. Englewood Cliffs, NJ: Prentice-Hall.
Smith, Dorothy E. (1979) A Sociology For Women', in J. Sherman and E. Torton Beck (eds) *The Prism of Sex: Essays in the Sociology of Knowledge*, pp. 135–87. Madison, WI: University of Wisconsin Press.
Stacey, Judith (1996) 'Can There Be a Feminist Ethnography?', in H. Gottfried (ed.) *Feminism and Social Change: Bridging Theory and Practice*, pp. 88–101. Urbana and Chicago, IL: University of Illinois Press.
Stacey, Judith (1997) 'Disloyal to the Disciplines: A Feminist Trajectory in the Borderlands', in B. Laslett and B. Thorne (eds) *Feminist Sociology: Life Histories of a Movement*, pp. 126–50. Rutgers, NJ: Rutgers University Press.
Stacey, Judith and Thorne, Barrie (1985) 'The Missing Feminist Revolution in Sociology', *Social Problems* 32(4): 301–16.
Thorne, Barrie (1980) 'You Still Takin' Notes?: Field Work and Problems of Informed Consent', *Social Problems* 27(3): 284–97.

E. Interpretation and Presentation of Observational Data

56

On Writing Fieldnotes: Collection Strategies and Background Expectancies

Nicholas H. Wolfinger

Introduction

At first glance, writing fieldnotes seems deceptively straightforward. Go to a research site, see what happens, then write it down. But this simple description raises a fundamental question: when typing up notes, exactly what does the researcher choose to annotate? Decisions made at this juncture of the research process may have a profound impact on the final ethnographic report. This article explores some of the taken-for-granted assumptions involved in writing fieldnotes.

Over time, ethnography has become increasingly self-conscious (Clifford, 1986; Emerson, 1983, 1987). Many scholars have examined the relationships between ethnographers and their subjects (Clifford, 1986; Woolgar, 1988) and ethnographic texts and their readers (Atkinson, 1990; Woolgar, 1988). Comparatively little attention has been paid to fieldnotes (Emerson et al., 1995; see also Garfinkel and Sacks, 1970). Most of the extant research has been either taxonomic (Ottenberg, 1990; Sanjek, 1990a) or philosophic (Clifford, 1990; Sanjek, 1990b), but not pragmatic. Fieldnote tutorials routinely neglect practical consideration of the issues involved in note-taking, a particularly noteworthy absence in light of ethnography's fundamentally subjective underpinnings. Van Maanen (1988) provides an eloquent statement of the problem:

Source: *Qualitative Research*, 2(1) (2002): 85–95.

> To put it bluntly, fieldnotes are gnomic, shorthand reconstructions of events, observations, and conversations that took place in the field. They are composed well after the fact as inexact notes to oneself and represent simply one of many levels of textualization set off by experience. To disentangle the interpretive procedures at work as one moves across levels is problematic to say the least. . . . Little wonder that fieldnotes are the secret papers of social research (pp. 223–4).

Given that *writing* is traditionally seen as the linchpin of ethnographic inquiry (see Geertz, 1973), it is important to understand how Van Maanen's 'many levels of textualization set off by experience' affect the note-taking process.

Emerson et al. (1995), the only monograph-length treatment of fieldnotes to date, provide much valuable information but treat the selection of experiences for annotation as a process guided by explicit instructions. As such, they gloss over the taken-for-granted mechanisms involved. I aim to show how these mechanisms, operating within formalized instructions for ethnographic scholars, affect the content of fieldnotes.

Data

I employ data from two sources. First, I use fieldnotes from my ethnography of public space interaction (Wolfinger, 1995); second, I draw upon ethnographic data produced by others, both published and from graduate classes.

Where Do Fieldnotes Come From?

When in the field, ethnographers are already deciding what to write about. What do they notice? What do they choose to focus their attention upon? What do they subsequently recall? Of what they remember, what do they choose to document in their notes? In what detail?

I begin with an example that reveals the complex relationship between a researcher's experiences in the field and his or her experiences at the typewriter or word processor. From a student ethnography of supermarket checkout queues:

> The most populous line unfortunately had only two people in it, one of whom was waiting for the last of her groceries to be bagged and so was not someone I felt I could rely on to contribute significantly to my 'setting'. I became slightly disappointed at the slimness of my pickin's [sic] but forged ahead, resolved to make the best of it and also a bit amazed at myself for the degree of preliminary planning and concern I was manifesting. Why didn't I just get in line and start watching people? A single minute would be more than enough time to produce material to fill pages if I simply gave that minute the opportunity to impress itself upon me.

This ethnographer spent a lot of time deciding what to watch. How much of his equivocation occurred at the time he conducted his observations and how much was recreated when typing up his notes? The last line in the excerpt suggests the latter: the concern of 'fill[ing] pages' is specific to the process of writing fieldnotes. How did his deliberation influence his data collection: did he treat the checkout queue he subsequently entered as a different phenomenon because it only had two people in it? Did this influence emerge when he went through the queue or did he just impute it afterwards when typing up his notes? Speculation could continue, but I think the point has been made: what the ethnographer does at the word processor and what he or she does in the field have a meaningful interaction. This idea is not lost on Emerson et al. (1995: 60–1), who describe how ethnographers may let understandings unfold in the course of writing fieldnotes. However, they focus on the rhetorical rather than the substantive implications. How does the process of textual recreation actually affect ethnographers' experiences, as depicted in their notes?

Ethnography places much discretion in the hands of the researcher. Symbolic interactionism (Blumer, 1969), social setting analysis (Lofland and Lofland, 1984) and grounded theory (Charmaz, 1983; Glaser, 1978; Glaser and Strauss, 1967) – three common theoretical warrants for ethnography – argue against the use of field research methods for testing preconceived hypotheses. Instead, these perspectives advocate letting one's experiences in the field guide a study's focus. While in the field, researchers presumably identify certain phenomena as interesting and worthy of annotation. They therefore must exercise discretion in deciding what should be documented in their fieldnotes. The sources of this discretion are explored in the following sections of this article.

Three practical considerations often shape the course of note-taking. First, a researcher will sometimes be able to take notes while in the field. Many fieldwork texts advocate this practice (Berg, 1989; Emerson et al., 1995; Goffman, 1989; Lofland and Lofland, 1984; Schatzman and Strauss, 1973). These preliminary notes generally form an outline when the researcher sits down at the end of the day to type out complete notes. Second, the focus of an ethnographic investigation typically narrows over time (Hammersley and Atkinson, 1983; Spradley, 1980), obviously influencing what a note-taker chooses to describe. Third, note-taking may be influenced by the perceived audience (Emerson et al., 1995). Within these broad constraints, however, an ethnographer will still have to decide exactly what should be annotated.

The Role of Tacit Knowledge in the Creation of Fieldnotes

I contend that tacit knowledge is perhaps the most important consideration in determining how particular observations are deemed worthy of annotation. Ethnographers use what Harold Garfinkel (1967: 78) calls the documentary method of interpretation when they choose to record an observation:

> Social objects . . . are the products of complicated judgements in which an 'underlying pattern' is built up from a temporally qualified succession of appearances. . . . Aspects of the process may be brought to consciousness, for example, when we struggle to 'make out' what somebody is doing or 'form hypotheses' about it (Heritage, 1984: 86).

In other words, all social actors actively strive to make informed decisions based upon their background knowledge, knowledge that is, in the words of Garfinkel (1967) 'seen but unnoticed' (p. 118). This includes ethnographic note-takers. Although it has been suggested that ethnographic writing depends upon taken-for-granted knowledge (Ely et al., 1991; Emerson, 1983; Kleinman et al., 1997; Pelto and Pelto, 1978), this idea has not been explored thoroughly or has been examined only in specific contexts such as conversation analysis (Watson, 1996).

The relationship between tacit knowledge and fieldnotes becomes apparent in considering an example. The following fieldnote describes Spradley's (1980) summons to a grand jury:

> There were rows of spectator benches, all made of heavy dark wood, oak or walnut, to match the paneled walls. The rows of benches went for more than twenty-five feet until they met a railing that seemed to neatly mark off a large area for 'official business'. . . . At the right of the area behind the railing were twelve high-back leather chairs behind another railing. A large oak table with massive chairs all faced toward a high lectern *which I took to be* the judge's bench (p. 74, emphasis added).

In the last line of this excerpt, Spradley identifies a structure as the judge's bench. How does he know this? He has background knowledge from three sources: he presumably has some previous exposure to courtrooms, he knows he has been summoned to the grand jury, and he has recorded other observations (of benches, railings, and a jurors' box) indicating the likely presence of a judge's bench. Thinking back to his time in the field, his observations resonate with his background knowledge to produce the identification of 'judge's bench'. This is a particularly useful example because Spradley himself assists in the argument I propose: by writing '*which I took to be* the judge's bench', he acknowledges the interpretive process involved in making an ethnographic observation. Furthermore, Spradley's analytic self-consciousness shows that the processes of interpretation and identification involved in any ethnographic study can be examined by the researcher. In this instance, it is curious that the judge's bench was apparently more difficult to identify than any other part of the courtroom. Did this reflect the importance Spradley attributed to the judge in the ensuing legal proceedings, his uncertainty regarding the judicial role in a grand jury hearing, an unexpectedly small lectern, or something else entirely? Presumably only Spradley knows for sure, but either way the answer might tell us more both about the grand jury and how he went about describing his experiences there.

Two Strategies for Writing Fieldnotes

When ethnographers type up their fieldnotes they inevitably must make decisions concerning the order in which they will document what happened in the field. Different techniques may be employed, and it is my position that the strategy adopted will influence the ensuing data by determining how tacit knowledge is used in the depiction of people, places and events.

Emerson et al. (1995: 48) describe two methods for writing fieldnotes. They are not mutually exclusive, but simply represent two practical approaches to the note-taking process. I will describe these two strategies, then suggest how each might produce different data depending upon an ethnographer's tacit knowledge and beliefs.

Strategy One: The Salience Hierarchy

In sitting down to record notes, ethnographers can start by describing whatever observations struck them as the most noteworthy, the most interesting, or the most telling. What might comprise one of these salient episodes? An example from an ethnography of phone sex 'chat line' employees:

> The phone monitors consistently refer to the callers as losers and sometimes as psychos. We answer the phones by releasing mute buttons for our headsets and some monitors use the mute buttons to ridicule the callers while they are still on the line with them. The most common practice is to talk to the caller and quickly press the mute button and say 'fucking loser' and then return to the caller.

This excerpt conveys an interesting piece of ethnographic information – I assume it is interesting because I found it to be so – so its author may well have been intrigued when she selected it for annotation. Ethnographers may be prepared to describe a number of such events when they sit down to write fieldnotes. These incipient data comprise what I will refer to as a *salience hierarchy*.[1] What makes an observation salient is highly subjective and depends upon the particular research context. Nevertheless, several rules of thumb apply and each demonstrates the role of tacit knowledge in writing fieldnotes.

Deviant cases often lead to salient data. Cases may be deviant in at least two ways. First, they may strike researchers as deviant with respect to their tacit expectations. For example, in his study of rural communards, Bennett Berger (1981) only mentions sexual practices that might be seen as deviant by society at large. He doesn't waste time describing the monogamous sex lives of cohabiting adults. Thus Berger's background knowledge influenced what he perceived as salient. The aforementioned description of phone sex workers ironically provides a subtler example of this point. The discovery that workers disparage their customers may have seemed noteworthy simply because the ethnographer presumably entered the setting fully aware of how important client relations are to the atmosphere of any retail business.

Tacit knowledge may also be developed in situ, so that a case may seem deviant just with respect to others observed. These other cases form the contrasts – in other words, the background knowledge – that lead to the identification of certain cases as deviant. This is a desirable occurrence because it can contribute to the development of theory (Katz, 1983). In my study of public space interaction, I found that pedestrians often looked at each other and exchanged pleasantries after narrowly averting a collision. One time a pedestrian addressed me without looking. I found this anomaly interesting and accorded it additional consideration in my fieldnotes:

> I looked at him (at his face, that is) and said 'sorry.' He did not look back at me in mumbling a response, which was something like 'alright.' His expression did not change when he said this. In retrospect, the apparent depth of his preoccupation or reverie rendered this exchange memorable – I was struck by it at the time.

This fieldnote may not have been written had the pedestrian manifested a more usual reaction to the near-collision.

An interaction's boundaries (temporal, spatial, or otherwise) or absence thereof might also render it memorable. In my study of public space interaction, a memory temporally and spatially contiguous to many others was often less memorable than one occurring on, for instance, an otherwise empty street. In the latter case, there's simply less mental fodder present to cloud one's memory when typing up fieldnotes.

Ethnographers frequently choose to record a particular observation because it stands out. Observations often stand out because they are deviant, either when compared to others or with respect to a researcher's existing knowledge and beliefs. Either way, background knowledge influences which cases are chosen for annotation.

Strategy Two: Comprehensive Note-taking

An alternate strategy for recording notes is to systematically and comprehensively describe everything that happened during a particular period of time, such as a single trip to the field. Systematic can mean a couple of different things. One place to start is with a generalized list of concerns, such as that provided by Lofland and Lofland (1984: 48):

Who is he?
What does he do?
What do you think she meant by that?
What are they supposed to do?
Why did she do that?
Why is that done?
What happens after ______?
What would happen if ______?
What do you think about ______?
Who is responsible if ______?

Spradley (1980: 78) provides a similar list:

1. *Space*: the physical place or places
2. *Actor*: the people involved
3. *Activity*: a set of related acts people do
4. *Object*: the physical things that are present
5. *Act*: single actions that people do
6. *Event*: a set of related activities that people carry out
7. *Time*: the sequencing that takes place over time
8. *Goal*: the things people are trying to accomplish
9. *Feeling*: the emotions felt and expressed

Another strategy is to organize one's note-taking temporally: start at the beginning and end at the end (Emerson et al., 1995). Many social settings have their own unique time-tables (Hammersley and Atkinson, 1983) and these can facilitate note-taking by serving as outlines. Note-taking that attends to the rhythms of its subjects may be more attentive to members' meanings, a generally desirable objective in ethnography (Emerson, 1983).

The *comprehensive* method of note-taking also has the advantage of forcing an ethnographer to recreate events in the order they really happened. This can aid in the recall of details that might otherwise have been forgotten. Cognitive psychology shows that people tend to retain information in specific 'knowledge structures' (Abelson, 1981; Schank and Abelson, 1977). It follows that remembering one aspect of an event might trigger recall of an entire sequence. One text on ethnographic method, cognizant of this, offers the following advice:

> *Concentrate on the first and last remarks in each conversation.* Conversations generally follow a logical sequence. That is, a certain question elicits a certain response; a certain remark provokes another; a certain topic leads to a related one. If you can remember how a conversation was initiated, you can frequently follow it through to the end in your own mind (Bogdan and Taylor, 1975: 62; original emphasis).

Similarly, Berg (1989: 73) encourages ethnographers to take notes on the sequence of events.

How does tacit knowledge affect note-taking when using the *comprehensive* method? In recounting entire segments of time spent in the field a researcher will often describe events that might otherwise seem too mundane to annotate. These data may later turn out to be valuable, because they can provide the contrasts that allow an ethnographer to identify deviant cases. Whether or not they are identified as such, these cases will form the background knowledge that guides subsequent note-taking. Methodological self-awareness increases because researchers can make their tacit knowledge explicit.

In my ethnography of public space I described several interactions that 'didn't happen'. This sort of material becomes useful when compared to interactions that did happen, because it allows a researcher to develop the criteria for defining an event in any given social setting. Here is an example of an interaction that didn't happen, that I would not have noted had I just been using the *salience hierarchy* method:

> I came to a halt at a red light, still on the curb. Across the street, I saw two studenty looking women. They were obviously poised to cross over to my side of the street when the light turned green. They were positioned far to my right. I looked over at them . . . to assess whether an interaction was forthcoming – whether either I or they would end up negotiating some sort of passage. . . . It quickly became apparent to me that no 'passage negotiating' was going to take place. We were too far apart laterally (somewhere between ten and fifteen feet), and this was very apparent to me as I stood still and saw them standing still. Thus, by being still I had the opportunity to assess their intended courses (which they followed), and they presumably had the opportunity to assess mine. The light soon turned green and we all passed, far apart, without exchanging glances. The two women had been talking to each other the entire time.

By describing non-interactions, I made explicit some of the criteria for defining an interaction in public space. Another example, from a student ethnography of a supermarket:

> I move along to the produce area where there are a number of people sorting through the apples, bananas and veggies. I am struck by how normal it all is. . . . No one is particularly interested in the doings of anyone else. Fruit and vegetables are occupying almost everyone's attention fully. Two young women, dressed in Melrose Ave. style, sweep through the scene, moving rapidly and talking in quiet but rushed voices. No one but me seems to even notice. The setting is almost languid.

By documenting omissions, ethnographers convey a more comprehensive depiction of a research site, allowing them to gain valuable insight into the background knowledge that guides subsequent note-taking. This in turn may prove useful in better understanding how and why events take place.

Conclusion

Fieldnotes are an oft-neglected yet fundamental part of ethnography. They serve the crucial role of connecting researchers and their subjects in the writing of an ethnographic report. Often an ethnographer spends enormous amounts of time writing notes. For these reasons field notes may take on a unique sacredness (Jackson, 1990a, 1990b; Sanjek, 1990b). As Freud might put it, we cathect our notes – they become part of us. Therefore it is important to understand the processes underlying their creation and how these processes can affect ethnographic texts.

The advent of reflexive ethnography has drawn attention to the relationship between researchers and their subjects, and that of texts and their readers. Moreover, it is well known that prose inevitably embodies the author's tacit convictions; consider, for example, Howard Becker's (1986) pithy remark in his book on writing for the social sciences: 'You have already made many choices when you sit down to write, but you probably don't know what they are' (p. 17). Fieldnotes, given their importance to ethnographic scholarship, deserve the same self-scrutiny that other prose is routinely subjected to.

Fieldnotes inevitably reflect the ethnographer's background knowledge, or tacit beliefs. In this article I have described some of the ways in which background knowledge might affect the construction of ethnographic notes, particularly subsequent to the decision to employ either the *salience hierarchy* or *comprehensive* strategies of note-taking. This decision represents one commonly taken-for-granted feature of ethnography that can easily be made explicit. Within each of these approaches reside more subtle choices. At this juncture the researcher's tacit knowledge affects which observations will be recorded in fieldnotes. By recognizing some of the assumptions involved, ethnographers can develop a more thorough understanding of their note-taking and ultimately of the social worlds they study.

Note

1. With apologies to Sheldon Stryker (1980), who gives this useful term an entirely unrelated meaning.

References

Abelson, Robert P. (1981) 'Psychological Status of the Script Concept', *American Psychologist* 36: 715–29.

Atkinson, Paul (1990) *The Ethnographic Imagination; Textual Constructions of Reality*. London: Routledge.

Becker, Howard S. (1986) *Writing for Social Sciences; How to Start and Finish Your Thesis, Book, or Article*. Chicago: University of Chicago Press.

Berg, Bruce L. (1989) *Qualitative Research Methods for the Social Sciences*. Needham Heights, MS: Allyn and Bacon.

Berger, Bennett M. (1981) *The Survival of a Counter Culture*. Berkeley, CA: University of California Press.

Bogdan, Robert and Taylor, Steven J. (1975) *Introduction to Qualitative Research Methods; A Phenomenological Approach to the Social Sciences*. New York: Wiley.

Blumer, Herbert (1969) *Symbolic Interactionism; Perspective and Method*. Englewood Cliffs, NJ: Prentice-Hall.

Charmaz, Kathy (1983) 'The Grounded Theory Method: An Explication and Interpretation', in R.M. Emerson (ed.) *Contemporary Field Research*, pp. 109–226. Prospect Heights, IL: Waveland Press.

Clifford, James (1986) 'Introduction: Partial Truths', in J. Clifford and G.E. Marcus (eds) *Writing Culture; The Poetics and Politics of Ethnography*, pp. 1–26. Berkeley: University of California Press.

Clifford, James (1990) 'Notes on Fieldnotes', in R. Sanjek (ed.) *Fieldnotes; The Makings of Anthropology*, pp. 47–70. Ithaca, NY: Cornell University Press.

Ely, Margot, Anzul, Margaret, Friedman, Teri, Garner, Diane and McCormack Steinmetz, Ann (1991) *Doing Qualitative Research; Circles within Circles*. London: Falmer.

Emerson, Robert M. (1983) 'Introduction', in R.M. Emerson (ed.) *Contemporary Field Research*, pp. 1–35. Prospect Heights, IL: Waveland Press.

Emerson, Robert M. (1987) 'Four Ways to Improve the Craft of Fieldwork', *Journal of Contemporary Ethnography* 16: 69–87.

Emerson, Robert M., Fretz, Rachel I. and Shaw, Linda L. (1995) *Writing Ethnographic Fieldnotes*. Chicago: University of Chicago Press.

Garfinkel, Harold (1967) *Studies in Ethnomethodology*. Oxford: Polity Press.

Garfinkel, Harold and Sacks, Harvey (1970) 'On Formal Structures of Practical Actions', in J.C. McKinney and E.A. Tiryakian (eds) *Theoretical Sociology*, pp. 337–66. New York: Appleton.

Geertz, Clifford (1973) *The Interpretation of Cultures*. New York: Basic Books.

Glaser, Barney G. (1978) *Theoretical Sensitivity*. Mill Valley, CA: Sociological Press.

Glaser, Barney G. and Strauss, Anselm L. (1967) *The Discovery of Grounded Theory; Strategies for Qualitative Research*. Chicago: Aldine.

Goffman, Erving. (1989) 'On Fieldwork', transcribed and edited by Lyn Lofland. *Journal of Contemporary Ethnography* 18: 123–32.

Hammersley, Martyn and Atkinson, Paul (1983) *Ethnography; Principles in Practice*. London: Tavistock.

Heritage, John (1984) *Garfinkel and Ethnomethodology*. Oxford: Polity Press.

Katz, Jack (1983) A Theory of Qualitative Methodology: The Social System of Analytic Fieldwork', in R.M. Emerson (ed.) *Contemporary Field Research*, pp. 127–18. Prospect Heights, IL: Waveland Press.

Kleinman, Sherryl, Copp, Martha A. and Henderson, Karla A. (1997) 'Qualitatively Different: Teaching Fieldwork to Graduate Students', *Journal of Contemporary Ethnography* 25: 469–99.

Jackson, Jean E. (1990a) '"Déjà Entendu" – the Liminal Qualities of Anthropological Fieldnotes', *Journal of Contemporary Ethnography* 19: 8–43.

Jackson, Jean E. (1990b) '"I am a Fieldnote": Fieldnotes as a Symbol of Professional Identity', in R. Sanjek (ed.) *Fieldnotes; The Makings of Anthropology*, pp. 3–33. Ithaca, NY: Cornell University Press.

Lofland, John and Lofland, Lyn (1984) *Analyzing Social Settings*, 2nd edn. Belmont, CA: Wadsworth.

Ottenberg, Simon (1990) 'Thirty Years of Fieldnotes: Changing Relationships to the Text', in R. Sanjek (ed.) *Fieldnotes; The Makings of Anthropology*, pp. 139–60. Ithaca, NY: Cornell University Press.

Pelto, Pertti J. and Pelto, Gretel H. (1978) *Anthropological Research; The Structure of Inquiry*, 2nd edn. London: Cambridge University Press.

Sanjek, Roger (1990a) A Vocabulary for Fieldnotes', in R. Sanjek (ed.) *Fieldnotes; The Makings of Anthropology*, pp. 92–121. Ithaca, NY: Cornell University Press.

Sanjek, Roger (1990b) 'Fire, Loss, and the Sorcerer's Apprentice', in R. Sanjek (ed.) *Fieldnotes; The Makings of Anthropology*, pp. 34–44. Ithaca, NY: Cornell University Press.

Schank, Roger C. and Abelson, Robert P. (1977) *Scripts, Plans, Goals, and Understanding*. Hillsdale, NJ: Erlbaum.

Schatzman, Leonard and Strauss, Anselm L. (1973) *Field Research; Strategies for a Natural Sociology*. Englewood Cliffs, NJ: Prentice-Hall.

Spradley, James P. (1980) *Participant Observation*. New York: Holt, Rinehart & Winston.

Stryker, Sheldon (1980) *Symbolic Interactionism; A Social Structural View*. Menlo Park, CA: Benjamin/Cummings.

Van Maanen, John (1988) *Tales of the Field; On Writing Ethnography*. Chicago: University of Chicago Press.

Watson, Graham (1996) 'Listening to the Native: The Non-Ironic Alternative to "Dialogic" Ethnography (as well as to Functionalism, Marxism, and Structuralism', *Canadian Review of Sociology and Anthropology* 33: 73–86.

Wolfinger, Nicholas H. (1995) 'Passing Moments: Some Social Dynamics of Pedestrian Interaction', *Journal of Contemporary Ethnography* 24: 323–40.

Woolgar, Steve (1988) 'Reflexivity is the Ethnographer of the Text', in S. Woolgar (ed.) *Knowledge and Reflexivity; New Frontiers in the Sociology of Knowledge*, pp. 14–34. London: Sage.

57

Thick Description: Toward an Interpretive Theory of Culture

Clifford Geertz

I

In her book, *Philosophy in a New Key*, Susanne Langer remarks that certain ideas burst upon the intellectual landscape with a tremendous force. They resolve so many fundamental problems at once that they seem also to promise that they will resolve all fundamental problems, clarify all obscure issues. Everyone snaps them up as the open sesame of some new positive science, the conceptual center-point around which a comprehensive system of analysis can be built. The sudden vogue of such a *grande idée*, crowding out almost everything else for a while, is due, she says, "to the fact that all sensitive and active minds turn at once to exploiting it. We try it in every connection, for every purpose, experiment with possible stretches of its strict meaning, with generalizations and derivatives."

After we have become familiar with the new idea, however, after it has become part of our general stock of theoretical concepts, our expectations are brought more into balance with its actual uses, and its excessive popularity is ended. A few zealots persist in the old key-to-the-universe view of it; but less driven thinkers settle down after a while to the problems the idea has really generated. They try to apply it and extend it where it applies and where it is capable of extension; and they desist where it does not apply or cannot be extended. It becomes, if it was, in truth, a seminal idea in the first place, a permanent and enduring part of our intellectual armory. But it no

Source: *The Interpretation of Cultures* (New York: Basic Books, 1973), pp. 3–30.

longer has the grandiose, all-promising scope, the infinite versatility of apparent application, it once had. The second law of thermodynamics, or the principle of natural selection, or the notion of unconscious motivation, or the organization of the means of production does not explain everything, not even everything human, but it still explains something; and our attention shifts to isolating just what that something is, to disentangling ourselves from a lot of pseudoscience to which, in the first flush of its celebrity, it has also given rise.

Whether or not this is, in fact, the way all centrally important scientific concepts develop, I don't know. But certainly this pattern fits the concept of culture, around which the whole discipline of anthropology arose, and whose domination that discipline has been increasingly concerned to limit, specify, focus, and contain. It is to this cutting of the culture concept down to size, therefore actually insuring its continued importance rather than undermining it, that the essays below are all, in their several ways and from their several directions, dedicated. They all argue, sometimes explicitly, more often merely through the particular analysis they develop, for a narrowed, specialized, and, so I imagine, theoretically more powerful concept of culture to replace E. B. Tylor's famous "most complex whole," which, its originative power not denied, seems to me to have reached the point where it obscures a good deal more than it reveals.

The conceptual morass into which the Tylorean kind of *pot-au-feu* theorizing about culture can lead, is evident in what is still one of the better general introductions to anthropology, Clyde Kluckhohn's *Mirror for Man*. In some twenty-seven pages of his chapter on the concept, Kluckhohn managed to define culture in turn as: (1) "the total way of life of a people"; (2) "the social legacy the individual acquires from his group"; (3) "a way of thinking, feeling, and believing"; (4) "an abstraction from behavior"; (5) a theory on the part of the anthropologist about the way in which a group of people in fact behave; (6) a "storehouse of pooled learning"; (7) "a set of standardized orientations to recurrent problems"; (8) "learned behavior"; (9) a mechanism for the normative regulation of behavior; (10) "a set of techniques for adjusting both to the external environment and to other men"; (11) "a precipitate of history"; and turning, perhaps in desperation, to similes, as a map, as a sieve, and as a matrix. In the face of this sort of theoretical diffusion, even a somewhat constricted and not entirely standard concept of culture, which is at least internally coherent and, more important, which has a definable argument to make is (as, to be fair, Kluckhohn himself keenly realized) an improvement. Eclecticism is self-defeating not because there is only one direction in which it is useful to move, but because there are so many: it is necessary to choose.

The concept of culture I espouse, and whose utility the essays below attempt to demonstrate, is essentially a semiotic one. Believing, with Max Weber, that man is an animal suspended in webs of significance he himself has spun, I take culture to be those webs, and the analysis of it to be therefore not an experimental science in search of law but an interpretive one in search of meaning. It is explication I am after, construing social expressions on their surface enigmatical. But this pronouncement, a doctrine in a clause, demands itself some explication.

II

Operationalism as a methodological dogma never made much sense so far as the social sciences are concerned, and except for a few rather too well-swept corners – Skinnerian behaviorism, intelligence testing, and so on – it is largely dead now. But it had, for all that, an important point to make, which, however we may feel about trying to define charisma or alienation in terms of operations, retains a certain force: if you want to understand what a science is, you should look in the first instance not at its theories or its findings, and certainly not at what its apologists say about it; you should look at what the practitioners of it do.

In anthropology, or anyway social anthropology, what the practioners do is ethnography. And it is in understanding what ethnography is, or more exactly *what doing ethnography is*, that a start can be made toward grasping what anthropological analysis amounts to as a form of knowledge. This, it must immediately be said, is not a matter of methods. From one point of view, that of the textbook, doing ethnography is establishing rapport, selecting informants, transcribing texts, taking genealogies, mapping fields, keeping a diary, and so on. But it is not these things, techniques and received procedures, that define the enterprise. What defines it is the kind of intellectual effort it is: an elaborate venture in, to borrow a notion from Gilbert Ryle, "thick description."

Ryle's discussion of "thick description" appears in two recent essays of his (now reprinted in the second volume of his *Collected Papers*) addressed to the general question of what, as he puts it, "*Le Penseur*" is doing: "Thinking and Reflecting" and "The Thinking of Thoughts." Consider, he says, two boys rapidly contracting the eyelids of their right eyes. In one, this is an involuntary twitch; in the other, a conspiratorial signal to a friend. The two movements are, as movements, identical; from an I-am-a-camera, "phenomenalistic" observation of them alone, one could not tell which was twitch and which was wink, or indeed whether both or either was twitch or wink. Yet the difference, however unphotographable, between a twitch and a wink is vast; as anyone unfortunate enough to have had the first taken for the second knows. The winker is communicating, and indeed communicating in a quite precise and special way; (1) deliberately, (2) to someone in particular, (3) to impart a particular message, (4) according to a socially established code, and (5) without cognizance of the rest of the company. As Ryle points out, the winker has not done two things, contracted his eyelids and winked, while the twitcher has done only one, contracted his eyelids. Contracting your eyelids on purpose when there exists a public code in which so doing counts as a conspiratorial signal *is* winking. That's all there is to it: a speck of behavior, a fleck of culture, and – *voilà*! – a gesture.

That, however, is just the beginning. Suppose, he continues, there is a third boy, who, "to give malicious amusement to his cronies," parodies the first boy's wink, as amateurish, clumsy, obvious, and so on. He, of course, does this in the same way the second boy winked and the first twitched: by contracting his right eyelids. Only this boy is neither winking nor twitching, he is parodying someone else's, as he takes it, laughable, attempt at winking. Here, too, a socially established code exists (he will "wink" laboriously, overobviously, perhaps adding a grimace – the usual artifices of the clown), and so also does a message. Only now it is not conspiracy but ridicule that is in the air. If the others think he is actually winking, his whole project misfires as completely, though with somewhat different results, as if they think he is twitching. One can go further: uncertain of his mimicking abilities, the would-be satirist may practice at home before the mirror, in which case he is not twitching, winking, or parodying, but rehearsing; though so far as what a camera, a radical behaviorist, or a believer in protocol sentences would record he is just rapidly contracting his right eyelids like all the others. Complexities are possible, if not practically without end, at least logically so. The original winker might, for example, actually have been fake-winking, say, to mislead outsiders into imagining there was a conspiracy afoot when there in fact was not, in which case our descriptions of what the parodist is parodying and the rehearser rehearsing of course shift accordingly. But the point is that between what Ryle calls the "thin description" of what the rehearser (parodist, winker, twitcher . . .) is doing ("rapidly contracting his right eyelids") and the "thick description" of what he is doing ("practicing a burlesque of a friend faking a wink to deceive an innocent into thinking a conspiracy is in motion") lies the object of ethnography: a stratified hierarchy of meaningful structures in terms of which twitches, winks, fake-winks, parodies, rehearsals of parodies are produced, perceived, and interpreted, and without which they would not (not even the zero-form twitches, which, *as a cultural category*, are as much nonwinks as winks are nontwitches) in fact exist, no matter what anyone did or didn't do with his eyelids.

Like so many of the little stories Oxford philosophers like to make up for themselves, all this winking, fake-winking, burlesque-fake-winking, rehearsed-burlesque-fake-winking, may seem a bit artificial. In way of adding a more empirical note, let me give, deliberately unpreceded by any prior explanatory comment at all, a not untypical excerpt from my own field journal to demonstrate that, however evened off for didactic purposes, Ryle's example presents an image only too exact of the sort of piled-up structures of inference and implication through which an ethnographer is continually trying to pick his way:

> The French [the informant said] had only just arrived. They set up twenty or so small forts between here, the town, and the Marmusha area up in the middle of the mountains, placing them on promontories so they could survey the countryside. But for all this they couldn't guarantee safety, especially at night, so although the *mezrag*, trade-pact, system was supposed to be legally abolished it in fact continued as before.

One night, when Cohen (who speaks fluent Berber), was up there, at Marmusha, two other Jews who were traders to a neighboring tribe came by to purchase some goods from him. Some Berbers, from yet another neighboring tribe, tried to break into Cohen's place, but he fired his rifle in the air. (Traditionally, Jews were not allowed to carry weapons; but at this period things were so unsettled many did so anyway.) This attracted the attention of the French and the marauders fled.

The next night, however, they came back, one of them disguised as a woman who knocked on the door with some sort of a story. Cohen was suspicious and didn't want to let "her" in, but the other Jews said, "oh, it's all right, it's only a woman." So they opened the door and the whole lot came pouring in. They killed the two visiting Jews, but Cohen managed to barricade himself in an adjoining room. He heard the robbers planning to burn him alive in the shop after they removed his goods, and so he opened the door and, laying about him wildly with a club, managed to escape through a window.

He went up to the fort, then, to have his wounds dressed, and complained to the local commandant, one Captain Dumari, saying he wanted his *'ar* – i.e., four or five times the value of the merchandise stolen from him. The robbers were from a tribe which had not yet submitted to French authority and were in open rebellion against it, and he wanted authorization to go with his *mezrag*-holder, the Marmusha tribal *sheikh*, to collect the indemnity that, under traditional rules, he had coming to him. Captain Dumari couldn't officially give him permission to do this, because of the French prohibition of the *mezrag* relationship, but he gave him verbal authorization, saying, "If you get killed, it's your problem."

So the *sheikh*, the Jew, and a small company of armed Marmushans went off ten or fifteen kilometers up into the rebellious area, where there were of course no French, and, sneaking up, captured the thief-tribe's shepherd and stole its herds. The other tribe soon came riding, out on horses after them, armed with rifles and ready to attack. But when they saw who the "sheep thieves" were, they thought better of it and said, "all right, we'll talk." They couldn't really deny what had happened – that some of their men had robbed Cohen and killed the two visitors – and they weren't prepared to start the serious feud with the Marmusha a scuffle with the invading party would bring on. So the two groups talked, and talked, and talked, there on the plain amid the thousands of sheep, and decided finally on five-hundred-sheep damages. The two armed Berber groups then lined up on their horses at opposite ends of the plain, with the sheep herded between them, and Cohen, in his black gown, pillbox hat, and flapping slippers, went out alone among the sheep, picking out, one by one and at his own good speed, the best ones for his payment.

So Cohen got his sheep and drove them back to Marmusha. The French, up in their fort, heard them coming from some distance ("Ba, ba, ba" said Cohen, happily, recalling the image) and said, "What the hell is that?" And Cohen said, "That is my *'ar*." The French couldn't believe he had actually done what he said he had done, and accused him of being a spy for the rebellious Berbers, put him in prison, and took his sheep. In the town, his family, not having heard from him in so long a time, thought he was dead. But after a while the French released him and he came back home, but without his sheep. He then went to the Colonel in the town, the Frenchman in charge of the whole region, to complain. But the Colonel said, "I can't do anything about the matter. It's not my problem."

Quoted raw, a note in a bottle, this passage conveys, as any similar one similarly presented would do, a fair sense of how much goes into ethnographic description of even the most elemental sort – how extraordinarily "thick" it is. In finished anthropological writings, including those collected here, this fact – that what we call our data are really our own constructions of other people's constructions of what they and their compatriots are up to – is obscured because most of what we need to comprehend a particular event, ritual, custom, idea, or whatever is insinuated as background information before the thing itself is directly examined. (Even to reveal that this little drama took place in the highlands of central Morocco in 1912 – and was recounted there in 1968 – is to determine much of our understanding of it.) There is nothing particularly wrong with this, and it is in any case inevitable. But it does lead to a view of anthropological research as rather more of an observational and rather less of an interpretive activity than it really is. Right down at the factual base, the hard rock, insofar as there is any, of the whole enterprise, we are already explicating: and worse, explicating explications. Winks upon winks upon winks.

Analysis, then, is sorting out the structures of signification – what Ryle called established codes, a somewhat misleading expression, for it makes the enterprise sound too much like that of the cipher clerk when it is much more like that of the literary critic – and determining their social ground and import. Here, in our text, such sorting would begin with distinguishing the three unlike frames of interpretation ingredient in the situation, Jewish, Berber, and French, and would then move on to show how (and why) at that time, in that place, their copresence produced a situation in which systematic misunderstanding reduced traditional form to social farce. What tripped Cohen up, and with him the whole, ancient pattern of social and economic relationships within which he functioned, was a confusion of tongues.

I shall come back to this too-compacted aphorism later, as well as to the details of the text itself. The point for now is only that ethnography is thick description. What the ethnographer is in fact faced with – except when (as, of course, he must do) he is pursuing the more automatized routines of data collection – is a multiplicity of complex conceptual structures, many of them superimposed upon or knotted into one another, which are at once strange, irregular, and inexplicit, and which he must contrive somehow first to grasp and then to render. And this is true at the most down-to-earth, jungle field work levels of his activity: interviewing informants, observing rituals, eliciting kin terms, tracing property lines, censusing households . . . writing his journal. Doing ethnography is like trying to read (in the sense of "construct a reading of") a manuscript – foreign, faded, full of ellipses, incoherencies, suspicious emendations, and tendentious commentaries, but written not in conventionalized graphs of sound but in transient examples of shaped behavior.

III

Culture, this acted document, thus is public, like a burlesqued wink or a mock sheep raid. Though ideational, it does not exist in someone's head; though unphysical, it is not an occult entity. The interminable, because unterminable, debate within anthropology as to whether culture is "subjective" or "objective," together with the mutual exchange of intellectual insults ("idealist!" – "materialist!"; "mentalist!" – "behaviorist!"; "impressionist!" – "positivist!") which accompanies it, is wholly misconceived. Once human behavior is seen as (most of the time; there *are* true twitches) symbolic action – action which, like phonation in speech, pigment in painting, line in writing, or sonance in music, signifies – the question as to whether culture is patterned conduct or a frame of mind, or even the two somehow mixed together, loses sense. The thing to ask about a burlesqued wink or a mock sheep raid is not what their ontological status is. It is the same as that of rocks on the one hand and dreams on the other – they are things of this world. The thing to ask is what their import is: what it is, ridicule or challenge, irony or anger, snobbery or pride, that, in their occurrence and through their agency, is getting said.

This may seem like an obvious truth, but there are a number of ways to obscure it. One is to imagine that culture is a self-contained "super-organic" reality with forces and purposes of its own; that is, to reify it. Another is to claim that it consists in the brute pattern of behavioral events we observe in fact to occur in some identifiable community or other; that is, to reduce it. But though both these confusions still exist, and doubtless will be always with us, the main source of theoretical muddlement in contemporary anthropology is a view which developed in reaction to them and is right now very widely held – namely, that, to quote Ward Goodenough, perhaps its leading proponent, "culture [is located] in the minds and hearts of men."

Variously called ethnoscience, componential analysis, or cognitive anthropology (a terminological wavering which reflects a deeper uncertainty), this school of thought holds that culture is composed of psychological structures by means of which individuals or groups of individuals guide their behavior. "A society's culture," to quote Goodenough again, this time in a passage which has become the *locus classicus* of the whole movement, "consists of whatever it is one has to know or believe in order to operate in a manner acceptable to its members." And from this view of what culture is follows a view, equally assured, of what describing it is – the writing out of systematic rules, an ethnographic algorithm, which, if followed, would make it possible so to operate, to pass (physical appearance aside) for a native. In such a way, extreme subjectivism is married to extreme formalism, with the expected result: an explosion of debate as to whether particular analyses (which come in the form of taxonomies, paradigms, tables, trees, and other ingenuities) reflect what the natives "really" think or are merely clever simulations, logically equivalent but substantively different, of what they think.

As, on first glance, this approach may look close enough to the one being developed here to be mistaken for it, it is useful to be explicit as to what divides them. If, leaving our winks and sheep behind for the moment, we take, say, a Beethoven quartet as an, admittedly rather special but, for these purposes, nicely illustrative, sample of culture, no one would, I think, identify it with its score, with the skills and knowledge needed to play it, with the understanding of it possessed by its performers or auditors, nor, to take care, *en passant*, of the reductionists and reifiers, with a particular performance of it or with some mysterious entity transcending material existence. The "no one" is perhaps too strong here, for there are always incorrigibles. But that a Beethoven quartet is a temporally developed tonal structure, a coherent sequence of modeled sound – in a word, music – and not anybody's knowledge of or belief about anything, including how to play it, is a proposition to which most people are, upon reflection, likely to assent.

To play the violin it is necessary to possess certain habits, skills, knowledge, and talents, to be in the mood to play, and (as the old joke goes) to have a violin. But violin playing is neither the habits, skills, knowledge, and so on, nor the mood, nor (the notion believers in "material culture" apparently embrace) the violin. To make a trade pact in Morocco, you have to do certain things in certain ways (among others, cut, while chanting Quranic Arabic, the throat of a lamb before the assembled, undeformed, adult male members of your tribe) and to be possessed of certain psychological characteristics (among others, a desire for distant things). But a trade pact is neither the throat cutting nor the desire, though it is real enough, as seven kinsmen of our Marmusha sheikh discovered when, on an earlier occasion, they were executed by him following the theft of one mangy, essentially valueless sheepskin from Cohen.

Culture is public because meaning is. You can't wink (or burlesque one) without knowing what counts as winking or how, physically, to contract your eyelids, and you can't conduct a sheep raid (or mimic one) without knowing what it is to steal a sheep and how practically to go about it. But to draw from such truths the conclusion that knowing how to wink is winking and knowing how to steal a sheep is sheep raiding is to betray as deep a confusion as, taking thin descriptions for thick, to identify winking with eyelid contractions or sheep raiding with chasing woolly animals out of pastures. The cognitivist fallacy – that culture consists (to quote another spokesman for the movement, Stephen Tyler) of "mental phenomena which can [he means "should"] be analyzed by formal methods similar to those of mathematics and logic" – is as destructive of an effective use of the concept as are the behaviorist and idealist fallacies to which it is a misdrawn correction. Perhaps, as its errors are more sophisticated and its distortions subtler, it is even more so.

The generalized attack on privacy theories of meaning is, since early Husserl and late Wittgenstein, so much a part of modern thought that it need not be developed once more here. What is necessary is to see to it that the news of it reaches anthropology; and in particular that it is made clear that to say that culture consists of socially established structures of meaning in terms of which people do such things as signal conspiracies and join them or perceive insults and answer them, is no more to say that it is a psychological phenomenon, a characteristic of someone's mind, personality, cognitive structure, or whatever, than to say that Tantrism, genetics, the progressive form of the verb, the classification of wines, the Common Law, or the notion of "a conditional curse" (as Westermarck defined the concept of *'ar* in terms of which Cohen pressed his claim to damages) is. What, in a place like Morocco, most prevents those of us who grew up winking other winks or attending other sheep from grasping what people are up to is not ignorance as to how cognition works (though, especially as, one assumes, it works the same among them as it does among us, it would greatly help to have less of that too) as a lack of familiarity with the imaginative universe within which their acts are signs. As Wittgenstein has been invoked, he may as well be quoted:

> We . . . say of some people that they are transparent to us. It is, however, important as regards this observation that one human being can be a complete enigma to another. We learn this when we come into a strange country with entirely strange traditions; and, what is more, even given a mastery of the country's language. We do not *understand* the people. (And not because of not knowing what they are saying to themselves.) We cannot find our feet with them.

IV

Finding our feet, an unnerving business which never more than distantly succeeds, is what ethnographic research consists of as a personal experience; trying to formulate the basis on which one imagines, always excessively, one has found them is what anthropological writing consists of as a scientific endeavor. We are not, or at least I am not, seeking either to become natives (a compromised word in any case) or to mimic them. Only romantics or spies would seem to find point in that. We are seeking, in the widened sense of the term in which it encompasses very much more than talk, to converse with them, a matter a great deal more difficult, and not only with strangers, than is commonly recognized. "If speaking *for* someone else seems to be a mysterious process," Stanley Cavell has remarked, "that may be because speaking *to* someone does not seem mysterious enough."

Looked at in this way, the aim of anthropology is the enlargement of the universe of human discourse. That is not, of course, its only aim – instruction, amusement, practical counsel, moral advance, and the discovery of natural order in human behavior are others; nor is anthropology the only discipline which pursues it. But it is an aim to which a semiotic concept of culture is peculiarly well adapted. As interworked systems of construable signs (what, ignoring provincial usages, I would call symbols), culture is not a power, something to which social events, behaviors, institutions, or processes can be causally attributed; it is a context, something within which they can be intelligibly – that is, thickly – described.

The famous anthropological absorption with the (to us) exotic – Berber horsemen, Jewish peddlers, French Legionnaires – is, thus, essentially a device for displacing the dulling sense of familiarity with which the mysteriousness of our own ability to relate perceptively to one another is concealed from us. Looking at the ordinary in places where it takes unaccustomed forms brings out not, as has so often been claimed, the arbitrariness of human behavior (there is nothing especially arbitrary about taking sheep theft for insolence in Morocco), but the degree to which its meaning varies according to the pattern of life by which it is informed. Understanding a people's culture exposes their normalness without reducing their particularity. (The more I manage to follow what the Moroccans are up to, the more logical, and the more singular, they seem.) It renders them accessible: setting them in the frame of their own banalities, it dissolves their opacity.

It is this maneuver, usually too casually referred to as "seeing things from the actor's point of view," too bookishly as "the *verstehen* approach," or too technically as "emic analysis," that so often leads to the notion that anthropology is a variety of either long-distance mind reading or cannibal-isle fantasizing, and which, for someone anxious to navigate past the wrecks of a dozen sunken philosophies, must therefore be executed with a great deal of care. Nothing is more necessary to comprehending what anthropological interpretation is, and the degree to which it *is* interpretation, than an exact understanding of what it means – and what it does not mean – to say that our formulations of other peoples' symbol systems must be actor-oriented.[1]

What it means is that descriptions of Berber, Jewish, or French culture must be cast in terms of the constructions we imagine Berbers. Jews, or

Frenchmen to place upon what they live through, the formulae they use to define what happens to them. What it does not mean is that such descriptions are themselves Berber, Jewish, or French – that is, part of the reality they are ostensibly describing; they are anthropological – that is, part of a developing system of scientific analysis. They must be cast in terms of the interpretations to which persons of a particular denomination subject their experience, because that is what they profess to be descriptions of; they are anthropological because it is, in fact, anthropologists who profess them. Normally, it is not necessary to point out quite so laboriously that the object of study is one thing and the study of it another. It is clear enough that the physical world is not physics and *A Skeleton Key to Finnegan's Wake* not *Finnegan's Wake*. But, as, in the study of culture, analysis penetrates into the very body of the object – that is, *we begin with our own interpretations of what our informants are up to, or think they are up to, and then systematize those* – the line between (Moroccan) culture as a natural fact and (Moroccan) culture as a theoretical entity tends to get blurred. All the more so, as the latter is presented in the form of an actor's-eye description of (Moroccan) conceptions of everything from violence, honor, divinity, and justice, to tribe, property, patronage, and chiefship.

In short, anthropological writings are themselves interpretations, and second and third order ones to boot. (By definition, only a "native" makes first order ones: it's *his* culture.)[2] They are, thus, fictions; fictions, in the sense that they are "something made," "something fashioned" – the original meaning of *fictiō* – not that they are false, unfactual, or merely "as if" thought experiments. To construct actor-oriented descriptions of the involvements of a Berber chieftain, a Jewish merchant, and a French soldier with one another in 1912 Morocco is clearly an imaginative act, not all that different from constructing similar descriptions of, say, the involvements with one another of a provincial French doctor, his silly, adulterous wife, and her feckless lover in nineteenth century France. In the latter case, the actors are represented as not having existed and the events as not having happened, while in the former they are represented as actual, or as having been so. This is a difference of no mean importance; indeed, precisely the one Madame Bovary had difficulty grasping. But the importance does not lie in the fact that her story was created while Cohen's was only noted. The conditions of their creation, and the point of it (to say nothing of the manner and the quality) differ. But the one is as much a *fictiō* – "a making" – as the other.

Anthropologists have not always been as aware as they might be of this fact: that although culture exists in the trading post, the hill fort, or the sheep run, anthropology exists in the book, the article, the lecture, the museum display, or, sometimes nowadays, the film. To become aware of it is to realize that the line between mode of representation and substantive content is as undrawable in cultural analysis as it is in painting; and that fact in turn seems to threaten the objective status of anthropological knowledge by suggesting that its source is not social reality but scholarly artifice.

It does threaten it, but the threat is hollow. The claim to attention of an ethnographic account does not rest on its author's ability to capture primitive facts in faraway places and carry them home like a mask or a carving, but on the degree to which he is able to clarify what goes on in such places, to reduce the puzzlement – what manner of men are these? – to which unfamiliar acts emerging out of unknown backgrounds naturally give rise. This raises some serious problems of verification, all right – or, if "verification" is too strong a word for so soft a science (I, myself, would prefer "appraisal"), of how you can tell a better account from a worse one. But that is precisely the virtue of it. If ethnography is thick description and ethnographers those who are doing the describing, then the determining question for any given example of it, whether a field journal squib or a Malinowski-sized monograph, is whether it sorts winks from twitches and real winks from mimicked ones. It is not against a body of uninterpreted data, radically thinned descriptions, that we must measure the cogency of our explications, but against the power of the scientific imagination to bring us into touch with the lives of strangers. It is not worth it, as Thoreau said, to go round the world to count the cats in Zanzibar.

V

Now, this proposition, that it is not in our interest to bleach human behavior of the very properties that interest us before we begin to examine it, has sometimes been escalated into a larger claim: namely, that as it is only those properties that interest us, we need not attend, save cursorily, to behavior at all. Culture is most effectively treated, the argument goes, purely as a symbolic system (the catch phrase is, "in its own terms"), by isolating its elements, specifying the internal relationships among those elements, and then characterizing the whole system in some general way – according to the core symbols around which it is organized, the underlying structures of which it is a surface expression, or the ideological principles upon which it is based. Though a distinct improvement over "learned behavior" and "mental phenomena" notions of what culture is, and the source of some of the most powerful theoretical ideas in contemporary anthropology, this hermetical approach to things seems to me to run the danger (and increasingly to have been overtaken by it) of locking cultural analysis away from its proper object, the informal logic of actual life. There is little profit in extricating a concept from the defects of psychologism only to plunge it immediately into those of schematicism.

Behavior must be attended to, and with some exactness, because it is through the flow of behavior – or, more precisely, social action – that cultural forms find articulation. They find it as well, of course, in various sorts of artifacts, and various states of consciousness; but these draw their meaning from the role they play (Wittgenstein would say their "use") in an ongoing pattern of life, not from any intrinsic relationships they bear to one another. It is what Cohen, the sheikh, and "Captain Dumari" were doing when they tripped over one another's purposes – pursuing trade, defending honor, establishing dominance – that created our pastoral drama, and that is what the drama is, therefore, "about." Whatever, or wherever, symbol systems "in their own terms" may be, we gain empirical access to them by inspecting events, not by arranging abstracted entities into unified patterns.

A further implication of this is that coherence cannot be the major test of validity for a cultural description. Cultural systems must have a minimal degree of coherence, else we would not call them systems; and, by observation, they normally have a great deal more. But there is nothing so coherent as a paranoid's delusion or a swindler's story. The force of our interpretations cannot rest, as they are now so often made to do, on the tightness with which they hold together, or the assurance with which they are argued. Nothing has done more, I think, to discredit cultural analysis than the construction of impeccable depictions of formal order in whose actual existence nobody can quite believe.

If anthropological interpretation is constructing a reading of what happens, then to divorce it from what happens – from what, in this time or that place, specific people say, what they do, what is done to them, from the whole vast business of the world – is to divorce it from its applications and render it vacant. A good interpretation of anything – a poem, a person, a history, a ritual, an institution, a society – takes us into the heart of that of which it is the interpretation. When it does not do that, but leads us instead somewhere else – into an admiration of its own elegance, of its author's cleverness, or of the beauties of Euclidean order – it may have its intrinsic charms; but it is something else than what the task at hand – figuring out what all that rigamarole with the sheep is about – calls for.

The rigamarole with the sheep – the sham theft of them, the reparative transfer of them, the political confiscation of them – is (or was) essentially a social discourse, even if, as I suggested earlier, one conducted in multiple tongues and as much in action as in words.

Claiming his *'ar*, Cohen invoked the trade pact; recognizing the claim, the sheikh challenged the offenders' tribe; accepting responsibility, the offenders' tribe paid the indemnity; anxious to make clear to sheikhs and peddlers alike who was now in charge here, the French showed the imperial hand. As in any discourse, code does not determine conduct, and what was actually said need not have been. Cohen might not have, given its illegitimacy in Protectorate eyes, chosen to press his claim. The sheikh might, for similar reasons, have rejected it. The offenders' tribe, still resisting French authority, might have decided to regard the raid as "real" and fight rather than negotiate. The French, were they more *habile* and less *dur* (as, under Mareschal Lyautey's seigniorial tutelage, they later in fact became), might have permitted Cohen to keep his sheep, winking – as we say – at the continuance of the trade pattern and its limitation to their authority. And there are other possibilities: the Marmushans might have regarded the French action as too great an insult to bear and gone into dissidence themselves; the French might have attempted not just to clamp down on Cohen but to bring the sheikh himself more closely to heel; and Cohen might have concluded that between renegade Berbers and Beau Geste soldiers, driving trade in the Atlas highlands was no longer worth the candle and retired to the better-governed confines of the town. This, indeed, is more or less what happened, somewhat further along, as the Protectorate moved toward genuine sovereignty. But the point here is not to describe what did or did not take place in Morocco. (From this simple incident one can widen out into enormous complexities of social experience.) It is to demonstrate what a piece of anthropological interpretation consists in: tracing the curve of a social discourse; fixing it into an inspectable form.

The ethnographer "inscribes" social discourse; *he writes it down*. In so doing, he turns it from a passing event, which exists only in its own moment of occurrence, into an account, which exists in its inscriptions and can be reconsulted. The sheikh is long dead, killed in the process of being, as the French called it, "pacified"; "Captain Dumari," his pacifier, lives, retired to his souvenirs, in the south of France; and Cohen went last year, part refugee, part pilgrim, part dying patriarch, "home" to Israel. But what they, in my extended sense, "said" to one another on an Atlas plateau sixty years ago is – very far from perfectly – preserved for study. "What," Paul Ricoeur, from whom this whole idea of the inscription of action is borrowed and somewhat twisted, asks, "what does writing fix?"

> Not the event of speaking, but the "said" of speaking, where we understand by the "said" of speaking that intentional exteriorization constitutive of the aim of discourse thanks to which the *sagen* – the saying – wants to become *Aus-sage* – the enunciation, the enunciated. In short, what we write is the *noema* ["thought," "content," "gist"] of the speaking. It is the meaning of the speech event, not the event as event.

This is not itself so very "said" – if Oxford philosophers run to little stories, phenomenological ones run to large sentences; but it brings us anyway to a more precise answer to our generative question, "What does the ethnographer do?" – he writes.[3] This, too, may seem a less than startling discovery, and to someone familiar with the current "literature," an implausible one. But as the standard answer to our question has been, "He observes, he records, he analyzes" – a kind of *veni, vidi, vici* conception of the matter – it may have more deep-going consequences than are at first apparent, not the least of which is that distinguishing these three phases of knowledge-seeking may not, as a matter of fact, normally be possible; and, indeed, as autonomous "operations" they may not in fact exist.

The situation is even more delicate, because, as already noted, what we inscribe (or try to) is not raw social discourse, to which, because, save very marginally or very specially, we are not actors, we do not have direct access, but only that small part of it which our informants can lead us into understanding.[4] This is not as fatal as it sounds, for, in fact, not all Cretans are liars, and it is not necessary to know everything in order to understand something. But it does make the view of anthropological analysis as the conceptual manipulation of discovered facts, a logical reconstruction of a mere reality, seem rather lame. To set forth symmetrical crystals of significance, purified of the material complexity in which they were located, and then attribute their existence to autogenous principles of order, universal properties of the human mind, or vast, a priori *weltanschauungen*, is to pretend a science that does not exist and imagine a reality that cannot be found. Cultural analysis is (or should be) guessing at meanings, assessing the guesses, and drawing explanatory conclusions from the better guesses, not discovering the Continent of Meaning and mapping out its bodiless landscape.

VI

So, there are three characteristics of ethnographic description; it is interpretive; what it is interpretive of is the flow of social discourse; and the interpreting involved consists in trying to rescue the "said" of such discourse from its perishing occasions and fix it in perusable terms. The *kula* is gone or altered; but, for better or worse, *The Argonauts of the Western Pacific* remains. But there is, in addition, a fourth characteristic of such description, at least as I practice it: it is microscopic.

This is not to say that there are no large-scale anthropological interpretations of whole societies, civilizations, world events, and so on. Indeed, it is such extension of our analyses to wider contexts that, along with their theoretical implications, recommends them to general attention and justifies our constructing them. No one really cares anymore, not even Cohen (well . . . maybe, Cohen), about those sheep as such. History may have its unobtrusive turning points, "great noises in a little room"; but this little go-round was surely not one of them.

It is merely to say that the anthropologist characteristically approaches such broader interpretations and more abstract analyses from the direction of exceedingly extended acquaintances with extremely small matters. He confronts the same grand realities that others – historians, economists, political scientists, sociologists – confront in more fateful settings: Power, Change, Faith, Oppression, Work, Passion, Authority, Beauty, Violence, Love, Prestige; but he confronts them in contexts obscure enough – places like Marmusha and lives like Cohen's – to take the capital letters off them. These all-too-human constancies, "those big words that make us all afraid," take a homely form in such homely contexts. But that is exactly the advantage. There are enough profundities in the world already.

Yet, the problem of how to get from a collection of ethnographic miniatures on the order of our sheep story – an assortment of remarks and anecdotes – to wall-sized culturescapes of the nation, the epoch, the continent, or the civilization is not so easily passed over with vague allusions to the virtues of concreteness and the down-to-earth mind. For a science born in Indian tribes, Pacific islands, and African lineages and subsequently seized with grander ambitions, this has come to be a major methodological problem, and for the most part a badly handled one. The models that anthropologists have themselves worked out to justify their moving from local truths to general visions have been, in fact, as responsible for undermining the effort as anything their critics – sociologists obsessed with sample sizes, psychologists with measures, or economists with aggregates – have been able to devise against them.

Of these, the two main ones have been: the Jonesville-is-the-USA "microcosmic" model; and the Easter-Island-is-a-testing-case "natural experiment" model. Either heaven in a grain of sand, or the farther shores of possibility.

The Jonesville-is-America writ small (or America-is-Jonesville writ large) fallacy is so obviously one that the only thing that needs explanation is how people have managed to believe it and expected others to believe it. The notion that one can find the essence of national societies, civilizations, great religions, or whatever summed up and simplified in so-called "typical" small towns and villages is palpable nonsense. What one finds in small towns and villages is (alas) small-town or village life. If localized, microscopic studies were really dependent for their greater relevance upon such a premise – that they captured the great world in the little – they wouldn't have any relevance.

But, of course, they are not. The locus of study is not the object of study. Anthropologists don't study villages (tribes, towns, neighborhoods . . .); they study *in* villages. You can study different things in different places, and some things – for example, what colonial domination does to established frames of moral expectation – you can best study in confined localities. But that doesn't make the place what it is you are studying. In the remoter provinces of Morocco and Indonesia I have wrestled with the same questions other social scientists have wrestled with in more central locations – for example, how comes it that men's most importunate claims to humanity are cast in the accents of group pride? – and with about the same conclusiveness. One can add a dimension – one much needed in the present climate of size-up-and-solve social science; but that is all. There is a certain value, if you are going to run on about the exploitation of the masses in having seen a Javanese sharecropper turning earth in a tropical downpour or a Moroccan tailor embroidering kaftans by the light of a twenty-watt bulb. But the notion that this gives you the thing entire (and elevates you to some moral vantage ground from which you can look down upon the ethically less privileged) is an idea which only someone too long in the bush could possibly entertain.

The "natural laboratory" notion has been equally pernicious, not only because the analogy is false – what kind of a laboratory is it where *none* of the parameters are manipulable? – but because it leads to a notion that the data derived from ethnographic studies are purer, or more fundamental, or more solid, or less conditioned (the most favored word is "elementary") than those derived from other sorts of social inquiry. The great natural variation of cultural forms is, of course, not only anthropology's great (and wasting) resource, but the ground of its deepest theoretical dilemma: how is such variation to be squared with the biological unity of the human species? But it is not, even metaphorically, experimental variation, because the context in which it occurs varies along with it, and it is not possible (though there are those who try) to isolate the *y*'s from *x*'s to write a proper function.

The famous studies purporting to show that the Oedipus complex was backwards in the Trobriands, sex roles were upside down in Tchambuli, and the Pueblo Indians lacked aggression (it is characteristic that they were all negative – "but not in the South"), are, whatever their empirical validity may or may not be, not "scientifically tested and approved" hypotheses. They are interpretations, or misinterpretations, like any others, arrived at in the same way as any others, and as inherently inconclusive as any others, and the attempt to invest them with the authority of physical experimentation is but methodological sleight of hand. Ethnographic findings are not privileged, just particular: another country heard from. To regard them as anything more (*or anything less*) than that distorts both them and their implications, which are far profounder than mere primitivity, for social theory.

Another country heard from: the reason that protracted descriptions of distant sheep raids (and a really good ethnographer would have gone into what kind of sheep they were) have general relevance is that they present the sociological mind with bodied stuff on which to feed. The important thing about the anthropologist's findings is their complex specificness, their circumstantiality. It is with the kind of material produced by long-term, mainly (though not exclusively) qualitative, highly participative, and almost obsessively fine-comb field study in confined contexts that the mega-concepts with which contemporary social science is afflicted – legitimacy, modernization, integration, conflict, charisma, structure, . . . meaning – can be given the sort of sensible actuality that makes it possible to think not only realistically and concretely *about* them, but, what is more important, creatively and imaginatively *with* them.

The methodological problem which the microscopic nature of ethnography presents is both real and critical. But it is not to be resolved by regarding a remote locality as the world in a teacup or as the sociological equivalent of a cloud chamber. It is to be resolved – or, anyway, decently kept at bay – by realizing that social actions are comments on more than themselves; that where an interpretation comes from does not determine where it can be impelled to go. Small facts speak to large issues, winks to epistemology, or sheep raids to revolution, because they are made to.

VII

Which brings us, finally, to theory. The besetting sin of interpretive approaches to anything – literature, dreams, symptoms, culture – is that they tend to resist, or to be permitted to resist, conceptual articulation and thus to escape systematic modes of assessment. You either grasp an interpretation or you do not, see the point of it or you do not, accept it or you do not. Imprisoned in the immediacy of its own detail, it is presented as self-validating, or, worse, as validated by the supposedly developed sensitivities of the person who presents it; any attempt to cast what it says in terms other than its own is regarded as a travesty – as, the anthropologist's severest term of moral abuse, ethnocentric.

For a field of study which, however timidly (though I, myself, am not timid about the matter at all), asserts itself to be a science, this just will not do. There is no reason why the conceptual structure of a cultural interpretation should be any less formulable, and thus less susceptible to explicit canons of appraisal, than that of, say, a biological observation or a physical experiment – no reason except that the terms in which such formulations can be cast are, if not wholly nonexistent, very nearly so. We are reduced to insinuating theories because we lack the power to state them.

At the same time, it must be admitted that there are a number of characteristics of cultural interpretation which make the theoretical development of it more than usually difficult. The first is the need for theory to stay rather closer to the ground than tends to be the case in sciences more able to give themselves over to imaginative abstraction. Only short flights of ratiocination tend to be effective in anthropology; longer ones tend to drift off into logical dreams, academic bemusements with formal symmetry. The whole point of a semiotic approach to culture is, as I have said, to aid us in gaining access to the conceptual world in which our subjects live so that we can, in some extended sense of the term, converse with them. The tension between the pull of this need to penetrate an unfamiliar universe of symbolic action and the requirements of technical advance in the theory of culture, between the need to grasp and the need to analyze, is, as a result, both necessarily great and essentially irremovable. Indeed, the further theoretical development goes, the deeper the tension gets. This is the first condition for cultural theory: it is not its own master. As it is unseverable from the immediacies thick description presents, its freedom to shape itself in terms of its internal logic is rather limited. What generality it contrives to achieve grows out of the delicacy of its distinctions, not the sweep of its abstractions.

And from this follows a peculiarity in the way, as a simple matter of empirical fact, our knowledge of culture . . . cultures . . . a culture . . . grows: in spurts. Rather than following a rising curve of cumulative findings, cultural analysis breaks up into a disconnected yet coherent sequence of bolder and bolder sorties. Studies do build on other studies, not in the sense that they take up where the others leave off, but in the sense that, better informed and better conceptualized, they plunge more deeply into the same things. Every serious cultural analysis starts from a sheer beginning and ends where it manages to get before exhausting its intellectual impulse. Previously discovered facts are mobilized, previously developed concepts used, previously formulated hypotheses tried out; but the movement is not from already proven theorems to newly proven ones, it is from an awkward fumbling for the most elementary understanding to a supported claim that one has achieved that and surpassed it. A study is an advance if it is more incisive – whatever that may mean – than those that preceded it; but it less stands on their shoulders than, challenged and challenging, runs by their side.

It is for this reason, among others, that the essay, whether of thirty pages or three hundred, has seemed the natural genre in which to present cultural interpretations and the theories sustaining them, and why, if one looks for systematic treatises in the field, one is so soon disappointed, the more so if one finds any. Even inventory articles are rare here, and anyway of hardly more than bibliographical interest. The major theoretical contributions not only lie in specific studies – that is true in almost any field – but they are very difficult to abstract from such studies and integrate into anything one might call "culture theory" as such. Theoretical formulations hover so low over the interpretations they govern that they don't make much sense or hold much interest apart from them. This is so, not because they are not general (if they are not general, they are not theoretical), but because, stated independently of their applications, they seem either commonplace or vacant. One can, and this in fact is how the field progresses conceptually, take a line of theoretical attack developed in connection with one exercise in ethnographic interpretation and employ it in another, pushing it forward to greater precision and broader relevance; but one cannot write a "General Theory of Cultural Interpretation." Or, rather, one can, but there appears to be little profit in it, because the essential task of theory building here is not to codify abstract regularities but to make thick description possible, not to generalize across cases but to generalize within them.

To generalize within cases is usually called, at least in medicine and depth psychology, clinical inference. Rather than beginning with a set of observations and attempting to subsume them under a governing law, such inference begins with a set of (presumptive) signifiers and attempts to place them within an intelligible frame. Measures are matched to theoretical predictions, but symptoms (even when they are measured) are scanned for theoretical peculiarities – that is, they are diagnosed. In the study of culture the signifiers are not symptoms or clusters of symptoms, but symbolic acts or clusters of symbolic acts, and the aim is not therapy but the analysis of social discourse. But the way in which theory is used – to ferret out the unapparent import of things – is the same.

Thus we are lead to the second condition of cultural theory: it is not, at least in the strict meaning of the term, predictive. The diagnostician doesn't predict measles; he decides that someone has them, or at the very most *anticipates* that someone is rather likely shortly to get them. But this limitation, which is real enough, has commonly been both misunderstood and exaggerated, because it has been taken to mean that cultural interpretation is merely post facto: that, like the peasant in the old story, we first shoot the holes in the fence and then paint the bull's-eyes around them. It is hardly to be denied that there is a good deal of that sort of thing around, some of it in prominent places. It is to be denied, however, that it is the inevitable outcome of a clinical approach to the use of theory.

It is true that in the clinical style of theoretical formulation, conceptualization is directed toward the task of generating interpretations of matters already in hand, not toward projecting outcomes of experimental manipulations or deducing future states of a determined system. But that does not mean that theory has only to fit (or, more carefully, to generate cogent interpretations of) realities past; it has also to survive – intellectually survive – realities to come. Although we formulate our interpretation of an outburst of winking or an instance of sheep-raiding after its occurrence, sometimes long after, the theoretical framework in terms of which such an interpretation is made must be capable of continuing to yield defensible interpretations as new social phenomena swim into view. Although one starts any effort at thick description, beyond the obvious and superficial, from a state of general bewilderment as to what the devil is going on – trying to find one's feet – one does not start (or ought not) intellectually empty-handed. Theoretical ideas are not created wholly anew in each study; as I have said, they are adopted from other, related studies, and, refined in the process, applied to new interpretive problems. If they cease being useful with respect to such problems, they tend to stop being used and are more or less abandoned. If they continue being useful, throwing up new understandings, they are further elaborated and go on being used.[5]

Such a view of how theory functions in an interpretive science suggests that the distinction, relative in any case, that appears in the experimental or observational sciences between "description" and "explanation" appears here as one, even more relative, between "inscription" ("thick description") and "specification" ("diagnosis") – between setting down the meaning particular social actions have for the actors whose actions they are, and stating, as explicitly as we can manage, what the knowledge thus attained demonstrates about the society in which it is found and, beyond that, about social life as such. Our double task is to uncover the conceptual structures that inform our subjects' acts, the "said" of social discourse, and to construct a system of analysis in whose terms what is generic to those structures, what belongs to them because they are what they are, will stand out against the other determinants of human behavior. In ethnography, the office of theory is to provide a vocabulary in which what symbolic action has to say about itself – that is about the role of culture in human life – can be expressed.

Aside from a couple of orienting pieces concerned with more foundational matters, it is in such a manner that theory operates in the essays collected here. A repertoire of very general, made-in-the-academy concepts and systems of concepts – "integration," "rationalization," "symbol," "ideology," "ethos," "revolution," "identity," "metaphor," "structure," "ritual," "world view," "actor," "function," "sacred," and, of course, "culture" itself – is woven into the body of thick-description ethnography in the hope of rendering mere occurrences scientifically eloquent.[6] The aim is to draw large conclusions from small, but very densely textured facts; to support broad assertions about the role of culture in the construction of collective life by engaging them exactly with complex specifics.

Thus it is not only interpretation that goes all the way down to the most immediate observational level: the theory upon which such interpretation conceptually depends does so also. My interest in Cohen's story, like Ryle's in winks, grew out of some very general notions indeed. The "confusion of tongues" model – the view that social conflict is not something that happens when, out of weakness, indefiniteness, obsolescence, or neglect, cultural forms cease to operate, but rather something which happens when, like burlesqued winks, such forms are pressed by unusual situations or unusual intentions to operate in unusual ways – is not an idea I got from Cohen's story. It is one, instructed by colleagues, students, and predecessors, I brought to it.

Our innocent-looking "note in a bottle" is more than a portrayal of the frames of meaning of Jewish peddlers, Berber warriors, and French proconsuls, or even of their mutual interference. It is an argument that to rework the pattern of social relationships is to rearrange the coordinates of the experienced world. Society's forms are culture's substance.

VIII

There is an Indian story – at least I heard it as an Indian story – about an Englishman who, having been told that the world rested on a platform which rested on the back of an elephant which rested in turn on the back of a turtle, asked (perhaps he was an ethnographer; it is the way they behave), what did the turtle rest on? Another turtle. And that turtle? "Ah, Sahib, after that it is turtles all the way down."

Such, indeed, is the condition of things. I do not know how long it would be profitable to meditate on the encounter of Cohen, the sheikh, and "Dumari" (the period has perhaps already been exceeded); but I do know that however long I did so I would not get anywhere near to the bottom of it. Nor have I ever gotten anywhere near to the bottom of anything I have ever written about, either in the essays below or elsewhere. Cultural analysis is intrinsically, incomplete. And, worse than that, the more deeply it goes the less complete it is. It is a strange science whose most telling assertions are its most tremulously based, in which to get somewhere with the matter at hand is to intensify the suspicion, both your own and that of others, that you are not quite getting it right. But that, along with plaguing subtle people with obtuse, questions, is what being an ethnographer is like.

There are a number of ways to escape this – turning culture into folklore and collecting it, turning it into traits and counting it, turning it into institutions and "classifying it, turning it into structures and toying with it. But they *are* escapes. The fact is that to commit oneself to a semiotic concept of culture and an interpretive approach to the study of it is to commit oneself to a view of ethnographic assertion as, to borrow W. B. Gallie's by now famous phrase, "essentially contestable." Anthropology, or at least interpretive anthropology, is a science whose progress is marked less by a perfection of consensus than by a refinement of debate. What gets better is the precision with which we vex each other.

This is very difficult to see when one's attention is being monopolized by a single party to the argument. Monologues are of little value here, because there are no conclusions to be reported; there is merely a discussion to be sustained. Insofar as the essays here collected have any importance, it is less in what they say than what they are witness to: an enormous increase in interest, not only in anthropology, but in social studies generally, in the role of symbolic forms in human life. Meaning, that elusive and ill-defined pseudoentity we were once more than content to leave philosophers and literary critics to fumble with, has now come back into the heart of our discipline. Even Marxists are quoting Cassirer; even positivists, Kenneth Burke.

My own position in the midst of all this has been to try to resist subjectivism on the one hand and cabbalism on the other, to try to keep the analysis of symbolic forms as closely tied as I could to concrete social events and occasions, the public world of common life, and to organize it in such a way that the connections between theoretical formulations and descriptive interpretations were unobscured by appeals to dark sciences. I have never been impressed by the argument that, as complete objectivity is impossible in these matters (as, of course, it is), one might as well let one's sentiments run loose. As Robert Solow has remarked, that is like saying that as a perfectly aseptic environment is impossible, one might as well conduct surgery in a sewer. Nor, on the other hand, have I been impressed with claims that structural linguistics, computer engineering, or some other advanced form of thought is going to enable us to understand men without knowing them. Nothing will discredit a semiotic approach to culture more quickly than allowing it to drift into a combination of intuitionism and alchemy, no matter how elegantly the intuitions are expressed or how modern the alchemy is made to look.

The danger that cultural analysis, in search of all-too-deep-lying turtles, will lose touch with the hard surfaces of life – with the political, economic, stratificatory realities within which men are everywhere contained – and with the biological and physical necessities on which those surfaces rest, is an ever-present one. The only defense against it, and against, thus, turning cultural analysis into a kind of sociological aestheticism, is to train such analysis on such realities and such necessities in the first place. It is thus that I have written about nationalism, about violence, about identity, about human nature, about legitimacy, about revolution, about ethnicity, about urbanization, about status, about death, about time, and most of all about particular attempts by particular peoples to place these things in some sort of comprehensible, meaningful frame.

To look at the symbolic dimensions of social action – art, religion, ideology, science, law, morality, common sense – is not to turn away from the existential dilemmas of life for some empyrean realm of de-emotionalized forms; it is to plunge into the midst of them. The essential vocation of interpretive anthropology is not to answer our deepest questions, but to make available to us answers that others, guarding other sheep in other valleys, have given, and thus to include them in the consultable record of what man has said.

Notes

1. Not only other peoples'; anthropology *can* be trained on the culture of which it is itself a part, and it increasingly is; a fact of profound importance, but which, as it raises a few tricky and rather special second order problems, I shall put to the side for the moment.
2. The order problem is, again, complex. Anthropological works based on other anthropological works (Lévi-Strauss', for example) may, of course, be fourth order or higher, and informants frequently, even habitually, make second order interpretations – what have come to be known as "native models." In literate cultures, where "native" interpretation can proceed to higher levels – in connection with the Maghreb, one has only to think of Ibn Khaldun; with the United States, Margaret Mead – these matters become intricate indeed.
3. Or, again, more exactly, "inscribes." Most ethnography is in fact to be found in books and articles, rather than in films, records, museum displays, or whatever; but even in them there are, of course, photographs, drawings, diagrams, tables, and so on. Self-consciousness about modes of representation (not to speak of experiments with them) has been very lacking in anthropology.
4. So far as it has reinforced the anthropologist's impulse to engage himself with his informants as persons rather than as objects, the notion of "participant observation" has been a valuable one. But, to the degree it has lead the anthropologist to block from

his view the very special, culturally bracketed nature of his own role and to imagine himself something more than an interested (in both senses of that word) sojourner, it has been our most powerful source of bad faith.

5. Admittedly, this is something of an idealization. Because theories are seldom if ever decisively disproved in clinical use but merely grow increasingly awkward, unproductive, strained, or vacuous, they often persist long after all but a handful of people (though *they* are often most passionate) have lost much interest in them. Indeed, so far as anthropology is concerned, it is almost more of a problem to get exhausted ideas out of the literature than it is to get productive ones in, and so a great deal more of theoretical discussion than one would prefer is critical rather than constructive, and whole careers have been devoted to hastening the demise of moribund notions. As the field advances one would hope that this sort of intellectual weed control would become a less prominent part of our activities. But, for the moment, it remains true that old theories tend less to die than to go into second editions.
6. The overwhelming bulk of the following chapters concern Indonesia rather than Morocco, for I have just begun to face up to the demands of my North African material which, for the most part, was gathered more recently. Field work in Indonesia was carried out in 1952–1954, 1957–1958, and 1971; in Morocco in 1964, 1965–1966, 1968–1969, and 1972.

58

Thinking through Fieldwork

Judith Okely

In this chapter, I describe the way in which I interpreted and wrote up my material from an intensive fieldwork study of Gypsies and aspects of government policy. The approach and methods which informed this work were those of social anthropology. I give details of how I recorded my material and, more relevant for this collection, how I made use of it along with the totality of my field experience for the ensuing publications. Since the term 'qualitative' has been applied to a range of different methodologies within the social sciences, it is important to outline the distinctive characteristics of social anthropological research. 'Qualitative' can refer to research using only a small sample of interviews, whether structured or unstructured. In either case the qualitative material is bounded by the cultural conventions of the interview. Paradoxically, the interview format is associated first and foremost with quantitative surveys whose positivistic conventions have set the agenda (cf. Oakley 1981). 'Qualitative' has also come to be used to describe the research which this chapter addresses, namely that which emerges from participant observation.

There are significant contrasts between anthropological and sociological empirical research, with implications for analysis. The two disciplines came from different contexts. Social anthropology was formerly associated with the study of non-western societies, mainly by westerners. Sociology's empirical work was concerned mainly with western societies of which the sociologist was a member. Unlike the sociologist, the anthropologist could not take much as given, he or she could not isolate one theme extracted from a

Source: Alan Bryman and Robert G. Burgess (eds), *Analyzing Qualitative Data* (London: Routledge, 1994), pp. 18–34.

wider context, since the society as a whole was largely unknown to the researcher, and undocumented. Rigidly formulated questionnaires were inappropriate. These and the interview mode are culture bound. The sociologist could be more presumptuous in knowledge of the wider social context. Whereas Durkheim (1897) could claim to identify and sub-classify suicide in France, Malinowski (1926) had first to discover and then redefine such a practice among the Trobrianders. He had no statistics to play with.

The way in which the anthropologist carries out fieldwork affects the sort of material produced, then analyzed and presented in the final texts. The anthropologist rarely commences research with an hypothesis to test. There are few pre-set, neatly honed questions, although there are multiple questions in the fieldworker's head. There are theories, themes, ideas and ethnographic details to discover, examine or dismiss. The anthropologist, despite months of literature reviews, possibly years of theoretical and comparative reading, will have to eject hypotheses like so much ballast. The people may not live as recorded, there could be famine, strife or civil war. Rituals may be missionized, nomads dispersed, leaders imprisoned, documentation a distortion or deflection from the outsider's gaze. The ethnographer must, like a surrealist, be *disponible* (cf. Breton 1937), and open to *objets trouvés*, after arriving in the field. This approach inevitably affects the subsequent interpretation and analysis.

Although early field anthropologists made claims for the scientific status of their work, they have been less vulnerable than empirical sociologists to demands for positivist legitimacy in methods. Formerly, it was considered sufficiently impressive that anthropologists actually uprooted themselves and went to live elsewhere for extended periods. Social scientists who stayed at home were not in a position to challenge the techniques of pioneers in the unknown fields of exotica. There has been greater freedom in the analysis of fieldwork research. A great deal is taken on trust about the way material is written up. There can be no easily replicable formulae. The notion of techniques to be applied uniformly across the globe is inappropriate. Granted, social anthropologists of the earlier school have been too cavalier in both preparing students for the field and in conveying advice about how to writeup. The lacunae are best filled by detailed autobiographical accounts of fieldwork and the ways in which interpretations are arrived at. These are relatively rare and split off from what are seen as the core concerns of the discipline (Okely 1992). The increasing bureaucratic and pedagogic demand for explicit methods 'training' has sometimes meant that social anthropologists have half-heartedly and inappropriately fallen back on textbooks devised for sociologists and others.

The historically divisive association of sociology with western societies and anthropology with non-western societies is no longer appropriate. Each discipline has strayed into the other's territory. While retaining its traditional methods, social anthropology can be used in the study of any group or society.

I have, for example, applied it in the study of transport, the elderly and planning in East Anglia (Okely 1991), in addition to the study of Gypsies and government policy in England. Others have adopted an anthropological approach in Britain for scrutinizing the police (Young 1991) and views of death (Hockey 1990). These studies show that the kind of qualitative material which anthropology's methods and theories generate is different from other disciplines within the social sciences.

Each discipline retains its different historical approaches to methodology. Within sociology there appears to be a widespread association of participant observation with the theoretical perspective of symbolic interactionism (Hammersley and Atkinson 1983, Silverman 1985).[1] By contrast, the research material gathered by anthropologists can be placed in as many of the theoretical perspectives as there exist in both the social sciences and, if relevant, the humanities.

It is the custom for the anthropologist to be both fieldworker and analyst-author. Division of research labour into discrete tasks, or between individuals, is at a minimum. The anthropologist fieldworker records, interprets and writes up his or her own material. For the anthropologist, the stages of knowledge as the research progresses are not sectioned between persons. So there is no need to formulate mechanical procedures and managerial-type instructions to ensure uniformity of perspective along some chain of command. The anthropologist does not have to check and double-check whether numerous assistants and interviewers have understood or even faked the collection of data. He or she has instead to look to his or her specific relationship with the people who are the subject of study. The anthropologist becomes the collector and a walking archive, with ever unfolding resources for interpretation. By contrast, a social scientist in a prestigious research centre asserted that in order to follow the correct social science procedure and to attain 'objectivity', ideally someone other than I, the fieldworker, should write up the final report with the aid of my field notes. The fact that I completed the task myself was seen uneasily as a form of intellectual cheating rather than a scientific necessity and standard anthropological practice. Such a division between collection and analysis might be possible in a research tradition where the researcher delegates the former to a reserve army of interviewers with pre-ordained questionnaire and clone-like application.[2] The pre-selected choice of answers gives material which can be mechanically classified as part of the analysis.

Agar, the anthropologist, has offered an alternative descriptive term for research and fieldwork which is not hypothesis bound. A somewhat mechanistic metaphor, which doubtless allays the worries of those wanting proof of 'tools' of research, is what he names the 'funnel approach' (1980: 13). From the outset of fieldwork, the anthropologist adopts an open-ended approach to the full range of information and to all manner of people. This is the essence of the holistic approach. The material and ethnographic concerns are not cut to size at the start. The people who are the subject of study are themselves free to volunteer their concerns in their own voice and context. All this has implications for the kind of material and field notes which the anthropologist is faced with when it comes to writing up.

Both during the fieldwork and after, themes gradually emerge. Patterns and priorities impose themselves upon the ethnographer. Voices and ideas are neither muffled nor dismissed. To the professional positivist this seems like chaos. The voices and material lead the researcher in unpredictable, uncontrollable directions. This is indeed not a controlled experiment. The fieldworker cannot separate the act of gathering material from that of its continuing interpretation. Ideas and hunches emerge during the encounter and are explored or eventually discarded as fieldwork progresses. Writing up involves a similar experience. The ensuing analysis is creative, demanding and all consuming. It cannot be fully comprehended at the early writing-up stages by someone other than the fieldworker.

Long-term participant experience helps to make sense of even the most detached survey data. Leach, in a critique (1967) of an extensive survey of landownership in Ceylon (Sarkar and Tambiah 1957), was able to draw on his fieldwork in just one village to counter some interpretations of the statistics for 57 villages. The survey had concluded that 335 households were landless. However, from his detailed first-hand observation of inheritance practice in the region, Leach was able to point out that over time, a considerable number of the younger informants would inherit land. The same applied to many sharecroppers who were in fact heirs of the owners. He also suggested that some of the interpretation of the apparently unproblematic survey data was convincing only because the main researchers, already familiar with the region, arrived inadvertently 'at their conclusions by intuitive methods. . . . The numerical apparatus in which these conclusions are embedded seems to me to be very largely a complicated piece of self deception' (1967: 76).

After fieldwork, the material found in notebooks, in transcripts and even in contemporary written sources is only a guide and trigger. The anthropologist-writer draws also on the totality of the experience, parts of which may not, cannot, be cerebrally written down at the time. It is recorded in memory, body and all the senses. Ideas and themes have worked through the whole being throughout the experience of fieldwork. They have gestated in dreams and the subconscious in both sleep and in waking hours, away from the field, at the anthropologist's desk, in libraries and in dialogue with the people on return visits. Photographs point to details hitherto unnoticed by the fieldworker in the midst of the action. They may also revive hidden memories. The anthropologist may notice ethnographic detail which photographers do not perceive. The photograph on the cover of my 1983 book shows the interior of a Gypsy trailer-caravan. The professional who provided it failed to notice both before and after the event that the kitchen area had no sink. This is a crucial clue to Gypsies' pollution taboos. Other sources have also to be carefully scrutinized. Snatches of music may conjure up images and forgotten or half-submerged insights. The act of interpretation and writing from past fieldwork may be as evocative and sensory as Proust's description of the tasting of the madeleine cake in *A la recherche du temps perdu* (1954).

The understanding and ways of making sense of the material and of writing cannot be routinized and streamlined as instructions for methodology textbooks. Nor can it be fully assessed at this stage by a non-participant. Instead, to admit to the vastness, unpredictability and creative turbulence in which the ethnographic writer is immersed can be a reassurance that positivism is no guide. The methods in which many social scientists have been instructed have been an intellectual carapace. The puzzled novice researcher may be contaminated by positivistic notions of 'contamination', 'detachment', 'prediction', 'operationalization' or 'typicality'. Since ethnographic openness or *disponibilité* have defied hypotheses, the material cannot be subjected to strict formulae. The problem is how to convince researchers from other traditions or those who are schooled in positivistic formulae.

Years after my intensive fieldwork on Gypsies, and after follow-up research, there are still reverberations, there are still things to write about or to reinterpret, especially in the light of shifts in the theoretical and substantive concerns of social anthropology. It is only now that I feel free to detail the way in which the material was written up for books published in 1975 and 1983. The reasons are both political and intellectual. The first book, *Gypsies and Government Policy in England* (Adams *et al.* 1975) was policy-oriented. It was believed by the research centre that overseered it that policy-makers would respond most effectively to quantitative data (Okely 1987). On the other hand, they had chosen to employ an anthropologist because this discipline was traditionally associated with the study of the exotic and non-literate 'other'. Thus there was a built-in contradiction which the centre had not properly confronted, if extensive quantitative data was to be the ideal. The details of the struggle for qualitative methods could not be easily published (Okely 1987). Although an anthropological methodology was eventually accepted, its description in any public sphere risked political controversy. A detailed account of participant observation among both local officials and a vulnerable ethnic minority, and the seemingly clinical way in which field notes were dissected, risked being misread and misrepresented. Simultaneously, an exploration of how cumulative field experience leads to imaginative and intuitive interpretation would have been unsettling or irrelevant to the bureaucrats and politicians whom the research was intended to influence. Their received notions of research were and are still drenched in scientistic clichés.

There are also intellectual reasons for the absence of earlier research descriptions. Few anthropologists by the late 1970s had devoted much space to a detailed description of fieldwork practice. Resisting pressure to put a description in an appendix, I included a chapter on my fieldwork (Okely 1983). This deliberately excluded references to note taking. An autobiographical account was not as free as I had wished. Subsequently, social anthropology has shown new interest in examining the construction of ethnographic texts. Key monographs have been scrutinized for the literary devices used in the presentation of non-fictional material (Clifford and Marcus 1986, Geertz 1988).

It is standard practice for an anthropologist to live alongside a group of people for at least a year. I lived in a trailer-caravan on Gypsy encampments and went out to work with them: calling for scrap metal and joining a potato-picking gang. Research entailed periods amounting to about two years, including return visits, as well as participant observation among government officials and others in regular contact with Gypsies. Nothing approximating to an interview was used in the research for which I was responsible. Accordingly, the material which informs my writing is very different from that gleaned from one-to-one interviews with individuals divorced from daily practice and context. When I conducted research among government officials and non-Gypsies or 'Gorgios', as Gypsies call them, in the area, free-flowing conversations and dialogue occurred as I accompanied them about their business and at their leisure. All together, the number of people, whether Gypsy or Gorgio, whom I encountered and from whom I gained information amounted to several hundred, but my approach still places me behind what is considered to be the qualitative divide. The number of 'informants' is not an adequate guide to the distinction between very different research approaches.

Some of the themes and subsequent chapters in my early publication on Gypsies were explicitly affected by the demands and brief of the policy-oriented centre. The aim was to examine the Gypsies' position and preferences in the light of recent legislation which presumed long-term sedentarization and assimilation of this ethnic group. The needs of this travelling people and their conflict with the dominant society were unavoidably a key focus in the sort of questions addressed by the ethnographer at all stages of the research. Thus the political context and funding proposal influenced the way fieldwork was conducted, the themes selected in analysis and the projected readership (Adams *et al.* 1975). Even without the policy subtext, however, it still would have been impossible to write of a 'self-contained community' (Okely 1983). The non-Gypsy or Gorgio made an appearance every day on the Gypsy camps in body or spirit. Thus the Gorgio appeared in field notes and published text.

With minimal success, I had combed the anthropological monographs and other literature for guidance and for reassurance in the face of increasing scepticism among my employers about non-questionnaire research. Then a chance meeting with the Africanist anthropologist, Malcolm Mcleod, afforded me the best and only detailed methodological advice I was to find at the outset of fieldwork. From his experience, he suggested, 'write down everything you hear, smell and see; even the colour of the carpets. . . .

Ideally you should fill an exercise book for each day.' So I jettisoned my earlier, increasingly unsatisfactory attempts at writing notes under prescribed headings. I had been prematurely deciding what was relevant and in the process omitting other details, possibly for ever. My notes took the form of a chronological journal. The only marker was the date on each page. Events were written up as soon as possible. The record for some days did indeed, when time allowed, stretch to exercise-book length. Ideas, tentative interpretations and dominant themes were also written into the text, as the field experience developed. However crude or speculative the ideas at that early stage, it did not matter. As both fieldworker and future author, I was free to allow the ideas to germinate in their own time and through my own thinking, not by proxy. Subsequent participant observation and extended contemplation would sift out the wheat from the chaff.

Given the Gypsies' resistance to direct questioning, especially by outsiders, the material was acquired through daily, informal communication, which I recorded in detail. Thus the seemingly trivial descriptions of domestic objects, trailer interiors, and stray remarks volunteered in conversation, all meticulously noted, assumed massive significance when scrutinized and sifted, months and sometimes years later. Often the description, if properly open-ended, can seem innocent of deeper meaning when it is first recorded. Later, it is up to the anthropologist to read for the meaning.

At the end of fieldwork, I was faced with thousands of words: the experience written in chronological form. How to analyze? How to write up? First, the notes were typed up with wide margins, but single spaced, each page with the date on top. The final amount filled over eight box files of foolscap paper. Already, themes had emerged and I made a provisional list of topics, with subheadings. The broad headings were consistent with the holistic traditions of classical anthropology. The fieldworker aims for the total context, which includes kinship, the economy, politics and religion. Nothing can be taken for granted nor rigidly prearranged. So the classification is made after, not before, fieldwork. Major examples include:

(1) *Gypsy – Gorgio relations* with (i) council officials, (ii) customers, (iii) the police;
(2) *Work* (i) occupations, e.g. scrap metal, tarmac, antiques, (ii) ideology of self-employment, (iii) links with geographical mobility;
(3) *Gypsy identity* (i) Gorgio categories, (ii) self-ascription;
(4) *Kinship and the family* (i) choice of spouse, (ii) weddings, (iii) child care, (iv) shared residence, (v) alliances;
(5) *Education and socialization*;
(6) *Finance* (i) earnings, (ii) visible assets, (iii) budgeting;
(7) *Gorgio law* (i) courts, (ii) evictions, (iii) violence;
(8) *Gypsy law and community* (i) exchange and reciprocity, (ii) disputes, (iii) alliances;

(9) *Travelling* (i) ideology, (ii) case-studies, (iii) housing;
(10) *Sites* (i) official, (ii) official and temporary, (iii) illegal, (iv) rented, (v) Gypsy owned;
(11) *Religious beliefs*.

As I re-read through the notes, I added to the subheadings and categories. Paragraph by paragraph was bracketed. In many instances there was overlap. For example, a passage on kinship alliances could also be relevant to work partnerships. A case-study of a move from a camp could overlap with the topic of a police eviction. Details of a group discussion would involve individuals from several family or domestic units. The entire piece would be relevant to each unit. This overlap was no technical problem, since twelve photocopies were made of each page and the paragraphs were cut up according to each heading, stapled chronologically, and placed into the relevant named file. Each extract was marked with its original date in order to locate and cross-reference it in the chronology. Today, all this can be done by computer, without the labour-intensive cutting and pasting. Nevertheless, no computer can stand in for the ethnographer's discovery of emergent themes as fieldwork progresses, nor the final thinking and analysis. No computer can think through the fieldwork.

Without a fixed ideal number, I had always grabbed the opportunity to meet new individuals or families. It was a matter of balancing this with the benefits of increasing rapport with core clusters of families in the camps where I lived and in the locality as a whole. I was not under pressure as is a pollster interviewer to find select individuals and then hurry on after each interview. In working through my notes, I found that I had been able to obtain detailed, quantifiable information on seventy-five Gypsy families or domestic units, covering, for example, occupations, shelter, motor vehicles, visible assets, relative wealth or poverty, offspring, past and current marriages, kinship connections, literacy, housing experience and travelling patterns. All this was obtained without clip-board, and its accuracy was assured by the multiple contexts in which it emerged. This information was collated in anonymized family folders and quantified, along with the information on fifty families collected by my three colleagues. Numerical tables were presented in the first publication (Adams *et al.* 1975). The material was satisfactorily interpreted with the aid of knowledge which comes from long-term participant observation. The representative of the computer company which prepared the numerical print-out was intrigued by the fact that this relatively poor group showed, in the 1970s, such an 'atypically' high percentage of motor vehicle ownership. He advised us to make something of this extraordinary fact. Of course this qualitative aspect was unsurprising for a semi-nomadic Traveller-Gypsy group. But such detail was unlikely to have become apparent in a mass survey with quantitative priorities.

The numerical material may have given apparent credibility for policy-makers, but it was only a clue to systematic patterns. Since anthropology has delighted in the specific counter-example to ethnocentric universalisms, the presentation of its research does not rest on numerical majorities. It may require only one remark, one individual's example to unravel the elusive intelligibility of the group or context. People's beliefs, values and actions are not necessarily revealed by head counting. Instead, these crucial revelations are much more likely to emerge from chance incidents, extended comments, and both informal and ceremonial gatherings. All this is the stuff found in field notes or the fieldworker's selective memory. It is worked and thought through from event to text.

Having identified themes with many sub-classifications, and having stapled them sequentially in category folders, I would read and re-read the minutiae of this material. The detail and specificity were material to think with, a starting-point for an elaborated interpretation. The resultant chapters were in no way a collage of these bits and pieces. The folders provided confirmation of hunches, or unravelled false leads. The germs of ideas jotted down in the field offered organizing principles which had continued to simmer away since they first surfaced in crude and tentative form. The newly classified notes contained extended examples. There were first-hand statements whose words had often been forgotten long after the event, now preserved in aspic on paper. These voices were appropriately reproduced in the final text.

Preliminary outlines and drafts of chapters drew also on a wealth of other material. For dominant themes and proposed chapters I made card indexes with references and quotations from wider anthropological reading, from historical and folklore writing on Gypsies and from detailed notes on available contemporary records. Contrary to a still popular stereotyped view of participant observation and the discipline as a whole, anthropology does not ignore other sources. Archival, historical and current written records such as official reports, newspapers and the visual media are all grist to the writer's mill. Before ever encountering Gypsies and local agencies, I was steeped in non-Gypsy writing about this non-literate group who had few of their own records. Outsiders' typologies of 'real' and 'counterfeit' Gypsies offered an important perspective for understanding how the dominant majority rationalizes its repression of an ethnic minority. The racist and sometimes sentimental typology pointed to the constraints or opportunities available to the group in its relations with outsiders. Later, I looked at the use of stereotypes by Gypsies in economic relations with non-Gypsies (Okely 1979). I suggested that the ethnicity of the Gypsy could be exoticized (+), or concealed (0), or degraded (–), or neutralized (+ –). I was not concerned with how many times and how many Gypsies adopted the different strategies. My analysis was intended to demonstrate the underlying principles, rather than any quantitative information.

Given social anthropology's fundamental concern with a people's self-perceptions and view of the world, I was inevitably concerned with how the group defined its own members. Here, first-hand quotes and both ambiguous and unambiguous examples of individuals accepted as authentic members indicated the Gypsies' priorities and significant criteria. A key chapter in the 1975 publication (Okely 1975a) and a more theoretically exploratory chapter in the subsequent monograph (Okely 1983: 66–76) were devoted to questions of Gypsy identity and outsiders' interlocking or contrasting classifications. The theoretical perspective of Barth (1969), who posited the notion of self-ascription for ethnic groups, was an illuminating organizing principle.

Not only is the question of the construction of identity a continuing intellectual concern of anthropology, it is also used as a political weapon by non-Gypsies. The rejection of visible local Gypsies as 'counterfeit' in contrast to a mythically 'real Romany' legitimates a policy of harrassment and oppression. So the selection of this theme in writing up was grounded in the historical and political context of the minority. The larger society has itself generated the question about real or counterfeit Gypsies, and invented, beyond the academy, answers to which the academic has to respond. Received views on the 'other' become even more sensitive when the anthropologist is writing about people who inhabit both his or her own geographical territory, and that of the reader. Readers both within and outside the academy have ready-made stereotypes about the 'real' Gypsy. I was never allowed to escape the matter in everyday encounters outside the field. Once my research interests were identified, outsiders invariably offered pronouncements on the popular typology, before seeking my comments.

Written documentation provided the major source for chapters on the history of local policy (Okely 1975c) and an extended overview (Okely 1983: 105–24) of local and national planning. However, its reading and interpretation depended heavily on participant observation both among non-Gypsy officials and voluntary groups. Intensive fieldwork in only a limited locality helped to make sense of wider patterns, just as Leach had once found (1967, see above). I had, like Agar (1980), approached the material with an open mind. The local authority where I completed the bulk of fieldwork had been represented as one of the most liberal. Temporary sites had been established in the teeth of ferocious opposition. I was, until a subsequent incident outlined below, allowed completely free access to all the county council records and the photocopying machine. I filled two box files with copies of what I deemed to be essential documents: committee minutes and reports, correspondence, ministry directives, an informal census, residents' complaining letters and the odd petition. In the midst of the numerous and bulging filing cabinets was just one intervention from a Gypsy; a non-literate man on one of the camps where I had lived. He must have asked a literate acquaintance to write the letter, in its neat old-fashioned style. It requested the installation of a tap on his temporary council site, something which temporary army camps always regarded as a minimum requirement. The request remained unheeded throughout the three years of the site run by the authority which collected a weekly 'service charge' for the Elsans and rubbish skip. Officially, Gypsies had no political voice except for that one request in the council files. The interpretation and writing must therefore also confront the absences in the material.

Other absences may occur in the published text, not for want of material and analysis but because of powerful controls on information. The investigation of elites and those with power, which Nader has called 'studying up' (1974), is more precarious. As unexpected as the Gypsy's letter was the discovery of correspondence which revealed secret, illegal plans by senior county officials to set up an emergency committee and action group to chase specified Gypsies out of the county. The plan explicitly invited the police to collaborate with council officials. Their proposed task was to identify 'non-local' Gypsies (a contradiction in terms when considering travelling groups, who can rarely be fixed within a single county), and arrange for their deportation. The documentation belied the council's public protestations that the police acted as an independent and impartial body and that only legal methods were used to evict Gypsies from public or private land. The police chief's written response in fact rejected the emergency proposal. He argued that his force had already devised perfectly adequate screening methods. This correspondence could not be published. When my interest in these newly found documents was realized, my access to files was curtailed. The quality and amount of data to be scrutinized and interpreted was thereby diminished. Although I could not publish the evidence, it none the less informed my overall analysis (Okely 1975c), and especially that in my subsequent single-authored publication (Okely 1983: ch. 7).

In some instances, there were no comparable studies and traditions within anthropology to inform the analysis. This is another feature of such research, whose material can produce new theoretical paradigms and insights. The open-ended approach allows space for the previously unimagined. I had the opportunity to challenge classical concepts and typologies in both economics and kinship. For example, the classical typology of nomads in economic anthropology includes only hunter-gatherers and pastoralists. There was nothing on the specific nomadic formation found among Gypsies, and which, taking off from the articles of Cotten (1954, 1955), I helped to formulate (Okely 1975b and 1983). Gypsies, I suggested, are a unique type of nomad whose economy is directly dependent on a wider, usually sedentary economy. They can never claim to have or approximate to self-sufficiency, as has been said of other nomads. Hitherto, writers about Gypsies had tended to identify their economic activity by a description of the content of their occupations; e.g. horse dealer, tinsmith, knife-grinder, fortune-teller, scrap-dealer, hop-picker. No one except Cotten had attempted to look for a common form or structure. My notes gave me the huge variety of Gypsy occupations. They could be fitted into a common schema which I identified as the 'occasional supply of goods, services and labour where demand and supply are irregular in time and place' (1975b: 114 and expanded 1983: 50–1). My fieldwork confirmed again and again the Gypsies' conscious rejection of wage labour, which they spoke of with contempt. Thus, the popular argument that Gypsies were excluded from wage labour employment solely because of prejudice was untenable. It also explained the failure of well-meaning attempts to 'train' them for 'ordinary' employment.

Similarly, there seemed to be no available categories for the kind of marriage exchanges found in my field material (1983: 175–80). Some anthropologists inaccurately labelled them sister exchange. Segalen, Zonabend (and others) have, I discovered much later, offered the term 'chaining or renewal of alliance' (1987: 114–5) for marriage patterns in the French peasantry and which are to some extent similar to those found among British Gypsies. My analysis was here again intimately bound up with the manner in which the material was acquired. Contrary to standard advice in anthropology, dating back at least to Malinowski, the collection of genealogical information was not a way into the group and a means of establishing rapport. Bits and pieces were conveyed in the most unexpected circumstances. Any information on relationships, even names, was considered explosive. My genealogy file was composed of scribbles on cigarette packets, paper hankies and food packaging. I could not ask Gypsies about rules and choices of marriage partners, as I was not supposed to know about, let alone systematize any such matters. My completed diagrams (all anonymized), and perhaps banal to other anthropologists, represented a triumph over all adversity (1983: 175–80). Only other Gypsiologists would understand this.

The opening chapter to *The Traveller-Gypsies* (1983) includes some avowed speculation on the history of Gypsies. The circumstances of contemporary Gypsies derived from my intensive fieldwork, and placed alongside scattered historical data, is used to question the popular diffusionist theories which postulate that the Gypsies' presence in Europe from about the fifteenth century is to be explained only by migration from India. My alternative and still controversial suggestion is that, after the collapse of feudalism, groups of travelling people were formed from landless serfs and indigenous people, as well as from individuals from along the oriental trade routes. Fieldwork had shown how over time Gypsies used consanguines and affinal relations for internal solidarity. Simultaneously, they could incorporate isolated individuals through marriage. They banded together to exploit occupations which required flexibility in space and time and which were outside the fixed wage labour system. More detailed investigation is required by historians.

Meanwhile, the use of the present to analyze the past should be permitted to anthropologists as much as it is to others. Historians such as Thomas (1971) and Ladurie (1981) have made creative use of anthropological insights from living societies to explain rituals, witchcraft and animal symbolism in cultures which flourished centuries ago. Others may do it less explicitly. Mayall (1988) for example, uses the anthropologists' conceptual description of the Gypsy economy, formulated from accumulative participant observation and economic anthropology, of the kind outlined above (Okely 1975b, 1983), to make sense of his nineteenth-century data on Gypsies. Whereas an historian is expected to footnote the written source of every *fact*, he may project *concepts* backwards, without always fully acknowledging their source in recent anthropological analysis and fieldwork (Mayall 1988: 46–54). The use of qualitative anthropological research by historians should discredit a still popular belief in sociology that findings from participant observation are only applicable to the fieldworker's single locality in both time and space.[3]

Chapter priorities emerge through the experiencing, thinking and rethinking. For example, at the outset I had no preconceived notion, let alone hypothesis, that pollution beliefs and animal classification among the Gypsies would turn out to be as significant as I realized they were at the time of the later publication (Okely 1983: chs. 6 and 12). Nor indeed did my policy-oriented research employers, who saw these aspects of Gypsy life as mere 'superstitions', that could only be of interest to denigrated folkloreists. Beliefs were not seen then as relevant to the prioritized and pre-defined issues of sedentarization and settlement. The Gypsies' use of pollution beliefs to construct ethnic boundaries became clearer to me through the repeated evidence in field notes of distinct practices; for example, the rejection of sinks in caravans, the use of multiple bowls, and explicit statements or just casual asides by Gypsies about the apparently dirty habits of non-Gypsies.

While unburdened by hypotheses, the anthropologist is of necessity steeped in the broader theoretical and ethnographic concerns of his or her time. A theoretical sensitivity to the subject of pollution or animal symbolism had also been formed from reading Douglas (1966), Leach (1972), Lévi-Strauss (1966) and Tambiah (1973). Once the groundwork has been done, chance plays its part with the open mind.

It was while I was browsing again through *The Journal of the Gypsy Lore Society* that signals I had been alerted to in the field and in anthropological studies were consolidated. A stray example in an article by Thompson, a 'gentleman scholar', gave a clue to the system of the Gypsies' animal classification, evidence of which I could see scattered throughout the field notes. Until then I had not found a pattern in the Gypsies' random comments in identifying one animal as clean and another as polluting. Thompson (1922: 23, cited in Okely 1983: 92) recorded a direct quote, without further investigation, from a Gypsy who denigrated cats and dogs as polluted because they licked themselves, whereas horses were considered clean because they did not have the same habit. Eureka! I saw the demarcation between the inside and outside of the body reflected in the Gypsies' selective focus on animal behaviour, and as consistent with their own view of the body. Ingestation of the 'outer body' broke the rules which governed the Gypsies' washing and eating practices. The bits and pieces, recorded in field notes and in the folklore literature, on dogs, cats, ponies and hedgehogs suddenly fell into a classification grid. In contrast to the work of Lévi-Strauss (1966) and Tambiah (1973), this classification also depended on a relationship with the perceived or actual classification of another, that of non-Gypsies. The analysis, while drawing on structuralists' codes, yet grounded the beliefs in the historical and materialist context of the Gypsies' relations with outsiders. Pollution beliefs among Gypsies elsewhere were to be found in studies carried out by anthropologists working contemporaneously (Kaminski 1980, Miller 1975 and Sutherland 1975). None of these studies had, however, cracked the animal code. As is sometimes the case, the material may be there, but not necessarily interpreted. Jane Dick Zatta (1989) has followed up my analysis of British Gypsy animal classification in a comparison with Italian Gypsies and with intriguing innovation. Qualitative research of this kind has implications beyond territorial boundaries.

Interpretations are attained not only through a combination of anthropological knowledge and textual scrutiny, but also through the memory of field experience, unwritten yet inscribed in the fieldworker's being. The ethnographer, as former participant observer, judges the authenticity of his or her conclusions and interpretations in terms of that total experience. Others cannot so easily do it for her. Anthropologists have not begun to articulate the way in which ideas and interpretations are arrived at. The self and its autobiographical experience are used to relive and rework the material, although many anthropologists are coy about explicitly discussing that self and its history (Okely 1992). The interpretation is intuitive (cf. Leach 1967 above), yet the intuition is a newly acquired one which has been formed from months of living in the different culture.

Ideas may emerge from only the most intangible link with recorded notes. They arise in part as a response to other theories and ideas, long after fieldwork. For example, when the first book had been sent to press, and I was no longer employed by the policy-oriented research centre but instead registered for a doctorate and in the very different atmosphere of a university anthropology department, new ways of thinking about my fieldwork and the Gypsies took shape. Again, the political as well as academic context had implications for the ensuing texts. An all-women seminar of anthropologists took place weekly. Banished by the professor from departmental buildings because men were discouraged, if not excluded,[4] it pursued its creative and co-operative course away from the usual academic cockfights. Suddenly I could see how my political feminism could link up with academic anthropological work and the rethinking of my field material. I had not previously placed women in a separate category when classifying and chopping up field notes. Gender was not consciously highlighted, except to analyze the division of labour. Although I was without a relevant file to think through, I nevertheless found ideas tumbling out. Perhaps the very absence of a file permitted lateral thinking. All the exotic and erotic images of the Gypsy woman in literature, opera and art assailed me. The ideas crystallized after a chance viewing of an old Hollywood movie on Gypsy Rose Lee, the stripper, who had no ethnic connections with the minority group. Key images pointed to the contradictions between the projected fantasy among Gorgios of the Gypsy woman which contrasted with the behaviour I had learned was required of the Gypsies I had encountered.

As a woman researcher, I learned also through personal experience about non-Gypsy projections. To outsiders learning of my research, I was sometimes seen as a 'Gypsy woman', with all the fantastic stereotypes. Vivid and contrasting examples of actual Gypsy women from my fieldwork were recalled. I pursued the contrast in a paper for the women's seminar. Thoughts came at unexpected times; on a walk, in the night, not necessarily when seated with pen and paper at a desk. After the broad schema of ideas was set down, I could look back for some exact details, incidents and statements in the chronological field notes. That is, the ideas and theories, having fermented in the subconscious, emerged by free association from unspecified experience. Only then was empirical evidence instrumentally sought as confirmation or elucidation. This form of analysis was the opposite to how I had worked on some of the other chapters; I now responded entirely to intuition and elusive memory before grounding myself in the recorded notes. There was a moment of relief and freedom when, after a discussion with Shirley Ardener, I felt it was right to trust my new intuitions, while being unable in any empiricist way to point to exact 'proof' in field notes that the ideas were sound. Since I had not and could not have predicted the paradoxes in Gypsy women's position either before or during fieldwork, it was logical that this would not be traceable in the notes. It was a matter of piecing together passing clues and seemingly neutral descriptions, while allowing an imagination, grounded in experience, its proper freedom. My approach rested absolutely on the initial fieldwork which itself helped to form the theories, in a mysterious and uncontrolled way.

In my text (Okely 1975d) I presented stereotypes from literature and opera as icons or as an epiphany to demonstrate the full range of possibilities in stereotypes, not as some numerically calculated evidence. This was misunderstood by some empiricists who complained that my 'proof' rested on just one opera (*Carmen*) and two or three quotes from novels or poems. Since these were labelled as 'only fiction', or 'numerically insignificant', they were not counted (literally) as sufficient evidence. These readers were inappropriately demanding a media content-analysis which entails mechanically counting the number of times a stereotype appears, for example, in newspapers.

To conclude, the interpretation of anthropological material is, like fieldwork, a continuing and creative experience. The research has combined action and contemplation. Scrutiny of the notes offers both empirical certainty and intuitive reminders. Insights emerge also from the subconscious and from bodily memories, never penned on paper. There are serendipitous connections to be made, if the writer is open to them. Writing and analysis comprise a movement between the tangible and intangible, between the cerebral and sensual, between the visible and invisible. Interpretation moves from evidence to ideas and theory, then back again. There can be no set formulae, only broad guidelines, sensitive to specific cases. The researcher is freed from a division of labour which splits fieldwork from analysis. The author is not alienated from the experience of participant observation, but draws upon it both precisely and amorphously for the resolution of the completed text.[5]

Notes

1. After eight years in a department of sociology, where I taught the core courses in 'qualitative methodology', I was astonished to be told by a theoretical Marxist colleague, unfamiliar with my publications, that I was a symbolic interactionist. His apparent proof was that I had done intensive fieldwork, including participant observation.
2. This tradition is often marked by a gender division of labour, with a named male researcher-author and unnamed female assistants.
3. Sociology undergraduates are routinely asked to write essays on the 'limitations' of participant observation. Textbooks inform them that its results may be 'valid', but 'unreliable' anywhere beyond the locality. In this positivistic exercise, quantitative criteria for generalizations predominate.
4. He did not see the inconsistency in being a member of All Souls, one of the richest Oxford colleges, which then excluded women.
5. The construction of anthropological texts has recently been scrutinized. My insertion of some reflexivity (see also Okely 1992), multiple quotations as voices and cultural references as ideology anticipated this. Simultaneously, my intention was to demonstrate that aspects of the classic monograph could be transposed to a western context.

References

Adams, B., Okely, J., Morgan, D. and Smith, D. (1975) *Gypsies and Government Policy in England*, London: Heinemann Educational.

Agar, M. (1980) *The Professional Stranger*, London: Academic Press.

Barth, F. (1969) 'Introduction', in F. Barth (ed.) *Ethnic Groups and Boundaries*, London: Allen & Unwin.

Breton, A. (1937) *L'Amour fou*, Paris: Gallimard.

Clifford, I. and Marcus, G. (1986) (eds) *Writing Culture*, Berkeley, Calif.: University of California Press.

Cotten, R.M. (1954) 'An anthropologist looks at gypsiology', *Journal of the Gypsy Lore Society* 3rd series, 33(3–4): 107–20.

——— (1955) 'An anthropologist looks at gypsiology", *Journal of the Gypsy Lore Society* 3rd series, 34 (1–2): 20–37.

Dick Zatta, J. (1989) '"The has tre mule!", tabous alimentaires et frontières ethniques', in P. Williams (ed.) *Tsiganes: Identité, évolution*, Paris: Syros Alternatives.

Douglas, M. (1966) *Purity and Danger*, London: Routledge & Kegan Paul.

Durkheim, E. (1897) *Le Suicide*, Paris: Alcan.

Geertz, C. (1988) *Works and Lives*, Oxford: Polity Press.
Hammersley, M. and Atkinson, P. (1983) *Ethnography: Principles in Practice*, London: Tavistock.
Hockey, J. (1990) *Experiencing Death: an Anthropological Account*, Edinburgh: Edinburgh University Press.
Kaminski, I.M. (1980) *The State of Ambiguity: Studies of Gypsy Refugees*, Gothenberg: Anthropological Research.
Ladurie, E. le Roy (1981) *Carnival in Romans*, Harmondsworth: Penguin Books.
Leach, E.R. (1967) 'An anthropologist's reflections on a social survey' in D. Jongmans and P. Gutkind (eds) *Anthropologists in the Field*, Assen: Van Gorcum.
________ (1972) 'Anthropological aspects of language: animal categories and verbal abuse', in P. Maranda (ed.), *Mythology*, Harmondsworth: Penguin Books.
Lévi-Strauss, C. (1966) *The Savage Mind*, trans, anon., London: Weidenfeld & Nicolson.
Malinowski, B. (1926) *Crime and Custom in Primitive Society*, London: Routlege & Kegan Paul.
Mayall, D. (1988) *The Gypsy-Travellers in Nineteenth Century Society*, Cambridge: Cambridge University Press.
Miller, C. (1975) 'American Rom and the ideology of defilement', in F. Rehfisch (ed.) *Gypsies, Tinkers and other Travellers*, London: Academic Press.
Nader, L. (1974) 'Up the anthropologists: perspectives gained from studying up', in D. Hymes (ed.) *Reinventing Anthropology*, New York: Vintage.
Oakley, A. (1981) 'Interviewing women: a contradiction in terms', in H. Roberts (ed.) *Doing Feminist Research*, London: Routledge.
Okely, J. (1975a) 'Gypsy identity', in B. Adams, J. Okely, D. Morgan and D. Smith *Gypsies and Government Policy in England*, London: Heinemann Educational.
________ (1975b) 'Work and travel', in B. Adams, J. Okely, D. Morgan and D. Smith *Gypsies and Government Policy in England*, London: Heinemann Educational.
________ (1975c) 'Conflict and accommodation', in B. Adams, J. Okely, D. Morgan and D. Smith *Gypsies and Government Policy in England*, London: Heinemann.
________ (1975d) 'Gypsy women: models in conflict', in S. Ardener (ed.) *Perceiving Women*, London: Malaby/Dent (part reproduced in Okely 1983).
________ (1979) 'Trading stereotypes', in S. Wallman (ed.) *Ethnicity at Work*, London: Macmillan.
________ (1983) *The Traveller-Gypsies*, Cambridge: Cambridge University Press.
________ (1987) 'Fieldwork up the M1: policy and political aspects', in A. Jackson (ed.) *Anthropology at Home*, London: Routledge.
________ (1991) 'The ethnographic method applied to rural transport, the elderly and planing', report to the Economic and Social Research Council.
________ (1992) 'Anthropology and autobiography: participant experience and embodied knowledge', in J. Okely and H. Callaway (eds) *Anthropology and Autobiography*, London: Routledge.
Proust, M. (1954) *A la recherche du temps perdu*, Paris: Gallimard.
Sarkar, N.K. and Tambiah, S.J. (1957) *The Disintegrating Village*, Colombo.
Segalen, M. and Zonabend, F. (1987) 'Social anthropology and the ethnology of France: the field of kinship and the family', in A. Jackson (ed.) *Anthropology at Home*, London: Routledge.
Silverman, D. (1985) *Qualitative Methodology and Sociology*, Aldershot: Gower.
Tambiah, S.J. (1973) 'Classification of animals in Thailand', in M. Douglas (ed.) *Rules and Meanings*, Harmondsworth: Penguin Books.
Thomas, K. (1971) *Religion and the Decline of Magic*, London: Weidenfeld & Nicolson.
Thompson, T.W. (1922) 'The uncleanness of women among English Gypsies', *Journal of Gypsy Lore Society* 3rd series, 2(3): 113–39.
Young, M. (1991) *An Inside Job*, Oxford: Oxford University Press.

59

On the Analysis of Observational Data: A Discussion of the Worth and Uses of Inductive Techniques and Respondent Validation[1]

M. Bloor

Introduction

It seems something of a commonplace among research sociologists that texts on methodology are only of very limited utility in study design, certainly they contain no templates which can be applied unproblematically for the resolution of particular research problems. There is an obvious reason for this namely that the appropriate methodology for any given study can only be chosen with reference to situational factors, factors specific to the study in question. Consequently, the methodological writings which most sociological researchers seem to find most useful tend to be those which are grounded in particular research projects rather than the general surveys of methodological techniques. For example, Whyte's description of his methodology in 'Street-corner Society' (Whyte, 1955, pp. 279–358) is justly more famous than more general surveys of participant observation.

In what follows therefore, I shall be discussing the problems and possibilities of inductive analysis and respondent validation for studies using observational data but I will attempt to ground the discussion in my own research in ENT out-patient clinics.

Source: *Sociology*, 12(3) (1978): 545–552.

Of necessity then, I must first say something about the nature of the research problem that concerned me and the data I collected. I was concerned to establish whether or not geographical differences in the incidence of adeno-tonsillectomy among children could be attributed to any marked degree to differences between ENT specialists in different geographical areas in their routine assessment practices for adeno-tonsillectomy cases (cf. Bloor, 1976a, Bloor, 1976b). In effect, I was following analogically Jack Douglas' argument that geographical differences in the incidence of officially designated suicides reflected similar geographical differences in the meanings of suicide employed by coroners and other official agents (cf. Douglas, 1967, Atkinson, 1971). The research task therefore involved the generation of sociological descriptions of a series of different specialists' assessment practices for comparative analysis. I tried to inductively reconstruct each specialist's own standard 'decision rules' which he normally used to decide on a disposal. Having arrived at the decision rules, I looked for each specialist's own standard search procedures whereby he generated the information necessary to put into use his decision rules. Then I was in a position to compare specialists' assessment practices by comparison of their individual decision rules and search procedures.

The data were collected as follows: I sat in on a series of clinics, the clinics being conducted by ten different specialists. I noted down the history taken for each relevant case, the examination findings and the results of any ancillary investigations (e.g. X-rays). I also noted down any comments the specialists made to me on cases. I asked questions on their reasons for taking particular disposal decisions whenever I got the chance. I also had some notes on more general conversations with specialists, for example during tea-breaks. Some use was also made of casenote materials. As in most observational studies then, the materials were not wholly or purely observational.[2]

I shall turn now to a description of the inductive method I adopted, its advantages and shortcomings, and its wider applications.

The Inductive Method

Above I stated that the initial research task was the accomplishment of descriptions of each specialist's assessment practices. These ethnographies cannot be merely compendia of specialists' actions and utterances in the clinic setting. I say 'cannot' both because such compendia would be very deficient since all observation is, by its very nature, selective, and because such compendia – devoid of any attribution of meaning to those actions and utterances – would be a mere babble, a senseless jumble. What is required is a description of specialists' assessment practices informed by a knowledge of specialists' definitions of the situation and of their practical purposes at hand.

Goodenough, writing of the so-called 'new ethnography' has neatly summed up the methodological requirements for producing such descriptions:

> 'Ethnographic description requires methods of processing observed phenomena such that we can inductively construct a theory of how our informants have organised the same phenomena'.
>
> (quoted by Psathas, 1968)

The analytical system that I adopted with this in mind was as follows:

1. For each specialist separately, cases were provisionally classified according to the disposal-category into which they fell.
2. The data on all a specialist's cases in a particular disposal-category were scrutinized in order to attempt to produce a provisional list of those case-features common to the cases in that category.
3. The 'deviant cases' (i.e. those cases where features common to many of the cases in the disposal-category were lacking) were scrutinized in order to ascertain whether (a) the provisional list of case-features common to a particular category could be so modified as to allow the inclusion of the deviant cases; or, (b) the classificatory system could be so modified as to allow the inclusion of the deviant cases within a modified category.
4. Having thus produced a list of case-features common to all cases in a particular category, cases in alternative categories were scrutinized to discover which case-features were shared with cases outside the first category considered. Such shared case-features were thus judged necessary rather than sufficient for the achievement of a particular disposal.
5. From the necessary and sufficient case-features associated with a particular category of cases sharing a common disposal, the specialist's relevant decision rules were derived.
6. For each decision rule the cases to which it applied were rescrutinized to derive the search procedures associated with that decision rule.
7. The above steps were then re-enacted for each of the other of the specialist's disposal-categories until a set of decision rules and associated search procedures for that specialist was derived which accounted for the disposal of all of that specialist's cases for which data had been gathered. The analysis was then repeated for each of the remaining specialists in turn.

Before I turn to assess the advantages and disadvantages of this procedure there are a number of points about the procedure that could usefully be amplified.

In the first place, the question of whether or not inductive thought is possible is, I am aware, philosophically contentious. I would like to sidestep this issue. Let me take up the position that Ayer has characterized as 'naive realism': it seemed to me that I was thinking inductively (what's good enough for Professor Ayer is good enough for me).

Second, I would not wish to claim that I derived my ethnographies of specialist practice *wholly* inductively. Specialists' comments, their answers to my questions, general impressions formed during data-collection, and so on, all meant that I had my views on what were the case features crucial to particular specialists' disposal decisions prior to the analysis. I did not attempt to suppress these views in the interests of purity of inductive thought, rather I attempted to incorporate them into the analysis. In effect, this introduced a degree of what Denzin (1970) calls 'triangulation' into the analysis: the same finding that was indicated from a specialist's comments was confirmed by comparative analysis of the transcripts. However, these analytical shortcuts were not always available: for certain specialists, and for certain aspects of other specialists' practices, I had to engage in induction.

Finally, it will be evident that the procedure I adopted had similarities with that technique that Robinson (1951) called 'analytic induction' and which was used in a number of criminological studies in the 1940s and 1950s, especially Lindesmith's (1947) study of opiate addiction and Cressey's (1953) study of embezzlers. The main difficulty with both studies was that the researchers were unable to distinguish between the necessary and sufficient causes of addiction or embezzlement because they lacked control groups in which necessary but not sufficient causes could be located. I was free of this difficulty because the cases in other disposal categories could stand as a control group for those cases in the disposal category that I was analysing.

However, in another respect I was at a disadvantage relative to Lindesmith and to Cressey insofar as, unlike them, I could not proceed with analysis contemporaneously with data-collection. Lindesmith and Cressey could discontinue data-collection to develop hypotheses, which could be tested out by further data-collection, which could be discontinued to modify the hypotheses, and so on. Logistically, it was impossible for me to break up my observations in specialists' clinics in this manner. Consequently, while Lindesmith's and Cressey's analytical technique indicated the appropriate cut-off point for data-collection automatically, I had to estimate in advance how many cases I would need to collect in order to have sufficient for analytical purposes. Inevitably, I tended to collect too many and the analyses were more cumbersome than they need have been as a result.

But here I am anticipating my discussion of the difficulties that went with my adopted analytic technique. Apart from this lack of an identifiable cut-off point, three major difficulties can be isolated.

Firstly, Schutz (1967, pp. 113–115) has pointed out that, although inductive analysis claims to be able to discover the features that an object has in common with others, it actually presupposes the existence of similar objects: the very act of comparison presupposes that the objects are comparable. Thus, in my study I was concerned with the description of a specialist's set of routine assessment practices, but the consistency of specialists' assessment practices – the existence of sets of routines – was implicitly assumed by the process of the analysis. Deviant cases were scrutinized for incorporation into a modified description of specialists' routines: peculiar case-features were isolated which accounted for their atypical disposal. But every case is to some extent unique: one can always find peculiar case-features which arguably crucially distinguish that case from other similar cases. In other words, the analytic method offers no means of distinguishing 'deviant' cases (which demand modifications to the analysis) from those cases where the specialist may simply be behaving inconsistently. Thus, the occurrence of an occasional inconsistent case can result in the analyst making spurious modifications to his description of a specialist's routines.

Another difficulty is the hoary old problem of drawing the correct inference from all the competing inferences that could be drawn from one's data. Always a difficulty, this is one which is most acutely present in observational studies because of the richness of the data. Inductive analysis can reduce, but not eliminate, the difficulty: some, but usually not all, competing inferences can be eliminated by comparison with other cases.

A third problem is that of the defeasibility of rules or routines (cf. Hart, 1951): routines are contingent and so subject to elaboration or modification where particular novel circumstances seem to demand it. Strictly speaking, one cannot derive routines which are sufficient for the disposal of all cases past, present, and future; one can only derive routines which are sufficient for the disposal of those cases on which the analysis is based.

Yet despite these difficulties, this analytic procedure does carry with it a number of fairly obvious advantages. Thus, it does introduce some sort of *system* into the analysis of qualitative data. Little appears to have been written on the actual logistics of analysing qualitative data, but it seems obvious that a systematic approach is necessary in order to break down the analytical task into manageable units of activity.

Again, the procedure forces the analyst to progressively elaborate his analysis since modifications or extensions to his initial analysis must be made so as to incorporate the 'deviant cases'. The analyst is discouraged from the selective use of data in the sense that he is discouraged from basing his analysis on certain cases while neglecting to fully consider other cases.

Nevertheless, because of the serious difficulties mentioned previously I could not be entirely happy with the analyses I produced. As a consequence I embarked on the validation exercise described below.

Respondent Validation

There has been some discussion of validation exercises among anthropologists, but the subject has not received much attention in sociological circles. There is a tendency among methodologists to equate validation with triangulation, to regard an analysis as validated when it is reproduced by the use of different methods. Triangulation seemed to offer no solution to my difficulties: in a sense, I had already tried it and pursued it as far as I was able. And triangulation is itself beset with problems of validity. For example, given the assumption that there is always one best method for tackling a particular research problem, what is the point of trying to reproduce one's findings by use of additional, inferior data generated by additional, inappropriate methods?

More important, perhaps, the very equation of validation with triangulation seems somewhat questionable. Garfinkel (1967) has criticized the view that the sociological enterprize should be seeking to produce sociological descriptions which are 'true' in some absolute and essentialist sense, since this leads to a position which sees collectivity members as 'cultural dopes' who are labouring under delusions about the nature of their social world, delusions which only the sociologist can dispel. The corollary of Garfinkel's argument is that the truth of our analyses, their validity, is constituted by establishing some sort of correspondence between an analyst's and collectivity members' views of their social world.

Let me be clearer about the sort of comparison that is required for establishing this correspondence. There is no question of comparing the sociologist's analysis with the member's analysis. If members could do sociological work there would be no point in sociologists doing it: collectivity members can only do sociological work if they are marginal to the collectivity. Kafka summed up the point neatly, albeit with characteristic gloom:

> 'Only he who is a party can really judge, but as a party he cannot judge. Hence it follows that there is no possibility of judgement in the world, only a glimmer of it'.
>
> (Kafka, 1954, p. 85.)

If collectivity members cannot themselves act as judges one can only establish a correspondence between the sociologist's and the member's view of the member's social world by exploring the extent to which members recognize, give assent to, the judgements of the sociologist. The task of getting collectivity members to assent to, or disagree with, one's analysis can be tackled in at least three different ways.

One such procedure is that practiced and recommended by the anthropologist Frake (1964), that of the prediction of members' classifications in the situations of their use in field settings. Frake was able to check his glossary of the diagnostic classifications of the Subudun people in this manner but there would be practical difficulties in attempting to predict specialists' disposal decisions in out-patient clinics. Roth and Eddy (1967) attempted to predict the selection of patients for a rehabilitation programme but abandoned the attempt because they had no way of ensuring that their predictions were based on the same information on patients as that used by the doctor

(Roth, personal communication). One might attempt to predict disposals in the clinic settings where one could ensure an identity of information with the doctor, but the doctor, being more practised in the task, would inevitably beat the researcher to the post.

An alternative procedure (also recommended by Frake) would be that of drawing up hypothetical cases and predicting one's respondents' responses to them. This procedure has been previously used in medical settings, albeit for the purposes of hypothesis-testing rather than prediction e.g. Haase (1964), Howie (1976). However, to systematically test my analyses of specialists' assessment practices would have entailed each specialist having to read and respond to a very large number of hypothetical cases; the procedure rapidly becomes a fatiguing one for respondents and so I eventually only made subsidiary use of this approach.

A third course, and the one I finally adopted, was that of simply writing out my analyses for my subjects and recording their reactions to them. For each separate specialist I wrote out a report describing his assessment practices, with each report I sent a covering letter requesting the specialist to read through the report to see how far it corresponded with his own impressions of his clinic practice. The discussion of the report then formed one of the themes of a tape-recorded interview. It is a commonplace for researchers to courteously report on their findings to their respondents, however the possibility of treating respondents' reactions to the reports as data seems not to have been considered.

What I was hoping for was a sort of self-recognition effect. This is well illustrated by the following specialist's reaction:

> 'I think it was a fair assessment, a fair summary. It put into words many of the things we do which are more second nature I think. You know, I think you've done quite well summarising (laughs) me: some of these things, "Well", I think, "you know, that *is* right", when I see it written down. And when you're doing it, well I suppose you have these things in the back of your mind but . . .'.

Not all the specialists reacted in such a gratifying fashion of course. Some of the specialists generally endorsed the report but queried a few specific points. Some of these points at issue were cleared up when it became evident that the specialist had misunderstood, or I had poorly expressed, my meaning. In other instances points of difference were fully discussed and I returned to my data to see if I had put a false construction on them and where the specialist's assertions fitted the data I adopted them for my own. For example, one specialist took issue with my finding that he never reviewed children presenting with histories of sore throats but always either listed them for surgery or discharged them at first clinic visit. We discussed the point and I asked him to describe the kind of sore throats cases that he *would* review. It turned out that such children exhibited a relatively narrow spectrum of symptomatology and so turned up at his clinics only comparatively infrequently. No such cases had presented during the period of my observations in his clinic, and I therefore modified my analysis accordingly.

There was one specialist, indeed, who substantially disagreed with my report. He claimed that I had over-emphasized the importance he attached to historical evidence compared to the examination findings. This seemed to me a fair point: unlike many of the specialists he had not verbalized his examination findings in the clinics for my benefit (I had obtained them later from the casenotes) amd so data-collection (and subsequent analysis) had possibly been biased towards the historical evidence. Again, we discussed the matter fully and I modified my analysis as a result.

On the face of things then, the validation exercise was a success – some respondents endorsed my descriptions of their practices, and where they did not the nature of the exercise was such as to enable me to correct the analysis so that this assent was no longer withheld. Nevertheless, I was not wholly satisfied with the exercise and my reservations grew as the exercise progressed.

There seemed to be two inter-related difficulties. In the first place, there is the difficulty of the problematic validity of one's validation exercise. How does one frame and present one's analysis in such a form and in such a setting that one can be confident that one's respondents will fully understand it, react to it in critical spirit and honestly and coherently tell one how it strikes them? A number of the criticisms from an ethnomethodological perspective of open-ended interviews (see especially Moore's 1972 analysis of interviewing) apply with almost equal force to the method used here.

In addition, I became aware of having made unwarranted ethno-centric assumptions when I planned the exercise: I had expected the specialists to respond to the reports in a manner similar to that of an academic colleague when one asks him to criticize a draft paper one has written. I became aware of having made this assumption when it was violated – I suspected that some of the specialists had not read the report in the expected critical spirit. They had read the report, I felt, in the way that we today might read a nineteenth century religious tract – with a modicum of detached, superficial interest, with a feeling that it displayed a certain peculiar charm perhaps, but without being so moved by its content as to feel the necessity to define one's own beliefs and practices in accordance with it or in contrast to it. They were unversed in the conventions of academic sociological criticism and they were perhaps only marginally interested in the content of the reports.[4]

As a consequence of these difficulties I began to conduct rough and ready checks on the validity of the validation exercise. There occurred instances in my analyses of almost every specialist's assessment practices where I found myself unable to choose between competing explanations for the disposal of certain cases. I therefore took the opportunity to fill in the resultant gaps in my analysis by the use of hypothetical cases or interview material. The rough and ready checks consisted of inserting here a few questions about some of the specialist's routine assessment practices, the answers to which were already incorporated in the written report. Where I found a specialist describing his practices in terms which contradicted the report that he had previously endorsed, then I asked more and more questions relating to practices described in the report. With two of the specialists I followed this procedure quite extensively: in effect, I ended up triangulating my validation exercise by the interview method.

Conclusion

Discussions of methodological techniques are best grounded in the context of particular research problems. This paper has shown that the inductive analytic method and the validation exercise that I employed in a study of medical assessment practices had a number of definite advantages that recommend their employment in other studies.

However, certain defects in both techniques meant that neither technique was wholly satisfactory and each had to be supplemented by the use of other methods: the inductive procedure was only of value in conjunction with the validation exercise, and the validation exercise needed triangulation by interviewing and the use of hypothetical cases.

Moreover, both techniques were situated devices developed in response to a particular research problem and particular situational constraints. All methodologies, to be successful, must exhibit this situational adaptability. Consequently, the two techniques may not be very broadly re-applicable even in a modified form. Some fundamental requirements for practicing the two techniques can be briefly outlined here.

The main requirement for inductive analysis is that sufficient data be available for the analyst to be able to examine all the potentially relevant aspects of the research problem. Methods of data-collection should thus ideally be open-ended, (like Lindesmith's and Cressey's), failing this the researcher must aim for blanket coverage. This also implies that the research problem (the focus of the analysis) should be precisely specified prior to data-collection. Despite the fact that there may be good grounds for not defining the research problem too closely too soon, unless one defines the focus of the analysis fairly precisely following piloting, then one may find oneself omitting some of the data crucial for the analytical task.

The main requirement for the validation exercise seems to be an adequate level of involvement among respondents in one's research or a degree of 'commitment' (cf. Becker, 1960) to the research topic. Inadequate involvement or commitment may result in uncritical evaluation of one's findings by respondents. Note however that a very high level of commitment may be equally dysfunctional insofar as reports of one's findings may clash with respondents' 'accounts' of their social world, accounts designed to serve particular organizational, political, or ideological purposes highly salient to the respondents: in such cases respondents may be too much a party not only to judge, but also to assent to the judgements of others.

Notes

1. The research where the methods related here were developed was funded by the Medical Research Council. This paper was first presented at a seminar in Aberdeen and has greatly benefited from the comments and criticisms of my colleagues at the Institute of Medical Sociology. I especially wish to thank Gordon Horobin and Philip Strong for their comments on an earlier draft.
2. I found it both impossible and undesirable to adopt a 'fly on-the-wall' posture: 'impossible' since the clinics are to a degree social events and all those present are constrained to engage in some purely social intercourse; and, 'undesirable' both because to ignore those constraints may cause other parties to withdraw their co-operation and because I wished to be in a position to be able to ask specialists clarificatory questions about cases.
3. The glimmer is presumably seen by he who takes up Cicourel's marginal position (Cicourel, 1964, pp. 40–49.)
4. This is perhaps a rather surprising assertion in view of the fact that the reports were detailed descriptions of their own assessment practices. However, it should be borne in mind that these specialists were surgeons, their work demanded manual dexterity rather than diagnostic flair: clinical assessment, especially for a routine surgical procedure like adeno-tonsillectomy, is not a central interest.

References

Atkinson, J. M. (1971). 'Societal Reactions to Suicide: the role of coroners' definitions', 165–191 in S. Cohen (ed.) *'Images of Deviance'*, Penguin, Harmondsworth.

Becker, H. S. (1960). 'Notes on the Concept of Commitment', *A.J.S.*, Vol. 66, pp. 32–40.

Bloor, M. J. (1976a). *'Decision-making in ENT Out-patient Clinics and Variation in the Hospitalisation of Children for Adeno-tonsillectomy'*, Ph.D. thesis, University of Aberdeen.

Bloor, M. J. (1976b). 'Bishop Berkeley and the Adeno-tonsillectomy Enigma, an exploration of the social construction of medical disposals', *Sociology*, Vol. 10, pp. 43–61.

Cicourel, A. V. (1964). *'Method and Measurement in Sociology'*, Free Press, New York.

Cressey, D. R. (1953). *'Other Peoples Money: a study of the social psychology of embezzlement'*, Free Press, Glencoe.

Denzin, N. K. (1970). *'The Research Act in Sociology'*, Butterworths, London.

Douglas, J. D. (1967). *'The Social Meanings of Suicide'*, Princeton University Press, Princeton.

Frake, C. O. (1964). 'Notes on Queries in Ethnography', *American Anthropologist*, Vol. 66, 132–145.

Garfinkel, H. (1967). *'Studies in Ethnomethodology'*, Prentice-Hall, Englewood Cliffs.

Haase, W. (1964). 'The Role of Socio-Economic Class in Examiner Bias', 241–247 in F. Riessman *et al.* (eds.) *'The Mental Health of the Poor'*, Free Press, New York.

Hart, H. L. A. (1951). 'The Ascription of Responsibility and Rights', in A. Flew (ed.) *'Logic and Language'*, Blackwell, Oxford.
Howie, J. G. R. (1976). 'Clinical judgement and antibiotic use in general practice', *B.M.J.*, part 2, 1061–1064.
Kafka, F. (1954). 'The Eight Octava Notebooks' in F. Kafka, *'Wedding Preparations in the Country and Other Posthumous Prose Writings*, Seeker and Warburg, London.
Lindesmith, A. L. (1947). *'Opiate Addiction'*, Principia Press, Bloomington.
Moore, M. (1972). *'The Negotiation of An Appropriate Occupational Role: a study of Progressive Church of England Clergymen'*. Ph.D. thesis, University of Kent.
Psathas, G. (1968). 'Ethnomethods and Phenomenology', *Social Research*, Vol. 35, pp. 500–520.
Robinson, W. S. (1951). 'The Logical Structure of Analytic Induction', *A.S.R.*, Vol. 16, pp. 812–818.
Roth, J. A. and Eddy, E. M. (1967). *'Rehabilitation for the Unwanted'*, Atherton Press, New York.
Schutz, A. (1967). 'Some Leading Concepts of Phenomenology', in M. Natanson (ed.), *'Alfred Schutz Collected Papers Volume 1'*, Martinus Nijhoff, The Hague.
Whyte, W. F. (1955). *'Street Corner Society'*, Second Edition, University of Chicago Press, Chicago.

60

The Presentation of Everyday Life: Some Textual Strategies for "Adequate Ethnography"

Kenneth Stoddart

Save for the organizational dimension (Lofland, 1974), sociologists have yet to give much serious analytic attention to the writing of ethnographic texts. Even as a practical matter the portrayal phase of research has been largely ignored: Indeed, with few exceptions (e.g., Lofland, 1976; Spradley, 1979, 1980), instructional materials leave the student with the impression that once the data have been gathered and analyzed, the distinctive aspects of the ethnographic task are complete – at this point, skills acquired in other quarters of the campus should properly come into play. As Marcus and Cushman (1982: 25) have noted, it is probably the case that this neglect is sustained by "conceiving of ethnography primarily as an activity that occurs in the field or by treating it as a method . . . of research."

Those anthropologists who have pioneered the study of ethnography as *text* (e.g., Clifford, 1980, 1981; Marcus, 1980; Marcus and Cushman, 1982) have rescued writing or portrayal as a topic by viewing ethnography solely as a *product*, that is, "simply an account resulting from having done fieldwork" (Marcus and Cushman, 1982: 27). The present work shares this view. As well, it shares their concern with writing as a solution to a very fundamental problem: establishing an ethnography's *authority* as a bona fide explication of a social world. Let me elaborate.

Source: *Journal of Contemporary Ethnography* (formerly *Urban Life*), 15(1) (1986): 103–121.

After coming to terms with the difficulties of data assembly and analysis, the ethnographer is faced with yet another problem: Given that his or her explication of everyday life in the domain of investigation is to stand as more than a personal document, how are findings to be presented so that they will be appreciated by colleagues with a confidence and certainty similar to his or her own? How will the ethnographer structure a textual account so that it achieves its effect as knowledge of "others," as an objective account about subjects among whom fieldwork was conducted, as *authoritative*? The answer, of course, is to be found by addressing ethnographies in terms of their textual strategies.

Although anthropologists have paid explicit attention to these strategies – that is, to the devices and conventions that "constantly reinforce unselfconsciously the reader's confidence in the author's knowledge as sufficient credibility for what the text states" (Marcus and Cushman, 1982: 38) – a similar trend has yet to develop in sociology. I offer the present work as a modest beginning.

My concerns here are twofold: (1) to specify some of the features of approved instances of the ethnographic genre and (2) to provide a preliminary inventory of textual strategies used by ethnographers to make their productions recognizable as approved ones. With regard to the latter, it will be seen that standardizing descriptions of research activities figure prominently in establishing adequacy. With regard to the former, readers will immediately note that only *classical* examples of the genre are addressed: More experimental ethnographic studies, such as those that incorporate and display the principle of reflexivity (e.g., Johnson, 1975; Stoddart, 1974a; Wieder, 1974), await later attention.

Some Received Criteria of Ethnographic Adequacy

Ethnographers know the program of their craft as a set of advisings relevant to the achievement of an account recognizably special in ways approved by their fellows. Here, an account so recognized shall be called an *adequate ethnography*. In the most general of terms, those ethnographies are received as adequate[1]

(1) that display the domain of investigation as it exists independently of the dimension added to it by the ethnographer's *presence* as an investigator;
(2) that display the features of a domain as received by its participants (i.e., that do not have a version of *ethnocentricism* as a property);
(3) that display the features of a domain as they exist independently of the techniques employed to assemble them (i.e., that do not have *methodogenesis* as a property); and

(4) if their accomplishment of 1 through 3 is founded in data gathered from participants knowledgeable in the ways of their domain (i.e., if 1 through 3 are not based on the productions of *unentitled informants*).

From the practitioner's point of view, an instance of a recognizably adequate ethnography may be achieved by treating these advisings as rules. These rules – descriptive of what might be called the good ethnographer's work – can be codified as follows:

(1) The good ethnographer regards his or her own presence in a domain as potentially tainting of its natural state.
(2) The good ethnographer seeks not to enforce an organizing schema upon encountered data but to surface from that data a schema local to the domain of investigation.
(3) The good ethnographer is attentive to the possibility that techniques of gathering data may create the data gathered.
(4) For the good ethnographer, any domain has as a feature the differential distribution of competence among its members.

That the ethnographer treats as rules the advisings presented above stands as an answer to the question, "How is adequate ethnography possible?" In its terms, insofar as ethnographers follow the maxims of the program they will produce an instance of adequate ethnography. Yet such an answer is suspect, for it appears to assume and neglect more than it reveals, as the following considerations suggest:

(1) How, short of not being there, can the ethnographer actually capture a domain as it exists independently of his or her presence?
(2) How, short of being a member and sharing in the totality of group members' experiences, including biographical ones, can the ethnographer actually know the features of their domain as they do?
(3) How, short of using no techniques at all, can the ethnographer actually know a domain's features as they exist independently of his or her tools of assembly?
(4) How, save via the investigation of the competence of all its members, can the ethnographer actually establish who in a domain is an entitled provisioner of data?

And the rule-following account of adequate ethnography neglects much of the work ethnographers must engage in in order to bring off their productions as adequate ones. I am speaking here of the deciding, recognizing, judging, and portraying work that is performed by ethnographers and the readers of their accounts. For example, the ethnographer must (1) decide that his or her research was conducted in accord with the above maxims and

(2) portray that research in such a way that it retains its character as research conducted in accordance with those maxims. The latter task – portraying the research as adequate – is of central interest here.

A leading feature of all research activities, including ethnographic ones, is their availability for description in an infinite number of ways, each one technically correct. Yet even the most limited examination of actual descriptions of research activities suggests that equal and serious attention is not extended to all candidates, despite their technical correctness as descriptions. Why is that the case? That question is partially answered by directing attention to what might be called – following Turner's (1970) exegesis of Austin's (1965) work – *the performative character of descriptions*. The view is taken that descriptions of research activities must be seen not merely as describing what the ethnographer did in the field but instead as doing the work of making what she or he did available as activities-conducted-in-accordance-with-the-advisings-of-good-ethnographic-work. Adequate ethnography, then, becomes a course of telling, of portrayal. The remainder of this article reviews and analyzes such tellings as ways of doing adequate ethnography. It inventories some sanctioned ways of portraying compliance to the rules of "good ethnographic work," of telling that the problems of presence, ethnocentrism, methodogenesis, and informants have been recognized and solved for all practical purposes.

The Problem of Presence

For those who do their sociological work via strategies requiring their interactional presence in the domain of concern, that very presence can become problematic in serious ways, perhaps even eroding the possibility of achieving the goal that occasioned their presence in the first place. Many practitioners hold that the ethnographer adds a new and perhaps necessarily tainting dimension to the domain under investigation. For example,

> Participant observation introduces a new person into the community – the participant observer, whose presence may change it to an unknown extent [Selltiz et al., 1960: 220].

> Reactive effects of observation are the most perplexing feature of participant observation, since the presence of an observer in any setting is often a "foreign object." The creation of the role of participant observer inevitably introduces some degree of reactivity into the field setting [Denzin, 1970: 203–204].

The intention here is to describe some of the ways in which competent and entitled practitioners of the ethnographic craft decide that their position in the field was unobtrusive and therefore that the data upon which they base their analyses are reflective of the "real" character of the domain they investigated. The data displayed here are taken from those sections of published ethnographies wherein the ethnographer provides for readers an account of the inconsequentiality of his or her presence.

It should be made clear that I am interested in speculating on neither the behavioral possibility of an unobtrusive position nor the issue of literal correspondence between researchers' portrayal of their position and their "actual" position. Rather, I deliberately suspend concern with those matters in favor of examining some of the ways in which adequate-for-all-practical-purposes solutions to the problem of being in a domain while at the same time attempting to retrieve its character independently of being there may be formulated within the research enterprise.

The literature reveals that a single theme is employed by ethnographers to assure their readers that they were able to capture their field in all its naturalness. This might be called *the theme of the invisible researcher*. The invisible researcher is the ideal researcher who sees without being observed and, consequently, captures the natural field without tainting it. Ethnographers regularly recommend their depiction of a domain as *the* depiction on the grounds that they were able to employ a practically possible, realizable version of that ideal. There appear to be at least six variations on this theme:

(1) Disattending: erosion of visibility by time. Some ethnographers claim that they were present in a domain as ethnographers for a period of time lengthy enough to erode the notice normal inhabitants initially took of them. This variation accomplishes the ethnographer's invisibility or unobtrusiveness via its portrayal of the researcher as having "faded into the background," his or her presence thereby coming to be "normalized" in the domain of investigation. Those persons whose activities are the topic of study eventually go about those activities as if an ethnographer were not present. Often this variation is presented in the form of a "natural history" of the ethnographer's visibility. For example,

> In doing fieldwork, investigating "death," I have been aware of the possible effect of my own presence on the behavior of participants in the scene, and my only claim to its minimal relevance is my feeling that because of my long stay in the hospital, particularly at County, staff came to take my presence for granted. While in the early days of my research at County I distinctly felt that, in many cases, staff members monitored their activities to show me what they thought I wanted to see, as I became well known to staff and they less suspect of my intentions, I felt sufficiently disregarded to be relatively secure that what I was witnessing would have gone on were I not there [Sudnow, 1967: 7].

(2) Disattending: erosion of visibility by display of no symbolic detachment. The second variation on the theme of the invisible researcher differs from the first in that it includes a specification of the conditions facilitating eventual

disattending. Often it is asserted that displaying no symbolic detachment from the ways of the domain of investigation promoted the "normalization" of the ethnographer's presence. The following example illustrates the use of this strategy:

> People often said, "People wouldn't talk to you when you first came here, but now they've adopted you everybody will talk to you." Of great aid toward this "adoption" were the facts that I drove a two-year old car, "know how to loaf and joke" on the street, wore old clothes, and otherwise imitated the average level of everyday life, and already know how to speak accurately whatever dialect level of the language was being spoken at any moment [West, 1945: xi].

> Blending in was facilitated by certain basic skills. One of the most important involved being "quick" . . . . Unless you are to be seen as something of a "divvy" you must be able to look after yourself in the verbal quickfire of the Corner and the pub. . . . Being able to kick and head a football reasonably accurately was also an important aspect of fitting into the scheme [Parker, 1974: 217–219].

(3) Disattending: erosion of visibility by display of symbolic attachment. In contrast to the second variation – wherein the ethnographer merely fits into the domain – the third variation depicts efforts at actively working at invisibility via display of attachment to the domain under research. Such attachment often involves participating with ordinary inhabitants in their routine activities. As he or she participates, the ethnographer's identity as ethnographer will become, it is asserted, more subsidiary and less focal. For example,

> Occasionally my aid was enlisted to assist in a particular procedure by passing an instrument, helping prop up a patient, tighten a tourniquet, and the like, all of which I did to make myself as much a part of the scene as possible. I felt that by helping the morgue attendant transfer a dead patient from the bed to the stretcher, I made him feel as though his actions were not being so distantly observed as when I stood in the background and silently watched him at work [Sudnow, 1967: 7–8].

Choice of a locally appropriate style may figure in here as well:

> I adopted the dress and sometimes the speech of the people with whom I was in most frequent contact, as best I could without looking silly or feeling uncomfortable. I came close in dress . . . with almost no effort at all. My vocabulary and diction changed, but not radically. . . . Thus, while remaining conspicuous in speech and perhaps in dress, I had dulled some of the characteristics of my background. I probably made myself more accessible to others [Liebow, 1967: 255].

(4) Disattending: erosion of visibility by personalizing the ethnographer-informant relationship. A fourth variation accomplishes the invisibility of the

ethnographer by portraying the relationship with normal inhabitants of the domain under investigation as a personal one: The ethnographer was invisible insofar as informants suspended concern with the "research" aspect of his or her identity and "liked him or her as a person." For example,

> I finally came to be considered a full-fledged community member, welcome in men's loafing centers, in church, and in many homes, able to talk to people in situations of my own devising, and free to record any private conversations I wanted to record, in the presence of my informants, without arousing any apparent embarrassment [West, 1945: xi].

This variation in use is evident as well in commonplace discussions of ethnographer-informant "rapport," and in the work of those whose claims to unobtrusiveness in a domain are based on their membership in that domain, independent of and prior to research interests (see Polsky, 1967; Scott, 1968).

(5) Misrepresentation: masking real research interests. This fifth variation is founded on the assumption that the activities most likely to be altered by the ethnographer's presence are those announced as objects of research interest. Thus, in the hope that the normal inhabitants of a domain will perform routinely the activities of real interest, the ethnographer provides a false account of the rationale underlying his or her presence. Unobtrusiveness can thus be recommended on the grounds that the ethnographer was not visible as ethnographer on those occasions when members were performing these research-relevant activities:

> In securing a place to live we spent three hours in Plainville and were in contact with a total of five people. We explained to each, in precisely the same way, that we were moving to Plainville to study changes in farming practices over the past fifteen to twenty years. . . . Our initial contacts seemed interested in the project, expressed the view that it made sense, and volunteered to help us with it. . . . For initial research emphasis we followed through with what we had told our initial contacts, and what we assumed was a relatively low affect interest area, that is, changes in agricultural technology since 1940. By using this as a device of entry we could expand gradually our emphasis to other areas of the culture as the research progressed and our rapport became more firmly established [Gallaher, 1964: 286–287].

(6) Misrepresentation: masking identity as ethnographer. A sixth variation consists of the ethnographer's claim to have represented him- or herself to informants in some way other than as ethnographer. Under its auspices the possibility of obtrusiveness is eliminated as the ethnographer cannot be formulated as such by the domain's normal inhabitants: The researcher is physically present but socially invisible. For example,

> I did not confide my research interests or the fact that I was recording observations to any of the patients or staff. They observed me writing frequently, but assumed that I was pursuing my academic studies. In general, my relationship with other patients and with staff members was similar to that of most patients [Roth, 1963: viii].

The Problem of Ethnocentrism

Ethnographers know the distinctiveness of their enterprise in its intended erosion of the ethnocentrism attendant to sociological work conducted by survey or experimental small group methods. Indeed, for ethnographers the difference between their own and other versions of the sociological enterprise consists in their commitment to addressing concerns native to the domain of interest (whereas practitioners of other sorts of sociology address concerns that derive from elsewhere – notably, from the culture of sociology itself). Ethnographers claim they investigate a domain for its relevancies. Their nonethnographer colleagues, it is said, do so in terms of their own. The difference is assertedly crucial: Whereas ethnography stands to reveal something of the investigated unit, it is held that other versions may reveal only things about their own culture.

One aspect of adequate ethnography consists in the portrayal of mundane or exotic social circumstances from the point of view of participants in those circumstances. A finding to the contrary is a finding against adequacy, as Matza (1969) and Tyler (1969: 11) note in their characterizations of early sociological and anthropological fieldwork. What does the enthnographer do to ensure readers that portrayed native practices and meanings are authentically the latter's own and not – to use the words of Marcus and Cushman (1982: 49) – "just the work of the ethnographic writer's imaginative bricolage from the knowledge that he has acquired from another form of life"? At least two strategies are in use, united by their promotion of natives in the "hierarchy of credibility" (Becker, 1967) from mere informant to "lay research assistant," "talent judge," and, in one version, "coauthor."

(1) Native as research assistant: the member test of validity. In the very way that ethnographers propose that they have captured the perspective native to a domain they propose as well a set of judges of the adequacy so captured – the natives of the domain under investigation. Ethnographers' use of native commentary on their work as a recommendation of its adequacy – called elsewhere (Douglas, 1970: 21) "the member test of validity" – is often encountered. Blumer, for example, recommends (1969: 41) and has used (1967) such commentary to enhance the received validity of ethnographic portrayals. And, as is well known, Whyte (1955) made extensive use of a major informant in the sorting and resorting of the findings that were eventually reported as *Street Corner Society*. Consider the following as further examples of this adequacy-recommending strategy in use:

> The description and analysis are based on twelve months of field work with the police departments of two large cities west of the Mississippi. . . . The observations were augmented by approximately one hundred interviews with police officers of all ranks. The formulations that will be proposed were discussed in these interviews. They were recognized by the respondents as elements of standard practice. The respondent's recognition was often accompanied by remarks indicating that they had never thought about things in this way and that they were not aware how standardized police work was [Bittner, 1967: 701].

> At Professor Herbert Blumer's suggestion, a final check on validity was attempted. We selected ten recreational users and ten heads who had been interviewed and in our judgement were very articulate observers of the scene in which they found themselves. We asked them to meet in a series of group interviews where we discussed what we concluded was the perspective of each of these groups. Our description of the Colony scene was corrected and extended as a result. This enabled us to separate individual perspectives from those more collective in character. . . . The group setting enabled us to raise points about which we were unclear and get the reaction of our informants on the authenticity of our description [Carey, 1968: 202].

(2) Native as talent judge: interaction effectiveness.[2] The idea of native as expert is further expressed in the work of those who – in the manner suggested by Frake (1964) – assert the adequacy of their recovery of local perspectives by claiming "interaction effectiveness" (Douglas, 1976: 123); that is, that their description can serve as *instructions* on how to act as a competent member of the unit that has been investigated. An example is provided by Wiseman (1970: 280–281):

> A set of instructions was given to . . . observers on how to dress and how to act so as to pass as a native at a given station on the loop – in this case, Skid Row and the Christian Missionaries. Successful passing (i.e., unquestioned acceptance) was seen as validating the instructions and the perspectives they reflected.

The Problem of Methodogenesis

Inquirers of whatever sort are concerned to offer the assurance that the tools of their inquiry explicated or retrieved a domain's actual features; a finding to the contrary is a finding against the inquiry's adequacy. Ethnographers recommend that what they have discovered about a domain was in fact discovered as opposed to "created" by (1) neutralizing or (2) invisibilizing their techniques of inquiry, and by (3) providing redundant demonstrations that what is reported was there independently of viewpoint. To consider briefly these strategies for accomplishing ethnographic adequacy:

(1) Neutralizing the techniques. Becker et al. (1961: 43) have directed researchers to assess the evidentiary value of members' statements on the basis of whether these have been "volunteered" by informants (or "merely overheard" in their talk) – that is, made independently – or produced in response to questions posed by the ethnographer. They portray the former statements as

> likely to reflect the observer's preoccupations and possible biases less than the one which is made in response to some action of the observer, for the observer's very question may direct the informant into giving an answer which might never occur to him otherwise [Becker, 1970: 30].

It is suggested that "the number of directed statements should be small in comparison to the volunteered statements" (Becker et al., 1961: 43) and implied that ethnographic adequacy is enhanced by such a ratio.

(2) Invisibilizing the techniques. This strategy erodes the possibility of methodogenesis by abandoning altogether the use of techniques producing "directed statements"; the ethnographer claims that the data are solely "overhearings" of talk among members. In his study of dance musicians, Becker (1963: 84) provides an approximation of this strategy in use:

> I seldom did any formal interviewing but concentrated rather on listening to and recording the ordinary kinds of talk that occurred amongst musicians.

(3) Redundant demonstrations. Ethnographers commonly detach an observation from a viewpoint by providing a sort of walking-around tour of the object or event in question. In this way they establish the independent "thereness" of the phenomenon by demonstrating over and over that variously located observers encountered it in an identical fashion. For example, in a discussion of policy changes that I claimed had occurred in a public drinking place (Stoddart, 1981), I present fieldnotes describing not only my own sense of such changes but also statements to this effect from waiters, bartenders, other patrons, and drug dealers, seeking to convey thereby that what I detected was there to be detected independently of my address. To paraphrase Stern's comments (1973, quoted in Marcus and Cushman, 1982: 29) on the novelist's use of descriptive diversion, such redundant demonstrations serve to add and superadd to the sense of assurance that any reader has about the work.

The Problem of Informants

Practitioners often formulate one goal of ethnography as the discovery and description of knowledge foundational to membership in a particular domain, that is, "the socially acquired and shared knowledge, or culture, that enables members of the society to behave in ways deemed appropriate by their fellows" (Frake, 1969: 123). Ethnographers commonly attempt to realize this goal by querying persons native to the domain in question, persons ethnographers have come to call informants.

In order to found portrayals on information general to the domain in question and not idiosyncratic to particular participants, cohorts, or social locations, practitioners insist in their technical writings that not just any native will "do" as an informant. What is needed, it is held, is a well-informed informant, to use Back's (1956) apt characterization. Consider the following:

> The critical issue . . . is whether or not the informant can be assumed to have the information that the fieldworker requires. . . . We are interested primarily in whether he is likely to know enough [Zelditch, 1969: 11–12].

> A good informant is one who knows the culture well [Spradley and McCurdy, 1972: 47].

> One should obviously seek participants in the sphere of life who are acute observers and who are well-informed [Blumer, 1969: 41].

By whatever means ethnographers locate natives entitled to speak as natives, the accomplishment of adequate ethnography is contingent on telling that they did so and on characterizing these natives as "well informed." Consider that for any identified informant the ethnographer has available a host of alternate descriptors, all potentially correct. Yet on any occasion of doing informant description, not "all" or "just any" descriptors are brought to bear. For example, in an account depicting what participants in a drug-using "community" understood as "the facts" about the effects of LSD (Stoddart, 1974b), informants' statements purportedly revealing of those facts were prefaced with such remarks as "as one experienced user put it." Other statements suggested that data were gathered from participants of sufficient tenure in the "community" to speak knowledgeably of its beliefs. A description of an informant, then, is not merely an assemblage of certain "facts" about the informant; rather, someone is described as "an experienced user," for example, not primarily because he or she is "in fact" an experienced user, but as a device for accomplishing his or her entitlement. Such a description formulates the informant as a competent provisioner of local knowledge and thereby accomplishes one aspect of adequate ethnography.

Conclusion

Despite the fact that all inquiries by their very nature must fall short of completeness, "perfect" objectivity, and the like (Garfinkel, 1967), ethnographers find with great regularity that their own and colleagues' inquiries are complete enough, objective enough, and, in a word, *adequate*. This article has identified certain received criteria of ethnographic adequacy and described, in a preliminary fashion, an ensemble of strategies employed by entitled practitioners of the craft to assert that adequacy.

Over the course of this examination of what has been called "the good ethnographer's work," two prominent themes have emerged. Both are suggestive of fundamental continuities between ethnographic inquiry and the inquiries undertaken by "ordinary" members of society.

First, it appears that for ethnographers – as for all inquirers – the "natural state" of any domain is that domain's paramount feature. In the formulation of their task and in their location of problems attendant to that task, ethnographers both presume this "natural state" and seek to preserve it in carrying out and reporting their research activities. Thus, for ethnographers the domain of their concern is possessed of features massively independent of whatever means are employed to make them visible; and the capture of those features as they are received by that domain's participants provides the ethnographic task. And, finally, ethnographers orient to any and all matters sub-sumable under the title "research activities" so as to provide for the defensible claim that they accomplished the task of capturing features of that domain in a practically adequate fashion.

Second, in their tellings of the inconsequentiality to the data of their own presence and technique, in their formulation of the entitlement of who provided the data, and so on, ethnographers reveal that their strategies for asserting adequacy are not particular to the ethnographic enterprise but are distinctive of tellings of adequate reportage in general – insofar as it is reportage of a referent with the character suggested in the preceding paragraph.

Notes

1. The list is obviously not exhaustive.
2. This section has benefited from a reading of Emerson (1981).

References

Austin, J. L. (1965) How to Do Things with Words. New York: Oxford Univ. Press.

Back, K. W. (1956) 'The well-informed informant." Human Organization 14: 30–33.

Becker, H. S. (1970) "Problems of inference and proof in participant observation," pp. 25–38 in H. S. Becker (ed.) Sociological Work: Method and Substance. Chicago: Aldine.

________ (1967) "Whose side are we on?" Social Problems 14: 239–248.

________ (1963) Outsiders: Studies in the Sociology of Deviance. New York: Free Press.

________ B. Geer, et al. (1961) Boys in White. Chicago: Univ. of Chicago Press.

Bittner, E. (1967) "The police on skid-row: a study of peace-keeping." Amer. Soc. Rev. 32: 699–715.

Bloor, M. (1978) "On the analysis of observational data: A discussion of the worth and uses of inductive techniques and respondent validation." Sociology 12: 545–552.
Blumer, H. (1969) Symbolic Interactionism: Perspective and Method. Englewood Cliffs, NJ: Prentice-Hall.
——— (1967) The World of Youthful Drug Use. Berkeley: Univ. of California School of Criminology.
Carey, J. (1968) The College Drug Scene. Englewood Cliffs, NJ: Prentice-Hall.
Clifford, J. (1981) "On ethnographic surrealism." Comparative Studies in Society and History 23: 539–564.
——— (1980) "Fieldwork, reciprocity, and the making of ethnographic texts: the example of Maurice Leenhardt." Man 15: 518–532.
Denzin, N. (1970) The Research Act. Chicago: Aldine.
Douglas, J. D. (1970) "Understanding everyday life," pp. 3–44 in J. D. Douglas (ed.) Understanding Everyday Life. Chicago: Aldine.
——— (1976) Investigative Social Research: Individual and Team Field Research. Beverly Hills, CA: Sage.
Emerson, R. M. (1981) "Observational field work." Annual Rev. of Sociology 7: 351–378.
Frake, C. O. (1969) "Notes on queries in ethnography," pp. 123–137 in S. A. Tyler (ed.) Cognitive Anthropology. New York: Holt, Rinehart & Winston.
——— (1964) "How to ask for a drink in subanun." Amer. Anthropologist 66: 132–145.
Gallaher, A., Jr. (1964) "Plainville: the twice-studied town," pp. 285–303 in A. J. Vidich et al. (eds.) Reflections on Community Studies. New York: John Wiley.
Garfinkel, H. (1967) Studies in Ethnomethodology. Englewood Cliffs, NJ: Prentice-Hall.
Johnson, J. M. (1975) Doing Field Research. New York: Free Press.
Liebow, E. (1967) Tally's Corner. Boston: Little, Brown.
Lofland, J. (1976) Doing Social Life: The Qualitative Study of Human Interaction in Natural Settings. New York: John Wiley.
——— (1974) "Styles of reporting qualitative field research." Amer. Sociologist 9: 101–111.
Marcus, G. (1980) "Rhetoric and the ethnographic genre in anthropological research." Current Anthropology 21: 507–510.
Marcus, G. and D. Cushman (1982) "Ethnographies as texts." Annual Rev. of Anthropology 11: 25–69.
Matza, D. (1969) Becoming Deviant. Englewood Cliffs, NJ: Prentice-Hall.
Parker, H. J. (1974) View from the Boys: A Sociology of Downtown Adolescents. London: David & Charles.
Polsky, N. (1967) Hustlers, Beats, and Others. Chicago: Aldine.
Roth, J. (1963) Timetables. Indianapolis: Bobbs-Merrill.
Schatzman, L. and A. L. Strauss (1973) Field Research: Strategies for a Natural Sociology. Englewood Cliffs, NJ: Prentice-Hall.
Scott, M. B. (1968) The Racing Game. Chicago: Aldine.
Selltiz, C., M. Jahoda, et al. (1960) Research Methods in Social Relations. New York: Holt, Rinehart & Winston.
Spradley, J. P. (1980) Participant Observation. New York: Holt, Rinehart & Winston.
——— (1979) The Ethnographic Interview. New York: Holt, Rinehart & Winston.
——— and D. W. McCurdy (1972) The Cultural Experience: Ethnography in Complex Society. Chicago: Science Research Associates.
Stern, J. P. (1973) On Realism. London: Routledge & Kegan Paul.
Stoddart, K. (1981) "As long as I can't see you do it: a case study of drug-related activities in public places." Canadian J. of Criminology 23: 391–406.

Stoddart, K. (1974a) "Pinched: notes on the ethnographer's location of argot," pp. 173–179 in R. Turner (ed.) Ethnomethodology. Harmondsworth: Penguin.

——— (1974b) "The facts of life about dope: observations of a local pharmacology." Urban Life and Culture 3: 179–204.

Sudnow, D. (1967) Passing On: The Social Organization of Dying. Englewood Cliffs, NJ: Prentice-Hall.

Turner, R. (1970) "Words, utterances, and activities," pp. 169–187 in J. D. Douglas (ed.) Understanding Everyday Life. Chicago: Aldine.

Tyler, S. A. [ed.] (1969) Cognitive Anthropology. New York: Holt, Rinehart & Winston.

West, J. (1945) Plainville, U.S.A. New York: Columbia Univ. Press.

Wieder, D. L. (1974) Language and Social Reality: The Case of Telling the Convict Code. The Hague: Mouton.

Whyte, W. F. (1955) Street Corner Society. Chicago: Univ. of Chicago Press.

Wiseman, J. P. (1970) Stations of the Lost: The Treatment of Skid Row Alcoholics. Englewood Cliffs, NJ: Prentice-Hall.

Zelditch, M., Jr. (1969) "Some methodological problems of field studies," pp. 5–18 in G. J. McCall and J. L. Simmons (eds.) Issues in Participant Observation. Reading, MA: Addison-Wesley.

61

Representation, Legitimation, and Autoethnography: An Autoethnographic Writing Story

Nicholas L. Holt

Introduction

The postmodern research movement has raised doubts about the privilege of any one method for obtaining authoritative knowledge about the social world (Agger, 1990). In the wake of the postmodern, traditional writing practices for qualitative research have been critiqued (Richardson, 2000). In particular, the ethnographic genre has been subjected to extensive scrutiny, and several types of ethnographic writing practices have become available to researchers (van Manen, 1988). One emergent ethnographic writing practice involves highly personalized accounts where authors draw on their own experiences to extend understanding of a particular discipline or culture. Such evocative writing practices have been labelled 'autoethnography' (Reed-Danahay, 1997). The movement toward personalized research reflects calls to place greater emphasis on the ways in which the ethnographer interacts with the culture being researched.

Autoethnography is a genre of writing and research that connects the personal to the cultural, placing the self within a social context (Reed-Danahay, 1997). These texts are usually written in the first person and feature dialogue, emotion, and self-consciousness as relational and institutional stories affected by history, social structure, and culture (Ellis & Bochner, 2000).

Source: *International Journal of Qualitative Methods*, 2(1) (2003): 18–28.

Reed-Danahay explained that autoethnographers may vary in their emphasis on *graphy* (i.e., the research process), *ethnos* (i.e., culture), or *auto* (i.e., self). Whatever the specific focus, authors use their own experiences in a culture reflexively to look more deeply at self-other interactions. By writing themselves into their own work as major characters, autoethnographers have challenged accepted views about silent authorship, where the researcher's voice is not included in the presentation of findings (e.g., Charmaz & Mitchell, 1997). This development may have liberated some researchers from the constraints of the dominant realist representations of empirical ethnography because, as Richardson (1995) argued, how researchers are expected to write influences what they can write about.

The use of self as the only data source in autoethnography has been questioned (see, for example, Denzin & Lincoln, 1994; Sparkes, 2000). Accordingly, autoethnographies have been criticized for being too self-indulgent and narcissistic (Coffey, 1999). Sparkes (2000) suggested that autoethnography is at the boundaries of academic research because such accounts do not sit comfortably with traditional criteria used to judge qualitative inquiries. Indeed, traditional criteria used to judge qualitative research in general may not be appropriate for autoethnography (Garratt & Hodkinson, 1999).

Denzin & Lincoln (1994, 2000) described key 'moments' in the history of qualitative research. The first moment was the traditional period (early 1900s), when qualitative researchers aspired to 'objective' accounts of field experiences. The second moment was the modernist phase (postwar years to 1970s), which was concerned with making qualitative research as rigorous as its quantitative counterpart. The third moment (1970–1986) was concerned with the blurring of genres. The fourth moment (mid-1980s) is characterized by crises of representation and legitimation. The fifth moment concerns experimental writing and participatory research. Additional stages include the sixth (postexperimental) and seventh (future) moments, whereby fictional ethnographies and ethnographic poetry become taken for granted. Whereas Denzin and Lincoln's moments characterize the historical development of North American qualitative literature, they do not generalize well across disciplines. For example, Sparkes (2002a) reflected that it is difficult to apply these moments to sport and physical activity research where qualitative research is a relatively new development.

This article is a writing story (i.e., a story about the process of conducting research [see Richardson, 1995]), concerned with publishing an autoethnographic account in the discipline of physical education pedagogy. Given that sport and physical activity research does not have well-developed qualitative research practices (Sparkes, 2002a), it may not be a surprise that the issues dealt with in this article are reflective of the fourth moment of qualitative research. Two crucial issues associated with the fourth moment of qualitative research are the dual crises of representation and legitimation (Denzin & Lincoln, 2000). The crisis of representation refers to the writing practices (i.e., how researchers write and represent the social world). Additionally, verification issues relating to methods and representation are (re)considered as problematic (Marcus & Fischer, 1986). The crisis of legitimation questions traditional criteria used for evaluating and interpreting qualitative research, involving a rethinking of terms such as validity, reliability, and objectivity.

The present article can be considered as a writing story that is intended to provide guidance for others wishing to produce evocative writing accounts. As such, the overall objective of this writing story is to critique an aspect of research culture (the peer review process experienced for an autoethnographic manuscript) pertaining to forms of representation and legitimation. In other words, this paper is an autoethnographic writing story about the process of getting an autoethnographic manuscript published. Specifically, reviewers' comments are used to highlight problematic issues relating to: (a) the use of verification strategies in autoethnographic studies; and (b) the use of self as the only data source.

The Origins of the Autoethnographic Account

Originally I developed an autoethnographic account depicting my experiences as a Ph.D. student trying to come to terms with teaching at the university level for the first time (Holt, 2001). The plot hinged on a series of clashes between my personal teaching history and the teaching ideology of the research institution I had recently joined. I drew guidance from an autoethnographic account produced by Sparkes (1996) and the associated discussion of the review process for his paper (Sparkes, 2000). Sparkes reported that he was criticized by the reviewers of his 1996 account for presenting a paper that was heavily theorized in the introduction and conclusion. With this in mind, I produced a manuscript in which I attempted to tell a story, rather than produce a heavily theoretical reflective piece, for, as Wolcott (1994) suggested, qualitative researchers need to be storytellers, and storytelling should be one of their distinguishing attributes.

The autoethnographic account produced by Sparkes (1996) that inspired my autoethnography concerned his experiences as a white, male, middle class, former elite athlete with a chronic back injury that ultimately curtailed his sporting career. Sparkes linked his personal experiences to social, sporting, medical, and academic discourses via a thorough sociological self-exploration. I was drawn to the manner in which he explored emotional dimensions and consequences of an interrupted body project, and attempted to bring the reader into his 'inner' world. Although the content and purposes of Sparkes' story differed from my own, I was attracted by the powerful and emotive way in which his experiences were communicated. I especially liked the connections he made between his personal experiences and the wider (sub) cultural settings in which he was located (i.e., sport, the medics who 'treated' him, his family, and his academic career). I thought autoethnography could be a useful way for examining my teaching experiences in a self-reflexive manner.

Autoethnography seemed to be the optimal choice for my article because, as Tierney (1998) asserted, "autoethnography confronts dominant forms of representation and power in an attempt to reclaim, through self-reflective response, representational spaces that have marginalized those of us at the borders" (p. 66). I tried to produce an account of a somewhat marginalized figure portraying the personal tensions I experienced integrating my pedagogical approach with the mandates of the university teaching program. The story seemed to develop into a discourse about the role of teaching in a research university, and the placelessness of a new graduate student in this subculture. In addition to these personal concerns, I wrote about how my actions and behaviours may have influenced the students in my classes. I drew on a reflective framework, pursued a methodological tradition (autoethnography) and linked my 'findings' back to relevant pedagogical research. Although I thought I had complied with the demands of rigorous science, three journal editors and their reviewers disagreed before the manuscript was finally accepted at the fourth attempt.

Deconstructing Reviewer Feedback: An Autoethnographic Writing Story

In the following section I have recreated the feedback provided by seven reviewers (and four editors) into one conversation between an author and two reviewers. I am the author, a Ph.D. student who feels desperately in need of publications to get to the first step of the career ladder. Although my feelings of desperation underpinned this writing process, publication for the sake of garnering another line on one's vita may be interpreted as self-serving. As such, I went through this writing process faced with a tension between my desire to communicate my pedagogical experiences with others (and thus to contribute deeper understandings about teaching), against my 'need' to publish.

The process of rejection and resubmission I experienced may be a familiar one to some readers: select a seemingly appropriate journal, submit the manuscript, receive rejection, integrate reviewer's feedback, send to another journal, and so on. For the purposes of context, some information about the journals to which the manuscript was submitted is required. All four journals were similar in that they published research and practice pertaining to physical education pedagogy. One journal was aimed more at practitioners, one journal mixed theory and practice, and two journals were more research-oriented. (The manuscript was eventually published in one of the more research-oriented journals.) In terms of the editorial policies, I deliberately chose journals that had a broad scope. For example, one editorial policy stated '*Research methodologies may be quantitative or qualitative and may use data gathered through historical analysis, surveys, field work, action research, participant observation, content analysis, simulations, or experiments*'. The aims of another were to *'facilitate research that enriches (the discipline) and disseminate findings to professionals and the public'*.

More specifically, the journals required that manuscripts adhere to the ethical and content guidelines of the American Psychological Association publication manual (APA, 2001). The APA criteria for evaluating the content of manuscripts are oriented toward quantitative research. For example, guidelines include: "Is the research question significant, and is the work original and important? Does the research design fully and unambiguously test the hypothesis? Is the research at an advanced enough stage to make the publication of results meaningful?" (APA, 2001, p. 6). However, because the journals I chose published both qualitative and quantitative research, there was also specific advice for judging qualitative manuscripts, such as assessments of the *'quality of analyses'* and providing evidence of *'saturation of data'* and *'trustworthiness'*.

In recreating the discourse presented below, I used the comments provided by journal reviewers as my 'data'. Initially, I simply collected and saved these reviews (after I had used the feedback to 'improve' my manuscript). After ultimately receiving an acceptance from the fourth journal, I had noticed an emergent pattern in the content of the reviews. In general, there was a set of reviewers who valued autoethnography but had concerns over its rigor. There was another set of reviewers who clearly did not think that authoethnography constituted scientific research. I 'analyzed' this data by categorizing the reviewer's comments into two main themes. The first theme included 'sympathetic' comments (i.e., the reviewers who thought that autoethnography was merited but were unsure of verification criteria). The second theme included 'skeptical' comments (i.e., the reviewers were unsure about the scientific merit of autoethnography). I then recreated these comments into a discourse that is intended to represent my personal experiences of the publication process. I created two reviewers to represent the two themes of comments. Reviewer A is presented as an amalgam of those who were sympathetic to autoethnography. In contrast, Reviewer B represents those who were more suspicious of the scientific merit of the genre. As far as possible, verbatim quotes from the reviewers' written comments have been crafted into a narrative dialogue to reflect on two central issues that arose from the review process: (a) The use of verification strategies in autoethnographic studies, and (b) the use of self as the only data source.

Recreating a Discourse I: The Use of Verification Strategies in Autoethnographic Studies

This first discourse is intended to convey a discussion of the verification strategies by which my autoethnographic manuscript was judged. The hypothetical setting is a university office, and the reviewers are settling down to discuss their perceptions of the manuscript with me.

Reviewer A: I believe that this methodology has great potential to contribute to our existing knowledge base. But, the problem is that this paper does not show the rigor that is necessary in a research-based article. Narrative research requires as much rigor as any other research design. . .

Author: This paper went through countless drafts and re-writes, and it was read and edited by several colleagues.

Reviewer B: Time does not equate to quality. There is no evidence of rigor in this narrative. . .

Author: [Defensively] And what do you mean by that? It was a rigorous process – in fact, this manuscript was subjected to more editing and critical appraisal from colleagues than anything else I've done.

Reviewer B: If I may interject. Qualitative investigators should address credibility, dependability, and trustworthiness of data. There must be some type of paper trail in which the results of the study can be supported with data.

Author: But these traditional forms of criteria do not apply to autoethnographic texts. Ellis and Bochner (2000) said so. . .

Reviewer B: Credibility, dependability, and trustworthiness are the cornerstones of qualitative rigor. Without these standards qualitative research becomes nothing more than journalism with a smattering of theory.

Reviewer A: How would you suggest we review this scientific contribution?

Author: I think the intrinsic interest and value of the story should not be dismissed. Laurel Richardson (1995, 2000) suggested that such work could be evaluated by questions like 'did the paper have an emotional or intellectual impact?'

Reviewer B: No, no, no. Certain scientific tenets must be adhered to. The manuscript should be grounded within a theoretical framework. We cannot publish good stories in an academic journal. Methodological procedures and data analysis must be clearly explained and supported with references. You failed to comply with these demands in such a manner that your work could be replicated.

Author: How could someone else replicate work based on my personal experiences?

Reviewer B: That is exactly why this material is not of publishable quality!

Reviewer A: Look, I am sympathetic to your plight, but your paper does not show clear relationships and patterns, does not have the completeness in the narrative, nor does it hold

> the phenomenon up to serious inspection. I am not able to find depth to your narrative. If you are interested in sharing your narrative, it is important that it have a clear focus, depth, and is believable to the reader. I would advise you to keep trying.

In analyzing this conversation, it is apparent that Reviewer B demanded the use of Lincoln and Guba's (1985) parallel perspective to establish the 'trustworthiness' of the investigation, which is understandable if one is not familiar with the criteria by which to judge an autoethnography. Reviewer A seemed more familiar with the appropriate criteria for this type of research, and echoed Ellis (1995), who argued a story could be considered valid if it evokes in the reader a feeling that the experience is authentic, believable, and possible. However, evaluation of autoethnographic research is problematic, as Josselson (1993) questioned: "What is a good story? Is just a good story enough? What must be added to a story to make it scholarship? How do we derive concepts from stories and then use these concepts to understand people?" (p. xi).

The reviewers called for several validity techniques that are normally associated with enhancing the rigor of qualitative investigations. However, some scholars have suggested that the criteria used to judge autoethnography should not necessarily be the same as traditional criteria used to judge other qualitative research investigations (Garratt & Hodkinson, 1999; Sparkes, 2000). Furthermore, the post-hoc use of techniques to ensure reliability and validity has been questioned recently (Morse, Barrett, Mayan, Olson, & Spiers, 2002). Morse et al suggested that more attention should be paid to constructive rather than evaluative (i.e., post-hoc) techniques. For example, describing investigator responsiveness during the research process would be a constructive approach to validity, as opposed to the inclusion of evaluative checks to establish the trustworthiness of completed research (e.g., an external audit). Both viewpoints question the application of 'traditional criteria' (i.e., post-hoc evaluative measures based largely on the parallel perspective) to judge contemporary qualitative investigations. There is common ground here because constructive approaches to validity and reliability would be more appropriate criteria to judge autoethnography than the post-hoc imposition of evaluative techniques associated with the parallel perspective. For example, reflexive techniques that examine the sensitivity of the researcher to the particular subculture under investigation may provide constructive approaches to validity in autoethnography.

Richardson (2000, pp. 15–16) described five factors she uses when reviewing personal narrative papers that includes analysis of both evaluative and constructive validity techniques. The criteria are: (a) Substantive contribution. Does the piece contribute to our understanding of social life? (b) Aesthetic merit. Does this piece succeed aesthetically? Is the text artistically shaped, satisfyingly complex, and not boring? (c) Reflexivity. How did the author come to write this text? How has the author's subjectivity been both a producer and a product of this text? (d) Impactfullness. Does this affect me emotionally and/or intellectually? Does it generate new questions or move me to action? (e) Expresses a reality. Does this text embody a fleshed out sense of lived experience? Autoethnographic manuscripts might include dramatic recall, unusual phrasing, and strong metaphors to invite the reader to 'relive' events with the author. These guidelines may provide a framework for directing investigators and reviewers alike.

The theoretical arguments outlined above seem to justify the use of different criteria to evaluate autoethnography. However, such theoretical arguments did little to alleviate my sense of helplessness as I continued to face reviewers who demanded 'traditional' verification criteria. I carefully addressed the reviewers' comments, and the next draft of my autoethnography complied with the traditional formula more readily as a result of the changes I had made. I did not realize I was to confront another set of issues that struck to the very heart of the autoethnographic genre.

Recreating a Discourse II: The Use of Self as the Only Data Source

The second dialogue of reviewer criticisms centers on demands for a more realist (and presumably acceptable) form of representation. Again, reviewer A represents those more credulous reviewers whereas reviewer B is representative of those with a skeptical view of autoethnography.

Reviewer B: It is generally not wise to conduct a study of the 'self.' There are always exceptions to this rule, however, a strong case must be presented. It would be difficult to classify this manuscript as 'true' research even after a revision.

Author: The genre of autoethnography is based on, and designed for, the use of self. Without the self there could be no autoethnography. You are dismissing the entire methodology rather than critiquing this particular investigation.

Reviewer A: Well, I personally support narrative as a form of qualitative research, but I am not able to find depth to this narrative. For example, including extracts from a research diary would have been useful. This would have allowed the investigator to incorporate direct quotes into the body of the manuscript.

Reviewer B: Yes, a good idea. But the manuscript is presented as a reflective synopsis of his teaching, and not a formal research investigation per se. . .

Author: Have you stopped to think that this paper may not have been presented or intended as a 'formal research investigation per se?'

Reviewer B: Well, why was a singular case the sole source of data and a narrative analysis used? I mean, you use 'I' ten times on page ten. This seems rather excessive. The final section of the manuscript seemed to be a revisit of personal experiences and emotions more so than substantial recommendations for teachers and administrators. These strategies are particularly contextual and unique, which makes applicability and transfer difficult.

Reviewer A: My comments are directed toward assisting the author in developing a manuscript that would fit the requirements of a research-oriented journal. The author does not convince me of the significance of the methodology for reflective purposes. In other words, elaboration concerning why this methodology is useful is needed, particularly for the reader naive to this type of methodology.

These comments left me at a loss. After all, I thought that I had complied with advice provided by the autoethnography methodology experts (i.e., Ellis & Bochner, 2000; Reed-Danahay, 1997) and demonstrated necessary methodological rigor. It seemed to me that the reviewers were leveling criticisms of my work that were more a criticism of the entire genre of autoethnographic research, which appeared to be because of their unfamiliarity with this form of representation and the use of self as the source of data. Reviewer B was critical of my expression of personal experiences and emotions, which is the essence of autoethnographic writing (Reed-Danahay, 1997). These reviewers seemed to agree with Charmaz and Mitchell (1997), who argued that scholarly writers are expected to stay on the sidelines and keep their voices out of their articles ("the proper voice is no voice at all" p. 194). The mistrust of the self as a research vehicle was evidenced by calls for 'empirical' data. Reviewer A wanted extracts from the research diary and direct quotes, which would require moving away from the tenets of autoethnography to produce a more realist version of the tale (Van Manen, 1988). Both reviewers questioned whether this type of manuscript constitutes proper research. I realized that establishing autoethnography as 'proper research' is problematic because such work is located "at the boundaries of disciplinary practices" (Sparkes, 2000, p. 21).

The problematic nature of self-as-the-only-data-source in autoethnography is further complicated because qualitative researchers have long been encouraged to consider how their personal subjectivity influences the investigative process. The investigator has been established as the research instrument *par excellence* in a range of qualitative research traditions (Lincoln & Guba, 1985). In ethnography, the subjectivity of the researcher is seen as a resource for understanding the problematic world they are investigating, as something to capitalize on rather than exorcise (Glesne & Peshkin, 1992). In the autoethnographic tradition the use of the investigator's *self* as the focus of the research is problematic. But, as Sparkes (2002b) questioned, why are personal narratives so offensive?

It appears that criticisms of narcissism are common because autoethnographies are considered too self-indulgent, introspective, and individualized (Sparkes, 2000). Coffey (1999) suggested that those who preach autoethnography are "in danger of gross self-indulgence" (p. 132). However, criticisms of narcissism and self-indulgence represent universal charges applied to all autoethnographies regardless of the qualities of the particular venture. Of course, autoethnographies can be self-indulgent, but the universal charge of self-indulgence labeled against autoethnographies is a result of misapprehensions of the genre due to a mistrust of the work of self (Sparkes, 2002b).

Even if reviewers do not trust the use of self, a good autoethnography should not be dismissed. Autoethnography is not necessarily limited to the self because people do not accumulate their experiences in a social vacuum (Stanley, 1993). Mykhalovskiy (1996) challenged reductive, dualistic views (i.e., self-other distinctions) of autoethnography, and suggested that to write individual experience is to write social experience. It seems that charges of narcissism and related criticisms function to reinforce ethnographic orthodoxy and resist change (Ellis & Bochner, 2000). Such criticisms, in turn, function to preserve the very types of dominant viewpoints that those using autoethnographic approaches may wish to question. If autoethnography is intended to confront dominant forms of representation and power in an attempt to reclaim marginalized representational spaces (Tierny, 1998), it is quite ironic that the method itself becomes marginalized by the academic review process.

Conclusion

The autoethnographic account featured here was eventually published after being examined by seven different reviewers and four journal editors. Much of the feedback provided a universal critique of autoethnography rather than a critique of the particular manuscript. The majority of the reviewers' comments were directed at making the autoethnography more realist, which would then enable them to evaluate it using more established, acceptable, and accessible criteria. Alternatively, feedback directed at improving the autoethnographic content was largely absent.

Autoethnography has been received with a significant degree of academic suspicion because it contravenes certain qualitative research traditions. The controversy surrounding autoethnography is in part related to the problematic exclusive use of the self to produce research (Denzin & Lincoln, 1994).

Additionally, it is apparent that traditional criteria used to pass judgement on qualitative research in general may not be suitable for autoethnography (Garratt & Hodkinson, 1999; Sparkes, 2000). These issues reflect the essence of the dual crises of representation and legitimation that characterize the fourth moment of qualitative inquiry (Denzin & Lincoln, 2000).

I could not help but wonder if the reviewers found themselves in a constrained and unenviable position. As a gatekeeper of research culture, the reviewer holds a powerful position. High quality reviews can provide valuable learning experiences, contribute to professional development, and even (indirectly) influence salary and promotion. Given that most reviewers take these responsibilities seriously, it must be a daunting experience to be confronted with an academic manuscript that veers so significantly with the tenets of a traditional review. Whereas I was frustrated by the peer-review process that I experienced for this particular manuscript, I had to acknowledge that the reviewers may have been frustrated by the lack of guidance from the wider research community concerning how to evaluate autoethnographic work. My experiences, and the sense of inadequacy I experienced as I was repeatedly unable to publish the account, left me pondering the worth of continuing in my attempts to get the article published. Eventually, I resolved to 'keep trying' because to admit defeat would be to accept the marginalization of *both* my pedagogical experiences *and* the method by which I chose to express them. To publish represented a small, but personally significant, victory over certain dominant perspectives.

In terms of considering the journals to which the manuscript was submitted, it must be noted that they were not 'pure' qualitative publications. However, I did select journals that published qualitative research and had a broad scope. Certainly the journals differed in terms of their focus on research versus practice, but the content of the reviews did not reflect research-practice concerns. Moreover, there was consistency in the reviews concerning the application of traditional criteria to judge autoethnographic research and the use of the self to produce research. Perhaps some of the problems reported upon here would not have been experienced if I had only submitted to purely qualitative journals. However, if autoethnographies are only submitted to purely qualitative journals then autoethnographic writers may be reinforcing the dominant practices of verification that emerged here. If autoethnography is to be justified as proper research then publication in broad 'mainstream' journals (i.e., that publish qualitative *and* quantitative research) is a necessary step.

Much like the autoethnographic texts themselves, the boundaries of research and their maintenance are socially constructed (Sparkes, 2000). In justifying autoethnography as proper research, it should be noted that ethnographers have acted autobiographically before, but in the past they may not have been aware of doing so, and taken their genre for granted (Coffey, 1999). Autoethnographies may leave reviewers in a perilous position. In the example examined here, the reviewers were not sure if the account was proper research (because of the style of representation), and the verification criteria they wished to judge this research by appeared to be inappropriate. Whereas the use of autoethnographic methods may be increasing, knowledge of how to evaluate and provide feedback to improve such accounts appears to be lagging. As reviewers begin to develop ways in which to judge autoethnography, they must resist the temptation to "seek universal foundational criteria lest one form of dogma simply replaces another" (Sparkes, 2002b, p. 223). However, criteria for evaluating personal writing have barely begun to develop (DeVault, 1997).

There are potential implications of the writing story presented here for investigators and reviewers. Caution is warranted in over-generalizing from this account to the general review criteria of all journals. This paper only provides a personal negative experience of the peer review process for one manuscript. It must also be noted that qualitative research in sport and physical activity may not be as well advanced as in other disciplines (Sparkes, 2002a). However, it is intended that this account be thought provoking for other reviewers and researchers. Researchers would be well advised to be persistent in their autoethnographic intentions, and be prepared to face rejection and critiques of their chosen genre. Resilience and conviction are required to pursue this methodology. Generally autoethnographic accounts have been limited to publication in qualitative journals or presentations at qualitative conferences. If autoethnography becomes more accepted as a form of scientific presentation, researchers must develop new avenues of criticism and support for such work.

Whereas those who produce autoethnography are at risk of being overly narcissistic and self-indulgent, there does seem to be a place for research that links the personal with the cultural. Autoethnography can encourage empathy and connection beyond the self of the author and contribute to sociological understandings (Sparkes, 2002b). In this article, I attempted to autoethnographically link personal experiences of a review process with pertinent issues reflective of research culture in order to contribute to understandings of the autoethnographic approach (of course it is up to the reader to judge if this has been achieved). In postmodernity, it is surely incumbent on the gatekeepers of research to share perspectives on a variety of research methodologies, styles of representation, and evaluative criteria, rather than privilege the authority of dominant viewpoints. It is essential that such discussions take place to ensure the future (or confirm the demise) of autoethnographic methodology.

References

Agger, B. (1990). *The decline of discourse: Reading, writing, and resistance in post-modern capitalism*. Lewes, UK: Falmer.

American Psychological Association. (2001). *Publication manual* (5th ed.). Washington: American Psychological Association.

Charmaz, K., & Mitchell, R. (1997). The myth of silent authorship: Self, substance, and style in ethnographic writing. In R. Hertz, (Ed.), *Reflexivity and voice* (pp. 193–215). London: Sage.

Coffey, P. (1999). *The ethnographic self.* London: Sage.

Denzin, N. K., & Lincoln, Y. S. (1994). Introduction: Entering the field of qualitative research. In N. K. Denzin & Y. S. Lincoln (Eds.), *Handbook of qualitative research* (pp. 1–17). Thousands Oaks, CA: Sage.

Denzin, N. K., & Lincoln, Y. S. (2000). The policies and practices of interpretation. In N. K. Denzin & Y. S. Lincoln (Eds.), *Handbook of qualitative research* (2nd ed., pp. 897–992). Thousand Oaks, CA: Sage.

DeVault, M. (1997). Personal writing in social research. In R. Hertz (Ed.), *Reflexivity and voice* (pp. 216–228). London: Sage.

Ellis, C. (1995). *Final negotiations.* Philadelphia: Temple University Press.

Ellis, C., & Bochner, A. (2000). Autoethnography, personal narrative, reflexivity: Researcher as subject. In N. K. Denzin & Y. S. Lincoln (Eds.), *Handbook of qualitative research* (2nd ed., pp. 733–768). Thousand Oaks, CA: Sage.

Garrett, D., & Hodkinson, P. (1999). Can there be criteria for selecting research criteria? A hermeneutical analysis of an inescapable dilemma. *Qualitative Inquiry, 4,* 515–539.

Glesne, C., & Peshkin, A. (1992). *Becoming qualitative researchers.* White Plains, NY: Longman.

Holt, N. L. (2001). Beyond technical reflection: Demonstrating the modification of teaching behaviors using three levels of reflection. *Avante, 7*(2), 66–76.

Josselson, R. (1993). A narrative introduction. In R. Josselson & A. Leiblich (Eds.), *The narrative study of lives* (pp. ix–xv). London: Sage.

Lincoln, Y. S., & Guba, E. G. (1985). *Naturalistic inquiry.* Newbury Park, CA: Sage.

Marcus, G., & Fischer, M. (Eds., 1986). *Anthropology as cultural critique.* Chicago: University of Chicago Press.

Morse, J. M., Barrett, M., Mayan, M., Olson, K., & Spiers, J. (2002). Verification strategies for establishing reliability and validity in qualitative research. *International Journal of Qualitative Methods, 1(2), Article 2.* Retrieved July 4 2002 from http://www.ualberta.ca/~ijqm/

Mykhalovskiy, E. (1996). Reconsidering table talk: Critical thoughts on the relationship between sociology, autobiography, and self-indulgence. *Qualitative Sociology, 19,* 131–151.

Reed-Danahay, D. (1997). *Auto/Ethnography.* New York: Berg.

Richardson, L. (1995). Writing-stories: Co-authoring 'The sea monster,' a writing story. *Qualitative Inquiry, 1,* 189–203.

Richardson, L. (2000). New writing practices in qualitative research. *Sociology of Sport Journal, 17,* 5–20.

Sparkes, A. C. (1996). The fatal flaw: A narrative of the fragile body-self. *Qualitative Inquiry, 2,* 463–294.

Sparkes, A. C. (2000). Autoethnography and narratives of self: Reflections on criteria in action. *Sociology of Sport Journal, 17,* 21–41.

Sparkes, A. C. (2002a). *Telling tales in sport and physical activity: A qualitative journey.* Champaign, IL: Human Kinetics.

Sparkes, A. C. (2002b). Autoethnography: Self-indulgence or something more? In A. Bochner & C. Ellis (Eds.), *Ethnographically speaking: Autoethnography, literature, and aesthetics.* New York: AltaMira.

Stanley, L. (1993). On auto/biography in sociology. *Sociology, 27,* 41–52.

Tierny, W. G. (1998). Life history's history: Subjects foretold. *Qualitative Inquiry, 4,* 49–70.

Van Manen, J. (1988). *Tales of the field: On writing ethnography.* Thousand Oaks, CA: Sage.

Wolcott, H. F. (1994). *Transforming qualitative data.* Thousand Oaks, CA: Sage.

F. Observational Screens: Photography, CCTV and Internet

62

Looking Emotionally: Photography, Racism and Intimacy in Research

Mónica G. Moreno Figueroa

Introduction: On Seeing and Looking

> I cannot reproduce the Winter Garden Photograph. It exists only for me. For you, it would be nothing but an indifferent picture, one of the thousand manifestations of the 'ordinary'; it cannot in any way constitute the visible object of a science; it cannot establish an objectivity, in the positive sense of the term; at most it would interest your *stadium*: period, clothes, photogeny; but in it, for you, no wound. (Barthes, 2000: 73)

Photographs inhabit the visible; they are at the same time a visual expression and a sign to be decoded; they move within 'chains of meanings', from the cultural practice of producing them to 'the cultural practice of looking and interpretation' (Hall, 1999: 310). As such, important aspects that come into play when reading an image are 'the subjective capacities of the viewer to make images signify' (ibid.). Barthes' reflection above on 'the' photograph, the motif of his *Camera Lucida*, is puzzling in its refusal to reproduce this image and in the implications this has for the audience of his text. The image dates to 1898 and shows his mother, when she was 5 years old, with her brother, then 7. They are 'standing together at the end of a little wooden bridge in a glassed-in conservatory, what was called a Winter Garden in those days' (Barthes, 2000: 67). Barthes found this picture in his mother's apartment just after her death when looking at her photographs: 'gradually moving back in time with her, looking for the truth of the

Source: *History of the Human Sciences*, 21(4) (2008): 68–85.

face I had loved. And I found it' (ibid.) In a loving and poetic manner he describes to us the detail of a 'just image' that could give him a sense of justice and accuracy in the midst of his grief. He doesn't want to run the risk of reducing his mother to the 'ordinary' and chooses not to show the picture. It seems as if he doesn't trust us – the viewers and readers of the text – and our subjective capacities, to make this image signify. It exists only for him, for us it would be 'indifferent'. While Barthes' resistance might be understood as honouring the 'wound' of love, longing and absence, throughout his book there is a sense that we do get to *look* at the Winter Garden photograph even if we don't *see* it, even if it is never shown to us. But still we don't see it: it is not physically printed on the page. Barthes' 'teasing' has triggered and engaged debates about the relation between words and image. He is emphasizing the context of interpretation where this image should, if shown, be understood. Hirsch (1997), for example, similarly to Mitchell's 'imagetext' (1994), describes how the Winter Garden photograph 'has been transformed into a "prose picture"' (1997: 3) since we only have the words Barthes uses and his reaction: the rejection of showing the image.

In this article, I shall discuss the implications of using images to exemplify and disseminate research analysis focusing on the dynamics of showing, seeing and looking in specific contexts. The benefits of showing and seeing specific images go beyond the mere function of illustration, pointing towards their pedagogical, political or 'raising-consciousness' advantages. There is an expectation that the production, collection or display of certain images will allow *understanding* or at least be the starting point of dialogue about the issues at stake (consider, for example, Back, 2004; Knowles, 2006). This position has not, however, gone unchallenged. A relevant example is Apel's (2003) excellent article about the touring exhibition *Without Sanctuary: Lynching Photographs in America* (2000–2). Apel discusses the heated debate generated in the United States, especially among African Americans, about whether to show a collection of photographs of lynchings that took place between 1880 and 1960. Mainly coming from the USA's 'Deep South', some of these images – as with Barthes' Winter Garden picture – were family photographs found in drawers, dusty corners, old chests in attics or family albums next to images of birthdays, weddings and vacations. In these photographs, lynch mobs smile proudly at the camera while cheering one of the most terrifying acts of racial violence. Overall Apel's article argues that the consensus was for showing these images. While some were worried about the commercialization of the suffering of 'black' people or that it would arouse 'black' rage and resentment against 'whites', the arguments for showing the images were more compelling. This is clearly summarized in the recognition that 'the reasons one might be disinclined to show such photographs "are outweighed by the importance of showing how people who otherwise believed in basic democratic principles turned into self-exonerating murderers"' (Apel, 2003: 466, citing Bauerlein). But Apel sensed an urge among the reviewers of the exhibition for witnessing, and explores how the exchange of looks is potent and even vital. For example, she quotes a *New York Times* analyst writing that 'we are a crowd looking at a crowd looking at a lynching. And we are looking at the lynching too. Again and again, a white mob looks back at us' (2003: 460). It is this exchange of gazes, the fluidity of subject (is it the mob, or the victim, or the gaze of the white mob looking back to the spectators?) and the possibilities of (dis)connection that gave this exhibition the potential to shake the subjective capacities of its spectators to make these images signify both politically and intimately.

The stark contrast between Barthes' intimate setting and these lynching photographs could imply that we are dealing with obviously different senses of 'gravity'. While both of these cases deal with images found in familial contexts that are about death and witnessing, the arguments for whether or not to show them seem to be based largely in a questioning of the audience's capacity to make such images signify; can we trust the viewer? This question raises at least two other ones: who are the 'we' deciding? And who are the viewers? We could say that it is understandable not to show a family picture as it 'enlarges' its dimensions and adds weight to the narrative of a man facing death and honouring his intimacy with his mother (apart from elaborating one of the most important and now classic works on photography). It is Barthes' prerogative. And, the 'not showing' makes this image poignant. We could also say it is understandable, then, that the showing of lynching photographs might be seen as necessary to humanize the victims 'through "bearing witness", that is, through the act of looking and seeing' as active engagement that condemns racial hatred (Apel, 2003: 467). But also, we might add, to shock the viewers, to open debate and perhaps to educate them. And here, the showing of gazes at play is what also makes these images poignant.

In this article, however, I propose a different approach to photographic images that extends both of these debates to a common ground. My interest is not in the explicit depiction of racism but in the subtle and everyday context of its reproduction and recognition through family photographs. Simultaneously, like Barthes' 'protection' of his mother's reduction to the 'ordinary', I am concerned with the need to honour the intimacy created in my research between the participants and myself, while being aware of the extent one can or should exercise this safeguard. The two examples above deal, among other things, with issues of representation and meaning in terms of being able to connect the viewer with, or disconnect the viewer from, say, death and justice (and of course issues of racial hatred). What I want to do here is extend the discussion about seeing and looking at photographic images to the context of research processes that investigate issues defined and produced by the same visibility, such as race and beauty. I am concerned with the ambivalence of making use of the visible to 'understand' while acknowledging the need to denounce that same visibility as core to the production of forms of racism.

Based on research that uses personal photographs as central to the exploration of contemporary practices of racism in Mexico, I shall discuss the implications of using such images to exemplify and disseminate the analysis. This research project was concerned with a group of Mexican *mestiza* (racially mixed) women and their experiences of racism from the particular perspective of the visible, that is, the ways they see and are seen; the elements that interact to inform their gaze; the meanings and values of the metaphor of their own image. The participants' social conditions as women and *mestizas* reproduce sexual and racial identities as stereotypes and prevailing postcolonial relationships: they are located within a context of being a visually fixed, stereotyped other where their bodies and skins are 'signifiers of discrimination [that] must be processed as visible' (Bhabha, 1999: 376). In this context the use of photographic images proved to be a poignant method in researching a social phenomenon marked by its visibility. It is my argument that showing and publishing these kinds of research-tool images as part of strategies of research dissemination has not been thoroughly problematized. I suggest that it is even more important to theorize issues of intimacy and trust when dealing with personal photographs than with other types of 'found images' that are 'out-in-the-world', or images produced through research processes by or with participants.

Here, when I speak of images, I refer specifically to the photographic image and to the 'images we imagine': those subjective images created mentally within specific socio-historical contexts and in dialogue with words, texts and narratives. Following Rose (2007) I also distinguish between vision (the capacity of the human eye to 'see') and visuality (the variety of social and cultural elements that construct, determine and structure such capacity of seeing) from the notion of visibility. Visibility can be understood both as 'the capacity of being seen'[1] but also as the political degree to which specific issues, identities, concerns, become publicly prominent, somehow 'exposed' and socially highlighted (Porfido, 2006; Puwar, 2004). The use of images in my research allowed both the participants and myself to grasp that the visibility of specific forms of racism was part of their everyday experience mainly in relation to their ideas of beauty and their family life. I will now turn to an example of this work to explore the 'common ground' where racism, visibility and research meet.

Sharing the Seeing of Beauty

> And then I looked at myself and I said 'I'm not that bad'. I was so ashamed, as if I was some kind of monster but that didn't correspond to what I saw . . . but can you see how powerful the idea is? More powerful than the evidence, because I was looking at myself but the idea was more powerful. (Patricia, 37 years old, Huajuapan, Mexico, July 2002, personal interview)

Here Patricia, one of my interviewees, is telling me what she used to think when looking at a particular set of photographs of herself as a child. She told me how, when she was maybe 7 or 8 years old, she would very often take out these pictures. She liked looking at them to cheer herself up when feeling uncomfortable with her family:

> I mean . . . how I must have felt . . . I felt in one way but saw myself in another way. I used to say: 'no, but I'm all right.' I liked my photos but there's this contradiction, that's why I'm showing them to you now. (Patricia, 37 years old, Huajuapan, Mexico, July 2002, personal interview)

In these two extracts the possibility of accessing the visible is redefined as an individual process that is never 'innocent' or a mere starting point in the process of looking. Patricia is talking about her capacity to make a fair and coherent judgement of her appearance and the struggle between sets of criteria that pull her perception in different directions. On the one hand, she sees her image in a photograph and is able to grant a value to her ability to see herself and to distinguish what she sees: 'I'm not that bad', 'I'm all right'. On the other, she has an idea of herself and she is also able to give a value to it: 'I was so ashamed, as if I was some kind of monster.' This type of tension between evidence and interpretation, between seeing and feeling, is clearly revealed in Patricia's extract, and has been thoroughly analysed by various scholars who reject the simplistic correlation of photography with 'truth' and 'reality' (Lalvani, 1996; Sekula, 1982, 1983; Tagg, 1988).

The struggle Patricia faces exemplifies the power of the configuration of the visible world and the different layers that determine what it is possible for her to see and the meanings she derives from her perception. These layers are the various aspects which, on the one hand, structure the visible world and, on the other, shape and organize not only the mere ability to see, but also the intricate process by which we give meaning to, and create a partial understanding of, what crosses our field of vision. Such aspects refer to the practicalities of the immediate social context and the emotional responses that the visible generates and they establish the criteria of what is beautiful, 'observable' and visible, and what is inadequate, shameful and invisible. The visual appreciation of beauty works simultaneously as a cause, a process of construction and an effect of a discourse of hegemonic beauty over the body, where 'regimes of difference' (Ahmed, 1998), such as femininity and notions of race, stand in internal (dis)agreement with each other, where they *collide*.

Does it make any difference to the reader, in terms of understanding Patricia's or Barthes' experience, to look at these photographs? Could we 'bear witness' without seeing the lynching photographs that Apel discusses? Do I, the researcher, or you, the reader, get a better insight of the experiences of racism, the feelings of shame and 'ugliness', the sheer pain of death and absence, by confirming visually what is already racially and aesthetically informed? Patricia's and the other women's accounts of their emotions and experiences of racism, alongside the sharing of their images, make poignant the need to be reflexive about such images' visual display. I argued earlier that Barthes' resistance might be understood as honouring the 'wound' after his mother's death, but also emphasized the relevance of honouring and humanizing the victims of lynching. Questioning the display or not of my participants' images and the implication of either decision, is also a way of thinking about how to honour the intimacies created with them in the space of a research project concerned with the experience of racism (the how-does-it-feel and what-does-it-do). One possible way forward in assessing the criteria to think about and through images is to explore what photographs do, how photographs inhabit the visible and the layers/chains of meanings they bear.

Analyses of photography have been readily concerned with issues of content and representation, of how images accrue value. I want to extend this understanding to include the problems of how to judge viewers' interpretations of images (Poole, 1997) and how photography and political claims of visibility meet. Gell's (1999) argument about artworks as traps is compelling and useful for the understanding and use of photography proposed here:

> A trap . . . communicates a deadly absence – the absence of the man [*sic*] who devised and set it, and the absence of the animal who will become the victim. Because of these marked absences, the trap, like all traps, functions like a powerful sign. Not designed to communicate or to function as a sign (in fact, designed to be hidden and escape notice) the trap nonetheless signifies far more intensely than most signs intended as such. . . . Since this is a sign that is not, officially, a sign at all, it escapes all censorship. (Gell, 1999: 200)

Arguing for a complex understanding of photography as traps means considering photography's capacity to illustrate, explain and evoke but also, simultaneously, to fascinate, give pleasure and ensnare (we would still like to see the Winter Garden Photograph; the lynching photographs compel the viewer to look; maybe Patricia is 'that bad' – let us see!). Such an understanding invites us to consider how we think about the use and content of photographs as research tools. But it also implies that the role of a photograph as part of a published text or a conference presentation has to be continuously critically assessed. (For Barthes it is straightforward: 'It exists only for me' whereas for Apel, the exchange of gazes needs exposing; Patricia's case is harder to pin down: should we see? What do we expect to find? What would our parameters of looking be? Would we dismiss her if we think she's wrong and just paranoid, or would we sympathize?)

If photographs are traps, it is paramount to be aware of their quality, despite seeming 'safe territory', to ensnare. In this respect, Poole (1997) observes in her analysis of the circulation of images between Europe and

Andean South America, that one important element – easily dismissed in our rush to reveal the complicity between power and representation, or in this case race and representation – is that 'we frequently forget that images are also about the pleasure of looking. Visual images fascinate us. They compel us to look at them, especially when the material they show us is unfamiliar or strange' (1997: 17).

Considering the history and the sometimes questionable social uses of photography can be useful to contextualize this ensnaring process. Since the invention of photography and its 'coincidence' with imperial expansion, scientific racism and changing forms of knowledge linked to logics of comparability and equivalency, images have symbolized a fascinating point of access to the 'other' (consider, for example, the works of Edwards, 1992; Pinney, 1992; Ryan, 1997; Williams, 2003). Poole argues that the wide circulation of photographic images played a key role in the 'formation of the racial culture of European modernity' since photographs were a core 'support to an emerging idea of race as a material, historical and biological fact' (1997: 15). She not only relates the historical coincidence of the emergence of 'race' with photography but also points towards the ambiguities of photographs' interpretation and the pleasure images produce:

> By phrasing identity in the rigorous methods and languages of the biological science of the time, nineteenth century racial theory translated the politics of colonial subjugation into the visual – and aesthetic – calculus of embodied 'natural' differences. In saying this, however, I do not want to imply that race operates only or even primarily through visual technologies and discourses. Nor do I want to say that either vision explains race, or race, vision. Rather, by looking at the historical intersections between visual and racial discourses, . . . vision and race [are approached] as autonomous but related features of a broad epistemic field in which knowledge was organized around principles of typification, comparability, and equivalency. (1997: 14–15)

Photographs, discourses of race and beauty, and experiences of racism can be easily entangled with the pleasure of looking, forgetting their histories of formation and the material and symbolic meanings they have accumulated. Simultaneously these processes of entanglement are further intertwined within the specific context of 'doing' and 'disseminating' research. When we approach images, what is implicit is an understanding that the visible world can be typified, compared and assessed with a series of parameters, such as racial and aesthetic ones, however shifting, that are already in place. So when listening to/reading Patricia's assertion 'I'm not that bad'/'I'm all right' we are compelled to see, to do an assessment of comparability ('she is not as bad as'/'of course she's not a monster!') and equivalency ('she looks like this or that'/'of course she's all right!'). Thus, looking and the perception of the visible world becomes a key element in the racist moment and its political (in)visibility. The act of looking reproduces notions of identification and belonging attached to the visible. Images are visual representations of the 'real' that are continuously struggling to make that 'real' believable: their meanings can be unstable, usually expressing more than they really 'want' to. But significantly, an image can be seen as the tip of the iceberg. It is open to multiple interpretations: 'the visible is defined by the invisible' (Phelan, 1996: 14) and at the same time it can never provide a total or complete meaning of the 'real' that it intends to signify (it is seen from a particular perspective/place/viewpoint/experience). Phelan (1996) argues that these qualities/defects of the image are precisely the space of creativity where identities can be subverted and contested. Such instability opens ruptures and gaps that can facilitate mechanisms of resistance and change. In this way, the instabilities of the act of looking might not only reproduce discrimination and reify racial stereotypes but also open, or keep open, opportunities to challenge practices of exclusion linked to the visible world. With this in mind, I will now turn to discuss the methodological dilemmas I faced when researching the visible through photographs.

Intimacy, Photographs and Research

The coupling of intimacy and sociological research can be instructive. They both rely on movements from private to public realms and vice versa though in no strict (pre)determined way or order. Intimacy has to do at the same time, with the most private and the most publicly regulated acts and practices. Berlant writes that 'the inwardness of the intimate is met by a corresponding publicness' (Berlant, 2000: 1). Often, doing and writing sociological research is the process of intimate listening, of developing the difficult 'art of listening', that is able to extend a story and its small, private minutiae to a 'bigger picture' of public issues (Back, 2007). So being intimate, *doing* intimacy, is what research is all about. Moreover, research practices usually involve facing decisions related to intimacy on a continuous basis.

There are at least three moments when these decisions become pressing. First, is when we have to determine what sorts of limits should be established between researcher and participants: what to do with the tension between seeking closeness and deep understanding while maintaining distance? Second, are the choices we face when trying to 'separate' the data from their narration in order to analyse them. And finally, once the data are analysed and ready to present or publish, we have to decide to what extent our conclusions are further adapted to suit both our theoretical allegiances and the particular audiences we talk to or write for. In this process the corresponding 'publicness' of the intimacy of research (which is by no means linear but rather convoluted), the loyalties that emerge through research relationships and intimate moments can be, and sometimes are, suspended.

Such moments of suspension in research refer to the understanding of what intimacy and intimate relations can be, and what their relationship with a sense of ethics is. The questions emerging from such a decision-making process are about the limits of loyalty (to participants, data and analysis) created through processes of intimacy and the various uses and understandings of ethics that cross the research process. What are the methodological arguments that underpin such decisions? How can we avoid a simplistic ethics argument that does not account for the complexity of the politics and commitment to intimacy involved in the research process?

I faced one such dilemma of intimacy in my research when I had to decide what to do with the personal photographs the participants shared with me. I must confess I did not think carefully about this before doing the interviews. Then, I only thought that the photographs from the participants' personal and family albums would somehow be at the centre of the development of my research. Indeed, that was what happened. The aim of this research project was to explore the *qualities* of the participants' experiences of racism and the importance of discourses of *mestizaje* (racial mixing) in everyday life (Moreno Figueroa, 2006). In 2002, I conducted six focus groups in three Mexican cities – Huajuapan, Oaxaca; Leon, Guanajuato; and Mexico City – with a total of 39 university-educated, working and middle-class women between the ages of 18 and 55. Sixteen participants agreed to additional life story interviews where they brought personal/family photographs to guide not only their life stories, but also discussions of how they saw themselves and their experiences of discrimination and/or racism. During the sessions, the participants displayed the images and used them as a guide for their life stories and to reflect on the *metaphors* of their own visual representation, as exemplified above by Patricia ('I wasn't that bad'/'I was so ashamed, as if I was some kind of monster').

The use of photographs was core for the participants to make what I have called 'contextual translations' of their images, which allowed them to articulate their life stories in relation to experiences of racism. The idea of 'contextual translations' refers to the process undertaken by the research participants of intimating with a particular image, of 'reading'/'interpreting' an image which is then shared in the context of research where 'looking emotionally' allows *understanding*. The notion of 'looking emotionally' refers to an engagement, of researchers and audiences of research, with participants' lived emotional experiences that explicitly confronts the historical and social legacies of the visible. I will come back to this below. Contextual translations and looking emotionally are processes based on the idea that photography has capacities for *readability* because of its association with what Sekula (1982) calls 'hidden or implicit text'. Here it is important to emphasize that the photographic image allows not one but multiple capacities for readability, as well as multiple and complex readings. On the one hand, there is the hidden or implicit text that moves within historical and cultural discourses of power. But on the other, there are multiple potential ways of reading a photographic image due to: the subject of the reading (Who is reading and from where?); the interpellation of the listener of such a subject (Who is listening and from where?); the sharing of the stories and meanings (What is being exchanged? What is it possible to tell at any given moment about an image? What is being translated?). So the process of readability undertaken by the women in this research was analysed mainly in terms of the contextualization – the experience – of the event reading a photographic image: it was in that moment that the participants became subjects of contextual translations and what I have called 'looking emotionally' was possible. The use of images created 'experiential events': the sessions became intimate moments/spaces that operated as experiences in themselves at the same time that the women's experiences were discussed. Those moments can be characterized as particularly intimate not only because they created closeness, familiarity and a sense of loyalty between the women, their images and me as researcher, but also because the narratives that disclosed the *qualities* of the racist experience were produced imparting a sense of understanding which informed my writing.

The use of photographs in this research was more than a 'tool' for memory, a mnemonic device or a way to elicit narratives. It was the key that opened up a possibility for the participants to exercise contextual translations of their images in order to articulate their life stories in relation to their experiences of racism. But using the images also raised the question of what to do with them next: to show or not to show? Regardless of the fact that most participants gave me a copy of their photographs, and most of them also authorized me to show and publish the images where pertinent to the research, I wondered whether displaying such images in the completed text would actually enhance the understanding of what these women shared with me. Would it make any difference? If so, what sort of difference?

In asking these questions, I was concerned with two main things: (1) a respect for the intimate relations created but also a questioning of the ways in which such intimacy is then redistributed and perhaps diluted in the process; and (2) the risk that the images would simply constitute a visual confirmation of the stereotypes that inform the discrimination these women experienced in the first place. I also wondered who would be looking? What contexts would inform their gaze? How could the privileges of intimacy be respected but also honoured? As Barthes says when looking at his mother's photographs after her death, 'if I were ever to show them to friends I could doubt that these photographs would *speak*' (2000: 64). I decided not to show them.

Let us look at a further example that unpacks this decision. When I was talking with Consuelo and exploring her relation to issues of beauty, appearance and shame, she looked at a portrait of herself and told me about the comments she had received from some of her relatives, mainly her uncles, while growing up:

> They used to call me: '*prieta* [dark], *cabezona* [big head], and *dientuda* [goofy]', and I used to say: 'Why do I have such big teeth?' 'Why am I so *morena* [dark]?' (Consuelo, 29 years old, Leon, Mexico, July 2002, personal interview)

Prieta and *morena* refer to a female with dark (or darker than) brown skin and can have a pejorative connotation depending on intonation. Within the context of these expressions and Consuelo's intonation in the interview, these words here have a pejorative intention. Consuelo approaches her understanding of beauty via what it is not. If 'they' call her those names with the intention to offend and bully her, then it seems that for her the contrary is the acceptable and adequate appearance. To be beautiful then means not to have dark brown skin, not to have a big head, and not to have big teeth. But then this belief leaves Consuelo in an upsetting situation where she asks herself: 'Why do I have such big teeth?' 'Why am I so *morena*?' These questions have no satisfactory answer. They seem to be asked as if an answer is both possible and impossible, making them poignant. Is there something to be explained? Is it just a rhetorical complaint resigned to silence? That she feels it is even possible to ask for a reason for her appearance seems to me to be the key question. Would then looking at her image give us – and her – an insight to the sources of such beliefs? Would you be able to 'get into' her experience if I show this image here? Could we then 'really' understand her terms of adequate and acceptable appearance?

Consuelo's questions are simultaneously directed towards the family and the nation. On the one hand, she is confronting the specific sets of relationships established among her family members that allow comments like the above to be shared openly and with no second thoughts (although I would argue that this is by no means a unique characteristic of Consuelo's family but points to a specific sense of cultural 'tolerance' where these utterances are acceptable). On the other, we can also locate her questions in the specific socio-historical context of Mexico and most of Latin America, where racial ideologies of *mestizaje* operate (De La Cadena, 2001; Knight, 1990; Lomnitz, 1992; Martinez-Echazabal, 1998; Wade, 2004). The process of *mestizaje* has offered the subjects of the nation a promise of a 'physical flexibility', of moulding, of a form of racial mixing that, contrary to the US 'one-drop rule', is able to enter in processes of whitening and therefore direct its legacy towards 'perfection'. This promise has stated that the 'race will be improved', and if it isn't, if people's appearances don't approximate to the 'white ideal', such 'failure' has involved the blaming of individuals for their lack of adequate planning. Consuelo's questions ('Why do I have such big teeth?' 'Why am I so *morena*?') seem to arise from the notion that 'things' can happen to the body. It seems that 'things' such as specific sizes, colours, heights and features work in favour of beauty or ugliness. But these questions also point to the idea that such 'things' could have been manipulated, planned, worked out previously giving depth to the lived experience of the racial discourse of *mestizaje* (racial mixing) that prevails in contemporary Mexico. Thus, resonances with the colonial understanding of *mestizaje* as a 'highly flexible' social identity within the caste system appear in this extract from Consuelo. The racist logic of 'improving the race' by 'whitening' the population towards a 'white-Europeanized' ideal is implicit in Consuelo's story. Her questioning of her appearance bears this racist logic and testifies to the non-spoken rules of a social hierarchy of distinctions and of the codes that circulate within the particular Mexican configuration of the visible world. The specific promise of *mestizaje* brings the element of chance as the only explanation when the 'race' is not improved. It is as if *mestizaje* has worked in favour of hazard, giving some Mexican women and men the benefit of beauty and others the benefit of dreaming what it could have been like to 'look' different. Would they perhaps have fair-coloured skin, 'coloured' eyes, 'fine' features; would they then be beautiful?

I question whether showing her image here would help the reader to become intimate with Consuelo's experience. Would the image help us 'truly' understand her questions and her notions of adequate and acceptable appearance? Would seeing the image help us deal with the nexus of intentionalities at play? It might be that showing that particular image – or indeed any image – interacts with the history of the emergence of the modern concept of race, within the ideological framework of positivism and scientific racism, and the (post)colonial function of photography, or with the weight of conventions that a photograph has as a cultural object and practice. These conventions, such as photography's illusion of truth and its expected revelation of the real, might in turn become conflated with prescribed and racialized notions of beauty. My concern in analysing the participants' accounts, however, was mainly to put forward their lived experiences of racism, how they came to confront their racialized and beautiful bodies, and in which ways and in what moments these were related and relevant to being Mexican, a national subject. So in this quest, I argue that the expression and articulation of the complexity of Consuelo's account, for example, comes into being through the use of her photographs during the interview. Nevertheless, the understanding of her experience goes further than social and cultural racialized visual codes. It is my argument that 'looking emotionally', as a specific way of looking, is a way forward to approach not only what Consuelo sees when in front of her portrait but also *how it feels, what it is like*. It is a way both to honour the intimacy of the process that generated this account and to question our own expectation of what a *morena* might look like. The proposal is to think of the visual without images, that is, access the configurations of the visible world by looking through images, beyond their immediate display to reach their sites of production.

Looking Emotionally

The backdrop of this discussion has been the notion of intimacy, both as closeness and depth, as demanding private communion, a sense of sharing and public responsibility (Berlant, 2000). I am not arguing just for 'showing' or 'not showing' photographs, but rather inviting viewers/readers to exercise their agency through reflexive caution. So, while the photos used in my research were core for looking at issues related to race discourses, notions of beauty and experiences of racism, the ambivalent processes that relate to showing and seeing these images have to be considered for their political implications. On the one hand, it is to turn back to the participants of research projects and honour the intimate space created between us by listening, engaging and 'looking emotionally' at their experience rather than taking for granted our (readers, students, researchers) informed gaze as altogether critical and beyond racist and racialized aesthetic prejudices. On the other, it is to critique the visual politics of beauty and racial stereotyping by engaging with the question of the visible, physical and imagined body and actively decide the sorts of engagements we, as audiences, want to have with the visible. It is to think about, for example, what to do with the ensnaring visual differentiations of skin colours, types of hair, facial features, heights and weights and how to deal with the risk of fixing body differences when aiming to expose the racially informed 'ways of seeing'. As Poole rightly asks, 'how it was that "race" [and beauty] came to be seen at all?' (Poole, 1997: 22).

As a way of approaching an image but also as an attitude towards it, 'looking emotionally' refers to the ongoing and open proposal of reflecting on visual research practices and assessing their accompanying processes. It also engages with recent debates about emotions, cultural practices and politics (Ahmed, 2001, 2004; Berlant, 2004, 2006; Sedgwick and Frank, 2003; Svasek and Milton, 2005; Zournazi, 2002). When using images, specifically personal photographs, as tools for research, these are not just devices for memory work or merely a technique to elicit responses. Rather, photographic images offer an intimate context of interaction between the researcher and the participant that demands a deep listening and, moreover, an emotional visual capacity that engages with all the layers of meaning being conveyed: the album, wall or box where the picture is kept; the pose of the sitter(s) and the detail in the background; the pain in remembering/reliving the shame felt; the doubt about what can be seen and how it feels to see it again; the pervasiveness of racism that infiltrates the minutiae of everyday life. In her 'sociality of emotions' model, Ahmed proposes that emotions are relational in that they *move* us and *attach* us to others: 'they involve (re)actions or relations of "towardness" or "awayness" in relation to such objects' (2004: 8). Moreover, emotions are also intentional, being 'about' things and involving directionality towards objects. Emotions are neither located *within* subjects, as a possession of the self, or *without*, as something affecting them. Instead, emotions are located in movement, circulating between objects: between images, participants, researchers and readers; between interpretations, contexts and moments of 'reading'. Emotions are a response (we give to objects) which do not come from 'inside' us as 'reaction', but are produced in-between ourselves and others as 'action'. Looking emotionally is located in such an 'in-between', in the movement and circulation: it is about engaging with the overall experience of selecting, showing, reading, seeing, interpreting and being affected by, a photographic image that is doubly entangled by its raw depiction and its charged 'telling' (by notions of 'race' and by the experience of racism). Looking emotionally means allowing that contact to shape the research interaction and establish a relationality between the researcher and the participant. It happens when the intimate moment of seeing an image is shared, a moment that deals with what is seen but also with how 'that' comes to be seen at all.

I have argued for a theoretical honouring of the intimate relations that produced particular narratives of photographic images bringing together the different elements of the research process: participants/researcher/analysis/data/audiences. Looking emotionally in the context of this discussion is a call to engage with experiences of racism and exclusion and to question our own (as readers, academics, researchers) expectations of racial and beauty prescriptions. I am not saying that photographs reproduce racism but that the stories that emerge from the experience of racism, while deeply and intimately visible, are larger than what the photographic images can ever produce. I have been confronted at conferences and seminars about choosing not to show these images and I am certainly not implying that the viewers' gaze translates in a racialized reading. I have also seen the 'benefits' of showing these photographs when trying to explain what *mestizaje*, mixedness, 'looks' like to European audiences. Observing a family with a widespread variety of skin colour tones visually shocks normalized ideas of what 'black' and 'white' bodies are expected to be, to look like, and where their borders are.

When talking about deeply painful experiences such as racism, what is at stake is not a challenge for readers to visually confirm what the participants said, but to see with them the ways in which 'a photograph is always invisible: it is not it that we see' (Barthes, 2000: 6). Ultimately, what the participants shared was a means for us (researchers/readers) to engage in the production of their identities, understandings and concerns by looking emotionally. Seeing a particular image – or indeed any image – interacts with the weight of conventions that a photograph has in itself, such as its illusion of truth and its expected revelation of the real. Nevertheless, this does not relieve us from the debate about 'bearing witness' by seeing and looking at images such as the lynching photographs. Rather, these different cases open the debate to engage with specific contexts and with reflexive spaces where issues of commitment, responsibility and political intervention are at stake. While I agree with Barthes that the person or thing depicted in an image can be said to 'have been there', in the case of the photographs of my interviewees, the understanding of someone's experience through the process of exercising a contextual translation of their photograph means looking beyond the photographs. The invitation is then to look emotionally, that is, to attend to webs of shared social understandings, individual perceptions of the past translated from the moment of the telling, and emotional reactions to the experience of reliving while narrating.

Note

1. 'visibility', the *Oxford English Dictionary*, 2nd edn, 1989: OED Online, Oxford University Press, 30 April 2008 [http://dictionary.oed.com/cgi/entry/50278233]

Bibliography

Ahmed, S. (1998) 'Animated Borders: Skin, Colour and Tanning', in M. Shildrick and J. Price (eds) *Vital Signs: Feminist Reconfigurations of the Bio/Logical Body*. Edinburgh: Edinburgh University Press, pp. 45–65.

Ahmed, S. (2001) 'Communities that Feel: Intensity, Difference and Attachment', in A. Koivunen and S. Paasonen (eds) *Proceedings of Conference on Affective Encounters: Rethinking Embodiment in Feminist Media Studies*. Turku: University of Turku, School of Art, Literature and Music, Media Studies.

Ahmed, S. (2004) *The Cultural Politics of Emotions*. London and New York: Routledge.

Apel, D. (2003) 'On Looking: Lynching Photographs and Legacies of Lynching After 9/11', *American Quarterly* 55(3): 457–78.

Back, L. (2004) 'Listening with Our Eyes: Portraiture as Urban Encounter', in C. Knowles and P. Sweetman (eds) *Picturing the Social Landscape: Visual Methods and the Sociological Imagination*. London: Routledge.

Back, L. (2007) *The Art of Listening*. Oxford and New York: Berg.

Barthes, R. (2000) *Camera Lucida*. London: Vintage.

Berlant, L. (2000) *Intimacy*. Chicago, IL: University of Chicago Press.

Berlant, L. (2004) *Compassion: The Culture and Politics of an Emotion*. New York and London: Routledge.

Berlant, L. (2006) 'Cruel Optimism', *Differences: A Journal of Feminist Cultural Studies* 17(3): 20–36.

Bhabha, H. (1999) 'The Other Question: The Stereotype and Colonial Discourse', in J. Evans and S. Hall (eds) *Visual Culture: The Reader*. London: Sage Publications and Milton Keynes, Bucks: the Open University, pp. 370–8.

De La Cadena, M. (2001) 'Reconstructing Race: Racism, Culture and Mestizaje in Latin America', *NACLA Report on the Americas* XXXIV(6): 16–23.

Edwards, E., ed. (1992) *Photography and Anthropology: 1860–1920*. London: Royal Anthropological Institute.

Gell, A. (1999) 'Vogel's Net: Traps as Artworks and Artworks as Traps', in E. Hirsch (ed.) *The Art of Anthropology*. London and New Brunswick, NJ: Athlone Press, pp. 187–214.

Hall, S. (1999) 'Introduction: Looking and Subjectivity', in J. Evans and S. Hall (eds) *Visual Culture: The Reader*. London: Sage Publications and Milton Keynes, Bucks: the Open University, pp. 309–14.

Hirsch, M. (1997) *Family Frames: Photography, Narrative and Postmemory*. Cambridge, MA and London: Harvard University Press.

Knight, A. (1990) 'Racism, Revolution, and Indigenismo: Mexico, 1910–1940', in R. Graham (ed.) *The Idea of Race in Latin America, 1870–1940*. Austin: University of Texas Press.

Knowles, C. (2006) 'Seeing Race through the Lens', *Ethnic and Racial Studies* 29(3): 512–29.

Lalvani, S. (1996) *Photography, Vision and the Production of Modern Bodies*. New York: State University of New York Press.

Lomnitz, C. (1992) *Exits from the Labyrinth: Culture and Ideology in the Mexican National Space*. Berkeley and Los Angeles: University of California Press.

Martinez-Echazabal, L. (1998) 'Mestizaje and the Discourse of National/Cultural Identity in Latin America, 1845–1959: Race and National Identity in the Americas', *Latin American Perspectives* 25(3): 21–42.

Mitchell, W. J. T. (1994) *Picture Theory: Essays on Verbal and Visual Representation*. Chicago, IL and London: University of Chicago Press.

Moreno Figueroa, M. G. (2006) 'The Complexities of the Visible: Mexican Women's Experiences of Racism, Mestizaje and National Identity', PhD thesis, Goldsmiths College, University of London.

Phelan, P. (1996) *Unmarked: The Politics of Performance*. London: Routledge.

Pinney, C. (1992) 'The Parallel Histories of Anthropology and Photography', in E. Edwards (ed.) *Anthropology and Photography, 1860–1920*. New Haven, CT and London: Yale University Press and Royal Anthropological Institute, pp. 74–95.

Poole, D. (1997) *Vision, Race, and Modernity: A Visual Economy of the Andean Image World*. Princeton, NJ: Princeton University Press.

Porfido, G. (2006) 'Becoming Visible: Gay Identity and Visual Justice', PhD thesis, London School of Economics and Political Science, University of London.

Puwar, N. (2004) *Space Invaders: Race, Gender and Bodies out of Place*. Oxford and New York: Berg.

Rose, G. (2007) *Visual Methodologies: An Introduction to the Interpretation of Visual Materials*. London, Thousand Oaks, CA and New Delhi: Sage.

Ryan, J. (1997) *Picturing Empire: Photography and the Visualization of the British Empire*. Chicago, IL: University of Chicago Press.

Sedgwick, E. K. and Frank, A. (2003) *Touching Feeling: Affect, Pedagogy, Performativity*. Durham, NC: Duke University Press.

Sekula, A. (1982) 'On the Invention of Photographic Meaning', in V. Burgin (ed.) *Thinking Photography*. London: Macmillan, pp. 84–109.

Sekula, A. (1983) 'Photography between Labour and Capital', in R. Wilkie (ed.) *Mining Photographs and Other Pictures: A Selection from the Negative Archives of Shedden Studio, Glace Bay*. Halifax: Nova Scotia College of Art and Design and University College of Cape Breton Press.

Svasek, M. and Milton, K., eds (2005) *Mixed Emotions: Anthropological Studies of Feeling*. Oxford and New York: Berg.

Tagg, J. (1988) *The Burden of Representation*. London: Macmillan.

Wade, P. (2004) 'Images of Latin America *Mestizaje* and the Politics of Comparison', *Bulletin of Latin American Research* 23(3): 355–66.

Williams, C. J. (2003) *Framing the West. Race, Gender, and the Photographic Frontier in the Pacific Northwest*. New York: Oxford University Press.

Zournazi, M. (2002) *Hope: New Philosophies for Change*. London: Lawrence & Wishart.

63

Using CCTV to Study Visitors in The New Art Gallery, Walsall, UK

Ela Beaumont

Introduction

Observation is a fundamental tool in visitor studies for researching both human activities and the physical setting in which such activities take place (Angrosino and de Perez, 1997). The use of ethnographic observational techniques represents one of the longest established traditions in social science research, and whilst early observational field work relied on notebook and pencil, more recently, cameras (still and movie) have augmented the ethnographer's tool box. I will argue that the routine use of CCTV surveillance in art galleries in the UK presents an opportunity for researchers to harness its potential as a powerful observational tool in visitor studies, and that recent developments in video technology have created new opportunities for observational research. The analysis of video data is in its infancy in the field of art gallery visitor studies, and this paper aims to contribute to the theoretical, ethical and practical debate by discussing a recent observational visitor study using in-house CCTV cameras in the New Art Gallery, Walsall.

There are differences between using CCTV filming as an observational method in visitor studies and the wider use of CCTV for the maintenance of public order. All surveillance tries to make visible the behaviours of people of interest to the agency in question (Lyon 2002). This study was interested in family group visitors, and CCTV was used within the strict boundaries of

Source: *Surveillance & Society,* 3(2/3) (2005): 251–269.

the terms of reference of the research, only coding specific information on the behaviours of the sample, ie. family group visitors to the New Art Gallery, Walsall. Other behaviours by other groups in the gallery, which were captured on CCTV during the research filming, for example groups of young people playing in the lifts or moving exhibits were not coded because they fell outside the scope of the investigation. The gallery is governed by Walsall Metropolitan Borough Council and formal permission was granted by Walsall MBC for CCTV footage to be used for research purposes, on the understanding that using the CCTV as an investigative tool for research was secondary to surveillance, which would take priority if necessary. Security staff monitored the CCTV concurrent with the research study, and there are occasional gaps in the video footage where the cameras swung round to focus on behaviour outside of the research study, that security staff considered inappropriate.

Background

Britain is one of the most watched nations in the world, with over 4 million surveillance cameras in use (Norris *et al.*, 2004). In 2004, The Council for Museums, Archives and Libraries in the UK, published guidelines for the use of CCTV in museums, suggesting that the main advantages that accrue from CCTV are: deterrence, an aid to invigilation, recording for post-incident investigation, entry control and site management out of hours (Resource, 2004). In this context, surveillance is understood as 'the purposeful watching of specific individuals who have aroused suspicion' (Lyon, 2001). New Art Gallery, Walsall, used CCTV as a security measure in the way described above, and we proposed to carry out a visitor study using the existing CCTV surveillance system within the New Art Gallery as an observational research tool. Utilising the New Art Gallery in-house CCTV system, this research made significant advances on previous observational visitor studies that have gathered 'covert observational data' (Office of Information Commissioners, 2002).

Definitions of surveillance include the '*close observation, especially of a suspected person*' (Concise Oxford Dictionary), yet much contemporary surveillance is 'not applied especially to a suspected person' (Marx, 2002). This visitor study embraced existing surveillance technology rather than setting up additional systems with extra cameras. Earlier video observational research has attempted to reduce the effect of video cameras on visitor's conduct by hiding them (Vom Lehn *et al.*, 2002), and in attempting to gather audio tapes of natural speech, previous studies have had problems obtaining good quality acoustic data (Allen, 2002). Solutions such as getting visitors to wear microphones and carry tape recorders bring their own difficulties in ensuring naturalistic responses. This study of visitors to the New Art Gallery,

Walsall, obtained good quality visual data from CCTV and used face-to-face interviews for verbal information.

There are a number of constraints upon using CCTV for research purposes, primarily relating to how data are collected and handled, including the legal requirements of 'informed consent'. However, we took the view that as the surveillance equipment was already constantly monitoring visitors to New Art Gallery, Walsall, whilst they were in the art gallery, this additional interrogation of their images for research purposes was a legitimate use of the CCTV footage. We modified the definition of surveillance offered above, to '*the purposeful watching of specific individuals*' For this study, the specific individuals were visitors coming to the New Art Gallery, Walsall.

Antecedents

A pioneering study of observational methods in the early days of visitor studies research highlighted the advantages of filming visitors. In 1942 at the University of Chicago, Neilson, an education research associate in the Museum of Science and Industry, identified the need for comprehensive visitor studies, suggesting that visitors themselves remain the least known, most unpredictable and most difficult aspect of museum work to study. He suggested that the method of determining how museum visitors react should be observation in the context of specific museum situations. However, he found there were four main limitations to traditional observation and tracking studies involving following visitors with notebook and stopwatch:

- High financial implications of training and maintaining staff
- Large number of hours necessary and the difficulty of the work
- Variable or unreliable data
- Inefficient yield of data related to hours of work

(Neilson, 1942)

Neilson proposed an alternative technique for observing visitors; that of time lapse photography. He mounted a movie camera equipped with single-exposure release high in one corner of the gallery, concealed as much as possible from visitor's view. The resulting film, a series of snapshots of the entire room, provided a record of all activities in the gallery, showing exactly:

- How many visitors entered the gallery
- How many there were at any given moment
- Where visitors went and in what sequence
- How they looked at the exhibits
- How long they stayed at each exhibit
- How long they stayed in the gallery as a whole

Neilson found that this data on film could be analysed and transcribed a number of times whilst looking for different aspects of visitor behaviour.

Current Video Observational Research

Observing visitors as a method in visitor studies research has its origins in anthropological and ethnographic investigations and observational and tracking studies have the great advantage that they can be administered unobtrusively. Visitors may be entirely unaware that they are being observed, though ethics demand that there be a sign somewhere prominent drawing visitors' attention to the fact that there is a study taking place.

There is a long-standing interest in using video recording for the observational study of visitor behaviour (Vom Lehn *et al.*, 2002). Neilson's early work in the 1940's has already been mentioned. Following Shettel's work in the 1960's and Hensel (1987) and Leichter *et al.* (1989) in the 1980's, researchers have used video to record visitors' navigation paths, how they look at exhibits and their patterns of behaviour (Falk, 1991; McManus, 1994; Phillips, 1995). The use of video technology principally replaces an in-situ observer with a camera and the great advantage of this method is the facility for the researcher to repeatedly view events away from the research site. As Vom Lehn (2002) has pointed out, the video camera augments the work of the observer, and, by recording at different times on different days, a substantial amount of data can be gathered and subsequently analysed. The technology makes it possible to record and analyse with a degree of particularity not available to the field researcher. In research of this type, the video camera does not supplant the observer, rather, the field observation is enhanced by the video data.

Recently video observation methods have been used by Heath (2002) in the Victoria and Albert Museum and the Science Museum in London, Gutwill-Wise (2002) at the Exploratorium in San Francisco, and Vom Lehn *et al.* (2002) at @Bristol in the UK and the Musee des Beaux Arts in Rouen, France. These studies have demonstrated how powerful film data can be, but have also shown the difficulties in operationalising studies that use these techniques. However, for some types of study, the benefits outweigh the difficulties[1] and the value of using video footage to investigate visitors is that it provides:

- a non-intrusive method of studying visitor behaviour
- an opportunity to examine visitor behaviour in different exhibitions in the same gallery
- stable evidence that can be viewed repeatedly
- data that can be replicated and re-examined

Video recording also captures, in real-time, visitor behaviour as it arises in a naturalistic setting. Details and patterns appear through intense and repeated scrutiny of the tapes which, as raw material, can be analysed by the research community in order to judge the reliability of findings based on the video data. But despite its potential as a research tool, collecting video data remains relatively unexplored within visitor studies for three reasons:

- The technical problems associated with obtaining good quality recordings in art gallery and museum settings (Phillips, 1995; Vom Lehn *et al.*, 2001)
- The analysis of video data is extremely time consuming (McManus, 2001; Allen, 2002)
- Visitors react to the camera and their behaviour becomes unnatural (Smith and Wolf, 1996)

In relation to the latter point, Borun and her colleagues (1997) audio and videotaped parents and children at the Franklin Institute in the United States as part of a family learning project, but if people realized they were being observed the observation was ended by the researcher and deleted from the sample because the observation was considered to be corrupted (Borun *et al.*, 1997). There are also potential analytical weaknesses associated with observational method, which are:

- Lack of rigorous design
- Large amounts of poorly collected and unanalysed data
- Superficial and hasty conclusions (Miles and Huberman, 1994)

Using CCTV for Research at the New Art Gallery, Walsall

All the studies referred to above involved setting up camera equipment specifically for the research project. This study in New Art Gallery, Walsall engaged with CCTV creatively as an observational tool to study visitors within art galleries by utilising the existing surveillance cameras. It opens up the prospect of a wider ideological debate about the use of CCTV in art galleries, and contributes to work in progress on a code of ethics for video observation in visitor studies (Gutwill-Wise, 2002).

The 'real time availability' of the observational data, which Marx (2002) suggests is a feature of the 'New Surveillance', was an advantage in this visitor study, and offered the opportunity to integrate data collection for research into the routine surveillance activity of the gallery (within the Office of the Information Commissioners CCTV Code of Practice guidelines). The pre-existing presence of cameras at the research site meant that gathering observational data through New Art Gallery, Walsall's internal CCTV security system was a distinct possibility. In consultation with the curatorial staff, front of house staff and security staff it was agreed that a CCTV camera in a gallery on the third floor would provide the most comprehensive data with least disturbance to the normal functioning of the CCTV cameras as front line security. There were 2 major advantages to using in-house CCTV in this visitor study:

- technical problems associated with obtaining good quality recordings had already been addressed and solved by the technical staff at the New Art Gallery
- CCTV cameras are discreet compared to bulky and often obtrusive video equipment

The case study at New Art Gallery, Walsall was tightly focused with the intention of gathering information about a particular group in a particular context within a particular time frame and analysis of the video observations used methods that provide an alternative to the extensive written notes that are the hallmark of traditional ethnographic inquiry (Nastasi, in Schensul *et al.*, 1999).

Ethical Considerations

The decision to use CCTV for the current research project was taken with the intention of minimising the intrusiveness of the filming without, however, compromising visitor's right to give their informed consent. The study took place in the '*real world of ethical tensions*' (Lyon, 2002), where, for example, the disclosure of images recorded by CCTV is restricted to uses compatible with the original reason for obtaining the images (France, 2000). In relation to videotaping visitors during observational and tracking studies, recent guidelines suggest that informed consent explaining the nature of the study is required. However, research studies have shown that informed consent can be undocumented or implied as long as signs are posted in the recorded areas (Heath, 2002; Gutwill-Wise, 2002).

The advantage for the current study in using CCTV film was, as suggested above, the prevalence of surveillance in public spaces in the UK (Figure 1). CCTV surveillance is generally perceived as a benign activity in the UK (Norris, 2004), and we took the view that by posting signs telling of the research, and in not disrupting the gallery with extra cameras or microphones, the CCTV footage would gather naturally occurring visitor activities in a highly structured way, yielding increasingly detailed, useful information.

Usually video observation relies on equipment that is very visible and therefore intrusive. Visitors are aware of being filmed when video equipment is set up at the research site, and the protocols for filmed observation

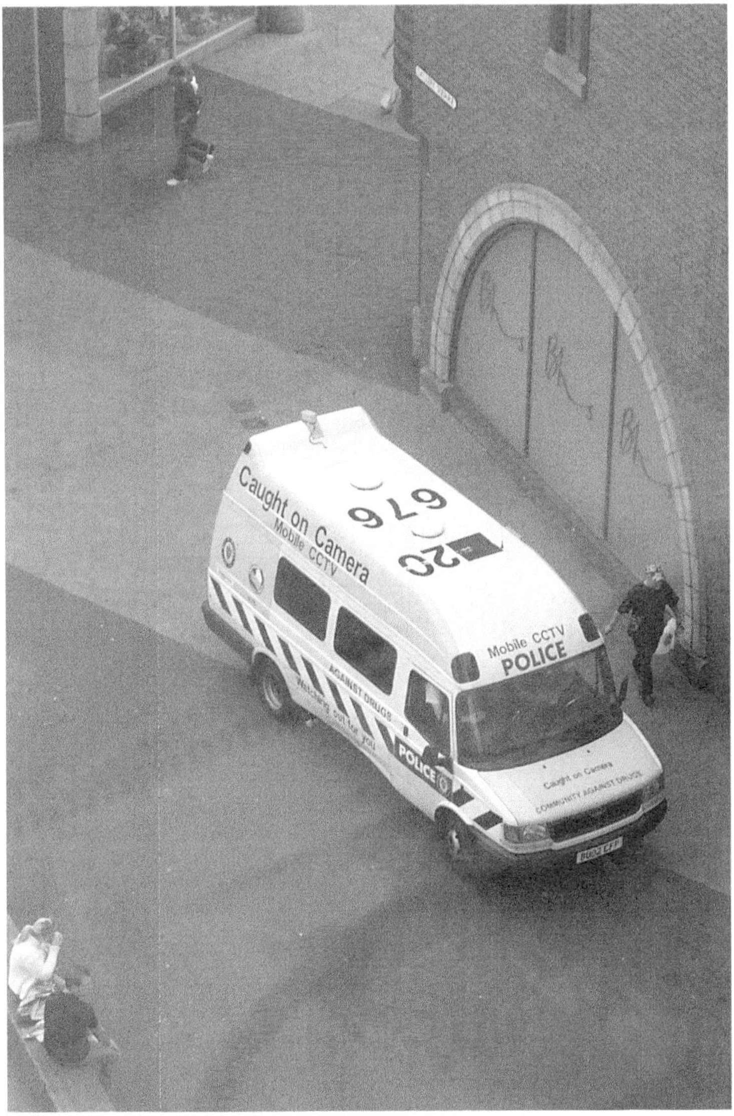

Figure 1: The view from the third floor of the New Art Gallery, Walsall. The police surveillance van was regularly parked directly outside the gallery entrance.

are constantly evolving and being refined through action research in light of emerging ethical considerations (Gutwill, 2002). The figures below show the posters devised for research into visitors at the Exploratorium in San Francisco. Figure 2 shows the original poster, figure 3 shows the poster modified because researchers found visitors had not noticed the first one.

Both in the United States and the UK, researchers need to consider the size and placing of signs which alert the public that they are entering a zone

Figure 2: First poster at the Exploratorium in San Francisco alerting visitors to the study. The posters are in Spanish as well as English because a large number of visitors have Spanish as their first language.

covered by surveillance equipment. Even large signs, with images of cameras, as shown above, are not always noticed by the public, and in any case, the First Data Protection Principle (UK) suggests that in certain circumstances, a written A4 sign at eye level may be enough (France 2000).

Ethical principles in 'action research' have been summarised in the context of visitors and visitor studies by Bicknell and Gammon (1995) through

Figure 3: Second poster devised after it was discovered the first poster was not noticed by visitors

their research in the Science Museum in London. This study in the New Art Gallery, Walsall, took account of the following principles:

- Observing protocol
- Including participants
- Negotiating with those affected
- Reporting progress

- Obtaining explicit authorisation before observing
- Obtaining explicit authorisation before examining files, correspondence or other documentation
- Negotiating descriptions of people's work
- Negotiating accounts from other points of view
- Obtaining explicit authorisation before using quotations
- Negotiating reports for various levels of release
- Accepting responsibility for maintaining confidentiality
- Retaining the right to report work
- Making principle and procedures known and binding

Extensive discussions with key personnel concerning exactly how visitors are informed and how to gain their cooperation will always be needed when filming is part of a research strategy. Particular protocols need to be developed in relation to the use of CCTV footage as the film is being used for a purpose other than that for which it was originally intended (France, 2000). The use of images derived from surveillance technology, albeit not applied to '*close observation of a suspected person*' (Concise Oxford Dictionary, 1998) needs careful treatment. The Data Protection Act of 1998, and subsequent interpretations of it, requires those who operate CCTV systems and who record images from which individuals can be identified through 'personal data' to register with the Information Commissioner (Taylor, 2002). 'Personal data' means:

> Any information relating to an identified or identifiable natural person; an identifiable person is one who can be identified, directly or indirectly, in particular by reference to an identification number or to one or more factors specific to his physical, physiological, mental, economic, cultural or social identity (European Directive 95/46/EC).

Evidence from other recent visitor studies suggests that researchers regard videotaping and the use of CCTV images for ethnographic methods for research purposes as ethically neutral (Angrosino and Perez, 1997). But three criteria should be applied to test for the probity of research which uses such methods:

- The means will not cause more harm than necessary to achieve the value
- No less harmful way exists at present to protect the value
- The means used to achieve the value (result) will not undermine it (Herrera, 1999)

In the context of studying human actions, the above criteria gives such research its moral meaning, 'in that sense proportionate refers to the specific value at stake and the limitations, the harm, or the inconvenience which will inevitably come about in trying to achieve that value' (Craig *et al.*, 2000).

Figure 4: Notice outside the gallery showing that CCTV cameras are in use

Practical Considerations

In consultation with security staff at the New Art Gallery, Walsall, we felt that the notice at the front of the gallery, informing visitors that CCTV cameras were in operation was not sufficient to gain implied consent (Figure 4), particularly as the notice was almost completely obscured by the front door as it opened to admit visitors (Figure 5, overleaf).

Figure 5: Door obscuring the CCTV camera notice

In discussion with senior gallery staff, the agreed procedure was to place a notice at the entrance to the gallery under investigation. The notice informed visitors about the research and invited those with reservations about the research to speak to an attendant. In practice, researchers have found that visitors do not object to the use of video in the conduct of research (Gutwill, 2002; Allen, 2002; Vom Lehn, 2002; Beaumont, 2004).

CCTV filming has not been used extensively as an observational method in visitor studies to date and had not been used before in the New Art

City Final

Express & Star

2 31 31 31 | Thursday, February 13, 2003 | Classifieds 01902 3

Gallery removes artwork from new Veil exhibition

Images portraying a veiled Statue of Liberty clutching the Koran and the Houses of Parliament converted into a mosque have been banned from an exhibition at Walsall's New Art Gallery.

Walsall Council took the step ahead of the opening to the public tomorrow of controversial exhibition Veil, which looks at history and current use of the concealing headdress particularly favoured by Muslim women.

Councillor Gary Perry, cabinet member for leisure and community services, said he had taken the decision to remove two works because of the current state of alert and tensions in the Middle East.

Blank

But the artists concerned - three Russian Jewish artists - have accused the council of censorship as visitors to the gallery will be faced with blank spaces where once the images would have been displayed.

The exhibition, which due also to tour Liverpool and Oxford galleries, is the first of its kind in Britain to explore the place of the veil worn by devout Muslim women.

Councillor Perry said that on viewing the exhibition he had, along with senior officers at the council, decided that two works in particular were inappropriate given current security alerts and tensions.

Report by Paul Hinton

One features the Statue of Liberty dressed in a full-length burkha and holding a Koran and the other is an image of the Houses of Parliament converted into a mosque with domes and minarets.

Both works, by the AES art group, were created in 1996 and confront fears of fundamentalist Islam in a satirical way.

But Councillor Perry said: "In view of the current climate nationally and the state of alert and the tension in the Middle East, we felt there could have been someone who would have gone into the gallery and made a statement which could have caused tension in the community."

Walsall Council spokesman Tony Moran said this afternoon: "Walsall is proud of the fact that its rich diversity of cultures and communities have always existed in harmony, understanding and peace.

"But during this period of heightened sensitivity and following the horrific events of September 11, it was felt that the two images could be viewed as reinforcing controversy, fear and prejudice."

The artwork with Parliament as a mosque

A veiled Statue of Liberty with the Koran

Figure 6: Two photographs were removed from the Veil show by Walsall MBC

Gallery, Walsall. The gallery is governed by Walsall Metropolitan Borough Council and a risk assessment was carried out with the Risk and Insurance Officer at Walsall MBC. Formal permission was granted by him on behalf of Walsall MBC for CCTV footage to be used in this way. Ironically perhaps, whilst granting permission for video research to take place in the gallery, Walsall MBC was also making the decision to withdraw photographs by the Moscow-based EAS group from the 'Veil' exhibition, fearing they were too controversial in the current climate, post September 11th (fig 6).

Significantly, some contemporary artists have responded to the implications of widespread CCTV surveillance, and their work shown in art galleries in the form of photographs, video and multimedia (Naldi and Kirkup, 1993; Stidworthy, 2004). Art galleries are 'laboratories for the study of value in art' (Ueki, 1998) and this study of the New Art Gallery, Walsall, contributes to the ongoing debate about the purpose of art galleries and the role of the art gallery in contemporary society, in particular the context in which contemporary art is viewed.

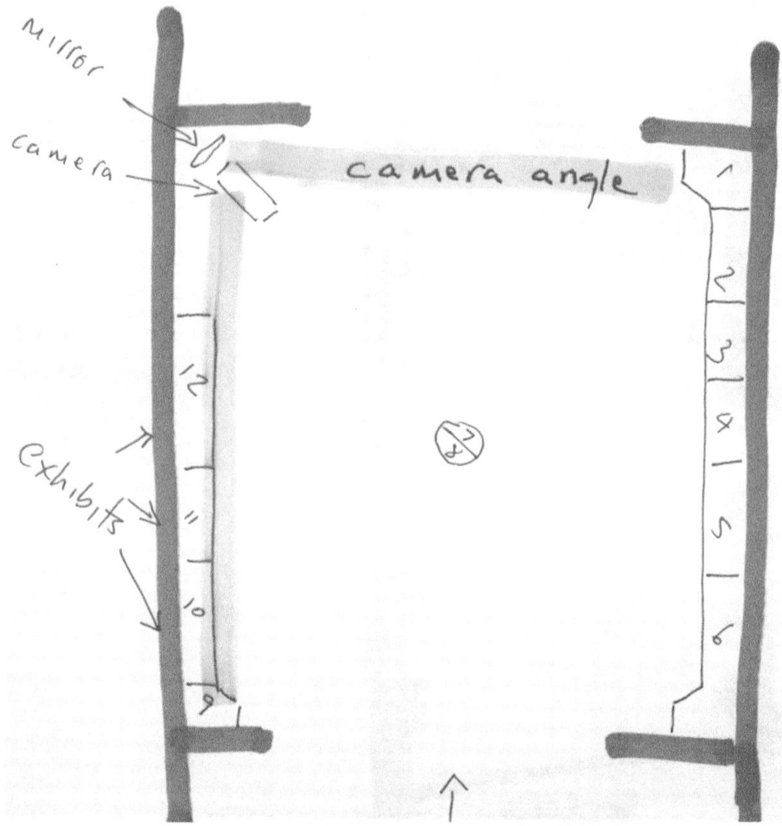

Figure 7: Diagram showing the camera angle in Neilson's study

Camera Position

Neilson, in his early study, was not able to obtain a wide angle camera, so he mounted the camera backwards, pointing towards a large convex mirror that reflected the image of the gallery (Figure 7). For the present study the CCTV camera coverage poses no such problem. The wide angle of the camera shows the whole gallery with the exception of a small area directly under the camera (Figure 8). Three hour VHS videotapes provide data that can be analysed time and again without being transcribed into tabular form.

Sample Size

Another potential problem with observational methods relates to stability and instability of observed behaviour. It is commonly assumed that behavioural observation provides a typical sample of behaviour, but the sample of

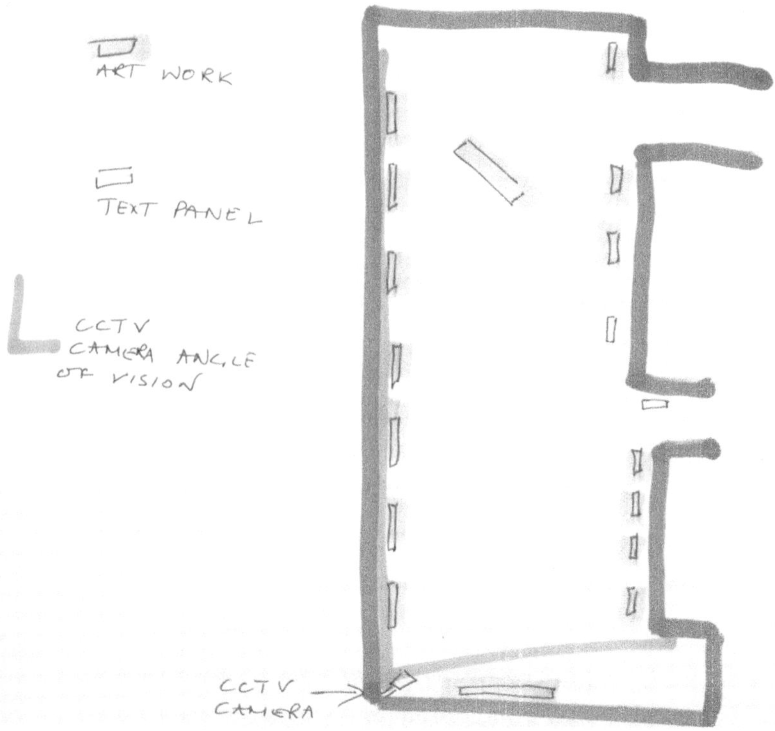

Figure 8: Diagram showing the CCTV camera angle at the cases study site, New Art Gallery, Walsall

behaviour may not be representative. In general, the larger the sample of behaviour, the more representative and accurate the sample of behaviour becomes (Holden and Edwards, 1989). Systematic studies of observational data on parent-child interactions, for example, have found that the more observations are combined, the more stability of behaviour increased (Touliatos *et al.*, 1990). In this study, a sample number was selected that would be sufficient to gather information about general trends, but small enough to analyse in great detail.

Coding Observational Data

In the New Art Gallery, Walsall study, observational video data used in conjunction with questionnaire surveys and in-depth interviews increased both reliability and validity. The overall methodological approach used in this study provided rich material, generating both quantitative and qualitative data. In applied fields such as visitor studies, there is often compatibility between qualitative and quantitative research in practice (Robson, 2002), and advantages to mixed method research design. There are many questions related to visitor behaviour and interaction where observation would be the preferred method (Lytton, 1971), but observational methods continue to be rare for two reasons:

- The approach requires more knowledge than other methods
- A considerable investment in time and training is needed in observation and coding data

Observing visitors usually involves an initial study to establish visitor behaviours that relate to the area of investigation, and the list of interesting visitor behaviours refined and often coded. Direct observation also includes the need to develop a set of observation methods that can accurately record the behaviours under investigation. Behavioural categories need to be defined carefully so that the method is reliable

Investigators usually conduct a series of preliminary observations during which all behaviours are recorded. These can be compared to lists of behaviours published in previous studies and a protocol and observation schedule developed accordingly.

The observation schedule consists of data sheets based on the floor plan of the study site. The purpose of this is to collect information in a systematic manner (Serrell, 1998). Observational data sheets are typically hand drawn and not to scale, but contain information relevant to the study. At New Art Gallery, Walsall, each observed visitor was allocated a unique number based on the video time code in the top left hand corner of the video image. For example, Figure 9 (overleaf) shows the observational data sheet relating to a grandmother and granddaughter visitor group; their unique number is 30.5.03, 13.00.51 which indicates the exact date and time they entered the gallery. Figure 10 (p.266) shows how the observations of the same group are recorded on video: further reviewing of the data involves simply locating the number on the videotape. This system prevents accidental transposing of information about visitors and allows for quality checks to be made, but crucially means that demographic and attitudinal data collected elsewhere can be correlated to the video data on visitor behaviour.

This method of coding data was particularly important because interviews took place concurrently with video observation. The time code enabled video observation to be matched with interview data on individual visitors without compromising their anonymity.

Conclusion

The contribution of this study of visitors to the New Art Gallery, Walsall, has been to focus on the behaviour of visitors to a newly built regional art gallery using CCTV video observation. Despite the growing body of research

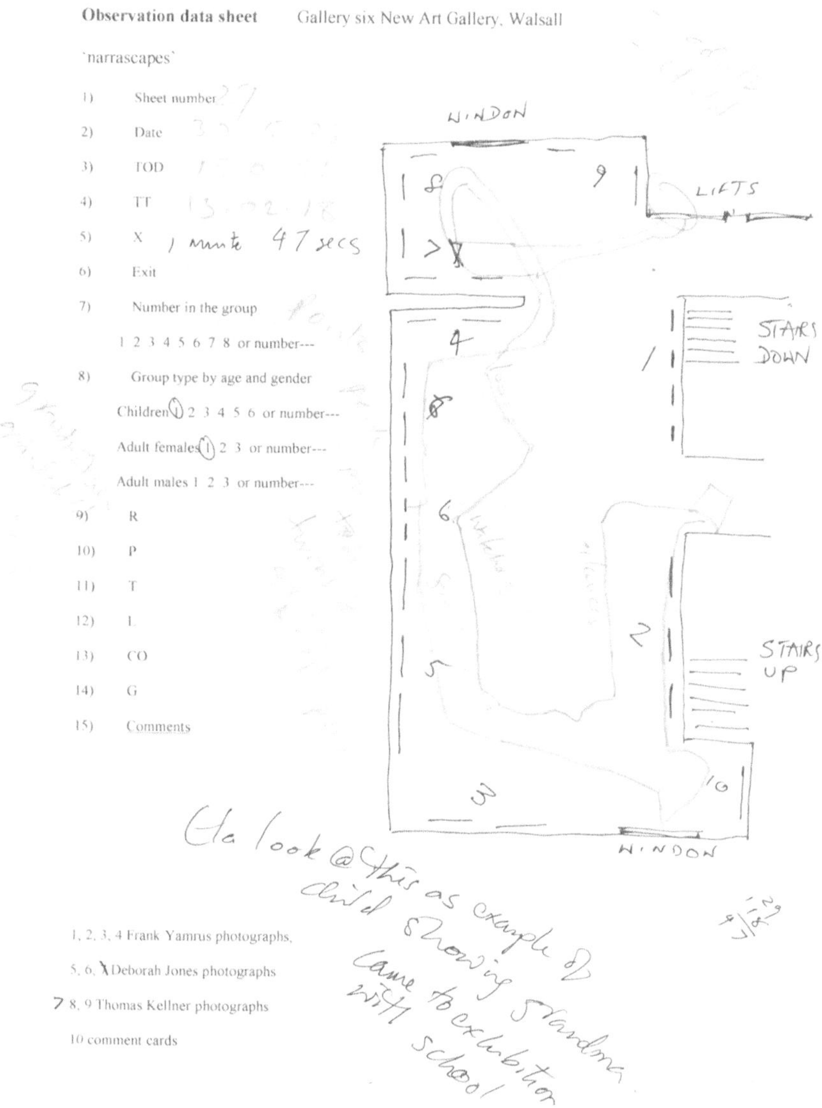

Observation data sheet Gallery six New Art Gallery, Walsall

'narrascapes'

1) Sheet number
2) Date
3) TOD
4) TT
5) X 1 minute 47 secs
6) Exit
7) Number in the group
   1 2 3 4 5 6 7 8 or number---
8) Group type by age and gender
   Children 1 2 3 4 5 6 or number---
   Adult females 1 2 3 or number---
   Adult males 1 2 3 or number---
9) R
10) P
11) T
12) L
13) CO
14) G
15) Comments

1, 2, 3, 4 Frank Yamrus photographs,
5, 6, 7 Deborah Jones photographs
7 8, 9 Thomas Kellner photographs
10 comment cards

Figure 9: Observation sheet No. 29 from the 'narrascape' show

into visitors, we still know relatively little about the action and interaction arising between visitors as they encounter exhibits (Vom Lehn, 2002), At the New Art Gallery, Walsall, the technological sophistication of pre-exisiting CCTV coverage was identified as having great potential as an observational tool for visitor research. Combined with other methods within a case study strategy, the observational CCTV information yielded valuable data about visitors to a significant new regional gallery. New galleries devoted to modern

Figure 10: CCTV video image of visitors, showing their unique reference number (which relates to the observation sheet)

and contemporary art have been established relatively easily, they are glamorous, they reflect a general shift of interest towards modern and contemporary art (Shubert, 2000) and they manifestly produce results by attracting large audiences in the short term. The period between 1994 and 2004 represents a remarkable cultural moment in the UK, and this detailed observational case study of visitors to the New Art Gallery, Walsall provides insights into visitor behaviour at an important regional millennium project.

Reeves (2002) recently suggested, in her wide-ranging review of current research initiatives for the Art Council in England, that there is scope for developing a national framework for visitor research which builds on the methodological strengths and practical experience of the best UK studies. Her report concluded that there is a need for more research that utilises both quantitative data and qualitative description, and a need for more in-depth case study research. This study contributes to the developing field of visitor studies in the UK and internationally by offering an original but replicable observational method using in-house CCTV.

References

Allen, S. (2002) Looking for Learning in Visitor Talk: A Methodological Exploration In Leinhardt, Crowley and Knutson (eds.) (2002) *Learning Conversations in Museums*. New Jersey: Lawrence Erlbaum.

Angrosino, M.V. and de Perez, K.A. (1997) Rethinking Observation: From Method to Content In Denzin and Lincoln (eds.) *Handbook of Qualitative Research* London: Sage.

Beaumont, E.T. (2004) An Empirical Study of Family Group Visitors to a Millennium Art Gallery. Unpublished PhD Thesis, University of Salford, UK.

Bicknell, S., and Gammon, B. (1995) Ethics and Visitor Studies- Or Not? *Visitor Studies: Theory, Research, and Practice*, 8(1):

Borun, M., M. Chambers *et al.* (1997) Enhancing family Learning through Exhibits. *Curator* 40(4): 279–295.

Craig, G., A. Corden and P. Thornton (2000) Safety in Social Research. *Social Research Update* 29.

Falk, J. (1991) Analysis of the Behaviour of Family Visitors in Natural History Museums. *Curator* 34(1): 44–50.

France, E. (2000) Office of the Information Commissioners CCTV Code of Practice Guidelines. London: OIC.

Gutwill-Wise, J. (2002) Audio/videotaping Museum Visitors: Methods for Obtaining Informed Consent. Presentation at the *Visitor Studies Conference* 2002, Cody, Wyoming.

Heath, C., P. Luff, *et al.* (2002) Crafting Participation: designing ecologies, configuring experience." *Visual Communication* 1(1): 9–34.

Hensel, K.A. (1987) Families in a museum: Interactions and conversations at displays. Unpublished PhD Thesis, Columbia University Teachers College, New York.

Herrera, C.D. (1999) Two Arguments for 'covert methods' in social research. *British Journal of Sociology* 50:

Holden, G.W. and Edwards, L. (1989) Parental attitudes towards childrearing: Instruments, issues and implications *Psychological Bulletin* 106: 29–58.

Leichter, H.J., Hensel, K. And Larsen, E. (1989) Families and Museums: Issues and Perspectives. In *Marriage and Family Review-Museum Visits and Activities for Family Life Enrichment* 1989, Hawarth Press, New York

Lyon, D. (2001) *Surveillance Society: Monitoring Everyday Life*. Buckingham: Open University Press.

Lytton, H. (1971) Observational Studies of Parent-Child Interaction: A Methodological Review *Child Development* 42(3): 651–684

Marx, G.T. (2002) What's New About the 'New Surveillance'? Classifying for Change and Continuity *Surveillance and Society* 1(1): 9–29. http://www.surveillance-and-society.org/articles1/whatsnew.pdf

McManus, P. (1994) Families in museums. In R. Miles and L. Zavala (eds.) (1994) *Towards The Museum of the Future* London: Routledge.

McManus, P. (2001) Getting to Know Your Visitors. Association for Heritage Interpretation. www.heritage-interpretation.org.uk

Miles, M.B. and A.M. Huberman (1994) *Qualitative Data Analysis*. Thousand Oaks, CA: Sage.

Naldi, P. and W. Kirkup (1993) Search Surveillance Still. Installation in the *Second Tyne International*, Newcastle.

Neilson, L.C. (1942) A Technique for Studying the Behaviour of Museum Visitors *Journal of Educational Psychology* 37: 103–110.

Norris, C., M. McCahill and D. Wood (2004) 'Editorial. The Growth of CCTV: a global perspective on the international diffusion of video surveillance in publicly accessible space,' *Surveillance & Society* 2(2/3): 110–135. http://www.surveillance-and-society.org/articles2(2)/editorial.pdf

Phillips, D. (1995) Evaluating Time-Lapse Video Evaluation *Museum Management and Curatorship*, 9: 19–29.

Reeves, M. (2002) Measuring the economic and social impact of the arts: a review, Arts Council of England, London.

Resource (2004) Guide to the Use of CCTV in Museums, Archives and Libraries, London http: //www. resource .gov.uk

Rounds, J. (2001) Is there a Core Literature in Museology? *Curator* 44(2): 194–206.

Schensul, S.L., J.J. Schensul and M.D. LeCompte (1999) *Essential Ethnographic Methods: Observations, Interviews and Questionnaires*. Walnut Creek CA: AltiMira Press.

Schubert, K. (2000) *The Curator's Egg-The Evolution of the Museum Concept from the French Revolution to the Present Day*. New York: One-Off Press.

Serrell, B. (1998) Paying Attention: Visitors and Museum Exhibitions. Washington: AAM.

Shettel, H. (1968) *Strategies for determining exhibit effectiveness* (AIR E95-4-68-FR) Pittsburgh, PA: American Institute for Research.

Smith, J. K. and L. Wolf (1996) Museum visitor preferences and intentions in constructing aesthetic experience. *Poetics* 24: 219–238

Stidworthy, I (2004) Artist's Talk, ICA London, April.

Taylor, N. (2002) State Surveillance and the Right to Privacy *Surveillance and Society* 1(1): 66–85. http://www.surveillance-and-society.org/articles1/statesurv.pdf

Touliatos, J., B.F. Perlmutter and M. Straus (1990) *Handbook of Family Measurement Techniques*. London: Sage.

Ueki, K. (1998) Shaping the Experience of Art: Art Museums and Curatorial Intentions. Unpublished MPhil Thesis, University of Manchester.

Vom Lehn, D., C. Heath and J. Hindmarsh (2002) Video-based Field studies in Museums and Galleries *Visitor Studies Today* Fall.

Vom Lehn., D., C. Heath and H. Knoblauch (2001) Configuring Exhibits: The Interactional Production of Experience in Museums and Galleries. In H. Knoblauch and H. Kotthoff (eds.) (2001) *Verbal Art Across Cultures: Aesthetics and Proto-Aesthetics of Communication* Tubingen: Gunter, Narr Verlag, pp. 281–297.

64

Ethnographic Approaches to the Internet and Computer-Mediated Communication

Angela Cora Garcia, Alecea I. Standlee, Jennifer Bechkoff and Yan Cui

Technologically mediated communication is being incorporated into ever more aspects of daily life (Clegg Smith 2004, 223; Mann and Stewart 2000, 209; Vayreda et al. 2002; Whitty 2002, 2003, and 2004). The distinction between online and offline worlds is therefore becoming less useful as activities in these realms become increasingly merged in our society, and as the two spaces interact with and transform each other (Bakardjieva 2005; Haythornthwaite and Kazmer 2002, 441; Suoranta and Lehtimäki 2004; Salaff 2002; Carter 2004).

While there exists a huge body of research on the Internet and computer-mediated communication (CMC), only some of this research is qualitative, and of this, an even smaller portion is ethnographic. With the exception of a relatively small group of ethnographers who focus research efforts on the Internet (for example Cherny 1999; Hampton and Wellman 1999 and 2001; Hine 2000; Kendall 2004; Kozinets 2001; Leung 2005; Lysloff 2003; Walstrom 2000a and 2000b), most ethnographers still conduct studies firmly situated in the "offline" social world. To continue to effectively explore some of the main and enduring concerns of ethnographic research (such as the nature of specific social worlds and subcultures; the construction of identity; the beliefs, values, and world views underlying human action and social life;

Source: *Journal of Contemporary Ethnography*, 38(1) (2009): 52–83.

and the experience of everyday life) ethnographers must incorporate the Internet and CMC into their research to adequately understand social life in contemporary society.

Our analysis of the ethnographic literature on technologically mediated phenomena suggests that ethnographers must alter their research techniques to accommodate these social changes. Specifically, we argue that the current blending of offline and online worlds requires ethnographers to incorporate CMC in their study design and procedures for approaching and interacting with research subjects:

1. Because online ethnographers are not physically co-present with their research subjects, they cannot use their interpersonal skills to access and interpret the social worlds they are studying. Instead, ethnographers must develop skills in the analysis of textual and visual data, and in the interactional organization of text-based CMC.
2. The process of gaining access to the setting and research subjects is different in online ethnography because of the lack of physical presence and the resulting anonymity provided by the medium. Ethnographers must therefore learn how to manage their identity and presentation of self in visual and textual media and to do impression management via CMC modalities such as e-mail, chat, and instant messaging.
3. The blurring of public and private in the online world raises ethical issues around access to data and techniques for the protection of privacy and confidentiality. Ethnographers must learn how to apply standard principles of human subject protection to a research environment which differs in fundamental ways from the face-to-face research contexts for which they were conceived and designed.

In this article we analyze the existing ethnographic literature on the Internet and CMC, as well as writings about conducting research online. We will first discuss methods of data collection and analysis, including participation observation, fieldnotes, and interviewing. The second section will address issues of obtaining access to the research setting and subjects. The final section will address ethical and human subject protection issues which arise from the unique characteristics of the online environment.

Defining the Setting in Online Ethnographic Research

When people are not interacting face-to-face, but only via chat rooms or by creating and reading Web pages, where is the ethnographer to put his or her feet? Some ethnographers argue that "virtual" sites of ethnographic research differ greatly from traditional "real world" settings (Hine 2000; Lysloff 2003). Forte (2004, 226) writes that "both involve the detached study of a "site" that pre-exists the ethnographer and which the ethnographer comes to "visit" as an "outsider." In Lysloff's (2003) study of an online music community she contrasts online "viewing images and listening to sounds and music, of reading and writing texts" (Lysloff 2003, 235) with her previous ethnographic experiences in Java which involved physical and social immersion in the day to day life of that community.

While some argue that the "virtual" world is a different "social space" than the "real world," we concur with those ethnographers who argue that there is one social world which contains both traditional and technologically advanced modes of communication and sites of social activity (e.g., Ruhleder 2000; Lyman and Wakeford 1999, 360). "Virtual reality" is not a reality separate from other aspects of human action and experience, but rather a part of it. Therefore, ethnographers should define the field or setting of their research on the basis of their research topic, rather than arbitrarily or prematurely excluding one arena or the other (see Hampton and Wellman 2001 and 1999; Campbell 2006, 277). The ethnographic research we reviewed reveals a variety of approaches to how the setting of online research is defined.

Social Phenomena Which Exist Solely Online

We found very few ethnographic studies of online settings in which members have no offline contact. Members of some online support groups have only online contact, as in Walstrom's (2000a, 2000b, and 2004a) study of people with eating disorders, Nelson and Otnes' (2005) study of brides planning international weddings, and LeBesco's (2004) study of the experience of being overweight. For some types of listservs all of the interaction between group members is online (e.g., Kleinman's [2004] ethnographic study of an international listserv for people interested in the topic of women in science and engineering). Because in these types of groups participants' only contact with each other is via CMC, ethnographers can study the social life of these support communities solely by examining their online behavior.

Social Phenomena Which Exist Primarily Online

More common are those settings in which members may have some offline contact, but the majority of their contacts and their primary experience of that setting are online. For example, Lysloff's (2003) ethnography of composers of computerized music involved a community whose activities and contacts were routinely computer mediated, with only occasional face-to-face encounters. Taylor's (1999, 437) study of "digital embodiment" investigated the online electronic medium used to create avatars for use in virtual worlds; the setting for this aspect of her research was therefore online behaviors, locations and technologies. Cybersex is by definition online behavior; Whitty (2004) defined the CMC through which these online relationships were conducted as the site of her study, while acknowledging that some of these online relationships do develop into offline relationships. In sum, for these types of locations it is feasible to limit the setting of the research to online/CMC phenomena.

Multimodal Social Worlds as Research Settings

When the social phenomena being studied are conducted through both CMC and face-to-face contacts, it is necessary to define the setting to include online and offline components. For example, in Hampton and Wellman's (2001) study of "Netville," an experimental wired community, one of the coauthors lived onsite and conducted both face-to-face and virtual participant observation of community life. Other ethnographic studies in which the phenomena observed had both online and offline components include Miller and Slater's (2000) study of Internet use by Trinidadians, Green's (1999) study of virtual reality technologies, Kozinets' (2001) study of Star Trek "fandom," Ruhleder's (2000) study of students in an online classroom, Kendall's (2002) study of masculinity in an online "pub," Clark's (2004) study of a Linux user's group, Correll's (1995) study of an "electronic bar," and Jones' (2005) study of gay men using chat rooms to find friends and sexual partners.

The ethnographies we examined varied in terms of how effectively they defined the setting of their research, with some studies erring in the direction of too great a reliance on online aspects of the social world. For example, Silver's (2003) study of a real city's "electronic village" relied primarily on online interactions (e.g. newsgroup postings and Web sites), although a small number of face-to-face interviews with participants were conducted as well. Offline participant observation would have strengthened this study and demonstrated how relationships, networks, and other connections within the community were formed or affected by their online dimensions.

The Offline Social World as a Research Setting

We found very few examples of ethnographic studies which dealt with the effect of CMC or the Internet on "offline" aspects of social life. Bakardjieva's (2005) ethnographic study of how having a computer in the home affected such aspects of everyday life as parental roles and tasks and family togetherness is one study which focuses on the offline impact of computer use.

Defining the Setting in Online Ethnographic Research

While some social phenomena exist solely online, we found that there are very few research topics that justify limiting the field to online phenomena. Therefore, while some researchers may perceive examining Web sites or CMC as a shortcut to data collection, the setting of the study should typically be defined to include relevant offline components of the social world as well as the CMC. Rather than deciding in advance to conduct an ethnography of an online site or community, the ethnographer should first choose their topic of interest, and then define the field in terms of whether and how that topic involves different modes of communication or technological locations.

Our review of existing research into the Internet and CMC suggests that "virtually all" ethnographies of contemporary society should include technologically mediated communication, behavior, or artifacts (e.g., Web sites) in their definition of the field or setting for the research. For example, much of the communication that used to take place face-to-face or over the phone is now routinely done electronically in workplaces, organizations, and institutions. In many organizations today, you can not understand how work is done without examining emails, instant messaging, hand held electronic devices, and company Web sites. Another example is ethnographic studies of the social worlds of children and teenagers. What would Adler and Adler's (1996) study of social stratification among school children reveal if the study were redone in the current era of cell phones, e-mail accounts, social networking Web sites, and text messaging? At the very least, the existence of enduring documents of the children's communication (e-mails, recordings of chat room conversations, Web site postings, etc.) would allow for more precise examination of their relationships and friendship networks. Even when participants in a social world do not utilize CMC, they are affected by their failure to use it, because the "traditional" modes of communication (face-to-face, telephone, and writing), are no longer the only, or in some settings, even the primary means of communication. The invisible effects of exclusion from CMC also need to be addressed (e.g., how does a homeless person without access to the Internet find a job, an apartment, or even social services?).

Once the ethnographer has defined the field or setting of his/er research, there are adjustments in data collection and analysis which must be made when CMC/online phenomena are studied. In the next sections we will address these issues.

Online Participant Observation

Our analysis of online ethnographic research suggests that the participation observation approach must be adjusted when working in online settings. First, since the ethnographer can not directly observe the people she or he is studying, the nature of observation changes. Second, the ability to technologically record events, interactions, and locations in online research settings changes the role of field notes and how findings are reported. Third, the nature of online data (e.g., textual and visual material rather than people speaking and acting) requires a different set of skills for understanding and analyzing it. Fourth, existing ethnographies of the Internet/CMC tend to privilege text-based phenomena at the expense of visual phenomena. And finally, the use of sound and movement in Web sites is under-analyzed in current online ethnographic work.

On "Being There": The Nature of Online Observation

Observation in online research involves watching text and images on a computer screen rather than watching people in offline settings. However, the technologically mediated environment still provides direct contact with the social world the ethnographer is studying, since participants in that setting communicate through online behavior. In her studies of online support groups, Walstrom (2004a and 2004b) uses the term "participant-experiencer" instead of "participant-observer" to characterize the nature of the researcher's role in the setting. The participant experiencer "entails the role of active contributor to the group being studied. This role specifically refers to a researcher who has personal experience with the central problem being discussed by group participants" (Walstrom, 2004a, 175). The use of the term "experiencer" instead of observer is helpful because in the online support group there is no opportunity to directly observe the other members of the group; the researcher can, however, experience what it is like to participate in the group by reading and posting messages to the group. Schaap (2002), writing about his ethnographic work in MUDs (online locations where participants interact via constructed characters) notes that:

> My observations are purely textual and I haven't met any of my informants face to face. While this poses some unorthodox problems, I believe that one learns to speak and listen, or rather write and read in this world just as one would in a particular physical locale. After a while one starts to discern what kind of conversation one is having, which clues to pick up on and when informants are reluctant to speak about a certain subject. (Schaap 2002, 29–30)

While in the offline world observation requires at least the minimal participation of "being there," many online settings provide the opportunity for completely unobtrusive observation: the researcher can observe by "lurking." In some online contexts the presence of the lurking researcher is undetectable, in others, the researcher's presence in the interaction may be detectable, but not his or her identity. We found that online ethnographers have a variety of perspectives on the advisability and ethics of lurking.

Some ethnographers advocate beginning a participant observation study of online phenomena by lurking first. For example, Kozinets and Handelman (1998) used lurking to gain information about consumer's boycotting behavior. This period of observation gave them information that they then used to create interview questions. In Shoham's (2004) ethnographic study of Israeli chat rooms he first lurked, then introduced himself as a newcomer to the community engaged in participant-observation research.

Other ethnographers advocate active engagement in the field as a superior data collection strategy (Bell 2001; Miller and Slater 2000). Bell (2001) writes "[l]urking is a one-way process, and one of the strengths of ethnography is its emphasis on *dialogue* with respondents – recasting research as collaboration rather than appropriation" (Bell 2001, 198). Heath et al. (1999) also argue that lurking is not the same as participating, and therefore does not produce the same benefits for the research as active participation.

> If we take seriously the imperative to locate ourselves within political, historical, and cultural processes of the practice of research, then we do not believe that lurking online, as a singular mode of ethnographic research, is a satisfactory means to understand and/or relate to our subject matter. (Heath et al. 1999, cited in Lyman and Wakeford 1999, 367)

While acknowledging the advantages of unobtrusive observation which lurking provides, Sveningsson (2004, 47) addresses the ethics of lurking: "What happens to the privacy and integrity of the people we study?" While there may be advantages to announcing one's presence as an online researcher (Clark 2004; Roberts, Smith, and Pollock 2004; Sveningsson 2004), in some cases disclosing one's presence to ask for consent may sacrifice participants' anonymity and disturb naturally occurring behavior (Soukup 1999). While some online participants prefer researchers to lurk first to gain knowledge about the setting before asking questions, LeBesco (2004) found that participants felt spied on when she was lurking, and would have preferred her to actively engage with them (see also Maczewski, Storey, and Hoskins 2004, 72; Kendall 2002). In their survey of listservs and newsgroups, Chen, Hall, and Johns (2004) found that while researchers may be encouraged to lurk first to familiarize themselves with the setting before asking questions, it was not acceptable to collect data from newsgroups and listservs without permission or without identifying oneself as a researcher (see also LeBesco 2004).

Conforming to the Norms of the Online Setting

Online ethnographers can gain access to a field setting and recruit potential research subjects by displaying cultural competence of the norms of the group they are studying (Walstrom 2004a and 2004b). For example, Clegg Smith (2004) discovered that posting a message announcing her study of an online listserv for doctors would probably have been badly received. In that online community any message off the topic of discussion was subject to negative comments by group members. Cherny (1999) also found her entrance to the field (MUDs – multiplayer online computer games) was facilitated by participating in the activity before beginning her research project. "Being a participant first, and sharing the community's response to much media coverage of MUDs, I no doubt had an easier entree as a researcher" (Cherny 1999, 301). Another technique is to identify oneself as a member, or at least a sympathizer of the group being studied. For example, in LeBesco's study of "fat positive" USENET discussion groups, she presented herself as "one of us:"

> I gained entree, organizational and individual, to these sites as a researcher, partly, I suspect, because I committed early on to a fat-positive perspective. Part of my introduction, aside from my academic credentials, explained my interest in studying online conversations about fat as stemming from my personal experience of corpulence. I positioned myself as someone who had lost weight and would no longer be considered "fat" by what I imagined to be their standards, but as someone who respected and still wished to participate in many of the struggles waged in fat communities. (LeBesco 2004, 66)

Catterall and Maclaran (2002) recommend using electronic resources to have a private conversation with a research participant to learn how best to conform to community norms in that particular online setting:

> Most chat rooms have a one-to-one facility for private conversations which enables one person to "whisper" to another member. This can be used to seek advice on how to comport oneself within the community, the most productive times to visit the room, and so forth. (Catterall and Maclaran (2002, 231)

As the examples above show, online settings present the ethnographic researcher with something of a dilemma as they try to learn the norms of the online research setting. While information about the setting can be learned by silently lurking, some groups will object to this behavior. Lurking first, if allowed by the site and by the IRB, is acceptable if that is how participants in that setting routinely participate. If not, ethnographers will get a more authentic experience of an online setting if they jump straight into participation. In short, the ethnographer should attempt to experience the online site the same way that actual participants routinely experience it.

The Nature of Data for Online Participant Observation

Technologically mediated environments profoundly change the nature of the information obtained through participant observation research. Soukup (2000) writes that most CMC research has a textual bias, focusing on the written word rather than on the full range of modalities available. While textual, visual, aural, and kinetic aspects of CMC are integrated in online environments, we will discuss them separately here.

Interpreting Textual Data

The textual data available for participant observation research include e-mail, chat room interactions, instant messaging, Web sites, and other online environments. There is a large body of research on emoticons in text-based CMC which ethnographers should be familiar with (see, for example, Mann and Stewart 2000 and 2002; see also Huffaker and Calvert 2005; Riva 2002). Campbell (2006) found emoticons and other aspects of online communication in a newsgroup for "skinheads" to be used by members to construct identity and form relationships.

> The participants made use of various techniques to convey physicality, emotion and feeling: colloquial vernacular (an indication of locality), the selection of specific words which expressed subtly in feeling, the use of uppercase letters to denote anger or shouting, and the more explicit use of emoticons. Methods for constructing individual idiosyncrasies included the use of quotes, automatically incorporated at the end of each message, lyrics from skinhead songs, or hypertext links to favorite or self-authored Web sites. These modes constructed and communicated versions of the "self," and they were read and interpreted by others as style and dress might be read offline. (Campbell 2006, 277–78)

Campbell (2006) argues that what participants write conveys important information about their identity, presentation of self, and how they define and perceive their world. However, it is not enough to simply "translate" emoticons or be fluent in their use; these and other aspects of participants' text-based interaction pose interpretive puzzles for the ethnographer in terms of their relationship to participants' presentation of self. Markham (2004) cautions against making assumptions about interviewees based on their use of emoticons, style of communication, or skill at writing, typing, or spelling, and so on. In one case a participant's textual performance led her to make assumptions about the person's gender, race, and level of education, which may not have been accurate or relevant to their interactions with other members of that online community.

While the use of slang terms, abbreviations, and emoticons in online communities should be of concern to ethnographers, what is not often as readily appreciated is how the medium changes the organization of online interactions. For example, the asynchronous nature of chat room talk changes the sequencing of messages which may make interpretation problematic. Others may post messages which intervene between a post and the message it responds to (Garcia and Jacobs 1998 and 1999; Markham 2004).

> Organizing interactions into retrospective linearity may be a natural way of making sense of what happened, but it does not allow us to view, investigate, and build our knowledge base of how fragmented and disorganized interactions construct identities and relationships. (Markham 2004, 153)

In short, a printout of a chat room conversation is not a substitute for observing the interactional process which produced it. Because of the effect of sequencing issues on the organization of chat room talk ethnographers should consider using a research setting which allows them to obtain screen save data from all participants, to determine how each party interpreted the conversation as it unfolded.

Integration of Visual Data

In offline participant observation, ethnographers routinely experience and analyze participants' verbal messages in conjunction with their facial expressions, tone of voice, and body language, along with the impression given by their appearance, clothing, and setting. Online ethnographers should also take care to integrate visual aspects of the data into their observations and analysis and treat visual data (e.g., the use of pictures, colors, page layout, and graphic design of Web sites) as a key aspect of the online location. This may require developing a new set of skills and data collection methods.[1] Ethnographers of the Internet/CMC have developed useful ways to incorporate visual data into their participant observation research:

Visual aspects of Web sites. Hine's (2000) analysis of Web sites devoted to a famous murder case included both the visual and textual components. She discussed the choice of photos that were used, the backgrounds for the photos, the types of layout or arrangement of objects and text on the page, and other aspects of the visual appearance of the sites in terms of how they conveyed an emotional message about the case.

Use of photos by online participants. Jones (2005) analyzed photos exchanged over the Internet in his study of gay men searching for online friends or sexual partners. He found that participants chose to display photos which conveyed information of interest to their co-interactant (e.g., physical appearance, body type), while at the same time making identification of the person from the photo difficult or impossible.

The use of avatars. Taylor (1999) studied how participants in virtual worlds create visual representations of their characters which they then control in interactions with the avatars of other participants. Participants' interactions are thus mediated by visual representations as well as by the characters that are created.

The use of Webcams by online participants. Keating and Mirus' (2003) study of Webcam Internet connections between deaf speakers of American Sign Language shows how the technology facilitated visual communication between the participants. In a case study by Lyons (2003) of a "wearable computer" expert, the researcher took screen shots from the participant's computer. When the participant was interviewed later, the screen shots were played back to jog his memory so he could better recount his experiences while wearing the mobile computer. The researcher found the screen shots to be an invaluable tool during the interview sessions. Wise (2004) describes the use of Webcams by individuals to broadcast their lives over the Internet (see also Andrejevic 2004; White 2003). Some people have installed video cameras in their houses, hooked them up to the Internet, and displayed their daily life to anyone who cares to watch. While these types of phenomena may make an interesting subject for ethnographic analysis, Wise notes that there are limitations to the amount of understanding they provide:

> [A]s a means of studying everyday life, Webcams can provide only a quite attenuated version of it. Despite what it can deliver (momentariness and longue durée), it cannot come close to presenting the thick description, if you will, the level of detail, density and embodiedness of everyday life usually sought by the myriad projects on everyday life since the turn of the last century (see Gardiner 2000; Highmore 2002). Though some Webcams provide sound, and a few allow for camera movement, many sensory and spatial dimensions are lost. They lack a sense of presence. (Wise 2004, 428)

White (2003) also notes that Webcam data may suffer from poor visibility or uncontrolled loss of the image, either of which may affect its usability for ethnographic research.

Videotaping computer users. In Ruhleder's (2000) study of video teleconferencing in a distributed workplace she used video cameras to record the participants interacting with each other and their geographically dispersed colleagues. Thus the visual and interactive data available to Ruhleder and her colleagues for analysis included much more than just the recording of the videoconference itself.

> The use of video cameras allowed for the creation of multisite video data. The resulting collection of videotapes not only captured various distributed interactions but also enabled us to retrospectively sync up the activities across multiple locations. This allowed us to pinpoint what was seen or heard locally and across the link. We captured reactions to local talk and gestures, to the other side's talk and gestures, and to materials shown on a computer or electronic whiteboard. (Ruhleder 2000, 7)

In sum, the analysis of visual material is likely to be an important component of online ethnographic studies.

Sound and Movement in Online Participant Observation Data

Sound and movement are becoming common aspects of Web site design as well as other types of online settings and communication. For example, while e-mails used to be simply typed text, they may now come on colored "stationery" and include pictures, sound, and moving graphics which provide enhanced resources for the presentation of self (Goffman 1959) in online environments. This evolutionary trend is not yet adequately dealt with by ethnographic research in online settings.

In addition, the increasing use of direct voice communication (e.g., via videoconferencing) provides data which more closely mirrors conversations in face-to-face settings. Thus while previous "generations" of online ethnography have been biased toward textual data, the next generation of online ethnographies must engage not only with textual and visual material, but also with sound and movement.

Recording and Communicating Online Participant Observation Data

The role of field notes. Both textual and visual data can be recorded by "screen save" programs (e.g., Camtasia or Hypercam[2]) providing a digital "videotape" of an online environment. An online ethnographer can digitally record their visit to a Web site and replay it at will, stopping, starting and moving around in the data as needed. Because of this ability to completely capture the field, the need to take field notes is diminished, or at least altered. However, field notes are still essential to help the researcher catalogue, describe, and develop theories from their observations, and to record their reactions and subjective experiences. Shoham (2004) took field notes in addition to saving printouts of chat windows. Schaap (2002) took field notes on his participant observation in online MUD fantasy role play games, and recorded logs of the interactions that he participated in.

Communicating field notes and recorded data. Researchers in both on and offline settings face tough decisions about how to represent observational data and field notes to their readers. Kendall (2004) notes that there is a cultural belief that online interaction is not "real." Kendall's (2004) audience tended to question her assumption that online interaction is really social interaction, or that the friendships formed online are really friendships; these attitudes betray our culture's ideas about what interaction and friendship mean. The ethnographer is therefore often trying to communicate with an audience which is not as well versed in CMC as is the researcher or the subjects of his or her study.

Because of the multimodal nature of online communication, ethnographers often find themselves trying to describe visual, aural, and kinetic observational data in verbal form. Ethnographers must also make decisions about how to represent participants' text-based actions for their readers. Markham (2004) argues that participants' writings should be conveyed to the reader exactly as they were written, without correcting spelling, grammatical or punctuation errors, or standardizing fonts.

> We literally reconfigure these people when we edit their sentences, because for many of them, these messages are a deliberate presentation of self. Even when they are not deliberate, texts construct the essence and meaning of the participant, as perceived and responded to by others. (Markham 2004, 153)

Schaap (2002) chose to correct typos to make quotes easier to read, but otherwise left the text exactly as written by the participants.

In sum, the online ethnographer must learn to engage in participant observation without a physical presence in the field, and must develop new skills and procedures for collecting and analyzing data, recording field notes and communicating the results of their analysis. Current online ethnographies tend to privilege textual aspects of the data and do not adequately integrate visual, aural, and kinetic phenomena. In particular, technological improvements are increasing the role of oral communication in online environments. If the Web continues to move in an aural direction ethnography may come full circle – e.g., with synchronous oral communication and visual access to research subjects (e.g., via Webcam), which (ironically) may make future Web-based experiences more similar to offline interaction than it currently is. The differences between textual and oral communication in online environments is also of critical importance for how ethnographers conduct interviews in online research.

Interviewing in Online/CMC Research

Ethnographers conducting online/CMC research can use online interviews, offline interviews, or both. For some topics, offline interviews may be more useful than online interviews. Offline interviews enable the researcher to verify information gathered online (Kozinets 1998; Turkle 1995) check the identities of online personas (Turkle 1995; Taylor 1999; see also Catterall and Maclaran 2002), and accommodate respondent preferences or characteristics [e.g., skill at expressing themselves in writing (Taylor 1999) or keyboarding skills (Mann and Stewart 2002)]. The goals of the research project should also be taken into account (Taylor 1999). Offline interviews can also fill gaps in online data collection and resolve ambiguities or queries (Leung 2005).

Online interviews are a useful resource for ethnographers (Fetterman 1998; see also Giesler and Pohlmann 2003; Schaap 2002; Shoham 2004; Taylor 1999). Williams and Copes (2005) used online interviews following participant observation and focused discussions to expand on themes that emerged from earlier analyses. Hampton and Wellman (1999) used online interviews in conjunction with offline interviews, online/offline participant observation, and surveys while studying Netville, an experimental wired community. Schaap (2002) used online interviews in his study of fantasy role play games in MUDs.

Most online interviewing is done via asynchronous text-based modes of CMC such as email or discussion boards. Respondents can choose to respond right away, or to deliberate over their responses (Catterall and Maclaran 2002; see also Kozinets 1998).[3]

For some research topics and participants, synchronous modes of CMC such as instant messaging, or "quasi-synchronous" modes such as chat will be helpful. These modes of CMC more closely replicate oral communication traditionally used by ethnographic researchers (Mann and Stewart 2002). In Williams and Copes' (2005) study of youth "straight edge" culture all their interviews "took place online using either an Instant Messaging (IM) or Internet Relay Chat (IRC) program, both of which are popular among young Internet users" (Williams and Copes 2005, 75). Instant messaging was an appropriate choice for this topic because it was the mode of online communication preferred by their research participants.

Because of the nature of text based interaction, online interviews may be structured differently than interviews conducted face to face (Jones 2005). Pauwels (2005, 605) notes that this "doesn't necessarily make it less valid (it might, conversely, be more thoughtful and denser information)." Kozinets and Handelman (1998) used asynchronous online interviews in their study of consumer boycotting behavior.

If the researcher conducting the interview is not particularly well versed in online communication conventions, he/she may need to hire someone else (such as a member of the community being studied) to conduct the interview (as did Jones 2005), spend time in the field to develop interpretive skill so as not to misrepresent the meanings behind the text (Kozinets 1998 and 2002), and/or use a synchronous mode of communication for the interview to be able to ask interviewees for clarification.

Because of the anonymity of CMC, respondents may be less concerned about the impressions they are making (Riva 2002). Thus, online interview responses may be more candid than those obtained from offline interviews. However, some argue that allowing respondents to put more thought into their responses may limit spontaneity (Catterall and Maclaran 2002). The anonymity factor in online interviews may balance power between interviewer and interviewee; interviewees may feel freer to challenge researchers than they would in a face-to-face interview (Catterall and Maclaran 2002).

However this ability to challenge the researcher may damage the results if interviewees respond with false information (Ayers 2004).

Videoconferencing systems or Webcams have been found to be useful technologies for conducting various types of interactions (e.g., Marziali and Donahue [2006] on support groups conducted via videoconferencing; Birkenmaier [2005] on videoconferencing technology in social work education; Keating and Mirus [2003] on the use of Webcams for Internet conversations between deaf speakers of American Sign Language; and Gilham and Moody [2001] on video conferencing for education and career counseling of incarcerated youth). Videoconferencing can also be a useful tool for conducting synchronous interviews online (Fetterman 1998 and 2002), enabling the interviewer and interviewee to observe each other's nonverbal behaviors.

Access and Identity in Online Environments

Gaining access to the research setting, building rapport with research subjects, and obtaining volunteers for interviews are issues for both traditional and online ethnography. In online ethnography, the resources and challenges involved in obtaining access differ because ethnographers can not rely on their physical presence, appearance, interactional style and conversational competence to help them gain access (Mann and Stewart 2002). In addition to the challenges of anonymity, members of some online communities may be more resistant to being studied than are people in analogous face-to-face venues. LeBesco (2004) states that during a single month, eight researchers tried to gain access to the online site she was studying, and all but one were rejected by group members. Bakardjieva (2005) reports that she had minimal success recruiting respondents through posting announcements on online newsgroups. She had to resort to creating a snowball sample from personal connections. Chen, Hall, and Johns (2004, 159) found that "[m]any list owners and newsgroup members deeply resent the presence of researchers and journalists in their groups" (see also Catterall and Maclaran 2002; Paccagnella 1997). In this section we consider some of the challenges, resources, and opportunities for obtaining access to sites and subjects for online ethnographic studies.

Anonymity and the Identity of Online Participants

Verifying authenticity. When interacting over the Internet the question of whether the person you are interacting with is actually who they say they are must be considered (Cherny 1999; Mann and Stewart 2000 and 2002). Ethnographers thus must consider whether and how they will attempt to verify the identity of their research subjects. In some online settings participants are very open about their identity and make no efforts to conceal it or deceive others. Koufaris (2001, 227) found this to be true in an online newsgroup on organ transplant recipients, who "always use their real names and talk about their real personal lives without hesitation." However, Nissenbaum (2003) reminds us that the anonymity of the Internet allows participants to conceal their offline identity. Some Internet spaces even require the use of pseudonyms. Livia (1999) writes that the French "Minitel," a text-based chat room type communication system, requires participants to use pseudonyms and forbids the use of identifiers such as phone numbers or addresses in these pseudonyms. This level of anonymity makes it difficult for researchers to verify information about participants.

Race, class, and gender. Personal characteristics which may be of critical importance in an ethnographic study (e.g., age, race, or gender) may also be concealed or altered in online interactions (Nissenbaum 2003; see also Mann and Stewart 2000, 58–9). For example, consider this excerpt from a detailed interview of "Su'ad," an engineering student in Kuwait (Wheeler 2001):

> I have a friend who is getting married to someone she met on the Internet. They only "chatted" for four months and now they're going to spend the rest of their lives together. I think she's stupid. It's possible to lie on the Internet. How does she know that he is really as good as he says he is on-line. One has to be careful . . . One time I was "chatting" with another engineer from Saudi Arabia. He kept asking are you a man or a woman. Finally I answered, "I'm a woman, is this important." He said, "Yes, I refuse to talk to you." (Wheeler 2001, 197)

This quote indicates that Internet users are well aware of the possibility of deception or misrepresentation over the Internet (see also Mann and Stewart 2000). The anonymity of online settings may enable research subjects to lie, either to each other or to researchers (Jones 2005; Kozinets 1998).

However, the mere possibility (or presence) of deception does not necessarily mean that the data are not useful for ethnographic analysis.

> In my own research, using face-to-face interviews, I have found that while chatrooms do provide opportunities for people to lie about themselves, this does not necessarily lead to the formation of shallow relationships. Paradoxically, it can open a space for a deeper level of engagement with others (Whitty and Gavin 2001). (Whitty 2004, 205)

Whitty, describing her research on the study of Internet relationships and sexuality, discusses how research participants may lie to each other:

> While Turkle (1996) (*sic* – 1995) suggested that it is fairly common for people to lie on the Internet, more recent empirical data suggests that

> experimentation with one's physical appearance online does not occur as frequently as one might expect (e.g., Cooper, Delmonico, & Burg, 2000; Roberts & Parks, 2001; Whitty, 2002a). Moreover, I have found that men tend to lie online more than women, typically exaggerating aspects of themselves, such as education, occupation and income, which are aspects men often tend to exaggerate offline in order to attract women (Whitty, 2002a). (Whitty 2004, 206)

Dealing with online identity issues. Ethnographic researchers have various ways of handling the possibility of deception online. Turkle (1995) decided "not to report on my own findings [from online interviews] unless I have met the Internet user in person rather than simply in persona" (Turkle 1995, 324). Researchers must also consider the possibility that some of their "online research participants" may actually be multiple personas or characters performed by the same offline individual. Taylor (1999) discusses the issue of "multiple embodiments" in her research on participant's use of avatars in virtual world play.

> Although some users maintain a consistency within a single avatar or character, many do not (by either having multiple bodies within a single space or through their use of multiple worlds). It is also worth noting that although some spaces provide information that lets you know, for example, that the characters Iona and Taylor have the same person at the other end of the keyboard, it is not always the case. Thus, it is quite possible to run into one of your informants in a form (literally a body) you do not recognize. (Taylor 1999, 439)

But Taylor (1999) reaches a different conclusion about the impact of anonymity on the research project than do some other researchers. She argues that even in the offline world, we create ourselves through action and display different selves in different social contexts. Therefore, the fact that participants' online personas are not the same as their offline personas, or that they may even conceal or misrepresent their offline identities, is not necessarily problematic for the researcher because these possibilities are part of social life, not just part of online life (see also Carter 2005). Kozinets (1998, 369) writes "[T]he same freedom which inspires people to mischievously construct deliberate falsehoods about themselves and their opinions also allows them and others the freedom to express aspects of themselves, their ambitions and inner conflicts, that they would otherwise keep deeply hidden." Thus online data may have value for ethnographers, even if participants' self representations are not, strictly speaking, "authentic." Just as the online environment allows the possibility of fluidity in how research participants present themselves, researchers also have choices about how to present themselves in the online context to facilitate access to settings, subjects, and data.

Anonymity and the Identity of the Researcher

The identity of the researcher (e.g., sex, race, or age) and how the researcher presents him or herself in the ethnographic field have long been a concern of offline ethnographers (e.g., Liebow 1967 [see also Mann and Stewart 2000]; Weissman 1970; Eder and Fingerson 2003; Schwalbe and Wolkomir 2003; Reinharz and Chase 2003; Dunbar, Rodriguez, and Parker 2003). The researcher's identity can affect how conspicuous they are in the setting and the likelihood that potential informants will be willing to talk to them.

In spite of the anonymity provided by the online environment, we found that the identity of the researcher may affect the data collection process in an ethnographic study. We will use the approaches taken by Kendall (2002), Hine (2000), and Ayers (2004) to illustrate some of the issues ethnographers face as they decide how to represent themselves in the online research environment. Both Kendall and Hine are female academics doing ethnographic research in an online setting. However, they made different decisions about how to present themselves to their research subjects, because the settings they were studying were different.

Lori Kendall's (2002) book is an ethnographic study of a "virtual pub" called "BlueSky." She first approached this online space as an anonymous guest, but quickly learned that anonymous guests were "not appreciated" on this MUD (p. 18). She describes her decision to choose the character name "Copperhead, whose aggressive and poisonous connotations might allow me both to fit in and to feel somewhat protected" (Kendall 2002, 18). She made this decision because she anticipated that BlueSky would be an "aggressive, male-oriented space."

Christine Hine (2000), in her ethnographic study of the Internet as used by supporters, producers, and consumers of news about a famous murder case, makes a very different decision about how to present herself to her research subjects. She decides to use the name "Christine" instead of Chris (what she is usually called), to present a less threatening image to potential research subjects and thus enable her to collect more data. The name Christine is clearly female, while "Chris" is gender neutral.[4]

Ayers (2004, 263) describes a research project on feminist activists comparing "two social movement groups: one that exists in cyberspace and one that exists in the physical world." Ayers conducted interviews in both settings, representing himself (accurately) as a male researcher. Ayers concluded that some subjects in the online study were not being authentic in their responses; some of the responses he received were fictitious or condescending. He concludes that these types of responses occurred because he had identified himself as a male researcher. He suggests that a male researcher working in this type of politicized female setting must take extra pains to create rapport with his research subjects before engaging in the study.

These three examples show that in an online research setting, the researcher's identity still has an impact even though the research participants can not see or hear the researcher. However, the resources for dealing with these identity issues are different, because the online research subjects will probably not be encountering the researcher in person. In the next section we discuss researcher's online resources for managing the impressions they give to potential research subjects.

The Researcher's Presentation of Self via Textual and Visual Material

In online research impressions about the researcher are communicated by the visual and textual material presented to potential subjects rather than the voice, mannerisms and physical presence of the researcher him or her self. Because of this, a different set of skills are required of the ethnographer to obtain access to the setting and to successfully recruit research subjects. Hine (2000) not only carefully considered how to construct the e-mail letter she sent to potential research subjects, she also referred potential research subjects to her university Web page to document her identity, to increase her response rates. Other researchers also use Web pages to establish researcher authenticity and recruit subjects (Bakardjieva 2005; Maczewski 1999; see also Catterall and Maclaran 2002; Mann and Stewart 2002). Mann and Stewart (2000, 82) suggest placing "strategic advertisements in appropriate newsgroups, mailing lists or BBSs" to recruit research participants. Taylor (1999) attached

> a URL (Universal Resource Locator, a pointer to my Web page) to my avatar as a way of directing potential participants to information about my research. Users could then simply click on my avatar, see the phrase "Avatar Research" and be automatically taken to my site . . . this proved a crucial method of soliciting participation, especially in spaces where researchers were looked on warily. (Taylor 1999, 446)

Similarly, Cherny (1999) attached a message about her research to her character description in the MUD she was studying.

Maczewski, Storey, and Hoskins (2004) explain why any Web site that is used to recruit subjects must be carefully designed:

> Instead of a voice on the phone, the web site now provided the first impression of the project and the researcher for the participants. Information design through text and images now conveyed the research project without personal contact and raised questions of inclusion and exclusion . . . if the researcher only chose clip art that represented males, it would be possible that females felt excluded. The web-site color choices

> may appeal to specific groups of people. The text style may attract or exclude certain groups of young people. (Maczewski Storey, and Hoskins 2004, 68)

Sometimes the impression made by the researcher, as in the case of Hine's (2000) research, is first formed by a recruitment letter sent to potential subjects. In online settings a formal letter on institutional e-letterhead may be more effective than a more casual approach, because of online participants' anxiety about interacting with strangers online, or concerns about the authenticity of the researcher.[5]

Lindlif and Shatzer (1998) found that mistakes made in how online research participants are approached may be difficult to repair:

> For example, Fisher, Margolis, and Resnick (1996) discuss several errors they made in distributing a survey to Usenet groups and Listserv lists, such as inadvertently cross-posting the instrument to all of the news groups, resulting in a flood of copies to each group and "considerable consternation and controversy on the Internet" for several days (p. 18). (Lindlif and Shatzer 1998, 189, fn 4)

Ethnographers of the Web would be well advised, therefore, to consider carefully their initial presentations of self to their research subjects.

In this section we have shown how the online environment creates new dilemmas and opportunities for ethnographic researchers as they seek access to research subjects, sites, and documents. Our review of the literature shows that there are no simple recipes for success; the choices ethnographers make must be tailored as closely as possible to the specific issues, participants, and technological modalities they are studying.

Ethical Dilemmas in Online Ethnographic Research

Public and Private Spheres in Online Ethnography

The boundaries between public and private "spaces" are drawn differently in online locations than they would be in comparable offline spaces. For example in Koufaris' (2001) ethnographic study of a newsgroup for organ transplant recipients she describes a Web site set up by a friend of a woman ("Susan") who was in need of a heart transplant.

> Her battle to live was followed by hundreds of people on the Internet . . . where one could find constant updates of Susan's health as well as pictures of her, anecdotes, poetry and prose related to their battle, and other links to sites with information on organ transplantation. The details of the last months of her life were shared with people from all over the world. At her funeral, friends from as far away as New Zealand were sending their regrets for a woman they only knew through their computers.

> And her story is available for anyone with a computer and a modem to see on the World Wide Web. (Koufaris 2001, 225)

In contrast, members of some online communities may define their "public" postings as private material (see Kozinets 1998).

> Waskul and Douglass (1996) described the private/public distinction in virtual environments in terms of "publicly private" and "privately public," noting that public and private are metaphorical labels based on the assumption that cyber-spaces are like physical places (p. 131). In cyberspace, the defining of public versus private "space" may be reduced to a matter of accessibility. A cyberspace may be viewed as a public space as it is publicly accessible, yet interactions that occur within that space may be deemed by the participants to be private. Waskul and Douglass (1996) argued that researchers do not have the right to define spaces as public or private to meet their own research needs. Rather, account should be taken of the size and nature of the online forum and the intrusiveness of the study. (Roberts, Smith, and Pollock 2004, 161)

In online research contexts, the lines between public and private may become blurred (see Catterall and Maclaran 2002; Chen, Hall, and Johns 2004; Barnes 2004; Bruckman 2004).

Ethnographers may also have to decide how to deal with explicitly private online sites to which the public does not have access. For example, in Jones' (2005) study of a chat room for gay men he found that participants rarely used the public chat rooms, almost always engaging in private one-on-one chats. Jones (2005) dealt with this data collection problem by soliciting participants to share their chat logs anonymously.

The practical and ethical issues around the boundaries between public and private online spaces are also relevant when collecting archived data. As Ruhleder (2000) and Taylor (1999) note, the use of archived materials enables ethnographic research on cultural artifacts (e.g., Web pages and listservs), and interactions (e.g., online discussions) which are examined after the fact rather than experienced by the researcher as they are unfolding (Maczewski et al. 2004, 65). Ruhleder (2000) used online archives to study interactional patterns in an online classroom. She found that the classroom archives were a useful source of data because they allowed deeper access to what it was like to participate in an online class. Students routinely used class archives to catch up with missed sessions and material (Ruhleder 2000, 10, 11).

The ethical landscape for the use of archival data are not yet clear. While some argue that Web sites are analogous to magazines or television shows, and hence are intentionally and inherently "public," others argue that some Internet locations are inherently private. For example, Döring (2002) argues that:

> Net forums are not addressed at a dispersed, broad audience, and instead fulfill an internal exchange aimed at only those people who are currently enlisted. Incognito logging of group interactions in the Net would thus

> be equivalent to the secret recording of a table conversation in a restaurant or a multiperson chat at a party, and would represent an unethical infringement of privacy laws. (Döring 2002, 343)

Ethnographers can increase their chances of making the right choice about how to gain access to archival and other online data by learning the norms of behavior in the specific environment they are studying.

Revealing or Concealing the Identity of the Online Ethnographer

Norms and practices about how participants portray their identity vary with the type of online setting. There is thus no single answer to the question of whether, when, or how the researcher should reveal his or her identity to research participants. In some online settings, members disapprove of identity deception. In a survey of online mailing lists and newsgroups, Chen, Hall, and Johns (2004, 166) found that "[w]hen respondents were asked if the information gatherers should disclose their identity and research intent in the groups, there is an unanimous 'yes.'" Roberts et al. (2004) had been participating in MOO activities for a period of time and needed to decide how to introduce their research roles to the MOO members (MOOs are types of MUDS). From their participation, they learned that MOO participants found deception to be strongly objectionable. For this reason, Roberts et al. (2004) decided to list their research identities on their MOO characters: "The open approach and linking of identities demonstrated respect for individuals, increasing the information they had in terms of giving informed consent to participate in the research" (Roberts, Smith, and Pollock 2004, 165). The measures Roberts, Smith, and Pollock (2004) took ensured the participants in the MOO always knew when they were interacting with researchers.

Livia's (1999, 430) field research on the French Minitel discusses "the ethical problems posed by the anonymizing effect of the new medium." Because the Minitel requires the use of pseudonyms, participants could not tell that Livia was a researcher until they had begun interacting with her. Livia found that when she revealed her researcher status at the beginning of the conversation co-interactants "hung up" (p. 430). The dilemma Livia experienced highlights the ethical and practical concerns connected to how to represent one self to participants in online studies.

Protecting the Privacy and Anonymity of Research Subjects Online

There is some concern among Internet researchers that the wiliness of the technology makes it difficult to guarantee anonymity for people studied online. Porr and Ployhart (2004, 142) state that "it is impossible to gain complete anonymity over the Internet because of technologies such as cookies, IP addresses and Internet server log files (Weckert, 2000)." One solution they suggest is to have online interviewees complete the research from a computer other than their own as a way of minimizing the chance that their interview could be matched with their identity (Porr and Ployhart 2004; see also Roberts, Smith, and Pollock 2004). Johns, Hall, and Crowell (2004) suggest keeping data in separate, locked computer files to protect the privacy of research subjects in online studies.

The publication of participants' online pseudonyms may also put the anonymity of research subjects at risk. A pseudonym may be (without the researcher knowing it) the subject's real name (Roberts, Smith, and Pollock 2004; Whitty 2004). Even if the pseudonym is not the participant's real name, it may be possible for members of the online community to identify the person from it. For example, Bruckman (2004) argues:

> [Users] may use the same pseudonym over an extended period of time and ultimately care about the reputation of that pseudonym. They may use that pseudonym on multiple sites. The pseudonym may in fact contain part or all of their real name. Furthermore, people engaging in serious conversation with one another online tend to continually reveal little bits of information. If the forum is archived, these accumulated small pieces of information often eventually begin to identify the individual. (Bruckman 2004, 102–3)

Because of these potential problems, Carter (2004 and 2005) changed the "nicknames" of her informants in her ethnographic study of "Cybertown" to protect their privacy. Mann and Stewart also raise the issue of anonymity and suggest that "with email (and asynchronous conferencing), real names, user names, domain names, signatures and even ISPs may all need to be adjusted" (Mann and Stewart 2000, 57; see also Turkle 1995; Paccagnella 1997). Cherny (1999) decided to change her research subjects' character names in her public reports to provide more privacy for her research participants.

Researchers can also decide to let participants edit or delete portions of the online messages/material to protect privacy. For example, in Lyons' (2003) study of a "wearable computer" expert, the research participant was able to voluntarily black out the screen to prevent sensitive information from being recorded for the study. These examples show that the ethnographer must have an understanding of how the technology operates to be aware of all of the potential threats to privacy and anonymity.

One can not assume that a subject that is considered sensitive or private in the offline world will necessarily be considered so online. As in offline ethnography, learning the norms and communicative practices of the people being studied will be helpful to the researcher, but because of the different boundaries between public and private, and the opportunities for unobtrusive observation provided online, it may be difficult to learn these norms and practices without taking the plunge into the online world. And finally, the ever-changing nature of the technology will continue to provide new challenges in terms of protecting the privacy and anonymity of research participants.

Conclusion

The increasing prevalent use of the Internet and CMC require ethnographers to adjust their traditional modes of research to the unique conditions in the online environment. The technologically mediated nature of online activities and artifacts provide researchers with choices about how to define the setting. How the research setting is defined will depend on the phenomena of interest, whether it is a Web site, chat room interaction, discussion group posting, MUD, MOO, virtual game, or other type of online activity. For some studies the research setting will be defined to provide access to the activities of producing and engaging in such online phenomena, or the relationship between the online activities and the individual's "real world" experiences and activities, or sense of self. Thus the research setting may be defined as an entirely online/technologically mediated phenomena, a combination of online and offline phenomena, or solely offline aspects of the problem.

The traditional methods for data collection in ethnographic studies (primarily participant observation and interviewing) are used differently in online research. We found that researchers are faced with the same problems as those of their research participants: how to communicate, present oneself, and interpret others' presentation of self in a technologically mediated interactional environment. While online ethnographers still must gain access to subjects and protect their privacy, the interactional resources available to do these things are different from face-to-face research settings. Thus ethnographers must learn how to translate observational, interviewing, ethical, and rapport-building skills to a largely text-based and visual virtual research environment. Furthermore, the nature of this online environment is in a continual state of transformation and development. As Kozinets (2005 and forthcoming) and others point out, Internet technologies (such as instant messaging) and cultural artifacts (such as blogs) have recently become more popular and important modes of communication; ethnographers will need to be alert to such shifts as they search for topics for research, define their research setting, choose and adjust methods of data collection, and use appropriate strategies for gaining access to research settings and subjects.

Notes

1. See Pink (2004) for a discussion of online resources for doing visual ethnography.
2. Camtasia (produced by TechSmith Corporation) produces digital copies of computer screens. Hypercam is produced by Hyperionics Technology, LLC.

3. See Mann and Stewart (2000, 126–129) for a discussion of the advantages and disadvantages of using various modalities of CMC for interviewing.
4. Livia (1999) also discusses her strategic choice of pseudonyms – one obviously female, the other gender neutral. See also Mann and Stewart (2000) on this topic.
5. See Mann and Stewart (2000) for a related discussion about formality in e-mail exchanges with research participants.

References

Adler, P., and P. Adler. 1996. Preadolescent clique stratification and the hierarchy of identity. *Sociological Inquiry* 66: 111–42.

Andrejevic, M. 2004. The Webcam subculture and the digital enclosure. In *Mediaspace: Place, scale and culture in a media age*, edited by N. Couldry and A. McCarthy, 193–208. London and New York: Routledge.

Ayers, M. D. 2004. Fact or fiction: Notes of a man interviewing women online. In *Virtual research ethics: Issues and controversies*, edited by E. A. Buchanan, 262–72. Hershey, PA: Information Science Publishing.

Bakardjieva, M. 2005. Internet society: The Internet in everyday life. London: Sage.

Barnes, Susan B. 2004. Issues of attribution and identification in online social research. In *Online social research: Methods, issues, & ethics*, edited by M. D. Johns, S. S. Chen, and G. J. Hall, 203–22. New York: Peter Lang.

Bell, D. 2001. *An introduction to cyberculture*. New York: Routledge.

Birkenmaier, J. 2005. Weaving a web: The use of Internet technology in field education. *Journal of Teaching in Social Work* 25 (1/2): 3–20.

Bruckman, A. 2004. Introduction: Opportunities and challenges in methodology and ethics. In *Online social research: Methods, issues, & ethics*, edited by M. D. Johns, S.-L. S. Chen, and G. J. Hall, 101–4. New York: Peter Lang.

Campbell, A. 2006. The search for authenticity: An exploration of an online skinhead newsgroup. *New Media & Society* 8 (2): 269–94.

Carter, D. M. 2004. Living in virtual communities: Making friends online. *Journal of Urban Technology* 11 (3): 109–25.

________. 2005. Living in virtual communities: An ethnography of human relationships in cyberspace. *Information, Communication & Society* 8 (2): 148–67.

Catterall, M., and Pauline M. 2002. Researching consumers in virtual worlds: A cyberspace odyssey. *Journal of Consumer Behavior* 1 (3): 228–38.

Chen, S.-L. S., G. J. Hall, and M. D. Johns. 2004. Research paparazzi in cyberspace: The voices of the researched. In *Online social research: Methods, issues, and ethics*, edited by M. D. Johns, S.-L. S. Chen, and G. J. Hall, 157–75. New York: Peter Lang.

Cherny, L. 1999. Conversation and community: Chat in a virtual world. Stanford, CA: CSLI Publications.

Clark, D. 2004. "What if you meet face to face?: A case study in virtual/material research ethics. In *Virtual research ethics: Issues and controversies*, edited by E. A. Buchanan, 246–61. Hershey, PA: Information Science Publishing.

Clegg Smith, K. M. 2004. Electronic eavesdropping: The ethical issues involved in conducting a virtual ethnography. In *Online social research: Methods, issues, and ethics*, edited by M. D. Johns, S.-L. S. Chen, and G. J. Hall, 223–38. New York: Peter Lang.

Cooper, A., D. L. Delmonico, and R. Burg. 2000. Cybersex users, abusers, and compulsives: New findings and implications. *Sexual Addiction & Compulsivity* 7 (2): 5–29.

Correll, S. 1995. The ethnography of an electronic bar: The lesbian cafe. *Journal of Contemporary Ethnography* 24 (3): 270–98.

Döring, N. 2002. Studying online love and cyber romance. In *Online social sciences*, edited by B. Batinic, U.-D. Reips, and M. Bosnjak, 233–356. Kirkland, WA: Hogrefe & Huber Publishers.
Dunbar, C. Jr., D. Rodriguez, and L. Parker. 2003. Race, subjectivity, and the interview process. In *Inside interviewing: New lenses, new concerns*, edited by J. A. Holstein and J. F. Gubrium, 33–53. Thousand Oaks, CA, London, New Delhi: Sage.
Eder, D. and L. Fingerson. 2003. Interviewing children and adolescents. In *Inside interviewing: New lenses, new concerns*, edited by J. A. Holstein and J. F. Gubrium, 33–53. Thousand Oaks, CA, London, New Delhi: Sage.
Fetterman, D. M. 1998. *Ethnography*, 2nd ed. Applied Social Research Methods Series, Volume 17. Thousand Oaks, CA: Sage.
———. 2002. Web surveys to digital movies: Technological tools of the trade. *Educational Researcher* 31 (6): 29–37.
Fisher, B., M. Margolis, and D. Resnick. 1996. Breaking ground on the virtual frontier: Surveying civic life on the Internet. *American Sociologist* 27 (1): 11–29.
Forte, M. 2004. Co-Construction and field creation: Web site development as both an instrument and relationship in action research. In *Virtual research ethics: Issues and controversies*, edited by E. A. Buchanan, 219–45. Hershey, PA: Information Science Publishing.
Garcia, A. C., and J. B. Jacobs. 1998. The interactional organization of computer-mediated communication in the college classroom. *Qualitative Sociology* 21 (3): 299–317.
———. 1999. Eyes of the beholder: Understanding the turn-taking system in quasi-synchronous computer-mediated communication. *Research on Language and Social Interaction* 32 (4): 337–369.
Gardiner, M. 2000. *Critiques of everyday life*. London and New York: Routledge.
Giesler, M., and M. Pohlmann. 2003. The anthropology of file sharing: Consuming Napster as a gift. *Advances in Consumer Research* 30: 273–79.
Gilham, C., and B. A. Moody. 2001. Face to-face: Video conferencing creates opportunities for incarcerated youth. *Journal of Correctional Education* 52 (1): 29–31.
Goffman, E. 1959. *The presentation of self in everyday life*. Garden City, New York: Doubleday.
Green, N. 1999. Disrupting the field: Virtual reality technologies and "multisited" ethnographic methods. *American Behavioral Scientist* 43 (3): 409–21.
Hampton, K. N., and B. Wellman. 1999. Netville online and offline: Observing and surveying a wired suburb. *American Behavioral Scientist* 43 (3): 475–492.
———. 2001. Long distance community in the network society: Contact and support beyond Netville. *The American Behavioral Scientist* 45 (3): 476–495.
Haythornthwaite, C. and M. M. Kazmer. 2002. Bringing the Internet home: Adult distance learners and their Internet, home and work worlds. In *The Internet in everyday life*, edited by B. Wellman and C. Haythornthwaite, 431–63. Oxford: Blackwell.
Heath, D., Koch, E., Ley B., and M. Montoya. 1999. Nodes and queries: Linking locations in networked fields of inquiry. *The American Behavioral Scientist* 43: 450–63.
Highmore, B. 2002. Everyday life and cultural theory: An introduction. New York: Routledge.
Hine, C. 2000. *Virtual ethnography*. London/Thousand Oaks/New Delhi: Sage.
Huffaker, D. A., and S. L. Calvert. 2005. Gender, identity, and language use in teenage blogs. *Journal of Computer-Mediated Communication* 10 (2) article 1.
Johns, M. D., G. J. Hall, and T. L. Crowell. 2004. Surviving the IRB review: Institutional guidelines and research strategies. In *Online social research: Methods, issues, & ethics*, edited by M. D. Johns, S.-L. S. Chen, and G. J. Hall, 105–24. New York: Peter Lang.
Jones, R. H. 2005. "You show me yours, I'll show you mine": The negotiation of shifts from textual to visual modes in computer-mediated interaction among gay men. *Visual Communication* 4 (1): 69–92.

Keating, E., and G. Mirus. 2003. American Sign Language in virtual space: Interactions between deaf users of computer-mediated video communication and the impact of technology on language practices. *Language in Society* 32: 693–714.

Kendall, L. 2002. *Hanging out in the virtual pub: Masculinities and relationships online*. Berkeley, CA: University of California Press.

________. 2004. Participants and observers in online ethnography: Five stories about identity. In *Online social research: Methods, issues, & ethics*, edited by M. D. Johns, S.-L. S. Chen, and G. J. Hall, 125–40. New York: Peter Lang.

Kleinman, S. S. 2004. Researching OURNET: A case study of a multiple methods approach. In *Online social research: Methods, issues, and ethics*, edited by M. D. Johns, S.-L. S. Chen, and G. J. Hall, 48–62. New York: Peter Lang.

Koufaris, M. 2001. The search for a new identity: Post-organ transplant recipients and the Internet. In *Our virtual world: The transformation of work, play and life via technology*, edited by L. Chidambaram and Z. Ilze, 291–33. Hershey, PA and London: Idea Group Publishing.

Kozinets, R. V. 1998. On netnography: Initial reflection on consumer research investigations of cyberculture. *Advances in Consumer Research* 25: 366–71.

________. 2001. Utopian enterprise: articulating the meanings of Star Trek's culture of consumption. *Journal of Consumer Research* 28: 67–88.

________. 2002. The field behind the screen: Using netnography for marketing research in online communities. *Journal of Marketing Research* 39 (1): 61–72.

________. 2005. Communal big bangs and the ever-expanding netnographic universe. *Thexis* 3: 38–41.

________. Forthcoming. Netnography 2.0. *Handbook of qualitative research methods in marketing*, edited by Russell W. Belk. Edward Elgar Publishing.

Kozinets, R. V., and J. Handelman. 1998. Ensouling consumption: A netnographic exploration of the meaning of boycotting behavior. *Advances in Consumer Research* 25: 475–80.

LeBesco, K. 2004. Managing visibility, intimacy, and focus in online critical ethnography. In *Online social research: Methods, issues, and ethics*, edited by M. D. Johns, S.-L. S. Chen, and G. J. Hall, 63–79. New York: Peter Lang.

Leung, Linda. 2005. *Virtual ethnicity: Race, resistance and the World Wide Web*. Burlington, VT: Ashgate Publishing Company.

Liebow, E. 1967. *Tally's corner: A study of Negro street corner men*. Boston, Toronto, London: Little, Brown and Company.

Lindlif, T. R., and M. J. Shatzer. 1998. Media ethnography in virtual space: Strategies, limits, and possibilities. *Journal of Broadcasting & Electronic Media* 42 (2): 170–90.

Livia, A. 1999. Doing sociolinguistic research on the French Minitel. *American Behavioral Scientist* 43 (3): 422–35.

Lyman, P., and N. Wakeford. 1999. Introduction: Going into the (virtual) field. *American Behavioral Scientist* 43 (3): 359–76.

Lyons, K. 2003. Everyday wearable computer use: A case study of an expert user. *Mobile HCI*: 61–75.

Lysloff, R. T. A. 2003. Musical community on the Internet: An on-line ethnography. *Cultural Anthropology* 18 (2): 233–63.

Maczewski, M. 1999. Interplay of online and onground realities: Internet research on youth experiences online. Master's thesis, University of Victoria, Victoria, British Columbia, Canada.

Maczewski, M., M. A. Storey, and M. Hoskins. 2004. Conducting congruent, ethical, qualitative research in Internet-mediated research environments. In *Virtual research ethics*, edited by E. Buchanan, 62–78. Information Science.

Mann, C., and F. Stewart. 2000. *Internet communication and qualitative research: A handbook for researching online*. London: Sage.

———. 2002. Internet interviewing. In *Handbook of interview research: Context and method*, edited by J. A. Holstein and J. F. Gubrium, 603–28. London: Sage.

Markham, A. N. 2004. Representation in online ethnographies: A matter of context sensitivity. In *Online social research: Methods, issues, & ethics*, edited by M. D. Johns, S.-L. S. Chen, and G. J. Hall, 141–55. New York: Peter Lang.

Marziali, E., and P. Donahue. 2006. Caring for others: Internet video-conferencing group intervention for family caregivers of older adults with neurodegenerative disease. *The Gerontologist* 46 (3): 398–403.

Miller, D., and D. Slater. 2000. *The Internet: An ethnographic approach*. Oxford & New York: Berg.

Nelson, M. R., and C. C. Otnes. 2005. Exploring cross-cultural ambivalence: A netnography of intercultural wedding message boards. *Journal of Business Research* 58: 89–95.

Nissenbaum, H. 2003. Securing trust online: Wisdom or oxymoron? In *Virtual publics: Policy and community in an electronic age*, edited by B. E. Kolko, 134–71. New York: Columbia University Press.

Paccagnella, L. 1997. Getting the seat of your pants dirty: Strategies for ethnographic research in virtual communities. *Journal of Computer-Mediated Communication* 3 (1) June.

Pauwels, L. 2005. Web sites as visual and multimodal cultural expressions: Opportunities and issues of online hybrid media research. *Media, Culture, and Society* 27 (4): 604–13.

Pink, S. 2004. Making links: On situating a new Web site. *Cambridge Journal of Education* 34 (2): 211–22.

Porr, B. W., and R. E. Ployhart. 2004. Organizational research over the Internet: Ethical challenges and opportunities. In *Virtual research ethics: Issues and controversies*, edited by E. A. Buchanan, 130–55. Hershey, PA: Information Science Publishing.

Reinharz, S., and S. E. Chase. 2003. Interviewing women. In *Inside interviewing: New lenses, new concerns*, edited by J. A. Holstein and J. F. Gubrium, 73–90. Thousand Oaks, CA, London, New Delhi: Sage.

Riva, G. 2002. The sociocognitive psychology of computer-mediated communication: The present and future of technology-based interactions. *CyberPsychology and Behavior* 5 (6): 581–98.

Roberts, L. D. and M. R. Parks. 2001. The social geography of gender-switching in virtual environments on the Internet. In *Virtual gender: Technology, consumption and identity*, edited by E. Green and A. Adam, 265–85. London: Routledge.

Roberts, L., L. Smith, and C. Pollock. 2004. Conducting ethical research online: Respect for individuals, identities and the ownership of words. In *Virtual research ethics: Issues and controversies*, edited by E. A. Buchanan, 156–73. Hershey, PA: Information Science Publishing.

Ruhleder, K. 2000. The virtual ethnographer: Fieldwork in distributed electronic environments. *Field Methods* 12 (1): 3–17.

Salaff, J. W. 2002. Where home is the office. In *The Internet in everyday life*, edited by B. Wellman and C. Haythornthwaite, 464–95. Malden, MA: Blackwell.

Schaap, Frank. 2002. *The words that took us there: Ethnography in virtual reality*. Amsterdam: Askant Academic Publishers.

Schwalbe, M. L. and M. Wolkomir. 2003. Interviewing men. In *Inside interviewing: New lenses, new concerns*, edited by J. A. Holstein and J. F. Gubrium, 33–53. Thousand Oaks, CA, London, New Delhi: Sage.

Shoham, Aviv. 2004. Flow experiences and image making: An online chat-room ethnography. *Psychology and Marketing* 21 (10): 855–82.

Silver, D. 2003. Communication, community, consumption: An ethnographic exploration of an online city. In *Virtual publics: Policy and community in an electronic age*, edited by B. E. Kolko, 327–54. New York: Columbia University Press.

Soukup, Charles. 1999. The gendered interactional patterns of computer-mediated chatrooms: A critical ethnographic study. *Information Society* 15 (3): 161–176.

________. 2000. Building a theory of multimedia CMC: An analysis, critique and integration of computer-mediated communication theory and research. *New Media & Society* 2 (4): 407–25.

Suoranta, J., and H. Lehtimäki. 2004. *Children in the information society*. New York: Peter Lang.

Sveningsson, M. 2004. Ethics in Internet ethnography. In *Virtual research ethics: Issues and controversies*, edited by E. A. Buchanan, 45–61. Hershey, PA: Information Science Publishing.

Taylor, T. L. 1999. Life in virtual worlds: Plural existence, multimodalities, and other online research challenges. *American Behavioral Scientist* 43 (3): 436–49.

Turkle, S. 1995. Life on the screen: Identity in the Age of the Internet. New York: Simon & Schuster.

Vayreda, A., A. Galvez, F. Nunez, and B. Callen. 2002. *Participating in an electronic forum: The difference gender makes*. Internet Research 3.0: Net/Work/Theory. Maastricht, The Netherlands.

Waskul, D., and M. Douglass. 1996. Considering the electronic participant: Some polemical observations on the ethics of on-line research. *The Information Society* 12: 129–39.

Walstrom, M. K. 2000a. The eating disorder is not YOU: Applying Bakhtin's theories in analyzing narrative co-construction in an Internet support group. In *Studies in Symbolic Interaction*, edited by N. Denzin, 241–60. Greenwich, CT: JAI Press.

________. 2000b. "You know, who's the thinnest?": Combating surveillance and creating safety in coping with eating disorders online. *CyberPsychology* 3 (5): 761–83.

________. 2004a. Ethics and engagement in communication scholarship: Analyzing public, online support groups as researcher/participant-experiencer. In *Virtual research ethics: Issues and controversies*, edited by E. A. Buchanan, 174–202. Hershey, PA: Information Science Publishing.

________. 2004b. "Seeing and sensing" online interaction: An interpretive interactionist approach to USENET support group research. In *Online social research: Methods, issues, & ethics*, edited by M. D. Johns, S.-L. S. Chen, and G. J. Hall, 81–97. New York: Peter Lang.

Weckert, J. 2000. What is new or unique about Internet activities? In *Internet Ethics*, edited by D. Langford, 47–64. New York: St. Martin's Press.

Weissman, Jacqueline P. 1970. *Stations of the lost: The treatment of skid row alcoholics*. Chicago and London: Chicago.

Wheeler, D. 2001. New technologies, old culture: A look at women, gender, and the Internet in Kuwait. In *Culture, technology, communication: Toward an intercultural global village*, edited by C. Ess, 187–212. Albany: SUNY.

White, M. 2003. Too close to see: Men, women, and Webcams. *New Media & Society* 5 (1): 7–28.

Whitty, M. 2002. Liar, liar! An examination of how open, supportive and honest people are in chat rooms. *Computers in Human Behavior* 18: 343–52.

________. 2003. Cyber-flirting: Playing at love. *Theory and Psychology* 13 (3): 339–57.

________. 2004. Peering into online bedroom windows: Considering the ethical implications of investigating Internet relationships and sexuality. In *Virtual research ethics: Issues and controversies*, edited by E. A. Buchanan, 203–18. Hershey, PA: Information Science Publishing.

Whitty, M., and J. Gavin. 2001. Age/sex/location: Uncovering the social cues in the developmint of online relationships. *CyberPsychology and Behavior* 4 (5): 623–30.

Williams, J. P., and H. Copes. 2005. How edge are you?: Constructing authentic identities and subcultural boundaries in a straightedge Internet forum. *Symbolic Interaction* 28 (1): 67–89.

Wise, J. M. 2004. An immense and unexpected field of action. *Cultural Studies* 18 (2/3): 424–42.

65

Ethnography, the Internet, and Youth Culture: Strategies for Examining Social Resistance and "Online-Offline" Relationships

Brian Wilson

> Virtual reality and cyberspace are commonly imagined in terms of reaction against, or opposition to, the real world. . . . In certain cases, these are presented as some kind of Utopian project. Virtual Reality is imagined as a nowhere-somewhere alternative to the dangerous conditions of contemporary social reality. . . . The mythology of cyberspace is preferred over its sociology. I have argued that it is time to re-locate virtual culture in the real world (the real world that virtual culturalists, seduced by their own metaphors, pronounce dead or dying). Through the development of new technologies, we are, indeed, more and more open to experiences of de-realisation and de-localisation. But we continue to have physical and localised existences. We must consider our state of suspension between these conditions. (Robins, 1996, pp. 16, 26)

In the years since Robins' (1996) critique, researchers have made important strides to better understand links between online and offline cultural life. Scholars like Burkhalter (1999), Ebo (1998) Harcourt (1999), and Stubbs (1999) have examined how race/ethnicity, class, and gender related identities are experienced offline and online. Other researchers have considered, for example, how youth subcultural life is a continuous virtual-real experience (Bennett, 2004; Wilson & Atkinson, 2005). Still others have

Source: *Canadian Journal of Education*, 29(1) (2006): 307–328.

examined online and offline addiction/recovery support group conventions (Pleace, Burrows, Loader, Muncer, & Nettleton, 2000).

Researchers are also considering how to best understand the experiences of those who navigate across online and offline spaces. In this context, Hine (2000), Markham (1998), Miller and Slater (2000), Mann and Stewart (2002, 2003), Kendall (1999), and others have emphasized the importance of an ethnographic approach to Internet research, and offered important insights into the use of (and relationships/differences between) computer mediated communication (CMC) and face to face (FTF) methods in interviewing and focus groups. Ethnographers who study Internet-related topics struggle to develop and apply novel approaches to their research, while remaining sensitive to still-useful elements of conventional techniques for qualitative inquiry. Markham (1998) called this the "the paradox of conducting a non-traditional ethnography in a non-traditional nonspace, with traditional sensibilities" (p. 62).

There remains much to explore in these areas given the various relationships between online and offline qualitative methodologies. For example, the Internet is a space where research subjects are recruited for offline interviews, documents produced by culture members are accessed for analysis, and experiential ethnographic explorations through cultural spaces and online environments take place. The area of research that specifically considers relationships between online and offline ethnographic methods remains especially rich for development because there is a relative scarcity of work that includes detailed reflections on ways that online and offline ethnographic techniques can be integrated to aid research focused on cultural groups and especially on cultural flows – a topic of particular interest for those who study the globalization of culture. The reason that this is important for those working in education is that understanding the dissemination of culture is a way of understanding the dissemination of a dominant form of knowledge for young people – cultural knowledge – and the sets of cultural knowledge that people acquire and possess inform their interpretations of the world around them (including their interpretations of forms of knowledge they are offered in formal education contexts).

The goal of this article is to contribute to existing literature around these topics by offering methodological reflections from my experiences conducting an ethnographic study of online and offline cultural life in a youth subculture; describing and outlining the rationale for the methodology for a recently designed study of the online and offline cultural lives of members of youth-driven social movement groups; and ultimately offering a set of suggestions for examining social resistance in a global age – an age where (youth) culture circulates globally and locally, and where collective action is increasingly transnational. The argument that underlies this article is that the integration of ethnographic methods, both traditional (offline and face-to-face) and virtual, can be helpful in developing rich and comprehensive understandings of relationships between online and offline cultural life, and for examining the diffuse character of youth culture and resistance. This research is particularly pertinent for those interested in the ways that young people interact with and through Internet technology in and out of educational settings, the ways that online and offline cultural lives of youth transcend educational settings, and for those concerned with the ways that knowledge of social issues is sometimes translated into social action.

In making this argument, I acknowledge that a combined online-offline approach is not always preferable to exclusively online or exclusively offline studies of Internet cultures and experiences. The choice of methods is largely dependent on the goal of the research and 'strands of experience' that the researcher is interested in studying (Eichhorn, 2001). As Hine (2000, p. 59) observes, even studies that include research conducted both online and offline should not be viewed as holistic, given that all ethnographic accounts are selective and partial. However, for research concerned with tracing connections/relationships between online forms of social organization and activism, and offline interactions and action, a *multi-site and multi-method approach* is sensible and desirable – although not without challenges and problems – as I intend to show.

Ethnography: Theory, Method, and Rationalizing an Online and Offline Approach

Ethnographic Methods and the Boundaries of Ethnographic Research

Although ethnography is a notoriously ambiguous term, some consensus occurs around the idea that ethnography includes some combination of participant and non-participant observation, informal and semi-structured interviews, and document analysis – and that the process of writing up research findings and (re)presenting life worlds is integrally related to the act of doing ethnography (Prus, 1996; Tedlock, 2000). Although oral interactions have traditionally been privileged "as part of the 'romantic legacy' of ethnography, that tends to treat speech as more authentic than writing," increasingly the written texts associated with cultures have become equally valued accounts of the realties of those being studied (Hine, 2000, p. 51, drawing on Hammersley & Atkinson, 1995). Hine (2000) elaborates on this more inclusive view of texts.

> Rather than being seen as more or less accurate portrayals of reality, texts should be seen as ethnographic material which tells us about the understanding which authors have of the reality which they inhabit. Texts are an important part of life in many of the settings which ethnographers now address, and to ignore them would be a highly impartial account of cultural

> practices. Rule books, manuals, biographies, scientific papers, official statistics and codes of practice can all be seen as ethnographic material in the ways in which they present and shape reality and are embedded in practice. (p. 51)

Including online and offline texts as part of a broader ethnographic analysis means properly contextualizing and situating the writing and reading practices in ways that make the texts meaningful, or as Hine (2000) argues "tying those texts to particular circumstances of production and consumption" (p. 52). Implicit to this argument is the view that ethnographic work around the Internet ideally takes place in multiple sites, a point pursued later in this paper.

Connecting Theory and Method Around a Critical Interactionist Approach

Although the roots of sociology-based ethnographic research lie in the symbolic interactionist tradition, ethnography (especially as it related to the study of urban youth cultures) came to be associated with more critically oriented writing and studies in the 1970s at the University of Birmingham's Centre for Contemporary Cultural Studies. The Centre, and especially centre-associate Paul Willis, advocated an approach known loosely as "critical interactionism" (Willis, 1977; see Kincheloe and McLaren, 2000 for an overview of this and related traditions). This approach – an approach that guided the research described in this article – is essentially an integration of a conventional, interpretive, micro-sociological approach to research and a structuralist approach to interpretation.

The critical interactionist position is aligned with a critical-realist stance – a positioning that is somewhat distinct from the more relativist stances adopted by those influenced by some strands of postmodern theory. The ontological position associated with critical realism is linked to the epistemological assumption that although multiple (even endless) interpretations of texts exist – whether these be images, webpage documents, interview transcripts – at some point a "relative anchorage" of meaning in interpretations of texts occurs (Hall, 1985, p. 93). Interpretations of media texts (e.g., webpages) made by media analysts/researchers can be useful in shedding light on how these texts might be used by audiences/users. In her textual analysis work on women in sport media, Margaret Duncan (1990) offers a succinct articulation of this position.

> Responsible textual analysis studies do not assert with absolute certainty how particular texts are interpreted. But they suggest the kinds of interpretations that may take place, based on available evidence, and likely interpretations of a particular text. Ultimately these interpretations must be judged on the basis of the persuasiveness and logic of the researcher's discussion. (p. 27)

This understanding is especially pertinent for textual analysis work focused on unveiling the potential meanings of webpages for users/audiences; and the likely goals of website producers. Textual analysis studies focused on the Internet are ideally complemented by ethnographic work that includes interviews with audiences and producers of online content, especially if insight into the relationships between online and offline cultural life are sought – a point that influenced the design of the second case study presented in this paper, described later.

This reflexive, critical-realist stance is inherently linked with an approach to studying and interpreting (youth) resistance that is grounded in the Marxist-related concepts hegemony and ideology. A neo-Marxist understanding of hegemony presumes a relationship between marginalized youth and a dominant group (e.g., moral entrepreneurs such as law-makers and media producers), a relationship that has been at least tentatively secured because the dominant group has been able to achieve and maintain consent to its dominance and because it has successfully allowed safety valve expressions of resistance amongst those who are marginalized (e.g., symbolic forms of resistance, like shocking hairstyles or musical forms, that might empower some young people, but seldom alter the social conditions that frame and reinforce the circumstances of their oppression – c.f., Hall & Jefferson, 1976; Wilson, 2002a, drawing on Gramsci, 1971). This understanding is inherently linked with Hall's and Duncan's view of interpreting texts (like webpages) because it is based on the assumption that dominant groups maintain their dominance through the use of ideological strategies (e.g., incorporating forms of alternative youth culture into mainstream culture) (Baron, 1989; Schissel, 1997). In a similar way, resistant (youth) groups, especially those who are (relatively) well-equipped and well-organized, often attempt to challenge the hegemony of dominant groups by undermining and disrupting the ideological messages and structures that have been created and disseminated to support their power positions. Succinct examples of this sort of undermining/disruption are evident in the work of those at the Vancouver-based magazine *Adbusters* who devise and publish counter-advertisements and write articles that are intended to unveil the contradictions that underlie the practices of multinational corporations.

Underlying critical analyses of dominant structures and media messages is the assumption that a real set of power relations privileges some groups and marginalizes other groups, and that behind the ideological messages that support and justify these relations (e.g., oversimplified, decontextualized messages/images about youth – see Acland, 1995) is a hidden reality, a reality that must remain hidden if the privilege of powerful groups is to remain relatively unquestioned and unchallenged. For critical interactionist researchers, a primary goal is to uncover contradictions that emerge when comparing the ideological fronts presented by power groups and the actual practices of these groups (Howell, Andrews, & Jackson, 2002). This is important context for subsequent parts of this article, especially the description of and rationale for methodologies that I adopted to study the resistance of youth who were guided by these critical interactionist principles.

Studying Youth Cultural Resistance in a Global Age

What is unique about the critical study of everyday experience and forms of cultural resistance in the age of Internet communication is that these experiences and forms often take place on a global level. The challenge for social theorists and methodologists has been to figure out and clearly articulate what it means to study experience on this level.

Guiding my research on this topic has been the theoretical work of Arjun Appadurai (2000). Appadurai's writings are uniquely focused on theorizing the dynamics of global cultural transmission, or what he called "five dimensions of global cultural flows," to demonstrate the various ways that culture moves across borders and around the world (p. 33). Appadurai outlined five dimensions, or "scapes," that need to be taken into account when examining global cultural flow: ethnoscapes, mediascapes, technoscapes, finanscapes, and ideoscapes. Ethnoscapes refers to the flow of people around the world (e.g., tourists, immigrants, or refugees). Technoscapes refers to the flow of technology (e.g., transnational business relocations or the hosting of mega-events like the Olympics that include the movement of new technologies to various countries). Finanscapes refers to the patterns of capital transfer on a global level. Mediascapes refers to the modes of mediated image distribution (e.g., electronic or print media), and to how these images allow viewers to gain access to other parts of the world. Ideoscapes refers to images that are invested with political-ideological meaning (e.g., propaganda images distributed to and through mass media outlets). At the core of Appadurai's framework is the assumption that the various disjunctures or interactions that occur between global cultural flows (as they relate to the various scapes) offer cultural analysts insight into the complex ways that local cultures relate to global forces, and the ways that culture circulates (Carrington & Wilson, 2002).

A variety of methodological strategies are used to study cultural flows and specifically the ways that culture flows to and from individuals around the world, although methods for studying cultural flow and the Internet (in ways that account for the complexities of global movement described by Appadurai) are only beginning to be considered. The body of work on the broad topic of globalization and cultural flows includes studies focused on the flow and impact of Americanizing forces, such as the global transmission of images of Michael Jordan (and the corporate values associated with Jordan's sponsor Nike [Andrews, Carrington, Mazur, & Jackson, 1996]) and the reactions that people have to these sorts of messages and images

(Wilson & Sparks, 2001). Methods used to examine such phenomena include the analysis of images, spaces, and discourses using textual analysis techniques drawn from media studies, historical analyses that are sensitized to the development of (and relationships between) political, economic, and cultural phenomena over time, and the use of focus group interviews as part of audience research projects that assess how individuals understand cultural messages (from abroad).

Michael Burawoy (Burawoy et. al., 1999) and his colleagues in *Global Ethnography: Forces, Connections, and Imaginations in a Postmodern World* offer a most encompassing approach to the study of global forces and power relations. These studies use what Burawoy termed an "extended case study" approach, an approach that was guided by the following four principles: that the researcher must enter the field to appreciate the experiences of individuals; that fieldwork must take place over time and space; that research must extend from micro-processes to macro-forces; and that theory is extended and challenged as due process when examining research findings (pp. 26–28). This is a useful departure point for considering globalization and the Internet because it requires analyses of both the micro-interactions that take place in the production of Internet content and across the Internet medium, as well as the macro-structures that frame these interactions (e.g., evident in analyses of who has access to the Internet and who offers Internet services). This approach is consistent with the critical interactionist position embodied in the classic work of Paul Willis (1977) – except that the "macro-forces" referred to by some of those in Burawoy et al.'s book include global phenomena not considered by Willis. What the following two case studies offer this literature is a way of thinking theoretically and methodologically about cultural flow as it relates specifically to the Internet communication medium, and with particular attention to flows through online and offline settings, and from local to global spaces.

Studying Youth Culture, Social Resistance, and Online-Offline Cultural Flow

In this section I have described and reflected on two studies of online and offline cultural life, social resistance, and youth. The first is a now-completed study of the rave youth subculture I conducted from 1995–1999 in Southern Ontario, Canada (Wilson, 2002a, 2006; Wilson & Atkinson, 2005). The second is an in-progress study of youth-driven social movement groups, with a focus on relationships between online organization/activism and offline collective action. In these contexts, the term youth is used loosely to refer to adolescents and young adults, with participants in the rave scene ranging from approximately 13–25 years old, although several ravers were older than this. Participants in the youth-driven activist groups range in age from approximately 13–30 years old.

Cultural Resistance, Globalization and Online-Offline Cultural Flow

Three interconnected arguments, derived from literature focused around youth, resistance, social movements, globalization, and communication technology, form the background for the discussion of the two studies. First, the subcultural lives of many Internet-using young people should not be understood as virtual *or* real because the online and offline experiences of youth are oftentimes continuous and interconnected. For example, in an ethnographic study of online-offline relationships embedded in cultural life in Trinidad and Tobago, Miller and Slater (2000) stated that the focus of so much research on "virtuality or separateness as the defining feature of the Internet may well have less to do with the characteristics of the Internet and more to do with the needs of these various intellectual projects" (p. 5). Extending Miller and Slater's point into the context of research on youth cultural life, I assert that it is important to consider not only how the division between online and offline is in many respects (for many youth) a theoretical one, but also how the study of connections between and flows through online and offline requires methodologies that are sensitive to this form of experience and interaction. This point is especially pertinent for those interested in better gauging the cultural experiences of young people in developing pertinent curricula in schools and other settings.

Second, a need exists to more adequately account for and theorize the increasingly global and political character of youth cultural life. Early theoretical explanation offered by those working in a classical American delinquency tradition (e.g., Cohen, 1955) helped subsequent researchers describe how young people react to feelings of marginalization and alienation by connecting with a group of similar others and creating an alternative (counter-middle class) value system. Researchers at the University of Birmingham (at the Centre for Contemporary Cultural Studies – the CCCS) in the 1970s theorized a reactive and proactive youth, a youth who assertively expressed their dissatisfaction with the dominant system through symbolic rituals and outlandish/shocking styles (Hall & Jefferson, 1976; Hebdige, 1979). More recently, theorists studying the most prominent late twentieth century subculture, the rave subculture, have argued that members of this supposedly postmodern youth group are less overtly political and confrontational than those of past subcultural generations (Malbon, 1998, 1999; McRobbie, 1993, 1994; Redhead, 1990; Redhead, O'Connor, & Wynn, 1997). In a similar way, classical social movement theorist Alberto Mellucci (1996) has commented on the apparent disappearance of political action among youth in the 1990s and beyond (compared to previous eras).

These approaches fail to account for more global and political forms of resistance and cultural dissemination that have emerged in recent years around a range of social issues, including the environment, globalization, poverty, and gender/racial/ethnic inequality (Barlow & Clarke, 2002; Klein, 2000; Niedzviecki, 2000; Sage, 1999; Wilson, 2002b), a point I elaborate on in my discussion of the methodology for the second case study. In the same way, the global flows of youth (culture), flows that occur through tourism, mass media, Internet media, migration, and other means, are only beginning to be accounted for in literature around youth resistance (Carrington & Wilson, 2002).

This observation is at the base of the third and final argument, which is, that the identified link between the rise of Internet communication and the emergence of various social movements (and the related potential for transnational collective action, [Downing, 2001; Dyer-Witheford, 2000; Fisher, 1998; Myers, 1994]) has not been investigated in any depth by those who study youth resistance (Wilson, 2002b) This lack of research focused on the global dissemination of youth cultural forms and on the enhanced potential for collective action in an age of Internet communication has also meant that methodologies aimed at examining online-offline culture, global flows of culture, and the impact of Internet communication on attempts at collective action remain underdeveloped and unexplored.

Case Study 1 – Rave Culture, Online and Offline

Rave culture, as it existed in Southern Ontario in the mid to late 1990s, was a largely middle class culture of youth whose members were renowned for their interest in computer-generated dance music, attendance at all-night dance parties and, in many cases, amphetamine drugs. Unlike previous youth subcultures that rejected mainstream progressions in communications and media, ravers embraced technology as part of their philosophy (Wilson, 2002a, 2006). My study did not initially aim to examine relationships between online and offline cultural life, although eventually this became an interest (and requirement) once it was clear that the Internet was a central meeting place, space of organization, and cultural reference point for youth ravers (Wilson & Atkinson, 2005).

I chose a variety of methods and data sources to study this group, many focused around the Internet. I joined three rave newsgroups: two Toronto-based newsgroups and one inhabited by ravers around the world. I read weekly and monthly online rave zines produced in Toronto and abroad. I attended online-offline raves that featured online video of DJs playing at a dance party, and a chatroom where those in attendance at the offline rave location could interact with online participants/viewers. I also, more conventionally, spent time at all-night rave dance parties in various Toronto locations, and conducted in-person and online interviews with rave DJs, rave promoters, and members of the rave subculture.

There were several instances where my work online and offline (especially my involvement on local rave newsgroups) was complementary. The following set of examples drawn from my research is evidence of this:

- information gleaned from reading newsgroup discussions and debates in the local rave scene provided a basis from which to develop early "sensitizing concepts." These experiences and early 'mini-hypotheses' informed the development of interview guides, guides designed to allow for explorations around these identified concepts.
- the local newsgroup was a forum through which I recruited interviewees for both offline and online interviews (although the majority were offline). I was especially aware of the need to be viewed as credible and trustworthy by potential interviewees because ravers were commonly being stigmatized in local and national media at the time (and for this reason, concerned about and suspicious of outsiders who might lurk on their newsgroups). At the same time, I was sensitive to the fact that posts that included too much depth and detail might not be read. For this reason, I adopted what I called a "two message" approach to recruitment. The first letter was a short summary of who I was, the nature of my involvement on the newsgroups, the reasons for the research, and my interest in interviewing ravers. In this letter, I directed those interested in the project to a second letter (i.e., newsgroup posting) where I provided greater detail about myself and the research. I received several responses from interested ravers. Perhaps surprisingly, there was no online negative response to the research request.
- the newsgroup was an excellent information source about upcoming events (i.e., online and offline events). My research schedule was heavily influenced by the daily information I received.

The online and offline qualitative approaches were harmonious in other ways. My online research, focused on websites that promoted the rave scene and offered insights into the rave philosophy, provided an excellent reference point for examining contradictions and tensions within the rave scene, and distinguishing between the official rhetoric about the scene, and the various (and often contradictory) practices and cultural behaviors embedded in it. In several instances, offline interviews included follow-up email conversations that allowed me to continue developing rapport and trust. In a general way, my early experiences in the rave scene (e.g., attending rave parties and reading about the scene on webpages and newsgroups), when referred to during interviews, allowed me to demonstrate to respondents that I was deeply interested in their culture, and was not looking to do a superficial, journalistic story about rave (and drugs) that would further discredit their culture.

The design of the second study, outlined in detail below, was heavily influenced by my realization that the research on rave was not only about a specific group and set of spaces, but also about the circuits through which information about the culture flowed and through which aspects of the culture were enabled. This is akin, retrospectively, to the position that Hine (2000) took in her online ethnographic research.

> By focusing on sites, locales and places, we may be missing out on other ways of understanding culture, based on connection, difference, heterogeneity and incoherence. We miss out on the opportunity to consider the role of space in structuring social relations. [It is simultaneously important to consider] the idea that a new form of space is increasingly important in structuring social relations. This space is the space of flows, which, in contrasts to the space of place, is organized around connection rather than location . . . [and that] the organization of social relations is not necessarily linked to local context in a straightforward way. By analogy, the field site of ethnography could become a field flow, which is organized around tracing connections rather than about location in a singular bounded way. (p. 61)

Case Study 2 – Connected Youth: A Study of Youth-Driven Social Movements, Globalization, and Community in the Age of the Internet

In recent years, youth-driven, social activist networks/organizations that address a variety of social, political, and cultural concerns have become abundant and at times prominent. Issues addressed within these networks are both youth-specific (e.g., school bullying) and more general (poverty, the environment, violence, human rights, Aboriginal issues), and are engaged on both a local and global level. Many of these organizations exist, and in some cases thrive, because they have access to and make strategic use of the far-reaching and relatively inexpensive Internet. In most cases, in fact, webpages are a central meeting point and basis for information dissemination and expression for these groups.

These developments are so notable because studies on youth, to date, have tended to focus on the symbolic, stylistic, apolitical, and local ways that young people respond to their feelings of marginalization and social concerns, a point elaborated on earlier in this article. Moreover, those who study youth resistance have not investigated in any depth the identified link between the rise of Internet communication and the emergence of various (transnational) social movements. For these reasons, several questions about youth, social action, identity, and globalization remain largely unanswered: What do these developments tell researchers about the nature of youth community and social cohesion at a time when young people are increasingly gaining access to and actively using the Internet as a forum for meeting similarly positioned peers? How might the development of these networks/movement-organizations alter how researchers explain the resistive efforts of young people? How might the emergence of transnational youth-driven efforts influence how young people's relationships to globalization are understood, relationships that have traditionally been understood in terms of youth being impacted by global forces? What is the relationship between young people's online (activist) activities and offline social action?

The study design was informed by traditional ethnographic principles, lessons derived from my experience conducting research on the rave subculture, and existing ethnographic work and methodological papers sensitive to online-offline relationships (especially Hine, 2000; Mann & Stewart, 2002, 2003; Sade-Beck, 2004) and to the study of globalization and cultural flows (Burawoy et al., 2000). The research focuses on youth-driven networks/organizations that use the Internet as a primary forum for promotion and communication. Organizations were chosen that had a mandate related to the engagement of local (i.e., Vancouver-based) issues and concerns and participation in and promotion of local networks. Others were chosen because of their engagement of global issues and participation in and promotion of global networks.

Representatives for thirty-seven organizations were interviewed over the course of study. The organizations/movements were focused on a range of issues/topics, including racial conflict, the environment, violence, native youth issues, native activism, globalization, sweatshops, genetic engineering, problems with mass media, war issues, and social development through sport. Because parts of the research are still in progress – and because this article is about the research methods and methodology adopted for the research (not the findings) – I will keep the organizations' identities anonymous.

Phase One. In the first phase, the content of these websites was monitored and analyzed, including an examination of articles on the websites, chatroom activities, and online materials designed to describe and promote the organization/movement. Following this stage, the websites remained a key information source (i.e., keeping the research team informed of upcoming events and emergent issues of interest to the group). The information also provided a basis to compare the values and goals of the group as they were formally described on their website to the informal and actual practices of the organization. In this way, the collected data acted as a foundation for subsequent interviews with movement organizers and website-producers, as well as for the ethnographic work at meetings and conferences (phases 2 and 3).

Phase Two. In phase two, team members conducted in-depth interviews with website producers and movement organizers. The interviews focused on the details of the movements' emergence and development, strategies underlying the movements' promotion (and the role of the Internet in this context), relationships between online and offline movement-activities, and the various identities and perspectives of those involved in the movement. Team members also considered relationships between the various participants in the movement (local relationships, global relationships), between the movement and other movements, and between the movement and others (e.g., mainstream press, politicians, government organizations, key figures relevant to the social concerns of the group). The key goal of this phase was to find out more about the main features of youth movement-organizations and attain a sense of the relationship between online writing/activism and offline culture/activism/action.

Phase Three. Phase three was based around observation and interviewing at formal and informal events organized by these movement-groups (e.g., rallies, cultural festivals, fundraisers, protests, demonstrations, or conventions) and around an analysis of the promotion of and media coverage surrounding higher profile events (e.g., meetings of these groups prior to the G8 Economic Summit or the Earth Summit – this part of the research is in-progress). These events are easily found on event calendars that can be accessed through the Internet. Large-scale events, such as Vancouver's *Rhyme and Resist* – a cultural festival attended by thousands of youth, include workshops on topics such as anti-imperialism and rainforest-defense – occur semi-regularly (while smaller events occur frequently). A combination of observation and informal interviewing followed by fieldnote-taking, transcription of recorded interviews, and ongoing (reflexive) analysis were central to this phase.

Rationale and Reflections

Phase one was designed with several ethnographic principles in mind. Perhaps most notably, the early explorations of the webpages informed the construction of the interview guide. At the same time, the website analysis could be viewed as 'casing the joint,' that is, using information from the website to make decisions about how research team members should present themselves to those the organizations, and for identifying potential gatekeepers and interviewees. I also consider the websites produced by the youth groups as documents to understand the official rules and stances underlying group-culture, acting as a key reference point during other parts of the study where the unofficial and informal rules, systems, and strategies will potentially be uncovered.

The task of assessing the ability of these youth-driven movement groups to organize and raise consciousness around certain issues (and accomplish other related goals they have set), while examining the Internet's role in this process, are being guided by Lemire's (2002) previous research on social movement groups. Lemire's research identified a list of known strategies for enabling social action through Internet communication. Factors identified in his research included using email and websites to mobilize the signing of petitions, using webpages to promote the ideologies/doctrine of the group/movement, and using the webpage as an alternative media source. The aim in considering these and other factors through interviews with key group members is to be in a position to comment on Internet-related strategies for collective action adopted by these groups, and the logic underlying these strategies. I also intend, with this background, to consider how these group members define the success of their Internet-related efforts, and ultimately reflect on the way that the Internet enabled or constrained in their efforts.

Conclusions, Concerns, and Future Directions

In this article, I have raised several issues pertinent to researchers interested in the study of youth cultural resistance in an age of Internet communication. I also consider in reflection how a critical interactionist/ethnographic approach, which is underscored by a theoretical sensitivity to everyday experiences and the power structures that frame these experiences, is a useful guide for understanding youth resistance in a global age and Internet-influenced culture. The Internet is a space where aspects of youth cultural resistance that are evident in community-formation and online production exist alongside and in the same virtual space as forms of culture operated by power groups associated with the Internet (e.g., AOL/TIME Warner). A critical ethnographic approach is useful in sensitizing scholars to this broader context that youth groups exist within, and for considering this struggle in relation to notions of hegemony and ideology described earlier – leading to questions about whether flows of youth cultural resistance are, in fact, effectively challenging power groups, and the extent to which youth cultural forms are consented to or incorporated. Underlying research on topics such as these is a commitment to understand 'whose knowledge counts' and understanding the Internet as a space where battles between different message producers and knowledge claims play out. This sort of research contributes to a body of work in education focused around the diffusion of knowledge from and through various media.

Ultimately, by considering how more conventional approaches to studying social groups' relationships with and use of the Internet (as an interactive, global medium) and the social-political context that Internet use takes place within, a better understanding of media, youth and resistance, and cultural experience into the twenty-first century can be approached.

References

Acland, C. (1995). *Youth, murder, spectacle: The cultural politics of "youth in crisis"* Boulder, CO: Westview Press.

Andrews, D., Carrington, B., Mazur, Z. & Jackson, S. (1996) Jordanscapes: A preliminary analysis of the global popular. *Sociology of Sport Journal*, 13(4), pp. 428–457.

Appadurai, A. (2000). *Modernity at large: Cultural dimensions of globalization*. Minneapolis, MN: University of Minnesota Press.

Barlow, M., & Clarke, T. (2002). *Global showdown: How the new activists are fighting global corporate rule*. Don Mills, ON: Stoddard.

Baron, S. (1989). Resistance and its consequences: The street culture of punks. *Youth & Society*, 21(2), 207–237.

Bennett, A. (2004). Virtual subculture? Youth identity and the Internet. In A. Bennett & K. Khan-Harris (Eds.), *After subculture: Critical studies in contemporary youth culture* (pp. 162–172). New York: Palgrave Macmillan.

Burawoy, M. Blum, J., George, S., Gille, Z., Gowan, T., Haney, L., Klawiter, M., Lopez, S., Riain, S., & Thayer, M. (2000). *Global ethnography: Forces, connections, and imaginations in a postmodern world*. Berkeley, CA. University of California Press.

Burkhalter, B. (1999). Reading race online: Discovering racial identity in Usenet discussions. In M. Smith & P. Kollack (Eds.), *Communities in cyberspace* (pp. 60–75). New York: Routledge.

Carrington, B., & Wilson, B. (2002). Global clubcultures: Cultural flows and 'late modern' dance music culture. In M. Cieslik & G. Pollock (Eds.), *Young people in a risk society: The restructuring of youth identities and transitions in late modernity* (pp. 74–99). Aldershot, Hampshire, UK: Ashgate Publishing.

Cohen, A. (1955). *Delinquent boys: The culture of the gang*. Glencoe IL: The Free Press of Glencoe.

Downing, J. (2001). *Radical media: Rebellious communication and social movements*. Thousand Oaks, CA: Sage.

Duncan, M. (1990). Sports photographs and sexual difference: Images of women and men in the 1984 and 1988 Olympic games. *Sociology of Sport Journal*, 7(1), 22–43.

Dyer-Witheford, N. (1999). *Cyber-Marx: Cycles and circuits of struggle in high technology capitalism*. Chicago: University of Illinois Press.

Ebo, B. (Ed.) (1998). *Cyberghetto or cybertopia: Race, class, and gender on the internet*. Westport, CT: Praeger.

Eichhorn, K. (2001). Sites unseen: Ethnographic research in a textual community. *Qualitative Studies in Education*, 14(4), 565–578.

Fisher, D. (1998). Rumoring theory and the internet: A framework for analyzing the grass roots. *Social Science Computer Review*, 16(2), 158–168.

Gramsci, A. (1971). *Selections from the prison notebooks* of Antonio Gramsci (Ed. & trans. Q. Hare & G. N. Smith). London, UK: Lawrence & Wishart.

Hall, S. (1985). Signification, representation, ideology: Althusser and post-structuralist debates. *Critical Studies in Mass Communication*, 2(2), 91–114.

Hall, S., & Jefferson, T. (Eds.). (1976). *Resistance through rituals: Youth subcultures in post-war Britain*. London, UK: Hutchison.

Hammersley, M., & Atkinson, P. (1995). *Ethnography: Principles in practice*. New York: Routledge.

Harcourt, W. (1999). *Women@Internet: Creating new cultures in cyberspace*. New York: Zed Books.

Hebdige, D. (1979). *Subculture: The meaning of style*. London, UK: Methuen.

Hine, C. (2000). *Virtual ethnography*. Thousand Oaks, CA: Sage.

Howell, J., Andrews, D., & Jackson, S. (2002). Cultural studies and sport studies: An interventionist practice. In J. Maguire & K. Young (Eds.), *Theory, sport & society* (pp. 151–177). New York: JAI.

Kendall, L. (1999). Recontexualizing "cyberspace": Methodological considerations for on-line research. In S. Jones (Ed.), *Doing Internet research: Critical issues and methods for examining the net* (pp. 57–74). Thousand Oaks, CA: Sage.

Kincheloe, J., & McLaren, P. (2000). Rethinking critical theory and qualitative research. In N. Denzin & Y. Lincoln (Eds.), *Handbook of qualitative research* (pp. 279–313). Thousand Oaks, CA: Sage.

Klein, N. (2000). *No Logo: Taking aim at brand bullies*. Toronto: Vintage Canada.

Lemire, M. (2002). Globalization, information society and social movement. In M. Pendakur & R. Harris (Eds.), *Citizenship and participation in the information age* (pp. 310–321). Aurora, ON: Garamond Press.

Malbon, B. (1998). Clubbing: Consumption, identity and the spatial practices of every-night life. In T. Skelton & G. Valentine (Eds.), *Cool places: Geographies of youth cultures* (pp. 266–286). New York: Routledge.

Malbon, B. (1999). *Clubbing: Dancing, ecstasy and vitality*. New York: Routledge.

Mann, C., & Stewart, F. (2002). *Internet communication and qualitative research: A handbook for researching online*. Thousand Oaks, CA: Sage.

Mann, C., & Stewart, F. (2003). Internet interviewing. In J. Gubrium & J. Holstein (Eds.), *Postmodern interviewing* (pp. 81–105). Thousand Oaks, CA: Sage.

Markham, A. (1998). *Life online: Researching real experience in virtual space*. Walnut Creek, CA: Altamira Press.

McRobbie, A. (1993). Shut up and dance: Youth culture and changing modes of femininity. *Cultural Studies*, 7(3), 406–426.

McRobbie, A. (1994). *Postmodernism and popular culture*. London, UK: Routledge.

Melucci, A. (1996). *Challenging codes: Collective action in the age of information*. New York: Cambridge University Press.

Miller, D., & Slater, D. (2000). *The Internet: An ethnographic approach*. New York: Berg.

Myers, D. (1994). Communication technology and social movements: Contributions of computer networks to activism. *Social Science Computer Review*, 12(2), 250–260.

Niedzviecki, H. (2000). *We want some too: Underground desire and the reinvention of mass culture*. Toronto: Penguin.

Pleace, N., Burrows, R., Loader, B., Muncer, S., & Nettleton, S. (2000). On-line with friends of Bill W: Social support and the Net. *Sociological Research On-line* 5(2).

Prus, R. (1996). *Symbolic interaction and ethnographic research: Intersubjectivity and the study of human lived experience*. Albany, NY: State University of New York.

Redhead, S. (1990). *End of the century party: Youth and pop towards 2000*. New York: St. Martin's Press.

Redhead, S., O'Connor, J., & Wynn, D. (Ed.), (1997). *The clubcultures reader: Readings in popular cultural studies*. Malden, MA: Blackwell.

Robins, K. (1996). Cyberspace and the world we live in. In J. Dovey (Ed.), *Fractal dreams: New media in social context* (pp. 1–30). London, UK: Lawrence and Wishard.

Sade-Beck, L. (2004). Internet ethnography: Online and offline. *International Journal of Qualitative Methods*, 3(2). Article 4. Retrieved Maarch 23, 2006, from http://www.ualberta.ca/~iiqm/backissues/3_2/pdf/sadebeck.pdf

Sage, G. (1999). Justice do it! The Nike transnational advocacy network: Organization, collective actions, and outcomes. *Sociology of Sport Journal*, 16, 206–235.

Schissel, B. (1997). *Blaming children: Youth crime, moral panics, and the politics of hate*. Halifax, NS: Fernwood Publishing.

Stubbs, P. (1999). Virtual diaspora? Imagining Croatia on-line. *Sociological Research On-line* 4(2). Retrieved March 23, 2006, from http://www.socresonline.org.uk/4/2/stubbs.html

Tanner, J. (1996). *Teenage troubles: Youth and deviance in Canada*. Toronto: Nelson Canada.

Tedlock, B. (2000). Ethnography and ethnographic representation. In N. Denzin & Y. Lincoln (Eds.), *Handbook of qualitative research* (second edition) (pp. 455–486). Thousand Oaks, CA: Sage.

Willis, P. (1977). *Learning to labour: How working class kids get working class jobs*. New York: Columbia University Press.

Wilson, B. (2002a). The Canadian rave scene and five theses on youth resistance. *Canadian Journal of Sociology*, 27(3), 373–412.

Wilson, B. (2002b). The "anti-jock" movement: Reconsidering youth resistance, masculinity and sport culture in the age of the Internet. *Sociology of Sport Journal*, 19(2), 207–234.

Wilson, B. (2006). *Fight, flight, or chill: Subcultures, youth, and rave into the 21st century*. Montreal and Kingston: McGill-Queen's University Press.

Wilson, B., & Atkinson, M. (2005). Rave and Straightedge, the virtual and the real: Exploring on-line and off-line experiences in Canadian youth subcultures. *Youth & Society*, 36(3), 276–311.

Wilson, B. & Sparks, R. (2001). Michael Jordan, sneaker commercials, and Canadian youth cultures. In D. Andrews (Ed.), *Michael Jordan Inc.: Corporate sport, media culture, and late modern America* (pp. 217–255). Albany, NY: State University of New York Press.

OBSERVATION METHODS

SAGE BENCHMARKS IN SOCIAL RESEARCH METHODS

OBSERVATION METHODS

VOLUME II

Edited by

Barry Smart, Kay Peggs and Joseph Burridge

SAGE

Los Angeles | London | New Delhi
Singapore | Washington DC

SAGE

Los Angeles | London | New Delhi
Singapore | Washington DC

SAGE Publications Ltd
1 Oliver's Yard
55 City Road
London EC1Y 1SP

SAGE Publications Inc.
2455 Teller Road
Thousand Oaks, California 91320

SAGE Publications India Pvt Ltd
B 1/I 1, Mohan Cooperative Industrial Area
Mathura Road
New Delhi 110 044

SAGE Publications Asia-Pacific Pte Ltd
3 Church Street
#10-04 Samsung Hub
Singapore 049483

First published 2013

Typeset by Arete Publishing Private Limited, Delhi

Printed on paper from sustainable resources

Printed by MPG Printgroup, UK

Library of Congress Control Number: 2012942348

British Library Cataloguing in Publication Data

A catalogue record for this book is available from the British Library

MIX
Paper from responsible sources
FSC
www.fsc.org
FSC® C018575

ISBN: 978-1-4462-0811-3 (set of four volumes)

Contents

Volume II

B. Reflections on the Practice of Observation (*Continued)*

C. Ethics, Risk and Observation

'The relation between what we see and what we know is never settled'.

Berger, J. (1972). *Ways of Seeing*. Harmondsworth: Penguin: 7.

B. Reflections on the Practice of Observation (*Continued*)

22

A Problem of Sociological Praxis: The Case for Interventive Observation in Field Work

Y. Michal Bodemann

Just as the economists are the scientific representatives of the bourgeoisie, so the socialists and communists are the theorists of the proletariat [. . .] But as history proceeds and as the struggle of the proletariat takes shape more clearly, (these theorists) need no longer look for science in their own minds; they only have to give account of what is happening before their eyes, and make themselves its voice. As long as they seek science and only create systems, as long as they are the beginning of the struggle, they see in misery nothing but misery, without recognising its revolutionary and subversive aspect which will overthrow the old society. But from this point on, science, produced by the historical movement and consciously associated with this movement, has ceased to be doctrinaire and has become revolutionary.

Karl Marx,
The Poverty of Philosophy.[1]

1. Pseudo-Praxis

Over the past few years, the skepticism against the stance of value neutrality, and the reception of the Frankfurt School in the wake of renewed attention to Marxism in a "Critical Sociology" or whatever, has once again focused attention on the problem of the integration of theory

Source: *Theory and Society*, 5(3) (1978): 387–420.

and praxis in sociology. Bearing in mind Marx's famous statement, "Philosophers have only interpreted the world in different ways, but the point is to change it," members of the Frankfurt School recognized the problem, and perhaps difficulty, of praxis in sociology from the very beginning,[2] and associates of Adorno and Horkheimer stated later:

> It is precisely a theory of society for which change does not merely represent a phrase piously invoked on Sundays only, which must work to incorporate within itself the entire force of resistant facticity, if it does not wish to remain an impotent dream – an impotence which can only be of advantage to the power of the status quo. The affinity of empirical social research to praxis, the negative moment of which one should certainly not evaluate lightly, includes also the potential for excluding self-deception and for intervening in reality with precision, and effectively.[3]

In North America, paradoxically perhaps due to the increasing professionalization of the discipline, the notion of praxis has largely been an alien concept or has been misunderstood, and the preoccupation has been since Alvin Gouldner's re-examination of Weber[4] with the problem of value neutrality. It is noteworthy, by the way, that Robert Lynd's[5] strong advocacy of value *commitment* has generally received a more gingerly treatment. Judging from the number of publications in which it has been discussed[6] and reprinted, Howard S. Becker's essay, "Whose Side Are We On?"[7] seems to provide some sort of consensus among the liberal majority in sociology: We are bound to be more sympathetic to the people we study than to their adversaries (or their superiors); we have to make our vantage point explicit and should hope for studies which would present the side of the other parties involved, so that "over the years each 'one sided' study will provoke further studies that gradually enlarge our grasp of all the relevant facets of an institution's operation."[8]

In other words, even though we may take sides "as our personal and political commitments dictate,"[9] someone else surely will come along who will provide the proper "balance," and the problem of our partisanship is thus taken care of in the general realm of sociological relativity where our value commitment experiences some sort of *Aufhebung*; it is neutralized by the contrary commitment of others. Not only, then, does Becker show us a way in which an approximation of value-free sociology is "naturally" assured, but the partisanship which he proposes never transcends that of some sort of "sympathy" towards whatever group happens to be under sociological observation:

> If the difficulty is that we gain sympathy with under-dogs by studying them, is it not also true that the superordinates in hierarchical relationship have their own superordinates with whom they must contend? Is it not true that we might study those superordinates or subordinates,

> presenting their point of view on their relations with their superiors and thus gaining a deeper sympathy with them and avoiding the bias of one-sided identification with those below them?[10]

In this sense, even if the standpoint of value neutrality, as Gouldner has pointed out in his excellent discussion of Becker,[11] has been abandoned with surprising ease, it has equally unsurprisingly been unable to face the problem of value *commitment* in social science.

Where Howard Becker and others are the new "mainliners," to use I. L. Horowitz's slightly tortured terminology[12] in that they have abused the critique on value neutrality to formulate a pseudo-partisanship which turns the sociological product with the same ease into a marketable entity as it happened in the era of professed value neutrality, there are also some "marginals," non-professionals mostly, who, fashionable as it may have been, saw in "action research" or "action anthropology"[13] a sincere attempt at praxis.

Most notable, in this respect, is the approach of Sol Tax[14] who, partly on the suggestion of his students who were shocked by the living conditions of Fox Indians in the American midwest, decided to contribute to a change of these conditions, and yet insisted that the Indians themselves knew best what was good for them.

> But to impose our choices on the assumption that "we know better than they what is good for them" not only restricts their freedom, but is likely to turn out to be empirically wrong. The point is that what is best for them involves what they want to be. Operationally this is knowable only by observing which alternatives they actually choose, and we defeat ourselves to the degree that we choose for them. Hence we find ourselves always discovering and not applying knowledge.[15]

While this lack of paternalism, degree of respect for the community, may be extraordinary for social scientists and particularly for "action" researchers, it appears that Tax and his associates have skirted what appalled his students in the first place: the misery of the community which is not overcome with the establishment of a program for the local tourist industry. Tax and his associates have failed to see and analyze the political roots of this misery, and to present this analysis in the form of options to a community which, on the basis of its demoralization, was unable to see these options.

Only too easily does this approach of "humanistic," unpolitical praxis merge with the type of research suggested by the *Vicos* project of the late Alan Holmberg of Cornell, where paternalism creeps back into action anthropology. Holmberg[16] apparently initiated significant economic changes in the community and, according to his own report, with great success. Here, the community is all too easily turned into a laboratory for the social scientist. What if these autocratically introduced changes had produced unanticipated negative results? When does pre-existing misery justify actions that have the potential of new misery? The sociologist or anthropologist, fortunately, does not have to live with the consequences of his work.

While action anthropology in general may have maintained a certain degree of "humanistic" sensitivity towards members of the setting, the sensitivities of action researchers in sociology, as represented, e.g. by Arthur Shostak,[17] seem to be generally more oriented to the desires of government bureaucracies. In safe distance from much maligned "armchair sociology" the discipline advertises under the label of "action research" or "applied sociology" its worth, its practical usefulness, its marketability to the sources of grants and the holders of power. It is the ideology of lower-echelon professionalism in sociology, and in essential accord with the relativistic advocacy of Howard Becker.

Others again, who claim the status of marginality in contrast to these "mainliners," vigorously reject the tainted label of action sociology. Henry Etzkowitz and his friends at the New School of Social Research confidently proclaimed "Sociology as Praxis" or as the "New Sociology of Institution Formation" in a "Manifesto for Sociologists"[18] (1968). Their point of departure is what even in 1968 was not exactly news in sociological debates: that "(a)lmost all of American sociology has abdicated the responsibility for finding solutions to the most urgent problems facing our society." The new problem-solvers, then, turn away from the "mere objective analysis" of both Structural-Functionalism and the "radical Nihilism" of "Coser, Vidich, and Stein,"[19] and begin solving problems: "American sociology" is returned "to a fundamental and basic institutional involvement"[20] with the establishment of a "Community Cooperative Center with a fifty-fifty white-Negro integrated Board of Directors" in Brooklyn.[21] While the Manifesto is silent about the reactions and sentiments of the second fifty percent to this cooperative, presumably due to the "the terrible void in verbal communication and interaction between Negroes and Whites,"[22] Etzkowitz and Schaflander did find "that when tape is used as an unobtrusive recorder of real-life interactions between people [. . .] the resources and materials are usually genuine and truthful," and "the taping is rarely a barrier or proscenium separating the sociologist from the respondent. In fact, most people tend to forget that the tape recorder is on." Thus, "the Heisenberg principle is overcome" with surprising ease: "because the research instrument (i.e., the long-forgotten tape recorder, or bug) rarely influences the process being studied."[23] Therefore, even though "laboratory manipulated studies" are denounced in the "Manifesto," Etzkowitz and Schaflander themselves establish a laboratory serving simultaneously as Utopia, with complete ignorance of the actual political situation and with contempt for the "young people [who] are bitterly screaming for 'black power' while they roam the streets."[24] Similarly, the vision of the Cooperative in Brooklyn as the starting point of radical transformation of American society blinds these engineers of Utopia against the fact that not alone the tape recorder but the entire cooperative is a research instrument in the hands of sociological experimenters. And the experiment appears to be of short duration, not for the laboratory rats perhaps, but rather for these two sociological engineers: a few months later, they have moved out of the adventure in Brooklyn and Bedford-Stuyvesant to pursue their academic careers in Boston. They may liken themselves to the old CIO organizers, but the difference is apparent: their interest and commitment, symbolically represented in the tape recorder, is different from that of the "slave-ridden Americans in the Ghetto;"[25] the institution they created smacks of paternalism – after all, we engineers know better – and *they* could withdraw at any time – an option not open to the ghetto dweller or the organizer in the CIO. And in fact, they *did* withdraw once the tape reels were filled.

The sociological exploits aside, this type of unreflected, misunderstood and mindless "praxis" is well potrayed in an earlier *Manifesto* of a different sort. Marx and Engels[26] describe these Utopian Socialists (if they can be described as socialists) as those who

> endeavor, and that consistently, to deaden the class struggle and to reconcile the class antagonisms. They still dream of experimental realization of their social Utopias, of founding isolated *phalanstères* [. . .] and to realize all these castles in the air they are compelled to appeal to the feelings and purses of the bourgeois.[27]

With these, in my view objectionable attempts at praxis in mind, it appears that today sociologists who see themselves in the Marxist tradition, are calling more and more emphatically for praxis in sociology, and yet with only the vaguest notions of what praxis may concretely mean, or with great reservations about the notion of praxis as in Norman Birnbaum's (1968) thorough discussion of the problem.[28] Being more specific and optimistic than most others, Hans Peter Dreitzel has stated recently:

> to develop a truly emancipatory social science is to bring about a methodological integration of theory and praxis, or, to be precise, an integration of the empirical research process with practical political involvement. . . . Only in this way can it be scientifically demonstrated that social research is not a methodologically directed focalization by the subject on a given object, but rather a symbolically mediated understanding – a methodologically directed communication between subjects, the goal of which is to bring about emancipatory change in all participants.[29]

In the following, I will examine the problem of praxis for the method of participant observation. It does not suffice to merely propose "praxis" if it remains uninformed by criticism. Contemporary approaches to field work will therefore be critically reviewed and important elements of praxis-oriented field work will be sought. From there, I will attempt to outline the shape of participant observation as praxis.

2. Field Work

The Participant Observer as Master of the Laboratory. The Problem of Participation

In contrast to other empirical research, especially survey studies, one of the most apparent features of participant observer studies are the lengthy methodological appendices and the tormented tone of ethical justifications regarding that particular field project. Their purpose is obvious: social scientists trained one way or another in the catechism of objectivity, replicability and quantitative methods feel a need to explain the validity and accuracy of their procedures to skeptical colleagues, and the ethical problem inherent in *all* empirical research where sociologists are in direct contact with individuals as objects of research, somehow appears to be most evident in the case of participant observation. These lengthy justifications, then, have two basic themes: first, how is participant observation, i.e., a type of "social espionage,"[30] justifiable, given the helplessness and misery which characterizes the members of most settings usually under sociological scrutiny and second, how is participant observation justifiable as a scientific method.

In his detailed and perceptive methodological appendices, Herbert Gans[31] distinguishes three types of participant observation.[32] As *total researcher*, the

> researcher is physically present at the event which he observes, but does not really participate in it. Indeed, his main function is to observe, and to abstain from participation so as not to affect the phenomenon being studied – or at least, to affect it no more than is absolutely unavoidable.[33]

In other words, Gans strikes here a methodologically dubious compromise. On the one hand, he adheres to the dogma that a vividly present observer is, due to his passivity, somehow irrelevant or neutral with regard to the interactions in the setting – the researcher in control of a "pure" laboratory experiment – on the other hand, he has to admit that his presence principally means he is a participant, and therefore has an effect upon the setting.[34] As an example, he mentions his allegedly unobtrusive attendance at meetings; but in another context, in contradiction to his alleged unobtrusiveness, Gans mentions how the organizer of one meeting praised him publicly for his regular attendance; of how others felt honored about his presence on various occasions.[35]

While the second type of participation, that of the "*researcher-participant*" remains rather vaguely defined – it is a hybrid between type one and type three – the definition of the third type, that of "*total participant*" comes more easily. Here, the researcher participates as a " 'real' participant."[36] At least in terms of social interaction, it remains a mystery, how I can participate "really" in one, and "unreally" in another role and setting. The sociological role is a perfectly real and natural role, especially so in contemporary urban settings.

One of the problems here is that Gans fails to introduce the notion of the stranger. Like Becker and others, Gans believes apparently that the sociologist *qua* researcher is exempt from the human affliction of subjectivity. The sociologist, if he only assumes the role of the "not really" participating researcher, is apparently able to have equal access to, and is able to enjoy equal informational yield from, all social settings. And yet, elsewhere Gans admits at least the most blatant cases which make this access impossible:[37] he was unable to gain equal access to the women in Levittown (or the West End in Boston) and his very role as researcher, as sociologist, his academic background, kept him, as he admits, from other sources of information. And further:

> Although being a real participant allows the researcher to understand the functioning of a group like no other method can, it also cuts him off from other parts of the society which are closed to its members. For example, it would be impossible for a participant studying one political party to study the opposition as well.[38]

and:

> [emotional involvement with the setting] allows the observer to understand the people with whom he is living, and to look at the world through their eyes. At the same time, it can also blind him to some of their behavior patterns, and thus distort the study.[39]

Two questions must be raised in this respect. First, is it not inherent in the definition of *participant* observation to make observations in a framework of interaction the purpose of which would be to gain in-depth understanding of the setting in which this interaction takes place? Is it the task of participant observation to produce, much in the style of 19th century novels,[40] omniscient analyses of settings outside the frame of face-to-face interaction? Successful field studies such as Liebow's *Tally's Corner* or Whyte's *Street Corner Society* seem to show the opposite. Second, much of contemporary field method literature seems to understand participant observation in terms of some sort of *model-adjustment*: members of the setting may be asked questions, but the setting is rarely *questionable*. The natural rebellion or deviance or resistance in which individuals are involved in everyday life – admirably characterized in Goffman's "The Underlife of a Public Institution"[41] – seems to be mysteriously absent from most participant-observer stances. The role of the observer, then, is more that of a benign guest than that of a normal member, or even normal stranger. When Gans argues that the full participant may be "blind" to some "behavior patterns" of the group,[42] the explanation for this deficiency must be sought in this "no-questions-asked" type of attitude in observation.[43]

I would argue that precisely the antagonism to the setting itself may provide more important insight into the setting. My conflicts with a bureaucracy, for example, will probably tell me more about this bureaucracy than my complacent submission to its directives.[44] In fact, the antagonism evident in Garfinkel's rather cynical experiments,[45] disruption of routines, questioning of commonsense assumptions etc. is more informative than any "real," but in fact *complacent*, participation elsewhere. Both anthropological and sociological studies are full of situations where most crucial facts became apparent in crises, confrontations and eruptions of conflict.

In short, Gans represents a field work approach, with deeply ingrained respect for value neutrality, which oscillates between various types of involvement and which lastly shies away from full, natural involvement in settings in return for an illusionary omniscience and objectivity. This approach discourages natural involvement in settings because one might develop undue sympathy for the setting (Becker), one might be blinded to some of the behavior patterns of the group (Gans), or, with "overrapport"[46] one might gain too much knowledge of one, too little knowledge of a second group in an observed setting.

Occasionally, however, Gans' scruples about participation and identification with the group and over-rapport become minimized. He writes:

> The identification is probably more intense if the people studied are suffering from deprivation, and if they are a low status group whose point of view is not being taken notice of in the world outside. In such a situation, the researcher feels a need to do something about the deprivation. [. . .] This reaction also befell me. [. . .] Indeed, this identification can be socially useful – at least from the liberal perspective – for the sociologist then becomes an informal spokesman for the groups who themselves lack the power to voice their demands in the larger society.[47]

But while Gans may have enjoyed the role of ombudsperson and advocate which expresses the power of the social scientist over others and which expresses the difference in social class, this pleasurable paternalistic role, and the sociologist's social conscience is sacrificed to the conscience of the detached scientist; in the case of the West End study, advocacy was abandoned in favor of the exigencies of the research project.[48]

This identification with the setting and its advocacy is not only in practice, but also in theory no deeply felt commitment, and is above all overruled by the demands of cool, objective science:

> Although identification can detract from the objectivity of the research, it need not do so – especially if the researcher knows what is happening to him. Moreover, the identification, likely to be strong at the beginning, decreases in intensity as the research proceeds. It is reduced even further in the time which elapses between the end of field work, the data analysis,

> and the writing of the report. Instances of over-identification in the field work can therefore be dealt with in later stages of the research.[49]

Since this type of field research does not demand any deep commitment to the setting – the members of the setting are *subjects* after all – the ethicality of this type of participant-observation is inevitably questionable. Thus, Gans has to admit, "(t)he fact that I was using friendly relationships for the collection of data, coupled with my feeling that I was thus exploiting these relationships, did create some guilt."[50]

The problem with this type of participant observation is then accurately stated: the "friendly relationships" provide data which lead to studies which lead to the researcher's professional advancement, while the benefit to the population, as in the case of the West End Study, and many other studies of its kind, remains demonstrably nil. I would therefore suggest that, while methodologically appropriate participant observation should involve full, natural participation in a setting, politically aware and reflexive, an ethically appropriate field methodology can only be a methodology the primary focus of which is the end of the misery which it observes.[51]

"Role" as Epistemological Harness

> "This event was, of course, a performance break in which I dropped my role as researcher and acted as a person."[52]

If Herbert Gans' view can be considered in the mainstream regarding participant observation, one has to conclude that the notion of value neutrality, the laboratory view of natural field settings and that ethical considerations, in a rather muddled way, all contribute to what amounts to a contrived style in much of contemporary field work, imposing serious constraints upon the stance of the participant observer. In the following, I would like to deal in more detail with a fourth self-imposed constraint: that which is generated in large part by role theory, a peculiar application of Weber's notion of interpretive understanding and a simplistic view of intersubjectivity.[53]

In a paper entitled "Objectivity and Realism in Sociology," Egon Bittner reiterates a point made by Max Weber:[54] "All knowledge of cultural reality, as may be seen, is always knowledge from *particular points of view*."[55] With Weber's apparent backing, Bittner finds that the basic problem of the field worker is that he is an outsider:

> It is not whether he observes well or poorly that matters, but the circumstance of his being an outsider observer with all the consequences issuing from it.[56]

And therefore:

> The more satisfied the field worker is about having done all in his power to solve procedural problems, the less likely he is to note, and if noted attribute significance to, the fact that the reality he has seen and is describing is of a special kind, no matter how closely it resembles the "real" reality.[57]

Bittner seems to argue, then, that there is a radical distinction between the reality experienced by the participant observer *qua* outsider, and that of the "natural," normal member of the setting. The particularity of the status of the outsider is emphasized further:

> Now, into this setting [the setting of practical everyday life; YMB] moves the field worker, a visitor whose main interest in things is to see them and to whom, accordingly, all things are primarily *exhibits*.[58] (his emphasis.)

Apparently, while "natural" members of the setting have practical concerns, the participant observer *qua* outsider has mere curiosity and is removed from the joys and sufferings of members:

> Thus, for example, he might learn to recognize the amusing without being amused by it; and even if he permits himself the joy of amusement, either indulgently or in remorseless exploitation of his own responses, he cannot appreciate it in its natural sense while remaining faithful to his aims as a field worker.[59]

Behind the façade of the detached outsider emerges thus the alienation of the object of research from the vocation of the social scientist; the scientist – even beyond mere detachment – is radically separated from the *interests* – the sufferings and joys – of the setting. The role of field worker, in fact, is characterized by the "special *epoché* of the observer of social reality."[60]

> The field work setting (reality bounded by an ethnographic field work project) is not the field worker's own, because he *cedes* it as not being his world. He has deliberately undertaken to view it as the world of others.[61]

But the contrived and ambivalent style in this type of field work becomes apparent when Bittner nevertheless requires that the "intellectual posture in this sort of field work [be] total immersion in the life studied."[62]

The radical hiatus between the researcher and the setting permits odd phenomenological support for the type of value pluralism proposed by Howard Becker. Becker's relativistic advocacy, however, is driven here to the extreme of passion less changing in and out of roles in the way in which one may change in and out of clothes:

> Since the field worker, as field worker, of course, always sees things from a freely chosen vantage point – chosen, to be sure from among actually taken vantage points – he tends to experience reality as being of subjective origin to a far greater extent than is typical in the natural attitude.

> Slipping in and out of points of view, he cannot avoid appreciating meanings of objects as more or less freely conjured.[63]

Although it may remain unclear how this serial role-swapping is reconciled with the status of the observer as outsider, this "colorful plurality" of the social world[64] calls to mind a very important part of Marx's eleventh thesis on Feuerbach, sometimes omitted in translation, namely, that "philosophers have only interpreted the world *in various ways* [*verschieden*], but the point is, to change it" (my emphasis).

Bittner acknowledges that this slipping in and out of roles, the awareness of cognitive plurality, in fact obfuscates the basic reality of the setting, "those traits of depth, stability and necessity that people recognize as actually inherent in the circumstances of their existence."[65] But the point is more reluctantly acknowledged, there is no remedy; at best, in Bittner's resigned attitude, it might re-assign the "proper relevance"[66] to the field worker's observations. It is my contention that it merely re-assigns the proper relevance to *this particular* observational stance and oblique notion of field work.

Raymond L. Gold probably agrees with Bittner on the point of "slipping in and out of roles":

> While a field worker cannot be all things to all men, he routinely tries to fit himself into as many roles as he can, as long as playing them helps him to develop relationships with informants in his master-role (i.e., participant-as-observer etc.).[67]

But peripatetic role-swapping is not the only way in which participant observers have shielded themselves from the setting, and distanced themselves from its members. Gold, who specifically advises the "protection of the (participant observer) role" in cases of possible emotional outbursts of the "self" in intolerable settings[68] – "performance break" in Gans' terms – insists that "role pretense" is a basic necessity of the participant observer. Much in the way in which Bittner seems to find full immersion compatible with "slipping in and out of roles" of the outsider-observer, Gold apparently sees no contradiction between complete participation and role pretence:

> The true identity and purpose of the complete participant in field research are not known to those whom he observes. He interacts with them as naturally as possible in whatever areas of their living interest him and are accessible to him as situations in which he can play, or learn to play, requisite day-to-day roles successfully.[69]

and:

> Role pretense is a basic theme in these activities. It matters little whether the complete participant in a factory situation has an upper-lower class background and perhaps some factory experience, or whether he has an

> upper-middle class background quite divorced from factory work and the norms of such workers. What really matters is that he knows he is pretending to be a colleague. [. . .] He must bind the mask of pretense to himself or stand the risk of exposure and research failure.[70]

It strikes me as completely mysterious why the field worker cannot "be himself" and why pretense is necessary in these types of circumstances.[71]

What is there to be pretended when the field worker with middle class background works in the setting of the factory? Furthermore, given the contradictory interests of the factory worker and the masked Observer-Sociologist – a type of social spy – does it still make a difference whether the informational yield has been gained through pretense or honesty? The methodologically fruitless introduction of masked observation, or pretense, merely shields and detaches the field worker from the setting *in the sense that* it neglects the promise of full, yet reflexive, participation.

Not unlike Gans and many others, Gold and Bittner hold resolutely to the view of the participant observer as an *outsider*, and disregard his potential as *stranger*: The outsider, here, maintains his shield of detachment from the setting throughout the process of observation; his *participation* is therefore compromised and indeed in serious question: his interaction with members of the setting is "not serious," it remains provisional and fragmentary, continuously subject to the constraints of a role which demands that he not be present "as a person".

The stranger, as newcomer to the group, however, behaves quite differently.[72] As a natural person, though quite aware of his alienness with respect to the group in which he is host, he interacts with people without this shield and artificial constraint. In the case of the stranger, the attribute of alienness is taken into account by the group and absorbed by the setting much in the way in which the attributes of "priest," "pupil," "poet," "deviant" or "shoemaker" are dealt with in a framework of interaction. Unlike the synthetic outsider, the novice, stranger, participates and intervenes in the joys and sufferings of the group, and takes account of members' interests.[73] The participant observer, whether as stranger or familiar member, however, is not satisfied with the facticity of the setting with which he is faced: he participates, and *intervenes*, and intervening, he breaks through the cognitive barrier of the taken-for-granted. I would suggest that this type of stance is more appropriate to the ethnographic observer, and the awareness that he remains – to whatever degree – a stranger to the setting indeed takes it for granted that his knowledge, too, is "always knowledge from a particular point of view".

Participation across Class Barriers: Anthropology

As I have noted earlier, for the case of *action anthropology*, anthropology had to concern itself particularly intensively with the problems of participation and the active, interventive role of the observer in field settings. Murray

Wax[74] has pointed out recently how Malinowski was the first anthropologist to engage in modern-type field work which meant already for him the researcher's intensive involvement in the field. In a demonstrative act, not only to the "savages" and the white administrators and missionaries, but particularly to fellow anthropologists, Malinowski decided to pitch his tent in the midst of a Melanesian village,[75] after he had found it impracticable to commute back and forth from the compound of the white administrators in the area. Around the time when sociologists of the Chicago School began to venture into local neighborhoods, Malinowski stipulated in the *Argonauts* that the proper conditions for ethnographic field work consisted "mainly in cutting oneself off from the company of other white men, and remaining in as close contact with the natives as possible, which really can only be achieved by camping right in their villages,"[76] and it also dawned on him, as Wax has pointed out that it is crucial for the anthropologist to live "intimately and for a prolonged period of time within a single native community."

Nevertheless: despite Malinowski's insistence that "savages" are "rational moral beings," Malinowski (and other anthropologists of his time) still treated their subject with a mixture of ambivalence and contempt. Malinowski may have been deeply involved in the setting, without any pretence of roles or masks or what not, but he rejected empathy with the "savages," saw them as *curiosa* and, above all, he "established himself as petty lord in the village."[77] He was obtrusive and exploitative of villagers much in the way of white colonialists from whom he sought to distance himself. As Wax points out further, Malinowski had to be tolerated by the villagers because he was "white and wealthy:"

> as they knew that I would thrust my nose into everything, even where a well-mannered native would not dream of intruding, they finished by regarding me as part and parcel of their life, to necessary evil or nuisance, mitigated by donations of tobacco.[78]

Unlike sociology which carried its teleological and critical objective along as it developed, from Saint-Simon and Comte to Marx, the Chicago and Frankfurt Schools, the Lynds and C. Wright Mills, the origins of *anthropology* were clearly of a different nature. And when anthropologists began to reflect upon their role, the legacy of Malinowski, among others, revealed itself also as a major burden. As Nash and Wintrob[79] have argued, the concern with the anthropologist's increasing personal involvement with subjects and with the field became a crucial epistemological issue with the famous re-studies: Dollard and Powdermaker on the Deep South,[80] Mead and Fortune on the Arapesh,[81] and most of all, Redfield and Lewis on Tepotztlan.[82] Furthermore, the character of anthropology became questionable with decolonization and the assertion of independence on the part of anthropological colonies: the issue of the anthropologist's interest in the field, the question, "Whose side are we on?" had to be faced.

Thus, J.J. Maquet[83] and others talked about "the disturbing correspondence between our discipline and the colonial regime."[84] Malinowski may have cut himself off "from the company of white men," but he was, financially and biographically, deeply tied to the colonizers, and anthropology has by and large remained that way. When anthropology became reflective, it may have become guilt-ridden, but it did not liberate itself from the class antagonism of anthropology and colony. As we saw, action anthropology, a maverick of sorts, tried to extricate itself either through the attempt to eliminate paternalism (Tax) or by some sort of paternalistic philanthropy, as in the case of Holmberg. Others again were still faced with the Malinowskian dilemma, and they attempted to bend it in various ways. I. C. Jarvie[85] speaks about the problem of "ethical integrity in participant observation" and sees it as an insoluble contradiction and "identity crisis" for the anthropologist:[86] as an observer, one must be a stranger, as a participant, one ought to be a friend; as anthropologist, one is observer, but the basic problem is that the anthropological method requires participation.[87]

Others felt guilt ridden like some sociologists, particularly where colonialism became intrinsic and thus a problem at home. Precisely in this situation, M. Shokeid[88] is tormented by his own insincerity in the field (among North-African immigrants to Israel) by hiding his true background and by exploiting the trust of members of a setting constituted of co-nationals. Unlike in much of sociology, however, the observer's effect on the field is readily acknowledged, for example by Jarvie. Although Raymond Apthorpe[89] finds that the effect of the observer on the field is "subversive of science" and generally unsettling, he dares to ask:

> Wouldn't it be better to accept it [viz., the observer's impact on setting] and to develop it into its logical conclusion, and perhaps even to make it a virtue? Most actions have effect, in the field situation as in the rest of life, and there can be no escaping this.[90]

While Apthorpe sees it as a useful methodological procedure, others find the prevalence of the participant observer role a commandment of philanthropy. Lodewijk Brunt speaks of the anthropologist's "ethical responsibility towards [. . .] humanity," (in Nash and Wintrop, 1972), but the limits of this activism are clearly drawn; anthropology and political activism seem fundamentally irreconcilable:

> Participant observation means *total immersion*. It may also mean action when called upon by his conscience and his hosts. But such action must be taken in integrity [. . .] I am not saying we should become activists, but our social responsibilities should be more happily married to our professional postures.[91]

As summary of this sketchy review of field work in anthropology, I would suggest the following points. First, anthropologists have, by and large, not developed elaborate roles which shielded them from the members of the setting – presumably because anthropologists in the field are naturally, and blatantly, strangers and therefore find the sociologist's artificial construct of the "outsider" dispensable.

Second, unlike sociologists, anthropologists have found role pretense unnecessary. In the past, at least, they have dealt with populations who lacked acquaintance with social scientists and were in this sense a less "sophisticated" population than people in slums and urban settings may be. Moreover, given that anthropologists usually lack all affinity to the economic system of their settings, unable or unwilling to pose as missionaries, traders, urban planners or psychiatric consultants, the question of mask and pretense is groundless. They are, pure and simple, anthropologists.

Third, despite these advantages, however, – lack of contrived roles – anthropologists have generally not overcome the gap in social class, as representatives of hostile interests. The occasional awareness of this gap, as, most notably, in the case of action anthropology, may have led at best to a "humanism" of various sorts, at worst to philanthropic paternalism. As such, however, anthropologists have not touched upon the colonial-oppressed status of their population.

This leaves us with two tasks. One is, to locate methodologically more adequate approaches to participant observation in sociology; the other, to present a model of participant observation *qua* political action: participant observation as *praxis*.

3. Steps towards Praxis: "Surrender," Self-Discovery and Intervention

> Field work is not merely one among several methods of social study, but it is paramount. It is, more than other methods of study, itself a practice, consciously undertaken, in sociology itself – in the perceiving and predicting of social roles, both one's own and those of others.[92]

As I have pointed out before, one very sincere step toward praxis, and yet politically and methodologically naive, was that taken by action anthropology as developed by Sol Tax and his students. Politically naive, because it operated on the level of some sort of do-it-yourself philanthropy; methodologically naive, since it did not concern itself in particular with the heuristic and epistemic consequences of this approach, even though Lisa Peattie and Alan Holmberg, for example, were in fact aware that an interventive approach would reveal the diachronicity, the dynamic elements of the setting usually closed to conventional structural-functionalist anthropology.[93]

In this direction, and remarkably enough, in virtual isolation from their contemporary colleagues, Morris and Charlotte Schwartz[94] have indeed pointed to the methodological characteristics and benefits of an interventive approach. Although still caught in the pseudoproblem of "objectivity" irreconcilable with "empathy," they do find that:

> The participant observer is an integral part of the situation he is observing. [. . .] together, the observer and the observed constitute a context which would be different if either participant were different or were eliminated. [. . .] Continued observed-observer transactions influence in many ways both the kinds of data that emerge (for the observer creates them to some degree) and the registering, interpreting and recording of them.[95]

On the basis of this cautiously formulated premise, M. Schwartz is able to conclude that "action"

> can lead to new insights, understandings and hypotheses about human behavior. For in initiating action on the basis of one's analyses one might uncover phenomena previously unrevealed and see connections heretofore not seen. In addition, action can provide support or lack of support for our hypotheses [. . .] We have the opportunity to verify [sic] [. . .] the relevance and accuracy of our interpretations. Finally, action research permits us to make a direct contribution to organizational goals and social ends we value.[96]

As can readily be seen, this approach is still overshadowed by the idea of the setting as laboratory, as sociological playground. On the other hand, their view has led the Schwartzes to the equally hazardous perhaps, yet very promising idea of field work as social psychiatry, an idea somewhat akin to the use of therapy groups.[97]

From my personal vantage point, I would see a major step toward praxis in the notion of participant observation as formulated by the early Chicago School, and by Everett C. Hughes as its contemporary representative. I would suggest that the major achievements of this School in terms of participant observation were these: first, the Chicagoans developed the 19th century-style "social survey" of Charles Booth and others – almost exclusively descriptive – into a *method:*[98] "People went out and did field observation designed to advance *theoretical* as well as *practical*, knowledge of modern, urban society."[99]

Secondly, this method was designed to *report* in Park's terms, "fully, freely and bitterly"[100] about the conditions in the urban slum, in the hope that this reportage would lead to change. Occasionally, reportage and writing in general was directly agitatorial, as in the case of Albion Small. This approach is notably different from the approach of say, Paul Goehre, a European contemporary of the Chicagoans. Goehre, a German theology student worked for three months in a factory in Chemitz to find out why the German working class was attracted by Social Democracy – and what could be done to stop it. (1891/1895).

But most importantly, perhaps the style of field work developed by Everett Hughes and some of his associates has always seen the participant observer as a *part* of the setting which is observed. This approach, quite correctly in my opinion, sees the participant observer as stranger – perhaps even, as Hughes puts it, "class stranger"[101] – to most settings, and yet:

> The observer, in greater or less degree is caught up in the very web of social interaction which he observes, analyzes and reports.[102]

In terms of his *stance*, he is a participant observer in the way of Malinowski; in fact, the Chicagoans called field work the "anthropological method." He is observer, stranger, but he can do without pretense and elaborate roles. His observing and reporting affects the setting, and may threaten the self-conception of the group:

> The hatred occasionally visited upon the debunking historian is visited almost daily upon the person who reports on the behavior of people he has lived among; and it is not so much the writing of the report as the very act of thinking in such objective terms that disturbs the people observed.[103]

Being "caught up in the web of social interaction" and, as observer, taking one's own role into account may, however, also prove "disturbing" to the observer.[104] I see this reflexive type of participation, "self-discovery," as Hughes has called it, also as the potential of liberation.[105] In another context,[106] a paper entitled, "Teaching as Field Work," Hughes hints in a similar sense at the emancipatory potential, of sociology *qua* self-discovery: he gives an account of students who, reflecting on their own tradition-bound background, are able to detach themselves from that background.[107] Somewhat introspectively perhaps, this type of participant observation becomes identical with critical self-examination and is *thus* a form of praxis. And in "Teaching as Field Work," finally, Hughes presents a viable model of participant observation in its critical, and praxis-sense: here, full, natural participation is reconciled with critical and self-critical observation.

This leaves us with the problem how the participant observer who is, in his stance, a member of the setting, who takes his own role critically into account, who does not succumb to the misery of the setting but wants to change it – how this participant observer can resolve the problem of his own identity vis-à-vis the setting. In a methodological – though not explicitly political sense – I see a solution in the concept of *surrender*, as formulated by Kurt H. Wolff.[108]

Wolff points to the military connotation of the term surrender. I find this connotation indicative and of crucial importance. Because as a military term, applied here to a sociological stance, it points to what sociological research has usually been: the invasion and "control" of alien communities.[109] For most of sociology and anthropology, the antagonism between the researcher and the community has been paramount. The term surrender, however, though military, carries within it also the negation of the military and of antagonism. Surrender (and more closely the German term *Hingebung* as opposed to *Ergebung*), furthermore transcends the antagonism between the researcher and the setting. Wolff can therefore point to its meaning as that of "cognitive love," and other meanings which follow from it, e.g. "total involvement, suspension of received notions, pertinence of everything, identification, and risk of being hurt."[110]

From there, I see two primary ideas in Wolff's notion of "surrender": first, again, that of "cognitive love," which I would understand as a composite of "love" and "understanding"; second, that of the observer's *epoché*: "suspension of received notions." Where the *epoché* implies "suspension of the very distinction between subject and object" and thus: cathectic identification of the researcher with the community, it appears of utmost importance. However, where the *epoché* refers to "suspension of received notions," suspension of notions "which they (political and philosophical radicals) can imagine *in any way* (my emphasis) to be pertinent to their undertakings,"[111] I find it a practical impossibility, much in terms of my critique of Bittner. The sociologist, (qua political and philosophical radical) despite his total involvement in the community, *must* maintain his reflexivity and his analytical commitment. Reflexivity and analysis are deeply grounded in sociology and in its philosophical heritage; in fact, stood at its very beginning: the reflexions of Herakleitos or Plato's analyses could not have been based on the facticity of one isolated polis alone; they were based on the awareness of options, of contrast and on comparability: the concern with other contemporary or historical communities. The field worker *in situ* remains analytical, precisely in the way in which the actor's role is *inseparably* one with his self, and which is furthermore a role which stays with the actor even after he has left his stage:[112] In the world as we experience and practice it, radical *epoché* is impossible. Finally, there may be a *choc*, but there is no *hiatus* between realities: They overlap or are congruent precisely in the way in which sociological praxis is the congruence of two only ideationally separable realities.

But the passage mentioned above suggests also a different interpretation: suspension of received notions "which (political and philosophical radicals) can imagine in any way to be pertinent to their undertakings" may also lead to a qualified suspension of received notions: It is ambiguous whether the passage implies that political radicals, for example, should bracket their political commitment and present themselves at a nullpoint, as faceless beings, to the community. In terms of an alternative interpretation, then, one may ask whether *epoché* here implies, e.g., the elimination of paternalistic radicalism; rejection of the idea of the "let's take a look, but we know of course already what your trouble is."[113]

Despite these reservations regarding the second point, the first point, i.e., the idea of cognitive love, is a crucial component of the stance of field work as praxis, and the sociologist tied to the middle class, "class stranger" to most of his settings is, with praxis, bound to overcome the gap in social class, and in this sense he must *surrender*.[114]

4. The Case for Praxis: Marx's *Enquête Ouvrière*

Most analyses dealing with questionnaires and social surveys at all historically usually mention in passing that Marx, *inter alia*, composed a questionnaire. In the following, I will try to show that Marx's *Enquête Ouvrière* occupies a unique status in social research and that it provides us with a lucid model of social research qua praxis.[115] Hilde Weiss[116] has supplied us with a thorough, penetrating study of the *Enquête*. Looking at Marx's *Enquête* in the context of other social surveys at that time in France, she finds that the *Enquêteurs* saw these surveys in a clearly defined framework: They found it necessary to examine the worker's attitudes and values and know about the subjective needs of the working class in order to satisfy these needs in the limits of the "just" and the "possible." The working class was seen in passive and manipulative terms. The *Dictionnaire de l'Economie Politique*[117] for example states explicitly: "Those who are to be questioned, may in no way participate in the survey". In this sense, there has been little change in the empirical methods of social science over the past century, Hilde Weiss concludes:

> The *enquêteurs* were faced on the one hand with the barrier of the "extent of the possible," i.e., a reform which the government could achieve; on the other hand, they wanted to study the cause of the workers' sympathy for socialist ideas, and the attractiveness of these ideas.[118]

As Hilde Weiss points out, unlike most other *Enquêtes* down to the 1880's, Marx addressed his *Enquête* directly to the workers themselves: not local bureaucrats, clergymen or "honourable citizens," but only the workers can "describe with full knowledge the evils which they endure; only they, and not providential saviours can energetically apply remedies to the social ills which they suffer."[119] As evident in the *Préambule* to the questionnaire and from the questionnaire itself, the *Enquête* had two primary goals: first, to present the workers' own description of their social condition to the public, and second, to increase the consciousness of the working class with the questionnaire *per se*: by means of a systematic catalogue of questions. This explains in part the wide publicity given to the *Enquête*: published in 1880 in the *Revue Socialiste*, 25,000 copies were distributed "to all the workers' societies, to the socialist and democratic groups and circles, to all French newspapers, and to all those individuals who asked for it."[120]

With the near polemical, propagandistic nature of its pointed questions, Marx tries to further shatter the confidence of the working class in the established political system. The *Préambule* states that no "French government dared to make a real investigation into the situation of the working class" since capitalist exploitation, as revealed in the social surveys of the English government, has aroused fear in the French bourgeoisie with regard to an impartial and systematic *Enquête*. And further: "Until the French government begins to make such an investigation, we will have to make the attempt, with the few means available to us, to begin such an investigations."

In other words: the *Préambule* (unlike introductions to most contemporary questionnaires) is fully explicit about the purpose of the questionnaire. The questionnaire itself seeks factual information, and at the same time, is didactic and political. Plain factual questions such as,

46. What kind of work contract do you have with your employer? Are you engaged by the day, by the week, by the month etc.?

are followed by such double-edged, near-cynical questions as #48 and #49:

48. In the event of the contract being broken, what penalty is imposed on the employer if it is his fault?
49. What penalty is imposed on the worker if it is his fault?

Many other questions are formulated in this mode. In #26 and #27, for example, Marx asks sarcastically:

26. In case of accidents, is the employer obliged *by law* to pay compensation to the worker or his family?
27. If not, has he ever paid compensation in any way to those who have met with an accident while working to enrich him?

As Rubel has pointed out,[121] the questionnaire reads in part like an outline of *Kapital*. In particular, the questionnaire covers quite systematically problems of absolute surplus value (the work day), relative surplus value (division of labor, rationalisation of work, factory legislation) and wages (piece rates, time rates, overtime etc.). Thus, while Marx asks factual questions, he simultaneously familiarises the worker with basic theoretical concepts by means of personal, everyday experience. In the second section of the questionnaire, for example, he alludes directly to the concept of relative surplus value:

43. Are the machines cleaned by workers specially employed for this work, or are they cleaned gratuitously by the workers who are employed on them during their ordinary working day?
45. How much time do you lose every day in getting to work and in returning home?

Similarly, he encourages the workers to observe fluctuations in the price of commodities in relation to their wages:

76. Indicate changes in the *price of the commodities* which you produce, or of the services which you provide, and indicate for comparison whether your wage has *changed at the same time*, or whether it remained the same. [his emphasis]

Marx's questionnaire, then, was not only intended for collection and presentation of data, it was also political and didactic. This is no dramatic conclusion. As I have tried to indicate earlier, all field work is intrinsically political. Questionnaires and opinion polls, despite presumptions to the contrary, are political. Opinion polls, for example, which present the voter with the alternative of two political parties or two candidates are political precisely because they exclude other options, and precisely because the findings which exclude these other options have a telescopic effect when they are being broadcast to a wider public.[122] The only dramatic finding may be that, as Werner Fuchs has pointed out, most sociologists seem either unaware or blind to the fact that the interaction of the researcher with the field is a political issue.[123] Furthermore, as Fuchs points out, while methodology textbooks may have recognized the influence of the researcher on the field, these same textbooks do not recognize this predicament as a potential. It appears that the increasing professionalization of the discipline simply discourages certain modes of research which are in conflict with the self-conception of the profession, while other methodological procedures, such as concealed observation, in general, are broadly labelled unethical[124] but again from a professional perspective, much like regulations which forbid physicians to advertise their services.

Where, then, does all this leave the sociologist who takes this Marxian praxis seriously, who has decided to throw in his lot with that of the oppressed and poor? First of all, it destroys the sociologist's arrogant self-image as "social engineer" in the style of Auguste Comte. As Marx states in the *Préambule*, the proletariat needs no providential saviours, "only the workers can describe with full knowledge the evils which they endure," and only *they* can apply the remedies. What the *Enquête* asks *us*, intellectuals, sociologists or what not, to do, is to gather "experiences," *and* to immediately feed back these experiences as *information*: ordered and presented in a theoretical framework. This is what Mao Tse-Tung means presumably by "teaching the masses clearly what we have received from them confusedly." As catalysts for theoretically grounded information, we can realise our educational and political responsibility; or, in the words of *The Poverty of Philosophy*, as theorists of the working class, we give account of what is happening before our eyes and make ourselves its voice.[125] In the *Préambule*, Marx made it clear that his *Enquête Ouvrière* is only a first attempt at a paradigm; it is up to us to take this paradigm seriously.

5. Conclusion

This paper is primarily an attempt to present a critique of current approaches to field work, and especially of participant observation. This particular focus is important for two reasons. First, ethnographic work has become of paramount importance given the sophisticated manipulation of our lives today: the intricate organisation of the work place, the divisive and paralysing structure especially of the urban setting (race and ethnicity, welfare, total institutions, leisure etc.). The second reason for such an extended critique is that it would guide us to politically and methodologically adequate approaches to field work.

I have also attempted to point to the importance of the *Enquête Ouvrière*. In its intent, the *Enquête* complements the *Communist Manifesto* in many ways: both documents should be seen more clearly as program; as guidelines to our work. While the *Enquête* has obvious bearing on how survey research may be conducted in the future, it has also direct implications for other types of field work, especially in conjunction with a critical review of field work approaches; the participant observer might note the following implications in particular:

1. He participates fully, freely and self-critically in the setting.
2. He observes and renders a description of the facts and "on-goings" of the setting, but in the context of his biographical position. (Paraphrasing Giambattista Vico, we could say that he understands the nature of the setting to the extent to which he *lives* – or in his study or in his mind, re-lives, re-creates – the setting.)
3. Given his commitment and the evidence, received and theoretically grounded, he can actively intervene: *first*, by presenting options to a community which has been deprived of options, divided and paralysed, (this may have didactical or organisational implications); *second*, by returning his findings (descriptive and theoretical) to the community, by sharing it with others who partake in a comparable predicament and with all those who identify with this predicament and who are willing to change it.

Committed to these objectives, I believe we fulfill the imperative of Vico's *New Science*: to reconnect his "philosophy" to his "philology"; to confront the "eternally true" value with the empirical fact; to measure what *is* on what actually *ought to be*.

Notes

1. K. Marx and F. Engels, *Werke* ("MEW"), vol. 4. Berlin: Dietz Verlag, p. 143.
2. See, e.g., Hilde Weiss' (1936) superb, yet fairly unknown examination of Marx's *Enquête Ouvrière*. This paper must be considered as a major contribution to the problem

of empirical research and praxis on the part of the Frankfurt School. For a recent, quite different analysis, see J. Habermas (1973).
3. *Frankfurt Institute For Social Research: Aspects of Sociology,* Boston: Beacon Press, 1972, p. 127.
4. A. W. Gouldner, 1961.
5. R. Lynd, *Knowledge for What?*, Princeton: Princeton University Press, 1939.
6. The *American Sociologist* in its February, 1971 issue made Becker's partisanship argument to something of a theme in papers by Peter L. Berger and Gresham Riley. In it, Berger postulates (to the comfort of the ASA, I assume) that sociology as discipline must be value free, while the "practitioner of sociology must not be value free" (to the comfort of Berger's liberal friends, I assume). Unfortunately, it remains unclear in his presentation how these two postulates may be reconciled, except perhaps in that the relativism on the individual level is, precisely as in Becker, transcended in social science at large, for he states: "There are plenty of passions available, and the sociologist may well participate in some of them." (1971:5).

   Responding to Gresham Riley, declared adversary of the sociology of knowledge and partisan of non-partisanship, or value-neutrality (1971, pp. 6–12), Becker recommits himself quite firmly into the ranks of objectivism.
7. H. S. Becker, *Social Problems*, 14, 1967, pp. 239–47.
8. *Ibid.*, p. 247.
9. *Ibid.*
10. *Ibid.*, p. 246.
11. A. W. Gouldner, "The Sociologist as Partisan: Sociology and the Welfare State," *The American Sociologist*, 3, 1968, pp. 103–16.
12. I. L. Horowitz, "Mainliners and Marginals: The Human Shape of Sociological Theory," in L. Gross (ed.), *Sociological Theory: Inquiries and Paradigms*, New York, Harper and Row, 1967, p. 359.
13. For a detailed discussion, see K. Wolff, *Versuch zu einer Wissenssoziologie*, Berlin und Neuwied: Luchterhand, 1968, esp. 237–49.
14. S. Tax, "The Fox Project," *Human Organization*, 17, 1958, pp. 17–19.
15. *Ibid.*, p. 18.
16. A. R. Holmberg, "The Research and Development Approach to the Study of Change," *Human Organization, op.cit.*, pp. 12–16; and A. R. Holmberg, "Participant Intervention in the Field," *Human Organization*, 14, 1955, pp. 23–8.
17. A. Shostak, (ed.), *Sociology in Action. Case Studies in Social Problems and Directed Social Change*, Homewood, Ill,: Dorsey, 1966.
18. A more recent statement by H. Etzkowitz, in quite parallel terms, but with a new label, is his paper, entitled, "Sociology and Praxis" (1971).
19. H. Etzkowitz, and G. J. Schaflander, "A Manifesto for Sociologists: Institution Formation – a New Sociology", *Social Problems*, 15, 1968, p. 401.
20. *Ibid.*, p. 399.
21. *Ibid.*, p. 399.
22. *Ibid.*, p. 406.
23. *Ibid.*, p. 400.
24. *Ibid.*, p. 404.
25. *Ibid.*, p. 399.
26. Marx, *op. cit.*, 1848.
27. Marx, *Ibid.*, Chapter III, p. 3.
28. N. Birnbaum, "The Crisis in Marxist Sociology," *Social Research,* 35, 1968, pp. 348–80.
29. H. P. Dreitzel, "Social Science and the Problem of Rationality: Notes on the Sociology of Technocrats," *Politics and Society*, 2, 1972, p. 181.

30. M. Nicolaus, "The Professional Organization of Sociology: A View from Below," *The Antioch Review,* 24, 1969.

    Coser (1959) was perhaps the first to characterize a particular type of concealed observation as "espionage." His attack contributed to a prolonged discussion about the ethics of concealed field work, especially in the framework of the ASA.
31. H. Gans, *The Urban Villagers; Group and Class in the Life of Italian Americans,* New York: The Free Press, 1962; and H. Gans, *The Levittowners*, New York: Random House, 1967.
32. Gans' typology is based, by and large, on the typology of R. Gold (1958).
33. Gans, *op cit.*, 1962, p. 338.
34. It is noteworthy that Habermas should be subject to a similar fallacy. He writes:

    > Access of data via understanding of meaning must be permitted. [. . .] In place of controlled observation, which generates the anonymity (exchangeability) of the observing subject and thus of the reproducibility of the observation, there arises a participatory relation of the understanding subject to the subject confronting him [11] (alter ego). The paradigm is no longer the observation but the dialogue – thus, a communication in which the *understanding subject must invest a part of his subjectivity, no matter in what manner this may be controllable,* in order to be able to meet confronting subjects at all on the intersubjective level which makes understanding possible. [. . .] The fashionable demand for a type of "action research," that is to combine political enlightenment with research, overlooks that the uncontrolled modification of the field is incompatible with the simultaneous gathering of data in that field, a condition which is also valid for the social sciences. (1967/1973: 10–11; my emphasis)

    In other words: on the one hand, Habermas finds, implicitly at least, that the observer influences the setting which he confronts; on the other hand, and interestingly enough, when it becomes explicitly a political issue, he warns against the observer's uncontrolled effects upon the setting. The promise of self-reflection, and in our context: the observer's possibility to take himself into account, which Habermas acknowledges elsewhere (see note 105 below and *Theorie und Praxis* German edition pp. 300 ff.) is neglected with regard to politically conscious, empirical research.
35. Gans, *op.cit.*, 1967, p. xxiii.
36. Gans, *op.cit.*, 1962, p. 339.
37. Gans, *op.cit.*, 1967, p. xxv.
38. Gans, *op.cit.*, 1962, p. 340.
39. *Ibid.*, p. 343.
40. About omniscience in Goethe's novels and later 19th-century works in Germany, cf. e.g., Eduard Spranger (1930) and Friedrich Beissner (1952). Spranger describes the novelist's omniscience as *Innensichtstandort* (perhaps "standpoint of inner view" or "inside view"); the story teller's position as being "evenly distributed" to all actors – to which Beissner quite rightly remarks that it is questionable whether we can here still speak of a *Standort* as a fixed position at all.
41. E. Goffman, *Asylums: Essays on the Social Situation of Mental Patients and Other Inmates*, 1959, p. 320.
42. Gans, *op.cit., loc. cit.* (1962).
43. In a footnote (1962:306), Gans himself points out, by the way, that on one occasion, "his research task" was facilitated when he made his "position clear": "[. . .] a number of people who previously refused to talk to me, changed their minds [. . .]." It is quite obvious that members of a community should distrust and stay away from an outsider who remains aloof and who does not make clear where he stands.

44. Some of this has recently been pointed out by Molotch and Lester (1973). T. R. Young's "conflict methodology" obviously points also in this direction. cf. Lehmann and Young (1974).
45. See, e.g., Garfinkel, 1967, p. 51.
46. S. M. Miller, "The Participant Observer and the 'Over-Rapport,'" *American Sociological Review*, 17, 1962, pp. 97–9.
47. Gans, *op.cit.*, 1962, p. 343.
48. *Ibid.*, p. 306 ff. The only remaining vestige of Gans' advocacy was a memorandum sent to one of the redevelopment commissions. As Gans puts it, the "memorandum was never acknowledged." Most of Chapter 14 of *The Urban Villagers* is an expanded version of this memorandum.
49. *Ibid.*, p. 343.
50. *Ibid.*, p. 345.
51. Or, as M. Nicolaus has put it in his paraphrase of Marx: "The point is not to re-interpret oppression, but to end it." (1969).
52. Gans, *op.cit.*, 1967, p. 443.
53. I find these elements represented, most notably, in Peter Berger and Thomas Luckmann's (1966) re-interpretation of Schutz and in Berger's (1963) presentation of role theory. For an early critique of Husserlian Phenomonology, cf., Theodor W. Adorno (1956/1972).
54. E. Bittner, in Psathas, 1973, pp. 109–25.
55. Weber's emphasis, quoted in Bittner, (118). The statement is taken from Weber's essay, " 'Objectivity' in Social Science and Social Policy."
56. Bittner, *op.cit.*, p. 119.
57. *Ibid.*
58. *Ibid.*, p. 121.
59. *Ibid.*
60. *Ibid.*, p. 123.
61. *Ibid.*, p. 121.
62. I would seriously doubt whether studies such as Laud Humphrey's (1970) *Tea Room Trade*, conducted in a non-participatory and concealed fashion, could be described as being the result of a posture of "total immersion," as Bittner argues. I also find, furthermore, that not enough attention is given to the problem of subcultural specificity. It appears that, to Bittner, "Those competences and resources [which researchers] possess as members of society" (p. 117) are sufficient competence of the study of societal subcultures. I would argue that any such competence is, strictly speaking, immaterial as long as the researcher is a stranger to the observed setting.
63. *Ibid.*, p. 121–2.
64. *Ibid.*, p. 64.
65. *Ibid.*, p. 123.
66. *Ibid.*
67. R. L. Gold, (1958), p. 219.
68. *Ibid.*, p. 219, fn. 4.
69. *Ibid.*, p. 219.
70. *Ibid., loc. cit.*
71. It must be pointed out that I am not dealing here with the problem of concealed observation in field work. In concealed observation (which, in specific circumstances – investigative reporting for one – I find ethically quite acceptable) pretense is obviously a necessity, but the character and objective of this type of research is quite a different matter. For an excellent example of conceded investigative reporting, cf. Günter Wallraff (1969).

72. We are neglecting here for the moment settings where researchers participate as non-strangers, e.g., the sociologist as mother or wife, as teacher (cf. Hughes, 1970/1971), as participant in a political movement (Thorne, 1971) etc.
73. As to the concept of the "stranger," cf. Simmel (1908/1971:143–149) who stresses that the stranger, unlike the "barbarian," is seen as member of a group and is simultaneously remote and near. On the other hand, Simmel also ascribes a not unambiguous objectivity to the stranger – an objectivity which he places in quotation marks: The stranger is not entangled in local quarrels, and is "not bound by ties which could prejudice his perception." (146) While I would admit that the stranger is, indeed, not entangled in local quarrels, and is thus able to judge certain relationships "objectively" the stranger is nevertheless not "free": he himself, upon entry into the group, is entangled in relationships of his own; the quality of his entry is furthermore variable and biographically determined (i.e., he may enter as pariah, as Court Jew etc.). Simmel, in contrast, believes that the stranger sees even "the close relationship (*das Nahverhaeltnis*) as though from a bird's eye view" – which comes close to Spranger's notion of omniscience. I would therefore conclude that, despite Simmel's vestige of realism in this description, Bittner stands in the tradition of Simmel when he formulates the outsider as one who – in my view – is endowed with superhuman qualities: that he stands *above* and *outside* group interaction and is able to look at relationships in a way in which no other human being can.
74. M. Wax, "Tenting With Malinowski," *American Sociological Review*, 37, 1972, pp. 1–13.
75. Malinowski is quite schoolmasterly here (1922:6): with two photographs, he shows his anthropological colleagues precisely how deeply one must pitch one's tent inside the community.
76. B. Malinowski, *Argonauts of the Western Pacific,* New York: E. P. Dutton, 1961, p. 6.
77. Wax, *op.cit.*, p. 11.
78. Malinowski, *op.cit.*, p. 7–8.
79. D. Nash and D. Wintrob, "The Emergence of Self-Consciousness in Ethnography," *Current Anthropology*, vol. 13, 5, 1972, pp. 527–32.
80. J. Dollard, *Caste and Class in a Southern Town*, New Haven: Yale University Press, 1937; and H. Powdermaker, *After Freedom: A Cultural Study in the Deep South*, New York: Viking Press, 1939.
81. M. Mead, *Sex and Temperament in Three Primitive Societies*, New York: W. W. Morrow, 1935; and R. Fortune, "Arapesh Warfare," *American Anthropologist,* 41, 1939, pp. 22–41.
82. R. Redfield, *Tepoztlán: A Mexican Village,* Chicago: University of Chicago Press, 1930; and O. Lewis, *Life in a Mexican Village: Tepoztlán Restudied,* Urbana, Ill.: University of Illinois Press, 1951.
83. J. J. Maquet, "Objectivity in Anthropology," *Current Anthropology*, 5, 1964, pp. 47–55.
84. *Ibid.*, p. 51.
85. I. C. Jarvie, "The Problem of Ethical Integrity in Participant Observation," *Current Anthropology*, 10, 1969, pp. 505–8.
86. *Ibid.*, p. 505.
87. Hughes (1960/1971:502f) makes a very similar point without, however, turning it into a sociological "identity crisis."
88. M. Shokeid, "Field Work as Predicament Rather than Spectacle," *Archives Européennes de Sociologie,* 11, 1970, pp. 11–22.
89. Jarvie, *op.cit.*
90. *Ibid.*, p. 513.
91. P. Gutkind, in Jarvie (1969), p. 516.

92. E. C. Hughes, *The Sociological Eye. Selected Papers on Work, Self, and the Study of Society,* Chicago and New York: Aldine Atherton, 1971, p. 498.
93. L. R. Peattie, "Intervention and Applied Science in Anthropology," *Human Organization*, 17, 1958, pp. 4–8.
94. M. S. Schwartz and C. Green Schwartz, "Problems in Participant Observation," *American Journal of Sociology*, 60, 1955, pp. 343–54; and M. S. Schwartz, "The Uses of Sociology in the Mental Hospital," *Social Problems*, 1963, pp. 219–27.
95. Schwartz & Schwartz, *op.cit.*, p. 346.
96. Schwartz, *op.cit.*, p. 227.
97. Personal communication; a book on the matter is in preparation.
98. Hughes, *op.cit.*, p. 501.
99. *Ibid.*, p. 500–1, my emphasis.
100. *Ibid.*, p. 499.
101. *Ibid.*, p. 502.
102. *Ibid.*, p. 505.
103. *Ibid.*, p. 505.
104. And Peter Berger gives us quite a scare regarding the disturbing effects of sociology in general which he believes is "subversive of established patterns of thought" (1971:2). So disturbing in fact, that it has driven at least one sociologist, Alfred Seidel by name, student of Weber, to suicide! Rather than suggesting this desperate consequence, I would hope that the disturbing effect created by some sociology would lead, not to passive resignation and desperate acts, but rather to liberation and change.
105. This interpretation would bring field work in proximity to Habermas' view of Marx and Freud. He writes in *Theory and Practice:*

    The critiques which Marx developed as a theory of society and Freud as metapsychology are distinguished precisely by incorporating in their consciousness an interest which directs knowledge, as interest in emancipation going beyond the technical and practical interest of knowledge. [. . .] self-reflection, which the analytic method has made possible and provoked, [. . .] can coincide with emancipation from unrecognized dependencies – that is, knowledge coincides with the fulfillment of the interest in liberation through knowledge.
106. Hughes, *op.cit.*, pp. 566–76.
107. *Ibid.*, p. 569.
108. K. Wolff, "Surrender and Community Study: The Study of Loma," in Vidich et al., 1964, pp. 233–63; "Toward Radicalism in Sociology and Every Day," in Psathas (ed.), 1973, pp. 47–58; and *op.cit.*, 1968.
109. *Ibid.*, (1964), p. 263.
110. *Ibid.*, p. 237. There are some striking parallels here to Paolo Freire's *Pedagogy of the Oppressed*, most notably in the existentialist tone of both postures. Freire speaks quite similarly of "love," "humility" and "sympathy" (1970:21); where Wolff speaks of the "risk of being hurt" in the act of *surrender* to the community, Freire talks about "risking acts of love" (35). Needless to say, Wolff's *surrender* is indeed radical; Freire – whose sincerity should certainly not be questioned – exhausts himself in admonitions of solidarity and dialogue with the "oppressed." Thus, where *surrender* risks being interpreted as a romanticised *unio mystica* with the community, Freire's *Pedagogy* is likely to be trivialised into some sort of dialogical benevolence towards the "oppressed." Furthermore, Freire's *Pedagogy* never quite loses the air of priestly humanism: his unceasing efforts at establishing relationships of trust with the oppressed are suspicious.

Ultimately, he relies upon bourgeois science: sociology, anthropology and social spychology become vehicles for "cultural change" and for the "modification of attitudes and values." Marx's "historical movement," in fact, historical forces in general, fail to play any significant role in his scheme. Revolution remains an abstract concept, political structure and organisation are degraded into "sectarianism" or "destructive fanaticism" or disappear in favor of the holy trinity of "leaders", "people" and "pedagogues." Freire's *conscientização* is directly *taught*, not rooted in the objective conditions of people's lives and then mediated by intellectuals – in the sense of Marx's "theoreticians" of the proletariat. Thus, even though Freire quotes Mao Tse-Tung at length – that cultural workers must "teach the masses clearly what they have received from them confusedly" (82), he fails to perceive the political presuppositions of such statements.

111. Wolff, 1973, *op.cit.*, p. 55.
112. This view is of course irreconcilable with Alfred Schutz's position.
113. It is true of course that the second interpretation is improbable: Wolff stresses repeatedly that *surrender* is more unconditional than phenomenological bracketing. "Catch," unlike the expected outcome of bracketing, may be uncognitive. This interpretation is also suggested by Franco Ferrarotti (1973:183) who argues that, while the idea of *surrender* is on the one hand congruent with a "sociologiacome partecipazione", it may entail, ultimately, the renunciation of the scientific discourse in favor of, possibly, poetry or mysticism.
114. This is e.g., the tenor of George Lukacs' early autobiographical sketch, "*Mein Weg zu Marx*", (My road to Marx) first published in 1933; reiterated, in similar fashion in the new (1967) preface to his *History and Class Consciousness*: there, he speaks of his changeover from one class to the other.

    I should emphasize that these "steps toward praxis" are primarily steps for me and from my point of view: they have autobiographical roots and I may be forgiven for these interpretations and idiosyncratic conclusions.
115. An English version of the *Enquête Ouvrière* is found in Bottomore and Rubel (1956:203–212). The German translation in MEW (vol. 19:230–237) from the English manuscript by Marx and the French version as it appeared in the *Revue Socialiste* lack the somewhat stilted tone of the otherwise careful translation by Bottomore. The original English manuscript version which is in part quite different from the French version, is evidently still not available in print. I have used the manuscript version wherever I have found significant discrepancies between the *Revue* version and the manuscript version.

    MEW seems to question, albeit implicitly, Marx's authorship of the *Préambule* to the questionnaire; it appears here only in the Appendix. I am fully convinced with Rubel that the *Preambule* was "sans doute redigé par Marx".
116. H. Weiss, "Die Enquête Ouvrière von Karl Marx," *Zeitschrift für Sozialforschung*, 1, fasc. 1 1936, pp. 76–98.
117. 1854, p. 706.
118. Weiss, *op.cit.*, p. 82.
119. MEW, *op.cit.*, 19, p. 570.
120. M. Rubel (ed.), *Karl Marx: Oeuvres*, 1, Paris: Gallimard, 1963, p. 1527.
121. M. Rubel, *Karl Marx: Essai de biographie intellectuelle*, Paris: Marcel Rivière, 1957, p. 423; and Rubel, *op.cit.*, p. 1734.
122. It is almost needless to point out that the *Revue Socialiste* lacked an effective retrieval system to ensure the return of completed questionnaires, and only a relatively small number of questionnaires seems in fact to have been returned to the editors. Furthermore, the length and the complexity of the questionnaire must have been trying the patience of most proletarians. In this respect at least, modern survey research offers a genuine advance.

123. Polling organizations are of course quite aware of this propagandistic influence, and the political role of surveys in general.
124. Most notably by Erikson (1967) who is particularly committed, it seems, to a rather rigid professional code of ethics.
125. The present discussion, of course, did not attempt to resolve the seemingly inextricable paradox of the role of intellectuals in the working class movement and the question of the *unity* of theory and practice, as pointed to by Gouldner (1974).

    I suggest, however, that the role of intellectuals over the past 30 years or so has changed quite radically. The discrepancy in social status and origin, so blatant in the time of Marx and Engels, has been reduced with the loss of status of the intellectual in Western Capitalism and the rising discontent by groups other than the working class proper. Given that only the oppressed groups are the ones who can "apply the remedies," we may be well advised to look at the relationship between the intellectuals and a revolutionary movement not in terms of domination, but in terms of division of labor and reciprocity.

References

Adorno, Theodor W., *Zur Metakritik der Erkenntnisstheorie. Studien uber Husserl und die phanomenologischen Antinomien,* Frankfurt: Suhrkamp, 1972.

Althusser, Louis, *Pour Marx*, Paris: Maspero, 1965.

Apthorpe, Raymond, "Comment," in Jarvie (1969).

Becker, Howard S. "Whose Side Are We On?", *Social Problems,* 14 (Winter), 1967, pp. 239–247.

________ "Reply to Riley's 'Partisanship and Objectivity,' " *American Sociologist*, 6 (February, 1971, p. 13.

Beissner, Friedrich, *Der Erzaehler Franz Kafka*, Stuttgart: Kohlhammer, 1952.

Berger, Peter L., *Invitation to Sociology,* Garden City, N.Y.: Doubleday, 1963.

________ "Sociology and Freedom". *American Sociologist*, 6 (February), 1971, pp. 1–5.

Berger, Peter L. and Thomas Luckmann, *The Social Construction of Reality*, Garden City, N.Y.: Doubleday, 1966.

Birnbaum Norman, "The Crisis in Marxist Sociology," *Social Research* 35 (Summer), 1968, pp. 348–380.

Bittner, Egon, "Objectivity and Realism in Sociology," in Psathas, (1973), pp. 109–125.

Bottomore, T. B. (ed.), *Karl Marx. Selected Writings in Sociology and Social Philosophy*, New York: McGraw-Hill, 1956.

Brunt, Lodewijk, "Comment" in Nash and Wintrob (1972).

Coser, Lewis A., "Participant Observation and the Military, An Exchange," *American Sociological Review*, 24. 1959, pp. 397–400.

Dictionnaire de l'Economie Politque, Paris, Quillanmin, 1854.

Dollard, John, *Caste and Class in a Southern Town*, New Haven: Yale University Press, 1937.

Dreitzel, Hans-Peter, "Social Science and the Problem of Rationality: Notes on the Sociology of Technocrats," *Politics and Society*, 2 (Winter), 1972, pp. 165–182.

Erikson, Kai T., "A Comment on Disguised Observation in Sociology," *Social Problems*, 14 (Spring), 1967, pp. 366–373.

Etzkowitz, Henry, "Sociology and Praxis," *Social Theory and Practice,* 1, 1971, pp. 1–9.

Etzkowits, Henry and Gerald M. Schaflander, "A Manifesto for Sociologists: Institution Formation – a New Sociology," *Social Problems*, 15 (Spring), 1968, pp. 399–408.

Ferrarotti, Franco, "A proposito Kurt Wolff in occasione del suo 60° compleanno," *La Critica Sociologia*, 25 (Spring), 1973, pp. 181–185.

Fortune, R., "Arapesh Warfare," in *American Anthropologist*, 41, 1939, pp. 22–41.
Frankfurt Institute for Social Research: Aspects of Sociology, Boston: Beacon Press, 1972. (First published as: *Soziologische Exkurse*, Frankfurt: Europaische Verlagsanstalt, 1956).
Freire, Paolo, *Pedagogy of the Oppressed,* New York: Seabury Press, 1970.
Fuchs, Werner, "Empirische Socialforschung als Politische Aktion," *Soziale Welt. Zeitschrift für Sozialwissenschaftliche Forschung und Praxis*, 1970, pp. 21–22 fasc. 1.
Galliher, John F., "The Protection of Human Subjects: A Reexamination of the Professional Code of Ethics," *The American Sociologist*, 8 (August), 1973, pp. 93–100.
Gans, Herbert J., *The Urban Villagers; Group and Class in the Life of Italian Americans*, New York: The Free Press, 1962.
——— *The Levittowners,* New York: Random House-Vintage Books, 1967.
Goffman, Erving, *Asylums: Essays on the Social Situation of Mental Patients and Other Immates,* Garden City: Anchor Books, 1961.
Göhre, Paul, *Three Months in a Workshop. A Practical Study*, London: Swan Sonnenschein & Co. 1895. (Originally as: *Drei Monate Fabrikarbeiter und Handwerksbursche. Eine praktische Studie,* Leipzig: F. W. Grunow, 1891.)
Gold, Raymond L., "Roles in Sociological Field Observations," *Social Forces*, 1958, pp. 217–223.
Gouldner, Alvin W., "Marxism and Social Theory," *Theory and Society*, 1, 1974, pp. 17–35.
Gouldner, Alvin W., "The Sociologist as Partisan: Sociology and the Welfare State," *The American Sociologist*, 3 (May), 1968, pp. 103–116.
Gutkind, Peter, "Comment" in Jarvie (1969).
Habermas, Jürgen, *Theory and Practice*, Boston: Beacon Press, 1973. (Partial translation of *Theorie und Praxis,* Frankfurt: Suhrkamp Verl., 1967.)
Hindess, Barry, "The 'Phenomenological' Sociology of Alfred Schutz," *Economy and Society,* 1 (February), 1972, pp. 1–27.
Holmberg, Alan R., "Participant Intervention in the Field," *Human Organization*, 14 (Spring), 1955, pp. 23–28.
Holmberg, Alan R., "The Research and Development Approach to the Study of Change," *Human Organization,* 17 (Spring), 1958, pp. 12–16.
Harowitz, Irving L., "Mainliners and Marginals: The Human Shape of Sociological Theory," in: Llewellyn Gross (ed.) *Sociological Theory: Inquiries and Paradigms,* New York: Harper and Row, 1967, pp. 358–383.
Hughes, Everett C., *The Sociological Eye. Selected Papers on Work, Self, and the Study of Society,* Chicago and New York: Aldine Atherton, 1971.
——— "The Place of Field Work in Social Science," in: Buford H, Junker, *Field Work: An Introduction to the Social Sciences*, Chicago: University of Chicago Press, 1960. (Reprinted in Hughes (1971) pp. 496–506.)
——— "Teaching as Field Work." *The American Sociologist*, 5, (February) 1970. (Reprinted in: Hughes (1971) pp. 566–576.
Humphreys, Laud, *Tea Room Trade,* Chicago: Aldine, 1970.
Hymes, Del (ed.), *Reinventing Anthropology,* New York: Random House, 1973.
Jarvie, I. C., "The Problem of Ethical Integrity in Participant Observation," *Current Anthropology,* 10 (December) 1969, 505–508.
Lanzardo, Dario, "Intervento socialista nella lotta operaia: l'inchiesta operaia di Marx," *Quaderni Rossi*, 5, 1965.
Levine, Donald N. (ed. and transl.), *Georg Simmel. On Individuality and Social Forms*, Chicago: University of Chicago Press, 1971 (Orig. publ. 1908.)
Lewis, Oscar, *Life in a Mexican Village: Tepoztlán Restudied*, Urbana, Ill.: University of Illinois Press, 1951.

Lukacs, Georg, *History and Class Consciousness*. Cambridge, Mass.: MIT Press, 1971. (First published as: *Geschichte und Klassenbewusstsein*, Berlin: Malik, 1923.)
Lynd, Robert, *Knowledge For What?* Princeton: Princeton University Press, 1939.
Macciocchi, Maria Antonietta, *Lettere dal Interno del P.C.I. a Louis Althusser*, Milano: Feltrinelli, 1969.
Malinowski, Bronislaw, *Argonauts of the Western Pacific*, New York: E. P. Dutton, 1961 (Reprint of work orig. publ. in 1922).
Maquet, Jacques J., "Objectivity in Anthropology," *Current Anthropology*, 5, 1964, pp. 47–55.
Marx, Karl and Engels, Friedrich. *Werke* ("MEW") vols. 4 and 19 Berlin: Dietz Verlag.
Mead, Margaret, *Sex and Temperament in Three Primitive Societies*, New York: W. W. Morrow, 1935.
Miller, S. M., "The Participant Observer and the 'Over-Rapport,'" *American Sociological Review*, 17, 1952, pp. 97–99.
Molotch, Harvey M. and Marilyn Lester, "Accidents, Scandals, and Routines: Resources for Insurgent Methodology," *The Insurgent Sociologist*, 3. (Summer), 1973, pp. 1–11.
Nash, Dennison and Ronald Wintrob, "The Emergence of Self-Consciousness in Ethnography," *Current Anthropology*, vol. 13, 5, 1972, pp. 527–532.
Nicolaus, Martin, "The Professional Organization of Sociology: A View from Below," *The Antioch Review*, 24, 1969.
Panzieri, Raniero, "Uso socialista dell 'inchiesta operaia,'" *Quaderni Rossi*, 5, 1965.
Peattie, Lisa R., "Interventionism and Applied Science in Anthropology," *Human Organization*, 17 (Spring), 1958, pp. 4–8.
Powdermaker, Hortense, *After Freedom: A Cultural Study in the Deep South*, New York: Viking Press, 1939.
Psathas, George (ed.), *Phenomenological Sociology: Issues and Applications*, New York, London: John Wiley and Sons, 1973.
Redfield, Robert, *Tepoztlán: A Mexican Village*, Chicago: University of Chicago Press, 1930.
Riley, Gresham, "Partisanship and Objectivity in the Social Sciences," *American Sociologist*, 6 (February), 1971 pp. 6–12.
Rubel, Maximilien (ed.), *Karl Marx. Oeuvres*, vol. 1. Paris: Gallimard, 1963.
Rubel, Maximilien, *Karl Marx: Essai de biographie intellectuelle*, Paris: Marcel Rivière, 1957.
Schwartz, Morris S. and Charlotte Green Schwartz, "Problems in Participant Observation," *American Journal of Sociology*, 60, 1955, pp. 343–354.
Schwartz, Morris S., "The Uses of Sociology in the Mental Hospital," *Social Problems* (Winter), 1963, pp. 219–227.
Shokeid, Moshe, "Field Work as Predicament Rather than Spectacle," *Archives Européennes de Sociologie*, 11, 1970, pp. 111–122.
Shostak, Arthur (ed.), *Sociology in Action. Case Studies in Social Problems and Directed Social Change*, Homewood, Ill.: Dorsey, 1966.
Sjoberg, Gideon (ed.), *Ethics, Politics and Social Research*, Cambridge, Mass.: Schenkman, 1967.
Spranger, Eduard, "*Der psychologische Perspektivismus im Roman. Eine Skizze zur Theorie des Romans erläutert an Goethes Hauptwerken*," Frankfurt A.M.: *Jahrbuch des Freien Deutschen Hochstifts*, 1930.
Tax Sol, "The Fox Project," *Human Organization*, 17 (Spring) 1958, pp. 17–19.
Thorne, Barrie, *Resisting the Draft: An Ethnography of the Draft Resistance Movement*, Unpublished Ph.D. Dissertation, Brandeis University, 1971.
Vidich, Arthur J., Joseph Bensman, Maurice R. Stein (eds.), *Reflections on Community Studies*, New York, London, Sydney: John Wiley and Sons, 1964.
Wallraff, Günter, *13 unerwünschte Reportagen*, Cologne: Kiepenheuer und Witsch, 1969.

Wax, Murray L., "Tenting With Malinowski," *American Sociological Review*, 37 (February), 1972, pp. 1–13.

Weiss, Hilde, "Die Enquête Ouvrière von Karl Marx," *Zeitschrift für Sozialforschung*. voll fasc. 1: 1936, pp. 76–98. (Reprinted in: Friedrich Fuerstenberg, *Industriesoziologie*. Neuwied: Luchterhand, 1959. pp. 81–101.

Wolff, Kurt H., "Values in Action: A Comment." *Human Organization* 17 (Spring) 1958, pp. 23–24.

——— "Surrender and Community Study: The Study of Loma," in: Vidich et al. (1964), pp. 233–263.

——— *Versuch zu einer Wissenssoziologie*, Berlin und Neuwied: Luchterhand, 1968.

——— "Toward Radicalism in Sociology and Every Day," in: Psathas, George (ed.) (1973), pp. 47–58.

23

Benefits of 'Observer Effects': Lessons from the Field

Torin Monahan and Jill A. Fisher

A frequent criticism of ethnographic research is that 'observer effects' will somehow bias and possibly invalidate research findings (LeCompte and Goetz, 1982; Spano, 2005). Put simply, critics assert that the presence of a researcher will influence the behavior of those being studied, making it impossible for ethnographers to ever really document social phenomena in any accurate, let alone objective, way (Wilson, 1977). Implicit in this negative evaluation of ethnographic methods is the assumption that other methods, particularly quantitative methods, are more objective or less prone to bias (Agar, 1980; Forsythe, 1999). This article is an initial response to that criticism.

Observer effects – also sometimes referred to as 'researcher effects,' 'reactivity,' or the 'Hawthorne effect' – are often understood to be so pervasive that ethnographers must make de facto explanations about how they will attempt to minimize them (McDonald, 2005; Shipman, 1997). By doing so, however, ethnographers effectively legitimize the concern. For instance, a key part of grant proposals is a description of the methods that ethnographers will mobilize to prevent their presence from becoming an intervention or changing the behaviors and activities of those whom they are studying (Agar, 1980).[1] Of course, the implication is that individuals will behave better (e.g., more ethically, more conscientiously, more efficiently) when being observed. In part this concern is a response to ethnographers' relative methodological indifference – unlike researchers using statistical methods – to

Source: *Qualitative Research,* 10(3) (2010): 357–376.

measuring the extent of any bias introduced or calculating the reliability and validity of their data (Atkinson and Hammersley, 2007).[2] Importantly, observer effects are framed as inevitably bad because they indicate a 'contamination' of the supposedly pure social environment being studied (Hunt, 1985). Some methodologists advise qualitative researchers to hone an awareness of possible observer effects, document them, and incorporate them as caveats into reports on fieldwork (Patton, 2002). Others encourage ethnographers to seek out explicitly evidence of observer effects to better understand – and then mitigate – 'researcher-induced distortions' (e.g., LeCompte and Goetz, 1982; Spano, 2006). The possibility that the ethnographer can both have an effect and *by doing so* tap into valuable and accurate data is seldom explored in contemporary literature on methods (e.g., Speer and Hutchby, 2003).

This article proposes that the capacity of ethnography to shape the discourses and practices being observed can be considered a strength of the method. While outsiders may see the data as 'biased,' ethnographers should be prepared to argue that informants' performances – however staged for or influenced by the observer – often reveal profound truths about social and/or cultural phenomena. To make this case, first we mobilize methodological insights from science studies to illustrate the contingency and partiality of all knowledge, from physics to ethnography, and to challenge the notion that ethnography is inherently less objective than other research methods. This section underscores the points that 'observer effects' (broadly conceived) are unavoidable in knowledge production regardless of the scientific field and that it is the different value that is placed on scientific disciplines that creates hierarchies – as well as criticisms – of particular methods. Second, we draw upon our ethnographic projects to illustrate the rich data that can be obtained from 'staged performances' by informants.[3] In addition, by detailing a few examples of questionable behavior on the part of informants, we challenge the fallacy that the presence of ethnographers will cause informants to self-censor. We are neither suggesting that ethnographers should strive to produce observer effects nor are we arguing that anything goes in qualitative research. Instead, our own experiences in the field convince us that observer effects are not in and of themselves liabilities to ethnographic inquiry. Observer effects can and do generate important data and critical insights.

Methodological Insights from Science Studies

Scholars working in the sociology of knowledge, science and technology studies (STS), and related fields know that knowledge is not simply made up of pre-existing, exogenous Truths about the natural or social worlds, but is instead produced through the interrelation of competing interests and values, organizational configurations, and material properties and constraints (Hess, 1995; Latour and Woolgar, 1986). In short, knowledge is socially constructed. As feminist scientist Ruth Hubbard (1988: 5) explains, 'Facts aren't just out there. Every fact has a factor, a maker . . . making facts is a social enterprise'. A crucial part of the scientific enterprise is the methods used in the creation of facts. These methods help to establish the rules that researchers follow within a field and enable distinctions between researchers in different disciplines. Not only do methods provide a blueprint for how to do science, but they also shape the types of questions researchers can pose and the types of facts that are legible within each discipline (Kaplan and Rogers, 1994).

All knowledge and methods for generating facts are not seen as equal. Professional research communities, government agencies, and private industry attach different values to the contribution of each type of science or knowledge, and this influences societal understandings of valuable and less valuable research. The rank ordering of disciples from the 'hardest' natural science (i.e., physics) to the 'softest' social science (i.e., anthropology) is indicative of the hierarchy imposed on fields and methods, with quantitative approaches ruling over qualitative ones (Porter, 1995). Hierarchies are not merely symbolic; in addition to differences in prestige bestowed on each field, practitioners are rewarded salaries and resources concomitant with each discipline's ranking.[4] Moreover, knowledge production in the humanities, especially literature and the arts, is often not even acknowledged as such by those in both the natural and social sciences.

In spite of popular conceptions to the contrary, laboratory and quantitative methods are not more objective or less biased than field and qualitative methods. Mathematics and physics are no less socially constructed than anthropology and sociology (Restivo, 1985; Restivo and Bauchspies, 2006). All knowledge is contingent on the interests of the scientists creating it, the tools and procedures they use to measure the phenomena under investigation, and the analytic frameworks they use to interpret their results (Knorr-Cetina, 1999; Latour and Woolgar, 1986). On one hand, this means that knowledge is always circumscribed by the ability to represent the natural or social worlds. On the other, it means that the beliefs and expectations of researchers or scientific communities have powerful effects on the ability to measure and interpret the world.

An interesting example of the latter can be found in the history of science. Before 19th century science established the basis for current understanding of the mechanics of reproduction, there were many models to explain the differentiation of the sexes (Beldecos et al., 1988; Tuana, 1988). One of the models established by Aristotle was that males have more 'heat' than do females, and it is the presence of heat that accounts for anatomical differences (Tuana, 1988). The question of why one sex would have more heat than the other was 'solved' in the 2nd century when Galen traced the differences to the anatomical structure of veins and arteries attributing heat to the flow of 'pure' blood through the arteries on the right side of the body and the lack of heat to 'impure' blood in veins on the left side of the body. Seed that developed on the right side (in the right testicle and right ovary for instance) would by consequence lead to male offspring, and seed that developed on the left side would become female offspring. This was an inventive anatomical explanation to support a widespread belief about sex differences. What is remarkable about Galen's claim is that it continued to be perpetuated for hundreds of years (until the 17th century) in spite of anatomical evidence to the contrary. As Nancy Tuana (1988: 49) explains:

> anatomists persistently held to the view that the female seed was defective because of the impurity of the blood that fed it. Although careful attention to the actual structure of the veins and arteries of the testicles and ovaries would refute this view, anatomists continued to overlook this error . . . It is perhaps not surprising that even an anatomist as careful as Vesalius would perpetuate such an error. The scientific theory he had inherited demanded this 'fact.' The belief that female seed arose from the 'serous, salty, and acrid' blood of the left testes was the only viable explanation of the perceived differences between women and men.

In other words, explanatory theories shape the perceptions of even the most careful scientists (Denzin, 1989). Observation is inflected with the values and beliefs of the observer in the laboratory and in the field whether the object of inquiry is a molecule, physiology, or human behavior.

This perspective is not unknown in the sciences. As Donna Haraway (1988) points out, 'The only people who end up actually *believing* and, goddess forbid, acting on the ideological doctrines of disembodied scientific objectivity enshrined in elementary textbooks and technoscience booster literature are nonscientists' (p. 576, original emphasis). The field of physics bears out Haraway's point. Specifically, Heisenberg's uncertainty principle acknowledges the impossibility of simultaneously observing, measuring, and representing nature (Lukacs, 2001). The uncertainty principle states that in an effort to gain precision about individual parts of a system, researchers lose precision in their knowledge of the whole system. For example, researchers can measure the momentum of a particle or its position, but both variables cannot be measured at once. The observer effect is the recognition that researchers are interacting with the system, usually through the instruments of measurement, and changing the phenomena being studied. In spite of these caveats, however, the field of physics is not generally thought of as biased. Nor should it be. The point here is that observer effects in physics share obvious similarities to those in ethnography, but ethnographers struggle more against *outsider claims* that the research is less valid, reliable, generalizable, and so forth than do other scientists.[5]

If all pursuits of knowledge invariably have observer effects of some sort, how then can unbiased, objective science ever be achieved? In its purest sense, it cannot. However, this is not the same thing as saying that science is doomed to be biased and subjective. Instead it means that there are limits to scientific representations of Truth. What *all* natural, behavioral, and social sciences do exceedingly well is represent (multiple) truths. Feminist philosophers of science have for decades been critiquing hegemonic conceptions of Truth and advocating for more nuanced theoretical approaches to knowledge production (e.g., Haraway, 1991, 1997; Harding, 1991, 1998; Longino, 1990). A primary component of this orientation to science is that all perspectives are partial.[6] Because knowledge is created through combinations of disciplinary, methodological, and theoretical approaches, there are limitations to claims that can be made about the natural or social phenomena under investigation. As Sandra Harding (1991) proposes in her model of 'strong objectivity,' knowledge production needs to be inclusive of more and different types of people with differing values and interests to create a more robust set of truth claims.

Just as measuring devices used by physicists provide useful ways of getting at particular truths about the world, so too does ethnography serve as an ideal vehicle for exploring truths about meaning-making practices, social relations, and power (Hess, 2001; Juris, 2008; Wall, 2009). Or more relevant to debates over quantitative and qualitative methods, the questions that can be answered with statistics are different from those to be answered with ethnography, but the limitations of each, not just of ethnography, need to be acknowledged. The entrenchment of hierarchies in evaluating methods is in part a result of professional norms and political and economic motivations that both influence which types of questions are considered most important to society and constrain possible answers to those questions (Woodhouse et al., 2002). Unfortunately, so-called disinterested research that does not disrupt the status quo often supports and reproduces modern social problems, from weapons proliferation to environmental degradation (Monahan, 2008; Restivo, 1988). Building upon this problematization of the possibility of non-biased research, the following sections draw upon empirical examples to respond more directly to the criticism of ethnographic methods as compromised by observer effects.

Benefits of Informants' 'Staged Performances'

Notwithstanding the inherent biases with all methods, a persistent critique of ethnography is that researchers can never obtain access to accurate representations of social phenomena (Wilson, 1977). The very act of mediation, of messy interactions with informants, of partial exposure to group practices somehow contaminates pure data that is always out there, just out of reach. A typical response of ethnographers, especially those operating in the anthropological tradition, is that with sufficient time, informants will become inured to the presence of the researcher, let down their guard, and behave 'normally' (Geertz, 1973; Stoddart, 1986). According to this defense, which we find mostly persuasive, it is simply too difficult for informants to maintain a façade for researchers for months or years at a stretch. Indeed, with time researchers become integrated into the communities they study so that no façade is necessary because the barriers between researchers and informants become less important than the social practices in which individuals and groups are engaged. This is not the same thing as saying that cultural differences are erased or power differentials neutralized. Instead, the ethnographer comes to realize that his or her project is – and always was – subordinate to the relations, functions, and logics of the community being studied, and that communities find places for researchers and assign meaning to their activities (Emerson et al., 1995). Put simply, informants have agency and will exercise it to make sense of and influence researchers and research results. The responses of communities to researchers are important data in and of themselves, revealing a great deal about the communities being studied.

From a positivist perspective, the very fact of informant responses to or engagement with ethnographers taints the data (Burawoy, 1991; Denzin and Lincoln, 2000). Thus, some researchers are haunted by the rich ethnographic material they have obtained and are plagued with existential misgivings that informants were simply staging 'false' performances for the researcher's benefit. For instance, Charles Bosk (2001: 204), reflecting on his research in US medical settings, confesses:

> I have been blessed and cursed with 'theatrical' natives with more than enough savvy to figure out my interests. Those disciplinings that modeled the surgical conscience . . . and those tortured, quasi-religious meditations on the ethics of genetics and its applications . . . – how genuine were they really? Might they have not taken place if I had been absent and subjects felt no obligation to show just how seriously they took those obligations in which I was most interested? I did not much entertain these doubts at the time of the fieldwork. The data were often too good to question in this way. But now I wonder if I need to reconsider how an attentive audience of one with a tape recorder or stenographer's notebook, taking down every word, might cause a self-conscious subject to play his or her social role 'over the top.'

Bosk's implication is that if performances are staged or scripted in response to the presence of an ethnographer, then the data are suspect and should probably be discarded. This is a troubling conclusion for us for a few reasons. It presumes that 'natives' are fixed in their social settings and that their beliefs are held constant, unless – and until – they are compelled to perform for the visiting researcher.[7] This underplays and therefore undervalues the *constant* reconstruction of cultural meanings and group identities that occurs through engagement with a steady stream of outsiders and insiders, policies and practices, technologies and symbols, and so on.

Meaning is not out there to be found by the researcher; it is continuously made and remade through social practice and the give-and-take of social interaction, including interaction with the researcher. As Emerson et al. (1995: 3–4) explain:

> The task of the ethnographer is not to determine 'the truth' but to reveal the multiple truths apparent in others' lives . . . Relationships between the field researcher and people in the setting do not so much disrupt or alter ongoing patterns of social interaction as reveal the terms and bases on which people form social ties in the first place . . . Through participation, the field researcher sees first-hand and up close how people grapple with uncertainty and confusion, how meanings emerge through talk and collective action, how understandings and interpretations change over time. In all these ways, the fieldworker's closeness to others' daily lives and activities heightens sensitivity to social life as process.

Therefore, some of the greatest strengths of ethnographic research lie in cultivating close ties with others and dispelling the illusion that robust data are best achieved through distance. As Casper Jensen and Peter Lauritsen (2005) elucidate, 'The traditional social scientist wants to keep his strength by staying distanced . . . these strategies are misplaced. Arguably, the problem of the social scientist is not that his connections are too many and too strong, but that they are too few and fragile' (p. 72).

'Staged performances' are important because they are deeply revealing of how individuals perceive themselves and would like to be perceived. Developing close relationships with informants, while not necessary for witnessing and interpreting performances, can certainly assist researchers in grasping the many intended – and sometimes contradictory – messages behind such plays. Rather than invalidate or cautiously tolerate data derived from staged performances, we embrace such data – not as a representation of any singular Truth, necessarily, but as rich symbolic texts that lend themselves to multiple interpretations and provide critical insights into the cultures being studied. It is important to remember that observations are data to be interpreted, not the 'results' themselves of the study, and as such, data need to be analysed by the ethnographer in light of the context in which they were generated.

Take for example the following description of a performance staged for one of us (TM) by transportation engineers in the US. This study was designed to document the surveillance capabilities of intelligent transportation systems (ITS), which include the integration of a complex array of technological systems, ranging from video cameras on streets and highways, to license-plate recognition systems, to sensors embedded in roads to detect the speed, flow, and density of automobile traffic (Monahan, 2007). The primary objective of such systems is to increase the throughput of vehicles on streets and highways, using smart technologies to rationalize traffic flow rather than building new lanes or roadways. While it is not the everyday interest of ITS engineers, these systems also enable a host of surveillance possibilities, from monitoring individuals and their movements, to collecting aggregate data on commuters, to assisting the police with criminal investigations. The performance staged by engineers was intended to alleviate any concerns that the researcher might have about the surveillance functions of the systems the engineers oversaw.

Near the end of an observational session and interview with three ITS engineers in their control room, one of the engineers prompted another to demonstrate for me something they had programmed into their video surveillance system. The engineer who was addressed said 'Oh yeah,' seemingly glad that his colleague reminded him, and then pulled up live video footage from an intersection. After punching a few buttons, he asked me to watch what happened when he turned the camera and zoomed-in to look into the window of a nearby apartment building. The other engineer explained:

> So the camera . . . it's on the southeast corner and that's an apartment complex that's kind of northeast there, and as he zooms in you'll notice that whole corner, the image just goes away. It just blurs it out.

When he got close to being able to see people in the apartment, a gray splotch emerged suddenly on the screen, obscuring the view into the apartment. They demonstrated for me what they could see both with and then without the software-based privacy protection. While providing claims that the camera had the software embedded in it, they still had the capability to alter the configuration remotely, turning it on or off. The reason they gave for programming the privacy patch was that they did not want to accidentally see people in their places of residence, which occasionally happened when the system lost power and the cameras moved into a default position – pointing right at people's windows.

This was clearly a performance for my benefit rather than something that the engineers do on their own when no outsiders are present. It was intended to communicate to me that they really care about privacy and are not interested in scrutinizing people, just analyzing and optimizing traffic flow. This shows, in part, that they know that people are concerned about potential threats to privacy, which is not something I even mentioned, and that they know that their systems look like surveillance systems that could compromise individual privacy. Thus, the performance also revealed that the engineers conceive of surveillance as being about scrutinizing individuals, especially in private places, instead of seeing surveillance as being more about the scrutiny and control of people more generally, perhaps even in groups. Finally, because they have the capability of activating and deactivating the privacy protections at will, and they were not trepidatious about letting me know that, I learned that they believe that they should be trusted and that they feel capable of shouldering that responsibility. Without the performance, it would have been much more difficult to obtain these data. Still, one cannot say that these findings are corrupt because they were given freely with a purpose in mind, namely of influencing the researcher and his interpretations. As with all data, qualitative or quantitative, these data require interpretation, which is the task of the researcher, the research community, and any other readers.

A second example of a staged performance was witnessed by one of us (JF) as part of research on the clinical trials industry. The purpose of the project was to explore the implications of the pharmaceutical industry's outsourcing of drug studies to private practices and for-profit research facilities in the US (Fisher, 2009). Part of the focus of the study was to better understand the process of informed consent, including how consent forms were explained and administered and how human subjects made decisions about participating in pharmaceutical clinical trials (Fisher, 2006). The staged performance occurred when a physician invited me to observe the informed consent visit of an elderly woman and her son who were considering enrolling the woman in a clinical trial to test the safety of an investigational treatment for Alzheimer's Disease.

After the woman and her son had left the physician's office (having signed the consent form), the physician turned to me and asked me to pretend that I was a clerk from the US Food and Drug Administration (FDA) and to evaluate the informed consent process. That this was the first thing he said to me following the session reveals the extent to which he was aware of my presence and likely performing, at least in part, for me. While one might question how the informed consent visit might have gone had I not been present, any observer effects that occurred do not, however, compromise the data. Because I was not representing the FDA, it did not matter how the informed consent visit went in a technical sense (which was more or less flawless); I was not there to evaluate the physician. Instead, my purpose was to understand issues pertaining to informed consent. The physician's staged performance did not diminish the importance of the insights I gleaned from the observation.

Specifically, the single most interesting finding from this interaction was the conflict between why the woman and her son had come to the clinic that day – in hopes of finding a miracle cure for the woman – and the purpose of the study – to determine if the treatment was safe enough to continue its clinical development. Scholars have discussed 'therapeutic misconceptions' that accompany participation in medical research (Appelbaum and Lidz, 2008), especially safety studies which are also known as Phase I trials. This occurs when potential human subjects believe that the clinical trial will have a therapeutic benefit for their condition when in fact the study is not designed to provide a benefit (as was the case with the Alzheimer's trial) or when the efficacy of the treatment is the very purpose of the study. There are often concerns that researchers or clinicians are not doing their part to minimize the occurrence of therapeutic misconceptions because it might be in their interest not to correct subjects' misunderstandings about the purpose of the studies. In the case of the informed consent visit I observed, the physician was very careful about explaining and reiterating that the study would not improve the woman's condition. Nonetheless, the son continued to search for some benefit to his mother if she were to participate. He settled in the end on the diagnostic benefits to the woman as a way to justify her enrollment in the study in spite of the lack of therapeutic benefit. In other words, what this interaction revealed was that even when observer effects could be said to be happening, much can still be learned from the observation.

What is important is that ethnographers interpret their data in light of the possibility that their informants are engaging in staged performances. In both of these examples, critical insights into the research questions were given. In the first case, ITS engineers demonstrated the ability of the technology to be used as a surveillance tool, their awareness of privacy concerns that are raised by this potential application of traffic cameras, and their conviction that they should be trusted. In the second case, the physician being on his best behavior could not mitigate the risk of potential subjects misunderstanding the purpose of clinical trials, which reveals how complex and delicate the informed consent process can be. Additionally, observer effects involving the putting forth of good behavior have the potential to expose the pervasiveness and even naturalization of less than desirable behaviors or interactions, which is the area we turn to next.

Refuting the Fallacy of Informant Self-Censoring

Often when US funding agencies receive grant proposals for projects that will use ethnographic methods, some reviewers will complain that informants will act differently in front of researchers, that they will censor themselves and be on their best behavior, therefore attenuating the validity of the data obtained (Agar, 1980).[8] Such comments, which are intended to influence study designs and funding outcomes, embody both explicit and implicit truth claims. The explicit claims are that observer effects will occur and that every effort should be taken to minimize them; the implicit claims are that this kind of bias is unique to ethnographic research, that it is bad because it prevents access to 'pure' data, and that other methods – particularly quantitative ones – are more scientific because they are not subject to these particular risks of data corruption.

Whereas we discussed the biases inherent in all research activities in the first section of this article, the goal of this section is to question the veracity of claims about self-censorship and good behavior by research subjects. Drawing upon a few examples from our research projects, we illustrate that informants tend to show little compunction about behaving unethically or making discriminatory statements in the presence of researchers. It may be the case, still, that they are trying to be on their best behavior, but their best behavior sometimes leaves a lot to be desired and, more importantly, provides insight into their understanding of their roles in relation to others. One could argue that questionable behavior in front of researchers is a *more valid* research finding than self-reporting or other measures because it is what cannot be contained by, or what subjects choose not to hide through, self-censorship.

For example, in a collaborative ethnographic research project of ours, we witnessed a medical procedure where a physician's failure to inform his patient – even in the presence of outside researchers observing the consent conference – may have endangered the patient. For this project, we were studying the use of new identification and tracking technologies in hospitals (Fisher and Monahan, 2008). One of the more exotic technologies being used is a radio-frequency identification (RFID) human implant, which is inserted into the triceps region of patients' arms. Once the chip is implanted, it can be scanned at close range by hand-held readers at hospitals to obtain a unique identifying number, which can then be used to access a patient's health records. The ostensible value of such a system is that if a patient with an RFID implant arrives at a hospital and is unable to communicate with hospital staff, then the number communicated by the implant will give staff access to a patient's name, medical history, allergies, and so on, so that rapid and appropriate treatment can be provided (Monahan and Wall, 2007). During a visit to one of our fieldsites, we observed and interviewed an elderly patient being implanted with one of these chips.

We were already present in the physician's office when the patient arrived. We requested and received permission to interview the patient and physician and observe the implant procedure, and we obtained informed consent from both the physician and patient. At the same time, the physician pushed a series of 'release' forms at the patient but did not explain them or tell the patient how the medical device worked or what its limitations were. The patient, who was an energetic and voluble man in his 90s, signed the release forms without reading them. When we asked why he was electing to be implanted, he told us a frightening story about collapsing in his home four months earlier and being unable to dial 911 or call for help. Someone finally did call paramedics, who came and took him to the hospital, where he eventually recovered. Medical staff at the hospital then convinced him to participate in a study on RFID implants. As the patient's narrative unfolded, all in the presence of the physician doing the procedure, it became clear that the patient believed that the implant would direct medical assistance to him in the future, should he ever collapse again and be unable to call for help. This is not how the technology works, however. It does not read vital signs; it does not have a locator device, such as a global positioning system unit; and it does not transmit information at all without the very close presence (less than 3 feet) of an RFID reader. Thus, because the patient thought that the technology would automatically send for help in the future, he might be less likely to struggle to reach a phone and dial 911. In short, he could be placed in greater danger with the implant than without it. The physician implanted the chip anyhow.

What is remarkable about this scene, especially for purposes of this discussion of self-censorship and best behavior on the part of those being studied, is that the physician heard from the patient everything that we did, but he chose not to correct the patient. For whatever reason, the physician was not particularly concerned about the fact that the patient had a profound misunderstanding about the capabilities of the technology or that the patient might be placed in greater physical danger later as a result. This is consistent with other evidence about 'normal' misbehavior that happens in the context of medical care or research (DeVries et al., 2006; Ziebland et al., 2007). It may be the case that self-censorship was occurring and that this was better-than-usual behavior for the physician, but one cannot say that observer effects prevented interesting and important data from being divulged.[9]

In another example from the clinical trials industry study described in the previous section, physicians and research staff were unguarded about their perceptions of human subjects based on gender, race, and class differences. Participation in pharmaceutical clinical trials is stratified with the poor and racial minorities enrolling in the riskiest drug studies and middle-class and whites enrolling in studies that have the most potential for personal benefit (Fisher, 2009). Given these inequalities, research staff rationalize the distribution of risk and burden by mapping it onto stereotyped characteristics of different groups. For example, poor women and women of color were often portrayed problematically as leading disorganized lives that are prone to crisis, being unreliable because of lack of transportation, and being irresponsible for having 'too many' children. I (JF) heard statements from research staff such as the following: 'Lots of times poor women have children attached to them, and it makes it a lot harder [to complete studies]' and

> I would love to be able to go to our county hospital which is overrun by women who have lots of unplanned pregnancies and say, 'Line up, come on over here, and let us do this for you [sterilize you].' . . . Maybe sometimes it's unfair [to say, but] you have to have women who can understand *their* commitment toward the clinical trial that they're participating in. I mean it's not all about what *we* can do for *you*.

Likewise, differences in the participation among racial groups in clinical trials were described in equally problematic ways. Compared to white participants, research staff argue that Asians are non-altruistic, blacks are suspicious, and Hispanics are very compliant (Fisher, 2009).

This indicates that the presence of an ethnographer does not automatically trigger informants to become politically correct. Yet, based on the body language of most of my informants who talked about perceived differences among potential human subjects and the care with which they seemed to be choosing their words, I got the strong impression that they were trying to be on their best, most PC behavior. What does this mean for observer effects and how one can interpret these data? On one hand, the views of my informants can be seen as simply biased – and sometimes classist or racist – statements they made that reveal something only about the person making these claims. On the other hand, however, these perceptions of potential subjects become normalized and perpetuate certain enrollment practices and by doing so have real effects. For example, research staff shared with me that they saw it as their duty to persuade people not to enroll in clinical trials when the staff did not think they were 'capable' of being committed, dependable study subjects. One staff member explained that she frequently used this technique on low-income women interested in participating in pharmaceutical studies:

> Usually you try to give them enough information that they [decide not to participate] themselves, you know? Maybe I'll say to you, 'Well, I don't know these hours are not going to coincide with your work or with your lifestyle, you know?' Or 'You've got such bad veins that I really don't recommend you get in this study, you know?' You look for something to kind of sway the thing [decision] to go in the other direction [of not participating] . . . They are not going to make it [through the end of the study] because they don't have the tools to do it. And you don't want to tell them [that] you don't feel that they're mentally capable.

Thus, perceptions of research staff determine who may be allowed to enroll in clinical trials. In spite of any tendency for informants to self-censor, research staff revealed what they believe is acceptable to say about different gender, racial, and class groups to an outsider and how these beliefs shape their practices.[10]

Negotiating observer effects such as the ones presented here requires interpretation on the part of ethnographers. In the example of the RFID implant, the question should not be how badly would the physician act without the ethnographers present (the answer to which might be that he would have acted the same). Instead, it presents an interesting moment in the relationship of the physician to the patient *because* the ethnographers were present. Had we not questioned the patient about why he was agreeing to get the implant, the physician may never have heard the extent to which the patient was confused about the implant's function. Nonetheless, the interaction illustrated a surprising lack of responsibility on the part of the physician to inform the patient and protect his safety. Likewise, in the clinical trials example, the task was for the ethnographer to map the larger trends in clinical trial participation onto the attitudes of research staff. What would be an ineffective – and certainly biased – approach would be to seek a definitive answer from informants about the stratified differences in drug study participation. Instead, the goal of ethnography is to interpret and make meaning out of the relationships among groups being studied. In both examples, if what we witnessed was informants being on their best behavior, we can interpret our observations accordingly to make claims about what informants' concerns are for how ethnographers or others might perceive them and the ways in which they may or may not be trying to make themselves look better.

Conclusion

There are politics involved whenever claims are made about the strengths or weaknesses of particular research methods. This can be seen when qualitative researchers question the practical significance of 'statistically significant' findings by quantitative researchers, or when they say that quantitative research is simply unable to access or understand the local meanings that subjects attach to social phenomena. Similarly, when researchers in the quantitative tradition refer to qualitative research as 'anecdotal,' 'without validity,' or subject to a host of biases, such as 'observer effects,' these may be legitimate concerns about the accuracy and generalizability of data, but they are also territorial efforts to discredit those working in qualitative methodological traditions (Forsythe, 1999; Wilson, 1977). Our position is that all research methods possess inherent biases, but that the hegemonic value system for social science research is that of scientism – a system that aspires toward value-free, objective findings while denying that this is a biased, partial, and exclusionary form of knowledge production. Qualitative research is often falsely assumed to be less objective and more prone to bias than are quantitative approaches.

Ironically, when ethnographers take seriously the injunction to minimize observer effects, they may be restricting their access to rich data in the field. If the keys to successful ethnographic research are gaining access to fieldsites, developing rapport with informants, and obtaining insider status (Fox, 2004), then the scientistic move toward distance and separation pulls ethnographers out of their element, thereby privileging modes of discrete data extraction over engaged, collaborative participant observation. Efforts by ethnographers to minimize observer effects not only lend validity to the critique that bias and data-corruption are occurring, but they also impose artificial constraints upon fieldwork and subsequently diminish the scientific endeavor.

Of course, we recognize that ethnographers are compelled to constrain their own practices in these ways to obtain funding, publish, and otherwise succeed in academia. Given the widespread nature of the biases against ethnographic methods, it would be unethical for us to advise researchers – especially junior researchers – to eschew these scientistic rituals that vouchsafe acceptance of ethnography while effectively subordinating it to quantitative modes of inquiry. This is further evidence, however, that there is still a long way to go in cultivating recognition across social science fields of the unique strengths, value, and validity of ethnographic methods. It is insufficient to see ethnography as an optional precursor to truly robust quantitative research that will be informed by ethnographic findings, which is a conciliatory move in several fields. This narrative further subordinates one method to another.[11] Ethnography is not simply a scout that is sent into a strange and possibly hostile field to spy on natives and report back to authorities so that 'real' research can follow. Even as a metaphor alone, this is a colonial and extractive approach to research that does not resonate well with ideals of engagement, participation, and collaboration that are present in the best – but unfortunately not all – ethnographic work.

Within this context, which is the context of the politics of knowledge production as well as research methods, this article has explored some salubrious outcomes of observer effects, especially when they take the form of staged performances, and questioned the extent to which self-censorship occurs in the field. Staged performances should be warmly accepted as gifts from informants; they are valuable treasures of meaning, abundantly wrapped in multiple layers of interest, assumption, and concern; they are alluring conceits overflowing with interpretive possibility. As anthropologists know, a gift is never merely a gift (Mauss, 1990). The giving of a gift creates and solidifies relationships and obligations, enmeshing people in shared social fabric through ritual. One does not discard the gift or scowl upon it (or its givers), no matter how ugly or useless the gift might seem. The obligation for ethnographers given the gift of a staged performance is to interpret it deeply. Ethnographers should reflect upon what staged performances communicate about what informants desire, how they would like to be seen, what they hold up as ideals, what they think might be important for an outsider to know, how they perceive researchers, and more. All of these messages are present in staged performances in a much more highly concentrated form than in routine interactions and mundane daily practices. This is a gift that must be accepted and passed on to others.

Self-censorship on the part of informants is another way that observer effects are purported to manifest. Undoubtedly, self-censorship or behavior modification do occur when informants feel scrutinized. Ethnographers have argued that over time self-censorship fades away, especially if the researcher becomes taken for granted by informants and/or integrated into the community being studied. There are also ways that ethnographers can perform 'validity checks' on data they suspect to be dubious, namely by comparing discourse to practice and looking for tensions, or by triangulating articulations from multiple informants and looking for inconsistencies. In employing such pragmatic validity checks, though, one should not lose sight of the fact that all data are open to interpretation and reinterpretation; it is too simplistic to think of some data as true and other data as false. Instead, there are many webs of signification that the ethnographer must navigate to construct a coherent story of cultural meanings, logics, and structures (Clarke, 2005; Geertz, 1973).

Another way of approaching the problem of self-censorship is by carefully noting instances where it does not appear to be happening. We related a few examples from our research projects where people acted against their professional code of ethics (i.e., physicians not obtaining informed consent from patients for voluntary medical procedures) or made racist, sexist, or classist statements seemingly without concern that their words would become data. In such instances, which are not at all rare, disturbing practices and articulations on the part of informants may actually be *more valid* for occurring in front of researchers. The reasoning behind this conclusion is that self-censorship may be occurring but that these unethical practices or prejudicial claims escaped that filter. In other words, informants might behave much worse when researchers are not present, but it is unlikely that they behave much better.

It is our hope that this serious discussion of observer effects will inform debates about the merits of ethnographic research more broadly. We are under no illusion that all ethnographic projects are equally robust or free from corrupting interests, but ethnographic methods should not be subordinated to other methods simply because they do not employ the same techniques, aspire to the same outcomes, or adhere to the same criteria. Additionally, staged performances and self-censorship, or lack thereof, are valuable data in and of themselves, not something to be dismissed. It is the ethnographer's close proximity to and interaction with informants, rather than distance and separation, that affords the transformation of observer effects from distasteful bias to serendipitous boon.

Notes

1. Both authors have received comments to this effect from grant reviewers on different projects submitted to the US National Science Foundation (NSF) and National Institutes of Health (NIH). In these instances, reviewers who self-identified as quantitative researchers expressed concern that ethnographic observation cannot help but act as a corrupting form of intervention.
2. This is not to say that there are no methods for measuring validity in qualitative research, rather these techniques are not as common as those used in quantitative research. For a discussion on formal approaches to validity in qualitative research, see Cho and Trent (2006).

3. Each project discussed mobilized qualitative methods that included ethnographic approaches to research. Detailed information about the specific methods for each of these projects can be found in the papers or books cited in the discussions of each example.
4. John Jackson Jr. (2008) reflected on the hierarchies created to rank the social sciences:

   There is a general pecking order in the social sciences. We all know that. It moves from economics down through the likes of political science and psychology, finally landing in the realm of sociology and anthropology. The closer one gets to serious mathematics as constituitive [*sic*] of the center of the discipline's exploits, the higher one's salary, the less diverse one's colleagues (in terms of categories such as race or gender), and the more powerful one's academic department. There are exceptions to this formulation, but it holds true quite a bit of the time, no?

5. It is common for quantitative researchers to criticize the 'anecdotal' nature of qualitative research, yet quite uncommon for qualitative research to deconstruct statistical methods. This likely stems from the misperception that qualitative methods – 'just talking to people' – is easier and requires less training than do quantitative methods (Forsythe, 1999: 131).
6. Building upon Haraway and others in science studies, Casper Jensen and Peter Lauritsen (2005) make the point that 'this partiality is not in itself a problem; indeed, our argument is that it is a *condition* of research. More problematic is the forgetfulness of this partiality' (p.68, original emphasis).
7. This view of informants and this approach to research is reminiscent of Orientalism. Edward Said (1979) criticized Orientalist scholars for their tendency to create the 'Other' by holding cultures as separate and incommensurable and romanticizing the presumed constancy of the cultures being studied.
8. See note 1 for additional information about reviews of grant proposals.
9. The question of what is the ethical obligation of ethnographers in cases like these is clearly an important one. Perhaps it is the ethnographers' ethical duty to intervene in order to protect their subject. In this particular case, we did not correct the patient during the consent conference. One of us (TM) did, however, attempt to talk with the patient afterwards, but the patient was ushered by the physician into another room after the procedure, and we were unable to speak with him later. This was an unfortunate and morally problematic outcome to the observation. Should we have the opportunity to re-live the experience, we would certainly have taken the opportunity we had with the patient to correct his misunderstanding.
10. Similarly, in her ethnographic study of maternity care, Bowler (1993) found that midwives in the UK perpetuated and mobilized racial stereotypes to shape the care they provided to women.
11. Lynne S. Giddings (2006) makes a compelling argument that the move toward 'mixed methods research' also effectively reduces robust and engaged qualitative inquiry to pragmatic instrumentality, usually for purposes of securing funding for research.

References

Agar, Michael (1980) 'Getting Better Quality Stuff: Methodological Competition in an Interdisciplinary Niche', *Urban Life* 9(1): 34–50.

Appelbaum, Paul S. and Lidz, Charles W. (2008) 'The Therapeutic Misconception', in E.J. Emanuel, R.A. Crouch, C. Grady, R. Lie, F. Miller and D. Wendler (eds) *The Oxford Textbook of Clinical Research Ethics*. New York: Oxford University Press.

Atkinson, Paul and Hammersley, Martyn (2007) *Ethnography: Principles in Practice* (3rd Edition). New York: Taylor & Francis.
Beldecos, Athena, Balley, Sarah, Gilbert, Scott, Hicks, Karen, Kenschaft, Lori, Niemczyk, Nancy, Rosenberg, Rebecca, Schaertel, Stephanie and Wedel, Andrew (1988) 'The Importance of Feminist Critique for Contemporary Cell Biology', *Hypatia* 3(1): 61–76.
Bosk, Charles L. (2001) 'Irony, Ethnography, and Informed Consent', in B. Hoffmaster (ed.) *Bioethics in Social Context,* pp. 199–220. Philadelphia, PA: Temple University Press.
Bowler, Isobel (1993) '"They're Not the Same as Us": Midwives' Stereotypes of South Asian Descent Maternity Patients', *Sociology of Health & Illness* 15(2): 157–78.
Burawoy, Michael (1991) 'Introduction', in M. Burawoy (ed.) *Ethnography Unbound: Power and Resistance in the Modern Metropolis*, pp. 1–7. Berkeley, CA: University of California Press.
Cho, Jeasik and Trent, Allen (2006) 'Validity in Qualitative Research Revisited', *Qualitative Research* 6(3): 319–40.
Clarke, Adele (2005) *Situational Analysis: Grounded Theory after the Postmodern Turn*. Thousand Oaks, CA: Sage.
Denzin, Norman K. (1989) *Interpretive Interactionism*. Newbury Park, CA: Sage.
Denzin, Norman K. and Lincoln, Yvonna S. (2000) 'The Discipline and Practice of Qualitative Research', in N.K. Denzin and Y.S. Lincoln (eds) *Handbook of Qualitative Research*, pp. 1–28. Thousand Oaks, CA: Sage.
DeVries, Raymond, Anderson, Melissa S. and Martinson, Brian C. (2006) 'Normal Misbehavior: Scientists Talk About the Ethics of Research', *Journal of Empirical Research on Human Research Ethics* 1(1): 43–50.
Emerson, Robert M., Fretz, Rachel I. and Shaw, Linda L. (1995) *Writing Ethnographic Fieldnotes*. Chicago: University of Chicago Press.
Fisher, Jill A. (2006) 'Procedural Misconceptions and Informed Consent: Insights from Empirical Research on the Clinical Trials Industry', *Kennedy Institute of Ethics Journal* 16(3): 251–68.
Fisher, Jill A. (2009) *Medical Research for Hire: The Political Economy of Pharmaceutical Clinical Trials*. New Brunswick, NJ: Rutgers University Press.
Fisher, Jill A. and Monahan, Torin (2008) 'Tracking the Social Dimensions of RFID Systems in Hospitals', *International Journal of Medical Informatics* 77(3): 176–83.
Forsythe, Diana E. (1999) '"It's Just a Matter of Common Sense": Ethnography as Invisible Work', *Computer Supported Cooperative Work* 8: 127–45.
Fox, Renée C. (2004) 'Observations and Reflections of a Perpetual Fieldworker', *The ANNALS of the American Academy of Political and Social Science* 595(1): 309–26.
Geertz, Clifford (1973) *The Interpretation of Cultures: Selected Essays*. New York: Basic Books.
Giddings, Lynne S. (2006) 'Mixed-methods Research: Positivism Dressed in Drag?', *Journal of Research in Nursing* 11(3): 195–203.
Haraway, Donna (1988) 'Situated Knowledges: The Science Question in Feminism and the Privilege of Partial Perspective', *Feminist Studies* 14(3): 575–99.
Haraway, Donna J. (1991) *Simians, Cyborgs, and Women: The Reinvention of Nature*. New York: Routledge.
Haraway, Donna J. (1997) *Modest_Witness@Second_Millennium.FemaleMan_Meets_OncoMouse: Feminism and Technoscience*. New York: Routledge.
Harding, Sandra (1991) *Whose Science? Whose Knowledge?: Thinking from Women's Lives*. Ithaca, NY: Cornell University Press.
Harding, Sandra (1998) *Is Science Multi-Cultural?: Postcolonialisms, Feminisms, and Epistemologies*. Bloomington, IN: Indiana University Press.

Hess, David J. (1995) *Science and Technology in a Multicultural World: The Cultural Politics of Facts and Artifacts*. New York: Columbia University Press.

Hess, David J. (2001) 'Ethnography and the Development of Science and Technology Studies', in P. Atkinson, A. Coffey, S. Delamont, J. Lofland and L. Lofland (eds) *Sage Handbook of Ethnography*, pp. 234–45. Thousand Oaks, CA: Sage.

Hubbard, Ruth (1988) 'Science, Facts, and Feminism', *Hypatia* 3(1): 5–17.

Hunt, Morton M. (1985) *Profiles of Social Research: The Scientific Study of Human Interactions*. New York: Russell Sage Foundation.

Jackson Jr., John L. (2008) 'Anthropology: The Softest Social Science?', *The Chronicle of Higher Education*, URL (consulted July 29 2008): http://chronicle.com/review/brainstorm/index.php?id=680

Jensen, Casper and Lauritsen, Peter (2005) 'Qualitative Research as Partial Connection: Bypassing the Power-Knowledge Nexus', *Qualitative Research* 5(1): 59–77.

Juris, Jeffrey S. (2008) *Networking Futures: The Movements Against Corporate Globalization*. Durham, MD: Duke University Press.

Kaplan, Gisela and Rogers, Lesley (1994) 'Race and Gender Fallacies: The Paucity of Biological Determinist Explanations of Difference', in E. Tobach and B. Rosoff (eds) *Challenging Racism and Sexism: Alternatives to Genetic Explanations of Difference*. New York: The Feminist Press at The City University of New York.

Knorr-Cetina, K. (1999) *Epistemic Cultures: How the Sciences Make Knowledge*. Cambridge, MA: Harvard University Press.

Latour, Bruno and Woolgar, Steve (1986) *Laboratory Life: The Construction of Scientific Facts*. Princeton, NJ: Princeton University Press.

LeCompte, Margaret D. and Goetz, Judith Preissle (1982) 'Problems of Reliability and Validity in Ethnographic Research', *Review of Educational Research* 52(1): 31–60.

Longino, Helen E. (1990) *Science as Social Knowledge: Values and Objectivity in Scientific Inquiry*. Princeton, NJ: Princeton University Press.

Lukacs, John (2001) 'Heisenberg's Recognitions: The End of the Scientific World View', in M. Lederman and I. Bartsch (eds) *The Gender and Science Reader*, pp. 225–30. New York: Routledge.

Mauss, Marcel (1990) *The Gift: Forms and Functions of Exchange in Archaic Societies*. London: Routledge.

McDonald, Seonaidh (2005) 'Studying Actions in Context: A Qualitative Shadowing Method for Organizational Research', *Qualitative Research* 5(4): 455–73.

Monahan, Torin (2007) '"War Rooms" of the Street: Surveillance Practices in Transportation Control Centers', *The Communication Review* 10(4): 367–89.

Monahan, Torin (2008) 'Editorial: Surveillance and Inequality', *Surveillance & Society* 5(3): 217–26.

Monahan, Torin and Wall, Tyler (2007) 'Somatic Surveillance: Corporeal Control through Information Networks', *Surveillance & Society* 4(3): 154–73.

Patton, Michael Quinn (2002) *Qualitative Research and Evaluation Methods* (3rd Edition). Thousand Oaks, CA: Sage.

Porter, Theodore M. (1995) *Trust in Numbers: The Pursuit of Objectivity in Science and Public Life*. Princeton, NJ: Princeton University Press.

Restivo, Sal (1985) *The Social Relations of Physics, Mysticism, and Mathematics: Studies in Social Structure, Interests, and Ideas*. New York: Springer.

Restivo, Sal (1988) 'Modern Science as a Social Problem', *Social Problems* 35(3): 206–25.

Restivo, Sal and Bauchspies, Wenda (2006) 'The Will to Mathematics: Minds, Morals, and Numbers', *Foundations of Science* 11(1–2): 197–215.

Said, Edward W. (1979) *Orientalism* (1st Vintage Books Edition). New York: Vintage Books.

Shipman, Marten D. (1997) *The Limitations of Social Research* (4th Edition). London: Longman.

Spano, Richard (2005) 'Potential Sources of Observer Bias in Police Observational Data', *Social Science Research* 34: 591–617.
Spano, Richard (2006) 'Observer Behavior as a Potential Source of Reactivity: Describing and Quantifying Observer Effects in a Large-Scale Observational Study of Police', *Sociological Methods & Research* 34(4): 521–53.
Speer, Susan A. and Hutchby Ian (2003) 'From Ethics to Analytics: Aspects of Participants' Orientations to the Presence and Relevance of Recording Devices', *Sociology* 37(2): 315–37.
Stoddart, Kenneth (1986) 'The Presentation of Everyday Life', *Journal of Contemporary Ethnography* 15(1): 103–21.
Tuana, Nancy (1988) 'The Weaker Seed The Sexist Bias of Reproductive Theory', *Hypatia* 3(1): 35–59.
Wall, Tyler (2009) 'The Fronts of War: Military Geographies, Local Logics, and the Rural Hoosier Heartland'. Doctoral Dissertation, Justice and Social Inquiry, Arizona State University, Tempe.
Wilson, Stephen (1977) 'The Use of Ethnographic Techniques in Educational Research', *Review of Educational Research* 47(2): 245–65.
Woodhouse, Edward, Hess, David, Breyman, Steve and Martin, Brian (2002) 'Science Studies and Activism: Possibilities and Problems for Reconstructivist Agendas', *Social Studies of Science* 32(2): 297–319.
Ziebland, S., Featherstone, K., Snowdon, C., Barker, K., Frost, H. and Fairbank, J. (2007) 'Does it Matter if Clinicians Recruiting for a Trial Don't Understand What the Trial is Really About? Qualitative Study of Surgeons' Experiences of Participation in a Pragmatic Multi-Centre RCT', *Trials* 8(1): 4.

24

Can There Be a Feminist Ethnography?

Judith Stacey

Most feminist researchers, committed, at a minimum, to redressing the sexist imbalances of masculinist scholarship, appear to select their research projects on substantive grounds. Personal interests and skills meld, often mysteriously, with collective feminist concerns to determine a particular topic of research, which, in turn, appears to guide the research methods employed in its service. Indeed, in such a fashion, I chose my last project, a study of patriarchy and revolution in China designed to address major theoretical questions about Western feminism and socialism. The nature of this subject, compounded by limitations in my training, necessitated the macro-structural, abstract approach based almost exclusively on library research that I adopted. And, as a consequence, *Patriarchy and Socialist Revolution in China*, its textual product, offered an analysis of socialism and patriarchy which, as several reviewers justly complained, left out stories about actual women or patriarchs (Stacey, 1983). My dissatisfaction with that kind of research process and outcome led me to privilege methodological considerations over substantive interests when I selected my current research project, a fieldwork study of family and gender relationships in California's Silicon Valley. I was eager for a "hands on," face-to-face research experience, which I also believed was more compatible with feminist principles.

Although there is no uniform canon of feminist research principles, and many lively debates about whether there should be, and, if so, what one should contain, still it is possible to characterize a dominant conception of feminist research currently prevailing among feminist scholars. Most view feminist research as primarily research on, by, and especially *for* women and draw

Source: *Women's Studies International Forum*, 11(1) (1988): 21–27.

sharp distinctions between the goals and methods of mainstream and feminist scholarship.[1] Feminist scholars evince widespread disenchantment with the dualisms, abstractions, and detachment of positivism, rejecting the separations between subject and object, thought and feeling, knower and known, and political and personal as well as their reflections in the arbitrary boundaries of traditional academic disciplines. Instead most feminist scholars advocate an integrative, trans-disciplinary approach to knowledge which grounds theory contextually in the concrete realm of women's everyday lives. The "actual experience and language of women is the central agenda for feminist social science and scholarship," asserts Barbara Du Bois in an essay advocating "Passionate Scholarship," and only a minority of feminist scholars would dissent (Du Bois, 1983: 108). Indeed feminists tend to celebrate "feeling, belief, and experientially based knowledge," which draw upon such traditionally feminine capacities as intuition, empathy, and relationship (Stanley and Wise, 1983a). Discussions of feminist methodology generally assault the hierarchical, exploitative relations of conventional research, urging feminist researchers to seek instead an egalitarian research process characterized by authenticity, reciprocity, and intersubjectivity between the researcher and her "subjects" (Duelli Klein, 1983; Du Bois, 1983; Mies, 1983; Reinharz, 1983; Stanley and Wise, 1983a, 1983b). "A methodology that allows for women studying women in an interactive process," Renate Duelli Klein argues, "will end the exploitation of women as research objects" (Duelli Klein, 1983: 95).

Judged by such criteria, the ethnographic method, by which I mean intensive participant-observation study which yields a synthetic cultural account, appears ideally suited to feminist research. That is why in "The Missing Feminist Revolution in Sociology," an essay reflecting on the limitations of feminist efforts to transform sociology, Barrie Thorne and I wondered with disappointment why so few feminist sociologists had turned to the ethnographic tradition of community studies within the discipline, a tradition that seemed to us far more compatible with feminist principles than are the more widely practiced positivist methods (Stacey and Thorne, 1985). Many other feminist scholars share the view that ethnography is particularly appropriate to feminist research (Duelli Klein, 1983; Mies, 1983; Reinharz, 1983; Stanley and Wise, 1983a, 1983b). Like a good deal of feminism, ethnography emphasizes the experiential. Its approach to knowledge is contextual and interpersonal, attentive like most women, therefore, to the concrete realm of everyday reality and human agency. Moreover, because in ethnographic studies the researcher herself is the primary medium, the "instrument" of research, this method draws on those resources of empathy, connection, and concern that many feminists consider to be women's special strengths and which they argue should be germinal in feminist research. Ethnographic method also appears to provide much greater respect for and power to one's research "subjects" who, some feminists propose, can and should become full collaborators in feminist research (Duelli Klein, 1983; Mies, 1983; Stanley and Wise, 1983a).

This, at least, is how ethnography appeared to me as I found myself unintentionally but irresistibly drawn to it in a study originally intended to be based on more conventional interview methods. An ethnographic approach seemed to resolve the "contradiction in terms" involved in interviewing women that Anne Oakley had identified in her critique of classical sociological interview methods (Oakley, 1981). Oakley rejected the hierarchical, objectifying, and falsely "objective" stance of the neutral, impersonal interviewer as neither possible nor desirable, arguing that meaningful and feminist research depends instead on empathy and mutuality. And I was reassured by Shulamit Reinharz's assertion that the problems of experiential fieldwork methodology "seem minor in comparison with the quality of relations that I develop with people involved in the study and the quality of the understanding that emerges from those relations" (Reinharz, 1983: 185).

But now after two and half years of fieldwork experience, I am less sanguine and more focused on the difficult contradictions between feminist principles and ethnographic method I have encountered than on their compatibility. Hence the question in my title which is modelled (but with a twist) on the implicit question in Oakley's "Interviewing Women: A Contradiction in Terms." The twist is that I now perceive the opposite contradiction between feminist ethics and methods than the one that Oakley discusses. I find myself wondering whether the appearance of greater respect for and equality with research subjects in the ethnographic approach masks a deeper, more dangerous form of exploitation.

There are two major areas of contradiction I wish to discuss. The first involves the ethnographic research process, the second its product. Precisely because ethnographic research depends upon human relationship, engagement, and attachment, it places research subjects at grave risk of manipulation and betrayal by the ethnographer, as the following vignette from my fieldwork illustrates. One of my key informants, now a married, fundamentalist Christian, was involved in a closeted lesbian relationship at the time of her conversion. I first learned of this relationship from her spurned lesbian lover, and this only six months after working in the field. Of course, this immediately placed me in an extremely awkward situation ethically, a situation of triangulation and potential betrayal in relation to these two women and of inauthenticity toward the more secretive one. Several months later (partly, I believe, in response to her perception of my inauthenticity) this informant "came out" to me about this affair, but she asked me to respect the confidentiality of this knowledge when relating to her relatives, friends, and co-workers. Moreover, she and her rejected lover began to compete for my allegiance, sympathy, and ultimately for my view of their shared history.

I could give numerous other examples (such as the case of a secret of paternity, of an illicit affair, and of illicit activities). All placed me in situations of inauthenticity, dissimilitude, and potential, perhaps inevitable betrayal, situations that I now believe are inherent in fieldwork method. For no matter how welcome, even enjoyable the fieldworker's presence may appear to "natives", fieldwork represents an intrusion and intervention into a system of relationships, a system of relationships that the researcher is far freer than the researched to leave. The inequality and potential treacherousness of this relationship seems inescapable.

So too does the exploitative aspect of ethnographic process seem unavoidable. The lives, loves, and tragedies that fieldwork informants share with a researcher are ultimately data, grist for the ethnographic mill, a mill that has a truly grinding power. More times than I would have liked, this study has placed me in a ghoulish, and structurally conflictual relationship to tragedy, a feature of ethnographic process that became particularly graphic during the recent death of another one of my key informants. My ethnographic role consigned me to experience this death both as friend and as researcher, and it presented me with numerous delicate, confusing dilemmas, such as whether or not, and to whom, to make a gift of the precious, but potentially hurtful tapes of an oral history I had once conducted with the deceased. I was confronted as well with the discomforting awareness that as researcher I stood to benefit from this tragedy. Not only would the funeral and family grieving process serve as further research "opportunity," but also the death may free me to include more of this family's "truths" in my ethnographic account than would have been possible had he lived. This and other fieldwork experiences forced my recognition that conflicts of interest and emotion between the ethnographer as authentic, related person (i.e. participant), and as exploiting researcher (i.e. observer) are also an inescapable feature of ethnographic method.

The second major area of contradiction between feminist principles and ethnographic method involves the dissonance between fieldwork practice and ethnographic product. Despite the aspects of intervention and exploitation I have described, ethnographic method appears to (and often does) place the researcher and her informants in a collaborative, reciprocal quest for understanding, but the research product is ultimately that of the researcher, however modified or influenced by informants. With very rare exceptions it is the researcher who narrates, who "authors" the ethnography. In the last instance an ethnography is a written document structured primarily by a researcher's purposes, offering a researcher's interpretations, registered in a researcher's voice.[2]

Here too, therefore, elements of inequality, exploitation, and even betrayal are endemic to ethnography. Perhaps even more than ethnographic process, the published ethnography represents an intervention into the lives and relationships of its subjects. As author an ethnographer cannot (and, I believe, should not) escape tasks of interpretation, evaluation, and judgement. It is possible (and most feminists might claim it is crucial) to discuss and negotiate one's final presentation of narrative with informants, but this does not eliminate the problem of authority, and it can raise a host of new contradictions for the feminist ethnographer.[3] For example after more than a year and a half and scores of hours of mutual reflections on the meaning of the lesbian relationship mentioned above, this "research collaborator" has asked me to leave this part of her history out of my ethnographic account. What feminist ethical principles can I invoke to guide me here? Principles of respect for research subjects and for a collaborative, egalitarian research relationship would suggest compliance, but this forces me to collude with the homophobic silencing of lesbian experience, as well as to consciously distort what I consider a crucial component of the ethnographic "truth" in my study. Whatever we decide, my ethnography will betray a feminist principle.

Indeed the irony I now perceive is that ethnographic method exposes subjects to far greater danger and exploitation than do more positivist, abstract, and "masculinist" research methods. The greater the intimacy, the apparent mutuality of the researcher/researched relationship, the greater is the danger.

The account I have just given of the paradoxes of feminist ethnography is itself falsely innocent. I have presented my methodological/ethical quandaries the way that I first conceptualized them as a feminist researcher, innocent as then I was of relevant methodological literature by ethnographers who long have grappled with related concerns. I am no longer so innocent and ignorant, but I retained this construction to help underscore a curious fact. There has been suprisingly little cross-fertilization between the discourses of feminist epistemology and methods and those of the critical traditions within anthropology and sociology.[4] Most pertinent is the dearth of dialogue between feminist scholarship and the contemporaneous developments in the literature referred to as the "new" or "postmodern" or "poststructuralist" ethnography.[5] This is curious, because the new or postmodern ethnography is concerned with quite similar issues as those that concern feminist scholars and, at first glance, it offers a potential resolution to the feminist ethnographic paradox.

Postmodern ethnography is critical and self-reflexive ethnography and a literature of meditation on the inherent, but often unacknowledged hierarchical and power-laden relations of ethnographic writing.[6] Like feminist scholars, critical ethnographers tear the veil from scientific pretensions of neutral observation or description. They attempt to bring to their research an awareness that ethnographic writing is not cultural reportage, but cultural construction, and always a construction of self as well as of the other. In James Clifford's words the "historical predicament of ethnography" is "the fact that it is always caught up in the intervention, not the representation of cultures" (Clifford, 1986: 2). And at rare moments, critical or "postmodern" ethnographers incorporate feminist insights into their reflexive critiques. Vincent Crapanzano, for example, suggests that "Interpretation has been understood as a phallic, a phallic-aggressive, a cruel and violent, a destructive act, and as a fertile, a fertilizing, a fruitful, and a creative one," and he self-consciously retains the male pronoun to refer to the ethnographer "despite his or her sexual identity, for I am writing of a stance and not of the person" (Crapanzano, 1986: 52).

As I understand it, the postmodern ethnographic solution to the anthropologist's predicament is to fully acknowledge the limitations of ethnographic process and product and to reduce their claims. Like feminists, critical ethnographers eschew a detached stance of neutral observation, and they perceive their subjects as collaborators in a project the researcher can never fully control. Moreover, they acknowledge the indispensably intrusive and unequal nature of their participation in the studied culture. Even more than most feminist scholars, I believe, critical ethnographers have been excrutiatingly self-conscious about the distortions and limitations of the textual products of their studies. Here they have attempted first to fully acknowledge and own the interpretive authorial self and second to experiment with dialogic forms of ethnographic representation that place more of the voices and perspectives of the researched into the narrative and that more authentically reflect the dissonance and particularity of the ethnographic research process.

Finally postmodern ethnographers, influenced by deconstructionist fashions, aim only for "Partial Truths" as James Clifford titled his introduction to a major collection of this genre:

> Ethnographic truths are thus inherently *partial* – committed and incomplete. This point is now widely asserted – and resisted at strategic points by those who fear the collapse of clear standards of verification. But once accepted and built into ethnographic art, a rigorous sense of partiality can be a source of representational tact (Clifford, 1986: 7).

This reflexivity and self-critique of "post-modern" ethnographic literature parallels and has much to contribute to feminist methodological reflections. Perhaps it unwittingly exploits some of the latter as well, as feminist social scientists have published similar reflections on matters of the self, commitment, and partiality in research (Krieger, 1985; Mies, 1983; Rosaldo, 1983; Stanley and Wise, 1983a, 1983b;). At the least it could temper feminist celebrations of ethnographic methods with a salutary note of humility about the limitations of cross-cultural and interpersonal understanding and representation. Certainly I favor much more dialogue and exchange between the two than has taken place to date.

Recently feminist anthropologist Marilyn Strathern also noted the surprising paucity of engagement between feminism and the new ethnography and, in an important contribution to such dialogue, offered an analysis of the grounds for mutual resistance that undergird what she termed the "awkward relationship" between the two (Strathern, 1987). Feminism and critical anthropology, Strathern claims, are mutually "vulnerable on the ethical grounds they hold to be so important": "each has a potential for undermining the other" because they rest upon incompatible constructions of the relationship between self and "Other" (Strathern, 1987: 289). Feminism, Strathern argues, presumes an antagonistic relationship to the male Other, a presumption which grounds its acute sensitivity to power inequalities and has the power to undermine those anthropological pretensions of alliance and collaboration with the Other upon which new ethnographic strategies for multiple authorship reside. Anthropology, in turn, from its cross-cultural vantage-point, suggests the illusory nature of feminist pretensions of actual separation from men of their own culture.

I view the resistances somewhat differently. Feminism's keen sensitivity to structural inequalities in research and to the irreconcilability of Otherness applies primarily, I believe, to its critique of research by men, particularly to research *by* men, but *about* women. The majority of *feminist* claims about feminist ethnographic and other forms of qualitative research, however, presume that such research occurs almost exclusively woman-to-woman. As such feminist researchers are apt to suffer the delusion of alliance more than the delusion of separateness and to suffer it more, I believe, than do most poststructuralist ethnographers. Recall the claims about empathy and identification between feminist researchers and the women they study and the calls by feminist scholars for an egalitarian research process, full collaboration, and even multiple authorship with which this essay began. Hence, it strikes me that a fruitful dialogue between feminism and critical ethnography might address their complementary sensitivities and naivetes about the inherent inequalities and the possibilities for relationships in the definition, study, and representation of the Other.

While I hope to further such a dialogue, in the end, I agree with Strathern that the relationship between feminism and ethnography is unavoidably ambivalent. I am less convinced than she of the virtues of this awkwardness, but I agree that while it can be mitigated, it cannot be effaced. Even an exhaustive, mutually beneficial exchange cannot resolve the feminist ethnographer's dilemma. The postmodern strategy is an inadequate response to the ethical issues endemic to ethnographic process and product that I have encountered and described. It acknowledges, but does little to ameliorate the problems of intervention, triangulation, or inherently unequal reciprocity with informants; nor can it resolve the feminist reporting quandries. For example, acknowledging partiality and taking responsibility for authorial construction cannot reduce my handling of the lesbian affair into a matter of "representational tact."

My current response to the question in my title is that while there cannot be a fully feminist ethnography, there can be (indeed there are) ethnographies that are partially feminist, accounts of culture enhanced by the application of feminist perspectives. There also can and should be feminist research that is rigorously self-aware and therefore humble about the partiality of its ethnographic vision and its capacity to represent self and other. Moreover, even after my loss of ethnographic innocence I believe the potential benefits of "partially" feminist ethnography seem worth the serious moral costs involved.

Indeed, as Carole Joffe has suggested to me, my assault on the ethical foundations of fieldwork may have been unduly harsh, a fairer measure, perhaps, of my prior illusions about ethnographic virtue than of ethnographic vice (Joffe, 1986). Certainly, as she and Shulamit Reinharz assert, fieldworkers can and do form valuable relationships with many of those we study, and some of our unsolicited interventions into the lives of our informants are constructive and deeply appreciated. Just last week, for example, a daughter of the informant whose death I mentioned above consoled me on the sudden death of my own father and thanked me for having allowed her to repair her hostile relationship with her father before he died by helping her to perceive his pride in and identification with her. In certain circumstances fieldwork research offers particular research subjects practical and emotional support and a form of loving attention, of comparatively nonjudgmental acceptance, that they come to value deeply.

But then again, beneficiaries of such attention may also come to depend upon it, and this suggests another ethical quandary in fieldwork, the potential for, indeed the likelihood of desertion by the researcher.[7] Yet rigorous self-awareness of the ethical pitfalls in the method enables one to monitor and then to mitigate some of the dangers to which ethnographers expose their informants. I conclude in this Talmudic fashion to leave the dialogue open, believing that an uneasy fusion of feminist and critical ethnographic consciousness may allow us to construct cultural accounts that, however partial and idiosyncratic, can achieve the contextuality, depth, and nuance I consider to be unattainable through less dangerous, but more remote research methods.

Notes

1. Perhaps the most comprehensive summary of the characteristic distinctions between these approaches which feminists draw appears in several pages of tables detailing contrasts between the two in Reinharz (1983: 168–72).
2. For just this reason Duelh Klein (1983), Mies (1983), and, to a lesser extent, Stanley and Wise (1983b) argue against this approach and for fuller collaboration between researcher and subjects, particularly for activist research in the tradition of Pablo Freire generated by and accountable to grass roots women's movement projects. But, as Carol Smart (n.d.) as well as Stanley and Wise recognize, such an approach places severe restraints on who and what can be studied and on what could be written, restraints that could seriously harm feminist interests.
3. Carol Smart (n.d.) offers important reflections on the adverse implications of this ethical principle when feminists study, as she believes we should, the powerful and the agents of social control rather than their targets.

4. Critical reflections on the ethics and politics of fieldwork have a long history in both disciplines, and by now the literature is vast. (For important examples from the past two decades, see Asad, 1973; Emerson, 1983; Haan, Bellah, Rabinow, and Sullivan, 1983; Hymes, 1974; Thorne, 1978, 1980.)
5. Howard Becker makes a similar point about the unfortunate paucity of exchange between critical traditions in sociology and poststructuralist anthropology in a review of *Writing Culture* (Clifford and Marcus, 1986), a major text on new ethnography (Becker, 1987).
6. A good sampler and bibliography of postmodern ethnographic criticism appears in Clifford and Marcus, 1986. Other important texts include Clifford (1983), Crapanzano (1977), and Marcus and Cushman (1982).
7. In her inimicable witty style, Arlene Kaplan Daniels discusses the etiquette of abandoning one's research subjects as well as other ethical questions in fieldwork (Daniels, 1983). I believe that the problem of desertion is more serious in longterm ethnographic studies than in those based on more limited contact characteristic of other forms of qualitative research.

References

Asad, Talal. 1973. *Anthropology and the Colonial Encounter*. Ithaca Press, London.

Becker, Howard. 1987. The writing of science. *Contemporary Sociology* **16**(1): 25–27.

Clifford, James. 1983. On ethnographic authority. *Representations* **1**(2): 118–46.

Clifford, James. 1986. Introduction: Partial truths. In Clifford, James and Marcus, George, eds, *Writing Culture. The Poetics and Politics of Ethnography.* University of California Press, Berkeley.

Clifford, James, and Marcus, George, eds. 1986. *Writing Culture: The Poetics and Politics of Ethnography*. University of California Press, Berkeley.

Crapanzano, Vincent. 1977 The writing of ethnography. *Dialectical Anthropology* **2**: 69–73.

Crapanzano, Vincent. 1986. Hermes' dilemma: The masking of subversion in ethnographic description. In Clifford, James and Marcus, George, eds, *Writing Culture. The Poetics and Politics of Ethnography*. University of California Press, Berkeley.

Daniels, Arlene Kaplan. 1983. Self-deception and self-discovery in fieldwork. *Qualitative Sociology* **6**(3): 195–214.

Du Bois, Barbara. 1983. Passionate scholarship: Notes on values, knowing and method in feminist social science. In Bowles, Gloria and Duelli Klein, Renate, eds, *Theories of Women's Studies.* Routledge and Kegan Paul, London.

Duelli Klein, Renate. 1983. How to do what we want to do: Thoughts about feminist methodology. In Bowles, Gloria and Duelli Klein, Renate, eds, *Theories of Women's Studies.* Routledge and Kegan Paul, London.

Emerson, Robert M. 1983. *Contemporary Field Research: A Collection of Readings*. Little-Brown, Boston.

Freire, Paulo. 1970. *Pedagogy of the Oppressed.* Seabury Press, New York.

Haan, Norma, Robert N. Bellah, Paul Rabinow, and William M. Sullivan, eds. 1983. *Social Science as Moral Inquiry*. Columbia University Press, New York.

Hymes, Dell, ed. 1974. *Reinventing Anthropology*. Vintage, New York.

Joffe, Carole. 1986. Personal Communication to Author.

Krieger, Susan. 1985. Beyond "subjectivity". The use of the self in social science. *Qualitative Sociology* **8**(4): 309–324.

Marcus, George E. and Cushman, Dick. 1982. Ethnographies as Texts. *Annual Reviews of Anthropology* **11**: 25–69.

Mies, Maria. 1983. Towards a methodology for feminist research. In Bowles, Gloria and Duelli Klein, Renate, eds. 1983. *Theories of Women's Studies.* Routledge and Kegan Paul, London.

Oakley, Anne. 1981. Interviewing women: A contradiction in terms. In Roberts, Helen, ed, *Doing Feminist Research.* Routledge and Kegan Paul, London.

Reinharz, Shulamit. 1983. Experiential analysis: A contribution to feminist research. In Bowles, Gloria and Duelli Klein, Renate, eds, *Theories of Women's Studies.* Routledge and Kegan Paul, London.

Rosaldo, Michelle Z. 1983. Moral/Analytic Dilemmas Posed By the Intersection of Feminism and Social Science. In Haan, Norma, Bellah, Robert N., Rabinow, Paul, and Sullivan, William M., eds, *Social Science as Moral Inquiry.* Columbia University Press, New York.

Smart, Carol, n.d. *Researching prostitution: Some problems for feminist research.* Unpublished paper, Institute of Psychiatry, London.

Stacey, Judith. 1983. *Patriarchy and Socialist Revolution in China.* University of California Press, Berkeley.

Stacey, Judith and Thorne, Barrie. 1985. The missing feminist revolution in sociology. *Social Problems* **32**(4): 301–316.

Stanley, Liz and Wise, Sue. 1983a. "Back into the personal" or: Our attempt to construct "feminist research." In Bowles, Gloria and Duelli Klein, Renate, eds, *Theories of Women's Studies.* Routledge and Kegan Paul, London.

Stanley, Liz and Wise, Sue. 1983b. *Breaking Out: Feminist Consciousness and Feminist Research.* Routledge and Kegan Paul, London.

Strathern, Marilyn. 1987. An awkward relationship: The case of feminism and anthropology. *Signs* **12**(2): 276–292.

Thorne, Barrie. 1978. Political activist as participant observer: Conflicts of commitment in a study of the Draft Resistance Movement of the 1960s. *Symbolic Interaction* **2**(1): 73–88.

Thorne, Barrie. 1980. "You still takin' notes?: Fieldwork and problems of informed consent. *Social Problems* **27**: 284–297.

25

On Tricky Ground: Researching the Native in the Age of Uncertainty

Linda Tuhiwai Smith

Introduction

In the spaces between research methodologies, ethical principles, institutional regulations, and human subjects as individuals and as socially organized actors and communities is tricky ground. The ground is tricky because it is complicated and changeable, and it is tricky also because it can play tricks on research and researchers. Qualitative researchers generally learn to recognize and negotiate this ground in a number of ways, such as through their graduate studies, their acquisition of deep theoretical and methodological understandings, apprenticeships, experiences and practices, conversations with colleagues, peer reviews, their teaching of others. The epistemological challenges to research – to its paradigms, practices, and impacts – play a significant role in making those spaces richly nuanced in terms of the diverse interests that occupy such spaces and at the same time much more dangerous for the unsuspecting qualitative traveler. For it is not just the noisy communities of difference "out there" in the margins of society who are moving into the research domain with new methodologies, epistemological approaches, and challenges to the way research is conducted. The neighbors are misbehaving as well. The pursuit of new scientific and technological knowledge, with biomedical research as a specific example, has presented new challenges to our understandings of what is scientifically

Source: Norman K. Denzin and Yvonna S. Lincoln (eds), *The Landscape of Qualitative Research* (3rd Edition) (Thousand Oaks, CA: SAGE, 2008), pp. 113–143.

possible and ethically acceptable. The turn back to the modernist and imperialist discourse of discovery, "hunting, racing, and gathering" across the globe to map the human genome or curing disease through the new science of genetic engineering, has an impact on the work of qualitative social science researchers. The discourse of discovery speaks through globalization and the marketplace of knowledge." Hunting, racing, and gathering" is without doubt about winning. But wait – there is more. Also lurking around the corners are countervailing conservative forces that seek to disrupt any agenda of social justice that may form on such tricky ground. These forces have little tolerance for public debate, have little patience for alternative views, and have no interest in qualitative richness or complexity. Rather, they are nostalgic for a return to a research paradigm that, like life in general, should be simple.

It is often at the level of specific communities in the margins of a society that these complex currents intersect and are experienced. Some indigenous communities are examples of groups that have been historically vulnerable to research and remain vulnerable in many ways, but also have been able to resist as a group and to attempt to reshape and engage in research around their own interests. This chapter applies indigenous perspectives to examine the intersecting challenges of methodologies, ethics, institutions, and communities. It is a chapter about arriving at and often departing from commonly accepted understandings about the relationships between methodology, ethics, institutional demands, and the communities in which we live and with whom we research. Rather than a story of how complex the world is and how powerless we are to change it, this chapter is framed within a sense of the possible, of what indigenous communities have struggled for, have tried to assert and have achieved.

Indigenous Research and the Spaces from Which It Speaks

Indigenous peoples can be defined as the assembly of those who have witnessed, been excluded from, and have survived modernity and imperialism. They are peoples who have experienced the imperialism and colonialism of the modern historical period beginning with the Enlightenment. They remain culturally distinct, some with their native languages and belief systems still alive. They are minorities in territories and states over which they once held sovereignty. Some indigenous peoples do hold sovereignty, but of such small states that they wield little power over their own lives because they are subject to the whims and anxieties of large and powerful states. Some indigenous communities survive outside their traditional lands because they were forcibly removed from their lands and connections. They carry many names and labels, being referred to as natives, indigenous, autochthonous, tribal peoples, or ethnic minorities. Many indigenous peoples come together at regional and international levels to argue for rights and recognition. In some countries, such as China, there are many different indigenous groups and languages. In other places, such as New Zealand, there is one indigenous group, known as Māori, with one common language but multiple ways of defining themselves.

There are, of course, other definitions of indigenous or native peoples, stemming in part from international agreements and understandings, national laws and regulations, popular discourses, and the self-defining identities of the peoples who have been colonized and oppressed (Burger, 1987; Pritchard, 1998; Wilmer, 1993). The category of the native Other is one that Fanon (1961/1963) and Memmi (1957/1967) have argued is implicated in the same category as the settler and the colonizer. As opposing identities, they constitute each other as much as they constitute themselves. Rey Chow (1993) reminds us, however, that the native did exist before the "gaze" of the settler and before the image of "native" came to be constituted by imperialism, and that the native does have an existence outside and predating the settler/native identity. Chow refers to the "fascination" with the native as a "labor with endangered authenticities." The identity of "the native" is regarded as complicated, ambiguous, and therefore troubling even for those who live the realities and contradictions of being native and of being a member of a colonized and minority community that still remembers other ways of being, of knowing, and of relating to the world. What is troubling to the dominant cultural group about the definition of "native" is not what necessarily troubles the "native" community. The desire for "pure," uncontaminated, and simple definitions of the native by the settler is often a desire to continue to know and define the Other, whereas the desire by the native to be self-defining and self-naming can be read as a desire to be free, to escape definition, to be complicated, to develop and change, and to be regarded as fully human. In between such desires are multiple and shifting identities and hybridities with much more nuanced positions about what constitutes native identities, native communities, and native knowledge in anti/postcolonial times. There are also the not-insignificant matters of disproportionately high levels of poverty and underdevelopment, high levels of sickness and early death from preventable illnesses, disproportionate levels of incarceration, and other indices of social marginalization experienced by most indigenous communities.

There are some cautionary notes to these definitions, as native communities are not homogeneous, do not agree on the same issues, and do not live in splendid isolation from the world. There are internal relations of power, as in any society, that exclude, marginalize, and silence some while empowering others. Issues of gender, economic class, age, language, and religion are also struggled over in contemporary indigenous communities. There are native indigenous communities in the developed and in the developing world, and although material conditions even for those who live in rich countries are often horrendous, people in those countries are still better off than those in developing countries. There are, however, still many native and indigenous families and communities who possess the ancient memories of another way of knowing that informs many of their contemporary practices. When the foundations of those memories are disturbed, space sometimes is created for alternative imaginings to be voiced, to be sung, and to be heard (again).

The genealogy of indigenous approaches to research and the fact that they can be reviewed in this chapter is important because they have not simply appeared overnight, nor do they exist – as with other critical research approaches – without a politics of support around them or a history of ideas. This chapter speaks from particular historical, political, and moral spaces, along with a set of relationships and connections between indigenous aspirations, political activism, scholarship, and other social justice movements and scholarly work. Indigenous communities and researchers from different parts of the globe have long and often voiced concern about the "problem of research" and represented themselves to be among the "most researched" peoples of the world. The critique of research came to be voiced in the public domain in the 1970s, when indigenous political activism was also reasserting itself (Eidheim, 1997; Humphery, 2000; Langton, 1981; L. T. Smith, 1999). The history of research from many indigenous perspectives is so deeply embedded in colonization that it has been regarded as a tool only of colonization and not as a potential tool for self-determination and development. For indigenous peoples, research has a significance that is embedded in our history as natives under the gaze of Western science and colonialism. It is framed by indigenous attempts to escape the penetration and surveillance of that gaze while simultaneously reordering, reconstituting, and redefining ourselves as peoples and communities in a state of ongoing crisis. Research is a site of contestation not simply at the level of epistemology or methodology but also in its broadest sense as an organized scholarly activity that is deeply connected to power. That resistance to research, however, is changing ever so slightly as more indigenous and minority scholars have engaged in research methodologies and debates about research with communities (Bishop, 1998; Cram, Keefe, Ormsby, & Ormsby, 1998; Humphery, 2000; Pidgeon & Hardy, 2002; Smith, 1985; Worby & Rigney, 2002). It is also changing as indigenous communities and nations have mobilized internationally and have engaged with issues related to globalization, education systems, sovereignty, and the development of new technologies.

Indigenous peoples are used to being studied by outsiders; indeed, many of the basic disciplines of knowledge are implicated in studying the Other and creating expert knowledge of the Other (Helu Thaman, 2003; Said, 1978; Minh-ha, 1989; Vidich & Lyman, 2000). More recently, however, indigenous researchers have been active in seeking ways to disrupt the "history of exploitation, suspicion, misunderstanding, and prejudice" of indigenous peoples in order to develop methodologies and approaches to research that privilege indigenous knowledges, voices, experiences, reflections, and analyses of their social, material, and spiritual conditions (Rigney, 1999, p. 117). This shift in position, from seeing ourselves as passive victims of all research to seeing ourselves as activists engaging in a counterhegemonic struggle over research, is significant. The story of that progression has been told elsewhere in more depth and is not unique to indigenous peoples; women, gay and lesbian communities, ethnic minorities, and other marginalized communities have made similar journeys of critical discovery of the role of research in their lives (Hill Collins, 1991; Ladson-Billings, 2000; Mies, 1983; Moraga & Anzaldúa, 1983; Sedgwick, 1991). There have been multiple challenges to the epistemic basis of the dominant scientific paradigm of research, and these have led to the development of approaches that have offered a promise of counterhegemonic work. Some broad examples of these include oral history as stories of the working class, the range of feminist methodologies in both quantitative and qualitative research, the development of cultural and anti/postcolonial studies, critical race theory, and other critical approaches within disciplines (Beverley, 2000; Ladson-Billings, 2000; McLaren, 1993; Mohanty, 1984; Reinharz, 1992; Spivak, 1987; Stanley & Wise, 1983). Critical theorists have held out the hope that research could lead to emancipation and social justice for oppressed groups if research understood and addressed unequal relations of power. Feminism has challenged the deep patriarchy of Western knowledge and opened up new spaces for the examination of epistemological difference. Third World women, African American women, black women, Chicanas, and other minority group women have added immensely to our understandings of the intersections of gender, race, class, and imperialism and have attempted to describe what that means for themselves as researchers choosing to research in the margins (Aldama, 2001; Elabor-Idemudia, 2002; Hill Collins, 1991; Ladson-Billings, 2000; Mohanty, 1984; Moraga & Anzaldúa, 1983; Te Awekotuku, 1999). Indigenous women have played important roles in exploring the intersections of gender, race, class, and difference through the lens of native people and against the frame of colonization and oppression (K. Anderson, 2000; Maracle, 1996; Moreton-Robinson, 2000; L. T. Smith, 1992; Te Awekotuku, 1991; Trask, 1986).

The decolonization project in research engages in multiple layers of struggle across multiple sites. It involves the unmasking and deconstruction of imperialism, and its aspect of colonialism, in its old and new formations alongside a search for sovereignty; for reclamation of knowledge, language, and culture; and for the social transformation of the colonial relations between the native and the settler. It has been argued elsewhere that indigenous research needs an agenda that situates approaches and programs of research in the decolonization politics of the indigenous peoples movement (L. T. Smith, 1999). I would emphasize the importance of retaining the connections between the academy of researchers, the diverse indigenous communities, and the larger political struggle of decolonization because the disconnection of that relationship reinforces the colonial approach to education as divisive and destructive. This is not to suggest that such a relationship is, has been, or ever will be harmonious and idyllic; rather, it suggests that the connections, for all their turbulence, offer the best possibility for a transformative agenda that moves indigenous communities to someplace better than where they are now. Research is not just a highly moral and civilized search for knowledge; it is a set of very human activities that reproduce particular social relations of power. Decolonizing research, then, is not simply about challenging or making refinements to qualitative research. It is a much broader but still purposeful agenda for transforming the institution of research, the deep underlying structures and taken-for-granted ways of organizing, conducting, and disseminating research and knowledge. To borrow from Edward Said (1978), research can also be described as "a corporate institution" that has made statements about indigenous peoples, "authorising views" of us, "describing [us], teaching about [us], settling [us] and ruling over [us]." It is the corporate institution of research, as well as the epistemological foundations from which it springs, that needs to be decolonized.

I name this research methodology as Indigenist.
– Lester Rigney (1999, p. 118)

Becoming an indigenous researcher is somewhat like Maxine Green's (2000) description of how artists from the margins come to re-imagine public spaces. "Through resistance in the course of their becoming – through naming what stood in their way, through coming together in efforts to overcome – people are likely to find out the kinds of selves they are creating"

Table 1: Corporate layers of research

- Foundations, genealogies, and disciplines of knowledge that define its methodologies and its systems of classification and representation
- Historical embeddedness in imperialism, the production of knowledge, and the development of science
- Cultures and subcultures of its institutions and infrastructures
- Communities of like-minded or trained scholars, disciplinary bodies, and research associations
- Ways in which research is regulated and inscribed through notions of ethics, ethical review boards, and codes of conduct
- Practices of reporting and publishing
- National and international funding agencies and their links to particular agendas
- Ways in which some forms of research legitimate dominant forms of knowledge and maintain hegemony or dominant myths
- Chain and distribution of benefits from research
- Intersection of research with policy and the design and implementation of interventions

(p. 301). Indigenous researchers are "becoming" a research community. They have connected with each other across borders and have sought dialogue and conversations with each other. They write in ways that deeply resonate shared histories and struggles. They also write about what indigenous research ought to be. Australian Aborigine scholar Lester Rigney (1999), emphasizing Ward Churchill's (1993) earlier declarations of indigenist positioning, has argued for an indigenist approach to research that is formed around the three principles of *resistance*, *political integrity*, and *privileging* indigenous voices. He, like other indigenous researchers, connects research to liberation and to the history of oppression and racism. Rigney argues that research must serve and inform the political liberation struggle of indigenous peoples. It is also a struggle for development, for rebuilding leadership and governance structures, for strengthening social and cultural institutions, for protecting and restoring environments, and for revitalizing language and culture. Some indigenous writers would argue that indigenous research is research that is carried out by indigenous researchers with indigenous communities for indigenous communities (Cram, 2001; Rigney, 1999). Implicit in such a definition is that indigenous researchers are committed to a platform for changing the status quo and see the engagement by indigenous researchers as an important lever for transforming institutions, communities, and society. Other writers state that purpose more explicitly in that they define indigenous research as being a transformative project that is active in pursuit of social and institutional change, that makes space for indigenous knowledge, and that has a critical view of power relations and inequality (Bishop, 1998; Brady, 1999; Pihama, 2001; L. T. Smith, 1991). Others emphasize the critical role of research in enabling peoples and communities to reclaim and tell their stories in their own ways and to give *testimonio* to their collective herstories and struggles (Battiste, 2000; Beverley, 2000; The Latina Feminist Group, 2001). Embedded in these stories are the ways of knowing, deep metaphors, and motivational drivers that inspire the transformative praxis that many indigenous researchers identify as a powerful agent for resistance and change. These approaches connect and draw from indigenous knowledge and privilege indigenous pedagogies in their practices, relationships, and methodologies. Most indigenous researchers would claim that their research validates an ethical and culturally defined approach that enables indigenous communities to theorize their own lives and that connects their past histories with their future lives (Marker, 2003). Indigenous approaches are also mindful of and sensitive to the audiences of research and therefore of the accountabilities of researchers as storytellers, documenters of culture, and witnesses of the realities of indigenous lives, of their ceremonies, their aspirations, their incarcerations, their deaths. (Pihama, 1994; Steinhauer, 2003; Te Hennepe, 1993; Warrior, 1995).

In New Zealand, Māori scholars have coined their research approach as Kaupapa Māori or Māori research rather than employing the term "indigenist."

There are strong reasons for such a naming, as the struggle has been seen as one over Māori language and the ability by Māori as Māori to name the world, to theorize the world, and to research back to power. The genealogy of indigenous research for Māori has one of its beginnings in the development of alternative Māori immersion-based schooling (Pihama, Cram, & Walker, 2002; G. H. Smith, 1990; L. T. Smith, 2000). Graham Smith (1990) has argued that the struggle to develop alternative schools known as Kura Kaupapa Māori helped produce a series of educational strategies that engaged with multiple levels of colonization and social inequality. These strategies included engagement with theory and research in new ways. Kaupapa Māori research has developed its own life, and as an approach or theory of research methodology, it has been applied across different disciplinary fields, including the sciences. It can be argued that researchers who employ a Kaupapa Māori approach are employing quite consciously a set of arguments, principles, and frameworks that relate to the purpose, ethics, analyses, and outcomes of research (Bishop & Glynn, 1999; Durie, 1992; Johnston, 2003; Pihama, 1993; L. T. Smith, 1991; Tomlins-Jahnke, 1997). It is a particular approach that sets out to make a positive difference for Māori, that incorporates a model of social change or transformation, that privileges Māori knowledge and ways of being, that sees the engagement in theory as well as empirical research as a significant task, and that sets out a framework for organizing, conducting, and evaluating Māori research (Jahnke & Taiapa, 1999; Pihama et al., 2002). It is also an approach that is active in building capacity and research infrastructure in order to sustain a sovereign research agenda that supports community aspirations and development (L. T. Smith, 1999). Those who work within this approach would argue that Kaupapa Māori research comes out of the practices, value systems, and social relations that are evident in the taken-for-granted ways that Māori people live their lives.

Indigenist research also includes a critique of the "rules of practice" regarding research, the way research projects are funded, and the development of strategies that address community concerns about the assumptions, ethics, purposes, procedures, and outcomes of research. These strategies often have led to innovative research questions, new methodologies, new research relationships, deep analyses of the researcher in context, and analyses, interpretations, and the making of meanings that have been enriched by indigenous concepts and language. To an extent, these strategies have encouraged nonindigenous researchers into a dialogue about research and, on occasion, to a reformulated and more constructive and collaborative research relationship with indigenous communities (Cram, 1997; Haig-Brown & Archibald, 1996; Simon & Smith, 2001; G. H. Smith, 1992). Critical and social justice approaches to qualitative research have provided academic space for much of the early work of indigenous research. Denzin and Lincoln (2000) describe a moment in the history of qualitative research (1970–1986) as the moment of "blurred genres" when local knowledge and lived realities became important, when a diversity of paradigms and methods developed, and when a theoretical and methodological blurring across boundaries occurred. Arguably, an indigenist research voice emerged in that blurred and liminal space as it paralleled the rise in indigenous political activism, especially in places like Australia, New Zealand, Norway, and North America. For indigenous activists, this moment was also one of recognition that decolonization needed a positive and more inclusive social vision and needed more tools for development and self-determination (as an alternative to violent campaigns of resistance). Research, like schooling, once the tool of colonization and oppression, is very gradually coming to be seen as a potential means to reclaim languages, histories, and knowledge, to find solutions to the negative impacts of colonialism and to give voice to an alternative way of knowing and of being. Indigenous research focuses and situates the broader indigenous agenda in the research domain. This domain is dominated by a history, by institutional practices, and by particular paradigms and approaches to research held by academic communities and disciplines. The spaces within the research domain through which indigenous research can operate are small spaces on a shifting ground. Negotiating and transforming institutional practices and research frameworks is as significant as the carrying out of actual research programs. This makes indigenous research a highly political activity that can be perceived as threatening, destabilizing, and privileging of indigeneity over the interests and experiences of other diverse groups. Decolonization is political and disruptive even when the strategies employed are pacifist because anything that requires a major change of worldview, that forces a society to confront its past and address it at a structural and institutional level that challenges the systems of power, is indeed political. Indigenous research presents a challenge to the corporate institution of research to change its worldview, to confront its past and make changes.

Indigenous research approaches, like feminist methodologies, have not emerged into a neutral context, although their arrival has been predicted by those working with silenced and marginalized communities. As Lincoln (1993) forewarned, however, social sciences cannot simply develop grand narratives of the silenced without including the voices and understandings of marginalized and silenced communities. There continues to be vigorous critique of indigenous approaches and claims to knowledge, and, indeed, the indigenous presence in the academy. In some cases, this critique is framed by the discourses of anti–affirmative action, such as calls for "color- and race-free" policies. In other cases, the critique is a very focused attack on the possibility that indigenous people have a knowledge that can be differentiated from dogma and witchcraft or is a very focused and personal attack on an individual (Trask, 1993). In other examples, the critique does reflect attempts by nonindigenous scholars to engage seriously with indigenous scholarship and understand its implications for the practices of nonindigenous scholars and their disciplines. In a limited sense, there has been an attempt at dialogue between indigenous and nonindigenous scholars, usually occurring after indigenous scholars have provided a critique of the discipline – for example Vine Deloria's (1995) critique of anthropology and Ngugi wa Thiong'o's (1981/1987) critique of what counted as African literature. Kenyan writer Ngugi wa Thiong'o viewed the language of the settler/colonizer as being implicated in the "colonization of the mind" and came to the decision that he would not write in the language of the colonizer but instead would write in his own language of Gikuyu or Ki-Swahili. Ngugi's stance helped create further space for debate about "postcolonial" literature and the role of literature in colonial education systems (Ashcroft, Griffiths, & Tiffin, 1989). Vine Deloria's sustained political critique of the place of the American Indian in the American system has created space for the further development of American Indian Studies and a dialogue with other disciplines (Biolsi & Zimmerman, 1997). Unfortunately, dialogue is often the solution to fractures created through lack of dialogue between those with power and marginalized groups. Similar debates have occurred and continue to occur in other fields, including literature (Cook-Lynn, 1996; Harjo & Bird, 1997; Womack, 1999), feminist studies (Maracle, 1996; Moraga & Anzaldúa, 1983; Moreton-Robinson, 2000), and multicultural and ethnic studies (Mihesuah, 1998). Some debates are very public media campaigns that invoke the prejudices and attitudes toward indigenous peoples held by the dominant social group.[1] In some of these campaigns, the ethnicity of the dominant group is masked behind such social categories as "the public," "the taxpayers," or "the rest of society." The fears and attitudes of the dominant social group, and of other minority social groups, are employed quite purposefully in public debates about indigenous knowledge as the arbiters of what indigenous people are permitted to do, of what they are allowed to know, and indeed of who they are.

Table 2: Strategies for building indigenous research capability

- The training of indigenous people as researchers
- The employment of indigenous people as researchers
- Participation by indigenous people in a wide range of research projects employing different kinds of approaches and methodologies
- The generating of research questions by communities
- Developing indigenous research methodologies
- Developing research protocols for working with communities
- The support by various individuals and communities of research-based decision making
- The establishment of indigenous research organizations
- Presentation of their research by indigenous researchers to other indigenous researchers
- Engagements and dialogue between indigenous and nonindigenous researchers and communities

An important task of indigenous research in "becoming" a community of researchers is about capacity building, developing and mentoring researchers, and creating the space and support for new approaches to research and new examinations of indigenous knowledge. That activity can now be seen in a range of strategies that are being applied by diverse communities across the world to build research capability. Conversations about indigenous methodologies – albeit in different historical, disciplinary, and institutional spaces – are being discussed and applied by a diverse range of indigenous scholars across the globe. These include Sami scholars in northern Norway, Finland, Sweden, and Russia (Keskitalo, 1997) and native scholars in the Pacific Islands (Helu Thaman, 2003; Kaomea, 2003). Sami literary scholar Harald Gaski (1997), for example, argues that "Ever since the world's various indigenous peoples began turning their efforts to co-operative endeavours in the 1970s, the Sami have participated actively in the struggle to make these peoples' and their own voice heard. Art and literature have always played an important role in this endeavour. Therefore, the time for Sami literature to join world literature is past due" (p. 6). Jan Henri Keskitalo (1997) points to a research agenda for Sami people that is "based on the freedom to define, initiate and organize research, and the possibility to prioritise what kind of research should be defined as Sami research, at least when using public funding" (p. 169). All these discussions represent cross-border conversations and activism, as the territorial boundaries of many indigenous communities have been intersected and overlaid by the formation of modern states. Some discussions occur through specific indigenous forums, or through feminist or environmentalist networks, and others occur through the diaspora of the Third World, the "developing world," and regional gatherings (Alexander & Mohanty, 1997; Saunders, 2002; Shiva, 1993; Spivak, 1987).

Researching the Native in the Knowledge Economy

Knowledge is a key commodity in the 21st century. We understand this at a commonsense level simply as an effect of living in the era of globalization, although it is also expressed as the consequences of life in the postindustrial age, the age of information and postmodernity. Knowledge as a commodity is a conception of knowledge (and curriculum) that is situated in the intersection of different visions of and alliances for globalization (Peters, 2003). Michael Apple (2001) refers to this alliance as one that brings together neoliberals, neoconservatives, authoritarian populists, and the new middle class. Apple defines neoliberals as those who are "deeply committed to markets and to freedom as 'individual choice,'" neoconservatives as ones who "want a return to discipline and traditional knowledge," authoritarian populists as ones who "want a return to (their) God in all of our institutions" (p. 11), and the new middle class as those who have created and stand to benefit most from this configuration of interests. The neoliberal economic vision of globalization is one in which the market shapes and determines most, if not all, human activities. Far from being simply an economic theory, neoliberal proponents have used their access to power to attempt to reform all aspects of society, including the relationships between the state and society. New Zealand is often used as a model, the "experiment" for how far this agenda can be pursued, because of the significant neoliberal reforms undertaken over the past 20 years (Kelsey, 1995). The reforms have included a "hollowing out" of the state; the reform and re-regulation of the welfare system – education, health, banking, and finance; and the removal of tariffs and other barriers to free trade (Moran, 1999). The reforms have been supported by a powerful ideological apparatus that has denied empirical evidence that groups were being marginalized further by policies and that the gaps between the rich and poor, the well and the sick, were widening under the reform regime. This ideological apparatus is most visible in its discursive strategies with rhetoric and slogans such as "user pays," privatization, increased competition, freedom of choice, and voucher education. It is also evident in the construction of new, idealized neoliberal subjects who are supposed to be "self-regulating selective choosers, highly competitive and autonomous individuals liberated from their locations in history, the economy, culture and community in order to become consumers in a global market" (L. T. Smith et al., 2002, p. 170).

The significance of the neoliberal agenda for social science research is that the "social," the "science," and the "research" have also been re-envisioned and re-regulated according to the neoliberal ideologies. One site where this re-envisioning and re-regulation of the social, the science, and the research intersects is in the economy of knowledge. As with other strategies of power, it is often the marginalized and silenced communities of society who experience the brunt and the cruelty of both the slogans and the material changes in their lives. The "knowledge economy" is a term used by businesspeople such as Thomas Stewart (1997) to define the ways in which changes in technology such as the Internet, the removal of barriers to travel and trade, and the shift to a postindustrial economy have created conditions in which the knowledge content of all goods and services will underpin wealth creation and determine competitive advantage. As a commodity, knowledge is produced under capitalist labor market conditions: it can be bought and sold, and it is private rather than public property. Researchers are knowledge workers who produce new knowledge. In this environment, new and unique knowledge products become highly prized objects of capitalist desire. Mapping the human genome and searching for cures to various diseases that will require the manufacturing of special products are just two examples of the "race" now on for "knowledge," the new El Dorado. Now, where can one discover new knowledge that is not already under private ownership? The laboratories? The rain forests? The human body? The knowledge and practices of those who have maintained their unique ways of living? The answer to all the above is "Yes," and there is more. Indigenous knowledge once denied by science as irrational and dogmatic is one of those new frontiers of knowledge. The efforts by indigenous peoples to reclaim and protect their traditional knowledge now coincides and converges with scientific interests in discovering how that knowledge can offer new possibilities for discovery (Stewart-Harawira, 1999).

One convergence of indigenous knowledge and science is in the field of ethnobotany, a field that has botanists and biologists working closely with indigenous communities in the collection and documentation of plants, medicinal remedies, and other practices. In doing science, ethnobotanists are also doing qualitative research, talking to community experts, observing practices, and developing word banks and other resources. The protocols that have been developed by the International Society of Ethnobotany will be discussed again later in this chapter. One use of the research that its members gather lies in the identification of medicinal properties that can be reproduced in the laboratory and developed for commercialization. The pharmaceutical industry has a keen hunger for such research, and there is real intensity in the hunt for new miracles to cure or alleviate both old and modern diseases. The search for new knowledge knows no borders. It is competitive and expensive, and only a few can participate. In the biomedical field, the rapid advances in knowledge and technology – for example, in reproductive birth technologies and in genetic engineering – present new challenges to what society thinks is ethically acceptable. Issues raised in relation to cloning a human being, new genetic therapies, and other remedies and practices stretch our understandings of what life is about. Although the science can develop the new knowledge, it is the *social* science that has an understanding of the nature of social change. Scientists, however, can also be powerful advocates of their own discoveries and fields of research, such that institutions and industries "buy into" the promise of new technologies and expect society to "catch up" to the ethical implications of the new knowledge. For qualitative research, new technologies present new vistas in a sense, new attitudes to examine and new dilemmas to resolve. For indigenous and other marginalized communities, the new vistas present new threats and risks in terms of their ability to protect their traditional knowledge and the likelihood of the benefits of research being distributed equitably to the poor rather than to the rich.

As Apple (2001) reminds, us, however, the neoliberal agenda also converges with the countervailing neoconservative and authoritarian tendencies that seek to protect and strengthen certain "traditional" forms of privilege. The "traditional" values and forms of knowledge being reified by these interest groups are not the same traditional values and ways of knowing that indigenous peoples speak of but are in fact the very antithesis of any form of non-Western, nonheterosexual, nonfeminist knowledge. Graham

Hingangaroa Smith (1994) argues that there are new types of colonization in the neoliberal version of globalization that enable dominant interests in society to be maintained. Smith further contends that in the global marketplace, where everything can be commodified, local communities, cultures, practices, and values are put at accelerated risk, with little room to maneuver or develop resistance. One analogy of how the global marketplace works to put local communities and knowledge at risk is the impact of the large multinational or national company that sets up its store or its mall in a town that has small and struggling businesses. There are powerful driving forces that shape the ways in which individual interests come to be either aligned with or marginalized from the new development. For example, some people may need employment and others may need access to cheaper products; some people need to retain their businesses or see their community as being defined by the "Main Street," not the Mall. Young people may see the Mall as presenting new social possibilities that would cater more to their tastes by providing access to more global brands. In the end, the community becomes divided by economic interests, although all may ultimately wish for a united community. In the end, the Mall wins: The small businesses either collapse or struggle on; Main Street looks even more depressing, driving more people to the Mall; and everyone in town begins wearing the global brands, just like the people on television and the people who live in the next community, the next state, the next country. Local products, if they are made, find their way to a boot sale or a market day, basically consigned to the margins of the economy and community consciousness. Some local or native products are selected as marketable in the Mall, such as native medicine wheels and small hanging crystals. These products are not produced locally, because that would cost too much, so the *image* is reproduced at a cheaper price in countries with poor labor market conditions and then sold in every mall in the world. Imagine this as a global process having an impact in every little community of the world. It is a very seductive process, but something gets lost, in this process, for the community. For indigenous communities, the "something lost" has been defined as indigenous knowledge and culture. In biological terms, the "something lost" is our diversity; in sociolinguistics, it is the diversity of minority languages; culturally, it is our uniqueness of stories and experiences and how they are expressed. These are the "endangered authenticities" of which Rey Chow (1993) speaks, ones that are being erased through the homogenization of culture.

The knowledge economy, as one theme of globalization, constitutes the new identities of the self-regulating and selective chooser, the consumer of knowledge products, the knowledge worker and knowledge manager, and the clients of knowledge organizations. McLaren (1993, p. 215) calls these *market identities* that reflect the corporate model of market education and educational consumption. One might think that this makes for a very educated and knowledgeable society – not so. The knowledge economy is about creating and processing knowledge, trading and using knowledge for competitive advantage – it is not about knowing or knowledge for its own sake, it is not about the pursuit of knowledge but about "creating" knowledge by turning knowledge into a commodity or product. Research plays an important role in the creation of knowledge and, as argued by Steven Jordan (2003) in an article he entitled "Who stole my methodology?," even the most participatory research models are being subjected to the processes of commodification "for the purposes of supporting and reproducing the social relations of accumulation in their multifarious forms" (p. 195). Jordan further suggests that the methodology of participatory research is being appropriated and reconstituted by neoliberal discourses of participation "in ways that are antithetical to both its founding principles and traditions" (p. 195).

The neoliberal version of globalization is not, however, the only ideology at work across the globe. There are other interests at work, some repressive and others progressive. Trafficking in drugs and people, catering to pedophilia, and other organized criminal activities also have gone global. More recently, global terrorism (recognizing that some communities have been terrorized for hundreds of years by various forms of colonialism) has heightened the impulses and fears of neoconservatives and authoritarian populists and simultaneously has created threats to the free operations of the global marketplace. The powerful nostalgia of neoconservatives and authoritarian populists for a curriculum of the right (Apple, 2001), a curriculum of simple "facts," and a reification of what Denzin (1991) refers to as "ancient narratives" augurs dangers for education, for educational research, and for any social justice research. Neoconservative and authoritarian interest groups seek to disrupt any agenda for social justice and already have been effective in peeling back gains in social justice programs, although Roman and Eyre (1997) caution us to see the dangers of "applying 'backlash' exclusively to Right-wing political reactions [that] fail to draw attention to reactionary and defensive politics within and across left-wing/progressive groups – whether feminist, critical multicultural/anti-racist, or anti-heterosexist" (p. 3). The neoliberal agenda crosses the left and right of the political spectrum, and to some extent the fellow travelers of neoliberalism manage to infiltrate a wide spectrum of politics.

Other, more progressive groups also have managed to go global and make use of knowledge in the pursuit of a social justice agenda. Nongovernmental organizations and communities of interest have managed to put up resistance to the powerful interests of wealthy nations and corporations. Some of these coalitions have brought together diverse interests and unusual bedfellows to contest free trade; others have organized important consciousness-raising activities to keep information about injustice in the public eye. Small communities still cling to their own schools and identities as they attempt to build democratic community consensus. One of the perspectives that indigenous research brings to an understanding of this moment in the history of globalization is that it is simply another historical moment (one of many that indigenous communities have survived) that reinscribes imperialism with new versions of old colonialisms. This is not as cynical as it may sound; rather, it comes from the wisdom of survival on the margins. This moment can be analyzed, understood, and disrupted by holding onto and rearticulating an alternative vision of life and society. It is also not the only defining moment: Other changes have occurred that make communities somewhat more prepared to act or resist. For example, more indigenous researchers are choosing to research alongside their own communities. There are more allies. There also are other imperatives that have driven an agenda of transformation; among them is language regeneration. Language regeneration programs have created a momentum, especially in New Zealand, that neoliberal reforms have not been able or willing to subvert, as these programs have a strong hold on the community's aspirations. Indigenous development is optimistic despite what often appear to be huge barriers.

The new subjectivities of the free market and the knowledge economy also include the re-envisioning and re-regulation of new native subjects, a reworked Other, still raced and gendered, idealized and demonized, but now in possession of "market potential." Some of these new subjectivities resonate with the global market, where evoking of "the image" is a powerful mechanism for distancing the material conditions of the people from the image itself. Other subjectivities are "turning the gaze" back onto the dominant settler society, reflecting the momentum of political, educational, and economic change that already has occurred in many indigenous communities. These identities are formed "in translation," in the constant negotiation for meaning in a changing context. New identities form and re-form in response to or as a consequence of other changes and other identities. New voices are expressed, new leaders emerge, new organizations form, and new narratives of identity get told.

One newly worked native identity is that of the native intellectual as scientist. This is a small, emerging group of native scientists with strong connections to their native knowledge and practices. These scientists represent a new type of translator or interlocutor, one who bridges different knowledge traditions in ways that Western scientists find difficult to dismiss and indigenous communities find acceptable (Little Bear, 2000; Thomas, 2001). The native scientist not only is the native healer, herbalist, or spiritual expert but also is someone who understands the philosophies, knowledge, and histories that underpin cultural practices and beliefs and who generates his or her science from these foundations. As Basso (1996) and Marker (2003) have suggested, these people are not in the academy to "play word and idea games" but intend to contribute to change for the benefit of communities, to ensure that science listens to, acknowledges, and benefits indigenous communities. The role of these indigenous professionals is similar to the role played by the first generation of indigenous teachers and nurses and by the first generation of medical doctors and social workers in native communities, a difficult role of translating, mediating, and negotiating values, beliefs, and practices from different worldviews in difficult political contexts.

Ethics and Research

One area of research being vigorously contested by indigenous communities is that of research ethics and the definitions and practices that exemplify ethical and respectful research. Indigenous researchers often situate discussions about ethics in the context of indigenous knowledge and values and in the context of imperialism, colonialism, and racism (Cram, 1993, 2001; Menzies, 2001; Rigney, 1999). Indigenous understandings of research ethics have often been informed by indigenous scholars' broad experience of research and other interactions with the media, health system, museums, schools, and government agencies. Increasingly, however, research ethics has come to be a focus of indigenous efforts to transform research and institutions (Worby & Rigney, 2002). Research ethics is often much more about institutional and professional regulations and codes of conduct than it is about the needs, aspirations, or worldviews of "marginalized and vulnerable" communities. Institutions are bound by ethical regulations designed to govern conduct within well-defined principles that have been embedded in international agreements and national laws. The Nuremberg Code (1949) was the first major international expression of principles that set out to protect the rights of people from research abuse, but there are other significant agreements, such as the World Medical Association Declaration of Helsinki Agreement of 1964 and the Belmont Report of 1979. National jurisdictions and professional societies have their own regulations that govern ethical conduct of research with human subjects. Increasingly, the challenges of new biotechnologies – for example, new birth technologies, genetic engineering, and issues related to cloning – also have given rise to ethical concerns, reviews, and revised guidelines.

For indigenous and other marginalized communities, research ethics is at a very basic level about establishing, maintaining, and nurturing reciprocal and respectful relationships, not just among people as individuals but also with people as individuals, as collectives, and as members of communities, and with humans who live in and with other entities in the environment. The abilities to enter preexisting relationships; to build, maintain, and nurture relationships; and to strengthen connectivity are important research skills in the indigenous arena. They require critical sensitivity and reciprocity of spirit by a researcher. Bishop (1998) refers to an example of relationship building in the Māori context as *whakawhanaungatanga*, "the process of establishing family (*whânau*) relationships, literally by means of identifying, through culturally appropriate means, your bodily linkage, your engagement, your

Table 3: "Community-up" approach to defining researcher conduct

Cultural values (Smith, 1999)	*Researcher guideline (Cram, 2001)*
Aroha ki te tangata	A respect for people – allow people to define their own space and meet on their own terms.
He kanohi kitea	It is important to meet people face to face, especially when introducing the idea of the research,"fronting up" to the community before sending out long, complicated letters and materials.
Titiro, whakarongo . . . kôrero	Looking and listening (and then maybe speaking). This value emphasizes the importance of looking/observing and listening in order to develop understandings and find a place from which to speak.
Manaaki ki te tangata	Sharing, hosting, being generous. This is a value that underpins a collaborative approach to research, one that enables knowledge to flow both ways and that acknowledges the researcher as a learner and not just a data gatherer or observer. It also facilitates the process of "giving back," of sharing results, and of bringing closure if that is required for a project but not to a relationship.
Kia tupato	Be cautious. This suggests that researchers need to be politically astute, culturally safe, and reflective about their insider/outsider status. It is also a caution to insiders and outsiders that in community research, things can come undone without the researcher being aware or being told directly.
Kaua e takahia te mana o te tangata	Do not trample on the "mana" or dignity of a person. This is about informing people and guarding against being paternalistic or impatient because people do not know what the researcher may know. It is also about simple things like the way Westerners use wit, sarcasm, and irony as discursive strategies or where one sits down. For example, Māori people are offended when someone sits on a table designed and used for food.
Kaua e mahaki	Do not flaunt your knowledge. This is about finding ways to share knowledge, to be generous with knowledge without being a "show-off" or being arrogant. Sharing knowledge is about empowering a process, but the community has to empower itself.

connectedness, and therefore, an unspoken but implicit commitment to other people" (p. 203). Worby and Rigney (2002) refer to the "Five Rs: Resources, Reputations, Relationships, Reconciliation and Research" (pp. 27–28) as informing the process of gaining ethical consent. They argue that "The dynamic relationship between givers and receivers of knowledge is a reminder that dealing with indigenous issues is one of the most sensitive and complex tasks facing teachers, learners and researchers at all levels . . ." (p. 27). Bishop and Glynn (1992) also make the point that relationships are not simply about making friends. They argue that researchers must be self-aware of their position within the relationship and aware of their need for engagement in power-sharing processes.

In *Decolonizing Methodologies* (L. T. Smith, 1999), I also gave some examples of the ways in which my communities may describe respect, respectful conduct, trustworthiness, and integrity at a day-to-day level of practice and community assessment. My concern was to show that community people, like everyone else, make assessments of character at every interaction. They assess people from the first time they see them, hear them, and engage with them. They assess them by the tone of a letter that is sent, as well as by the way they eat, dress, and speak. These are applied to strangers as well as insiders. We all do it. Different cultures, societies, and groups have ways of masking, revealing, and managing how much of the assessment is actually conveyed to the other person and, when it is communicated, in what form and for what purpose. A colleague, Fiona Cram (2001), has translated how the selected value statements in *Decolonizing Methodologies* could be applied by researchers to reflect on their own codes of conduct. This could be described as an exercise of "bottom-up" or "community-up" defining of ethical behaviors that create opportunities to discuss and negotiate what is meant by the term "respect." Other colleagues have elaborated on the values, adding more and reframing some to incorporate other cultural expressions. One point to make is that most ethical codes are top down, in the sense of "moral" philosophy framing the meanings of ethics and in the sense that the powerful still make decisions for the powerless. The discussions, dialogues, and conversations about what ethical research conduct looks like are conducted in the meeting rooms of the powerful.

No one would dispute the principle of *respect*; indeed, it is embedded in all the major ethical protocols for researching with human subjects. However, what is *respect*, and how do we know when researchers are behaving respectfully? What does *respect* entail at a day-to-day level of interaction? To be respectful, what else does a researcher need to understand? It is when we ask questions about the apparently universal value of respect that things come undone, because the basic premise of that value is quintessentially Euro-American. What at first appears a simple matter of *respect* can end up as a complicated matter of cultural protocols, languages of respect, rituals of respect, dress codes: in short, the "p's and q's" of etiquette specific to cultural, gender, and class groups and subgroups. *Respect*, like other social values, embraces quite complex social norms, behaviors, and meanings, as one of many competing and active values in any given social situation. As an ethical principle, *respect* is constructed as universal partly through the process of defining what it means in philosophical and moral terms, partly through a process of distancing the social value and practice of *respect* from the messiness of any particular set of social interactions, and partly through a process of wrapping up the principle in a legal and procedural framework. The practice of *respect* in research is interpreted and expressed in very different ways on the basis of methodology, theoretical paradigms, institutional preparation, and individual idiosyncrasies and "manners."

Similarly, the principle and practice of *informed consent* presents real-world problems for researchers and for the researched. Fine, Weis, Weseen, and Wong (2000) already have discussed the ways in which "the consent form sits at the contradictory base of the institutionalisation of research" (p. 113). The form itself can be, as they argue, a "crude tool – a conscience – to remind us of our accountability and position" (p. 113). They argue that a consent form makes the power relations between researchers and researched concrete, and this can present challenges to researchers and researched alike, with some participants *wanting* to share their stories while others may feel *compelled* to share. The form itself can be the basis of dialogue and mediation, but the individual person who is participating in the research still must sign it. The principle of *informed consent* is based on the right of individuals to give consent to participation once they have been informed about the project and believe that they understand the project. In some jurisdictions, this right does not necessarily apply to children, prisoners, or people who have a mental illness. Nevertheless, the right is an individual one. However, what if participating in a research project, unwittingly or wittingly, reveals collective information to researchers – for example, providing DNA, sharing the making of a medicine, or revealing secret women's or men's business as may occur in societies like Aboriginal Australian communities, where men's knowledge and women's knowledge is strictly differentiated? Researching with children already has opened up the possibility that family secrets, especially stories of abuse, require actions to be taken beyond the simple gathering of data. One concern of indigenous communities about the *informed consent* principle is about the bleeding of knowledge away from collective protection through individual participation in research, with knowledge moving to scientists and organizations in the world at large. This process weakens indigenous collectively shared knowledge and is especially risky in an era of knowledge hunting and gathering. Another concern is about the nature of what it really means to be informed for people who may not be literate or well educated, who may not speak the language of the researcher, and who may not be able to differentiate the *invitation* to participate in research from the enforced compliance in signing official forms for welfare and social service agencies.

The claim to universal principles is one of the difficulties with ethical codes of conduct for research. It is not just that the concepts of respect, beneficence, and justice have been defined through Western eyes; there are other principles that inform ethical codes that can be problematic under certain conditions. In some indigenous contexts, the issue is framed more around the concept of *human rights* rather than principles or values. However, whether it is about principles, values, or rights, there is a common underpinning. Ethics codes are for the most part about protecting the individual, not the collective. Individuals can be "picked off" by researchers even when a community signals it does not approve of a project. Similarly, the claim to beneficence, the "save mankind" claim made even before research has been completed, is used to provide a moral imperative that certain forms of research must be supported at the expense of either individual or community consent. Research is often assumed to be beneficial simply because it is framed as research; its benefits are regarded as "self-evident" because the intentions of the researcher are "good." In a review of health research literature reporting on research involving indigenous Australians, I. Anderson, Griew, and McAullay (2003) suggest that very little attention is paid to the concept of benefit by researchers, and even less attention is paid to the assessment of research benefit. A consequence of the lack of guidelines in this area, they argue, is that "in the absence of any other guidelines the values that guide such a judgement will reflect those of the ethics committee as opposed to those of the Indigenous community in which research is proposed" (p. 26).

A more significant difficulty, already alluded to, can be expressed more in terms of "who" governs, regulates, interprets, sanctions, and monitors ethical codes of conduct." "Who" is responsible if things go terribly wrong? And "who" really governs and regulates the behaviors of scientists outside institutions and voluntary professional societies? For example, rogue scientists and quirky religious groups are already competing for the glory of cloning human beings with those whose research is at least held to an acceptable standard because of their employment in recognized institutions. From an indigenous perspective, the "who" on ethical review boards is representative of narrow class, religious, academic, and ethnic groups rather than reflecting the diversity of society. Because these boards are fundamentally supportive of research for advancing knowledge and other high-level aims, their main task is to advance research, not to limit it. In other words, their purpose is not neutral; it is to assist institutions to undertake research – within acceptable standards. These boards are not where larger questions about society's interests in research ought to be discussed; they generally are the place where already determined views about research are processed, primarily to protect institutions. Marginalized and vulnerable groups are not, by and large, represented on such boards. If a marginalized group is represented, its voice is muted as one of many voices of equal weight but not of equal power. Hence, even if a representative of a marginalized group is included on a review board, the individual may not have the support, the knowledge, or the language to debate the issue among those who accept the dominant Western view of ethics and society. These are difficult concerns to resolve but need to be discussed in an ongoing way, as ethical challenges will always exist in societies.

King, Henderson, and Stein (1999) suggest that there are two paradigms of ethics, the one we know as principalist and a potentially new one in process that is about relationships. King, Henderson, and Stein argue that the ethics regulations that researchers currently work under are based on three factors:

- Balancing principles: autonomy, beneficence, justice, informed consent, and confidentiality
- Ethical universalism (not moral relativism): truth (not stories)
- Atomistic focus: small frame, centered on individuals.

In the case of the International Society of Ethnobiology (ISE), a society of scientists whose work involves indigenous communities, the Code of Ethics that was developed with indigenous participation identifies 15 principles upon which ethical conduct rests. These principles include such things as the principles of self-determination, inalienability, traditional guardianship, and active participation. The ISE Code of Ethics suggests that research needs to be built on meaningful partnerships and collaboration with indigenous communities. Similarly, the Australian Institute of Aboriginal and Torres Strait Islander Studies published the *Guidelines for Ethical Research in Indigenous Studies* (2000) after conducting workshops with indigenous studies researchers. The Guidelines connect the notion of ethical principles with human rights and seek to "embody the best standards of ethical research and human rights" (p. 4). The *Guidelines* propose three major principles, inside of which are fuller explanations of the principles and practical applications. The three main principles are

- Consultation, negotiation, and mutual understanding
- Respect, recognition, and involvement
- Benefits, outcomes, and agreement.

Within the principles of the *Guidelines* are further subprinciples, such as respect for indigenous knowledge systems and processes, recognition of the diversity and uniqueness of peoples and individuals, and respect for intellectual and cultural property rights and involvement of indigenous individuals and communities as research collaborators.

Principles are balancing factors that still rest upon the assumption that the principles are understood as meaning the same thing to all people under all circumstances. As Denzin (2003) argues, this approach implies a singular approach to all forms of inquiry that oversimplifies and dehumanizes the human subject. Indigenous communities and other marginalized groups may not understand the history of the ethical code of conduct or its basis in Western moral philosophy, but they do understand breaches of respect and negative impacts from research such as the removal of their rights and lands. Qualitative researchers also know that emerging methodologies and emerging researchers have a difficult time making their way through the review process to gain approval. Kathleen M. Cumiskey (1998) narrates her experiences in dealing with her institutional review board as ones that came down to a reminder that graduate students would not be indemnified if she happened to be arrested or her work subpoenaed. The emphasis on procedural issues, including the balancing of risks and benefits, inhibits or limits the potential for institutions and society to examine ethics against a much broader social and epistemological framework.

What does an indigenous approach to research contribute to a discussion about ethical standards? Indigenous perspectives challenge researchers to reflect upon two significant contributions. In the first instance, indigenous communities share with other marginalized and vulnerable communities a collective and historically sustained experience of research as the Object. They also share the use of a "research as expert" representation of who they are. It is an experience indigenous communities associate with colonialism and racism, with inequality and injustice. More important, indigenous communities hold an alternative way of knowing about themselves and the environment that has managed to survive the assaults of colonization and its impacts. This alternative way of knowing may be different from what was known several hundred years ago by a community, but it is still a way of knowing that provides access to a different epistemology, an alternative vision of society, an alternative ethics for human conduct. It is not, therefore, a question of whether the knowledge is "pure" and authentic but whether it has been the means through which people have made sense of their lives and circumstances, that has sustained them and their cultural practices over time, that forms the basis for their understanding of human conduct, that enriches their creative spirit and fuels their determination to be free. The first contribution of an indigenous perspective to any discussion about research ethics is one that challenges those of us who teach about research ethics, who participate in approving and monitoring ethics proposals, to understand the historical development of research as a corporate, deeply colonial institution that is structurally embedded in society and its institutions. It is not just about training and then policing individual researchers, nor about ensuring that research with human subjects is an ethical activity. One thing we must have learned from the past is that when research subjects are not regarded as human to begin with, when they have been dehumanized, when they have been marginalized from "normal" human society, the human researcher does not see human subjects. To unravel the story of research ethics with human subjects, teachers and students must understand that research ethics is not just a body of historical "hiccups" and their legal solutions. It is a study of how societies, institutions, disciplines, and individuals *authorize, describe, settle, and rule*. It is a study of historical imperialism, racism, and patriarchy and the new formations of these systems in contemporary relations of power. It is a study of how humans fail and succeed at treating each other with respect.

Just as important, the second contribution indigenous research offers is a rich, deep, and diverse resource of alternative ways of knowing and thinking about ethics, research relationships, personal conduct, and researcher integrity. There are other ways to think about ethics that are unique to each culture. There are other ways to guide researcher conduct and ensure the integrity of research and the pursuit of knowledge. In New Zealand, as one example, Māori are discussing ethics in relation to *tikanga*, defined briefly by Mead (2003) as "A body of knowledge and customary practices carried out characteristically by communities" (p. 15). Mead argues that Tikanga has three main aspects, of knowledge, practices, and actors, and that among other things, tikanga provides guidelines about moral and behavioral issues and informs ethical matters. He proposes five "tests" that can be applied to an ethical dilemma from a *tikanga Māori* perspective. These "tests" draw on Māori values to provide a framework for arriving at a Māori position on a specific ethical issue. The "tests" include the following:

- Applying cultural understandings of knowledge (for example applying *mauri*, the view that every living thing has a *mauri or life force*)
- Genealogical stories (such as those that explain how living things were created)
- Precedents in history
- Relationships
- Cultural values (such as the value of looking after people).

Mead suggests that examining an ethical issue against each of the five "tests" provides a framework that enables the dilemma created by new technologies to be thought through in a way that meets cultural and ethical scrutiny while remaining open to new possibilities. It is also a way to build a cultural and community body of knowledge about new discoveries, technologies, and research ethics.

It may be that these and other explorations connect with King, Henderson, and Stein's (1999) conception of a relationships paradigm that includes the following elements:

Layering of relevant relationships – individuals and groups
Context based – what are the relevant contexts? Culture, gender, race/ethnicity, community, place, others

Crosscutting issues, wider frame of reference
Narrative focus
Continuity – issues arise before and continue after projects
Change – in relationships over time

It may also be a way that connects with Denzin's (2003) call for a more inclusive and flexible model that would apply to all forms of inquiry. Also, as suggested by I. Anderson, Griew, and McAullay (2003), there is a tension between the regulations of practice and the development of ethical relationships. They argue that there is a need to develop at least two layers of responsiveness, one involving institutional collaborations with communities and the other involving researcher relationships with communities that are also mediated by reformed research structures. Indigenous research offers access to a range of epistemic alternatives. I would not want to suggest that such ways are simply out there waiting to be discovered, but certainly there are people and communities willing to engage in a meaningful dialogue, and there is much to talk about.

Qualitative Travelers on Tricky Ground

Qualitative research in an age of terrorism, in a time of uncertainty, and in an era when knowledge as power is reinscribed through its value as a commodity in the global market place presents tricky ground for researchers. It is often at the local level of marginalized communities that these complex currents intersect and are experienced as material conditions of poverty, injustice, and oppression. It is also at this level that responses to such currents are created on the ground, for seemingly pragmatic reasons. Sometimes this approach may indeed be a reasonable solution, but at other times it draws into question the taken-for-granted understandings that are being applied to decisions made under pressure. What maps should qualitative researchers study before venturing onto such terrain? This is not a trick question but rather one that suggests that we do have some maps. We can begin with all the maps of qualitative research we currently have, then draw some new maps that enrich and extend the boundaries of our understandings beyond the margins. We need to draw on all our maps of understanding. Even those tired and retired maps of qualitative research may hold important clues such as the origin stories or genealogical beginnings of certain trends and sticking points in qualitative research.

Qualitative researchers, however, must be more than either travelers or cultural tourists. Qualitative research is an important tool for indigenous communities because it is the tool that seems most able to *wage the battle of representation* (Fine et al., 2000); to weave and unravel competing *storylines* (Bishop, 1998); to situate, place, and contextualize; to create spaces for decolonizing (Aldama, 2001; Tierney, 2000); to provide frameworks for hearing silence and listening to the voices of the silenced (LeCompte, 1993; L. T. Smith, 2001); to create spaces for dialogue across difference; to analyze and make sense of complex and shifting experiences, identities, and realities; and to understand little and big changes that affect our lives. Qualitative research approaches have the potential to respond to epistemic challenges and crises, to unravel and weave, to fold in and unmask the layers of the social life and depth of human experience. This is not an argument for reducing qualitative research to social activism, nor is it an argument that suggests that quantitative research cannot also do some of the same things, but rather an argument for the tools, strategies, insights, and expert knowledge that can come with having a focused mind trained on the qualitative experience of people.

Qualitative research has an expanding set of tools that enable finer-grained interpretations of social life. Expanding the understandings and tools of qualitative researchers is important in an era when the diversity of human experience in social groups and communities, with languages and epistemologies, is undergoing profound cultural and political shifts. Although it could be argued that this has always been the case because societies always are dynamic, there is an argument to be made about the rapid loss of languages and cultures, the homogenization of cultures through globalization, and the significance for many communities of the impact of human beings on the environment. Indigenous communities live with the urgency that these challenges present to the world and have sought, through international mobilization, to call attention to these concerns. It is considered a sign of success when the Western world, through one of its institutions, pauses even momentarily to consider an alternative possibility. Indigenous research actively seeks to extend that momentary pause into genuine engagement with indigenous communities and alternative ways of seeking to live with and in the world.

Note

1. For example, in January, 2004, a series of speeches was made in New Zealand by a conservative political leader that attacked the role of the Treaty of Waitangi in legislation, that claimed Māori had extra holiday entitlements, that Māori with academic qualifications had lower standards because of affirmative action entry practices, and that purported to represent a "race free" vision for New Zealand. The speeches were quickly taken up as a populist message even though they were based on information later found to be incorrect and exaggerated and were clearly underpinned by an understanding of race and ethnicity that resonated with the racist messages of Australia's One Nation Leader Pauline Hanson.

References

Aldama, A. J. (2001). *Disrupting savagism: Intersecting Chicana/o, Mexican immigrant, and Native American struggles for self-representation*. Durham, NC: Duke University Press.

Alexander, M. J., & Mohanty, C. T. (1997). *Feminist genealogies, colonial legacies, democratic futures*. New York: Routledge.

Anderson, I., Griew, R., & McAullay, D. (2003). Ethics guidelines, health research and indigenous Australians. *New Zealand Bioethics Journal, 4*(1), 20–29.

Anderson, K. (2000). *A recognition of being: Reconstructing native womanhood*. Toronto: Sumach Press.

Apple, M. (2001). *Educating the "right" way: Markets, standards, God, and inequality*. New York: RoutledgeFalmer.

Ashcroft, B., Griffiths, G., & Tiffin, H. (1989). *The empire writes back: Theory and practice in post-colonial literatures*. London: Routledge.

Australian Institute of Aboriginal and Torres Strait Islander Studies. (2000). *Guidelines for ethical research in indigenous studies*. Canberra: Author.

Basso, K. H. (1996). *Wisdom sits in places: Landscape and language among the Western Apache*. Albuquerque: University of New Mexico Press.

Battiste, M. (Ed.). (2000). *Reclaiming indigenous voice and vision*. Vancouver: University of British Columbia Press.

Beverley, J. (2000). Testimonio, subalternity, and narrative authority. In N. Denzin & Y. S. Lincoln (Eds.), *Handbook of qualitative research* (2nd ed., pp. 555–566). Thousand Oaks: Sage.

Biolsi, T., & Zimmerman, L. J. (Eds.). (1997). *Indians and anthropologists: Vine Deloria and the critique of anthropology.* Tucson: University of Arizona Press.
Bishop, R. (1998). Freeing ourselves from neo-colonial domination in research: A Māori approach to creating knowledge. *Qualitative Studies in Education, 11*(2), 199–219.
Bishop, R., & Glynn, T. (1992). He kanohi kitea: Conducting and evaluating educational research. *New Zealand Journal of Educational Studies, 27*(2), 125–135.
Bishop, R., & Glynn, T. (1999). Researching in Māori contexts: An interpretation of participatory consciousness. *Journal of Intercultural Studies, 20*(2), 167–182.
Brady, W. (1999). Observing the Other. *Eureka Street, 9*(1), 28–30.
Burger, J. (1987). *Report from the frontier: The state of the world's indigenous peoples.* London: Zed Books.
Chow, R. (1993). *Writing diaspora: Tactics of intervention in contemporary cultural studies.* Bloomington: Indiana University Press.
Churchill, W. (1993). I am indigenist. In W. Churchill (Ed.), *Struggle for the land: Indigenous resistance to genocide, ecocide, and expropriation in contemporary North America* (pp. 403–451). Monroe, ME: Common Courage Press.
Cook-Lynn, E. (1996). *Why I can't read Wallace Stegner and other essays: A tribal voice.* Madison: University of Wisconsin Press.
Cram, F. (1993). Ethics in Māori research. In L. Nikora (Ed.), *Cultural justice and ethics.* [Proceedings of the Cultural Justice and Ethics Symposium held as part of the New Zealand Psychological Society's annual conference]. Wellington, New Zealand: Victoria University.
Cram, F. (1997). Developing partnerships in research: Pâkehâ researchers and Māori research. *Sites, 35,* 44–63.
Cram, F. (2001). Rangahau Māori: Tona tika, tona pono – The validity and integrity of Māori research. In M. Tolich (Ed.), *Research ethics in Aotearoa New Zealand* (pp. 35–52). Auckland, New Zealand: Pearson Education.
Cram, F., Keefe, V., Ormsby, C., & Ormsby, W. (1998). Memory work and Māori health research: Discussion of a qualitative method. *He Pukenga Kôrero: A Journal of Māori Studies,* 37–45.
Cumiskey, K. M. (1998). (De)facing the Institutional Review Board: Wrangling the fear and fantasy of ethical dilemmas and research on "at-risk" youth. In J. Ayala et al. (Eds.), *Speed bumps: Reflections on the politics and methods of qualitative work* (pp. 28–31). New York: State University of New York, Graduate School of Education.
Deloria, V., Jr. (1995). *Red earth, white lies: Native Americans and the myth of scientific fact.* New York: Scribner.
Denzin, N. (1991). *Images of postmodern society: Social theory and contemporary cinema.* Newbury Park, CA: Sage.
Denzin, N., & Lincoln, Y. S. (2000). The discipline and practice of qualitative research. In N. Denzin & Y. S. Lincoln (Eds.), *Handbook of qualitative research* (2nd ed., pp. 128). Thousand Oaks, CA: Sage.
Durie, A. (1992). *Whaia te Ara Tika: Research methodologies and Māori.* Seminar on Māori research at Massey University, Palmerston North, New Zealand.
Eidheim, H. (1997). Ethno-political development among the Sami after World War II: The invention of self-hood. In H. Gaski (Ed.), *Sami culture in a new era: The Norwegian Sami experience* (pp. 29–61). Kárásjohka, Norway: Davvi Girji.
Elabor-Idemudia, P. (2002). Participatory research: A tool in the production of knowledge in development discourse. In K. Saunders (Ed.), *Feminist development and thought: Rethinking modernity, post-colonialism and representation* (pp. 227–242). London: Zed Books.
Fanon, F. (1963). *Wretched of the earth* (C. Farrington, Trans.). New York: Grove Press. (Original work published 1961)

Fine, M., Weis, L., Weseen, S., & Wong, L. (2000). For whom? Qualitative research, representations, and social responsibilities. In N. Denzin & Y. S. Lincoln (Eds.), *Handbook of qualitative research* (2nd ed., pp. 107–132). Thousand Oaks, CA: Sage.

Gaski, H. (Ed.). (1997). *In the shadow of the midnight sun: Contemporary Sami prose and literature*. Kárásjohka, Norway: Davvi Girji.

Green, M. (2000). Lived spaces, shared spaces, public spaces. In L. Weis & M. Fine (Eds.), *Construction sites: Excavating race, class, and gender among urban youth* (pp. 293–304). New York: Teachers College Press.

Haig-Brown, C., & Archibald, J. (1996). Transforming First Nations research with respect and power. *Qualitative Studies in Education, 9*(3), 245–267.

Harjo, J., & Bird, G. (1997). *Reinventing the enemy's language. Contemporary native women's writing of North America*. New York: W. W. Norton and Company.

Helu Thaman, K. (2003). *Re-presenting and re-searching Oceania: A suggestion for synthesis*. Keynote address to the Pacific Health Research Fono, Health Research Council of New Zealand, Auckland.

Hill Colllins, P. (1991). Learning from the outsider within. In M. Fonow & J. A. Cook (Eds.), *Beyond methodology: Feminist scholarship as lived research* (pp. 35–57). Bloomington: Indiana University Press.

Humphery, K. (2000). *Indigenous health and "Western research"* (Discussion paper for VicHealth Koori Health Research and Community Development Unit). Melbourne: Centre for the Study of Health and Society, University of Melbourne.

Humphery, K. (2002). Dirty questions: Indigenous health and "Western research." *Australian and New Zealand Journal of Public Health, 25*(3), 197–202.

Jahnke, H., & Taiapa, J. (1999). Māori research. In C. Davidson & M. Tolich (Eds.), *Social science research in New Zealand: Many paths to understanding* (pp. 39–50). Auckland, New Zealand: Longman Pearson Education.

Johnston, P. M. (2003). Research in a bicultural context: The case in Aotearoa/New Zealand. In J. Swann & J. Pratt (Eds.), *Educational research practice: Making sense of methodology* (pp. 98–110). London: Continuum.

Jordan, S. (2003). Who stole my methodology? Co-opting PAR. *Globalisation, Societies and Education, 1*(2), 185–200.

Kaomea, J. (2003). Reading erasures and making the familiar strange: Defamiliarizing methods for research in formerly colonized and historically oppressed communities. *Educational Researcher, 32*(2), 14–25.

Kelsey, J. (1995). *The New Zealand experiment*. Auckland, New Zealand: Auckland University Press.

Keskitalo, J. H. (1997). Sami post-secondary education – Ideals and realities. In H. Gaski (Ed.), *Sami culture in a new era: The Norwegian Sami experience* (pp. 155–171). Kárásjohka, Norway: Davvi Girji.

King, N., Henderson, G. E., & Stein, J. E. (1999). *Beyond regulations. Ethics in human subjects research*. Chapel Hill: University of North Carolina Press.

Ladson-Billings, G. (2000). Racialized discourses and ethnic epistemologies. In N. K. Denzin & Y. S. Lincoln (Eds.), *Handbook of qualitative research* (2nd ed., pp. 257–278). Thousand Oaks, CA: Sage.

Langton, M. (1981). Anthropologists must change. *Identity, 4*(4), 11.

The Latina Feminist Group. (2001). *Telling to live: Latina feminist testimonios*. Durham, NC: Duke University Press.

LeCompte, M. (1993). A framework for hearing silence: What does telling stories mean when we are supposed to be doing science? In D. McLaughlin & W. G. Tierney (Eds.), *Naming silenced lives* (pp. 9–28). New York: Routledge.

Lincoln, Y. S. (1993). I and thou: Method, voice, and roles in research with the silenced. In D. McLaughlin & W. G. Tierney (Eds.), *Naming silenced lives* (pp. 29–50). New York: Routledge.

Little Bear, L. (2000). Jagged worldviews colliding. In M. Battiste (Ed.), *Reclaiming indigenous voice and vision* (pp. 77–85). Vancouver: University of British Columbia Press.

Maracle, L. (1996). *I am woman: A native perspective on sociology and feminism*. Vancouver: Press Gang Publishers.

Marker, M. (2003). Indigenous voice, community, and epistemic violence: The ethnographer's "interests" and what "interests" the ethnographer. *Qualitative Studies in Education, 16*(3), 361–375.

McLaren, P. (1993). Border disputes: Multicultural narrative, identity formation, and critical pedagogy in postmodern America. In D. McLaughlin & W. G. Tierney (Eds.), *Naming silenced lives* (pp. 201–236). New York: Routledge.

Mead, H. M. (2003). *Tikanga Māori: Living by Māori values*. Wellington, New Zealand: Huia Publications and Te Whare Wananga o Awanuiarangi Press.

Memmi, A. (1967). *The colonizer and the colonized*. Boston: Beacon Press. (Original work published 1957)

Menzies, C. (2001). Researching with, for and among indigenous peoples. *Canadian Journal of Native Education, 25*(1), 19–36.

Mies, M. (1983). Towards a methodology for feminist research. In G. Bowles & R. D. Klein (Eds.), *Theories of women's studies* (pp. 117–139). New York: Routledge.

Minh-ha, T. T. (1989). *Woman, native, other: Writing, postcoloniality and feminism*. Bloomington: Indiana University Press.

Mohanty, C. (1984). Under Western eyes: Feminist scholarship and colonial discourses. *Boundary, 12*(3) and *13*(1), 338–358.

Moraga, C., & Anzaldúa, G. (Eds.). (1983). *This bridge called my back*. New York: Kitchen Table Press.

Moran, W. (1999). Democracy and geography in the reregulation of New Zealand. In D. B. Knight & A. E. Joseph (Eds.), *Restructuring societies: Insights from the social sciences* (pp. 33–58). Ottawa: Carleton University Press.

Moreton-Robinson, A. (2000). *Talkin'up to the white woman: Indigenous women and feminism*. St. Lucia: University of Queensland Press.

Ngugi Wa Thiong'o. (1997). *Writers in politics: A re-engagement with issues of literature and society*. Oxford, UK: James Currey. (Original work published 1981)

Peters, M. A. (2003). Classical political economy and the role of universities in the new knowledge economy. *Globalization, Societies and Education, 1*(2), 153–168.

Pidgeon, M., & Hardy, C. (2002). Researching with Aboriginal peoples: Practices and principles. *Canadian Journal of Native Education, 26*(2), 96–106.

Pihama, L. (1993). *Tungia te ururua kia tupu whakarirorito te tupu o te harakeke*. Unpublished master's thesis, University of Auckland.

Pihama, L. (1994). Are films dangerous?: A Māori woman's perspective on *The Piano*. *Hecate, 20*(2), 239–242.

Pihama, L. (2001). *Tihei Mauriora: Honouring our voices – mana wahine as a kaupapa Māori theoretical framework*. Unpublished doctoral thesis, University of Auckland.

Pihama, L., Cram, F., & Walker, S. (2002). Creating methodological space: A literature review of Kaupapa Māori research. *Canadian Journal of Native Education, 26*(1), 30–43.

Pritchard, S. (Ed.). (1998). *Indigenous peoples, the United Nations and human rights*. London: Zed Books.

Reinharz, S. (1992). *Feminist methods in social research*. New York: Oxford University Press.

Rigney, L. (1999). Internationalization of an indigenous anticolonial cultural critique of research methodologies. A guide to indigenist research methodology and its principles. *Wicazo SA Journal of Native American Studies Review, 14*(2), 109–121.

Roman, L. G., & Eyre, L. (1997). *Dangerous territories: Struggles for difference and equality in education*. New York: Routledge.

Said, E. (1978). *Orientalism*. London: Vintage Books.
Saunders, K. (Ed.). (2002). *Feminist development and thought: Rethinking modernity, post-colonialism and representation*. London. Zed Books.
Sedgwick, E. K. (1991). *Epistemology of the closet*. New York: Harvester Wheatsheaf.
Shiva, V. (1993). *Monocultures of the mind*. London: Zed Books.
Simon, J., & Smith, L. T. (Eds.). (2001). *A civilising mission? Perceptions and representations of the New Zealand Native Schools system*. Auckland, New Zealand: Auckland University Press.
Smith, G. H. (1990). The politics of reforming Māori education: The transforming potential of kura kaupapa Māori. In H. Lauder & C. Wylie (Eds.), *Towards successful schooling* (pp. 73–89). Basingstoke: Falmer.
Smith, G. H. (1992). *Research issues related to Māori education*. Auckland, New Zealand: Research Unit for Māori Education, The University of Auckland.
Smith, G. H. (1994). For sale: Indigenous language, knowledge and culture. *Polemic: A Journal of the University of Sydney Law School, 4*(3).
Smith, L. T. (1991). Te rapunga i te ao marama (the search for the world of light): Māori perspectives on research in education. In T. Linzey & J. Morss (Eds.), *Growing up: The politics of human learning* (pp. 46–55). Auckland, New Zealand: Longman Paul.
Smith, L. T. (1992). Māori women: Discourses, projects and mana wahine. In S. Middleton & A. Jones (Eds.), *Women and education in Aotearoa 2* (pp. 33–51). Wellington, New Zealand Bridget Williams Books.
Smith, L. T. (1999). *Decolonizing methodologies: Research and indigenous peoples*. London: Zed Books.
Smith, L. T. (2000). Kaupapa Māori research. In M. Battiste (Ed.), *Reclaiming indigenous voice and vision* (pp. 225–247). Vancouver: University of British Columbia Press.
Smith, L. T. (2001). Troubling spaces. *Journal of Critical Psychology, 4*, 175–182.
Smith, L. T., Smith, G. H., Boler, M., Kempton, M., Ormond, A., Chueh, H. C., et al. (2002). "Do you guys hate Aucklanders too?" Youth: Voicing difference from the rural heartland. *Journal of Rural Studies, 18*, 169–178.
Spivak, G. (1987). *In other worlds: Essays in cultural politics*. New York: Methuen.
Stanley, L., & Wise, S. (1983). *Breaking out: Feminist consciousness and feminist research*. London: Routledge & Kegan Paul.
Steinhauer, E. (2003). Thoughts on an indigenous research methodology. *Canadian Journal of Native Education, 26*(2), 69–81.
Stewart, T. A. (1997). *Intellectual capital: The new wealth of organizations*. New York. Doubleday/Currency.
Stewart-Harawira, M. (1999, October). Neo-imperialism and the (mis)appropriation of indigenousness. *Pacific World, 54*, pp. 10–15.
Te Awekotuku, N. (1991). *Mana wahine Māori*. Auckland, New Zealand: New Women's Press.
Te Awekotuku, N. (1999). Māori women and research: Researching ourselves. In *Māori psychology: Research and practice* (pp. 57–63) [Proceedings of a symposium sponsored by the Māori and Psychology Research Unit, University of Waikato]. Hamilton, New Zealand: University of Waikato.
Te Hennepe, S. (1993). Issues of respect: Reflections of First Nations students' experiences in post-secondary anthropology classrooms. *Canadian Journal of Native Education, 20*, 193–260.
Thomas, G. (2001). The value of scientific engineering training for Indian communities. In K. James (Ed.), *Science and Native American communities* (pp. 149–154). Lincoln: University of Nebraska Press.
Thompson, P. (1978). *The voices of the past*. London: Oxford University Press.

Tierney, W. G. (2000). Undaunted courage: Life history and the postmodern challenge. In N. K. Denzin & Y. S. Lincoln (Eds.), *Handbook of qualitative research* (2nd ed., pp. 537–554). Thousand Oaks, CA: Sage.

Tomlins-Jahnke, H. (1997). Towards a theory of mana wahine. *He Pukenga Kôrero: A Journal of Māori Studies, 3*(1), 27–36.

Trask, H.-K. (1986). *Eros and power: The promise of feminist theory.* Philadelphia: University of Pennsylvania Press.

Trask, H.-K. (1993). *From a native daughter: Colonialism and sovereignty in Hawai'i.* Monroe, ME: Common Courage Press.

Vidich, A. J., & Lyman, S. M. (2000). Qualitative methods: Their history in sociology and anthropology. In N. K. Denzin & Y. S. Lincoln (Eds.), *Handbook of qualitative research* (pp. 37–84). Thousand Oaks, CA: Sage.

Warrior, R. A. (1995). *Tribal secrets: Recovering American Indian intellectual traditions.* Minneapolis: University of Minnesota Press.

Wilmer, F. (1993). *The indigenous voice in world politics.* Newbury Park, CA: Sage.

Womack, C. S. (1999). *Red on red: Native American literary separatism.* Minneapolis: University of Minnesota Press.

Worby, G., & Rigney, D. (2002). Approaching ethical issues: Institutional management of indigenous research. *Australian Universities Review, 45*(1), 24–33.

26

Ethnographic Showcases, 1870–1930

Raymond Corbey

> Il n'y a ni bourg, ni hameau, ni maison séparée dans l'île; Zamé a voulu que toutes les possessions d'une province fussent réunies dans une même enceinte afin que l'oeil vigilant du commandant de la ville pût s' étendre avec moins de peine sur tous les sujets de la contrée
>
> – D. A. F. de Sade, *Aline et Valcour*

"To see is to know" – this motto was attached to the anthropological exhibits of the World's Columbian Exposition of 1893, one of the many world fairs during the era of imperialism and colonialism (Rydell 1984:44). At these gigantic exhibitions, staged by the principal colonial powers, the world was collected and displayed. Natives from a wide range of colonized cultures quickly became a standard part of most manifestations of this kind. Together with their artifacts, houses, and even complete villages, so-called savages or primitives were made available for visual inspection by millions of strolling and staring Western citizens. Comparable places of spectacle such as zoos, botanical gardens, circuses, temporary or permanent exhibitions staged by missionary societies and museums of natural history, all exhibited other races and/or other species and testified to the imperialism of 19th-century nation-states.

In this article I will put these ethnographic exhibits into the wider context of the collecting, measuring, classifying, picturing, filing, and narrating of colonial Others during the heyday of colonialism. All these modes of dealing with the exotic, with colonial otherness, functioned in a context of European hegemony, testifying to the successful imperialist expansion of 19th-century

Source: *Cultural Anthropology*, 8(3) (1993): 338–369.

nation-states and to the intricate connections that developed between scientific and political practices. Of course, I cannot bypass the historical changes and national differences in exhibitionary practices in the period under study – the last decades of the 19th and the first decades of the 20th century – but I will concentrate on the similarities, which in my view are predominant, arguing that it is possible to have a wide range of seemingly divergent modes of dealing with the Other within one single analytic field.

The World on Show

World fairs or international expositions (*exposition universelle, Weltausstellung*) were very large-scale happenings that combined features of trade and industrial fairs, carnival, music festivals, political manifestations, museums, and art galleries. But primarily they were "pilgrimage sites of commodity fetishism," as Walter Benjamin (1984:441) put it rather pointedly. From 1851 onward, when the first international exposition took place in London, an enormous variety of industrial and technological products were exhibited, including steam machines, lawn mowers, elevators, photographic cameras, mechanized weaving looms, and household appliances. In addition, colonial raw materials and products were displayed, along with archaeological artifacts. Various architectural styles were presented, and after 1885 the arts became a recurrent theme. The idea was to show progress in all fields – not only in industry, trade, and transportation, but also in the arts, the sciences, and culture. Meanwhile, there was no mention of poverty, sickness and oppression, or social and international conflicts.

World fairs have been compared to gigantic potlatches, joyous ritual displays of richness and power, where possessions were given away and even destroyed in great numbers in order to gain prestige and to outdo others, as occurred among the Kwakiutl and other Amerindian societies of the North American Northwest Coast (Benedict 1983:7 ff). In both cases – world fairs and potlatches – ritualized competition, gaining prestige, and keeping up reciprocity between parties of comparable caliber played important roles. Both kinds of manifestations were large-scale, expensive festivals that – having social, economic, political, juridical, moral, and aesthetic aspects – displayed the character of a *fait social total* in the sense of M. Mauss, and both regulated relationships between rival groups (such as nations and large cities). Economic interests went hand in hand with cultural ones, and nationalistic ambitions were apparent with the international character of the fairs, where each country built its own monumental pavilion in its particular national style. The architecture was meant to impress; it could never be large, imposing, or unusual enough: the Crystal Place (London), the Eiffel Tower (Paris), the Atomium (Brussels). The first world fair, the Great Exhibition of the Works of Industry of all Nations, in the Crystal Palace in

London in 1851, attracted 6 million visitors; the world fair in Paris in 1878, 16 million; the 1900 Paris fair, still before the era of the cinema and television, 50 million.

So-called colonial exhibitions – such as the Colonial and Indian Exhibition in London in 1886, the British Empire Exhibition in 1924–25, or the Exposition Coloniale in Paris in 1931 – were even more closely associated with the idea and ideal of empire. But even for the 1851 Great Exhibition, the Society of Arts had fanatically developed scenarios for, as one of its spokesmen put it, "promoting and spreading Christianity, civilization and commerce among peoples still steeped in barbarity and idolatry – in terms of quantity nearly half of humanity" (quoted in Haltern 1971:314). World fairs quickly became inseparable from imperialism and nationalism. For the British Exhibition at Wembley in 1924, a historical pageant was staged with fifteen thousand participants that took three days to pass (Benedict 1983:47). With this gigantic spectacle, accompanied by music and texts written for the occasion, and structured like a heroic epos governed by a fundamental opposition between the civilized and the barbaric, the English people, self-consciously and full of pride, presented and represented their newly created world empire. Such manifestations are of interest in the present context, not only because colonial natives had a role to play too, but also because they express the civilizatory idiom that formed the basis of contemporary views of "primitives," tied up with imperialist ideology and social Darwinism.

In covering the Great Exhibition, the German newspaper *Allgemeine Zeitung* stressed the prevalent "spirit of encyclopaedism" (Haltern 1971:352). In the course of their development, the encyclopedic character of world exhibits – an explicitly stated goal – became increasingly prominent. The history of subjects like the home, labor, and transportation was shown, while an equally panoramic variety of cultures was displayed. An inventory and census of the whole world and the whole history of humankind was constructed in a way reminiscent of medieval maps of the world – *mappae mundi*, offering an encyclopedic survey of the world as creation – or of cabinets of curiosity during the Renaissance. During the same period, journals like the French *Le Tour du Monde*, with their synoptic and panoptic illustrations usually copied from photographs, had a comparable function, as did photography as such, not least as put to use at the world fairs (Favrod 1989; Lederbogen 1986; Theye 1989).

In all cultures, political and religious elites tend to accumulate and flaunt their rare and precious objects from faraway places in order to gain prestige and to display their knowledgeability. Everyone from Renaissance princes and cardinals to Chinese emperors owned collections of exotic animals, objects, and even people. Something similar happened at world fairs, where not individual collectors but states, metropoles, and their elites were involved. Mary Helms offers the analogy of the tribal shaman's medicine pouch holding a collection of strange objects. The more rare and exotic these objects, the more effective they were and the more they contributed to the shaman's prestige and power.

> The emperor's zoo and botanical gardens, like the shaman's pouch, contained bits and pieces of the animate cosmos, power-filled natural wonders, examples of the rare, the curious, the strange, and the precious – all expressions of the unusual and the different attesting to the forces of the dynamic universe that by definition lies outside the (again by definition) controlled, socialized, civilized heartland. [Helms 1988:166]

In each of these cases, a microcosm sampled and presented the macrocosm. The world fair can also be read as a microcosm, created by the Promethean Western middle classes, with their unlimited trust in Enlightenment ideas and the rational constructibility of the world – a world made after their own image, to European standards. Here, nature was of only secondary importance, appearing only in the shape of cultivated crops and painted backdrops; here, everything had been fabricated by man. Marx and Engels referred to the Great Exhibition as "a pantheon in modern Rome" where the bourgeoisie class of this world "exhibits with self-congratulatory pride the gods it has created for itself" (quoted in Haltern 1971:314). In high euphoria, the bourgeoisie celebrated progress, the attainment of world power, and the creation of Western middle-class culture by its own efforts – which, in its own eyes, was the purpose to which world history (and indeed cosmic evolution) had been directed from its earliest beginnings. Progress and civilization were the key concepts behind these large-scale representations of middle-class Selves and savage Others. Another prominent theme was the Enlightenment ideal of universal brotherhood, connected with the Christian ideals of peace and love.

World fairs, as Carol Breckenridge (1989:196) remarks, were part of a unitary, though not necessarily uniform, landscape of discourse and practice, providing a cultural technology for situating metropole and colony within a single analytic field, thus creating an imagined ecumene. The fairs told the story of mankind, the very same narrative that accompanied and legitimized colonial expansion. In this epic, staged by themselves, white, rational, civilized European citizens cast themselves in the role of the hero.

Savages on Show

Placed alongside all kinds of objects and products, colonial natives quickly became a standard part of world fairs, for the education and entertainment of Western citizens. Not only the citizens themselves but also the natives figured as categories in Western representations of Self, as characters in the story of the ascent to civilization, depicted as the inevitable triumph of higher races over lower ones and as progress through science and imperial conquest.

Often ethnologists were ahead of their times concerning interpretations of other cultures, but Charles Rau, for one, who created the ethnological exhibits at the Philadelphia Centennial Exhibition in 1876 on behalf of the Smithsonian Institution, stated that

> the extreme lowness of our remote ancestors cannot be a source of humiliation; on the contrary, we should glory in our having advanced so far above them, and recognize the great truth that progress is the law that governs the development of mankind. [Quoted in Rydell 1984:24]

Two years later, the Paris world fair of 1878 was the first one in which many people from non-Western cultures were exhibited, in specially constructed pavilions and "native villages" (*Village indigène*). The display of 400 natives from the French colonies Indochina, Senegal, and Tahiti met with huge success, as did the exhibits of indigenous peoples from Java, Samoa, Dahomey, Egypt, and North America itself at the World's Columbian Exposition in 1893.

Native villages were a standard part of world fairs from 1878 onward. Equally popular were the "foreign streets" such as the "Rue de Caire." Around the turn of the century, the International Anthropological Exhibit Company commercially exploited exhibitions of non-Western people in the United States in several settings, including world fairs. At the Dutch "Internationale Koloniale en Uitvoerhandel Tentoonstelling" at Amsterdam in 1883, natives from the Dutch East Indies and West Indies were shown. The Greater Britain Exhibition of 1899 included a "Kaffir Kraal – A Vivid Representation of Life in the Wilds of the Dark Continent," an exhibit featuring African animals and 174 natives from several South African peoples brought under control only shortly before. They were divided into four native villages, showing their crafts, performing "war dances," and riding on ponies. Among them were San, who characteristically were exhibited as part of the natural history of Africa, together with baboons (MacKenzie 1984:104). Often the European impresarios traveled from one world fair to another with the same group of people – the Senegalese who constituted the well-known "Senegalese village" (Figure 1), for example – and had them perform at other venues and on other occasions as well.

A brochure commenting on the "Village from Dahomey" at the Imperial International Exhibition of 1909 stressed the violent brutality of indigenous Africa, especially Dahomey with its "bloodthirsty potentates" and woman warriors or "Amazons" who were one of the main attractions of the village; it praised the French intervention in 1892 with the following words:

> Order and decency, trade and civilization have taken the place of rule by fear of the sword. France has placed its hand on the blackest spot in West Africa, and wiped out some of the red stain that made Dahomey a byword in the world. . . . Today . . . (the) days of savagery are passing away. [Quoted in MacKenzie 1984:116]

Figure 1: "Senegalese Village" in the Brussels world fair of 1910, the setting of physical confrontation of European citizens with exotic Africans, for amusement and edification. Over the entry – a semiotically highly charged threshold between two worlds, one primitive and one civilized – religion, trades, songs, dances, tam-tam, and a harem are announced. Postcard, collection of P. Faber, Amsterdam.

It was light against dark, order against violence, and a European nation as the bringer of civilization. The exhibited "Amazons" – depicted as both barbarous and alluring, true personifications of the Dark Continent – performed throughout Europe. When they appeared in the Moskauer Panoptikum in Frankfurt in 1899, they were introduced as "wild females" – *wilde Weiber*. A group of women from Samoa (Figure 2), however, was described by the press and in brochures as a breathtakingly beautiful, always cheery, erotically permissive, and lazy people from the paradisiacal Pacific Ocean (Plakate 1880–1914:257). North American Indians were similarly idealized and romanticized.

The 1909 world fair that featured the Amazons also included a native village of nomadic Kalmuks from Central Asia, brought under the control of the Russian empire shortly before. At the Berliner Gewerbe-Ausstellung of 1896, which led to the foundation of the Deutsches Kolonialmuseum, over a hundred natives from the German colonies were present, each group in its own carefully imitated cultural and natural setting. They had to call "hurrah" at set times in praise of emperor and *Reich* (G. Schneider 1982:167). Governments were keenly aware of the opportunity to publicize their colonial policies and to manipulate public attitudes toward the newly acquired territories. German, Dutch, and Irish villages, among others, with native people in traditional clothing were also (re)presented at the world fairs as

Figure 2: A direct consequence of German colonial expansion. "Our new fellow-countrymen." *Schautruppe* from Samoa performing in the Berliner Tiergarten in 1900 or 1901. Cabinet photograph, Marquardt Brothers, collection of the Museum voor Volkenkunde, Rotterdam.

part of the national exhibits, staged in this case, however, by the exhibited peoples themselves, not their colonizers.

It seems to have been quite usual for visitors to throw money to the performing natives and for the natives to beg for it. The exhibited peoples' behavior and movement was strictly controlled. They were presented as "different" and forced to behave that way. At most – though not all – manifestations, it was unthinkable that they should mingle spontaneously with the visitors, and usually there were few possibilities for contact between parties. The living exhibits had to stay within a precisely circumscribed part of the exhibition space, which represented their world; the boundary between this world and that of the citizens visiting and inspecting them (Figure 3), between wildness and civility, nature and culture, had to be respected unconditionally. All signs of acculturation were avoided as long as the natives were on show. One of my colleagues who grew up in postwar Berlin told me of his astonishment when, as a boy, he came across an African man whom he had seen only hours before in native attire at Castan's Panoptikum, now in European clothes on a tramcar, smoking a cigarette. Primitivity was staged in minute detail.

The desirability of "civilizing" North American Indian peoples was an important theme at the St. Louis fair, as it was at other world fairs in the United States. Their "dull-minded and self-centered tribal existence" had to be replaced by "[the] active and constructive and broadminded life of modern humanity" (Francis 1913:529, quoted in Benedict 1983:50). The

Figure 3: An imperial encounter. "African Village: Women eating their meal," scrutinized by visitors to the colonial exposition at Antwerp in 1930. The gaze of the visitors is not returned. Postcard, collection of S. Wachlin.

imposition of cultural assimilation, as typical of internal colonialism in the United States as it was of French colonialism, was less important in the English colonial regime. English visitors of French world fairs and colonial exhibitions, therefore, often expressed their astonishment when confronted with indigenous people in European clothing. In St. Louis, the living exhibits were typically organized on a scale from civilized to barbaric. The lower a people or race was deemed to be, the further removed it was from the "Indian school" that marked one pole of the scale, that of civilization. Philippine Igorots and African Pygmies were situated near the pole of barbarity at the other end of the scale.[1] At the World's Columbian Exposition of 1893 in Chicago, the opposition between wild and civilized and the desirability of civilizing peoples that were still wild was expressed by showing an old Indian in shabby traditional clothing next to his son in a neat new suit (Benedict 1983:49 ff.).

Hagenbeck's *Völkerschau*

Persons from non-Western cultures appeared not only at world and colonial exhibitions, but also at special ethnographic shows called *Völkerschau* in Germany, where this type of manifestation had proliferated since 1874. In that year Carl Hagenbeck, a dealer in wild animals in Hamburg and later

Figure 4: Nomadic Samen or "Laplanders" with reindeers, sleighs, and tents, probably in the zoo of Halle, Germany, c. 1927. Photograph, collection of the Museum voor Volkenkunde, Rotterdam.

director of a zoo and a circus, began exhibiting Samen – Lapps – as "purely natural people" (*reine Naturmenschen*, Lehmann 1955) in several German cities, together with their tents, tools, weapons, and other possessions, as well as reindeer (Figure 4). In 1876, he sent one of his collaborators to Egyptian Sudan in order to bring back Nubians and indigenous animals. This group of savages with their wild personalities, as Hagenbeck describes them in his autobiography (Hagenbeck 1909), was scrutinized by over thirty thousand German visitors on the first day of their appearance in Breslau. Subsequently, the Nubians were exhibited in other European cities, including Paris and London.

When this venture became a success, Hagenbeck extended his profitable activities to include North American Indians, Inuit, people from India, and Zulus (*Zulukaffern*). He and other entrepreneurs also exhibited Sudanese, Bushmen, and *Somalinegerknaben* riding on ostriches; later, Dinka, Maasai, and Ashanti were recruited in Africa and brought to Europe (cf. Thode-Arora 1989:168–178). The supply of natives closely followed the colonial conquests. Tuareg, for instance, were on exhibit in Paris within months of the French capture of Timbuktu in 1894, and natives from Madagascar appeared a year after the French occupation of that island (W. Schneider 1977:101). The new genre quickly caught on, and ethnographic exhibits stayed very popular well into the present century. Hagenbeck presented them to the middle classes with the stated intention of promoting the

Bildung – the knowledge and culture a civilized person should possess – and stimulating the German people's nationalistic zest for colonial expansion. Often the members of *Naturvölker* ("natural peoples"), more closely associated with living nature than with civilization, were exhibited in local zoos behind bars or wire fences; fairgrounds and public parks served as settings, too. In France, the Paris Jardin d'Acclimatation, created in 1859 for the study and popularization of exotic animals and plants, became a popular setting for ethnological exhibits similar to the German ones.

For decades, the German press wrote about the appearance, behavior, and nature of the foreign visitors in a very negative tone, expressing disgust and contempt for them. The general reaction of the public visiting the exhibitions seems to have been the same; but near the turn of the century, press coverage began to change for the better, and more attention was given to ethnographic detail. Another reason for negative reactions to black Africans in particular, apart from deeply ingrained stereotypes of barbarity and primitivity, was the stubborn and often bloody resistance of several African peoples toward the European expansion in Africa, which was covered extensively by the European press. In the eyes of many Germans, a black African was some sort of savage monster. France consciously played on such fears in the French-German war of 1870–71 by putting black *tirailleurs indigènes* trained in Algeria to use against German troops (Goldmann 1985:258). In general, the more an indigenous people resisted colonization, the more ferocity its representatives had to display when staged.

Fear was but one of the mixed feelings German citizens experienced when visiting ethnological exhibitions. Another reaction was sexual fascination and curiosity, as is clear from contemporary press coverage and from preserved posters. Admiration of the supposedly great sexual potency of the scarcely clothed primitives competed with depreciation because of their alleged bestial lust (Goldmann 1985:263–264; Thode-Arora 1989:115–119). Disgust alternated with exalted attention, wonder, and enchantment when Western citizens were confronted with picturesque scenes from savage life.

In the Netherlands, too, ethnological exhibits took place. In the year 1900, for instance, the *Groote Achantees Karavanen* ("Large Ashanti Caravans") attracted much attention in Amsterdam, Rotterdam, The Hague, Utrecht, and Nijmegen. The Ashanti, usually shown at the Jardin d'Acclimatation in Paris, now toured the rest of Western Europe.[2] In the Netherlands, they were described on a poster as "old natives from the Gold Coasts of Africa . . . Warriors, Fetish, Priests, Snake-charmer, Women, Girls and Children. The most uncommon human race that has ever been seen in Europe. Most interesting for everyone." (Municipal Archives, Rotterdam). A few years earlier, *De Boschmannen of wilden van Afrika* ("The Bushmen or savages of Africa"), as the title of the accompanying brochure reads, were on tour. Judging from their appearance, this brochure states, "they show more similarity to Apes than to people. . . . Notwithstanding their ferocity these Bushmen are nearly harmless, and even the most fearful person can approach and feel all over them with the greatest confidence" (Municipal Archives, Rotterdam). The suggestion that they could be touched indicates how close the attitude toward these people was to the attitude toward animals. That the exhibited people were similar – metonymically, metaphorically, qua appearance and behavior – to animals, especially apes, was indeed a common perception, fed by contemporary scientific theory. In recent decades, in contrast, Bushmen once again came to play a positive role in Western imagination, similar to the one they played in the 18th century – that of noble savages, spontaneously and innocently enjoying a pure, natural, paradisiacal existence.

The natives performed in several roles. The American firm William Foote & Co. African American Characters exploited a show with African-Americans – as the letterhead of the firm stated – appearing as "Savages, Slaves, Soldiers and Citizens" (Thode-Arora 1989:41). Crafts, hunting techniques, rituals, dances, and songs were among the activities staged, as well as stereotypical "authentic" performances like warfare, cannibalistic acts, and head-hunting. At the 1904 Louisiana Purchase Exposition in St. Louis, Igorots from the Philippines could be seen eating dog meat, a food taboo in the West, while African Pygmies illustrated decapitation. The above-mentioned Dahomey "Amazons," heavily armed, simulated fights. Aborigines from Queensland, Australia, presented as *Austral Neger*, on exhibit at the Frankfurt Zoo and elsewhere in May 1885, were described on posters as cannibals and bloodthirsty monsters – "*wirklich blutdürstige Ungeheuer*" (Figure 5). Another poster, printed for their appearance in England, continued a European iconographical tradition reaching back to De Bry's late 16th-century *Grands Voyages* and earlier, by depicting them engaged in a ferocious cannibal ritual, with the following text:

> Male and female Australian cannibals/R. A. Cunningham, Director/The first and only obtained colony of these strange, savage, disfigured and most brutal race ever lured from the remote interior wilds, where they indulge in ceaseless bloody feuds and forays, to feast upon each other's flesh/The very lowest order of mankind, and beyond conception most curious to look upon. [Plakate 1880–1914:228]

How did the exhibited individuals themselves, often more or less coerced into participation, experience and cope with the exhibits? Many had to battle with homesickness, emotional confusion, difficulties of adjustment to the European climate and food, and vicious infections. Often they actively resisted the roles that were forced on them, for instance by running away, and they could be put back in harness only by force. Now and then bad treatment led to court cases. In the *Zeitschrift für Ethnologie* of 1880, Rudolf Virchow, the prominent anthropologist and politician, describes in detail how an Inuit woman whose measurements he wished to take literally ran into the walls of the room in total panic (quoted by Thode-Arora 1989:129–130),

Figure 5: A troupe of Aborigines from Queensland, Australia. Exploited by the ruthless A. R. Cunningham, they arrived in Germany in 1883, probably from England, where they performed at the Crystal Palace, Sydenham, a popular venue for ethnological exhibits. This picture was bought in Berlin in 1884. In 1896, only three individuals out of eight were still alive. Cabinet photograph, Negretti and Zambra, London, collection of the Museum voor Volkenkunde, Rotterdam.

an incident that was by no means exceptional. The percentage of those who died soon after their arrival in Europe was considerable. When the aforementioned group of Aborigines (Figure 5) arrived in Germany in 1883, there were eight of them; in May 1885, when they appeared in the Frankfurt Zoo, five were still alive; in October 1896, when they were examined by members of the Berliner Gesellschaft für Anthropologic, Ethnologie und Urgeschichte, there were only three survivors of the original group. The members of two Inuit families from Labrador, on exhibit in several German cities during the year 1880 with their dogs and kayaks (Figure 6), all succumbed to an infection in January 1881 (Plakate 1880–1914:236). At the cemetery of Tervuren, Belgium, close to the site of the 1897 colonial exhibition, a number of Africans lie buried who were part of the "Kongolese Village" and rowed canoes on the Tervuren ponds. In some cases, however, the exhibited natives were paid very well, treated with warmth and care, and offered sightseeing tours and dinners with local prominents.

How the individuals on display – confronted with an alien world, unfamiliar food, strange customs, and a different climate – experienced and handled their situation is an important question. The relative scarcity of sources concerning their intentions and feelings should not lead us to underestimate their subjectivity or to leave it understudied. Who were they? How did they

Figure 6: Some members of two Inuit families, on exhibit in 1880 at Berlin, Frankfurt, and several other German cities. In January 1881, both families succumbed to an infectious disease. The diary of one of the male members of the group is one of the few preserved testimonies from the perspective of the exhibited individuals (Taylor 1981). This carte-de-visite, with the printed text "Hagenbeck's Thierpark, Hamburg/Photographie von J. M. Jacobsen, St. Pauli" on its reverse, was bought during the month of October 1880 in Berlin by a Dutch visitor. Collection of the Museum voor Volkenkunde, Rotterdam.

think and feel? How did they negotiate their identities? How did they creatively exploit the roles they were cast into to realize their intentions and express their reactions?[3] How did these roles influence their conceptions of themselves and of the Western people they met? How, for instance, did Penobscot Indian Frank Loring (alias Chief Big Thunder), Micmac Indian Jeremy Bartlett (alias Doctor Lone Cloud), and Maliseet Indian Henry Perley (alias Red Eagle) feel about their roles as itinerant Native American performers (Prins 1991)?[4]

The diary of one of the males from the Inuit troupe touring Germany has been preserved (Taylor 1981). Much is known about Jefke, a black African boy on exhibit at the Antwerp (Belgium) Zoo, as well as about Ota Benga, a Pygmy boy displayed in a cage with a chimpanzee and later with an orangutan at the Bronx Zoo (Bieder 1991; Bradford and Blume 1992). Less is known about Klikko, "The Wild Dancing Bushman," a Khoi-San performing in London around 1913 (Parsons 1988). Mitchell (1988, 1989) quotes reactions of Egyptian scholars visiting the "Rue de Caire" at the Paris Exposition universelle of 1889. How did Black Elk conceive of his appearances in Buffalo Bill's Wild West Show (Figure 7; see Neihardt 1988; Rice 1991), or

Figure 7: Members of the successful "Buffalo Bill's Wild West Show," visiting London in 1887 on the occasion of the 50th jubilee of Queen Victoria's succession to the throne. In subsequent years, the show toured Europe twice, including an appearance at the 1889 world fair in Paris. This photograph was bought at London in July 1887 by a Dutch visitor. The handwritten numbers correspond to the handwritten names and tribes of the pictured Indians on the reverse. Cabinet photograph, Elliot and Fry, London, collection of the Museum voor Volkenkunde, Rotterdam.

Mary Alice Nelson (alias Molly Spotted Elk), a Penobscot Indian, of her dance performances in Paris, where she thrilled the public with her floor-length eagle feather headdress (McBride 1989)?[5] What was on the minds of the Indians from Surinam performing at the 1883 colonial exhibition in Amsterdam (see Figures 8 and 9), who seem to have been wildly enthusiastic during the first days of exposure to their new world, but soon had to cope with boredom, sickness, and the disappointment of not meeting their Dutch king as promised?

Commerce and Science

Hagenbeck was certainly not the first to take such an initiative. Although *Völkerschauen* became very popular during the period 1870–1930 and took place on a larger scale than ever before, the phenomenon as such was by no means new. During the 1820s, Captain Samual Hadlock from Maine toured Europe with a troupe of Inuit, which was exhibited in London, Hamburg, Berlin, Leipzig, Dresden, Prague, and Vienna. Part of his company was a "Maori chieftain" whom he had come across in England. When this Maori suddenly died, his head was preserved chemically and fixed to a model of

Figure 8: "Natives from Surinam," probably with their impresario, at the Amsterdam 1883 colonial exhibition that, besides these Indians (and Marrons) from the Dutch West Indies, featured colonial natives from the Dutch East Indies (cf. Figure 9). Cabinet photograph, Photographie Française, Amsterdam, Collection of the Tropenmuseum, Amsterdam.

his body. "We do not even need the Captain's word for it to be convinced," the Austrian *Allgemeine Theaterzeitung und Unterhaltungsblatt* wrote on the occasion of the exhibition of the reconstructed Maori at Vienna, "that this man from New Zealand, before he was taken aboard, really has eaten other people, because that's indeed the way he looks" (Goldmann 1985:256).

In London, to give but a few more examples, "Red Indians" could be seen in 1844; in 1845 and 1847, Bushmen; in 1850 and 1853, south African "Kaffirs," in the Egyptian hall – reflecting England's renewed interest in that part of the world during those years. In London, as in Berlin or Paris, learned societies such as the Ethnological Society and the competing Anthropological Society showed great interest in the ethnological exhibitions. Here too, as in Germany, the general attitudes of the public were rather negative, not least under the influence of contemporary missionary propaganda "In appearance," the *Times* wrote on the Bushmen in 1847,

> they are little above the monkey tribe, and scarcely better than the mere brutes of the field. . . . They are sullen, silent and savage – mere animals in propensity, and worse than animals in appearance. . . . In short, a more miserable set of human beings – for human they are, nevertheless – was never seen. [Quoted in Altick 1978:281]

Figure 9: Two gamelan players and two women dancers who performed in the "Javanese Kampong" (Javanese native village) at the 1883 "International Colonial and Exportation Exhibition" in Amsterdam. Together with others, they were recruited in the Dutch East Indies by the Dutch Missionary Society (Nederlands Zendingsgenootschap). Cabinet photograph, F. D. van Rosmalen, collection of the Museum voor Volkenkunde, Rotterdam.

And Charles Dickens wondered, "Is it idiosyncratic in me to abhor, detest, abominate, and abjure that noble savage?" adding that he hoped something would happen to the stove these people slept around, so that they would suffocate. In his eyes, and in the eyes of most of his contemporaries, the savage was

> a prodigious nuisance and a gross superstition . . . cruel, false, thievish, murderish; addicted more or less to grease, entrails, and beastly customs; a wild animal with the questionable gift of boasting; a conceited, tiresome, bloodthirsty, monotonous humbug . . . if we have anything to learn from the Noble Savage, it is what to avoid. His virtues are a fable; his happiness is a delusion; his nobility, nonsense. [Dickens 1938:133–138, quoted in Altick 1978]

We should add, however, that in later years Dickens traded his ideas on Anglo-Saxon superiority and progress for a form of cultural relativism implying much milder views on foraging and tribal peoples.

Columbus and Hernán Cortés had already brought back Indians and Aztecs from the New World. European princes, such as the Medici at Florence, had scores of aliens at their courts as curiosities and for purposes of prestige. During the age of European expansion, virtually every generation of Europeans could see Nubians, Inuit, Saami, North American Indians, and Pygmies at fairs, in inns – like the Amsterdam "Blaauw Jan," precursor to the Artis Zoo – and theaters or, together with exotic animals, in zoos and princely menageries. An analogous practice was that of exhibiting the insane – usually presented in cages, with an admissions fee. In 18th-century France, insanity was seen as a decline to a state of wildness and unruly animality, associated traditionally with all that was wicked and unnatural (Foucault 1961), while at the same time there existed a whole body of publications theorizing on similarities of physical appearance between particular types of insanity and particular animal species. "What was presented here," Dörner writes,

> was wild and indomitable nature, <beastliness>, absolute and destructive unruliness, social danger, which, behind the bars installed by reason, could be staged the more dramatically for showing at the same time to the public reason as the necessity of controlling nature, as a constraint upon unlimited freedom and as securing the order of the state. [1984:22]

Order was contrasted with chaos, reason with wildness.

Also comparable were the "monsters" and "freaks" shown at fairs, in circus side-shows, and in the amusement zones of world fairs: individuals with clubfeet, Siamese twins, bearded women, giants, dwarves, and so on (cf. Bogdan 1988). A famous case is that of Merrick, the "elephant-man," an intelligent and sensitive person who was terribly maimed by sickness, exhibited in a freak show in London at the end of the 19th century until a prominent surgeon took him into his custody (Montagu 1971). The so-called panopticums in large cities, like Castan's Panoptikum and the Passagen-Panoptikum in Berlin, combined features of a museum of anatomy, a cabinet of curiosities, and a horror cabinet. The way exotic animals were – and still are – shown and handled in circus performances elucidates practices of discipline and the concomitant idiom of wildness and taming that were present more implicitly in many exhibits involving people. P. T. Barnum's shows and, somewhat later, the German Circus Sarrasani for decades had ethnological acts on their program, often combined with acrobatics.

During the 18th and 19th centuries, exhibits of life specimens were increasingly reframed in terms of science, especially physical anthropology and natural history. Aside from their entertainment and curiosity value, their educational value came to be stressed more and more. Hagenbeck, for instance, advertised his manifestations as "anthropological-zoological exhibitions" (*Anthropologisch-Zoologische Ausstellung*). In many ways, exhibitions of human individuals were related to scientific practices and purposes anyhow. The lunatic asylums where the insane were put on show were in the process of being medicalized; what was monstrous or exotic was often as interesting from a scientific point of view as it was shocking or fascinating to the general public. At the beginning of the 19th century, the Khoi-Khoi ("Hottentot") woman Saartjie Baartmann was put on show when alive, and dissected by the famous Cuvier when dead. A mold of her body was then exhibited – in fact, until a few years ago, at the Paris Musée de l'Homme – as was, half a century later, the skeleton of the Tasmanian woman Truganini, which was cremated and committed to the waves only in the 1970s. Anthropologists used to be represented on the committees heading the anthropological sections of world fairs, often quarreling with those who wished to cater more to commercial than to scientific or educational interests.

In anthropometric and psychometric laboratories at the world fairs, visitors could witness and even take part in scientific research on racial characteristics. Phrenology, craniology, physiognomy, and anthropometry shared the assumption that in the outward shape and physical appearance of the body, the inner character – of different races, but also of criminals, prostitutes, and deviants – was manifest. The outward shape, therefore, had to be measured and mapped meticulously (cf. Sekula 1986). Particularly manifest in this context are the interconnections between exhibiting colonial natives and scientifically collecting, measuring, classifying, and filing them. At the same time, anthropological societies and museums of natural history accumulated tens of thousands of native skulls.[6] Saartjie Baartmann and Truganini are typical cases. According to contemporary views of the genetic variation of mankind, Khoi-Khoi, Tasmanians, Australian Aborgines (Figure 5), and several native peoples from Tierra del Fuego (Figures 10 and 11) stood

Figure 10: A number of Indians from Tierra del Fuego, probably Selk'nam, with their impresario, on exhibit in Paris in 1889. During the first decades of the present century, missionaries protested sharply against the systematic murder of the natives of Tierra del Fuego by white colonists and their bounty hunters. The skulls of murdered natives were often sold to museums in the Western world, fetching high prices. In addition, infectious diseases played havoc, so by around 1940, only a hundred individuals remained from an original population of many thousands. The anthropologist Anne Chapman (1982) could locate only ten genetically unmixed Selk'nam during three fieldwork stays between 1964 and 1974. From Gusinde (1931).

Figure 11: Another picture of "Patagonians," possibly Techuelches, taken in a Hamburg studio and bought in May 1879 on the occasion of a *Völkerschau* at Hamburg – as written by hand on the reverse. While North American Indians usually were admired and romanticized, as were the Samoans in Figure 2, the fascination with most other groups was tinged with fear and disgust. Cabinet photograph, J. M. Jacobsen, Hamburg, collection of the Museum voor Volkenkunde, Rotterdam.

closest to the apes and prehistoric ape-men in the racial hierarchy, and therefore were outstanding examples of "contemporary ancestors" and "missing links."

Exhibited colonial natives typically had to appear before anthropological societies such as those of Oxford and London or the Société d'Anthropologié in Paris. The Berliner Gesellschaft für Anthropologic, Ethnologie und Urgeschichte used to organize special sessions for this purpose, often at the same locations as the ethnological shows. On March 28, 1896, for instance, a number of members of the Berlin anthropological society studied an "authentic Arabic harem" at the above-mentioned Passagen-Panoptikum (Theye 1989:103). Usually anthropometric measurements and photographs were taken on such occasions, and it will come as no surprise that the objects of research, who were urged to undress, were often uneasy or even reluctant. The manufacture of plaster casts of different parts of the body was another research activity, as were linguistic and ethnomusicological observations. The learned societies provided certificates of authenticity of the people exhibited, and suggested new target groups that would serve the financial needs of the impresarios as fully as their own scientific interests.

So science, commerce, and imperialism went hand in hand. When Buffalo Bill's Wild West Show (Figure 7) performed in Paris during the summer of

1889, Prince Roland Bonaparte, ethnologist and anthropologist, was present continuously in order to question and measure the Cheyenne and Sioux who took part in the show. J. A. Jacobsen, a Norwegian and one of the most important collaborators of the Hagenbeck family, also collected for the Berlin Museum of Ethnology and supervised the manufacture of ethnographical dioramas featuring wax, plaster, or papiermâché mannequins. These were fabricated with the help of life casts and photographs, and they showed dramatic scenes from ritual and daily life. The Hagenbecks maintained close relations with the Hamburg Museum für Völkerkunde, to the benefit of both parties. And museums of ethnography or so-called colonial museums often originated from world fairs that, despite their many aspects, were first and foremost commercial happenings: the Musée d'Ethnographie du Trocadéro – now called the Musée de l'Homme – in Paris was created on the occasion of the 1878 world fair (Dias 1991), while the Koninklijk Museum voor Midden-Afrika at Tervuren, Belgium, resulted from the colonial exhibition of 1897.

William Schneider (1977:98–99) signals a certain shift back to the traditional amusement-oriented character of European ethnological exhibits, which he situates at about 1890. In order to increase profits, the organizers began to stress the unusual and the bizarre and to add spectacular performances, such as mock battles or cannibalistic rituals. What had begun to develop into a means of scientifically educating and edifying the public about faraway peoples and their customs turned into a form of amusement again, yet without impeding the persistence of a third function: that of political and imperialistic propaganda. Similar differences existed between the official ethnographic exhibits and the less scholarly orchestrated midway amusement zones at North American world fairs, which also served as venues for ethnographic types and spectacles.[7]

Science and imperialism went hand in hand; and so, of course, did missionary activities and imperialism. I do not know of any human showcases within the context of missionary propaganda, but other practices of categorizing, picturing, and exhibiting uncivilized, "heathen" peoples had much in common with more profane happenings. During the first half of the twentieth century, many European missionary museums and countless missionary exhibitions, Roman Catholic as well as Protestant, tried to persuade their visitors to take certain views and certain actions concerning colonially dominated non-Western, non-Christian peoples. Such exhibitions, permanent or temporary, were staged by missionary societies with the help of objects, photographs, maps, and sometimes dressed mannequins or busts. On such occasions, the well-known narrative plots and metaphors, slightly modified, return: civilized or Christian whites bringing the light of civilization or religion to savages or pagans, in the name of some higher instance, be it progress or God. Missionary photography for propagandistic goals showed characters and scenes from such narratives.

On December 24, 1924, Pope Pius XI opened the World Missionary Exhibition at the Vatican, which attracted over 750,000 visitors during the following year. There was a Hall of Propaganda, a Hall of the Holy Land as the Cradle of Christianity, one of the History of the Missions, and one dedicated to quantitative data concerning the missions. All of these halls were crammed with ethnographical objects showing the customs and morals of the heathen cultures that were the target of the Roman Catholic civilizatory and religious offensive, neatly arranged according to ethnologist Father Wilhelm Schmidt's theory of *Kulturkreise* and their historical development (Kilger 1925). Rome was represented as the center of the world – a role that on medieval world maps was still reserved for Jerusalem and in imperialist discourse was given to the metropoles of the leading imperialist nation-states. Roman Catholic missionary pageants in Belgium, the southern Netherlands, and elsewhere in Europe during the first decades of this century, with locals made up and dressed to represent black Africans, are reminiscent of Indian painter George Catlin's Indian Show in England during the 1840s, where native English people dressed up as North American Indians, or of the show in honor of King Henry II in 1550 near Rouen, France, where French sailors played Brazilian Indians.

The 1930s witnessed the decline of the ethnological exhibition, at least in the specific form it had taken until then. Criticism of imperialism and racism increased, and ethnographic shows were found objectionable on moral grounds.[8] Ethnographic films and numerous scientific, semi-scientific, and pseudo-scientific anthropological treatises, abundantly illustrated with photographs, took over much of the function of ethnological exhibits, as did colonial and missionary propaganda films. The increasing acculturation of colonial natives thwarted the creation of romantic or depreciating scenes of their natural lives. Recruiting exhibit groups in the German colonies became very difficult after 1901 because of new laws and regulations, and then World War I and its aftermath complicated things even more. During the 1930s, the *Völkerschau* was prohibited by the National Socialists, who feared they would increase the sympathy of the German people for other races.

Nonetheless, manifestations that are quite similar in several respects have persisted until today; for instance, the presence of Maori at the Te Maori exhibition at New York or of Sulawesi Toradja building a rice barn at the Toradja exhibition in the London Museum of Mankind during the 1980s. As a counterpart to exotic native villages at late 19th-century world fairs, a "Dutch city" in Nagasaki was opened in the spring of 1992 featuring four miles of canals with full-scale replicas of well-known Dutch buildings such as the Utrecht Cathedral. The cinema, television, and tourism also share certain functions with the Völkerschau, which was often recommended as an opportunity to visit faraway cultures.

The Story and the Gaze

It is not difficult to show the pivotal role of narrative structures in 19th- and early 20th-century world fairs, museums, or missionary exhibitions. Narrative plots are as pervasive in the civilizatory, imperialist, missionary, and scientific discourses of the period as in the three-dimensional spectacles that, to a considerable degree, were governed by these discursive activities. As many contemporary book titles suggest, the history of mankind was narrated essentially as a heroic ascent toward the natural and ultimate goal of cosmic evolution: the industrial civilization of white, European, middle-class citizens of the 19th century. Other races followed the same path, it was postulated – especially in evolutionist ethnology, which was a scientific manifestation of the discourse on progress – but lagged behind culturally and physically. Imperialist expansion was represented in terms of a social Darwinist natural history, and European hegemony as a natural and therefore desirable development. There has been some controversy over the question of whether the master narrative of progress and civilization is essentially a secularized avatar of the Christian idea of world history as God's working, but in any case, it is not formulated in religious terms. The implied development is from lack of civilization to civilized state, from wildness to civility, achieved heroically by the white, Caucasian race under its own power, and by the other races with the help of the Caucasian one, insofar at least as their constitutions allowed them to progress. The stagewise development from savagery through barbarism to civilization was suggested by organizing museum and world fair exhibits into evolutionary sequences.

Sekula's (1986:58) stress on the spirit of optical empiricism and encyclopedism of pictorial archives, with their purely iterative character, is heuristically useful and certainly justified to a certain degree; but in many contexts of collecting, filing, and exhibiting, an order was imposed on the data that went beyond mere iteration and taxonomy. In many cases, all essential ingredients of the story, or at least of a certain type of story, are present: a beginning where some desirable good is lacking; an end that is somehow implied by that beginning teleologically; acting subjects; strife and struggle; and other plot elements. World fairs and museums not only categorized peoples, races, cultures, species, and artifacts by creating taxonomies, but also ordered them syntagmatically, creating the well-known plots of civilized/Christian whites bringing light to the savage/heathen in the name of some higher instance. The same goes for many photographs from colonial contexts, showing moments from the story they presuppose and illustrate (Corbey 1988, 1989, 1990). Those well-known plots – flexible and capable of incorporating disparate elements, of outdoing alternative readings – are as pervasive in 19th-century civilizatory, imperialist, missionary, and scientific discourses as in the spectacles and pictures that were governed by these discursive activities. This century saw the proliferation of historicized, evolutionary

Figure 12: Hagenbeck's "Troop of Malabars," probably 1925. Often, as in this case, the exhibited persons on request signed postcards bearing their picture, which were sold at the venue. Postcard, collection of S. Wachlin, Amsterdam.

frameworks of representation – of artifacts and natural history specimens, of human, racial, and national origins.

One aspect of these spectacles, pictures, and narratives was that they neutralized the cognitive dissonance and the threat to Western middle-class identity constituted by the baffling cultural difference of new peoples. Colonial Others were incorporated narratively. In a *mise-en-intrigue*, they were assigned their roles in the stories told by museum exhibitions, world fairs, and colonial postcards (Figure 12). They were cast as contemporary ancestors, receivers of true civilization and true religion. The radical difference of the Other was made sense of and thus warded off by a narrative <discordant concordance> between "civilized" and "savage." Money, trade, and exchange mediated between peoples (cf. Hinsley 1991:362), but on another level stories were created in order to mediate the basic contradiction between the two states of mankind. Here I concur with Lévi-Strauss's interpretation of myth as a struggle with contradictions or paradoxes, as a syntagmatic mediation of paradigmatic oppositions.[9] Carol Breckenridge (1989:211) points out the analogy between the building of private collections by colonial officials, creating an illusion of cognitive control over a colonial experience that might otherwise have been disturbingly chaotic, and the world fair as a reminder of the orderliness of empire, which consolidated the sense of imperial knowledge and control in the imagined Victorian ecumene.

But the resultant concordance will never be complete; the attempt to harmonize is ultimately bound to fail. For as the plot develops, the initial discordance between civilized and primitive, white and black, Christian and heathen, is slowly but never totally overcome. Struggle develops into contracts, but some antagonism and difference is necessary all the way, to keep the story going. The story familiarizes and exoticizes at the same time.[10] Also on a different level – that of the citizen's personal experience – the Other seems to preserve an elusive quality; he or she never yields completely to incorporation within the framework of the familiar, stubbornly resisting a textualizing closure of spontaneous experience, of fascination, of wonder.[11]

To return now to the 1893 world fair motto we began with – "To see is to know" – of course we do not know how things are by simply looking. The eye is not innocent. The motto succinctly expresses an underlying ideology that is at work in a range of seemingly disparate practices in colonial times: photography, colonialist discourse, missionary discourse, anthropometry, collecting and exhibiting, and so on. What people saw, rather than reality as it is, was, to a considerable extent, reality as perceived, as actively constructed by images, conceptions, native taxonomies, stories, and motivational attitudes on the spectator's mind (cf. Mason 1990). The perceived order was an imposed one; the citizen's gaze on alien people was determined to a considerable degree by stories and stereotypes in his or her mind.

"Visualization and spatialization," Fabian writes in his study of the central role of the gaze and the visual in the history of anthropology,

> [became] a program for the new discipline of anthropology. There was a time when this meant, above all, the exhibition of the exotic in illustrated travelogues, museums, fairs, and expositions. These early ethnological practices established seldom articulated but firm convictions that presentations of knowledge through visual and spatial images, maps, diagrams, trees, and tables are particularly well suited to the description of primitive cultures. [Fabian 1983:121]

In this context, completely in line with our analysis of ethnological exhibitions, he stresses the ideological effects of *visualism* as a cognitive style. What is seen, the objectified Other, is looked on as coming not only from faraway places, but also, and more importantly, from a different, allochronic time. Fabian shows how a temporal gap is constructed between citizens and their "contemporary ancestors," "how anthropology has managed to maintain distance, mostly by manipulating temporal coexistence through the denial of coevalness" (Fabian 1983:121). This occurred in anthropology, just as it did in the political ideology of imperialism, in Christian discourse on heathens and the mission, at world fairs, and in certain photographic practices.

> When modern anthropology began to construct its Other in terms of topoi implying distance, difference, and opposition, its intent was above all, but at least also, to construct ordered Space and Time – a cosmos – for Western society to inhabit, rather than "understanding other cultures," its ostensible vocation. [Fabian 1983:111–112]

Besides plotting, there is a second aspect of the storied nature of the imperial imagination that has relevance to a proper understanding of ethnological exhibitions. It has to do with the encyclopedic character of world fairs. "As with other dimensions of the show," notes Greenhalgh (1988:87), "the imperial dimension was underpinned by the belief that it was possible to present a complete knowledge, to create a physical encyclopedia capable of capturing and explaining a total world view." What is interesting here is the cognitive position ascribed to the visitors: they are assumed to be situated high above the world they gaze upon. On a very basic level, activities such as narrating, taking pictures, or just plain looking create the illusion of the surveyability or transparency of reality, connected with the suggestion that those who narrate, take pictures, or look, find themselves in the privileged position of a panoptic spectator. In spontaneous visual perception, we already tend to experience ourselves as the natural center of the world we see – a world that seems to be organized around the onlooker as its pivot; in other words, perspective is more than a neutral mathematical projection on external space. All three activities – looking, taking pictures, narrating – are perspectivistic, not only in a literal sense, as in the case of looking or photographing, but also cognitively, emotionally, and ideologically. External reality is constructed from and around a central position, that of the onlooker, the photographer, the narrator – and also that of the citizen roaming the world fair. These spectators by their very activity seem to be panoptic, omniscient; their point of view is or seems to be panoramic, that of a bird's eye rather than a frog's eye. The narrator, hovering high above the plot, oversees time and space to a considerable degree, having relatively free access to what narrative characters are doing at different times and places. This panoptic gaze is very wide, and might indeed be coextensive with story or narrative as a highly general discourse structure, operating not only in fiction and literature, but also in the sciences and in philosophy, in religious practices, and in everyday conversation.

The position of the narrator is usually external to the story, somewhere above the narrated events, permitting an overview of space and time, although it may be more limited, tied to one of the characters in the represented events and part of these.[12] External narrators find themselves in an excellent position to assign significance and value to the events and characters, including, in the case of a typical 19th-century master narrative, their own heroic role, natural superiority, and unshakable moral and cognitive orientation. "The panoramic approach lays out the whole world conceptually in a Linnean classification or evolutionary scheme, or experientially in a scenic effect. . . . The view is comprehensive, extensive, commanding, aggrandizing. As a prospect, it holds in it scenarios for future action" (Kirshenblatt-Gimblett 1991:413). Panoramic landscape descriptions in novels or travel accounts, as well as landscape photographs and turn-of-the-century postcards, often embody a discourse of empire and domination, with the seer as a "monarch-of-all-I-survey" (Pratt 1988; cf. Fabian 1983:118–123), in firm control of the seen. The same gaze – self-confident, panoptic, voyeuristic – is to be found in many photos of colonial natives. "There is an aggression implicit in every use of the camera," Susan Sontag (1984:7) has remarked, and taking pictures was indeed another means of taking possession of native peoples and their lands, as was narration.

A fine example of the encyclopedic urge in an imperialist context was an ambitious project the Asiatic Society of Bengal tried to realize in Calcutta in 1865: a synoptic exhibition of living representatives of the races of the Old World (or at least of India), to be visited primarily by scientists (Falconer 1984–86). The idea was that of a kind of panopticon, as we know it from the writings of Jeremy Bentham and Michel Foucault. The spirit of the project once again was one of optical empiricism, as expressed by the Chicago motto "To see is to know."[13] The effort failed. A somewhat meager but similar outcome realized instead of the exhibit of life specimens was constituted by E. T. Dalton's *Descriptive Ethnology of Bengal* (1872), illustrated with lithographs based on photographs, and the eight-volume *The People of India: A Series of Photographic Illustrations with Descriptive Letterpress of the Races and Tribes of Hindustan* (Watson and Kaye 1868–75), commissioned by the British government, both covering the native peoples from this part of the British empire. In a comparable way, richly illustrated publications like Karl Ernst von Baer's *Types principaux des différentes raçes humaines dans les cinq parties du monde* (1861) proffered the "photographic museum of the human races" that had been postulated theoretically 20 years before by E. R. A. Serres (1845:243; cf. Theye 1989) in France and had quickly become a quite common ideal, all the more so because it was felt that soon it would be too late. In the year 1872, the German ethnologist Adolf Bastian also formulated the project of "a photographic museum of the human races," which was indeed created later by the Berliner Gesellschaft für Ethnologie, Anthropologie und Urgeschichte; tens of thousands of accumulated photographs were destroyed during the Second World War (Theye 1989).

Apart from illustrating the synoptic theme, the project of a photographic census of humanity also brings out nicely the role of photography in the complex of practices regarding colonial subjects: collecting, scrutinizing, measuring, categorizing, filing, controlling, narrating. The second half of the 19th century witnessed the quick rise of photography as another machinery of capturing and displaying the world. Here, too, we come across the illusion of authenticity, of unmediated encounter. Time and again the unbiased, true character of the photographic picture was stressed; photos were seen as windows on the world, as unmediated and therefore unbiased copies of nature itself. Photography was applied on a large scale in many scientific disciplines, in a spirit of optical empiricism. While the prospering middle classes of Western industrial societies presented themselves honorifically in self-congratulatory studio portraits, hundreds of thousands of photos of their Others – other races, criminals, prostitutes, the insane, deviants – functioned in the context of repression. Publications like *The People of India*, Carl Damman's *Anthropologisch-Ethnologisches Album in Photographien* (c. 1872), or, somewhat later, the scenes-and-types postcard genre, were matched by photographic albums with "types" of criminals. In this context, the needs of nation-states went hand in hand with scientific purposes. Breckenridge (1989:195–196) points out that agencies such as archives, libraries, surveys, revenue bureaucracies, folklore and ethnographic agencies, censuses, and museums provided a context for surveilling, recording, classifying, and evaluating called for by the new order of 19th-century nation-states with their imperializing and disciplinary bureaucracies – whether it concerned colonies abroad or criminals and slums at home.

Tony Bennett, in an argument that in many ways parallels and in other ways complements the one developed here, has unraveled relations between power and knowledge in the development of what he appropriately calls an "exhibitionary complex." It encompasses museums of art, history, and natural science; dioramas and panoramas; national and international exhibitions; arcades and department stores, serving as "linked sites for the development and articulation of new disciplines (history, biology, art history, anthropology) and their discursive formations (the past, evolution, aesthetics, man) as well as for the development of new technologies of vision" (Bennett 1988:73).

Conclusion

I have explored some of the complex interdependencies between the colonialist, scientific, and visual appropriation of cultural others in the context of world fairs and ethnographic exhibits. Persons from tribal cultures, on show in the West, were commodified, labeled (Bouquet and Branco 1988), scripted, objectified, essentialized, decontextualized, aestheticized, and fetishized. They were cast in the role of backward, allochronic contemporary ancestors, receivers of true civilization and true religion in the stories told by museums, world fairs, and imperialist ideologies, thus becoming narrative characters in the citizen's articulation of identity – of Self and Other. Their own voices and views – ironically often as ethnocentric and omniscient as Western ones – were neutralized. Fitting cultural others into narrative plots, we suggested, was a way the citizen's panoptic eye/I dealt with their wondrous, disturbing difference without annihilating it completely. These plots came with the illusion of the panoptic position of an omniscient spectator, functioning as another strategy of power – the illusion that "to see is to know."

Over the last centuries, the "we" – group, as an emic category of Western middle classes characterized by true humanity, has been expanding continuously to include many categories that were formerly excluded or considered ambiguous: women, slaves, peasants, the poor, and non-Western peoples.[14] In this article, occasional reference was made to analogies between how other races and species were thought of and treated in the late 19th and early 20th centuries. By now, the boundary of the human species has been reached and, in fact, is being questioned – not least as to its moral significance – and transgressed. The discussion is now shifting toward zoos, circuses, dolphin shows, bioindustry, and animal experiments; toward "simian Orientalism" (Haraway 1989) and other forms of anthropocentrism. It would seem that our observations on ethno-/eurocentric ethnographic exhibits during the heyday of colonialism are in many ways readily extendable to present-day forms, in theory and practice, of anthropocentrism and speciesism.

Notes

1. In this context, Ivan Karp's analytical distinction between exoticizing and assimilating strategies of exhibition is relevant (in Karp and Lavine 1991:375 ff.); both strategies are examples of the cognitive strategy of assimilation – as opposed to accommodation – in the sense of Piagetian developmental psychology.
2. For a survey of ethnographic exhibits in the Jardin d'Acclimatation, see Schneider 1977.
3. A fruitful line of analysis not exploited here might pursue the infamous character of similar trades and professions; cf. Blok 1990.
4. Prins (1991) argues that these early showmen functioned as cultural mediators between dominant Euro-American society and their own respective native communities, and that their performances inspired the imagery currently used to express ethnic self-identity among tribespeople in the Northeast.
5. McBride 1989 provides some biographical information on Molly Spotted Elk.
6. The Smithsonian Institution in Washington, D.C., for instance, harbours the remains of several tens of thousands non-Western individuals.
7. As remarked before, I am trying to bring out the general character of ethnographic exhibits in the period under consideration more than differences between nations, over time, or between types of manifestation. For an analysis of national differences, see Benedict 1991.
8. One would also expect conflicts between the scholarly and the popular imagination to have increased toward the end of the period 1870–1930 with the advent, in anthropology, of paradigms critical of evolutionist anthropology.
9. At the same time, as may be clear to insiders, we take some inspiration from the structuralist narratology of A. J. Greimas and the Paris School – without necessarily subscribing to all its presuppostions, however. For an as powerful but more radical, poststructuralist, analytical approach to exhibitionary practices, see Bal 1992.
10. Here again, Karp's distinction between exoticizing and assimilating strategies is relevant; see note 1.
11. Stephen Greenblatt's (1991) stress on the enduring fascination with "the marvellous" complements the narratologically inspired perspective developed here.
12. In contemporary narratology, the narrator's perspective is analyzed in terms of internal and external "focalization"; compare Genette 1972 and Bal 1991, who stress the

ideological effects of focalization not only in texts but also in pictures and museums, developing the theory of focalization into a powerful tool of cultural criticism.

13. I am here indeed indirectly and loosely drawing on certain ideas from the work of Michel Foucault, which has proved to be inspiring and heuristically useful in the field of cultural studies (cf. Mitchell 1988, 1989), despite its lack of historical precision, its too-sweeping generalizations, and its tendency toward sometimes rather obscure rhetoric.
14. Even early hominids may be included in this group. A poster announcing the 1989 exhibition "*Archéologie de la France: 30 ans de découvertes*" at the Grand Palais in Paris shows a *Homo erectus* man with a modern baby on his knee, the caption reading "*Nous avons tous 400,000 ans,*" which only 20 years ago would still have been unthinkable.

References Cited

Altick, Richard D. 1978 The Shows of London. A Panoramic History of Exhibitions, 1600–1862. Cambridge, MA: Belknap.

Bal, Mieke 1991 On Story-Telling: Essays in Narratology. Sonoma: Polebridge Press.

——— 1992 Telling, Showing, Showing-off. Critical Inquiry 18:556–594.

Benedict, Burton 1983 The Anthropology of World's Fairs: San Francisco's Panama Pacific International Exposition of 1915. London: Scolar Press, in association with the Lowie Museum of Anthropology.

——— 1991 International Exhibitions and National Identity. Anthropology Today 7(3):5–9.

Benjamin, Walter 1984 Paris, die Hauptstadt des XIX. Jahrhunderts. *In* Allegorien kultureller Erfahrung. Ausgewählte Schriften 1920–1940. Leipzig: Reclam.

Bennett, Tony 1988 The Exhibitionary Complex. New Formations 4 (Spring):74–102.

Bieder, Robert E. 1991 Zoo: Race, Sex and Power. Bloomington, IN, unpublished MS.

Blok, Anton 1990 Infamy. London: Polity.

Bogdan, R. 1988 Freaks Show: Presenting Human Oddities for Amusement and Profit. Chicago: University of Chicago Press.

Bouquet, Mary, and Jorge Branco 1988 Melanesian Artifacts – Postmodernist Reflections. Lisbon: Museu de Etnologia.

Bradford, Phillips, and Harvey Blume 1992 Ota Benga: The Pygmy at the Zoo. New York: St. Martin's Press.

Breckenridge, Carol A. 1989 The Aesthetics and Politics of Colonial Collecting: India at World Fairs. Comparative Studies in Society and History 31:195–216.

Chapman, Anne 1982 Drama and Power in a Hunting Society. The Selk'nam of Tierra del Fuego. New York: Cambridge University Press.

Corbey, Raymond 1988 Alterity: The Colonial Nude. Critique of Anthropology 8(3):75–92.

——— 1989 Wildheid en beschaving. De Europese verbeelding van Afrika. Baarn: Ambo.

——— 1990 Der Missionar, die Heiden und das Photo. Zeitschrift für Kulturaustausch 40:460–465.

Dalton, Edward T. 1872 Descriptive Ethnology of Bengal. Calcutta: Office of the Superintendent of Government Printing.

Damman, Carl n.d.[c 1872] Anthropologisches-Ethnologisches Album in Photographien. Berlin: Wiegard, Hempel, & Parey.

de Sade, D. A. F. 1990 Oeuvres. Paris: Bibliothèque de la Pleiade.

Dias, Nelia 1991 Le Musée d'ethnographie du Trocadéro 1878–1908. Anthropologie et Muséologie en France. Paris: CNRS.
Dickens, Charles 1938 The Noble Savage. Reprinted Pieces. London: Bloomsbury Press.
Dömer, Klaus 1984 Bürger und Irre. Zur Sozialgeschichte und Wissenschaftssoziologie der Psychiatrie Frankfurt am Main: Europäische Verlagsanstalt.
Fabian, Johannes 1983 Time and the Other. How Anthropology Makes its Object. New York: Columbia University Press.
Falconer, John 1984–86 Ethnographical Photography in India. Photographic Collector 5:16–46.
Favrod, Charles-Henri 1989 Voir les autres autrement. *In* Étranges étrangères: Photographie et exotisme, 1850/1910. Paris: Centre National de la Photographie.
Foucault, Michel 1961 Folie et déraison. Histoire de la folie à l'âge classique. Paris: Plon.
Francis, D. R. 1913 The Universal Exposition of 1904. St. Louis: Louisiana Purchase Exposition Company.
Genette, Gerard 1972 Figures III. Paris: Seuil.
Goldmann, Stefan 1985 Wilde in Europa. *In* Wir und die Wilden. Einblicke in eine kannibalische Beziehung. Thomas Theye, ed. Pp. 243–269. Reinbek: Rowohlt.
Greenblatt, Stephen 1991 Resonance and Wonder. *In* Exhibiting Cultures: The Politics and Poetics of Museum Display. Ivan Karp and Steven D. Lavine, eds. Pp. 42–56. Washington, DC: Smithsonian Institution Press.
Greenhalgh, Paul 1988 Ephemeral Vistas. The Expositions Universelles, Great Exhibitions and Worlds Fairs, 1851–1939. Manchester. Manchester University Press.
Gusinde, Martin 1931 Die Feuerland-Indianer. *In* Die Selk'nam. Vom Leben und Denken eines Jägervolkes auf der grossen Feuerlandinsel, Mödling, Austria: S.V.D.
Hagenbeck, Karl 1909 Von Tieren und Menschen. Erlebnisse und Erfahrungen. Berlin-Charlottenburg.
Haltem, Ulrich 1971 Die Londoner Weltausstellung von 1851. Ein Beitrag zur Geschichte der bürgerlich-industriellen Gesellschaft im 19. Jahrhundert. Münsten Verlag Aschendorf.
Haraway, Donna 1989 Primate Visions: Gender, Race and Nature in the World of Modem Science. New York: Routledge.
Helms, Mary W. 1988 Ulysses Sail: An Ethnographic Odessey of Power, Knowledge and Geographical Distance. Princeton: Princeton University Press.
Hinsley, Curtis M. 1991 The World as Marketplace: Commodification of the Exotic at the World's Columbian Exposition. *In* Exhibiting Cultures: The Politics and Poetics of Museum Display. Ivan Karp and Steven D. Lavine, eds. Pp. 344–365. Washington: Smithsonian Institution Press.
Karp, Ivan, and Steven D. Lavine, eds. 1991 Exhibiting Cultures: The Politics and Poetics of Museum Display. Washington and London: Smithsonian Institution Press.
Kilger, Laurenz 1925 Die vatikanische Missions-Ausstellung 1925. Die Katholischen Missionen 53(6):167–179.
Kirshenblatt-Gimblett, Barbara 1991 Objects of Ethnography. *In* Exhibiting Cultures: The Politics and Poetics of Museum Display. Ivan Karp and Steven D. Lavine, eds. Pp. 386–443. Washington, D.C.: Smithsonian Institution Press.
Lederbogen, Jan 1986 Fotografie als Völkerschau. Fotogeschichte: Beiträge zur Geschichte und Ästhetikder Photographie 6:47–64.
Lehmann, Alfred 1955 Zeitgenössische Bilder der ersten Völkerschauen. *In* Von fremden Völker und Kulturen. Beiträge zur Völkerkunde. Hans Plischke zum 65. Geburtstag. W. Lang, W. Nippold, and G. Spannaus eds. Pp. 31–38. Dusseldorf: Droste Verlag.
Mackenzie, John M. 1984 Propaganda and Empire: The Manipulation of British Public Opinion 1860–1960. Manchester Manchester University Press.
Mason, Peter 1990 Deconstructing America. Representations of the Other. New York: Routledge.

Mason, Peter, ed. 1992 Indianen en Nederlanders, 1492–1992. Wampum 11, special issue.
McBride, Bunny 1989 A Penobscot in Paris. Down East August: 63–65, 80–81.
Mitchell, Timothy 1988 Colonising Egypt. Cambridge: Cambridge University Press.
——— 1989 The World as Exhibition. Comparative Studies in Society and History 31:217–236.
Montagu, Ashley 1971 The Elephant Man: A Study in Human Dignity. New York: Outerbridge and Dienstfrey.
Neihardt, John 1988 Black Elk Speaks. Lincoln: University of Nebraska Press.
Parsons, Q. N. 1988 Frantz or Klikko, The Wild Dancing Bushman: A Case Study in Khoisan Stereotyping. Botswana Notes and Records 20:71–76.
Pieterse, Evelien 1992 Amerika binnen handbereik. In Indianen en Nederlanders, 1492–1992. Peter Mason, ed. Wampum 11:16–39.
Plakate 1880–1914 n.d. Exhibition catalogue, Historisches Museum, Frankfurt.
Pratt, Mary Louise 1988 Conventions of Representation: When Discourse and Ideology Meet. *In* The Taming of the Text: Explorations in Language, Literature, and Culture. Willie van Peer, ed. Pp. 15–34. London: Routledge and Kegan Paul.
Prins, Harald E. L. 1991 Public Performance and Ethnic Identity: Chief Big Thunder and the Peddling of 'Indian' Culture. Paper presented at the American Society for Ethnohistory, Annual Meeting, 7–10 November.
Rice, Julian 1991 Black Elk's Story: Distinguishing its Lakota Purpose. Albuquerque: University of New Mexico Press.
Rydell, Robert W. 1984 All the World's a Fair: Visions of Empire at American International Expositions, 1876–1916. Chicago: University of Chicago Press.
Schneider, G. 1982 Das deutsche Kolonialmuseum in Berlin und seine Bedeutung im Rahmen der preussischen Schulreform um die Jahrhundertwende. *In* Die Zukunft beginnt in der Vergangenheit. Schriften des Historischen Museums XVI. Pp. 155 ff. Frankfurt: Historisches Museum.
Schneider, W. 1977 Race and Empire: The Rise of Popular Ethnography in the Late Nineteenth Century. Journal of Popular Culture 11:98–109.
Sekula, Alan 1986 The Body and the Archive. October 39(Winter):3–64.
Serres, Etienne-Renaud-Augustin 1845 Observations sur l'application de la photographie à l'étude des raçes humaines. Comptes Rendues de l'Académie des Sciences 21:243 ff.
Sontag, Susan 1984 On Photography. Harmondsworth: Penguin Books.
Taylor, J. Garth 1981 An Eskimo Abroad, 1880: His Diary and Death. Canadian Geographic Oct./Nov.
Theye, Thomas 1989 <Wir wollen nicht glauben, sondern schauen>: Zur Geschichte der ethnographischen Fotografie im 19. Jahrhundert. *In* Idem (red.), Der geraubte Schatten: Die Photographie als ethnographisches Dokument. Pp. 60–119. München: Münchner Stadtmuseum.
Thode-Arora, Hilke 1989 Für fünfzig Pfenning um die Welt: Die Hagenbeckschen Völkerschauen. New York: Campus Verlag.
Von Baer, Karl Ernst 1861 Types principaux des différentes raçes humaines dans les cinq parties du monde. 2nd ed. St. Petersburg.
Watson, J. F., and J. W. Kaye 1868–75 The People of India: A Series of Photographic Illustrations with Descriptive Letterpress of the Races and Tribes of Hindustan. 8 vols. London: India Museum.

27

Why Look at Animals?

John Berger

In the accompanying ideology, animals are always the observed. The fact that they can observe us has lost all significance. They are the objects of our ever-extending knowledge. What we know about them is an index of our power, and thus an index of what separates us from them. The more we know, the further away they are.

Yet in the same ideology, as Lukacs points out in *History and Class Consciousness*, nature is also a value concept. A value opposed to the social institutions which strip man of his natural essence and imprison him. 'Nature thereby acquires the meaning of what has grown organically, what was not created by man, in contrast to the artificial structures of human civilization. At the same time, it can be understood as that aspect of human inwardness which has remained natural, or at least tends or longs to become natural once more.' According to this view of nature, the life of a wild animal becomes an ideal, an ideal internalized as a feeling surrounding a repressed desire. The image of a wild animal becomes the starting-point of a daydream: a point from which the daydreamer departs with his back turned.

The degree of confusion involved is illustrated by the following news story: 'London housewife Barbara Carter won a "grant a wish" charity contest, and said she wanted to kiss and cuddle a lion. Wednesday night she was in a hospital in shock and with throat wounds. Mrs Carter, 46, was taken to the lions' compound of the safari park at Bewdley, Wednesday. As she bent forward to stroke the lioness, Suki, it pounced and dragged her to

Source: *About Looking* (NY: Pantheon Books, 1980), pp. 27–37.: *About Looking* (London: Bloomsbury Publishing Plc, 2009), pp. 16–28.

the ground. Wardens later said, "We seem to have made a bad error of judgment. We have always regarded the lioness as perfectly safe."'

The treatment of animals in 19th-century romantic painting was already an acknowledgement of their impending disappearance. The images are of animals *receding* into a wildness that existed only in the imagination. There was, however, one 19th-century artist who was obsessed by the transformation about to take place, and whose work was an uncanny illustration of it. Grandville published his *Public and Private Life of Animals* in instalments between 1840 and 1842.

At first sight, Grandville's animals, dressed up and performing as men and women, appear to belong to the old tradition, whereby a person is portrayed as an animal so as to reveal more clearly an aspect of his or her character. The device was like putting on a mask, but its function was to unmask. The animal represents the apogee of the character trait in question: the lion, absolute courage: the hare, lechery. The animal once lived near the origin of the quality. It was through the animal that the quality first became recognizable. And so the animal lends it his name.

But as one goes on looking at Grandville's engravings, one becomes aware that the shock which they convey derives, in fact, from the opposite movement to that which one first assumed. These animals are not being 'borrowed' to explain people, nothing is being unmasked; on the contrary. These animals have become prisoners of a human/social situation into which they have been press-ganged. The vulture as landlord is more dreadfully rapacious than he is as a bird. The crocodiles at dinner are greedier at the table than they are in the river.

Here animals are not being used as reminders of origin, or as moral metaphors, they are being used *en masse* to 'people' situations. The movement that ends with the banality of Disney began as a disturbing, prophetic dream in the work of Grandville.

The dogs in Grandville's engraving of the dog-pound are in no way canine; they have dogs' faces, but what they are suffering is imprisonment *like men*.

The bear is a good father shows a bear dejectedly pulling a pram like any other human bread-winner. Grandville's first volume ends with the words 'Goodnight then, dear reader. Go home, lock your cage well, sleep tight and have pleasant dreams. Until tomorrow.' Animals and populace are becoming synonymous, which is to say the animals are fading away.

A later Grandville drawing, entitled *The animals entering the steam ark*, is explicit. In the Judaeo-Christian tradition, Noah's Ark was the first ordered assembly of animals and man. The assembly is now over. Grandville shows us the great departure. On a quayside a long queue of different species is filing slowly away, their backs towards us. Their postures suggest all the last-minute doubts of emigrants. In the distance is a ramp by which the first have already entered the 19th-century ark, which is like an

American steamboat. The bear. The lion. The donkey. The camel. The cock. The fox. Exeunt.

'About 1867', according to the *London Zoo Guide*, 'a music hall artist called the Great Vance sang a song called *Walking in the zoo is the OK thing to do*, and the word "zoo" came into everyday use. London Zoo also brought the word "jumbo" into the English language. Jumbo was an African elephant of mammoth size, who lived at the zoo between 1865 and 1882. Queen Victoria took an interest in him and eventually he ended his days as the star of the famous Barnum circus which travelled through America – his name living on to describe things of giant proportions.'

Public zoos came into existence at the beginning of the period which was to see the disappearance of animals from daily life. The zoo to which people go to meet animals, to observe them, to see them, is, in fact, a monument to the impossibility of such encounters. Modern zoos are an epitaph to a relationship which was as old as man. They are not seen as such because the wrong questions have been addressed to zoos.

When they were founded – the London Zoo in 1828, the Jardin des Plantes in 1793, the Berlin Zoo in 1844 – they brought considerable prestige to the national capitals. The prestige was not so different from that which had accrued to the private royal menageries. These menageries, along with gold plate, architecture, orchestras, players, furnishings, dwarfs, acrobats, uniforms, horses, art and food, had been demonstrations of an emperor's or king's power and wealth. Likewise in the 19th century, public zoos were an endorsement of modern colonial power. The capturing of the animals was a symbolic representation of the conquest of all distant and exotic lands. 'Explorers' proved their patriotism by sending home a tiger or an elephant. The gift of an exotic animal to the metropolitan zoo became a token in subservient diplomatic relations.

Yet, like every other 19th-century public institution, the zoo, however supportive of the ideology of imperialism, had to claim an independent and civic function. The claim was that it was another kind of museum, whose I purpose was to further knowledge and public enlightenment. And so the first questions asked of zoos belonged to natural history; it was then thought possible to study the natural life of animals even in such unnatural conditions. A century later, more sophisticated zoologists such as Konrad Lorenz asked behaviouristic and ethological questions, the claimed purpose of which was to discover more about the springs of human action through the study of animals under experimental conditions.

Meanwhile, millions visited the zoos each year out of a curiosity which was both so large, so vague and so personal that it is hard to express in a single question. Today in France 22 million people visit the 200 zoos each year. A high proportion of the visitors were and are children.

Children in the industrialized world are surrounded by animal imagery: toys, cartoons, pictures, decorations of every sort. No other source of imagery can begin to compete with that of animals. The apparently spontaneous interest that children have in animals might lead one to suppose that this has always been the case. Certainly some of the earliest toys (when toys were unknown to the vast majority of the population) were animal. Equally, children's games, all over the world, include real or pretended animals. Yet it was not until the 19th century that reproductions of animals became a regular part of the decor of middle-class childhoods – and then, in this century, with the advent of vast display and selling systems like Disney's – of all childhoods.

In the preceding centuries, the proportion of toys which were animal was small. And these did not pretend to realism, but were symbolic. The difference was that between a traditional hobby horse and a rocking horse: the first was merely a stick with a rudimentary head which children rode like a broom handle: the second was an elaborate 'reproduction' of a horse, painted realistically, with real reins of leather, a real mane of hair, and designed movement to resemble that of a horse galloping. The rocking horse was a 19th-century invention.

This new demand for verisimilitude in animal toys led to different methods of manufacture. The first stuffed animals were produced, and the most expensive were covered with real animal skin – usually the skin of still-born calves. The same period saw the appearance of soft animals – bears, tigers, rabbits – such as children take to bed with them. Thus the manufacture of realistic animal toys coincides, more or less, with the establishment of public zoos.

The family visit to the zoo is often a more sentimental occasion than a visit to a fair or a football match. Adults take children to the zoo to show them the originals of their 'reproductions', and also perhaps in the hope of re-finding some of the innocence of that reproduced animal world which they remember from their own childhood.

The animals seldom live up to the adults' memories, whilst to the children they appear, for the most part, unexpectedly lethargic and dull. (As frequent as the calls of animals in a zoo are the cries of children demanding: Where is he? Why doesn't he move? Is he dead?) And so one might summarize the felt, but not necessarily expressed, question of most visitors as: Why are these animals less than I believed?

And this unprofessional, unexpressed question is the one worth answering.

A zoo is a place where as many species and varieties of animal as possible are collected in order that they can be seen, observed, studied. In principle, each cage is a frame round the animal inside it. Visitors visit the zoo to look at animals. They proceed from cage to cage, not unlike visitors in an art gallery who stop in front of one painting, and then move on to the next or the one after next. Yet in the zoo the view is always wrong. Like an image out of focus. One is so accustomed to this that one scarcely notices it any more; or, rather, the apology habitually anticipates the disappointment, so that the latter is not felt. And the apology runs like this: What do you expect? It's not a dead object you have come to look at, it's alive. It's leading its own life. Why should this coincide with its being properly visible? Yet the reasoning of this apology is inadequate, The truth is more startling.

However you look at these animals, even if the animal is up against the bars, less than a foot from you, looking outwards in the public direction, *you are looking at something that has been rendered absolutely marginal*; and all the concentration you can muster will never be enough to centralize it. Why is this?

Within limits, the animals are free, but both they themselves, and their spectators, presume on their close confinement. The visibility through the glass, the spaces between the bars, or the empty air above the moat, are not what they seem – if they were, then everything would be changed. Thus visibility, space, air, have been reduced to tokens.

The decor, accepting these elements as tokens, sometimes reproduces them to create pure illusion – as in the case of painted prairies or painted rock pools at the back of the boxes for small animals. Sometimes it merely adds further tokens to suggest something of the animal's original landscape – the dead branches of a tree for monkeys, artificial rocks for bears, pebbles and shallow water for crocodiles. These added tokens serve two distinct purposes: for the spectator they are like theatre props: for the animal they constitute the bare minimum of an environment in which they can physically exist.

The animals, isolated from each other and without interaction between species, have become utterly dependent upon their keepers. Consequently most of their responses have been changed. What was central to their interest has been replaced by a passive waiting for a series of arbitrary outside interventions. The events they perceive occurring around them have become as illusory in terms of their natural responses as the painted prairies. At the same time this very isolation (usually) guarantees their longevity as specimens and facilitates their taxonomic arrangement.

All this is what makes them marginal. The space which they inhabit is artificial. Hence their tendency to bundle towards the edge of it. (Beyond its edges there may be real space.) In some cages the light is equally artificial. In all cases the environment is illusory. Nothing surrounds them except their own lethargy or hyperactivity. They have nothing to act upon – except, briefly, supplied food and – very occasionally – a supplied mate. (Hence their perennial actions become marginal actions without an object.) Lastly, their dependence and isolation have so conditioned their responses that they treat any event which takes place around them – usually it is in front of them, where the public is – as marginal. (Hence their assumption of an otherwise exclusively human attitude – indifference.)

Zoos, realistic animal toys and the widespread commercial diffusion of animal imagery, all began as animals started to be withdrawn from daily life. One could suppose that such innovations were compensatory. Yet in reality the innovations themselves belonged to the same remorseless movement as was dispersing the animals. The zoos, with their theatrical decor for display, were in fact demonstrations of how animals had been rendered absolutely marginal. The realistic toys increased the demand for the new animal puppet: the urban pet. The reproduction of animals in images – as their biological reproduction in birth becomes a rarer and rarer sight – was competitively forced to make animals ever more exotic and remote.

Everywhere animals disappear. In zoos they constitute the living monument to their own disappearance. And in doing so, they provoked their last metaphor. *The Naked Ape, The Human Zoo*, are titles of world bestsellers. In these books the zoologist, Desmond Morris, proposes that the unnatural behaviour of animals in captivity can help us to understand, accept and overcome the stresses involved in living in consumer societies.

All sites of enforced marginalization – ghettos, shanty towns, prisons, madhouses, concentration camps – have something in common with zoos. But it is both too easy and too evasive to use the zoo as a symbol. The zoo is a demonstration of the relations between man and animals; nothing else. The marginalization of animals is today being followed by the marginalization and disposal of the only class who, throughout history, has remained familiar with animals and maintained the wisdom which accompanies that familiarity: the middle and small peasant. The basis of this wisdom is an acceptance of the dualism at the very origin of the relation between man and animal. The rejection of this dualism is probably an important factor in opening the way to modern totalitarianism. But I do not wish to go beyond the limits of that unprofessional, unexpressed but fundamental question asked of the zoo.

The zoo cannot but disappoint. The public purpose of zoos is to offer visitors the opportunity of looking at animals. Yet nowhere in a zoo can a stranger encounter the look of an animal. At the most, the animal's gaze flickers and passes on. They look sideways. They look blindly beyond. They scan mechanically. They have been immunized to encounter, because nothing can any more occupy a *central* place in their attention.

Therein lies the ultimate consequence of their marginalization. That look between animal and man, which may have played a crucial role in the development of human society, and with which, in any case, all men had always lived until less than a century ago, has been extinguished. Looking at each animal, the unaccompanied zoo visitor is alone. As for the crowds, they belong to a species which has at last been isolated.

This historic loss, to which zoos are a monument, is now irredeemable for the culture of capitalism.

C. Ethics, Risk and Observation

28

Ethical Challenges in Participant Observation: A Reflection on Ethnographic Fieldwork

Jun Li

Introduction

After completing my doctoral study in education, I received a one-year postdoctoral fellowship from the Ontario Problem Gambling Research Center to investigate female gambling culture at the department of sociology. In order to understand how women come to gamble and develop gambling problems, I employed two integrated ethnographic data collection methods, participant observation, and in-depth interviewing. While participant observation aimed to uncover the world of female gamblers with first-hand exploration of naturalistic gambling settings, in-depth interviewing intended to reveal the meanings of female gambling by attending to women's personal perspectives and interpretations. This labour-intensive ethnographic work was paid off with rich and informative narrative data. However, due to the disturbing emotions of ethical dilemmas I experienced in the field as an embedded participant researcher, in preparing the final report, I did not include some of my field encounters.

Although the ethical pitfalls inherent in participant observation are a well-recognized concern in ethnographic research, confessional tales (Van Maanen, 1988), especially unexpected mistakes occurring in the field, are less addressed in the literature. A few years have passed, but I still feel that

Source: *The Qualitative Report,* 13(1) (2008): 100–115.

this gambling research experience should be told to re-inform ethnographers, particularly the novice researchers, about potential and often unforeseen contingencies of such fieldwork.

In this reflective essay I first outline the necessity of using ethnographic methods to investigate sensitive topics involving vulnerable individuals or groups; then I detail ethical dilemmas and socio-emotional discomfort I encountered in the field; and finally I reflect on the lessons I learned from conducting this gambling study.

Ethnography and Sensitive Research

It has long been acknowledged that, when studying non-mainstream groups in society such as the marginalized and the stigmatized, researchers must tailor their data collection methods to both the sensitivity of the research topic and the vulnerability of research subjects (Goffman, 1963; Hobbs, 2002; Lee, 1993). Because observational research does not intervene in the activities of the people being studied (Alder & Alder, 2000), ethnography is, in particular, suitable to investigating sensitive issues because such work can provide rich, detailed descriptions about the unknown or the little known. As the only field method that allows researchers to observe what people do in "real life" contexts, not what they say what they do, ethnographic participant observation can supply detailed, authentic information unattainable by any other research method (Homan, 1980; Humphreys, 1970; Gans, 1999).

Since participant observation has the greatest potential to uncover contextualized, honest data, otherwise inaccessible, it ontologically and epistemologically underpins human quests for understanding multiple realities of life in context (Rossman & Rallis, 2003). Furthermore, in revealing what lies beneath, ethnographic research can empower the very people being studied, transforming the "public consciousness" and "common sense" about the disadvantaged in society (Fine, Weis, Weseen, & Wong, 2003).

Ethical Dilemmas in Participant Observation

In order to discover, describe, and represent the world of the researched, ethnographers such as cultural anthropologists and symbolic interactionist sociologists have traditionally committed themselves to the first-hand exploration of research settings with naturalistic field methods (Atkinson, Coffey, Delamont, Lofland, & Lofland, 2002; Hammersley & Atkinson, 1995; Hume & Mulcock, 2004). Ethnographic participant observation could be overt or covert, with or without revealing research purpose and research identity to the researched. Although covert participant observation is more likely to provide detailed portraits of contextualized social realities, it stirred much controversy and debate on research ethics, mainly regarding the deception and the absence of informed consent from the people being studied (Bulmer, 1980, 1982; Dingwall, 1980; M. L. Wax, 1979). With increasing limitations imposed on covert research, recent writings on this topic remain rare.

Nevertheless, researchers with an advocacy and emancipatory paradigm, especially those who aim to experience and represent social lives of the disadvantaged people, continue to challenge ethical restrictions on covert research methods. For example, in a feminist analysis of doing research with homeless women, the researcher contends that the unique context of the lives of these women demands a re-definition of the conventional ethical constructs to uncover unheard stories of troubled lives, so as to prevent a perpetuation of the stereotyping, stigmatization, and marginalization they face on a day-to-day basis (Paradis, 2000). Since ethnographic research consistently yields more detailed, contextualized findings than consent interviews and other overt qualitative data collection methods, the distinctive contribution of this embedded fieldwork should be recognized (Charmaz & Olesen, 1997), and ethnographic research should receive more support from the ethics review board (Tope, Chamberlain, Crowley, & Hodson, 2005). The position taken by these researchers has significant implications for sensitive research involving with vulnerable participants, as I explain next.

Gambling as a Sensitive Research Topic

Although many social and cultural phenomena can be considered as "sensitive" in general terms, sensitive research refers to the study of secretive, stigmatized, or deviant human activity and behaviour involving vulnerable research subjects. Gambling is a sensitive research topic because it is characterized by a mixture of popularity, glamour, secrecy, and stigma in contemporary context.

In the past decade legal gambling in Canada has been increasingly accepted as one of life's legitimate pleasures, emerging as national pastime. According to the data released by Statistics Canada, net revenue from the government-run lotteries, video lottery terminals, and casinos rose from 2.7 billion in 1992 to 12.4 billion in 2004 (Statistics Canada, 2005). The economic impact of the gambling industry is evident in measure, as numbers do not lie. However, the social, psychological, and health costs of gambling addiction remain largely unknown because stories behind such staggering gambling revenue growth are rarely told by gamblers themselves. Although public attitudes towards recreational or social gambling in the West have been rapidly shifting in parallel with the expansion of legal gambling, excessive or problem gambling remains a sensitive topic people are afraid of talking about. This reality has posed enormous challenges to gambling studies in general and to treatment research in particular (Toneatto, 2005).

With accelerating gambling prevalence, more and more women are lured to games of chance, investing their time and money in this previously male-dominant activity. Compared to male gamblers, female gamblers are more likely to use gambling to get away from their personal and family problems (Brown & Coventry, 1998). Not surprisingly, women who struggle with adverse life circumstances are particularly vulnerable to gamble out of control (Boughton & Falenchuk, 2007). Perceiving unspoken stigma attached to female gambling and negative consequences of gambling exposure, women tend to keep their gambling experiences personal and private (Hraba & Lee, 1996; King, 1999). They often safeguard their gambling behaviour from being known to outsiders or non-gamblers, including family and friends. Needless to say, the chance for women to willingly share their private gambling experience with a researcher is very slim because a researcher is perceived as in the position of public knowledge production (Lee, 1993; McCorkel & Myers, 2003). Such psychological tendency not only subjects female gamblers to negative consequences of gambling alone, in silence, but also prevents them from seeking help and participating in research.

Although female gambling is becoming increasingly popular, women's ways of gambling remain under-researched and poorly understood due to the sensitivity of the topic, the vulnerability of female gamblers, and the lack of alternative paradigms and methodologies in gender specific gambling research (Hing & Breen, 2001; Lesieur, 1989; Mark & Lesieur, 1992). Because it is difficult to recruit female participants who are willing to provide honest and accurate information commonly shared backstage among gamblers, a covert data collection method is needed to help gain research access to female gambling culture.

Participant Observation as Embedded Gambling Research

According to Alder and Alder (1987), when studying a subculture particularly a deviant group, "researchers must assume social roles that fit into the worlds they are studying." (p. 8). Generally speaking, the more secretive and amorphous are the activities of the researched, the more necessary it is for the researcher to participate in their activities to learn about their culture. In order to have an insider's experience and point of view, I set out to explore everyday realities of female gambling by immersing myself into the culture I studied. Entering into the field as a participant researcher, I intended to take the perspectives of female gamblers, to see the world through their eyes, to feel what they feel, and to experience what they experience.

Because this study was designed to reveal backstage realities of female gambling largely unknown to public, using covert participant observation was well justified in this gambling research. First, since the data collection took place in constant changing public social settings with a flow of gamblers from all walks of life, the disclosure of my research identity was neither necessary nor practical because doing so would interrupt gamblers' routine activities. Second, given the sensitivity of female gambling, revealing my research identity could potentially cause gamblers' unnatural reactions such as fleeing from the scene or acting as "good participants". Third, since I intended to record physical gambling settings and activities without identifying the people I observed and interacted with, this ethnographic work would pose minimal risks on the anonymous informants I came across in the field.

Participating in female gambling culture as an embedded researcher, I frequently reminded myself to walk in the shoes of female gamblers. However, as a non-gambler, who gambled with the budgeted research money, I found myself incapable of becoming a complete participant, or "going native" to join them as a real gambler. In the absence of shared values and objectives, I could not live through the same emotional ups and downs experienced by real gamblers, such as the excitement of winning and the fear of losing. However, by taking part in their activities and bearing witness to their joys and sorrows, I learned that the lives of female gamblers were complex realities, different from the statistical results generated by quantitative studies. From my perspective, women's gambling choice, participation, and addiction cannot be fully understood without taking account of the socioeconomic and personal circumstances in which their lives evolved.

The naturalistic observational technique I employed in this study allowed me to observe and document both physical settings and social activities of female gamblers as the flow of gambling scenes naturally unfolded. However, the field practice of ethnography was much more complex and ethically challenging than what I had originally anticipated. I learned first-hand that socially sensitive research could pose special ethical problems for the researcher (Sieber & Stanley, 1988). On a number of occasions my disguised social interactions with female gamblers caused disturbing feelings of awkwardness and uneasiness in both parties. I became concerned about harmful psychological consequences of my covered research work, wondering to what extent I could adjust my research role to the complexity and the dynamics of field social relationships. This concern eventually resulted in the shifting of membership roles I assumed as a participant researcher, as I detail next.

Fieldwork Encounters in Gambling Settings

Undertaking this gambling study as a participant researcher the first time, I was more curious than apprehensive at the beginning of the fieldwork. Although I was aware of the embarrassing and frustrating situations reported by other ethnographers, without first-hand field experience I was psychologically unprepared for ethical challenges embedded in ethnography until encountering a parade of problems in the field.

After obtaining ethical clearance from the Research Ethics Board at the university with which I was affiliated, I started my solo research endeavour. I made about 40 field trips to the three largest casinos in southeast Ontario, regularly taking shuttle buses with casino-goers. The round-trip bus ride to each casino took about two to three hours. For each trip I was out around 10 am and back home around 7 pm. My routine fieldwork as a participant researcher included, but was not limited to, observing female gambling activities, playing slot machines with the budget research money ($20 per trip), and interacting with female gamblers whenever possible. To avoid sceptical reactions from gamblers, I did not take field notes on the spot. Usually I recorded what I had seen and heard when I was alone in casino cafeterias or outdoor gardens. Some notes were condensed and some expanded (Spradley, 1980), depending on the research relevance of the situations and occurrences I observed. I also kept regular journal entries to reflect on positive and negative experiences I had encountered during each trip.

Since anyone over the age of 18 is legally eligible to play in Ontario casinos, I obtained a "player's card" from each casino I visited, identifying myself as a beginning player who was curious about recreational gambling entertainment. Once I entered the field of public space, riding buses and playing in casinos with other gamblers, I had no control over mobile social settings such as what would happen and whom I would meet. Maybe because of my petite Asian appearance, or my ponytail and casual student outfit, or I simply looked out of place, at the very beginning I was stopped many times by casino gatekeepers to check my identity card. Some female gamblers I met in the field also mistook me as a young university student. A few of them even initiated to lecture me about the addictive nature of casino gambling. Three dramatic field encounters led me to adjust my field membership roles from covert to overt and finally to peripheral participation.

First Field Scenario: Encounter with Three Chinese Female Gamblers

In my first visit to one of the casinos, I became acquainted with three Chinese women, two seniors and one middle-aged. This acquaintance came naturally because we are all recent immigrants from Mainland China who speak Mandarin. While walking back to the slot machines from an outdoor chit-chat, one of the seniors drew me aside, asking me why I came to gamble and what did I do for a living. I told her I worked at a university and it was my first trip to this casino. She was very worried about me and initiated a serious discussion about the consequence of my gambling participation. She warned me that casino gambling was dangerously addictive, and that it would ruin my life if I came here to gamble regularly. At that moment I felt really guilty and uneasy about my research deception, so I told her not to worry about me because I was there for the purpose of research. She was quite taken back by this unexpected disclosure, but in the meantime she looked appeased because obviously I would not become a gambling victim. After discovering my research role, these Chinese women became curious in my study. I briefly explained my project to them and also gave them some of the research flyers I prepared in the Chinese language. The flyers clearly outlined the research purpose, the interview procedure, the eligibility and confidentiality of participation, and also the monetary compensation of each interview session. Initially, two of them were interested in attending interviews, but later cancelled their appointments due to firm family objection. They explained that their family members objected to their gambling participation in the first place, let alone to talk about it to a researcher from the same cultural background.

Second Field Scenario: Encounter with a White Female Gambler

On my return bus trip from the casino, I sat down by the window looking out the greenery landscape. After a while, a middle-aged White woman left her seat, politely asking if she could sit beside me. We started to chat together. Later I figured out that she came to talk to me for the sole purpose of educating me about the consequence of casino gambling. After relaying her gambling stories, especially how she was lured to slot machines by "beginner's luck" and gradually developed a powerful urge to casino gambling, she asked me why I came to gamble. I simply responded that it was my first casino trip. She became very concerned and warned me right then and there.

> It is ok for people like me, but you are too young. You should not come to play those games. You can learn lots of things elsewhere but you will learn nothing here. Trying once is more than enough. Forget it! Don't come again!

Once again I felt pressured to reveal my research role to ease her worries, so I did. Upon my disclosure, she was somehow embarrassed for mistaking me as a university student, a potential gambling victim she felt obligated to rescue. I expressed my appreciation for her good intentions and then briefly talked about my gambling study, hoping that she would be willing to participate in consent interviewing. At that moment I wanted to recruit her as a participant because from our conversation I saw a potentially rich informant who would have a lot to tell from her two decades of gambling experience. Furthermore, she was very open, articulate, and persuasive; a strong woman who was not afraid of speaking her mind. I gave her a flyer of my research advertisement when she was about to get off the bus. However, I never heard from her.

Third Field Scenario: Encounter with a South Asian Female Gambler

Another day I was chatting with a South Asian woman who sat beside me on the return bus trip. She asked me if I had any luck, and I told her I lost $20 (the research money) in the casino. She commented that it was nothing compared to her loss of the day, "you know how much I lost today? $300! I have lost about $50,000 in three years." I asked if she ever tried to quit, she responded, "I want to stop but I keep coming back . . ." By this time her cell phone was ringing, the driver asked her why she did not answer the phone, her voice dropped lower, "I know who is calling, it is my husband. . . . I have three sons . . . they are very sad that their mom is a gambler. . ." I was very empathetic and felt compelled to help her. I suggested that she could call gambling helpline or go to get free professional counselling on gambling addiction. I also told her about my gambling study, suggesting that the confidential research interviewing could also serve as a potential outlet for the release of repressed psychological and emotional distress. Despite my good intentions, she stopped talking about her gambling stories right after I disclosed my role as a researcher. She did take a research flyer, but never contacted me.

Since these women did not attend my interviews, I did not know who they were, nor did I know what circumstances and motivational forces had driven them to gamble, and gamble compulsively. However, to this day I still vividly remember these field trip encounters. On the one hand, I treasured this fieldwork experience and believed that these casual conversations that occurred in the field truly reflected the hearts and minds of these female gamblers; on the other hand, I struggled with ethical dilemmas of my research deception because these women volunteered their information before knowing who I was, but with my research role shifting from concealment to disclosure, their attitudes towards me accordingly changed from sharing to silence.

Responding to Ethical Challenges and Adjusting Field Membership Roles

Performing as an undercover researcher, I was emotionally disturbed by these unexpected awkward situations that occurred in the field. I felt that my disguised interaction with these female gamblers had unintentionally infringed their right to privacy, and also subjected myself to psychological pressure and inner conflicts. I also felt that in order to carry out an ethical research, I should not use the end to justify the means, even though the potential research benefits would outweigh psychological risks. To avoid ethical dilemmas inherent in research concealment, in the subsequent field trip I was determined to make my research overt. It was out of question for me to unveil my research inside casinos, but nevertheless I decided to try it on the bus. After obtaining permission from the driver, I announced that I was doing research on women's gambling experiences and I would appreciate their information and participation. Female gamblers generally took my flyers, but they rarely offered to participate in research. A woman expressed her concerns, "I don't want to talk to you about gambling because you work for the government." Another woman explained further,

> You know why I don't want to come to your study? I would feel guilty if I come to your interviews because I know I would lie to you. I don't want to talk because I don't want my private life going public. Believe me – the gamblers who come to you will lie to you. You won't get the truth.

From this woman's honest remarks on the discrepancy between public accounts and private lives of female gamblers, I started to understand why it was difficult for women to tell their gambling stories. Regardless of my efforts, the overt approach I attempted was unsuccessful because my research disclosure immediately changed social relationships and silenced voices. Once female gamblers were made known of my research role, they started to view me differently, treating me as a suspicious outsider who should not be entrusted because I did not share their experiences. This perception and suspicion immediately created a more awkward social space between me and them, the researcher and the researched.

As the overt approach did not work in the field, I decided to make another change, adjusting the level of my involvement and participation from active to moderate (Spradley, 1980). I still took part in women's routine gambling activities, but intentionally limited personal contacts in the field. Instead of actively engaging personal interactions or substantial conversations with female gamblers, I mainly recorded the gambling activities I observed and the spontaneous conversations I overheard during my field trips. In other words, by primarily assuming a covert, peripheral membership role (Alder & Alder, 1987), I became a detached insider.

The peripheral membership I assumed in the field turned out to be a good fit with the uniqueness of female gambling culture. This adjustment not only helped me avoid similar ethical dilemmas I experienced at the beginning of this fieldwork, but also gave me a much-needed psychological space to participate in women's gambling activity as an insider and observe it as an outsider. For the rest of my fieldwork, I did not encounter further awkward situations or experience emotional uneasiness and discomfort.

Participant Observation and Consent Interviewing as Complementary Research Methods

Although none of the female gamblers I met in the field offered to come to my interviews, I received about two dozen responses to my research advertisement placed in public places such as community centers and grocery stores.

After screening all responses by telephone, I successfully conducted two sessions of consent interviews with seven women from five different cultural backgrounds (Eastern Indian, English, French, Native Canadian, and Pakistani), who regularly played various games of chance, such as slots, bingo, and poker.

When the woman I met on the bus informed me that the female gamblers who came to my interviews would not tell the truth, I worried if the data I painstakingly collected would be valid at all. In order to establish trust and rapport with the vulnerable participants, and also elicit reliable and authentic information from them, I conducted all interviews in a respectful, friendly, and non-judgmental manner. Before interviewing I repeatedly explained the purpose of the study to make sure these female gamblers understood the benefits and risks of their participation, particularly their right to privacy and confidentiality. As the interviews progressed from simply telling gambling stories to reflectively interpreting the meanings of gambling, these women became more and more comfortable sharing their private gambling experiences. Their openness yielded detailed and in-depth information on how they came to gamble and what gambling meant to them. The findings showed that these women progressed from gambling entertainment to gambling addiction by emotionally relying on games of chance as their hope, escape, therapy, reward, and social life (Li, 2007).

Regardless of the ethical dilemmas I encountered in the field, two integrated ethnographic data collection methods made a complementary contribution to this study because they helped me sketch a fuller picture of female gambling culture. Looking back over this fieldwork experience, I feel that in sensitive research, ethnographic participant observation should not be simply measured and judged by the same ethical guidelines for consent interviews or other overt data collection methods. Rather, a great emphasis should be placed on how to handle unavoidable ethical challenges and unexpected situations in the field.

Reflections on Ethical Challenges of Participant Observation

From this ethnographic gambling study I have learned that in sensitive research the revelation of hidden realities and the pitfall of ethical dilemmas are co-existed realities that are not easy to reconcile. As I have illustrated, in a constant-shifting gambling environment, I came across different female gamblers in every field trip, thus it would be awkward for me to identify my research before starting a conversation with female gamblers. However, without informing them about my study, the conversations and interactions I engaged in the field for the purpose of research unintentionally violated the privacy right of the informants, and also subjected myself to psychological conflicts.

From this ethnographic work I have also learned that the characteristics of the researcher can shape field social interactions in unexpected ways, which often gives rise to problems of relational ethics.

Researcher's Personal Characteristics in Sensitive Research

It has been noted that the evolving researcher-participant relationship in field research is often conditioned by the researcher's personal characteristics such as gender, race, and age (Dewalt & Dewalt, 2002; Ladson-Billings, 2003; R. H. Wax, 1979).

Firstly, gender mattered in this study. As a female researcher participating in women's gambling activities, I had easy access to female gamblers as I was not perceived as threatening to them. This gender advantage in field research has lent support to the position that in order to understand women's life situations and experiences, more studies should be conducted by women and for women (Ettorre, 1989).

Secondly, culture mattered in this study. Having the same racial and cultural background as the researched is usually advantageous in fieldwork. However, in this gambling study my cultural background did not help me gain access to Chinese female gamblers' private experiences. During my field trips I witnessed a significant Asian presence in all three casinos I visited. Needless to say, behind such a popular scene there would be countless stories to be told by Asian female gamblers. Although I made an effort to recruit Chinese female participants, none of them were willing to attend or consent to interviews. Two interrelated cultural accounts may provide explanations to this paradox. First, in Chinese culture gambling is historically viewed as one of five vices and institutionally opposed by the government. Second, as a result of the Confucian prescription on morality, Chinese people tend to hold strong feelings of shame and honour about cultural undesirable or desirable behaviours. Perhaps because exposing a culturally undesirable behaviour to a researcher from the same culture would activate strong negative emotions, Chinese female gamblers chose to restrain themselves from participating in my study.

Thirdly, age mattered in this study. Some teacher-student relationships accidentally occurred in the field, which were obviously triggered by my age; the appearance of age being perceived by the public. The Chinese senior took the initiative to talk to me because it was culturally considered a noble social practice for an elder to educate the young about the negative consequences of their choices. Similarly, the middle-aged White woman felt compelled to save me from the devastating consequences of gambling addiction because she perceived me as a young university student who ventured out in a wrong place.

Although these women did not know who I was, they made an effort to prevent me from becoming a victim of gambling addiction. Having been trapped in a vicious gambling cycle and suffered negative consequences of addiction, they did not want to see the young falling into the same pitfalls. Contemplating these dramatic field encounters, I have realized what motivated these women to open their hearts and minds to a complete stranger was deeply rooted humanity that unites all of us. By taking actions, these women sent out two important messages: (a) prevention is the best treatment for gambling addiction and (b) the general public, especially the young, should be educated that their gambling choices are not free from negative consequences. I believe the revelation of these stories can counteract public stereotypes as well as research biases, which often simplify complex realities of female gambling world, blaming victims for their irresponsible choices.

Unforeseen Field Circumstances in Sensitive Research

As qualitative researchers we tend to connect ourselves to the researched individuals on both cognitive and emotional levels (Mitchell, 1991), therefore our lives are often embedded within our fieldwork experiences in such a way that all of our interactions in the field involve moral choices (Tedlock, 2000). Everyday social realities are fluid, unpredictable, and fragile because they are largely shaped by situated and often unforeseen circumstances produced by the people who come into contact with each other (O'Brien, 2006). As such, ethnographers who engaged in distant settings have no choice but to situationally solve many on-site relational ethical problems they encountered in the research process (Clarke, 1975; Ellis, 2007; Etherington, 2007).

Doing ethnography in sensitive research is like walking a tightrope, as such field work requires ethnographers to keep a mindful awareness of ongoing relationships and to make frequent adjustments accordingly. In embedded research, good intentions do not always result in smooth social interaction but sometimes produces awkward outcomes (O'Nell, 2002). As this gambling study has revealed, in conducting ethnographic sensitive research with vulnerable individuals, social relationships and interactions between the researcher and the researched are extremely fragile, subject to break downs. However, such disruptions can always be repaired with appropriate field strategies. Certain skills, such as the psychological preparedness for the unforeseen occurrences of ethical dilemmas and the flexibility of adjusting to different field circumstances, are crucial in successful fieldwork because "an understanding of a number of potential difficulties and capacity to be flexible may sooth the way to dealing with some pitfalls that revolve around sensitive topics, regions, and methods" (Laine, 2000, p. 93).

At the beginning, I made field entry as a covert participant researcher to learn about the world of female gambling, but only discovered that the very people I intended to study wanted to protect me from the ravages of gambling addiction. When the female gamblers expressed their concerns about my gambling participation, I was pressured to unveil my research because I wanted to ease their worries. I also intended to recruit them for consent interviews because from their actions I saw the richness of gambling experiences they would be able to provide. However, due to my inexperience in handling the pressure of unpredictable field situations and my eagerness to recruit potential participants, the solution to one problem naturally caused another; my disclosure took away their worries, but simultaneously silenced their voices.

Similarly, when I shared my gambling experience with the South Asian woman, she promptly offered her personal information in exchange. This exchange also ended abruptly after I informed her about my research. Obviously her initial openness was activated by an understanding that she was sharing her stories with a like-minded peer, a misperception led by my concealment. When this woman perceived me as an insider, she felt psychologically safe to share because a fellow gambler would not judge her from the perspective of an outsider. Most likely, if she was made aware of my research in the first place, she would not have volunteered the same information.

My active gambling participation not only created the awkward situations and dramatically changed the behaviour of the female gamblers, but also subjected me as a researcher to psychological risks. I felt uneasy with my disguised research role because, by pretending to be one of them, I successfully obtained the information for the purpose of the research, but the data was obtained at the expense of the autonomy of the female gamblers. To avoid such disturbing inner struggles and conflicts, in later field trips I adjusted my membership role and reduced the level of involvement. There might be other possible solutions for this ethical dilemma, but I believe I made a good field decision given the field circumstances in which I was situated. With my ongoing adjustment, in the later stage of my fieldwork I was able to document the female gambling culture by moderately or peripherally participating in their activities as a detached insider, without experiencing further ethical problems.

Lessons Drawn from Fieldwork Reflections

In revealing private lives and telling others' stories, field researchers often face ethical dilemmas and moral choices that cannot be easily resolved with general ethical guidelines. At the field entry, although my interactions with female gamblers ended in uncomfortable halts, those encounters helped reveal gambling stories different from public perceptions and stereotypes. Given the pros and cons of participant observation, it is difficult to draw a clear line between the ethics and the politics of ethnographic fieldwork

(Murphy & Dingwall, 2002, Punch, 2000). This is because when a study aims to supply both evidence and ground to contest against social prejudice, the research process itself becomes praxis and advocacy (Lather, 2004). From this standpoint, I feel that in order to conduct an ethical study, researchers must always prepare themselves psychologically and technically for the unexpected, willing to make adaptive changes in mobile field settings. The ethical and moral responsibilities of ethnographic research should not be simply aimed at eliminating covert research to avoid ethical dilemmas, but to take full consideration of the sensitivity of the research topic, the vulnerability of the researched population, and the plasticity of field membership roles.

First, it is important for researchers to tailor their data collection methods to the sensitivity of the research topic. Although every effort must be made to avoid research concealment, under circumstances that an overt fieldwork is unpractical, or likely to trigger emotional reactions from the researched, the covert approach should be considered because it can help reveal what lies beneath without altering the nature of reality. As this gambling study illustrated, a covert researcher should take caution about performing such a role and set limits on personal interaction in the field. If done skillfully, the potential social benefits of covert work often outweigh its ethical risks, as the data collected can help push the boundaries of our current understanding of disadvantaged individuals and groups in society. Such understanding can serve as advocacy to inform better policy and practice, ultimately bringing positive changes for people being studied.

Second, it is important for researchers to adapt their data collection methods to the vulnerability of the researched. Because the existing social order and classifications systematically marginalize and stigmatize certain groups, the individuals being assigned such labels are virtually made invisible and voiceless in our society. As this fieldwork clearly showed, female gamblers did not want to seek help or participate in research because they were afraid of exposure and judgment. Researchers undertaking sensitive ethnographic studies should always make effort to fulfill their social responsibilities in field interactions, protecting the vulnerable from further stigmatization or marginalization.

Third, it is important for embedded researchers to psychologically prepare for the plasticity of field membership roles. Given the uncertainty and complexities of field social interactions, participant researchers should not assume certain fixed positions at the field entry and then cling to them from start to finish. They must adjust their levels of involvement and participation, should the ethical issues arise in the field. In this gambling study, after experiencing the awkward situations at the beginning of my research trips, I subsequently adjusted my research participation from covert to overt and then to peripheral. These field adaptations not only prevented negative and disturbing emotions the informants would feel about deceitful interactions in which they were unconsciously involved, but also helped me avoid experiencing further uneasy emotions and inner conflicts derived from ethical dilemmas.

From this fieldwork I have also learned two practical techniques. One is that when conducting covert fieldwork, the concealment of research role should be maintained from start to finish, so as to minimize psychological risks of such work posed on both the researcher and the researched. Another technique is that in sensitive research an embedded researcher should not attempt to recruit participants in a constantly changing environment because the possibility of establishing trust and rapport with the informants in mobile social settings is rather minimal, if not impossible.

Conclusion

As an embedded researcher participating in female gambling culture, I have learned first-hand that the challenges of ethnographic fieldwork are centred on the issues of quest for knowledge and ethical dilemmas. By assuming a covert research role, I was able to observe natural occurrences in gambling settings, but unable to make peace with disturbing feelings of my research concealment. By uncovering my research identity, I was able to fulfill ethical obligations as a researcher, but unable to get the female gamblers to speak their minds. This fieldwork has illustrated that ethnography as a research methodology has the greatest strength and the greatest weakness (Hume & Mulcock, 2004). The mastery of this naturalistic data collection method comes with the lessons learned and experiences gained from the field. Given the fact that in covert participant observation the search for truth is often at odds with the conformity to conventional research ethics, field researchers must be psychologically prepared to negotiate two opposing forces that co-exist in constant tension.

In retrospection, this fieldwork not only drew me closer to the world of female gamblers, but also offered me an excellent opportunity for reflexive research practice. On the one hand, my participation in women's gambling activities helped me understand their subjective experiences; on the other hand, it allowed me to enrich the data by adding personal and emotional depth to my work (Smith & Kornblum, 1996). Although the research process of ethnography is challenging and difficult, the outcome of such work has proven its worth and benefit.

Ethnography is both a process and a product. As I reflect on the experience of participant observation, especially on how to balance the search for truth and the ethical challenges that often co-exist in parallel in ethnographic sensitive research, I come to realize that there may be no easy or universal solutions for the inherent dilemma of this unique field method. Given the active voices I heard in concealment and the immediate silence I encountered after my research disclosure, I contend that in sensitive studies research ethics must go beyond the simple avoidance of research covertness to a mindful consideration of the well-being of marginalized individuals and communities being studied because as researchers, not only should we aim to uncover and interpret the voices, but also to understand the silence.

References

Alder, P. A., & Alder, P. (1987). *Membership roles in field research*. Newbury Park, CA: Sage.

Alder, P. A., & Alder, P. (2000). Observational techniques. In N. K. Denzin & Y. S. Lincoln (Eds.), *Handbook of qualitative research* (pp. 377–392). Thousand Oaks, CA: Sage.

Atkinson, P., Coffey, A., Delamont, S., Lofland, J., & Lofland, L. (Eds.). (2002). *Handbook of ethnography*. London: Sage.

Boughton, R., & Falenchuk, O. (2007). Vulnerability and comorbidity factors of female problem gambling. *Journal of Gambling Studies, 23*, 323–334.

Brown, S., & Coventry, L. (1998). *Queen of hearts: The needs of women with gambling problems*. Melbourne, Australia: Victorian Women's Trust.

Bulmer, M. (1980). Comment on the ethics of covert methods. *British Journal of Sociology, 31*, 59–65.

Bulmer, M. (1982). The merits and demerits of covert participant observation. In M. Bulmer (Ed.), *Social research ethics* (pp. 217–251). London: Macmillan.

Charmaz, K., & Olesen, V. (1997). Ethnographic research in medical sociology: Its foci and distinctive contributions. *Sociological Methods and Research, 25*, 452–494.

Clarke, M. (1975). Survival in the field: Implications of personal experience in fieldwork. *Theory and Society, 2*, 95–123.

Dewalt, K. M., & Dewalt, B. R. (2002). *Participant observation: A guide for fieldworkers*. New York: Altamira Press.

Dingwall, R. (1980). Ethics and ethnography. *The Sociological Review, 28*, 871–891.

Ellis, C. (2007). Telling secrets, revealing lives: Relational ethics in research with intimate others. *Qualitative Inquiry, 13*, 3–29.

Etherington, K. (2007). Ethical research in reflexive relationships. *Qualitative Inquiry, 13*, 599–616.

Ettorre, B. (1989). Women and substance use/abuse: Towards a feminist perspective or how to make dust fly. *Women's Studies International Forum, 12*, 593–602.

Fine, M., Weis, L., Weseen, S., & Wong, L. (2003). For whom? Qualitative research, representations, and social responsibilities. In N. K. Denzin & Y. S. Lincoln (Eds.), *The landscape of qualitative research: Theories and issues* (pp. 167–207). Thousand Oaks, CA: Sage.

Gans, H. J. (1999). Participant observation in the era of "ethnography". *Journal of Contemporary Ethnography, 28*, 540–548.

Goffman, E. (1963). *Stigma*. Englewood Cliffs, NJ: Prentice Hall.

Hammersley, M., & Atkinson, P. (1995). *Ethnography: Principles in practice*. London: Routledge.

Hing, N., & Breen, H. (2001). Profiling lady luck: An empirical study of gambling and problem gambling amongst female club members. *Journal of Gambling Studies, 17*, 47–69.

Hobbs, D. (2002). Ethnography and the study of deviance. In P. Atkinson, A. Coffey, S. Delamont, J. Lofland, & L. Lofland (Eds.), *Handbook of ethnography* (pp. 204–219). London: Sage.

Homan, R. (1980). The ethics of covert methods. *British Journal of Sociology, 31*, 46–59.

Hraba, J., & Lee, G. (1996). Gender, gambling, and problem gambling. *Journal of Gambling Studies, 12*, 83–101.

Hume, L., & Mulcock, J. (Eds.). (2004). *Anthropologists in the field: Cases in participant observation*. New York: Columbia University Press.

Humphreys, L. (1970). *Tearoom trade*. Chicago: Aldine.

King, A. (1999). *Diary of a powerful addiction*. Tyndall, Manitoba, Canada: Crown.

Ladson-Billings, G. (2003). Racialized discourses and ethnic epistemologies. In N. K. Denzin & Y. S. Lincoln (Eds.), *The landscape of qualitative research: Theories and issues* (pp. 398–432). Thousand Oaks, CA: Sage.

Laine, M. (2000). *Fieldwork, participation, and practice: Ethics and dilemmas in qualitative research*. London: Sage.

Lather, P. (2004). Research as praxis. In R. A. Caztambide-Fernandez, H. A. Harding, & T. Sorde-Marti (Eds.), *Cultural studies and education: Perspectives on theory, methodology, and practice* (Reprint Series No. 38, pp. 41–60). Cambridge, MA: Harvard Educational Review.

Lee, R. M. (1993). *Doing research on sensitive topics*. London: Sage.

Lesieur, H. R. (1989). Current research into pathological gambling and gaps in the literature. In H. Shaffer (Ed.), Compulsive gambling: Theory, research, and practice (pp. 225–248). Toronto: Lexington Books.

Li, J. (2007). Women's ways of gambling and gender-specific research. *Sociological Inquiry, 77*, 626–636.

Mark, M. E., & Lesieur, H. R. (1992). A feminist critique of problem gambling research. *British Journal of Addiction, 8*, 549–565.

McCorkel, J. A., & Myers, K. (2003). What difference does difference make? Position and privilege in the field. *Qualitative Sociology, 26*, 199–231.

Mitchell, R. G., Jr. (1991). Secrecy and disclosure in the fieldwork. In W. B. Shaffir & R. A. Stebbins (Eds.), *Experiencing fieldwork: An inside view of qualitative research* (pp. 97–108). Newbury Park, CA: Sage.

Murphy, E., & Dingwall, R. (2002). The ethics of ethnography. In P. Atkinson, A. Coffey, S. Delamont, J. Lofland, & L. Lofland (Eds.), *Handbook of ethnography* (pp. 339–351). London: Sage.

O'Brien, J. (2006). Building and breaching reality. In J. O'Brien (Ed.), *The production of reality: Essays and readings on social interaction* (pp. 334–352). Thousand Oaks, CA: Pine Forge.

O'Nell, P. (2002). Good intentions and awkward outcomes: Ethical gatekeeping in field research. In W. C. Hoonaard (Ed.), *Walking the tightrope: Ethical issues for qualitative researchers* (pp. 17–25). Toronto: University of Toronto.

Paradis, E. K. (2000). Feminist and community psychology ethics in research with homeless women. *American Journal of Community Psychology, 28*, 839–858.

Punch, M. (2000). Politics and ethics in qualitative research. In N. K. Denzin & Y. S. Lincoln (Eds.), *Handbook of Qualitative research* (pp. 83–97). Thousand Oaks, CA: Sage.

Rossman, G. B., & Rallis, S. F. (2003). *Learning in the field: An introduction to qualitative research*. London: Sage.

Sieber, J. E., & Stanley, B. (1988). Ethical and professional dimensions of socially sensitive research. *American Psychologist, 43*, 49–55.

Smith, C. D., & Kornblum, W. (Eds.). (1996). *In the field: Readings on the field research experience*. Westport, CT: Praeger.

Spradley, J. P. (1980). *Participant observation*. New York : Holt, Rinehart, and Winston.

Statistics Canada. (2005). *Perspectives on labour and income: Fact-sheet on gambling*. Retrieved January 5, 2007, from http://www.statcan.ca/english/studies/75-001/comm/2005_01.pdf

Tedlock, B. (2000). Ethnography and ethnographic representation. In N. K. Denzin & Y. S. Lincoln (Eds.), *Handbook of qualitative research* (pp. 455–486). Thousand Oaks, CA: Sage.

Toneatto, T. (2005). A perspective on problem gambling treatment: Issues and challenges. *Journal of Gambling Studies, 21*, 73–78.

Tope, D., Chamberlain, L. J., Crowley, M., & Hodson, R. (2005). The benefits of being there: Evidence from the literature on work. *Journal of Contemporary Ethnography, 34*, 470–493.

Van Maanen, J. (1988). *Tales of the field: On writing ethnography*. Chicago: The University of Chicago.

Wax, M. L. (1979). On the presentation of self in fieldwork: The dialectic of mutual deception and disclosure. *Humanity & Society, 3*, 248–259.

Wax, R. H. (1979). Gender and age in fieldwork and fieldwork education: No good training is done by any man alone. *Social Problems, 26*, 509–523.

29

The Risk of 'Going Observationalist': Negotiating the Hidden Dilemmas of Being an Insider Participant Observer

Robert V. Labaree

As a faculty member active in my university's Academic Senate during the past three years, I initiated my preliminary dissertation research with both a professional as well as a personal motivation to explore the limitations of pseudo-legislative forms of faculty involvement in shared governance. Entering a doctoral program in higher education leadership offered a means for exploring various aspects of shared governance from the dual perspective of a senate faculty member and a researcher pursuing advanced studies grounded in theories of distributive politics, organizational behavior and educational leadership in a large research university setting.

A review of the literature on shared governance reveals a measurable reliance on 'perceptions', 'opinions' and 'judgments' concerning the faculty's role in academic decision-making. While these perspectives can be helpful in revealing the opportunities and challenges of senate-based faculty involvement in institutional governance, this literature largely fails to provide an intimate view of the internal dynamics of shared governance. Personal observations by authors who participated in faculty senates were rarely documented in a systematic manner or conducted over a long period of time. Observations were predominantly reflective and rarely occurred while the researcher was deeply imbedded in the 'real-time' governance process and connected to the complex web of committees, task-force groups, and joint

Source: *Qualitative Research*, 2(1) (2002): 97–122.

administrative–faculty units that comprise the true mechanisms of policy formulation in shared decision-making. In addition, the literature rarely discusses the role of informal arenas of shared governance and bargaining, such as luncheons and senate retreats, even though I have observed that these forums represent a significant avenue for policy creation and negotiation between the faculty and administration. Finally, a reliance on documentary analysis of minutes of meetings is inadequate in providing the complete picture of shared governance. Although minutes of meetings and other materials are considered official records of senate activities, they most often represent a filtered 'truth' of what happens during the policy formulation process. Minutes of meetings rarely contain the details necessary for analyzing the complex dimensions of faculty involvement in shared governance; they do not document the informal forms of negotiation and debate mentioned earlier, and are unevenly collected in committees.

With these issues in mind, I sought a new approach to studying the role of faculty senates in shared governance – one that is: (1) researched systematically over time; (2) takes into account informal methods of negotiation among constituents; and (3) is conducted within the borders of a faculty senate during debates over official university policy: in other words, a study conducted by an insider. More recently, I have participated in senate activities with a deliberate dual purpose. On the one hand, I perform the seen activities of moderating the university-wide faculty listserv, helping to develop the senate's web site, and participating on a variety of senate committees and task-force groups, while at the same time acting as one of the library faculty's senate representatives. On the other hand, I also engage in the unseen purpose of laying the groundwork for forging possible links with informants once I begin the data-gathering aspect of my doctoral research, of learning as much as I can about the structures of influence and distributive politics within the organization and of contextualizing this knowledge within the cultural framework of the institution. During this exploratory stage of research, I approached this mission with a desire to protect my 'insiderness', sharing my ulterior motives as an insider participant observer with only a few senate colleagues.

I consider my insiderness as the key to delving into the hidden crevices of the organization and, as a result of this case study, developing a deeper understanding of the complex dynamics of faculty influence on decision-making that is rarely seen in current research on shared governance. Standing before me, however, are a number of dilemmas about my insiderness that raise several critical issues for the qualitative researcher. On a broad level, there are issues concerning the conceptual definition of insiderness and its relationship to outsiderness, and the search to understand why insiderness is considered revealing in an epistemological manner that is considered inaccessible to an outsider. Related to this are questions concerning whether insiderness can provide a unique perspective that can never be penetrated by an outside participant observer and whether insiderness actually diminishes a broader, unbiased understanding that an outsider could bring to the study (Merton, 1972; Surra and Ridley, 1991). On an introspective level, there are questions of motivation and the motivations of others and the careful consideration of ethical issues. Finally, at the center of these questions lies the general issue of whether the outcomes and interpretive conclusions would differ significantly if the study were to be conducted by an insider participant observer rather than an outsider.

A significant volume of literature is devoted to the insider–outsider debate in qualitative research. Many contemporary researchers are focusing on their desire to look beyond this methodological dichotomy. However, these studies are generally reflective narratives and they are diffused among many fields of study, including anthropology (Burke, 1989; Haniff, 1985; Headland et al., 1990; Jones, 1970; Narayan, 1993; Ohnuki-Tierney, 1984), education (Arnold, 1994), ethnic studies (Beoku-Betts, 1994; Brewer, 1986; Cerroni-Long, 1994; De Andrade, 2000; Wilson, 1974; Zinn, 1979), family research (Christensen and Dahl, 1997; Daly, 1992; Farnsworth, 1996; Olson, 1977; Surra and Ridley, 1991), feminist studies (Acker and Esseveld, 1991; DeVault, 1996; England, 1994; Hawkesworth, 1989; Naples, 1996; Reay, 1995, 1996; Riessman, 1987; Smith, 1987), folklore research (Parades, 1977; Scheiberg, 1990; Sherman, 1986; Yocum, 1980), geography (Clark, 1998; Mullings, 1999; Rose, 1997), nursing (Kauffman, 1994; Thorne and Paterson, 2000), social work (Kanuha, 2000; MacDonald, 1990) and sociology (Blauner, 1987; Griffith, 1998; Merton, 1972; Wilson, 1974).

The purpose of this article is to review the literature devoted to analyzing the insider–outsider issue in qualitative inquiry and to draw upon my preliminary research of the faculty senate in order to build a framework for exploring the hidden dilemmas of being an insider participant observer. Methodological and ethical dilemmas faced by the insider participant observer are often hidden because the insider may assume there is a measurable advantage inherent in seeing things from the inside and because the achievement of insiderness is, at least in part, a product of the researcher's own perceptions of achieved access and the interpretations of those being investigated. As this analysis hopes to illustrate, the assumption that insiderness provides greater access can be true, but only to a degree, and it is situational depending on a number of critical factors that are determined by the circumstances of the moment.

This article is divided into five parts. Part one examines the concept of insiderness and compares and contrasts it with the idea of outsiderness. Part two discusses the perceived value of insiderness and the possible motivations for transition into the role of an insider participant observer. Part three provides an overview of the issue of insiderness in participatory observation as it relates to the degree to which the researcher should be foregrounded or shadowed in the text. Part four expands on these discussions by carefully analyzing the hidden dilemmas of entering the field, positioning and disclosure, shared relationships and disengagement. In this section, particular emphasis is placed on ethical issues. The article concludes by summarizing the conceptualizations of insiderness and identifying what I believe are issues that need greater clarification and analysis within the field of qualitative inquiry.

Part I: Mapping the Boundaries of Insiderness

Merton (1972) defines the insider–outsider position as an epistemological principle centered on the issue of access. The issue of access can take two forms. It is either a monopolistic access, in which the researcher possesses exclusive knowledge of the community and its members, or the researcher has privileged access, in which he or she has a claim to the hidden knowledge of the group that an outsider must otherwise acquire at greater risk. Merton cites Simmel's idea of the outsider as a 'professional stranger' who is detached from the commitments of the group under study, can more readily acquire objectivity, and 'is unrestricted by prejudged practice and theory, and therefore, can raise questions unlikely to be raised by Insiders' (Kauffman, 1994). Merton thus frames access in terms of an insider doctrine and an outsider doctrine. In this framework, the insider is an individual who possesses intimate knowledge of the community and its members due to previous and ongoing association with that community and its members. The general assumption is that this intimate knowledge offers insights that are difficult or impossible to access by an outsider.

Olson (1977), writing about family research, described perceptions of insiderness and outsiderness in similar terms. Insiderness was gained from family members most often utilizing self-reporting methods. Researchers provide an outsider view of families based on observations. Olson (1977) saw the dichotomy of insiderness and outsiderness as 'two mutually exclusive frames of reference'. Surra and Ridley (1991) expanded this conceptualization by arguing that insiderness and outsiderness are points on a continuum. A researcher's place on that continuum is circumscribed by the relationship between the observer and the phenomena being observed.

Christensen and Dahl (1997) view Olson's 'mutually exclusive frames of reference' as problematic and imply that the perspective of insiderness or outsiderness resting on a continuum is incomplete. This conceptualization of research creates a polarizing effect that undermines the ability of the researcher to move beyond understanding insiderness and outsiderness as an either/or proposition. In this sense, Christensen and Dahl point out that 'We must also consider how "outside" outsiders really are in regard to what they are researching' (p. 273). For example, researchers conducting studies on families in an identifiable society need to recognize that they too are often enculturated members of this society. The purpose of this introspective analysis is to bring forth more accurate accounts of the phenomenon being studied and to encourage multiple perspectives that 'bring flexibility and wholeness to the research process' (p. 274).

Merton's insider–outsider doctrine and Olson's mutually exclusive perspective frames of reference, however, contribute to an 'insider–outsider' dichotomy that ignores the fact that researchers are multiple insiders and outsiders (Deutsch, 1981). In the case of my research, I am an insider of the academic senate due to my role as a faculty member and, therefore, most closely aligned to the community of faculty members who comprise the majority of the senate's membership. However, the organization also comprises student research assistants, an administrative assistant and ex officio administrators who may not be granted the authority to vote on resolutions or appoint committee members, but who, nevertheless, influence the process of shared governance in the senate through the work they perform in support of senate activities. Although I am an insider faculty member, I am at the same time an outsider to these other group members and their 'truths' (as well as fictions) about the organization. In this sense, the boundaries of insiderness are situational and defined by the perceptions of those being researched.

Haniff (1985) concludes that insiderness and being 'native' are difficult concepts to define because there is so little 'distance' between the researcher and the researched. As she states: 'an insider may be of more detriment than an outsider and a native more foreign than a foreigner' (p. 112). Nevertheless, Haniff argues that insiderness is crucial in research because outsiderness is, by its very nature, limiting in terms of understanding hidden meanings and achieving a deep level of trust with the informant. She summarizes her position by saying:

> It is only when we are perceived and accepted as an insider that we can truly understand the meaning of the lives we study. An insider or native must take this status seriously. Its methodological implications are profound, for it is this group who can either do the most harm or the most good (pp. 112–13).

The concepts of what it means to be an insider participant observer have evolved over time. Merton initiated much of the debate in 1970 when he articulated insiderness and outsiderness as exclusionary domains of understanding within the realm of the sociology of knowledge (Zinn, 1979). Olson (1977), Surra and Ridley (1991), Christensen and Dahl (1997), and others later sought to explain the dichotomy of insiderness and outsiderness as two extremes on a continuum that represents the reporter's frame of reference. Deutsch (1981) viewed this form of debate as inadequate in explaining the complexities of the insider and outsider dichotomy by noting that 'we are all multiple insiders and outsiders' (p. 174). She argued that only by moving beyond a dichotomous cause and effect framework that focuses on which position is more important, can researchers begin to explore the more critical issue of how the positional status of the researcher may influence the phenomenon being studied. This perspective attempts to create a framework for understanding the researcher's positionality of insiderness as a process of achievement rather than simply ascribed or bestowed upon the investigator.

Most recently, De Andrade (2000) has called for a rejection of the idea that insiderness is either achieved or ascribed, but is instead a process of ongoing evaluation. Griffith (1998) argues that 'different knowledges are imbedded in both the researcher's biography and the social relations of power and privilege in which the researcher is located' (p. 363). Insiderness is therefore, a result of the person's biographical profile, political activities, research agenda and the relationship with the community under study. Griffith describes the researcher's social location and knowledges as 'always located somewhere', yet continuously moving back and forth between the positional boundaries of insiderness and outsiderness.

Part II: The Value and Motivating Factors of Insiderness

The status of insiderness must hold some value or there would be little incentive for researchers to study those subjects or situations with which they are most intimately associated. Haniff (1985) attaches an almost noble attribute to the pursuit of insiderness. However, defining the advantages of insider participant observation and native anthropology in this way is overly simplistic. In this section, I examine the reasons why native anthropologists, feminist researchers and others utilizing an insider perspective, place unique value on insiderness, based in part on the concept of epistemological privilege, and the possible motivations of transition from a community member to a researcher of that community.

One can easily cite a number of examples, both of a non-scholarly and scholarly nature, that illustrate the perceived value of being on the inside. For example, a measurable segment of historical research is devoted to describing hidden truths that the public is unaware of, due to the fact that the author was 'an insider' with privileged access to information not otherwise obtainable. There are also numerous examples in popular culture which reflect a near obsession with seeking an unseen 'reality' in virtually every corner of American society – the professional taboo of the magician revealing trade secrets, television shows that place the viewer 'on the inside', the proliferation of reality programming that shows the viewer 'what really happens' during a drug raid or in the emergency room of a hospital. Whether the concept of being on the inside is related to non-academic work or not, the overall implication is that there is a 'definite advantage' (Ohnuki-Tierney, 1984) associated with insiderness. The idea is that the relative position of the observer participant observer can reveal a new perspective, a hidden meaning, or a unique understanding that is not otherwise achievable by an outsider. In short, most researchers argue that, either implicitly or explicitly, insiderness privileges the researcher, although the form of this secured positionality differs depending on the objectives of the research project and the level of access to key informants. Researchers cite a variety of interrelated advantages to insiderness, which I have categorized into four broad values: the value of shared experiences; the value of greater access; the value of cultural interpretation; and the value of deeper understanding and clarity of thought for the researcher. Each of these advantages also has concurrent challenges that the insider participant observer must negotiate and come to terms with.

Early feminist researchers, such as Oakley (1981), were drawn to qualitative inquiry and ethnographic research, in part because of the idea that women possess a common set of interests, concerns and social values, and that these shared experiences therefore formed the basis 'for dismantling the hierarchies, fictions, and avoidances of research based on positivist frameworks' and facilitated more complete interpretations of data based on a shared understanding of women's experiences in society (DeVault, 1996: 37). This assumption was similar to that of Zinn (1979) who argued that minority scholars had special lenses through which to understand the social reality of their community. Zinn also pointed out that minority scholars were more likely to grasp certain aspects of racial phenomena that would be inaccessible to members of the dominant culture (outsiders) and that this enhanced perception would allow the insider researcher to break down 'self-protective behaviors' formed as a result of interactions with outsiders from the dominant culture.

More contemporary feminist researchers, however, have argued that this assumption is overly simplistic and does not take into account the complex nature of relationships between the researcher and the informant. In other words, insider participant observation practiced within one's own community, however that community is defined, remains a process of negotiated achievement. Riessman (1987) encapsulated this idea when she stated that 'gender is not enough' in establishing a harmonious relationship with informants. This realization has subsequently been applied to the idea that being 'black is not enough' (Beoku-Betts, 1994) and in studying other under-represented communities, such as Zavella's (1993) research with Chicano informants and earlier by Styles (1979) in his research on gay bathhouses ('I did not possess any special access to the life of the baths merely because I am gay', p. 148).

Feminist literature, in particular, held the idea that shared experiences provided a privileged means of establishing a rapport between the researcher and the informant. This assumption has recently been modified and a balance appears to have been struck. Few contemporary authors totally reject the idea that there is no value in shared experiences among women and other groups because, by simply being an insider, there are experiential commonalities that can form the basis for building trust and forging a strong relationship. However, these experiences are not to be confused with the need to work toward achieving insiderness and nurturing an ongoing level of trust.

Another categorical advantage cited in the literature concerns the idea that insiderness facilitates greater access at the start of one's research, to special groups within the community, and to critical information. This is perhaps the most universally accepted advantage given to being an insider. Ohnuki-Tierney (1984) argues that there exists 'a definite advantage' to being part of a society at the beginning of a research project because this positionality significantly reduces the need for preliminary negotiation that an outsider must conduct to gain access to the community and to work with key informants. However, she cautions that this does not mean that the participant observer has somehow achieved an inside understanding of the community's culture. Related to this form of access is the assertion by Haniff (1985) and Hsiung (1996) that being an insider contributes to the establishment of initial levels of trust and can lead to more open exchanges with informants.

In addition, access relates to the ability of the insider to obtain certain types of information (Jones, 1970). The idea is that an insider would possess a better understanding of how information is gathered, synthesized, stored and disseminated within the community. Insiderness may also contribute clues as to why information is possessed or distributed by some stakeholders but withheld from others. As a member of the academic senate, for example, I have a comprehensive understanding of how information is collected and distributed by and about the senate. I also have relatively easy access to the textual artifacts of the organization. Outsiders, on the other hand, must be very explicit in explaining why they want access to certain documents held in senate files or stored electronically. Access to information can also be facilitated by the fact that insiders, as unobtrusive observers, are more capable of taking advantage of 'privileged eavesdropping' (Burke, 1989).

The literature also describes insiderness as holding special value in terms of being able to interpret the culture of the community. Ohnuki-Tierney (1984) stated, for example, that 'native anthropologists are in a far more advantageous position in understanding the emotive dimensions of behavior' (p. 584), which she argues is difficult for outsiders to interpret. Emotive and sensory dimensions of behavior and their relationship to culture are perhaps the least understood area of anthropological study, according to Ohnuki-Tierney. She cautions, however, that the intensity by which an insider may identify with these emotional dimensions can undermine the ability of the research to find patterns of emotions.

On a related level, insiderness also provides an avenue for interpreting culture-based cues in an interview (Haniff, 1985). Haniff used these cues as a means of placing the respondent at ease in her study of Caribbean women because there was a shared understanding of the normative rules of the community and this implicit understanding contributed to minimizing class differences. Also, the ability to utilize insiderness to create a rapport free from tension, according to Haniff, contributes to the legitimacy of the researcher in the eyes of the informants. Although not mentioned by Haniff, by extension, the interpretation of cultural cues can also relate to other forms of communication that are facilitated by a shared knowledge of normative rules, values and belief systems, such as non-verbal cues and the use of unique terminology.

Finally, several authors describe the value of insiderness based upon an ability of the insider participant observer to utilize their position of insiderness to discover greater clarity of purpose and understanding of one's own work. This position is largely based on the idea that insiderness can inform a deeper understanding and knowledge of the experiences of those being researched. For example, Zinn (1979) found that her insiderness allowed her to formulate hypotheses and the discovery of data that extended beyond traditional frameworks of understanding and that were unique from the experiences of outsiders. In commenting on the study of Italian-American ethnicity, Cerroni-Long (1994) noted achieving a greater understanding of the fact that there is no 'real truth' imbedded in the privilege of insiderness and that the understanding of group dynamics is dependent on the perspective of the researcher, whether an insider or not.

The values of shared experience, greater access, cultural interpretation, and deeper understanding and clarity of thought, are closely tied together and inform one another in a variety of ways. For example, insider access to cultural norms and attitudes supports a means for developing deeper clarity about one's own purpose in studying the group. The advantages claimed by insider researchers do not necessarily inform our understanding of their motivations, however. So what does motivate individuals to become engaged in researching their own social identity groups?

Kanuha (2000) describes three reasons for studying 'my own kind' that represent common themes found in the literature of native anthropologists and feminist scholars. First, the nature of her identity and life experiences as a marginalized lesbian and woman of color motivated Kanuha to seek more meaning about her own social identity. As she puts it, to find 'more knowledge, more analysis, and ultimately more understanding of others whose life experiences were similar to mine' (p. 441). A second motivation for Kanuha was the sense of being constricted within the boundaries of traditional theoretical explanations and conceptual frameworks that rarely spoke to the stigmatized social identity and social experiences of her life. Her scholarly pursuit in seeking a greater understanding represented a journey to discover new modes of analysis concerning the community with which she identified most closely. The third motivation summarized by Kanuha was rooted in the desire to make a contribution to the practice of social work in a way that could lead to more meaningful service to socially stigmatized groups in society, such as gays and people of color.

Kanuha also alludes to two other forms of motivation that are more closely aligned with methodological choices than historical, social and ideological relationships. Kahuna speculates that there may be a motivation derived from 'the construction and meaningfulness of the researcher as subject–object' (p. 441). Although reflexivity is a primary motivational component of both outsider and insider research, the insider occupies a dualistic position as both the object of the study and as the subject of that study. As Kanuha argues in citing the work of Fine (1994), this duality places the insider participant observer, not within the domain of the subject or the object, but at the intersection of both. The other form of motivation from a methodological standpoint alluded to by Kanuha relates to her work with people who rely on narrative traditions, oral communication and cooperative relationships. This is often the primary means of transmitting culture within marginalized groups and, therefore, is most easily accessible by an insider of that group. This motivation is closely related to the value of being an insider to interpret and understand socially constructed cues and patterns of communication.

Unlike Kanuha, my own decision to become an insider participant observer is not grounded in the idea of seeking a greater understanding of my social identity in a marginalized community. For native researchers, seeking social justice and satisfying one's own identity quest (Narayan, 1993) are primary motivations for defining the social purpose of their research as an insider and nurturing a sense of self-actualized legitimacy. Although an abstract similarity exists in terms of my initial research in the sense that our senate is marginalized in terms of its ultimate influence of university-wide policy-making, the implications are certainly not transferable. However, like Kanuha, I did decide on the transition from academic senate member to an insider participant observer because I saw a need to improve my community (the senate) and my practice. I want to stand as an insider at the intersection of being both a subject of the study and as a participant object of that study (Kanuha, 2000: 441), and in this way make a unique contribution to expanding the theoretical and practice-based understandings of shared governance in tertiary education. In the next section, I build on the assumed values of insiderness by outlining some of the challenges related to the introspective negotiations of locating the insider participant observer in the text, maintaining objectivity, rejecting familiarity, and formulating presumptions of truthfulness.

Part III: Negotiating the Insider's Position

In its broadest sense, the struggle to come to terms with the issue of insiderness in participatory observation is directly related to the degree to which the researcher should be illuminated or shadowed in the text (Mullings, 1999). This should not imply that the author is totally absent from the text. The positionality of insiderness commits researcher–participants to showing their place in the setting that they are investigating. This acknowledgment can take many forms and parallels a growing movement in qualitative inquiry to describe the participating self in a symbiotic relationship with the observed. The text, therefore, is intended to reveal the interconnectedness of both (see, for example, Wolf, 1996; Okely and Callaway, 1992; Sparkes, 1995). However, a clear distinction is emerging in the literature concerning the effects of insiderness on the depiction of reality (De Andrade, 2000).

According to Allum (1991), qualitative researchers must come to terms with at least four interrelated, introspective negotiations regarding the position of insiderness. The first introspective negotiation concerns the struggle to properly locate the author within the text. Several factors may determine when, where, and to what degree the insider enters the text. This can include: (1) the capacity of the insider to penetrate into the deepest recesses of the community and to show at what levels access is denied despite the insider status; (2) the ability of the insider to nurture a trustworthy rapport with an informant; (3) the disclosure of any advanced knowledge about the community's culture, its political and power structure, and its normative rules and belief systems; and, (4) acknowledgment of any problems that arise as a result of being an insider. The central issue encompassing these points noted by Allum concerns the crisis of representation in the text and whether or not it is necessary to tell a story about yourself if the research is about someone else. If what Kanuha states is accurate concerning the insider's role as both the instrument of inquiry as well as the subject of that inquiry due to their sharing the same experiential spaces, then the insider ethnographer should be made explicit in the text, as Kondo (1986) argues. However, this is a complex endeavor that is governed in large part by the remaining three introspective negotiations.

The second issue participant observers must negotiate as an insider is the maintenance of objectivity and accuracy. This, of course, rests on the assumption that some manner of objectivity is possible. Even if the author totally rejects the pretense of objectivity, there remain points in a story where accuracy in portraying the community and its membership is important to the researcher. For the outsider, this may involve methods of triangulation, interviewing multiple members of the community, or document analysis and the study of material culture to gain as accurate a portrait of reality as possible. The insider, already existing within the community, but re-entering the setting as an observer, possesses a considerable amount of pre-constructed assumptions and knowledge about the community (Beoku-Betts, 1994; Narayan, 1993). This advanced knowledge can serve as a source of understanding that informs the insider as he or she attempts to develop a moment of objectivity in the text or an accurate depiction of reality. However, the researcher should resist the 'comfortable perquisites' of insiderness (Deutsch, 1981: 190). Possession of advanced knowledge should not lead to a disregard for questioning one's own insider knowledge. Advanced knowledge only provides guideposts for approaching the setting as an insider participant observer. Insiders must negotiate objectivity and accuracy before they enter the setting with the same rigor as outside participant observers (Miller, 1997).

Closely related to the maintenance of objectivity and accuracy is the desire of the participant observer to reject the sense of familiarity that comes with the intimacy of insiderness. The experience of Ohnuki-Tierney (1984) illustrates this point. She returned to Kobe, Japan, to conduct an anthropological study there after living in the United States since 1958, visiting Japan on many occasions, but only to study the Ainu people of Hokkaido. Upon her return to Japan, the 'distancing' that had taken place from her own people in Kobe accentuated a sense of awareness about the 'strange [and] . . . intriguing behavioral patterns and thought processes' of the people there. However, after only a month and a half, the clarity with which she observed these differences began to fade, subsumed by the familiar and of 'becoming more like them' (p. 584). Ohnuki-Tierney noted it was only after she returned to the United States that she was able to regain her perspective and return to Kobe with a renewed focus. The experience of Ohnuki-Tierney perhaps illustrates the point that, while insiderness should be made explicit in the text as both an object and a subject, the process of personal distancing is still required to bring clarity to the research endeavor.

For the insider participant observer, familiarity with the community and its people wraps the person in a consciousness of comfort (Maynard and Purvis, 1994) that hides the opportunity for the mundane and the ordinary to inform the study. Just as it is dangerous to assume an ability to gain an objective and accurate picture of the insider's community by simply 'stepping back' from the phenomenon, the literature also reveals the need for the insider participant observer to actively question familiarity (Bulmer, 1982). This is due, at least in part, to a recognition by a growing number of social scientists that important insights may be revealed by investigating what Brekhus (1998) calls 'unmarked' phenomena – ordinary actions and events that are unnoticed during the course of 'everyday' social interaction.

The fourth introspective negotiation the insider must come to terms with is the issue of constructing and deconstructing presumptions of truthfulness in the text. As with the concepts of objectivity, the achievement of a consensus ethnographic 'truth' recognized by both the observer and the observed may be an impossible goal (Angrosino and Mays de Perez, 2000). Nevertheless, what the insider accepts as reality is likely to be governed to some degree by advanced, intimate knowledge of the setting and, thus, constructed quite differently from the view of reality of the outsider, who is not privileged with advanced knowledge (Beoku-Betts, 1994).

There are complex issues concerning the manner in which insider participant observers should be situated in the text and how insiders represent themselves, their interpretations, and their research settings and informants (Tierney and Lincoln, 1997). The positionality of insiderness demands a set of introspective negotiations that determine the location of the insider in a way that differs from the outsider's location in the text. This is due, in part, to the insider's dual role as both object and subject. The goal of locating the author/insider appropriately in the text intersects with many of the hidden ethical and methodological dilemmas of being an insider participant observer which shape the text in a particular way. The next section outlines some of these dilemmas in more detail.

Part IV: The Hidden Ethical and Methodological Dilemmas of Insiderness

Each of the introspective negotiations that a researcher must address contain hidden dilemmas involving the assumption that insiderness holds a tangible advantage in understanding the community and its members and that this reasoned awareness is unattainable for the outside participant observer. There is a diverse volume of literature devoted to highlighting the challenges of being on the inside. Much of this literature is composed of critical reflections and after-the-fact narratives concerning the personal identity work that comes with prolonged intimate knowledge (Beoku-Betts, 1994; Brewer, 1986; Cerroni-Long, 1994; Haniff, 1985; Kanuha, 2000; Styles, 1979; Zinn, 1979). To gain a clearer understanding of the ethical and methodological dilemmas cited by insider participant observers, I have subdivided the research process into four segments – entering the field, positioning and disclosure, shared and significant relationships, and disengagement from the study. I describe each segment within the context of how they are negotiated by the insider participant observer.

The Hidden Dilemmas of Entering the Field

Entry for the insider participant observer involves negotiating through what Mullings (1999) refers to as 'positional spaces', or 'areas where the situated knowledge of both parties in the interview encounter, engender a level of trust and cooperation' (p. 340). The situatedness of the insider, which in some respects is a journey that first requires the insider to step outside in order to gain a new understanding of the inside ('distancing'), speaks to the possession of previous knowledge about the community, acceptance within, and culturally defined forms of responsibility. Responsibility to one's own community is a particularly vexing issue because any false representations of the phenomena, either real or perceived, could lead to feelings of betrayal on the part of the participants. This is one of the ethical dilemmas I face as I enter into a study of my own faculty senate. In the politically charged arena of shared governance at my institution, where negotiation, bargaining and jockeying for position are commonplace, I must be very explicit in explaining my reasons for engaging in a study of the organization; given the distributive politics of the situation, I must be clear in my research agenda and unambiguous about my long-term purpose.

Beoku-Betts (1994) found that the observer brings multiple identities into the setting and that insiderness only facilitates knowledge about particular lives and events. Yet this knowledge is not contextualized through the experiences of the respondent. Beoku-Betts also found that her African background was an advantage in establishing a rapport with the Gullah women whom she interviewed. However, this rapport also came with the 'special responsibilities' of being an unmarried black woman entering the community. For example, cultural norms dictated that she must consciously avoid associating with the local black men on a regular basis. She points out that researchers who are not black are not likely to be affected by these behavioral expectations, although they may not have access to, or even knowledge of, certain avenues of access that were made available to the author.

The multiple identities alluded to by Beoku-Betts in her study of the African-American communities of the Sea Islands of South Carolina illustrate an important ethical and methodological dilemma related to entering the field. Beoku-Betts appears to ascribe insiderness to the fact that her racial background helped establish a trusted rapport with the black women whom she was studying. This insider relationship appeared to offer Beoku-Betts a greater level of information and access than would usually be granted to an outsider. However, is Beoku-Betts truly an insider or is her insiderness more a reflection of the situational circumstances of the moment and based on the perceptions of her respondents rather than her own place in the community? The challenge is to take advantage of the privileged access of being an insider, while at the same time resisting confinement within a given role as researcher. According to Burke (1989), the unintended result would be working with epistemological tunnel vision, confined to understanding social reality only from the perspective of the ascribed role granted by the respondents.

This position supports Allum's (1991) observations about the hidden dilemmas of entry. Allum argues 'that entry begins long before one physically arrives' (p. 5) and that the relationships forged between the observer and the observed is one of constant negotiation and renegotiation. In other words, there is not one form of entry, but many forms depending on the situatedness of both the observer and the observed. Previous knowledge should not typecast the observer into being in one form of acceptance, but in multiple forms that must fit the particular social setting and identity issues of the respondents. In this sense, the researcher is both an insider and an outsider simultaneously. The insider/outsider boundaries are transitory depending upon the methodological strategies utilized in the study (Griffith, 1998).

Negotiating the hidden dilemmas of entry, therefore, requires a hyper sensitivity to one's own previous knowledge assumptions and the positional space one occupies during entry. Entry does not occur once, but is a process of continual negotiation as new actors enter the field. The negotiated space often is defined by obligations and responsibilities to those being studied that an outsider may never encounter. On the other hand, insiderness can facilitate access to, and knowledge of, key entry points that may be unavailable to the outside participant observer, at least in the beginning of the research endeavor.

The Hidden Dilemmas of Positioning and Disclosure

The responsibilities of Beoku-Betts within the Gullah community, due to her race, gender and marital status, are related to the ways in which participant observers are positioned within a particular social setting by those whom they interview. The insider as outsider status 'requires that one look both from the outside in and from the insider out and to understand both' (Merton, 1972, quoted in Beoku-Betts, 1994: 419). This is an important feature of insiderness, not only because of expected behavioral and social responsibilities within the community, but it is critical to conceptualizing how observers as insiders are positioned by the respondents due to their status. For example, the status of Beoku-Betts as an educated professional and a university researcher proved to be a challenge to establishing trust. Some of the respondents wanted to know, after learning that she was going to write a book about her experiences with the Gullah, if they were going to share in the publication royalties. Beoku-Betts discovered, in particular among the women in her own age group, that they were highly suspicious of her motives and thought she was exploiting their information for financial gain.

Her findings illustrate two critical issues concerning the hidden dilemma of positioning and disclosure. First, insiderness is no guarantee to avoiding unintended and unanticipated positioning by others and ensuring the establishment of trust with respondents. Mullings (1999) highlights this issue in her research of managers and employers in Jamaican information processing companies. Mullings found herself 'caught in the middle of the quiet conflict between' the two groups and, as such, was forced to convince both sides that she was not representing the interests of the other. Any perception of falling within one particular camp would compromise the trust needed to research the other. In establishing trust with each group, Mullings resorted to conducting 'clandestine meetings' with workers outside of the office building to establish interview times (p. 347).

Second, status in any number of forms can exude power and this can be perceived as threatening by certain respondents. The research of Mullings illustrates this point. Hsiung (1996) demonstrated in her research, though, that power relations in the field are more complex than simply the unequal relationship between the researcher and the respondent. She argues that this binary power relationship, common in recent discussions of feminist methodology, is inadequate in understanding the multiple dimensions of power relationships in the field. Hsiung studied small factories in Taiwan composed primarily of no more than 30 female workers and supervised exclusively by men, representing the gender-defined and rigid social structure of Taiwanese society. Hsiung's status was simultaneously an insider and an outsider. She was an outsider in the factory due to her status as a researcher studying in America but, as an indigenous woman, she was also an insider. While Hsiung's insider status gained her access to the women in the factories, her status as a Taiwanese woman subjected her to the prejudices and stereotypes of women by men in Taiwanese society. Hsiung demonstrates that power relationships are complex and multidimensional and often based on the unequal distribution of power in a patriarchal social system.

For Hsiung, the dilemma of positionality and disclosure is complicated in several ways. She was positioned by the fact that she inserted herself into the unequal social relationship between female workers and the male supervisor. Similar to the experience of Mullings, she was careful not to appear to be aligning with either the workers or the supervisors for fear of losing trust with the other. Her status as an indigenous women, however, left her vulnerable to subjugation by the male supervisors. As she stated, 'My goodwill in establishing reciprocal relationships with [the male supervisors] in the field left me in a vulnerable position that was subject to engulfment and potential exploitation' (Hsiung, 1996: 127). Finally, Hsiung was also positioned as an outsider by the female workers in the factory who expressed genuine concern over the fact that she was childless, illustrating how difficult it was for them to step out of the cultural norms and rules of society.

The work of Mullings and Hsiung provide clear evidence that the hidden dilemmas of positionality and disclosure, in addition to complex and multidimensional variables such as gender and the degree of insiderness and outsiderness in relation to the respondent, present challenges that must be carefully negotiated. In the faculty senate, my control over unintended positioning and my location within the power structure of the organization are complicated by my relative youth (inexperience) and the fact that I am a librarian.

The Hidden Dilemmas of Shared and Significant Relationships

Another hidden dilemma related to the issue of unintended positioning, such as the challenge Beoku-Betts faced, owing to her status as an unmarried woman in the Gullah community, is the potential meaningfulness of field-work relationships that are forged between the researcher and the respondent. Amanda Coffey (1999) analyzes this issue in her book, *The Ethnographic Self: Fieldwork and the Representation of Identity*. Coffey notes that the interactional quality of ethnographic inquiry, especially if it is conducted in an atmosphere of duress, can often lead to relationships that are both professional and personal. This is not necessarily an undesirable situation. Coffey argues that the ethnographic investigation is not a straightforward process, but requires 'a level of rapport and sharing' (p. 40). These qualities of rapport and sharing are conducted within a framework of reciprocity and are often negotiated emotionally in the social setting. In other words, participant observers cannot be immunized from their respondents, acting like a detached recording instrument that merely synthesizes the data and disseminates the findings. Both the insider and outsider participant observers are dependent upon social interaction to develop meaningful relationships with informants.

Coffey brings the researcher closer to understanding the roles of the informant as a friend or a colleague, and in some cases, as a family member (Farnsworth, 1996; Miller, 1997; Scheiberg, 1990). However, the perspective Coffey presents is that of an outsider entering the field and negotiating between the potential benefits and challenges that may arise from developing a friendship with an informant. This perspective has been detailed in several ethnographic studies (Crick, 1992; El-Or, 1997; Hendry, 1992). The issue of shared or significant relationships between the observer and the observed is complicated by insiderness in at least two critical ways. First, the insider brings more to the social setting than an understanding of power structures, knowledge of key events, access to informants and other advanced knowledge. The insider as observer also brings previous friendships and collegial relationships into the community as a participant observer. The process of qualitative investigation is likely to require a newly negotiated bond between the observer and informant that ties the two individuals together – one based on the previous friendship and another based on the perspective of 'I'm the researcher and you are one of my informants.' The possible ramifications of damaging this first relationship in pursuit of the second must be clearly understood by the insider.

Second, previous friendship with informants requires an upfront and clearly stated research agenda to avoid any ambiguity regarding the intent of the insider–researcher. For example, Beoku-Betts (1994) found that her association with one informant in particular 'meant that some people were less willing to cooperate than they might otherwise have been for reasons pertaining to personal or local community politics' (p. 424). The previous or negotiated personal relationships one forges with particular individuals are not isolated from the interpersonal dynamics experienced between one's own informant-friend and other members of the community. The friendships that insiders create prior to becoming participant observers are likely to be open for all to see within the community. However, when the transition moves from community member to achieving the role of being a researcher of that community, suspicions of illegal partnerships or hidden purposes can arise, undermining access to others. In my own research of the faculty senate, for example, I have chosen to ameliorate some of the historical mistrust between the senate and the central administrators by making it clear that the primary purpose of my research is to discover ways to improve the decision-making processes of shared governance and to apply this knowledge to my own institution.

While the benefits of forging close relationships between outsider participant observers and respondents have been explored in qualitative research literature and ethnography, the possible influence of previous relationships and friendships between the insider participant observer and informants is underdeveloped. The issue is a critical one, however. The possibility of 'going observationalist' implies negotiating a different relationship between the informant-friend and the participant observer that may threaten the personal bond between the two because the change in status orientation of the insider fundamentally redefines their mutual relationship. In addition, the self-interest of the informant would also imply a need for assessing relationships with others in the community that may lead to an outcome that is not beneficial to the informant.

The Hidden Dilemmas of Disengagement

Maines et al. (1980) state the following about leaving the research setting:

> . . . leaving cannot be seen as an isolated act. Rather, it is a process whose essential character is conditioned by the manner in which researchers enter the field, the nature of social relationships they form while conducting the research, the bargains they make, and the kinds of persons they become to their subject (p. 263).

The decision to leave the community often involves choosing whether to do so abruptly or gradually over time. The process of disengagement also involves addressing ethical questions about the researchers' obligations to their informants after leaving the setting (Zinn, 1979). Although every participant observer must consider the most appropriate manner for leaving the community they are investigating, the issue is not given the same consideration among qualitative researchers as entering the field, negotiating social relations, and other thresholds of ethnographic observation (Allum, 1991: 30).

Snow (1980) observed that an exploration of disengagement in participant observation 'has received scant attention' in the literature. This trend has generally continued, although the lack of attention is particularly pronounced when applied to the process of disengagement and to the positionality and situatedness of the insider participant observer. One possible explanation for this is the assumption that leaving is not relevant to the insider. As opposed to the positionality of the outsider, the nature of insiderness implies that the researcher has never had to 'enter' the setting in the first place, except perhaps in terms of introspective 'distancing' noted by some native anthropologists and feminist researchers (Kanuha, 2000; Ohnuki-Tierney, 1984). The lack of focus on insider disengagement implies that leaving the field is conceptualized more as a form of introspective transitioning from the insider role back to being a community member again, rather than as part of a rigorously negotiated and achieved methodological process. However, this is a false assumption because the idea that insider disengagement is only an introspective act implies that the insider's perceptions of social reality are not transformed by the experience of participant observation, either personally or through the eyes of informants. New knowledge acquired during the study, including that which arises from an examination of the mundane and unmarked of the culture, in all likelihood provides the insider participant observer with a broader sense of the community's hidden complexities that were largely invisible to the individual before assuming the role of a researcher. Informants whom the insider interacted with during the study are likely to continue to harbor residual definitions of the insider not just as a friend, colleague or peer group member, but also as an investigator.

Disengagement from the setting involves careful consideration of responsibilities and obligations to the research subjects after the investigation is completed. Allum (1991) notes that disengagement is made more difficult if false identity or strategic deceit is employed during the social interactions with informants. At issue for many ethnographers is the degree to which the researcher shares the results of the study. Allum noted that, in order to avoid being asked to mediate a conflict or provide advice that would alienate the researcher from those who are in disagreement with the outcome, he was very calculated in his selection of who would read the final report (Bogdan and Taylor, 1984[1975]). This is a particularly challenging issue for insiders because they remain in the community and may suffer possible repercussions from those who may feel they were misrepresented in the study or who challenge the interpretations or conclusions made by the insider.

Based on the work of McCormick (1973) and McCormick and Ramsey (1978), Angrosino and Mays de Perez (2000) offer three criteria for measuring whether or not 'a proper relationship' exists between the value of the research and the other elements of the endeavor: the means used will not cause more harm than necessary to achieve the value; no less harmful way exists at present to protect the value; and the means used to achieve the value will not undermine it. These ethical judgments are appropriate to consider for all forms of participant observation (i.e. insider–outsider, passive–active, etc.). However, the three criteria must be considered within the context of knowing that the methodological nature of insiderness implies that there is no possibility of total disengagement from the community once the study is completed. Practices of strategic deceit, the tactical use of withholding information, and making conscious decisions about limiting who will read about the study's findings can follow the insider participant observer in the community long after an outsider has moved on to other research projects. These are risks that should be negotiated and carefully calculated by the insider participant observer before the study begins.

Part V: Conclusion

The idea of 'going native' is most often attributed to Bronislaw Malinowski (1922). He argued that, in order to gain the most complete understanding of 'the native's point of view', anthropologists must emphasize their role as participants within the social milieu of the culture. For the insider, who is already a 'native', the corollary presumption is that true understanding can only be attained by 'going observationalist'. This concept speaks to the need to distance oneself introspectively from the phenomena so that a more complete understanding of the community's social reality can be obtained. In this article, I have attempted to describe the hidden dilemmas encountered by the insider participant observer related to this conceptualization. The literature shows that insiderness does hold value for the researcher due to possession of previous knowledge that informs multiple points of access to informants and provides a head-start in interpreting the meaning of organization's language and the unsaid and unmarked features of a community's culture and belief systems. However, as Beoku-Betts, Styles, Zinn and many others demonstrate when discussing their experiences as an insider, these advantages are not absolute. They contain hidden dilemmas related to issues of unintended positioning, shared relationships and disclosure, and the often conflicting negotiations concerning the process of 'entering' and disengagement.

On a philosophical level, the conceptualization of insiderness reflects a fragmented and often conflicting domain due, in large part, to a significant number of critical questions that so far have resisted clear answers. Among the questions still open for debate are: Who is more entitled to interpret the 'correct meaning' (Surra and Ridley, 1991) of cultural phenomena? Should the representations of the 'native' be vested in one identifiable group based on the assumption that insiders have a more complete understanding of the culture? Does insiderness reveal a more sophisticated understanding of social realities? At what point could insiderness possibly diminish access to the culture and the trusted rapport with respondents? To what degree is insiderness an ascribed and achieved status, or, as De Andrade (2000) argues, is insiderness neither of these, but rather 'created through an ongoing process of evaluation that is dependent on the performance of the group membership by researchers and participants at multiple levels?' (p. 283). Is there a way to account for a possible derivative Heisenberg effect associated with the transition from community member to insider participant observer? Under what conditions should researchers seek objectivity by distancing themselves from their own privileged insiderness? At what points should insiders seek to camouflage their positionality and to what extent could this affect the outcome of their studies? In short, a common thread tying together literature that reflects the experiences of insider participant observers, also supported by observations made during the preliminary stages of my own research, is the struggle to come to terms with what it means to be an insider and the ability to conduct sound research while an insider.

From a methodological standpoint, a review of insiderness literature over the past 30 years provides some indications that qualitative research is edging closer to delineating the advantages and disadvantages of insiderness, at least in terms of how insiderness applies to issues of gaining access, discerning cultural cues, and obtaining a clarity of purpose for the participant observer. For example, the severe insider–outsider dichotomy first articulated by Merton in the early 1970s was largely rejected and replaced by those who saw insiderness and outsiderness as two endpoints of a continuum (Olson, 1977). More contemporary insider participant observers point to evidence based on their own experiences in the field that a researcher is simultaneously both an insider and outsider as dictated by the circumstances of positionality and the achievement of trusted insiderness in relation to those being studied (Deutsch, 1981).

However, as I confront my own research as an insider participant observer of my university's faculty senate, I see at least two issues that represent underdeveloped areas of analysis. First, there remains a need to develop a more sophisticated understanding of the insider as both object and subject within qualitative inquiry by continuing to move away from locating insiderness and outsiderness on a continuum or within a conceptual matrix. These formulations are inadequate in explaining the location of hidden ethical and methodological dilemmas of 'going observationalist'. From the vantage point of my research, the dilemmas of insider research are hidden because they reside primarily on a Z axis from this perceived continuum between insiderness and outsiderness. This Z axis represents the *degree* to which a researcher has gained insiderness at any given moment in time and space. For example, I would have a high degree of insiderness when I examine a five-member faculty senate task force of which I am also a member, but

I would possess a lesser degree of insiderness in relation to studying the Executive Board of the Senate of which I am not a member. I would have a low degree of insiderness in trying to discover the ways in which the work of the task force affects the course of decision-making within the university administration, since my access to this group would almost parallel that of an outsider. Therefore, similar to what De Andrade (2000) articulates, it is more epistemologically beneficial to view insiderness not in terms of an ascribed or achieved status, which gives the impression that one can reach a final level of understanding, but as a continual process of introspective inquiry that researchers can use to monitor their access to the community. This is closely tied to Cerroni-Long's (1994) idea that no particular position is privileged to see the 'real truth', but that 'social experience and its perception are continuously "created" by the social actors' (p. 135).

A second common theme found in the literature devoted to examining insiderness in qualitative research is the emphasis on the role of insiderness in interpreting the experiences of under-represented groups in native anthropology and feminist literature. There are a number of reasons for this. One explanation is the feeling among 'native' groups that they must extend analysis beyond the pre-determined boundaries of prevailing frames of discourse. They are distanced from society as a whole due to factors such as ethnicity, class, culture and educational attainment (Jones, 1970; Narayan, 1993). A related explanation for the focus in insider research on describing the social realities of under-represented groups is the desire of African-Americans, Native Americans, feminists, and others, to create a representation of their social reality that outsiders cannot penetrate due to their inability 'to transcend their own cultural biases' (Munro, 1994).

The defining characteristics between the suppressed subculture and the dominant culture are usually quite distinct. However, other forms of insiderness can be completely invisible to the other members of a community (Mullings, 1999). This is often the case within the organizational life of professional settings, such as a hospital or university, where the distinctions are less apparent and positionality is defined more by what you do than who you are (Herndon and Kreps, 2001; Symon and Cassell, 1998). Although the recent work edited by Gellner and Hirsch (2001) offers an excellent starting point – in particular the three essays by Chapman, Mascarenhaus-Keyes, and O'Neill – there remain few examples where the perceived privilege of insiderness is applied to research of professional work cultures. I found it particularly difficult in my review of the literature to find a transferable model of insiderness equivalent to a faculty senate, a pseudo-legislative advisory organization in an academic setting. Beoku-Betts (1994) noted that her professional status as a researcher posed 'a major stumbling block in establishing trust' (p. 429), but this discussion was examined only in relation to her being positioned unintendedly as a black woman professional in a community where most people had never attended college. Mullings (1999) examined the relationship between workers and managers in Jamaica, but the prevailing undercurrent of the study related to negotiating class struggles.

The professions (i.e. doctors, accountants, lawyers, teachers, etc.) create distinctive occupational subcultures that interpret reality on the basis of status and work identity. The structural, political and symbolic frames of organizational life (Bolman and Deal, 1997) are discrete from those described in sociological and anthropological studies, especially as they apply to the initial invisibility of the insider participant observer. The examination of insiderness within the professional domain and the ways in which this domain is distinct from, although certainly not mutually exclusive of, the more common forms of insiderness analysis related to race, ethnicity, and gender, for example, has been largely unexplored. The challenge of unmarked insiderness within the professions is an issue in need of further analysis in qualitative inquiry.

References

Acker, J., Barry, K. and Esseveld, J. (1991) 'Objectivity and Truth: Problems in Doing Feminist Research', in Mary Margaret Fonow and Judith A. Cook (eds) *Beyond Methodology: Feminist Scholarship as Lived Research*, pp. 133–53. Indianapolis: Indiana University Press.

Allum, Keith F. (1991) 'On Conducting Educational Research in the Field: The Evolution of an Ethnographic Experience from Passive to Active Participant Observer', paper presented at the Annual Meeting of the American Educational Research Association, Chicago, 5 April (ERIC Document ED334224).

Angrosino, Michael V. and Mays de Perez, Kimberly A. (2000) 'Rethinking Observation: From Method to Context', in Norman K. Denzin and Yvonna S. Lincoln (eds) *Handbook of Qualitative Research*, pp. 673–702. Thousand Oaks, CA: Sage.

Arnold, Fiona (1994) 'Bullying, a Tale of Everyday Life: Reflections on Insider Research', *Educational Action Research* 2: 183–93.

Beoku-Betts, Josephine (1994) 'When Black Is Not Enough: Doing Field Research among Gullah Women', *NWSA Journal* 6 (Fall): 413–33.

Blauner, Robert (1987) 'Problems of Editing "First Person" Sociology', *Qualitative Sociology* 10: 46–64.

Bogdan, Robert C. and Taylor, Steven J. (1984[1975]) *Introduction to Qualitative Research Methods: The Search for Meanings*, 2nd edn. New York: Wiley.

Bolman, Lee G. and Deal, Terrence E. (1997) *Reframing Organizations: Artistry, Choice, and Leadership*, 2nd edn. San Francisco, CA: Jossey-Bass.

Brekhus, W. (1998) 'Sociology of the Unmarked: Redirecting our Focus', *Sociological Theory* 16 (March): 34–51.

Brewer, Rose M. (1986) 'Research for Whom and by Whom: A Reconsideration of the Black Social Scientist as Insider', *Wisconsin Sociologist* 23 (Winter): 19–28.

Bulmer, Martin (1982) 'When Is Disguise Justified? Alternatives to Covert Participant Observation', *Qualitative Sociology* 5 (Summer): 251–64.

Burke, Joan E. (1989) 'Becoming an "Inside-Outsider"', *JASO: Journal of the Anthropological Society of Oxford* 20: 219–27.

Cerroni-Long, E.L. (1994) 'Insider and Outsider Perspectives in the Study of Ethnicity: The Case of the Italian-American', *Ethnic Forum* 14 (Spring): 121–38.

Christensen, Donna Hendrickson and Dahl, Carla M. (1997) 'Rethinking Research Dichotomies', *Family and Consumer Sciences Research Journal* 25 (March): 269–85.
Clark, G. (1998) 'Stylized Facts and Close Dialogue: Methodology in Economic Geography', *Annals of the Association of American Geographers* 88: 73–87.
Coffey, Amanda (1999) *The Ethnographic Self: Fieldwork and the Representation of Identity*. Thousand Oaks, CA: Sage.
Crick, M. (1992) 'Ali and Me: An Essay in Street-Corner Anthropology', in Judith Okely and Helen Callaway (eds) *Anthropology and Autobiography*, pp. 175–92. London: Routledge.
Daly, Kerry (1992) 'Parenthood as Problematic: Insider Interviews with Couples Seeking to Adopt', in Jane F. Gilgun, Kerry Daly and Gerald Handel (eds) *Qualitative Methods in Family Research*, pp. 103–25. Newbury Park, CA: Sage.
De Andrade, Lelia Lomba (2000) 'Negotiating from the Insider: Constructing Racial and Ethnic Identity in Qualitative Research', *Journal of Contemporary Ethnography* 29 (June): 268–90.
Deutsch, Cynthia P. (1981) 'The Behavioral Scientists: Insider and Outsider', *Journal of Social Issues* 37: 172–91.
DeVault, Marjorie L. (1990) 'Talking and Listening from Women's Standpoint: Feminist Strategies for Interviewing and Analysis', *Social Problems* 37 (February): 96–116.
DeVault, Marjorie L. (1996) 'Talking Back to Sociology: Distinctive Contributions of Feminist Methodology', *Annual Review of Sociology* 22: 29–50.
El-Or, Tamar (1997) 'Do You Really Know How They Make Love? The Limits on Intimacy with Ethnographic Informants', in Rosanna Hertz (ed.) *Reflexivity and Voice: Reports and Interpretations*, pp. 169–89. Thousand Oaks, CA: Sage.
England, K. (1994) 'Getting Personal: Reflexivity, Positionality, and Feminist Research', *Professional Geographer* 46: 80–9.
Farnsworth, Elizabeth B. (1996) 'Reflexivity and Qualitative Family Research: Insider's Perspectives in Bereaving the Loss of a Child', *Marriage and Family Review* 24: 399–415.
Fine, Michelle (1994) 'Working the Hyphens: Reinventing Self and Other in Qualitative Research', in Norman K. Denzin and Yvonna S. Lincoln (eds) *Handbook of Qualitative Research*, pp. 70–82. Thousand Oaks, CA: Sage.
Gellner, David N. and Hirsch, Eric (eds) (2001) *Inside Organizations: Anthropologists at Work*. Oxford: Berg.
Griffith, Alison I. (1998) 'Insider/Outsider: Epistemological Privilege and Mothering Work', *Human Studies* 21: 361–76.
Haniff, Nesha Z. (1985) 'Toward a Native Anthropology: Methodological Notes on a Study of Successful Caribbean Women by an Insider', *Anthropology and Humanism Quarterly* 10: 107–13.
Hawkesworth, Mary E. (1989) 'Knowers, Knowing, Known: Feminist Theory and Claims of Truth', *Signs: Journal of Women in Culture and Society* (Spring): 533–57.
Headland, Thomas N., Pike, Kenneth L. and Harris, Marvin (eds) (1990) *Emics and Etics: The Insider/Outsider Debate. Frontiers of Anthropology*, Vol. 7. Newbury Park, CA: Sage.
Hendry, Joy (1992) 'The Paradox of Friendship in the Field: Analysis of a Long-term Anglo-Japanese Relationship', in Judith Okely and Helen Callaway (eds) *Anthropology and Autobiography*, pp. 163–74. London: Routledge.
Herndon, Sandra L. and Kreps, Gary L. (eds) (2001) *Qualitative Research: Applications in Organizational Communication*, 2nd edn. Cresskill, NJ: Hampton Press.
Hsiung, Ping-Chen (1996) 'Between Bosses and Workers: The Dilemmas of a Keen Observer and a Vocal Feminist', in Diane L. Wolf (ed.) *Feminist Dilemmas in Fieldwork*, pp. 122–37. Boulder, CO: Westview Press.

Jones, Delmos J. (1970) 'Toward a Native Anthropology', *Human Organization: A Scientific Quarterly for the Study of Development Change* 29 (Winter): 251–9.

Kanuha, Valli Kalei (2000) '"Being" Native versus "Going Native": Conducting Social Work Research as an Insider', *Social Work* 45 (October): 439–47.

Kauffman, Karen S. (1994) 'The Insider/Outsider Dilemma: Field Experience of a White Researcher "Getting In" a Poor Black Community', *Nursing Research* 43 (May–June): 179–83.

Kondo, Dorinne K. (1986) 'Dissolution and Reconstitution of Self: Implications for Anthropological Epistemology', *Cultural Anthropology* 1: 74–88.

MacDonald, Frederick F. (1990) 'A Case Study of Insider–Outsider Dynamics in Rural Community Development', *Human Services in the Rural Environment* 14 (Summer): 15–20.

Maines, David R., Shaffir, William and Turowetz, Allan (1980) 'Leaving the Field in Ethnographic Research: Reflections on the Entrance–Exit Hypothesis', in William Shaffir, Robert A. Stebbins and Allan Turowetz (eds) *Fieldwork Experience: Qualitative Approaches to Social Research*, pp. 261–81. New York: St Martin's Press.

Malinowski, Bronislaw (1922) *Argonauts of the Western Pacific: An Account of Native Enterprise and Adventure in the Archipelagoes of Melanesian New Guinea*. New York: Holt, Rinehart & Winston.

Maynard, M. and Purvis, J. (eds) (1994) *Researching Women's Lives from a Feminist Perspective*. London: Taylor and Francis.

McCormick, Richard A. (1973) *Ambiguity and Moral Choice*. Milwaukee, WI: Marquette University Press.

McCormick, Richard A. and Ramsey, Paul (1978) *Doing Evil to Achieve Good*. Chicago: Loyola University Press.

Merton, Robert (1972) 'Insiders and Outsiders: A Chapter in the Sociology of Knowledge', *American Journal of Sociology* 78 (July): 9–47.

Miller, Kim (1997) 'All in the Family: Family Folklore, Objectivity, and Self-Censorship', *Western Folklore* 56 (Summer): 331–46.

Mullings, Beverly (1999) 'Insider or Outsider, Both or Neither: Some Dilemmas of Interviewing in a Cross-Cultural Setting', *GeoForum* 30: 337–50.

Munro, Doug (1994) 'Who "Owns" Pacific History? Reflections on the Insider/Outsider Dichotomy', *The Journal of Pacific History* 29 (December): 232–7.

Naples, Nancy A. (1996) 'A Feminist Revisiting of the Insider/Outsider Debate: The "Outsider Phenomenon" in Rural Iowa', *Qualitative Sociology* 19: 83–106.

Narayan, Kirin (1993) 'How Native is the "Native" Anthropologist?', *American Anthropologist* 95 (1993): 671–86.

Oakley, Ann (1981) 'Interviewing Women: A Contradiction in Terms', in Helen Roberts (ed.) *Doing Feminist Research*, pp. 30–61. London: Routledge & Kegan Paul.

Ohnuki-Tierney, Emiko (1984) '"Native" Anthropologists', *American Ethnologist* 11: 584–6.

Okely, Judith and Callaway, Helen (eds) (1992) *Anthropology and Autobiography*. London: Routledge.

Olson, David H. (1977) 'Insiders' and Outsiders' Views of Relationships: Research Studies', in George Levinger and Harold L. Raush (eds) *Close Relationships: Perspectives on the Meaning of Intimacy*, pp. 115–35. Amhurst: University of Massachusetts Press.

Parades, Americo (1977) 'On Ethnographic Work Among Minority Groups: A Folklorist's Perspective', *New Scholar* 6: 1–53.

Reay, Diane (1995) 'Feminist Research: The Fallacy of Easy Access', *Women's Studies International Forum* 18 (March–April): 205–13.

Reay, Diane (1996) 'Insider Perspectives or Stealing the Words out of Women's Mouths: Interpretation in the Research Process', *Feminist Review* 53 (Summer): 57–73.

Riessman, Catherine K. (1987) 'When Gender Is Not Enough: Women Interviewing Women', *Gender and Society* 1 (June): 172–207.
Rose, G. (1997) 'Situated Knowledge: Positionality Reflexivities and Other Tactics', *Progress in Human Geography* 21: 305–20.
Scheiberg, Susan L. (1990) 'A Folklorist in the Family: On the Process of Fieldwork Among Intimates', *Western Folklore* 49: 208–14.
Sherman, Sharon (1986) 'That's How the Seder Looks: A Fieldwork Account of Videotaping Family Folklore', *Journal of Folklore Research* 23: 53–70.
Smith, Dorothy (1987) *The Everyday World as Problematic: A Feminist Sociology*. Boston, MA: Northeastern University Press.
Snow, David A. (1980) 'The Disengagement Process: A Neglected Problem in Participant Observation Research', *Qualitative Sociology* 3 (Summer): 100–22.
Sparkes, Andrew C. (1995) 'Writing People: Reflections on the Dual Crisis of Representation and Legitimation in Qualitative Inquiry', *Quest* 47 (March): 158–95.
Styles, Joseph (1979) 'Outsider/Insider: Researching Gay Baths', *Urban Life: A Journal of Ethnographic Research* 8 (July): 135–52.
Surra, Catherine A. and Ridley, Carl A. (1991) 'Multiple Perspectives on Interaction: Participants, Peers, and Observers', in Barbara M. Montgomery and Steve Duck (eds) *Studying Interpersonal Interaction*, pp. 25–55. New York: Guilford Press.
Symon, Gillian and Cassell, Catherine (eds) (1998) *Qualitative Methods and Analysis in Organizational Research: A Practical Guide*. Thousand Oaks, CA: Sage.
Thorne, Sally E. and Paterson, Barbara L. (2000) 'Two Decades of Insider Research: What We Know and Don't Know About Chronic Illness Experience', *Annual Review of Nursing Research* 18: 3–25.
Tierney, William G. and Lincoln, Yvonna S. (eds) (1997) *Representation and the Text: Reframing the Narrative Voice*. Albany: State University of New York Press.
Wilson, William J. (1974) 'The New Black Sociology: Reflections on the "Insiders" and "Outsiders" Controversy', in James E. Blackwell and Morris Janowitz (eds) *Black Sociologists: Historical and Contemporary Perspectives*, pp. 322–38. Chicago: University of Chicago Press.
Wolf, Diane L. (ed.) (1996) *Feminist Dilemmas in Fieldwork*. Boulder, CO: Westview Press.
Yocum, Margaret S. (1980) 'Fieldwork in Family Folklore and Oral History: A Study in Methodology', PhD dissertation, University of Massachusetts.
Zavella, Patricia (1993) 'Feminist Insider Dilemmas: Constructing Ethnic Identity with "Chicana"', *Frontiers: A Journal of Women's Studies* 13: 53–79.
Zinn, Maxine Baca (1979) 'Field Research in Minority Communities: Ethical, Methodological and Political Observations by an Insider', *Social Problems* 27 (December): 209–19.

30

Informed Consent, Anticipatory Regulation and Ethnographic Practice

Elizabeth Murphy and Robert Dingwall

Introduction

This paper discusses the impact of the concept of informed consent, adopted in ethical review by most anticipatory regulatory regimes, on qualitative research in sociology, with particular reference to ethnography. Drawing upon our own experiences of such research over the last 35 years, we consider the relevance of these regimes to the ethical dilemmas which, in practice, confront ethnographers. We also consider the threat of such bureaucratic regulation to ethnography's contribution to improving the efficiency, effectiveness, equity and humanity of health services.

Ethnography typically involves researchers spending extended periods of time (sometimes a year or longer) in one or more settings (for example, a hospital ward, a general practice surgery, a laboratory or an out-patient clinic) observing what goes on, talking to members of the setting, collecting documents and, on occasion, interviewing. The researcher's objective is to recover the 'situated rationality of action' – the ways in which, in context, people's actions make sense, even when they seem, to others outside the situation, to be inappropriate or counter-productive.

Studies of this kind have a long and distinguished history in sociological research on health and health care. Classic examples include Goffman (1961) on mental hospitals, Glaser and Strauss (1965) on death and dying, Becker, Geer, Hughes, and Strauss (1961) on the professional socialisation of medical

Source: *Social Science & Medicine*, 65(11) (2007): 2223–2234.

students and Davis (1963) on the experiences of polio victims and their families. More recently ethnographies of hospital wards (Allen, 1997), responses to medical error (Bosk, 1979), paediatric clinics (Strong, 1979), genetic counselling in a children's hospital (Bosk, 1992), care in private nursing homes (Diamond, 1992), paediatric intensive care (Anspach, 1992), emergency rooms (Timmermans, 1999), the construction of disease (Mol, 2002) and medical examiners (Timmermans, 2006) have all made important contributions to our understanding of health care organization.

Ethnographies have a key role to play in creating a more efficient, more effective, more equitable and more humane health care system, particularly in illuminating the organizational and interactional processes through which health care is delivered. They offer important information, to policy makers and practitioners, about factors that compromise or promote high-quality care, particularly the ways in which well-intentioned actions may have unanticipated negative consequences.

The elaborate, bureaucratised, systems of ethical review currently operating in the US, Canada, UK and elsewhere threaten the survival of ethnographic research with little gain in protecting research participants. These anticipatory regulatory regimes were developed for governance of clinical and biomedical research, primarily in response to abusive experimentation in Nazi Germany revealed in the Nuremberg Trials (Annas & Grodin, 1992; Hazelgrove, 2002; Lifton, 1986; Schmidt, 2004; Weindling, 2004). The subsequent uncovering of highly questionable clinical and biomedical research practices between the 1920s and 1950s, in both the US and the UK, strengthened the case for the regulatory regimes that began to develop from the mid-1960s (Beecher, 1959, 1966; Hazelgrove, 2002; Jones, 1981; Papworth, 1967; Rothman, 1993). However, these regimes are ill-suited to assessing ethnographic study proposals (Murphy & Dingwall, 2003). This is neither because ethnographic research is free from ethical challenges, nor because such challenges are unusually difficult or intransigent. It is, rather, because anticipatory regulatory regimes are based on assumptions derived from the model of clinical trials or biomedical experimentation, with prior specification of hypotheses, design, instruments and implementation in protocols that are finalised before the study begins. In principle, this model is readily transferable to survey research, where it is possible to specify, in advance, exactly what a study will involve. This is not true of ethnographic research. In the rest of this paper, we explore the ways in which ethnographic research in sociology departs from the assumptions underlying these regimes before concluding by considering the implications for designing appropriate regulation.

We should, though, first be clear that we have no quarrel with the notion that research participation should be free from coercion. The concept of informed consent is deeply embedded in liberal individualist assumptions about the virtue of autonomy and the priority of individual over community rights (D'Agostino, 1998; Wolpe, 1998). However, respect for persons, as operationalised through the practice of informed consent, is only one of the three principles articulated in *The Belmont Report* (National Commission for the Protection of Human Subjects of Biomedical and Behavioral Research, 1979) which underpins the subsequent development of Institutional Review Boards (US), Research Ethics Boards (Canada) and Ethics Committees (UK). The other principles, of justice and beneficence, are equally relevant and may, in some instances, compete with the principle of autonomy (Wolpe, 1998). Nevertheless, our critique in this paper is neither of the *principle* of autonomy, nor of its *operationalisation* in the requirement that research should only be carried out with consenting participants. Our challenge is to the identification of the principle with a particular set of *bureaucratic practices* more suited to biomedical experimentation.

Ethnographers do not universally share our commitment to openness in research. Some reject the principle of informed consent altogether, defending covert methods and deliberately misleading people about their research purposes (Douglas, 1976). Others defend deceit in terms of the value of the research findings. For example, Diamond argues that the value of his exposé of the exploitation of nursing assistants and the erosion of patient autonomy, privacy and dignity in three Chicago nursing homes justifies any wrong done to unwitting participants in his 'undercover' research study (Diamond, 1992). If his employers had found out that he was doing research, he would have been dismissed and unable to complete the study. Given the choice between abandoning his research and misleading staff and patients, he chose the latter. The compromise of autonomy was, in his view, outweighed by the beneficence of the findings for residents and workers.

Diamond is, however, in the minority among contemporary ethnographers. Most, including ourselves, interpret the principle of respect for persons as requiring us to conduct research as overtly as possible consistent with avoiding distress or disruption to the settings we study. While deception may be 'shaded in many different ways' (Haggerty, 2004) and, in research as in everyday life, the boundary between truth and deceit is often blurred, we are wary of the argument that the ends justify the means. Such judgements often depend too heavily on individual researchers' estimation of the value of ends in which they have a heavy personal investment. We also believe that the difficulties of obtaining valid data using overt methods in even very challenging settings can be exaggerated. There are numerous examples of ethnographers being granted access to highly sensitive settings including the UK night-time economy (Monaghan, 2004), the British Civil Service (Heclo & Wildavsky, 1974), the Mafia (Ianni & Ianni, 1972), professional 'fences' Klockars (1974), professional criminals (Polsky, 1971) and drug barons (Adler, 1985) (see also the discussions in Lee-Treweek and Linkogle (2000)). Moreover, covert ethnographers may underestimate harm caused to those they study, if and when they discover they have been misled about the researcher's real interest (Bulmer, 1980). Subjects of covert research may be harmed as well as wronged.

Our argument is not that informed consent is trivial or irrelevant. It is, rather, that informed consent in ethnography is neither achievable nor demonstrable in the terms set by anticipatory regulatory regimes that take clinical research or biomedical experimentation as their paradigm cases. This is because of fundamental differences in the practices of ethnographic and experimental research that we now examine.

Extended Periods of Involvement in Research Settings

In ethnography, researchers spend extended periods of time, often a year or more, in their research setting. The practice of obtaining prior informed consent was developed in relation to the discrete episodic interventions typical of clinical trials or biomedical experimentation. Such interventions lend themselves to the legalistic, contractual approach embodied in obtaining signed consent from research subjects prior to the start of research. In ethnography, consent is more likely to be negotiated and renegotiated over time as the relationship between the ethnographer and the research hosts develops (Adler & Adler, 2002). Although ethnographers have traditionally described the people involved in their research as participants or informants, we prefer the terminology of 'hosts' here because it better captures the nature of the relationship: ethnographers are guests in someone else's setting and, like guests everywhere, there are clear, if not always explicitly articulated, expectations of proper behaviour. The status of guests is always fragile and depends on appropriate conduct towards their hosts, although this is not necessarily reciprocated.

Typically, at the start of such research, consent is both tentative and limited and the researcher's access to sensitive aspects of the setting may be restricted. Over time, as the trust between researcher and hosts develops, access may be granted to previously restricted areas or interactions. However, such access is rarely guaranteed and depends upon the ethnographers' capacity for sustaining the goodwill and co-operation of their hosts. Such cooperation can be, and sometimes is, withdrawn. At times, this may lead to the researcher having to leave the setting completely. In Dingwall and Murray's (1983) work on child protection, for example, fieldwork in the A&E department of a major university hospital was terminated when the junior doctors realised that Dingwall had observed their, officially deviant, informal re-organization of their shifts to allow more of them to attend the hospital ball for at least some part of the evening. More often, the researcher finds access to particular aspects of the setting or its activities is restricted, as when, for example, the researcher is refused access to particularly sensitive meetings or areas within the research setting. In his Ph.D. study of the training of health visitors (public health nurses), for example, Dingwall (1974, 1977) was unable to observe one-to-one tutorial meetings between the course staff and students, which clearly played an important part in the way that staff searched for, and generated evidence of, student competence, because of the tutors' concern for the students' privacy in receipt of criticism.

Ethnographic consent is a relational and sequential process rather than a contractual agreement and lasts throughout the period of research (Katz & Fox, 2004). It is based on trust between researcher and researched and is a matter over which research hosts exercise ongoing judgement. The relationship between the ethnographer and their hosts is more akin to that of a patient with a general practitioner than with a surgeon. The patient gives tentative and limited consent before contact with a general practitioner. Over time, as the relationship develops, the patient may agree to or solicit further contact, accept or reject advice as it is offered and, from time to time, agree to specific interventions. Mostly, consent is implied rather than stated and operates within a context where the patient may withdraw cooperation at any time. Unlike the anaesthetised patient, the general practice patient surrenders relatively little autonomy. The same is true of most ethnographic research, where successful execution depends upon establishing and maintaining a long-term relationship of trust with research hosts (Bosk, 2004; Katz & Fox, 2004).

Of course ethnographers do face ethical dilemmas around consent. Indeed, some dilemmas are heightened by the extended periods that researchers spend in the field and may be more nuanced and subtle than in the case of the discrete, episodic interventions that characterise biomedicine (Corrigan, 2003). It is in the nature of ethnography that researchers seek to minimise their own impact on the settings they study. This reflects a concern, on the one hand, to reduce any disturbance or inconvenience to hosts and, on the other, to avoid distorting the setting and hence rendering findings unrepresentative. As researchers become integrated into the setting, they form relationships with hosts. Over time, these hosts may come to overlook the research purpose and the researcher's identity *qua* researcher begins, for the hosts at least, to fade into the background. The line between researcher and friend or confidante becomes blurred and hosts may disclose information that they do not recognise as relevant to the research but which the researcher considers to be so. This raises questions about the extent to which the hosts' prior consent justifies the use of such disclosures as data. In Dingwall's (1974) Ph.D. work, for example, he was invited to parties thrown by members of the student group. Observation at these events was an important source of data on the lateral life roles of the students, particularly a number of conversations with women in their late 20s and early 30s about their prospects of marriage or whether they would end up settling for a career in nursing. The study was strongly influenced by the US work of Olesen and Whittaker (1968) on baccalaureate student nurses and of Psathas (1968) on diploma students, where this tension had been identified as a significant issue for understanding student commitment to their courses and intended occupation and these were clearly relevant data. However, these were 'off-duty' occasions where some tongues were undoubtedly loosened by alcohol. In the end, Dingwall decided not to use specific quotes, except for one instance where a student used a line from a popular song to describe her feelings about the impact of a number of unsuccessful relationships. However, he did make general references to these observations and the sorts of conversations that occurred between him and the students, which seemed important evidence for his discussion of the interactions between personal and professional lives.

Emergent Research Focus and Design

Since Nuremberg, there has been general agreement between ethicists, regulators and scientists that research participation must not only be voluntary but also based on adequate knowledge and understanding of the nature, duration, purpose, methods and potential hazards of the study. This has, in practice, been translated into the requirement that the research participant give written consent, at the outset, based on a clear specification of the research. Clearly, it is not straightforward to determine what is to count as adequate knowledge in either biomedical or social research. In both, the research is often embedded in highly technical and theoretical concepts that may be opaque to potential participants. There are practical difficulties in making explanations accessible to participants without distorting the true nature of the research or, as Brewster Smith (1979) observed, 'sending informants and cohabitants to graduate school' (p. 14). Accounts that both biomedical and ethnographic researchers give to potential research participants are inevitably partial without necessarily reflecting any desire to deceive. Such accounts can never represent the simple, unmediated truth. Of necessity, they will be designed for the audiences for which they are intended.[1] Otherwise, they might be true but incomprehensible. The adequacy of such accounts is always a matter of judgement and debate.

However, ethnographic research confronts an additional difficulty. In biomedical research, it is normally possible to specify in advance precisely what the study will involve. In ethnography, both the research focus and the research design typically emerge during the course of the research (O'Neill, 2002). Consequently, the researcher is rarely able to give a full account at the outset of what the research will involve, simply because she or he may not yet know. Flexibility of research design, and capacity to respond to insights emerging from early empirical work, are distinctive contributions of ethnography, but complicate the process of obtaining prior informed consent. For example, Murphy is currently researching the transition from children's to adult services for young people with moderate to severe intellectual disabilities. Gaining initial consent to carry out this longitudinal research was an elaborate process involving the young people, their parents, social workers and health care providers as well as managers and supervisors. While the team was able to negotiate initial consent to the study with each of these groups, it was not always obvious from the outset exactly what it would be useful to observe or whom it would be appropriate to interview. Indeed, in some cases, it was only once the research was under way that they were able to identify the key meetings, facilities and personnel. Moreover, these changed over the period of the fieldwork as new policies and systems were introduced and new appointments made. This necessitated frequent negotiations and renegotiations, which cannot be accommodated within a model that relies on fixing consent in writing at the outset.

Analysis of ethnographic data is an iterative process where, typically, the researcher sets out with a broad question to be investigated and the precise lines of analysis emerge as the data accumulate. Again, this can be difficult to reconcile with the requirement that participants are fully informed about the purpose of the research before it commences. Some authors (e.g. Richards & Schwartz, 2002) suggest returning to participants at the end of the study to seek consent to use findings once the shape of the analysis is clear. This approach risks undermining the credibility and impartiality of ethnographic findings. There is always the possibility that honest inquiry will produce uncomfortable findings. To grant participants the power of veto invites them to suppress findings they find unpalatable.

We believe that what is required is caution when explaining the research at the outset. For example, in a study of mothers' experiences of feeding their babies and young children, Murphy described the purpose of the study as 'improving our understanding of the choices which mothers make about what, when and how to feed their babies'. This inclusive formulation gave her team freedom to develop lines of analysis that had not been fully anticipated at the outset (Murphy, 1999, 2000, 2003, 2004, 2007). Similarly, Dingwall's child protection study (Dingwall, Eekelaar, & Murray, 1983) was described as 'an attempt to understand the workings of a system that was widely considered at the time to be problematic so that its members and stakeholders could consider how it might be improved'. This second example highlights the importance of not being drawn into promising what we are unlikely to be able to produce. Not unreasonably, research participants frequently base their decision about whether to participate in research on an evaluation of the likely benefits to themselves or others. There is often a temptation to over-claim the potential contribution of a piece of research to solving participants' current problems. In this case, the researchers avoided suggesting that the findings would prescribe improvements in the child protection system. They made it clear that the contribution of the research would be to illuminate how the current system works as a resource that policy-makers and practitioners might use to identify possible reforms. Thus, they maintained the boundary between the function of the researcher and that of the policy-maker and clarified what participants could reasonably expect.

Nature, Degree and Positioning of Risk in Ethnographic Research

Giving potential participants 'information on all known foreseeable risks' (Corrigan, 2003) is central to the operationalisation of informed consent. The application of this principle to ethnographic research is more complex than in biomedical experimentation, where it is normally possible to specify and quantify risks in advance with reasonable precision. While ethnographic research is not risk-free, for either hosts or researchers (Lee-Treweek & Linkogle, 2000; Punch, 1994; Renzetti & Lee, 1993), potential harm to hosts is usually different in nature, degree and positioning from that faced by the subjects of biomedical experiments (Lincoln & Tierney, 2004). Most commonly it involves psychological or social harm. Participants may be upset, worried or offended during the fieldwork. For example, the close relationships which develop during ethnographic research sometimes lead hosts to share information they find distressing or embarrassing to talk about, raising concerns about the possible re-awakening of memories and consequent re-victimisation of those who have experienced difficult or traumatic circumstances. Like most ethnographers, we have experienced such interactions in research, as in our everyday lives. Sometimes such disclosures appear to have been therapeutic. At others, they may have been experienced, in retrospect, as embarrassing or regrettable. However, it is unclear how any conceivable application of the principle of informed consent could protect against such harm. In all cases, they appear to have chosen freely to share their troubles with us even though they had not, we assume, anticipated doing so when they first agreed to our research. Bureaucratic practices are largely irrelevant to such situations. The ethical challenge lies in ensuring that, in such circumstances, hosts are treated with decency, discretion and respect, and have access to whatever support they may subsequently need.

Mostly, however, the point of greatest risk occurs when research findings are published (Murphy & Dingwall, 2001). The reputations and relationships of hosts may be disturbed or damaged by publication. Hosts may be wounded by what is written about them, sometimes in ways that surprise well-meaning researchers (Ellis, 1995). In a few extreme cases, publication may put participants at risk of legal sanction, which is particularly problematic because, in many jurisdictions, researchers have no legal protection from the obligation to disclose their sources (Haggerty, 2004). However careful and respectful researchers are in their reports, they have minimal control over how these are taken up and transformed by others, including tabloid journalists. Murphy had the experience of seeing the observation that a significant proportion of parents reported giving their babies token tastes of alcohol, for example by allowing them to suck a finger dipped in the Christening champagne, sensationalised under headlines such as 'Tiddly tots' and 'Alcoholic babies' (Murphy, Parker, & Phipps (1998)).

The risks that exist in ethnographic research are highly indeterminate, subjective and not always easy to specify and communicate in advance to hosts (Bosk, 2004). In practice, there is often real difficulty in persuading research participants to engage meaningfully with such hypothetical risks.[2] Their pre-conceptions influence their interpretations of whatever information the researcher offers. In biomedical experimentation, it is usually possible to offer some actuarial assessment of the probability that risks will occur, but this is not the case in ethnographic research. Many potential hazards mirror those normally encountered in everyday life, and for which formal consent is not normally considered necessary (American Association of University Professors, 2000; Haggerty, 2004). As Haggerty argues, there is a danger that the unpredictability and indeterminacy of risk may lead to ethics committees applying the precautionary principle in ways that make ethnographic research impossible, without any gain in protection for participants. In such cases, as Alberti (2000) suggests, fulfilling one function of ethics committees (protecting participants from harm) is achieved at the expense of the other (encouraging research that will improve health care). Nicholl (2000) argues that when regulatory mechanisms become a barrier to research that is ethical, this is itself unethical.

In practice, ethnographers are most likely to identify risks to their hosts only in the course of carrying out the project. Guillemin and Gillam (2004) observe that it is in the day-to-day practice of research that 'ethically important moments' occur and ethnographers are faced with the decision about how they should respond. In such circumstances, hosts are likely to be served less well by procedural ethics embodied in regulations requiring fully informed advance consent than by researchers' conscientious commitment to avoiding exploitative and potentially harmful relationships with those they study.

Power in Researcher/Participant Relationships

Regulatory practices around biomedical research were born out of concerns about the possible exploitation of power differentials between researcher and researched. To a large extent, the resulting regulations model the researcher–researched relationship on that believed to prevail between doctors and patients. The research participant is taken to be passive, vulnerable and in need of protection and the researcher is understood to be powerful and capable of exploiting dependent patients. Written informed consent in advance is designed to mitigate this unequal power relationship and protect the subjects' rights during experimental interventions where researcher control is intrinsic to the research design.

This understanding of an asymmetrical relationship between a powerful researcher and a vulnerable research subject may be appropriate in relation to experimental interventions, where research subjects submit to the researcher by surrendering their right to self-determination for the duration of the experiment (but see Morris & Balmer, 2006). Its relevance to ethnographic research, where power is located differently, is less clear-cut. As Haggerty (2004) observes, while the imagined subjects of social science research are the 'poor and marginalised' (p. 409) who require special protection, this is not always the case. Ethnographers often study powerful people in publicly funded settings where informed consent could be interpreted as a self-serving requirement directed at protecting the *status quo* from interrogation and powerful members from scrutiny. Research hosts are not invariably the vulnerable and passive individuals imagined in many ethics texts. There are real questions about whether power brings with it an obligation to contribute to improving knowledge, particularly in relation to the organisation and management of public services. Ethnography is, in many instances, one aspect of public accountability for the use of public resources or the impact of private interests on the public sphere.

Even where research hosts have less overt power than those discussed above, it would be highly unusual for the researcher to be able to exercise the kind of control that characterises experimenter-subject relations. In ethnography, researchers typically adopt a passive rather than an active role. They are highly dependent upon the willingness of hosts to act and speak within their sight and hearing. Continuing data collection depends crucially upon on-going cooperation from hosts and hosts who are uneasy about the research can obstruct or even terminate it in numerous ways. Negotiating consent in ethnography is continuous and constantly vulnerable to withdrawal. Participants have considerable capacity for controlling researchers leading, in extreme cases, to ethnographers abandoning the research altogether. Dingwall (2001), for example, recently sought to write a case study of a major scientific scandal as a problem in organisational deviance and its management. Although much documentation was in the public domain, it soon became clear that the major research university where many of the protagonists worked was unlikely to co-operate, placing an insuperable barrier in the way of funding the project. While the results would contribute to debates about research governance, and means for implementing this without stifling innovation, the university's leaders were more concerned with resolving their reputational problems and avoiding further publicity, whether positive or negative, that might draw attention to their experiences.

Public versus Private Settings

A further uncertainty concerns the application of the requirement for informed consent to public as opposed to private settings. Some ethnographies are conducted in settings which are open to the public and where there is no expectation that presence or participation requires prior negotiation. Such settings are typically highly complex and mobile and this makes obtaining written, or even oral, informed consent from all who pass through impractical. Sometimes strict application of the requirement that all those are the subject of research must give individual, fully informed consent would mean that the ethnographer was so occupied in negotiating consent that she would have no time to carry out the research and would destroy the setting itself as an object of study. A number of authors (Adler & Adler, 2002; Haggerty, 2004; van den Hoonaard, 2002) report how, in such circumstances, ethical review boards' preoccupation with obtaining informed consent has made conducting otherwise unexceptionable studies impossible.

Such restrictions threaten to extinguish research in public spaces, with consequent loss of important data about social interaction and practices. It is unclear why researchers' observations of behaviour in public spaces should be controlled in this way or how such restrictions protect those observed. People in public spaces knowingly make their behaviour available for scrutiny by anonymous others – British cities are full of CCTV systems continuously scrutinized by a variety of law enforcers, for example. In that sense, people's consent to being observed is implied by their simple presence. Why should ethnographers be subject to more stringent requirements in this respect than other citizens or surveillance agencies? That said, even public settings have private moments and researchers have an obligation to treat such moments with respect and decency, which may well require them to adopt a self-denying ordinance when faced with what might otherwise be relevant data. Once again, this requires the kind of ethical sensitivity and situational judgement that cannot be enforced by anticipatory regulatory regimes.

However, many settings ethnographers study are semi-public rather than public and, it may be unclear from whom consent is required. Whereas biomedical and interview research focus on individuals, ethnographies typically involve extended observations, for example in hospital wards or outpatient departments. The population is often highly transient and it is not always obvious who should be classed as research participants. If, for example, the focus of the research is on nurse behaviour on a particular ward, one clearly needs to seek consent from the nurses. But what about all the other people who are based on or pass through that ward? How far does that extend: to doctors? to cleaners? to patients? to maintenance staff who make a one-off visit to the ward? During extended periods of observation, ethnographers will have many casual encounters with people who pass through. Once again, if the setting is not to be disrupted, and the research is not to be impractical, one must distinguish between those for whom the research is likely to be consequential and those who are tangential. Often the risk of harm is so minimal that it is not clear whose interests obtaining consent actually serves. Elsewhere, Dingwall (1980) has distinguished between 'principals' and 'spear-carriers'. In the child protection study, for example, consent was not sought from families involved in the candidate cases that were tracked through the network of agencies in the main study site. For this study, they were 'spear carriers' because the project's focus was on how cases were co-produced by professionals and agencies. Parents' accounts were irrelevant to this and they were not interviewed as a result of a deliberate decision not to generate ironic data. Agency decisions were based on what staff deemed to constitute evidence and their mutual assessments of its adequacy for justifying various courses of action. These data were embodied in official records, internal and inter-agency review meetings between professionals and court hearings. The people involved in these were the 'principals' whose consent was carefully negotiated. Clearly, one could envisage a project where these roles are reversed, where parents' stories are the object of interest and child protection professionals are the spear carriers. In such a case, the negotiation of consent would necessarily be inverted.

Discussion

Anticipatory regulatory regimes that transfer models of consent from clinical or biomedical research to ethnography are highly problematic. Approaches currently adopted in the UK and other developed countries are erecting insuperable barriers to ethnographic work or creating such disincentives that scholars are choosing not to do it. This presents us with a stark choice – a choice that hinges upon whether or not we believe that ethnography has anything to offer to the creation of a better society, or, at least, to the more efficient, effective, equitable and humane use of resources in the one that we have. Do we accept the disappearance of ethnographic work or do we revisit the models of regulation that have been imposed on the social sciences?

We happen to believe that ethnography has a vital role in examining the extent to which institutions do what they claim to do, fairly and civilly and with an appropriate mobilisation of resources. It is the gold standard for the study of processes, a study that is as important for policymakers and practitioners as the study of outcomes (Murphy & Dingwall, 2003). If this is accepted, then a regulatory regime that is inimical to such research will replace systematic knowledge with systematic ignorance and undermine the democratic accountability of major institutions operating in or impinging on the public sphere. Neither seems a particularly ethical consequence of ethical regulation. In effect, autonomy once again trumps both justice and beneficence as well-intentioned actions bring about negative outcomes.

There are two further considerations. The first is a reasonable assessment of the actual risks involved in observational research. Faced with the need to estimate, in advance, potential harm, researchers and regulators find themselves anticipating hypothetical worst case scenarios with little attention to the empirical evidence of the likelihood of the worst case actually occurring. In reality the risks to participants of observational research do not begin to approach the risks inherent in most forms of clinical or biomedical experimentation.

The second is that, if academic researchers are no longer able to conduct observational research, this does not mean that such investigations will not take place. They will simply be done by our rivals as cultural commentators (Strong, 1983), other writers and journalists, working with different kinds of motive and subject to different ethical principles – relatively free of regulatory constraints. Although journalists can be sued after the event, prior restraint is difficult to achieve. There are many examples of such investigations – and not only by tabloid journalists. Serious writers and investigative journalists undertake what amounts to covert research, justified by appeals to the public interest on the one hand and to the freedom of the press on the other (e.g. Ehrenreich, 2002; Orwell, 2003; Toynbee, 2003). Regulation will not stop observational work, but simply ensure that it is done by people outside its jurisdiction and unconstrained by the standards of scientific rigour expected of academic researchers. A few academics may use their own life experiences in the form of auto-ethnographic accounts, but these are notoriously unsystematic and unrepresentative, although they may usefully identify potential areas for further work ("Locus Congressi", 2006). Policymakers, professionals and citizens will be forced to rely upon less robust findings as a basis for improving the public sphere, since rigorous research will have been forced out by the perverse effects of anticipatory ethical regulation.

Ethical conduct of ethnographic research ultimately depends upon the personal integrity and ethical education of the researcher. It is simply not possible to write an unambiguous set of rules for conducting such research or fully to evaluate proposals in an anticipatory process. During extended fieldwork, unanticipated ethical dilemmas will arise and require immediate resolution. Increasing bureaucratisation of such research through what Haggerty has called 'ethics creep' risks undermining the moral and ethical responsibility of the very researchers it seeks to control. When ethics becomes institutionalised, rule-following replaces a commitment to working out the 'right thing to do' as researchers negotiate the complex moral territory of fieldwork. The danger is that researchers will come to see satisfying the demands of ethics committees as the be all and end all of ethical practice.

None of this is to suggest that we can simply ignore ethics or leave it to the consciences of individual researchers. The good reputation of social scientists is critical to the institutional legitimacy and public standing of our disciplines: unethical behaviour by one member of our community may compromise society's support for us all. Given the ever-increasing institutional pressures on individual researchers to increase productivity, the temptations to ride rough-shod over ethical practice are very real. Some kind of regulation is both inevitable and appropriate. However, the current system of anticipatory, bureaucratic regulation is not an effective means of ensuring ethical research practice in ethnography.

It is time to reclaim research ethics from the bureaucrats. We need to develop and strengthen 'professional' models of regulation – albeit in an environment that is, at least temporarily, hostile to professions. Such models would emphasise education, training and mutual accountability. They would place more weight on the promotion of sound ethical behaviour than upon pre-empting unacceptable behaviour through anticipatory bureaucracy. Such professional models might include, for example, a stronger emphasis on ethical discussion in basic research training. We might consider a system of accrediting or licensing principal investigators and research student supervisors based on the completion of training in ethics and subsequent evidence of continued ethical engagement and awareness. Accredited individuals could be exempted from external review except in cases where there was significant risk of harm to research participants or a proposed project raised particularly contentious issues.

There are a number of potential strategies for accomplishing this. One would place institutional ethics committees at the centre of the accreditation process. Freed from the excessive workload arising from the current bureaucratic model of prospective consideration of all projects, ethics committees might also find time to monitor research practice in institutions, possibly through a 'dip-stick' approach to reviewing projects that are currently underway or recently completed. They could also consider complaints from those who believe that they have been the butt of unethical research practice. Ultimately, ethics committees could have the power to suspend the accreditation of investigators and supervisors who are deemed to have engaged in unethical research practices.

However, given their record of bureaucratic excess, it may be better to explore more traditional models that would place accreditation in the hands of the relevant professional communities. Ferlie and Geraghty (2005) have noted the failure of confrontational regulation to achieve bureaucratic dominance over professionals, resulting as it does in implementation deficits and, often, the delegitimation of the regulators in the face of public trust in the professionals. Nevertheless, the problem of accountability for expertise remains. In the light of experience, they argue that traditional forms of peer review and self-regulation may have more to contribute than has been acknowledged in recent years. What may be required is a process of re-professionalization, where the first tier of regulation influencing skills and values remains under the control of the expert group, even if there is a certain amount of external oversight.

Journal editors and those examining doctoral theses could also have an enhanced role in promoting sound ethical practice. As editors ourselves, we are concerned by an increased reliance among referees on authors' reports that the project has been granted ethical clearance by the relevant ethics committee. This appears to be at the expense of the referees' own judgement about the acceptability of the research practices which generated the findings reported. A similar degrading of the assessment of the ethics underpinning doctoral submissions, by the substitution of ethics committee approval for a professional judgment about the acceptability of the research practice reported in the thesis, can also be observed. Reasserting the professional responsibilities of reviewers and examiners could be expected to increase the awareness and sensitivity of researchers to ethical dimensions of their practice by highlighting their mutual accountability to their peers.

There is, of course, no reason why any one strategy for regulation should necessarily have a monopoly. Researchers might choose to seek accreditation through disciplinary associations where these have a strong identity, as in psychology, or to opt for an institutional licence, particularly if they are working in a new or interdisciplinary field. To some extent, competition between accrediting bodies would serve to keep over-reaching in check, since a licensing regime that was too burdensome would simply lose clients to one taking a more liberal approach.

Current approaches to the governance of ethnographic research are simply not fit for the purposes for which they are intended. We have identified a number of characteristics of ethnography that make anticipatory regulation inappropriate. This is not a problem that can be solved simply by tinkering with the current approach to such research by ethics committees. It requires a root and branch rethinking of ethical regulation, and a reassertion of the continuing value of professional self-regulation and mutual accountability.

Notes

1. The situation is further complicated by the multiple audiences for such accounts. Such audiences sometimes include, for example, lawyers who review consent documentation on behalf of universities or publishers to identify any possible grounds for legal action (Adler & Adler, 2002). As a result, there may be a bias against plain English and in favour of legal dialects, which may be as impenetrable as scientific ones.
2. This is a difficulty that has also been reported in relation to biomedical research (Morris & Balmer, 2006).

References

Adler, P. (1985). *Wheeling and dealing: An ethnography of an upper-level drug dealing and smuggling community*. New York: Columbia University Press.

Adler, P. A., & Adler, P. (2002). Do university lawyers and the police define research values? In W. C. van den Hoonaard (Ed.), *Walking the tightrope: Ethical issues for qualitative researchers* (pp. 34–42). Toronto: University of Toronto Press.

Alberti, G. (2000). Multicentre research ethics committees: Has the cure been worse than the disease? *British Medical Journal, 3201*, 1157–1158.

Allen, D. (1997). The nursing-medical boundary: A negotiated order? *Sociology of Health and Illness, 19*(4), 498–520.

American Association of University Professors (2000). Protecting Human Beings: Institutional Review Boards and Social Science Research. <http://www.aaup.org/statements/Redbook/repirb.htm>. Accessed 17 February 2006.

Annas, G. J., & Grodin, M. A. (1992). *Nazi doctors and the Nuremberg code: Human rights in human experiments.* Oxford: Oxford University Press.

Anspach, R. R. (1992). *Deciding who lives: Fateful choices in the intensive care nursery.* Berkeley: University of California Press.

Becker, H. S., Geer, B., Hughes, E. C., & Strauss, A. L. (1961). *Boys in white.* Chicago: University of Chicago Press.

Beecher, H. (1959). Experimentation in man. *Journal of the American Medical Association, 169*, 461–478.

Beecher, H. (1966). Ethics and clinical research. *New England Journal of Medicine, 24*, 1345–1360.

Bosk, C. L. (1979). *Forgive and remember: Managing medical failure*. Chicago: University of Chicago Press.

Bosk, C. L. (1992). *All God's mistakes: Genetic counseling in a pediatric hospital.* Chicago: University of Chicago Press.

Bosk, C. (2004). The ethnographer and the IRB: Comment on Kevin D. Haggerty, "Ethics creep: governing social science research in the name of ethics. *Qualitative Sociology, 27*(4), 417–420.

Brewster Smith, M. (1979). Some perspectives on ethical/political issues in social science research. In M. Wax, & J. Cassell (Eds.), *Federal regulations: Ethical issues and social research* (pp. 11–22). Boulder. Colorado: Westview Press.

Bulmer, M. (1980). Comment on the ethics of covert methods. *British Journal of Sociology, 31*(1), 59–65.

Corrigan, O. (2003). Empty ethics: The problem with informed consent. *Sociology of Health and Illness, 25*(3), 768–792.

D'Agostino, F. (1998). Two conceptions of autonomy. *Economy and Society, 27*(1), 28–49.

Davis, F. (1963). *Passage through crisis: Polio victims and their families.* Indianapolis: Bobbs Merrill.

Diamond, T. (1992). *Making gray gold.* Chicago: University of Chicago Press.

Dingwall, R. (1974). *The Social Organisation of Health Visitor Training.* Ph.D. thesis, University of Aberdeen.

Dingwall, R. (1977). *The Social Organisation of Health Visitor Training*. London: Croom Helm.

Dingwall, R. (1980). Ethics and ethnography. *Sociological Review, 28*(4), 871–891.

Dingwall, R. (2001). Scientific misconduct as organisational deviance. *Zeitschrift für Rechtssoziologie, 22*(2), 245–258.

Dingwall, R., Eekelaar, J., & Murray, T. (1983). *The protection of children: State intervention and family life*. Oxford: Blackwell.

Dingwall, R., & Murray, T. (1983). Categorisation in accident departments: "Good" patients, "bad" patients and "children". *Sociology of Health and Illness, 5*(2), 127–148.

Douglas, J. D. (1976). *Investigative social research: Individual and team field research.* Beverly Hills: Sage.

Ehrenreich, B. (2002). *Nickel and dimed: Undercover in low-wage America*. London: Granta Books.

Ellis, C. (1995). Emotional and ethical quagmires in returning to the field. *Journal of Contemporary Ethnography, 24*(1), 68–98.

Ferlie, E., & Geraghty, K. J. (2005). Professionals in public service organizations: implications for public sector 'reforming'. In E. Ferlie, L. E. Lynn Jr., & C. Pollitt (Eds.), *The Oxford handbook of public management* (pp. 422–445). Oxford: Oxford University Press.

Glaser, B. G., & Strauss, A. L. (1965). *Awareness of dying*. Chicago: Aldine.

Goffman, E. (1961). *Asylums*. New York: Doubleday.

Guillemin, M., & Gillam, L. (2004). Ethics, reflexivity and "ethically important moments" in research. *Qualitative Inquiry, 10*(2), 261–280.

Haggerty, K. D. (2004). Ethics creep: Governing social science research in the name of ethics. *Qualitative Sociology, 27*(4), 391–414.

Hazelgrove, J. (2002). The old faith and the new science: The Nuremberg code and human experimentation ethics in Britain 1946–73. *Social History of Medicine, 15*, 109–135.

Heclo, H., & Wildavsky. (1974). *The private government of public money: Community and policy inside British politics*. London: Macmillan.

Ianni, F., & Ianni, E. (1972). *A family business: Kinship and social control in organised crime*. London: Routledge.

Jones, J. H. (1981). *Bad blood: The tuskegee syphilis experiment*. New York: Free Press.

Katz, A. M., & Fox, K. (2004). *The process of informed consent. What's at stake*. Boston: Children's Hospital Boston.

Klockars, C. B. (1974). *The professional fence*. New York: Free Press.

Lee-Treweek, G., & Linkogle, S. (2000). *Danger in the field: Risks and ethics in social research*. London: Routledge.

Lifton, R. J. (1986). *The Nazi doctors: Medical killing and the psychology of genocide*. New York: Basic Books.

Lincoln, Y. S., & Tierney, W. G. (2004). Qualitative research and institutional review boards. *Qualitative Inquiry, 10*(2), 219–234.

"Locus Congressi". (2006). Perspective: "Four days in a strange place". *Journal of Health Services Research and Policy, 11*(2), 122–124.

Mol, A.-M. (2002). *The body multiple: Ontology in medical practice*. Durham, NC: Duke University Press.

Monaghan, L. F. (2004). Doorwork and legal risk: Observations from an embodied ethnography. *Social and Legal Studies, 13*(4), 453–480.

Morris, N., & Balmer, B. (2006). Volunteer human subjects' understandings of their participation in a biomedical research experiment. *Social Science & Medicine, 62*, 998–1008.

Murphy, E. (2007). Images of childhood in mothers' accounts of contemporary childrearing. *Childhood, 14*(1), 105–127.

Murphy, E. A. (1999). 'Breast is Best': Infant feeding and maternal deviance. *Sociology of Health and Illness, 21*(2), 187–208.

Murphy, E. A. (2000). Risk, responsibility and rhetoric in infant feeding. *Journal of Contemporary Ethnography, 29*(3), 291–325.

Murphy, E. (2003). Expertise and forms of knowledge in the government of families. *The Sociological Review, 51*(4), 433–462.

Murphy, E. (2004). Anticipatory accounts. *Symbolic Interaction, 27*(2), 129–154.

Murphy, E., & Dingwall, R. (2001). The ethics of ethnography. In P. Atkinson, A. Coffey, S. Delamont, J. Lofland, & L. Lofland (Eds.), *Handbook of ethnography* (pp. 339–351). London: Sage.

Murphy, E., & Dingwall, R. (2003). *Qualitative methods and health policy research*. New York: Aldine de Gruyter.
Murphy, E. A., Parker, S., & Phipps, C. (1998). Competing agendas in infant feeding. *British Food Journal, 100*(3), 128–132.
National Commission for the Protection of Human Subjects of Biomedical and Behavioral Research (1979). *The Belmont report: Ethical principles and guidelines for the protection of human subjects of research*, US Department of Health, Education and Welfare.
Nicholl, J. (2000). The ethics of research ethics committees. *British Medical Journal, 3201*, 217.
O'Neill, P. (2002). Good intentions and awkward outcomes: Ethical gatekeeping in field research. In W. C. van den Hoonaard (Ed.), *Walking the tightrope: Ethical issues for qualitative researchers* (pp. 17–25). Toronto: University of Toronto Press.
Olesen, V. L., & Whittaker, E. W. (1968). *The silent dialogue: A study in the social psychology of professional socialization*. San Francisco: Jossey Bass.
Orwell, G. (2003). *Down and out in London and Paris*. London: Penguin.
Papworth, M. H. (1967). *Human guinea pigs*. London: Routledge and Kegan Paul.
Polsky, N. (1971). *Hustlers, beats and others*. Harmondsworth: Penguin.
Psathas, G. (1968). *The student nurse in the Diploma School of Nursing*. New York: Springer Publishing Company.
Punch, M. (1994). Politics and ethics in qualitative research. In N. Denzin, & Y. Lincoln (Eds.), *Handbook of qualitative research* (pp. 83–97). Thousand Oaks: Sage.
Renzetti, C. M., & Lee, R. M. (Eds.) (1993). *Researching sensitive topics*. Newbury Park CA: Sage.
Richards, H. M., & Schwartz, L. J. (2002). Ethics of qualitative research: Are there special issues for health services research? *Family Practice, 19*, 135–139.
Rothman, D. J. (1993). *Strangers at the bedside: A history of how law and bioethics transformed medical decision making*. New York: Basic Books.
Schmidt, U. (2004). *Justice at Nuremburg: Leo Alexander and the Nazi doctors' trial*. London: Palgrave.
Strong, P. M. (1979). *The ceremonial order of the clinic*. London: Routledge and Kegan Paul.
Strong, P. M. (1983). The rivals: An essay on the sociological trades. In R. Dingwall, & P. Lewis (Eds.), *The sociology of the professions: Lawyers, doctors and others* (pp. 59–77). London: Macmillan.
Timmermans, S. (1999). *Sudden death and the myth of CPR*. Philadelphia: Temple University Press.
Timmermans, S. (2006). *Postmortem: How medical examiners explain suspicious deaths*. Chicago: University of Chicago Press.
Toynbee, P. (2003). *Hard work: Life in low-pay Britain*. London: Bloomsbury.
Van den Hoonaard, W. C. (2002). Introduction: Ethical norming in qualitative research. In W. C. van den Hoonaard (Ed.), *Walking the tightrope: Ethical issues for qualitative researchers* (pp. 3–16). Toronto: University of Toronto Press.
Weindling, P. (2004). *Nazi medicine and the Nuremberg trials: From medical war crimes to informed consent*. London: Palgrave.
Wolpe, P. R. (1998). The triumph of autonomy in American bioethics. In R. DeVries, & J. Subedi (Eds.), *Bioethics and society: Constructing the ethical enterprise*. New Jersey: Prentice-Hall.

31

The Art and Politics of Covert Research: Doing 'Situated Ethics' in the Field

David Calvey

Introduction

This article discusses a covert participant observational study of door supervisors, or 'bouncers' as they are more traditionally called, in the night-time economy (NTE) of Manchester, England. The primary fieldwork was conducted over six months between January and June 1996 with types of secondary data being collected about media reactions and post-fieldwork management of the bouncer role until the present day.

This article is organized into six broad sections. The first section is concerned with outlining the traditional case against covert research, including the governance and regulation of research ethics. The second section critiques informed consent, followed by an examination of the submerged tradition of covert research. The next two sections, based on my case study, explore the 'covert self' and 'situated ethics' in the field. The concluding section proposes a creative and imaginative role for covert research.

The Opposition to Covert Research

The guidance that professional associations and bodies[1] offer their members amounts to a moral and methodological 'frowning upon' covert research. This 'standard position' or 'professional consensus' is strongly tied up with

Source: *Sociology*, 42(5) (2008): 905–918.

the principle of informed consent and, related to this, the public image of the discipline and the legitimacy of its knowledge.

Homan (1980, 1991), who is regularly quoted in the literature, lists thirteen 'objections' to covert research, including flouting the principle of informed consent; the erosion of personal liberty; betraying trust; pollution of the research environment; producing a negative reputation of social research; discrimination against the defenceless and powerless; damage to the behaviour or interests of subjects; covert methods may become habitual in the everyday life of the person doing the research; the habit of deception may spread to other spheres of human interaction; covert methods are invisibly reactive; covert methods are seldom necessary; covert methods have the effect of confining the scope of research; and that the covert researcher suffers excessive strain in maintaining cover.

Some of the objections raised by Homan (1980, 1991) and the discussion by Bulmer (1982) on the merits of covert research are very useful descriptions of the conditions of doing covert research and are certainly important considerations to be aware of. However, for me, they do not amount to a sustained rejection of covert research.

Within what I call the 'standard view' on covert research, when there is recognition that informed consent is not always achievable in its absolute form, there is often, a call for retrospective explicit debriefing of subjects, so that at least they know that research has been taking place even if they did not know it beforehand. This is an idealized view of the research setting and, again, contributes to covert work being fettered and stifled. The most famous case, which is regularly raised as a clear example of 'bad ethics' in the social sciences, is Laud Humphreys' Tearoom Trade (1970). This has received much academic criticism, mainly on the grounds of deception. Alternatively, Mitchell (1993) feels that 'secrecy' is an integral and unacknowledged part of robust fieldwork. More recently, the place of covert research in criminological research has been debated. Interestingly, Wells (2004) argues that covert research should be discussed in terms of research necessity and quality of data, rather than the emotive debates about morality and ethics which have traditionally framed it.

The opposition to covert research is articulated further in the discourse on the governance and regulation of research ethics. In a statement on ethical practice, the British Sociological Association (BSA) state:

> . . .covert methods violate the principles of informed consent and may invade the privacy of those being studied. Participant or non-participant observation in non-public spaces or experimental manipulation of research participants without their knowledge should be resorted to only where it is impossible to use other methods to obtain essential data. (2002: 6)

The ESRC has demonstrated an increased concern about the regulation of ethical conduct in their development of the Research Ethics Framework (REF). It states:

> Covert research may be undertaken when it may provide unique forms of evidence or where overt observation might alter the phenomenon being studied. The broad principle should be that covert research must not be undertaken lightly or routinely. It is only justified if important issues are being addressed and if matters of social significance which cannot be uncovered in other ways are likely to be discovered. (ESRC, 2005: 21)

Although the ESRC is supporting some grounds for covert research by asking for additional justification, it is my contention that covert research is effectively stigmatized in the research world.

In 2005 Max Travers, in a short statement within the newsletter of the British Sociological Association, makes some very useful observations about these developments:

> The rise of ethical review should be understood as the latest in a series of legislative and institutional measures in which the state has restricted academic freedom and exerted greater control over social scientific research. . . It will also result in a massive amount of new administrative work, and a whole industry around these institutions. Justified on the entirely spurious grounds that sociological research harms society unless this is properly regulated. (Travers, 2005: 20)

The Critique of the Doctrine of Informed Consent: The Blurred Reality of Fieldwork

One of the issues here is what I refer to as the 'consent to what' problem, in that social research is often contingent and all probabilities cannot be covered by the consent form. Indeed, some research involves a semi-covert element when there is no obvious source of consent, or when some but not all the participants are informed (Burgess, 1985). Concerns have also been raised about the ethicality of ethics committees (Hammersley, 2006) and the ritualism around completing consent forms (Sin, 2005). Richardson and McMullan (2007) argue that overly rigid ethics committees can be counterproductive, although their aim is to suggest procedures to ensure the highest ethical standards for sociology by exploring lessons learnt from the NHS research ethics process. Part of the problem is the continued adoption of inappropriate medical models of doing social research, yet we still continue to turn to that mentality in terms of research governance (Israel, 2004).

The general problem I have is that there is a sort of codification about the qualitative research journey, which is not always open to risk assessment, review and evaluation. There has been a rise in the bureaucratization around ethics, field research and risk taking and attempts at regulation, most commonly through University Research Ethics Committees, which, if wrongly managed and centre driven from above, can deter, fetter and discourage creative covert research. What these ethics codes articulate is ultimately a species of both protectionism and privileging.

Research is a situated business and not open to rationalistic planning. It is in the particular cases of the here and now with participants that ethics are situationally accomplished. Let me stress that I do not disagree with the concept of meaningful consent (Williams, 2006) in that covert research would definitely not be appropriate for certain settings or groups. It is neither an 'anything goes' nor a 'one size fits all' policy but what is appropriate in that setting. Lugosi (2006), in his ethnographic study of commercial hospitality, argues that concealment and disclosure are negotiated throughout the fieldwork period. He argues that it is the relationships with specific informants that determine overtness or covertness in the research. For him, concealment is sometimes necessary and often unavoidable.

In the research setting, even when doing overt research, you can find yourself in a blurred situation with regard to confidentiality and consent. Howard Parker's (1974) ambivalent position of receiving stolen goods from the boys and his selective publication policy mirror my own feelings, with what I eventually published from the fieldwork data. Punch (1986) comments that, in large organizations, it is difficult to get the full and informed consent of everybody involved. Indeed, Punch (1986) stresses that trying to gain absolute informed consent can ironically end up terminating some research projects. What the researcher is trying to capture by investigative methods is often the unofficial view of an organization – what Shulman (1994) describes as 'dirty data'.

What consent forms cover the remit to observe and document all social behaviour and conversations? Anderson and Bissell (2004) have drawn attention to the fact that overt and covert research is a moral continuum, where the boundaries can become blurred in the doing. Bourgois offers some nuanced reflections on doing fieldwork:

> . . . we are taught in our courses preparatory to fieldwork that the gifted researcher must break the boundaries between outsider and insider. We are supposed to 'build rapport' and develop such a level of trust and acceptance in our host societies that we do not distort social interaction. Anything less leads to the collection of skewed or superficial data. How can we reconcile effective participant/observation with truly informed consent? Is rapport building a covert way of saying 'encourage people to forget that you are constantly observing them and registering everything they are saying and doing?' (Bourgois, 2007: 296–7)

Engagement with the ethics of research is not a ritualistic tick box process that once done at the beginning of the project can then be obviated, but runs throughout the lifetime of a project. Although such ethical reflexivity is more pronounced in covert research it applies clearly to overt research contexts as well. A proportion of covert practices are routinely glossed over in sanitized overt accounts. Moreover, it is difficult to plan, or indeed cleanse, research so that ethical dilemmas don't arise. In certain research contexts it is difficult to maintain a strict either/or division between overt and covert hence the messy reality is more akin to a continuum. As Punch states:

> . . .The semi-conscious tactics of the field – eavesdropping, fudging over one's purpose, simulating friendship, surreptitiously reading documents, etc. – make for good data but bad consciences. (Punch, 1986: 73)

The Submerged Tradition of Covert Research

This is clearly not an exhaustive list but a range of work exists from different fields and disciplines that have used covert methods, either deliberately or more emergently, to study a given area. These include Festinger et al.'s (1956) study of an apocalyptic religious cult, Lofland and Lejeune's (1960) study of Alcoholics Anonymous, Rosenhahn's (1973) work on mental health hospitals, Parker's (1974) work on juvenile gangs, Wallis's (1976) study of scientology, Ditton's (1977) work on fiddling and pilfering by bread salesmen, Homan's (1978) work on pentecostalists, Fielding's (1982) work on extreme right-wing organizations, Holdaway's (1983) study of the police, Burgess's (1985) work on school culture, Thompson's (1988) work on the Ku Klux Klan, Taylor's (1991) work on mental health wards, Fountain's (1993) study of cannabis dealers, Goode's (1996) use of bogus personal ads to investigate courtship, Hobbs et al. (2003), Monaghan (2002) and Sanders' (2005) work on bouncers, Scheper-Hughes's (2004) study of organ-traffickers, Lugosi's (2006) work on commercial hospitality and Shulman's (2007) work on lying in the workplace.

Partly, what is required is to both recognize and treat this covert tradition in a dedicated manner and recover it from the index status that it has in much of the methodology literature. Despite the growing literature on danger and risk in fieldwork (Bloor et al., 2007; Lee, 1993, 1995, 2000; Lee-Treweek and Linkogle, 2000) and a general move to more reflexive auto ethnography, the covert research experience is often subsumed in ethics chapters and mentioned more momentarily in methodological appendices and disregarded footnotes.

If we look at the methodology of some of the classics, covert research has radically shaped sociology and criminology, if in unintended ways. Goffman undertook a covert ethnographic study of social life in mental hospitals – *Asylums* (1961) – in order to investigate how the inmates were being treated. He justified this approach by claiming that the findings of his research, whilst he was employed as a physical therapist's assistant, would contribute to the well-being of the inmates. Ultimately, Goffman's study increased our knowledge about the process of institutionalization. Garfinkel's public breaching experiments, which informed his development of ethnomethodology as regards the normative order of ordinary life, were done without seeking informed consent and had a radical influence on generations of sociologists. Melville Dalton was employed as an administrative assistant, wherein he collected data on the informal culture of organizational life in *Men Who Manage* (1959), and thereby influenced many in occupational sociology. Many of the Chicago School studies were done under a broad banner of participant observation, but included elements of covert research. The strong critics of covert research seem to have amnesia over such studies. In more contemporary times, the development of club studies and recreational drug cultures has been driven by accounts seeking insider knowledge, which often employ covert dimensions (Measham and Moore, 2006).

The 'Covert Self': Going Bouncer and Nomadic Ethnography

My ethnography was grounded in the 'lived experiences' of the participants and the commitment was to doing an ethnography, which attempts to provide thick (Geertz, 1973) and faithful (Bittner, 1973) descriptions of the natural and routine setting of door work by engagement with those studied. The context of the study was the night-time economy and in particular the world of bouncers and bouncing, to use the traditional term. There is an emergent literature on bouncers (Calvey, 2000; Hobbs et al., 2003; Monaghan, 2002; Sanders, 2005; Winlow, 2001; Winlow and Hall, 2006). In the UK, the bouncer trade is being professionalized by various agencies which are distancing it from the shadow economy and links to criminality. This project was a very small-scale one, which was partly self-funded and had some departmental assistance. The autonomy of the project didn't produce any problems of bounded censorship or overly policy-driven enquiry, although it did result in some press attention, which compounded the management of the post-fieldwork biography.

The benefits of this covert study were to contribute in some way to a more nuanced understanding of the still under-researched night-time economy and, in particular, the associated theories about the commodification of violence. For many of the commentators in this embryonic field, the standard view is that the world of bouncers is one that is saturated in violence and often equated with forms of masculinity and cultural capital. For me, violence certainly is a typical feature of the door environment and forms part of their 'dirty work'. However, I also feel, from my covert study, that the role and use of violence is overplayed in certain accounts of the work of bouncers. Violence is often an exaggerated part of 'war stories' told amongst bouncers about their work world as part of their occupational worth. Thus, what is appealed to is an exotic account, with the more mundane and routine features of door work being rather submerged. For me, like the other authors in this field, bouncers play the role of a much maligned private police force. The continual drive is to resist the temptation to paint a picture that fits with one's theoretical framework and becomes a vehicle to tell certain academic stories about society. The goal is about getting a proportionate and authentic picture of that cultural world, which I feel the covert aspect of the study has helped me to do, although my version is one among several.

My interests in studying this area grew out of being a regular consumer of pubs and clubs for many years, where bouncers were classic figures of both fear and fascination. I had trained in martial arts for several years prior to undertaking the study, which gave me the hardness passport (Patrick, 1973) for entry. This was a type of embodied ethnography, displaying the relevant body and cultural capital (Monaghan, 2002; Wacquant, 1995; Winlow, 2001) and interaction rituals (Goffman, 1967) to pass (Garfinkel, 1967) as a doorman. It was the art and politics of 'faking friendship' (Duncombe and Jessop, 2002) as I constructed a door career. After I had completed a local city council door staff training course I had made contact with a door agency that supplied door staff to pubs and clubs in Manchester city centre.

It was the very strong sense of a door community that produced difficult dynamics for me in the post-fieldwork management of self, as many of the door people I had briefly encountered identified me as a doorman for some time afterwards and regularly offered privileges in the shape of free entry to clubs and pubs in the city. After all, to them, I had merely finished doing their door but not necessarily other doors. Consequently, I would regularly be asked which door I was working on. For me, the research was not only about getting on and staying on the door but also staying out. My 'bouncer's story' of who I was still had to be sustained well after the study had finished and I was often, unexpectedly, called upon to 'get back into character', if you will, as I bumped into former bouncers. The usual reply would be that I was in 'early retirement' or 'I couldn't stand the pace any more'. As I was no longer just a punter, but an ex-bouncer with a certain status, I had to distance myself strategically from the field. My partner commented to me, during the fieldwork, 'when you put that jacket on, your bouncer head comes on'. This summed up the duality of the role that I was performing.

This type of embedded ethnography was emotionally demanding. In this sense, it was a form of what Blackman (2007) refers to as 'hidden ethnography', wherein emotional relations develop between the researcher and the researched, but are left unexplained and often glossed over. It was a very open-ended research project with no finishing date. The experience and management of the post-fieldwork self is not new but can be a typical feature of covert work, which is marginalized in most methodological accounts, mainly because it is not a conventional sort of problem.

The particular nomadic role that I adopted arose partly from my comparative desire to understand different door modes but also as a sort of ethnographic risk management. As I developed rapport and closer bonds in doing door work, the covert role became more of an emotional risk in that I felt a sort of 'guilt clause' in the sustained deception that I was involved in. In short, the paradox was getting close to them without them getting close to me. It was definitely a type of edgework (Lyng, 2005) or what I described as 'sub-aqua ethnography'. In this role, I constantly engineered appropriate exits such as pay or personality conflicts with management, as I moved around various doors. In the event that my cover was blown I had rehearsed certain stories but I was not sure if they would be accepted. Merely stating that I was an academic would offer little or no protection. When asked where I lived I would never give precise locations but had standard responses like 'not far away', 'local' or, as a last resort, the 'city centre'. Also, if I gave a phone number, it would be a mobile one and never my home one. Thus, communication was kept on a first name basis or nickname with payment being done on a cash basis.

I covered 10 door sites, including pubs, clubs and café bars, which exhibit both their door order and status hierarchy. In this sense, I constructed a 'door career' as part of my fieldwork credibility strategy as I was often asked where I had worked and with whom. So when asked if I knew the local 'heads' or people who were 'connected' to criminal fraternities, or at least had aspirations to be, I lied and emphatically said 'Yes, of course' on any occasion that I was asked. This line of work is a type of emotional labour which, although often temporary, develops shifting relationships with door people. Door work typically features a lot of bonding and comradeship as articulated in the practitioner's literature (Thompson, 1994, 1996; Twemlow, 1980).

A journalist who interviewed me after the study stated that when I first sat down he thought I was a bouncer who had become an academic and not the reverse. I take his reaction as a sign of successful 'disguised observation' (Denzin, 1968).

Situated Ethics in the Field

Ethical codes, and various ethnographic accounts, offer a sanitized picture of social research. They offer little or at best limited nuanced understanding of the emotional, biographical and shifting character of fieldwork where ethical decisions are occasioned practices. I am not seeking to abolish professional codes nor add extensive regulations to them. Clearly, they are case contingent.

The move I wish to make is to view ethics as contingent, dynamic, temporal, occasioned and situated affairs. Thus, particularly in my covert case, one is involved in a web of shifting and mixed connections, tactics, identities and motives (Lovatt and Purkis, 1996), which is deeply biographical (Roberts, 2000) and brings a new dimension to 'taking a side' (Becker, 1967).

I wanted to resist what I considered to be 'analytic exotica' throughout my project. Hence, I abstained from value judgements about the participants or, put another way, was indifferent to them. The management of situated ethics is not only about adopting a theoretically reflexive attitude but also about a whole series of practical manoeuvres and tactics. Jean Rafferty, an investigative journalist doing a covert study of a homeless hostel in Glasgow, stresses:

> The main problem with the use of subterfuge, it seems to me, is not its morality but the practicality of carrying it out. Sustaining a role – that is remembering the details of a fabricated life – is not easy. (2004, 128)

In many ways, it is more of a 'blurred self' that is managed in the covert field setting. Donal MacIntyre, a popular television journalist, has some useful comments to make on his covert role:

> The essence of the technique is getting people to tell their stories in their own words, as they would to one of their own. But this can only happen if they believe I am who I say I am. . . .There is no blueprint, no Scotland Yard course or City University module to prepare you for this kind of work. . . .The job goes beyond normal health and safety regulations and is outside every EU working directive. (1999: 8)

During the course of my research I experienced, both in the sense of witnessing and participating in, various 'ethical dilemmas' around drug taking, violence towards and from bouncers, withholding information from the police and taking cuts from door money, but these were all occasioned features of the setting, which I 'geared into' as a member of the setting, not an academic zookeeper or moral guardian. When a senior doorman of a famous Manchester nightclub, which has since closed down, told me 'I was in the firm' I simultaneously felt pleased with the credibility of my deception but also troubled about the future consequences of this status. Some would view my role as problematic in terms of collusion, which is only an issue if one retains a traditional conception of fieldworker objectivity. Obviously, these encounters could have put me in an ambiguous legal position, as I had acquired a type of deviant knowledge (Walters, 2003) but fortunately that never happened. In a way, it was a type of 'fingers crossed ethnography' where my luck might have run out.

My role could place me in a 'moral fix' (Van Maanen, 1983). This was made very clear when one former female student recognized me on the door and I had to deny this and assertively accused her of being drunk, which caused her some rather public confusion and distress. I didn't need sociology to tell me to 'turn the tape off' when a bouncer in the field that I had developed trust with began to open up to me about his emotional relationship to his estranged wife and son and the brutalization of door work. In this sense, my ethical conduct is 'self regulated'.

My research dreams were driven by realism and naturalism and, thus, trying to gain naturally occurring data. In some ways, part of the dream is to reduce artificiality and covert research is interesting in obviating that, although it is not a panacea and routinely suffers from the instigation tactics of field researchers. That is to say, researchers can effectively create or alter situations to catalyse certain types of behaviours or responses they are more interested in. In this sense, covert research is not free of active probing, or other methods of getting the story, which is at odds with the picture of a more purist capturing of data that it trades on.

Conclusions: Covert Research as Part of the Sociological Imagination

My aim has not been to offer formulaic advice for ethnographic research but rather to reflect on some particular problems I encountered which led me to reflect more widely on the issue of research ethics and to critique what I consider to be a 'fetish on ethics'. Another researcher with a different intellectual diet and gaze might arrive at a different and even opposing set of conclusions and reflections. For me, the discourse on ethics trades on an exaggerated idea of the role and importance of the academic. In this climate of hyper-sensitivity, covert research is frowned upon and many researchers, apart from a small minority, are effectively deterred from doing it and I do not see a dramatic shift in that. Obviously, we must be sensitized to the complex set of ethical problems in doing covert research, but it is worth pursuing. Ultimately, for me, the analytic understanding gained from the membership that a covert role can bring to the setting outweighs its ethical disadvantages. The decision to do it was driven by neither sensationalism nor romanticism. Erving Goffman concludes his presidential address to the American Sociological Association in 1982 by stressing that sociology must sustain a 'spirit of unfettered, unsponsored inquiry' (Goffman, 1983). For me, covert research clearly demonstrates that spirit.

I see covert research as part of a wider process of disruptive thinking in sociology and the social sciences, where one's normal status and privilege in the setting is removed. Covert research is part of a somewhat submerged tradition that needs to be recovered for future usage in its own right rather than being treated correctively as teaching material for cases of 'failed or bad ethics'. In many cases, covert research is an informed choice of research style rather than an enforced one. Moreover, research in this mould is a tradition that has significantly shaped, often in controversial ways, debates about the research relationship. My deep concern is that, in the present context of governance, we develop forms of 'methodological hypochondria'. This is not a belligerent stance nor a heroic portrayal of the covert researcher as, quite clearly, covert research is not appropriate for certain sensitive topics.

What is interesting is the popular public voyeuristic appetite for covert documentaries while, simultaneously, covert work is being more regulated in the professional social science community. What has been, and still is, routinely used in practitioner fields over the years, for different purposes, has been under utilized in the social science community. For me, covert research has a potentially creative and imaginative part to play and a voice to be heard in the sociological community. Covert research is not to everyone's taste and neither should it be. For me it can be artful, and it does have a place in the history of the research relationship and, I hope, an interesting, if controversial, future as well.

Note

1. There are various ethics codes and policy statements from professional and associations. These include the British Sociological Association's *Statement of Ethical Practice* (2002); the Social Research Association's *Ethical Guidelines* (2003); the Market Research Society's *Code of Conduct* (1999); the National Union of Journalists' *Code of Conduct* (1990); the British Society of Criminology *Code of Research Ethics* (2002); and the American Anthropological Code of Ethics (1997).

References

Anderson, C. and P. Bissell (2004) 'Using Semi Covert Research to Evaluate an Emergency Hormonal Contraception Service', *Pharmacy World and Science* 26(2): 102–6.

Becker, H.S. (1967) 'Whose Side Are We On?', *Social Problems* 14(3): 239–47.

Bittner, E. (1973) 'Objectivity and Realism in Sociology', in G. Psathas (ed.) *Phenomenological Sociology*, pp. 109–25. New York: Wiley.

Blackman, S. (2007) 'Hidden Ethnography: Crossing Emotional Borders in Qualitative Accounts of Young People's lives', *Sociology* 41(4): 699–716.

Bloor, M., B. Fincham and H. Sampson (2007) *Qualiti (NCRM) Commissioned Inquiry into the Risk to Well-Being of Researchers in Qualitative Research*. Cardiff: ESRC National Centre for Research Methods.

Bourgois, P. (2007) 'Confronting the Ethics of Ethnography; Lessons from Fieldwork in Central America', in A.C.G.M. Robben and J.A. Sluka (eds) *Ethnographic Fieldwork; An Anthropological Reader*, pp. 288–97. Oxford: Blackwell.

British Sociological Association (2002) *Statement of Ethical Practice*, URL (consulted May 2008): http://www.sociology.org.uk/as4bsoce.pdf

Bulmer, M. (ed.) (1982) *Social Research Ethics: An Examination of the Merits of Covert Participant Observation*. London: Macmillan.

Burgess, R.G. (1985) 'The Whole Truth? Some Ethical Problems of Research in a Comprehensive School', in R.G. Burgess (ed.) *Field Methods in the Study of Education*, pp. 139–62. Lewes: Falmer.

Calvey, D. (2000) 'Getting on the Door and Staying There: A Covert Participant Observational Study of Bouncers', in G. Lee-Treweek and S. Linkogle (eds) (2000) *Danger in the Field: Risk and Ethics in Social Research*, pp. 43–61. London: Routledge.

Dalton, M. (1959) *Men Who Manage*. New York: John Wiley and Sons.

Denzin, N. (1968) 'On the Ethics of Disguised Observation', *Social Problems* 15(4): 502–4.

Ditton, J. (1977) *Part Time Crime: An Ethnography of Fiddling and Pilferage*. London: Macmillan.
Duncombe, J. and J. Jessop (2002) 'Rapport and the Ethics of "Faking Friendship"', in M. Mauthner and M. Birch (eds) *Ethics in Qualitative Research*, pp. 107–22. London: SAGE.
ESRC (2005) *Research Ethics Framework*, URL (consulted May 2008): http://www.esrcsocietytoday.ac.uk/ESRCInfoCentre/opportunities/research_ethics_framework/index.aspx?ComponentId=11292&SourcePageId=19165
Festinger, L., H. Riecken and S. Schachter (1956) *When Prophecy Fails*. St Paul, MN: University of Minnesota Press.
Fielding, N. (1982) 'Observational Research on the National Front', in M. Bulmer (ed.) *Social Research Ethics: An Examination of the Merits of Covert Participant Observation*, pp. 80–104. London: Macmillan.
Fountain, J. (1993) 'Dealing with Data', in T. May and D. Hobbs (eds) *Interpreting the Field: Accounts of Ethnography*, pp. 145–73. Oxford: Clarendon Press.
Garfinkel, H. (1967) *Studies in Ethnomethodology*. Englewood Cliffs, NJ: Prentice Hall.
Geertz, C. (1973) *The Interpretation of Cultures*. New York: Basic Books.
Goffman, E. (1961) *Asylums*. New York: Doubleday.
Goffman, E. (1967) *Interaction Rituals*. Chicago, IL: Aldine.
Goffman, E. (1983) 'The Interaction Order', 1982 Presidential Address to the American Sociological Association, *American Sociological Review* 48(February): 1–17.
Goode, E. (1996) 'The Ethics of Deception in Social Research: A Case Study', *Qualitative Sociology* 10(1): 11–33.
Hammersley, M. (2006) 'Are Ethics Committees Ethical?', *Qualitative Researcher* 2(Spring): 4–8.
Hobbs, D., P. Hadfield, S. Lister and S. Winlow (2003) *Bouncers: Violence and Governance in the Night-time Economy*. Oxford: Oxford University Press.
Holdaway, S. (1983) *Inside the British Police: A Force at Work*. Oxford: Blackwell.
Homan, R. (1978) 'Interpersonal Communications in Pentecostal Meetings', *Sociological Review* 26(3): 499–518.
Homan, R. (1980) 'The Ethics of Covert Methods', *British Journal of Sociology* 31(1): 36–59.
Homan, R. (1991) *The Ethics of Social Research*. London: Macmillan.
Humphreys, L. (1970) *Tearoom Trade: Impersonal Sex in Public Places*. Chicago, IL: Aldine.
Israel, M. (2004) *Ethics and the Governance of Criminological Research in Australia*. Sydney: New South Wales Bureau of Crime Statistics and Research.
Lee, R.M. (1993) *Doing Research on Sensitive Topics*. London: SAGE.
Lee, R.M. (1995) *Dangerous Fieldwork*. London: SAGE.
Lee, R.M. (2000) *Unobtrusive Methods in Social Research*. Buckingham: Open University Press.
Lee-Treweek, G. and S. Linkogle (2000) *Danger in the Field: Risk and Ethics in Social Research*. London: Routledge.
Lofland, J. and R.A. Lejeune (1960) 'Initial Interactions of Newcomers in Alcoholics Anonymous: A field experiment in class symbols and socialization', *Social Problems* 8(2): 102–11.
Lovatt, A. and J. Purkis (1996) 'Shouting in the Street: Popular Culture, Values and Ethnography', in J. O'Conner and D. Wynne (eds) *From the Margins to the Centre*, pp. 249–57. Aldershot: Arena.
Lugosi, P. (2006) 'Between Overt and Covert Research: Concealment and Disclosure in an Ethnographic Study of Commercial Hospitality', *Qualitative Inquiry* 12(3): 541–61.
Lyng, S. (ed.) (2005) *Edgework: The Sociology of Risk-Taking*. London: Routledge.
MacIntyre, D. (1999) *One Man, Four Lives Undercover*. London: BBC Worldwide.

Measham, F. and K. Moore (2006) 'Reluctant Reflexivity, Implicit Insider Knowledge and the Development of Club Studies', in B. Sanders (ed.) *Drugs, Clubs and Young People*, pp. 13–26. Aldershot: Ashgate.
Mitchell, R. (1993) *Secrecy and Fieldwork*. London: SAGE.
Monaghan, L. (2002) 'Regulating "Unruly" Bodies: Work Tasks, Conflict and Violence in Britain's Night-Time Economy', *British Journal of Sociology* 53(3): 403–31.
Parker, H. (1974) *A View from the Boys: A Sociology of Downtown Adolescents*. London: David & Charles.
Patrick, J. (1973) *A Glasgow Gang Observed*. London: Eyre Methuen.
Punch, M. (1986) *The Politics and Ethics of Fieldwork*. London: SAGE.
Rafferty, J. (2004) 'Interviewing: The Unspoken Compact', in E. Williamson and M. Smyth (eds) *Researchers and their 'Subjects'*, pp. 121–36. Bristol: Policy.
Richardson, S. and M. McMullan (2007) 'Research Ethics in the UK: What Can Sociology Learn from Health?', *Sociology* 41(6): 1115–32.
Roberts, B. (2000) *Biographical Research*. Buckingham: Open University Press.
Rosenhahn, D.L. (1973) 'On Being Sane in Insane Places', *Science* 179(4070): 250–8.
Sanders, B. (2005) 'In the Club: Ecstasy Use and Supply in a London Nightclub', *Sociology* 39(2): 241–59.
Scheper-Hughes, N. (2004) 'Parts Unknown: Undercover Ethnography of the Organs-Trafficking Underworld', *Ethnography* 5(1): 29–73.
Shulman, D. (1994) 'Dirty Data and Investigative Methods', *Journal of Contemporary Ethnography* 3(2): 214–53.
Shulman, D. (2007) *From Hire to Liar: The Role of Deception in the Workplace*. New York: Cornell University Press.
Sin, Chih Hoong (2005) 'Seeking Informed Consent', *Sociology* 39(2): 277–95.
Taylor, S.J. (1991) 'Leaving the Field: Research Relationships and Responsibilities', in W.B. Shaffir and R. Stebbins (eds) *Experiencing Fieldwork: An Inside View of Qualitative Research*, pp. 238–47. Newbury Park, CA: SAGE.
Thompson, G. (1994) *Watch My Back: A Bouncer's Story*. Sussex: Summersdale.
Thompson, G. (1996) *On the Door: The Continuing Adventures of a Nightclub Doorman*. Sussex: Summersdale.
Thompson, J. (1988) *My Life in the Klan*. Nashville, TN: Rutledge Hill Press.
Travers, M. (2005) 'A Question of Ethics', *Network* 90: 20.
Twemlow, C. (1980) *The Tuxedo Warrior*. Chichester: Summersdale.
Van Maanen, J. (1983) 'The Moral Fix: On the Ethics of Fieldwork', in R. Emerson (ed.) *Contemporary Field Research: A Collection of Readings*. Boston: Little, Brown.
Wacquant, L.J.D. (1995) 'Pugs at Work: Bodily Capital and Bodily Labour among Professional Boxers', *Body and Society* 1(1): 65–93.
Wallis, R. (1976) *The Road to Total Freedom: A Sociological Analysis of Scientology*. London: Heinemann Educational.
Walters, R. (2003) *Deviant Knowledge: Criminology, Politics and Policy*. London: Willan.
Wells, H.M. (2004) 'Is There a Place for Covert Research Methods in Criminology?', *Graduate Journal of Social Science* 1(1): 1–19.
Williams, B. (2006) 'Meaningful Consent to Participate in Social Research on the Part of People under Eighteen', *Research Ethics Review* 2(1): 19–24.
Winlow, S. (2001) *Badfellas: Crime, Tradition and New Masculinities*. Oxford: Berg.
Winlow, S. and S. Hall (2006) *Violent Night: Urban Leisure and Contemporary Culture*. Oxford: Berg.

32

Covert Participant Observation: On Its Nature and Practice

Richard A. Hilbert

The aim of this article is to address a number of issues pertinent to covert participant observation not ordinarily discussed in other work. The article is concerned neither with the advantages and disadvantages of covert research nor with ethical questions related to relative secrecy.[1] My concern lies mainly with the nature of covert participant observation per se, that is, what it is that covert analysts are actually doing when they engage in covert activity. I shall argue that analysts' practical concerns with protecting their identities closely coincide with other methodological strategies of ethnographic investigation.

I should add, however, that much of what follows is of direct relevance to traditional concerns, and I shall indicate some of these relevancies toward the conclusion of this discussion. In anticipation of these relevancies, I might state that the distinction between secret and nonsecret observation may challenge the assumption that analysts explicitly and consciously "choose" to be covert, overt, or somewhere in between. As a consequence, the question of ethics, though of considerable importance as an issue, will have to be settled according to standards other than "how covert" somebody's research is. Finally, the question of the advantages and disadvantages of covert research becomes moot, as such matters are situational, being essentially decided within the terms of research settings themselves.

Source: *Journal of Contemporary Ethnography* (formerly *Urban Life*), 9(1) (1980): 51–77.

Covert Participant Observation as an Unstudied Phenomenon

Traditional concerns with methodological advantages of covert participant observation and the ethical problems of secrecy have failed to address the nature of the research activity itself and the general strategies located within it. This is the topic I am undertaking here. The ideas developed below were inspired by a one-year ethnographic investigation, related to the completion of my doctoral dissertation, of an elementary teaching credentials program. My chosen strategy was enrollment in the program with a decided preference for not revealing my research interests. Though I successfully completed the program (securing certification), these research interests were paramount. Until quite late in the academic year, my research interests were not known by most of the program personnel, staff, students, or teachers with whom I worked in the public schools. My intended public identity, for the most part, was that of student, student teacher, and prospective schoolteacher.[2]

In this article, I intend not to report findings relevant to teacher education or to other theoretical issues to which my study was directed. Rather what concerns me is the nature of the research itself, particularly the congruence of secrecy-guarding with practical, methodological strategies. The claim will be made that my experience was not accidental or idiosyncratic but was a consequence of the essential nature of the research activity itself. As such, it is of interest both theoretically as a sociological phenomenon and practically as an explication of methodological strategies for analysts about to embark on similar research ventures.

The sort of research I wish to generalize about involves full participation in a social gathering, group, or organization. I am further discussing research in which it is the analyst's task to come to terms with what it takes to sustain competent membership covertly in the group in question. Thus I am not concerned with simple spying or deception for purposes of access to secret files of "facts and figures", or of gaining the confidence of key officials who subsequently "tell all." The research I am discussing focuses on the mundane realities of day-to-day social existence and the problem of how the perceived stabilities of such realities are produced and managed. Conventional solutions to this problem recommend discovering the "subjective realities" or internalized rules governing proper conduct within social settings. Such solutions have no doubt led to traditional discussions of validity when it comes to the use of "informants." Informants can ostensibly "tell" researchers about these subjective realities, and it is therefore incumbent upon researchers to ensure that their informants are being truthful, represent the judgments of "typical" group members, and so forth. It will be argued here, however, that covert participant observation is participation par excellence and as such is itself "typical" – if the term is at all meaningful – of whatever it is that actors do to sustain competent membership.

The Secret in Covert Participant Observation

Insofar as analysts have determined that full disclosure would be injurious to their research interests, an abiding concern must be for maintaining a secret. We might initially think of the analyst's research role as the "reality" behind a manufactured or projected social identity. In short, we have the sociologist trying to pass himself or herself off as "something else."

In a classic investigation into the life of an intersexed person, Garfinkel (1967: 116–185) characterizes this sort of activity as "passing," i.e., behavior directed toward achieving a social identity other than what one might achieve without this behavior. What is instructive in this article is that the subject of the study, known only as Agnes, artfully secured her social right to be natural, normal member of a female population within a wider population consisting only of men and women whose ascribed sexual status, as Agnes knew, was beyond their control, just as hers was for her. However, throughout her passing performance, Agnes protected a secret which if revealed would have brought certain ruin. Her secret consisted of (a) her past upbringing as a boy until the age of seventeen, (b) her plans to secure surgically the anatomy that "nature intended," and (c) her present male genitals. While these three items were equally important to her secret, it was the last of them that proved most problematic to Agnes in her day-to-day encounters.

If Agnes's secret can be trichotomized, so can that of the covert analyst. The covert researcher guarding the secret must generally attend to aspects of his or her past, present, and future, which if revealed could lead to the breakdown of the projected identity. (a) The fact that he or she has been trained as a sociologist, has written literature reviews relevant to the setting, and possibly has written a research proposal are all potentially tricky areas to manage should they become public, as any research-motivated method of access to the setting might be. (b) Similarly, the fact that the analyst plans to write about the setting generally has to be withheld from other participants. (c) Most problematic, however, are certain aspects of analysts' present, day-to-day professional lives. A number of things could be involved, such as routine discussions with colleagues about the progress of the study, but the most likely, and probably most basic thing concerns the fact that analysts take extensive field notes and spend a great amount of time away from the field (possibly while other members are sleeping or watching television) poring over the notes, typing them, filing them, keeping records of conversations which to members are trivial and forgotten, and so forth. One might wonder, while typing these conversations alone at night, what it would be like for participants to these conversations to walk in unexpectedly or otherwise see the notes.[3]

Some might say that despite the strategic necessity of guarding the secret, guarding the secret is after all secondary to the actual purposes of the research.

Guarding the secret, so the argument might go, is a methodological device making it possible to conduct the research itself, namely discovering shared attitudes and norms through observation and interaction. Such characterizations overlook the sense in which guarding the secret *is* the research, *is* the method through which analysts discover whatever it is about group membership that is "typical." This is so, at least insofar as the secret is managed and successful passing is accomplished.

Sociologists electing to use covert field methods are aware that full disclosure will alter their participant status and will as a consequence affect their abilities to participate "typically."[4] By the same token, the way in which they guard the secret must be consistent with their membership activity. That is, they must make use of their knowledge of "what typical members do" in order to protect the secret; they must protect the secret by doing those things. Not to do so risks disclosure. At the same time, if this is successfully managed, analysts succeed in finding out "what typical members do," i.e., what they, the analysts, themselves were actually doing vis-à-vis a relevant social community. Their dual interests – that of managing the secret and that of doing research – fundamentally coincide. They are inseparable.[5]

We can take a few tips from Agnes on this point. Throughout her daily life, she continually relied upon her vague, incomplete, and unspecifiable background knowledge concerning feminine behavior to conceal her secret. She used "feminine modesty," for example, to avoid situations that might jeopardize the integrity of her performance. However, part of her very problem consisted of her lack of experience at being feminine. This lack of experience she "filled in" in ways that were to her always unpredictable. She was also able to make use of others' formulations of what she was doing, what she had done, and what she ought to do in discovering further and further aspects of naturally feminine behavior and in discovering what femininity consisted of "all along." Thus her concerns with guarding her secret coincided precisely with her methods of establishing, maintaining, and sustaining her "after all" femininity within a moral order of naturally sexed persons.

Agnes's passing was of immense importance to her, since through it she achieved her social identity as a woman. For Agnes, "nature's mistake," her "growth," her misspent youth, and her eventual need to turn to medical science to "fix things" were aspects of her life that she knew others would not understand. They were not, for Agnes, signs which contradicted her natural femininity, but she shared societal notions of sexuality in such a way as to render anyone's potential horror upon learning her secret understandable, normal, realistic. Indeed, she herself shared in this horror in her claims that she was not "one of those"; i.e., she had nothing in common with transsexuals, transvestites, or homosexuals. Furthermore, on at least one occasion, when part of her secret had been revealed to her boyfriend, she wondered if he in fact was normal, given the ease with which he continued to accept her "after all" status as a woman. Thus for Agnes, managing the secret had nothing to do with concealing her identity in favor of a false one; rather the necessity of such management derived from certain features of her biography which would have been incomprehensible to others. When confronted with these facts, others might mistakenly have identified her as "one of those."

Agnes successfully did female-behavior, and as such her *behavior* is formally indistinguishable from what "typical" women do. This is to say that what for normals is nonproblematic was for Agnes problematic every step of the way. In short, Agnes viewed organized feminine behavior much as an ethnographer would. Thus Garfinkel's (1967) study is not about deviant behavior or about con-artistry; it is about what it takes to sustain natural femininity. The study was possible through a case study of one person who had to attend to things that for most people go unnoticed.

Covert participant observers share in Agnes's problem. In order to learn what constitutes membership in a social setting, they have to assure themselves that what they are witnessing, what they are doing, is not colored by the mere *fact* that they are, in addition to participating in group processes, doing research. Simultaneously they seek to notice features of group membership that for other members may go unnoticed or unthematized. These are not contradictory objectives; but as indicated above they are synonymous. Successfully managing the secret *requires* attention to areas which to other members may be nonproblematic. Successfully managing the secret is also what we mean when we say successfully maintaining competent, "typical" membership.

Some General Features of Social Settings Which Covert Participant Observers Can Depend On

To restate: Covert research oriented toward discovering features of interaction which constitute typical membership does not depend upon guarding the secret simply for purposes of establishing a perspective otherwise inaccessible. Rather, guarding the secret is the process through which typical membership is realized. Simultaneously, the analyst uses his or her already established, albeit incomplete, knowledge concerning what typical membership consists of in order to manage the secret successfully. Thus, managing the secret and sustaining typical group membership are mutually elaborating activities which together constitute covert participant observation research activity.

What follows is a list of features general enough to apply to nearly any social setting. Items in the list are neither exhaustive nor mutually exclusive. They represent setting features that covert analysts can initially depend upon in their research activity. When I say analysts can depend on them, I am not indicating their literal or nonproblematic status, nor am I proclaiming them to be "true" in any precise sense. Rather, I am saying that they can be useful to analysts in managing the secret and sustaining group membership; they represent idealized resources whose specific sense awaits discovery from setting to setting and from occasion to occasion. I have also included examples from my own experience to fill in some of the shadowy detail.

(1) First and most useful is the tendency for people to view others as what they claim to be (see Goffman, 1959) – in cases relevant to this discussion, as fellow members. People are not therefore *inclined* to see anyone as doing sociological research. This may hardly need mentioning, but to sociologists entering a setting for covert research purposes, the likelihood of exposure may be drastically overestimated. In my own case, the setting itself included an academic department in which the art of ethnography is taught to graduate students. To an astonishing degree, the possibility that I was simultaneously doing research while getting a teaching credential was not entertained, even though program personnel knew of my identity as a sociology graduate student. This was so, despite the fact that numerous unannounced research projects were conducted throughout the year using us, students in the program, as unsuspecting subjects. Thus, even within a setting where both ethnography and "secret research" were available to staff and personnel as day-to-day concepts, the possibility that the setting itself was simultaneously being systematically noticed for research purposes was apparently so remote as to escape any consideration whatsoever. I eventually came to feel that I had been overprotective of the secret, almost as though no one would have believed it if invited to do so. This was extremely helpful and is relevant to the paragraphs below.

(2) All participants to the setting have past biographies and future intentions which conform to patterns only "more or less." Their pasts and futures are "typical," but this is an occasioned typicality accomplished through talk. It is possible, then, for the analyst to engage in such talk without pretext. Talk which for members is essentially equivocal is, for the analyst, purposefully vague. On many occasions, I had the opportunity to discuss past "experiences in academia," doubts that I would ever be a schoolteacher, and "alternative career lines" (including sociology) in ways that conformed to other students' talk about disquiet and doubt concerning their present purposes in the credentials program.

(3) Similarly, membership itself is "typical" by virtue of membership activity. In this regard, most social settings allow for eccentricity (see Lofland, 1971: 100–101). The degree to which this is so depends upon unspecifiable, situationally specific features of the setting which can only be revealed in the course of the research. The *fact* that this is so, however, makes it possible for analysts to probe areas that others may find ludicrous without surrendering their status as "typical" members.

The extent to which it is advantageous for a sociologist to accept and cash in on the social label "a little bit odd" is something that has to be decided by individual analysts. The point is that it can be done and it can be useful.

(4) In keeping with the above, there often will be other participants in the setting who are labeled eccentric (see Vidich, 1955: 357). There will also be those judged marginally competent or incompetent altogether. These people should be watched. They take risks the analysts may be unwilling to take. Sometimes their activity will be of the sort the analyst would indeed like to engage in, and with similar results. One of my colleagues, upon reading an early draft of my study, thought I was referring to myself sometimes in the first person and sometimes as "a student" and recommended that I standardize my writing style. What other students did, asked, and said often so closely resembled the sort of thing I would do for research purposes that my reader initially thought that I had been doing all of it. *This* was the point that needed clarification in subsequent drafts.

(5) Nothing counts as "the action that was taken" independent of talk about action. What is typical behavior within a social setting is typical by virtue of its linguistically accomplished typicality. Thus within indeterminate limitations, the analyst can engage in research activity without direct attention to covering it up. There is nothing about that activity that is self-evidently research activity. In fact, insofar as members attend to that activity at all, they will not "see" research activity; they will see some sort of membership activity instead. This is to say that they will tend to see it as somehow in conformity with notions of normal membership activity.[6] It is therefore in the analyst's interest at times like these to allow members to inform him or her of what he or she is actually doing or has done vis-à-vis the setting. That is, vis-à-vis the setting, the activity is not research activity. By noting members' compliments, observations, complaints, and so on, the analyst can turn necessary practicalities of research activity into opportunities to generate data – allowing members to provide accounts of how that activity conforms to notions of typical membership.

(6) Talk is irremediably vague (see Garfinkel, 1967: 1–11). This enables the analyst to talk about issues, ask questions, and so forth; to members, such talk appears less monumental than it does to the analyst; i.e., small talk to members can be significant talk for the researcher. Many times I asked questions of instructors, supervising teachers, and other program personnel regarding specific course requirements, practical questions for getting the assignments done. On occasion, these question masqueraded my intent to explore sociological themes interesting to me at the moment; more often these questions simultaneously served mundane and research ends. In still other cases, I asked questions

for purely pragmatic purposes and the sociological import of the exchange did not strike me until months later when I was reviewing my notes.

(7) The equivocal nature of language also allows analysts to enter, more or less freely, into talk about past training, future plans, and present sociological interests. When directly asked about matters that might affect the integrity of the secret, analysts can follow Agnes's lead and select characterizations that are likely answers to questions typically asked of members of the setting. They can, furthermore, state their responses in such a way that despite essential ambiguities will not require further information or invite further inquiry. While no clear recipe for doing this will be forthcoming, analysts can nevertheless depend upon their abilities to improvise answers and any further clarification with the certain knowledge that routine interaction will not require them to disclose more than what remains under their control. This knowledge rests on a fundamental feature of talk in any setting: Absolute clarification, where all is known and brought to light, is an essential impossibility (see Garfinkel, 1967). Put differently, nothing counts as "telling all," and members both know that and count on it. To quote Wittgenstein (1953: 1) "Explanations come to an end."

Several occasions arose in which my background as a sociologist became thematized. Students knew me from TA sections, had friends I "should know," or had heard through various channels that I had a master's degree. It was relatively effortless for me to allow students to reveal their practical concerns with these matters: Could I "do" anything with an M.A.? (No.) Didn't I "just hate sociology?" (At times.) Did I think labeling theory was just another form of Marxism? (I hadn't thought of that.) Often students' interests revolved around the issue of my M.A. either with respect to my starting salary or my very employability as a schoolteacher. Sometimes I concertedly turned conversations in these directions, talking about whether I would ever "go back" to sociology, get a Ph.D. and the rest of it. In fact, by emphasizing the disadvantages of a Ph.D. in the school-teaching job market, I was able to speculate quite freely on the possibility of using my experiences in the credentials program to write a dissertation.[7] I might add, parenthetically, that several students encouraged me to do just that, saying that "somebody should get *something* out of all this!"

(8) When new information arises, members have at their disposal procedures for viewing that information not as information withheld but as what-could-have-been-said-earlier-but-never-came-up. Thus as parts of the secret are disclosed, or perhaps as the secret itself is being revealed (as indeed it often must be), analysts can trade on members' inclination not to view themselves as having been "fooled." That is, this inclination is an invaluable resource for the analyst in managing either inadvertent slip-ups or the ultimate debriefing of setting participants. If

successfully managed, debriefing can itself become an occasion for analysts to display typical membership and to learn how members normalize such revelations as insufficient to warrant doubting analysts' "after all" status as members.

I began formal debriefing of key personnel during the last eight weeks of the academic year. Insofar as I had been established as a competent student and student teacher, I found full disclosure to be an unrealizable ideal; i.e., my student role, in each case, assimilated my revelation. The secret was received in most cases with mild lack of interest. Students appeared not to believe me or thought I was fantasizing because of the poor job market. One former supervising teacher said: "So why are you telling *me* this? It seems you'd have to get somebody in sociology to help you on that." The supervising teacher I was working with at the time was quite interested and even wanted to know some of the specifics; however, after a lengthy discussion of my dissertation and my upcoming advancement to candidacy (for which I would need time off), she shifted the topic to "practical" matters – my need to write "plans of some sort" for an all-class takeover I was about to conduct. The fact that she did not want to see these written plans herself, her acknowledgement that I "might not really *need* them," and her assertion that writing them anyway would be "good practice" for the types of things I would be doing in the future ("Sometimes a principal may walk in and want to see them . . . you just never know . . .") all contributed to a hunch that my revelations were somehow assimilated by my "after all" status as student teacher. This hunch subsequently proved correct.

Two Examples

In this section we take up two general research strategies which depend upon setting features enumerated above. The congruence of managing the secret and membership activity will be illustrated by emphasizing first how normal membership activity can be turned to research advantage (secret interview) and subsequently how necessary research procedures (secret note-taking) can be construed within social settings to be normal membership activity. To some degree, the specifics of what follows are consequences of my own research setting; however, their central import should not be lost on those generally interested in participant observation.

Participation-as-research: the secret interview. Analysts concerned with normal membership can learn from other members' knowledge concerning this activity. Thus it is proper to ask questions (see Vidich, 1955: 354–360). In so doing, covert analysts are using their knowledge that requesting and providing information are features of almost any social setting; analysts can begin there. Asking for information accomplishes at least three things:

(a) It provides responses which to the analyst serve as examples, models, and indicators of proper talk within the setting under study. These reponses are data in the same way that any observed talk is data. Here, however, the analyst controls the topic, intervenes as it were, and solicits talk about areas particular to his or her research interests – talk which otherwise would not take place.
(b) Simultaneously, the analysts, in asking questions, is practicing the art of talk within the setting. Certain sorts of questions are not taken seriously. They may, for example, appear to go beyond the bounds of "what everybody knows" – and instead of responses may only draw bewildered looks (see Stoddart, 1974). Thus the analyst can be an apprentice in members' talk in the very process of attempting to solicit it.
(c) Combining these facts, it becomes possible to identify certain parameters of members' talk, that is, boundaries which competent members do not ordinarily cross. Such boundaries have been referred to as "idealizations" in the ethnomethodological literature – unstated premises, assumptions, presuppositions, and background expectancies which provide for the perceived reasonableness of all members' talk while at the same time being elaborated and documented as "after all correct" whenever such talk takes place.

Whenever analysts talk within the setting, they could be conducting secret interviews. A concerted interview, however, aimed at the objectives outlined in the last paragraph, will generally require some degree of risk. The analyst may want to approach the boundaries of perceived reasonableness or cross them temporarily for purposes of satisfying himself or herself as to "where" those boundaries are; the line of demarcation is routinely avoided by members, and it may be possible to identify it only by crossing it – with caution. Such practices, if engaged in continuously, will be costly in terms of role credibility. Much can be gained, though, from using these devices strategically and sparingly. "How much" to use them can only be determined in the context of doing the research itself; such determinations simultaneously become research data.

In my own research, a tremendous resource resided in the fact that students in the program were *expected* to ask questions. We were supposed to be learning how to teach. In framing questions as directed toward the presumed objective, I was able to generate data concerning how to be a student within the credentials porgram learning how to teach. Both in the responses provided and in the way questions were taken, I learned something about schooling that is so taken for granted that it should only seldom be brought up for review, and even then only in theory or in jest. I learned the sorts of questions students can ask. Sometimes I was told in college classes that such-and-such was a "good" question and that it deserved an answer, but that I should ask it somewhere else. Students sometimes groaned when I began flirting with such issues as how bureaucratically mandated "behavioral objectives" can tell teachers what is going on inside a child's mind or by what justification a child not "conforming to school expectations" or working up to "grade level" can be called a problem child. At one point late in the first term, a staff member called me into her office to relay some complaints made by students and instructors about some of my questions. She was unable to tell me what kinds of questions I should ask, but she suggested I keep to questions like: "When is it due?" or "how many pages should it be?"[8]

As already suggested, analysts cannot overly engage in such risk-taking if they are to maintain typical membership. This is not to say that the secret would be revealed, but significant tendencies on the part of members to see analysts as incompetent or otherwise obstructionist will probably not be beneficial to ongoing research concerns. It should be noted in this regard, however, that several of the teachers in one of my student teaching assignments commented on my questioning, saying that it was my "strongest point" and that it indicated I was a "really conscientious student."

Research-as-participation: taking field notes. Field notes are possibly the most universal ingredient of participant observation and of ethnography in general. Most textbooks on field methods provide suggestions for taking and processing notes (see Forcese and Richer, 1973: 149–151; Lofland, 1971: 101–113, 117–133). What is not addressed, possibly because little has been said about covert field methods, is the sense in which note-taking for research purposes can be systematically concealed and made to conform with membership activity. Analysts can use whatever it is about the setting that transforms "paper and ink" into tools of membership in order to accomplish their practical ends of note-taking while simultaneously discovering further aspects of what they are doing vis-à-vis the same setting.

The credentials program was particularly amenable to this sort of practice, since students were expected to take notes and even to keep a "journal" of day-to-day activities and observations. An observation guide was provided listing 38 areas with which students were to become familiar. Many of these areas closely approximated my own research interests. A program coordinator elaborated on the list, suggesting that we avoid being overly conspicuous in our note-taking in the public schools. We might, she said, want to "remember" things and write them down later.

All told, I collected 450 pages of typed, single-spaced notes by the end of the year. By far the easiest to manage notes were those that doubled as "class notes." In college classes I wrote nonstop as instructor lectures and student-instructor dialogue took place. My seemingly fanatical note-taking served, within the setting, to document my "seriousness" with respect to my student role. It also served to document my eccentricity. Upon occasions in which students broke into small groups for various purposes and were to choose "recorders" to take notes and report back to the class, I was chosen as the "recorder" on the grounds that "he takes notes on everything anyway."

I made it a point in all of my note-taking to reveal my illegible handwriting in order to discourage people from attempting to read my notes.

Taking notes was not as easy or straightforward in other settings within the program, such as "professional seminars" in which students shared mutual problems and successes encountered while student teaching, special topics seminars, informal get-togethers, potluck dinners, and so forth. In these settings, I was able to vary the degree to which my note-taking was in the open by monitoring other students' note-taking activity. Even where little note-taking was taking place, paper and notebooks were usually highly visible. In such cases, I was able to participate in group processes while periodically turning to a piece of paper to jot down some bit of information to be interpreted later in the evening. I also engaged in such practices as writing while looking elsewhere, for example at a book or blackboard, giving the impression that I was copying something down and was not paying attention to the discussion which was the object of my attention.

Note-taking was most difficult to manage in student teaching assignments, which lasted most of the year. It was here that my participant role was all-consuming. I had practical concerns, bureaucratic responsibilities, and problems to resolve. The most useful resource in this setting was the recommendation by program personnel that we take notes, because supervising teachers expected us to take notes. I initially asked one of these supervising teachers if such note-taking made her nervous. Her response was: "Are you kidding? After how used to being observed I am?" She said she was glad to see that I was a "conscientious student" and that she herself had taken "notes like that" as a student teacher. She also said that it made her feel important to see me taking notes; "Like it makes me think I'm doing something worth writing down."

Occasionally, however, problems arose. For example, during one student teaching assignment, my note-taking became potentially troublesome when the supervising teacher wanted to read my notes. "I probably ought to look at them to see what I'm doing that I might not even know about." I was not sure how to handle this, and I stalled for a few days. Later, when she was taking notes on my student teaching performance, she remarked that she did not want me to read her notes for the same reason you don't want me to read yours." In the course of this conversation, I learned that it was routine for evaluators to take notes on teachers, notes which if read by the teachers being evaluated, might be misinterpreted as overly negative. Thus teachers routinely withhold their written observations, and vis-à-vis the setting, this is what I appeared to be doing.

The point I wish to emphasize here is that taking notes need not be seen simply as the recording of events and occurrences within the setting in which the analyst is also a participant. Rather the note-taking ought properly to be viewed as a constituent feature of that very setting.[9] Thus the analyst's technical research work is part of his or her very role as a participant. Members view it that way; that they view it that way is part of competent membership, and as such it is a topic for analysis. Analysts can then use this acquired information as resources for further managing the problem of discreet note-taking. Throughout the year I was able to take notes while waiting to see what I was doing. Taking notes in the teachers' lunchroom was drawing up "lesson plans"; taking notes on teachers' talk concerning "control devices" of the classroom was the accumulation of such devices for future use as a schoolteacher. The point here is that what was an unfortunately necessary component of field work was simultaneously competent student-teacher activity. Thus my analyst activity and my membership activity were congruent.

As suggested above, the specifics of my research setting may provide for some exaggerated claims regarding the possibility of open and unabashed note-taking. Nevertheless, it is always in the analyst's interest to discover the taken for granted corpus of knowledge operating within a setting regarding the normal, routine use of paper and ink, or whatever facilities are available for taking notes. This corpus of knowledge can be simultaneously discovered in the actual course of taking notes and used to manage note-taking, allowing the analyst both to discover and to display typical membership activity in and through necessary research procedure.

Some Objections

The central point I have been making is that for covert participant observers, managing the secret is not incidental to the research itself, nor is it incidental to the display of typical membership. It is not as though managing the secret is hiding one's "real" identity for purposes of gaining access to a setting or a participant's "perspective" from which it then becomes possible to do research. Managing the secret *is* the research, *is* the display of typical membership. These activities are not independent of one another, nor can they be analytically segregated. They coincide, they are congruent, they are inseparable.

Objections can be raised to these arguments, and I will attempt to deal with some of them in this section. I have been discussing these matters as though everything analysts do in their capacities as researchers consists of managing the secret, as though that were all they do. One might ask whether analysts need constantly be on guard in this way. Certainly during most of the time spent in a setting, the secret is in no immediate danger; there simply is no foreseeable thematic opening for the topic to arise. At times like these, however, it is the analyst, and only the analyst, who has made the determination to relax his or her guard. And even within these contexts, the analyst cannot know precisely when and how guarding the secret might reemerge as relevant. It is in this sense, then, that the analyst must continually monitor and manage the secret, pay close attention, be ever vigilant.

I also might be misunderstood on the following point: I do not mean to suggest that all of the material gathered in the process of managing the secret will be relevant to the analyst's immediate theoretical concerns. When I say that managing the secret is simultaneously membership activity and research activity, I mean that it is through such practices that analysts can assure themselves of typical membership. Such assurances are crucial for analysts in determining the validity of their data. That is, without these assurances, their possibly flaws participant status may undercut the credibility of their research activity, whatever their immediate theoretical research interests might be.

A more serious objection might be that the secret is, after all, a secret, and the analyst could be discredited; this in and of itself renders the analyst atypical of group members. Such an objection presupposes several things about social life, most notably that all "typical" members are "up-front," have "nothing to hide," or in some objective way fit a "mold" of typicality. Such a "mold" is conventionally characterized as conformity to a set of internalized group norms, such conformity "causing" the observed typicality.[10] The arguments presented here, however, are informed by a perspective which regards "typicality" as an accomplishment of occasioned, ongoing membership activity which includes the presentation of self through an array of practices of selecting relevancies, glossing incidentals, ignoring or often concealing idiosyncratic biographical specifics which otherwise might call competent membership into question, and disregarding an unlimited number of alternative ways of constituting what "really is going on" or "what I'm really doing, what my interests really are" (see Garfinkel, 1967). Nothing counts as "typical membership" beyond a working consensus (see Goffman, 1959) of the setting in question; beyond the terms of that setting, "typicality-within-that-setting" is a meaningless expression. Thus, just as Agnes's behavior was sufficient for her socially recognized right to be a woman, successful passing on the part of analysts is sufficient to establish their typicality.

A related objection might be raised: Is it not the case that researchers engaged in such analyses are depending for their data upon case studies of one, namely themselves? Worse, are they not overly dependent upon their own subjective impressions regarding "what it takes" to sustain membership where other members' impressions might significantly differ? Answers to these questions revolve around an emphasis upon what it is that analysts interested in what it takes to sustain membership are concerned with: *doing*. They are not concerned with subjective formulations, attitudes, or opinions.[11] What it takes to sustain membership is not inside anyone's head but consists of socially accountable activity. Furthermore, analysts, in engaging in such activity, are not doing so in isolation. Indeed, successful passing on their part is not possible without the active cooperation of other members of the setting; other members facilitate analysts' performance. Thus successful passing does not mean coming to view things from a "typical perspective," but rather depends upon full participation in the setting with other members who also participate and whose participation renders analysts' participation typical. In the final analysis, then, it can be stated: The analyst *is* a typical member by virtue of successful passing.

Implications for Traditional Issues in Participant Observation

The above observations suggest a mutually determining relationship between secret and open observation; i.e., the techniques that it takes to protect secrecy are those very techniques that are used to display public membership. This indicates a fundamental error in attempting to characterize a research role as secret *or* open; i.e., the dichotomy is spurious in its essentials. Nor, given this, can we characterize various research strategies as positions along a continuum, since a continuum by necessity retains an analytic distinction between secret and open participant observation (see Junker, 1960: 144–147; Schatzman and Strauss, 1973: 58–63). Roth (1962: 283) is close to the point in his assertion that the ends of a secrecy-openness continuum "probably do not exist" and that "it is the non-dichotomous part of this continuum which is of most interest." But what is suggested here is an interdependence or mutual identity of covert/overt which renders any attempt to locate "points" along a continuum, even one without ends, an exercise in futility.

This being the case, it should not be inferred that it is impossible to characterize certain research roles as more or less secret in certain ways and more or less open in others. Nor is it impossible to compare research roles, say in different studies, in terms of their relative degrees of secrecy. We do this all the time. But the fact that such characterizations and comparisons are routinely made should not lead us to believe that "degrees of secrecy" can be formalized or can be found to transcend research roles from one study to another. Such determinations by necessity are made on an ad hoc basis through the use of unthematized folk theories about the dichotomous and "obvious" distinction between secret and open observation. If, as I have argued, this radical distinction between covert and overt research cannot be sustained analytically,[12] then questions regarding "how covert" a given study is or "how many" degrees of secrecy are available in participant observation are not amenable to analytic scrutiny, at least not in the way these questions have been conveniently addressed.

It is also likely that rather than making "methodological decisions" about "how covert" to be in advance of an investigation, researchers simply do their studies with some sort of understanding that they will be able to characterize their research methods should they be required to account for themselves in the future. When that time comes, analysts invoke commonsense methods of making such determinations, possibly with no intent of considering these determinations as part and parcel of their sociological theorizing or analysis. Further, it should be no surprise that methodological appendices can account for "just that amount of secrecy" as the amount that was required to do the job, since presumably the analyst was already in the course of the research itself using whatever methodological strategies that were deemed appropriate for getting the job done.

Finally, the question of research ethics often seems to turn on the secrecy issue. The above observations suggest that it is going to have to turn on something else; i.e., while "full disclosure" is meaningless as an ethical ideal, "full concealment" is equally meaningless as something to avoid. Were we to take the secrecy criterion at face value, we would be led to state that the more covert analysts are the more unethical they are, which would lead us to the objectionable conclusion that all participant observers are unethical to a degree.[13] I would suggest, in fact, that while secrecy is often raised as a red herring, literature concerning ethics in participant observation (see especially Erikson, 1967) backs away from secrecy in favor of a paramount moral maxim to participant observers which should be hard for no one to swallow: People should not be hurt.[14] Secrecy in this context is often associated with broader moral questions, for example, to what extent sociologists should enter exotic, religious, or deviant subcultures where they may be in no position to judge the damage their work might cause natives (see Fichter and Kolb, 1953). While proposed general criteria for ethical decisions are beyond the scope of this article, it should be clear that such determinations cannot be made (and therefore in fact *are* not made) on the basis of degree of secrecy *per se*.

Concluding Remarks

The ideas presented in this article were inspired by a covert participant observation study within an elementary school teaching credentials program. These experiences suggest that the characterization "covert research" can often be misleading. While my research required the preservation of secrecy and warranted the label covert, it indicated the congruence between the *covert* and *public* natures of any research role. Thus this discussion can be extended to participant observation in general.

A participant observation is irremediably covert to some degree, because of the essential glossing practices any member engages in to produce typicality. But for participant observers, these practices normally have a special significance. Not only is it seldom desirable for research-motivated behavior to emerge as dominant in the eyes of fellow participants, it is probably difficult for analysts to reveal the entire character of their research interests and goals, even if this were their research policy; indeed it is doubtful they could articulate it fully even to themselves (see Forceses and Richer, 1973: 146). Yet it is through guarding the integrity of such concerns that analysts both achieve acceptance within a social community and accomplish their research ends, assuring that whatever their research interests might be, they are after all participating typically within the setting. In short, all participant observers must, in some sense and to some degree, pass.

Notes

1. For discussions of relative advantages and disadvantages of secrecy, see Forcese and Richer, 1973: 144; Lofland, 1971: 93–94; Helmstadter, 1970: 78; Gold, 1958. Ethical considerations are found in Roth, 1962; Erikson, 1967; Shils, 1959; Denzin; 1970: 333–341; Fichter and Kolb, 1953; Riesman and Watson, 1964: 260–269. Finally, attempts to identify specific degrees of secrecy on an overt-covert continuum include Bogdan and Taylor, 1975: 28–30; Lofland, 1971: 93–99; Junker, 1960: 144–147; Schatzman and Strauss, 1973: 58–63; Gold, 1958; and Roth, 1962.
2. So as not to leave the impression that I ignored or glossed over ethical considerations, let me say that these issues, as they related directly to my study, were of constant concern to me and are covered in the dissertation. Briefly, my own solution to the ethical problem turned on (a) my unwillingness to represent myself as something I was not (I was enrolled in the credentials program), (b) my lack of interest in "backstage activity" concomitant with my unwillingness to use my public role as access to group secrets, (c) my lack of interest in anyone's personal life, (d) my data source, which was "anybody's" data source, and (e) the nature of the setting itself, i.e., neither sacred nor deviant, a tax-supported institution which is publicly accountable. I am not, therefore, recommending my sort of research activity as ethically pure by abstract criteria, nor is relative secrecy methodologically desirable in every case.
3. Lofland (1971: 108–109) notes that such musings can often lead to "a sense of betrayal" on the part of analysts.
4. Whether the analyst is correct in this or not is not in question. The point is simply that this is usually a working assumption in covert research. In that capacity, it is part and parcel of the decision not to disclose one's intent to do research, write papers, and so on.
5. This stands as opposed to characterizations of the "scientific role" and the "social role" as merely "interdependent" (Bruyn, 1966: 18–20).
6. Whether and to what degree such membership activity is judged to be *competent* is a research problem. But even the grossest form of incompetence is nevertheless so judged from members' perspective; i.e., it is by and through such a perspective that the action is determined to be what it is and is judged incompetent. Thus even incompetent activity "fits in" with the overall intended gestalt of possible membership activity, documenting what that gestalt is and "was all along." For a treatment of how members allow perceived irreconcilable contradictions to count as evidence of a presupposed reality that contains no contradictions, see Hilbert, 1977.
7. The use of the term "speculate" is neither idle nor Machiavellian. During most of the year, doubts about the very feasibility of the project often suppressed any sense of direction I might have been more comfortable pursuing. These doubts, which were never fully allayed until the completion of the first draft of the dissertation, are something field researchers often experience, but they also suggest a similarity between me and other students in the credentials program: uncertainty about the future and present purposes for being in the program.
8. It is of no small interest that I was only one of several students sanctioned for asking the wrong kinds of questions. I was not the only "eccentric."

9. This should not be confused with the ethnomethodological notion of reflexivity or the sense in which a description of a setting is an essential part of the very setting it organizes and describes (see Garfinkel, 1967; Wieder, 1974a, 1974b). That is another matter entirely.
10. Shared rules for behavior are Parsons's solution to the problem of the "factual" social order; identifying these internalized rules in such a way as to explain behavior meets Weber's criterion of adequate causal analysis.
11. This is not to say that such studies cannot be conducted or be beneficial; attitudes and other subjective phenomena have been dealt with quite successfully (see for example Becker et al., 1961). However, it is unwarranted to use such information to account for "typical membership" or to discover what it takes to sustain competent membership within a social setting. Such an attempt would depend upon the notion that internalized norms and values are sufficient for causing certain kinds of objectively observable behavior. The inadequacies of this model have been dealt with extensively (see Wilson, 1970; Garfinkel, 1967; Wieder, 1970, 1974a, 1974b; Zimmerman, 1970; Zimmerman and Wieder, 1970).
12. A colleague of mine tells an interesting story. He was hired as a consultant to a biology department; he was to provide qualitative material regarding social networks, interpersonal relations, and departmental politics. He was to roam freely, interviewing whom he chose, asking whatever he thought pertinent. All department members knew of the study and supported it. Though this is about as "open" as you can get, he soon found himself privy to an array of secrets and interpersonal confidences he had not purposely sought. Suggestions were made about whom to interview and why; assessments about other faculty were made in confidence; "plausible" hypotheses, research findings, and conclusions were constantly being provided. In short, the researcher had become assimilated by the very community he wished to study, complete with the simultaneity of (a) the public display of proper research behavior within that setting and (b) the practice of concealing whatever information that was necessary to conceal in order to accomplish the research task.
13. It would be no great leap from here to argue that because of the essential glossing practices described by Garfinkel (1967), members themselves are irremediably unethical to some degree.
14. The issue of relative secrecy as an ethical consideration was raised partly as a result of such disguised observations as Lofland and Lejeune's (1960) penetration of Alcoholics Anonymous. Appropriately, the defense they offered (Lofland, 1960–1961) in response to a critique by Davis (1960–1961) was just that they had not hurt anybody.

References

Becker, H. et al. (1961) Boys in White: Student Culture in Medical School. Chicago: Univ. of Chicago Press.

Bogdan, R., and S. J. Taylor (1975) Introduction to Qualitative Research Methods. New York: John Wiley.

Bruyn, S. T. (1966) The Human Perspective in Sociology: The Methodology of Participant Observation. Englewood Cliffs, NJ: Prentice-Hall.

Davis, F. (1960–1961) "Comment on 'Initial interaction of newcomers in Alcoholics Anonymous.'" Social Problems: 364–365.

Denzin, N. K. (1970) The Research Act: A Theoretical Introduction to Sociological Methods. Chicago: Aldine.

Erikson, K. T. (1967) "A comment on disguised observation in sociology." Social Problems 14: 366–373.

Fichter, J. H. and W. L. Kolb (1953) "Ethical limitations on sociological reporting." Amer. Soc. Rev. 18: 544–550.
Goffman, E. (1959) The Presentation of Self in Everyday Life. Garden City, NY: Doubleday.
Forcese, D. P. and S. Richer (1973) Social Research Methods. Englewood Cliffs, NJ: Prentice-Hall.
Garfinkel, H. (1967) Studies in Ethnomethodology. Englewood Cliffs, NJ: Prentice-Hall.
Gold, R. L. (1958) "Rules in sociological field observations." Social Forces 36: 217–223.
Helmstadter, G. C. (1970) Research Concepts in Human Behavior: Education, Psychology, Sociology. Englewood Cliffs, NJ: Prentice-Hall.
Hilbert, R. A. (1977) "Approaching reason's edge: 'nonsense' as the final solution to the problem of meaning." Soc. Inquiry 47: 25–31.
Junker, B. H. (1960) Field Work: An Introduction to the Social Sciences. Chicago: Univ. of Chicago Press.
Lofland, J. (1971) Analyzing Social Settings: A Guide to Qualitative Observation and Analysis. Belmont, CA: Wadsworth.
________ (1960–1961) "Reply to Davis." Social Problems 8: 365–367.
________ and R. A. Lejeune (1960) "Initial interaction of newcomers in Alcoholics Anonymous: A field experiment in class symbols and socialization." Social Problems 8: 102–111.
Riesman, D. and J. Watson (1964) "The sociability project: a chronicle of frustration and achievement," in P. E. Hammond (ed.) Sociologists At Work. New York: Basic.
Roth, J. (1962) "Comments on 'secret observation'." Social Problems 9: 283–284.
Schatzman, L. and A. Strauss (1973) Field Research: Strategies for a Natural Sociology. Englewood Cliffs, NJ: Prentice-Hall.
Shils, E. A. (1959) "Social inquiry and the autonomy of the individual," pp. 114–157 in D. Lerner (ed.) The Human Meaning of the Social Sciences. New York: Meridian.
Stoddart, K. (1974) "Pinched: notes on the ethnographer's location of argot," pp. 173–179 in R. Turner (ed.) Ethnomethodology. New York: Viking.
Vidich, A. J. (1955) "Participant observation and the collection and interpretation data." Amer. J. of Sociology 60: 354–360.
Wieder, D. L. (1974a) Language and Social Reality: The Case of Telling the Convict Code. The Hague: Mouton.
________ (1974b) "Telling the Code," pp. 21–26 in R. Turner (ed.) Ethnomethodology. New York: Viking.
________ (1970) "On meaning by rule," pp. 107–135 in J. D. Douglas (ed.) Understanding Everyday Life. Chicago: Aldine.
Wilson, T. P. (1970) "Normative and interpretive paradigms in sociology," pp. 57–79 in J. D. Douglas (ed.) Understanding Everyday Life. Chicago: Aldine.
Wittgenstein, L. (1953) Philosophical Investigations. New York: Macmillan.
Zimmerman, D. H. (1970) "The practicalities of rule use," pp. 221–238 in J. D. Douglas (ed.) Understanding Everyday Life. Chicago: Aldine.
________ and D. L. Wieder (1970) "Ethnomethodology and the problem of order: comment on Denzin", pp. 285–295 in J. D. Douglas (ed.) Understanding Everyday Life. Chicago: Aldine.

33

Between Overt and Covert Research: Concealment and Disclosure in an Ethnographic Study of Commercial Hospitality

Peter Lugosi

Covert research and the use of covert methods have always been contentious issues among social scientists. Advocates have argued that covert methods offer researchers access to information that is otherwise denied to them (Calvey, 2000; Lauder, 2003; Miller, 2001), while critics have denounced covert methods as ethically and professionally unsound and vilified all those who engaged in "duplicitous" research (Beauchamp, Faden, Wallace, & Walters, 1982; Bulmer, 1982a; Herrera, 1999; Warwick, 1982). This article suggests that instead of labeling research "covert" and dismissing it on the grounds of ethical irresponsibility, it is more useful to consider how the nature of the study, the character of the fieldwork context, and the relationships between informants and ethnographer determine overtness or covertness in the field. I argue that prevailing critiques of covert methods instill a sense of ethical hypersensitivity that does not help ethnographers to resolve problems of covertness in their fieldwork. Consequently, this article offers a more nuanced understanding of the processes of concealment and disclosure in fieldwork that helps to reassess the usefulness of existing critiques of covert research and covert methods.

I begin by briefly reviewing the historical treatment of covert research and suggest that this has created a professional climate in which all forms

Source: *Qualitative Inquiry*, 12(3) (2006): 541–561.

of concealment are treated as inherently transgressive. In the next section, I provide a brief description of my study and the research context and then proceed to discuss the techniques I used to recruit informants and elicit information from them. I focus on the factors that determined when and how details of the research were disclosed and, also, on those factors that determined how my identity as researcher could be presented.

The case study demonstrates why concealment was pervasive throughout the research and why there was constant ambiguity regarding my identity and motivations. Concealment of this kind is often considered ethically problematic, but I maintain that ethnographers in similar situations should not be hampered by all the criticisms leveled against covert research. I am not advocating covert research or covert methods per se; however, I do suggest that examining the contextual nature of fieldwork helps ethnographers and their potential critics to understand why concealment is a necessary and often-inevitable part of research. This article thus seeks to provide an alternative point of reference that informs debates surrounding the emergence of concealment and covertness in research.

The Historical Treatment of Covert Research and Covert Methods

Critics have continually questioned the necessity and the usefulness of researchers disguising their identities and concealing their research agendas (cf. Erikson, 1995; Herrera, 1999; Warwick, 1982). Shils (1982), for example, argued that covert methods were inexcusable forms of civil betrayal that violated the individual's right not to be studied. He felt it was morally obnoxious of scientists to assume that the search for truth granted them a license to disregard the rights of those being studied. In contrast to some authors, who argued that the knowledge gained through covert research justified the use of deception (e.g., Denzin & Erikson, 1982; Goode, 2001; Miller, 2001), Shils rejected the claim that the value of such knowledge outweighed the problems caused by the infringement on individual rights.

Bok (1986) pointed to the emotional and psychological stress that covert research causes to those deceived and to those who continually have to deceive. The tensions caused by concealment are often found in the fieldwork accounts of ethnographers (cf. de Laine, 2000; Diamond, 1992; Wong, 1998). The pressure to produce open, reflexive ethnographies has meant that researchers are obliged to account for their indiscretions in the field. On one hand, this provides a richer understanding of how the relationships in the field affected the nature of the data, while on the other hand ethnographers are forced into a perpetual cycle of critical self-analysis in which every aspect of their relationships with informants is problematized.

Evidence of stress among informants caused by concealment is harder to find, although it is generally assumed that they would be offended or troubled by being misled. Kimmel (1996) presented numerous cases where deceptive research, particularly within psychological experiments, had caused stress for participants. However, some authors have noted that the majority of criticisms of covert methods are based on deductive reasoning and that there is little actual evidence to suggest that discovery of the researcher's hidden identity or intention always causes distress among informants (e.g., Mitchell, 1993; Oakes, 2002). Nevertheless, Beauchamp et al., (1982), Bulmer (1982a), and Warwick (1982) concluded that knowledge of concealment in fieldwork would raise self-doubt and suspiciousness among informants and make them reluctant to participate in future research.

This legacy of stigmatization, coupled to today's culture of litigation and ethical hypersensitivity (cf. Nelson, 2004; Wright, 2004), has certainly made the use of covert methods a perilous endeavor. Funding bodies and host organizations, already dismissive of exploratory research that does not use conventional, "safe" methodologies (Lincoln & Tierney, 2004), are thus far more inclined to restrict studies using covert methods. The problem is that the historical criticisms of covert research have established a series of professional benchmarks that are used to evaluate any and all forms of concealment in research. Many contemporary research methods texts continue to use studies such as Milgram's (1974) experiments on obedience and Humphreys's (1970) study of gay men's sexual activities as common reference points in their discussions on covert methods (e.g., Bryman, 2004; Crow, 2000; Esterberg, 2002; Gomm, 2004; May, 2001; Oakley, 2000).[1] The notion of covert research has come to represent a distinct and reprehensible strategy where researchers consciously obscure their motives, purposively deceive their informants, and in the case of participant observers disguise their identities. In short, covert research is often treated as an antithesis to open and overt research. The danger is that understanding all forms of concealment in fieldwork through this frame of reference means any research program involving covertness becomes vulnerable to censure.

Many authors have recognized that there is not a clear divide between overt and covert research (Agar, 1996; Bulmer, 1982b; Gomm, 2004; Herrera, 1999; Hilbert, 1980). A researcher may be able to maintain the subterfuge and conduct fully covert research, but completely overt research can never be guaranteed. It is certainly a mistake to assume that ethnographic fieldwork can ever be fully open and overt, with all the relevant participants giving their continued support based on a consistent understanding of the research. Consequently, any critique of covertness must question whether all occurrences of concealment should be considered universally unethical.

Most practiced ethnographers concede that fieldwork relationships inevitably involve some covertness (see, e.g., Grills, 1998; Shaffir & Stebbins, 1991; Smith & Kornblum, 1996). Relationships with informants are often contrived, and despite many developing into genuine friendships, the ulterior motives of ethnographers are entangled with all social encounters during the course of the research, however informal or incidental (Coffey, 1999). It is interesting to note that, even for some of the most hardened critics of covert research, certain forms of deception were still an accepted part of fieldwork. Cassell (1982), for example, treated the severity of deception as a semantic issue: Clear transgressions such as Humphreys's were unforgivable, but insincere compliance with social etiquette or exaggerated empathy was deemed to be appropriate when building rapport with informants. According to Cassell, these kinds of "social lies" (p. 18) were often part of everyday interaction and not ethically problematic.

In contrast, many writers have questioned the ethical implications of "rapport management" (Birch & Miller, 2000; Duncombe & Jessop, 2002; Finch, 1984; Luff, 1999). Finch (1984) argued that the carefully chosen gestures that make informants feel at ease and encourage them to be more candid in their responses were not benign actions but instrumental and potentially exploitative tactics. For Duncombe and Jessop (2002) rapport management reflected the power of researchers to control field relationships, and the relative powerlessness of informants to either detect this insincerity or to formulate their own strategies of resistance. Duncombe and Jessop consequently suggested that researchers "should continue to worry about these issues as they emerge in each piece of research and each individual interview" (p. 121). Such critiques encourage researchers to accentuate the ethical questions surrounding their every action. As a result, all attempts by researchers to engage with informants are potentially treated as a series of lies that (a) mask the researcher's true intent and (b) distort informants' ability to make appropriate decisions about the information they divulge.

This highlights the perpetual tension between the moral and professional expectations of academia and the contextual factors that are pervasive in ethnographic research. These critiques instill an intense ethical awareness, but such awareness merely problematizes our relationships in the field, while concealment and covertness may remain a constant part of ethnographic fieldwork. The key issue is how ethnographers and their critics attempt to address this inherent conflict. Institutional Review Boards and ethics committees expect researchers to predict where and when issues of concealment will emerge, assess the implications of such concealment, and construct bureaucratized forms of disclosure and consent that demonstrably negate any potential risk. It is assumed that concealment is thus avoidable because the contractual agreements between researchers and informants clearly define their mutual rights, roles, and obligations, and consensual participation is based on an appropriate understanding of the research.

Advocates of a more participatory approach to research have suggested that the only way to avoid concealment in fieldwork is to develop inquiries with the full collaboration of informants (Christians, 2000; Guba & Lincoln, 1989; Lincoln & Guba, 1989). At the heart of this participative model is an "ethics of care" "rooted in reciprocity, relatedness, and responsiveness" (Noddings, 2003, p. 2), and authors such as Denzin (1997) argued that research should be driven by a moral imperative to empower informants and encourage them to be coauthors of the research. It is presumed that building and nurturing open relationships in fieldwork allows for a constructive dialogue between ethnographers and informants about the development of the study. Consequently, "because the research-subject relation is reciprocal, invasion of privacy, informed consent, and deception are [assumed to be] nonissues" (Christians, 2000, p. 149).

The fundamental problem with the institutional review system and the participatory approach is the assumptions they perpetuate about relationships with informants. The rationalizing tendencies of the institutional review system conceptualize field relationships as coherent, formal processes and, in doing so, seem to ignore or deny the ambiguities intrinsic to relationships in exploratory research. Advocates of participatory research acknowledge the existence of ambiguity and suggest that emerging problems can be addressed through open dialogue (e.g., Guba & Lincoln, 1989), but the expectations surrounding how relationships are built and maintained are equally unreasonable. In establishing contacts, building relationships with individuals or gaining entry into social networks, concealment is still likely to pose moral dilemmas. Within participant observational studies of fragmented groups, with discontinuous membership, the sporadic and ephemeral nature of encounters make these problems particularly intense (see, e.g., Adler, 1993; Adler & Adler, 2002, 2004). The social context impedes the ethnographer's ability to maintain open, interactive relationships with all the potential informants throughout the research, and certain aspects of the ethnographer's identity and motivations inevitably remain concealed.

These models continue to treat all forms of concealment as inherently corrupt, which in turn has the potential to transform every encounter in the field into a moral quagmire. Therefore, what is needed is a pragmatic examination of why concealment arises in research and how related ethical issues may be addressed. Exposition of case studies such as the one in this article will not resolve the moral and professional crises that surround covert methods, but they will inform debates surrounding the practicalities of concealment in research. My aim is to encourage researchers, academic review boards, and ethics committees to develop a more critical understanding of the research process and the real-world contexts in which it takes place. Funding bodies and academic host organizations can thus avoid treating all forms of concealment as essentially flawed and cease to restrict research involving covert methods.

The Research

Aim

The aim of the current study was to consider how commonality and identification was articulated through hospitality exchange, and I sought to understand how the social aspects of hospitality were entangled with commercial agendas and the business of hospitality. I could not rely on secondary data or on survey methods to explore these issues, and I assumed that the most useful way to understand how commonality may (or may not) operate was by observing social interaction in specific contexts and by discussing people's experiences with them. From the beginning of the research, I attempted to identify potential research sites by making exploratory visits to bars, restaurants, and clubs and examining the nature of interaction between "hosts" and "guests." Consequently, any visit to a commercial hospitality venue became a potential source of information.

Context

The "Freelands" was a small bar located in a peripheral district of "Compton," a suburban town in England.[2] It was largely patronized by gay and lesbian consumers, although it is misleading to call the Freelands a "gay bar." Heterosexual men and women from the local neighborhood also frequented the venue, and the gay and "straight" clientele continually mixed. Patrons regularly came alone, in pairs, or in small groups of three to five and then formed larger groups that incorporated strangers. The size and location of the Freelands meant the operators could not rely exclusively on gay or straight consumers, and the bar thrived because the managers actively encouraged mixed patronage. My initial experience of the Freelands typified the inclusive culture that the managers sought to create. During my first visit, which was with a female friend, the bar staff were very friendly toward us and invited us to a birthday celebration taking place on the following Saturday. On the night of the party, the bar was officially closed at 11 p.m. but the managers encouraged us to stay for "after hours" drinks, and we eventually left the bar after 4 a.m. Being heterosexual did not seem to impede us from engaging with gay or lesbian patrons or exclude us from the Freelands.

Covertness in Research

The inclusive culture of the Freelands encouraged me to visit the bar as a consumer, but it quickly became apparent that the character of the social space made it relevant to my studies. During my subsequent visits, the interactions with consumers and service staff were increasingly driven by ethnographic curiosity. My cursory observations highlighted the need for a more intensive examination of who came to the bar, why they consumed there, and how their relationships were negotiated; however, being a consumer limited the time I could spend in the bar to short periods, ranging from 1 to 4 hours per visit. To understand how the consumer profile changed throughout the day, week, and month, it was necessary to spend extended periods in the bar. I, therefore, considered applying for a job at the Freelands so I could gain access to the consumers and staff.

I expected to have to engage in emotionally stressful negotiations with the managers about conducting my research, but gaining access proved to be relatively easy. Only three people worked at the Freelands, and one of the managers, Adam, told me that they needed more staff. I expressed interest in working at the Freelands so when the second manager, Shawn, went on holiday for a week, Adam asked if I wanted to replace him temporarily. After 2 weeks, Shawn phoned and said he had lost his passport so I continued to work there in his absence. When he eventually returned to Britain, he did not come back to work, so Adam, a third member of staff and I were left to manage the Freelands.

Like many researchers in organizational settings, I found it difficult to be open about my research because I feared that the bar's operators would be reluctant to expose themselves to scrutiny. Diamond (1992), for example, had to pursue the majority of his research on nursing homes covertly because he felt that the doctors and senior staff would feel threatened by his study and deny him access. Similar to Diamond, I anticipated that defining myself as a researcher early in our relationship would be counterproductive, and my sociological curiosity remained veiled in my enthusiasm for working at the Freelands. However, in contrast to him, I did not intend to conduct my research covertly. I assumed that once the managers and I developed a closer relationship, in which I was seen as a sympathetic "active member" (Adler & Adler, 1987) within their social milieu, they would not be as threatened by my ethnographic intent. During my 2nd week, I revealed to the managers my interest in the bar and its patrons. I explained that I wanted to write about the relationship between people's sense of identities and their experiences of drinking venues and, that I wished to interview customers and bar staff. The managers said they would be happy to contribute and from then on they regularly introduced me to customers as the "straight man writing a book about us." I did not gain entry to a group simply because there was no coherent group as such. Nevertheless, working at the Freelands provided me with the opportunity to interact with the staff and customers.

I worked at the bar for 27 months, working nearly every night of the week for the first 8 months, and 1 or 2 nights per week after this period. In addition, I usually came to the Freelands once or twice during the days and on my nights off. Having got to know some of the customers, I also participated in a number of social events with people outside the bar. I regularly went to house parties and to other gay and straight venues in and around London with people from the Freelands. Meeting people outside of the Freelands were sincerely enjoyable social activities, despite the fact that I often commented on these events in one of my diaries. There was an effective collapse of the work-leisure divide, and it was often impossible to separate the moments when I assumed the role of ethnographer and when I resumed being a "civilian."

It is interesting to note that Shils (1982, pp. 131–132) drew a simple divide between situations where observation was part of everyday life, where it was deemed healthy, and situations where observation was academically motivated and made possible through some kind of intentional manipulation on the part of the researcher. However, although the basic principle of Shils's argument is sound, such criticism does not adequately recognize the duality and multiplicity involved in all social encounters. Awareness of the social and physical surroundings is obviously an essential quality for an ethnographer, and this sense of awareness becomes instinctive. Furthermore, the context blurred the divide between a social encounter and a sociologically useful encounter.

The ethnographic intent was inseparably entangled with my social life, and I continually appropriated idle gossip, conversations, and comments during my visits to the Freelands and other hospitality venues. Patrons constantly provided useful information during momentary encounters, but most of the people I saw and heard did not formally consent to share their experiences. For example, in one incident, two people were dancing in the Freelands – being loud and drawing considerable attention from others. Another (older) patron rolled his eyes and said the words *bloody queens* as a passing comment to some of the others and me in the bar. Publicly broadcast declarations such as this did not necessarily warrant elaborate clarification of my reception or potential interpretation. Writers generally agree that informing people in public places of the researcher's intent is unnecessary (Bulmer, 1982b; Dingwall, 1980; Lee, 2000; Lofland & Lofland, 1995; Punch, 1986, 1994; Roth, 1962). During momentary encounters such as this, stopping the person to explain who I was, what I intended to do, and then asking for him or her to repeat his opinion was certainly impractical. However, as the diary extract below demonstrates, supposedly public space could easily be transformed into private space, which, in turn, radically changed the ethical implications of the encounter.[3]

> A man came in around 8 p.m. . . . I'd never seen him before, and he did not seem like the sociable type. He sat at the bar alone drinking, smoking and staring into empty space. I was really nervous, but I really wanted to find out who he was and what he was doing here. As I was cleaning the bar, I casually mentioned that I had never seen him before. He said

> he did not come here very often because the last time he came people got a "bit funny" with him. I asked what he meant, but he seemed reluctant to tell me about it. He said he used to come here before [the Freelands started to attract gay and lesbian consumers] and when he read about [the bar's new customer policy in a local newspaper] he came in again. He said he was "surprised at how ordinary it was." I felt the ice was breaking and asked if he went anywhere else. He responded bluntly with "I am not gay!" When I told him that I was not gay either, he seemed to relax a lot more. He told me about his job as well as his identity crisis as a bisexual man. [During the next 15 minutes, Mike talked about his past encounters, sexual preferences, his homophobic family, and his views on bisexuality.] Whenever he mentioned his sexual experiences, he lowered his voice and looked around. Even here, he was so conscious about what he was saying. I felt terrible because he confided in me, and I knew his comments would make it into the diary. (April 02, 2001)

When customers revealed personal or sensitive information it instantly turned relatively unproblematic conversations into more intimate exchanges. Working at the Freelands facilitated this kind of shift in interaction, and it, therefore, became necessary to continually question the context of the encounter and the personal factors that determined when and how I could provide credible explanations of the research. Within the following sections, I examine the issues that affected my relationship with informants and then proceed to illustrate the techniques I used to communicate my research intentions. I maintain that while these techniques contain elements of covertness, it is wrong to suggest that they are unethical.

Negotiating Covertness

The encounters in the Freelands were often ephemeral, which made it difficult to build rapport, but levels of education and specific issues surrounding sexuality and gender were also critical in determining the relationships I formed with informants. It became evident that those with postsecondary school education, especially those who had been to university, were often much more interested in my research.[4] Others, usually those without degree-level education, tended to pay less attention to my well-rehearsed explanations and appeared less interested in my work. During these encounters, I was forced to either abridge my accounts or abandon explanations altogether and concentrate on developing informal relationships and building rapport.

Recruiting lesbian consumers to participate in the research was also difficult because male clientele outnumbered females, and lesbian patrons were frequently less integrated into other social networks. Lesbian women tended to drink in couples or small groups and mixed less with other male or female customers. Being a heterosexual male meant I knew less about lesbian women; and, because I had fewer opportunities to interact with them, there were fewer opportunities to find out more about them. Approaching couples or small groups of women was often impossible to do casually, and my attempts to engage members of such groups spontaneously were often fruitless. Only one woman refused outright to participate in the study, but seven other women who initially agreed to take part in interviews did not come to the arranged meetings and never contacted me again.

In contrast, heterosexual women were easier to interact with, although recruitment was still problematic. The vast majority of straight women treated the Freelands as a liberating space where they could interact with men without the danger of heterosexual male objectification (cf. Moran, Skeggs, Tyrer, & Corteen, 2001; Skeggs, 1999). They were often very communicative, although on at least three occasions women misunderstood my motivations and assumed that my conversations about research were part of an elaborate mating ritual. Consequently, I began to emphasize the relationship with my girlfriend early in conversations to avoid misunderstandings.

Abrupt and Incremental Disclosure

Each encounter in the Freelands brought with it different opportunities and tensions, but the strategies and tactics I used to negotiate these encounters became increasingly repeated. In short, the communication of my research occurred either abruptly or incrementally. Within the abrupt method, I approached people unexpectedly, introduced myself, and explained my work before asking a series of questions. It was often necessary to use this approach with lesbian women and other infrequent customers because there were fewer opportunities to obtain their opinions. In many cases, especially in the beginning of my research, people saw this as an intrusion into their leisure time and space and were reluctant to participate. Not everyone reacted negatively, but when this was the outcome of an encounter, withdrawal became the only alternative. I often approached people opportunistically either because they were on their own or because they were part of a larger, boisterous, and friendly looking group with three or more members.[5] However, as I explain in more detail below, in the later stages of the research my relationship with existing patrons helped to legitimize my work, and emphasizing that other patrons had contributed to the study usually made new informants more comfortable about discussing their opinions with me.

Within the incremental approach, I established an informal contact and developed it into a research relationship. I often neglected to highlight my research intentions during initial meetings and then introduced my work casually during subsequent encounters. For example, I purposively talked to others about books and about how my work was progressing in front of Mike for over a year so he could hear us. I used these opportunities to determine whether it would be reasonable to ask him to participate formally in the research. He did not understand, or seem to want to understand, and in general he was apathetic toward these conversations. Despite this, we regularly talked about relationships, holidays, families, and the Freelands. In certain situations, our relationship seemed more like friendship, although I found it difficult to ask him to participate in a recorded interview. After much deliberation, I eventually told Mike about my work and that I had written about him. I felt I owed him an explanation, and it certainly felt good to be honest with him. He was unsure about it at first, but after I explained my research in more detail he seemed positive about it. He said he would like to read my work although we were skeptical about giving him extracts in case members of his family found them.

The incremental approach was most effective with the regular male clients, gay and straight, and their female acquaintances. Throughout my research, I tried both variations depending on the situation and my courage at the time, although I tended to use the incremental approach. People were slowly eased into the research relationship and were given time to adjust: The interaction was longitudinal, and mutual understanding of our roles and obligations developed over time. For example, most of my 26 key informants continued to share "gossip" and offer suggestions about who I should interview and what I should ask, without solicitation from me; and in return, I reciprocated with small gifts, drinks, and chauffeured people in my car.

The Language of Concealment and Disclosure

The incremental approach usually involved subtle, often covert methods to elicit information from people in the beginning of the relationship. For example, instead of asking informants to comment on specific people, which implied surveillance, I casually encouraged them to look around the room and comment on what they thought about the customers, the decorations, or the venue in general. Observation no longer focused on the individual but shifted, semantically, to the general, and this transformed a potentially alien activity into a mundane form of voyeurism.

This method was useful during brief encounters, but within longer interactions it was often more practical to offer broader explanations about my work. Being more overt allowed me to ask direct questions not only about people's opinions of the bar and its consumers but also about their perceptions of other venues and their patrons. However, even in my attempts to be more open, elements of concealment were still present. For example, like many other ethnographers, I usually avoided using terms such as *sociology* or *ethnography* in my initial introductions and explanations of my work (cf. Agar, 1996; Pryce, 1986; Shaffir, 1991; Whyte, 1981). I used terms such as *writing* instead of *researching* because I assumed that the latter would imply invasive surveillance and a separation between the researcher and researched (as the object of study). The term *writing* was deliberately meant to inject ambiguity to help build rapport between potential informants and me. For the same reason, I highlighted the dominance of male perspectives in my study when trying to recruit lesbian informants. I encouraged them to contribute so they can make their narratives a more overt part of the Freelands's legacy. The explanations of the work I gave to lesbian informants deemphasized my interest in the business of hospitality and, instead, stressed my sociological curiosity about their lives and experiences.

The prevailing critiques of covert methods and concealment led me to question whether such partial and incomplete explanations offered reasonable enough clarification of the study for informants. However, the fieldwork made it apparent that esoteric notions such as identity performance, behavioral patterns, or social networks were alien and meaningless to most of the people in the bar. I reshaped the descriptions of my research to suit the audience, and specific interpretations of my work were purposive attempts to create empathetic relationships between informants and me. The claims that I was writing about the places where people drink, or questions about where else they drank, or what their opinion was of another venue were not meant to mislead informants. To another sociologist these statements and questions may seem like distorted interpretations of my work, but these were attempts to present the research in terms informants found meaningful. My informants' understanding of my work was certainly not as intensive or as nuanced as mine, although it is misleading to suggest that they were deceived. I gave elaborate explanations to individuals who were prepared to listen, but I inevitably abridged my accounts to people for whom clarification appeared superfluous.

Visibility and Change in Concealment and Disclosure

Working at the Freelands allowed me to be increasingly open about my work: Explanations of my studies were gradually woven into casual conversations, which then filtered through to other clients. Existing informants and patrons who knew about my research began to discuss it with others and, similarly to Mann's study of "Brady's bar," news of my research reached many of the regular clientele through "bar talk" (Mann, 1976; Spradley & Mann, 1975). For example, one of my key informants, Dave, found out about my work after talking to the manager. He approached me and asked about my research, and after a brief discussion we arranged to meet for an informal interview. Following this initial meeting, I regularly talked with him in the bar about my work, and I was often able to turn to others present and ask for their contributions. This ability to discuss the research openly became an effective way to make my work and my dual role in the social setting more visible. For instance, my first interview with Dave was conducted in the Freelands. Three other customers saw us and then amused each other by walking past our table and talking into the tape recorder. After a brief explanation, they joined us and three more people sat down with us within the next hour. The interview proved to be very productive, and the majority of the initial interviews with other informants were subsequently conducted in the bar.[6] Visibility thus helped make the research seem more broadly accepted and encouraged new people to take an interest.

The research became increasingly overt during the 27 months, and existing informants supported this progressive disclosure. As I noted previously, informants continued to offer information openly throughout the study, and many of the key informants subsequently helped to recruit new participants. However, the research never became completely overt: Some of the patrons remained unaware of the study, while others continued to have only partial knowledge or understanding of it. Consequently, the problems of concealment persistently surfaced throughout the study, and disclosure was always incomplete.

These undisclosed aspects of the study continue to raise moral and professional dilemmas at the publication stage. Feelings of disillusionment or betrayal are common among informants once they encounter textual representations of their lives (cf. Boelen, 1992; Bosk, 2001; Morgan, 1972; Scheper-Hughes, 2000; Vidich & Bensman, 2000; Whyte, 1981).[7] The problems may be greater here because many patrons were unaware that they would be included in such accounts. Those championing a more inclusive research agenda have argued that such problems can be avoided by persuading informants to take an active part in the writing and publication process (e.g., Denzin, 1997; Guba & Lincoln, 1989; Lincoln, 1995). This approach, however, rests on the assumption that informants are as interested in or committed to the enterprise as the researcher. It was certainly difficult to expect this sort of commitment or interest from most of the informants in this research context. As I have argued throughout this article, many relationships in the Freelands were ephemeral and tenuous, while others such as the one I built with Mike, though longer and more intensive, still made it largely impossible to engage in meaningful dialogue about my research.

To date, only three informants have seen my written conclusions, and only one, Damien, has read them in detail. He spent several weeks reading drafts of my thesis, and his critical comments were an invaluable help. The vast majority of those who contributed openly to the study no longer patronize the Freelands, and I remain hesitant about distributing my work among the people who continue to consume there. My research arises occasionally in conversations with patrons who know of my study but did not participate formally, although I have only pursued the idea of sharing my conclusions with a small number of them. This is partly because the thesis that emerged from the fieldwork is esoteric and written for an academic audience, but principally because it contains a series of candid observations that I fear would place many people in compromising positions.[8]

The conclusions of my research will most likely appear in heavily edited academic articles. I have agreed to give copies of articles emerging from the study to one of my key informants, and I plan to provide copies to two others. They will undoubtedly pass these articles on to others and, in doing so, will help those unfamiliar with the study to gain a broader understand of it; but I also anticipate that many of my future encounters with patrons will be devoted to mediating between them and the representations of their social milieu that come to light. I suspect that some of these will be tense exchanges, but writing and publishing this article before any others will, I hope, create a clear point of reference that informs any potential dialogue.

Conclusion

Conceptualizing field research as an ethical enterprise reflects the maturity of ethnographic inquiry. However, when we attempt to address the ethical implications of our work, our frames of reference draw on a culture of denigration that treats all our untruths as professional misconduct. My fieldwork experiences demonstrate that our inquiry, and the ethical critiques of our inquiry, can never ignore the context in which the research takes place. By understanding the consequences that specific contextual factors have on ethnographic research, we can accept that the problems of concealment are likely to resurface in our inquiries. It is, therefore, essential that researchers and their critics understand why the relationships between ethnographers and informants are entangled with concealed truths.

Within my research, the social context was critical in determining the level and the nature of concealment. The venue was patronized by a diverse range of consumers who often spent short periods in the bar. The discontinuous nature of the community and the transitory encounters with informants meant these relationships were inherently based on partial knowledge of my intentions. Although this may seem unique to this social context, ethnographers will inevitably encounter fragmented communities in which their roles as researchers remain veiled because they have limited opportunities to disclose information about their work. It is a truism to suggest that certain contexts do not lend themselves to simplistic acts of disclosure or consent; however, I illustrated how the context fundamentally influenced the social relationships, which in turn were critical in determining the overt or covert status of the study.

The Freelands was part of my own personal "leisure geographies," and this consequently eroded the distinction between work and leisure. It was impossible to disentangle those moments when awareness and interpretation was purely sociological from when it was social, and my academic motivations were, therefore, inherently veiled. Furthermore, I illustrated how such social factors as sexuality, gender, and differences in education influenced the nature of my relationships with informants, which in turn were critical in determining how my work and I could be presented. Certain descriptions of my work, those which abandoned references to esoteric academic concepts, were often more meaningful to informants and, therefore, more useful in encouraging them to participate in the research. It is misleading, however, to suggest that abridged reconstructions of my research were unethical. In contrast, these incomplete explanations of my work and me were used to create and nurture social bonds that could be developed into open relationships in which I could be more explicit about my study.

The earlier stages of my research were characterized by covertness regarding my identity. During the later stages of my fieldwork, when I had begun to develop a network of acquaintances and informants who could legitimize my multiple statuses as consumer, employee, and researcher, I was able to disseminate information about my work more overtly. This did not necessarily mean my explanations were more elaborate and therefore more candid: The problems I outlined above continually made it difficult to form relationships with certain patrons, and many were not interested in my accounts. Nevertheless, I purposefully attempted to make my roles more transparent, and the research became more overt. This demonstrates that instead of focusing on any individual act of concealment, it is important to understand how relationships may change and develop throughout the fieldwork.

However, the most important question to emerge from my research is not simply why concealment was evident or how I negotiated these problems: The key question is how we treat the strategies and tactics I used in the fieldwork. The literature on research ethics discouraged me from conducting my study totally covertly. At the same time, the critiques of covert methods and concealment, particularly those that problematized rapport building and maintenance, made all relationships vulnerable to the charge of being exploitative. In the Freelands, every encounter was tainted with the potential to be sociologically fruitful, which made it subject to these critiques of rapport. Within every encounter, however mundane or trivial, I was acutely aware that the people I encountered were not equipped with comprehensive information about my work or me. There was constant pressure to compensate for this perpetual sense of dishonesty, but despite my efforts to make my research more overt, the fieldwork context made it impossible to eliminate concealment from the study.

The critiques of covert research and concealment did not eliminate these problems, but they did help to nurture a paralyzing state of reflexive self-criticism where every act was open to perpetual debate. I am not encouraging others to engage in subterfuge, and I do not claim that researchers should feel comfortable about misleading their informants. Nor do I suggest that researchers stop engaging in reflexive criticism of their own actions in the field. However, I do urge ethnographers and their critics to look to this and other accounts of concealment in fieldwork and to reassess the appropriateness of existing critiques of concealment and covert methods (cf. Agar, 1996; Leo, 1995, 1996; Shaffir, 1991). If all encounters in the field are treated as inherently unethical, and the researcher's every act is treated as a potential source of ethical crisis, we risk jeopardizing the future of all ethnographic inquiry. Regardless of whether universities or funding organizations restrict covert or semicovert research because of legal or moral reasons, the fact remains that if all concealment is considered universally unethical, any ethnographic research is potentially at risk of being suppressed because the problems surrounding concealment and disclosure I described here are likely to reemerge.

Therefore, the critical questions for ethnographers do not simply concern how they engage with informants but also how they distinguish between those concealments that are necessary or unavoidable in these relationships and those that represent dangerous or irresponsible moral transgressions. The divide between these different kinds of untruths will always be ambiguous, but understanding the realities of fieldwork can help researchers to avoid agonizing over all their duplicities. Furthermore, reflecting on covert methods without treating them as inherently transgressive can help to avoid polarizing researchers and academic review boards or ethics committees in their interpretations of concealment and disclosure.

Notes

1. Stanley Milgram (1974) conducted a series of experiments in which subjects were asked to administer electric shocks to respondents if they answered questions incorrectly. The electric shocks were not real and the respondents were research confederates pretending to feel pain, but the experience caused considerable emotional stress for the subjects. Laud Humphreys studied the behavior of men who engaged in sex with other men in public toilets. Humphreys observed the men and then recorded their car number plates, which he then used to trace their identities. He subsequently disguised himself and interviewed a number of the men under the pretence that he was conducting a public health survey.
2. The names of all the people and places have been changed to try to maintain anonymity.
3. Denzin (1997) has argued that the distinction between private and public are no longer appropriate and everything that was once imagined to be individual should instead be thought of as "public and part of the local and moral community" (p. 278). He consequently suggested that all knowledge should be considered equally sacred and that representations of people's lives had to be constructed through open and honest dialogue with informants. I agree with Denzin that the distinction between public and private is inevitably blurred, and also that the knowledge emerging from our interactions with informants should be treated respectfully; but, as this case study demonstrates, the sort of pluralistic dialogue he championed is not always feasible.
4. Secondary school is the mandatory level of education in the United Kingdom.
5. Approaching couples abruptly often resulted in short, awkward interactions and it was usually better to avoid disturbing two people unless they started the conversation with me or showed active willingness to interact.

6. These public interviews were then followed by private interviews with individual informants away from the bar.
7. I appreciate that offending or upsetting individuals are not the only risks surrounding publication. Revealing the illegal activities that took place in the bar may invite police scrutiny, descriptions of people may undermine their wish to keep their sexuality secret, and my portrayal of individual gay, lesbian, or bisexual consumers may serve various political or commercial interests, but the lack of space prevents me from addressing these issues here.
8. When Damien read drafts of the thesis he easily recognized certain characters in spite of my attempts to disguise their identities, and distributing the whole manuscript in such a parochial social context would compromise many of my informants' anonymity.

References

Adler, P. A. (1993). *Wheeling and dealing: An ethnography of an upper-level drug dealing and smuggling community* (2nd ed.). New York: Columbia University Press.

Adler, P. A., & Adler, P. (1987). *Membership roles in field settings*. Newbury Park, CA: Sage.

Adler, P. A., & Adler, P. (2002). Do university lawyers and the police define research values? In W. C. van den Hoonaard (Ed.), *Walking the tightrope: Ethical issues for qualitative researchers* (pp. 34–42). Toronto, Canada: University of Toronto Press.

Adler, P. A., & Adler, P. (2004). *Paradise laborers: Hotel work in the global economy*. Ithaca, NY: Cornell University Press.

Agar, M. (1996). *The professional stranger: An informal introduction to ethnography* (2nd ed.). San Diego, CA: Academic Press.

Beauchamp, T., Faden, R., Wallace, J., & Walters, L. (Eds.). (1982). *Ethical issues in social scientific research*. Baltimore: John Hopkins University Press.

Birch, M., & Miller, T. (2000). Inviting intimacy: The interview as "therapeutic opportunity." *Social Research Methodology, Theory and Practice, 3*, 189–202.

Boelen, W. A. M. (1992). Street corner society: Cornerville revisited. *Journal of Contemporary Ethnography, 21*(1), 11–51.

Bok, S. (1986). *Secrets: On the ethics of concealment and revelation*. Oxford, UK: Oxford University Press.

Bosk, C. L. (2001). Irony, ethnography, and informed consent. In B. Hoffmaster (Ed.), *Bioethics in social context* (pp. 199–220). Philadelphia: Temple University Press.

Bryman, A. (2004). *Social research methods*. Oxford, UK: Oxford University Press.

Bulmer, M. (1982a). Ethical problems in social research: The case of covert participant observation. In M. Bulmer (Ed.), *Social research ethics: An examination of the merits of covert participant observation* (pp. 3–12). London: Macmillan.

Bulmer, M. (1982b). The merits and demerits of covert participant observation. In M. Bulmer (Ed.), *Social research ethics: An examination of the merits of covert participant observation* (pp. 217–251). London: Macmillan.

Calvey, D. (2000). Getting on the door and staying there: A covert participant observational study of bouncers. In G. Lee-Treweek & S. Linkogle (Eds.), *Danger in the field: Risks and ethics in social research* (pp. 43–60). London: Routledge.

Cassell, J. (1982). Harms, benefits, wrongs, and rights in fieldwork. In J. E. Sieber (Ed.), *The ethics of social research: Fieldwork, regulation, and publication* (pp. 7–31). New York: Springer.

Christians, C. G. (2000). Ethics and politics in qualitative research. In N. K. Denzin & Y. S. Lincoln (Eds.), *Handbook of qualitative research* (2nd ed., pp. 133–155). Thousand Oaks, CA: Sage.

Coffey, A. (1999). *The ethnographic self: Fieldwork and the representation of identity*. London: Sage.

Crow, I. (2000). The power of research. In D. Burton (Ed.), *Research training for social scientists: A handbook for postgraduate researchers* (pp. 68–80). London: Sage.

de Laine, M. (2000). *Fieldwork, participation and practice: Ethics and dilemmas in qualitative research*. London: Sage.

Denzin, N. K. (1997). *Interpretative ethnography: Ethnographic practices for the 21st century*. Thousand Oaks, CA: Sage.

Denzin, N. K., & Erikson, K. T. (1982). On the ethics of disguised observation: An exchange. In M. Bulmer (Ed.), *Social research ethics: An examination of the merits of covert participant observation* (pp. 142–151). London: Macmillan.

Diamond, T. (1992). *Making gray gold: Narratives of nursing home care*. Chicago: University of Chicago Press.

Dingwall, R. (1980). Ethics and ethnography. *Sociological Review, 28*(4), 871–891.

Duncombe, J., & Jessop, J. (2002). "Doing rapport" and ethics of "faking friendship." In M. Mauthner, M. Birch, J. Jessop, & T. Miller (Eds.), *Ethics in qualitative research* (pp. 107–122). London: Sage.

Erikson, K. (1995). Commentary. *American Sociologist, 26*(2), 4–11.

Esterberg, K. G. (2002). *Qualitative methods in social research*. London: McGraw-Hill.

Finch, J. (1984). It's great to have someone to talk to: The ethics and politics of interviewing women. In C. Bell & H. Roberts (Eds.), *Social researching: Politics, problems, practice* (pp. 70–87). London: Routledge.

Gomm, R. (2004). *Social research methodology: A critical introduction*. Basingstoke, UK: Palgrave Macmillan.

Goode, E. (2001). The ethics of deception in social research: A case study. In J. M. Miller & R. Tewksbury (Eds.), *Extreme methods: Innovative approaches to social science research* (pp. 239–257). Boston: Allyn & Bacon.

Grills, S. (Ed.). (1998). *Doing ethnographic research: Fieldwork settings*. Thousand Oaks, CA: Sage.

Guba, E. G., & Lincoln, Y. S. (1989). *Fourth generation evaluation*. Newbury Park, CA: Sage.

Herrera, C. D. (1999). Two arguments for "covert research" in social research. *British Journal of Sociology, 50*(2), 331–341.

Hilbert, R. A. (1980). Covert participant observation: "On its nature and practice." *Urban Life, 9*(1), 51–78.

Humphreys, L. (1970). *Tearoom trade: Impersonal sex in public places*. London: Duckworth.

Kimmel, A. J. (1996). *Ethical issues in behavioral research: A survey*. Oxford, UK: Blackwell.

Lauder, M. (2003). Covert participant observation: Justifying the use of deception. *Journal of Contemporary Religion, 18*(2), 185–196.

Lee, R. M. (2000). *Unobtrusive methods in social research*. Buckingham, UK: Open University Press.

Leo, R. A. (1995). Trial and tribulations: Courts, ethnography, and the need for an evidentiary privilege for academic researchers. *American Sociologist, 26*(1), 113–134.

Leo, R. A. (1996). The ethics of deceptive research roles reconsidered: A response to Kai Erikson. *American Sociologist, 27*(1), 122–128.

Lincoln, Y. S. (1995). Emerging criteria for quality in qualitative and interpretative research. *Qualitative Inquiry, 1*(3), 275–289.

Lincoln, Y. S., & Guba, E. S. (1989). Ethics: The twin failure of positivist science. *Review of Higher Education, 12*(3), 221–240.

Lincoln, Y. S., & Tierney, W. G. (2004). Qualitative research and institutional review boards. *Qualitative Inquiry, 10*(2), 219–234.

Lofland, J., & Lofland, L. H. (1995). *Analyzing social settings: A guide to qualitative observation and analysis* (3rd ed.). London: Wadsworth.

Luff, D. (1999). Dialogue across the divides: "Moments of rapport" and power in feminist research with anti-feminist women. *Sociology, 33*(4), 707–731.

Mann, B. J. (1976). The ethics of fieldwork in an urban bar. In A. Rynkiewich & J. P. Spradley (Eds.), *Ethics and anthropology: Dilemmas in fieldwork* (pp. 95–109). New York: John Wiley.

May, T. (2001). *Social research: Issues, methods and process* (3rd ed.). Buckingham, UK: Open University Press.

Milgram, S. (1974). *Obedience to authority: An experimental view*. London: Tavistock.

Miller, J. M. (2001). Covert participant observation: Reconsidering the least used method. In J. M. Miller & R. Tewksbury (Eds.), *Extreme methods: Innovative approaches to social science research* (pp. 13–21). Boston: Allyn & Bacon.

Mitchell, R. G. (1993). *Secrecy and fieldwork*. Newbury Park, CA: Sage.

Moran, L., Skeggs, B., Tyrer, P., & Corteen, K. (2001). Property, boundary, exclusion: Making sense of hetero-violence in safer places. *Social and Cultural Geography, 2*(4), 407–420.

Morgan, D. H. J. (1972). The British Association scandal: The effect of publicity on a sociological investigation. *Sociological Review, 20*(2), 185–206.

Nelson, C. (2004). The brave new world of research surveillance. *Qualitative Inquiry, 10*(2), 207–218.

Noddings, N. (2003). *Caring: A feminine approach to ethics and moral education* (2nd ed.). Berkeley: University of California Press.

Oakes, J. M. (2002). Risks and wrongs in social science research: An evaluator's guide to the IRB. *Evaluation Review, 26*(5), 443–479.

Oakley, A. (2000). *Experiments in knowing: Gender and method in the social sciences*. Cambridge, UK: Polity.

Pryce, K. (1986). *Endless pressure* (2nd ed.). Bristol, UK: Bristol Classical Press.

Punch, M. (1986). *The politics and ethics of fieldwork*. Newbury Park, CA: Sage.

Punch, M. (1994). Politics and ethics in qualitative research. In N. K. Denzin & Y. S. Lincoln (Eds.), *Handbook of qualitative research* (pp. 83–97). Thousand Oaks, CA: Sage.

Roth, J. A. (1962). Comments on "secret observation." *Social Problems, 9*(3), 283–284.

Scheper-Hughes, N. (2000). Ire in Ireland. *Ethnography, 1*(1), 117–140.

Shaffir, W. B. (1991). Managing a convincing self presentation: Some personal reflections on entering the field. In W. B. Shaffir & R. A. Stebbins (Eds.), *Experiencing fieldwork: An inside view of qualitative research* (pp. 72–81). Newbury Park, CA: Sage.

Shaffir, W. B., & Stebbins, R. A. (Eds.). (1991). *Experiencing fieldwork: An inside view of qualitative research*. Newbury Park, CA: Sage.

Shils, E. (1982). Social inquiry and the autonomy of the individual. In M. Bulmer (Ed.), *Social research ethics: An examination of the merits of covert participant observation* (pp. 125–141). London: Macmillan.

Skeggs, B. (1999). Matter out of place: Visibility and sexualities in leisure spaces. *Leisure Studies, 18*, 213–232.

Smith, C. D., & Kornblum, W. (Eds.). (1996). *In the field: Readings on the field research experience* (2nd ed.). Westport, CT: Praeger.

Spradley, J. P., & Mann, B. J. (1975). *The cocktail waitress: Woman's work in a man's world*. New York: John Wiley.

Vidich, A. J., & Bensman, J. (2000). *Small town in mass society: Class, power, and religion in a rural community* (rev. ed.). Chicago: University of Chicago Press.

Warwick, D. P. (1982). Tearoom trade: Means and end in social research. In M. Bulmer (Ed.), *Social research ethics: An examination of the merits of covert participant observation* (pp. 38–58). London: Macmillan.

Whyte, W. F. (1981). *Street corner society: The social structure of an Italian slum* (3rd ed.). London: University of Chicago Press.
Wong, L. M. (1998). The ethics of rapport: Institutional safeguards, resistance, and betrayal. *Qualitative Inquiry, 4*(2), 178–199.
Wright, D. (2004). Creative nonfiction and the academy: A cautionary tale. *Qualitative Inquiry, 10*(2), 202–206.

34

Ethical Covert Research

Paul Spicker

Covert research has been heavily criticized on ethical grounds; in some discussions of ethical principles in research, it is treated as one of the principal issues (e.g. Bulmer, 1982; Punch, 1986). Students are being advised not to undertake such research. This, for example, comes from a textbook on research for social workers:

> Some of the more famous participant observation studies (such as Whyte's study of street corner gangs) collected rich data because they did not disclose to the observed that their behaviour was being researched. Nowadays, such studies are viewed as ethically untenable and we are no longer able to conduct such research. (Alston and Bowles, 2003: 196)

This reflects an atmosphere in which research is becoming increasingly regulated and hemmed in by ethical constraints (Haggerty, 2004). In an environment where institutions are actively scrutinizing the ethical implications of research proposals, and where there tends to be a presumption that covert research is problematic, it is becoming increasingly difficult to plan and execute research if covert elements have been identified (e.g. Iphofen, 2006; Tysome, 2006).

Covert research is often seen as intrinsically illegitimate.

> As a general principle, deception or concealment of the purposes of a study from, or covert observation of, identifiable participants are not considered ethical because they are contrary to the principle of respect for persons in that free and fully informed consent cannot be given. (Australian NHMRC, 1999; cf. Government of South Africa, 2004)

Source: *Sociology*, 45(1) (2011): 118–133.

The identification here of covert research with deception is very common (e.g. Bulmer, 1982; Herrera, 1999; Homan, 1991). Little in the article that follows refers to deception, but to avoid confusion I need to make a clear distinction between them at the outset. Covert research is research which is not disclosed to the subject – where the researcher does not reveal that research is taking place. It is covert research, for example, if a researcher simply stands and watches what people are doing, like checking whether motorists are using mobile phones (Walker et al., 2006). It is covert research when a researcher attends a public event, like a trial, a football match or a political meeting, and writes about it. It is covert research if a researcher accumulates information about subjects who do not know it – for example, someone writing an unauthorized biography, or someone undertaking secondary analysis of data about human subjects. It is covert research if a researcher sees and records something when people have not realized that research is going on (cf. Power, 1989). The term 'limited disclosure', favoured by the Australian National Health and Medical Research Council, helps to make two points about covert action. The first is that disclosure is a positive act; a researcher may not actively be concealing the research, but unless there are steps taken to disclose what is happening, the activity may still be considered covert. The second is that disclosure is not a dichotomous concept, something that is done or not done, but a spectrum of activity; researchers can reveal nothing, a little, or a lot, and still not be making a full disclosure (Australian NHMRC, 2007: 2.3). As many writers have recognized, overt research often has covert elements (Lugosi, 2006)

Deception, by contrast, occurs where the nature of a researcher's action is misrepresented to the research subject. This can be done at the same time as covert work, but deception is not necessarily, or even usually, done in the form of covert research. Researchers who engage in deception mainly say they are doing one thing when they are actually doing another (as in Milgram's famous experiments [1974]). In a review of the use of deception in social psychology in nearly 1200 studies over 20 years, participants were unaware that a study was going on in only one study in seven (cited Kimmel, 1996: 73). In other words, although some deceptive research is also covert, most is not.

Many of the objections which are made to covert research are objections to deception rather than covert activity. Homan's lengthy review of objections to covert activity (1991) includes the concerns that:

- deception betrays trust;
- deception 'pollutes the research environment', leading people to be suspicious of researchers;
- the habit of deception infects the researcher's behaviour – it 'becomes a way of life';
- the strain of maintaining a deception may be damaging to the researcher
- deception is bad for the reputation of research; and
- use of the method may legitimate deception more widely.

Homan finds good reasons to qualify them: the concerns are over-generalized, and there are many circumstances in which they do not apply. Whether or not they are justifiable, however, I do not think that they have much to do with the use of covert research. They apply, rather, to research where deception is practised, whether it is covert or overt.

The widespread confusion between deception and covert activity means, unfortunately, that a conventional literature review does little to clarify the ethical issues; I have consequently avoided referring to the usual litany of deceptive studies and the papers that discuss them. In this article, I want to focus instead on the arguments about covert research in itself – that is, research done without the knowledge, and so without the informed consent, of the people who are its subjects.

Why Not Disclose?

The main reasons for limiting disclosure are practical and methodological. The methodological argument is a strong one. Every researcher knows, or should know, that the experience of being observed is likely to affect the behaviour of the person who is being observed; the Hawthorne experiments are a familiar staple of basic textbooks (Olsen et al., 2004). It has been argued, for example, that:

> . . . in participant observation studies it is virtually impossible to obtain consent from all observed individuals . . . Obtaining consent would interfere with the strength of the 'naturalist' approach of ethnography. Seeking consent from participants in these situations may lead to behavioural changes that would invalidate the research. (Australian NHMRC, 2001: E130)

There are different ways of trying to deal with the problem, but if the objective of a study calls for observation of people's behaviour in a particular context, the research has to minimize the effect of the process of observation on the way that people behave. Limiting disclosure by the researcher is the obvious way to counter that problem.

The main argument against this is that it is often impossible not to affect the subject under study. Attendance at events, for example, is often taken to be implicit support of those events, and participant observation in particular often involves engagement in the activity being studied. Homan argues that covert research is 'invisibly' reactive, whereas overt research tends to be explicit about it (1991). This limits the scope to claim that research has not affected the behaviour being studied, but it does not vitiate the process. In the same way as it is impossible to be perfectly clean, it may be impossible for a researcher to be completely unobtrusive. That does not mean that it is not worth trying to do.

There are also some basic practical reasons why disclosure may be limited. It may not be possible to get consent because observation is fleeting or anonymous, because the process of gaining consent is burdensome or intrusive, or because the research draws on information gained before the research project began. There may not have been an opportunity to disclose. Observations might take place before there has been any verbal interaction; in my own work I have turned up for a pre-arranged interview, but observed interactions in the reception area while I was waiting. Disclosure to one party is not disclosure to everyone; wherever there are incidental social interactions with third parties, other people may be unwittingly involved. And research has a tendency to twist in the researcher's hand; it is possible that the researcher does not realize that something ought to be recorded till after it has happened (cf. Whyte, 1955). Researchers who work exclusively to fixed schedules and a deductive methodology may feel they are able to work within the confines of a strictly prescribed approach, but for those who work in flexible, inductive, qualitative modes, this is neither possible nor desirable. In ethical terms, even if it is difficult for researchers not to see and not to be aware of the things that are happening around them, there are still choices to be made – researchers do not necessarily have to use the information available to them. It is not uncommon, consequently, for ethical decisions to be made after the event – whether or not to refer to material that has been obtained without adequate disclosure.

Homan (1991) reviews three main arguments against the effectiveness of covert research as a method:

- that covert research affects the behaviour of research subjects, regardless of its intention;
- that covert research is unnecessary; and
- that covert research limits the scope of research, by preventing overt interaction.

The same kind of reservations could be made about most research methods: the third objection, for example, could be made to any fixed-form questionnaire with closed questions. What is right and appropriate for one research problem may not be right and appropriate for another. Whether covert research is 'necessary' equally depends on the context – in field research of any kind, some failure to disclose may be unavoidable. How effective it is, by comparison with overt research, depends on the design and purpose of the research within its particular situation.

Practical and methodological arguments, whatever their merits, do not outweigh ethical considerations. If there are ethical objections to using material obtained covertly, they establish a presumption against either conducting research without disclosure, or using research materials obtained in that way.

The Rights of Research Subjects

The core of the ethical objection to covert research is that it infringes the moral rights of research subjects. What are those rights?

There are two main classes of moral rights which ought to be considered: particular rights, which are specific to individuals, and general rights, which apply to everyone (e.g. Hart, 1955). Particular rights are dependent on the relationships between parties; examples are the right to have a promise kept, or to hold someone to a contract. Particular rights in research are generated by the relationship between the researcher and others involved in the research – usually, participants rather than subjects in general, because most particular rights in research are held by people who are actively engaged with the researcher. Researchers should behave towards research participants with integrity; that is, their actions should be consistent with the principles they affirm. If they promise confidentiality, they should maintain it. By the same token, if researchers are working in circumstances where confidentiality cannot be maintained, they should not be promising anything different. Israel records a substantial literature in which promises of confidentiality in the field of criminology are liable to be compromised by the actions of law enforcement agencies (2004a).

General rights are not dependent on the relationships between the parties; they are held by anyone in the same position (Raz, 1986). There are rights of citizenship, like the rights guaranteed in the US constitution, general legal rights, and human rights, including rights to privacy where appropriate, and the right not to be exploited. The general duties of researchers include beneficence – the promotion of welfare and the avoidance of harm and respect for persons. Those duties apply not just to research participants – those who knowingly take part in a study – but to research subjects, to third parties and potentially to wider social groups. There may be conflicting rights and interests. If a research participant is abusing the rights of another person, it is not necessarily the case that the researcher is bound to respect or protect the participant's position. Research on issues like violence, drug dealing or sex abuse has to consider the rights of the victims, not just the perpetrators.

Some researchers present the rights of research participants as if they take priority over other ethical considerations. In Pearson's research on football hooliganism, for example, he comments that he 'perceived that the duty he owed to his unwitting research subjects overrode any "moral" obligation to society or authority' (Ferdinand et al., 2007: 539; see also Pearson, 2009). The Association of Social Anthropologists suggests as a general principle that:

> . . . most anthropologists would maintain that their paramount obligation is to their research participants and that where there is conflict, the interests and rights of those studied should come first. (ASA, 1999: s 1.1)

This is difficult to defend. The rights of research subjects are not unlimited, and they certainly do not pre-empt other ethical issues. It is more appropriate to think of the process as striking a balance between competing principles. The British Sociological Association emphasizes the need for sociologists to 'protect the rights of those they study', but qualifies this by saying:

> In some cases, where the public interest dictates otherwise and particularly where power is being abused, obligations of trust and protection may weigh less heavily. Nevertheless, these obligations should not be discarded lightly. (BSA, 2002: para 15)

I would differ here in the order of priority. Particular rights generated in the process of research are not more important than human rights. The initial presumption should be that researchers will respect people's human rights, and there have to be extremely strong reasons to qualify that position.

Because covert research is done without making any commitment about a research relationship, it is difficult for research subjects to claim particular rights in respect of the research done. This may be undesirable, because making things explicit is usually a better way to protect the interests of all parties, but that is not a fatal objection. Whether it is problematic depends on the context in which the research takes place. The obligation on researchers to respect general rights applies to covert research in the same way as to overt research.

Privacy

Privacy is not the only general right governing research, but it has come to dominate the debate on covert research. Privacy refers in the first instance to non-interference; respecting privacy means that people are not inappropriately observed, inconvenienced or reported on. The private sphere is a 'zone of protected activity' (Faden and Beauchamp, 1986). However, it has come to stand beyond that for a much stronger concept, based on 'the claim of individuals, groups or institutions to determine for themselves when, how and to what extent information about them is communicated to others' (Westin, cited in Kimmel, 1988). This goes well beyond non-interference; it asserts that people have elements of their lives which is for them to control, and no-one else (Australian NHMRC, 1999: 18; cf. Rössler, 2005: 72). It follows that information obtained in research lies in the control of the research subject or participant, and the use of such information without the person's consent is unethical.

The strong interpretation of privacy is contentious. The idea that all information remains under the control of the people it relates to would establish a presumption against the free circulation of information – a position fundamentally opposed to democratic thought as well as to scientific discourse. For example, if taken literally, it should imply that most of the personal data currently held in data archives, regardless of anonymity, should be expunged. Faden and Beauchamp comment that the principle came to occupy its prominent role without being thought through. The goals are 'imperfectly distinguished and poorly formulated . . . We can claim little beyond the indisputable fact that there has developed a general inchoate societal demand for the protection of patients' and subjects' rights' (1986: 223). The consequences of that uncritical adoption are now widely experienced in the research community.

Whenever research intrudes into the private sphere, it is taken as axiomatic that it must respect people's right to control information about themselves, and consent is essential to legitimate any such intervention – a necessary element, even if it is not a sufficient one. Research subjects have the right to be informed about research, to consent, or to withdraw from it if they are not content. So, for example, the Social Policy Association suggests that 'Research participation should wherever possible be based on individuals' freely given informed consent' and that:

> Consent to participate in a research study should be regarded as an ongoing process and it should be made clear to participants that they are free to withdraw from the study or withhold information at any point. (SPA, 2009)

If research is covert, people are not informed, they are not able to consent, and they cannot effectively withdraw. The Socio-Legal Studies Association takes the view that:

> Despite the technical advantages of covert research methods they clearly violate the principle of consent and invade the privacy of those studied. (SLSA, n.d.)

That, in a nutshell, is the principal ethical objection to covert research in practice.

The Public Sphere

There is a substantial qualification to make about the application of privacy as a general right. The concept of privacy hinges on a basic distinction between the private and the public spheres. Even if individuals are considered to control personal information, that control does not relate to information in the public sphere. Shils, whose passionate criticism of covert observation established one of the most influential arguments against it, writes:

> Observation which takes place in public or in settings in which the participants conventionally or knowingly accept the responsibility for the public character of their actions and expressions . . . is different from observation which seeks to enter the private sphere unknown to the actor. . . . The open sphere – the sphere in which the individual has committed himself to publicity – is a legitimate object of observation, as it is of interviewing. (1982 [1959]: 132)

Different norms apply to public actions than to private ones. Public actions can be publicly observed. There is nothing wrong with a researcher attending a trial as a member of the public, and reporting on it, even if the effect of reporting a 'guilty' verdict is to expose the subjects to public disapproval. More than not being wrong, it is ethically desirable. The public nature of the act makes the ability to report and comment essential to the functioning of the system of criminal justice. Where a function is public, researchers can legitimately put themselves in the place of service users, 'mystery shoppers' or citizens. Research in a democracy has an important public function, which is to scrutinize social processes and to bring people to account. Observation, recording, the investigation of records and public criticism are not only permissible, but it may be unethical to put stumbling blocks in their way (Spicker, 2007) – as some Research Ethics Committees have tried to do (Israel, 2004b). And there are contexts where undisclosed research may be the method of choice, shining a light on the conduct of public services or issues of public interest which people, organizations, businesses or governments may wish to conceal.

Some actions are public in their very nature. The formal actions of governments and public agencies, even in closed rooms, are intrinsically public. A political demonstration is a public action. So is standing for elected office. Information which has moved into the public domain, like census returns or research reports, can legitimately be used without further reference to the subjects of the information. I am not sure that action in the public sphere has 'implicit' consent; what happens, rather, is that consent becomes morally irrelevant, because the information is beyond the rights of the individual to control. This is the main answer to the allegation that covert research denies research subjects the opportunity to be informed, to consent, or to withdraw from research. If they are in the public domain, they have no such rights.

Disputes about ethics in these contexts focus necessarily, then, not on the application of a general presumption of consent, but on the question of whether the circumstances are truly public. The boundaries between public and private spheres are complex, and they can be difficult to determine. The treatment in research of several critical issues, like the study of domestic violence or liability to personal taxation, have hinged on the question of whether they should be classified as public or private. (There is a growing sentiment that domestic violence should be thought of as public: see Schneider, 1994.) The Australian National Health and Medical Research Council has argued:

> Covert observation is usually regarded as acceptable if undertaken in a public place . . . it is implicit in sanctioning unconsented observation of individuals in a public place that they are aware that any of their actions will be visible to anyone simultaneously present. (1999)

This seems to be an argument that doing things in a public place is public because it is visible. This is not quite right. The distinctions are conventional, and people can do private things in public spaces. What defines something as public is not the geographical location where it happens, but the nature of the act. It can be difficult to separate the private and public spheres distinctly. A bank, a restaurant, a park or a public convenience may be a public place, but some of the things that people do there are private. Clubs and meetings in hired rooms in pubs are not self-evidently in one sphere or the other, and some of the controversies about research in these settings are about whether or not they should be seen as intrusive. The study of policing is particularly problematic, because the nature of police action often involves intervention in the private sphere in order to examine whether it ought to be considered public. Punch writes:

> . . . during my research with the police in Amsterdam, the patrol car was directed to a fight and the two constables jumped out and started wrestling with the combatants. Was I supposed to step up to this writhing shindig and shout 'freeze!' and then, inserting my head between the entangled limbs, whip out my code and, Miranda-like, chant out the rights of the participants? (1986: 36)

I think he can rest easy: fights in public places which are rowdy enough to draw the police into a brawl have clearly stepped into the public domain.

Autonomy

Faden and Beauchamp argue that the central principle behind the idea of consent is not so much privacy, as autonomy or self-determination – the ability of people to decide for themselves what happens to them (1986). The principle of consent to research is derived from issues in medical ethics, because medical intervention undertaken without consent can be held to constitute a battery. It was well established in US law that:

> . . . every human being of adult years in sound mind has a right to determine what shall be done with his own body; and a surgeon who performs an operation without his patient's consent commits a battery for which he is liable in damages. (*Schloendorff*, cited in Faden and Beauchamp, 1986: 127)

That principle was directly extended into the governance of research through the Nuremberg code, developed to address the abuses in bio-medical experimentation perpetrated under the Nazis. The experiments conducted by Nazi doctors violated that principle. The decision in *United States of America v Karl Brandt* consequently specified that 'The voluntary consent of the human subject is absolutely essential' and that 'During the course of the experiment the human subject should be at liberty to bring the experiment to an end . . .' (Nuremberg Code, 1947). These principles are commonly translated into a presumption that consent must always be obtained for research to be legitimate. They have spread out from bio-medical research to cover most fields involving research with human subjects.

One of the key objections to covert research, then, is that 'covert methods help erode personal liberty' (Homan, 1991). In so far as the actions of the researcher limit the behaviour of the research subject without the subject's consent, they could be argued to limit that person's scope for self-determination. If the process of the research is burdensome, intrusive or oppressive, that may be true; but it is not clear why it should be more true of covert research than it is of other kinds, undertaken with or without the subject's consent. The process of disclosure may be intrusive and burdensome; in some circumstances it may be alarming.

There seems to be an assumption in the emphasis on consent that if people are able to consent to the research openly, they are more likely to be able to protect their interests. This is probably true, but it is only true up to a point. Wiles et al. found that even experienced researchers, when asked to become research participants, tend to misjudge the degree to which they wish to reveal information (2006: 293–4). Some subjects are vulnerable, some may not see the consequences, and damage done to subjects by a researcher is not excused by saying 'well, I warned them'. Corrigan lambasts reliance on formal processes of consent, and the model of participants as rational actors, as an example of 'empty ethics' (2003).

The Association of Social Anthropologists comments, I think rightly, that:

> Consent from subjects does not absolve anthropologists from their obligation to protect research participants as far as possible against the potentially harmful effects of research. (ASA, 1999: s 1.2)

The Australian NHMRC suggests that covert research might be undertaken in certain cases where disclosure is limited in practice (Australian NHMRC, 2007). However:

> The covertness of the method is unlikely to be regarded as preventing the duty of confidentiality from arising. It is probable that researchers who conceal the fact that they are researchers when they gain information from potential participants still owe those participants the same obligation

> to respect their confidence as they would have had if their researcher status had been disclosed. (Australian NHMRC, 2001: L21)

The argument here implies that there are particular rights, as well as general ones, in covert research. If the objection to covert research is that it is done without sufficient safeguards, it seems to follow that covert research might be possible, in cases where the research is methodologically justifiable, provided that other safeguards are strengthened. Public exposure is commonly avoided through the use of anonymity – not, in itself, an ethical principle, but a device to try to limit the degree to which information gained in research can be attributed to the circumstances of particular individuals or groups.

An example of circumstances where alternative safeguards are appropriate is the case where a researcher is writing about personal experience or past events. It is not necessarily the case that research begins with the identification of a problem or issue; often researchers systematize information from their previous experience. Many researchers choose areas of work where they have some prior knowledge or familiarity with the setting. (I began my own career working in housing management, and my first published work was related to that experience.) Knowledge of this sort, however, is invariably obtained for a different purpose than research. Does that mean that it cannot be used? I do not think that such a constraint is ethically required, or even reasonable. The limitations that researchers recognize in such circumstances relate to the context where they have acquired the information, and sensitivity to the position of the people affected. That would typically lead to anonymization where appropriate, and perhaps altering details which might otherwise breach confidentiality. That is not the same as giving control over information to the research subjects.

This assumes that the principle of consent is defeasible, and the safeguards it offers may reasonably be offered in other ways. If it is accepted that this is possible, it implies that the strong concept of privacy, which treats private information as being solely under the control of the subject, is not decisive. It also follows that covert research may be undertaken without obtaining consent, or at least with limited disclosure, in some circumstances within the private sphere. There is an obligation to consider appropriate safeguards to protect participants, subjects and others from undesirable consequences. If covert research is undertaken, it is subject to such safeguards.

The Avoidance of Harm

Researchers have a duty of beneficence – or at least of non-maleficence, the avoidance of harm – to those affected by research, going beyond the issues of privacy and consent. No researcher should be blind to the potential consequences of research. If a research report is effective, it may well heighten awareness, change the way that people respond to the issues raised, or inform policy. This may, in some cases, increase risk to those involved in research. There is no rule, however, to say that harm must never occur. For example, evaluations of policy or service delivery always carry the risk that activities may be curtailed or funding withdrawn; that does not mean that the research is not legitimate.

Possibly the most damning criticism made of covert research has been that it has a distinct potential to harm the people who are being studied (cf. Homan, 1991), and that consequently it is inconsistent with the obligation to protect the position of research subjects. The criticism is paradoxical. One of the main arguments for doing things covertly is to avoid changing the nature of the behaviour that is being observed. In most cases, if it is done successfully, covert action means that the subjects of research are doing exactly what they did before, and they are exposed to no risk from the research process beyond the risk already associated with the activity. There may be a problem if their action should not be reported – typically, because it is genuinely private – but it should not be a problem otherwise. The test of beneficence does not imply that research subjects should not be subject to changes in their circumstances, but that research should not expose subjects to harm beyond the consequences of their own actions.

The issue here is not that covert research does not have the potential to have undesirable effects. Many forms of research do. A judgement about whether research is ethical depends on an engagement with the ethical considerations. At root, the tests that need to be applied to covert research are the same as the tests that apply to other forms of research – whether there is a potential for harm, whether people's rights are infringed, whether the research is intrusive and what safeguards are needed. Ethical conduct in covert research depends, not on the automatic application of programmatic rules, but on examining the ethical implications of process and outcomes in the context where the research is done.

The hardest test of this principle has been criminal activity. Covert research is often favoured as an approach for the exploration of criminal activity, because of the concern that researchers will otherwise not be able to get access, because subjects will disguise their behaviour and invalidate the findings, or because the researchers will themselves be put at risk. The argument that covert research is harmful in this context relates crucially to the fact that their action is being reported – that attention is being drawn to it. Wiles and her colleagues report that social researchers are likely to break confidence only in the most serious cases:

> Researchers appear to feel a responsibility to report, with their participants' permission, cases where their participants are at risk of harm but not cases of illegal activity. There is a clear consensus that the abuse of

> vulnerable people should always be reported with participants' permission. (2008)

There are reasonable concerns about the potential harm that might be done when a person is not willing to make a complaint, but two parts of this seem to me misplaced. One is the qualification of both sentences with the proviso that this should only be done 'with the participants' permission'. That elevates the particular rights to participants to a status where they take priority over both the risk of harm and the protection of vulnerable people. The other is the suggestion that researchers will not report illegal activity. Some researchers have certainly taken the view that it is unethical to report a crime revealed in the process of the research (e.g. Australian NHMRC, 2001: 130; Loxley et al., 1997; Pearson, 2009; and see Israel, 2004a for further examples). I think they are profoundly mistaken. There are 'rules of recognition' in any system of law (Hart, 1961); a criminal action is one which has been defined as criminal through a recognized public process. The investigation of the circumstances of people engaged in criminal acts can always be seen as public rather than private. There is consequently no immediate ethical bar to using covert research in these contexts, because there is no implicit duty of confidentiality about criminal activity.

The Social Research Association informs us that:

> The US Office for Protection from Research Risks allows observational research to be exempt from consent unless . . . 'any disclosure of the human subjects' responses outside the research could reasonably place the subjects at risk of criminal or civil liability . . . (SRA, 2003: 33)

This seems to say that researchers can observe people without consent, going about their private business, only until they do something that *prima facie* falls into the public sphere. That reasoning is perverse. It is precisely at this point that consent from the subject is no longer required. There is always an additional individual responsibility in ethics to deliberate whether the law is indeed moral (for example, laws restricting certain sexual behaviours). A researcher who wants to take a different view about the legitimacy of the law has both to justify that and to accept the consequences. That, however, is the exception rather than the rule.

The duties which the researcher has are not subject to the consent of the offender. There is nothing intrinsic in the rights of the research subject that implies that the researcher should maintain silence or confidentiality, or that the researcher must become complicit in the crime. There is no right which allows a research subject to commit crimes with impunity, and if a person's actions place them at risk, they have to accept the consequences. The main rights that people have in those circumstances are not to be arraigned for crimes without the process of law – which means that researchers need to exercise considerable caution in publishing details of crimes, even in circumstances where the research participant is apparently confessing. Researchers have no general duty to protect people from the law. On the contrary, if the test of rights is that there is due process of law, the principle may require ethical researchers to cooperate actively with the authorities.

Other Objections to Covert Research

The kind of issue that the literature focuses on most clearly is a special case: the use of covert participant observation, or 'undercover' work, where researchers either pretend to be something different, like Rosenhan's classic study of psychiatric hospitals (1973), or become something different, as Calvey became a security guard and Pearson became a football hooligan (Calvey, 2008; Pearson, 2009). There are many problems with this kind of research, but its covert nature is the least of them. If covert participant observation is undertaken in a legitimate public setting, without deception – for example, Homan's participation in religious services (cited in Homan, 1991) – there is nothing evidently wrong with it. There are four other more significant reservations about undercover work. The first reservation concerns what kind of activity the researcher is participating in. Covert participant observation in criminal activity is likely itself to be criminal, and the advancement of knowledge is unlikely to be a legitimate justification for that. Pearson describes his participation in football hooliganism:

> I found myself both witnessing criminal offences and being put under pressure to commit them personally . . . I committed 'minor' offences (which I tentatively defined as those which would not cause direct physical harm to a research subject) on a weekly basis as part of the research routine. (2009: 245, 246–7)

Participation in criminal offences, such as offences against public order, seems to me to require a much stronger justification than completion of a doctoral thesis.

Second, there is the extent to which the observation can be said to be intrusive. Many private actions are of general interest – issues like nutrition, family care, sexuality or physical capacity have implications for public policy – but it is difficult to think of circumstances in these fields where research in non-public settings, on aspects of the issues which do not impinge on the rights of others, and undertaken without the consent of the subject, would not breach the privacy of those observed. Garfinkel's famous 'experiments with trust', where students were asked to disrupt the everyday lives of their unsuspecting parents and siblings, seems to me to fall foul of that principle. As one student's sister commented: 'Please, no more experiments. We're not rats, you know.' (Garfinkel, 1984 [1967]: 48–9.)

The third objection is the potential risk to the researcher. Sometimes the role implies exposure to physical violence: Calvey's work depended in part on his abilities in martial arts (2008). Sometimes the risk is to physical integrity; the risks to Rosenhan's researchers of permanent physical or mental damage through inappropriate medical intervention were, in retrospect, staggering.

Fourth, there is the issue of deception. Some researchers do deliberately misrepresent their situation. They need to consider whether it does indeed represent a betrayal – a breach of undertakings given to the research subjects.

Homan lists 13 key objections to covert research, but six, as noted earlier, are objections to deception rather than covert research, and three are about the effectiveness of covert research as a method. That leaves four main ethical objections:

- covert procedures flout the principle of informed consent . . .
- covert methods help erode personal liberty . . .
- covert methods discriminate against the defenceless and powerless . . .
- Covert methods may damage the behaviour or interests of subjects. (Homan, 1991)

I have reviewed three of these arguments in the course of this article. The argument that remains is the suggestion that the process is intrinsically biased against people who are disadvantaged. This seems to rest in the view that people who are vulnerable and disadvantaged are less able to defend their privacy than people who are wealthy. That may be true, but it does not follow that a procedure which does not go through the process of disclosure puts them more at risk. It is no less true that privacy has also been used to assert disadvantage, and the effect of defending people from intrusion has been to deflect potential criticism of the way people behave in the domestic sphere – for example, in marital violence and child abuse (Rössler, 2005; Schneider, 1994).

Covert Research in Practice

I gave earlier some basic practical and methodological reasons why researchers might limit disclosure in the process of research. There is, though, another, more subtle, and more substantial, argument for non-disclosure. In the course of the argument so far, I have given several instances of research where disclosure is limited – such as the researcher who is drawing on, reflecting and systematizing previous experience; the researcher who decides to use information obtained fortuitously; or the researcher whose agenda changes in the course of the research. Most academic researchers will, I think, recognize that kind of process. Undeclared, incremental learning and discovery is part of the normal process of academic learning and investigation.

I work as an advisor, consultant and researcher, undertaking applied policy research to meet the needs of various agencies. The things I learn from each project add to the next. I use examples from previous research, practice and informal enquiries liberally in my teaching and writing. In the course of the last year, I have visited or discussed issues with government officials, benefit administrators, voluntary sector workers, employment advisers, welfare rights workers, community workers, campaigners, politicians, disability activists, journalists and of course academics. None of those discussions has been part of a formally constituted research project. Finding out about things – having conversations, forming networks, learning from people, and piecing information together – is part of my work. I do not generally do 'covert research' in the sense of deliberately constructing research that is intended to be concealed from view; but equally, there may be no point at which it would be opportune or appropriate to make an explicit disclosure about the process of research and how the information I glean might be used. That method of working, in the current climate of ethical governance, has become problematic. Any researcher who came forward with a proposal to fish unsystematically for serendipitous material would find it difficult to justify the proposal to a Research Ethics Committee.

The problem here, I think, does not lie with the practice, but with a misconceived view of research. Ethical review procedures have often been based in a stereotypical, restrictive understanding of research activity. The kind of research they relate to is a closely defined, project-based, time-bounded programme of work, with a clear plan, statement of objectives and method of working. The idea of 'research', however, includes much more than project-based investigation. It takes in scholarship, exploration, discovery, and practice. Research is often done informally, and informal research is rarely overt. If the process of acquiring and systematizing knowledge cannot be done without the full, explicit knowledge and cooperation of research subjects, in many cases it will not happen at all.

The rules which are being applied to covert research are based in concerns about a marginal set of special cases, they are unduly influenced by concerns within deception, which is another issue entirely, and there is widespread confusion about the duties that researchers owe to their subjects. The importance of the issue stretches far beyond the narrow issue of covert participant observation. The ethical principles which apply to formal research have to be applicable to informal research – any other position would be incoherent. We cannot sensibly say, for example, that researchers have a duty of beneficence in research that begins with a project proposal and do not have such a duty in research that does not. Currently we seem to imagine that formal research is somehow morally different from other approaches to the acquisition of knowledge – scholarship, discovery, experience, engagement with practice, or the kind of networking and interaction that is central to learning. This is not sustainable. The rules which are applied to covert research necessarily apply to any equivalent situation where people are not aware that research is taking place. These restrictions threaten the very nature of academic enquiry.

References

Alston M and Bowles W (2003) *Research for Social Workers*. London: Routledge.

ASA (1999) *Ethical Guidelines for Good Research Practice*. London: Association of Social Anthropologists of the UK and the Commonwealth.

Australian National Health and Medical Research Council (NHMRC) (1999) *National Statement on Ethical Conduct in Research Involving Humans*. Canberra: Commonwealth of Australia. Available at: http://www.nhmrc.gov.au/publications/_files/e35.pdf

Australian NHMRC (2001) *Human Research Ethics Handbook*. Canberra: Commonwealth of Australia. Available at: http://www.nhmrc.gov.au/_files_nhmrc/file/publications/hrecbook/hrechand.pdf

Australian NHMRC (2007) *National Statement of Ethical Conduct in Human Research*. Available at: http://www.nhmrc.gov.au/publications/ethics/2007_humans/contents.htm

BSA (2002) *Statement of Ethical Practice for the British Sociological Association*. Available at: http://www.sociology.org.uk/as4bsoce.pdf

Bulmer M (ed.) (1982) *Social Research Ethics*. London: Macmillan.

Calvey D (2008) The art and politics of covert research. *Sociology* 42(5): 905–18.

Corrigan O (2003) Empty ethics: The problem with informed consent. *Sociology of Health and Illness* 25(3): 769–92.

Faden R and Beauchamp T (1986) *A History and Theory of Informed Consent*. Oxford: Oxford University Press.

Ferdinand J, Pearson G, Rowe M and Worthington F (2007) A different kind of ethics. *Ethnography* 8(4): 519–43.

Garfinkel H (1984 [1967]) *Studies in Ethnomethodology*. Cambridge: Polity.

Government of South Africa (2004) *Ethics in Health Research: Principles, Structures and Processes*. Available at: http://www.doh.gov.za/docs/factsheets/guidelines/ethnics/sec10.pdf#search=%22covert%20observation%20informed%20consent%22

Haggerty K (2004) Ethics creep. *Qualitative Sociology* 27(4): 391–414.

Hart H (1955) Are there any natural rights? *Philosophical Review* 64: 175–91.

Hart H (1961) *The Concept of Law*. Oxford: Oxford University Press.

Herrera C (1999) Two arguments for 'covert methods' in social research. *British Journal of Sociology* 5(2): 331–43.

Homan R (1991) *The Ethics of Social Research*. London: Macmillan.

Iphofen R (2006) Ethical scrutiny. *Times Higher Education Supplement*, 8 September.

Israel M (2004a) Strictly confidential? *British Journal of Criminology* 44(5): 715–40.

Israel M (2004b) *Ethics and the Governance of Criminological Research in Australia*. Sydney: New South Wales Bureau of Crime Statistics Research. Available at: http://www.bocsar.nsw.gov.au/lawlink/bocsar/ll_bocsar.nsf/vwFiles/R55.pdf/$file/R55.pdf

Kimmel A (1988) *Ethics and Values in Social Research*. London: Sage.

Kimmel A (1996) *Ethical Issues in Behavioural Research*. Oxford: Blackwell.

Loxley W, Hawks D and Bevan J (1997) Protecting the interests of participants in research into illicit drug use *Addiction* 92(9): 1081–5.

Lugosi P (2006) Between overt and covert research. *Qualitative Inquiry* 12(3): 541–61.

Milgram S (1974) *Obedience to Authority*. London: Tavistock.

Olson R, Verley J, Santos L and Salas C (2004) What we teach students about the Hawthorne studies: A review of content within a sample of introductory I-O and OB textbooks. *The Industrial-Organizational Psychologist* 41(3): 23–39.
Nuremberg Code (1947) Available at: http://ohsr.od.nih.gov/guidleines/nuremberg.html
Pearson G (2009) The researcher as hooligan. *International Journal of Social Research Methodology* 12(3): 244–55.
Power R (1989) Participant observation and its place in the study of illicit drug abuse. *Addiction* 84(1): 43–52.
Punch M (1986) *The Politics and Ethics of Fieldwork*. London: Sage.
Raz J (1986) Rights based moralities. In: *The Morality of Freedom*. Oxford: Clarendon Press, 193–216.
Rosenhan D (1973) On being sane in insane places. *Science* 179(4070): 250–8.
Rössler B (2005) *The Value of Privacy*. Cambridge: Polity.
Schneider E (1994) The violence of privacy. In: Fineman M and Mykitiuk R *The Public Nature of Private Violence*. New York: Routledge, 49–54.
Shils E (1982 [1959]) Social inquiry and the autonomy of the individual. In: M Bulmer (ed.) *Social Research Ethics*. London: Macmillan.
SLSA (n.d.) *First Re-statement of Research Ethics*. Available at: http://www.kent.ac.uk/slsa/download/ethics_drft2.pdf
SPA (2009) *Social Policy Association Guidelines on Research Ethics*. Available at: http://www.social-policy.com/documents/SPA_code_ethics_jan09.pdf
SRA (2003) *Ethical Guidelines*. Available at: http://www.the-sra.org.uk/documents/pdfs/ethics03.pdf
Spicker P (2007) The ethics of policy research. *Evidence and Policy* 3(1): 99–118.
Tysome T (2006) Ethics guards are 'stifling' creativity. *Times Higher Education Supplement*, 25 August.
Walker L, Williams J and Jamrozik L (2006) Unsafe driving behaviour and four wheel drive vehicles. *British Medical Journal* 331(8 July): 71.
Whyte W (1955) *Street Corner Society*. Chicago: University of Chicago.
Wiles R, Crow G, Heath S and Charles V (2006) Researching researchers. *Qualitative Research* 6(3): 283–99.
Wiles R, Crow G, Heath S and Charles V (2008) The management of confidentiality and anonymity in social research. *International Journal of Social Research Methodology* 11(5): 417–28.

35

Lone Researchers at Sea: Gender, Risk and Responsibility

Helen Sampson and Michelle Thomas

> Perhaps my imagination . . . ran away with me, because it suddenly came into my head that the crew might start throwing all of us overboard one after another . . . . True, eleven crewmen were up against sixteen officers and cadets, but the crew would have the advantage of surprise. Perhaps in 1985 such thoughts seem melodramatic to landsmen, but the dockside bars of the world abound with stories of recent acts of piracy, violence and mutiny on the high seas.
>
> (Young, 1986: 376)

Introduction

As Young (1986) illustrates so vividly in his account of his experiences aboard the *Piranha*, life at sea can be 'worlds apart' from life ashore for both seafarers and 'passengers'. As a male passenger he experienced many moments of trepidation and fear in the course of his relatively short but ill-fated voyage from Rio which ended prematurely with the grounding of the ship. This may be surprising to some who have never been at sea but, to most of those who have, Young's fears are easy to relate to and indeed they reflect the realities of being in the midst of the ocean, alone, vulnerable, and reliant on the goodwill of shipmates.

If ships are dangerous environments for seafarers and passengers, so too do they contain risks for researchers and this article specifically considers

Source: *Qualitative Research*, 3(2) (2003): 165–189.

the ways in which ambient risks amplify situational risk using the example of women researchers aboard cargo ships. Research in many fieldwork settings can be rife with hazards, and yet these issues continue to receive little attention despite the various remote and dangerous settings that researchers enter in the course of their work (Lee-Treweek and Linkogle, 2000a). Within sociology, accounts of fear, risk and danger have tended to be limited to ethnography's oral tradition of shared anecdotes and feature less commonly in written accounts (Coffey, 1999). However recent exceptions can be found in the work of Lee (1995), Lee-Treweek and Linkogle (2000a), and Nordstrom and Robben (1995), which draw together 'tales from the field' and explicitly consider the dangers and risks inherent in fieldwork situations.

In doing this Lee considers a wide range of fieldwork settings and, importantly, distinguishes between ambient and situational risk. He characterizes ambient risks as those encountered in dangerous research settings, for example working with police officers on the streets of Northern Ireland. These differ from 'situational risks', which are defined as those that are evoked by a researcher's presence or actions in a specific research setting. As research sites, ships are settings carrying both ambient and situational risk. The ship is a potentially hazardous environment in terms of the ambient risks associated with the potential for on-board explosions, collisions, groundings, founderings, and injuries linked to the seafaring labour process. These risks are clearly demonstrated by available data on incidents and mortality rates at sea. For example, fatal injuries and drownings among Danish seafarers have been found to be 11 times higher than average rates among the Danish workforce ashore (Hansen, 1996). In 1998 according to Lloyds Casualty Report figures 175 ships were lost around the world (ICONS, 2000). In addition, on certain routes and trades, piracy, often involving lethal force, is rife (Reederei Nord, 2000). Manifestly these risks are as present for researchers at sea as they are for seafarers. In addition, researchers, and particularly women researchers, are likely to be faced with situational risks aboard. Seafarers aboard merchant cargo vessels are inculcated with a strong occupational culture that, in part, is defined in opposition to that of 'landlubbers'.[1] Ships are additionally overwhelmingly male environments which may increase the hostility expressed towards female researchers who may be regarded as 'outsiders' to an even greater extent than their male counterparts.[2]

Researchers' own biographies inevitably impact upon research settings and populations (Coffey, 1999). In some cases they may effect the nature and level of risk that a researcher is exposed to during the course of his/her study. Gender is one aspect of researcher identity that has recently received attention (Warren, 1988) and, similarly, sexual harassment in the field has been documented by some female researchers (Gearing, 1995; Gurney, 1985; Moreno, 1995) and does not appear to be uncommon (Stanko, 1992,[3] Whitehead and Price, 1986).

This article will focus on dangerous fieldwork, considering the specific example of a hazardous occupation. It will explore the ways in which the ambient risks of organizational settings combine with the situational risks associated with being female researchers exposed to sexism and sexual harassment on-board ships. The article traces the range of issues and problems encountered in the course of the research, from the moment of selecting the research site, through to the negotiation of access and actual fieldwork experience.

Fieldwork Context and Scope

The article draws on the research experiences of the authors whilst working on an ESRC/Seafarers International Research Centre funded project (ESRC ref: L214252036) on Transnational Seafarer Communities (see Kahveci et al., 2002). The project sought to consider the extent to which mixed nationality crews could be considered as groups of transnationals and required researchers to interview seafarers aboard, subject to their consent, and maintain detailed fieldnotes of crew interaction and associated observations. The initial researcher appointed to work on the project on a contract basis was a man, resulting in an exclusively male research team. However, in 1999 Sampson (who had five years experience of private sector qualitative research and was registered for a part-time PhD), was requested to assist with the shipboard fieldwork and undertook one voyage on a consultancy basis. Subsequently the appointed researcher left the project and Sampson was given a two-year research contract as his replacement in the course of which she undertook five voyages totalling 118 days at sea. Thomas was appointed to SIRC in 1999 to work on matters relating to seafarers' occupational health and safety. Thomas had completed a PhD in 1999 and had five years' experience working on various university research projects utilizing quantitative and qualitative research methods. As part of her orientation and to assist with the project she agreed to undertake a voyage aboard a Gas Carrier and subsequently returned to the ship for follow-up research. She spent a total of seven weeks at sea and a further two weeks aboard the ship in dry dock. In the course of the project we therefore undertook a total of seven fieldtrips, aboard six different cargo ships of different types, in the period 1999–2001. A further seven voyages were conducted by other members of the established research team (Kahveci, Lane) and SIRC staff. All the other researchers involved in the project were male. Fieldwork was conducted by lone researchers over periods of between two and seven weeks. Ships were ocean-going cargo vessels with long sea-passages and few port calls. All were run by multinational crews of between 17 and 29 people and on only one ship were female officers, or seafarers' female partners, present (see Table 1).

Table 1: Research voyage details

Ship pseudonym	*Ship type*	*Length of voyage (days)*	*Crew size*	*Nationality of crew*	*Women aboard*
Eclipse	Refrigerated cargo	42	25	Swedish Filipino	None
Santos Sunset	Bulk carrier	29	25	Indian Bangladeshi Filipino	None
Singapore Orchid (1st trip)	Gas carrier	27	29	British Filipino Australian	One officer One partner
Norwegian Imp	Oil tanker	21	25	Filipino Canadian British Croatian German Indian	None
Singapore Orchid (2nd trip)	Gas Carrier	32	29	British Filipino Australian	Two partners
Qui Aura	Refrigerated cargo	16	17	German Spanish Chilean Peruvian	None
Calmex	Oil Tanker	21	26	Croatian Pakistani Filipino	None

The ship is a relatively unusual setting for fieldwork. It is an isolated environment with limited possibilities for communication with others beyond the field. In some senses the isolation of fieldworkers at sea may be likened to that experienced by anthropologists researching geographically remote populations. However the ship is different in as much as, during deep sea voyages, researchers do not have the opportunity to leave vessels, no matter how dire the circumstances or how desperate their need.

For a variety of reasons modern cargo ships are predominantly male environments characterized by rigid occupational hierarchies. Crews are divided into officers and ratings and within each of these groups further ranks exist ranging from Chief Engineer to 4th engineer, from Captain to 3rd Officer and from Bosun to Ordinary Seaman (OS) (see Table 2). So important is occupational hierarchy that it is quite usual for crew members of modern cargo ships to introduce, and refer to, themselves and others by rank rather than name. Nicknames may also derive from rank and are often passed from one crewmember to his/her reliever 'outliving' the tour of duty of any individual. Thus a pumpman may commonly be 'named' 'Pumpie', a radio officer 'Sparks', an electrician 'Lecie' and so on.

In terms of spatial use and layout, ships vary according to age, trade (refrigerated cargo, bulk carrier or oil tanker, for example), ownership, and

Table 2: Positions on-board[4]

Deck department	*Engine department*	*Catering department*
Captain/Master	Chief Engineer	Chief cook
Chief Officer/Mate	1st Engineer (infrequent)	2nd Cook
1st Officer (rare)	2nd Engineer	Messman/Steward
2nd Officer/2nd Mate	3rd Engineer	
3rd Officer/3rd Mate	4th Engineer	
Cadet	(Junior Engineer) Cadet/Engineering Trainee	
Radio/Admin. Officer	Reefer engineer*	
Bosun	Electrician*	
Pumpman	Electrical Technical Officer*	
Able Seaman (AB)	Cargo Engineer (infrequent)*	
Ordinary Seaman (OS)	Fitter Oiler/Motorman Wiper	

*Equivalent in status to 2nd/3rd Engineer

place of construction. Accommodation and work spaces include the bridge (where navigation and watchkeeping takes place), engine room, galley (kitchen), bars/recreation/mess rooms (usually segregated by rank) and cargo control rooms/deck offices (where cargo operations, loading and discharging are coordinated). Some vessels may also have additional facilities such as a library, gym, sauna, cinema, or swimming pool. Cabins are usually arranged on decks separated by rank, with the highest ranking officers occupying cabins on the higher decks of accommodation blocks. On all the ships where we conducted fieldwork crewmembers each had their own cabins and we were provided with our own en-suite accommodation. However this is not always the case and it is not uncommon for ratings to share bathrooms and toilets (as was the case on two of the vessels included here).

Opportunities to communicate with people ashore vary by ship and route. Many modern ships have email; however, this is not always confidential and emails may have to be sent and received via the ship's captain. Most modern ships have satellite phones from which (for some considerable expense) calls may be made to land from almost any part of the ocean. However one of the vessels included in this research did not have a satellite telephone and (whilst underway) serving seafarers (and the researcher) relied on VHS radio communication on passing 'posts' such as 'Las Palmas Radio' for radiotelephone contact with families and friends. Access to any of a ship's communication systems requires agreement from the Captain for both researchers and seafarers. On coastal routes personal mobile phones can be used if a ship is close enough to land. However opportunities to communicate ashore are generally rather limited and dependent on permission from the Captain.

The length of time a ship is in port depends on a variety of factors including its type, cargo, design, and the facilities of the port itself. In general, container vessels spend little time in port and bulk carriers tend to take longer to load or discharge cargoes. Reefers (refrigerated cargo ships) often visit remote ports with poor shore-side facilities and may therefore spend longer in some ports than others. Tankers may load/discharge cargoes out at sea, moored at specially designed buoys via deep sea pipelines, or may berth at remote refineries with few facilities for seafarers. However, regardless of ship type or port capacity it is true to say that in general the length of time ships spend alongside is declining and they are therefore becoming increasingly isolated workplaces (Kahveci, 1999).

Research Issues

Selecting the Site

One of the first steps for an ethnographic researcher is to identify the particular settings to which they would like to gain access. In the course of our research we deliberately selected a high proportion of ships with long sea passages and few port calls. This was a result of our finding that seafarers tend to have little free time in port and some, for example chief officers, have many demands made on their time in port and therefore become very fatigued. Seafarers with time 'off' generally go ashore and may be unavailable to researchers who may find that research opportunities are limited aboard ships engaged in what are termed short-sea trades, where port calls are very frequent and close together. Our choice of ship was thus driven by our assessment of the optimum conditions for data collection, rather than any consideration of ambient or situational risk.

In retrospect, we realized that from the outset we neglected our needs as researchers. In particular, in prioritizing research requirements over our own needs, we ourselves, our institutions, and our funding body, failed to recognize the full implications of the isolation of ships engaged in long haul voyages and the attendant risks posed to the health and safety of our team of researchers. Deep sea voyages may be of considerable duration (two weeks or more) during which there is no possibility of leaving a vessel (no matter what the circumstances) and where opportunities to communicate ashore may be limited or completely absent. Such conditions may not present difficulties if the fieldwork is relatively unproblematic; however, in the event of finding oneself in a hostile or threatening research environment, the consequences can be potentially disastrous. In addition there are more prosaic risks to consider. For example, cargo ships carry limited medical supplies and crews do not generally include qualified medical professionals. The outcome of an injury or sudden illness is therefore likely to be more serious than if it occurred on, or near, land (Lane et al., 2001).

Prioritizing research requirements over our own health and safety was perhaps a consequence of our preoccupation with the more general risks entailed in such research. Ships are notoriously unsafe work environments and indeed one vessel, in a total of 14 involved in our study, sank less than 12 months after a researcher had been aboard. Perhaps an awareness of the general *ambient*, as opposed to *situational*, dangers involved in undertaking this research led to oversights in relation to the specific (situational) risks entailed in researching such remote environments alone. We believe however that it is also symptomatic of the research culture in which we operate. This in our view under-emphasizes the personal health and safety of researchers (Lee-Treweek and Linkogle, 2000b) whilst fostering an atmosphere in which there are pressures to venture into ever more remote, exotic, bizarre, and, importantly, unresearched settings. It is not unheard of for researchers to go missing or indeed be assaulted (Thompson, 1967) or even murdered in the course of, or as a result of, their research (Oglesby, 1995). Equally there are risks relating to researchers' mental and emotional health when engaged in fieldwork (Lee-Treweek, 2000) which can have serious consequences including suicide (Howell, 1990). These are frequently overlooked or ignored in the design and contemplation of fieldwork. Certainly, despite our knowledge of ships as environments in which isolation is common, (they have been likened by some to prisons [Lane, 1998]), we did not pay sufficient attention to either the psychological or practical preparation necessary for entry into such challenging fieldwork settings.

Encountering and Minimizing Situational Risk

Keeping on the Right Side of the Gatekeepers

Once we had identified individual ships as potential research sites we made concerted efforts to gain access. The process of negotiation for research access is usually staged with access agreed by both formal and informal gatekeepers (Brewer, 2000). In this, shipboard research is no exception. Access to a ship must first be negotiated with a shipowning or managing company. However, the *key* gatekeeper is invariably the captain of a vessel. As noted earlier, merchant cargo ships are strongly hierarchical work environments and are one of a small number of unique occupational settings where divisions of occupational status extend to dominate not only work but also leisure time. Within these settings, and bearing in mind the common usage of the term 'Master' as synonymous with 'Captain', the maxim 'Captain is King' seems appropriate. At sea, Captains are ultimately responsible not only for the operation of their vessels but for the safety of each individual aboard. With such responsibilities come significant powers, and Captains are generally in a position to have crew members 'sent home' (i.e. sacked) from any port, at any time, without being asked to substantially justify their decisions. In such circumstances, seafarers may be given another opportunity on another ship if employed directly by a sympathetic owner/operator; however, if they are employed by a crewing agent or in a country where they may be 'blacklisted', their career in shipping may be, effectively, at an end.

The position of researchers in these settings is somewhat ambiguous. In a system that upholds the Captain's authority as paramount, they have little genuine autonomy despite being, at least nominally, independent. Without the Captain's blessing and support research cannot flourish and, as a result, having struggled up a gangway, laden with baggage, the first priority for a newly arrived researcher is generally meeting, and trying to negotiate access with, the Captain.

Such first meetings are not always encouraging. Berthed ships are generally at their busiest and most commercially vulnerable. Cargoes are discharged or loaded as fast as possible to minimize berthing fees and in this period of transition they, and their ships, are at risk of damage. These are tense times when seafarers may work long hours in stressful contexts and Captains are often somewhat preoccupied. They are also tense and important moments in the research process as the wrong responses or actions so early in the research may compromise its entire success and even, and more importantly, the safety of the researcher. The following fieldnotes are indicative of the kinds of reception we met on joining vessels:

> It was clear that the Chief Officer . . . was very surprised by my arrival. He had no idea where I would be sleeping and hadn't been expecting me. The Captain was asleep and the C.O. did not want to wake him. . . .
> (Fieldnotes)

> I am also taken to meet the Captain who looks very severe and formal dressed in uniform. . . . The Captain says hello briefly and then tells me that he is busy today but will show me around and talk to me tomorrow.
> (Fieldnotes)

Such receptions are of less consequence in other research settings where a researcher may simply go home to return at a better time. Aboard ships, berthed in remote ports and subject to sudden shifts in timetabled departures, researchers cannot leave and may be somewhat lost given the institutional setting in which they find themselves and its associated formal and informal rules and regulations. Nevertheless we found that it was vital in this early stage of the research encounter to sublimate our own needs, however pressing, and remain in the background as far as possible and await the Captain's convenience.

Whilst Captains are the main gatekeepers and grant access to staff, work spaces such as 'the bridge', and even the social side of shipboard life, negotiating access is something of a full-time occupation in a shipboard context. It involves a number of gatekeepers other than Captains, including Chief

Engineers who may determine access to engineers and engine rooms, and both Bosuns and Chief Officers, who generally mediate access to deck ratings and their work environments. Researchers are consequently engaged in a constant round of requests for access throughout the research process as the following fieldnote examples demonstrate:

> After breakfast I talked with the Chief Officer about the way I wanted to work the interviews. . . . The Chief Officer said he was going to hold a meeting with all the deck crew and I asked if I could sit in. He agreed. (Fieldnotes)

> Met Bosun, [Chief Mate] told him it was okay for me to walk around the deck and talk to anyone I wanted to – the Bosun agreed. Tour also included a brief visit to crew quarters. (Fieldnotes)

Research in such hierarchical settings is particularly demanding, requiring as it does a continuous process of negotiation and re-negotiation. The process on-board ship is comparable to that found in other hierarchical organizations such as police forces (Brewer, 1990, 1991; Fox and Lundman, 1974). The level of power held by the Captain is such that there is a fundamental need for a researcher to maintain a good relationship with him or her and to remain 'in favour'. Aboard however, it is also necessary to maintain good relationships with other less formally powerful gatekeepers. These requirements can put researchers in situations that they find problematic or that make them feel uncomfortable, and they additionally create ideal conditions for sexual harassment.[5] The research requirements and the research setting combine to place Captains and indeed other gatekeepers in a position where they may be encouraged to exert unwelcome sexual pressure on researchers or subject them to varying levels of harassment. In the context of a ship where situations can be very emotionally charged and volatile, this may pose a considerable challenge to researchers, a challenge that is amplified by the extent to which fieldworkers on ships are not so much immersed (Coffey, 1999) as 'trapped' in the field. In handling such issues 'getting it wrong' can have serious consequences, not only for the research process but also for the safety of researchers who rely on Captains for a considerable degree of 'protection'. This applies not only in terms of physical safety in the context of a dangerous working environment but also in the wider sense of protection against anti-social behaviour.

Building Good Relationships with Crew Members: Balancing Demands

A surprising feature of conducting research on-board ships was that, on all but one vessel, rather than finding ourselves bored, as we had anticipated, we tended to find ourselves over-extended. There never seemed to be enough hours in the 24 hour lifecycle of the ship to see all that we wanted to see, hear all we wanted to hear, and do all that we wanted to do. We were reminded of Lareau's admission that:

> Like a greedy child on Christmas day who keeps opening package after package without stopping to play with them and then asks for more, I kept going to the field, didn't write it up, but went back to the field anyway for fear of missing something important.
>
> (Lareau, 1996: 218)

In the relatively unique context of a ship the first days aboard tend to be filled with new sights, sounds, and experiences. As a result, when sea-staff were welcoming[6] we found it difficult to set our own agendas in terms of how, and with whom, we spent our time. On a ship, whilst crew members have clearly defined work times and tasks researchers have the appearance of being entirely 'free' from such constraints and thus at the 'disposal' of seafarers aboard. In pursuit of cordial relations and unfettered access we were generally reluctant to decline offers to be shown particular procedures or events or, indeed, to join in social activities; however we found that this was frequently problematic and incrementally stressful involving, as it often did, negotiating 'rejections'. The following fieldnote is illustrative:

> I was invited by the Third Engineer to join the officers at the ABC bar and I would have liked to have accepted unequivocally but the agent had already said he would be going with the Captain for a pizza and had invited me as well.
>
> (Fieldnotes)

Being in a fieldwork setting and gaining initial access to a site is no guarantee of acceptance much less trust or even popularity. Hard won trust and rapport can be quickly lost in the face of a perceived rejection or 'social snub' and this may have particularly negative implications in hierarchical multinational settings. Indeed our experience suggests that maintaining relationships is just as demanding and requires just as much skill as gaining initial access to a research site (Adler and Adler, 1991). This can take a toll on a researcher's stamina and it demands considerable tact and patience. In the context of a ship it is not unduly melodramatic to state that a researcher cannot afford to alienate seafarers, not only in order to preserve the integrity of their research but also to minimize the degree of risk that they are exposed to as individuals. There are many incidents involving seafarer 'disappearance' even in today's modern merchant navy and a number of more explicit examples of homicide at sea. In addition there are many alternative methods by which an individual or group might seek to retaliate against a lone, isolated, researcher at sea. Such 'retaliation' against researchers is not unknown as the tragic example of the murder of Myrna Mack serves to highlight (Oglesby, 1995).

In the course of a research encounter, social management is further complicated by the presence of 'cliques' within groups or communities (Warren and Rasmussen, 1977). In the hierarchical structure of a ship these are encouraged by the divisions between ratings and officers, senior and junior officers, deck and engine crew, and members of different nationalities. In this context time and attention allocated to both individuals and on-board social groupings has to be carefully balanced in order to avoid close association with some individuals at the expense of others. The following fieldnotes are illustrative of our increasing awareness of subgroups within the ship's crew and the problems of managing our time between such groups:

> [A senior officer] talked and talked and talked and showed me countless files which he kindly agreed I could come back and read at my leisure. They were a mixture of reports concerning the ship. . . . At 15.00 [he] was still talking and I was both exhausted and keen to go and meet the crew during their coffee break. I was unable to manoeuvre us out in time to get to the crew break and instead settled for persuading him to go and get a coffee in the deck office.
>
> (Fieldnotes)

> . . . the conversation turned to the engineers/deckies divide. The CE was bemoaning the fact that the Captain got paid slightly more than he did and there were the now familiar comments about all they do is steer and if anything goes wrong we have to deal with it. In fact this divide is nearly as apparent as the crew/officer one.
>
> (Fieldnotes)

> Whilst I was talking [one of the officers] passed by and we started talking and actually ended up talking for about an hour (thus blowing my intentions of talking to crew). I'm going to get better at extricating myself from these conversations, later on same day went to see Captain about emails and ended up talking to him for some time. Its not that I find it easier to talk to the officers rather than the crew it's just I find it hard to get out of these conversations without appearing rude.
>
> (Fieldnotes)

In addition to the temporal demands associated with a ship as a research environment, we also found that seafarers' living and working conditions predisposed them to making personal disclosures that could cause problems for a researcher. It seemed that the lonely and socially isolated environment in which seafarers operate increased the likelihood that, once accepted, we would be likely to be subjected to both emotional tales and demands for emotional support. The following fieldnotes are illustrative:

> I don't know whether this was stress or an indication of what a nice man he is but he told me about a Bosun who was killed in an accident on a [ship type] and he started to cry.
>
> (Fieldnotes)

> C . . . had been telling me that his father had died just before he joined the ship and he had had to fly on the morning of the day of the funeral and so had been unable to attend (he had just recovered from a dislocated shoulder which had required him to be sent home – he fell downstairs in rough weather – so he didn't dare ask for more leave). He clearly hadn't had much chance to get over his father's death (he was 66 and died of a stroke) and told me that every night he dreamt of his dad and had only just begun to be able to sleep with the light off.
>
> (Fieldnotes)

Many seafarers have been through traumatic experiences either as a direct consequence of their work (for example, witnessing the death of a colleague) or in circumstances where their working conditions have increased the long-term emotional impact of such events (for example, being absent from home when a child dies). Seafarers rarely discuss such issues with colleagues:

> People steer clear of particular topics on the ship especially personal matters. The Fitter on the ship has experienced two of his children dying in the last three years (the last one in September) but no one but me knows. There must be times when he wants to tell people and it seems incredible that such an enormous thing remains hidden.
>
> (Fieldnotes)

When researchers take an interest in the lives of seafarers aboard ship they invite the rare opportunity for seafarers to emotionally unload. In this respect seafarers' isolation and the extent to which they have a need for a sympathetic 'ear' mirrors, perhaps surprisingly, the situation of many house-bound female research participants. For example Finch (1993) found that the women she interviewed who were married to clergymen, and in many ways isolated from their communities, similarly appreciated the opportunity to talk intimately with a complete stranger in the guise of a researcher. This puts researchers in situations of heightened emotional tension where the importance of a satisfactory response (from the seafarer's, or indeed clergy-man's wife's, perspective) is critical. In failing to make such a response researchers may become exposed to increased situational risk as unintentional offence may be caused and future interaction may break down resulting in hostility of a potentially serious nature. It is worth noting the delicate balance between researcher versus counsellor and the extent to which getting such a balance 'wrong' can land researchers in difficult emotional territory as well as being potentially damaging for the seafarer. This has been discussed in the context of interviews on 'sensitive' or 'private' subjects such as couple relationships (see Duncombe and Marsden, 1996); however, clearly, emotional and personal disclosures can occur in research in the most seemingly innocuous settings (see Farquhar and Das, 1999, discussion of the concept of 'sensitive research subjects'). Thus we were frequently privy to deeply emotional disclosures even in the course of formal tape-recorded interviews on-board. The following response from a rating who had not seen his mother for decades and was planning to try to find her when he returned home demonstrates the kind of response that seemingly straightforward questions about background might elicit:

> My father has already died. My Mum, I don't know. I was also part of a broken family. They were separated when I was eight or ten years old. I stayed with my Dad until he died. My mother is in Mindanao. That's why when I go home this year, God help me, I can try to find her. . . .
>
> (Seafarer interview)

These kinds of disclosures and conversations may reflect gendered perceptions of researchers. Whilst an interested and sympathetic male researcher is also likely to be subject to unsolicited emotional disclosures, gender stereotypes may encourage the perception that women are 'good listeners' and are likely to provide a 'sympathetic ear'. Padfield and Procter have highlighted the extent to which women are more likely to confide in a woman researcher than a man (1996). Our discussions with our colleagues and consideration of all of our fieldnotes suggest that men similarly see women as a repository of confidences. As others have observed this may stem from a desire to develop closer personal relationships with researchers and in some cases may represent a form of sexual hustle (Brewer, 2000). Whatever the motivation, however, this has implications for researchers' time and for their emotional wellbeing. Such conversations are emotionally draining and researchers, like seafarers, are operating away from their normal sources of emotional support – friends, families, colleagues. Indeed in common with other researchers (Karim, 1993) they are likely to experience feelings of isolation, and loneliness, exacerbated by their inability to leave the 'field' for even short periods of time, during long sea-passages, or to communicate privately with people beyond the fieldwork setting.

> I have learned that the only way to communicate with the shore is to have the Captain type a message for you and send it via the computer. This does not make for much privacy! The VHF telephone can be used after passing Las Palmas.
>
> (Fieldnotes)

> The nature of the ship is such that people notice if they do not see you and there is an expectation to be sociable (particularly amongst crew). This can get pretty exhausting. Also now I am beginning to experience a feeling not exactly like claustrophobia but just the sense that there is no means of getting away from it all and this can make it difficult to switch off.
>
> (Fieldnotes)

Such demands expose researchers to a different set of work-related risks relating to their emotional and mental well-being. In ethnographic fieldwork the twenty-four hour a day nature of the fieldwork, coupled with emotional demands and sleep deprivation can leave researchers physically and emotionally exhausted. Ethnographic fieldwork can be intense and poorly bounded. This situation is exacerbated when conducting fieldwork in residential settings where events occur around the clock and may negatively impact upon the attendant level of situational risk. Such behaviour is not sustainable long term and can result in 'under-performing'. Others have pointed to the need for researchers to engage in activities that allow them to 'distance' themselves from the field (Lee, 1995); however aboard ships we found making the time and space for such pursuits difficult and at times impossible. Nevertheless we may have been better at 'allowing' ourselves 'time off' had we been more effectively prepared for the likelihood of being drawn into regimes of almost twenty-four hour working.

Encountering, Avoiding, and Deflecting Sexism

As noted earlier, merchant cargo ships are highly gender-segregated spaces with very few women working aboard even those ships owned, or managed, by the most 'progressive' shipping companies. Women who do join as cadets frequently leave the merchant navy prematurely and low rates of female labour retention are likely to relate, in large part, to the hostile environment they find themselves in once at sea (Belcher et al., 2001; Zhao, 1998). As researchers have found in other fieldwork settings (Easterday et al., 1977; Gurney, 1985), women in the context of the ship are typically accorded low status when they are aboard as either a seafarer's permanent or 'temporary' partner. The 'low status' of 'wives' can be illustrated by the following account where one of us was mistaken by a visiting superintendent as a seafarers' partner:

> In the morning a [company name] inspector came aboard and went around the ship with the Captain. . . . He later apologized to me [for not responding to my efforts to engage him in conversation at lunch] saying that he thought I was someone's wife. The implication being that I was insignificant as 'a wife' but worth talking to as 'a researcher'.
>
> (Fieldnotes)

The low status of partners on-board reflects, in part, the low status accorded to women in general aboard ship. Apart from seafarers' partners, the majority of women encountered by most seafarers in the course of their tours of duty are commercial sex-workers. Seafarers have little on-board exposure to women working in other fields (although there are a small number of women superintendents, customs officials, seafarer mission personnel etc. who come and go whilst ships are in port). This general absence of women onboard can work to both the advantage and disadvantage of the female researcher. On the one hand some seafarers welcome 'female' company and particularly in the guise of women who are interested in listening to details of their work histories, personal lives, and emotional issues. Indeed some commented that talking to us was a bit like talking to their wives, mothers, or sisters. On the other hand, a number of seafarers were suspicious of women who were not in the category of either 'partner' or 'prostitute'. Aboard one ship where the Captain was excessively hostile to our research a fieldnote recorded the following observation:

> I have no doubt whatsoever that most of the Captain's attitude relates to my sex. In fact I imagine that he would bully female cadets or officers in exactly the way he is trying to bully me. Perhaps some seafarers get so used to only dealing with women as wives or prostitutes that they can't handle them in other relationships. . . It seems he had been expecting a man until they had left Brasil when he obviously discovered I was going to be female.
>
> (Fieldnotes)

When women aboard are from an establishment such as a university seafarers may feel challenged or even threatened, and this can result in a somewhat hostile reception for the researcher. In these cases it appears that gendered stereotypes and a general limited exposure to women on-board contribute to situational risk. Researchers may attempt to overcome this by cultivating a non-threatening temporary persona (Gurney, 1985; Lofland and Lofland, 1984) and seafarers may 'deal with' the perceived threat they feel by subordinating the researcher to status of 'student', 'the girl', or even the 'girlfriend' (Easterday et al., 1977) as the following examples demonstrate:

> The Captain told me he had told the agent in Spain that I was his (i.e. the Captain's) girlfriend. He said the agent had said 'oh I understand. I understand Captain' in a manner suggesting that his secret was safe with him.
>
> (Fieldnotes)

> Since Sunday I have also been told that he had been heard – prior to meeting me – making mutterings in the general vein of 'bloody students'. Not exactly encouraging but I felt determined to strike up a conversation with him and try and get him to see beyond this prejudice . . . [some days later] . . . I have obviously made progress with [name] . . . who now instead of calling me a 'bloody student' is referring to me as 'girlie'.
>
> (Fieldnotes)

In the main seafaring remains a bastion of male endeavour. Where significant numbers of women are present in the labour force, for example aboard cruise ships, they encounter relatively high levels of sexism from their colleagues. Zhao identifies several examples of such sexism in her study of the cruise industry. One respondent suggested to her that:

> Ships are not a place for women – they should stay with their husbands and children (quoted in Zhao, 2001).

The role of researcher does not insulate women from such sexism which tends to increase associated levels of situational risk. In establishing rapport our experiences confirmed accounts by others of being 'challenged' by our respondents' use of sexist comments or remarks (Gurney, 1985). We were conscious of being 'tested' on many occasions and tended to feel that to pass the test we had to both register the comment, perhaps with a sardonic glance or a raised eyebrow, whilst resisting the temptation to 'rise to the bait'. Other 'tests' were less good-humoured and made use of sexualized references to other women aboard, ashore, or serving on other ships. They tended to be made by officers sharing the same ethnic status as the researcher and were not so prevalent in cross-cultural encounters. These references seemed best ignored and when made we tended to try to steer the conversation back to the issues at hand – for example, was the seafarer in question a good engineer? Dealing with sexism and other prejudices can be a source of strain to researchers (Warren and Rasmussen, 1977) and ignoring comments and behaviours that conflict with researchers' own values is not something which is always undertaken without a sense of moral cost (Keith, 1992). However, as Warren notes, there is a 'trade-off between accepting sexism on the one hand and the acquisition of knowledge [. . .] on the other' (1988: 36). Failing to challenge such sexist behaviours is not, we would contend, a betrayal of our beliefs (Gurney, 1985 quoted in Warren, 1988) but is rather the only practical and, in the context of ships, safe way to conduct research. In the course of a research career many offensive things may be heard, including racist and sexist remarks; however we were assisted by our view of the role of the researcher as noting and interpreting such comments rather than seeking to achieve the impossible task of 'reforming' individuals who make them.

Whereas other researchers have adopted 'traditional feminine' roles as dictated by the context of their research (see, for example, Gurney, 1985) in minimizing the impact of our gender on the levels of situational risk we encountered, we attempted to distance ourselves from our gendered identities. We achieved this through conscious efforts to monitor and 'manage' our images (Green et al., 1993) and, in this, cross-cultural differences and ambiguities may have assisted us. We resisted undertaking 'feminine tasks' such as helping with cleaning, administration, and catering whilst engaging in counter-stereotyped activities such as painting, chipping,[7] taking an active part in life boat drills[8] etc. This strategy had the further benefit of allowing us increased access to the crew and work of the ship and additional opportunities to observe all aspects of life on-board. It was interesting to note that in the confined and rather public context of a ship our image management tended to extend beyond our physical selves and our actions to management of our physical environment and personal space. For example, we took care over the appearance of our cabins and what was on display there, and even carefully 'managed' the drying of laundry in communal laundry rooms.

Dealing with Specific Situational Risks: Sexual Harassment and Sexual Hustling

Attempting to distance oneself from traditional female roles within the research setting and avoiding a gendered (and feminized) identity does not provide adequate protection from sexual hustling and harassment. With the best will in the world the extent to which one may go 'genderless' within the field is debatable (see Moreno, 1995) and in a male-dominated environment such as a ship, gender identity may be particularly significant. Seafarers report that the arrival of any new woman on-board is an event typically preceded by much speculation, interest and excitement. However, despite the sexist and remote environment of the ship the level of sexual harassment we encountered as researchers did not compare with the experiences of working female seafarers (Belcher et al., 2001; Zhao, 1998). This is likely to be because as 'guests' of shipping companies we were afforded some 'protection' from heavy-handed physical advances. Gurney (1985) in describing the problems that may be encountered by female researchers defines sexual hustling as ranging from 'flirtatious behaviour and sexually suggestive remarks to overt sexual propositioning'. Using this definition there were undoubtedly occasions when we were subjected to hustling and harassment. In dealing with such behaviours, whilst intent on maintaining good on-board relationships, we minimized their significance and only on writing this article properly reflected on the unacceptability of some of the behaviours we encountered. Field relations are delicate and concern to maintain rapport can make rejections to sexual overtures difficult for the researcher. In the close (indeed almost 'panoptic') environment of a ship or similar institution, there may be considerable reluctance to 'report' an incident to others for fear of social disruption. It may additionally be difficult to determine who to report an incident to given that it could, for example, involve the behaviour of a senior officer and even a Captain or Chief Engineer. Furthermore, when working with economically vulnerable groups (as many ratings are) researchers may wish to take account of the consequences for an 'offender' of being reported for sexual harassment. We were fortunate in managing to deflect sexual interest directed towards us aboard ship. However, we noted that efforts to develop close and 'friendly' relationships with informants (Fetterman, 1998) in order to further the research had to be carefully balanced against maintaining appropriate distance with individuals who were often isolated and lonely. We found that reference to 'relationship' status was of some help in rejecting emotional/sexual advances tactfully; however we did not find that being married/in a relationship and publicly referencing our status was sufficient to deter all such advances. Being childless appeared to undermine our 'attached' status on occasion and in common with some other researchers (Berik, 1996; Kenna, 1992) we found that some informants regarded childlessness as an indication of a poor marriage or partnership.

Whilst sexual/emotional advances had to be handled carefully we did not experience a situation when an individual withdrew or threatened to withdraw from the research as a result of 'rejection'. However, managing such interaction enhanced our awareness of the situational and ambient risks with which we were faced.

In considering these issues others have noted that preparation prior to entering the field is important. Knowledge of the role and position of women within research settings may be helpful whilst prior consideration of how to deal with incidents should they occur seems imperative (Gurney, 1991). Nevertheless these were not steps that we were encouraged to take by more experienced or knowledgeable researchers prior to entering 'the field'. In our view this was a particularly serious omission as in the context of a ship researchers cannot decide whether to 'stick it out and salvage a study, if possible, or abandon a research site and begin anew' (Gurney, 1985: 59; see also Sluka, 1995). Abandoning ship is clearly *not* a viable option (although on one ship it appeared an attractive one!). However, despite the inherent barriers to leaving the field when aboard ship, we consider that it is of value to discuss alternative action in the event that unpleasant or personally risky dangerous situations occur in such, or similar, settings. Open discussion of possible strategies to cope in such situations should be encouraged within research institutions amongst both male and female researchers. In the event of sexual harassment it is also important that researchers feel supported by their institutions. In our case we were both aware of feeling that acknowledging problems of sexual harassment back at 'base' would undermine our perceived 'success' as fieldworkers. We were particularly concerned to avoid reinforcing a view that shipboard research should not be conducted by women and as a result significantly underplayed and under-reported the situations with which we were, at times, confronted. However, following discussions with each other and at times with other colleagues we concluded that sexual harassment in the field is too serious a subject to be overlooked or brushed under the carpet. Fears that problems of sexual harassment may contribute to beliefs about 'appropriate' research topics and environments for women need to be broken down and indeed discussion and support concerning these issues should be in place for male and female researchers alike. Sexual harassment can be problematic for any researcher[9] and the effects are likely to be amplified where there is no one within the field with whom the fieldworker has established a particular rapport allowing for the sharing of confidences. Despite a growing awareness of such issues, within and beyond research institutions (Lee-Treweek and Linkogle, 2000a), in common with others (Green et al., 1993) our experiences led us to the view that institutions do not currently pay sufficient attention to the safety of researchers (male or female) and risk thrusting young and inexperienced researchers unprepared, and unprotected, into the field.

Ambient Risk as an Amplifier

The risk encountered by researchers in fieldwork settings can result in various undesirable and serious outcomes ranging from the onset of mental health problems, through to physical assault, suicide, 'unexplained' disappearance, and even murder (Lee, 1995; Nordstrom and Robben, 1995). In the course of our shipboard fieldwork, a situation arose aboard one particular vessel that was tense, dangerous, and extremely difficult to manage. This served to highlight the extent to which we, and our institution, had neglected to properly consider the range of risks associated with shipboard research and the ways in which the ambient risk attached to ships as research sites serves to amplify the situational risk that may be encountered aboard. On the particular ship involved, the researcher was subject to considerable verbal harassment and physical intimidation by the Captain. In the course of the 16 day sea passage, he attempted to prevent her from talking with any member of the crew by instructing her not to talk with them, and telling them not to talk with her unless they wanted to 'end up in gaol'. She was accused of being a 'spy' (first commercial and later trade-union) and was banned from all areas of the ship except her cabin and the messroom where she was allocated a specific table alone. The situation became very worrying as the Captain went so far as to refuse her access to the satellite telephone or any alternative form of shipshore communication, effectively isolating her from the outside world. The following fieldnote is illustrative of her feelings:

> I was so upset after my encounter with the Captain that I found myself shaking. I'm not sure if it is possible to convey how it feels to be treated with such hostility in this particular environment. Although I try to avoid the Captain [. . .] there are some things which I just can't do without his permission (like make a telephone call) and I can't understand why he should want to stop me from doing something as reasonable as using the telephone to call home anyway.
>
> (Fieldnotes)

Whilst gender was felt to be a factor here, it was not considered to be the only one, and we are aware that male researchers have also experienced considerable hostility in the field. As Van Maanen notes, 'Disagreeable and unapproachable people are sure to be among those with whom the researcher must deal' and indeed, as he found in his fieldwork with police agencies, 'unambiguous rejection' is not an impossible event (1991: 36):

> 'Stay out of my life, Van Maanen. I don't have nothing to say to you and you don't have nothing to say to me. I'm putting in my time . . . I don't know what you want and I wouldn't give a shit even if I did. You mind your business and I'll mind mine.' (Van Maanen, 1991: 36)

Accusations of spying have been reported elsewhere; indeed, Sluka (1995) suggests they are commonplace in anthropological research. However, when such accusations and the hostility they ferment are encountered aboard ship there are limits to what can be done. Researchers cannot leave the field to allow for a 'cooling off' period, they cannot abandon the research, and they may not even be able to communicate with the outside world. In such situations there can be considerable personal risk and researchers may find they have frighteningly little control over their situation. There is no way to ascertain the likelihood of such events occurring. However, our experiences served to highlight the amplifying effect of the ambient risk associated with research aboard ship and the dangers inherent in adopting a 'cavalier' attitude to fieldwork in this setting, and similar settings of attendant risk such as conflict zones (Sluka, 1995).

Conclusion

Traditionally social science publications have tended to focus more on discussions of the findings and practicalities of undertaking research than on researchers' own personal experiences. However, the last 20 years have seen a change of emphasis in the work of many ethnographers who have taken a more consciously reflexive approach in reporting research findings. Nevertheless, issues relating to the personal health, safety, and wellbeing of social researchers continue to be under-emphasized in the work of many sociologists.

In developing accounts of risk in research settings, those who have attended to issues of researcher health and safety have found it useful to distinguish between the ambient risks attached to specific research sites and situational risks to which researchers become exposed in the course of their work. In this article we have sought to highlight the ways in which gender can add to situational risk in some research contexts and, importantly, the extent to which ambient risks, and particularly those associated with remote research sites, can act as an amplifier of situational risk.

The research settings we have focused on here, merchant navy ships, provide environments in which both ambient and situational risks are high, particularly for female researchers. Whilst accepting that some risks associated with on-board research in a male-dominated and remote environment are immutable, our experiences suggest that there are greater steps that can be taken at both an institutional and a personal level to minimize such risk in the course of the research process (Lankshear, 2000).

In most dangerous fieldwork settings there are steps that can be taken to minimize personal risk. Attending to the minimization of such ambient risk has the additional benefit of reducing the amplifier effect it tends to have on the situational risk to which researchers are invariably exposed. In terms of the ship, one of the major sources of ambient risk is the remoteness of deep sea vessels once at sea, and the consequent isolation of researchers aboard. The identification of a means of independent communication between researchers and the 'outside' world goes some considerable way to reducing such remoteness. Whilst expensive, the dangerous situation experienced aboard one vessel convinced us and, fortunately, our institution, that a mobile satellite telephone should be purchased and carried by all researchers joining vessels in order to facilitate contact with shore-based colleagues and families. This has served to significantly reduce the extent to which researchers are endangered by their isolated status whilst undertaking work on-board, and it has helped to erode the impact of the situational risks they may encounter in the course of their work. However, as has been noted elsewhere (Jamieson, 2000) the proliferation of contract research and current mechanisms for research funding militate against such provision in many cases.

Other measures may also be beneficial in terms of minimizing the ambient risks attached to voyage-based research. For example ships may be selected with closer consideration to voyage cycles and those without an initial 'short-sea' passage may be deemed unsuitable as research sites. Companies with poor health and safety track records may be researched using alternative methods that do not involve ethnography aboard, and researchers may be encouraged to undertake pre-sea survival courses to ensure that in the event of a disaster they can take steps to minimize personal injury and the chances of mortality. In addition, a fundamental requirement in the conduct of future fieldwork aboard ships should be that research institutions maintain regular formal contact with researchers in the field and monitor and log details of their vessels, location, and contact details which may alter as they move between satellite telephone 'code zones'.

In terms of reducing and counteracting situational risks on-board we implemented a number of different strategies, some of which have been usefully employed by other researchers in alternative field settings. Many of the situational risks we encountered related to our gender and these may have been exacerbated by the male-dominated nature of the research setting. Whilst it is not our view that one can become 'genderless', we found that certain strategies were useful in reducing the impact of our gender on the research process and experience. Specifically these included the avoidance of stereotypically 'feminine' tasks and behaviours and general image management. Despite these efforts, our experience suggests that in the environment of a cargo ship female researchers are more likely to be confronted by sexism, sexual harassment and 'sexual hustling' than their male colleagues. Rather than minimizing the extent to which such issues arise and their impact on both the fieldwork process and researchers as individuals, we believe researchers would benefit from formally discussing such issues, along with potential responses and coping strategies, with colleagues prior to entering the field. In order to facilitate such discussion research institutions should provide a supportive environment for their researchers and in addition should encourage open discussion of, and 'debriefing' on, fieldwork experiences (Jamieson, 2000; Sluka, 1995).

Our experiences of conducting fieldwork, relating to this and to other projects, indicate that institutions and funding bodies[10] have yet to take issues relating to researcher health and safety sufficiently seriously. This can, in part, be attributed to the prevalence of a male-dominated and competitive research culture in institutions of higher education, a culture which is often reflected in accounts of risk and bravery in the field (Patrick, 1973). In order to 'fit in' to this culture, and to enjoy academic success, men and women may find themselves complicit in compromising their own health and safety. Certainly we feel that we have, to some extent, been guilty of this. However, the tendency to overlook researcher health and safety also reflects a more general failure to fully appreciate the extent to which ambient risks amplify the more usual 'situational' risks encountered by researchers. When faced with entry to a research site in which the ambient risk is high, researchers, their colleagues, families, and friends may be so concerned with the obvious associated dangers, such as catching a bullet in a war zone (Olujic, 1995) or sinking having boarded a rusty old cargo ship, that they fail (as we did) to properly consider the ways in which such dangers amplify the everyday situational risks attached to research. As a result they also fail to consider strategies for dealing with challenging situations and minimising such risks, a situation which Sluka reminds us can be a '*matter of life and death*' (Sluka, 1995: 291).

Notes

1. This is an old term used lightheartedly at sea to refer to land-people as opposed to seafarers.
2. Ships in the merchant navy have traditionally sailed with male crews and recent estimates indicate that women continue to be underrepresented on-board. Furthermore, recent research suggests that a considerable degree of hostility towards women seafarers exists amongst male crew, who have difficulties accepting women on-board as colleagues (SIRC/ILO report 2002).
3. A survey of female criminologists found almost one in three reporting some form of sexual harassment during fieldwork (Stanko, 1992).
4. This table is not intended to be comprehensive but rather lists the occupational ranks we encountered aboard. Companies may use different titles for the same occupational rank, e.g. in the UK the chief officer may be referred to as the 'chief mate' or 'mate' and in addition some jobs are specific to particular ship types (e.g. reefer engineers). We have not attempted to cover all such variations.

5. Male researchers may also be sexually harassed in the fieldwork setting, as has happened to a male researcher from our centre whilst aboard ship and see also McKeganey and Bloor, 1991.
6. Aboard *Qui Auora* the reception of the Captain, and as a result the sea-staff, was hostile and uncooperative.
7. Chipping rusting paint with jet hammers is one of the less pleasant routine maintenance tasks aboard ship.
8. Lifeboat drills are recognized within the industry as being extremely hazardous and seafarers are often reluctant to take part in them. The UK MAIB recently noted that in the last 10 years 12 fatalities have occurred during lifeboat drills whilst lifeboats have not saved the life of a single seafarer in British waters (MAIB, 2001).
9. In addition to experiencing sexual harassment (McKeganey and Bloor, 1991), male researchers conducting fieldwork in traditionally female domains have reported problems where efforts to secure interviews have been perceived as 'sexual advances' or attempts to set up 'dates' (see Sparke, 1996).
10. Funding bodies occasionally utilize unrealistic daily subsistence rates that may force researchers to reside in unsafe living conditions whilst conducting research.

References

Adler, P.A. and Adler, P. (1991) 'Stability and Flexibility: Maintaining Relations Within Organised and Unorganised Groups', in W.B. Shaffir and R.A. Stebbins (eds) *Experiencing Fieldwork: An Inside View of Qualitative Research*, pp. 173–83. London: Sage.

Belcher, P., Lane A.D., Sampson, H., Thomas, M., Veiga, J. and Zhao, M., (2001) *Women Seafarers: Global Employment Policies and Practices*, SIRC report for ILO, December.

Berik, G. (1996) 'Understanding the Gender System in Rural Turkey: Fieldwork Dilemmas of Conformity and Intervention' in D.L. Wolf (ed.) *Feminist Dilemmas in Fieldwork*, pp. 56–71. Boulder, CO: Westview.

Brewer, J.D. (1990) 'Sensitivity as a Problem in Field Research', *American Behavioural Scientist* 33: 578–93.

Brewer, J.D. (1991) *Inside the RUC: Routine Policing in a Divided Society*. Oxford: Clarendon Press.

Brewer, J.D. (2000) *Ethnography*. Buckingham: Open University Press.

Coffey, A. (1999) *The Ethnographic Self: Fieldwork and the Representation of Identity*. London: Sage.

Duncombe, J. and Marsden, D. (1996) 'Can We Research the Private Sphere? Methodological and Ethical Problems in the Study of the Role of Intimate Emotion in Personal Relationships', in L. Morris and E. Stina Lyon (eds) *Gender Relations in Public and Private: New Research Perspectives*, pp. 141–55. London: Macmillan.

Easterday, L., Papademas, D., Schorr, L. and Valentine, C. (1977) 'The Making of a Female Researcher: Role Problems in Field Work', *Urban Life* 6(3): 333–48.

Farquhar, C. and Das, R. (1999) 'Are Focus Groups Suitable For "Sensitive" Topics?', in R.S. Barbour and J. Kitzinger (eds) *Developing Focus Group Research: Politics, Theory and Practice*, pp. 47–63. London: Sage.

Fetterman, D. (1998) *Ethnography*. London: Sage.

Finch, J. (1993) '"It's Great to Have Someone to Talk To": Ethics and Politics of Interviewing Women', in M. Hammersley (ed) *Social Research: Philosophy, Politics and Practice*, pp. 166–80. London: Sage.

Fox, J. and Lundman, R., (1974) 'Problems and Strategies in Gaining Access to Police Organisation', *Criminology* 12: 52–69.

Gearing, J. (1995) 'Fear and Loving in the West Indies: Research from the Heart (As Well as the Head)' in D. Kulick and M. Willson (eds) *Taboo: Sex, Identity and Erotic Subjectivity in Anthropological Fieldwork*, pp. 186–218. London and New York: Routledge.

Green, G., Barbour, R.S., Barnard, M., Kitzinger, J. (1993) 'Who Wears the Trousers? Sexual Harassment in Research Settings', *Women's Studies International Forum* 16(6): 627–37.

Gurney, J.N. (1985) 'Not One of the Guys: The Female Researcher in a Male-Dominated Setting', *Qualitative Sociology* 8: 42–60.

Gurney, J.N. (1991) 'Female Researchers in Male Dominated Settings' in W.B. Shaffir and R.A. Stebbins (eds) *Experiencing Fieldwork: An Inside View of Qualitative Research*, pp. 53–61. London: Sage.

Hansen H.L. (1996) 'Surveillance of Deaths On-board Danish Merchant Ships 1986–93: Implications for Prevention', *Occupational & Environmental Medicine* 53(4): 269–75.

Howell, N. (1990) *Surviving Fieldwork: A Report of the Advisory Panel on Health and Safety in Fieldwork*. Washington, DC: American Anthropological Association.

ICONS (2000) *Information Paper International Shipping*. [http://www.ICONS.hl.com.au]

Jamieson, J. (2000) 'Negotiating Danger in Fieldwork on Crime: A Researcher's Tale', in G. Lee-Treweek and S. Linkogle (eds) *Danger in the Field: Risk and Ethics in Social Research*, pp. 61–71. London: Routledge.

Kahveci, E. (1999) *Fast Turnaround Ships and their Impact on Crews*. Cardiff: SIRC.

Kahveci, E., Lane, T. and Sampson, H. (2002) *Transnational Seafarer Communities*. Cardiff: SIRC.

Karim, J.W. (1993) 'With Moyang Melu in Carey Island: More Endangered, More Engendered' in D. Bell, P. Caplan and W. Jahan Karim (eds) *Gendered Fields: Women, Men and Ethnography*, pp. 78–92. London: Routledge.

Keith, M. (1992) 'Angry Writing: (Re)presenting the Unethical World of the Ethnographer', *Environment and Planning D: Society and Space* 10: 551–68.

Kenna, M.E. (1992) 'Changing Places and Altered Perspectives: Research on a Greek Island in the 1960s and in the 1980s', in J. Okely and H. Callaway (eds) *Anthropology and Autobiography*, pp. 147–62. London: Routledge.

Lane, A.D. (1998) 'Being On a Ship is Being in Jail', *Naftika Kronika*, November: 88–9.

Lane A.D., Alderton, T., Bloor, M., Kahveci, E., Obando-Rojas, B., Sampson, H., Thomas, M., Winchester, N. and Zhao, M. (2001) *The Impact on Seafarers' Living and Working Conditions of Changes in the Structure of the Shipping Industry*. International Labour Organisation Sectoral Activities Programme, Geneva: International Labour Office, JMC/29/2001/3.

Lankshear, G. (2000) 'Bacteria and Babies: A Personal Reflection on Researcher Risk in a Hospital', in G. Lee-Treweek and S. Linkogle, (eds) *Danger in the Field: Risk and Ethics in Social Research*, pp. 72–90. London: Routledge.

Lareau, A. (1996) 'Common Problems in Fieldwork: A Personal Essay', in A. Lareau and J. Shultz (eds) *Journeys through Ethnography: Realistic Accounts of Fieldwork*, pp. 195–236. Oxford: Westview Press.

Lee, R.M. (1995) *Dangerous Fieldwork*. London: Sage.

Lee-Treweek, G. (2000) 'The Insight of Emotional Danger: Research Experience in a Home for Older People', in G. Lee-Treweek, and S. Linkogle (eds) *Danger in the Field: Risk and Ethics in Social Research*, pp. 114–31. London: Routledge.

Lee-Treweek, G. and Linkogle, S. (2000a) *Danger in the Field: Risk and Ethics in Social Research*. London: Routledge.

Lee-Treweek, G. and Linkogle, S. (2000b) 'Putting Danger in the Frame', in G. Lee-Treweek and S. Linkogle (eds) *Danger in the Field: Risk and Ethics in Social Research*, pp. 8–25. London: Routledge.

Lofland, J. and Lofland, L.H. (1984) *Analysing Social Settings: A Guide to Qualitative Observation and Analysis*. Belmont: Wadsworth.

MAIB (2001) *Review of Lifeboat and Launching Systems' Accidents*. Safety Study 1/2001. Southampton: Department of Environment, Transport and Regions (DETR), Marine Accident Investigation Branch (MAIB).

McKeganey, N. and Bloor, M. (1991) 'Spotting the Invisible Man: The Influence of Male Gender on Fieldwork Relations', *British Journal of Sociology* 42(2): 195–210.

Moreno, E. (1995) 'Rape in the Field: Reflections from a Survivor', in D. Kulick and M. Willson (eds) *Taboo: Sex, Identity and Erotic Subjectivity in Anthropological Fieldwork*, pp. 219–50. London and New York: Routledge.

Nordstrom, C. and Robben, A.C.G.M. (1995) *Fieldwork Under Fire: Contemporary Studies of Violence and Culture*. California: University of California Press.

Oglesby, E. (1995) 'Myrna Mack', in C. Nordstrom and A.C.G.M. Robben (eds) *Fieldwork Under Fire: Contemporary Studies of Violence and Culture*. California: University of California Press.

Olujic, M.B. (1995) 'The Croatian War Experience', in C. Nordstrom and A.C.G.M. Robben (eds) *Fieldwork Under Fire: Contemporary Studies of Violence and Culture*. California: University of California Press.

Padfield, M. and Procter, I. (1996) 'The Effect of Interviewer's Gender on the Interviewing Process: A Comparative Enquiry', *Sociology* 30(2): 355–66.

Patrick, J. (1973) *A Glasgow Gang Observed*. London: Eyre Methuen.

Reederei Nord (2000) 'The Number of Pirate Attacks Increased Again in 2000', *Nord News* (Winter Issue), Cyprus: Reederei 'NORD' Klaus E. Oldendorff Ltd.

Sluka, J.A. (1995) 'Reflections on Managing Danger in Fieldwork: Dangerous Anthropology in Belfast', in C. Nordstrom and A.C.G.M. Robben (eds) *Fieldwork Under Fire: Contemporary Studies of Violence and Culture*, pp. 276–94. California: University of California Press.

Sparke, M. (1996) 'Displacing the Field in Fieldwork: Masculinity, Metaphor and Space', in N. Duncan (ed.) *Bodyspace: Destabilizing Geographies of Gender and Sexuality*, pp. 212–33. London and New York: Routledge.

Stanko, E.A. (1992) 'Intimidating Education: Sexual Harassment in Criminology', *Journal of Criminal Justice Education* 3: 331–40.

Thompson, H. (1967) *Hell's Angels*. Harmondsworth: Penguin.

Van Maanen, J. (1991) 'Playing Back the Tape: Early Days in the Field', in W.B. Shaffir and R.A. Stebbins (eds) *Experiencing Fieldwork: An Inside View of Qualitative Research*, pp. 31–42. London: Sage.

Warren, C.A.B. (1988) *Gender Issues in Field Research*. Newbury Park, CA: Sage.

Warren, C.A.B. and Rasmussen, P.K. (1977) 'Sex and Gender in Field Research', *Urban Life* 6(3): 349–69.

Whitehead, T.L. and Price, L. (1986) 'Summary: Sex and the Fieldwork Experience', in T.L. Whitehead and M.E. Conaway (eds) *Self, Sex and Gender in Cross-Cultural Fieldwork*, pp. 289–304. Urbana, IL: University of Illinois Press.

Young, G. (1986) *Slow Boats Home*. London: Penguin.

Zhao, M. (1998) *Women Seafarers in the EC: A Preliminary Report Based On German and UK Case Studies*. Cardiff: SIRC, Cardiff University.

Zhao, M. (2001) 'The Role of Female Crew Members On-Board Cruise Ships'. Unpublished paper.

36

When Is Disguise Justified? Alternatives to Covert Participant Observation

Martin Bulmer

The question of whether or to what extent particular research methods are ethically or practically justified has been given new point by the progressive extensions of federal regulation of research into the field of the social sciences (Duster et al., 1979). Old debates about the ethics of covert participation observation (e.g. Coser, 1959; Davis, 1961; Denzin, 1968; Erikson, 1967; Galliher, 1973; Lofland and Lejune, 1960; Roth, 1962; Sullivan et al. 1958) are being revived on both sides of the Atlantic at a time when sociology is increasingly in the public eye and under public scrutiny (cf. Bulmer, 1980; Dingwall, 1980; Douglas, 1976; Hilbert, 1980; Homan, 1980; Krockars & O'Connor, 1980; Whyte, 1979). This article argues that the classic contrast drawn between the participant-as-observer (playing an open role) and the complete participant (whose role is disguised) (Gold, 1958) is an inadequate formulation of the alternatives available, which stultifies debate and hinders methodological innovation.

Covert, secret or disguised participant observation may be defined as research situations where the real identity of the observer as a social researcher remains secret and entirely unknown to those with whom he or she is in contact. The investigator purports to be a complete participant and is in fact something else. Classic instances of such research include Caudill's study of a psychiatric hospital (Caudill, Redlich, Gilmore and Brody, 1952), Festinger's study of a group who believed the end of the world was coming (Festinger et al., 1956), Humphries' study of homosexual encounters in public toilets

Source: *Qualitative Sociology*, 5(4) (1982): 251–264.

(1970) and Rosenhan's study using pseudo-patients of the labelling process in psychiatry (1973). Several of these studies have yielded knowledge not otherwise available, which could probably not have been obtained if a direct, overt, observational strategy had been adopted.

Controversy over Covert Observational Methods

Nevertheless, such research methods have aroused fierce controversy and stinging criticism. On ethical grounds they have been criticized for violating the principle of informed consent, invading subjects' privacy, using deception, and betraying trust, possibly harming the interests of the group studied, and having unmeasurable effects on the investigator him or herself and the data collected. If subjected to scrutiny by Institutional Review Boards, covert observational methods are particularly likely to be the object of criticism and disapproval for these reasons.

There are also practical and methodological reasons for doubting their value. Hiding completely one's research intent may affect the quality of the data gathered in unknown ways. The observer may be unaware of his or her own biases, and may be misled about how he or she is perceived by those being studied. A rare comparison is provided by William Caudill, who studied the same psychiatric hospital using first covert and later overt observational methods. He concluded that the overt approach provided data equal in richness to that yielded by the covert approach, and gave access to a wider range of data over which it was possible to exercise a greater degree of control. Caudill was more satisfied, both intellectually and emotionally, with the results of the second, overt, piece of fieldwork (1958: xv).

It is also the case that the need or appropriateness of covert research may be exaggerated. Many accounts of observational research stress that its success depends more upon the acceptance of the investigator as a person to be trusted by those being studied, than on elaborate role pretense. In studies of professional criminals, for example, the techniques of espionage or undercover police work are inadvisable and unnecessary (Klockars, 1975, 1977; Polsky, 1969: 124–35). Apparently closed or inaccessible milieux such as the Mafia (Ianni and Ianni, 1972) or the British civil service (Heclo & Wildovsky, 1974) may be successfully studied by observational methods without resort to disguises and concealment of research purpose. "Powerful people seem so convinced of their own righteousness and so sure of their power that they usually talk freely" (Galliher, 1980, p. 303).

Sometimes covert research is defended on the grounds that it is only different in degree in the extent of deception used from "open" research. Rather than being polar alternatives, there is a continuum of greater or lesser openness or secrecy on the part of the investigator (Roth, 1962). For example, there are a number of studies where the identity and purpose of the researcher was known to one or two members of a group but kept secret from the other members (cf. Patrick, 1973; Pryce, 1979). *Total* honesty is characteristic of very few types of social relationship (cf. Bok, 1978) and all field research involves giving misinformation, less than full information or even mild deceit *to some extent*. This is particularly true in the negotiation of access to an institution, association or informal group, and in conveying to those being studied what exactly it is that the obeserver is doing there.

Despite such caveats, the debate still tends to polarize between those who adopt highly moral and condemnatory attitudes against concealment and exaggerated general justifications which would assimilate sociological field research to the paradigm of investigative journalism (cf. Douglas, 1976). There is a need for subtler distinctions to be made, and the wider variety of observational research strategies used in practice to be recognised in methodological debate (Bulmer, 1982). Four such strategies will now be discussed, which cut across the "either/or" character of covert versus overt field research.

Retrospective Participant Observation

If poetry is "emotion recollected in tranquility," one mode of participant observation might be described as "experience recollected in academia." A number of distinguished observational studies have been conducted (at least partially) retrospectively, after the event. After World War II, a number of sociologists wrote studies of units in the armed forces, based on their own experience of war-time service. An example is Homans' study of the small warship (1946). Slightly later, Howard Becker's study of dance musicians was begun in 1948–49 when he himself was working as a musician and also going through college.

> At the time I made the study I had played the piano professionally for several years and was active in musical circles in Chicago. . . . I worked with many different orchestras and many different kinds of orchestras during that period and kept extensive notes on the events that occurred while I was with other musicians. Most of the people I observed did not know that I was making a study of musicians. I seldom did any formal interviewing, but concentrated rather on listening to and recording the ordinary kinds of conversation that occurred among musicians. (1963: 84)

Most of the observations were made on the job, even while playing, or in local "job markets" where musicians found work.

Ned Polsky's study of poolroom hustling, though, "carried out" over eight months in 1962–63, was grounded in and inseparable from his own experience in poolrooms. He had frequented them for over 20 years, since billiard playing was his chief recreation. Prior to the time of the study he had played billiards an average of six hours per week, mainly in the major poolhall of New York. He knew and had played regularly with several of the hustlers who were the subject of the research (1969: 46–48).

In conventional covert research the sequence of research roles is:

academic sociologist	→	total participant (sociologist)	→	academic sociologist again

In retrospective participant observation this sequence is reversed:

total participant nonsociologist)	→	academic sociologist	→	retrospective observer (sociologist)

As a total participant in this role, the "observer" was completely immersed in the setting which he or she was studying, but had not yet developed a research interest in it. No deception of those studied was involved, no breach of informed consent occurred, since no scientific study was in view at the time the study was carried out.

The Native as Stranger

Discsussion of covert versus overt methods tend to assume that in most studies, with the exception of retrospective observation, the observer is an outsider who learns to become an insider of a "native." This is reflected in the perceived danger of "going native." The observer has to maintain the necessary detachment or distance, and the personal strains which this sets up when doing covert research are likely to be a particular problem (cf. Redlich, 1973). The possibility that an insider or native might become an observer is rarely considered. "An outsider could be made into an insider, but never the reverse" (Heilman, 1980: 101).

Nevertheless, the "native-as-stranger" approach to observational research can circumvent some of the problems faced in deciding when to do an overt or covert study as an outsider. Rather than the sociologist being a stranger who enters an unfamiliar setting purporting to be a native, someone indigenous to that setting is trained as a sociologist and then observes it further. Unlike retrospective observation, the observer returns to the milieu but enters under different auspices, that of a legitimate member of the culture who has sought the necessary degree of distance and detachment in order to study it. Several notable covert studies have been done of small religious groups by pretending to be full members. Heilman's account (1980) of his study of his own orthodox Jewish synagogue is of particular interest as a contrast to this, since he was studying it both as a sociologist and as a "native" insider.

Another interesting example is provided by Simon Holdaway's (1977) research upon the British police. The police in Britain have in the past recruited few university graduates. In recent years, in order to train a cadre of future senior officers, a few outstanding serving policemen have been selected competitively and sent for three years (on full pay) to follow an undergraduate degree course at university, on condition that they return and serve for at least a minimum period afterwards. The expectation of course is that they will remain in the police force. Most study law, but Holdaway was selected for the scheme and elected to study sociology, having previously spent nine years in the police force. Upon graduation, he returned to the police force and spent a further two years as a serving police officer. During this time he was also registered" for part-time study for a higher degree, on the subject of informal practices within police stations when charging suspects (Holdaway, 1977). This research interest was not explicitly acknowledged to his superiors or fellow policemen (Holdaway, 1982). Subsequently, Holdaway obtained a university post teaching sociology, from which he has continued to write about the police (cf. 1979) and write up his observational data.

This type of research, like Heilman's, follows the sequence:

total participant (non sociologist) → academic sociologist → total participant (non sociologist *and* sociologist) → academic sociologist

There can thus be different research "careers" from the conventional one of the sociologist-stranger trying to turn him or herself into a native for the purposes of research. When such research strategies are being pursued, different questions are raised about the extent and legitimacy of concealing one's research intentions. In the most extreme (though admittedly quite atypical) situations, as in Bettelheim's study of the concentration camp (1943), the "native" investigator may even adopt the research role as a strategy for survival.

The Covert Outsider

The roles of retrospective observer or native-as-stranger are open only to those whose backgrounds and experiences are congruent with their research interests. In the case of the broader class of observational studies where the observer does not have special prior experience to capitalize upon, there are nevertheless alternatives open which are not captured by a simple dichotomy between a decision to adopt a covert or overt approach. A distinction between the "insider" and the "outsider" is helpful at this point. Research which is more or less covert in character may vary in the character of the role assumed by the sociological observer.

Debates on covert methods usually assume that the investigator seeks to become an accepted insider – whether as sect member, mental patient, watchqueen or whatever. An extreme example in the literature is the infiltration of the Ku Klux Klan as a pretended believer. In fact, however, many observational studies are carried out in settings where a variety of roles exist, and it is possible for the sociologist to adopt one of these roles as a convenient cover for research. The sociology of education provides a clear example. Adoption of the role of teacher has been a standard strategy for carrying out observational research in schools (e.g. Hargreaves, 1967; Lacey, 1970). The sociologist who is not explicit about his or her research interests can use the teacher role as a cover for research work, without becoming completely identified with those being studied (Burgess, 1980). It is a legitimate professional role which does not require complete immersion in a particular milieu.

The teacher role can and has been used as a means of entree to other settings to gain access and conduct semicovert research. Gypsies are a tightly knit and closed group, suspicious of outsiders and resistant to being the subjects of field work. Anne Sutherland found in her study of Californian gypsies that they did not distinguish between "social anthropologist" and "journalist" or "detective," each a category of people whom they greatly distrusted. Nevertheless she was able to carry out an ethnographic study principally because she acted for nearly a year as principal of a specially established gypsy school, which was highly valued by the local Rom and brought her into close contact with them. In this case, acting as a teacher gave her access to a group which otherwise would have been closed (Sutherland, 1975: 30, 20–25.).

A notable British prison study, Cohen and Taylor's study of the effects of long-term imprisonment (1972) was also carried out through their playing the role of teacher. They visited "E Wing" of Durham prison, where Category "A" prisoners serving very long-term prison sentences were housed, as adult education lecturers, giving regular seminars to a small group of prisoners over a period of some months:

> The classes were more stimulating than those we took at the university; the men actually read the recommended books and had plenty of practical examples with which to refute our crasser generalizations. But much of the time it was very like any other extramural (extension) class. . . . (A)n academic ethos was a welcome relief from the usual relationships which (prisoners) encountered in the prison: the vulgar authoritarianism of the "screws" (i.e. prison officers) and the patronizing condescension of social workers, psychiatrists and chaplains. (Cohen and Taylor, 1977: 68).

The idea of doing the research emerged slowly in the course of the teaching situation and came as much from the prisoners as from the sociologists. It developed when the latter ran out of teaching material and turned to talking more conversationally about all sorts of subjects. Cohen and Taylor used various methods including "literary identification" to try to structure the collection of data in the classes. Research continued under the guise of the classes for a considerable period, while formal application was made through the hierarchy of the prison and the British Home Office for permission to conduct research. This was eventually refused, but the authors were able to reply that most of it had already been done in their adult education class. Cohen and Taylor's work may be criticized on methodological grounds, and they tended to take an "oppositional" stance towards the prison authorities which eventually undermined their attempts to mount a further, more systematic study (Cohen and Taylor, 1977: 78–86). However, their adoption of a semicovert role in the prison as outsiders in a nonresearch role made possible a study of a special group of prisoners. It was, moreover, distinctively different in strategy from the Morrises' overt observational study of another British prison ten years earlier (1962), where they were present in the prison for an extended period but recognized throughout that time as researchers.

The teacher role is not the only one which can be adopted by what may be termed the "covert outsider" in a research setting, but it provides a good example of the potential for being present in a nonsociological guise without becoming an insider. Moreover, such a nonresearch outsider role can provide an opportunity to carry out research which would be denied to the outsider-who-is-researcher.

Thus in addition to the usual alternatives of the outsider-as-researcher participating in order to look in (as in classic studies such as Whyte's *Street Corner Society* 1955) or the covert insider (the sociologist as pseudopatient, or gang member, or sect member), a third role has been recognized, that of covert outsider where the researcher plays some other role, such as that of teacher.

The Overt Insider

There is, however, an interesting case which is the reverse of the latter, that of the sociologist who does become an insider, but does so overtly and with the knowledge and agreement of members of the setting being studied. Thus Jonathan Rubinstein, after working for a year as a police reporter on a Philadelphia newspaper, negotiated with the director of the Philadelphia Police Department that he actually work with police on the street. In September 1969, he entered the Philadelphia Police Academy, going through the full training of a police officer. On graduation he joined a patrol unit, working with different units full-time for one year until September 1970 and at weekends for a further year until September 1971. He never spent less that two months with any unit.

Rubinstein describes his role as follows:

> Legal complications prevented me from becoming a sworn police officer, although I had fulfilled all the requirements at the academy. But I was permitted to go on the street as an armed observer (travelling in patrol cars). Since I worked mostly with uniformed policemen but had to wear plainclothes, I carefully chose some that would give me the appearance of a detective or superior officer. Many of the men I worked with knew that I was not a policeman (though some believed I was a federal agent or an undercover police operative), but policemen from other units and districts whom we encountered in the course of work did not. Only on a few occasions did my companions inform these men of my identity. On no occasion did any policeman I worked with inform a private citizen of my status in my presence, and only once did anyone question whether I was a police officer. . . . I worked only with those men who willingly accepted me as a worker and not just as a passenger. Everywhere I went there were men who granted me their company and introduced me to the police craft by allowing me to share their work. (Rubinstein, 1974: xii–xiii)

Yet this is written by a social science researcher, attached to the University of Pennsylvania and supervised by Erving Goffman, not by a policeman. Rubinstein's work was done with the full knowledge and cooperation of the police department. Rubinstein's immediate associates knew he was not a policeman, and most knew he was a researcher. But in the course of the research some other policemen, and all members of the public, were misled as to his identity. The role played by the "overt insider" indeed differs relatively little from the role played by someone just "hanging around" in a particular setting doing research, except that the deliberate adoption of (and training for) the role gives the researcher access to settings from which the "hanger-around" might be excluded. It is not a panacea. For example, McCall (1978: 85–90) suggests that in observational studies of the police, observers have a particular tendency to "go native" and adopt a police perspective. This is likely to be greatly enhanced if the observer has undergone police training.

There is considerably more scope for the adoption of this type of approach in sociological research, the taking on of a new occupational role specifically for research purposes, if necessary being trained for the work to quite a high standard. It might be described as "prospective" participant observation. Instead of exploiting past social experience, the sociologist deliberately seeks out new social experience, as a legitimate member of an occupational group within an organization, in order to understand better the life of the group and the workings of the organization.

Conclusion

Discussion and formalization of research methods – particularly, perhaps, qualitative methods – tends to lag behind practice. It is clear from a number of monographs that a simple polarization between completely secret and entirely overt participant observation does not adequately capture the range of research practice, even though such a polarization is found n recent debates on the merits of covert methods (for some recent examples see Bulmer, 1982).

In particular, the variety of research roles which may be adopted are more various than is commonly recognized in methodology texts. In the first place, the recognition of a time dimension and of the possibility of retrospective or "prospective" participant observation adds fresh complexity to the simple "either/or" polarity between secret and open research. Movement back and forth between different stages of a research career points up the difficulties of formulating simple ethical prescriptions.

More formally, the variety of roles available to the participant observer – as shown in Figure 1 – is not adequately captured in the conventional debates. These tend to contrast the open (top right – Whyte) and the covert (bottom center – Humphries) as polar alternatives, without recognizing the distinction between different types of "insider" role, particularly between the native-as-stranger and the stranger coming from outside the setting. Moreover, the scope for covert research within organizations in an outsider role, and for overt "insider" research as a socialized stranger, suggests that the argument over the merits of covert participant observation is not a matter of white and black, from whatever perspective one views it.

	INSIDER ROLE		OUTSIDER ROLE
	Native as stranger	Stranger	
Overt	For example: Samuel Heilman, Synagogue Life	For example: Jonathan Rubinstein, City Police	For example: William F. Whyte, Street Corner Society
Covert	For example: Simon Holdaway, "An inside job"	For example: Laud Humphreys, Tearoom Trade	For example: Stanley Cohen and Laurie Taylor, Psychological Survival

Figure 1: Varieties of research role played by field researchers using participant observation

References

Becker, Howard S. 1963 Outsiders. New York: The Free Press of Glencoe.
Bettelheim, Bruno 1943 "Individual and Mass Behaviour in Extreme Situations." Journal of Abnormal and Social Psychology 38:417–452.
Bok, Sisella 1978 Lying: Moral Choice in Public and Private Life. New York: Random House.
Bulmer, Martin 1980 "Comment on 'The Ethics of Covert Methods' (Homan, 1980)." British Journal of Sociology 31:59–65.
Burgess, Robert 1980 "Some Fieldwork Problems in Teacher-Based Research." British Educational Research Journal 6:165–173.
——— (ed.) 1982 Social Research Ethics: An Examination of the Merits of Covert Participant Observation. New York: Holmes and Meier.
Caudill, William, Frederick Redlich, Helen Gilmore and Eugene Brody. 1952 "Social Structure and Interaction Processes on a Psychiatric Ward." American Journal of Orthopsychiatry 22:314–334.
Cohen, Stanley and Laurie Taylor 1972 Psychological Survival: The Experience of Long-Term Imprisonment. Harmondsworth: Penguin.
——— 1977 "Talking About Prison Blues." Pp. 67–86 in Colin Bell and Howard Newby (eds.), Doing Sociological Research. London: Allen & Unwin.
Coser, Lewis et al. 1959 "Participant Observation and the Military: An Exchange." American Sociological Review 24:397–400.
Davis, Fred 1961 "Comment on Initial Interactions of Newcomers in Alcoholics Anonymous' (Lofland and Lejeune, 1960)." Social Problems 8:364–365.
Dinzin, Norman 1968 "On the Ethics of Disguised Observation." Social Problems 15: 502–504.
Dingwall, Robert 1980 "Ethics and Ethnography." The Sociological Review 28: 871–891.
Douglas, Jack D. 1976 Investigative Social Research. Beverly Hills: Sage.
Duster, Troy et al. 1979 "Fieldwork and the Protection of Human Subjects." The American Sociologist 14:136–142.
Erickson, Kai T. 1967 "A Comment on Disguised Observation in Sociology." Social Problems 14:366–373.
Festinger, Leo 1956 When Prophecy Fails. Minneapolis: University of Minnesota Press.
Galliher, John F. 1973 "The Protection of Human Subjects: A Reexamination of the Professional Code of Ethics." The American Sociologist 8:93–100.
——— 1980 "Social Scientists' Ethical Responsibilities to Superordinates: Looking Upward Meekly," Social Problems 27:298–308.
Gold, Raymond L. 1958 "Roles in Sociological Field Observations." Social Forces 36:217–233.
Hargreaves, David H. 1967 Social Relations in a Secondary School. London: Routledge.
Heclo, Hugh and Aaron Wildovsky 1974 The Private Government of Public Money. London: Macmillan.
Heilman, Samuel C. 1980 "Jewish Sociologist: Native-as-Stranger." The American Sociologist 15:100–108.
Hilbert, Richard A. 1980 "Covert Participant Observation: On Its Nature and Practice." Urban Life 9:51–76.
Holdaway, Simon 1977 "Changes in Urban Policing." British Journal of Sociology 28:119–137.
——— 1979 The British Police. London: Edward Arnold.
——— 1982 "An Inside Job; A Case Study of Covert Research on the Police." Pp. 59–79 in M. Bulmer (ed.), Social Research Ethics. London: Macmillan.

Homan, Roger 1980 "The Ethics of Covert Methods." British Journal of Sociology 31:46–59.

Homans, George C. 1946 "The Small Warship." American Sociological Review 11:294–300.

Humphreys, Laud 1970 Tearoom Trade: Impersonal Sex in Public Places. Chicago, Aldine (Revised edition, 1975, with additional appendices).

Ianni, Francis A.J. and Elizabeth Ianni 1972 A Family Business: Kinship and Social Control in Organized Crime. London: Routledge.

Klockars, Carl 1975 The Professional Fence. London: Tavistock.

________ 1977 "Field Ethics for the Life History." Pp. 201–206 in Robert S. Weppner (ed.), Street Ethnography. Beverly Hills: Sage Annual Review of Drug and Alcohol Abuse, Vol. 1.

Klockars, Carl and Finbar W. O'Connor (eds.) 1980 Deviance and Decency: The Ethics of Research with Human Subjects. Beverly Hills: Sage.

Lacey, Colin 1970 Hightown Grammar. Manchester: Manchester University Press.

Lofland, John and R.A. LeJeune 1960 "initial Interaction of Newcomers in Alcoholics Anonymous: A Field Experiment in Class Symbols and Socialization." Social Problems 8:102–111.

McCall, George J. 1978 Observing the Law: Field Methods in the Study of Crime and the Criminal Justice System. New York: Free Press.

Morris, Terence P. and Pauline Morris 1962 Pentonville. London: Routledge.

Patrick, James 1973 A Glasgow Gang Observed. London: Eyre-Methuen.

Polsky, Ned 1969 Hustlers, Beats and Others. Harmondsworth: Penguin.

Pryce, Ken 1979 Endless Pressure. Harmondsworth: Penguin.

Redlich, Frederick 1973 "Ethical Aspects of Clinical Observation of Behaviour." Journal of Nervous and Mental Disease 157 (November):313–319.

Rosenhan, D.L. 1973 "On Being Sane in Insane Places." Science 179:350–358.

Roth, Julius 1962 "Comments on 'Secret Observation.'" Social Problems 9:283–284.

Rubinstein, Jonathan 1974 City Police. New York: Ballantine (first published 1973).

Sullivan, Mortimer A. et al. 1958 "Participant Observation as Employed in the Study of a Military Training Program." American Sociological Review 23:660–667.

Sutherland, Anne 1975 Gypsies: The Hidden Americans. London: Tavistock.

Whyte, William F. 1955 Street Corner Society. Chicago: University of Chicago Press (Revised Edition: first published 1943).

________ 1979 "On Making the Most of Participant Observation." The American Sociologist 14:56–66.

37

A Comment on Disguised Observation in Sociology

Kai T. Erikson

At the beginning of their excellent paper on the subject, Howard S. Becker and Blanche Geer define participant observation as "that method in which the observer participates in the daily life of the people under study, either openly in the role of researcher or covertly in some disguised role . . ."[1]

The purpose of this paper is to argue that the research strategy mentioned in the last few words of that description represents a significant ethical problem in the field of sociology. In point of sheer volume, of course, the problem is relatively small, for disguised" participant observation is probably one of the rarest research techniques in use among sociologists. But in point of general importance, the problem is far more serious – partly because the use of disguises seems to attract a disproportionate amount of interest both inside and outside the field, and partly because it offers a natural starting point for dealing with other ethical issues in the profession.

In recent years, a handful of studies have been reported in the literature based on the work of observers who deliberately misrepresented their identity in order to enter an otherwise inaccessible social situation. Some of these studies have already provoked a good deal of comment – among them, for instance, the cases of the anthropologist who posed as a mental patient by complaining of symptoms he did not feel,[2] the sociologists who joined a gathering of religious mystics by professing convictions they did not share,[3] the Air Force officer who borrowed a new name, a new birth date, a new

Source: *Social Problems*, 14(4) (1967): 366–373.

personal history, a new set of mannerisms and even a new physical appearance in order to impersonate an enlisted man,[4] and the group of graduate students who ventured into a meeting of Alcoholics Anonymous wearing the clothes of men from other social classes than their own and the facial expressions of men suffering from an unfortunate disability.[5]

In taking the position that this kind of masquerading is unethical, I am naturally going to say many things that are only matters of personal opinion; and thus the following remarks are apt to have a more editorial flavor than is usual for papers read at professional meetings. But a good deal more is at stake here than the sensitivities of any particular person, and my excuse for dealing with an issue that seems to have so many subjective overtones is that the use of disguises in social research affects the professional climate in which all of us work and raises a number of methodological questions that should be discussed more widely.

I am assuming here that "personal morality" and "professional ethics" are not the same thing. Personal morality has something to do with the way an individual conducts himself across the range of his human contacts; it is not local to a particular group of persons or to a particular set of occupational interests. Professional ethics, on the other hand, refer to the way a group of associates define their special responsibility to one another and to the rest of the social order in which they work. In this sense, professional ethics often deal with issues that are practical in their application and limited in their scope: they are the terms of a covenant among people gathered together into a given occupational group. For instance, it may or may not be ethical for an espionage agent or a journalist to represent himself as someone he is not in the course of gathering information, but it certainly does not follow that the conduct of a sociologist should be judged in the same terms; for the sociologist has a different relationship to the rest of the community, operates under a different warrant, and has a different set of professional and scientific interests to protect. In this sense, the ethics governing a particular discipline are in many ways local to the transactions that discipline has with the larger world.

The argument to be presented here, then, is that the practice of using masks in social research compromises both the people who wear them and the people for whom they are worn, and in doing so, violates the terms of a contract which the sociologist should be ready to honor in his dealings with others. There are many respects in which this is true, but I will be dealing here in particular with the relationship between the sociologist and a) the subjects of his research, b) the colleagues with whom he works, c) the students he agrees to teach, and d) the data he takes as his subject matter.

The first of these points has to do with the responsibilities a sociologist should accept toward other institutions and other people in the social order. It may seem a little cranky to insist that disguised observation constitutes an ugly invasion of privacy and is, on that ground alone, objectionable. But it is a matter of cold calculation to point out that this particular research strategy can injure people in ways we can neither anticipate in advance nor compensate for afterward. For one thing, the sheer act of entering a human transaction on the basis of deliberate fraud may be painful to the people who are thereby misled; and even if that were not the case, there are countless ways in which a stranger who pretends to be something else can disturb others by failing to understand the conditions of intimacy that prevail in the group he has tried to invade. Nor does it matter very much how sympathetic the observer is toward the persons whose lives he is studying: the fact of the matter is that he does not *know* which of his actions are apt to hurt other people, and it is highly presumptuous of him to act as if he does – particularly when, as is ordinarily the case, he has elected to wear a disguise exactly because he is entering a social sphere so far from his own experience.

So the sheer act of wearing disguises in someone else's world may cause discomfort, no matter what we later write in our reports; and this possibility raises two questions. The first, of course, is whether we have the right to inflict pain at all when we are aware of these risks and the subjects of the study are not. The second, however, is perhaps more important from the narrow point of view of the profession itself: so long as we suspect that a method we use has at least *some* potential for harming others, we are in the extremely awkward position of having to weigh the scientific and social benefits of that procedure against its possible cost in human discomfort, and this is a difficult business under the best of circumstances. If we happen to harm people who have agreed to act as subjects, we can at least argue that they knew something of the risks involved and were willing to contribute to that vague program called the "advance of knowledge". But when we do so with people who have expressed no readiness to participate in our researches (indeed, people who would presumably have refused if asked directly), we are in very much the same ethical position as a physician who carries out medical experiments on human subjects without their consent. The only conceivable argument in favor of such experimentation is that the knowledge derived from it is worth the discomfort it may cause. And the difficulties here are that we do not know how to measure the value of the work we do or the methods we employ in this way, and, moreover, that we might be doing an extraordinary disservice to the idea of detached scholarship if we tried. Sociologists cannot protect their freedom of inquiry if they owe the rest of the community (not to mention themselves) an accounting for the distress they may have inadvertently imposed on people who have not volunteered to take that risk.

The second problem with disguised observation to be considered here has to do with the sociologist's responsibilities to his colleagues. It probably goes without saying that research of this sort is liable to damage the reputation of sociology in the larger society and close off promising areas of research for future investigators. This is true in the limited sense that a particular agency –

say, for example, Alcoholics Anonymous – may decide that its integrity and perhaps even its effectiveness was violated by the appearance of sociologists pretending to be someone else and deny access to other students who propose to use an altogether different approach. And it is also true in the wider sense that any research tactic which attracts unfavorable notice may help diminish the general climate of trust toward sociology in the community as a whole. So long as this remains a serious possibility, the practice of disguised observation becomes a problem for everyone in the profession; and to this extent, it is wholly within the bounds of professional etiquette for one sociologist to challenge the work of another on this score.

This objection has been raised several times before, and the answer most often given to it is that the people who are studied in this fashion – alcoholics or spiritualists or mental patients, for example – are not likely to read what we say about them anyway. Now this argument has the advantage of being correct a good deal of the time, but this fact does not prevent it from being altogether irrelevant. To begin with, the experience of the past few years should surely have informed us that the press is more than ready to translate our technical reports into news copy, and this means that we can no longer provide shelter for other people behind the walls of our own anonymity. But even if that were not the case, it is a little absurd for us to claim that we derive some measure of protection from the narrowness of our audience when we devote so much time trying to broaden it. The fact is that we are increasingly reaching audiences whose confidence we cannot afford to jeopardize, and we have every right to be afraid that such people may close their doors to sociological research if they learn to become too suspicious of our methods and intentions.

The third objection to be raised here, if only as a note in passing, concerns the responsibilities the profession should accept toward its students. The division of labor in contemporary sociology is such that a considerable proportion of the data we use in our work is gathered by graduate students or other apprentices, and this proportion is even higher for research procedures that require the amount of energy and time necessary for participant observation. Of the dozen or more observers who took part in the studies I have cited, for example, all but one was a graduate student. Now a number of sociologists who have engaged in disguised observation have reported that it is apt to pose serious moral problems and a good deal of personal discomfort, and I think one might well argue that this is a heavy burden to place on any person who is, by our own explicit standards, not yet ready for professional life. I am not suggesting here that students are too immature to make a seasoned choice in the matter. I am suggesting that they should not be asked to make what one defender of the method has called "real and excruciating moral decisions" while they are still students and presumably protected from the various dilemmas and contentions which occupy us in meetings like this – particularly since they are so likely to be academically, economically, and even psychologically dependent upon those elders who ask them to choose.[6]

The fourth objection I would like to raise here about the use of undercover observation is probably the most important – and yet the most remote from what is usually meant by the term "ethics." It seems to me that any attempt to use masquerades in social research betrays an extraordinary disrespect for the complexities of human interaction, and for this reason can only lead to bad science. Perhaps the most important responsibility of any sociologist is to appreciate how little he really knows about his intricate and elusive subject matter. We have at best a poor understanding of the human mind, of the communication signals that link one mind to another, or the social structures that emerge from those linkages – and it is the most arrant kind of over-simplification for us to think that we can assess the effect which a clever costume or a few studied gestures have on the social setting. The pose might "work" in the sense that the observer is admitted into the situation; but once this passage has been accomplished, how is he to judge his own influence on the lives of the people he is studying? This is a serious problem in every department of science, of course, and a good deal of time has been devoted to its solution. But the only way to cope with the problem in even a preliminary way is to have as clear a picture as possible of the social properties that the observer is introducing into the situation, and this is altogether impossible if we ourselves are not sure who he is. We can *impersonate* other modes of behavior with varying degrees of insight and skill, but we cannot *reproduce* them; and since this is the case, it seems a little irresponsible for a sociologist to assume that he can enter social life in any masquerade that suits his purpose without seriously disrupting the scene he hopes to study.

When people interact, they relate to one another at many different levels at once, and only a fraction of the messages communicated during that interchange are registered in the conscious mind of the participant. It may be possible for someone to mimic the conventional gestures of fear, but it is impossible for him to reproduce the small postural and chemical changes which go with it. It may be possible for a middle-class speaker to imitate the broader accents of lower-class speech, but his vocal equipment is simply not conditioned to do so without arousing at least a subliminal suspicion. It may be possible for a trained person to rearrange the slant of his body and reset his facial muscles to approximate the bearing of someone else, but his performance will never be anything more than a rough imposture. Now we know that these various physiological, linguistic, and kinetic cues play an important part in the context of human interaction, but we have no idea how to simulate them – and what is probably more to the point, we never will. For one thing, we cannot expect to learn in a matter of hours what others have been practicing throughout a lifetime. For another, to imitate always means to parody, to caricature, to exaggerate certain details of behavior at the expense of others, and to that extent any person who selects a disguise will naturally emphasize those details which *he* assumes are most important to the character he is portraying. In doing so, of course, he is really only portraying a piece of himself. It is interesting to speculate, for example, why the Air Force lieutenant mentioned earlier thought he needed to present himself as a near-delinquent youth with a visible layer of personal problems in order to pose as an enlisted man. Whatever the reasoning behind this articular charade, it would certainly be reasonable for someone to suspect that it tells us more about the investigators' impression of enlisted men than it does about the men themselves – and since we have no way of learning whether this is true or not, we have lost rather than gained an edge of control over the situation we are hoping to understand. What the investigators had introduced into the situation was a creature of their own invention, and it would be hardly surprising if the results of their inquiry corresponded to some image they had in advance of the enlisted man's condition. (It is perhaps worth noting here that impersonation always seems easier for people looking down rather than up the status ladder. We find it reasonable to assume that officers "know how" to portray enlisted men or that sociologists have the technical capacity to pose as drunks or religious mystics, but it is not at all clear that the reverse would be equally true.)

This, then, is the problem. If we provide observers with special masks and coach them in the "ways" of the private world they are hoping to enter, how can we learn what is happening to the people who meet them in this disguise? What information is registered in the unconscious minds of the other people who live in that world? How does the social structure accommodate to this peculiar invasion?

It is clear, I think, that something happens – something over which we have no control. Let me relate two incidents drawn from the studies mentioned earlier. The first has to do with the Air Force officer who posed as an enlisted man. In their report of the study, the investigators used several pages of a short paper to describe the elaborate masquerade they had fashioned for the observer and the coaching he had received in the ways of the adolescent sub-culture. "So successful was the tutoring," reads the brief report, "that when the time for 'enlistment' arrived, the recruiting sergeant suggested that the observer not be accepted by the Air Force because by all appearances he was a juvenile delinquent."[7] And later, during an interview with a service psychologist, the observer was recommended for re-classification on the grounds that he appeared quite anxious over the death of his father. Now these events may indeed suggest that the pose was successful, for the observer *was* trying to look somewhat delinquent and *did* have a story memorized about the death of his father in an auto accident. But who would care to argue that the diagnosis of the sergeant and the psychologist were inaccurate? Surely something was wrong, and if they perceived an edge of uneasiness which reminded them of anxiety or detected a note of furtiveness which looked to them like delinquency, they may only have been responding to the presence of a real conflict between the observer and his mask. We may leave it to the psychoanalysts to ask whether vague anxieties about "killing" one's father are an unlikely impression for someone to leave behind when he is parading around with a new name, a new background, a new history, and, of course, a new set of parents. The authors of the article tell us that the observer "did have something of a problem to transform himself from a 27-year-old, college trained, commissioned officer into a 19-year-old, near-delinquent high school graduate," and this is certainly easy to believe.[8] What is more difficult to believe is that such a transformation is possible at all – and if it is not, we can have very little confidence in the information gathered by the observer. Since we do not know to what kind of creature the enlisted men were responding, we do not know what sense to make of what they said and did.

The second example comes from the study of the apocalyptic religious group. At one point in the study, two observers arrived at one of the group's meeting places under instructions to tell quite ordinary stories about their experience in spiritualism in order to create as little commotion as possible. A few days afterwards, however, the leader of the group was overheard explaining that the two observers had appeared upset, excited, confused, and unsure of their errand at the time of their original visit, all of which helped confirm her suspicion that they had somehow been "sent" from another planet. In one sense, of course, this incident offered the observers an intriguing view of the belief structure of the cult, but in another sense, the leader's assessment of the situation was very shrewd: after all, the observers *had* been sent from another world, if not another planet, and she may have been quite right to sense that they were a bit confused and unsure of their errand during their early moments in the new job. "In both cases," the report informs us, the visits of the observers "were given as illustrations that 'strange things are happening.'"[9] Indeed, strange things *were* happening; yet we have no idea how strange they really were. It is almost impossible to evaluate the reaction of the group to the appearance of the pair of observers because we do not know whether they were seen as ordinary converts or as extraordinary beings. And it makes a difference, for in the first instance the investigators would be observing a response which fell within the normal range of the group's experience, while in the second instance they would be observing a response which would never have taken place had the life of the group been allowed to run its own course.

My point in raising these two examples, it should be clear, is not to insist on the accuracy of these or any other interpretations, but to point out that a wide variety of such interpretations is possible so long as one has no control over the effects introduced by the observer. A company of recruits with a disguised officer in its midst is simply a different kind of organization than one without the same ingredient; a group of spiritualists which numbers as many as eight observers among its twenty or so members has a wholly different character than one which does not – and so long as we remain unable to account for such differences, we cannot know the meaning of the information we collect.

In one of the most sensible pieces written on the subject, Julius Roth has reminded us that all social research is disguised in one respect or another and that the range of ethical questions which bear on the issue must be visualized as falling on a continuum.[10] Thus, it is all very well for someone to argue that deliberate disguises are improper for sociologists, but it is quite another matter for him to specify what varieties of research activity fall within the range of that principle. Every ethical statement seems to lose its crisp authority the moment it is carried over into marginal situations where the conditions governing research are not so clearly stipulated. For instance, some of the richest material in the social sciences has been gathered by sociologists who were true participants in the group under study but who did not announce to other members that they were employing this opportunity to collect research data. Sociologists live careers in which they occasionally become patients, occasionally take jobs as steel workers or taxi drivers, and frequently find themselves in social settings where their trained eye begins to look for data even though their presence in the situation was not engineered for that purpose. It would be absurd, then, to insist as a point of ethics that sociologists should always introduce themselves as investigators everywhere they go and should inform every person who figures in their thinking exactly what their research is all about.

But I do think we can find a place to begin. If disguised observation sits somewhere on a continuum and is not easily defined, this only suggests that we will have to seek further for a relevant ethic and recognize that any line we draw on that continuum will be a little artificial. What I propose, then, at least as a beginning, is the following: first, that it is unethical for a sociologist to *deliberately misrepresent* his identity for the purpose of entering a private domain *to which he is not otherwise eligible*; and second, that it is unethical for a sociologist to *deliberately misrepresent* the character of the research in which he is engaged. Now these negative sanctions leave us a good deal of leeway – more, perhaps, than we will eventually want. But they have the effect of establishing a stable point of reference in an otherwise hazy territory, and from such an anchored position as this we can move out into more important questions about invasion of privacy as an ethical issue.

In the meantime, the time has probably come for us to assume a general posture on the question of disguised participant observation even if we are not yet ready to state a specific ethic, and a logical first step in this direction would be to assess how most members of the profession feel about the matter. I am not suggesting that we poll one another on the merits of adopting a formal code, but that we take some kind of unofficial reading to learn what we can about the prevailing climate of opinion in the field. If we discover that a substantial number of sociologists are uncomfortable about the practice, then those who continue to employ it will at least know where they stand in respect to the "collective conscience" of their discipline. And if we discover that only a scattering of sociologists are concerned about the matter, we will at least have the satisfaction of knowing that the profession – as a profession – has accepted the responsibility of knowing its own mind.

Notes

1. Howard S. Becker and Blanche Geer, "Participant Observation and Interviewing: A Comparison," *Human Organization*, 16 (1957), pp. 28–32.
2. William C. Caudill *et al.*, "Social Structure and Interaction Processes on a Psychiatric Ward," *American Journal of Orthopsychiatry*, 22 (1952), pp. 314–334.
3. Leon Festinger, Henry W. Riecken, and Stanley Schacter, *When Prophecy Fails*. Minneapolis: University of Minnesota Press, 1956.
4. Mortimer A. Sullivan, Stuart A. Queen, and Ralph C. Patrick, Jr., "Participant Observation as Employed in the Study of a Military Training Program," *American Sociological Review*, 23 (1958), pp. 660–667.
5. John F. Lofland and Robert A. Lejeune, "Initial Interaction of Newcomers in Alcoholics Anonymous: A Field Experiment in Class Symbols and Socialization," *Social Problems*, 8 (1960), pp. 102–111.
6. To keep the record straight, I might add that I first became interested in these matters when I was a graduate student and applied for one of the observer posts mentioned here.
7. Sullivan, Queen, and Patrick, *op. cit.*, p. 663.
8. Stuart A. Queen, "Comment," *American Sociological Review*, 24 (1959), pp. 399–400.
9. Festinger, Riecken, and Schacter, *op. cit.*, p. 241–242.
10. Julius A. Roth, "Comments on 'Secert Observation,'" *Social Problems*, 9 (1962), pp. 283–284.

38

Methods: The Sociologist as Voyeur

Laud Humphreys

In the summer of 1965, I wrote a research paper on the subject of homosexuality. After reading the paper, my graduate adviser raised a question, the answer to which was not available from my data or from the literature on sexual deviance: "But where does the average guy go just to get a blow job? That's where you should do your research." I suspected that the answer was "to the tearooms," but this was little more than a hunch. We decided that this area of covert deviant behavior, tangential to the subculture, was one that needed study.

Stories of tearoom adventures and raids constantly pass along the homosexual grapevine. There is a great deal of talk (usually of a pejorative nature) about these facilities in gay circles. Most men with whom I had conversed at any length during my brief research admitted to "tricking" (engaging in one-time sexual relationships) in tearooms from time to time.

Sociologists had studied bar operations[1] and the practice of male prostitution by teen-age gang members,[2] but no one had tackled the scenes of impersonal sex where most arrests are made. Literature on the subject indicates that, up to now, the police and other law enforcement agents have been the only systematic observers of homosexual action in public restrooms. In some localities, these agents have been very busy with such observations. For example, of the 493 charges of felony for supposed homosexual conduct made during a recent four-year period in Los Angeles County, California, 56 per cent were against persons arrested in public restrooms.[3]

Social scientists have avoided this area of deviant behavior, perhaps due to the many emotional and methodological problems it presents – some of

Source: *Tearoom Trade* (Enlarged Edition) (New Jersey: Transaction, 1975), pp.16–44.

which are common to any study of a deviant population. Ethical and emotional problems, I suspect, provide the more serious obstacles for most prospective researchers. As Hooker points out, "Gaining access to secret worlds of homosexuals, and maintaining rapport while conducting an ethnographic field study, requires the development of a non-evaluative attitude toward all forms of sexual behavior."[4] Such an attitude, involving divorce from one's socialization, is not easy to come by. No amount of intellectual exercise alone can enable the ethnographer to make such emotional adjustments, and ethical concerns (see the Postscript for a full discussion) serve to complicate the task.

I am inclined to agree with Polsky when he says: "Most difficulties that one meets and solves in doing field research on criminals are simply the difficulties one meets and solves in doing field research."[5] The obstructions encountered in the course of this study – insurmountable as some appeared at the time – are, for the most part, shared with other ethnographers, particularly those who take on deviant populations.

Unless one intends to study only that beleaguered, captive population, the students in our college classrooms, the first problem is one of locating subjects for research. In his study of the Negro street corner man, Elliot Liebow found that "he is no more at home to the researcher than he is to the case worker or the census taker."[6] The refreshingly human account of Whyte's "first efforts" at research in *Street Corner Society* underscores this point.[7] Masters and Johnson turned to prostitutes for respondents who were "knowledgeable, cooperative, and available for study" in the early stages of their investigation of human sexual response.[8]

As indicated in the preceding chapter, my initial problem was one of locating the more popular tearooms. Once I could find where the "action" was, I knew that potential research subjects would be involved. This is the advantage of studying a population defined only by their participation in a specific sort of interaction. To observe those involved in race track society, the scientist goes to the race tracks.[9] To study "homosexuals" or "schizophrenics," however, one must first overcome the vague, stereotypical generalizations to which even social science falls victim in order to define (much less isolate and sample) the population.

This is not a study of "homosexuals" but of participants in homosexual acts. The subjects of this study have but one thing in common: each has been observed by me in the course of a homosexual act in a public park restroom. This is the activity, and these are the actors, that I set out to study in 1965. The physical traces that helped lead me to such performers and performances are discussed in Chapter 1. When a researcher is able to engage participants in conversation outside the tearooms, he may be directed to some of the more active spots of a city. In the early stages of research, however, such measures as location, surroundings, and the number of cars parked in front are very helpful in locating the more profitable places for research.

Other Tearooms – Other Variables

There are, of course, other tearooms, not located in parks, that might have been studied. Those in the Y's and transportation facilities have received the greatest publicity.[10] My study, however, has been focused upon the park facilities for two reasons. First, the latter have the greatest notoriety in the homosexual subculture. Second, I wanted to control the ecological and demographic variables as much as possible. All but two of the restrooms in which I conducted systematic observations were of the same floor plan, and all shared common environmental conditions. Of greater importance is the "democratic" nature of outdoor facilities. Parks are much more apt to draw a representative sample of the population.

In the same city, there is a well-known tearoom in a courthouse, another in a large department store, and a third in the basement of a class B movie theater. Each caters to a different clientele, is subject to different influences from the physical surroundings, and is supervised by different forces of social control. In the department store tearoom, most of the men wear neckties. Participants venture there during lunch hour from their nearby offices. This is a white collar facility, patrolled by the store's detectives. Word has it that an apprehended offender is taken to the office of the store manager, who administers reprimands and threats and then pronounces sentence: he revokes the guilty man's credit card! I once spent an hour counselling a distraught participant who was contemplating suicide in apprehension of what his wife might be told if she tried to charge anything at this popular store, where her husband had been caught in an act of fellatio.

Although I have made informal observations of tearoom activity in New York, Chicago, St. Louis, Kansas City, Des Moines, Tulsa, Denver, Los Angeles, and San Francisco, the greater part of the research was concentrated in one metropolitan area. Admittedly, there are a number of factors that cast doubt on the applicability of these findings to other regions of the United States, much less to other nations and cultures.

One feature of the toilet stalls in the city where my research was concentrated constitutes an important variable: there are no doors on the stalls in the public parks. Signals from the stalls, therefore, are all of the bodily motion variety – gestures of the head or hands. One social scientist questioned an earlier paper of mine that had omitted reference to other types of signaling:

> You don't say anything about men who go into a stall, close the door and then make contact with the chap in the next stall by means such as foot-tapping or passing notes. I presume that, given the layout of the restrooms you surveyed, this was not common practice, but it is certainly done widely here. As a matter of fact, the University recently removed the doors from *every other* stall in several of the larger men's rooms on campus in order to cut down on this activity.[11]

In tearooms where there were doors on the stalls, I *have* observed the use of foot-tapping as a means of communication. What the university authorities mentioned above apparently failed to realize, however, is that doors on the stalls serve as hindrances rather than aids to homoerotic activity. Certainly, the passing of notes would cause inconvenience and place the actor in greater jeopardy.

Other variables such as climate, availability of parks, the nature of police surveillance, amount of newspaper publicity accorded offenders, or relative popularity of other sexual outlets could result in wide variations in the volume of tearoom activity.[12] My contention, however, is that the basic rules of the game – and the profile of the players – are applicable to any place in the United States.

This much may be said with certainty: there is probably no major city in the nation without its tearooms in current operation. These facilities constitute a major part of the free sex market for those in the homosexual subculture – and for millions who might never identify with the gay society. For the social scientist, these public toilets provide a means for direct observation of the dynamics of sexual encounters *in situ*; moreover, as will be seen, they facilitate the gathering of a representative sample of secret deviants, for most of whom association with the deviant subculture is minimal.

Neatness versus Accuracy

I employed the methods described herein not because they are the most accurate in the sense of "neatness" or "cleanness" but because they promised the greatest accuracy in terms of faithfulness to people and actions as they live and happen. These are strategies that I judged to be the least obtrusive measures available – the least likely to distort the real world.

My biases are those that Bruyn attributes to the participant observer, who "is interested in people as they are, not as he thinks they ought to be according to some standard of his own."[13] To employ, therefore, any strategies that might distort either the activity observed or the profile of those who engage in it would be foreign to my scientific philosophy and inimical to my purposes.

Some methods, then, have grown quite naturally from the chromosomal messages of a particular "school" of sociology. Others are mutations resulting from interaction with my research environment. As obstacles developed, means were devised to circumvent them. Unusual difficulties call for unusual strategies. Although I have employed a number of "oddball measures," as they are called by Webb and his associates, these research methods are actually only uncommon applications of such tested measures as physical traces, the running record, and simple observation.[14]

My concern in this study has been with the description of a specific style of deviant behavior and of the population who engage in that activity. Beyond such systematic, descriptive analyses, I have tried to offer, in the light of deviance theory, some explanation as to why and how these people participate in the particular form of behavior described. I have not attempted to test any pre-stated hypotheses. Such an approach tends to limit sociological research to the imagery of the physical sciences. It seems to me equally valid to apply a number of measures to one population or one type of social interaction for the purpose of describing that encounter and its participants.

Hypotheses should develop *out of* such ethnographic work, rather than provide restrictions and distortions from its inception. Where my data have called for a conceptual framework, I have tried to supply it, sometimes with the help of other social scientists. In those cases where data were strong enough to generate new theoretical approaches, I have attempted to be a willing medium. The descriptive study is important, not only in obtaining objective and systematic knowledge of behavior that is either unknown or taken for granted, but in providing the groundwork for new theoretical development. If the social scientist is to move back and forth between his data and the body of social theory, the path of that movement should not be restricted to a set of predestined hypotheses.[15]

The research in which I engaged, from the summer of 1965 through the winter of 1967–68, may be broken down into two distinct stages, each with its subcategories. The first was an ethnographic or participant-observation stage. This part of the research extended over two years on a part-time basis (I was also involved in graduate study at the time).

The second half involved six months of full-time work in administering interview schedules to more than one hundred respondents and in attempting to interview another twenty-seven. Another year has been devoted to analysis of the resulting data.

Preparing for the Field

As an ethnographer, my first task was to acquaint myself with the homosexual subculture. Because of my pastoral experience, I was no total stranger to those circles. While a seminarian, I was employed for two years in a parish that was known in the homosexual world as Chicago's "queen parish" – a place to which the homosexuals could turn for counsel, understanding priests, good music, and worship with an aesthetic emphasis. I soon came to know the gay parishioners and to speak their language. Seminarians who worked there called on people who lived in unbelievable squalor and in houses of prostitution, so it was nothing for us to seek the flock in gay bars as well. One of the faithful churchmen was bartender and part-time entertainer in one of the more popular spots, and he always looked after the seminarians and warned us of impending raids.

This particular part of my education was supplemented in the summer of 1953, when I spent three months in clinical training at the State University of Iowa's psychiatric hospital. This was a model institution, operated primarily for research and training purposes, and (in line with research interests of the Head of Staff) was well stocked that summer with male homosexual patients. That training provided me with a background in psychoanalytic theory regarding homosexuality.

From 1955 to 1965, I served parishes in Oklahoma, Colorado, and Kansas, twice serving as Episcopal campus chaplain on a part-time basis. Because I was considered "wise" and did not attempt to "reform" them, hundreds of homosexuals of all sorts and conditions came to me during those years for counselling. Having joined me in counselling parishioners over the coffee pot for many a night, my wife provided much understanding assistance in this area of my ministry.

The problem, at the beginning of my research, was threefold: to become acquainted with the sociological literature on sexual deviance; to gain entry to a deviant subculture in a strange city where I no longer had pastoral, and only part-time priestly, duties; and to begin to listen to sexual deviants with a scientist's rather than a pastor's ear.

Passing as Deviant

Like any deviant group, homosexuals have developed defenses against outsiders: secrecy about their true identity, symbolic gestures and the use of the eyes for communication, unwillingness to expose the whereabouts of their meeting places, extraordinary caution with strangers, and admission to certain places only in the company of a recognized person. Shorn of pastoral contacts and unwilling to use professional credentials, I had to enter the subculture as would any newcomer and to make contact with respondents under the guise of being another gay guy.[16]

Such entry is not difficult to accomplish. Almost any taxi driver can tell a customer where to find a gay bar. A guide to such gathering places[17] may be purchased for five dollars. The real problem is not one of making contact with the subculture but of making the contact "stick." Acceptance does not come easy, and it is extremely difficult to move beyond superficial contact in public places to acceptance by the group and invitations to private and semiprivate parties. This problem has been well expressed by a team engaged in homosexual research at the University of Michigan:

> An outsider – be he a novitiate deviant, police officer, or sociologist – finds it necessary to cope with a kind of double closure one confronts around many kinds of subcultural deviance; to wit, one may gain entrance into the deviant enterprise only if he has had previous connection with it, but he can gain such connections only if he has them.[18]

On one occasion, for instance, tickets to an after-hours party were sold to the man next to me at a bar. When I asked to buy one, I was told that they were "full up." Following the tip of another customer, I showed up anyway and walked right in. No one questioned my being there. Since my purpose at this point of the field study was simply to "get the feel" of the deviant community rather than to study methods of penetrating its boundaries, I finally tired of the long method and told a friendly potential respondent who I was and what I was doing. He then got me invited to cocktail parties before the annual "drag ball," and my survey of the subculture neared completion.

During those first months, I made the rounds of ten gay bars then operating in the metropolitan area, attended private gatherings and the annual ball, covered the scene where male prostitutes operate out of a coffee house, observed pick-up operations in the parks and streets, and had dozens of informal interviews with participants in the gay society. I also visited the locales where "instant sex" was to be had: the local bathhouse, certain movie theaters, and the tearooms.

From the beginning, my decision was to continue the practice of the field study in passing as deviant. Although this raises questions of scientific ethics, which will be dealt with later, there are good reasons for following this method of participant observation.

In the first place, I am convinced that there is only *one* way to watch highly discreditable behavior and that is to pretend to be in the same boat with those engaging in it. To wear a button that says "I Am a Watchbird, Watching You" into a tearoom, would instantly eliminate all action except the flushing of toilets and the exiting of all present. Polsky has done excellent observation of pool hustlers because he is experienced and welcome in their game. He is accepted as one of them. He might also do well, as he suggests, in interviewing a jewel thief or a fence in his tavern hangout. But it should be noted that he does not propose watching them steal, whereas my research required observation of criminal acts.[19]

The second reason is to prevent distortion. Hypothetically, let us assume that a few men could be found to continue their sexual activity while under observation. How "normal" could that activity be? How could the researcher separate the "show" and the "cover" from standard procedures of the encounter? Masters and Johnson might gather clinical data in a clinical setting without distortion, but a stage is a suitable research site only for those who wish to study the "onstage" behavior of actors.

Serving as Watchqueen

In *Unobtrusive Measures*, the authors refer to the participant observation method as one of "simple observation."[20] This is something of a misnomer for the study of sexually deviant behavior. Observation of the tearoom encounters –

far from being simple – became, at some stages of the research, almost impossibly complex.

Observation is made doubly difficult when the observer is an object of suspicion. Any man who remains in a public washroom for more than five minutes is apt to be either a member of the vice squad or someone on the make. As yet, he is not suspected as being a social scientist. The researcher, concerned as he is in uncovering information, is unavoidably at variance with the secretive interests of the deviant population. Because his behavior is both criminal[21] and the object of much social derision, the tearoom customer is exceptionally sensitive to the intrusion of all strangers.

Bruyn points out three difficulties attendant to participant observation: "how to become a natural part of the life of the observed," "how to maintain scientific integrity," and "problems of ethical integrity."[22] Each of these problems is intensified in the observation of homosexual activity. When the focus of an encounter is specifically sexual, it is very difficult for the observer to take a "natural part" in the action without actual involvement of a sexual nature. Such involvement would, of course, raise serious questions of both scientific and ethical integrity on his part. His central problem, then, is one of maintaining both objectivity and participation (the old theological question of how to be *in*, but not *of*, the world).

In their excellent and comprehensive paper on the subject of passing as deviant, Black and Mileski outline ways "by which the social organization itself can be mobilized by the investigator in the interests of his research."[23] Unfortunately, this paper had not been written when I needed it; nevertheless, my preliminary observations of tearoom encounters led to the discovery of an essential strategy – the real methodological breakthrough of this research – that involved such mobilization of the social organization being observed.

The very fear and suspicion encountered in the restrooms produces a participant role, the sexuality of which is optional. This is the role of the lookout ("watchqueen" in the argot), a man who is situated at the door or windows from which he may observe the means of access to the restroom. When someone approaches, he coughs. He nods when the coast is clear or if he recognizes an entering party as a regular.

The lookouts fall into three main types. The most common of these are the "waiters," men who are waiting for someone with whom they have made an appointment or whom they expect to find at this spot, for a particular type of "trick," or for a chance to get in on the action. The others are the masturbaters, who engage in autoerotic behavior (either overtly or beneath their clothing) while observing sexual acts, and the voyeurs, who appear to derive sexual stimulation and pleasure from watching the others. Waiters sometimes masturbate while waiting – and I have no evidence to prove some are not also voyeurs. The point is that the primary purpose of their presence differs from that of the pure mastubater or voyeur: the waiters expect to become players. In a sense, the masturbaters are all voyeurs, while the reverse is not true.

In terms of appearances, I assumed the role of the voyeur – a role superbly suited for sociologists and the only lookout role that is not overtly sexual. On those occasions when there was only one other man present in the room, I have taken a role that is even less sexual than that of the voyeur-lookout: the straight person who has come to the facility for purposes of elimination. Although it avoids sexual pressure, this role is problematic for the researcher: it is short-lived and invariably disrupts the action he has set out to observe. (See Chapter 3 for discussion of this role and others.)

Before being alerted to the role of lookout by a cooperating respondent, I tried first the role of the straight and then that of the waiter. As the former, I disrupted the action and frustrated my research. As the latter – glancing at my watch and pacing nervously from window to door to peer out – I could not stay long without being invited to enter the action and could only make furtive observation of the encounters. As it was, the waiter and voyeur roles are subject to blurring and I was often mistaken for the former.

By serving as a voyeur-lookout, I was able to move around the room at will, from window to window, and to observe all that went on without alarming my respondents or otherwise disturbing the action. I found this role much more amenable and profitable than the limited roles assumed in the earlier stages of research. Not only has being a watchqueen enabled me to gather data on the behavioral patterns, it has facilitated the linking of participants in homosexual acts with particular automobiles.

During the first year of observations – from April of 1966 to April 1967 – my field research notes were made with the aid of a portable tape recorder, concealed under a pasteboard carton on the front seat of my automobile. Research efforts during this time were directed toward comprehensiveness. I attempted to survey all of the active tearooms in one city and to extend my observations, whenever possible, to other communities across the country. My concern was to observe the activity across a representative range of times and places.

The park restrooms first become active as sexual outlets between 7:30 and 8:30 A.M., when the park attendants arrive to unlock them. The early customers are men who meet on their way to work. After 9:00, the activity drops off sharply until lunch time. During the first two hours of the afternoon, there is another abrupt increase in activity from those who spend their lunch breaks in the park, followed by a leveling-off until about 4:00 P.M. From this time until about 7:00 in the evening, the great bulk of tearoom participants arrive. Most participants stop off in the park restrooms while driving home from work. As one respondent stated, he tries to make it to a tearoom "nearly every evening about 5:30 for a quick job on the way home."

A few of these facilities remain open until as late as 9:00 P.M., but most are locked up by 7:30. On Saturdays and Sundays, the over-all volume of activity is much greater, reaching its peak between 4:00 and 4:30 P.M. I have observed a drop in tearoom action on the weekends immediately after lunch time, a period that coincides with the greatest amount of picnic activity. Otherwise, the curve is roughly bell-shaped, rising to the late afternoon peak. There are, of course, variations in these patterns from park to park and during different seasons of the year. The "hunting season" is described in Chapter 1, but I estimate that the months of greatest sexual volume in the park restrooms are from July through October.

My interest, then, was in distributing observation time throughout these periods of varying activity – and in different parks and different seasons. In all, during this first year, I observed some 120 sexual acts in nineteen different men's rooms in five parks of the one city. Not including the time spent outdoors, in driving, or engaged in informal interviews with participants, I spent close to sixty hours in the tearooms in this first stage of the observations. This time was broken into segments of an hour or less (averaging about twenty minutes), between which periods I drove to other tearooms or parks, sat in my automobile, talked with the few men I could involve in conversations outdoors, or simply stood outside.

This excerpt from one of the tapes I made in October, 1966, may communicate to the reader an idea of my observational techniques at that stage of the research:

> I was in this facility about five minutes, during that time the Negro in his thirties, neatly dressed, who had the [Ford], stood constantly at the urinal and masturbated, making no attempt to hide what he was doing. There was a young Negro in there at the same time – very neat, very well dressed, in his late teens I would gather – with glasses, student type. He stood at the window throughout the time and said nothing. I stood near him at the window, and he made no approach. I went to the other window, and he made no move. As I was leaving, the man from the white Chevrolet left his car and went in. . . . Now, as I describe the two Negroes, I know the one man was alone in his car. The younger man obviously came on foot to this facility. All right, on to some other places. It is now 4:47, traffic is very heavy, much distraction. . . . I'm now approaching the facility again and there's not much point going there, because there are no cars out front – so I'll move on to Hillside.

My purpose in this "time and place sampling" was to avoid the research errors outlined by Webb and others – particularly the danger "that the timing of the data collection may be such that a selective population periodically appears before the observer, while another population, equally periodically, engages in the same behavior, but comes along only when the observer is absent. Similarly, the individual's behavior may shift as the hours or days of the week change."[24]

Sampling Covert Deviants

Hooker has noted that homosexuals who lead secret lives are usually available for study "only when caught by law enforcement agents, or when seeking psychiatric help."[25] To my knowledge, no one has yet attempted to acquire a *representative* sample of *covert* deviants for any sort of research. Polsky's efforts to secure a representative sample of "beats," in order to effect a survey of drug use among them, is a possible exception to this generalization, although I question that Village beats could be called *covert* deviants.

Following Rainwater's suggestion, I gathered a sample of the tearoom participants by tracing the license plates of the autos they drive to the parks. I have already indicated ways in which automobiles provide observable traces of their drivers' movements (see Chapter 1). Operation of one's car is a form of self-presentation that tells the observant sociologist a great deal about the operator.

For several months, I had noted fluctuations in the number of automobiles that remained more than fifteen minutes in front of the sampled tearooms. My observations had indicated that, with the sole exception of police cars, autos that parked in front of these public restrooms (which, as has been mentioned, are usually isolated from other park facilities) for a quarter of an hour or more invariably belonged to participants in the homosexual encounters. The same is true for cars that appeared in front of two or more such facilities in the course of an hour.

These variations in frequency were recorded for half-hour periods from 11:00 A.M. to 7:00 P.M., each day of the week, for each of four parks observed during the summer months. Averages of these volumes were calculated for each thirty-minute time period, the weekdays and weekends being separated for control purposes. Although the original calculations were made separately for each park, no major differences were observed in the over-all traffic pattern thus recorded. These data were then collapsed into the graphs that appear as Figures 1 and 2.

In September of 1966, then, I set about to gather a sample in as systematic a manner as possible under the circumstances. With the help of the tape recorder, I took the license numbers of as many cars during each half-hour period as equalled approximately 10 per cent of the average volume of "likely" autos at that time on that day of the week. At least for the largest park (which represents roughly half of the observed homosexual activity of this sort in the city), the results were fairly representative on a time basis. Random selection cannot be claimed for this sample: because of the pressures of time and possible detection, I was able to record only a portion of the license plates of participating men I saw at any one time, the choice of which to record being determined by the volume and flow of traffic and the position in which the autos were parked.

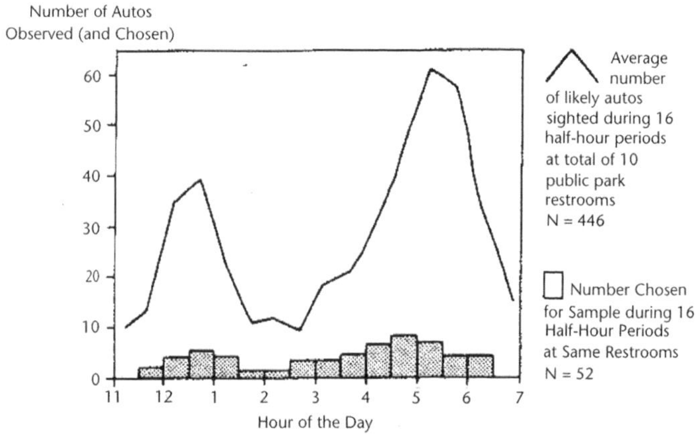

Figure 1: Illustrating volume of autos observed at restrooms and sampling method (averaged from data gathered on weekdays)

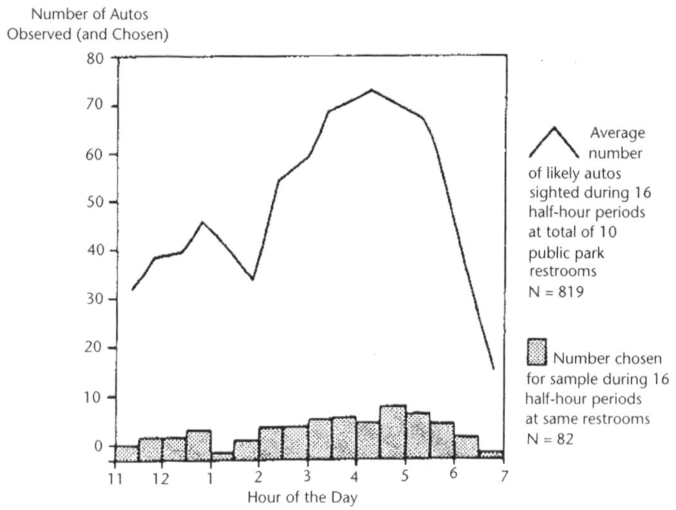

Figure 2: Illustrating volume of autos observed at restrooms and sampling method (averaged from data gathered on weekends)

I also noted, whenever possible, a brief description of both the car and its driver. By means of frequent sorties into the tearooms for observation, each recorded license number was verified as belonging to a man actually observed in homosexual activity inside the facilities. Sometimes the numbers were taped prior to my entrance, in anticipation of what I would find inside. In most cases, however, I observed the activity, left the tearoom, waited in my car for the participants to enter their autos – then recorded the plate numbers and brief descriptions. For each of these men but one I added to the data the role he took in the sexual encounter.

The original sample thus gained was of 134 license numbers, carefully linked to persons involved in the homosexual encounters, gathered from the environs of ten public restrooms in four different parks of a metropolitan area of two million people. With attrition and additions that will be described later, one hundred participants in the tearoom game were included in the final sample.

Systematic Observation

Before leaving the account of my observation strategies to consider the archival measures employed during the first half of my research, I want to describe the techniques employed in "tightening up" my data. Following the preliminary observations, I developed a "Systematic Observation Sheet" on which to record my observations. This form – used by myself in describing fifty encounters and by a cooperating participant in the recording of thirty others – helped to assure consistent and thorough recording of the observed encounters.

Figure 3 is a reproduction of an actual Systematic Observation Sheet, as filled out by me immediately upon return to my office one summer afternoon. Only the date, place, and description of an auto have been blanked out in order to avoid incrimination. This was the first, and briefest, of a series of three successive encounters observed in that popular restroom in the course of thirty-five minutes. After they were concluded, I drove to another part of the park and, with the use of a clipboard, filled in the diagrams and made written notes. As a left-hander, I find writing difficult; so I waited until I could use a typewriter to add the running commentary at the bottom.

As may be seen, this report sheet includes places for recording the time and place involved; a description of the participants (their age, attire, auto, and role in the encounter); a description of weather and other environmental conditions; a diagram on which movments of the participants could be plotted, along with location of the places of contract and fellatio; as well as a complete description of the progress of the encounters and reactions of the observer.

Such care was taken for several reasons. My first concern has been for objective validity – to avoid distortion of the data either by my presence or my presuppositions. I have also desired to make future replications and comparative studies possible, by being as systematic as possible in recording and gathering data.

Finally, I wanted to make the best of a rather unique opportunity for participant observation. The tearooms are challenging, not only because they present unusual problems for the researcher but because they provide an extraordinary opportunity for detailed observation. Due to the lack of verbal

SYSTEMATIC OBSERVATION SHEET   Date: Day: Sat

O = Observor (1) (2)   General Conditions: Weather & temp. 80° –

X = Principal Aggressor   # & type of people in parks:

Y = Principal Passive Participant   est. volume of gay activity:

A - N = Other Participants   Place:

Z = Law Enforcement Personnel

Time Began: 3:20 (Encounter A)

Time Ended: 3:30

Participants: [include symbol, est. age, attire, other distinguishing characteristics, type of auto driven]

X: 40 –

Y: 45 –

Others: A – 20 –

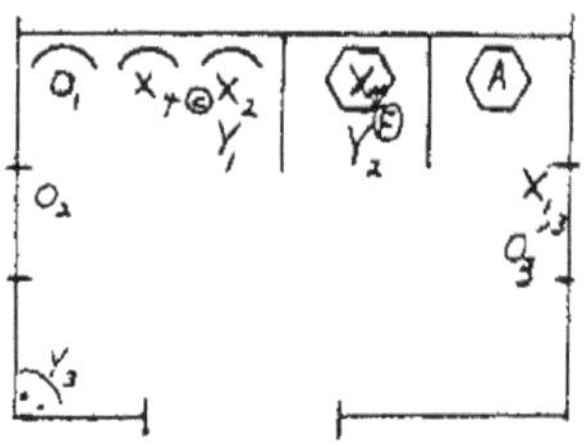

(F) = Fellatio performed

(C) = Contract made

Description of Action: [note: when possible, indicate: delays in autos, etc., before entering tearoom...manner of approaches...types of sexual roles taken... nature of interruptions and reaction to them...ANYTHING WHICH MAY BE SPOKEN... any masturbation going on...actions of lookout(s)...REACTION TO TEENAGERS AND ANY PARTICIPATION BY THEM...reactions to observor...length of time of sexual acts...spitting, washing of hands, wiping, etc.]

X was standing at right window & A seated on stool when O entered. O went to first urinal. X moved into third urinal, unzipped pants but did not urinate. He began to look at me. I zipped pants and moved in front of left window. X went back to opposite window. I saw Y get out of car and approach tearoom. He went to third urinal as soon as entered. In about 2 minutes, X moved into center urinal and began to cruise Y. By then, Y had an erection – X reached over and began to masturbate him with right hand, himself with left. I moved over to the far window. A looked at me. I smiled and nodded. X and Y walked together to stal # 1. X lowered pants and sat down. Y, standing in front with pants still unzipped and erection showing, continued to masturbate for another minute, then he inserted penis in X's mouth. He reached clime-x in about 3 minutes, having clasped hands behind X"s neck. He then went to basin, washed hands & left.

Figure 3

communication and the consistency of the physical settings, a type of laboratory is provided by these facilities – one in which human behavior may be observed with the control of a number of variables.

The analysis of the encounters that follows in Chapters 3 and 4 is based, primarily, on the fifty systematic observations I made between March and August, 1967. The numerous informal observations I made previously – and the thirty systematic observations made by a cooperating respondent – have served mainly as checks against this systematic portion of my research. Although I can think of no way in which my earlier observations run counter to these detailed ones, their greatest value was as preparation for this stage of the participant observation. Those of the respondent were also in general agreement with mine. Perhaps because he was frequently a sexual participant in the encounters he observed, however, he tended to concentrate more on the details of the sexual acts and less on the interaction leading to them. It is also interesting to note that his estimate of the age of participants was lower than mine. It is not possible to say whether this discrepancy resulted from a tendency on his part to view others as potential sexual partners or from a parallel difference in our own ages. Anyway, it is doubtful that the populations we observed differed greatly.

The Talk Outside

A sociologist without verbal communication is like a doctor without a stethoscope. The silence of these sexual encounters confounded such research problems as legitimation of the observer and identification of roles. As indicated above, however, it has certain advantages in limiting the number of variables that must be observed, recorded, and evaluated. When action alone is being observed and analyzed, the patterns of behavior themselves acquire meaning independent of verbalization. "The method of participant observation," Bruyn states, "is a research procedure which can provide the basis for establishing adequacy at the level of meaning."[26] What verbal research is possible through outside interviews then becomes an independent means of verifying the observations.

Despite the almost inviolate silence within the restroom setting, tearoom participants are neither mute nor particularly taciturn. Away from the scenes where their sexual deviance is exposed – outside what I shall later discuss as the "interaction membrane" – conversation is again possible. Once my car and face had become familiar, I was able to enter into verbal relationships with twelve of the participants, whom I refer to as the "intensive dozen." Eight of these men are included in the final sample. Four others, although engaged in dialogue near the tearooms where I observed them, were not included in the sample. Of the eight in the sample, five (including the two "walkers," who had walked rather than driven to the tearooms) were contacted after leaving the scene of an encounter, and three became cooperating respondents as a result of relationships that developed from the formal interviews.

After the initial contacts with this intensive dozen, I told them of my research, disclosing my real purpose for being in the tearooms. With the help of some meals together and a number of drinks, all agreed to cooperate in subsequent interviewing sessions. A few of these interviews were taped (only two men were sufficiently unafraid to allow their voices to be recorded on tape – and I don't blame the others) but most were later reconstructed from notes. Apart from the systematic observations themselves, these conversations constitute the richest source of data in the study.

Some may ask why, if nine of these cooperating respondents were obtained without the formal interviews, I bothered with the seemingly endless task of acquiring a sample and administering questionnaires – particularly when interviews with the intensive dozen provided such depth to the data. The answer is simple: these men are not representative of the tearoom population. I could engage them in conversation only because they are more overt, less defensive, and better educated than the average participant.

This suggests a problem for all research that relies on willing respondents. Their very willingness to cooperate sets them apart from those they are meant to represent. *Tally's Corner* and *Street Corner Society* stand high among the classics of social science – and rightly so – but I wonder sometimes how well Tally and Doc represent the apathetic, alienated, uninvolved men of the street corners. When authors such as Liebow and Whyte strive to compensate for this by extending their research throughout the friendship networks, great ethnography results. But the saddest works in the name of social science are those that barrage the reader with endless individual case studies and small samples from private psychiatric practices, few of which can be representative of the vast numbers of human beings who are supposed to be "understood" in terms of these deviant deviants.

Archival Evidence

The unobtrusive measures of participant observation and physical traces, combined with a limited use of open-ended interviews for purposes of correction and validation, enabled me to describe the previously unexplored area of tearoom encounters. The preliminary description of the participant population, however, began only after the establishment of a verified sample. For this stage of the study, I turned to archival measures, "the running record."[27]

Identification of the sample was made by using the automobile license registers of the states in which my respondents lived. Fortunately, friendly policemen gave me access to the license registers, without asking to see the numbers or becoming too inquisitive about the type of "market research" in which I was engaged. These registers provided the names and addresses of those in the sample, as well as the brand name and year of the automobiles thus registered. The make of the car, as recorded in the registers, was checked against my transcribed description of each car. In the two cases where these descriptions were contradictory, the numbers were rejected from the sample. Names and addresses were then checked in the directories of the metropolitan area, from which volumes I also acquired marital and occupational data for most of the sample.

Geographic mobility and data gaps plague the researcher who attempts to use the city directory as a source of information. Fortunately, however, new directories had been issued just prior to my need for them. Somewhat to my surprise, I had another advantage due to residential stability on the part of the population under study. Only 17 per cent of the men in the sample were not listed in these directories. Occupational data were not given for 37 per cent of the men (including those not in the directories).

In those few cases where addresses in license registers did not correspond with those in the city and county directories, I took advantage of still another archival source: the telephone company's index of numbers by street addresses, which had been published more recently than either of the other archival sources. By the time my sample had been verified and identified, none of the archival measures employed was over a year old, and the most recent had been published only the week before completion of this stage of the research.[28]

For fear of eliminating variables that might profitably be studied at a later date, I did not scrub from my sample those for whom the archives provided no marital or occupational data. These men, I felt, might represent either a transient or secretive portion of tearoom participants, the exclusion of which would have distorted the population.

Other biases were not avoidable; however, where possible, I have attempted to compensate for them. In the first place, I did not record the license numbers of automobiles from states not represented in the metropolitan area. My estimate is that about 5 per cent of the cars driven by participants in these homosexual encounters bore such out-of-state plates. The majority of these autos also had stickers that identified the owners as armed forces personnel from nearby military installations. This fact is important for indicating (*a*) that a very small percentage of the participants are tourists or salesmen who travel from a great distance, and (*b*) that the military should have a larger representation in my sample than the 2 per cent indicated by the available occupational data. If the archives and local license plates are biased against any one segment of the population, it is members of the armed forces. When compensation for this bias has been made, the indications (including those gained from interviews) are that some 10 per cent of tearoom participants will be armed forces personnel. This factor, however, should vary with the proximity of parks to large military installations. There are no such bases within a twenty-mile radius of the parks in this study.

Other license numbers were excluded from the sample because they precluded research. Because these may have produced a bias, they will be mentioned. First, eighteen were eliminated because they were not listed in the license registers. (The police attribute such gaps in their data to clerical error.) Seven men resided outside the metropolitan area under study. Another five were driving leased or company cars. For two, the automobile description in the registers did not fit my description. Finally, the license registers listed two nonexistent addresses. Other adjustments were made before I reached the final sample of 100. Two men were dropped from the sample for identity reasons. To replace them, I added two young men who had walked to the tearooms and were included in my intensive dozen. I had estimated that a small percentage of the participants walked to their favorite tearooms and thought it important to have representation from the pedestrian population.

If one may disregard for a moment the strong possibility that most of these exclusions may have resulted from errors on the part of clerks, printers, or myself, one other factor may be hidden behind these omissions. Is it not possible that some of these unidentifiable plates may have been switched, counterfeited, or acquired by other illegitimate means? Some men might have leased cars or given false addresses in order to preserve their anonymity. I know of no unobtrusive way to test the degree of deliberate deception. If deception were the motive in these instances, it was successful.

A View from the Streets

Like archives, park restrooms, and automobiles, the streets of our cities are subject to public regulation and scrutiny. They are thus good places for nonreactive research (nonreactive in that it requires no response from the research subjects). Having gained addresses for every person in my sample, I spent a Christmas vacation on the streets and highways. By recording a description of every residence and neighborhood represented in the sample, I was able to gain further data on my research subjects.

The first purpose of this survey of homes was to acquire descriptions of the house types and dwelling areas that, when combined with occupational data gleaned from the archives, would enable me to use Warner's Index of Status Characteristics (I. S. C.) for a socioeconomic profile of my population.[29] Generally speaking, this attempt was not successful: job classifications were too vague and large city housing units too difficult to rank by Warner's criteria.

As physical evidence, however, homes provide a source of data about a population that outweighs any failure they may have as a status index. Swing sets and bicycles in the yards indicate that a family is not childless. A shrine to Saint Mary suggests that the resident is Roman Catholic in religious identification. Christmas decorations bespeak at least a nominal Christian preference. A boat or trailer in the driveway suggests love of the outdoor life. "For Rent" signs may indicate the size of an average apartment and, in some cases, the price. The most important sign, however, was the relative "neatness" of the house and grounds. Some implications of this information are pointed out in Chapter 7.

Obtrusive Measures

Realizing that the majority of my participant sample were married – and nearly all of them quite secretive about their deviant activity – I was faced with the problem of how to interview more than the nine willing respondents.

Formal interviews of the sample were part of the original research design. The little I knew about these covert deviants made me want to know a great deal more. Here was a unique population just waiting to be studied – but I had no way to approach them. Clearly, I could not knock on the door of a suburban residence and say, "Excuse me, I saw you engaging in a homosexual act in a tearoom last year, and I wonder if I might ask you a few questions." Having already been jailed, locked in a restroom, and attacked by a group of ruffians, I had no desire to conclude my research with a series of beatings.

Perhaps I might have had some success by contacting these men at their work, granted that I could obtain their business addresses. This strategy would have precluded the possibility of seeing their homes and meeting their wives, however, and I believed these confrontations to be important.

About this time, fortunately, I was asked to develop a questionnaire for a social health survey of men in the community, which was being conducted by a research center with which I had been a research associate. Based on such interview schedules already in use in Michigan and New York, the product would provide nearly all the information I would want on the men in my sample: family background, socioeconomic factors, personal health and social histories, religious and employment data, a few questions on social and political attitudes, a survey of friendship networks, and information on marital relationships and sex.

With the permission of the director of the research project, I added my deviant sample to the over-all sample of the survey, making certain that only one trusted, mature graduate student and I made all the interviews of my respondents. Thus legitimized, we set out to interview. Using a table of random numbers, I randomized my sample, so that its representativeness would not be lost in the event that we should be unable to complete all 100 interviews.

More will be written later of the measures taken to safeguard respondents; the important thing to note here is that none of the respondents was threatened by the interviews. My master list was kept in a safe-deposit box. Each interview card, kept under lock and key, was destroyed with completion of the schedule. No names or other identifying tags were allowed to appear on the questionnaires. Although I recognized each of the men interviewed from observation of them in the tearooms, there was no indication that they remembered me. I was careful to change my appearance, dress, and automobile from the days when I had passed as deviant. I also allowed at least a year's time to lapse between the original sampling procedure and the interviews.

This strategy was most important – both from the standpoint of research validity and ethics – because it enabled me to approach my respondents as normal people, answering normal questions, as part of a normal survey. They *are* part of a larger sample. Their being interviewed is not stigmatizing, because they comprise but a small portion of a much larger sample of the population in their area. They were not put on the spot about their deviance, because they were not interviewed as deviants.

The attrition rate for these interviews was high, but not discouragingly so. Attempts were made at securing seventy-five interviews, fifty of which were completed. Thirty-five per cent were lost by attrition, including 13 per cent who refused to cooperate in the interviews. In addition to the fifty completed schedules, three fathers of participants consented to interviews on the social health survey, as did two fathers of the control sample.

Because of the preinterview data obtained by the archival and observational research previously described, it was possible to learn a great deal even from the losses. As should be expected, the residue of men with whom interviews were completed are slightly overrepresentative of the middle and upper classes; they are suburbanites, more highly educated men. Those who were lost represent a more transient group (the most common reason for loss was that the subject had moved and left no forwarding address), employed in manual jobs. From preinterview information it was learned that the largest single occupational class in the sample was the truck drivers. Only two members of this class remained among those interviewed.

The refusals also indicated some biases. These men, when pinpointed on a map, clustered around the Italian and working class German areas of the city. Of the ten lost in this manner, three had Italian names and five bore names of distinctly Germanic origin.

Once these interviews were completed, preparations could be made for the final step of the research design. From names appearing in the randomly selected sample of the over-all social health survey, fifty men were selected, matched with the completed questionnaires on the following four characteristics: I. S. C. occupational category, race, area of the metropolitan region in which the party resided, and marital status. The loss here was not from refusals or lost addresses but from those who, when interviewed, failed to correspond with the expected characteristics for matching. Our precedure, in those cases, was simply to move on to another name in the larger sample.

These last fifty interviews, then, enabled me to compare characteristics of two samples – one deviant, one control – matched on the basis of certain socioeconomic characteristics, race, and marital status. Although I made a large proportion of these interviews, and nearly all of the deviant interviews, I found it necessary to hire and train two graduate students to assist with interviewing the control sample. A meeting was held with the assistant interviewers immediately following the completion of each schedule – and all coding of the questionnaires was done by us in conference.

There were a number of open-ended questions in the interview schedules, but the majority included a wide range of precoded answers, for the sake of ease in interviewing and economy in analysis. In addition, the interviewers were trained to make copious marginal notes and required to submit a postinterview questionnaire with each schedule. The median time required for administering the interview schedules did not differ greatly between the two samples: one hour for the deviants, fifty-five minutes for the "straights." Even the days of the week when respondents were interviewed showed little variation between the two samples: Sunday, Tuesday, and Saturday, in that order, being the more popular days.

Summary

From a methodological standpoint, the value of this research is that it has employed a variety of methods, each testing a different outcropping of the research population and their sexual encounters. It has united the systematic use of participant observation strategies with other nonreactive measures such as physical traces and archives. The exigencies of research in a socially sensitive area demanded such approaches; and the application of unobtrusive measures yielded data that call, in turn, for reactive methods.

Research strategies do not develop *ex nihilo*. In part, they are the outgrowth of the researcher's basic assumptions. Special conditions of the research problem itself also exercise a determining influence upon the methods used. This chapter has been an attempt to indicate how my ethnographic assumptions, coupled with the difficulties inhering in the study of covert deviants and their behavior, have given rise to a set of strategies.

With the help of "oddball" measures, the outlines of the portrait of participants in the homosexual encounters of the tearooms appeared. Reactive strategies were needed to fill in the distinguishing features. They are human, socially patterned features; and it is doubtful that any one method could have given them the expressive description they deserve.

Notes

1. In this connection, Evelyn Hooker, "The Homosexual Community," in *Personality Research* (Copenhagen: Monksgaard, 1962) is important. See also Sherri Cavan, *Liquor License: An Ethnography of Bar Behavior* (Chicago: Aldine, 1966), especially pp. 211–226.
2. Albert J. Reiss, Jr., "The Social Integration of Queers and Peers," *Social Problems*, Vol. 9, No. 2 (Fall, 1961), pp. 102–120.
3. These arrests for felonious homosexual offenses in Los Angeles County, from 1962–64, were analyzed in Jon J. Gallo and others, "The Consenting Adult Homosexual and the Law: An Empirical Study of Enforcement and Administration in Los Angeles County," *UCLA Law Review*, 13 (March, 1966), p. 804. A number of references to public restrooms in this excellent study constitute nearly all the literature on this subject. Outside of occasional comments in gay novels and other works on the subject of homosexuality that tearooms provide "one of the known sexual outlets" for the homosexual, there is no other mention to be found.
4. Hooker, "The Homosexual Community," p. 40.

5. Ned Polsky, *Hustlers, Beats, and Others* (Chicago: Aldine, 1967), p. 126.
6. Elliot Liebow, *Tally's Corner* (Boston: Little, Brown, 1967), p. 7.
7. William Foote Whyte, *Street Corner Society* (Chicago: University of Chicago Press, 1943), pp. 288–298.
8. William H. Masters and Virginia E. Johnson, *Human Sexual Response* (Boston: Little, Brown, 1966), p. 10.
9. An interesting example of precisely this sort of study is Marvin B. Scott, *The Racing Game* (Chicago: Aldine, 1968).
10. Perhaps the most famous tearoom arrest in America was that of a presidential assistant in the restroom of a Washington YMCA in 1964.
11. In a letter from Martin Hoffman, M.D., School of Criminology, University of California, Berkeley, August 2, 1967.
12. For instance, informants tell me that the policy of Denver's daily newspapers to publish names, addresses, and places of employment of all men arrested on charges of homosexual behavior has caused a decrease of tearoom activity in the city – with a corresponding rise in the popularity of other homosexual outlets.
13. Severyn T. Bruyn, *The Human Perspective in Sociology* (Englewood Cliffs, N.J.: Prentice-Hall, 1966), p. 18.
14. Eugene J. Webb and others, *Unobtrusive Measures: Nonreactive Research in the Social Sciences* (Chicago: Rand McNally, 1966). All of these measures are described in some detail in this work.
15. For a description of theory development at its best, see C. Wright Mills, *The Sociological Imagination* (New York: Grove Press, 1959), p. 73.
16. My reticence at admitting I was a sociologist resulted, in part, from the cautioning of a gay friend who warned me that homosexuals in the community are particularly wary of sociologists. This is supposedly the result of the failure of a graduate student at another university to disguise the names of bars and respondents in a master's thesis on this subject.
17. *Guild Guide* (Washington: Guild Press, 1968).
18. Donald J. Black and Maureen A. Mileski, "Passing as Deviant: Methodological Problems and Tactics," unpublished working paper available through the Department of Sociology, University of Michigan, Ann Arbor, pp. 4–5.
19. Polsky, *Hustlers, Beats, and Others*, p. 127.
20. Webb and others, *Unobtrusive Measures*, p. 49.
21. The Revised Statutes of the state under study, for instance, read on this wise: H563.230. *The abominable and detestable crime against nature–penalty.* – Every person who shall be convicted of the detestable and abominable crime against nature, committed with mankind or with beast, with the sexual organs or with the mouth, shall be punished by imprisonment in the penitentiary not less than two years.
22. Bruyn, *Human Perspective.*
23. Black and Mileski, "Passing as Deviant," p. 2.
24. Webb and others, *op. cit.*, p. 136.
25. Hooker, *op. cit.*, p. 169.
26. Bruyn, *op. cit.*, p. 179.
27. Webb and others, *op. cit.*, pp. 53–87.
28. Because identification of the city in which this research was conducted might result in pressure being brought to bear on law enforcement agencies or respondents, it has been necessary for me to omit references to the archival volumes used. The name of the city, county, or state appears in the title of each of these sources.
29. See W. Lloyd Warner and others, *Social Class in America* (Chicago: Science Research Associates, 1949).

39

Controversies Surrounding Laud Humphreys' Tearoom Trade: An Unsettling Example of Politics and Power in Methodological Critiques

Michael Lenza

Laud Humphreys' *Tea Room Trade: Impersonal Sex In Public Places* (1970) is commonly presented in many sociological methods texts as an example of covert and deceptive research methods that endangered subjects without their consent. The following excerpt is reasonably representative of how this work tends to be presented:

> "In Laud Humphreys' tearoom trade study (a study of male homosexual encounters in public restrooms), about 100 men were observed engaging in sexual acts as Humphrey's pretended to be a "watchqueen" (a voyeur and lookout). Subjects were followed to their cars, and their license numbers were secretly recorded. Names and addresses were obtained from police registers when Humphreys posed as a market researcher. A year later, in disguise, Humphreys used a deceptive story about a health survey to interview the subjects in their homes. Humphreys was careful to keep names in a safety deposit boxes, and identifiers with subject names were burned. He significantly advanced knowledge of homosexuals who frequent "tearooms" and overturned previous false beliefs about them. There has been controversy over the study: The subjects never consented; deception was used; and the names could have been used to blackmail subjects, to end marriages, or to initiate criminal prosecutions" (Neuman 1997: 447).

Source: *International Journal of Sociology and Social Policy*, 24(3/4/5) (2004): 20–31.

Regarding these issues surrounding research on human subjects, this paper will argue that this dominant view of Humphreys' tearoom trade study, focusing primarily upon respect for autonomy (informed consent), misinforms the reader as much as it informs of the underlying moral and ethical foundations for research with human subjects. Three moral and ethical principles provide the foundations for most medical, scientific, and social research methodologies: beneficence, justice, and respect for autonomy (informed consent) (Faden and Beauchamp 1986: 5). I will first review each of these three ethical foundations for sociological research, then, I will examine Humphreys' Tearoom Trade study through the vantage point of each and the historical facts surrounding the controversies.

Humphreys' study in casual public sex, when historically situated and examined through each of the three ethical and moral pillars, will provide a more balanced framework from which to view his work. In addition, this more balanced view may promote better understanding of how ethical and methodological dilemmas arise when the underlying principles for human research come into conflict with each other.

The Pillar of Beneficence

Beneficence carries within it perhaps the most basic ethical and moral consideration before a social researcher. It requires the researcher avoid harming the subject. *Beneficence* also carries with it a stipulation that research activities decidedly benefit others:

> "The principle of beneficence includes the following four elements, all linked through the common theme of promoting the welfare of others: (1) one ought not to inflict evil or harm; (2) one ought to prevent evil or harm; (3) one ought to remove evil or harm; (4) one ought to do or promote good. (Faden and Beauchamp 1986: 10)

These four elements are not so much a hierarchical ordering of considerations as an interconnected pattern of considerations that should be weighed and present in all stages of research. A small potential for harm can be acceptable if it serves to prevent, remove, or correct a larger harm or evil likely to be experienced by the subjects. There is an unavoidable element of paternalism in *beneficence* that can, and often does, create conflicts with respect for autonomy.

The Pillar of Respect for Autonomy

Rooted in our principle of respect for autonomy, *informed consent* is believed to protect a subject's right to make autonomous decisions so as not to infringe upon self-determination (Faden and Beauchamp 1986: 8–9,

27–28). Although research involving human subjects is as old as medicine itself, institutional concerns and formal requirements for *informed consent* for human subjects is a relatively recent development that grew out of accounts of misconduct uncovered at the Nuremberg trials of Nazi physician's experiments with human subjects.

It is interesting to note that it was actually Germany, in 1931, who where the first modern nation to enact strict regulations on the use of human subjects in medical research. These regulations specifically required researchers obtain the willing consent "in a clear and undebatable manner" before human subjects could be used (Sass 1983; Faden and Beauchamp 1986: 154). This fact is omitted in historical revisionist reviews of *informed consent* that present the Nuremberg Code of 1947 as the first major code addressing this issue (Belmont Report 1979).

It is important to recognize that horrific human experimentations, conducted by the Nazis, were carried out in the presence of strict legal doctrine prohibiting such crimes. History should inform and temper researcher's beliefs that bureaucratic structures function independent of the social and political events surrounding them. As history indicates, Internal Review Boards and Professional Codes of Ethics, though helpful, cannot serve to replace the ethics of individual researchers.

The Pillar of Justice

The concept of *justice*, as addressed in the literature on human subjects, is problematic for sociologist. It is approached principally from an economic and rational choice model for distributive justice. For example, the Belmont Report (1979) discusses Justice in Section B, Part 3:

> "**Justice.** – Who ought to receive the benefits of research and bear its burdens? This is a question of justice, in the sense of "fairness in distribution" or "what is deserved." An injustice occurs when some benefit to which a person is entitled is denied without good reason or when some burden is imposed unduly" (Belmont Report 1979).

The subsequent discussion following the above quote is relevant to medical, biological, and social scientific research projects that are developing new products or procedures, but has limited value in guiding social research. The American Sociological Association (ASA) code of Ethics does not have an explicit section on justice or beneficence, though it does address informed consent at considerable length (ASA 2003).

In a search of the JSTOR database of 24 sociological journals, for articles with the keyword *justice*, 135 articles were returned representing all major paradigms in sociology. The bulk of publications prove to be concerned with social justice. However, *justice*, as defined above is not a primary concern of sociology's acknowledged governing professional body. It is therefore both curious and disheartening to see an explicit emphasis on *informed consent* within the ASA Code of Ethics, which fulfills federal funding requirements, while discussion of justice and beneficence, arguably the heart and soul of the sociological endeavor, is conspicuously absent. Although there are numerous differences and debates about what social justice entails, too numerous for this writing, it is possible to recognize what should be a modest point of agreement: social justice entails an understanding of what is or was, combined with a view to what could be, for human benefit.

An Examination of Humphreys' Tearoom Trade through the Three Ethical Pillars of Social Science Research

An examination Humphreys' study through all three of the moral pillars underlying social research will establish that his research did not violate any premise of either beneficence or the sociological interest in social justice. His methods in interviewing subjects within their homes under the guise of conducting a health care survey did violate respect for autonomy. However, it is necessary to recognize that there were no federal or professional guidelines requiring informed consent in human research at that time.

Research that conducted a review of studies published in the *Journal of Personality and Social Psychology* in 1971 established that deception was used in 47% of the articles published that year (Menges 1973). Exemplary work in sociology commonly utilized deception. Goffman (1961) posed as an athletic director in his fieldwork for *Asylums*. Gusfield (1955) in *Symbolic Crusade*, informed the Woman's Christian Temperance Union (WCTU) he was a disinterested investigator of social movements (Gusfield 1955: 29; Mitchell 1993: 3). There was nothing exceptional in Laud Humphreys' use of deception in his fieldwork. It fell well within the accepted methodologies of that time period. Further, a more recent in-depth review of secrecy and fieldwork (Mitchell 1993), firmly establishes that secrecy remains to this day an essential element of most research involving human subjects. A belief that informed consent fully informs human subjects and in practice protects their right to autonomy is more a methodological myth than fact:

The following five elements have been identified as the concepts analytical components:

1. Disclosure
2. Comprehension
3. Voluntariness
4. Competence
5. Consent (Faden and Beauchamp 1986: 274)

Research indicates that what *informed consent* really means to subjects is closer to a release, "by letting the doctor do whatever is necessary, best, or whatever he sees fit" (Faden and Beauchamp 1986: 277). While the actual practice of gaining inform consent from patients tended to boil down to merely a ritual of "obtaining a signature on a consent form" (Zussman 1997: 178; Faden and Beauchamp 1986: 277; Mitchell 1993). *Informed consent* implies that subjects understand and agree to every facet of an experiment. This is clearly not true.

The underlying focus and function of *informed consent* is based in its legalistic importance as a signed contract establishing proof of voluntary cooperation protecting the researcher and the sponsoring institution from liability, aligning sociological research with the current funding requirements (Mitchell 1993: 35, 28).

Despite a recent institutional focus upon informed consent as the principle means to protect subjects, beneficence remains the most achievable and strongest moral principle protecting human subjects in social research: (1) one ought not to inflict harm; (2) one ought to prevent harm; (3) one ought to remove harm; (4) one ought to do or promote good.

The issue of inflicting harm on his subjects revolves around Humphreys' commitment and ability to protect their identity. Humphreys kept the list of his subject's identities in a locked box, at a secret location, 1000 miles away from where his research was conducted. He destroyed the list when a possibility arose that there could be outside interests in the list. He took the identity of his subjects with him to the grave (Reiss 1978: 175; Mitchell 1993: 53). The facts clearly show Humphreys was absolutely committed to the protection of his subjects and did not allow harm to befall them through his actions. Regardless of one's position on the issues surrounding his methodology, one has to admit he admirably upheld that tenant, "one ought not to harm," (Faden and Beauchamp 1986: 10) and successfully protected his subjects' identities.

It was the publicity and homophobic hysteria surrounding Humphreys' dissertation work at Washington University that created a dangerous situation for the subjects. One should ask whether Alvin Gouldner's revelations to Nicholas von Hoffman and the resulting public scandal created a greater potential for harm to subjects than actions by Humphreys (see Goodwin et al. 1991). Metaphorically, if one were to become aware of possible fire hazards and lack of exits in a given building, it would be highly inappropriate to then run through the building screaming "Fire," and thereby actually creating a more immediately dangerous situation. Gouldner's actions in communicating his opinion of Humphreys's study outside of department to Von Hoffman, away from structures of confidentiality for subjects, intentionally created a public scandal where no protections for subjects existed. This could have brought serious harm to Humphreys' participants, as newspapers are exempt from libel if reporting is absent malice. In reality, the publicity generated by Gouldner and Von Hoffman posed a greater danger to participants than Humphreys' observations and follow-up interviews ever did.

Warwick (1982) notes the importance of recognizing the difference between "harms that are *intrinsic* to the research process and those that are *extrinsic* to that process" (Warwick 1982: 121) and even directly quotes Humphreys on the extrinsic publicity intentionally brought by others:

> "In the wake of front-page publicity, *fostered by members of the administration and faculty* at Washington University soon after the completion of the research, I am surprised that no such investigation followed" (emphasis my own; Warwick 1982: 108; Humphreys 1975: 229).

Despite this recognition of these extrinsic forces in play, Warwick, who engages in unfettered exposition of imagined harms, does not address or discuss the evidence of actual harms that resulted from these intentional extrinsic acts by others. In fact, the only substantive evidence we have of any harm befalling Humphreys' subjects are the anxious phone calls he received from several subjects after the extrinsic sensationalizing this study in the public sphere (Glazer 1972; Humphreys 1975: 215).

Should there not be some reasonable constraints upon critics? Is it proper for professionals to publicly accuse other researchers of possible crimes or harms to subjects with no more substantive proof than their mere imaginations or fanciful what if stories that never occurred? If a professional knowingly presents fabricated claims within the public sphere should they then not be responsible for:

1. The anxiety, fears, and harms to subjects that result from the contrived accusations.
2. The harms they create for the accused researcher's reputation and future.
3. The harm and fears brought to the field of study that arises from these unsubstantiated claims.

Intentionally placing someone in a false light in the public eye is unethical (Pickard 1982: 261). Many ethical critics of sociological research would be well served to turn their imaginative critical musings upon themselves.

If there is a lesson to be learned from this it would be that our first and foremost ethical requirement as practitioners should be the protection of subjects, which would include proper procedures for the bringing about and settling of ethical and methodological disputes. This would be accomplished in a manner that insures the confidentiality and protection of subjects, apart from media attention and personal conflicts.

There is another dilemma in the discussions and presentations of Humphreys' study. Should a possible harm that did not occur take precedence over what actually did occur?

> "The subjects never consented; deception was used; and the names could have been used to blackmail subjects, to end marriages, or to initiate criminal prosecutions" (Neuman 1997: 447).

This presentation, as do most methodological reviews of Humphreys tearoom trade study, fits what Best (1990) calls "the atrocity tale" which is then tied to a particular policy position (Best 1990: 132–137; Toulmin et al. 1979: 43–56; Mitchell 1993: 27). This presentation clearly ties a violation of informed consent to the atrocity tales of destroyed marriages, blackmail, and criminal prosecutions. Neuman fails to acknowledge that if *informed consent* forms were used in Humphreys' project, it would have constituted a grave danger to the subjects as those records could be subject to subpoena (Bond 1978: 146; Mitchell 1993: 34). It is not that atrocity tales are not useful. They can inform us of all the many things that could possibly go wrong when conducting sociological research. However, critics can themselves make substantive and ethical errors if they present their arguments in a manner that intentionally substitutes their imaginative musings for the actual substantive reality.

On the way to the office yesterday, one could have become distracted, run over, horribly mangled and killed a neighbor's child, destroyed a family, and caused a terrible grief. An interesting element of "what ifs" is that such statements are very difficult to defend against. By the mere telling of the tale one creates and assigns guilt for events that never occurred. Therein lies the power of the "what if" scenario. *The burden of proof shifts from the teller of the tale to the accused.* These types of arguments, substantively based upon nonevents can be inherently unfair, particularly, as in Humphreys' case, when the facts speak to the care and consideration that was undeniably taken to insure that damage to subjects would not occur.

These what if stories surrounding Humphreys work also provide us with a view to politics and power within our profession. Humphreys had little status or power as a PhD candidate completing his dissertation relative to the status and power of established professionals such as Alvin Gouldner and Nicholas von Hoffman. This difference in power and status allowed his critics to merely imaginatively create what if tales, void of any substantive proofs, to forever caste the specter of ethical impropriety around this work. There is no exoneration from what if. Even when established as untrue the specter of the horror remains. It haunted Humphreys throughout his life (Humphreys 1975: 223–232; Nardi 1995). These fabricated "what if" tales did in reality create more harm to subjects, damage to Laud Humphreys' reputation, and did more to undermine public faith in sociological research than any act by Humphreys in conducting the research itself.

More importantly, from the vantage point of beneficence as well as from a concern with social justice, Humphreys' tearoom study made significant positive contributions to the population he studied. His thorough fieldwork exploded the myth that the populations of men engaging in these casual sexual encounters were criminals, transients, or predatory pedophiles that communities needed to hunt down and criminally prosecute. His study firmly established that this casual sexual behavior was limited to consenting adults that transpired in a manner that made it highly unlikely that youths or disinterested parties would be approached or harassed. Furthermore, he established that this population principally consisted of individuals of good standing within the community, many of which were married, employed, and former members of the armed forces. Humphreys' work seriously questioned the received view that homosexual contact should be subjected to the retributive harms of imprisonment and ultimately created doubt in the dichotomous categories of gender that closely paralleled sexual behavior (Humphreys 1975: 64, and chapters 6 and 7). His conceptualization of the *breastplate of righteousness* was a significant contribution to our understanding that expressed religious and political views sometimes serve as purposeful acts of misdirection. Few studies in sociology have accomplished as much in a single work. Laud Humphreys' *Tearoom Trade* (1970) did not violate the deeper ethical and social concerns of sociology, expressed in the concepts of *beneficence* and *justice*, and it is necessary to recognize both its historical and continuing importance in understanding human sexuality.

References

American Sociological Association Code of Ethics. 2003. http://www.asanet.org/members/ecostand2.html.

Belmont Report. 1979. http://ohrp.osophs.dhhs.gov/humansubjects/guidance/belmont.htm.

Best, J. 1990. *Threatened Children: Rhetoric and Concern about Child Victims*. Chicago: University of Chicago Press.

Bond, K. 1978. "Confidentiality and the Protection of Human Subjects in Social Science Research: A Report on Recent Developments." *American Sociologist* 13: 14–152.

Faden, Ruth R. and Beauchamp, Tom L. 1986. *A History and Theory of Informed Consent*. New York: Oxford University Press.

Glazer, Myron. 1975. "Impersonal Sex." In Humphreys, L. *Tearoom Trade: Impersonal Sex in Public Places*. pp. 213–222. Chicago: Aldine Publishing Company.

Goffman, Erving. 1961. *Asylums: Essays on the Social Situation of Mental Patients and Other Inmates*. Garden City, N.Y.: Anchor Books.

Goodwin, Glen A., Horowitz, Irving Louis, and Nardi, Peter, M. 1991. "Laud Humphreys: A Pioneer in the Practice of Social Science." *Sociological Inquiry* 61: 139–147.

Gusfield, J. 1955. "Fieldwork Reciprocities in Studying a Social Movement." *Human Organization* 14: 29–33.

Humphreys, Laud. 1970. *Tearoom Trade: Impersonal Sex in Public Places*. Chicago: Aldine Publishing Company.

———. 1975. *Tearoom Trade: Impersonal Sex in Public Places*. Chicago: Aldine Publishing Company.

Menges, Robert J. 1973. "Openness and Honesty Versus Coercion and Deception in Psychological Research." *American Psychologist* 28: 1030–34.

Mitchell, Richard G. 1993. *Secrecy and Fieldwork*. Newbury Park: Sage Publications.
Nardi, Peter, M. 1995. "The Breastplate of Righteousness: Twenty-Five Years After Laud Humphreys' Tearoom Trade Impersonal Sex in Public Places." *Journal of Homosexuality* 30: pp. 1–10.
Neuman, W. Lawrence. 1997. *Social Research Methods: Qualitative and Quantitative Approaches*. 3rd ed. Boston: Allyn and Bacon.
Pickard, Terry. 1982. "Invasions of Privacy in Social Science Research." In Beauchamp, Tom L., R. Faden, R. Wallace, and L. Walters (eds.) *Ethical Issues in Social Science Research*. pp. 257–273. Baltimore: John Hopkins University Press.
Reiss, A.J. 1978. "Conditions and Consequences of Consent in Human Subject Research," In K. Wulff (Ed.) *Regulation of Scientific Inquiry*. pp. 161–184. Boulder: Westview Press.
Sass, Hans Martin. 1983. "Reichsrundschreiben 1931: Pre Nuremberg German Regulations Concerning New Therapy and Human Experimentation." *Journal of Medicine and Philosophy* 8: pp. 99–111.
Toulmin, S., Rieke, R., and Janik, A. 1979. *An Introduction to Reasoning*. New York: Macmillan.
Warwick, Donald P. 1982. "Types of Harm in Social Research." In Beauchamp, Tom L., R. Faden, R. Wallace, and L. Walters (eds.) *Ethical Issues in Social Science Research*. pp. 101–124. Baltimore: John Hopkins University Press.
Zussman, Robert. 1997. "Sociological Perspectives on Medical Ethics and Decision Making." *Annual Review of Sociology* 23: pp. 171–189.

40

Working in Hostile Environments

Nigel Fielding

Defining Hostile Environments

British diplomats formerly received an allowance when assigned a difficult posting. The city where I spent my teenage years – Washington, D.C. – was such a posting. It wasn't the troublesome natives, though. The embassy crew got the money because of the summer heat and yards-deep winter snows. Washington was a 'hostile environment'. So is the deck of a North Atlantic trawler, the cockpit of a Formula One racing car and the entrance to many British nightclubs after dark. Human cultures recognize hostile environments by paying people 'danger money' to work in them. The irony is that after acclimatization these payments are 'money for old rope'. People who work in hostile environments get used to the conditions. They develop a culture around them. The immediate sense of danger subsides (Pollnac et al., 1995; Winlow et al., 2001).

Researchers work in hostile environments, too. They seldom get paid danger money, though they sometimes get a bit of respect from other researchers. But while they don't get the money, researchers who do this kind of work do develop a culture, like any other group sharing an experience which isn't universal. The culture expresses practices, beliefs and justifications that people sharing the experience are drawn towards, and, like other cultures, it often expresses these things by contrast with the practices, beliefs and justifications held by people who haven't shared the experience. This chapter's purpose is to tell you about some of the working practices, unwritten beliefs

Source: Clive Seale, Giampietro Gobo, Jaber F. Gubrium and David Silverman (eds), *Qualitative Research Practice* (London: SAGE, 2004), pp. 236–248.

and personal justifications of people who research in hostile environments. A particular theme is that research in such settings involves a dialectic of cooperation and resistance, and the tension between these qualities is a potential analytic resource as well as an important way to gauge the adequacy of the data and monitor threats to the fieldworker's welfare.

Of course, I should have written '*a person* who researches in hostile environments'. Like other qualitative researchers I have an eye for the particular, but it does not stop me generalizing from my own experience. I will support some of my assertions with those published authorities that agree with me, but much of what I say reflects my own circumstances and decisions. I should say what that experience is. My first exposure to researching in hostile environments was also my first experience of field research. I didn't expect hostility. With hair down to my waist and membership of a hard-left political group, I expected to be welcomed as a brother in the US communes I set out to study for my undergraduate dissertation. Arriving in my mother's sports car with an interview schedule that revealed I was almost completely ignorant of the practical realities of communal living, I came to appreciate what feminist methodologists later deemed a 'second wave' feminist insight: shared characteristics do not automatically make for rapport. I learned that nearly any research environment can be hostile if you are incompetent, but also that the emergency repair techniques we use to negotiate awkward situations in ordinary social life can be effective in awkward field situations too.

My PhD's analytic agenda, comparing extremist groups as cases of political deviance, led me back to the field. I discovered that there was no empirical research on the leading party on the British extreme Right, the National Front. It was the early 1970s, with the economy in decline and bitter class conflict. The NF was making headway, and in the two general elections that took place in 1974 it fielded enough candidates to get electoral airtime on national TV. United only by xenophobia, the NF was an uneasy coalition of traditionalist ex-Conservatives, ex-socialists who attributed high unemployment to excessive immigration, straightforward fascists, and 'skinheads' (back then a shaved male head was not a fashion statement but signified a belief in assault as the solution to most social problems). There were frequent violent confrontations with opponents, prominent officials had served prison sentences for violence and weapons-carrying, and the party had active links to Loyalist paramilitary groups in Northern Ireland. As well as the blacks, the Jews and the communists, the party scapegoated sociologists. My study of the NF combined covert and overt observation with interviews, about which more later.

Events at the end of fieldwork and following publication of my research persuaded me that a new research field might be desirable. Before my PhD I had taught at the Metropolitan Police College. It occurred to me – for largely biographical reasons – that one of the things that interested me most was the way people change as they move into new occupations, new relationships and new cultures. After all, I could see how I'd changed, as I moved from a student's critical perspective on 'the system' to a job teaching police recruits. I connected my experience of teaching in police college with my interest in change, and came up with the idea of researching police recruit socialization. Since then I have been engaged in police research. Policing can be a hostile research environment. It must be foremost among working environments that can brutalize its practitioners (constables have told me theirs is 'the only job in which you have to wash your hands *before* you go to the toilet'). It is also hostile in a more straightforward way, as I realized when one tour of observational fieldwork ended with an enraged citizen attempting to use a still plugged-in electric fire to lasso me and the officer I was accompanying.

Hostility to Research

Since almost any research environment can be hostile (remembering my blundering non-entry into the world of the commune), we must acknowledge that it isn't only qualitative researchers who negotiate hostile research environments. It is really a matter for anyone who treads 'the field'. Because it was recognized that if fieldworkers faced obstacles in the field it would affect response rates and data quality, survey research had an early concern with fieldworker safety and security. From this area we derive much of our understanding not just of personal security and access issues but how people respond to being asked about 'sensitive' topics, and how ongoing dealings with sponsors or other interested parties during analysis and publication can be hostile. Since survey research has often dealt with social problems, this literature has been directed to negotiating sensitive topics, ethical safeguards, and protecting the rights and welfare of respondents and researchers.

However, these themes only partially illuminate working in hostile environments. Working in hostile environments should be differentiated from working on sensitive topics. The latter suggests a rather passive respondent group who are getting by quite comfortably until a researcher comes along and upsets them by asking about that class of things that are best left unsaid – drunkenness, drug-taking, masturbation, not wanting to live near people unlike themselves. There is a difference between that and the circumstances we face when the research environment – like the physical environment on the deck of a deep-sea trawler – is actively hostile to one's presence. Hostile research environments are those where the research population is actively resistant to research. It isn't simply indifferent, uninformed, or susceptible to being upset by certain questions or poor technique. It does not want research done, and if research nevertheless takes place, it seeks to control the research and the researcher.

This said, there is great variation among hostile environments. The spectrum covers settings like the National Front, which explicitly repudiates social research, to those such as the police and other occupational groups, some of whose practices are sensitive and in which careless fieldwork can provoke resistance. The latter types of setting are more numerous, where a broadly sympathetic field or setting can become inhospitable in response to specific actions by fieldworkers that are insensitive, irresponsible or unwittingly inappropriate.

Such circumstances pose ethical as well as practical challenges. The highly developed literature on research ethics touches at many points on my concerns (e.g. Homan, 1991). But it seems to me that this literature both reflects something noble – sociology's aspiration to be a moral discipline that falls over backwards to respect its human objects of study – and something less impressive: timidity about entering the more difficult reaches of the social world. It is the latter that has produced a situation where we know a good deal more about the poor – those who are as powerless to resist research as they are to resist every other imposition – than we do about the powerful. Despite years of refinement, no code of ethics can satisfactorily anticipate the messiness of field relations. Those who accept this messiness justify going where their analytic agenda leads by reference to 'situation ethics' (Fielding, 1992). I will discuss situation ethics later, but won't attempt to justify the research topics I have pursued en route.

Rather, I take the view that work in hostile environments is not only a topic for paralysing ethical conundrums but an analytic resource. From work in such circumstances comes distinctive insights. Resistance to research is not only an ethical problem, much less a simply technical one (even if 'recipes' to overcome resistance were available). Resistance to research tells us much about organizations and their members, as Dalton (1959) long ago recognized. Qualitative researchers are familiar with the idea that they are 'the research instrument'. Especially in participant observation, but in intensive interviewing too, understanding flows from interrogating our own field experience. Dalton's discussion of obstacles to access suggests that much can be learned about organizations from who is willing or unwilling to speak to researchers, a theme in several other works of Chicago qualitative sociology by students of Burgess or Hughes (see Fine, 1995).

Obstruction, evasion, refusals and other troubles can in themselves be significant sources of data. Resistance is an important way to understand the culture being researched, and researchers can derive insight into belief systems and other facets of organizational culture from a reflexive stance towards field relations, from being alert to how people interact with them as researchers. Examples abound in elite studies, a fine instance of which is Smigel's (1958) study of Wall Street lawyers. The reasons these powerful professionals gave for why research was out of the question – constant time pressure, corporate clients needing to be assured of confidentiality, and previous research on law having 'gone wrong' – reveal the fetishization of secrecy by which legal elites avoid accountability. Resistance is not, of course, unambiguously an analytic resource. Sometimes researchers may accept that it is legitimate, in which case they should back off.

Like cooperation, resistance is not an invariant quality. Cooperation and resistance are continually renegotiated during fieldwork. In qualitative research we are *always* negotiating access. Van Maanen (1982: 138) captures the unpredictability well, remarking that

> in the abstract, relations in the field are such that the researcher is provided with trusted information of the sort necessary to both understand and empathize with the observed, but the researcher's presence itself creates little change or disturbance . . . concretely, however, such relations wax and wane over the course of a study, approach or exceed the upper and lower limits with different individuals on the scene, and vary according to the practical situation.

Sometimes it is hard to gauge whether refusal represents momentary inconvenience or enduring unwillingness (something survey researchers also ponder; Couper, 1997, found that when people say they are too busy or not interested in the topic they generally are indeed busy or uninterested, and it is seldom worth pressing them).

The signs of resistance or hostility are not necessarily gross. Research participants can use subtle means to regulate data collection and exert influence over data analysis. After all, even our relations with intimates are not entirely explicit or predictable, and the social world is rich in resources with which to deny candour, rescind confidences and excuse missed appointments. When Ostrander (1993) studied a network of wealthy women active in charities she learned much about the way the powerful apparently 'open up' while actually talking past the research question and use such things as invitations to meetings in expensive restaurants to suggest intimacy while engendering a sense of being beholden on the part of the researcher. Ostrander offers tactics useful in meeting these subtle forms of resistance, though none was proof against the discovery that members of her sample sat on the board of the organization from which she was seeking her next job, a discovery that obliged her to delay for several years publication of her book about the research.

That resistance is something we must anticipate, gauging how it may differently affect each element of fieldwork and group of research participants, is illustrated in Silverman's (1997) account of conducting research in a sensitive field of professional practice, HIV counselling. Silverman's conversation-analytic research required precise capture of speech in the setting; recording was essential. 'We ended up with a method of recruitment whereby [HIV] counsellors themselves explained the research to patients (often with the aid of written materials) and invited them to participate . . . the audiotapes were simply sent to me by each of the centres for analysis' (Silverman, 1997: 227). Access to clients was ultimately available only on the basis that the practitioners would decide whether to pass on to Silverman any tapes they judged to be usable. As Silverman notes, being in control of the tape recorder had some appeal to the practitioners.

Nespor (1998) encountered innocent resistance to research among primary schoolchildren. His topic simply did not register as important with them and they had nothing to say. Nespor decided to trust the children. He gave them the recorder and let them record what they wished after conducting a classroom project. Though not squarely relating to his topic, some of what they recorded illuminated a dimension of the issue he had not considered. Sometimes, then, resistance is not loss but value-added. As these examples suggest, field research is never something one does to research subjects but something one does *with* research subjects. It is co-produced by researcher and subjects. Both our methodological and analytic practice must be sensitive to the fact that, for example, interview data involve co-construction rather than plucking respondents' 'true answer' from them like fruit from the bough (Dean and Whyte, 1958). Anticipating and satisfying subjects' desire to influence the research is not just a matter of smooth field relations or a factor in weighing responses but can help us understand what they find objectionable and why, and contribute fresh analytic themes.

Protecting Rights and Welfare

Research with the reluctant or obstructive raises concerns about fieldworker security. Before each episode of fieldwork it is sensible to think it through, identifying potential problems and how to respond to them. Where we study those who may be a direct threat to our physical safety and security, the guidelines developed by professional groups in regular contact with dangerous others may be useful. These include going accompanied, and if this is not possible, notifying others where you are, when you aim to return, and times to check in by calling a colleague at intervals. It may also be sensible to meet with research subjects in public places and avoid night-time fieldwork. Social workers, police and benefits staff are trained to recognize signs that the people they are dealing with are becoming threatening, and in counter-measures; it may be worth obtaining such guidance. Professional associations and researchers' organizations also produce such guidance. The methodological literature includes useful treatments of these concerns, particularly from the survey tradition. Surveys tend not to be conducted by the researcher who designs the study but by subcontracted fieldworkers employed by large research organizations alert to their duty of care for employees. Gwiasda et al. (1997) offer practical guidance on safety and security in fieldwork in risky situations, based on their work in high-crime ghettoes.

Every year since 1980 I have sent between 15 and 60 students into the field to conduct observation and interviewing exercises (and my postgraduate students have used these methods more extensively for their dissertations/theses), and it is perhaps reassuring to say that none has suffered physical harm in the course of fieldwork. However, there has been occasional unanticipated resistance. A postgraduate who received permission to observe visitors to a stately home was asked to leave by the senior caretaker because he had been informally interviewing visitors, although she had agreed he could do so. It emerged that the issue was not interviewing but what the respondents had said. All reported that their visit was marred by the policy of keeping rooms in near-darkness to protect antique tapestries. While irrelevant to the student's topic, the overheard criticisms of 'their' stately home had upset the caretakers, who had passed the complaints, with their own view that the visitors were right, to the senior caretaker who had imposed the policy. It revealed an interesting theme concerning the pride the working-class caretakers felt about a property owned by the very rich, but one could not have anticipated that the fieldwork would expose conflict between the staff. One is left with few resources beyond tact and clarity about one's own rights. In this case, it was right for the student to refuse to censor or surrender his fieldnotes, as he had made it clear that the fieldwork involved asking visitors for their impressions.

Resistant Fields of Study

I have suggested that any setting can prove resistant or hostile to research, but particular research fields do display characteristic qualities. Here I am largely considering settings that are not hostile to research per se but may display resistance to research on particular aspects of their work. Those who study political groups enjoy the advantage that, because such groups seek new adherents, events that would be closed in other organizations are open to scrutiny. Also, politicians talk for a living and are often willing to be interviewed. However, since they strike positions on almost any topic, and ensure that all party representatives espouse it, one tends to hear the same line from every interviewee. Getting genuine engagement with analytically driven topics is not easy. Political organizations are also guarded about the business dimension: party finances, ownership of facilities, the party's backers. They are also cagey about alliances they might form and other tactical aspects of particular electoral scenarios. Researchers should remember that political organizations have massive experience of managing information flows to journalists, and researchers are likely to be seen as a branch of that species.

There are some similarities in researching cults, sects and subcultural groups. While these generally also wish to recruit new members, that is an advantage only if one is prepared to conduct covert participant observation, although it may be possible in an overt study to present the research as an opportunity to tell the 'outside world' what the group is 'really like'. Cults, sects and subcultures may also require converts to have particular characteristics that cannot be faked or fudged, like a particular ethnicity, tangible enjoyment of a particular substance or activity, or willingness to commit criminal acts. Where observation is the preferred method, and whether covert or overt, fieldworkers should be prepared for tests either of their commitment (if passing as a pseudo-member) or of their adherence to conditions of access. Perhaps the best-known problem associated with research on cults arose in Festinger's (1956) study of an end-of-the-world cult. The observers significantly increased the apparent number of the group's believers, strengthening the commitment of genuine members, especially when they learned that some of the new 'converts' were flying in from other cities to participate. One senses that these researchers never fully intended to conduct covert observation, and blundered into a fieldwork relationship that was at the 'fudging' rather than the 'faking' end of the spectrum. It is difficult to tell from the defensive published accounts of the fieldwork whether they experienced much resistance, as opposed to embarrassment, but the study is generally taken as a warning against changing the object of research by the methods used to research it. Very small sects, cults or subcultural groups are probably best studied by interview.

Field research on crime and criminals faces a large obstacle: the behaviours of interest are illegal. This is why 'criminology' is misnamed – it is largely the study of criminal justice, not crime. Nevertheless there is a small but valuable genre of studies of 'active criminals' (ones who are at large; the more numerous studies of criminals in confinement tell us only about unsuccessful criminals, those who have been caught). Naturally most active criminals are resistant to research and the group is therefore an example of one that is generally hostile to fieldwork. The lengths needed to access it are illustrated by Wright and Decker (1994), who persuaded a fringe member of a network of burglars to vouch for them on several conditions, including payment for interviews, interviews conducted in locations decided by respondents, and all data being anonymized and de-contextualized; in other studies, data have been lodged outside national jurisdiction in a country whose laws bar releasing such documents to agents of a foreign government (see Feenan, 2002). Such lengths are justified by the yield of original, high-quality information from studies of criminals at large. I discuss studies of agents and agencies of criminal justice later.

With an increasing range of social institutions playing a regulatory role, research interest has grown in 'informal' social control. Some private-sector regulation is itself dubious, and the profit motive adds another sensitivity that makes for resistance to research. Prior experience of the informal control organization is useful in overcoming it. Indeed, Winlow et al. (2001) suggest it was essential to their study of nightclub bouncers and the people whose behaviour they regulate. Winlow had previously worked as a bouncer and 'looked the part'. The violent and bloody incident forming the article's centrepiece indicates that, as well as looking the part, one must be prepared to play it.

Some argue that the resistance to research that has increasingly been expressed by deprived communities and groups such as asylum-seekers is legitimate and should be respected. They ask why researchers should join other bodies that prosecute their own interests on the backs of the deprived. It is hard to challenge this. Many researchers feel that potential research subjects should have the right to refuse to be researched. Fieldworkers may justify their intention by the good that may come of it. For example, where welfare programmes are at risk because they are abused or ineffective, qualitative research can reveal success stories that disappear in the aggregate of formal programme evaluation methodologies (Ong, 1999). Another response is to pay. Indeed, some deprived communities have formalized a rate set as a percentage of the research budget that is built into the grant application. Researchers proceeding thus should satisfy themselves about the means of distribution into the researched community.

Resistance to research remains less frequent among the powerless than the powerful. Like other resources, privacy flows most directly to the stakeholders in society. Among these are 'the professions'. The concept of social closure was itself developed in the sociology of the professions. Professions have elaborate and effective means to avoid oversight. The insulation from oversight that marks the 'great professions' may, though, be the researcher's best resource. Professionals are aware of the need to grant a measure of accountability if full public oversight is to be avoided. They also often feel their contribution to society is considerable and the world should know about it. Among the professions exercising strong social closure (by which access to professionals and information about their practices is granted only under carefully regulated conditions, prime among which is as a client) are medical and health-related trades, the law, and workers in religion. Elite studies have given rise to an insightful literature on negotiating resistance to research, as discussed earlier. The point to draw out is that research in almost all substantive fields, including that on social institutions having broadly benign functions, will encounter offstage areas and practices that those who inhabit them will prefer to shield from scrutiny.

Hostility in Two Field Studies

I have suggested that neither 'cooperation' nor 'resistance' should be regarded as invariant throughout the fieldwork process, that the means by which these qualities are made manifest to the researcher are not necessarily gross, and that the nature of cooperation or resistance can tell us much analytically. While fieldworkers have long identified the importance of rapport, the several interactions represented by this concept require particularly close attention in fieldwork in hostile environments. I will elaborate these themes in two examples: the first from the National Front research and the second from my community policing research. The National Front research mentioned earlier examined links between extremist politics and political violence, originally comparing extreme Left and extreme Right. However, I soon found there was no empirically based literature on the contemporary extreme Right and the research was refocused on the NF. The community policing research attempted to gauge and document the practical means community officers used to achieve social control sensitive to the community they policed.

The National Front Fieldwork

It was fortunate for the National Front research that there were two general elections in 1974. In that time of political turbulence, extremist politics of all hues gained unusual prominence. As the main extreme Right party, the National Front fielded candidates in both general elections, and enjoyed its first party political broadcasts. The NF was a coalition of smaller factions. Significant not in numbers but for their links to the Conservative Party, two factions, the Racial Preservation Society (RPS) and the League of Empire Loyalists (LEL), had particular strength in large South Coast towns like Brighton, where I lived, while membership based on former Labour supporters featured in cities like London, where my PhD was registered.

I conducted my first interviews with branch officers with no definite research design in mind. It seemed best to see whether I could gain access rather than fix a research design ahead of time that might prove unrealistic. The only certainty was that the study would of necessity be qualitative. The NF was virulently hostile to research, and specifically castigated sociology. There was no prospect of getting a membership list and, even if my interests could be expressed in questionnaire items, a general survey would require a large over-sample to collect enough NF voters for statistical analysis. While the status of qualitative method was more subordinate then than now, one warrant for qualitative work was exploratory research, which was what I proposed.

My first interview was successful in the sense that I was not thrown out. Despite my long hair and status as a student sociologist (I presented myself as an undergraduate, thinking this may be less threatening), I was politely received. My respondent warmed somewhat when I deliberately asked a series of questions requiring him to do no more than recite the party ideology. I saw this as a quid pro quo: I would be taken seriously if I took my respondent seriously. It needs to be said that taking my respondent seriously involved not choking with rage or laughter at the expression of views that were both repellent and ludicrous.

Like a police officer's first arrest, this first interview prompted much reflection. I decided that to obtain the information I wanted it was best to avoid antagonizing NF members. If this involved not only reserving my own view but actively misrepresenting it by acquiescing to the views they expressed, I would do so. I had noticed that, when my respondent had delivered himself of a strongly expressed opinion, he momentarily scanned my expression, presumably for signs of condemnation. By simply acknowledging his statement I could encourage him to say more, and these further statements went beyond the recital of standard ideology. It was not hard to offer mildly encouraging or positive acknowledgements because I was familiar with the ideology and current position on contentious policies from party publications.

Another reflection on the first interview was also consequential. The respondent was a candidate for the local council. The interview ranged from the 'big' national issues to the minutiae of local politics. On local matters his views were barely distinguishable from those of mainstream candidates. I was forming an uneasy awareness of the banality of evil, that the danger of extremist politics lies in its ability to cast itself as 'normal', as the real voice of the ordinary people, and that murderous things can be done matter-of-factly. It came to me that it was essential that my research design balanced the interviews with field observation. Reliance on interviews might leave me with a corpus of data that merely invested the ideology in the party literature with the particularities of those I interviewed. I wanted to know how the people who laid claim to this ideology deployed it.

Observation method demanded more complicated decisions regarding my role than those involved in seeking interviews. I wanted to see how the people I was interviewing conducted their party's business in branch activities, and since I would be known to them from the interviews, this could only be done overtly. But my main interest was in political violence, and this was unlikely even to be discussed in the presence of a known observer. I decided to conduct overt observation in branches where I was known from interviews and covert observation in branches and at events where I was not known. Even at a time of lower awareness of fieldworker welfare and security concerns, this was probably foolhardy. I do not commend it as an example, I do not challenge those whose eloquent ethical critiques have condemned what I did, but I do say there was no other way to obtain the data.

There was exhilaration in finding out what the NF's insiders and hardcore opponents knew and I would not gainsay its effect on my motivation. But since I was able to compare what I would have 'known' if I had rested with interviews and what I learned from direct observation and informant testimony, I was also able to assess the 'scientific' justification for what I was doing. The interview response proved consistently misleading about the party's sources of finance, links with paramilitary groups, the way it monitored its opponents and conducted physical interventions against them.

I have already mentioned that I used what amounts to deception in the front management I adopted during interviews. Once I moved towards overt observation I needed to use the interviews not only to collect data but to negotiate access to branch activities. I needed to observe more than one branch so I could compare the branches where ex-RPS and LEL members predominated with those where ex-Labour supporters predominated. I also wanted to observe branches more than once for reasons of validity. At one branch, my attendance at branch meetings was readily agreed. At another, I sensed during the interview that without an other-than-neutral tack I was not going to get any further than this one interview. For these reasons a change in my approach was necessary.

I decided that I would have to do more than encourage further responses by mildly positive acknowledgements of what was said. I began interjecting statements that implied that I was becoming tantalized and even persuaded by what I was hearing. These statements, which seemed crude and shameless to me, were warmly received. I had arrived at the stance I took in the rest of the overt fieldwork. I would be the researcher-turned-convert. Like Festinger's pseudo-converts, I always fudged exactly where I stood. I was also aware I was under test. However, non-researchers displaying availability-for-conversion are also under test: each event one attends is an earnest of ongoing interest and an occasion for members to work at securing the final commitment. Testing was frequent but not continual; I never felt there was a point of no return and I worked to ensure that members did not think so either.

Cooperation and resistance were thus related to my evolving role. My tactics secured an enhanced but conditional degree of cooperation. Cooperation also varied between different members. I found that it was not hard to get on the wavelength of those that one journalist labelled 'the seaside landladies of the NF'. I had heard some of the views they expressed from my relatives. But these were not the activists in whom I was most interested. Where working-class and young male members predominated, aggressive action was more squarely on the agenda, and my limited experience of this group and lack of interest in football made relations harder to establish.

This hiatus was resolved by a coincidence. As well as the branch fieldwork, I had approached headquarters for interviews with senior officials. Negotiations were slow, being conducted by post (I wanted a record of our exchanges but, more to the point, as a student I could not afford the telephone). When I was finally granted an interview with a relatively senior official it was tense and largely repeated the too-familiar ideology, interspersed with statements like 'that information is not released' in answer to the more interesting questions. I learned more from observing the interview setting than from the interview (for example, party HQ, an ordinary end-of-terrace house, was encased in corrugated metal, and two-way mirrors protected a weapons safe under the stairs). However, I dutifully wrote down the 'no comments' and my respondent's parade of racist clichés. When I contacted HQ after the interview using the foot-in-door technique that I had stupidly forgotten to ask a few questions and would like to come back, the request was accepted. This time I would be seen by a less elevated official, the National Front Student Association organizer. He turned out to be a sociology graduate and ex-Maoist from Belfast. I would lie if I said other than that we got on extremely well, something that gave me pause at the time and has ever since. Through him I gained an interview with the party's deputy chairman, a meeting with the party leader, and access to a branch actively involved in confronting opponents.[1] I did not inform him that, as well as the overt fieldwork, I was attending other branches and events

I was reaching the final stage of fieldwork when I decided to attend the NF Remembrance Day rally. Branches held their own commemoration at the Cenotaph in London following the official ceremony and then marched to a rally. The event was regarded as the high point of the NF calendar. I had an invitation to attend with the Brighton branch, the day beginning early in the morning and ending in the early hours of the next day as the coach made its final drop-off. A sustained fieldwork occasion like that makes it very apparent that relations with research participants vary from the cooperative to the resistant. Indeed, they varied from the embarrassingly warm to the disturbingly hostile. To show this can be useful analytically, I recount a segment from the morning coach trip.

Sat next to a spry older woman, I was asked how I came to be involved as I did not look like a 'typical National Fronter'. I retailed my line about it having begun as research but having become a personal interest, explaining the NF beliefs that particularly attracted me. When I finished she asked 'Well, what do we believe, dear?' in relation to one of the planks of party belief that I had taken to be most firmly grounded of all. Was this an innocent question? Perhaps the lady was ex-RPS and did not know the NF policy. Or was it a test? I offered a careful statement of the party line on the topic, concluding that 'When all was said and done Britain is for the British'. The round of applause from the surrounding seats showed me that, witting or not, the lady's question had functioned as a test. When we got off the coach the male activists were deployed to form the edge of the column and point of contact with anti-Nazi protesters. The branch was left without able-bodied men to carry the branch banner. I was asked to carry it, the branch chairman taking the other pole. Analytically the value of this incident came in showing variation in awareness of key party ideology, and in displaying how members achieve a sense of shared purpose and commitment.

As branches arrived at the rally, members milled around behind police cordons waiting for the platform speakers to begin. I stood with the branch committee. I felt something prod into my back and the phrase 'stick 'em up' said in a Belfast accent into my ear. Instinct told me to respond very slowly and treat it lightly. It was the NFSA organizer. He asked whether I was 'here in an objective capacity or just out of interest today'. With the branch group present, I was grateful for the option offered by that construction. He proceeded to offer a connoisseur's critique of the proceedings. This suggested that I had passed another test, by noncommittal statements and because it would have been difficult to interrogate me about my purposes with the branch committee present.

That I was on test that day seemed to be confirmed when the branch broke its journey at an isolated rural pub. The venue was clearly pre-arranged as we were shown to a back room. A burly branch member who was regarded as the branch's most reliable lynchpin in confrontations with opponents said he would like to talk about branch security. While I was pleased that one of my key interests had come up, it rapidly became clear that the discussion was a barely concealed device to warn me about what happened to people who pretended to affiliation in order to obtain information and then disappear. Particular attention was paid to injuries received by those involved in previous episodes of 'entryism'. I blandly shared in condemning such subterfuge and made further noises about my growing interest in membership. Discussion closed in a dark corner of the room next to an exit well away from the rest of the branch. I was pointedly handed a membership application form and told that after the enjoyable day in which I had participated I would no doubt want to complete the form immediately.

I had always offered to provide NF HQ with a copy of any book I wrote as a result of the research. That the NF wanted to know what I was writing about it became apparent after my thesis was lodged in the university library. The library soon contacted me requesting a further copy, as the middle 200 pages had been cut out. This happened a second time, after which the thesis was withdrawn from public access. Although my book was ready soon after completing the thesis, my publisher's consultations with lawyers took two years, during which the NF referred others wishing to research the party to me. The day the book was published, copies were sent to the NF chairman, deputy chairman, and the NFSA organizer. The deputy chairman was a physically imposing figure who dealt with the practical running of the party while the chairman concentrated on ideology and presenting a statesmanlike image (despite convictions for arms offences). From the chairman I received a polite but cold letter of a few lines some weeks after publication. The evening of publication I received from the deputy chairman a telephone call in which I was threatened with death. The telephone call was marked by alternate bellowing of abuse and bitterly expressed criticism of points in the book. Fortunately I had recently obtained a permanent job and was leaving London, though I did obtain an unlisted telephone number when I moved.

Much of the NF fieldwork had the feel of shadow-boxing. While 'immersion' in the field and intense field relationships feature in the literature about observation method, these were difficult standards when applied to work with a violent racist group. Indeed, the one relationship that came close to these qualities, that with the NFSA organizer, caused me much soul-searching. This sense of shadow-boxing and near-relationships made it hard to gauge the calibre of some of the data. Analytically I did get much of the information I wanted, but some lines of inquiry were closed because of the uncertain status of the data. Cooperation and resistance not only vary but it can be hard for researchers even to tell if they are receiving one or the other. Drawing on her work with US white supremacists, Blee (1998) suggests that in such circumstances we monitor the strategic use of emotion by informants. This can both expose the emotional content of social movements and help us to see how interpretations are jointly shaped in the fieldworker/informant relationship. In this and several other respects there is strong affinity between our respective understandings of fieldwork in extremist political groups, suggesting the enduring character of field relations in such settings.

The Police Fieldwork

The National Front research contrasts with my second example in having been relatively short-term. While my friends joked about the inevitability of my move from researching racists to researching the police, my interest was not due to the connection with an occupation widely thought to attract authoritarian racists, but precisely because the police I encountered so little fitted the stereotype. Following my appointment to a permanent academic post, and drawing on my pre-doctorate experience lecturing at Hendon Police College, I took the opportunity to explore how people changed when they joined the police. That was in the late 1970s and I have since remained in police research. Long-term presence in a research field makes for different resonances of the cooperative/resistant dimension. There are challenges that could rapidly make it a hostile environment if wrongly handled. A 'long-term research presence' is enabled by careful professional work, of course, but also requires responsiveness to the needs of the research area. It requires skill in seeing how what 'the system' wants can also be made to serve one's own analytic interests.

As well as studies of police socialization I have researched the police investigation of child sexual abuse, community policing, the recruitment of ethnic minority police officers, 'intelligence-led policing', and police efforts to manage the public's fear of crime. The work on community policing combined observation with interviews in an effort to identify factors that promoted successful community policing. We prepared both qualitative fieldnotes and quantitative observation forms, which enabled us to compare the working style of community officers and regular patrol officers, and to assess whether citizens of different ethnicity, status or gender were dealt with differently. We decided that each fieldworker should be attached to one group of officers throughout the fieldwork. I spent a year on fieldwork, so I got to know 'my' officers quite well, not just on patrol but in the canteen and after hours.

One of their number was manning a patrol car when I accompanied him on a night shift from a Friday evening into the early hours of Saturday morning. He had been a participant in the administration of 'street justice' to a black male who had assaulted an undercover female decoy during a bag snatch, which I have discussed in Fielding (1990) as an instance of the need to look at incidents from the participants' perspective to gain a holistic understanding. Earlier in the fieldwork we had discussed the topic and method of the research. Although the research team kept it to a minimum to avoid reminding officers that their work was being observed, we periodically jotted notes in the officers' presence. On this occasion the shift had started quietly, as it was raining. In the late evening we received a call. A man was acting erratically and dodging back and forth from sidewalk to roadway near a major intersection. We found the man easily. He was a heavily built Afro-Caribbean and appeared to be one of the area's large number of vagrants. The constable recognized him. He was 'basically all right' but could be aggressive and hard to handle when drunk. Tonight it looked like he had had 'a skinful'. We approached without blue lamps and parked a few feet behind the man. The constable approached the man, calling him by name in an entirely matter-of-fact tone of voice: 'Hey John, it's not a nice night is it.' The man muttered something, pushed the constable's chest and then hit out at his shoulder. The constable sidestepped the blow and in doing so, steadied John from the waist and, rather as if strolling with a lover, guided him through 180 degrees and into the car. The patrol car was a battered two-door model. Its cramped interior was not designed for maintaining aplomb when steering a 250 lb drunk on to the back seat, but the constable took elaborate and subtle care (not at all attended to by the drunk, who was trying to remember the officer's name) to get him on to the seat without hitting his head on the bodywork.

These manoeuvres were performed with a seamless balletic grace that sticks in my memory as an instance of the unsung but significant portion of policework that is not about verbal or negotiating skill but about body control, handholds and balance. To capture the moment I jotted a note, trying to record the movement as well as the few words that were said. The station was nearby so we took the drunk there to sober up in a cell. Back in the car the constable said he had noticed me writing a note and wondered what I had seen that was worth writing down. Officers seldom displayed interest in what we were doing, especially after the initial novelty wore off. This, and the fact that I felt I had a friendly relationship with this officer, made me say that I had taken the note because I was impressed with how he had handled the incident. I added that his movements had struck me as nearing on the graceful and that I felt his skills of coordination had disarmed the man and largely accounted for a surprisingly quick and easy resolution. 'You mean you're writing things down about *me*?!', the constable asked. The tone was one of near-incredulity, but behind it was an unmistakable shading of betrayal. I said that he'd be welcome to read what I had written. There was nothing to hide and my commentary was the opposite of adverse. 'No, I *don't* bloody want to read it!' I attempted repair by emphasizing that I was seriously impressed by his physical performance. Whether this sounded uncomfortably close to the kind of appreciation a gay man might offer or my flattery was too unctuous I do not know, but the attempted repair failed. The rest of the shift passed in virtual silence. Those who have experienced the deadliness of the early morning hours of a quiet night shift will know that I was being punished.

The episode suggests that cooperation/resistance is a stilted way to express the nature of field relations in long-term fieldwork with small working groups. A sense of relief at being tolerated sometimes excites not only gratitude but a misplaced sense on the part of fieldworkers that they have achieved genuine friendship with those they observe. For all our talk about reflexivity, we are sometimes lazy about taking the role of the other. Perhaps I would not have been shocked by the tone of betrayal the constable expressed if I had troubled to think what it is like to have someone with a notepad staring at you while you do your job, the job on which your mortgage, self-image and friendship network depend. Perhaps if I had bothered seeing it from the constable's perspective I would have recognized how a statement about my 'balletic grace', when I thought all I was doing was stopping a drunken tramp punching me and puking on my wreck of a patrol car, would not only be unbelievably pompous but would highlight the distance between my world and the observer's. Perhaps, indeed, I would remember that these researchers had said they were concentrating on members of the public experiencing community policing interventions rather than watching the officers conducting the interventions and evaluating not only what they did but even how they moved.

For my part I had overestimated how much the officer understood about what we were doing, what type of action our access warranted, and the bounds of our informal relationship. Offering sight of the jotted note was particularly inadvisable. My jotted fieldnotes largely consist of abbreviations and short phrases spoken by field participants that trigger recall when writing full fieldnotes. Even if the constable had read them he would have been unable to see in them the mental picture I had formed. The significance of offering sight of the notes was not that I expected the offer to be accepted or, if accepted, that it would help. Its significance lay in something more primordial than any fieldworker/participant relationship. It was about owning the guilt of betrayal. It was about self-abasement. I had wronged a friend and something should be given to make good the hurt. It did not work.

From such circumstances there is no return to normal. The gift relationship lies at the very heart of sociality (Douglas, 1970). When the gift is refused, and there is no other gift to give, the relationship withers.

This was just one field contact and I was just one fieldworker of three. All of us at times felt the bind of personal relationships, particularly when the depth of our contact gave us insight into what lay behind callous, brutal acts of policing, forcing us to consider whether, similarly placed, we could have done any better. Such feelings intrude particularly sharply when 'policy relevance' is criterial. Expensive work like this is seldom funded without some such claim and this encourages the tendency to evaluate rather than simply interpret. Operating with long-term commitment to a substantive field like policing, politics or health gives us another set of interests to negotiate beyond ethical practice, personal feelings and our analytic agenda. It is seldom as crass as pressure to produce a particular analysis or cover up inconvenient data, although this certainly does happen. More insidious, better concealed, and more frequent, is self-censorship.

When a researcher commits to a given field they have to balance the nature of the contribution they wish to make to the discipline against what they have to compromise to operate in that field. That is not something about which principles can be laid down by others. However, one can 'speak truth to power' and survive in one's field, provided one takes the long view. A highly regarded criminal justice researcher conducted in the 1970s a methodologically rigorous study that produced an extremely unflattering picture of a major criminal justice institution. The study refuted government policy in respect of this institution. The researcher was banned from ever entering again a branch of this institution. The researcher turned attention to an unrelated criminal justice institution, making key contributions to that field. Recently the researcher was appointed by the government department dealing with criminal justice to the top national role in respect of a third area of criminal justice, an appointment that has to be approved by a top politician. The point is that the relationship between the policy and research worlds is one marked by competing interests on the part of both policy-makers and researchers (Fielding, 1999). If a controversial researcher is rehabilitated it is because policy-makers have seen an interest in doing so, bearing in mind that a reputation for independence can lend credibility to contested policies and that, since governments change, the thorn in one government's side does not prick for ever.

Since operating in long-term substantive fields that are policy-relevant produces field-specific expertise, and since it is often hard to predict how one's work will be received since one cannot tell which policy faction will predominate in the response to it, it seems best to let integrity decide whether the research environment is hostile or hospitable. But we do need to recognize that all this is out there, affecting our field relations as much as the interpersonal dimensions discussed in my example.

Conclusion

The tone of this chapter has run from the jaunty and cavalier to the portentous and perhaps maudlin. It has seemed horribly self-referential in the writing, but we say often enough that in qualitative research the researcher is the research instrument and we must therefore accept being a little more open about our practices than seems comfortable. It is time to extract some ideas from all this autobiography. In terms of practical procedures there are several measures that help protect the fieldworker's physical and mental security and welfare. The heightened sensitivity to the cooperative/hostile dimension of rapport that has been highlighted in this chapter is not just an analytic but a practical resource. Close and reflexive monitoring of reactions to our presence, actions and the topics we raise helps us negotiate the field. It is important to anticipate what challenges may arise so that appropriate precautions are used: a tense situation can be inflamed if respondents realize that fieldworkers regard contact with them as requiring precautions. After each fieldwork event it is worth keeping note of any aspect relating to security or which made the fieldworker uncomfortable; Bruyn's (1966) schema for recording subjective aspects of fieldwork is recommended.

Other security measures are no different from those that people ordinarily take when meeting strangers, with the exception of fieldwork with people engaged in criminal or unethical practices. There thought needs to be given to the legal aspects of holding data in which the authorities have an interest. The sparse case law in this area is based on researchers being analogous to investigative journalists. Conducting research is not a legally acceptable reason to refuse the authorities access to research materials. In the few documented cases, researchers have refused access rather than incriminate their sources, and been sentenced for contempt of court (for a discussion, see Van Maanen, 1982).

In discussing work in hostile research environments I have emphasized the importance of interpersonal skills, especially observation and negotiation. Sensitivity to the others' behavioural display involves monitoring paralinguistic elements as well as verbalizations. The early discussions of interview method made much of its necessary combination with observation, and attention to paralinguistics can be useful both in managing the fieldwork situation and analysing the data. Fieldwork can be seen as a series of negotiations, with the researcher and the researched continually testing the bounds of the relationship. Flexibility is needed, and this means we need to be prepared in difficult research environments to make decisions in the field. We cannot anticipate everything and sometimes we will be asked to do things that give us serious pause. This is why it is important to be clear about our ethical precepts ahead of entry to the field, because when we are in the field they will be tested and our effort will be directed to deciding how to apply them.

I have referred to the utility of play-acting, of being comfortable with playing roles that do not reflect how we are when not in the field. In rehearsing how each fieldwork event may go, we need to consider the demands that may arise which are only solvable by deception, and to decide what degree of deception we are prepared to accept. The practical application of research ethics has less to do with formal codes than with situated practice. Field situations vary so much that blanket injunctions are not helpful, and, in any case, when we think seriously about the methods we use it is difficult to maintain some of the assumptions on which they are founded. For example, we make a distinction between overt and covert observation, but precisely when do sociologists stop observing? If my subject is policing, I may have carefully negotiated with Scotland Yard the threshold above which I would feel obliged to notify superior officers of malpractice rather than use it as data (a quite usual negotiation), but what happens if by chance I witness such an occurrence while walking home after fieldwork? Should I report it? Use it in my analysis? Even if I don't, can I erase it from my awareness so effectively that it won't affect the analysis? Should I approach those I have observed and tell them I plan to use what they did in my research?

Decisions about what is right (or least wrong) in such instances do not involve pulling the correct principle off the shelf but an awareness of all the principles that must be balanced, a determination of their degree of applicability, and a judgement about their balance. It is similar to the judge's procedure in sentencing: in light of *actus rea* (the thing done) and *mens rea* (the doer's intention), decide the ceiling of penalty attaching to the act, take account of mitigating factors and set the most appropriate penalty. That is the kind of judgement we face in research ethics, not the simple bipolar choice – guilty or not – that we permit the untrained jury. Professional judgement is about the precise application of principles and that is the kind of work that situational ethics requires (Duster et al., 1979). Such ethics rule little out on principle but require a more acute awareness of principles in practice than any code of ethics offers.

One of the hardest judgements is when we consider the gain in knowledge (the 'scientific justification') to warrant means we regard as dubious. What makes it so difficult is that we usually do not know precisely what knowledge will be gained. Agreeing to conceal in one's apartment a stolen gun entrusted to us by a field contact expecting a police search (Polsky, 1971) may enhance our standing with our contact or their contempt for our pliability. Enhanced standing may encourage our contact to share more information with us, but we cannot know if the information will be useful. Refusing to secrete the gun may put us at risk too – declining to do this favour may make our contact indifferent to our security when we operate in their milieu.

For all our care over situational ethics, decisions like this rely at least partly on how our individual personalities respond to risk. Those who enjoy the kind of risk involved in suspending familiar identities have an advantage in working in hostile environments. When deception is involved it may be a matter of degree rather than an absolute, as in cases where one allows others to draw conclusions one does not correct. More than a willingness to use what means are necessary to gather the information we want, working in hostile research environments requires thoughtful planning, anticipating the things that may happen on each fieldwork occasion, interpersonal sensitivity in the field, and flexibility. Above all, it demands awareness that every field decision, including the decision not to go further, must be treated by reference to the *practical* application of ethical *prinicipIes*.

Note

1. There is no implication that the branch or its members precipitated such actions. Details of the circumstances are in Fielding (1981) and Fielding and Fielding (1986).

References

Blee, K. (1998) 'White-knuckle research: emotional dynamics in fieldwork with racist activists', *Qualitative Sociology*, 21(4): 381–99.

Bruyn, S.T. (1966) *The Human Perspective in Sociology*. Englewood Cliffs, NJ: Prentice-Hall.

Couper, M. (1997) 'Survey introductions and data quality', *Public Opinion Quarterly*, 61(2): 317–38.

Dalton, M. (1959) *Men Who Manage: Fusions of Feeling and Theory in Administration*. New York: Wiley.

Dean, J. and Whyte, W. (1958) 'How do you know if the informant is telling the truth?', *Human Organization*, 17: 34–8.

Douglas, M. (1970) *Natural Symbols*. London: Barrie.

Duster, T., Matza, D. and Wellman, D. (1979) 'Fieldwork and the protection of human subjects', *American Sociologist*, 14(3): 136–42.

Feenan, D. (2002) 'Legal issues in acquiring information about illegal behaviour through criminological research', *British Journal of Criminology*, 42: 762–81.

Festinger, L. (1956) *When Prophecy Fails*. Minneapolis: University of Minnesota Press.

Fielding, N. (1981) *The National Front*. London: Routledge.

Fielding, N. (1990) 'Mediating the message: the co-production of field research', *American Behavioral Scientist*, 33(5): 608–20.

Fielding, N. (1992) 'Affinity and hostility in research on sensitive topics', in C. Renzetti and R.M. Lee (eds), *Researching Sensitive Topics*. London: Sage, pp. 146–59.

Fielding, N. (1999) 'Research and practice in policing: a view from Europe', *Police Practice and Research*, 1(1): 1–29.

Fielding, N. and Fielding, J. (1986) *Linking Data*. Beverly Hills, CA: Sage.

Fine, G.A. (ed.) (1995) *A Second Chicago School*? Chicago: University of Chicago Press.

Gwiasda, V., Taluc, N. and Popkin, S. (1997) 'Data collection in dangerous neighborhoods' *Evaluation Review*, 21(1): 77–93.

Homan, R. (1991) *Ethics of Social Research*. London: Longman.
Nespor, J. (1998) 'The meanings of research: kids as subjects and kids as inquirers', *Qualitative Inquiry*, 4(3): 369–88.
Ong, P. (ed.) (1999) *Impacts of Affirmative Action*. Los Angeles: AltaMira Press.
Ostrander, S. (1993) 'Surely you're not in this just to be helpful: access, rapport and interviewing in three studies of elites', *Journal of Contemporary Ethnography*, 22: 7–27.
Pollnac, R., Poggie J. and Vandusen, C. (1995) 'Cultural adaptation to danger and the safety of commercial oceanic fishermen', *Human Organization*, 54(2): 355–79.
Polsky, N. (1971) *Hustlers, Beats and Others*. Harmondsworth: Penguin.
Silverman, D. (1997) *Discourses of Counselling*, Appendix 1: 'The history of the research project'. London: Sage.
Smigel, E. (1958) 'Interviewing a legal elite: the Wall Street lawyers', *American Journal of Sociology*, 64: 159–64.
Van Maanen, J. (1982) *Varieties of Qualitative Research*. London: Sage.
Winlow, S., Hobbs, D., Lister, S. and Hadfield, P. (2001) 'Get ready to duck: bouncers and the realities of ethnographic research on violent groups', *British Journal of Criminology*, 41: 536–48.
Wright, R. and Decker, S. (1994) *Burglars on the Job*. Boston: Northeastern University Press.

41

'Dangerous Fieldwork' Re-examined: The Question of Researcher Subject Position

Pamela Nilan

Introduction

This article considers the problem of conducting ethnographic fieldwork in dangerous and difficult circumstances in the course of a sociological research project. Although 'dangerous fieldwork' is not a new topic – one must acknowledge the excellent work of Van Maanen (1988) and the valuable working paper by Lee (1995) among many others – this article takes a different slant. The specific difficulties this researcher found in conducting ethnographic fieldwork in risky and trying circumstances were further complicated by the constant sense that one ought to be more 'in control' of the situation than was really possible. Looking back at field journal entries which refer to this anxiety over control, brought the realization that this was probably not just an idiosyncratic or isolated fieldwork phenomenon, but was pertinent to contemporary debates in qualitative research more generally. So this article, despite its somewhat sensationalist title, actually grows out of some 13 years of conducting sociological research with young people on gender, education and youth culture. Across a variety of projects in Australia and in the Southeast Asian region, this researcher has struggled to reconcile the combination of two different approaches in qualitative social science research: 'formal' methods (interviews, surveys,

Source: *Qualitative Research,* 2(3) (2002): 363–386.

focus group work), and ethnographic fieldwork – participant observation. The major problem has been effectively moving between the two subject positions of researcher implied in the contrasting research paradigms here. The first is constituted within a discourse of control, objectivity, even emotional detachment. The second is constituted within a discourse of immersion, reflexivity and rapport. I find the former researcher position easier to assume, but the latter, although much more uncomfortable, always yields rich and interesting data, which the first sometimes does not. It is in the latter of these researcher subject positions that the possibility of danger and risk most commonly arises, yet this is the position of least control and enhanced emotional vulnerability.

The project from which the examples in this article are taken used both approaches – a common feature of contemporary qualitative research. For example, Eriksen advises that even anthropologists depend on a combination of formal and unstructured methods in their fieldwork (1995: 15). Elsewhere in the social sciences there exists a 'carnivalesque profusion' of methodological approaches (Atkinson et al., 2001: 4; see also Hammersley, 1998: 7). This is so, not only in 'foreign' fieldwork settings such as I describe here, but in institutional settings such as hospitals, welfare agencies and schools (Street, 2001). Using the more formal approach, the social science researcher obtains some 'hard' data through surveys, interviews, focus groups, then, using the second approach, seeks to support or interrogate it through ethnographic data gained through immersion in the local scene or culture. Taken together, these qualitative methods constitute a valuable tool for investigating the complex relationship between social structures and human agency (Goodman, 1998). At the level of the local scene, phenomena are both mapped using formal empirical methods, and deeply interrogated, using ethnographic methods. Inevitably the more formal approach seems much more 'scientific' than participant observation. The design and delivery of surveys, the design and execution of one-to-one interviews, the assembly and conducting of focus groups, all imply control of the research context by the researcher. He or she enters the research context and sets about deliberately constructing a 'controlled' set of research circumstances – controlled in the sense of deriving primarily from the intentions and interventions of the researcher, rather than primarily from local participants. The research subjects assemble themselves (physically) and their accounts of phenomena relative to the set of research circumstances constructed by the researcher.

Yet quasi-scientific 'control' is the very thing which must be set aside for effective participant observation to take place. The aim of the ethnographic researcher is to observe 'naturally' occurring phenomena relative to the research question. The term 'naturally' here refers to the fact that events and accounts derive primarily from the intentions and actions of local participants, rather than from researcher interventions (Hammersley, 1998: 2). The researcher who is employing both survey/interview data collection methods and ethnographic methods in the same field context at the same time therefore has to move rapidly between different subject positions *qua* research. Sometimes she/he is in a very 'controlling' position (recruiting subjects, distributing surveys, conducting interviews at a specific time and place), and sometimes occupies the very opposite data collection position, with little or no control at all over the research context, attempting to blend in and be as unobtrusive as possible in order to better explore the taken-for-grantedness of local routines, rituals, procedures and assumptions. I did not find these rapid shifts in the subject position of researcher easy, especially in the foreign, exotic or risky fieldwork context, where one is beset by confusion and anxiety. In unfamiliar settings which carry an element of threat, the subject position of the controlling researcher seems far more secure and reassuring.

In qualitative research projects where ethnographic fieldwork is not used, the emotionally detached researcher invites subjects to present themselves in a formal research context, has minimal contact with them (especially in the case of surveys), then retreats back to the reassuring safety and familiarity of home, hotel or cultural enclave with neat stacks of satisfyingly ordered data. By contrast, observational and incidental data gathered while the researcher is living in, or spending extended time in the field, is inevitably far less neat and logically ordered. While Hammersley (1998) politely labels the process 'unstructured' (p. 2), Kumar (1992) describes it as a 'brash, awkward, hit-and-run encounter' (p. 1). Although potentially rich, fieldwork yields up messy notes, personal reflections, photographs, sketches, assemblages of documents and even locally collected items. Writing up fieldnotes relies on the frequently faulty device of memory. The researcher tends to collect as much data as possible, in part because it is often not clear what may turn out to be relevant in the end. The methodological mandate to create ethnographic 'thick description' (Geertz, 1973), and the fear of missing something which may later turn out to be important, usually results in the gathering of large amounts of data which cannot be easily categorized and analysed. The researcher may suffer from isolation, anxiety, stress and depression (Lee, 1995: 13), even in relatively straightforward fieldwork. The research accounts produced from ethnographic data can often seem like descriptive narratives, and inevitably, in the course of producing a research account which will be taken seriously by the academy, a lot of detail must be set aside. Finally, a range of ethical issues seem to emerge in the ethnographic component of qualitative research which were not considered at the time of planning and obtaining permission for the research. This is especially the case in difficult or dangerous fieldwork situations.

The issue I take up here is the positional shift of the researcher who moves between very different forms of data collection in these combined method studies which take place in the same setting. My argument is that the two subject positions are not easily separated. Willson describes this as resulting in a qualitative research situation where the ethnographer is expected to be 'simultaneously *detached* and yet intensely *engaged*' (1995: 255, my emphasis). In my experience, the ethically enshrined precept of researcher 'control' tends to flow from the context of formal data collection over into the ethnographic context of data collection (see Geertz, 1983 – 'blurred genres'), both at the actual time of data collection and during the later phase of analysis and interpretation. This article looks at some extreme examples of ethnographic research in dangerous and unpredictable fieldwork situations. A combination of qualitative methods involved a move from the subject position of emotionally detached researcher, which implies control, to the subject position of emotionally immersed researcher, which implies vulnerability. The key themes of this article resonate with a question recently posed by Lerum, 'how do researchers' emotional engagements affect the kind of knowledge that is created?' (2001: 470). Since this is about my own recent research experiences, I acknowledge that the problem of moving from one research position to the other may be dismissed as a personal and idiosyncratic one, but accounts from doctoral students employing similar combinations of methods in field research indicate that this is a wider experience. The fact that I refer to extreme examples is not intended to glamorize the experiences. Rather the marked lack of researcher control at key moments in the situations described, points up the positional dichotomy issue I wish to address in a broader sense.

Evidence from Ethics Committee Approval

We can initially examine this dichotomy through an examination of the research ethics application in which I requested permission to conduct the original case study. The reader should note that permission to conduct the case study was granted, and the project proceeded as planned.

The 1999 ethics approval application form at my university required the applicant to state:

9. Scientific or Educational Aims and Scientific Value of Project
The applicant must immediately take up the available subject position of researcher within the discourse of scientific (controlled, emotionally-detached) research, just in responding to the question. Trying to frame an adequate response, I wondered about the ethnographic component. Embedded participant ethnography seeks to understand the meaning/making of members of a specific culture (for example, a remote African village, a workplace, an urban street gang and so on). Since the fieldwork context would not normally be considered as known in advance, or predictable, but something to be directly experienced by the researcher before any decoding or explanation is possible, the ultimate value of ethnographic enquiry cannot be known in advance.

Framing a response was therefore a source of tension. In the meantime, I stated the aims:

1. To identify to what extent 'culture', in the contemporary Balinese example, is being transformed through education, development, and political change, and to what extent new forms of identity for young people are being created through a synthesis of traditional and modernist discourses.
2. To investigate whether it is indeed the case that students in North Bali are coming to understand themselves primarily as individuals, or see themselves primarily as members of extended family, caste, clan or village.
3. To ascertain whether students, who were involved in recent political activism to support the cause of Megawati Soekarnoputri's PDI Perjuangan party, understand themselves as attempting to ensure a particular kind of imagined future, through their active support for her as a leader.
4. To chart student, community and employer dissatisfaction in North Bali with the Indonesian education system as it currently exists.

Suitable means of obtaining data which could fulfil these research aims were then stated:

11. Research Methodology

(a) Observations of student life and activities recorded in a daily journal of my teaching and involvement with the teachers' college and the attached demonstration school – this will not involve any identification of students or teachers. It will be a record of events without names or identifying labels.
(b) Fieldwork notes on student conversations, disputes and resolutions. Also field notes on any student demonstrations or political rallies. Once again, this will not involve any identification of the actual people involved.
(c) Surveys and selected interviews with tertiary students, teachers, parents and local employers (165 subjects).
(d) Focus group interviews with political activists (5 groups of 5 persons).

The application form demanded the submission of all documentation relating to data collection items (c) and (d) above, but (a) and (b), the ethnographic components, elicited no such demands. This differential emphasis indicates the dichotomy between researcher subject positions I wish to consider in this article. De Laine (2000: 21), with reference to Denzin (1997), argues that the researcher considers ethical issues of proposed ethnographic research within a dualistic configuration of moral concerns ranging from lurking positivism to overtly feminist concerns, and this is why fieldworkers are confused and anxious. However, this has not been my experience.

Nevertheless, in supplying the necessary documentation for (c) and (d) above, the researcher is certainly constituted within the recognizable 'scientific' discourse of doing research on humans. Yet (a) and (b) also refer explicitly to research on humans, some of whom the researcher will obviously get to know well in the course of prolonged participant observation (which was specified as four months elsewhere in the document). So what exactly constitutes the subject position of fieldworker researcher if he or she is not required to give much indication at all of what they will be doing? This silence is not just a fault in the form. There has been no publicly available 'rhetoric of method' (Marcus, 2001: 527) for fieldwork practice (see also Anderson, 1989: 263; Hammersley, 1998: 18). Perhaps this suits some researchers well. Nevertheless, the prior demand that the scientific value of the project be demonstrated is a powerful discourse, and one within which all proposed data collection practices obviously need to be considered. For this researcher it meant filling the empty semantic data collection space of (a) and (b) with sufficient ethics and rigour (or anxiety about these things) to ensure that the whole project would eventually turn out to have some 'scientific value' in the end, such as 'reflexive' techniques (Anderson, 1989: 254 – see discussion below), and focusing the ethnographic data collection very tightly on the aims of research. However, tight focus in participant observation carries the obvious danger of only noting down what you consider to be relevant at the time, which carries the risk of missing something which may later prove relevant.

Furthermore, set up to deal primarily with bio-medical experimental research, ethics committees require applications from researchers to sound as though all research subjects will be treated dispassionately and objectively while adhering to detailed protocols (Madjar and Higgins, 1996: 131). Implicitly this applies to the full range of methods used, including any ethnographic component. In this way, ethics committees manage to convey the expectation that, if approval to conduct research is granted, objectivity and impartiality will prevail in gathering data from human subjects. These concerns imply high levels of control over the research context and over the degree of researcher emotional involvement in the research setting (Willson, 1995: 255). However, while the researcher may have a great, if perhaps illusionary, sense of control, certainty of practice and distance when using formal methods such as (c) and (d) above, the same cannot be said of ethnographic immersion to collect participant observation data. Popkewitz (1984) argues that immersed ethnographic research has always treated the social world as in 'flux', teaming with complexity, contradiction and human agency (p. 50). This research 'world' is not an orderly place and the humans in it are not orderly either. Far from being fixed, the ethnographic human research subject resembles the poststructuralist subject, imagined as constantly in process. Davies (1997) claims the human subject only exists 'as process' (p. 274).

While Haraway claims that human subjectivity is 'partial in all its guises . . . always constructed and stitched together imperfectly' (1991: 193). If this is so for research subjects in ethnographic research, then the researcher as human subject is similarly in flux, dealing constantly with shifting realities and contradictions.

With such slippery human subjects as its epistemological basis, there are relatively few 'scientific' rules and guidelines in ethnographic research, except for the researcher to find willing participants, record as much data as possible while trying not to affect the 'natural' setting, and to stay immersed for a reasonably long period of time (Hammersley, 1998: 8). Beyond this advice, it has been claimed that:

> There is as yet no alternative modality of method or articulation of a set of regulative ideals governing fieldwork that gives professional legitimacy to what is in fact happening to fieldwork. (Marcus, 2001: 527)

Yet reflexivity in ethnographic research is often cited as a means to achieve a methodological rigour which at least parallels the 'scientific' rigour of quantitative social science research. However, there did not seem to be anywhere to insert reflexive considerations and practices in the ethics application form. Although a term subject to various interpretations, reflexivity refers on the one hand to the cultural attitude and approach of the ethnographic researcher, and on the other hand to certain practices, derived from the latter, which achieve rigour in the fieldwork process. The appropriate reflexive approach for an outsider entering a fieldwork situation is the constant acknowledgment of the social and cultural position of the researcher in relation to locals, for example, his or her relative privilege, ingrained cultural assumptions, language and communication problems, tendency to impose simplistic theoretical interpretations and so on. This intensive 'locating' of the researcher relative to the local scene and the data collected is usually realized in the form of a reflexive diary or journal in which all kinds of personal jottings are recorded. The purpose is to remind, and clarify for the researcher, exactly what kind of subject position he or she occupies at all stages of the fieldwork – to record the viewing of oneself (and the data collection) in a 'growing variety of local mirrors' (Herzfeld, 2001: 260). Yet despite the comforting reassurance of rigorous daily reflections and journal-keeping, reflexive practice did not soothe my preoccupation with considerations of 'scientific' control. In part this is because reflexivity belongs far more to the tradition of anthropological research than to modes of sociological enquiry.

Ethnography – participant observation – is the traditional method of anthropology (see for example, Eriksen, 1995; Malinowski, 1961), while surveys, interviews and focus groups as research methods prevail in sociology, education, human geography, economics, psychology and market research.

If, as in my case, the researcher is not trained as an anthropologist, but was apprenticed within the latter set of social science disciplines, then the push for 'scientific' rigour is almost inescapable. The practice of fieldwork – while relatively conventional and assumed for anthropologists – may constitute an unusual or even a political act for those from other disciplines (Wolf, 1996: 8). As a sociologist, I undertake fieldwork because I have found it to yield exceptionally rich and politically significant data, yet the sociological researcher subject position remains contingent upon practices which embody 'objectivity, impartiality and disinterested observation' (Madjar and Higgins, 1996: 131), and I find this troubling for the practice of immersion. Hammersley (1998) maintains that within sociology the question still remains as to whether ethnography is scientific (p. 6).

The implication of 'scientific' control in fieldwork continues to be exemplified in published social science research accounts. Most research accounts in journals within the field of sociology analyse ethnographic data sets as if they were calmly gathered in a rational and orderly context. Indeed, I do it myself. The implicit concepts of experiencing the world are set up in such accounts so as to make certain things disappear – confusion, threats, danger, the unpredictable, the non-event. A universality of control, objectivity, things going according to plan is conveyed in neutral, academic prose – 'a measured intellectual style' (Van Maanen, 1988: 47). As Lerum points out in relationship to sociology, 'at the core of this objective ideology is the rule of *emotional detachment*' (2001: 472, emphasis in original).

In this article I want to explore what happens when things do go wrong and the idea of *emotional detachment* is a bad joke. The first example is a sudden civil emergency – a riot – which Lee (1995: 3) would describe as *situational* danger – unexpected crisis. The second describes an event when the researcher enters a local nightclub where drugs and sex are sold and violence threatens. Lee's term for this kind of danger is *ambient* (1995: 3) – goes with the setting. In both situations, gender increased the danger (see Warren, 1988). A middle-aged female researcher is trying to collect data in an environment dominated by potentially violent young men. Yet one does not seek to over-dramatize these events. Obviously I survived to tell the tale. The value of the examples lies in their utility to examine some pivotal qualitative research methodology questions which pertain to control and anxiety about the lack of it:

1. To what extent can ethnographic fieldwork data collected in dangerous and chaotic settings constitute orderly and coherent accounts of the phenomena?
2. What kinds of ethical and methodological issues emerge from the practice of a researcher collecting fieldwork data in dangerous and chaotic situations, which are, nonetheless, relevant to the major research question?

The Research Context

It might surprise the reader to learn that things got dangerous and unpredictable in Bali, one of the most popular tourist destinations in the world and generally counted as very safe. Yet beyond the tourist enclaves Bali seethes with many of the same kinds of political resentments as elsewhere in the archipelago, and there is a thriving underworld trade in sex and drugs which does not exist just for the overseas tourist market. My research project was entitled 'Young People, Education, Politics and Radical Change in North Bali'. Bali seemed to offer an Indonesian context in which identity frames of tradition and modernity were more or less equally available as cultural resources to be taken up within a modern notion of the self by young people. The research was carried out over four months in Singaraja, North Bali, a large provincial town in economic decline. Two important sources of local revenue were the teachers' college (where I taught English and research methods) and the tourist enclave of Lovina, about ten kilometres away on the coast.

I arrived in August 1999 and spent two months conducting both the large scale survey and ethnographic research. I lived alone in a guest house on a main street near the college and close to the centre of Singaraja. The teaching was enjoyable, but everything else was extremely difficult, including getting out of town. This could only be achieved during daylight hours and involved three transfers between *bemos*, small, rusting, cramped vehicles which often broke down. Responses given on ethics applications tend to assume the target research setting can be relatively easily entered and exited by the researcher from his/her privileged position of professionally trained outsider/observer, even during the ethnographic component of research. We might summarize this as the unproblematized belief that privilege confers remoteness from consequences. However, any assumption of this kind is confounded by the methodological imperative to live simply in the local community in order to be 'in a far better position to be really in touch with the natives than any other white resident', as Malinowski (1961: 18) puts it. The ethnographic emphasis on investigating the subjective dimensions of human experience, and the contextual nature of those experiences, demand researcher immersion. And, logically, the immersed researcher is likely to suffer the same consequences as the local inhabitants or workers if anything goes wrong. Ethics committees are concerned about danger and risk, but not in reference to the researcher. Madjar and Higgins (1996) note that ethics committees tend to focus on the 'worst case scenarios' and 'what if' situations (p. 130), but this is only about participants, not researchers (see Lee, 1995). I had been in the field for about six weeks when East Timor went to the polls. Australia and Indonesia moved rapidly into a foreign relations crisis and local anti-Australian feeling was running high. I was verbally harassed in the street, which I duly recorded (keeping control). I felt extremely isolated, anxious and stressed (which I noted in my reflexive journal). None of this seemed very 'scientific', just personal and subjective.

By October 1999, when local political unrest began, I had three exercise books full of participant observation fieldnotes, interspersed with reflexive journal entries. I had become familiar with the student leaders of local political groups (whom I intended to interview in focus groups), and had gained some idea of regional Indonesian politics from the inside. At a deeper level I had become aware of the strong resentment of Balinese Hindus (a minority) towards the Muslim majority in the country. Bitter feelings against the political hegemony of Java embodied by Suharto and his cronies (and the military), and fear of a rising tide of Islamic fundamentalism actively fuelled the almost fanatical support by Balinese for the political party PDI-Perjuangan (Indonesian Democratic Party of Struggle) (Surya, 1999), and for its leader, Megawati Sukarnoputri.

The Riots

1999 was a landmark year in the growth of Indonesian democracy. Following the fall of Suharto in the previous year, early 1999 saw the first open democratic elections. PDI-Perjuangan, secured the most primary votes and a parliament was assembled in Jakarta. On 20th October 1999, election for President of the Republic was held among the members of parliament. It was widely expected that Megawati Sukarnoputri, as leader of the majority party, would take the presidency. However, this was not the case. Instead, Abdurrahman Wahid, known as Gus Dur, the moderate head of the Islamic party NU, was elected president by parliamentary vote. In reaction there were riots by PDI Perjuangan supporters in many parts of the country. Commentators afterwards agreed that the worst riots and property destruction took place in Bali (*Nusa Tenggara*, 1999: 1). In Singaraja some damage was done in small attacks during the night of the 20th, but the well-organized mass riots began early on the morning of the 21st, which was afterwards always referred to as *Kamis Kelabu* – Black Thursday (*Nusa Tenggara*, 1999: 3).

On the morning of the 21st October I was scheduled to give a lecture commencing at 7.00 am at the teachers' college. The campus was curiously deserted when I arrived, and only about 20 students came for the lecture. About 15 minutes after I began speaking I received a call on my cellular phone from one of my co-researchers, warning of violent riots up near the *kantor bupati* (mayoral office), about three kilometres away. He told me to go home as quickly as possible and stay inside. Having warned and dismissed the students I ran outside to the street but all public transport had vanished and I had to walk and run with several of my female students, up the long street to our houses. As we ran past the closed and shuttered, silent houses and government offices, a series of loud explosions shook the ground and two large plumes of smoke streamed up a hundred metres into the sky from the direction of the mayoral office. The roar of a thousand voices could be heard. All three of the students with me began to cry. I felt an intense sense of foreboding, but asked the students what they were feeling (researcher in control and getting field data). '*Takut!*' (frightened) they whispered, as they ran into their houses. Obvious really.

Over the next eight hours, nearly all the public buildings in Singaraja were burnt out, as well as many homes of government officials. Seven of them were in the street near my home or in the street behind. 'The mob just went by. They have set huge burning pyres and hurled explosive missiles at the Department of Justice, and all the other government departments in my street' (Fieldnotes, 21/10/99). Around 20 cars of government officials were torched, as well as innumerable motorbikes of family members. The town centre where I lived was blockaded on all sides. No-one could get out. Power and telephone lines were cut, including mobile phone networks. The mob occupied the central business district, throwing fire-bombs, cutting down trees, throwing rocks, bottles, shouting slogans and beating on fences with sticks and iron bars. The smoke and noise were intense, especially during the two-hour period when my street was under attack. For most of this time I was lying on the floor of my house with a wet handkerchief over my mouth and nose, eyes streaming from the toxic smoke, in momentary expectation of a fire bomb coming through the front window, since mine was a typical government resident's house. There was nothing to do but endure. '*Emotional detachment*' (Lerum, 2001: 472) was entirely lost.

The kind of fear you experience at such times is hard to describe. I thought about my children, partner and family back in Australia and wondered if I would ever see them again. However, I was also drawn, in the subject position of researcher, to peep through the curtains at the front window, despite the danger, whenever the mob raged by, to see whether there were faces I knew in the crowd. There were familiar faces, and I jotted their details down, while gasping for breath – an action which now seems ridiculous. I even considered, once or twice, running out to the front gate and trying to ask some questions of the young rioters. But instead I filled pages and pages of a notepad with commentary – sure that I should be both gathering data and practising reflexivity at this historically and politically significant juncture. Two of the entries made early in the riot period read: 'I sneaked to the fence and asked my neighbour where the police were. She replied '*pulang*' (gone home). Similarly it appears that the local army contingent has locked itself up in barracks', and 'Lots of the neighbours are out on the footpath watching. They express great solidarity with the demonstrators' (Fieldnotes, 21/10/99). However, after this the fire-bombers arrived and I did not dare venture out.

As mentioned, I wrote pages of commentary during the riot experience. The journal comments can be best understood like this:

ME
FAMILY
NEIGHBOURHOOD, COLLEAGUES, STUDENTS
POLITICAL SITUATION IN BALI
POLITICAL SITUATION IN INDONESIA
INDONESIA/AUSTRALIA RELATIONS (EAST TIMOR)
GLOBAL ORDER (RESENTMENT OF DEVELOPING WORLD CITIZENS TOWARDS DEVELOPED WORLD CITIZENS)

The hierarchy in this notebook schema is of arenas of fear and concern, moving downward from the intensely personal level to the global level. So, first I was fearful for my own safety – an older woman, alone and unprotected in a flimsy house, with a mob of a thousand or so violently angry young men outside. Second, I was intensely aware of my family back in Australia and my responsibilities and feelings for them. Beyond that, I was concerned for my neighbours, my colleagues, the mayor (whom I had interviewed two days before) and above all my students, including those involved in the riots, those cowering at home, and the Muslim students who surely felt deeply threatened. Beyond that, I was concerned that the political situation in Bali had spiralled out of control and the local economy would suffer drastically in lost tourism revenue, which would impact adversely on people I knew. In the national context, I worried that the level of local civil unrest gave the clue to the long-term viability of Indonesia's fledgling democracy, and the country might lurch back into a 'balkanized' state characterized by small warring enclaves and ethno-religious primordialism. At the regional level, I was well aware that Australian troops were in East Timor, heading up the United Nations post-election peace-keeping force. Massive Indonesian resentment was simmering about this, fanned by media and political rhetoric in both countries. I had already been verbally harassed in the local neighbourhood as an Australian visitor and told to go home. It was not unthinkable that someone would choose to target the local Australian in the rage of riot fever. Finally, I was aware that some level of resentment is always present when a westerner from an affluent country is living in any developing country, no matter what the philanthropic orientations of specific individuals might be.

The schema above is an example of the self-reflexivity techniques advocated by many critical ethnographers, and it was comforting at the time to have to consider the crisis with some level of objectivity. Yet, like Hunt (1989), after the harrowing incident in the field, it first appeared I had written down a lot of strikingly irrelevant information. The urge to write it all down as it happened was driven by the recurrent fieldwork question, what kind of 'scientific value' for the project overall will this data prove to have? Every moment was supposed to count, so it had to be recorded. Yet there was nothing to record but explosions, smoke, shouting, and the thoughts of the researcher. In retrospect though, the reflexivity exercised in the act of writing was in itself valuable because it signified the researcher as a cultural presence outside, and yet within, the social and cultural context of a street riot in the developing world. So while it was not instantly utilizable data, it was a valuable reflexive moment because it shone a brilliant spotlight upon the epistemological problem of shifting between researcher subject positions. The writing made explicit the contrast between pitifully huddling alone on the floor, choking on smoke, (the ethnographic experience) and sitting down in comfortable surroundings to interview someone according to a carefully planned set of questions (formal qualitative method). If I had not chosen to live locally and conduct participant observation as a component of the project I would have been safe in a heavily-guarded hotel in the tourist enclave.

The rioting and property destruction went on until late afternoon, when the power mysteriously came back on. The rioters all vanished to watch the vote for Vice-President, broadcast live from Jakarta. Megawati Sukarnoputri was elected Vice-President, and there was no more civil unrest. Of course it was after the immediate danger had passed that some excellent data collection prospects arose, because I was then theoretically able to move around the neighbourhood asking people how they felt. However, I asked no questions and wrote down no observations. I had rushed over to my neighbour's house and was glued to the television like everyone else, watching the voting count, holding the hand of my elderly neighbour who was sobbing and repeating 'Bu Mega' (Mother Megawati). Her husband, and dozens of relatives and neighbours, were all quite affected emotionally as we watched the outcome together. We were all in the same situation. If Megawati lost the vote for Vice-President, there was no telling what the rioting mobs might do.

This was a subject position far from the idealized position of neutral participant observer. However, it was a moment of deep rapport, problematic as that term is currently (see Marcus, 2001). Shared extreme experiences are psychologically bonding for those involved, as Geertz (1983) found so many years ago, and allow unique researcher insights. So, to return to my first question regarding this context of data collection; to what extent can fieldwork data collected in dangerous and chaotic settings be formed into orderly, coherent accounts of the phenomena? This article is the only attempt I have made to form the journal data into a navigable account of the phenomena as it actually happened. There is an article written around the theoretical notion of embodiment in which I include some of the empirical material (Nilan, 2001), but it deals with the manipulation of the young people in the rioting mob by political agitators, information which I did not piece together until later that week. So there are two answers to the question. The facts and explanations that allow a researcher to make sense of chaotic and dangerous research situations often are not available at the time, but have to be collected later. Second, the emotional experience of the researcher in chaotic and dangerous research situations is such a large part of the data that one cannot rely on it alone to create an orderly and coherent account of the phenomena.

If we turn to the second question of the ethical and methodological issues which emerge from the phenomenon of a researcher collecting data in dangerous and chaotic situations, then the duty of care of the researcher towards participants can be singled out as a most important ethical issue, although personal safety is also an issue. Clearly it is not appropriate, as I indicated above, to ask participants silly questions when there is a crisis going on. It does seem ethically appropriate to offer sympathy and consolation to informants and associates who have suffered in the crisis, such as my weeping neighbour, or her husband, whose consultancy firm had lost everything (few Indonesian businesses have any insurance), or my Muslim students, whom I telephoned later to see if they were safe. The methodological issue is more complex. As I have shown by example, 'data' collected in chaotic and threatening circumstances not only lacks objectivity and neutrality, if these are indeed always desirable qualities, but often lacks factual content, not to mention coherence. The notes written by this researcher were not about the people I was supposed to be studying, but were mostly about me and how I was feeling. I concur with Probyn (1993) that reflexive writings are often self-centred and ethnographically sterile (p. 80). On their own, they serve the purpose of anchoring the researcher in a reflexive sense, while he/she is dealing with 'post-facto' explanatory data. Nevertheless 'crisis' data is packed with emotional immediacy, and this should dissuade any researcher from dismissing it out of hand (see Van Maanen, 1988). As Carspecken asks, 'how does an outsider gain an insider's view?' (1996: 17). In this instance, the shared experience allowed this researcher to sympathize more fully with the fears and anxieties of locals in the crisis, even though the two sets of responses could not be imagined as exactly congruent due to historical and cultural differences. Finally, it became clear to me after the riot that it was important I had witnessed these dangerous events myself from the same vulnerable position as my neighbours. Local people, and even some of the rioters themselves, were very prepared to talk to me openly about the riot. After all, I had been through it too, and they knew that. So, although I certainly experienced a sense of 'loss of control' at the time of greatest danger, this was an aspect of the situation that was more or less true of everyone who experienced the riot. The real research 'value' of the experience came afterwards, when I was able to achieve high levels of rapport with locals and obtain frank and detailed data about the events from all kinds of points of view.

The Nightclub

The second example of trying to collect data in difficult and risky (rather than dangerous and chaotic) circumstances was the result of a decision on the part of the researcher, unlike the previous example. In the course of collecting background ethnographic data about groups of young people in Singaraja, I was told that some of them frequented a nightclub with a reputation for commercial sex and drugs (see *Bali Post*, 1999a; Fajans et al., 1995) in the tourist enclave about ten kilometres away. Since my research project (see above) implied the assembly of information about the available meaning/identity frames of the young people in the local area, I determined to go to the nightclub after midnight and observe. I was aware, generally speaking, of the potential dangers in fieldwork on illegal drug use (Lee, 1995: 39), but I did not anticipate any problems in Bali.

Although the neon sign outside was impressive, the California Club was in poor repair and did not improve during the time of my observations (I visited three times). There was a dance floor in the middle and a raised platform where the band was warming up. To the left were two large pool rooms, dimly-lit and half-partitioned. To my surprise, drug dealing was going on quite openly in these spaces. Tables and chairs stood around the brightly-lit dance floor. A few tables were occupied by young (and some very young) prostitutes of both sexes, and 'guides' (see Lette, 1996) – who were not actually selling sex, but were looking for romance with a tourist. They were carefully dressed in western clothes, with styled hair and make-up. On both sides, at tables more dimly lit, were western tourists and backpackers, eyeing the colourful group near the dance floor. Everybody was waiting for the band to begin playing. I sat down at the bar. With little idea of how to get information, I started talking to one of the barmen about my purpose. Warmed by my use of the local language, he offered me some advice – pay one of the older 'guides' to act as bodyguard and informant. The place was dangerous and I might get robbed or worse. He knew more than I did in his acknowledgment of the dangers to personal security which arise for researchers in the drug subculture' (Lee, 1995: 40). While I was still thinking about his advice, a fight broke out among the drug dealers. At the same time I was approached by a woman offering in English to sell me *shabu-shabu* (crystal methylamphetamine). When I refused in Indonesian she glared and walked back to her friends, who all stared and muttered. Distinct qualms about my own personal safety started to register (Van Maanen, 1988: 86). I realized that it was going to be difficult, if not impossible in this potentially violent context to collect data if I was taken for a spy or a possible robbery target (Lee, 1995). It would be much more productive for data collection if I could 'pass', look like someone who had the usual reasons for being there. There was an older Australian man sitting at the bar. Earlier I had greeted him and he told me to 'f – off'. Even drunker than before, he was now calling out and trying to touch the very young *cowok jegeg* (pretty boys) who were flirting with a much older German woman. A glum-looking young Balinese man was sitting next to the older Australian. I leaned over and quietly asked him to explain what was going on (in Indonesian). He told me that he had an arrangement to go home eventually with the Australian but he didn't want to. He said that he much preferred romance with tourist women but that since he was no longer so young, and a heavy drug user, it was easier to find old men who wanted to pay him for sex. How old was he? Twenty. I explained my purpose in being there and offered him 20 dollars to walk around with me and explain things, a practice which is not uncommon in some kinds of fieldwork (see Lee, 1995: 41). He readily agreed. The older Australian was not pleased. He threatened to punch me but he was too drunk.

It was remarkable that as soon as I paired up with Wayan (not his real name), I felt much less conspicuous, and much more in control of the data collection possibilities. It was an uncomfortable thought that this sense of 'control' came primarily from my financial control over Wayan. However, the issue of passing was very relevant. There were quite a few older western women paired up with young 'guides' (Jennaway, 1996, see also Bagus, 1997 and Lette, 1996), which seems to be an accepted pattern in Bali, so I did not stand out in the crowd. Wayan proved most helpful. As a local lad and long-time guide, he knew almost everyone and their local history. He helped me recognize the nightclub situation in a 'culturally typified' way (Carspecken, 1996: 19) which I could not do alone. For example, he confirmed that most of the young men and women offering sex were locals from poor backgrounds like himself, as were those selling drugs in the poolrooms. However, the group of very young *bencong* (transvestite men) (Oetomo, 1996) who had appeared earlier were from Denpasar, the capital of Bali. Wayan also assured me that the '*boss-boss*' (pimps and drug distributors) who managed the local trade in sex and drugs were from Java or Madura. It may have been so, since most North Balinese are poor and would probably lack the resources to make the necessary capital investment. The California Club venue was reportedly run by a local man, but owned by a consortium of Denpasar businessmen.

Wayan was keen to show me as much as possible about the sex and drugs trade. He pointed out that the tourist 'action' was at the front near the band and the dance floor, while the local 'action' was going on out the back. Far in the right hand back corner was a flashing sign – DiscoHouse – above swinging doors. In the far left hand corner was a door which lead out to the toilets through an untidy courtyard garden. It had not been obvious to me before, but those areas were busy with the local 'action' as Wayan called it. It was at this point in the evening, when I moved from the tourist domain where I 'belonged', to the local domain backstage, as it were, that the data collection context became most difficult. While tourists are catered for, much of the sex trade in venues like the California Club involves local male clients. An article in the *Bali Post* in 1999 reported that most of the customers in a Buleleng (North Bali) *Seks Kafe* in the eastern part of Lovina were local 'civil servants or members of the military'. Young female sex workers were charging them 25,000 rupiah for a 'short time' (*Bali Post*, 1999b).

Few tourists ever seemed to make their way through the gloom at the back of the California Club into DiscoHouse, which was the local scene. I noticed it was one-way traffic. Those who went in did not usually re-appear. Wayan explained; those who went into DiscoHouse continued out beyond the dance space into the courtyard outside, then went upstairs via a back staircase. We went through and stood at the bottom of the staircase. He pointed upstairs and said '*tempat tidur*' (beds), then started to ascend, urging me to follow. It was not a comfortable situation. I declined (keeping control?). As Lerum puts it, 'the study of sex work is a politically and emotionally tricky endeavour' (2001: 468). This was particularly so for me as a woman. I felt a strong sense of risk and distaste at the thought of being a sexual 'voyeur' (Kulick, 1995: 9). Yet afterwards I thought perhaps I should have gone up the staircase. The scientifically rigorous pursuit of ethnographic data probably demanded it.

We walked around to the area outside the toilets, talking about the local drug trade. It appeared magic mushrooms, marijuana, ecstasy and shabu-shabu were readily obtainable. Heroin was also apparently available from the right source. Wayan discreetly pointed out some groups of young men, glimpsed only dimly in the bushes, who were injecting drugs, using shared needles. Did I want to meet them? No. I was embarrassed by both of Wayan's invitations. It did not seem respectful, sensitive to context, or indeed ethically sound, to go in and visually observe young people doing high-risk things they usually did in relative privacy. Besides, I did not want to draw too much attention to myself in such a high risk situation. Yet what important data might I be missing through such evasion? Moments like these convinced me that, not only did I find foreign culture fieldwork uncomfortable, but that I belonged less to the 'Indiana Jones' class of ethnographer, and more to the 'timid and retiring' class (Lee, 1995: vii). I fled to the women's toilets, only to find myself unwittingly listening in (see Humphreys, 1970) to two young local women arguing over money they had just obtained for a threesome with an Indonesian client. One wanted equal shares but the other maintained that she had done more in the encounter and deserved a higher proportion of the payment. The other flounced off with her lesser share. Just a little while later she left by the front door on the arm of a German backpacker. While this provided some data about the prevalence of multiple sexual partners, it was opportunistically obtained, in the sense that the girls probably assumed that, as a foreigner, I wouldn't be able to understand their conversation. It also worried me that these local girls were so young, and taking such high risks in sex work. I shied away, however, from getting involved. I was concerned that I would, as it were, blow my cover (Lee, 1995), and draw attention to myself, which would compromise my ability to freely observe.

Some time was then spent near the dance floor, observing the interaction between tourists and local young people. Once again, Wayan was able to confirm what I thought I was observing. Relations between young female 'guides' and tourist women are not mentioned much at all in accounts of tourist phenomena in the region, but it was happening in the California Club:

> Two Indonesian girls came in together. Late. One was very pretty, with long curly hair and wore a dress. The other had very short hair and was frankly butch. The pretty one in the dress was dirty dancing centre-stage with a very butch Danish lesbian I had been talking to earlier. They were wrapped around each other. Her '*cowok*' girlfriend seemed to be trying to chat up some older French women. (Fieldnotes, 23/10/99)

These girls were described to me by Wayan as '*cowok*' (the boy) and '*cewek*' (the girl) respectively. Blackwood (1995) concurs that these labels are commonly applied to butch-femme lesbian partners in Indonesia (see also Gayatri, 1993). Wayan confirmed that there was quite a lesbian 'scene' operating in the local tourist enclave, which survived on sex and romance with German and Dutch female tourists who 'knew' where to find it. Murray (1995) maintains that in Java open lesbianism is generally found only in lower class butch-femme subculture. Furthermore, she notes that,

> Lower class lesbian subcultures are . . . part of an urban milieu which includes elements such as crime and prostitution, and my observations are that a disproportionate number of lesbians are involved in the sex industry. (p. 23)

I include these references and quotes to indicate the relative accuracy of the data I collected in the California Club on this, and on other occasions, with the assistance of Wayan. There is no doubt that engaging him as a paid informant enabled the collection of significant data which I have written about elsewhere (Nilan, 2001), as well as endowing the data collection process with a satisfying sense of control. I mention this specifically because it is now appropriate to consider again the first of the questions which guide this article: to what extent can fieldwork data collected in difficult and risky settings be formed into orderly, coherent accounts of the phenomena?

First, in both the examples I have used in this article, the data I ultimately obtained was the most politically significant and volatile of that I gathered across the project as a whole. This is an important point as Lee (1995) points out. Often difficult and dangerous fieldwork is worth the experience because of the highly politically potent nature of the data, which is almost impossible to collect by any other means. Second, as I think I have demonstrated here, there are some difficult and risky fieldwork settings in which it is improbable that the researcher who is very much an outsider will be able to collect useful data. In the setting of the California Club, had I not passed as a client and engaged a paid informant, I would have been left conjecturing about almost everything. There are a lot of other data collected in the California Club, some of it relevant to the groups I was studying back in Singaraja, which I would never have obtained but for my paid informant. Because I did have access to an insider view during my observations in the venue, it has been relatively easy to form that data into orderly, coherent accounts. Had that not been the case, the data would have been far more disorderly, personally oriented and speculative in nature. However, it is notable that I did not mention the use of paid informants in my application to the ethics committee, and my practice in this regard was not pre-approved. Like many other qualitative researchers I was 'personally challenged' in trying to adhere to granted research protocols (Madjar and Higgins, 1996: 133).

This leads, obviously, to the second question in relation to this data: What ethical and methodological issues emerge from the phenomenon of a researcher collecting fieldwork data in difficult and risky situations? The first thing to note is that the collection of data in challenging participant observation situations demands considerable ingenuity on the part of the researcher (Lee, 1995). Ingenuity practised at the time can obviously lead to a range of ethical and methodological issues to be considered (and even suffered, see below) after the event. Another matter to settle here is whether the research questions justified the collection of data in such a venue using a paid informant. There is no doubt in my mind that they did, given the valuable data I was able to collect on the range of possible identity/meaning frames available to local young people, as well as the cultural constitution of health risks. However, the moral dilemma of having engaged a young male sex-worker as my paid informant remains, as does the question of whether it is ethical practice to merely observe young people engaged in criminal and high-risk behaviour without warning them in any way, or notifying anyone about it. Or, indeed, whether it is ethical to eavesdrop on other people's private conversations, without letting them know you can understand what they are saying. These are demanding ethical questions, but I argue that there would be few qualitative researchers who do not come up against similar dilemmas, since collecting ethnographic data never takes place in an orderly context in which all ethical issues can be foreseen, and guaranteed through specific protocols. Furthermore, as Lee advises, ethnographers studying drug scenes generally avoid drawing any official attention to local people or settings (1995: 47). Inciardi makes the point that a researcher's ability to intervene directly can be quite strictly limited (1993: 155). As described below, there were negative social consequences to my fieldwork at the California Club at a local level, and these would have been even more dire if I had chosen to draw any official attention to the high-risk behaviour of the young people involved.

I went back to the nightclub two weeks later and engaged Wayan again, then a week later as well. I do not consider that any damage was done by Wayan acting as a paid informant for me in the venue, and this is probably the major ethical consideration. Marcus (2001) acknowledges that the ethnographer routinely establishes these kinds of expedient fieldwork links as working partnerships for which no 'regulative ideal or modality' exists in documented fieldwork practice (p. 524). Unfortunately however, while Wayan's local reputation presumably remained intact, mine did not. Gossip went around Singaraja that I had a young 'boyfriend'. News travels very fast in Bali and a whole range of local people (both men and women) became suspicious of my motives in the research and were less than co-operative with me after that. I did explain myself most heartily, but I feel I was not believed by a lot of people. I keenly felt their disappointment at my apparent sexual misconduct (Kulick, 1995: 7), which was constituted within a complicated gendered discourse of the older academic western woman who, though foreign, had seemed to fulfil the ideal of the married 'modest woman' (Parameswaran, 2001: 82; see also Warren, 1988). Methodologically then, I solved one data collection problem, only to create another. Fortunately, this did not happen until just before I was due to leave for home. It is ironic that a distinctly non-rapport relationship compromised a range of carefully negotiated 'rapport' relationships. This example of rapport failure certainly bears out Coffey's (1999) claim that the 'self' of the ethnographer has an effect on every aspect of the research process (p. 6), especially the gendered aspect of the self (Kelsky, 2001; Sherif, 2001; Venkateswar 2001). It also demonstrates how the expediency of problem-solving in the difficult and risky participant observation context can have unforeseen consequences. Reaching for greater control in one way may lead to an unexpected loss of it elsewhere. As Lerum (2001) observes, engaging fully 'in the moment' with one's subjects '*often ruins one's sense of control*' (p. 481, my emphasis).

Conclusion: The Issue of Writing Accounts from Data Collected during Difficult Circumstances

In conclusion it must be stated again that, although composed of difficult and unsettling experiences, the ethnographic fieldwork described above provided immensely rich and valuable, as well as politically significant, sources of data. However, this potential could only be realized in the production of appropriate research accounts. As indicated earlier, there is very little difficulty in writing up accounts from data collected by formal methods within qualitative research. The data can easily be reduced to summary, and various analytical procedures, for example, descriptive statistics and content analysis can be carried out if needed. The subject position of writer, similarly to the subject position of the 'scientific' researcher, is constituted through the production of texts which speak with the voice of 'objectivity' (Lerum, 2001: 472), 'rationality' (Denzin, 1997: iv) and 'certainty' (Goodman, 1998: 63). However, as argued previously, data collected through ethnographic fieldwork is notoriously messy and chaotic. So, to what extent can data collected in dangerous and chaotic settings constitute orderly and coherent accounts of the phenomena which speak with a 'voice' which will be found acceptable within the academy?

The problem is that, if we put aside the accepted 'scientific' criteria for acceptable research accounts which conform to a quantitative research paradigm, then the criteria by which thesis examiners, journal editors and book publishers judge qualitative research accounts have to be taken into account. So, if it had not occurred before, then certainly at the stage of writing up data and publishing research accounts, the precept of researcher 'control' tends to blur across into the analysis and interpretation of ethnographic data.

Detachment and distancing from human informants secures a safe jurisdiction for academic work, and this protective boundary successfully facilitates the production of a specific brand of knowledge' (Lerum, 2001: 475; see also Willson, 1995: 255).

In a competitive publishing environment, fieldwork researchers who wish to see their work reviewed seriously need to put a satisfyingly neat presentation of apparently empirically valid 'facts' about local phenomena out into the public arena, no matter how poorly this may reflect the context of data collection (see Van Maanen, 1988). Any attempt to do otherwise implies the troubled power relations that underlie production of academic knowledge (Parameswaran, 2001: 93) and usually results in rejection. This dilemma is routinely experienced by anthropologists:

> The tidy, systematic and well-rounded texts written by anthropologists are more often than not the end-product of long periods in the field characterized by boredom, illness, personal privations, disappointments and frustration. (Eriksen, 1995: 15)

In my own successful publishing from the ethnographic component of the project so far, I have eliminated any reference to collecting data in difficult or chaotic circumstances, instead constituting the research account text primarily in terms of a particular theoretical 'take' on the data, which is expected within my discipline. Perhaps this can be reduced to a 'desire for mastery' over the data which is unlikely to be challenged (Alcoff, 1991: 22) by the academic review process. The texts certainly do not openly admit of any problems of representation (Lather, 1993: 674) relative to the data.

We might well then ask, what type of ethnographic subject is produced through data analysis, since it can be claimed that the very analytic texts we produce are themselves constitutive effects of discourse (see for example, Honan et al., 2000: 30). I would argue that the descriptive and analytic texts habitually written up from ethnographic fieldwork data constitute human research subjects which seem to be stable, orderly, rational, consistent characters in a coherent and compelling narrative. This is so because of the necessity to write within the acceptable discourses of academic prose and disciplinary genre. The human research subject 'constantly in process' (Davies, 1997: 274) – only existing in process, complexity and contradiction – is almost impossible to depict accurately (but see Honan et al., 2000). So what of the researcher/writer constituted as human subject in descriptive and analytic texts habitually written up from ethnographic fieldwork data? It might be just possible to depict research participants as constantly in process, complex and contradictory, but this is a luxury the writer/researcher cannot afford with regard to their own authorial voice in the text. Although as Goodman (1998) says, even though we know nothing 'for sure', we have to give the impression of 'certainty' (p. 63).

Turning to the directly practical matter of writing accounts, I suggest the strategy for the writer/researcher is not to identify the most emotionally intense moments in fieldwork research as the pre-eminent moments of data collection which must somehow find their way into the eventual textual accounts. It is obvious that such events stick in the memory long after more mundane experiences have faded, yet it is often impossible to collect really informative and useful data at such times. As I have tried to illustrate through the examples in this article, it is often data collected after the event or in quieter moments which turn out to be the most evocative, in the end. In the example of the riots, it was really everything I found out afterwards about the events which produced the most satisfying explanatory account of the phenomenon. In the example of the California Club, while the first visit was emotionally challenging, subsequent visits with Wayan as informant yielded much more analytically rich data. This is the final argument for control as a significant component of the ethnographic researcher subject position. The researcher/writer needs to stay on top of the data, in the field and at the computer, in order to write respectful and well-informed accounts of people and their lives. In my discipline, like many others, there is a noted disdain for personal narratives in accounts (Kulick, 1995: 3). So while personally-oriented confessional pieces about difficult fieldwork (like this article) have their place, at their best ethnographic accounts permit the reader to experience some of the 'activity, creativity and human agency' (Willies, 1977: 3) of people and events in the field. Good ethnography creates meaning out of the fieldwork experience, in carefully crafted accounts which attempt to capture the richness and complexity of the original actors and events, while keeping within the accepted limits of reader comprehension and publishing criteria of the discipline.

References

Alcoff, L. (1991) 'The Problem of Speaking for Others', *Cultural Critique* 20(1): 5–32.

Anderson, G. (1989) 'Critical Ethnography in Education: Origins, Current Status, and New Directions', *Review of Educational Research* 59(3): 249–70.

Atkinson, P., Coffey, A. and Delamont, S. (2001) *Handbook of Ethnography*. London: Sage.

Bagus, M.I. (1997) 'Sows on Heat and Disoriented Boars – Straight Sex from Bali to Melbourne', *Melbourne Journal of Politics* 24: 69–93.

Bali Post, (1999a) 'Meningkat drastis penderita narkoba di Bali', 9/11/99, p. 3.

Bali Post, (1999b) 'Kafe di Buleleng dijadikan ajang bisnis seks', 14/10/99, p. 3.

Blackwood, E. (1995) 'Falling in Love with an-Other Lesbian', pp. 30–51 in D. Kulick and M. Wilson (eds) *Taboo: Sex, Identity and Erotic Subjectivity in Anthropological Fieldwork*. London & New York: Routledge.

Carspecken, P.F. (1996) *Critical Ethnography in Educational Research*. New York & London: Routledge.

Davies, B. (1997) 'The Subject of Poststructuralism: A Reply to Alison Jones', *Gender and Education* 9(3): 271–83.

De Laine, M. (2000) *Fieldwork Participation and Practice: Ethics and Dilemmas in Qualitative Research*. London: Sage.

Denzin, N. (1997) *Interpretive Ethnography: Ethnographic Practice for the 21st Century*. Thousand Oaks: Sage Publications.

Eriksen, T.H. (1995) *Small Places, Large Issues*. London: Pluto Press.

Fajans, P., Ford, K. and Wirawan, D.W. (1995) 'AIDS Knowledge and Risk Behaviours of Domestic Clients of Sex Workers in Bali, Indonesia', *Social Science and Medicine* 41(3): 409–17.

Gayatri, B.J.D (1993) 'Coming Out But Remaining Hidden: A Portrait of Lesbians in Java'. Paper presented at the International Congress of Anthropological and Ethnological Sciences, Mexico City, Mexico.

Geertz, C. (1973) *The Interpretation of Cultures*. New York: Basic Books.

Geertz, C. (1983) *Local Knowledge*. New York: Basic Books.

Goodman, J. (1998) 'Ideology and Critical Ethnography', pp. 50–66 in G. Shacklock and J. Smyth (eds) *Being Reflexive in Critical Educational and Social Research*. London: Falmer Press.

Hammersley, M. (1998) *Reading Ethnographic Research: A Critical Guide* (2nd edition). London & New York: Longman.

Haraway, D. (1991) *Simians, Cyborgs and Women: The Reinvention of Nature*. London & New York: Routledge.

Herzfeld, M. (2001) 'Performing Comparisons: Ethnography, Globetrotting and the Spaces of Social Knowledge', *Journal of Anthropological Research* 57(3): 259–76.

Honan, E., Knobel, M., Baker, C. and Davies, B. (2000) 'Producing Possible Hannahs: Theory and the Subject of Research', *Qualitative Inquiry* 6(1): 9–32.

Humphreys, L. (1970) *Tearoom Trade: A Study of Homosexual Encounters in Public Places*. London: Duckworth.

Hunt, J.C. (1989) *Psychoanalytic Aspects of Fieldwork*. Newbury Park, CA: Sage.

Inciardi, J. (1993) 'Some Considerations on the Methods, Dangers and Ethics of Crack-House Research', pp. 153–72 in J. Inciardi, D. Lockwood and A.E. Pottinger (eds) *Women and Crack-Cocaine*. New York: Macmillan.

Jennaway, M. (1996) 'Sweet Breath and Bitter Honey: HIV/AIDS and the Embodiment of Desire Among North Balinese Women', unpublished PhD Thesis, The University of Queensland, Australia.

Kelsky, K. (2001) 'Who Sleeps With Whom, or How (Not) to Want the West in Japan', *Qualitative Inquiry* 7(4): 418–35.

Kulick, D. (1995) 'The Sexual Life of Anthropologists: Erotic Subjectivity and Ethnographic Work', pp. 1–28 in D. Kulick and M. Wilson (eds) *Taboo: Sex, Identity, and Erotic Subjectivity in Anthropological Fieldwork*. London & New York: Routledge.

Kumar, N. (1992) *Friends, Brothers and Informants: Fieldwork Memoirs of Banaras*. Berkeley: University of California Press.

Lather, P. (1993) 'Fertile Obsession: Validity After Poststructuralism', *The Sociological Quarterly* 34(4): 673–93.
Lee, R.M. (1995) *Dangerous Fieldwork*. Qualitative Research Methods Series 34. London: Sage Publications.
Lerum, K. (2001) 'Subjects of Desire: Academic Armor, Intimate Ethnography, and the Production of Critical Knowledge', *Qualitative Inquiry* 7(4): 466–83.
Lette, H. (1996) '"Changing My Thinking With a Western Woman": Javanese Youths' Construction of Masculinity', *Asia-Pacific Viewpoint* 37(2): 195–207.
Madjar, I. and Higgins, I. (1996) 'Of Ethics Committees, Protocols and Behaving Ethically in the Field: A Case Study of Research With Elderly Residents in a Nursing Home', *Nursing Enquiry* 3(3): 130–37.
Malinowski, B. (1961) *Argonauts of the Western Pacific*. New York: Dutton Press.
Marcus, G.E. (2001) 'From Rapport Under Erasure to Theaters of Complicit Reflexivity', *Qualitative Enquiry* 7(4): 519–28.
Murray, A. (1995) 'Where Are the Indonesian Lesbians?', *Inside Indonesia* 42(1): 22–3.
Nilan, P. (2001) 'Young People as Embodied Subjects in Bali: Ritual, Politics and Risk/Profit', Working Papers: Centre for Asia-Pacific Social Transformation (CAP-STRANS), http://www.uow.edu.au/research/centres/capstans. University of Wollongong and University of Newcastle, Australia.
Nusa Tenggara, (1999) Amuk massa itu seperti orang ngunying. 28/10/99: 1–3.
Oetomo, D. (1996) 'Gender and Sexual Orientation in Indonesia', pp. 259–69 in L.J. Sears (ed.) *Fantasizing the Feminine in Indonesia*. Durham & London: Duke University Press.
Parameswaran, R. (2001) 'Feminist Media Ethnography in India: Exploring Power, Gender and Culture in the Field', *Qualitative Inquiry* 7(1): 69–103.
Popkewitz, T. (1984) *Paradigm and Ideology in Educational Research*. New York: Falmer Press.
Probyn, E. (1993) *Sexing the Self: Gendered Positions in Cultural Studies*. London & New York: Routledge.
Sherif, B. (2001) 'The Ambiguity of Boundaries in the Fieldwork Experience: Establishing Rapport and Negotiating Insider/Outsider Status', *Qualitative Inquiry* 7(4): 437–47.
Street, A. (2001) 'Critical Ethnography', Paper distributed at Critical Ethnography Workshop, University of Newcastle: 21–22 September.
Surya, I.M. (1999) Bali election results, 19/6/99.
Van Maanen, J. (1988) *Tales of the Field*. Chicago: Chicago University Press.
Venkateswar, S. (2001) 'Strategies of Power: An Analysis of an Encounter in the Andaman Islands', *Qualitative Inquiry* 7(4): 448–65.
Warren, C.A.B. (1988) *Gender Issues in Field Research*. Newbury Park, CA: Sage.
Willson, M. (1995) 'Perspective and Difference: Sexualization, the Field and the Ethnographer', pp. 251–75 in D. Kulick and M. Wilson (eds) *Taboo: Sex, Identity, and Erotic Subjectivity in Anthropological Fieldwork*. London & New York: Routledge.
Wolf, D.L. (1996) 'Situating Feminist Dilemmas in Fieldwork', pp. 1–55 in D.L. Wolf (ed.) *Feminist Dilemmas in Fieldwork*. Boulder Colorado: Westview Press.

42

Doing Participant Observation in a Psychiatric Hospital – Research Ethics Resumed

Christine Oeye, Anne Karen Bjelland and Aina Skorpen

Introduction

In health research, there is a growing tendency among social scientists and researchers in medical professions to conduct participant observation research within psychiatric institutions (Barrett, 1996; Hem & Heggen, 2003; Hummelvoll & Severinsson, 2001; van Dongen, 2004). Social scientists entering the medical field are being called on to comply with ethical regulations, usually enforced by medical ethical review boards, to ensure that ethical guidelines are followed. Social scientists are then held accountable for their research methods, and are questioned with specific reference to medical research guidelines (Fluehr-Lobban, 2003a). The consideration of participant observation research proposals by medical review boards often has uncomfortable consequences for social scientists. The review board's interpretation and proposals challenge the methodological "habitus" of participant observation (Mills, 2003). Social scientists such as anthropologists have questioned the role of medical ethics boards, and report uneasiness when trying to follow medical research ethics guidelines (Angrosino & Mays de Pèrez, 2000; Mills, 2003; Pels, 1999). Angrosino and Mays de Pèrez (2000) claim ethical boards within psychology and biomedicine do not have experience in participant observation research. For that reason, social scientists

Source: *Social Science & Medicine*, 65(11) (2007): 2296–2306.

employing participant observation are caught between two different models of research: the experimental, hypothesis-testing, clinical model, and participant observation, which is intrusive in character (Angrosino & Mays de Pèrez, 2000; Hoeyer, Dahlager, & Lynöe, 2005). Wax (1999) claimed that ethical guidelines and the institutional review system are prejudiced in favor of research methodologies that mimic epidemiology, and kindred "objective" approaches that deal with humans as detached social atoms. These approaches do not address the needs or practices of participant observation.

Earlier in this journal, Hoeyer et al. (2005) outlined two different research ethics traditions: medical research ethics and research ethics in anthropology. They showed how differences in central ethical issues can lead to clashes between social scientists and medical staff. Their article addressed the difficulties of measuring one ethical tradition by the standards of the other. In their concluding remarks they suggest building bridges between the two ethical traditions in order to prevent or cope with research conflicts. Their suggestions were made without looking thoroughly at the epistemological paradoxes inherited by both research ethics traditions, or at the practical research problems that arise in participant observation when trying to follow medical research ethics. This article is therefore inspired by Hoeyer et al., and based on cross-disciplinary specific study in a psychiatric setting in Norway by the first and third authors (C.O. and A.S.). This paper traces an ambiguous and paradoxical notion of vulnerability in medical research ethics. Further, the article presents the implications of this notion of vulnerability for participant observation in a psychiatric hospital.

The Research Case

The main aim of the research was to examine how milieu therapy is practised, organized and experienced by milieu therapists, and patients. In short, milieu therapy is about how daily life activities are organized in a psychiatric ward for the purpose of gaining therapeutic effects (Gunderson, 1978). Activities also include interactions between staff and patients (Ammon, 1993). We planned to use participant observation to observe both milieu therapists and patients, and observe the interaction between them in different settings in public rooms in two different psychiatric wards. We included patients and milieu therapists for different purposes in our study. Patient informants were recruited as recipients of such therapy, to shed light on milieu therapy and to answer the following question: what are their experiences with daily life routines and activities? Milieu therapists were recruited so that researchers could observe how they employ this therapeutic tool. Milieu therapists could be psychiatric nurses, nurses, ergonomists, enrolled nurses, activators, or unskilled staff. For interviews, we planned to recruit primarily milieu therapists who had some experience with milieu therapy to explain the practice of milieu therapy. Included patients would be of any type, independent of diagnosis, with some daily life experience (e.g., routines and activities) within mental health institutions. We planned to use individual-based informed consent forms for interviews with both groups.

The Hospital Management Board, in conjunction with the senior nursing officer of two different wards, approved our research plan. In contrast, obtaining research approval from the research ethics committee proved difficult. The committee, both in writing and orally, strongly recommended that we request written informed consent from each psychiatric patient to observe milieu therapeutic practices. The arguments for this demand were various. Firstly, psychiatric patients are regarded as being vulnerable, with impaired decision-making capacity as regards giving informed consent. Therefore, we were to cooperate with the medical staff in charge to decide whether or not each psychiatric patient was capable of providing informed consent. The second argument followed the first one: their vulnerability increases the potential for the study to do harm and creates doubt over how psychiatric patients will benefit from the research. Thirdly, because psychiatric patients are often involuntarily hospitalized, they are vulnerable and not capable of voluntarily giving informed consent. The committee claimed that psychiatric patients who were not willing to participate in the study were not able to withdraw from the psychiatric ward whenever the ethnographers were present to observe milieu therapeutic activities. Participant observation as a research method, they claimed, challenges the ethical principle of voluntariness. They suggested two different solutions to this problem: (1) non-participating patients could be transferred to another ward or (2) we could leave the ward and do observations in another ward where everyone agreed to be observed. We discussed different solutions with the committee before, and at the beginning of the fieldwork, because we found the suggestions difficult to comply with. The first suggestion we found very unethical; it would have drastically altered the treatment of patients, and by doing so, "punished" patients who did not agree. The committee's second suggestion was followed, and most patients were asked for written informed consent to allow us to observe them in the public rooms of the ward. The process of asking patients for informed consent involved asking staff in charge for advice and permission about which patients to ask for informed consent. Most patients gave their consent, and the ones who were never asked or did not give their consent, were not followed as closely as the ones who gave their consent. It was not practically possible to avoid observing situations in which patients who had not given their consent were involved. The committee was not worried about us engaging with milieu therapists, and informed consent to observe them was not requested.

In order to address the committee's recommendations and follow the medical research ethical guidelines, we were forced to examine and revise our research plan. Following these guidelines, we experienced some obstacles and difficulties when employing participant observation. The obstacles experienced were mainly due to: (a) the opposition of an individual-based informed consent process and a collective-based participant observation method; (b) opposition of the principle of voluntariness and elimination of all influential circumstances and participant observation as an influential-based method; and (c) being able to predict benefits and avoid all sorts of harm, which is a principle that is difficult to follow when the purpose of the research is to scrutinize a therapeutic tool anchored in professional identity. The underlying premise behind these ethical guidelines is the notion of vulnerability among psychiatric patients as a participant group.

Medical Research Ethics Committees in Norway, Europe and the US

The World Medical Association's Declaration of Helsinki in 1964 stressed the importance of establishing Research Ethics Committees (RECs) (called Institutional Review Boards in the US), and stipulated that researchers should not take ethical responsibilities entirely upon themselves (Hoeyer et al., 2005). In Norway, a Regional Medical Research Ethics Committee (REK) was set up in 1985, for each of Norway's five health regions, in accordance with the guidelines laid down by the Norwegian Ministry of Education and Research (REK, 2003). The committees give advice and guidance based on generally accepted principles of research ethics established and regulated by the Norwegian Medical Research Ethics Committee (NEM) (1990) and international bodies (e.g., the Declaration of Helsinki) (http://www.etikkom.no/REK/english/reference). In addition to the medical committees, there are two independent but coordinated national committees for research ethics to cover scientific disciplines such as Social Sciences and Humanities, and Science and Technology (http://www.etikkom.no/English/about). If the research is carried out in a medical institution, the researchers must submit a research application for ethical review to the REK in their region. Therefore, our research, despite its social science foundation, had to be submitted to a medical research ethics committee. Furthermore, the committee for medical research ethics operates with a descending scale of permission. It suggests that the researcher uses healthy adults as research participants as their first preference, before vulnerable groups. These patient groups are listed hierarchically, with healthy adults at the top of the list, followed by increasingly more vulnerable patient groups: the mentally ill, intellectually challenged, demented, drugged and intoxicated, acutely confused, unconscious, exhausted or marked by a serious disease, and children (http://www.etikkom.no/Etikk-torget/omForskningsEtikk). This does not imply that every person in the latter groups has impaired decision-making capacity. The committee claims that some people with mental illness will be capable of giving informed consent to participate in research. Recently, there has been a discussion within the NEM on how to handle participant observation studies when vulnerable participant groups are involved, mainly focused on impaired decision-making capacity among groups of patients, the risk of disturbing health personnel in their work, and difficulties in complying with professional confidentiality. Despite these problematic issues, the NEM (2005b) emphasizes the importance of gaining knowledge on how health personnel function when treating vulnerable groups. Furthermore, the NEM has recognized problematic issues when studying vulnerable groups, and therefore suggested in 2005 that researchers should consult user/patient councils (The National Committees for Research Ethics in Norway (NEM), 2005a).

The Norwegian model departs from the more common Research Ethics Committees (RECs) in Europe and the US. Throughout Europe, there are different practices on how RECs are organized. For instance, in the UK, the era of self-regulation ended in May 2004, and RECs became legally accountable to a new government body, the UK Ethics Committee Authority (UKECA). Ethical reviews in social science research depart from medical research, which are still regulated at institutions such as universities (Kerrison & Pollock, 2005). In the US, Institutional Review Boards (IRBs) are independent or commercial, and based at academic or medical institutions, but are regulated by the same federal regulations (Gilbert, 2006). Ethnographic research is governed by federal regulations because human participants are involved, but not all research projects require full IRB reviews – for example, those involving observations of public behavior where subjects cannot be identified (National Science Foundation, 2006). In contrast, projects that focus on sensitive information, where the disclosure of respondents could harm the respondents require full review. This includes research on vulnerable groups such as the mentally ill (Kennedy, 2005). The National Science Foundation (NSF) has questioned IRBs' reviews and social science methodologies, especially ethnography and the use of written informed consent (NSF, 2006). A request for written, formal consent could seem suspicious, inappropriate, rude and perhaps even fraught with danger in many societies and communities. Therefore, the NSF suggests, the use of "group consent" in such societies (e.g., presenting the project in an open meeting, and allowing questions to be raised and answered publicly) in addition to individual-based informed consent (NSF, 2006).

Participant Observation

Despite REK's ethical suggestions and demands, participant observation seemed to us to be a suitable and necessary research method to use, that would allow us to understand how milieu therapy is organized, how it takes place, and how milieu therapists and patients interact, as it is a method used to study daily life activities (Alvesson & Sköldberg, 2000) – also a principal focus of milieu therapy. The method is a mixture of participation, observation, and interviewing (Delamont, 2004), where the purpose is to observe how participants interact with, and influence each other (Atkinson & Hammersley, 1994). Participant observation is a deep investigation of social life and life experience. This process entails becoming involved in social life in order to see and experience how daily life activities are practised (Lincoln & Denzin, 2003; Lofland, Snow, Anderson, & Lofland, 2006). It is undertaken to understand how the studied culture works and what the world looks like for the people involved in the study. Because the anthropologist more or less lives and participates in the studied community, he or she is in a position to interpret the meanings and possible consequences of actors' behaviors within that specific cultural context (Chambers, 2000). Participant observation then becomes an ongoing attempt to place specific encounters, events and understanding into a fuller, more meaningful context (Tedlock, 2000). Research participants and their knowledge are not essentially unaffected by the research or its circumstances; the research is constructed in a social, cultural and historical context rather than discovered and revealed irrespective of circumstantial factors (Klee, 1997).

Medical Research Ethics and the Assumption of Vulnerability

The World Medical Association has developed the Declaration of Helsinki (1964) as a statement of ethical principles that provides guidance to physicians and other participants in medical research involving human subjects (WMA, 2002). Medical research ethics is anchored in the Hippocratic Oath, in doing no harm, and tends to take the protection of the individual as its main objective (Hoeyer et al., 2005). Therefore, according to the Declaration of Helsinki, studies should be designed in the safest manner possible and every medical research project involving human subjects should be preceded by careful assessment of predictable risks and burdens weighed against foreseeable benefits to the subject or to others. Furthermore, the participants in research must be volunteers and informed of the research project to freely give informed consent, preferably in writing. Accordingly, the participants must provide entirely voluntary agreement to participate, with no physical or psychological coercion (Christians, 2003).

The notion of vulnerability among psychiatric patients seems to underlie restrictions on ethical proposals. According to the Declaration of Helsinki, some research populations are vulnerable and in need of special protection. The World Medical Association is concerned about the possibility of exploiting vulnerable groups (see WMA, 2002) for these reasons:

(1) By virtue of their illnesses and common psychiatric symptoms, such as ambivalence, apathy, paranoia, self-destructiveness, and impulsivity, those with mental illness may be prevented from meaningfully participating in the consent process (Osborn & Fulford, 2003). Psychiatric patients are also believed to have impaired cognitive and emotional functioning, as well as a lack of awareness and insight into illness, which implies lack of comprehension and rational reasoning when given information about the research project (DuVal, 2004; Eichelman, Wikler, & Hartwig, 1984).

(2) Institutional constraints apply, which means that in psychiatric hospitals, patients are "captive" and often involuntarily detained, indicating a lack of freedom and autonomy (Regehr, Edwardh, & Bradford, 2000; Roberts, 2002). Patients are often deprived of the facilities required for exercising autonomy (Radden, 2002).

(3) Patients are subject to social factors that arise from stigma, dependence, fractured relationships, fear and discomfort displayed by others, unemployment, homelessness, poverty, loneliness, and other factors primarily brought on by their illness (DuVal, 2004). Therefore, the psychiatric patient is believed to be in a vulnerable position in research, because mental illness impairs their capacity to understand, reason, choose or act (Lützén & Nordin, 1994). Accordingly, psychiatric patients lack the capacity to make acceptable decisions about their participation (Appelbaum, 1998).

All the reasons for seeing psychiatric patients as vulnerable make the informed consent process a challenge. Informed consent means that the researcher gives the participant the fullest possible information about the study and collects a written (or sometimes oral) consent. In recent years, the focus of discussions on informed consent has shifted from the researcher's obligation to disclose information, to the quality of a research participant's understanding and decision-making capacity (Leino-Kilpi et al., 2002). Participants should not only be informed, but their capacity to understand, in order to give informed consent, should also be checked.

The medical research ethics tradition draws on different traditions of knowledge in understanding psychiatric patients as being vulnerable with impaired decision-making capacity. First, medical research ethics draws on biological knowledge, such that psychiatric patients are to be regarded as vulnerable due to certain innate biological traits. They have less capacity to understand the information given due to illness, symptoms and medication. Second, medical research ethics draws on an understanding of research participants as basically autonomous, which implies self-determination free of any kind of coercion. The psychiatric patient will, in most cases, not be able to fulfill such an ethical demand, or the ideal of the perfect research participant, due to the influence of illness, the coercive circumstances of institutionalization and by social factors such as stigma, poverty, and dependence. Accordingly, medical research ethics is based on the idea that, as far as possible, research participants should not be under any influence of any kind, including illness, coercion, and institutional, social or relational constraints. The more a research participant is influenced by these constraints, the more he/she is regarded as vulnerable with impaired decision-making capacity and in need of protection by others, including the research committee and medical staff.

Research Ethics in Social Science and Anthropology

The standards for ethical conduct do not vary fundamentally between medical and anthropological research, even though medical research has been more regulated than social behavioral research (Fluehr-Lobban, 2003a). In the field of social anthropology, it used to be quite common to do covert participant observation (Atkinson & Hammersley, 1994). In psychiatric hospitals, the study of Caudill (1958) is an example of covert participant observation. Today, neither the American Anthropologist Association (AAA) nor the Association of Social Anthropologists (ASA) permit covert participant observation.

Social scientists continue to discuss ethical standards that will guide their research, but in anthropology, achieving agreement has never been easy (Caplan, 2003). There is currently no common agreement on what such standards should be, although many proposals have been made from various social scientific communities and researchers (Lincoln & Denzin, 2003). Both the AAA, in 1971 (AAA, 1998), and the ASA, in 1987 (ASA, 1999), formulated ethical guidelines for participant observation research. These guidelines are based on the Nuremberg Trials and include obligations of openness and informed consent (ASA, 1999). For years, anthropologists have dismissed informed consent as being applicable only to biomedical research. However, in the last decade, realization has been growing that ethnographers are not exempt from obtaining informed consent in theory and practice (Fluehr-Lobban, 2003a), and that anthropologists should offer the fullest possible information and disclosure for the goals and potential uses of research before it is undertaken (Fluehr-Lobban, 2003b). Furthermore, anthropologists should be sensitive to harmful effects (AAA, 1998; ASA, 1999), especially when it comes to vulnerable research populations (ASA, 1999). These vulnerable groups are often listed as ethnic or religious minorities in danger of being hurt by research due to an uncertain political situation in the country in which they live. In addition to ethnic and religious minorities, Fluehr-Lobban (2003a) regards children, the incarcerated, and the mentally incompetent to be vulnerable, which prevents these groups from giving full informed consent.

Paradoxes in Research Ethics

The roots of protection lie in the Hippocratic Oath and paternalism; the patient does not know what is good for them (Leino-Kilpi et al., 2002). The risk/benefit ratio has its roots in utilitarian ethics, a calculus of presumed benefits and risks, whereas informed consent becomes a moral obligation (Wax, 1999). Medical research ethics is, then, anchored in different ethical traditions. On the one hand, the protection of psychiatric patients is based on the Hippocratic Oath and paternalism, but on the other hand, the medical research ethics is anchored in absolute respect of patient autonomy and self-determination. We may ask whether the medical ethics doctrine places itself in a dim epistemological landscape, between the human right to voluntarily participate in scientific research, and the human right to protection. The recognition of these paradoxes is deeply rooted in the different humanist-liberal and paternalistic traditions of epistemology. Both traditions imply that participants are, by nature, either fundamentally autonomous or innately vulnerable due to certain biological traits, and therefore in need of protection. The ethical doctrine of informed consent is derived from the recognition of the patient's vulnerability, as well as respect for the patient's autonomy, and the enlightenment ideals of humanism, liberalism and human rights thinking (Leino-Kilpi et al., 2002). A mixture of different epistemological traditions in medical research ethics implies paradoxical guidelines when it comes to doing participant observation research. Thus, ethical guidelines, such as those about informed consent, the principle of voluntariness, and estimating the risk/benefit ratio, become substantial challenges when employing participant observation.

Informed Consent and Participant Observation – a Challenge

According to medical ethical guidelines, psychiatric patients are potentially considered mentally incapable, with impaired decision-making capacity for giving informed consent due to their assumed vulnerability. Tancredi (1995) claimed that the concept of a patient's ability to understand is somehow problematic because the notion of what it is "to understand" is unclear. We can further ask: "What is rationality?" "How should we define it?" "What should be the criteria for assessing what is rational?" The concept of vulnerability presupposes the notion that psychiatric patients are incapable of expressing their values, meanings and thoughts of relevance to research (Lützén & Nordin, 1994). Our experiences of conducting participant observation research in a psychiatric hospital question this notion. In fact, the majority of psychiatric patients in the wards were expecting us and wanted to tell their stories. Psychiatric patients had several relevant questions and comments regarding research practices. Most were willing to participate, and those who were in doubt provided understandable arguments for not giving their consent. Those who were unsure of participating were either afraid of offending staff members by gossiping, afraid of sanctions by staff members or concerned that the researchers would violate professional confidentiality.

Challenges were presented by the necessity of collecting informed consent forms when doing participant observation from one group of participants (psychiatric patients) and not the other group (milieu therapists), when individuals of both groups were present in the study. In doing so, we communicated to psychiatric patients that we must be more careful in our interaction with them, be more considerate, show more respect, and be more careful than in our interactions with milieu therapists. Our interaction with psychiatric patients became in a way "unnatural" – being very considerate and careful, asking again and again for permission to stay with them, and being overly pleasant. In doing so, we undermined their mental capacity, and, in a sense, communicated distrust in their capacity to make their own decisions. After we had informed one of our participants several times about our study and asked for his written informed consent signature on two different forms, he commented: "You know, I am not stupid, I understand the purpose of the study." If, as researchers, we do not give psychiatric patients the opportunity to participate in the research, we may cause or maintain their feeling of unworthiness and confirm their stigma and lack of self-determination. Furthermore, researchers are in danger of maintaining the view of psychiatric patients as vulnerable, and further stigmatizing this specific group. Such research is in danger of representing the professional's point of view at the expense of the patient's point of view (Van der Geest & Finkler, 2004).

Punch (1994) was strongly critical of recommendations of informed consent and stated: ". . . gaining consent is quite inappropriate, because activity is taking place that cannot be interrupted. In much fieldwork there seems to be no way around the predicament that informed consent – divulging one's own identity and research purpose to all and sundry – will kill many a project stone dead" (Punch, 1994, p. 90). Therefore, obtaining informed consent becomes inappropriate, because activity that cannot be interrupted is taking place, even though the mere presence of a researcher does change the situation. This is known as the "observer's paradox."

The Principle of Voluntariness and Participant Observation – a Challenge

The principle of voluntariness and the elimination of all extenuating circumstances seem to rub uneasily against participant observation, which is an influence-based method. As we have seen, according to medical research ethics, all participants must voluntarily agree to participate without physical or psychological coercion. Accordingly, one of the main reasons for assuring voluntariness is to avoid exploitation and harm (Kottow, 2003).

Psychiatric patients are not wholly free to choose when it comes to activities such as leaving the hospital, and cannot choose when visitors enter or leave the ward (these are some reasons for regarding this group as vulnerable). Nor are milieu therapists wholly free to choose when it comes to being participants in a participant observation study, despite reading information sheets and signing informed consent sheets. Some members of staff did not consider themselves to be entirely free to make a choice, which was been recognized by others (Moore & Savage, 2002).

Medical research ethics guidelines have not incorporated an understanding of participants as members of a social community, where participants are shaped and reshaped in social interaction. On the contrary, participants are seen as solitary individuals who are supposed to choose freely with minimal constraints or influence of any sort. Participant observation has been understood as an intrusive method (Angrosino & Mays de Pèrez, 2000), because participants can hardly escape the context of discovery. As mentioned earlier, in the social sciences, participants and their knowledge are understood as being created through social interactions and settings, rather than being discovered and revealed in isolation from circumstantial factors. Therefore, employing participant observation means interacting with, getting to know about, and being involved in events and participants' lives to learn about their social life and cultural ideas. Only by being involved, can the researcher gain knowledge about daily life matters in different social settings.

Estimating the Risk/Benefit Ratio When Employing Participant Observation – a Challenge

As mentioned earlier, according to medical research ethics and the Helsinki Declaration, researchers should estimate the potential risk/benefit ratio. Social scientists employing participant observation methods have expressed concern over how to predict possible risks and benefits (Akeroyd, 1984; Christians, 2003). First, any benefits that the informants should expect from the research and the researcher, as well as the obligations of the researcher to informants in a participant observation study, are not clear (Jokinen, Lappalainen, Meriläinen, & Pelkonen, 2002). Second, the ethical principle of avoiding all sorts of harm and minimizing risk seems somehow vague; what kind of harm and risks are relevant when doing participant observation? According to Labott and Johnson (2004), we are talking about psychological and social risks in daily life research, even though what might be understood as minimal social and psychological risks is a bit unclear. Du Toit (1980) suggested that possible risks include the danger of embarrassing or causing pain to once-trusted participant friends. According to Angrosino and Mays de Pèrez (2000), participant observation implies being involved in the lives of the participants and thus is fraught with all sorts of possibilities for harm and risks. In our study, the potential for harm is unclear, but we have observed in the field how participants of both groups have reacted when we have been present doing participant observation in terms of expectations, ambivalence, embarrassment, worries and unpleasantness.

We have identified three different conditions that cause what some would call emotional or social harm or possible benefits to participant individuals, either patients or milieu therapists, when employing participant observation as a research method. First, the very purpose of this study is to scrutinize a therapeutic tool through the method of participant observation, which can be experienced by milieu therapists as an unpleasant form of surveillance. As ethnographers, we followed milieu therapists in their daily work and life. Doing participant observation in a psychiatric hospital means that the milieu therapists allow access to their professional life and work. This includes access to formal and informal meetings, where they discuss patients, routines and different forms of milieu therapeutic efforts. In these settings, the milieu therapists discuss matters that are delicate and difficult, sometimes with a great deal of despair and distaste. They also let us follow them in different forms of interactive activities with patients, which included attempts to calm patients, set limits, deliver unpopular messages, and be strict or harsh in some way – tasks that many of the milieu therapists find uncomfortable. Afterwards, milieu therapists tried to explain or make excuses for not coping. We did not only observe and map interactions, activities and episodes, but we also requested professional explanations and recorded these in writing. Participant observation then became an inquiry in which milieu therapists' professional identities were scrutinized.

Second, when doing participant observation, the researcher becomes a mediator between or benefactor to different groups, which is a well-known experience in the field (Atkinson & Hammersley, 1994; Wax, 1999). Patient participants, in most cases, expected us and were often anxiously waiting for days in the smoking room to tell their stories about daily life matters inside and outside the institution. Most patients saw an opportunity to be seen, to be listened to, to have someone to whom they could explain their misery, and not least, to use us as mediators between themselves and staff or the rest of society. Some milieu therapists also saw the opportunity to use us as mediators between themselves and hospital management. There were also those who wanted the study to map and document their practice, which they were confident in and proud of, for the professional community. On the other hand, milieu therapists worried about how we interpreted their practice, whether we liked their practice, and how we would write about it. Some were afraid of being presented in an unfavorable way or being tested, e.g., "Are you here to test how much knowledge we have about milieu therapy?".

Third, in participant observation studies, participants of both groups obtain glimpses of the translation process by means of curious ethnographers observing and asking questions. The principle of participant observation is obtaining the "native's point of view," whilst at the same time keeping a distance so as not to "go native," or fully acting as a participant, in order to keep one's eyes open (Behar, 1996). Participant observation is then deeply paradoxical. Acting as natives, we would communicate understanding and friendship to participants in our bid for a trustworthy relationship. Simultaneously, they knew "we had our eyes open" in order to record information about them. Potential emotional harm inflicted on participants can be understood as anchored in this paradox. Here lies our power as researchers, of potentially causing harm to professional souls.

The Question of Vulnerability and Research Ethics Resumed

An important mission of research ethics, either medical or anthropological, is to avoid the exploitation of, to protect, and to do no harm to, research participants, as researchers learned from the experiments by Nazi physicians. Therefore, we will argue that research ethics is an important issue to be formulated, reformulated and discussed for the appropriateness of the research methods used and for the research situation at hand. However, in Norway, social scientists are being questioned with specific reference to medical research guidelines when doing participant observation in a medical setting.

Participant observation in a psychiatric hospital will necessarily involve informants of different categories, in our case, they are patients and milieu therapists. Medical research ethics has classified the participant category of psychiatric patients as vulnerable, without looking at the reasons for including this group in the study. The purpose of including psychiatric patients in our study differs from the purpose of including milieu therapists in the study. While psychiatric patients are primarily included to shed light on a therapeutic tool, milieu therapists are included to observe how they "use" and explain this therapeutic tool. In this case, who is to be regarded as vulnerable is predefined. Medical research ethics, then, assume that psychiatric patients are more exposed to potential harm than other participant categories.

This article has tried to challenge such an assumption by drawing on participant observation experience where participants of two different groups, patients and milieu therapists, are included. Participant observation often affects the lives of those studied, because of the researchers' continual presence. This can lead participants to see themselves in ways that they feel are disrespectful, or in other ways they do not like, through the interrogation process, and so, in these respects, either produce or invoke vulnerabilities or harm. Participants representing both groups showed glimpses of doubt, ambivalence and unpleasantness on the one hand, and expectations and trust on the other. In our study, it seemed that more was at stake for milieu therapists, than for psychiatric patients. We have not argued that milieu therapists should be regarded as vulnerable *per se*, but we have argued that participant observation is capable of producing vulnerabilities and hurting professionals when the purpose of the study is to scrutinize a therapeutic tool anchored in professional identity. Therefore, researchers employing participant observation should give an account of what is at stake. What is at stake is not an exact branch of knowledge that fits a utilitarian model – to predict harm and benefit ratios – in any exact way, but more a way of giving an account of what might be at stake for the different participant categories involved. Predefined notions of who is to be regarded as vulnerable, due to different traits or capacities, should be replaced by providing accounts of what might be at stake for those involved to an ethics board or committee.

Participant observation creates a problem for obtaining informed consent, because informed consent is an individual-based ethical guideline and participant observation is based on observing interaction between participants, which makes it a collective approach. When doing participant observation we will often observe participants interacting when one or more participants are believed to have impaired decision-making capacity, or have not yet given their consent to be observed. In our study, we were bound to follow medical ethical restrictions. Participant observation became a practical, as well as a scientific, research problem of informing, inviting and collecting consent on entering the field for the purpose of observing human activities. In doing so, there is a risk of radically altering human activities and interaction.

As informants in participant observation research, both groups were subjected to the continual presence of the researcher, which is the very core of participant observation. Neither milieu therapists nor patients could escape the context of discovery. Being informants in a participant observation research project implies it is almost impossible to hide from the eyes of a curious ethnographer. Both groups are in difficult circumstances, created by the institution. While patients can withdraw to their private rooms, staff members do not have any private rooms that are inaccessible to the researchers.

We suggest that the individual-based ethical principle of informed consent and voluntariness seems utopian when employing participant observation, which is a method based on how informants and researchers influence and construct each other. Therefore, we concur with the NEM (National Committees for Research Ethics in Norway, 2005a), which recently suggested that researchers should consult user/patient councils. Furthermore, we suggest that the user/patient council should be composed of potential participants at the institution where the research is carried out and should also include members of staff and patients. Permission to carry out participant observation should be given by a research council at the institution, consisting of researchers, staff members with different occupational backgrounds, and patients.

Conclusion

In Norway, medical research committees question social scientists, with specific reference to medical research guidelines, when the social scientists are doing participant observation in a medical setting. In our study, the regional medical research committee recommended that we ask each psychiatric patient to sign individual-based informed consent forms for permission to observe them. The committee's suggestions were attempted, but proved difficult. It was not practically possible to avoid observing situations that involved patients who had not given their consent. The underlying premise for the committees' recommendations is a predefined notion of psychiatric patients as vulnerable.

The notion of vulnerability is anchored in different epistemological traditions in medical research ethics, which leads to paradoxical guidelines when it comes to doing participant observation research. Therefore, ethical guidelines, such as informed consent, the principle of voluntariness, and estimating the risk/benefit ratio, become substantial challenges when employing participant observation. As long as participant observation projects are carried out in medical institutions, informed consent, the principle of voluntariness, and predicting risk/benefit ratios in medical ethical guidelines need to be reviewed. We have therefore argued that the ethical question of vulnerability in participation observation research at psychiatric institutions requires a nuanced reformulation. Predefined notions of who is to be regarded as vulnerable are not suitable when doing participant observation. However, that vulnerability is a condition that must be understood as a construct of the purpose and context of the study.

References

Akeroyd, A. V. (1984). Ethics in relation to informants, the profession and governments. In R. F. Ellen (Ed.), *Ethnographic research: A guide to general conduct* (pp. 133–154). London: Academic Press.

Alvesson, M., & Sköldberg, K. (2000). *Reflexive methodology: New vistas for qualitative research*. London: Sage Publications.

American Anthropological Association. (1998). *Code of Ethics of the American Anthropological Association*. <http://www.aaanet.org/committees/ethics/ethicscode.pdf>.

Ammon, G. (1993). Milieu therapy. Paper presented at the Ninth World Congress of the World (WPA) (pp. 24–30), Rio de Janeiro, Brazil.

Angrosino, M. V., & Mays de Pèrez, K. A. (2000). Rethinking observation: From method to context. In K. Denzin, & Y. S. Lincoln (Eds.), *Handbook of qualitative research* (pp. 673–702). Thousand Oaks, CA: Sage.

Appelbaum, P. S. (Ed.). (1998). Missing the boat: Competence and consent in psychiatric research. Editorial. *American Journal of Psychiatry, 155*(11), 1486.

Association of Social Anthropologists. (1999). *Ethical guidelines for good research practice*. <http://www.theasa.org/ethics/ethics_guidelines.htm>.

Atkinson, P., & Hammersley, M. (1994). Ethnography and participant observation. In K. Denzin, & Y. S. Lincoln (Eds.), *Handbook of qualitative research* (pp. 83–99). Thousand Oaks, CA: Sage.

Barrett, R. (1996). *The psychiatric team and the social definition of schizophrenia: An anthropological study of person and illness*. New York: Cambridge University Press.

Behar, R. (1996). *The vulnerable observer: Anthropology that breaks your heart*. Boston: Beacon Press.

Caplan, P. (2003). Introduction: Anthropology and ethics. In P. Caplan (Ed.), *The ethics of anthropology: Debates and dilemmas* (pp. 1–35). London: Routledge.

Caudill, W. (1958). *The psychiatric hospital as a small society*. Cambridge, MA: Harvard University Press.

Chambers, E. (2000). Applied ethnography. In K. Denzin, & Y. S. Lincoln (Eds.), *Handbook of qualitative research* (pp. 673–702). Thousand Oaks, CA: Sage.

Christians, C. G. (2003). Ethics and politics in qualitative research. In K. Denzin, & Y. S. Lincoln (Eds.), *The landscape of qualitative research: Theories and issues* (pp. 208–243). Thousand Oaks, CA: Sage.

Delamont, S. (2004). Ethnography and participant observation. In C. Seale, G. Gobo, J. F. Gubrium, & D. Silverman (Eds.), *Qualitative Research Practice* (pp. 217–230). London: Sage.

Du Toit, B. (1980). Ethics, informed consent, and fieldwork. *Journal of Anthropology Research, 36*(3), 274–285.

DuVal, G. (2004). Ethics in psychiatric research: Study design issues. *Canadian Journal of Psychiatry, 49*(1), 55–59.

Eichelman, B., Wikler, D., & Hartwig, A. (1984). Ethics and psychiatric research: Problems and justification. *American Journal of Psychiatry, 141*(3), 400–405.

Fluehr-Lobban, C. (2003a). Darkness in El Dorado: Research ethics, then and now. In C. Fluehr-Lobban (Ed.), *Ethics and the profession of anthropology: Dialogue for ethically conscious practice* (pp. 85–106). Walnut Creek: AltaMira.

Fluehr-Lobban, C. (2003b). Informed consent in anthropological research: We are not exempt. In C. Fluehr-Lobban (Ed.), *Ethics and the profession of anthropology: Dialogue for ethically conscious practice* (pp. 159–177). Walnut Creek: AltaMira.

Gilbert, S. G. (2006). Supplementing the traditional Institutional Review Board with an Environmental Health and Community Review Board. *Environmental Health Perspectives, 114*(10), 1626–1629.

Gunderson, J. G. (1978). Defining the therapeutic processes in psychiatric milieus. *Psychiatry, 41*, 327–335.

Hem, M. H., & Heggen, K. (2003). Being professional and being human: One nurse's relationship with a psychiatric patient. *Journal of Advanced Nursing, 43*(1), 101–108.

Hoeyer, K., Dahlager, L., & Lynöe, N. (2005). Conflicting notions of research ethics. The mutually challenging traditions of social scientists and medical researchers. *Social Science & Medicine, 61*(8), 1741–1749.

Hummelvoll, J. K., & Severinsson, E. (2001). Coping with everyday reality: Mental health professionals? Reflections on the care provided in an acute psychiatric ward. *Australian and New Zealand Journal of Mental Health Nursing, 10*(3), 156–166.

Jokinen, P., Lappalainen, M., Meriläinen, P., & Pelkonen, M. (2002). Ethical issues in ethnographic nursing research with children and elderly people. *Scandinavian Journal of Caring Sciences, 16*(2), 165–170.

Kennedy, J. M. (2005). Institutional Review Boards and institutional researchers. *New Directions for Institutional Research, 127*, 17–31.

Kerrison, S., & Pollock, A. M. (2005). The reform of UK research ethics committees: throwing the baby out with the bath water? *Journal of Medical Ethics, 31*, 487–489.

Klee, R. (1997). *Introduction to the philosophy of science – Cutting nature at its seams*. Oxford: Oxford University Press.

Kottow, M. H. (2003). The vulnerable and the susceptible. *Bioethics, 17*(5–6), 460–471.

Labott, S. M., & Johnson, T. P. (2004). Psychological and social risks of behavioral research. *IRB: Ethics & Human Research, 26*(3), 11–15.

Leino-Kilpi, H., Välimaki, M., Arndt, M., Dassen, T., Gasull, M., Lemonidou, C., et al. (2002). *Patient's autonomy, privacy and informed consent.* Amsterdam: IOS Press Ohmsha.

Lincoln, Y. S., & Denzin, N. K. (2003). The ethical revolution. In Y. S. Lincoln, & N. K. Denzin (Eds.), *Turning points in qualitative research: Tying knots in a handkerchief* (pp. 217–219). Walnut Creek: AltaMira.

Lofland, J., Snow, D., Anderson, L., & Lofland, L. H. (2006). *Analyzing social settings – A guide to qualitative observation and analysis.* Belmont, CA: Thomson Wadsworth.

Lützén, K., & Nordin, C. (1994). Modifying autonomy – A concept grounded in nurses' experiences of moral decision-making in psychiatric practice. *Journal of Medical Ethics, 20*(2), 101–107.

Mills, D. (2003). 'Like a horse in blinkers?' A political history of anthropology's research ethics. In P. Caplan (Ed.), *The ethics of anthropology: Debates and dilemmas* (pp. 37–54). London: Routledge.

Moore, L., & Savage, J. (2002). Participant observation, informed consent and ethical approval. *Nurse Researcher, 9*(4), 58–69.

National Science Foundation. (2006). <http://www.nsf.gov/bfa/dias/policy/hsfaqs.jsp>.

Osborn, D. P. J., & Fulford, K. W. M. (2003). Psychiatric research: What ethical concerns do LRECs encounter? A postal survey. *Journal of Medical Ethics, 29*(1), 55–56.

Pels, P. (1999). Professions of duplexity: A prehistory of ethical codes in anthropology. *Current Anthropology, 40*(2), 101–136.

Punch, M. (1994). Politics and ethics in qualitative research. In K. Denzin, & Y. S. Lincoln (Eds.), *Handbook of qualitative research* (pp. 83–99). Thousand Oaks, CA: Sage Publications.

Radden, J. (2002). Psychiatric ethics. *Bioethics, 16*(5), 397–411.

Regehr, C., Edwardh, M., & Bradford, J. (2000). Research ethics and forensic patients. *Canadian Journal of Psychiatry, 45*(10), 892–899.

Roberts, L. W. (2002). Ethics and mental illness research. *The Psychiatric Clinics of North America, 25*(3), 525–545.

Tancredi, L. R. (1995). The limits of empirical studies on research ethics. *Ethics & Behavior, 5*(3), 217–236.

Tedlock, B. (2000). Ethnography and ethnographic representation. In K. Denzin, & Y. S. Lincoln (Eds.), *Handbook of qualitative research* (pp. 455–486). Thousand Oaks, CA: Sage.

The National Committees for Research Ethics in Norway (NEM). (2005a). *Redusert samtykke kompetanse i helsefaglig forskning.* <http://www.etikkom.no/retningslinjer/samtykke>.

National Committees for Research Ethics in Norway (NEM). (2005b). <http://www.etikkom.no/HvaGjorVi/Uttalelser/NEM/2005/200605/view>.

The National Model for Research Ethics in Norway. (1990). <http://www.etikkom.no/REK/english/reference>.

The Regional Committees for Medical Research Ethics in Norway (REK). (2003). <http://www.etikkom.no/REK/english/reference>.

Van der Geest, S., & Finkler, K. (2004). Hospital ethnography: Introduction. *Social Science & Medicine, 59*(10), 1995–2001.

Van Dongen, E. (2004). *Worlds of psychotic people, wanderers, 'bricoleurs' and strategists.* London: Routledge.

Wax, M. L. (1999). Comments to P. Pels: Professions of duplexity: A prehistory of ethical codes in anthropology. *Current Anthropology, 40*(2), 129–130.

World Medical Association. (2002). *Declaration of Helsinki.* <http://www.wma.net/e/policy/d1.htm>.

43

The Researcher as Hooligan: Where 'Participant' Observation Means Breaking the Law

Geoff Pearson

Introduction

This article looks at the unique problems that researchers face carrying out participant observation (PO) in fields where criminal activity is commonplace and where 'participation' can often mean breaking the law. It does not intend to revisit the well-worn arguments concerning the ethics of covert participant observation (CPO), although as we will see, many of the same debates about proportionality and justifiability arise. The merits, dilemmas and ethical problems arising from CPO have already provoked much academic debate (e.g. Bulmer, 1982; Crow, Wiles, Heath, & Charles, 2006; Erikson, 1967), and bodies such as the British and American Sociological Associations and the ESRC have felt obliged to draft codes and frameworks stating the desired objectives and minimum standards of social research, resulting in increased ethical regulation (Crow et al., 2006, p. 85). This guidance suggests that academic researchers in the social sciences should adhere to ethical standards based around principles such as informed consent, anonymity of research subjects and an adherence to standards of conduct that will not damage the reputation of academia or endanger subsequent researchers. However, whilst this is desirable, some have argued that in exceptional situations, non-adherence to these guidelines to allow

Source: *International Journal of Social Research Methodology*, 12(3) (2009): 243–255.

covert research may be *justifiable* in situations where the field would otherwise be closed to research, where overt techniques would unduly distort the field leading to inaccurate results, or where the safety of the researcher is at stake (e.g. Denzin & Erikson, 1982).

This position is supported to some extent by the codes themselves. For example, the Statement of Ethical Practice for the British Sociological Association states that 'the use of covert methods may be justified in certain circumstances'.[1] Similarly, the ESRC Research Ethics Framework notes that where researchers pursue projects where informed consent is not gained this must be justified: 'Deception (i.e. research without consent) should only be used as a last resort when no other approach is possible'.[2] The British Society of Criminology Code of Ethics is even more stringent, noting that the 'exceptional circumstances' in which research without informed consent could be justified relate 'to exceptional importance of the topic rather than difficulty of gaining access'.[3]

These limited exceptions to the general ethical standards suggest that where it is impossible to pursue avenues of 'ethical' research, methods such as CPO can be undertaken, provided every effort is undertaken to prevent harm befalling the research subjects (e.g. by preserving anonymity). The codes point to a balance between the beneficial outcomes of research on the one hand, and the potential 'harm' caused by methods such as CPO on the other. Essentially the issue is one of proportionality: where the 'harm' is minimal (e.g. a minor infringement of the right to privacy) and the research beneficial, the research is justifiable, but where methods cause serious harm (e.g. the reporting of the research subject to the police), it is more difficult for the research to be justified by its outcomes. A research method that causes harm but delivers little public benefit, we can assume, would be considered disproportionate and therefore unjustifiable.

However, do these arguments about proportionality and justifiability apply to the dilemma of researchers being pressurised to commit criminal offences in the field in the course of obtaining research findings? This is a predicament that researchers studying criminal or quasi-criminal social groups frequently face, particularly those involved in PO. In these fields, research subjects are often mistrustful of academics, whom they may believe will report them to the authorities (see Armstrong, 1993, especially pp. 23–25; Stott, Hutchinson, & Drury, 2001, p. 364). This can lead the observer (particularly one who possesses no previous connection with those under observation) to commit criminal offences during participation in order to integrate into the field and gain the trust of the subjects. Unlike the utilisation of covert methods, the issue of criminality is not expressly referred to in the ethical codes, although it can be presumed through reading them that academics are expected to abide by the law whilst carrying out their research.[4]

The question I wish to address is whether the committal of criminal offences is ever justifiable in social research and what guidance should be given to researchers operating in these fields. Should we assume that unlawful methods are always unethical? If so, do the same arguments of proportionality apply to this problem as are applied to the use of covert research methods in the codes? After all, the admission that covert methods are sometimes justifiable is controversial. In private research spheres in particular, the covert researcher essentially wilfully misleads research subjects who have not consented to being observed, invading their privacy. It also potentially puts non-consenting subjects in danger, for example, should the authorities compel the researcher to surrender research data or testify in a court of law (see Coomber, 2002). Furthermore, the requirement in the codes that the primary concern when covert methods are used should be to avoid harming the subjects places researchers under pressure not to report crimes that are being planned or committed (e.g. Feenan, 2002; Norris, 1993), making the observer complicit in these offences. If deception and the hiding of criminal offences from the authorities can be justified where the research field is otherwise inaccessible and/or the findings are in the public interest, then should the same principles apply to the actual committal of offences in the field?

The dilemma of whether or not to commit offences in the course of research arose in my own socio-legal ethnographic study into legislative and policing responses to crowds of 'risk' English football supporters (commonly referred to as 'hooligans'), which was carried out between 1995 and 2007. This research was not exclusively limited to the activities of hardcore 'hooligan' firms in the manner of, for example, Armstrong's work with 'the Blades' (Armstrong, 1998). However, early attempts at interviews proved to be unreliable; often non-violent fans would exaggerate their involvement in disorder whilst serious 'hooligans' tended to 'play-down' their involvement for fear of being reported. Direct observations were therefore required, but my desire first not to distort a field, partly populated by groups of fans committing criminal offences, and second to remain anonymous to police and ground stewards (combined with the general distrust of the fans towards researchers) meant that CPO was adopted.

In order to gain entrance to the field and then maintain a useful research position within it, I found myself both witnessing criminal offences and being put under pressure to commit them personally. This article does not intend to justify my own methodology. Indeed, there were a number of strategies adopted in my earlier research that I would now consider very difficult to defend. I would not claim that my method of approaching the dilemma was the 'right way'. The contradictions and inconsistencies of my own approach will become apparent. Instead, I seek to problematise my own research practice in retrospect to draw out the dilemmas and ethical conundrums that researchers face whilst operating within these kinds of fields. Examples from this fieldwork are included to illustrate the discussion of where the committal of criminal offences fits into the general debate on ethics in fieldwork. Reference will also be made to a number of other ethnographers who have worked in criminal or quasi-criminal fields, for example, Armstrong (football 'hooligans' 1998), Drury and Stott (anti-roads protesters, 2001), Hobbs (petty criminals, 1996; armed robbers, 1995), Hobbs, Hadfield, Lister, and Winlow (bouncers, 2003), Patrick (gangs, 1973) and Williams, Dunlap, Johnson, and Ansley (crack users and dealers, 1992).

Criminal Offences in Football Crowds

It is inevitable that carrying out CPO in groups involved in criminal activity means the researcher will face pressure to participate in the committal of offences. If the true identity and purpose of the researcher is unknown, and s/he is gaining acceptance from the research subjects through participation in their activities then a choice presents itself. Should the researcher commit offences in line with their participation in earlier non-criminal activities of the group? If the researcher follows this path then their position in the research field will be maintained (or even enhanced should suspicions remain about the researcher). This may lead to the subjects being more forthcoming and honest to the researcher, allowing the collection of richer and more accurate data. Furthermore, where the researcher commits an offence that the research subjects are participating in, s/he may also be preventing undue distortion of the field and the damaging of observational data. However, if the researcher refuses to commit the offence, then their 'undercover' position may become threatened where the subjects become suspicious as to their motive, which could lead to a loss of the research position. In some cases, this could also result in a threat to the researcher's safety. Even if the research position is maintained, the refusal to participate in the offence has the potential to distort the field by altering the dynamic of the group and making some research subjects 'think twice' about committing the crime.[5]

A number of ethical and practical issues need to be addressed when making this decision. First, should an academic researcher pursuing knowledge 'to serve the public good'[6] commit criminal offences that in the short term would not be conducive to this aim? If the researcher breaks the law, is s/he bringing their reputation, that of their employer/funder and even the academic profession itself, into disrepute? The second issue is the practicality of committing criminal offences whilst carrying out research, most notably the risk of arrest, prosecution and negative publicity for the researcher's institution/funder.

The committal of offences by researchers embarking on PO (including overt observations) is nothing new, and a number of academics have admitted committing crimes whilst carrying out research. Armstrong, for example, admitted 'two feeble attempts at punching' and running and 'chasing' with the research group whilst carrying out ethnographic research with a 'hooligan firm' (Armstrong, 1993, p. 17), and Drury was amongst a crowd of 12 protesters arrested for breach of the peace whilst carrying out PO at an anti-roads protest (Drury & Stott, 2001, p. 61). Other researchers have decided not to commit offences. Patrick, for example, felt forced to withdraw from the field (a study of gangs in Glasgow) because he resisted being drawn into escalating violence (Patrick, 1973, pp. 135–139). Polsky also argues that the researcher must 'draw a line' between her/himself and the 'criminal':

> . . . in doing field research on criminals you damned well better *not* pretend to be 'one of them', because they will test this claim out and one of two things will happen: either you will (. . .) get sucking into 'participant' observation of the sort you would rather not undertake, or you will be exposed, with still greater negative consequences. (1969, p. 124)[7]

However, despite the dilemma not being a new one, little guidance is provided by the codes and frameworks on ethical research conduct. Initially my own research adopted a phenomenological position, stating that the only way I could gain an understanding of the 'lifeworld' of the subjects was to fully immerse myself in the group's activities through PO (see Bruyn, 1970, pp. 69–70, p. 204; Sellitz et al., 1965, p. 283), which would include participation in any offences they may commit. Such a course of action would have put me, the research subjects and the reputation of my institution in serious jeopardy and I felt a compromise had to be found. In the field of football-crowd disorder, research subjects frequently commit a wide range of mainly lower level criminal offences, the most typical being: drink and drug offences, ticket touting, breaches of the Football (Offences) Act 1990 (invading the pitch, throwing missiles, indecent or racialist chanting), disorderly and threatening behaviour, assault and actual bodily harm, theft and criminal damage. Additionally, the groups would on occasion commit more serious offences such as riot, affray, violent disorder, robbery and assault occasioning grievous bodily harm.

Whilst it was possible to avoid committing some individual offences, a refusal to commit crimes on a regular basis would have aroused suspicions and reduced research opportunities. As a result, I committed 'minor' offences (which I tentatively defined as those which would not cause direct physical harm to a research subject) on a weekly basis as part of the research routine. My strategy was to commit only the offences which the majority of the research subjects were committing and that I considered necessary to carry out the research. Furthermore, whilst I would commit lesser offences with regularity I would, if possible, avoid more serious ones. In practice, this strategy was not as simple to follow as first envisaged, and on numerous occasions I found myself straying from it either on the basis of my personal morals or to maintain or enhance my position in the field. There follows three examples taken from a three-year period (ending in 1998) carrying out covert research with fans of an English football club for my PhD thesis that illustrate some of the aforementioned dilemmas. The work was initially self-funded and at no stage was I required to gain ethical clearance, although I was encouraged to be ethically reflexive about my methods. These examples demonstrate the rationale for my decisions – i.e. to enable me to gather rich, undistorted empirical data that would otherwise have been inaccessible.

'Accessing the Field' in a Literal Sense

During my period in the field, the research subjects frequently invaded the football pitch in my presence. Legislation makes any infringement onto the pitch an arrestable criminal offence and under certain circumstances it can be construed as threatening behaviour under the Public Order Act. One extreme example of this practice occurred when approximately 400 fans (including the 'risk supporters' under observation) scaled perimeter fencing and massed on the side of the pitch as a match drew to a close. At the other end of the stadium, rival fans were also climbing the fences to invade the pitch and the police were preparing a large-scale deployment. It was clear that a confrontation was going to occur and that the best place to observe the intricacies of the confrontation and policing would be on the pitch. Therefore, I joined the supporters on their charge across the pitch at the final whistle that resulted in them chasing the home supporters (who simultaneously invaded the playing field) back into their own stand. My positioning on the pitch, rather than watching lawfully from the stands, enabled me to pick up the mood of the supporters involved in the 'charge' (many fans, for example, became distressed when they realised the celebratory charge was turning disorderly). It also enabled me to witness first-hand the actions of the police officers one of whom seized me and threatened me with arrest if I did not return to the stands immediately. In line with the actions of research subjects I had observed in similar interactions, I followed the officer's advice, on this occasion preventing possible embarrassment for myself and my institution.

'This Train Has Been Designated Alcohol-free'

Another football-specific criminal offence I committed was taking and consuming alcohol on 'football specials' (i.e. transport organised to carry visiting supporters to matches).[8] This is a low-level offence that is frequently committed by fans and when I was travelling with 'risk supporters', it was expected that alcohol would be taken onto transport. Smuggling alcohol past the police and drinking on football specials proved to be an important method of encouraging social interaction, which in turn led to increased opportunities for data collection. Finally, the process of attempting to smuggle alcohol onto football specials proved an excellent way of testing the effectiveness of this public order tactic, demonstrating that where supporters were determined to breach the legislation, they would usually be successful. Furthermore, I approached the research from a socio-legal perspective which, in contrast to positivist legal approaches, encourages critical analysis of the basis and utilisation of laws such as this. From an early stage of my research, I became highly critical of the provisions of the Sporting Events (Control of Alcohol etc.) Act 1985, considering it badly drafted, rushed legislation that caused more public order problems than it solved (Pearson, 2000). As a result, I had no moral qualms about breaching this statute, and found little criticism from my academic peers. However, that is not to say that such activity was therefore justifiable, or would have been considered acceptable by my institution.

'You're a Bit of a Nutter!'

The most serious dilemma I encountered was where I considered it necessary to commit an offence that the research subjects were not participating in at the relevant time. This is difficult to justify because the 'distorting the field' argument cannot be used. For individual observations and non-covert work, it is difficult to think of occasions when this would be necessary but my own covert research with a particular group of supporters over an extended period provided new challenges. For short periods of covert research, it is easier to avoid committing offences, or to 'pick and choose' which offences to commit. Issues of trust between the subjects and the researcher are less complex and researchers can either avoid committing offences, or are able to risk 'pretending' to commit them (e.g. miming offensive chanting in a crowd).

However, during longer periods of covert research, these tactics for avoiding the committal of individual offences are more difficult to apply. Whilst avoiding participation in disorder in which research subjects are involved on a few occasions may not endanger the field position, acting in this way over a number of months will inevitably lead to questions and suspicions arising that could lead to a loss of the research position. In groups of 'risk supporters', this can be particularly problematic in the aftermath of undercover police operations and the recent indiscreet revelations of covert journalists with hidden cameras.[9] As a result, the covert researcher needs to act in line with research subject norms over the entire period of research if s/he wishes to retain trust and access. This obviously puts the researcher under greater pressure to commit criminal acts, but at the same time provides the opportunity to 'opt out' of some actions, provided that s/he has already gained the trust of the research subjects.

The pressure I faced to commit minor offences such as those noted above was always present, but I was usually able to avoid committing the more serious offences provided that trust was maintained when my behaviour was viewed over a longer period. As a result, on several occasions, I committed offences that, whilst recognisable to the sub-culture of the research subjects, were not always in keeping with what was occurring in the field at that given moment. I took this action when I was concerned that there were suspicions about my conduct that could risk my research position particularly where I had 'opted out' of crimes committed by the subjects (or where I intended to in the future), usually because I considered them unacceptable risks to the safety of myself or other research subjects, or because I feared identification by the authorities.[10] On one occasion, for example, when I believed it necessary to prove my reliability to the subjects, I individually confronted a small group of rival supporters in a public house. The attempt was purely 'for show' as I predicted the group would intervene and prevent any serious physical confrontation. Nonetheless, the action was both criminal (threatening behaviour) and in the short term seriously distorted the field.[11] My justification for this action at the time was that it enhanced my position in the field and I was accepted for the remainder of the season as one of the 'hardcore' despite my continual 'opting out' of more serious offences. In fact, towards the end of this research period, I was informed by one of the research subjects that I was considered to be 'a bit of a nutter' by some of the group. This was despite the fact that during my entire time in the field, I had not 'connected' with a single punch nor been arrested, in contrast to many of the research subjects.

This example demonstrates the extremes of the dilemma of the covert participant observer working for long periods of time in 'criminal' fields. A methodology which required the researcher to carry out such actions may well fail to get ethical clearance by a University or funding body. However, in closing the door to this type of research strategy, the possibility of carrying out an ethnographic analysis of an undistorted field without a highly increased risk of physical harm (and arrest) would be closed. Without taking such actions, future observations would have needed to be external and of an overt nature. This, in turn, may have distorted the field and increased the chances of observer misinterpretation of events and the motives/actions of the research subjects. I considered that my actions were justifiable in terms of research data gained from a closed field which over the years has been much misunderstood, and from research subjects who have been subject to serious injustices and maltreatment by those in authority (see Armstrong & Hobbs, 1994; Giulianotti, 1994; Greenfield & Osborn, 1996; James & Pearson, 2006; Pearson, 1998, 1999, 2006; Scraton, Jemphrey, & Coleman, 1995; Stott & Pearson, 2006). The data gathered fed into findings that were publicised in academic publications, books, newspaper articles, radio and television, highlighting these injustices. Later, work in the area as part of a larger project relying in part on similar methods fed directly into police practice which resulted in a significant reduction in both disorder and arrests at international football tournaments (Stott & Pearson, 2007). However, in terms of arguing that 'the end justified the means', it should be kept in mind that the extent of this impact could not have been predicted when I initially embarked upon the research. My initial overriding intention was not to publicise injustices or change things 'for the better' but merely to gain suitable data for my PhD thesis! So attempting to justify research methods on utilitarian grounds of harm and benefit is problematic as we can never be sure of the positive impact our research will have (nor, indeed, of the scale of the harm that may arise).

Consistency, Guidance and Academic Freedom

Even if we accept that the type of strategies I followed could be justified (either by their necessity or their outcomes), my approach could still be criticised due to a severe lack of consistency. At no stage was I able to determine and justify hard-and-fast rules about what offences I would commit and when. My justifications were initially based on the 'distortion of the field' argument, that I would do everything that was necessary to reduce distortion of the field by my own presence. However, personal ideas of ethics and morality impinged upon this and there were some specific offences that I refused to commit purely for these reasons. For example, I refused to indulge in racist chanting, even though in the eyes of the law this is considered less serious than other offences I committed. Practical pressures meant that I also avoided other offences, particularly when my crowd observations were combined with interviews with police officers and football officials who were present on the ground. My decision not to follow the 'distortion of the field' argument through to its logical conclusion therefore endangered both the quality of my data and my position in the field. It also damaged my ability to justify some of the unethical methods used during my research as it weakened the claim that committing specific offences was absolutely necessary to carry out the research.

There are a number of approaches that could be adopted when dealing with these issues. The first is to claim that guidance, regulation and consistency about what offences can be committed and under what circumstances is unnecessary, and that decisions about conduct in the field should depend upon the conscience of the individual researcher. It can be argued that hard-and-fast rules unduly hamper research and that only the researcher her/himself knows the unique problems and dilemmas that will arise in their field. Reynolds (1982), for example, argues that it is up to the researcher to determine how s/he wishes to 'serve society' (i.e. whether knowledge itself is an end or whether personal morality is of greater importance), suggesting that a Kantian view of following 'universal' moral rules is not the only way to determine the researcher's conduct.[12]

This argument suggests that it should be the researcher who determines: (1) the importance of their own research to academia and society, and (2) whether any ethical infringements are justified in this light. With the current ethical codes proving to be ineffective in providing clear and unambiguous guidance to researchers embarking on research into criminal fields, this is essentially the current situation (except in cases where applications for funding are carefully vetted and/or where ethical clearance is required). However, this position provides little protection for the individual researcher when it comes to criticism of their methods by their institution, potential funders or the media. It also provides no support whatsoever in the event of an investigation or prosecution by the authorities. Moreover, such an approach invites criticism where it is perceived that a researcher has an inflated opinion of the value of their own work or where s/he considers 'unethical' behaviour to be less problematic than colleagues (e.g. Warwick's (1982) criticism of Humphreys, especially p. 56).

An alternative approach would be to adopt the ethical norms of the research field, in other words, to follow the lead of research subjects in terms of what offences to commit and what to avoid. This stance is supported by Hobbs (1996) who adopted it in his participant observational study of 'wheelers and dealers' in London's East End. Hobbs participated in the field even when this meant involvement in criminal activity, but argued this was not unethical because the ethics adopted were those of the community. He argued that his research necessitated '. . . a willingness to abide by the ethics of the researched culture and not the normative ethical constraints of sociological research' and that 'a failure to adhere to these norms would have closed the research field' (1996, p. 7). This stance could easily be applied in the field of football-crowd disorder because those involved typically follow sub-cultural norms that the committal only of certain crimes, or in certain contexts, is acceptable.[13]

However, this approach also has problems in that it could reduce the opportunity for covert research in areas of more serious criminality, which it can be argued (due to their severity) are in greater need of academic exposure and analysis. Adhering to Hobbs's line in the research of, for example, a 'paedophile ring' would be unlikely to be seen as justifiable by third parties (particularly the media), however beneficial the research outcome.

A third approach is for better guidelines to be laid down by research bodies and institutions about how work in controversial and potentially dangerous fields should be conducted, most obviously by adding to the current codes of practice. Academic peers have a legitimate interest in how research is carried out by others; unethical conduct by colleagues may harm participants, damage the reputation of a department, organisation or institution and make future research in similar areas more difficult, so it is easy to argue that such research should be more heavily regulated. If codes could be re-drafted to include more guidance on research in criminal fields this could theoretically open up research funding to fields that are currently under-researched due to the difficulty of obtaining ethical clearance from funders relying on the current codes. Another advantage of set rules is that this could provide some protection for researchers, particularly against investigation or punishment for the committal of offences in the course of research. Whilst it is not suggested that additions to the codes could – or should – grant any form of guaranteed 'immunity' from prosecution, it would provide some guidance and protection for researchers in this area that would be beneficial in the event of threats of litigation, employment disputes with institutions or investigations by the media.

Unfortunately in the current climate such a development looks unrealistic and here lies a fundamental 'Catch-22'. There are serious problems with the current codes that make the addition of more 'hard-and-fast' rules problematic. The codes are not universally accepted as appropriate frameworks for researchers in the social sciences; Homan, for example, criticised them as being more likely to close discussion on ethical issues rather than stimulate it (1991, p. 38). More recently Hammersley has expressed concern about the role of current codes and ethical committees, noting their tendency to rely upon medical and psychological models that are not always appropriate for social sciences (2006, p. 5; see also Coomber, 2002 with specific reference to informed consent) and arguing that ethical committees are not always in the best position to determine what research should proceed. It is also argued that 'hard-and-fast' rules are not always the best way to approach questions of methodology and ethics, particularly in areas where access is impossible by overt methods (Crow et al., 2006, p. 88) and that as a result '. . . the rigidity of standardised regulation will need to be tempered by a degree of flexibility according to the characteristics of specific research contexts' (p. 95). Increased regulation of covert fieldwork in such fields would almost certainly see *reduced* freedom for academics carrying out such research. Funders and institutions wary of their legal and public relations position would be unlikely to give approval to proposals that could see a researcher risk prosecution or media expose. Additionally, legal obligations under Health and Safety Legislation would make it difficult for an institution to sanction the taking of risks in the field that may lead to physical injury.

In light of these legal and practical realities, the solution may be not increased regulation through codes and ethics committees, but increased non-binding guidance for researchers considering utilising such methods. New researchers in particular could be made aware of the ethical and practical problems with such methods (as well as their potential technical advantages) and more comprehensive research training modules could provide guidance and 'lessons from the field' from those who have used such methods (in the manner of Williams et al.'s (1992) work on personal safety in research). Researchers should be encouraged to adopt strategies to minimise threats such as by carrying out risk assessments, and made aware of the pressures on institutions and funders. They should also be encouraged to be self-aware of the limitations of their own research in terms of justifying contentious methods and encouraged to be forward thinking and open about the actual potential value their research will have above and beyond their acquisition of a PhD and/or publications in high-ranking academic journals. However, such guidance must fall short of preventing such methods being adopted where no other 'ethical' method could be employed to achieve the same results and senior colleagues and research committees should be wary of determining that projects are not 'worthy' of utilising contentious methods.

Conclusion

Little formal guidance is provided to researchers in the social sciences who wish to carry out ethnographic research within 'criminal' fields. If current ethical guidelines, relying on principles such as informed consent, are applied to research proposals in such areas, the research methods that can be utilised are dramatically reduced. However, in areas of criminality, where distrust of authority figures is commonplace, often methods such as CPO can be the only way of gaining meaningful and undistorted data. Ethical codes in the social sciences acknowledge that such methods can be used 'as a last resort' but give little guidance either to individual researchers or ethics committees on how such methods can be used in practice. In particular, how should covert observers act when faced with the pressure to commit offences themselves?

If we accept that the purpose of academic research is to advance knowledge, rather than to serve any particular dominant moral standpoint, then covert methods should be justifiable if they achieve the aim of enhancing understanding of otherwise 'hidden' areas. Furthermore, where we accept that this legitimate aim can justify covert methods, we also need to acknowledge that in many cases, this will lead to the researcher acting illegally within these fields. It seems illogical to permit covert activity in fields of criminality but not accept that the researcher may have to commit occasional unlawful acts. Without the freedom to make these choices, CPO in these areas will be hamstrung. There are some areas of society to which the overt researcher will not be permitted and to which the covert researcher will never be accepted unless s/he participates in activities which would be considered unacceptable according to traditional ethical standards. However, these areas should not be out of bounds for the researcher and, if properly managed, the potential value of research into these areas can offset the risks to both individual and institution.

'Criminal' research fields are deserving of serious academic analysis and the adoption of the most appropriate research method to provide the best data. Without researchers who are willing to embark upon covert research, and are sometimes willing to break the law in order to gather this data, some aspects of society will remain hidden or misunderstood, with negative consequences for those in, or affected by, those areas. In the long term, 'protecting' the research subjects by prohibiting covert methods may have the opposite effect, keeping harm and injustices hidden from the public. It will also reduce the value of academic research generally in terms of 'making a difference', as researchers are steered towards topics that are as 'mundane' as possible (Crow et al., 2006, p. 89).

However, the current situation, whereby researchers operating in these fields often 'go it alone' by justifying their own methods results in difficulties and dangers. Such a situation grants important academic freedom to researchers who are not bound by ethical clearance procedures, but the risks that come with operating in this way should not be underestimated by either researcher or institution. As a result, further guidance for those who consider it is necessary to adopt such strategies is important in order to guide their ethical decisions. This guidance could be used to provide some protection from critical employers or funders and in some cases, police forces, state prosecution services and the courts. However, such guidance should not in the present climate materialise in the form of extended frameworks or codes of practice, which have been criticised for being overly restrictive on methods and closing debate on contentious subjects. There is a difference between guidance and increased regulation. The latter, in the current climate, would almost certainly reduce the freedom to carry out research in areas where fieldwork is required to understand phenomena and drive forward social change. Until the issues and dangers of such methods are further debated and illuminated, some research areas will remain closed and other researchers will continue to operate in potentially dangerous research fields without adequate risk assessment or guidance.

Notes

1. 2002, s. 31. Similar statements are contained within the Socio-Legal Studies Association First Re-Statement of Research Ethics (s. 6.6(2)) and the American Sociological Association's Code of Ethics 1999 (s. 12.01(b)).
2. Para 3.2.2.
3. 2006, s. 4(iii).
4. For example with reference to the reporting of offences against children and the right to privacy.
5. The participant observer must, of course, also be careful that their participation does not encourage offences that may otherwise have not been committed.
6. American Sociological Association Code of Ethics 1999 Principle E.

7. Although Homan points out that Polsky removed symbols of the social researcher and suggests he may not have been as open as he claimed (1991, p. 120).
8. Prohibited by the Sporting Events (Control of Alcohol etc.) Act 1985.
9. Such as the 'McIntyre Investigates' (BBC1, 9 November 1999) and 'Hooligans: A Panorama Special' (BBC1, 2 August 2006).
10. This was a particular concern as my research increasingly saw covert observation combined with interviews/observations with the police, Home Office and football authorities.
11. Although conversely in doing so it led to rich data about the ability of groups of risk supporters to 'self police'.
12. Suggested other methods that could be used include cost/benefit analyses and a balancing of the individual rights of research subjects against the rights of society.
13. For example, the 'unwritten rule' that fighting fellow 'risk supporters' is acceptable, but attacking ordinary fans and members of the public is not.

References

Armstrong, G. (1993). Like that Desmond Morris? In D. Hobbs & T. May (Eds.), *Interpreting the field – Accounts of ethnography* (pp. 3–44). Oxford: Oxford University Press.

Armstrong, G. (1998). *Football hooligans: Knowing the score*. Oxford: Berg.

Armstrong, G., & Hobbs, D. (1994). Tackled from behind. In R. Giulianotti, N. Bonney, & M. Hepworth (Eds.), *Football, violence and social identity* (pp. 196–228). London: Routledge.

Bruyn, S. (1970). The new empiricists: The participant observer and the phenomenologist. In W. Filstead (Ed.), *Qualitative methodology: Firsthand involvement with the social world* (pp. 283–287). Chicago, IL: Markham.

Bulmer, M. (1982). The merits and demerits of covert participant observation. In M. Bulmer (Ed.), *Social research ethics* (pp. 217–251). New York: Holmes & Meier.

Coomber, R. (2002). Protecting our research subjects, our data and ourselves from respective prosecution, seizure and summons/subpoena. *Addiction Research and Theory, 10*(1), 1–5.

Crow, G., Wiles, R., Heath, S., & Charles, V. (2006). Research ethics and data quality: The implications of informed consent. *International Journal of Social Research Methodology, 9*(2), 83–95.

Denzin, N., & Erikson, K. (1982). On the ethics of disguised observation. In M. Bulmer (Ed.), *Social research ethics* (pp. 142–151). New York: Holmes & Meier.

Drury, J., & Stott, C. (2001). Bias as a research strategy: The case of intergroup conflict. *Field Methods, 13*(1), 47–67.

Erikson, K. (1967). A comment on disguise observation in sociology. *Social Problems, 14*, 366–373.

Feenan, D. (2002). Researching paramilitary violence in northern Ireland. *International Journal of Social Research Methodology, 5*(2), 147–163.

Giulianotti, R. (1994). Taking liberties: Hibs casuals and Scottish law. In R. Giulianotti, N. Bonney, & M. Hepworth (Eds.), *Football, violence and social identity* (pp. 229–261). London: Routledge.

Greenfield, S., & Osborn, G. (1996). After the act: The (re)construction of football fandom. *Journal of Civil Liberties, 1*(1), 7–28.

Hammersley, M. (2006). Are ethical committees ethical? *Qualitative Researcher, 2*, 4–8.

Hobbs, D. (1995). *Bad business: Professional crime in modern Britain*. Oxford: Oxford University Press.

Hobbs, D. (1996). *Doing the business: Entrepreneurship, detectives and the working class in the east end of London*. Oxford: Clarendon.
Hobbs, D., Hadfield, P., Lister, S., & Winlow, S. (2003). *Bouncers: Violence and governance in the night-time economy*. Oxford: Oxford University Press.
Homan, R. (1991). *The ethics of social research*. London: Longman.
James, M., & Pearson, G. (2006). Football banning orders: Analysing their use in court. *Journal of Criminal Law, 70*(6), 509–530.
Norris, C. (1993). Some ethical considerations on field-work with the police. In D. Hobbs & T. May (Eds.), *Interpreting the field – Accounts of ethnography* (pp. 122–144). Oxford: Oxford University Press.
Patrick, J. (1973). *A Glasgow gang observed*. London: Eyre-Methuen.
Pearson, G. (1998). The English disease? The socio-legal construction of football hooliganism. *Youth and Policy – The Journal of Critical Analysis, (60)*, 1–15.
Pearson, G. (1999). Legitimate targets? The civil liberties of football fans. *Journal of Civil Liberties, 4*(1), 28–47.
Pearson, G. (2000). Legislating for the football hooligan: A case for reform. In S. Greenfield & G. Osborn (Eds.), *Law and Sport in Contemporary Society* (pp. 182–200). London: Frank Cass.
Pearson, G. (2006). Contextualising the football disorder act: Proportionality under the hammer. In S. Greenfield & G. Osborn (Eds.), *Readings in law and popular culture* (pp. 91–118). London: Routledge.
Polsky, N. (1969). *Hustlers, beats and others*. Harmondsworth: Penguin.
Reynolds, P. (1982). Moral judgements: Strategies for analysis with application to covert participant observation. In M. Bulmer (Ed.), *Social research ethics* (pp. 185–216). New York: Holmes & Meier.
Scraton, P., Jemphrey, A., & Coleman, S. (1995). *No last rights: The denial of justice and the promotion of myth in the aftermath of the Hillsborough disaster*. Liverpool: LCC/Alden Press.
Sellitz, C., Jahoba, H., Deutsch, M., & Cook, S. W. (1965). *Research methods in social relations*. London: Methuen.
Stott, C., Hutchinson, P., & Drury, J. (2001). 'Hooligans' abroad? Inter-group dynamics, social identity and participation in collective 'disorder' at the 1998 World Cup finals. *British Journal of Social Psychology, 40*, 359–384.
Stott, C., & Pearson, G. (2006). Football banning orders, proportionality and public order. *Howard Journal of Criminal Justice, 45*(3), 241–254.
Stott, C., & Pearson, G. (2007). *Football 'hooliganism', policing and the war on the 'English disease'*. London: Pennant Books.
Warwick, D. (1982). Tearoom trade: Means and ends in social research. In M. Bulmer (Ed.), *Social research ethics* (pp. 38–58). New York: Holmes & Meier.
Williams, T., Dunlap, E., Johnson, B., & Ansley, H. (1992). Personal safety in dangerous places. *Journal of Contemporary Ethnography, 21*(3), 343–374.

44

Ethnographic Intimacy: Thinking through the Ethics of Social Research in Sex Worlds

Maria Pérez-y-Pérez and Tony Stanley

Introduction

Ethnography enables researchers to spend extended periods of time in one or more settings, develop close relationships with participants, and look for links between the diverse pieces of that world and between that world and other social worlds (Agar, 1996; Calestani, 2009; Mattingly, 2005; Strathern, 1999). The value of ethnographic methodology is that we can enter the world of the people that interest us (Pérez-y-Pérez and Stanley, 2005), or 'climb inside it and spend time with its inhabitants' (Agar, 2006a, p. 18). The principles of ethnography are central to following and describing the actions, activities and relationships of the worlds we wish to understand or *peer into* (Latour, 2005; Law, 2004; Mol, 2002). The objective is to record the 'situated rationality of action' (Murphy and Dingwall, 2007). Key to building such intimate knowledge, participant observation is a data collection method (informal conversations, casual observations and other unobtrusive interactions) that establishes rapport and forges personal relationships, thus weaving the researcher into the diverse worlds that constitute the research field (Bolton, 1995, 1996; Pérez-y-Pérez and Stanley, 2005; Rambo Ronai, 1998; Weseley, 2006).

Source: *Sociological Research Online,* 16(2) 13 (2011). http://www.socresonline.org.uk/16/2/13.html

Ethnographic methodology brings into question what participant observation entails and means. For some researchers, 'going native', in sex worlds, for example, may jeopardise their professional status (Hart, 1998). For others, complete participation becomes the key to insider status. Bolton (1995, 1996) immersed himself in a gay community to address the processes of adaptation and change towards issues such as AIDS, albeit using unorthodox methods (see Wengle, 1988). Other researchers of sex worlds have also chosen participatory roles with less contentious pathways (McKeganey and Barnard, 1996; O'Connell Davidson, 1998; Sanders, 2005). As Hodgson contends, 'What makes ethnographically based cultural critique so much richer – that is, our engagement with peoples as well as "texts" – is what from another perspective makes it so problematic, in terms of the potential repercussions of our work for those we study' (Hodgson, 1999, p. 202). Personal and disciplinary ethics make us accountable in unique ways to the people we study for the possible consequences of our work and writing, however intended or unintended. Research subjects consent to an intimate relationship within which they are going to be used as objects. The research process is one of objectification and appropriation, it does not become an ethical activity simply because researchers have followed agreed rules and guidelines (O'Connell Davidson, 2008).

Using our experiences of ethnographic research, we consider some of the ethical challenges we encountered and introduce some of the strategies we assembled as researchers involved in *intimate* research. The sticky ethical issues we faced from the outset and over the duration of our studies were largely unanticipated, and required us to make ongoing quick but sound ethical decisions. Our participant observer positions, Pérez-y-Pérez as a receptionist/shift manager in a massage parlour, and Stanley as a social worker in a non-profit organisation providing HIV prevention services, allowed us entry into intimate aspects of the lives of the people who inhabit these worlds. We use examples from our own studies to discuss how we entered and positioned ourselves within these worlds; reflect on the level of intimacy developed by researcher and researched and how we managed these relationships; and reflect on our 'hybridity' and the *sticky* issues as they presented themselves: informed consent, confidentiality and criminal activities. We discuss these issues further within the context of ethics committee policies, processes and guidelines and the lack of *fit* between these and ethnographic research.

Choosing Ethnography

We chose ethnography as a methodology to avoid the standard 'question and answer' approach. We wanted to enter our fields of research to better understand the meanings and experiences of the people we talked to. In 'entering' their worlds and 'following' their lives (Latour, 1999), the discovery of connections (Agar, 2006b) and the stories people shared became crucial to the meanings and the findings we were able to describe (Pérez-y-Pérez, 2003; Stanley, 1999, 2000). By entering the sex world, Pérez-y-Pérez was able to explore the complex social organisation of prostitution (O'Neill, 1997), and consider the configuration of networked actors (social and technical) that produces sex markets (Pérez-y-Pérez and Stanley, 2005), at a time when legislative changes were seeking to decriminalise prostitution (Prostitution Reform Act, 2003). As a social worker and active within the local gay community, Stanley was able to enter spaces where sex was negotiated. This was crucial to sparking talk and stories in the context of risk and HIV, when police activity and media coverage added further complexity to the ethical considerations of ethnographic research.

'Doing' Ethical Research: Issues *for* Ethnographers

Ethics committee guidelines shape the performance of research and the relationship between researcher and researched. Professional codes of conduct establish principles of justice, safety, truthfulness, confidentiality and respect that the researcher is expected to follow (O'Brien, 2001). Ethical codes and guidelines are generally informed by legislation (Dawson et al., 2003), which also stipulates sanctions that follow failure to meet ethical requirements. In practice, researchers often develop a set of protocols and procedures to suit their research setting or context based on ethical guidelines. Rather than set their own agendas, researchers are required to undertake active and careful discussion and negotiation with ethics committees, and cautious consideration when attending to potentially difficult ethical decisions once fieldwork is underway (O'Brien, 2001).

In considering applications for research approval, ethics committees may raise methodological concerns where, rather than traditional approaches, qualitative data collection relies on observation and participation, and informal, conversational-style interviews (Mattingly, 2005; Murphy and Dingwall, 2007; Sanders, 2006). The validity of such methods, the amount and accuracy of information to be collected, how analysis will be undertaken and the subsequent quality of the data and findings may be questioned (Harper and Corsín Jiménez, 2005; Strathern, 2000). In addition, methods such as covert data collection, in sex work research, for example (Pérez-y-Pérez, 2003; Sanders, 2006), where not all inhabitants of the research setting can be informed that they are being observed,[1] can appear unethical and potentially dangerous for the researcher. Misunderstandings about the research setting, such as a 'red light district', massage parlour, apartment, or illegal brothel, often become the focus of concern, particularly where activities are considered illegal. Stereotypes of *marginal sexualities*, and the link with criminal behaviour, risk and disease, raise questions of researcher safety.

Informed Consent

The concept of informed consent is a central tenet of research ethics policies where the transparency of a social and psychological reality is assumed that enables researchers to provide full and accurate information about the research and all 'foreseeable risks' to autonomous subjects able to make informed choices is assumed (Wilkinson, 2001). The protocol for informed consent involves giving participants an information sheet or letter with comprehensive and accurate details about the research, including the demands, risks, inconveniences, discomforts and benefits that might be involved, and the participant's right to withdraw from the study (Dawson et al., 2003; O'Brien, 2001). Researchers must ensure participation is voluntary (no coercion or inducement beyond reasonable compensation), participants are adequately informed of all aspects of the research process, informed consent is gained (departures from this must be fully justified) and confidentiality and anonymity is assured (where possible).

Ethnographers, however, may encounter a lack of *fit* between ethnographic practice and ethics committee requirements (Hodgson, 1999; Mattingly, 2005; Murphy and Dingwall, 2007). In the practical reality of an ethnographic study, relationships within the field can be unpredictable and precarious, subject to individual whims, organisational necessity, and so forth; these may restrict access to sensitive aspects of the setting (Aagaard-Hansen and Vang Johansen, 2008; Meskell and Pels, 2005; Strathern, 2000). Equally, new opportunities may arise or be sought as the study progresses. Access to settings, interactions and activities are rarely guaranteed, can be blocked as quickly as given, and depend upon a researcher's capacity for sustaining the goodwill and cooperation of the participants (Murphy and Dingwall, 2007; Punch, 1986). If access is withdrawn, the researcher must renegotiate and re-establish goodwill and trust, and seek new ways of *getting back in*. The initial informed consent can be both tentative and limited. Over time, as trust is established between researcher and researched, access may be granted to previously restricted areas or interactions. Consent should, therefore, be viewed as a relational, sequential and continual process, rather than a contractual agreement per se (Murphy and Dingwall, 2007).

Risk: Relationships, Law, and Personal Boundaries

In ethnographic research, issues of accountability and consequences are more personalised and, therefore, precarious than in other research methodologies (Castañeda, 2006; Hodgson, 1999; Mattingly, 2005; Pels, 1999). Indeed, the risks can be highly indeterminate, subjective, and not easily specified or communicated in advance to participants (O'Connell Davidson, 2008; Roberts and Sanders, 2005). Persuading potential participants to engage in discussions of hypothetical risks, may compromise entrance to a research group (O'Connell Davidson, 2008). The intimate relationships conscientiously built, often over extended periods of time, may make participants vulnerable to hurt or even harm (Stacey, 1988). Many potential risks, however, mirror those encountered in everyday life, for which formal consent is unnecessary (Sanders, 2006). A practical solution, therefore, is to base consent on the trust between researcher and researched to exercise ongoing judgement (Murphy and Dingwall, 2007). This is particularly pertinent when a researcher is firmly embedded as a *participant* and becomes inextricably socially and emotionally tied with the inhabitants of the field (see Pérez-y-Pérez, 2003; Allison, 1994; Flowers, 1998a; Weseley, 2006).

In addition to the issue of informed consent, ethnographic researchers may need to consider matters of illegal or criminal behaviour, as well as personal and institutional moral boundaries (Pérez-y-Pérez, 2003; O'Connell Davidson, 1998, 2008; Punch, 1993; Sanders, 2005; Van Maanen, 1992). When Pérez-y-Pérez undertook her study, sex work was still regarded as a criminal activity. As a receptionist/shift manager, she negotiated the sale of sexual services performed by women who operated from the massage parlour. She also took part in a number of prostitution-related activities (Pérez-y-Pérez, 2009). Similarly, Stanley found himself in an ethical 'minefield' as participants discussed sexual negotiation while police executed a search warrant as part of an unfolding criminal investigation into the activities of HIV positive men knowingly transmitting HIV. Ethical considerations, therefore, were negotiated in spaces where research practice was taking place.

Associations between HIV/AIDS and disease, drug-injecting sex workers, potential violence within research and sex settings, sexual deviance, and criminal underworlds all surface in institutional ethics discourses relating to ethnographic research involving sex worlds, fuelled by assumptions about the type of people who organise and inhabit these worlds. A central concern for ethics committees can be the danger that the researcher is exposed to (and, possibly, the reputation of the institution responsible for the researcher) (Murphy and Dingwall, 2007). Indeed, the routines the researcher engages in may entail long hours, in fairly unknown or secluded locations, often late at night, and sometimes alone, mirroring the lives of participants (Pérez-y-Pérez, 2003; Rambo Ronai, 1998; Weseley, 2006).

Ethnographers confront a number of sticky questions when negotiating the 'fit' between ethics committee expectations and the challenges of research methodology: What constitutes consent? What becomes coercive? How does one handle the complex and often conflicting requests for confidentiality, on one hand, and authorship and ownership, on the other? How much of one's interpretations should be shared with informants? (Adams, 1981; Fluehr-Lobban, 2003; Jorgensen, 1971; Mattingly, 2005; Silverman, 2003).

Intimacy and Ethnography

The nature of ethnography requires researchers to minimise their impact on the research setting, reduce any disturbance or inconvenience to participants and avoid disrupting the setting, thereby rendering the findings unrepresentative. Thus, researchers become integrated into the setting, form relationships with participants and participate as a member of their social worlds. When working with *marginalised* groups of people, building relationships is a necessary part of gaining trust and access (Agustín, 2008; Boynton, 2002; Wahab, 2003). Over time, participants may come to overlook the research purpose and the researcher's identity. As the line between researcher, friend, confidante or colleague becomes blurred, participants may disclose information that they do not consider relevant but which the researcher considers is relevant to the research (O'Connell Davidson, 2008). The dilemma then is in reconciling effective participation with truly informed consent. Bourjois questions the expectation that the researcher will 'build rapport' without distorting the 'host' setting as a means of encouraging people to forget that they are being observed and recorded. He argues, 'To maintain truly informed consent we should interrupt controversial conversations and activities to re-announce our presence and to make sure everyone is aware of the implications of what they are saying or doing' (Bourjois, 1990, p. 52).

Reflexivity is needed for researchers to consider how to manage the intimacies that develop within the field, how these come into being, and to what extent they are considered data. For example, whether the participants' prior consent justifies the use of disclosures, activities and personal settings as data (Murphy and Dingwall, 2007; O'Connell Davidson, 2008; Roberts and Sanders, 2005); or whether people who consent to the research intimacies implied by an ethnographic study, consent to a relationship that is time-limited and ultimately leads to their objectification (O'Connell Davidson, 2008). What then does it mean to consent to such an arrangement? How do we, at the onset of the research, explain the process and anticipate what will occur in the field? How 'native' will we go to build such relationships? What risks are we prepared to take in terms of our own safety and that of our participants in building intimate relationships?

The ethical or moral base of the relationship between researcher and researched cannot be adequately captured by the language of contractual consent or of care. In other contexts, when we contract or ask someone for help or assistance, we think in terms of gratitude and obligation rather than informed consent. Roberts and Sanders (2005) see the willingness of participants to be researched as a *gift*, one of immense value for the studies and subsequent careers of researchers; there may be little direct or immediate benefit for the subjects nor financial or material gain (see Agustín, 2005; O'Connell Davidson, 2008; Roberts and Sanders, 2005; Van Maanen, 1983).

O'Connell Davidson (2008) suggests that such *gift-giving* can lead us into a symbolic domain that is non-contractual, a terrain where people come together in a 'complex web of mutual dependency and obligation', and are usually careful about whom to approach for help. To conduct social research it is often necessary to ask for help indiscriminately, or to ask for help from those whom we would not usually accept gifts – perhaps because we perceive them to be too vulnerable – but sometimes because we find them morally, politically or personally objectionable and would wish to avoid any on-going obligation to them (O'Connell Davidson, 2008, p. 21).

The relationship between researcher and researched in ethnography can resemble but also differ from normal social relations, such as mutual friendships, therapist and client, employer and employee, service provider and consumer. A research relationship, however, is, inevitably, a social and emotional relationship. This complicates our understanding of 'informed consent': not all outcomes of a social relationship can be predicted at the outset; nor can the relationship's development be controlled. In our research, we also inhabited the worlds of our participants, crafting bodies and language that allowed us to be enrolled as members of these worlds. This meant that we were frequently at odds with our hybrid selves.

In sex world research, issues of strong stereotypes and potential dangers must be confronted before the project begins. Although research designs are, therefore, perhaps over scrutinised and charged with queries that expect more insight from the novice researcher, this also encourages reflexivity, even at this initial stage. Thinking ahead to foreseeable problems, having a set of plan 'B's' if the initial methods fail, and taking time out of the fieldwork to reassess 'good practice' in an intense and volatile environment are crucial and highlight the importance of collegial relationships in the field. We argue that these considerations go beyond those of ethics committee and institutional guidelines.

The following sections discuss these ethical concerns of intimacy and ethnographic research. Examples from our own ethnographies illustrate how we managed some of the foreseeable matters and how we grappled with some unplanned encounters. First, Pérez-y-Pérez examines the ethical issues that emerged in her sex work research. Stanley then outlines the ethical dilemmas he experienced in his research on sexual negotiations/risk and HIV.

Researching Sex Work: Fitting *in* and Staying *in*

I entered the sex work industry and embarked upon my fieldwork armed with information sheets, consent forms and the good intentions of professional ethics informed by my institutional guidelines and accounts of ethnographic fieldwork (see Gilbert, 1993; Oakley, 1981; Punch, 1986; Stanley and Wise, 1991; Wolcott, 1995). However, at my massage parlour *job interview* (for the receptionist/shift manager position) the senior manager made it clear that my 'research presence' was insignificant compared to my 'paid employee presence'. As I began to explain how I would assure safety, confidentiality and anonymity, I was silenced by a wave of the hand. The manager listed the responsibilities of a shift manager, then reiterated that my research should not interfere with my employee role. Roberts and Sanders (2005) see this as an example of how the researcher needs to pre-empt how an audience will construct the researcher in relation to their own lives.

Despite the lack of interest in my research, gaining informed consent from potential participants (formal interviews) and inhabitants of the research field (participant observation) remained vital to me as I was bound by my institutional ethics committee guidelines. However, I also entered the field as a participant, and realised that the issue of consent was not a linear or transparent process. To devise an ethical research strategy, therefore, I needed to be guided by the practices of my research world. Sex work networks are organised around verbal agreements and unspoken understandings. Sex work practice is learned through oral storytelling and by the piecing together of shared knowledge, experiences and ideas of good and bad sex work practice that serve as a vehicle of community memory (Orr, 1990) or a knowledge base for workers (Pérez-y-Pérez, forthcoming; George, 2008; Gherardi and Nicolini, 2002). The implications of a signed consent form in this context soon became clear. The illegality, stigma, and 'hidden' nature of sex work are such that the disclosure of real identities is problematic for workers. For many, sex work is a means of supplementing income, such as unemployment and child benefits. Any documents that require real names and signatures may be perceived as a potential risk to this income (Pérez-y-Pérez, 2003; Phoenix, 1999; Sanders et al., 2009). Documentation of personal details has also been associated with the surveillance of prostitution and the assignation of the prostitute label (Pérez-y-Pérez, 2003; Palmer and Reed, 2001; Prostitution Law Review Committee, 2005).

Following the traditional route and ethics guidelines, I drafted information sheets for the women within the massage parlour. These explained the proposed study, and contained sex work topics that were regularly discussed around the *staff table* in the massage parlour during 'down time'. In the printed document, however, these topics no longer seemed familiar, disrupting accepted conventions of sex work, and rendering the building of relationships between myself and the women problematic. The initial four participants commented at the end of their interviews that they felt uneasy about documenting their participation in the study. They did not want a copy of their consent form or information sheet, quickly reading before handing them back to me. Two of the women refused to sign the consent form in spite of my assurances of anonymity and confidentiality, commenting that this 'paper work' made it seem 'too formal' or 'official-like'.

A reasonable and practical solution, therefore, was to convey information and receive consent verbally. Thus, I followed Agar's recommendation: 'To begin your work honestly by presenting yourself and your task in some way that will make sense to the group members' (Agar, 1996, p. 111). I needed to explain, at regular intervals, to the women, management and workers, that my role as shift manager would be a form of data collection. Moreover, rather than information sheets, the gossip grapevine within the massage parlour was extremely effective. By the time I began my first shift, most of the women were aware that I was a researcher and student studying the sex industry. I did not, therefore, have the awkward situation of going in 'cold' to explain who I was and what I was doing.

What to say and who to tell in ethnographic studies is often a tricky issue. Norris (1993) suggests that explanations constructed for the research are conditional upon the audience being addressed. Flowers (1998a), in her study of phone sex operators, saw no need to disclose her graduate *student* status. This information was not well received by her co-workers, who felt uncomfortable and distanced themselves, when initially there had been camaraderie. To increase their comfort level, Flowers responded with her own stories of harassment and agony and thus gained entry to the setting and relative acceptance (Flowers, 1998b).

My participation in the massage parlour world required me to learn the role of shift manager in order to understand the organisation of sex work. Rather than the 'passive observed', the women I worked with were 'experts' in relation to me. Like Flowers, I intended my presence to be perceived as non-threatening, and to become the student in order to understand, observe and participate (Agar, 1996). This 'recasting of the self' in the role of 'apprentice' as part of the democratising process can be critical in terms of group acceptance and the establishment of trust (Agar, 1996; Punch, 1993). The peculiarity of my 'apprenticeship' was that I entered the massage parlour as a member of management.

Like other ethnographers (Hastrup, 1995; Kondo, 1990), I observed and learned the cultural competencies of how to act and behave and the nuances of interaction and engagement in the field. The culture of my research subjects was 'written over me' as I became just another 'damn shift manager' or 'one of the girls'. How I presented my(self) body was critical and required an element of authenticity (see Hochschild, 1983); seemingly a contradiction when prostitution is predicated on pseudonyms, pseudo-bodies, pseudo-personalities and identities. I responded to the norms of the massage parlour, where bodies are displayed, sexualised, disciplined, desired, shaped, touched and talked about, 'crafting my body as part of the crafting of the field' (Coffey, 1999, p. 68). Fieldwork is an embodied experience; how our body looks, and is used can impact upon access, field roles and relationships (Coffey, 1999). This I intuitively understood as necessary for the successful accomplishment of fieldwork (see Allison, 1994; Hobbs, 1988; Punch, 1993).

Part of this 'field self' is the ability to learn physical actions and articulations that fit the context (Hochschild, 1983). I did this through watching and being guided by the women I worked with, piecing together my own body management, just as other workers do (Pérez-y-Pérez, forthcoming). I also used my real name, despite the use of pseudonyms to meet the need for confidentiality. This decision was an attempt to present myself as 'honestly' as possible in order to gain the trust of workers and management, and non-threatening in terms of competition. However, using my real name could also be read as an attempt to separate myself from sex work *per se*.

The shifts I worked were long, often beginning at 6.30pm and finishing at 6.00am. During the night there would be blocks of 'down time' (Pérez-y-Pérez, forthcoming), filled by talking, sharing stories, or partaking in 'bonding activities'. Sharing intimate personal life experiences around the *gossip table*, in an area not accessed by clients, or in small, intimate groups, became the key way in which workers could make decisions concerning me, not as a sociologist or researcher, but as another woman. At first, I was the subject of research, regularly interrogated and asked key questions: 'What do you think so far?' 'What does your partner think of you working in a massage parlour?' I also established friendships with women who shared similar interests and life experiences, and accepted invitations to join women who worked together regularly on the same shift when they socialised outside work. Through these informal meetings, I learned a great deal about the research questions I was interested in, such as their personal lives and their work. I was also introduced to and socialised with their families and friends; most had no idea that these women worked in the sex work industry, and I took care to use their real rather than fictional working names.

The nature of ethnography means that not all field inhabitants are aware of their participation. Late into the night on busy shifts, I often found myself 'entertaining' one or two clients at the bar; some divulged personal details and history to me without prompting. To inform everyone who walked through the door of my research would have altered the dynamics of the setting, and, potentially, risked the loss of business clientele (see Agar, 1996; Fountain, 1993; Hobbs, 1993; and Norris, 1993). By default and design, clients often believed, or were led to believe, that I was simply a massage parlour receptionist or shift manager. Under such pretences, I organised their sexual encounters with sex workers. Thus, the practice of participant observation can be seen as interactionally deceitful (Bulmer, 1982; Norris, 1993). However, I was also actively participating in the field as a paid employee, and recording my experience of work, of which clients were a small part.

My hybrid status as a worker/researcher provided me with understandings of the business of sex work that had not been detailed by other researchers. As shift manager, my job was to coordinate the work of a parlour and present it to clients and the police in such a way that it ran as a 'normal' business. This provided me with a perspective on the sex work world that differed from that of the sex workers but, at the same time, placed me in proximity to them.

Researcher Intimacy with HIV Positive 'Subjects'

As a social worker employed within a non-profit organisation providing HIV prevention services, my research interest stemmed from the relative silence within social work research around sexual practices connected with HIV/AIDS (Stanley, 2000). In particular, I wanted to explore issues of sexual risk and pleasure and the contexts within which sexual behaviour is embedded (Stanley, 1999). With the introduction of protease inhibitor drug treatments in the late 1990s, new sets of social issues required increased social work assistance. I wanted to answer questions about how sexual negotiation is organised when one lives with HIV or AIDS. From a legal perspective, the case law relating to HIV positive sex practices was unfolding while I was in the field. More importantly, the research literature did not cover all the ethical dilemmas I faced as a researcher (Bolton, 1996). Like Pérez-y-Pérez, I negotiated a range of ethical boundaries and transgressions to enter the spaces where talk of sexual negotiation happened.

With my knowledge of Foucauldian notions of power and understanding (Foucault, 1972), I established an ethnographic project and immersed myself in the field. As a qualified social worker, sociologist, queer man and knowing the sexual spaces of interest, I applied to two institutional bodies, health and academic, for ethics approval. In preparing my applications, I considered a number of questions: How much do we comfortably reveal in these processes? How much does the ethics committee need to know? And, paradoxically, is keeping back 'the gay sexual self' an easier means of gaining ethics approval? From a professional perspective, I was faced with the work of HIV positive sexual negotiation on a day-to-day basis. From the perspective of a researcher, I was guided by professional codes of ethics. These became the dominant models that guided my fieldwork. Ethics committee guidelines also seek to keep researchers and institutions free from risk. As other ethnographers have discovered, the ethical risks are complex, complicated and ever present: What would we do if a subject advised us of knowingly infecting another man? Do we ask participants to stop just before they disclose crimes? And if we do, in what way do we complicate the co-construction of the 'research subject' for the purposes of institutional categorisation and, therefore, audit and risk management?

To gain an understanding of the complex issues around how HIV positive men negotiated, organised and managed sex with men, I entered spaces where HIV was visible and hidden, and changed how people responded to each other with regard to sexual negotiation. I went 'inside' the bathhouses, backrooms, parks, and bedrooms inhabited by people living with HIV. At the time I was 'loitering with intent', however, there was extensive local media coverage of the arrests of a small number of sexually active, HIV positive men. My concern was that participants in my study would be hesitant to talk about their sexual practices because of this. But quite the opposite happened. The media coverage provided a focus for the ten men who readily shared their thoughts, feelings, views, ideas and sexual experiences.

In the context of increased public attention and blame, talking to participants in their homes and spaces where they enjoyed sex was an important strategy in generating rich stories about having and managing sex while living with HIV/AIDS. At the beginning of each interview, a process of negotiated trust opened between the participant and me in my professionally visible self (social worker) and visible private self (gay man in the club scene). Participants were interested in discussing how I would ensure their confidentiality, where data would be stored, and who would have access to it. In one way, the media and police activity encouraged the participants to exercise their research subjectivity. The men renegotiated how they would set up the ethical protections they felt necessary. Some requested further levels of anonymity: where arrangements had been made for work to be posted, meetings were set up to verbally discuss the analysis of the data, thus removing paper-based notes from production. As police attention increased, the management of confidentiality and anonymity become vital. A number of participants also asked for clarification about the legal repercussions of police action.

From an ethical perspective, the events at that time could not have been foreseen. This highlighted the precarious position of the researcher within the network of required ethics approval. Maintaining fluidity around the use of names and identifiers to ensure audit trails meet ethics requirements, concomitantly offers potential criminal investigation audit trails. In the final stages of the fieldwork, and quite unrelated to it, the police, as part of their investigation into the sexual actions of two HIV positive clients, who were subsequently arrested, executed a search warrant on the office of the organisation where I worked. Acting under a court order, the police seized a small number of my client files and client notes belonging to another counsellor.[2] The police obtained names and addresses of sexual partners of one of the arrested men, via these files, and the police investigation widened. As a social worker *and* researcher, this raised a number of as yet non-traversed ethical issues. At that time, I was still entering the homes of HIV positive men to enquire about social and sexual practices. This led to questions such as: Where does the social worker and researcher begin and end? Do they? Should they?

When client files were seized by the police, I collected interview transcripts and identifying information about participants, and after consultation with my academic supervisors, shredded everything. Protecting the identities of participants is ethical research practice. Building knowledge with clients about their experiences is ethical social work practice, from which research and publications can facilitate wider access to information about this sensitive, complex and 'hard to reach' group (Stanley, 1999, 2000). Following the police activities, the organisation I worked for published recommendations for HIV positive people. Increased awareness of disclosure issues and the new definition of non-safe sexual activity as a criminal offence also introduced a further perspective to my study and brought results that might not be found in other studies. The experiences of the men who took part, therefore, cannot be generalised for all people with HIV.

Being an Ethical and *Intimate* Researcher

Our experience of ethnographic research suggests that ethics policies and processes provide guidance but not definitive solutions to questions about ethical research and moral behaviour. Trying to craft research with emerging ways of 'knowing', or 'doing research' to fit ethics practices, particularly where these are rigid and dogmatic, creates a number of dilemmas. Anticipatory regulatory regimes based on models of clinical or biomedical research, where risk is more readily assessed, are highly problematic (Aagaard-Hansen and Vang Johansen, 2008; Meskell and Pels, 2005; Murphy and Dingwall, 2007; Strathern, 2000). The dialogue between researchers and ethics committees becomes constrained when there is a lack of *fit* between traditional research models and those of the researcher. In such a climate, researchers may find themselves committed to processes that are conceptually flawed or impossible to implement. On the other hand, the ethics process may foster deception and cultures of counterfeit practice (Gordon, 2003). Furthermore, researchers may be requested to assess all risks, and perhaps to anticipate hypothetical worst-case scenarios, Pérez-y-Pérez , for example, was asked to consider the possibility of arrest for not disclosing the real identities of her co-workers. This may create disincentives to undertake research.

In the intimacy of ethnographic research, the application of the principle of informed consent is more complex than intimated by institutional ethics guidelines and practice. Establishing *close relationships* in the field is not a risk-free endeavour for participants or researchers (Pérez-y-Pérez, 2003; Punch, 1994). Such relationships could lead participants to share information they find distressing. Such disclosures may appear therapeutic; others, in retrospect, may be embarrassing or regrettable, possibly reawakening memories of trauma, and result in participants becoming upset, worried or offended. How the principle of informed consent could protect against such harm, particularly as participants may choose to share troubles in ways they did not anticipate when they gave their consent, is unclear. The ethical challenge lies in ensuring that participants are treated with decency, discretion and respect, and have access to whatever support they may need. Given the difficulty of gaining fully informed consent, researchers suggest returning to participants at the end of the study to reaffirm that the findings can be used (Liamputtong and Ezzy, 2005). This approach, however, risks undermining the credibility and impartiality of the findings. To grant the right to veto may invite participants to suppress findings they find unpalatable. A further alternative is to apply caution when explaining the research process, rather than making assurances that we, as researchers, are unlikely to sustain.

The ethnographic approach provides the opportunity to make choices and decisions within a reflexive standpoint that allows the researcher to contemplate wider structural processes driving and influencing data collection. Becker contends: 'No matter how carefully one plans in advance the research is designed in the course of its execution. Indeed as we experienced it, final research findings are the result of hundreds of decisions, large and small, whilst the research is underway' (Becker, 1965, p. 602). While the literature offers novice ethnographers/researchers information about the importance of following ethical guidelines, this did not prepare us for the intimate and challenging questions that participants presented, or the extent to which we needed to engage in self-disclosure in order to develop effective relationships on which to build trust. Given the personal investment in maintaining relationships, self-disclosure and emotional involvement, ethnography could be described as being 'situated between autobiography and anthropology' (Hastrup, 1992, p. 117).

Conclusion

Our management of intimate ethnographic research was guided and restricted by the ethics policies of the academic institutions where we were located. The purpose of such policies is to summon into being ethical research and researchers (Murphy and Dingwall, 2007; Wilkinson, 2001), yet new research methodologies and strategies for generating knowledge can lead to conflict with the social, organisational and cultural practices of ethics processes. When prescribed ethical protocols fail to engage with the complex lives and work of participants and researchers, the policies and practices of research ethics function to construct both researchers and participants as contrary to, different from, and *other* to its aims. This can also result in ethics committees and researchers being positioned as binary opposites: with or without power; rigid versus flexible; or methodologically conservative or methodologically explorative. In underestimating the extent to which we knowingly or unknowingly position ourselves within these spaces, we become complicit in preserving the very things that our work seeks to erode. Ultimately, in the process of ethnographic research, ethical conduct depends on the personal integrity and ethical education of the researcher.

The process of informed consent also has implications for reifying the research subject. This raises questions about what is and what is not ethical. We conclude, however, by reflecting on Roberts and Sanders' (2005) notion of the relationship of researched to researcher as a form of 'gifting'. The willingness of participants to be part of our studies we considered a gift to us, of immense value to our careers as researchers and academics. We took their willingness to engage with us as we wove ourselves into their worlds and private spaces as their consent to be objectified; in return we gave them our respect and acknowledged the contribution, as inhabitants of the worlds that interested us, they made to our world.

Notes

1. This may occur for the sake of access to the research location and at the request of sex workers and managers to avoid disruption of business activities.
2. There is no legal privilege in New Zealand to protect the social worker/counsellor–client relationship.

References

Aagaard-Hansen, J. and M. Vang Johansen (2008) 'Research Ethics Across Disciplines', *Anthropology Today* 24(3): 15–19. [doi:10.1111/j.1467-8322.2008.00585.x]

Adams, R. (1981) 'Ethical Principles in Anthropological Research: One or Many?' *Human Organization* 40: 155–60.

Agar, M. (1996) *The Professional Stranger: An Informal Introduction to Ethnography*. San Diego: Academic Press.

Agar, M. (2006a) *Dope Double Agent: The Naked Emperor on Drugs*. Lulubooks.

Agar, M. (2006b) 'An Ethnography by Any Other Name . . .', *Forum: Qualitative Social Research* 7(4).

Agustín, L. (2005) 'New Research Directions: The Cultural Study of Commercial Sex', *Sexualities* 8: 618–631.

Agustín, L. (2008) *Sex at the Margins: Migration, Labour Markets and the Rescue Industry*. London: Zed Books.

Allison, A. (1994) *Nightwork: Sexuality, Pleasure, and Corporate Masculinity in a Tokyo Hostess Club*. Chicago: Chicago University Press.

Becker, H. (1965) 'Review of Sociologists at Work: Essays on the Craft of Social Research', *American Sociological Review* 30: 602–603.

Bolton, R. (1995) 'Tricks, Friends, and Lovers', in D. Nulick and M. Willson (eds) *Taboo: Sex, Identity and Erotic Subjectivity in Anthropological Fieldwork*. London: Routledge.

Bolton, R. (1996) 'Coming Home: The Journey of a Gay Ethnographer in the Years of the Plague', in E. Lewin and W.C. Leap (eds) *Out in the Field: Reflections of Lesbian and Gay Anthropologists*. Chicago: University of Illinois Press.

Bourjois, P. (1990) 'Confronting Anthropological Ethics: Ethnographic Lessons from Central America', *Journal of Peace Research* 27(1): 43–54. [doi:10.1177/0022343390027001005]

Boynton, P.M. (2002) 'Life on the Streets: The Experiences of Community Researchers in a Study of Prostitution', *Journal of Community & Applied Social Psychology* 12(1): 1–12. [doi:10.1002/casp.657]

Bulmer, M. (1982) *The Uses of Social Research*. London: Allen and Unwin.
Calestani, M. (2009) 'An Anthropology of "The Good Life" in the Bolivian Plateau', *Soc Indic Res* 90: 141–153. [doi :10.1007/s 11205-008-9317-5]
Castañeda, Q.E. (2006) 'Ethnography in the Forest: An Analysis of Ethics in the Morals of Anthropology', *Cultural Anthropology* 21(1): 121–145.
Coffey, A. (1999) *The Ethnographic Self: Fieldwork and the Representation of Identity*. London: Sage.
Dawson, J., M. Foley and N. Peart (2003) 'Research Ethics Committees', in J. Dawson and N. Peart (eds) *The Law of Research: A Guide*. Dunedin: University of Otago Press.
Fluehr-Lobban, C. (2003) *Ethics and the Profession of Anthropology: Dialogue for Ethically Conscious Practice*, 2nd edn. Walnut Creek, CA: Alta Mira Press.
Flowers, A. (1998a) *The Fantasy Factory: An Insider's View of the Phone Sex Industry*. Philadelphia, PA: Pennsylvania University Press.
Flowers, A. (1998b) 'Research from Within: Participant Observation in the Phone-Sex Workplace', in J.E. Elias, V.L. Bullough, V. Elias and G. Brewer (eds) *Prostitution: On Whores, Hustlers, and Johns*. New York: Prometheus Books.
Foucault, M. (1972) *The Archeology of Knowledge*. London: Tavistock.
Fountain, J. (1993) 'Dealing with Data', in D. Hobbs and T. May (eds) *Interpreting the Field: Accounts of Ethnography*. Oxford: Clarendon Press.
George, M. (2008) 'Interactions in Expert Service Work: Demonstrating Professionalism in Personal Training', *Journal of Contemporary Ethnography*, 37(1), 108–131. [doi:10.1177/0891241607309498]
Gherardi, S. and D. Nicolini (2002) 'Learning the Trade: A Culture of Safety in Practice', *Organization* 9: 191–223 [doi:10.1177/1350508402009002264]
Gilbert, N. (1993) *Researching Social Life*. London: Sage.
Gordon, E. (2003) 'Trials and Tribulations of Navigating IRBs: Anthropological and Biomedical Perspectives of "Risk" in Conducting Human Subjects Research', *Anthropological Quarterly* 76: 299–320. [doi:10.1353/anq.2003.0023]
Harper, I. and A. Corsín Jiménez (2005) 'Towards Interactive Professional Ethics', *Anthropology Today* 21(6): 10–12. [doi:10.1111/j.1467-8322.2005.00394.x]
Hart, A. (1998) *Buying and Selling Power: Anthropological Reflections on Prostitution in Spain*. Oxford: Westview Press.
Hastrup, K. (1992) 'Writing Ethnography: State of Art', in J. Oakley and H. Callaway (eds) *Anthropology and Ethnography*. New York: Routledge. [doi:10.4324/9780203450536_chapter_7]
Hastrup, K. (1995) *A Passage to Anthropology: Between Experience and Theory*. London: Routledge.
Hobbs, D. (1988) *Doing the Business: Entrepreneurship, the Working Class and Detectives in the East End of London*. Oxford: Oxford University Press.
Hobbs, D. (1993) 'Peers, Careers, and Academic Fears: Writing as Field-Work', in D. Hobbs and T. May (eds) *Interpreting the Field: Accounts of Ethnography*. Oxford: Clarendon Press.
Hochschild, A.R. (1983) *The Managed Heart*. Berkley, CA: University of California Press.
Hodgson, D.L. (1999) 'Critical Interventions: Dilemmas of Accountability in Contemporary Ethnographic Research', *Identities* 6(2–3): 201–224. [doi:10.1080/1070289X.1999.9962643]
Jorgensen, J. (1971) 'On Ethics and Anthropology', *Current Anthropology* 12: 321–34. [doi:10.1086/201209]
Kondo, D.K. (1990) *Crafting Selves: Power, Gender, and Discourses of Identity in a Japanese Workplace*. Chicago: University of Chicago Press.
Latour, B. (1999) *Pandora's Hope: Essays on the Reality of Science Studies*. Cambridge, MA: Harvard University Press.

Latour, B. (2005) *Reassembling the Social: An Introduction to Actor-Network-Theory*. Oxford: Oxford University Press.

Law, J. (2004) *After Method: Mess in Social Sciences Research*. London, New York: Routledge.

Liamputtong, P. and D. Ezzy (2005) *Qualitative Research Methods*. Oxford: Oxford University Press.

Mattingly, C. (2005) 'Toward a Vulnerable Ethics of Research Practice', *Health: An Interdisciplinary Journal for the Social Study of Health, Illness and Medicine* 9(4): 453–471.

McKeganey, N. and M. Barnard (1996) *Sex Work on the Streets: Prostitutes and their Clients*. Buckingham: Open University Press.

Meskell, L. and P. Pels (2005) 'Introduction: Embedding Ethics', in L. Meskell and P. Pels (eds) *Embedding Ethics*. Oxford and New York: Berg.

Mol, A. (2002) *The Body Multiple: Ontology in Medical Practice*. Durham, NC: Duke University Press.

Murphy, E. and R. Dingwall (2007) 'Informed Consent, Anticipatory Regulation and Ethnographic Practice', *Social Science & Medicine* 65: 2223–2234. [doi:10.1016/j.socscimed.2007.08.008]

Norris, C. (1993) 'Some Ethical Considerations on Field-work with the Police', in D. Hobbs and T. May (eds) *Interpreting the Field: Accounts of Ethnography*. Oxford: Clarendon Press.

Oakley, A. (1981) 'Interviewing Women: A Contradiction in Terms', in H. Roberts (ed) *Doing Feminist Research*. London: Routledge.

O'Brien, M. (2001) 'Doing Ethical Research Legally: Research Ethics and the Law', in M. Tolich (ed) *Research Ethics in Aotearoa New Zealand*. Auckland: Longman.

O'Connell Davidson, J. (1998) *Prostitution, Power and Freedom*. Cambridge: Polity Press.

O'Connell Davidson, J. (2008) 'If No Means No, Does Yes Mean Yes? Consenting to Research Intimacies', *History of the Human Sciences* 21(4): 49–67. [doi:10.1177/0952695108095511]

O'Neill, M. (1997) 'Prostitute Women Now', in G.S.A. Scambler (ed) *Rethinking Prostitution: Purchasing Sex in the 1990s*. London: Routledge.

Orr, J.E. (1990) 'Sharing Knowledge, Celebrating Identity: Community Memory in a Service Culture', in D.M. Edwards (ed) *Collective Remembering*. London: Sage.

Palmer, T. and A. Reed (2001) *Study of the Attitudes of Christchurch Sex Workers to the Police Registration Scheme*. Christchurch: New Zealand Prostitutes Collective.

Payne, M. (2006) *Narrative Therapy: An Introduction for Counsellors*, 2nd edn. Los Angeles: Sage.

Pels, P. (1999) 'Professions of Duplexity: A Prehistory of Ethical Codes in Anthropology', *Current Anthropology* 40(2): 101–136. [doi:10.1086/200001]

Pérez-y-Pérez, M. (2003) 'Discipline, Autonomy and Ambiguity: Organisations, Markets and Work in the Sex Industry', PhD thesis, Department of Sociology, University of Canterbury, Christchurch.

Pérez-y-Pérez, M. (2009) 'Doing it with the Lights On: The Decriminalisation of New Zealand Sex Work', *The International Journal of Interdisciplinary Social Sciences* 4(8): 139–152.

Pérez-y-Pérez, M. (forthcoming) 'Down Time: Circulating Knowledge in a Massage Parlour.'

Pérez-y-Pérez, M. and Stanley, T. (2005) 'Researching Sex: Getting Our Hands Dirty', *Social Work Review* 17(4): 30–38.

Phoenix, J. (1999) *Making Sense of Prostitution*. Basingstoke: Palgrave. [doi:10.1057/9780333985472]

Prostitution Law Review Committee (2005) *The Nature and Extent of the Sex Industry in New Zealand: An Estimation*. Wellington: Ministry of Justice.

Prostitution Reform Act, 2003.

Punch, M. (1986) *The Politics and Ethics of Fieldwork*. Beverley Hills, CA: Sage.

Punch, M. (1993) 'Observation and the Police: The Research Experience', in M. Hammersley (ed) *Social Research: Philosophy, Politics and Practice*. London: Sage/The Open University Press.

Punch, M. (1994) 'Politics and Ethics in Qualitative Research', in N. Denzin and Y. Lincoln (eds) *Handbook of Qualitative Research*. Thousand Oaks, CA: Sage.

Rambo Ronai, C. (1998) 'Sketching with Derrida: An Ethnography of a Researcher/Erotic Dancer', *Qualitative Inquiry* 4: 405–420. [doi:10.1177/107780049800400306]

Roberts, J.M. and T. Sanders (2005) 'Before, During and After: Realism, Reflexivity and Ethnography', *The Sociological Review* 53(2): 294–313. [doi:10.1111/j.1467-954X.2005.00515.x]

Sanders, T. (2005) *Sex Work: A Risky Business*. Cullompton: Willan.

Sanders, T. (2006) 'Sexing Up the Subject: Methodological Nuances in Researching the Female Sex Industry', *Sexualities* 9: 449–468. [doi:10.1177/1363460706068044]

Sanders, T., M. O'Neill and J. Pitcher (2009) *Prostitution: Sex Work, Policy and Politics*. London: Sage.

Silverman, M. (2003) 'Everyday Ethics: A Personal Journey in Rural Ireland, 1980–2001', in P. Caplan (ed) *The Ethics of Anthropology: Debates and Dilemmas*. London: Routledge.

Stacey, J. (1988) 'Can There be a Feminist Ethnography?', *Women's Studies International Forum* 11(1): 21–27. [doi :10.1016/0277-5395(88)90004-0]

Stanley, T. (1999) 'Positive Sex and Risk: Social and Sexual Negotiation with HIV', MA thesis, Department of Sociology, University of Canterbury, Christchurch.

Stanley, T. (2000) 'Living with HIV: Managing the Social Tensions', *Social Work Review* 12(3): 29–31.

Stanley, L. and S. Wise (1991) 'Feminist Research, Feminist Consciousness and Experiences of Sexism', in M.M. Fonow and J.A. Cook (eds) *Beyond Methodology: Feminist Scholarship as Lived Research*. Bloomington, IN: Indiana University Press.

Strathern, M. (1999) *Property, Substance and Effects: Essays on Persons and Things*. New Brunswick, NJ: Athlone Press.

Strathern, M. (2000) *Audit Cultures: Anthropological Studies in Accountability, Ethics and the Academy*. London and New York: Routledge. [doi:10.4324/9780203449721]

Van Maanen, J. (1983) 'The Fact of Fiction in Organizational Ethnography', in J. Van Maanen (ed) *Qualitative Methodology*. London: Sage.

Van Maanen, J. (1992) 'Drinking Our Troubles Away: Managing Conflict in a British Policy Agency', in D.M. Kolb and J.M. Bartunek (eds) *Hidden Conflict in Organizations: Uncovering the Behind-the-Scenes Disputes*. London: Sage.

Wahab, S. (2003) 'Creating Knowledge Collaboratively with Female Sex Workers: Insights from a Qualitative, Feminist, and Participatory Study', *Qualitative Inquiry* 9: 625–642. [doi:10.1177/1077800403252734]

Wengle, J.L. (1988) *Ethnographers in the Field: The Psychology of Research*. Tuscaloosa, AL: University of Alabama.

Wesely, J.K. (2006) 'Negotiating Myself: The Impact of Studying Female Exotic Dancers on a Feminist Researcher', *Qualitative Inquiry* 12(1): 146–162. [doi:10.1177/1077800405282791]

Wilkinson, T. (2001) 'The Core Ideas of Research Ethics', in M. Tolich (ed) *Research Ethics in Aotearoa New Zealand*. Auckland: Longman.

Wolcott, H.F. (1995) 'Fieldwork: The Basic Arts', in *The Art of Fieldwork*. Walnut Creek, CA: Altamira Press.